ASTRONOMY AND ASTROPHYSICS ABSTRACTS

A Publication of the Astronomisches Rechen-Institut Heidelberg
Member of the Abstracting Board
of the International Council of Scientific Unions

Volume 28
Literature 1980, Part 2

Edited by
S. Böhme W. Fricke I. Heinrich W. Hofmann
D. Krahn D. Rosa L. D. Schmadel G. Zech

Springer-Verlag Berlin Heidelberg GmbH 1981

Astronomisches Rechen-Institut Heidelberg
Director: Professor Dr. Walter Fricke

Astronomy and Astrophysics Abstracts
Editors-in-Chief: Inge Heinrich, Dr. Lutz D. Schmadel

Astronomy and Astrophysics Abstracts is prepared
under the auspices of the International Astronomical Union

ISBN 978-3-662-12327-0 ISBN 978-3-662-12325-6 (eBook)
DOI 10.1007/978-3-662-12325-6

Originally published by Springer-Verlag Berlin Heidelberg New York in 1981
Softcover reprint of the hardcover 1st edition 1981

Library of Congress Catalog Card Number 72-104650.
2153/3130-543210

Preface

Astronomy and Astrophysics Abstracts, which has appeared in semi-annual volumes since 1969, is devoted to the recording, summarizing and indexing of astronomical publications throughout the world. It is prepared under the auspices of the International Astronomical Union (according to a resolution adopted at the 14th General Assembly in 1970).

Astronomy and Astrophysics Abstracts aims to present a comprehensive documentation of literature in all fields of astronomy and astrophysics. Every effort will be made to ensure that the average time interval between the date of receipt of the original literature and publication of the abstracts will not exceed eight months. This time interval is near to that achieved by monthly abstracting journals, compared to which our system of accumulating abstracts for about six months offers the advantage of greater convenience for the user.

Volume 28 contains literature published in 1980 and received before February 10, 1981; some older literature which was received late and which is not recorded in earlier volumes is also included.

We acknowledge with thanks contributions to this volume by Dr. J. Bouška, Prague, who surveyed journals and publications in Czech and supplied us with abstracts in English.

It is a pleasure to thank Ms. Dipl.-Math. Ute Esser, our colleague on leave, and Dr. Vladimír R. Matas, who joined our staff in January 1981, for their valuable contributions.

We express our warmest thanks again to Ms. Helga Ballmann, Ms. Nadja Bentele, Ms. Mona El-Choura, Ms. Monika Kohl, Ms. Sylvia Matyssek and Ms. Angelika Meßmer for typing the text of this volume on IBM 72 Composers, for compiling the pages from abstract slips in a perfect form for offset reproduction, and for punching material for the author index and for the subject index, which finally were printed with a TN chain on a 1403 IBM high-speed printer. Finally, we have to thank Mr. Claus Leitherer and Mr. Werner Sanns who supported our task by careful proofreading.

Heidelberg, April 1981

Siegfried Böhme
Walter Fricke
Inge Heinrich
Wilfried Hofmann
Dietlinde Krahn
Dorothea Rosa
Lutz D. Schmadel
Gert Zech

Contents

Positional Astronomy, Celestial Mechanics

Space Research

Theoretical Astrophysics

Sun

Earth

Planetary System

Stars

Interstellar Matter, Nebulae

Radio Sources, X-ray Sources, Cosmic Radiation

Stellar Systems, Galaxy, Extragalactic Objects, Cosmology

Introduction

Astronomical bibliographies

Astronomy and Astrophysics Abstracts begins documentation and abstracting from the year 1969. For information on astronomical literature before this date consultation of one of the following bibliographies is suggested:
(1) J. J. de Lalande, Bibliographie Astronomique, Paris 1803 (this work covers the time from 480 B. C. to the year 1803, VIII + 966 pages).
(2) J. C. Houzeau, A. Lancaster, Bibliographie générale de l'astronomie, Volume I (in two parts), Bruxelles 1882, 1887, Volume II, Bruxelles 1889. The complete title of Volume II is "Bibliographie générale de l'astronomie ou catalogue méthodique des ouvrages, des mémoires et des observations astronomiques, publiés depuis l'origine de l'imprimerie jusqu'en 1880". A new edition of these volumes was prepared by D. W. Dewhirst in 1964.
(3) Bibliography of Astronomy, 1881 - 1898. The literature of this period was recorded on standard slips by the Observatoire Royal de Belgique. From the material (some 52,000 items) a microfilm version was produced by University Microfilms Limited, Tylers Green, High Wycombe, Buckinghamshire, England, in 1970.
(4) Astronomischer Jahresbericht, 1899 gegründet von Walter Wislicenus, herausgegeben vom Astronomischen Rechen-Institut in Heidelberg (formerly in Berlin), Verlag W. de Gruyter, Berlin. For the period from 1899 to 1968 sixty-eight volumes were published, each of which, in general, covers the literature of one year.
(5) Bulletin Signalétique – Section 120: Astronomie, Physique Spatiale, Géophysique. Published by Centre de Documentation du Centre National de la Recherche Scientifique, Paris. This publication is a continuation of "Bibliographie Mensuelle de l'Astronomie" founded in 1933 by the Société Astronomique de France. The publication is continued.
(6) Referativnyj Zhurnal. Founded in 1953 and published by Vsesoyuznyj Institut Nauchnoj i Tekhnicheskoj Informatsii, Akademiya Nauk, Moskva. The publication is continued.

Concept of Astronomy and Astrophysics Abstracts

This abstracting service aims to present a comprehensive documentation of the literature in all fields of astronomy and astrophysics and their border fields. It appears in semi-annual volumes. Two of these volumes cover the literature of one calendar year. The half-yearly period of issue is regarded as an optimal period for summarizing papers into subject categories and for the presentation of abstracts as quickly as possible after the publication of the original literature.

The recording summarizing and indexing of astronomical publications of the year 1980 received from July 1980 to February 1981 are subjects of **Volume 28**. It also records a number of papers issued before 1980 but received within this period.
The main characteristics of the concept of Astronomy and Astrophysics Abstracts may be summarized as follows:
(1) The subdivision of astronomy and its border fields into subject categories is facilitated by the fact that the astronomical objects appear to be particularly well suited for the formation of categories. It may be assumed that such subdivisions can be maintained for a long period. Experience shows, however, that progress in research might imply minor changes in the classification scheme.
(2) Each paper has been classified into one of 108 numbered subject categories and given a serial number within the category. In this way each item is numbered by six figures: the first three indicate the number of the category, the following three the serial number within the category. Reference to an abstract in Volume 1 is indicated by "01" before the number of the category; for example: 01.074.028, denotes Volume 1, category 074, abstract 028.
A paper might be classified into more than one category. In this case, its abstract is placed only in one category, whereas in the other categories only cross references are given. These are listed at the end of each category.
(3) Authors' abstracts are used whenever possible. Popular articles are not abstracted.
(4) If possible, titles of papers and abstracts are given in English. A special reference is made to titles which we have not taken in the original language.

Transliteration scheme for the Russian alphabet

The transliteration of the Russian alphabet in use in Astronomy and Astrophysics Abstracts is presented here.

А	а	a	П	п	p
Б	б	b	Р	р	r
В	в	v	С	с	s
Г	г	g	Т	т	t
Д	д	d	У	у	u
Е	е	e	Ф	ф	f
Ё	ё	e	Х	х	kh
Ж	ж	zh	Ц	ц	ts
З	з	z	Ч	ч	ch
И	и	i	Ш	ш	sh
Й	й	j	Щ	щ	shch
К	к	k	Ы	ы	y
Л	л	l	Ь	ь	'
М	м	m	Э	э	eh
Н	н	n	Ю	ю	yu
О	о	o	Я	я	ya

This transliteration was recommended by the Abstracting Board of the International Council of Scientific Unions in 1969. It corresponds essentially to the transliteration proposed by the Academy of Sciences, Moscow, which is used by the Referativnyj Zhurnal. In this case the letters can be read and printed by usual data processing machines.
If the names of Russian authors in the literature are transliterated in a different scheme, we present the names as they are given in the references cited and in addition in brackets according to our transliteration table.

Sources of information

The majority of sources of information for this volume is given in section **001 Periodicals** and in section **008 Observatories, Institutes**. Section 001 records 800 periodicals indicating full titles and publishers. It may be noted that the titles of the periodicals are given in the original languages, and that Russian titles have been transliterated applying the transliteration scheme given above. Section 008 records 96 periodicals; these are publication series of observatories and astronomical institutes. Titles of the periodicals have been given following the recommendations of the "International List of Periodical Title Word Abbreviations" and its additions (see also **Abbreviations**, p. 10). In most cases they permit recognition of the full title without recourse to the key in section 001.

If other abstracting journals have been consulted in order to examine the degree of completeness of our service, we cite these papers and give reference to the abstracting service.

Author index and subject index

The subject category and the serial number have been used as a reference both in the author index and the subject index. These references are more precise than page references. They offer considerable advantages in indexing by means of data processing machines, and are more convenient for the user.

The author index of this volume contains 9246 names. A complete reference comprises six figures, three for the subject category and three for the serial number within the category. In the case of more than one reference to abstracts in one category, the number of the category is given only once and not repeated in the immediately following references. The total number of papers (some do not give names of authors) recorded in this volume amounts to 7835.
We consider the subject index as an approximation to an optimal index covering all fields of astronomy and astrophysics and their border fields. The assigning of one or more key words to a paper is, undoubtedly, a difficult task. Some journals have started giving key words together with the titles of papers. These key words are chosen by the authors themselves. Starting with Volume 18, the subject index was enlarged to a certain extent in order to provide a thesaurus of astronomical and astrophysical terms. This is done not only for the users' convenience, but also with the intention to propose the use of special key words to authors and publishers.

While each volume is scheduled to contain an author index and a subject index, the magnetic tapes containing the index information will be used to produce separate index volumes (authors and subjects) at intervals of five years.

The sorting program for the author and subject indexes is based on the IBM SORT/MERGE Program. This program sorts blank before hyphen (–) and before letters. Apostrophes are ignored by a special routine. The computations and printing were carried out on an IBM 360/44.

The two most common and widely used classification systems in astronomy and astrophysics are given by Class 9 of the International Classification System for Physics, published by the International Council of Scientific Unions Abstracting Board (Second edition 1978. ICSU-AB, 17 Rue Mirabeau, 75017 Paris, France, ISSN 0305-9618), and the Astronomy and Astrophysics Abstracts classification. In order to facililitate literature searches, we introduce a concordance relation between these two very different systems. This solution is only a unilateral one. Starting from the fourth hierarchical level of the ICSU-AB system, the appropriate Astronomy and Astrophysics Abstracts chapter numbers are listed. This cannot imply an identical content of the respective chapters in both systems. In many cases there is only a rather partial concordance, and therefore the Astronomy and Astrophysics Abstracts numbers are enclosed in parentheses. Due to the fact that our service only aims to present a comprehensive documentation of the literature in all fields of astronomy and astrophysics, only a certain part of Class 9 of the ICSU-AB scheme is covered.

The users are requested to inform us on spelling errors within the author and subject indexes in order to assist us in eliminating mistakes in future cumulative indexes.

Concordance Relation

between the ICSU-AB International Classification System for Physics and the Astronomy and Astrophysics Abstracts Classification Scheme

ICSU-AB International Classification System for Physics	Astronomy and Astrophysics Abstracts Classification Scheme
0 General	
01.10 Announcements, news, and organizational activities	
01.10.C	006 (010)
01.10.F	011
01.10.H	013 (010)
01.30 Physics literature and publications	
01.30.B	012 (014)
01.30.C	012
01.30.E	003
01.30.K	002 (003)
01.30.M	003
01.30.P	003
01.30.R	014
01.30.T	002
01.40 Education	014
01.50 Educational aids	014
01.60 Biographical, historical, and personal notes	004 (005, 006, 007)
01.65 History of science	004
01.90 Other topics of general interest	015

ICSU-AB International Classification System for Physics	Astronomy and Astrophysics Abstracts Classification Scheme
9 Geophysics, Astronomy, and Astrophysics	
91.10 Geodesy and gravity	
91.10.B	046
91.10.N	044 (045)
91.10.Q	081
91.25 Geomagnetism and paleomagnetism; geoelectricity	084
91.35 Earth's interior structure and properties	081
91.90 Other topics in solid Earth physics	081
92.60 Meteorology	082
92.65 Atmospheric optics	082
94.10 Physics of the neutral atmosphere	
94.10.B	082
94.10.D	082
94.10.F	082
94.10.G	082 (063)
94.10.H	082
94.10.L	082
94.10.N	106
94.10.Q	082
94.10.S	084
94.20 Physics of the ionosphere	
94.20.B	083
94.20.D	083
94.20.M	083 (084)
94.20.P	083 (084)
94.20.W	083 (062)
94.20.Y	083 (084)
94.30 Physics of the magnetosphere	
94.30.C	084
94.30.D	084

ICSU-AB International Classification System for Physics	Astronomy and Astrophysics Abstracts Classification Scheme
94.30.E	084
94.30.F	084 (062)
94.30.G	084 (062)
94.30.H	084
94.30.L	084
94.30.M	084
94.30.S	084
94.30.V	084 (074)
94.30.W	084 (078, 143)
94.40 Cosmic rays	
94.40.C	143
94.40.E	143 (078, 106)
94.40.H	078
94.40.K	143 (085)
94.40.L	143 (078)
94.40.V	105 (143)
94.60 Interplanetary space	
94.60.D	074
94.60.F	074
94.60.G	074 (062)
94.60.K	106
94.60.M	106
94.60.Q	074 (091, 094.0)
94.60.R	106 (062)
94.80 Aerospace facilities and techniques, space research	
94.80.P	053 (051)
94.80.R	054 (051)
94.80.W	032.5
95.10 Fundamental astronomy	
95.10.C	042 (043, 052)
95.10.E	042 (052)
95.10.G	041 (079, 095, 096)
95.10.J	041

ICSU-AB International Classification System for Physics	Astronomy and Astrophysics Abstracts Classification Scheme
95.30 Fundamental aspects of astrophysics	
95.30.C	061 (022)
95.30.E	022
95.30.G	022 (061)
95.30.J	063
95.30.L	062
95.30.Q	062
95.30.S	066.0 (162)
95.45 Observatories	008 (009)
95.55 Astronomical instruments	
95.55.B	032.0
95.55.C	032.0 (031.0)
95.55.E	032.0
95.55.J	033
95.55.L	032.5
95.65 Auxiliary and recording instruments	034
95.70 Other instrumentation and techniques (including clocks, frequency standards, etc.)	035 (031.5, 034, 036)
95.75 Techniques of observation and reduction	
95.75.D	031.5
95.75.F	031.5
95.75.H	031.5
95.75.K	031.5
95.75.M	031.5 (021)
95.75.P	021
95.80 Catalogues, atlases, etc.	002 (047)
96.10 General, solar nebula, and cosmogony	107 (091)
96.20 Moon	
96.20.B	094.0
96.20.D	094.0 (094.5)
96.20.J	094.0

ICSU-AB International Classification System for Physics	Astronomy and Astrophysics Abstracts Classification Scheme
96.30 Planets and satellites (excluding the moon)	
96.30.D	092
96.30.E	093
96.30.G	097
96.30.H	098
96.30.K	099
96.30.M	100
96.30.T	101
96.50 Other objects in the planetary system	
96.50.D	106
96.50.G	102 (103)
96.50.K	104
96.50.M	105
96.60 Solar physics	
96.60.C	080 (075)
96.60.F	071 (080)
96.60.K	080
96.60.M	071
96.60.N	073 (074)
96.60.P	074
96.60.Q	072
96.60.R	073 (076, 077)
96.60.S	073
96.60.V	078 (074)
97.10 Stellar characteristics and properties	
97.10.B	131
97.10.C	065 (061)
97.10.E	064 (063)
97.10.F	112 (064)
97.10.H	064 (112)
97.10.K	116 (065)
97.10.L	116 (065)
97.10.N	115
97.10.Q	115
97.10.R	115 (113, 114).
97.10.T	114
97.10.V	111
97.10.W	111

ICSU-AB International Classification System for Physics	Astronomy and Astrophysics Abstracts Classification Scheme
97.20 Normal stars (by class): general or individual	
97.20.D	121
97.20.R	126
97.30 Variable and peculiar stars (including novae)	
97.30.E	114
97.30.F	116
97.30.G	122 (123)
97.30.J	122 (123)
97.30.K	122 (123)
97.30.N	122 (123)
97.30.Q	124 (122, 123)
97.30.S	122 (123)
97.60 Late stages of stellar evolution (including black holes)	
97.60.B	125
97.60.G	141.5
97.60.J	066.5
97.60.L	066.0
97.60.S	066.0 (065)
97.80 Binary and multiple stars (including extrasolar planetary systems)	
97.80.D	118 (002)
97.80.F	120
97.80.H	119
97.80.J	142.0 (117)
97.80.K	118
97.80.M	118
98.10 Stellar dynamics	151
98.20 Stellar clusters and associations	
98.20.C	152
98.20.E	153
98.20.H	154

ICSU-AB International Classification System for Physics	Astronomy and Astrophysics Abstracts Classification Scheme
98.40 Interstellar matter and nebulae	
98.40.B	131
98.40.C	131
98.40.F	132
98.40.H	132 (134)
98.40.J	133 (112)
98.40.K	134 (131)
98.40.M	134
98.40.N	125
98.50 The Galaxy; extragalactic objects and systems	
98.50.C	158
98.50.E	151 (158)
98.50.H	158 (162)
98.50.K	160
98.50.L	155 (156, 157)
98.50.M	160 (159)
98.50.R	158
98.50.T	161
98.70 Other objects and background radiations of unknown origin or distances	
98.70.D	141.0
98.70.J	141.0
98.70.L	133
98.70.Q	142.0 (142.5)
98.70.S	143
98.70.V	066.0 (142.0, 142.5, 162)
98.80 Cosmology	
98.80.B	162
98.80.D	162 (066.0)
98.80.F	061 (162)

Abbreviations

Abbreviations used in Astronomy and Astrophysics Abstracts are primarily based on the 'International List of Periodical Title Word Abbreviations', prepared for the UNISIST/ICSU-AB Working Group on Bibliographic Descriptions (1970).

A.A.B.	Associazione Astrofili Bolognesi
Aarg.	Aargang
AAS	American Astronomical Society
AAVSO	American Association of Variable Star Observers
Abh.	Abhandlung–
Abstr.	Abstract–
Abt.	Abteilung–
Acad.	Academi–, Academy
Accad.	Accademi–
Act.	Active, Activit–
Adm.	Administr–
Adv.	Advanc–
Aehron.	Aehronomi–
Aeron.	Aeronom–
Aeronaut.	Aeronauti–
Aerosp.	Aerospace
AG	Astronomische Gesellschaft
AIAA	American Institute of Aeronautics and Astronautics
AJB	Astronomischer Jahresbericht
Akad.	Akadem–
Ala.	Alabama
Alm.	Almanac–, Almanak–
Amat.	Amateur–
An.	Anais, Anale–, Anali–, Anals
Anal.	Analis–, Analit–, Analys–, Analyt–
Angew.	Angewandt–
Ann.	Annaes, Annal–
Annu.	Annu–
Anst.	Anstalt
Anu.	Anual–, Anuar–
Anz.	Anzeiger
Appl.	Applied
Arb.	Arbeit
Arch.	Archiv–
Årg.	Årgang
Ariz.	Arizona
Ark.	Arkiv–
Arkh.	Arkhiv–
Artif.	Artifici–
ASA	Astronomical Society of Australia
Asoc.	Asocia–
ASP	Astronomical Society of the Pacific
ASSA	Astronomical Society of Southern Africa
Assem.	Assembl–
Assoc.	Associ–
Assoz.	Assozi–
Astrofis.	Astrofisic–
Astrofiz.	Astrofizi–
Astrometr.	Astrometr–
Astron.	Astronom–
Astronaut.	Astronauti–, Astronauty–
Astrophys.	Astrophys–
ASV	Astronomical Society of Victoria
ASWA	Astronomical Society of Western Australia
At.	Atom–
Atmos.	Atmosf–, Atmosph–
BAA	British Astronomical Association
Bayer.	Bayerisch–
Beitr.	Beitrag, Beiträge
Beob.	Beobacht–
Ber.	Bericht–
Bibl.	Bibliot–
Bibliogr.	Bibliograf–, Bibliograph–
BIH	Bureau International de l'Heure (Paris)
Bimest.	Bimestr–
Bl.	Blatt, Blätter
Bol.	Boletin
Boll.	Bolletino
Bul.	Buleten–, Buletin–, Bulten
Bull.	Bulletin–, Bullettino
Bur.	Bureau–
Byul.	Byuleten–, Byuletin–
Byull.	Byulleten–
C.R.	Comptes Rendus
Cah.	Cahier–
Calif.	California
Cas.	Casopis
Cent.	Center–, Central, Centrale, Centrally, Centre
Cercet.	Cercetary
Chem.	Chemi–
Chim.	Chimi–
Chron.	Chronic–, Chronik, Chronique
Chronom.	Chronometr–
Cie.	Compagnie
Cienc.	Ciencia–
Cient.	Cientific–
Circ.	Circolar–, Circolo, Circolaire–, Circular–, Circulo
Cirk.	Cirkulaer–
Cl.	Clasa, Classe–
Co.	Companies, Company
Coll.	College
Collect.	Collect–
Colloq.	Colloqui–
Colo.	Colorado
Comet.	Cometary
Commentat.	Commentat–
Commun.	Communica–
Comput.	Computation, Computer–, Computing
Comun.	Comunica–
Conf.	Conferen–
Congr.	Congres–
Conn.	Connecticut
Contract.	Contract–
Contrib.	Contribu–
Cosm.	Cosmic–
Cosmochim.	Cosmochimi–
COSPAR	Committee on Space Research
Crystallogr.	Crystallograph–
CSIRO	Commonwealth Scientific and Industrial Research Organization
Cult.	Cultur–, Cultuur
Curr.	Current
D.C.	District of Columbia
DDR	Deutsche Demokratische Republik
Del.	Delaware
Dep.	Departament, Département, Department
Dev.	Development–, Développement–
Diss.	Disserta–
Div.	Divis–
Doc.	Document–
Dok.	Dokument–
Dokl.	Doklad–
Ehksp.	Ehksperiment–
Eidg.	Eidgenössisch–
Eksp.	Eksperiment–
Electron.	Electroni–

Eng.	Engineer–
Environ.	Environment–
Equip.	Equipement, Equipment
Ergeb.	Ergebnis–
ESA	European Space Agency
ESO	European Southern Observatory
ESRO	European Space Research Organization
Eval.	Evaluation–
Exp.	Experiment–
Extraterr.	Extraterrestr–
F. R. Germany	Federal Republic of Germany
Fac.	Facolt–, Faculd–, Facult–
Fak.	Fakult
Fasc.	Fascicul–
Fenn.	Fenni–
Fis.	Fisic–, Fisik–
Fiz.	Fizic–, Fizik–, Fizyk–
Fla.	Florida
Fluid.	Fluidi–
Fond.	Fondation–, Fondazione
Fortschr.	Fortschritt–
Fotogr.	Fotograf–
Found.	Foundation–
Freq.	Frequen–
Fundam.	Fundamenta–
Fys.	Fysik–, Fysisch, Fysisk–
Fyz.	Fyzik–
G.	Giornale
Ga.	Georgia
Gaz.	Gazeta, Gazette
Gazz.	Gazzetta
Gen.	General
Geochem.	Geochem–
Geochim.	Geochim–
Geod.	Geodaes–, Geodaet–, Geodes–, Geodet–, Geodez–
Geofis.	Geofis–
Geofiz.	Geofiz–
Geofys.	Geofys–
Geogr.	Geograf–, Geograph–
Geokhim.	Geokhim–
Geol.	Geolog–, Geolosk–
Geomagn.	Geomagneti–
Geophys.	Geophys–
Ges.	Gesellschaft
Gesch.	Geschichte
Gl.	Glavno–
Glas.	Glasnik
Gos.	Gosudarst–
Gov.	Government–
Grenzgeb.	Grenzgebiet–
GSFC	Goddard Space Flight Center
H. M.	Her Majesty's, His Majesty's
Handb.	Handbook, Handbuch
Her.	Herald–
Hist.	History
Hochsch.	Hochschule
Hoegsk.	Hoegskol–
HR-diagram	Hertzsprung-Russell diagram
Hydrogr.	Hydrograf–, Hydrograph–
IAF	International Astronautical Federation
IAU	International Astronomical Union
IBM	International Business Machines Corporation
ICSU	International Council of Scientific Unions
ICSU-AB	International Council of Scientific Unions–Abstracting Board
IEEE	Institute of Electrical and Electronics Engineers
Ill.	Illinois
Inc.	Incorporated
Ind.	Industr–

Inf.	Informat–, Informaz–, Informe–
Ing.	Ingenieur
INIS	International Nuclear Information System
INSPEC	International Information Services for the Physics and Engineering Communities
Inst.	Institut–, Instytut–
Instn.	Institution
Instrum.	Instrument–
Int.	Internationa–, Internazional–
Inter.	Intérieur–, Interior
Interplanet.	Interplanetary
Intez.	Intezet–
Invest.	Investiga–
Ionos.	Ionosfer–, Ionospher–
Iskusstv.	Iskusstvenn–
Issled.	Issledovan–
Ist.	Istitut–
Izd.	Izdatel–
Izv.	Izvesti–
J.	Joernaal–, Jornal–, Journal–
Jaarb.	Jaarboek–
Jahrb.	Jahrbuch, Jahrbücher
Jahresber.	Jahresbericht–
Jahresschr.	Jahresschrift
Jahrg.	Jahrgang
JPL	Jet Propulsion Laboratory
K.	Königlich–, Koninklijk–, Kunglig–
Kans.	Kansas
Kartogr.	Kartograf–
Kernforsch.	Kernforschung
Kernphys.	Kernphysik–
Khem.	Khemyi–
Khim.	Khimi–
Kim.	Kimija–, Kimya
Kl.	Klass–
Kolloq.	Kolloquium–
Komet.	Kometnyj
Komm.	Kommission–
Konf.	Konfer–
Kongr.	Kongress
Kosm.	Kosmich–
Kosmog.	Kosmogon–
Kozp.	Kozponti
KPNO	Kitt Peak National Observatory
Kut.	Kutato
Ky.	Kentucky
La.	Louisiana
Lab.	Laborato–
Lett.	Letter–, Lettra, Lettre
Libr.	Librair–, Librar–
Mag.	Magasin, Magazin–
Magn.	Magneti–, Magnitn–
Mass.	Massachusetts
Mat.	Matemaat–, Matemat–
Mater.	Material–
Math.	Mathemat–
Md.	Maryland
Meas.	Measur–
Mec.	Mecani–
Mech.	Mechani–
Medd.	Meddelande–, Meddelelse
Meded.	Mededeeling–, Mededeling–
Mekh.	Mekhani–
Mem.	Memento–, Memoir–, Memori–, Memory–, Memuary
Memo.	Memorand–
Mens.	Mensile, Mensual–, Mensuel–
Messtech.	Messtechni–
Meteorol.	Meteorolog–
Mich.	Michigan
Micromec.	Micromecaniq–
Miner.	Mineral, Minerale–, Minerali–

Mineral.	Mineralog–
Minn.	Minnesota
Miss.	Mississippi
MIT	Massachusetts Institute of Technology
Mitt.	Mitteilung–
Mo.	Missouri
Mod.	Modern–
Mol.	Molecul–, Molekul–
Mon.	Monat, Monatlich–, Month–
Monogr.	Monograph–
Mont.	Montana
MPI	Max-Planck-Institut
Mus.	Museum
N. C.	North Carolina
N. D.	North Dakota
N. H.	New Hampshire
N. J.	New Jersey
N. M.	New Mexico
N. Y.	New York
Nablyud.	Nablyudeni–
Nac.	Nacion–
Nachr.	Nachricht–
NASA	National Aeronautics and Space Administration
Nat.	Natur–
Natl.	National–
Naturforsch.	Naturforsch–
Naturwiss.	Naturwissenschaft–
Natuurkd.	Natuurkunde
Nauchn.	Nauchny–
Nauk.	Nauka, Naukite, Naukov–, Naukow–
Naut.	Nautic–
Nav.	Naval–
Navig.	Navigat–
Naz.	Nazion–
Nebr.	Nebraska
Nev.	Nevada
Newsl.	Newsletter–
Not.	Notationes, Notic–, Notise–, Notizi–
Nouv.	Nouveau–, Nouvell–
Nov.	Novoe
Nucl.	Nucléaire–, Nuclear–, Nucl–
Nukl.	Nukle–
Numer.	Numeri–
O-va	Obshchestva
O-vo	Obshchestvo
Obs.	Observ–
Obz.	Obzor–
Okla.	Oklahoma
Opt.	Optic–, Optik–, Optique
Oreg.	Oregon
Oss.	Osserva–
Pa.	Pennsylvania
Paleontol.	Paleontolog–
Pap.	Paper–, Papier
Part.	Particle
Perem.	Peremenn–
Period.	Periodi–
Petrol.	Petrolog–
Philos.	Philosoph–
Photogr.	Photograf–, Photograph–
Photogramm.	Photogrammetr–
Photom.	Photometr–
Phys.	Physic–, Physik–, Physique–, Physisch–
Pict.	Picture–
Planet.	Planetary
Pr.	Prac–
Prelim.	Prelimin–
Prepr.	Preprint
Prib.	Pribor–
Prikl.	Prikladnoj
Prir.	Prirodn–

Prirodoved.	Prirodoved–
Probl.	Problem–
Proc.	Proceedings
Prod.	Prodott–, Produc–, Produkt,
Prog.	Progres–
Propag.	Propagation
Prov.	Provinc–, Provints–, Provinz–
Pubbl.	Pubblicazion–
Publ.	Publicac–, Publicas–, Publicat–, Publikas–, Publikat–
Q.	Quarterly
Quant.	Quantit–
R.	Royal
R. I.	Rhode Island
Radiat.	Radiati–
Radioact.	Radioactiv–, Radioaktiv–
Radioisot.	Radioisotop–
Rap.	Raport–
Rapp.	Rapport–
RAS	Royal Astronomical Society
Rec.	Record–
Rech.	Recherche–
Ref.	Referat–, Reference–, Referieren
Relat.	Related, Relation–
Relativ.	Relativit–
Rend.	Rendicont–
Rep.	Report–
Repr.	Reprint–
Repub.	Republi–
Res.	Research–
Result.	Resultad–, Resultat–
Rev.	Review–, Revisio, Revista, Revue–
Rezul't.	Rezul'tat–
Ric.	Ricerca, Ricerche
Riv.	Rivist–
Rundsch.	Rundschau
S. C.	South Carolina
S. D.	South Dakota
SAF	Société Astronomique de France
SAI	Società Astronomica Italiana
Samml.	Sammlung–
SAO	Smithsonian Astrophysical Observatory
SAS	Société Astronomique de Suisse
Satell.	Satellite
Sb.	Sbornik–
Schr.	Schrift–
Schriftenr.	Schriftenreihe
Sci.	Scienc–, Scient–, Scienz–
Scr.	Scripta, Scritt–
Secc.	Seccion–
Sect.	Secti–
Seer.	Seeria
Sekc.	Sekci–, Sekcj–
Sekt.	Sektion–, Sektor–
Sekts.	Sektsi–
Sel.	Seleccion–, Select–, Selek–, Selezione
Selsk.	Selskab–, Selskap–
Semin.	Séminair–, Seminar–
Sep.	Separat–
Ser.	Seria–, Serie–, Seriya
Serv.	Servic–, Serviz–
Sess.	Sessi–
Signal.	Signalétique–
Simp.	Simpoz–
Sitzungsber.	Sitzungsbericht–
Skr.	Skrift–
Soc.	Sociedad–, Societ–
Sol.	Solar
Soln.	Solnechn–
Sonderdr.	Sonderdruck–
Soobshch.	Soobshchen–
South.	Southern

Spacecr.	Spacecraft
Spat.	Spatial–
Spec.	Special–
Spectrosc.	Spectroscop–
Spectrosk.	Spectroskop–
Spets.	Spetsial–
Spez.	Spezial–, Speziell–
SSR	Sovetskaya Sotsialisticheskaya Respublika
SSSR	Soyuz Sovetskikh Sotsialisticheskikh Respublik
St.	Saint–, Sankt–, Sant
– St.	– Straße, Street
Stand.	Standard–, Standart
Sternw.	Sternwarte–
Stiint.	Stiintific–
Stn.	Station, Stazione
Stud.	Studia, Studie–, Studii
Supl.	Suplement–, Supliment–
Suppl.	Supplement–
Surv.	Survey–
Symp.	Sympos–, Sympoz-
Syst.	System–
Sz.	Szemle
Teach.	Teacher–, Teaching
Tec.	Tecni–
Tech.	Techni–
Technol.	Technolog–
Tecnol.	Tecnolog–
Teh.	Tehnic–, Tehnika, Tehnisk–
Tehnol.	Tehnolog–, Tehnolosk–
Tek.	Tekni–
Tekh.	Tekhni–
Tekhnol.	Tekhnolog–
Teknol.	Teknolog–
Telesc.	Telescop–
Telev.	Television–
Tenn.	Tennessee
Teor.	Teoret–, Teori–
Terr.	Terrestr–
Test.	Testing
Tex.	Texas
TH	Technische Hochschule
Theor.	Theoret–, Theori–
Tidschr.	Tidschrift–
Tidskr.	Tidskrift–
Tidsskr.	Tidsskrift–
Top.	Topic–
Tr.	Trudy
Trans.	Transactions, Transazione
Tsentr.	Tsentral–
Tsirk.	Tsirkulyar–
TU	Technical University
Uch.	Uchen–
Uchebn.	Uchebn–
UK	United Kingdom
Umsch.	Umschau
UN	United Nations
Univ.	Universidad–, Universit–, Univerzitet–
US	United States
USA	United States of America
USSR	Union of Soviet Socialist Republics
Va.	Virginia
Var.	Various
Ver.	Verein–, Verenig–
Veränderl.	Veränderlich–
Verh.	Verhandl–
Vermess.	Vermessung–
Vermessungswes.	Vermessungswesen
Veröff.	Veröffentlich–
Vesn.	Vesnik
Vestn.	Vestnik
Vetensk.	Vetenskap–
Vidensk.	Videnskab–, Videnskap
Vierteljahresschr.	Vierteljahresschrift–
Vierteljahrsschr.	Vierteljahrsschrift–
VLB	Very Long Baseline
Volcanol.	Volcanolog–
Vopr.	Vopros–
Vortr.	Vorträge
Vses.	Vsesoyuzn–
Vt.	Vermont
Vyp.	Vypusk–
Vyssh.	Vyssh–
Vyzk.	Vyzkum–
W. Va.	West Virginia
Wash.	Washington
West.	Western
Wet.	Wetenschap–, Wetenskap–
Wis.	Wisconsin
Wiss.	Wissenschaft–
Wyo.	Wyoming
Yad.	Yadern–
Z.	Zeitschrift–
ZA	Zero Age
ZAED	Zentralstelle für Atomkernenergie-Dokumentation
Zap.	Zapisk–, Zapyisk–
Zaved.	Zaveden–
Zent.	Zentral
Zentralbl.	Zentralblatt
Zesz.	Zeszyt
Zh.	Zhurnal–
Zirk.	Zirkular

Periodicals, Proceedings, Books, Activities

001 Periodicals

A. A. O. Newsl.
Anglo-Australian Observatory Newsletter. Published by the Anglo-Australian Observatory, PO Box 296, Epping, NSW 2121, Australia.

AAS Photo-Bull.
AAS (American Astronomical Society) Photo-Bulletin. Published by the Working Group on Photographic Materials. Produced by Eastman Kodak Co., Rochester, N.Y.

AAVSO Bull.
Bulletin of the American Association of Variable Star Observers, 187 Concord Avenue, Cambridge, Mass., 02138, U.S.A.

Acad. **R. Belgique, Bull. Cl. Sci.**
Académie Royale de Belgique, Bulletin de la Classe des Sciences (Koninklijke Academie van België, Mededelingen van de Klasse der Wetenschappen). 5ᵉ Série, Palais des Académies, Bruxelles.

Acta Astron.
Acta Astronomica. An international quarterly journal. Publisher: Polska Akademia Nauk, Komitet Astronomii (Polish Academy of Sciences, Committee of Astronomy), Warszawa – Wrocław.

Acta Astron. Sinica
Acta Astronomica Sinica. Published by Purple Mountain Observatory, Academia Sinica, Nanking, China.

Acta Astronaut.
Acta Astronautica. Journal of the International Academy of Astronautics. Publisher: Pergamon Press Inc., Elmsford, New York, U.S.A.; Pergamon Press Ltd., Oxford, England.

Acta Cosmologica
Acta Cosmologica. Published by Obserwatorium Astronomiczne Universytetu Jagiellońskiego, Kraków, Poland.

Acta Crystallogr. A
Acta Crystallographica, Section A: Crystal Physics, Diffraction, Theoretical and General Crystallography. Munksgaard International Booksellers and Publishers Ltd., 35 Norre Sogade, DK 1370 Kobenhavn K, Denmark.

Acta Geod. Geophys. Montan.
Acta Geodaetica, Geophysica et Montanistica. Akademiai Kiado, 1054 Budapest, Alkotmany utca 21, Hungary.

Acta Geophys. Polonica
Acta Geophysica Polonica. ARS Polona-Ruch, 00-068 Warszawa, Krakowskie Przedmiescie 7, P.O. Box 1001, Poland.

Acta Geophys. Sinica
Acta Geophysica Sinica. Chinese Academy of Sciences, Department of Geophysical Research. Published by Science Press, Peking, People's Republic of China.

Acta Mech. Sinica
Acta Mechanica Sinica, Science Press, Peking, People's Republic of China. Subscription address: Guozi Shudian, P.O. Box 399, Peking.

Acta Phys. Acad. Sci. Hungaricae
Acta Physica Academiae Scientiarum Hungaricae. Postafiok 24, Budapest 502, Hungary.

Acta Phys. Austriaca
Acta Physica Austriaca. Springer-Verlag, A-1011 Wien, Molkerbastei 5, Postfach 367, Austria.

Acta Phys. Polonica B
Acta Physica Polonica B. ARS Polona-Ruch, Warszawa 1, P.O. Box 154, Poland.

Acta Phys. Sinica
Acta Physica Sinica. Chinese Academy of Sciences, Institute of Physics, Peking, People's Republic of China. [English translation in: Chinese J. Phys. (*USA*)].

Acta Phys. Slovaca
Acta Physica Slovaca. VEDA Publishing House of the Slovak Academy of Sciences, 895 30 Bratislava, Klemensova 27, Czechoslovakia.

Acta Polytech. III
Acta Polytechnica. Series III. Elektrotechnická fakulta ČVUT v Praze, Technická ul. 2, Praha 6-Dejvice, Czechoslovakia.

Acta Sci. Nat. Univ. Pekinensis
Acta Scientiarum Naturalium Universitatis Pekinensis. Peking, People's Republic of China.

Acta Tech. CSAV
Acta Technica Československá Akademie Věd. Academia, Publishing House of the Czechoslovak Academy of Sciences, Vodickova 40, 112 29 Praha 1, Czechoslovakia. John Benjamins N.V., Periodical Trade, Warmoesstraat 54, Amsterdam, Netherlands.

Acta Univ. Carolinae Math. Phys.
Acta Universitatis Carolinae, Mathematica et Physica. Administrace: Matematicko-fyzikální fakulta University Karlovy, Praha.

Adv. Astron. Astrophys.
Advances in Astronomy and Astrophysics. Publisher: Academic Press, New York–London.

Adv. Colloid Interface Sci.
Advances in Colloid and Interface Science. Elsevier Scientific Publishing Co., P.O. Box 211, Amsterdam, Netherlands.

Adv. Phys.
Advances in Physics. Taylor & Francis Ltd., 10–14 Macklin Street London, WC2B 5NF, England.

AIAA J.
AIAA Journal. A Publication of the American Institute of Aeronautics and Astronautics devoted to Aerospace Research and Development. Published by the American Institute of Aeronautics and Astronautics, New York, N.Y.

AIP Conf. Proc.
AIP Conference Proceedings. American Institute of Physics, 335 East 45th Street, New York, N.Y. 10017, USA.

Alta Freq.
Alta Frequenza. Ufficio Centrale AEI-CEI, Viale Monza 259, 20126 Milano, Italy.

American J. Phys.
American Journal of Physics. Published for the American Association of Physics Teachers by the American Institute of Physics, 335 East 47th Street, New York, N.Y. 10017, USA.

American Mineral.
American Mineralogist. Mineralogical Society of America, 1707 L Street, N.W., Washington, DC 20036, USA.

American Sci.
American Scientist. Society of Sigma XI, 345 Whitney Avenue, New Haven, CT 06510, USA.

An. Acad. Brasil. Cienc.
Anais da Academia Brasileira de Ciencias. Caixa Postal 229, ZC-00 Rio de Janeiro gb, Brazil.

An. Fac. Cienc.
Anais da Faculdade de Ciências. Universidade do Porto, Portugal.

An. Fis.
Anales de Física. Real Sociedad Española de Física y Química (Facultad de Ciencias), Ciudad Universitaria, Madrid-3, Spain.

An. Stiint. 'Al. I. Cuza' Iasi (Ser. Noua) I
Analele Stiintifice ale Universitatu 'Al. I. Cuza' din Iasi (Serie Noua), Sectiunea I Fizica. Calea 23 August, Iasi, Rumania.

An. Univ. Bucuresti Fiz.
Analele Universitatii Bucuresti Fizica. Biblioteca Centrala Universitara, Serviciul Schimb de Publicatii, Bucuresti, Str. Onesti 1, Rumania.

Ann. Acad. Sci. Fennicae, Ser. A. VI
Annales Academiae Scientiarum Fennicae, Series A VI (Physica). Snellmaninkato 9-11, 00170 Helsinki-17, Finnland.

Ann. Geofis.
Annali di Geofisica. Istituto Nazionale di Geofisica, Citta Universitaria, Via Ruggero Bonghi 11/B, 00184 Roma, Italy.

Ann. Géophys.
Annales de Géophysique. Service des Publications du CNRS, 15 Quai Anatole-France, 75700 Paris, France.

Ann. Inst. Henri Poincaré A
Annales de l'Institut Henri Poincaré, Section A (Physique Theorique). 11 Rue Pierre-Curie, Paris 5, France.

Ann. Nucl. Energy
Annals of Nuclear Energy. Pergamon Press Ltd., Headington Hill Hall Oxford, OX3 0BW, England.

Ann. Obs. Astron. Météorol. Toulouse
Annales de l'Observatoire Astronomique et Météorologique de Toulouse. Publisher: Gauthier-Villars, Paris.

Ann. Physics
Annals of Physics. Academic Press Inc., 111 Fifth Avenue, New York, NY 10003, USA.

Ann. Physik
Annalen der Physik. 7. Folge. Publisher: Johann Ambrosius Barth, Salomonstr. 18B, Leipzig 701, German Democratic Republic.

Ann. Physique
Annales de Physique. Publisher: Masson et Cie., 120 Boulevard Saint-Germain, Paris 6, France.

Ann. Sci.
Annals of Science. Taylor & Francis Ltd., 10-14 Macklin Street, London, WC2B 5NF, England.

Ann. Soc. Sci. Bruxelles I
Annales de la Société Scientifique de Bruxelles. Série I: Sciences Mathématiques, Astronomiques et Physiques. Rue de Bruxelles 61, B 5000 Namur, Belgium.

Ann. Télécommun.
Annales des Télécommunications. Centre National d'Études des Télécommunications, 38 rue du Général Leclerc, 92 Issy-les-Moulineaux, France.

Ann. Tokyo Astron. Obs.
Annals of the Tokyo Astronomical Observatory. University of Tokyo, Mitaka, Tokyo, Japan.

Ann. Univ.-Sternw. Wien
Annalen der Universitäts-Sternwarte Wien. In Kommission bei Ferd. Dümmlers Verlag, Bonn.

Annu. Rep. Astron. Inst. Greece
Annual Reports of the Astronomical Institutes of Greece. Published by the Greek National Committee for Astronomy. Academy of Athens, Research Center for Astronomy and Applied Mathematics.

Annu. Rev. Astron. Astrophys.
Annual Review of Astronomy and Astrophysics. Publisher: Annual Reviews Inc., Palo Alto, California.

Annu. Rev. Earth Planet. Sci.
Annual Review of Earth and Planetary Sciences. Annual Reviews Inc., 4139 El Camino Way, Palo Alto, Calif. 94306, USA. ISSN 0084-6597.

Annu. Univ. Sofia Fac. Phys.
Annuaire de l'Université de Sofia Faculté de Physique, Sofiya, Bulgaria.

Antenna
L'Antenna. Via Monte Generoso 6/a, 20155 Milano, Italy.

Anz. Österreich. Akad. Wiss. Math.-Naturwiss. Kl.
Anzeiger. Österreichische Akademie der Wissenschaften. Mathematisch-Naturwissenschaftliche Klasse. Publisher: Springer-Verlag, Wien.

APL Tech. Dig.
APL Technical Digest. Applied Physics Laboratory, The John Hopkins University, 8621 Georgia Avenue, Silver Spring, MD 20910, USA.

Appl. Opt.
Applied Optics. A monthly publication of the Optical Society of America. Published for the Optical Society of America by the American Institute of Physics, 335 East 45th Street, New York, NY 10017, USA.

Appl. Phys.
Applied Physics. Springer-Verlag, Heidelberger Platz 3, D-1000 Berlin 33, F. R. Germany.

Appl. Phys. Lett.
Applied Physics Letters. American Institute of Physics, 335 East 45th Street, New York, N.Y. 10017, USA.

Appl. Sci. Res.
Applied Scientific Research. Martinus Nijhoff, Lange Voorhout 9, Den Haag, Netherlands.

Appl. Spectrosc.
Applied Spectroscopy. 428 East Preston Street, Baltimore, MD 21202, USA.

Appl. Spectrosc. Rev.
Applied Spectroscopy Reviews. Marcel Dekker Inc., 95 Madison Avenue, New York, NY 10016, USA.

Arch. Hist. Exact Sci.
Archive for History of Exact Sciences. Springer-Verlag, Berlin - Heidelberg - New York. ISSN 0003 - 9519.

Arch. Mech.
Archives of Mechanics (Archiwum Mechaniki Stosowanej). Polish Scientific Publishers, Swietokrzyska 21, Warszawa, Poland.

Arch. Sci.
Archives des Sciences, éditées par la Société de Physique et d'Histoire Naturelle de Genève. Publisher: Imprimerie Kundig, Genève. Subscription address: Librairie Payot, Genève.

Archaeoastronomy
Archaeoastronomy. Supplement to Journal for the History of Astronomy. Published by Science History Publications Ltd, Halfpenny Furze, Mill Lane, Chalfont St Giles, Bucks, England, HP8 4NR. ISSN 0142–7253.

Archaeometry
Archaeometry. Cambridge University Press, P.O. Box 92, London, NW1 2DB, England.

Ark. Mat.
Arkiv för Matematik. Published by Institut Mittag-Leffler, Auravägen 17, S-182 62 Djursholm, Sweden.

Artif. Satell.
Artificial Satellites. Publication of Polish Scientific Institutions. Polish Academy of Sciences, National Committee of Geophysics and Geodesy, National Committee for Space Research, Warsaw. Space Research Centre, Pałac Kultury i Nauki 2301, 00-901 Warszawa, Poland.

Asoc. Argentina Astron. Bol.
Asociación Argentina de Astronomía. Boletin. Editor: Instituto Argentino de Radioastronomía, Provincia de Buenos Aires, Argentina. Printer: Talleres Gráficos "Renovación", La Plata, República Argentina.

Assoc. Veneta Oss. Stelle Variabili Bull.
Associazione Veneta Osservatori di Stelle Variabili, Bulletin. c/o Gruppo Astrofili di Padova, Corso Garibaldi 41, 35100 Padova, Italy.

Astrofiz. Issled. Izv. Spets. Astrofiz. Obs.
Astrofizicheskie Issledovaniya. Izvestiya Spetsial'noj Astrofizicheskoj Observatorii. Akademiya Nauk SSSR. Publishers: Izdatel'stvo "Nauka", Leningradskoe Otdelenie, Leningrad.

Astrofizika
Astrofizika. Izdatel'stvo Akademii Nauk Armyanskoj SSR, Erevan. [An English translation is published in "Astrophysics"].

Astrometr. Astrofiz.
Astrometriya i Astrofizika. Respublikanskij Mezhvedomstvennyj Sbornik. Akademiya Nauk Ukrainskoj SSR, Glavnaya Astronomicheskaya Observatoriya. Naukova Dumka, Kiev.

Astron. Astrophys.
Astronomy and Astrophysics. A European Journal. Published by Springer-Verlag, Berlin–Heidelberg–New York.

Astron. Astrophys., Suppl. Ser.
Astronomy and Astrophysics. Supplement Series. A European Journal. Published by les Editions de Physique, Orsay, France, on behalf of the Board of Directors of the European Southern Observatory. ISSN 0365-0138.

Astron. Circ.
Astronomical Circular. Compiled by the editor section of Acta Astronomica Sinica, Purple Mountain Observatory, Nanking, China.
Edited by the Chinese Astronomical Society.

Astron. Data Cent. Bull.
Astronomical Data Center Bulletin. National Space Science Data Center/World Data Center A for Rockets and Satellites. National Aeronautics and Space Administration, Goddard Space Flight Center, Greenbelt, Maryland 20771, USA.

Astron. Her.
Astronomical Herald. Astronomical Society of Japan, Tokyo Astronomical Observatory, Oosawa Mitaka, Tokyo, Japan.

Astron. J.
The Astronomical Journal. Published for the American Astronomical Society by the American Institute of Physics, New York, N. Y. Editorial Office: Department of Astronomy, Columbia University, New York, N. Y.

Astron. Mitt. Eidg. Sternw. Zürich
Astronomische Mitteilungen der Eidgenössischen Sternwarte Zürich, Switzerland.

Astron. Nachr.
Astronomische Nachrichten. Publisher: Akademie-Verlag, Berlin.

Astron. Pap.
Astronomical Papers prepared for the use of the American Ephemeris and Nautical Almanac. Published by the

Nautical Almanac Office, U.S. Naval Observatory by direction of the Secretary of the Navy and under the authority of Congress. U.S. Government Printing Office, Washington, D.C.

Astron. Rep.
The Astronomical Reports. Polish Amateur Astronomical Society. Polskie Towarzystwo Miłośników Astronomii, Kraków, Poland.

Astron. Schule
Astronomie in der Schule. Zeitschrift für die Hand des Astronomielehrers. Herausgegeben vom Verlag Volk und Wissen, Berlin. Redaktion: Sternwarte Bautzen.

Astron. Soc. Western Australia, Circ.
The Astronomical Society of Western Australia (Inc.), Circular.

Astron. Tidsskr.
Astronomisk Tidsskrift. Edited by Astronomisk Selskab, København; Norsk Astronomisk Selskap, Oslo; Svenska Astronomiska Sällskapet, Stockholm. Printed by John Griegs Boktrykkeri, Bergen.

Astron. Tsirk.
Astronomicheskij Tsirkulyar, izdavaemyj Byuro Astronomicheskikh Soobshchenij Akademii Nauk SSSR. Tipografiya Astrosoveta AN SSSR, Moskva.

Astron. Vestn.
Astronomicheskij Vestnik. Publishers: Izdatel'stvo "Nauka", Moskva.

Astron. Zh. Akad. Nauk SSSR
Astronomicheskij Zhurnal. Akademiya Nauk SSSR. Publishers: Izdatel'stvo "Nauka", Moskva. [An English translation is published in "Soviet Astronomy"].

Astronomia
Astronomia. Periodico trimestrale dell'Unione Astrofili Italiani.

Astronomie
L'Astronomie et Bulletin de la Société Astronomique de France. Société Astronomique de France, Paris.

Astronomy
Astronomy. AstroMedia Corp., 757 North Broadway, Suite 204, Milwaukee, WI 53202, USA.

Astrophys. J.
The Astrophysical Journal. Published for the American Astronomical Society by the University of Chicago Press, Chicago, Illinois.

Astrophys. J.,Lett.
The Astrophysical Journal. Letters to the Editors. Published for the American Astronomical Society by the University of Chicago Press, Chicago, Illinois.

Astrophys. J., Suppl. Ser.
The Astrophysical Journal. Supplement Series. Published for the American Astronomical Society by the University of Chicago Press, Chicago, Illinois.

Astrophys. Lett.
Astrophysical Letters. Published by NASA-Goddard Space Flight Center. Gordon and Breach Science Publishers Ltd., New York–London–Paris.

Astrophys. Space Sci.
Astrophysics and Space Science. An International Journal of Cosmic Physics. Published by D. Reidel Publishing Company, Dordrecht, Holland.

Astrophysics
Astrophysics. A cover-to-cover translation of Astrofizika (USSR). Consultants Bureau, New York, N. Y.

At. Data Nucl. Data Tables
Atomic Data and Nuclear Data Tables. Academic Press Inc., 111 Fifth Avenue, New York, NY 10003, USA.

At. Energy Rev.
Atomic Energy Review. International Atomic Energy, Kärntner Ring 11, P.O. Box 590, A-1011 Vienna, Austria.

Atmos. Environ.
Atmospheric Environment. Pergamon Press Ltd., Headington Hill Hall, Oxford, OX3 OBW, England.

Atomkernenergie
Atomkernenergie. Verlag Karl Thiemig, Pilgersheimerstrasse 38, 8 München 90, Postfach 900740, F.R. Germany.

Atti Accad. Ligure Sci. Lett.
Atti della Accademia Ligure di Scienze e Lettere. Palazzo Reale, Via Balbi 10, 16126 Genova, Italy.

Atti Accad. Naz. Lincei, Mem. Ser. Ottava
Atti della Accademia Nazionale dei Lincei. Serie Ottava. Memorie. Classe di Scienze fisiche, matematiche e naturali. Sezione I: Matematica, Meccanica, Astronomia, Geodesia e Geofisica. Published by Accademia Nazionale dei Lincei, Roma.

Atti Accad. Naz. Lincei, Rend. Ser. Ottava
Atti della Accademia Nazionale dei Lincei. Serie Ottava. Rendiconti. Classe di Scienze fisiche, matematiche e naturali. Published by Accademia Nazionale dei Lincei, Roma.

Atti Accad. Sci. Torino I
Atti della Accademia delle Scienze di Torino. I. Classe di Scienze Fisiche, Mathematiche e Naturali. Via Accademia delle Scienze 6, Via Maria Vittoria 3, Torino (208), Italy.

Atti Fond. Giorgio Ronchi
Atti della Fondazione Giorgio Ronchi. Largo Enrico Fermi 1, 50125 Arcetri-Firenze, Italy.

Australian J. Phys.
Australian Journal of Physics. Published by the Commonwealth Scientific and Industrial Research Organization, 372 Albert Street, East Melbourne, Victoria 3002, Australia.

Australian J. Phys., Astrophys. Suppl.
Australian Journal of Physics, Astrophysical Supplement. Published by Commonwealth Scientific and Industrial Research Organization, 372 Albert Street, East Melbourne, Victoria 3002, Australia.

Autom. Strum.
Automazione e Strumentazione. Associazione Nazionale Italiana per l'Automazione, Via Le Premuda 2, 21029 Milano, Italy.

B. I. H., Paris, Circ.
Bureau International de l'Heure, B.I.H., Paris, Circulars. 61, Avenue de l'Observatoire, 75014-Paris.

BAV Rundbrief
BAV Rundbrief. Mitteilungsblatt der Berliner Arbeitsgemeinschaft für Veränderliche Sterne. Editor: BAV Berliner Arbeitsgemeinschaft für Veränderliche Sterne eV., Berlin.

BBSAG Bull.
Bedeckungsveränderlichen Beobachter der Schweizerischen Astronomischen Gesellschaft, [Swiss Astronomical Society's Eclipsing Variable Observers], Bulletin. To be obtained from R. Diethelm, Winterthur, Switzerland.

Bell Lab. Rec.
Bell Laboratories Record. Bell Telephone Laboratories, Mountain Avenue, Murray Hill, NJ 07974, USA.

Bell Syst. Tech. J.
Bell System Technical Journal. American Telephone and Telegraph Co., 195 Broadway, New York, NY 10007, USA.

Ber. Bunsenges. Phys. Chem.
Berichte der Bunsengesellschaft für Physikalische Chemie. Verlag Chemie, 6940 Weinheim/Bergstrasse, Postfach 1260/1280, Germany.

Blick Weltall
Blick in das Weltall. Monatsprogramm und Mitteilungen für Sternfreunde. Archenhold-Sternwarte, Berlin-Treptow.

BMR J. Australian Geol. Geophys.
BMR Journal of Australian Geology & Geophysics. Bureau of Mineral Resources, Geology & Geophysics, P. O. Box 378, Canberra, A. C. T. 2601, Australia.

Bol. Acad. Cienc. Fis. Mat. Nat.
Boletin de la Academia de Ciencias Fisicas Matematicas y Naturales. Printed by Italgrafica, S.R.L. Republica de Venezuela.

Bol. Astron.
Boletin Astronômico. Observatório do Capricórnio, Prefeitura Municipal de Campinas–SP, Brazil.

Bol. Astron. Obs. Madrid
Boletín Astronómico del Observatorio de Madrid. Instituto Geografico Nacional, General Ibáñez de Ibero, 3. Madrid 3. Spain.

Bol. Inst. Mat., Astron. Fis., Univ. Nac. Córdoba
Boletin del Instituto de Matematica, Astronomia y Fisica, Universidad Nacional de Córdoba (R. A.). Dirección General de Publicaciones, Córdoba (Argentina).

Bol. Inst. Tonantzintla
Boletin del Instituto de Tonantzintla. Instituto Nacional de Astrofisica, Optica y Electronica, Apartados Postales Nos. 216 y 51, Puebla, Pue, Mexico.

Bol. Liga Latinoamericana Astron.
Boletin de la Liga Latinoamericana de Astronomia. Publicado por la Asociacion Argentina Amigos de la Astronomia, Buenos Aires, Argentina.

Bol. Obs. Ebro
Boletín del Observatorio del Ebro, Tortosa. Printed by Cooperativa Gráfica Dertosense, Tortosa.

Boll. Geod. Sci. Affini
Bolletino di Geodesia e Scienze Affini. Pubblicazione dell'Istituto Geografico Militare, Firenze.

Boll. Geofis. Teor. Appl.
Bollettino di Geofisica Teorica ed Applicada. Osservatorio Geofisico Sperimentale, 34123 Trieste, Italy.

Boundary-Layer Meteorol.
Boundary-Layer Meteorology. D. Reidel Publishing Co., P.O. Box 17, Dordrecht, Netherlands.

British Astron. Assoc. Circ.
British Astronomical Association, Circular. Editorial Office: S. W. Milbourn, Brookhill Road, Copthorne Bank, Crawley, West Sussex, RH10 3QJ.

British J. Philos. Sci.
British Journal for the Philosophy of Science. Cambridge University Press, Bentley House, 200 Euston Road, London, NW1 2DB, England.

British J. Photogr.
British Journal of Photography. Henry Greenwood & Co., 24 Wellington Street, London, WC2E 7DH, England.

Bul. Inst. Politeh. 'Gheorghe Gheorghiu-Dej' Bucuresti.
Buletinul Institutului Politehnic 'Gheorghe Gheorghiu-Dej' Bucuresti.Calea Grivitei 132, Bucuresti, Rumania. Journal split into three series, Bul. Inst. Politeh. 'Gheorghe Gheorghiu-Dej' Bucuresti Ser. Chim. – Metal., Ser. Electroteh. and Ser. Mec. (Rumania).

Bul. Inst. Politeh. Iasi I
Buletinul Institutului Politehnic din Iasi. Sectia I Matematica, Mecanica, Teoretica, Fizica. Polytechnic Institute, Iasi, Rumania.

Bulgarian J. Phys.
Bulgarian Journal of Physics. Bulgarian Academy of Sciences, Faculty of Physics, 5 Anton Ivanov Blvd., 1126 Sofia, Bulgaria.

Bull. Acad. Polonaise Sci., Ser. Sci. Tech.
Bulletin de l'Académie Polonaise des Sciences. Série des Sciences Techniques. 00-901 Warszawa, Palac Kultury i Nauki, P. O. Box 20, Poland.

Bull. AFOEV
Bulletin de l'Association Française des Observateurs d'Etoiles Variables. Rédaction et publication: E. Schweitzer, "La Moineaudière", 16, rue de Plobsheim, 67100 Strasbourg–Neudorf, France.

Bull. American Astron. Soc.
Bulletin of the American Astronomical Society. Published for the American Astronomical Society by the American Institute of Physics, 335 East 45th Street, New York, N.Y. 10017, USA.

Bull. American Meteorol. Soc.
Bulletin of the American Meteorological Society. 45 Beacon Street, Boston, MA 02108, USA.

Bull. Assoc. Suisse Electr.
Bulletin de l'Association Suisse des Electriciens (Organe Commun de l'Association Suisse des Electriciens (ASE) et de l'Union des Centrales Suisses d'Electricité (UCS)). Seefeldstrasse 301, Case Postale 229, 8008 Zürich, Switzerland.

Bull. Astron. Inst. Czechoslovakia
Bulletin of the Astronomical Institutes of Czechoslovakia. Published under the auspices of the Czechoslovak Academy of Sciences by Academia, Praha. Editor: Astronomical Institute of the Czechoslovak Academy of Sciences, Praha.

Bull. Astron., Obs. R. Belgique
Bulletin Astronomique, Observatoire Royal de Belgique. (Astronomisch Bulletin, Koninklijke Sterrenwacht van België).

Bull. Astron. Soc. India
Bulletin of the Astronomical Society of India. Edited and published by M. S. Vardya, Tata Institute of Fundamental Research, Bombay on behalf of the Astronomical Society of India, Osmania University, Hyderabad.

Bull. Earthquake Res. Inst. Univ. Tokyo
Bulletin of the Earthquake Research Institute, University of Tokyo. 1-1 Yayoi 1-chome, Bunkyo-ku, Tokyo, Japan.

Bull. Géod.
Bulletin Géodésique. The Journal of the International Association of Geodesy. Publié par le Bureau Central de l'Association Internationale de Géodésie, 39 Rue Gay-Lussac, 75005 Paris, France.

Bull. Geogr. Surv. Inst.
Bulletin of the Geographical Survey Institute. Published by the Geographical Survey Institute, Ministry of Construction, Tokyo, Japan.

Bull. Groupe Rech. Géod. Spat.
Groupe de Recherches de Géodésie Spatiale. Bulletin. CNES/Toulouse, France.

Bull. Inf. Cent. Données Stellaires
Bulletin d'Information du Centre de Données Stellaires. Compiled at Observatoire de Strasbourg, 11, rue de l'Université, 67000–Strasbourg, France.

Bull. Inst. Chem. Res. Kyoto Univ.
Bulletin of the Institute for Chemical Research, Kyoto University. Kyoto, Japan.

Bull. Inst. Space Aeronaut. Sci. Univ. Tokyo A
Bulletin of the Institute of Space and Aeronautical Science, University of Tokyo A. Tokyo, Japan.

Bull. Inst. Space Aeronaut. Sci., Univ. Tokyo, B
Bulletin of the Institute of Space and Aeronautical Science, University of Tokyo, B. Tokyo, Japan.

Bull. Minéral.
Bulletin de Minéralogie. Masson, 120 Boulevard Saint-Germain, 75280 Paris Cedex 06, France.

Bull. Obs. Astron. Belgrade
Bulletin de l'Observatoire Astronomique de Belgrade. Editor: Observatoire Astronomique de Belgrade. Printed by Naucna delo, Belgrade.

Bull. Res. Inst. Sci. Meas. Tôhoku Univ.
Bulletin of the Research Institute for Scientific Measurements, Tôhoku University. Sendai, Japan.

Bull. Sci. Yougoslavie
Bulletin Scientifique. Conseil des Academies des Sciences et des Arts de la RSF de Yougoslavie. Section A: Sciences Naturelles, Techniques et Médicales. Rédaction et Administration: Opatička ul. 18/II, Zagreb, Yougoslavie.

Bull. Seismol. Soc. America
Bulletin of the Seismological Society of America. Seismological Society of America, 2907 Claremont Avenue, Berkeley, CA 94705, USA.

Bull. Signal.
Bulletin Signalétique. Section 120: Astronomie, physique spatiale, geophysique. Centre Nationale de la Recherche Scientifique, Informascience, Centre de Documentation Scientifique et Technique, 26, rue Boyer, 75971 Paris. ISSN 0007-5337.

Bull. Soc. R. Sci. Liège
Bulletin de la Société Royale des Sciences de Liège. L'Université, 15 Avenue des Tilleurs, Liège, Belgium.

Bull. Tokyo Gakugei Univ., Ser. IV
Bulletin of Tokyo Gakugei University. Series IV (Mathematics and Natural Sciences) 4-1-1 Nukui-kita-machi, Koganei, Tokyo, Japan.

Bull. Yamagata Univ. (Nat. Sci.)
Bulletin of the Yamagata University (Natural Science). Yamagata, Japan.

Byull. Abastumanskaya Astrofiz. Obs.
Abastumanskaya Astrofizicheskaya Observatoriya, Gora Kanobili. Byulleten'. Akademiya Nauk Gruzinskoj SSR. Publishers: Izdatel'stvo "Metsniereba", Tbilisi.

Byull. Inst. Astrofiz.
Byulleten' Instituta Astrofiziki, Akademiya Nauk Tadzhikskoj SSR. Izdatel'stvo Donish, Dushanbe.

Byull. Inst. Teor. Astron.
Byulleten' Instituta Teoreticheskoj Astronomii. Izdatel'stvo Nauka, Leningradskoe Otdelenie, Leningrad.

C. R. Acad. Bulgare Sci.
Comptes Rendus de l'Académie Bulgare des Sciences. (Doklady Bolgarskoj Akademii Nauk). Sofiya, Bulgaria.

C. R. Acad. Sci. Paris
Comptes Rendus hebdomadaires des Séances de l'Académie des Sciences, publiés par MM. les Secrétaires Perpétuels. Imprimerie: Gauthier-Villars, Montreuil, France.

Canadian Aeronaut. Space J.
Canadian Aeronautics and Space Journal. Canadian Aeronautics and Space Institute, Saxe Building, 60 - 75 Sparks Street, Ottawa, Ontario K1P 5A5, Canada.

Canadian Electr. Eng. J.
Canadian Electrical Engineering Journal. Engineering Institute of Canada, Suite 700, EIC Building, 2050 Mansfield Street, Montreal, Canada.

Canadian J. Earth Sci.
Canadian Journal of Earth Sciences. National Research Council of Canada, Ottawa KIA OR6, Canada.

Canadian J. Phys.
Canadian Journal of Physics. Published by the National Research Council of Canada, Ottawa. Printed in Canada by the University of Toronto Press, Toronto, Ont.

Carter Obs., Astron. Bull.
Carter Observatory, Astronomical Bulletin. Carter Observatory, P.O. Box 2909, Wellington 1, New Zealand.

Celestial Mech.
Celestial Mechanics. An International Journal of Space Dynamics. Publishers: D. Reidel Publishing Company, Dordrecht, Holland.

Cent. Astrophys. Prepr. Ser.
Center for Astrophysics, Preprint Series. Harvard College Observatory, Smithsonian Astrophysical Observatory. Center for Astrophysics, 60 Garden St., Cambridge, Mass. 02138.

Centaurus
Centaurus. International magazine of the history of mathematics, science, and technology. Munksgaard, Copenhagen.

Ceskoslovensky Cas. Fyz., A.
Československý časopis pro fyziku. Sekce A. Academia Publishing House of the Czechoslovak Academy of Sciences, Vodičkova 40, 112 29 Praha 1, Czechoslovakia.

Chem. Phys. Lett.
Chemical Physics Letters. North-Holland Publishing Co., P.O. Box 211, Amsterdam-C, Netherlands.

Chinese Astron.
Chinese Astronomy. A cover-to-cover translation of Acta Astron. Sinica and Stud. Astron. Sinica. Published by Pergamon Press, Headington Hill Hall, Oxford, OX3 0BW, England – Maxwell House, Fairview Park, Elmsford, N.Y. 10523, USA.

Chinese J. Phys.
Chinese Journal of Physics. Physical Society of the Republic of China, Physics Department, National Taiwan University, Taipei, Taiwan, China.

Ciel
Le Ciel. Bulletin de la Société Astronomique de Liège. Éditeur responsable: Françoise Rameau, 19 - 21, rue des genêts, 4310 St. Nicolas.

Ciel Terre
Ciel et Terre. Bulletin de la Société Belge d'Astronomie, de Météorologie et de Physique du Globe. Administration: Avenue Circulaire, 3, Bruxelles. Printed by Imprimerie R. Louis, Bruxelles.

Circ. Czechoslovak Obs., Time and Latitude
Circular of the Czechoslovak Observatories, Time and Latitude. Czechoslovak Academy of Sciences, Astronomical Institute, Prague, Czechoslovakia.

Circ. Inf.
Circulaire d'Information. Union Astronomique Internationale. Commission des Etoiles Doubles. Address: Observatoire de Meudon, Meudon, France.

Circ. Stn. Astron. Int. Latitudine, Carloforte-Cagliari
Circolari della Stazione Astronomica Internazionale di Latitudine, Carloforte-Cagliari. Serie A printed by Tipo-Offset "3T", Cagliari. Serie B printed by Multi Copy, Milano.

Circ. Time and Latitude Serv.
Circular Time and Latitude Service. Polish Academy of Sciences, Astronomical Latitude Observatory, Borowiec, Poland.

Clim. Change
Climatic Change. D. Reidel Publishing Co., P.O. Box 17, 3300 AA Dordrecht, Netherlands.

Coelum
Coelum. Periodico bimestrale per la Divulgazione dell' Astronomia. Editor: Osservatorio Astronomico Universitario di Bologna.

Collect. Phenom.
Collective Phenomena. Gordon and Breach Science Publishers Ltd., 41 and 42 William IV Street, London, WC2, England.

Commentat. Phys.-Math.
Commentationes Physico-Mathematicae. Societas Scientiarum Fennica, Helsinki-Helsingfors. Printed by Keskuskirjapaino-Centraltryckeriet, Helsinki-Helsingfors.

Comments Astrophys.
Comments on Astrophysics. A Journal of Critical Discussion of the Current Literature. Comments on Modern Physics: Part C. Publishers: Gordon and Breach, Science Publishers Ltd., 42 William IV Street, London WC2, England.

Comments At. Mol. Phys.
Comments on Atomic and Molecular Physics. Gordon & Breach Science Publishers Ltd., 41 and 42 William IV Street, London, WC2, England.

Comments Nucl. Part. Phys.
Comments on Nuclear and Particle Physics. Gordon & Breach Science Publishers Ltd., 41 and 42 William IV Street, London, WC2, England.

Comments Plasma Phys. Controlled Fusion
Comments on Plasma Physics and Controlled Fusion. Gordon & Breach Science Publishers Ltd., 41 and 42 William IV Street, London, WC2, England.

Commun. Math. Phys.
Communications in Mathematical Physics, Springer-Verlag, Postfach 105280, 6900 Heidelberg 1, F.R. Germany.

Commun. Stat. Simulation Comput.
Communications in Statistics – Simulation and Computation. Marcel Dekker Inc., 270 Madison Avenue, New York, NY 10016, USA.

Comput. Geosci.
Computers & Geosciences. Pergamon Press Ltd., Headington Hill Hall, Oxford OX3 0BW, England.

Comput. Math. with Appl.
Computers & Mathematics with Applications. Pergamon Press Ltd., Headington Hill Hall, Oxford, OX3 0BW, England.

Comput. Phys. Commun.
Computer Physics Communications. North-Holland Publishing Co., P. O. Box 211 Amsterdam, Netherlands.

Comun. Obs. Astron. Univ. Coimbra
Comunicações do Observatório Astronomico da Universidade de Coimbra, Portugal.

Comunicaciones
Comunicaciones. Centro de Informacion de Comunicaciones, Ministerio de Comunicaciones, Habana, Cuba.

Contemp. Phys.
Contemporary Physics.Taylor and Francis Ltd., 10 - 14 Macklin Street, London, WC2B 5NF, England.

Contrib. Astron. Obs. Skalnaté Pleso.
Contributions of the Astronomical Observatory Skalnaté Pleso. VEDA, vydavatel'stvo Slovenskej akadémie vied, Bratislava, Czechoslovakia.

Contrib. Atmos. Phys.
Contributions to Atmospheric Physics – Beiträge zur Physik der Atmosphäre. Publisher: Friedrich Vieweg & Sohn, Braunschweig.

Contrib. Geophys. Inst. Slovak Acad. Sci. Ser. Meteorol.
Contributions of the Geophysical Institute of the Slovak Academy of Sciences, Series of Meteorology. Dubravska cesta, 899 30 Bratislava, Czechoslovakia.

Contrib. Obs. New Mexico State Univ.
Contributions of the Observatory of New Mexico State University. Published by the Astronomy Department, Box 4500, New Mexico State University, Las Cruces, New Mexico 88003.

COSPAR Inform. Bull.
COSPAR. Information Bulletin. Address: COSPAR Secretariat, Paris.

CQ Radio Amat. J.
CQ Radio Amateur's Journal. 14 Vanderventer Avenue, Port Washington, Long Island, NY 11050, USA.

Cryogenics
Cryogenics. IPC Science and Technology Press Ltd., IPC House 32 High Street, Guildford, Surrey GU1 3EW, England.

CSELT Rapp. Tec.
CSELT Rapporti Tecnici. CSELT - Centro Studi e Laboratori Telecomunicazioni, Torino, Via Guglielmo Reiss Romoli 274, Italy.

CSIO Commun.
CSIO Communications. Central Scientific Instruments Organisation, Sector 30, Chandigarh-160020, India.

Curr. Sci.
Current Science, Current Science Association, Raman Research Institute, Bangalore 6, India.

Czechoslovak J. Phys. B
Czechoslovak Journal of Physics, Section B. Czechoslovak Academy of Science, Akademia, Vodičkova 40, 112 29 Praha 1, Czechoslovakia.

Data Rep. Hydrogr. Obs., Ser. Astron. Geod., Tokyo
Data Report of Hydrographic Observations. Series of Astronomy and Geodesy. Maritime Safety Agency, Hydrographic Department Tsukiji-5, Chuo-ku, Tokyo, 104 Japan.

Def. Sci. J.
Defence Science Journal.Metcalf House, New Delhi 6, India.

Deutsche Geod. Komm. Bayerisch. Akad. Wiss.
Deutsche Geodätische Kommission bei der Bayerischen Akademie der Wissenschaften. Reihe A: Höhere Geodäsie; Reihe B: Angewandte Geodäsie; Reihe C: Dissertationen; Reihe D: Tafelwerke; Reihe E: Geschichte und Entwicklung der Geodäsie. Published by Verlag der Bayerischen Akademie der Wissenschaften, München.

Dimensions NBS.
Dimensions NBS.U.S. Department of Commerce, Washington, DC 20234, USA.

Dokl. Akad. Nauk SSSR
Doklady Akademii Nauk SSSR. Seriya Matematika, Fizika. Publishers: Izdatel'stvo "Nauka", Moskva.

Dudley Obs. Rep.
Dudley Observatory Reports. Dudley Observatory, Albany, N.Y., USA.

Dunsink Obs. Publ.
Dunsink Observatory Publications. The Observatory of the School of Cosmic Physics, Dublin Institute for Advanced Studies, Dublin.

Earth Extraterr. Sci.
Earth and Extraterrestrial Sciences. Gordon & Breach Science Publishers Ltd., 41 and 42 William IV Street, London, WC2, England.

Earth Planet. Sci. Lett.
Earth and Planetary Science Letters. A Letter Journal devoted to the Development in Time of the Earth and Planetary System. Publisher: North-Holland Publishing Company, Amsterdam, Netherlands.

El Universo
El Universo. Organo de la Sociedad Astronomica de Mexico, Mexico, D.F.

Electricidade
Electricidade. Empresa Editorial Electrotecnica Edel, Rua de Dona Estefania, 48, 3., Esq., Lisboa 1, Portugal.

Electron. Lett.
Electronics Letters. Institution of Electrical Engineers, Savoy Place, London, WC2R OBL, England.

Electron. Today Int.
Electronics Today International. Modmags Ltd., 25 - 27 Oxford Street, London, W1R 1RF, England.

Electronics
Electronics. McGraw-Hill Publishing Co., 1221 Avenue of the Americas, New York, N.Y. 10020, USA.

Elektron. Anz.
Elektronik Anzeiger, Verlag W.Girardet, 43 Essen 1, Girardetstrasse 2 - 38, Postfach 9, Germany.

Elektrotech. Z. B
Elektrotechnische Zeitschrift. Ausgabe B: Der Elektrotechniker. Verband Deutscher Elektrotechniker Publication address: VDE-Verlag, Bismarckstrasse 33, 1000 Berlin 12, F. R. Germany.

EOS Trans. American Geophys. Union
EOS Transactions of the American Geophysical Union. 1707 L Street, N.W., Washington, DC 20036, USA.

ESA Bull.
ESA Bulletin. Editorial Office: ESA Scientific and Technical Publications Branch, ESTEC, Noordwijk, The Netherlands. ISSN 0376-4265.

ESA IUE Newsl.
ESA IUE Newsletter. Published by The ESA IUE Observatory, Villafranca Satellite Tracking Station, Apartado 54065, Madrid, Spain.

ESA J.
ESA Journal. Editorial Office: ESA Scientific and Technical Publications Branch, ESTEC, Noordwijk, The Netherlands. ISSN 0379-2285.

ESO Sci. Prepr.
European Southern Observatory, Scientific Preprints. Available from Preprints Service, ESO-Library c/o CERN, 1211 Geneva 23, Switzerland.

ESO Tech. Rep.
European Southern Observatory, (ESO), Technical Report. Published by the European Southern Observatory Telescope Project Division, CERN, Geneva, Switzerland.

Exp. Tech. Phys.
Experimentelle Technik der Physik, VEB Deutscher Verlag der Wissenschaften, Traubenstrasse 10, 108 Berlin 8, German Democratic Republic.

Feingerätetechnik
Feingerätetechnik. VEB Verlag Technik, Oranienburger Strasse 13/14, 1020 Berlin, DDR.

Feinwerktech. Messtech.
F & M. Feinwerktechnik und Messtechnik. Fusion of "Feinwerktechnik" and "Messtechnik" (formerly Zeitschrift für Instrumentenkunde) beginning with Jahrgang 82, No. 5 (1974). Publishers: Karl Hanser Verlag, Kolbergerstr. 22, D-8000 München 80. F.R.Germany.

Fiz. Sz.
Fizikai Szemle. Kiadja a Lapkiado Vallalat, Budapest VII, Lenin korut 9–11, Hungary.

Fizika
Fizika. 'Mladost' Export-Import, Zagreb, Ilica 30, Yugoslavia.

Fortschr. Phys.
Fortschritte der Physik. Akademie-Verlag. DDR-108 Berlin, Leipzigerstrasse 3 - 4, Germany.

Found. Phys.
Foundations of Physics. Plenum Publishing Co., 8 Scrubs Lane, Harlesden, London, NW10 6SE, England.

Fra Fys. Verden
Fra Fysikkens Verden. Fysisk Institutt, Universitetet i Trondheim, Norges Laererhogskole, 7000 Trondheim, Norway.

Fudan J.
Fudan Journal. Fudan University, Shanghai, People's Republic of China.

Fundam. Cosmic Phys.
Fundamentals of Cosmic Physics. Gordon and Breach Science Publishers Ltd., New York–London–Paris.

Funkschau
Funkschau. Francis-Verlag, 8 München 37, Postfach 37 01 20, Karlstrasse 37, Germany.

Fys. Tidsskr.
Fysisk Tidsskrift. Subscription address: Jul. Gjellerups Boghandel, Solvgade 87, 1307 Kobenhavn, Denmark.

G. A.A.B.
Giornale dell'A.A.B. Notiziario trimestrale delle attività culturali e scientifiche della Associazione Astrofili Bolognesi, Bologna, Italy.

G. Astron.
Giornale di Astronomia, Pubblicazione della Società Astronomica Italiana. Printed by Tipolitografia Lodigraf S.p.A. Lodi (MI).

Gen. Relativ. Gravitation
General Relativity and Gravitation. Published under the auspices of the International Committee on General Relativity and Gravitation GRG. Publishing Office: Plenum Publishing Corporation, 227 West 17th Street, New York, N.Y. 10011, USA.

Geochim. Cosmochim. Acta
Geochimica et Cosmochimica Acta. Journal of the Geochemical Society. Publishing House: Pergamon Press, Ltd., Oxford.

Geod. Geophys. Veröff., Reihe III
Geodätische und Geophysikalische Veröffentlichungen. Reihe III: Physik der festen Erde. Herausgegeben vom Nationalkomitee für Geodäsie und Geophysik bei der Akademie der Wissenschaften der Deutschen Demokratischen Republik.

Geod. Kartogr.
Geodezja i Kartografia. Komitet Geodezji Polskiej Akademii Nauk. Publisher: Państwowe Wydawnictwo Naukowe, Warszawa.

Geoexploration
Geoexploration. Elsevier Publishing Co., P.O. Box 211, Amsterdam, Netherlands.

Geomagn. Aehron.
Geomagnetizm i Aehronomiya. Akademiya Nauk SSSR. Izdatel'stvo "Nauka", Moskva [An English translation is published in "Geomagnetism and Aeronomy", American Geophysical Union, Washington, D.C.].

Geomagn. Ser. Earth Phys. Branch
Geomagnetic Series, Earth Physics Branch. Energy, Mines and Resources Canada, 1 Observatory Crescent, Ottawa K1A OE4, Canada.

Geophys. Astrophys. Fluid Dyn.
Geophysical and Astrophysical Fluid Dynamics. Gordon and Breach Science Publishers Ltd., 41/42 William IV Street, London, WC2, England.

Geophys. J. R. Astron. Soc.
The Geophysical Journal of the Royal Astronomical Society. Published for the Royal Astronomical Society by Blackwell Scientific Publications, Oxford–Edinburgh. American Office of the Geophys. J., US Geological Survey, Stop 967, Box 25046, Federal Center, Denver, Colorado 80225, USA.

Geophys. Res. Lett.
Geophysical Research Letters. Published monthly by the American Geophysical Union, Washington, D.C., U.S.A.

Geophys. Surv.
Geophysical Surveys. D. Reidel Publishing Co., P.O. Box 17, Dordrecht, Netherlands.

Geophysics
Geophysics. Society of Exploration Geophysicists, P.O. Box 3098, Tulsa, OK 74101, USA.

GEOS
GEOS. Department of Energy, Mines and Resources, Ottawa, Canada.

GEOS Circ.
GEOS (Groupe: Etude et Observation Stellaire and Gruppo Europeo di Osservazione Stellare) Circulars, Series: RR (RR Lyrae type variables), SR (red variables), EB (eclipsing binaries). Published by A. Figer, GEOS, 12 rue Bezout, 75014 Paris, France.

Gerlands Beitr. Geophys.
Gerlands Beiträge zur Geophysik. Publisher: Akademische Verlagsgesellschaft Geest & Portig K.-G., Leipzig.

Glasnik Mat.
Glasnik Matematicki. Published by the Society of Mathematicians and Physicists of the S. R. of Croatia. Publisher: Drustvo Matematicara i Fizicara S. R. Hrvatske, Zagreb.

GSFC Doc.
Goddard Space Flight Center, Greenbelt, Maryland. Available from Technical Information Division, Code 250, Goddard Space Flight Center, Greenbelt, Maryland 20771.

Hadronic J.
Hadronic Journal. Hadronic Press Inc., Nonantum, MA 02195, USA.

Heavens
The Heavens. The Oriental Astronomical Association, Otsu-shi, Shiga-ken, Japan. In Japanese.

Helvetica Phys. Acta
Helvetica Physica Acta. Schweizerische Physikalische Gesellschaft. Publisher: E. Birkhäuser, Elisabethenstr. 19, CH-4000 Basel 10, Switzerland.

HHI Sol. Data
HHI Solar Data. Heinrich-Hertz-Institut, Solare Beobachtungsergebnisse. Akademie der Wissenschaften der DDR, Zentralinstitut für Solar-Terrestrische Physik (Heinrich-Hertz-Institut), DDR-1199 Berlin-Adlershof.

Hvar Obs. Bull.
Hvar Observatory Bulletin. Faculty of Geodesy. 41000 Zagreb, Kačićeva 26, Yugoslavia.

I.U.A.A. Bull.
I.U.A.A. Bulletin. International Union of Amateur Astronomers, Contributions. I.U.A.A. c/o Achille Leani, via Bertesi 15, 26100 Cremona, Italy.

IAU Circ.
International Astronomical Union, Circular. Central Bureau for Astronomical Telegrams, Smithsonian Astrophysical Observatory, Cambridge, Mass.

IBM Tech. Disclosure Bull.
IBM Technical Disclosure Bulletin. International Business Machines Co., Armonk, New York, NY 10504, USA.

Icarus
Icarus. International Journal of Solar System Studies. Publisher: Academic Press, New York – London.

ICSU Bull.
ICSU Bulletin. International Council of Scientific Unions. Secretariat: 51, Bd de Montmorency, Paris, France.

IEE J. Microwave Opt. Acoust.
IEE Journal on Microwave, Optics and Acoustics. Institution of Electrical Engineers, Publishing Department, P.O. Box 8, Southgate House, Stevenage, Herts. SG1 1HQ, England.

IEEE Spectrum
IEEE Spectrum. Published monthly by the Institute of Electrical and Electronics Engineers, 345 East 47th Street, New York, N.Y. 10017, USA.

IEEE Trans. Aerosp. Electron. Syst.
IEEE Transactions on Aerospace and Electronic Systems. Published by the Institute of Electrical and Electronics Engineers, 345 East 47th Street, New York, N.Y. 10017, USA.

IEEE Trans. Antennas Propag.
IEEE Transactions on Antennas and Propagation. Published by the Institute of Electrical and Electronics Engineers, 345 East 47th Street, New York, N.Y. 10017, USA.

IEEE Trans. Commun.
IEEE Transactions on Communications. Institute of Electrical and Electronics Engineers, 345 East 47th Street, New York, NY 10017, USA.

IEEE Trans. Consum. Electron.
IEEE Transactions on Consumer Electronics. Institute of Electrical and Electronics Engineers, 345 East 47th Street, New York, N. Y. 10017, USA.

IEEE Trans. Electromagn. Compat.
IEEE Transactions on Electromagnetic Compatibility. Institute of Electrical and Electronics Engineers, 345 East 47th Street, New York, NY 10017, USA.

IEEE Trans. Electron Devices
IEEE Transactions on Electron Devices. Published by the Institute of Electrical and Electronics Engineers, 345 East 47th Street, New York, N. Y. 10017, USA.

IEEE Trans. Geosci. Electron.
IEEE Transactions on Geoscience Electronics. Published by the Institute of Electrical and Electronics Engineers, 345 East 47th Street, New York, N.Y. 10017, USA.

IEEE Trans. Instrum. Meas.
IEEE Transactions on Instrumentation and Measurement. Published by the Institute of Electrical and Electronics Engineers, 345 East 47th Street, New York, N.Y. 10017, USA.

IEEE Trans. Magn.
IEEE Transactions on Magnetics. Institute of Electrical and Electronics Engineers, 345 East 47th Street, New York, N. Y. 10017, USA.

IEEE Trans. Microwave Theory Tech.
IEEE Transactions on Microwave Theory and Techniques. Published by the Institute of Electrical and Electronics Engineers, 345 East 47th Street, New York, N.Y. 10017, USA.

IEEE Trans. Nucl. Sci.
IEEE Transactions on Nuclear Science. Institute of Electrical and Electronics Engineers, 345 East 47th Street, New York, N.Y. 10017, USA.

IEEE Trans. Plasma Sci.
IEEE Transactions on Plasma Science. Institute of

Electrical and Electronics Engineers, 345 East 47th Street, New York, NY 10017, USA.

Ind. At. Spat.
Industries Atomiques & Spatiales. SERTNA S. A., 29 avenue R. Poincaré, Paris 16, France.

Ind. Math.
Industrial Mathematics. Industrial Mathematics Society, P.O. Box 159, Roseville, MI 48066, USA.

Ind. Res./Dev.
Industrial Research/Development. Technical Publishing Co., 222 S. Riverside Plaza, Chicago, IL 60606, USA.

Indian East. Eng.
Indian and Eastern Engineer. 'Piramal Mansion', 235 Dr. D. Naoroji Road, Bombay 400001, India.

Indian J. Hist. Sci.
Indian Journal of History of Science. Published and printed by Indian National Science Academy, Bahadur Shah Zafar Marg, New Delhi 110002, at Mudranika, 13-A Bepin Pal Road, Calcutta 700026.

Indian J. Meteorol. Hydrol. Geophys.
Indian Journal of Meteorology, Hydrology & Geophysics. Formerly: Indian J. Meteorol. Geophys. Indian Meteorological Department, Civil Lines, Delhi 110006, India.

Indian J. Phys. Part B
Indian Journal of Physics Part B. Indian Association for the Cultivation of Science, 2 & 3 Raja Subodh Chandra Mallik Road, Calcutta 700032, India.

Indian J. Pure Appl. Math.
Indian Journal of Pure and Applied Mathematics. National Institute of Sciences India, Bahadur Shah Zafar Marg, New Delhi 1, India.

Indian J. Pure Appl. Phys.
Indian Journal of Pure and Applied Physics. Council of Scientific and Industrial Research, Hillside Road, New Delhi 110012, India.

Indian J. Radio Space Phys.
Indian Journal of Radio & Space Physics. Council of Scientific & Industrial Research. Editorial address: Publications & Information Directorate, Hillside Road, New Delhi 110012, India.

Inf. Bull. South. Hemisphere
Information Bulletin for the Southern Hemisphere. Editorial Office: Observatorio Astronómico, La Plata, Argentina.

Inf. Bull. Variable Stars
Commission 27 of the I.A.U. Information Bulletin on Variable Stars. Konkoly Observatory, Budapest.

Informeto Astron. Obs. Univ. Turku
Informeto Astronomia Observatorio Universitato de Turku, Finnlando.

Infrared Phys.
Infrared Physics. An International Research Journal. Publisher: Pergamon Press Ltd., Oxford, England.

Ingenieur
De Ingenieur. Koninklijk Institut van Ingenieurs. Editorial address; 23 Prinsessegracht, Den Haag, Netherlands.

Inst. Theor. Astrophys., Blindern–Oslo, Rep.
Institute of Theoretical Astrophysics, Blindern–Oslo, Report. Universitetsforlagets trykningssentral, Oslo.

Int. At. Energy Agency Bull.
International Atomic Energy Agency Bulletin. Kärntnerring 11, P.O. Box. 590, A-1011 Wien, Austria.

Int. Comet Q.
The International Comet Quarterly, Physics Department, Appalachian State University, Boone, NC 28608.

Int. J. Electron.
International Journal of Electronics. Taylor and Francis Ltd., 10–14 Macklin Street, London, WC2B 5BF, England.

Int. J. Eng. Sci.
International Journal of Engineering Science. Pergamon Press Ltd., Headington Hill Hall, Oxford, OX3 0BW, England.

Int. J. Heat Mass Transfer.
International Journal of Heat and Mass Transfer. Pergamon Press Ltd., Headington Hill Hall, Oxford, OX3 OBW, England.

Int. J. Mass Spectrom. Ion Phys.
International Journal of Mass Spectrometry and Ion Physics. Elsevier Scientific Publishing Co., P.O. Box 211 Amsterdam, Netherlands.

Int. J. Theor. Phys.
International Journal of Theoretical Physics. Plenum Publishing Co. Ltd., Davis House, 8 Scrubs Lane, London, NW10 6SE, England.

Interavia
Interavia (English Edition). Interavia S.A., 86 Avenue Louis-Casai, P.O. Box 162, 1216 Cointrin, Geneva, Switzerland.

Interdisciplinary Sci. Rev.
Interdisciplinary Science Reviews. Heyden & Son Ltd., Spectrum House, Alderton Crescent, London NW4 3XX, England.

Irish Astron. J.
The Irish Astronomical Journal. A Quarterly Publication under the auspices of the Observatories of Armagh and Dunsink. Armagh Observatory, Northern Ireland.

ISIS
ISIS. An international review devoted to the history of science and its cultural influences. Publication and Editorial Office, Department of History and Sociology of Science, University of Pennsylvania, Philadelphia 19104.

Israel J. Technol.
Israel Journal of Technology. Weizmann Science Press, P.O. Box 801, Jerusalem 91000, Israel.

Izv. Akad. Nauk Armyan. SSR
Izvestiya Akademii Nauk Armyanskoj SSR. Fizika. Publisher: Izdatel'stvo AN Armyanskoj SSR, Erevan.

Izv. Astron. Ehngel'gardt. Obs.
Izvestiya Astronomicheskoj Ehngel'gardtovskoj Observatorii. Izdatel'stvo Kazanskogo Universiteta, Kazan.

Izv. Glav. Astron. Obs. Pulkovo
Izvestiya Glavnoj Astronomicheskoj Observatorii v Pulkove. Akademiya Nauk SSSR. Izdanie Glavnoj astronomicheskoj observatorii v Pulkove, Leningrad.

Izv. Krymskoj Astrofiz. Obs.
Izvestiya Krymskoj Astrofizicheskoj Observatorii. Akademiya Nauk SSR. Publishers: Izdatel'stvo "Nauka", Moskva.

J. American Assoc. Variable Star Obs.
The Journal of the American Association of Variable Star Observers. Published by The American Association of Variable Star Observers, 187 Concord Avenue, Cambridge, Mass. 02138, USA.

J. Appl. Meteorol.
Journal of Applied Meteorology. American Meteorological Society, 45 Beacon Street, Boston, MA 02108, USA.

J. Appl. Photogr. Eng.
Journal of Applied Photographic Engineering. Society of Photographic Scientists and Engineers, Suite 204, 1330 Massachusetts Avenue, N.W. Washington, D.C. 20005, USA.

J. Appl. Phys.
Journal of Applied Physics. American Institute of Physics, 335 East 45th Street, New York, NY 10017, USA.

J. Astron. Soc. Egypt
Journal of the Astronomical Society of Egypt. Published by Helwan Observatory, Helwan, Egypt.

J. Astron. Soc. Western Australia
The Journal of the Astronomical Society of Western Australia. Edited by the Astronomical Society of Western Australia, Perth, W. A.

J. Astronaut. Sci.
Journal of the Astronautical Sciences. American Astronautical Society, 6060 Duke Street, Alexandria, VA 22304, USA.

J. Astrophys. Astron.
Journal of Astrophysics and Astronomy. Published by Indian Academy of Sciences, Post Box No. 8005, Bangalore 560080, India.

J. Atmos. Sci.
Journal of the Atmospheric Sciences. American Meteorological Society, 45 Beacon Street, Boston, MA 02108, USA.

J. Atmos. Terr. Phys.
Journal of Atmospheric and Terrestrial Physics. Pergamon Press Ltd., Oxford, England.

J. Australian Math. Soc., Ser. B
Journal of the Australian Mathematical Society, Series B (Applied Mathematics). Editorial address: Department of Mathematics, University of Western Australia, Nedlands, Western Australia 6009, Australia.

J. British Astron. Assoc.
Journal of the British Astronomical Association. Burlington House, Piccadilly, London, W1V ONL, England.

J. British Interplanet. Soc.
Journal of the British Interplanetary Society. British Interplanetary Society, 12 Bessborough Gardens, London, SW1V 2JJ, England.

J. Chem. Educ.
Journal of Chemical Education. Publications Office, 119 West 24th Street, New York, NY 10011, USA.

J. Colloid Interface Sci.
Journal of Colloid and Interface Science. Academic Press Inc., 111 Fifth Avenue, New York, N.Y. 10003, USA.

J. Comput. Phys.
Journal of Computational Physics. Academic Press Inc., 111 Fifth Avenue, New York, NY 10003, USA.

J. Fluid Mech.
Journal of Fluid Mechanics. Cambridge University Press, Bentley House, 200 Euston Road, London, NW1 2DB, England.

J. Funct. Anal.
Journal of Functional Analysis. Academic Press Inc., 111 Fifth Avenue, New York, NY 10003, USA.

J. Geomagn. Geoelectr.
Journal of Geomagnetism and Geoelectricity. Society of Terrestrial Magnetism and Electricity of Japan, Geophysical Institute, Tokyo University, Tokyo 113, Japan.

J. Geophys.
Journal of Geophysics / Zeitschrift für Geophysik. Springer Verlag, D-6900 Heidelberg 1, Postfach 105280, F. R. Germany.

J. Geophys. Res.
Journal of Geophysical Research. Published by American Geophysical Union, 1909K Street, N.W. Washington D.C. First section: Space physics; Second section: Physics and chemistry of the solid earth, planetology, geodesy; Third section: Oceans and atmospheres.

J. Guid. Control
Journal of Guidance and Control. American Institute of Aeronautics and Astronautics, 1290 Avenue of the Americas, New York, NY 10019, USA.

J. Hist. Arabic Sci.
Journal for the History of Arabic Science. Institute for the History of Arabic Science, University of Aleppo, Aleppo, Syria.

J. Hist. Astron.
Journal for the History of Astronomy. Published by Science History Publications Ltd., Halfpenny Furze, Mill Lane, Chalfont St Giles, Buckinghamshire, England.

J. Illum. Eng. Inst. Japan
Journal of the Illuminating Engineering Institute of Japan. 3 Yurakucho, 1-chome, Chiyodaku, Tokyo 100, Japan.

J. Illum. Eng. Soc.
Journal of the Illuminating Engineering Society. 345 East 47th Street, New York, NY 10017, USA.

J. Indian Inst. Sci.
Journal of the Indian Institute of Science. Bangalore 560012, India.

J. Inst. Elektron. Commun. Eng. Japan
Journal of the Institute of Electronics and Communication Engineers of Japan. Denshi Tsushin Gakkai,

Kikai-Shinko-Kaikan, 5 - 8, Shibakoen 3 Chome Minato-ku, Tokyo 105, Japan.

J. Inst. Math. Appl.
Journal of the Institute of Mathematics and its Applications. Academic Press Inc. (London) Ltd., 24 - 28 Oval Road, London NW1 7DX, England.

J. Inst. Telev. Eng. Japan
Journal of the Institute of Television Engineers of Japan. Kikai-Shinko Building, 3-5-8 Shiba Park, Minato-ku, Shiba PO, Tokyo, Japan.

J. Instn. Electron. Telecommun. Eng.
Journal of the Institution of Electronics and Telecommunication Engineers. 2 Lodi Road, Institutional Area, New Delhi 110003, India.

J. Low Temp. Phys.
Journal of Low Temperature Physics. Plenum Publishing Corp., 227 West 17th Street, New York, NY 10011, USA.

J. Magn. Magn. Mater.
Journal of Magnetism and Magnetic Materials. North-Holland Publishing Co., P.O. Box 211, Amsterdam, Netherlands.

J. Math. Phys.
Journal of Mathematical Physics. American Institute of Physics, 335 East 45th Street, New York, N.Y. 10017, USA.

J. Math. Phys. Sci.
Journal of Mathematical and Physical Sciences. Indian Institute of Technology, Madras 600036, India.

J. Mech. Eng. Lab.
Journal of Mechanical Engineering Laboratory. Agency of Industrial Science and Technology, Igusa Suginami-ku, Tokyo, Japan.

J. Mol. Spectrosc.
Journal of Molecular Spectroscopy. Academic Press Inc., 111 Fifth Avenue, New York, NY 10003, USA.

J. Nanjing Univ.
Journal of Nanjing University. Nanjing Daxue Xuebao. (Natural Science Edition). Nanking University, Nanking, China.

J. Navig.
The Journal of Navigation. The Royal Institute of Navigation at the Royal Geographical Society, Kensington Gore, London, SW7 2AT. Scottish Academic Press Ltd., 33 Montgomery Street, Edinburgh EH7 5JX.

J. Opt. *(France)*
Journal of Optics. Formerly: Nuov. Rev. Opt. Masson Editeur, 120 Boulevard Saint-Germain, 75280 Paris Cedex 06, France.

J. Opt. *(India)*
Journal of Optics. Optical Society of India, Department of Applied Physics, University of Calcutta, 92 Acharya Prafulla Chandra Road, Calcutta-9, India.

J. Opt. Soc. America
Journal of the Optical Society of America. American Institute of Physics, 335 East 45th Street, New York, N.Y. 10017, USA.

J. Photogr. Sci.
Journal of Photographic Science. Royal Photographic Society, 14 South Audley Street, London, W1Y 5DP, England.

J. Phys. A
Journal of Physics A, (Mathematical, Nuclear and General). Institute of Physics, 47 Belgrave Square, London, SW1X 8QX, England.

J. Phys. B
Journal of Physics B, (Atomic and Molecular Physics). Institute of Physics, 47 Belgrave Square, London, SW1X 8QX, England.

J. Phys. Chem. Ref. Data
Journal of Physical and Chemical Reference Data. American Chemical Society, 1155 Sixteenth Street, N.W., Washington, DC 20036, USA.

J. Phys. Colloq.
Journal de Physique Colloque. Société Française de Physique, 87 bis Avenue du Général Leclerc, 75014 Paris, France.

J. Phys. E
Journal of Physics E, (Scientific Instruments). Formerly: J. Sci. Instrum. (GB). Institute of Physics, 47 Belgrave Square, London, SW1X 8QX, England.

J. Phys. F
Journal of Physics F, (Metal Physics). Institute of Physics, 47 Belgrave Square, London, SW1X 8QX, England.

J. Phys. G
Journal of Physics G, (Nuclear Physics). Institute of Physics, 47 Belgrave Square, London, SW1X 8QX, England.

J. Phys. Soc. Japan
Journal of the Physical Society of Japan. Room 211, Kikai Shinko Building, Shiba Koen, Minato-ku, Tokyo 105, Japan.

J. Physique
Journal de Physique. Z. I. de Courtaboeuf, B. P. 112, 91402 Orsay, France.

J. Plasma Phys.
Journal of Plasma Physics. Cambridge University Press, Bentley House, 200 Euston Road, London, NW1 2DB, England.

J. Proc. R. Soc. New South Wales
Journal and Proceedings of the Royal Society of New South Wales. Science Centre, 35 Clarence Street, Sydney, N.S.W. 2000, Australia

J. Quant. Spectrosc. Radiat. Transfer
Journal of Quantitative Spectroscopy & Radiative Transfer. Pergamon Press Ltd., Headington Hill Hall, Oxford, OX3 OBW, England.

J. R. Astron. Soc. Canada
The Journal of the Royal Astronomical Society of Canada, devoted to the advancement of astronomy and allied sciences. The Royal Astronomical Society of Canada, 124 Merten Street, Toronto, Ontario, Canada.

J. Radio Res. Lab.
Journal of the Radio Research Laboratories. Chief Planning Section, Radio Research Laboratories, Ministry of

Posts & Telecommunications, Nukui-Kitamachi, Koganei-shi, Tokyo 184, Japan.

J. Radioanal. Chem.
Journal of Radioanalytical Chemistry. Elsevier Sequoia S.A., P.O. Box 851, 1001 Lausanne 1, Switzerland. Kultura, H-1389 Budapest, 62, P.O. Box 149, Hungary.

J. Res. Natl. Bur. Stand. B
Journal of Research of the National Bureau of Standards. Section B (Mathematics and Mathematical Physics). US Government Printing Office, Division of Public Documents, Washington, DC 20402, USA.

J. Sci. Ind. Res.
Journal of Scientific and Industrial Research. Sales & Distribution Office, Publications & Information Directorate, Hillside Road, New Delhi 110012, India.

J. Sci. Res. Banaras Hindu Univ.
Journal of Scientific Research of the Banaras Hindu University. PO-Banaras Hindu University, India.

J. Spacecr. Rockets
Journal of Spacecraft and Rockets. American Institute of Aeronautics and Astronautics, 1290 Avenue of the Americas, New York, N. Y. 10019, USA.

J. Spectrosc. Soc. Japan
Journal of the Spectroscopical Society of Japan. (Bunkyo Kenkyu). 2-15-1, Nakai, Shinjuku-ku, Tokyo 161, Japan.

J. Stat. Phys.
Journal of Statistical Physics. Plenum Publishing Corp., 227 West 17th Street, New York, N. Y. 10011, USA.

Japanese J. Appl. Phys.
Japanese Journal of Applied Physics. Publication Office, 2nd Toya Kaiji Building, 24-8 Shinbashi, Minato-ku, Tokyo 105, Japan.

Jenaer Rundsch. (Jena Rev.)
Jenaer Rundschau (Jena Review). Publisher: VEB Verlag Technik, Berlin, German Democratic Republic.

JETP Lett.
JETP Letters. A translation of JETP Pis'ma v Redaktsiyu of the Academy of Sciences in the USSR. American Institute of Physics, 335 East 45th Street, New York, NY 10017, USA.

JPL Tech. Memo.
Jet Propulsion Laboratory, California Institute of Technology, Pasadena, California. National Aeronautics and Space Administration. Technical Memorandum.

JPL Tech. Rep.
Jet Propulsion Laboratory, California Institute of Technology, Pasadena, California. National Aeronautics and Space Administration. Technical Report.

Kexue Tongbao
Kexue Tongbao. Academia Sinica, Peking, People's Republic of China [English translation in: Kexue Tongbao (Scientia)(USA)].

Kodaikanal Obs. Bull.
Kodaikanal Observatory Bulletins, Series A, Indian Institute of Astrophysics, Bangalore, India.

Komet. Tsirk.
Kometnyj Tsirkulyar. Gruppa po Issledovaniyu Komet Astrosoveta i Mezhduvedomstvennyj Geofizicheskij Komitet Akademii Nauk SSSR. Kievskij Universitet im. T. G. Shevchenko.

Komety i Meteory
Komety i Meteory. Akademiya Nauk Tadzhikskoj SSR. Astronomicheskij Sovet Akademii Nauk SSSR. Publishers: Izdatel'stvo "Donish", Dushanbe.

Kosm. Issled.
Kosmicheskie Issledovaniya. Akademiya Nauk SSSR. Publishers: Izdatel'stvo "Nauka", Moskva [An English translation is published as "Cosmic Research", Consultants Bureau, New York, N. Y.].

Kozmos
Kozmos. Popular Astronomical Journal of the Slovak Central Observatory in Hurbanovo. Publisher: Slovenská ústredná hvezdáren v Hurbanove.

Lett. Math. Phys.
Letters in Mathematical Physics. D. Reidel Publishing Co., P.O. Box 17, Dordrecht, Netherlands.

L'Universo
L'Universo. Rivista dell'Istituto Geografico Militare. Direzione, Redazione e Amministrazione: Istituto Geografico Militare, Firenze.

Magn. Polya Soln. Pyaten
Magnitnye Polya Solnechnykh Pyaten. (Supplements to Solnechnye Dannye. Byulleten' (*Solar Data*). Publishers: Izdatel'stvo "Nauka", Leningrad.

Manuscr. Geod.
Manuscripta Geodetica. H. Kremers Verlag, Postfach 31 0801, D-1000 Berlin (West) 31. ISSN 0340-8825.

Marconi Rev.
Marconi Review. Marconi Co., Marconi House, Chelmsford, Essex, England.

Mat.-Fys. Medd. K. Danske Vidensk. Selsk.
Matematisk-Fysiske Meddelelser Konglige Danske Videnskabernes Selskab. Dantes Plads 5, DK-1556 Copenhagen V. Publisher: Munksgaard Ltd., 6 Norregade, DK-1165 Copenhagen K, Denmark.

Math. Intelligencer
The Mathematical Intelligencer. Springer-Verlag, Berlin–Heidelberg–New York, 175 Fifth Avenue, New York, NY 10010, USA.

Math. Proc. Cambridge Philos. Soc.
Mathematical Proceedings of the Cambridge Philosophical Society. Formerly: Proceedings of the Cambridge, Philosophical Society (Mathematical and Physical Sciences). Cambridge University Press, Bentley House, 200 Euston Road, London, NW1 2DB, England.

Meccanica
Meccanica, Tamburini Editore, s.p.a., Via G. Pascoli, 55-Milano, Italy.

Mech. Res. Commun.
Mechanics Research Communications. Pergamon Press Ltd., Headington Hill Hall, Oxford, OX3 OBW, England.

Meded. K. Acad. Wet. Lett. Schone Kunsten Belgie
Mededelingen van de Koninklijke Academie voor

Wetenschappen, Letteren en Schone Kunsten van Belgie. Paleis der Academien, Hertogsstraat 1, Bruxelles, Belgium.

Mem. Astron. Soc. India
Memoirs of the Astronomical Society of India. Edited and published by M.S. Vardya, Tata Institute of Fundamental Research, Bombay 400005 on behalf of the Astronomical Society of India, Osmania University, Hyderabad 500007.

Mem. Fac. Eng. Kyoto Univ.
Memoirs of the Faculty of Engineering, Kyoto University, Kyoto, Japan.

Mem. Fac. Eng. Osaka City Univ.
Memoirs of the Faculty of Engineering, Osaka City University. 459 Sugimoto-cho, Sumi Yoshi-kum, Osaka, Japan.

Mem. Fac. Sci. Kyoto Univ.
Memoirs of the Faculty of Science, Kyoto University. Series of Physics, Astrophysics, Geophysics, and Chemistry. Printed by Yamashiro Printing Publishing Co. Ltd., Kamigyo, Kyoto.

Mem. Japan Astron. Study Assoc.
Memoirs of the Japan Astronomical Study Association. c/o National Science Museum, Ueno Park, Taito-ku, Tokyo, Japan.

Mém. Sci. Rev. Metall.
Mémoires Scientifiques de la Revue de Metallurgie. 5 Rue Paul Cézanne, 75008 Paris, France.

Mem. Soc. Astron. Italiana
Memorie della Società Astronomica Italiana. Presso Laboratorio di Astrofisica Spaziale, Castella Postale 67, 00044 Frascati, Italy.

Mercury
Mercury. The Journal of the Astronomical Society of the Pacific. Published by the Astronomical Society of the Pacific, 1290 24th Avenue, San Francisco, California 94122, USA. (415) 661 - 8660.

Messenger
The Messenger, El Mensajero. Edited by European Southern Observatory, Schleißheimer Straße 17, D-8046 Garching bei München, F. R. Germany.

Meteor Sect. Rep.
Netherlands Association for Astronomy and Meteorology. Meteor Section Report. Meteor Section NVWS, De Sitterlaan 37, 2313 TK Leiden, Netherlands.

Meteoritics
Meteoritics. The Journal of the Meteoritical Society. Published quarterly by The Meteoritical Society and Arizona State University Bureau of Publications. Editorial address: Center for Meteorite Studies, The Arizona State University, Tempe, Ariz. 85281, USA.

Meteoritika
Akademiya Nauk SSSR. Komitet po Meteoritam. Publishers: Izdatel'stvo "Nauka", Moskva.

Meteorol. Mag.
Meteorological Magazine. Director General Meteorological Office, London Road, Bracknell, Berks RG12 2SZ, England.

Meteorol. Rundsch.
Meteorologische Rundschau. Springer-Verlag, D-1000 Berlin 33, Heidelberger Platz 3, Germany.

Metrologia
Metrologia. Springer-Verlag, Heidelberger Platz 3, D-1000 Berlin 33, F. R. Germany.

Microwave J.
Microwave Journal. To be obtained from 610 Washington Street, Dedham Plaza, Dedham, Massachusetts, U.S.A.

Microwave Syst. News
Microwave Systems News. 3975 East Bayshore Road, Palo Alto, CA 94303, USA.

Microwaves
Microwaves. Hayden Publishing Co., 50 Essex Street, Rochelle Park, NJ 07662, USA.

Minor Planet Bull.
The Minor Planet Bulletin. Bulletin of the Minor Planets Section of the Association of Lunar and Planetary Observers. Editorial Office: R. G. Hodgson, Dordt College, Sioux Center, Iowa, U.S.A.

Minor Planet Circ., (M. P. C.)
The Minor Planet Circulars/Minor Planets and Comets. Edited under the supervision of B. G. Marsden. Published by Minor Planet Center, Smithsonian Astrophysical Observatory, Cambridge, Mass. 02138, USA.

Mitt. Astron. Ges.
Mitteilungen der Astronomischen Gesellschaft, Hamburg. Available from Astron. Instit. Univ. Bochum, Postfach 10 21 48, D-4630 Bochum. ISSN 0172–5483.

Mitt. Inst. Theor. Geod. Univ. Bonn
Mitteilungen aus dem Institut für Theoretische Geodäsie der Universität Bonn, Nußallee 17, 5300 Bonn 1, F. R. Germany.

Mitt. Karl-Schwarzschild-Obs. Tautenburg
Mitteilungen des Karl-Schwarzschild-Observatoriums Tautenburg der Deutschen Akademie der Wissenschaften der DDR. Zentralinstitut für Astrophysik.

Mitt. Satell.-Beobachtungsstn. Zimmerwald
Mitteilungen der Satelliten-Beobachtungsstation Zimmerwald. Hausdruckerei Institut für Exakte Wissenschaften, Bern, Switzerland.

Mitt. Sternw. Ungar. Akad. Wiss.
Mitteilungen der Sternwarte der Ungarischen Akademie der Wissenschaften. A Magyar Tudományos Akadémia Csillagvizsgáló Intézetének Közleményei, Budapest–Szabadsághegy, HU ISSN 0324-2234.

Mitt. Veränderl. Sterne (MVS)
Mitteilungen über Veränderliche Sterne. Herausgegeben von der Sternwarte Sonneberg der Akademie der Wissenschaften der DDR, Sonneberg, German Democratic Republic.

Mod. Geol.
Modern Geology. Gordon & Breach Science Publishers Ltd., 41 and 42 William IV Street, London WC2, England.

Mon. Not. R. Astron. Soc.
Monthly Notices of the Royal Astronomical Society. Published for the Royal Astronomical Society by Black-

well Scientific Publications, Oxford – London – Edinburgh – Melbourne.

Mon. Notes Astron. Soc. South. Africa
Monthly Notes of the Astronomical Society of Southern Africa. Published by the Astronomical Society of Southern Africa, S. A. Astronomical Observatory, Cape Province, South Africa.

Mon. Notes Int. Polar Motion Serv.
Monthly Notes of the International Polar Motion Service. Published by the Central Bureau, International Latitude Observatory of Mizusawa, Mizusawa-shi, Iwate-ken, Japan.

Monitor
Monitor. Formerly: Proc. Instn. Radio Electron. Eng. Australia. Institution of Radio and Electronics Engineers Australia. Science House, 157 Gloucester Street, Sydney, N.S.W. 2000, Australia.

Moon Planets
The Moon and the Planets. An International Journal of Comparative Planetology. Publisher: D. Reidel Publishing Company, Dordrecht, Holland – Boston, USA. Formerly: Moon.

Nablyud. Iskusstv. Nebesn. Tel
Nablyudeniya Isskusstvennykh Nebesnykh Tel. Published by Astronomicheskij Sovet Akademii Nauk SSSR, Moskva.

Nachr. Akad. Wiss. Göttingen II
Nachrichten der Akademie der Wissenschaften in Göttingen. II. Mathematisch-Physikalische Klasse. Vandenhoeck & Ruprecht, Göttingen.

Nachr. Elektron.
Nachrichten Elektronik. Formerly: Int. Elektron. Rundsch. Elektro-Welt-Verlag Dr. Hüthig, D - 6900 Heidelberg 1, Postfach 102869, Im Weiher 10, F. R. Germany.

Nachr. Karten-, Vermessungswesen
Nachrichten aus dem Karten- und Vermessungswesen. Editor: Institut für Angewandte Geodäsie (Abt. II des Deutschen Geodätischen Forschungsinstituts). Published by Verlag des Instituts für Angewandte Geodäsie, Frankfurt a. M.

Nachr. Olbers-Ges. Bremen
Nachrichten der Olbers-Gesellschaft Bremen. Werderstraße 73, Bremen.

Nachrichtentech. Elektron.
Nachrichtentechnik Elektronik. VEB Verlag Technik, DDR 102 Berlin, Oranienburger Strasse 13/14.

NASA Conf. Publ.
NASA Conference Publication. National Aeronautics and Space Administration. Scientific and Technical Information Branch, Washington, D.C. For sale by the National Technical Information Service, Springfield, Virginia 22161.

NASA Contract. Rep.
NASA Contractor Report. National Aeronautics and Space Administration, Washington, D.C. For sale by the National Technical Information Service, Springfield, Virginia 22161.

NASA Ref. Publ.
NASA Reference Publication. National Aeronautics and Space Administration. Scientific and Technical Information Office. Washington, D.C. 20546. For sale by the National Technical Information Service, Springfield, Virginia 22161.

NASA Tech. Memo.
NASA Technical Memorandum. National Aeronautics and Space Administration, Washington, D.C. For sale by the National Technical Information Service, Springfield, Virginia 22161.

NASA Tech. Note
NASA Technical Note. National Aeronautics and Space Administration, Washington, D.C. For sale by the National Technical Information Service, Springfield, Virginia 22161.

NASA Tech. Pap.
NASA Technical Paper. National Aeronautics and Space Administration, Washington, D.C. For sale by the National Technical Information Service, Springfield, Virginia 22161.

Natl. Geogr.
National Geographic. Official Journal of the National Geographic Society, Washington, D.C. 17th and M Sts. N.W., Washington, D.C. 20036.

Nature
Nature. Editorial and Publishing Offices: Macmillan Journals Limited, 4 Little Essex Street, London WC2R 3LF, 711 National Press Building, Washington, D.C. 20045.

Naturwiss. Rundsch.
Naturwissenschaftliche Rundschau. Wissenschaftliche Verlagsgesellschaft mbH. Birkenwaldstraße 44, Stuttgart. F. R. Germany.

Naturwissenschaften
Die Naturwissenschaften. Publisher: Springer-Verlag, Berlin – Heidelberg – New York.

Nauchn. Inf.
Nauchnye Informatsii. Astronomicheskij Sovet Akademii Nauk SSSR, Moskva.

Naučna Misao
Naučna Misao. Društvo za Unapredivanje i Širenje Nauke, Zagreb, Baboničeva 54, Yugoslavia. Scientific Idea. Society for Promotion and Propagation of Science.

Navigation *(France)*
Navigation. Institut Française de Navigation, 3 avenue Octave-Greard, Paris 7, France.

Navigation *(USA)*
Navigation. Journal of the Institute of Navigation, Institute of Navigation, Suite 832, 815 15th Street, N. W., Washington, DC 20005, USA.

NBS Monogr.
National Bureau of Standards Monograph. U.S. Government Printing Office, Washington, D.C. 20402.

Nederlands Tijdschr. Natuurkd. A
Nederlands Tijdschrift voor Natuurkunde, Publisher: Martinus Nijhoff, Lange Voorhout 9, Den Haag, Netherlands.

New Phys. (Korean Phys. Soc.)
New Physics (Korean Physical Society). Seoul, Korea.

New Scientist
New Scientist. New Science Publications, 128 Long Acre, London, WC2E 9QH, England.

New Zealand Energy J.
New Zealand Energy Journal. Formerly: New Zealand Electr. J. Technical Publications Ltd., 127 Molesworth Street, P. O. Box 3047, Wellington, New Zealand.

New Zealand J. Sci.
New Zealand Journal of Science. Department of Scientific and Industrial Research, Private Bag, Wellington, New Zealand.

News Lett. Astron. Soc. N.Y.
News Letter of the Astronomical Society of New York. A. G. D. Philip (Editor). Astronomical Society of New York, Dudley Observatory, 69 Union Avenue, Schenectady, New York 12308.

NHK Lab. Note
NHK Laboratories Note. Nippon Hoso Kyokai, 1-10-11 Kinuta, Setagaya-ku, Tokyo, Japan.

Nouv. Autom.
Nouvel Automatisme. Formerly: Automatisme (France). AFCET, 41 rue de la Grange-aux-Belles, 75483 Paris Cedex 10, France.

Nucl. Instrum. Methods
Nuclear Instruments and Methods. North-Holland Publishing Co., P. O. Box 211, Amsterdam, Netherlands.

Nucl. Phys. A
Nuclear Physics, Volume A. North-Holland Publishing Co., P. O. Box 211, Amsterdam, Netherlands.

Nucl. Tracks Methods Instrum. Appl.
Nuclear Tracks, Methods, Instruments and Application. Formerly: Nucl. Track Detect. (GB) Pergamon Press Ltd., Headington Hill Hall, Oxford OX3 0BW, England.

Nukleonika
Nukleonika. Polska Akademia Nauk, 00-901 Warszawa, Polac Kultury i Nauki, Poland.

Numer. Math.
Numerische Mathematik. Springer-Verlag, Berlin–Heidelberg–New York.

Nuovo Cimento A
Il Nuovo Cimento. A. Societa Italiana di Fisica, vialle XII Guigno 1, 40124 Bologna, Italy.

Nuovo Cimento B
Il Nuovo Cimento. B. Societa Italiana di Fisica, vialle XII Guigno 1, 40124 Bologna, Italy.

Nuovo Cimento C
Il Nuovo Cimento. C. Societa Italiana di Fisica, vialle XII Guigno 1, 40124 Bologna, Italy.

Nuovo Cimento, Lett.
Lettere al Nuovo Cimento, a Cura della Società Italiana di Fiscia. Via Degli Andalo 2, 40124 Bologna, Italy.

Nuovo Cimento, Riv.
Rivista del Nuovo Cimento. Società Italiana di Fisica, Via Degli Andalo 2, Bologna 40124, Italy.

Nuovo Cimento, Suppl.
Supplemento al Nuovo Cimento. Società Italiana di Fisica. Via Degli Andalo 2, 40124 Bologna, Italy.

Obs. Artif. Earth Satell.
Observations of Artificial Satellites of the Earth (Nablyudeniya Iskusstvennykh Sputnikov Zemli). Magyar Tudományos Akadémia Csillagvizsgáló Intézete. Budapest.

Obs. Astrophys. Lab., Univ. Helsinki.Rep.
Observatory and Astrophysics Laboratory, University of Helsinki. Report. Helsinki, Finland.

Observatory
The Observatory. A Review of Astronomy. Publishers: The Editors of 'The Observatory', Royal Greenwich Observatory, Herstmonceux Castle, Hailsham, Sussex, England, BN27 1RP.

Occas. Rep. R. Obs. Edinburgh
Occasional Reports of the Royal Observatory, Edinburgh, Blackford Hill, Edinburgh EH9 3HJ, Scotland.

Occultation Newsl.
Occultation Newsletter. Published by the International Occultation Timing Association (I.O.T.A.). 6N106 White Oak Lane, St. Charles, Ill. 60174, USA.

Österreich. Z. Vermessungswes. Photogramm.
Österreichische Zeitschrift für Vermessungswesen und Photogrammetrie. Editor and Publisher: Österreichischer Verein für Vermessungswesen und Photogrammetrie, Wien, Austria.

Opt. Acta
Optica Acta. Taylor and Francis Ltd., 10 - 14 Macklin Street, London,WC2B 5NF, England.

Opt. Commun.
Optics Communications. North-Holland Publishing Co., P.O. Box 211, Amsterdam, Netherlands.

Opt. Eng.
Optical Engineering. Society of Photo-Optical Instrumentation Engineers, 337 Tejon Place, Palos Verdes Estates, CA 90274, USA.

Opt. Lett.
Optics Letters. A publication of the Optical Society of America. American Institute of Physics, 335 East 45th Street, New York, N. Y. 10017, USA.

Opt. Pura Apl.
Optica Pura y Aplicada. Serrano 121, Madrid 6, Spain.

Opt. Spectra
Optical Spectra. The Magazine of Optical/Electro-Optical/Laser Technology. Published by The Optical Publishing Co., Inc., 59 Bartlett Ave., P.O. Box 1146, Pittsfield, Mass. 01201, USA.

Optik
Optik. Zeitschrift für das gesamte Gebiet der Licht- und Elektronenoptik. Publishers: Wissenschaftliche Verlagsgesellschaft mbH, Postfach 40, D-7000 Stuttgart, F. R. Germany.

Org. Geochem.
Organic Geochemistry. A publication of the International Association of Geochemistry and Cosmochemistry. Pergamon Press, Headington Hill Hall, Oxford OX3 0BW.

Maxwell House, Fairview Park, Elmsford, N.Y. 10523, USA.

Origins of Life
Origins of Life (Formerly Space Life Sciences). An International Journal. Publisher: D. Reidel Publishing Company, Dordrecht, Holland.

Orion
Orion. Zeitschrift der Schweizerischen Astronomischen Gesellschaft (SAG). Revue de la Société Astronomique de Suisse (SAS). Printed by A. Schudel & Co. AG, 4125 Riehen, Switzerland.

Orione
Orione. Rivista Trimestrale di Divulgazione Astronomica. Via Roma 6, 10025 Pino Torinese (Torino).

Oss. Astrofis. Catania, Pubbl.
Osservatorio Astrofisico di Catania, Pubblicazione. Printed by Scuola Salesiana del Libro, Catania.

Oss. Mem. Oss. Astrofis. Arcetri
Osservazioni e Memorie dell'Osservatorio Astrofisico di Arcetri. Università Degli Studi di Firenze, Firenze, Italy.

Oyo Buturi
Oyo Buturi. Japan Society of Applied Physics, Room No. 209-2, Kikai-Shinko Building, 21 Shiba-Koen Minato-ku, Tokyo, Japan.

Perem. Zvezdy, Byull.
Peremennye Zvezdy, Byulleten', izdavaemyj Astronomicheskim Sovetom Akademii Nauk SSSR. Published by Astronomicheskij Sovet Akademii Nauk SSSR, Moskva.

Perem. Zvezdy, Prilozhenie
Peremennye Zvezdy, Prilozhenie (The Variable Stars, Supplement). Astronomicheskij Sovet Akademii Nauk SSSR, Moskva.

Philos. Mag.
The Philosophical Magazine. A Journal of Theoretical, Experimental and Applied Physics. Eighth Series. Publisher: Taylor & Francis, Ltd., 10 - 14 Macklin Street, London, WC2B 5NF, England.

Philos. Trans. R. Soc. London, Ser. A
Philosophical Transactions of the Royal Society of London. Series A, Mathematical and Physical Sciences. Carlton House Terrace, London, SW1Y 5 AG England.

Photogr. J. Sun
Photographic Journal of the Sun. Supplement to Monthly Bulletin, Solar Phenomena. Osservatorio Astronomico di Roma.

Photogr. Sci. Eng.
Photographic Science and Engineering. Society of Photographic Scientists and Engineers, Suite 204, 1330 Massachusetts Avenue N.W., Washington, DC 20005, USA.

Photogramm. Eng. Remote Sensing
Photogrammetric Engineering and Remote Sensing. Formerly: Photogramm. Eng. American Society of Photogrammetry, 105 North Virginia Avenue, Falls Church, VA 22046, USA.

Phys. Abstr.
Physics Abstracts. Science Abstracts, Series A. An INSPEC Publication, published by The Institution of Electrical Engineers in Association with the Institute of Electrical and Electronics Engineers Inc. Printed by Pindar & Son Ltd., Scarborough, N. Yorkshire, England.

Phys. Bl.
Physikalische Blätter. Physik-Verlag GmbH, Pappelallee 3, Postfach 1260/1280, D-6940 Weinheim, F. R. Germany.

Phys. Briefs
Physics Briefs. Physikalische Berichte. Edited by Deutsche Physikalische Gesellschaft and Fachinformationszentrum Energie, Physik, Mathematik in cooperation with American Institute of Physics. Published by Physik Verlag GmbH, Postfach 1260/1280, D-6940 Weinheim, F. R. Germany. ISSN 0170-7434.

Phys. Bull.
Physics Bulletin. Published by the Institute of Physics, 47 Belgrave Square, London, SW1X 8QX, England.

Phys. Canada
Physics in Canada. The Bulletin of The Canadian Association of Physicists. Suite 903, 151 Slater Street, Ottawa, Ontario K1P 5H3.

Phys. Earth Planet. Inter.
Physics of the Earth and Planetary Interiors. A journal devoted to observational and experimental studies of the Earth and Planetary interiors and their theoretical interpretation by the physical sciences. Publisher: North-Holland Publishing Company, Amsterdam, Netherlands.

Phys. Educ.
Physics Education. Institute of Physics, 47 Belgrave Square, London, SW1X 8QX, England.

Phys. Energ. Fortis Phys. Nucl.
Physica Energiae Fortis et Physica Nuclearis Science Press, Peking, People's Republic of China. Subscription address: Guozo Shudian. P.P. Box 399, Peking.

Phys. Fluids
The Physics of Fluids. Published by the American Institute of Physics, 335 East 45th Street, New York, NY 10017, USA.

Phys. Lett.
Physics Letters. Volumes A and B. Publisher: North-Holland Publishing Company, Amsterdam.

Phys. Norvegica
Physica Norvegica. Institute of Physics, University of Oslo, Blindern, per Oslo, Norway.

Phys. Rep.
Physics Reports. Formerly: Phys. Rev. Phys. Lett. C. North-Holland Publishing Co., P.O. Box 211, Amsterdam, Netherlands.

Phys. Rev. A
Physical Review A, General Physics. Published for the American Physical Society by the American Institute of Physics, 335 East 45th Street, New York, N. Y. 10017, USA.

Phys. Rev. B
Physical Review B, Solid State. Published for the American Physical Society by the American Institute of Physics, 335 East 45th Street, New York, NY 10017, USA.

Phys. Rev. C
Physical Review C, Nuclear Physics. Published for the

American Physical Society by the American Institute of Physics, 335 East 45th Street, New York, NY 10017, USA.

Phys. Rev. D
Physical Review D, Particles and Fields. Published for the American Physical Society by the American Institute of Physics, 335 East 45th Street, New York, NY 10017, USA.

Phys. Rev. Lett.
Physical Review Letters. Published for the American Physical Society by the American Institute of Physics, 335 East 45th Street, New York, NY 10017, USA.

Phys. Scr.
Physica Scripta. (Formerly Arkiv för Fysik). Published by the Royal Swedish Academy of Sciences, S-104 05 Stockholm 50, Sweden.

Phys. Teach.
Physics Teacher. American Institute of Physics, 335 East 45th Street, New York, NY 10017, USA.

Phys. Today
Physics Today. Published by the American Institute of Physics, 335 East 45th Street, New York, NY 10017, USA.

Phys. unserer Zeit
Physik in unserer Zeit. Verlag Chemie GmbH, Pappelallee 3, Postfach 1260/1280, D-6940 Weinheim, F. R. Germany.

Physica B, C
Physica B & C. Subscription address: North-Holland Publishing Co., P.O. Box 211, Amsterdam, Netherlands.

Pis'ma v Astron. Zhurn.
Pis'ma v Astronomicheskij Zhurnal. Akademiya Nauk SSSR. Publishers: Izdatel'stvo 'Nauka', Moskva. Translation in Soviet Astron. Lett.

Planet. Astron.
Planetary Astronomy. Published by the James-Mims Observatory, Inc., P. O. Box 15854, Baton Rouge, La. 70895. ISSN 0195-8038.

Planet. Space Sci.
Planetary and Space Science. Pergamon Press Ltd., Headington Hill Hall, Oxford, OX3 OBW, England.

Plasma Phys.
Plasma Physics. Pergamon Press Ltd., Headington Hill Hall, Oxford, OX3 0BW, England.

Pokroky
Pokroky matematiky, fyziky a astronomie. Editor: Jednota čs. matematiků a fyziků. Publisher: Academia, Praha.

Postępy Astron.
Postępy Astronomii. Czasopismo Poświecone Upowszechnianiu Wiedzy Astronomicznej. Polskie Towarzystwo Astronomiczne, Warszawa. Printed in Poland by Pánstwowe Wydawnictwo Naukowe, Lódź.

Postępy Fiz.
Postępy Fizyki. Polskie Towarzystwo Fizykczne, 00-681 Warszawa, ul. Hoza 69, Poland.

Pramāṇa
Pramāṇa. Indian Academy of Sciences, Bangalore 560006, India.

Priroda
Priroda. Publishers: Izdatel'stvo "Nauka", Moskva.

Probl. Kosm. Fiz.
Problemy Kosmichskoj Fiziki. Mezhvedomstvennyj Nauchnoj Sbornik. Izdatel'skoe Obedinenie Vishcha Shkola. Izdatel'stvo pri Kievskom Universitete, Kiev.

Proc. Astron. Soc. Australia
Proceedings of the Astronomical Society of Australia. Published for the Society by Sydney University Press, Sydney.

Proc. IEEE
Proceedings of the IEEE. Published by the Institute of Electrical and Electronics Engineers, 345 East 47th Street, New York, NY 10017, USA.

Proc. Indian Acad. Sci., Sect. A
Proceedings of the Indian Academy of Sciences, Section A. Bangalore 560006, India.

Proc. Indian Acad. Sci., Sect. C
Proceedings of the Indian Academy of Sciences, Section C: Engineering Sciences, Bangalore 560006, India.

Proc. Indian Natl. Sci. Acad., Part A.
Proceedings of the Indian National Science Academy, Part A, Bahadur Shah Zafar Marg, New Delhi 1, India.

Proc. Int. Latitude Obs. Mizusawa
Proceedings of the International Latitude Observatory of Mizusawa. Published by the International Latitude Observatory of Mizusawa, Japan.

Proc. Japan Acad., Ser. B
Proceedings of the Japan Academy. Series B, Physical and Biological Sciences, Ueno Park, Tokyo 110, Japan.

Proc. K. Nederlandse Akad. Wet. B
Koninklijke Nederlandse Akademie van Wetenschappen. Proceedings. Series B, Physical Sciences. Publisher: North-Holland Publishing Company, Amsterdam, Netherlands.

Proc. Natl. Acad. Sci. India, Sect. A
Proceedings of the National Academy of Sciences of India. Section A (Physical Sciences). Lajpatrai Road, Allahabad-2, India

Proc. Natl. Acad. Sci. U.S.A.
Proceedings of the National Academy of Sciences of the United States of America. National Academy of Sciences, 2101 Constitution Avenue, Washington, DC 20418, USA.

Proc. R. Instn. Great Britain
Proceedings of the Royal Institution of Great Britain. 21 Albemarle Street, London W1, England.

Proc. R. Irish Acad., Sect. A
Proceedings of the Royal Irish Academy, Section A (Mathematical, Astronomical and Physical Science). 19 Dawson Street, Dublin 2, Ireland.

Proc. R. Soc. Edinburgh, Sect. A
Proceedings of the Royal Society of Edinburgh, Section A (Mathematical and Physical Sciences). 22 George Street, Edinburgh, EH2 2PQ, Scotland.

Proc. R. Soc. London, Ser. A
Proceedings of the Royal Society of London. Series A:

Mathematical and Physical Sciences. Published by the Royal Society, 6 Carlton House Terrace, London, SW1Y 5AG, England.

Proc. Res. Inst. Atmos. Nagoya Univ.
Proceedings of the Research Institute of Atmospherics Nagoya University. Nagoya University, 13 Honohara, 3 Chrome, Toyokawa 442, Japan.

Proc. Soc. Photo-Opt. Instrum. Eng.
Proceedings of the Society of Photo-Optical Instrumentation Engineers, Bellingham, WA 98225, USA.

Prog. Aerosp. Sci.
Progress in Aerospace Sciences. Pergamon Press Ltd., Headington Hill Hall, Oxford, OX3 0BW, England.

Prog. Part.Nucl. Phys.
Progress in Particle and Nuclear Physics. Pergamon Press Ltd., Headington Hill Hall, Oxford OX3 0BW, England.

Prog. Theor. Phys.
Progress of Theoretical Physics. Published for the Research Institute for Fundamental Physics and the Physical Society of Japan. Publication Office: Progress of Theoretical Physics, Yukawa Hall, Kyoto University, 606 Kyoto, Japan.

Prog. Theor. Phys. Suppl.
Supplement of the Progress of Theoretical Physics. Published for the Research Institute for Fundamental Physics and The Physical Society of Japan. Publication Office: Progress of Theoretical Physics. Yukawa Hall, Kyoto University, 606 Kyoto, Japan.

PTB Mitt.
PTB Mitteilungen. Forschen + Prüfen. Fachorgan für Wirtschaft und Wissenschaft. Amts- und Mitteilungsblatt der Physikalisch-Technischen Bundesanstalt. Braunschweig–Berlin. Deutscher Eichverlag, Postfach 3367, D-3300 Braunschweig, F.R. Germany.

Publ. Astron. Soc. Japan
Publications of the Astronomical Society of Japan. Published by the Astronomical Society of Japan. Office of the Society: Tokyo Astronomical Observatory, Mitaka, Tokyo. Agent: Maruzen Co. Ltd. (Export Department), Nihonbashi, Tokyo, Japan.

Publ. Astron. Soc. Pacific
Publications of the Astronomical Society of the Pacific. Published by the Astronomical Society of the Pacific, 1290 24th Avenue, San Francisco, California 94122, USA. (415) 661 - 8660.

Publ. Beijing Astron. Obs.
Publications of the Beijing Astronomical Observatory. Beijing Astronomical Observatory, Academia Sinica, China.

Publ. Bosscha Obs.
Publications of the Bosscha Observatory. Bandung Institute of Technology, Department of Science. Lembang, Indonesia.

Publ. Dep. Astron., Univ. Beograd
Publications of the Department of Astronomy, University of Beograd, Faculty of Sciences (Publications de la Chaire d'Astronomie, Universitè de Beograd, Facultè des Sciences) Beograd. YU ISSN 0350-3283.

Publ. Dep. Geod. Astron., Univ. Thessaloniki
Publications of the Department of Geodetic Astronomy, University of Thessaloniki.

Publ. Dominion Astrophys. Obs.
Publications of the Dominion Astrophysical Observatory, Victoria, B. C. National Research Council of Canada.

Publ. Eidg. Sternw. Zürich
Publikationen der Eidgenössischen Sternwarte Zürich. Schulthess Polygraphischer Verlag, Zürich.

Publ. Inst. Geophys., Polish Acad. Sci.
Publications of the Institute of Geophysics, Polish Academy of Sciences. Państwowe Wydawnictwo Naukowe, Warszawa-Łódź. ISBN 83-01-02210-8. ISSN 0138-0214.

Publ. Inst. R. Meteorol. Belgique A
Publications, Institut Royal Meteorologique de Belgique. Serie A. 3 Avenue Circulaire, Uccle-Bruxelles 1180, Belgium.

Publ. Int. Latitude Obs. Mizusawa
Publications of the International Latitude Observatory of Mizusawa. Published by the International Latitude Observatory of Mizusawa, Japan.

Publ. Korean Natl. Astron. Obs.
Publications of the Korean National Astronomical Observatory. Published by the Korean National Astronomical Observatory, Seoul, Korea.

Publ. Obs. Astron. Beograd
Publications de l'Observatoire Astronomique de Beograd. Editeur: Observatoire Astronomique de Belgrade, 11050 Beograd, Volgina 7, Yougoslavie.

Publ. R. Obs. Edinburgh
Publications of the Royal Observatory, Edinburgh. Published by The Royal Observatory, Edinburgh, Scotland.

Publ. Secc. Mat. Univ. Autònoma Barcelona
Publicacions de la Secció de Matemàtiques, Universitat Autònoma de Barcelona. (Bellaterra) Spain.

Publ. Shensi Astron. Obs.
Publications of the Shensi Astronomical Observatory. Academia Sinica, P.O. Box 18, Lintong, near Sian, China.

Publ. Tartu Astrofiz. Obs.
W. Struve nimelise Tartu Astrofüüsika Observatooriumi, Publikatsioonid. Eesti NSV Teaduste Akadeemia, Tartu.

Publ. United States Naval Obs.
Publications of the United States Naval Observatory. Department of the Navy, U.S. Naval Observatory, Washington. U.S. Government Printing Office, Washington, D.C.

Publ. Variable Star Sect.,R. Astron. Soc. New Zealand
Publications of the Variable Star Section, Royal Astronomical Society of New Zealand. Director: F. M. Bateson, Greerton, Tauranga, New Zealand.

Publ. Warner Swasey Obs.
Publications of the Warner and Swasey Observatory, Case Western Reserve University, Cleveland, Ohio 44106.

Pure Appl. Geophys.
Pure and Applied Geophysics. Birkhäuser Verlag, CH-4010 Basel, Elizabethenstrasse 19, Switzerland.

Q. Appl. Math.
Quarterly of Applied Mathematics. American Mathematical Society, P. O. Box 1571, Providence, RI 02901, USA.

Q. Bull. Sol. Act.
International Astronomical Union, Quarterly Bulletin on Solar Activity. Published by the Tokyo Astronomical Observatory. Beginning with No. 197; formerly Zürich.

Q. J. R. Astron. Soc.
Quarterly Journal of the Royal Astronomical Society. Burlington House, London, W1V ONL, England.

Q. J. R. Meteorol. Soc.
Quarterly Journal of the Royal Meteorological Society. Cromwell House, High Street, Bracknell, Berks, England.

Quaternary Res.
Quaternary Research. Academic Press Inc., 111 Fifth Avenue, New York, NY 10003, USA.

R. Greenwich Obs., Time Latitude Serv.
Royal Greenwich Observatory, Time and Latitude Service Royal Greenwich Observatory, Herstmonceux Castle, Hailsham, East Sussex BN27 1RP, England.

R. Obs. Ann.
Royal Observatory Annals, Royal Greenwich Observatory, Herstmonceux Castle, Hailsham, East Sussex, BN 27 1RP, England.

Radiat. Eff.
Radiation Effects. Gordon & Breach Science Publishers Ltd., 41 and 42 William IV Street, London, WC2, England.

Radio Commun.
Radio Communication. Radio Society of Great Britain, 35 Doughty Street, London, WC1N 2AE, England.

Radio Sci.
Radio Science, American Geophysical Union, 2901 Byrdhill Road, Richmond, VA 23228, USA.

Radiochem. Radioanal. Lett.
Radiochemical and Radioanalytical Letters. Elsevier Sequoia S. A., P. O. Box 851, CH-1001 Lausanne, Switzerland. Akademiai Kiado, Alkotmany U 21, Budapest 5, Hungary.

Radiotekh. Ehlektron.
Radiotekhnika i Ehlektronika. Moskva TSP-3, Pr. Karl Marx 18, USSR.

Rech. Aerosp.
Recherche Aerospatiale. Office National d'Études et de Recherches Aerospatiales, 29 Avenue de la Division Leclerc, 92320-Chatillon, France.

Recherche
Recherche, 4 Place de l'Odéon, Paris 6, France.

Ref. Zh. 51. Astron.
Referativnyj Zhurnal. 51. Astronomiya. Vsesoyuznyj Institut Nachnoj i Tekhnicheskoj Informatsii. Moskva.

Ref. Zh. 52. Geod. i Aehrosemka
Referativnyj Zhurnal. 52. Geodeziya i Aehrosemka. Vsesoyuznyj Institut Nachnoj i Tekhnicheskoj Informatsii. Moskva.

Ref. Zh. 62. Issled. kosm. prostranstva
Referativnyj Zhurnal. 62. Issledovanie Kosmicheskogo Prostranstva. Vsesoyuznyj Institut Nauchnoj i Tekhnicheskoj Informatsii. Moskva.

Rep. Finnish Geod. Inst.
Reports of the Finnish Geodetic Institute. Suomen Geodeettisen Laitoksen Tiedonantoja. Helsinki, Finland.

Rep. Math. Phys.
Reports on Mathematical Physics, Pergamon Press Ltd., Headington Hill Hall, Oxford OX3 0BW, England. Subscription address: ARS Polona-Ruch Foreign Trade Enterprise, Krakowskie Przedmiescie 7, 00-068 Warszawa.

Rep. NRL Prog.
Report of NRL Progress. National Technical Information Service, U. S. Department of Commerce, Springfield, VA 22151, USA.

Rep. Obs. Lund
Reports from the Observatory of Lund. Lunds Universitet, Institutionen för Astronomie, S-222 24 Lund, Sweden.

Rep. Prog. Phys.
Reports on Progress in Physics. Published by the Institute of Physics, 47 Belgrave Square, London, SW1X 8QX, England.

Rep. Ser. Dep. Phys. Sci., Univ. Turku
Report Series, Department of Physical Sciences, Institute of Astronomy, University of Turku, SF-20500 Turku 50, Finland.

Rep. Univ. Electro-Commun.
Reports of the University of Electro-Communications. University of Electro-Communications, 1-5-1 Chofugaoka, Chofu-shi, Tokyo, Japan.

Res. Lab. Electron. Onsala Space Obs.
Research Laboratory of Electronics and Onsala Space Observatory, Chalmers University of Technology, Gothenburg, Sweden. Research Report.

Rev. Acad. Cienc. Zaragoza
Revista de la Academia de Ciencias Zaragoza. Academia de Ciencias Exactas, Fisico-Quimicas y Naturales de Zaragoza, Plaza de Paraiso 1, Zaragoza, Spain.

Rev. Astron.
Revista Astronomica. Organo de la Asociación Argentina Amigos de la Astronomia, Avenida Patricias Argentinas 550, Buenos Aires 5, Argentina.

Rev. Brasil. Fis.
Revista Brasileira de Fisica. Sociedade Brasileira de Fisica, Cx. Postal 20553, Sao Paolo SP, Brazil.

Rev. Espanola Electron.
Revista Espanola de Electronica. Ediciones Tecnicas 'Rede', Apartado 5252, Barcelona, Spain.

Rev. Fac. Sci. Univ. Istanbul C
Revue de la Faculté des Sciences de l'Université d'Istanbul (Istanbul Universitesi Fen Fakultesi Mecmuasi) Série C (Astronomie, Physique, Chimie). Beyazit, Istanbul, Turkey.

Rev. Geofis.
Revista de Geofisica, Instituto Nacional de Geofisica, Serrano 123, Madrid 2, Spain.

Rev. Geophys. Space Phys.
Reviews of Geophysics and Space Physics (formerly Reviews of Geophysics). Published by the American Geophysical Union, 1909 K Street, N.W., Washington, DC 20006, USA.

Rev. Hist. Sci.
Revue d'Histoire des Sciences. Revue trimestrielle publiée avec le concours du C.N.R.S. Centre International de Synthèse, Section d'Histoire des Sciences.

Rev. Mexicana Astron. Astrofis.
Revista Mexicana de Astronomia y Astrofisica. Dirección: Instituto de Astronomia, Universidad Nacional Autónoma de México, Apartado Postal 70-264, Mexico 20, D. F., Mexico.

Rev. Mexicana Fis.
Revista Mexicana de Fisica. Sociedad Mexicana de Fisica, Apartado Postal No. 20-364, Mexico 20, D. F. Mexico.

Rev. Mod. Phys.
Reviews of Modern Physics. Published for the American Physical Society by the American Institute of Physics, 335 East 35th Street, New York, NY 10017, USA.

Rev. Phys. Appl.
Revue de Physique Appliquée. Société Française de Physique. 87 bis, Avenue du Général-Leclerc, 75014 Paris, France.

Rev. Polytech.
Revue Polytechnique. Chemin de la Caroline 22, 1213 Petit-Lancy, Genève, Switzerland.

Rev, Radio Res. Lab.
Review of the Radio Research Laboratories. Ministry of Posts & Telecommunications, Nukui-Kitamachi, Konganei-shi, Tokyo 184, Japan.

Rev. Roumaine Phys.
Revue Roumaine de Physique. Academie Republicii Populare Romine, Boite Postale 134 - 135, Bucuresti, Rumania.

Rev. Sci. Instrum.
Review of Scientific Instruments.American Institute of Physics, 335 East 45th Street, New York, NY 10017, USA.

Rezul't. Nablyud. Iskusstv. Sputnikov Zemli
Rezul'taty Nablyudenij Iskusstvennykh Sputnikov Zemli. Published by Astronomicheskij Sovet Akademii Nauk SSSR, Ryazanskij Gosudarstvennyj Pedagogicheskij Institut, Ryazan'.

Ric. Astron.
Ricerche Astronomiche. Specola Vaticana, Città del Vaticano.

Ric. Spettrosc.
Ricerche Spettroscopiche. Specola Vaticana, Città del Vaticano.

Říše hvězd
Říše hvězd. Czech popular astronomical journal. Publisher: Panorama, Praha.

Rumanian Sci. Abstr.
Rumanian Scientific Abstracts. Natural Sciences. Publishers: The Scientific Documentation Centre of the Academy of the Socialist Republic of Romania, Bucureşti.

Sci. American
Scientific American. 415 Madison Avenue, New York, NY 10017, USA.

Sci. Atmos. Sinica
Scientia Atmospherica Sinica. Science Press, Peking. Subscription address: Guozi Shudian, P.O. Box 399, Peking, Peoples's Republic of China.

Sci. Dimension
Science Dimension. National Research Council of Canada, Ottawa K1A 0R6, Canada.

Sci. Light
Science of Light. Published by the Institute for Optical Research, Tokyo University of Education (in collaboration with the Spectroscopical Society of Japan). 400, Hyakunintyo-4, Sinyuku, Tokyo 160, Japan.

Sci. Pap. Inst. Phys. Chem. Res.
Scientific Papers of the Institute of Physical and Chemical Research. Rikagaku Kenkyusho, Wako-shi, Saitama 351, Japan.

Sci. Prog.
Science Progress. Blackwell Scientific Publications, Oxford, England.

Sci. Prog. Découverte
Science Progrès Découverte (formerly Science Progrès, La Nature). Revue publiée avec la participation du Palais de la Découverte. Published by Dunod, Editeur, Paris. Imprimerie Bayeusaine, Bayeux.

Sci. Rep. Tôhoku Univ. Ser. 1.
The Science Reports of the Tôhoku University. First Series (Physics, Chemistry, Astronomy). Published by the Faculty of Science, Tôhoku University, Sendai, Japan. ISSN 0040-8778.

Sci. Rep. Tôhoku Univ., Ser. 8.
The Science Reports of the Tôhoku University. Eighth Series (Physics and Astronomy). Published by the Faculty of Science, Tôhoku University, Sendai, Japan. ISSN 0388-5607.

Sci. Rev.
Scienca Revuo. Prof. B. Popovic, Ognjena Price 80, Beograd, Yugoslavia.

Sci. Sinica
Scientia Sinica. Science Press, Peking. Subscription address: Guozi Shudian, P.O. Box 399, Peking, China.

Science
Science. American Association for the Advancement of Science, 1515 Massachusetts Avenue, N. W. Washington, D. C. 20005, USA.

Scr. Fac. Sci. Nat. Univ. Purkynianae Brunensis Phys.
Scripta Facultatis Scientiarum Naturalium Universitatis Purkynianae Brunensis, Physica. University J. E. Purkyne, 61137 Brno-Kotlarská 2, Czechoslovakia.

Sdelovaci Tech.
Sdělovací technika. Publishers of Technical Literature, Spálená 51, 11302 Praha 1, Czechoslovakia.

Shensi Astron. Obs. Repr.
Shensi Astronomical Observatory Reprints. Academia Sinica, P.O. Box 18 Lintong, near Sian, China.

SIAM J. Appl. Math.
SIAM Journal on Applied Mathematics. Society for Industrial and Applied Mathematics, 33 South 17th Street, Philadelphia, PA 19103, USA.

Siemens Rev.
Siemens Review. Siemens-Aktiengesellschaft, Postfach 325, 8520 Erlangen 2, F. R. Germany.

Sitzungsber. Akad. Wiss. DDR
Sitzungsberichte der Akademie der Wissenschaften der DDR. Mathematik-Naturwissenschaften-Technik. Akademie-Verlag, Berlin.

Sitzungsber. Bayerische Akad. Wiss.
Bayerische Akademie der Wissenschaften. Mathematisch-Naturwissenschaftliche Klasse. Sitzungsberichte. Publisher: Verlag der Bayerischen Akademie der Wissenschaften, München.

Sitzungsber. Heidelberger Akad. Wiss.
Sitzungsberichte der Heidelberger Akademie der Wissenschaften. Mathematisch-Naturwissenschaftliche Klasse. Publisher: Springer-Verlag, Heidelberg.

Sitzungsber. Österreich. Akad. Wiss.
Sitzungsberichte. Österreichische Akademie der Wissenschaften. Mathematisch-Naturwissenschaftliche Klasse. Abteilung II: Mathematik, Astronomie, Meteorologie und Technik. Publisher: Springer-Verlag, Wien.

Sky Telesc.
Sky and Telescope. Published by Sky Publishing Corporation, 49-50-51 Bay State Road, Cambridge, Mass. 02138, USA.

Smithsonian Astrophys. Obs., Spec. Rep.
Smithsonian Astrophysical Observatory, Special Report. Available from the Publications Division, Distribution Section, Smithsonian Astrophysical Observatory, Cambridge, Mass. 02138.

Smithsonian Contrib. Astrophys.
Smithsonian Contributions to Astrophysics. Smithsonian Institution Astrophysical Observatory, Cambridge, Mass. Printed by Smithsonian Institution Press, City of Washington. For sale by the Superintendent of Documents, U.S. Government Printing Office, Washington, D.C.

Sol. Energy
Solar Energy, Pergamon Press, Maxwell House, Fairview Park, Elmsford, NY 10523, USA.

Sol. Phenom.
Solar Phenomena. Osservatorio Astronomico di Roma.

Sol. Phys.
Solar Physics. A Journal for Solar Research and the Study of Solar Terrestrial Physics. Publisher: D. Reidel Publishing Company, Dordrecht, Holland.

Sol. Terr. Environ. Res. Japan
Solar Terrestrial Environmental Research in Japan. Formerly: Rep. Ionos. Space Res. Japan. Institute of Space and Aeronautical Science, University of Tokyo, 4-6-1, Komaba, Meguro-ku, Tokyo 153, Japan.

Soln. Dannye, Byull.
Solnechnye Dannye. Byulleten'. (*Solar Data*). Publishers: Izdatel'stvo "Nauka", Leningradskoe Otdelenie, Leningrad.

Sonne
Sonne. Mitteilungsblatt der Amateursonnenbeobachter. Peter Völker, c/o Wilhelm-Foerster-Sternwarte, Munsterdamm 90, 1000 Berlin 41.

Soobshch. Byurakan. Obs.
Soobshcheniya Byurakanskoj Observatorii. Akademiya Nauk Armyanskoj SSR, Erevan.

Soobshch. Gos. Astron. Inst. Shternberg
Soobshcheniya Gosudarstvennogo Astronomicheskogo Instituta im. P. K. Shternberga. Publishers: Izdatel'stvo Moskovskogo Universiteta, Moskva.

Soobshch. Spets. Astrofiz. Obs.
Soobshcheniya Spetsial'noj Astrofizicheskoj Observatorii. Izdanie Spetsial'noj Astrofizicheskoj Observatorii AN SSSR.

South African Astron. Obs. Circ.
South African Astronomical Observatory, Circulars. S.A. Astronomical Observatory, Observatory, Cape.

South African J. Phys.
South African Journal of Physics (Suid-Afrikaanse Tydskrif vir Fisika). Bureau for Scientific Publications, P.O. Box 1758, Pretoria 0001, South Africa.

South. Stars
Southern Stars. The Journal of the Royal Astronomical Society of New Zealand (Inc.). Address of the Society: P.O. Box 3181, Wellington C1, New Zealand.

Soviet Astron.
Soviet Astronomy. A translation of Astronomicheskij Zhurnal (Astronomical Journal). Published by the American Institute of Physics, New York, N.Y.

Soviet Astron. Lett.
Soviet Astronomy Letters. A translation of "Pis'ma v Astronomicheskij Zhurnal". Published by the American Institute of Physics.

Space Sci. Instrum.
Space Science Instrumentation. An International Journal of Scientific Instruments for Aircraft, Balloons, Sounding Rockets, and Spacecraft. Published by D. Reidel Publishing Company, Dordrecht, Holland.

Space Sci. Rev.
Space Science Reviews. Publishers: D. Reidel Publishing Company, Dordrecht, Holland.

Spaceflight
Spaceflight. Published by the British Interplanetary Society. Printed by Unwin Brothers Ltd., at the Gresham Press, Old Woking, England.

Spaceworld
Spaceworld. Palmer Publications Inc., Amherst, WI 54406, USA.

Spectrosc. Lett.
Spectroscopy Letters. Marcel Dekker Inc., 270 Madison Avenue, New York, NY 10016, USA.

Sterne
Die Sterne. Zeitschrift für alle Gebiete der Himmelskunde. Johann Ambrosius Barth, Leipzig, German Democratic Republic.

Sterne Weltraum
Sterne und Weltraum. Astronomische Monatsschrift. Publisher: Verlag Sterne und Weltraum Dr. Vehrenberg, Düsseldorf, F. R. Germany.

Sternenbote
Sternenbote. Monatsschrift für Österreichs Amateurastronomen. Publisher: Astronomisches Büro. Hermann Mucke, Wien, Austria.

Stockholms Obs. Ann.
Stockholms Observatorium Annaler. Printed by Almquist & Wiksell, Stockholm, Sweden.

Stockholms Obs. Rep.
Stockholms Observatorium, Saltsjöbaden, Sweden, Report.

Strolling Astron.
The Strolling Astronomer. The Journal of The Association of Lunar and Planetary Observers, Publication Office: The Strolling Astronomer, Box 3 AZ, University Park, New Mexico, USA.

Stud. Appl. Math.
Studies in Applied Mathematics. American Elsevier Publishing Co., 52 Vanderbilt Avenue, New York, NY 10017, USA.

Stud. Astron. Sinica
Studia Astronomica Sinica. Published by the Purple Mountain Observatory, Academia Sinica, Nanking, People's Republic of China.

Stud. Cercet. Astron.
Studii şi Cercetări de Astronomie. Editura Academiei Republicii Socialiste România. Editorial Office: Observatorul Astronomic, Bucureşti, Rumania.

Stud. Cercet. Fiz.
Studii şi Cercetări de Fizica. Academia Republicii Populare Romine. P.O. Box 134-5, Calca Victoriei 126, Bucureşti, Rumania.

Stud. Geophys. Geod.
Studia geophysica et geodaetica. Published for the Geophysical Institute of the Czechoslovak Academy of Sciences by Academia, Praha.

Stud. Hist. Philos. Sci.
Studies in History and Philosophy of Science. Pergamon Press Ltd., Headington Hill Hall, Oxford OX3 0BW, England.

Stud. Soc. Sci. Torunensis
Studia Societatis Scientiarum Torunensis, Toruń – Polonia. Sectio F (Astronomia).

Stud. Univ. Babeş-Bolyai
Studia Universitatis Babeş-Bolyai. Series Mathematica-Physica. Publishers: Intreprinderea Poligrafica, Cluj.

Tartu Astron. Obs. Teated
Tartu Astronoomia Observatoorium Teated. Eesti NSV Teaduste Akadeemia W. Struve nim. Tartu Astrofüüsika Observatoorium, Tartu.

Tec. Regul. Mando Autom.
Tecnica de la Regulacion y Mando Automatico. Compania Espanola de Editoriales Tecnologicas Internacionales SA. (CETISA), Concepcion Arenal 5 y 7, Barcelona (16), Spain.

Tech. Rep. Inst. At. Energy, Kyoto Univ.
Technical Reports of the Institute of Atomic Energy, Kyoto University. Kyoto, Japan.

Technica
Technica. E. Birkhäuser, CH-4010 Basel 10, Switzerland.

Tectonophysics
Tectonophysics. Elsevier Scientific Publishing Co., P. O. Box 211, Amsterdam, Netherlands.

Tehnika
Tehnika. Beograd, Kneza Milosa 7/11, Yugoslavia.

Telecommun. J.
Telecommunication Journal.(English Edition). International Telecommunications Union, Place des Nations, 1211 Genève 20, Switzerland.

Tellus
Tellus, a bi-monthly Journal of Geophysics. Svenska Geofysiska Foreningen, Arrhenius laboratoriet, Fack, S - 104 05 Stockholm, Sweden.

Texas J. Sci.
Texas Journal of Science. The Talley Press, San Angelo, Texas, USA.

Time Freq. Serv. Bull.
Time and Frequency Services Bulletin. Shensi Astronomical Observatory, Chinese Academy of Sciences, Lintong, Sian, China.

Tokyo Astron. Bull., Second Ser.
Tokyo Astronomical Observatory, Japan. Tokyo Astronomical Bulletin, Second Series.

Tokyo Astron. Obs. Rep.
University of Tokyo, Tokyo Astronomical Observatory, Japan. Report. ISSN 0374-4639.

Tokyo Astron. Obs., Time and Latitude Bull.
Tokyo Astronomical Observatory, Time and Latitude Bulletins. Mitaka, Tokyo, Japan.

Tr. Astrofiz. Inst. Alma-Ata
Trudy Astrofizicheskogo Instituta, Alma-Ata. Akademiya Nauk Kazakhskoj SSR. Publishers: Izdatel'stvo "Nauka" Kazakhskoj SSR, Alma-Ata.

Tr. Astron. Obs., ***Leningrad***
Uchenye Zapiski Gosudarstvennogo Universiteta im. A. A. Zhdanova, Seriya matematicheskikh nauk = Trudy Astronomicheskoj Observatorii. Izdatel'stvo Leningradskogo Universiteta, Leningrad.

Tr. Glav. Astron. Obs. Pulkovo
Trudy Glavnoj Astronomicheskoj Observatorii v Pulkove. Akademiya Nauk SSR. Izdanie Glavnoj astronomicheskoj observatorii v Pulkove, Leningrad.

Tr. Inst. Teor. Astron., *Leningrad*
Trudy Instituta Teoreticheskoj Astronomii. Akademiya Nauk SSSR. Publishers: Izdatel'stvo "Nauka", Leningrad.

Tr. Kazan. Gorod. Astron. Obs.
Trudy Kazanskoj Gorodskoj Astronomicheskoj Observatorii. Izdatel'stvo Kazanskogo Universiteta, Kazan.

Tr. Tashkent. Astron. Obs.
Trudy Tashkentskoj Astronomicheskoj Observatorii. Akademiya Nauk Uzbekskoj SSR. Publishers: Izdatel'stvo "FAN" Uzbekskoj SSR, Tashkent.

Trans. American Nucl. Soc.
Transactions of the American Nuclear Society. 244 East Ogden Avenue, Hinsdale, IL 60521, USA.

Trans. ASME. J. Heat Transfer
Transactions of the ASME. Journal of Heat Transfer. Formerly: Trans. ASME Ser. C, J. Heat Transfer (USA). American Society of Mechanical Engineers, 345 East 47th Street, New York, NY 10017, USA.

Trans. Astron. Obs. Yale Univ.
Transactions of the Astronomical Observatory of Yale University. Published by the Observatory, New Haven.

Trans. IAU
Transactions of the International Astronomical Union. Published and distributed for the IAU (UAI) by D. Reidel Publishing Company, Dordrecht, Holland – Boston, U.S.A.

Trans. Inst. Electron. Commun. Eng. Japan E
Transactions of the Institute of Electronics and Communication Engineers of Japan, Section E. Denshi Tsushin Gakkai, Kikai-Shinko-kaikan Bldg., Shiba Park 21-1-5, Minatoku, Tokyo, Japan. English translation of selected articles appear in Electron & Commun. Jap. (USA) and Syst. Comput. Control (USA).

Trans. R. Soc. Canada
Transactions of the Royal Society of Canada (Mémoires de la Société Royale du Canada). National Library, 395 Wellington Street, Ottawa 4, Canada. University of Toronto Press, Toronto 5, Ontario.

Trans. South African Inst. Electr. Eng.
Transactions of the South African Institute of Electrical Engineers. Kelvin Publications, Loveday Street, P. O. Box 2988, 2000 Johannesburg, South Africa.

Tsirk. Astron. Inst. Tashkent
Tsirkulyar Astronomicheskogo Instituta. Akademiya Nauk Uzbekskoj SSR. Izdatel'stvo "FAN" Uzbekskoj SSR, Tashkent.

Tsirk. Astron. Obs. L'vov
Tsirkulyar. Astronomicheskaya Observatoriya. L'vovskij Ordena Lenina Gosudarstvennyj Universitet imeni Ivana Franko. Publisher: Izdatel'stvo L'vovskogo Universiteta, L'vov.

Umschau
Umschau in Wissenschaft und Technik. Umschau Verlag, Stuttgarter Str. 18 - 24, D-6000 Frankfurt/M., F. R. Germany.

United States Naval Obs., Circ.
United States Naval Observatory, Circular. U.S. Naval Observatory, Washington, D.C. 20390.

Univ. Chile, Dep. Astron., Publ.
Universidad de Chile, Facultad de Ciencias Fisicas y Matematicas, Departamento de Astronomía, Publicaciones. Observatorio Astronómico Nacional, Cerro Calán, Santiago de Chile.

Urania Barcelona
Urania. Revista de Astronomía y Ciencias Afines. Órgano de la Sociedad Astronómica de España y América, Barcelona; Unión Nacional de Astronomía y Ciencias Afines, Madrid, Spain.

Urania Kraków
Urania. Miesięcznik Polskiego Towarzystwa Miłośników Astronomii, Kraków. Publisher: Krakowska Drukarnia Prasowa, Kraków, Poland.

Vasiona
Vasiona. Revue d'Astronomie et d'Astronautique. Bulletin de la Société Astronomique "R. Bošković", Beograd.

Vatican Obs. Publ.
Vatican Observatory Publications, Specola Vaticana, Città del Vaticano.

Veröff. Astron. Rechen-Inst. Heidelberg
Veröffentlichungen des Astronomischen Rechen-Instituts Heidelberg. Verlag G. Braun, Karlsruhe, F.R. Germany.

Veröff. Bayer. Komm. Int. Erdmessung, Bayer. Akad. Wiss., Astron.-Geod. Arb.
Veröffentlichungen der Bayerischen Kommission für die Internationale Erdmessung der Bayerischen Akademie der Wissenschaften. Astronomisch-Geodätische Arbeiten. Published by Verlag der Bayerischen Akademie der Wissenschaften, München, F. R. Germany. ISSN 0340-7691, ISBN 3-7696-9782-0.

Veröff. Remeis-Sternw. Bamberg
Veröffentlichungen der Remeis-Sternwarte Bamberg, Astronomisches Institut der Universität Erlangen-Nürnberg.

Veröff. Sternw. Sonneberg
Akademie der Wissenschaften der DDR, Zentralinstitut für Astrophysik, Veröffentlichungen der Sternwarte in Sonneberg. Publisher: Akademie-Verlag, Berlin, German Democratic Republic.

Veröff. Zentralinst. Phys. Erde
Akademie der Wissenschaften der DDR, Forschungsbereich Geo- und Kosmowissenschaften. Veröffentlichungen des Zentralinstituts für Physik der Erde, Potsdam, German Democratic Republic.

Vesmír
Vesmír. Přírodovědecký časopis Čs. akademie věd. Publisher: Academia, Praha.

Vestn. Khar'kov. Univ.
Vestnik Khar'kovskogo Universiteta. Seriya Astronomicheskaya. Publishers: Izdatel'stvo Khar'kovskogo Universiteta, Khar'kov.

Vestn. Kiev. Univ.
Vestnik Kievskogo Universiteta. Seriya Astronomii. Publishers: Izdatel'stvo Kievskogo Universiteta, Kiev.

Vierteljahrsschr. Naturforsch. Ges. Zürich
Vierteljahrsschrift der Naturforschenden Gesellschaft in Zürich. Printer and Publisher: Leeman AG, Zürich, Switzerland.

Vistas Astron.
Vistas in Astronomy. An international review journal. Pergamon Press, Oxford – New York – Braunschweig.

Weather
Weather. James Glaisher House, Grenville Place, Bracknell, Berks RG12 1BX, England.

Wiad. Telekomun.
Wiadomosci Telekomunikacyjne. Editorial address: Warszawa, Kazimierzowska 52, Poland.

Wiss. Z. Friedrich-Schiller-Univ. Jena
Wissenschaftliche Zeitschrift der Friedrich-Schiller-Universität. Jena. Mathematisch-Naturwissenschaftliche Reihe; Edited by the Rektor der Friedrich-Schiller-Universität Jena, Am Anger 24, Jena, German Democratic Republic.

Wiss. Z. Humboldt-Univ. Berlin
Wissenschaftliche Zeitschrift der Humboldt-Universität zu Berlin. Mathematisch-Naturwissenschaftliche Reihe. Edited by the Rektor der Humboldt-Universität Berlin, Unter den Linden 6, 108 Berlin, German Democratic Republic.

Wuli
Wuli. Science Press, Peking, People's Republic of China. Subscription address: Guozi Shudian, P.O. Box 399, Peking.

Yamamoto Circ.
Yamamoto Circular. Published by the Yamamoto Observatory, Kamitanakami-Kiryutyo, Otu, Siga-Ken, [520-21] Japan.

Z. angew. Math. Mech.
Zeitschrift für angewandte Mathematik und Mechanik. Akademie-Verlag GmbH, 108 Berlin, Leipziger Strasse 3–4, German Democratic Republic.

Z. angew. Math. Phys.
Zeitschrift für angewandte Mathematik und Physik. Verlag Birkhauser, Postfach 4000, Basel 24, Switzerland.

Z. Elektr. Inf. - Energietech.
Zeitschrift für Elektrische Informations- und Energietechnik. Akademische Verlagsgesellschaft Geest & Portig K. G., DDR 701 Leipzig, Sternwartenstrasse 8, Germany. Formerly: Hochfrequenztech. & Elektroakust. (Germany) and Wiss. Z. Elektrotech. (Germany).

Z. Flugwiss. Weltraumforsch.
Zeitschrift für Flugwissenschaften und Weltraumforschung. Formerly: Z. Flugwiss., Postfach 3267, D-3000 Braunschweig, Germany.

Z. Naturforsch.
Zeitschrift für Naturforschung. Teil A. A Europhysics Journal. Physik–Physikalische Chemie–Kosmophysik. Verlag der Zeitschrift für Naturforschung, Tübingen. P. O. Box 2645, D–7400 Tübingen, F. R. Germany.

Z. Phys. A
Zeitschrift für Physik A. Atoms and Nuclei. Springer-Verlag, Berlin–Heidelberg–New York.

Z. Phys. B
Zeitschrift für Physik B. Condensed Matter and Quanta. Springer-Verlag, Berlin–Heidelberg–New York.

Z. Phys. C
Zeitschrift für Physik C. Particles and Fields. Springer-Verlag, P.O. Box 105280, D-6900 Heidelberg 1, Germany.

Z. Vermessungswes.
Zeitschrift für Vermessungswesen. Verlag Konrad Wittwer, 7000 Stuttgart 1, Nordbahnhofstrasse 16, Postfach 147, F. R. Germany.

Zeiss Inf.
Zeiss Information. Carl Zeiss, Oberkochen. F. R. Germany.

Zemlya i Vselennaya
Zemlya i Vselennaya. Astronomiya, Geofizika. Issledovaniya Kosmicheskogo Prostranstva, Nauchno-Populyarnyj Zhurnal Akademii Nauk SSSR. Publishers: Izdatel'stvo "Nauka", Moskva.

Zenit
Populair wetenschappelijk maandblad over sterrenkunde/weerkunde/ruimtevaart/ruimte-onderzoek/aanverwante wetenschappen en technieken. Bureau: Stichting De Koepel, Utrecht.

Zentralbl. Math. Grenzgeb. – Math. Abstr.
Zentralblatt für Mathematik und ihre Grenzgebiete – Mathematics Abstracts. Publisher: Springer-Verlag, Berlin–Heidelberg–New York.

Zvaigžņota Debess
Latvijas PSR Zinātņu Akadēmijas Radioastrofizikas Observatorijas Populārzinatnisks Gadalaiku Izdevums. Izdevnieciba "Zinātne", Riga.

Journals abstracted completely.

A selected number of periodicals listed in category 001 are central to the subject scope of *Astronomy and Astrophysics Abstracts.* Depending on their relevance, almost all papers of the journals listed below are abstracted in our service.

AAVSO Bull.
Acta Astron.
Acta Astron. Sinica
Acta Cosmologica
Adv. Astron. Astrophys.
Ann. Tokyo Astron. Obs.
Ann. Univ.-Sternw. Wien
Annu. Rep. Astron. Inst. Greece
Annu. Rev. Astron. Astrophys.
Assoc. Veneta Oss. Stelle Variabili Bull.
Astrofiz. Issled. Izv. Spets. Astrofiz. Obs.
Astrofizika
Astrometr. Astrofiz.
Astron. Astrophys.
Astron. Astrophys., Suppl. Ser.
Astron. J.
Astron. Nachr.
Astron. Tidsskr.
Astron. Tsirk.
Astron. Vestn.
Astron. Zh. Akad. Nauk SSSR
Astronomia
Astronomie
Astrophys. J.
Astrophys. J., Lett.
Astrophys. J., Suppl. Ser.
Astrophys. Lett.
Astrophys. Space Sci.
Australian J. Phys., Astrophys. Suppl.
BAV Rundbrief
BBSAG Bull.
Bol. Astron. Obs. Madrid
Bol. Inst. Tonantzintla
British Astron. Assoc. Circ.
Bull. AFOEV
Bull. American Astron. Soc.
Bull. Astron. Inst. Czechoslovakia
Bull. Astron., Obs. R. Belgique
Bull. Astron. Soc. India
Bull. Inf. Cent. Données Stellaires
Bull. Obs. Astron. Belgrade
Byull. Abastuman. Astrofiz. Obs.
Byull. Inst. Astrofiz.
Byull. Inst. Teor. Astron.
Carter Obs., Astron. Bull.
Celestial Mech.
Circ. Inf.
Circ. Stn. Astron. Int. Latitudine, Carloforte-Cagliari
Coelum
Comments Astrophys.
Comun. Obs. Astron. Univ. Coimbra
Contrib. Obs. New Mexico State Univ.
Dudley Obs. Rep.
Dunsink Obs. Publ.
El Universo
ESO Sci. Prepr.
ESO Tech. Rep.
G. Astron.
IAU Circ.
Icarus
Inf. Bull. Variable Stars
Inst. Theor. Astrophys., Blindern–Oslo, Rep.
Irish Astron. J.
Izv. Astron. Ehngel'gardt. Obs.
Izv. Glav. Astron. Obs. Pulkovo
Izv. Krymskoj Astrofiz. Obs.
J. American Assoc. Variable Star Obs.
J. British Astron. Assoc.
J. Hist. Astron.
J. R. Astron. Soc. Canada
Komet. Tsirk.
Komety i Meteory
Mem. Soc. Astron. Italiana
Mercury
Messenger
Meteoritics
Meteoritika
Minor Planet Bull.
Minor Planet Circ.
Mitt. Astron. Ges.
Mitt. Veränderl. Sterne (MVS)
Mon. Not. R. Astron. Soc.
Mon. Notes Astron. Soc. South Africa
Mon. Notes Int. Polar Motion Serv.
Moon Planets
Nablyud. Iskusstv. Nebesn. Tel
Nauchn. Inf.
News Lett. Astron. Soc. N.Y.
Obs. Astrophys. Lab., Univ. Helsinki. Rep.
Observatory
Occas. Rep. R. Obs. Edinburgh
Occultation Newsl.
Orion
Oss. Astrofis. Catania, Pubbl.
Oss. Mem. Oss. Astrofis. Arcetri
Perem. Zvezdy, Byull.
Perem. Zvezdy, Prilozhenie
Pis'ma v Astron. Zhurn.
Postępy Astron.
Probl. Kosm. Fiz.
Proc. Astron. Soc. Australia
Proc. Int. Latitude Obs. Mizusawa
Publ. Astron. Soc. Japan
Publ. Astron. Soc. Pacific
Publ. Dominion Astrophys. Obs.
Publ. Eidg. Sternw. Zürich
Publ. Int. Latitude Obs. Mizusawa
Publ. R. Obs. Edinburgh
Publ. Tartu Astrofiz. Obs.
Publ. United States Naval Obs.
Publ. Variable Star Sect. R. Astron. Soc. New Zealand
Q. Bull. Sol. Act.
Q. J. R. Astron. Soc.
R. Obs. Ann.
Rep. Obs. Lund
Rev. Mexicana Astron. Astrofis.
Rezul't. Nablyud. Iskusstv. Sputnikov Zemli
Sky Telesc.
Smithsonian Astrophys. Obs., Spec. Rep.
Smithsonian Contrib. Astrophys.
Sol. Phys.
Soln. Dannye, Byull.
Soobshch. Byurakan. Obs.
Soobshch. Gos. Astron. Inst. Shternberg
Soobshch. Spets. Astrofiz. Obs.
South African Astron. Obs. Circ.
South. Stars
Space Sci. Instrum.
Space Sci. Rev.
Sterne
Sterne Weltraum
Stockholms Obs. Ann.

Stockholms Obs. Rep.
Strolling Astron.
Tartu Astron. Obs. Teated
Tokyo Astron. Bull.
Tokyo Astron. Obs. Rep.
Tr. Astrofiz. Inst. Alma-Ata
Tr. Astron. Obs., *Leningrad*
Tr. Glav. Astron. Obs. Pulkovo
Tr. Inst. Teor. Astron., *Leningrad*
Tr. Kazan. Gorod. Astron. Obs.
Tr. Tashkent. Astron. Obs.
Trans. IAU
Tsirk. Astron. Inst. Tashkent
Tsirk. Astron. Obs. L'vov
United States Naval Obs., Circ.
Urania Barcelona
Urania Kraków
Vatican Obs. Publ.
Veröff. Astron. Rechen-Inst. Heidelberg
Veröff. Sternw. Sonneberg
Vestn. Khar'kov. Univ.
Vestn. Kiev. Univ.
Vistas Astron.
Yamamoto Circ.

002 Bibliographical Publications, Catalogues, Atlases

002.001 **North American theses and dissertations on the history of astronomy.**
J. Hist. Astron., Vol. 11, 148 - 151 (1980).

002.002 **Vilnius photometric catalogue (magnetic tape).** P. North.
Astron. Astrophys., Suppl. Ser., Vol. 41, 395 - 396 (1980).

This catalogue on magnetic tape contains the data in the *UPXYZVTS* system of Vilnius Observatory published until the end of 1979. There are 2 095 measurements concerning 1 879 stars.

002.003 **Viking Orbiter Stereo Imaging Catalog.** K. R. Blasius, A. V. Vetrone, M. D. Martin.
NASA Contract. Rep., NASA CR 3277, 302 pp. Price $ 11.75 (1980).

The extremely long missions of the two Viking Orbiter spacecraft have produced a wealth of photos of surface features. Many of these photos can be used to form stereo images allowing the earth-bound student of Mars to examine his subject in 3-D. This catalog is a technical guide to the use of stereo coverage within the complex Viking imaging data set.

002.004 **Astronomical infrared data base.** M. Schmitz, D. Y. Gezari, J. M. Mead.
Astron. Data Cent. Bull., Vol. 1, 12 - 13 (1980).

002.005 **Dearborn Observatory Catalogue of Faint Red Stars.** T. A. Nagy, R. S. Hill.
Astron. Data Cent. Bull., Vol. 1, 14 - 16 (1980).

002.006 **Machine-readable Durchmusterung Catalogues.** T. A. Nagy, J. M. Mead, W. H. Warren, Jr.
Astron. Data Cent. Bull., Vol. 1, 17 - 18 (1980).

002.007 **Catalog of supplemental Bonner Durchmusterung stars.** W. H. Warren, Jr., K. Kress.
Astron. Data Cent. Bull., Vol. 1, 19 - 20 (1980).

002.008 **Zone statistics for the Durchmusterung Catalogues.** W. H. Warren, Jr., T. A. Nagy, R. S. Hill.
Astron. Data Cent. Bull., Vol. 1, 21 - 24 (1980).

002.009 **A revised magnetic tape of the Boss General Catalogue of 33342 stars for the epoch 1950.**
T. A. Nagy, J. M. Mead.
Astron. Data Cent. Bull., Vol. 1, 25 - 27 (1980).

002.010 **Status report on machine-readable astronomical catalogues. Astronomical Data Center, NASA–Goddard Space Flight Center.**
W. H. Warren, Jr., T. A. Nagy, J. M. Mead.
Astron. Data Cent. Bull., Vol. 1, 32 - 50 (1980).

002.011 **Catalog of the Washington scale pairs observed with the Moscow zenith telescope.**
L. P. Basurmanova-Gribko.
Tr. Gos. Astron. Inst. Shternberga, Tom 49, 186 - 195 (1980). In Russian.

This article considers a catalog of 20 scale pairs compiled from 34 stars of the program of the Washington PZT. The catalog of differences of declinations of scale pairs obtained in the period 1968.2 to 1970.7 is given in a table.

002.012 **Preliminary catalogue of the declinations of the latitude programme stars obtained by the micrometer method.** I. M. Kalinina.
Tr. Gos. Astron. Inst. Shternberga, Tom 49, 196 - 212 (1980). In Russian.

The continuation of the Moscow declination catalogue stars of the latitude programmes contains 823 stars in declination zones 24°45′ to 36°30′ and 73°24′ to 86°39′.

002.013 **A revised optical catalog of quasi-stellar objects.** A. Hewitt, G. Burbidge.
Astrophys. J., Suppl. Ser., Vol. 43, 57 - 158 (1980).

A catalog containing the basic optical information on all QSOs and BL Lacertae objects which have been certainly identified, meaning, in the case of QSOs, that redshifts have been measured (complete to 1979 September), is given. Included are the coordinate designations, other names, equatorial coordinates, visual magnitudes, $(B-V)$ and $(U-B)$ colors, the emission-line redshifts, emission lines detected in the optical spectrum, absorption-line redshifts, and references for finding charts, absorption- and emission-line redshifts, and photometry and polarimetry for 1549 QSOs and BL Lacertae objects. Also given are tables containing the list of QSOs which lie in the fields of bright galaxies and QSOs which form close pairs of multiple systems.

002.014 **Bulletin GRG, No. 40. List of publications.** Gen. Relativ. Gravitation, Vol. 12, 305 - 345 (1980).

002.015 **Photometric catalogues: present and future.** B. Hauck.
Celestial Mech., Vol. 22, (see 012.002), 135 - 138 (1980).

Current sources of stellar photometric data are described, together with a summary of the quality of such data. The main future problem is the adequate dissemination of the information collected.

002.016 **Biographies of Einstein.** D. C. Cassidy. Einstein Symposion Berlin, (see 012.003), p. 490 - 500 (1979).

002.017 **On-line computer access to the Bibliographical Star Index.** J. M. Mead, T. A. Nagy, R. S. Hill.
Bull. American Astron. Soc., Vol. 12, 459 (1980). – Abstract.

002.018 **Infrared astronomical data base and Master Catalogue.** M. Schmitz, D. Y. Gezari, J. M. Mead.
Bull. American Astron. Soc., Vol. 12, 524 (1980). – Abstract.

002.019 **Catalogues of individual corrections to declinations of FK4 stars.** A. G. Oborneva.
Tr. Gos. Astron. Inst. Shternberga, Tom 50, 73 - 78 (1980). In Russian.

The FK4 stars were observed as reference stars for differential observations of the FKSZ. The two catalogues contain 93 and 283 individual corrections which are given for the epoch of observations. Internal mean square errors of one correction are ±0″.28 and 0″.38.

002.020 **Catalogue of the declinations of stars in the PZT program zone of Washington and Mizusawa obtained with the Moscow ZTL-180 telescope.** I. M. Kalinina.
Tr. Gos. Astron. Inst. Shternberga, Tom 50, 95 - 104 (1980). In Russian.

The catalogue contains 327 stars in declination zones 38°00′ to 39°30′ and 71°54′ to 73°24′ symmetric to the zenith. The mean epoch of the observations is 1975.2; the mean error of one observation is ± 0″.21.

002.021 **NLTT Catalogue. Volume IV. –30° to –90°.** W. J. Luyten.

Published by University of Minnesota, Minneapolis, Minnesota, USA. 3 + 151 pp. (1980).

The fourth volume of the catalogue contains about half the number of stars of that in the other three volumes. It is divided into two zones, viz. from −30 to −45, and from −45 to −90, each of which contains somewhat more than 4,000 stars. In the first of these the positions are given to 1^s, and 0'.1, and the motions to 0''.001, though for more than half the stars these data are merely copied from the LTT and BPM Catalogues, and do not have the accuracy of those in the preceding volumes. The final zone, however, is not much more than a repeat of the LTT and BPM Catalogues with the addition of a small number of stars from the other catalogues mentioned above. Therefore, in this zone positions are given to only $0^m.1$ and 1', and the motions to 0''.01.

002.022 **The nearest stars.** V. A. Zakhozhaj.
Vestn. Khar'k. Univ., No. 190. Fiz. Luny Planet. Fundam. Astrometr., Vyp. 14, p. 52 - 77 (1979). In Russian.

002.023 **Catalogue général d'étoiles de type O. Données spectroscopiques et photométriques. Bande magnétique et listage.** G. Goy.
Astron. Astrophys., Suppl. Ser., Vol. 42, 91 - 102 (1980).

A tape was prepared with the following contents: File 1: Data for 971 O type stars (3 card images of 80 characters each for 1 star). HD and BD number or other designation, 1900 position, galactic coordinates, polarization data, H II region membership, 7 colors in Geneva system, *V* magnitude and detailed spectral types are given. File 2: Monographs. All peculiarities are given for 385 stars. File 3: Abbreviations, content of columns, description, bibliography.

002.024 **On the derivation of a catalogue of radio source positions from interferometric observations.**
H. G. Walter.
Astron. Astrophys., Vol. 89, 198 - 203 (1980).

A catalogue of radio source positions is compiled from currently available high precision interferometric observations of compact extragalactic objects. There are twenty-eight objects constituting a reference coordinate system whose systematic error is approximately 0''.01 in both coordinates. The individual errors of the radio source positions amount to 0''.03 in right ascension and 0''.02 in declination. Systematic errors are derived for the various independent surveys contributing to the catalogue formation, and assessment of their weights is attempted.

002.025 **A catalog of nonstellar water maser sources.**
A. St. Clair Dinger, D. F. Dickinson.
Astron. J., Vol. 85, 1247 - 1254 (1980).

A list of 195 nonstellar galactic sources of water maser emission has been compiled. This list is not a comprehensive catalog of all the available information; it is intended to be used as an index. As such, it contains only coordinates, velocities, and selected references. The extragalactic sources are listed in a separate table.

002.026 **Distribution of astronomical sources in the second Equatorial Infrared Catalogue.** T. A. Nagy, L. H. Sweeney, J. R. Lesh, J. M. Mead, S. P. Maran, T. F. Heinsheimer. F. F. Yates.
Proc. Soc. Photo-Opt. Instrum. Eng., Vol. 197, (see 012.017) 217 - 224 (1979). – Abstr. in Phys. Abstr., Vol. 83, Abstr. 90346 (1980).

002.027 **A search for carbon stars in the AFGL Catalogue.**
A. Altamore, F. Smriglio, E. Bussoletti, C. E. Corsi, L. Rossi.
Astrophys. Space Sci., Vol. 72, 159 - 165 (1980).

A comparison between the General Catalogue of Cool Carbon Stars (CCS) and the AFGL Catalogue has been performed. Eighty-five stars have been found in common between the two lists. Eighty-four stars which were present in Baumert's comparison between CCS and the 2 μ Sky Survey have no counterpart in the AFGL. Four new tentative identifications are given. The analysis of the two colours diagrams K-[4.2 μ] vs. I−K and I-[4.2 μ] vs. [4.2 μ]−[11 μ] led to the conclusion that all the infrared emission from the sources seems to come from a single circumstellar shell.

002.028 **International Astronomical Union, Commission 42. Bibliography and Program Notes on Close Binaries, Nos. 33, 34.** (Material received by September 15, 1980).
G. Larsson-Leander (Editor).
Published by Lund Observatory, Lund, Sweden. 29 + 18 pp. (1980).

002.029 **NSSDC Data Listing.**
National Space Science Data Center/World Data Center A for Rockets and Satellites, National Aeronautics and Space Administration, Goddard Space Flight Center, Greenbelt, Maryland, NSSDC/WDC-A-R & S 80-10, 58 pp. (1980).

The NSSDC Data Listing provides a convenient reference to space science and supportive data available from the National Space Science Data Center (NSSDC). The first part of this listing, Satellite Data, is in an abbreviated form compared to the data catalogs normally published by NSSDC/WDC-A-R & S (World Data Center A for Rockets and Satellites). The second part, Non-Satellite Data, contains a listing of ground-based data, models computer routines, and composite spacecraft data that are available from NSSDC.

002.030 **Charts for southern variables. Series No. 12.**
F. M. Bateson, M. Morel, B. Sumner, R. Winnett.
Published by Astronomical Research Ltd., P. O. Box 3093, Greerton, Tauranga, New Zealand. 9 pp. + charts 501 - 551 (1980).

002.031 **Catalogs of proper motion stars. II. Stars brighter than visual magnitude 15 and south of declination +30° with annual proper motion between 0''.7 and 1''.0.**
O. J. Eggen.
Astrophys. J., Suppl. Ser., Vol. 43, 457 - 468 (1980).

Photometry is presented for stars with annual proper motion between 0''.7 and 1''.0, brighter than visual magnitude 15, and south of declination +30°. Available astrometric data for these objects are also summarized.

002.032 **First supplement to the catalogue of modern light curve synthesis photometric solutions of close binary systems.** B. Cester, G. Giuricin, F. Mardirossian, M. Mezzetti, F. Predolin.
Mem. Soc. Astron. Italiana, Vol. 51, 197 - 232 (1980).

Modern light curve synthesis photometric solutions of 16 close binary systems have been listed, which can be regarded as a supplement to the catalogue of photometric solutions by Cester et al. (1979).

002.033 **Bibliography of Jan Hendrik Oort 1922 - 1979.**
Compiled by D. Ondei, J. Ekers, W. N. Brouw, H. van Woerden, with a subject index.
Oort and the universe, (see 003.005), p. 175 - 195, 197 - 200 (1980).

002.034 **Results of radar observations of faint meteors. Catalogue of orbits to $+12^m$.**
B. L. Kashcheev, A. A. Tkachuk.
Mezhduved. geofiz. kom. pri Prezidiume AN SSSR. 232 pp. (1980). In Russian. – Abstr. in Ref. zh., 51. Astron., 10.51.61 (1980).

002.035 **Preliminary catalogue of 337 stars observed with the photoelectric astrolabe GDII No. 2 of Beijing**

Observatory. L.-z. Lu, D.-j. Luo, Z.-z. Wang.
Acta Astron. Sinica, Vol. 21, 305 - 313 (1980). In Chinese.
The residuals of 355 fundamental stars and 367 catalogue stars observed during 1976- 1978 with GDII No. 2 have been reduced and analysed. The relations between $2K$ and magnitude as well as spectrum type of FK4 stars observed have been discussed.

002.036 **The results of a treatment of the tape version of the [Fe/H] catalogue by Morel et al.**
S. V. Vereshchagin.
Bull. Inf. Cent. Données Stellaires, No. 19, p. 39 - 40 (1980).
The magnetic tape version of the catalogue by Morel et al. (1976) contains 517 stars with metallicity indices [Fe/H] obtained from the curves of growth. The author has investigated 225 dwarfs out of them. The aim of this investigation is to derive some conclusions on the evolution of the galactic disk.

002.037 **Discordances between SAO and HD numbers for bright stars.** D. Hoffleit.
Bull. Inf. Cent. Données Stellaires, No. 19, p. 41 - 51 (1980).

002.038 **Corrections to the catalog of open clusters (7022).** G. Lyngå.
Bull. Inf. Cent. Données Stellaires, No. 19, p. 52 (1980).

002.039 **Catalogue of extinction data of 12547 O....F stars, galactic clusters and δ-Cephei stars.**
T. Neckel, G. Klare, M. Sarcander.
Bull. Inf. Cent. Données Stellaires, No. 19, p. 61 - 62 (1980).

002.040 **UBV data 1976 - 1979.** J.-C. Mermilliod.
Bull. Inf. Cent. Données Stellaires, No. 19, p. 63 - 64 (1980).

002.041 **A compilation of Balmer lines photometric data.**
J.-C. Mermilliod, M. Mermilliod.
Bull. Inf. Cent. Données Stellaires, No. 19, p. 65 - 66 (1980).

002.042 **Catalogue of masses and ages of stars in 68 open clusters.** A. Piskunov.
Bull. Inf. Cent. Données Stellaires, No. 19, p. 67 - 70 (1980).
In the present catalogue the evolutionary masses and ages of about 7000 stars in 68 open clusters as derived from their positions in the theoretical H-R diagram are listed.

002.043 **Catalogue of physical parameters of spectroscopic binary stars.** Z. Kraitcheva (*Krajcheva*), E. Popova, A. Tutukov, L. Yungelson (*Yungel'son*).
Bull. Inf. Cent. Données Stellaires, No. 19, p. 71 - 73 (1980).

002.044 **The SAO catalogue with astrophysical data.**
F. Ochsenbein.
Bull. Inf. Cent. Données Stellaires, No. 19, p. 74 - 81 (1980).
The SAO catalogue with astrophysical data is now available at the Centre de Données Stellaires, both on microfiche and on magnetic tape. A description of the contents of this catalogue is given, and some statistics are provided.

002.045 **Catalogs recently published, to be published or in preparation. List XIII.** IAU Commission 45 Working Group on Spectroscopic and Photometric Data.
Bull. Inf. Cent. Données Stellaires, No. 19, p. 82 - 84 (1980).

002.046 **Catalogs currently available on microfiche.**
Bull. Inf. Cent. Données Stellaires, No. 19, p. 85 - 91 (1980).

002.047 **Star catalogs available at the C. D. S.: additions and corrections.**
Bull. Inf. Cent. Données Stellaires, No. 19, p. 92 - 93 (1980).

002.048 **Spectroscopy – what are the needs once space astrometry has given us a new data base.**
C. Jaschek.
Highlights of Astronomy, Vol. 5, (see 012.029), 795 - 798 (1980).

002.049 **Articles and notes on methods and instruments of modern astronomy published in "Zemlya i Vselennaya" from 1965 to 1979.**
Zemlya Vselennaya, 1980, No. 5, p. 78 - 79. In Russian.

002.050 **New USNO Zodiacal Star Catalog.**
R. E. Schmidt, T. E. Corbin, T. C. Van Flandern.
Bull. American Astron. Soc., Vol. 12, 740 (1980). – Abstract.

002.051 **Data information subsystem – Upper Atmosphere Data Bank.** M. Kramarski.
Artif. Satell., Vol. 15, No. 2, p. 19 - 25 (1980).
Assumption and description of the Upper Atmosphere Data Bank are presented, containing physical parameters of the troposphere, stratosphere and mesosphere, measured with rocket and balloon sounding. The Bank contains current data for the years 1964 - 1970 for the world network and data for the years 1960 - 1970 achieved from the measuring network in USSR. Programs and files used in subsystem with respect of Data Bank are described.

002.052 **Bibliography of dissertations on the history of astronomy in the USSR after the Second World War.**
J. Hist. Astron., Vol. 11, 221 - 224 (1980).

002.053 **Spectrophotometry with a self-scanned silicon diode array. III. A catalogue of energy distributions.**
A. L. Cochran.
Univ. Texas, Publ. Astron., No. 16, 282 pp. (1980).
A self-scanned silicon photodiode array was used to obtain spectrophotometry on a sample of 98 stars covering the nominal spectral range from 4600–10200 Å. The sample includes stars of known angular diameters and photometric standard stars. Tables and plots of the energy distributions are given.

002.054 **Sample spectral atlas for Sirius.**
R. L. Kurucz, I. Furenlid.
Smithsonian Astrophys. Obs., Spec. Rep. No. 387, 142 pp. (1979).
The authors present a sample spectral atlas of Sirius for the region 354 to 440 nm in which they show a photographic spectrum that resolves the line profiles and has a small-scale signal to noise better than 300. The atlas also shows a calculated spectrum that demonstrates its capabilities in spectrum synthesis. A roughly photometric spectrum at 10 nm/panel, then at 5 nm/panel with a comparison rotationally broadened calculated spectrum, then at 0.8 nm/panel with comparison unbroadened and rotationally broadened calculated spectra and line identifications is given.

002.055 **Lowell proper motions XIX.** Proper Motion Survey in the southern hemisphere with the 13-inch photographic telescope of the Lowell Observatory.
H. L. Giclas, R. Burnham, Jr., N. G. Thomas.
Lowell Obs. Bull. No. 165, Vol. 8, 145 - 154 (1979).

002.056 **Lowell Proper Motion Survey.** Summary catalog of GD and GR stars. Proper Motion Survey with the 13-inch photographic telescope of the Lowell Observatory.
H. L. Giclas, R. Burnham, Jr., N. G. Thomas.
Lowell Obs. Bull. No. 166, Vol. 8, 157 - 206 (1980).

002.057 **The ESO/Uppsala survey of the ESO (B) atlas of the southern sky – VIII.**
A. Lauberts, E. B. Holmberg, H.-E. Schuster, R. M. West.

ESO Sci. Prepr. No. 105, 83 pp. (1980). – Submitted to Astron. Astrophys., Suppl. Ser.

002.058 **Bibliographie der Mitarbeiter des Zentralinstituts für Physik der Erde 1969 - 1978.** L. Lerbs.
Veröff. Zentralinst. Phys. Erde, Nr. 55, 88 pp. (1980).

002.059 **Astronomy and Astrophysics Abstracts.** Vol. 27. Literature 1980, Part 1.
S. Böhme, U. Esser, W. Fricke, I. Heinrich, W. Hofmann, D. Krahn, D. Rosa, L. D. Schmadel, G. Zech (Editors).
Published for Astronomisches Rechen-Institut by Springer-Verlag, Berlin–Heidelberg–New York, 10 + 939 pp. Price DM 118.00; ca. US $ 64.90 [Subscription price DM 94.40; ca. US $ 52.00] (1980). ISBN 3-540-10134-9. ISBN 0-387-10134-9.

002.060 **Bibliography of the papers carried out at the Satellite Tracking Station No. 1132 Cluj-Napoca between 1957-1977.**
Visual Obs. Artif. Earth Satell., p. 135 - 141 (1977) = Publ. Obs. Astron. Univ. Cluj-Napoca, Romania, Nr. B 46-58.

002.061 **A bibliography of planetary geology principal investigators and their associates, 1979 - 1980.**
Compiled by E. Lettvin, J. M. Boyce.
NASA Tech. Memo., NASA TM 82180, 85 pp. (1980).

A compilation of selected bibliographic data specifically relating to recent publications (May 1979 through May 1980). Serves as a companion piece to NASA TM 81776, "Reports of Planetary Geology Programs, 1979 - 1980".

002.062 **Mapping the southern sky with the ESO 1 m Schmidt telescope.** H.-E. Schuster.
Messenger, No. 22, p. 7 - 10 (1980).

002.063 **A catalogue of masses, radii, and luminosities of 71 double-lined spectrum eclipsing binaries.**
B. Cester, G. Giuricin, F. Mardirossian, F. Predolin.
Mem. Soc. Astron. Italiana, Vol. 51, 539 - 586 (1980).

The masses, radii, and luminosities of 71 double-lined spectrum eclipsing binaries having lightcurves analyzed by means of modern lightcurve synthesis numerical models have been listed in a catalogue.

002.064 **Milestones in globular cluster research: a guide to the literature.** D. A. Hanes.
Globular clusters, (see 003.008), p. 1 - 20 (1980).

002.065 **Catalogue of proper motions of stars in selected regions of the sky with galaxies. II.**
N. V. Kharchenko.
Glav. astron. obs. AN USSR, Kiev, 1980. 151 pp. In Russian. Abstr. in Ref. zh., 51. Astron., 11.51.639 (1980).

002.066 **An update of the status of the Revised 3C Catalog of Radio Sources; 22 new galaxy redshifts.**
H. E. Smith, H. Spinrad.
Publ. Astron. Soc. Pacific, Vol. 92, 553 - 569 (1980).

The authors present an updated catalog of the important optical data for the radio sources in the Revised Third Cambridge Catalog. Optical positions, redshifts, magnitudes, and identifications have been included as well as radio flux densities and spectral indices for the sample of 297 extragalactic 3 CR sources. A bibliography of optical/radio data is also given. In addition to the literature compilation the authors present new spectroscopic observations of 22 3CR galaxies with previously unpublished redshifts.

002.067 **Making a photographic star atlas with a 5.5-inch Schmidt camera.** D. Lemay.
J. R. Astron. Soc. Canada, Vol. 74, 366 (1980). – Abstract.

002.068 **A union catalogue of antiquarian astronomy books.** R. C. Brooks.
J. R. Astron. Soc. Canada, Vol. 74, 367 (1980). – Abstract.

002.069 **List of dissertations defended at sessions of specialized councils in the field of astronomy and geophysics at the Moscow State University.**
L. N. Bondarenko.
Astron. Tsirk., No. 1108, p. 7 - 8 (1980). In Russian.

002.070 **Dissertations at the Odessa State University in 1979.** E. N. Makarenko.
Astron. Tsirk., No. 1116, p. 8 (1980). In Russian.

002.071 **Viking Orbiter views of Mars.**
M. H. Carr, W. A. Baum, K. R. Blasius, G. A. Briggs, J. A. Cutts, T. C. Duxbury, R. Greeley, J. Guest, H. Masursky, B. A. Smith, L. A. Soderblom, J. Veverka, J. B. Wellman, C. R. Spitzer (Editor), with a foreword by T. A. Mutch.
National Aeronautics and Space Administration, Washington, D.C., Scientific and Technical Information Branch, NASA SP-441, 8 + 182 pp. (1980). For sale by the Superintendent of Documents, U.S. Government Printing Office, Washington, D.C. 20402.

002.072 **The AAVSO Variable Star Atlas.**
Prepared by C. E. Scovil.
Published by Sky Publishing Corporation, 49 Bay State Road, Cambridge, Mass. 02238. 4 pp. + 178 charts, Price $ 39.95 (1980). ISBN 0-933346-28-X.

The purpose of the AAVSO Variable Star Atlas is to indicate the locations of all the brighter variable stars whose magnitude range exceeds 0.5. Also included are all other fainter variables in the current observing program of the American Association of Variable Star Observers and the Variable Star Section of the Royal Astronomical Society of New Zealand. The AAVSO atlas also contains all galaxies in Harvard Annals, Vol. 88, No. 2 (the so-called Shapley-Ames catalogue), all Messier objects, and many other deep-sky objects (open and globular clusters, as well as diffuse and planetary nebulae).

002.073 **Catalogue of cosmic gamma-ray bursts from data of the "Konus" experiment. Part 3.**
E. P. Mazets, S. V. Golenetskij, V. N. Il'inskij, V. N. Panov, R. L. Aptekar', Yu. A. Gur'yan, A. V. D'yachkov, M. P. Proskura, I. A. Sokolov, Z. Ya. Sokolova, N. G. Khavenson, T. V. Kharitonova.
Fiz.-tekh. inst. AN SSSR. Prepr., 1980, No. 662, 39 pp. In Russian. – Abstr. in Ref. zh., 51. Astron., 12.51.485 (1980).

002.074 **Annotations on papers on geomagnetism and aeronomy published in "News of Universities, Radiophysics", Vol. 12, Nos. 5 - 12 (1979). In Russian.**
Geomagn. Aehron., Tom 20, 1132 - 1134 (1980). In Russian.

002.075 **A master list of nonstellar optical astronomical objects.** R. S. Dixon, G. Sonneborn (Editors).
Ohio State University, Columbus, Ohio. 885 pp. Price $ 30.00 (1980). – From Phys. Today, Vol. 33, No. 9, p. 77 (1980).

002.076 **Atlas of the planets.** P. Doherty.
Hamlyn/McGraw Hill, Feltham, UK. 143 pp. Price £ 6.95, $ 16.95 (1980). ISBN 0-600-30439-6. – Review in Nature, Vol. 288, 514; 1980 (*D. W. Hughes*).

002.077 **Dictionary of scientific biography, Vol. 16: index.**
C. C. Gillispie (Editor).
Scribner's, New York. 510 pp. Price $ 65.00 (1980). – Review in Sky Telesc., Vol. 60, 321 (1980).

002.078 **New star chart.** L. Hoffmann.
Published with accompanying booklet by the author. Rio de Janeiro, Price $ 10.00 (1979). – Review in J. British Astron. Assoc., Vol. 91, 99 - 100; 1980 (*S. Dunlop*).

002.079 **Handbook for astronomical societies, 1980.**
B. Jones (Editor).
Federation of Astronomical Societies, Alcyone, 28 High House Avenue, Bradford, West Yorkshire. 78 pp. Price £ 1.90. Review in J. British Astron. Assoc., Vol. 90, 590 - 591; 1980 (*H. Hatfield*).

002.080 **A complete lunar map** ("Polnaya Karta Lunyi").
Yu. N. Lipskij (Scientific Supervisor).
Sternberg Astronomical Institute, Moscow, "Nauka" ("Science") Publishers, Moscow (1979). – Review in Irish Astron. J., Vol. 14, 33; 1979 (*E. Öpik*).

002.081 **True visual magnitude photographic star atlas. Vol. 5. Northern stars.** C. Papadopoulos, C. Scovil.
Pergamon Press, Oxford. Price $ 225.00 (1980). ISBN 0-08-021626-9. – Review in Phys. Abstr., Vol. 83, Abstr. 109376 (1980).

002.082 **The illustrated encyclopedia of astronomy and space.** I. Ridpath (Editor).
Revised edition. Macmillan, London, 240 pp. Price £ 9.95 (1979). – Review in J. British Astron. Assoc., Vol. 90, 588; 1980 (*H. Miles*).

002.083 **Near infrared photographic sky survey: a field index.** G. S. Rossano, E. R. Craine.
Astronomy and Astrophysics Series, Vol. 8. Pachart Publ. House, Tucson, Arizona. 202 pp. Price $ 38.00 (1980). ISBN 0-912918-11-X. – Review in Astrophys. Lett., Vol. 21, 79; 1980 (*R. W. Capps*).

002.084 **Atlas van Maan, Mars, Venus.** A. Rükl.
La Rivière & Voorhoeve, Zwolle. Price f 21.50 (1980). ISBN 90-6084-409-2. – Review in Zenit, 7e Jaarg., 237; 1980 (*G. Schilling*).

Webb Society deep-sky observer's handbook, Vol. 3: Open and globular clusters. See Abstr. 003.061.

The Astronomical Data Center at Goddard Space Flight Center. See Abstr. 013.004.

On-going and future astronomical data base activities at the Goddard Space Flight Center – February 1980. See Abstr. 013.025.

Automated plate assignment. See Abstr. 021.007.

On the construction of a comprehensive general catalogue of star positions. See Abstr. 041.001.

Systematic errors of the plate measurements of the Yale zone +9° to +20°. See Abstr. 041.002.

Astrometric radio source catalogs. See Abstr. 041.003.

Catalogue of positions of 15,000 stars in the zone 60° to 70° south declination. See Abstr. 041.005.

Fundamental Catalog – past and future. See Abstr. 041.011.

The FK5, an improved fundamental reference system extended to fainter stars. See Abstr. 041.012.

A comparison of the Smithsonian Astrophysical Observatory catalogue (SAO) with AGK3R and Perth 70. See Abstr. 041.033.

On the determination of the equinox and equator of the new fundamental reference coordinate system, the FK5. See Abstr. 043.001.

A catalogue of fine structures in type IV solar radio bursts. See Abstr. 077.026.

A catalogue of [Fe/H] determinations. See Abstr. 114.010.

Micrometer measures of 1,980 double stars. See Abstr. 118.027.

Binarité des étoiles supergéantes. See Abstr. 120.012.

High dynamic range observations in the fields of strong extragalactic radio sources. See Abstr. 141.128.

A survey of the sky between 0° to +2° declination. See Abstr. 141.172.

A Westerbork survey of clusters of galaxies. XIII. Deep 610 MHz source counts from the Cancer cluster field. See Abstr. 160.028.

003 Books

003.001 **Annual review of nuclear and particle science, Vol. 29.** J. D. Jackson, H. E. Gove, R. F. Schwitters (Editors). Annual Reviews Inc., Palo Alto, Calif., USA. 477 pp. (1979). ISBN 0-8243-1529-4. – Review in Phys. Abstr., Vol. 83, Abstr. 67940 (1980). – See abstracts 022.027, 061.008.

003.002 **Advances in planetary geology.** A. Woronow (Editor), with a foreword by J. M. Boyce. NASA Tech. Memo., NASA TM-81979. 5 + 332 pp. Price $ 12.00 (1980). – The individual contributions (dissertations) are included in their corresponding subject categories – see abstracts 097.013 - 097.015.

003.003 **Problems of cosmic physics. Vypusk 15.** S. K. Vsekhsvyatskij (Editor). Respublikanskij Mezhvedomstvennyj Nauchnyj Sbornik. Izdatel'stvo pri Kievskom Gosudarstvennom Universitete Izdatel'skogo Obedineniya "Vishcha Shkola", Kiev. 160 pp. Price 1 Rbl. 60 Kop. (1980). In Russian. – The individual contributions are included in their corresponding subject categories – see abstracts 034.021, 054.005, 061.020, 064.040, 074.038, 078.007, 080.030, 082.032, 084.039, 102.011 - 102.013, 103.002, 103.104, 103.121, 103.141, 104.019 - 104.021, 116.011.

003.004 **Annual Review of Astronomy and Astrophysics. Volume 18.** G. Burbidge, D. Layzer, J. G. Phillips (Editors). Annual Review Inc., 4139 El Camino Way, Palo Alto, Calif. 94306, USA. 9 + 591 pp. Price $ 21.00 (1980). ISBN 0-8243-0918-9. – The individual contributions are included in their corresponding categories – see abstracts 013.012, 064.045, 066.108, 091.017, 107.013, 112.017, 115.004, 122.102, 126.024, 131.128, 131.129, 141.098, 141.099, 143.029, 162.060.

003.005 **Oort and the universe. A sketch of Oort's research and person.** Liber amicorum presented to Jan Hendrik Oort on the occasion of his 80th birthday 28 April 1980. H. van Woerden, W. N. Brouw, H. C. van de Hulst (Editors). D. Reidel Publishing Company, Dordrecht, Holland – Boston, U. S. A. – London, England. 8 + 210 pp. Price Dfl. 55.00, US$ 29.00 cloth; Dfl. 25.00, US$ 12.95 paper (1980). ISBN 90-277-1180-1, ISBN 90-277-1209-3 pbk. – The individual contributions are included in their corresponding subject categories – see abstracts 002.033, 005.009 - 005.023, 013.013 - 013.017, 134.029.

003.006 **Highlights of Astronomy, Vol. 5**, as presented at the XVIIth General Assembly of the IAU, Montreal, 1979. P. A. Wayman (Editor). D. Reidel Publishing Company, Dordrecht, Holland–Boston, U.S.A.–London, England. 8 + 868 pp. Price Dfl. 160.00, US $ 84.00 cloth, Dfl 75.00, US $ 39.50 paper (1980). ISBN 90-277-1146-1 cloth, ISBN 90-277-1147-X paper. – See abstracts 012.022 - 012.031, 022.106, 051.021, 066.122, 117.027.

003.007 **Asymptotic analysis.** From theory to application. F. Verhulst (Editor). Lecture Notes in Mathematics. Vol. 711. Springer Verlag, Berlin – Heidelberg – New York. 240 pp. (1979). ISBN 3-540-09245-5. ISBN 0-387-09245-5. – The individual contributions within the subject scope of Astronomy and Astrophysics Abstracts are included in their corresponding categories – see abstracts 042.062, 042.063.

003.008 **Globular clusters.** Based on the proceedings of a Nato Advanced Study Institute held at the Institute of Astronomy, University of Cambridge, August 1978. D. A. Hanes, B. F. Madore (Editors). Cambridge University Press, Cambridge - London - New York - New Rochelle - Melbourne - Sydney. 7 + 390 pp. Price DM 142.80 (1980). ISBN 0-521-22861-1. – The individual contributions are included in their corresponding subject categories – see abstracts 002.064, 142.117, 142.118, 151.086, 151.087, 154.028 - 154.041, 158.224.

003.009 **Helios solar probes, science summaries.** J. H. Trainor (Editor). NASA Tech. Memo., NASA TM 82005, 105 pp. (1980). The individual contributions are included in their corresponding subject categories – see abstracts 032.555 - 032.566, 051.036 - 051.038, 052.048.

003.010 **Solar system plasma physics.** E. N. Parker, C. F. Kennel, L. J. Lanzerotti (Editors), with an epilogue by E. N. Parker. Vol. 1: Solar and solar wind plasma physics. Vol. 2: Magnetospheres. Vol. 3: Solar system plasma processes. North-Holland Publishing Company, Amsterdam – New York– Oxford. 344 + 49 + 402 + 49 + 378 + 49 pp. Price DM 479.80 (1979). ISBN Vol. 1:0444851151, ISBN Vol. 2:0444852662, ISBN Vol. 3:0444852670, ISBN Set no.:0444852689. The individual contributions within the subject scope of Astronomy and Astrophysics Abstracts are included in their corresponding categories – see abstracts 062.126, 062.127, 073.116, 074.106 - 074.108, 080.075, 083.048, 083.049, 084.097 - 084.102, 091.052, 091.053, 092.005, 097.079, 099.150, 102.038, 143.045.

003.011 **Geodynamics and astrometry. Foundations, methods, results.** Collected articles dedicated to the 100th anniversary of the birthday of A. Ya. Orlov. E. P. Fedorov (Editor), and with introductory remarks. Akademiya Nauk Ukrainskoj SSR, Glavnaya Astronomicheskaya Observatoriya, Poltavskaya Gravimetricheskaya Observatoriya Instituta Geofiziki im. S. I. Subbotina. Naukova Dumka, Kiev. 167 pp. Price 2 Rbl. 50 Kop. (1980). In Russian. – The individual papers are included in their corresponding subject categories – see abstracts 004.080, 005.044, 005.045, 031.610, 041.037 - 041.040, 044.038, 045.011, 081.033, 081.034.

003.012 **Solar flares.** A monograph from Skylab Solar Workshop II. P. A. Sturrock (Editor). Colorado Associated University Press, Boulder, Colo., USA. 10+513 pp. Price $17.50 (1980). ISBN 0-87081-076-6. – The individual contributions are included in their corresponding subject categories – see abstracts 011.041, 073.128 - 073.137.

003.013 **The state of the Universe.** Wolfson College Lectures 1979. G. T. Bath (Editor). Clarendon Press, Oxford. 7+199 pp. Price £8.95, $24.95 (1980). ISBN 0-19-857549-1. – Reviews in J. British Astron. Assoc., Vol. 91, 94 - 95; 1980 (*J. Mitton*); Nature, Vol. 288, 34 - 35; 1980 (*J. Silk*); Observatory, Vol. 100, 135; 1980 (*R. C. Smith*). – The individual contributions are included in their corresponding categories – see abstracts 031.622, 066.191, 091.068, 114.168, 142.140, 158.319, 161.045, 162.118.

003.014 **Cosmic X-ray astronomy.** D. J. Adams. Monographs on astronomical subjects: 6. Adam Hilger Ltd., Bristol, England. 10 + 150 pp. Price £ 13.50

(1980). ISBN 0-85274-253-3. – Reviews in Astron. Tidsskr., Årg. 13, 190 - 191; 1980 (*K. Gyldenkerne*); Ciel Terre, Vol. 96, 405; 1980 (*M. Gabriel*); J. British Astron. Assoc., Vol. 91, 97 - 98; 1980 (*G. T. Bath*); J. British Interplanet. Soc., Vol. 33, 440 (1980).

003.015 **The planet Saturn.** A history of observation, theory and discovery. A. F. O'D. Alexander.
Reprint of the 1962 edition. Dover, New York. 474 pp. Price $ 8.00 (1980). – Review in Sky Telesc., Vol. 60, 321 (1980).

003.016 **Nel cosmo alla ricerca della vita.** P. Angela.
Ed. Garzanti, Milano. 293 pp. Price L. 9,000 (1980). Review in Orione, Vol. 2, 126 - 127; 1980 (*S. Baroni*).

003.017 **The Universe.** I. Asimov.
Third edition. Walker & Co., 720 Fifth Ave., New York, N.Y. 10019. 321 pp. Price $ 15.95 (1980). – Review in Sky Telesc., Vol. 60, 418 (1980).

003.018 **Charge-coupled devices and their applications.** J. D. E. Beynon, D. R. Lamb (Editors).
McGraw-Hill, London. 275 pp. Price $ 20.00 (1980). ISBN 0-07-084522-0. – Review in Astrophys. Lett., Vol. 21, 79; 1980 (*G. E. Danielson*).

003.019 **Project Orion.** A design study of a system for detecting extrasolar planets. D. C. Black (Editor).
National Aeronautics and Space Administration, NASA SP-436. For sale by the Superintendent of Documents, U.S. Government Printing Office, Washington, D. C. 20402. 13 + 204 pp. (1980).

003.020 **Schumann resonances in the earth-ionosphere cavity.** P. V. Bliokh, A. P. Nikolaenko, Yu. F. Fillippov.
Peter Peregrinus, Stevenage, England. 10 + 166 pp. (1980). ISBN 0-906048-33-8. – Review in Phys. Abstr., Vol. 83, Abstr. 86050 (1980).

003.021 **Astronomy illustrated.**
B. L. Bonneau, B. A. Smith.
Third edition. Kendall/Hunt Publishing Co., 2460 Kerper Blvd. Dubuque, Iowa 52001. 248 pp. Price $ 13.95 (1980). Review in Sky Telesc., Vol. 60, 519 (1980).

003.022 **Calcul astronomique pour amateurs** adapté à l'emploi d'un calculateur électronique de poche.
S. Bouiges.
Second edition. Masson, Paris–New York–Barcelone–Milan. 144 pp. (1980). ISBN 2-225-66803-5.
Contents: Rappel de notions élémentaires d'astronomie. Le calcul des positions planétaires. Les relations Ciel-Terre. La programmation sur calculatrice de poche. Exemples de programmes. Note sur les fonctions trigonométriques inverses.

003.023 **Basic steps in astronomy.** J. Boulton.
Sterling Publishing Co., 2 Park Ave., New York, N.Y. 10016. 144 pp. Price $ 9.95 (1979). – Review in Sky Telesc., Vol. 60, 519 (1980).

003.024 **The galactic club: intelligent life in outer space.** R. N. Bracewell.
W. W. Norton, 141 pp. Price £ 2.50 (1980). ISBN 0-393-95022-0. – From Nature, Vol. 287, No. 5785, p. XVII (1980).

003.025 **Nordlyset.** A. Brekke, A. Egeland.
Grøndahl & Søn Forlag, Oslo, 143 pp. Price N. kr. 148.00 (1979). – Review in Astron. Tidsskr., Årg. 13, 140 - 141; 1980 (*T. S. Ringnes*).

003.026 **Analytical algorithms of celestial mechanics.** V. A. Brumberg.
Nauka. Glavnaya redaktsiya fiziko-matematicheskoj literatury, Moskva. 208 pp. Price 1 Rbl. 70 Kop. (1980). In Russian.

003.027 **The age of Stonehenge.** C. Burgess.
Dent/Biblio. 402 pp. Price £ 12.00, $ 25.00 (1980). Review in Nature, Vol. 288, 39 - 40; 1980 (*R. J. C. Atkinson*).

003.028 **Spacetime, geometry, cosmology.** W. L. Burke.
University Science Books, 20 Edgehill Rd., Mill Valley, Calif. 94941, U.S.A. 329 pp. Price $ 22.50 (1980). ISBN 0-935702-10-6. – Review in Sky Telesc., Vol. 60, 419 (1980).

003.029 **Rings of stone.** The prehistoric stone circles of Britain and Ireland. A. Burl.
Frances Lincoln, London. 280 pp. Price £ 9.95 (1979). Review in Archaeoastronomy, No. 2, p. S103 - S104; 1980 (*O. Gingerich*).

003.030 **Cosmos, earth, and man: a short history of the universe.** P. Cloud.
Yale University Press. 372 pp. Price £ 5.00 (1980). ISBN 0-300-02594-7. – From Nature, Vol. 287, No. 5785, p. XVII (1980).

003.031 **Interiors of the planets.** A. H. Cook.
Cambridge University Press, Cambridge–London–New York–New Rochelle–Melbourne–Sydney.
11 + 348 pp. Price DM 129.80 (1980). ISBN 0-521-23214-7.
Contents: The internal structure of the earth. Methods for the determination of the dynamical properties of planets. Equations of state of terrestrial materials. The moon. Mars, Venus and Mercury. High pressure metals. Jupiter and Saturn, Uranus and Neptune. Departures from the hydrostatic state.

003.032 **La conquista dello spazio.**
M. Coradini, M. Fulchignoni.
Ed. Newton Compton, Rome. 96 pp. Price L 5,000 (1980). Review in Orione, Vol. 2, 126; 1980 (*V. Zappalà*).

003.033 **Theory of stellar pulsation.** J. P. Cox.
Princeton University Press, Princeton, N.J. 14 + 382 pp. Price $ 40.00 cloth; $ 13.50 paper (1980). – Reviews in Nature, Vol. 287, 876; 1980 (*R. J. Tayler*); Science, Vol. 210, 1119 - 1120; 1980 (*J.-P. Zahn*); Sky Telesc., Vol. 60, 418 (1980).

003.034 **Other worlds: space, superspace and the quantum universe.** P. Davies.
J. M. Dent, London–Toronto–Melbourne. 207 pp. Price £ 7.50 (1980). ISBN 0-460-04400-1. – Reviews in J. British Astron. Assoc., Vol. 91, 97; 1980 (*C. A. Ronan*); J. British Interplanet. Soc., Vol. 33, 440 (1980); Nature, Vol. 287, 567 - 568; 1980 (*J. S. Bell*).

003.035 **Stardoom.** P. Davies.
Paperpack edition of "The runaway universe", 1978. Fontana. 219 pp. Price £ 0.95 (1979). – Review in J. British Astron. Assoc., Vol. 90, 492; 1980 (*I. Nicolson*).

003.036 **The search for gravity waves.** P. C. W. Davies.
Cambridge University Press, Cambridge–London–New York–New Rochelle–Melbourne–Sydney.
8 + 144 pp. Price DM 36.80 (1980). ISBN 0-521-23197-3.
Contents: Electromagnetic waves. What are gravity waves? Sources of gravity waves. Gravity wave detectors. Have they been seen? Theory of gravity wave detectors.

003.037 **Astronomia e poesia.** R. R. de Freitas Mourão.
Difel/Difusão Editorial S. A., Rio de Janeiro–São Paulo, Brazil. 149 pp. (1977).

003.038 **Buracos negros.** Universos em colapso.
R. R. de Freitas Mourao.
Second edition. Editora Vozes Ltda., rua Frei Luís 100, 25.600 Petrópolis, Brasil. 123 pp. (1980).

003.039 **Röntgenastronomie.** Monografie over Astronomie en Astrofysica. T. Dethier.
Available from Volkssterrenwacht Urania, Mattheessensstraat 62, 2540 Hove, Belgium. Price BF 150.00. – Review in Zenit, 7e Jaarg., 169 (1980).

003.040 **Galileo.** S. Drake.
Oxford University Press. 100 pp. Price £ 4.50 hbk; 95 p. pbk. (1980). ISBN 0-19-287527-2 hbk.; ISBN 0-19-287526-4 pbk. – From Nature, Vol. 287, No. 5785, p. XVII (1980).

003.041 **Physics of the interstellar medium.**
J. E. Dyson, D. A. Williams.
Manchester University Press. 194 pp. Price £ 11.95 (1980). ISBN 0-7190-0797-6. – From Nature, Vol. 287, 373 (1980).

003.042 **Who goes there?** E. Edelson.
McGraw-Hill. 196 pp. Price $ 4.95 (1980). – Review in Sky Telesc., Vol. 60, 417 (1980).

003.043 **Life beyond earth: the intelligent earthling's guide to life in the universe.**
G. Feinberg, R. Shapiro.
William Morrow, New York. 480 pp. Price $ 14.95 hbk., $ 7.95 pbk. (1980). ISBN 0-688-03642-2 hbk.; ISBN 0-688-08642-X pbk. – Review in Nature, Vol. 288, 34; 1980 (*P. Davies*).

003.044 **Galaxies.** T. Ferris.
Thames & Hudson, Sierra Club. 192 pp. Price £ 20.00, $ 75.00 (1980). – Review in Nature, Vol. 288, 515; 1980 (*P. Campbell*).

003.045 **Galileo and the art of reasoning. Rhetorical foundations of logic and scientific method.**
M. A. Finocchiaro.
Boston Studies in the Philosophy of Science, Vol. 61. D. Reidel Publishing Company, Dordrecht, Holland–Boston, USA, London–England. 478 pp. Price Dfl. 80.00, $ 42.00 hbk, Dfl. 40.00, $ 21.00 pbk. (1980). ISBN 90-277-1094-5 hbk., ISBN 90-277-1095-3 pbk. – From Nature, Vol. 288, 523 (1980).

003.046 **A theory of cosmology.** The small bang theory and an introduction to the new physics. D. Fiske.
Vantage Press, New York. 14 + 138 pp. Price $ 13.00 (1980). From Science, Vol. 210, 1158 (1980).

003.047 **The meteorite crater Zhamanshin.**
P. V. Florenskij, A. I. Dabizha.
Nauka, Moskva. 127 pp. (1980). In Russian. – Review in Ref. zh., 51. Astron., 1.51.45 (1981).

003.048 **The enigma of Stonehenge.**
J. Fowles, B. Brukoff.
Cape/Summit. 128 pp. Price £ 6.95, $ 19.95 (1980). – Review in Nature, Vol. 288, 39 - 40; 1980 (*R. J. C. Atkinson*).

003.049 **The ambidextrous universe: mirror asymmetry and time-reversed worlds.** M. Gardner.
Second edition. Scribner's, New York. 293 pp. Price $ 4.95 (1979). – From Phys. Today, Vol. 33, No. 10, p. 81 (1980).

003.050 **Guide de l'astronome amateur.** D. Godillon.
Maloine S. A. éditeur, 27 rue de l'école de médecine, 75006 Paris, France. 610 pp. Price F 148.00 (1980). – Review in Sky Telesc., Vol. 60, 418 (1980).

003.051 **The strangest star.** A scientific account of the life and death of the sun. J. Gribbin.
Athlone Press, London. 184 pp. Price £ 6.95 (1980). ISBN 0-485-11207-8. – Review in Astron. Tidsskr., Årg. 13, 147 (1980).

003.052 **Geodesy and the earth's gravity field. Vol. II: Geodynamics and advanced methods.** E. Groten.
Dümmlerbuch 7838. F. Dümmler, Bonn. 5 pp. + p. 411 - 724 (1980). – Review in Gerlands Beitr. Geophys., Band 89, 534; 1980 (*K. Arnold*).

003.053 **Flare stars.** G. A. Gurzadyan.
International Series in Natural Philosophy, Vol. 101. Pergamon Press, Oxford–New York. 14 + 344 pp. Price $ 50.00, £ 25.00 (1980). ISBN 0-08-0230350. – Review in Sky Telesc., Vol. 60, 419 (1980).

003.054 **Non-linear methods of spectral analysis.**
S. Haykin (Editor).
Topics in Applied Physics, Vol. 34. Springer-Verlag, Berlin–Heidelberg–New York. 10 + 247 pp. Price DM 89.00, $ 49.00 (1979). – Review in Space Sci. Rev., Vol. 26, 454; 1980 (*C. de Jager*).

003.055 **Relativistic cosmology.** An introduction.
J. Heidmann. Translated by S. Mitton, J. Mitton.
Springer-Verlag, Berlin–Heidelberg–New York. 15 + 168 pp. Price DM 48.00, US $ 28.40 (1980). ISBN 3-540-10138-1.

Contents: The metagalaxy out to a distance of one gigaparsec: Distance scale. The distribution of galaxies in space. The expansion of the universe. Nearby intergalactic matter. The density of the universe. The age of the universe. – Spaces with constant curvature: Locally Euclidean spaces. Locally non-Euclidean spaces. Spherical and hyperbolic spaces. – Model universes: Uniform relativistic model universes. Theory of observations in the relativistic zone. The cosmological constant. Cosmological horizons. – The metagalaxy in the relativistic zone: The Hubble diagram for galaxies. Distant intergalactic material. Radio galaxies and quasars. The cosmic microwave background.

003.056 **General relativity and gravitation. Vol. 1 and 2.** One hundred years after the birth of Albert Einstein.
A. Held (Editor).
Plenum Press, New York. Vol. 1: 18 + 540 pp. Price $ 49.75; Vol. 2: 20 + 598 pp. Price $ 49.75 (1980). – Reviews in Science, Vol. 209, 1234; 1980 (*S. A. Bludman*); Sky Telesc., Vol. 60, 419 (1980).

003.057 **Greenwich time, and the discovery of longitude.**
D. Howse.
Oxford University Press, Oxford. 18 + 254 pp. Price £ 7.95, $ 24.95 (1980). ISBN 0-19-215948-8. – Reviews in J. British Astron. Assoc., Vol. 90, 489; 1980 (*P. Moore*); Nature, Vol. 288, 516; 1980 (*D. de Solla Price*); Sci. American, Vol. 243, No. 3, p. 46 - 48 (1980); Sky Telesc., Vol. 60, 233 (1980); Sky Telesc., Vol. 60, 513 - 514; 1980 (*M. M. Thomson*).

003.058 **Cosmogonia del sistema solare.** F. Hoyle.
Mondadori. Price L. 6,500 (1979). – Review in G. A. A. B., No. 59 - 60, p. 13 - 14; 1980 (*S. Ghedini*).

003.059 **Steady-state cosmology re-visited.** F. Hoyle.
University College Cardiff Press, Cardiff, U. K. 72 pp. (1980). ISBN 0-906449-22-7.

Contents: The SS theory, according to F. H. The SS theory, according to Bondi and Gold. The steady-state concept. The redshift-magnitude relation. The counting of radio sources. The microwave background in anticipation. The conformal invariance and the conservation (or otherwise) of baryons. Thermalising the microwave background. The hot big-bang does not produce galaxies. Helium again. White holes in general and a white hole of galactic mass in particular. White holes in the SS theory. The microwave background again. The helium abundance (for the last time) and the deuterium problem. An hierarchical picture. Condensations within a galactic white hole. The information content of the universe and its relevance for cosmology. An historical note.

003.060 **The brightest stars.** C. de Jager.
Geophysics and Astrophysics Monographs, Vol. 19. D. Reidel Publishing Company, Dordrecht, Holland–Boston, USA–London, England. 12+458 pp. Price Dfl. 140.00, $ 73.50 cloth, Dfl. 60.00, $ 31.50 paper (1980). ISBN 90-277-1109-7, ISBN 90-277-1110-0 (pbk.).

Contents: The upper boundaries of the Hertzsprung-Russell diagram. The main observational characteristics of the most luminous stars. Spectral characteristics and stellar parameters for the main groups of luminous stars; some prototypes. The evolution of massive stars. The structure of very tenuous stellar atmospheres. Chromospheres, coronae, gas and dust around luminous stars. Mass-loss from bright stars. Variability of supergiant atmospheres. Stars of transient extreme brightness; novae, supernovae.

003.061 **Webb Society deep-sky observer's handbook, Vol. 3: Open and globular clusters.**
K. G. Jones (Editor).
Lutterworth Press, Guildford, England; Enslow Publishers, Hillside, N. J. 14 + 206 pp. Price £ 4.95, $ 8.95 (1980). Reviews in J. British Astron. Assoc., Vol. 90, 587 - 588; 1980 (*H. Couper*); Sky Telesc., Vol. 60, 519 (1980).

003.062 **Earth and cosmos.**
R. S. Kandel, with a preface by J.-C. Pecker.
Pergamon Press, Oxford–New York–Toronto–Sydney–Paris–Frankfurt. 12 + 254 pp. Price $ 30.00, £ 13.50 hardcover, $ 14.90, £ 6.95 flexicover (1980). ISBN 0-08-025016-5 hardcover, ISBN 0-08-023086-5 flexicover. – Review in Nature, Vol. 288, 34; 1980 (*P. Davies*).

Contents: Overview. Matter, radiation, and the basic forces of nature. The earth and the universe. The earth in the Galaxy. The stability of the terrestrial environment. Sun and earth. The energy balance of the atmosphere. The astronomical rhythms: day and night, the seasons, and tides. The global circulation of the atmosphere. Continents, oceans, and climate. Life and the earth. The impact of man. The future of humanity.

003.063 **L'Etrenne ou la Neige Sexangulaire.**
J. Kepler. Traduction critique par R. Halleux.
J. Vrin, Paris. 4 + 169 pp. (1975). – Review in J. Hist. Astron., Vol. 11, 143 (1980).

003.064 **The state of the planet.** A report prepared by the International Federation of Institutes for Advanced Study. A. King.
Pergamon Press, Oxford–New York. 11 + 130 pp. Price $ 22.00, £ 10.00 hbk., $ 11.00, £ 5.00 flexi. (1980). ISBN 0-08-024717-2 hbk, ISBN 0-08-024716-4 flexi. – From Nature, Vol. 286, 917 (1980).

003.065 **The history of the telescope.** H. C. King.
Reprint of the 1955 edition. Dover, New York. 18 + 456 pp. Price $ 8.95 (1979). – From Science, Vol. 210, 456 (1980).

003.066 **Hundert Milliarden Sonnen.** Geburt, Leben und Tod der Sterne. R. Kippenhahn.
Piper Verlag, München. 276 pp. Price DM 48.00 (1980). Review in Sterne Weltraum, Jahrg. 19, 352; 1980 (*I. Appenzeller*).

003.067 **Coherence and correlation in atomic collisions.**
H. Kleinpoppen, J. F. Williams.
Plenum Press, New York. 14 + 806 pp. (1980). ISBN 0-306-40250-5. – Review in Phys. Abstr., Vol. 83, Abstr. 90615 (1980).

003.068 **K. E. Tsiolkovskij. Textbook for students.**
A. A. Kosmodem'yanskij.
Prosveshchenie, Moskva. 144 pp. (1980). In Russian. – Review in Ref. zh., 51. Astron., 8.51.26 (1980).

003.069 **Heralds of the universe.** L. M. Kuznetsova.
Znanie, Moskva. 160 pp. (1980). In Russian.
From Ref. zh., 51. Astron., 11.51.60 (1980).

003.070 **The earth's variable rotation: geophysical causes and consequences.** K. Lambeck.
Cambridge Monographs on Mechanics and Applied Mathematics, Cambridge University Press, Cambridge. 11 + 449 pp. Price £ 35.00 (1980). ISBN 0-521-22769-0. – Review in Phys. Abstr., Vol. 83, Abstr. 76902 (1980).

003.071 **Star maps for beginners.**
I. M. Levitt, R. K. Marshall.
Simon and Schuster, New York. 64 pp. Price $ 4.95, £ 2.95 (1980). ISBN 0-671-25422-7. – Reviews in Nature, Vol. 288, 514; 1980 (*D. W. Hughes*); Sky Telesc., Vol. 60, 417 (1980).

003.072 **Atmospheric phenomena.** Reprinted Scientific American articles. D. K. Lynch (Compiler).
W. H. Freeman. 175 pp. Price $ 15.95 cloth, $ 8.95 paper (1980). Review in Sky Telesc., Vol. 60, 519 (1980).

003.073 **Die erste Sternwarte Europas mit ihren Instrumenten und Uhren. 400 Jahre Jost Bürgi in Kassel.**
L. von Mackensen.
Callwey Verlag, Munich. 158 pp. Price DM 36.00 (1979). Review in J. Hist. Astron., Vol. 11, 212 - 213; 1980 (*O. Gingerich*).

003.074 **Special report to Celestron telescope owners.**
R. L. Mansfield.
Astronomical Data Service, 3922 Leisure Lane, Colorado Springs, Colo. 80917. 24 pp. Price $ 4.75 (1980). – Review in Sky Telesc., Vol. 60, 321 (1980).

003.075 **Physics and the physical universe.** J. B. Marion.
Third edition. Wiley, New York. 455 pp. Price $ 19.95 (1980). – From Phys. Today, Vol. 33, No. 10, p. 81 (1980).

003.076 **Plasma astrophysics. Nonthermal processes in diffuse magnetized plasmas. Vol. 1: The emission, absorption and transfer of waves in plasmas. Vol. 2: Astrophysical applications.** D. B. Melrose.
Gordon & Breach, New York–London–Paris. Vol. 1: 9 + 269 pp. Price DM 114.85 (1980). Vol. 2: 8 + 423 pp. Price DM 159.75 (1980). ISBN 0-667-03490-3. – Review in Phys. Abstr., Vol. 83, Abstr. 105446 (1980).

Contents: Waves in plasmas. Spontaneous emission. Gyromagnetic, synchrotron and inverse Compton emissions. The induced processes and quasi-linear theory. The absorption and transfer of radiation. Scattering of fast particles. Acceleration of fast particles. Interpretation of synchrotron spectra. The plasma emission processes. Solar radio bursts. Radiation into the magnetoionic mode and mode coupling. Radiation from rotating magnetospheres.

003.077 **Solar eclipse of July 31, 1981 and its observation.** A. A. Mikhajlov (Editor).
Nauka, Moskva. 159 pp. (1980). In Russian. – Review in Ref. zh., 51. Astron., 12.51.36 (1980).

003.078 **Physical processes in the ionosphere and magnetosphere.** E. V. Mishin (Editor).
Inst. zemn. magn., ionos. i rasprostr. radiovoln, Moskva. 150 pp. Price 1 Rbl. 20 Kop. (1979). In Russian. – Review in Ref. zh., 62. Issled. kosm. prostranstva, 8.62.34 (1980).

003.079 **Life in space.** D. L. Moche.
A & W Publ. Inc., New York. 160 pp. Price $ 10.95 paper, $ 25.00 cloth (1980). – From J. R. Astron. Soc. Canada, Vol. 74, 316 (1980).

003.080 **Magnetic field generation in electrically conducting fluids.** H. K. Moffatt.
Translated from the English edition. Mir, Moskva. 339 pp. (1980). In Russian. – Review in Ref. zh., 51. Astron., 9.51.38 (1980).

003.081 **1981 Yearbook of Astronomy.** P. Moore (Editor).
Sidgwick & Jackson Ltd., London. 248 pp. Price £ 6.95 hbk., £ 3.95 pbk. (1980). – Review in J. British Astron. Assoc., Vol. 91, 93 - 94; 1980 (*R. D. Prout*).

003.082 **Astronomy with binoculars.** J. Muirden.
Crowell Publishing Co., 10 East 53rd Street, New York, N. Y. 10022. 150 pp. Price $ 8.95 (1979). – Review in Strolling Astron., Vol. 28, 156; 1980 (*E. F. Bailey*).

003.083 **The cloudy night book.** G. S. Mumford.
Sky Publishing Corp., Cambridge, Mass. 115 pp. Price $ 4.95 (1979). – Review in J. R. Astron. Soc. Canada, Vol. 74, 374; 1980 (*G. L. H. Harris*).

003.084 **Space developments for the future of mankind.** L. G. Napolitano (Editor).
Pergamon Press. 600 pp. Price £ 45.00 (1980). – Review in J. British Interplanet. Soc., Vol. 33, 440 (1980).

003.085 **Främmande världar.** P. Nilson.
Raben & Sjögren, Stockholm. 207 pp. Price Sv. kr. 75.00 (1980). – Review in Astron. Tidsskr., Årg. 13, 147 (1980).

003.086 **The image of eternity: roots of time in the physical world.** D. Park.
University of Massachusetts Press. 149 pp. Price $ 14.50 (1980). – Review in Nature, Vol. 288, 305 - 306; 1980 (*G. E. Uhlenbeck*).

003.087 **Frequency analysis of astronomical time series.** Ya. Pel't.
Valgus, Tallin. 135 pp. (1980). In Russian. – Review in Ref. zh., 51. Astron., 1.51.51 (1981).

003.088 **Starlight nights.** L. C. Peltier.
Sky Publishing Corporation, Cambridge, Mass. 236 pp. Price $ 7.95 paperbound (1980). – Review in Sky Telesc., Vol. 60, 517 - 518; 1980 (*D. Hoffman*).

003.089 **Cosmogenic structures on the earth.** L. L. Perchuk (Editor).
Nauka, Moskva. 81 pp. (1980). In Russian. – From Ref. zh., 51. Astron., 10.51.59 (1980).

003.090 **Geomagnetic field and interior structure of the earth.** V. I. Pochtarev (Editor).
Moskva. 304 pp. (1980). In Russian. – From Ref. zh., 51. Astron., 11.51.58 (1980).

003.091 **The total solar eclipse of July 31, 1981.** D. D. Polozhentsev.
Nauka, Leningrad. 27 pp. (1980). In Russian. – Review in Ref. zh., 51. Astron., 12.51.37 (1980).

003.092 **Aspheric surfaces in astronomical optics.** G. M. Popov.
Nauka, Moskva. 159 pp. (1980). In Russian. – Review in Ref. zh., 51. Astron., 1.51.53 (1981).

003.093 **Theoretical cosmology.** A. K. Raychaudhuri.
Clarendon (Oxford University Press), New York. 10 + 298 pp. Price $ 26.50 (1980). – Review in Science, Vol. 210, 568 (1980).

003.094 **Mitt första teleskop.** S. O. Rehnlund.
Bokförlaget INOVA, Bromma. 176 pp. Price Sv. kr. 35.00 (1980). – Review in Astron. Tidsskr., Årg. 13, 147 (1980).

003.095 **Astronomy data book.** J. H. Robinson, J. Muirden.
Second edition. David and Charles. 272 pp. Price £ 6.95 (1979). From J. British Astron. Assoc., Vol. 90, 494 (1980).

003.096 **Lectures on the density wave theory.** K. Rohlfs.
Translated from the English edition. Mir, Moskva. 208 pp. (1980). In Russian. – Review in Ref. zh., 51. Astron., 1.51.50 (1981).

003.097 **Motion of bodies in general relativity.** A. P. Ryabushko.
Vyshehjshaya Shkola, Minsk. 240 pp. Price 2 Rbl. 60 Kop. (1979). In Russian.

003.098 **Cosmos.** C. Sagan.
Random House. 416 pp. Price $ 19.95, £ 12.50 (1980). Review in Nature, Vol. 288, 514 - 515; 1980 (*J. Maddox*).

003.099 **Problems of optical astronomy.** P. V. Shcheglov.
Nauka, Moskva. 271 pp. (1980). In Russian. – Review in Ref. zh., 51. Astron., 8.51.28 (1980).

003.100 **Universe, life, intelligence.** I. S. Shklovskij.
5th revised and completed edition. Nauka, Moskva. 352 pp. (1980). In Russian. – From Ref. zh., 51. Astron., 10.51.54 (1980).

003.101 **The Martian surface.** A. V. Sidorenko (Editor).
Nauka, Moskva. 238 pp. (1980). In Russian. – Review in Ref. zh., 51. Astron., 11.51.59 (1980).

003.102 **Astronomie astrophysique de notre temps, 1980.** J. Thurnheer.
Available from Robert Rumley, Rue du Temple 7, 1020 Rennes, France. 300 pp. Price Fr. 25.00 (1980). – Review in Orion, 38. Jahrg., 172; 1980 (*M. Roud*).

003.103 **Out of the darkness: the planet Pluto.** C. W. Tombaugh, P. Moore.
Stackpole Books, Cameron and Kelker Sts., Harrisburg, Pa. 17105. 221 pp. Price $ 14.95, £ 7.95 (1980). – Reviews in Nature, Vol. 288, 514; 1980 (*D. W. Hughes*); Sky Telesc., Vol. 60, 519 (1980).

003.104 **Antique scientific instruments.** G. L'E. Turner. Blandford/Sterling. 168 pp. Price £ 3.95, $ 12.95 hbk., $ 6.95 pbk. (1980). – Review in Nature, Vol. 288, 516; 1980 (*S. A Bedini*).

003.105 **Buoyancy effects in fluids.** J. S. Turner. Cambridge University Press. 368 pp. Price £ 7.95 (1979). – Review in Observatory, Vol. 100, 174; 1980 (*R. C. Smith*).

003.106 **Foundations of cosmology.** A. Tursunov. Mysl', Moskva. 238 pp. (1979). In Russian. – Review in Priroda, No. 9, p. 124 - 126; 1980 (*N. F. Ovchinnikov*).

003.107 **Course of geodetic astronomy. Textbook for students of universities teaching astronomical geodesy.** S. S. Uralov. Nedra, Moskva. 592 pp. Price 1 Rbl. 60 Kop. (1980). In Russian. – Review in Ref. zh., 52. Geod. Aehrosemka, 10.52.25 (1980).

003.108 **Methods of investigation in astronomical optics.** Eh. A. Vitrichenko. Nauka, Moskva. 152 pp. (1980). In Russian. – Review in Ref. zh., 51. Astron., 10.51.55 (1980).

003.109 **Adaptive optics. Collected articles.** A. Vlasenko, A. Kuksenko (Editors). Translated from the English edition. Mir, Moskva. 456 pp. (1980). In Russian. – Review in Ref. zh., 51. Astron. 1.51.52 (1981).

003.110 **Light from the sky**. Readings from Scientific American. With an introduction by J. Walker. W. H. Freeman. 78 pp. Price £ 3.30, £ 7.10 hbk. (1980). ISBN 0-7167-1222-9 flexi., ISBN 0-7167-1221-0 hbk. – From Nature, Vol. 287, 94 (1980).

003.111 **Gravity, particles, and astrophysics.** A review of modern theories of gravity and G-variability, and their relation to elementary particle physics and astrophysics. P. S. Wesson. Astrophysics and Space Science Library, Vol. 79. D. Reidel Publishing Company, Dordrecht, Holland–Boston, USA–London, England. 8 + 188 pp. Price Dfl. 65.00. $ 34.00 (1980). ISBN 90-277-1083-X.

The book deals with the relationship between gravitation and elementary particle physics, and the implications of these subjects for astrophysics. The main theories of gravity involved are those of Dirac, Hoyle/Narlikar and Canuto et al., in which gravity becomes weaker with time. The main theories of particle physics involved are the gauge theories of Weinberg, Salam and others, which hold out the hope of unifying some of the apparently different forces of Nature. These new theories can be tested by observations of astrophysical systems.

003.112 **The planet Pluto.** A. J. Whyte. Pergamon Press, 147 pp. Price £ 8.75 (1980). Review in J. British Interplanet. Soc., Vol. 33, 440 (1980).

003.113 **The book of time.** C. Wilson (Editor). Westbridge Books/David and Charles. 320 pp. Price £ 10.50, $ 24.95 (1980). – Review in Nature, Vol. 288, 516 - 517; 1980 (*D. de Solla Price*).

003.114 **Starseekers.** C. Wilson. Hodder and Stoughton/Doubleday. 271 pp. Price £ 10.95, $ 15.95 (1980). – Reviews in J. British Astron. Assoc., Vol. 91, 95; 1980 (*P. Moore*); Nature, Vol. 288, 35 - 36; 1980 (*D. W. Hughes*).

003.115 **The infrared handbook.** W. L. Wolfe, G. J. Zissis. Published by the Office of Naval Research, USA. 1720 pp. (1980). – Review in Infrared Phys., Vol. 20, 351 - 352; 1980 (*W. D. Lawson*).

003.116 **Sun, moon and standing stones.** J. E. Wood. Oxford University Press. 217 pp. Price £ 3.50 (1980). – Review in Observatory, Vol. 100, 173 - 174; 1980 (*L. V. Morrison*).

003.117 **"Sojus-22" erforscht die Erde.** Gemeinsame Ausgabe der Akademie der Wissenschaften der DDR and der Akademie der Wissenschaften der UdSSR. H. Stiller, R. Joachim, K. Müller, A. Zickler, R. Z. Sagdeev, Ya. L. Ziman, A. S. Eliseev, Yu. P. Semenov. (Editors). Akademie-Verlag, Berlin, GDR, 283 pp. Price M 28.00, DM 29.90 (1980).

Contents: Wissenschaftlich-methodische Grundlagen des Experimentes "Raduga". Das Raumschiff "Sojus-22". Die Multispektralgeräte. Vorbereitung der MKF-6 auf den Flug. Flugbedingungen. Das wissenschaftliche Programm des Experimentes "Raduga" und seine Durchführung. Bearbeitung der Fotomaterialien. Erste Ergebnisse der thematischen Bildinterpretation.

003.118 **Astronomi.** Gyldendal, Nordisk Forlag, Köpenhamn. 495 pp. Price D. kr. 525.00 (1979). – Review in Astron. Tidsskr., Årg. 13, 146; 1980 (*G. Larsson-Leander*).

003.119 **Tafeln zur Astronomie aus der Encyclopédie.** Recueil de planches, sur les sciences, les arts libéraux, et les arts méchaniques, Paris 1763. Nachdruck. Diderot – d'Alembert. Verlag POOL, E. D. Lang, Ländlestr. 6, 7760 Radolfzell 18. 35 Tafeln, Price DM 210.00 (1980). – Review in Sterne Weltraum, Jahrg. 19, 352, 354, 356 (1980).

003.120 **Cosmic investigations made in the USSR in 1979.** Nauka, Moskva. 43 pp. Price 30 Kop. (1980). In Russian. – Review in Ref. zh., 62. Issled. kosm. prostranstva, 9.62.69 (1980).

003.121 **Burning and explosion in cosmos and on the earth. Collected articles. All-Union Astronomical-Geodetical Society of the USSR Academy of Sciences.** Moskva. 155 pp. (1980). In Russian. – Abstr. in Ref. zh., 51. Astron., 10.51.58 (1980).

003.122 **Dynamics of a continuum in space and on the earth. Collected articles. All-Union Astronomical and Geodetical Society of the USSR Academy of Sciences.** Moskva. 165 pp. (1978). In Russian. – Abstr. in Ref. zh., 51. Astron., 8.51.32 (1980).

004 History of Astronomy

004.001 **The Japanese record of the guest-star of 1408.**
K. Imaeda, T. Kiang.
J. Hist. Astron., Vol. 11, 77 - 80 (1980).

004.002 **"The great laboratories of the universe": William Herschel on matter theory and planetary life.**
S. Schaffer.
J. Hist. Astron., Vol. 11, 81 - 111 (1980).

004.003 **Johann Schroeter on the light of the nebulae.**
W. B. Forbush III.
J. Hist. Astron., Vol. 11, 111 - 113 (1980).

004.004 **"Sicut nodus in tabula": de la rotation propre du soleil au seizième siècle.**
M.-P. Lerner.
J. Hist. Astron., Vol.11, 114 - 129 (1980).

004.005 **On writing the history of modern astronomy.**
M. Hoskin, O. Gingerich.
J. Hist. Astron., Vol. 11, 145 - 146 (1980).

004.006 **The Huggins archives at Wellesley College.**
J. Morgan.
J. Hist. Astron., Vol. 11, 147 (1980).

004.007 **Wer hat die Relativitätstheorie geschaffen?**
F. Hund.
Phys. Bl., 36. Jahrg., 237 - 240 (1980).

004.008 **A Dollond–Wollaston telescope.**
H. E. Dall, E. J. Hysom, C. A. Ronan.
J. British Astron. Assoc., Vol. 90, 422 - 428 (1980).

004.009 **Comet (sometimes Newton) P/1680.**
C. M. Botley.
J. British Astron. Assoc., Vol. 90, 449 - 450 (1980).

004.010 **La grande Meridiana in San Petronio.**
D. Maccaferri.
Coelum, Vol. 49, 115 - 126 (1980).

004.011 **Hipparchus's determination of the length of the tropical year and the rate of precession.**
N. M. Swerdlow.
Arch. Hist. Exact Sci., Vol. 21, 291 - 309 (1980).

004.012 **Sirius B and the gravitational redshift: an historical review.** N. S. Hetherington.
Q. J. R. Astron. Soc., Vol. 21, 246 - 252 (1980).

004.013 **Was Ptolemy a fraud?** O. Gingerich.
Q. J. R. Astron. Soc., Vol. 21, 253 - 266 (1980).

004.014 **Galileo saw Neptune in 1612.** D. W. Hughes.
Nature, Vol. 287, 277 - 278 (1980).

004.015 **Galileo's observations of Neptune.**
C. T. Kowal, S. Drake.
Nature, Vol. 287, 311 - 313 (1980).

The authors have found that Galileo observed the planet Neptune on 28 December 1612 and 28 January 1613. The latter observation may be of astrometric value, and differs by 1 arc min from the predicted position of Neptune. Galileo also detected the motion of Neptune.

004.016 **Solar rotation in 1612: Galileo vs. Harriot.**
R. B. Herr.
Bull. American Astron. Soc., Vol. 12, 504 (1980). – Abstract.

004.017 **First achromatic telescopes.**
V. A. Gurikov.
Zemlya Vselennaya, 1980, No. 4, p. 68 - 71. In Russian.

004.018 **The motion of the sun in the Vasistha Siddhānta.**
G. Abraham.
Arch. Hist. Exact Sci., Vol. 22, 1 - 3 (1980).

004.019 **Perturbations and solar tables from Lacaille to Delambre: the rapprochement of observation and theory, part I.** C. A. Wilson.
Arch. Hist. Exact Sci., Vol. 22, 53 - 188 (1980).

004.020 **Perturbations and solar tables from Lacaille to Delambre: the rapprochement of observation and theory, part II.** C. A. Wilson.
Arch. Hist. Exact Sci., Vol. 22, 189 - 304 (1980).

004.021 **The paradox of the dark night sky.**
E. R. Harrison.
Mercury, Vol. 9, 83 - 93, 101 (1980).

004.022 **Tycho's supernova.** K. W. Kamper.
Mercury, Vol. 9, 97 - 98, 103 (1980).

004.023 **John Tebbutt and the astronomy at Windsor Observatory.** G. L. White.
Proc. Astron. Soc. Australia, Vol. 3, 408 - 411 (1979).

004.024 **The history of the fundamental catalogues.**
F. Schmeidler.
Mitt. Astron. Ges., Nr. 48, (see 012.015), p. 11 - 23 (1980). In German.

After a discussion of the conception of a fundamental catalogue of stellar positions, it is pointed out that problems connected with the construction of such a catalogue were different in antiquity and in the middle ages from those of our times. The procedure of the astronomers of the renaissance epoch in measuring star positions is outlined. The first fundamental catalogue according to modern definitions was that of Maskelyne. The history of the first Fundamental Catalogue of the Astronomische Gesellschaft and of its successors is described and compared to that of the fundamental catalogues constructed in the United States during the last one hundred years. The paper is concluded by some remarks on the accuracy of astronomical observations, and on the importance of astrometric measurements for astronomy.

004.025 **Antike Sternsagen.** E. Knobloch.
Sterne Weltraum, Jahrg. 19, 338 - 343 (1980).

004.026 **Henry Tappan, Franz Brünnow, and the founding of the Ann Arbor school of astronomers, 1852 - 1863.** H. Plotkin.
Ann. Sci., Vol. 37, 287 - 302 (1980).

004.027 **Treatise "On the improvement of the largest declination and latitude of a town" by Abu Mahmud Khugandi.** Kh. F. Abdulla-zade.
Izv. AN TadzhSSR. Otd. fiz.-mat., khim. i geol. nauk, 1980, No. 1, p. 17 - 22. In Russian. – Abstr. in Ref. zh., 51. Astron., 8.51.1 (1980).

004.028 **Le zodiaque dans l'astronomie grecque.**
G. Aujac.
Rev. Hist. Sci., Tome 33, No. 1, p. 3 - 32 (1980).

004.029 **Astronomie in Gotha. II. Die Sternwarte in der Jägerstraße.** M. Strumpf, T. Marold.
Sterne, 56. Band, 227 - 236 (1980).

004.030 **Spherical trigonometry in the work of mathematicians and astronomers of the medieval East.**
G. P. Matvievskaya.
Nauka i tekh. Vopr. istorii i teor., Moskva-Leningrad, 1979, No. 10, p. 81 - 82. In Russian. – Abstr. in Ref. zh., 51. Astron., 9.51.2 (1980).

004.031 **Notes on the book of Copernicus.** O. Gingerich.
Vopr. istorii estestvozn. i tekh., 1980, No. 2, p. 103 - 106. In Russian. – Abstr. in Ref. zh., 51. Astron., 9.51.6 (1980).

004.032 **N. A. Morozov's contribution to the development of Russian natural sciences.** Yu. I. Solov'ev.
Vopr. istorii estestvozn. i tekh., 1980, No. 2, p. 67 - 72. In Russian. – From Ref. zh., 51. Astron., 9.51.9 (1980).

004.033 **Ole Rømers meridiankreds og Peder Horrebows behandling af meridian-observationer.**
N. T. Jørgensen.
Astron. Tidsskr., Årg. 13, 97 - 108 (1980).

004.034 **Albert Einstein – Ein Jahr danach.** A. Unsöld.
Phys. Bl., 36. Jahrg., 337 - 339 (1980).

004.035 **Acerca de un original grabado en madera.**
J. Ashbrook.
Rev. Astron., No. 212, p. 12 - 13 (1980).

004.036 **The history of the chronometer.** I. S. Pandul.
Zemlya Vselennaya, 1980, No. 5, p. 58 - 60. In Russian.

004.037 **Herschel's determination of the solar apex.**
M. Hoskin.
J. Hist. Astron., Vol. 11, 153 - 163 (1980).

004.038 **The solar activity in the time of Galileo.**
K. Sakurai.
J. Hist. Astron., Vol. 11, 164 - 173 (1980).

004.039 **Bernard Walther's astronomical observations.**
R. L. Kremer.
J. Hist. Astron., Vol. 11, 174 - 191 (1980).

004.040 **Two star tables from Muslim Spain.**
P. Kunitzsch.
J. Hist. Astron., Vol. 11, 192 - 201 (1980).

004.041 **Astronomical identities of Mesoamerican gods.**
D. H. Kelley.
Archaeoastronomy, No. 2, p. S1 - S54 (1980).

004.042 **Astronomical origin of the offset street grid at Teotihuacan.** B. C. Chiu, P. Morrison.
Archaeoastronomy, No. 2, p. S55 - S64 (1980).

004.043 **On the analysis of megalithic lunar sightlines in Scotland.** L. V. Morrison.
Archaeoastronomy, No. 2, p. S65 - S77 (1980).

004.044 **A new study of all megalithic lunar lines.**
A. Thom, A. S. Thom.
Archaeoastronomy, No. 2, p. S78 - S89 (1980).

004.045 **Astronomical foresights used by megalithic man.**
A. Thom, A. S. Thom.
Archaeoastronomy, No. 2, p. S90 - S94 (1980).

004.046 **A solstitial site near Peterborough?**
A. S. Thom.
Archaeoastronomy, No. 2, p. S95 (1980).

004.047 **Tycho Brahe en het verdwenen hemelpaleis.**
G. W. E. Beekman.
Zenit, 7e Jaarg., 2 - 7 (1980).

004.048 **Intihuatana: een gnomon?** W. Buijze.
Zenit, 7e Jaarg., 192 - 193 (1980).

004.049 **Op zoek naar Observatorium Tusculanum.**
G. W. E. Beekman.
Zenit, 7e Jaarg., 430 - 431 (1980).

004.050 **Parameter disposition in pre-Newtonian planetary theories.** R. C. Riddell.
Arch. Hist. Exact Sci., Vol. 23, 87 - 157 (1980).
Contents: Introduction. Phenomena, models and explanations. The modelling process. Ptolemy. Copernicus. Kepler. From Hipparchus to Newton.

004.051 **Discovery of the Earth's core.** S. G. Brush.
American J. Phys., Vol. 48, 705 - 724 (1980).
Abstr. in Phys. Abstr., Vol. 83, Abstr. 105486 (1980).

004.052 **Galileo's sighting of Neptune.**
S. Drake, C. T. Kowal.
Sci. American, Vol. 243, No. 6, p. 52 - 59 (1980).
He first observed it in 1612 and thought it was a fixed star, some 234 years before it was discovered to be a planet. His data cast doubt on the accuracy of modern orbital calculations for Neptune.

004.053 **Centenaries for 1981.** B. Hetherington.
J. British Astron. Assoc., Vol. 91, 79 - 80 (1980).

004.054 **A unique Greek sundial recently discovered in Central Asia.** R. R.-J. Rohr.
J. R. Astron. Soc. Canada, Vol. 74, 271 - 278 (1980).
In 1975 a French team of archaeologists brought to light a Greek sundial of the time of Alexander the Great, of so unexpected and mathematically correct a pattern that a new chapter seems to have opened in the story of early gnomonics.

004.055 **Basis Geographiae recentioris Astronomica.**
A. Hanle, O. Mittelstaedt.
Sterne Weltraum, 19. Jahrg., 407 - 410 (1980).

004.056 **How the U. S. Naval Observatory began, 1830 - 65.**
S. J. Dick.
Sky Telesc., Vol. 60, 466 - 471 (1980).

004.057 **The history of astronomy in the University of Durham from 1835 to 1939.**
G. D. Rochester.
Q. J. R. Astron. Soc., Vol. 21, 369 - 378 (1980).

004.058 **The sources of Eratosthenes' measurement of the Earth.** R. R. Newton.
Q. J. R. Astron. Soc., Vol. 21, 379 - 387 (1980).
Ancient Greek geographers expressed the north-south coordinate of a point in at least two different ways before the use of latitude became standard. Coordinates expressed in these ways were naturally rounded to convenient values. Later, when latitude was adopted, its values were often calculated from these rounded values instead of being measured

afresh. This process may be the origin of the angular data that Eratosthenes used to estimate the size of the Earth.

004.059 **Comments on 'Was Ptolemy a fraud?' by Owen Gingerich.** R. R. Newton.
Q. J. R. Astron. Soc., Vol. 21, 388 - 399 (1980).
In the paper called 'Was Ptolemy a fraud?', (see 004.013) Gingerich concludes that he was not. He concludes instead that Ptolemy's Syntaxis is the 'greatest surviving astronomical work from antiquity' and that Ptolemy was 'the greatest astronomer of antiquity'. In this paper the author reviews his arguments and concludes that the answer to his question is 'yes'.

004.060 **History of the mass of Mercury.** R. A. Lyttleton.
Q. J. R. Astron. Soc., Vol. 21, 400 - 413 (1980).

004.061 **Graduate astronomy education in the early days of Lick Observatory.** D. E. Osterbrock.
Mercury, Vol. 9, 151 - 156 (1980).

004.062 **A survey of historical astrolabes of Delhi.** K. Behari, V. Govind.
Indian J. Hist. Sci., Vol. 15, 94 - 104 (1980).
An attempt has been made to highlight the following: (a) to trace out the history and utility of the astrolabes during the medieval period; (b) the main components of astrolabes; (c) the typological variations of astrolabes with reference to their material, size, longitudinal and latitudinal positions and utility; (d) the significant astrolabes of Delhi with their salient features; (e) the astrolabe makers and astrolabe making centres in India; and (f) problems and some suggestions relating to the study, conservation and deciphering of the damaged historical astrolabes.

004.063 **Observation and theory in Babylonian astronomy.** A. Aaboe.
Centaurus, Vol. 24, 14 - 35 (1980).

004.064 **Some investigations on the ephemerides of the Babylonian moon tests, system A.**
L. Brack-Bernsen.
Centaurus, Vol. 24, 36 - 50 (1980).

004.065 **The full moon serpent. A foundation stone of ancient astronomy?** K. P. Moesgaard.
Centaurus, Vol. 24, 51 - 96 (1980).

004.066 **Hipparchus' empirical basis for his lunar mean motions.** A historical footnote to Olaf Pedersen, A survey of the Almagest, 161-64. G. J. Toomer.
Centaurus, Vol. 24, 97 - 109 (1980).

004.067 **The conjunction of 3102 B.C.** B. L. van der Waerden.
Centaurus, Vol. 24, 117 - 131 (1980).

004.068 **The status of models in ancient and medieval astronomy.** B. R. Goldstein.
Centaurus, Vol. 24, 132 - 147 (1980).

004.069 **An unusual value for the length of the meridian degree: 66 1/2 miles, in Ibn Yunus' Hakimitic Zij.**
W. Hartner.
Centaurus, Vol. 24, 148 - 152 (1980).

004.070 **The sphera solida and related instruments.** R. Lorch.
Centaurus, Vol. 24, 153 - 161 (1980).

004.071 **Astronomical events from a Persian astrological manuscript.** E. S. Kennedy.
Centaurus, Vol. 24, 162 - 177 (1980).

004.072 **Appendix to E. S. Kennedy "Astronomical events from a Persian astrological manuscript".**
O. Gingerich.
Centaurus, Vol. 24, 178 - 180 (1980).

004.073 **Long-period motions of the earth in De revolutionibus.** N. M. Swerdlow.
Centaurus, Vol. 24, 212 - 245 (1980).

004.074 **Newton and Huygens' explanation of the 22° halo.** A. E. Shapiro.
Centaurus, Vol. 24, 273 - 287 (1980).

004.075 **Kepler, Newton and Flamsteed on refraction through a 'regular Aire': the mathematical and the practical.** D. T. Whiteside.
Centaurus, Vol. 24, 288 - 315 (1980).

004.076 **Roger Joseph Boscovich and John Robison on terrestrial aberration.** K. Møller Pedersen.
Centaurus, Vol. 24, 335 - 345 (1980).

004.077 **On the date used in Chinese historical annals when recording observations made during the latter half of the night.** T. Kiang.
Acta Astron. Sinica, Vol. 21, 323 - 333 (1980). In Chinese.

004.078 **The hexagram "Feng" in the book of changes is not a sunspot record.** Z.-r. Xu.
Acta Astron. Sinica, Vol. 21, 340 - 341 (1980). In Chinese.

004.079 **The roots of Stonehenge.** J. Olson.
Publ. Astron. Soc. Pacific, Vol. 92, 548 (1980).
Abstract.

004.080 **Division of the history of astrometry into periods.** I. G. Kolchinskij.
Geodynamics and astrometry, (see 003.011), p. 120 - 130 (1980). In Russian.

004.081 **Castle Frederick Observatory: location and design.** R. L. Bishop.
J. R. Astron. Soc. Canada, Vol. 74, 364 (1980). – Abstract.

004.082 **Review of "A letter concerning earthquakes".** J. E. Kennedy.
J. R. Astron. Soc. Canada, Vol. 74, 364 (1980). – Abstract.

004.083 **Astronomical tests for the determination of the age of megalithic circles at Odry.** M. Kubiak.
Urania Kraków, Vol. 51, 322 - 328 (1980). In Polish.

004.084 **Odry – Tucholian "Stonehenge" and Capella – Pomeranian Sirius.** L. Zajdler.
Urania Kraków, Vol. 51, 329 - 335 (1980). In Polish.

North American theses and dissertations on the history of astronomy. See Abstr. 002.001.

Bibliography of dissertations on the history of astronomy in the USSR after the Second World War.
See Abstr. 002.052.

A union catalogue of antiquarian astronomy books.
See Abstr. 002.068.

The age of Stonehenge. See Abstr. 003.027.

Rings of stone. See Abstr. 003.029.

The enigma of Stonehenge. See Abstr. 003.048.

Greenwich time, and the discovery of longitude. See Abstr. 003.057.

L'Etrenne ou la Neige Sexangulaire. See Abstr. 003.063.

Die erste Sternwarte Europas mit ihren Instrumenten und Uhren. 400 Jahre Jost Bürgi in Kassel. See Abstr. 003.073.

Antique scientific instruments. See Abstr. 003.104.

The Herschel dynasty – Part II: John Herschel. See Abstr. 005.005.

Dr John William Draper. See Abstr. 005.006.

Johannes Kepler – Die Ganzheit von Mensch, Natur und Gott. See Abstr. 005.008.

Die Welt als Uhr - ein Sprach- und Denkbild. See Abstr. 009.014.

Aristarchus of Samos symposium, held on Samos, Greece 1980 June 17 - 19. See Abstr. 011.013.

Symposium on the history of Chinese astronomy was held in Chengdu. See Abstr. 011.021.

L'astronomie dans l'antiquité classique. See Abstr. 012.056.

L'astronomia in Jugoslavia. See Abstr. 013.009.

Tobias Mayers Meridianbeobachtungen der Sonne und die Frage der Veränderlichkeit des Sonnendurchmessers. See Abstr. 080.029.

The necessity of new dynamical theories of the planets. See Abstr. 091.007.

Contributions to the study of Babylonian lunar theory. See Abstr. 094.022.

Op zoek naar Pluto. See Abstr. 101.024.

One hundred and fifteen years of meteor spectroscopy. See Abstr. 104.028.

Erratum

004.901 **Erratum: "Johann Heinrich Lambert and the determination of orbits for planets and comets"** [Celestial Mech., Vol. 21, 237 - 250 (1980)]. O. Volk.
Celestial Mech., Vol. 22, 415 (1980). – See Abstr. 27.004.012.

005 Biography

005.001 **P. Angelo Secchi, S.J., 1818 - 1878.**
H. A. Brück.
Irish Astron. J., Vol. 14, 9 - 13 (1979).

005.002 **The diaries of T. G. Elger.**
J. Burnett.
J. British Astron. Assoc., Vol. 90, 444 - 448 (1980).

005.003 **Einstein und Deutschland.** A. Hermann.
Einstein Symposion Berlin, (see 012.003), p. 537 - 550 (1979).

005.004 **Charles Messier,** on the occasion of the 250th anniversary of his birthday (1730 - 1817).
A. I. Eremeeva.
Zemlya Vselennaya, 1980, No. 4, p. 48 - 51. In Russian.

005.005 **The Herschel dynasty – Part II: John Herschel.**
P. M. Millman.
J. R. Astron. Soc. Canada, Vol. 74, 203 - 215 (1980).

005.006 **Dr John William Draper.** D. Trombino.
J. British Astron. Assoc., Vol. 90, 565 - 571 (1980).
John William Draper (1811 - 1882) was first to photograph the Moon successfully; and later to apply photography to spectroscopic analysis. Draper anticipated, by years, the work of Kirchhoff and others. This paper describes, briefly, some of these achievements.

005.007 **Morozov as revolutionary and scientist.**
V. A. Tvardovskaya.
Vopr. istorii estestvozn. i tekh., 1980, No. 2, p. 58 - 66. In Russian. – From Ref. zh., 51. Astron., 9.51.8 (1980).

005.008 **Johannes Kepler – Die Ganzheit von Mensch, Natur und Gott.** Zum 350. Todestag des großen Astronomen und Naturphilosophen. F. Burdecki.
Phys. Bl., 36. Jahrg., 323 - 327 (1980).

005.009 **Jan H. Oort's work.** As told to the youngsters at Leiden – and elsewhere. A. Blaauw.
Oort and the universe, (see 003.005), p. 1 - 19 (1980).

005.010 **Meritus Emeritus.** The first decade of Jan Oort's retirement. H. van der Laan.
Oort and the universe, (see 003.005), p. 21 - 29 (1980).

005.011 **Jan Hendrik Oort and Dutch astronomy.**
H. G. van Bueren.
Oort and the universe, (see 003.005), p. 31 - 37 (1980).

005.012 **Oort's scientific importance on a world-wide scale.**
B. Strömgren.
Oort and the universe, (see 003.005), p. 39 - 44 (1980).

005.013 **Oort and international co-operation in astronomy.**
D. H. Sadler.
Oort and the universe, (see 003.005), p. 45 - 50 (1980).

005.014 **Reminiscences of the early nineteen-twenties.** P. van de Kamp.
Oort and the universe, (see 003.005), p. 51 - 54 (1980).

005.015 **The first five years of Jan Oort at Leiden, 1924 to 1929.** B. J. Bok.
Oort and the universe, (see 003.005), p. 55 - 58 (1980).

005.016 **Early galactic structure.** P. O. Lindblad.
Oort and the universe, (see 003.005), p. 59 - 64 (1980).

005.017 **Oort's work on comets.** M. Schmidt.
Oort and the universe, (see 003.005), p. 111 - 115 (1980).

005.018 **Oort and extragalactic astronomy.** M. Burbidge, G. Burbidge.
Oort and the universe, (see 003.005), p. 141 - 150 (1980).

005.019 **The earth and the universe.** A. H. Oort.
Oort and the universe, (see 003.005), p. 153 - 156 (1980).

005.020 **The challenge of Jan Oort.** J. H. Bannier.
Oort and the universe, (see 003.005), p. 157 - 160 (1980).

005.021 **Jan Oort at the telescope.** F. Walraven.
Oort and the universe, (see 003.005), p. 161 (1980).

005.022 **Personal recollections.** To Jan Oort, on his eightieth birthday. G. Westerhout.
Oort and the universe, (see 003.005), p. 163 - 164 (1980).

005.023 **Style of research.** H. van de Hulst.
Oort and the universe, (see 003.005), p. 165 - 171 (1980).

005.024 **Joseph Leclerc et la Société Astronomique de France.** M. Laffineur.
Astronomie, Vol. 94, 513 - 515 (1980).

005.025 **A. Ya. Orlov**, on the occasion of the 100th anniversary of his birthday. E. P. Fedorov.
Zemlya Vselennaya, 1980, No. 5, p. 42 - 44. In Russian.

005.026 **Magellan's scientific deed.** On the occasion of the 500th anniversary of his birthday.
A. F. Plakhotnik.
Zemlya Vselennaya, 1980, No. 6, p. 52 - 56. In Russian.

005.027 **Wendelen: een vaak vergeten astronoom.**
Zenit, 7e Jaarg., 136 (1980).

005.028 **Die astronomischen Arbeiten von Peter Anich.** J. Fuchs.
Tiroler Wirtschaftsstud., Vol. 32, 211 - 220 (1976) = Mitt. Sternw. Innsbruck Nr. 30. – Abstract.

005.029 **The Herschel dynasty – Part III: Alexander Stewart Herschel.** P. M. Millman.
J. R. Astron. Soc. Canada, Vol. 74, 279 - 290 (1980).
The astronomical research of Professor Alexander Stewart Herschel, grandson of Sir William Herschel, is described with particular reference to his pioneer work in meteor spectroscopy. Brief notes on other members of the Herschel family are included.

005.030 **Le centenaire de A. J. Orlov 1880 - 1954.** A. Stoyko.
Astronomie, Vol. 94, 537 - 538 (1980).

005.031 **Couder and coma.** P. B. Fellgett.
Q. J. R. Astron. Soc., Vol. 21, 491 - 492 (1980).

005.032 **Alfred Wegener und die Himmelskunde. Ein Beitrag zum 100. Geburtstag des bedeutenden Naturforschers.** G. Ehmke.
Sterne, 56. Band, 331 - 340 (1980).

005.033 **Citation of Dr. J. Paul Wild as recipient of the G. E. Hale Prize awarded by the Solar Physics Division of the American Astronomical Society, 17 June 1980.**
Sol. Phys., Vol. 68, 421 - 422 (1980).

005.034 **Le rôle éminent de Bernard Lyot dans le développement de l'Astronomie française.**
C. Fehrenbach.
C. R. Acad. Sci. Paris, Vie Acad., Tome 291, 76 - 79 (1980).

005.035 **Le principe et le développement par Bernard Lyot du coronographe.** J.-C. Pecker.
C. R. Acad. Sci. Paris, Vie Acad., Tome 291, 79 - 83 (1980).

005.036 **Johannes Kepler.** Z. Horsky.
Vesmír, Vol. 59, 357 - 359 (1980). In Czech.

005.037 **Johannes Kepler and Weil der Stadt.** J. Bouška.
Říše hvězd, Vol. 61, 236 - 237 (1980). In Czech.

005.038 **W. Hohmann, 1880 - 1945.**
Vesmír, Vol. 59, 350 (1980). In Czech.

005.039 **B. V. Kukarkin, 1909 - 1977.** J. Bouška.
Vesmír, Vol. 59, 318 (1980). In Czech.

005.040 **V. Láska, 1862 - 1943.** A. Zátopek.
Vesmír, Vol. 59, 286 (1980). In Czech.

005.041 **Einstein and the problems of optics.**
I. M. Frank.
Astron. Nachr., Band 301, 261 - 275 (1980). In German.
A survey is given about the early Russian editions of papers of Einstein, strongly influencing the progress of theoretical physics in the Soviet Union. The question of faster-than-light velocities as well as problems of the Doppler and the Wawilow-Tscherenkow effects are considered extensively. Results obtained by Einstein are confronted with later developments, showing their great influence on modern optics.

005.042 **Albert Einstein.** P. L. Kapitsa.
Vestn. AN SSSR, 1980, No. 7, p. 37 - 40. In Russian. Abstr. in Ref. zh., 51. Astron., 11.51.20 (1980).

005.043 **A. Einstein and modern science.**
Ya. B. Zel'dovich.
Vestn. AN SSSR, 1980, No. 7, p. 40 - 46. In Russian. – Abstr. in Ref. zh., 51. Astron., 11.51.21 (1980).

005.044 **Alexander Yakovlevich Orlov: Life, work, scientific heritage.** E. P. Fedorov.
Geodynamics and astrometry, (see 003.011), p. 7 - 24 (1980). In Russian.

005.045 **Activities of A. Ya. Orlov in Odessa.**
V. P. Tsesevich.
Geodynamics and astrometry, (see 003.011), p. 24 - 27 (1980). In Russian.

005.046 **Jan Śniadecki.** P. Rybka.
Urania Kraków, Vol. 51, 310 - 315 (1980). In Polish.

005.047 **Johann Kepler.** S. R. Brzostkiewicz.
Urania Kraków, Vol. 51, 336 - 343 (1980). In Polish.

005.048 **Giant of the golden age of physics.**
R. Popić.
Vasiona, Année 28, 49 - 51 (1980). In Serbo-Croatian.
A review is given of Einstein's ideas and contributions to physics.

Biographies of Einstein. See Abstr. 002.016.

Bibliography of Jan Hendrik Oort 1922 - 1979.
See Abstr. 002.033.

Dictionary of scientific biography, Vol. 16: index.
See Abstr. 002.077.

John Tebbutt and the astronomy at Windsor Observatory. See Abstr. 004.023.

006 Personal Notes

Frank Bateson received the Amateur Achievement Award. L. Kuhi, A. Fraknoi.
Mercury, Vol. 9, 162 (1980).

Roger Bonnet received the Prix Deslandres.
J.-C. Pecker.
C. R. Acad. Sci. Paris, Vie Acad., Tome 291, 122 (1980).

André Brahic received the Prix Lalande-Valz.
J.-C. Pecker.
C. R. Acad. Sci. Paris, Vie Acad., Tome 291, 122 (1980).

Paul Couteau received the Prix Henri de Parville.
J.-C. Pecker.
C. R. Acad. Sci. Paris, Vie Acad., Tome 291, 129 (1980).

G. de Vaucouleurs received the Herschel Medal.
Phys. Today, Vol. 33, No. 8, p. 66 (1980).

René Dumont received the Prix Antoine d'Abbadie.
G. Wlérick.
C. R. Acad. Sci. Paris, Vie Acad., Tome 291, 122 (1980).

Sir David Gill received the Catherine Wolfe Bruce Medal.
Mercury, Vol. 9, 124 - 126, 130, 132, 134 (1980).

George H. Herbig received the Bruce Medal.
L. Kuhi, A. Fraknoi.
Mercury, Vol. 9, 159 - 160 (1980).

G. H. Herbig received the Bruce Medal by the Astronomical Society of the Pacific.
Observatory, Vol. 100, 176 (1980).

H. G. L. M. Lamers received the Pastoor Schmeits-prijs.
Zenit, 7e Jaarg., 258 (1980).

James Liebert received the Trumpler Award.
L. Kuhi, A. Fraknoi.
Mercury, Vol. 9, 160 - 161 (1980).

Aden B. Meinel received the 1980 Frederic Ives Medal.
J. Opt. Soc. America, Vol. 70, 1167 (1980).

P. M. Millman received the Gold Medal of the Czechoslovak Academy of Sciences.
Říše hvězd, Vol. 61, 170 (1980). In Czech.

Chaim Leib Pekeris received the Gold Medal of the Royal Astronomical Society.
Q. J. R. Astron. Soc., Vol. 21, 245 (1980).

V. Ptáček, 60th birthday.
Říše hvězd, Vol. 61, 192 - 193 (1980). In Czech.

Luis Rodriguez received the Trumpler Award.
L. Kuhi, A. Fraknoi.
Mercury, Vol. 9, 160 - 161 (1980).

G. Schrutka-Rechtenstamm, 70 Jahre.
H. Haupt.
Sternenbote, 23. Jahrg., 123 (1980).

Walter Sullivan received the Dorothea Klumpke-Roberts Award. L. Kuhi, A. Fraknoi.
Mercury, Vol. 9, 161 - 162 (1980).

J. Paul Wild received the George Ellery Hale Prize.
Phys. Today, Vol. 33, No. 9, p. 81 - 82 (1980).

J. Paul Wild received the Royal Medal.
Phys. Today, Vol. 33, No. 9, p. 82 (1980).

J. Paul Wild received the George Ellery Hale Prize.
Sol. Phys., Vol. 68, 419 (1980).

007 Obituaries

Joseph Ashbrook, 1918 - 1980 August 4.
M. A. Hoskin.
J. Hist. Astron., Vol. 11, 201 (1980).

Joseph Ashbrook: Renaissance man, 1918 April 4 - 1980 August 4. With contributions by L. J. Robinson, B. J. Bok, C. R. Chapman, C. A. Federer, Jr., O. Gingerich, W. H. Haas, L. G. Jacchia, B. G. Marsden, M. W. Mayall, J. White.
Sky Telesc., Vol. 60, 281 - 284 (1980).

M. Bloch, 1902 July 26 - 1979 August.
G. Paturel.
Bull. AFOEV, Tome 14, 37 (1980).

E. Buchar died 1979 September 20.
Stud. Geophys. Geod., Vol. 24, 104 - 105 (1980).

Sir Edward Bullard, 1907 September 21 - 1980 April 3. A. H. Cook.
Q. J. R. Astron. Soc., Vol. 21, 483 - 486 (1980).

Daniel Chalonge, 1895 January 21 - 1977 November 28. J.-C. Pecker.
Q. J. R. Astron. Soc., Vol. 21, 481 - 483 (1980).

Charles Arthur Cross, 1922 - 1980.
P. Moore.
J. British Astron. Assoc., Vol. 90, 469 (1980).

Umberto Dall'Olmo, 1925 - 1980 January 11.
J. Hist. Astron., Vol. 11, 145 (1980).

Herbert Dingle, 1890 August 2 - 1978 September 4.
G. J. Whitrow.
Q. J. R, Astron. Soc., Vol. 21, 333 - 337 (1980).

V. V. Fedynskij, 1908, May 1 - 1978, June 17.
Meteoritika, Vyp. 38, p. 157 - 158 (1979). In Russian.

Joachim Otto Fleckenstein, 1914 July 7 - 1980 February 21.
J. Hist. Astron., Vol. 11, 144 - 145 (1980).

Joachim Otto Fleckenstein, 1914 July 7 - 1980 February 21. E. Proverbio.
Mem. Soc. Astron. Italiana, Vol. 51, 379 - 382 (1980).

Leo John Gleeson.
K. C. Westfold.
Proc. Astron. Soc. Australia, Vol. 3, 406 - 407 (1979).

Edward Aubrey Glennie, 1889 July 18 - 1980 February 15. G. Bomford.
Q. J. R. Astron. Soc., Vol. 21, 337 - 338 (1980).

Vladimir Guth, 1905, February 3 - 1980, June 24.
Z. Ceplecha.
Bull. Astron. Inst. Czechoslovakia, Vol. 31, 256 (1980).

V. Guth died 1980 June 24. Ľ. Kresák.
Kozmos, Vol. 11, 153 (1980). In Slovak.

V. Guth died 1980 June 24.
Říše hvězd, Vol. 61, 169 - 170 (1980). In Czech.

V. Guth died 1980 June 24. Z. Ceplecha.
Vesmír, Vol. 59, 284 (1980). In Czech.

Harold L. Johnson died 1980 April 2.
R. I. Thompson.
Phys. Today, Vol. 33, No. 9, p. 83 - 84 (1980).

H. L. Johnson, April 17, 1921 - April 2, 1980.
E. E. Mendoza V.
Rev. Mexicana Astron. Astrofis., Vol. 5, 3 (1980).

Harold L. Johnson, 1921 - 1980 April 2.
Sky Telesc., Vol. 60, 273 (1980).

Paul Labitzke, 5 January 1890 - 31 October 1979.
F. Schmeidler.
Mitt. Astron. Ges., Nr. 50, p. 9 - 10 (1980).

Thomas R. McGetchin, July 31, 1936 - October 22, 1979.
Proc. Tenth Lunar Planet. Sci. Conf., (see 012.050), p. VII - VIII (1979).

A. S. Partridge, 1907 - 1979 June 10.
L. Warner.
South. Stars, Vol. 28, 118 - 120 (1980).

H. H. Plaskett, 1893 July 5 - 1980 January 26.
P. M. Millman.
J. R. Astron. Soc. Canada, Vol. 74, 234 - 236 (1980).

Harry Hemley Plaskett, 1893 - 1980 January 26.
M. G. Adam.
Q. J. R. Astron. Soc., Vol. 21, 486 - 488 (1980).

Karl Reinmuth, 4 April 1892 - 6 May 1979.
J. Schubart.
Mitt. Astron. Ges., Nr. 50, p. 7 - 8 (1980).

Stephen Frederick Smerd, 1916 - 1978.
J. P. Wild.
Proc. Astron. Soc. Australia, Vol. 3, 406 (1979).

S. I. Syrovatskii, 1925 March 2 - 1979 September 26.
S. L. Mandel'stam (*Mandel'shtam*), B. V. Somov.
Sol. Phys., Vol. 67, 3 - 4 (1980).

008 Observatories, Institutes

Reports, communications and publications of observatories and astronomical institutes are recorded in this section; included are numbered series of reprints. Whenever possible, the numbers of the abstracts referring to the publications are given. Observatories and institutes are listed in alphabetical order of their towns. In some cases observatory publications do not give the name of the town; the following list which gives names and towns of some institutions may serve as an aid in such cases.

Aarne Karjalainen Observatory	**Oulu**, Finland
Algonquin Radio Observatory	**Lake Traverse**, Ontario, Canada
Allegheny Observatory	**Pittsburgh**, Pennsylvania, USA
Anglo-Australian Observatory	**Epping**, N. S. W., Australia
Archenhold-Sternwarte	**Berlin-Treptow**, German Democratic Republic
Argentine Radioastronomy Institute	**Pereyra Iraola**, Argentina
Arizona State University	**Tempe**, Arizona, USA
Arthur J. Dyer Observatory	**Nashville**, Tennessee, USA
Astronomical Latitude Station, Polish Academy of Sciences	**Borowiec**, Poland
Astronomisches Rechen-Institut	**Heidelberg**, F. R. Germany
Bell Laboratories	**Murray Hill**, New Jersey, USA
Bell Telephone Laboratories	**Holmdel**, New Jersey, USA
Bosscha Observatory	**Lembang**, Indonesia
Boyden Observatory	**Bloemfontein**, South Africa
Bureau International de l'Heure	**Paris**, France
Cajigal Observatory	**Caracas**, Venezuela
California Institute of Technology	**Pasadena**, California, USA
Carter Observatory	**Wellington**, New Zealand
Catalina Station	**Tucson**, Arizona, USA
Cavendish Laboratory	**Cambridge**, England
Centre de Données Stellaires	**Strasbourg**, France
Cerro Tololo Interamerican Observatory	**La Serena**, Chile
Československá akademie věd, Astronimický ústav	**Prague**, Czechoslovakia
Chamberlin Observatory, University of Denver	**Denver**, Colorado, USA
Columbia University, Department of Astronomy	**New York**, New York, USA
Commonwealth Observatory	**Canberra**, Australia
Cornell University, Center for Radiophysics and Space Research	**Ithaca**, New York, USA
Corralitos Observatory	**Las Cruces**, New Mexico, USA
Crawford Hill Laboratory	**Holmdel**, New Jersey, USA
David Dunlap Observatory, University of Toronto	**Richmond Hill**, Ontario, Canada
Dearborn Observatory	**Evanston**, Illinois, USA
Department of Astronomy and Observatory, Univ. California	**Los Angeles**, California, USA
Department of Astronomy Swarthmore College	**Swarthmore**, Pennsylvania, USA
Department of Astronomy, University of Texas	**Austin**, Texas, USA
Deutsches Hydrographisches Institut (DHI)	**Hamburg**, F. R. Germany
Division Radiophysics, C.S.I.R.O. University Grounds	**Sydney**, Australia
Dominion Astrophysical Observatory	**Victoria**, B.C., Canada
Dominion Observatory	**Ottawa**, Ontario, Canada
Dominion Radio Astrophysical Observatory	**Penticton**, B. C., Canada,
Dudley Observatory	**Albany**, New York, USA
Dunsink Observatory	**Dublin**, Ireland
Dyer Observatory, Vanderbilt University	**Nashville**, Tennessee, USA
Ege University Observatory	**Izmir**, Turkey
Engelhardt Observatory	**Kazan**, USSR
Erwin W. Fick Observatory, Iowa State University	**Ames**, Iowa, USA
European Southern Observatory	**La Silla**, Chile
Felix Aguilar Observatory	**San Juan**, Argentina
Fernbank Observatory	**Atlanta**, Georgia, USA
Five College Observatories	**Amherst**, Massachusetts, USA
Florida State University Radio Observatory	**Tallahassee**, Florida, USA
Flower and Cook Observatories, University of Pennsylvania	**Philadelphia**, Pennsylvania, USA
George R. Wallace Jr. Astrophysical Observatory	**Cambridge**, Massachusetts, USA
Georgetown Observatory	**Washington**, D.C., USA
Glavnaya Astronomicheskaya Observatoriya AN SSSR	**Pulkovo**, USSR
Goddard Space Flight Center	**Greenbelt**, Maryland, USA
Goethe Link Observatory, Indiana University	**Bloomington**, Indiana, USA
"Guido Horn d'Arturo" Observatory	**Bologna**, Italy
H. M. Nautical Almanac Office, Royal Greenwich Observatory	**Greenwich**, England
Hale Observatories	**Pasadena**, California, USA
Harvard College Observatory	**Cambridge**, Massachusetts, USA
Harvard Radio Astronomy Station	**Cambridge**, Massachusetts, USA
Haute Provence Observatory	**Saint Michel**, France
Haystack Observatory	**Westford**, Massachusetts, USA
Heinrich-Hertz Institut	**Berlin-Adlershof**, German Democratic Republic
Herzberg Institute of Astrophysics	**Victoria**, B. C.; **Ottawa**, Canada
High Altitude Observatory, University of Colorado	**Boulder**, Colorado, USA
Hopkins Observatory	**Williamstown**, Massachusetts, USA
Horn d'Arturo Observatory	**Bologna**, Italy
Hvar Observatory	**Zagreb**, Yugoslavia
IBM Thomas J. Watson Research Center	**Yorktown Heights**, New York, USA
Indian Institute of Astrophysics	**Bangalore**, India
Institute for Astronomy, University of Hawaii	**Honolulu**, Hawaii, USA
Institute for Theoretical Astronomy (Institut Teoreticheskoj Astronomii)	**Leningrad**, USSR
Institute of Astronomy and Space Science, University of British Columbia	**Vancouver**, B. C., Canada
Institute of Theoretical Astrophysics, Blindern	**Oslo**, Norway

Instituto Argentino de Radioastronomía	**Villa Elisa**, Provincia de Buenos Aires, Argentina
Instituto de Astronomía y Física del Espacio (IAFE)	**Buenos Aires**, Argentina
Instituto Venezolano de Astronomia	**Merida**, Venezuela
Instituto y Observatorio de Marina	**San Fernando (Cádiz)**, Spain
Inter-American Observatory	**Cerro Tololo**, La Serena, Chile
International Latitude Observatory	**Mizusawa**, Japan
IUE Observatory, European Space Agency	**Villafranca**, Madrid, Spain
Jet Propulsion Laboratory, California Institute of Technology	**Pasadena**, California
Joint Institute for Laboratory Astrophysics (JILA)	**Boulder**, Colorado, USA
Judson B. Coit Observatory	**Boston**, Massachusetts, USA
Kandilli Observatory	**Istanbul**, Turkey
Kansas University Observatory	**Lawrence**, Kansas, USA
Kapteyn Astronomical Laboratory	**Groningen**, Netherlands
Karl-Schwarzschild-Observatorium	**Tautenburg**, German Democratic Republic
Kenneth Mees Observatory	**Rochester**, New York, USA
Kiepenheuer-Institut für Sonnenphysik, formerly Fraunhofer-Institut	**Freiburg**, F. R. Germany
Kitt Peak National Observatory	**Tucson**, Arizona, USA
Kodaikanal Observatory	Bangalore, India
Korean National Astronomical Observatory	**Seoul**, Korea
Kwasan and Hida Observatories	**Kyoto**, Japan
Lamont-Hussey Observatory	**Bloemfontein**, South Africa
Landessternwarte Heidelberg-Königstuhl	**Heidelberg**, F. R. Germany
Las Campanas Observatory	**Pasadena**, California, USA
Lawrence Livermore Laboratory, University of California	**Livermore**, California, USA
Leander McCormick Observatory, University of Virginia	**Charlottesville**, Virginia, USA
Lee Observatory	**Beirut**, Lebanon
Leopold-Figl-Observatorium	**Vienna**, Austria
Leuschner Observatory	**Berkeley**, California, USA
Lick Observatory	**Santa Cruz**, (Mount Hamilton), California, USA
Lindheimer Astronomical Research Center	**Evanston**, Illinois, USA
Lockheed Palo Alto Research Laboratory	**Palo Alto**, California, USA
Lockheed Solar Observatory	**Saugus**, California, USA
Lohrmann-Observatorium der Technischen Universität Dresden	**Dresden**, German Democratic Republic
Louisiana State University Observatory	**Baton Rouge**, Louisiana, USA
Lowell Observatory	**Flagstaff**, Arizona, USA
Lunar and Planetary Laboratory	**Tucson**, Arizona, USA
Max-Planck-Institut für Astronomie	**Heidelberg**, F. R. Germany
Max-Planck-Institut für Physik und Astrophysik	**Munich**, F. R. Germany
Max-Planck-Institut für Radioastronomie	**Bonn**, F. R. Germany
McDonald Observatory at Mount Locke	**Austin**, Texas, USA
McGraw-Hill Observatory	**Kitt Peak**, Arizona, USA
McMath Hulbert Observatory	**Pontiac**, Michigan, USA

C.E.K. Mees Observatory, University of Rochester	**Rochester**, New York, USA
Michigan State University Observatory	**East Lansing**, Michigan, USA
Molonglo Radio Observatory, University of Sydney	**Sydney**, Australia
Monterey Institute for Research in Astronomy	**Carmel Valley**, California, USA
Mount Cuba Observatory	**Wilmington**, Delaware, USA
Mount John Observatory	**Lake Tekapo**, New Zealand
Mount Stromlo Observatory	**Canberra**, Australia
Mount Wilson Observatory	**Pasadena**, California, USA
Mt. Laguna Observatory	**San Diego**, California, USA.
Mullard Radio Astronomy Observatory	**Cambridge**, England
Mullard Space Science Laboratory	**London**, England
Narrabri Observatory, University of Sydney	**Sydney**, Australia
National Bureau of Standards	**Washington**, D.C., USA
National Observatory, USA	**Kitt Peak**, Arizona, USA
National Radio Astronomy Observatory	**Charlottesville**, Virginia, USA
	Green Bank, West Virginia, USA
	Socorro, New Mexico, USA
	Tucson, Arizona, USA
National Research Council of Canada	**Ottawa**, Ontario, Canada
New Mexico State University Observatory	**Las Cruces**, New Mexico, USA
Nicholas Copernicus Observatory and Planetarium	**Brno**, Czechoslovakia
Nizamiah & Rangapur Observatories	**Hyderabad**, India
Nuffield Radio Astronomy Laboratories, Jodrell Bank University of Manchester	**Manchester**, England
Observatoire Royal de Belgique	**Uccle**, Belgium
Observatories of the University of Western Ontario	**London**, Canada
Observatório Astronômico do Instituto de Física da Universidade Federal do Rio Grande do Sul	**Porto Alegre**, Rio Grande do Sul, Brazil
Observatorio de Cartuja	**Granada**, Spain
Observatorio del Ebro	**Tortosa**, Spain
Observatorio Fabra	**Barcelona**, Spain
Observatorio Nacional	**Rio de Janeiro**, Brazil
Observatorio Nacional de Física Cósmica	**San Miguel**, Argentina
Observatory, University of Michigan	**Ann Arbor**, Michigan, USA
Ohio State University Radio Observatory	**Columbus**, Ohio, USA
Ole Roemer-Observatoriet	**Aarhus**, Denmark
Onsala Space Observatory	**Göteborg**, Sweden
Owens Valley Radio Observatory	**Big Pine**, California, USA
Palomar Observatory	**Pasadena**, California, USA
Perkins Observatory, Ohio State and Wesleyan Universities	**Delaware**, Ohio, USA
Physical Research Laboratory	**Ahmedabad**, India
Purple Mountain Observatory	**Nanking**, China
Radcliffe Observatory	**Pretoria**, South Africa
Raman Research Institute	**Bangalore**, India
Rattlesnake Mountain Observatory	**Richland**, Washington, USA
Remeis-Sternwarte	**Bamberg**, F. R. Germany

Ritter Astrophysical Research Center of the University of Toledo	**Toledo**, Ohio, USA
Rosemary Hill Observatory	**Gainesville**, Florida, USA
Rothney Astrophysical Observatory	**Calgary**, Canada
Royal Aircraft Establishment, Geophysical Studies in Space Department	**Farnborough**, England
Royal Radar Establishment, Radio Astronomy Division	**Malvern**, England
Sacramento Peak Observatory	**Sunspot**, New Mexico, USA
Sagamore Hill Radio Observatory	**Hamilton**, Massachusetts, USA
San Fernando Observatory	**El Segundo**, California, USA
Shensi Astronomical Observatory	**Lintong**, Sian, China
Siding Spring Observatory	**Siding Spring**, New South Wales
Smithsonian Astrophysical Observatory	**Cambridge**, Massachusetts, USA
Sonnenobservatorium Kanzelhöhe	**Graz**, Austria
South African Astronomical Observatory	**Cape Town**, South Africa
Specola Astronomica Vaticana	**Castel Gandolfo**, Vatican
Specola di Padova	**Asiago**, Italy
Sproul Observatory	**Swarthmore**, Pennsylvania, USA
Stellar Data Center	**Strasbourg**, France
Sternberg Astronomical Institute	**Moscow**, USSR
Steward Observatory, University of Arizona	**Tucson**, Arizona, USA
W. Struve Tartu Astrophysical Observatory	**Tartu**, USSR
Tata Institute of Fundamental Research	**Bombay**, India
United States Naval Observatory	**Washington**, D. C., USA
University of Alabama	**University**, Alabama, USA
University of California	**Berkeley**, California, USA
University of Florida Observatories	**Bronson**, Florida, USA
University of Florida, Radio Observatory	**Old Town**, Florida, USA
University of Hawaii	**Honolulu**, Hawaii, USA
University of Illinois Observatory	**Urbana**, Illinois, USA
University of Kansas Observatory	**Lawrence**, Kansas, USA
University of Maryland	**College Park**, Maryland, USA
University of Michigan Observatories	**Ann Arbor**, Michigan, USA
University of Minnesota	**Minneapolis**, Minnesota, USA
University of South Florida Observatory	**Tampa**, Florida, USA
University of Texas, Department of Astronomy	**Austin**, Texas, USA
University of Washington, Astronomy Department	**Seattle**, Washington, USA
Uttar Pradesh State Observatory	**Nainital**, India
Van Vleck Observatory	**Middletown**, Connecticut, USA
Vatican Observatory	**Castel Gandolfo**, Vatican
Venezuelan Astronomical Institute	**Merida**, Venezuela
Wallace Astrophysical Observatory	**Cambridge**, Massachusetts, USA
Warner and Swasey Observatory	**Cleveland**, Ohio, USA
Washburn Observatory, University of Wisconsin	**Madison**, Wisconsin, USA
West Melton Observatory	**Christchurch**, New Zealand
Wilhelm-Förster Sternwarte	**Berlin**, F. R. Germany
Yale University Observatory	**New Haven**, Connecticut, USA
Yerkes Observatory	**Williams Bay**, Wisconsin, USA
Zentralinstitut für Astrophysik, Sternwarte Babelsberg, (Fachbereich Kosmische Physik)	**Potsdam-Babelsberg**, German Democratic Republic
Zentralinstitut für Astrophysik, Sternwarte in Sonneberg	**Sonneberg**, German Democratic Republic
Zentralinstitut für solar-terrestrische Physik	**Berlin-Adlershof**, German Democratic Republic

008.001 **Aarhus**

Institute of Astronomy, University of Aarhus.
Annual report 1979.
Published by Inst. Astron. Univ. Aarhus, Denmark. 11 pp. (1980).

008.002 **Abastumani**

Abastumanskaya Astrofizicheskaya Observatoriya, Gora Kanobili, Byulleten', No. 52 (28.012.034).

008.003 **Armagh**

Armagh Observatory in 1978/79. Annual report covering the period 1 April, 1978 - 31 March, 1979.
M. de Groot.
Irish Astron. J., Vol. 14, 1 - 8 (1979).

Armagh Observatory Contribution, Nos. 102 (26.114.116), 103 (28.131.003), 104 (28.122.076).

Armagh Observatory, Leaflet, Nos. 138 (25.008.009), 139, 140 (26.015.031), 141, 142 (26.015.032), 143 (26.011.026), 144 (28.008.003), 145.

008.004 **Austin, Tex.**

Department of Astronomy and McDonald Observatory of the University of Texas, Austin, Texas, Reprints Nos. 733, 734, 751 (26.141.032), 752 (25.123.043), 753 (22.123.041), 754 (22.123.039), 755 (25.123.037), 756 (22.117.049), 757 (26.099.002), 758 (26.158.079), 759 (26.124.002), 760 (26.101.026), 761 (26.122.035), 762 (26.141.011), 763 (26.113.004), 764 (26.118.006), 765 (26.158.057), 766 (26.124.202), 767 (26.122.012), 768 (26.117.005), 769 (26.141.023), 770 (26.122.014), 771 (26.158.028), 773 (26.141.050), 774 (26.131.051), 775 (26.124.601), 776 (26.099.124), 777 (26.122.036), 778 (26.155.027), 779 (22.115.027), 780 (26.155.066), 781 (26.065.016), 782 (26.162.028), 783 (26.064.016), 784

(26.106.044), 785 (26.151.033), 786 (27.141.149), 787 (26.100.048), 788 (26.100.033), 789 (26.041.018), 790 (26.132.031), 791 (26.155.060), 792 (26.091.068), 793 (26.158.148), 794 (26.120.015), 795 (26.131.069), 796 (26.125.401), 797 (26.158.209), 798 (26.099.168), 799 (26.122.091), 800 (26.022.138), 801 (26.114.077), 802 (27.158.031), 803 (26.122.094), 804 (26.114.095), 805 (26.122.096), 806 (27.141.029), 807 (26.126.031), 808 (26.131.163), 809 (27.061.005), 810 (27.116.021), 811 (27.034.015), 812 (27.114.091), 813 (27.122.001), 814 (27.141.006), 815 (27.114.006), 816 (27.158.068), 817 (27.021.016), 818 (26.131.158), 819 (27.114.004), 820 (27.120.008), 821 (27.114.015), 822 (26.063.039), 823 (27.034.001), 824 (27.131.029), 825 (27.131.030), 826 (27.022.019), 827 (27.103.402), 828 (27.096.004), 829 (27.096.003), 830 (27.124.402), 831 (27.162.010), 832 (27.114.052), 833 (27.141.113), 834 (27.122.041), 835 (27.122.042), 836 (27.158.191), 837 (27.101.019), 838 (27.101.017), 839 (08.142.082), 840 (08.142.083), 841 (27.158.213), 842 (27.122.101), 843 (27.122.149), 844 (27.099.073), 845 (27.099.074), 846 (27.141.165), 847 (22.122.163), 848 (22.064.092), 849 (26.031.607), 850 (26.122.095).

University of Texas Publications in Astronomy, No. 16 (28.002.053).

008.005 **Bamberg**

Dr.-Remeis-Sternwarte Bamberg, Astronomisches Institut der Universität Erlangen-Nürnberg. – Report for 1979. I. Bues.
Mitt. Astron. Ges., No. 49, p. 7 - 12 (1980).

008.006 **Basel**

Astronomisches Institut der Universität Basel. Report for 1979. G. A. Tammann.
Mitt. Astron. Ges., Nr. 49, p. 13 - 18 (1980).

008.007 **Berlin**

Institut für Astronomie und Astrophysik der Technischen Universität Berlin. – Report for 1979. R. Wielen.
Mitt. Astron. Ges., Nr. 49, p. 19 - 26 (1980).

Heinrich-Hertz-Institut. Solare Beobachtungsergebnisse. Akademie der Wissenschaften der DDR, Zentralinstitut für Solar-Terrestrische Physik (Heinrich-Hertz-Institut), Berlin-Adlershof, HHI Solar Data, Vol. 31. 1980 February - June (28.072.038).

008.008 **Bloemfontein**

University of the Orange Free State: Boyden Observatory, Department of Astronomy. A. H. Jarrett.
Mon. Notes Astron. Soc. South. Africa, Vol. 39, 7 - 9 (1980).

008.009 **Bochum**

Astronomisches Institut der Ruhr-Universität Bochum. – Report for 1979. T. Schmidt-Kaler.
Mitt. Astron. Ges., Nr. 49, p. 27 - 38 (1980).

Bereich Extraterrestrische Physik in der Abteilung XII der Ruhr-Universität Bochum. – Report for 1979. R. H. Giese.
Mitt. Astron. Ges., Nr. 49, p. 38 - 40 (1980).

Institut für Theoretische Physik, Lehrstuhl IV der Ruhr-Universität Bochum. – Report for 1979. K. Schindler.
Mitt. Astron. Ges., Nr. 49, p. 41 - 42 (1980).

008.010 **Bogota**

Publicaciones del Observatorio Astronomico Nacional, Universidad Nacional de Colombia, Facultad de Ciencias, Bogotá, No. 6 (28.079.401).

008.011 **Bonn**

Astronomische Institute der Universität Bonn. Sternwarte mit Observatorium Hoher List; Radioastronomisches Institut mit Radioobservatorium Stockert; Institut für Astrophysik und Extraterrestrische Forschung; Sonderforschungsbereich Radioastronomie. – Reports for 1979. O. Hachenberg, W. Priester, H. Schmidt.
Mitt. Astron. Ges., Nr. 49, p. 43 - 75 (1980).

Veröffentlichungen der Astronomischen Institute Bonn, Nos. 92 - 93 (28.120.019; 28.112.041).

Max-Planck-Institut für Radioastronomie. – Report for 1979. R. Schwartz.
Mitt. Astron. Ges., Nr. 49, p. 77 - 106 (1980).

Max-Planck-Gesellschaft zur Förderung der Wissenschaften, M. P. I. für Radioastronomie, Bonn, Sonderdruck, Ser. A, Nos. 376 (27.125.060), 384 (27.033.001), 385 (26.132.039), 386 (27.071.026), 387 (26.141.083), 388 (27.156.007), 389 (27.141.143), 390 (27.141.028), 391 (27.131.036), 392 (28.022.148), 393 (27.158.014), 394 (27.141.174), 395 (28.131.122), 396 (27.151.076), 397 (27.022.031), 398 (27.125.026), 399 (27.132.044), 400 (27.132.023), 401 (27.132.027), 402 (27.141.552), 403 (27.131.183), 404 (27.141.058), 405 (27.141.516), 406 (27.041.011), 407 (28.132.004).

008.012 **Borowiec**

Circular, Time and Latitude Service, Nos. 153 - 154 (28.044.016).

008.013 **Boulder, Colo.**

The High Altitude Observatory of the National Center for Atmospheric Research, Boulder, Colorado. Report from Solar Institute. R. M. MacQueen.
Sol. Phys., Vol. 68, 411 - 418 (1980).

008.014 **Bucharest**

Observations solaires. Rotations 1677 - 1689 (28. 072.039).

008.015 **Budapest**

A Magyar Tudományos Akadémia Csillagvizsgáló Intézetének Közleményei – Mitteilungen der Sternwarte der Ungarischen Akademie der Wissenschaften, Budapest – Szabadsághegy, Nos. 74 (28.122.135), 76 (28.122.136).

008.016 **Buenos Aires**

Instituto de Astronomia y Fisica del Espacio (IAFE), Buenos Aires, Argentina, Serie Publicaciones Científicas, No. PC-1-79 (28.082.061).

Instituto de Astronomia y Fisica del Espacio, (IAFE), Buenos Aires, Argentina, Tirada Aparte Nos. 24 (28.022.154), 25 (25.119.008), 26 (25.120.018), 27 (26.135.005), 28 (25.119.053), 29 (27.117.009), 30 (27.117.010), 31 (22. 114.552), 32 (27.120.028).

008.017 **Cambridge, England**

Institute of Astronomy. Annual report 1979 - 1980. M. J. Rees.
Published by Univ. Cambridge, The Observatories, Madingley Road, Cambridge CB3 OHA. 41 pp. (1980).

Mullard Radio Astronomy Observatory, Cavendish Laboratory, University of Cambridge. – Report for the period 1978 October 1 - 1979 September 30.
M. Ryle.
Q. J. R. Astron. Soc., Vol. 21, 293 - 306 (1980).

008.018 **Cambridge, Mass.**

Smithsonian Astrophysical Observatory, Special Report, Nos. 387 (28.002.054), 388 (28.032.549), 389 (28.012.038).

IAU Circulars, Nos. 3491 - 3557 (1980).

Minor Planet Circulars, (M.P.C.), Nos. 5391 - 5658 (1980).

008.019 **Cape Town**

South African Astronomical Observatory. Report for the year ending 1 January 1980.
Council for Scientific and Industrial Research (South Africa); Science Research Council (United Kingdom). 36 pp. (1980).

Notes from SAAO. I. S. Glass.
Mon. Notes Astron. Soc. South. Africa, Vol. 39, 2 - 3 (1980).

South African Astronomical Observatory Circulars, Vol. 1, No. 5 (28.113.032; 28.122.143; 28.113.033; 28.122.144; 28.113.034; 28.114.113; 28.113.035; 28.122. 145).

University of Cape Town: Department of Astronomy. B. Warner.
Mon. Notes Astron. Soc. South. Africa, Vol. 39, 4 - 7 (1980).

University of Cape Town: Department of Applied Mathematics. G. F. R. Ellis.
Mon. Notes Astron. Soc. South. Africa, Vol. 39, 7 (1980).

008.020 **Carloforte-Cagliari**

Rapporto annuale per l'anno 1977. E. Proverbio.
Pubbl. Stn. Astron. Int. Latitudine, Carloforte-Cagliari, Nuova Ser., N. 69, 22 pp.

Rapporto annuale per l'anno 1978. E. Proverbio.
Pubbl. Stn. Astron. Int. Latitudine, Carloforte-Cagliari, Nuova Ser., N. 74, 22 pp.

Circolari della Stazione Astronomia Internazionale di Latitudine, Carloforte-Cagliari, Serie A(5), N. 14 (28.044. 017).

Pubblicazioni della Stazione Astronomica Internazionale di Latitudine, Carloforte-Cagliari, Nuova Serie, Nos. 38 (28.021.037), 55 (22.035.014), 58 (22.052.015), 60 (21.032.045), 61 (22.045.008), 63 (22.082.142), 64 (26.045. 009), 65 (28.162.087), 66 (25.080.058), 69 (28.008.020), 70 (28.046.011), 72 (28.046.012), 73 (28.046.013), 74 (28.008.020).

008.021 **Cluj-Napoca**

Contributions of the Astronomical Observatory, Universitatea "Babes-Bolyai", Cluj-Napoca, 1978 (28.012.009).

Publicaţiile Observatorului Astronomio al Universităţii din Cluj, Nr. B 46-58 (28.052.039; 28.052,040; 28.052.041; 28.052.042; 28.054.008; 28.052.043; 28.052.044; 28.052.045; 28.052.046; 28.054.009; 28.054.010; 28.031.575; 28.054.011; 28.002.060).

008.022 **Coimbra**

Comunicações do Observatório Astronómico da Universidade de Coimbra, Nos. 21, 22, 23 (28.031.016).

008.023 **Córdoba, Argentina**

Observatorio Astronómico (Universidad Nacional de Córdoba, Córdoba, Argentina), Tirada Aparte, Nos. 228 (27.158.062), 229 (28.113.026), 230 (26.158.168).

008.024 **Cracow**

Cracow Observatory, Reprint No. 125 (28.106.074).

008.025 **Crimea**

Chronicle.
Izv. Krymskoj Astrofiz. Obs., Tom 62, 202 - 203 (1980). In Russian.

Izvestiya Ordena Trudovogo Krasnogo Znameni Krymskoj Astrofizicheskoj Observatorii, Akademiya Nauk SSSR, Tom 61 (28.075.023, 28.021.021, 28.071.033, 28.071.034, 28.073.090, 28.077.041, 28.077.042, 28.142.513, 28.142.075, 28.142.076, 28.141.112, 28.114.087, 28.122.077, 28.122.078, 28.116.013, 28.122.079, 28.122.080, 28.158.129, 28.158.130, 28.032.014, 28.031.012, 28.031.013, 28.032.537)
Tom 62 (28.114.145; 28.122.204; 28.066.185; 28.114.146; 28.158.237; 28.064.092; 28.114.147; 28.114.148; 28.022.175; 28.134.039; 28.142.124; 28.142.125; 28.141.168; 28.075.025; 28.072.059; 28.072.060; 28.072.061; 28.073.121; 28.073.122; 28.077.058; 28.072.062; 28.031.062; 28.031.063; 28.021.047; 28.021.048; 28.008.025).

008.026 **Dresden**

Technische Universität Dresden, Lohrmann-Observatorium, Zirkular, Nos. 91 - 92 (28.045.007).

008.027 **Edinburgh**

Royal Observatory Edinburgh. Annual report for the year ended 30 September 1979. V. C. Reddish.
Published by R. Obs. Edinburgh. 36 pp. (1980). ISSN 0309-0108. ISBN 0-902553-23-2.

Communications from the Royal Observatory, Edinburgh, Nos. 340 (27.120.004), 341 (27.114.018), 346 (27.131.044), 347 (27.158.047), 348 (27.122.027), 349 (27.141.036), 350 (27.154.004), 352 (27.158.051), 355 (27.034.045), 356 (27.159.027), 357 (27.159.001), 358 (27.122.037), 359 (27.158.188), 360 (27.160.083), 361 (28.021.033), 362 (27.117.017), 363 (27.034.035), 364 (27.011.031), 365 (28.122.007), 366 (27.031.566), 367 (27.118.022), 368 (27.031.577), 370 (28.114.003), 371 375 (27.113.015), 380, 381 (27.160.082).

008.028 **Epping**

Anglo-Australian Telescope 1979/80. Report of the Anglo-Australian Telescope Board 1 July 1979 to 30 June 1980. H. Massey.
Australian Government Publishing Service, Canberra. 6 + 39 pp (1980).

Anglo-Australian Observatory, Preprint Nos. 134 (28.141.531), 135 (28.031.571), 136 (28.112.042), 137 (28.124.021), 138 (28.122.137), 139 (28.158.183), 140 (28.162.085), 141 (28.114.111), 142 (28.141.142), 143 (28.159.017), 144 (28.158.196).

A. A. O. Newsletter, No. 15 (1980).

008.029 **Erlangen**

Physikalisches Institut, Angewandte Optik.
Report for 1979. G. Weigelt.
Mitt. Astron. Ges., Nr. 49, p. 107 - 108 (1980).

008.030 **Farnborough**

Royal Aircraft Establishment, Farnborough. Geophysical Studies in Space Department. Report for the year ending 1980 March 31.
Q. J. R. Astron. Soc., Vol. 21, 460 - 463 (1980).

008.031 **Flagstaff, Ariz.**

Lowell Observatory, Flagstaff, Arizona, Bulletin, Vol. 8, Nos. 5 (28.002.055), 6 (28.002.056).

008.032 **Frankfurt**

Institut für Theoretische Physik/Astrophysik.
Report for 1979. W. H. Kegel.
Mitt. Astron. Ges., Nr. 49, p. 109 - 110 (1980).

008.033 **Freiburg**

Kiepenheuer-Institut für Sonnenphysik. – Report for 1979. E. H. Schröter.
Mitt. Astron. Ges., Nr. 49, p. 111 - 122 (1980).

008.034 **Göteborg**

Research Laboratory of Electronics and Onsala Space Observatory, Chalmers University of Technology, Gothenburg, Sweden. Research Report, Nos. 140(28.131.261), 141 (28.082.060).

008.035 **Göttingen**

Universitäts-Sternwarte Göttingen und Institut für Sonnenforschung Locarno-Orselina (Tessin). – Report for 1979. H. H. Voigt, K. J. Fricke.
Mitt. Astron. Ges., Nr. 49, p. 123 - 132 (1980).

Solar observations at the Göttingen University Observatory. (Report from Solar Institute).
E. Wiehr, A. Wittmann, H. Wöhl.
Sol. Phys., Vol. 68, 207 - 212 (1980).

008.036 **Graz**

Institut für Astronomie der Universität Graz; Institut für Astronomie (Universitäts-Sternwarte); Observatorium Lustbühel; Sonnenobservatorion Kanzelhöhe. – Report for 1979. H. Haupt.
Mitt. Astron. Ges., Nr. 49, p. 133 - 137 (1980).

008.037 **Green Bank, W. Va.**

National Radio Astronomy Observatory, Green Bank, Reprints, Series A, Nos. 1080 (27.141.007), 1081 (27.131.010), 1082 (27.131.011), 1083 (27.008.026), 1084 (27.141.045), 1085 (27.160.007), 1086 (27.125.009), 1087 (27.064.009), 1088 (27.141.110), 1089 (27.141.112), 1090 (27.155.012), 1091 (27.132.043), 1092 (27.141.530), 1093 (27.131.143), 1094 (27.158.076), 1095 (27.158.211), 1096 (27.141.145), 1097 (27.141.146), 1098 (26.158.188), 1099 (27.131.152), 1100 (27.122.099), 1101 (27.141.157), 1102 (27.141.158), 1103 (28.033.019), 1104 (27.077.011), 1105 (27.125.058), 1106 (27.132.068), 1107 (27.141.053), 1108 (27.141.054), 1109 (27.160.060), 1110 (27.131.151), 1111 (27.158.236), 1112 (27.112.017), 1113 (27.116.012), 1114 (27.160.080), 1115 (27.141.193), 1116 (27.141.194), 1117 (27.141.195), 1118 (27.131.179), 1119 (27.134.042), 1120 (28.141.001), 1121 (28.158.003), 1122 (27.158.328), 1123 (27.064.091), 1124 (27.159.028), 1125 (28.160.001), 1126 (28.141.029), 1127 (28.131.036), 1128 (28.141.028), 1129 (28.131.029), 1130 (28.158.008), 1131 (28.158.009), 1132 (28.158.011), 1133 (28.031.573).

National Radio Astronomy Observatory, Green Bank, Reprints, Series B, Nos. 506 (27.132.024), 507 (27.131.068), 508 (27.132.031), 509 (28.141.139), 510 (27.031.656), 511 (27.116.028), 512 (28.141.002).

008.038 **Greenwich**

The report of the Royal Greenwich Observatory. For the period 1978 October 1 to 1979 September 30.
F. G. Smith (Director), G. A. Wilkins, B. D. Yallop (Editors). Published by Royal Greenwich Observatory, Herstmonceux Castle, Hailsham, East Sussex, 55 pp. Price £ 2.50 (1980). ISSN 0308-3322.

The Royal Greenwich Observatory. Report for the period 1978 October 1 to 1979 September 30.
F. G. Smith.
Q. J. R. Astron. Soc., Vol. 21, 432 - 441 (1980).

Royal Observatory Annals. Royal Greenwich Observatory Herstmonceux, No. 13 (28.072.040).

Greenwich Time Report. Royal Greenwich Observatory, Time and Latitude Service, 1979 October - December (28.044.018).

H. M. Nautical Almanac Office, Library Reprint, Nos. 348 (27.043.004), 349 (27.045.037), 350 (28.044.035).

008.039 **Groningen**

Nederlandse Vereniging voor Weer- en Sterrenkunde. Kapteyn Astronomical Laboratory, Groningen, Netherlands. Observations of Variable Stars. Report. No. 34 (28.123.014).

008.040 **Hamburg**

Hamburger Sternwarte. – Report for 1979.
A. Weigert.
Mitt. Astron. Ges., Nr. 49, p. 139 - 145 (1980).

Deutsches Hydrographisches Institut. 33./34. Jahresbericht 1978/79. G. Zickwolff.
Deutsches Hydrographisches Institut, Hamburg. 137 pp. (1980).

Deutsches Hydrographisches Institut, Hamburg. Zeit- und Breitendienst, 1979 January - December (28.044.019).

008.041 **Hannover**

Astronomische Station am Institut für Theoretische Geodäsie der Universität. – Report for 1979.
G. Seeber.
Mitt. Astron. Ges., Nr. 49, p. 147 - 149 (1980).

008.042 **Heidelberg**

Astronomisches Rechen-Institut. – Report for 1979.
W. Fricke.
Mitt. Astron. Ges., Nr. 49, p. 151 - 161 (1980).

Astronomy and Astrophysics Abstracts, Vol. 27 (28.002.059).

Astronomische Grundlagen für den Kalender 1982 (28.047.029).

Institut für Theoretische Astrophysik. – Report for 1979. B. Baschek, G. Traving.
Mitt. Astron. Ges., Nr. 49, p. 163 - 165 (1980).

Landessternwarte.– Report for 1979.
I. Appenzeller.
Mitt. Astron. Ges., Nr. 49, p. 177 - 184 (1980).

Max-Planck-Institut für Astronomie. Deutsch-Spanisches Astronomisches Zentrum. – Report for 1979.
H. Elsässer.
Mitt. Astron. Ges., Nr. 49, p. 185 - 209 (1980).

Max-Planck-Institut für Kernphysik. Abteilung Kosmochemie. – Report for 1979.
H. Fechtig, H. Völk.
Mitt. Astron. Ges., Nr. 49, p. 167 - 175 (1980).

008.043 **Innsbruck**

Institut für Astronomie. – Report for 1979.
J. Pfleiderer.
Mitt. Astron. Ges., Nr. 49, p. 211 - 214 (1980).

Mitteilungen der Sternwarte Innsbruck, Nos. 19 (14.008.048), 20 (14.063.008), 21 (14.113.007), 22 (13.063.014), 23 (14.155.008), 24 (17.106.013), 25 (17.031.213), 26 (17.034.009), 27 (17.031.403), 28 (18.008.046), 29 (18.116.026), 30 (28.005.028), 31 (18.031.420), 32 (18.158.242), 33 (18.031.320), 34 (22.113.069), 35 (19.008.070), 36 (19.158.117), 37 (20.135.020), 38 (20.135.021), 39 (20.160.065), 40 (22.031.213), 41 (22.160.013), 42 (22.158.056), 43 (22.135.011), 44 (22.113.007), 45 (22.008.050), 46 (22.153.030), 47, 48 (28.158.184), 49 (26.153.009), 50 (26.135.013), 51 (26.008.038), 52 (25.135.035).

008.044 **Ithaca, N.Y.**

National Astronomy and Ionosphere Center, Cornell University, Ithaca, Astronomy Publications, Nos. A80-1 (27.125.009), A80-2 (27.141.518), A80-3 (27.158.027), A80-4 (26.141.043), A80-5 (27.112.013), A80-6 (26.141.544), A80-7 (27.141.531), A80-8 (27.015.007), A80-9 (27.032.077), A80-10, A80-11 (27.158.071), A80-12 (27.141.530), A80-13 (27.100.049), A80-14 (27.141.013), A80-15 (27.135.044), A80-16 (27.135.012).

008.045 **Kazan**

Izvestiya Astronomicheskoj Ehngel'gardtovskoj Observatorii, Kazan, Izdatel'stvo Kazanskogo Universiteta, No. 45 (28.082.076; 28.100.125; 28.096.015; 28.119.098; 28.119.099).

008.046 **Kiel**

Institut für Theoretische Physik und Sternwarte der Universität. – Report for 1979. K. Hunger.
Mitt. Astron. Ges., Nr. 49, p. 215 - 218 (1980).

Institut für Reine und Angewandte Kernphysik, Arbeitsgruppe Mathematische Physik. – Report for 1979. K. O. Thielheim.
Mitt. Astron. Ges., Nr. 49, p. 219 - 220 (1980).

008.047 **Kiev**

Astrometriya i Astrofizika, Akademiya Nauk Ukrainskoj SSR, Glavnaya Astronomicheskaya Observatoriya, Vyp. (No.) 41, 42 (1980).

008.048 **Kyoto**

Contributions from the Kwasan and Hida Observatories, University of Kyoto, Nos. 242 (28.097.075), 243 (27.097.041).

008.049 **La Plata**

Observatorio Astronómico de la Universidad Nacional de La Plata, Serie Especial, No. 25 (28.031.017).

008.050 **La Serena**

Cerro Tololo Interamerican Observatory, La Serena, Chile. – Report for the period 1 January 1978 - 30 September 1979. M. M. Phillips, J. Muñoz.
Bull. American Astron. Soc., Vol. 12, 570 - 583 (1980).

008.051 **La Silla**

European Southern Observatory. Annual report 1979. L. Woltjer.
Published by European South. Obs., Garching, Germany. 56 pp. (1980). ISSN 0531-4496.

ESO Scientific Preprint Nos. 99 (28.158.185), 100 (28.064.068), 101 (28.158.186), 102 (28.131.262), 103 (28.133.012), 104 (28.125.053), 105 (28.002.057), 106 (28.131.263), 107 (28.151.072), 108 (28.159.015), 109 (28.158.187), 110 (28.160.052), 111 (28.134.034), 112 (28.158.188), 113 (28.158.189), 114 (28.158.190), 115 (28.141.140), 116 (28.159.016), 117 (28.158.191), 118 (28.114.112), 119 (28.158.192), 120 (28.151.073), 121 (28.125.054), 122 (28.141.141), 123 (28.160.053), 124 (28.122.138), 125 (28.122.139), 126 (28.103.210), 127 (28.158.220), 128 (28.117.071), 129 (28.141.158), 130 (28.113.041), 131 (28.142.115).

The Messenger – El Mensajero, Nos. 21 - 22 (1980).

008.052 **Lake Tekapo**

Mt John University Observatory. Annual report for 1980. J. B. Hearnshaw.
South. Stars, Vol. 28, 137 - 139 (1980).

008.053 **Leicester**

University of Leicester X-ray Astronomy Group, Department of Physics. Report for the period 1978 October to 1980 March. K. A. Pounds.
Q. J. R. Astron. Soc., Vol. 21, 464 - 480 (1980).

008.054 **Leningrad**

Chronicle.
Byull. Inst. Teor. Astron., Tom 14, 655 (1980). In Russian.

Byulleten' Instituta Teoreticheskoj Astronomii, Akademiya Nauk SSSR, Tom 14, Nos. 9 (28.031.600; 28.041.034; 28.081.029; 28.041.035; 28.098.086), 10 (28.021.046; 28.042.068; 28.081.030; 28.052.050; 28.052.051; 28.008.054).

008.055 **Lintong**

Publications of the Shensi Astronomical Observatory, 1980, No. 1 (28.044.033; 28.035.003; 28.035.004; 28.035.005; 28.041.029; 28.041.030; 28.041.031).

Time and Frequency Services Bulletin, Nos. 4 - 6 (28.044.020).

008.056 **Lund**

Reports from the Observatory of Lund, Nos. 14 (28.009.021), 15 (28.034.040), 16 (28.153.021), 17 (28.063.052).

008.057 **Madrid**

Boletín Astronómico del Observatorio de Madrid, Vol. 10, No. 3 (28.032.021; 28.034.041; 28.072.041).

Universidad Complutense – Facultad de Ciencias, Madrid. Seminario de Astronomia y Geodesia, Publicación, Nos. 97 (25.002.027), 98, 99, 100.

008.058 **Manchester**

University of Manchester, Nuffield Radio Astronomy Laboratories, Jodrell Bank. Report for the year ending 1979 September 30. B. Lovell.
Q. J. R. Astron. Soc., Vol. 21, 317 - 331 (1980).

008.059 **Manila**

Manila Observatory, Solar Division, Solar maps and activity, 1980 May - September (28.072.042).

008.060 **Mizusawa**

Monthly Notes of the International Polar Motion Service, 1980 Nos. 5 - 10 (28.045.008).

Bulletins, Time Service of the Mizusawa Observatory, Vol. 23 (28.044.021).

Publications of the International Latitude Observatory of Mizusawa, Vol. 13, No. 1 (28.046.009; 28.083.043; 28.041.025).

008.061 **Mons**

Communications du Département d'Astrophysique de la Faculté des Sciences de Mons. Mons Astrophysical Papers. Nos. 80 (27.062.086), 81 (27.114.130), 82 (27.098.066).

008.062 **Moscow**

Soobshcheniya Gosudarstvennogo Astronomicheskogo Instituta im. P. K. Shternberga, Izdatel'stvo Moskovskogo Universiteta, Nos. 218 (28.042.017; 28.042.018), 219 (28.114.020).

Trudy Gosudarstvennogo Astronomicheskogo Instituta im. P. K. Shternberga, Izdatel'stvo Moskovskogo Universiteta, Tom 49 (28. 042.007; 28.042.008; 28.042.009; 28.042.010; 28.042.011; 28.052.003; 28.052.004; 28.052.005; 28.052.006; 28.052.007; 28.052.008; 28.002.011; 28.002.012; 28.093.009) 50 (28.041.009; 28.002.019; 28.041.010; 28.002.020; 28.094.009; 28.094.010; 28.094.011; 28.094.012).

008.063 **Münster**

Astronomisches Institut der Universität. Außenstation Schalkenmehren. – Report for 1979.
W. C. Seitter.
Mitt. Astron. Ges., Nr. 49, p. 257 - 262 (1980).

008.064 **Munich**

Institut für Astronomie und Astrophysik der Universität München. Unitersitäts-Sternwarte. – Report for 1979. P. Wellmann.
Mitt. Astron. Ges., Nr. 49, p. 251 - 255 (1980).

Max-Planck-Institut für Physik und Astrophysik. Institut für Astrophysik und Institut für extraterrestrische Physik. – Reports for 1979.
R. Kippenhahn, H. Rothermel.
Mitt. Astron. Ges., Nr. 49, p. 221 - 249 (1980).

008.065 **Paris**

Rapport d'activité de l'Observatoire de Paris.
1er janvier - 31 décembre 1979.
61, Avenue de l'Observatoire, 75014 Paris. 106 pp. (1980).

Observatoire de Paris, Section d'Astrophysique, à Meudon, **Cartes synoptiques,** Vol. 6, Fasc. 3, anneés 1978 - 1979 (28.072.043).

Bureau International de l'Heure, (B. I. H.), Circular A (28.044.022), D165 - 169 (28.044.023).

008.066 **Pittsburgh, Penn.**

Allegheny Observatory, University of Pittsburgh, Pittsburgh, Pennsylvania 15214. – Report for the period 1 July 1976 to 30 June 1978.
Bull. American Astron. Soc., Vol. 12, 567 - 569 (1980).

008.067 **Potsdam**

Zentralinstitut für Astrophysik, Sternwarte Babelsberg, Mitteilungen, Neue Folge, Nos. 196 (19.042.002), 197 (19.160.001), 198 (19.160.002), 199 (19.066.060), 200 (19.066.061), 201 (19.066.062), 202 (19.160.032), 203 (19.034.085), 204 (19.155.055), 205 (20.066.042), 206 (20.158.051), 207 (20.162.092), 208 (20.141.142), 209 (20.160.058), 210 (21.141.097), 211 (21.062.057), 212 (21.158.151), 213 (21.160.052), 214 (21.160.054), 215 (28.066.143), 216 (28.158.194), 217 (28.158.195), 218 (28.160.055), 219 (28.160.056), 220 (22.062.025), 225 (22.031.297), 226 (22.160.034), 227 (22.031.298), 228 (25.141.035), 229 (25.141.036), 231 (25.066.042), 232 (25.158.101), 233 (25.141.064), 250 (28.066.142).

Zeit- und Breitenbestimmungen, Zeitsysteme, Präzisionszeitvergleiche, Jahrg. 1979, Nos. 5 - 6 (28.044.024)

Veröffentlichungen des Zentralinstituts für Physik der Erde, Potsdam, Nr. 55 (28.002.058).

008.068 **San Diego, Calif.**

Mt. Laguna Observatory, San Diego, California 92182. – Report. M. S. Snowden. Bull. American Astron. Soc., Vol. 12, 622 - 623 (1980).

008.069 **San Fernando (Cadiz)**

Memoria de las actividades en 1979. Inst. Obs. Marina, San Fernando (Cadiz), Espania. 22 pp. (1980).

008.070 **Santa Cruz, Calif.**

Lick Observatory Bulletin, Nos. 680 (10.141.078), 791 (21.131.150), 809 (22.135.030), 821 (26.135.010), 822 (26.141.054), 823 (27.158.007), 826 (26.158.022), 827 (26.158.211), 828 (26.158.172), 829 (26.158.193), 830 (26.132.016), 833 (27.131.015), 834 (25.114.144), 835 (26.158.032), 836 (26.134.007), 837 (26.141.051), 838 (26.141.021), 839 (26.034.009), 840 (26.158.044), 841 (26.158.048), 842 (26.122.025), 843 (26.131.117), 845 (26.118.018), 846 (27.062.005), 847 (27.062.004), 849 (27.124.003), 850 (26.121.013), 851 (27.122.064), 852 (26.121.015; 27.121.901), 853 (27.158.072), 855 (27.135.034), 858 (27.134.012), 859 (27.121.019), 860 (27.154.025), 861 (27.158.159), 864 (27.158.265).

Lick Observatory Contributions Nos. 415 (25.153.007), 418 (26.154.001), 419 (25.141.106), 421 (27.107.004).

008.071 **Sendai**

Annual report of research activities in 1979. Astronomy group. Sci. Rep. Tôhoku Univ., Ser. 8, Vol. 1, No. 2, p. 71 - 76 (1980).

Sendai Astronomiaj Raportoj, Nos. 206 (26.151.063), 207 (26.075.017), 208 (26.103.403), 209 (26.064.055), 210 (27.062.011), 211 (26.155.016), 212 (27.155.046), 213 (28.122.140), 214 (28.062.098), 215 (28.065.064), 216 (28.125.055), 217 (28.008.071), 218 (28.131.025), 221 (28.122.019).

008.072 **Shanghai**

Time Service Annual Report 1978 (28.044.025).

008.073 **Skalnaté Pleso**

Contributions of the Astronomical Observatory Skalnaté Pleso, Vol. 9 (28.074.007, 28.074.008, 28.104.009, 28.104.010, 28.104.011, 28.098.005, 28.119.017).

008.074 **Sonneberg**

Zentralinstitut für Astrophysik, Sonneberg. Mitteilungen über Veränderliche Sterne, Band 8, Heft 9 (28.113.030; 28.122.141; 28.123.015; 28.123.016; 28.123.017; 28.123.018; 28.122.142; 28.119.085; 28.119.086).

008.075 **St. Andrews**

Communications from the University Observatory, St. Andrews, Nos. 20 (20.113.027), 21 (21.113.026), 22 (21.121.024), 23 (25.113.026), 24 (25.158.039), 25 (25.158.040), 26 (25.119.022), 27 (25.115.011), 28 (25.119.042), 29 (25.158.182), 30 (27.113.002), 31 (27.158.045), 32 (27.158.152).

University Observatory, St. Andrews. Reprint, Nos. 75 (20.064.014), 76 (21.131.171), 77 (21.155.038), 78 (26.113.034), 79 (27.113.015), 80 (27.155.004), 81 (27.117.060), 82 (25.155.013), 83 (25.114.017).

008.076 **Stockholm**

Stockholms Observatorium, Saltsjöbaden, Sweden, Report Nos. 16 (28.121.005), 17 (28.021.032).

008.077 **Strasbourg**

Centre de Données Stellaires. Annual report of the director for 1979. C. Jaschek. Bull. Inf. Cent. Données Stellaires, No. 19, p. 53 - 56 (1980).

Bulletin d'Information du Centre de Données Stellaires, No. 19 (1980).

008.078 **Sydney**

Sydney Observatory. Report for the year ending 1979 December 31. W. H. Robertson. Q. J. R. Astron. Soc., Vol. 21, 332 (1980).

008.079 **Tartu**

W. Struve nimeline Tartu Astrofüüsika Observatoorium, Teated Nr. 60 (28.114.046).

ENSV Teaduste Akadeemia, Prepr. A-1 (28.114.046); A-3 (28.064.014); A-4 (28.064.015).

008.080 **Thessaloniki**

Department of Geodetic Astronomy, University of Thessaloniki. – Annual report 1979. L. N. Mavridis. Annu. Rep. Astron. Inst. Greece, 1979, p. 20 - 24 (1980).

008.081 **Tokyo**

Annals of the Tokyo Astronomical Observatory, University of Tokyo, Second Series, Vol. 18, No. 1 (28.063.053; 28.113.031).

University of Tokyo, Tokyo Astronomical Observatory, Report (No. 73), Vol. 19, No. 2 (28.031.018; 28.072.044; 28.034.042; 28.031.572; 28.041.026; 28.034.043; 28.074.089; 28.047.030; 28.047.031).

Tokyo Astronomical Observatory, Reprints Nos. 575 (28.158.030), 576 (28.131.026), 577 (28.064.007), 578 (28.033.001), 579 (28.094.004), 580 (27.155.032), 581 (27.081.070), 582 (28.134.030), 583 (28.120.010), 584 (28.126.026), 585 (28.077.043), 586 (28.142.007).

Tokyo Astronomical Observatory, Time and Latitude Bulletins, Vol. 53, No. 4, Vol. 54, Nos. 1 - 2 (28.044.026).

Tokyo Astronomical Observatory, Kiso Information Bulletin, Vol. 1, Nos. 1 - 3 (1979/80).

Tokyo Astronomical Bulletin, Second Series, No. 263 (28.124.022).

008.082 **Toruń**

Bulletin of the Astronomical Observatory of N. Copernicus University in Toruń, No. 61 (25.042.024; 25.114.165; 26.135.023; 25.153.022; 27.114.061; 27.120.010; 27.114.060).

008.083 **Tucson, Ariz.**

Kitt Peak National Observatory, Tucson, Arizona. Report for the period 1 January - 30 September 1979.
Bull. American Astron. Soc., Vol. 12, 584 - 621 (1980).

008.084 **Tübingen**

Astronomisches Institut der Universität und Lehrstuhl für Theoretische Astrophysik. – Report for 1979.
G. Elwert, M. Grewing.
Mitt. Astron. Ges., Nr. 49, p. 263 - 271 (1980).

008.085 **Uccle**

Observatoire Royal de Belgique, Communications. Koninklijke Sterrenwacht van België, Mededelingen, Série A, No. 58 (28.081.001).

Observatoire Royal de Belgique, Communications. Koninklijke Sterrenwacht van België, Mededelingen, Série B, No. 116 (27.045.029).

008.086 **Utrecht**

Utrechtse Sterrekundige Overdrukken, Nos. 503 (27.142.019), 504 (27.114.039), 505 (27.114.040), 506 (27.032.518), 507 (27.071.004), 508 (27.062.028), 509 (27.122.021), 510, 511 (27.132.020), 512 (27.074.035), 513 (27.074.036), 514 (27.126.019), 515 (27.076.015), 516 (27.073.034), 517 (27.077.066), 518 (27.073.079), 519 (27.125.062), 520 (28.063.043), 521 (27.115.012).

008.087 **Victoria**

Dominion Astrophysical Observatory, Victoria, British Columbia. Herzberg Institute of Astrophysics National Research Council, Canada. Report for the year 1979 April 1 to 1980 March 31. S. van den Bergh.
Q. J. R. Astron. Soc., Vol. 21, 442 - 459 (1980).

Dominion Astrophysical Observatory, Victoria, B. C.
J. B. Hutchings.
J. R. Astron. Soc. Canada, Vol. 74, 302 - 303 (1980).

008.088 **Vienna**

Institut für Astronomie der Universität Wien.
Report for 1979. K. Rakos.
Mitt. Astron. Ges., Nr. 49, p. 273 - 287 (1980).

Astronomische Mitteilungen Wien, Nr. 24 (28.046.010).

008.089 **Villa Elisa, Provincia de Buenos Aires**

Contribuciones del Instituto Argentino de Radioastronomia, Nos. 83 (26.131.039), 84 (26.153.027), 85 (27.155.064), 86 (27.002.059), 87 (27.155.063), 88 (27.158.337), 89 (27.152.013), 90 (27.153.081), 91 (27.125.098), 92 (27.125.025), 93 (27.155.008).

008.090 **Vilnius**

Vilniaus Astronomijos Observatorijos Biuletenis (Bulletin of the Vilnius Astronomical Observatory), Nr. 53 (28.034.077; 28.034.078; 28.034.079; 28.064.105), 55 (28.082.087; 28.082.088; 28.113.048; 28.114.167).

008.091 **Warsaw**

Warsaw University Observatory and Polish Academy of Sciences, N. Copernicus Astronomical Center, Reprint Nos. 417 (27.142.047), 418 (27.119.028), 419 (27.119.030), 420 (27.119.031), 421 (27.119.032), 422 (27.066.245), 423 (27.066.246), 424 (27.122.136), 425 (28.122.014), 426 (28.119.002), 427 (28.142.005), 428 (28.066.501).

Polish Academy of Sciences, N. Copernicus Astronomical Center, Preprint No. 111 (28.064.069).

Publications of the Institute of Geophysics, Polish Academy of Sciences, F-5 (136), F-6 (137) (28.044.031; 28.044.032; 28.031.574; 28.045.009; 28.032.022; 28.045.010).

008.092 **Washington, D.C.**

Astronomical Papers prepared for the use of the American Ephemeris and Nautical Almanac, Vol. 21, Part 3 (28.041.028).

United States Naval Observatory, Washington, D.C., Circular, Nos. 159 (28.041.027), 160 (28.079.501).

Publications of the United States Naval Observatory, Washington, Second Series, Vol. 24, Part VI (28.118.027).

U.S. Naval Observatory, Washington, D.C. Time Service Publications. Series 4, Nos. 700 - 725; Series 6, Nos. 53 - 58; Series 7, Nos. 653 - 678; Series 14, No. 28 (28.044.027 - 28.044.030).

008.093 **Wellington**

Annual report of the Carter Observatory Board for the year ended 1980 March 31. J. B. Mackie.
Carter Obs., Astron. Bull., No. 95, 5 pp. (1980).

Report of the director for the year ended 1980 March 31. B. M. Lewis.
Carter Obs., Astron. Bull., No. 95, 4 pp. (1980).

Carter Observatory, Astronomical Bulletin, Nos. 93 (28.047.032), 95 (28.008.093).

Carter Observatory, Wellington, Reprint Series 2, No. 12 (28.160.054).

008.094 **Würzburg**

Institut für Astronomie und Astrophysik, Lehrstuhl Astronomie. – Report for 1979. F.-L. Deubner.
Mitt. Astron. Ges., Nr. 49, p. 289 - 293 (1980).

009 Notes on Observatories, Planetaria, Exhibitions

009.001 **Rhodes University: Department of Physics and Electronics.** E. E. Baart.
Mon. Notes Astron. Soc. South. Africa, Vol. 39, 9 - 10 (1980).

009.002 **University of South Africa: Department of Mathematics, Applied Mathematics and Astronomy.**
P. D. Bennewith, J. Wolterbeek.
Mon. Notes Astron. Soc. South. Africa, Vol. 39, 10 - 11 (1980).

009.003 **La Silla, die südliche Beobachtungsstation der europäischen Astronomie. Teil I: Geschichte und Ausstattung des Observatoriums.**
N. Vogt.
Sterne Weltraum, Jahrg. 19, 284 - 289 (1980).

009.004 **Astronomy Centre, University of Sussex.** Report for the year ending 1979 September 30.
L. Mestel.
Q. J. R. Astron. Soc., Vol. 21, 307 - 316 (1980).

009.005 **L'Observatoire Flammarion de Juvisy.**
A. Dollfus.
Astronomie, Vol. 94, 363 - 366 (1980).

009.006 **L'Observatoire de Juvisy: souvenir et avenir.**
J.-M. Huin.
Astronomie, Vol. 94, 367 - 374 (1980).

009.007 **L'Observatoire de la Roque de los Muchachos.**
N. O'Hora.
Astronomie, Vol. 94, 375 - 384 (1980).

009.008 **Alpine Observatory in Central Asia.**
S. B. Novikov.
Zemlya Vselennaya, 1980, No. 4, p. 52 - 58. In Russian.
Report on the construction of the Alpine Observatory of the Sternberg Astronomical Institute in South Uzbekistan.

009.009 **Annual report for magnetic observatories – 1977.**
E. I. Loomer.
Geomagn. Ser. Earth Phys. Branch, No. 15, p. 1 - 139 (1979).
In French and English. – Abstr. in Phys. Abstr., Vol. 83, Abstr. 72418 (1980).

009.010 **Australian magnetic observatories.**
P. M. McGregor.
BMR J. Australian Geol. Geophys., Vol. 4, 361 - 371 (1979).
Abstr. in Phys. Abstr., Vol. 83, Abstr. 72632 (1980).

009.011 **Geodetic work and projects at the Observatory of Graz-Lustbühel.** K. Rinner.
Acta Geod. Geophys. Montan., Vol. 14, No. 1 - 2, p. 17 - 29 (1979). – Abstr. in Phys. Abstr., Vol. 83, Abstr. 77117 (1980).

009.012 **La Silla, die südliche Beobachtungsstation der europäischen Astronomie. Teil II: Der Arbeitsalltag bei ESO.** N. Vogt.
Sterne Weltraum, Jahrg. 19, 324 - 330 (1980).

009.013 **A new observatory for Mexico.**
C. Allen, J. de la Herran, H. Johnson.
Sky Telesc., Vol. 60, 270 - 273 (1980).

009.014 **Die Welt als Uhr - ein Sprach- und Denkbild.**
K. Maurice.
Sterne Weltraum, Jahrg. 19, 364 - 367 (1980).
Das Bayerische Nationalmuseum veranstaltet zu seinem 125jährigen Gründungsjubiläum die Ausstellung "Die Welt als Uhr – deutsche Uhren und Automaten 1550 - 1650".

009.015 **'Holländische Sternwarte' in Zwitserland.**
S. van Heck, F. Cornelis.
Zenit, 7e Jaarg., 96 - 98 (1980).

009.016 **Europese sterrenwacht op La Palma.**
Zenit, 7e Jaarg., 99 (1980).

009.017 **Reuzentelescopen op Mauna Kea in gebruik.**
G. W. E. Beekman.
Zenit, 7e Jaarg., 99 (1980).

009.018 **Experiment in het planetarium van Brussel 'Sterrenkunde voor de lagere school'.**

C. de Loore, L. van Itterbeeck, A. Pien, J. Keuppens.
Zenit, 7e Jaarg., 148 - 149 (1980).

009.019 **Nieuw planetarium in Apeldoorn.**
G. W. E. Beekman.
Zenit, 7e Jaarg., 198 - 199 (1980).

009.020 **Grote Europese sterrenwacht in wording.**
C. de Jager.
Zenit, 7e Jaarg., 354 - 356 (1980).

009.021 **The Lund Centre for Automated Measurements of Astronomical Plates.** T. Elvius, H. Lindgren, G. Lyngå, N. Wihlborg.
Rep. Obs. Lund, No. 14, 14 pp. (1978).

009.022 **Ausbau der Steuerung des Stuttgarter Planetariums zur Vollautomatik.**
G. Bräuninger, H.-U. Keller.
Sterne Weltraum, 19. Jahrg., 427 - 429 (1980).

009.023 **NASA's new infrared eye.** C. M. Telesco, E. E. Becklin, R. W. Capps, A. T. Tokunaga.
Sky Telesc., Vol. 60, 462 - 465 (1980).
Concerning the Mauna Kea Observatory.

009.024 **Opening of the Astronomical Center of the Polish Academy of Sciences.**
Zh. Pol'sk. A. N., Vol. 23, No. 4, p. 109 - 111 (1978, 1980).
Abstr. in Ref. zh., 51. Astron., 11.51.32 (1980).

009.025 **A planetarium for London, Ontario.**
P. Jedicke.
J. R. Astron. Soc. Canada, Vol. 74, 364 - 365 (1980).
Abstract.

009.026 **How Belgrade got its planetarium.**
P. Đurković, B. Ševarlić.
Vasiona, Année 28, 4 - 5 (1980). In Serbo-Croatian.

009.027 **A decade of the Belgrade Planetarium.**
M. Jeličić.
Vasiona, Année 28, 5 - 7 (1980). In Croatian.

Astronomie in Gotha. II. Die Sternwarte in der Jägerstraße. See Abstr. 004.029.

How the U. S. Naval Observatory began, 1830 - 65. See Abstr. 004.056.

010 Societies, Associations, Organizations

010.001 American Association of Variable Star Observers (AAVSO)

Variable stars and the AAVSO.
See Abstr. 122.017.

010.002 American Astronomical Society (AAS)

The 156th meeting of the American Astronomical Society, held 15 - 18 June 1980 at College Park, Maryland. Abstracts of papers presented.
Bull. American Astron. Soc., Vol. 12, 443 - 565 (1980).

Report of the Committee on the Status of Women. American Astronomical Society. I. King.
Bull. American Astron. Soc., Vol. 12, 624 - 635 (1980).

12th annual meeting of the Division for Planetary Sciences of the American Astronomical Society held 14 - 17 October 1980 at Tucson, Arizona. Abstracts of papers presented.
Bull. American Astron. Soc., Vol. 12, 639 - 737 (1980).

Abstracts of papers presented at the Dynamical Astronomy Division Meeting held 18 - 20 June 1980 at College Park, Maryland.
Bull. American Astron. Soc., Vol. 12, 738 - 746 (1980).

Late-paper abstracts from the 156th meeting of the American Astronomical Society held 15 - 18 June 1980 at College Park, Maryland.
Bull. American Astron. Soc., Vol. 12, 747 - 752 (1980).

Bulletin of the American Astronomical Society, Vol. 12, Nos. 2 - 3 (1980).

AAS Photo-Bulletin, Issue 23, No. 1 (1980).

010.003 Association Française des Observateurs d'Etoiles Variables (A. F. O. E. V.)

La vie de l'Association. E. Schweitzer.
Bull. AFOEV, Tome 14, 50, 74 (1980).

Activité de l'A. F. O. E. V. E. Schweitzer.
Astronomie, Vol. 94, 516 - 517 (1980).

Bulletin de l'Association Française des Observateurs d'Etoiles Variables, Tome 14, Nos. 2, 3 (1980).

010.004 Association of Lunar and Planetary Observers (A. L. P. O.)

Minor Planets Section news.
Minor Planet Bull., Vol. 7, 32 (1980).

The Minor Planet Bulletin. Bulletin of the Minor Planets Section of the Association of Lunar and Planetary Observers, Vol. 7, No. 4, Part 1 - 2 (1980).

The Strolling Astronomer. The Journal of the Association of Lunar and Planetary Observers, Vol. 28, Nos. 7 - 8 (1980).

010.005 Astronomical Society of Australia (ASA)

Meetings of the Society.
Proc. Astron. Soc. Australia, Vol. 3, 289 (1978), Vol. 3, 403 (1979).

Proceedings of the Astronomical Society of Australia, Vol. 3, Nos. 3+4 (1978), Nos. 5+6 (1979).

010.006 Astronomical Society of Egypt

Journal of the Astronomical Society of Egypt, Vol. 1 (1979).

010.007 Astronomical Society of India

Bulletin of the Astronomical Society of India, Vol. 8, Nos. 1 - 3 (1980).

Memoirs of the Astronomical Society of India, Vol. 1, August (1980).

010.008 Astronomical Society of Japan

Publications of the Astronomical Society of Japan, Vol. 32, Nos. 2 - 4 (1980).

010.009 Astronomical Society of New York

Abstracts of papers given at the Spring Meeting at Cornell University, Ithaca, N. Y., 1980 April 26.
News Lett. Astron. Soc. N. Y., Vol. 1, No. 7, p. 3 - 24 (1980).

News Letter of the Astronomical Society of New York, Vol. 1, No. 7 (1980).

010.010 Astronomical Society of the Pacific (ASP)

Abstracts of papers presented at the University of Arizona Meeting of the Astronomical Society of the Pacific, 7 - 12 July 1980.
Publ. Astron. Soc. Pacific, Vol. 92, 546 - 552 (1980).

Publications of the Astronomical Society of the Pacific, Vol. 92, Nos. 547 - 549 (1980).

Mercury. The Journal of the Astronomical Society of the Pacific, Vol. 9, Nos. 4 - 6 (1980).

010.011 **Astronomical Society of Southern Africa (ASSA)**

Notices.
Mon. Notes Astron. Soc. South. Africa, Vol. 39, 1 (1980).

Monthly Notes of the Astronomical Society of Southern Africa, Vol. 39, Nos. 1 - 4 (1980).

010.012 **Astronomical Society of Western Australia (ASWA)**

Report on proceedings of ordinary meetings.
J. Astron. Soc. Western Australia, Vol. 27, No. 12, Vol. 28, Nos. 1 - 4 (1980).

The Journal of the Astronomical Society of Western Australia. Vol. 27, No. 12, Vol. 28, Nos. 1 - 4 (1980).

010.013 **Astronomische Gesellschaft (AG)**

Berichte über die wissenschaftlichen Tagungen Bochum, 12. - 14. März 1979, Mainz, 5. - 7. März 1980.
Mitt. Astron. Ges., Nr. 50, p. 13 - 168 (1980).

Mitteilungen der Astronomischen Gesellschaft, Nr. 48 - 50 (1980).

010.014 **British Astronomical Association (BAA)**

Meetings and activities of the Association.
J. British Astron. Assoc., Vol. 90, 403 - 410, 470 - 471, 501 - 523, 526 - 528, Vol. 91, 2 - 8, 12 - 13 (1980).

Notices.
J. British Astron. Assoc., Vol. 90, 401 - 402, 524 - 525, Vol. 91, 1 (1980).

Section reports.
J. British Astron. Assoc., Vol. 90, 451 - 463, 560 - 564, Vol. 91, 8 - 12 (1980).

Journal of the British Astronomical Association, Vol. 90, Nos. 5 - 6, Vol. 91, No. 1 (1980).

British Astronomical Association Circular, Nos. 607 - 612 (1980).

010.015 **British Interplanetary Society (BIS)**

JBIS. Journal of the British Interplanetary Society, Vol. 33, Nos. 9 - 12 (1980).

Spaceflight. A publication of the British Interplanetary Society, Vol. 22, Nos. 9 - 10, 11 - 12 (1980).

010.016 **European Space Agency (ESA)**

Programmes under development and operations (*ESA*).
ESA Bull., No. 21, p. 49 - 58 (1980). – Abstr. in Phys. Abstr., Vol. 83, Abstr. 72742 (1980).

ESA IUE Newsletter, Nos. 7 - 8 (1980).

010.017 **International Association of Geodesy (IAG)**

The Geodesist's Handbook 1980.
I. I. Mueller (Editor).
Bull. Geod., Vol. 54, 243 - 502 (1980).

010.018 **International Astronomical Union (IAU)**

Transactions of the International Astronomical Union, Volume XVIIB: Proceedings of the Seventeenth General Assembly, Montreal 1979. See Abstr. 012.042.

L'Union Astronomique Internationale vit...
See Abstr. 011.030.

Circulaire d'Information, No. 82 (1980).

IAU Circulars, Nos. 3491 - 3557 (1980).

Information Bulletin on Variable Stars, Commission 27 of the I.A.U., Nos. 1810 - 1902 (1980).

Minor Planet Circulars, (M.P.C.), Nos. 5391 - 5658 (1980).

010.019 **Meteoritical Society**

Meteoritics. The Journal of the Meteoritical Society, Vol. 15, Nos. 2 - 3 (1980).

010.020 **Nantucket Maria Mitchell Association**

The Nantucket Maria Mitchell Association.
Seventy-eighth annual report for the year ending December 31, 1979.
Edited by the Nantucket Maria Mitchell Assoc., Nantucket, Mass., 56 pp. (1980).

010.021 **Oriental Astronomical Association**

The Heavens, Nos. 661 - 666, Vol. 61, Nos. 6 - 11 (1980).

010.022 **Royal Astronomical Society (RAS)**

Meetings of the Society.
Observatory, Vol. 100, 89 - 106, 137 - 148 (1980).

Meetings and activities of the Society.
Q. J. R. Astron. Soc., Vol. 21, 340 - 365, 489 - 490 (1980).

RAS discussion meeting on 'The early solar system', held on 1980 April 15 at Cardiff.
Observatory, Vol. 100, 175 - 176 (1980).

Star formation. Summaries of papers presented at the RAS specialist discussion, held 1980 April 15 at Cardiff.
Observatory, Vol. 100, 177 - 181 (1980).

RAS specialist discussion on results from the *Voyager* encounters with Jupiter and its satellites, held 1980 January 11. See Abstr. 099.003.

Geophysical Journal of the Royal Astronomical Society, Vol. 62, Nos. 2 - 3, Vol. 63, Nos. 1 - 3 (1980).

Monthly Notices of the Royal Astronomical Society, Vol. 192, Nos. 2 - 3, Vol. 193, Nos. 1 - 3 (1980).

The Quarterly Journal of the Royal Astronomical Society, Vol. 21, Nos. 3 - 4 (1980).

010.023 **Royal Astronomical Society of Canada (RAS Canada)**

Meetings of the Society.
J. R. Astron. Soc. Canada, Vol. 74, 305 - 309, 354 - 368 (1980).

Eleventh meeting of the Canadian Astronomical Society held at Saint Mary's University, Halifax, Nova Scotia, June 25 - 28, 1980. Abstracts of papers presented.
J. R. Astron. Soc. Canada, Vol. 74, 354 - 368 (1980).

The Journal of the Royal Astronomical Society of Canada, Vol. 74, Nos. 4 - 6 (1980).

National Newsletter. Supplement to the Journal of the Royal Astronomical Society of Canada, Vol. 74, Nos. 4 - 6 (1980).

The Observer's Handbook 1981 (28.047.027).

010.024 **Royal Astronomical Society of New Zealand (RAS New Zealand)**

Southern Stars. Journal of the Royal Astronomical Society of New Zealand, Vol. 28, Nos. 5 - 6 (1980).

010.025 **Schweizerische Astronomische Gesellschaft (SAG)**

Mitteilungen.
Orion, 38. Jahrg., 123 - 126, 157 - 160, 185 - 188 (1980).

BBSAG Bulletin, Nos. 49 - 51 (1980).

Orion. Zeitschrift der Schweizerischen Astronomischen Gesellschaft. Revue de la Société Astronomique de Suisse, 38. Jahrg., Nr. 179 - 181, Sondernummer (1980).

010.026 **Società Astronomica Italiana (S. A. It.)**

Giornale di Astronomia, Vol. 6, N. 3 (1980).

Memorie della Società Astronomica Italiana, Vol. 51, Nos. 2 - 3 (1980).

010.027 **Société Astronomique de France (SAF)**

Seances, commissions, activités de la Société.
Astronomie, Vol. 94, 550 - 568 (1980).

L'Astronomie et Bulletin de la Société Astronomique de France, Vol. 94, septembre - décembre (1980).

010.028 **Société Astronomique de Liège**

Le Ciel, Vol. 42, 151 - 258, septembre - décembre (1980).

010.029 **Société Belge d'Astronomie, de Météorologie et de Physique du Globe**

Ciel et Terre. Bulletin de la Société Belge d'Astronomie, de Météorologie et de Physique du Globe, Vol. 96, Nos. 4 - 6 (1980).

010.030 **VAGO (Astronomical-Geodetical Society of the USSR)**

Crimean plenary session of the VAGO Central Council. V. A. Bronshtehn.
Zemlya Vselennaya, 1980, No. 5, p. 49 - 50. In Russian.

010.031 **Vereinigung der Sternfreunde e. V.**

Nachrichten der Vereinigung der Sternfreunde e. V.
Sterne Weltraum, Jahrg. 19, 431 - 434 (1980).

Sonne. Mitteilungsblatt der Amateursonnenbeobachter. Jahrg. 4, Nos. 15 - 16 (1980).

011 Reports on Colloquia, Congresses, Meetings, Symposia, Expeditions

011.001 **What is the mass of Mercury?** Report of a seminar at Cambridge 1980 May 2. C. A. Ronan.
J. British Astron. Assoc., Vol. 90, 438 - 441 (1980).

011.002 **Some new results in X-ray astronomy presented at the annual meeting of the High Energy Astrophysics Division of AAS**, held at Cambridge, USA, January 28 - 30, 1980. P. C. Agrawal.
Bull. Astron. Soc. India, Vol. 8, 31 - 34 (1980).

011.003 **Probable duplicity of Pallas from speckle interferometry and solar shrinkage from eclipse data reported at A.A.S. Meeting.** D. W. Dunham.
Occultation Newsl., Vol. 2, 102 - 103 (1980).

011.004 **Autumn 1979 MIST (*Magnetosphere, Ionosphere and Solar-Terrestrial Environment*) meeting with discussion of a proposed cometary mission.**
P. A. Hadjiry, M. J. Laird, M. K. Wallis.
Q. J. R. Astron. Soc., Vol. 21, 282 - 287 (1980).

011.005 **Professor A. T. Price memorial meeting.**
P. A. Hadjiry, M. J. Laird.
Q. J. R. Astron. Soc., Vol. 21, 288 - 292 (1980).

011.006 **The elementary universe.** M. G. Edmunds.
Nature, Vol. 287, 103 - 104 (1980).
The author reports the Santa Cruz Workshop in astronomy and astrophysics on 'The origin and distribution of the elements' held July 6 - July 18 at the University of California, Santa Cruz.

011.007 **Radio astronomy at the 1979 WARC (World Administrative Radio Conference).**
V. Pankonin, W. C. Erickson.
Bull. American Astron. Soc., Vol. 12, 502 (1980). – Abstract.

011.008 **Problems of astrophysical investigations of the moon.** Report on the conference in Abastumani, 1979, Sept. 25 - 29. V. V. Shevchenko.
Zemlya Vselennaya, 1980, No. 4, p. 64 - 66. In Russian.

011.009 **Infrared astronomy comes of age.**
R. D. Wolstencroft.
Nature, Vol. 287, 386 - 387 (1980).
The International Astronomical Union symposium on infrared astronomy, held in Kailua-Kona, Hawaii, June 23 - 27, 1980.

011.010 **Variable stars and galaxies.**
J. Osborne, R. Cole.
Nature, Vol. 287, 484 - 486 (1980).
Fifth European Regional Meeting of the IAU "Variability in stars and galaxies", held at the University of Liège, Belgium, on 28 July 1980.

011.011 **Colloquium No. 42 of the IAU "The interaction of variable stars with their environment".**
V. Ureche.
Proceedings of the colloquium of astronomy, Cluj-Napoca, (see 012.009), p. 73 - 75 (1978).

011.012 **Bericht über die wissenschaftliche Tagung der Astronomischen Gesellschaft in Heidelberg am 9. - 10. Oktober 1979.** W. Fricke.
Mitt. Astron. Ges., Nr. 48, (see 012.015), p. 5 - 6 (1980).

011.013 **Aristarchus of Samos symposium**, held on Samos, Greece 1980 June 17 - 19. O. Gingerich.
Sky Telesc., Vol. 60, 376 - 377 (1980).

011.014 **XVII. Generalversammlung der Internationalen Astronomischen Union.**
H. Lorenz, G. Ruben.
Sterne, 56. Band, 237 - 244 (1980).

011.015 **Report on the Second European IUE Conference, Tübingen (Germany), 26 - 28 March 1980.**
ESA J., Vol. 4, 180 (1980).

011.016 **Report on the Fifth European Symposium on European Sounding Rocket and Balloon Programmes and Related Research, Bournemouth (UK), 14 - 18 April 1980.**
ESA J., Vol. 4, 181 (1980).

011.017 **Joint Discussion Meeting of IAU Comissions 10, 12, 14, and 44 on the Solar Maximum Year, held on August 15th, 1979 in Montreal, Canada. Brief report of the meeting.**
Prepared by Z. Švestka, with extended summaries of talks given by G. van Hoven, P. Hoyng, M. Kuperus.
Sol Phys., Vol. 67, 379 - 384 (1980).

011.018 **All-Union conference "Structure and physics of galaxies".** Borzhomi, 1980, June 17 - 20.
V. G. Surdin.
Astron. Zh., Tom 57, 1348 - 1350 (1980). In Russian.
English translation in Soviet Astron., Vol. 24, No. 6.

011.019 **I. A. P. P. P. International Amateur and Professional Photoelectric Photometry.**
South. Stars, Vol. 28, 140 - 143 (1980).

011.020 **Plasma Physics Division meets in San Diego,** 1980 November 10 - 14.
Phys. Today, Vol. 33, No. 10, p. 55 - 57 (1980).

011.021 **Symposium on the history of Chinese astronomy was held in Chengdu.** S.-z. Zhang.
Acta Astron. Sinica, Vol. 21, 320 - 322 (1980). In Chinese.

011.022 **XVII th General Assembly of the International Astronomical Union, Montreal, August 14 - 23, 1979.** C. Iwaniszewska.
Postępy Astron., Tom 28, 243 - 245 (1980).

011.023 **Space astrometry – its impact on astronomy and astrophysics – introductory comments.**
G. Westerhout.
Highlights of Astronomy, Vol. 5, (see 012.029), 779 - 781 (1980).

011.024 **Sun – climate links.** T. M. L. Wigley.
Nature, Vol. 288, 317 - 319 (1980).
Report on the conference "Sun and climate", held at Toulouse, France, 30 September - 3 October 1980.

011.025 **Where do cosmic rays come from?** Report of the VIIth European Cosmic Ray Symposium held at Leningrad, September 15 - 19.
Nature, Vol. 288, 438 (1980).

011.026 **The WPGM (*AAS Working Group on Photographic Materials*) meets in San Francisco.** B. Schoening.
AAS Photo-Bull., Issue 23, No. 1, p. 12 (1980).

011.027 **International cooperation in the field of investigation of the cosmos. XXIIIth COSPAR congress.**
Budapest, 1980, June.
Priroda, 1980, No. 11, p. 67 - 69. In Russian.

011.028 **Vrijwel alle sterren verliezen massa.**
P. R. Wesselius.
Zenit, 7e Jaarg., 117 (1980).

011.029 **De I. A. U. te Montréal.** P. C. van der Kruit.
Zenit, 7e Jaarg., 256 - 258 (1980).

011.030 **L'Union Astronomique Internationale vit...**
J.-C. Pecker.
Astronomie, Vol. 94, 527 - 534 (1980).

011.031 **9. Internationale Konferenz über Allgemeine Relativitätstheorie und Gravitation (GR 9).**
E. Schmutzer.
Astron. Schule, 17. Jahrg., 98 (1980).

011.032 **Solar and interplanetary dynamics.** Report of IAU Symposium 91, 27 - 31 August 1979, Cambridge, Massachusetts, USA. M. Dryer, E. Tandberg-Hanssen.
Comments Astrophys., Vol. 9, 51 - 59 (1980).

011.033 **Review of the COSPAR/IAU/IUPAC symposium on non solar gamma rays (>30 MeV).**
G. F. Bignami.
Space Research, Vol. XX, (see 012.043), 277 - 282 (1980).

011.034 **Summary of the Czechoslovak solar corona observations.** J. Sýkora.
Bull. Astron. Soc. India, Vol. 8, 73 - 75, 77, 79 (1980).

011.035 **Planets and their systems at the XXIIIth plenary meeting of COSPAR.** M. Burša.
Říše hvězd, Vol. 61, 177 - 183 (1980). In Czech.

011.036 **"The Europhysics Study Conference on Gamma-Ray Astronomy after COS-B" Erice, Italy, 17-23 May 1979.** G. F. Bignami.
Non-solar gamma-rays, (see 012.047), p. 227 - 237 (1980).
The Study Conference was organized under the auspices of the EPS at the Centro di Cultura Scientifica "Ettore Majorana" in Erice, Italy and was held from May 17 to 23, 1979. One of the highlights of the Conference has been the presentation of the latest set of data of the COS-B mission.

011.037 **Scientific session of the Department of General Physics and Astronomy and the Department of Nuclear Physics of the USSR Academy of Sciences, 21 - 22 November 1979.**
Usp. fiz. nauk, Vol. 131, 511 - 513 (1980). In Russian.
Abstr. in Ref. zh., 51. Astron., 11.51.26 (1980).

011.038 **Scientific session of the Department of General Physics and Astronomy of the USSR Academy of Sciences, 20 - 21 February 1980.**
Usp. fiz. nauk, Vol. 131, 721 - 725 (1980). In Russian. – From Ref. zh., 51. Astron., 11.51.27 (1980).

011.039 **Bericht über die Tagung in Bochum.** Frühjahrstagung der Arbeitsgemeinschaft Extraterrestrische Physik, der Astronomischen Gesellschaft und der Fachausschüsse Kurzzeitphysik und Plasma- und Gasentladungsphysik der Deutschen Physikalischen Gesellschaft vom 12. - 14. März 1979. T. Schmidt-Kaler.
Mitt. Astron. Ges., Nr. 50, p. 11 - 12 (1980).

011.040 **Bericht über die Tagung in Mainz.** Die Frühjahrstagung der Arbeitsgemeinschaft für Extraterrestrische Physik, der Astronomischen Gesellschaft und des Paneth Kolloquiums für Kosmochemie vom 5. - 7. März 1980.
H. J. Fahr.
Mitt. Astron. Ges., Nr. 50, p. 81 (1980).

011.041 **Introduction to "Solar flares. A monograph from Skylab Solar Workshop II".**
P. A. Sturrock.
Solar flares, (see 003.012), p. 1 - 16 (1980).
Contents: Historical background. Goals and functioning of the workshop. Results of the workshop.

011.042 **Symposium dedicated to the centenary of Albert Einstein's birth.** B. Jovanović.
Vasiona, Année 28, 61 - 63 (1980). In Croatian.

011.043 **All-Union seminar "Evolution of orbits of celestial bodies".** Tomsk, 1980, May 19 - 20.
I. N. Potapov.
Komet. Tsirk., Kiev, No. 264 (1980). In Russian.

011.044 **All-Union seminar on physics and dynamics of small bodies.** Dushanbe, 1980, November 18 - 23.
N. A. Belyaev, K. I. Churyumov.
Komet. Tsirk., Kiev, No. 270 (1980). In Russian.

012 Proceedings of Colloquia, Congresses, Meetings, Symposia

012.001 **Star formation.** Tenth advanced course of the Swiss Society of Astronomy and Astrophysics, held in Saas-Fee, Switzerland, March 24 - 29, 1980.
A. Maeder, L. Martinet (Editors).
Published and sold by Geneva Observatory, CH–1290 Sauverny, Switzerland. 7 + 222 pp. Price SFr. 30.00 (1980). – The individual contributions are included in their corresponding categories – see abstracts 131.021 - 131.023.

012.002 **Star catalogues, positional astronomy and celestial mechanics.** Proceedings of a symposium held in honor of Paul Herget at the U.S. Naval Observatory, Washington, November 30, 1978. Parts III, IV. (Parts I, II see Abstr. 27.012.038).
Celestial Mech., Vol. 22, No. 2, p. 113 - 202 (1980). – The individual contributions are included in their corresponding subject categories – see abstracts 002.015, 013.005, 031.514, 032.503 - 032.507, 041.001 - 041.003, 043.001.

012.003 **Einstein Symposion Berlin.** A symposium held at the Technische Universität Berlin, March 25 - 30, 1979.
H. Nelkowski, A. Hermann, H. Poser, R. Schrader, R. Seiler (Editors).
Lecture Notes in Physics, Vol. 100. Springer-Verlag, Berlin–Heidelberg–New York. 8 + 550 pp. Price DM 49.50 (1979). ISBN 3-540-09718-X, ISBN 0-387-09718-X. – The individual contributions within the subject scope of Astronomy and Astrophysics Abstracts are included in their corresponding categories – see abstracts 002.016, 005.003, 015.004, 015.005, 066.013 - 066.015.

012.004 **Isolating gravitating systems in general relativity.** Conference held at Varenna on Lake Como, Italy, 28 June - 10 July 1976. J. Ehlers (Editor).
Proceedings of the International School of Physics "Enrico Fermi", Course 67. North-Holland Publishing Company, Amsterdam, Netherlands. 16 + 500 pp. Price $ 85.25 (1979). ISBN 0-444-85329-4. – Review in Phys. Abstr., Vol. 83, Abstr. 73089 (1980). – See abstracts 066.032 - 066.037, 066.039 - 066.041, 141.515.

012.005 **Physics of dense matter.** Conference held at Paris, France, 17 - 21 September 1979.
J. Phys. Colloq., Vol. 41, No. C-2 (1980). – Review in Phys. Abstr., Vol. 83, Abstr. 73094 (1980). – See abstracts 022.031, 022.032, 062.024 - 062.026, 065.026 - 065.029, 065.037, 066.508 - 066.512, 080.017, 091.003, 125.027, 126.022.

012.006 **Proceedings of the Los Alamos Conference on Optics '79.** Conference held at Los Alamos, N. Mex., USA, 23 - 25 May 1979.
Proceedings of the Society of Photo-Optical Instrumentation Engineers, Vol. 190 (1979). – Review in Phys. Abstr., Vol. 83, Abstr. 73095 (1980).

012.007 **Interferometry.** Conference held at San Diego, Calif., USA, 29 - 30 August 1979.
Proceedings of the Society of Photo-Optical Instrumentation Engineers, Vol. 192 (1979). – Review in Phys. Abstr., Vol. 83, Abstr. 73096 (1980).

012.008 **Stellar turbulence.** Proceedings of Colloquium 51 of the International Astronomical Union, held at the University of Western Ontario, London, Ontario, Canada, August 27 - 30, 1979.
D. F. Gray, J. L. Linsky (Editors), with a summary by E. Böhm-Vitense.
Lecture Notes in Physics. Vol. 114. Springer-Verlag, Berlin – Heidelberg – New-York, 10 + 308 pp. Price DM 37.60 (1980). ISBN 3-540-09737-6, ISBN 0-387-09737-6. The individual contributions are included in their corresponding subject categories – see abstracts 031.528, 031.529, 063.025 - 063.028, 064.019 - 064.036, 071.017 - 071.025, 073.050 - 073.052, 080.025 - 080.028, 112.013, 114.057 - 114.061, 116.005, 119.025.

012.009 **Proceedings of the colloquium of astronomy. Section astrophysics.** Conference held at Cluj-Napoca, November 1977.
Contributions of the Astronomical Observatory, Universitatea Babes-Bolyai Cluj-Napoca, Facultatea de Matematica. 96 pp. (1978). – The individual contributions are included in their corresponding subject categories. – See abstracts 011.011, 021.015, 031.530, 062.038, 064.037, 117.017, 117.018, 118.012, 118.013, 119.029, 122.061.

012.010 **Congrès Général de la Société Française de Physique (General Congress of the French Physics Society).** Conference held at Toulouse, France, 25 - 30 June 1979.
J. Phys. Colloq., Vol. 41, No. C-3 (1980). – Review in Phys. Abstr., Vol. 83, Abstr. 77536 (1980). – See abstracts 062.046, 100.014, 131.099.

012.011 **Optical systems engineering.** Conference held at San Diego, Calif., USA, 27 - 28 August 1979.
Proc. Soc. Photo-Opt. Instrum. Eng., Vol. 193 (1979). Review in Phys. Abstr., Vol. 83, Abstr. 77541 (1980). – See abstracts 031.008, 031.009, 032.520.

012.012 **Optica hoy y mañana. (Optics present and future).** Proceedings of the Eleventh Congress of the International Commission for Optics ICO-11, held at Madrid, Spain, 10 - 17 September 1978.
J. Bescos, A. Hidalgo, L. Plaza, J. Santamaria (Editors).
Published by Instituto de Optica "Daza de Valdes", CSIC, Sociedad Española de Optica, Madrid, Spain. 15 + 839 pp. (1978). ISBN 84-600-1317-0. – Review in Phys. Abstr., Vol. 83, Abstr. 77547 (1980). – See abstracts 031.010, 031.570, 032.007.

012.013 **Symposium on the bow shock.** Conference held at Strasbourg, France. 31 August - 1 September 1978.
Nuovo Cimento C, Ser. 1, Vol. 2C, No. 6 (1979). – Review in Phys. Abstr., Vol. 83, Abstr. 82573 (1980). – See abstracts 022.059, 051.008, 062.052, 062.053, 084.033 - 084.037, 106.015 - 106.017, 106.050.

012.014 **Recent advances in TV sensors and systems.** Conference held at San Diego, Calif., USA, 27 - 28 August 1979.
Proc. Soc. Photo-Opt. Instrum. Eng., Vol. 203 (1979). – See abstracts 032.521 - 032.525.

012.015 **Astrometrie und dynamische Astronomie.** Tagung der Astronomischen Gesellschaft in Heidelberg, 9. - 10. Oktober 1979 aus Anlass des hundertjährigen Jubiläums des Fundamentalkatalogs der Astronomischen Gesellschaft.
W. Fricke, T. Schmidt-Kaler, W. Seggewiss (Editors).
Mitt. Astron. Ges., Nr. 48, 189 pp. (1980). The individual contributions are included in their corresponding subject categories – see abstracts 004.024, 011.012, 031.535, 032.008, 041.011 - 041.021, 042.022, 042.023, 043.002, 051.007, 066.065, 066.066, 080.029, 091.007, 091.008, 098.012, 098.013, 107.007, 111.011 - 111.013, 151.038, 151.039, 154.013, 155.016, 158.106, 159.010.

012.016 **Measurements of optical radiations.** Conference held at San Diego, Calif., USA, 29 - 30 August 1979.
Proc. Soc. Photo-Opt. Instrum. Eng., Vol. 196 (1979).
Review in Phys. Abstr., Vol. 83, Abstr. 86036 (1980). – See abstracts 032.010, 034.019.

012.017 **Modern utilization of infrared technology V.** Conference held at San Diego, Calif., USA, 29 - 30 August 1979.
Proc. Soc. Photo-Opt. Instrum. Eng., Vol. 197 (1979).
Review in Phys. Abstr., Vol. 83, Abstr. 86037 (1980). – See abstracts 032.526, 002.026 .

012.018 **Proceedings of the 13th Lunar and Planetary Symposium,** held at Tokyo, July 7 - 9, 1980.
Published by the Institute of Space and Aeronautical Science, University of Tokyo. 6 + 423 pp. (1980). – The individual contributions within the subject scope of Astronomy and Astrophysics Abstracts are included in their corresponding categories – see abstracts 015.013, 022.078 - 022.082, 053.005, 063.043, 066.104, 075.014, 075.015, 082.040, 085.014, 091.015, 091.016, 093.025, 094.029, 094.519, 094.520, 097.026 - 097.028, 098.020, 099.033 - 099.041, 100.026, 100.027, 102.019, 103.107 - 103.109, 105.064 - 105.070, 106.056, 106.057, 107.009 - 107.011, 131.126.

012.019 **Solid particles in the solar system.** International Astronomical Union. Symposium No. 90, held at Ottawa, Canada, August 27 - 30, 1979.
I. Halliday, B. A. McIntosh (Editors), with a summary by P. M. Millman.
D. Reidel Publishing Company, Dordrecht, Holland - Boston, U.S.A. - London, England. 16 + 441 pp. Price Dfl. 102.50 (1980). ISBN 90-277-1164-X, ISBN 90-277-1165-8 (pbk.). The individual contributions are included in their corresponding subject categories – see abstracts 022.069 - 022.073, 031.541 - 031.544, 032.528 - 032.532, 042.031, 051.012, 063.041, 091.012, 094.027, 094.028, 099.027, 102.015 - 102.018, 103.105, 103.106, 103.901, 104.023 - 104.042, 105.057 - 105.059, 106.021 - 106.049, 131.116 - 131.117.

012.020 **Solar and interplanetary dynamics.** International Astronomical Union. Symposium No. 91, held in Cambridge, Mass., U.S.A., August 27 - 31, 1979.
M. Dryer, E. Tandberg-Hanssen (Editors), with a symposium summary by M. Kuperus.
D. Reidel Publishing Company, Dordrecht, Holland–Boston, U.S.A.–London, England. 19 + 558 pp. Price Dfl. 134.50 (1980). ISBN 90-277-1162-3, ISBN 90-277-1163-1 (pbk.). The individual contributions are included in their corresponding subject categories – see abstracts 033.012, 051.016, 051.017, 062.070, 064.050, 072.028, 072.029, 073.075 - 073.088, 074.044 - 074.072, 075.017 - 075.022, 076.021 - 076.023, 077.036 - 077.040, 080.036 - 080.040, 084.047, 085.015, 101.007, 106.059 - 106.067.

012.021 **ESO workshop on methods of abundance determination for stars.** Geneva, 24 - 26 March 1980.
P. E. Nissen, K. Kjär (Editors), with an introduction by P. O. Lindblad, P. E. Nissen.
Published by European Southern Observatory, Munich, F. R. Germany. 9+56 pp. (1980). – The individual contributions are included in their corresponding subject categories – see abstracts 022.174, 034.067, 034.068, 113.045, 114.141 - 114.144.

012.022 **Large-scale velocity fields on the sun.** Joint Discussion No. 1 at the XVIIth General Assembly of the IAU, Montreal, 1979. M. Stix (Editor).
Highlights of Astronomy, Vol. 5, (see 003.006), 71 - 127 (1980). – The individual contributions are included in their corresponding subject categories – see abstracts 062.083, 072.032, 080.044, 080.045.

012.023 **Nuclei of normal galaxies.** Joint Discussion No. 3 at the XVIIth General Assembly of the IAU, Montreal, 1979. W. B. Burton (Editor).
Highlights of Astronomy, Vol. 5, (see 003.006), 129 - 225 (1980). – The individual contributions are included in their corresponding subject categories – see abstracts 151.058, 151.059, 155.023 - 155.026, 158.135 - 158.143.

012.024 **Ultraviolet astronomy – new results from recent space experiments.** Joint Discussion No. 4 at the XVIIth General Assembly of the IAU, Montreal, 1979.
R. Davis (Editor).
Highlights of Astronomy, Vol. 5, (see 003.006), 227 - 329 (1980). – The individual contributions are included in their corresponding subject categories – see abstracts 114.089, 114.090, 124.011, 126.027, 131.152, 131.153, 135.027, 142.080, 158.144 - 158.146.

012.025 **Very hot plasmas in circumstellar, interstellar, and intergalactic space.** Joint Discussion No. 5 at the XVIIth General Assembly of the IAU, Montreal, 1979.
W. I. Axford (Editor).
Highlights of Astronomy, Vol. 5, (see 003.006), 331 - 428 (1980). – The individual contributions are included in their corresponding subject categories – see abstracts 062.084, 062.085, 064.053, 073.094, 074.077, 125.043, 131.154, 158.147, 161.003, 161.004.

012.026 **Stellar instabilities.** Joint Discussion No. 6 at the XVIIth General Assembly of the IAU, Montreal, 1979. A. N. Cox (Editor).
Highlights of Astronomy, Vol. 5, (see 003.006), 429 - 519 (1980). – The individual contributions are included in their corresponding subject categories – see abstracts 065.053 - 065.057, 080.046 - 080.048, 122.086 - 122.097, 124.012, 125.044, 125.045, 135.028.

012.027 **Physics of the chromosphere – corona-wind complex and mass loss in stellar atmospheres.** Joint Discussion No. 7 at the XVIIth General Assembly of the IAU, Montreal, 1979. D. H. Mihalas (Editor).
Highlights of Astronomy, Vol. 5, (see 003.006), 521 - 613 (1980). – The individual contributions are included in their corresponding subject categories – see abstracts 064.054 - 064.058, 073.095, 074.078, 080.049, 142.081.

012.028 **Extragalactic high energy astrophysics.** Joint Discussion No. 8 at the XVIIth General Assembly of the IAU, Montreal, 1979. H. van der Laan (Editor).
Highlights of Astronomy, Vol. 5, (see 003.006), 615 - 774 (1980). – The individual contributions are included in their corresponding subject categories – see abstracts 141.117 - 141.120, 142.082, 158.148 - 158.154, 160.038 - 160.044, 161.005, 162.069.

012.029 **Space astrometry.** Joint Commission Meetings of Commissions 8, 20, 25, 26, 30, 33, 37, 44, 45 at the XVIIth General Assembly of the IAU, Montreal, 1979.
G. Westerhout (Chairman).
Highlights of Astronomy, Vol. 5, (see 003.006), 777 - 807 (1980). – The individual contributions are included in their corresponding subject categories – see abstracts 002.048, 011.023, 032.540, 032.541, 111.015, 111.016.

012.030 **Stellar abundances and stellar rotation.** Joint Commission Meetings of Commissions 29, 36, 45 at the XVIIth General Assembly of the IAU, Montreal, 1979.
M. Hack (Chairman).
Highlights of Astronomy, Vol. 5, (see 003.006), 809 - 837

(1980). – The individual contributions are included in their corresponding subject categories – see abstracts 114.091, 116.014, 116.015, 117.028, 154.020.

012.031 **Close binaries and stellar activity.** Joint Commission Meetings of Commissions 10, 27, 40, 42, 44, 48 at the XVIIth General Assembly of the IAU, Montreal, 1979.
G. Larsson-Leander (Chairman).
Highlights of Astronomy, Vol. 5, (see 003.006), 839 - 865 (1980). – The individual contributions are included in their corresponding subject categories –see abstracts 064.059, 064.060, 116.016, 117.029 - 117.031, 122.098, 142.083.

012.032 **Evolution der Planetenatmosphären und des Lebens.** 2. DFG - Kolloquium über Planetenforschung, held in Schliersee, F. R. Germany, October 2 - 5, 1979.
P. Blum (Editor), with a panel discussion by K. Rawer and a closing discussion by F. M. Neubauer.
Published by Deutsche Forschungsgemeinschaft. 10 + 214 pp. (1980). – The individual contributions within the subject scope of Astronomy and Astrophysics Abstracts are included in their corresponding categories – see abstracts 082.047, 085.017, 091.021, 091.022, 102.022, 105.076, 131.155.

012.033 **Interstellar molecules.** International Astronomical Union. Symposium No. 87, held at Mont Tremblant, Quėbec, Canada, August 6 - 10, 1979.
B. H. Andrew (Editor).
D. Reidel Publishing Company, Dordrecht, Holland – Boston, U.S. A. – London, England. 40 + 704 pp. Price Dfl. 145.00, $ 76.50 cloth, Dfl. 65.00, $ 34.00 paper (1980).
ISBN 90-277-1160-7, ISBN 90-277-1161-5 (pbk).
The individual contributions are included in their corresponding subject categories – see abstracts 015.027, 022.107 - 022.127, 031.560, 032.542, 033.014 - 033.017, 034.028, 034.029, 061.030, 112.020 - 112.035, 131.156 - 131.243, 134.031, 155.027, 158.158 - 158.160.

012.034 **Proceedings of the conference "Spiral structure of the Galaxy", Kiev, 21 - 23 November 1977.**
E. K. Kharadze, V. I. Voroshilov (Editors).
Abastumanskaya Astrofizicheskaya Observatoriya, Gora Kanobili, Byulleten', No. 52, 120 pp. (1980). In Russian.
The individual contributions are included in their corresponding subject categories – see abstracts 131.247- 131.249 151.069, 151.070, 155.028 - 155.034, 158.162 , 158.163.

012.035 **XIth seminar on cosmophysics in Leningrad. Interaction of cosmic rays with matter, Leningrad, 1979, November 30 - December 2.**
G. E. Kocharov (Editor).
Fiz.-tekh. inst. AN SSSR, Leningrad. 398 pp. (1979). In Russian. – From Ref. zh., 51. Astron., 8.51.31 (1980). – See abstracts 022.065, 061.022, 062.056, 062.057, 071.026, 072.025, 075.012, 078.008, 080.031, 097.021, 106.020, 143.021 - 143.024.

012.036 **Applications of digital image processing. III.** Conference held at San Diego, Calif., USA, 27 - 29 August 1979.
Proc. Soc. Photo-Opt. Instrum. Eng., Vol. 207 (1979).
Review in Phys. Abstr., Vol. 83, Abstr. 105424 (1980).

012.037 **Space Shuttle: dawn of an era.**
Proceedings of the 26th AAS Annual Conference, Los Angeles, Calif., USA, 29 October - 1 November 1979.
W. F. Rector, P. A. Penzo (Editors).
Advances in the Astronomical Sciences, Vol. 41, Part I, 452 pp. Price $ 40.00 hardcover, $ 30.00 softcover; Part II 428 pp. Price $ 45.00 hardcover, $ 35.00 softcover (1980).
See abstracts 27.032.558, 27.032.559, 27.051.035, 27.081.040, 032.546, 032.547, 051.029, 051.030, 084.071.

012.038 **Cool stars, stellar systems, and the sun.** Proceedings of a workshop held January 31, 1980 at the Harvard-Smithsonian Center for Astrophysics, Cambridge, Mass. A. K. Dupree (Editor).
Smithsonian Astrophys. Obs., Spec. Rep. 389, 235 pp. (1980). The individual contributions are included in their corresponding subject categories – see abstracts 064.070 - 064.076, 080.057, 112.043, 113.036, 114.114 - 114.116, 116.021, 117.053 - 117.056, 120.020, 142.105, 153.022.

012.039 **Origin and distribution of the elements.**
Proceedings of the second symposium, Paris, May 1977. L. H. Ahrens, G. Protas (Editors).
International Series in Earth Sciences, Vol. 34.
Pergamon Press, Oxford – New York – Toronto – Sidney – Paris –Frankfurt. 11 + 909 pp. Price DM 298.80 (1979).
ISBN 0-08-022947-6. – The individual contributions within the subject scope of Astronomy and Astrophysics Abstracts are included in their corresponding categories – see abstracts 061.034 - 061.036, 071.042, 094.532 - 094.537, 099.123, 105.093 - 105.095, 107.022.

012.040 **3rd meeting on methodologies of information processing,** held in Rome, 27 and 28 May, 1980.
G. A. De Biase, G. Sedmak (Editors).
Mem. Soc. Astron. Italiana, Vol. 51, 387 - 535 (1980). – The individual contributions are included in their corresponding subject categories – see abstracts 013.037, 021.039 - 021.043, 031.582 - 031.592, 122.152.

012.041 **Magnetospheric boundary layers.** A Sydney Chapman Conference. Proceedings of an international conference held in Alpbach, Austria, 11 - 15 June 1979.
B. Battrick, J. Mort (Editors), with a foreword by G. Haerendel and J. Ortner.
ESA SP-148. ESA Scientific and Technical Publications Branch, ESTEC, Noordwijk, The Netherlands. 16 + 411 pp. Price FF 120.00 (1979). – The individual contributions within the subject scope of Astronomy and Astrophysics Abstracts are included in their corresponding categories – see abstracts 062.108 - 062.111, 066.525, 066.526, 084.081 - 084.090, 093.063, 099.130 - 099.133, 141.534.

012.042 **Transactions of the International Astronomical Union, Volume XVIIB: Proceedings of the Seventeenth General Assembly, Montreal 1979.**
P. A. Wayman (Editor).
D. Reidel Publishing Company, Dordrecht, Holland - Boston, USA - London, England. 10 + 527 pp. Price Dfl. 130.00, US $ 68.50 (1980). ISBN 90-277-1159-3.
This volume includes information on the finances of the Union and a proposed budget for the forthcoming triennium 1980-82. It also contains the actual proceedings of the General Assembly at Montreal as well as the reports of the Finance Committee, the Resolutions Committee, the Special Nominating Committee, and an alphabetical list of IAU members.
Contents: Preface. Part 1: Report of the Executive Committee. Part 2: Inaugural Ceremony. Report of the XVIIth General Assembly (Resolutions included). Part 3: Reports of Meetings of Commissions. Part 4: Astronomer's Handbook. Appendix.

012.043 **Space Research, Vol. XX.**
Proceedings of the open meetings of the working groups on physical sciences of the twenty-second plenary meeting of COSPAR, Bangalore, India, 29 May - 9 June 1979.
M. J. Rycroft (Editor), with a welcome address by Y. Pal.
Pergamon Press, Oxford - New York - Toronto - Sydney - Paris - Frankfurt. 14 + 285 pp. Price $ 50.00, £ 22.20, DM 160.80 (1980). ISBN 0-08-024437-8. – The individual contributions within the subject scope of Astronomy and Astrophysics Abstracts are included in their corresponding subject categories – see abstracts 011.033, 013.038, 032.554,

073.104, 076.027, 082.066 - 082.068, 093.064 - 093.071, 094.042, 102.033, 143.038, 143.039.

012.044 **Variability in stars and galaxies**. Proceedings of the fifth European Regional meeting in Astronomy, held in Liège, Belgium, 28 July - 1 August 1980.
With a preface by P. Ledoux.
Sold and distributed by Institut d'Astrophysique, avenue de Cointe, 5, B-4200 Cointe-Ougrée. 19 + 439 pp. (1980).
The individual contributions are included in their corresponding subject categories – see abstracts 015.026, 064.081, 064.082, 080.065 - 080.067, 112.046, 114.130, 117.070, 121.011, 122.156 - 122.162, 125.063, 131.286, 141.151 - 141.155, 142.114, 158.211 - 158.214, 162.057, 162.097.

012.045 **Jupiter**. Astronomisches Seminar, Universität Heidelberg, Wintersemester 79/80.
Leitung: H. Fechtig, C. Leinert, G. Morfill.
Published by Max-Planck-Institut für Kernphysik, Heidelberg, F. R. Germany. 2 + 162 pp. (1980). – The individual contributions are included in their corresponding categories – see abstracts 042.067, 051.035, 099.138 - 099.144.

012.046 **Stellar hydrodynamics**. Proceedings of the 58th Colloquium of the International Astronomical Union held in Los Alamos, U. S. A., August 12 - 15, 1980.
A. N. Cox, D. S. King (Editors), with an introduction by R. S. Stobie.
Space Sci. Rev., Vol. 27, Nos 3/4, p. 219 - 686 (1980).
The individual contributions are included in their corresponding subject categories – see abstracts 062.122 - 062.124, 064.088 - 064.090, 065.077 - 065.086, 066.532, 071.045, 071.046, 080.073, 122.172 - 122.201, 124.024, 125.071 - 125.075, 131.305 - 131.309, 135.038, 142.123.

012.047 **Non-solar gamma-rays**. Proceedings of a Symposium of the Twenty-second Plenary Meeting of COSPAR, Bangalore, India 29 May - 9 June 1979. R. Cowsik, R. D. Wills (Editors), with opening remarks by B. V. Sreekantan.
Advances in Space Exploration, COSPAR Symposium Series Vol. 7. Pergamon Press,Oxford - New York - Toronto - Sydney - Paris - Frankfurt. 12 + 276 pp. Price $ 50.00 (1980). ISBN 0-08-024440-8. – The individual contributions are included in their corresponding subject categories – see abstracts 011.036, 032.572 - 032.575, 125.077, 142.520 - 142.532, 151.090, 157.011 - 157.016, 158.233.

012.048 **Cosmophysical aspects of cosmic ray investigations. International seminar.**
E. V. Kolomeets et al. (Editors).
Kazakhsk. univ. Alma-Ata. 103 pp. (1980). In Russian.
From Ref. zh., 51. Astron., 11.51.57 (1980).

012.049 **Comet Halley micrometeoroid hazard workshop.** Proceedings of an international workshop held in ESTEC, Noordwijk, The Netherlands, 18 - 19 April 1979.
N. Longdon (Editor).
ESA SP-153, ESA Scientific and Technical Publications Branch, ESTEC, Noordwijk, The Netherlands. 4 + 147 pp. (1979). – The individual contributions within the subject scope of Astronomy and Astrophysics Abstracts are included in their corresponding categories – see abstracts 051.040 - 051.043, 053.011, 053.012, 102.040, 102.041, 103.503, 103.504.

012.050 **Proceedings of the Tenth Lunar and Planetary Science Conference**, Houston, Texas, March 19 - 23, 1979.
Compiled by the Lunar and Planetary Institute, Houston, Texas. R. B. Merrill (Managing Editor), D. D. Bogard, F. Hörz, D. S. McKay (Editors).
Vol. 1: Meteorites and lunar rocks. Vol. 2: Early solar system and lunar regolith. Vol. 3: Planetary interiors and surfaces. Geochim. Cosmochim. Acta, Suppl. 11. Pergamon Press, New York–Oxford–Toronto–Sydney–Frankfurt–Paris. 23 + 11 + 11 + 3077 + 23 + 23 + 27 pp. Price $ 220.00 (1979). ISBN 0-08-025128-5. – The individual contributions are included in their corresponding subject categories – see abstracts 022.176 - 022.186, 051.039, 061.044, 081.032, 091.057 - 091.063, 092.006, 092.007, 093.075, 094.045 - 094.058, 094.540 - 094.635, 097.086 - 097.093, 099.151, 104.049, 105.122 - 105.145, 107.028 - 107.031.

012.051 **Photometry, kinematics and dynamics of galaxies.** Proceedings of a conference held at the University of Texas at Austin, August 6 - 8,1979.
D. S. Evans (Editor).
Published by the Department of Astronomy, University of Texas, Austin, Texas, USA. 11 + 492 pp. Price $ 16.00 (1979). ISBN 0-96-037960-6. Review in Nature, Vol. 287, 259; 1980 (*C. Pritchet*). The individual contributions are included in their corresponding subject categories – see abstracts 031.611 - 031.614, 034.070, 132.042, 151.092 - 151.115, 155.048, 158.243 - 158.299, 159.023, 160.072 - 160.074, 161.007.

012.052 **Limits of life**. Proceedings of the Fourth College Park Colloquium on chemical evolution, University of Maryland, College Park, Maryland, U.S.A., October 18 - 20, 1978. C. Ponnamperuma, L. Margulis (Editors).
D. Reidel Publishing Company, Dordrecht, Holland–Boston, U.S.A.–London, England. 12 + 199 pp. Price Dfl. 50.00, US $ 26.50 (1980). ISBN 90-277-1155-0. – The individual contributions within the subject scope of Astronomy and Astrophysics Abstracts are included in their corresponding categories – see abstracts 080.080, 082.078.

012.053 **Strategies for the search for life in the universe.** Proceedings of a joint session of Commissions 16, 40, and 44, held in Montreal, Canada, during the IAU General Assembly, 15 - 16 August, 1979.
M. D. Papagiannis (Editor) with a foreword by L. Goldberg. Astrophysics and space science library. Vol. 83. D. Reidel Publishing Company, Dordrecht, Holland–Boston, U.S.A.–London, England. 15 + 253 pp. Price Dfl. 70.00, $ 37.00 cloth, ISBN 90-277-1181-X, Dfl. 30.00, $ 14.95 pbk., ISBN 90-277-1226-3 (1980). – The individual contributions are included in their corresponding subject categories – see abstracts 015.035 - 015.051, 031.617 - 031.619, 118.036.

012.054 **X-ray and gamma-ray astronomy in the 1980's.** Results of an international gathering of scientists at Spitzingsee near Munich, October 9 - 13, 1978.
B. Aschenbach, K. Pinkau.(Editors), with a foreword by A. G. W. Cameron, H. S. W. Massey.
Published by European Science Foundation, 1, quai Lezay-Marnésia, F - 67000 Strasbourg, France. 54 pp. (1979). – The individual contributions are included in their corresponding subject categories – see abstracts 032.579, 064.101, 131.329, 142.129 - 142.131, 142.534, 143.048, 157.017, 158.309, 160.075.

012.055 **X-ray astronomy**. Proceedings of the NATO Advanced Study Institute held at Erice, Sicily, July 1 - 14, 1979. R. Giacconi, G. Setti (Editors).
NATO Advanced Study Institutes Series C, Vol. 60.
D. Reidel Publishing Company, Dordrecht, Holland - Boston, U.S.A. - London, England. 8 + 406 pp.
Price Dfl. 90.00, $ 47.50 (1980). ISBN 90-277-1156-9.
The individual contributions are included in their corresponding subject categories – see abstracts 013.041, 064.104,

066.533, 125.085 - 125.088, 141.177, 142.132 - 142.138, 158.311 - 158.315, 160.077 - 160.083.

012.056 **L'astronomie dans l'antiquité classique.** Colloquium held at Toulouse in 1977.
G. Aujac, J. Soubiran (Editors).
Collection d'Etudes Anciennes, Les Belles Lettres, Paris. 260 pp. (1979). – Review in J. Hist. Astron., Vol. 11, 204 - 205; 1980 (*G. E. R. Lloyd*).

012.057 **Proceedings of the 2nd European IUE Conference.** Tübingen, Germany, 26 - 28 March 1980.
Compiled by B. Battrick, J. Mort.
European Space Agency, Paris. 29 + 368 pp. Price F 125.00 (1980). ISBN 0379-6566. – Review in Phys. Abstr., Vol. 83, Abstr. 82584 (1980).

012.058 **Fifth ESA Symposium on European rocket & balloon programmes & related research.** Conference held at Bournemouth, England, 14 - 18 April 1980.
T. D. Guyenne, G. Levy (Editors).
ESA Sci. Tech. Publication Branch, ESTEC, Noordwijk, Netherlands. 29 + 526 pp. (1980). – Review in Phys. Abstr., Vol. 83, Abstr. 105433 (1980).

012.059 **Space – New opportunities for international ventures.** Based on papers presented at the 17th Goddard Memorial Symposium, held March, 1979, in Washington, D.C. W. C. Hayes, Jr. (Editor).
Univelt Inc., P. O. Box 28130, San Diego, Calif. 92128. 290 pp. Price $ 35.00 cloth, $ 25.00 paper (1980). – Review in Sky Telesc., Vol. 60, 419 (1980).

012.060 **Life sciences and space research, Vol. XVIII.** Proceedings of the open meeting of the working group on space biology of the 22nd plenary meeting of COSPAR, Bangalore, India, May 1979.
R. Holmquist (Editor).
Pergamon Press. 220 pp. Price £ 20.00 (1980). ISBN 0-08-024436-X. – From Nature, Vol. 287, 373 (1980).

012.061 **Remember the future – the Apollo legacy.** S. Kent (Editor).
Univelt Inc., P.O. Box 28130, San Diego, Calif. 92128. 202 pp. Price $ 25.00 cloth, $ 15.00 paper (1980). – Review in Sky Telesc., Vol. 60, 417 (1980).

012.062 **Space and development. Vol. 6. Advances in space exploration.** Y. Pal (Editor).
COSPAR Symposium Series. Pergamon Press. 97 pp. Price $ 18.00, £ 8.00 (1980). ISBN 0-08-024441-6. – From Nature, Vol. 286, 916 (1980).

012.063 **Light scattering by irregularly shaped particles.** Conference held at Albany, N.Y., USA, 5 - 7 June 1979. D. W. Schuerman (Editor).
Plenum Press. 10 + 334 pp. Price £ 39.50 (1980). ISBN 0-306-40421-4. – Reviews in Geophys. J. R. Astron. Soc., Vol. 63, 798; 1980 (*F. W. Taylor*); Phys. Abstr., Vol. 84, Abstr. 5028 (1981).

012.064 **Giant molecular clouds in the Galaxy.** Third Gregynog Astrophysics Workshop.
P. M. Solomon, M. G. Edmunds (Editors).
Pergamon Press, Oxford–New York. 13 + 344 pp. Price £ 18.00, $ 41.00 (1980). ISBN 0-08-023068-7. – Reviews in Nature, Vol. 287, 176; 1980 (*B. E. Turner*); Sky Telesc., Vol. 60, 321 (1980).

012.065 **Israel Physical Society 1980 Annual Meeting.** Conference held at Rehovot, Israel, 9 - 10 April 1980.
Bull. Israel Phys. Soc., Vol. 26 (1980). – Review in Phys. Abstr., Vol. 83, Abstr. 94909 (1980).

012.066 **Fifth Annual Radio and Space Sciences Symposium.** Conference held at New Delhi, India, 22 - 25 Jan. 1979.
Indian J. Radio Space Phys., Vol. 8, No. 5 - 6 (1979). Review in Phys. Abstr., Vol. 83, Abstr. 94912 (1980).

012.067 **Sun and the heliosphere.** Conference held at London, England, 3 - 4 April 1979.
Philos. Trans. R. Soc. London, Ser. A, Vol. 297, No. 1433 (1980). – Review in Phys. Abstr., Vol. 83, Abstr. 82574 (1980).

013 Reports on Astronomy in Various Countries and Particular Fields, International Cooperation

013.001 **Ways of development and progress of Soviet meteoritics.** E. L. Krinov.
Meteoritika, Vyp. 38, p. 3 - 11 (1979). In Russian.

013.002 **Progress in investigation of meteoritic matter in the USSR.** A. A. Yavnel'.
Meteoritika, Vyp. 38, p. 12 - 18 (1979). In Russian.

013.003 **NASA–CDS cooperative agreement.**
J. M. Mead.
Astron. Data Cent. Bull., Vol. 1, 2 (1980).

013.004 **The Astronomical Data Center at Goddard Space Flight Center.**
T. A. Nagy, J. M. Mead, W. H. Warren, Jr.
Astron. Data Cent. Bull. Vol. 1, 3 - 11 (1980).

013.005 **Positional astronomy at the Royal Greenwich Observatory.** C. A. Murray.
Celestial Mech., Vol. 22, (see 012.002), 139 (1980).

013.006 **The Astronomical Data Center at Goddard Space Flight Center.**
W. H. Warren, Jr., J. M. Mead, T. A. Nagy.
Bull. American Astron. Soc., Vol. 12, 459 (1980). – Abstract.

013.007 **Ultraviolet astronomy enters the eighties.**
S. P. Maran, A. Boggess III.
Phys. Today, Vol. 33, No. 9, p. 40 - 46 (1980).
A decade of observations by earth-orbiting satellites has led to the discovery of compact, hot components in many stellar systems and extended coronas surrounding our Galaxy and others.

013.008 **Die Weiterführung der Zürcher Sonnenflecken-Statistik.**
Orion, 38. Jahrg., 149 (1980).

013.009 **L'astronomia in Jugoslavia.** Z. Knežević.
Orione, Vol. 2, 104 - 110 (1980).

013.010 **Some results from the Space Science Department research programme.** D. E. Page.
ESA Bull., No. 23, p. 19 - 25 (1980).

013.011 **ESA's Science Programme: the present situation and future perspectives.** V. Manno.
ESA Bull., No. 23, p. 34 - 36 (1980).

013.012 **On some trends in the development of astrophysics.**
V. A. Ambartsumian (*Ambartsumyan*).
Annu. Rev. Astron. Astrophys., Vol. 18, (see 003.004), 1 - 13 (1980).

013.013 **Early galactic radio astronomy at Kootwijk.**
C. A. Muller.
Oort and the universe, (see 003.005), p. 65 - 70 (1980).

013.014 **Oort and his large radiotelescope.**
W. N. Christiansen.
Oort and the universe, (see 003.005), p. 71 - 78 (1980).

013.015 **Ten years of discovery with Oort's synthesis radio telescope.** R. J. Allen, R. D. Ekers.
Oort and the universe, (see 003.005), p. 79 - 110 (1980).

013.016 **Oort's work reflected in current studies of galactic CO.** W. B. Burton.
Oort and the universe, (see 003.005), p. 123 - 128 (1980).

013.017 **On high-energy astrophysics.** V. L. Ginzburg.
Oort and the universe, (see 003.005), p. 129 - 140 (1980).

013.018 **Investigations in the field of mathematical methods of celestial mechanics in the USSR.**
V. K. Abalakin.
Narisi z istor. prirodozn. i tekh. Kiev, 1980, No. 26, p. 30 - 37, 126. In Ukrainian. – Abstr. in Ref. zh., 51. Astron., 10.51.4 (1980).

013.019 **Development of mathematical methods in astrophysics in the USSR.** O. F. Bogorods'kij.
Narisi z istor. prirodozn. i tekh. Kiev, 1980, No. 26, p. 38 - 44, 126. In Ukrainian. – Abstr. in Ref. zh., 51. Astron., 10.51.5 (1980).

013.020 **Scientific astronomical centres in Poland.**
Postępy Astron., Tom 28, 241 (1980). In Polish.

013.021 **Moyens que l'Administration Française des Télécommunications peut mettre à disposition pour la transmission des données.** M. Meyer.
Bull. Inf. Cent. Données Stellaires, No. 19, p. 14 - 19 (1980).

013.022 **Le raccordement à la banque de données du C. D. S. au Centre de Calcul de Strasbourg-Cronenbourg (CCSC).** M. Wenger.
Bull. Inf. Cent. Données Stellaires, No. 19, p. 20 - 23 (1980).

013.023 **EURONET- DIANE: l'information en Europe.**
Y. Salle.
Bull. Inf. Cent. Données Stellaires, No. 19, p. 24 - 31 (1980).

013.024 **Les programmes de la bibliothèque du Centre de Calcul du C. N. R. S. et interuniversitaire.**
J. Thomann.
Bull. Inf. Cent. Données Stellaires, No. 19, p. 32 - 34 (1980).
A description is given of the different programs available at the Computing Center, in the field of statistics, numerical computation and graphic representation.

013.025 **On-going and future astronomical data base activities at the Goddard Space Flight Center – February 1980.** J. Mead.
Bull. Inf. Cent. Données Stellaires, No. 19, p. 57 - 59 (1980).

013.026 **News from the Soviet Astronomical Data Center.**
O. Dluzhnevskaya.
Bull. Inf. Cent. Données Stellaires, No. 19, p. 60 (1980).

013.027 **Space research in Poland during 1978.**
J. Kryński.
Artif. Satell., Vol. 15, No. 1, p. 3 - 28 (1980).
Space research in Poland in the fields space physics, satellite geodesy, cosmic biology and medicine, remote sensing and space meteorology during 1978 is reviewed.

013.028 **Continuation of the series of relative sunspot numbers.**
Yamamoto Circ., No. 1942 (1980).

013.029 **Sterrenkunde in Brazilië.** B. Audenaert.
Zenit, 7e Jaarg., 106 - 108 (1980).

013.030 **Sterrenkunde in Oostenrijk.** B. Audenaert.
Zenit, 7e Jaarg., 200 - 202 (1980).

013.031 **Sterrenkunde in Denemarken.** G. W. E. Beekman.
Zenit, 7e Jaarg., 282 - 286 (1980).

013.032 **Sterrenkunde in Griekenland.** B. Audenaert.
Zenit, 7e Jaarg., 340 - 343 (1980).

013.033 **Zürich stopt bepaling Wolf-getal.** H. Potters.
Zenit, 7e Jaarg., 446 - 447 (1980).

013.034 **Sterrenkunde in Canada.** B. Audenaert.
Zenit, 7e Jaarg., 474 - 478 (1980).

013.035 **UK nova/supernova search programme.**
British Astron. Assoc., Circ., No. 612 (1980).

013.036 **Low-frequency radio astronomy in Antarctica.**
A. Fant, Jr.
Sky Telesc., Vol. 60, 488 - 489 (1980).

013.037 **The Image Processing Center of the Faculty of Sciences of Rome University.** G. A. De Biase.
Mem. Soc. Astron. Italiana, Vol. 51, (see 012.040), 397 - 398 (1980). – Abstract.

013.038 **Infrared astronomical research in India.**
P. V. Kulkarni.
Space Research, Vol. XX, (see 012.043), 271 - 275 (1980).

013.039 **Aden and Marjorie Meinel's China trip October-November 1979.** A. B. Meinel, M. P. Meinel.
Appl. Opt., Vol. 19, 2666 - 2669 (1980).

013.040 **An amateur observing program in southern Arizona.** D. H. Levy.
J. R. Astron. Soc. Canada, Vol. 74, 366 (1980). – Abstract.

013.041 **X-ray astronomy.** R. Giacconi.
X-ray astronomy, (see 012.055), p. 1 - 13 (1980).
Contents: Introduction and historical background. Early development of X-ray astronomy. Development of focusing X-ray optics. Einstein.

Reminiscences of the early nineteen-twenties.
See Abstr. 005.014.

Report of the IAU Working Group on Cartographic Coordinates and Rotational Elements of the Planets and Satellites. See Abstr. 091.013.

014 Teaching in Astronomy

014.001 **Laboratory exercises in astronomy – the orbit of a visual binary.** A. Evans.
Sky Telesc., Vol. 60, 195 - 197 (1980).

014.002 **A worthwile observational project for non-science astronomy classes.** W. J. Bisard.
Bull. American Astron. Soc., Vol. 12, 451 (1980). – Abstract.

014.003 **The interactive computer in undergraduate astronomy.** R. J. Dukes, Jr.
Bull. American Astron. Soc., Vol. 12, 451 - 452 (1980).
Abstract.

014.004 **Erkenntnisinteresse, Erkenntnismotiv und Aktivierung im Astronomieunterricht.**
H. Kühnhold.
Astron. Schule, 17. Jahrg., 83 - 86 (1980).

014.005 **Theoretische Untersuchungen zum Begriffssystem der Astronomie im Astronomieunterricht.**
M. Schukowski.
Astron. Schule, 17. Jahrg., 90 - 92 (1980).

014.006 **Compact solar camera.** A. Juergens.
American J. Phys., Vol. 48, 273 - 274 (1980).
Abstr. in Phys. Abstr., Vol. 83, Abstr. 67975 (1980).

014.007 **A solar calendar.** G. Ménard.
J. R. Astron. Soc. Canada, Vol. 74, 242 - 246, with a correction p. 370 (1980).
This note describes how a solar calendar based on the analemma can be designed.

014.008 **Plenary session of SPAK in the Georgien SSR.**
V. A. Hagen-Thorn.
Zemlya Vselennaya, 1980, No. 5, p. 51 - 52. In Russian.

014.009 **Astronomy as a specific discipline and the problems of teaching it.** V. Kourganoff.
Vistas Astron., Vol. 24, 239 - 244 (1980).

014.010 **Het Hertzsprung-Russell-Diagram.**
J. Brand, T. de Zeeuw.
Zenit, 7e Jaarg., 483 - 490 (1980).

014.011 **Solar eclipse predictions.** J. Mottmann.
American J. Phys., Vol. 48, 626 - 628 (1980).
Abstr. in Phys. Abstr., Vol. 83, Abstr. 105452 (1980).

014.012 **On measuring the earth with a hose.**
D. Rawlins.
J. R. Astron. Soc. Canada, Vol. 74, 299 - 301 (1980).

014.013 **Unterrichtstheorie und Astronomieunterricht.**
O. Mader.
Astron. Schule, 17. Jahrg., 104 - 108 (1980).

014.014 **Empfehlungen zur Nutzung von Unterrichtsmitteln in der AG (R) „Astronomie und Raumfahrt".**
N. Franke.
Astron. Schule, 17. Jahrg., 108 - 111 (1980).

014.015 **Erfahrungen mit neu zusammengestellten Dia-Reihen.** V. Kluge.
Astron. Schule, 17. Jahrg., 111 - 113 (1980).

014.016 **Astronomische Themen im Schulunterricht.** H. Kaiser.
Orion, 38. Jahrg., 191 - 192 (1980).

014.017 **Zu den Funktionen der obligatorischen Schülerbeobachtungen.** U. Walther.
Astron. Schule, 17. Jahrg., 131 - 133 (1980).

014.018 **Veranschaulichung astronomischer Strukturen.** M. Schukowski.
Astron. Schule, 17. Jahrg., 133 - 137 (1980).

014.019 **Zur unterrichtlichen Behandlung der nichtoptischen Astronomie.** K. Lindner.
Astron. Schule, 17. Jahrg., 137 - 139 (1980).

014.020 **Schulkommission der Astronomischen Gesellschaft. Abschlußbericht.** T. Schmidt-Kaler.
Mitt. Astron. Ges., Nr. 50, p. 169 - 172 (1980).

Graduate astronomy education in the early days of Lick Observatory. See Abstr. 004.061.

Experiment in het planetarium van Brussel 'Sterrenkunde voor de lagere school'. See Abstr. 009.018.

015 Miscellaneous Papers (Philosophical Aspects, Extraterrestrial Civilizations, etc.)

015.001 **The cost-effectiveness in terms of publications and citations of various optical telescopes at the Kitt Peak National Observatory.** H. A. Abt.
Publ. Astron. Soc. Pacific, Vol. 92, 249 - 254 (1980).

This study is based upon (1) the 445 papers published in 1973–76 by visiting and staff astronomers from observations with the two 0.4-m, two 0.9-m, 1.3-m, and 2.1-m telescopes at the Kitt Peak National Observatory, (2) the 4179 references (citations) made in 1973–78 to those papers, and (3) the relative annual operating costs for those telescopes. The intermediate results are that the numbers of published papers vary as the 1.1-power of the aperture, the numbers of citations vary as the 1.5-power of the aperture, and the initial costs and annual operating expenses vary as the 2.37- and 2.1- powers of the aperture, respectively. These indicate that smaller telescopes, properly equipped, are several times more productive of publications and citations than the largest one.

015.002 **Carl Sagan's "Cosmos": prime-time astronomy.** J. K. Beatty.
Sky Telesc., Vol. 60, 191 - 194 (1980).

015.003 **Extraterrestrial intelligent beings do not exist.** F. J. Tipler.
Q. J. R. Astron. Soc., Vol. 21, 267 - 281 (1980).

015.004 **Mathematical theories and philosophical insights in cosmology.** R. Torretti.
Einstein Symposion Berlin, (see 012.003), p. 320 - 335 (1979).

015.005 **Die philosophische Relevanz der Kosmologie.** B. Kanitscheider.
Einstein Symposion Berlin, (see 012.003), p. 336 - 357 (1979).

015.006 **The impact of space technology on astronomy – the political process.** R. A. Williamson.
Bull. American Astron. Soc., Vol. 12, 502 (1980). – Abstract.

015.007 **Some advantages of wide over narrow band signals in the search for extraterrestrial intelligence (SETI).** P. F. Clancy.
J. British Interplanet. Soc., Vol. 33, 391 - 395 (1980).

015.008 **Ice ages on the earth and their astronomical implications.** Z. Kopal.
Moon Planets, Vol. 23, 253 - 258 (1980).

It is pointed out that while the long-periodic variations of the elements of the terrestrial orbit around the sun are probably sufficient to account for the frequency-spectrum of recurrent ice ages established from the geological record of climatic changes experiences by the earth in the course of the past half a million years, such kinematic phenomena cannot account naturally for the sudden onset of the ice age at the end of the Tertiary epoch. Other astronomical phenomena (such as the galactic orbit of the solar system, which may cause our earth temporarily to pass through different types of galactic climate; or temporary fluctuations in the energy output of the sun), as well as geophysical phenomena (changes in atmospheric chemistry, and consequent fluctuations of the 'greenhouse effect'), may have to be invoked to account for the geological facts by their combined effects.

015.009 **Are we alone after all?** R. Sheaffer.
Spaceflight, Vol. 22, 334 - 337 (1980).

015.010 **The search for intelligence.** E. J. Coffey.
Spaceflight, Vol. 22, 338 - 339 (1980).

015.011 **Does the radio spectrum have room for radio astronomy?** B. J. Robinson, J. B. Whiteoak.
Proc. Astron. Soc. Australia, Vol. 3, 396 - 400 (1979).

015.012 **Autographs of famous astronomers.** N. I. Nevskaya.
Nauka i tekh. Vopr. istorii i teor., Moskva-Leningrad, 1979, No. 10, p. 83 - 86. In Russian. – Abstr. in Ref. zh., 51. Astron., 9.51.5 (1980).

015.013 **The genetic code as fossil vestiges of the primitive atmosphere.** M. Shimizu.
13th Lunar and Planetary Symposium, (see 012.018), p. 394 - 402 (1980).

015.014 **Reflections on the earth and cosmos.** K. F. Ogorodnikov.
Zemlya Vselennaya, 1980, No. 5, p. 20 - 24. In Russian.

015.015 **Signal of an extraterrestrial civilization?** G. M. Rudnitskij.
Zemlya Vselennaya, 1980, No. 5, p. 37. In Russian.

015.016 **Asteroid theory of extinctions strengthened.** R. A. Kerr.
Science, Vol. 210, 514 - 517 (1980).
An asteroid may well have hit Earth at the close of the dinosaur age, but how that impact might have affected life is still obscure.

015.017 **De hemel van Wega.** H. Feijth.
Zenit, 7e Jaarg., 290 - 291 (1980).

015.018 **Star names in western astronomy.** J. P. R. Engledew.
J. British Astron. Assoc., Vol. 91, 63 - 67 (1980).

015.019 **L'étoile de Noël.** D. Proust, avec des notes complémentaires par E. Schweitzer.
Bull. AFOEV, Tome 14, 67 - 69 (1980).

015.020 **Neuere Betrachtungen zum Stern von Bethlehem.** U. Lemmer.
Sterne Weltraum, 19. Jahrg., 404 - 406 (1980).

015.021 **Comets and the origin of life.** N. J. McNaughton, C. T. Pillinger.
Nature, Vol. 288, 540 (1980).

015.022 **The ecosphere of the Sun and the planets. Emigration of life between the planets.** L. Boni.
Atti Fond. Giorgo Ronchi, Vol. 35, 357 - 384 (1980). In Italian. – Abstr. in Phys. Abstr., Vol. 84, Abstr. 4403 (1981).

015.023 **Life from an orderly cosmos.** S. W. Fox.
Naturwissenschaften, 67. Jahrg., 576 - 581 (1980).
Recent astrophysical studies suggest a high degree of order in the inanimate universe, stemming from cosmic beginnings. This state is consistent with the nonrandomness observed experimentally in the thermal polymers of amino acids that figure as an early inanimate stage in organic evolution. The various stages in inanimate matter, protocells, and evolved cells and the degree of order that they represent comport with the second law of thermodynamics on a cosmic scale.

015.024 **Die Aussprache der arabischen Sternnamen und der arabisch-persischen Namen von Mondobjekten.**
P. Kunitzsch.
Sterne, 56. Band, 358 - 363 (1980).

015.025 **Questions concerning the social status of astronomy.** S. Vaghi, E. Schatzman.
Fundam. Sci., Vol. 1, 275 - 293 (1980).

015.026 **L'apparition de la vie dans l'univers.** A. H. Delsemme.
Variability in stars and galaxies, (see 012.044), p. C.P.1 - 26 (1980).
The nature of the phenomena that have lead to the appearance of life on Earth, suggests that these phenomena may have been duplicated many times elsewhere; the scenario for the appearance of life, as it has been partially reconstructed, does not seem to call for unlikely or outstanding coincidences and, everywhere in the universe, the required chemicals are among the most abundant available.

015.027 **Detection of interstellar BS in the Cirrus dark cloud of the Numbbum association. I. An intuitive model and its subsequent observation.** J. J. Charfman.
Interstellar molecules, (see 012.033), p. 645 - 648 (1980).
The editors of Astronomy and Astrophysics Abstracts read this article of the well known astronomer J. J. Charfman with great pleasure.

015.028 **The scientific case against astrology.** I. Kelly.
Mercury, Vol. 9, 135 - 142 (1980).

015.029 **Some tests of astrology.** J. Pasachoff.
Mercury, Vol. 9, 137 (1980).

015.030 **Säkularer Tiefpunkt der Sonnenaktivität – Ursache einer Kälteperiode um das Jahr 2000?**
T. Landscheidt.
Jahrbuch der Wittheit zu Bremen, Band 24, 189 - 220 (1980).

015.031 **Astrology and the fortunes of churches.** J. D. North.
Centaurus, Vol. 24, 181 - 211 (1980).

015.032 **Cosmic factors of origin and development of life and intellect in the universe.** F. A. Tsitsin.
Zh. Vses. khim. o-va im. D. I. Mendeleeva, Vol. 25, 435 - 439 (1980). In Russian. – Abstr. in Ref. zh., 51. Astron., 11.51.2 (1980).

015.033 **On the strategy of search for extraterrestrial civilizations.** N. S. Kardashev.
Zh. Vses. khim. o-va im. D. I. Mendeleeva, Vol. 25, 455 - 460 (1980). In Russian. – Abstr. in Ref. zh., 51. Astron., 11.51.3 (1980).

015.034 **Cosmic vacuum prevents spontaneous migration of microorganisms in the universe.**
M. D. Nusinov, S. V. Lysenko.
Zh. Vses. khim. o-va im. D. I. Mendeleeva, Vol. 25, 452 - 455 (1980). In Russian. – Abstr. in Ref. zh., 51. Astron., 11.51.4 (1980).

015.035 **Strategies for the search for life in the universe. Highlights of the proceedings.**
M. D. Papagiannis.
Strategies for the search for life in the universe, (see 012.053), p. 3 - 12 (1980).

015.036 **The number N of advanced civilizations in our Galaxy and the question of galactic colonization. An introduction.** P. Morrison.
Strategies for the search for life in the universe, (see 012.053), p. 15 - 18 (1980).

015.037 **N is very small.** M. H. Hart.
Strategies for the search for life in the universe, (see 012.053), p. 19 - 25 (1980).
If N, the number of advanced technological civilizations in a typical galaxy the size of our own, were a large number, then the solar system would probably have been colonized long ago. The author infers that N is small. The most likely cause of N being small is that f_{life} (i.e., the fraction of suitable planets on which life actually arises) is very small; reasons are given why this should be so.

015.038 **N is neither very small nor very large.** F. D. Drake.
Strategies for the search for life in the universe, (see 012.053), p. 27 - 34 (1980).
The laws of physics, biology, and interstellar distances make interstellar colonization unreasonable, and moderate values of N continue to be most plausible.

015.039 **Galactic-scale civilization.** T. B. H. Kuiper.
Strategies for the search for life in the universe, (see 012.053), p. 35 - 43 (1980).

015.040 **The number N of galactic civilizations must be either very large or very small.**
M. D. Papagiannis.

Strategies for the search for life in the universe, (see 012.053), p. 45 - 57 (1980).

The author concludes that either the colonization of the galaxy has already taken place, in which case N must be very large ($10^{10} - 10^{11}$), or that the colonization has not yet occurred because very few advanced civilizations capable of initiating it have appeared in the galaxy over its past history, in which case N must be very small ($10^{0} - 10^{1}$). The probability that the galaxy is in the process of being colonized now is very low ($\simeq 0.1\%$).

015.041 **Uncertainty in estimates of the number of extraterrestrial civilizations.** P. A. Sturrock.
Strategies for the search for life in the universe, (see 012.053), p. 59 - 72 (1980).

015.042 **A new approach to the number N of advanced civilizations in the Galaxy.** V. S. Troitskij.
Strategies for the search for life in the universe, (see 012.053), p. 73 - 76 (1980).

The nearly simultaneous origin of life everywhere in our galaxy, and possibly in the entire Universe, is proposed as an alternative to the generally accepted concept of a continuous appearance of life. The impulsive appearance of life provides a better explanation to the possible absence of advanced civilizations in our galaxy.

015.043 **Strategies for SETI through radio waves. An introduction.** B. M. Oliver.
Strategies for the search for life in the universe, (see 012.053), p. 79 - 80 (1980).

015.044 **Microwave searches in the U.S.A. and Canada.** B. Zuckerman, J. Tarter.
Strategies for the search for life in the universe, (see 012.053), p. 81 - 92 (1980).

015.045 **A bimodal search strategy for SETI.** S. Gulkis, E. T. Olsen, J. Tarter.
Strategies for the search for life in the universe, (see 012.053), p. 93 - 105 (1980).

015.046 **Search for planets and early life in other solar systems. An introduction.** J. L. Greenstein.
Strategies for the search for life in the universe, (see 012.053), p. 109 - 110 (1980).

015.047 **The search for early forms of life in other planetary systems: future possibilities afforded by spectroscopic techniques.** T. Owen.
Strategies for the search for life in the universe, (see 012.053), p. 177 - 185 (1980).

It is possible to establish criteria for habitable planets in terms of their sizes and distances from their stars. If such planets can be found in other solar systems, and observed separately from their stars, simple spectrophotometry can reveal whether or not their atmospheres contain gases such as oxygen, methane and water vapor in concentrations and/or combinations that would indicate the presence of life.

015.048 **Manifestations of advanced cosmic civilizations. An introduction.** S. von Hoerner.
Strategies for the search for life in the universe, (see 012.053), p. 189 - 196 (1980).

015.049 **Starships and their detectability.** A. R. Martin, A. Bond.
Strategies for the search for life in the universe, (see 012.053), p. 197 - 226 (1980).

The feasibility of interstellar vehicles journeying among the stars is considered. The discussion is then extended to consideration of fast starships, relatively small vehicles which travel at a reasonable percentage of the speed of light, and much slower, much larger world ships which spend millenia on their journey. Some of the characteristics of such vehicles are commented upon.

015.050 **Radio leakage and eavesdropping.** W. T. Sullivan, III.
Strategies for the search for life in the universe, (see 012.053), p. 227 - 239 (1980).

In addition to searches for purposeful signals, those attempting interstellar communication should also consider the possibility of eavesdropping on radio emissions inadvertently "leaking" from other technical civilizations. It is concluded that, given the present modest understanding of the cultural and technical evolution of civilizations, any initial interstellar radio contact has a priori even chances of placing us in the role of eavesdropper or of intended recipient.

015.051 **Conclusions and recommendations from the joint session on strategies for the search for life in the universe.** M. D. Papagiannis.
Strategies for the search for life in the universe, (see 012.053), p. 243 - 245 (1980).

The galactic club: intelligent life in outer space. See Abstr. 003.024.

Who goes there? See Abstr. 003.042.

Life beyond earth: the intelligent earthling's guide to life in the universe. See Abstr. 003.043.

Life in space. See Abstr. 003.079.

Applied Mathematics, Physics

021 Mathematical Papers Related to Astronomy and Astrophysics, Computing, Data Processing

021.001 **A processor for compression of multi-spectral image data on-board remote sensing satellites.**
E. Mattsson.
J. British Interplanet. Soc., Vol. 33, 327 - 332 (1980).

021.002 **The CLIP4 array processor.** A. Wood.
J. British Interplanet. Soc., Vol. 33, 338 - 344 (1980).

021.003 **Making the VICAR image processing system portable.**
M. D. Lawden, D. Pearce.
J. British Interplanet. Soc., Vol. 33, 369 - 376 (1980).

021.004 **Basis formulas to convert astronomical calendars by computer programming.** M. Gossler.
Astron. Nachr., Band 301, 191 - 194 (1980). In German.

Direct formulas for calculating Julian Days out of calendar dates and vice versa are given for several secular astronomical calendars.

021.005 **Error analysis of the recurrent algorithm of least squares estimate in satellite geodesy.** P.-z. Jia.
Acta Astron. Sinica, Vol. 21, 112 - 121 (1980). In Chinese.

An error analysis of the recurrent algorithm of least squares estimate for satellite geodesy is given. The analysis has shown that this method can be applied to the current satellite geodesy.

021.006 **Calculs astronomiques pour amateurs.**
J. Meeus.
Astronomie, Vol. 94, 402 - 407 (1980).

021.007 **Automated plate assignment.**
T. A. Nagy.
Astron. Data Cent. Bull., Vol. 1, 28 - 31 (1980).

021.008 **The Mark III software system – an integrated system for the scheduling, acquisition, processing, and analysis of VLBI experiments.**
J. W. Ryan, N. R. Vandenberg.
Bull. American Astron. Soc., Vol. 12, 457 (1980). – Abstract.

021.009 **Iteration method for reduction of measurements on the basis of the regularized process of the Newton-Kantorovich type.** S. K. Arutyunov, A. Eh. Brzhozovskij.
Kosm. Issled., Tom 18, 518 - 526 (1980). In Russian.

021.010 **Systems of formulae for calculating the ephemerides of the sun.** H. Lichtenegger.
Acta Geod. Geophys. Montan., Vol. 14, No. 1 - 2, p. 125 - 133 (1979). In German. – Abstr. in Phys. Abstr., Vol. 83, Abstr. 77367 (1980).

021.011 **MHD-calculations for cometary plasmas.**
H. U. Schmidt, R. Wegmann.
Comput. Phys. Commun., Vol. 19, 309 - 326 (1980).

The authors present a relatively simple method for the numerical solution of the MHD-equations on a curvilinear grid. They consider the full 3D equations as well as a modified axisymmetric version. The difference scheme is based on the principle of forming backward differences along characteristics in the spatial variables. This method is used to calculate the interaction of the interplanetary magnetic field with the plasma flow around a comet. The MHD-equations are modified by source terms, which describe the transfer of mass, momentum and energy from a given background (the comet) to the plasma.

021.012 **A new method of deconvolution and its application to lunar occultations.** C. R. Subrahmanya.
Astron. Astrophys., Vol. 89, 132 - 139 (1980).

A new method of deconvolution is described which uses our prior knowledge about the solution to derive some of the information obscured in the data because of the smoothing nature of convolution and the presence of noise. It uses a regularised least-squares criterion of agreement with the data, according to which the computed solution will lead to a minimum variance of noise and also be smooth in the sense of minimum variance of its second differences. The new algorithm is a rapidly converging sequence of iterations for minimising a weighted sum of squares of the deviations of the solution from the specified bounds.

021.013 **Restoration of lunar occultation scans.**
K. von der Heide.
Astron. Astrophys., Vol. 89, 220 - 222 (1980).

A very general least squares algorithm for the retrieval of the shape of an object from images blurred by an arbitrary, and only partly known, point spread function was applied to a simulated scan of an optical lunar occultation. Although a dynamical smoothing is included in the algorithm to suppress the noise, an excellent resolution is achieved.

021.014 **Tests of two different Maximum Likelihood algorithms for determining statistical parallaxes.**
D. H. P. Jones, A. Heck, J. Dawe, S. V. M. Clube.
Astron. Astrophys., Vol. 89, 225 - 226 (1980).

Synthetic data have been used to test two Maximum Likelihood algorithms for statistical parallax. Notwithstanding their different formulations, both give the correct solution.

021.015 **A method of numerical integration of differential equations of order II and its application to the determination of a white dwarf model.** I. H. A. Sass.
Proceedings of the colloquium of astronomy, Cluj-Napoca, (see 012.009), p. 91 - 95 (1978).

The article contains a rapid method of numerical integration which can be successfully applied to differential equations of order two. To prove this method the author integrated the equilibrium equation in the case of complete degeneration.

021.016 **An efficient implementation of the algorithm "CLEAN".** B. G. Clark.
Astron. Astrophys., Vol. 89, 377 - 378 (1980).

The computational effort of the CLEAN algorithm can be substantially reduced by selecting many components and subtracting them in a single operation, rather than separately.

021.017 **Appunti sull'elaborazione dei dati sperimentali (III parte).** E. Tamburini.
Astronomia, N. 3, p. 3 - 9 (1980).

021.018 **How to obtain the true correlation from a 3-level digital correlator.** S. R. Kulkarni, C. Heiles.
Astron. J., Vol. 85, 1413 - 1420 (1980).

A set of explicit approximation formulas to relate correlation as measured by a 3-level digital autocorrelator to the true correlation has been developed. These formulas are accurate to much less than 0.5%, and are also applicable to the more general case of a cross correlator. With the help of numerical simulations, the effects of the drifts of the transition levels on the power spectrum and the inaccuracy of the formulas have been studied. In the cases studied, which are representative of practical applications, the deviations from the scaling are less than 0.5% of the peak signal. The authors conclude that the inaccuracies in the reduction methods and drifts in transition levels do not give rise to any significant errors in observed power spectra.

021.019 **LODAS – a lunar occultation data acquisition system.** D. Hall
South. Stars, Vol. 28, 101 - 108 (1980).

This paper is concerned with the design and construction of a high-speed photometric data-acquisition system which is primarily intended for use in lunar occultation work.

021.020 **A new statistical test with application to globular cluster X-ray source masses.**
A. P. Lightman, P. Hertz, J. E. Grindlay.
Astrophys. J., Vol. 241, 367 - 373 (1980).

The authors develop an absolute likelihood method to statistically determine the mass of the typical globular cluster X-ray source, given a data set of measured positions and measurement errors for a number of sources. This method employs the Kolmogorov-Smirnov test, which the authors generalize to incorporate varying relative measurement errors in the observations. The authors give minimal criteria required of a data set to usefully distinguish between hypotheses. They illustrate their methods, for the X-ray source mass problem, with simulated data sets generated by single-component King potentials, by two-component "core-within-a-core" King potentials, and by a mixture of X-ray source masses.

021.021 **Effects of discontinuity of time series on computed periods.** D. N. Rachkovskij.
Izv. Krymskoj Astrofiz. Obs., Tom 61, 12 - 19 (1980). In Russian.

The method of superposition of epochs for long discontinuous time series is considered. Importance of a preliminary examination of the power spectrum of the discontinuity distribution function is mentioned. Some results concerning the treatment of solar observations by the method under consideration are discussed. The periods of $134^{m}_{.}503$ and $148^{m}_{.}361$ could be the satellites of an unobservable line $141^{m}_{.}092$, when assuming a daily shift phase of the period $T = 141^{m}_{.}092$ equal to π.

021.022 **Mathematical approximations of the ephemerides and the almanac for computers.**
X.-h. Di, D.-z. Xian, Z.-k. Yu.
Acta Astron. Sinica, Vol. 21, 287 - 292 (1980). In Chinese.

Methods using polynomials to approximate the functions in the astronomical ephemerides are discussed. For equally spaced data the algorithms for calculating the coefficients of Chebyshev series and Chebyshev best-fit polynomials are given. Normal algorithms are used to produce the almanac for computers.

021.023 **Algorithms for the combined determination of proper motions of the earth's surface, of variations of the gravitational field, of polar motion and irregularities of the earth's rotation.** V. M. Panin.
Novosib. inst. inzh. geod., aehrofotosemki i kartogr. Novosibirsk, 1980. 18 pp. In Russian. – Abstr. in Ref. zh., 52. Geod. Aehrosemka, 10.52.74 (1980).

021.024 **An efficient, automated method of finding asteroid occultations of catalog stars.**
L. Wasserman, E. Bowell.
Bull. American Astron. Soc., Vol. 12, 664 (1980). – Abstract.

021.025 **A differential correction theory for the determination of atmospheric functions.** J. W. Siry.
Bull. American Astron. Soc., Vol. 12, 741 (1980). – Abstract.

021.026 **Matrix formulation of the Picard method for array processing machines.**
P. E. Nacozy, T. W. Feagin.
Bull. American Astron. Soc., Vol. 12, 742 (1980). – Abstract.

021.027 **On covariance matrices arising in least squares adjustments.** W. H. Jefferys.
Bull. American Astron. Soc., Vol. 12, 742 - 743 (1980). Abstract.

021.028 **A low order m-fold Runge-Kutta algorithm.**
D. G. Bettis.
Bull. American Astron. Soc., Vol. 12, 743 (1980). – Abstract.

021.029 **Real time asteroid identification.**
L. G. Taff.
Bull. American Astron. Soc., Vol. 12, 743 (1980). – Abstract.

021.030 **Least-squares solution of ill-conditioned systems. II.**
R. L. Branham, Jr.
Astron. J., Vol. 85, 1520 - 1527 (1980).

A singular-value analysis of normal equations from observations of minor planets 6, 7, 8, 9, and 15 is undertaken to determine corrections to a number of astronomical parameters, particularly the equinox correction for the FK4. In a previous investigation the test for small singular values was criticized because it resulted in discordant equinox determinations. Here it is shown that none of the test employed by singular-value analysis leads to solutions superior to those given by classical least squares. Some suggestions are made regarding the desirability of planning observational programs in such a way that the observations do not lead to extremely ill-conditioned systems.

021.031 **Kaartprojekties in de sterrenkunde (I), (II).**
K. Velt.
Zenit, 7e Jaarg., 62 - 71, 150 - 155 (1980).

021.032 **Fortran programmes for reducing uvby-Hβ photometry and for deriving physical properties of stars.**
K. P. Lindroos.
Stockholms Obs. Rep. No. 17, 68 pp. (1980).

021.033 **An algorithm for the real time analysis of digitised images.** R. K. Lutz.
Comput. J., Vol. 23, 262 - 269 (1980) = Commun. R. Obs. Edinburgh, No. 361.

021.034 **Calculs astronomiques pour amateurs.**
J. Meeus.
Astronomie, Vol. 94, 541 - 546 (1980).

021.035 **1. Koordinatensysteme der Astronomie.**
H. Schilt.
Orion, 38. Jahrg., Sondernr., p. 14 - 17 (1980).

021.036 **2. Koordinaten-Transformationen.**
H. Schilt.
Orion, 38. Jahrg., Sondernr., p. 17 - 23 (1980).

021.037 **Analisi dello spettro di metodi di perequazione.**
E. Proverbio, S. Uras.
Ist. Lombardo (Rend. Sc.) A 112, 150 - 162 (1978) = Pubbl. Stn. Astron. Int. Latitudine, Carloforte-Cagliari, Nuova Ser., N. 38.

021.038 **A digital correlator for real time spectral analysis of radio astronomy signals.** G. F. W. Woodhouse.
Trans. South African Inst. Electr. Eng., Vol. 71, Part 7, 188 - 192 (1980). – Abstr. in Phys. Abstr., Vol. 84, Abstr. 9385 (1981).

021.039 **Situation of information processing in astronomy.**
G. A. De Biase.
Mem. Soc. Astron. Italiana, Vol. 51, (see 012.040), 391 - 394 (1980).

021.040 **The Frascati RPCNET node of CNR computer network.** I. Mazzitelli.
Mem. Soc. Astron. Italiana, Vol. 51, (see 012.040), 395 - 396 (1980).
The purposes and the present structure of the RPCNET pole of the CNR computers net are described. Also, some hints on the possible development of the pole are provided.

021.041 **The ELSPEC/4 interactive spectra processing package.**
F. Pasian, L. Rusconi, G. Sedmak.
Mem. Soc. Astron. Italiana, Vol. 51, (see 012.040), 405 - 421 (1980).
The package accepts prism and grating spectra digitized on a PDS 1010A microdensitometer and runs on a DEC PDP 11 computer under RSX-11M operating system. The package performs spectra linearization, background removal, continuum normalization, and several lines parameters analysis procedures.

021.042 **ICL: a user oriented interactive command language for data processing in astrophysics.**
M. Pucillo, P. Santin.
Mem. Soc. Astron. Italiana, Vol. 51, (see 012.040), 437 - 441 (1980).
A description is given of an interactive command language, designed and built on the data processing system of the Astronomical Observatory of Trieste. A particular attention is given to the language syntax and to that feature that allows the user to implement new functions to be added to the system.

021.043 **Automated graphical plots for the study of the gravitational N-body problem.**
P. Carnevali, P. Santangelo.
Mem. Soc. Astron. Italiana, Vol. 51, (see 012.040), 529 - 535 (1980).
The authors' code for the numerical integration of the gravitational N-body problem produces a great amount of data (typically some Mbytes). It is shown that automated graphical plots obtained from these data are a powerful tool for the comprehension of the physics of the N-body problem.

021.044 **Inverting x, y grid coordinates to obtain latitude and longitude in the van der Grinten projection.**
D. P. Rubincam.
NASA Tech. Memo., NASA TM 81998, 3 + 9 pp. (1980).
The latitude and longitude of a point on the earth's surface are found from its x, y grid coordinates in the van der Grinten projection. The latitude is a solution of a cubic equation and the longitude a solution of a quadratic equation. Also, the x, y grid coordinates of a point on the earth's surface can be found if its latitude and longitude are known by solving two simultaneous quadratic equations.

021.045 **Nomograph and table for Doppler linewidths.**
S. O. Kastner.
J. Opt. Soc. America, Vol. 70, 1371 - 1372 (1980).
A nomograph for finding the Doppler linewidth in terms of a given wavelength, temperature, and atomic mass is described. For a more accurate evaluation in terms of elemental atomic number, Z, a table of constants is provided.

021.046 **Investigation of the efficiency of numerical algorithms using stabilizing transformations.**
T. V. Bordovitsyna, L. E. Sukhoplyueva.
Byull. Inst. Teor. Astron., Tom 14, 591 - 596 (1980). In Russian.
The efficiency of three methods of stabilization for the integration of differential equations by Runge – Kutta and Adams – Moulton methods has been investigated. The method of the stabilizing factor, the method of Nacozy and the K–S transformation have been considered.

021.047 **Multichannel information-measuring system.**
L. V. Granitskij, A. B. Bukach, Yu. V. Kaplin, N. I. Bukach.
Izv. Krymskoj Astrofiz. Obs., Tom 62, 193 - 197 (1980). In Russian.

021.048 **The on-line radioastronomical data processing system based on computer M-6000 in the process of observations.** S. L. Domnin, V. A. Efanov, V. A. Korsenskij, E. S. Korsenskaya, I. G. Moiseev, N. S. Nesterov, P. S. Nikitin.
Izv. Krymskoj Astrofiz. Obs., Tom 62, 198 - 201 (1980). In Russian.

021.049 **ASAAD (Signal-Acquisition-Averaging-Display) system.** M. MacDonell.
J. R. Astron. Soc. Canada, Vol. 74, 365 (1980). – Abstract.

021.050 **On the possibility of using Laplace series for the gravitational potential of the surface of a planet. I.**
V. A. Antonov, K. V. Kholshevnikov.
Astron. Zh., Tom 57, 1323 - 1330 (1980). In Russian. English translation in Soviet Astron., Vol. 24, No. 6.

021.051 **On a variant of the implicit method of Runge-Kutta.**
M. S. Yarov-Yarovoj.
Tr. MVTU im. N. Eh. Baumana, 1980, No. 336, p. 3 - 11. In Russian. – Abstr. in Ref. zh., 62. Issled. kosm. prostranstva, 1.62.154 (1981).

021.052 **New criterion of selection of rough errors in astronomical observations.** S. F. Shaporev.
Komet. Tsirk., Kiev, No. 264 (1980). In Russian.

021.053 **Automatic calculation of the moonless period in observations of comets.**
N. A. Konovalova, A. N. Pushkarev.
Komet. Tsirk., Kiev, No. 270 (1980). In Russian.

Status report on machine-readable astronomical catalogues. Astronomical Data Center, NASA–Goddard Space Flight Center. See Abstr. 002.010.

On-line computer access to the Bibliographical Star Index. See Abstr. 002.017.

Calcul astronomique pour amateurs adapté à l'emploi d'un calculateur électronique de poche.
See Abstr. 003.022.

Frequency analysis of astronomical time series. See Abstr. 003.087.

The Lund Centre for Automated Measurements of Astronomical Plates. See Abstr. 009.021.

Les programmes de la bibliothèque du Centre de Calcul du C. N. R. S. et interuniversitaire. See Abstr. 013.024.

The Image Processing Center of the Faculty of Sciences of Rome University. See Abstr. 013.037.

The interactive computer in undergraduate astronomy. See Abstr. 014.003.

Solution of block-structured least-squares problems. See Abstr. 031.050.

HP-41C program to trace skew rays through systems with conic surfaces. See Abstr. 031.058.

General ray tracing with a pocket calculator. See Abstr. 031.059.

BASIC language meridional ray trace. See Abstr. 031.060.

The application of a coherent optical data processing system to photographically recorded astronomical spectra. See Abstr. 031.550.

Iterative method applied to image reconstruction and to computer-generated holograms. See Abstr. 031.557.

Iterative method applied to image reconstruction and to computer-generated holograms. See Abstr. 031.569.

An interactive method for the reduction of echelle spectrograms. See Abstr. 031.582.

Interactive methods for feature analysis in astronomical images. See Abstr. 031.584.

The "ELIA" package with some applications to astronomical image processing. See Abstr. 031.585.

Data recording system for 60 cm reflector at KSC. See Abstr. 034.020.

Almanac for computers 1981. See Abstr. 047.034.

Of computing and astronauts. See Abstr. 051.013.

Numerical treatment of the unsteady hydromagnetic thermal boundary layer problem. See Abstr. 062.010.

Spurious solutions of the Navier-Stokes equations. See Abstr. 062.069.

Exact solution of a simple time-dependent integro-differential equation by the method of Laplace transform and the theory of linear singular operators. See Abstr. 063.010.

Computation of the evolution model of normal stars. See Abstr. 065.050.

Flux linkages of bipolar sunspot groups: a computer study. See Abstr. 072.027.

A "fast" model of the solar convection zone. See Abstr. 080.053.

Coordinated ionospheric and magnetospheric observations from the ISIS 2 satellite by the ISIS 2 experimenters. Volume 1. Optical auroral images and related direct measurements. See Abstr. 084.044.

Coordinated ionospheric and magnetospheric observations from the ISIS 2 satellite by the ISIS 2 experimenters. Volume 3. High-latitude charged particle, magnetic field, and ionospheric plasma observations during northern summer. See Abstr. 084.109.

Properties of solutions to integral master equations for closed and open stellar systems. See Abstr. 151.053.

022 Physical Papers Related to Astronomy and Astrophysics

022.001 **Junction conditions across perturbed contact discontinuities.** B. F. Schutz.
Mon. Not. R. Astron. Soc., Vol. 192, 503 - 504 (1980).

Near an oscillating 'contact discontinuity' – a discontinuity which particles do not cross – the appropriate junction condition is that the Lagrangian change in the pressure be continuous. This corrects earlier claims that an inequivalent condition, continuity of the Eulerian pressure change, should be used when the usual linearized equations for the Eulerian perturbation are employed.

022.002 **Diameter to depth dependence of impact craters.** K. Nagel, H. Fechtig.
Planet. Space Sci., Vol. 28, 567 - 573 (1980).

Laboratory impact experiments in the micron to millimeter projectile size range in silicate and metal targets have been performed in order to clarify the still ambigously interpreted velocity dependence of the crater diameter to depth ratios (D/T). The experimental results clearly show the independence of the D/T ratio of velocities above a threshold velocity of 3 - 4 km s^{-1}. The D/T ratio is a function of target properties and of projectile density ρ. For a given target, the resulting approximate relation is $D/T \sim \rho^{-\alpha}$ with α varying between 1/2 and 1/5.

022.003 **On the explosive formation of macroscopic hypervelocity projectiles for use in the study of planetary cratering.** J. W. Bond, R. J. Keyse, G. Newton.
Planet. Space Sci., Vol. 28, 599 - 608 (1980).

The authors show that by modifying a conventional lined hollow charge, such that the explosive is asymmetrically distributed, it is possible to form projectiles of controllable mass (up to ~ 1 g) moving in the forward direction with speeds of the order of 10 km s^{-1}. Measurements of the projectile speed and estimates of its mass are found to agree well with the predictions of the one dimensional theory of Carleone and Chou (1974). A comparison is made of the crater formation obtained by firing both modified and unmodified lined hollow charges at basalt. The authors indicate how it should be possible to increase the velocity of the projectile up to ~ 14 km s^{-1} and the application that such a technique would have to problems in both cratering physics and planetary studies.

022.004 **A line parameter list for the ν_2 and ν_4 bands of $^{12}CH_4$ and $^{13}CH_4$, extended to $J' = 25$ and its application to planetary atmospheres.**
G. S. Orton, A. G. Robiette.
J. Quant. Spectrosc. Radiat. Transfer, Vol. 24, 81 - 95 (1980).

Line parameters (transition frequencies, line strengths, line widths, ground state energies and quantum identifications) for the ν_2 and ν_4 bands of $^{12}CH_4$ and $^{13}CH_4$ have been calculated for $J' \leqslant 25$ using the simultaneous coupled fitting procedure of Gray and Robiette. Molecular constants for the ν_2 band of $^{13}CH_4$ were estimated from isotopic shifts from $^{12}CH_4$ values. Agreement with laboratory spectra, where available, is always well within 1 cm^{-1} over the entire spectral range covered by the list. Applications of the parameter list are demonstrated for remote sounding of the Jovian atmosphere. The list is available on magnetic tape from the authors.

022.005 **An accurate representation of the transmission functions of the H_2O and CO_2 infrared bands.**
T. Aoki.
J. Quant. Spectrosc. Radiat. Transfer, Vol. 24, 191 - 202 (1980).

A simple but accurate formula is given that describes the characteristics of the transmission function of gaseous absorption bands which are not well described by the Mayer-Goody random band model. The temperature dependence of the transmission function is also well described. The parameters for the model have been determined for infrared bands of water vapor and carbon dioxide by fitting values of transmittance calculated from line parameter data of McClatchey et al. The results have been compared with experimental measurements.

022.006 **Quantitative laboratory spectra and spectral line parameters for the ν_2 and ν_4 bands of PH_3 applicable to spectral radiative models of the atmosphere of Jupiter.**
A. Goldman, G. R. Cook, F. S. Bonomo.
J. Quant. Spectrosc. Radiat. Transfer, Vol. 24, 211 - 218 (1980).

Quantitative laboratory PH_3 absorption spectra were obtained in the 800 - 1350 cm^{-1} region, at ~0.05 cm^{-1} resolution, with gas amounts corresponding to observed PH_3 absorptions in the atmosphere of Jupiter. A compilation of spectral line positions, intensities and ground state energies has been generated for the ν_2 and ν_4 bands of PH_3. Line-by-line calculations have been compared with the experimental spectra.

022.007 **Probabilities for electronic transitions of molecular systems of high-temperature air components – II. The γ and β systems of NO and the (4+) system of CO.**
N. E. Kuz'menko, L. A. Kuznetsova, A. P. Monyakin, Yu. Ya. Kuzyakov.
J. Quant. Spectrosc. Radiat. Transfer, Vol. 24, 219 - 227 (1980).

022.008 **Line broadening and oscillator strength measurements for the nitric oxide γ (0,0) band.**
L. G. Dodge, J. Dusek, M. F. Zabielski.
J. Quant. Spectrosc. Radiat. Transfer, Vol. 24, 237 - 249 (1980).

The effective collision diameters for line broadening by several foreign gases and the oscillator strength for the NO γ(0,0) band have been determined with the use of high-resolution spectra and a detailed computer model.

022.009 **Statistical inference of four Ar_{II} transition probabilities.** C. B. Shaw, Jr.
J. Quant. Spectrosc. Radiat. Transfer, Vol. 24, 259 - 261 (1980).

Using the multi-line method, several hundred temperature measurements were made in a partially stabilized high pressure argon arc. Statistically significant evidence enquires revision of transition probabilities for the Ar_{II} lines at 476.489, 480.607, 484.790 and 487.990 nm.

022.010 **About the relation between picocosmical and megacosmical quantities.** M. Schubert.
Astron. Nachr., Band 301, 153 - 154 (1980). In German.

The cosmological meaning of the microscopically determined Dirac number hc/fm^2 is discussed under the assumption $M/m = b(R/r)^2$. The subsequent derivation of the expression for the cosmological baryon number is compared with the derivation following from the inertia-free mechanics.

022.011 **Actinide crystal–liquid partitioning for clinopyroxene and $Ca_3(PO_4)_2$.**
T. Benjamin, W. R. Heuser, D. S. Burnett, M. G. Seitz.
Geochim. Cosmochim. Acta, Vol. 44, 1251 - 1264 (1980).

022.012 **Reflectance spectrophotometry extended to u.v. for terrestrial, lunar and meteoritic samples.**
A. Dollfus, A. Cailleux, B. Cervelle, C. T. Hua, J. C. Mandeville.
Geochim. Cosmochim. Acta, Vol. 44, 1293 - 1310 (1980).

Extension of remote sensing of planetary bodies to the ultraviolet is now feasible up to 2000 Å from earth-orbiting telescopes and spacecraft. The benefits of this extension is analysed on the basis of laboratory spectra taken on a large variety of terrestrial, lunar and meteoritic samples. Knowledge of the albedo for two wavelengths at 2300 and 6500 Å permits classification of a surface into one of the following types: lunar, carbonaceous chondrites, ordinary chondrites, achondrites or acidic rocks, basaltic rocks, irons. For asteroids and non-icy satellites, rock-type classification and constraints in chemical abundances of Si, Al, Fe and Ti can be derived from photometry at 2300 and 6500 Å.

022.013 **Carbon in solid solution in forsterite – a key to the untractable nature of reduced carbon in terrestrial and cosmogenic rocks.**
F. Freund, H. Kathrein, H. Wengeler, R. Knobel, H. J. Heinen.
Geochim. Cosmochim. Acta, Vol. 44, 1319 - 1333 (1980).

Recently it has been shown that carbon can dissolve in refractory oxides, like MgO and CaO, in the form of carbon atoms.The experimental results obtained with carbonaceous MgO are reviewed and new results are presented which demonstrate that synthetic forsterite and natural olivines can also take up atomic carbon in solid solution. The incorporation of the carbon atoms is treated thermodynamically.

022.014 **On the $2\,s^2S{-}2\,p^2P^\circ$ resonance lines of O VI and the $2\,s^2\ {}^1S_0{-}2\,s\,2\,p^3P_1^\circ$ intercombination line of O V.**
C. M. Brown.
Astron. Astrophys., Vol. 88, 273 - 274 (1980).

The wavelengths of the O VI resonance lines have been measured in the laboratory, and compared with several other measurements and an isoelectronic interpolation. Wavelengths of 1031.929±0.005 Å and 1037.617±0.005 Å are reported for the two components $2\,s^2S_{1/2}{-}2p^2P^\circ_{3/2,1/2}$, respectively. These values are in good agreement with two other recent measurements, and with the isoelectronic interpolation. The measured wavelength of the O V $2\,s^2\ {}^1S_0{-}2\,s\,2\,p^3P_1^\circ$ intercombination line is 1218.344±0.010 Å.

022.015 **Lagrangian and Hamiltonian for the motion of charged particles up to the fourth-order terms.**
D. A. Vaiopoulos.
Astrophys. Space Sci., Vol. 71, 239 - 247 (1980).

It is known that a special case of the Lagrangian of two identical charged particles up to fourth order terms has been found. The generalized Lagrangian is derived by a scheme which is somewhat different from the one known. On the other hand one can then proceed to give the Hamiltonian form up to the same order.

022.016 **Lifetime of levels belonging to the $3p^5\ 4p$ configuration of Ar(I).**
M. J. G. Borge, J. Campos.
J. Quant. Spectrosc. Radiat. Transfer, Vol. 24, 263 - 268 (1980).

022.017 **Curves of growth for van der Waals broadened spectral lines.** C. Park.
J. Quant. Spectrosc. Radiat. Transfer, Vol. 24, 289 - 292 (1980).

The author describes the functional dependence of the equivalent line widths on the van der Waals force which may have application in future line-by-line studies of high gas pressure environments.

022.018 **Oscillator strengths of ionized chromium lines in the 2413–2718 Å wavelength range.**
A. Goly, S. Weniger.
J. Quant. Spectrosc. Radiat. Transfer, Vol. 24,335 -340 (1980).

022.019 **Intensities and pressure-broadened widths of CO_2 R-branch lines at 15 μm from tunable laser measurements.** G. L. Tettemer, W. G. Planet
J. Quant. Spectrosc. Radiat. Transfer, Vol. 24, 343 - 345 (1980).

022.020 **Stark broadening trends in homologous ions.**
M. H. Miller, A. Lesage, J. Purić.
Astrophys. J., Vol. 239, 410 - 413 (1980).

Regularities are noted in the experimental Stark widths of nonresonance lines belonging to the homologous sequences C II, Si II, Ge II, Sn II, Pb II, and N II, P II . . . Bi II. Similarities between structure-related trends in line strengths and in Stark widths are examined. Distinctions are seen between broadening trends for resonance lines and nonresonance lines.

022.021 **Large enhancements of electron impact produced OH $A{-}X$ emission in the presence of certain catalyzer gases.** P. Erman, M. Larsson.
Phys. Scr., Vol. 22, 348 - 352 (1980).

The properties of the OH $A{-}X$ emission resulting from electron impact on H_2O has been studied using time resolved techniques. It is found that the addition of various catalyzers such as rare gases, N_2, N_2O, etc. cause a strong enhancement of this emission and a drastic change of the decay curves.

022.022 **Transition probabilities and their accuracy.**
M. C. E. Huber, R. J. Sandeman.
Phys. Scr., Vol. 22, 373 - 385 (1980).

The authors review the classical experimental methods for determining transition probabilities, viz. the method of determining lifetimes and branching ratios, as well as the absorption, hook, and emission techniques. They conclude by mentioning old and new techniques – magneto-rotation and nonlinear optics – which bear a hitherto unexploited potential for deriving accurate oscillator strengths.

022.023 **A kinetic theory of grain formation.**
B. Donn.
Bull. American Astron. Soc., Vol. 12, 447 (1980). – Abstract.

022.024 **The periodic system of the free diatomic molecules.**
R. A. Hefferlin, H. Kuhlman.
Bull. American Astron. Soc., Vol. 12, 459 (1980). – Abstract.

022.025 **Circular polarization of molecular spectra.**
R. M. E. Illing.
Bull. American Astron. Soc., Vol. 12, 477 (1980). – Abstract.

022.026 **Vibrational disequilibrium in astrophysical systems.**
J. Nuth, B. Donn.
Bull. American Astron. Soc., Vol. 12, 530 (1980). – Abstract.

022.027 **Cosmology confronts particle physics.**
G. Steigman.
Annual review of nuclear and particle science, Vol. 29, (see 003.001), 313 - 338 (1979). – Abstr. in Phys. Abstr., Vol. 83, Abstr. 68440 (1980).

022.028 **Thick-target measurements and astrophysical thermonuclear reaction rates. Proton-induced reactions.** N. A. Roughton, M. R. Fritts, R. J. Peterson, C. S. Zaidins, C. J. Hansen.
At. Data Nucl. Data Tables, Vol. 23, No. 2, p. 177 - 194 (1979). – Abstr. in Phys. Abstr., Vol. 83, Abstr. 73108 (1980).

022.029 **On relativistic kinetic theory of transport processes, in particular of neutrino systems.** S. R. de Groot.
Ann. Inst. Henri Poincaré Sect. A, Vol. 31, 377 - 386 (1979). Abstr. in Phys. Abstr., Vol. 83, Abstr. 73283 (1980).

022.030 **Rotational analysis of the (2,0) band of ${}^{13}C^{16}O^+$ molecule of comet-tail system.**
B. R. Vujisic, D. S. Pesic, S. Weniger, D. Rakotorijimy.

Indian J. Pure Appl. Phys., Vol. 18, 370 - 372 (1980).
Abstr. in Phys. Abstr., Vol. 83, Abstr. 74157 (1980).

022.031 **Dense matter in laser driven fusion: laboratory experiments.** R. L. McCrory, J. Wilson.
J. Phys. Colloq., Vol. 41, No. C-2, (see 012.005), p. C2/165 - 177 (1980). – Abstr. in Phys. Abstr., Vol. 83, Abstr. 75077 (1980).

022.032 **Phase separation of ionic mixtures.**
J. P. Hansen.
J. Phys. Colloq., Vol. 41, No. C-2, (see 012.005), p. C2/43 - 52 (1980). – Abstr. in Phys. Abstr., Vol. 83, Abstr. 77279 (1980).

022.033 **Cosmological limits on photon splitting.**
D. C. Wilkins.
Phys. Rev. D, Vol. 21, 2122 - 2136 (1980). – Abstr. in Phys. Abstr., Vol. 83, Abstr. 77284 (1980).

022.034 **Nuclear and neutron matter with isobars – a transition potential model.**
T. Ainsworth, R. A. Smith.
Phys. Lett. B, Vol. 91B, 317 - 320 (1980). – Abstr. in Phys. Abstr., Vol. 83, Abstr. 68573 (1980).

022.035 **Microwave spectrum of water in the ν_2 excited vibrational state.** H. Kuze.
Astrophys. J., Vol. 239, 1131 - 1133 (1980).

Four microwave lines in the ν_2 excited vibrational state of $H_2{}^{16}O$ have been newly observed. The frequencies of the lines are (in MHz) $4_{2,3}-3_{3,0}$, 12008.80 ± 0.03; $5_{3,2}-4_{4,1}$, 26834.27 ± 0.03; $4_{4,0}-5_{3,3}$, 96261.16 ± 0.1; and $2_{2,0}-3_{1,3}$, 119995.94 ± 0.1. An improved set of molecular constants is obtained through a combined analysis of microwave and infrared data.

022.036 **On the $e\ {}^1\Pi - X\ {}^1\Sigma^+$ and ${}^1\Sigma^+ - X\ {}^1\Sigma^+$ transitions of ZrO related to S star spectra.** P. S. Murty.
Astrophys. J., Vol. 240, 363 - 367 (1980).

In view of the prevalent interest in the singlet band systems of ZrO, which are present in the spectra of pure S stars, values of band intensity factors (FCFs) and *r*-centroids are evaluated and displayed for the $e\ {}^1\Pi - X\ {}^1\Sigma^+$ and ${}^1\Sigma^+ - X\ {}^1\Sigma^+$ transitions. The results suggest the possible presence of new bands in both the systems and warrant a reinvestigation of the spectrum of ZrO in the 5500–6600 Å region. Predicted band head positions of the new bands are presented to aid both laboratory and stellar spectral studies.

022.037 **Oscillator strengths and collision strengths for S IV.**
K. Bhadra, R. J. W. Henry.
Astrophys. J., Vol. 240, 368 - 373 (1980).

Collision strengths for S IV for excitation between the fine structure levels of the $3s^2\,3p$ for ground state and excited states $3s3p^2\ {}^4P, {}^2D, {}^2P$, and ${}^2S$ are calculated in two-state and five-state close-coupling approximations for the energy range $1.3 < k^2 < 6.0\ R$. Configuration-interaction target wave functions which gave oscillator strengths accurate to 10% are used in the expansion. Comparison with beam-foil experimental data is good. It is concluded that use of elaborate wave functions is more important than the choice of the scattering approximation.

022.038 **A laboratory study on the dissociative recombination of vibrationally excited O_2^+ ions.**
E. C. Zipf.
J. Geophys. Res., Vol. 85, 4232 - 4236 (1980).

022.039 **Population inversion and suprathermal excitation in carbon monoxide.** J. Köppen, W. H. Kegel.
Astron. Astrophys., Suppl. Ser., Vol. 42, 59 - 67 (1980).

The authors investigate under which physical conditions (kinetic temperature, H_2 density, CO column density) the CO molecule shows suprathermal excitation and population inversion. For the $J = 1 - 0$ line they find for $T_{kin} \gtrsim 20$ K suprathermal excitation and for $T_{kin} \gtrsim 50$ K population inversion for H_2 densities about $10^4\ cm^{-3}$ and not too large column densities. However, only weak masers are found. Most of the observed clouds seem to have physical conditions which allow the application of the LTE analysis. However, the authors expect that in clouds of lower density the LTE analysis will underestimate the CO column density significantly.

022.040 **Lyman- and Balmer-like transitions for the hydrogen atom in strong magnetic fields.**
G. Wunner, H. Ruder.
Astron. Astrophys., Vol. 89, 241 - 245 (1980).

For two values of the magnetic field strength ($B = 2.35 \times 10^{11}$ Gauss, and $B = 4.70 \times 10^{12}$ Gauss) the authors calculate with high computational accuracy the wavelengths and intensities of transitions between hydrogen-like states, and compare the resulting spectra with the ones of the field-free hydrogen atom.

022.041 **An analysis of collisional vibrational excitations and the astrophysical SiO maser phenomenon.**
W. D. Watson, M. Elitzur, R. J. Bieniek.
Astrophys. J., Vol. 240, 547 - 552 (1980).

Because of the likely conclusion that astrophysical SiO masers are pumped by collisions, the authors have analyzed the properties of collisional pump rates for vibrational-rotational excitations of SiO at high temperatures. They find that the pump rates from the ground vibrational state, and probably from excited vibrational states, are independent of the angular momentum of the final state. The analysis is based on general properties of the cross sections as described by the sudden approximation for the rotational part of the excitation. A detailed knowledge of vibration-rotation cross sections for SiO is thus unnecessary for establishing the primary excitation processes and for progress toward a quantitative understanding of the SiO maser phenomenon.

022.042 **Theoretical profiles for the 1–0 $S(1)$ H_2 line in carbon stars.**
D. Goorvitch, J. H. Goebel, G. C. Augason.
Astrophys. J., Vol. 240, 588 - 596 (1980).

The spectra of carbon stars have been synthesized from the models of Querci, Querci and Tsuji and Querci and Querci in the region of the 1–0 $S(1)$ vibration-rotation quadrupole line of H_2. The line is shown to be sufficiently strong to be seen against the numerous lines of the CN red system for models with effective temperatures less than about 2700 K. The usefulness of the line as a diagnostic of the atmosphere is discussed, and a comparison with the measured spectra of UU Aur and S Cep is made. It is concluded that the outer atmospheric layers of carbon stars are significantly warmer than the models predict. An additional opacity source in the outer layers is required.

022.043 **Stark effect at the Si I series limit.**
C. Jordan, J.-D, F. Bartoe, G. E. Brueckner.
Astrophys. J., Vol. 240, 702 - 708, plates 6 - 8 (1980).

In the present paper the authors discuss some small redshifts observed to occur in a few of the high members of the Si I series. They propose that these are due to quadratic Stark effect, although the precise physical origin is not yet certain, because not enough is known about the polarizability of the states concerned.

022.044 **Proton collisional excitation in the ground ${}^3P$ terms of oxygen and sulfur isoelectronic ions.**
D. A. Landman.
Astrophys. J., Vol. 240, 709 - 717 (1980).

Proton collisional excitation cross sections and rate constants are presented for transitions between the fine-structure levels of the p^4 ^3P ground terms for ions belonging to the oxygen (F II–S IX) and sulphur (Cl II–Ni XIII) isoelectronic sequences. A symmetrized semiclassical impact parameter method was used involving direct integration of the Schrödinger equation for electric quadrupole excitation.

022.045 **The measurement of *f*-values for lines of astrophysical interest by beam-foil spectroscopy.**
J. E. Ross, B. J. O'Mara.
Proc. Astron. Soc. Australia, Vol. 3, 263 - 264 (1978).

022.046 **Vacuum instability, cosmology and constraints on particle masses in the Weinberg-Salam model.**
A. D. Linde.
Phys. Lett. B, Vol. 92B, 119 - 122 (1980). – Abstr. in Phys. Abstr., Vol. 83, Abstr. 78007 (1980).

022.047 **Collisional and infrared radiative pumping of molecular vibrational states: the carbon monoxide infrared bands.**
N. Z. Scoville, R. Krotkov, D. Wang.
Astrophys. J., Vol. 240, 929 - 939 (1980).

Excitation of observable infrared emission from molecular vibrational states may arise from collisions, direct pumping by infrared radiation, and ultraviolet fluorescence. The authors analyze and compare the non-local thermodynamic equilibrium (NLTE) excitation resulting from all three processes to identify the observational "signatures" of each and to estimate their efficiency in production of vibrational line photons. The authors treat the direct vibrational excitation resulting from collisions and IR absorption. For all modes of excitation they find it common that the relative populations of the vibrational levels are characterized by a single excitation temperature. This result is borne out by solution of the full rate equations governing statistical equilibrium of the CO level populations taking account of line photon trapping when transitions become optically thick. A specific example to which this treatment is applied is the emission at $\lambda = 2.35$ μm in the CO $\Delta\nu = 2$ overtone bands recently discovered in Orion.

022.048 **Ultraviolet pumping of molecular vibrational states: the CO infrared bands.**
R. Krotkov, D. Wang, N. Z. Scoville.
Astrophys. J., Vol. 240, 940 - 949 (1980).

Observable infrared emission from molecular clouds may arise as fluorescence pumped by absorption of ultraviolet photons from a nearby star. General features of such infrared emission are discussed for various diatomic molecules. It is found that the populations of the vibrational levels are commonly described by a single vibrational temperature which is easily estimated for a number of limiting cases. The recent observations of CO overtone bands in Orion are considered in detail to see if the emission could be produced by UV pumping only. It is concluded that both the observed relative intensities of the three bands and also their absolute strengths are inconsistent with such a model.

022.049 **A new, dielectronic-like recombination process for low temperatures and the radio recombination lines of carbon.**
W. D. Watson, L. R. Western, R. B. Christensen.
Astrophys. J., Vol. 240, 956 - 961 (1980).

A dielectronic-like recombination and autoionization process involving fine-structure states is considered in relation to the predicted strengths of radio recombination lines from ionized carbon. Unlike ordinary dielectronic recombination in which the autoionizing complex is stabilized by the radiative de-excitation of the core, the recombination proposed in this paper is temporarily stabilized through collisions that change the angular momentum of the highly excited electron. Although the stabilization is only temporary, it lasts sufficiently long to contribute significantly to the populations involved in radio recombination lines. As a result, the populations of states of high principal quantum number for carbon atoms are not the same as for hydrogen atoms in the neutral interstellar gas.

022.050 **Theoretical microwave spectral constants for C_2N, C_2N^+, and C_3H.** S. Green.
Astrophys. J., Vol. 240, 962 - 967 (1980).

Theoretical microwave spectral constants have been computed for C_2N, C_3H, and C_2N^+. For C_2N these are compared with values obtained from optical data. Calculated hyperfine constants are also presented for HNC, DNC, and $HCNH^+$. The possibility of observing these species in dense interstellar clouds is discussed.

022.051 **Theoretical microwave spectral constants for C_3H^+ and C_4H^+.** S. Wilson, S. Green.
Astrophys. J., Vol. 240, 968 - 970 (1980).

Theoretical bond lengths and rotation constants are presented for C_3H^+ and C_4H^+. Calculations for C_3 are used to assess the accuracy of the former. Recent results for C_2H^+ are also discussed.

022.052 **Transition probabilities for the hydrogen atom in intense magnetic fields using the adiabatic approximation: comparison with variational results.**
G. Wunner.
Astrophys. J., Vol. 240, 971 - 973 (1980).

Using accurate numerical wave functions within the adiabatic approximation, the author calculates transition rates for the hydrogen atom in strong magnetic fields (of order 10^{10}–10^{12} gauss, which are assumed for neutron stars), and compares with variational results of Wadehra. It is found that using numerical wave functions within the adiabatic approximation leads to values for the transition rates which are larger than the ones of the variational calculation by about 15%.

022.053 **An ab initio prediction of the $J = 1 \leftarrow 0$ transition frequency of $HNCH^+$.** P. S. Dardi, C. E. Dykstra.
Astrophys. J., Lett., Vol. 240, L171 - L173 (1980).

Well-correlated molecular electronic wave functions have been used to determine the equilibrium structure of $HNCH^+$. The ion is found to be linear and the equilibrium rotational constant, B_e, is 37.21 GHz with an estimated accuracy of better than 0.2%. Allowing for anharmonicity and nonrigidity corrections leads to a $J = 1 \leftarrow 0$ transition frequency of 74.07 GHz with a likely error range of ±0.15 GHz.

022.054 **Atomic data for Fe II.**
H. Nussbaumer, P. J. Storey.
Astron. Astrophys., Vol. 89, 308 - 313 (1980).

Electron-ion collision strengths for the lowest four terms of Fe^+ are calculated in the close coupling approximation with a 2 configuration basis. Transition probabilities for the corresponding forbidden transitions and for UV1, UV2, UV3 are obtained in a 4 configuration basis. With these data the authors derive relative population densities for a^6D, a^4F, a^4D, a^4P for a range of electron temperatures and densities.

022.055 **Calculated Stark widths of oxygen ion lines.**
J. D. Hey, P. Breger.
J. Quant. Spectrosc. Radiat. Transfer, Vol. 24, 349 - 364 (1980).

Calculations have been performed on the electron impact broadening of isolated lines from singly-ionized and doubly-ionized oxygen emitted from a plasma of electron density 10^{17} cm^{-3} and temperature about 2eV. These have been compared with results of measurements performed by Platiša, Popović, and Konjević on a plasma produced by a low pressure pulsed arc. Good overall agreement has been obtained for both ionization stages, which the authors interpret as strong support for a recently derived expression for the effective Gaunt factor in line broadening calculations.

022.056 **Rotational variation of predissociation linewidths for the Schumann–Runge bands of molecular oxygen.** B. R. Lewis, J. H. Carver, T. I. Hobbs, D. G. McCoy, H. P. F. Gies.
J. Quant. Spectrosc. Radiat. Transfer, Vol. 24, 365 - 369 (1980).

022.057 **Vacuum structure in gauge theories. The problem of strong CP violation and cosmology.** A. D. Linde.
Phys. Lett. B, Vol. 93B, 327 - 330 (1980). – Abstr. in Phys. Abstr., Vol. 83, Abstr. 86578 (1980).

022.058 **The potential energy of an infinite system of nucleons and delta resonances.** N. H. Goodwin.
J. Phys. G, Vol. 6, 815 - 839 (1980). – Abstr. in Phys. Abstr., Vol. 83, Abstr. 86836 (1980).

022.059 **Collisionless shocks: simulation and laboratory experiments.** I. M. Podgorny (*Podgornyj*).
Nuovo Cimento C, Ser. 1, Vol. 2C, (see 012.013), 834 - 853 (1979). – Abstr. in Phys. Abstr., Vol. 83, Abstr. 87893 (1980).

022.060 **Coriolis phenomenon in nuclei and planets: rotation alignment.** H. A. Smith, Jr.
American J. Phys., Vol. 48, 577 - 578 (1980). – Abstr. in Phys. Abstr., Vol. 83, Abstr. 90636 (1980).

022.061 **Conformal-symmetry breaking and cosmological particle creation in $\lambda\phi^4$ theory.**
N. D. Birrell, P. C. W. Davies.
Phys. Rev. D, Vol. 22, 322 - 329 (1980). – Abstr. in Phys. Abstr., Vol. 83, Abstr. 90998 (1980).

022.062 **Lifetime measurements of stepwise collisional and radiative excited Mn I levels.**
U. Becker, M. Kwiatkowski, U. Teppner, P. Zimmermann.
J. Phys. B, Vol. 13, 2505 - 2516 (1980). – Abstr. in Phys. Abstr., Vol. 83, Abstr. 91501 (1980).

022.063 **Microwave spectra of molecules of astrophysical interest. XVIII. Formic acid.**
E. Willemot, D. Dangoisse, N. Monnanteuil, J. Bellet.
J. Phys. Chem. Ref. Data, Vol. 9, No. 1, p. 59 - 160 (1980). Abstr. in Phys. Abstr., Vol. 83, Abstr. 91550 (1980).

022.064 **Experimental simulation on solar flare – experiment on merging of two current-carrying plasma columns.**
S. Besshou, N. Kawashima.
Bull. Inst. Space Aeronaut. Sci., Univ. Tokyo B, Vol. 15, 555 - 571 (1979). In Japanese. – Abstr. in Phys. Abstr., Vol. 83, Abstr. 94775 (1980).

022.065 **Experiments for studying the dynamics of magnetic reconnection in a current layer.**
S. Yu. Bogdanov, V. S. Markov, S. I. Syrovatskij, A. G. Frank, A. Z. Khodzhaev.
XIth seminar on cosmophysics, (see 012.035), p. 79 - 82 (1979). In Russian. – Abstr. in Ref. zh., 51. Astron., 8.51.386 (1980).

022.066 **Physical regularities and numerical values of fundamental constants.** I. L. Rozental'.
Inst. kosm. issled. AN SSSR. Prepr., 1979, No. 524, 24 pp. In Russian. – Abstr. in Ref. zh., 51. Astron., 8.51.766 (1980).

022.067 **Theoretical oscillator strengths of some transitions of Y II and Zr II.** V. Pirronello, G. Strazzulla.
Astrophys. Space Sci., Vol. 72, 55 - 59 (1980).

The authors calculate the *gf*-values of some transitions of Y II and Zr II to test the reliability of the commonly used values for astrophysical applications. For Zr II they show the inadequacy of the values used at the present time.

022.068 **Amorphous interstellar grains: wavelength dependence of far-infrared emission efficiency.**
J. Seki, T. Yamamoto.
Astrophys. Space Sci., Vol. 72, 79 - 86 (1980).

It is shown theoretically that the emission efficiency of amorphous grains with radii smaller than ~ 100 Å has a λ^{-1}-dependence in the wavelength region longer than ~ 100 μ, whereas those of crystalline, metallic, and larger amorphous grains are proportional to λ^{-2} in the far infrared. Astrophysical implications of the amorphous grains are discussed.

022.069 **Experiments on dust collection for a cometary mission.** J. Kissel, B. C. Clark, D. Clair.
Solid particles in the solar system, (see 012.019), p. 271 (1980). – Abstract.

022.070 **Laboratory measurements on the infrared features of interstellar silicate grains.** W. Krätschmer.
Solid particles in the solar system, (see 012.019), p. 351 - 354 (1980).

Thin layers of an amorphous silicate coating were obtained by ion sputtering of crystalline olivine on KBr substrates. The dielectric functions of the coating were determined from reflection and transmission measurements in the wavelength range between 5 and 25 μ. The small particle extinction of the amorphous silicate shows a strong feature at 9.7 μ and a weaker one at 17.5 μ. In wavelength position both absorption bands agree with astronomically observed features.

022.071 **Cosmic dust synthesized in reducing environments.**
B. N. Khare, C. Sagan.
Solid particles in the solar system, (see 012.019), p. 355 - 356 (1980). – Abstract.

022.072 **Collisional processes of iron and steel projectiles on targets of different densities.**
H. Fechtig, K. Nagel, N. Pailer, E. Schneider.
Solid particles in the solar system, (see 012.019), p. 357 - 364 (1980).

Cratering experiments for μm- and mm- sized iron- and steel-projectiles on various target materials show that crater depths and the ratios of crater diameter to crater depth D/T depend on the densities of the projectile- and target-material and on the ductility of the target material. Cratering experiments into low density material (Saffile, ρ = 0.28 g/cm^3) have produced elongated impact "craters". For low target densities the "crater" depth is up to 100 times the projectile diameter, depending on its impact speed. This impact process leads to a complete accretion of the projectile mass within the target.

022.073 **Determination of particle densities by penetration studies.** N. Pailer, E. Grün.
Solid particles in the solar system, (see 012.019), p. 365 - 370 (1980).

022.074 **Absolute transition probabilities of neutral titanium lines.** A. Holys, J. R. Fuhr.
Astron. Astrophys., Vol. 90, 14 - 17 (1980).

Transition probabilities for 21 lines of Ti I have been measured in emission with a wall-stabilized arc. These lines cover a wavelength range of 3600-6300 Å and originate from either the $y^3 G°$ or the $y^5 G°$ upper level. Weak as well as strong lines were measured, and the resulting relative *A*-values were converted to the absolute scale of Bell et al. (1975). The uncertainties of the absolute transition probabilities are estimated to be within 20-30%. Comparisons of this work with other recent experiments show excellent agreement.

022.075 **Proton excitation rates for fine structure transitions in C III, O V, and Ne VII in the sun.**
J. G. Doyle, A. E. Kingston, R. H. G. Reid.
Astron. Astrophys., Vol. 90, 97 - 101 (1980).

Proton excitation rates are calculated for the fine structure transitions in the $(2s, 2p)^3P$ and $(2p)^2\ ^3P$ levels in C III, O V, and Ne VII and the importance of these transitions in determining population densities and line ratios is considered.

022.076 **The case against UV photostimulated oxidation of magnetite.** R. V. Morris, H. V. Lauer, Jr.
Geophys. Res. Lett., Vol. 7, 605 - 608 (1980).

The authors have studied the kinetics of magnetite oxidation in O_2- bearing atmospheres in the presence of electromagnetic radiation. No perceptible oxidation of magnetite by ultraviolet (UV) photostimulation occurred. Therefore, although the authors cannot totally rule out the possibility that the process actually occurs, they conclude that there is not yet a basis in laboratory experiments for inferring that UV photostimulated oxidation of magnetite occurs naturally on the surface of Mars.

022.077 **An infrared band system of the ZrCl molecule.**
J. G. Phillips, S. P. Davis, D. C. Galehouse.
Astrophys. J., Suppl. Ser., Vol. 43, 417 - 434 (1980).

Ten bands discovered in the 0.97 μm to 1.15 μm region have been found to be produced by the ZrCl molecule. Resultant molecular constants are compared with theoretical predictions.

022.078 **Mid and far infrared spectra of hydrous silicates.**
C. Koike, H. Hasegawa.
13th Lunar and Planetary Symposium, (see 012.018), p. 210 - 216 (1980).

022.079 **Evaporation experiment: implication to the origin of cosmic materials.** H. Nagasawa, K. Yamakoshi.
13th Lunar and Planetary Symposium, (see 012.018), p. 217 - 222 (1980).

The authors present some results of evaporation experiments and their application to the study of the origin of the silicate spherules collected from deep-sea sediments and the Ca, Al-rich inclusions in the Allende meteorite.

022.080 **Low temperature sintering of fine powders – an application to the consolidation process of meteorites.** K. Yomogida, T. Matsui, M. Honda.
13th Lunar and Planetary Symposium, (see 012.018), p. 223 - 231 (1980).

022.081 **The significance of low melting-temperature materials in consolidation of planetesimals.**
M. Miyamoto, K. Ito, N. Fujii, Y. Kobayashi.
13th Lunar and Planetary Symposium, (see 012.018), p. 232 - 238 (1980).

The authors carried out experiments on low-velocity impacts in order to study both the mechanics of fragmentation and rebound, and the mechanical properties of loosely consolidated aggregates containing the small amount of the materials which have the low melting-temperature or different mechanical properties from the host material.

022.082 **Model experiment of collisional breakup of asteroids: on the velocities of fragments.**
A. Fujiwara, A. Tsukamoto.
13th Lunar and Planetary Symposium, (see 012.018), p. 290 (1980). – Abstract.

022.083 **Calculated Stark widths of isolated S(III) and S(IV) lines.** J. D. Hey, P. Breger.
J. Quant. Spectrosc. Radiat. Transfer, Vol. 24, 427 - 439 (1980).

Calculations have been performed on the electron-impact broadening of isolated lines from doubly-ionized and triply-ionized sulphur emitted from a plasma of electron density 10^{17} cm^{-3} and temperature 28,500°K. These have been compared with results of measurements performed by Platiša, Popović, Dimitrijević, and Konjević on a low-pressure, pulsed arc. Good overall agreement has been obtained for both ionization stages, in confirmation of earlier conclusions based on a similar comparison for oxygen ion lines.

022.084 **Stark widths of doubly- and triply-ionized atom lines.** M. S. Dimitrijević, N. Konjević.
J. Quant. Spectros. Radiat. Transfer, Vol. 24, 451 - 459 (1980).

The authors report modifications of well known semi-empirical and semiclassical approximation formulas for Stark line-width calculations. Comparisons with experiments for doubly ionized atoms yield as an average ratio of measured to calculated widths 1.06 ± 0.31 for a modified semiempirical formula and 0.96 ± 0.24 for a modified semiclassical formula. For triply ionized atoms these ratios are 0.91 ± 0.42 and 1.08 ± 0.41, respectively. Comparison with other theoretical calculations have also been made.

022.085 **Tunable infrared diode laser measurements of line strengths and collision widths of $^{12}C^{16}O$ at room temperature.** P. L. Varghese, R. K. Hanson.
J. Quant. Spectrosc. Radiat. Transfer, Vol. 24, 479 - 489 (1980).

A tunable diode laser was used to scan 33 vibration-rotation lines in the fundamental band of CO in room-temperature CO-N_2 and CO-Ar mixtures. Each absorption record was fitted with a Voigt profile from which the line strength and collision width were determined. The fundamental band strength of $^{12}C^{16}O$ at 273.2°K was determined to be 283 ± 4 cm^{-2} atm^{-1}.

022.086 **Dielectronic recombination rate for the Be-sequence.**
Y. Hahn, J. N. Gau, R. Luddy, M. Dube, N. Shkolnik.
J. Quant. Spectrosc. Radiat. Transfer, Vol. 24, 505 - 515 (1980).

The dielectronic recombination (DR) rate coefficient α^{DR} is explicitly calculated for the Mo, Fe, Ar, and Ox target ions of the Be-sequence with four electrons, in the isolated resonance approximation. This work extends a previous study of the Mo^{38+} ions at 1.4, 2.8, and 5.6 keV electron temperatures. Both $\Delta n \neq 0$ and $\Delta n = 0$ transitions are considered in detail.

022.087 **Absorption of *H* and *D* Ly-α radiation by O_2 molecules at high temperatures.**
C.-C. Chiang, G. B. Skinner.
J. Quant. Spectrosc. Radiat. Transfer, Vol. 24, 525 - 528 (1980).

022.088 **Can lines of C_2^- be observed in carbon stars?**
M. S. Vardya, K. S. Krishna Swamy.
Chem. Phys. Lett., Vol. 73, 616 - 617 (1980). – Abstr. in Phys. Abstr., Vol. 83, Abstr. 98661 (1980).

022.089 **Partition function and dissociation equilibrium constant of the $^{12}C\ ^{16}O$ molecule.** A. P. Sarychev.
Astron. Zh., Tom 57, 1020 - 1022 (1980). In Russian.
English translation in Soviet Astron., Vol. 24, No. 5.

The partition function and the dissociation equilibrium constant in the range of temperatures 1000 - 10000 K with a step of 100 K are evaluated on the basis of the new spectroscopic constants of the $^{12}C\ ^{16}O$ molecule in the ground state $X^1\Sigma^+$. Empirical formulae approximating the given quantities in the interval of 3000 - 8000 K with a high degree of accuracy are obtained.

022.090 **Laboratory simulation of the induced magnetospheres of comets and Venus.** I. M. Podgorny (*Podgornyj*), E. (*Eh.*) M. Dubinin, P. L. Israelevich (*Izrajlevich*).
Moon Planets, Vol. 23, 323 - 338 (1980).

The comparison of data obtained in laboratory experiments on the solar wind interaction with a body endowed with a plasma shell, the observations of comet type I tails and the direct measurements near Venus show that an induced magnetosphere is formed with an extended magnetic tail. This magnetosphere appears due to currents associated with unipolar induction. The distribution of electrodynamical forces associated with the formation of the induced magnetosphere makes it possible to explain the acceleration of matter towards the tail as in the motion across the tail observed in comets and Venus. The analysis of the condensation motion in Halley's comet yields an estimate of tail magnetic field of 30 to 50 γ. A three-dimensional model of the induced magnetospheres of Venus and comets is developed.

022.091 **Entropies need not be concave.**
P. T. Landsberg, D. Tranah.
Phys. Lett. A, Vol. 78 A, 219 - 220 (1980). – Abstr. in Phys. Abstr., Vol. 83, Abstr. 98867 (1980).

022.092 **Stellar reaction rate of $^{26}Mg(p,\gamma)^{27}Al$.**
L. Buchmann, H. W. Becker, K. U. Kettner, W. E. Kieser, P. Schmalbrock, C. Rolfs.
Z. Phys. A, Vol. 296, 273 - 280 (1980). – Abstr. in Phys. Abstr., Vol. 83, Abstr. 99132 (1980).

022.093 **Structures in the beta strength function and consequences for nuclear physics and astrophysics.**
H. V. Klapdor, C. O. Wene.
J. Phys. G, Vol. 6, 1061 -1104 (1980). – Abstr. in Phys. Abstr., Vol. 83, Abstr. 99152 (1980).

022.094 **The relative band strengths of a_2-X_2, b_2-X_2 systems and the dissociation energy of the CeO molecule.** V. M. Rao, M. L. P. Rao, P. T. Rao.
Acta Phys. Polonica A, Vol. A58, 237 - 240 (1980). – Abstr. in Phys. Abstr., Vol. 83, Abstr. 99495 (1980).

022.095 **The microwave spectrum of CH_2DOH.**
C. R. Quade, R. D. Suenram.
J. Chem. Phys., Vol. 73, 1127 - 1131 (1980). – Abstr. in Phys. Abstr., Vol. 83, Abstr. 99498 (1980).

022.096 **An additional uncertainty in calculated radiative association rates of molecular formation at low temperatures.** E. Herbst.
Astrophys. J., Vol. 241, 197 - 199 (1980).

The effect of introducing a centrifugal barrier to a previously published theory of radiative association rate coefficients for polyatomic products is demonstrated. Discrepancies of up to one order of magnitude from earlier calculated rate coefficients are found.

022.097 **Charge transfer of multiply charged ions with hydrogen and helium: quantal calculations.**
S. E. Butler, T. G. Heil, A. Dalgarno.
Astrophys. J., Vol. 241, 442 - 447 (1980).

The rate coefficients are presented for a number of charge transfer reactions of astrophysical importance involving two and three times ionized C, N, O, and Ne interacting with H and He at thermal energies. Complete quantal calculations are reported which utilize potential curves and matrix elements calculated with large configuration-interaction wave functions in a close-coupled formulation of the scattering. Astrophysical applications are discussed.

022.098 **Laboratory production of candidates for the diffuse interstellar bands.** T. J. Wdowiak.
Astrophys. J., Lett., Vol. 241, L55 - L58 (1980).

A method of synthesis via plasma discharge and rare gas cryogenic matrix isolation has produced large carbon-containing molecules from a 0.5% methane in argon mixture, giving rise to a reproducible pattern of features corresponding to diffuse interstellar bands (DIBs), including the strongest and widest 4430 (4428) Å diffuse interstellar band, the second strongest and wide 6283 (6284) Å DIB along with its narrower and shorter wavelength companion, the 6269 Å DIB, and the 5778 - 5780 Å DIBs.

022.099 **Oscillator strengths for O III and the Bowen fluorescent mechanism.** H. E. Saraph, M. J. Seaton.
Mon. Not. R. Astron. Soc., Vol. 193, 617 - 629 (1980).

Oscillator strengths are calculated for transitions between O III triplet terms using a frozen cores approximation. The O III Bowen mechanism involves excitation of O III $2p^2\ {}^3P_2 - 2p3d\ {}^3P_2^o$ by He II Ly α; $P(\lambda)$ is the probability of obtaining a quantum in an O III line of wavelength λ, following excitation of $2p3d\ {}^3P_2^o$. Values of $P(\lambda)$ are calculated. From optical and UV observations of the planetary nebula NGC 7662, intensities are obtained for the six O III multiples $2p3d\ {}^3P^o - 2p3p\ {}^3L$ and $2p3p\ {}^3L - 2p3s\ {}^3P^o$, with $L = S, P$ and D, and for He II λ 4686.

022.100 **Recombination of ions and electrons on grains and the ionization degree in dense interstellar clouds.**
T. Umebayashi, T. Nakano.
Publ. Astron. Soc. Japan, Vol. 32, 405 - 421 (1980).

The authors investigate the processes relevant to the recombination of ions and electrons on grains. The sticking probability of thermal electrons on grains is found to be between 0.3 and 1.0, and that of the ions about 1.0. The charge-state distribution of grains is determined as a function of the grain radius and the gas temperature, and it is found that most grains have one excess electron. By using a simplified reaction scheme in the gas phase together with the recombination of ions and electrons on grains, the ionization degree of the gas in the steady state is investigated.

022.101 **Restricted quantum-mechanical three-body problems. II. A general theory of helium-like ions.**
S. Barcza.
Astrophys. Space Sci., Vol. 72, 497 - 507 (1980).

A new method is presented in a general form to solve the Schrödinger equation of helium-like ions. The wave function is expanded in terms of the eigenfunctions of a moving electron in the field of two Coulombic ions which are fixed in space. This makes the method similar to the Dirac perturbation theory (perturbation theory for time-dependent problems). In the present method an infinitely coupled system of infinitely many second-order ordinary differential equations must be solved instead of one second-order partial differential equation of three variables. The nature of the singular points and boundary conditions are discussed and some general relations are given which are useful for the numerical treatment.

022.102 **Consequences of structures in the beta strength function on nuclear physics and astrophysics.**
H.-V. Klapdor, C. O. Wene.
Nukleonika, Vol. 25, No. 1, p. 105 - 131 (1980). – Abstr. in Phys. Abstr., Vol. 83, Abstr. 102149 (1980).

022.103 **Cross-section measurements and thermonuclear reaction rates for $^{49}Ti(p,\gamma)^{50}V$ and $^{49}Ti(p,n)^{49}V$.**
S. R. Kennett, M. R. Anderson, Z. E. Switkowski, D. G. Sargood.
Nucl. Phys. A, Vol. A344, 351 - 360 (1980). – Abstr. in Phys. Abstr., Vol. 83, Abstr. 102187 (1980).

022.104 **Microwave properties of solid CO_2.**
R. A. Simpson, B. C. Fair, H. T. Howard.
J. Geophys. Res., Vol. 85, 5481 - 5484 (1980).

Laboratory measurements on the electrical properties of solid carbon dioxide (dry ice) were made over the frequency range 2.2 to 12 GHz. These give a dielectric constant which varies with density according to the Rayleigh mixing formula and is independent of frequency; at 1 g/cm^3 the dielectric constant is 1.7. An upper limit of 0.005 for the loss tangent was found.

022.105 **Abiotic organic synthesis in space.** M. R. Bloch, H. L. Wirth.
Naturwissenschaften, 67. Jahrg., 562 - 564 (1980).

022.106 **The interplay of molecular spectroscopy and astronomy.** G. Herzberg.
Highlights of Astronomy, Vol. 5, (see 003.006), 3 - 26 (1980).
The author shows that the frequent and intimate interplay between laboratory spectroscopy and astronomical observation is responsible for a great deal of progress both in molecular spectroscopy and in astronomical problems. One problem in which this interplay is especially important is the problem of the diffuse interstellar lines.

022.107 **On the identification of mm-wavelength U-lines.** B. E. Turner.
Interstellar molecules, (see 012.033), p. 45 - 46 (1980).

022.108 **Long carbon chain molecules in the laboratory and in space.** G. Winnewisser, F. Toelle, H. Ungerechts, C. M. Walmsley.
Interstellar molecules, (see 012.033), p. 59 - 65 (1980).

022.109 **Population inversion and suprathermal excitation in carbon monoxide.**
J. Köppen, W. H. Kegel.
Interstellar molecules, (see 012.033), p. 127 (1980).

022.110 **The hydrogen molecule as a collision partner.** T. Oka.
Interstellar molecules, (see 012.033), p. 221 - 229 (1980).

022.111 **Recent laboratory work on molecules of possible importance for interstellar studies.**
G. Herzberg.
Interstellar molecules, (see 012.033), p. 231 - 238 (1980) = NRCC 18508.

022.112 **Far infrared laser magnetic resonance spectroscopy.** K. M. Evenson, R. J. Saykally.
Interstellar molecules, (see 012.033), p. 239 - 245 (1980).

022.113 **Laboratory measurements of oscillator strengths of ultraviolet molecular lines of HCl and H_2O and column densities of these molecules in the Zeta Ophiuchi cloud.** P. L. Smith, K. Yoshino, W. H. Parkinson.
Interstellar molecules, (see 012.033), p. 269 - 270 (1980).

022.114 **Charge exchange and fine structure excitation in O–H^+ collisions.** G. Chambaud, J. M. Launay, B. Lévy, P. Millie, E. Roueff, F. Tran Minh.
Interstellar molecules, (see 012.033), p. 287 - 288 (1980).

022.115 **Experimental measurements of ion-molecule reactions.** F. C. Fehsenfeld.
Interstellar molecules, (see 012.033), p. 291 - 296 (1980).
The applicability of ion-molecule reaction rate constants measured at room temperature to simulation of interstellar cloud chemistry is discussed.

022.116 **An ICR study of ion-molecule reactions in the C_2H_2/HCN system.** M. J. McEwan, V. G. Anicich, W. T. Huntress, Jr.
Interstellar molecules, (see 012.033), p. 299 - 303 (1980).

022.117 **An ICR study of an association reaction at low pressure.** M. J. McEwan, V. G. Anicich, W. T. Huntress, Jr., P. R. Kemperer, M. T. Bowers.
Interstellar molecules, (see 012.033), p. 305 - 306 (1980).

022.118 **Laboratory studies of interstellar carbon/nitrogen ion chemistry.** H. I. Schiff, G. I. Mackay, G. D. Vlachos, D. K. Bohme.
Interstellar molecules, (see 012.033), p. 307 - 310 (1980).
Laboratory measurements are reported for ion-molecule reactions involving CN^+, HCN^+, C_2N^+ and HCN, and their implications for interstellar synthesis for C/N compounds are discussed.

022.119 **The formation of complex interstellar molecules by radiative association.** E. Herbst.
Interstellar molecules, (see 012.033), p. 317 - 321 (1980).

022.120 **The formation of interstellar molecules via radiative association reactions.** D. Smith, N. G. Adams.
Interstellar molecules, (see 012.033), p. 323 - 324 (1980).
The radiative association rate coefficients and their temperature dependences have been estimated for several likely interstellar ion-molecule reactions from laboratory collisional association rate data.

022.121 **On the formation of interstellar linear molecules.** A. Sakata.
Interstellar molecules, (see 012.033), p. 325 - 329 (1980).
A possible mechanism for the formation of interstellar linear molecules is studied experimentally by means of synthesis apparatus. C_2 and CN radicals are abundantly formed from plasmas containing C, N and H atoms. The C_2 and CN radicals survive electron bombardment. They collide with each other and recombine to form linear molecules.

022.122 **Laboratory and modeling studies of chemistry in dense molecular clouds.** W. T. Huntress, Jr., S. S. Prasad, G. F. Mitchell.
Interstellar molecules, (see 012.033), p. 331 - 336 (1980).
A chemical evolutionary model with a large number of species and a large chemical library is used to examine the principal chemical processes in interstellar clouds. Simple chemical equilibrium arguments show the potential for synthesis of very complex organic species by ion-molecule radiative association reactions.

022.123 **Laboratory and theoretical results on interstellar molecule production by grains in molecular clouds.**
J. M. Greenberg, L. J. Allamandola, W. Hagen, C. E. P. van de Bult, F. Baas.
Interstellar molecules, (see 012.033), p. 355 - 363 (1980).
Laboratory and theoretical studies have been made of the effects of ultraviolet photolysis of interstellar grain mantles. It has been shown that grain photolysis should be important even in dense clouds.

022.124 **The formation of hydrocarbons and iron-hydrides on cold interstellar grains - experimental studies.**
A. Bar-Nun, M. Litman, M. Pasternak, M. L. Rappaport.
Interstellar molecules, (see 012.033), p. 367 - 371 (1980).

022.125 **Reproduction of the interstellar ice band by grain mantle analogs.** W. Hagen, A. G. G. M. Tielens, J. M. Greenberg.
Interstellar molecules, (see 012.033), p. 387 - 388 (1980).

022.126 **Λ-doublet population inversion in collisions of OH, OD, CH, CD and NH^+.**
R. N. Dixon, D. Field.
Interstellar molecules, (see 012.033), p. 583 - 587 (1980).
The results of a new approach to the problem of the collisional step in the pumping cycle for OH and CH masers are reported.

022.127 **Collisional inversion of the populations of Λ-doublets in CH and OH: a critical study.**
D. R. Flower.
Interstellar molecules, (see 012.033), p. 589 - 590 (1980).

022.128 **Theoretical studies of impact cratering.**
D. Roddy, K. Kreyenhagen, S. Schuster, D. Orphal.
Bull. American Astron. Soc., Vol. 12, 661 - 662 (1980). Abstract.

022.129 **Regions of weak methane absorptions in the near infrared: laboratory measurements and planetary applications.** C. de Bergh, J. P. Maillard, J. Brault, J. C. Buriez, B. Lutz, T. Owen.
Bull. American Astron. Soc., Vol. 12, 668 (1980). – Abstract.

022.130 **Oscillator strengths and collision strengths for SV.**
R. J. W. Henry, W. L. van Wyngaarden.
Bull. American Astron. Soc., Vol. 12, 675 (1980). – Abstract.

022.131 **RE-LAB (*Reflectance Experiment Laboratory*): first results for MOPL *(Mean Optical Path Length)* measurements.** C. M. Pieters.
Bull. American Astron. Soc., Vol. 12, 681 (1980). – Abstract.

022.132 **Estimations of mineral proportions in particulate samples from albedo and absorption measurements.**
T. V. V. King, C. M. Pieters, D. L. Sandlin.
Bull. American Astron. Soc., Vol. 12, 681 - 682 (1980). Abstract.

022.133 **Effects of low temperature and pressure on the reflectance spectra of clay minerals.**
T. G. Farr, B. Smith, J. B. Adams, D. B. Wenner.
Bull. American Astron. Soc., Vol. 12, 682 (1980). – Abstract.

022.134 **Altered basaltic glass – a Martian soil analog.**
C. C. Allen, J. L. Gooding, M. Jercinovic, K. Keil.
Bull. American Astron. Soc., Vol. 12, 682 (1980). – Abstract.

022.135 **The formation of the Martian soil by hydrothermal alteration of impact melt sheets.** H. E. Newsom.
Bull. American Astron. Soc., Vol. 12, 682 - 683 (1980). Abstract.

022.136 **Pressure dependence of the absolute rate constant for the reaction OH + C_2H_2 from 228 – 413 K: implications for CO modeling of the Jovian atmosphere.**
D. F. Nava, J. V. Michael, R. P. Borkowski, W. A. Payne, L. J. Stief.
Bull. American Astron. Soc., Vol. 12, 687 (1980). – Abstract.

022.137 **The decomposition of methane by hot hydrogen atoms: a different approach.**
S. Aronowitz, J. Flores, T. Scattergood, S. Chang.
Bull. American Astron. Soc., Vol. 12, 693 (1980). – Abstract.

022.138 **Analysis of infrared laboratory CH_4 spectra.**
D. C. Benner, U. Fink.
Bull. American Astron. Soc., Vol. 12, 697 (1980). – Abstract.

022.139 **Experimental and theoretical investigation of the 2–0 vibrational absorption spectrum of hydrogen at below ambient temperature.**
D. Goorvitch, P. M. Silvaggio, R. W. Boese.
Bull. American Astron. Soc., Vol. 12, 706 (1980). – Abstract.

022.140 **Band modelling of laboratory methane spectra from 5500 to 6200 cm^{-1} at 110, 190, and 273 K.**
P. M. Silvaggio, R. W. Boese. L. P. Giver, D. Goorvitch.
Bull. American Astron. Soc., Vol. 12, 706 (1980). – Abstract.

022.141 **Laboratory simulations of central pit craters.**
J. Fink, R. Greeley, D. Gault.
Bull. American Astron. Soc., Vol. 12, 712 (1980). – Abstract.

022.142 **Charge transfer of multiply charged ions with hydrogen and helium: Landau-Zener calculations.**
S. E. Butler, A. Dalgarno.
Astrophys. J., Vol. 241, 838 - 843 (1980).

Estimates are presented of the rate coefficients for several astrophysically important multiply charged ionic systems undergoing charge transfer recombination with atomic hydrogen and helium at thermal energies, and for some potentially significant reverse ionization reactions. Quantal calculations are analyzed to develop procedures for estimating the collision parameters, and the Landau-Zener approximation is used to compute the rate coefficients. The emission lines produced in the decay of the reaction products are identified.

022.143 **SO_2 frost: UV-visible reflectivity and Io surface coverage.** D. B. Nash, F. P. Fanale, R. M. Nelson.
Geophys. Res. Lett., Vol. 7, 665 - 668 (1980).

The spectral reflectance of laboratory SO_2 frost in the range 0.24 - 0.85 μm was measured. The results, when compared with Io's full-disk, earth-based reflectance spectrum, suggest that SO_2 frost is not the dominant component of Io's optical surface layer. Various calculated UV-visible spectra for mixtures of SO_2 frost with other plausible surface components (sulfur allotropes and sodium sulfide) provide a good match to Io's spectrum and limit optically thick SO_2 frost coverage to less than about 20% of Io's average projected surface area.

022.144 **A laboratory study of the λ2145 Å auroral mystery feature.** P. W. Erdman, P. J. Espy, E. C. Zipf.
Geophys. Res. Lett., Vol. 7, 761 - 764 (1980).

022.145 **Beam-foil lifetimes of Be-like ions of elements from Mg to S.** E. Träbert, P. H. Heckmann.
Phys. Scr., Vol. 22, 489 - 492 (1980).

022.146 **Resonant interaction between an active molecular medium and a free electron laser.** H. Wilhelmsson.
Phys. Scr., Vol. 22, 503 - 506 (1980).

The system of a relativistic electron beam and an active molecular (maser) medium in the presence of a ripple magnetic field is studied theoretically.

022.147 **Effective O_2 absorption cross sections for solar absorption data from Lyman α ion chambers.**
M. Ilyas.
J. Geophys. Res., Vol. 85, 5113 - 5118 (1980).

Effective absorption cross sections applicable to total signal from Lyman α ion chambers at all altitudes are presented as a function of normalized signal as well as altitude (and solar zenith angle) for four solar H Lyman α flux conditions (2, 3, 4, and 6 ergs $cm^{-2} s^{-1}$) using Carver et al.'s recent σ_{O_2} data (over Lyman α) at 195 K and 294 K.

022.148 **Ground state spectroscopic constants of $H^{15}NCS$, $HN^{13}CS$, and $HNC^{34}S$, and the molecular structure of isothiocyanic acid.**
K. Yamada, M. Winnewisser, G. Winnewisser, L. B. Szalanski, M. C. L. Gerry.
J. Mol. Spectrosc., Vol. 79, 295 - 313 (1980) = Max-Planck-Gesellschaft zur Förderung der Wissenschaften, M. P. I. für Radioastronomie, Bonn, Sonderdrucke, Ser. A, No. 392.

022.149 **Die Rolle des Laboratoriumsexperiments bei der Erforschung des kosmischen Staubes.**
J. Gürtler, J. Dorschner.
Sterne, 56. Band, 300 - 314 (1980).

022.150 **Possible production of 'collapsed' hadronic matter in very-high-energy nucleon-nucleon collisions.**
A. K. Mann, H. Primakoff.
Phys. Rev. D, Vol. 22, 1115 - 1119 (1980). – Abstr. in Phys. Abstr., Vol. 84, Abstr. 586 (1981).

022.151 **Similarity solutions of a strong shock wave propagation in a mixture of a gas and dusty particles.**
S. I. Pai, S. Menon, Z. Q. Fan.
Int. J. Eng. Sci., Vol. 18, 1365 - 1373 (1980). – Abstr. in Phys. Abstr., Vol. 84, Abstr. 2024 (1981).

022.152 **A possible role for triplet H_2CN^+ isomers in the formation of HCN and HNC in interstellar clouds.**
T. L. Allen, J. D. Goddard, H. F. Schaefer III.
J. Chem. Phys., Vol. 73, 3255 - 3263 (1980). – Abstr. in Phys. Abstr., Vol. 84, Abstr. 4066 (1981).

022.153 **Relative intensities and predicted new bands of the A (E – X) and B (D – A) systems of ZrO.**
P. S. Murty.
Astrophys. Lett., Vol. 21, 17 - 20 (1980).

022.154 **Unpolarised and spin-change collisions between He^+ and H at low energies.**
C. Falcón, L. Opradolce, R. D. Piacentini.
J. Phys. B., Vol. 11, 3033 - 3038 (1978) = Inst. Astron. Fis. Espacio, Buenos Aires, Tirada Aparte No. 24.

022.155 **The O (^1S) quantum yield from O_2^+ dissociative recombination.** D. R. Bates, E. C. Zipf.
Planet. Space Sci., Vol. 28, 1081 - 1086 (1980).

022.156 **H_α and H_β emission cross sections for low-energy H and H^+ collisions with N_2 and O_2.**
B. Van Zyl, H. Neumann.
J. Geophys. Res., Vol. 85, 6006 - 6010 (1980).

Absolute cross sections for the emission of Balmer alpha and Balmer beta radiations from H and H^+ collisions with N_2 and O_2 are reported. The H and H^+ projectile energies ranged between about 50 eV and 2.5 keV. For projectile energies above 300 eV, the contributions to these radiations from decay of the long-lived 3*s* and 4*s* excited states of hydrogen were resolved. The emission cross sections for the case of H atom impact were found to be substantially larger than those for H^+ impact. The results are discussed within the context of their application to analysis of the hydrogen aurora.

022.157 **Charge transfer of doubly charged oxygen ions in helium.** A. Dalgarno, S. E. Butler, T. G. Heil.
J. Geophys. Res., Vol. 85, 6047 - 6048 (1980).

022.158 **On the derivation of electron density and temperature from [S II] and [O II] line intensity ratios.**
J. Cantó, K. H. Elliott, J. Meaburn, A. C. Theokas.
Mon. Not. R. Astron. Soc., Vol. 193, 911 - 919 (1980).

Line intensity ratios for [S II] and [O II] due to collisional de-excitation are briefly discussed. [O II] and [S II] contour plots are presented, which allow effective electron temperatures and densities to be estimated from pairs of line ratios.

022.159 **Behavior of molecules on interstellar grains: application of the Langevin equation and iterative extended Hückel.** S. Aronowitz, S. Chang.
Astrophys. J., Vol. 242, 149 - 164 (1980)

The Langevin equation was used to explore an adsorbate desorption mechanism. Calculations were performed using iterative extended Hückel on a silica model site with various small adsorbates, e. g., H, CH, OH, NO, CO. It was found that barriers to free traversal from one site to another are substantial (~3 - 10 eV). The nature of the silica grain and that of the "cold" desorption mechanism, when considered together, suggest that the abundance of very small grains might be less common than anticipated.

022.160 **Laboratory studies of isotope fractionation in the reactions of C^+ and HCO^+ with CO: interstellar implications.** D. Smith, N. G. Adams.
Astrophys. J., Vol. 242, 424 - 431 (1980).

The authors' experiments have shown that fraction of the ^{13}C and ^{18}O isotopes can occur in specific ion-molecule reactions at the lowest temperatures at present accessible in their apparatus (~ 80 K). Extrapolation of the data to the temperatures of dark and dense interstellar clouds indicates that this phenomenon can account rather well for the observed enhanced abundance of ^{13}CO and to some extent that of $H^{13}CO^+$ in these clouds.

022.161 **Intermediate coupling collision strengths for $\Delta n=0$ transitions produced by electron impact on highly charged ions. I. Theoretical development and application to $n=2$ levels in helium-like ions.** D. H. Sampson, R. E. H. Clark.
Astrophys. J., Suppl. Ser., Vol. 44, 169 - 191 (1980).

A Coulomb-Born exchange method has been extended to apply to transitions involving no change in principal quantum number n with intermediate coupling and configuration mixing effects included. The theory is used to calculate the collision strengths and electric dipole line strengths for all fine-structure transitions with $\Delta n=0$ and $n=2$ in helium-like ions with nuclear charge number Z in the range $6 \leq Z \leq 74$.

022.162 **Intermediate coupling collision strengths for $\Delta n=0$ transitions produced by electron impact on highly charged ions. II. Transitions between states of the $1s^22s^2$ and $1s^22p^2$ configurations and those of the $1s^22s2p$ configuration in beryllium-like ions.**
D. H. Sampson, R. E. H. Clark, L. B. Golden.
Astrophys. J., Suppl. Ser., Vol. 44, 193 - 213 (1980).

The theory given in the preceding paper is used to calculate the collision strengths and line strengths for all fine-structure transitions between states of the $1s^22s^2$ and $1s^22p^2$ configurations and those of the $1s^22s2p$ configuration in beryllium-like ions with nuclear charge number Z in the range $14 \leq Z \leq 74$. The results are found to be in mostly very good agreement with the data available by other means. They are expected to be useful in application to high temperature plasmas such as occur in the coronal regions of the Sun and in controlled nuclear fusion research.

022.163 **Intermediate coupling collision strengths for $\Delta n=0$ transitions produced by electron impact on highly charged ions. III. Transitions within the $1s^22s2p$ and $1s^22p^2$ configurations and between the $1s^22s^2$ and $1s^22p^2$ configurations in beryllium-like ions.** R. E. H. Clark, D. H. Sampson, A. D. Parks.
Astrophys. J., Suppl. Ser., Vol. 44, 215 - 222 (1980).

Intermediate coupling collision strengths are obtained for all fine-structure transitions within the $1s^22s2p$ and $1s^22p^2$ configurations and between the $1s^22s^2$ and $1s^22p^2$ configurations in highly charged beryllium-like ions with Z in the range $14 \leq Z \leq 74$. The results are expected to be useful in the study of high temperature plasmas such as occur in the solar coronal regions and in fusion research.

022.164 **Further analysis of the $B^2\Sigma^+ \rightarrow X^2\Sigma^+$ system of the YO molecule.** A. Bernard, R. Gravina.
Astrophys. J., Suppl. Ser., Vol. 44, 223 - 239 (1980).

The $B \rightarrow X$ system of the YO molecule has been reinvestigated. A rotational analysis is reported for seven new emission bands. Accurate rotational constants are obtained in both states, mainly from a simultaneous multiband fitting. A catalogue is given in which are listed the experimental line wavenumbers together with the differences between observed and calculated values.

022.165 **Electromagnetic transitions for the hydrogen atom in strong magnetic fields.**
G. Wunner, H. Ruder.
Astrophys. J., Vol. 242, 828 - 842 (1980).
The authors provide the relevant formulae for electromagnetic transitions in strong magnetic fields, and apply them to the hydrogen atom. For magnetic field strengths in the range 2.35×10^{11} to 4.7×10^{12} gauss, which are characteristic of neutron stars, they calculate dipole strengths, oscillator strengths, sum rules, asymptotic formulae, transition probabilities, lifetimes, and intensities.

022.166 **The spectrum of lanthanum oxide: a reanalysis of the rotational data.** A. Bernard, A. M. Sibaï.
Z. Naturforsch., Band 35a, 1313 - 1316 (1980).

022.167 **Series formulas for the spectrum of atomic sodium (Na I).** W. C. Martin.
J. Opt. Soc. America, Vol. 70, 784 - 788 (1980).

022.168 **High resolution spectrum of the N_2^+ Meinel system to 11250 Å.**
W. Benesch, D. Rivers, J. Moore.
J. Opt. Soc. America, Vol. 70, 792 - 799 (1980).

022.169 **Cr, Co, and Ni transitions isoelectronic to the Fe XXIV – Fe XVII lines around 11 Å in laser-produced plasma.**
N. Spector, A. Zigler, H. Zmora, J. L. Schwob.
J. Opt. Soc. America, Vol. 70, 857 - 861 (1980).
The laser-produced plasma spectra of Cr, Co, and Ni observed between 9 and 14 Å isoelectronic to the iron lines in the wavelength region 10.5–12 Å are measured. Spectral lines belonging to many ionization stages fall in this wavelength region. Classification of most of these lines is given, based on isoelectronic extrapolation with comparison to other works on iron.

022.170 **Oscillator strengths of neutral yttrium (Y I) from hook-method measurements in a furnace.**
B. L. Cardon, W. H. Parkinson, F. S. Tomkins.
J. Opt. Soc. America, Vol. 70, 1372 - 1375 (1980).

022.171 **Transition probabilities for allowed and forbidden transitions from the 1s2p ^{3}P levels in He-like O VII and F VIII.** L. Engström, C. Jupén, B. Denne, S. Huldt, W. T. Meng, P. Kaijser, J. O. Ekberg, U. Litzén, I. Martinson.
Phys. Scr., Vol. 22, 570 - 574 (1980).

022.172 **A critical compilation of energy levels in the configurations $2s^m 2p^k$ (m = 2, 1, 0) of F I-, O I- and Be I-like spectra.** B. Edlén.
Phys. Scr., Vol. 22, 593 - 602 (1980).

022.173 **Hyperfine splitting in the rotational levels of the C_2H molecule.** A. A. Rejtblat.
Pis'ma Astron. Zh., Tom 6, 768 - 773 (1980). In Russian. English translation in Soviet Astron. Lett., Vol. 6.
The hyperfine splittings of the rotational levels of the C_2H molecule are calculated up to $N = 10$. Line strengths and transition probabilities are given for allowed transitions with $\Delta N = 1$, $\Delta F = 0, \pm 1$.

022.174 **Atomic and molecular data.** A. Omont.
ESO workshop on methods of abundance determination for stars, (see 012.021), p. 25 - 27 (1980).

022.175 **Relative intensities of the hydrogen lines in moving media.** V. P. Grinin, N. A. Katysheva.
Izv. Krymskoj Astrofiz. Obs., Tom 62, 66 - 78 (1980). In Russian.
A calculation program for hydrogen spectra formed in a non-LTE gas with non-zero boundary conditions has been made. The limited case of collisional excitation and ionization is discussed in detail . For this there have been calculated the Lyman and Balmer decrements, the ratio $I_{L\alpha}/I_{H\alpha}$ on a large range of change of gas parameters. The influence of diffuse L_c- radiation on the ionization degree is also considered.

022.176 **Radiative cooling experiments on lunar glass analogues.** J. Arndt, K. Flad, M. Feth.
Proc. Tenth Lunar Planet. Sci. Conf., (see 012.050), p. 355 - 373 (1979).

022.177 **A simplified model for glass formation.**
D. R. Uhlmann, P. I. K. Onorato, G. W. Scherer.
Proc. Tenth Lunar Planet. Sci. Conf., (see 012.050), p. 375 - 381 (1979).

022.178 **The olivine-ilmenite thermometer.**
D. J. Andersen, D. H. Lindsley.
Proc. Tenth Lunar Planet. Sci. Conf., (see 012.050), p. 493 - 507 (1979).

022.179 **Experimental studies of nickel and chromium partitioning into olivine from synthetic basaltic melts.** H. D. Schreiber.
Proc. Tenth Lunar Planet. Sci. Conf., (see 012.050), p. 509 - 516 (1979).

022.180 **Spectra of Fe-Ti silicate glasses: implications to remote-sensing of planetary surfaces.**
D. A. Nolet, R. G. Burns, S. L. Flamm, J. R. Besancon.
Proc. Tenth Lunar Planet. Sci. Conf., (see 012.050), p. 1775 - 1786 (1979).

022.181 **Comparison of single crack propagation in lunar analogue glass and the failure strength of rocks.**
N. Soga, H. Spetzler, H. Mizutani.
Proc. Tenth Lunar Planet. Sci. Conf., (see 012.050), p. 2165 - 2173 (1979).

022.182 **Paleointensity determinations at elevated temperatures: sample preparation technique.**
L. A. Taylor.
Proc. Tenth Lunar Planet. Sci. Conf., (see 012.050), p. 2183 - 2187 (1979).

022.183 **Experimental shock metamorphism of mono- and polycrystalline olivine: a comparative study.**
J. F. Bauer.
Proc. Tenth Lunar Planet. Sci. Conf., (see 012.050), p. 2573 - 2596 (1979).

022.184 **Calculational investigation of impact cratering dynamics: early time material motions.**
F. M. Thomsen, M. G. Austin, S. F. Ruhl, P. H. Schultz, D. L. Orphal.
Proc. Tenth Lunar Planet. Sci. Conf., (see 012.050), p. 2741 - 2756 (1979).

022.185 **A material-strength model for apparent crater volume.** K. A. Holsapple, R. M. Schmidt.
Proc. Tenth Lunar Planet. Sci. Conf., (see 012.050), p. 2757 - 2777 (1979).

022.186 **Effects of lift force on ejecta transport.**
S. I. Pai, S. Menon, P. H. Schultz.
Proc. Tenth Lunar Planet. Sci. Conf., (see 012.050), p. 2779 - 2797 (1979).

022.187 **Hydrogenic emission and recombination coefficients for a wide range of temperature and wavelength.**
G. J. Ferland.
Publ. Astron. Soc. Pacific, Vol. 92, 596 - 602 (1980).
Hydrogenic emission and recombination coefficients are presented which cover extremes of both wavelength ($912 Å \leqslant \lambda \leqslant 20 \mu$) and temperature ($500 K \leqslant T_e \leqslant 2 \times 10^6 K$). These tables should facilitate interpretation of both satellite and ground-based observations of exotic gaseous nebulae.

022.188 **Übergangswahrscheinlichkeiten für das Wasserstoffatom im starken Magnetfeld.**
G. Wunner, H. Ruder.
Mitt. Astron. Ges., Nr. 50, p. 129 - 130 (1980).

022.189 **Positronium im starken Magnetfeld: Elektromagnetische Übergänge und Zerfall.**
G. Wunner, H. Herold, H. Ruder.
Mitt. Astron. Ges., Nr. 50, p. 131 - 132 (1980).

022.190 **Sind Laserübergänge in chemisch erzeugtem OH möglich?** F. J. Comes, K. H. Gericke.
Mitt. Astron. Ges., Nr. 50, p. 152 - 155 (1980).

022.191 **Light as seen by Einstein.** P. Grujić.
Vasiona, Année 28, 51 - 52 (1980). In Serbo-Croatian.

022.192 **Photodetachment and radiative attachment involving the $2p^2\ {}^3P^e$ state of H^-.**
V. L. Jacobs, A. K. Bhatia, A. Temkin.
Astrophys. J., Vol. 242, 1278 - 1281 (1980).
Cross sections for the photodetachment processes $H^-(2p^2\ {}^3P^e) + \hbar\omega \rightarrow H(2s \text{ or } 2p) + e^-$ are calculated and used in the evaluation of the rate coefficients for the inverse radiative attachment processes. The rate coefficients are compared with those for radiative attachment into the $1s^2\ {}^1S$ ground state of H^-, and it is found that the latter processes are more probable for all electron temperatures. However, radiative attachment into (and decay of) the $H^-(2p^2\ {}^3P^e)$ state may be an important source of infrared emission.

Non-linear methods of spectral analysis.
See Abstr. 003.054.

Coherence and correlation in atomic collisions.
See Abstr. 003.067.

Dielectronic recombination rates, ionization equilibrium, and radiative emission rates for calcium and nickel ions in low-density high-temperature plasmas.
See Abstr. 063.023.

High resolution IR balloon-borne solar spectra and laboratory spectra in the HNO_3 1720-cm^{-1} region: an analysis.
See Abstr. 071.047.

Construction of fundamental systems of oscillator strengths and abundance scales in the solar photosphere. Fe I.
See Abstr. 071.049.

Comparison of Viking Orbiter multispectral images and laboratory reflectance spectra of terrestrial samples.
See Abstr. 097.045.

Surface texture of Vesta from optical polarimetry.
See Abstr. 098.032.

Physical characteristics of cometary dust from optical studies. See Abstr. 102.015.

Visible spectrum of π^1 Gruis: identification of new $e\ {}^1\Pi - X\ {}^1\Sigma^+$ and ${}^1\Sigma^+ - X\ {}^1\Sigma^+$ bands of ZrO.
See Abstr. 114.055.

Collisional excitation of interstellar molecules due to H_2: linear molecules CO, OCS, SiO, HCN and HC_3N in ${}^1\Sigma$ state. See Abstr. 131.018.

Redox reactions and the optical properties of interstellar grains. See Abstr. 131.103.

The predicted $1s^2$ - $1s6p$ H^- auto-ionization resonance observed as a diffuse interstellar line at 7581 Å.
See Abstr. 131.143.

The photodissociation of interstellar CH^+.
See Abstr. 131.211.

Molecular synthesis in interstellar clouds: the radiative association reaction $H + OH \rightarrow H_2O + h\nu$.
See Abstr. 131.212.

Gas phase synthesis of amino-, cyano- and nitroso-compounds in interstellar clouds. See Abstr. 131.214.

Molecule formation in cool, dense interstellar clouds.
See Abstr. 131.217.

Charge transfer of Ne^{2+} with helium.
See Abstr. 135.021.

Erratum

022.901 **Erratum: 'Radiative lifetimes and hyperfine induced decay at the $1s2p\ 3p_2$ and $3p_0$ levels in Al XII'** [Phys. Scr., Vol. 22, 45 - 48 (1980)].
B. Denne, S. Huldt, J. Pihl, R. Hallin.
Phys. Scr., Vol. 22, 666 (1980). – See Abstr. 27.022.166.

Astronomical Instruments and Techniques

031 Astronomical Optics, Methods of Observation and Reduction

Astronomical Optics

031.001 **A simple null test for a Schmidt camera aspheric corrector.** R. V. Willstrop.
Mon. Not. R. Astron. Soc., Vol. 192, 455 - 466 (1980).

Existing accurate optical tests of the aspheric corrector of a Schmidt camera generally require auxiliary optics of good quality and large diameter. The test described here, which is believed to be original, requires only the spherical mirror which is to be used in the camera, and is accurate to the diffraction limit if the camera is not so fast that the profile of the corrector requires a term in r^6 in addition to the usual terms in r^2 and r^4. Another attractive feature of the test is that it may be carried out in yellow or red light although the corrector is to be figured for use in the blue or ultraviolet spectrum.

031.002 **Optical designs for large specialized telescopes.** A. Baranne, G. Lemaître.
C. R. Acad. Sci., Paris, Tome 291, Sér. B, 39 - 41 (1980). In French.

The authors show that with a fixed corrector at the Cassegrain focus of a telescope, it is possible to use a spherical primary. For certain specialized telescopes this leads to a low cost.

031.003 **Spotdiagramme astronomischer Spiegelsysteme.** H. Jungbluth.
Sterne Weltraum, Jahrg. 19, 305 - 307 (1980).

031.004 **Hartmann test for the 6-meter mirror of BTA *(Large Azimuthal Telescope).*** L. I. Snezhko.
Astron. Zh. Tom 57, 869 - 877 (1980). In Russian.
English translation in Soviet Astron., Vol. 24, No. 4.

The method of simultaneous processing of two hartmannograms applied in the Hartmann test of BTA is described. The results of the investigation of thermal deformations of the BTA main mirror are presented.

031.005 **X-ray mirror replication.** M. P. Kowalski, M. P. Ulmer.
Bull. American Astron. Soc., Vol. 12, 527 (1980). – Abstract.

031.006 **Scattered-light measurements of optical surfaces.** R. N. Smartt.
Proceedings of the Los Alamos Conference on Optics '79, (see 012.006), p. 58 - 66 (1979). – Abstr. in Phys. Abstr., Vol. 83, Abstr. 74463 (1980).

031.007 **Field correction of a Ritchey–Chrétien telescope at several focal ratios.** C. G. Wynne.
Mon. Not. R. Astron. Soc., Vol. 193, 7 - 13 (1980).

It is shown that both coma and astigmatism can be corrected by the addition of an afocal thin doublet lens between the secondary mirror and the focus, together with a modification to the secondary. The analysis leads to a discriminant that determines the range of two-mirror systems for which such correction is possible. Two numerical examples are given, for conversion of the 2.5-m du Pont $f/7.5$ telescope to work at $f/15$ and $f/30$.

031.008 **Optical system engineering approach to Cassegrain telescope selection, design, and tolerancing.**
J. H. Oberheuser.
Proc. Soc. Photo-Opt. Instrum. Eng., Vol. 193, (see 012.011), 27 - 33 (1979). – Abstr. in Phys. Abstr., Vol. 83, Abstr. 82199 (1980).

031.009 **Diffraction limit in coronagraphs.** A. E. DeCew, Jr.
Proc. Soc. Photo-Opt. Instrum. Eng., Vol. 193, (see 012.011), 47 - 52 (1979). – Abstr. in Phys. Abstr., Vol. 83, Abstr. 82200 (1980).

031.010 **'Chromotron', a spectral image analysis device.** M. Detaille.
Optics present and future, (see 012.012), p. 165 - 168 (1978). Abstr. in Phys. Abstr., Vol. 83, Abstr. 83683 (1980).

031.011 **Hartmann test results of the BTA second main mirror.** L. I. Snezhko.
Pis'ma Astron. Zh., Tom 6, 667 - 670 (1980). In Russian. English translation in Soviet Astron. Lett., Vol. 6.

031.012 **Ritchey-Chrétien systems with a lens corrector before the image.** G. M. Popov, M. B. Popova.
Izv. Krymskoj Astrofiz. Obs., Tom 61, 160 - 167 (1980). In Russian.

A modified Ritchey-Chrétien system with a lens corrector before the image is described. The original method of designing this system by a computer is discussed. Some systems with a plate corrector with an aspheric surface are designed.

031.013 **A universal Offner null corrector.** G. M. Popov.
Izv. Krymskoj Astrofiz. Obs., Tom 61, 168 - 172 (1980). In Russian.

The special kind of the Offner null corrector is discussed. It can be designed to test concave paraboloidal or hyperboloidal mirrors. It is possible to fit this corrector to testing mirrors with different curvatures and eccentricities.

031.014 **Hartmann tests on large telescopes carried out with a small screen in a pupil image.** B. Loibl.
Astron. Astrophys., Vol. 91, 265 - 268 (1980).

For Hartmann tests on new large telescopes an apparatus was developed and tested where a small screen is located in an image plane of the entrance pupil.

031.015 **Negen telescoopsystemen.** Vergeleken door middel van optische doorrekening.
H. G. J. Rutten, M. A. M. van Venrooij.
Zenit, 7e Jaarg., 396 - 402 (1980).

031.016 **Alguns espelhos e telescopios reflectores e um Cassegrain de 40 cm (traçagem automática).**
I. D. Simões.
Comun. Obs. Astron. Univ. Coimbra, No. 23, 50 pp. (1980).

031.017 **Tolerancias en el montaje de telescopios de dos espejos.** E. J. Campitelli.
Obs. Astron. Univ. Nac. La Plata, Ser. Especial, No. 25, 59 pp. (1979). ISSN 0325-3015.

031.018 **On a Schmidt camera with a modified collimator for use in spectrographs.** Y. Yamashita.
Tokyo Astron. Obs. Rep. (No. 73), Vol. 19, 171 - 178 (1980). In Japanese.

031.019 **On the use of transmission gratings for astronomical optics.** A. Baranne.
C. R. Acad. Sci. Paris, Tome 291, Sér. B, 205 - 207 (1980). In French.

Are transmission gratings really less luminous than reflexion gratings? A grism can be twice as luminous as an ordinary transmission grating and for certain applications can challenge the best reflexion gratings.

031.020 **Optiken für die Amateur-Astronomie.** E. Wiedemann.
Sterne Weltraum, 19. Jahrg., 411 - 418 (1980).

031.021 **Ein Vergleichstest: "Immersionsobjektiv" von Wolfgang Busch – Zeiss-B-Objektiv.**
B. Wedel.
Sterne Weltraum, 19. Jahrg., 422 - 423 (1980).

031.022 **Comparison of Wolter I and Kirkpatrick-Baez X-ray optics for a Spacelab LAMAR facility.**
R. C. Catura, W. A. Brown, L. W. Acton.
Opt. Eng., Vol. 19, 602 - 609 (1980). – Abstr. in Phys. Abstr., Vol. 84, Abstr. 4771 (1981).

031.023 **Stressed mirror polishing. 1: A technique for producing nonaxisymmetric mirrors.**
J. Lubliner, J. E. Nelson.
Appl. Opt., Vol. 19, 2332 - 2340 (1980).

The theoretical basis is developed for a technique to fabricate nonaxisymmetric mirrors. Stresses are applied to a mirror blank that would have the effect of elastically deforming a desired surface into a sphere. A sphere is then polished into the blank, and upon release of the applied stress, the spherical surface deforms into the desired one.

031.024 **Stressed mirror polishing. 2: Fabrication of an off-axis section of a paraboloid.** J. E. Nelson, G. Gabor, L. K. Hunt, J. Lubliner, T. S. Mast.
Appl. Opt., Vol. 19, 2341 - 2352 (1980).

031.025 **Adjusting spatially separated plane mirrors to coplanarity.** R. Q. Twiss, W. T. Welford.
Appl. Opt., Vol. 19, 2416 - 2418 (1980).

An interferometric technique based on a white light fringe setting was used to set mirrors defining a base plane in a Michelson stellar interferometer to coplanarity. The method is in principle capable of extension to very large separations, and the precision of visual setting is of the order of the wavelength of light.

031.026 **Centering the secondary mirror of the Tirgo telescope.** S. Guidarelli.
Appl. Opt., Vol. 19, 2520 - 2523 (1980).

A procedure for centering the secondary mirror of a Cassegrainian telescope on its mechanical mount has been devised, using the classical principle of observing the movement of a reflected image when the mirror is rotated. The centering is obtained by iteration, keeping the reflecting surface in contact with a circular centered edge. The achievable accuracy, accounting for the change in the hyperboloid shape of the mirror from a sphere, has been calculated; the theoretical values were checked using the procedure for centering the secondary mirror of the Tirgo infrared telescope.

031.027 **Testing large telescope mirrors in the optical shop by an autocollimation method with multiple pendulum flat mirrors.** N. Hu.
Appl. Opt., Vol. 19, 2680 - 2682 (1980).

Design and a compensation technique are presented to enable slope error measurements to be made via use of a pendulum having a small flat mirror mounted on the lower end. A multiplicity of such pendulum mirrors can act as a substitute for a full diameter flat for testing large astronomical primary and secondary mirrors.

031.028 **Design considerations for a 10-m optical table telescope.** A. B. Meinel, M. P. Meinel.
Appl. Opt., Vol. 19, 2683 - 2687 (1980).

A study of a telescope configuration for a 10-m segmented-mirror telescope done at National Central Universtiy, Chung-li,Taiwan, indicates that the application of the principle employed in optical tables can result in a telescope cost lying well below the scaling law for conventional telescopes. A similar design configuration could be used for a large monolithic primary mirror. The optical table geometry is relevant only to altazimuth mountings where the gravity vector remains in a vertical plane through the structure.

031.029 **Strehl number degradation by large-scale systematic surface deviations.** A. Greve.
Appl. Opt., Vol. 19, 2948 - 2951 (1980).

Experience shows that optical and radio reflector surfaces may have systematic residual deviations to which the statistical analysis is not necessarily applicable. For large-scale systematic deviations the author derives general expressions for calculating the degraded on-axis intensity, i. e., the Strehl number. For this type of deviation he introduces a quasi-rms value, which has the same optical significance as the rms value of random surface deviations. Numerical examples of realistic cases are presented.

031.030 **Design and optimization technique for three-mirror telescopes.** D. Korsch.
Appl. Opt., Vol. 19, 3640 - 3645 (1980).

A high performance design and optimization technique for three-mirror telescopes, based on the generation of rigorously aplanatic configurations, has been developed. The theory and five sample designs are presented. The work was done using an HP-System 35 desk-top computer.

031.031 **Automatic computation of optical aberration coefficients.** T. B. Andersen.
Appl. Opt., Vol. 19, 3800 - 3816 (1980).

A method based on ray tracing equations is presented facilitating the computation of polynomial functions that predict the x and y intercepts, direction tangents, and optical path length of an arbitrary skew ray on a defocused image surface. The optical systems considered may contain surfaces of general nonrotationally symmetric figure and orientation, refracting as well as reflecting, including plane diffraction gratings. The classical aberration coefficient equations are a subset of these polynomials. The method can, without hard programming work, be extended to arbitrarily high orders.

031.032 **Generalized ray tracing, caustic surfaces, generalized bending, and the construction of a novel merit function for optical design.** R.-S. Chang, O. N. Stavroudis.
J. Opt. Soc. America, Vol. 70, 976 - 985 (1980).

The authors suggest that caustic surfaces can be used as a merit function in the optical design process, that the merit functions can be calculated by means of generalized ray tracing, and that generalized bending provides an effective means of optimizing the design when included in a feedback loop.

031.033 **Distributed computing for optical design.** W. G. Peck.
J. Opt. Soc. America, Vol. 70, 1032 (1980). – Abstract.

031.034 **Current status of the SYNOPSYS lens design program.** D. C. Dilworth.
J. Opt. Soc. America, Vol. 70, 1032 (1980). – Abstract.

031.035 **New developments in code V™.** T. I. Harris.
J. Opt. Soc. America, Vol. 70, 1032 (1980).
Abstract.

031.036 **Optical design programs at Imperial College.**
P. M. J. H. Wormell, M. J. Kidger.
J. Opt. Soc. America, Vol. 70, 1032 (1980). – Abstract.

031.037 **Uniform scheme for aberration calculations and image assessment for general optical systems.**
H. H. Hopkins.
J. Opt. Soc. America, Vol. 70, 1032 - 1033 (1980). – Abstract

031.038 **Optical design with desktop computers.**
M. J. Kidger.
J. Opt. Soc. America, Vol. 70, 1033 (1980). – Abstract.

031.039 **Optical design using small dedicated computers.**
D. C. Sinclair.
J. Opt. Soc. America, Vol. 70, 1033 (1980). – Abstract.

031.040 **Dedicated minicomputers in optical design.**
D. E. Gustafson.
J. Opt. Soc. America, Vol. 70, 1033 (1980). – Abstract.

031.041 **Considerations for minicomputer application to large optical design programs.**
R. A. Arnold, R. E. Casas, J. W. Figoski.
J. Opt. Soc. America, Vol. 70, 1033 (1980). – Abstract.

031.042 **Microcomputers and optical design.** A. Cox.
J. Opt. Soc. America, Vol. 70, 1033 - 1034 (1980).
Abstract.

031.043 **TRS-80 as an optical design computer.**
R. R. Willey.
J. Opt. Soc. America, Vol. 70, 1034 (1980). – Abstract.

031.044 **Easy-to-use optical program for Commodore "PET" computer.** J. A. Gibson.
J. Opt. Soc. America, Vol. 70, 1034 (1980). – Abstract.

031.045 **New approach to the optimization of lens systems.**
M. Harrigan.
J. Opt. Soc. America, Vol. 70, 1035 (1980). – Abstract.

031.046 **Ten years of lens design with Glatzel's adaptive method.** J. L. Rayces.
J. Opt. Soc. America, Vol. 70, 1035 (1980). – Abstract.

031.047 **Orthogonal polynomials as lens-aberration coefficients.** D. S. Grey.
J. Opt. Soc. America, Vol. 70, 1035 (1980). – Abstract.

031.048 **Aberration balancing criterion for non-diffraction-limited lenses.** B. Tatian.
J. Opt. Soc. America, Vol. 70, 1035 (1980). – Abstract.

031.049 **Application of Zernike polynomial lens sensitivity program ZEST to "optimizing" a lens design.**
D. Friedman.
J. Opt. Soc. America, Vol. 70, 1035 (1980). – Abstract.

031.050 **Solution of block-structured least-squares problems.**
C. R. Crawford.
J. Opt. Soc. America, Vol. 70, 1035 - 1036 (1980). – Abstract.

031.051 **Lens design using optical aberration coefficients.**
P. N. Robb.
J. Opt. Soc. America, Vol. 70, 1036 (1980). – Abstract.

031.052 **Method of lens design utilizing Seidel's coefficients.**
F. Kondoh.
J. Opt. Soc. America, Vol. 70, 1036 (1980). – Abstract.

031.053 **Pursuit of symmetry in wide-angle reflective optical designs.** I. R. Abel, M. R. Hatch.
J. Opt. Soc. America, Vol. 70, 1040 (1980). – Abstract.

031.054 **Family of flat-field meniscus corrector catadioptric lenses of the Maksutov type.** J. J. Villa.
J. Opt. Soc. America, Vol. 70, 1040 - 1041 (1980). – Abstract.

031.055 **Review of lens design using gradient index materials.**
D. T. Moore.
J. Opt. Soc. America, Vol. 70, 1042 (1980). – Abstract.

031.056 **Nonstatistical method to evaluate tolerances of an optical system.** G. Pinto.
J. Opt. Soc. America, Vol. 70, 1043 (1980). – Abstract.

031.057 **Monochromator designs with aberration-corrected gratings.** R. Chipman.
J. Opt. Soc. America, Vol. 70, 1043 (1980). – Abstract.

031.058 **HP-41C program to trace skew rays through systems with conic surfaces.** D. S. Nicholson.
J. Opt. Soc. America, Vol. 70, 1058 (1980). – Abstract.

031.059 **General ray tracing with a pocket calculator.**
D. E. Stoltzmann.
J. Opt. Soc. America, Vol. 70, 1060 (1980). – Abstract.

031.060 **BASIC language meridional ray trace.** L. Larks.
J. Opt. Soc. America, Vol. 70, 1063 (1980).
Abstract.

031.061 **Optical system of a new chromospheric telescope.**
Yu. A. Klevtsov, V. D. Trifonov.
Issled. po geomagn., aehron. i fiz. Solntsa, Moskva, 1980, No. 52, p. 71 - 75. In Russian. – Abstr. in Ref. zh., 51. Astron., 11.51.141 (1980).

031.062 **Two-mirror aberration-free systems with one mirror aspherical surface known.** G. M. Popov.
Izv. Krymskoj Astrofiz. Obs., Tom 62, 183 - 189 (1980). In Russian.

Described is a novel and exact method of determining the figure of one of the mirrors of an aberration-free two-mirror system with known figure of the other mirror.

031.063 **On the dependence of the coefficients of reflection for mirrors coated with a layer of $Al+Al_2O_3$ on the Al_2O_3 layer thickness in the spectral range 1700 - 10000 Å.**
S. V. Fedulov, V. A. Volosnov, G. S. Dyatlova.
Izv. Krymskoj Astrofiz. Obs., Tom 62, 190 - 192 (1980). In Russian.

031.064 **Active mirrors for ground-based telescopes.**
J. W. Hardy.
Publ. Astron. Soc. Pacific, Vol. 92, 547 (1980). – Abstract.

031.065 **Design of optical mirror elements of Fresnel.**
Yu. K. Vifanskij.
Opt.-mekh. prom-st', 1980, No. 6, p. 17 - 20. In Russian. Abstr. in Ref. zh., 51. Astron., 12.51.85 (1980).

031.066 **Investigation of the distribution of light in the focal plane of the AFU-75 objective.**

Yu. V. Sizonenko.
Astrometr. Astrofiz., Vyp. (No.) 42, p. 99 - 102 (1980). In Russian.

Aspheric surfaces in astronomical optics.
See Abstr. 003.092.

Methods of investigation in astronomical optics.
See Abstr. 003.108.

Adaptive optics. Collected articles.
See Abstr. 003.109.

A Dollond–Wollaston telescope.
See Abstr. 004.008.

Minimum-cost 4-m telescope developed at October 1979 Nanjing study of telescope design and construction.
See Abstr. 032.027.

Improved Wadsworth mounting with aspherical holographic grating. See Abstr. 034.052.

Methods of Observation and Reduction

031.501 **Voyager image processing at the Image Processing Laboratory.** P. L. Jepsen, J. A. Mosher, G. M. Yagi, C. C. Avis, J. J. Lorre, G. W. Garneau.
J. British Interplanet. Soc., Vol. 33, 315 - 322 (1980).

031.502 **Analysis of astronomical images using moments.** R. S. Stobie.
J. British Interplanet. Soc., Vol. 33, 323 - 326 (1980).

031.503 **Quantitative analysis of autoradiographic image intensification using Thiourea-S^{35}.**
B. S. Askins, C. R. O'Dell.
Publ. Astron. Soc. Pacific, Vol. 92, 375 - 381 (1980).
Photographic images enhanced by the method of Thiourea-S^{35} autoradiography are evaluated in terms of signal-to-noise ratio, detective quantum efficiency (DQE), and Wiener spectrum analysis using digitized images. It is determined that the original signal-to-noise ratio is not degraded by the intensification process which allows an increase in the practical working DQE as a function of density. These results apply at all spatial frequencies that were tested. The advantage given by autoradiography is the ability to produce usable images from emulsions originally exposed to the low densities corresponding to the maximum DQE and movement of faint image densities above the level of the threshold for detection.

031.504 **Full aperture solar photography for reflectors.** S. D. Ringwood.
J. British Astron. Assoc., Vol. 90, 429 - 433 (1980).

031.505 **The use of colour filters in visual planetary observation.** J. H. Robinson.
J. British Astron. Assoc., Vol. 90, 434 - 437 (1980).

031.506 **Image reconstruction methods in radio astronomy.** C. R. Subrahmanya.
Bull. Astron.Soc. India, Vol. 8, 5 - 13 (1980).
An observation of a celestial source generally results in data which are insufficient for a unique reconstruction of the desired brightness profiles of the source. Two typical situations are considered here which pertain to the reconstruction of an object from the measurements of (A) an incomplete set of its Fourier components, and (B) of a diffraction limited image of the object.

031.507 **Processing of ground-based astronomical images obtained using photon-counting techniques.**
D. D. Walker, M. L. Kendall.
J. British Interplanet. Soc., Vol. 33, 347 - 351 (1980).
Digital images of faint astronomical objects have been obtained with the UCL Image Photon Counting System. This paper describes the manner in which the images were processed, using the image processing facilities at the Appleton Laboratory, and the interactive data reduction facility at UCL.

031.508 **Image processing for the ESA Faint Object Camera.** P. Norris.
J. British Interplanet. Soc., Vol. 33, 352 - 356 (1980).
The Faint Object Camera (FOC) is Europe's contribution to the scientific instrument complement on NASA's Space Telescope (ST), which is currently scheduled for a late-1983 Shuttle launch. After a brief description of the instrument and its operating modes, its scientific objectives are summarised. The basic data processing required to remove instrument signature effects from the FOC images is then discussed. More subtle errors in the data are then presented and the techniques for their removal identified. In general, these secondary errors require interaction between the observer and the computer for their complete elimination.

031.509 **Calibration and image processing software for the Space Telescope Faint Object Camera.** E. Golton.
J. British Interplanet. Soc., Vol. 33, 357 - 360 (1980).
Software is being written for ESA in order to process calibration data and perform routine corrections on images from the Space Telescope Faint Object Camera. Methods to be used for calibrating the detector and also the complete camera are described. An explanation is given of the methods proposed for analysing calibration data, and the image processing software techniques developed for this task and for routine image correction are described.

031.510 **Completeness and reliability of star counts from microphotometer scannings.** G. M. Richter.
Astron. Nachr., Band 301, 157 - 163 (1980).
If three plates of the same field are scanned and star catalogues are compiled from these scans then completeness and reliability of the catalogues and an estimate of the true star numbers can be derived as functions of the brightness or other parameters down to the plate limit. An application is described.

031.511 **Possibilities of photometric investigation of globular clusters by means of photographic equidensity curves.** A. A. Strugatskaya, W. Högner, R. Ziener.
Astron. Nachr., Band 301, 189 - 190 (1980). In German.
A method of estimating the surface-brightness of globular clusters with equidensity curves is described.

031.512 **Die Beobachtung des Siriusbegleiters mit einem Spezial-Kameraansatz.** B. Wedel.
Sterne Weltraum, Jahrg. 19, 303 - 304 (1980).

031.513 **A New Zealander makes WOPAI (*Williams Occultation Position Angle Indicator*).** H. O. Williams.
Occultation Newsl., Vol. 2, 98 (1980).

031.514 **Global astrometry by space techniques.** J. Kovalevsky.
Celestial Mech., Vol. 22, (see 012.002), 153 - 163 (1980).

The European Space Agency project of an astrometric satellite – HIPPARCOS – is shortly described. It will measure the angles between stars situated in fields separated by about 70°. The precision of the elementary measurements is expected to be of the order of 0".005. A similar accuracy is found to apply to the basic reduction giving the abscissae of stars referred to great circles on the sky. The final overall reduction should yield accuracies better than 0".002 in position and parallaxes and 0".002 per year in proper motions. The main features of the final catalogue are described and some possible consequences for fundamental astronomy are given.

031.515 **A new method to restore the spatial distribution of stars in globular clusters and its application to flare stars in the Pleiades.** E. L. Kosarev.
Pis'ma Astron. Zh., Tom 6, 408 - 413 (1980). In Russian. English translation in Soviet Astron. Lett., Vol. 6.

A new method to restore the spatial star distribution in globular clusters is presented. The method gives both an estimate of the unknown spatial distribution and the restoring error. This error has statistical origin and depends only on the number of stars in the cluster. The method is applied to restore the spatial density of 441 flare stars in the Pleiades.

031.516 **Results of basic improvements to the extraction of spectra from IUE images.** D. Lindler, R. Bohlin.
Bull. American Astron. Soc., Vol. 12, 459 - 460 (1980). Abstract.

031.517 **A new method for measuring the solar radius.** T. L. Duvall, Jr., H. P. Jones.
Bull. American Astron. Soc., Vol. 12, 474 (1980). – Abstract.

031.518 **Critique of singular value analysis.** R. L. Branham, Jr.
Bull. American Astron. Soc., Vol. 12, 510 (1980). – Abstract.

031.519 **A method for determining the distribution of plasma loop parameters in solar coronal active regions.** R. G. Teske, E. B. Mayfield.
Bull. American Astron. Soc., Vol. 12, 519 (1980). – Abstract.

031.520 **Picture processing of the globular cluster NGC 6712.** D. H. Martins, C. A. Harvel.
Bull. American Astron. Soc., Vol. 12, 523 (1980). – Abstract.

031.521 **Application of digital image processing techniques to faint solar flare phenomena.**
D. L. Glackin, S. F. Martin.
Bull. American Astron. Soc., Vol. 12, 525 (1980). – Abstract.

031.522 **Two dimensional Fourier analysis of galactic images.** W. Krakow, P. E. Seiden, J. M. Huntley.
Bull. American Astron. Soc., Vol. 12, 528 (1980). – Abstract.

031.523 **Stellar velocity fields and the spectroscopic detectability of planetary systems.** D. Deming.
Bull. American Astron. Soc., Vol. 12, 540 (1980). – Abstract.

031.524 **Low-cost high-resolution spectral measurements of solar and atmospheric lines using a commercially available scanning interferometer.**
R. E. Bruce, Jr., J. Elon Graves.
Interferometry, (see 012.007), p. 235 - 238 (1979). – Abstr. in Phys. Abstr., Vol. 83, Abstr. 77298 (1980).

031.525 **Estimate of the spectral density of images with the help of a coherent-optical spectrum analyser.**
V. N. Dudinov, V. S. Tsvetkova, S. G. Kuz'menkov, V. V. Konichek.
Vestn. Khar'k. Univ., No. 190. Fiz. Luny Planet. Fundam. Astrometr., Vyp. 14, p. 16 - 24 (1979). In Russian.

031.526 **On an approximation of photoelectric profiles of star images.** D. V. Dimitrov.
Vestn. Khar'k. Univ., No. 190. Fiz. Luny Planet. Fundam. Astrometr., Vyp. 14, p. 24 - 27 (1979). In Russian.

031.527 **Magnetic field measurements on stellar sources: a new method.** R. D. Robinson, Jr.
Astrophys. J., Vol. 239, 961 - 967 (1980).

The standard method for measuring stellar magnetic fields involves the deduction of Zeeman splitting through the interpretation of circular polarization measurements of magnetically sensitive spectral lines. This technique has two serious drawbacks. (*a*) Effects from oppositely directed field elements will cancel, so that only residual fields are measured, and (*b*) dilution of the signal by nonmagnetic areas on the stellar surface greatly reduces the measured field strengths. To overcome these problems the author has developed a technique for measuring field strengths without using polarization measurements. Basically, this method deduces Zeeman splitting by accurately measuring changes in line shape caused by magnetic fields. This is done by comparing the profile from a magnetically sensitive line to that of a magnetically insensitive line having similar spectral characteristics. The advantage of this method is that it measures both absolute values of field strength as well as the fraction of the visible hemisphere of the star covered by the fields. The principal disadvantage is that the technique is insensitive to small fields. In most instances, only field strengths greater than 1000 gauss can be detected.

031.528 **The determination of stellar turbulence by low resolution techniques.**
R. Głębocki, A. Stawikowski.
Stellar turbulence, (see 012.008), p. 55 - 74 (1980).

031.529 **Analysis of high resolution stellar line profiles.** D. F. Gray.
Stellar turbulence, (see 012.008), p. 75 - 84 (1980).

031.530 **The determination of the own photometric system and the passing at the international system.**
G. D. Chiş.
Proceedings of the colloquium of astronomy, Cluj-Napoca, (see 012.009), p. 86 - 90 (1978).

031.531 **An inversion technique for the determination of velocity fields from spectral line profiles.**
A. Kavetsky, B. J. O'Mara.
Proc. Astron. Soc. Australia, Vol. 3, 325 - 326 (1979).

031.532 **Methods for calculating circular polarisation in magnetic white dwarfs.** B. Martin, D. T. Wickramasinghe.
Proc. Astron. Soc. Australia, Vol. 3, 351 - 352 (1979).

031.533 **Analysis of multiple imagery at Jet Propulsion Laboratory's Image Processing Laboratory.**
W. B. Green, N. A. Bryant, P. L. Jepsen, R. G. McLeod, J. A. Mosher, R. H. Selzer, W. D. Stromberg, G. M. Yagi, A. L. Zobrist.

Opt. Eng., Vol. 19, 168 - 179 (1980). – Abstr. in Phys. Abstr., Vol. 83, Abstr. 82219 (1980).

031.534 **Recent results in high-resolution astronomical imaging.** R. L. Frost.
Proceedings of the 1980 IEEE International Symposium on Circuits and Systems, Part II, held at Houston, Tex., USA, 28 - 30 April 1980. IEEE, New York, USA (1980). p. 505 - 509. – Abstr. in Phys. Abstr., Vol. 83, Abstr. 82225 (1980).

031.535 **Rigorous treatment of stellar aberration.** P. Stumpff.
Mitt. Astron. Ges., Nr. 48, (see 012.015), p. 87 (1980).

031.536 **The use of a large-aperture radio system for meteor studies.** R. W. Herring, P. A. Forsyth.
Canadian J. Phys., Vol. 58, 779 - 787 (1980). – Abstr. in Phys. Abstr., Vol. 83, Abstr. 85870 (1980).

031.537 **A short discourse on planetary photography.** J. Sanford.
Strolling Astron., Vol. 28, 129 - 134 (1980).

031.538 **A general theoretical method for star pointing with the coelostat of the Meudon solar tower.**
R. Freire.
J. Opt. (*France*), Vol. 11, 195 - 200 (1980). – Abstr. in Phys. Abstr., Vol. 83, Abstr. 90338 (1980).

031.539 **Adjusting the orientation of a polar axis.** R. L. Waterfield.
J. British Astron. Assoc., Vol. 90, 529 - 538 (1980).

The author describes some modifications and additions to E. S. King's method for adjusting the orientation of a polar axis by measuring the lengths and directions of star-trails near the pole.

031.540 **An astrometric test of the doublet prime-focus corrector on the Anglo-Australian Telescope, and the optical position of OR 103.** A. N. Argue, C. Sullivan.
Observatory, Vol. 100, 152 - 154 (1980).

031.541 **Method for the determination of density and phase functions of interplanetary dust.**
A. Mujica, G. López, F. Sánchez.
Solid particles in the solar system, (see 012.019), p. 55 - 60 (1980).

A method for the determination of the scattered light intensity by a unit-volume of interplanetary space is presented. From ground based zodiacal light measurements and experimental results of Pioneer X the density and phase functions are obtained without any previous assumptions about them.

031.542 **Method of scattering plane scanning.** J. Buitrago, P. Alvarez, G. López, A. Mujica, F. Sánchez.
Solid particles in the solar system, (see 012.019), p. 61 - 65 (1980).

The number density of interplanetary dust particles as a function of altitude above the symmetry plane can be determined from ground-based measurements by a new method.

031.543 **Inversion of the zodiacal brightness integral: a new geometric approach suitable for the out-of-ecliptic zodiacal light programme.**
R. Dumont, A.-C. Levasseur-Regourd.
Solid particles in the solar system, (see 012.019), p. 67 - 70 (1980).

The information provided by measurements of the zodiacal brightness, Z, is discussed as a function of the looking direction, with respect to the velocity vector of the observer through the solar system. In addition to a few directions which allow inversion of the brightness integral completely free of assumptions, it is shown that the orbital plane gives inverting possibilities, based on a single assumption on the heliocentric dependences of the volume scattering efficiency. Important features of the zodiacal cloud distribution, very difficult to obtain from earthbound observations, will easily be derived from the data of the out-of-ecliptic zodiacal light experiment.

031.544 **The analysis of meteor data.** W. J. Baggaley.
Solid particles in the solar system, (see 012.019), p. 109 - 110 (1980).

A problem commonly encountered in meteor studies is the need for a realistic interpretation of statistical data which contains uncertainties in two variables. A simple solution is provided and for illustration applied to diffusion-height data.

031.545 **Methods for accurate photographic stellar spectrophotometry using the solar spectrum as calibration.**
J. Lind, D. Dravins.
Astron. Astrophys., Vol. 90, 151 - 162 (1980).

Methods for photographic spectrophotometry using single-pass spectrographs are developed with the aim of obtaining stellar spectra of sufficiently high quality to permit detailed spectral line studies over extended wavelength regions. The spectrograph instrumental profile and photographic development effects are studied and the corresponding modulation transfer functions (*MTF*) determined by measuring the modulation undergone by a calibration spectrum of sky- or moonlight which is exposed side by side with the stellar spectrum on each plate.

031.546 **Multitemperature analysis of solar X-ray line emission.**
J. Sylwester, J. Schrijver, R. Mewe.
Sol. Phys., Vol. 67, 285 - 309 (1980).

The authors propose and test a new method of multitemperature analysis of solar X-ray spectra. The method is based on a technique developed by Withbroe (1975). Various tests of the method on simulated temperature models establish its usefulness, generality and stability. The possibilities of deriving the relative element abundances are analysed.

031.547 **Method of registration of the gravitational radiation from extraterrestrial sources in the kHz frequency band.** A. I. Gusak, V. M. Sidorov.
Astron. Zh., Tom 57, 1082 - 1093 (1980). In Russian. English translation in Soviet Astron., Vol. 24, No. 5.

The possibility of detecting gravitational radiation from an extraterrestrial source with the system "earth and high-frequency gravimetre on its surface" is considered.

031.548 **Radial velocities of faint stars from objective prism plates.** J. Stock, W. Osborn.
Astron. J., Vol. 85, 1366 - 1378 (1980).

A simple method by which the approximate radial velocity of a star may be obtained from objective prism plates is described in detail. The method has been used to derive the velocities of 41 faint stars, most of which have metal-weak spectra and are shown to be high-velocity objects. It is shown that this method may be generalized so that the radial velocities for all measurable stars on an objective prism plate can be obtained.

031.549 **Astrometric analyses of 14 Sproul plate series.** J. L. Hershey.
Astron. J., Vol. 85, 1399 - 1402 (1980).

Parallaxes are presented for 15 stars of spectral type G, K, and M, with internal errors typically one-third to nearly one-fifth of the errors given in current catalogs. Two central control stars suggest that the external errors are nearly as small as the internal errors. Most of the plate material is from the past 15 yr and shows high telescope stability with yearly mean residuals as low as 5 milliarcsec (s. e.) in some series.

031.550 **The application of a coherent optical data processing system to photographically recorded astronomical spectra.** B. Bates, D. L. Giaretta, P. J. P. Sweeney.
Astron. Astrophys., Vol. 90, 318 - 323 (1980).
An outline is given of a simple optical data processing system which has been used to provide image enhancement in a photographic astronomical spectrum. The technique has been applied in particular to the examination of the velocity structure of the interstellar Mg I line towards κ Ori recorded during a balloon flight.

031.551 **Measuring stellar temperatures. (Part I).** J. B. Hearnshaw.
South. Stars, Vol. 28, 109 - 117 (1980).

031.552 **La observación del Sol.** Complemento de la 5ª parte, filtros de objetivos. 6ª parte, evaluación de la actividad del Sol. A. Papetti.
Rev. Astron., No. 212, p. 9 - 11 (1980).

031.553 **Automatic technique for accurately locating planet centers in Voyager images.** J. J. Lorre.
Image understanding systems. II. Conference held at San Diego, Calif., USA, 29 - 30 August 1979. Proc. Soc. Photo-Opt. Instrum. Eng., Vol. 205, 175 - 179 (1979). – Abstr. in Phys. Abstr., Vol. 83, Abstr. 101534 (1980).

031.554 **Statistical reconstruction of a Martian scene; G-mode cluster analysis results from multispectral data population.** M. Poscolieri.
Mem. Soc. Astron. Italiana, Vol. 51, 309 - 328 (1980).

031.555 **The use of objective prism plates from the UK Schmidt telescope for low resolution spectrophotometry of quasars.** R. G. Clowes, D. Emerson, M. G. Smith, P. T. Wallace, R. D. Cannon, A. Savage, A. Boksenberg.
Mon. Not. R. Astron. Soc., Vol. 193, 415 - 426 (1980).
The paper describes a simple technique for obtaining low-resolution, conventional spectra of quasars from microdensitometer tracings of their objective prism spectra. Comparison with IPCS spectra indicates the capabilities of the technique. Redshifts and equivalent widths can be determined with typical discrepancies of 1 and 40 per cent respectively.

031.556 **Method of single parameter to search for and identify artificial satellites in the atmosphere.**
P.-x. Xu.
Acta Astron. Sinica, Vol. 21, 293 - 296 (1980). In Chinese.
In this method the indefinite influence of air drag on the motion of an artificial satellite is simply represented with one parameter, which enables us to search for and identify the satellite in a simple way, when the satellite is missed owing to the influence of air drag.

031.557 **Iterative method applied to image reconstruction and to computer-generated holograms.**
J. R. Fienup.
Opt. Eng., Vol. 19, 297 - 305 (1980). – Abstr. in Phys. Abstr., Vol. 83, Abstr. 102634 (1980).

031.558 **Problems of single-frequency solar observations.** K. M. Borkowski.
Postępy Astron., Tom 28, 175 - 183 (1980). In Polish.
Besides a short historical background of the narrow-band radio observations of the Sun the article deals with observational and data reduction problems met primarily in metric wavelenghts observations.

031.559 **The effect of turbulence on stellar interferometer operation.** A. A. Tokovinin.
Pis'ma Astron. Zh., Tom 6, 730 - 734 (1980). In Russian. English translation in Soviet Astron. Lett., Vol. 6.
An estimate of the average fringe visibility under Kolmogorov turbulence of the atmosphere is derived.

031.560 **New experimental possibilities and the future at far IR wavelengths.** C. H. Townes.
Interstellar molecules, (see 012.033), p. 637 - 644 (1980).
The observation of molecular phenomena in interstellar clouds by spectroscopy in the longer wavelength region of the infrared is discussed. This is a region where sensitivity is almost universally affected by background continua, appropriate techniques may be either coherent or incoherent, and work above most of the atmosphere is required for much of the spectral range.

031.561 **The search for the evolute flash at Jupiter.** J. M. Martin.
Bull. American Astron. Soc., Vol. 12, 684 (1980). – Abstract.

031.562 **Turbulence in deep occultations.** B. S. Haugstad.
Bull. American Astron. Soc., Vol. 12, 684 (1980). – Abstract.

031.563 **Correlation spectrometry by pressure modulation: a technique for very high spectral resolution remote sensing of planetary atmospheres from spacecraft.**
D. J. McCleese, J. V. Martonchik.
Bull. American Astron. Soc., Vol. 12, 702 (1980). – Abstract.

031.564 **A comparison of relative astrometric techniques as applied to minor planets.**
P. D. Hemenway, R. L. Duncombe.
Bull. American Astron. Soc., Vol. 12, 743 (1980). – Abstract.

031.565 **Method for calculating an object's position using two reference stars.** C. McEldery.
Minor Planet Bull., Vol. 7, 29 - 30 (1980).

031.566 **Fourier transform enhanced photography of the M 51 system.** M. S. Burkhead, W. Matuska.
AAS Photo-Bull., Issue 23, No. 1, p. 13 - 15 (1980).
Fourier transform image enhancement techniques and results for digitized photographic data are presented and compared with photographic techniques.

031.567 **X-ray astronomical spectroscopy.** S. S. Holt.
Vistas Astron., Vol. 24, 301 - 317 (1980).

031.568 **Radioastronomische waarneemtechnieken: 1950 - 1980.** J. van Nieuwkoop.
Zenit, 7e Jaarg., 416 - 422 (1980).

031.569 **Iterative method applied to image reconstruction and to computer-generated holograms.**
J. R. Fienup.
Proc. Soc. Photo-Opt. Instrum. Eng., Vol. 207, (see 012.036), 2 - 13 (1979). – Abstr. in Phys. Abstr., Vol. 83, Abstr. 106475 (1980).

031.570 **Astronomical speckle interferometry and speckle holography.** G. P. Weigelt.
Optics present and future, (see 012.012), p. 551 - 554 (1978). Abstr. in Phys. Abstr., Vol. 83, Abstr. 109373 (1980).

031.571 **An approach to the reduction of digital spectral data.** J. Straede.
Anglo-Australian Obs. Prepr. No. 135, 8 pp. (1980).
Reduction of digital spectral data can be facilitated by the provision of a system which isolates, as far as possible, the

user from the non-astronomical aspects of computer programming. A successful system, implemented at the Anglo-Australian Observatory, is described and its short-comings discussed with suggestions as to how these might be avoided in future implementations.

031.572 **Basic experiments for absolute intensity measurements in the VUV region by the photon counting method.** K. Higashi, K. Nishi, A. Yamaguchi.
Tokyo Astron. Obs. Rep. (No. 73), Vol. 19, 232 - 268 (1980).
In Japanese.

031.573 **Improved correction for millimeter-wavelength atmospheric attenuation.** B. L. Ulich.
Astrophys. Lett., Vol. 21, 21 - 28 (1980).

The standard chopper-wheel method of calibrating millimeter-wavelength corrected antenna temperature data has several deficiencies which require significant corrections to produce repeatable measurements. Reducing the chopper-wheel brightness temperature below ambient temperature can result in a simplified calibration equation and a more accurate correction for atmospheric attenuation. An empirical calibration procedure is described to determine the effective temperature of a cooled chopper which is nearly at the mean atmospheric temperature.

031.574 **Wind influence on determination of clock correction.** M. Lehmann.
Publ. Inst. Geophys., Polish Acad. Sci., F-6 (137), p. 47 - 57 (1980).

One of the factors influencing results of astrometric observations is the wind effect. Occurrence of the wind effect in time observations performed by a visual transit instrument for the period 1968-1971 in Borowiec was studied. An influence of the wind direction was observed. The order of values of the wind influence is below 0^s012.

031.575 **Note sur la réduction des observations photographiques des satellites artificiels à l'aide des ordinateurs.** M. Trifu.
Visual Obs. Artif. Earth Satell., p. 126 - 129 (1977) = Publ. Obs. Astron. Univ. Cluj-Napoca, Romania, Nr. B 46-58.

031.576 **Astrophotography for beginners: III.**
G. S. Pearce, R. Scagell.
J. British Astron. Assoc., Vol. 91, 26 - 40 (1980).

031.577 **Measurement of areas on the solar disk.**
L. M. Dougherty.
J. British Astron. Assoc., Vol. 91, 75 - 78 (1980).

A simple technique using a prepared graticule, scaled to an associated projected solar disk diameter, is presented. Some comments on the precision of the method are offered. Instructions for the construction of a suitable graticule are given.

031.578 **Astrophotographie mit einfachen Adaptern.**
J. Dubiel, J. Emmert.
Sterne Weltraum, 19. Jahrg., 424 (1980).

031.579 **Fast photometry – new facilities at La Silla.**
H. Pedersen.
Messenger, No. 22, p. 16 - 17 (1980).

031.580 **Die Beobachtung der veränderlichen Sterne.**
A. Gautschy.
Orion, 38. Jahrg., Sondernr., p. 5 (1980).

031.581 **Astrophotography.** R. Malmstrom.
British J. Photogr., Vol. 127, 680 - 683 (1980).
Abstr. in Phys. Abstr., Vol. 84, Abstr. 9381 (1981).

031.582 **An interactive method for the reduction of echelle spectrograms.**
G. A. De Biase, S. Gaudenzi, R. Lombardi, C. Rossi.
Mem. Soc. Astron. Italiana, Vol. 51, (see 012.040), 399 - 404 (1980).

An interactive method for the analysis and reduction of echellograms is described. The method is sufficiently independent of the spectrum shape and allows the employment of almost all information recorded on the data array.

031.583 **Interactive reduction of solar flare spectra.**
E. Acampa, R. Falciani, A. M. Sambuco, L. A. Smaldone.
Mem. Soc. Astron. Italiana, Vol. 51, (see 012.040), 423 - 436 (1980).

031.584 **Interactive methods for feature analysis in astronomical images.**
M. L. Malagnini, G. L. Sicuranza.
Mem. Soc. Astron. Italiana, Vol. 51, (see 012.040), 443 - 450 (1980).

The paper describes some interactive procedures used for morphological analysis and for the identification of structural characteristics which help to discriminate different categories of images.

031.585 **The "ELIA" package with some applications to astronomical image processing.**
G. Buscema, G. Iannicola, D. Nanni, G. Pittella, D. Trevese, A. Vignato.
Mem. Soc. Astron. Italiana, Vol. 51, (see 012.040), 451 - 461 (1980).

The "ELIA" (Elaborazione Immagini Astronomiche) software package is described. This package was written at the Observatory of Rome and is devoted to astronomical image processing, with special regard to the automated multiobjects photometry.

031.586 **Preliminary estimates and corrections in the two-dimensional spectroscopy of the solar chromospheric active regions.** F. Caroti Ghelli.
Mem. Soc. Astron. Italiana, Vol. 51, (see 012.040), 463 - 472 (1980).

031.587 **Interactive approach to integrated photometry of globular clusters.**
R. Buonanno, G. Buscema, G. A. De Biase, G. Iannicola, R. Lombardi.
Mem. Soc. Astron. Italiana, Vol. 51, (see 012.040), 473 - 481 (1980).

The contribution of the resolved most luminous stars is eliminated in order to study the colour gradient of the less luminous stars as a whole.

031.588 **Automatic stellar photometry on photographic plates.**
R. Buonanno, G. Buscema, C. Corsi, G. Iannicola.
Mem. Soc. Astron. Italiana, Vol. 51, (see 012.040), 483 - 489 (1980).

031.589 **On the shape of the innermost isophotes of galaxies.**
M. Capaccioli, R. Rampazzo.
Mem. Soc. Astron. Italiana, Vol. 51, (see 012.040), 491 - 496 (1980).

The problem of the effect of seeing on the shape of extended sources is discussed. Results of numerical simulations of seeing convolution on a simple model of an elliptical galaxy are presented. They show that at least part of the observed rounding-off of the innermost isophotes of galaxies can be accounted for by seeing convolution. A specific application to the case of the elliptical galaxy NGC 3379 is also given.

031.590 **An analytical approximation to the problem of spatial deprojection of rotation curves of elliptical galaxies.** M. Capaccioli.
Mem. Soc. Astron. Italiana, Vol. 51, (see 012.040), 497 - 501 (1980).

A simple analytical recipe is provided to perform the line of sight deprojection of the rotation curve of an elliptical galaxy under the assumptions that the galaxy is axially symmetric, seen edge-on and fully transparent.

031.591 **Analysis of ellipticity and twisting of the isophotes of some bright galaxies in Virgo using the INMP interactive numerical mapping package.**
R. Barbon, L. Benacchio, M. Capaccioli, G. A. De Biase, P. Santin, G. Sedmak.
Mem. Soc. Astron. Italiana, Vol. 51, (see 012.040), 503 - 504 (1980). – Abstract.

031.592 **A "clustering" method applied to the analysis of sky maps in gamma-ray astronomy.**
V. Di Gesu, B. Sacco, G. Tobia.
Mem. Soc. Astron. Italiana, Vol. 51, (see 012.040), 517 - 528 (1980).

031.593 **On the linearity of electronography.** S. Jeffers.
Astron. Astrophys., Vol. 92, 196 - 199 (1980).

Absolute sensitometry of the electronographic emulsions XM, G_5, and L_4 is reported for a range of incident electron energies (5 - 35 kV). All these emulsions are found to exhibit nonlinear response to incident electron flux. For a given electron energy, the emulsion characteristic curve can be fitted to an equation of the form $D = D_s(1-e^{-AE})$ where D = density, D_s = saturation density, A = average area of developed grain, E = exposure.

031.594 **Zur Reduktion photographischer Positionen von kleinen Planeten und Kometen.**
M. Gressmann.
Sterne, 56. Band, 364 - 370 (1980).

031.595 **Effects influencing the exactness of visual observations of variable stars.** A. Pliska.
Říše hvězd, Vol. 61, 187 - 191 (1980). In Czech.

031.596 **Speckle processing gives diffraction-limited true images from severely aberrated instruments.**
F. M. Cady, R. H. T. Bates.
Opt. Lett., Vol. 5, 438 - 440 (1980).

Speckle images of spatially incoherent objects viewed under severe seeing conditions are formed in the optical laboratory (organized for simulations of stellar speckle interferometry). The brightest pixel in each speckle image is shifted to the center of image space, and the translated image is added to all other speckle images that have been similarly processed. A recognizable diffraction-limited version of the true image of the object results, even when the imaging instrument is defocused such that, under perfect seeing conditions, the Airy disk is spread over an area comparable with that covered by a typical speckle image.

031.597 **Incoherent optical 1-bit cross-correlators for radio antenna arrays.** T. W. Cole.
Appl. Opt., Vol. 19, 2169 - 2173 (1980).

A new incoherent optical approach to an important digital signal processing problem in radio astronomy is presented. It greatly simplifies the many thousands of 1-bit digital cross-correlations that can be required by using a digitally controlled light panel and photodiode array.

031.598 **Field-of-view constraints on actively controlled long-baseline stellar interferometry.** W. Waller.
J. Opt. Soc. America, Vol. 70, 1096 - 1100 (1980).

The field-of-view limitations to actively controlled long-baseline interferometry are investigated. Without resorting to any specific atmospheric model, it is shown that the deleterious effects of non-isoplanatism do not grow with further baseline separation (higher spatial frequency sampled). Numerical calculations, which assume an atmospheric model, plus recently acquired data on the vertical profile of turbulence, suggest that the atmosphere induced field-of view limitation is about 2 arcsec at optical wavelengths and about 50 arcsec at infrared wavelengths.

031.599 **Space-time analysis of photon-limited stellar speckle interferometry.** K. A. O'Donnell, J. C. Dainty.
J. Opt. Soc. America, Vol. 70, 1354 - 1361 (1980).

The standard method of stellar speckle interferometry, in which short exposure photographs are individually analyzed, is not the most general method of extracting object information from the time-varying image intensity. The authors introduce a space-time analysis in which both spatial and temporal fluctuations are taken into account; the aim is to measure the power spectrum of the image with an increased signal to noise ratio. This more general space-time analysis does not yield an improved signal to noise ratio at very low light levels.

031.600 **Specialized complex using of three very-long-base radiointerferometers.** I. D. Zhongolovich, V. I. Valyaev, A. A. Malkov, T. B. Sabanina.
Byull. Inst. Teor. Astron., Tom 14, 529 - 551 (1980). In Russian.

Specialized complex using of three VLBI opens the possibility to improve accurate determining of some astrometric and geodynamic parameters. The corresponding theory is outlined and numerical examples are given.

031.601 **Use of two-beam interference in an apparatus for investigation of the dynamics of the solar atmosphere.** S. M. Gorskij, I. E. Kozhevatov, E. Kh. Kulikova, V. P. Lebedev, N. P. Cheragin.
Issled. po geomagn., aehron. i. fiz. Solntsa, Moskva, 1980, No. 52, p. 116 - 119. In Russian. – Abstr. in Ref. zh., 51. Astron., 11.51.150 (1980).

031.602 **Threshold contrasts at photographic recording of stars on the sky background.** I. I. Brejdo.
Zh. nauch. i prikl. fotogr. i kinematogr., Vol. 25, 285 - 289 (1980). In Russian. – Abstr. in Ref. zh., 51. Astron., 11.51. 259 (1980).

031.603 **Territorial networks of optico-electronic means for observations of astronomical objects (present state and problems of development).**
S. I. Vereshchagin, A. G. Sergeev.
Opt.-mekh. prom-st', 1980, No. 3, p. 6 - 10. In Russian. Abstr. in Ref. zh., 51. Astron., 11.51.264 (1980).

031.604 **Input actions of an optico-electronic system for measuring the angular coordinates of astronomical objects.**
Yu. A. Seliverstov, B. N. Avsenin, S. A. Likhomanov, S. I. Vereshchagin.
Opt.-mekh. prom-st', 1980, No. 3, p. 10 - 13. In Russian. Abstr. in Ref. zh., 51. Astron., 11.51.265 (1980).

031.605 **On the accuracy of determination of the angle of astronomical refraction by the visual method with the help of universal field instruments.**
A. E. Filippov, D. I. Maslich, M. N. Rusin, L. I. Tsmykal, N. F. Nelyubin.
Soveshchanie po atmos. opt. Tez. dokl. Ch. 2. Tomsk, 1980, p. 151 - 154. In Russian. – Abstr. in Ref. zh., 52. Geod. Aehrosemka, 11. 52.68 (1980).

031.606 **Additional regime of photographic artificial earth satellites for the AFU-75 camera.** S. V. Tolbin.
Geod. i kartogr., 1980, No. 7, p. 37 - 39. In Russian. – Abstr. in Ref. zh., 52. Geod. Aehrosemka, 11.52.71 (1980).

031.607 **Automatic identification of reference stars in photographic observations of artificial earth satellites.** S. M. Dzhudzhev, G. G. Muzdrakov.
Geod., kartogr., zemeustr., Vol. 20, No. 2, p. 10 - 14 (1980). In Bulgarian. – Abstr. in Ref. zh., 52. Geod. Aehrosemka, 11.52.72 (1980).

031.608 **High-resolution infrared spectroscopy.** D. N. B. Hall.
Publ. Astron. Soc. Pacific, Vol. 92, 546 (1980). – Abstract.

031.609 **An improved method to derive periods of cyclic phenomena.** H. G. Marraco, J. C. Muzzio.
Publ. Astron. Soc. Pacific, Vol. 92, 700 - 701 (1980).
The authors present two simple improvements to the method described by Jurkevich (1971) to derive periods of cyclic phenomena and which is particularly useful when large numbers of observations are available. An example of the results of both the original and the improved methods is given.

031.610 **Radio interferometry as a means of solving basic problems of astrometry and astrophysics.**
A. F. Dravskikh, A. M. Finkel'shtejn.
Geodynamics and astrometry, (see 003.011), p. 137 - 158 (1980). In Russian.

031.611 **The measurement of faint galaxy magnitudes.** M. R. S. Hawkins.
Photometry, kinematics and dynamics of galaxies, (see 012.051), p. 105 - 107 (1979).

031.612 **A new technique for eliminating errors in computerised surface photometry.** C. P. Blackman.
Photometry, kinematics and dynamics of galaxies, (see 012.051), p. 135 - 137 (1979).

031.613 **Velocity fields in late-type galaxies by Hα Fabry-Perot interferometry.** G. de Vaucouleurs.
Photometry, kinematics and dynamics of galaxies, (see 012.051), p. 249 - 251 (1979).

031.614 **Color photography of galaxies.** J. D. Wray.
Photometry, kinematics and dynamics of galaxies, (see 012.051), p. 311 - 320 (1979).

031.615 **CCD images obtained with the Canada-France-Hawaii Telescope.** P. Hickson, J. R. Auman, G. G. Fahlman, T. K. Menon, G. A. H. Walker, R. Johnson, T. Lester.
J. R. Astron. Soc. Canada, Vol. 74, 364 (1980). – Abstract.

031.616 **A least-squares determination of cepheid maxima.** R. G. McCallum, J. M. Roney.
J. R. Astron. Soc. Canada, Vol. 74, 367 (1980). – Abstract.

031.617 **Search for planets by spectroscopic methods.** K. Serkowski.
Strategies for the search for life in the universe, (see 012.053), p. 155 - 161 (1980).
To detect a Jupiter-like planet around a solar-type star by spectroscopic methods, the annual average of measured velocity should have a mean error not larger than about ±5m/s. This means that accuracy should be almost two orders of magnitude better than commonly achieved for stellar radial velocities.

031.618 **The search for planets in other solar systems through use of the Space Telescope.**
W. A. Baum.
Strategies for the search for life in the universe, (see 012.053), p. 163 - 166 (1980).
The Space Telescope can make important contributions to the search for extrasolar planetary systems. This note particularly calls attention to the capability of the CCD camera system and discusses criteria for the selection of candidate stars.

031.619 **A comparison of alternative methods for detecting other planetary systems.** D. C. Black.
Strategies for the search for life in the universe, (see 012.053), p. 167 - 175 (1980).

031.620 **Quantitative Analyse von Skylab-Röntgenbildern der Sonne mit iterativer Entfaltung.**
K. Maute, G. Elwert.
Mitt. Astron. Ges., Nr. 50, p. 98 - 100 (1980).

031.621 **Tracking velocity as source of systematic errors in observations with a contact micrometer.**
A. Khadzhijski.
Vissha geod., 1980, No. 5, p. 58 - 64. In Bulgarian. – Abstr. in Ref. zh., 51. Astron., 12.51.118 (1980).

031.622 **New ways of seeing the Universe.** F. G. Smith.
The state of the Universe, (see 003.013), p. 181 - 195 (1980).
The author discusses the spectacular extension of observational techniques which we have seen in astronomy over the last 25 years. These techniques now extend through practically the whole electromagnetic spectrum involving radio telescopes, radio interferometers with intercontinental baselines, balloon-borne gamma-ray telescopes, and X-ray telescopes on satellites. He describes the relation between the old and the new astronomies, and shows that the new has not led to the abandonment of the old, but rather to its renewal.

031.623 **On the reduction of observational results of star tremors.**
A. V. Arkhangel'skij, S. D. Volzhanin.
Soveshchanie po atmos. opt. Tez. dokl. Ch. 2. Tomsk, 1980, p. 169 - 171. In Russian. – Abstr. in Ref. zh., 51. Astron., 1.51.97 (1980).

031.624 **Taking into account refraction effects in determination of the position of artificial celestial bodies.**
M. V. Bratijchuk, A. G. Kirichenko, I. I. Motrunich, I. V. Shvalagin.
Vliyanie atmos. na astron. nablyudeniya v optich. i radiodiapazonakh. Tez. dokl. soveshch. probl.-temat. gruppy po teor. astrometr. sekts. astrometr. AS AN SSSR, 1980. Irkutsk, 1980, p. 26 - 28. In Russian. – Abstr. in Ref. zh., 51. Astron., 1.51.131 (1981).

031.625 **Taking into account the influence of refraction of the atmosphere on Doppler measurements.**
O. M. Bulygina.
Vliyanie atmos. na astron. nablyudeniya v optich. i radiodiapazonakh. Tez. dokl. soveshch. probl.-temat. gruppy po teor. astrometr. sekts. astrometr. AS AN SSSR, 1980. Irkutsk, 1980, p. 55 - 57. In Russian. – Abstr. in Ref. zh., 51. Astron., 1.51.134 (1981).

031.626 **Estimate of and allowance for the influence of atmospheric conditions on the accuracy of measurement of displacement of the effective center of solar radio radiation.** A. A. Kurbanov.
Dokl. AN TadzhSSR, Vol. 23, 242 - 246 (1980). In Russian. Abstr. in Ref. zh., 51. Astron., 1.51.761 (1981).

031.627 **On the possibility of receiving photoelectric information with given error and minimum consumption of observational time.**
L. V. Granitskij, A. B. Bukach, N. I. Bukach.
Krym. astrofiz. obs., AN SSSR. P. Nauchnyj, 1980, 22 pp. In Russian. – Abstr. in Ref. zh., 51. Astron., 1.51.793 (1981).

The Lund Centre for Automated Measurements of Astronomical Plates. See Abstr. 009.021.

Automated plate assignment.
See Abstr. 021.007.

Fortran programmes for reducing uvby-Hβ photometry and for deriving physical properties of stars.
See Abstr. 021.032.

A digital correlator for real time spectral analysis of radio astronomy signals. See Abstr. 021.038.

'Chromotron' a spectral image analysis device.
See Abstr. 031.010.

Astrometry with the Space Telescope.
See Abstr. 032.504.

The ability of the Space Telescope to detect extrasolar planetary systems. See Abstr. 032.505.

Applications of interferometers in space to astrometry and planetary detection. See Abstr. 032.507.

The MSFC solar correlation tracker.
See Abstr. 034.008.

Optimal sampling of charge-coupled devices.
See Abstr. 034.027.

Velocity fields in late-type galaxies from Hα Fabry-Perot interferometry. I. Instrumentation and data reduction. See Abstr. 034.047.

Anwendung der Fernsehtechnik in der Astrometrie.
See Abstr. 041.020.

Die Plate Overlap-Methode: Steifheit des Systems und Lösungsverfahren. See Abstr. 041.021.

On the application of the general theory of methods of geodetic astronomy to the determination of coordinates in points of experimental measurements of refraction angles.
See Abstr. 046.015.

Theory of scan plane flux anisotropies.
See Abstr. 062.106.

On the gravitational wave induced oscillations of viscoelastic bodies. See Abstr. 066.194.

Interpretation of filter magnetograph results including solar magneto-optical effects: observations.
See Abstr. 075.005.

Interpretations of filter magnetograph results including solar magneto-optical effects: theory.
See Abstr. 075.006.

Problems of single-frequency meter-wavelength solar observations. See Abstr. 077.035.

Atmospheric limitations of narrow-field optical astrometry. See Abstr. 082.028.

Theoretical infrared stratospheric emission spectra from balloons and satellites. See Abstr. 082.041.

On the restoration of perturbations of the vertical density profile in the atmosphere from photographs of the sun received on an artificial earth satellite.
See Abstr. 082.081.

Experimental determination of the astronomical refraction from observations on large zenith distances.
See Abstr. 082.083.

Synchronous television recording of active forms and spectra of aurorae. See Abstr. 084.054.

Application of the regularization method to the analysis of lunar occultation observations of close double stars and radio sources. See Abstr. 096.007.

The mosaics of Mars. As seen by the Viking Lander Cameras. See Abstr. 097.025.

Lyman-alpha observations in the vicinity of Saturn with Copernicus. See Abstr. 100.120.

Saturn satellite observations with reflecting coronagraph optics. See Abstr. 100.128.

Speckle interferometry with the CFHT 3.60 m. I. Resolution of the system Pluto-Charon.
See Abstr. 101.032.

Method for the determination of density and phase functions of interplanetary dust. See Abstr. 106.002.

An analysis of parallaxes determined in two coordinates. See Abstr. 111.014.

Measurement of the absolute flux from Vega in the *K* band (2.2 μm). See Abstr. 113.009.

Vector space methods of photometric analysis. II. Refinement of the MK grid for B stars. See Abstr. 113.043.

An automatic procedure for a determination of the effective temperature and gravity. Application to 100 O-type stars. See Abstr. 114.008.

Angular diameters by the lunar occultation technique. III. See Abstr. 115.008.

The astrometric search for neighboring planetary systems. See Abstr. 118.036.

An interactive procedure to determine the mean radius of Cepheids. See Abstr. 122.152.

On the accuracy and precision of extinctions derived from general star counts. See Abstr. 131.107.

Spectroscopic observations of galactic nebulae and galaxies with the Imaging Photon Counting System (IPCS).
See Abstr. 135.024.

Membership in the open cluster NGC 6494. Astrometry with a PDS microdensitometer. See Abstr. 153.004.

The first half century of galaxy photometry: methods and results. See Abstr. 158.243.

Computer enhanced photography of M 51.
See Abstr. 158.292.

032 Astronomical Instruments, Space Instrumentation

Astronomical Instruments

032.001 **Das Multiple Mirror Telescope MMT auf Mount Hopkins.** Prototyp einer neuen Fernrohr-Generation.
A. Tarnutzer.
Orion, 38. Jahrg., 110 - 115 (1980).

032.002 **Multiple-mirror telescope systems in astronomy.**
J. F. Grainger.
Astrophys. Space Sci., Vol. 71, 229 - 234 (1980).

It is far from obvious that the conventional large telescope is the optimum fore-optical system for looking through a turbulent atmosphere. There are considerable scientific and financial advantages in going to multiple-imaging-elements and several groups in different parts of the world are working in this direction.

032.003 **Vibrations of a telescope support under wind load.**
N. G. Bochkarev, D. N. Spitsyna, E. V. Sosnovskij.
Astron. Zh., Tom 57, 864 - 868 (1980). In Russian.
English translation in Soviet Astron., Vol. 24, No. 4.

The results of a theoretical analysis of vibrations of a frame-type telescope support are given. The main frequencies of natural vibrations are found and the evaluation of the wind load dynamical influence is given, the wind load being taken in the form of a random function which depends on the wind velocity.

032.004 **Study of the errors of 2′ divisions in selected limb regions of the Wanschaff vertical circle.**
N. F. Minyajlo.
Astrometr. Astrofiz., Vyp. (No.) 41, p. 92 - 98 (1980). In Russian.

032.005 **Investigation of diameter corrections of the Brorfelde transit circle.**
C. Fabricius, L. Helmer, H. J. Fogh Olsen.
Astron. Astrophys., Vol. 89, 57 - 60 (1980).

At the transit circle in Brorfelde a comparison has been made between several determinations of diameter corrections of which the latest has been performed with automatic telescope setting and circle reading. The difference between new and old diameters is around 0.″09. The internal mean error of one diameter correction is 0.″035. Using six diameters in observing a star, the remaining systematic error arising from errors in the diameter corrections, rarely exceeds 0.″1 in declination.

032.006 **Recent work with the UK Schmidt Telescope.**
S. B. Tritton.
Proc. Astron. Soc. Australia, Vol. 3, 206 - 207 (1978).

032.007 **Some design concepts for optical and infrared telescopes of 25-meter equivalent aperture.**
I. Ghozeil.
Optics present and future, (see 012.012), p. 169 - 172 (1978).
Abstr. in Phys. Abstr., Vol. 83, Abstr. 82203 (1980).

032.008 **Automatic photoelectric transit circles.**
Y. Requieme.
Mitt. Astron. Ges., Nr. 48, (see 012.015), p. 109 - 125 (1980).

On the threshold of the space astrometry era, at least six automatic meridian circles are in or will shortly come into operation, offering remarkable possibilities in magnitude, accuracy and observation speed. The expected results will give by themselves a new start to classical astrometry, even if these promises seem at present eclipsed by the HIPPARCOS satellite project. An urgent task is to show clearly the complementation of ground-based and space astrometry and to define consequently the observation programmes.

032.009 **The Canada-France-Hawaii telescope.** R. Cayrel.
Messenger, No. 21, p. 1 - 3 (1980).

032.010 **High-resolution optical telescope for ultraviolet (UV) radiation field.** W. W. Karayan.
Proc. Soc. Photo-Opt. Instrum. Eng., Vol. 196, (see 012.016), 40 - 50 (1979). – Abstr. in Phys. Abstr., Vol. 83, Abstr. 90335 (1980).

032.011 **Big eyes.** M. J. Disney.
Nature, Vol. 287, 679 - 680 (1980).

032.012 **Große optische Teleskope.**
J. Reiche, H.-J. Teske.
Sterne, 56. Band, 203 - 219 (1980).

032.013 **Determination of natural vibration frequencies of a telescope frame.**
N. G. Bochkarev, D. N. Spitsyna, E. V. Sosnovskij.
Astron. Zh., Tom 57, 1078 - 1081 (1980). In Russian.
English translation in Soviet Astron., Vol. 24, No. 5.

032.014 **On the theory of the Popov-Bowen camera.**
B. A. Burnasheva, R. E. Gershberg, V. I. Pronik.
Izv. Krymskoj Astrofiz. Obs., Tom 61, 150 - 159 (1980). In Russian.

The Popov-Bowen two-mirror camera is systematically analysed. The fastest attainable systems are found and the possible correction of high orders of spherical aberration is determined both for pure mirror systems and for similar systems with a plane-parallel plate at the exit. Possibilities to obtain a flat field up to 9° in diameter with a simple concave-flat lens placed just before the focus of the systems are considered in detail. For several focal ratios optimal parameters of such systems are given in a table; proper spot diagrams are displayed in a figure.

032.015 **Solar astrolabe construction at C.E.R.G.A.**
F. Laclare, J. Demarcq, F. Chollet.
C. R. Acad. Sci. Paris, Tome 291, Sér. B, 189 - 192 (1980). In French.

In 1974 a Danjon astrolabe was used at C.E.R.G.A. for solar observations. Since then some modifications have permitted solar observations at three zenith distances and thus to cover an orbital arc of 9 months. The accuracy of the observations obtained thus far allows to determinate the orbital elements of the earth.

032.016 **Optical infrared sky survey instrumentation.**
E. R. Craine.
Opt. Eng., Vol. 19, 397 - 403 (1980). – Abstr. in Phys. Abstr., Vol. 83, Abstr. 105167 (1980).

032.017 **Une petite merveille: le 6 pouces de Strasbourg.**
P. Muller.
Astronomie, Vol. 94, 509 - 512 (1980).

032.018 **U of Md two-color refractometry.**
D. D. Wellnitz, D. G. Currie.
Bull. American Astron. Soc., Vol. 12, 748 (1980). – Abstract.

032.019 **Ervaringen met de Celestron 8.** K. Teuwen.
Zenit, 7e Jaarg., 32 - 35 (1980).

032.020 **Ervaringen met een 115 mm Newtontelescoop.** E. Hartsuiker.
Zenit, 7e Jaarg., 306 - 307, 318 - 319 (1980).

032.021 **El telescopio de 1.5 metros del Observatorio Astronomico Nacional.** J. F. Lahulla.
Bol. Astron. Obs. Madrid, Vol. 10, No. 3, p. 3 - 7 (1978).

032.022 **Analysis of instrument stability of a screw step in observations of latitude variations at Borowiec.** W. Jakś.
Publ. Inst. Geophys., Polish Acad. Sci., F-6 (137), p. 67 - 70 (1980).

The micrometer screw of a zenith telescope has been investigated using the scale pairs. The paper contains the results of analysing the temperature coefficient of the screw step.

032.023 **L'observation dans un grand instrument – avantages – difficultés.** M. Verdenet.
Bull. AFOEV, Tome 14, 65 - 66 (1980).

032.024 **Pointing of the 3.6 m telescope.** A. B. Muller.
Messenger, No. 22, p. 17 - 18 (1980).

032.025 **The application of the cross-correlation function to the photoelectric time registration of stellar transits.** N. Solaric.
Zb. Rad. Jurema, Vol. 25, Part 2, p. 125 - 127 (1980).
Abstr. in Phys. Abstr., Vol. 84, Abstr. 4775 (1981).

032.026 **Astronomiske instrumenter til privatbrug.** P. Darnell.
Astron. Tidsskr., Årg. 13, 149 - 155 (1980).

032.027 **Minimum-cost 4-m telescope developed at October 1979 Nanjing study of telescope design and construction.** A. B. Meinel, M. P. Meinel, N. Hu, Q. Hu, C. Pan.
Appl. Opt., Vol. 19, 2670 - 2679 (1980).

A lightweight 4-m telescope with a 6400-kg primary mirror of *f*/1.5 was developed during a two-week workshop at the Nanjing Astronomical Instruments Factory. A central column supports the secondary mirror, thus eliminating all structures around the periphery of the primary mirror. The altazimuth mounting has the elevation axis behind the primary mirror and cell, requiring a counterweight. The Cassegrain focal position coincides with the elevation axis. A single secondary mirror and appropriate field correctors enable operations at the Harland Epps-Dan Schulte, Cassegrain, Nasmyth, and coudé foci.

032.028 **On the history of the construction of the 6-m telescope.** N. N. Mikhel'son.
Nauka i tekh. Vopr. istor. i teor., Moskva-Leningrad, 1979, No. 10, p. 87 - 90. In Russian. – Abstr. in Ref. zh., 51. Astron., 11.51.79 (1980).

032.029 **Solar telescopes.** V. G. Banin.
Issled. po geomagn., aehron. i fiz. Solntsa, Moskva, 1980, No. 52, p. 60 - 70. In Russian. – Abstr. in Ref. zh., 51. Astron., 11.51.140 (1980).

032.030 **The polar siderostat of the large solar vacuum telescope.** V. I. Kruglov, M. A. Shamsutdinov, A. K. Kitov, Yu. A. Kuznetsov, V. M. Grigor'ev, V. Ya. Govorukhin.
Issled. po geomagn., aehron. i fiz. Solntsa, Moskva, 1980, No. 52, p. 76 - 81. In Russian. – Abstr. in Ref. zh., 51. Astron., 11.51.142 (1980).

032.031 **On the decrease of the vibration level of the experimental model of the large solar vacuum telescope using a dynamic shock absorber.** A. K. Kitov.
Issled. po geomagn., aehron. i fiz. Solntsa, Moskva, 1980, No. 52, p. 82 - 84. In Russian. – Abstr. in Ref. zh., 51. Astron., 11.51.143 (1980).

032.032 **On a possible vibration of the image during vibrations of the experimental model of the large solar vacuum telescope.** A. K. Kitov.
Issled. po geomagn., aehron. i fiz. Solntsa, Moskva, 1980, No. 52, p. 85 - 89. In Russian. – Abstr. in Ref. zh., 51. Astron., 11.51.144 (1980).

032.033 **Some investigations of the effect of astronomical pneumatic unloading systems on the image quality.** V. M. Dul'kin, A. A. Fadeev, V. P. Zhokhov.
Issled. po geomagn., aehron. i fiz. Solntsa, Moskva, 1980, No. 52, p. 90 - 91. In Russian. – Abstr. in Ref. zh., 51. Astron., 11.51.145 (1980).

032.034 **A technique of choosing an astronomical mirror unloading system.** V. M. Dul'kin.
Issled. po geomagn., aehron. i fiz. Solntsa, Moskva, 1980, No. 52, p. 92 - 93. In Russian. – Abstr. in Ref. zh., 51. Astron., 11.51.146 (1980).

032.035 **In praise of smaller telescopes.** J. R. Percy.
J. R. Astron. Soc. Canada, Vol. 74, 334 - 341 (1980).

032.036 **Gear selection, considering gear-load balancing, for astrophotography.** M. P. Edwards.
J. R. Astron. Soc. Canada, Vol. 74, 366 - 367 (1980). Abstract.

032.037 **On the efficiency of taking into account 2′ circle division errors in determination of declinations.** N. F. Minyajlo.
Astron. Tsirk., No. 1108, p. 3 - 5 (1980). In Russian.

032.038 **Investigation of the aerodynamics of astronomical domes (preliminary results).** V. F. Umarov.
Astron. Tsirk., No. 1110, p. 5 - 8 (1980). In Russian.

032.039 **Influence of the ellipticity of pivots of the horizontal axis of an instrument during azimuthal astronomical observations.** Ts. Darakchiev.
Vissha geod., 1980, No. 5, p. 49 - 57. In Bulgarian. – Abstr. in Ref. zh., 51. Astron., 12.51.117 (1980).

032.040 **Investigation of the variation of the azimuth of a transit instrument.** D. Lkhagvasurehn, B. Bal'zhinova, G. Davaakhuu.
Ehrdehm shinzhilgehehnij butehehl. Fiz.- tekh. khurehehlehn. Mat. khurehehlehn. BNMAU shinzhlekh ukhaany Asad., 1980, No. 17, p. 107 - 111. In Russian. – Abstr. in Ref. zh., 51. Astron., 12.51.120 (1980).

032.041 **What is worth to observe using a very large telescope?** W. Kluźniak.
Urania Kraków, Vol. 51, 290 - 299 (1980). In Polish.

032.042 **On the compensation of the inclination of the axis of a transit instrument.** J. Dittrich, H. Jochmann.
Astrometr. Astrofiz., Vyp. (No.) 42, p. 96 - 99 (1980). In Russian.

032.043 **Influence of the surrounding medium on the temperature change of parts of an instrument.** G. M. Blank.

Vliyanie atmos. na astron. nablyudeniya v optich. i radiodiapazonakh. Tez. dokl. soveshch. probl.-temat. gruppy po teor. astrometr. sekts. astrometr. AS AN SSSR, 1980. Irkutsk, 1980, p. 72 - 73. In Russian. – Abstr. in Ref. zh., 51. Astron., 1.51.136 (1981).

032.044 **Examination of the Talcott levels of the ZTL-180 zenith telescope of the Poltava Observatory.**
N. A. Popov, A. P. Stehpa.
Vrashchenie i priliv. deformatsii Zemli, Kiev, 1980, No. 12, p. 68 - 74. In Russian. – Abstr. in Ref. zh., 51. Astron., 1.51.771 (1981).

A Dollond–Wollaston telescope.
See Abstr. 004.008.

First achromatic telescopes.
See Abstr. 004.017.

Compact solar camera. See Abstr. 014.006.

The cost-effectiveness in terms of publications and citations of various optical telescopes at the Kitt Peak National Observatory. See Abstr. 015.001.

Hartmann tests on large telescopes carried out with a small screen in a pupil image.
See Abstr. 031.014.

Centering the secondary mirror of the Tirgo telescope. See Abstr. 031.026.

Design considerations for a 10-m optical table telescope. See Abstr. 031.028.

Adjusting the orientation of a polar axis.
See Abstr. 031.539.

Stellar images of the Mizusawa PZT.
See Abstr. 041.025.

Laser displacement sensor with application to gravitational radiation detection. See Abstr. 066.178.

Space Instrumentation

032.501 **An imaging gas scintillation counter for X-ray astronomy.** J. Davelaar, G. Manzo, A. Peacock, B. G. Taylor, R. D. Andresen, J. A. M. Bleeker.
Astron. Astrophys., Vol. 87, 276 - 281 (1980).

A position sensitive gas scintillation proportional counter suitable as a focal plane detector for an X-ray imaging telescope is presented. The detector has an on-axis position and energy resolution of ~ 2 mm and ~ 10% respectively for 6 keV X-rays with a useful aperture diameter of 7 cm. Further design improvements are outlined which should improve the position resolution to much better than 1 mm at 6 keV for a useful aperture of 8 cm.

032.502 **Imaging of cosmic X-ray sources using coded mask telescopes.** G. K. Skinner.
J. British Interplanet. Soc., Vol. 33, 333 - 337 (1980).

032.503 **Space Telescope, a versatile new instrument.** N. Roman.
Celestial Mech., Vol. 22, (see 012.002), 165 - 174 (1980).

The design, in-orbit functioning, and projected performance of the Space Telescope are discussed.

032.504 **Astrometry with the Space Telescope.** W. H. Jefferys.
Celestial Mech., Vol. 22, (see 012.002), 175 - 181 (1980).

The Space Telescope is described, along with its initial complement of instruments. Particular attention is paid to the fine guidance sensors, which will be the principal astrometric instrument. The responsibilities of the Space Telescope astrometry team towards the Space Telescope astrometry project are described. A description of the scientific results to be expected from Space Telescope astrometry is given. The impact of other proposed Space Astrometry projects, such as the European Astrometry Satellite, on Space Telescope astrometry is also discussed.

032.505 **The ability of the Space Telescope to detect extra-solar planetary systems.** W. A. Baum.
Celestial Mech., Vol. 22, (see 012.002, 183 - 190 (1980).

The Space Telescope can play a key role in searching for and investigating the contents of extra-solar planetary systems. For about 90 nearby stars, positional variations due to major planets would be well within the astrometric capability of the wide-field/planetary camera system. Since the centroids of star images will be determined to within a milliarcsecond down to 22nd magnitude, there will be an abundance of reference stars at very small angular distances from each planetary system candidate, and they will have small enough motions of their own to provide a reference frame of the stability required.

032.506 **Space Telescope astrometry from CCD images.** L.-T. G. Chiu.
Celestial Mech., Vol. 22, (see 012.002), 191 - 196 (1980).

The astrometric application of the wide field camera and the planetary camera is reviewed. It is shown that the digital image centering algorithms can yield a positional accuracy of 0.1 milli-arcsecond. Deconvolution of CCD's sensitivity, non-flatness of the filters, and crinkling of the CCD surface may limit the positional accuracy to 1 milli-arcsecond.

032.507 **Applications of interferometers in space to astrometry and planetary detection.**
S. H. Knowles, D. L. Thacker.
Celestial Mech., Vol. 22, (see 012.002), 197 - 202 (1980).

Optical and infrared interferometers located outside the Earth's atmosphere should attain fully coherent performance over baselines as long as desired. This new capability should lead, if problems of compensating for angular stability can be solved, to extremely precise measurements of the angular distance between a star and one or more references. This property can be used for astrometric planetary detection, as well as for parallax and proper motion measurements. In addition, the interferometer's high rejection of stray radiation allows the construction of a sensitive direct planetary detection device.

032.508 **Effects of temperature fluctuations on IUE data quality.** R. Thompson, B. Turnrose, R. Bohlin.
Bull. American Astron. Soc., Vol. 12, 461 (1980). – Abstract.

032.509 **Improvements to the accuracy of the IUE wavelength scales.** B. Turnrose, C. Harvel, R. Bohlin.
Bull. American Astron. Soc., Vol. 12, 461 (1980). – Abstract.

032.510 **A solar flare X-ray polarimeter for OSS-1.**
G. A. Chanan, J. P. Hughes, J. R. Lemen, R. Novick, I. Rochwarger, M. Sackson, L. J. Tramiel.
Bull. American Astron. Soc., Vol. 12, 475 (1980). – Abstract.

032.511 **A solar extreme ultraviolet telescope and spectrograph for Space Shuttle.**
W. M. Neupert, R. J. Thomas, G. L. Epstein.
Bull. American Astron. Soc., Vol. 12, 475 (1980). – Abstract.

032.512 **High resolution spectrograph for the Space Telescope.** J. C. Brandt, A. Boggess, S. R. Heap, S. P. Maran, A. M. Smith, E. A. Beaver, M. Bottema, J. B. Hutchings, M. A. Jura, J. L. Linsky, B. D. Savage, L. M. Trafton, R. J. Weymann.
Bull. American Astron. Soc., Vol. 12, 488 (1980). – Abstract.

032.513 **SMM (*Solar Maximum Mission*) orbiting coronagraph – early results.**
C. Sawyer, W. J. Wagner, E. Hildner, L. L. House.
Bull. Amerioan Astron. Soc., Vol. 12, 531 (1980). – Abstract.

032.514 **Observations from the Flat Crystal Spectrometer on the Solar Maximum Mission.** K. T. Strong.
Bull. American Astron. Soc., Vol. 12, 533 (1980). – Abstract.

032.515 **The Ultraviolet Spectrometer and Polarimeter (UVSP) on the Solar Maximum Mission and initial results in polarimetry.**
E. Tandberg-Hanssen, R. G. Athay, E. C. Bruner, J. M. Beckers, J. C. Brandt, R. D. Chapman, C. C. Cheng, J. Gurman, W. Henze, C. L. Hyder, A. G. Michalitsianos, R. A. Shine, S. A. Schoolman, B. E. Woodgate.
Bull. American Astron. Soc., Vol. 12, 534 (1980). – Abstract.

032.516 **SMM (*Solar Maximum Mission*) Coronagraph/Polarimeter observations of coronal transient manifestations correlated with flare data.** L. L. House, A. Csoeke-Poeckh, C. Sawyer, W. J. Wagner, E. Hildner.
Bull. American Astron. Soc., Vol. 12, 535 (1980). – Abstract.

032.517 **First radio telescope in space.**
M. B. Zakson, N. S. Kardashev, A. I. Savin, A. G. Sokolov, K. P. Feoktistov.
Zemlya Vselennaya, 1980, No. 4, p. 2 - 9. In Russian.

032.518 **Wolter-Schwarzschild optics for the extreme ultraviolet: the Berkeley stellar spectrometer and the EUV Explorer.** R. F. Malina, S. Bowyer, D. Finley, W. Cash.
Opt. Eng., Vol. 19, 211 - 218 (1980). – Abstr. in Phys. Abstr., Vol. 83, Abstr. 82197 (1980).

032.519 **The Pioneer-Venus Solar Flux Radiometer.**
J. M. Palmer.
Opt. Eng., Vol. 19, 224 - 228 (1980). – Abstr. in Phys. Abstr., Vol. 83, Abstr. 82198 (1980).

032.520 **Space Telescope fine guidance – a design for manufacturability.** I. Friedman.
Proc. Soc. Photo-Opt. Instrum. Eng., Vol. 193, (see 012.011), 53 - 62 (1979). – Abstr. in Phys. Abstr., Vol. 83, Abstr. 82201 (1980).

032.521 **Space Telescope Digicon.** H. R. Alting-Mees, H. A. Wenzel, E. A. Beaver, J. L. Shannon.
Proc. Soc. Photo-Opt. Instrum. Eng., Vol. 203, (see 012.014), 12 - 20 (1979). – Abstr. in Phys. Abstr., Vol. 83, Abstr. 85858 (1980).

032.522 **800 × 800 charge-coupled device (CCD) camera for the Galileo Jupiter Orbiter Mission.**
M. C. Clary, K. P. Klaasen, L. M. Snyder, P. K. Wang.
Proc. Soc. Photo-Opt. Instrum. Eng., Vol. 203, (see 012.014), 98 - 108 (1979). – Abstr. in Phys. Abstr., Vol. 83, Abstr. 85859 (1980).

032.523 **Integrated infrared detector arrays for low-background astronomy.** C. R. McCreight.
Proc. Soc. Photo-Opt. Instrum. Eng., Vol. 203, (see 012.014), 109 - 116 (1979). – Abstr. in Phys. Abstr., Vol. 83, Abstr. 85860 (1980).

032.524 **The charge injection device (CID) as a stellar tracking sensor.**
H. K. Burke, D. M. Brown, A. Grafinger, G. J. Michon, H. W. Tomlinson, T. L. Vogelsong, R. Wilson.
Proc. Soc. Photo-Opt. Instrum. Eng., Vol. 203, (see 012.014), 124 - 129 (1979). – Abstr. in Phys. Abstr., Vol. 83, Abstr. 85861 (1980).

032.525 **Infrared focal plane arrays for planetary missions.**
J. B. Wellman.
Proc. Soc. Photo-Opt. Instrum. Eng., Vol. 203, (see 012.014), 166 - 172 (1979). – Abstr. in Phys. Abstr., Vol. 83, Abstr. 85862 (1980).

032.526 **Design and test of the near infrared mapping spectrometer (NIMS) focal plane for the Galileo Jupiter Orbiter Mission.** G. Bailey.
Proc. Soc. Photo-Opt. Instrum. Eng., Vol. 197, (see 012.017), 210 - 216 (1979). – Abstr. in Phys. Abstr., Vol. 83, Abstr. 90336 (1980).

032.527 **Hadamard transform X-ray telescope.**
S. Miyamoto, H. Tsunemi, K. Tsuno.
Bull. Inst. Space Aeronaut. Sci., Univ. Tokyo B, Vol. 15, 585 - 610 (1979). In Japanese. – Abstr. in Phys. Abstr., Vol. 83, Abstr. 94714 (1980).

032.528 **Planned observations of the diffuse sky radiation during Shuttle mission STS–4.**
J. L. Weinberg, R. C. Hahn, F. Giovane, D. W. Schuerman.
Solid particles in the solar system, (see 012.019), p. 25 - 28 (1980).

The skylab flight spare 10-color photopolarimeter is being refurbished for use in Space Shuttle mission STS–4, a test flight currently scheduled for October 1980. Observations will be made of zodiacal light, background starlight, and the Shuttle-induced atmosphere (spacecraft corona), with emphasis on regions of sky closer than 90° to the sun.

032.529 **An attempt to observe zodiacal light at 5μ with a balloon experiment.** E. Pitz.
Solid particles in the solar system, (see 012.019), p. 29 - 32 (1980).

A liquid nitrogen cooled photometer with indium antimonide detectors is described which has been developed for balloon-borne observations of the zodiacal light. The provision of both difference and absolute flux measurement modes will allow the observation of the zodiacal light, and the presumably much more intense emission of the high atmosphere, respectively.

032.530 **Dust experiment for a rendezvous cometary mission.** B.-K. Dalmann, D. Bahr, H. Fechtig, J. Kissel.
Solid particles in the solar system, (see 012.019), p. 273 (1980). – Abstract.

032.531 **An impact mass-spectrometer for the Halley-probe.** G. Braun, E. Grün, J. Kissel, N. Pailer.
Solid particles in the solar system, (see 012.019), p. 275 - 276 (1980).

032.532 **Micrometeoroid multiple foil penetration and particle recovery experiments on board Space Shuttle's Long Duration Exposure Facility (LDEF).**
J. C. Mandeville, J. A. M. McDonnell.
Solid particles in the solar system, (see 012.019), p. 395 - 400 (1980).

Experiments designed for the investigation of the near-earth micrometeoroid flux on the Space Shuttle Long Duration Exposure Facility (LDEF) are described. The paper examines, in particular, how two of the experiments deploy a series of multiple layer foil arrays to investigate the physical properties of incident meteoroids and lead to partial recovery of the micrometeoroids for laboratory analysis. Several thousand penetrations are expected to be returned after 12 months' exposure in space.

032.533 **Fishing on Saturn with the KAO (*Kuiper Airborne Observatory*).** D. DiCicco.
Sky Telesc., Vol. 60, 367 - 371 (1980).

032.534 **Light from electron avalanches and background rejection in X-ray astronomy.** O. H. W. Siegmund, P. W. Sanford, I. M. Mason, J. L. Culhane, R. Cockshott.
Nature, Vol. 287, 808 - 810 (1980).

The authors report that a modified version of the parallel plate imaging proportional counter has been constructed to investigate the application of risetime discrimination to the scintillation pulses caused by the electron avalanche process. They show that efficient background event rejection (>90%) is achieved and discuss an application of this system for X-ray astronomy.

032.535 **Universal multichannel spectrometer for investigation of energetic distributions of electron and proton streams in the ionosphere and magnetosphere.**
N. M. Shyutte, A. I. Puolokajnen, G. I. Volkov, A. I. Belyashin, L. I. Denshchikova, L. E. Gasilin, V. F. Kopylov, Yu. M. Grashin, V. I. Dvoretskij, O. A. Tyurikov.
Inst. kosm. issled. AN SSSR. Prepr., 1980, No. 540, 59 pp. In Russian. – Abstr. in Ref. zh., 62. Issled. kosm. prostranstva, 9.62.118 (1980).

032.536 **Helium refrigerator for cooling the receivers of the submillimeter telescope of the orbital station Salyut 6.** A. A. Nikonov, V. A. Maslakov, V. N. Kurkin, E. I. Klimenko, A. E. Salomonovich, T. M. Sidyakin, A. S. Khajkin, V. N. Bakun.
Fiz. inst. AN SSSR. Prepr., 1980, No. 12, 42 pp. In Russian. Abstr. in Ref. zh., 62. Issled. kosm. prostranstva, 9.62.119 (1980).

032.537 **Thermal regime and protection of mirrors on the Orbiting Solar Telescope OST-1 against degradation.**
A. V. Bruns, V. V. Benyukh.
Izv. Krymskoj Astrofiz. Obs., Tom 61, 173 - 181 (1980). In Russian.

032.538 **High pressure gas scintillation spectrometers for X-ray astronomy.** G. Manzo, A. Peacock, R. D. Andresen, B. G. Taylor.
Nucl. Instrum. Methods, Vol. 174, 301 - 315 (1980). – Abstr. in Phys. Abstr., Vol. 83, Abstr. 105166 (1980).

032.539 **X-ray spectrometer spectrograph telescope system.** E. C. Bruner, Jr., L. W. Acton, W. A. Brown, S. W. Salat, A. Franks, G. Schmidtke, W. Schweizer, R. J. Speer.
Opt. Eng., Vol. 19, 433 - 437 (1980). – Abstr. in Phys. Abstr., Vol. 83, Abstr. 105168 (1980).

032.540 **The European astrometry satellite, Hipparcos.** E. Høg.
Highlights of Astronomy, Vol. 5, (see 012.029), 783 - 787 (1980).

A description of the instrument is given. It is expected to obtain positions, annual proper motions and parallaxes for 100 000 stars in a 2.5 years mission. The accuracy (s.e.) will be $<\pm0''002$ for stars of $m_B < 11$ and about $\pm0''006$ at $m_B = 14$. An extension of the mission to 3.5 years would improve the annual proper motion to $\pm0''0010$ for stars brighter than $m_B = 9$.

032.541 **The U.S. Space Telescope: astrometric capabilities.** W. H. Jefferys.
Highlights of Astronomy, Vol. 5, (see 012.029), 789 - 794 (1980).

This paper describes the Space Telescope and the instruments it will carry which will be used for astrometry. Particular attention is paid to the two imaging cameras and to the fine guidance sensors. A brief outline of the kinds of programs which are to be carried out is also given.

032.542 **Future possibilities for ultraviolet observations of interstellar molecules.** G. R. Carruthers.
Interstellar molecules, (see 012.033), p. 613 - 614 (1980).

032.543 **Overview of the Voyager ultraviolet spectrometry results through Jupiter encounter.**
A. L. Broadfoot, B. R. Sandel, D. E. Shemansky, J. C. McConnell, G. R. Smith, J. B. Holberg, S. K. Atreya, T. M. Donahue, D. F. Strobel, J. L. Bertaux.
Bull. American Astron. Soc., Vol. 12, 685 - 686 (1980). Abstract.

032.544 **Viking lander camera calibration files.** S. Wall, L. Cullen.
Bull. American Astron. Soc., Vol. 12, 689 (1980). – Abstract.

032.545 **A new instrument for cometary dust studies.** M. Hanner, H. Patashnick, G. Rupprecht.
Bull. American Astron. Soc., Vol. 12, 732 (1980). – Abstract.

032.546 **X-ray astronomy from the Space Shuttle.** G. P. Garmire.
Space Shuttle: dawn of an era. Part II, (see 012.037), p. 537 - 548 (1980). – Abstr. in Phys. Abstr., Vol. 83, Abstr. 109363 (1980).

032.547 **X-ray astronomy with the advanced X-ray Astrophysics Facility (AXAF).** L. Van Speybroeck.
Space Shuttle:dawn of an era. Part II, (see 012.037), p. 623 (1980). – Abstr. in Phys. Abstr., Vol. 83, Abstr. 109366 (1980).

032.548 **An airborne far-infrared spectrometer for astronomical observations.** J. W. V. Storey, D. M. Watson, C. H. Townes.
Int. J. Infrared Millim. Waves, Vol. 1, No. 1, p. 15 - 25 (1980). Abstr. in Phys. Abstr., Vol. 83, Abstr. 109371 (1980).

032.549 **Wave and particle phenomena induced by an electrodynamic tether.** M. Dobrowolny.
Smithsonian Astrophys. Obs., Spec. Rep. No. 388, 30 pp. (1979).

032.550 **The Faint Object Camera for the Space Telescope.** F. Macchetto, H. C. van de Hulst, S. di Serego Alighieri, M. A. C. Perryman.
European Space Agency, ESA SP-1028, 48 pp. (1980).
The FOC as now being built is described, which also provides the information needed to make a first appraisal of the feasibility of a particular observing programme. An idea of the complexity and novelty of the types of programmes that are possible can be gained from the examples reviewed in two chapters.

032.551 **Television optics for the Voyager mission to Jupiter and Saturn.** L. Snyder.
Opt. Eng., Vol. 19, 566 - 576 (1980). – Abstr. in Phys. Abstr., Vol. 84, Abstr. 4743 (1981).

032.552 **An MgF_2-NO ion chamber with O_2 gas filter as a detector of solar H Lyman-α radiation.**
N. Watanabe, I. Higashino, T. Oshio.
J. Phys. E, Vol. 13, 1164 - 1165 (1980). – Abstr. in Phys. Abstr., Vol. 84, Abstr. 4768 (1981).

032.553 **State-of-the-art Space Telescope digicon performance data.** R. O. Ginaven, J. P. Choisser, L. Acton, W. Wysoczanski, H. R. Alting-Mees, R. D. Smith, II, E. A. Beaver, H. J. Eck, A. Delamere, J. L. Shannon.
Proc. Soc. Photo-Opt. Instrum. Eng., Vol. 217, 55 - 68 (1980). Abstr. in Phys. Abstr., Vol. 84, Abstr. 9370 (1981).

032.554 **Ion states of low energy cosmic rays: the Indian experiment on the first Space Shuttle-Spacelab mission.** S. Biswas.
Space Research, Vol. XX, (see 012.043), 267 - 270 (1980).

032.555 **The plasma experiment on Helios (E 1).**
H. Rosenbauer, R. Schwenn, B. Meyer, H. Miggenrieder, J. Wolfe.
Helios solar probes, science summaries, (see 003.009), p. 6 - 11 (1980).

032.556 **Scientific results obtained by the Helios Technical University of Braunschweig flux-gate (E 2) and search-coil (E 4) magnetometer experiments.**
F. B. Neubauer, G. Dehmel, G. Musmann, A. Maier, E. Lammers.
Helios solar probes, science summaries, (see 003.009), p. 17 - 22 (1980).

032.557 **Rome-GSFC magnetic field experiment (E 3).**
N. F. Ness, F. Mariani, B. Bavassano, L. F. Burlaga, U. Villante,
Helios solar probes, science summaries, (see 003.009), p. 23 - 26 (1980).

032.558 **Electric field experiment (E 5b).** P. J. Kellogg.
Helios solar probes, science summaries, (see 003.009), p. 30 - 31 (1980).

032.559 **Radio astronomy experiment (E 5c).**
R. Stone.
Helios solar probes, science summaries, (see 003.009), p. 32 - 35 (1980).

032.560 **Cosmic ray experiment (E 6).**
H. Kunow, G. Wibberenz, G. Green, R. Müller-Mellin, M. Witte, H. Hempe.
Helios solar probes, science summaries, (see 003.009), p. 36 - 51 (1980).

032.561 **Cosmic ray experiment (E 7).**
J. H. Trainor, F. B. McDonald, T. L. Cline, U. D. Desai, B. J. Teegarden, K. G. McCracken, M. Van Hollebeke.
Helios solar probes, science summaries, (see 003.009), p. 52 - 61 (1980).

032.562 **Spectrometer for measurements of low-energy electrons and ions (E 8).** E. Keppler, A. K. Richter, K. Richter, G. Umlauft, B. Wilken, D. J. Williams.
Helios solar probes, science summaries, (see 003.009), p. 62 - 67 (1980).

032.563 **Zodiacal light experiment (E 9).**
C. Leinert, E. Pitz, H. Link, I. Richter, M. Hanner.
Helios solar probes, science summaries, (see 003.009), p. 68 - 74 (1980).

032.564 **The micrometeoroid analyzer (E 10). Micrometeoroid experiment – data analysis.**
E. Grün, H. Fechtig, J. Kissel, P. Gammelin, R. H. Giese, K. D. Schmidt, G. Schwehm.
Helios solar probes, science summaries, (see 003.009), p. 75 - 82 (1980).

032.565 **Faraday rotation experiment (E 12).**
G. S. Levy, H. Volland, M. K. Bird, C. T. Stelzried, B. L. Seidel.
Helios solar probes, science summaries, (see 003.009), p. 85 - 94 (1980).

032.566 **Helios occultation experiment – time delay measurements (E OC).**
P. Edenhofer, P. B. Esposito, E. Lüneburg.
Helios solar probes, science summaries, (see 003.009), p. 95 - 102 (1980).

032.567 **Large-aperture high-resolution X-ray collimator for the Solar Maximum Mission.** R. A. Nobles, L. W. Acton, E. G. Joki, J. W. Leibacher, R. C. Peterson.
Appl. Opt., Vol. 19, 2957 - 2966 (1980).

032.568 **Spherical crystal imaging spectrometer (SCIS) for cosmic X-ray spectroscopy.**
H. W. Schnopper, P. O. Taylor.
Appl. Opt., Vol. 19, 3306 - 3312 (1980).
The application of a spherically bent crystal X-ray spectrometer to cosmic X-ray problems is discussed. This is the only geometry whose diffraction properties are preserved under all rotations of the spacecraft. The combination of Bragg reflection and spherical aberration provides for stigmatic imaging of extended sources and minimum spatial and/or spectral resolution loss arising from source extent and spacecraft pointing errors. The sensitivity of the instrument is discussed in the context of a Spacelab mission.

032.569 **Holographic cylindrical grating for cosmic X-ray and XUV spectroscopy in grazing incidence.**
M. Singh, S. Singh.
Appl. Opt., Vol. 19, 3313 - 3317 (1980).
The theory of the holographically prepared grating on a cylindrical surface has been developed, to be applicable in grazing incidence, in a convergent beam of light and in transmission in the objective mode for the design of X-ray and XUV spectrographic systems. The design parameters for a grating with minimum astigmatism for the entire wavelength range up to 0.050μm and zero astigmatism at one wavelength in this region have been calculated. Coma type aberration and resolution at different wavelengths have also been estimated.

032.570 **Solar physics in the VUV: the importance of high resolution observations.** G. E. Brueckner.
Appl. Opt., Vol. 19, 3994 - 4001 (1980).

032.571 **Ultraviolet spectroscopy of interstellar and intergalactic matter.** J. M. Shull.
Appl. Opt., Vol. 19, 4002 - 4006 (1980).

032.572 **The IHXR80 hard X-ray experiment.** P. Ubertini, L. Boccaccini, C. La Padula, M. Mastropietro, G. Medici, R. Patriarca, F. V. Polcaro.
Non-solar gamma-rays, (see 012.047), p. 241 - 245 (1980).
Large area Multiwire Proportional Chambers for hard X-ray Astronomy have been succesfully employed in the HXR76 and HXR78 balloon borne experiments flown from the CNR Milo Base (Sicily) respectively on-board transatlantic and transmediterranean flights.

032.573 **The NASA gamma ray observatory (G.R.O.).** P. Durouchoux.
Non-solar gamma-rays, (see 012.047), p. 247 - 264 (1980).
The gamma-ray observatory (G.R.O.) is a project to carry gamma-ray astronomy forward in the 1980's. The G.R.O. is planned as a free-flying satellite to be launched from the space shuttle, carrying 5 large gamma-ray experiments aimed at providing data for a major advance in gamma-ray astronomy. These experiments, with their scientific objectives and sensitivities, are described in this paper.

032.574 **The India/Soviet Gamma Ray Astronomy Programme.** S. V. Damle.
Non-solar gamma-rays, (see 012.047), p. 265 - 270 (1980).
The Indo-Soviet Collaborative Programme in Gamma Ray Astronomy (GRISP) was launched in 1976. A series of balloon flights for the Anna-6 and Natalya-I oriented gamma ray telescopes were conducted in India by the National Balloon Facility of the Tata Institute of Fundamental Research, from Hyderabad during 1977 autumn and 1979 spring seasons. This paper reviews the status of the GRISP programme.

032.575 **Background effects between 250 km and 1100 km.** D. R. Parsignault.
Non-solar gamma-rays, (see 012.047), p. 271 - 274 (1980).
The author presents data on the cosmic-ray-induced and electron-induced background obtained with a pair of large-area, argon-filled, proportional counters, part of the Hard X-ray Experiment, flown on board the Astronomical Netherlands Satellite. These data can be used to estimate the expected total background counting rates in the energy range of 1.4 - 7.2 keV of any proposed proportional counter experiment in Earth orbit.

032.576 **Operating body of two components for a swimming probe in the Venus atmosphere.**
G. M. Moskalenko.
Kosm. Issled., Tom 18, 914 - 917 (1980). In Russian.

032.577 **Entwicklung eines 1m-Teleskops für den extremen Ultraviolett-Bereich.**
M. Grewing, C. Wulf-Mathies, G. Krämer.
Mitt. Astron. Ges., Nr. 50, p. 53 - 56 (1980).

032.578 **ASTRO-HEL, ein Raketenexperiment zur Beobachtung der Heliumhintergrundstrahlung bei 30,4 nm und 58,4 nm.**
C. Wulf-Mathies, H. J. Fahr, G. Lay.
Mitt. Astron. Ges., Nr. 50, p. 146 - 151 (1980).

032.579 **Description of the EXUV, GRO and AXAF satellites.** X-ray and gamma-ray astronomy in the 1980's, (see 012.054), p. 40 - 41 (1979).
Contents: The extreme ultraviolet and the extreme ultraviolet/soft X-ray sky surveys (EUV and EXUV). The gamma ray observatory (GRO). The advanced X-ray astronomy facility (AXAF).

"Sojus-22" erforscht die Erde.
See Abstr. 003.117.

Introduction to "Solar flares. A monograph from Skylab Solar Workshop II". See Abstr. 011.041.

X-ray astronomy. See Abstr. 013.041.

A processor for compression of multi-spectral image data on-board remote sensing satellites.
See Abstr. 021.001.

Diffraction limit in coronagraphs.
See Abstr. 031.009.

Comparison of Wolter I and Kirkpatrick-Baez X-ray optics for a Spacelab LAMAR facility. See Abstr. 031.022.

Image processing for the ESA faint object camera.
See Abstr. 031.508.

Calibration and image processing software for the Space Telescope Faint Object Camera. See Abstr. 031.509.

Global astrometry by space techniques.
See Abstr. 031.514.

New experimental possibilities and the future at far IR wavelengths. See Abstr. 031.560.

Deuteriumlampen und Wolframbandlampen als raumfahrttaugliche Kontinuumstrahler im Spektralbereich 160 nm - 3,5 μm. See Abstr. 034.074.

The Einstein Observatory: New perspectives in astronomy. See Abstr. 051.002.

The astrometry satellite Hipparcos.
See Abstr. 051.004.

Scientific projects under development.
See Abstr. 051.015.

Infrared Space Observatory (ISO). Pre-phase A study.
See Abstr. 051.031.

The contribution of OSO-8 to solar physics from data analyzed as of May 1979. See Abstr. 073.104.

Una medicion de lineas de radiacion γ atmosferica a bajas latitudes. See Abstr. 082.061.

Coordinated ionospheric and magnetospheric observations from the ISIS 2 satellite by the ISIS 2 experimenters. Volume 1. Optical auroral images and related direct measurements. See Abstr. 084.044.

Coordinated ionospheric and magnetospheric observations from the ISIS 2 satellite by the ISIS 2 experimenters. Volume 3. High-latitude charged particle, magnetic field, and ionospheric plasma observations during northern summer. See Abstr. 084.109.

Infrared remote sensing of the atmosphere of Venus from the Pioneer 12 orbiter. See Abstr. 093.070.

Extra-atmospheric cometary observations. I. Instruments. See Abstr. 103.002.

Two-dimensional photographic photometry of the zodiacal light from spatial observations. See Abstr. 106.025.

Description of ten investigations. See Abstr. 142.131.

The vertical component of 1–20 MeV gamma rays at balloon altitudes. See Abstr. 142.510.

Energy spectra of cosmic-ray nuclei to above 100 GeV per nucleon. See Abstr. 143.006.

Flugzeugbeobachtungen von Galaxien im fernen IR-Bereich. See Abstr. 158.307.

Erratum

032.901 **Erratum: "Die Justierung parallaktischer Montierungen mit Hilfe von Taschenrechnern"** [Orion, 38. Jahrg., 100 - 102 (1980)]. A. H. Kleyn, H. U. Fuchs. Orion, 38. Jahrg., 122 (1980). – See Abstr. 27.032.043.

033 Radio Telescopes and Equipment

033.001 **Nobeyama radiospectrographs for solar observations.** K. Kai, M. Sawa, Y. Shiomi, S. Aiba, H. Sekiguchi, N. Shibuya, T. Kosugi, H. Nakajima.
Publ. Astron. Soc. Japan, Vol. 32, 371 - 376 (1980).

Descriptions are given of high-sensitivity radiospectrographs for monitoring solar activities at decimeter- and meter-wavelengths. The radio spectrographs have been operational since September 1977 at the Nobeyama Solar Radio Observatory, cooperating with both the 160-MHz and 17-GHz interferometers. The radiospectrographs consist of a multichannel spectrograph in the 70–220-MHz band and an acousto-optical spectrograph in the 200–600-MHz band. Both of them are able to measure the right handed- and left handed-circular polarizations on a time-sharing basis. Some dynamic spectra of solar bursts recorded with the radiospectrographs are presented.

033.002 **VLB interferometer sensitivity and the number of compact radio sources.** D. B. Shaffer.
Bull. American Astron. Soc., Vol. 12, 494 - 495 (1980). Abstract.

033.003 **Initial observations with the 4-metre millimetre-wave telescope at Epping.**
F. F. Gardner, R. A. Batchelor, M. G. McCulloch, L. W. Simons, J. B. Whiteoak.
Proc. Astron. Soc. Australia, Vol. 3, 264 - 266 (1978).

033.004 **Molecular line observations at Parkes with a 3 GHz maser.** B. Höglund, J. B. Whiteoak, F. F. Gardner.
Proc. Astron. Soc. Australia, Vol. 3, 321 - 323 (1979).

033.005 **Initial observations with the Tidbinbilla 64-m telescope at λ = 13.5 mm.**
N. Fourikis, D. L. Jauncey.
Proc. Astron. Soc. Australia, Vol. 3, 353 - 354 (1979).

033.006 **A proposed correlator backend for the Culgoora radioheliograph.** D. J. McLean, M. Beard, A. Bos.
Proc. Astron. Soc. Australia, Vol. 3, 371 - 375 (1979).

033.007 **Radio astronomy: achievements and prospects.** N. Kardashev.
Sky Telesc., Vol. 60, 268 - 269 (1980).

033.008 **A computer simulation for evaluating the array performance of the 10-mϕ 5-element super-synthesis telescope.** K.- I. Morita, M. Ishiguro.
Proc. Res. Inst. Atmos. Nagoya Univ., Vol. 27, 49 - 60 (1980). Abstr. in Phys. Abstr., Vol. 83, Abstr. 98548 (1980).

033.009 **A new control system for the improved λ 8-cm radioheliograph.** M. Nishio.
Proc. Res. Inst. Atmos. Nagoya Univ., Vol. 27, 61 - 77 (1980). Abstr. in Phys. Abstr., Vol. 83, Abstr. 98549 (1980).

033.010 **The Jodrell Bank radio-linked interferometer network.**
J. G. Davies, B. Anderson, I. Morison.
Nature, Vol. 288, 64 - 66 (1980).

The Multi Telescope Radio Linked Interferometer (MTRLI) has just been brought into operation at Jodrell Bank and here the authors introduce the instrument, describe its capabilities, and present some of the first maps to be made with it. MTRLI produces high quality maps of radio sources with resolutions varying from ~1 arc s to ~0.02 arc s depending on the frequency of operation.

033.011 **Radiointerferometer Crimea-Pushchino.**
L. I. Matveenko, L. R. Kogan, L. S. Chesalin, V. I. Kostenko, A. Kh. Papatsenko, G. D. Kopelyanskij. I. G. Moiseev, V. A. Efanov, R. L. Sorochenko.
Pis'ma Astron. Zh., Tom 6, 662 - 666 (1980). In Russian. English translation in Soviet Astron. Lett., Vol. 6.

The main parameters of the Crimea-Pushchino radiointerferometer and results of observation of the H_2O maser sources W 49, W 51 and Orion A are given.

033.012 **Interplanetary scintillation – preliminary observations at 103 MHz.** S. K. Alurkar, R. V. Bhonsle.
Solar and interplanetary dynamics, (see 012.020), p. 405 - 408 (1980).

A 3-station interplanetary scintillation (IPS) observatory is being developed mainly with a view to study the solar wind plasma.

033.013 **The 8mm wavelength solar radio telescope of Purple Mountain Observatory.**
Acta Astron. Sinica, Vol. 21, 314 - 316 (1980). In Chinese.

A radio telescope operating at 35 GHz has been constructed to observe solar radiation and to measure atmospheric absorption. The paraboloid dish and the pyramid horn have been used for relative and absolute measurements respectively.

033.014 **Maser amplifiers.** E. Kollberg.
Interstellar molecules, (see 012.033), p. 615 - 617 (1980).

033.015 **Wideband spectrometers for millimetre wavelengths.** B. J. Robinson.
Interstellar molecules, (see 012.033), p. 619 - 623 (1980).

033.016 **Acousto-optic radiospectrometers for mm-wave spectroscopy.** Y. Chikada, N. Ukita, J. Inatani, N. Kaifu, S. Kodaira.
Interstellar molecules, (see 012.033), p. 625 - 626 (1980).

033.017 **Future spectral line research with the VLA.** K. J. Johnston.
Interstellar molecules, (see 012.033), p. 627 - 629 (1980).

033.018 **Resultaten met amateur radioteleskoop.** R. Peelen.
Zenit, 7e Jaarg., 392 - 393 (1980).

033.019 **Absolute brightness temperature measurements at 3.5-mm wavelength.**
B. L. Ulich, J. H. Davis, P. J. Rhodes, J. M. Hollis.
IEEE Trans. Antennas Propag., Vol. AP-28, 367 - 377 (1980) = Natl. Radio Astron. Obs., Green Bank, Repr. Ser. A, No. 1103.

Observations have been made at 86.1 GHz to derive the absolute brightness temperatures of the Sun (7914±192 K), Venus (357.5±13.1 K), Jupiter (179.4±4.7 K), and Saturn (153.4±4.8 K) with a standard error of about three percent. A stable transmitter and novel superheterodyne receiver were constructed and used to determine the effective collecting area of the Millimeter Wave Observatory (MWO) 4.9-m antenna relative to a previously calibrated standard gain horn. The brightness temperatures may be used to establish an absolute calibration scale and to determine the antenna aperture and beam efficiencies of other radio telescopes at 3.5-mm wavelength.

033.020 **The VLA: ears on the universe.** C. A. Federer, Jr.
Sky Telesc., Vol. 60, 472 - 473 (1980).

033.021 **Strong coma lobes from small gravitational deformations.** S. von Hoerner.
IEEE Trans. Antennas Propag., Vol. AP-28, 652 - 657 (1980). Abstr. in Phys. Abstr., Vol. 84, Abstr. 4767 (1981).

033.022 **A new 17-GHz solar radio interferometer at Nobeyama.** H. Nakajima, H. Sekiguchi, S. Aiba, Y. Shiomi, T. Kuwabara, M. Sawa, H. Hirabayashi, T. Kosugi, K. Kai.
Publ. Astron. Soc. Japan, Vol. 32, 639 - 650 (1980).

The design of the interferometer is based particularly on the following two requirements: to achieve a time resolution high enough to follow rapidly changing phenomena such as impulsive bursts, and to make adjustment of the phase and gain as accurately and frequently as possible. The interferometer is able to produce east-west images of the whole Sun with a time resolution up to 0.8 s for two circular polarizations. The spatial resolution is ~45". The sidelobe levels can be reduced to ≤ 3% under a good condition.

033.023 **The Very Large Array.** A. R. Thompson, B. G. Clark, C. M. Wade, P. J. Napier.
Astrophys. J., Suppl. Ser., Vol. 44, 151 - 167, plates 1 - 2 (1980).

The VLA, located on the plains of San Augustin in west-central New Mexico, provides the capability of mapping the sky at centimeter and decimeter wavelengths with resolution down to tenths of arcseconds and sensitivity of approximately 100 microjanskys. The 27 antennas, each of 25 m diameter, are arranged on a three-armed array. The paper describes the array at the completion of construction at the end of 1980.

033.024 **A new high-gain, broadband, steerable array to study Jovian decametric emission.**
A. Boischot, C. Rosolen, M. G. Aubier, G. Daigne, F. Genova, Y. Leblanc, A. Lecacheux, J. De La Noë, B. Møller-Pedersen.
Icarus, Vol. 43, 399 - 407 (1980).

A large array of antennae has been built at the Radioastronomy Observatory, Nançay, France, to study solar and planetary decametric emissions. This array has a high gain (25db) in a broad range of frequencies and is steerable through a large part of the sky. The authors present the main characteristics of this array, and the receivers which are used to show the importance of the equipment for Jovian studies. The authors summarize the results already obtained and describe some topics which are presently being studied.

033.025 **The Canadian Very-Long-Baseline Array.** T. H. Legg.
J. R. Astron. Soc. Canada, Vol. 74, 363 (1980). – Abstract.

033.026 **A new spectroscopic facility at millimetre wavelengths.** A. Baudry, J. Brillet, J. M. Desbats, J. Lacroix, G. Montignac, P. Encrenaz, R. Lucas, G. Beaudin, P. Dierich, A. Germont, P. Landry, G. Rerat.
J. Astrophys. Astron., Vol. 1, 193 - 196 (1980).

A new millimetre-wave facility is in operation at the Bordeaux Observatory for spectroscopic observations of interstellar and stratospheric molecules. A cooled receiver has been installed on a 2.5-m radio telescope. The overall system temperature is in the range 400 to 600 K (single side band) in the operating frequency range 75 to 115 GHz.

Low-frequency radio astronomy in Antarctica. See Abstr. 013.036.

Radioastronomische waarneemtechnieken: 1950 - 1980. See Abstr. 031.568.

Incoherent optical 1-bit cross-correlators for radio antenna arrays. See Abstr. 031.597.

Some recent explorations of the solar corona from the Culgoora Solar Radio Observatory. See Abstr. 077.024.

Radio studies of stellar activity with the Arecibo interferometer. See Abstr. 116.002.

Radio observations with a wide fractional bandwidth. See Abstr. 141.521.

034 Auxiliary Instrumentation

034.001 **SEC Vidicon photometry of the main sequence of ω Centauri.** A. W. Rodgers, E. B. Newell, T. Stapinski, P. Harding, J. Norris.
Publ. Astron. Soc. Pacific, Vol. 92, 288 - 299 (1980).

A television area-photometer based on a computer controlled SEC Vidicon is described. The performance of this photometer is discussed in terms of the results obtained during tests on laboratory point sources and on a stellar *UBV* sequence. The authors apply the detector to a study of the main-sequence turnoff stars in ω Cen.

034.002 **Near infrared photography with a vacuum-cold camera.**
G. S. Rossano, R. W. Russell, R. H. Cornett.
Publ. Astron. Soc. Pacific, Vol. 92, 357 - 361 (1980).

Sensitized cooled plates have been obtained of the Orion nebula region and of Sh2-149 in the wavelength ranges 8000 Å–9000 Å and 9000 Å–11,000 Å with a recently designed and constructed vacuum-cold camera. Sensitization procedures are described and the camera design is presented. The Orion plates reveal a very red object, demonstrating one application of the system.

034.003 **Thermal background subtraction in photodiode detectors.** J. W. Percival, K. H. Nordsieck.
Publ. Astron. Soc. Pacific, Vol. 92, 362 - 367 (1980).

The nature of the thermal diode background in integrating photodiode detectors is investigated. It is found that a naive linear subtraction of a scan of background (thermal plus light) from a scan of signal plus background introduces a small multiplicative error in the apparent gain of the detector. The fractional gain error is approximately equal to the fraction of diode saturation due to integrated dark current and does not depend on the amount of illumination. A simple linear correction algorithm is described. Application of this algorithm allows the use of integration times for which diode background becomes a large fraction of diode saturation, thus allowing detection of lower light levels.

034.004 **An integrating preamplifier for indium antimonide infrared detectors.** J. R. Barton, D. A. Allen.
Publ. Astron. Soc. Pacific, Vol. 92, 368 - 374 (1980).

The authors describe electronics developed at the Anglo-Australian Observatory for use with InSb infrared detectors. The circuit avoids some of the disadvantages of conventional preamplifiers by eliminating the feedback resistor, and offers a sensitive and versatile infrared system.

034.005 **High-resolution solar spectroscopy with double-pass solar spectrometers.** Eh. A. Gurtovenko.
Astrometr. Astrofiz., Vyp. (No.) 41, p. 15 - 28 (1980). In Russian.

A short description is given for the main characteristics and properties of double-pass systems used for high-resolution solar spectroscopy. The decisive influence of the intermediate slit upon the wings of the apparatus function is emphasized. The pass of beams in the system equipped with a right-angle mirror is described and the formulas for the apparatus function are deduced. The main errors of the double-pass solar spectrometers are considered and some suggestions on the adjustment of the apparatus are given.

034.006 **Implementation of a fiber coupled spectrograph.** S. C. Barden, L. W. Ramsey, R. J. Truax.
Bull. American Astron. Soc., Vol. 12, 460 - 461 (1980). Abstract.

034.007 **A polarimeter for a Fourier Transform Spectrometer and initial solar observations.**
J. Harvey, J. Brault, J. Stenflo, C. Zwaan.
Bull. American Astron. Soc., Vol. 12, 476 (1980). – Abstract.

034.008 **The MSFC solar correlation tracker.** N. P. Cumings, W. R. McIntosh.
Bull. American Astron. Soc., Vol. 12, 476 (1980). – Abstract.

034.009 **A CCD imaging polarimeter.** J. A. Tyson. Bull. American Astron. Soc., Vol. 12, 488 (1980). Abstract.

034.010 **Photon-counting CCD camera systems for astronomy.**
W. C. Wickes, D. G. Currie, J. L. Johnson, L. A. Buennagel.
Bull. American Astron. Soc., Vol. 12, 488 - 489 (1980). Abstract.

034.011 **Differential camera.** D. Socker, B. Woodgate, L. Brown, R. W. Hobbs.
Bull. American Astron. Soc., Vol. 12, 489 (1980). – Abstract.

034.012 **A CCD camera for astrometric observations.** D. G. Currie, J. L. Hershey, L. A. Buennagel, W. C. Wickes.
Bull. American Astron. Soc., Vol. 12, 499 (1980). – Abstract.

034.013 **A new multichannel polarimeter.** L. Tomaszewski, J. Landstreet, G. Symonds.
Bull. American Astron. Soc., Vol. 12, 499 (1980). – Abstract.

034.014 **Acousto-optic spectral line receiver for molecular astronomy.** G. Chin, D. Buhl, J. M. Florez.
Bull. American Astron. Soc., Vol. 12, 506 - 507 (1980). Abstract.

034.015 **1 mm continuum observations.** T. L. Roellig, J. R. Houck.
Bull. American Astron. Soc., Vol. 12, 526 (1980). – Abstract.

034.016 **Automatic digital micrometer.** V. A. Psarev. Vestn. Khar'k. Univ., No. 190. Fiz. Luny Planet. Fundam. Astrometr., Vyp. 14, p. 27 - 38 (1979). In Russian.

034.017 **Astronomical spectrometer using a charge coupled device detector.** S. S. Meyer.
Rev. Sci. Instrum., Vol. 51, 638 - 641 (1980). – Abstr. in Phys. Abstr., Vol. 83, Abstr. 82207 (1980).

034.018 **Differential photometer controlled by microprocessor.** D. Dubet, M. Rouxel.
Tec. Regul. Mando Autom., No. 94, p. 38 - 40, 42 - 45 (1979). In Spanish. – Abstr. in Phys. Abstr., Vol. 83, Abstr. 82222 (1980).

034.019 **Narrow-field radiometry in a quasi-isotropic atmosphere.** A. Holmes, J. M. Palmer, M. G. Tomasko.
Proc. Soc. Photo-Opt. Instrum. Eng., Vol. 196, (see 012.016), 27 - 32 (1979). – Abstr. in Phys. Abstr., Vol. 83, Abstr. 90345 (1980).

034.020 **Data recording system for 60 cm reflector at KSC.** K. Takagishi, M. Matsuoka.
Bull. Inst. Space Aeronaut. Sci., Univ. Tokyo B, Vol. 15, 611 - 625 (1979). In Japanese. – Abstr. in Phys. Abstr., Vol. 83, Abstr. 94717 (1980).

034.021 **Calculation of illumination in the shadow cast by an external occulting shield of a coronograph. III. Double disk.** A. V. Lenskij.
Problems of cosmic physics. Vyp. 15, (see 003.003), p. 29 - 32 (1980). In Russian.

034.022 **A low-cost interferometrically controlled grid-generator.** J. F. Grainger, J. A. Job.
Astrophys. Space Sci., Vol. 72, 237 - 240 (1980).
A low-cost easily constructed instrument for the manufacture of repeated element grids is described. Such grids may be used as entrance or exit apertures in spectrometers. The instrument is interferometrically controlled and can produce grids with regular spacing or in accordance with some desired function. The shape of the repeated element is in the operator's control.

034.023 **Protuberanzfotos mit Kurzprotuberanzansatz.** J. Salami.
Orion, 38. Jahrg., 154 - 155 (1980).

034.024 **Due filtri in aiuto degli astrofili.** P. Bianucci.
Orione, Vol. 2, 111 - 115 (1980).

034.025 **Measurements of low intensities by means of a silicon-intensified-target vidicon.** R. Schielicke.
Feingerätetechnik, Vol. 29, 168 - 170 (1980). In German. Abstr. in Phys. Abstr., Vol. 83, Abstr. 98545 (1980).

034.026 **Michelson spectrometer system.** H. L. Johnson, T. D. Fay, W. Z. Wisniewski.
System aspects of electro-optics. Conference held at Huntsville, Ala., USA, 22 - 23 May 1979. Proc. Soc. Photo-Opt. Instrum. Eng., Vol. 187, 141 - 147 (1979). – Abstr. in Phys. Abstr., Vol. 83, Abstr. 98552 (1980).

034.027 **Optimal sampling of charge-coupled devices.** D. J. Hegyi, A. Burrows.
Astron. J., Vol. 85, 1421 - 1424 (1980).
The authors show that when the output noise spectrum of a charge-coupled device is flat, the maximum signal-to-noise ratio is obtained using a differential-averaging sampling function (positive when negative square wave sampling). However, in astronomical applications, where image quality rather than readout speed is more desirable, $1/f$ flicker noise of the on-chip amplifier dominates the noise spectrum. They show that even when $1/f$ noise dominates, the same differential-averaging sampling function yields a signal-to-noise ratio which is never more than 36% from optimum. Their results were obtained using a matched-filter analysis.

034.028 **New experimental possibilities and future prospects for $1-5\mu m$ infrared spectroscopy of interstellar molecules.** D. N. B. Hall.
Interstellar molecules, (see 012.033), p. 631 - 632 (1980).

034.029 **A 10 micron heterodyne receiver for ultra high resolution astronomical spectroscopy.**
D. Buhl, G. Chin, J. Faris, T. Kostiuk, M. J. Mumma, D. Zipoy.
Interstellar molecules, (see 012.033), p. 633 - 636 (1980).

034.030 **A portable photometer and data recording system.** E. Dunham, R. Baron, J. L. Elliot.
Bull. American Astron. Soc., Vol. 12, 691 (1980). – Abstract.

034.031 **An important nonlinearity in the counting properties of intensified dissector scanners.**
P. M. Rybski.
Bull. American Astron. Soc., Vol. 12, 751 - 752 (1980). Abstract.

034.032 **An RCA charge-coupled device used as the detector in a Cassegrain spectrophotometer.**
P. M. Rybski, K. Darden.
Bull. American Astron. Soc., Vol. 12, 752 (1980). – Abstract.

034.033 **A high-accuracy optical polarization analyser.** M. Semel.
Astron. Astrophys., Vol. 91, 369 - 371 (1980). In French.
A system of optical analyser suitable for the precise measurement of polarization (down to 10^{-4} or less) is described. Two states of polarization in the solar spectrum for a wide spectral range and for many solar points can be observed simultaneously. The analysis of polarization is obtained by using a pair of quartz crystals ensuring the same optical path for the two beams, whose separation is very small.

034.034 **Ervaringen met een H-alpha filter.** A. Mak.
Zenit, 7e Jaarg., 76 - 78 (1980).

034.035 **Universeel besturingsapparaat voor meteoorkamera's.** P. B. van der Wal.
Zenit, 7e Jaarg., 240 - 242 (1980).

034.036 **Kruisdraadoculair.** H. J. de Visser.
Zenit, 7e Jaarg., 507 (1980).

034.037 **The Digicon system of Cima Ekar Observatory.** C. Barbieri, F. Bortoletto, S. Di Serego Alighieri.
Astrophys. Space Sci., Vol. 73, 199 - 206 (1980).
This paper describes the Digicon system in current operation for spectroscopic acquisition at the 182-cm telescope of Cima Ekar (Asiago). An example of observation of the emission-line galaxy VV 565 provides evidence of the quality of the system.

034.038 **Spectropolarimetry with an intensifier-dissector-scanner.** L. Tomaszewski, G. R. Symonds, J. D. Landstreet.
Publ. Astron. Soc. Pacific, Vol. 92, 518 - 527 (1980).
A spectropolarimeter has been constructed which combines a Pockels cell polarization modulator, a standard Cassegrain spectrograph, and an intensifier-dissector-scanner-type detector. The resulting instrument is capable of measuring a wavelength dependence of polarization over a 2000 Å band with 15 Å-20 Å resolution. This paper describes the distinctive design features of the spectropolarimeter and number of tests of its performance.

034.039 **Astrometric stability of the Texas Mark II electronographic camera.** J. D. Mulholland, P. J. Griboval.
Published by Univ. Texas, Austin. 12 pp. (1980). Paper submitted to Astron. J.

034.040 **A spectrum comparator for astronomical purposes.** A. Ardeberg, N. Hansson, E. Olsen.
Rep. Obs. Lund, No. 15, 8 pp. (1979).
An outline is given of a spectrum comparator for astronomical purposes. The design is adequate for low-cost construction in a well-equipped workshop. Features include equally-imaging light paths, projection, individual illumination and magnification and provision for rough radial-velocity measurements. The construction is discussed to some extent.

034.041 **El espectrografo "Coudé" del telescopio de 1.5 metros.** J. Pensado.
Bol. Astron. Obs. Madrid, Vol. 10, No. 3, p. 9 - 27 (1978).

034.042 **A multichannel polarimeter. III. Test observations.** S. Kikuchi, M. Konno, Y. Mikami.
Tokyo Astron. Obs. Rep. (No. 73), Vol. 19, 217 - 231 (1980). In Japanese.

034.043 **PMT shutter for satellite laser ranging systems.** H. Sato.
Tokyo Astron. Obs. Rep. (No. 73), Vol. 19, 278 - 287 (1980). In Japanese.

034.044 **The ESO 1 m Schmidt telescope equipped with a Racine wedge.** A. B. Muller.
Messenger, No. 22, p. 18 - 19 (1980).

034.045 **Analogical device for a rough localization of gravitational-wave sources.** S. Frasca.
Nuovo Cimento C, Ser. 1, Vol. 3C, 237 - 255 (1980). – Abstr. in Phys. Abstr., Vol. 84, Abstr. 4770 (1981).

034.046 **A photometer for infrared astronomy.** M. Roth, L. Carrasco, J. Franco, G. Resendiz.
Rev. Mexicaña Fis., Vol. 27, 39 - 54 (1980). In Spanish. Abstr. in Phys. Abstr., Vol. 84, Abstr. 9375 (1981).

034.047 **Velocity fields in late-type galaxies from Hα Fabry-Perot interferometry. I. Instrumentation and data reduction.** G. de Vaucouleurs, W. D. Pence.
Astrophys. J., Vol. 242, 18 - 29 (1980).

An image-tube Hα Fabry-Perot interferometer used for the mapping of velocity fields in late-type galaxies with the Galaxymeter at McDonald Observatory is described. The semiautomated method of reduction leading to the construction of detailed velocity maps at 4″ to 10″ resolution from 500 to 5000 velocities per galaxy is outlined. Accidental and systematic errors are discussed.

034.048 **Polarization study of the solar corona using a double polarigraph.** K. A. Raju.
Bull. Astron. Soc. India, Vol. 8, 65 - 68 (1980).

The double polarigraph, used in the total solar eclipse of 16th Feb. 1980 to study the polarization of the corona at two wavelengths, is described briefly. The negatives are being studied.

034.049 **Multiple object spectroscopy: the Medusa spectrograph.**
J. M. Hill, J. R. P. Angel, J. S. Scott, D. Lindley, P. Hintzen.
Astrophys. J., Lett., Vol. 242, L69 - L72, plates L4 - L7 (1980).

The authors have built and tested an instrument to obtain simultaneous spectra of many objects in the field of view of the Steward 90 inch (2.29 m) telescope. The present device, while already giving a sixfold reduction in the mean telescope time per galaxy, has significant light losses because it is not ideally matched to the telescope. An instrument being designed for the prime focus will transmit light from each object as efficiently as a conventional spectrograph.

034.050 **Bandpass interference filters for very far infrared astronomy.** J. E. Davis.
Infrared Phys., Vol. 20, 287 - 290 (1980).

Three bandpass filters have been constructed for use with a 0.3K bolometer. The center frequencies and bandwidths of the filters are chosen to correspond with the 1, 2 and 3 mm atmospheric windows. The filters are formed of resonant metallic grids separated by air spacers. Peak transmissivities of about 90% at 3 mm and 65% at 1 mm have been achieved.

034.051 **An infrared hygrometer for astronomical site testing.** E. Büscher, D. Lemke.
Infrared Phys., Vol. 20, 321 - 325 (1980).

The paper describes an i.r. hygrometer which measures the absolute water vapour content in the atmosphere. The instrument is easy to handle, it can be used with either the sun or the moon as a radiation source, and its design minimizes errors due to atmospheric aerosols. The instrument has already been used for astronomical site testing.

034.052 **Improved Wadsworth mounting with aspherical holographic grating.** M. Duban.
Appl. Opt., Vol. 19, 2488 - 2489 (1980).

034.053 **Matched tandem etalon camera–MATEC–and its application to auroral observations.**
E. R. Young, K. C. Clark.
Appl. Opt., Vol. 19, 2631 - 2637 (1980).

034.054 **Echelle efficiencies: theory and experiment.** D. J. Schroeder, R. L. Hilliard.
Appl. Opt., Vol. 19, 2833 - 2841 (1980).

Comparisons of theoretical calculations and experimental measurements of echelle grating efficiencies are given for $R2$ echelles used in three possible configurations: $\alpha > \beta$; $\alpha < \beta$; and the quasi-Littrow mode. The throughput-resolution products for these various cases are also compared.

034.055 **An integrating photometer with automatic data processing.** C.-j. Wang, X.-z. Shi, C.-j. Shen.
Acta Astron. Sinica, Vol. 21, 409 - 414 (1980). In Chinese.

The photometer is based on dc integrating circuit. It makes use of an A/D converter to transfer the analog data into digital ones. Through an interface these data are punched on paper tape. The data are processed and final results are obtained automatically. Two kinds of monitors are provided for real-time check.

034.056 **Autocollimation interferometer on diffraction gratings.** I. I. Dukhopel. A. G. Seregin,
G. N. Rassudova, L. N. Orlova, Yu. A. Bliznyuk.
Issled. po geomagn., aehron. i fiz. Solntsa, Moskva, 1980, No. 52, p. 94 - 95. In Russian. – Abstr. in Ref. zh., 51. Astron., 11.51.147 (1980).

034.057 **Solar spectrographs.** Eh. A. Gurtovenko.
Issled. po geomagn., aehron. i fiz. Solntsa, Moskva, 1980, No. 52, p. 96 - 112. In Russian. – Abstr. in Ref. zh., 51. Astron., 11.51.148 (1980).

034.058 **Interference solar spectrograph.** E. S. Kulagin.
Issled. po geomagn., aehron. i fiz. Solntsa, Moskva, 1980, No. 52, p. 113 - 115. In Russian. – Abstr. in Ref. zh., 51. Astron., 11.51.149 (1980).

034.059 **Investigation of the "vibration of the spectrum" in spectrographs.** S. M. Gorskij, I. E. Kozhevatov,
E. Kh. Kulikova, V. P. Lebedev, T. M. Soldatkina.
Issled. po geomagn., aehron. i fiz. Solntsa, Moskva, 1980, No. 52, p. 120 - 121. In Russian. – Abstr. in Ref. zh., 51. Astron., 11.51.151 (1980).

034.060 **Choosing the scheme of the spectrograph of the large solar vacuum telescope.** N. M. Firstova.
Issled. po geomagn., aehron. i fiz. Solntsa, Moskva, 1980, No. 52, p. 122 - 127. In Russian. – Abstr. in Ref. zh., 51. Astron., 11.51.152 (1980).

034.061 **Monochromatic filters for solar observations.** V. I. Skomorovskij, S. B. Ioffe.
Issled. po geomagn., aehron. i fiz. Solntsa, Moskva, 1980, No. 52, p. 128 - 149. In Russian. – Abstr. in Ref. zh., 51. Astron., 11.51.153 (1980).

034.062 **Narrow-band tunable Fabry-Perot filter.** Eh. V. Kononovich, O. B. Smirnova,
V. V. Chichmar'.
Issled. po geomagn., aehron. i fiz. Solntsa, Moskva, 1980, No. 52, p. 150 - 152. In Russian. – Abstr. in Ref. zh., 51. Astron., 11.51.154 (1980).

034.063 **Improvement of the characteristics of birefringent filters by applying different types of crystals.**
T. A. Vinogradova.
Issled. po geomagn., aehron. i fiz. Solntsa, Moskva, 1980, No. 52, p. 153 - 154. In Russian. – Abstr. in Ref. zh., 51. Astron., 11.51.155 (1980).

034.064 **Solar magnetographs.**
V. M. Grigor'ev, N. I. Kobanov.
Issled. po geomagn., aehron. i fiz. Solntsa, Moskva, 1980, No. 52, p. 155 - 176. In Russian. – Abstr. in Ref. zh., 51. Astron., 11.51.156 (1980).

034.065 **Cosmic ray spectrograph based on the effect of local generation of neutrons.**
V. L. Yanchukovskij, A. L. Yanchukovskij.
Issled. po geomagn., aehron. i fiz. Solntsa, Moskva, 1980, No. 52, p. 52 - 56. In Russian. – Abstr. in Ref. zh., 51. Astron., 11.51.257 (1980).

034.066 **On a possibility of making a cosmic ray spectrograph with controllable coupling coefficients.**
A. L. Yanchukovskij, V. L. Yanchukovskij.
Issled. po geomagn., aehron. i fiz. Solntsa, Moskva, 1980, No. 52, p. 57 - 59. In Russian. – Abstr. in Ref. zh., 51. Astron., 11.51.258 (1980).

034.067 **The two new ESO tools for stellar abundance work.**
J. Andersen.
ESO workshop on methods of abundance determination for stars, (see 012.021), p. 11 - 17 (1980).

034.068 **Instrumentation – existing and forthcoming.**
P. E. Nissen.
ESO workshop on methods of abundance determination for stars, (see 012.021), p. 19 - 23 (1980).
The aim of this paper is to inform about two new ESO instru ments, the Cassegrain Echelle Spectrograph, CASPEC, and the Coudè Echelle Spectrometer, CES, and to discuss these instruments from the user's point of view. Similar instruments already in operation at other observatories were described also.

034.069 **The remote-controlled spectrograph, area scanner, and spectropolarimeter for the Lick 3-m telescope.**
J. S. Miller, L. B. Robinson, G. D. Schmidt.
Publ. Astron. Soc. Pacific, Vol. 92, 702 - 712 (1980) = Lick Obs. Bull., No. 870.
The purpose of this paper is to present a description of the spectrograph used in conjunction with the image-tube, image-dissector scanner (ITS). In addition to the spectrograph the authors discuss the area scanner modification to the original ITS system which allows raster scanning and limited picture taking. They present the design of a new addition to the system, à polarization analyzer, which allows spectropolarimetry at the full spectral resolution of the spectrograph-ITS. With high through-put and simultaneous sky and sky-plus-star beams, the spectropolarimeter is a powerful device for faint-object study.

034.070 **Galaxy electrography at McDonald Observatory.**
P. Griboval.
Photometry, kinematics and dynamics of galaxies, (see 012.051), p. 321 - 323 (1979).

034.071 **Protective filters for solar observation.**
B. R. Chou.
J. R. Astron. Soc. Canada, Vol. 74, 365 (1980). – Abstract.

034.072 **Polarimeter designed for the investigation of elliptical polarization of solar radio emission.**
L. I. Tsvetkov.
Astron. Tsirk., No. 1113, p. 1 - 2 (1980). In Russian.

034.073 **Use of tilted interference filters in astronomical observations.**
V. G. Kornilov, A. M. Cherepashchuk.
Astron. Zh., Tom 57, 1331 - 1338 (1980). In Russian. English translation in Soviet Astron., Vol. 24, No. 6.

034.074 **Deuteriumlampen und Wolframbandlampen als raumfahrttaugliche Kontinuumstrahler im Spektralbereich 160 nm - 3,5 μm.**
U. Finkenzeller, W. Schwarz, D. Labs.
Mitt. Astron. Ges., Nr. 50, p. 51 - 53 (1980).

034.075 **Ein Infrarot-Absorptions-Hygrometer.**
E. Büscher, D. Lemke.
Mitt. Astron. Ges., Nr. 50, p. 56 (1980). – Abstract.

034.076 **Dünnschichtfilter für astronomische Messungen im extremen UV.** C. Wulf-Mathies, M. Grewing.
Mitt. Astron. Ges., Nr. 50, p. 141 - 146 (1980).

034.077 **Informational criterion for quality estimate of paralysable photon counters.**
D. Ralys, R. Kalytis.
Bull. Vilnius Astron. Obs., Nr. 53, p. 3 - 12 (1980). In Russian.
A statical analysis of the signal transfer in paralysable photon counters is made. The mean and variance of the output signal are expressed in terms of photon counter parameters. The informational criterion for a quality estimate (Kalytis and Ralys, 1978) was used in order to compare quantitatively paralysable and nonparalysable photon counters within their common dynamic range.

034.078 **A gradational criterion for photon counter quality estimate.** D. Ralys.
Bull. Vilnius Astron. Obs., Nr. 53, p. 13 - 21 (1980).

034.079 **Influence of operational conditions on the stability of the FEU-79 and FEU-106 photomultipliers.**
R. Kalytis.
Bull. Vilnius Astron. Obs., Nr. 53, p. 22 - 32 (1980). In Russian.
The influence of high voltage and temperature on the stability of the Soviet-made photomultipliers FEU-79 and FEU-106 is investigated.

034.080 **Superconductive bolometers in IR-astronomical technique. I. Integral sensitivity. Operating conditions.**
V. P. Kuz'kov.
Astrometr. Astrofiz., Vyp. (No.) 42, p. 90 - 96 (1980). In Russian.

034.081 **Integral thermo-sensor.**
G. M. Blank, V. S. Grigor'ev.
Vliyanie atmos. na astron. nablyudeniya v opt. i radiodiapazonakh. Tez. dokl. soveshch. probl.-temat. gruppy po teor. astrometr. sekts. astrometr. AS AN SSSR, 1980. Irkutsk, 1980, p. 71. In Russian. – Abstr. in Ref. zh., 51. Astron., 1.51.773 (1981).

034.082 **Two-channel photometer for submillimeter and infrared observations at the BTA 6-m telescope of the USSR Academy of Sciences.**
G. B. Sholomitskij, I. A. Maslov, N. A. Pankratov, Yu. V. Kulikov, V. G. Malyarov, S. A. Ignatenko, G. G. Popov, Yu. V. Nikol'skij, V. A. Soglasnova.
Inst. kosm. issled. AN SSSR. Prepr., 1980, No. 572, 22 pp. In Russian. – Abstr. in Ref. zh., 51. Astron., 1.51.807 (1981).

Charge-coupled devices and their applications.
See Abstr. 003.018.

The infrared handbook. See Abstr. 003.115.

Adjusting spatially separated plane mirrors to coplanarity. See Abstr. 031.025.

New experimental possibilities and the future at far IR wavelengths. See Abstr. 031.560.

Mesures absolues de la température apparente des mers lunaires en infra-rouge. See Abstr. 094.506.

The astrometric search for neighboring planetary systems. See Abstr. 118.036.

035 Clocks and Frequency Standards

035.001 **WWV synchronized microprocessor clock.**
G. P. McCook, F. P. Maloney, J. C. Lochner, C. C. Harris.
Bull. American Astron. Soc., Vol. 12, 462 - 463 (1980). Abstract.

035.002 **Een eenvoudig astronomisch uurwerk.**
H. A. J. Rynja.
Zenit, 7e Jaarg., 160 - 163 (1980).

035.003 **Environment effect on portable rubidium clocks in flight.** H.-q. Chen.
Publ. Shensi Astron. Obs., No. 1, p. 15 - 26 (1980).
According to the experiment of the time synchronization by flying rubidium atomic clocks, the environment effects on the clocks are discussed. To improve the synchronous precision, the modified coefficients to the environment are given.

035.004 **Predicting dissemination delay of the first hop of 100 kHz sky wave.**
Y.-r. Miao, K.-j. Yang, H.-q. Zheng.
Publ. Shensi Astron. Obs., No. 1, p. 27 - 34 (1980).

035.005 **The effect of solar activity on long wave propagation.** Y.-r. Miao, K.-j. Yang, L.-d. Pan.
Publ. Shensi Astron. Obs., No. 1, p. 35 - 42 (1980).

035.006 **A converter for receipt of time signals.**
R. Fangor.
Urania Kraków, Vol. 51, 305 - 310 (1980). In Polish.

035.007 **The system of time indication in astronomical pavilions.** V. K. Budz'ko, B. I. Brodskij.
Vrashchenie i priliv. deformatsii Zemli, Kiev, 1980, No. 12, p. 89 - 95. In Russian. – Abstr. in Ref. zh., 51. Astron., 1.51.774 (1981).

The history of the chronometer.
See Abstr. 004.036.

A unique Greek sundial recently discovered in Central Asia. See Abstr. 004.054.

Comparison and coordination of time scales.
See Abstr. 044.006.

Research of timekeeping in setting up atomic time scale. See Abstr. 044.033.

036 Photographic Materials and Techniques

036.001 **Vapor-deposited graded-thickness films.**
H. Herzig, R. S. Spencer.
NASA Tech. Briefs, Vol. 4, No. 1, p. 142 (1979). – Abstr. in Phys. Abstr., Vol. 83, Abstr. 76164 (1980).

036.002 **L'effetto Sabattier: l'applicazione della teoria delle isodense nella fotografia astronomica di oggetti estesi.** A. Buzzoni.
Coelum, Anno 50, 195 - 210 (1980).

036.003 **Techniques of astronomical photography.**
O. D. Dokuchaeva.
Zemlya Vselennaya, 1980, No. 5, p. 26 - 30. In Russian.

036.004 **Photographic reproduction of large astronomical glass plates: some problems and pitfalls.**
R. M. West, B. Dumoulin.
AAS Photo-Bull., Issue 23, No. 1, p. 3 - 8 (1980).

The authors describe a number of problems and their solutions in connection with on-glass contact copying of large astronomical glass-based photographs. The problems concern cleaning of the original and copy plates, the establishment of physical contact between the plates before exposure, separation of the plates after exposure, and processing of the copy plates – in particular, development and drying. The recommendations given are based on extensive experience at the ESO Sky Atlas Laboratory with the production of the ESO (B) and ESO/SRC Atlases of the Southern Sky.

036.005 **Comparison of Kodak plates, type IIa-O, with ORWO ZU-2 astronomical plates: the influence of baking in dry nitrogen.** O. Zichová, V. Kraitschev.
AAS Photo-Bull., Issue 23, No. 1, p. 9 - 11 (1980).

Two batches of Kodak spectroscopic plates, type IIa-O, were compared with two batches of type ZU-2 plates made by ORWO for their response to baking in dry nitrogen. Sensitivity, granularity, gamma, and fog were evaluated for all samples.

036.006 **Colour photography in astronomy.** D. F. Malin.
Vistas Astron., Vol. 24, 219 - 238 (1980).

036.007 **Astrophotographie mit dem Kodak Ektachrome 400.**
P. Riepe.
Sterne Weltraum, 19. Jahrg., 426 (1980).

036.008 **Hypersensitization of Kodak type IIIaJ plates using vapour of technical fluid nitrogen.**
M. M. Fedorov.
Astron. Tsirk., No. 1109, p. 6 - 8 (1980). In Russian.

The WGPM (*AAS Working Group on Photographic Materials*) meets in San Francisco. See Abstr. 011.026.

Full aperture solar photography for reflectors.
See Abstr. 031.504.

A short discourse on planetary photography.
See Abstr. 031.537.

Fourier transform enhanced photography of the M 51 system. See Abstr. 031.566.

Astrophotography. See Abstr. 031.581.

Color photography of galaxies.
See Abstr. 031.614.

Recent work with the UK Schmidt Telescope.
See Abstr. 032.006.

Positional Astronomy, Celestial Mechanics

041 Astrometry

041.001 **On the construction of a comprehensive general catalogue of star positions.** H. Eichhorn.
Celestial Mech., Vol. 22, (see 012.002), 127 - 134 (1980).

The procedures that would lead to the construction of a comprehensive general catalogue of star positions are outlined and discussed. Attention is drawn to the problems faced by a definitive reduction of the Astrographic Catalogue.

041.002 **Systematic errors of the plate measurements of the Yale zone +9° to +20°.** C. A. Smith.
Celestial Mech., Vol. 22, (see 012.002), 141 - 142 (1980). Summary.

041.003 **Astrometric radio source catalogs.** K. J. Johnston, J. H. Spencer, G. H. Kaplan, W. J. Klepczynski, D. D. McCarthy.
Celestial Mech., Vol. 22, (see 012.002), 143 - 151 (1980).

A review of the accuracy of radio interferometric positions of extragalactic sources is given. A mean catalog of radio positions is presented. With the accuracy currently attainable, radio observations can contribute significantly to the determination of astronomical constants.

041.004 **Comparison of the accuracy of measurements of images of stars and galaxies on plates taken with the wide-angle and long-focus astrographs of the Main Astronomical Observatory (Kiev).** S. P. Rybka, A. I. Yatsenko.
Astrometr. Astrofiz., Vyp. (No.) 41, p. 64 - 67 (1980). In Russian.

041.005 **Catalogue of positions of 15,000 stars in the zone 60° to 70° south declination.** F. W. Fallon.
Bull. American Astron. Soc., Vol. 12, 454 (1980). – Abstract.

041.006 **An assessment of the astrometric quality of the SRC J Survey.** G. F. Benedict, T. C. Talley.
Bull. American Astron. Soc., Vol. 12, 454 (1980). – Abstract.

041.007 **High-precision astrometry.** G. D. Gatewood, J. W. Stein.
Bull. American Astron. Soc., Vol. 12, 455 (1980). – Abstract.

041.008 **Improving planetary occultation astrometry.** W. S. Penhallow.
Bull. American Astron. Soc., Vol. 12, 500 (1980). – Abstract.

041.009 **Results of observations of right ascensions of PZT stars with the Moscow meridian circle. II.** O. A. Kozina, V. A. Korobova, L. M. Khommik.
Tr. Gos. Astron. Inst. Shternberga, Tom 50, 47 - 72 (1980). In Russian.

The observations of the second part of the Catalogue consist of ten PZT programs. The Catalogue was made in the FK4 system. The standard error is $\pm 0^{s}0198 \sec \delta$. The fundamental stars observations are given as individual corrections to the FK4 ascensions.

041.010 **Accurate positions of stars in SA 18.** Yu. A. Shokin.
Tr. Gos. Astron. Inst. Shternberga, Tom 50, 79 - 94 (1980). In Russian.

The method of determination of accurate star positions in SA 18 is described. The catalogue of positions of 86 stars in SA 18 proposed as a standard for estimation of the television positional method's precision is given.

041.011 **Fundamental Catalog – past and future.** T. Schmidt-Kaler.
Mitt. Astron. Ges., Nr. 48, (see 012.015), p. 7 - 9 (1980). In German.

Some remarks are made on the history of the Fundamental Catalog and his author, Arthur Auwers, and on future developments of fundamental astronomy.

041.012 **The FK5, an improved fundamental reference system extended to fainter stars.** W. Fricke.
Mitt. Astron. Ges., Nr. 48, (see 012.015), p. 29 - 41 (1980).

Presented is a status report on work on the FK5 combined with a review of the principles underlying the formulation of a dynamical reference system. More detailed information is given on the determinations of the equinox and equator of the FK5. At present it can be said that all α_{FK4} require the correction $E = +0^{s}.031$ at 1950 and that all FK4 proper motions must be corrected by $\Delta\mu_{\alpha} = \dot{E} = +0^{s}.085$ simultaneously with the application of the corrections due to the new value of the general precession in order to ensure that the FK5 equinox is as closely as possible identical to the dynamical equinox. These values may not yet be the final ones, but it is certain that they will not change by more than a few milliseconds. No significant correction to the FK4 equator was found.

041.013 **On the improvement of the FK4 equinox and equator by means of photographic observations of asteroids.** V. I. Orelskaya (*Orel'skaya*).
Mitt. Astron. Ges., Nr. 48, (see 012.015), p. 43 - 49 (1980).

The astronomical coordinate system represented by the FK4 is now in the process of further improvement. Corrections to the zero points of this Catalogue using selected minor planet observations are determined at the Institute of Theoretical Astronomy of the USSR Academy of Sciences. The determinations are presented and discussed.

041.014 **Equator and equinox of FK4 by 2240 observations of (51) Nemausa.** L. K. Kristensen.
Mitt. Astron. Ges., Nr. 48, (see 012.015), p. 50 - 55 (1980).

Corrections to the equator and equinox of FK4 have been determined by 2240 photographic observations of minor planet (51) Nemausa to respectively $\Delta\delta_{\bigcirc} = +''.03$ and $E = +^{s}.044$ at the epoch 1956.

041.015 **Systematische Bewegungen von Fundamental-Sternen.** P. Brosche, H. Schwan.
Mitt. Astron. Ges., Nr. 48, (see 012.015), p. 55 (1980).

041.016 **Vorschlag für die Ableitung eines Fundamentalkatalogs aus Beobachtungen mit Zenitkameras.** K. Pilowski.
Mitt. Astron. Ges., Nr. 48, (see 012.015), p. 56 - 58 (1980).

041.017 **The International Reference Star programs.** C. Smith.
Mitt. Astron. Ges., Nr. 48, (see 012.015), p. 95 - 108 (1980).

Observations of FK4 stars made during the AGK3R/SRS programs will lead to substantial improvement of the FK4 system and individual positions and proper motions. Improvement of the uniformity of the distribution of AGK3R/SRS on the celestial sphere could be realized by adding about 1700 stars in the declination zone 0° to −30° to bring the density closer to 1 star per square degree. A serious deficiency in the selection of B-, A-, and F-type stars in the declination zone −30° to −90° has occurred. Revised lists of AGK3R and SRS should be prepared and discussed soon to aid in planning for the next coordinated efforts at observing the International Reference Star lists.

041.018 **Atmospheric limitations to narrow-field optical astrometry.** L. Lindegren.
Mitt. Astron. Ges., Nr. 48, (see 012.015), p. 147 (1980).

041.019 **Optical and radio positions of 3C 273B in the FK4-system.** C. de Vegt, U. K. Gehlich.
Mitt. Astron. Ges., Nr. 48, (see 012.015), p. 147 (1980).

041.020 **Anwendung der Fernsehtechnik in der Astrometrie.** B. Wedel.
Mitt. Astron. Ges., Nr. 48, (see 012.015), p. 150 (1980).

041.021 **Die Plate Overlap-Methode: Steifheit des Systems und Lösungsverfahren.** K. von der Heide.
Mitt. Astron. Ges., Nr. 48, (see 012.015), p. 150 (1980).

041.022 **Construction of minimum catalogues of reference stars from ephemerides.**
N. A. Rodionova, V. I. Troitskij.
Avtomatiz. protsessov upr. i obrab. inf. Leningrad, 1979, p. 81 - 83. In Russian. – Abstr. in Ref. zh., 51. Astron., 8.51.138 (1980).

041.023 **On the systematic errors $\Delta\alpha_a$ and $\Delta\mu_a$ of right ascensions and proper motions of the General Catalogue.** E. V. Vityazeva.
Vestn. LGU, 1980, No. 7, p. 102 - 109. In Russian. – Abstr. in Ref. zh., 51. Astron., 10.51.151 (1980).

041.024 **The Photoelectric Astrolabe Catalogue of the Shanghai Observatory.**
T.-q. Xu, P.-z. Lu.
Acta Astron. Sinica, Vol. 21, 297 - 304 (1980). In Chinese.
Observational results with the photoelectric astrolabe type II during the period 1975 September to 1978 are analysed. Using the data obtained during this period, the corrections of star position of 103 stars are calculated and the systematic corrections of the catalogue $(\Delta\alpha)_\delta$, $(\Delta\delta)_\delta$ are given.

041.025 **Stellar images of the Mizusawa PZT.**
G. Murakami.
Publ. Int. Latitude Obs. Mizusawa, Vol. 13, 33 - 42 (1979).
Stellar images on the PZT plates are affected by the very existence and the motion of the carriage, and by other functions of the PZT. Diameters of stellar images of the Mizusawa PZT are measured along the diurnal motion and the meridian direction, respectively, by a microdensitometer. It is found that the diameter along the meridian direction is larger than that along the diurnal motion. The most predominant cause of increasing image diameter along the meridian can be explained by diffraction of light through the carriage rod.

041.026 **Short period terms in the observations of right ascensions with the meridian circle.** R. Fukaya.
Tokyo Astron. Obs. Rep. (No. 73), Vol. 19, 269 - 277 (1980). In Japanese.

041.027 **Observations of the sun, moon, and planets. Six-Inch Transit Circle results.** F. S. Gauss.
United States Naval Obs. Circ., No. 159, 43 pp. (1979).
This Circular contains positions of the sun, moon and planets observed with the Six-Inch Transit Circle between 6 May 1975 and 4 July 1977. These results are provisional. Definitive results will be published later in a volume of Publications of the U. S. Naval Observatory.

041.028 **The orbits of five minor planets and corrections to the FK4 equator and equinox.**
R. L. Branham, Jr.
Astron. Pap., Vol. 21, Part 3, 167 - 459 (1979).
Minor planets (6), (7), (8), (9), and (15) were used to determine the equator and equinox errors of the FK4 catalog system. Other quantities derived were: the secular variations of the equator and equinox; the elements of the Earth's orbit; the secular variation of the obliquity; and a correction to Newcomb's general precession in longitude. The analysis was based on a total of 14,560 observations in right ascension and 14,333 in declination. It was possible to reduce seventy-four per cent of these to the FK4. The main results of the study are: the correction to the precession is 1.″318 ± 0.″238; the equator and equinox corrections at epoch 1950 are, respectively, −0.″036 ± 0.″037 and 0.″283 ± 0.″342; the motion of the equinox is 0.″784 ± 0.″208; the secular variation of the obliquity caused by a motion of the ecliptic is −0.″295 ± 0.″163; and the secular variation of the obliquity caused by a motion of the equator is −0.″201 ± 0.″104. But evidence is presented to show that these secular variations of the obliquity are more fictitious than real.

041.029 **Theoretical stellar spectral deviations in observations using Danjon astrolabes.** Z.-g. Li.
Publ. Shensi Astron. Obs., No. 1, p. 43 - 49 (1980).

041.030 **Stellar spectral deviations on the model-1 photo-astrolabe and Danjon astrolabe.**
J.-y. Xu, T.-g. Yang.
Publ. Shensi Astron. Obs., No. 1, p. 51 - 60 (1980).

041.031 **Development of space astrometry and the importance of classical observations from the earth.**
J.-y. Xu.
Publ. Shensi Astron. Obs., No. 1, p. 61 - 66 (1980).

041.032 **Astronomia di posizione. Effemeridi e almanacchi. VII. Effemeridi della luna.** M. Veltri.
G. Astron. Vol. 6, 265 - 287 (1980).

041.033 **A comparison of the Smithsonian Astrophysical Observatory catalogue (SAO) with AGK3R and Perth 70.** C. Sullivan, A. N. Argue.
Mon. Not. R. Astron. Soc., Vol. 193, 921 - 929 (1980).
The SAO catalogue has been compared with AGK3R for 19 596 stars having $\delta > 0°$, and with Perth 70 for 18 775 stars having $\delta < 0°$. Plots are given showing the comparisons for 10° zones in δ. Systematic differences can exceed 0.5 arcsec in certain areas. Recommendations are made for the use of SAO in the optical identification of radio sources.

041.034 **Introduction of the corrections in the calculated coordinates of a satellite and their influence on the accuracy of determination of the observational station.**
A. G. Kirichenko.
Byull. Inst. Teor. Astron., Tom 14, 552 - 556 (1980). In Russian.
The paper deals with the estimation of the values of corrections introduced in the calculated coordinates of a satellite for the phase angle of a satellite, the refractional parallax, and the reduction of the observations to the common centre. It is shown that the correction for the phase angle of a satellite may cause an error in the determination of the position of the observation station about 15 m and that for the refractional parallax about 3 - 18 m.

041.035 **Comparison of solar observations made in Washington from 1911 to 1970 with Newcomb's theory.** M. L. Sveshnikov.
Byull. Inst. Teor. Astron., Tom 14, 571 - 588 (1980). In Russian.
Corrections to Newcomb's elements of the sun's orbit, the obliquity of the ecliptic, the equinox and equator of the FK4 catalogue system have been determined from solar observations. It is shown that the observed declinations of the sun are affected by systematic errors with secular trend.

041.036 **On the connection of the radio and optical systems of positions and proper motions.** P. Brosche.
Astron. J., Vol. 85, 1674 - 1675 (1980).
This paper emphasizes the importance of determining fictitious proper motions of optical counterparts of compact radio sources, especially with regard to the planned astrometry satellite Hipparcos. A list of objects is given.

041.037 **A general look at astrometry.** E. P. Fedorov.
Geodynamics and astrometry, (see 003.011), p. 74 - 109 (1980). In Russian.

041.038 **On basic coordinate systems used in astrometry and geodynamics.** Ya. S. Yatskiv, V. S. Gubanov.
Geodynamics and astrometry, (see 003.011), p. 110 - 120 (1980). In Russian.

041.039 **Astrometry and celestial mechanics.**
V. K. Abalakin.
Geodynamics and astrometry, (see 003.011), p. 130 - 136 (1980). In Russian.

041.040 **Astrometrical investigations with a satellite of ESA.**
E. Høg.
Geodynamics and astrometry, (see 003.011), p. 158 - 164 (1980). In Russian.

041.041 **Individual corrections of the declinations of 76 FK4 stars.** N. F. Minyajlo.
Astron. Tsirk., No. 1097, p. 6 - 8 (1980). In Russian.

041.042 **Observations of the moon and outer planets obtained in Nikolaev in 1969 - 1972.**
Ya. E. Gordon, L. F. Gorel', E. V. Khrutskaya.
Nikolaev. otd. GAO AN SSSR. Nikolaev, 1980. 42 pp. In Russian. – Abstr. in Ref. zh., 51. Astron., 12.51.135 (1980).

041.043 **Influence of random bursts on the registration of star transit moments.** D. Ojdov.
Ehrdem shinzhilgehehnij butehehl. Fiz.-tekh. khurehehlehn. Mat. khurehehlehn. BNMAU shinzhlekh ukhaany Asad., 1980, No. 17, p. 134 - 139. In Russian. – Abstr. in Ref. zh., 51. Astron., 12.51.141 (1980).

Catalog of the Washington scale pairs observed with the Moscow zenith telescope. See Abstr. 002.011.

Preliminary catalogue of the declinations of the latitude programme stars obtained by the micrometer method. See Abstr. 002.012.

Catalogues of individual corrections to declinations of FK4 stars. See Abstr. 002.019.

Catalogue of the declinations of stars in the PZT program zone of Washington and Mizusawa obtained with the Moscow ZTL-180 telescope. See Abstr. 002.020.

New USNO Zodiacal Star Catalog.
See Abstr. 002.050.

Calcul astronomique pour amateurs adapté à l'emploi d'un calculateur électronique de poche.
See Abstr. 003.022.

The history of the fundamental catalogues.
See Abstr. 004.024.

Division of the history of astrometry into periods.
See Abstr. 004.080.

Space astrometry – its impact on astronomy and astrophysics – introductory comments.
See Abstr. 011.023.

Positional astronomy at the Royal Greenwich Observatory. See Abstr. 013.005.

Kaartprojekties in de sterrenkunde (I), (II).
See Abstr. 021.031.

Global astrometry by space techniques.
See Abstr. 031.514.

Critique of singular value analysis.
See Abstr. 031.518.

Rigorous treatment of stellar aberration.
See Abstr. 031.535.

Wind influence on determination of clock correction. See Abstr. 031.574.

Zur Reduktion photographischer Positionen von kleinen Planeten und Kometen. See Abstr. 031.594.

Input actions of an optico-electronic system for measuring the angular coordinates of astronomical objects.
See Abstr. 031.604.

Automatic identification of reference stars in photographic observations of artificial earth satellites.
See Abstr. 031.607.

Taking into account refraction effects in determination of the position of artificial celestial bodies.
See Abstr. 031.624.

Investigation of diameter corrections of the Brorfelde transit circle. See Abstr. 032.005.

Solar astrolabe construction at C.E.R.G.A.
See Abstr. 032.015.

Astrometry with the Space Telescope.
See Abstr. 032.504.

Space Telescope astrometry from CCD images.
See Abstr. 032.506.

Applications of interferometers in space to astrometry and planetary detection. See Abstr. 032.507.

The European astrometry satellite, Hipparcos.
See Abstr. 032.540.

The U.S. Space Telescope: astrometric capabilities.
See Abstr. 032.541.

On the determination of the equinox and equator of the new fundamental reference coordinate system, the FK5.
See Abstr. 043.001.

The IAU (1976) System of Astronomical Constants. See Abstr. 043.002.

Astronomisch-geodätische Messungen mit einem Zenitteleskop. See Abstr. 046.010.

The astrometry satellite Hipparcos. See Abstr. 051.004.

Space astrometry as NASA and ESA projects. See Abstr. 051.007.

Relativistic corrections in astrometric observations of planets. See Abstr. 066.180.

Tobias Mayers Meridianbeobachtungen der Sonne und die Frage der Veränderlichkeit des Sonnendurchmessers. See Abstr. 080.029.

Atmospheric limitations of narrow-field optical astrometry. See Abstr. 082.028.

On the character of influence and on the method of calculation of the chromatic refraction in determination of declinations of stars with the methods of position astronomy. See Abstr. 082.096.

Catalogue of the positions of Venus obtained with the wide-angle astrograph of the Sternberg Astronomical Institute. See Abstr. 093.009.

An analysis of total lunar occultations made in the years 1943 to 1974. See Abstr. 096.011.

Astrometric observations of Saturn. See Abstr. 100.072.

Photographic observations of Saturn satellites. See Abstr. 100.130.

Astrometric study of the Uranus satellite Miranda. See Abstr. 101.003.

The needs in the radial velocity area in view of impending space astrometry projects. See Abstr. 111.016.

Classification of stellar populations and luminosity classes from accurate proper motions. See Abstr. 111.019.

The triple system Gliese 815 (Furuhjelm 54): an astrometric study. See Abstr. 118.015.

Extragalactic radio source positions determined with VLBI. See Abstr. 141.035.

Astrometry and high-speed photometry of an optical candidate for PSR 1913+16. See Abstr. 141.531.

042 Celestial Mechanics, Figure of Celestial Bodies

042.001 **The triple collision manifold in the isosceles case of the planar three-body problem.**
J. M. Irigoyen.
C. R. Acad. Sci. Paris, Tome 290, Sér. B, 489 - 492 (1980).
In French.

042.002 **The resonance overlap criterion and the onset of stochastic behavior in the restricted three-body problem.** J. Wisdom.
Astron. J., Vol. 85, 1122 - 1133 (1980) = Contrib. No. 3390, Div. Geol. Planet. Sci., Calif. Inst. Technol., Pasadena.

The resonance overlap criterion for the onset of stochastic behavior is applied to the planar circular-restricted three-body problem with small mass ratio (μ). Its predictions for $\mu = 10^{-3}$, 10^{-4}, and 10^{-5} are compared to the transitions observed in the numerically determined Kolmogorov-Sinai entropy and found to be in remarkably good agreement. In addition, an approximate scaling law for the onset of stochastic behavior is derived.

042.003 **On the solving of Hill's differential equation through the method of operational calculus.**
J. Meffroy.
Moon Planets, Vol. 23, 73 - 97 (1980).

Using a method previously applied to the treatment of the Mathieu differential equation, the author solves Hill's differential equation of lunar theory through the way of operational calculus, which avoids the cumbersome infinite determinants of the classical procedure.

042.004 **Further analysis about the regions of motion of the general three-body problem.** Y.-s. Sun, D.-j. Luo.
Acta Astron. Sinica, Vol. 21, 96 - 103 (1980). In Chinese.

042.005 **Numerical simulation of a three-dimensional gravitating system of colliding masses.**
J.-q. Zheng, H.-n. Zhou.
Acta Astron. Sinica, Vol. 21, 104 - 111 (1980). In Chinese.

The authors have investigated numerically a model of a three-dimensional gravitating system of colliding particles in which the particles are moving in the gravitational field of a central rotational ellipsoid body and interact through inelastic collisions.

042.006 **A statistical theory for the disruption of four-body systems.** P. E. Nash, J. J. Monaghan.
Mon. Not. R. Astron. Soc., Vol. 192, 809 - 820 (1980).

A statistical theory based on phase space averaging is used to describe the disruption of bound four-body systems. The dominant final states are assumed to be either two binaries or one binary and two unbound particles. The relative frequency of the two final states is, however, sensitive to the choice of cut-off. The mass distributions of the final states are calculated.

042.007 **Sur le problème généralisé restreint des trois corps solides.** G. N. Duboshin.
Tr. Gos. Astron. Inst. Shternberga, Tom 49, 3 - 31 (1980).
In Russian.

042.008 **Le problème spécial généralisé restreint des trois corps solides.** G. N. Duboshin.
Tr. Gos. Astron. Inst. Shternberga, Tom 49, 32 - 56 (1980).
In Russian.

042.009 **On the 4th order short-period solar perturbations in the motion of the satellites of the major planets.**
A. A. Orlov, V. M. Chepurova.
Tr. Gos. Astron. Inst. Shternberga, Tom 49, 57 - 68 (1980).
In Russian.

With the aid of von Zeipel's method in the limits of the restricted elliptic problem of three bodies short-period terms of the fourth order in the characteristic function expansion in powers of the sun and satellite mean motions ratio are obtained. Design formulae for the calculation of the short-period solar perturbations in the motion of the satellites to terms of the fourth order inclusive are derived.

042.010 **On intermediate orbits of some triple stellar systems.** A. A. Orlov, N. A. Solovaya.
Tr. Gos. Astron. Inst. Shternberga, Tom 49, 69 - 81 (1980).
In Russian.

042.011 **On some varieties of intermediate orbits in the theory of satellite motion.**
A. A. Orlov, N. A. Solovaya.
Tr. Gos. Astron. Inst. Shternberga, Tom 49, 82 - 89 (1980).
In Russian.

042.012 **An analytic solution of the planar averaged restricted three-body problem in the case of circulation of the pericentre of the orbit of a particle.** F. Veres.
Astron. Zh., Tom 57, 824 - 832 (1980). In Russian.
English translation in Soviet Astron., Vol. 24, No. 4.

The problem considered can be solved by quadratures.

042.013 **Formal integrals of motion in the planar restricted three-body problem.** E. I. Timoshkova.
Astron. Zh., Tom 57, 833 - 840 (1980). In Russian.
English translation in Soviet Astron., Vol. 24, No. 4.

042.014 **Construction of conditionally periodic solutions of the restricted circular three-body problem in the three-dimensional resonant case. I. Elimination of the short-period terms.** S. G. Zhuravlev.
Astron. Zh., Tom 57, 841 - 850 (1980). In Russian.
English translation in Soviet Astron., Vol. 24, No. 4.

A practical construction of conditionally periodic solutions of the restricted circular three-body problem in a three-dimensional case, when a sharp commensurability between frequencies is taken into account, is realized. In Part 1, the procedure of selection of secular, long-period and resonant parts of the Hamiltonian and the procedure of elimination of short-period terms of the Hamiltonian are described.

042.015 **New u-type families of periodic orbits in the restricted three-body problem.** M. K. Zikides.
Astron. Astrophys., Vol. 88, 298 - 301 (1980).

New characteristic curves of u-type families of periodic orbits of the restricted three-body problem are found. The evolution of these families is numerically studied for various values of the mass ratio μ of the primary bodies.

042.016 **On the stability of Lagrangian solutions of the elliptic restricted three-body problem.**
A. P. Ivanov, S. R. Karimov, A. G. Sokol'skij.
Pis'ma Astron. Zh., Tom 6, 442 - 448 (1980). In Russian.
English translation in Soviet Astron. Lett., Vol. 6.

The question of the stability of the Lagrangian solutions of the planar elliptic restricted three-body problem for the values of parameters (mass ratio of the main attracting bodies and the eccentricity of their Kepler ellipses) corresponding to boundaries of the stability regions of the linearized system is solved in non-linear treatment.

042.017 **A special case of series expansion of the general force function of two rigid bodies.**
G. N. Duboshin.

Soobshch. Gos. Astron. Inst. Shternberga, No. 218, p. 3 - 10 (1980). In Russian.

A force function of interaction of two rigid bodies is considered provided that elementary particles of these bodies are attracted or repelled, the force of interaction being proportional to the mass product and some power of their mutual distance. It is supposed that one of the bodies is fully contained inside the boundary surface of the other.

042.018 **A partial case of the restricted problem of three rigid bodies.** G. N. Duboshin.
Soobshch. Gos. Astron. Inst. Shternberga, No. 218, p. 11 - 33 (1980). In Russian.

The problem of three bodies is considered, one of the bodies being proposed "passive" while the other two "active". Each of the bodies is assumed to be an axially symmetric rigid body having the plane of symmetry normal to the axis of symmetry. The elementary particles of the two different bodies attract each other with a force proportional to some function of the mutual distance. It is supposed that at the initial moment of time all three planes of symmetry coincide and that the initial conditions are chosen so that the mass centres of the bodies always remain in this common plane and each body rotates regularly around the axis of symmetry. Besides it is supposed that the orbit of the mass centre of one active body around that of the other is circular.

042.019 **Solution of the variational equations in the problem of motion of a point in a central field of forces.** Yu. S. Aleksandrov.
Kosm. Issled., Tom 18, 483 - 489 (1980). In Russian.

042.020 **On the two triaxial rigid body problem.** M. Šidlichovský.
Bull. Astron. Inst. Czechoslovakia, Vol. 31, 240 - 253 (1980).

A system of two triaxial rigid spheroids, the gravitational field of which includes terms with J_2 and J_{22} is investigated. The first-order Lie-Hori perturbation method is employed with Delaunay's variables for the orbital motion and the action variables, based on Andoyer's variables, for the rotational motion of the bodies. The relation between the averaged energy of perturbation and the constants of integration is found. The configurations of the three angular momenta (two rotational and one orbital) of the state with minimum energy are investigated.

042.021 **A contribution to the stability of the triangular points in the elliptic restricted three-body problem.**
R. Meire.
Bull. Astron. Inst. Czechoslovakia, Vol. 31, 312 - 316 (1980).

Hill's equations are derived in a more compact form and this will be of great importance for the truncation errors in the numerical stability analysis. The obtained transition curves in the parameterplane (μ, e) confirm Danby's results, but the new results are more accurate.

042.022 **Modern approaches to research in dynamical astronomy.** J. Kovalevsky.
Mitt. Astron. Ges., Nr. 48, (see 012.015), p. 67 - 80 (1980).

The main scientific objectives of the dynamics of the solar system are described. They include the kinematics and dynamics of the Earth and the Earth-Moon system, the determination of the reference system, the motions and the evolution in the solar system and the structure of planets and satellites. The theoretical aspects in support to these objectives are dealt with using celestial mechanics. Many new observational techniques of dynamical astronomy have appeared during the last 20 years. The relevance of these techniques to the scientific objectives is analyzed.

042.023 **Die Bestimmung der Kleinen Divisoren mit Hilfe der exzentrischen Anomalie: Ein Beitrag zur modernen Planetentheorie.** R. Dvorak.
Mitt. Astron. Ges., Nr. 48, (see 012.015), p. 89 - 91 (1980).

042.024 **Keplerian representation of a non-Keplerian orbit.** B. E. Baxter.
J. Guid.Control, Vol. 3, 151 - 153 (1980). – Abstr. in Phys. Abstr., Vol. 83, Abstr. 90324 (1980).

042.025 **Periodic solutions of Schwarzschild type in the relativistic restricted problem of three bodies.**
V. Singh.
Indian J. Pure Appl. Math., Vol. 7, 1351 - 1357 (1976). Abstr. in Phys. Abstr., Vol. 83, Abstr. 90730 (1980).

042.026 **On the two body problem in invariant mechanics.** I. Mihaila.
Rev. Roumaine Phys., Vol. 25, 289 - 293 (1980). In French. Abstr. in Phys. Abstr., Vol. 83, Abstr. 94696 (1980).

042.027 **Construction of the second intermediate Hill orbit in the restricted circular three-body problem.**
M. D. Shinibaev.
Redkol. zh. "Vestn. AN KazSSR." Alma-Ata, 1980. 24 pp. In Russian. – Abstr. in Ref. zh., 51. Astron., 8.51.110 (1980).

042.028 **Gravitationally-inertial potential force field.** G. G. Polyakov.
Astrakhan. gos. ped. inst. Astrakhan', 1980. 17 pp. In Russian. Abstr. in Ref. zh., 62. Issled. kosm. prostranstva, 8.62.465 (1980).

042.029 **On the character of evolution of orbits in the vicinity of the 1 : 3 Kirkwood gap.**
A. N. Simonenko, V. G. Kruchinenko, L. M. Sherbaum.
Meteoritika, Vyp. 39, p. 121 - 133 (1980). In Russian.

042.030 **Notes on the central force r^n.** R. Broucke.
Astrophys. Space Sci., Vol. 72, 33 - 53 (1980).

The author collects several results related to the classical problem of two-dimensional motion of a particle in the field of a central force proportional to a real power of the distance r. At first he generalizes Whittaker's result of the fourteen powers of r which lead to integrability with elliptic functions. Next, he studies the stability of the circular solutions, which are the singular solutions of the problem, in Whittaker's terminology. Finally, it is shown that the stable singular circular solutions of the central force problem generalize to stable singular elliptic solutions of the two-fixed-center problem. The stability and the bifurcations with other families of periodic solutions of the two-fixed-center problem are also described.

042.031 **The effect of radiation pressure on the restricted three-body problem.** D. W. Schuerman.
Solid particles in the solar system, (see 012.019), p. 285 - 288 (1980).

The classical restricted three-body problem is generalized to include the force of radiation pressure and the Poynting-Robertson effect. Implications for space colonization and a mechanism for producing asymmetries in the interplanetary dust complex are discussed.

042.032 **The family i 1 ν of the three-dimensional general three-body problem.**
K. Katopodis, S. Ichtiaroglou, M. Michalodimitrakis.
Astron. Astrophys., Vol. 90, 102 - 105 (1980).

The authors computed the biparametric family i 1 ν of symmetric periodic orbits of the three-dimensional general three-body problem. The periodic orbits refer to a suitably defined rotating frame.

042.033 **Periodic orbits and ergodic components of a resonant dynamical system.**
G. Contopoulos, M. Zikides.
Astron. Astrophys., Vol. 90, 198 - 203 (1980).
The authors study a dynamical system with two equal frequencies as the energy h increases. For small h the system is close to an integrable one while for h approaching the escape energy it has several ergodic components that show a hierarchical structure. This structure is related to the bifurcation of new families of periodic orbits.

042.034 **Complete solution for a problem on translatory-rotational motion of two absolute solid bodies.**
V. V. Vidyakin.
Arkhang. gos. ped. inst. Arkhangel'sk, 1980. 19 pp. In Russian. Abstr. in Ref. zh., 51. Astron., 9.51.102 (1980).

042.035 **On the convergence of series representing a class of periodic solutions of the averaged equations of the outer variant of the restricted elliptic three-body problem.**
V. P. Evteev, B. M. Nagorev.
Dokl. AN TadzhSSR, Vol. 23, 143 - 146 (1980). In Russian. Abstr. in Ref. zh., 51. Astron., 9.51.106 (1980).

042.036 **A note on a Lagrangian formulation for motion about the collinear points.** D. L. Richardson.
Celestial Mech., Vol. 22, 231 - 236 (1980).
A Lagrangian formulation for the three-dimensional motion of a satellite in the vicinity of the collinear points of the circular-restricted problem is reconsidered. It is shown that the influence of the primaries can be expressed in the form of two third-body disturbing functions. By use of this approach, the equations for the Lagrangian and for the motion itself are readily developed into highly compact expressions.

042.037 **On the variational equations associated with a Lagrangian.** A. Hennawi.
Celestial Mech., Vol. 22, 237 - 240 (1980).
Broucke (1976) has studied the symplectic properties of the variational equations of a Lagrangian of a very particular form, with constant coefficients. In this article, the author generalizes his results to the case of an arbitrary Lagrangian. He shows that the characteristic exponents of a periodic solution can be computed in Lagrangian formulation as well as in the more usual Hamiltonian formulation.

042.038 **Analytic construction of periodic orbits about the collinear points.** D. L. Richardson.
Celestial Mech., Vol. 22, 241 - 253 (1980).
A third-order analytical solution for halo-type periodic motion about the collinear points of the circular-restricted problem is presented. The three-dimensional equations of motion are obtained by a Lagrangian formulation. The solution is constructed using the method of successive approximations in conjunction with a technique similar to the Lindstedt-Poincaré method. The theory is applied to the Sun-Earth system.

042.039 **Application of Hamilton's law of varying action to the restricted three-body problem.**
D. L. Hitzl, D. A. Levinson.
Celestial Mech., Vol. 22, 255 - 266 (1980).
Hamilton's law is developed in its most general form and is used to produce series solutions of the restricted three-body problem. Finally, for illustrative purposes, numerical results are presented for several symmetric periodic orbits.

042.040 **Theory of the Trojan asteroids. Part III.**
B. Garfinkel.
Celestial Mech., Vol. 22, 267 - 287 (1980).
In the previously published Parts I and II of the paper, the author has constructed a formal long-periodic solution for the case of 1 : 1 resonance in the restricted problem of three bodies to $0(m^{3/2})$, where m is the small mass parameter of the system. The time-dependence $t(\lambda; \alpha, m)$, where λ is the mean synodic longitude and α is related to the Jacobi constant, has been expressed by a hyperelliptic integral. It is shown here that with the approximation $m = 0$ in the integrand, the function $t(\lambda, \alpha, 0)$ can be expanded in a series involving standard elliptic functions. Then the problem of inversion can be formally solved, yielding the function $\lambda(t, \alpha, 0)$.

042.041 **Properties of linearized mappings associated with periodic orbits in the three-body problem.**
P. C. Kammeyer.
Celestial Mech., Vol. 22, 289 - 296 (1980).
Three results on the linearized mapping associated with the plane three body problem near a periodic orbit are established. It is first shown that linear stability of such an orbit is independent of initial position on the orbit and of coordinate system. Second, the relation of Hénon connecting the rates of change of rotation angle and period of an isoenergetic family of periodic orbits is proved, together with a similar relation for families of orbits closing exactly in a rotating coordinate system. Finally, a condition for a critical orbit is given which is applicable to any family of periodic orbits.

042.042 **On the origin of the Kirkwood Gaps.**
T. A. Heppenheimer.
Celestial Mech., Vol. 22, 297 - 304 (1980).
It is proposed that the Kirkwood Gaps are primordial, representing regions where asteroids failed to form by accretion. A brief scenario is presented to indicate the main features of a model for the early history of the asteroids. A discussion is given of two problems: the origin of asteroidal eccentricities and inclinations, and the likelihood that Jupiter suffered major changes in its semimajor axis during its formation.

042.043 **On the analogy between orbital dynamics and rigid body dynamics.** J. L. Junkins, J. D. Turner.
J. Astronaut. Sci., Vol. 27, 345 - 348 (1979). – Abstr. in Phys. Abstr., Vol. 83, Abstr. 98537 (1980).

042.044 **Non-linear axisymmetric oscillations of a homogeneous sphere.** A. S. Baranov.
Astron. Zh., Tom 57, 968 - 974 (1980). In Russian. – English translation in Soviet Astron., Vol. 24, No. 5.
Non-linear periodic oscillations of an ellipsoid of revolution with respect to a sphere are considered. The figure is assumed to consist of an ideal fluid described in terms of constant density and isotropic pressure. It is shown that with increasing amplitude the oscillations of the body are slowing down.

042.045 **Periodic solution of the generalized Hill problem.**
I. N. Latyshev.
Astron. Zh., Tom 57, 1063 - 1069 (1980). In Russian. English translation in Soviet Astron., Vol. 24, No. 5.
The Galaxy – star – satellite case of the planar limited Hill problem is studied. The acceleration of the Galaxy's gravitational force is considered to be proportional to the k-th power of the satellite's distance from the centre of the Galaxy provided that $-2 \leqslant k < +1$. A periodic solution of the problem is obtained in the form of a double series in increasing powers of m and $\Phi = 1 - k$, when m is a small parameter. For $|m| \leqslant 0.18$ the convergence of the series is shown. With the help of numerical integration periodic orbits of the problem are found for values of m which exceed $m = 0.18$.

042.046 **Two particular kinds of solution of the planar averaged restricted three-body problem.**
F. Veres.
Astron. Zh., Tom 57, 1070 - 1077 (1980). In Russian.

English translation in Soviet Astron., Vol. 24, No. 5.

Under small eccentricities a libration-type solution appears in which both the pericentre angle and the eccentricity vary periodically around some mean values. Under large eccentricities an unstable solution appears in which the eccentricity reaches unity and the particle falls onto the central body. A method for reversing the quadratures is proposed for the cases of libration and of the unstable solution. Analytical expressions for the orbital elements as explicit functions of time are obtained.

042.047 **Solution of a problem of translatory-rotational motion of two axisymmetrical bodies.**
V. V. Vidyakin.
Pis'ma Astron. Zh., Tom 6, 651 - 653 (1980). In Russian. English translation in Soviet Astron. Lett., Vol. 6.

A particular case of the two-body problem is considered which is referred to as the "two floats" case according to Duboshin's (1961) terminology. The problem is proved to have a solution in quadratures.

042.048 **On the problem of rotational motion of an axisymmetric satellite in a resonance case.**
Yu. G. Markov.
Pis'ma Astron. Zh., Tom 6, 654 - 658 (1980). In Russian. English translation in Soviet Astron. Lett., Vol. 6.

By means of the perturbation theory method the first approximation for an axisymmetric satellite in a resonance case is obtained on the basis of intermediate motion. The satellite is supposed to be situated at the triangular libration point L_4 of the restricted circular three-body problem.

042.049 **Restricted problem: families of vertical critical periodic orbits.**
S. Ichtiaroglou, K. Katopodis, M. Michalodimitrakis.
Astron. Astrophys., Vol. 90, 324 - 326 (1980).

The authors give four families of vertical critical periodic orbits of the circular planar restricted three-body problem, found by numerical continuation, with respect to μ, of the vertical critical periodic orbits b1v, b2v, c1v, c2v of the Copenhagen problem (with μ =0.5).

042.050 **The existence and stability of the libration points of an axisymmetric body moving around another axisymmetric body.** K. B. Bhatnagar, U. Gupta.
Astron. Astrophys., Vol. 91, 194 - 201 (1980).

The motion of two mutually attracting rigid bodies, both of which are axisymmetric with axisymmetric density distribution has been considered. Nine particular solutions corresponding to the libration points and analogous to the points Spoke, Arrow, and Float (Duboshin, 1959) have been found. The stability of these libration points has been discussed in two categories of cases. In the first category, different shapes of the bodies have been taken and in the second category, the mass and the linear dimensions of one of the bodies have been taken small in comparison to the other. Lastly, one of the bodies has been taken a rod which is considered as a limiting case of a prolate spheroid.

042.051 **What does the mean motion mean?**
J. D. Mulholland.
Bull. American Astron. Soc., Vol. 12, 743 (1980). – Abstract.

042.052 **A note on the tidal evolution of eccentric orbits.**
A. W. Harris.
Bull. American Astron. Soc., Vol. 12, 744 (1980). – Abstract.

042.053 **Dynamics of particle orbits for narrow eccentric rings maintained by small satellites.**
C. D. Murray, S. F. Dermott.
Bull. American Astron. Soc., Vol. 12, 744 - 745 (1980). Abstract.

042.054 **Resonant periodic orbits in the problem of three bodies.** J. H. Kwok, P. E. Nacozy.
Bull. American Astron. Soc., Vol. 12, 745 (1980). – Abstract.

042.055 **The calculation of the Trojan period.**
B. Garfinkel.
Bull. American Astron. Soc., Vol. 12, 745 (1980). – Abstract.

042.056 **A solution to the Painlevé-Wintner problem and its expanding gravitational system analog.**
N. D. Hulkower, D. G. Saari.
Bull. American Astron. Soc., Vol. 12, 745 (1980). – Abstract.

042.057 **The inclination changes in the problem of two triaxial rigid spheroids.** M. Šidlichovský.
Celestial Mech., Vol. 22, 343 - 355 (1980).

The first-order perturbations of a system of two triaxial rigid spheroids under Hori-Lie transformation are investigated. The time dependence of the configuration of the three angular momentum vectors, two rotational and one orbital, is studied. The motion of the relative configuration of the angular momentum vectors is periodical except in a special aperiodic case. The expressions for the periods are given.

042.058 **Stability criteria in many-body systems. I. An empirical stability criterion for co-rotational three-body systems.**
I. W. Walker, A. G. Emslie, A. E. Roy.
Celestial Mech., Vol. 22, 371 - 402 (1980).

An expansion of the force function of n-body dynamical systems, where the equations of motion are expressed in the Jacobian coordinate system, is shown to give rise naturally to a set of $(n-1)(n-2)$ dimensionless parameters ϵ^{ki}, representative of the size of the disturbances on the Keplerian orbits of the various bodies. The expansion is particularized to the case $n = 3$ which involves the consideration of only two parameters, ϵ^{23} and ϵ_{32}. Treating a system of n bodies as a set of disturbed three-body systems the authors use existing data from the solar system, known triple systems and numerical experiments in the many-body problem to plot a large number of triple systems in the ϵ^{23}, ϵ_{32} plane and show the results agree well with the ϵ^{23}, ϵ_{32} analysis above (eccentricities and inclinations as appropriate to most real systems being negligible) They deal briefly with the extension of the ϵ criteria to many-body systems where $n>4$, and discuss several interesting cases of dynamical systems.

042.059 **Bifurcations of triple-periodic orbits.**
G. Contopoulos, P. Michaelidis.
Celestial Mech., Vol. 22, 403 - 413 (1980).

The authors consider families of periodic orbits in potentials symmetric with respect to the x-axis. The characteristics of triple-periodic orbits (i.e. orbits intersecting the x-axis three times) that bifurcate from the central characteristic do not have their maximum or minimum energy (or perturbation) at the point of intersection. The authors explain theoretically that this happens only for triple-periodic orbits and not for any other type of resonant periodic orbits and verify this fact by numerical calculations.

042.060 **Concave hamburger equilibrium of rotating bodies.**
T. Fukushima, Y. Eriguchi, D. Sugimoto, G. S. Bisnovatyi-Kogan (*Bisnovatyj-Kogan*).
Prog. Theor. Phys., Vol. 63, 1957 - 1970 (1980). – Abstr. in Phys. Abstr., Vol. 83, Abstr. 109422 (1980).

042.061 **Lagrangian solutions to the three-body problem with forces r^{-p} (p integer).**
R. De. A. Campos, P. L. Ferreira.
Rev. Brasil. Fis., Vol. 10, No. 1, p. 59 - 75 (1980). – Abstr. in Phys. Abstr., Vol. 84, Abstr. 9351 (1981).

042.062 **The 1 : 2 : 1-resonance, its periodic orbits and integrals.** E. van der Aa, J. A. Sanders.
Asymptotic analysis, (see 003.007), p. 187 - 208 (1979).

042.063 **Approximations of higher order resonances with an application to Contopoulos' model problem.**
J. A. Sanders, F. Verhulšt.
Asymptotic analysis, (see 003.007), p. 209 - 228 (1979).

042.064 **Elliptic Hill's problem: the continuation of periodic orbits.** S. Ichtiaroglou.
Astron. Astrophys., Vol. 92, 139 - 141 (1980).

The existence of families of periodic orbits in the case of Hill's problem ($\mu=0$), when the body P_1 describes elliptic orbits around the massive body P_2 is provéd. The periodic orbits of P_3 can be continued from the orbits of the circular Hill problem with period equal to $2k\pi$ with respect to the eccentricity of the orbit of P_1.

042.065 **Theory of the motions of the Moon and of the satellites with Laplace's variables.** J.-F. Lestrade.
Astron. Astrophys., Vol. 92, 302 - 314 (1980). In French.

Laplace emphasizes the choice of the independent variable taken in the differential equations of the dynamics to treat them successfully by approximations. Choosing the true longitude ν which identifies itself with the geometrical positions of the celestial body all along its trajectory must be more advantageous, according to Laplace, than the mean anomaly, or in another word the time, deviating too much from it. The arguments formed by combinations of the true longitudes of the Moon ν and of the Sun ν' make the series representing the motion more convergent than the classical ones. Series concerning the Moon and Jupiter VI are obtained.

042.066 **General three-body problem: families of vertical critical periodic orbits.**
S. Ichtiaroglou, K. Katopodis, M. Michalodimitrakis.
Astrophys. Space Sci., Vol. 73, 445 - 451 (1980).

In this paper the authors present four families of vertical critical periodic orbits found by continuation, with respect to the small mass m_3, of the vertical critical periodic orbits $l1v$, $i1v$, $m1v$, $c3v$ of the circular restricted problem. The periodic orbits refer to a suitably defined rotating frame of reference.

042.067 **Der Einfluß des Jupiter im Sonnensystem.** W. Sanns.
Jupiter, (see 012.045), p. 1 - 17 (1980).

Four methods for the investigation of the disturbing influence of Jupiter in the dynamical system sun – Jupiter – asteroid are discussed: Hill's method; search for quasi-periodic orbits; search for periodic orbits; numerical methods. Their application to families of asteroids, Kirkwood gaps, and cases of resonance are shown.

042.068 **The numerical power series method for comets.** V. F. Myachin, O. A. Sizova.
Byull. Inst. Teor. Astron., Tom 14, 597 - 607 (1980). In Russian.

The regularized numerical integration method based on the Taylor–Steffensen power series is applied to the n-body problem when the masses of some considered bodies may be neglected.

042.069 **Fast non-resonance rotations of a satellite in the restricted three-body problem.**
P. S. Krasil'nikov.
Pis'ma Astron. Zh., Tom 6, 774 - 777 (1980). In Russian. English translation in Soviet Astron. Lett., Vol. 6.

Relative motions of a satellite on an arbitrary quasi-periodic orbit in the restricted three-body problem, when the ratio of the satellite's angular velocity to the mean motion of the primary bodies is small, are investigated.

042.070 **Motion about the stable libration points in the linearized, restricted three-body problem.**
D. Mittleman.
NASA Ref. Publ., NASA RP 1065, 7 + 151 pp. Price $ 6.75 (1980).

The motion of a point particle in the neighborhood of a triangular libration point (L_4 or L_5) in the linearized, restricted problem of three bodies in the plane is described. The derivation of the equations of motion is standard. From these equations, three invariants of the motion are obtained; the Jacobi integral is expressed linearly in terms of two of these. The trajectories for varied initial conditions are drawn, and a complete geometric description of the particle motion is given in elementary terms.

042.071 **On long-periodic perturbations of Trojan asteroids.** B. Érdi, W. H. Presler.
Astron. J., Vol. 85, 1670 - 1673 (1980).

Previous analytical results (Érdi 1979) concerning the long-periodic perturbations of the eccentricity and of the longitude of the perihelion of Trojan asteroids are compared with numerical integration of the equations of motion of the corresponding plane elliptic restricted three-body problem. Thirty asteroids around the Lagrangian points L_4 and L_5 are tested. The periodicity of the eccentricity is approximately 3600 yr. The perihelion circulates (20 asteroids) or librates (ten asteroids) with the same period. There is good agreement between the analytical and numerical results.

042.072 **Periodic solution of the generalized Hill problem.** I. N. Latyshev.
Astron. Tsirk., No. 1100, p. 7 - 8 (1980). In Russian.

042.073 **Differential equations of the perturbed translatory-rotational motion of a two-body system in Delaunay-Andoyer elements.** D. Z. Koenov.
Dokl. AN TadzhSSR, Vol. 23, 238 - 241 (1980). In Russian. Abstr. in Ref. zh., 51. Astron., 12.51.101 (1980).

042.074 **On the stability of the second intermediate Hill orbit.** M. D. Shinîbaev.
Probl. mekh. upravlyaem. dvizheniya. Perm', 1980, p. 167 - 169. In Russian. – Abstr. in Ref. zh., 51. Astron., 12.51.103 (1980).

042.075 **Mutual gravitational influence of two protoplanets in the planar three-body problem in the case of initially circular orbits.** S. I. Ipatov.
Inst. prikl. mat. AN SSSR. Prepr. 1979, No. 183, 32 pp. In Russian.

Numerical solutions of the planar three-body problem are obtained in order to investigate the motion of two protoplanets (material points) moving around a central massive body (the sun) in initially circular orbits. The cases in which the masses of the protoplanets equal each other and in which the mass of one protoplanet is considerably larger than the mass of the other one are investigated. The mass of the larger protoplanet is varied from $10^{-9} M_{\odot}$ to $10^{-3} M_{\odot}$. Formulae are obtained allowing to estimate the limits of changes of the semi-major axes and eccentricities during evolution and some additional characteristics of motion if the particles move almost periodically.

042.076 **Particular solutions of the problem obtained from Huang's model by averaging according to Fatou's scheme.** G. I. Shirmin.
Vestn. MGU. Fiz. astron., Vol. 21, No. 4, p. 60 - 66 (1980). In Russian. – Abstr. in Ref. zh., 51. Astron., 1.51.68 (1981).

042.077 **On the distribution of mean motions of bodies of small masses in a system of two gravitating centres with variable separation.** B. B. D'yakov, B. I. Reznikov. Fiz.-tekh. inst. AN SSSR. Prepr., 1980, No. 672, 15 pp. In Russian. – Abstr. in Ref. zh., 51. Astron., 1.51.69 (1981).

042.078 **Quasi-periodic solutions of oscillating systems with small parameter.** I. V. Tupikova. Inst. teor. astron. AN SSSR, Leningrad, 1980. 56 pp. In Russian. – Abstr. in Ref. zh., 51. Astron., 1.51.70 (1981).

042.079 **On the construction of a universal algorithm of calculation of the second and higher orders perturbations from the moon.** V. D. Petelina. Tr. MVTU im. N. Eh. Baumana, 1980, No. 336, p. 37 - 44. In Russian. – Abstr. in Ref. zh., 62. Issled. kosm. prostranstva, 1.62.153 (1981).

Analytical algorithms of celestial mechanics. See Abstr. 003.026.

Parameter disposition in pre-Newtonian planetary theories. See Abstr. 004.050.

Automated graphical plots for the study of the gravitational N-body problem. See Abstr. 021.043.

Astrometry and celestial mechanics. See Abstr. 041.039.

Spatial periodic oscillations of a satellite relative to the mass center. See Abstr. 052.015.

On the problem of stability of regular precessions of a symmetric satellite. See Abstr. 052.016.

Rotational dynamics of a deformable medium: further generalization. See Abstr. 062.118.

On general-relativistic celestial mechanics. See Abstr. 066.065.

Relativistische Bewegung ausgedehnter Körper. See Abstr. 066.066.

Life near the Roche limit: behavior of ejecta from satellites close to planets. See Abstr. 091.006.

The necessity of new dynamical theories of the planets. See Abstr. 091.007.

The main problem of lunar theory solved by the method of Brown. See Abstr. 094.017.

The evolution of the lunar orbit revisited, II. See Abstr. 094.018.

Planetary perturbations of the moon. Comparison of ELP-1900 with Brown's theory. See Abstr. 094.030.

Perturbations due to the shape of the Moon in lunar theory. See Abstr. 094.036.

Direct perturbations of the planets on the Moon's motion. See Abstr. 094.037.

On the asymmetry of distribution of asteroids near exact commensurabilities. See Abstr. 098.022.

Graphical measurement of Saturn's oblateness and the radius of the Encke gap. See Abstr. 100.111.

Interior structure of Saturn inferred from Pioneer 11 gravity data. See Abstr. 100.112.

Evolution of comet orbits under the perturbing influence of the giant planets and nearby stars. See Abstr. 102.010.

Numerische Untersuchungen zur Entwicklung des Planetensystems. See Abstr. 107.007.

Stability of tidal equilibrium. See Abstr. 117.069.

043 Astronomical Constants

043.001 **On the determination of the equinox and equator of the new fundamental reference coordinate system, the FK5.** W. Fricke.
Celestial Mech., Vol. 22, (see 012.002), 113 - 125 (1980).

According to IAU recommendations for the improvement of the specification of the fundamental reference system, the positions and centennial variations in the FK5 shall correspond as closely as possible to the dynamical reference frame. Described here is the status of work on the determination of a correction to the FK4 equinox and a correction to the proper motions in right ascension of the FK4 such that the equinox of the FK5 corresponds at all times as closely as possible to the dynamical equinox. Evidence is presented for a correction to the right ascensions of the FK4 (equinox correction) of about $E = +0\overset{s}{.}050$ at 1960, and for a correction to the FK4 proper motions of about $\Delta\mu_a = \dot{E} = +1\overset{''}{.}25$ per century. This investigation has given new findings on the deficiencies of older equinox determinations that have given rise to confusion for a long time. Concerning the determination of the equator point of the FK5 it appears that available new data do not support a significant correction to the FK4. The newly derived expression for the correction to the FK4 equinox is $E = 0\overset{s}{.}035 \pm 0\overset{s}{.}003 + (0\overset{s}{.}085 \pm 0\overset{s}{.}10)(T - 19.50)$, where T is counted in centuries and the errors are standard deviations. It will be applied together with the new value of the general precession in constructing the FK5.

043.002 **The IAU (1976) System of Astronomical Constants.** T. Lederle.
Mitt. Astron. Ges., Nr. 48, (see 012.015), p. 59 - 65 (1980).

A system of astronomical constants can be defined as a set of parameters whose adopted numerical values are needed for the reduction of observations. It should be a consistent set, i. e. the theoretical relations known between the constants have to be exactly fulfilled. The IAU (1976) System of Astronomical Constants will be introduced in 1984 together with the FK5. Finally a detailed explanation for the change of the coefficients in the conventional formula for GMST at 0^h UT1 is given.

043.003 **Kinematics of stars and determination of precessional corrections.** A. N. Balakirev.
Astron. Zh., Tom 57, 1102 - 1104 (1980). In Russian. English translation in Soviet Astron., Vol. 24, No. 5.

It is shown that the sample of 512 distant stars in FK4/FK4 Sup used by Fricke to determine the precessional corrections consists of at least two different kinematic groups, B-stars and A–M-stars. The precessional corrections found separately for each of these groups are different, thus Fricke's values of corrections are questionable.

043.004 **On the motion of the FK4 equinox.** V. I. Orel'skaya.
Pis'ma Astron. Zh., Tom 6, 659 - 661 (1980). In Russian. English translation in Soviet Astron. Lett., Vol. 6.

Possible causes of a false non-precessional motion of the equinox and the way of its establishment for the Fundamental Catalogue are described. Special observations of 10 selected minor planets are used to determine the corrections of the FK4 equinox and equator positions. The results obtained are in good agreement with other independent investigations.

043.005 **The effect of corrections to nutation on astrometric VLBI.** C. Ma.
Bull. American Astron. Soc., Vol. 12, 742 (1980). – Abstract.

043.006 **Geodetic Reference System 1980.** H. Moritz.
Bull. Géod., Vol. 54, 395 - 405 (1980).

The 17. General Assembly of the IUGG in Canberra, December 1979, has adopted a new "Geodetic Reference System 1980" defined by the following constants of the Earth: equatorial radius, geocentric gravitational constant, dynamic form factor, and angular velocity.

043.007 **Non-precessional motion of the vernal equinox and Fricke's precessional corrections.**
A. N. Balakirev.
Astron. Tsirk., No. 1100, p. 4 - 7 (1980). In Russian.

043.008 **Nutation in the IAU system of astronomical constants.** Ya. S. Yatskiv.
Inst. teor. fiz. AN USSR. Prepr., 1980, No. 95, 59 pp. In Russian. – Abstr. in Ref. zh., 51. Astron., 1.51.137 (1981).

Least-squares solution of ill-conditioned systems. II. See Abstr. 021.030.

Astrometric radio source catalogs. See Abstr. 041.003.

The FK5, an improved fundamental reference system extended to fainter stars. See Abstr. 041.012.

On the improvement of the FK4 equinox and equator by means of photographic observations of asteroids. See Abstr. 041.013.

Equator and equinox of FK4 by 2240 observations of (51) Nemausa. See Abstr. 041.014.

The orbits of five minor planets and corrections to the FK4 equator and equinox. See Abstr. 041.028.

New nutation theories. See Abstr. 044.004.

044 Time, Rotation of the Earth

044.001 **Zonal tides and changes in the length of day.**
J. B. Merriam.
Geophys. J. R. Astron. Soc., Vol. 62, 551 - 561 (1980).
The theory of changes in the length of day produced by the long period zonal tides is examined in an attempt to resolve the discrepancy between theoretical and observational results. It is argued here that as a result of the lack of coupling between core and mantle each conserves angular momentum separately with the result that only the changes in mantle moment of inertia influence the changes in rotation rate.

044.002 **Improvement of the Universal Time Service by means of an AR series model.**
D.- w. Zheng, H.- y. Huang, D.- c. Liao, S.- f. Luo.
Acta Astron. Sinica, Vol. 21, 122 - 130 (1980). In Chinese.

044.003 **On the secular change of the angular velocity of the earth's rotation.** V. P. Dolgachev, E. P. Kalinina.
Astron. Zh., Tom 57, 851 - 858 (1980). In Russian.
English translation in Soviet Astron., Vol. 24, No. 4.
The secular acceleration of the earth's rotation is studied with the aid of numerical integration of an 18th-order system of differential equations of motion at a 200-year time interval. The angular rate of rotation of the earth increases by 0.0002 sec per century.

044.004 **New nutation theories.**
P. K. Seidelmann, G. H. Kaplan.
Bull. American Astron. Soc., Vol. 12, 509 (1980). – Abstract.

044.005 **Results of observations of Universal Time at the Astronomical Observatory of the Khar'kov State University and KhGNIIM in the year 1972 - 1977.**
V. I. Turenko, N. G. Litkevich.
Vestn. Khar'k. Univ., No. 190. Fiz. Luny Planet. Fundam. Astrometr., Vyp. 14, p. 3 - 7 (1979). In Russian.

044.006 **Comparison and coordination of time scales.**
J. McK. Luck.
Proc. Astron. Soc. Australia, Vol. 3, 357 - 363 (1979).

044.007 **Time and latitude results of observations made at Merate Observatory with the astrolabe for the year 1978.** L. Buffoni, F. Chlistovsky, A. Manara, F. Mazzoleni.
Astron. Astrophys., Suppl. Ser., Vol. 42, 177 - 178 (1980).
Results of the observations made with the astrolabe Danjon OPL n° are given. These results are in the FK4 system.

044.008 **On determination of time at high latitudes.**
A. N. Kuznetsov, V. N. Baranov.
Izv. vuzov. Geod. i aehrofotosemka, 1980, No. 2, p. 10 - 75. In Russian. – Abstr. in Ref. zh., 52. Geod. Aehrosemka, 8.52.69 (1980).

044.009 **On the irregularity of the earth's rotation.**
A. I. Rybakov, E. P. Kalinina.
Astron. Zh., Tom 57, 1099 - 1101 (1980). In Russian.
English translation in Soviet Astron., Vol. 24, No. 5.
Differential equations of motion for the earth – sun – moon system were integrated numerically on a time interval of 200 years. The table of values of the terrestrial angle of proper rotation obtained in this way was treated with the aid of the least squares method provided that the angular acceleration is constant. The value of the latter was found to be equal to $-0.271522 \times 10^{-11}$ (day)$^{-2}$.

044.010 **Om jordens rotation, månens avstånd och tidvattnet.**
N. Hansson.
Astron. Tidsskr., Årg. 13, 109 - 118 (1980).

044.011 **The earth's rotation and ΔT, 1820 - 1980.**
T. C. Van Flandern, M. R. Lukac.
Bull. American Astron. Soc., Vol. 12, 740 (1980). – Abstract.

044.012 **Monitoring the earth's nutation with radio interferometry.** G. H. Kaplan, F. J. Josties.
Bull. American Astron. Soc., Vol. 12, 741 (1980). – Abstract.

044.013 **Doppler satellite observations of high frequency variations in UT1 - UTC.**
R. J. Anderle, W. L. Stein.
Bull. American Astron. Soc., Vol. 12, 741 (1980). – Abstract.

044.014 **Tidal variations of earth rotation.**
C. F. Yoder, J. G. Williams, W. S. Sinclair, M. E. Parke.
Bull. American Astron. Soc., Vol. 12, 744 (1980). – Abstract.

044.015 **Wie entstand die astronomische Weltzeit?**
M. Gossler.
Sternenbote, 23. Jahrg., 171 - 174 (1980).

044.016 **Time and Latitude Service.** 1980 January - June.
Circ. Time Latitude Serv., Nos. 153 - 154 (1980).

044.017 **Time service for the years 1976-1977-1978.**
Circ. Stn. Astron. Int. Latitudine, Carloforte-Cagliari, Ser. A(5), N. 14, 80 pp. (1979).

044.018 **Greenwich Time Report.** 1979 October - December.
F. G. Smith.
R. Greenwich Obs., Time Latitude Serv., p. 171 - 184 (1980).

044.019 **Zeit- und Breitendienst.** January - December 1979.
Deutsches Hydrogr. Inst., Hamburg (1980).

044.020 **Time and Frequency Services Bulletin,** Nos. 4 - 6.
Published by Shaanxi Astron. Obs., Chinese Acad. Sci., Lintong, Xian, China (1980).

044.021 **Time service of the Mizusawa Observatory.**
Bulletins, Vol. 23, 1978. S. Yumi.
Published by the Int. Latitude Obs. Mizusawa, Mizusawa-Shi, Iwate-Ken, Japan. 37 pp. (1979). ISSN 0580-6585.
This Bulletin contains the results of time service and astronomical observations made at the Mizusawa Observatory from 1 January to 31 December 1978.

044.022 **UT2-UT1 for 1981.**
B. I. H., Paris, Circ. A (1980).

044.023 **Bureau International de l'Heure (B. I. H.) Circular D.**
August - December, 1980.
B. I. H., Paris, Circ. D165 - D169 (1980).
Universal time and coordinates of the pole. (1979 BIH System), Coordinates Universal Time. International Atomic Time.

044.024 **Zeit- und Breitenbestimmungen, Zeitsysteme, Präzisionszeitvergleiche.** September - Dezember 1979.
Published by Akademie der Wissenschaften der DDR, Zentralinstitut für Physik der Erde, Potsdam, Geodätisch-astronomisches Observatorium, Jahrg. 1979, Nos. 5 - 6 (1980).

044.025 **Time Service Annual Report, 1978.**
Xu-Jia-Hui Section, Shanghai Obs., Acad. Sinica, Shanghai, China, 128 pp. (1980).

044.026 **Time and Latitude Bulletins.** October 1979 - June 1980. Z. Suemoto.
Tokyo Astron. Obs., Time and Latitude Bull., Vol. 53, 45 - 59 (1979), Vol. 54, 1 - 29 (1980).

044.027 **Daily time differences and relative phase values.** 1980 July - December.
U. S. Naval Obs., Washington, D.C. Time Serv. Publ., Ser. 4, Nos. 700 - 725 (1980).

044.028 **A.1-UT1 data.** 1980 July - December .
U. S. Naval Obs., Washington, D.C. Time Serv. Publ., Ser. 6, Nos. 53 - 58 (1980).

044.029 **Preliminary times and coordinates of the pole.** 1980 July - December.
U. S. Naval Obs., Washington, D.C. Time Serv. Publ., Ser. 7, Nos. 653 - 678 (1980).

044.030 **Time service announcement. UTC time scale.** G. M. R. Winkler.
U. S. Naval Obs., Washington, D.C. Time Serv. Publ., Ser. 14, No. 28 (1980).

044.031 **Lokalna skala czasu UTC_{BO}.** I. Domiński.
Publ. Inst. Geophys., Polish Acad. Sci., F-6 (137), p. 3 - 27 (1980).

044.032 **Determination of the Borowiec-Potsdam longitude difference.** S. Schillak.
Publ. Inst. Geophys., Polish Acad. Sci., F-6 (137), p. 29 - 46 (1980).

The paper presents results of the longitude observations between the Institute of Terrestrial Physics in Potsdam and Astronomical Latitude Observatory in Borowiec from September 1, 1971 to December 31, 1974. The difference of the Borowiec-Potsdam longitude did not show any essential variations during the campaign and was in agreement with determinations in the previous years.

044.033 **Research of timekeeping in setting up atomic time scale.** X.-p. Pan, H. Yang.
Publ. Shensi Astron. Obs., No. 1, p. 1 - 13 (1980).

The method of the measurement and the design of the system especially the data comparison are discussed. An automatical comparison system has been built with the precision of $\pm 2 \times 10^{-14}$/day or $\pm 5 \times 10^{-13}$/hour in frequency calibration, the precision of ± 1 ns in timing and the precision of a few parts in 10^{12}/day in international comparison. Also, the expression of the characteristics of the clocks and the selection of the timekeeping clocks are discussed. The mathematical model of the rubidium atomic clocks is set up.

044.034 **The detection of short term fluctuations in the length of day.** M. Feissel, D. Gambis.
C. R. Acad. Sci. Paris, Tome 291, Sér. B, 271 - 273 (1980). In French.

The comparison of independent series of measurements of the Earth's rotation allows, for the first time, the detection of transient fluctuations in the length of day with a total amplitude of 0.35 ms and a recurrence of 55 days.

044.035 **Atmospheric angular momentum fluctuations and changes in the length of the day.**
R. Hide, N. T. Birch, L. V. Morrison, D. J. Shea, A. A. White.
Nature, Vol. 286, 114 - 117 (1980) = H. M. Naut. Alm. Off., Libr. Repr. No. 350.

Fluctuations in the angular momentum of the Earth's atmosphere correspond fairly well with equal and opposite fluctuations in the angular momentum of the Earth's 'solid' mantle. A decrease of 20% of the relative angular momentum of the atmosphere during the second half of May 1979 was accompanied by a speeding up of the rotation of the mantle causing the length of the day (l.o.d.) to decrease by 0.6×10^{-3} s. Although motions in the Earth's liquid core produce the more pronounced 'decade' variations in the l.o.d. there is no evidence that core motions produce l.o.d. variations on much shorter time-scales.

044.036 **Changes in length-of-day and atmospheric circulation.** K. Lambeck.
Nature, Vol. 286, 104 - 105 (1980).

044.037 **Investigation of the results of determination of clock corrections obtained from observations of star pairs.** M. N. Pyshnenko.
Khabarovsk. gos. ped. inst. Khabarovsk, 1980, 12 pp. In Russian. – Abstr. in Ref. zh., 51. Astron., 11.51.183 (1980).

044.038 **Study of the earth's rotation: a complex problem of geodynamics.** Ya. S. Yatskiv.
Geodynamics and astrometry, (see 003.011), p. 63 - 73 (1980). In Russian.

044.039 **Influence of the change of the collimation on the accuracy of determination of clock corrections.**
G. M. Blank.
Astron. Tsirk., No. 1112, p. 6 - 7 (1980). In Russian.

044.040 **On the influence of wind on the results of time service astronomical observations.**
D. Lkhagvasurehn, G. Davaakhuu.
Ehrdehm shinzhilgehehnij butehehl. Fiz.-tekh. khurehehlehn. Mat. khurehehlehn. BNMAU shinzhlekh ukhaany Asad., 1980, No. 17, p. 112 - 114. In Russian. – Abstr. in Ref. zh., 51. Astron., 12.51.121 (1980).

044.041 **Investigation of the influence of the temperature field in a pavilion on the observations with a transit instrument.**
G. M. Blank, V. S. Grigor'ev, V. L. Molchanova, E. N. Fedoseev.
Vliyanie atmos. na astron. nablyudeniya v opt. i radiodiapazonakh. Tez. dokl. soveshch. probl.-temat. gruppy po teor. astrometr. sekts. astrometr. AS AN SSSR 1980. Irkutsk, 1980, p. 32 - 34. In Russian. – Abstr. in Ref. zh., 51. Astron., 1.51.769 (1981).

Greenwich time and the discovery of longitude. See Abstr. 003.057.

The earth's variable rotation: geophysical causes and consequences. See Abstr. 003.070.

Algorithms for the combined determination of proper motions of the earth's surface, of variations of the gravitational field, of polar motion and irregularities of the earth's rotation. See Abstr. 021.023.

Variation der Chandler-Periode aus den am Observatorium Potsdam beobachteten Breiten. See Abstr. 045.004.

Luni-solar nutation tables and the liquid core of the earth. See Abstr. 081.001.

On separating indirect effects from variations of the gravitational field, of the figure and rotation of the earth depending on time determined from data of repeated measurements. See Abstr. 081.019.

On the mechanism of the influence of solar activity on the velocity of the earth's daily rotation.
See Abstr. 085.007.

An analysis of total lunar occultations made in the years 1943 to 1974. See Abstr. 096.011.

045 Latitude Determination, Polar Motion

045.001 **An analysis of the homogeneous ILS polar motion series.** C. R. Wilson, R. O. Vicente.
Geophys. J. R. Astron. Soc., Vol. 62, 605 - 616 (1980).

A new reduction of the International Latitude Service Observations for the years 1899 - 1977 has recently been completed under the direction of S. Yumi and the International Astronomical Union Commission 19. This paper examines the annual, Chandler frequency, and long period motion of the Earth's pole implied by this new homogeneous data set. Analysis of the long period motion shows an apparent drift and, in agreement with earlier studies using other data, there is some evidence of an approximately 30-yr oscillation in the pole position.

045.002 **Analysis on the stability of the JYD system of the pole coordinates.** P.-r. Yi, B.-t. Zhou.
Acta Astron. Sinica, Vol. 21, 131 - 135 (1980). In Chinese.

The accuracy or stability of the system of the pole coordinates referred to the JYD (1968.0) is investigated. These results are compared with those of some systems of coordinates of the pole. It is shown that the accuracy of the JYD (1968.0) system is better and less than ±0".01.

045.003 **Bestimmung astronomischer Längendifferenzen für das europäische Längennetz in den Jahren 1977 bis 1979.** K. Kaniuth, W. Wende.
Deutsche Geod. Komm. Bayer. Akad. Wiss., München, Reihe B: Angew. Geod., Heft Nr. 250, 44 pp. (1980).

As a contribution to the establishment of a modern astronomical longitude reference system in Europe three series of longitude difference measurements between national reference stations have been performed by the authors with a Danjon-Astrolabe. The evaluation of the observations in two modes, i.e. the usual adjustment in groups as well as a common adjustment of all star transits, is described in detail. The internal accuracies of the derived longitude differences are in the order of ±0ˢ.0010 to ±0ˢ.0016.

045.004 **Variation der Chandler-Periode aus den am Observatorium Potsdam beobachteten Breiten.**
J. Höpfner.
Gerlands Beitr. Geophys., Band 89, 182 - 186 (1980) = Mitt. Zentralinst. Phys. Erde, Nr. 854.

On the base of the significant variation of the phase of the Chandler wobble, studies were made with consideration being given to a variable Chandler period. Using the simple and modified harmonic analysis, the results were derived for periods of treatment of six years at intervals of three months. According to this, the Chandler period varies between 1.185 and 1.193 years from 1957.8 to 1978.0.

045.005 **Über Variationen der Periode der freien Polbewegung (Chandler-Periode).** H. Jochmann.
Gerlands Beitr. Geophys., Band 89, 187 - 194 (1980) = Mitt. Zentralinst. Phys. Erde, Nr. 853.

The period of Chandler wobble, calculated by suitable modified harmonic analysis, seems to vary in short periods of time with remarkable amounts (0.04 a). These variations cannot be proved by the physical theory of polar motion. By an input-output analysis of the system of polar motion it was found that nonsecular variations of the period of Chandler wobble do not amount to the value obtained by harmonic analysis.

045.006 **On changes in the kinetic energy and frequency of the Chandler nutation.**
H. J. M. Abraham.
Proc. Astron. Soc. Australia, Vol. 3, 355 - 356 (1979).

045.007 **Breitenbestimmung. 1979 June - 1980 June.**
Tech. Univ. Dresden, Lohrmann-Obs., Zirk. Nr. 91 - 92 (1980).

045.008 **Monthly Notes of the International Polar Motion Service.**
Mon. Notes Int. Polar Motion Serv., Nos. 5 - 10, p. 43 - 104 (1980).

Announces the values of latitudes observed at the collaborating stations during May - October 1980.

045.009 **Analysis of accuracy of latitude determination at Borowiec.** W. Jakś.
Publ. Inst. Geophys., Polish Acad. Sci., F-6 (137), p. 59 - 66 (1980).

Corrections to the star declinations were obtained and the smoothed curve of the latitude variations for the period 1970-1974 was determined analytically.

045.010 **Analysis of periodical variations of the Ottawa latitude.** W. Jakś, M. Lehmann.
Publ. Inst. Geophys., Polish Acad. Sci., F-6 (137), p. 71 - 79 (1980).

045.011 **Some results of observations of latitude variations at the Poltava Gravimetrical Observatory.**
N. I. Panchenko.
Geodynamics and astrometry, (see 003.011), p. 59 - 63 (1980). In Russian.

045.012 **On a possible reason of personal errors of latitude observations.** V. V. Lapaeva.
Astron. Tsirk., No. 1099, p. 6 - 8 (1980). In Russian.

Algorithms for the combined determination of proper motions of the earth's surface, of variations of the gravitational field, of polar motion and irregularities of the earth's rotation. See Abstr. 021.023.

Time and latitude results of observations made at Merate Observatory with the astrolabe for the year 1978.
See Abstr. 044.007.

Om jordens rotation, månens avstånd och tidvattnet.
See Abstr. 044.010.

Monitoring the earth's nutation with radio interferometry. See Abstr. 044.012.

Time and Latitude Service. 1980 January - June. See Abstr. 044.016.

Greenwich Time Report. See Abstr. 044.018.

Zeit- und Breitendienst. January - December 1979. See Abstr. 044.019.

Zeit- und Breitenbestimmungen, Zeitsysteme, Präzisionszeitvergleiche. September - Dezember 1979. See Abstr. 044.024.

Time and Latitude Bulletins. October 1979 - June 1980. See Abstr. 044.026.

Astrofix en coordonnées rectangulaires: méthode de Bhattacharji, méthode de Döllen et méthode de Döllen géneralisée. See Abstr. 046.001.

Simultanbestimmungen der Lotabweichungskomponenten ξ und η mit dem Prismenastrolabium. See Abstr. 046.002.

046 Astronomical Geodesy, Satellite Geodesy, Navigation

046.001 **Astrofix en coordonnées rectangulaires: méthode de Bhattacharji, méthode de Döllen et méthode de Döllen géneralisée.** A. Vassallo.
Bull. Géod., Vol. 54, 213 - 220 (1980).

046.002 **Simultanbestimmungen der Lotabweichungskomponenten ξ und η mit dem Prismenastrolabium. XI.** Beobachtungen auf den Hauptdreieckspunkten Breungeshain, Burgholdinghausen, Damscheid, Gerolzhofen (Wiebelsberg), Montabaur, Muxerath, Niederreifenberg, Nindorf, Stade und Tünsdorf im Jahre 1976. A. Rödde.
Deutsche Geod. Komm. Bayer. Akad. Wiss., München, Reihe B: Angew. Geod., Heft Nr. 248, 70 pp. (1980) = Mitt. Nr. 157 Inst. Angew. Geod., Frankfurt.

The author reports on the astronomical determinations of longitudes and latitudes simultaneously performed on 9 first-order triangulation stations pertaining to the West German part of the European Triangulation Net. The components ξ and η of the deviation of the vertical in the system RE 1950 have been computed.

046.003 **On variations of geocentric positions with time.** E. Groten.
Acta Geod. Geophys. Montan., Vol. 14, No. 1 - 2, p. 31 - 57 (1979). – Abstr. in Phys. Abstr., Vol. 83, Abstr. 76894 (1980).

046.004 **Zur lokalen Geoidbestimmung aus terrestrischen Messungen vertikaler Schweregradienten.** B. Heck.
Deutsche Geod. Komm. Bayer. Akad. Wiss., München, Reihe C: Diss., Heft Nr. 259, 95 pp. (1979).

046.005 **Die Arbeiten des Sonderforschungsbereiches 78 Satellitengeodäsie der Technischen Universität München im Jahre 1979.** M. Schneider.
Veröff. Bayer. Komm. Int. Erdmessung, Bayer. Akad. Wiss., Astron.-Geod. Arb., Heft Nr. 40, 192 pp. (1980).

046.006 **Automatische Laser-Satellitenentfernungsmessung in Potsdam.**
H. Fischer, L. Grunwaldt, R. Neubert.
Gerlands Beitr. Geophys., Band 89, 177 - 181 (1980) = Mitt. Zentralinst. Phys. Erde, Nr. 868.

046.007 **Direct reduction of astronomical sights without logarithms or calculator.** D. H. Sadler.
J. Navig., Vol. 33, 430 - 434 (1980).

046.008 **Surface perspective azimuthal cartographic projections of an ellipsoid with positive image (projections of space photographs of surfaces of celestial bodies).**
L. M. Bugaevskij.
Izv. vuzov. Geod. i aehrofotosemka, 1980, No. 2, p. 85 - 92. In Russian. – Abstr. in Ref. zh., 52. Geod. Aehrosemka, 10.52.171 (1980).

046.009 **Doppler positioning obtained by MX 1502 with the aid of precise ephemeris.**
S. Yumi, C. Kakuta, K. Sato, M. Aihara.
Publ. Int. Latitude Obs. Mizusawa, Vol. 13, 1 - 18 (1979).

046.010 **Astronomisch-geodätische Messungen mit einem Zenitteleskop.** P. Jackson.
Sitzungsber. Österreichische Akad. Wiss., Math.-Naturwiss. Kl., Abt. II, 188. Band, 201 - 217 (1979) = Astron. Mitt. Wien, Nr. 24.

This is a report on tests with a zenith telescope 135/1750 used as an astrolabe. The Horrebow levels of the instrument provide an exact reproduction, and preservation for a sufficiently long time, of any zenith distance chosen at will (i.e. 10° in this test series). Furthermore, a photographic method of observation was combined with a special procedure to record the time of star transits through the almucantar. This way of use of the zenith telescope is intermediate between the photographic zenith telescope proper (PZT) and the French astrolabe à prisme.

046.011 **The satellite laser ranging system at Cagliari Observatory.** L. Cugusi.
Laser workshop. Third international workshop on laser ranging instrumentation. Proceedings, Lagonissi, May 23 - 27, 1978 = Pubbl. Stn. Astron. Int. Latitudine, Carloforte-Cagliari, Nuova Ser., N. 70, 5 pp.

046.012 **Il progetto di telemetria laser e Doppler.** E. Proverbio.
Pubbl. Stn. Astron. Int. Latitudine, Carloforte-Cagliari, Nuova Ser., N. 72, 13 pp. (1979).

046.013 **Il programma spaziale ed i progetti di telemetria laser e Doppler.** E. Proverbio.
Pubbl. Stn. Astron. Int. Latitudine, Carloforte-Cagliari, Nuova Ser., N. 73, 10 pp. (1979).

046.014 **The determination of geocentric coordinates by means of satellite Doppler method on a single station.** W.-y. Zhu, H.-g. Xu, S.-d. Xu, G.-l. Zhu, Z.-y. Chu.
Acta Astron. Sinica, Vol. 21, 399 - 403 (1980). In Chinese.

046.015 **On the application of the general theory of methods of geodetic astronomy to the determination of coordinates in points of experimental measurements of refrac-**

tion angles. N. N. Redichkin.
Soveshchanie po atmos. opt. Tez. dokl. Ch. 2. Tomsk, 1980, p. 162 - 165. In Russian. – Abstr. in Ref. zh., 52. Geod. Aehrosemka, 11. 52. 69 (1980).

046.016 **Navigation vs astronomy.** R. Auclair.
J. R. Astron. Soc. Canada, Vol. 74, 367 (1980).
Abstract.

046.017 **On the calculation of the azimuth and corrections of a chronometer according to the heights of the sun with averaged data.** A. V. Butkevich, A. S. Lavnikevich.
Geod., kartogr. i aehrofotosemka, L'vov, 1980, No. 32, p. 3 - 8. In Russian. – Abstr. in Ref. zh., 52. Geod. Aehrosemka, 1.52.97 (1981).

046.018 **On the determination of the height of objects above the earth's surface from one station.**
B. Gezeman, P. Milojković.
Vasiona, Année 28, 9 - 23 (1980). In Croatian.

Geodesy and the earth's gravity field. Vol. II: Geodynamics and advanced methods. See Abstr. 003.052.

Course of geodetic astronomy. Textbook for students of universities teaching astronomical geodesy.
See Abstr. 003.107.

Geodetic work and projects at the Observatory of Graz-Lustbühel. See Abstr. 009.011.

Error analysis of the recurrent algorithm of least squares estimate in satellite geodesy. See Abstr. 021.005.

PMT shutter for satellite laser ranging systems.
See Abstr. 034.043.

Least squares combination of satellite harmonics and integral formulas in physical geodesy.
See Abstr. 081.023.

Ionospheric refraction in Doppler satellite observation. See Abstr. 083.043.

047 Ephemerides, Almanacs, Calendars, Chronology

047.001 **Efemérides Astronómicas para o ano de 1980.**
Published by Observatório Astronómico da Universidade de Coimbra, Coimbra. 17 + 256 pp. (1979).

047.002 **Astronomical Phenomena for the year 1982.**
Prepared by The Nautical Almanac Office, United States Naval Observatory and Her Majesty's Nautical Almanac Office, Royal Greenwich Observatory. U.S. Government Printing Office, Washington; Her Majesty's Stationery Office, London. 71 pp. (1979). ISBN 0-11-886903-5.

047.003 **The Handbook of the British Astronomical Association 1981.**
Prepared by the Computing Section of the Association under the supervision of G. E. Taylor.
Office of the Association: Burlington House, Piccadilly, London, W1V 0NL. 106 pp. Price £ 2.50 (1980).

047.004 **Almanac for Geodetic Engineers 1980.**
Prepared by the Astronomical Observation Division, National Geophysical and Astronomical Office, PAGASA, under the supervision of R. L. Kintanar, W. A. Miñoza.
For sale by the PAGASA (Weather Bureau), Quezon City. 10 + 26 pp. Price P. 5.00 (1979).

047.005 **Philippine Astronomical Handbook 1980.**
Prepared by the Astronomical Observation Division of the National Geophysical and Astronomical Office under the supervision of R. L. Kintanar, W. A. Miñoza.
Published by the Philippine Atmospheric, Geophysical, and Astronomical Services Administration. 12 + 63 pp. (1979). ISSN 0115-1207.

047.006 **The Astronomical Almanac for the year 1981.**
Data for astronomy, space sciences, geodesy, surveying, navigation and other applications.
Issued by the Nautical Almanac Office, United States Naval Observatory, Washington; and Her Majesty's Nautical Almanac Office, London. For sale by the Superintendent of Documents, U.S. Government Printing Office, Washington, D.C., 20402 and Her Majesty's Stationery Office, London. 9 + 539 pp. (1980).

047.007 **Annuaire de l'Observatoire Royal de Belgique** [Jaarboek van de Koninklijke Sterrenwacht van België] **1981.**
Imprimerie Hayez, Rue Fin 4, 1080 Bruxelles. 148e année, 231 pp. (1980).

047.008 **The Indian Astronomical Ephemeris for the year 1980.**
Prepared by Nautical Almanac Unit, Regional Meteorological Centre, New Alipore, Calcutta-700053 under the supervision of S. A. Bandyopadhyay, P. K. Das (Director-General of Meteorology).
Published by the Controller of Publications, Civil Lines, Delhi. 18 + 490 pp. Price Rs. 325.00, £ 37.89, $ 117.00 (1979).

047.009 **Almanaque Nautico, 1981. Con suplemento para la navegacion aerea.**
Published by Instituto y Observatorio de Marina, San Fernando (Cádiz). 418 + 30 + 5 pp. (1980). ISBN 84-7469-005-6; ISSN 0210-735X.

047.010 **1981 Nautical Almanac.** Pub. No. 681.
Published by Hydrographic Office of Japan, Maritime Safety Agency, Tokyo, Japan. 6 + 466 + 7 pp. (1980)

047.011 **1981 Abridged Nautical Almanac.** Pub. No. 683.
Published by Hydrographic Office of Japan, Maritime Safety Agency, Tokyo, Japan. 4 + 242 + 7 pp. (1980)

047.012 **Éphémérides Astronomiques pour l'An 1981. Connaissance des Temps.** Nouvelle série.
B. Morando (Chef du Service des Calculs).
Bureau des Longitudes, 77 Avenue Denfert Rochereau, 75014 Paris. Diffusé par les Editions Gauthier-Villars. Available from l'Etablissement Principal du Service Hydrographique et

Océanographique de la Marine, 29275 Brest. 39 + 126 pp. (1980). ISBN 2-11-080316-9.

047.013 **Éphémérides Nautiques pour l'an 1981.** Ouvrage publié par le Bureau des Longitudes spécialement a l'usage des marins.
Gauthier-Villars, Paris. 485 pp. (1980). ISBN 2-04-010678-2.

047.014 **Nautisches Jahrbuch 1981.**
Edited by Seehydrographischer Dienst der Deutschen Demokratischen Republik, Rostock. 31. Jahrg. 45 + 366 pp. (1980).

047.015 **The Air Almanac 1981, January - June.**
Air Publication 1602. Issued by Her Majesty's Nautical Almanac Office, London; and Nautical Almanac Office, United States Naval Observatory, Washington. Her Majesty's Stationery Office, London; US Government Printing Office, Washington, D.C. 20402. p. 1 - 364, A 105 + F 4 pp. Price £ 11.50 (1980). ISBN 0-11-772264-2.

047.016 **Kalender für Sternfreunde 1981.** Kleines astronomisches Jahrbuch. P. Ahnert.
Johann Ambrosius Barth, Leipzig, German Democratic Republic. 188 pp. Price M 4.80, DM 8.00 (1980).

047.017 **Astronomical Yearbook of the USSR for the year 1983.** V. K. Abalakin (Editor).
Institut Teoreticheskoj Astronomii Akademii Nauk SSSR. Izdatel'stvo "Nauka", Leningradskoe Otdelenie, Leningrad. 720 pp. Price 10 Rbl. 40 Kop. (1980). In Russian.

047.018 **Extended studies to the reform of the Coptic calendar.** J. S. Mikhail.
J. Astron. Soc. Egypt, Vol. 1, 41 - 52 (1979).
Comments on the claim that the civil year used in the Coptic calendar is sidereal are given. Explanations to the computations for determining the age of the Moon used in the ecclesiastical feasts are given. Proposals to adjust the lunar side of the calendar are given.

047.019 **Rocznik Astronomiczny Obserwatorium Krakowskiego 1981.** International Supplement Nr. 52.
K. Rudnicki (Editor).
Państwowe Wydawnictwo Naukowe, Kraków. 5 + 137 pp. Price zł 72.00 (1980). ISBN 83-01-02585-9. – See abstracts 047.020, 119.041, 119.042, 122.081.

047.020 **Geocentric ephemerides for the year 1981 of the libration points L_4 and L_5 in the Earth-Moon system and in the Sun-Venus system.**
K. Kordylewski, R. Szafraniec.
Rocznik Astronomiczny Obserwatorium Krakowskiego 1981, (see 047.019), p. 131 - 136 (1980).

047.021 **1981 Polaris Almanac for Azimuth Determination.** Pub. No. 685.
Published by Hydrographic Office of Japan. Maritime Safety Agency, Tokyo, Japan. 15 pp. (1980).

047.022 **Ephémérides astronomiques ou calendrier des événements célestes pour 1981.**
A. Koeckelenbergh.
Ciel Terre, Vol. 96, 265 - 334 (1980).

047.023 **The preparation of new planetary ephemerides.**
E. Santoro, K. F. Pulkkinen, G. H. Kaplan, P. Espenschied, T. C. Van Flandern, P. K. Seidelmann.
Bull. American Astron. Soc., Vol. 12, 740 (1980). – Abstract.

047.024 **On the use of numerical ephemerides.**
J. D. Mulholland, P. J. Shelus.
Bull. American Astron. Soc., Vol. 12, 742 (1980). – Abstract.

047.025 **Anuarul Astronomic 1981.**
G. Stănilă, M. Stavinschi, S. Dinulescu (Editors).
Centrul de astronomie şi ştiinţe spaţiale, Bucureşti. Editura Academiei Republicii Socialiste România, R. 79717 Bucureşti, Calea Victoriei 125. 275 pp. Price Lei 10.00 (1980).

047.026 **Anuario del Observatorio Astronomico Nacional 1980.**
Published by Observatorio Astronomico Nacional, Universidad Nacional de Colombia, Facultad de Ciencias, Bogotá, Colombia, S. A. 112 pp. (1980). ISSN 0120-2758.

047.027 **The Observer's Handbook 1981.**
J. R. Percy (Editor).
Royal Astronomical Society of Canada, 124 Merton Street, Toronto, Canada M4S 2Z2. 144 pp (1980). ISSN 0080-4193.

047.028 **Almanacco Astronomico della Rivista Coelum per l'anno 1981.** P. Battistini.
Coelum, Suppl. al fasc. 11-12, 43 + 38 pp. Price L 6,000 (1980).

047.029 **Astronomische Grundlagen für den Kalender 1982.**
Compiled by T. Lederle, edited by Astronomisches Rechen-Institut, Heidelberg.
Verlag G. Braun, Karlsruhe. 92 pp. Price DM 42.00 (1980).

047.030 **Public documents on calendar and the Tokyo Astronomical Observatory appearing in the "Hooki Bunrui Taizen, Part II".** S. Ito.
Tokyo Astron. Obs. Rep. (No. 73), Vol. 19, 295 - 304 (1980). In Japanese.

047.031 **A list of documents related to the history of calendars in Japan.** M. Utida.
Tokyo Astron. Obs. Rep. (No. 73), Vol. 19, 305 - 322 (1980). In Japanese.

047.032 **Astronomical Handbook for 1980.**
G. A. Eiby.
Carter Obs., Astron. Bull., No. 93, 32 pp. (1979).

047.033 **Astronomiskais Kalendārs 1981.**
J. Bikše, I. Daube, M. Diriķis, J. Francmanis, V. Freijs, J. Miezis (Editors).
Latvijas PSR Zinātņu Akadēmija Radioastrofizikas Observatorija, Vissavienības Astronomijas un Ģeodēzijas Biedrības Latvijas Nodaļa. Izdevniecība "Zinātne", Riga. 195 pp. Price 50 Kop. (1980).

047.034 **Almanac for computers 1981.**
Nautical Almanac Office, United States Naval Observatory, Washington, D. C. 20390. A14 + B17 + C23 + D33 + E10 pp. (1980).

047.035 **Annuaire du Bureau des Longitudes. Éphémérides 1981.** Calendriers–soleil–lune–planètes–satellites–étoiles–marées–déclinaison magnétique.
Gauthier-Villars, Paris. 15 + 259 pp. (1980). ISBN 2-04-011416-5 = Supplément à l'Astronomie de Janvier 1981.

047.036 **Efemérides Astronómicas 1981.**
Published by Instituto y Observatorio de Marina, San Fernando (Cádiz), Spain. Vol. 190, 16 + 477 pp. (1980). ISBN 84-7469-006-4. ISSN 0080-5971.

047.037 **Dados Astronómicos para os Almanaques de 1980 para Portugal.** E. M. L. Cabrita.

Published by Observatório Astronómico de Lisboa, 59 pp. (1979).

047.038 **Dados Astronómicos para os Almanaques de 1981 para Portugal.** E. M. L. Cabrita.
Published by Observatório Astronómico de Lisboa. 52 pp. (1980).

047.039 **Chinese Astronomical Ephemeris 1981.** Published by Purple Mountain Observatory, Academia Sinica, Nanking, China. 539 pp. (1980).

047.040 **Astronomical Calendar of the Sofia Observatory for the year 1981.** D. Rajkova, Z. Krajcheva, Z. Ivanova, A. Antov; edited by A. Bonov.
Izdatelstvo na Blgarskata Akademiya na Naukite, Sofiya. 101 pp. Price 1.16 Lv. (1980). In Bulgarian.

047.041 **Hvězdářská ročenka 1981.** V. Guth, B. Onderlička, P. Příhoda, J. Ruprecht (Editors).
Ročník 57, svazek 1. Tabulky efemerid. Academia, nakladatelství Československé akademie věd, Praha. 151 pp. Price 21.00 Kčs (1980).

047.042 **Hvězdářská ročenka 1980.** V. Guth, B. Onderlička, P. Příhoda, J. Ruprecht (Editors).
Ročník 56, svazek 2: Přehled pokroků v astronomii. Academia, Nakladatelství Československé akademie věd, Praha. 160 pp. Price 22.00 Kčs (1980).

047.043 **Anuario del Observatorio Astronómico de Madrid para 1981.**
Published by Instituto Geografico Nacional, Madrid. 441 pp. Price 300 pesetas (1980). ISBN 84-500-3470-1, ISSN 0373-5125.

047.044 **Satellites galiléens de Jupiter. Phénomènes et configurations pour 1981.**
B. Morando (Directeur du Service des Calculs), J.-E. Arlot, Y. Jannot, W. Thuillot, D. T. Vu (Rédaction et calculs).
Supplément à la Connaissance des Temps, Bureau des Longitudes, Paris, 61 pp. (1980).

047.045 **1982 Japanese Ephemeris.** Pub. No. 684. Compiled by A. Yamazaki, T. Mori, Y. Tano, A. Senda, Y. Kubo, Y. Ganeko, F. Ono, T. Takemura, Y. Harada, K. Inoue, M. Sasaki.
Published by Hydrographic Department, Maritime Safety Agency, Tokyo, Japan. 6 + 474 + 29 pp. (1980).

047.046 **Almanacco di Astronomia 1981 dell'Unione Astrofili Italiani.**
Suppl. al Num. 4 della Riv. Astronomia, 88 pp. (1980).

047.047 **Sterrengids 1980.** W. Gielingh, J. Meeus.
Uitgegeven door de Stichting 'de Koepel' in opdracht van de NVWS. 152 pp. Price f 24.50. – Review in Zenit, 7e Jaarg., 124; 1980 (*G. P. Können*).

047.048 **Astronomisk Årsbok 1979/80.** A. Larsson, J. Schildt.
Bokförlaget INOVA, Stockholm. 96 pp. Price S. kr. 15.00 (1980). – Review in Astron. Tidsskr., Årg. 13, 193; 1980 (*S. Söderhjelm*).

047.049 **Himmelskalender 1981.** Astronomisches Jahrbuch für Österreich. H. Mucke.
Astronom. Büro. Hasenwartg. 32, A-1238 Wien. 105 pp. Price öS 50.00 (1980). – Review in Sternenbote, 23. Jahrg., 198 - 199 (1980).

Basis formulas to convert astronomical calendars by computer programming. See Abstr. 021.004.

Systems of formulae for calculating the ephemerides of the sun. See Abstr. 021.010.

Mathematical approximations of the ephemerides and the almanac for computers. See Abstr. 021.022.

Calculs astronomiques pour amateurs.
See Abstr. 021.034.

Changement de système photométrique dans l'Annuaire du Bureau des Longitudes.
See Abstr. 113.019.

Space Research

051 Extraterrestrial Research, Spaceflight Related to Astronomy and Astrophysics

051.001 **The stormy sun.** B. O'Leary.
Sky Telesc., Vol. 60, 199 - 201 (1980).

051.002 **The Einstein Observatory: New perspectives in astronomy.** R. Giacconi, H. Tananbaum.
Science, Vol. 209, 865 - 876 (1980).

051.003 **The distant geomagnetic tail observing satellite Geos-3.** K. Knott.
ESA J., Vol. 4, 1 - 14 (1980). – Abstr. in Phys. Abstr., Vol. 83, Abstr. 82163 (1980).

051.004 **The astrometry satellite Hipparcos.** F. Beeckmans.
ESA J., Vol. 4, 15 - 30 (1980). – Abstr. in Phys. Abstr., Vol. 83, Abstr. 82164 (1980).

051.005 **The extreme-ultraviolet and soft X-ray sky-survey project EXUV.** A. Peacock.
ESA J., Vol. 4, 31 - 40 (1980). – Abstr. in Phys. Abstr., Vol. 83, Abstr. 82165 (1980).

051.006 **The International Comet Mission (ICM).** R. Reinhard.
ESA J., Vol. 4, 41 - 58 (1980). – Abstr. in Phys. Abstr., Vol. 83, Abstr. 82166 (1980).

051.007 **Space astrometry as NASA and ESA projects.** E. Høg.
Mitt. Astron. Ges., Nr. 48, (see 012.015), p. 127 - 146 (1980).

Two projects for astrometry from earth satellites are described and their impact on astrometry, astrophysics and stellar astronomy are outlined. The NASA Space Telescope will provide very accurate relative parallaxes and proper motions, primarily of a few hundred faint stars. The European astrometry satellite HIPPARCOS will obtain absolute positions, proper motions and parallaxes for 100 000 stars, particularly of stars brighter than m = 11, and with an accuracy of ±0.″002. It is discussed how the HIPPARCOS project has evolved during the recent years of ESA study. Finally, the data analysis for deriving the astrometric star catalogue is outlined.

051.008 **The International Sun-Earth Explorer mission.** A. C. Durney.
Nuovo Cimento C, Ser. 1, Vol. 2C, (see 012.013), 722 - 736 (1979). – Abstr. in Phys. Abstr., Vol. 83, Abstr. 90320 (1980).

051.009 **Planet-A project on a magnetic measurement in the interplanetary space during the Halley-Venus mission.** T. Saito, S. Kokubun, I. Aoyama, M. Seto, H. Fukunishi, A. Nishida.
Bull. Inst. Space Aeronaut. Sci., Univ. Tokyo B, Vol. 15, 487 - 500 (1979). In Japanese. – Abstr. in Phys. Abstr., Vol. 83, Abstr. 94672 (1980).

051.010 **Summary of terrestrial and planetary radio and plasma waves observed by the initial phase of JIKIKEN (EXOS-B) observations.**
H. Oya, A. Morioka.
Bull. Inst. Space Aeronaut. Sci., Univ. Tokyo B, Vol. 15, 637 - 648 (1979). In Japanese. – Abstr. in Phys. Abstr., Vol. 83, Abstr. 94675 (1980).

051.011 **Hipparcos. A new astronomical base line.** C. Turon Lacarrieu.
Recherche, Vol. 11, 709 - 711 (1980). In French. – Abstr. in Phys. Abstr., Vol. 83, Abstr. 94688 (1980).

051.012 **The International Solar Polar Mission: zodiacal light/background starlight experiment (ISPM–ZLE).**
G. H. Schwehm.
Solid particles in the solar system, (see 012.019), p. 23 - 24 (1980).

The International Solar Polar 1983 Mission will for the first time provide the opportunity to perform measurements of the zodiacal light out of the ecliptic and above the solar poles.

051.013 **Of computing and astronauts.** P. Herget.
Sky Telesc., Vol. 60, 373 - 375 (1980).

051.014 **Recent scientific achievements of ESA spacecraft.** D. E. Page.
ESA Bull., No. 23, p. 10 - 18 (1980).

This article concentrates on results obtained more recently from the Cos-B, Geos, ISEE and IUE spacecraft.

051.015 **Scientific projects under development.** M. Delahais.
ESA Bull., No. 23, p. 26 - 33 (1980).

The author presents a brief recapitulation of the scientific aims of five missions (European X-Ray Observatory Satellite, Sled, International Solar-Polar Mission, Space Telescope, Hipparcos), together with the attendant spacecraft designs, the essential features of the operational phases, and the status of development.

051.016 **A program for the observations of the sun and heliosphere from space 1980 - 1995.**
J. D. Bohlin, E. G. Chipman.
Solar and interplanetary dynamics, (see 012.020), p. 523 - 539 (1980).

Solar and heliospheric physics will be studied by essentially every solar space mission either now approved, or in the planning stage, for the period of 1980 to 1995. These missions include traditional earth-orbiting satellites; Shuttle/ Spacelab sortie missions, free flyers that transit the solar polar caps and probe the innermost corona (both frontier regions of the heliosphere), and finally possible semi-permanent orbiting platforms for advanced solar /heliosphere observations.

051.017 **Proposal for an interplanetary mission to sound the outer regions of the solar corona.**
H. Porsche, H. Volland, K. Bird, P. Edenhofer.
Solar and interplanetary dynamics, (see 012.020), p. 541 - 545 (1980).

The mission of HELIOS had been started in order to investigate in situ the innermost regions of the interplanetary space.

051.018 **Verändertes Jupiter-Programm Galileo.**
H. W. Köhler.
Sterne Weltraum, Jahrg. 19, 374 - 375 (1980).

051.019 **Direct and reverse satellite space lifts on Mercury and Venus.** G. G. Polyakov.
Astrakhan. gos. ped. inst., Astrakhan'. 1980. 12 pp. In Russian. Abstr. in Ref. zh., 62. Issled. kosm. prostranstva, 10.62.421 (1980).

051.020 **Continuation of the expedition on Salyut 6.**
S. A. Nikitin.
Priroda, 1980, No. 11, p. 104 - 105. In Russian.

051.021 **Scientific need for space astronomy.**
L. Goldberg.
Highlights of Astronomy, Vol. 5, (see 003.006), 63 - 68 (1980).

051.022 **A search of the solar neighborhood for sub-stellar objects.** J. Tarter, R. Reynolds, R. Walker.
Bull. American Astron. Soc., Vol. 12, 707 (1980). – Abstract.

051.023 **Ruimte-onderzoek en sterrekunde (1).** (UV-astronomie). W. de Graaff.
Zenit, 7e Jaarg., 28 - 31 (1980).

051.024 **Het eerste jaar van IUE.** A. J. Willis, C. de Loore, E. L. van Dessel, M. Burger.
Zenit, 7e Jaarg., 138 - 141 (1980).

051.025 **SMM (*Solar Maximum Mission*) bespiedt zonnevlammen.** G. W. E. Beekman.
Zenit, 7e Jaarg., 184 - 185 (1980).

051.026 **Ruimte-onderzoek en sterrekunde (2).** Röntgen- en gamma-astronomie. W. de Graaff.
Zenit, 7e Jaarg., 194 - 197 (1980).

051.027 **De IUE röntgenveertiendaagse.** M. Burger, E. L. van Dessel, C. de Loore, A. J. Willis.
Zenit, 7e Jaarg., 218 - 221 (1980).

051.028 **In de keuken van het SMM-team.** C. Titulaer.
Zenit, 7e Jaarg., 471 - 472 (1980).

051.029 **The International Solar Polar Mission.**
M. Delahais, D. Eaton.
Space Shuttle: dawn of an era. Part II, (see 012.037), p. 829 - 843 (1980). – Abstr. in Phys. Abstr., Vol. 83, Abstr. 109335 (1980).

051.030 **The Space Shuttle and deep space missions.**
J. C. Beckman.
Space Shuttle: dawn of an era. Part II, (see 012.037), p. 857 - 865 (1980). – Abstr. in Phys. Abstr., Vol. 83, Abstr. 109336 (1980).

051.031 **Infrared Space Observatory (ISO).** Pre-phase A study.
European Space Agency, SCI (80) 9, Paris. 4+67 pp. (1980).

The crucial step into space, which removes the restrictions imposed by the atmosphere, will be taken first by IRAS, which will be launched into orbit in 1982. ISO is proposed to follow IRAS in exploiting the space environment for IR observations and will use a cooled telescope with cooled instruments. The main purpose for ISO, however, is to provide spectroscopic studies of sources which have been identified to be of interest from photometric measurements. An important facility will also be to make detailed photometric studies of small areas of sky, using long integration times to achieve the highest levels of sensitivity.

051.032 **To encounter a star – the Solar Probe mission.**
J. E. Randolph.
J. Astronaut. Sci., Vol. 28, No. 1, p. 1 - 13 (1980). – Abstr. in Phys. Abstr., Vol. 84, Abstr. 4741 (1981).

051.033 **Microwave communications from interplanetary space.** J. Buj.
Rev. Española Electron. Vol. 27, No. 309 - 310, p. 24 - 29 (1980). In Spanish. – Abstr. in Phys. Abstr., Vol. 84, Abstr. 4744 (1981).

051.034 **MAGSAT – a new satellite to survey the Earth's magnetic field.**
F. F. Mobley, L. D. Eckard, G. H. Fountain, G. W. Ousley.
IEEE Trans. Magn.,Vol. MAG-16, 758 - 760 (1980). – Abstr. in Phys. Abstr., Vol. 84, Abstr. 9346 (1981).

051.035 **Die Galileo - Raumflugmission.**
H. Sarcander.
Jupiter, (see 012.045), p. 141 - 162 (1980).

051.036 **A short review of the overall scientific tasks and of some general principles of Helios.**
H. Porsche.
Helios solar probes, science summaries, (see 003.009), p. 1 - 5 (1980).

051.037 **Calculation of the disturbances of the low energy electron measurements (E 1 - Z).**
G.-H. Voigt, U. Isensee, M. Maassberg.
Helios solar probes, science summaries, (see 003.009), p. 12 - 16 (1980).

051.038 **A summary of progress in space physics made with Helios plasma wave instrument data (E 5a).**
D. A. Gurnett, R. R. Anderson.
Helios solar probes, science summaries, (see 003.009), p. 27 - 29 (1980).

051.039 **Space science: retrospect and prospect.**
N. W. Hinners.
Proc. Tenth Lunar Planet. Sci. Conf., (see 012.050), p. XVII - XXIII (1979).

051.040 **The joint NASA/ESA cometary mission to comets Halley and Tempel 2.**
D. Dale, R. Pacault, R. Reinhard.
Comet Halley micrometeoroid hazard workshop, (see 012.049), p. 3 - 5 (1979).

051.041 **The micrometeoroid hazard to a space probe in the vicinity of the nucleus of Halley's comet.**
D. W. Hughes.
Comet Halley micrometeoroid hazard workshop, (see 012.049), p. 51 - 56 (1979).

051.042 **Dust hazard of the cometary probe.** H. Fechtig.
Comet Halley micrometeoroid hazard workshop, (see 012.049), p. 63 - 65 (1979).

051.043 **The treatment of the problems due to hypervelocity impact during a fast fly-by (57 km/s) of Halley's comet.** M. J. v. d. Hoek.
Comet Halley micrometeoroid hazard workshop, (see 012.049), p. 121 - 129 (1979).

051.044 **Space report.**
Spaceflight, Vol. 22, 321 - 327 (1980).

La conquista dello spazio. See Abstr. 003.032.

Space developments for the future of mankind. See Abstr. 003.084.

"Sojus-22" erforscht die Erde. See Abstr. 003.117.

Programmes under development and operations (*ESA*). See Abstr. 010.016

Autumn 1979 MIST (*Magnetosphere, Ionosphere and Solar-Terrestrial Environment*) meeting with discussion of a proposed cometary mission. See Abstr. 011.004.

Fifth ESA Symposium on European rocket & balloon programmes & related research. Conference held at Bournemouth, England, 14 - 18 April 1980. See Abstr. 012.058.

Space – New opportunities for international ventures. See Abstr. 012.059.

Life sciences and space research. Vol. XVIII. See Abstr. 012.060.

Space and development. Vol. 6. Advances in space exploration. See Abstr. 012.062.

Some results from the Space Science Department research programme. See Abstr. 013.010.

ESA's Science Programme: the present situation and future perspectives. See Abstr. 013.011.

The European astrometry satellite, Hipparcos. See Abstr. 032.540.

Cinderella spacecraft: deep-space probes face funding starvation, despite brilliant achievements in 1979. See Abstr. 053.003.

The OPEN program: an example of the scientific rationale for future solar-terrestrial research programs. See Abstr. 085.015.

Apollo asteroids. See Abstr. 098.089.

Comets and cometary missions. See Abstr. 102.033.

The needs in the radial velocity area in view of impending space astrometry projects. See Abstr. 111.016.

052 Astrodynamics, Navigation of Space Vehicles

052.001 **Long term evolution of the LAGEOS orbit.**
D. E. Smith, P. J. Dunn.
Geophys. Res. Lett., Vol. 7, 437 - 440 (1980).

The authors report on the observed evolution of the LAGEOS orbit over the first thirty-two months after launch and assess the present understanding and capability to determine the LAGEOS orbit.

052.002 **A method for improving the orbit of an artificial satellite with sparse observations.** P.-x. Xu.
Acta Astron. Sinica, Vol. 21, 93 - 95 (1980). In Chinese.

In this method, mean anomaly is separated from the other orbital elements, so that sparse observations, distributed over a long time-interval and covering a large section of the orbit, may be used in improving the orbit of an artificial satellite.

052.003 **Determination of the orbital elements of artificial earth satellites from photographic and laser observations.** E. P. Aksenov, S. N. Vashkov'yak, N. V. Emel'yanov.
Tr. Gos. Astron. Inst. Shternberga, Tom 49, 90 - 115 (1980). In Russian.

A computer programme of orbit improvement by means of photographic and laser observations is presented. The orbit derived from the solution of the problem of two fixed centres has been selected as an intermediate orbit. The position of the satellite is determined taking into consideration gravitational and luni-solar perturbations.

052.004 **Construction of the conditional equations when improving the intermediate orbit of an artificial earth satellite.**
E. P. Aksenov, S. N. Vashkov'yak, N. V. Emel'yanov.
Tr. Gos. Astron. Inst. Shternberga, Tom 49, 116 - 121 (1980). In Russian.

Formulae for the derivatives of the rectangular coordinates of an artificial satellite with respect to the elements of the intermediate orbit are given. As an intermediate is chosen the solution of the problem of two fixed centres. These orbits take into account second and third harmonics of the geopotential.

052.005 **A method of calculation of luni-solar perturbations of the orbital elements of artificial earth satellites.**
N. V. Emel'yanov.
Tr. Gos. Astron. Inst. Shternberga, Tom 49, 122 - 129 (1980). In Russian.

052.006 **Long-period perturbations in satellite motion by the oblateness of the earth's atmosphere.**
B. N. Noskov.
Tr. Gos. Astron. Inst. Shternberga, Tom 49, 130 - 155 (1980). In Russian.

The long-period perturbations in the motion of an artificial earth satellite caused by the earth atmosphere oblateness are investigated. The analytic expressions derived for these perturbations hold for all values of the eccentricity less than unity.

052.007 **Determination of the coordinates of intermediate motion of an artificial earth satellite by means of power series according to regularized time.** N. Georgiev.
Tr. Gos. Astron. Inst. Shternberga, Tom 49, 156 - 171 (1980). In Russian.

A method of determination of parameters of an intermediate motion of a satellite on short arcs is proposed. This method makes it possible to represent the variable parameters by polynomials on the regularized time.

052.008 **Power series of the perturbations of the coordinates of intermediate motion due to zonal harmonics of the geopotential.** N. Georgiev.

Tr. Gos. Astron. Inst. Shternberga, Tom 49, 172 - 185 (1980). In Russian.

052.009 **Halo-orbit formulation for the ISEE-3 mission.** D. L. Richardson.
Astrodynamics Conference, (see 027.012.026), Paper 79-127, 16 pp. (1980).
The path to which the ISEE-3 satellite is to be controlled is a periodic solution to the three-dimensional equations for motion about the interior libration point of the Sun-Earth-satellite three-body system. These particular periodic "halo" orbits are constructed using analytical and numerical methods.

052.010 **A note on stable halo orbits.** J. V. Breakwell. Astrodynamics Conference, (see 27.012.026), Paper 79-130, 2 pp. (1980).

052.011 **Determination of the initial entry conditions of a space vehicle into the Martian atmosphere using a detached navigation probe.** N. M. Ivanov, V. D. Belykh.
Kosm. Issled., Tom 18, 507 - 517 (1980). In Russian.

052.012 **On a class of particular solutions of the variation problem of flight in a noncentral field.**
A. G. Azizov, N. A. Korshunova.
Kosm. Issled., Tom 18, 643 - 646 (1980). In Russian.

052.013 **Rotation motions of a gyrostatic satellite in a Kepler orbit.** M. C. Nucci.
Z. Angew. Math. Mech., Vol. 60, 113 - 114 (1980). – Abstr. in Phys. Abstr., Vol. 83, Abstr. 86080 (1980).

052.014 **Earth satellite orbits with resonant lunisolar perturbations. I. Resonances dependent only on inclination.** S. Hughes.
Proc. R. Soc. London, Ser. A, Vol. 372, 243 - 264 (1980). Abstr. in Phys. Abstr., Vol. 83, Abstr. 90326 (1980).

052.015 **Spatial periodic oscillations of a satellite relative to the mass center.**
V. A. Sarychev, V. V. Sazonov, N. V. Mel'nik.
Kosm. Issled., Tom 18, 659 - 677 (1980). In Russian.

052.016 **On the problem of stability of regular precessions of a symmetric satellite.** A. G. Sokol'skij.
Kosm. Issled., Tom 18, 698 - 706 (1980). In Russian.

052.017 **Influence of different approximations of the geomagnetic field on the accuracy of model parameters of the rotational motion of an artificial earth satellite.**
S. M. Zabluda, A. M. Yanshin.
Kosm. Issled., Tom 18, 715 - 721 (1980). In Russian.

052.018 **On a non-iteration method of calculation of a trajectory with fixed end points.**
V. N. Borovenko.
Kosm. Issled., Tom 18, 793 - 796 (1980). In Russian.

052.019 **Resonance rotations of a satellite with interaction between a magnetic and a gravitational field.**
V. V. Beletskij, A. N. Shlyakhtin.
Inst. prikl. mat. AN SSSR. Prepr., 1980, No. 46, 30 pp. In Russian. – Abstr. in Ref. zh., 62. Issled. kosm. prostranstva, 8.62.297 (1980).

052.020 **Uniaxial gravitational orientation of artificial satellites.** V. A. Sarychev, V. V. Sazonov.
Inst. prikl. mat. AN SSSR. Prepr., 1980, No. 49, 27 pp. In Russian. – Abstr. in Ref. zh., 62. Issled. kosm. prostranstva, 8.62.298 (1980).

052.021 **Gravitational orientation of a rotating satellite.** V. A. Sarychev, V. V. Sazonov.
Inst. prikl. mat. AN SSSR. Prepr., 1980, No. 72, 28 pp. In Russian. – Abstr. in Ref. zh., 62. Issled kosm. prostranstva, 8.62.299 (1980).

052.022 **Geostationary-orbit determination by single-ground-station tracking.** M. Soop.
ESA J., Vol. 4, 159 - 169 (1980).
The two types of tracking data that can be obtained from a single ground station for a geostationary spacecraft are ground-antenna pointing direction and range. This paper analyses how the spacecraft orbital parameters determined from that single station depend on the various tracking inputs and their accuracies.

052.023 **Solar-synchronous and multiple orbits.**
P. M. Vingardt, P. E. Ehl'yasberg.
Inst. kosm. issled. AN SSSR. Prepr., 1980, No. 542, 26 pp. In Russian. – Abstr. in Ref. zh., 62. Issled. kosm. prostranstva, 9.62.330 (1980).

052.024 **Investigation of the mutual position of the orbit of a satellite with high apogee and the magnetopause, of the shock wave and the shadow of the earth by means of a machine diagram.** V. I. Prokhorenko.
Inst. kosm. issled. AN SSSR. Prepr., 1980, No. 541, 69 pp. In Russian. – Abstr. in Ref. zh., 62. Issled. kosm. prostranstva, 9.62.341 (1980).

052.025 **Resonance motions of a space vehicle relative to the mass center situated in the triangular libration point of the earth – moon system.**
Yu. V. Barkin, Yu. G. Markov.
Prikl. mat. i mekh., Vol. 44, 569 - 573 (1980). In Russian. Abstr. in Ref. zh., 62. Issled. kosm. prostranstva, 9.62.349 (1980).

052.026 **On an integrable case of satellite motion in an approximate gravitational field of the earth.**
E. L. Lukashevich.
Nauchn. chteniya po aviats. i kosmonavt., 1979. Moskva, 1980, p. 158. In Russian. – Abstr. in Ref. zh., 62. Issled. kosm. prostranstva, 8.62.270 (1980).

052.027 **Optimum expectation orbits.** B. L. Zhurin. Nauchn. chteniya po aviats. i kosmonavt., 1979. Moskva, 1980, p. 172. In Russian. – Abstr. in Ref. zh., 62. Issled. kosm. prostranstva, 8.62.278 (1980).

052.028 **Forecast of interplanetary flights to Jupiter and Saturn.**
V. A. Kotin, A. V. Leshchenko, O. V. Papkov.
Nauchn. chteniya po aviats. i kosmonavt., 1979. Moskva, 1980, p. 168 - 169. In Russian. – Abstr. in Ref. zh., 62. Issled. kosm. prostranstva, 8.62.281 (1980).

052.029 **Characteristics of trajectories to the comets of the Jovian group.** A. K. Platonov, R. K. Kazakova.
Inst. prikl. mat. AN SSSR. Prepr., 1980, No. 74, 29 pp. In Russian. – Abstr. in Ref. zh., 62. Issled. kosm. prostranstva, 8.62.282 (1980).

052.030 **Improvement of the second order perturbative solution of artificial earth satellites.**
L. Liu, D.-z. Zhao.
Acta Astron. Sinica, Vol. 21, 278 - 286 (1980). In Chinese.

052.031 **On periodic motions of artificial earth satellites relative to the mass center in a gravitational field.**
I. G. Azova.
Redkol. zh. "Vestn. LGU. Mat., mekh., astron." Leningrad,

1980, 16 pp. In Russian. – Abstr. in Ref. zh., 62. Issled. kosm. prostranstva, 10.62.296 (1980).

052.032 **The probability of collisions on the geostationary ring.** M. Hechler, J. C. van der Ha.
ESA J., Vol. 4, 277 - 286 (1980).
Fundamental data are derived via deterministic orbit calculations for a representative sample of uncontrolled objects and an intersection process with the geostationary ring described by a probability distribution for the active satellites.

052.033 **Doubly periodic orbits in the sun-earth-moon system.** D. W. Dunham, D. P. Muhonen.
Bull. American Astron. Soc., Vol. 12, 738 (1980). – Abstract.

052.034 **Poynting-Robertson drag on satellites near synchronous altitude.** V. J. Slabinski.
Bull. American Astron. Soc., Vol. 12, 741 (1980). – Abstract.

052.035 **Natural intermediaries.** A. Deprit.
Bull. American Astron. Soc., Vol. 12, 745 (1980). Abstract.

052.036 **Improvement of ephemeris by observations of one station.** R. Dietrich.
Artif. Satell., Vol. 15, No. 1, p. 63 - 65 (1980).
Investigations have been made to improve orbital elements by observations of one station. The numerical difficulties of almost singular normal equations can be avoided by using fictive worldwide observations. These fictive observations are calculated by the given orbital elements. Some calculations show a sufficient accuracy for blind positioning for short prediction periods.

052.037 **The algorithm for numerical calculations of an artificial satellite free motion around its center of mass.** M. Banaszkiewicz.
Artif. Satell., Vol. 15, No. 2, p. 3 - 11 (1980).
In the paper it is shown how to use quick numerical procedures computing elliptic functions and integrals to determine the free motion of an artificial satellite around its center of mass. The algorithm presented here gives the values of the parameters determining the motion for an arbitrary moment t if initial values of the parameters for the moment t_0 are known. The obtained formulae enable to calculate the motion in each of three cases: the motion around the minor, the intermediary, and the major main axis of the moment of inertia tensor.

052.038 **System of orbital computation (ORBIT-1). Part 2.: improvement of satellites orbits.**
A. Drożyner.
Artif. Satell., Vol. 15, No. 2, p. 13 - 18 (1980).
In this paper methods of satellites orbits improvement, used in system ORBIT-1, are described.

052.039 **On the determination of quasi-osculating orbital elements of an artificial satellite by using optical observations.**
A. Pál, T. Oproiu, V. Mioc, E. Radu, B. Pârv.
Visual Obs. Artif. Earth Satell., p. 7 - 21 (1977) = Publ. Obs. Astron. Univ. Cluj-Napoca, Romania, Nr. B 46-58.
The paper deals with the determination of quasi-osculating orbital elements of an artificial satellite. The determination of geocentric positions and orbital elements, by means of optical observations, is shown. Two procedures for these determinations are presented.

052.040 **On the improvement of the orbital elements of an artificial satellite.** A. Pál, T. Oproiu, H. Alexandrescu, V. Mioc, E. Radu, B. Pârv.
Visual Obs. Artif. Earth Satell., p. 22 - 33 (1977) = Publ. Obs. Astron. Univ. Cluj-Napoca, Romania, Nr. B 46-58.
A method for the differential improvement of the orbital elements of an artificial Earth satellite is presented. An improvement program in FORTRAN IV for the computer FELIX C-256 is described.

052.041 **Comparative study of some formulae approximating the atmospheric density and their utilization in the theory of artificial satellites motion.** L. Burs, A. Pál.
Visual Obs. Artif. Earth Satell., p. 34 - 60 (1977) = Publ. Obs. Astron. Univ. Cluj-Napoca, Romania, Nr. B 46-58.
Two new general formulae for atmospheric density are proposed. The equations of the motion of an artificial satellite, using a new empirical formula for air density, are integrated and the first order secular perturbations of semi-major axis and eccentricity are deduced. A computer program for the calculation of the secular perturbations of semi-major axis and eccentricity is briefly presented.

052.042 **L'influence du quatrième harmonique zonal du potentiel gravitationnel terrestre sur la période nodale des satellites artificiels.** T. Oproiu, V. Mioc.
Visual Obs. Artif. Earth Satell., p. 61 - 71 (1977) = Publ. Obs. Astron. Univ. Cluj-Napoca, Romania, Nr. B 46-58.
Dans une application numérique, cette influence est comparée à celle du second harmonique zonal du potentiel gravitationnel.

052.043 **Lorentz force influence on a charged satellite motion.** V. Mioc, E. Radu.
Visual Obs. Artif. Earth Satell., p. 86 - 94 (1977) = Publ. Obs. Astron. Univ. Cluj-Napoca, Romania, Nr. B 46-58.
The paper deals with the influence of the main geomagnetic field on the nodal period of a charged artificial satellite moving in a quasi-circular orbit.

052.044 **Determination of preliminary orbits of artificial Earth satellites by using the algorithm of Gauss-Banachiewicz.** I. Zongor.
Visual Obs. Artif. Earth Satell., p. 95 - 102 (1977) = Publ. Obs. Astron. Univ. Cluj-Napoca, Romania, Nr. B 46-58.
A computer program for these determinations is briefly described. A numerical example is given.

052.045 **Sur l'influence des harmoniques tesséraux du potentiel gravitationnel terrestre sur le rayon-vecteur des satellites artificiels.** T. Zlaczki.
Visual Obs. Artif. Earth Satell., p. 103 - 110 (1977) = Publ. Obs. Astron. Univ. Cluj-Napoca, Romania, Nr. B 46-58.
On présente une forme quelque peu modifiée de l'algorythme proposé par Zieliński (1968) pour évaluer l'influence des harmoniques tesséraux sur le rayon-vecteur des satellites artificiels. Un programme en FORTRAN IV pour ordinateur FELIX C-256 a été élaboré afin d'évaluer l'influence des harmoniques tesséraux C_{32} et S_{32}. Une application numérique est présentée.

052.046 **Sur la variation du demi-grand axe de l'orbite d'un satellite artificiel, causée par la résistence atmosphérique.** V. Mureşan.
Visual Obs. Artif. Earth Satell., p. 111 - 115 (1977) = Publ. Obs. Astron. Univ. Cluj-Napoca, Romania, Nr. B 46-58.
La possibilité d'étudier la variation du demi-grand axe de l'orbite d'un satellite artificiel à l'aide de l'intégration numérique est montrée. Pour déterminer la densité de la haute atmosphère, on a employé le sous-programme élaboré par Jacchia (1972). On a fait une application au satellite Explorer 19.

052.047 **Samos 2 (1961α1): orbit determination and analysis at 31 : 2 resonance.** D. M. C. Walker.
Planet. Space Sci., Vol. 28, 1059 - 1072 (1980).

052.048 **Celestial mechanics (E 11).** W. Kundt, W. G. Melbourne, J. D. Anderson.
Helios solar probes, science summaries, (see 003.009), p. 83 - 84 (1980).

052.049 **The difference between the nodal and keplerian periods of artificial satellites as an effect of atmospheric drag.** V. Mioc.
Astron. Nachr., Band 301, 311 - 315 (1980).
An analytical expression for the difference caused by the atmospheric drag between the nodal and keplerian periods of artificial satellites with quasi-circular orbits, moving in a spherically symmetrical atmosphere, is given. Some corresponding equations found by different authors are obtained as particular cases of the given equation.

052.050 **On symmetric trajectories of a satellite of a non-spherical planet.** V. G. Sokolov.
Byull. Inst. Teor. Astron., Tom 14, 617 - 620 (1980). In Russian.
The conditions for the existence of symmetric satellite trajectories about a planet with longitude and latitude dependent gravitational potential have been established.

052.051 **Motion of a satellite of the earth. I. Linear perturbations.** A. M. Fominov.
Byull. Inst. Teor. Astron., Tom 14, 621 - 654 (1980). In Russian.
The problem of the construction of an artificial earth satellite motion theory under the action of the principal perturbing forces is investigated. The perturbing forces are presented as expansions in terms of Legendre polynomials. Similar method with use of well known functions of the inclination and functions of the eccentricity have been used for the derivation of the perturbations of the orbital elements. Formulae for the linear perturbations of the elements of the orbit due to the earth's non-central gravitational field, lunar and solar gravity, tidal deformation of the earth, solar and terrestrial radiation pressure, air drag and motion of the equatorial plane of the earth have been derived.

052.052 **The third order solution of Vinti's problem and the Poisson's brackets of elements.**
L.- d. Wu, F. Tong.
Acta Astron. Sinica, Vol. 21, 389 - 398 (1980). In Chinese.
The third order solution of Vinti's problem with J_3 has been derived here concretely. The a, e, s, M_s, ψ_s and O_s are used as the elements of orbit, and their Poisson's brackets are given in this paper and may be used in process of construction of the third order perturbations of an artificial satellite.

052.053 **Resonance and periodic motions of a solid satellite relative to its mass center situated in a triangular libration point.** Yu. V. Barkin, Yu. G. Markov.
Kosm. Issled., Tom 18, 832 - 843 (1980). In Russian.

052.054 **On the stability of stationary motions of a gyrostat in a central field.** S. Kh. Gevorkyan.
Kosm. Issled., Tom 18, 933 - 935 (1980). In Russian.

052.055 **Improvement of the estimate of the damping decrement of the angular velocity of rotation of the SEO artificial earth satellite based on an analysis of information from the Ariabata satellite.**
V. I. Dranovskij, A. M. Yanshin.
Kosm. Issled., Tom 18, 935 - 937 (1980). In Russian.

052.056 **Influence of fluctuations of the atmosphere's density on the accuracy of determination and forecast of orbits of artificial earth satellites.** P. E. Ehl'yasberg.
Inst. kosm. issled. AN SSSR. Prepr., 1980, No. 286, 21 pp. In Russian. – Abstr. in Ref. zh., 62. Issled. kosm. prostranstva, 1.62.156 (1981).

Bibliography of the papers carried out at the Satellite Tracking Station No. 1132 Cluj-Napoca between 1957-1977. See Abstr. 002.060.

Method of single parameter to search for and identify artificial satellites in the atmosphere.
See Abstr. 031.556.

Note sur la réduction des observations photographiques des satellites artificiels à l'aide des ordinateurs.
See Abstr. 031.575.

A note on a Lagrangian formulation for motion about the collinear points. See Abstr. 042.036.

Motion about the stable libration points in the linearized, restricted three-body problem.
See Abstr. 042.070.

Variations in the quasi-nodal period of the satellite Explorer 19 caused by the atmospheric drag.
See Abstr. 054.009.

On the ballistic coefficient of Cosmos 378 rocket.
See Abstr. 054.010.

On the secular decrease in the semimajor axis of Lageos's orbit. See Abstr. 054.012.

The Earth's gravitational field and orbital resonances of the Intercosmos satellites.
See Abstr. 081.028.

053 Lunar and Planetary Probes and Satellites

053.001 **Pioneer 11 close to Saturn.**
G. A. Burba.
Priroda, 1980, No. 7, p. 88 - 90. In Russian.

053.002 **At the boundaries of the solar system.**
D. Yu. Gol'dovskij.
Zemlya Vselennaya, 1980, No. 4, p. 9. In Russian.

053.003 **Cinderella spacecraft: deep-space probes face funding starvation, despite brilliant achievements in 1979.** C. Bulloch.
Interavia, Vol. 35, 231 - 234 (1980). – Abstr. in Phys. Abstr., Vol. 83, Abstr. 72743 (1980).

053.004 **Microwave communications from outer planets: the Voyager project.** A. G. Brejcha.
Microwave J., Vol. 23, No. 1, p. 25 - 30, 34 - 35, 44 (1980). Abstr. in Phys. Abstr., Vol. 83, Abstr. 77263 (1980).

053.005 **Future transportation systems for planetary missions.** R. Akiba.
13th Lunar and Planetary Symposium, (see 012.018), p. 416 - 423 (1980).

053.006 **Voyager 1 bei Saturn angelangt.**
H. W. Köhler.
Sterne Weltraum, Jahrg. 19, 375 - 376 (1980).

053.007 **Planetary exploration by spacecraft.**
Z. Kopal.
Contemp. Phys., Vol. 21, 359 - 380 (1980). – Abstr. in Phys. Abstr., Vol. 83, Abstr. 105146 (1980).

053.008 **L'exploration de Saturne par la sonde spatiale Pioneer.** T. Gehrels.
Astronomie, Vol. 94, 467 - 499 (1980).
Une sonde spatiale Pioneer a survolé Saturne le 1er septembre 1979. L'auteur présente ici une synthèse des résultats obtenus sur la magnétosphere, les satellites, l'atmosphère de Titan, les anneaux, ainsi que sur l'intérieur et l'atmosphère de Saturne.

053.009 **Voyager 1 bereikt Saturnus.** O. Namba.
Zenit, 7e Jaarg., 414 - 415 (1980).

053.010 **Pioneer 11 Saturn encounter.**
T. G. Northrop, A. G. Opp, J. H. Wolfe.
J. Geophys. Res., Vol. 85, 5651 - 5652 (1980).

053.011 **Preliminary design concepts for the Halley probe shield.** R. Reinhard.
Comet Halley micrometeoroid hazard workshop, (see 012.049), p. 7 - 15 (1979).

053.012 **A particle impact detector for the Halley probe.**
E. Grün, J. A. M. McDonnell.
Comet Halley micrometeoroid hazard workshop, (see 012.049), p. 17 - 21 (1979).

NSSDC Data Listing. See Abstr. 002.029.

Voyager 1 and Saturn. See Abstr. 100.025.

054 Artificial Earth Satellites

054.001 **The Satellite Power System (SPS) Reference System.** G. M. Stokes, K. C. Davis.
Bull. American Astron. Soc., Vol. 12, 501 (1980). – Abstract.

054.002 **Satellite Power System – effects on optical astronomy.** P. B. Boyce.
Bull. American Astron. Soc., Vol. 12, 501 (1980). – Abstract.

054.003 **The effects of a Satellite Power System on ground-based radio astronomy.**
W. C. Erickson, A. R. Thompson.
Bull. American Astron. Soc., Vol. 12, 501 - 502 (1980). Abstract.

054.004 **Les satellites et les manifestations de l'activité solaire dans l'espace interplanétaire.**
J.- L. Steinberg.
Astronomie, Vol. 94, 453 - 461 (1980).

054.005 **Orbital elements of satellite GEOS-C obtained from photographic observations at one station.**
A. G. Kirichenko, K. A. Kudak.
Problems of cosmic physics. Vyp. 15, (see 033.033), p. 47 - 53 (1980). In Russian.

054.006 **Direct solar and earth-albedo radiation pressure effects on the orbit of Pageos 1.** S. Zerbini.
Celestial Mech., Vol. 22, 307 - 334 (1980).

The orbit of the Pageos 1 balloon satellite has been investigated in detail over the early part of the balloon's lifetime. The analysis herein focuses on how Pageos's orbit was affected by direct solar and albedo radiation pressure. The author has evaluated the near-earth micrometeoroid-particle flux, which turns out to be $5 \times 10^{-8}\ cm^{-2}\ sec^{-1}$. With the assumptions for the satellite's area-to-mass ratio and reflection coefficient, one would need a solar constant of 1.95 cal $cm^{-2}\ min^{-1}$ to give a best-fit to the data.

054.007 **Visual observations of artificial earth satellites in Finland 1979.** A. Tuominen.
Observations of Satellites, No. 20. Published by the Finnish Meteorological Institute, Helsinki. 61 pp. (1980). ISSN 0355-2004. ISBN 951-697-136-9.

054.008 **Photometric observations of artificial Earth satellites.** V. Mioc.
Visual Obs. Artif. Earth Satell., p. 72 - 85 (1977) = Publ. Obs. Astron. Univ. Cluj-Napoca, Romania, Nr. B 46-58.

The results obtainable from photometric observations of artificial Earth satellites are presented. Different formulae for atmospheric density determination from photometric data are shown. The activity in this field at the Satellite Tracking Station No. 1132 Cluj-Napoca is pointed out.

054.009 **Variations in the quasi-nodal period of the satellite Explorer 19 caused by the atmospheric drag.**
E. Radu, V. Mioc.
Visual Obs. Artif. Earth Satell., p. 116 - 121 (1977) = Publ. Obs. Astron. Univ. Cluj-Napoca, Romania, Nr. B 46-58.

The variations in the quasi-nodal period of the satellite Explorer 19 due to the air drag between January 22 - February 12, 1973 are determined using visual observations performed by 9 tracking stations. A justification of the formula used is given.

054.010 **On the ballistic coefficient of Cosmos 378 rocket.** A. Pál, L. Burs.
Visual Obs. Artif. Earth Satell., p. 122 - 125 (1977) = Publ. Obs. Astron. Univ. Cluj-Napoca, Romania, Nr. B 46-58.

054.011 **Note on the satellite visibility prediction.** M. Roşca. Visual Obs. Artif. Earth Satell., p. 130 - 134 (1977) = Publ. Obs. Astron. Univ. Cluj-Napoca, Romania, Nr. B 46-58.

A computer program for satellite visibility prediction is described.

054.012 **On the secular decrease in the semimajor axis of Lageos's orbit.** D. P. Rubincam.
NASA Tech. Memo., NASA TM 80734, 41 pp. (1980).

The semimajor axis of the Lageos orbit is decreasing secularly at the rate of -1.1 mm day^{-1} due to an unknown force. Nine possible mechanisms are investigated here to discover which one, if any, might be the force. Five of the mechanisms, resonance with the earth's gravitational field, gravitational radiation, the Poynting-Robertson effect, transfer of spin angular momentum to the orbital angular momentum, and drag from near-earth dust are ruled out because they are too small or require unacceptable assumptions to account for the observed rate. Three other mechanisms, the Yarkovsky effect, the Schach effect, and terrestrial radiation pressure could possibly give the proper order-of-magnitude for the decay rate, but the characteristic signatures of these perturbations do not agree with the observed secular decrease. Atmospheric drag from a combination of charged and neutral particles is the most likely cause for the orbital decay.

054.013 **Computation of ephemerides of artificial earth satellites for the laser ranging system of the first generation.** N. T. Mironov, K. A. Bogatyrev.
Astrometr. Astrofiz., Vyp. (No.) 42, p. 83 - 89 (1980). In Russian.

Different procedures of ephemeris computation for artificial earth satellites are considered. The suggested algorithm of ephemeris computation permits improving the procedure of deriving the visible passes of satellite orbits and to save computer time.

054.014 **Satellite digest – 140, 141.** R. D. Christy. Spaceflight, Vol. 22, 318 - 319, 349 - 350, 363 - 364 (1980).

NSSDC Data Listing. See Abstr. 002.029.

Bibliography of the papers carried out at the Satellite Tracking Station No. 1132 Cluj-Napoca between 1957 - 1977. See Abstr. 002.060.

Note sur la réduction des observations photographiques des satellites artificiels à l'aide des ordinateurs. See Abstr. 031.575.

Long term evolution of the LAGEOS orbit. See Abstr. 052.001.

Improvement of ephemeris by observations of one station. See Abstr. 052.036.

The algorithm for numerical calculations of an artificial satellite free motion around its center of mass. See Abstr. 052.037.

Coordinated ionospheric and magnetospheric observations from the ISIS 2 satellite by the ISIS 2 experimenters. Volume 1. Optical auroral images and related direct measurements. See Abstr. 084.044.

Theoretical Astrophysics

061 General Aspects (Nucleosynthesis, Neutrino Astronomy, etc.)

061.001 **In defense of anti-matter.**
S. Rogers, W. B. Thompson.
Astrophys. Space Sci., Vol. 71, 257 - 260 (1980).

There appears to be a prejudice in the astronomical world against an obvious high-energy source – the mutual annihilation of matter and anti-matter. In favour of this prejudice is the lack of any convincing evidence of the presence of naturally occuring anti-matter. Only recently have cosmic-ray anti-protons been detected and then in numbers consistent with secondary production in flight, while annihilation X-rays have also been detected, but again in circumstances where they might well be attributed to secondary effects of some other high-energy process.

061.002 **Astronomical consequences of the neutrino rest mass. III. The nonlinear stage of evolution of perturbations and the hidden mass.** A. G. Doroshkevich, Ya. B. Zel'dovich, R. A. Syunyaev, M. Yu. Khlopov.
Pis'ma Astron. Zh., Tom 6, 465 - 469 (1980). In Russian. English translation in Soviet Astron. Lett., Vol. 6.

The influence of the existence of the neutrino rest mass on the phenomenon of "hidden mass" of galaxies and clusters of galaxies, on the nonlinear stage of evolution of inhomogeneities and on the problem of H I observations in the spectra of quasars with large redshifts are discussed.

061.003 **Gamma ray energy deposition in expanding nebulae.**
J. T. Kriese, A. G. Petschek.
Bull. American Astron. Soc., Vol. 12, 448 - 449 (1980). Abstract.

061.004 **Astrophysical implications of apparent neutrino oscillations.** M. M. Shapiro, R. Silberberg.
Bull. American Astron. Soc., Vol. 12, 450 (1980). – Abstract.

061.005 **Helium synthesis, neutrino flavors, and cosmological implications.** F. W. Stecker.
Bull. American Astron. Soc., Vol. 12, 489 (1980). – Abstract.

061.006 **Constraints for r-process nucleosynthesis from solar system isotopic anomalies.** T. Lee.
Bull. American Astron. Soc., Vol. 12, 543 (1980). – Abstract.

061.007 **Recent advances in infrared astronomy.**
E. I. Robson.
Phys. Educ., Vol. 15, No. 2, p. 68 - 73 (1980). – Abstr. in Phys. Abstr., Vol. 83, Abstr. 67958 (1980).

061.008 **Experimental neutrino astrophysics.** K. Lande.
Annual review of nuclear and particle science, Vol. 29, (see 003.001), 394 - 410 (1979). – Abstr. in Phys. Abstr., Vol. 83, Abstr. 72769 (1980).

061.009 **Possible astrophysical sites for the production of galactic nuclei.** T. Ohnishi.
Tech. Rep. Inst. At. Energy Kyoto Univ., No. 179, p. 1 - 11 (1979). – Abstr. in Phys. Abstr., Vol. 83, Abstr. 72932 (1980).

061.010 **The origin of matter and stellar evolution.**
A. Maedov.
C. R. Séances Soc. Phys. Hist. Nat. Genève, Vol. 14, No. 1, p. 26 - 42 (1979). In French. – Abstr. in Phys. Abstr., Vol. 83, Abstr. 72933 (1980).

061.011 **On the propagation of electromagnetic waves in a cosmological neutrino sea.**
V. De Sabbata, M. Gasperini.
Nuovo Cimento, Lett., Ser. 2, Vol. 28, 181 - 185 (1980). Abstr. in Phys. Abstr., Vol. 83, Abstr. 77283 (1980).

061.012 **Synthesis of light metals in the Galaxy. Aluminium abundances in cool halo stars.**
M. Spite, F. Spite.
Astron. Astrophys., Vol. 89, 118 - 122 (1980).

The theories of pure explosive nucleosynthesis of carbon burning predict that the supernovae with low metal content produce light metals with a strongly enhanced odd-even effect. Echelle spectra of halo stars, dwarfs and giants, were obtained, using an electronic camera. The results are not compatible with the predictions of the pure explosive nucleosynthesis theories. They appear to disagree with results about hotter stars: either hotter dwarfs (Peterson, 1978) or hotter horizontal branch stars (Kodaira et al., 1969). The most likely explanation for this disagreement is that, in hotter stars, diffusion is efficient and modified the relative abundance of the elements. The ratios Al/Mg and Al/Fe thus could be lowered.

061.013 **Relations between nucleosynthesis rates and the metal abundance.** B. M. Tinsley.
Astron. Astrophys., Vol. 89, 246 - 248 (1980).

The metal abundance resulting from a given rate of nucleosynthesis is often expressed as the mass of metals ever ejected from stars divided by the total mass of the system. For this expression to be valid, the "total mass" must include only material that was enriched during the appropriate period of galactic evolution, and must exclude any objects that condensed at an earlier epoch. This obvious requirement seems to have been overlooked in the literature. When a consistent "total mass" is used, the above expression is essentially equal to the yield, which is known to be a useful estimate of the metal abundance.

061.014 **On the neutrino number and isotropy of the Universe in grand unified theories.**
D. V. Nanopoulos, D. Sutherland, A. Yildiz.
Nuovo Cimento, Lett., Ser. 2, Vol. 28, 205 - 208 (1980). Abstr. in Phys. Abstr., Vol. 83, Abstr. 85852 (1980).

061.015 **A new equation of state of supernova matter.**
M. F. El Eid, W. Hillebrandt.
Astron. Astrophys., Suppl. Ser., Vol. 42, 215 - 226 (1980).

An equation of state of matter at high densities ($\rho \geq 10^{11}$ g cm^{-3}) and temperatures ($T \geq 5 \times 10^9$ K) is presented, which includes several important effects such as neutron and proton degeneracy as well as nucleon-nucleon interactions and nuclear excitations. Results are given for matter in both nuclear statistical equilibrium and β-equilibrium for different values of entropy, density and temperature.

061.016 **Hydrogen burning of ^{24,25,26}Mg in explosive carbon burning.** J. Keinonen, S. Brandenburg.
Nucl. Phys. A, Vol. A341, 345 - 364 (1980). – Abstr. in Phys. Abstr., Vol. 83, Abstr. 86875 (1980).

061.017 **Neutrinos.** H. Fraas.
Phys. unserer Zeit, 11. Jahrg., 136 - 144 (1980).

061.018 **Infrarotastronomie.** S. Drapatz.
Phys. unserer Zeit, 11. Jahrg., 145 - 156 (1980).

061.019 **Escape of neutrinos from within cylindrically symmetric cluster of particles.** K. P. Singh.
Indian J. Pure Appl. Math., Vol. 7, 1400 - 1404 (1976).
Abstr. in Phys. Abstr., Vol. 83, Abstr. 94707 (1980).

061.020 **Absolute stable state of matter in the interior of cold superdense strongly magnetized astrophysical objects.**
A. R. Bakhalbashyan, G. M. Nedyalkova, V. S. Sekerzhitskij, G. A. Shul'man.
Problems of cosmic physics. Vyp. 15, (see 003.003), p. 137 - 142 (1980). In Russian.

The most stable of Ae- and Aen-phases of cold matter with frozen magnetic field is studied. The quantitative assessments of the appearance threshold of free neutrons in the matter of a cold superdense star for different magnetic field intensities on its surface are given.

061.021 **Modelling physical conditions in stars by ultra-high compression.**
V. A. Belokon', V. S. Imshennik, Ya. M. Kazhdan, N. I. Leonova, L. A. Pliner, M. V. Stepanova, Z. V. Suraeva.
Inst. prikl. mat. AN SSSR. Prepr., 1980, No. 29, 35 pp. In Russian. – Abstr. in Ref. zh., 51. Astron., 8.51.167 (1980).

061.022 **On a mechanism of generation of X-ray flares.**
Yu. E. Charikov.
XIth seminar on cosmophysics, (see 012.035), p. 60 - 68 (1979). In Russian. – Abstr. in Ref. zh., 51. Astron., 8.51.385 (1980).

061.023 **Compressibility of cold catalyzed matter.**
P. Haensel.
Astron. Astrophys., Vol. 90, 70 - 72 (1980).

The compression modulus, $\partial p/\partial \rho$ of electrically neutral, spatially homogeneous mixture of electrons, neutrons and protons at $T = 0$ K is studied for nucleon number density $0.15\ \mathrm{fm}^{-3} < \rho < 0.4\ \mathrm{fm}^{-3}$. The effects of nuclear interactions are included using the methods of Landau theory of Fermi liquids. The (negative) contribution to $\partial p/\partial \rho$ resulting from the change of composition of matter implied by compression is small. To a very good approximation, the compression modulus of such a model of dense cold catalyzed matter is given by the nucleon contribution to $\partial p/\partial \rho$, calculated at fixed composition of matter.

061.024 **L'evoluzione chimica nell'universo.** P. Galeotti.
Orione, Vol. 2, 97 - 103 (1980).

061.025 **Astrophysical constraints on the mass of heavy stable neutral leptons.** Ya. B. Zel'dovich,
A. A. Klypin, M. Yu. Khlopov, V. M. Chechetkin.
Inst. prikl. mat. AN SSSR. Prepr., 1980, No. 44, 22 pp. In Russian. – Abstr. in Ref. zh., 51. Astron., 9.51.141 (1980).

061.026 **Oszillierende Neutrinos und die Materiedichte des Weltalls.** B. Kröger.
Phys. Bl., 36. Jahrg., 339 - 340 (1980).

061.027 **On the limiting mass of a dense stellar matter.**
J. P. Sharma.
Acta Astronaut., Vol. 7, 209 - 217 (1980). – Abstr. in Phys. Abstr., Vol. 83, Abstr. 101611 (1980).

061.028 **Neutrino emission from a supernova shock.**
H. A. Bethe, J. H. Applegate, G. E. Brown.
Astrophys. J., Vol. 241, 343 - 354 (1980).

In the shock resulting from the "bounce" of a supernova, the neutrinos are prevented from leaving by the high neutrino opacity of the material just outside the shock, until the density of that material has decreased to less than 10^{11} g cm^{-3}. The number of neutrinos then emitted is moderate. Neutrino pairs are formed, mainly from electron pairs, and their emission again is limited by the opacity. The energy lost from the shock by neutrino pairs is estimated, and is from 20 to 50% for reasonable initial shock energies.

061.029 **If the rest mass of a neutrino is not zero ......**
G. S. Bisnovatyj-Kogan.
Zemlya Vselennaya, 1980, No. 5, p. 24 - 25. In Russian.

061.030 **Interpretation of isotopic abundances in interstellar clouds.** M. Guélin, J. Lequeux.
Interstellar molecules, (see 012.033), p. 427 - 438 (1980).

061.031 **Stellar rates for the ^{24}Mg(α, n)^{27}Si, ^{25}Mg(p, n)^{25}Al, ^{27}Al(p, n)^{27}Si, and ^{28}Si(α, n)^{31}S reactions.**
C. W. Cheng, J. D. King.
Astrophys. J., Vol. 241, 844 - 847 (1980).

Ground-state reaction rates have been deduced from recent cross section measurements for the endoergic ^{24}Mg(α, n)^{27}Si, ^{25}Mg(p, n)^{25}Al, ^{27}Al(p, n)^{27}Si, and ^{28}Si(α, n)^{31}S reactions, supplemented by an extrapolation from the first experimental point to threshold using the statistical model predictions of Woosley and his colleagues. These rates are in very good agreement with those calculated from the statistical model. Stellar rates have been derived from the ground-state rates by multiplying by the ratio of stellar to ground-state rates given by the statistical model. Both ground-state and stellar rates have been represented by analytical functions of the temperature.

061.032 **Thermodynamic laws of neutrino and photon emission.** P. J. Walsh, C. F. Gallo.
American J. Phys., Vol. 48, 599 - 603 (1980). – Abstr. in Phys. Abstr., Vol. 83, Abstr. 105451 (1980).

061.033 **Astrophysical bounds on very-low-mass axions.**
D. A. Dicus, E. W. Kolb, V. L. Teplitz, R. V. Wagoner.
Phys. Rev. D, Vol. 22, 839 - 845 (1980). – Abstr. in Phys. Abstr., Vol. 83, Abstr. 109353 (1980).

061.034 **Study of some processes of nucleosynthesis in the evolving galaxy.** A. K. Lavrukhina,
R. I. Kuznetsova.
Origin and distribution of the elements, (see 012.039), p. 3 - 9 (1979).

Calculations of the rate of the isotope formations of ^{9}Be, ^{40}K, ^{50}V and "passing" nuclides arising from the solar system matter irradiated by cosmic rays were made. The calculations were carried out at the various irradiation conditions corresponding to the astrophysical processes arising in all stages of evolution of the solar system matter. The agreement between observed abundances of the nuclides mentioned above and their production in the fast particle reactions was received for the first time in the frames of the model of exploded supernova (or galactic nucleus).

061.035 **Some long-lived and stable nuclides produced by nuclear reactions.** R. Gensho, O. Nitoh,
T. Makino, M. Honda.
Origin and distribution of the elements, (see 012.039), p. 11 - 18 (1979).

061.036 **The chemical evolution of the Galaxy and isotopic ratios in the solar system.** J. Audouze.
Origin and distribution of the elements, (see 012.039), p. 83 - 90 (1979).

The relationship between the present knowledge of the isotopic ratios in the solar system and that of the chemical evolution of the galaxies is tentatively described.

061.037 **On a system of self-gravitating Fermi-degenerate neutrinos.** R. Ruffini.
Nuovo Cimento,Lett., Ser. 2, Vol. 29, 161 - 162 (1980).
Abstr. in Phys. Abstr., Vol. 84, Abstr. 9362 (1981).

061.038 **Nuclear reaction rate in Debye-Hückel electrolytic plasma in stellar interiors.**
H. L. Duorah, A. E. Md Khairozzaman.
Pramāṇa, Vol. 15, 145 - 151 (1980). – Abstr. in Phys. Abstr., Vol. 84, Abstr. 9363 (1981).

061.039 **Seed abundances for *r*-processing in the helium shells of supernovae.**
J. J. Cowan, A. G. W. Cameron, J. W. Truran.
Astrophys. J., Vol. 241, 1090 - 1093 (1980).

The production of *r*-process nuclei as a consequence of shock heating of the helium shells in supernovae is examined. The assumption of an initial (seed) distribution which is enriched in heavy *s*-process elements relative to solar system matter is found to allow production of *r*-process abundance features compatible with observations. With such enrichments, the production of *r*-process nuclei in solar proportions in the helium shells of supernovae becomes possible.

061.040 **Microscopic calculation of the beta-decay of medium-mass and heavy nuclei up to the *r*-process path.** H. V. Klapdor, T. Oda.
Astrophys. J., Lett., Vol. 242, L49 - L52 (1980).

Beta-decay half-lives are calculated microscopically for the first time up to the *r*-process path (for $^{89}Rb - ^{119}Rb$). The calculated lifetimes are shorter by an order of magnitude than those predicted by the gross theory used at present in *r*-process calculations. This could solve the longstanding problem of reproducing the positions of the *r*-process abundance peaks. Calculations of the beta strength function S_β for the $^{250,252,256}Np$ isotopes show that the shape of S_β will strongly affect the decay back from the *r*-process path to the valley of β-stability by beta-delayed fission.

061.041 **Heavy neutrinos with nonzero rest mass and astrophysical consequences of their discovery.**
M. Šolc.
Říše hvězd, Vol. 61, 161 - 164 (1980). In Czech.

061.042 **Cosmological and astrophysical implications of heavy Majorana particles.**
T. Yanagida, M. Yoshimura.
Phys. Rev. Lett., Vol. 45, 71 - 74, with a correction p. 498 (1980).

It is pointed out that heavy Majorana particles contemplated in certain grand unified theories may explain the cosmological baryon excess. Induced neutrino mixing can give oscillation lengths consistent with reactor and accelerator experiments, but capable of explaining the solar-neutrino "puzzle".

061.043 **Have massive cosmological neutrinos already been detected?** F. W. Stecker.
Phys. Rev. Lett., Vol. 45, 1460 - 1462 (1980).

The possibility is investigated that the decay of massive cosmological neutrinos may have produced a spectral signature which has already been detected in observations of the ultraviolet background radiation. Various implications are discussed including a possible implied neutrino mass of 13.8–14.8 eV. A lower limit is also placed on the lifetime of heavy neutrinos ν_H with respect to the decay $\nu_H \rightarrow \nu_L + \gamma$ based on the cosmic UV observations.

061.044 **Xenon from intermediate zones of supernovae.**
D. Heymann, M. Dziczkaniec.
Proc. Tenth Lunar Planet. Sci. Conf., (see 012.050), p. 1943 - 1959 (1979).

061.045 **The origin of the elements.** R. J. Tayler.
The state of the Universe, (see 003.013), p. 40 - 67 (1980).

The author has described the progress that has been made in trying to understand the present chemical composition of the Universe in terms of its initial composition and subsequent modifications produced by nuclear reactions during its life history.

061.046 **New horizons in neutrino astrophysics.**
S. Đorgovski.
Vasiona, Année 28, 38 - 43 (1980). In Croatian.

061.047 **The chemical abundances of the Cassiopeia A fast-moving knots: explosive nucleosynthesis on a minicomputer.** M. D. Johnston, P. C. Joss.
Astrophys. J., Vol. 242, 1124 - 1132 (1980).

The authors describe a simplified nuclear reaction network for explosive nucleosynthesis calculations, in which only the most abundant nuclear species and the most important reactions linking these species are considered. They obtain good agreement with previous calculations employing more complex reaction networks. They apply this scheme to the observed chemical abundances of the fast-moving knots in the supernova remnant Cassiopeia A and find that a wide range of initial conditions could yield the observed abundances. The agreement between the calculated and observed chemical abundances in Cas A and similar supernova remnants depends primarily upon the relevant nuclear physics and does not provide strong evidence in favor of any particular model of the supernova event.

061.048 **Astrophysical nuclear partition function.**
S. S. M. Wong, E. Zhao, X. Zhu.
Astrophys. J., Lett., Vol. 242, L173 - L177 (1980).

The asymptotic divergences of astrophysical nuclear partition function and mean excitation energy are eliminated by performing the calculations in a finite spectroscopy space using the statistical spectroscopy method. Results for ^{56}Fe and ^{56}Ni are presented and compared with those of existing models.

Thick-target measurements and astrophysical thermonuclear reaction rates. Proton-induced reactions.
See Abstr. 022.028.

On relativistic kinetic theory of transport processes, in particular of neutrino systems. See Abstr. 022.029.

Phase separation of ionic mixtures.
See Abstr. 022.032.

Stellar reaction rate of $^{26}Mg(p,\gamma)^{27}Al$.
See Abstr. 022.092.

Structures in the beta strength function and consequences for nuclear physics and astrophysics.
See Abstr. 022.093.

Consequences of structures in the beta strength function on nuclear physics and astrophysics.
See Abstr. 022.102.

Cross-section measurements and thermonuclear reaction rates for $^{49}Ti(p,\gamma)^{50}V$ and $^{49}Ti(p,n)^{49}V$. See Abstr. 022.103.

On the behavior of the short-range pair correlation function in a mixture of two-component plasma. See Abstr. 062.034.

Neutrino, gamma-ray, electron, and positron production in an ultrarelativistic plasma. See Abstr. 062.114.

Mass loss from red giants: some effects on the interstellar medium. See Abstr. 064.082.

Time-dependent neutrino transport out of dense stellar cores. See Abstr. 065.010.

Neutrino energy equilibration models. See Abstr. 065.017.

A new flash mixing. See Abstr. 065.047.

New solar-neutrino flux calculations and implications regarding neutrino oscillations. See Abstr. 080.074.

A revision of the meteorite based cosmic abundance of boron. See Abstr. 105.081.

Noble gas anomalies and synthesis of the chemical elements. See Abstr. 106.012.

Nucleosynthesis in evolved globular clusters. See Abstr. 154.033.

Astronomical consequences of the neutrino rest mass. I. The universe. See Abstr. 162.012.

Astronomical consequences of the neutrino rest mass. II. Spectrum of density perturbations and small-scale fluctuations of the microwave background. See Abstr. 162.013.

Primordial nucleosynthesis and Dirac's Large Numbers Hypothesis. See Abstr. 162.019.

Neutrinos in cosmological models. See Abstr. 162.034.

Spatially homogeneous neutrino cosmologies. See Abstr. 162.045.

Neutrinos and the age of the Universe. See Abstr. 162.061.

Cosmological implications of the primeval nucleosynthesis. See Abstr. 162.068.

Neutrinos and the universe. See Abstr. 162.099.

Neutrinos and the properties of the universe. See Abstr. 162.100.

Do neutrino rest masses affect cosmological helium production? See Abstr. 162.104.

Galactic neutrinos and UV astronomy. See Abstr. 162.105.

Some consequences of the theory of interacting fields in cosmology and astrophysics. See Abstr. 162.116.

Erratum

061.901 **Erratum: 'Enhancement of thermonuclear reaction rate due to strong screening. II. Ionic mixtures'** [Astrophys. J., Vol. 234, 1079 - 1084 (1979)].
N. Itoh, H. Totsuji, S. Ichimaru, H. E. DeWitt.
Astrophys. J., Vol. 239, 415 (1980). – See Abstr. 26.061.032.

062 Hydrodynamics, Magnetohydrodynamics, Plasma

062.001 **On the equilibrium structures of self-gravitating masses of gas containing axisymmetric magnetic fields.** I. Lerche, B. C. Low.
Mon. Not. R. Astron. Soc., Vol. 192, 611 - 620 (1980).

The authors give the general equations describing the equilibrium shapes of self-gravitating gas clouds containing axisymmetric magnetic fields. The general equations admit of a large class of solutions. It is shown that if the one additional *(ad hoc)* assumption is made that the mass be spherically symmetrically distributed then the gas pressure and the boundary conditions are sufficiently constraining that the general topological structure of the solution is effectively determined. The further assumption of isothermal conditions for this case demands that all solutions possess force-free axisymmetric magnetic fields. The authors also outline how the construction of aspherical (but axisymmetric) configurations can be achieved in some special cases and they show that the detailed form of the possible equilibrium shapes depends upon the arbitrary choice of the functional form of the variation of the gas pressure along the field lines.

062.002 **Analysis of X-ray line spectra from a transient plasma under solar flare conditions. I. General outline.** R. Mewe, J. Schrijver.
Astron. Astrophys., Vol. 87, 261 - 268 (1980).

For the set-up of the observing programs to study the impulsive phase of a solar flare with the X-ray Polychromator (XRP) in the NASA Solar Maximum Mission (SMM), a detailed understanding of the time behaviour of the X-ray spectra is needed. The authors have performed various model calculations for the wavelength regions observable by the XRP bent crystal spectrometer. Resulting time-varying spectra are given for plasma models in which the electron temperature either increases with a steep jump or gradually rises in about one minute and declines afterwards.

062.003 **Cellular convection in a stratified atmosphere.** J. M. Massaguer, J.-P. Zahn.
Astron. Astrophys., Vol. 87, 315 - 327 (1980).

The authors examine the properties of thermal convection in an atmosphere spanning several density scale heights. Both the conductivity and the shear viscosity are assumed to be constant. The inelastic approximation is used, together with a modal expansion procedure in the horizontal directions: this expansion is severely truncated, keeping one or two modes. Only steady solutions are built, for Rayleigh numbers up to 10^5 times critical and Prandtl numbers of 1, 10^{-2} and 10^{-4}.

062.004 **Rays and foci in a magneto-ionic medium with linearly varying magnetic field.**
R. White, M. J. Laird.
Planet. Space Sci., Vol. 28, 837 - 846 (1980).

An analytical study is carried out of rays from a source in a magneto-ionic medium in which the magnetic field is linear in the position coordinates. A general solution is given, and special cases, including that in which the magnetic field vector is confined to a plane, are examined. In the latter case, the specific sub-case of null current density, in which the field lines are rectangular hyperbolas, has been the subject of detailed numerical calculations. It is shown that focusing of rays can occur, though perhaps less readily than in the case of parabolic field lines considered by previous workers. The authors relate their parametrisation quantitatively to the terrestrial magnetic field.

062.005 **On the distribution of the field of electromagnetic waves emitted by a dipole in a homogeneous magnetoactive plasma.** Ya. L. Al'pert, B. S. Moiseyev.
J. Atmos. Terr. Phys., Vol. 42, 521 - 528 (1980).

The paper deals with the emission of an electric dipole in a homogeneous magnetoactive multicomponent plasma in all resonance frequency ranges ω which are lower than the Langmuir electron frequency ω_0.

062.006 **Rapidly rotating α^2-dynamo models.** G. Rüdiger.
Astron. Nachr., Band 301, 181 - 187 (1980).

062.007 **Exact static equilibrium of vertically oriented magnetic flux tubes. I. The Schlüter-Temesváry sunspot.**
B. C. Low.
Sol. Phys., Vol. 67, 57 - 77 (1980).

A method is prescribed for generating exact solutions of magnetostatic equilibrium describing a cylindrically symmetric magnetic fluxtube oriented vertically in a stratified medium. A particular solution is obtained by this method for a medium sized sunspot whose magnetic field obeys the similarity law of Schlüter and Temesváry (1958). With this solution, it is possible for the first time to illustrate explicitly the confinement of the magnetic field of the cool sunspot by the hotter external plasma in an exact relationship involving both magnetic pressure and field tension as well as the support of the weight of the plasma by pressure gradients. It is found that the cool region of the sunspot is not likely to extend much more than a few density scale heights below the photosphere.

062.008 **On absolute and convective instabilities.** D. L. Giaretta.
Astron. Astrophys., Vol. 88, 113 - 116 (1980).

The importance of distinguishing between convective and absolute instabilities is discussed and a method for distinguishing between the two is briefly described. This method is applied to several instabilities which may arise in interstellar space.

062.009 **Non-linear interaction of Alfvén waves with compressive fast magnetosonic waves.**
C. Lacombe, A. Mangeney.
Astron. Astrophys., Vol. 88, 277 - 281 (1980).

This paper is devoted to the study of the non-linear interactions between oblique fast m.h.d. waves, which are observed to be almost always present in the solar wind, and the Alfvén waves, which form the most important component of the m.h.d. turbulence in the solar wind. The main "weak turbulence" process is the so-called non-linear Landau effect; it damps the compressible modes, heats, but not strongly, the thermal protons, and may enhance the high-frequency Alfvén waves; the anisotropy between backward and forward Alfvén waves is always increased by this process. However, this non-linear process is not very efficient in the solar wind, so that Alfvén waves can be considered as decoupled from compressive waves in the major part of the m.h.d. spectral range.

062.010 **Numerical treatment of the unsteady hydromagnetic thermal boundary layer problem.**
M. A. Drymonitou, V. S. Geroyannis, C. L. Goudas.
Astrophys. Space Sci., Vol. 71, 87 - 100 (1980).

The paper presents a suitable numerical method for the treatment of the unsteady hydromagnetic thermal boundary layer problem for flows past an infinite porous flat plate, the motion of which is governed by a general time-dependent law, under the influence of a transverse externally set magnetic field. The normal velocity of suction/injection at the plate is also assumed to be time-dependent. Analytical approximations are given for the cases of a plate (1) generally accelerated and (2) harmonically oscillating. This problem is closely related to the motions and heat transfer occurring locally on the surfaces of stars.

062.011 **Suspended particles and the gravitational instability of a rotating plasma.** R. C. Sharma, K. C. Sharma.
Astrophys. Space Sci., Vol. 71, 325 - 332 (1980).

The gravitational instability of an infinite homogeneous self-gravitating and finitely conducting, rotating gas-particle medium, in the presence of a uniform vertical magnetic field, is studied to include finite Larmor radius and suspended particles effects. The particular cases of the effects of rotation, finite conductivity, finite Larmor radius and suspended particles on the waves propagated along and perpendicular to the magnetic field have been discussed. Jeans's criterion determines the gravitational instability.

062.012 **Free convection effects on the oscillatory flow in the Stokes problem past an infinite porous vertical limiting surface with constant suction, II.**
N. G. Kafousias, A. A. Raptis, G. A. Georgantopoulos, C. V. Massalas.
Astrophys. Space Sci., Vol. 71, 337 - 352 (1980).

The authors present the two-dimensional free convection flow of an incompressible viscous fluid past an infinite vertical limiting surface (porous wall) for the Stokes problem when the fluid is subjected to a constant suction velocity. The flow is normal to the porous wall and the free stream oscillates about a mean value. As the mean steady flow has been presented in Part I, only the solutions for the transient velocity profiles, transient temperature profiles, the amplitude and the phase of the skin friction and the rate of heat transfer are presented. The influence of the Grashof number G and Eckert number E on the unsteady flow field is discussed for air ($P = 0.71$) and water ($P = 7$) and for the cases of externally heating and cooling the porous limiting surface by free convection currents.

062.013 **Slow convection of a magnetized plasma and the earth plasma sheet.** A. Hruška.
Astrophys. Space Sci., Vol. 71, 459 - 474 (1980).

Stationary convection of an isotropic, infinitely conducting plasma in a magnetic field with non-trivial geometry is discussed under the assumption that the inertial term in the equation of motion may be ignored. The energy gained or lost by a volume element of plasma per unit time does not vary along the field-lines. Simple relations between the components of the current density, depending on the field-line geometry, exist. Similar relations hold for the components of the plasma velocity. The theoretical analysis is applied to the geomagnetically-quiet plasma sheet and a qualitative physical picture of the sheet is suggested.

062.014 **Systematic trends in the radiative emission rates for low- and medium-Z elements in high-temperature plasmas.** J. Davis, V. L. Jacobs.
J. Quant. Spectrosc. Radiat. Transfer, Vol. 24, 283 - 288 (1980).

The power radiated by an optically thin, low-density ($N_e \leq 10^{14}$ electrons/cm^3) plasma has been calculated for the electron temperature range 1-10^6 eV taking into account resonance line emission, direct recombination radiation, dielectronic recombination radiation, and bremsstrahlung from the ions of a given element. The ionization structure has been determined by using a corona equilibrium model in which collisional ionization and inner-shelled excitation followed by autoionization are balanced by direct radiative and dielectronic recombination. Based on the results for representative elements from carbon through nickel, graphs are presented of the maximum radiated power, the maximum emission temperature, and the mean charge at the maximum for each shell as functions of the atomic number Z. Assuming that the maximum emission temperature can be achieved, aluminum and iron are predicted to be the most efficient K-shell radiators for $Z \leq 28$.

062.015 **Electron temperature measurements in low-density plasmas by helium spectroscopy.**
N. Brenning.
J. Quant. Spectrosc. Radiat. Transfer, Vol. 24, 293 - 318 (1980).

062.016 **Spontaneous formation of knots in relativistic flows: a model for variability in compact synchrotron sources.** A. P. Marscher.
Astrophys. J., Vol. 239, 296 - 304 (1980).

A model is proposed in which regions of enhanced synchrotron emission are formed via radiative thermal instabilities in relativistic flows. A perturbed volume in which the magnetic field is enhanced by a few tens of a percent relative to the steady value collapses due to the increased cooling rate and consequent loss in pressure. The collapse further increases the magnetic field, leading to even shorter radiative lifetimes. The instability progresses in this way until either the external electron pressure is balanced by the perturbed magnetic pressure or the overall expansion of the flow becomes important. A fluid-dynamical treatment of the instability is performed, and the results are shown to apply to flux variations and formation of knots in the relativistic flows which are thought to occur in quasars and active galactic nuclei. The knots possess the same kinematic properties as the collimated flow and can thus be responsible for the apparent superluminal motions observed in compact radio sources.

062.017 **Higher order fluid equations for multicomponent nonequilibrium stellar (plasma) atmospheres and star clusters.** S. Cuperman, I. Weiss, M. Dryer.
Astrophys. J., Vol. 239, 345 - 359 (1980).

A generalized fluid theory which is required for the description of time-dependent, spatially nonhomogeneous, anisotropic, multispecies, spherically symmetric systems of particles obeying an inverse-square law of interactions is developed. The resulting equations apply to the expansion of stellar atmospheres (and in particular for the case of the solar wind), stellar systems, as well as controlled thermonuclear devices based on spherical inertial confinement (e.g., laser-pellet interaction, etc.).

062.018 **The heating of a thermally conducting stratified medium. I. Self-gravitating gas threaded by magnetic fields.** I. Lerche, B. C. Low.
Astrophys. J., Vol. 239, 360 - 376 (1980).

The authors present two theoretical problems to illustrate how the magnetic field induced anisotropy in thermal conduction influences the equilibrium structure of gravitating gas. Magnetic forces alone can drastically alter the equilibrium structures of a gas. To remove the effect of magnetic forces, the two problems presented involve either nearly force-free or potential magnetic fields. The magnetic field, therefore, does not participate directly in the force balance and its role is confined to channeling the thermal conduction flux along field lines.

062.019 **Evolving force-free magnetic fields. III. States of nonequilibrium and the preflare stage.**
B. C. Low.
Astrophys. J., Vol. 239, 377 - 388 (1980).

The author considers whether a neighboring magnetostatic equilibrium exists to allow a magnetic field initially in a force-free configuration to accommodate any imposed weak pressure. The following problem is treated. Suppose the foot points of the field are fixed and the plasma is frozen into the field lines under the approximation of infinite electrical conductivity. Introduce a weak pressure and ask whether infinitesimal plasma displacements exist to adjust the field lines to a new equilibrium without changing the field line connectivity. The analysis is carried out for the bipolar force-free fields forming one of two evolutionary sequences modeling the development of the preflare stage as presented in the first paper of this series.

062.020 **Fast plasma heating by anomalous current dissipation.** A. Duijveman, P. Hoyng, J. A. Ionson.
Bull. American Astron. Soc., Vol. 12, 481 (1980). – Abstract.

062.021 **Further development of numerical MHD model of coronal dynamics.**
S. T. Wu, S. M. Han, Y. Nakagawa.
Bull. American Astron. Soc., Vol. 12, 516 (1980). – Abstract.

062.022 **A new method of solving the force-free magnetic field equations and its application to solar physics.**
H.-m. Chang.
Bull. American Astron. Soc., Vol. 12, 516 - 517 (1980). Abstract.

062.023 **Particle acceleration by a shock wave in a turbulent medium.**
V. N. Vasil'ev, I. N. Toptygin, A. G. Chirkov.
Kosm. Issled., Tom 18, 556 - 566 (1980). In Russian.

062.024 **Transport coefficients of dense plasmas.** M. Baus.
J. Phys. Colloq., Vol. 41, No. C-2, (see 012.005), p. C2/69 - 76 (1980). – Abstr. in Phys. Abstr., Vol. 83, Abstr. 77280 (1980).

062.025 **Dense plasmas, nuclear reactions and astrophysics.** E. Schatzman.
J. Phys. Colloq., Vol. 41, No. C-2, (see 012.005), p. C2/89 - 96 (1980). – Abstr. in Phys. Abstr., Vol. 83, Abstr. 77281 (1980).

062.026 **Infinitely magnetized hydrogen atom and pulsar crust.** C. Angelie, C. Deutsch, M. Signore.
J. Phys. Colloq., Vol. 41, No. C-2, (see 012.005), p. C2/133 - 138 (1980). – Abstr. in Phys. Abstr., Vol. 83, Abstr. 77282 (1980).

062.027 **Pulsar radio emission from beam plasma instability.** E. Asseo, R. Pellat, M. Rosado.
Astrophys. J., Vol. 239, 661 - 670 (1980).

A two step maser mechanism has been recently proposed for the pulsar radio emissions: a spatial modulation of the density of relativistic particles results from an electrostatic beam plasma instability; the bunches of particles flowing along curved magnetic field lines coherently radiate electromagnetic radio waves. The authors show that such a mechanism is not correct. For this purpose they reconsider the relativistic beam plasma instability, including in the analysis the curvature of magnetic field lines and finite transverse beam dimensions. The main result of this work is that the finite beam dimensions provide a natural coupling between the most unstable quasi-electrostatic beam plasma wave and a fully electromagnetic vacuum wave.

062.028 **Sunspots and the physics of magnetic flux tubes. X. On the hydrodynamic instability of buoyant fields.**
K. C. Tsinganos.
Astrophys. J., Vol. 239, 746 - 760 (1980).

The hydrodynamics of cylindrical buoyant bubbles (magnetic flux tubes) in an atmosphere is examined. The author considers their dynamical stability, with particular attention to the Rayleigh-Taylor and Kelvin-Helmholtz instabilities. The instabilities are sufficient to cause fragmentation in the absence of some strong stabilizing effect such as surface tension or an internal azimuthal magnetic field. The results of the stability analysis are first applied to rising gas bubbles in liquids. Then, since agreement with the experimental data is found, the results are applied to the similar problem of the rising magnetic flux in the Sun.

062.029 **A new calculation on rotating protostar collapse.** M. L. Norman, J. R. Wilson, R. T. Barton.
Astrophys. J., Vol. 239, 968 - 981 (1980).

The authors calculate the isothermal collapse of a rotating axisymmetric gas cloud. The cloud is slowly rotating so that the collapse is deep, severely testing the angular momentum advection. Special care is taken to improve local conservation of angular momentum, which is monitored by a mass versus specific angular momentum spectrum. The authors find the collapse is a runaway yielding central disklike regions of increasing mass density and flatness. Although the numerical solution cannot become singular they show that a possible final state for the cloud consistent with its mass versus specific angular momentum spectrum is an equilibrium thin disk which is singular in surface density and angular velocity at the rotation axis.

062.030 **Flow past a massive object and the gravitational drag.** Y. Rephaeli, E. E. Salpeter.
Astrophys. J., Vol. 240, 20 - 24 (1980).

The gravitational interaction between a massive object and ambient gas is considered. The authors give a description of subsonic and supersonic flows past a massive, mass ejecting object in nonviscous gas, including an estimate of the maximum density enhancements near the shock-front. The drag and the associated gas heating rates are calculated and compared with results for motion through collisionless gas.

062.031 **Hydromagnetic rotational braking of magnetic stars.** R. C. Fleck, Jr.
Astrophys. J., Vol. 240, 218 - 222 (1980).

It is suggested that the magnetic Ap stars can be rotationally decelerated to long periods by the braking action of the associated magnetic field on time scales of order 10^7 - 10^{10} years depending on whether the star's dipole field is aligned perpendicular or parallel to the rotation axis. Rotation includes a toroidal magnetic field in the plasma surrounding a star, and the accompanying magnetic stresses produce a net torque acting to despin the star. These results indicate that it is not necessary to postulate mass loss or mass accretion for this purely hydromagnetic braking effect.

062.032 **Stability of the Primakoff-Sedov blast wave and its generalizations.** I. B. Bernstein, D. L. Book.
Astrophys. J., Vol. 240, 223 - 234 (1980).

The linear stability of those Sedov blast wave similarity solutions for which the flow is homologous behind the shock, of which the best-known example is the Primakoff point blast model, along with that of their two-dimensional counterparts, is proved analytically. This conclusion in the three-dimensional case applies to all perturbations, and in the two-dimensional case to flutelike ($k_z = 0$) modes. Implications for supernova remnant evolution are discussed.

062.033 **Flows along magnetic flux tubes. I. Equilibrium and buoyancy of a slender magnetic loop in the interior of a star.** M. Schüssler.
Astron. Astrophys., Vol. 89, 26 - 32 (1980).

It has been speculated that loop formation may enhance the buoyant loss of magnetic flux from stellar interiors. In this paper, the effect on magnetic buoyancy of a downflow along the arms of a slender, isolated magnetic loop formed from an initially horizontal flux tube is investigated excluding the uppermost sub-surface layers of the star. For convective regions as the envelope of the sun, only very broad loops with a radius of curvature of the order of their distance to the stellar centre may grow and do not enhance the buoyant loss significantly.

062.034 **On the behavior of the short-range pair correlation function in a mixture of two-component plasma.**
R. M. Singh, R. P. Singh, Y. Singh.
Astrophys. J., Vol. 240, 672 - 679 (1980).
The short-range behavior of the pair correlation function in a mixture of two-component plasma is investigated. The method which has been adopted involves the evaluation of the contribution to a pair correlation function of interactions between particles 1 and 2, using a quantum statistical mechanical treatment, and the contribution of all other interactions by using the Wigner-Kirkwood expansion in powers of $\hbar^2$.

062.035 **Charge transfer in astrophysical shocks.**
S. E. Butler, J. C. Raymond.
Astrophys. J., Vol. 240, 680 - 684 (1980).
The effects of charge transfer reactions upon astrophysical shocks are investigated. The column densities, line intensities, and line ratios of the important ionic species of N, O, Si, and S are substantially modified for both low- and high-velocity shocks. Some general implications for shock models of astronomical objects are discussed.

062.036 **Comments on the dissipation of hydromagnetic surface waves.** M. A. Lee.
Astrophys. J., Vol. 240, 693 - 695 (1980).
A recent paper by Wentzel, which claims to calculate a plasma heating rate due to dissipation of surface waves in an ideal magnetohydrodynamic (MHD) fluid, is found to be in error in interpretation. A well-established general theorem pertaining to the conservative ideal MHD fluid requires that the normal mode calculated by Wentzel be oscillatory in time. Within ideal MHD, dissipation and plasma heating are therefore impossible.

062.037 **Subcritical solutions in a special nonlinear $\alpha\omega$-dynamo model.** H. Bräuer.
Astron. Nachr., Band 301, 203 - 206 (1980). Elaboration of a talk presented at an international workshop on "Dynamo theory and the generation of the Earth's magnetic field", Alšovice (ČSSR), 15 - 20 October 1979.
It is shown in an idealized $\alpha\omega$-dynamo model that a finite flux density not only is able to prevail against its own back-reaction on the generating mechanism but can even make the dynamo more responsive to excitation and enable it to run subcritically.

062.038 **Sur un modèle magnétogravitationnelle d'un plasma cosmique doué d'une conductivité électrique finie. L'equation de dispersion.** M. Vasiu.
Proceedings of the colloquium of astronomy, Cluj-Napoca, (see 012.009), p. 76 - 85 (1978).
L'auteur établit l'équation de dispersion pour un modèle de plasma cosmique compressible, nonvisqueux, doué d'une conductivité électrique finie, en mouvement de rotation, sous l'action d'un champ magnétique uniforme et axial et sous l'action de son propre champ gravifique. Les équations magnétohydrodynamiques utilisées sont écrites en coordonées cylindriques.

062.039 **Structure within a magnetic flux tube.**
P. R. Wilson.
Proc. Astron. Soc. Australia, Vol. 3, 225 - 226 (1978).

062.040 **The tearing mode instability in a partially ionized plasma.** N. F. Cramer, I. J. Donnelly.
Proc. Astron. Soc. Australia, Vol. 3, 367 - 368 (1979).

062.041 **Necessary conditions for the magnetohydrodynamic dynamo.** M. R. E. Proctor.
Geophys. Astrophys. Fluid Dyn., Vol. 14, 127 - 145 (1979).
Abstr. in Phys. Abstr., Vol. 83, Abstr. 81807 (1980).

062.042 **A self-consistent treatment of simple dynamo systems.** H. K. Moffatt.
Geophys. Astrophys. Fluid Dyn., Vol. 14, 147 - 166 (1979).
Abstr. in Phys. Abstr., Vol. 83, Abstr. 81808 (1980).

062.043 **Spherical dynamos with anisotropic α-effect.**
F. H. Busse, S. W. Miin.
Geophys. Astrophys. Fluid Dyn., Vol. 14, 167 - 181 (1979).
Abstr. in Phys. Abstr., Vol. 83, Abstr. 81809 (1980).

062.044 **Dynamos theorems.** R. Hide.
Geophys. Astrophys. Fluid Dyn., Vol. 14, 183 - 186 (1979). – Abstr. in Phys. Abstr., Vol. 83, Abstr. 81810 (1980).

062.045 **Thermal and magnetic instabilities in a rapidly rotating fluid sphere.** D. R. Fearn.
Geophys. Astrophys. Fluid Dyn., Vol. 14, 103 - 126 (1979).
Abstr. in Phys. Abstr., Vol. 83, Abstr. 81850 (1980).

062.046 **Dielectronic recombination. Applications in the study of certain astrophysical plasmas.** J. Dabau.
J. Phys. Colloq., Vol. 41, No. C-3, (see 012.010), p. 181 - 184 (1980). – Abstr. in Phys. Abstr., Vol. 83, Abstr. 82188 (1980).

062.047 **Electron trapping in the solar magnetic field and emission of decimetric continuum radio bursts.**
A. O. Benz.
Astrophys. J., Vol. 240, 892 - 907 (1980).
The conditions for confinement of energetic electrons in the solar corona are studied with respect to the influence of electrostatic waves, which have been suggested to cause type IV_{dm} decimetric continuum emission. A hydrodynamic approach is taken for simplicity. The unstable growth of these waves is found to be effectively limited by a change of the particle gyroperiod in the electric field of the wave detuning the resonance. Saturation of wave energy density occurs at a low level, which is proportional to the fraction of energetic particles. The low wave level excludes induced scattering on thermal ions for the hydrodynamic instability. A new model is proposed based on conversion by interaction with low-frequency waves, in particular, lower hybrid waves, which are known to exist in loss-cone situations of the magnetosphere.

062.048 **Two basis sets for the *g*- and *p*-modes of self gravitating fluids.**
V. V. Dixit, B. Sarath, Y. Sobouti.
Astron. Astrophys., Vol. 89, 259 - 263 (1980) = Biruni Obs., Contrib. No. 8.
Completeness of the sets of trial functions used in the numerical calculation of the eigenvalues and eigenvectors of the *g*- and *p*-modes of convectively neutral fluids is explicitly shown. Also, in the case of *g*-modes, it is proved that the eigenvectors obtained by the Rayleigh-Ritz variational method do converge to the actual eigenvectors.

062.049 **Normal modes of rotating fluids.** Y. Sobouti.
Astron. Astrophys., Vol. 89, 314 - 335 (1980) = Biruni Obs., Contrib. No. 7.
The normal modes of oscillations of a rotating fluid have been expressed in terms of those of a non-rotating and convectively neutral fluid. The *p*-modes accept a double perturbation expansion in which the rotation and deviation of the fluid from convective neutrality are considered as two perturbation parameters. The *g*-modes do not yield to such a treatment. Their strong interaction with toroidal displacements of the fluid violates the criteria for perturbation expansions. Axisymmetric and non-axisymmetric modes of the fluid are treated in their full generality. Some numerical values of the *p*-eigenvalues and eigenvectors in different perturbation orders and for different spherical harmonic numbers are presented.

062.050 **Some effects of magnetic fields in spatially homogeneous universes with conductivity.** A. J. Fennelly.
Phys. Rev. D, Vol. 21, 2107 - 2118 (1980). – Abstr. in Phys. Abstr., Vol. 83, Abstr. 83864 (1980).

062.051 **Dynamical accumulation of an infinitely conducting plasma towards the neutral line of a magnetic field.**
T. Kadonaga, A. Tomimatsu.
Prog. Theor. Phys., Vol. 63, 1202 - 1212 (1980). – Abstr. in Phys. Abstr., Vol. 83, Abstr. 83977 (1980).

062.052 **The theory of magnetic shocks in collisionless plasma.** N. A. Krall.
Nuovo Cimento C, Ser. 1, Vol. 2C, (see 012.013), 693 - 711 (1979). – Abstr. in Phys. Abstr., Vol. 83, Abstr. 85854 (1980).

062.053 **Kinetic models of shocks in collisionless plasmas.** R. A. Cairns.
Nuovo Cimento C, Ser. 1, Vol. 2C, (see 012.013), 712 - 721 (1979). – Abstr. in Phys. Abstr., Vol. 83, Abstr. 90330 (1980).

062.054 **On exact equilibrium states in external gravitational fields of heated, self-gravitating gas clouds cooling by conduction and radiation.** I. Lerche, B. C. Low.
Physica D, Vol. 10, 203 - 218 (1980). – Abstr. in Phys. Abstr., Vol. 83, Abstr. 94862 (1980).

062.055 **Dynamo problems in astrophysics.**
Ya. B. Zel'dovich, A. A. Ruzmajkin.
Inst. prikl. mat. AN SSSR. Prepr., 1980, No. 52, 70 pp. In Russian. – Abstr. in Ref. zh., 51. Astron., 8.51.168 (1980).

062.056 **Specific lines of a force-free magnetic field.**
N. A. Bobrova, S. I. Syrovatskij.
XIth seminar on cosmophysics, (see 012.035), p. 101 - 120 (1979). In Russian. – Abstr. in Ref. zh., 51. Astron., 8.51.370 (1980).

062.057 **Magnetic reconnection and acceleration of particles.**
S. V. Bulanov, P. V. Sasorov, S. I. Syrovatskij.
XIth seminar on cosmophysics, (see 012.035), p. 83 - 100 (1979). In Russian. – Abstr. in Ref. zh., 51. Astron., 8.51.384 (1980).

062.058 **Adiabatic-drift-loss modification of the electromagnetic loss-cone instability for anisotropic plasma.**
B. Juhl, R. A. Treumann.
Astrophys. Space Sci., Vol. 72, 97 - 110 (1980).
Observations of the adiabatic behaviour of energetic particle pitch-angle distributions in the magnetosphere in the past indicated the development of pronounced minima or drift-loss cones on the pitch-angle distributions centred at $\alpha \simeq 90°$ in connection with storm-time changes in magnetospheric convection and magnetic field. Using a model of a drift-modified loss-cone distribution of the butterfly type, the linear stability of electromagnetic whistler or ion-cyclotron waves propagating parallel to the magnetic field has been investigated.

062.059 **Magnetogasdynamic converging spherical detonation waves.** B. G. Verma, J. B. Singh.
Astrophys. Space Sci., Vol. 72, 133 - 142 (1980).
The problems of magnetogasdynamic collapsing detonation waves having spherical symmetry near the point through a gas with varying initial density, is studied. The effects of varying density and magnetic field on the front velocity and other variables have been considered.

062.060 **The quiet aligned rotator.** F. C. Michel.
Astrophys. Space Sci., Vol. 72, 175 - 181 (1980).
The author concludes that the magnetospheric structure around an aligned rotating magnetized neutron star, in the case of a completely charge-separated plasma, consists of a dome of charge about the polar caps and an equatorial disk of opposite charge which, together, entirely envelope the surface but do not fill the magnetosphere. Although the aligned rotator is a 'standard model' for analyzing pulsar emission, the magnetosphere obtained needs not emit particles and needs not generate a stellar wind, contrary to previous expectations. Pair production only seems to modify the detailed shape of the dome and disk, such modification serving to shut off further pair production. Such a magnetohydrodynamically inactive object may have difficulty simulating pulsar emission.

062.061 **The influence of the plasma inhomogeneity on the critical velocity phenomenon.** A. Piel,
E. Möbius, G. Himmel.
Astrophys. Space Sci., Vol. 72, 211 - 221 (1980).
The phenomenon of a critical ionization velocity is discussed under the aspect of experimentally measured inhomogeneities. The difficulties inherent in homogeneous plasma models on the phenomenon are shown. A simple sheath model is presented which can be compared with the local plasma parameters measured in a rotating plasma device. The origin of the observed turbulent heating is attributed to a modified two-stream instability occurring in the sheath under discussion.

062.062 **On the equilibrium of a cylindrical plasma supported horizontally by magnetic fields in uniform gravity.** I. Lerche, B. C. Low.
Sol. Phys., Vol. 67, 229 - 243 (1980).
The authors consider the mechanical equilibrium of a cylinder of plasma suspended horizontally by magnetic fields in uniform gravity. This configuration is what may be expected if a quiescent prominence were to condense in a region initially filled with a uniform magnetic field. A set of exact solutions describing the equilibrium situation is presented. The set of solutions covers a particular case of a uniform temperature as well as cases where the temperature rises from zero at the center of the plasma cylinder to rapidly reach a constant asymptotic value outside the cylinder. The physical properties of these solutions are described.

062.063 **Magnetohydrodynamic equilibrium and stability of pre-flare loops. Constant pitch field.** S. S. Hasan.
Sol. Phys., Vol. 67, 267 - 283 (1980).
The purpose of this study is to demonstrate that a stable configuration is possible for a loop in which adequate magnetic energy can be stored prior to a solar flare. It is shown that a cylindrically symmetric force-free field with a constant pitch is unsuitable as it is unstable for any degree of twist. However, the author finds that the constant pitch field in the presence of a small transverse pressure gradient can be stable and hence provides an acceptable configuration for energy storage.

062.064 **Asymptotic estimates for an axisymmetric rotating fluid (*star*).** A. Friedman, B. Turkington.
J. Funct. Anal., Vol. 37, 136 - 163 (1980). – Abstr. in Phys. Abstr., Vol. 83, Abstr. 98542 (1980).

062.065 **On shift instability of accretion disks.**
R. S. Iroshnikov.
Astron. Zh., Tom 57, 985 - 990 (1980). In Russian. – English translation in Soviet Astron., Vol. 24, No. 5.
The question is investigated whether the turbulization of disks with Keplerian law of rotation is possible.

062.066 **On the influence of neutral current sheets in a cosmic plasma on the frequency spectrum of propagating radiation.** V. V. Zheleznyakov, E. Ya. Zlotnik.
Astron. Zh., Tom 57, 1038 - 1046 (1980). In Russian.
English translation in Soviet Astron., Vol. 24, No. 5.
The conditions for the propagation of electromagnetic

waves through neutral current sheets in a cosmic plasma (in particular, in the solar corona) are considered. The character of changes of the frequency spectrum of the radio emission having crossed a neutral sheet is outlined.

062.067 **The nonaxisymmetric configurations of uniformly rotating polytropes.** P. O. Vandervoort.
Astrophys. J., Vol. 241, 316 - 333 (1980).

This paper is devoted to the study of the nonaxisymmetric configurations of a uniformly rotating polytrope. The nonaxisymmetric configurations occur only for values of the polytropic index less than 0.808 according to James. An analytic method, valid for small values of the polytropic index, is formulated for the approximate solution of the governing equations. Sequences of axisymmetric and nonaxisymmetric configurations are constructed and delineated. It is shown how centrifugal effects restrict the occurrence of nonaxisymmetric configurations to cases in which the polytropic index is small.

062.068 **The heating of a thermally conducting stratified medium. II. A simple plane model of an atmosphere.**
I. Lerche, B. C. Low.
Astrophys. J., Vol. 241, 459 - 467 (1980).

Exact solutions of the following theoretical problem are presented: A plane atmosphere is in hydrostatic equilibrium with a uniform gravity. The ideal gas law is assumed. Heat is generated elsewhere at a rate proportional to the local density. The atmosphere is maintained in a steady state through cooling by thermal conduction and radiation. The solutions reproduce the macroscopic ordering of a hot "corona" separated from a "photosphere" by a layer of temperature minimum. The analytic solutions allow direct illustration of the interplay between steady energy transport and the requirements of hydrostatic equilibrium.

062.069 **Spurious solutions of the Navier-Stokes equations.**
D. Summers.
Astrophys. J., Vol. 241, 468 - 473 (1980).

Previous authors, notably Scarf and Noble, have applied the Navier-Stokes equations to the solar wind and have found that viscous forces are very significant in the vicinity of, and downstream from, the sonic point and that there is a resulting great enhancement of flow speed in these regions. It is shown here that such conclusions and solutions are erroneous. A method of integration of the Navier-Stokes equations is presented which involves upstream integration from a great heliocentric distance, at which point the flow variables are evaluated by appropriate asymptotic series. The resulting critical viscous solutions are found to be well behaved and generally involve a slight decrease in flow speed and increase in temperature, as expected on physical grounds.

062.070 **A model for impulsive electron acceleration to energies of tens of kT_e.** P. Hoyng,
A. Duijveman, T. F. J. van Grunsven, D. R. Nicholson.
Solar and interplanetary dynamics, (see 012.020), p. 299 - 302 (1980).

The authors describe a model for first stage electron acceleration based on quasi-linear interaction with Langmuir waves. The acceleration takes place in a MHD-unstable plasma region that contains many microscopically unstable current layers, which act as (quasi-stationary) sources of Langmuir waves.

062.071 **Solution of the equation of heat flow.**
S. H. Margolis, E. Knobloch.
Mon. Not. R. Astron. Soc., Vol. 193, 345 - 351 (1980).

The geometry of sunspots has been used to suggest a problem in heat flow. The equation of heat transport is solved for the case of a cylinder with a given thermal conductivity imbedded in an otherwise uniform medium with different conductivity. The surface of this region radiates heat with flux proportional to temperature. The variations in temperature along the radiating surface have been determined. The analysis is used to set limits on the ratio of diameter to depth for cases which preserve the sharp surface temperature transition across the cylinder.

062.072 **Magnetohydrodynamic Kelvin-Helmholtz instabilities in astrophysics – I. Relativistic flows – plane boundary layer in vortex sheet approximation.**
A. Ferrari, E. Trussoni, L. Zaninetti.
Mon. Not. R. Astron. Soc., Vol. 193, 469 - 486 (1980).

The authors re-examine some unresolved problems of the linear MHD Kelvin-Helmholtz instability, starting from the analysis of relativistic (and non-relativistic) flows in the approximation of a plane vortex sheet, for the contact layer between the fluids in relative motion. Results are discussed for a range of physical parameters in specific connection with application to models of jets in extragalactic radio sources.

062.073 **Upper limit on the electric field along a magnetic O line.** V. M. Vasyliunas.
J. Geophys. Res., Vol. 85, 4616 - 4620 (1980).

062.074 **Acceleration of relativistic charged particles by supersonic hydromagnetic turbulence.**
L. E. (*L. Eh.*) Gurevich, A. A. Rumyantsev.
Astrophys. Space Sci., Vol. 72, 261 - 270 (1980).

The acceleration of relativistic particles is considered during their intersection with hydromagnetic shock fronts in the presence of randomly distributed large-scale magnetic fields. In a series of astronomical objects, the Larmor radius of the relativistic particles exceeds the width of the shock front. In this case there is a change in the adiabatic invariant which results in an increase in the energy of the particle when it crosses the front in any direction.

062.075 **Propagation of axi-symmetric relativistic shock waves.**
G. Deb Ray, T. K. Chakraborty, S. N. Banerjee.
Astrophys. Space Sci., Vol. 72, 323 - 345 (1980).

Solutions in series for the propagation of relativistic shock waves with axial symmetry are obtained. The authors assume that the gaseous elements move almost radially and that the disturbance moves through a cold gas at rest wherein the nucleon number density and the energy density obey an exponential law of distance from a given plane. The motion is sustained by continuous explosions in the central region liberating energy varying as the cube of time. Also, the authors assume the equation of state of the moving elements as that of photonic gas.

062.076 **Relativistic hydrodynamics of a free expansion and a shock wave in one-dimension. Super-light expansion of extragalactic radio sources.**
M. Yokosawa, S. Sakashita.
Astrophys. Space Sci., Vol. 72, 447 - 475 (1980).

Dynamical evolution of a relativistic explosion resulting from a large amount of energy release in a homogeneous medium is studied using the Khalatnikov equation describing relativistic, hydrodynamic, planar flow. The early phase of the explosion is idealized to two stages: a free expansion and a shock wave stage. By the hodograph transformation inverting the dependent and independent variables, the hydrodynamic equations for the relativistic flow are reduced to second-order linear equations in a velocity-enthalpy space and they are solved by the method of Laplace transformation. The propagation laws and flow structures of the relativistic expansion are obtained at each stage. The transition time from a free expansion to a shock wave stage suggests that the super-light expansion observed in extragalactic radio sources has no spherical geometry but must be confined to a narrow cone.

062.077 **Magnetic (electric) strings in relativistic magnetohydrodynamics.** G. Prasad, B. B. Sinha.

C. R. Acad. Sci. Paris, Tome 291, Sér. A, 439 - 442 (1980). In French.

A number of lemmas describing the existence of magnetic (electric) strings and their dual strings has been established. Finally, a theorem describing the frozen-in property of magnetic fields has been established.

062.078 **Weak turbulence noise propagation in a celestial environment.** B. Chen.
Acta Astron. Sinica, Vol. 21, 257 - 261 (1980). In Chinese.

The author derives a useful integral formula for the computation of weak turbulence noise propagation in a nearly isothermal celestial environment under gravity. He shows that the noise propagation is severely limited by the gravity factor both in frequency and in propagation distance.

062.079 **Strong Langmuir wave turbulence: some results with selfconsistent Landau damping.**
T. F. J. van Grunsven, P. Hoyng, D. R. Nicholson.
Astron. Astrophys., Vol. 91, 7 - 16 (1980).

The coupled mode equation for the ion density and the envelope of the high frequency Langmuir wave electric field are solved in two spatial dimensions by the split-timestep Fourier method, together with the quasi-linear equation for the evolution of the tail of the electron velocity distribution (only the isotropic part of the velocity distribution is considered). In this way, an approximate selfconsistent time evolution for waves and particles is obtained; no justification is given for the use of quasi-linear theory for the case of strong turbulence. The authors study an initial value problem, and at $t = 0$ two equally strong pumps at $\pm \boldsymbol{k}_0$ are initialized in $\boldsymbol{k}$-space. The early time evolution always shows the familiar parametric growth of daughter waves. Two cases are studied in detail.

062.080 **Impulsive electron acceleration to energies of tens of kT_e by Langmuir wave turbulence.**
P. Hoyng, A. Duijveman, T. F. J. van Grunsven, D. R. Nicholson.
Astron. Astrophys., Vol. 91, 17 - 24 (1980).

Electron acceleration to energies of tens of kT_e is believed to take place in a source region that contains many microscopically unstable current layers, which generate Langmuir waves. It is argued that the shape of the tail of the electron velocity distribution is determined in the first place by two effects: (1) the spatial inhomogeneity of the Langmuir wave distribution and (2) escape of fast electrons from the acceleration region. An equation describing the evolution of the tail of the electron distribution is presented; its stationary solution $f(v)$ and the resulting flux density of escaping electrons is studied analytically and numerically. Applications for solar radio bursts are suggested.

062.081 **Propagation of waves in an atmosphere in the presence of a magnetic field. II. The reflection of Alfvén waves.** B. Leroy.
Astron. Astrophys., Vol. 91, 136 - 146 (1980).

The propagation of linear Alfvén waves in an isothermal atmosphere is studied without invoking the WKBJ approximation. In order to clarify existing work, the treatment of the wave reflection induced by the density gradient is fully revisited. A special emphasis is put on the coupled modes representation. As an application, the reflectance of an atmosphere with respect to Alfvén waves is derived.

062.082 **An analytical version of the free-energy-minimization method for the equation of state of stellar plasmas.** W. Däppen.
Astron. Astrophys., Vol. 91, 212 - 220 (1980).

In the free energy method statistical mechanical models are used to construct a free energy function of the plasma. The equilibrium composition for given temperature and density is found where the free energy is a minimum. Further simplifications are made to obtain an analytic expression for the free energy. Thus the minimum is rapidly found using a second order algorithm, whereas until now numerical first order derivatives and a steepest-descent method had to be used. Consequently time-consuming computations are avoided and the analytical version of the free energy method has successfully been incorporated into the stellar evolution programmes at Geneva Observatory.

062.083 **A formalism for differential rotation.**
B. R. Durney, H. C. Spruit.
Highlights of Astronomy, Vol. 5, (see 012.022), 121 - 127 (1980).

062.084 **Very hot plasmas in the solar system.** W. I. Axford.
Highlights of Astronomy, Vol. 5, (see 012.025), 351 - 359 (1980).

A great deal of understanding of the origin and behaviour of the very hot plasmas which exist in the solar system has been achieved during the 20 years in which in situ measurements have been possible. In retrospect, perhaps the biggest surprises have been the diverse nature of these plasmas and the wide range of processes which heat them and cause some particles to be accelerated to very high energies. These processes must surely also be important in more general astrophysical contexts.

062.085 **Acceleration mechanisms, flares, magnetic reconnection and shock waves.** S. A. Colgate.
Highlights of Astronomy, Vol. 5, (see 012.025), 397 - 410 (1980).

There are three general classifications of acceleration processes. The first is hydrodynamic, second coherent electromagnetic, and third stochastic electromagnetic processes. The author discusses each of these briefly in terms of the spectrum of accelerated particles.

062.086 **Hydromagnetic planetary waves in a vertically sheared zonal flow and a transverse magnetic field.**
O. M. El Mekki.
Sol. Phys., Vol. 68, 3 - 15 (1980).

Hydromagnetic planetary waves propagating through a zonal flow and a transverse magnetic field both of which are sheared in the vertical direction are studied. It is found that the effect of the transverse magnetic field is to make planetary waves, which characteristically propagate westwards, propagate eastwards in both westerly and easterly zonal flows. It is also shown that at a critical level the rays are guided by the zonal flow only and that the waves are either attenuated or escalated by an exponential factor as they cross a critical level.

062.087 **Vertical motions in an intense magnetic flux tube. IV: Radiative relaxation in a uniform medium.**
A. R. Webb, B. Roberts.
Sol. Phys., Vol. 68, 71 - 85 (1980).

Radiative damping of waves is important in the upper photosphere. It is thus of interest to examine the effect of radiative relaxation on the propagation of waves in an intense magnetic flux tube embedded in a uniform atmosphere. Assuming Newton's law of cooling, it is shown that the radiative energy loss leads to wave damping. Both the 'damping per wavelength' and the 'damping per period' reach maximum value when the sound and radiative timescales are comparable. The stronger the magnetic field, the greater is the damping.

062.088 **Vertical motions in an intense magnetic flux tube. V: Radiative relaxation in a stratified medium.**
A. R. Webb, B. Roberts.
Sol. Phys., Vol. 68, 87 - 102 (1980).

It is of interest to examine the effect of radiative relaxation on the propagation of waves in an intense magnetic flux tube embedded in a stratified atmosphere. The radiative energy loss (assuming Newton's law of cooling) leads to a decrease in the vertical phase-velocity of the waves, and to a damping of the amplitude for those waves with frequencies greater than the adiabatic value of the tube cut-off frequency. The phase-shift between velocity oscillations at two different levels and the phase-difference between temperature and velocity perturbations are compared with the available observations. Radiative dissipation of waves propagating along an intense flux tube may be the cause of the high temperature (and excess brightness) observed in the network.

062.089 **Energy balance in current sheets: from Petschek to gravity driven reconnection?**
C. Mercier, J. Heyvaerts.
Sol. Phys., Vol. 68, 151 - 176 (1980).

Energy balance processes play a very important role in the determination of the reconnection regime in the central diffusive region of a steady Petschek flow: as a consequence of the plasma thermal properties, abrupt transitions in the reconnection regime may occur for special external conditions. The regime becomes then a dynamical one. The authors reexamine the problem of onset of such a dynamical transition and conclude that plasma microturbulence does not appear in a straightforward way. However it is possible that the canonical Petschek regime may evolute into a new one in which the dissipative sheet is no longer infinitesimal with respect to the dimensions of the structure, and in which gravity plays an important role. Flare triggering, if related to the reconnection regime, must then proceed by more complex processes, possibly related to tearing mode dynamics, or to more global properties of the magnetic structure of the active region.

062.090 **Axisymmetric convection driven by latitudinal temperature gradients in rotating spherical shells.**
D. H. Hathaway, P. A. Gilman, J. Miller, J. Toomre.
Bull. American Astron. Soc., Vol. 12, 686 (1980). – Abstract.

062.091 **On the supersonic dynamics of magnetized jets of thermal gas in radio galaxies.**
K. L. Chan, R. N. Henriksen.
Astrophys. J., Vol. 241, 534 - 551 (1980).

A stellar-wind type formulation is developed to study axisymmetric MHD jet flows in radio galaxies. When the magnetic energy is small compared to the total energy of the jet flow and the azimuthal field has a pinching configuration, similarity methods can be employed to reduce the two-dimensional partial differential equations of the problem to a system of ordinary differential equations. These equations allow some approximate analytical solutions in the outer jet region where many parameters approach their asymptotic values, but numerical methods have to be used in the inner jet region where variables change rapidly. The results of the numerical experiments to study the effects of a pinching external pressure and a pinching magnetic field are discussed, and the observed width of the northern radio jet in 3C 449 will be fitted by each of the two pinching models. The asymptotic analytic expressions are tested numerically.

062.092 **Magneetvelden in de astrofysika.**
W. van Rensbergen.
Zenit, 7e Jaarg., 292 - 295 (1980).

062.093 **Quantum effects in cyclotron plasma absorption.**
G. G. Pavlov, Yu. A. Shibanov, D. G. Yakovlev.
Astrophys. Space Sci., Vol. 73, 33 - 62 (1980).

It is shown that X-ray radiation of neutron stars with magnetic fields $B = 10^{11}$ - 10^{13} G near cyclotron resonances is deeply affected by such quantum effects as electron-positron vacuum polarization, the quantizing character of the magnetic field, the non-harmonic character of the Landau levels, and the quantum recoil of electrons. These effects should be taken into account for an interpretation of observational data on X-ray pulsars (e.g., Her X-1) and other X-ray sources associated with neutron stars.

062.094 **The radiation of a hot magnetized plasma accreted by degenerate stars with a strong magnetic field.**
G. G. Pavlov, I. G. Mitrofanov, Yu. A. Shibanov.
Astrophys. Space Sci., Vol. 73, 63 - 82 (1980).

The absorption coefficients for extraordinary and ordinary electromagnetic modes are found for a tenuous hot magnetized plasma, taking into account the collisions between plasma particles and the scattering of photons. The intensity, Stokes parameters and polarization of radiation of a homogeneous plasma slab are calculated for conditions which may be realized in the heated regions of accreted plasma in an AM Herculis-type system. The large difference between the absorption coefficient of extra-ordinary and ordinary modes near the cyclotron harmonics may result in the emission of the broad polarized continuum together with the narrow cyclotron lines. The polarization of these lines has a complicated spectral dependence. The results obtained are shown to be useful for explaining the main properties of AM Herculis-type objects.

062.095 **Absorption of intense electromagnetic beams in a magnetoplasma.** N. Gopalswamy, V. Krishan.
Astrophys. Space Sci., Vol. 73, 179 - 186 (1980).

The multiphoton inverse bremsstrahlung absorption of two intense electromagnetic beams passing through a magnetized plasma is studied. The rate of absorption of electromagnetic energy by the electrons is calculated by deriving a kinetic equation for the electrons. It is found that the absorption enhances when the frequency of one electromagnetic beam is more, and that of the other electromagnetic beam is less, than the electron-cyclotron frequency. A possible application to extragalactic radio sources is discussed.

062.096 **High-frequency transverse Fresnel drag in a moving magneto-active plasma.** N. Meyer-Vernet.
Astrophys. Space Sci., Vol. 73, 207 - 212 (1980).

The bending of a beam of radiation normally incident on an anisotropic cold plasma moving transversely is calculated. There are generally two emergent beams – refracted by different amounts – which may be of both signs, depending on the parameters. As was previously found for an isotropic plasma, the bending is generally small except in a few particular cases.

062.097 **Equilibrium of self-gravitating polytropic cylinders with a magnetic field.** N. K. Sood.
Astrophys. Space Sci., Vol. 73, 213 - 225 (1980).

The effect of a prevalent magnetic field on static and uniformly rotating self-gravitating cylinders of infinite length is examined. A variety of magnetic-field configurations are shown to be admissible solutions of equations of motion, from which some feasible cases are presented. The homogeneous case is also studied and shows interesting magnetic-field patterns.

062.098 **Spherically symmetric flows by the dissipative finite difference scheme.** Y. Nakamura.
Sci. Rep. Tôhoku Univ., Ser. 1, Vol. 62, 121 - 133 (1980).

062.099 **Expansions into magnetized media.**
A. R. Garlick.
Observatory, Vol. 100, 181 (1980). – Abstract.

062.100 **On the α-effect for slow and fast rotation.**
G. Rüdiger.
Astron. Nachr., Band 299, 217 - 222 (1978).

This paper deals with the influence of global rotation on a special inhomogeneous field of incompressible turbulence. All its deviations from homogeneity and isotropy may be described only by a constant gradient of turbulence intensity. Using a

Fourier representation the author is able to determine the α-effect for any given rotational rates and material constants.

062.101 **Convective instability when the temperature gradient and rotation vector are oblique to gravity. II. Real fluids with effects of diffusion.**
D. H. Hathaway, J. Toomre, P. A. Gilman.
Geophys. Astrophys. Fluid Dyn., Vol. 15, 7 - 37 (1980).
Abstr. in Phys. Abstr., Vol. 84, Abstr. 4760 (1981).

062.102 **The Cowling anti-dynamo theorem.**
R. W. James, P. H. Roberts, D. E. Winch.
Geophys. Astrophys. Fluid Dyn., Vol. 15, 149 - 160 (1980).
Abstr. in Phys. Abstr., Vol. 84, Abstr. 4761 (1981).

062.103 **Eigenvalue bounds in magnetoatmospheric shear flow.** J. A. Adam.
J. Phys. A, Vol. 13, 3325 - 3338 (1980). – Abstr. in Phys. Abstr., Vol. 84, Abstr. 6898 (1981).

062.104 **On plasma-neutral gas interaction.**
N. Venkataramani, S. K. Mattoo.
Pramāna, Vol. 15, 117 - 136 (1980). – Abstr. in Phys. Abstr., Vol. 84, Abstr. 7006 (1981).

062.105 **Rays in magneto-ionic theory – II.**
K. G. Budden, G. F. Stott.
J. Atmos. Terr. Phys., Vol. 42, 791 - 800 (1980).

In a previous paper, for a plasma in which the only effective carriers are electrons, a formula was given for finding the refractive indices when the ray direction is given. In this paper the formula is extended to allow for the presence of heavy ions. It is an equation of degree six whose solutions are associated with six different waves, although not all of them are necessarily excited by a given source. The properties of the equation are illustrated by discussion of its solutions at cut-off frequencies and resonance frequencies, and by showing its connection with resonance cones, Storey cones, and window points. A specific example of a plasma containing protons and He^+ ions is studied as an illustration.

062.106 **Theory of scan plane flux anisotropies.**
T. G. Northrop, M. F. Thomsen.
J. Geophys. Res., Vol. 85, 5719 - 5724 (1980).

When a spacecraft detector measures particle flux as a function of look direction in a plane (the scan plane) anisotropy is often seen. This anisotropy is caused by spatial gradients, by **E** × **B** particle drift, and by various spectral and geometric effects. This paper treats all of these effects systematically, starting from the nonrelativistic Vlasov equation. The general analysis is applied to a simple model of an anisotropic distribution to give a relation between the **E** × **B** drift, the gradient and the experimentally observed first, second, and third harmonics of the flux as a function of angle in the scan plane.

062.107 **Properties of the longitudinal dielectric function: an application to the auroral plasma.**
P. B. Dusenbery, R. L. Kaufmann.
J. Geophys. Res., Vol. 85, 5969 - 5976 (1980).

The longitudinal dielectric function is used to study purely electrostatic plasma waves. The paper is concerned with linear waves in a spatially homogeneous, collisionless, uniformly magnetized plasma. The plasma is described by a two temperature, streaming Maxwellian distribution. Sample plasma parameters are selected to model the auroral ionosphere at rocket altitudes, and wave numbers are selected to produce resonance with energetic auroral electrons.

062.108 **Spindown of rotating magnets.** W. Kundt.
Magnetospheric boundary layers, (see 012.041), p. 265 - 267 (1979).

The torque exerted on a rotating magnet by impinging plasma has been derived from certain general assumptions plus conservation laws. Its relevance to binary evolution (neutron stars, white dwarfs, in particular: binary X-ray sources, AM Her, 4U 0115+63, SS 433), to the generation of cosmic rays, and to the (early) planets and their satellites is explained.

062.109 **The magnetospheres of magnetic A stars and of pulsars.** J. Arons.
Magnetospheric boundary layers, (see 012.041), p. 271 - 279 (1979).

Aspects of the hydromagnetic structure of the magnetospheres surrounding magnetic A stars are considered. The magnetopause radius is estimated. The supply of mass to the magnetosphere from the star is considered; a new model for formation of a corona is suggested. Plasma supply from planets contained within the magnetosphere is considered as well as the possibility of discovering these planets from photons emitted as a consequence of the planet-magnetosphere interaction. A number of mechanisms for rotational spin down of magnetic A stars are considered. A new spin down theory, based on emission of interstellar magnetosonic modes from the shear unstable magnetopause, is suggested. A comparison is made to certain pulsar model classes.

062.110 **Kinetic theory of the boundary layer between a flowing isotropic plasma and a magnetic field.**
L. R. O. Storey, L. Cairo.
Magnetospheric boundary layers, (see 012.041), p. 289 - 293 (1979).

062.111 **A microscopic description of interpenetrated plasma regions.** M. A. Roth.
Magnetospheric boundary layers, (see 012.041), p. 295 - 309 (1979).

A model of steady-state tangential discontinuities has been developed using the kinetic theory of multi-components collisionless plasmas. The author considers the magnetopause layer and the inner edge of the plasma boundary layer as regions where two different hydrogen plasmas are interpenetrated and the model is used to describe their microstructures.

062.112 **Interstellar shock waves with magnetic precursors.**
B. T. Draine.
Astrophys. J., Vol. 241, 1021 - 1038 (1980).

The structure of steady, radiative, one-dimensional shock waves in partially ionized gas with a transverse magnetic field B_0 is investigated. It is found that such shocks may be preceded by a "magnetic precursor" which heats and compresses the medium ahead of the front where the neutral gas undergoes a discontinuous change of state. The physical processes operative in such shocks are examined, including the effects of charged dust grains in dense molecular clouds. Numerical examples are shown for $v_s = 10\,km\,s^{-1}$ shocks propagating into diffuse H I or H_2.

062.113 **Cyclotron absorption in accreting magnetic white dwarfs.** G. Chanmugam.
Astrophys. J., Vol. 241, 1122 - 1130 (1980).

The cyclotron absorption coefficient is calculated using a three-dimensional Maxwellian distribution for the electrons for a wide range of temperature and frequencies. The results are applied to a plasma slab which is perpendicular to the magnetic field, and it is shown that there are deviations from the Rayleigh-Jeans spectrum predicted in earlier works. Comparison with observations for AM Herculis suggests that its magnetic field is about 5×10^7 gauss (or less).

062.114 **Neutrino, gamma-ray, electron, and positron production in an ultrarelativistic plasma.**
A. P. Marscher, W. T. Vestrand, J. S. Scott.
Astrophys. J., Vol. 241, 1166 - 1174 (1980).

The authors examine neutrino, γ-ray, electron, and

positron production resulting from inelastic proton-proton collisions in a highly relativistic plasma. Analytic expressions for the production spectra are obtained in the case of a power law relativistic proton distribution. Numerical results are presented for a relativistic Maxwellian proton distribution.

062.115 **Stability of strong linearly polarized electromagnetic waves in dense plasmas.** A. Che, W. H. Kegel.
Astron. Astrophys., Vol. 92, 204 - 211 (1980).

It is shown that strong linearly polarized transverse electromagnetic waves are unstable to a certain type of density fluctuation in the sense that transverse wave energy is transferred into the longitudinal mode. Since the strictly periodic solutions of the non-linear wave equation are of a singular nature, one would expect that more general perturbations lead to the excitation of other modes. The analysis has been carried out for a plane wave only. It appears, however, that the conclusions drawn hold qualitatively also for other geometries. From this the authors conclude for pulsars that a periodic electromagnetic field can be expected only in their close vicinity.

062.116 **Two dimensional magnetic merging and reconnection.** A. F. Cheng.
Astrophys. J., Vol. 242, 326 - 335 (1980).

Magnetic energy release in merging flows is investigated by numerical integration of unsteady resistive MHD equations in two dimensions. No explosive magnetic energy release is found in two topologically distinct merging flow geometries, in the absence of imposed strong plasma inflows. The results indicate that two dimensional merging without strong forcing is not promising as a mechanism for explosive magnetic energy release, but is more promising for gradual plasma heating and magnetic field dissipation.

062.117 **Plasma at densities above the nuclear.** G. S. Sahakian (*Saakyan*), L. S. Grigorian (*L. Sh. Grigoryan*).
Astrophys. Space Sci., Vol. 73, 307 - 318 (1980).

The state of degenerate plasma is investigated at densities above the nuclear density.

062.118 **Rotational dynamics of a deformable medium: further generalization.** V. S. Geroyannis.
Astrophys. Space Sci., Vol. 73, 453 - 467 (1980).

Cauchy's fundamental first law of continuum mechanics is integrated over the whole mass of a self-gravitating deformable finite material continuum, viscolinear (i.e., Newtonian), not necessarily constrained to obey Stokes' condition, with viscosity coefficients given as arbitrary functions of the coordinates. The general Eulerian equation is derived, governing generalized rotation on which certain other cooperating deformations are superimposed. Finally, the explicit form of this equation is given for the case of a viscous gaseous polytrope.

062.119 **An explosion model in magnetogasdynamics.** B. G. Verma, J. B. Singh.
Astrophys. Space Sci., Vol. 73, 469 - 479 (1980).

By taking into account the interaction with magnetic field, in an axially symmetric inhomogeneous medium and assuming Gaussian density profile in radial direction instead of the cusped exponential law, a point explosion has been investigated by generalising the method suggested by Laumbach and Probstein (1969). The shock envelope becomes increasingly elongated along the axis of rotation, until it finally breaks through into the intergalactic space before it can spread considerably in equatorial directions as in ordinary gasdynamics. A comparison has been made between the authors' results and those obtained in ordinary gasdynamics. Explosion models of double radio sources and related objects are suggested.

062.120 **Plasma acceleration by ion-acoustic turbulence.** V. Krishan.
Sol. Phys., Vol. 68, 343 - 350 (1980).

An energetic proton beam passing through a stationary ionized medium excites ion-acoustic turbulence. The ion-acoustic instability saturates due to the non-linear indirect wave-particle scattering. The electric field associated with the ion-acoustic waves accelerates the plasma particles. Applicability of the results to cometary tails is discussed.

062.121 **Collapse and equilibrium of rotating, adiabatic clouds.** A. P. Boss.
Astrophys. J., Vol. 242, 699 - 709 (1980).

A numerical hydrodynamics computer code has been used to follow the collapse and establishment of equilibrium of adiabatic gas clouds restricted to axial symmetry. The clouds are initially uniform in density and rotation, with adiabatic exponents $\gamma = {}^5/_3$ and ${}^7/_5$. The numerical technique allows, for the first time, a direct comparison to be made between the dynamic collapse and approach to equilibrium of unconstrained clouds on the one hand, and the results for incompressible, uniformly rotating equilibrium clouds, and the equilibrium structures of differentially rotating polytropes, on the other hand. Models with differing initial values of α (ratio of thermal to gravitational energy) and β (ratio of rotational to gravitational energy) have been calculated with both values of γ to ascertain the possible end states of adiabatic clouds.

062.122 **One dimensional hydrodynamics of asteroid-neutron star collisions.**
M. J. Newman, A. N. Cox.
Space Sci. Rev., Vol. 27, (see 012.046), 591 - 594 (1980).

It has been suggested by several authors that the observed cosmic gamma-ray bursts might be produced by the collision of comet or asteroid-sized bodies with a compact object. The authors present the results of simplified one-dimensional hydrodynamic-radiation diffusion calculations of such an occurrence.

062.123 **Thermodynamic transport properties in dense stars.** T. W. Edwards.
Space Sci. Rev., Vol. 27, (see 012.046), 627 - 633 (1980).

The thermodynamic transport properties of special relativistic imperfect fluids, as found in dense stars, are investigated. These properties, which include thermal and electrical conductivities, electrothermal coefficients, and bulk and shear viscosities may be formulated in terms of the momentum distribution functions obtained from the solution of the Boltzmann transport equation. Spherical harmonic solutions of the relaxation form of the relativistic magnetic Boltzmann transport equation have also been obtained which give the non-equilibrium momentum distribution function perturbation $f - f^{(0)} = \Delta f(p)$ in terms of electromagnetic and thermal fields.

062.124 **σ-stability analysis of hydromagnetic instabilities in stars with toroidal magnetic fields.**
M. Goossens, D. Biront.
Space Sci. Rev., Vol. 27, (see 012.046), 667 - 672 (1980).

σ-stability analysis is used to investigate the adiabatic stability of a star containing an axisymmetric toroidal magnetic field. Necessary and sufficient conditions for σ-stability are derived. Special attention is devoted to the typical hydromagnetic instabilities that can be introduced by a weak toroidal magnetic field in a star that is stably stratified in the absence of any magnetic field. An expression for the maximum growth rate of instability is derived and the basic properties of the displacement fields associated with the instabilities are indicated.

062.125 **Fully developed anisotropic hydromagnetic turbulence in interplanetary space.**

M. Dobrowolny, A. Mangeney, P. Veltri.
Phys. Rev. Lett., Vol. 45, 144 - 147 (1980).

The solar-wind magnetohydrodynamic turbulence is observed to be mainly made of Alfvénic fluctuations propagating away from the sun. It is shown that such an asymmetric state is a general consequence of the evolution of developed magnetohydrodynamic turbulence, which, starting from an initial asymmetry between modes with cross helicity +1 and −1, tends, as a consequence of nonlinear interactions, towards a state where the only modes left are those initially prevailing (with either cross helicity +1 or −1).

062.126 **Shock systems in collisionless space plasmas.** E. W. Greenstadt, R. W. Fredricks.
Solar system plasma physics, Vol. 3, (see 003.010), 3 - 43 (1979).

Contents: Introduction. Brief history of collisionless shock experience. Earth's bow shock system. Planetary shocks. Astrogenic shocks. Discussion. Recommendations.

062.127 **Magnetic field reconnection.** B. U. Ö. Sonnerup.
Solar system plasma physics, Vol. 3, (see 003.010), 45 - 108 (1979).

It is the purpose of this paper to provide a concise qualitative summary of the present state of reconnection theory and observations, with special reference to the earth's magnetosphere, and to bring into focus a number of specific problems and questions concerning the reconnection process in its magnetospheric application which should be studied both theoretically and observationally.

062.128 **Phenomenological description of a non-equilibrium magnetospheric plasma in adiabatic approximation.**
V. D. Pletnev, G. A. Skuridin.
Kosm. Issled., Tom 18, 851 - 876 (1980). In Russian.

062.129 **Simple method of determination of a gas velocity field in the envelopes of active objects.**
S. N. Fabrika.
Astron. Tsirk., No. 1109, p. 1 - 2 (1980). In Russian.

062.130 **Freie Energie und Stabilität magnetosphärischer und solarer Plasmastrukturen.** K. Schindler.
Mitt. Astron. Ges., Nr. 50, p. 13 (1980). – Abstract.

062.131 **Investigation of the stability and non-uniqueness of flows in rotating spherical layers.**
Yu. N. Belyaev, A. A. Monakhov, G. N. Khlebutin, I. M. Yavorskaya.
Inst. kosm. issled. AN SSSR. Prepr., 1980, No. 567, 71 pp. In Russian. – Abstr. in Ref. zh., 51. Astron., 12.51.161 (1980).

062.132 **Spectra of multicharged ions in laboratory and astrophysical plasma.**
M. A. Mazing, S. L. Mandel'shtam.
8 Nats. konf. po atom. spektroskop. s mezhdunar. uchastiem, Varna, 1978. Dokl. p. 9 - 37. In Russian. – Abstr. in Ref. zh., 51. Astron., 1.51.145 (1981).

062.133 **On the equilibrium of heated self-gravitating masses: cooling by conduction.** I. Lerche, B. C. Low.
Astrophys. J., Vol. 242, 1144 - 1155 (1980).

An investigation is given of the equilibrium states available to a self-gravitating mass of gas, cooling by conduction, and being heated at a rate proportional to the local gas density. For a constant thermal conductivity it is shown that the gas density has either a central maximum or a central minimum, depending on the ratio of the thermal conductivity to a parameter taken to be a measure of the rate of heating. For a thermal conductivity which is a positive power of the temperature, it is shown that the gas density always has a central minimum and a maximum at the outer boundary of the configuration. For cylindrical and spherical geometrical configurations the same general properties are obtained. The physical origin of this behavior is discussed.

062.134 **On equilibrium states of heated self-gravitating gas clouds cooling by conduction in an external gravitational field.** I. Lerche, B. C. Low.
Astrophys. J., Vol. 242, 1156 - 1165 (1980).

Exact, analytic solutions are presented for equilibrium states of a self-gravitating, one-dimensional cloud of gas, embedded in an external gravitational field due to a plane of "stars", being heated at a rate proportional to the local gas density, and cooling by thermal conduction. The authors have done the calculations to illustrate that (1) the role of thermal conduction in determining the equilibrium distributions of material in gas clouds is considerably more complicated than simple dimensional arguments might otherwise indicate and (2) the phenomenon of density "inversion" obtaining in the absence of an external field is a rugged topological property obtaining also for any strength of the external gravitational field. The authors suggest that these calculations may be of interest in investigations of interstellar molecular clouds and of filamentary structure in supernova remnants and may also aid in the modeling of more realistic gas distributions around "cocoon" protostars than have heretofore been available.

062.135 **Grundmechanismen der Resistiven Tearing Instabilität.** L. Janicke.
Mitt. Astron. Ges., Nr. 50, p. 80 (1980). – Abstract.

Plasma astrophysics. Nonthermal processes in diffuse magnetized plasmas. Vol. 1: The emission, absorption and transfer of waves in plasmas. Vol. 2: Astrophysical applications. See Abstr. 003.076.

Magnetic field generation in electrically conducting fluids. See Abstr. 003.080.

Buoyancy effects in fluids. See Abstr. 003.105.

Non-thermal electron scattering in a homogeneous plasma. See Abstr. 063.017.

Dielectronic recombination rates, ionization equilibrium, and radiative emission rates for calcium and nickel ions in low-density high-temperature plasmas.
See Abstr. 063.023.

Mesoturbulence. See Abstr. 063.025.

Thermal effects on the cyclotron line formation process in X-ray pulsars. See Abstr. 063.057.

Stellar convection theory. See Abstr. 064.019.

Hydrodynamic modeling of mass loss from cataclysmic variable secondaries. See Abstr. 064.090.

Convective stability in magnetic stars.
See Abstr. 065.009.

Differential rotation of magnetic stars.
See Abstr. 065.024.

Neutronization, lepton escape and stellar hydrodynamics. See Abstr. 065.026.

Stellar convection. II. A multimode numerical solution for convection in spheres. See Abstr. 065.030.

Stellar convection. III. Convection at large Rayleigh numbers. See Abstr. 065.033.

Astrophysical problems of condensed matter in huge magnetic fields. See Abstr. 065.037.

Numerical solution of the 1 D spherical non stationary radiating shock using characteristics. Application to protostars. See Abstr. 065.043.

The time scale of secularly unstable stellar rotation. See Abstr. 065.052.

σ-stability analysis of a toroidal magnetic field in stars. See Abstr. 065.065.

The effect of rotation on the hydrodynamics of stellar collapse. See Abstr. 065.083.

The effect of a magnetic field on stellar pulsations as a singular perturbation problem. See Abstr. 065.086.

Time-dependent, optically thick accretion onto a black hole. See Abstr. 066.047.

Gravitational radiation from colliding, compact stars: hydrodynamic calculations in one dimension. See Abstr. 066.050.

Hamiltonian formalism for perfect fluids in general relativity. See Abstr. 066.070.

Radiative heat transfer in surface layers of neutron stars with a magnetic field. See Abstr. 066.504.

Magnetohydrodynamic shock propagation in the vicinity of a magnetic neutral sheet. See Abstr. 074.094.

The evolution of active region loop plasma. See Abstr. 074.095.

Hydromagnetic waves and turbulence in the solar wind. See Abstr. 074.107.

Evolution of solar magnetic fields: a new approach to MHD initial-boundary value problems by the method of nearcharacteristics. See Abstr. 075.011.

The pulses as a diagnostic technique in the sun. See Abstr. 080.003.

Earth's magnetosphere: global problems in magnetospheric plasma physics. See Abstr. 084.097.

Towards a comparative theory of magnetospheres. See Abstr. 091.053.

Hydrodynamic studies of the nova outburst. See Abstr. 124.024.

Hot plasmas in supernova remnants. See Abstr. 125.043.

Supernova explosions – the role of a Rayleigh-Taylor instability. See Abstr. 125.074.

The effect of stellar structure on supernova remnant evolution. See Abstr. 125.075.

Wave-wave interactions in a rotating gravitating gas cloud. See Abstr. 131.031.

Numerical calculations of the collapse of non-rotating, magnetic gas clouds. See Abstr. 131.033.

Nonlinear hydrodynamics of acoustic instabilities in diffuse clouds. See Abstr. 131.089.

H II bubbles and disruption of molecular clouds. See Abstr. 131.118.

On the motion and destruction of grains in interstellar clouds. See Abstr. 131.120.

Amplification of protostellar magnetic fields. See Abstr. 131.266.

Criteria for fragmentation in a collapsing rotating cloud. See Abstr. 131.290.

Fluid jets in radio sources. See Abstr. 141.053.

Turbulent generation of magnetic fields in extended extragalactic radio sources. See Abstr. 141.103.

Turbulence-related morphology in extragalactic radio sources. See Abstr. 141.104.

Alfvén-driven cyclotron corona as a model for quasar infrared. See Abstr. 141.116.

On the self-consistent description of axisymmetric pulsar magnetospheres – II. A method of solution. See Abstr. 141.501.

Pulsating X-ray sources: the oblique dipole configuration. See Abstr. 142.014.

Implications of the high-state iron-like feature and soft X-ray emission of Hercules X-1. See Abstr. 142.042.

Comments on stochastic acceleration of cosmic rays. See Abstr. 143.040.

Galactic spiral shocks: vertical structure, thermal phase effects, and self-gravity. See Abstr. 151.026.

Recent developments in the mathematical investigation of the initial value problem of stellar dynamics and plasma physics. See Abstr. 151.077.

063 Radiative Transfer, Scattering

063.001 **Scaling laws for resonance line photons in an absorbing medium.** H. Frisch.
Astron. Astrophys., Vol. 87, 357 - 360 (1980).

Resonance line scattering in the presence of a source of continuous absorption is studied for very small values of the ratio of the continuous to line opacity, β. Scaling laws for the thermalization length, the thermalization frequency, the mean number of scatterings and the mean path length are extracted from an asymptotic analysis of the equation of transfer in the limit $\beta \to 0$. An interpretation is given for asymptotic scaling laws inferred from numerical data by Hummer and Kunasz (1978) and Bonilha et al. (1979).

063.002 **The multiband method in radiative transfer calculations.** D. E. Cullen, G. C. Pomraning.
J. Quant. Spectrosc. Radiat. Transfer, Vol. 24, 97 - 117 (1980).

The multiband method is developed for use in radiative transfer problems. The essence of the method is to divide the frequency range into both energy groups and cross section bands. Included in the formulation are the effects of continuously varying (in space and time) opacities, which leads to band-to-band streaming transfer terms. Several numerical examples indicate the increased accuracy possible by using a combination of groups and bands, as contrasted to groups alone.

063.003 **Scattering and absorption properties of CO_2 ice spheres in the region 360 - 4000 cm^{-1}.**
G. E. Hunt, E. A. Mitchell, H. H. Kieffer, R. Ditteon.
J. Quant. Spectrosc. Radiat. Transfer, Vol. 24, 141 - 146 (1980).

The authors present the results of Mie calculations for CO_2 ice particles of radius $a = 0.1$, 1 and 10 μm in the region 360 - 4000 cm^{-1}. Since CO_2 possesses a very small imaginary refractive index, the larger particles behave as conservative scatters. This is quite different from the equivalent water ice particles. The results presented in this study will be useful for investigation of the Martian atmosphere and surface.

063.004 **The F_N method for polarization studies – II. Numerical results.** J. R. Maiorino, C. E. Siewert.
J. Quant. Spectrosc. Radiat. Transfer, Vol. 24, 159 - 165 (1980).

The F_N method is used to establish numerical results basic to polarization studies in plane parallel atmospheres.

063.005 **Radiation transfer with synthetic scattering phase function – II.** S. A. El Wakil.
J. Quant. Spectrosc. Radiat. Transfer, Vol. 24, 179 - 183 (1980).

Equations connecting the transmission and reflection functions, from a finite medium to a semi-infinite medium, are used to reproduce recently published numerical results. The approach yields quite good agreement, especially for large τ_0 and $|1 - c| \ll 1$.

063.006 **On the numerical characteristics of an inverse solution for three-term radiative transfer.**
W. L. Dunn, J. R. Maiorino.
J. Quant. Spectrosc. Radiat. Transfer, Vol. 24, 203 - 209 (1980).

Certain numerical characteristics of an inverse formulation for three-term scattering radiative transfer are investigated. Specifically, approximate solutions to the direct problem are constructed by the F_N and Monte Carlo methods, allowing approximation of the various surface angular moments and related quantities needed for the inverse calculation. Several numerical schemes are employed in order to demonstrate the computational characteristics for some specific phase functions. The numerical results indicate that the single-scatter albedo can be calculated fairly consistently and accurately, but the higher order coefficients of the scattering law are more difficult to obtain by this method.

063.007 **On the transfer of line radiation in random magnetic fields.** R. Faulstich.
J. Quant. Spectrosc. Radiat. Transfer, Vol. 24, 229 - 236 (1980).

063.008 **On the transfer of polarized radiation in inhomogeneous media.** D. B. Wilson.
Mon. Not. R. Astron. Soc., Vol. 192, 787 - 797 (1980).

It is demonstrated that the equation for the transfer of polarized radiation in inhomogeneous media expressed in terms of Stokes parameters proportional to energy flux can be taken to be the same as that in homogeneous media provided that the generalized Faraday rotation length-scale and the scale-length of the plasma are much greater than the wavelength of the radiation. A simple example is discussed to illustrate the behaviour.

063.009 **Exact solution of a basic equation in finite atmosphere by the method of Laplace transform and linear singular operators.** R. N. Das.
Astrophys. Space Sci., Vol. 71, 25 - 35 (1980).

A finite atmosphere having distribution of intensity at both surfaces with definite form of scattering function and source function is considered. The basic integro-differential equation for the intensity distribution at any optical depth is subjected to the finite Laplace transform to have linear integral equations for the surface quantities under interest. These linear integral equations are transformed into linear singular integral equations by use of the Plemelj's formulae. The solution of these linear singular integral equations are obtained in terms of the X-Y equations of Chandrasekhar by use of the theory of linear singular operators which is applied in Das (1978).

063.010 **Exact solution of a simple time-dependent integro-differential equation by the method of Laplace transform and the theory of linear singular operators.**
R. N. Das.
Astrophys. Space Sci., Vol. 71, 37 - 43 (1980).

The simplest form of the equation of transfer for a time dependent radiation field in finite atmosphere is considered. This equation of transfer is an integro-differential equation, the solution of this equation is based on the theory of separation of variables, the Laplace transform and the theory of linear singular operators. The emergent intensities from the bounding faces of the finite atmosphere are determined in terms of X-Y equations of Chandrasekhar.

063.011 **Determination of the single-scattering albedo of a dense Rayleigh-scattering atmosphere with true absorption.** N. J. McCormick.
Astrophys. Space Sci., Vol. 71, 235 - 238 (1980).

A procedure is developed to determine the single-scattering albedo from polarization measurements of the angle-dependent intensity at two locations within, or on the boundaries of, a homogeneous finite atmosphere which scatters radiation according to Rayleigh's law with true absorption. The density of the atmosphere need not be known.

063.012 **Formation of universal and diffusion regions of non-linear spectra of relativistic electrons in spatially limited sources.** V. M. Kontorovich, A. E. Kochanov.
Astrophys. Space Sci., Vol. 71, 265 - 293, 295 - 324 (1980).
In English and Russian.

It is demonstrated that in the case of hard injection of relativistic electrons accompanied by the joint action of synchrotron (Compton) losses and energy-dependent spatial diffusion, a spectrum with 'breaks' is formed containing universal (with index $\gamma = 2$) and diffusion regions, both independent of the injection spectrum. The effect from non-linearity of the electron spectrum is considered in averaged electromagnetic spectra for various geometries of sources (sphere, disk, arm). It is shown that a universal region (with index $\alpha = 0.5$) can occur in the radiation spectrum.

063.013 **Light scattering by an optically thin inhomogeneous, spherically-symmetric planetary atmosphere: brightness at the zenith near the terminator.**
S. J. Wilson, K. K. Sen.
Astrophys. Space Sci., Vol. 71, 405 - 410 (1980).

The approximate method of representing the intensity by a three-stream angular division is used to compute the brightness at the zenith near the terminator of an optically thin, spherically-symmetric planetary atmosphere illuminated by parallel solar radiation. The results are compared with those of Sobolev obtained under classical Eddington approximation.

063.014 **Resonance-line polarization. VI. Line wing transfer calculations including excited state interference.**
L. H. Auer, D. E. Rees, J. O. Stenflo.
Astron. Astrophys., Vol. 88, 302 - 308 (1980).

A heuristic theory of polarized radiative transfer is developed for the wings of solar resonance lines. Magnetic fields are neglected. The theory includes quantum mechanical interference between $j = 1/2$ and $3/2$ excited states of line transitions sharing a common $j = 1/2$ ground state. Examples of such lines are Ca II H and K, Na I D_1 and D_2, and Mg II h and k. Calculations are made with the HSRA solar model for these lines as well as the dipole-type transition Ca I 4227 which is not affected by interference. The results for Ca I 4227, Ca II H and K and Na I D_1 and D_2 compare very well with recent observations, lending support to the authors' theory. The polarization predicted in the Mg II h and k lines is the largest of all indicating these lines to be prime candidates for linear polarization observations in the UV spectrum.

063.015 **Coaxial radiative and convective heat transfer in gray and nongray gases.** A. T. Mattick.
J. Quant. Spectrosc. Radiat. Transfer, Vol. 24, 323 - 334 (1980).

Coupled radiative and convective heat transfer is investigated for an absorbing gas flowing in a finite length channel and heated by blackbody radiation directed along the flow axis. The problem is formulated in one dimension and numerical solutions are obtained for the temperature profile of the gas and for the radiation escaping the channel entrance, assuming both gray and nongray absorption spectra.

063.016 **On path lengths of integration in the computation of the Voigt function.** T. B. Andersen.
J. Quant. Spectrosc. Radiat. Transfer, Vol. 24, 341 - 342 (1980).

Path lengths are discussed for the computation of the Voigt function in a stellar atmosphere using numerical integration and quadrature.

063.017 **Non-thermal electron scattering in a homogeneous plasma.** T. Bai.
Bull. American Astron. Soc., Vol. 12, 482 (1980). – Abstract.

063.018 **Thermal and vacuum effects on the cyclotron process in X-ray pulsars.** J. Kirk, P. Mészaros.
Bull. American Astron. Soc., Vol. 12, 514 (1980). – Abstract.

063.019 **Comoving frame calculations of spectral lines formed in rapidly expanding media with the partial frequency redistribution function for zero natural line width.**
A. Peraiah.
J. Astrophys. Astron., Vol. 1, 3 - 16 (1980).

Comoving frame calculations have been performed by using the angle-averaged partial frequency redistribution function to obtain flux profiles which can be compared with observations. The profiles calculated with line emission ($\epsilon = 10^{-3}$) resemble those observed in some quasars.

063.020 **Lines formed in rotating and expanding atmospheres.** A. Peraiah.
J. Astrophys. Astron., Vol. 1, 17 - 23 (1980).

The effects of rotational velocities on the formation of spectral lines formed in an atmosphere with radial velocities are investigated. Radial motion of the gases introduces a P Cygni type-shape and rotational motion increases the emission on either side of the centre of the line although the line remains asymmetric.

063.021 **On scattering of light by small nearly spherical particles.** Yu. V. Aleksandrov, V. P. Tishkovets.
Vestn. Khar'k. Univ., No. 190. Fiz. Luny Planet. Fundam. Astrometr., Vyp. 14, p. 9 - 15 (1979). In Russian.

063.022 **A probabilistic approach to radiative energy loss calculations for optically thick atmospheres: hydrogen lines and continua.** R. C. Canfield, P. J. Ricchiazzi.
Astrophys. J., Vol. 239, 1036 - 1044 (1980).

The authors simultaneously solve an approximate probabilistic radiative transfer equation and the statistical equilibrium equations for a model hydrogen atom consisting of three bound levels and ionization continuum. They explicitly solve the transfer equation for Lα, Lβ, Hα, and the Lyman continuum, assuming complete redistribution. The authors have tested the accuracy of this approach by comparing source functions and radiative loss rates to values obtained with a method that solves the exact transfer equation. Two recent model solar-flare chromospheres are used for this test.

063.023 **Dielectronic recombination rates, ionization equilibrium, and radiative emission rates for calcium and nickel ions in low-density high-temperature plasmas.**
V. L. Jacobs, J. Davis, J. E. Rogerson, M. Blaha, J. Cain, M. Davis.
Astrophys. J., Vol. 239, 1119 - 1130 (1980).

First the authors outline the extended version of the theory of dielectronic recombination and present the results of their calculations of the total dielectronic recombination rates for Ca and Ni ions. Particular emphasis is given to the effect of autoionization to excited states on the relative importance of $\Delta n = 0$ and $\Delta n \neq 0$ stabilizing radiative transitions of the recombining ion core. They present the results of their calculations for the corona ionization equilibrium abundances which were obtained by using the new dielectronic recombination rates. Then the authors give their results for the radiative emission rates for resonance line radiation, direct recombination radiation, dielectronic recombination radiation, and bremsstrahlung. Results are also presented for the temperature dependence of some relative line intensities in Ca and conclusions are discussed.

063.024 **Angle-dependent frequency redistribution in a plane-parallel medium: external source case.**
J.-S. Lee, R. R. Meier.
Astrophys. J., Vol. 240, 185 - 195 (1980).

A model of partial frequency redistribution for resonant scattering in a plane-parallel, optically thick atmosphere has been developed, which exactly takes into account the angular dependence of the redistribution function. The excitation source is assumed to be a continuum, external to the medium. A Monte Carlo method is employed to simulate the scattering of photons by two-level Maxwell-Boltzmann atoms with discrete lower atomic states and broadened upper states. Media with line center optical depths of up to 10^6 are considered. An unexpected depletion of photons in the near wings of back-scattered emission lines from very optically thick media, which the authors found in earlier work, has been verified. Its presence suggests the possibility of direct measurement of the redistribution function.

063.025 **Mesoturbulence.** G. Traving.
Stellar turbulence, (see 012.008), p. 172 - 182 (1980).

The influence of a stochastic velocity field with a finite scale length l on the transfer of line radiation is described by means of a generalization of the transfer equation. Micro- and macroturbulence are contained in this mesoturbulence approach as limiting cases $l \to 0$ and $l \to \infty$ respectively.

063.026 **Stochastic approach.** H.-P. Gail.
Stellar turbulence, (see 012.008), p. 183 - 194 (1980).

A general formalism for describing the radiation transfer in a medium with arbitrary velocity fields is presented. It is demonstrated that classical microturbulence and mesoturbulent models based on Markov processes can be considered as the two lowest order members within a hierarchy of model equations with an increasing degree of approximation to reality. Some preliminary results concerning the relevance of low order model equations are presented.

063.027 **The application of mesoturbulence to stellar atmospheres.** E. Sedlmayr.
Stellar turbulence, (see 012.008), p. 195 - 210 (1980).

For realistic stellar atmospheres the equations describing mesoturbulent line formation are solved numerically. The general dependence of theoretical line profiles and equivalent widths on the correlation length l and the mean square turbulent velocity σ is demonstrated. Also empirical relations between the basic parameters of the micro-macroturbulence description (v_{mic}, v_{mac}) and the fundamental mesoturbulence parameters (l, σ) are derived.

063.028 **Effects of acoustic waves on spectral line profiles.** L. E. Cram.
Stellar turbulence, (see 012.008), p. 211 (1980). – Abstract.

063.029 **Non-thermal emission mechanisms in astronomy.** K. C. Westfold.
Proc. Astron. Soc. Australia, Vol. 3, 195 - 199 (1978).

063.030 **Boundary effects and the circular polarization of synchrotron sources.** D. B. Melrose.
Proc. Astron. Soc. Australia, Vol. 3, 229 - 231 (1978).

063.031 **Radiative transfer through an arbitrarily thick, scattering atmosphere.**
A. H. Karp, J. Greenstadt, J. A. Fillmore.
J. Quant. Spectrosc. Radiat. Transfer, Vol. 24, 391 - 406 (1980).

A method is presented for solving the equation of radiative transfer in a vertically inhomogeneous, planetary atmosphere. The method, based on the spherical harmonics expansion, can be used to compute models with an arbitrarily large optical thickness and any scattering phase function. It is extremely efficient, requiring the equivalent of only two matrix multiplications per layer. This efficiency combined with its stability makes the method useful for computing realistic models for planetary atmospheres. To illustrate the range of validity of this method the authors compute the plane albedo from model atmospheres containing clouds with optical thicknesses ranging from 0 to 10^6.

063.032 **Generalization of the Curtis-Godson approximation to inhomogeneous scattering atmospheres.**
J. C. Buriez, Y. Fouquart.
J. Quant. Spectrosc. Radiat. Transfer, Vol. 24, 407 - 419 (1980).

The Curtis-Godson approximation, which was initially derived to compute the transmission through clear inhomogeneous atmospheres, has been generalized to any cloudy atmosphere by means of the scaled amount distribution. For most of the realistic atmospheres, the accuracy of this generalized approximation is comparable to or even better than for the clear case.

063.033 **A fast, exact code for scattered thermal radiation compared with a two-stream approximation.**
A. C. Cogley, D. K. Pandey, R. W. Bergstrom.
Icarus, Vol. 43, 96 - 101 (1980).

For radiatve transfer in plane-parallel emitting, absorbing, and scattering media, the two-stream approximation, and its various modifications or related methods, is probably mathematically the most simple to use. Unfortunately this physical approximation produces errors that are neither analytically known nor controllable. For externally (Sun) driven problems, many error studies exist for reflectivity, transmissivity, and certain defined albedos. A two-stream accuracy study for internally (thermal) driven problems is presented in this paper by comparison with a recently developed "exact" adding/doubling method.

063.034 **Vacuum polarization effects on Compton scattering by unmagnetized electrons in an ambient magnetic field.** R. J. Stoneham.
Opt. Acta, Vol. 27, 537 - 544 (1980). – Abstr. in Phys. Abstr., Vol. 83, Abstr. 86631 (1980).

063.035 **Vacuum polarization effects on Thomson scattering in a strong magnetic field.** R. J. Stoneham.
Opt. Acta, Vol. 27, 545 - 548 (1980). – Abstr. in Phys. Abstr., Vol. 83, Abstr. 86632 (1980).

063.036 **Scattering by nonspherical particles of size comparable to a wavelength: a new semi-empirical theory and its application to tropospheric aerosols.**
J. B. Pollack, J. N. Cuzzi.
J. Atmos. Sci., Vol. 37, 868 - 881 (1980). – Abstr. in Phys. Abstr., Vol. 83, Abstr. 90220 (1980).

063.037 **Monte Carlo simulation of relativistic Comptonization.** F. Takahara.
Prog. Theor. Phys., Vol. 63, 1551 - 1566 (1980). – Abstr. in Phys. Abstr., Vol. 83, Abstr. 90566 (1980).

063.038 **On the numerical evaluation of the *H*-functions of transfer problems in multiplying media.**
S. R. Das Gupta, Z. Islam, B. Majee.
Astrophys. Space Sci., Vol. 72, 71 - 78 (1980).

Extensive tables of the values of *H*-functions $H^0(z,\omega)$ and $H_1^0(z,\omega)$ appropriate for the problems of radiative transfer in multiplying media characterized by $\omega > 1$, have been constructed correctly to the sixth decimal place for values of ω in the range 1.05–10. This accuracy has been attained with the aid of a 32-point Gaussian quadrature.

063.039 **The coherence of radiation by fast-moving charges.** J. Katz.
Astrophys. Space Sci., Vol. 72, 117 - 125 (1980).

The coherence of the radiation from a cubic lattice of Q charges is evaluated at all wavelengths. The novelty of a second characteristic length, the distance between charges, adds a domain of wavelength in which enormous amplifications of power may exist ($Q^{1/3}$ times that of incoherent radiation). This amplification may be significant even if the charges are only poorly ordered because Q is extremely high in astrophysically relevant conditions.

063.040 **The complete solution for the scattering of polarized light in a Rayleigh and isotropically scattering atmosphere.** C. E. Siewert, J. R. Maiorino.
Astrophys. Space Sci., Vol. 72, 189 - 201 (1980).

The F_N method is used to solve, in a concise manner, the complete problem concerning the diffusion of polarized light in a plane-parallel Rayleigh and isotropically scattering atmosphere.

063.041 **A simple derivation of the radiation forces felt by scattering particles.** J. A. Burns, S. Soter.
Solid particles in the solar system, (see 012.019), p. 281 - 284 (1980).

The radiation pressure (RP) felt by a perfectly absorbing particle is due to the momentum withdrawn each second from the beam. The Poynting-Robertson (PR) drag is produced since the particle continually absorbs mass in the form of radiation, which, upon re-emission, has the same mean momentum density as the particle itself. The authors find that, relative to the force felt by a perfectly absorbing particle, the RP+PR forces felt by a scattering particle must be multipled by Q_{pr}, the radiation pressure coefficient, which can be evaluated from Mie theory.

063.042 **Spectral line formation in an atmosphere with exponential distribution of sources.**
G. M. Arutyunyan.
Dokl. AN ArmSSR, Vol. 70, 41 - 45 (1980). In Russian.
Abstr. in Ref. zh., 51. Astron., 9.51.152 (1980).

063.043 **Initial-value solution of the searchlight problem with anisotropic scattering: diffuse reflection.** S. Ueno.
13th Lunar and Planetary Symposium, (see 012.018), p. 1 - 10 (1980).

An initial-value solution of the searchlight problem is presented for finite, vertically inhomogeneous atmospheres with anisotropic scattering. With the aid of the integral operator method, the Riccati-type of nonlinear integro-differential equation for the scattering function is found in the case of the absorbing underlying surface.

063.044 **The diffusivity factor re-examined.**
J. P. Apruzese.
J. Quant. Spectrosc. Radiat. Transfer, Vol. 24, 461 - 470 (1980).

The issue of the optimum diffusivity factor for use in atmospheric heating calculations is re-examined in detail. It is shown that the value of this factor depends on the method by which the heating is calculated.

063.045 **A comparison of fast codes for the evaluation of the Voigt profile function.**
J. T. Twitty, P. L. Rarig, R. E. Thompson.
J. Quant. Spectrosc. Radiat. Transfer, Vol. 24, 529 - 532 (1980).

Several fast codes for the evaluation of individual Voigt profile functions are compared for accuracy and speed. An apparent error in one of these codes is discussed, along with a suggested correction. A general comment regarding the use of such codes in line-by-line radiative transfer programs is made.

063.046 **On the formation of Fe II lines in stellar spectra. I. Solar spatial intensity variation of λ3969.4.**
L. E. Cram, R. J. Rutten, B. W. Lites.
Astrophys. J., Vol. 241, 374 - 384, plate 2 (1980).

The authors employ high-spatial-resolution solar observations of the weak Fe II λ3969.4 line to study non-local thermodynamic equilibrium effects in Fe II line formation. Observed profiles of the Fe II resonance lines in the UV are used to define formation parameters in a 15-level atomic model computation, which shows that Fe II subordinate lines are generally formed out of local thermodynamic equilibrium as a result of pumping by UV line-wing photons from the deep photosphere. For the λ3969.4 line, this pumping results in large sensitivity to the atmospheric structure in layers deeper than the layer of formation of the H-wing background intensity. The authors discuss the absence of intense emission cores in the Fe II resonance lines, the effects of partially coherent scattering, and the effects of chromospheric and photospheric inhomogeneities. They find that emission of λ3969.4 provides a diagnostic of the inhomogeneous structure of the deep photosphere, for the Sun and for late-type stars.

063.047 **Two-dimensional radiative transfer. II. The wings of Ca K and Mg *k*.** S. P. Owocki, L. H. Auer.
Astrophys. J., Vol. 241, 448 - 458 (1980).

The effect of horizontal radiative transfer on the Ca K and Mg *k* line wing intensities in two-component models of the solar atmosphere is investigated. No significant influence on the spatially unresolved wing profiles of either line was found even for models in which the lateral variation was extreme over distances approaching a vertical scale height. The major conclusion of this work is that the Ca K wing intensity can, in principle, retain a temperature-induced lateral contrast down to horizontal scales approaching an opacity scale height.

063.048 **Broadening of Non-LTE lines by a turbulent velocity field with a finite correlation length.**
C. Froeschlé, H. Frisch.
Astron. Astrophys., Vol. 91, 202 - 211 (1980).

Profiles of Non-LTE lines broadened by a turbulent velocity field with a finite correlation length are calculated numerically with the "effective source function" method introduced in Frisch and Frisch (1976). This method allows to reformulate, for lines formed under complete frequency redistribution, stochastic transfer as a standard Non-LTE problem with an effective escape probability given by the average over all realizations of the velocity field of the usual escape probability. Effective source functions and mean emergent profiles are calculated for a two-level atom with two choices of the thermal source (isothermal and exponential mimicking a chromospheric temperature rise) and two choices of the line profile (Doppler and Voigt). It is also shown that a micro-macro-turbulent model cannot reproduce satisfactorily a turbulent velocity field with a finite correlation length.

063.049 **Small-scale velocity fields and mean line profiles.**
C. J. Durrant.
Astron. Astrophys., Vol. 91, 251 - 253 (1980).

The mean equation of transfer for a line formed in the presence of a small-scale velocity field is derived. Approximate expressions for the effects resulting from fluctuating quantities are illustrated by the case of short-period waves. Correlations between fluctuations cannot be neglected. It is concluded that mean line profiles are best generated by averaging the time variation of accurate individual profile sequences.

063.050 **The Cerenkov line radiation and the emission-line spectra of QSOs.** J.-h. You, F.-h. Cheng.
Acta Phys. Sinica, Vol. 29, 927 - 936 (1980). In Chinese.
Abstr. in Phys. Abstr., Vol. 83, Abstr. 105378 (1980).

063.051 **Light scattering by crystals of NH_3 and H_2O.**
A. Holmes, R. Paxman, H. P Stahl, M. Tomasko.
Bull. American Astron. Soc., Vol. 12, 705 - 706 (1980).
Abstract.

063.052 **Scattering by ensembles of small particles. Experiment, theory and application.**
B. Aa. S. Gustafson.
Rep. Obs. Lund, No. 17, 133 pp. (1980). ISSN 0349-4217.

A hypothetical selfconsistent picture of evolution of

prestellar interstellar dust through a comet phase leads to predictions about the composition of the circum-solar dust cloud. Scattering properties of thus resulting conglomerates with a "bird's-nest" type of structure are investigated using a micro-wave analogue technique. Approximate theoretical methods of general interest are developed which compare favorably with the experimental results. The principal features of scattering of visible radiation by zodiacal light particles are reasonably reproduced. A component which is suggestive of β-meteoroids is also predicted.

063.053 **Light scattering by spheroidal grains.** T. Onaka.
Ann. Tokyo Astron. Obs., Second Ser., Vol. 18, 1 - 54 (1980).

Light scattering by a spheroid with an impurity band is calculated for a wide range of grain parameters on the basis of the embedded-cavity model. The degree of asymmetry and the polarization change relative to the extinction change within the impurity band provide a strong constraint on the grain parameters, such as shape, size orientation and thickness of coating.

063.054 **On computing eigenvalues in radiative transfer.** C. E. Siewert.
J. Math. Phys., Vol. 21, 2468 - 2470 (1980). – Abstr. in Phys. Abstr., Vol. 84, Abstr. 4763 (1981).

063.055 **The similarity principle in the non-stationary radiation field.** M. Matsumoto.
Publ. Astron. Soc. Japan, Vol. 32, 629 - 638 (1980).

With the aid of the uniqueness of the solution to the non-stationary equation of transfer in an arbitrary geometry, the similarity principle that gives a functional relation for internal intensity in an absorbing and a non-absorbing homogeneous atmosphere is derived with allowance for quite arbitrary time parameters; one is the mean time spent by a photon in an absorbed state, and the other is the mean free time of the photon.

063.056 **Resonance radiative transfer for cyclotron line emission with recoil.**
I. Wasserman, E. Salpeter.
Astrophys. J., Vol. 241, 1107 - 1121 (1980).

The authors discuss the radiative transfer of cyclotron line photons through an optically thick atmosphere of resonant scattering electrons. Because of electron recoil, which is found to be important for cyclotron line emission from accreting neutron stars, most line photons escape in the red wing of the line, forming a single sharp spectral feature rather than the twin-peaked absorption profile characteristic of resonance radiative transfer without recoil. Implications of their results for the observed cyclotron emission line in the pulsed X-ray spectrum of Her X-1 are discussed.

063.057 **Thermal effects on the cyclotron line formation process in X-ray pulsars.**
J. G. Kirk, P. Mészáros.
Astrophys. J., Vol. 241, 1153 - 1160 (1980).

The authors derive expressions for the scattering and absorption cross sections of photons in a hot plasma including the effects of vacuum polarization in a strong magnetic field. Near the cyclotron resonance, these expressions depart significantly from previous cold plasma calculations. An approximate calculation of the radiative transfer is presented, and cyclotron absorption and emission-line fits are compared. This simplified transfer model suggests that an emission-line interpretation is slightly more probable at this stage.

063.058 **A reinvestigation of the redistribution functions R_{III} and R_{IV}.** S. J. McKenna.
Astrophys. J., Vol. 242, 283 - 293 (1980).

The behavior of the Hummer partial redistribution function R_{III} as a function of the scattering angle, and incoming and outgoing frequencies is investigated. It is found that, when scattering angle tends to 0 or π, the photons do not scatter coherently, as was implied by Reichel and Vardavas, but are partially redistributed in frequency. A quantum-mechanically consistent counterpart to R_{IV} is derived. The form of the redistribution function is more complicated, and its behavior as a function of scattering angle, and incoming and outgoing frequencies is discussed.

063.059 **A new approximation for the high magnetic field Compton cross-section.** R. Lieu.
Astrophys. Space Sci., Vol. 73, 481 - 498 (1980).

A mildly relativistic quantum-mechanical treatment of the Compton scattering cross-section in a strong magnetic field regime appropriate to observed electron temperature and field strengths in a neutron star magnetosphere is provided. The approximations used in the evaluation of the electron propagator are different from those employed previously and lead finally to simpler and more usable expressions for the cross-section. Various conceptual difficulties with the quantum cyclotron line mechanism are also discussed and the problem of the translational invariance of the Hamiltonian commonly used is addressed.

063.060 **Anisotropic light scattering in an inhomogeneous atmosphere. Asymptotic separation of angular variables in an optically thick layer.** Eh. G. Yanovitskij.
Astron. Zh., Tom 57, 1277 - 1286 (1980). In Russian. English translation in Soviet Astron., Vol. 24, No. 6.

Separation of angular variables in the problem of light scattering in a plane optically thick inhomogeneous atmosphere is studied. Relations connecting the solution of the Milne problem with the problem of diffuse reflection are found. The cases of a nearly conservative atmosphere and of an atmosphere overlaying a reflecting surface are considered separately.

063.061 **R_{II} partial frequency redistribution function and its effects on the formation of lines in expanding spherical atmospheres.** A. Peraiah.
Kodaikanal Obs. Bull., Ser. A, Vol. 2, 203 - 212 (1979).

The author has compared the profiles calculated with complete redistribution (CRD) and partial redistribution (PRD) function of R_{II} in a spherically symmetric expanding medium. Those profiles formed in CRD show more emission and deeper absorption wherever these occur compared to those formed by the PRD function. Profiles formed in a spherically symmetric medium show larger wing emission due to curvature scattering. The extended and expanding medium with emission in the continuum exhibits P Cygni type profiles.

063.062 **Solution of the radiative transfer equation in spherically symmetric media with spherical harmonic approximation.** A. Peraiah.
Kodaikanal Obs. Bull., Ser. A, Vol. 2, 230 - 239 (1979).

A numerical method for obtaining a solution of the radiative transfer equation in spherically symmetric media with spherical harmonic approximation is presented. The angle derivative is approximated by an orthonormal polynomial and this is represented by a matrix called curvature matrix, for a given beam of rays. An error analysis of the curvature matrix and results using the solution for few representative cases have been presented.

063.063 **Comoving frame calculations with Lorentz profiles in radially expanding media.**
A. Peraiah, G. Raghunath.
Kodaikanal Obs. Bull., Ser. A, Vol. 2, 240 - 251 (1979).

063.064 **Radiation pressure in resonance lines.**
A. Peraiah.

Kodaikanal Obs. Bull., Ser. A, Vol. 2, 260 - 262 (1979).

The effects of large-scale gas motions on the radiation pressure in a resonance line are investigated.

063.065 **Comments on the source function equality in (Zeeman)-multiplets.** L. G. Stenholm, R. Wehrse.
J. Astrophys. Astron., Vol. 1, 97 - 100 (1980).

The conditions for the source functions of a multiplet to be equal are studied for plasmas with and without magnetic fields. It is found that the source function equality holds – in addition to the case of collisional predominance – only when the redistribution functions are all identical and no interlocking with other lines occurs. When magnetic fields are present, the assumption of source function equality leads to a violation of the invariance conditions of the scattering matrix and should therefore not be made.

063.066 **An iterative simultaneous solution of the equations of statistical equilibrium and radiative transfer in the comoving frame.** A. Peraiah.
J. Astrophys. Astron., Vol. 1, 101 - 111 (1980).

A direct iteration has been performed to obtain a simultaneous solution of the equations of line transfer in an expanding spherically symmetric atmosphere in the comoving frame with statistical equilibrium for a non-LTE, two-level atom. The solution converges in three or four iterations to an accuracy of 1 per cent of the ratio of the population densities of the two levels. As initial values, the upper level population was set equal to zero or to LTE densities. The final solution on convergence indicates enhanced population of these levels over the initial values assumed. Large velocity gradients enhance this effect whereas large geometrical sizes of the atmospheres tend to reduce it.

063.067 **Effects of high velocities on photoionization lines.** A. Peraiah, G. Raghunath.
J. Astrophys. Astron., Vol. 1, 113 - 117 (1980).

The authors have treated the formation of spectral lines in a comoving frame where photoionization is predominant over collisional processes. The calculated line profiles are those seen by an observer at infinity. *P* Cygni-type profiles are observed in the case of a medium with no continuum absorption. For a medium with continuum absorption double peaked asymmetric profiles are noticed when the velocities are small; the two emission peaks merge into a single asymmetric peak for larger velocities.

063.068 **Radiative transfer in dust clouds. I. Hot-centered clouds associated with regions of massive star formation.** M. Rowan-Robinson.
Astrophys. J., Suppl. Ser., Vol. 44, 403 - 426 (1980).

The author describes a program he has developed to solve the equation of radiative transfer in spherically symmetric dust clouds with as few assumptions as possible. He compares his results with those from various approximate solutions and presents detailed models for hot-centered dust and molecular clouds associated with regions of massive star formation.

Plasma astrophysics. Nonthermal processes in diffuse magnetized plasmas. Vol. 1: The emission, absorption and transfer of waves in plasmas. Vol. 2: Astrophysical applications. See Abstr. 003.076.

Light scattering by irregularly shaped particles. See Abstr. 012.063.

Radiative diffusion in stellar atmospheres. See Abstr. 064.007.

Formation of resonance doublet profiles in rapidly expanding envelopes. See Abstr. 064.066.

Formation of chromospheric resonance line profiles in supergiants. See Abstr. 064.106.

The theory of radiatively driven stellar wind. I. A physical interpretation. See Abstr. 064.107.

Radiative heat transfer in surface layers of neutron stars with a magnetic field. See Abstr. 066.504.

Radiative-dominated cooling of the flare corona and transition region. See Abstr. 074.042.

Two-stream approximations to radiative transfer in planetary atmospheres: a unified description of existing methods and a new improvement. See Abstr. 091.041.

Optical investigation of dust in the solar system. See Abstr. 106.021.

The outer atmospheres of cool stars. VII. High resolution, absolute flux profiles of the Mg II *h* and *k* lines in stars of spectral types F8 to M5. See Abstr. 114.165.

Elektronenstreuung in der einfallenden Hülle von S CrA. See Abstr. 121.019.

Line-formation – Rechnungen in der Supernovahülle 1969*l*. See Abstr. 125.501.

Analysis and interpretation of H I self-absorption lines. I. See Abstr. 132.037.

A Monte Carlo model for light scattering by dark nebulae. See Abstr. 134.022.

Polarization in reflection nebulae. III. How good a grain diagnostic? See Abstr. 134.025.

Radiative transfer in dusty nebulae. III. The effects of dust albedo. See Abstr. 134.035.

Astrophysical gamma-ray production by inverse Compton interactions of relativistic electrons. III. Cutoff effect for inverse Compton spectra applied to the case of the hard X-ray and gamma-ray emission of NGC 4151. See Abstr. 158.095.

The Penrose Photoproduction Scenario for NGC 4151; (PCS–SSC). A black hole γ-ray emission mechanism for active galactic nuclei and Seyfert galaxies. See Abstr. 158.105.

Gas close to the radiative continuum source(s) in active nuclei. See Abstr. 158.315.

064 Stellar Atmospheres, Stellar Envelopes, Mass Loss, Accretion

064.001 **On the structure and composition of the Wolf-Rayet atmospheres.** J. Sahade.
Astron. Astrophys., Vol. 87, L7 - L9 (1980).

It is suggested that the Wolf-Rayet stars are not H-deficient objects, that they probably have a normal chemical composition but that the Balmer lines are not observed because the physical conditions of the "extended envelopes" of the Wolf-Rayet stars are such that the photospheric lines cannot be "seen" – except apparently in late WN's – nor H lines are formed in the layers of the "extended envelope" that can be investigated in the conventional region of the spectrum. It is further suggested that the "extended envelope" in a Wolf-Rayet star is a sort of stellar corona with two regions of maximum T_e and that the transition zone with the photosphere, if any, must be extremely narrow.

064.002 **Chromospheres of F, G, K type stars. IV. The zone model of heating.**
Z. Musielak, J. Sikorski.
Acta Astron., Vol. 30, 167 - 182 (1980).

The authors propose the so called "zone model" of heating in which the chromosphere is divided into three regions corresponding to dissipation of energy by acoustic, fast, and slow plus Alfvén waves, respectively. They also introduce a phenomenological χ parameter simulating a development of wave into shock on a finite distance. The influence of magnetic field, wave period, and χ parameter on the amount of dissipated energy is investigated for known semiempirical chromospheric models.

064.003 **Self accreting stellar winds.** R. L. T. Wolfson.
Mon. Not. R. Astron. Soc., Vol. 192, 881 - 889 (1980).

A two-dimensional theory of axisymmetric accretion about a gravitating mass point is extended to cover outflows as well. Symmetry properties and asymptotic solutions are established which suggest the possibility that outflow and accretion may occur in different regions of the same flow field. The simplest solution includes an outflow column coupled to an accretion column, with the two columns extending in opposite directions from the mass point. Asymptotic and numerical analyses suggest that the solutions displayed are the first in a family of increasingly complex outflow/accretion situations.

064.004 **Thick accretion disks and supercritical luminosities.**
B. Paczyński, P. J. Wiita.
Astron. Astrophys., Vol. 88, 23 - 31 (1980).

The authors present the basic equations they use in studying stationary accretion disks of arbitrary thickness. The assumptions made and the specialized equations needed by their approach in obtaining the shape and luminosity of thick disks are given. Specific cases are calculated and their results are summarized. The authors draw conclusions and point out modifications that should be included in more refined calculations.

064.005 **On penetrative instabilities.** S. K. Pandey.
Astrophys. Space Sci., Vol. 71, 499 - 506 (1980).

The investigation of instabilities in a penetrative atmosphere discussed in an earlier paper (Pandey et al., 1979) is extended to include thermal dissipation in a more realistic approximation. It is shown that the convective modes are not very sensitive to the presence of an overlying stable layer. The overstabilization of gravity modes is, however, found to occur under very stringent conditions, while the acoustic overstability is enhanced by the presence of an overlying stable layer.

064.006 **An envelope model of Phi Persei.**
M. Suzuki.
Publ. Astron. Soc. Japan, Vol. 32, 331 - 340 (1980).

Hynek's (1940, 1944) data on the radial velocity of φ Persei are reinterpreted under the assumption that the circumstellar envelope of the star is a gas ring in a stable periodic orbit of the restricted three-body problem. The masses of the primary and the secondary stars are deduced to be $20M_\odot$ and $4M_\odot$, respectively, with the separation of 1.42 AU. The effective gas ring with a radius of about 0.6 AU is also found to revolve around the primary B-type star in the orbital plane of the binary system.

064.007 **Radiative diffusion in stellar atmospheres.**
K. Nariai.
Publ. Astron. Soc. Japan, Vol. 32, 347 - 357 (1980).

The author obtains a static flux-constant model of stellar atmospheres solving the equation of radiative diffusion with an appropriate boundary condition at the bottom. The convergence of the iteration procedure is uniform but slow.

064.008 **Limb-darkening on distorted stars.**
E. J. Devinney, Jr.
Bull. American Astron. Soc., Vol. 12, 501 (1980). – Abstract.

064.009 **The structure of X-ray illuminated stellar atmospheres.** R. London, R. McCray, L. H. Auer.
Bull. American Astron. Soc., Vol. 12, 520 (1980). – Abstract.

064.010 **Mass-loss from the central star of NGC 6543.**
S. R. Heap.
Bull. American Astron. Soc., Vol. 12, 540 (1980). – Abstract.

064.011 **Atmospheres for hot, high-gravity stars. I. Pure hydrogen models.**
F. Wesemael, L. H. Auer, H. M. Van Horn, M. P. Savedoff.
Astrophys. J., Suppl. Ser., Vol. 43, 159 - 303 (1980).

An extensive grid of pure hydrogen model atmospheres for hot, high-gravity stars is presented. The models are intended to aid the analysis of visual, ultraviolet, and soft X-ray spectra of hot DA white dwarfs and EUV sources. The grid extends from $\log g = 4.0(1.0)9.0$ and T_{eff} ranges from 20,000 K up to the Eddington limit for most surface gravities. Most of the models are LTE unblanketed calculations, but selected NLTE models and blanketed LTE models have also been computed in order to assess the importance of these effects. For each model, continuum fluxes are tabulated covering the entire range of wavelengths for which there is significant flux. Strömgren colors, *UBV* colors, and bolometric corrections are also given. Profiles and equivalent widths of the $L\alpha$, $L\beta$, $L\gamma$, $H\alpha$, $H\beta$, $H\gamma$, and $H\delta$ lines are given as well.

064.012 **Chemical structure of circumstellar shells.**
J. M. Scalo, D. B. Slavsky.
Astrophys. J., Lett., Vol. 239, L73 - L77 (1980).

The authors discuss nonequilibrium chemistry in isotropic models of expanding circumstellar shells surrounding oxygen-rich red giant stars, using time-scale arguments and chemical kinetic calculations. The authors examine a simplified shell model to show that the abundances of many interesting molecules in oxygen-rich shells should exhibit rapid increases exterior to the radius at which OH is formed through photodissociation of H_2O by the interstellar ultraviolet radiation field if chromospheric radiation and shocks are unimportant. Results of chemical kinetic calculations for oxygen-rich shells are also reported.

064.013 **Relativistic accretion: the optically thick case.**
A. W. Gillman, R. F. Stellingwerf.
Astrophys. J., Vol. 240, 235 - 241 (1980).

The authors investigate the process of steady-state spherical accretion onto black holes when the infalling gas is optically thick to the outgoing radiation. The transfer of radiation is treated in the diffusion approximation, and the calculation is done relativistically so that the region near the Schwarzschild radius may be included. The luminosity is taken as a free parameter. It is found that, in cases of sufficiently high luminosity and mass flux, a radiation-broadened shock must form and prevent steady supersonic flow through the Schwarzschild radius. Fluctuations on time scales of microseconds to milliseconds could result.

064.014 **On the structure of Be star envelopes.** L. Luud. Tartu Inst. Astrofiz. Fiz. atmos., Prepr. A-3, 34 pp. (1980). In Russian.

From comparison of the calculated and observed intensities of Balmer, Paschen and Brackett emission lines it has been found that in the envelopes of Be stars nearly 10% of their volume consists of high density and optically thick matter and nearly 90% of low density and optically thin matter. It has been shown that the rapid variations of Hα and Hβ line profiles in Be stars with the characteristic time of the order of 10 minutes are real.

064.015 **Models of the P Cygni envelope. An analysis of the formation of hydrogen spectral lines.** I. Kolka. Tartu Inst. Astrofiz. Fiz. atmos., Prepr. A-4, 59 pp. (1980). In Russian.

Six kinematic models of the P Cygni envelope have been analyzed. It appears that only the model with three zones can explain all the observed features of Balmer-line profiles. The dependence of the kinetic temperature on the radial distance in this model can explain also the existence of low excitation metallic (Fe II, Ni II, Mn II) spectral lines found mainly in the ultraviolet spectral region.

064.016 **Star dust, mass loss and the late stages of stellar evolution.** S. Kwok. J. R. Astron. Soc. Canada, Vol. 74, 216 - 233 (1980).

The author discusses the role of dust grains in the thermal and dynamical structure of the circumstellar envelopes of late-type stars. It is suggested that radiation pressure on grains is responsible for the large-scale mass loss. Since the conditions necessary for grains to form are dependent on the chemical composition of the star, metallicity can be an important parameter in determining the end point of stellar evolution. The association of dust grains with stellar winds also has a significant effect on the interstellar medium. The observed amount of interstellar grains can be supplied by two processes: the ejection of silicate grains from M stars, and the ejection of carbon-based grains from the ionized winds of planetary nebulae and novae.

064.017 **Buoyancy effects in spherical accretion.** A. R. Garlick. Astron. Astrophys., Vol. 89, 48 - 56 (1980).

The effects considered here are those of buoyancy which may lead to gravity waves or convection. The work is guided by the methods used in stationary atmospheres and some discussion is devoted to the two approximations which are made: firstly an adiabatic condition and secondly a sound wave filtering approximation. The resulting equations are then investigated in three ways: for small scale perturbations, using a quasi-adiabatic approach and the exchange of stabilities method respectively. The results can be summed up as follows. Both gravity waves and convection tend to be inhibited by the spherical accretion motions, probably by advection of energy out of the region of interest. The consequences of these results for physical models of accretion are discussed briefly.

064.018 **The nature of SiO masers in late-type stars.** M. Elitzur. Astrophys. J., Vol. 240, 553 - 566 (1980).

It is shown that currently available theoretical models for SiO masers cannot explain all the recent observations, especially the maser emission from $\nu = 2$ and $\nu = 3$ states and the fact that the masers are observed in stars which cover a wide range of parameters, in particular mass loss rates. A new model for the SiO masers in late-type stars is therefore developed.

064.019 **Stellar convection theory.** J.-P. Zahn. Stellar turbulence, (see 012.008), p. 1 - 14 (1980).

064.020 **Instabilities in a polytropic atmosphere.** H. M. Antia, S. M. Chitre. Stellar turbulence, (see 012.008), p. 15 (1980). – Abstract.

064.021 **Thermal and continuum driven convection in B-stars.** G. D. Nelson. Stellar turbulence, (see 012.008), p. 16 (1980). – Abstract.

064.022 **Differential rotation in stars with convection zones.** P. A. Gilman. Stellar turbulence, (see 012.008), p. 19 - 37 (1980).

064.023 **Generation of oscillatory motions in the stellar atmosphere.** Y. Osaki. Stellar turbulence, (see 012. 008), p. 38 - 50 (1980).

Thermal overstability of non-radial eigenmodes of stars is discussed as one of possible causes for generating non-thermal motions in the stellar atmosphere. The nature of oscillatory motions in stars is first considered both in the local and the global standpoints. Then, the excitation of eigen-oscillations is discussed and results of numerical studies so far made are reviewed for the vibrational stability of various stellar models against non-radial oscillations. It is found that many of non-radial p-modes of high tesseral harmonics are likely excited in various stars of the HR diagram and that they possibly manifest themselves as non-thermal velocity fields in the stellar atmosphere.

064.024 **Turbulence in main sequence stars.** T. Gehren. Stellar turbulence, (see 012.008), p. 103 - 112 (1980).

064.025 **Observational aspects of macroturbulence in early type stars.** D. Ebbets. Stellar turbulence, (see 012.008), p. 113 - 125 (1980).

064.026 **Photospheric macroturbulence in late-type stars.** M. A. Smith. Stellar turbulence, (see 012.008), p. 126 - 135 (1980).

064.027 **Depth-dependence of turbulence in stellar atmospheres.** R. E. Stencel. Stellar turbulence, (see 012.008), p. 136 - 143 (1980).

064.028 **Microturbulence: age dependences.** R. Foy. Stellar turbulence, (see 012.008), p. 164 - 169 (1980).

064.029 **Turbulence in the atmosphere of B-type stars.** K. Kodaira. Stellar turbulence, (see 012.008), p. 170 - 171 (1980). Abstract.

064.030 **Mechanical energy transport.** R. F. Stein, J. W. Leibacher. Stellar turbulence, (see 012.008), p. 225 - 247 (1980).

064.031 **Stellar chromospheres.** J. L. Linsky. Stellar turbulence, (see 012.008), p. 248 - 277 (1980).

064.032 **Observations of the outer atmospheric regions of α Orionis.** A. P. Bernat, L. Goldberg.
Stellar turbulence, (see 012.008), p. 278 (1980). – Abstract.

064.033 **Stellar winds and coronae in cool stars.** A. K. Dupree, L. Hartmann.
Stellar turbulence, (see 012.008), p. 279 - 291 (1980).

Recent observational and theoretical results are reviewed that pertain to the presence and characteristics of stellar coronae and winds in late-type stars. It is found that stars – principally dwarfs – exist with "hot" coronae similar to the Sun with thermally driven winds. For stars at the lowest effective temperatures, and gravities characteristic of supergiant and giant stars, high temperature ($\sim 10^5$ K) atmospheres are absent (or if present are substantially weaker than in the dwarf stars), and massive winds are present. There also exist "hybrid" examples – luminous stars possessing both a "hot" corona and a supersonic stellar wind. Constraints for theoretical models are discussed.

064.034 **Relationship between envelope structure and energy source of non-thermal motions.** H. Ando.
Stellar turbulence, (see 012.008), p. 292 (1980). – Abstract.

064.035 **An analysis of microturbulence in the atmosphere of the F-type supergiant Gamma Cygni.**
A. A. Boyarchuk, L. S. Lyubimkov.
Stellar turbulence, (see 012.008), p. 292 - 293 (1980). Abstract.

064.036 **Excitation dependent gf-values and depth dependent microturbulences.** T. Hasegawa.
Stellar turbulence, (see 012.008), p. 294 (1980). – Abstract.

064.037 **Sur la stabilisation des oscillations d'une étoile dans le champ magnétique vertical et horizontal.**
N. Lungu.
Proceedings of the colloquium of astronomy, Cluj-Napoca, (see 012.009), p. 46 - 51 (1978).

The equations of the adiabatic pulsations for a magnetic star with slow rotation are considered. The condition of the stability is $P_o > P_m$. The case when only a certain superficial envelope of the star pulsates and the temperature gradient is constant is considered.

064.038 **Atmospheric shock effects in early B stars.** P. A. Stamford, R. D. Watson.
Proc. Astron. Soc. Australia, Vol. 3, 273 - 275 (1978).

064.039 **A suggestion concerning the generation of the physical state of stellar mantles.** A. B. Underhill.
Astrophys. J., Lett., Vol. 240, L153 - L156 (1980).

The physical state of the winds of B supergiants varies from one of a high level of ionization and rapid outflow to one of relatively low ionization and a moderate rate of outflow as one goes from type B0 to type A0. It is proposed that this is the result of magnetodynamic energy being released in the mantles of these stars, much as occurs in the sun. Data are presented to show that this idea is plausible. It is noted that the release of magnetodynamic energy as the result of interaction between magnetic fields and differential motion in an inhomogeneous structure is probably the cause of the heating in all outer stellar atmospheres (mantles).

064.040 **Calculation of the parameters on a strong shock wave front moving in a stellar envelope.**
I. A. Klimishin, B. I. Gnatyk.
Problems of cosmic physics. Vyp. 15, (see 003.003), p. 142 - 146 (1980). In Russian.

The properties of shock-wave adiabats are discussed taking into account the radiation pressure and density. A nomogram is proposed for the calculation of parameter jumps on the shock wave front moving in a pure hydrogen stellar envelope.

064.041 **Fluctuation theory of the mass flux from the stars: the refutations.** C. D. Andriesse.
Astrophys. Space Sci., Vol. 72, 167 - 173 (1980).

The third paper on the stellar mass loss as a fluctuation phenomenon addresses the refutations. It discusses the instability of the subphotosphere to outward mass flows, which develop into stochastic surges. By applying the principle of minimum free energy, the average outward velocity to which these flows are amplified is found to be just below the atmospheric escape velocity. Small fluctuations above this average value will therefore lead to mass loss. This justifies the Langevin equation, introduced previously to describe the mass loss, and a formal derivation of this equation is presented. Finally, the paper discusses deviations of the stellar relaxation time from GM^2/RL, when a strong concentration of matter is at the centre. These deviations explain the anomalous behaviour of the mass loss rate of red (super)giants, so that the essential prediction of the fluctuation theory for the loss rate is not refuted.

064.042 **Atmospheric structure of a carbon star with thin circumstellar shell.** M. Hirai.
Bull. Fukuoka Univ. Educ., Vol. 29, Part III, 99 - 104 (1979).

Results of a spectroscopic analysis for cool stars are discussed in view of the atmospheric structure. The overabundances of some metals are related to the ionization degree in the gas mixture. Assuming a simple model the effects of the chemical abundances are estimated on the outer layer of cool stars with circumstellar shell. As a result the circumstellar shell may significantly affect both the atmospheric structures and the chemical abundances of late type stars.

064.043 **Near-unstable supergiants.** C. de Jager.
ESA J., Vol. 4, 123 - 128 (1980).

064.044 **Thermochemical equilibrium in the atmospheres of cool stars. Atoms and ions.**
N. S. Komarov, V. V. Tsymbal.
Astron. Zh., Tom 57, 1010 - 1015 (1980). In Russian.
English translation in Soviet Astron., Vol. 24, No. 5.

A method of solving the equations of state equilibrium in atmospheres of late-type stars is described. The influence of abundances of chemical elements on the structure of the atmospheres has been studied as well as that of fundamental characteristics of the atmosphere on the concentration of atoms and ions as function of the optical depth.

064.045 **Stellar chromospheres.** J. L. Linsky.
Annu. Rev. Astron. Astrophys., Vol. 18, (see 003.004), 439 - 488 (1980).

Contents: What is a stellar chromosphere? In what regions of the H-R diagram do chromospheres exist? What trends are emerging from semiempirical chromospheric models of single stars? Are theoretical models of chromospheres becoming realistic? Why does the Wilson-Bappu relation work? Are there systematic flow patterns in stellar chromospheres? How are chromospheres in close binary systems different from chromospheres in single stars? Future prospects.

064.046 **Stability of accretion column flows.**
J. M. Hameury, S. Bonazzola, J. Heyvaerts.
Astron. Astrophys., Vol. 90, 359 - 365 (1980).

The authors show that under optically thick conditions, radiation-pressure dominated flows near the Eddington limit are, under the usual physical conditions, unstable to a Raleigh-Taylor type of instability. On the other hand, optically thin flows may remain homogeneous even on a large scale. It is concluded that realistic accretion models must take into account strong micro-inhomogeneity likely to be present in accretion columns.

064.047 **Spicular downflows and the transition to high mass-loss rates in G and K giants.**
S. G. Wallenhorst.
Astrophys. J., Vol. 241, 229 - 234 (1980).

The mass-loss rates and coronal base pressures of stars that have evolved up to and beyond the so-called high mass-loss rate transition locus have been examined, using an energy-balance model in which the coronal energy losses due to downflowing spicular material are included. The minimum flux assumption of A. G. Hearn is used to evaluate the jump in mass-loss rate expected when the sonic point of the stellar wind reaches the coronal base. In addition, this jump in mass-loss rate is estimated, using a temperature-height profile for a transition region dominated by spicular downflows to estimate the amount of downflowing material. Both methods give similar results, leading to the conclusion that as a star evolves across the transition locus, an order-of-magnitude increase in stellar mass-loss rate is to be expected.

064.048 **X-ray emission from the winds of hot stars.**
L. B. Lucy, R. L. White.
Astrophys. J., Vol. 241, 300 - 305 (1980) = Columbia Astrophys. Lab. Contrib. No. 184.

A phenomenological theory is proposed for the structure of the unstable line-driven winds of early-type stars. These winds are conjectured to break up into a population of blobs that are being radiatively driven through, and confined by ram pressure of an ambient gas that, because of shadowing by the blobs, is not itself being radiatively driven. With plausible choices for the theory's two free parameters, radiation from the bow shocks preceding the blobs can account for the X-ray luminosity of ζ Puppis. The theory breaks down, however, when used to model the much lower density wind of τ Scorpii, for then the blobs are destroyed by heat conduction from shocked gas.

064.049 **Nuclear burning in massive accretion disks.**
S. C. Vila.
Astrophys. J., Vol. 241, 355 - 357 (1980).

The author has constructed models for the vertical structure of disks around a black hole of $10^6 M_\odot$ at distances from it smaller than the tidal breakup radius. These disks are massive enough that nuclear burning occurs in their central layers. It has been found that convection occurs for disks with $\log T_{eff} > 4.25$, and the viscosity effect associated with it causes matter to spiral inward to the black hole.

064.050 **Reflexion and transmission of Alfvén waves in an atmosphere.** N. Bel, B. Leroy.
Solar and interplanetary dynamics, (see 012.020), p. 131 - 133 (1980).

064.051 **A comparison between the observed and predicted UV line blocking for blanketed model atmospheres of early type stars.**
F. Castelli, H. J. G. L. M. Lamers, F. Llorente de Andrés, E. A. Müller.
Astron. Astrophys., Vol. 91, 32 - 35 (1980).

For early B main-sequence and giant stars the observed near UV blocking factors are larger than those predicted by blanketed model atmospheres with a micro-turbulence of 2 km s^{-1}. Better agreement would be reached if a micro-turbulence of 4 to 5 km s^{-1} were adopted. Evidence was found that the line opacities due to doubly ionized metals are unsufficiently accounted for in the model atmospheres. For late B main-sequence and giant stars the observed near UV blocking is smaller than predicted which points to an overestimation of the line opacities due to singly ionized metals. The observed blocking in early B supergiants is much larger than predicted and would require a micro-turbulence of 10 km s^{-1} at least. If this large blocking extends to the far UV, the energy distribution given by the models might be changed drastically.

064.052 **Simulation of variable ultraviolet line blanketing in Ap Si stars.** J. Borsenberger, C. Jamar.
Astron. Astrophys., Vol. 91, 247 - 250 (1980).

The strong absorption which depresses the flux around 1400 Å in hot Ap Si stars relative to normal stars has been grossly simulated in NLTE models at T_{eff} = 13,000 and 15,000 K. The redistribution of the flux begins in the Balmer continuum where it is very strong on the red side of the null wavelength region; it renders the Paschen slope bluer than in the normal stars of the same T_{eff}. For the chosen opacity law, the difference in T_{eff} between normal models and blanketed models of same Paschen slope is 900 K. The Balmer discontinuity is systematically smaller in the blanketed models than in the comparison models of same Paschen slope.

064.053 **Stellar coronae.** G. S. Vaiana.
Highlights of Astronomy, Vol. 5, (see 012.025), 419 - 428 (1980).

064.054 **Chromospheres, coronae, and mass loss in stars hotter than the sun.** T. P. Snow, Jr.
Highlights of Astronomy, Vol. 5, (see 012.027), 525 - 531 (1980).

For the F stars, chromospheric indicators are present in the form of emission lines, seen in visible and ultraviolet wavelengths. Winds are present in A supergiants, but not in main sequence stars, although at least a few of the latter are X-ray sources, indicating the possible existence of coronae. Most OB supergiants are X-ray sources as well, indicating, along with the presence of super-ionization, that these stars have coronae. On the main sequence, the O stars and some B stars have mass loss with highly-ionized species in the wind. The winds in the O and B stars are commonly variable.

064.055 **Chromospheres, coronae and mass loss in solar and late-type stars.** C. Jordan.
Highlights of Astronomy, Vol. 5, (see 012.027), 533 - 540 (1980).

064.056 **Small-scale dissipative processes in stellar atmospheres.** J. W. Leibacher, R. F. Stein.
Highlights of Astronomy, Vol. 5, (see 012.027), 581 - 590 (1980).

The outer atmospheres of stars must be heated by some non-thermal energy flux to produce chromospheres and coronae. The authors discuss processes which convert the non-thermal energy flux or organized, macroscopic motions into random, microscopic (thermal) motions. Recent advances in the description of the chromosphere velocity field suggest that the acoustic waves observed there transmit very little energy, and hence are probably incapable of heating the upper chromosphere and corona. The apparent failure of this long held mechanism and the growing appreciation of the importance of strong magnetic fields in the chromosphere and corona have led to hypotheses of heating by the dissipation of currents (both oscillatory and quasi-steady).

064.057 **Stellar wind theories.** A. G. Hearn.
Highlights of Astronomy, Vol. 5, (see 012.027), 591 - 600 (1980).

064.058 **Mass-loss from spinning stars.**
S. R. Sreenivasan.
Highlights of Astronomy, Vol. 5, (see 012.027), 601 - 613 (1980).

The effects of mass-loss and angular momentum loss on the evolution of massive stars are discussed bringing out the main results as well as the limitations of recent studies. It is pointed out that an acceptable theory of stellar winds in early as well as late type stars is needed as well as a satisfactory assessment of a number of instabilities in these contexts for an adequate understanding of the evolutionary consequences for a wide variety of population I and population II stars, which are affected by mass-loss.

064.059 **On the differences at chromospheric levels between RS CVn-type binaries, active and quiet chromosphere single stars, and active and quiet regions in the sun.**
J. L. Linsky.
Highlights of Astronomy, Vol. 5, (see 012.031), 861 - 862 (1980).

064.060 **Mass transfer between binary stars.**
J. L. Modisette.
Highlights of Astronomy, Vol. 5, (see 012.031), 863 - 865 (1980).

064.061 **On the possible cause of observable underabundance of He in the atmospheres of Bp stars.**
V. L. Khokhlova.
Pis'ma Astron. Zh., Tom 6, 723 - 729 (1980). In Russian. English translation in Soviet Astron. Lett., Vol. 6.

The possibility of non-LTE excitation of He I in the atmospheres of hot chemically peculiar stars due to the interaction of UV lines of Fe III, Fe IV and Mn IV with the He I resonance line λ 584 is discussed. Some intensity anomalies of certain UV lines of these ions are predicted.

064.062 **Winds from supergiants.** S. F. Nerney.
Bull. American Astron. Soc., Vol. 12, 747 (1980). Abstract.

064.063 **Giant flares on supermassive accretion disks: polarization properties and energetics.**
S. Pineault.
Astrophys. J., Vol. 241, 528 - 533 (1980).

Some aspects of hot spots or giant flares on supermassive accretion disks are investigated, with particular emphasis on the circumstances under which the orbital motion of such a spot can lead to a quasi-linear rate of rotation of the polarization position angle. A tentative interpretation of recent polarization observations of AO 0235+164 is given, and some of the consequences are briefly discussed.

064.064 **Dipole confined by a disk.**
W. Kundt, M. Robnik.
Astron. Astrophys., Vol. 91, 305 - 310 (1980).

Exact solutions are given for a static magnetic dipole field in 2-dim and 3-dim space confined by an infinitesimally thin diamagnetic concentric plain disk. Whereas the 3-dim problem is relevant for accretion problems onto compact stars, the 2-dim problem is far easier to solve, and turns out to be a fair approximation. Computer drawings illustrate the field line geometry.

064.065 **Summary of solutions to the spherically symmetric accretion and stellar wind problems.** D. Summers.
Astrophys. Space Sci., Vol. 73, 3 - 9 (1980).

The stellar wind and accretion problems described by the steady, spherically symmetric continuum equations incorporating thermal conduction and viscosity are studied. A summary of solutions, including some new solutions, is presented and the solution properties are briefly examined.

064.066 **Formation of resonance doublet profiles in rapidly expanding envelopes.** J. Surdej.
Astrophys. Space Sci., Vol. 73, 101 - 158 (1980).

Generalization of the escape probability method introduced by Sobolev allows the author to study the transfer of spectral line radiation for a resonance doublet in rapidly expanding envelopes. For the cases of outward-accelerating (or equivalently inward-decelerating) and outward-decelerating (or equivalently inward-accelerating) envelopes he derives, in the frame of a three-level atom model, the expressions for the spectral radiation fields for the resulting radiative force exerted per atom and for the resonance doublet profile. For various physical and geometrical conditions prevailing in the expanding media, he discusses the behaviours of those quantities as well as their dependence on the parameters of the model. The author also stresses the importance of treating a resonance doublet as being formed by two distinct resonance transitions when evaluating the resulting radiative force acting on an atom. It is indeed shown that if one uses a two-level atom model to represent a resonance doublet – i.e., assigning to it an oscillator strength equal to the sum of the oscillator strengths of both resonance transitions – the amplitude of the resulting radiative force can be underestimated by factors reaching 100% and more in the regions of the expanding envelope which are optically thick to the spectral line radiation. In this context, it would be essential to revise the previous models of radiation-driven winds developed for early-type stars in which the lines belonging to any multiplet were treated as a single line.

064.067 **Lyman continuum photons emission from hot stars.**
H. A. Dottori.
Astrophys. Space Sci., Vol. 73, 175 - 177 (1980).

The number of Lyman continuum photons emitted from stars with temperatures between 15 000 K and 50 000 K for several values of the surface gravity are calculated on the basis of Kurucz's new models of stellar atmospheres. Results are compared with previous data.

064.068 **Profile of a line emitted by an accretion disk. Influence of the geometry upon its shape parameters.**
D. Gerbal, D. Pelat.
ESO Sci. Prepr. No. 100, 21 pp. (1980). – Submitted to Astron. Astrophys.

064.069 **Test computations of model stellar atmospheres with a new computer program.** J. Madej.
Acta Astron., Vol. 30, 249 - 258 (1980).

The paper gives a short description of an iterative FORTRAN program for calculating model stellar atmospheres of early type stars, including Balmer line blanketing. LTE equation of state and no convection is assumed. The program presented here can be run with small size digital computer (64 K of central memory required).

064.070 **Coronal dichotomies.** T. Ayres.
Smithsonian Astrophys. Obs., Spec. Rep. 389, (see 012.038), p. 65 - 78 (1980).

064.071 **Stellar coronae: interpretation and modeling of stellar activity.** R. Rosner.
Smithsonian Astrophys. Obs., Spec. Rep. 389, (see 012.038), p. 79 - 96 (1980).

064.072 **Theory of the solar wind and winds from late-type stars.** T. E. Holzer.
Smithsonian Astrophys. Obs., Spec. Rep. 389, (see 012.038), p. 153 - 181 (1980).

064.073 **Stellar wind energetics and mass loss in late-type supergiant stars.** R. E. Stencel.
Smithsonian Astrophys. Obs., Spec. Rep. 389, (see 012.038), p. 183 - 188 (1980).

064.074 **Non-thermal stellar winds in cool stars.**
D. J. Mullan.
Smithsonian Astrophys. Obs., Spec. Rep. 389, (see 012.038), p. 189 - 192 (1980).

The author proposes that magnetic reconnection drives

rapid mass loss in stars which lie above the STL (*Supersonic Transition Locus*).

064.075 **Stellar coronae: overview of the Einstein/CFA (*Harvard-Smithsonian Center for Astrophysics*) stellar survey.** G. S. Vaiana.
Smithsonian Astrophys. Obs., Spec. Rep.389, (see 012.038), p. 195 - 215 (1980).

064.076 **Theory of stellar coronae: an interpretation of X-ray emission from non-degenerate stellar sources.**
J. L. Linsky.
Smithsonian Astrophys. Obs., Spec. Rep. 389, (see 012.038), p. 217 - 235 (1980).

064.077 **The end of accretion onto early-type stars, and the onset of the stellar wind.** F. D. Kahn.
Observatory, Vol. 100, 181 (1980). – Abstract.

064.078 **Nonthermal structure of stellar atmospheres.**
L. E. Cram.
Comments Astrophys., Vol. 9, 25 - 50 (1980).

The discussion is a prelude to a series of monographs titled *Nonthermal Structure of Stellar Atmospheres* currently being prepared under the auspices of the National Aeronautics and Space Administration and le Centre National de la Recherche Scientifique.

064.079 **Chromospheres of F, G, K type stars. V. Radiative losses in spectral lines.**
R. Głębocki, Z. Musielak, J. Sikorski.
Acta Astron., Vol. 30, 259 - 266 (1980).

Radiative losses in weak and strong lines are compared with losses in continua. In spite of large numbers of weak lines (more than 20000 in visible regions) their total energy losses are small in comparison to losses in continua.

064.080 **The instability of radiation-driven stellar winds.**
R. G. Carlberg.
Astrophys. J., Vol. 241, 1131 - 1140 (1980).

The stability of a radiation-driven stellar wind is examined by applying a completely linearized stability analysis to a physically realistic wind model. Three types of instability are found. Both the line-shape and the radiation-driven sound wave instabilities are purely vertical motions which have been previously discovered. A new "gradient" instability has motion in both the vertical and horizontal directions.

064.081 **Stellar atmospheres with magnetic fields.**
K. Stępień.
Variability in stars and galaxies, (see 012.044), B.5.6 - 5.19 (1980).

Several possible effects of the magnetic field on the atmosphere of a magnetic star are discussed. The diffusion of electrons and ions in the magnetic field is discussed and it is shown that the slow-down of the diffusion of electrons across the magnetic field can influence the diffusion of ions, thus producing chemical inhomogeneities connected with the geometry of the magnetic field. Some results concerning the influence of chemical peculiarities on the energy distributions of Ap stars are presented.

064.082 **Mass loss from red giants: some effects on the interstellar medium.** A. Renzini.
Variability in stars and galaxies, (see 012.044), p. E.3.1 - 3.11 (1980).

This review focuses on intermediate-mass stars ($1 < M < 8M_\odot$), and on some effects of the red-giant mass-loss processes on the ISM. In particular, the relevance of these mass-loss processes for the galactic nucleosynthesis, for the production of the interstellar grains, and for planetary nebulae is discussed.

064.083 **Momentum and energy deposition in late-type stellar atmospheres and winds.**
L. Hartmann, K. B. MacGregor.
Astrophys. J., Vol. 242, 260 - 282 (1980).

The authors investigate the possibility that universal mechanisms of mechanical energy generation and deposition occur in all late-type stars, and that their specific effects on chromospheric temperature structure and mass loss are modified principally by surface gravity. The discussion focuses on the behavior of acoustic and magnetic wave modes known to exist on the Sun. These modes can be much more efficient than radiation pressure in driving mass loss.

064.084 **On the possibility of using metal unblanketed model atmospheres for analysis of population II stars.**
M. Moretti, L. Rossi.
Astrophys. Space Sci., Vol. 73, 379 - 387 (1980).

A comparison between model atmosphere grids with and without metal blanketing has been performed in the low-temperature range (T_{eff} = 6000 K to 10 000 K) and for different metal contents ($\log A = 0, -1, -2$; A = scaled solar abundance). For $A \leqslant 10^{-1}$ and $T_{eff} \geqslant 7000$ K, the Johnson colour indices $U\!-\!B$ and $B\!-\!V$, together with the Strömgren $u\!-\!b$, $b\!-\!y$ and the bolometric correction are little affected by metal blanketing. On the other hand, the trend of the physical quantities and the m_1 and c_1 colour indices reflect the inadequateness of the models even with $A = 10^{-2}$ and T_{eff} = 9500 K, where hydrogen line blanketing is expected to dominate. This fact discourages once and for all the use of metal unblanketed atmospheres other than for comparison of colour indices or the calculation of bolometric corrections for population II A–F spectral types.

064.085 **Stellar winds, fast rotators, and magnetic acceleration.** S. Nerney.
Astrophys. J., Vol. 242, 723 - 737 (1980).

The assumption that observed mass outflow from a star is due to a magnetically driven wind implies an upper bound on the surface magnetic field strength from regions where the wind originates. Upper bounds for the surface fields in supergiants and Be stars are derived from fast magnetic rotator theory. The author reports evidence that corroborates Rosendhal's observation of an abrupt change in the velocity-gradient-luminosity relationship for B8 and later supergiants. Finally, he considers the effect of magnetic acceleration on the dispersal of the solar nebula.

064.086 **Thick accretion disks with super-Eddington luminosities.**
M. A. Abramowicz, M. Calvani, L. Nobili.
Astrophys. J., Vol. 242, 772 - 788 (1980).

The authors describe a Newtonian version of the theory of thick accretion disks orbiting black holes. Assuming surface distributions of angular momentum and radiation flux, one can construct a model of a thick, non-Keplerian accretion disk with no knowledge of the viscosity mechanism. Thick accretion disks can have luminosities 100 times above the Eddington limit. The authors study the self-consistency constraints for the theory of thick accretion disks with super-Eddington luminosities.

064.087 **A connection between macroturbulence and non-radial oscillations in late-type stars.** M. A. Smith.
Astrophys. J., Lett., Vol. 242, L115 - L118 (1980).

The behavior of the mean amplitude of nonradial oscillations in late-type stellar atmospheres is considered. It is concluded that a large component of macroturbulence in the atmospheres of late-type stars is probably intimately tied to, if not directly caused by, 5-minute-like nonradial oscillations. Estimates are made of the likelihood of observing these modes as a function of luminosity class.

064.088 **Nonradial oscillations: the cause of macroturbulence in late-type stars?** M. A. Smith.
Space Sci. Rev., Vol. 27, (see 012.046), 307 - 312 (1980).

The author examines published theoretical estimates for relative nonradial pulsation amplitudes in order to see whether this mechanism can account for observed macroturbulences in late-type stars.

064.089 **The critical frequency in the stellar pulsation theory.** M. Takeuti.
Space Sci. Rev., Vol. 27, (see 012.046), 413 - 417 (1980).

The standing oscillation in an isothermal semi-infinite plane-parallel atmosphere has a critical frequency. The significance of critical frequency in the stellar pulsation is discussed briefly. The critical period is calculated for the cepheid instability strip. The pulsational instability is reduced by the linear running wave.

064.090 **Hydrodynamic modeling of mass loss from cataclysmic variable secondaries.** R. L. Gilliland.
Space Sci. Rev., Vol. 27, (see 012.046), 641 - 642 (1980). Abstract.

064.091 **The continuous absorption coefficients of stellar atmospheres.** J.-h. Jin.
Acta Astron. Sinica, Vol. 21, 342 - 348 (1980). In Chinese.

This paper shows that it is profitable to use the continuous absorption coefficient $\alpha_c(T, \lg P_e)$ per particle of the most abundant element instead of the commonly used continuous absorption coefficient per gram.

064.092 **Using model atmospheres for investigations of B - G stars.** L. S. Lyubimkov.
Izv. Krymskoj Astrofiz. Obs., Tom 62, 44 - 53 (1980). In Russian.

A version of model atmosphere analysis is described which is intended for investigations of B - G stars. The calculations of the continuum, profiles of hydrogen lines, and equivalent widths and depths of formation of metal lines are considered. Tables of the Hα - Hδ profiles calculated for the model atmospheres of Parsons on the basis of the unified theory of Vidal, Cooper, and Smith are given. Criteria which are used for the determination of the effective temperature and the surface gravity are discussed. A method of investigation of depth dependence of the microturbulence parameter in the atmosphere is described.

064.093 **A connection between macroturbulence and nonradial oscillations in late-type stars.** M. A. Smith.
Publ. Astron. Soc. Pacific, Vol. 92, 550 (1980). – Abstract.

064.094 **Column densities of opacity particles above the photospheres of cool stars.** V. V. Tsymbal.
Astron. Tsirk., No. 1103, p. 7 - 8 (1980). In Russian.

064.095 **The source of heating in the mantles of early-type supergiants.** A. B. Underhill.
J. R. Astron. Soc. Canada, Vol. 74, 356 (1980). – Abstract.

064.096 **Nonradial accretion onto magnetized neutron stars.** V. M. Lipunov.
Astron. Zh., Tom 57, 1253 - 1265 (1980). In Russian.
English translation in Soviet Astron., Vol. 24, No. 6.

Accretion onto magnetized neutron stars in close binary systems is considered. The shape of the magnetosphere in the two-stream model of nonradial accretion has been found. It is shown that the magnetospheres of long-period X-ray pulsars (with periods ~10 sec) are closed. The magnetospheres of pulsars with shorter periods can be open.

064.097 **Balmerlinienprofile von Protosternhüllen.** U. Bastian.
Diss. Naturwiss.-Math. Gesamtfak. Univ. Heidelberg. 104 pp. (1980).

064.098 **Einfluß der kinematischen Zähigkeit auf stellare Akkretionsraten.** M. Hosseini, H. J. Fahr.
Mitt. Astron. Ges., Nr. 50, p. 57 - 64 (1980).

064.099 **Spherical accretion onto a hydrostatic protostellar core.** H. Zinnecker, W. Tscharnuter.
Mitt. Astron. Ges., Nr. 50, p. 109 (1980). – Abstract.

064.100 **Modellrechnungen zur Struktur der Schockkühlschicht von YY-Orionis-Sternen.** R. Mundt.
Mitt. Astron. Ges., Nr. 50, p. 123 - 125 (1980).

064.101 **Stars and their coronae.** C. de Jager.
X-ray and gamma-ray astronomy in the 1980's, (see 012.054), p. 17 - 20 (1979).

064.102 **On the structure of Be star envelopes.** L. S. Luud.
Otd-nie fiz.-mat. i tekh. nauk AN EhstSSR. Prepr., 1980, No. A-3, 34 pp. In Russian. – Abstr. in Ref. zh., 51. Astron., 12.51.421 (1980).

064.103 **Models of the envelope of P Cyg. Analysis from hydrogen spectral lines.** I. R. Kolka.
Otd-nie fiz.-mat. i tekh. nauk AN EhstSSR. Prepr., 1980, A-4, 59 pp. In Russian. – From Ref. zh., 51. Astron., 12.51.423 (1980).

064.104 **Stellar coronae from Einstein: observations and theory.** R. Rosner, G. S. Vaiana.
X-ray astronomy, (see 012.055), p. 129 - 151 (1980).

Stellar X-ray emission appears to be the rule rather than the exception. In most cases the levels of observed emission far exceed predictions based upon the "standard" theories of coronal formation and heating, including the most recent modification of such theories. One is therefore faced with the tasks of rethinking the problems of coronal formation and reexamining its role in stellar evolution. In order to place these problems in perspective, the authors will first outline some of the larger issues, and then discuss how the new X-ray observations impinge on the questions that are raised.

064.105 **Taking into account convection in computation of model stellar atmospheres by the method of total linearization.** A. Galdikas.
Bull. Vilnius Astron. Obs., Nr. 53, p. 33 - 39 (1980). In Russian.

A method of temperature correction for convective model stellar atmospheres based on zero divergence of the total flux is proposed.

064.106 **Formation of chromospheric resonance line profiles in supergiants.** G. S. Basri.
Astrophys. J., Vol. 242, 1133 - 1143 (1980).

The formation of chromospheric resonance line profiles is examined for the case of the relatively low-density atmospheres appropriate to late-type supergiants. The effects of partial frequency redistribution control the emergent line profiles, to the extent that even an isothermal atmosphere can give rise to apparent emission features. Schematic model chromospheres are used to demonstrate the effects of different velocity fields and temperature-density structures on the line profiles, as a first step toward a clearer understanding of the chromospheric line profiles and why supergiants have broad emission lines (the Wilson-Bappu effect).

064.107 **The theory of radiatively driven stellar winds. I. A physical interpretation.** D. C. Abbott.
Astrophys. J., Vol. 242, 1183 - 1207 (1980).

This series of papers extends the line-driven wind theory of Castor, Abbott, and Klein (1975). The present paper develops a physical interpretation of line-driven flows using analytic methods.

The brightest stars. See Abstr. 003.060.

Vrijwel alle sterren verliezen massa. See Abstr. 011.028.

The interplay of molecular spectroscopy and astronomy. See Abstr. 022.106.

The determination of stellar turbulence by low resolution techniques. See Abstr. 031.528.

Cellular convection in a stratified atmosphere. See Abstr. 062.003.

Higher order fluid equations for multicomponent nonequilibrium stellar (plasma) atmospheres and star clusters. See Abstr. 062.017.

The heating of a thermally conducting stratified medium. II. A simple plane model of an atmosphere. See Abstr. 062.068.

Propagation of waves in an atmosphere in the presence of a magnetic field. II. The reflection of Alfvén waves. See Abstr. 062.081.

Cyclotron absorption in accreting magnetic white dwarfs. See Abstr. 062.113.

On path lengths of integration in the computation of the Voigt function. See Abstr. 063.016.

Lines formed in rotating and expanding atmospheres. See Abstr. 063.020.

Mesoturbulence. See Abstr. 063.025.

Stochastic approach. See Abstr. 063.026.

The application of mesoturbulence to stellar atmospheres. See Abstr. 063.027.

Effects of acoustic waves on spectral line profiles. See Abstr. 063.028.

Evolution of young stars with mass accretion. See Abstr. 065.005.

On the binary frequency distribution and evolution of Wolf-Rayet stars. See Abstr. 065.007.

Evolution of massive stars losing mass and angular momentum. V: The origin of Wolf-Rayet stars. See Abstr. 065.023.

Influence of a stellar wind on the evolution of a star of 30 $M_\odot$. See Abstr. 065.038.

Surface composition changes in massive star evolution with mass loss. See Abstr. 065.071.

Convective mixing in extended horizontal branch envelope models: the sdB/sdO transition. See Abstr. 065.093.

Time-dependent, optically thick accretion onto a black hole. See Abstr. 066.047.

Transonic disk accretion onto black holes. See Abstr. 066.051.

Outer parts of accreting discs around supermassive black holes in the nuclei of galaxies and quasars. See Abstr. 066.123.

SS 433: a high-energy neutrino source? See Abstr. 066.506.

Iron K photons from weakly magnetized neutron stars in X-ray binaries. See Abstr. 066.515.

Gamma-ray lines from accreting neutron stars. See Abstr. 066.517.

X-ray bursts from thermonuclear runaways on accreting neutron stars. See Abstr. 066.520.

Instabilities of accretion disk – magnetosphere interfaces. See Abstr. 066.525.

Interaction of accretion disk and rotating magnetic field of a neutron star. See Abstr. 066.526.

Thermal limit cycle oscillations on the surface of accreting neutron stars – X-ray bursters. See Abstr. 066.535.

Some more effects of waves on spectral line analysis. See Abstr. 071.016.

Effects of flux tubes on conventional chromospheric diagnostics. See Abstr. 073.052.

On the structural and stochastic motions in the solar and stellar atmospheres. See Abstr. 080.027.

Overshooting motions from the convection zone and their role in atmospheric heating. See Abstr. 080.049.

Mass loss from planetary protoatmospheres and from the protoplanetary nebula. See Abstr. 107.024.

IUE observations of circumstellar lines and mass loss from B-stars. See Abstr. 112.013.

Circumstellar chlorine chemistry and a search for AlCl. See Abstr. 112.014.

Envelopes around late-type giant stars. See Abstr. 112.017.

Behavior and significance of circumstellar clouds. See Abstr. 112.020.

Observations of stellar winds from cool stars. See Abstr. 112.043.

Detection of CO emission at 1.3 millimeters from the Betelgeuse circumstellar shell. See Abstr. 112.047.

Empirical ionization fractions in the winds and the determination of mass-loss rates for early-type stars. See Abstr. 112.049.

Line blanketed model atmospheres of Ap stars. IV. *UBV* colors and effective temperatures of Ap stars. See Abstr. 113.039.

Mercury in a thin layer in HgMn stars: a test of a diffusion model. See Abstr. 114.021.

Variations in the K-line emission of Arcturus. See Abstr. 114.042.

The spectrum variations of HD 153919. See Abstr. 114.043.

The ultraviolet flux of HD 122563. See Abstr. 114.067.

Outer atmospheres of cool stars. V. IUE observations of Capella: the rotation-activity connection. See Abstr. 114.084.

A survey of chromospheric Ca II H and K emission in field stars of the solar neighborhood. See Abstr. 114.107.

Comparison of activity cycles in old and young main-sequence stars. See Abstr. 114.108.

Solar-type phenomena on late-type stars. See Abstr. 114.114.

An estimate of active region filling factors for dMe and dM stars. See Abstr. 114.115.

Chromospheric and coronal activity in F, G and K type stars. See Abstr. 114.116.

Spectroscopic analysis of Pollux relative to the Sun with special reference to Arcturus. See Abstr. 114.126.

Variability in early-type stars. See Abstr. 114.130.

How could we improve abundance analyses of late-type stars – and should we? See Abstr. 114.143.

The Balmer decrement in the spectra of moving envelopes of stars. See Abstr. 114.148.

Transition region and chromospheric models of 24 UMa based on IUE ultraviolet spectrograms. See Abstr. 114.150.

The outer atmospheres of cool stars. VII. High resolution, absolute flux profiles of the Mg II *h* and *k* lines in stars of spectral types F8 to M5. See Abstr. 114.165.

Differential rotation and magnetic activity of the lower main sequence stars. See Abstr. 116.005.

Rotational modulation of chromospheric variations of main-sequence stars. See Abstr. 116.026.

Radio superflares in RS Canum Venaticorum binaries. See Abstr. 116.031.

Evolution of a blue supergiant with a neutron star companion immersed in its envelope. See Abstr. 117.001.

Accretion disk radii in cataclysmic variables. See Abstr. 117.035.

Turbulence in the atmospheres of eclipsing binary stars. See Abstr. 119.025.

Accretion disks in cataclysmic variables. I. The eclipse-related phase shifts in DQ Herculis and UX Ursae Majoris. See Abstr. 119.039.

P Cygni-Profile im UV Spektrum von UW Canis Majoris. See Abstr. 119.103.

The chromospheric explanation of T Tauri spectra. See Abstr. 121.004.

Elektronenstreuung in der einfallenden Hülle von S CrA. See Abstr. 121.019.

Polarization models for hot nonradial pulsators. See Abstr. 122.016.

HR 1099 and the starspot hypothesis for RS Canum Venaticorum binaries. See Abstr. 122.050.

A wave model for dwarf novae. See Abstr. 122.201.

Radial accretion of H-rich material onto a He white dwarf. See Abstr. 124.010.

The basic structure of hot white dwarfs atmospheres as a function of composition. See Abstr. 126.001.

Theory and observations of the optical continuum and line spectra of accretion discs around white dwarfs. See Abstr. 126.030.

Acoustic fluxes in white dwarfs. See Abstr. 126.039.

Comments on the existence of circumstellar clouds derived from interstellar Mg observations. See Abstr. 131.017.

Observations of CO in L1551: evidence for stellar wind driven shocks. See Abstr. 131.036.

A stellar-wind focusing mechanism as an explanation for Herbig-Haro objects. See Abstr. 131.070.

High energy phenomena in molecular clouds. See Abstr. 131.286.

Simultaneous far-infrared, near-infrared, and radio observations of OH/IR stars. See Abstr. 133.007.

The calculation of the optical spectra of NGC 6888. See Abstr. 134.004.

Pulsating X-ray sources: the oblique dipole configuration. See Abstr. 142.014.

The coronae of 40 Eridani. See Abstr. 142.019.

A discussion of the eccentric binary hypothesis for transient X-ray sources. II. Gradual acceleration stellar wind model. See Abstr. 142.065.

Mass loss and stellar wind in massive X-ray binaries. See Abstr. 142.081.

X-ray burst sources – a simple two zone model. See Abstr. 142.123.

065 Stellar Structure and Evolution

065.001 **L'évolution des étoiles et l'origine de l'hélium dans l'Univers en expansion.** A. Maeder.
Orion, 38. Jahrg., 116 - 117 (1980).

065.002 **Theoretical surface abundances of Population I giants.** S. J. Shadick, H. J. Falk, R. Mitalas.
Mon. Not. R. Astron. Soc., Vol. 192, 493 - 502 (1980).

The abundance predictions of standard models of stellar evolution are not in agreement with the observed CNO and other abundance ratios in all giants. The authors consider the effects of variation of initial composition and mass loss on the abundance ratios of $^{12}C/^{13}C$, $^{12}C/^{14}N$, $^{16}O/^{12}C$, $^{7}Li/H$, and $^{9}Be/H$ in giants. To test these effects, stellar models of 1.0 $M_\odot$, 1.5 $M_\odot$ and 2.5 $M_\odot$ are evolved with and without mass loss from the zero age main sequence (ZAMS) to the giant branch. Mass loss is treated using a scaled Reimers' mass loss algorithm and an acoustic luminosity algorithm. Sufficient mass loss, if present before the giant branch and/or variation of initial composition could be possible explanations for the anomalous abundances in giants.

065.003 **Deviations from the standard model of SS433.** M. Milgrom.
Astron. Astrophys., Vol. 87, L15 - L17 (1980).

Deviations from the standard model of SS433 are considered. The author considers mainly deviations in which the symmetry between the two line emitting regions is broken (such as a misalignment of the two velocity vectors). Even if the velocity vectors are constant in the rotating frame, the average wavelength of the two satellites $(\lambda^+ + \lambda^-)/2$ is no more constant with time but varies with the rotation period.

065.004 **Stability of the stellar structure in non-equilibrium thermodynamics.**
Q.-h. Peng, K.-l. Huang, K.-h. Zhan, X.-t. He.
Acta Astron. Sinica, Vol. 21, 172 - 179 (1980). In Chinese.

065.005 **Evolution of young stars with mass accretion.** S. K. Bhattacharjee, I. P. Williams.
Mon. Not. R. Astron. Soc., Vol. 192, 841 - 846 (1980).

The effects of gas accretion on the pre-main-sequence evolution of stars have been discussed. It is found that the evolution is dominated by the accretion rate rather than by initial conditions, and that the past evolutionary behaviour of a star cannot be determined from a knowledge of its present evolutionary state. However, the masses of the pre-main-sequence stars can be determined uniquely irrespective of the existence of accretion while the determination of their ages is very model dependent.

065.006 **The relationship between the envelope composition of a 6 $M_\odot$ red-giant model and its future evolution.**
D. Prialnik, G. Shaviv.
Astron. Astrophys., Vol. 88, 127 - 134 (1980).

The questions regarding the maximum mass of a main-sequence star that becomes a white-dwarf; the mass of single white dwarfs; the enhanced abundance ratios of N/O and He/H in planetary nebulae and the He^3 abundance in planetaries and in the interstellar medium are shown to be interrelated. The authors find that a model star of 6 $M_\odot$ on the main sequence develops prior to the double-shell source phase a C-O core of 0.82 $M_\odot$, an enrichment of 27% in the He/H ratio and a N/O ratio 5.5 times higher than the solar value. These results indicate that a 6 $M_\odot$ star is a plausible planetary-nebula progenitor. They also show that intermediate mass stars, ejecting most of their mass during the red-giant stage, are important contributors of C and O isotopes and of He^3 to the interstellar medium.

065.007 **On the binary frequency distribution and evolution of Wolf-Rayet stars.**
D. Vanbeveren, P. S. Conti.
Astron. Astrophys., Vol. 88, 230 - 239 (1980).

First the authors have summarized briefly some new information available since I. A. U. Symposium 49 in 1973 concerning the effective temperature, the luminosity and the chemical composition of WR stars. Kuhi (1973) stated that ~ 73% WR stars appear in binary systems with an OB companion. He only considered stars with an apparent magnitude $m_v \leqq 10^m$ based on the reasonable supposition that spectroscopic data were poor for faint stars. It is shown that this limitation is statistically biased. Newly obtained spectroscopic data are used to rediscuss the frequency of WR stars with "detected" O-type companions (e.g. absorption lines present). The authors suggest that a considerable fraction of the "single" line WR stars (i.e. a WR star where no OB companion is seen) could be WR stars with a compact companion but that others are truly single.

065.008 **Differential rotation along the lower main sequence: a theoretical investigation.**
G. Belvedere, L. Paternò, M. Stix.
Astron. Astrophys., Vol. 88, 240 - 247 (1980) = Mitt. Kiepenheuer - Inst. Nr. 183.

In this paper the authors extend to the lower main sequence stars the analysis of convection interacting with rotation in a compressible medium, already done for the solar case. As well known, among main sequence stars only those belonging to the lower half – say spectral types from F5 down to M – are allowed to possess convective envelopes and are then supposed to develop differential rotation, if one accepts the main mechanism generating differential rotation to be the coupling between convective transport and rotation itself.

065.009 **Convective stability in magnetic stars.** R. Simon.
Astrophys. Space Sci., Vol. 71, 111 - 121 (1980).

Dynamical stability of a static axisymmetrical magnetic star with respect to high-order modes of oscillation is investigated by means of the energy method, neglecting the Eulerian perturbation of gravity. The magnetic field is assumed to be continuous across the surface of the star and its first-order spatial derivatives, but it may have both toroidal and poloidal components. The second variation of the potential energy is written in a way which, in the case of a purely toroidal field, and for axisymmetrical and non-axisymmetrical modes, yields Tayler's local stability criteria which are necessary and sufficient conditions for convective stability, and in the case of a general field yields a single local stability criterion, which is a sufficient condition for convective stability.

065.010 **Time-dependent neutrino transport out of dense stellar cores.**
I. Lichtenstadt, A. Ron, N. Sack, J. J. Wagschal, S. A. Bludman.
Astrophys. Space Sci., Vol.71, 219 - 227 (1980).

Time-dependent neutrino transport out of an optically thick neutronized stellar core is calculated to study the effects of neutrino degeneracy and of source depletion. Neutrino trapping inhibits further neutrino emission until neutrinos peel out of the outer zones of the core, exposing successively inner zones. This inwardly propagating neutrino rarefaction wave can lead to $e^- + p \rightleftharpoons \nu + n$ oscillations in chemical composition. The effect of neutrino Fermi statistics is to retard considerably and disperse neutrino leakage out of the core, making neutrino transport insignificant during fast stages of core collapse.

065.011 **Emden–Chandrasekhar axisymmetric, solid-body rotating polytropes. I: Exact solutions for the special cases n = 0,1 and 5.** R. Caimmi.
Astrophys. Space Sci., Vol. 71, 415 - 457 (1980).

The basic theory on polytropes is revisited and EC polytropes are defined. The first-order approximation theory of Chandrasekhar (1933) and Chandrasekhar and Lebovitz (1962) is reviewed, refined and extended in such a way that better results are obtained without involving hard analytical or numerical techniques. A more precise equation is given in defining non-outer equipotential surfaces, and a new method is adopted in determining the explicit expression of the gravitational potential. Detailed results are given for the special cases n = 0, 1 and 5.

065.012 **Supernova triggered by electron captures.**
S. Miyaji, K. Nomoto, K. Yokoi, D. Sugimoto.
Publ. Astron. Soc. Japan, Vol. 32, 303 - 329 (1980).

Final evolution of the stars in the mass range of $8-12M_\odot$ has been investigated fully taking account of the effects of electron captures and oxygen deflagration. Only an O–Ne–Mg core has been treated numerically from the stages well in advance of electron captures and of oxygen burning.

065.013 **Vibrational stability of rotating stars.**
H. Shibahashi.
Publ. Astron. Soc. Japan, Vol. 32, 341 - 346 (1980).

The vibrational stability of rotating stars against axisymmetric perturbations is studied in the local theory using the Boussinesq approximation. The sufficient conditions for stability are explicitly given in terms of the rotation law of a star. In the case of a chemically homogeneous zone, the conditions are reduced to that (1) the angular velocity of rotation Ω is cylindrical and that (2) the temperature gradient is subadiabatic. The physical cause of the overstability induced by breakdown of the former condition is discussed, and is compared with that of the baroclinic instability in the dynamic meteorology.

065.014 **Evolutionary status of stars with $M \gtrsim 50\ M_\odot$.**
A. V. Tutukov, L. R. Yungel'son.
Pis'ma Astron. Zh., Tom 6, 491 - 494 (1980). In Russian. English translation in Soviet Astron. Lett., Vol. 6.

Observed masses and spatial velocities of main-sequence stars with $M \gtrsim 50\ M_\odot$ and of part of most luminous Wolf-Rayet stars (WN7/WN8) are explained by mass exchange and supernova explosions in close binary systems. Within the same notions blue supergiants that actively lose mass may represent common envelope stars with one component of the nucleus being a relativistic object.

065.015 **Numerical investigations of collapsing protostars. Physical and mathematical formulation of the problem. II. Numerical algorithms.** I. G. Kolesnik.
Astrometr. Astrofiz., Vyp. (No.) 41, p. 40 - 58 (1980). In Russian.

Hydrodynamic difference equations and an implicit code for their solution are considered. Equations for the internal structure of a quasistatic starlike core are described. The conditions of the core conjugation with the dynamical envelope as well as the methods of a time step choice and numerical algorithms modifications are discussed.

065.016 **Magnetic mixing and the Arcturus problem.**
E. N. Hubbard, D. S. P. Dearborn.
Astrophys. J., Vol. 239, 248 - 252 (1980).

A model for main-sequence mixing driven by the buoyancy of magnetic flux tubes is proposed, and the effect of such a process on the CNO isotopes is investigated. Such a mixing scheme applied to a $2.0M_\odot$ model is used to predict the ratios $^{14}N/^{15}N$, $^{16}O/^{17}O$, $^{16}O/^{18}O$, and C/O for any observed combination of $^{12}C/^{13}C$ and C/N.

065.017 **Neutrino energy equilibration models.**
D. L. Tubbs, T. A. Weaver, R. L. Bowers, J. R. Wilson, D. N. Schramm.
Astrophys. J., Vol. 239, 271 - 283 (1980).

Neutrino energy equilibration by neutrino-electron scattering is studied in four physical models, whose temperatures and densities are representative of stellar collapse conditions when electron scattering may be important. Results of Monte Carlo simulations are presented as data against which approximate transport methods may be tested.

065.018 **The core helium flash with two-dimensional convection.** P. W. Cole, R. G. Deupree.
Astrophys. J., Vol. 239, 284 - 291 (1980).

The dynamical conservation equations are integrated in two spatial dimensions and time near the peak of the core helium flash. The two-dimensional nature eliminates the need for a phenomenological theory of convection, as convective flow patterns are followed. The calculations indicate that the core helium flash is nonviolent, but that the relation between convection and the thermal runaway is considerably more complex than heretofore believed. The authors also find that the region interior to the flash peak is significantly heated by downward convective energy transport, and presumably mixed as well. Subsidiary calculations indicate that this heating will begin well before the peak of the thermal runaway. Effects of composition variation are discussed.

065.019 **The equilibria of rotating stars and stellar systems.**
P. O. Vandervoort, D. E. Welty.
Bull. American Astron. Soc., Vol. 12, 449 (1980). – Abstract.

065.020 **Progress towards a quasistatic two dimensional rotation code.** M. P. Savedoff.
Bull. American Astron. Soc., Vol. 12, 449 - 450 (1980). Abstract.

065.021 **Improved secular stability limits for differentially rotating polytropes and degenerate dwarfs.**
J. N. Imamura, R. H. Durisen.
Bull. American Astron. Soc., Vol. 12, 465 (1980). – Abstract.

065.022 **Evolution with mass loss during core helium burning of a massive star.** H. J. Falk, R. Mitalas.
Bull. American Astron. Soc., Vol. 12, 466 (1980). – Abstract.

065.023 **Evolution of massive stars losing mass and angular momentum.V: The origin of Wolf-Rayet stars.**
S. R. Sreenivasan, W. J. F. Wilson.
Bull. American Astron. Soc., Vol. 12, 539 (1980). – Abstract.

065.024 **Differential rotation of magnetic stars.**
K. Maezawa.
Bull. Yamagata Univ. (Nat. Sci.), Vol. 10, No. 1, p. 51 - 55 (1980). – Abstr. in Phys. Abstr., Vol. 83, Abstr. 72928 (1980).

065.025 **Stars on the wane: stellar geriatrics.**
Sci. Dimension, Vol. 12, No. 2, p. 26, 28 (1980). Abstr. in Phys. Abstr., Vol. 83, Abstr. 72931 (1980).

065.026 **Neutronization, lepton escape and stellar hydrodynamics.** W. D. Arnett.
J. Phys. Colloq., Vol. 41, No. C-2, (see 012.005), p. C2/25 - 29 (1980). – Abstr. in Phys. Abstr., Vol. 83, Abstr. 77388 (1980).

065.027 **Stellar collapse.** R. Canal.
J. Phys. Colloq., Vol. 41, No. C-2, (see 012.005), p. C2/105 - 110 (1980). – Abstr. in Phys. Abstr., Vol. 83, Abstr. 77389 (1980).

065.028 **The afterclap of degenerate carbon ignition revisited.** J. R. Buchler, S. A. Colgate, T. J. Mazurek.
J. Phys. Colloq., Vol. 41, No. C-2, (see 012.005), p. C2/259 - 263 (1980). – Abstr. in Phys. Abstr., Vol. 83, Abstr. 77390 (1980).

065.029 **A eutectic in carbon-oxygen white dwarfs?** D. J. Stevenson.
J. Phys. Colloq., Vol. 41, No. C-2, (see 012.005), p. C2/61 - 64 (1980). – Abstr. in Phys. Abstr., Vol. 83, Abstr. 77405 (1980).

065.030 **Stellar convection. II. A multimode numerical solution for convection in spheres.**
P. S. Marcus.
Astrophys. J., Vol. 239, 622 - 639 (1980).

The author has computed a stable, equilibrium solution for convection in a self-gravitating sphere of Boussinesq fluid by using a modal analysis in which the θ, ϕ dependence of the fluid is expanded in the set of 168 spherical harmonics, $Y^{l,m}$, with $l \leq 12$. To compute the numerical solution of his hierarchy of nonlinearly coupled equations, he has developed a new relaxation method. The temperature, velocity, and convective flux of the fluid as well as the kinetic and thermal energy spectra as functions of wavelength are computed. The author examines the dynamics of the energy cascade by computing the ratio of the amount of energy dissipation at a particular wavelength to the amount of energy produced at that same wavelength. The fraction of the convective flux that is carried by each wavelength and the degree of anisotropy associated with each length scale are also determined.

065.031 **History of the stellar birthrate from lithium abundances in red giants.**
J. M. Scalo, G. E. Miller.
Astrophys. J., Vol. 239, 953 - 960 (1980).

The authors discuss the adopted theoretical scenario for Li burning on the main sequence and Li dilution during the red giant phase, and the resulting reasons for the sensitivity of Li in red giants to the birthrate history. The details of the calculations are given, the observations are examined, and the comparison of theory with observation is presented along with a discussion of the possibility of uncovering a nonmonotonic time dependence of the birthrate.

065.032 **On the dependence of some helium shell flash characteristics on core mass.**
D. Havazelet, Z. Barkat.
Astrophys. J., Vol. 240, 196 - 202 (1980).

A theoretical derivation of the intershell mass-core mass relation of Paczyński is attempted. Formulae developed by Sugimoto and Fujimoto are extended to less massive ($m_c \lesssim 1$) cores.

065.033 **Stellar convection. III. Convection at large Rayleigh numbers.** P. S. Marcus.
Astrophys. J., Vol. 240, 203 - 217 (1980).

The author presents the results of a numerical, three-dimensional study of convection in a self-gravitating sphere of Boussinesq fluid with a Rayleigh number of 10^{10} and a Prandtl number of 1. The velocity and temperature are computed by using spectral methods in the horizontal and finite-differencing in the radial directions. An eddy viscosity and diffusivity are needed to model the subresolution flow. The amplitudes of the eddy viscosity and diffusivity are determined. By computing the energy spectra as well as the detailed energy budgets as a function of wavenumber, the author shows that for Rs = 10^{10} there is an inertial range for the modes corresponding to spherical harmonics with $l > 6$. He also shows that the velocity field becomes nearly isotropic at small wavelengths. Intermittent bursts of convective flux and energy propagate inward from the outer boundary layer. The bursts are shown to cascade from large to small wavelengths.

065.034 **On the calculation of the frequency splitting of adiabatic nonradial stellar oscillations by slow differential rotation.** J. Cuypers.
Astron. Astrophys., Vol. 89, 207 - 208 (1980).

The integrals with respect to the angular coordinates that appear in the expressions for the splitting of adiabatic nonradial modes of stellar oscillations by slow differential rotation are evaluated analytically for a general class of differential rotation laws. Only numerical integrations with respect to the radial coordinate r are required to obtain the amount of splitting.

065.035 **Adiabatic stellar core collapse: propagation of the shock.** K. A. Van Riper.
Astrophys. J., Vol. 240, 658 - 671 (1980).

Lichtenstadt, Sack, and Bludman have pointed out the importance of thermal stiffness, the postshock adiabatic index which determines how much shock-dissipated energy goes into pressure, in determining whether or not core collapse of massive stars will result in an explosion. The author shows how the thermal stiffness enters into the equation of state and how it effects the shock propagation. His numerical models confirm the findings of Lichtenstadt et al. that, with a large thermal stiffness, an explosion can occur without an inner core rebound following bounce.

065.036 **Stability of nonradial g^+ -mode pulsations in 1 $M_\odot$ models.** H. Saio.
Astrophys. J., Vol. 240, 685 - 692 (1980).

In this paper the author assumes that the amplitudes of the g^+ modes do not grow large enough to mix the core, so that he computes evolutionary models up to 4.5×10^9 yr without mixing and examines the stability of these models. The author finds that a combination of the destabilizing mechanisms due to the hydrogen ionization in the envelope and the $^3He + {}^3He$ reaction in the core may excite the $g_2{}^+$ mode with $l = 1$ (the period is about 80 minutes) in a model of age 4.5×10^9 yr, where l is the "angular momentum quantum number."

065.037 **Astrophysical problems of condensed matter in huge magnetic fields.** M. Ruderman.
J. Phys. Colloq., Vol. 41, No. C-2, (see 012.005), p. 125 - 131 (1980). – Abstr. in Phys. Abstr., Vol. 83, Abstr. 82187 (1980).

065.038 **Influence of a stellar wind on the evolution of a star of 30 $M_\odot$.** R. Stothers, C.-w. Chin.
Astrophys. J., Vol. 240, 885 - 891 (1980).

A coarse grid of theoretical evolutionary tracks has been computed for a star of 30 $M_\odot$, in an attempt to delineate the role of mass loss in the star's evolution during core helium burning. For all of the tracks, Cox-Stewart opacities have been adopted, and the free parameters have included the rate of mass loss, criterion for convection, and initial chemical composition. A grid of evolutionary tracks based on the Schwarzschild criterion for convection is presented, as well as a grid based on the Ledoux criterion is examined. A wide variety of possible evolutionary tracks can be obtained for an initial mass of 30 $M_\odot$ with reasonable choices of the free parameters.

065.039 **General relativistic collapse of an axially symmetric star. I. The formulation and the initial value equations.** T. Nakamura, K. Maeda, S. Miyama, M. Sasaki.
Prog. Theor. Phys., Vol. 63, 1229 - 1244 (1980). – Abstr. in Phys. Abstr., Vol. 83, Abstr. 85960 (1980).

065.040 **Differential rotation set up by latitude-dependent heat transport (*stellar rotation*).**
G. Belvedere, L. Paterno, M. Stix.
Geophys. Astrophys. Fluid Dyn., Vol. 14, 209 - 224 (1980). Abstr. in Phys. Abstr., Vol. 83, Abstr. 90411 (1980).

065.041 **Structure of relativistic stellar configurations. Linear stellar model in GRT.** V. Ureche.
Rev. Roumaine Phys., Vol. 25, 301 - 310 (1980). – Abstr. in Phys. Abstr., Vol. 83, Abstr. 94709 (1980).

065.042 **Effect of accretion on the pre-main sequence evolution of low-mass stars.** A. V. Fedorova.
Astron. Zh., Tom 57, 1105 - 1107 (1980). In Russian. English translation in Soviet Astron., Vol. 24, No. 5.

The pre-main sequence evolution of a star with initial mass 0.05 $M_\odot$ with accretion of protostellar cloud matter at a rate $10^{-7} M_\odot$/year is calculated. The resulting mass of the star is 1) 0.5 $M_\odot$, 2) 1 $M_\odot$. After the stop of accretion, the stars 0.5 and 1 $M_\odot$ can be situated $1^m - 1^m.5$ and $0^m.25$ below the main sequence. The hypothesis is suggested that the stars which lie below the main sequence in young clusters could strongly accrete matter at the pre-main sequence stage.

065.043 **Numerical solution of the 1 D spherical non stationary radiating shock using characteristics. Application to protostars.** P. J. Morel, A. Baglin.
Astron. Astrophys., Vol. 90, 327 - 337 (1980).

An implicit numerical code for solving, by means of characteristics,the spherically symmetric hydrodynamical equations of stellar evolution is presented. The spherically symmetric transfer equation (grey and time independent approximations), with a discontinuous source function, is formulated as a two points boundary-value problem and solved simultaneously with hydrodynamics by means of a variable factor of anisotropy (instead of the variable Eddington factor). As an example the procedure is applied to the hydrodynamical collapse of protostars. Comments on the advantages of characteristics and on the validity of the models with pseudo-viscosity are given.

065.044 **Asymptotic approximations for stellar nonradial pulsations.** M. Tassoul.
Astrophys. J., Suppl. Ser., Vol. 43, 469 - 490 (1980).

The problem of linear nonradial oscillations in stars is reconsidered using sound asymptotic expansions. High-order *p*- and *g*-modes are investigated. In the case of *g*-modes, the possibility of resonance occurs when the model possesses more than one zone in which spatial oscillations may occur.

065.045 **Turbulent convection and pulsational stability of variable stars.** D. Xiong.
Sci. Sinica, Vol. 23, 1139 - 1149 (1980).

According to the time-dependent convection theory the coupling between radial pulsation and convection is studied. The calculations of linear nonadiabatic pulsations of RR Lyrae and Delta Cephei models show that when the coupling between pulsations and convection is considered one can specify the red edge of the instability strip, and it may be expected to account for the light variability of red giants and supergiants.

065.046 **The collapse of carbon-oxygen white dwarfs.** R. Canal, J. Isern, J. Labay.
Astrophys. J., Lett., Vol. 241, L33 - L36 (1980).

Carbon-oxygen white dwarfs formed in close binary systems may become unstable by mass accretion. Recent results concerning carbon-oxygen separation at the freezing point during the phase of cooling may have very important consequences for the problem of neutron star formation. The central, high-density regions of the star are then made of pure oxygen, the carbon being rejected to lower-density layers. When the star is compressed, carbon ignition can only happen after neutronization of the central (oxygen) regions. The authors show that, in this case, the chances of collapse to a neutron star are independent from the rate of mass accretion, in contrast with previous studies.

065.047 **A new flash mixing.** I.-J. Sackmann.
Astrophys. J., Lett., Vol. 241, L37 - L40 (1980).

It was found that even for stars evolved away from the red giant branch, a new mixing of nucleosynthesis products from the hydrogen- and helium-burning shells into surface layers was possible, from the penetration of the contaminated intershell region into the H- and He-ionization convection zones. This is due to the helium shell flash driving an immense expansion of an inner carbon pocket. The surface would be enriched in carbon (^{12}C), helium (^{4}He), and *s*-process elements, but not significantly in nitrogen (^{14}N), oxygen (^{16}O), or the isotope ^{13}C. This new type of mixing might provide the missing clue for FG Sagittae.

065.048 **Turbulent surface layer of rotating B-type stars.** K. Kodaira.
Publ. Astron. Soc. Japan, Vol. 32, 435 - 444 (1980).

An order-of-magnitude quantitative study is carried out about the turbulence field which is energized via the shear flow driven by the thermal imbalance in the surface layer of rotating B-type stars, using the one-zone approximation. For an early to middle B-type star with the equatorial rotational velocity of about 100-300 km s^{-1}, the present model predicts macro- and micro-turbulence velocities of 0.1 - 1 km s^{-1} and a differential rotation of about 0.1 percent in the surface layer. It is pointed out that horizontal two-dimensional turbulence may also contribute to the observed macroturbulence.

065.049 **Equilibrium structure of stars with radiation pressure in the presence of a magnetic field.**
M. K. Das, J. Kar, J. N. Tandon.
Astrophys. Space Sci., Vol. 72, 397 - 410 (1980).

Equilibrium configuration of the upper main-sequence stars, with significant radiation pressure and having an interior magnetic field (matching with and external dipole field) has been considered. The structural parameters have been calculated for low and high magnetic fields by using a first-order perturbation method and a modified pertubation technique respectively. With the increase of radiation pressure, the star is seen to become more centrally condensed.

065.050 **Computation of the evolution model of normal stars.** R.-q. Huang, E.-r. Lou, G.-x. Chen.
Acta Astron. Sinica, Vol. 21, 243 - 251 (1980). In Chinese.

A method of computing the evolution model of normal stars is worked out. It is developed after having given an improvement to the computing method generally used. The programme for computation is made up by means of the Algol-60 language, and the computation is performed for a star with a mass of 2.82 $M_\odot$. The various evolution characteristics given by the results of computation have shown that this method and the programme worked out are correct and feasible.

065.051 **The time scale of thermohaline mixing in stars.** R. Kippenhahn, G. Ruschenplatt, H.-C. Thomas.
Astron. Astrophys., Vol. 91, 175 - 180 (1980).

The time scale of mixing due to thermohaline convection is estimated in the case of a gas in which the molecular weight decreases in the direction of gravity. The mass elements of higher molecular weight which sink into the regions below are hotter in the new surroundings. They therefore radiate into their neighborhood and create a circulation system which mixes the mass element with the surroundings before it has moved over a distance much bigger than its size. The "self destruction" reduces the mean free path of the sinking elements and therefore lengthens the time scale during which a (secularly) stable stratification is reached. The results are applied to a main sequence star which during mass exchange has received helium from a helium star companion. It is also applied to the case of the noncentral helium flash of a star of 1.3 $M_\odot$.

065.052 **The time scale of secularly unstable stellar rotation.** R. Kippenhahn, G. Ruschenplatt, H.-C. Thomas.
Astron. Astrophys., Vol. 91, 181 - 185 (1980).

Time scales for mixing due to the instabilities are estimated which occur in a rotating star when the conditions $\partial(s^2\omega)/\partial s \geq 0$, $\partial\omega/\partial z = 0$ are violated. It is argued that in the fully developed instability, eddies transporting angular momentum are destroyed before they can move through a distance greater than their thickness. This reduction of the "mean free path" of the eddies makes the time scale for redistribution of angular momentum through the star comparable with the Eddington-Vogt time scale.

065.053 **Types of stellar instabilities.** P. Ledoux.
Highlights of Astronomy, Vol. 5, (see 012.026), 433 - 436 (1980).

065.054 **Secular instabilities.** B. Paczyński.
Highlights of Astronomy, Vol. 5, (see 012.026), 437 - 439 (1980).

The relation between a linear series and thermal (i.e. secular) stability of stellar models is discussed. Models of accreting degenerate dwarfs are considered as an example of transition from instability to stability. This transition occurs while the accretion rate increases. Such models are relevant for novae and symbiotic stars.

065.055 **Shell flashes.** I.-J. Sackmann.
Highlights of Astronomy, Vol. 5, (see 012.026), 445 - 448 (1980).

Shell flashes are discussed with regard to three different stages of evolution: 1. normal stars, 2. white dwarfs accreting H-rich matter and 3. neutron stars accreting matter.

065.056 **Non-radial oscillations in red giants.** D. Keeley.
Highlights of Astronomy, Vol. 5, (see 012.026), 497 - 500 (1980).

The structure of red giant stars allows non-radial oscillation modes which propagate as p-modes near the surface, to propagate below the convection zone as g-modes with very high radial wave number. The conditions required for a clean separation of interior g-modes from surface p-modes may be well satisfied for extreme giant branch models. However, for somewhat hotter stars it is apparent that the interaction of the g-mode and p-mode regions may have interesting observational consequences.

065.057 **Dynamic catastrophes in stars.** J. Perdang.
Highlights of Astronomy, Vol. 5, (see 012.026), 501 - 504 (1980).

065.058 **The evolution of protostars. I. Global formulation and results.** S. W. Stahler, F. H. Shu, R. E. Taam.
Astrophys. J., Vol. 241, 637 - 654 (1980).

The authors review the controversy and give a new formulation for the problem of the evolution of protostars. Their method entails the division of the global problem into a set of more manageable subproblems. They derive the jump conditions of the radiative accretion shock which joins the hydrostatic mass-gaining core to the dynamic inner cloud envelope. Different approximations hold with high accuracy in the regions that are called the dust envelope, the opacity gap, the radiative precursor, the accretion shock, and the hydrostatic core. All equations to be solved are ordinary differential equations. Thus, standard integration schemes yield the high accuracy needed to resolve complex spatial structures which span many orders of magnitude in density and temperature. A $M_\odot$ protostar ends its main accretion phase moderately high up in the H-R diagram. This star begins its pre-main-sequence phase of quasi-static contraction on a convective Hayashi track.

065.059 **On dynamos in the cores of magnetic stars.** D. Moss.
Astron. Astrophys., Vol. 91, 319 - 321 (1980).

If the fields of the magnetic stars that lie on or near the upper main sequence are built by a contemporary dynamo operating in the core, it is suggested that for larger values of the angular velocity the dynamo may be oscillatory, and consequently the amplitude of the field that diffuses to the surface, and is therefore observable, may be negligible.

065.060 **Magnetic cycles of lower main sequence stars.** G. Belvedere, L. Paternò, M. Stix.
Astron. Astrophys., Vol. 91, 328 - 330 (1980) = Mitt. Kiepenheuer-Inst., Freiburg, Nr. 184.

The analysis of solar dynamo models, based on differential rotation and an α-effect (a mean current parallel to the mean magnetic field), is extended to main sequence stars. The authors compute oscillatory fields for five stellar models representing F5, G5, K0, K5, and M0 stars. The oscillation periods are approximately inversely proportional to the turbulent Prandtl number, *Pr*. For *Pr* = 0.01 the periods range from about 1 yr for F5 stars to about 40 yr for M0 stars. The results are discussed in the framework of the observational evidence of stellar activity.

065.061 **Current problems on horizontal-branch (HB) stars. IV. The influence of rotation on the expected properties of RR Lyrae pulsators.**
V. Castellani, G. Ponte, A. Tornambè.
Astrophys. Space Sci., Vol. 73, 11 - 25 (1980).

The influence of main sequence stellar rotation on horizontal branch stars and on RR Lyrae pulsational properties is studied. Synthetic HB's are computed for a wide range of original chemical compositions taking into account evolutionary effects based on Fusi Pecci-Renzini's evolutionary scheme. Theoretical properties of RR Lyrae pulsators are discussed: it is shown that main sequence rotation is able to induce in HB structures sensitive variations, so that some fundamental relations obtained in the classical frame (i. e., no rotation) have to be released. Consequences on the decodification of observational properties in galactic globular clusters are discussed.

065.062 **A new method for determining the internal rotational angular velocity of the stars.** H. Ando.
Astrophys. Space Sci., Vol. 73, 159 - 174 (1980).

Some physical aspects of the effect of rotation on the nonradial oscillations are discussed by the perturbation method. A new method to estimate the inner rotational angular velocity in stars with use of the rotational frequency splitting of the nonradial oscillations is proposed, and the "effective depth" for a specified mode is defined for practical procedure. The method is applied to three stars; B-type variable star, 53 Per, β Cep star, 12 Lac and δ Sct star, 1 Mon. It is shown that the rotational angular velocities obtained seem reasonable. However, whether or not the differential rotation in the stars is actually real cannot be confirmed in this paper because of the difficulty of getting exact information about the stellar mass and evolutionary stage.

065.063 **Power-series solutions of the Lane-Emden equation.** C. Mohan, A. R. Al-Bayaty.
Astrophys. Space Sci.,Vol. 73, 227 - 239 (1980).

The problem of representing the numerical solutions of the Lane-Emden equation analytically by means of a convergent power series has been considered. The results show that it is possible to represent the numerical solutions of the Lane-Emden equation by means of a power series which can be convergent in the whole interior of a polytropic model.

065.064 **Formation of the inhomogeneous zone in the convective core of a horizontal branch star.**
N. Arimoto.

Sci. Rep. Tôhoku Univ., Ser. 1, Vol. 62, 134 - 157 (1980).

Following the scheme of a method proposed by Suda and Uchida to treat the evolving convective core of a massive main sequence star, the author investigates the evolutionary character of the convective core in the early stages of the horizontal branch evolution. In spite of the different behaviour of the opacity, the situation is essentially identical with those in the early stages of the upper main sequence evolution. The chemically inhomogeneous zone resembling to the Sakashita and Hayashi's semiconvection zone will be formed and develop in these early stages. It has been shown that the early stages of the horizontal branch evolution would be characterized not by the overshooting from the convective core but by the formation of the Sakashita and Hayashi type inhomogeneous convection zone.

065.065 **σ-stability analysis of a toroidal magnetic field in stars.** M. Goossens.
Geophys. Astrophys. Fluid Dyn., Vol. 15, 123 - 147 (1980). Abstr. in Phys. Abstr., Vol. 84, Abstr. 4907 (1981).

065.066 **Newtonian spherical gravitational collapse.** E. N. Glass.
J. Phys. A, Vol. 13, 3097 - 3104 (1980). – Abstr. in Phys. Abstr., Vol. 84, Abstr. 9540 (1981).

065.067 **On unstably stratified toroidal magnetic fields in stars.** M. Goossens, R. J. Tayler.
Mon. Not. R. Astron. Soc., Vol. 193, 833 - 848 (1980).

A previous discussion of the stability of stars containing an axisymmetric toroidal magnetic field is extended and it is shown that there exist continuous perturbations which give a negative value of δW^{σ} when one of the conditions for stability is violated. The main characteristics of the displacement fields that are associated with the instabilities in the special, but astrophysically relevant, case of a weak toroidal field are indicated. In particular, axisymmetric toroidal magnetic fields with a strength of 10^8G in the cores of solar-type stars produce instabilities with *e*-folding times of the order of a few hours.

065.068 **The most massive stars in the Galaxy and the LMC: quasi-homogeneous evolution, time-averaged mass loss rates and mass limits.** A. Maeder.
Astron. Astrophys., Vol. 92, 101 - 110 (1980).

The evolution of massive stars in the range of 15 $M_{\odot}$ to 240 $M_{\odot}$ with two chemical compositions appropriate for the Galaxy and the Large Magellanic Cloud and various mass loss rates is studied on the basis of numerical models. In particular it is found that stars with masses $\gtrsim$ 100 $M_{\odot}$ and time-averaged mass loss rates $\gtrsim 2 \times 10^{-5}$ $M_{\odot}$ yr^{-1} evolve, after a stay near the Main-Sequence, directly towards the left in the HR diagram. The time-averaged mass loss rates of the most massive stars in the Galaxy and the LMC are estimated. For $M_{bol} > -11$, $\langle \dot{M} \rangle$ appears to be clearly inferior in the LMC with respect to the Galaxy. On the basis of the luminosities, the initial stellar mass spectrum is estimated to extend up to a maximum mass of about 150 - 200 $M_{\odot}$ with no significant differences between the Galaxy and the LMC. Both the high mass loss rates and stellar masses make rather likely the picture of quasi-homogeneous evolution for stars with $M \geqslant 100\,M_{\odot}$.

065.069 **CH subgiants and the mixing hypothesis.** J. A. Smith, P. Demarque.
Astron. Astrophys., Vol. 92, 163 - 166 (1980).

Some consequences of internal mixing at the helium core flash have been investigated for low mass, Population II stars. On the basis of the luminosity criterion, it seems unlikely that the internal mixing hypothesis can explain the CH subgiants, though it cannot be ruled out for the CH giants or other chemically peculiar giants. The authors propose a binary origin for the CH subgiants.

065.070 **The evolution of protostars. II. The hydrostatic core.** S. W. Stahler, F. H. Shu, R. E. Taam.
Astrophys. J., Vol. 242, 226 - 241 (1980).

The authors compute the structure and evolution of a protostellar core as a problem of stellar interiors. Novel techniques are introduced to account for the dynamical addition of mass and to follow the strikingly disparate thermal responses in different parts of the star. The authors present numerical results from two evolutionary calculations that begin with very different distributions of the specific entropy. The two sequences converge rapidly in time, justifying a posteriori the assumption that the final results are insensitive to the detailed mechanism for the formation of the core. The authors follow one model until the core mass reaches 1 $M_{\odot}$.

065.071 **Surface composition changes in massive star evolution with mass loss.**
A. Noels, P. S. Conti, M. Gabriel, J.-M. Vreux.
Astron. Astrophys., Vol. 92, 242 - 245 (1980).

A series of evolutionary models of 40-100 $M_{\odot}$ objects undergoing mass loss is constructed with the explicit inclusion of the surface composition of H, He, C, N, O elements. Mass loss rates similar to those observed in Of stars, 4 to 7×10^{-6} $M_{\odot}$ yr^{-1}, result in an appearance at the surface of "equilibrium" CNO products, i.e. enhanced nitrogen and diminished carbon, while the star is still burning hydrogen in the core. The authors suggest that these objects might reasonably be identified as those luminous "late" type WN stars still containing surface hydrogen.

065.072 **Stability of superdense stars in the bimetric theory of gravitation.**
V. Bálek, E. V. Chubarian (*Eh. V. Chubaryan*).
Astrophys. Space Sci., Vol. 73, 333 - 336 (1980).

Calculations of superdense star models in the Rosen bimetric theory of gravitation are presented. There is a family of 'quark stars' whose members are probably stable even if their masses do not exceed the maximum mass of neutron stars.

065.073 **Stellar age and mass determinations from kinematic data.** W. Iwanowska.
Astrophys. Space Sci., Vol. 73, 435 - 443 (1980).

Assuming that some exchange of energy between stars in the galactic disk may be at work, the author has generalized the formulae expressing the growth of the velocity variance with time by adding a mass term $\sim (M/M_{\odot})^{-\alpha}$ with $\alpha = 0.35$, which is a good fit to the kinematic data for young stars. Generalized formulae and the evolutionary mass-age relation (Iben, 1967) were used jointly to derive mass and age values for different stellar species.

065.074 **The surface chemistry of stars. III. The electric field of a chemically inhomogeneous star.** C. Alcock.
Astrophys. J., Vol. 242, 710 - 722 (1980).

The distribution of the electric field and chemical composition inside a star in diffusive equilibrium are derived. Hydrogen floats on the surface of a helium-rich star; a boundary layer of thickness about one pressure scale height produces a smooth transition in the electric field. Electrically driven mass loss does not occur unless the hydrogen content is extremely small. The electric field is dynamically unimportant outside the star, unless the star has a thermally driven wind. Arguments are given that suggest that white dwarf stars do not have such winds. It is also shown that if the star has a modest magnetic field, and if it rotates, the exterior electric field is dominated by the quadrupole contribution of the rotating magnetosphere.

065.075 **Axial rotation, tangled magnetic fields, and theoretical models of very massive stars.**
R. Stothers.

Astrophys. J., Vol. 242, 756 - 764 (1980).

A simple method of computing theoretical models of very massive stars endowed with fast axial rotation and tangled magnetic fields is described. Both of the two perturbing (non-gravitational) forces induce changes in the luminosity and radius that are studied as functions of zero-age chemical composition, opacity, and evolutionary state of the interior. The central condensation of the star is found to have a significant influence on shifts of the upper main-sequence band in the H-R diagram if the perturbing force is concentrated in the stellar envelope. It is shown that fast uniform rotation and intense envelope magnetic fields lead to probably the largest possible shifts of the main-sequence band in the H-R diagram that rotation and magnetic fields can induce.

065.076 **Evolution of the stars.** M. Vetešník.
Kozmos, Vol. 11, 131 - 132 (1980). In Czech.

065.077 **A numerical study of the fission hypothesis for rotating polytropes.** R. H. Durisen, J. E. Tohline.
Space Sci. Rev., Vol. 27, (see 012.046), 267 - 273 (1980).

065.078 **Mode discrimination in nonradially oscillating stars from light, colour and velocity observations.**
L. A. Balona, R. S. Stobie.
Space Sci. Rev., Vol. 27, (see 012.046), 371 - 376 (1980).

Expressions for the amplitudes and phases of the light, colour and radial velocity variations are derived for a star in nonradial oscillation. For stars in the cepheid instability strip the spherical harmonic mode of the oscillation can be obtained from the phase difference between the light and colour variations. For β Cep stars the mode can be estimated from the amplitude ratio of the light and colour variations.

065.079 **An extended work integral for pulsating stars.**
N. R. Simon.
Space Sci. Rev., Vol. 27, (see 012.046), 437 - 442 (1980).

The paper contains a preliminary report on research aimed at extending the pulsational work integral to next highest order beyond the linear regime.

065.080 **Spherical oscillation patterns.**
W. D. Pesnell, J. M. Coggins.
Space Sci. Rev., Vol. 27, (see 012.046), 473 - 474 (1980). – Abstract.

065.081 **Low-mass evolution from He ignition to beyond the horizontal branch.** K. Despain.
Space Sci. Rev., Vol. 27, (see 012.046), 483 - 489 (1980).

The evolution of an 0.6 $M_{\odot}$ stellar model during core helium burning is presented. Following the off-center ignition of helium in the "core" flash, the star remains on the red giant branch for $> 10^6$ years, undergoing twelve additional flashes. After leaving the giant branch, the star evolves on the horizontal branch for 8.15×10^7 years before returning to the giant branch and undergoing strong helium-shell flashes. The implications for horizontal branch and RR Lyrae stars are discussed.

065.082 **The core helium flash.** P. W. Cole, R. G. Deupree.
Space Sci. Rev., Vol. 27, (see 012.046), 491 - 493 (1980).

The role of convection in the core helium flash is simulated by two-dimensional eddies interacting with the thermonuclear runaway. The results predict that a considerable amount of helium in the core will be burned before the horizontal branch is reached and that some envelope mass loss is likely.

065.083 **The effect of rotation on the hydrodynamics of stellar collapse.** J. E. Tohline, J. M. Schombert, A. P. Boss.
Space Sci. Rev., Vol. 27, (see 012.046), 555 - 561 (1980).

065.084 **Rotational and tidal perturbations of nonradial oscillations in a polytropic star.** H. Saio.
Space Sci. Rev., Vol. 27, (see 012.046), 649 - 652 (1980).

065.085 **Rotational modes in a uniformly rotating star.**
P. Smeyers.
Space Sci. Rev., Vol. 27, (see 012.046), 653 - 659 (1980).

The linear adiabatic oscillations of a uniformly rotating star are examined with respect to a co-rotating frame of reference. It is shown that time-independent displacement fields are allowed and can be represented as toroidal fields. The time-dependent oscillation modes are governed by a system of differential equations of the fourth order in time for axisymmetric perturbations, and of the fifth order for non-axisymmetric perturbations. Therefore, in comparison to a non-rotating spherical star, a rotating star allows a new class of non-axisymmetric oscillation modes with non-zero frequencies. These modes correspond to the r-modes, which originate from purely toroidal displacement fields in a non-rotating spherical star.

065.086 **The effect of a magnetic field on stellar pulsations as a singular perturbation problem.**
M. Goossens, D. Biront.
Space Sci. Rev., Vol. 27, (see 012.046), 661 - 666 (1980).

The perturbation problem that describes the effect of a weak magnetic field on stellar adiabatic oscillation is considered. This perturbation problem is singular when the magnetic field does not vanish at the stellar surface, and a regular perturbation scheme fails where the magnetic pressure is comparable to the thermodynamic pressure. The application of the Method of Matched Asymptotic Expansion is used to obtain expressions for the eigenfunctions and the eigenfrequencies.

065.087 **On the nature of Wolf-Rayet stars.**
A. N. Tikhonov, A. V. Goncharskij,
A. M. Cherepashchuk, A. G. Yagola.
Dokl. AN SSSR, Vol. 253, 572 - 576 (1980). In Russian.
Abstr. in Ref. zh., 51. Astron., 11.51.556 (1980).

065.088 **Some properties of thermal equilibrium models of stars.** A. V. Tutukov, L. R. Yungel'son.
Astron. Zh., Tom 57, 1266 - 1272 (1980). In Russian.
English translation in Soviet Astron. Vol. 24, No. 6.

Thermal equilibrium models of 6, 10, 16, 32, 64 $M_{\odot}$ stars with evolutionary conditioned distributions of hydrogen and helium are computed. The models that correspond to the Schwarzschild criterion of convective stability cannot explain the position of the blue border of the region occupied by supergiants in the Hertzsprung-Russell diagram. If the star is irradiated by an external flux that exceeds the own flux of radiation of the star, the temperature of the stellar envelope increases practically without any changes of the stellar radius and luminosity. Increase of the external pressure under a constant temperature gives rise to convection in the stellar envelope.

065.089 **The effect of rotation and magnetic field on the minimum mass of a main-sequence star.**
A. V. Fedorova.
Astron. Zh., Tom 57, 1273 - 1276 (1980). In Russian.
English translation in Soviet Astron., Vol. 24, No. 6.

The effect of uniform and differential rotation and the effect of a small-scale random magnetic field and a poloidal magnetic field on the minimum mass of a main-sequence star are investigated. The conclusion is made that rotation and magnetic fields whose energy is 10 - 20% of the gravitational energy of a star can increase the minimum mass of a main-sequence star by a factor of 1.5 - 2.

065.090 **Qualitative analysis of star flare models.**
O. I. Bogoyavlenskij.
Usp. mat. nauk, Vol. 35, No. 4, p. 155 (1980). In Russian.
Abstr. in Ref. zh., 51. Astron., 12.51.410 (1980).

065.091 **Final stages of stellar evolution and supernova outbursts.** V. S. Imshennik, D. K. Nadezhin.
Inst. teor. i ehksperim. fiz. Moskva, Prepr., 1980, No. 91.
61 pp. In Russian. – Abstr. in Ref. zh., 51. Astron., 1.51.401 (1981).

065.092 **Final stages of stellar evolution and the hydrodynamic theory of supernova outbursts. Detailed review.** V. S. Imshennik, D. K. Nadezhin.
Inst. teor. i ehksperim. fiz. Moskva, Prepr., 1980, No. 98.
66 pp. In Russian. – Abstr. in Ref. zh., 51. Astron., 1.51.402 (1981).

065.093 **Convective mixing in extended horizontal branch envelope models: the sdB/sdO transition.**
D. E. Winget, W. Cabot.
Astrophys. J., Vol. 242, 1166 - 1175 (1980).

Static model envelopes are calculated from a $0.6\,M_\odot$ pre-white dwarf evolutionary sequence passing through the extended horizontal branch (EHB) region of the H-R diagram. The possibility of convective mixing of compositionally stratified (H-rich layer/He layer) models is investigated. It is found that mixing does occur in the range $\log T_{\rm eff} \sim 4.1-4.3$. It is found that an upper layer mass of $\leqslant 10^{-8} M_*$ is required for mixing to occur within the EHB. The presence of some initial He in the upper H-rich layer is required for mixing to occur; simple analytical calculations suggest a lower limit on the helium mass fraction of $Y \sim 0.1$.

065.094 **The evolution of mixed long-lived stars.**
H. Saio, J. C. Wheeler.
Astrophys. J., Vol. 242, 1176 - 1182 (1980).

The authors have computed evolutionary models of partially mixed stars by treating the extent of mixing and the degree of nonthermal pressure support as parameters. Unmixed and mixed models are compared to the distribution of blue stragglers in the old open cluster NGC 7789. Standard models fail to account for the apparent frequency of occurrence of blue stragglers redward of the normal main-sequence band. Theoretical time-constant loci for mixed models indicate that the distribution of blue stragglers can be understood in this way. The authors conclude that extensive internal mixing remains a viable hypothesis to account for the existence of blue stragglers.

The brightest stars. See Abstr. 003.060.

Hundert Milliarden Sonnen. See Abstr. 003.066.

Concave hamburger equilibrium of rotating bodies. See Abstr. 042.060.

Possible astrophysical sites for the production of galactic-nuclei. See Abstr. 061.009.

The origin of matter and stellar evolution. See Abstr. 061.010.

Relations between nucleosynthesis rates and the metal abundance. See Abstr. 061.013.

Nuclear reaction rate in Debye-Huckel electrolytic plasma in stellar interiors. See Abstr. 061.038.

Dense plasmas, nuclear reactions and astrophysics. See Abstr. 062.025.

Hydromagnetic rotational braking of magnetic stars. See Abstr. 062.031.

Flows along magnetic flux tubes. I. Equilibrium and buoyancy of a slender magnetic loop in the interior of a star. See Abstr. 062.033.

Asymptotic estimates for an axisymmetric rotating fluid (*star*). See Abstr. 062.064.

An analytical version of the free-energy-minimization method for the equation of state of stellar plasmas. See Abstr. 062.082.

Equilibrium of self-gravitating polytropic cylinders with a magnetic field. See Abstr. 062.097.

Thermodynamic transport properties in dense stars. See Abstr. 062.123.

σ-stability analysis of hydromagnetic instabilities in stars with toroidal magnetic fields. See Abstr. 062.124.

Stellar convection theory. See Abstr. 064.019.

Differential rotation in stars with convection zones. See Abstr. 064.022.

Sur la stabilisation des oscillations d'une étoile dans le champ magnétique vertical et horizontal. See Abstr. 064.037.

Near-unstable supergiants. See Abstr. 064.043.

Mass-loss from spinning stars. See Abstr. 064.058.

Spherical accretion onto a hydrostatic protostellar core. See Abstr. 064.099.

Experimental bounds on the coupling strength of torsion potentials. See Abstr. 066.042.

The collapsed state of a star. See Abstr. 066.049.

Some properties of static general relativistic stellar models. See Abstr. 066.084.

Hydrogen burning on accreting neutron star. See Abstr. 066.501.

Neutron star formation by collapse of white dwarfs. See Abstr. 066.532.

The solar wind and related astrophysical phenomena. See Abstr. 074.106.

The stars as suns. See Abstr. 114.168.

An observational H-R diagram for early-type stars. See Abstr. 115.002.

Supergiant variability: amplitudes and pulsation constants in relation with mass loss and convection. See Abstr. 115.005.

Evolution of a blue supergiant with a neutron star companion immersed in its envelope. See Abstr. 117.001.

Evolution of low-mass, distorted stars in a very close binary system with the inclusion of gravitational-radiation losses. See Abstr. 117.077.

A model of the subdwarf binary system LB 3459. See Abstr. 119.002.

Three-mode resonances in double-mode cepheids. See Abstr. 122.019.

The convection and stellar activity in Cepheid variables. See Abstr. 122.140.

Nonlinear calculations for BL Her stars. See Abstr. 122.193.

R Coronae Borealis pulsations. See Abstr. 122.196.

Pulsations of the R Coronae Borealis stars. See Abstr. 122.197.

Core collapse, bounce and shock propagation. See Abstr. 125.071.

Supernova explosions – the role of a Rayleigh-Taylor instability. See Abstr. 125.074.

The effect of stellar structure on supernova remnant evolution. See Abstr. 125.075.

Nucleosynthesis during nova explosions and grain formation in ejected envelopes. See Abstr. 126.025.

The long-term evolution of accreting carbon white dwarfs. See Abstr. 126.034.

Protostars and pre-main sequence objects. See Abstr. 131.021.

Star-formation and star-rotation. See Abstr. 131.268.

On solar type protostars. See Abstr. 131.308.

X-ray sources and stellar evolution. See Abstr. 142.134.

Galactic origin of cosmic gamma-ray bursts. See Abstr. 142.512.

Stellar evolution and globular clusters. See Abstr. 154.029.

Erratum

065.901 **Erratum: "Maximization of surface redshift"** [Bull. Astron. Soc. India, Vol. 7, 115 (1979)].
M. C. Durgapal, P. S. Rawat, K. Pandey.
Bull. Astron. Soc. India, Vol. 8, 30 (1980). – See Abstr. 27.065.051.

066 Relativistic Astrophysics, Gravitation Theory, Background Radiation, Black Holes, Neutron Stars

066.001 **Primordial baryon generation by black holes.**
J. D. Barrow.
Mon. Not. R. Astron. Soc., Vol. 192, 427 - 438 (1980).

The production of baryons by black holes small enough to evaporate X, $\bar{X}$ bosons is compared with the Weinberg–Wilczek model for baryon production which appeals only to the free decay of X-bosons in the SU(5), grand unified gauge theory. A detailed model of the baryon generation is given in an isotropically expanding Universe containing a fraction of its energy density in black holes small enough to evaporate X, $\bar{X}$ bosons by the Hawking effect.

066.002 **Observational tests of the cosmic turbulence theory.**
L. Danese, G. De Zotti.
Astron. Astrophys., Vol. 87, 303 - 306 (1980).

The authors discuss the constraints imposed to the basic parameter W of cosmological turbulence by observations of the microwave background spectrum and isotropy, and of the dynamics of galaxies. They show that those values of W which could meet the isotropy constraint entail excessive distortions of the background spectrum and galactic properties inconsistent with observations.

066.003 **Indirekter Nachweis von Gravitationswellen gelungen.**
H. Ritter.
Umschau, 80. Jahrg., 537 - 538 (1980).

The astronomical observations of the binary pulsar and of the binary Z Cha, which both present indirect evidence for the existence of gravitational radiation are briefly discussed.

066.004 **The nature of gravitation.** A. H. Cook.
Nature, Vol. 287, 189 - 190 (1980).

066.005 **Diffraction of electromagnetic waves in Schwarzschild's space-time.** T. Elster.
Astrophys. Space Sci., Vol. 71, 171 - 194 (1980).

In order to study the gravitational lens effect in detail the paper investigates the electromagnetic radiation generated by an electric dipole oscillating with high frequency in the Schwarzschild metric outside the horizon. Expressions for the Newman-Penrose tetrad components characterizing the radiation field are derived in a suitable approximation. The Poynting vector is discussed in the region behind the deflecting mass (black hole, neutron star) in the fully coherent case for large distances of source and observer from the centre of the lens. Intensity and gain in the focal region coincide with the corresponding values applying to scalar radiation. An observer sees double images of the source outside the focal region. The results (positions of the images, relative intensities) are compared with those occurring in the case of an incident plane wave and with the properties of the images in the (geometrical optics) limit of incoherent radiation.

066.006 **Primary black holes.**
I. D. Novikov, A. G. Polnarev.
Priroda, 1980, No. 7, p. 12 - 18. In Russian.

066.007 **Shear hell holes and anisotropic universes.**
B. J. Carr, J. D. Barrow.
Gen. Relativ. Gravitation, Vol. 11, 383 - 389 (1979).

If the early universe was highly anisotropic, primordial black holes may have formed prolifically even if the initial density fluctuations were small. However, the holes would initially be endowed with an immense amount of shear, so it is not obvious that they would evolve into the conventional type of stationary black hole envisaged by the "no hair" theorem. In principle there might exist soliton-type solutions which represent holes with shear which persists indefinitely. Such "shear hell holes", as we term them, could have even more dramatic properties than the usual stationary holes: in particular, they might be prolific generators of gravitational radiation and they could be associated with interesting quantum effects.

066.008 **On the electromagnetic detection of gravitational waves.** V. B. Braginsky (*Braginskij*), L. P. Grishchuk, A. G. Doroshkevich, M. B. Mensky, I. D. Novikov, M. V. Sazhin, Ya. B. Zeldovich (*Zel'dovich*).
Gen. Relativ. Gravitation, Vol. 11, 407 - 409 (1979).

The general principles of the electromagnetic detection of gravitational waves are discussed. A critical comment on a previous paper by Baierlein devoted to the same problem is given.

066.009 **High-energy gravity and the very early universe.**
H. Terazawa.
Gen. Relativ. Gravitation, Vol. 12, 93 - 97 (1980).

It is suggested that gravity may not be asymptotically free at short distances because of the interaction of the graviton with matter. If gravity indeed becomes strong at high energies, a revolutionary change of our present theory on the early universe would seem to be necessary. During the first extremely small fraction of a second in the big-bang universe, gravity would have been so strong that it might not have been described by Einstein's theory of general relativity. The possibility of abnormally strong gravity at high energies or short distances is discussed in some detail. A possible explanation is proposed for the nonvanishing mean baryon number density of the universe. It is also pointed out that the universe may well escape from the catastrophic singularity of Penrose and Hawking.

066.010 **Gravitational collapse with charge and small asymmetries. II. Interacting electromagnetic and gravitational perturbations.** J. Bičák.
Gen. Relativ. Gravitation, Vol. 12, 195 - 204 (1980).

Paper I (see 09.065.165) analyzed the evolution of nonspherical scalar-field pertubations of an electrically charged, collapsing star; this paper treats coupled electromagnetic and gravitational perturbations. It employs the results of recent detailed work in which coupled perturbations were studied in a gauge-invariant manner by using the Hamiltonian (Moncrief's) approach and the Newman-Penrose formalism, and the relations between the fundamental quantities of these two methods were obtained.

066.011 **Integrability conditions for a gravitational theory with nonmetric connection.**
M. Francaviglia, J. Kijowski.
Gen. Relativ. Gravitation, Vol. 12, 279 - 286 (1980).

In the framework of a new gravitational theory with nonmetric connection it is shown that the matter stress tensors satisfy a certain identity, which, via the contracted Bianchi identities, turns out to be a formal integrability condition for the gravitational field equations. The conservation law for the Hilbert tensor is also discussed.

066.012 **Radiation and the structure of space-time.**
M. Soffel, B. Müller, W. Greiner.
Gen. Relativ. Gravitation, Vol. 12, 287 - 303 (1980).

The classical bremsstrahlung problem is discussed with respect to the global structure of space-time. Various physical phenomena (such as the Hawking effect, etc.) connected with the large-scale structure of space-time are mentioned.

066.013 **Einstein's theory of gravitation.** J. Ehlers.
Einstein Symposion Berlin, (see 012.003), p. 10 - 35 (1979).
Contents: Basic assumptions about spacetime and local, nongravitational physics. Frames, orientations, bundles and spinor fields. Classical descriptions of matter and radiation. The Einstein-Hilbert field equation of gravitation. Some consequences and problems of GR (*general theory of relativity*). Isolated systems. The n-body problem in GR. Connection between GR and observations.

066.014 **Recent advances in global general relativity: a brief survey.** R. Penrose.
Einstein Symposion Berlin, (see 012.003), p. 36 - 45 (1979).

066.015 **Der Dualismus von Feld und Materie in der allgemeinen Relativitätstheorie.** P. Mittelstaedt.
Einstein Symposion Berlin, (see 012.003), p. 308 - 319 (1979).

066.016 **Microwave background radiation in the directions to clusters of galaxies.** R. A. Syunyaev.
Pis'ma Astron. Zh., Tom 6, 387 - 393 (1980). In Russian. English translation in Soviet Astron. Lett., Vol. 6.
Clusters of galaxies containing hot intergalactic gas may be considered as powerful sources of submillimeter radiation and "negative" sources at millimeter wavelengths due to scattering of the microwave background photons on hot electrons. The radiation flux density from the cluster, its spectrum and the "luminosity" dependence on redshift are computed. The microwave background brightness in the direction to the cluster is evaluated in the single scattering approximation.

066.017 **Black holes and gravitational waves. III. The resonant frequencies of rotating holes.** S. Detweiler.
Astrophys. J., Vol. 239, 292 - 295 (1980).
The free oscillations of rotating black holes are studied. Values of the complex resonant frequencies are given as functions of the Kerr parameter a for a variety of spherical harmonic indices l and m. The appendix uses analytic methods to show that the maximally rotating Kerr solution is, in some sense, marginally unstable.

066.018 **Gravitational focusing by a slowly rotating, relativistic, spherical mass.** J. K. Lawrence.
Astrophys. J., Vol. 239, 305 - 309 (1980).
Null rays emitted inside a relativistic, spherical mass can, depending on the location of the emission point, emerge in highly focused patterns. In order to increase the realism of this result, the author allows slow rotation of the mass. Depending on the location of the emission point, the effect of the rotation may be nil, the beam direction may be shifted in the direction of the rotation, the radiation pattern may be stretched in the direction opposite to the rotation, or the beams may be defocused by an amount proportional to the specific angular momentum.

066.019 **Quantum mechanics and gravitational waves.** B. Marx.
Nature, Vol. 287, 276 - 277 (1980).

066.020 **Dipole anisotropy in the 2.7 K background.** E. S. Cheng, S. Boughn, D. T. Wilkinson.
Bull. American Astron. Soc., Vol. 12, 489 (1980). – Abstract.

066.021 **The role of general relativity in astronomy: retrospect and prospect.** S. Chandrasekhar.
J. Astrophys. Astron., Vol. 1, 33 - 45 (1980). – Invited discourse delivered at the Seventeenth General Assembly of the International Astronomical Union held at Montreal, Canada.

066.022 **Symmetric vectors and algebraic classification.** E. Leibowitz.
J. Math. Phys., Vol. 21, 1141 - 1148 (1980). – Abstr. in Phys. Abstr., Vol. 83, Abstr. 68094 (1980).

066.023 **A new formalism of the Einstein equations for relativistic rotating systems.**
K. Maeda, M. Sasaki, T. Nakamura, S. Miyama.
Prog. Theor. Phys., Vol. 63, 719 - 721 (1980). – Abstr. in Phys. Abstr., Vol. 83, Abstr. 68101 (1980).

066.024 **Conformal coupling of gravitational wave field to curvature.** L. P. Grishchuk, V. M. Yudin.
J. Math. Phys., Vol. 21, 1168 - 1175 (1980). – Abstr. in Phys. Abstr., Vol. 83, Abstr. 68105 (1980).

066.025 **Physical and mathematical consequences of a 'gravitational' radiation which was attenuated upon its passage through matter.** I. A. Adamut.
Lucr. ICPE, No. 31, p. 23 - 32. In French. – Abstr. in Phys. Abstr., Vol. 83, Abstr. 68106 (1980).

066.026 **On the shielding of the gravitational field. An electrothermodynamical theory of gravitation.**
I. A. Adamut.
Lucr. ICPE, No. 31, p. 35 - 41. In Rumanian. – Abstr. in Phys. Abstr., Vol. 83, Abstr. 68107 (1980).

066.027 **Eleven-dimensional supergravity on the mass shell in superspace.** L. Brink, P. Howe.
Phys. Lett. B, Vol. 91B, 384 - 386 (1980). – Abstr. in Phys. Abstr., Vol. 83, Abstr. 68116 (1980).

066.028 **Detecting gravitational waves.** S. Boughn.
American Sci., Vol. 68, 174 - 183 (1980). – Abstr. in Phys. Abstr., Vol. 83, Abstr. 68122 (1980).

066.029 **Perturbations of spherically symmetric black holes.** B. C. Xanthopoulos.
Phys. Lett. A, Vol. 77A, 7 - 8 (1980). – Abstr. in Phys. Abstr., Vol. 83, Abstr. 72992 (1980).

066.030 **On the possibility of general relativistic oscillations.** H. Knutsen, R. Stabell.
Ann. Inst. Henri Poincaré Sect. A, Vol. 31, 339 - 353 (1979). Abstr. in Phys. Abstr., Vol. 83, Abstr. 73217 (1980).

066.031 **Generalized radial observers and the Reissner-Nordström field.** C. Ftaclas, J. M. Cohen.
Phys. Rev. D, Vol. 21, 2103 - 2106 (1980). – Abstr. in Phys. Abstr., Vol. 83, Abstr. 73223 (1980).

066.032 **Asymptotic structure of isolated systems.** B. G. Schmidt.
Isolating gravitating systems in general relativity, (see 012.004), p. 11 - 49 (1979). – Abstr. in Phys. Abstr., Vol. 83, Abstr. 73230 (1980).

066.033 **Asymptotic symmetries, energy-momentum and angular momentum at future null infinity.**
M. Walker.
Isolating gravitating systems in general relativity, (see 012.004), p. 50 - 60 (1979). – Abstr. in Phys. Abstr., Vol. 83, Abstr. 73231 (1980).

066.034 **Equations of motion and radiation reaction in the special and general theory of relativity.** P. Havas.
Isolating gravitating systems in general relativity, (see 012.004). p. 74 - 155 (1979). – From Phys. Abstr., Vol. 83, Abstr. 73232 (1980).

066.035 **Extended bodies in general relativity: their description and motion.** W. G. Dixon.
Isolating gravitating systems in general relativity, (see 012.004), p. 156 - 219 (1979). – From Phys. Abstr., Vol. 83, Abstr. 73233 (1980).

066.036 **Perturbation methods for interactions between strongly self-gravitating systems.** P. D. D'Eath.
Isolating gravitating systems in general relativity, (see 012.004), p. 249 - 288 (1979). – From Phys. Abstr., Vol. 83, Abstr. 73234 (1980).

066.037 **Topics in the dynamics of general relativity.** A. E. Fischer, J. E. Marsden.
Isolating gravitating systems in general relativity, (see 012.004), p. 322 - 395 (1979). – Abstr. in Phys. Abstr., Vol. 83, Abstr. 73236 (1980).

066.038 **Multipole expansions of gravitational radiation.** K. S. Thorne.
Rev. Mod. Phys., Vol. 52, No. 1, Part 1, p. 299 - 339 (1980). Abstr. in Phys. Abstr., Vol. 83, Abstr. 73240 (1980).

066.039 **Remark on Trautman's radiation condition.** M. Walker.
Isolating gravitating systems in general relativity, (see 012.004), p. 61 - 62 (1979). – Abstr. in Phys. Abstr., Vol. 83, Abstr. 73241 (1980).

066.040 **The slow-motion approximation in radiation problems.** W. L. Burke.
Isolating gravitating systems in general relativity, (see 012.004), p. 220 - 248 (1979). – Abstr. in Phys. Abstr., Vol. 83, Abstr. 73242 (1980).

066.041 **Gravitational energy loss in scattering problems.** A. Rosenblum.
Isolating gravitating systems in general relativity, (see 012.004), p. 313 - 317 (1979). – Abstr. in Phys. Abstr., Vol. 83, Abstr. 73244 (1980).

066.042 **Experimental bounds on the coupling strength of torsion potentials.** D. E. Neville.
Phys. Rev. D, Vol. 21, 2075 - 2080 (1980). – Abstr. in Phys. Abstr., Vol. 83, Abstr. 73249 (1980).

066.043 **Massless limit of vector multiplets in supergravity.** S. Deser.
Phys. Rev. D, Vol. 21, 2436 - 2437 (1980). – Abstr. in Phys. Abstr., Vol. 83, Abstr. 73250 (1980).

066.044 **Gravitational-wave research: current status and future prospects.** K. S. Thorne.
Rev. Mod. Phys., Vol. 52, No. 1, Part 1, p. 285 - 297 (1980). Abstr. in Phys. Abstr., Vol. 83, Abstr. 73254 (1980).

066.045 **Search for the intermediate-scale anisotropy of the cosmological background radiation.**
R. Fabbri, B. Melchiorri, F. Melchiorri, V. Natale, N. Caderni, K. Shivanandan.
Phys. Rev. D. Vol. 21, 2095 - 2102 (1980). – Abstr. in Phys. Abstr., Vol. 83, Abstr. 77524 (1980).

066.046 **From conformal supergravity in ordinary space to its superspace constraints.**
P. Van Nieuwenhuizen, P. C. West.
Nucl. Phys. B, Vol. B169, 501 - 513 (1980). – Abstr. in Phys. Abstr., Vol. 83, Abstr. 98844 (1980).

066.047 **Time-dependent, optically thick accretion onto a black hole.** D. L. Gilden, J. C. Wheeler.
Astrophys. J., Vol. 239, 705 - 711 (1980).

The authors have used a fully relativistic hydrodynamics code which incorporates diffusive radiation transport to study time-dependent, spherically symmetric, optically thick accretion onto a black hole. They find that matter free-falls into the hole regardless of whether the diffusion time scale is longer or shorter than the dynamical time.

066.048 **Particle field in bimetric general relativity.** D. Falik, N. Rosen.
Astrophys. J., Vol. 239, 1024 - 1031 (1980).

The field equations of the bimetric general relativity theory proposed recently by one of the authors (N. Rosen) are put into a static form. The equations are solved near the Schwarzschild sphere, and it is found that the field differs from that of the Einstein general relativity theory: instead of a black hole, one has an impenetrable sphere. For larger distances the field is found to agree with that of ordinary general relativity, so that solar system observations cannot distinguish between the two theories. For very large distances one gets a cosmic contribution to the field which may affect the dynamics of clusters of galaxies.

066.049 **The collapsed state of a star.** N. Rosen.
Astrophys. J., Vol. 239, 1032 - 1035 (1980).

From the field equations of the bimetric general relativity theory one finds that, as the end product of gravitational collapse, there exists a collapsed state of a star in which the Schwarzschild sphere is completely filled with matter of uniform density and pressure. The mass is given by $8\pi m^2 = 1/(\rho - P)$. In the case of pressureless matter the metric is continuous on the stellar surface. In the case of nonzero pressure there is a discontinuity of the metric across this surface. Inside the collapsed object the physical metric has the same form as that of an Einstein closed universe.

066.050 **Gravitational radiation from colliding, compact stars: hydrodynamic calculations in one dimension.**
S. L. Shapiro.
Astrophys. J., Vol. 240, 246 - 263 (1980).

Gravitational radiation from the collision of two, identical, uncollapsed compact stars which collide head-on at free-fall velocity is calculated. The hydrodynamic treatment is Newtonian and adiabatic. The collision is analyzed by restricting all motion to be parallel to the collision axis. This one-dimensional approximation, though naive, should yield reasonable estimates provided the dominant contribution to the burst of radiation following impact results from the initial compression, deceleration, and reexpansion of gas along the collision axis. Two hydrodynamic models are considered. The first model ignores gravity and can essentially be solved analytically. This model facilitates the formulation of reliable numerical procedures for calculating gravitational wave luminosities from raw, hydrodynamic data on a spacetime lattice. The second model incorporates gravity by utilizing "planar polytropes". These plane-symmetric configurations model stars which, prior to collision, are in hydrostatic equilibrium along the collision axis. Planar polytropes are used to analyze collisions between neutron stars and supermassive stars. Possible implications of the results for the search for gravitational waves are explored briefly.

066.051 **Transonic disk accretion onto black holes.** E. P. T. Liang, K. A. Thompson.
Astrophys. J., Vol. 240, 271 - 274 (1980).

The solution for the radial drift velocity of thin disk accretion onto black holes must be transonic, and is analogous to the critical solution in spherical Bondi accretion, except for the presence of angular momentum. The transonic requirement yields a correct treatment of the inner region of the disk not found in the conventional Keplerian models and may lead to significantly different overall disk structures. Possible observational consequences, relevant to the black hole hypothesis for Cyg X-1 and other candidates, are discussed.

066.052 **Angular distribution of the microwave background and its intensity in the directions of clusters of galaxies.** Ya. B. Zel'dovich, R. A. Syunyaev.
Pis'ma Astron. Zh., Tom 6, 545 - 547 (1980). In Russian. English translation in Soviet Astron. Lett., Vol. 6.

Observations of the microwave background intensity and polarization in the directions of clusters and superclusters of galaxies permit to obtain information on the angular distribution of the background and in particular on its quadrupole component.

066.053 **Electrodynamics in scale-covariant gravity theory.** V. N. Mansfield, S. Malin.
Astron. Astrophys., Vol. 89, 70 - 73 (1980).

Utilizing the inherent scale-invariance of Maxwell's Equations, classical electrodynamics is incorporated into the theory of scale-invariant gravity. In this incorporation the gravitational constant G is shown to transform like β^{-2} (β is the gauge function), the generalized Lorentz Force Law is derived, the electric charge is shown to be invariant under gauge transformation, and matter creation is shown to be a necessity. In all nontrivial gauges a modified version of QED is obtained. The deviation from standard QED, however, is shown to be beyond the range of experimental detection when $G\alpha\beta^{-2}$.

066.054 **Gravitational radiation from slowly rotating collapse: an exact Green function.**
B. D. Gaiser, R. V. Wagoner.
Astrophys. J., Vol. 240, 648 - 657 (1980).

The production of gravitational radiation by pressureless, axisymmetric, slowly rotating bodies is analyzed within general relativity. The calculation is carried out at first order in the angular velocity, in which case the density remains spherically symmetric, but with arbitrary initial radial structure. The authors obtain a Green function which relates the radiation received far from the star to boundary conditions on the surface of the star. They find that at late times, the Green function is dominated by a few quasi-normal modes of the Schwarzschild geometry.

066.055 **A new test of the theory of general relativity.** D. F. Crawford.
Proc. Astron. Soc. Australia, Vol. 3, 364 (1979).

066.056 **On caustics and singularities in general relativity.** H.-H. von Borzeszkowski, H. Paul.
Ann. Physik, Vol. 37, 102 - 108 (1980). – Abstr. in Phys. Abstr., Vol. 83, Abstr. 77654 (1980).

066.057 **Relativistic two-body interactions: a Hamiltonian formulation.** R. Giachetti, E. Sorace.
Nuovo Cimento B, Ser. 11, Vol. 56B, 263 - 301 (1980). Abstr. in Phys. Abstr., Vol. 83, Abstr. 77666 (1980).

066.058 **Propagation equations for test bodies with spin and rotation in theories of gravity with torsion.**
P. B. Yasskin, W. R. Stoeger.
Phys. Rev. D, Vol. 21, 2081 - 2094 (1980). – Abstr. in Phys. Abstr., Vol. 83, Abstr. 77670 (1980).

066.059 **Where has the fifth dimension gone?** A. Chodos, S. Detweiler.
Phys. Rev. D, Vol. 21, 2167 - 2170 (1980). – Abstr. in Phys. Abstr., Vol. 83, Abstr. 77679 (1980).

066.060 **Absorption of gravitational waves by water.** J. Chiba.
Nuovo Cimento C, Ser. 1, Vol. 2C, 549 - 568 (1979). – Abstr. in Phys. Abstr., Vol. 83, Abstr. 82189 (1980).

066.061 **The motion of a charged black hole in an electromagnetic field.** J. Bičák.
Proc. R. Soc. London, Ser. A, Vol. 371, 429 - 438 (1980). Abstr. in Phys. Abstr., Vol. 83, Abstr. 82407 (1980).

066.062 **Time-asymmetric initial data for black holes and black-hole collisions.** J. M. Bowen, W. York, Jr.
Phys. Rev. D, Vol. 21, 2047 - 2056 (1980). – Abstr. in Phys. Abstr., Vol. 83, Abstr. 82408 (1980).

066.063 **Quantum-mechanical instability of the Kerr-Newman black hole interior.** W. A. Hiscock.
Phys. Rev. D, Vol. 21, 2057 - 2063 (1980). – Abstr. in Phys. Abstr., Vol. 83, Abstr. 82409 (1980).

066.064 **Vacuum polarization in Schwarzschild spacetime.** P. Candelas.
Phys. Rev. D, Vol. 21, 2185 - 2202 (1980). – Abstr. in Phys. Abstr., Vol. 83, Abstr. 82410 (1980).

066.065 **On general-relativistic celestial mechanics.** H.-J. Treder.
Mitt. Astron. Ges., Nr. 48, (see 012.015), p. 177 - 188 (1980).

The fundamental principles of general relativistic dynamics are deduced from Einstein's field equations and the one- and two-particle problems in the relativistic celestial mechanics are discussed. The author demonstrates the determination of the Riemannian space curvature by the motions of twins of spinning satellites. Further he discusses the different opinions about the physical meaning of the calculations on gravitation radiation for double stars and so on. The author proves that these different opinions are founded by different interpretations of Einstein's gravitation equations as generally covariant determinations of the spacetime metrics or as gauge-invariant tensor field equations in a given space-time background. Gravitation radiation becomes physically meaningful by a recent point of view.

066.066 **Relativistische Bewegung ausgedehnter Körper.** A. Caporali.
Mitt. Astron. Ges., Nr. 48, (see 012.015), p. 189 (1980).

066.067 **Causal paradoxes implied by the hypothetical co-existence of positive- and negative-mass matter.**
R. A. de Martins.
Nuovo Cimento, Lett., Ser. 2, Vol. 28, 265 - 268 (1980). Abstr. in Phys. Abstr., Vol. 83, Abstr. 82685 (1980).

066.068 **On the problem of the singularities in the general cosmological solution of the Einstein equations.**
V. A. Belinskii *(Belinskij)*, I. M. Khalatnikov, E. M. Lifshitz *(Lifshits)*.
Phys. Lett. A, Vol. 77A, 214 - 216 (1980). – Abstr. in Phys. Abstr., Vol. 83, Abstr. 82686 (1980).

066.069 **On rotating charged dust in general relativity. III.** J. N. Islam.
Proc. R. Soc. London, Ser. A, Vol. 372, 111 - 115 (1980). Abstr. in Phys. Abstr., Vol. 83, Abstr. 82688 (1980).

066.070 **Hamiltonian formalism for perfect fluids in general relativity.** J. Demaret, V. Moncrief.
Phys. Rev. D, Vol. 21, 2785 - 2793 (1980). – Abstr. in Phys. Abstr., Vol. 83, Abstr. 82689 (1980).

066.071 **Electromagnetic phenomena induced by weak gravitational fields. Foundations for a possible gravitational wave detector.**
J. M. Codina, J. Graells, C. Martin.
Phys. Rev. D, Vol. 21, 2731 - 2735 (1980). – Abstr. in Phys. Abstr., Vol. 83, Abstr. 82692 (1980).

066.072 **Gravitational wave perturbations and gauge conditions.** J. Centrella.
Phys. Rev. D, Vol. 21, 2776 - 2784 (1980). – Abstr. in Phys. Abstr., Vol. 83, Abstr. 82693 (1980).

066.073 **Electromagnetic fields in space-times with local rotational symmetry.**
S. V. Dhurandhar, C. V. Vishveshwara, J. M. Cohen.
Phys. Rev. D, Vol. 21, 2794 - 2804 (1980). – Abstr. in Phys. Abstr., Vol. 83, Abstr. 82694 (1980).

066.074 **No black holes: a gravitational gauge theory possibility.** D. B. Chang, H. H. Johnson.
Phys. Lett. A, Vol. 77A, 411 - 415 (1980). – Abstr. in Phys. Abstr., Vol. 83, Abstr. 82702 (1980).

066.075 **Do black holes exist?** N. Rosen.
Nuovo Cimento, Lett., Ser. 2, Vol. 28, 221 - 224 (1980). – Abstr. in Phys. Abstr., Vol. 83, Abstr. 85853 (1980).

066.076 **Surfaces of infinite red-shift around a uniformly accelerating and rotating particle.**
H. Farhoosh, R. L. Zimmerman.
Phys. Rev. D, Vol. 21, 2064 - 2074 (1980). – Abstr. in Phys. Abstr., Vol. 83, Abstr. 85855 (1980).

066.077 **Stable circular orbits of test particles moving in the equatorial plane of a black hole.**
M. M. Kumar, S. S. Prasad.
Nuovo Cimento, Lett., Ser. 2, Vol. 28, 269 - 274 (1980). Abstr. in Phys. Abstr., Vol. 83, Abstr. 85970 (1980).

066.078 **Stochastic evolution of Schwarzschild black hole in a radiation heat bath.** W. H. Zurek.
Phys. Lett. A, Vol. 77A, 399 - 403 (1980). – Abstr. in Phys. Abstr., Vol. 83, Abstr. 85971 (1980).

066.079 **Conserved energy flux for the spherically symmetric system and the back reaction problem in black hole evaporation.** H. Kodama.
Prog. Theor. Phys.,Vol. 63, 1217 - 1228 (1980). – Abstr. in Phys. Abstr., Vol. 83, Abstr. 85972 (1980).

066.080 **Cosmic microwave background spectrum and G-varying cosmology.** J. V. Narlikar, N. C. Rana.
Phys. Lett. A, Vol. 77A, 219 - 220 (1980). – Abstr. in Phys. Abstr., Vol. 83, Abstr. 86020 (1980).

066.081 **The rigidly rotating relativistic dust cylinder.** W. B. Bonnor.
J. Phys. A, Vol. 13, 2121 - 2132 (1980). – Abstr. in Phys. Abstr., Vol. 83, Abstr. 86158 (1980).

066.082 **Space-times with distribution valued curvature tensors.** A. H. Taub.
J. Math. Phys., Vol. 21, 1423 - 1431 (1980). – Abstr. in Phys. Abstr., Vol. 83, Abstr. 86163 (1980).

066.083 **Matching of the plane symmetric static and homogeneous vacuum solutions of the Einstein field equation.** G. T. Carlson, Jr., J. L. Safko.
J. Math. Phys., Vol. 21, 1442 - 1448 (1980). – Abstr. in Phys. Abstr., Vol. 83, Abstr. 86164 (1980).

066.084 **Some properties of static general relativistic stellar models.** L. Lindblom.
J. Math. Phys., Vol. 21, 1455 - 1459 (1980). – Abstr. in Phys. Abstr., Vol. 83, Abstr. 86165 (1980).

066.085 **On gravitational Lagrangians in flat space-time.** M. Castagnino, L. Chimento.
Nuovo Cimento, Lett., Ser. 2, Vol. 28, 471 - 475 (1980). Abstr. in Phys. Abstr., Vol. 83, Abstr. 86169 (1980).

066.086 **Propagation of a shock wave in general relativity. Stationary approximation.** T. Ishizuka.
Prog. Theor. Phys., Vol. 63, 1541 - 1550 (1980). – Abstr. in Phys. Abstr., Vol. 83, Abstr. 86172 (1980).

066.087 **Can spin thermodynamics explain superradiance in rotating black holes?** A. Curir, M. Francaviglia.
Nuovo Cimento, Lett., Vol. 28, 426 - 428 (1980). – Abstr. in Phys. Abstr., Vol. 83, Abstr. 86179 (1980).

066.088 **Design of a tunable detector for gravitational radiation.** M. Karim.
J. Phys. A, Vol. 13, 2133 - 2142 (1980). – Abstr. in Phys. Abstr.,Vol. 83, Abstr. 86195 (1980).

066.089 **Does the geodesic equation contradict Hawking's black hole area theorem?** O. Gron.
Phys. Lett. A, Vol. 78A, 31 - 32 (1980). – Abstr. in Phys. Abstr., Vol. 83, Abstr. 90468 (1980).

066.090 **Relativistic corrections to the gravitational radiation of a binary system and the fine structure of the spectrum.**
D. V. Galtsov (*Gal'tsov*), A. A. Matiukhin, V. I. Petukhov.
Phys. Lett. A, Vol. 77 A, 387 - 390 (1980). – Abstr. in Phys. Abstr., Vol. 83, Abstr. 90479 (1980).

066.091 **Gauge theories, time dependence of the gravitational constant and antigravity in the early Universe.**
A. D. Linde.
Phys. Lett. B, Vol. 93B, 394 - 396 (1980). – Abstr. in Phys. Abstr., Vol. 83, Abstr. 90591 (1980).

066.092 **Conformal covariance of general relativity.** N. Ionescu-Pallas, I. Gottlieb.
Rev. Roumaine Phys., Vol. 25, 233 - 244 (1980). – Abstr. in Phys. Abstr., Vol. 83, Abstr. 90736 (1980).

066.093 **Light deviation in invariant mechanics.** M. Agop.
Rev. Roumaine Phys., Vol. 25, 295 - 299 (1980). In French. – Abstr. in Phys. Abstr., Vol. 83, Abstr. 90740 (1980).

066.094 **Statistical formulation of gravitational radiation reaction.** B. F. Schutz.
Phys. Rev. D, Vol. 22, 249 - 259 (1980). – Abstr. in Phys. Abstr., Vol. 83, Abstr. 90749 (1980).

066.095 **Geometric quantization and gravitational collapse.** M. J. Gotay, J. A. Isenberg.
Phys. Rev. D, Vol. 22, 235 - 248 (1980). – Abstr. in Phys. Abstr., Vol. 83, Abstr. 90752 (1980).

066.096 **The gravitational mirage.** M. Lachieze-Rey, J. Schneider.
Recherche, Vol. 11, 827 - 828 (1980). In French. – Abstr. in Phys. Abstr., Vol. 83, Abstr. 94710 (1980).

066.097 **Giant black holes.** B. Carter, J. P. Luminet.
Recherche, Vol. 11, 694 - 701 (1980). In French. Abstr. in Phys. Abstr., Vol. 83, Abstr. 94828 (1980).

066.098 **Gravitational lens in the universe?** V. F. Mukhanov.
Priroda, 1980, No. 10, p. 107 - 108. In Russian.

066.099 **Emission of massive particles from a black hole.** A. B. Gaina, I. A. Obukhov.
MGU, Moskva, 1980, 11 pp. In Russian. – Abstr. in Ref. zh., 51. Astron., 8.51.776 (1980).

066.100 **Scattering and absorption of scalar particles and fermions in a Reissner-Nordstrøm field.**
A. B. Gaina.
MGU, Moskva, 1980, 20 pp. In Russian. – Abstr. in Ref. zh., 51. Astron., 8.51.777 (1980).

066.101 **Gibt es Schwarze Löcher?** H.-E. Fröhlich.
Sterne, 56. Band, 220 - 226 (1980).

066.102 **Supergravitation.** J. Wess.
Naturwissenschaften, 67. Jahrg., 484 - 487 (1980).

066.103 **"Abnormal" temperature profile solutions of accretion disc around black holes.**
L. Fang, J. Zhang, S. Jiang.
Kexue Tongbao, Vol. 25, 851 - 853 (1980).

066.104 **Planet lander and verification of general relativity.**
N. Kawashima.
13th Lunar and Planetary Symposium, (see 012.018), p. 407 - 415 (1980).

066.105 **A spherical cavity in an Einstein universe.**
N. K. Kofinti.
Int. J. Theor. Phys., Vol. 19, 177 - 183 (1980). – Abstr. in Phys. Abstr., Vol. 83, Abstr. 95013 (1980).

066.106 **Isotropy of the velocity of light.** K. Ruebenbauer.
Int. J. Theor. Phys., Vol. 19, 217 - 219 (1980).
Abstr. in Phys. Abstr., Vol. 83, Abstr. 95026 (1980).

066.107 **Quantum many-particle systems in curved spacetime.**
I. Ichonose.
Phys. Lett. B, Vol. 94B, 269 - 271 (1980). – Abstr. in Phys. Abstr., Vol. 83, Abstr. 98751 (1980).

066.108 **Measurements of the cosmic background radiation.**
R. Weiss.
Annu. Rev. Astron. Astrophys., Vol. 18, (see 003.004), 489 - 535 (1980).

066.109 **On the gravitational radiation formula.**
G. Schäfer, H. Dehnen.
J. Phys. A, Vol. 13, 2703 - 2722 (1980). – Abstr. in Phys. Abstr., Vol. 83, Abstr. 98838 (1980).

066.110 **Another geometrical representation of supergravity.**
M. Pilati.
Nuovo Cimento A, Ser. 11, Vol. 57A, 361 - 376 (1980).
Abstr. in Phys. Abstr., Vol. 83, Abstr. 98841 (1980).

066.111 **The material vacuum.** S. V. M. Clube.
Mon. Not. R. Astron. Soc., Vol. 193, 385 - 397 (1980).

Observations of galactic nuclei may be interpreted as evidence of temporary hypermassive states ($\gamma \gg 1$) adopted by ordinary matter under conditions of extreme gravitational collapse. By analysing the observations in terms of an horizon-free flat space-time theory of gravity, one is guided towards a pre-Maxwellian understanding of gravitational fields and the need for a non-relativistic Lorentz invariant material vacuum. Such a medium has to be cold and provides a natural explanation of the microwave background. One of the pillars of the big bang theory is thus dissolved, and a particular kind of steady state theory is described involving recurrent activity in galaxies caused by hypermassive nuclei. Possible time variations of the non-cosmological component of quasar redshifts become one of the crucial tests of the theory.

066.112 **Primordial black holes and the deuterium abundance.**
D. Lindley.
Mon. Not. R. Astron. Soc., Vol. 193, 593 - 601 (1980).

The destruction of deuterium as a result of high energy photon emission from evaporating primordial black holes is examined. The decrease in abundance is found to be comparable in magnitude with the increase predicted by Zeldovich et al. in their consideration of nucleon emission by black holes, indicating that the fate of deuterium in a universe containing small black holes needs further examination.

066.113 **Trapped radial oscillations of gaseous disks around a black hole.** S. Kato, J. Fukue.
Publ. Astron. Soc. Japan, Vol. 32, 377 - 388 (1980).

In gaseous disks around a black hole, the epicyclic frequency does not increase monotonically inward in the radial direction. The effects of general relativity make it the maximum at $4a$, decrease it inward and eventually make it zero at $3a$, where a is the Schwarzschild radius. This implies that a low-frequency wave excited in the inner disk is trapped and cannot propagate outward much beyond the region of radius $4a$. Characteristics of this trapped oscillation are examined under some approximations. For typical values of parameters, the oscillation period of 100d, which is a typical observed period of time variations of QSOs and Seyfert galaxies, is realized when the mass of the black hole is $10^9 - 10^{10} M_\odot$.

066.114 **Computer experiment for calculating light trajectories.** V. G. Boltyanskij, L. K. Nikolaev.
Differ. uravn., Vol. 16, 548 - 550 (1980). In Russian. – Abstr. in Ref. zh., 51. Astron., 10.51.810 (1980).

066.115 **Post-Newtonian quantum effects of particle motion in a gravitational field. Fine and superfine structure of the levels of massive particles in Kerr-Newman fields.**
A. B. Gaina. G. A. Chizhov.
MGU. Moskva, 1980, 28 pp. In Russian. – Abstr. in Ref. zh., 51. Astron., 10.51.811 (1980).

066.116 **The perihelion advance of a charged test body with magnetic dipole moment in an electromagnetic and spherically symmetric gravitational field of a neutron star.**
D. Z. Taipov.
Vopr. Teor. otnositel'nosti. Alma-Ata, 1979, p. 55 - 60. In Russian. – Abstr. in Ref. zh., 51. Astron., 10.51.815 (1980).

066.117 **On the motion of a test body in an outer gravitational field. II. Equations of deviation of geodetic lines, maximum dimensions, state of tension and deformation in a Schwarzschild field.** V. Mileva, J. Popov.
Godishn. vissh. ucheb. zaved. Tekh. fiz., 1976 (1980), Vol. 13, No. 2, p. 7 - 22. In Bulgarian. – Abstr. in Ref. zh., 51. Astron., 10.51.817 (1980).

066.118 **Gravitational effects in the field theory of gravitation.**
A. A. Vlasov, V. I. Denisov, A. A. Logunov, M. A. Mestvirishvili.
Teor. i mat. fiz., Vol. 43, 147 - 186 (1980). In Russian.
Abstr. in Ref. zh., 51. Astron., 10.51.820 (1980).

066.119 **Relativistic transformation of solid angle.**
J. M. McKinley.
American J. Phys., Vol. 48, 612 - 614 (1980). – Abstr. in Phys. Abstr., Vol. 83, Abstr. 101683 (1980).

066.120 **Colliding plane gravitational waves.**
D. Ray.
Phys. Lett. A, Vol. 78A, 315 - 316 (1980). – Abstr. in Phys. Abstr., Vol. 83, Abstr. 101741 (1980).

066.121 **Primordial black holes and baryon production in grand unified theories.** A. F. Grillo.
Phys. Lett. B, Vol. 94B, 364 - 366 (1980). – Abstr. in Phys. Abstr., Vol. 83, Abstr. 101970 (1980).

066.122 **The role of general relativity in astronomy: retrospect and prospect.** S. Chandrasekhar.
Highlights of Astronomy, Vol. 5, (see 003.006), 45 - 61 (1980).

066.123 **Outer parts of accreting discs around supermassive black holes in the nuclei of galaxies and quasars.**
P. I. Kolykhalov, R. A. Syunyaev.
Pis'ma Astron. Zh., Tom 6, 680 - 686 (1980). In Russian. English translation in Soviet Astron. Lett., Vol. 6.

If the mass of a solitary black hole and the accretion rate are sufficiently great the outer parts of the accreting disc become "selfgravitating" and may desintegrate into fragments. This fragmentation forms around the black hole a ring consisting of stars and gas. The ring continues to exist even after the accretion has ceased. The orbits of stars in the ring are approximately of Kepler type and their velocity dispersion is small. The mass of the ring can reach some per cent of the mass of the black hole.

066.124 **Bimetric general relativity and cosmology.** N. Rosen.
Gen. Relativ. Gravitation, Vol. 12, 493 - 510 (1980).

A modification of the general relativity theory is proposed (bimetric general relativity) in which, in addition to the usual metric tensor $g_{\mu\nu}$ describing the space-time geometry and gravitation, there exists also a background metric tensor $\gamma_{\mu\nu}$. The latter describes the spacetime of the universe if no matter were present. One can set up simple isotropic closed models of the universe which first contract and then expand without going through a singular state. It is suggested that the maximum density of the universe was of the order of $\sim 10^{93}$ g/cm^3. The expansion from such a high-density state is similar to that from the singular state ("big bang") of the general relativity models. In the case of the dust-filled model one can fit the parameters to present cosmological data. Using the radiation-filled model to describe the early history of the universe, one can account for the cosmic abundance of helium and other light elements in the same way as in ordinary general relativity.

066.125 **On the motion of test particles in the field of a plane gravitational wave.**
N. V. Mitskievic, S. N. Pandey.
Gen. Relativ. Gravitation, Vol. 12, 581 - 583 (1980).

The motion of test particles in the field of a plane gravitational wave is studied in order to derive some general properties of such motion, especially the possibilities of repeated meetings of two inertially moving particles.

066.126 **Canonical quantization of gravity and a problem of scattering.** V. A. Rubakov.
Gen. Relativ. Gravitation, Vol. 12, 585 - 596 (1980).

Linearized theory of gravity is quantized both in a naive way and as a proper limit of the Dirac-Wheeler-De Witt approach to the quantization of the full theory. The equivalence between the two approaches is established. The problem of scattering in the canonically quantized theory of gravitation is investigated. The concept of the background metric naturally appears in the canonical formalism for this case. The equivalence between canonical and path-integral approaches is established for the problem of scattering. Some kinematical properties of functionals in Wheeler superspace are studied in an appendix.

066.127 **A Newman-Penrose-type formalism for space-times with torsion.** S. Jogia, J. B. Griffiths.
Gen. Relativ. Gravitation, Vol. 12, 597 - 617 (1980).

066.128 **Causes and cures for the infinities in slow-motion expansions in general relativity.**
J. L. Anderson, L. S. Kegeles.
Gen. Relativ. Gravitation, Vol. 12, 633 - 647 (1980).

The causes of the divergent integrals arising in slow - motion expansions of the general relativistic field equations are studied and a remedy for them is suggested. This is done within the context of a model problem involving a coupled nonlinear scalar field and isotropic oscillator. The model is shown to give rise to divergent integrals directly attributable to the nonlinearity when the field is assumed to be analytic in a slowness parameter. Application of a nonregular perturbation approach which includes the method of matched asymptotic expansions is shown to eliminate the infinite contributions.

066.129 **An approximation scheme for scalar waves in a Kerr geometry.** W. E. Couch.
Gen. Relativ. Gravitation, Vol. 12, 665 - 673 (1980).

An approximation scheme, analogous to ones previously considered for the Schwarzschild and Reissner-Nordström geometries, is applied to scalar waves in a Kerr geometry. Spectral curves of transmission coefficient for monochromatic, axially symmetric scalar waves imploding on a rotating black hole are calculated using the approximation scheme.

066.130 **Kosmische lenzen en quasars.** G. W. E. Beekman.
Zenit, 7e Jaarg., 222 - 224 (1980).

066.131 **Afbuiging van straling door de zon. 1. De zonsverduistering van 1919.** R. J. Rutten.
Zenit, 7e Jaarg., 276 - 281 (1980).

066.132 **Afbuiging van straling door de zon. 2. Nieuwe metingen.** R. J. Rutten.
Zenit, 7e Jaarg., 372 - 378 (1980).

066.133 **Gravitationswellen von Doppelsternsystem?**
H.-J. Blome.
Phys. unserer Zeit, 11. Jahrg., 179 - 190 (1980).

066.134 **Another simple model for energy emission by black holes.** L. M. Celnikier.
American J. Phys., Vol. 48, 725 - 727 (1980). – Abstr. in Phys. Abstr., Vol. 83, Abstr. 105454 (1980).

066.135 **Canonical forms for axial symmetric space-times.**
G. T. Carlson, Jr., J. L. Safko.
Ann. Physics., Vol. 128, 131 - 153 (1980). – Abstr. in Phys. Abstr., Vol. 83, Abstr. 105545 (1980).

066.136 **Propagation of the general relativistic blast wave.**
T. Ishizuka, S. Sakashita.
Prog. Theor. Phys., Vol. 63, 1945 - 1949 (1980). – Abstr. in Phys. Abstr., Vol. 83, Abstr. 105549 (1980).

066.137 **Cosmic censorship and test particles.**
T. Needham.
Phys. Rev. D, Vol. 22, 791 - 796 (1980). – Abstr. in Phys. Abstr., Vol. 83, Abstr. 105552 (1980).

066.138 **Killing horizons around a uniformly accelerating and rotating particle.** H. Farhoosh,
R. L. Zimmerman.
Phys. Rev. D, Vol. 22, 797 - 801 (1980). – Abstr. in Phys. Abstr., Vol. 83, Abstr. 105557 (1980).

066.139 **Supergravity with and without superspace.**
S. Ferrara, P. Van Nieuwenhuizen.
Ann. Physics, Vol. 127, 274 - 288 (1980). – Abstr. in Phys. Abstr., Vol. 83, Abstr. 105558 (1980).

066.140 **The gravitational Doppler effect explored by means of a geostationary satellite.** O. Gron.
Found. Phys., Vol. 10, 567 - 579 (1980). – Abstr. in Phys. Abstr., Vol. 83, Abstr. 109351 (1980).

066.141 **Relativity of temperature and the Hawking effect.** G. L. Sewell.
Phys. Lett. A, Vol. 79A, 23 - 24 (1980). – Abstr. in Phys. Abstr., Vol. 83, Abstr. 109456 (1980).

066.142 **On the non-significance of super-gravitation for classical field theories.** H.-J. Treder.
Ann. Physik, Band 36, 399 - 400 (1980) = Zentralinst. Astrophys. Sternw. Babelsberg, Mitt. Neue Folge Nr. 250.

066.143 **Die Asymmetrie der kosmischen Zeit und Riemanns Gravitationstheorie.** H.-J. Treder.
Astron. Nachr., Band 299, 165 - 170 (1978).

In a general-relativistic approach Riemann's ansatz means that in empty space-time domains the world-geometry is the purely metrical "Riemannian" geometry. However, in domains with a non-vanishing matter tensor the geometry becomes "non-Riemannian" affine connecting and is of the type of Weyl's geometry or of the "Einstein-Cartan theories of gravitation".

066.144 **Einstein's field theory with teleparallelism and Dirac's electrodynamics. III. Static, spherically symmetric solution of the approximated field equations.** E. Kreisel.
Ann. Physik, Band 37, 301 - 311 (1980). – Abstr. in Phys. Abstr., Vol. 84, Abstr. 123 (1981).

066.145 **Gravitational bounce.** K. Lake, L. A. Nelson.
Phys. Rev. D, Vol. 22, 1266 - 1269 (1980).
Abstr. in Phys. Abstr., Vol. 84, Abstr. 138 (1981).

066.146 **New method for extracting static equilibrium configurations in general relativity.**
J. J. Matese, P. G. Whitman.
Phys. Rev. D, Vol. 22, 1270 - 1275 (1980). – Abstr. in Phys. Abstr., Vol. 84, Abstr. 139 (1981).

066.147 **Production of gravitational waves through electromagnetic radiation.** K. Buchner, R. Rosca.
J. Phys. A, Vol. 13, L371 - L374 (1980). In German.
Abstr. in Phys. Abstr., Vol. 84, Abstr. 141 (1981).

066.148 **Two-soliton waves in anisotropic cosmology.** V. Belinsky (*Belinskij*), D. Fargion.
Nuovo Cimento B, Ser. 11, Vol. 59B, 143 - 162 (1980).
Abstr. in Phys. Abstr., Vol. 84, Abstr. 142 (1981).

066.149 **Force on a static charge outside a Schwarzschild black hole.** A. G. Smith, C. M. Will.
Phys. Rev. D, Vol. 22, 1276 - 1284 (1980). – Abstr. in Phys. Abstr., Vol. 84, Abstr. 146 (1981).

066.150 **Einstein's Hermitian theory of relativity as unification of gravo- and chromodynamics.**
H.-J. Treder.
Ann. Physik, Band 37, 250 - 258 (1980). In German.
Abstr. in Phys. Abstr., Vol. 84, Abstr. 147 (1981).

066.151 **Massive shell models in the gravitational theories with higher derivatives.**
H.-H. von Borzeszkowski, V. P. Frolov.
Ann. Physik, Band 37, 285 - 293 (1980). – Abstr. in Phys. Abstr., Vol. 84, Abstr. 152 (1981).

066.152 **The gravitational perturbations of the Kerr black hole. IV. The completion of the solution.**
S. Chandrasekhar.
Proc. R. Soc. London, Ser. A, Vol. 372, 475 - 484 (1980).
Abstr. in Phys. Abstr., Vol. 84, Abstr. 4936 (1981).

066.153 **Non-existence of equilibrium configurations of charged black holes.** G. W. Gibbons.
Proc. R. Soc. London, Ser. A, Vol. 372, 535 - 538 (1980).
Abstr. in Phys. Abstr., Vol. 84, Abstr. 4937 (1981).

066.154 **Introduction to black holes and pulsars.**
I. M. L. Dass, V. N. Rai.
Sci. Cult., Vol. 45, 199 - 202 (1979). – Abstr. in Phys. Abstr., Vol. 84, Abstr. 4939 (1981).

066.155 **The gravitational constant at time zero.**
L. S. Levitt.
Nuovo Cimento, Lett., Ser. 2, Vol. 29, 23 - 24 (1980).
Abstr. in Phys. Abstr., Vol. 84, Abstr. 5006 (1981).

066.156 **Photon trajectories in the Kerr-Newman metric.**
M. Calvani, F. de Felice, L. Nobili.
J. Phys. A, Vol. 13, 3213 - 3219 (1980). – Abstr. in Phys. Abstr., Vol. 84, Abstr. 5185 (1981).

066.157 **Gauge-invariant coupled gravitational, acoustical, and electromagnetic modes on most general spherical space-times.** U. H. Gerlach, U. K. Sengupta.
Phys. Rev. D, Vol. 22, 1300-1312 (1980). – Abstr. in Phys. Abstr., Vol. 84, Abstr. 5192 (1981).

066.158 **Conformal invariance, microscopic physics, and the nature of gravitation.** J. D. Bekenstein.
Phys. Rev. D, Vol. 22, 1313 - 1324 (1980). – Abstr. in Phys. Abstr., Vol. 84, Abstr. 5193 (1981).

066.159 **Curvature collineations of non-expanding and twist-free vacuum type-N metrics in general relativity.** W. D. Halford, C. B. G. McIntosh, E. H. Van Leeuwen.
J. Phys. A, Vol. 13, 2995 - 3000 (1980). – Abstr. in Phys. Abstr., Vol. 84, Abstr. 5194 (1981).

066.160 **Gravitational radiation in Szekere's quasi-spherical space-times.** G. M. Covarrubias.
J. Phys. A, Vol. 13, 3023 - 3028 (1980). – Abstr. in Phys. Abstr., Vol. 84, Abstr. 5195 (1981).

066.161 **Gravitational and sound waves in stiff matter.**
K. A. Bronnikov.
J. Phys. A, Vol. 13, 3455 - 3463 (1980). – Abstr. in Phys. Abstr., Vol. 84, Abstr. 5196 (1981).

066.162 **Scalar field generalizations of electromagnetic Bianchi models of types II, VIII and IX.**
D. Lorenz.
Nuovo Cimento, Lett., Ser. 2, Vol. 29, 238 - 240 (1980).
Abstr. in Phys. Abstr., Vol. 84, Abstr. 5200 (1981).

066.163 **N=8 supergravity in various dimensions and the implications for four dimensions.** J. H. Schwarz.
Phys. Lett. B, Vol. 95B, 219 - 221 (1980). – Abstr. in Phys. Abstr., Vol. 84, Abstr. 5202 (1981).

066.164 **A formula for the induced gravitational constant.**
S. L. Adler.
Phys. Lett. B, Vol. 95B, 241 - 243 (1980). – Abstr. in Phys. Abstr., Vol. 84, Abstr. 5203 (1981).

066.165 **Particle transmutations in quantum gravity.**
D. N. Page.
Phys. Lett. B, Vol. 95B, 244 - 246 (1980). – Abstr. in Phys. Abstr., Vol. 84, Abstr. 5213 (1981).

066.166 **The massless scalar field around a static black hole.**
V. P. Frolov, A. I. Zel'nikov.

J. Phys. A, Vol. 13, L345 - L347 (1980). – Abstr. in Phys. Abstr., Vol. 84, Abstr. 9560 (1981).

066.167 **Imaginary-frequency interior modes of black holes.** R. A. Matzner, N. A. Zamorano.
Proc. R. Soc. London, Ser. A, Vol. 373, 223 - 233 (1980). Abstr. in Phys. Abstr., Vol. 84, Abstr. 9561 (1981).

066.168 **On the determination of the degree of cosmological Compton distortions and the temperature of the cosmic blackbody radiation.** Y. Rephaeli.
Astrophys. J., Vol. 241, 858 - 863 (1980).

The author describes a method for a determination of the cosmic blackbody radiation temperature and its degree of cosmological Compton distortions. Differential measurements at only three frequencies of the additional Compton distortions in a direction to a gas-rich cluster of galaxies are sufficient to determine both of these quantities. The method is essentially independent of the gas properties and, it is argued, can significantly improve on previous determinations of the radiation temperature. The method is implicitly based on the universality of the radiation and thus, if successfully applied, can give strong evidence for the universal nature of the radiation.

066.169 **Interacting gravitational shocks in vacuum plane-symmetric cosmologies.** J. Centrella.
Astrophys. J., Vol. 241, 875 - 885 (1980).

Gravitational shocks in vacuum cosmological models are studied numerically. The author follows the free propagation of the shocks, including the formation of tails behind the shocks and their collisional interactions.

066.170 **Inhomogeneous spherical accretion onto massive black holes as a model for active galactic nuclei.**
L. Maraschi, G. C. Perola, A. Treves.
Astrophys. J., Vol. 241, 910 - 914 (1980).

The authors have discussed the characteristics of inhomogeneities consistent with the accretion flow onto a massive black hole assuming that the external pressure is of the order of the gravitational energy density and that the internal pressure is mainly thermal.

066.171 **The Jeans instability in an expanding medium.** J. P. Baptista, D. Gerbal.
Astrophys. Space Sci., Vol. 73, 349 - 353 (1980).

In this paper the authors solve, by the two time-scale method, the equation governing the evolution of a density perturbation in any expanding medium. The surprizingly simple result they obtained allows an accurate insight into the role played by cumulative but contradictory effects such as gravitational attraction forces and the expanding velocity field. Although the existence of a critical length is confirmed, there is no catastrophic growth of 'instability'.

066.172 **Stellar tidal disruption by a massive binary black hole.** G. W. Collins II.
Astrophys. Space Sci., Vol. 73, 355 - 378 (1980).

The effects of formation of a binary black hole in a dense star cluster are found to have significant effects on the dynamics of the cluster. Tidal destruction of stars captured into bound orbits during the formation of the black hole binary provide a sizeable source of very high temperature thermal radiation as well as a source of radially outward moving clouds of gas. The efficiency of subsequently accreted matter onto the binary components as an energy source is investigated and suggestive evolutionary models of the dynamics of the binary system are presented. Lifetimes of the system are shown to be compatible with contemporary estimates. It is suggested that the high-density cluster core provides a suitable environment for the operation of a number of models for the core of active galactic nuclei.

066.173 **The measurement of the gravitational constant in an orbiting laboratory.**
P. Farinella, A. Milani, A. M. Nobili.
Astrophys. Space Sci., Vol. 73, 417 - 433 (1980).

The authors propose to measure the gravitational constant G by putting in an orbiting laboratory a known mass of very high density and by tracking the motion of a small test mass under the gravitational influence of the primary mass. They analyse the different sources of perturbation. In order to maximize the time of interaction it is proposed to put the test mass as close as possible to the stable manifold of a collinear equilibrium point of the system earth-primary mass. This method will allow the determination of the value of G within a few parts over 10^5.

066.174 **Red and blue shifts near compact objects.** J. M. Cohen, M. F. Struble.
Astrophys. Space Sci., Vol. 73, 507 - 511 (1980).

The authors show that radiation emitted from material falling toward a black hole or neutron star can be blue-shifted as well as red-shifted. Although the red shift can be arbitrarily large near a black hole, there is an upper limit for the blue shift of 1/2. Material incident toward the poles of a magnetic neutron star can simultaneously radiate red and blue-shifted lines. Near an oblique magnetic rotator, the red and blue shifts will show a sinusoidal variation. Such spectral variations are associated with SS 433.

066.175 **A new upper limit on optical bursts from primordial black hole explosions.**
C. L. Bhat, H. Razdan, M. L. Sapru.
Astrophys. Space Sci., Vol. 73, 513 - 516 (1980).

Two wide-angle photomultiplier systems were simultaneously operated over a baseline of nearly 30 km in a search for cosmic optical bursts of fractional microsecond time-scale. In 74 hours of overlapping observation, one event was recorded coincident to 1 ms as against the corresponding accidental coincidence value of 0.14 from Čerenkov light pulses from unrelated cosmic ray showers. The possible cosmic origin of this event, including that from primordial black-hole explosions, is discussed and the corresponding upper limits derived.

066.176 **A reinvestigation of the standard model for the dynamics of a massive black hole in a globular cluster.** D. N. C. Lin, S. Tremaine.
Astrophys. J., Vol. 242, 789 - 798 (1980).

The authors investigate the validity of two of the standard assumptions used in calculating the dynamics of a massive black hole in the center of a dense stellar system such as a globular cluster. First, the motion of the hole due to perturbations from its associated cusp of bound stars is usually neglected. The authors find that this assumption introduces substantial errors in the relaxation rate in the cusp; however, these errors are less than the statistical uncertainties due to the small number of cusp stars. The authors confirm that close encounters have only a small effect on the stellar density distribution in the cusp, but they find that ejection by close encounters is the dominant hole-induced source of mass loss from the cluster core: a star entering the cusp is more likely to be ejected from the core than to be swallowed by the hole. The results show that close encounters play a major role in the evolution of a hole plus cluster system.

066.177 **The approximation of radiative effects in relativistic gravity: gravitational radiation reaction and energy loss in nearly Newtonian systems.** M. Walker, C. M. Will.
Astrophys. J.,Lett., Vol. 242, L129 - L133 (1980).

An argument is presented to determine the accuracy with which a solution of Einstein's field equations of gravitation must be approximated in order to describe the dominant effects of gravitational radiation emission from weak-field systems. Several previous calculations are compared in the light of this argument, and some apparent discrepancies among them are resolved. The majority of these calculations support the "quadrupole formulae" for gravitational radiation energy loss and radiation reaction.

066.178 **Laser displacement sensor with application to gravitational radiation detection.**
M. Weksler, Z. Vager, G. Neumann.
Appl. Opt., Vol. 19, 2717 - 2725 (1980).

066.179 **On theories of gravitation in which the dynamical equations do not follow from the field equations and the Birkhoff theorem.** U. Bleyer, J. P. Mücket.
Astron. Nachr., Band 301, 285 - 295 (1980).

In general the Birkhoff theorem is violated in non-Einsteinian theories of gravitation. The authors show for theories in which the dynamical equations do not follow from the field equations that time-dependent vacuum solutions are needed in order to join nonstatic spherically symmetric incoherent matter distributions. It is shown for Treder's tetrad theories that such vacuum solutions exist and a continuous and unique junction is possible. In generalization of these results the authors consider the problem in what theories of gravitation the dynamical equations do not follow from the field equations. This consideration leads to non-Einsteinian theories like bimetric theories or Treder's tetrad theories containing supplemetary geometrical quantities which are not dynamical variables of the theory.

066.180 **Relativistic corrections in astrometric observations of planets.** M. D. Kislik.
Pis'ma Astron. Zh., Tom 6, 778 - 784 (1980). In Russian. English translation in Soviet Astron. Lett., Vol. 6.

Formulas for relativistic corrections to the parameters measured in precise astrometric observations of planets are presented. They account for relativistic disturbances in the motion of planets and in the propagation of light. The coordinate independence of results of orbit determination in the space of measured parameters is demonstrated.

066.181 **Do black holes really evaporate thermally?**
F. J. Tipler.
Phys. Rev. Lett., Vol. 45, 949 - 951 (1980).

The Raychaudhuri equation is used to analyze the effect of the Hawking radiation back reaction upon a black-hole event horizon. It is found that if the effective stress-energy tensor of the Hawking radiation has negative energy density as expected, then an evaporating black hole initially a solar mass in size must disappear in less than a second. This implies that either the evaporation process, if it occurs at all, must be quite different from what is commonly supposed, or else black-hole event horizons – and hence black holes – do not exist.

066.182 **Gravitational radiation quadrupole formula is valid for gravitationally interacting systems.**
M. Walker, C. M. Will.
Phys. Rev. Lett., Vol. 45, 1741 - 1744 (1980).

An argument is presented for the validity of the quadrupole formula for gravitational radiation energy loss in the far field of nearly Newtonian (e.g., binary stellar) systems. This argument differs from earlier ones in that it determines beforehand the formal accuracy of approximation required to describe gravitationally self-interacting systems, uses the corresponding approximate equation of motion explicitly, and evaluates the appropriate asymptotic quantities by matching along the correct space-time light cones.

066.183 **Test of relativistic gravitation with a space-borne hydrogen maser.** R. F. C. Vessot, M. W. Levine, E. M. Mattison, E. L. Blomberg, T. E. Hoffman, G. U. Nystrom, B. F. Farrel, R. Decher, P. B. Eby, C. R. Baugher, J. W. Watts, D. L. Teuber, F. D. Wills.
Phys. Rev. Lett., Vol. 45, 2081 - 2084 (1980).

The results of a test of general relativity with use of a hydrogen-maser frequency standard in a spacecraft launched nearly vertically upward to 10000 km are reported. The agreement of the observed relativistic frequency shift with prediction is at the 70×10^{-6} level.

066.184 **On the Einstein-Maxwell field equations.**
D. D. Dionysiou.
Astrophys. Space Sci., Vol. 73, 295 - 305 (1980).

Starting with the Einstein-Maxwell field equations in general relativity the author constructs the general differential equations governing the components of the metric tensor. These equations allow to find h_{ij} in various orders. An answer up to the first relativistic corrections is a computational work to find ${}_4h_{00}$, since the other terms ${}_2h_{00}$, ${}_2h_{\alpha\beta}$ and ${}_3h_{0\alpha}$ are as in a pure gravitational case. On the other hand, using the defined Einstein-Maxwell tensor, the author gives the equations of motion of two charged particles in the 0th order; also, the generalization is given in the case of n particles.

066.185 **Method of definition of the sky background brightness from data of the experiment "Galaktika".**
A. M. Zvereva.
Izv. Krymskoj Astrofiz. Obs., Tom 62, 27 - 33 (1980). In Russian.

A method of useful signal isolation (from the sky background) based on the space experiment "Galaktika" on sputnik "Prognoz 6" data in the spectral region $1100 < \lambda < 1900$ Å has been discussed. The highly elongated orbit of this sputnik allows to receive spectral scans with different instrumental scattered light contribution, which is the result of intensive diffuse Lα emission of the geocorona and interstellar hydrogen. Records analysis at different distances from the earth are giving straight dependence between Lα records and other wavelengths. Extrapolation of this dependence to zero Lα emission gives the value of the true background radiation.

066.186 **On the criteria of observation of the gravitational lens effect.** N. V. Mitskievič.
Astron. Zh., Tom 57, 1339 - 1340 (1980). In Russian. English translation in Soviet Astron., Vol. 24, No. 6.

In connection with a possible interpretation of the observations of the double quasar 0957 + 561 as a manifestation of the gravitational lens effect the author considers the problems of checking the coherence of the radiation of the two sources in the optical and radio ranges and of possible distortions in the interference picture due to this coherence. The importance of observations of temporal variations in the appearance of the sources is also pointed out.

066.187 **Motion of a particle in the background of a white hole.** R. C. Kapoor.
Kodaikanal Obs. Bull., Ser. A, Vol. 2, 252 - 259 (1979).

A study has been made of the motion of a test particle in the background of a white hole. It is shown that radial as well as non-radial photons from such a particle can leak through the Schwarzschild radius even when the particle and the white hole boundary have not yet burst through. Implications of the results when applied to the case of a grey hole are also discussed.

066.188 **On a generalized theory of gravitation.**
P. B. Abramyan, G. S. Saakyan.
Uch. zap. Erevan. univ. Estestv. nauk, 1980, No. 1, p. 69 - 77. In Russian. – Abstr. in Ref. zh., 51. Astron., 12.51.678 (1980).

066.189 **On the description of a black hole in the expanding universe.** N. F. Dandash, N. V. Mitskievič.
Mater. 3-j konf. mol. uchen. Univ. druzhby narodov (mat. fiz., khim.), 1980. Moskva, 1980, p. 48 - 52. In Russian. – From Ref. zh., 51. Astron., 12.51.696 (1980).

066.190 **External gravitational field of rotating configurations (Ω^3 approximation).** M. O. Minasyan.
Uch. zap. Erevan. univ. Estestv. nauk, 1980, No. 1, p. 63 - 68. In Russian. – Abstr. in Ref. zh., 51. Astron., 12.51.697 (1980).

066.191 **Black holes.** R. Penrose.
The state of the Universe, (see 003.013), p. 121 - 143 (1980).

066.192 **Plane waves in gauge theories of gravitation.** W. Adamowicz.
Gen. Relativ. Gravitation, Vol. 12, 677 - 691 (1980).
Exact solutions representing pp waves are found in a wide class of gauge theories of gravitation. Algebraic and symmetry properties are investigated and a special case of plane waves is discussed.

066.193 **Notes on recent U_4 theories of gravitation.** R. P. Wallner.
Gen. Relativ. Gravitation, Vol. 12, 719 - 732 (1980).
Some of the proposed Lagrangians and their corresponding field equations for a gravitational theory based on a Riemann-Cartan space with metric-compatible connection (U_4 theory) are compared and a new one is suggested.

066.194 **On the gravitational wave induced oscillations of viscoelastic bodies.** J. G. Papastavridis, G. Lianis.
Gen. Relativ. Gravitation, Vol. 12, 743 - 766 (1980).
An equation for the propagation of oscillations in a viscoelastic solid, induced by gravitational waves, is derived here. A linearized version of a relativistically invariant constitutive equation of integral type is employed in connection with the appropriately linearized field equations of general relativity. This theory could be applied for a more realistic design of gravitational wave detectors.

066.195 **Algebraic isometric embeddings of charged spherically symmetric space-times.**
M. Ferraris, M. Francaviglia.
Gen. Relativ. Gravitation, Vol. 12, 791 - 804 (1980).
Recent results concerning isometric embeddings of charged spherically symmetric space-times as algebraic 4-surfaces of a pseudo-Euclidean R^n are given. Several problems arising in algebraic isometric embeddings of space-times with horizons are analyzed by studying the Reissner-Nordström solution. The connection between the images of radial null geodesics and the possibility of extending the algebraic isometric embedding through a horizon is studied in detail.

066.196 **Two theorems on flat space-time gravitational theories.** M. Castagnino, L. Chimento.
Gen. Relativ. Gravitation, Vol. 12, 825 - 835 (1980).

066.197 **A new unified field theory based on the conformal group.** E. Pessa.
Gen. Relativ. Gravitation, Vol. 12, 857 - 862 (1980).
The author develops here a new unified theory of the electromagnetic and gravitational field, based on a six-dimensional generalization of Maxwell's equations; additional space-time coordinates are interpreted only as mathematical tools in order to obtain a linear realization of the four-dimensional conformal group.

066.198 **General solutions for a static isotropic metric in the Brans-Dicke gravitational theory.**
N. Van Den Bergh.
Gen. Relativ. Gravitation, Vol. 12, 863 - 869 (1980).

066.199 **Motion of primordial black holes in the early universe and their likely distribution today.**
J. N. Islam, B. F. Schutz.
Gen. Relativ. Gravitation, Vol. 12, 881 - 893 (1980).
The authors discuss in detail those effects which slow down black holes of mass $\sim 10^{15}$ g and affect their spatial distribution today. In particular they treat effects caused by the charge fluctuations of the hole which result from quantum-mechanical processes. The dominant energy-loss mechanism for the holes is the expansion of the universe, which leaves them virtually at rest at the time of galaxy formation. The resultant violent relaxation should concentrate roughly half of them in present-day galaxies and their halos.

066.200 **Newtonian analogs of Szekeres' space-times.** G. Lawitzky.
Gen. Relativ. Gravitation, Vol. 12, 903 - 912 (1980).
By means of a covariant formulation of Newton's theory of gravitation Newtonian analogs of Szekeres' space-times are found.

066.201 **Apparent violation of the principle of equivalence and Killing horizons.**
R. L. Zimmerman, H. Farhoosh.
Gen. Relativ. Gravitation, Vol. 12, 935 - 943 (1980).
By means of the principle of equivalence the authors deduce the qualitative behavior of the Schwarzschild horizon about a uniformly accelerating particle. This result is confirmed for an exact solution of a uniformly accelerating object in the limit of small accelerations. For large accelerations the Schwarzschild horizon appears to violate the qualitative behavior established via the principle of equivalence. When similar arguments are extended to an observable such as the redshift between two observers, there is no departure from the results expected from the principle of equivalence. The resolution of the paradox is brought about by a compensating effect due to the Rindler horizon.

066.202 **Post-Newtonian generation of gravitational waves in a theory of gravity with torsion.**
M. Schweizer, N. Straumann, A Wipf.
Gen. Relativ. Gravitation, Vol. 12, 951 - 961 (1980).
The authors adapt the post-Newtonian gravitational-radiation methods developed within general relativity by Epstein and Wagoner to the gravitation theory with torsion and show that the two theories predict in this approximation the same gravitational radiation losses. Since they agree also on the first post-Newtonian level, they are at the present time–observationally–indistinguishable.

066.203 **A comment on "Cosmic censorship, black holes, and particle orbits".** N. T. Bishop.
Gen. Relativ. Gravitation, Vol. 12, 971 - 972 (1980).
A paper by Hiscock (see also 25.066.068) implies that the cosmic censorship hypothesis is false. It is shown here that this result is not proven and, indeed, is probably wrong.

066.204 **E = mc².** V. Zlatarov.
Vasiona, Année 28, 55 - 57 (1980). In Croatian.

066.205 **Space, time and the universe.** M. Blagojević.
Vasiona, Année 28, 60 - 61 (1980). In Croatian.

066.206 **Einstein's theory of relativity.**
Z. Ivanović, M. Mijić.
Vasiona, Année 28, 63 - 68 (1980). In Serbo-Croatian.

Bulletin GRG, No. 40. List of publications.
See Abstr. 002.014.

Spacetime, geometry, cosmology.
See Abstr. 003.028.

The search for gravity waves.
See Abstr. 003.036.

Buracos negros. See Abstr. 003.038.

Relativistic cosmology. See Abstr. 003.055.

General relativity and gravitation, Vol. 1 and 2. See Abstr. 003.056.

Motion of bodies in general relativity. See Abstr. 003.097.

Gravity, particles and astrophysics. See Abstr. 003.111.

Sirius B and the gravitational redshift: an historical review. See Abstr. 004.012.

9. Internationale Konferenz über Allgemeine Relativitätstheorie und Gravitation (GR 9). See Abstr. 011.031.

Cosmological limits on photon splitting. See Abstr. 022.033.

Entropies need not be concave. See Abstr. 022.091.

Method of registration of the gravitational radiation from extraterrestrial sources in the kHz frequency band. See Abstr. 031.547.

Analogical device for a rough localization of gravitational-wave sources. See Abstr. 034.045.

Periodic solutions of Schwarzschild type in the relativistic restricted problem of three bodies. See Abstr. 042.025.

Have massive cosmological neutrinos already been detected? See Abstr. 061.043.

Acceleration of relativistic charged particles by supersonic hydromagnetic turbulence. See Abstr. 062.074.

Magnetic (electric) strings in relativistic magnetohydrodynamics. See Abstr. 062.077.

Neutrino, gamma-ray, electron, and positron production in an ultrarelativistic plasma. See Abstr. 062.114.

Thick accretion disks and supercritical luminosities. See Abstr. 064.004.

Relativistic accretion: the optically thick case. See Abstr. 064.013.

Nuclear burning in massive accretion disks. See Abstr. 064.049.

Giant flares on supermassive accretion disks: polarization properties and energetics. See Abstr. 064.063.

Thick accretion disks with super-Eddington luminosities. See Abstr. 064.086.

General relativistic collapse of an axially symmetric star. I. The formulation and the initial value equations. See Abstr. 065.039.

Structure of relativistic stellar configurations. Linear stellar model in GRT. See Abstr. 065.041.

Newtonian spherical gravitational collapse. See Abstr. 065.066.

Stability of superdense stars in the bimetric theory of gravitation. See Abstr. 065.072.

Observations of sidereal motions of the earth's surface. See Abstr. 081.015.

Gravitational radiation and the evolution of cataclysmic binaries. See Abstr. 117.012.

The slaved disc model for SS 433. See Abstr. 117.059.

On the relativistic motion of the periastron in the eclipsing binary system DI Herculis. See Abstr. 119.009.

The interaction between the relativistic jets of SS433 and the interstellar medium. See Abstr. 125.005.

A new gravitational redshift for the white dwarf o^2 Eri B. See Abstr. 126.023.

Multiple quasar may indicate another gravitational lens. See Abstr. 141.078.

On the role of relativistic effects in the ejection and expansion of components of extragalactic double radio sources. See Abstr. 141.113.

The discovery of a gravitational lens. See Abstr. 141.122.

The double quasar Q0957 + 561 A, B: a gravitational lens image formed by a galaxy at $z = 0.39$. See Abstr. 141.129.

Multiple quasars and gravitational lenses. See Abstr. 141.143.

A range of time delays for the double quasar 0957 + 561 A, B. See Abstr. 141.148.

The effect of undetected gravitational lenses on statistical measures of quasar evolution. See Abstr. 141.180.

The lens galaxy of the twin QSO 0957+561. See Abstr. 141.181.

Relativistic observable effects in the binary pulsar PSR 1913 + 16. See Abstr. 141.515.

On observations of the cosmic radiation background. See Abstr. 142.109.

The 1979 March 5 gamma ray transient reviewed: its source location in N49 within the LMC and its characteristics as evidence for a vibrating neutron star. See Abstr. 142.504.

On the origin of the March 5, 1979 gamma ray transient: a vibrating neutron star in the Large Magellanic Cloud. See Abstr. 142.507.

Discrete gamma-ray sources: theory. See Abstr. 142.522.

Cosmic gamma-ray lines: theory. See Abstr. 142.528.

Star clusters containing massive, central black holes. III. Evolution calculations. See Abstr. 151.024.

Relativistic star clusters with high central redshift. See Abstr. 151.057.

Numerical models of star clusters with a central black hole. I. Adiabatic models. See Abstr. 151.119.

The primary source and the fates of galactic positrons. See Abstr. 155.002.

Massive black hole binaries in active galactic nuclei. See Abstr. 158.035.

Models of central sources in active galactic nuclei. See Abstr. 158.040.

Evolutionary implications of a rotating Kerr black hole model of active Seyfert galaxies. See Abstr. 158.061.

The Penrose Photoproduction Scenario for NGC 4151; (PCS–SSC). A black hole γ-ray emission mechanism for active galactic nuclei and Seyfert galaxies. See Abstr. 158.105.

The black hole in the nucleus of galaxies. See Abstr. 158.201.

Absorption effects due to intergalactic long whiskers of pyrolytic graphite and the cosmic microwave background. See Abstr. 161.001.

Fluctuations in the cosmic background radiation produced by evolving hierarchical cosmologies. See Abstr. 162.005.

Spontaneous symmetry breaking and the expansion rate of the early universe. See Abstr. 162.018.

Cosmological gravitational waves: their origin and consequences. See Abstr. 162.023.

Local inhomogeneities in a Robertson-Walker background. I. General framework. See Abstr. 162.031.

Cosmological solution with matter in a new theory of gravitation. See Abstr. 162.032.

Classical predictive electrodynamics in a conformally flat universe. The case of the Einstein-de Sitter universe. See Abstr. 162.037.

Cosmological model with gravitational, electromagnetic, and scalar waves. See Abstr. 162.038.

Cosmic matter-antimatter asymmetry and gravitational force. See Abstr. 162.046.

Polarization effects in cosmological models with anisotropic curvature. See Abstr. 162.051.

Microwave background radiation as a probe of the contemporary structure and history of the Universe. See Abstr. 162.060.

Cosmological spaces and closed conformal infinitesimal transformations. See Abstr. 162.066.

Gravitational radiation dominated cosmologies. See Abstr. 162.072.

Dipole and quadrupole anisotropies in homogeneous cosmological models. See Abstr. 162.082.

Cosmogenesis and the origin of the fundamental length scale. See Abstr. 162.090.

Space-time singularities and microwave background radiation. See Abstr. 162.093.

On the effect of a variable vacuum energy density upon the spectrum of the cosmological microwave background radiation. See Abstr. 162.094.

Possible determination of $\bar{q}_0$ using lunar occultations and laser ranging observations. See Abstr. 162.095.

Poisson's equation in de Sitter space-time. See Abstr. 162.120.

Local inhomogeneities in a Robertson-Walker background. II. Flux conditions at boundary surfaces. See Abstr. 162.126.

Neutron Stars

066.501 **Hydrogen burning on accreting neutron star.** M. Czerny, M. Jaroszyński.
Acta Astron., Vol. 30, 157 - 166 (1980).

Stationary burning of hydrogen accreted onto the surface of a neutron star of mass $1.4 M_\odot$ and radius 10 km is investigated. The nuclear reaction network includes 20 atomic species. General-relativistic effects are taken into account. There is some evidence that for low accretion rates hydrogen burning is unstable. For high accretion rates helium burning begins in the layer where hydrogen has not been exhausted. This phenomenon may qualitatively change the picture of helium flashes on neutron stars.

066.502 **Non-rigid massive spheres in general relativity.** M. C. Durgapal, P. S. Rawat.
Mon. Not. R. Astron. Soc., Vol. 192, 659 - 662 (1980).

Solutions corresponding to spheres with a slowly varying density ($\rho \propto 1 - r^2/K^2$) are evaluated. This has been applied to construct a model of a neutron star. The mass of a neutron star comes out to be $3.34 M_\odot$, when everywhere the speed of sound remains less than the speed of light. Under extreme conditions the mass of a neutron star approaches the limit of $8 M_\odot$. The mass/radius ratio for the physical condition $P = (1/3)\rho$ comes out to be more than that for a constant density sphere.

066.503 **Pair formation and electric field boundary conditions at neutron star magnetic poles.** P. B. Jones.
Mon. Not. R. Astron. Soc., Vol. 192, 847 - 860 (1980).

The paper gives the results of detailed calculations of pair formation in the inner magnetosphere of a neutron star made under the assumption that the surface electric field satisfies $\mathbf{E} \cdot \mathbf{B} = 0$. It is assumed that high multipole components in the magnetic field increase the curvature of open flux lines and allow pair formation at the rotation period of the typical radio pulsar. No dense pair plasma is formed. The outward moving plasma consists largely of a single stream of electrons so that for this boundary condition, there is no promising source of radio emission in the inner magnetosphere.

066.504 **Radiative heat transfer in surface layers of neutron stars with a magnetic field.**
N. A. Silant'ev, D. G. Yakovlev.
Astrophys. Space Sci., Vol. 71, 45 - 50 (1980).

Thermal conductivity due to Thomson scattering and free-free absorption of photons is numerically evaluated for a non-relativistic non-degenerate plasma in a magnetic field for a number of values of $b = \hbar\omega_B/kT \leqslant 1000$. In the case of pure scattering, simple fitting formulae are derived. At $b \gg 6$, the magnetic field is shown to decrease (by about one order) the characteristic densities above which heat transfer is mainly determined by free-free transitions.

066.505 **Thermogalvanomagnetic phenomena in neutron stars and white dwarfs.**
V. A. Urpin, D. G. Yakovlev.
Astron. Zh., Tom 57, 738 - 748 (1980). In Russian.
English translation in Soviet Astron., Vol. 24, No. 4.

The electron thermal and electrical conductivities and thermoelectric coefficient are found and the induction and heat transfer equations are presented for degenerate cores of white dwarfs and degenerate layers of the envelopes of neutron stars in the presence of a nonquantizing magnetic field. The thermogalvanomagnetic phenomena which can take place in neutron stars and white dwarfs are discussed.

066.506 **SS 433: a high-energy neutrino source?** D. S. Eichler.
Bull. American Astron. Soc., Vol. 12, 541 (1980). – Abstract.

066.507 **The magnetic fields of neutron stars and some related effects.** P. Meszaros.
J. Magn. Magn. Mater., Vol. 15 - 18, Part 3, 1551 - 1554 (1980). – Abstr. in Phys. Abstr., Vol. 83, Abstr. 72991 (1980).

066.508 **Dense neutron matter.** M. Rho.
J. Phys. Colloq., Vol. 41, No. C-2, (see 012.005), p. C2/1 - 8 (1980). – Abstr. in Phys. Abstr., Vol. 83, Abstr. 77277 (1980).

066.509 **Superfluidity in neutron stars.** J. Shaham.
J. Phys. Colloq., Vol. 41, No. C-2, (see 012.005), p. C2/9 - 23 (1980). – Abstr. in Phys. Abstr., Vol. 83, Abstr. 77441 (1980).

066.510 **Pulsars and compact X-ray sources: cosmic laboratories for the study of neutron stars and hadron matter.** D. Pines.
J. Phys. Colloq., Vol. 41, No. C-2, (see 012.005), p. C2/111 - 124 (1980). – Abstr. in Phys. Abstr., Vol. 83, Abstr. 77442 (1980).

066.511 **The cooling of neutron stars.** M. Soyeur.
J. Phys. Colloq., Vol. 41, No. C-2, (see 012.005), p. C2/139 - 146 (1980). – Abstr. in Phys. Abstr., Vol. 83, Abstr. 77443 (1980).

066.512 **On the mass-radius relationship of neutron stars.** J. Van Paradijs.
J. Phys. Colloq., Vol. 41, No. C-2, (see 012.005), p. C2/147 - 152 (1980). – Abstr. in Phys. Abstr., Vol. 83, Abstr. 77444 (1980).

066.513 **On the cooling of neutron stars.** G. Glen, P. Sutherland.
Astrophys. J., Vol. 239, 671 - 684 (1980).

The authors present results of detailed calculations of neutron star cooling curves, employing the best available information on equations of state, opacities, conductivities, and neutrino emissivities. General relativistic effects are included, and, for the more massive and compact neutron stars, these can have a significant impact on the thermal radiation as detected "at infinity".

066.514 **An upper limit to the rate of formation of neutron stars in the Galaxy. II.** J. G. Hills.
Astrophys. J., Vol. 240, 242 - 245 (1980).

In a recent paper Endal found the rate of formation of neutron stars in the Galaxy to be about one every 4 years, which is substantially larger than the maximum rate of one every 27 years which the author found in an earlier paper. Using more recently published observational data, the author finds that the maximum possible rate of neutron-star formation in the Galaxy which is consistent with energy conservation is one every 21 years, and the probable rate is roughly one every 48 years.

066.515 **Iron K photons from weakly magnetized neutron stars in X-ray binaries.** T. Bai.
Astrophys. J., Vol. 240, 264 - 270 (1980).

Hot plasmas that produce continuum X-rays from X-ray binaries are found to produce negligible amounts of iron K photons. In contrast, the atmosphere of the neutron star in an X-ray binary might be an important source of iron K photons ($\epsilon \approx 6.5$ keV), because it is bombarded by a large number of hard X-rays capable of photo-ejecting K-shell electrons from iron atoms. The author discusses the astrophysical information that is obtainable from the observations of iron K photons from the neutron star atmosphere.

066.516 **On the surface states sustained by the electric field of rotating neutron stars.** L. Z. Fang.
Mon. Not. R. Astron. Soc., Vol. 193, 107 - 110 (1980).

It is shown that, in an electric field typical of rotating neutron stars, electronic surface states can exist. The wavelengths corresponding to transitions between surface states lie in the ultraviolet or soft X-ray wavebands. The author proposes that detection of emission or absorption lines from surface states could become a new probe into the surface of neutron stars.

066.517 **Gamma-ray lines from accreting neutron stars.** K. Brecher, A. Burrows.
Astrophys. J., Vol. 240, 642 - 647 (1980).

The detection of γ-ray lines produced near the surface of accreting neutron stars can serve to test the strong, weak, electromagnetic, and gravitational interactions under extreme physical conditions unavailable in terrestrial laboratories. The authors have therefore computed the γ-ray line fluxes to be expected from (known) accreting neutron stars in binary X-ray sources. The strongest lines are most likely to be the 2.22 MeV deuterium formation line and the 0.511 MeV positron annihilation line, though γ-ray lines from excited states of carbon, oxygen, and other light nuclei may also contribute observable lines.

066.518 **Massive spheres with an isothermal core.** M. C. Durgapal, K. Pandey, R. Banerjee, A. K. Pande.
J. Phys. A, Vol. 13, 1729 - 1736 (1980). – Abstr. in Phys. Abstr., Vol. 83, Abstr. 82405 (1980).

066.519 **Experimental determination of the mass of a neutron star.** C. Yu.
Kexue Tongbao, Vol. 25, 886 - 887 (1980).

066.520 **X-ray bursts from thermonuclear runaways on accreting neutron stars.** R. E. Taam.
Astrophys. J., Vol. 241, 358 - 366 (1980).

The author has followed the thermonuclear runaways which develop in the hydrogen-rich envelopes of 0.476 and 1.41 $M_{\odot}$ neutron stars. Variations in the neutron star luminosity, CNO abundance of the accreted material, and mass accretion rate have been explored. In all cases, the thermonuclear shell flash led to high envelope temperatures ($T \gtrsim 7 \times 10^8$ K) and to the emission of an X-ray burst. Large variations in burst characteristics are found. Due to the uncertainty associated with the thermal state of the accreted envelope, the mass determination of the neutron star based on a comparison of temporal profile of computed and observed X-ray bursts is inconclusive.

066.521 **Thermische Strahlung von Neutronensternen.** W. Brinkmann.
Sterne Weltraum, Jahrg. 19, 378 - 383 (1980).

066.522 **On hydrogen and helium burning under conditions of accreting neutron stellar envelopes.**
A. D. Kudryashov, E. Ergma.
Pis'ma Astron. Zh., Tom 6, 712 - 716 (1980). In Russian. English translation in Soviet Astron. Lett., Vol. 6.

The burning of hydrogen-rich matter at high temperature and density is investigated.

066.523 **Numerical models of hydrogen and helium burning flashes on the surface of neutron stars.**
E. V. Ergma, A. D. Kudryashov.
Astrophys. Lett., Vol. 21, 13 - 16 (1980).

Results of numerical calculation of hydrogen and helium flashes on the surface of neutron stars are presented. It is shown that if the density at the bottom of the envelope is $3 \cdot 10^6 \mathrm{g/cm^3}$, then hydrogen burning always leads to helium ignition. Depending on the mass of the accreted envelope of a neutron star, the surface luminosity may have one or two maxima. The influence of the hydrogen burning network is discussed. The computational results and observational data for bursters and fast transients are compared.

066.524 **Quark matter core in neutron stars.** J. D. Anand, P. Bhattacharjee, S. N. Biswas.
J. Phys. A, Vol. 13, 3105 - 3112 (1980). – Abstr. in Phys. Abstr., Vol. 84, Abstr. 9555 (1981).

066.525 **Instabilities of accretion disk – magnetosphere interfaces.** U. Anzer, G. Börner.
Magnetospheric boundary layers, (see 012.041), p. 259 - 261 (1979).

Configurations which consist of a Keplerian accretion disk and a magnetosphere with solid rotation are investigated. Because of the velocity difference between disk and magnetosphere the boundary between the two is unstable to the Kelvin-Helmholtz instability. This instability grows strongest within a ring around the radius of corotation. It will lead to turbulent diffusion of disk material into the magnetosphere, and thus give a sufficiently high accretion rate.

066.526 **Interaction of accretion disk and rotating magnetic field of a neutron star.** U. Anzer, G. Börner.
Magnetospheric boundary layers, (see 012.041), p. 263 - 264 (1979).

Utilizing the instabilities of the boundary between disk and magnetosphere, a model for the flow of disk material into the neutron star's magnetosphere is constructed. The authors show that within a thin ring around the corotation radius gas flows over. The gas is partly accelerated towards the neutron star – producing the accretion – partly outwards. In terms of this model the authors explain certain observed properties of pulsating X-ray sources.

066.527 **The effects of crystallization on neutron star evolution.** M. B. Richardson.
News Lett. Astron. Soc. N. Y., Vol. 1, No. 7, p. 9 (1980). Abstract.

066.528 **Thermal emission from neutron stars.** D. Helfand.
News Lett. Astron. Soc. N. Y., Vol. 1, No. 7, p. 21 (1980). Abstract.

066.529 **Massive spheres with an isothermal core and $e^{\nu} \propto r^{2n}$ envelope.** M. C. Durgapal, A. K. Pande, R. Banerjee, K. Pandey.
Mon. Not. R. Astron. Soc., Vol. 193, 641 - 644 (1980).

Exact solutions for massive spherical structures with an isothermal core ($P = K\rho$) and an envelope in which $e^{\nu} \propto r^{2n}$, have been obtained. It is seen that only the case $K \leqslant 0.6$ corresponds to a physically real situation. It has been shown that such structures are pulsationally stable. The surface and central redshifts come out to be 0.73 and 8.9 respectively. Suitable neutron star models based upon such configurations have been constructed and their relevant parameters have been obtained. It is found that for a realistic case the mass and size of a neutron star comes out to be 2.4 $M_{\odot}$ and 16.2 km respectively.

066.530 **Thermonuclear burning in the envelope of a neutron star: effects of initial density, temperature and chemical composition.** A. D. Kudryashov, E. V. Ergma.
Acta Astron., Vol. 30, 453 - 467 (1980).

Characteristics of hydrogen (and helium) burning as depending on the density at the envelope bottom and the envelope temperature are studied. The influence of chemical composition on the behaviour of the flash is investigated. It is shown that if the density at the bottom of the envelope is greater than 10^6 g/cm^3 the hydrogen burning almost immediately leads to helium ignition. If $3 \times 10^5 \leqslant \rho_c \leqslant 10^6$ then helium burning begins after the exhaustion of a considerable part of hydrogen. The influence of the hydrogen burning kinetics on the character of the flash is discussed.

066.531 **Forced precession of neutron stars.**
W. M. DeCampli.
Astrophys. J., Vol. 242, 306 - 318 (1980).

The dynamics of a neutron star subject to an external torque is investigated. The neutron star is treated as a two-component body, one component being the solid crust together with the interior plasma coupled to it by a magnetic field, the other component being the interior neutron superfluid. The latter is treated with classical hydrodynamics. Two stresses are assumed to act between the crust and interior – a viscous stress and a normal or "inertial" stress. The equations of motion are solved by a perturbation method in the limit that the spin rate changes slowly over a precession period. The strength of the inertial couple is found to be crucial in determining the decay rate of forced precessional motion. Its presence greatly lengthens the viscous decay time of precessional motion. Its absence places severe constraints on models for SS 433.

066.532 **Neutron star formation by collapse of white dwarfs.**
R. Canal, J. Isern, J. Labay.
Space Sci. Rev., Vol. 27, (see 012.046), 595 - 600 (1980).

066.533 **X-rays from the surface of neutron stars.**
S. Tsuruta.
X-ray astronomy, (see 012.055), p. 73 - 87 (1980).

The author reports on the results of the attempts to observe point sources in five supernova remnants, as well as nine radio pulsars, which have already been made with the Einstein Observatory. He discusses what could come out of careful point source investigations of the Crab pulsar and the Vela pulsar with the Einstein Observatory and other X-ray space programs scheduled for the near future. Emphasis is placed on the possible theoretical implication of the outcome of such investigations. The author summarises the observational status as of October 1979, and explains possible theoretical implications of these observations.

066.534 **Thermonuclear bursts on neutron stars.**
E. Ergma, A. D. Kudryashov.
Usp. fiz. nauk, Vol. 132, 391 - 392 (1980). In Russian.
Abstr. in Ref. zh., 51. Astron., 1.51.517 (1981).

066.535 **Thermal limit cycle oscillations on the surface of accreting neutron stars – X-ray bursters.**
M. Barranco, J. R. Buchler, M. Livio.
Astrophys. J., Vol. 242, 1226 - 1231 (1980).

With the help of a very simple two-zone model the authors demonstrate the possibility of periodic thermal relaxation (limit cycle) oscillations in the helium burning envelope of accreting neutron stars. Physically reasonable model parameters can be chosen which yield agreement with the observed features of X-ray bursts. The authors suggest that this limit cycle is operative in neutron stars which have an accretion rate in a specific range. For hydrogen burning a similar cycle is possible, but it operates at such high temperatures that an unrealistically large accretion rate would be required.

Nuclear and neutron matter with isobars – a transition potential model. See Abstr. 022.034.

The potential energy of an infinite system of nucleons and delta resonances. See Abstr. 022.058.

Electromagnetic transitions for the hydrogen atom in strong magnetic fields. See Abstr. 022.165.

Übergangswahrscheinlichkeiten für das Wasserstoffatom im starken Magnetfeld. See Abstr. 022.188.

Positronium im starken Magnetfeld: Elektromagnetische Übergänge und Zerfall. See Abstr. 022.189.

Compressibility of cold catalyzed matter.
See Abstr. 061.023.

On the limiting mass of a dense stellar matter.
See Abstr. 061.027.

Quantum effects in cyclotron plasma absorption.
See Abstr. 062.093

Spindown of rotating magnets.
See Abstr. 062.108.

One dimensional hydrodynamics of asteroid-neutron star collisions. See Abstr. 062.122.

Resonance radiative transfer for cyclotron line emission with recoil. See Abstr. 063.056.

A new approximation for the high magnetic field Compton cross-section. See Abstr. 063.059.

Dipole confined by a disk.
See Abstr. 064.064.

Nonradial accretion onto magnetized neutron stars.
See Abstr. 064.096.

Supernova triggered by electron captures.
See Abstr. 065.012.

Astrophysical problems of condensed matter in huge magnetic fields. See Abstr. 065.037.

Shell flashes. See Abstr. 065.055.

New method for extracting static equilibrium configurations in general relativity. See Abstr. 066.146.

Red and blue shifts near compact objects.
See Abstr. 066.174.

Evolution of a blue supergiant with a neutron star companion immersed in its envelope. See Abstr. 117.001.

Pulsar activity and the morphology of supernova remnants. See Abstr. 125.025.

The superfluid phase transition in pulsars.
See Abstr. 141.530.

Pulsars and compact X-ray sources – cosmic laboratories for the study of neutron stars and hadron matter.
See Abstr. 141.539.

Characteristics of the Cen X-3 neutron star from correlated spin-up and X-ray luminosity measurements.
See Abstr. 142.013.

X-ray burst sources – a simple two zone model.
See Abstr. 142.123.

Origin of the 5 March 1979 γ-ray transient: a vibrating neutron star. See Abstr. 142.502.

Synchrotron and annihilation radiations of $e^+ - e^-$ pairs in the magnetosphere of a vibrating neutron star: the radiation mechanism for the March 5, 1979 gamma ray transient. See Abstr. 142.506.

On the origin of the March 5, 1979 gamma ray transient: a vibrating neutron star in the Large Magellanic Cloud. See Abstr. 142.507.

Lines in gamma-burst energy spectra. See Abstr. 142.514.

Collisions of asteroids on neutron stars as a cause of cosmic gamma-ray bursts. See Abstr. 142.518.

Discrete gamma-ray sources: theory. See Abstr. 142.522.

Cosmic gamma-ray lines: theory. See Abstr. 142.528.

Gamma-ray emission under gas accretion onto a neutron star. See Abstr. 142.533.

Errata

066.901 **Erratum: "Bimetric gravitation theory on a cosmological basis"** [Gen. Relativ. Gravitation, Vol. 9, 339 - 351 (1978)]. N. Rosen.
Gen. Relativ. Gravitation, Vol. 12, 265 (1980). – See Abstr. 21.066.073.

066.902 **Erratum: "Interplanetary phase scintillation and the search for very low frequency gravitational radiation"** [Astrophys. J., Vol. 230, 570 - 574 (1979)].
J. W. Armstrong, R. Woo, F. B. Estabrook.
Astrophys. J., Vol. 240, 719 (1980). – See Abstr. 25.066.080.

066.903 **Erratum: "Seismic detection of gravitational radiation"** [Rev. Geophys. Space Phys., Vol. 17, 2057 - 2069 (1979)]. O. G. Jensen.
Rev. Geophys. Space Phys., Vol. 18, 887 (1980). – See Abstr. 27.066.060.

Sun

071 Photosphere, Spectrum

071.001 **Solar granulation and bright dots of the sunspot umbra: power spectra analysis.**
A. Adjabshirzadeh.
C. R. Acad. Sci. Paris, Tome 290, Sér. B, 541 - 544 (1980). In French.

The author used a picture of the solar granulation and a picture of the "Bright Umbral-dots" of a sunspot to deduce separately by a Fourier Transform Power Spectrum method the distributions of their spectral densities.

071.002 **Centre to limb variation of rotational temperature of the CH molecules in the solar photosphere.**
M. C. Pande, V. P. Gaur, K. R. Bondal, R. C. Dubey.
Bull. Astron. Soc. India, Vol. 8, 27 - 29 (1980).

Spectrophotoelectric observations of the CH lines are presented to determine the centre-to-limb behaviour of rotational temperatures for the (0–0) and (1–1) bands of the electronic transition $A^2\Delta - X^2\Pi$ of the molecule. Taking into account the large probable errors in measurements, the authors find that the rotational temperature from centre to limb appears practically constant. Further accurate observations up to the extreme limb with a larger set of data appear necessary.

071.003 **On the determination of damping constants for Fe I lines in the solar photosphere.**
W. Van Rensbergen, G. Deridder.
Sol. Phys., Vol. 67, 5 - 7 (1980).

The line broadening calculated with a Smirnow–Roueff potential yields a fairly good description of the intensity profile in the wings of 31 solar Fe I lines.

071.004 **The solar granulation. I. Two dimensional power-spectrum analysis using optical data processing methods.** S. Koutchmy, A. Legait.
Astron. Astrophys., Vol. 88, 345 - 349 (1980).

The authors used optical data processing methods to analyse a good time sequence of selected high resolution solar granulation pictures obtained at the Vacuum solar telescope of Sacramento Peak Observatory. Several two-dimensional power-spectra were composited and averaged by rotation to deduce a statistically significant distribution. No regularities appear on the deduced spectra. The speckles of partly instrumental origin seen on individual 2-dimensional power spectra are stable, their characteristic lifetime being up to 7 min, indicating a similar lifetime for solar granulation.

071.005 **Iron abundance reduced to the Kurucz-Peytremann oscillator strengths scale.** G. L. Fedorchenko.
Astrometr. Astrofiz., Vyp. (No.) 41, p. 12 - 15 (1980). In Russian.

The solar iron abundance determined in some recent papers (since 1970) is recalculated in the Kurucz-Peytremann system of oscillator strengths. The mean value of the abundance is equal to 7.63 ± 0.05.

071.006 **Photospheric subrotation, differential rotation and zonal wind bands: a reverse pirouette.**
K. H. Schatten, H. G. Mayr.
Bull. American Astron. Soc., Vol. 12, 473 (1980). – Abstract.

071.007 **Solar photospheric evolution during the lifetime of a granule.** R. C. Altrock.
Bull. American Astron. Soc., Vol. 12, 474 - 475 (1980). Abstract.

071.008 **Are gravity waves trapped in the solar photosphere?** T. M. Brown.
Bull. American Astron. Soc., Vol. 12, 475 (1980). – Abstract.

071.009 **Two dimensional radiative transfer in resonance line wings.** S. P. Owocki.
Bull. American Astron. Soc., Vol. 12, 517 (1980). – Abstract.

071.010 **Solar limb brightening in the continuum from photoelectric eclipse measurements.**
W. A. Rosen, H. L. Poss.
Bull. American Astron. Soc., Vol. 12, 519 - 520 (1980). Abstract.

071.011 **Solar plages and the interpretation of stellar Ca II H and K line variations in late type dwarfs.**
J. W. Cook.
Bull. American Astron. Soc., Vol. 12, 520 (1980). – Abstract.

071.012 **ATM observations of the solar C I multiplets at λ1560 and λ1657.**
D. Roussel-Dupré, J. Heasley.
Bull. American Astron. Soc., Vol. 12, 526 (1980). – Abstract.

071.013 **A morphological model of the fine structure of the photospheric brightness field.** V. N. Karpinskij.
Soln. Dannye 1980 Byull., No. 2, p. 91 - 102 (1980). In Russian.

A quantitative method for an investigation of outbursts of the photospheric brightness field was used for an analysis of its fine structure and a construction of a morphological model.

071.014 **On the redshift of wavelength of telluric O_2 spectral lines with increase of the effective pressure.**
Kh. I. Abdusamatov, A. G. Zlatopol'skij.
Soln. Dannye 1980 Byull., No. 3, p. 90 - 96 (1980). In Russian.

The detected redshift of the telluric O_2 spectral lines during the day correlates well with the growth of the zenith distance of the sun. The shift is explained by an increase of the effective pressure along the path of the solar ray with growth of the zenith distance of the sun.

071.015 **Radial oscillations of the solar photosphere near the disk centre.** N. A. Drake, Yu. A. Solonskij.
Soln. Dannye 1980 Byull., No. 3, p. 97 - 102 (1980). In Russian.

The parameters of vertical oscillations of the solar photosphere are investigated by means of quiet region spectrograms being obtained during 195 minutes every 20 seconds. In the calculated power spectrum of the oscillations the main maximum with a period of about 5 minutes dominates and also there are peaks corresponding to 15, 20 and 50 minutes.

071.016 **Some more effects of waves on spectral line analysis.** C. J. Durrant.
Astron. Astrophys., Vol. 89, 80 - 87 (1980).

The effects on spectral line shifts due to velocity-correlated opacity and source function fluctuations in waves are discussed in the context of a simple exponential atmospheric model. These correlations contribute a steady net line shift and, if the velocity amplitude is small enough, a small correction to the fluctuating component of the shift. The results and limitations of the model are discussed with reference to the large-amplitude calculations of Cram et al. (1979) and the estimates of the high frequency power in the sun made by Deubner (1976).

071.017 **The height dependence of granular motion.**
A. Nesis.
Stellar turbulence, (see 012.008), p. 17 (1980). – Abstract.

071.018 **Numerical simulations of the solar granulation.**
Å. Nordlund.
Stellar turbulence, (see 012.008), p. 17 - 18 (1980). Abstract.

071.019 **The evolution of an average solar granule.**
R. C. Altrock.
Stellar turbulence, (see 012.008), p. 51 (1980). – Abstract.

071.020 **Observed solar spectral line asymmetries and wavelength shifts due to convection.** D. Dravins.
Stellar turbulence, (see 012. 008), p. 51 - 52 (1980). Abstract.

071.021 **Temporal and spatial fluctuations in widths of solar EUV lines.** R. G. Athay, O. R. White.
Stellar turbulence, (see 012.008), p. 53 (1980). – Abstract.

071.022 **Formation of the profiles of absorption lines in the inhomogeneous medium.**
R. I. Kostik (*Kostyk*).
Stellar turbulence, (see 012.008), p. 53 - 54 (1980). Abstract.

071.023 **Diagnostic use of Fe II H & K wing emission lines.**
L. E. Cram, R. J. Rutten, B. W. Lites.
Stellar turbulence, (see 012.008), p. 102 (1980). – Abstract.

071.024 **Numerical simulation of granular convection: effects on photospheric spectral line profiles.**
Å. Nordlund.
Stellar turbulence, (see 012.008), p. 213 - 224 (1980).

The results of numerical simulations of the solar granulation are used to investigate the effects on photospheric spectral lines of the correlated velocity and temperature fluctuations of the convective granular motions.

071.025 **On the establishment of internally consistent abundance-oscillator strength scales.**
E. A. Gurtovenko (*Eh. A. Gurtovenko*), R. I. Kostik (*Kostyk*).
Stellar turbulence, (see 012.008), p. 296 (1980). – Abstract.

071.026 **Energy transfer by Alfvén waves.**
Yu. D. Zhugzhda, V. A. Lotsans.
XIth seminar on cosmophysics, (see 012.035), p. 28 - 36 (1979). In Russian. – Abstr. in Ref. zh., 51. Astron., 8.51.363 (1980).

071.027 **The supergranule velocity field.**
R. G. Giovanelli.
Sol. Phys., Vol. 67, 211 - 228 (1980).

A study of supergranule motions confirms horizontal velocities with peak values of typically 0.36 km s^{-1} as observed in Fe I 8688 Å. Near disk center, supergranule vertical velocities in Fe I 8688 have rms values $\leqslant \pm 0.01$ km s^{-1}, after allowance for the residual effects of the line-of-sight component of the horizontal supergranule motions, the five-minute oscillations, granule motions, and detector drift. There is a marginally-significant association of magnetic elements, and hence of cell boundaries, with downward motions.

071.028 **The solar scandium abundance.**
N. H. Youssef.
J. Astron. Soc. Egypt, Vol. 1, 96 - 99 (1979).

The Sc abundance is derived from six Sc I lines in the photospheric spectrum using synthetic profile calculations. Equivalent widths are obtained from intensity measurements of Sc I lines. The abundance of scandium is found to be $\log \Sigma_{Sc} = 3.080 \pm 0.03$ in the $\log \Sigma_H = 12.00$ scale.

071.029 **The abundance determination of molybdenum and yttrium in the solar atmosphere.**
N. H. Youssef, A. Abdel-Azim.
J. Astron. Soc. Egypt, Vol. 1, 100 - 104 (1979).

The solar Mo and Y abundance is derived from six Mo lines and three Y lines in the photospheric spectrum. By using the synthetic profile calculations the authors derived the abundance values: $A_{Mo} = 2.37 \pm 0.25$ and $A_Y = 2.71 \pm 0.15$ in the logarithmic $A_H = 12.00$ scale.

071.030 **Multiple methods of the analysis of the profiles of Fraunhofer lines.** A. A. Galal.
J. Astron. Soc. Egypt, Vol. 1, 105 - 120 (1979).

An empirical technique for the interpretation of the observed profiles of multiplet lines is suggested. The method gives also the possibility to carry out the analysis of the observed profiles of multiplet lines without the assumption of LTE. In the present study the method was used to evaluate the function of Mg I b-lines in the solar atmosphere.

071.031 **Absolute fluxes, equivalent width and centre-to-limb profiles of the solar Mg II resonance lines (I).**
A. Greve, C. D. McKeith.
Astron. Astrophys., Vol. 90, 224 - 230 (1980).

For the average quiet Sun the authors derive from high resolution Fabry Perot-echelle spectrograms profiles of the Mg II resonance lines in the wavelength region 2760 Å $\lesssim \lambda \lesssim$ 2820 Å. They also derive the Mg II flux profile in the wavelength region 2660 Å $\lesssim \lambda \lesssim$ 2940 Å. For the region 2770 Å $\lesssim \lambda \lesssim$ 2820 Å a wavelength averaged limb darkening curve is derived. The Mg II profiles are calibrated using a combination of this limb darkening curve and the low spectral resolution flux profile.

071.032 **The formation of the Mg II resonance line wings in the solar atmosphere (II).** A. Greve.
Astron. Astrophys., Vol. 90, 231 - 238 (1980).

Synthetic LTE profiles of the Mg II resonance lines, calculated for the HSRA, VAL and the so-called "Ca II and Mg II" model atmospheres introduced by Ayres and Linsky (1976), are compared with profiles of the average quiet Sun observed in the wavelength region Å $\lesssim \lambda \lesssim$ 2820 Å. As a prerequisite for calculating profiles the author derives empirically for the model atmospheres under consideration the total opacity as a function of the height. These opacity functions reproduce at λ 2660 Å and λ 2940 Å the observed local continuum intensities and the adopted wavelength averaged limb darkening curve. It is found that the line wings, at distances $|\Delta\lambda| \gtrsim 2$ Å from either line centre, are formed under LTE conditions. The comparison of the calculated profiles and the observed profiles seems to indicate that none of the one-component, static model atmospheres for the average quiet Sun reproduces the centre-to-limb variation for the central parts of the Mg II profiles.

071.033 **Profiles of the H and K Ca II lines in the continuous emission bands.** A. N. Koval'.
Izv. Krymskoj Astrofiz. Obs., Tom 61, 20 - 25 (1980). In Russian.

Profiles of the H and K Ca II lines were obtained in 35

continuous emission bands H and K Ca II lines (14 profiles for the continuous spectrum of flares, 7 profiles for moustaches and 14 for the emission grains). A comparison of the line profiles in the continuous emission threads with those in the spectrum of the undisturbed photosphere has been made.

071.034 **The sun as a possible energy standard in the ultraviolet spectral region with λ < 3000 Å.**
A. M. Zvereva.
Izv. Krymskoj Astrofiz. Obs., Tom 61, 26 - 33 (1980). In Russian.

The ultraviolet spectrum of the sun as a star according to all available data of extra atmospheric observations is analysed

071.035 **The empirical determination of damping constants in the solar photosphere. I: Preliminary results for Fe I lines..** E. A. (*Eh. A.*) Gurtovenko, N. N. Kondrashova.
Sol. Phys., Vol. 68, 17 - 29 (1980).

The evaluation of damping constants is discussed and the results of their empirical determination by various authors are listed. The results of the determination of damping constants by analysing the wings of 38 strong Fe I lines are discussed and compared with previous results obtained by the same method from 27 moderate Fe I lines. The van der Waals damping constant γ_6 multiplied by an enhancement factor $E = 2.5$ is suggested for the interpretation of Fe I Fraunhofer lines.

071.036 **Coherence analysis of granular intensity.**
F. J. Kneer, W. Mattig, A. Nesis, W. Werner,
Sol. Phys., Vol. 68, 31 - 39 (1980) = Mitt. Kiepenheuer Inst. Nr. 166.

A high resolution spectrogram of the Mg b_2 line from the quiet Sun disc centre is subjected to a coherence analysis. The authors find that the coherence between intensity fluctuations in the continuum and the wings of the line breaks down at a distance $\Delta\lambda = 0.35$ Å from line centre. From this and the r.m.s. intensity contrast as a function of $\Delta\lambda$ the authors are led to the following simple model of temperature fluctuation δT in the solar photosphere: A lower part (below 50 km) with strongly inward increasing δT and an upper part (above 50 km) with constant $\delta T = 75$ K. The two parts are supposed to fluctuate incoherently.

071.037 **Some comments on the limb shift of solar lines. III. Variation of limb shift with solar latitude, across plages, and across supergranules.** J. M. Beckers, W. R. Taylor.
Sol. Phys., Vol. 68, 41 - 47 (1980).

The authors searched for a variation with heliographic latitude of the solar limb effect by comparing the relative wavelengths of weak and strong Fraunhofer lines. The blue shifts associated with the limb effect appear 9% ± 5% larger in the polar radius vector than in an equatorial radius vector at $\cos\theta = 0.5$. This should perhaps be interpreted as an increase with latitude of either solar convection or of convective overshoot. Recent observations of poleward meridional flows should be corrected for this limb effect variation. A search for a similar variation in plages and in network boundaries had negative results.

071.038 **Latitude variations of photospheric activity areas with particular reference to solar faculae.**
G. M. Brown, D. R. Evans.
Sol. Phys., Vol. 68, 141 - 149 (1980).

Detailed studies of the development of photospheric activity centres for two solar cycles show that Spörer's Law holds in a very similar form to that applying to sunspots for the faculae which inhabit the sunspot zones. Similar differences between the two solar hemispheres can arise, and it seems to be confirmed that the average latitude of faculae tends to be a few degrees poleward of that of sunspots throughout a given cycle. It is shown that the normal averaging process involved in deriving Spörer's Law obscures a detail which is revealed in a breakdown into the variations within successive narrow latitude strips. These show the existence within a cycle of three separate maxima of activity occurring at different epochs and with different preferred latitudes. The main properties of these maxima are discussed.

071.039 **Dynamical models of convective penetration and high-spatial resolution observations.** S. L. Keil.
Bull. American Astron. Soc., Vol. 12, 747 (1980). – Abstract.

071.040 **Solar and meteoritic abundance of silicon.**
U. Becker, P. Zimmermann, H. Holweger.
Geochim. Cosmochim. Acta, Vol. 44, 2145 - 2149 (1980).

Recent lifetime measurements on excited electronic states of neutral silicon lead to reassessment of widely used experimental transition probabilities of Si I lines. This translates into a 25% downward revision of the Si abundance determined from the solar spectrum. A solar atomic ratio, Si/Ca = 15.5 is inferred. This value coincides with that found in carbonaceous chondrites, but contrasts with ordinary and enstatite chondrites.

071.041 **A comment on the character of the photospheric granular net.** L. Hejna.
Bull. Astron. Inst. Czechoslovakia, Vol. 31, 362 - 364 (1980).

On the basis of the statistics of the number of elements immediately neighbouring on randomly selected granules, the degree of similarity of the real photospheric granular structure with a purely hexagonal structure has been determined. It was found that these two types of structures differ to a considerable extent, although a certain "quasi-hexagonalization" of the granular structure can be observed; this is expressed by a relatively expressive maximum in the distribution of the occurrence frequency of the number of neighbouring elements.

071.042 **Solar abundances. A new table (October 1976).**
B. E. J. Pagel.
Origin and distribution of the elements, (see 012.039), p. 79 - 80 (1979).

071.043 **An approximate calculation of the effect of opacity in the solar spectral lines of C III.**
J. G. Doyle, R. W. P. McWhirter.
Mon. Not. R. Astron. Soc., Vol. 193, 947 - 955 (1980).

By introducing the measured abundance of carbon and the results of ionization balance calculations an estimate is made of the line-of-sight physical thickness of the regions emitting C III lines at the disc centre.

071.044 **The formation of Na I spectral lines in the solar atmosphere.**
B. Caccin, M. T. Gomez, G. Roberti.
Astron. Astrophys., Vol. 92, 63 - 69 (1980).

The kinetic equilibrium of Na I in the solar atmosphere is studied for different atomic models (up to 9 bound levels) and the results are discussed. A satisfactory agreement with the observations can be reached, for the whole spectrum, with the HSRA and a suitable choice of the micro and macroturbulence, when the currently accepted values of the atomic parameters are used.

071.045 **Numerical simulation of the solar granulation.**
L. D. Cloutman.
Space Sci. Rev., Vol. 27, (see 012.046), 293 - 299 (1980).

The solar granulation has been simulated by numerical solution of the multidimensional, time-dependent, nonlinear Navier-Stokes equations applied to the solar atmosphere. Granules may be explained as buoyantly rising bubbles created at the level where T = 8000 K, and which have collapsed into vortex rings. The calculation is in quantitative agreement with observations and has a number of implications for solar physics and convection theory.

071.046 **Radiative damping of gravity waves in the solar atmosphere.** J. D. Logan, H. A. Hill.
Space Sci. Rev., Vol. 27, (see 012.046), 301 - 306 (1980).

The nonlocal character of the radiation field significantly modifies the radiative damping of perturbations in the solar photosphere. Gravity waves are not usually considered to exist in the solar photosphere because the radiative damping time, when based on the Newtonian approximation, is too short. However, this restriction does not apply to low order gravity waves. In fact, with the inclusion of nonlocal effects, the radiative damping for low order gravity waves becomes negative for some region in the photosphere and thus acts as a driving mechanism for gravity waves there.

071.047 **High resolution IR balloon-borne solar spectra and laboratory spectra in the HNO_3 1720-cm^{-1} region: an analysis.** A. Goldman, D. G. Murcray, F. J. Murcray, E. Niple.
Appl. Opt., Vol. 19, 3721 - 3724 (1980).

071.048 **Morphological elements and characteristics of the fine structure of the photospheric brightness field near the solar disc center.** V. N. Karpinskij.
Soln. Dannye 1980 Byull., No. 7, p. 94 - 103 (1980). In Russian.

The fine structure is considered as an ensemble of the following local morphological elements: the starting level, downward and upward excursion at the starting level, granule, porule, intergranular lane, subgranular structures, peculiar formations. A morphological map is given.

071.049 **Construction of fundamental systems of oscillator strengths and abundance scales in the solar photosphere. Fe I.** Eh. A. Gurtovenko, R. I. Kostyk.
Inst. teor. fiz. AN USSR. Prepr., 1979 (1980), No. 138, 45 pp. In Russian. – Abstr. in Ref. zh., 51. Astron., 11.51.428 (1980)

071.050 **Improved identification of some lines in the solar spectrum from λλ 4145 to 4190 Å.**
G. A. Porfir'eva.
Astron. Tsirk., No. 1101, p. 2 - 4 (1980). In Russian.

071.051 **Determination of the damping constant from the wings of strong Fraunhofer lines of the Liège Atlas of the solar spectrum.**
N. N. Kondrashova, Eh. A. Gurtovenko.
Astrometr. Astrofiz., Vyp. (No.) 42, p. 14 - 25 (1980). In Russian.

A modified method of Fraunhofer line wings analysis is applied for 38 Fe I lines, 3 Mg I lines, 3 Na I lines and 7 Ca I lines of the Liège Atlas of the solar spectrum. The damping constant for Fe I lines exceeds the van der Waals damping constant 2 - 2.5 times on the average. The results of other previous investigations of the damping constants for Fe I, Na I, Mg I and Ca I lines are discussed briefly.

Stark effect at the Si I series limit.
See Abstr. 022.043.

Lifetime measurements of stepwise collisional and radiative excited Mn I levels. See Abstr. 022.062.

An inversion technique for the determination of velocity fields from spectral line profiles.
See Abstr. 031.531.

High-resolution solar spectroscopy with double-pass solar spectrometers. See Abstr. 034.005.

A polarimeter for a Fourier Transform Spectrometer and initial solar observations. See Abstr. 034.007.

On the formation of Fe II lines in stellar spectra. I. Solar spatial intensity variation of λ3969.4.
See Abstr. 063.046.

Two-dimensional radiative transfer. II. The wings of Ca K and Mg *k*. See Abstr. 063.047.

Variations in Fraunhofer spectra of solar active regions and their connection with flares.
See Abstr. 072.070.

Characteristics of plage fragments with photospheric network properties. See Abstr. 073.002.

Comment on 'Average photospheric poloidal and toroidal magnetic field components near solar minimum' by Duvall et al. See Abstr. 075.001.

Analysis of changes in photospheric magnetic fields within a flare-productive active region. See Abstr. 075.007.

Five minute microwave solar oscillations.
See Abstr. 077.002.

Examples of non-thermal motions as seen on the sun.
See Abstr. 080.026.

Convective instability in the solar envelope.
See Abstr. 080.085.

Thorium in Arcturus, Pollux, Procyon and the sun.
See Abstr. 114.070.

Methods for the analysis of stellar spectra veiled by lines (III). See Abstr. 114.081.

The sun among the stars. III. Energy distributions of 16 northern G-type stars and the solar flux calibration.
See Abstr. 114.088.

072 Sunspots, Faculae, Activity Cycles, Solar Patrol

072.001 **A possible interpretation of retard evolutions of solar active regions.** Q. Yin, L. Fang.
Kexue Tongbao, Vol. 25, 656 - 659 (1980).

072.002 **Theory of force-free magnetic fields for unipolar sunspots and the energy of solar flares.**
H.-s. Yang, H.-m. Zhang.
Acta Astron. Sinica, Vol. 21, 136 - 142 (1980). In Chinese.

072.003 **Theory of force-free magnetic fields for bipolar sunspots and the energy of solar flares.**
H.-s. Yang, H.-m. Zhang, W.-b. Li.
Acta Astron. Sinica, Vol. 21, 143 - 151 (1980). In Chinese.

072.004 **A continuum bright point at the penumbral edge.**
H. Zirin, R. L. Moore.
Sol. Phys., Vol. 67, 79 - 82 (1980).
A small continuum bright point, observed at the outer edge of the penumbra of a small spot in a large complex spot group, is related to an occurrence beneath the Sun's surface. The characteristics of the point appear to be unique, and the name 'penumbra-periphery bright point' is proposed.

072.005 **Structure of the penumbra.** R. L. Moore.
Bull. American Astron. Soc., Vol. 12, 476 - 477 (1980). – Abstract.

072.006 **Resonant modes of umbral oscillation in sunspots.**
J. H. Thomas, M. A. Scheuer.
Bull. American Astron. Soc., Vol. 12, 477 (1980). – Abstract.

072.007 **A comparison of solar cycle 21 with previous solar cycles.** W. H. Marquette, S. F. Martin.
Bull. American Astron. Soc., Vol. 12, 508 (1980). – Abstract.

072.008 **Comparison of slowly and rapidly evolving magnetic structures in active regions seen in Hα and EUV.**
E. J. Schmahl, Z. Mouradian, M.-J. Martres, I. Soru-Escaut.
Bull. American Astron. Soc., Vol. 12, 526 (1980). – Abstract.

072.009 **Active region morphology and evolution-images from the Ultraviolet Spectrometer and Polarimeter.**
R. A. Shine, J. C. Brandt, R. D. Chapman, P. J. Kenny, A. G. Michalitsianos, B. E. Woodgate, E. C. Bruner, R. Rehse, S. A. Schoolman, C. C. Cheng, E. A. Tandberg-Hanssen, G. R. Athay, J. M. Beckers, J. Gurman, W. Henze, C. L. Hyder.
Bull. American Astron. Soc., Vol. 12, 531 (1980). – Abstract.

072.010 **Morphology of active regions and flares.**
R. D. Bentley.
Bull. American Astron. Soc., Vol. 12, 533 (1980). – Abstract.

072.011 **Density diagnostic of solar active region and flare plasmas from Si IV/O IV line ratio as observed from SMM (*Solar Maximum Mission*).**
E. C. Bruner, R. Rehse, S. A. Schoolman, J. C. Brandt, R. D. Chapman, P. J. Kenny, A. G. Michalitsianos, R. A. Shine, B. E. Woodgate, C. C. Cheng, E. A. Tandberg-Hanssen, G. R. Athay, J. M. Beckers, J. Gurman, W. Henze, C. L. Hyder.
Bull. American Astron. Soc., Vol. 12, 534 - 535 (1980). Abstract.

072.012 **Sunspot observations with the Ultraviolet Spectrometer and Polarimeter experiment on the Solar Maximum Mission.**
J. B. Gurman, B. E. Woodgate, R. A. Shine, J. C. Brandt, R. D. Chapman, A. G. Michalitsianos, P. J. Kenny, E. C. Bruner, R. Rehse, S. A. Schoolman, C. C. Cheng, E. A. Tandberg-Hanssen, G. R. Athay, J. M. Beckers, W. Henze, C. L. Hyder.
Bull. American Astron. Soc., Vol. 12, 535 (1980). – Abstract.

072.013 **Year of solar maximum.**
V. E. Stepanov, V. V. Kasinskij, V. M. Tomozov.
Zemlya Vselennaya, 1980, No. 4, p. 33 - 37. In Russian.

072.014 **On the use of the method of regressions for forecasting smoothed monthly Wolf numbers.**
Yu. I. Vitinskij.
Soln. Dannye 1980 Byull., No. 2, p. 103 - 106 (1980). In Russian.
Equations of linear regressions are given for forecasting smoothed monthly Wolf numbers separately for the ascending and descending branches of the 11-year solar cycle. It is shown that the method permits a prognosis of Wolf numbers a year and a half in advance for the end of the ascending branch and three years in advance for the beginning of then descending branch.

072.015 **Photometric analysis of the sunspot umbral dots. I. Dynamical and structural behaviour.**
A. Adjabshirzadeh, S. Koutchmy.
Astron. Astrophys., Vol. 89, 88 - 94 (1980).
A sequence of selected high resolution white light pictures of a unipolar sunspot was studied in order to measure the main characteristics of umbral dots size, contrast value, and lifetime. Using a statistical analysis, the behaviour of a typical dot was deduced. The large effect of smearing on the values of umbral intensities and sizes of umbral dots is discussed. An analytic function is proposed to describe the time variation of some of the umbral dots.

072.016 **Seasonal variation of oriental sunspot sightings.**
D. M. Willis, M. G. Easterbrook, F. R. Stephenson.
Nature, Vol. 287, 617 - 619 (1980).
The authors discuss what information the dates of the oriental sunspot sightings provide on past atmospheric conditions in the Orient and hence on past solar activity.

072.017 **On two populations of sunspot groups.**
G. V. Kuklin.
Bull. Astron. Inst. Czechoslovakia, Vol. 31, 224 - 232 (1980).
The principal component method was applied studying the sunspot groups distribution in respect to the maximum area for the individual 11-year cycles 12–19 (Lopez Arroya and Lahulla, 1974) and for the years 1900 - 1964 (Mandrykina, 1974). The existence of two populations of sunspot groups is confirmed. The characteristic distinction between populations I and II is apparently the magnetic structure of the groups belonging to them (bipolar and unipolar ones).

072.018 **Some pecularities in the development of the large August 1972 sunspot group.**
V. Bumba, L. Hejna.
Bull. Astron. Inst. Czechoslovakia, Vol. 31, 257 - 267 (1980).
On the basis of a large series of good quality sunspot photographs and with the aid of daily Mt. Wilson magnetograms the individual phases in the August 1972 proton-flare sunspots group development are found and their relation to the changes of magnetic field topology are estimated. A detailed description of two types of sunspot light-bridge evolution and their dependence on the magnetic field polarity distribution is given. A brief summary of the investigation of the chromospheric fine structure morphology determination by the underlying photospheric as well as magnetic field details in the group is presented.

072.019 **On the relative inhomogeneity of long-term series of sunspot indices.**
M. Kopecký, G. V. Kuklin, B. Růžičková-Topolová.
Bull. Astron. Inst. Czechoslovakia, Vol. 31, 267 - 283 (1980).

The fundamental observation series of sunspots, in particular the Greenwich, Zurich and Pulkovo, were compared. The runs of the relative numbers, total areas, numbers of groups of various types, the sunspot groups frequency distribution according to their importance, etc. were compared. It was found that these fundamental observational series are not homogeneous with respect to one another; some inhomogeneities within the series themselves are also pointed out. All this has to be taken into account in using them.

072.020 **Evidence for extreme divergence of open field lines from solar active regions.**
G. A. Dulk, D. B. Melrose, S. Suzuki.
Proc. Astron. Soc. Australia, Vol. 3, 375 - 379 (1979).

072. 021 **Neues zur Theorie der Sonnenaktivität.**
M. Schüssler.
Sterne Weltraum, Jahrg. 19, 331 - 337 (1980).

072.022 **Identification of the CrH molecule in a sunspot spectrum.** O. Engvold, H. Wöhl, J. W. Brault.
Astron. Astrophys., Suppl. Ser., Vol. 42, 209 - 213 (1980).

The $^6\Sigma^+ - {}^6\Sigma^+$ infrared system of the CrH molecule has been identified in the spectrum of a large sunspot.

072.023 **Le cycle d'activité solaire.** P. Simon.
Astronomie, Vol. 94, 417 - 430 (1980).

072.024 **Champ magnétique et champ de vitesse dans les régions actives solaires.** J. Rayrole.
Astronomie, Vol. 94, 431 - 437 (1980).

072.025 **Some peculiarities of solar activity for 1958 - 1976.**
G. A. Bazilevskaya, E. S. Vernova, G. F. Krymskij, M. I. Tyasto.
XIth seminar on cosmophysics, (see 012.035), p. 305 - 315 (1979). In Russian. – Abstr. in Ref. zh., 51. Astron., 8.51.399 (1980).

072.026 **L'activité solaire et les atmosphères planétaires.**
M. Grenon.
Orion, 38. Jahrg., 147 - 149 (1980).

072.027 **Flux linkages of bipolar sunspot groups: a computer study.** P. J. Baum, A. Bratenahl.
Sol. Phys., Vol. 67, 245 - 258 (1980).

Past studies of the structure of solar magnetic fields have used magnetograph data to compute selected field lines for comparison with the morphology of structures seen in various spectral wavelengths. While those analyses examine one of the integral properties of magnetic fields (field lines), they are not complete since they fail to determine the other important integral property: the boundaries of the flux of field lines of given connectivity. The authors determine such a system of boundaries, called separatrices, for the current free field of two p-f spot pairs so as to exhibit the line of self-intersection, called the separator. The analysis is compared with previous analytical work.

072.028 **Solar polar field reversals and secular variation of cosmic ray intensity.** H. S. Ahluwalia.
Solar and interplanetary dynamics, (see 012.020), p. 79 - 86 (1980).

The profile of the well-known 11-year variation of the cosmic ray intensity appears to depend upon the emerging solar polar magnetic field regime in a very characteristic manner. The author suggests two model configurations for the heliosphere. He believes that an "open" heliosphere model applies to solar activity cycles 18 and 20. A "closed" heliosphere model is obtainable during solar activity cycles 17 and 19. These results are discussed.

072.029 **Evidence for open field lines from active regions: short communication.** K. V. Sheridan.
Solar and interplanetary dynamics, (see 012.020), p. 261 (1980). – See 072.020.

072.030 **Änderungen der differentiellen Rotation und meridionale Bewegungen von Sonnenflecken 1940 bis 1968.** H. Balthasar, H. Wöhl.
Sterne Weltraum, Jahrg. 19, 385 - 388 (1980).

072.031 **The year of solar maximum.** V. E. Stepanov.
Vestn. AN SSSR, 1980, No. 6, p. 94 - 98. In Russian. – Abstr. in Ref. zh., 51. Astron., 10.51.37 (1980).

072.032 **The uniqueness of the geographic South Pole for the observations of large scale velocity fields on the sun: some experiences from a 120 hours continuous H-alpha patrol during January 1979.** U. Kusoffsky, M. A. Pomerantz.
Highlights of Astronomy, Vol. 5, (see 012.022), 89 - 90 (1980).

072.033 **Relationship between sunspot numbers during years of sunspot maximum and sunspot minimum.**
R. P. Kane, N. B. Trivedi.
Sol. Phys., Vol. 68, 135 - 139 (1980).

A correlation analysis shows that the peaks of the last eight solar cycles are well-correlated with the sunspot numbers in heliolatitudes 20°- 40° (specially in the southern hemisphere) occurring in the solar minimum years immediately preceding the solar maximum years.

072.034 **On the prediction of anomalous monthly Wolf numbers for each year of an 11-year solar cycle.**
Yu. I. Vitinskij.
Soln. Dannye 1980 Byull., No. 4, p. 104 - 107 (1980). In Russian.

Regression equations are given for the forecast of maximum and minimum values of monthly Wolf numbers of each year in an 11-year solar cycle for the descending and ascending branches separately. The reliability of the forecast is estimated. These indices are predicted for 1980 - 1982.

072.035 **On the origin of the Maunder minimum.**
P. R. Romanchuk.
Soln. Dannye 1980 Byull., No. 5, p. 103 - 107 (1980). In Russian.

The author's theory of the solar cycles was applied for explaining the origin of the Maunder minimum of solar activity. The Maunder minimum is explained in terms of fluctuations of resonance curves. Some arguments are proposed for explaining the origin of fluctuations.

072.036 **On the absorption coefficient and brightness temperature in sunspots.** G. F. Sitnik.
Soln. Dannye 1980 Byull., No. 6, p. 99 - 107 (1980). In Russian.

From observations the umbra brightness temperature has been determined and the following purely observational facts have been established. The sunspot does not radiate as a black body, since in the opposite case its brightness temperature would be constant for all wavelengths. In the spectral region $4800\ \text{Å} < \lambda < 21000\ \text{Å}$ the continuous absorption due to the existence of negative hydrogen ions really dominates in the umbra, but here it is decreased relative to the photosphere. In the spectral region $\lambda \leqslant 4800\ \text{Å}$ a new continuous absorption agent exists.

072.037 **Provisional sunspot-numbers for June - November 1980.**
Yamamoto Circ., Nos. 1938, 1940, 1942, 1944, 1946, 1948 (1980).

072.038 **Solare Beobachtungsergebnisse. Solar data.** Solar radio emission. 1980 February - June.
HHI Sol. Data, Vol. 31, 13 - 94 (1980).

072.039 **Observations solaires. Rotations 1677 - 1689, 8 janvier - 27 décembre 1979.**
E. Tifrea, V. Dinulescu, S. Dinulescu, G. Mariş, I. D. Niţă.
Cent. Astron. Sci. Spat., Acad. Repub. Socialiste România, Bucarest. 84 pp. Price Lei 3.75 (1980).

072.040 **Photoheliographic results 1972 - 1976.**
R. Obs. Ann., No. 13, 123 pp. Price £ 8.00 (1980).
ISSN 0080-4371.

072.041 **Actividad solar en 1978.** M. Lopez Arroyo.
Bol. Astron. Obs. Madrid, Vol. 10, No. 3, p. 29 - 93 (1978).

072.042 **Solar maps and activity**, 1980 May - September.
F. J. Heyden, V. L. Badillo.
Manila Obs., Sol. Div. (1980).

072.043 **Cartes synoptiques de la chromosphère solaire et catalogues des filaments et des centres d'activité.**
M.-J. Martres, G. Zlicaric.
Obs. Paris, Sect. Astrophys. Meudon, Vol. 6, Fasc. 3, Années 1978 - 1979 (1980). – Rotations Nos. 1663 à 1689, 1977 December 21 - 1979 December 27.

072.044 **Polar faculae of the sun.** Y. Tanaka.
Tokyo Astron. Obs. Rep. (No. 73), Vol. 19, 179 - 216 (1980). In Japanese.

072.045 **Twenty years of solar observing.**
R. J. Livesey.
J. British Astron. Assoc., Vol. 91, 68 - 74 (1980).

072.046 **1978 data on proton flare prediction by means of spiral spots.** Y. Ding, B. Luo, B. Zhang, W. Li.
Kexue Tongbao, Vol. 25, 928 - 932 (1980).

072.047 **„A" Sonnenfleckenbeobachtungen von blossem Auge.** H. U. Keller.
Orion, 38. Jahrg., 180 - 184 (1980).

072.048 **A prediction method for sunspot activity.**
M. Ichinose, R. Maeda, S. Ito.
Rev. Radio Res. Lab., Vol. 25, 443 - 448 (1979). In Japanese.
Abstr. in Phys. Abstr., Vol. 84, Abstr. 4846 (1981).

072.049 **La observación del sol. 7ª Parte: Evaluación de las áreas y diámetros de las manchas y fáculas.**
A. Papetti.
Rev. Astron., No. 213, p. 13 - 14 (1980).

072.050 **El aficionado y el sistema solar. Observaciones de manchas solares.**
Rev. Astron., No. 213, p. 19 - 23 (1980).

072.051 **Le maximum de l'activité solaire s'est-il produit en décembre 1979?** A. Koeckelenbergh.
Ciel Terre, Vol. 96, 355 - 358 (1980).

This note briefly recalls the nature and the methods used for the determination of the Wolf numbers. The evolution of the solar activity during the 21st cycle is presented and a prediction for the next six months (from 1980 November to 1981 April) is suggested.

072.052 **Differential rotation and meridional motions of sunspots in the years 1940 - 1968.**
H. Balthasar, H. Wöhl.
Astron. Astrophys., Vol. 92, 111 - 116 (1980).

Using positions of sunspots from the Greenwich Photoheliographic Results from 1940 to 1968 equations for the differential rotation and meridional motions of sunspot groups are determined. The differential rotation depends on the phase in the solar cycle and on the type of the groups. The meridional motions show a general southdrift of the spots. The significance of the equatorward motion near the equator found by Ward (1965) seems to be due to an effect of selection.

072.053 **Alfvén waves in sunspots.**
A. H. Nye, J. V. Hollweg.
Sol. Phys., Vol. 68, 279 - 295 (1980).

The propagation of Alfvén waves in a simple model of a sunspot is considered. The authors find that the observations of non-thermal motions near the temperature minimum (Beckers, 1976) and in the corona (Beckers and Schneeberger, 1977) are both consistent with an upward-propagating Alfvénic energy flux density of a few times 10^7 erg cm^{-2} s^{-1}. This flux density is too small to cool the sunspot, but it is large enough to supply the energy requirements of the transition region and corona above a sunspot. This conclusion depends on the assumptions that the observed motions are indeed Alfvénic with periods near 180 s.

072.054 **A note on permissible values of the vertical gradient of the sunspot magnetic field.**
V. A. Osherovitch (*Osherovich*).
Sol. Phys., Vol. 68, 297 - 302 (1980).

Some estimations for the vertical gradient of the sunspot magnetic field have been obtained under the similarity assumption. It is proved that the assumption is incompatible with large values of the vertical gradient. The greater the size of the sunspot is the smaller is its permissible value of dH/dz.

072.055 **Distribution of sunspots 1874–1976.**
B. D. Yallop, C. Y. Hohenkerk.
Sol. Phys., Vol. 68, 303 - 305 (1980).

A diagram of the distribution in latitude of sunspot groups during the period 1874–1976 is given. As an indication of sunspot activity, a diagram of the mean total daily sunspot areas for each synodic rotation is also given.

072.056 **Visual observations of the sun in Czechoslovakia in the year 1979.** L. Schmied.
Říše hvězd, Vol. 61, 201 - 203 (1980). In Czech.

072.057 **On the fine structure of light rings of sunspots.**
L. D. Parfinenko.
Soln. Dannye 1980 Byull., No. 7, p. 85 - 88 (1980). In Russian.

Using high-quality stratospheric and ground observations of sunspots the fine structure of light rings was studied with TV methods. The outer light ring is formed by a relative clustering of bright photospheric granules of the photosphere, bordering with the sunspot penumbra. The inner light ring is formed by ends of light filaments of the penumbra through a brightening of filaments from the quiet photosphere towards the sunspot umbra. Light rings are observed in all developed sunspots.

072.058 **Using the method of the heliolatitude-longitude index of solar activity for an analysis of the 27-day cosmic ray variations.** L. F. Churunova.
Ionos. i soln.-zemn. svyazi. Alma-Ata, 1980, p. 105 - 113. In Russian. – Abstr. in Ref. zh., 51. Astron., 11.51.476 (1980).

072.059 **Change of the physical conditions in sunspots of the active McMath 14943 region in September 12 - 19, 1977.** Eh. A. Baranovskij, N. N. Stepanyan.

Izv. Krymskoj Astrofiz. Obs., Tom 62, 125 - 130 (1980). In Russian.

Empirical models are obtained for 20 sunspot umbras. Five of these models represent the development of one spot for the 8-day period. The models are derived from the wings of K Ca II and D_2 Na I lines. It is shown that the method of sunspot models calculation using the wings of the above mentioned lines permits the derivation of spot density with agreeable accuracy. For the developing spot it is found that at first the temperature is scarcely changing, but the density is decreasing; then both temperature and density are increasing.

072.060 **On a relationship of brightness variations of Hα plages in active regions on the sun.** M. B. Ogir'.
Izv. Krymskoj Astrofiz. Obs., Tom 62, 131 - 141 (1980). In Russian.

Variations of brightness in seven spot groups belonging to five active regions are discussed. A correlation in the brightness variations of plages situated in the regions of a growing magnetic field is obtained.

072.061 **Sunspot motion and the flare on 4 July 1974.** A. N. Babin.
Izv. Krymskoj Astrofiz. Obs., Tom 62, 142 - 147 (1980). In Russian.

The pecularities of umbra motion within the δ magnetic configuration of McMath region 13043 before the flare on 4 July 1974 with white light emission are considered. A few hours before the flare one of the umbrae started to emerge from the δ configuration at velocity > 300 m/s. After the flare appeared, the process stopped. The exact position of the maxima of Hα and (Hα ± 1 Å) flare knots and the brightest core of the white light flare are compared.

072.062 **The behaviour of active region McMath 14822 at 13.5 mm wavelength.**
S. L. Domnin, V. A. Efanov, I. G. Moiseev, N. S. Nesterov.
Izv. Krymskoj Astrofiz. Obs., Tom 62, 176 - 182 (1980). In Russian.

The degrees of left- and right-handed polarizations of S-component sources, related to spots, show dependence on the strengths of appropriate spots' magnetic fields, and the sign of polarization corresponds to dominating of an extraordinary wave in emission. The main parameters of 10 bursts observed in the time of monitoring the region are presented. All bursts were circularly polarized with degree reaching 60 per cent.

072.063 **Study of faculae from the wings of the Ca II K line.** T. M. Minasyants.
Astron. Tsirk., No. 1102, p. 6 - 8 (1980). In Russian.

072.064 **Combined radio-optical observations of active solar regions associated with the S-component of solar microwave emission.** V. Gaizauskas, K. F. Tapping.
J. R. Astron. Soc. Canada, Vol. 74, 358 (1980). – Abstract.

072.065 **On a connection between polar faculae and sunspot formation activity.** V. V. Makarova.
Astron. Tsirk., No. 1107, p. 3 - 4 (1980). In Russian.

072.066 **The two components of the solar activity cycle as a consequence of the shock transition model of the solar magnetic cycle.** M. H. Gokhale.
Kodaikanal Obs. Bull., Ser. A, Vol. 2, 217 - 221 (1979).

Justification is given for comparing the consequences of the 'shock transition model' of the solar magnetic cycle with important observed properties of the solar activity cycle without waiting for the basic postulates of the model to be mathematically established. Such a comparison shows that the creation and the evolution of the two topologically distinct families of magnetic flux tubes and their different spatial distributions can account for the 'two component nature' of the solar activity cycle and also for the main qualitative differences in the intensity and distribution of activity in the two components.

072.067 **Why is geomagnetic activity during the ending years of a solar cycle well-correlated to the maximum of the next cycle?** M. H. Gokhale.
Kodaikanal Obs. Bull., Ser. A, Vol. 2, 222 - 223 (1979).

The 'two-component model' of the activity cycle provides a basic physical relation between successive solar cycles which might account for the observed correlation between the geomagnetic activity during the 'ending years' of one cycle and the maximum sunspot number in the next cycle.

072.068 **Periodic structure of solar activity and cosmic ray intensity at the 1964 - 1965 minimum.**
E. S. Vernova, A. A. Petrova, N. G. Ptitsyna, M. I. Tyasto.
Geomagn. pole i vnutr. stroenie Zemli. Moskva, 1980, p. 275 - 284. In Russian. – Abstr. in Ref. zh., 51. Astron., 12.51.342 (1980).

072.069 **Display of solar activity on the branch of growth of the 21st cycle in cosmic rays from data of the artificial earth satellite Meteor.**
N. K. Pereyaslova, M. N. Nazarova, I. E. Petrenko.
Geomagn. Aehron., Tom 20, 977 - 981 (1980). In Russian.

072.070 **Variations in Fraunhofer spectra of solar active regions and their connection with flares.**
K. V. Alikaeva, S. I. Gandzha, N. N. Kondrashova, P. N. Polupan.
Astrometr. Astrofiz., Vyp. (No.) 42, p. 3 - 14 (1980). In Russian.

Results of automatized processing of solar active region (AR) spectra are considered. It is found that the variations of the line profiles take place both in the different structural elements of AR and during the evolution of AR. In most cases the cores of lines are less deep and somewhat narrower and the wings are weaker in comparison with the undisturbed photosphere. The deviations of the central intensities from those in the undisturbed photosphere depend upon the optical depths of formation of the considered lines. The results obtained show close interactions of the processes occurring at chromospheric and photospheric levels of AR.

072.071 **Daily maps of the sun and magnetic fields of sunspots.**
Soln. Dannye 1980 Byull., No. 1, p. 1 - 87; No. 2, p. 1 - 84; No. 3, p. 1 - 83; No. 4, p. 1 - 94; No. 5, p. 1 - 102; No. 6, p. 1 - 98; No. 7, p. 1 - 84 (1980). In Russian,

072.072 **L'activité solaire.** M.-J. Martres, G. Zňicaric.
Astronomie, Vol. 94, 410 - 411, 462 - 463, 518 - 519, 570 (1980).

072.073 **Sunspot numbers.**
Sky Telesc., Vol. 60, 257, 347, 451, 547 (1980).

072.074 **Zürcher Sonnenfleckenrelativzahlen.**
Sterne Weltraum, Jahrg. 19, 318, 358, 396, 438 (1980).

The solar activity in the time of Galileo.
See Abstr. 004.038.

Die Weiterführung der Zürcher Sonnenflecken-Statistik. See Abstr. 013.008.

Continuation of the series of relative sunspots numbers. See Abstr. 013.028.

Zürich stopt bepaling Wolf-getal.
See Abstr. 013.033.

Circular polarization of molecular spectra. See Abstr. 022.025.

La observación del Sol. See Abstr. 031.552.

Measurement of areas on the solar disk. See Abstr. 031.577.

The effect of solar activity on long wave propagation. See Abstr. 035.005.

Exact static equilibrium of vertically oriented magnetic flux tubes. I. The Schlüter-Temesváry sunspot. See Abstr. 062.007.

A new method of solving the force-free magnetic field equations and its application to solar physics. See Abstr. 062.022.

Sunspots and the physics of magnetic flux tubes. X. On the hydrodynamic instability of buoyant fields. See Abstr. 062.028.

Solutions of the equation of heat flow. See Abstr. 062.071.

Solar granulation and bright dots of the sunspot umbra: power spectra analysis. See Abstr. 071.001.

Latitude variations of photospheric activity areas with particular reference to solar faculae. See Abstr. 071.038.

On the relationship between the eruption of quiescent filaments and the development of new active centers. See Abstr. 073.012.

Motions in the solar atmosphere associated with the white light flare of 11 July 1978. See Abstr. 073.065.

Vector magnetic field measurements at flare locations. See Abstr. 075.008.

Comparative magnetospherology, part 10. Two-hemisphere model on a reversal process of the heliomagnetosphere. See Abstr. 075.015.

Flux tube dynamo approach to the solar cycle. See Abstr. 075.016.

Morphology and spatial distribution of XUV and X-ray emissions in an active region observed from Skylab. See Abstr. 076.019.

Thermal radio emission from solar active regions. See Abstr. 077.021.

Observation with the VLA of a stationary loop structure on the sun at 6 centimeter wavelength. See Abstr. 077.025.

Polarized radio emission of an intense active region on the sun in July 1974 at wavelengths 1.9, 2.5 and 3.5 cm. See Abstr. 077.041.

Radio emission of solar active regions in the centimetric wavelength range. See Abstr. 077.047.

Comment on "Variability of the far-infrared solar temperature minimum with the solar cycle". See Abstr. 077.055.

Development of local radio sources and burst activity on the sun at short centimeter wavelengths. See Abstr. 077.058.

The sun is observed to be a torsional oscillator with a period of 11 years. See Abstr. 080.008.

Physical conditions in the solar atmosphere above an active region. See Abstr. 080.020.

Magnetic activity and variations in solar luminosity. See Abstr. 080.024.

Variations in the solar constant due to solar active regions. See Abstr. 080.068.

Solar rotation and activity in the past and their possible influence upon the evolution of life. See Abstr. 080.081.

Differential rotation of solar features and its variation as deduced from the 'shock-transition model' of the solar cycle. See Abstr. 080.082.

Differential rotation of the sun and the Maunder minimum of solar activity. See Abstr. 080.087.

On the location of the ionospheric current systems responsible for the lunar and solar magnetic variations. See Abstr. 083.042.

Evidence in the auroral record for secular solar variability. See Abstr. 084.041.

Secular variation of the solar activity during the geological age and its associated effect on the planetary environments. See Abstr. 085.014.

Lyman alpha albedo of Jupiter and solar activity. See Abstr. 099.002.

Titan: aerosol photochemistry and variations related to the sunspot cycle. See Abstr. 100.122.

Disintegration of comets and solar activity. See Abstr. 102.012.

The effect of solar-cycle ultraviolet flux variations on cometary gas. See Abstr. 102.031.

Four years of zodiacal light observations from the Helios space probes: evidence for a smooth distribution of interplanetary dust. See Abstr. 106.022.

Change in the zodiacal light with solar activity. See Abstr. 106.024.

Variations of interplanetary parameters and cosmic-ray intensities. See Abstr. 106.063.

Erratum

072.901 **Erratum: "Microwave, EUV, and X-ray observations of active region loops: evidence for gyroresonance absorption in the corona"** [Astron. Astrophys., Vol. 82, 265 - 271 (1980)].
M. R. Kundu, E. J. Schmahl, M. Gerassimenko.
Astron. Astrophys., Vol. 91, 377 (1980). – See Abstr. 27.072.010.

073 Chromosphere, Flares, Prominences

073.001 **Flare build-up in preflare low-lying loops and non-linear force-free magnetic field.** Q.-r. Su.
Acta Astron. Sinica, Vol. 21, 152 - 157 (1980). In Chinese.

073.002 **Characteristics of plage fragments with photospheric network properties.**
D. J. Nauer, R. G. Teske, G. E. Elste.
Sol. Phys., Vol. 67, 23 - 28 (1980).

Using data taken with the multi-channel magnetograph at KPNO, the authors demonstrate that plage regions surrounding a sunspot have thermal properties found in the photospheric network. These network-like regions existed up to the edge of the penumbra of the sunspot. Temperature gradients inferred from equivalent width fluctuations in the authors' data do not conflict with the requirements of the theory (Parker, 1978) for flux tubes to exist at subphotospheric levels.

073.003 **Measurements of the magnetic field and the gradient of temperature in the solar atmosphere above a flocculus using radio observations.**
V. M. Bogod, G. B. Gelfreikh (*Gel'frejkh*).
Sol. Phys., Vol. 67, 29 - 46 (1980).

The authors develop a method of measuring magnetic fields in the solar atmosphere from thermal bremsstrahlung and demonstrate it, using observations of a flocculus (plage) during August 1–3, 1977. The observations show that the flocculus under investigation possessed bipolar magnetic structure with peak to peak amplitude of magnetic field strength of about 40 G at the level of the upper chromosphere and the transition region. An analysis of the spectra of polarized radio emission gives an opportunity to determine the temperature gradient in the chromosphere-corona transition region.

073.004 **The flare of September 7, 1973: a typical example of a newly recognized class of solar transients.**
R. Pallavicini, G. S. Vaiana.
Sol. Phys., Vol. 67, 127 - 142 (1980).

X-ray, extreme-ultraviolet and optical observations of a solar flare are discussed. It is shown that the flare exemplifies a class of transient events characterized by long duration and long decay time and by the development of high systems of loops, generally brighter at the top.

073.005 **Helium ionization of solar radiation at $\lambda \leqslant 504$ Å by hydrogen absorption.**
Ch. Lkhagvazhav, N. A. Yakovkin, M. Yu. Zel'dina.
Soln. Dannye 1980 Byull., No. 1, p. 88 - 93 (1980).
In Russian.

The degree of helium ionization by solar radiation $\lambda \leqslant 504$ Å for a prominence in the form of a vertical slab is calculated. Hydrogen absorption was taken into account. The results obtained for a various probability of L_c (He I) quantum survival < I are given.

073.006 **Balmer line formation in prominence spectra. Evaluation of the accuracy of approximate solutions and comparison of theory with observations.**
K. I. Selyakov.
Soln. Dannye 1980 Byull., No. 1, p. 93 - 101 (1980).
In Russian.

A table and graphs are presented which give an evaluation of accuracy of approximate solutions used in computations of Balmer spectra and permitting to reduce their errors. A simple method is proposed for determinations of the prominence optical depth at the Hα line center and the Doppler width of Balmer lines.

073.007 **Measurements of line-of-sight velocities in eruptive prominences using a Wollaston prism.**
A. B. Delone, E. A. Makarova, G. S. Mikhalev.
Soln. Dannye 1980 Byull., No. 1, p. 102 - 106 (1980).
In Russian.

An "Opton" Hα filter passband 0.25 Å supplemented with a Wollaston prism was used to obtain two images of the same object at $\lambda_0 \pm 0.12$ Å. The comparison of the two images gives the line-of-sight velocities. The velocities from 30 to 70 km/s were measured in some knots of the eruptive prominence of August 27, 1977.

073.008 **The optical thickness of quiescent prominences in the Ca II K line and the central reversal in the spectral lines.** J. Kubota.
Publ. Astron. Soc. Japan, Vol. 32, 359 - 369 (1980).

Among 164 prominences observed by the author, 20 showed significant central reversal in their spectral lines. Most of these 20 prominences were located near plage regions and/or the position of changing magnetic neutral lines near the active regions. Some characteristics of central reversal in the Hα and the Ca II H and K lines are consistent with the hypothesis that the appearance of the central reversal is due to the absorption of light from the prominence body by a thin absorbing cloud or layer near the surface of the prominence.

073.009 **Spectral features of filamentary physically inhomogeneous prominences. I. Hydrogen (Excitation).**
N. N. Morozhenko, V. V. Zharkova.
Astrometr. Astrofiz., Vyp. (No.) 41, p. 3 - 11 (1980). In Russian.

Models of quiescent prominences with physical and structural inhomogeneities are considered. The steady state equation for the 2 - 5 hydrogen levels together with the transfer equation for the intensity in the Hα line are solved for various models with $\tau^0_{23} = 10$, 50 and 100. Variation of the source function with depth along the line of sight was found as well as the profiles and intensities of the Hα, Hβ and Hγ lines.

073.010 **The EUV continuum of neutral helium in solar prominences.** F. Q. Orrall, E. J. Schmahl.
Bull. American Astron. Soc., Vol. 12, 477 (1980). – Abstract.

073.011 **Exact magnetostatic models of filament prominences.** B. C. Low.
Bull. American Astron. Soc., Vol. 12, 477 (1980). – Abstract.

073.012 **On the relationship between the eruption of quiescent filaments and the development of new active centers.** L. M. Hermans, S. F. Martin.
Bull. American Astron. Soc., Vol. 12, 477 (1980). – Abstract.

073.013 **Size of the X-ray kernel of the large 20 August 1979 solar flare.** P. B. Landecker, D. L. McKenzie.
Bull. American Astron. Soc., Vol. 12, 478 (1980). – Abstract.

073.014 **Multifrequency observations of a solar flare at up to 0.2″ spatial resolution.**
K. A. Marsh, H. Zirin, G. J. Hurford.
Bull. American Astron. Soc., Vol. 12, 478 (1980). – Abstract.

073.015 **Transition-zone observations of rapid flare events as observed by OSO-8.** B. W. Lites.
Bull. American Astron. Soc., Vol. 12, 479 (1980). – Abstract.

073.016 **XUV observations of a dense compact flare.**
K. G. Widing, D. S. Spicer.
Bull. American Astron. Soc., Vol. 12, 479 (1980). – Abstract.

073.017 **On the high densities observed in solar flare plasmas.**
C. C. Cheng.
Bull. American Astron. Soc., Vol. 12, 479 - 480 (1980). Abstract.

073.018 **The necessary conditions for white-light flaring from proton bombardment.**
H. S. Hudson, B. N. Dwivedi.
Bull. American Astron. Soc., Vol. 12, 480 (1980). – Abstract.

073.019 **Extra AR flare brightenings and type III reverse slope bursts.** F. Tang.
Bull. American Astron. Soc., Vol. 12, 480 - 481 (1980). Abstract.

073.020 **Electrostatic ion-cyclotron heat flux instability of a flaring plasma.** P. J. Morrison, J. A. Ionson.
Bull. American Astron. Soc., Vol. 12, 481 (1980). – Abstract.

073.021 **Helium I 10830 Å filament – associated structures.**
M. K. McCabe, D. L. Mickey.
Bull. American Astron. Soc., Vol. 12, 503 - 504 (1980). Abstract.

073.022 **Flare-like events in quiescent prominences.**
G. D. Toot, J. M. Malville.
Bull. American Astron. Soc., Vol. 12, 504 (1980). – Abstract.

073.023 **Measurements of the Na D lines in quiescent prominences.** D. A. Landman.
Bull. American Astron. Soc., Vol. 12, 504 (1980). – Abstract.

073.024 **X-ray spectroscopy during the decay phase of a solar flare.** D. L. McKenzie, P. B. Landecker.
Bull. American Astron. Soc., Vol. 12, 506 (1980). – Abstract.

073.025 **Observations of solar filaments at 8, 15, 22, and 43 GHz.**
E. J. Schmahl, M. Bobrowsky, M. R. Kundu.
Bull. American Astron. Soc., Vol. 12, 507 (1980). – Abstract.

073.026 **Observations of transient magnetic fields in two solar flares.** H. Zirin, A. Patterson.
Bull. American Astron. Soc., Vol. 12, 515 (1980). – Abstract.

073.027 **Lα/Hα in solar flares and QSOs.**
R. C. Canfield, R. C. Puetter.
Bull. American Astron. Soc., Vol. 12, 517 (1980). – Abstract.

073.028 **Origin of the soft X-ray emission from impulsive solar flares.**
C. J. Crannell, J. T. Karpen, R. J. Thomas.
Bull. American Astron Soc., Vol. 12, 527 - 528 (1980). Abstract.

073.029 **High resolution solar flare X-ray spectra.**
U. Feldman, G. A. Doschek, R. W. Kreplin.
Bull. American Astron. Soc., Vol. 12, 529 - 530 (1980). Abstract.

073.030 **Density and temperature measurements for short lived transition zone phenomena.**
K. R. Nicolas, K. P. Dere, J.- D. F. Bartoe, G. E. Brueckner.
Bull. American Astron. Soc., Vol. 12, 530 - 531 (1980). Abstract.

073.031 **Dynamics of the high temperature flare.**
J. W. Leibacher.
Bull. American Astron. Soc., Vol. 12, 531 (1980). – Abstract.

073.032 **Solar flare and surge image sequences as seen by the Ultraviolet Spectrometer and Polarimeter on SMM (*Solar Maximum Mission*).**
W. Henze, J. C. Brandt, R. D. Chapman, P. J. Kenny, A. G. Michalitsianos, R. A. Shine, B. E. Woodgate, E. C. Bruner, R. Rehse, S. A. Schoolman, C. C. Cheng, E. A. Tandberg-Hanssen, G. R. Athay, J. M. Beckers, J. Gurman, C. L. Hyder.
Bull. American Astron. Soc., Vol. 12, 532 (1980). – Abstract.

073.033 **SMM (*Solar Maximum Mission*) joint observations of solar flares.** K. J. Frost.
Bull. American Astron. Soc., Vol. 12, 532 (1980). – Abstract.

073.034 **Low energy gamma ray continuum emission from solar flares of March 29, 1980.**
J. M. Ryan, D. J. Forrest, E. L. Chupp, C. Reppin, E. Rieger, K. Pinkau, G. Kanbach, G. Share, R. L. Kinzer, M. Strickman,
Bull. American Astron. Soc., Vol. 12, 532 (1980). – Abstract.

073.035 **High energy X-ray spectral changes during the solar flare of March 29, 1980.**
A. Kiplinger, B. R. Dennis, K. J. Frost, L. E. Orwig.
Bull. American Astron. Soc., Vol. 12, 533 (1980). – Abstract.

073.036 **Preliminary results from calcium and iron solar flare spectra from a Bent Crystal Spectrometer.**
E. Antonucci.
Bull. American Astron. Soc., Vol. 12, 533 (1980). – Abstract.

073.037 **Search for flare non-thermal electrons in iron and calcium BCS (*Bent Crystal Spectrometer*) spectra.**
M. A. Kayat, A. H. Gabriel, K. J. H. Phillips.
Bull. American Astron. Soc., Vol. 12, 533 - 534 (1980). Abstract.

073.038 **Temporal comparison of Fe Kα and hard X-ray emission during several solar flares.**
C. J. Wolfson.
Bull. American Astron. Soc., Vol. 12, 534 (1980). – Abstract.

073.039 **Inner-shell ionization (Kα) lines in flare impulsive phases.** K. J. H. Phillips.
Bull. American Astron. Soc., Vol. 12, 534 (1980). – Abstract.

073.040 **The dynamics of solar flares and surges as seen at the solar limb in the transition zone.**
B. E. Woodgate, J. C. Brandt, R. D. Chapman, P. J. Kenny, A. G. Michalitsianos, R. A. Shine, E. C. Bruner, R. Rehse, S. A. Schoolman, C. C. Cheng, E. A. Tandberg-Hanssen, G. R. Athay, J. M. Beckers, J. Gurman, W. Henze, C. L. Hyder.
Bull. American Astron. Soc., Vol. 12, 535 (1980). – Abstract.

073.041 **High-resolution X-ray spectra of solar flares. III. General spectral properties of X1–X5 type flares.**
G. A. Doschek, U. Feldman, R. W. Kreplin, L. Cohen.
Astrophys., J., Vol. 239, 725 - 737 (1980).

High-resolution X-ray spectra of six class X1–X5 flares are discussed. The wavelength ranges are: 1.82–1.97 Å, 2.98–3.07 Å, and 3.14–3.24 Å. Electron temperatures are derived from dielectronic-satellite-line–to–resonance-line ratios as a function of time for each flare.

073.042 **Directivity of 50–100 keV X-ray emission from impulsive solar flares.** S. R. Kane,
K. A. Anderson, W. D. Evans, R. W. Klebesadel, J. G. Laros.
Astrophys. J., Lett., Vol. 239, L85 - L88 (1980).

"Stereoscopic" observations of 50–100 keV solar X-rays,

made with two spacecraft in heliocentric orbit, have been used to measure the directivity of impulse solar X-ray bursts. The observations are consistent with an essentially isotropic emission of 50–100 keV X-rays over observation angles between 13° and 79°.

073.043 **A survey of ~1 MeV nucleon^{-1} solar flare particle abundances, $1 \leq Z \leq 26$, during the 1973–1977 solar minimum period.**
G. M. Mason, L. A. Fisk, D. Hovestadt, G. Gloeckler.
Astrophys. J., Vol. 239, 1070 - 1088 (1980).

The authors have surveyed the abundances of the major elements over the range H–Fe in solar flare energetic particles near 1 MeV nucleon^{-1} for a large number of flares during the period 1973–1977. The observations were carried out in interplanetary space using the University of Maryland/Max-Planck-Institut instrumentation on the *IMP 8* spacecraft. The survey considered two types of solar flare events: (1) large events from which the average abundances were deduced, and (2) events which had significant abundance differences from the average.

073.044 **The role of high energetic levels of helium in statistical equilibrium equations for prominences.**
Ch. Lkhagvazhav.
Soln. Dannye 1980 Byull., No. 2, p. 85 - 91 (1980). In Russian.

In the case of the helium atom formulae for evaluation of the radiative and electron impact interaction between higher and lower levels in the statistical equilibrium equations are obtained. Calculations have been made for prominences. It is shown that one must take into account the interaction of higher and lower levels in those cases, when radiative and triple recombinations are important.

073.045 **On peculiarities of limb surges.**
M. N. Stoyanova.
Soln. Dannye 1980 Byull., No. 3, p. 84 - 90 (1980). In Russian.

The structure and development of several surges are studied. Hα spectra and Hα spectroheliograms were obtained simultaneously. A recurring surge was of special interest. The conclusion is made that the occurrence of surges is due to wave processes in the complex structure of the magnetic field which lead to condensation of the coronal matter.

073.046 **On hydrodynamic models of the effect of flares on the solar chromosphere.** B. V. Somov.
Pis'ma Astron. Zh., Tom 6, 597 - 601 (1980). In Russian. English translation in Soviet Astron. Lett., Vol. 6.

The current state of the theory of nonsteady hydrodynamic phenomena caused by flare energy release in the solar atmosphere is discussed. It is emphasized that an impulsive optical continuum radiation (a white flare) can be due to thermal instability of heated chromospheric plasma.

073.047 **Impulsive heating of the solar chromosphere during flares.**
B. Ya. Sermulinya, B. V. Somov, A. R. Spektor.
Pis'ma Astron. Zh., Tom 6, 602 - 605 (1980). In Russian. English translation in Soviet Astron. Lett., Vol. 6.

The problem of nonstationary hydrodynamic plasma flow under impulsive heating of the chromosphere by large thermal fluxes and accelerated particles is solved numerically. Taking into account all essential dissipative processes as well as heat flux saturation, the electron and ion temperature, the density and the hot flare plasma velocity are calculated.

073.048 **Microturbulence near the edge of a solar plage.**
G. Simon, S. Dumont, Z. Mouradian, J. C. Pecker, G. Artzner, J. C. Vial.
Astron. Astrophys., Vol. 89, L8 - L9 (1980).

Observations of the Ca II K line at the edge of a solar plage show enhanced separation of the K2 peaks with respect to the measured value inside the plage and in the quiet sun. This effect may be interpreted as a variation of microturbulent motions at the height of formation of K2.

073.049 **Reviews of solar activities, coronal holes and geomagnetic storms in 1978.**
K. Marubashi, Y. Miyamoto, T. Ishii.
Sol. Terr. Environ. Res. Japan, Vol. 3, 107 - 115 (1979).
Abstr. in Phys. Abstr., Vol. 83, Abstr. 72919 (1980).

073.050 **Some effects of strong acoustic waves on strong spectral lines.** P. Gouttebroze, J. Leibacher.
Stellar turbulence, (see 012.008), p. 212 (1980). – Abstract.

073.051 **The solar chromospheric microturbulence and the emission observed at eclipse.** Y. Cuny.
Stellar turbulence, (see 012.008), p. 293 (1980). – Abstract.

073.052 **Effects of flux tubes on conventional chromospheric diagnostics.** T. R. Ayres.
Stellar turbulence, (see 012.008), p. 299 (1980). – Abstract.

073.053 **Time variation of the E–W asymmetry of flare numbers with respect to the phase of solar cycles.**
V. Letfus, B. Růžičková-Topolová.
Bull. Astron. Inst. Czechoslovakia, Vol. 31, 232 - 239 (1980).

The time variation of the annual values of the index of the E–W asymmetry in the occurrence of flares on the solar disk in the years 1959 - 1976 was studied. The analysis of the period 1935 - 1976 indicates that the characteristic run of the asymmetry was preserved in the 19th cycle, however, in the 20th it was not proved. On the average, there exists a persistent real asymmetry in favour of the E half of the disk. The hypothesis about the external effect on the longitudinal distribution of activity on the Sun, which should be reflected in the annual or semi-annual variation of the asymmetry index, was not substantiated.

073.054 **On proton and electron acceleration by shock waves during large solar flares.**
V. M. Gubchenko, V. V. Zaitsev *(Zajtsev)*.
Proc. Astron. Soc. Australia, Vol. 3, 236 - 238 (1978).

073.055 **Radio evidence on the particle distribution functions in the corona following flares.**
G. A. Dulk, D. B. Melrose, S. F. Smerd.
Proc. Astron. Soc. Australia, Vol. 3, 243 - 247 (1978).

073.056 **On the relative importance of radiative, mechanical and magnetic energy release of flares.**
D. J. McLean, G. A. Dulk.
Proc. Astron. Soc. Australia, Vol. 3, 251 - 252 (1978).

073.057 **Evidence on chromospheric structure from observations of solar brightness distribution at millimetre wavelengths.** N. R. Labrum.
Proc. Astron. Soc. Australia, Vol. 3, 256 - 259 (1978).

073.058 **The H I Lyman continuum in solar prominences and its interpretation in the presence of inhomogeneities.** F. Q. Orrall, E. J. Schmahl.
Astrophys. J., Vol. 240, 908 - 922 (1980).

Observations of the H I Lyman continuum are presented for nine hedgerow prominences observed at the limb with the Harvard EUV spectrometer on Skylab. Methods of analysis for both the continuum emission and absorption are developed for two contrasting models for the prominence fine structure, namely multiple resolved slabs, and multiple unresolved cylindrical threads. The absorption data set a lower limit on the total optical thickness at λ912 and on the number of structures in the line of sight. The emission data yield a mean electron temperature of 7524 ± 739 K and a mean brightness temperature of 6316 ± 75 K. Taken together, the emission and absorption data set firm constraints on the ionization, density, and fractional volume.

073.059 **Particle acceleration by shock waves in solar flares.** A. Achterberg, C. A. Norman.
Astron. Astrophys., Vol. 89, 353 - 362 (1980).

The authors propose a detailed shock wave model for second phase acceleration associated with solar flares. The acceleration of protons and electrons by solar flare generated shock waves is analyzed. A number of key observational consequences of the theory are described.

073.060 **Relativistic protons in the solar flare of 1977, September 24.** V. A. Blyudov, N. N. Volodichev, G. Ya. Kolesov, O. Yu. Nechaev, A. N. Podorol'skij, I. A. Savenko, A. A. Suslov.
Kosm. Issled., Tom 18, 801 - 804 (1980). In Russian.

073.061 **Continuous emission of solar and stellar flares.** M. A. Livshits, O. G. Badalyan, A. G. Kosovichev, M. M. Katsova.
Inst. zemn. magn., ionos. i rasprostr. radiovoln AN SSSR. Prepr., 1980, No. 7, 30 pp. In Russian. – Abstr. in Ref. zh., 51. Astron., 8.51.378 (1980).

073.062 **Upper limits on the power in solar oscillations at 1.2 mm, 9 mm, 3.7 cm, and 11.1 cm wavelengths.** M. R. Kundu, E. J. Schmahl.
Astron. Astrophys., Vol. 90, 192 - 197 (1980).

A search for solar oscillations has been made using the NRAO 36 ft telescope at 1.2 mm, the NRL 85 ft telescope at 9 mm, and the NRAO four-element interferometer at 3.7 and 11.1 cm wavelengths. After corrections for the small coherence length of the optically observed oscillations, for their known spectral bandwidth, and for the visibility function of the interferometer, upper limits have been placed on the fluctuation power at oscillation frequencies near 3 mHz. The interferometric observations at 3.7 cm and the single-dish observations at 1.2 and 9 mm imply that less than 0.3%, 0.04% and 0.1% respectively, of the bremsstrahlung photons emitted from the chromosphere show periodic fluctuations.

073.063 **Weißlichtflares.** M. Waldmeier.
Sonne, Jahrg. 4, 110 (1980).

073.064 **Na-light flare observations: McMath 13043–July 1974.** A. Cacciani, T. Fortini, M. Torelli.
Sol. Phys., Vol. 67, 311 - 316 (1980).

The authors extract a temporal sequence of the 13:55 UT 4 July, 1974 event from monochromatic filtergrams in Na light on the McMath region No. 13043–July 1974. Due to the properties of Na filtergrams the authors derive the exact relative position among sunspots, magnetic fields and flare-knots.

073.065 **Motions in the solar atmosphere associated with the white light flare of 11 July 1978.** L. Dezsö, L. Gesztelyi, L. Kondás, Á. Kovács, S. Rostás.
Sol. Phys., Vol. 67, 317 - 338 (1980).

The two main results of the present analysis of the major flare event of 11 July 1978 are that the solar phenomenon called Hα flare was predominantly a bright loop prominence system and that a filament, which appeared to directly cause this flare by its sudden disruption, was re-formed from below during the late phase of the flare.

073.066 **Observed Lα profiles for two solar flares: 14:12 UT 15 June, 1973 and 23:16 UT 21 January, 1974.** R. C. Canfield, M. E. Van Hoosier.
Sol. Phys., Vol. 67, 339 - 350 (1980).

Photographic observations of the time development of the profile of the Lα line of hydrogen during flares were obtained. The profiles for the 15 June, 1973 and 21 January, 1974 flares cover both core and wings of the line. The time sequences begin before flare maximum, and continue well into the decay phase. The authors discuss core symmetry and shift, and show that the observations imply integrated flare Lα/Hα intensity ratios within a factor of two of unity for these two flares.

073.067 **Thermodynamic models and fine structure of prominences.** O. Engvold.
Sol. Phys., Vol. 67, 351 - 355 (1980).

Observed Hα brightness versus size of emission substructures of quiescent prominences are compared with values predicted from thermodynamical models. The measured size of an emission element of a given brightness is substantially less than the theoretical value. Two possible causes for the discrepancy are suggested: (1) The partial filling of a recording aperture, due to the prominence fine structure, may affect the measurements seriously. (2) Changes of individual fine structure elements on a time scale of a few minutes implies that the prominence plasma may be in a non-stationary radiative state.

073.068 **A classification scheme for solar flare models.** D. S. Spicer, J. C. Brown.
Sol. Phys., Vol. 67, 385 - 392 (1980). Based on invited talks given at the SERF Workshop (Aug. 13, 1979) and the IAU Meeting of Commission 10 (Aug. 15, 1979) held in Montreal.

The authors present a classification scheme for solar flare models that utilize magnetic free energy-currents. The classification scheme is geometry independent and delineates models into two categories: those models utilizing currents flowing parallel to **B** and those utilizing currents flowing perpendicular to **B**. This delineation of drivers allows to specify what kinds of plasma-magnetic field configurations should be expected for a given current driver. Further, the delineation of drivers allows to identify both the strengths and the weaknesses of the various models.

073.069 **On discrete components of large solar flares.** V. N. Ishkov, Eh. I. Mogilevskij.
Fiz. protsessy v ionos. i magnitosfere. Moskva, 1979, p. 5 - 22. In Russian. – Abstr. in Ref. zh., 51. Astron., 9.51.371 (1980).

073.070 **On the study of emission lines of prominences.** G. F. Sitnik.
Astron. Zh., Tom 57, 1016 - 1019 (1980). In Russian. English translation in Soviet Astron., Vol. 24, No. 5.

A new method of investigation of emission lines of solar prominences, based on the use of triplet lines, is suggested. The main results of its application to the study of the magnesium green triplet lines and the Balmer lines H_{10}, H_{11}, and H_{12} in the quiescent prominence of October 15, 1960, are presented.

073.071 **A successful observation of the flash spectrum at the total solar eclipse of February 16, 1980.**
Astron. Circ., No. 7, p. 1 - 2, 9 (1980). – In Chinese and English.

073.072 **On the relation between spectra of electrons and bremsstrahlung X-radiation in solar flares.** E. I. Dajbog.
Pis'ma Astron. Zh., Tom 6, 641 - 644 (1980). In Russian. English translation in Soviet Astron. Lett., Vol. 6.

The spectrum of bremsstrahlung X-quanta is calculated for an exponential injection spectrum of radiating electrons in solar flares. It is shown that in the thick-target model the spectrum of quanta is steeper than the primary electron spectrum in contrast to the case of a power electron injection spectrum.

073.073 **Coronal holes, the height of the chromosphere, and the origin of spicules.** D. Rabin, R. L. Moore.
Astrophys. J., Vol. 241, 394 - 401, plate 3 (1980).

Hα observations led the authors to the hypothesis that spicules are taller under coronal holes, and that the smooth latitude trend found by Lippincott was an average over upward jumps at hole boundaries which were located mainly at high latitudes. The authors present the results of measurements of the height of the Hα chromosphere designed to test this hypothesis.

073.074 **Nonequilibrium ionization due to thermal diffusion and mass flows.** R. Roussel-Dupré.
Astrophys. J., Vol. 241, 402 - 408 (1980).

Recent calculations of diffusion coefficients are used in the continuity equation to compute ion populations of carbon in the solar transition region. Thermal diffusion causes strong departures from ionization equilibrium in the region where the temperature gradient is steepest. Mass-conserving flows are also included in the calculations. These dominate over thermal diffusion depending on the magnitude of the flows and also lead to departures from ionization equilibrium. These results have important implications for the interpretation of EUV line emission.

073.075 **Energy and mass injected by flares and eruptive prominences.** O. Engvold.
Solar and interplanetary dynamics, (see 012.020), p. 173 - 188 (1980).

The author describes and discusses various types of dynamical events at chromospheric levels and in the lower corona, and also attempts to evaluate their significance with regard to coronal disturbances. There are strong evidences that ascending prominences and flare sprays are essential in the process which brings about observable coronal transients.

073.076 **On a peculiar type of filament activation.** A. Bruzek.
Solar and interplanetary dynamics, (see 012.020), p. 203 - 206 (1980).

Hα observations of a peculiar type of filament activation are presented and discussed. Ejected filament material was stopped at a distant position and formed a new stable filament for 1/2 - 1 hour until it returned to its source or faded in situ. A similar event has been observed at the solar limb.

073.077 **The filament eruption in the 3B flare of July 29, 1973: onset and magnetic field configuration.**
R. L. Moore, B. J. LaBonte.
Solar and interplanetary dynamics, (see 012.020), p. 207 - 211 (1980).

073.078 **Dynamics of a quiescent filament.** B. Schmieder, M.-J. Martres, P. Mein, I. Soru-Escaut.
Solar and interplanetary dynamics, (see 012.020), p. 213 - 215 (1980).

073.079 **Particle acceleration in the process of eruptive opening and reconnection of magnetic fields.**
Z. Švestka, S. F. Martin, R. A. Kopp.
Solar and interplanetary dynamics, (see 012.020), p. 217 - 221 (1980).

073.080 **Recent observations of energetic electrons in solar flares.** S. R. Kane.
Solar and interplanetary dynamics, (see 012.020), p. 227 - 230 (1980).

073.081 **An energy storage process and energy budget of solar flares.** K. Tanaka, Z. Smith, M. Dryer.
Solar and interplanetary dynamics, (see 012.020), p. 231 - 234 (1980).

073.082 **Flare associated eruptive prominence activity of February 1, 1979.** A. Bhatnagar, R. M. Jain, D. B. Jadhav, R. N. Shelke, R. V. Bhonsle.
Solar and interplanetary dynamics, (see 012.020), p. 235 - 240 (1980).

Observations and analysis of solar flare activated ascending "Fountain type" prominence of 1 February 1979 are presented. It is shown that, as the "Fountain" prominence rises, it carries along with it the complex magnetic field which unfolds as the prominence material expands into distinct magnetic field lines.

073.083 **Flare model with force-free fields and helical symmetry.** D. K. Callebaut.
Solar and interplanetary dynamics, (see 012.020), p. 279 - 282 (1980).

Physical arguments are given indicating that solar flare magnetic energy storage may happen through force-free fields with helical symmetry ($\partial_z + \lambda(r)\,\partial_\varphi = 0$). The mathematical results turn out simple for helical fields whether general, in equilibrium or force-free. A preliminary stability analysis points to appropriate properties.

073.084 **The filament instability in a sheared field.** C. Chiuderi, G. Van Hoven.
Solar and interplanetary dynamics, (see 012.020), p. 295 - 298 (1980).

The authors summarize the results of a self-consistent calculation of thermal instability in a non-uniform field, showing how the dynamic response of density and temperature to the competing effects of optically-thin radiation and field-collimated thermal conduction leads to the formation of characteristic "knife-blade" filaments.

073.085 **A model flare and the continued post-flare mass release from the flare region.** Y. Uchida.
Solar and interplanetary dynamics, (see 012.020), p. 303 - 306 (1980).

A mechanism for slow and enduring mass release well after the flare onset, which is inferred from the enduring enhancement in the interplanetary mass flux after flare activity, is discussed in the modified scheme of a "neutral sheet" model of flares.

073.086 **A model of surge.** G. Noci.
Solar and interplanetary dynamics, (see 012.020), p. 307 - 311 (1980).

This paper reports on the application of a siphon flow model to the late stage of a surge observed in the UV radiation by the S055 experiment on board Skylab.

073.087 **Radiative hydrodynamics of flares: preliminary results and numerical treatment of the transition region.** A. N. McClymont, R. C. Canfield.
Solar and interplanetary dynamics, (see 012.020), p. 313 - 316 (1980).

The authors report on a comprehensive numerical simulation of flare dynamics, encompassing the corona, transition region and chromosphere. A coronal loop geometry, whose magnetic pressure dominates gas pressure, is assumed. The authors discuss difficulties in modelling the transition region under flare conditions, and suggest tentative solutions.

073.088 **Classification and investigation of solar flare situations conformably to interplanetary and magnetospheric disturbances.** K. G. Ivanov, N. V. Mikerina, L. V. Evdokimova.
Solar and interplanetary dynamics, (see 012.020), p. 421 - 424 (1980).

A new classification of large solar flares is presented.

073.089 **X-radiation and charged particles in the solar event in 1977, November 22.**
B. I. Valnicek, O. B. Likin, E. I. Morozova, N. F. Pisarenko, F. Farnik, I. V. Ehstulin.
Geomagn. Aehron., Tom 20, 777 - 784 (1980). In Russian.

073.090 **On the chromospheric network structure around developed active regions.** L. G. Kartashova.
Izv. Krymskoj Astrofiz. Obs., Tom 61, 34 - 36 (1980). In Russian.

The chromospheric network structure around several developed active regions was studied on the basis of observations in the Hα line.

073.091 **Spectral analysis of solar flares. I. Observational results.** Q.-V. Dinh.
Publ. Astron. Soc. Japan, Vol. 32, 495 - 514 (1980).

The spectra of three chromospheric flares observed on June 30 and July 6, 1974 and January 4, 1969 are analyzed. Electron densities are determined. The metal lines of the flare of June 30, 1974 at the maximum phase are studied. It is found that the emission of these metal lines comes from deeper layers than that of hydrogen. The ratio of the intensity of helium to that of hydrogen lines in the observed flares is found to be smaller than it is in the chromosphere and prominences. It is concluded that the observed flares can be interpreted with a homogeneous stratified model in which helium is emitted at the top, hydrogen in the middle, and metal lines in the lower layer.

073.092 **Spectral analysis of solar flares. II. Non-LTE analysis of hydrogen spectra.**
Q.-V. Dinh.
Publ. Astron. Soc. Japan, Vol. 32, 515 - 532 (1980).

Non-LTE calculations have been carried out on a hydrogen model atom in semi-empirical models of the solar flare by solving simultaneously the equations of radiative transfer and of statistical equilibrium. These flare models are homogeneous plane-parallel and hydrostatic and have large gas pressures in the chromosphere-corona transition region. A three-level then a four-level plus continuum representations of the hydrogen atom are assumed. The influence of the temperature-height distribution, turbulence motion, and atomic model on hydrogen emission in the Hα line and the Lyman continuum and on physical parameters of the flare models is investigated. The computational results are compared with the observations of the 30 June and 6 July, 1974 flares.

073.093 **Steady flow models of dark filaments.**
E. Ribes, W. Unno.
Astron. Astrophys., Vol. 91, 129 - 135 (1980).
Analytic models representing characteristics of filaments are obtained. Geometries and physical quantities are assumed to be independent of one horizontal direction (x-direction); but the magnetic and the flow vectors may have the x-components. General formulation of the problem is given in an appendix. Simplifying geometrical assumptions are made in order to obtain analytic solutions, and the energy equation is ignored in the derivation of analytic expressions for physical quantities and geometries. An August 7, 1975 filament is shown to be well simulated by the present model with appropriate choice of parameters.

073.094 **Plasma energetics in solar flares.** G. Van Hoven.
Highlights of Astronomy, Vol. 5, (see 012.025), 343 - 350 (1980).

The author concentrates on the high-temperature and quasi-thermal aspects of a flare, and on the basic physical mechanisms connected with the primary energization and dissipation processes. He treats the reconnection of the magnetic field, the bulk acceleration of particles, the thermalization and the ultimate radiation of the energy.

073.095 **A high-resolution view of the solar chromosphere and corona.** G. E. Brueckner.
Highlights of Astronomy, Vol. 5, (see 012.027), 557 - 569 (1980).

The observation of loops in all layers of the solar atmosphere together with realization of very rapid time changes and supersonic motions make it very likely that the heating of the outer solar atmosphere as well as the energy needed for the propulsion of the solar wind, is caused by magneto-acoustic waves or electric currents.

073.096 **Density dependence of solar emission lines of boron-like ions.** B. N. Dwivedi, P. K. Raju.
Sol. Phys., Vol. 68, 111 - 123 (1980).

Steady state level density of 20 of boron-like ions, Mg VIII and Si X, have been computed as a function of electron density and temperature. Using the computed level density, line intensities have been obtained as a function of electron density and temperature. A number of line intensity ratios in the Mg VIII and Si X line emission spectrum are found to be density sensitive. Absolute line fluxes from these ions at earth distance have been computed and are found to be comparable with values obtained, using various satellite and rocket measurements.

073.097 **Does H^- truly cool the solar chromosphere?**
T. R. Ayres.
Sol. Phys., Vol. 68, 125 - 133 (1980).

The author examines the controversial problem of H^- radiative cooling in the solar chromosphere. He finds, in agreement with Praderie and Thomas, that H^- is a substantial source of radiative heating in the outer atmosphere, especially when departures from LTE are important. The role of H^- as a chromospheric heating agent must be considered carefully before net radiative cooling rates can be assessed from empirical chromospheric models, or calculations of nonradiative heating, for example by acoustic waves, can be pursed meaningfully.

073.098 **O VI (λ = 1032 Å) profiles in and above an active region prominence, compared to the quiet Sun center and limb profiles.**
J. C. Vial, P. Lemaire, G. Artzner, P. Gouttebroze.
Sol. Phys., Vol. 68, 187 - 206 (1980).

O VI (λ = 1032 Å) profiles have been measured in and above a filament at the limb. They are compared to profiles measured at the quiet Sun center and at the quiet Sun limb. Absolute intensities are found to be about 1.55 times larger than above the quiet limb at the same height (3"); at the top of the prominence (15" above the limb) one finds a maximum blue shift and a minimum line width. The inferred non-thermal velocity (29 km s^{-1}) is about the same as in cooler lines while the approaching line-of-sight velocity (8 km s^{-1}) is lower than in Ca II lines. The O VI profile recorded 30" above the limb outside the filament is wider (FWHM = 0.33 Å). It can be interpreted as a coronal emission of O VI ions with a temperature of about 10^6 K, and a non-thermal velocity (NTV) of 49 km s^{-1}. This NTV is twice the NTV of quiet Sun center O VI profiles.

073.099 **Spatial and temporal correlation of high and low temperature solar flare emissions.**
D. M. Rust, R. W. Buhmann, B. R. Dennis, R. D. Robinson, R. R. Willson, M. Simon.
Bull. American Astron. Soc., Vol. 12, 752 (1980). – Abstract.

073.100 **Horizontal distribution of the X-ray energy deposit in the chromosphere and Hα two ribbon flares.**
J. C. Hénoux, D. Rust.
Astron. Astrophys., Vol. 91, 322 - 327 (1980).

The two-ribbon Hα brightening and the X-ray emitting coronal loop arcade during the very late phase of the 29 July 1973 flare are examined. By means of a simple geometrical model of the X-ray emitting structures, the horizontal distribution of X-radiation is computed for several different levels in the chromosphere. It is found that an arcade of X-ray emitting loops, commonly found over large two-ribbon flares, gives an energy deposit pattern in the chromosphere similar in shape to the Hα flare emission. The maximum X-ray energy deposit rate of 1.5×10^6 erg $cm^{-2} s^{-1}$ is slightly higher than the peak Hα emission rate. It is concluded that X-ray radiation is an important source of energy for Hα flares.

073.101 **De chromosfeer boven de rustige zon.**
W. van Tend.
Zenit, 7e Jaarg., 190 - 191 (1980).

073.102 **August 1972 proton-flare region and different phases of its background magnetic field development.** V. Bumba.
Bull. Astron. Inst. Czechoslovakia, Vol. 31, 351 - 362 (1980).

On the basis of magnetic synoptic charts as well as daily maps preceding and following the development of the August 1972 proton-flare region in an extended time interval, the author demonstrates that the evolution of such a complex solar process may be divided into more simple evolutionary phases. The establishment of larger-scale activity phenomena, connected with the forming of a proton-flare region, is very slow till the occurrence of proton-flares, and then it decreases very fast. With the forming of a proton-flare situation, the inflow of new magnetic flux into the studied extended solar region stops and during a few rotations the magnetic field practically disappears over a large area of the solar surface.

073.103 **On time variations of the chromospheric network.**
H. C. Dara, C. J. Macris.
Bull. Astron. Inst. Czechoslovakia, Vol. 31, 364 - 368 (1980).

The authors present a study of the variation of the chromospheric network during the 19th solar cycle. They find a change of the relative density of the small emission centres of the K_{232}, solar chromosphere (flocculi) with a maximum near the time of the solar activity maximum and a minimum about one year after the solar activity minimum.

073.104 **The contribution of OSO-8 to solar physics from data analyzed as of May 1979.**
R. M. Bonnet.
Space Research, Vol. XX, (see 012.043), 239 - 253 (1980).

The NASA Orbiting Solar Observatory-8 was launched on June 1975 and was operated for 39 months. The main results obtained to date with the four solar instruments on board the spacecraft are reviewed and presented in an synthetic form. The main contribution of OSO-8 to solar physics is in the area of wave propagation, dynamics of the upper chromosphere, models of the chromosphere and transition region and solar activity.

073.105 **High-resolution X-ray spectra of solar flares. IV. General spectral properties of M type flares.**
U. Feldman, G. A. Doschek, R. W. Kreplin, J. T. Mariska.
Astrophys. J., Vol. 241, 1175 - 1185 (1980).

High-resolution X-ray spectra of class M flares have been recorded by four Bragg crystal spectrometers flown by NRL on an Air Force spacecraft. The wavelength ranges are 1.82 - 1.97 Å, 2.98 - 3.07 Å, 3.14 - 3.24 Å, and 8.26 - 8.53 Å. Electron temperatures are derived from dielectronic satellite-line-to-resonance-line ratios as a function of time for several typical flares. For two of the M flares, blueshifted components to the resonance lines of Ca XIX and Fe XXV were observed that indicate motions along the line of sight of about 400 km s^{-1}. Random nonthermal motions in M flare spectra are largest during the soft X-ray rise phase.

073.106 **Size of the X-ray kernel of the large 1979 August 20 solar flare.** P. B. Landecker, D. L. McKenzie.
Astrophys. J., Lett., Vol. 241, L175 - L178 (1980).

At the time of the intense class X5 solar flare on 1979 August 20, monochromatic small raster maps near 7 Å were recorded of the flaring active region with 20″ resolution (FWHM) by the SOLEX Solar X-ray Spectrometer/Spectroheliograph. From these maps, the spatial extent (FWHM) of the X-ray kernel during the rise of this flare at about 0912 UT was determined to be less than 28″. The electron density in the hot flare plasma was estimated to be above 10^{11} cm^{-3}. This is the first soft X-ray measurement of the angular size of such an intense flare.

073.107 **Semiempirical models of chromospheric flare regions.**
M. E. Machado, E. H. Avrett, J. E. Vernazza, R. W. Noyes.
Astrophys. J., Vol. 242, 336 - 351 (1980).

The authors present homogeneous plane-parallel semi-empirical flare atmospheres that approximately reproduce observations in lines and continua of H I, Si I, C I, Ca II, and Mg II. The models have a thin transition zone at the top of an enhanced chromosphere, and they imply a substantial amount of heating from the transition zone down to the temperature minimum level. The authors' results are compared with theoretical models that invoke different modes of energy transfer from the corona into the chromosphere. Significant discrepancies are found in their comparison with the particle-heated models. The authors' radiative transfer solution shows that there is substantial Lα radiative heating in the upper chromosphere. The source of this heating may be the conductive energy flux in the transition zone where the Lα line cools the gas.

073.108 **VLA observations of impulsive solar flares at 4.9 GHz.**
K. A. Marsh, G. J. Hurford, H. Zirin, R. M. Hjellming.
Astrophys. J., Vol. 242, 352 - 358, plate 5 (1980).

Five impulsive solar microwave bursts (peak flux 2 - 1200 sfu) were observed during 1978 July 15 - 18, using the Very Large Array (VLA) at 4.9 GHz. It was found that, at the times of peak emission, all five flares were dominated by a single source whose size was in the range 12″ – 18″. Comparison with high-resolution Hα photographs and magnetograms showed that in each case the source was located between the Hα kernels, close to the magnetic neutral line. Analysis of spectral data from Sagamore Hill (1.4 - 35 GHz) indicated that the spectral turnover at low frequencies was due to absorption by the ambient active region. These results, together with the observed magnetic configuration, suggest a model in which the evolution in source structure at 4.9 GHz could be understood in terms of magnetic reconnection.

073.109 **Radiation signatures from a locally energized flaring loop.** A. G. Emslie, L. Vlahos.
Astrophys. J., Vol. 242, 359 - 373 (1980).

The authors calculate the radiation signatures from a locally energized solar flare loop, at a variety of wavelengths. The calculations depend strongly on the physical properties of the energy release mechanism which the authors qualitatively discuss. The model is found to be consistent with hard X-ray, microwave, and EUV observations for plausible source parameters. An important prediction of the model is that the intrinsic polarization of the hard X-ray burst should increase significantly over the photon energy range 20 keV $\lesssim \epsilon \lesssim$ 100 keV. This feature is not present in any other flare model hitherto presented, and the use of this, and other suggested observational diagnostics, to determine the various model parameters is discussed.

073.110 **Preflare conditions, changes and events.**
S. F. Martin.
Sol. Phys., Vol. 68, 217 - 236 (1980).

Preflare conditions, changes and events are loosely categorized as distinct, evolutionary or statistical. Distinct preflare phenomena are those for which direct physical associations with flares are implied. Evolutionary preflare changes are considered to be any long-term effect that may be related to the flare build-up even though the same changes may occur in the absence of flares. Statistical preflare changes logically include both distinct and evolutionary preflare changes.

073.111 **On the large scale brightness fluctuations in the solar atmosphere.**
V. A. Krat, V. I. Makarov, K. S. Tavastsherna.
Sol. Phys., Vol. 68, 237 - 242 (1980).

October 1976 spectroscopic observations of the solar chromosphere were analysed. It was found that the brightness distribution in the chromosphere is essentially bimodal and has characteristic scales of 1.2×10^4 km and 2.4×10^4 km. The supergranulation (3.5×10^4 km) in the brightness field was found to show up more faintly at all heights.

073.112 **Comments on the mechanism for the spicule support.**
P. G. Papushev.
Sol. Phys., Vol. 68, 275 - 278 (1980).

A one-dimensional model of non-viscous gas in an isothermal flow is considered, using the theory of Abramovich (1976). A mechanism of spicule flow is proposed which successfully explains the observed spicule height. The model further predicts inhomogeneities in the magnetic field topology, and in the direction and magnitude of the velocity flow.

073.113 **Steady flows in the chromosphere and transition-zone above active regions as observed by OSO-8.**
B. W. Lites.
Sol. Phys., Vol. 68, 327 - 337 (1980).

The author has examined more than two years of data from the University of Colorado instrument aboard OSO-8 for observational evidence of steady flows in order to determine their patterns of occurrence and their magnitude. The results show that steady flows occur frequently in the transition-zone near sunspots.

073.114 **A model of hot loops associated with solar flares. I. Gasdynamics in the loops.** F. Nagai.
Sol. Phys., Vol. 68, 351 - 379 (1980).

A dynamical model is proposed for the formation of soft X-ray emitting hot loops in solar flares. It is examined by numerical simulations how a solar model atmosphere in a magnetic loop changes its state and forms a hot loop when the flare energy is released in the form of heat liberation either at the top part or around the transition region in the loop.

073.115 **Turbulent velocities and kinetic temperatures in the chromosphere as calculated using the results of the total solar eclipse observations on June 30, 1973. I.**
V. M. Sobolev, G. F. Vyal'shin, Yu. A. Nagovitsyn.
Soln. Dannye 1980 Byull., No. 7, p. 88 - 93 (1980). In Russian.

Methods are proposed for calculating halfwidths of chromospheric lines from spectrograms obtained with a slitless spectrograph during the total solar eclipse on 30 June, 1973. The distribution of turbulent velocities through the lower chromosphere was obtained. Even in the lower chromosphere (h = 450 km) simultaneously with "cool" regions "hot" regions are present with $T_{\text{kin}} > 20000°$ emitting in He lines.

073.116 **Solar flares.** D. M. Rust.
Solar system plasma physics, Vol. 1, (see 003.010), 51 - 98 (1979).

Contents: 1. Introduction: flares and plasma physics, flare emissions, literature. 2. The preflare state: magnetic fields, Skylab observations of flare precursors. 3. Energy release: impulsive phase energy release, production of solar cosmic rays, electron beams, main phase energy release, post main-phase energy release, overview of energy release. 4. Flare models: emerging flux model, difficulties, unstable loop model, conclusions.

073.117 **A quasi-CA distribution for the regions of flare activity.** T.-j. Cao, S.-y. Yin, A.-a. Xu.
Acta Astron. Sinica, Vol. 21, 349 - 353 (1980). In Chinese.

By tracing the positions of the regions of flare activity from Jan. 1970 to Dec. 1975, it is proposed that these regions exhibit a regional distribution of quasi- CA (Complex of Activity) on the surface. 95% active regions were distributed in 21 quasi- CA districts.

073.118 **On the orientation of penumbral filaments.**
Eh. B. Kandrashov, R. B. Teplitskaya.
Issled. po geomagn., aehron. i fiz. Solntsa, Moskva, 1980, No. 52, p. 25 - 31. In Russian. – Abstr. in Ref. zh., 51. Astron., 11.51.446 (1980).

073.119 **Calcium flocculi and modulation of solar wind flow concentration with solar semirotational period.**
N. N. Lyakhov, V. V. Kasinskij.
Issled. po geomagn., aehron. i fiz. Solntsa, Moskva, 1980, No. 52, p. 39 - 45. In Russian. – Abstr. in Ref. zh., 51. Astron., 11.51.467 (1980).

073.120 **Catalogue of solar proton flares for 1970 - 1976.**
I. E. Pogodin.
Redkol. zh. Vestn. LGU. Fiz., khim. Leningrad, 1980. 27 pp. In Russian. – Abstr. in Ref. zh., 51. Astron., 11.51.474 (1980).

073.121 **Distribution of chromospheric flares relative to the sector boundaries of the interplanetary magnetic field extrapolated to the sun.** L. S. Levitskij.
Izv. Krymskoj Astrofiz. Obs., Tom 62, 148 - 153 (1980). In Russian.

The distribution of ~7000 flares of importance ⩾ 1 was studied around the sector boundaries of the interplanetary magnetic field (+–) and (–+) extrapolated to the sun. The data obtained for the time period July 1955 - December 1961 were used. The distributions obtained were analysed jointly with the same distributions for 1964 - 1974. It is shown that a stable concentration of flares is observed only near the boundaries (–+) for both hemispheres of the sun during increase of the activity and near the maximum cycles No. 19 and 20. There is no difference between "Hale" and "non-Hale" boundaries for these flares.

073.122 **Formation of hydrogen lines in quiescent prominences.** Ch. Tsovookhuu.
Izv. Krymskoj Astrofiz. Obs., Tom 62, 154 - 165 (1980). In Russian.

Simultaneous solutions of the radiative transfer and statistical equilibrium equations have been carried out for the hydrogen atom with eight bound levels and a continuum in the plane-parallel model of a prominence with the total optical thickness in the Hα-line-center equal to 1, 5, 10, 25, 50, 75, 100. The source functions, the total energy and the profiles of the lines Hα - Hη are calculated. The theoretical curve of growth and the curve of half-width are constructed.

073.123 **Magnetohydrodynamic simulation of surge events.**
V. Dermendzhiev, P. Dukhlev, G. Zakhariev.
Astron. Tsirk., No. 1107, p. 5 - 6 (1980). In Russian.

073.124 **Radial velocity field in the solar flare of August 4, 1972.** V. A. Abgaryan, Eh. V. Kononovich, I. F. Nikulin, S. L. Ovchinnikov, S. V. Startsev.
Astron. Tsirk., No. 1110, p. 2 - 5 (1980). In Russian.

073.125 **Rayleigh-Taylor instability in the plasma of solar prominences.** A. Z. Dolginov, V. M. Ostryakov.
Astron. Zh., Tom 57, 1302 - 1309 (1980). In Russian. English translation in Soviet Astron., Vol. 24, No. 6.

The instability range of tangential discontinuities in a stratified plasma with magnetic field is determined. The values of concentrations, temperatures and magnetic fields at which the tangential discontinuity confining a solar prominence becomes unstable are found. Disturbances can be both large- and small-scale. In the latter case, their development can be observed as jets of cold gas flowing into the corona. The sizes of prominences are possibly restricted by the Rayleigh-Taylor instability.

073.126 **Berechnungen der Energie- und Winkelverteilung von Flare-Elektronen und der von ihnen emittierten Röntgenstrahlung.** M. Wälder, G. Elwert.
Mitt. Astron. Ges., Nr. 50, p. 91 - 95 (1980).

073.127 **A study of chromospheric response to heating by accelerated electrons with numerical methods.**
A. G. Kosovichev, M. A. Livshits, Yu. P. Popov.
Inst. prikl. mat. AN SSSR. Prepr., 1980, No. 68, 39 pp. In Russian. – Abstr. in Ref. zh., 51. Astron., 12.51.308 (1980).

073.128 **The preflare state.** G. Van Hoven, U. Anzer, D. D. Barbosa, J. Birn, C.-C. Cheng, R. T. Hansen, B. V. Jackson, S. F. Martin, P. S. McIntosh, Y. Nakagawa, E. R. Priest, E. M. Reeves, E. J. Reichmann, E. J. Schmahl, J. B. Smith, C. V. Solodyna, R. J. Thomas, Y. Uchida, A. B. C. Walker.
Solar flares, (see 003.012), p. 17 - 81 (1980).

Contents: Background of the study. Precursors and evolution. The flare of 5 September 1973. Theorems of the preflare state. Results of the preflare study.

073.129 **Primary energy release.** S. Kahler, D. Spicer, Y. Uchida, H. Zirin.
Solar flares, (see 003.012), p. 83 - 116 (1980).

The authors first review those flare observations which seem to bear most directly on the primary energy release mechanism and follow that with an analysis of the observations contributed by team members. They then review some of the relevant theories of primary energy release and discuss the models of Spicer and Uchida.

073.130 **Energetic particles in solar flares.**
R. Ramaty, S. A. Colgate, G. A. Dulk, P. Hoyng, J. W. Knight, R. P. Lin, D. B. Melrose, F. Orrall, C. Paizis, P. R. Shapiro, D. F. Smith, M. Van Hollebeke.
Solar flares, (see 003.012), p. 117 - 185 (1980).

Contents: Introduction. Energy contained in the 10 - 100 keV electrons. On the existence and consequences of reverse currents. The rapid heating of coronal plasma during flares: nonequilibrium ionization diagnostics. Acceleration and energization mechanisms for the 10 - 100 keV electrons. Radio evidence on the particle distribution functions in the corona following flares. Energetic solar particles at 1 AU. Solar gamma rays. Other manifestations of particle acceleration. Second phase acceleration. Summary.

073.131 **Impulsive phase of solar flares.**
S. R. Kane, C. J. Crannell, D. Datlowe, U. Feldman, A. Gabriel, H. S. Hudson, M. R. Kundu, C. Mätzler, D. Neidig, V. Petrosian, N. R. Sheeley, Jr.
Solar flares, (see 003.012), p. 187 - 229 (1980).

This review of our present understanding of the impulsive phase begins with a brief background and an identification of the key questions related to this phenomenon. These are followed by an analysis of the various impulsive emissions and the models proposed to explain their origin. Finally the authors consider the role of the impulsive phase in the overall flare process and suggest future observational and theoretical studies which will bring us closer to an understanding of solar flares.

073.132 **The chromosphere and transition region.**
R. C. Canfield, J. C. Brown, G. E. Brueckner, J. W. Cook, I. J. D. Craig, G. A. Doschek, A. G. Emslie, J.-C. Henoux. B. W. Lites, M. E. Machado, J. H. Underwood.
Solar flares, (see 003.012), p. 231 - 271 (1980).

The authors confine themselves to studies of physical processes that involve the low-temperature part of the flare, principally the transition region and the chromosphere. The fundamental questions they address concern the role of conduction, radiation, fast particles, and mass motion in the chromospheric-coronal interaction. The data used in their approach to these problems are intensities of lines and continua in the EUV and XUV, broad-band soft X-ray and radio fluxes, spectral line profiles and Doppler shifts in the EUV and visible, and the variation of these quantities with space and time. The authors consider energetic particles, the impulsive phase, and the thermal X-ray plasma, but only insofar as they interact with the chromosphere and photosphere or play a role in model flare atmospheres.

073.133 **Mass ejections.** D. M. Rust, E. Hildner, M. Dryer, R. T. Hansen, A. N. McClymont, S. M. P. McKenna Lawlor, D. J. McLean, E. J. Schmahl, R. S. Steinolfson, E. Tandberg-Hanssen, R. Tousey, D. F. Webb, S. T. Wu.
Solar flares, (see 003.012), p. 273 - 339 (1980).

The purpose of the authors' study was to discover what role mass ejections play in the flare phenomenon. They sought to determine the energy and mass of the ejecta and to discover the forces that propel matter from the flare site and constrain it as it moves. They tried to determine the physical conditions in the ejecta and in the disturbed atmosphere. The study of ejecta provides insight into the origins of the flare plasma, the height and duration of energy release, and the large-scale magnetic field configuration before and after flare onset.

073.134 **The thermal X-ray flare plasma.**
R. Moore, D. L. McKenzie, Z. Švestka, K. G. Widing, S. K. Antiochos, K. P. Dere, H. W. Dodson-Prince, E. Hiei, K. R. Krall, A. S. Krieger, H. E. Mason, R. D. Petrasso, G. W. Pneuman, J. K. Silk, J. A. Vorpahl, G. L. Withbroe.
Solar flares, (see 003.012), p. 341 - 409 (1980).

The purpose of this paper is to outline the state of observational knowledge of the X-ray plasma and its physical interpretation as of the beginning of the Flare Workshop. First, the observed characteristics and the physical properties derived from the observations are summarized. The basic physical questions posed by the observational results are then considered, and current physical ideas and models which have been proposed in answer to these questions are briefly reviewed.

073.135 **Flare models.** P. A. Sturrock.
Solar flares, (see 003.012), p. 411 - 449 (1980).

The purpose of this paper is to review the current status of the modeling of solar flares. The author proposes requirements of flare models which are set by the observational data, and discusses models. A distinction has been made between "primary" requirements and models, and "secondary" requirements and models. A number of models are compared against the primary requirements; there is no corresponding comparison of models against the secondary requirements.

073.136 **Radiative energy output of the 5 September 1973 flare.** R. C. Canfield, C.-C. Cheng, K. P. Dere, G. A. Dulk, D. J. McLean, R. D. Robinson, Jr., E. J. Schmahl, S. A. Schoolman.
Solar flares, (see 003.012), p. 451 - 469 (1980).

The authors present results of a unique study: for the first time, they have measured the radiative energy output of a single flare over a range of more than ten decades in wavelength, from below one Ångstrom to above one meter. Their data permit them to determine the absolute intensity of radiative energy output over this entire range at the time of flare maximum (1831 UT, 5 September 1973).

073.137 **Mechanical energy output of the 5 September 1973 flare.**
D. F. Webb, C.-C. Cheng, G. A. Dulk, S. J. Edberg, S. F. Martin, S. McKenna Lawlor, D. J. McLean.
Solar flares, (see 003.012), p. 471 - 499 (1980).

The paper gives the results of the observational energy estimates in five categories: the cool Hα eruptive material in the eruptive filament, or spray, and the large surge; the emission front observed in Hα and Ca K; the assumed piston-driven shock wave; the assumed coronal transient; and flare core motions.

073.138 **XUV observations of a dense, compact flare.**
K. G. Widing, D. S. Spicer.
Astrophys. J., Vol. 242, 1243 - 1256, plates 25 - 26 (1980).

The purpose of the present paper is to present a comparison study of the 1973 December 17 flare based on the flare images photographed in the 170–630 Å spectral region by the NRL objective grating spectroheliograph on Skylab. In particular, the present study concentrates on the bright kernel of the December 17 flare, and it is compared with the August 9 compact flare.

Introduction to "Solar flares. A monograph from Skylab Solar Workshop II". See Abstr. 011.041.

Experimental simulation on solar flare – experiment on merging of two current-carrying plasma columns. See Abstr. 022.064.

Cr, Co, and Ni transitions isoelectronic to the Fe XXIV – Fe XVII lines around 11 Å in laser-produced plasma. See Abstr. 022.169.

Application of digital image processing techniques to faint solar flare phenomena. See Abstr. 031.521.

Interactive reduction of solar flare spectra. See Abstr. 031.583.

Preliminary estimates and corrections in the two-dimensional spectroscopy of the solar chromospheric active regions. See Abstr. 031.586.

Protuberanzfotos mit Kurzprotuberanzansatz. See Abstr. 034.023.

Analysis of X-ray line spectra from a transient plasma under solar flare conditions. I. General outline. See Abstr. 062.002.

Evolving force-free magnetic fields. III. States of nonequilibrium and the preflare stage. See Abstr. 062.019.

Fast plasma heating by anomalous current dissipation. See Abstr. 062.020.

On the equilibrium of a cylindrical plasma supported horizontally by magnetic fields in uniform gravity. See Abstr. 062.062.

Magnetohydrodynamic equilibrium and stability of pre-flare loops. Constant pitch field. See Abstr. 062.063.

Energy balance in current sheets: from Petschek to gravity driven reconnection? See Abstr. 062.089.

Non-thermal electron scattering in a homogeneous plasma. See Abstr. 063.017.

A probabilistic approach to radiative energy loss calculations for optically thick atmospheres: hydrogen lines and continua. See Abstr. 063.022.

Stellar chromospheres. See Abstr. 064.031.

On the differences at chromospheric levels between RS CVn-type binaries, active and quiet chromosphere single stars, and active and quiet regions in the sun. See Abstr. 064.059.

Chromospheres of F, G, K, type stars. V. Radiative losses in spectral lines. See Abstr. 064.079.

Theory of force-free magnetic fields for unipolar sunspots and the energy of solar flares. See Abstr. 072.002.

Theory of force-free magnetic fields for bipolar sunspots and the energy of solar flares. See Abstr. 072.003.

Comparison of slowly and rapidly evolving magnetic structures in active regions seen in Hα and EUV. See Abstr. 072.008.

Morphology of active regions and flares. See Abstr. 072.010.

Density diagnostic of solar active region and flare plasmas from Si IV/O IV line ratio as observed from SMM *(Solar Maximum Mission)*. See Abstr. 072.011.

1978 data on proton flare prediction by means of spiral spots. See Abstr. 072.046.

Sunspot motion and the flare on 4 July 1974. See Abstr. 072.061.

Mass motions in the transition zone of coronal holes. See Abstr. 074.011.

Radiative-dominated cooling of the flare corona and transition region. See Abstr. 074.042.

X-ray evidence of coronal preflare emission. See Abstr. 074.057.

Spicules and macrospicules. See Abstr. 074.058.

Radio data and computer simulations for shock waves generated by solar flares. See Abstr. 074.061.

Reconnection driven coronal transients. See Abstr. 074.063.

Transient disturbances of the outer corona. See Abstr. 074.065.

Monochromatic observations of the solar corona and prominences. See Abstr. 074.089.

Evidence for an X-type neutral sheet producing chromospheric activity. See Abstr. 075.002.

A new approach to understanding preflare buildup of magnetic energy. See Abstr. 075.009.

Measurements of impulsive EUV and hard X-ray solar flare emission. See Abstr. 076.003.

A presentation of some BCS (*Bent Crystal Spectrometer*) spectra. See Abstr. 076.008.

Early results from the Hard X-ray Burst Spectrometer on the Solar Maximum Mission. See Abstr. 076.009.

Temporal and spatial fluctuations in strengths and widths of C IV and Si II lines observed with OSO 8. See Abstr. 076.013.

Solar flare X-ray spectra between 7.8 and 23.0 Angstroms. See Abstr. 076.020.

UV emitting spicules. See Abstr. 076.022.

Analysis of X-ray spectra emitted by the 24 October 1970 flare. See Abstr. 076.025.

Solar flare X-ray spectra. III. Initial and final phase. See Abstr. 076.028.

On Doppler shifts of the Fe XXV ion resonance line in solar flare X-ray spectra. See Abstr. 076.029.

On the type of spectra of S-component sources and their correlation with flare occurence. See Abstr. 077.003.

Sharp-cutoff short-cm wavelength bursts from proton activity centers. See Abstr. 077.015.

Double ribbon events observed in He I 10830 Å associated with filament disappearances. See Abstr. 077.016.

Two-dimensional VLA maps of solar bursts at 15 and 23 GHz with arcsec resolution. See Abstr. 077.027.

Evidence for a peak in the number of isolated type III bursts prior to large solar flares. See Abstr. 077.033.

Location of compact microwave sources with respect to concentrations of magnetic field in active solar regions. See Abstr. 077.036.

Is there a limit on solar flare protons fluxes? See Abstr. 078.001.

Solar flare increase of cosmic ray intensity on November 22, 1977. See Abstr. 078.005.

The spatial anisotropy, rigidity spectrum, and propagation characteristics of the relativistic solar particles during the event on May 7, 1978. See Abstr. 078.013.

Energy balance from the chromosphere-corona transition region. See Abstr. 080.021.

Hot downflows above supergranular boundaries. See Abstr. 080.022.

Ionospheric currents associated with solar flares: a short review. See Abstr. 083.001.

Variations of ion composition and electron density accompanying solar flares. See Abstr. 083.039.

Magalert: August 27, 1978. See Abstr. 084.047.

Estimate of typical cross dimensions of interplanetary shock waves from powerful isolated solar flares. See Abstr. 106.058.

Transient phenomena originating at the sun – an interplanetary view. See Abstr. 106.062.

IPS observations of flare-generated disturbances. See Abstr. 106.064.

074 Corona, Solar Wind

074.001 **La corona solare e l'origine del vento solare.** G. Di Giovanni.
Coelum, Vol. 49, 93 - 114, 146 - 167 (1980).

074.002 **An estimation of the amount of heating for some solar coronal loops.** G. Elwert, U. Narain.
Bull. Astron.Soc. India, Vol. 8, 21 - 26 (1980).

Based on the standard idea that short cooling times imply continued heating, a method of estimation of the amount of heating is developed and applied to the five examples of Krieger (1978) assuming conduction to be the dominant cooling mechanism in plane parallel and line dipole geometries. As expected, one requires more heating in former geometry than in the latter. The required total energy supplied by heating is found to be comparable to the total thermal energy of the events under consideration.

074.003 **A low β coronal loop model. I: Kink instabilities in the $\beta = 0$ limit.** R. M. J. Sillen, A. Kattenberg.
Sol. Phys., Vol. 67, 47 - 56 (1980).

A localized force-free current is proposed as a model for the observed coronal loops. An upper limit for the growth rate of kink instabilities in this model is found by solving numerically, in cylinder symmetry, the MHD equation of motion, with the boundary condition $\beta = 0$ outside the loop. For various current densities a spectrum of kinks is found. These instabilities will disrupt the loops that are long or strongly twisted, on a time scale of a few seconds.

074.004 **A density model for the north polar coronal hole at the 1973 eclipse.**
F. Crifo-Magnant, J. P. Picat.
Astron. Astrophys., Vol. 88, 97 - 101 (1980).

From white-light polarized pictures taken by Koutchmy during the 1973 eclipse, brightness curves could be recorded in the north polar hole, in the darkest region between the polar plumes, between 1.3 and 3.2 $R_\odot$. This region is assumed to represent the homogeneous, spherically-symmetric background of the hole. Brightness curves are extended up to 5 $R_\odot$ by those of Munro and Jackson, which lie 1.5 higher in intensity and are thus divided by this factor. A classical inversion gives the electron density and the brightness of the *F*-corona. Adopting the geometry of Munro and Jackson for the extension of the hole allows the solar wind speed to be calculated.

074.005 **Energy source of the solar wind.** P. Carlqvist, H. Alfvén.
Astrophys. Space Sci., Vol. 71, 203 - 209 (1980).

A direct transfer of energy from photospheric activity to the solar wind by means of electric currents is discussed. Currents are assumed to flow in quiescent prominences which occasionally erupt and give rise to expanding loop-like structures in the corona, as observed from Skylab. It is proposed that energy is transferred from photospheric activity to the solar wind in the following ways: (1) as kinetic energy of the ejected loop matter; (2) as electric power directly fed into the extended loops; and (3) as torsional waves produced by fluctuations in the loop currents.

074.006 **Irregular structure and velocity of solar wind plasma according to Venera 10 data.**
O. I. Yakovlev, A. I. Efimov, V. M. Razmanov, V. K. Shtrykov.
Astron. Zh., Tom 57, 790 - 798 (1980). In Russian.
English translation in Soviet Astron., Vol. 24, No. 4.

The solar wind electron density power spectra, velocity and inner scale of the turbulence are inferred from observations of amplitude and frequency scintillations and spectral broadening made with the Venera 10 spacecraft.

074.007 **Intensities of the coronal emission line 530.3 nm observed at Lomnický Štit in the years 1971 - 1976.**
V. Rušin.
Contrib. Astron. Obs. Skalnaté Pleso, Vol. 9, 7 - 35 (1980).

The intensities of the emission coronal line 530.3 nm which were obtained at Lomnický Štit in the years 1971-1976 are given in a table. They are supplemented to Rybanský's paper (1977) for the whole solar cycle No. 20.

074.008 **Coronal index of the solar activity IIa, 1972 and 1973.** M. Rybanský.
Contrib. Astron. Obs. Skalnaté Pleso, Vol. 9, 37 - 113 (1980).

This paper contains tables of values which were obtained from measurements of the emission coronal line 530.3 nm by all corona stations. They are the continuation of tables which were published by the author earlier (Rybanský, 1977).

074.009 **The lateral expansion of the August 14, 1979 coronal transient.**
M. J. Koomen, D. J. Michels, R. A. Howard, N. R. Sheeley, Jr.
Bull. American Astron. Soc., Vol. 12, 515 (1980). – Abstract.

074.010 **Coronal heating by stochastic magnetic pumping.** P. A. Sturrock, Y. Uchida.
Bull. American Astron. Soc., Vol. 12, 516 (1980). – Abstract.

074.011 **Mass motions in the transition zone of coronal holes.** G. A. Doschek, J. T. Mariska, U. Feldman.
Bull. American Astron. Soc., Vol. 12, 518 (1980). – Abstract.

074.012 **Observations linking X-ray bright points with the source of the mass input to the solar wind.**
J. M. Davis.
Bull. American Astron. Soc., Vol. 12, 518 (1980). – Abstract.

074.013 **On the acceleration of thermal coronal ions by flare induced shock waves.** M. E. Pesses.
Bull. American Astron. Soc., Vol. 12, 518 (1980). – Abstract.

074.014 **On the thermal stability of coronal loop plasma.** S. K. Antiochos, A. G. Emslie.
Bull. American Astron. Soc., Vol. 12, 519 (1980). – Abstract.

074.015 **Kinematical analysis of a spray observed in the corona.** D. F. Webb, B. V. Jackson.
Bull. American Astron. Soc., Vol. 12, 527 (1980). – Abstract.

074.016 **The frequency, locations, sizes, and speeds of coronal mass ejections at the peak of solar cycle 21 – early results from SMM (*Solar Maximum Mission*).**
E. Hildner, L. L. House, C. B. Sawyer, W. J. Wagner.
Bull. American Astron. Soc., Vol. 12, 535 (1980). – Abstract.

074.017 **Diffuse density fronts moving through the outer corona observed by the HAO Coronagraph/Polarimeter on the Solar Maximum Mission spacecraft.**
W. J. Wagner, E. Hildner, L. L. House, C. Sawyer.
Bull. American Astron. Soc., Vol. 12, 535 - 536 (1980). Abstract.

074.018 **The non-spherical source surface magnetic model: comparison with coronal data.**
E. N. Frazier, M. Schulz, R. H. Levine.
Bull. American Astron. Soc., Vol. 12, 544 - 545 (1980). Abstract.

074.019 **Coronal holes, solar wind streams, and geomagnetic disturbances during 1978 and 1979.**
N. R. Sheeley, Jr., J. W. Harvey.
Bull. American Astron. Soc., Vol. 12, 545 (1980). – Abstract.

074.020 **Dynamic simulation of coronal mass ejections.**
R. S. Steinolfson, S. T. Wu.
Bull. American Astron. Soc., Vol. 12, 545 (1980). – Abstract.

074.021 **Simultaneous measurement of coronal Faraday rotation and total electron content during solar occultation of PSR 0525+21.**
R. A. Howard, M. K. Bird, M. J. Koomen, D. J. Michels, N. R. Sheeley, Jr.
Bull. American Astron. Soc., Vol. 12, 545 - 546 (1980). Abstract.

074.022 **On the nature of obstacles decelerating the solar wind near Venus and Mars and on peculiarities of interaction between the solar wind and the atmospheres of these planets.** T. K. Breus, K. I. Gringauz.
Kosm. Issled., Tom 18, 587 - 599 (1980). In Russian.

074.023 **The neutral component of the solar wind near the earth's orbit.** M. A. Gruntman.
Kosm. Issled., Tom 18, 649 - 651 (1980). In Russian.

074.024 **Thermal instability of a current sheet as the cause for formation of cold loops in the solar corona.**
B. V. Somov, S. I. Syrovatskij.
Pis'ma Astron. Zh., Tom 6, 592 - 596 (1980). In Russian. English translation in Soviet Astron. Lett., Vol. 6.

In MHD approximation the stability problem is solved for a neutral current sheet and small disturbances propagating along the current. It is shown that a radiative energy loss can lead to a break-up of the current sheet into a system of more cold and dense filaments which are parallel to the magnetic field lines.

074.025 **Geometry of expansion of the solar corona and parameters of the solar wind.** M. B. Krajnev.
Geomagn. Aehron., Tom 20, 577 - 582 (1980). In Russian.

074.026 **Solar corona electron density distribution.**
P. B. Esposito, P. Edenhofer, E. Lueneburg.
J. Geophys. Res., Vol. 85, 3414 - 3418 (1980).

Three and one-half months of single-frequency time delay data (earth-to-spacecraft and return signal travel time) were acquired from the Helios 2 spacecraft around the time of its solar occultation (May 16, 1976). Following the determination of the spacecraft trajectory the excess time delay due to the integrated effect of free electrons along the signal's ray path could be separated and modeled. An average solar corona, equatorial, electron density profile, during solar minimum, was deduced from time delay measurements acquired within 5-60 solar radii of the sun. The Helios electron density model is compared with similar models deduced from a variety of different experimental techniques.

074.027 **The low-frequency continuum as observed in the solar wind from ISEE 3: thermal electrostatic noise.**
S. Hoang, J. -L. Steinberg, G. Epstein, P. Tilloles, J. Fainberg, R. G. Stone.
J. Geophys. Res., Vol. 85, 3419 - 3430 (1980).

The low-frequency continuum, observed in the solar wind by ISEE 3, is shown to be generated by local electrostatic (thermal) plasma waves and not by electromagnetic waves.

074.028 **Observations of large fluxes of He^+ in the solar wind following an interplanetary shock.**
J. T. Gosling, J. R. Asbridge, S. J. Bame, W. C. Feldman, R. D. Zwickl.
J. Geophys. Res., Vol. 85, 3431 - 3434 (1980).

Los Alamos Scientific Laboratory instrumentation on Imp 7 has detected large fluxes of He^+ within that volume of solar wind plasma believed to be the solar ejecta driving the interplanetary shock wave disturbance of July 29, 1977. The very high He^+/He^{++} abundance ratio of 0.3 measured during this event suggests that this was solar prominence material only partially ionized by its passage through the corona.

074.029 **Normals of non-WKB Alfvén waves in the solar wind.** M. Heinemann.
J. Geophys. Res., Vol. 85, 3435 - 3441 (1980).

A theoretical treatment of wave normals of monochromatic non-WKB Alfvén waves in the solar wind is developed. The method is based on the one-fluid ideal MHD equations governing small-amplitude, undamped, poloidal Alfvén waves propagating in the solar equatorial plane of a Weber and Davis spiral field. Numerical results for waves propagating from the sun to 1 AU are presented.

074.030 **Backstreaming ions outside the earth's bow shock and their interaction with the solar wind.**
C. Bonifazi, A. Egidi, G. Moreno, S. Orsini.
J. Geophys. Res., Vol. 85, 3461 - 3472 (1980).

074.031 **Der Sonnenwind reicht weit in den Weltraum hinaus. Die Heliosphäre im Lichte der Plasmatheorie.**
E. Keppler.
Umschau, 80. Jahrg., 586 - 591 (1980).

The heliosphere is described as being defined by the domain controlled by the interplanetary magnetic field. The heliopause which ultimately surrounds the heliosphere towards interstellar space is probably hidden behind a shock wave towards interstellar winds. The heliosphere is sustained by the outflowing solar wind, which is described as a solar phenomenon, released from coronal holes and propagating radially outwards. Interactions of the solar wind with planets are briefly mentioned.

074.032 **Photometry of the monochromatic corona 530.3 nm observed at Lomnický Štít on Feb. 16, 1980.**
M. Rybanský.
Bull. Astron. Inst. Czechoslovakia, Vol. 31, 316 - 318 (1980).

Observations made at the coronal station of Lomnický Štít during the total solar eclipse of Feb. 16, 1980 are described. The results of processing the observational material are the isophotes in the Fe XIV – 530.3 nm line and drawings of prominences.

074.033 **A gyro-synchrotron maser in the solar corona?**
D. B. Melrose, S. M. White.
Proc. Astron. Soc. Australia, Vol. 3, 231 - 233 (1978).

074.034 **Comparison of radioheliograph, coronagraph and K-coronameter observations of a coronal streamer.**
B. V. Jackson, K. V. Sheridan, G. A. Dulk.
Proc. Astron. Soc. Australia, Vol. 3, 387 - 389 (1979).

074.035 **An observational picture of solar-wind MHD turbulence.** P. Veltri.
Nuovo Cimento C, Ser. 1, Vol. 3C, 45 - 55 (1980). – Abstr. in Phys. Abstr., Vol. 83, Abstr. 85834 (1980).

074.036 **Alfvén-wave acceleration of the solar wind.**
E. Leer, T. Fla, T. E. Holzer.
Nuovo Cimento C, Ser. 1, Vol. 3C, 114 - 122 (1980). – Abstr. in Phys. Abstr., Vol. 83, Abstr. 85923 (1980).

074.037 **Observations of heavy ions in the solar wind from data of the Prognoz 7 satellite.**
O. L. Vajsberg, Yu. I. Ermolaev, G. N. Zastenker,

A. N. Omel'chenko.
Kosm. Issled., Tom 18, 761 - 765 (1980). In Russian.

074.038 **Structure and dynamical peculiarities of the solar corona on June 30, 1973.**
N. I. Dzyubenko, V. I. Ivanchuk, O. S. Popov, G. A. Rubo.
Problems of cosmic physics. Vyp. 15, (see 003.003), p. 3 - 23 (1980). In Russian.

From corona negatives obtained during the solar eclipse of June 30, 1973 in Africa as well as 10-m coronagraphs the general, detailed and fine corona structure is studied up to $r \approx 3{,}0-4{,}0\ R_\odot$. The structure of outer corona rays up to $12\ R_\odot$ was studied from published reproductions of the enhanced solar corona image. The effects are discussed of quasi-continuous dynamic disturbances occurring in small coronal holes and in the vicinity of them. Results are obtained concerning fine filament structure of the inner and outer corona, their relation to magnetic fields and neutral (current) sheets, spatial structure of large helmet-like rays, relation of general corona structure to N–S asymmetry of activity at the 20th cycle and so on.

074.039 **Variation of the average 'freezing-in' temperature of oxygen ions with solar wind speed.**
K. W. Ogilvie, C. Vogt.
Geophys. Res. Lett., Vol. 7, 577 - 580 (1980).

Observations of the average oxygen ionization equilibrium as a function of speed of the solar wind are presented. At low solar wind speeds they indicate a coronal temperature at the freezing-in point of $1.6 \pm 0.2 \times 10^6$ °K. At speeds above 450 km sec^{-1} the apparent temperature starts to rise rapidly. This rise is tentatively interpreted in terms of a lack of thermodynamic equilibrium in the source region.

074.040 **On the stabilization of electron flows in an inhomogeneous plasma of the solar corona.**
V. I. Vigdorchik.
Izv. vuzov. Radiofiz., Vol. 23, 38 - 41 (1980). In Russian.
Abstr. in Ref. zh., 51. Astron., 9.51.357 (1980).

074.041 **On the physical significance of white light polar plumes in the solar corona.**
B. Sornette, B. Fort, J. P. Picat, M. Cailloux.
Astron. Astrophys., Vol. 90, 344 - 349 (1980).

Using white light photographs of the 30 June 1973 solar eclipse, it is shown that the plumes observed on the south pole are the upper parts of jets originating at the edge of the polar coronal hole (filament channel). A density and temperature model is derived which confirms this interpretation. It is deduced that the XUV plumes associated with bright points do not generally correspond to the white light polar plumes.

074.042 **Radiative-dominated cooling of the flare corona and transition region.** S. K. Antiochos.
Astrophys. J., Vol. 241, 385 - 393 (1980).

The author develops models for the cooling of coronal flare plasma in which radiation dominates, and in particular he investigates the emission measure profiles that such models predict.

074.043 **Global properties of the solar wind. I. The invariance of the momentum flux density.**
R. Steinitz, M. Eyni.
Astrophys. J., Vol. 241, 417 - 424 (1980).

From a statistical analysis of Helios 1 and Mariner 2 data the authors show that the momentum flux density carried by the solar wind is invariant with respect to velocity structures. The distribution of the momentum flux shows a dispersion which diminishes as the Sun is approached. This suggests that the momentum flux invariance is due to initial constraints determining the evolution of the solar wind. Thus, in modeling the solar wind, the density cannot initially be chosen independently of other quantities, such as flow velocity.

074.044 **Radio observations of coronal holes.**
K. V. Sheridan, G. A. Dulk.
Solar and interplanetary dynamics, (see 012.020), p. 37 - 43 (1980).

Coronal holes have been observed on several occasions with the 80 and 160 MHz radioheliograph at Culgoora. At 160 MHz the holes invariably appear as areas of low brightness, either on the disk or at the limb. At 80 MHz holes on the limb always appear less bright than their surroundings but on the disk frequently appear brighter. The simplest interpretation is that the coronal temperature in holes near the 80 MHz critical density (8×10^7 cm^{-3}) is higher than in normal quiet regions, but that the density at this level is lower.

074.045 **A model for the north coronal hole observed at the 1973 eclipse, between 1.3 and 3.2 $R_\odot$.**
F. Crifo, J.-P. Picat.
Solar and interplanetary dynamics, (see 012.020), p. 45 - 48 (1980).

074.046 **On the possibility of identifying coronal holes on synoptic maps of the green corona.**
V. Letfus, L. Kulčár, J. Sýkora.
Solar and interplanetary dynamics, (see 012.020), p. 49 - 53 (1980).

The coronal holes of the Skylab period are treated to identify them with the low-brightness regions in our synoptic maps of the λ 530.3 nm emission corona. Possibilities and difficulties of this identification are discussed.

074.047 **Solar observations with a new earth-orbiting coronagraph.**
N. R. Sheeley, Jr., R. A. Howard, D. J. Michels, M. J. Koomen.
Solar and interplanetary dynamics, (see 012.020), p. 55 - 59 (1980).

Since March 28, 1979, the Solwind coronagraph has been observing the sun's white light corona (2.6 - 10.0 $R_\odot$) routinely with a spatial resolution of approximately 1.25 arc min and a repetition rate of 10 minutes during the one-hour sunlit portion of each 97-minute satellite orbital period. These are the first satellite observations of the outer corona near the peak of a sunspot cycle when coronal transients and high-latitude streamers are common.

074.048 **Coronal structure and solar wind.** J. N. Tandon.
Solar and interplanetary dynamics, (see 012.020), p. 73 - 78 (1980).

Recent observations of large scale coronal structures and solar wind have been studied. The intercorrelation of the two have qualitatively explained through the focussing of solar-ion streams taking account of the local and general solar magnetic fields. This explains the association of coronal holes with weak, diverging open magnetic field lines and envisages the transfer of hydromagnetic wave energy from nearby active centers to account for the enhanced outflow of solar wind associated with coronal holes.

074.049 **The coronal responses to the large-scale and long-term phenomena of the lower layers of the sun.**
J. Sýkora.
Solar and interplanetary dynamics, (see 012.020), p. 87 - 104 (1980).

Based on the assumption, generally accepted over the past decade, that all the forms of solar and interplanetary activity are responses to the magnetic fields generated initially in the subphotosphere, some characteristics of the large-scale and long-term behaviour of the solar corona during the last three solar cycles are presented.

074.050 **Stellar mass flux and coronal heating by shock waves.** P. Couturier, A. Mangeney, P. Souffrin.
Solar and interplanetary dynamics, (see 012.020), p. 127 - 130 (1980).

The authors study a self-consistent oversimplified model which maintains the global balance of energy sources and sinks from the chromospheric level to the interplanetary medium. The heating mechanism chosen is the shock wave dissipation.

074.051 **Mode-coupled MHD waves in the corona and solar wind.** M. Heinemann, S. Olbert.
Solar and interplanetary dynamics, (see 012.020), p. 139 - 141 (1980).

A model is outlined of mode-coupled MHD compressional waves in the corona and solar wind.

074.052 **Properties of magnetohydrodynamic turbulence in the solar wind.**
M. Dobrowolny, A. Mangeney, P. L. Veltri.
Solar and interplanetary dynamics, (see 012.020), p. 143 - 146 (1980).

074.053 **An empirical relation between density, flow velocity and heliocentric distance in the solar wind.**
M. Eyni, R. Steinitz.
Solar and interplanetary dynamics, (see 012.020), p. 147 - 150 (1980).

074.054 **Are solar wind measurements of different spacecraft consistent?** R. Steinitz, M. Eyni,
Solar and interplanetary dynamics, (see 012.020), p. 151 - 154 (1980).

Results of solar wind measurements by different spacecraft are not always in full accord. Such measurements are in general not from one and the same distance r from the sun, nor are they taken at the same phase of the solar activity cycle. The authors examine the possibility of reconciling the apparent discrepancies.

074.055 **Observation of dust generated hydrogen in the solar vicinity.** H. J. Fahr, H. W. Ripken, G. Lay.
Solar and interplanetary dynamics, (see 012.020), p. 155 - 158 (1980).

Solar wind protons impinging on interplanetary dust grains are trapped, deionized, and subsequently desorbed. The steady state distribution of desorbed neutral hydrogen inside of 0.4 AU can be deduced by observation of resonantly scattered solar 121.6 nm radiation.

074.056 **Model calculations of solar wind expansion including an enhanced fraction of ionizing electrons.**
E. F. Petelski, H. J. Fahr, H. W. Ripken.
Solar and interplanetary dynamics, (see 012.020), p. 159 - 162 (1980).

Implications for solar wind and interstellar gas dynamics are calculated by simultaneously solving continuity equations for solar wind protons, interstellar hydrogen atoms, and energetic electrons.

074.057 **X-ray evidence of coronal preflare emission.**
D. F. Webb.
Solar and interplanetary dynamics, (see 012.020), p. 189 - 193 (1980).

074.058 **Spicules and macrospicules.** W. van Tend.
Solar and interplanetary dynamics, (see 012.020), p. 195 - 197 (1980).

The transition zone overheating model and the melon seed model are compared with observations of spicules, macrospicules, surges and sprays.

074.059 **The disruption of EUV coronal loops following a mass ejection transient.** E. J. Schmahl.
Solar and interplanetary dynamics, (see 012.020), p. 241 - 244 (1980).

074.060 **Decameter radio and white light observations of the 21 August 1973 coronal transient.**
T. E. Gergely, M. R. Kundu.
Solar and interplanetary dynamics, (see 012.020), p. 245 - 249 (1980).

074.061 **Radio data and computer simulations for shock waves generated by solar flares.**
A. Maxwell, M. Dryer.
Solar and interplanetary dynamics, (see 012.020), p. 251 - 255 (1980).

The authors have compared radio data on the shocks with computer simulations for the propagation of fast-mode MHD shocks through the solar corona.

074.062 **MHD aspects of coronal transients.** U. Anzer.
Solar and interplanetary dynamics, (see 012.020), p. 263 - 277 (1980).

The author reports about X-ray transients and white light transients. Models with single structures and continuum models are discussed.

074.063 **Reconnection driven coronal transients.**
G. W. Pneuman.
Solar and interplanetary dynamics, (see 012.020), p. 317 - 321 (1980).

The author considers flare associated phenomena in the corona.

074.064 **Two-fluid theory of interplanetary shock waves.**
P. Rosenau.
Solar and interplanetary dynamics, (see 012.020), p. 327 - 331 (1980).

A 2-fluid time-dependent analytical model of the perturbed solar wind is presented. The expansion of newly emitted material, caused, for instance, by the outburst of a solar flare, is simulated by a spherical piston.

074.065 **Transient disturbances of the outer corona.**
R. T. Stewart.
Solar and interplanetary dynamics, (see 012.020), p. 333 - 355 (1980).

The author gives a brief historical review of coronal transient disturbances, followed by a detailed description of three events observed at radio and whitelight wavelengths, concentrating mainly on the radio evidence.

074.066 **Measurements of mass flow in the transition region and inner corona.** G. J. Rottman.
Solar and interplanetary dynamics, (see 012.020), p. 375 - 378 (1980).

A recent sounding rocket experiment has provided high spectral resolution line profiles across the solar disk. The objective of this experiment is to provide information on the systematic velocity fields at the base of the corona by observing the displacement, width and shape of EUV emission lines.

074.067 **The solar mass ejection of 8 May 1979.**
D. J. Michels, R. A. Howard, M. J. Koomen, N. R. Sheeley, Jr., B. Rompolt.
Solar and interplanetary dynamics, (see 012.020), p. 387 - 391 (1980).

This paper describes the main features of the 8 May 1979 solar mass ejection, including the eruption of a polar crown filament to 1.5 $R_\odot$ and the passage of material through the outer corona, from 2.6 to 10.0 $R_\odot$.

074.068 **Two classes of fast solar wind streams: their origin and influence on the galactic cosmic ray intensity.**
N. Iucci, M. Parisi, M. Storini, G. Villoresi.
Solar and interplanetary dynamics, (see 012.020), p. 399 - 401 (1980).

074.069 **A large decametric wavelength antenna array for IPS (*interplanetary scintillation*) observations of** radio sources. C. V. Sastry.
Solar and interplanetary dynamics, (see 012.020), p. 403 (1980).

074.070 **The cross sectional magnetic profile of a coronal transient.**
M. K. Bird, H. Volland, B. L. Seidel, C. T. Stelzried.
Solar and interplanetary dynamics, (see 012.020), p. 475 - 481 (1980).

(I.) Coronal transients in white light and Faraday rotation and (II.) Faraday rotation from a transient flux tube are considered.

074.071 **Dynamics of coronal transients: two-dimensional non-plane MHD models.**
Y. Nakagawa, S. T. Wu, S. M. Han.
Solar and interplanetary dynamics, (see 012.020), p. 495 - 498 (1980).

Numerical results are obtained for non-plane MHD responses to a sudden energy release in a stratified model atmosphere. In agreement with observations, it is shown that after the energy release, the magnetic field affected by the energy release relaxes toward the potential configuration while the outer atmospheric fields increase its shear. Additional results suggest a new way of interpreting the energy storage and release in repeated flares.

074.072 **Observations of interplanetary scintillation and a theory of high-speed solar wind.**
H. Washimi, T. Kakinuma, M. Kojima.
Solar and interplanetary dynamics, (see 012.020), p. 499 - 502 (1980).

The observations of interplanetary scintillation show the existence of high-speed solar wind flows of 800 km/s out of the polar coronal regions and the well-coincidence to the model of coronal holes extending from the polar regions.

074.073 **Kinetic theory of weak disturbances in the solar wind.** I. S. Veselovskij.
Geomagn. Aehron., Tom 20, 769 - 776 (1980). In Russian.

074.074 **Conductive solar wind models in rapidly diverging flow geometries.** T. E. Holzer, E. Leer.
J. Geophys. Res., Vol. 85, 4665 - 4679 (1980).

A detailed parameter study of conductive models of the solar wind has been carried out, extending the previous similar studies of Durney (1972) and Durney and Hundhausen (1974) by considering collisionless inhibition of thermal conduction, rapidly diverging flow geometries, and the structure of solutions for the entire n_0-T_0 plane (n_0 and T_0 are the coronal base density and temperature). Primary emphasis is placed on understanding the complex effects of the physical processes operative in conductive solar wind models.

074.075 **Energy addition in the solar wind.**
E. Leer, T. E. Holzer.
J. Geophys. Res., Vol. 85, 4681 - 4688 (1980).

A general study of energy addition, energy loss, and energy redistribution in the solar wind, for both spherically symmetric and rapidly diverging flow geometries, is presented. It is found that energy addition in the region of subsonic flow increases the solar wind mass flux but either has little effect on (for heat addition) or significantly reduces (for momentum addition) the solar wind flow speed at 1 AU. In contrast, energy addition in the region of supersonic flow has no effect on the solar wind mass flux but significantly increases the flow speed at 1 AU. It is also found that both momentum loss in the subsonic region and energy exchange (involving loss in the subsonic region and gain in the supersonic region) can lead to an increase in the asymptotic flow speed.

074.076 **Energization of solar wind ions by reflection from the earth's bow shock.**
G. Paschmann, N. Sckopke, J. R. Asbridge, S. J. Bame, J. T. Gosling.
J. Geophys. Res., Vol. 85, 4689 - 4693 (1980).

The existence of ion beams with energies a few times the solar wind energy and streaming outward from the earth's bow shock has been known for some time. To explain the observed ion energies, a simple reflection model has been proposed in which the particles gain energy by displacement parallel to the interplanetary electric field. In this model the energy gained in the reflection can be described as a function of the angles between the interplanetary magnetic field, the solar wind velocity, and the local shock normal. Ion beams under widely varying conditions have been observed with Los Alamos Scientific Laboratory/Max-Planck-Institut instrumentation on Isee 1 and 2. For 18 cases, with beam energies ranging from ~ 1.4 to ~ 30 times the solar wind energy, a comparison between the observed and the predicted beam energies has been made. Good agreement between the reflection theory and the observations has been found. Thus simple reflection can account for the several keV ion beams observed upstream from the earth's bow shock.

074.077 **The solar corona.** C. Chiuderi.
Highlights of Astronomy, Vol. 5, (see 012.025), 335 - 341 (1980).

Recent observations from space have shown that the solar corona is spatially a very structured medium and temporally a very dynamic one. The consequent changes in the current theoretical ideas about coronal physics are reviewed. The role of the magnetic fields in shaping and heating the coronal structures is especially underlined.

074.078 **The large-scale structure of the corona.**
J. B. Zirker.
Highlights of Astronomy, Vol. 5, (see 012.027), 549 - 556 (1980).

Considerable progress has been made recently in understanding how the large scale structure of the solar corona controls the genesis of the solar wind and the distribution of slow and fast wind streams throughout the three-dimensional space surrounding the sun. The author discusses some of the progress made in this field during the last few years. He emphasizes the observational data and the inferences that can be made more or less directly from them.

074.079 **The thermal statics of coronal loops.**
B. Roberts, S. Frankenthal.
Sol. Phys., Vol. 68, 103 - 109 (1980).

The thermal statics of constant pressure coronal loops is discussed, with particular emphasis on non-equilibrium and scaling relations. An analytical solution showing explicitly the occurrence of non-equilibrium in radiation dominated loops is presented. In addition, the general scaling law for hot loops is given. However, in view of the uncertainties in the coronal heating function and the observational determined loop parameters, it is suggested that scaling laws are currently of limited value.

074.080 **The orientation of pre-transient coronal magnetic fields.** G. Trottet, R. M. MacQueen.
Sol. Phys., Vol. 68, 177 - 186 (1980).

From the group of transient events studied by Munro et al. (1979) for which definite surface associations exist, the authors find loop transients are strongly correlated with filament regions where the filament axis was oriented north-south. From direct soft X-ray observations of an expanding arch, the possible identification of the soft X-ray signature of footpoints of transient loops, and monochromatic observations of low coronal loops, the authors infer that loop-like coronal transients have their origin in low-lying coronal loops nearly co-planar with the north-south aligned filament axis. The situation with respect to non-loop events is less clear. Possible reasons for the preference of transients to arise from north-south filament-oriented regions are discussed.

074.081 **K-coronameter observations of the solar corona, 15 - 16 February 1980.** R. Fisher, A. Poland.
Bull. American Astron. Soc., Vol. 12, 750 (1980). – Abstract.

074.082 **Solar corona spectra observations in active, quiescent and hole regions.**
L. B. Gilliam, R. N. Smartt, J. B. Zirker.
Bull. American Astron. Soc., Vol. 12, 751 (1980). – Abstract.

074.083 **The variable nature of the solar wind interaction with cometary atmospheres.**
H. L. F. Houpis, D. A. Mendis.
Bull. American Astron. Soc., Vol. 12, 751 (1980). – Abstract.

074.084 **Measurements of coronal kinetic temperatures from 1.5 to 3 solar radii.** J. L. Kohl, H. Weiser, G. L. Withbroe, R. W. Noyes, W. H. Parkinson, E. M. Reeves, R. H. Munro, R. M. MacQueen.
Astrophys. J., Lett., Vol. 241, L117 - L121 (1980).

A rocket-borne Lα coronagraph has been used to make the first measurements of the spectral line profile of resonantly scattered hydrogen Lα coronal radiation between 1.5 and 3 solar radii. These data provide, for the first time, direct measurements of coronal temperatures above 1.5 solar radii. Data were obtained in a coronal hole, quiet region, and streamer.

074.085 **Thermal iron ions in high speed solar wind streams: detection by the IMP 7/8 energetic particle experiments.** D. G. Mitchell, E. C. Roelof.
Geophys. Res. Lett., Vol. 7, 661 - 664 (1980).

The first measurements of the abundance of iron ions in high speed (> 600 km s^{-1}) solar wind streams have been made. The identification of iron ions is quantitatively established. Preliminary estimates of the Fe/H ratio are within a factor of 2 of the adopted coronal abundance (5×10^{-5}), and there is some evidence that Fe/H may remain approximately constant within a given stream. In the peaks of fast streams ($700 < V < 800$ km s^{-1}), about 50 iron ion counts are obtained every 20 s, offering the possibility of studying the Fe/H ratio with ~ 1 m time resolution in high speed streams.

074.086 **On temperature and speed of He^{++} and O^{6+} ions in the solar wind.**
W. K. H. Schmidt, H. Rosenbauer, E. G. Shelly, J. Geiss.
Geophys. Res. Lett., Vol. 7, 697 - 700 (1980).

First results on fluid parameters of heavy ions in the solar wind from ISEE-1 mass spectrometer data are presented. Temperatures and speeds of the He^{++} and O^{6+} ions have been determined for about 150 hours of data in late 1977. It is found that the temperatures are roughly proportional to atomic mass of the ion species. When temperatures are comparatively low, this proportionality does not hold for H^+. At high speeds and high temperatures the minor ions stream somewhat faster than H^+, but there is on average practically no difference between speeds of He^{++} and O^{6+}.

074.087 **An empirical polytrope law for solar wind thermal electrons between 0.45 and 4.76 AU: Voyager 2 and Mariner 10.** E. C. Sittler, Jr., J. D. Scudder.
J. Geophys. Res., Vol. 85, 5131 - 5137 (1980).

Empirical evidence is presented that solar wind thermal electrons obey a polytrope law of the form $P = \Sigma n^{\gamma}$ with polytrope index $\gamma = 1.175 \pm 0.03$ (3σ). The Voyager 2 and Mariner 10 data used span the radial range from 0.45 to 4.76 AU and have a large dynamic range in density (four decades), and in temperature (over one decade), which is crucial for an unambiguous determination of γ. The polytrope index γ was found to be insensitive to systematic dependence on the bulk velocity.

074.088 **Uitzonderlijke vliegtuigwaarnemingen van de zonnekorona.** C. de Jager.
Zenit, 7e Jaarg., 226 - 227 (1980).

074.089 **Monochromatic observations of the solar corona and prominences.** K. Kumagai.
Tokyo Astron. Obs. Rep. (No. 73), Vol. 19, 288 - 294 (1980). In Japanese.

074.090 **Abundance ratios $^4He^{++}/^3He^{++}$ in the solar wind.**
K. W. Ogilvie, M. A. Coplan, P. Bochsler, J. Geiss.
J. Geophys. Res., Vol. 85, 6021 - 6024 (1980).

The ion composition experiment aboard the ISEE 3 spacecraft has been used to measure the $^4He^{++}/^3He^{++}$ ratio in the solar wind. The average ratio $^4He^{++}/^3He^{++}$ is found to be $2.1 \pm 0.2 \times 10^3$, in agreement with previous foil measurements. Individual observations vary widely from the average and the low $^4He^{++}/^3He^{++}$ ratios occur at times of low $^4He^{++}$ flux.

074.091 **Microturbulence in solar wind streams.**
S. Ananthakrishnan, W. A. Coles, J. J. Kaufman.
J. Geophys. Res., Vol. 85, 6025 - 6030 (1980).

The solar wind velocity estimated from observations of interplanetary scintillations (IPS) can be compared with spacecraft measurements during 1973 - 75 when the large scale structure was very stable. Coles et al. (1978) made such a comparison both to 'calibrate' the IPS data of 1973 and to infer the distribution of the electron density fluctuations that cause IPS. In this paper the authors have estimated the distribution of this 'microturbulence' more precisely by using a refined technique, and they have extended the comparison through 1975.

074.092 **Deceleration of the solar wind in the earth's foreshock region: ISEE 2 and Imp 8 observations.**
C. Bonifazi, G. Moreno, A. J. Lazarus, J. D. Sullivan.
J. Geophys. Res., Vol. 85, 6031 - 6038 (1980).

The deceleration of the solar wind, in the region of the interplanetary space filled by ions backstreaming from the earth's bow shock and associated waves, is studied using a two spacecraft technique. This deceleration, which is correlated with the "diffuse" but not with the "reflected" ion population, depends on the solar wind bulk velocity: at low velocities (below 300 km/s) the velocity decrease is ~ 5 km/s, while at higher velocities (above 400 km/s) the decrease may be as large as 30 km/s. Along with this deceleration the solar wind undergoes a deflection by ~ 1° away from the direction of the earth's bow shock. The energy balance shows that the kinetic energy loss far exceeds the thermal energy which is possibly gained by the solar wind; therefore at least part of this energy must go into waves and/or into the backstreaming ions.

074.093 **Observations of the velocity distribution of solar wind ions.**
K. W. Ogilvie, P. Bochsler, J. Geiss, M. A. Coplan.
J. Geophys. Res., Vol. 85, 6069 - 6074 (1980).

Measurements made by the ISEE 3 ion composition experiment have been used to determine the kinetic temperatures of $^3He^{++}$, $^4He^{++}$, $^{16}O^{6+}$, and $^{16}O^{7+}$ in the solar wind. It is found that these temperatures generally obey the relation that T_i/m_i = const, but fluctuations, some of which are caused by dynamical effects in the flow, are observed. The temperature of oxygen sometimes rises above 10^6 °K, which is very strong evidence for heating outside the collisional region of the corona. The velocity distribution function of helium is observed to be non-Maxwellian, with a pronounced high velocity tail. As this is one condition for heating by wave dissipation, this mechanism must still be considered as a heating mechanism.

074.094 **Magnetohydrodynamic shock propagation in the vicinity of a magnetic neutral sheet.**
R. S. Steinolfson, D. J. Mullan.
Astrophys. J., Vol. 241, 1186 - 1194 (1980).

This paper reports a numerical investigation of the propagation of magnetohydrodynamic (MHD) shocks in the vicinity of magnetic neutral sheets. The attenuation of a shock after passing through a neutral sheet has been evaluated. In a parameter study, the authors have examined values of shock speed, polytropic index, plasma beta, and neutral-sheet thickness which are representative of solar coronal conditions. If solar cosmic rays are accelerated in association with a flare-induced shock (as seems most likely), then their results suggest that the spatial structure of solar particle sources will be influenced by helmet streamers. Such streamers are most readily detectable by Hα filaments in the underlying chromosphere.

074.095 **The evolution of active region loop plasma.**
K. R. Krall, S. K. Antiochos.
Astrophys. J., Vol. 242, 374 - 382 (1980).

The authors investigate numerically the adjustment of coronal active-region loops to changes in their heating rate. The calculated evolution of physical parameters suggests that (1) mass supplied during chromospheric evaporation is much more effective in moderating coronal temperature excursions than when downward heat flux is dissipated by a static chromosphere, and (2) the method by which the chromosphere responds to changing coronal conditions can significantly influence coronal readjustment time scales. Observations are cited which illustrate the range of possible fluctuations in the heating rates.

074.096 **The Kitt Peak coronal velocity experiment.**
W. Livingston, J. Harvey, L. A. Doe, B. Gillespie, G. Ladd.
Bull. Astron. Soc. India, Vol. 8, 43 - 45, 47 (1980).

Multi-slit spectra of (Fe XIV 5303) plus neon comparisons were successfully obtained on February 16, 1980. The objective is to confirm previous results of March 1970 and June 1973, viz. that the corona is remarkably quiescent and that the majority of detected flows at 0.3 - 1.0 $R_\odot$ are directed toward the Sun.

074.097 **Coronal emission line measurements of the Feburary 16, 1980 total solar eclipse.**
D. H. Liebenberg, W. M. Sanders, E. A. Brown, H. S. Murray, R. N. Kennedy.
Bull. Astron. Soc. India, Vol. 8, 49 - 51, 53, 55, 57 (1980).

The authors have observed coronal emission line profiles at the February 16, 1980 total solar eclipse from an NC-135 aircraft. The data from Fe XIV and Ca XV emission lines can be analysed to obtain detailed ion temperature and density with 4 arc-s spatial resolution and extending to beyond 2 solar radii in equatorial regions. This paper describes the observations and methods of data analysis.

074.098 **Interferometric eclipse observations of coronal Fe XIV emission.**
R. N. Smartt, J. B. Zirker, H. A. Mauter.
Bull. Astron. Soc. India, Vol. 8, 59 - 61 (1980).

An experiment carried out during the February 16, 1980 total solar eclipse to measure velocities and intensity distributions associated with coronal loops is described. A preliminary discussion of the data is presented.

074.099 **Search for optical modulation of the solar corona during the February 16, 1980 total solar eclipse.**
E. J. Seykora.
Bull. Astron. Soc. India, Vol. 8, 63 - 64 (1980).

The optical modulation and solar scintillations in the frequency range from 20 Hz - 15 kHz and 5 - 15 MHz were recorded during the February 16, 1980 total solar eclipse. Preliminary results are discussed concerning both the coronal modulation and atmospheric modulation of the light. Recordings of the shadow band phenomena are presented.

074.100 **Coronal interferogram in 5303 Å obtained during the total solar eclipse of February 16, 1980 and coronal temperatures.** T. Chandrasekhar, N. M. Ashok. J. N. Desai, D. B. Vaidya, P. D. Angreji.
Bull. Astron. Soc. India, Vol. 8, 87 - 89 (1980).

074.101 **A semi-analytical approach to time-dependent coronal expansion.** R. A. Kopp.
Sol. Phys., Vol. 68, 307 - 316 (1980).

The author points out the existence of a special class of solutions to the nonlinear hydrodynamic equations describing the time-dependent solar wind, namely that for which the velocity profile is time-invariant but the density at each point of the corona changes exponentially with time. Theoretical velocity curves are calculated for the case of isothermal expansion and compared with the Parker model for steady-state expansion. These solutions can be used to obtain quantitative estimates for the degree of departure from the latter of a real corona undergoing evolution on a finite time scale.

074.102 **Thermal cyclotron radio emission of neutral current sheets in the solar corona.**
V. V. Zheleznyakov, E. Ya. Zlotnik.
Sol. Phys., Vol. 68, 317 - 326 (1980).

Cyclotron radiation (its frequency spectrum and polarization) of thermal electrons in a neutral current sheet is considered. It is shown that cyclotron radiation is able to escape from a thin edge of the sheet where the magnetic field is practically homogeneous. Due to this, the frequency spectrum has the form of comparatively narrow lines with integer ratio of frequencies. This fact enables one to recognize neutral current sheets in the solar corona by their radio emission.

074.103 **Estimation of the coronal magnetic field from the Razin effect in a solar decametric continuum burst.**
R. V. Bhonsle, S. S. Degaonkar.
Sol. Phys., Vol. 68, 339 - 342 (1980).

From the observed low frequency cut-off of a type IV continuum burst in the decameter range, the authors have obtained estimates of the coronal magnetic field (~6 G) around $2R_\odot$ from the center of the sun and the magnitude of its perturbation (1 to 30%) as a result of an MHD wave propagating through that region.

074.104 **On dielectronic recombination in some coronal ions.**
H. P. Mital, U. Narain.
Z. Naturforsch., Band 35a, 1325 - 1329 (1980).

Dielectronic recombination rate coefficients for some ions found in the solar corona have been investigated in the temperature range 20–1000 eV using two different expressions. A comparison has also been made with the corresponding radiative recombination rate coefficients.

074.105 **Coronal holes.** L. Kulčár, P. Prykryl.
Kozmos, Vol. 11, 103 - 105 (1980). In Slovak.

074.106 **The solar wind and related astrophysical phenomena.**
T. E. Holzer.
Solar system plasma physics, Vol. 1, (see 003.010), 101 - 176 (1979).

The author reviews the current state of understanding of

the large-scale dynamics of the solar wind and suggests some directions that research in this area is likely to follow in the future. He also considers the relationship between physical processes in the solar wind and in various astrophysical plasmas, with an emphasis being given to the study of stellar winds. The paper is concluded with a detailed summary and a tentative timetable for solar wind research over the next 25 years.

074.107 **Hydromagnetic waves and turbulence in the solar wind.** A. Barnes.
Solar system plasma physics, Vol. 1, (see 003.010), 249 - 319 (1979).
The author reviews the current state of knowledge of interplanetary fluctuations, their origins and their effects on the solar wind, and suggests directions for future inquiry. He discusses some astrophysical problems for which solar-wind hydromagnetic-wave studies have relevance.

074.108 **Kinetic processes in the solar wind.** W. C. Feldman.
Solar system plasma physics, Vol. 1, (see 003.010), 321 - 344 (1979).
The author reviews some of the recent work on kinetic plasma processes active in the solar wind near 1 AU. The scope of this review is confined to processes with wavelengths shorter than about 10 proton gyroradii which are driven by non-Maxwellian particle velocity distributions.

074.109 **Effect of temporal variations of solar wind flux velocity on the structure of the interplanetary magnetic field.**
V. A. Kovalenko, V. I. Mordvinov, M. A. Filippov.
Issled. po geomagn., aehron. i fiz. Solntsa, Moskva, 1980, No. 52, p. 32 - 38. In Russian. – Abstr. in Ref. zh., 51. Astron., 11.51.468 (1980).

074.110 **Coronal features in the λ 284 Å line and polarity of photospheric magnetic fields.**
M. Rybanskij, N. F. Tyagun.
Issled. po geomagn., aehron. i fiz. Solntsa, Moskva, 1980, No. 52, p. 14, 15. In Russian. – Abstr. in Ref. zh., 51. Astron., 11.51.473 (1980).

074.111 **Coronal scattering of interplanetary-spacecraft radio signals.** H. M. Bradford, D. Routledge.
J. R. Astron. Soc. Canada, Vol. 74, 358 (1980). – Abstract.

074.112 **Breitenvariation des Sonnenwindflusses und der Sonnenwindgeschwindigkeit berechnet aus UV-Beobachtungen von Mariner 10.**
N. Witt, J. M. Ajello, P. W. Blum.
Mitt. Astron. Ges., Nr. 50, p. 33 - 36 (1980).

074.113 **Einflüsse hoher interstellarer Gasdichten auf die Sonnenwind-Expansion.**
H. W. Ripken, H. J. Fahr.
Mitt. Astron. Ges., Nr. 50, p. 37 - 41 (1980).

074.114 **Indirekte Beobachtungen magnetohydrodynamischer Wellenaktivität in der Sonnenkorona.**
P. Edenhofer, M. K. Bird, H. Volland, J. V. Hollweg.
Mitt. Astron. Ges., Nr. 50, p. 42 - 45 (1980).

074.115 **Radiale Intensitätsgradienten der koronalen Emissionslinie λ 5303.** C. Spannagl.
Mitt. Astron. Ges., Nr. 50, p. 45 - 46 (1980).

074.116 **On the role of Alfvénic fluctuations in the inner solar system.** U. Villante.
J. Geophys. Res., Vol. 85, 6869 - 6873 (1980).
The author carefully examines the experimental implications of the current theoretical models of the Alfvénic fluctuations associated with the recurrent high-velocity solar wind streams and argues that under typical solar wind conditions the saturated waves might be possibly distinguished from the undamped ones at heliocentric distances smaller than 1 AU. He also shows that the experimental observations performed by Helios 1 and 2 may be tentatively considered to be consistent with a saturated propagation in the inner solar system. If this is the case, the contribution of the Alfvénic fluctuations to the energetics of the solar wind expansion may be appreciably larger than was previously estimated.

Summary of the Czechoslovak solar corona observations. See Abstr. 011.034.

A method for determining the distribution of plasma loop parameters in solar coronal active regions. See Abstr. 031.519.

SMM (*Solar Maximum Mission*) orbiting coronagraph – early results. See Abstr. 032.513.

SMM (*Solar Maximum Mission*) Coronagraph/Polarimeter observations of coronal transient manifestations correlated with flare data. See Abstr. 032.516.

The plasma experiment on Helios (E 1). See Abstr.032.555.

Scientific results obtained by the Helios Technical University of Braunschweig flux-gate (E 2) and search-coil (E 4) magnetometer experiments. See Abstr. 032.556.

Faraday rotation experiment (E 12). See Abstr. 032.565.

Calculation of the disturbances of the low energy electron measurements (E1 - Z). See Abstr. 051.037.

A summary of progress in space physics made with Helios plasma wave instrument data (E 5a). See Abstr. 051.038.

Investigation of the mutual position of the orbit of a satellite with high apogee and the magnetopause, of the shock wave and the shadow of the earth by means of a machine diagram. See Abstr. 052.024.

Les satellites et les manifestations de l'activité solaire dans l'espace interplanétaire. See Abstr. 054.004.

Non-linear interaction of Alfven waves with compressive fast magnetosonic waves. See Abstr. 062.009.

Higher order fluid equations for multicomponent nonequilibrium stellar (plasma) atmospheres and star clusters. See Abstr. 062.017.

Further development of numerical MHD model of coronal dynamics. See Abstr. 062.021.

Comments on the dissipation of hydromagnetic surface waves. See Abstr. 062.036.

Electron trapping in the solar magnetic field and emission of decimetric continuum radio bursts. See Abstr. 062.047.

On the influence of neutral current sheets in a cosmic plasma on the frequency spectrum of propagating radiation. See Abstr. 062.066.

Spurious solutions of the Navier-Stokes equations. See Abstr. 062.069.

Very hot plasmas in the solar system. See Abstr. 062.084.

Fully developed anisotropic hydromagnetic turbulence in interplanetary space. See Abstr. 062.125.

Stellar coronae. See Abstr. 064.053.

Stellar wind theories. See Abstr. 064.057.

Theory of the solar wind and winds from late-type stars. See Abstr. 064.072.

Energy transfer by Alfvén waves. See Abstr. 071.026.

Density and temperature measurements for short lived transition zone phenomena. See Abstr. 073.030.

Reviews of solar activities, coronal holes and geomagnetic storms in 1978. See Abstr. 073.049.

A successful observation of the flash spectrum at the total solar eclipse of February 16, 1980. See Abstr. 073.071.

Coronal holes, the height of the chromosphere, and the origin of spicules. See Abstr. 073.073.

Nonequilibrium ionization due to thermal diffusion and mass flows. See Abstr. 073.074.

Energy and mass injected by flares and eruptive prominences. See Abstr. 073.075.

Radiative hydrodynamics of flares: preliminary results and numerical treatment of the transition region. See Abstr. 073.087.

A high-resolution view of the solar chromosphere and corona. See Abstr. 073.095.

Density dependence of solar emission lines of boron-like ions. See Abstr. 073.096.

Steady flows in the chromosphere and transition-zone above active regions as observed by OSO-8. See Abstr. 073.113.

Calcium flocculi and modulation of solar wind flow concentration with solar semirotational period. See Abstr. 073.119.

The chromosphere and transition region. See Abstr. 073.132.

Mass ejections. See Abstr. 073.133.

Forecast of the AE index of magnetic activity one hour ahead from solar wind parameters. See Abstr. 075.013.

Evolution of coronal magnetic structures. See Abstr. 075.021.

Magnetically driven motions in solar corona. See Abstr. 075.022.

Morphology and spatial distribution of XUV and X-ray emissions in an active region observed from Skylab. See Abstr. 076.019.

X-ray structures associated with disappearing Hα filaments in active regions. See Abstr. 076.021.

UV emitting spicules. See Abstr. 076.022.

Initial results from a 3 to 25 Å solar X-ray spectrometer/spectroheliograph experiment. See Abstr. 076.027.

Synoptic charts of solar 9.1 cm and coronal hole data. See Abstr. 077.001.

Some recent explorations of the solar corona from the Culgoora Solar Radio Observatory. See Abstr. 077.024.

Radio observations of a massive, slow-moving ejection of coronal material. See Abstr. 077.030.

Solar radar observations. See Abstr. 077.037.

The association of type III bursts and coronal transient activity. See Abstr. 077.039.

Synchronous variation of the fluxes of distant type I sources and connection between the different activity centers. See Abstr. 077.046.

Radioverschijnselen als boodschappers uit de zonnecorona (I). Het decor der radioverschijnselen. See Abstr. 077.052.

Radioverschijnselen als boodschappers uit de zonnecorona (2). Kennismaking met enkele radioverschijnselen. See Abstr. 077.053.

Solar cosmic rays and isotopic composition of the solar wind. See Abstr. 078.008.

Energy balance from the chromosphere-corona transition region. See Abstr. 080.021.

Physical driving forces and models of coronal responses. See Abstr. 080.039.

A comparison of solar wind and ionospheric ion acoustic waves. See Abstr. 083.041.

The energy coupling function and the power generated by the solar wind–magnetosphere dynamo. See Abstr. 084.003.

Mechanism of magnetospheric substorms and the turbulence of solar wind. See Abstr. 084.031.

Observations of energetic electrons of magnetospheric origin in the magnetosheath and in the solar wind. See Abstr. 084.037.

Study of individual geomagnetic storms in terms of the solar wind. See Abstr. 084.043.

Dependence of the probability of substorm display on solar wind velocity and on the vertical component of the interplanetary magnetic field. See Abstr. 084.055.

The effects on the earth's magnetotail from shocks in the solar wind. See Abstr. 084.064.

Solar wind evolution and recurrent geomagnetic disturbances. See Abstr. 084.073.

Rates of mass, momentum, and energy transfer at the magnetopause. See Abstr. 084.088.

Impulsive penetration of solar wind plasma and its effects on the upper atmosphere. See Abstr. 084.089.

A new predictive model for determining solar wind-terrestrial planet interactions. See Abstr. 091.067.

Solar wind nitrogen and indigenous nitrogen in lunar material. See Abstr. 094.532.

Solar wind carbon chemistry as revealed by lunar sample analysis. See Abstr. 094.533.

The interaction of the solar wind with Mars, Venus and Mercury. See Abstr. 097.079.

Correlated variations of planetary albedos and solar-interplanetary parameters. See Abstr. 101.007.

Interplanetary gas. XXV. A solar wind and interplanetary magnetic field interpretation of cometary light outbursts. See Abstr. 102.029.

Cometary atmospheres. I. Solar wind modification of the outer ion coma. See Abstr. 102.034.

The interaction of the solar wind with comets. See Abstr. 102.038.

A very rapid turning of the plasma-tail axis of comet Bradfield 1979*l* on 1980 February 6. See Abstr. 103.206.

Propagation directions of hydromagnetic waves in interplanetary space: Pioneer 10 and 11. See Abstr. 106.011.

Observations of nonlinear turbulence in the upstream solar wind. See Abstr. 106.014.

Observations of backstreaming protons in the solar wind near the Earth's bow shock. See Abstr. 106.015

Energetic particle, solar wind plasma and magnetic field measurements on board Prognoz-6 during the large scale interplanetary disturbance of Jan. 3 - 4, 1978. See Abstr. 106.019.

The origin of interplanetary sectors. See Abstr. 106.059.

Interplanetary response to solar long time-scale phenomena. See Abstr. 106.060.

Transient phenomena originating at the sun – an interplanetary view. See Abstr. 106.062.

Theoretical interpretation of traveling interplanetary phenomena and their solar origins. See Abstr. 106.065.

Global modeling of disturbances in the corona–interplanetary space. See Abstr. 106.067.

On the deformation of the plane of the interplanetary current layer by solar wind streams. See Abstr. 106.068.

The thickness of interplanetary collisionless shock waves. See Abstr. 106.077.

Wechselwirkungsprozesse zwischen Sonnenwind und interplanetarem Staub. See Abstr. 106.084.

Solar wind interaction with interstellar helium. See Abstr. 131.004.

Particle trapping and acceleration during the August 1972 event. See Abstr. 143.002.

Expected fluctuations of the heliosphere and long-term cosmic ray variations. See Abstr. 143.023.

Galactic cosmic ray currents in high-velocity recurrent fluxes of the solar wind and long-term changes of cosmic ray anisotropy. See Abstr. 143.025.

The interactions of energetic particles with the solar wind. See Abstr. 143.045.

Solar polar coronal holes and galactic cosmic ray intensities. See Abstr. 143.052.

On the three-dimensional nature of the modulation of galactic cosmic rays. See Abstr. 143.053.

Erratum

074.901 **Erratum: 'Temperature distribution in the transition region and inner corona'** [Sol. Phys., Vol. 62, 93 - 105 (1979)]. B. Alam, S. M. R. Ansari, A. Qaiyum. Sol. Phys., Vol. 67, 207 (1980). – See Abstr. 25.074.082.

075 Magnetic Fields

075.001 **Comment on 'Average photospheric poloidal and toroidal magnetic field components near solar minimum' by Duvall et al.** P. Foukal, T. L. Duvall, Jr.
Sol. Phys., Vol. 67, 9 - 12 (1980).

The authors discuss the dynamical interpretation of evidence for an azimuthal tilt of the global magnetic field from the radial direction at the photosphere. They point out that the Reynolds stresses of supergranular convective motions might produce the required small tilt of intense flux tubes, without implying an unacceptably large momentum flux across the photospheric surface into the solar wind. The authors conclude that there is little reason, at present, to infer (Duvall et al., 1979) a separate low intensity constituent of the global magnetic field, from the observational evidence for an azimuthal tilt.

075.002 **Evidence for an X-type neutral sheet producing chromospheric activity.** N. Seehafer, J. Staude.
Sol. Phys., Vol. 67, 121 - 125 (1980).

Force-free magnetic-field extrapolation for the region McMath 12417 on 4 July 1973 corroborates a suggestion by Roy and Michalitsanos (1974): A large moving magnetic feature presses together opposite fluxes to form an X-type neutral sheet; the supposed geometry of the field as derived from chromospheric activity (subflares, ejections) is confirmed by the calculated lines of force.

075.003 **Magnetic fields and the solar constant.**
D. S. P. Dearborn, J. B. Blake.
Nature, Vol. 287, 365 - 366 (1980).

075.004 **A two-sector solar magnetic structure with 29 day rotation.**
J. T. Hoeksema, P. H. Scherrer, J. M. Wilcox.
Bull. American Astron. Soc., Vol. 12, 474 (1980). – Abstract.

075.005 **Interpretation of filter magnetograph results including solar magneto-optical effects: observations.**
E. A. West, M. J. Hagyard, J. E. Smith.
Bull. American Astron. Soc., Vol. 12, 476 (1980). – Abstract.

075.006 **Interpretations of filter magnetograph results including solar magneto-optical effects: theory.**
M. J. Hagyard, E. A. West, J. Smith.
Bull. American Astron. Soc., Vol. 12, 476 (1980). – Abstract.

075.007 **Analysis of changes in photospheric magnetic fields within a flare-productive active region.**
K. R. Krall, S. T. Wu, M. J. Hagyard, E. A. West, N. P. Cumings, J. B. Smith.
Bull. American Astron. Soc., Vol. 12, 514 (1980). – Abstract.

075.008 **Vector magnetic field measurements at flare locations.** J. B. Smith, Jr., N. P. Cumings, M. J. Hagyard, J. E. Smith, E. A. West, K. R. Krall.
Bull. American Astron. Soc., Vol. 12, 514 (1980). – Abstract.

075.009 **A new approach to understanding preflare buildup of magnetic energy.** D. S. Spicer, J. Ionson.
Bull. American Astron. Soc., Vol. 12, 515 (1980). – Abstract.

075.010 **Evolution of solar magnetic fields, a new method of analysis.** Y. Nakagawa.
Bull. American Astron. Soc., Vol. 12, 516 (1980). – Abstract.

075.011 **Evolution of solar magnetic fields: a new approach to MHD initial-boundary value problems by the method of nearcharacteristics.** Y. Nakagawa.
Astrophys. J., Vol. 240, 275 - 299 (1980).

Observations indicate that the magnetic field in the solar atmosphere is subject to continuous agitations by motions and flux changes at the photospheric level. The proper treatment of this subject, namely the evolution of a non-force-free magnetic field including the atmospheric responses, poses a mathematically complex MHD initial-boundary value problem. A new approach to this MHD initial-boundary problem is presented in the paper. The formulation is based on the method of nearcharacteristics developed recently by Werner and Shin and Kot. The physical validity of the method is demonstrated with examples, and discussion is given of their significance in interpreting observations.

075.012 **Changes of the magnetic field as a cause of non-stationary phenomena in the solar atmosphere.**
B. V. Somov, S. I. Syrovatskij.
XIth seminar on cosmophysics, (see 012.035), p. 12 - 14 (1979). In Russian. – Abstr. in Ref. zh., 51. Astron., 8.51.376 (1980).

075.013 **Forecast of the AE index of magnetic activity one hour ahead from solar wind parameters.**
N. I. Dvinskikh, B. G. Dolgoarshinnykh, N. M. Rudneva.
Issled. po geomagn., aehron. i fiz. Solntsa (Moskva), 1980, No. 50, p. 99 - 105. In Russian. – Abstr. in Ref. zh., 51. Astron. 8.51.405 (1980).

075.014 **Comparative magnetospherology, part 9. Solar-terrestrial phenomena as explained by heliomagnetic excursion in 1974.** M. Seto, T. Hayasaka, T. Saito.
13th Lunar and Planetary Symposium, (see 012.018), p. 116 - 123 (1980).

075.015 **Comparative magnetospherology, part 10. Two-hemisphere model on a reversal process of the heliomagnetosphere.** T. Saito, K. Yumoto, A. Eitoku, M. Yamauchi, M. Seto.
13th Lunar and Planetary Symposium, (see 012.018), p. 124 - 135 (1980).

075.016 **Flux tube dynamo approach to the solar cycle.**
M. Schüssler.
Nature, Vol. 288, 150 - 152 (1980).

A calculation similar to Leighton's magneto-kinematic model (1969) shows that a flux tube dynamo model can operate and reproduce the essential features of the solar cycle. It can also explain the cyclic variation of sunspot brightness and ephemeral active regions.

075.017 **Evolution of coronal and interplanetary magnetic fields.** R. H. Levine.
Solar and interplanetary dynamics, (see 012.020), p. 1 - 20 (1980).

The author points out the latest techniques and studies of the global solar magnetic field and its relation to the interplanetary field.

075.018 **Dynamics of large-scale magnetic field evolution during solar cycle 20.** P. S. McIntosh.
Solar and interplanetary dynamics, (see 012.020), p. 25 - 28 (1980).

The evolution of large-scale solar magnetic fields has been studied for a complete solar cycle using the atlas of H-alpha synoptic charts for 1964–1974. The results include: a unique magnetic pattern coinciding with major coronal holes: variations in the rate of solar rotation through the solar cycle; discovery of convergence and divergence among long-lived magnetic patterns; periodic discontinuities in the organization of large-scale magnetic fields; and a new cause for coronal transients.

075.019 **The false equilibrium of a force-free magnetic field.** B. C. Low.
Solar and interplanetary dynamics, (see 012.020), p. 283 - 289 (1980).

It has been a customary assumption that any force-free magnetic field represents an equilibrium field in the solar atmosphere under the extreme condition $8\pi p/B^2 << 1$. An example of a force-free magnetic field is presented for which this assumption fails in the sense that no equilibrium is possible for the magnetic field if imposed with an arbitrary ambient pressure, however weak the pressure is. A simple mechanism is proposed for the onset of eruption in the course of otherwise quasi-static evolution of magnetic fields in the solar atmosphere.

075.020 **Energy storage and instability in magnetic flux tubes.** T. Sakurai.
Solar and interplanetary dynamics, (see 012.020), p. 291 - 294 (1980).

Now it is known that the solar corona consists of many loops which are believed to represent the structure of the magnetic field. Since the plasma is very tenuous in the corona, the equilibrium of the magnetic field is approximated by the force-free field: rot $\underset{\sim}{B} \times \underset{\sim}{B} = 0$. In this paper the author will propose a method of solution for this equation and will discuss on the energy build up and instability in the magnetic flux tubes.

075.021 **Evolution of coronal magnetic structures.** R. S. Steinolfson, S. T. Wu.
Solar and interplanetary dynamics, (see 012.020), p. 483 - 486 (1980).

Numerical solutions of the time-dependent, two-dimensional, dissipationless, MHD equations of motion are used to examine the formation of a coronal streamer magnetic structure and the evolution of the streamer following an explosive solar event in the closed-field region.

075.022 **Magnetically driven motions in solar corona.** B. V. Somov, S. I. Syrovatskii (*Syrovatskij*).
Solar and interplanetary dynamics, (see 012.020), p. 487 - 489 (1980).

Solution of the nonlinear MHD problem of plasma flow in an increasing dipolar magnetic field is obtained in the approximation of a strong field. The distributions of plasma velocity, displacement, and density are calculated. The situation when the magnetic dipole is 'increased' by rapid process of magnetic reconnection or current sheet rupture is illustrated. Possible applications are discussed in connection with plasma ejections from chromosphere in corona.

075.023 **Magnetic field of the sun as a star, 1969 - 1976.** V. A. Kotov, M. L. Demidov.
Izv. Krymskoj Astrofiz. Obs., Tom 61, 3 - 11 (1980). In Russian.

A superposed epoch analysis of the solar mean magnetic field measured in Crimea during 1969 - 1976 was performed using 215 sector boundaries of the interplanetary magnetic field. The most principal recurrence period found in the data is $27^d.04 \pm 0^d.06$.

075.024 **An exploratory two-dimensional study of the coarse structure of network magnetic fields.**
R. G. Giovanelli.
Sol. Phys., Vol. 68, 49 - 69 (1980).

Analysis of a Mg b_2 magnetogram reveals that, in active regions (and, hence, wherever the magnetic network is well developed) fields cover associated supergranules completely at heights mostly below 500 - 600 km (zero height is at $\tau_{5000} = 1$) but possibly up to 700 - 800 km at great distances (e.g. $> 10^4$ km) from the network. These lie much lower than previously believed, mostly around the solar average temperature minimum. Near plagettes, the low-lying field has been measured out to ~6000 - 7000 km. One consequence is that in active regions and plagettes, the chromosphere-corona transition region probably penetrates below 600 km; another is that potential theory is inapplicable at coronal heights below about 15 000 km.

075.025 **Some characteristics of large-scale magnetic fields on the sun.** V. A. Kotov, N. N. Stepanyan.
Izv. Krymskoj Astrofiz. Obs., Tom 62, 117 - 124 (1980). In Russian.

Some properties of the solar large-scale magnetic fields were determined by comparing the solar mean magnetic field measured by magnetographs and the background fields inferred from H-alpha maps.

Magnetic field measurements on stellar sources: a new method. See Abstr. 031.527.

Flows along magnetic flux tubes. I. Equilibrium and buoyancy of a slender magnetic loop in the interior of a star. See Abstr. 062.033.

Structure within a magnetic flux tube. See Abstr. 062.039.

Electron trapping in the solar magnetic field and emission of decimetric continuum radio bursts. See Abstr. 062.047.

Magnetohydrodynamic equilibrium and stability of pre-flare loops. Constant pitch field. See Abstr. 062.063.

Vertical motions in an intense magnetic flux tube. IV: Radiative relaxation in a uniform medium. See Abstr. 062.087.

Vertical motions in an intense magnetic flux tube. V: Radiative relaxation in a stratified medium. See Abstr. 062.088.

Evidence for extreme divergence of open field lines from solar active regions. See Abstr. 072.020.

Champ magnétique et champ de vitesse dans les régions actives solaires. See Abstr. 072.024.

Evidence for open field lines from active regions: short communication. See Abstr. 072.029.

1978 data on proton flare prediction by means of spiral spots. See Abstr. 072.046.

A note on permissible values of the vertical gradient of the sunspot magnetic field. See Abstr. 072.054.

The two components of the solar activity cycle as a consequence of the shock transition model of the solar magnetic cycle. See Abstr. 072.066.

Measurements of the magnetic field and the gradient of temperature in the solar atmosphere above a flocculus using radio observations. See Abstr. 073.003.

Observations of transient magnetic fields in two solar flares. See Abstr. 073.026.

Particle acceleration in the process of eruptive opening and reconnection of magnetic fields.
See Abstr. 073.079.

Flare model with force-free fields and helical symmetry. See Abstr. 073.083.

Preflare conditions, changes and events.
See Abstr. 073.110.

The cross sectional magnetic profile of a coronal transient. See Abstr. 074.070.

Estimation of the coronal magnetic field from the Razin effect in a solar decametric continuum burst.
See Abstr. 074.103.

Coronal features in the λ 284 Å line and polarity of photospheric magnetic fields. See Abstr. 074.110.

A comparison of type III metric radio bursts and global solar potential field models. See Abstr. 077.019.

Observation with the VLA of a stationary loop structure on the sun at 6 centimeter wavelength.
See Abstr. 077.025.

Polarized solar type III bursts between 2.3 and 4.9 MHz. See Abstr. 077.051.

Torsional oscillations of the sun and magnetic flux eruption. See Abstr. 080.009.

Two-level model of the solar dynamo.
See Abstr. 080.031.

Search for giant cells in the solar convection zone.
See Abstr. 080.036.

Propagation of an MHD shock in the vicinity of a magnetic neutral sheet. See Abstr. 080.038.

Physical driving forces and models of coronal responses. See Abstr. 080.039.

Differential rotation of solar features and its variation as deduced from the 'shock-transition model' of the solar cycle. See Abstr. 080.082.

A model of the heliospheric magnetic field configuration. See Abstr.106.001

On the origin of the B_x and B_z components of the interplanetary magnetic field. See Abstr. 106.005.

Cosmic ray effects due to the general magnetic field of the sun. See Abstr. 143.021.

Twenty-two year modulation of cosmic rays associated with polarity reversal of polar magnetic field of the Sun.
See Abstr. 143.036.

Erratum

075.901 **Erratum: "Evidence for a lower limit of solar magnetic field strengths"** [Astron. Astrophys., Vol. 73, L19 - L20 (1979)]. E. Wiehr.
Astron. Astrophys., Vol. 91, 377 (1980). – See Abstr. 25.075.008.

076 UV, X, Gamma Radiation

076.001 **Dynamic spectral characteristics of thermal models for solar hard X-ray bursts.**
J. C. Brown, I. J. D. Craig, J. T. Karpen.
Sol. Phys., Vol. 67, 143 - 162 (1980).
The dynamic spectral characteristics of the thermal model for solar hard X-ray bursts recently proposed by Brown et al. (1979)(BMS) are investigated. The authors show that a conductively cooled single kernel is incompatible with the observed spectral decay of even a simple spike event. They show that the dissipative thermal model is consistent with the data if the source comprises not a single BMS-type kernel but rather an assembly of numerous, small, dissipative kernels, each very short-lived, which are continuously produced throughout the spike burst.

076.002 **Upper limits to the solar hard X-ray burst variability at 32 milliseconds time resolution.**
D. W. Datlowe, W. L. Imhof, J. R. Kilner, G. H. Nakano, J. B. Reagan.
Bull. American Astron. Soc., Vol. 12, 479 (1980). – Abstract.

076.003 **Measurements of impulsive EUV and hard X-ray solar flare emission.**
D. M. Horan, R. W. Kreplin, G. G. Fritz.
Bull. American Astron. Soc., Vol. 12, 479 (1980). – Abstract.

076.004 **Adiabatic and non-adiabatic processes in thermal models of solar hard X-ray bursts.** A. G. Emslie.
Bull. American Astron. Soc., Vol. 12, 481 (1980). – Abstract.

076.005 **Limits on the streaming and escape of electrons in thermal models for solar hard X-ray bursts.**
D. F. Smith, J. C. Brown.
Bull. American Astron. Soc., Vol. 12, 481 (1980). – Abstract.

076.006 **Measured variation in solar EUV emission.**
R. W. Kreplin, D. M. Horan, R. G. Taylor.
Bull. American Astron. Soc., Vol. 12, 517 (1980). – Abstract.

076.007 **The effect of diffusion and mass flows on C IV and Si IV EUV line profiles computed for various solar features.** R. A. Roussel-Dupré, C. Beerman.
Bull. American Astron. Soc., Vol. 12, 525 - 526 (1980). Abstract.

076.008 **A presentation of some BCS (*Bent Crystal Spectrometer*) spectra.** A. N. Parmar.
Bull. American Astron. Soc., Vol. 12, 531 (1980). – Abstract.

076.009 **Early results from the Hard X-ray Burst Spectrometer on the Solar Maximum Mission.**
L. E. Orwig, K. J. Frost, B. R. Dennis.
Bull. American Astron. Soc., Vol. 12, 532 (1980). – Abstract.

076.010 **Continuum emission observed with the BCS (*Bent Crystal Spectrometer*).** C. G. Rapley.
Bull. American Astron. Soc., Vol. 12, 534 (1980). – Abstract.

076.011 **Latitudinal anisotropy of the solar far ultraviolet flux: effect on the Lα sky background.**
J. W. Cook, R. R. Meier, G. E. Brueckner, M. E. Vanhoosier.
Bull. American Astron. Soc., Vol. 12, 544 (1980). – Abstract.

076.012 **The solar ultraviolet continuum.**
J. N. Dragon, J. P. Mutschlecner.
Astrophys. J., Vol. 239, 1045 - 1069 (1980).
Predictions have been made of the ultraviolet continuum of the Sun in an effort to improve upon previous predictions and to understand the nature of the apparent opacity deficiency in this region. Central intensity and flux were predicted for two standard solar models and were compared with selected observations over the wavelength range from 4500 Å to 1500 Å. Following a detailed study, all known important extinction sources have been included in the calculations. Bound-free agents include magnesium, aluminium, silicon, and iron. Molecular bands of CH, NH, OH, CN, and SiO have been included by a smeared-line opacity procedure adapted for this study. Line absorption was included by an opacity distribution function method for the 4th positive band system of CO and for a total of 13 elements including the iron group.

076.013 **Temporal and spatial fluctuations in strengths and widths of C IV and Si II lines observed with OSO 8.**
R. G. Athay, O. R. White.
Astrophys. J., Vol. 240, 306 - 321 (1980).
Time series of profiles for lines of C IV λ1548, and Si II λ1816.93, 1817.45, observed with an effective aperture of 2″ × 20″ and with time resolution less than 30 s are analyzed to determine the statistical properties of temporal and spatial fluctuations in line widths and strengths. Three classes of fluctuations with substantial amplitudes are evident in the data: (1) short term fluctuations with a characteristic fluctuation time near 5 minutes, (2) intermediate term fluctuations with a characteristic time of 30 minutes or longer, and (3) large scale spatial fluctuations associated with supergranule cell, network, and plage structure. For C IV the dominant fluctuations in line width are short term, whereas for Si II the dominant width fluctuations are intermediate term. Fluctuations in line strength for both C IV and Si II are appreciable on both short and intermediate term but are dominated by the large scale spatial structure.

076.014 **Measurements of solar EUV and soft X ray emission during sudden frequency deviations.**
D. M. Horan, R. W. Kreplin.
J. Geophys. Res., Vol. 85, 4257 - 4269 (1980).
Broadband sensors aboard the Solrad 11B satellite measured solar emission in the 0.5- to 3-Å, 1- to 8-Å, 8- to 20-Å, 100- to 500-Å, and 700- to 1030-Å bands during six sudden frequency deviation (SFD) events in 1977. Simultaneous analysis of the solar measurements and the SFD sonograms led to a proposed extension of an existing model to provide a qualitative description of the influence of solar flare emission on SFD development throughout the course of an SFD event. The proposed model clearly reflects the observed greater importance of the rate of change, rather than the magnitude, of solar emission during the early phases of an SFD and then a shift of dominance to solar emission magnitude as the SFD progresses. The observations also demonstrate that solar emission at wavelengths less than 20Å does not cause the positive frequency deviation peaks in larger SFD events.

076.015 **The solar spectral irradiance 1200–2550Å at solar maximum.**
G. H. Mount, G. J. Rottman, J. G. Timothy.
J. Geophys. Res., Vol. 85, 4271 - 4274 (1980).
Full-disk solar spectral irradiances at solar maximum were obtained in the spectral range 1200–2550Å at a spectral resolution of approximately 1Å from rocket observations above White Sands, New Mexico, on June 5, 1979. Comparison with measurements made near solar minimum indicates approximately a factor of 2.5 increase in the irradiance at 1200Å, a 22% increase near 1800Å, and no increase within the measurement errors (±15%) above 2100Å. Irradiances in the range 1800–2100Å are in excellent agreement with previous measurements, but those in the 2100- to 2550-Å range

are significantly lower. The intensities of strong emission lines at wavelengths below 1850Å are also reported.

076.016 **Effect of asymmetry on a trap model for solar hard X-ray bursts.** D. B. Melrose, S. M. White.
Proc. Astron. Soc. Australia, Vol. 3, 369 - 371 (1979).

076.017 **An observed correlation between the flux densities of extended hard X-ray and microwave solar bursts.**
R. T. Stewart, G. J. Nelson.
Proc. Astron. Soc. Australia, Vol. 3, 390 - 392 (1979).

076.018 **Interpretation of a correlation between the flux densities of extended hard X-ray and microwave solar bursts.** G. J. Nelson, R. T. Stewart.
Proc. Astron. Soc. Australia, Vol. 3, 392 - 395 (1979).

076.019 **Morphology and spatial distribution of XUV and X-ray emissions in an active region observed from Skylab.** C.-C. Cheng, J. B. Smith, Jr., E. Tandberg-Hanssen.
Sol. Phys., Vol. 67, 259 - 265 (1980).

The authors studied the morphology and spatial distribution of loops in an active region. The active region loops fall basically into two distinctive groups: the hot loops with temperatures $2 - 3 \times 10^6$ K as observed in coronal lines and X-rays, and the relatively cool loops with temperature $5 \times 10^5 - 1 \times 10^6$ K as observed in transition-zone lines (Ne VII, Mg IX). The brightest hot coronal loops in the active region are mostly low-lying, compact, closely-packed, and show greater stability than the transition-zone loops, which are fewer in number, large, and slender. The observed aspect ratio of the hot coronal loops is in the range of 0.1 and 0.2, which are almost two orders of magnitude larger than those for the Ne VII loops.

076.020 **Solar flare X-ray spectra between 7.8 and 23.0 Angstroms.** D. L. McKenzie, P. B. Landecker, R. M. Broussard, H. R. Rugge, R. M. Young, U. Feldman, G. A. Doschek.
Astrophys. J., Vol. 241, 409 - 416 (1980).

The authors present high-resolution X-ray spectra taken during a large solar flare on 1979 June 10. Many lines of highly ionized iron are resolved and identified for the first time in solar spectra. Lines with a wide range of excitation temperatures are found to have similar time development during the flare's rapid rise phase. The authors discuss density sensitive line ratios in Fe XXI and Fe XXII.

076.021 **X-ray structures associated with disappearing Hα filaments in active regions.** S. W. Kahler.
Solar and interplanetary dynamics, (see 012.020), p. 61 - 65 (1980).

This study examines in detail the relationship between active region disappearing Hα filaments and the associated coronal X-ray structures observed both before the disappearance event and afterwards.

076.022 **UV emitting spicules.** G. Poletto.
Solar and interplanetary dynamics, (see 012.020), p. 199 - 201 (1980).

Extreme ultraviolet observations of the chromospheric network in a coronal hole obtained in 1973 by the Harvard College Observatory experiment aboard Skylab are analyzed. Upper and lower limits to the actual emission measure in UV spicules have been obtained, and the consistency of the derived values with the hypothesis that UV spicules are Hα spicules falling back after being heated is discussed.

076.023 **On the thermalisation of flare-time energetic electrons observed at radio and X-ray wavelengths.**
S. S. Degaonkar, H. S. Sawant, R. V. Bhonsle.
Solar and interplanetary dynamics, (see 012.020), p. 223 - 226 (1980).

076.024 **Electron concentration as an indicator of variations of the spectrum of solar X-rays.**
L. I. Dorman, B. T. Zhumabaev, I. D. Kozin, B. A. Turkeeva, B. M. Rubinshtejn.
Geomagn. Aehron., Tom 20, 945 - 946 (1980). In Russian.

076.025 **Analysis of X-ray spectra emitted by the 24 October 1970 flare.** B. Sylwester, J. Sylwester, J. Jakimiec, V. V. Korneev, S. L. Mandelstam (*Mandel'shtam*), I. A. Zhitnik, B. Valnicek.
Postępy Astron., Tom 28, 237 - 240 (1980). In Polish.

The analysis deals with high resolution X-ray spectra measured by means of the Bragg quartz-crystal spectrometer on board the Intercosmos 4 satellite. On the basis of the fluxes measured in the X-ray lines, models of the temperature distribution of the emission measure were calculated. Comparison of the models corresponding to the rise and maximum phase of the flare was performed. Simultaneous measurements of the hard X-ray flux confirm the results of the model calculations.

076.026 **Solar flux variability in the Schumann-Runge continuum as a function of solar cycle 21.**
M. R. Torr, D. G. Torr, H. E. Hinteregger.
J. Geophys. Res., Vol. 85, 6063 - 6068 (1980).

Measurements of the solar flux in the Schumann-Runge continuum (1350 - 1750 Å) by the Atmosphere Explorer satellites reveal a strong dependence on solar activity. Solar intensities over the rising phase of cycle 21, increase by more than a factor of two at the shorter wavelengths (1350 Å), with a smaller change (~10%) at 1750 Å. A significant 27 day variability is found to exist superimposed on the solar cycle variation.

076.027 **Initial results from a 3 to 25 Å solar X-ray spectrometer/spectroheliograph experiment.**
P. B. Landecker, D. L. McKenzie, H. R. Rugge.
Space Research, Vol. XX, (see 012.043), 255 - 258 (1980).

A solar X-ray spectrometer/spectroheliograph experiment, designated CRLS-229, was launched in the solar pointed section of the U.S. Air Force P78-1 satellite in February 1979. Initial results of coronal X-ray spectra and maps obtained with this instrument, with emphasis on solar flares, are presented.

076.028 **Solar flare X-ray spectra. III. Initial and final phase.** V. V. Korneev, V. V. Krutov, S. L. Mandelstam (*Mandel'shtam*), B. Sylwester, I. P. Tindo, A. M. Urnov, B. Valníček, I. A. Zhitnik.
Sol. Phys., Vol. 68, 381 - 389 (1980).

Spectra of 3 large flares on 24 Oct., 5 Nov. and 16 Nov. 1970 in the region $\lambda = 1.75 - 1.95$ Å, obtained with the 'Intercosmos-4' satellite during solar activity maximum are given. The physical conditions at the initial and final (decaying) phases are mainly studied. The line spectra are compared with hard continuum in the region 8–80 keV and results of polarization measurements, obtained simultaneously aboard the same satellite.

076.029 **On Doppler shifts of the Fe XXV ion resonance line in solar flare X-ray spectra.** V. V. Korneev, I. A. Zhitnik, S. L. Mandelstam (*Mandel'shtam*), A. M. Urnov.
Sol. Phys., Vol. 68, 391 - 392 (1980).

Doppler shifts of the Fe XXV line in three solar flares show prevalently downward motions with velocities up to 200 km s^{-1}.

076.030 **Limits on the streaming and escape of electrons in thermal models for solar hard X-ray emission.**

D. F. Smith, J. C. Brown.
Astrophys. J., Vol. 242, 799 - 805 (1980).

Upper limits on the number of fast electrons streaming through and escaping from a plasma whose electrons have been heated to $\sim 10^8$ K and confined by a collisionless ion-acoustic thermal conduction front are determined. It is shown that such a front is fairly transparent to fast electrons with velocities much larger than the thermal velocity because the anisotropic ion-acoustic waves cannot scatter them, making them collisionless on a scale much larger than the thickness of the front. The collisionless analog of the collisional thermoelectric field is derived self-consistently and shown to offer a significant impediment to fast electrons because they must climb over a larger potential barrier than in the collisional case. The rate of production is determined for the case of a Maxwellian whose tail is being filled collisionally. Requirements for the stability of these electrons in the hot source plasma and conduction front are given. Methods of refining these limits are discussed.

076.031 **Gibt es Cluster von hellen Punkten?**
G. Kittelberger, G. Elwert.
Mitt. Astron. Ges., Nr. 50, p. 96 - 97 (1980).

076.032 **Solar spectral radiance and irradiance at 225.2–319.6 nanometers.**
J. L. Kohl, W. H. Parkinson, C. A. Zapata.
Astrophys. J., Suppl. Ser., Vol. 44, 295 - 317 (1980).

Mean absolute intensities (spectral radiance) over 0.1 nm intervals between 225.2 nm and 319.6 nm at disk center and near the limb of the Sun ($\mu = 0.23 \pm 0.04$) are derived from the high spectral resolution measurements published by Kohl, Parkinson, and Kurucz. The corresponding limb-to-center ratios and spectral irradiance values are provided. A comparison with existing measurements of solar spectral radiance and spectral irradiance for the most part shows agreement within the estimated error limits, although some narrow band variations may be outside experimental errors. The contribution to the solar constant of the 230–305 nm band is derived to be 19.7 Wm^{-2}, ± 12%.

Multitemperature analysis of solar X-ray line emission. See Abstr. 031.546.

Quantitative Analyse von Skylab-Röntgenbildern der Sonne mit iterativer Entfaltung. See Abstr. 031.620.

A solar flare X-ray polarimeter for OSS-1. See Abstr. 032.510.

Observations from the Flat Crystal Spectrometer on the Solar Maximum Mission. See Abstr. 032.514.

The Ultraviolet Spectrometer and Polarimeter (UVSP) on the Solar Maximum Mission and initial results in polarimetry. See Abstr. 032.515.

Solar physics in the VUV: the importance of high resolution observations. See Abstr. 032.570.

Analysis of X-ray line spectra from a transient plasma under solar flare conditions. I. General outline. See Abstr. 062.002.

Temporal and spatial fluctuations in widths of solar EUV lines. See Abstr. 071.021.

Sunspot observations with the Ultraviolet Spectrometer and Polarimeter experiment on the Solar Maximum Mission. See Abstr. 072.012.

Size of the X-ray kernel of the large 20 August 1979 solar flare. See Abstr. 073.013.

XUV observations of a dense compact flare. See Abstr. 073.016.

Extra AR flare brightenings and type III reverse slope bursts. See Abstr. 073.019.

Electrostatic ion-cyclotron heat flux instability of a flaring plasma. See Abstr. 073.020.

X-ray spectroscopy during the decay phase of a solar flare. See Abstr. 073.024.

Origin of the soft X-ray emission from impulsive solar flares. See Abstr. 073.028.

High resolution solar flare X-ray spectra. See Abstr. 073.029.

SMM (*Solar Maximum Mission*) joint observations of solar flares. See Abstr. 073.033.

Low energy gamma ray continuum emission from solar flares of March 29, 1980. See Abstr. 073.034.

High energy X-ray spectral changes during the solar flare of March 29, 1980. See Abstr. 073.035.

Preliminary results from calcium and iron solar flare spectra from a Bent Crystal Spectrometer. See Abstr. 073.036.

Search for flare non-thermal electrons in iron and calcium BCS (*Bent Crystal Spectrometer*) spectra. See Abstr. 073.037.

Temporal comparison of Fe Kα and hard X-ray emission during several solar flares. See Abstr. 073.038.

Inner-shell ionization (Kα) lines in flare impulsive phases. See Abstr. 073.039.

Directivity of 50–100 keV X-ray emission from impulsive solar flares. See Abstr. 073.042.

Recent observations of energetic electrons in solar flares. See Abstr. 073.080.

Horizontal distribution of the X-ray energy deposit in the chromosphere and Hα two ribbon flares. See Abstr. 073.100.

High-resolution X-ray spectra of solar flares. IV. General spectral properties of M type flares. See Abstr. 073.105.

Size of the X-ray kernel of the large 1979 August 20 solar flare. See Abstr. 073.106.

Radiation signatures from a locally energized flaring loop. See Abstr. 073.109.

Energetic particles in solar flares. See Abstr. 073.130.

Impulsive phase of solar flares. See Abstr. 073.131.

The thermal X-ray flare plasma. See Abstr. 073.134.

XUV observations of a dense, compact flare.
See Abstr. 073.138.

Observations linking X-ray bright points with the source of the mass input to the solar wind.
See Abstr. 074.012.

X-ray evidence of coronal preflare emission.
See Abstr. 074.057.

The disruption of EUV coronal loops following a mass ejection transient. See Abstr. 074.059.

Nature of meter-decameter bursts associated with hard X-ray bursts. See Abstr. 077.014.

Correlated observations of a spatially resolved type III solar radio burst group and the associated hard X-ray emission. See Abstr. 077.049.

Microwave time delays as evidence for a collisionless conduction front. See Abstr. 077.056.

Physical conditions in the solar atmosphere above an active region. See Abstr. 080.020.

Non-Maxwellian velocity distribution functions associated with steep temperature gradients in the solar transition region. I. Estimate of the electron velocity distribution functions. See Abstr. 080.069.

Solar ultraviolet continuum radiation: the photosphere, the low chromosphere, and the temperature-minimum region. See Abstr. 080.086.

Una medicion de lineas de radiacion γ atmosferica a bajas latitudes. See Abstr. 082.061.

Statistical evaluation of gamma-ray line observations.
See Abstr. 142.515.

Cosmic γ ray line observations.
See Abstr. 142.527.

077 Radio, Infrared Radiation

077.001 **Synoptic charts of solar 9.1 cm and coronal hole data.** F. L. Wefer, M. D. Papagiannis.
Sol. Phys., Vol. 67, 13 - 21 (1980).

Synoptic charts for Carrington rotations 1601 - 1605 (May–August, 1973) were prepared using the central meridian column of the daily 9.1 cm Stanford solar radio maps. Synoptic charts of coronal holes from the ATM-Skylab were superimposed on the radio data to investigate the ability of the radio charts to show coronal holes. The conclusion reached is that in spite of certain problems due to active regions, side-lobe effects and a rather large beamwidth, the 9.1 cm synoptic charts can be of substantial value in identifying large coronal holes, especially during periods of low solar activity.

077.002 **Five minute microwave solar oscillations.** F. M. Strauss, P. Kaufmann, R. Opher.
Sol. Phys., Vol. 67, 83 - 87 (1980).

Oscillations with a period of 5.6 min. were observed on 10 July, 1978 while tracking at 22 GHz the active region McMath 15403. The oscillations were strong, clearly defined, had no damping, and lasted for about two hours. The rarity of the phenomenon is indicated by the fact that it occurred only once in more than 250 hr of solar observations. The possibility that these oscillations are due to a standing Alfvén wave driven by the photospheric velocity field is discussed.

077.003 **On the type of spectra of S-component sources and their correlation with flare occurrence.**
P. Steffen.
Sol. Phys., Vol. 67, 89 - 100 (1980).

From solar maps at 8.6 mm wavelength and total flux measurements at wavelengths of 1.7 cm to 122 cm, various spectra of the slowly varying component have been studied. The main distinction between these various types of spectra is the slope of the spectra toward wavelengths of less than 2 cm. It has been shown that the probability of flare occurrence is correlated with the type of the source spectra.

077.004 **On the distribution of magnitudes of solar microwave events.** A. D. Fokker.
Sol. Phys., Vol. 67, 101 - 108 (1980).

A microwave magnitude is defined as a logarithmic measure of the energy content of a microwave event. The author studies the distribution of the magnitudes of microwave events. He indicates how this distribution can be reproduced by a phenomenological model for the flare build-up process.

077.005 **Wide-band average spectra of solar radio bursts.** M. K. Das Gupta, T. K. Das, S. K. Sarkar.
Sol. Phys., Vol. 67, 109 - 120 (1980).

Peak flux spectra of solar radio bursts in a wide frequency band have been statistically determined for different morphological types of bursts, for various ranges of magnetic field of the burst-associated sunspots and also for the bursts occurring in the central and limb region of the solar disk. The interpretations of the obtained spectra give an insight into the possible generation mechanisms, pointing to the location of the source region in the solar atmosphere.

077.006 **The position and polarization of type III solar bursts.** G. A. Dulk, S. Suzuki.
Astron. Astrophys., Vol. 88, 203 - 217 (1980).
The authors study the position and polarization of type III bursts in the range 24–220 MHz, concentrating on bursts that continue to frequencies lower than 24 MHz – i.e. bursts occurring on field lines extending from active regions into interplanetary space. The 997 bursts studied fall into two classes: fundamental–harmonic (*F–H*) pairs and "structureless" bursts with no visible *F–H* structure.

077.007 **The position and polarization of type V solar bursts.** G. A. Dulk, S. Suzuki, D. E. Gary.
Astron. Astrophys., Vol. 88, 218 - 229 (1980).

In the present study the authors concentrate on the polarization of type V bursts as revealed by the spectropolarimeter, supplemented by spectrograph measurements of intensity and frequency range and by heliographic information on source positions, sizes and brightnesses. They describe the equipment and the procedure for selection of bursts and give the observational results: first, of some examples and case studies; second, of the statistics of burst polarization; third, of the relation between polarization and source position; and fourth, of the brightness temperatures and source sizes. The authors discuss the interpretation of their results in terms of the coronal magnetic field and the plasma physical processes involved in the emission. Finally, they summarize and conclude the study.

077.008 **Fine structure in radio emission spectra of local sources on the sun in the 5.0 - 12.0 GHz range and current sheets of active regions.** N. S. Kaverin, M. M. Kobrin, A. I. Korshunov, V. V. Shushunov.
Astron. Zh., Tom 57, 767 - 770 (1980). In Russian.
English translation in Soviet Astron., Vol. 24, No. 4.

Frequency radio emission spectra of three local sources in which a flux density jump in the narrow frequency interval is present are given. It is shown that this type of structure in the spectra may be due to the influence of a current sheet on the conditions of radio wave propagation in the solar corona.

077.009 **On the fine structure of microwave radio emission from solar activity centres.**
V. V. Zheleznyakov, E. Ya. Zlotnik.
Astron. Zh., Tom 57, 778 - 789 (1980). In Russian.
English translation in Soviet Astron., Vol. 24, No. 4.

Different types of frequency spectra of thermal cyclotron radio emission from active regions of the solar atmosphere which have a fine structure are discussed. The conditions in the activity centre that provide formation of frequency spectra containing a system of cyclotron lines and high-frequency cutoffs are indicated.

077.010 **Brightness distribution on the quiet sun in the region of 2 - 4 cm from observations with RATAN-600.**
V. N. Borovik.
Pis'ma Astron. Zh., Tom 6, 426 - 431 (1980). In Russian.
English translation in Soviet Astron. Lett., Vol. 6.

Solar observations made with RATAN-600 at five wavelengths in the region of 2 - 4 cm have shown that the brightness temperature is distributed evenly over the disk up to 0.95 $R_{\odot}$. There is a limb brightening $\lesssim$ 10% at 4.0 cm, which decreases at shorter wavelengths.

077.011 **Solar limb brightening at submillimeter wavelengths.** C. Lindsey, R. Hildebrand, T. de Graauw.
Bull. American Astron. Soc., Vol. 12, 474 (1980). – Abstract.

077.012 **Two-dimensional snapshot maps with the VLA of a solar burst at 20 cm wavelength.**
T. Velusamy, M. R. Kundu.
Bull. American Astron. Soc., Vol. 12, 478 (1980). – Abstract.

077.013 **Microwave images of solar bursts with arcsecond resolution.** G. J. Hurford, K. A. Marsh, H. Zirin.
Bull. American Astron. Soc., Vol. 12, 478 (1980). – Abstract.

077.014 **Nature of meter-decameter bursts associated with hard X-ray bursts.**
M. R. Kundu, T. E. Gergely, S. R. Kane.
Bull. American Astron. Soc., Vol. 12, 478 - 479 (1980). Abstract.

077.015 **Sharp-cutoff short-cm wavelength bursts from proton activity centers.**
E. W. Cliver, D. A. Guidice.
Bull. American Astron. Soc., Vol. 12, 480 (1980). – Abstract.

077.016 **Double ribbon events observed in He I 10830 Å associated with filament disappearances.**
K. Harvey, N. Sheeley, Jr., J. Harvey.
Bull. American Astron. Soc., Vol. 12, 503 (1980). – Abstract.

077.017 **Time variability and structure of quiet sun sources at 6 cm wavelength.** F. T. Erskine, M. R. Kundu.
Bull. American Astron. Soc., Vol. 12, 504 (1980). – Abstract.

077.018 **A new look at James' solar radar results.** D. G. Wentzel.
Bull. American Astron. Soc., Vol. 12, 505 (1980). – Abstract.

077.019 **A comparison of type III metric radio bursts and global solar potential field models.**
B. V. Jackson, R. H. Levine.
Bull. American Astron. Soc., Vol. 12, 515 (1980). – Abstract.

077.020 **Persistent quasi-periodic microwave pulsations from a non-flaring compact source in a complex active region.** V. Gaizauskas, K. F. Tapping.
Bull. American Astron. Soc., Vol. 12, 515 (1980). – Abstract.

077.021 **Thermal radio emission from solar active regions.** G. D. Holman, M. R. Kundu.
Bull. American Astron. Soc., Vol. 12, 517 (1980). – Abstract.

077.022 **Observation with the VLA of a stationary loop structure on the sun at 6 cm wavelength.**
M. R. Kundu, T. Velusamy.
Bull. American Astron. Soc., Vol. 12, 519 (1980). – Abstract.

077.023 **Kilometer wavelength type II solar radio bursts.** H. V. Cane, R. G. Stone, J. Fainberg, J. L. Steinberg.
Bull. American Astron. Soc., Vol. 12, 546 (1980). – Abstract.

077.024 **Some recent explorations of the solar corona from the Culgoora Solar Radio Observatory.**
K. V. Sheridan.
Proc. Astron. Soc. Australia, Vol. 3, 185 - 194 (1978).

The paper begins with a brief description of the instruments now in use at Culgoora for solar radio observations. Some of the more recent results obtained from Culgoora observations and their interpretation are then discussed, together with their relationship to other observations.

077.025 **Observation with the VLA of a stationary loop structure on the sun at 6 centimeter wavelength.**
M. R. Kundu, T. Velusamy.
Astrophys. J., Lett., Vol. 240, L63 - L67, plates L1 - L3 (1980).

A looplike structure connecting two sunspots of opposite polarity in an active region has been observed at 6 cm with a resolution of 3″.5, using the Very Large Array. This loop structure is reminiscent of the X-ray loops, as observed, for example, from Skylab. The brightness temperature in the "loop" is ~ 10^6 K and ~ 5×10^6 K near its foot points. Most of the bright peaks in the "loop" are well aligned with a long neutral line. Several compact, highly circularly polarized emission peaks were observed over emerging flux regions near one of the spots. Some of these sources appear to be associated with arch filament systems (AFS). The low brightness emission in the "loop" is attributed to optically thin thermal bremsstrahlung. The emission at the foot points of the "loop" and that associated with the emerging flux regions are believed to be due to gyroresonance process.

077.026 **A catalogue of fine structures in type IV solar radio bursts.** T. Bernold.
Astron. Astrophys., Suppl. Ser., Vol. 42, 43 - 58 (1980).

Fine structures often appear superimposed on the continuum of type IV bursts. This catalogue presents a selection of the most typical ones. A set of characteristics of the different structures has been defined, and based on these, the observed structures were classified.

077.027 **Two-dimensional VLA maps of solar bursts at 15 and 23 GHz with arcsec resolution.**
K. A. Marsh, G. J. Hurford.
Astrophys. J., Lett., Vol. 240, L111 - L114, L6 - L8 (1980).

The purpose of the present study was to map solar bursts in two dimensions with high spatial resolution (~ 1″) at frequencies sufficiently high that obscuring effects by the ambient active region would be unimportant.

077.028 **A possible association of solar type III bursts and white light transients.**
B. V. Jackson, K. V. Sheridan, G. A. Dulk. D. J. McLean.
Proc. Astron. Soc. Australia, Vol. 3, 241 - 242 (1978).

077.029 **Observations of high brightness temperatures in moving type IV solar radio bursts.**
R. T. Stewart, R. A. Duncan, S. Suzuki, G. J. Nelson.
Proc. Astron. Soc. Australia, Vol. 3, 247 - 249 (1978).

077.030 **Radio observations of a massive, slow-moving ejection of coronal material.**
K. V. Sheridan, B. V. Jackson, D. J. McLean, G. A. Dulk.
Proc. Astron. Soc. Australia, Vol. 3, 249 - 250 (1978).

077.031 **Heliograph observations of both fast drift (type III) and continuum (type IV) solar emission from the one source high above the solar limb.** R. A. Duncan.
Proc. Astron. Soc. Australia, Vol. 3, 253 - 256 (1978).

077.032 **Position and polarization of solar drift pair bursts.** S. Suzuki, D. E. Gary.
Proc. Astron. Soc. Australia, Vol. 3, 379 - 383 (1979).

077.033 **Evidence for a peak in the number of isolated type III bursts prior to large solar flares.**
B. V. Jackson, K. V. Sheridan.
Proc. Astron. Soc. Australia, Vol. 3, 383 - 386 (1979).

077.034 **A plasma-emission mechanism for type I solar radio emission.** D. B. Melrose.
Sol. Phys., Vol. 67, 357 - 375 (1980).

A theory for type I emission is developed based on fundamental plasma emission due to coalescence of Langmuir waves with low-frequency waves. The Langmuir waves are attributed to energetic electrons trapped in a magnetic loop over an active region. The continuum can be explained in terms of Langmuir waves generated by a 'gap' distribution formed through collisional losses over a timescale of several tens of minutes. Bursts are attributed to local enhancements in the Langmuir turbulence associated with a loss-cone instability. No triggering mechanism for the bursts is identified.

077.035 **Problems of single-frequency meter-wavelength solar observations.**
K. M. Borkowski, P. Zlobec, C. A. Zanelli.
Mem. Soc. Astron. Italiana, Vol. 51, 247 - 261 (1980).

Methods and problems of solar radio data recordings at

different observing stations are given. Improvements and standardization of data are suggested.

077.036 **Location of compact microwave sources with respect to concentrations of magnetic field in active solar regions.** V. Gaizauskas, K. F. Tapping.
Solar and interplanetary dynamics, (see 012.020), p. 33 - 36 (1980).

The combined radio-optical investigation reported here was initiated to explore, in regions at various stages of evolution, both the variability of compact microwave sources and their association with specific optical features.

077.037 **Solar radar observations.** A. O. Benz.
Solar and interplanetary dynamics, (see 012.020), p. 135 - 138 (1980).

077.038 **Estimation of shock thickness from dynamic spectra of type II bursts.** H. S. Sawant, S. S. Degaonkar, S. K. Alurkar, R. V. Bhonsle.
Solar and interplanetary dynamics, (see 012.020), p. 257 - 259 (1980).

Twenty type II solar radio bursts were observed during the period 1968 to 1972 by a solar radio spectroscope (240-40 MHz) at Ahmadebad. Intensity variations in type II bursts as a function of frequency and time are sometimes observed in their dynamic spectra. This fine structure enables determination of the shock thickness of the order of a few hundred to a few thousand kilometers.

077.039 **The association of type III bursts and coronal transient activity.**
B. V. Jackson, G. A. Dulk, K. V. Sheridan.
Solar and interplanetary dynamics, (see 012.020), p. 379 - 380 (1980).

077.040 **Recent very bright type IV solar metre-wave radio emissions.**
R. A. Duncan, R. T. Stewart, G. J. Nelson.
Solar and interplanetary dynamics, (see 012.020), p. 381 - 385 (1980).

The authors report Culgoora radioheliograph observations of four type IV radio sources, some moving, some stationary, but all with brightness temperatures above 10^9 K, and one with a brightness temperature above 10^{13}K. They also describe one previously reported event (that of 1977 September 20) in more detail.

077.041 **Polarized radio emission of an intense active region on the sun in July 1974 at wavelengths 1.9, 2.5 and 3.5 cm.**
A. F. Bachurin, A. S. Dvoryashin, N. N. Eryushev, L. I. Tsvetkov.
Izv. Krymskoj Astrofiz. Obs., Tom 61, 37 - 51 (1980). In Russian.

The circularly polarized radio emission of an intense active region on the sun (spot group N 96, McMath region 13043) was considered at the beginning of July 1974. The observations were performed with the 22 m radio telescope of the Crimean Astrophysical Observatory at wavelengths 1.9, 2.5 and 3.5 cm simultaneously. The results are presented.

077.042 **Observations of high brightness regions near the poles of the sun at wavelengths 3.5, 8 and 13.5 mm.**
V. A. Efanov, N. Labrum, I. G. Moiseev, N. S. Nesterov, R. Stewart.
Izv. Krymskoj Astrofiz. Obs., Tom 61, 52 - 54 (1980). In Russian.

The results of observations of solar polar regions at waves 3.5, 8 and 13.5 mm, carried out in October and December 1977, are presented.

077.043 **Observations of the quiet sun during the solar minimum (cycles 20 - 21) with the Toyokawa λ8-cm radioheliograph.** M. Ishiguro, S. Énomé, K. Shibasaki, H. Tanaka.
Publ. Astron. Soc. Japan, Vol. 32, 533 - 541 (1980).

Daily radio maps were statistically processed to obtain the monthly-averaged quiet sun. In spite of the presence of the seasonal variation in these quiet-sun maps, the differences between the E–W and N–S center-to-limb variations of the brightness temperature showed a relatively coherent feature. This center-to-limb effect of the quiet sun can be explained by a simple coronal model with arches and holes.

077.044 **A model of radiation for the solar radio SVC (*slowly varying component*).** R.-y. Zhao, S.-j. Qian.
Acta Astron. Sinica, Vol. 21, 262 - 271 (1980). In Chinese.

The authors have studied the radiation mechanism of the SVC for a unipolar sunspot – the combined radiation mechanism consisting of a bremsstrahlung and cyclotron resonance radiation mechanism. The calculated results show that the maxima both of the flux density spectrum and the polarization degree spectrum are respectively at $\lambda = 6$ cm and $\lambda = 3$ cm. Those are essentially consistent with observational facts. Some interesting results regarding the distribution characteristics, which correspond to a resolving power as 0.″3–4.″8, of the SVC radiation over the sunspot are given. It is indicated that the cyclotron resonance radiation is overwhelmingly dominant in the SVC radiation.

077.045 **Nonlinear scattering mechanism of the spectrum of type IV solar decimetric radio bursts.**
J.-x. Yao.
Acta Astron. Sinica, Vol. 21, 272 - 277 (1980). In Chinese.

A mechanism of coherent plasma radiation of a type IV_{dm} solar radio burst and a method of estimating its spectrum theoretically are suggested.

077.046 **Synchronous variation of the fluxes of distant type I sources and connection between the different activity centers.** X.-z. Liu.
Acta Astron. Sinica, Vol. 21, 317 - 319 (1980). In Chinese.

Two examples which show the close relation between the fluxes of distant type I sources at meter waves are discussed. The coronal structure and the connection between the different activity centers are investigated from these examples.

077.047 **Radio emission of solar active regions in the centimetric wavelength range.** M. Siarkowski.
Postępy Astron., Tom 28, 233 - 235 (1980). In Polish.

Examples are presented of the brightness temperature and polarization variations of the radio emission with the distance from the spot axis. The calculations are performed using the number of different models describing the height variation of the electron density, temperature and magnetic field.

077.048 **Long-lived microwave pulsations observed in a complex solar active region.**
V. Gaizauskas, K. F. Tapping.
Astrophys. J., Vol. 241, 804 - 810, plate 16 (1980).

Microwave pulsations were detected on 1977 September 13 in the intense emission from a compact microwave source associated with the large, slowly rotating, and magnetically complex solar active region, McMath 14943. These pulsations persisted over 5 1/2 hours, with the dominant repetition rate remaining close to 0.4 Hz; they were not associated with flare activity. The core of the microwave emission was located over a plage rather than over the major spot in the region. A mechanism for the pulsating source is proposed in which radial oscillations in an arched magnetic flux tube modulate the gyrosynchrotron emission from high energy electrons trapped in the tube.

077.049 **Correlated observations of a spatially resolved type III solar radio burst group and the associated hard X-ray emission.** S. R. Kane, M. Pick, A. Raoult. Astrophys. J., Lett., Vol. 241, L113 - L116 (1980).

The first measurements of the spatial structure of a group of type III solar radio bursts associated with an impulsive hard X-ray burst are presented. At 169 MHz the radio source has been found to consist of two principal regions separated by $\sim 3 \times 10^5$ km. The two regions together produced a total of four component bursts in good time correlation with spikes in the hard X-ray emission. The observations indicate that electron acceleration/injection occurs over a region which covers a wide range of magnetic field lines.

077.050 **On the factors influencing the circular polarization sign of radio bursts at $\lambda = 3.2$ cm.**
G. N. Zubkova, V. P. Nefed'ev.
Soln. Dannye 1980 Byull., No. 4, p. 95 - 98 (1980). In Russian.

The solar disk burst distribution is investigated with account for the sign of polarization. The observed asymmetry of the burst distribution can be explained with the change of flare positions relative to sunspot magnetic fields.

077.051 **Polarized solar type III bursts between 2.3 and 4.9 MHz.**
J. Hanasz, R. Schreiber, V. I. Aksenov.
Astron. Astrophys., Vol. 91, 311 - 318 (1980).

The authors have observed from spacecraft 3 highly polarized type III bursts at frequency range from 2.3 to 4.9 MHz, using the ionosphere as the polaroid. From high polarization degree they infer on a considerable contribution of emission at the fundamental of plasma frequency in the bursts observed. Applying the theoretical formula for completely polarized fundamental emission the authors calculated that the lower limit of the magnetic field strength is 1.2×10^{-3} Gauss for the burst generation region at 4 MHz coronal level (heliocentric distance of about 8.5 solar radii).

077.052 **Radioverschijnselen als boodschappers uit de zonnecorona (I). Het decor der radioverschijnselen.**
A. D. Fokker, J. M. E. Kuijpers.
Zenit, 7e Jaarg., 334 - 338 (1980).

077.053 **Radioverschijnselen als boodschappers uit de zonnecorona (2). Kennismaking met enkele radioverschijnselen.** A. D. Fokker, J. M. E. Kuypers.
Zenit, 7e Jaarg., 436 - 440 (1980).

077.054 **Zonnewaarnemingen te Westerbork.**
A. Kattenberg.
Zenit, 7e Jaarg., 470 (1980).

077.055 **Comment on "Variability of the far-infrared solar temperature minimum with the solar cycle".**
J. W. Cook, G. E. Brueckner, M. E. VanHoosier.
Astron. Astrophys., Vol. 92, L7 - L8 (1980).

The authors compare the solar cycle variation of the full disk brightness temperature minimum in the far-infrared reported by Müller et al. (1980) (see Abstract 27.077.084) to the variation found in the far-ultraviolet T_{min} continuum. The far-ultraviolet observations suggest that the far-infrared variability should be nearer 40-60 K than the value of 200 K reported by Müller et al., whose solar cycle variation is comparable to their measurement error.

077.056 **Microwave time delays as evidence for a collisionless conduction front.**
H. J. Wiehl, W. A. Schöchlin, A. Magun.
Astron. Astrophys., Vol. 92, 260 - 266 (1980).

Nine simple impulsive microwave bursts were investigated for time delays in their times of maximum emission at high and low microwave frequencies. In seven of them the high frequency time profile preceded the low frequency time profile by 8–49 s. Using the fact that the microwave radiation is severely attenuated by free-free absorption and that the electron density decreases towards the observer the heights of the emitting volumes for each frequency were calculated. From the differences of the emission heights the time delays could be derived by assuming a mean ion-sound speed of 900 km s^{-1}.

077.057 **Dynamics of the power spectrum of quasi-periodic fluctuations of solar radio emission at 3.2 cm during chromospheric flares.** A. A. Bezotosnyj.
Ionos. i soln.-zemn. svyazi. Alma-Ata, 1980, p. 132 - 138. In Russian. – Abstr. in Ref. zh., 51. Astron., 11.51.460 (1980).

077.058 **Development of local radio sources and burst activity on the sun at short centimeter wavelengths.**
A. F. Bachurin, A. S. Dvoryashin, N. N. Eryushev, L. I. Tsvetkov.
Izv. Krymskoj Astrofiz. Obs., Tom 62, 166 - 175 (1980). In Russian.

The radio emission of proton and non-proton regions on the sun in the wavelength range 1.9 - 3.5 cm and burst activity of the proton sunspot group (McMath region 13043) in July 1974 at wavelength 1.9 cm are considered.

Plasma astrophysics. Nonthermal processes in diffuse magnetized plasmas. Vol. 1: The emission, absorption and transfer of waves in plasmas. Vol. 2: Astrophysical applications. See Abstr. 003.076.

Problems of single-frequency solar observations.
See Abstr. 031.558.

Estimate of and allowance for the influence of atmospheric conditions on the accuracy of measurement of displacement of the effective center of solar radio radiation.
See Abstr. 031.626.

Electric field experiment (E 5b).
See Abstr. 032.558.

Radio astronomy experiment (E 5c).
See Abstr. 032.559.

Nobeyama radiospectrographs for solar observations.
See Abstr. 033.001.

Absolute brightness temperature measurements at 3.5-mm wavelength. See Abstr. 033.019

A new 17-GHz solar radio interferometer at Nobeyama. See Abstr. 033.022.

Polarimeter designed for the investigation of elliptical polarization of solar radio emission.
See Abstr. 034.072.

A summary of progress in space physics made with Helios plasma wave instrument data (E 5a).
See Abstr. 051.038.

Electron trapping in the solar magnetic field and emission of decimetric continuum radio bursts.
See Abstr. 062.047.

A model for impulsive electron acceleration to energies of tens of kT_e. See Abstr. 062.070.

Impulsive electron acceleration to energies of tens of kT_e by Langmuir wave turbulence.
See Abstr. 062.080.

Solare Beobachtungsergebnisse. Solar data. Solar radio emission. 1980 February - June.
See Abstr. 072.038.

The behaviour of active region McMath 14822 at 13.5 mm wavelength. See Abstr. 072.062.

Combined radio-optical observations of active solar regions associated with the S-component of solar microwave emission. See Abstr. 072.064.

Measurements of the magnetic field and the gradient of temperature in the solar atmosphere above a flocculus using radio observations. See Abstr. 073.003.

Multifrequency observations of a solar flare at up to 0.2″ spatial resolution. See Abstr. 073.014.

Helium I 10830 Å filament – associated structures. See Abstr. 073.021.

Observations of solar filaments at 8, 15, 22, and 43 GHz. See Abstr. 073.025.

Radio evidence on the particle distribution functions in the corona following flares. See Abstr. 073.055.

Evidence on chromospheric structure from observations of solar brightness distribution at millimetre wavelengths. See Abstr. 073.057.

VLA observations of impulsive solar flares at 4.9 GHz. See Abstr. 073.108.

Impulsive phase of solar flares. See Abstr. 073.131.

Comparison of radioheliograph, coronagraph and K-coronameter observations of a coronal streamer. See Abstr. 074.034.

Radio observations of coronal holes. See Abstr. 074.044.

Radio data and computer simulations for shock waves generated by solar flares. See Abstr. 074.061.

Transient disturbances of the outer corona. See Abstr. 074.065.

Search for optical modulation of the solar corona during the February 16, 1980 total solar eclipse. See Abstr. 074.099.

Thermal cyclotron radio emission of neutral current sheets in the solar corona. See Abstr. 074.102.

Estimation of the coronal magnetic field from the Razin effect in a solar decametric continuum burst. See Abstr. 074.103.

An observed correlation between the flux densities of extended hard X-ray and microwave solar bursts. See Abstr. 076.017.

Interpretation of a correlation between the flux densities of extended hard X-ray and microwave solar bursts. See Abstr. 076.018.

On the thermalisation of flare-time energetic electrons observed at radio and X-ray wavelengths. See Abstr. 076.023.

Investigations of the sporadic radio radiation of the sun and of the parameters of the earth's ionosphere aboard Intercosmos-Copernicus 500. 5. Results of investigations of the electron concentration in the ionosphere. See Abstr. 083.022.

Interplanetary radio manifestations of large flare events. See Abstr. 106.007.

078 Cosmic Radiation

078.001 **Is there a limit on solar flare proton fluxes?**
D. D. Barbosa.
Sol. Phys., Vol. 67, 181 - 188 (1980).
The author explores the possibility that solar flare proton fluxes are limited in magnitude by saturation effects inherent to the acceleration mechanism. If cyclotron damping of Alfvén waves acts to accelerate protons, the criterion that the damping time is comparable to the acceleration time provides a fast particle number density at which protons load the wave spectrum. The limiting flux at 1 AU is obtained by a volume integration over the acceleration region and redistribution into an interplanetary emission cone.

078.002 **Radiation measurements aboard Cosmos 900. 3. Spectrometry of protons and α-particles with energies > 1 MeV/nucleon.**
T. A. Ivanova, V. D. Maslov, M. I. Panasyuk, Eh. N. Sosnovets.
Kosm. Issled., Tom 18, 567 - 571 (1980). In Russian.

078.003 **Connection of characteristics of the interplanetary medium with the intensity variations of charged particles from the flare of 1973, September 7.**
M. S. Kazaryan, B. M. Kuzhevskij, V. L. Maduev, Yu. V. Mineev, E. S. Spir'kova, I. P. Shestopalov.
Kosm. Issled., Tom 18, 572 - 579 (1980). In Russian.

078.004 **Spectral characteristics of solar protons in the initial anisotropic stage of their propagation.**
N. K. Pereyaslova, N. A. Mikirova.
Geomagn. Aehron., Tom 20, 583 - 587 (1980). In Russian.

078.005 **Solar flare increase of cosmic ray intensity on November 22, 1977.**
A. G. Fenton, K. B. Fenton, J. E. Humble.
Proc. Astron. Soc. Australia, Vol. 3, 238 -241 (1978).

078.006 **On the nature of electron increase with energies of 0.3 - 3 MeV from measurements aboard the Prognoz 4 satellite.** Yu. V. Mineev, E. S. Spir'kova.
Kosm. Issled., Tom 18, 805 - 807 (1980). In Russian.

078.007 **Determination of the kinematic characteristics of solar corpuscular streams from outbursts of comets.**
M. Z. Markovich, R. S. Osherov.
Problems of cosmic physics. Vyp. 15, (see 003.003), p. 98 - 103 (1980). In Russian.
The method proposed for determination of a corpuscular stream's kinematic characteristics is based on the assumption that brightness outbursts occur at the comet's passage through the region of its orbit crossing the trajectory of solar particles. The kinematic characteristics of the streams are calculated which caused the outbursts of comets 1951 VII and 1956 II.

078.008 **Solar cosmic rays and isotopic composition of the solar wind.** G. A. Koval'tsov.
XIth seminar on cosmophysics, (see 012.035), p. 237 - 256 (1979). In Russian. – Abstr. in Ref. zh., 51. Astron., 8.51.398 (1980).

078.009 **Gradient drift of solar cosmic radiation in dipole-type magnetic traps.** Yu. I. Okulov.
Fiz. protsessy v ionos. i magnitosfere, Moskva, 1979, p. 29 - 34. In Russian. – Abstr. in Ref. zh., 62. Issled. kosm. prostranstva, 9.62.172 (1980).

078.010 **Coherent effect in the preferential acceleration of relativistic solar heavy cosmic ray nuclei.**
S. Yousef.
J. Astron. Soc. Egypt, Vol. 1, 67 - 74 (1979).
A coherent mechanism is developed for the preferential acceleration of heavy solar cosmic ray nuclei. In this mechanism, jets of relativistic electrons serve as vehicles for the acceleration of trapped positive ions over very short distances.

078.011 **The interplanetary transport of solar cosmic rays.**
T. I. Gombosi, A. J. Owens.
Astrophys. J., Lett., Vol. 241, L129 - L132 (1980).
Numerical solutions are presented for the propagation of solar cosmic rays in interplanetary space, including the effects of pitch-angle scattering adiabatic focusing. The intensity-time profiles can be well fitted by a simple radial spatial diffusion equation with scattering mean-free path λ_{fit}. The radial mean-free path so obtained is significantly larger than the true scattering mean-free path for low-rigidity particles due to both adiabatic focusing and the inapplicability of the diffusive approximation early in the event. The well-known discrepancy between λ_{fit} and the theoretical predictions may be resolved by these calculations.

078.012 **Low-energy proton spectra in recurrent increases from Prognoz 4 measurements.**
M. A. Zel'dovich, Yu. I. Logachev, S. P. Ryumin, V. G. Stolpovskij.
Kosm. Issled., Tom 18, 943 - 946 (1980). In Russian.

078.013 **The spatial anisotropy, rigidity spectrum, and propagation characteristics of the relativistic solar particles during the event on May 7, 1978.**
H. Debrunner, J. A. Lockwood.
J. Geophys. Res., Vol. 85, 6853 - 6860 (1980).
The data from the worldwide network of neutron monitors have been used to deduce the spatial distribution, rigidity spectrum, and propagation characteristics of the GLE on May 7, 1978. The large magnitude of the GLE and hard rigidity spectrum enabled an analysis to be made utilizing both high-latitude and mid-latitude stations.

Cosmic ray experiment (E 6). See Abstr. 032.560.

Cosmic ray experiment (E 7). See Abstr. 032.561.

A survey of ~1 MeV nucleon^{-1} solar flare particle abundances, $1 \leq Z \leq 26$, during the 1973–1977 solar minimum period. See Abstr. 073.043.

Energetic particles in solar flares. See Abstr. 073.130.

Magnetohydrodynamic shock propagation in the vicinity of a magnetic neutral sheet. See Abstr. 074.094.

If you've seen one magnetosphere, you haven't seen them all: energetic particle observations in the Saturn magnetosphere. See Abstr. 100.103.

Brightness variations of comets and solar corpuscular activity. See Abstr. 102.011.

Cometary brightness outbursts and solar corpuscular activity. See Abstr. 102.045.

Grain disruption by collisions with solar energetic particles. See Abstr. 106.048.

Energetic particles in space. See Abstr. 106.053.

On a connection of the configuration of interplanetary shock waves from powerful isolated flares with proton events. See Abstr. 106.085.

Cosmic ray effects in solar system objects. See Abstr. 143.039.

The interactions of energetic particles with the solar wind. See Abstr. 143.045.

079 Solar Eclipses

079.001 **Total eclipses of the sun.** J. B. Zirker.
Science, Vol. 210, 1313 - 1319 (1980).
Some of the advances in solar physics resulting from eclipse observations are discussed. Experiments at the total eclipse of 16 February 1980 in India are also described. These included a test of general relativity, studies in coronal physics, investigations of solar prominences, diameter measurements, a search for interplanetary dust, a study of the gravity waves in the earth's atmosphere, and experiments on the biological effects on animals and humans.

079.002 **On the calculation of the second derivative of lunar elongation.** A. T. Fomenko.
Probl. mekh. upravlyaem. dvizheniya. Perm', 1980, p. 161 - 166. In Russian. – Abstr. in Ref. zh., 51. Astron., 1.51.86 (1981).

Solar eclipse predictions. See Abstr. 014.011.

Solar eclipse 1980 February 16

079.101 **Die Sonnenfinsternis vom 16. Februar 1980.** Beobachtungen der ETH-Sternwarte in Indien.
J. Dürst.
Orion, 38. Jahrg., 118 - 121 (1980).

079.102 **Die indische Sonnenfinsternis.**
M. Waldmeier.
Sterne Weltraum, Jahrg. 19, 290 - 294 (1980).

079.103 **De zonsverduistering van 16 februari 1980.**
C. Deetman, A. J. Deetman.
Zenit, 7e Jaarg., 308 - 311 (1980).

079.104 **De zon tussen de wolken.** H. Nieuwenhuis.
Zenit, 7e Jaarg., 312 - 314 (1980).

079.105 **Die totale Sonnenfinsternis vom 16. Februar 1980.**
H. Bernhard.
Sterne, 56. Band, 315 - 317 (1980).

079.106 **L'éclipse totale de Soleil du 16 février 1980.**
Ciel Terre, Vol. 96, 394 (1980).

079.107 **Observation of Baily's beads during the total solar eclipse of 1980 February 16.**
A. D. Fiala, D. W. Dunham, J. B. Dunham.
Bull. Astron. Soc. India, Vol. 8, 81 - 82 (1980).
Precise timings of Baily's beads activity were made from stations in India near the limits of the path of totality. The technique and planned use are outlined.

079.108 **Observations of the total solar eclipse of February 16, 1980.**
V. P. Gaur, K. R. Bondal, K. Sinha, G. C. Joshi, M. C. Pande.
Bull. Astron. Soc. India, Vol. 8, 83 - 85 (1980).
The instrumental set-up and preliminary results of the experiments performed during the eclipse are briefly reported.

079.109 **Intensity variation of photons during the solar eclipse.** S. V. Chandole, S. S. Shah.
Bull. Astron. Soc. India, Vol. 8, 91 - 92 (1980).
A decrease in the intensity of photons is found to be of the order of two.

079.110 **Preliminary report of solar eclipse observations.**
K. R. Sivaraman.
Bull. Astron. Soc. India, Vol. 8, 93 (1980).

079.111 **Solförmörkelseresa till Kenya februari 1980 som också blev en resa i tiden.**
P.-Å. Björklund.
Astron. Tidsskr., Årg. 13, 156 - 165 (1980).

079.112 **Observation of the total solar eclipse of 16 February 1980.** V. Rušin, Š. Knoška.
Říše hvězd, Vol. 61, 157 - 161, 165 - 168 (1980). In Slovak.

079.113 **To India for the solar eclipse of 16 February 1980.**
J. Sýkora.
Kozmos, Vol. 11, 107 - 116 (1980). In Slovak.

Summary of the Czechoslovak solar corona observations. See Abstr. 011.034.

Polarization study of the solar corona using a double polarigraph. See Abstr. 034.048.

A succesful observation of the flash spectrum at the total solar eclipse of February 16, 1980. See Abstr. 073.071.

The Kitt Peak coronal velocity experiment. See Abstr. 074.096.

Coronal emission line measurements of the February 16, 1980 total solar eclipse. See Abstr. 074.097.

Interferometric eclipse observations of coronal Fe XIV emission. See Abstr. 074.098.

Search for optical modulation of the solar corona during the February 16, 1980 total solar eclipse. See Abstr. 074.099.

Coronal interferogram in 5303 Å obtained during the total solar eclipse of February 16, 1980 and coronal temperatures. See Abstr. 074.100.

Total eclipses of the sun. See Abstr. 079.001.

Preliminary results from observations made near the edges of the path of the 1980 February 16 total solar eclipse. See Abstr. 080.012.

Satellite radio beacon study of the ionospheric variations at Hyderabad during the total solar eclipse of February 16, 1980. See Abstr. 083.046.

Solar eclipse 1973 June 30

A density model for the north polar coronal hole at the 1973 eclipse. See Abstr. 074.004.

Solar eclipse 1979 February 26

079.301 **Submillimeter airborne observation of the total solar eclipse of February 26th, 1979.**
T. A. Clark, R. T. Boreiko.
Bull. American Astron. Soc., Vol. 12, 750 (1980). – Abstract.

Solar eclipse 1980 August 10

079.401 **Eclipse anular de sol del 10 de Agosto de 1980.**
W. Cepeda.
Publ. Obs. Astron. Nac., Univ. Nac. Colombia, Bogotá, No. 6, 14 pp. (1980).

079.402 **Der Sonnenring von Calamarca.** F. Dorst.
Sterne Weltraum, 19. Jahrg., 420 - 421 (1980).

079.403 **Ringförmige Sonnenfinsternis in Südamerika.**
H. Salm.
Orion, 38. Jahrg., 189 - 190 (1980).

Solar eclipse 1981 July 31

079.501 **Total solar eclipse of 31 July 1981.**
A. D. Fiala, M. R. Lukac.
United States Naval Obs. Circ., No. 160, 55 pp. (1980).

Solar eclipse of July 31, 1981 and its observation. See Abstr. 003.077.

The total solar eclipse of July 31, 1981. See Abstr. 003.091.

080 Atmosphere, Figure, Internal Constitution, Neutrinos, Rotation, etc.

080.001 **Solar size variation.** D. W. Hughes.
Nature, Vol. 286, 439 - 440 (1980).

080.002 **Solar luminosity variations.** R. L. Gilliland.
Nature, Vol. 286, 838 - 839 (1980).

080.003 **The pulses as a diagnostic technique in the sun.**
G. C. Das.
Astrophys. Space Sci., Vol. 71, 353 - 361 (1980).

The author discusses a method of finding physical parameters by studying the pulses in the sun. For the sake of a mathematical approach, he considers an ideal, highly relevant model which could exist in the sun with the effects of ionization, due to which there will be a continuous formation of ionized particles. It is observed that the pulse originated at the centre of a dipole field propagates along the magnetic field. The author derives a dispersion relation for these types of pulses, propagating from the centre to the solar surface. The time taken by the pulse from its source to the solar surface is also estimated, with due account of the ionization effects on the pulse. Temporal and spatial damping of the pulses lead to estimates of the velocity distribution of the ionized particles and of the amplitude of the magnetic field of the wave in pulse.

080.004 **Model of the layer-cellular motions in the convective zone of the sun. I.** V. P. Savchenko.
Astron. Zh., Tom 57, 771 - 777 (1980). In Russian.
English translation in Soviet Astron., Vol. 24, No. 4.

080.005 **Detection of 160-min solar oscillations and atmospheric extinction.**
A. B. Severny (*Severnyj*), V. A. Kotov, T. T. Tsap.
Astron. Astrophys., Vol. 88, 317 - 319 (1980).

The influence of the Earth's atmosphere on the search for global solar oscillations with a period of 160^m is studied. The authors show that neither the amplitude nor the phase behaviour of the oscillations persisting during the last 5 years (1974 - 1978) can be explained in terms of the terrestrial atmospheric influences or by the statistical method of treatment. The observational evidence strongly suggests that the 160^m periodicity is of solar origin.

080.006 **The sun among the stars. II. Solar color, Hyades metal content, and distance.** J. Hardorp.
Astron. Astrophys., Vol. 88, 334 - 344 (1980).

The solar color, Hyades distance and metal content are not independent of each other. This is shown here by using results of model interior calculations and of model stellar atmospheres, in a differential way only, to interpret new scanner observations of 9 Hyades dwarfs. This interpretation removes the long-standing discrepancy between metal abundances determined spectroscopically, and from the position of the main sequence. For a distance of 45 pc, masses of visual and spectroscopic binaries are no longer found to be too small. Thus, for the first time, the theory of stellar evolution correctly predicts the Hyades' main sequence, as well as the mass luminosity relation, with solar values of Y and Z.

080.007 **Effect of the equation of time in the 160 min solar oscillations.**
V. A. Kotov, S. Koutchmy, A. B. Severnyj, T. T. Tsap.
Pis'ma Astron. Zh., Tom 6, 421 - 425 (1980). In Russian.
English translation in Soviet Astron. Lett., Vol. 6.

It is suggested that if the solar pulsation with the $160^m\!.010$ period, which is close to 1/9 of a day, where of the earth atmosphere in origin, the power spectrum and the exact value of the period inferred from the Crimean and Stanford measurements would depend on the equation of time. An analysis of the Crimean velocity measurements for 1974 - 1979 do not show such dependence thus reinforcing a solar interpretation of the $160^m\!.010$ period.

080.008 **The sun is observed to be a torsional oscillator with a period of 11 years.** R. Howard, B. J. LaBonte.
Astrophys. J., Lett., Vol. 239, L33 - L36 (1980).

Twelve years of full-disk Mount Wilson velocity data have been analyzed to study horizontal east-west motions. The authors find a torsional wave pattern with alternating latitude zones of slow and fast rotation, after subtracting a differentially rotating frame. Amplitudes of the flow pattern average about 3 m s^{-1}. It requires about 22 years for zones to drift from the poles, where they originate, to the equator, where they disappear. The pattern is symmetric about the equator. This pattern evidently represents a deep-seated circulation pattern and is the first evidence of the association of mass motions with large-scale characteristics of the solar activity cycle.

080.009 **Torsional oscillations of the sun and magnetic flux eruption.** B. J. LaBonte, R. Howard.
Bull. American Astron. Soc., Vol. 12, 473 (1980). – Abstract.

080.010 **Further investigation of the solar torsional oscillator with a period of 11 years.**
P. H. Scherrer, J. M. Wilcox.
Bull. American Astron. Soc., Vol. 12, 473 (1980). – Abstract.

080.011 **Large scale solar velocity features.**
J. R. Kuhn, S. P. Worden.
Bull. American Astron. Soc., Vol. 12, 473 (1980). – Abstract.

080.012 **Preliminary results from observations made near the edges of the path of the 1980 February 16 total solar eclipse.** A. D. Fiala, D. W. Dunham, J. B. Dunham.
Bull. American Astron. Soc., Vol. 12, 474 (1980). – Abstract.

080.013 **Radial and low-order non-radial p-mode oscillations of the sun.** E. J. Rhodes, Jr., R. K. Ulrich.
Bull. American Astron. Soc., Vol. 12, 475 (1980). – Abstract.

080.014 **Transmission line approach to global electrodynamic coupling of the solar atmosphere.**
J. A. Ionson, P. J. Morrison, D. S. Spicer.
Bull. American Astron. Soc., Vol. 12, 516 (1980). – Abstract.

080.015 **Transient plasmas in the solar transition zone.**
K. P. Dere, K. R. Nicolas, G. E. Brueckner.
Bull. American Astron. Soc., Vol. 12, 518 (1980). – Abstract.

080.016 **The dynamics of brightened interconnecting loops.**
R. Howard, Z. Švestka.
Bull. American Astron. Soc., Vol. 12, 519 (1980). – Abstract.

080.017 **Coalescence of iron in the centre of the sun.**
C. Deutsch, M. M. Gombert, H. Minoo.
J. Phys. Colloq., Vol. 41, No. C-2, (see 012.005), p. C2/65 - 68 (1980). – Abstr. in Phys. Abstr., Vol. 83, Abstr. 77373 (1980).

080.018 **A search for large-scale convection cells in the solar atmosphere.** R. Howard, B. J. LaBonte.
Astrophys. J., Vol. 239, 738 - 745 (1980).

Mount Wilson magnetograph velocity observations are used to search for east-west motions resulting from hypothetical cellular patterns extending over one or two hemispheres in the latitude direction. No such solar patterns were found. Upper limits established by this analysis depend on the cell lifetime and the pattern stability, but in all cases they are no more than about 10 m s^{-1}.

080.019 **Doppler observations of solar rotation.**
P. H. Scherrer, J. M. Wilcox.
Astrophys. J., Lett., Vol. 239, L89 - L90 (1980).
Daily observations of the photospheric equatorial rotation rate using the Doppler effect are made at the Stanford Solar Observatory. The observations show no variations in the rotation rate that exceed the observational error of about 1%. The average rotation rate is indistinguishable from that of sunspots and large-scale magnetic field structures.

080.020 **Physical conditions in the solar atmosphere above an active region.**
J. T. Mariska, U. Feldman, G. A. Doschek.
Astrophys. J., Vol. 240, 300 - 305 (1980).
The authors analyze EUV emission line observations obtained above an active region at the solar limb. These observations and recent atomic physics calculations are used to determine the volume emission measure as a function of temperature at different heights over the active region and the electron density at coronal temperatures at two locations over the active region. The results are compared with the predictions of recent models of the structure of active region loops. Because there are emission lines from more than one ion in each temperature interval of interest, the analysis is also a test of recent atomic physics calculations.

080.021 **Energy balance from the chromosphere-corona transition region.** M. Alvarez.
Astrophys. J., Vol. 240, 322 - 326 (1980).
A detailed energy balance of the chromosphere-corona transition region and lower corona is computed starting with the multicomponent temperature model developed to study the EUV emission lines. It becomes clear that in order to keep the energy balance, a "mechanical" energy source and sink must be included in the computations. The shape and detailed study of this "mechanical" energy source and sink allow to search for the physically plausible mechanism responsible for this behavior. Heat conduction is a very efficient mechanism to heat the dense regions of the chromosphere, and a "sudden" liberation of energy has to be invoked in order to maintain equilibrium.

080.022 **Hot downflows above supergranular boundaries.**
G. Poletto.
Astrophys. J., Lett., Vol. 240, L69 - L71 (1980).
In recent years, several UV observations have detected, above supergranular boundaries, hot downflows, which have been interpreted as cool spicular matter falling back after being heated to coronal temperatures. An examination of possible mechanisms leads the author to discard the process of conductive or radiative heating from the hot corona. An alternative process which involves the compressive heating of the cool spicular material is proposed and shown to be able to reproduce the required temperature rise over the correct time scale.

080.023 **The phenomenon of turbulent diffusion and the solar neutrino problem.**
E. Schatzman, A. Maeder.
C. R. Acad. Sci. Paris, Tome 291, Sér. B, 81 - 83 (1980). In French.

080.024 **Magnetic activity and variations in solar luminosity.**
E. A. Spiegel, N. O. Weiss.
Nature, Vol. 287, 616 - 617 (1980).
The authors indicate how strong magnetic fields at the base of the convective zone of the sun can alter the local convection. The resulting changes in thermal energy are large enough to produce variations of order 0.1% in the solar luminosity over the 11-yr sunspot cycle.

080.025 **On the dynamics of the solar convection zone.**
B. R. Durney, H. C. Spruit.
Stellar turbulence, (see 012.008), p. 15 - 16 (1980).
Abstract.

080.026 **Examples of non-thermal motions as seen on the sun.**
J. M. Beckers.
Stellar turbulence, (see 012.008), p. 85 - 101 (1980).
On the sun we can identify many of the motions derived from stellar spectral analysis. A summary is given of the observed solar velocity phenomena. Many of these (e. g. meridional flow, giant cells, solar differential rotation, supergranulation) are of great interest in astrophysics especially for interior structure and chromospheric and coronal structuring but contribute virtually nothing to the velocities derived from a solar irradiance spectrum analysis. Others (granulation, very small scale motions and to a lesser extent, oscillations) do contribute substantially to the integrated sun velocity analysis. Some of the properties of these motion fields are described.

080.027 **On the structural and stochastic motions in the solar and stellar atmospheres.**
E. I. Mogilevsky (*Eh. I. Mogilevskij*).
Stellar turbulence, (see 012.008), p. 294 - 295 (1980).
Abstract.

080.028 **Small-scale versus large-scale motions in the solar atmosphere derived from a non-LTE calculation of multiplet 38 of Ti I.** R. Cayrel, S. Dumont, P. Martin.
Stellar turbulence, (see 012.008), p. 298 (1980). – Abstract.

080.029 **Tobias Mayers Meridianbeobachtungen der Sonne und die Frage der Veränderlichkeit des Sonnendurchmessers.** A. Wittmann.
Mitt. Astron. Ges., Nr. 48, (see 012.015), p. 25 - 28 (1980).

080.030 **Simultaneous determination of the damping constant and iron abundance in the solar atmosphere.**
B. T. Babij, M. M. Koval'chuk.
Problems of cosmic physics. Vyp. 15, (see 003.003), p. 23 - 29 (1980). In Russian.
From the analysis of 40 iron lines of different strengths the change of the damping constant with depth and the abundance of iron in the solar atmosphere are determined simultaneously. Corrections to the system of oscillator strengths of Kurucz and Peytremann are proposed.

080.031 **Two-level model of the solar dynamo.**
V. A. Dogel', S. I. Syrovatskij.
XIth seminar on cosmophysics, (see 012.035), p. 15 - 27 (1979). In Russian. – Abstr. in Ref. zh., 51. Astron., 8.51.348 (1980).

080.032 **Model of the differential rotation of the sun.**
A. S. Monin, L. M. Simuni.
Dokl. AN SSSR, Vol. 250, 1348 - 1351 (1980). In Russian.
Abstr. in Ref. zh., 51. Astron., 8.51.349 (1980).

080.033 **Solar neutrinos and nonradial pulsations of the sun.**
G. T. Zatsepin, E. A. Gavryuseva, Yu. S. Kopysov.
Dokl. AN SSSR, Vol. 251, 1342 - 1345 (1980). In Russian.
Abstr. in Ref. zh., 8.51.425 (1980).

080.034 **Model of layer-cellular motions in the convective zone of the sun. II.** V. P. Savchenko.
Astron. Zh., Tom 57, 1023 - 1032 (1980). In Russian.
English translation in Soviet Astron., Vol. 24, No. 5.
A simple semi-empirical model of the sun's convective zone structure for the range of depths 3 - 200 thousand kilometres is calculated.

080.035 **Detection of 160 min solar intensity variations: sampling effect.**
S. Koutchmy, O. Koutchmy, V. A. Kotov.
Astron. Astrophys., Vol. 90, 372 - 376 (1980).

The hypothesis that the 2h 40m period in the Crimean and Stanford global solar velocity oscillation measurements, which is close to a 1/9 of a day, might be produced partly by a 1-day sampling regularity in observations, is checked using the IR center-limb intensity variations measurements. The authors find: (1) The power spectrum of the series of these data does show a peak near the 2h 40m period. (2) Power spectra of the series with observing windows "filled" with a constant do not exhibit a significant peak at the 9th harmonic of a day, nor do the power spectra of the set with observing windows "filled" with randomly generated numbers.

080.036 **Search for giant cells in the solar convection zone.** B. J. LaBonte, R. Howard.
Solar and interplanetary dynamics, (see 012.020), p. 21 - 23 (1980).

The authors have searched for large scale persistent east-west horizontal flows.

080.037 **A two-level solar dynamo based on solar activity, convection, and differential rotation.**
A. Bratenahl, P. J. Baum, W. M. Adams.
Solar and interplanetary dynamics, (see 012.020), p. 29 - 32 (1980).

080.038 **Propagation of an MHD shock in the vicinity of a magnetic neutral sheet.**
D. J. Mullan, R. S. Steinolfson.
Solar and interplanetary dynamics, (see 012.020), p. 323 - 326 (1980).

080.039 **Physical driving forces and models of coronal responses.**
S. I. Syrovatskii (*Syrovatskij*), B. V. Somov.
Solar and interplanetary dynamics, (see 012.020), p. 425 - 441 (1980). – Invited paper.

The reasons for nonstationary hydrodynamic flow in the solar atmosphere are reviewed. It is emphasized that a rapid local heating of the corona or the upper chromosphere can scarcely provide a very large mass of solar plasma ejections observed in the corona and the interplanetary space. The authors suggest that coronal transients and interplanetary ejections are produced by magnetic field evolving in the solar atmosphere. Magnetic reconnection in current sheets can play an essential role in this process. A suitable approximation of the strong magnetic field is formulated. Some solutions of MHD equations in this approximation are demonstrated. Their applications to coronal conditions are discussed.

080.040 **Gas dynamics of impulsive heated solar plasma.** B. J. Sermulina *(B. Ya. Sermulinya)*, B. V. Somov, A. R. Spektor, S. I. Syrovatskii *(Syrovatskij)*.
Solar and interplanetary dynamics, (see 012.020), p. 491 - 494 (1980).

Numerical solutions for the problem of hydrodynamic response of the inhomogeneous (exponential) atmosphere on impulsive heating by energetic electrons or by very high-temperature thermal fluxes are discussed.

080.041 **On the statistical significance of the reported phase coherence for solar oscillations.**
T. P. Caudell, H. A. Hill.
Mon. Not. R. Astron. Soc., Vol. 193, 381 - 383 (1980).

The statistical significance of the reported phase coherence in the observed solar oscillations is estimated more clearly through the use of Monte Carlo techniques. These simulations strongly support the global normal mode interpretation of the phenomenon.

080.042 **Quasi-biennial periodicity in the solar neutrino flux: a further result.** K. Sakurai.
Publ. Astron. Soc. Japan, Vol. 32, 547 - 549 (1980).

On the basis of the analysis of the up-dated experimental results obtained by Davis et al. (1978), it is shown that the observed flux of the solar neutrinos has been varying quasi-biennially since 1970. It is suggested that this variation may be generated by some physical processes inside the sun.

080.043 **Polytropic zone models of the sun and neutrino fluxes.** E. A. Gavryuseva, Yu. S. Kopysov.
Inst. yader. issled. AN SSSR. Prepr., 1980, No. 0157, 19 pp. In Russian. – Abstr. in Ref. zh., 51. Astron., 10.51.508 (1980).

080.044 **Global oscillations of the sun.** F.-L. Deubner. Highlights of Astronomy, Vol. 5, (see 012.022), 75 - 87 (1980).

080.045 **Global circulation of the sun: where are we and where are we going?** P. A. Gilman.
Highlights of Astronomy, Vol. 5, (see 012.022), 91 - 119 (1980).

The author presents an up-to-date picture of observation and theory of global circulation of the sun. He connects this subject to neighboring fields of interest, particularly the solar dynamo and solar variability as well as stellar rotation. He gives some opinions as to where the action should be in the near future.

080.046 **Theoretical review of secular instabilities in the sun.** M. Gabriel.
Highlights of Astronomy, Vol. 5, (see 012.026), 441 - 444 (1980).

The author discusses the problems raised by the discovery that the sun was, in the past, unstable towards non-radial oscillations.

080.047 **Observations of solar pulsations.** H. A. Hill. Highlights of Astronomy, Vol. 5, (see 012.026), 449 - 452 (1980).

Observational evidence for global oscillations of the sun with periods $\lesssim 1$ hr and > 5 min is presented.

080.048 **Solar pulsations.** A. B. Severny (*Severnyj*), V. A. Kotov, T. T. Tsap.
Highlights of Astronomy, Vol. 5, (see 012.026), 453 - 456 (1980).

The authors point out the following interesting features of the oscillations: 1. Sometimes the 160^m oscillations disappear or almost disappear (in May - June 1976, etc.) and reappear again with nearly the same phase. 2. The oscillations can offer the possibility of energy transport different from a radiative one, especially if one takes into account that at higher modes and high amplitudes near the center of a star the Reynolds stresses at such wave motions might be very large.

080.049 **Overshooting motions from the convection zone and their role in atmospheric heating.** J. Toomre.
Highlights of Astronomy, Vol. 5, (see 012.027), 571 - 580 (1980).

The author concentrates on the overshooting motions, for they may be the most effective in revealing details of the structure of the convection zone. Such structure is seen to influence most of the coupling mechanisms between the inside of the Sun and the atmosphere, and thus plays a vital role in all attempts to understand how chromospheres and coronae are maintained.

080.050 **Effects of certain analysis procedures on solar global velocity signals.** P. A. Gilman, G. A. Glatzmaier.
Astrophys. J., Vol. 241, 793 - 803 (1980).

The authors examine the data reduction procedures used by Howard and colleagues to deduce global solar velocities from the original Mount Wilson Doppler-magnetograph record. They conclude that substantially more "giant cell" velocity signal could be present in the Mount Wilson data than has been estimated by Howard and LaBonte. They suggest some more sensitive techniques which do not unnecessarily attenuate the signal.

080.051 **The rotation of the sun: observations at Stanford.** P. H. Scherrer, J. M. Wilcox, L. Svalgaard.
Astrophys. J., Vol. 241, 811 - 819 (1980).

Daily observations of the photospheric rotation rate using the Doppler effect have been made at the Stanford Solar Observatory since 1976 May. These observations show no daily or long-period variations in the rotation rate that exceed the observational error of about 1%. The average rotation rate is the same as that of the sunspots and the large-scale magnetic field structures.

080.052 **Structure in the 5 minute oscillations of integral sunlight.** A. Claverie, G. R. Isaak, C. P. McLeod, H. B. van der Raay, T. Roca Cortes.
Astron. Astrophys., Vol. 91, L9 - L10 (1980).

Additional information on the discrete structure in the 5 minute velocity oscillation observed in integral sunlight is presented. Measurements taken during 1976, 77, 78 and 79 show good consistency and some 25 discrete frequencies, established to an accuracy of $\sim 1:10^3$, are tabulated.

080.053 **A "fast" model of the solar convection zone.** G. Belvedere, L. Paterno, I. W. Roxburgh.
Astron. Astrophys., Vol. 91, 356 - 359 (1980).

A model of the solar convection zone is proposed, which is a good compromise between accuracy and the execution-time in a computer. The model incorporates the mixing-length theory in the simple scheme developed by Faulkner et al. (1965) which gives an accurate representation of the super-adiabatic layers. The authors estimate that the program runs approximately 20 times faster than the programs of more sophisticated models.

080.054 **Blijft de zon altijd schijnen?** A. J. M. Wanders.
Zenit, 7e Jaarg., 17 - 20 (1980).

080.055 **Krimpt de zon?** H. Potters.
Zenit, 7e Jaarg., 238 - 239 (1980).

080.056 **Wringende trillingen van het zonsoppervlak.** C. de Jager.
Zenit, 7e Jaarg., 468 - 469 (1980).

080.057 **The sun as a star.** R. W. Noyes.
Smithsonian Astrophys. Obs., Spec. Rep. 389, (see 012.038), p. 3 - 13 (1980).

080.058 **Solar neutrino production of long-lived isotopes and secular variations in the sun.**
W. C. Haxton, G. A. Cowan.
Science, Vol. 210, 897 - 899 (1980).

Long-lived isotopes produced in the earth's crust by solar neutrinos may provide a method of probing secular variations in the rate of energy production in the sun's core. Only one isotope, calcium-41, appears to be suitable from the dual standpoints of reliable nuclear physics and manageable backgrounds. The proposed measurement also may be interesting in view of recent evidence for neutrino oscillations.

080.059 **Observations of a probable change in the solar radius between 1715 and 1979.** D. W. Dunham, S. Sofia, A. D. Fiala, D. Herald, P. M. Muller.
Science, Vol. 210, 1243 - 1245 (1980).

Solar eclipses were observed from locations near both edges of the paths of totality in England in 1715, in Australia in 1976, and in North America in 1979. Analysis of these observations shows that the solar radius has contracted by 0.34 ± 0.2 arc second in 264 years.

080.060 **Solar oscillations.** H. B. van der Raay.
Nature, Vol. 288, 535 - 536 (1980).

080.061 **Solar oscillations: full disk observations from the geographic South Pole.**
G. Grec, E. Fossat, M. Pomerantz.
Nature, Vol. 288, 541 - 544 (1980).

Observing conditions at the geographic South Pole enable modes of global solar oscillations and theoretical models of the internal solar structure to be identified.

080.062 **Is the sun helium-deficient?**
J. Christensen-Dalsgaard, D. O. Gough.
Nature, Vol. 288, 544 - 547 (1980).

The recent observations of solar 5-min oscillations of low degree agree approximately with the predictions of a standard solar model with normal abundances of helium and heavy elements. Much of the apparent discrepancy noticed when the observations were first announced was a result of having neglected the influence of the sun's atmosphere in the normal mode analysis of the theoretical models. Standard solar models are not in perfect agreement with observation, but it seems that major modifications will not be necessary to remove the remaining small discrepancies.

080.063 **The constancy of the solar diameter over the past 250 years.**
J. H. Parkinson, L. V. Morrison, F. R. Stephenson.
Nature, Vol. 288, 548 - 551 (1980).

Reconsideration of meridian circle observations with analysis of transits of Mercury and durations of total solar eclipses indicate that there has been no detectable secular change in the solar diameter during the past 250 yr.

080.064 **The evolution of the solar 'constant'.**
M. J. Newman.
Origins of Life, Vol. 10, 105 - 110 (1980). – Abstr. in Phys. Abstr., Vol. 84, Abstr. 9462 (1981).

080.065 **Solar variability.** J. O. Stenflo.
Variability in stars and galaxies, (see 012.044), p. GL.1.1 - 1.20 (1980).

The author gives an overview of solar variability, and discusses in some detail progress in the fields of solar oscillations, magnetic activity, and global energy output.

080.066 **Solar oscillations (theory).**
J. Christensen-Dalsgaard.
Variability in stars and galaxies, (see 012.044), p. A.1.1 - 1.28 (1980).

Recent results of theoretical studies of solar oscillations are presented. Furthermore the interpretation of the different types of observed solar oscillations, and the potential of these oscillations as probes of the solar interior, are discussed.

080.067 **Observations of the hydrodynamic wave mode spectrum of the sun.** F.-L. Deubner.
Variability in stars and galaxies, (see 012.044), p. A.2.1 - 2.16 (1980).

The spectrum of periods of hydrodynamic oscillations and waves presently investigated in the sun ranges from seconds to several hours. The significance, and even the solar origin of some of the signals observed in the outer ranges of this spectrum is still being hotly debated, whereas the p-mode character of the so called 5-min oscillations is now well established. The recent development in the field of longer period oscillations observed at the solar limb as well as in the integrated light of the disk is discussed. Fluctuations observed spectroscopically in the outer layers of the solar atmosphere are less regular, and the energy flux connected to them appears insufficient for upper chromospheric and coronal heating.

080.068 **Variations in the solar constant due to solar active regions.** G. A. Chapman.
Astrophys. J., Lett., Vol. 242, L45 - L48 (1980).

Solar activity is expected to affect the solar constant at some level. Recent observations and data analysis show the amount of variation to be expected from active regions, faculae, and sunspots on the apparent solar brightness. It is concluded that the maximum effect is about 20 times greater for sunspots than for faculae per unit area.

080.069 **Non-Maxwellian velocity distribution functions associated with steep temperature gradients in the solar transition region. I. Estimate of the electron velocity distribution functions.** R. Roussel-Dupré.
Sol. Phys., Vol. 68, 243 - 263 (1980).

The author shows that, in the presence of the steep temperature gradients characteristic of EUV models of the solar transition region, the electron and proton velocity distribution functions are non-Maxwellian and are characterized by high energy tails. He estimates the magnitude of these tails for a model of the transition region and computes the heat flux.

080.070 **Non-Maxwellian velocity distribution functions associated with steep temperature gradients in the solar transition region. II. The effect of non-Maxwellian electron distribution functions on ionization equilibrium calculations for carbon, nitrogen and oxygen.**
R. Roussel-Dupré.
Sol. Phys., Vol. 68, 265 - 274 (1980).

Non-Maxwellian electron velocity distribution functions, computed for Dupree's (1972) model of the solar transition region, are used to calculate ionization rates for ions of carbon, nitrogen, and oxygen. Ionization equilibrium populations for these ions are then computed and compared with similar calculations assuming Maxwellian distribution functions for the electrons. The results show that the ion populations change (compared to the values computed with a Maxwellian) in some cases by several orders of magnitude depending on the ion and its temperature of formation.

080.071 **Über die Entwicklung und den inneren Aufbau der Sonne.** H. J. Haubold, R. W. John.
Astron. Schule, 17. Jahrg., 126 - 131 (1980).

080.072 **Solar neutrinos.** J. Grygar.
Vesmír, Vol. 59, 270 - 272 (1980). In Czech.

080.073 **Solar pulsations.** H. A. Hill.
Space Sci. Rev., Vol. 27, (see 012.046), 283 - 291 (1980).

Recent observational evidence on solar oscillations is reviewed; this evidence strongly favors the global interpretation for much of the observed spectrum. Implications of these observations for the study of the solar interior and atmosphere are discussed.

080.074 **New solar-neutrino flux calculations and implications regarding neutrino oscillations.**
J. N. Bahcall, S. H. Lubow, W. F. Huebner, N. H. Magee, Jr., A. L. Merts, M. F. Argo, P. D. Parker, B. Rozsnyai, R. K. Ulrich.
Phys. Rev. Lett., Vol. 45, 945 - 948 (1980).

The results of new calculations of solar-neutrino fluxes are presented; the fluxes are obtained from detailed solar models that make use of improved opacities and nuclear-physics cross sections. By evaluating known uncertainties in the predicted capture rate for the ^{37}Cl solar-neutrino experiment, the authors find that the ratio of theoretical to (best-estimate) observed capture rate lies in the range 4.0 to 2.6. These results constitute a strong constraint on models of neutrino oscillations if the entire discrepancy is ascribed to neutrino oscillations.

080.075 **A broad look at solar physics.** Adapted from the solar physics study of August 1975.
E. Parker, J. Beckers, A. Hundhausen, M. R. Kundu, C. E. Leith, R. Lin, J. Linsky, F. B. MacDonald, R. Noyes, F. Q. Orrall, L. E. Peterson, D. M. Rust, P. Sturrock, A. B. C. Walker, Jr., A. Timothy, K. A. Janes, R. Hart.
Solar system plasma physics, Vol. 1, (see 003.010), 3 - 49 (1979).

Contents: Introduction. The role of solar physics. Solar research – progress and prospects. Proposed solar research.

080.076 **Some approaches to the problem of determination of spectral line formation depth in the solar atmosphere.** M. L. Demidov.
Issled. po geomagn., aehron. i fiz. Solntsa, Moskva, 1980, No. 52, p. 3 - 13. In Russian. – Abstr. in Ref. zh., 51. Astron., 11.51.427 (1980).

080.077 **Deviation from the local thermodynamical equilibrium in the solar atmosphere. Methodology of the problem. The line source function.** N. G. Shchukina.
Inst. teor. fiz. AN USSR. Prepr., 1980, No. 33, 46 pp. In Russian. – Abstr. in Ref. zh., 51. Astron., 11.51.433 (1980).

080.078 **Temperature, number of molecules and turbulent velocity in the CO layer on the sun.**
A. P. Sarychev.
Astron. Tsirk., No. 1104, p. 5 - 6 (1980). In Russian.

080.079 **Model of the layer-cellular motion in the solar convective zone. III.** V. P. Savchenko.
Astron. Zh., Tom 57, 1295 - 1301 (1980). In Russian.
English translation in Soviet Astron., Vol. 24, No. 6.

Large-scale layer-cellular motion parameters for Spruit's model of the convection zone structure (1974) are calculated. Some indirect applications of the results to the prediction of solar and geomagnetic activity are considered.

080.080 **The evolution of the solar 'constant'.**
M. J. Newman.
Limits of life, (see 012.052), p. 99 - 104 (1980).

Has the solar 'constant', the rate at which energy is received by the earth from the sun per unit area per unit time, been constant at its present level since Archean times? Three mechanisms by which it has been suggested that the solar energy output can vary with time are discussed, characterized by long ($\sim 10^9$ yr), intermediate ($\sim 10^8$ yr), and short ($\sim$ years to decades) time scales.

080.081 **Solar rotation and activity in the past and their possible influence upon the evolution of life.**
E. H. Geyer.
Mitt. Astron. Ges., Nr. 50, p. 101 - 104 (1980).

080.082 **Differential rotation of solar features and its variation as deduced from the 'shock-transition model' of the solar cycle.** M. H. Gokhale.
Kodaikanal Obs. Bull., Ser. A, Vol. 2, 224 - 229 (1979).

From the topology of the two families of flux tubes in the shock-transition model of the solar cycle the author predicts the general nature of the differential rotation of: (1) sunspots and small-scale emission features in active regions and

(2) large-scale magnetic and emission features which have major contributions from the quiet regions. He predicts the manner in which the differential rotation of features in these categories will vary with the solar cycle. These predictions are in good agreement with the so far observed rotation characteristics of various features in each category.

080.083 **Investigation of the structure of the velocity field in the solar atmosphere.** I. E. Kozhevatov.
Issled. po geomagn., aehron. i fiz. Solntsa. Moskva, 1980, No. 52, p. 118 - 119. In Russian. – Abstr. in Ref. zh., 51. Astron., 12.51.303 (1980).

080.084 **Solar energetics.** I. A. Vlasov.
Atom. ehnerg., Vol. 49, No. 3, p. 155 - 161 (1980). In Russian. – Abstr. in Ref. zh., 51. Astron., 12.51.348 (1980).

080.085 **Convective instability in the solar envelope.** D. Narasimha, S. K. Pandey, S. M. Chitre.
J. Astrophys. Astron., Vol. 1, 165 - 175 (1980).

The characteristics of the most unstable fundamental mode and the first harmonic excited in the convection zone of a variety of solar envelope models are shown to be in reasonable agreement with the observed features of granulation and supergranulation.

080.086 **Solar ultraviolet continuum radiation: the photosphere, the low chromosphere, and the temperature-minimum region.** D. Samain.
Astrophys. J., Suppl. Ser., Vol. 44, 273 - 294 (1980).

A comparison of solar disk-center intensity measurements with theoretical values calculated for atmospheric models derived from the temperature distributions found by J. Vernazza and his colleagues indicates that generally good agreement is found with an atmospheric model having a minimum temperature of about 4150 K or possibly higher. Empirical opacity values including LTE departures and absorption coefficients which best represent the radiation field in the range 1460 Å–2100 Å are given. Precise values are obtained for the required opacity distribution, presumably due to lines, longward of 1682 Å.

080.087 **Differential rotation of the sun and the Maunder minimum of solar activity.**
R. N. Ikhsanov, Yu. I. Vitinskij.
Dokl. AN SSSR, Vol. 254, 577 - 580 (1980). In Russian. Abstr. in Ref. zh., 51. Astron., 1.51.325 (1981).

The strangest star. See Abstr. 003.051.

Solar rotation in 1612: Galileo vs. Harriot. See Abstr. 004.016.

Effects of discontinuity of time series on computed periods. See Abstr. 021.021.

Proton excitation rates for fine structure transitions in C III, O V, and Ne VII in the sun. See Abstr. 022.075.

A new method for measuring the solar radius. See Abstr. 031.517.

New horizons in neutrino astrophysics. See Abstr. 061.046.

The heating of a thermally conducting stratified medium. II. A simple plane model of an atmosphere. See Abstr. 062.068.

A formalism for differential rotation. See Abstr. 062.083.

Vertical motions in an intense magnetic flux tube. IV: Radiative relaxation in a uniform medium. See Abstr. 062.087.

Vertical motions in an intense magnetic flux tube. V: Radiative relaxation in a stratified medium. See Abstr. 062.088.

Resonance-line polarization. VI. Line wing transfer calculations including excited state interference. See Abstr. 063.014.

Two-dimensional radiative transfer. II. The wings of Ca K and Mg *k*. See Abstr. 063.047.

Mechanical energy transport. See Abstr. 064.030.

Small-scale dissipative processes in stellar atmospheres. See Abstr. 064.056.

Stability of nonradial g^+ -mode pulsations in 1 $M_\odot$ models. See Abstr. 065.036.

Differential rotation set up by latitude-dependent heat transport (*stellar rotation*). See Abstr. 065.040.

Afbuiging van straling door de zon. 1. De zonsverduistering van 1919. See Abstr. 066.131.

Afbuiging van straling door de zon. 2. Nieuwe metingen. See Abstr. 066.132.

Numerical simulation of granular convection: effects on photospheric spectral line profiles. See Abstr. 071.024.

The formation of Na I spectral lines in the solar atmosphere. See Abstr. 071.044.

Änderungen der differentiellen Rotation und meridionale Bewegungen von Sonnenflecken 1940 bis 1968. See Abstr. 072.030.

Differential rotation and meridional motions of sunspots in the years 1940 - 1968. See Abstr. 072.052.

Upper limits on the power in solar oscillations at 1.2 mm, 9 mm, 3.7 cm, and 11.1 cm wavelengths. See Abstr. 073.062.

A model of hot loops associated with solar flares. I. Gasdynamics in the loops. See Abstr. 073.114.

Magnetic fields and the solar constant. See Abstr. 075.003.

The solar ultraviolet continuum. See Abstr. 076.012.

Solar spectral radiance and irradiance at 225.2–319.6 nanometers. See Abstr. 076.032.

Correlated variations of planetary albedos and coincident solar-interplanetary variations. See Abstr. 091.050.

A possible detection of solar variability from photometry of Io, Europa, Callisto, and Rhea, 1976–1979. See Abstr. 099.001.

Hydromagnetics and period changes in RR Lyrae stars. See Abstr. 122.108.

Earth

081 Structure, Figure, Gravity, Orbit, etc.

081.001 **Luni-solar nutation tables and the liquid core of the earth.** P. Melchior.
Astron. Astrophys., Vol. 87, 365 - 368 (1980).

Recent experimental measurements performed by new astronomical techniques confirm that there was an urgent need to adopt a new luni-solar nutations table taking into account the dynamical effects of the liquid core of the earth. The author presents in this paper the most recent results obtained in this respect from earth tide measurements and a discussion about the nutation table to be adopted.

081.002 **The earth and planetary sciences.** G. W. Wetherill, C. L. Drake.
Science, Vol. 209, 96 - 104 (1980).

081.003 **Angular dislocation of the earth principal axes of inertia.**
W. C. C. Silva, H. U. Pilchowski, L. D. D. Ferreira.
Bull. Géod., Vol. 54, 181 - 189 (1980).

The coefficients C_{21} and S_{21} are usually considered to be zero in theory. However, satellite measurements show that C_{21} as well as S_{21} are non-zero, indicating that the axis of rotation of the earth is not the principal axis of inertia. This work determines the angles by which the principal axes of inertia must be rotated in relation to the geocentric coordinate system (where the Z axis coincides with the rotating axis of the earth).

081.004 **The cooling earth.** H. N. Pollack.
Nature, Vol. 286, 655 - 656 (1980).

081.005 **Palaeomagnetism of beach ridges in South Australia and the Milankovitch theory of ice ages.**
M. Idnurm, P. J. Cook.
Nature, Vol. 286, 699 - 702 (1980).

The authors discuss evidence for the Milankovitch theory from a South Australian sea-level record.

081.006 **Geoid anomalies in the vicinity of subduction zones.** D. C. McAdoo.
NASA Tech. Memo., NASA TM 80678, 45 pp. (1980).

081.007 **Geophysical parameters of the earth-moon system.** A. J. Ferrari, W. S. Sinclair, W. L. Sjogren,
J. G. Williams, C. F. Yoder.
J. Geophys. Res., Vol. 85, 3939 - 3951 (1980).

Doppler tracking data from Lunar Orbiter 4 have been combined with laser ranging data from lunar retroreflectors to yield a number of geophysical and geodetic parameters for the earth and moon. This joint solution gives values of (1) the lunar principal polar moment $C/MR^2 = 0.3905 \pm 0.0023$, (2) $MG_E = 398600.461 \pm 0.026$ km^3/s^2, and (3) an earth/moon mass ratio at 81.300587 ± 0.000049. Also determined are the harmonics of a complete lunar gravity field through degree and order 5, the obliquity of the lunar pole, selenocentric coordinates of the lunar retroreflectors, geocentric coordinates of the McDonald Observatory, and the lunar secular acceleration. The lunar potential Love number is weakly determined at 0.022 ± 0.013, and a surprisingly large dissipation of rotational energy is inferred, though either solid body tidal dissipation or liquid core mantle interactions could be causes.

081.008 **A ring around the earth during the Pleistocene?** J. A. O'Keefe.
Bull. American Astron. Soc., Vol. 12, 510 (1980). – Abstract.

081.009 **Dynamics of a fluid core with inward growing boundaries.** H. H. Schloessin, J. A. Jacobs.
Canadian J. Earth Sci., Vol. 17, 72 - 89 (1980). – Abstr. in Phys. Abstr., Vol. 83, Abstr. 76934 (1980).

081.010 **Does the whole of the earth's core convect?** K. A. Whaler.
Nature, Vol. 287, 528 - 530 (1980).

Higgins and Kennedy (1971) used data on the behaviour of iron at high temperatures and pressures to infer that the top of the earth's liquid core is stably stratified. Attempts to confirm this result both thermodynamically and using seismological data have been inconclusive. The author presents here geomagnetic results which may resolve the controversy.

081.011 **The chemical composition of the interior shells of the Earth.** J. M. Herndon.
Proc. R. Soc. London, Ser. A, Vol. 372, 149 - 154 (1980).
Abstr. in Phys. Abstr., Vol. 83, Abstr. 85624 (1980).

081.012 **Magnetic waves in the core of the Earth. II.** S. I. Braginsky (*Braginskij*).
Geophys. Astrophys. Fluid Dyn., Vol. 14, 189 - 208 (1980).
Abstr. in Phys. Abstr., Vol. 83, Abstr. 89993 (1980).

081.013 **A model of Earth's structure inferred from eigenperiods of torsional oscillation. II.** H. Oda.
Sci. Rep. Tokohu Univ. Fifth Ser. Geophys., Vol. 26, No. 2, p. 45 - 66 (1979). – Abstr. in Phys. Abstr., Vol. 83, Abstr. 90044 (1980).

081.014 **Diamond mantle – product of the early evolution of the earth.** B. A. Mal'kov.
Dokl. AN SSSR, Vol. 252, 175 - 177 (1980). In Russian.
Abstr. in Ref. zh., 51. Astron., 8.51.251 (1980).

081.015 **Observations of sidereal motions of the earth's surface.** L. Knopoff, P. A. Rydelek.
Rev. Geophys. Space Phys., Vol. 18, 723 - 724 (1980).

Vertical gravity observations from the geographic south pole are analyzed in the semidiurnal tidal frequency band and show the presence of a sidereal semidiurnal spectral component. This component is not unexpected and probably does not provide evidence for the existence of gravitational waves of extraterrestrial origin.

081.016 **Lateral density variations in elastic earth models from an extended minimum energy approach.**
B. V. Sanchez.
NASA Tech. Memo., NASA TM 80742, 6 + 23 pp. (1980).

Kaula's minimum energy approach has been extended to include the nonhydrostatic gravitational potential energy and the density perturbation field has been obtained to degree and order 8. The results indicate that the addition of the gravitational potential energy in the minimization process does not change significantly the conclusions based on results for the minimum shear strain energy case, concerning the inability of the mantle to withstand the lateral loading elastically.

081.017 **Satellite determination of short wavelength gravity variations.** J. V. Breakwell.
J. Astronaut. Sci., Vol. 27, 329 - 344 (1979). – Abstr. in Phys. Abstr., Vol. 83, Abstr. 98216 (1980).

081.018 **Marées terrestres.** P. Melchior (Editor).
Bull. Inf., (Obs. R. Belgique, Bruxelles), No. 84, p. 5321 - 5412 (1980).

081.019 **On separating indirect effects from variations of the gravitational field, of the figure and rotation of the earth depending on time determined from data of repeated measurements.** V. V. Buzuk, V. M. Panin.
Novosib. inst. inzh. geod., aehrofotosemki i kartogr. Novosibirsk, 1980. 15 pp. In Russian. – Abstr. in Ref. zh., 52. Geod. Aehrosemka, 10.52.73 (1980).

081.020 **Determination of the geopotential from satellite-to-satellite tracking data.**
B. C. Douglas, C. C. Goad, F. F. Morrison.
J. Geophys. Res., Vol. 85, 5471 - 5480 (1980).

The authors present the results of error analyses of possible satellite-to-satellite tracking missions to determine the geopotential at a resolution of 1° × 1°. To achieve an accuracy of a few milligals at this resolution requires a satellite altitude at or near 150 km and measurement of intersatellite speed to 10^{-6} m/s.

081.021 **Grand unification magnetic monopoles inside the Earth.** R. A. Carrigan, Jr.
Nature, Vol. 288, 348 - 350 (1980).

Elementary particle grand unification theories admit the possibility of very massive magnetic monopoles, but standard cosmology gives too many poles. Massive monopoles are affected both by gravity and magnetism, requiring reassessment of existing monopole searches. Monopoles inside the Earth would be able to annihilate with each field reversal, thereby giving rise to additional internal heat. This perspective is used to estimate an upper limit for the monopole-to-baryon ratio at accretion of 10^{-28}.

081.022 **Solid earth tides determined with VLBI.**
B. Schupler.
Bull. American Astron. Soc., Vol. 12, 742 (1980). – Abstract.

081.023 **Least squares combination of satellite harmonics and integral formulas in physical geodesy.**
L. Sjöberg.
Gerlands Beitr. Geophys., Band 89, 371 - 377 (1980).

Satellite harmonics and terrestrial gravity data are combined to least squares solutions of gravity field components. In a first approach the gravity field is treated as a homogeneous and isotropic stochastic process. Then it is replaced by a non-stochastic signal with random noise. It is shown that the minimum variance solution in the first approach equals the global least squares solution in the second approach.

081.024 **Residual currents, tidal ellipses and bottom friction.** P. Brosche, W. Hövel.
Deutsche Hydrogr. Z., Jahrg. 33, 110 - 123 (1980). In German.

For the case that oceanic tidal motions consist only of a residual current and of the tidal ellipse, the average frictional force corresponding to the average torque is given explicitly.

081.025 **Global gravity field to degree and order 30 from Geos 3 satellite altimetry and other data.**
E. M. Gaposchkin.
J. Geophys. Res., Vol. 85, 7221 - 7234 (1980).

A description of the earth's gravity field in spherical harmonics through degree and order 30 is obtained by combining satellite-to-sea-surface altimetry data from Geos 3, terrestrial gravity measurements, and satellite perturbation analysis. Reference orbits and a tide model are used with the altimetry data to calculate sea-surface heights, and these heights are assumed to be the geoid.

081.026 **Geoid height versus age for symmetric spreading ridges.** D. Sandwell, G. Schubert.
J. Geophys. Res., Vol. 85, 7235 - 7241 (1980).

Geoid height-age relations have been extracted from Geos 3 altimeter data for large areas in the North Atlantic, South Atlantic, southeast Indian, and southeast Pacific oceans. Except for the southeast Pacific area, geoid height decreases approximately linearly with the age of the ocean floor for ages less than about 80 m.y. in agreement with the prediction of an isostatically compensated thermal boundary layer model.

081.027 **An earth dynamo with anisotropic resistivity.**
E. W. Cowan.
J. Geophys. Res., Vol. 85, 7242 - 7246 (1980).

A dynamo mechanism generating a self-sustaining dipole field as a result of the motion of a fluid with anisotropic resistivity is a possibility for the dynamo generating the dipole field of the earth. The anisotropy appears as a large-scale average effect resulting from small-scale difference in the resistivity of the materials in ascending and descending convection currents within the liquid core, which are deflected in the required form by the Coriolis acceleration. A numerical analysis of time-evolving magnetic fields in a model with anisotropic resistivity supports the basic mechanism of the dynamo effect. The differences in resistivity required to give regeneration for the core of the earth are too large to be accounted for by temperature differences but could conceivably result from differences in chemical composition.

081.028 **The Earth's gravitational field and orbital resonances of the Intercosmos satellites.**
J. Klokočník.
Říše hvězd, Vol. 61, 227 - 229 (1980). In Czech.

081.029 **Solution of continuous and discrete boundary value problems.** M. S. Petrovskaya.
Byull. Inst. Teor. Astron., Tom 14, 557 - 570 (1980). In Russian.

The possibility of improving the convergence of Molodensky's series is considered. Then new formulas are derived for the solution of the geodetic boundary value problem. The strict relation has been derived between Stokes' constants and the surface anomaly. Closed formulas are derived for Stokes' constants and the anomalous potential in terms of the surface anomalies and heights.

081.030 **Determination of Stokes' constants and the disturbing potential in terms of gravity anomalies.**
M. S. Petrovskaya.
Byull. Inst. Teor. Astron., Tom 14, 608 - 616 (1980). In Russian.

An expansion of the disturbing potential is provided as a series in surface harmonics which is valid both in the space and on the earth's surface. The principal terms in it correspond to the solid harmonics of the external potential expansion, the coefficients being Stokes' constants. As a whole the potential on the surface is presented in terms of only Stokes' constants with the accuracy of the fourth order of the heights. Simple expressions are found to present Stokes' constants as explicit functions of the surface anomalies. A formula for the disturbing potential on the surface is derived in terms of the surface anomalies.

081.031 **On the classification of annular structures of the earth.**
Ya. G. Kats, V. L. Avdeev, V. P. Belov.
Kosmogen. struktury Zemli. Mater. semin. Moskva, 1980,

p. 22 - 26. In Russian. – Abstr. in Ref. zh., 51. Astron., 11.51.411 (1980).

081.032 **The abundances of major, minor and trace elements in the earth's mantle as derived from primitive ultramafic nodules.** E. Jagoutz, H. Palme, H. Baddenhausen, K. Blum, M. Cendales, G. Dreibus, B. Spettel, V. Lorenz, H. Wänke.
Proc. Tenth Lunar Planet. Sci. Conf., (see 012.050), p. 2031 - 2050 (1979).

081.033 **Development of A. Ya. Orlov's ideas on studying the tidal tilts of the earth's surface in works of the Poltava Gravimetrical Observatory.** P. S. Matveev.
Geodynamics and astrometry, (see 003.011), p. 27 - 52 (1980). In Russian.

081.034 **The works of A. Ya. Orlov on gravimetry and their development at the Poltava Observatory.**
I. A. Dychko, V. G. Bulatsen, V. G. Balenko.
Geodynamics and astrometry, (see 003.011), p. 52 - 58 (1980). In Russian.

Interiors of the planets. See Abstr. 003.031.

Geodesy and the earth's gravity field. Vol. II: Geodynamics and advanced methods. See Abstr. 003.052.

Earth and cosmos. See Abstr. 003.062.

The state of the planet. See Abstr. 003.064.

Geomagnetic field and interior structure of the earth. See Abstr. 003.090.

"Sojus-22" erforscht die Erde. See Abstr. 003.117.

Discovery of the Earth's core. See Abstr. 004.051.

Algorithms for the combined determination of proper motions of the earth's surface, of variations of the gravitational field, of polar motion and irregularities of the earth's rotation. See Abstr. 021.023.

Method of registration of the gravitational radiation from extraterrestrial sources in the kHz frequency band. See Abstr. 031.547.

Field-of-view constraints on actively controlled long-baseline stellar interferometry. See Abstr. 031.598.

Zonal tides and changes in the length of day. See Abstr. 044.001.

New nutation theories. See Abstr. 044.004.

Om jordens rotation, månens avstånd och tidvattnet. See Abstr. 044.010.

Atmospheric angular momentum fluctuations and changes in the length of the day. See Abstr. 044.035.

Changes in length-of-day and atmospheric circulation. See Abstr. 044.036.

On variations of geocentric positions with time. See Abstr. 046.003.

Zur lokalen Geoidbestimmung aus terrestrischen Messungen vertikaler Schweregradienten. See Abstr. 046.004.

On an integrable case of satellite motion in an approximate gravitational field of the earth. See Abstr. 052.026.

L'influence du quatrième harmonique zonal du potentiel gravitationnel terrestre sur la période nodale des satellites artificiels. See Abstr. 052.042.

Sur l'influence des harmoniques tesséraux du potentiel gravitationnel terrestre sur le rayon-vecteur des satellites artificiels. See Abstr. 052.045.

Thermal and magnetic instabilities in a rapidly rotating fluid sphere. See Abstr. 062:045.

The late 1960's secular variation impulse: further constraints on deep mantle conductivity. See Abstr. 084.094.

Spatial power spectra of the crustal geomagnetic field and core geomagnetic field. See Abstr. 084.095.

Tri-axiality of the Earth, the Moon and Mars. See Abstr. 091.051.

Some comments on anorthosites and basalts of the moon and the earth in connection with the evolution of these planets. See Abstr. 094.023.

Scientific achievements from ten years of lunar laser ranging. See Abstr. 094.026.

Moon and Earth: compositional differences inferred from siderophiles, volatiles, and alkalis in basalts. See Abstr. 094.529.

Volcano-ice interactions on the Earth and Mars. See Abstr. 097.014.

Dissolution of the primordial rare gases into the molten Earth's material. See Abstr. 107.026.

082 Atmosphere (Refraction, Scintillation, Extinction, Airglow, Site Testing)

082.001 **The evolution of atmospheric ozone.**
J. F. Kasting, T. M. Donahue.
J. Geophys. Res., Vol. 85, 3255 - 3263 (1980).

A one-dimensional coupled chemistry and flow model of the earth's atmosphere is used to study the relationship between ozone content and oxygen level. The effects of the biogenic trace gases methane and nitrous oxide are considered, as well as the production of odd nitrogen in lightning discharges.

082.002 **Energetics and thermal structure of the middle atmosphere.** S. Chandra.
Planet. Space Sci., Vol. 28, 585 - 593 (1980).

A study of a large number of temperature measurements in the middle atmosphere shows a much more complex thermal structure of this region than described in the U.S. Standard Atmosphere,1976. The mesopause height which is generally assumed to be at 80 km varies between 70 - 100 km, often with two minima in temperature at about 70 and 100 km and a maximum between 80 - 85 km. By solving the energy balance equation and the equations of continuity, the physical significance of the observed thermal structure is discussed in terms of the energetics of the various regions of the middle atmosphere.

082.003 **Supernovae effects on the terrestrial atmosphere.**
A. C. Aikin, S. Chandra, T. P. Stecher.
Planet. Space Sci., Vol. 28, 639 - 644 (1980).

082.004 **Stratospheric H_2O.**
H. W. Ellsaesser, J. E. Harries, D. Kley, R. Penndorf.
Planet. Space Sci., Vol. 28, 827 - 835 (1980).

The present state of our knowledge and understanding of H_2O in the stratosphere is reviewed. This reveals continuing discrepancies between observations and expectations following from the Brewer-Dobson hypothesis of stratospheric circulation. In particular, available observations indicate unexplained upward and poleward directed H_2O gradients immediately downstream from the tropical tropopause and variable vertical gradients above 20 km which generally disagree with those expected from oxidation of CH_4.

082.005 **Emission line shapes produced by dissociative excitation of atmospheric gases.**
E. C. Zipf, W. C. Wells.
Planet. Space Sci., Vol. 28, 859 - 866 (1980).

082.006 **An Antarctic NO density profile deduced from the gamma band airglow.** N. Iwagami, T. Ogawa.
Planet. Space Sci., Vol. 28, 867 - 873 (1980).

082.007 **Lunar tides in the middle atmosphere from NIMBUS 5 data.** D. M. Schlapp.
J. Atmos. Terr. Phys., Vol. 42, 529 - 532 (1980).

Lunar tides in temperature have been determined at stratospheric heights from about 2 yr of radiance measurements by the NIMBUS 5 satellite. The tides have an amplitude of order 0.1 K, and results are presented for the variations with height and latitude.

082.008 **Simultaneous mixing ratio profiles of stratospheric NO and NO_2 as derived from balloon-borne infrared solar spectra.**
R. D. Blatherwick, A. Goldman, D. G. Murcray, F. J. Murcray, G. R. Cook, J. W. VanAllen.
Geophys. Res. Lett., Vol. 7, 471 - 473 (1980).

Balloon-borne infrared solar spectra at ~0.02 cm^{-1} resolution obtained during sunset are used for the derivation of simultaneous vertical mixing ratio profiles of NO and NO_2. A simplified photochemical model for the diurnal variation of NO is included in the analysis.

082.009 **Excitation of the Na D-doublet of the nightglow.**
D. R. Bates, P. C. Ojha.
Nature, Vol. 286, 790 - 791 (1980).

The authors infer from the absolute intensity of the nocturnal D-doublet emission, the abundance of free Na and other relevant information that there is very little possibility of an alternative to Chapman's theory (1939) that the excited Na atoms involved are released by NaO–O collisions. The nature of the collision process is discussed.

082.010 **A comparison of two microthermometric methods capable to estimate astronomical image degradation in the turbulent atmosphere.**
A. Eh. Gur'yanov, S. L. Zubkovskij.
Astron. Zh., Tom 57, 859 - 863 (1980). In Russian.
English translation in Soviet Astron., Vol. 24, No. 4.

Results of the comparison of structural and spectral measurements of the temperature structure coefficient C_T^2 under different conditions are reported.

082.011 **On the values of the refraction constant accepted in different tables.** A. I. Nefed'eva.
Astron. Zh., Tom 57, 878 - 880 (1980). In Russian.
English translation in Soviet Astron., Vol. 24, No. 4.

The light wavelength and the values of the refraction constant used in compiling different tables of astronomical refraction are calculated. The results are given in a table.

082.012 **Stratospheric NO_2 and H_2O mixing ratio profiles from high resolution infrared solar spectra using nonlinear least squares.** E. Niple, W. G. Mankin, A. Goldman, D. G. Murcray, F. J. Murcray.
Geophys. Res. Lett., Vol. 7, 489 - 492 (1980).

082.013 **Comparison of near coincident LRIR and OAO–3 measurements of equatorial night ozone profiles.**
J. C. Gille, G. P. Anderson, P. L. Bailey.
Geophys. Res. Lett., Vol. 7, 525 - 528 (1980).

082.014 **The UV dayglow 2, Lyα and Lyβ emissions and the H distribution in the mesosphere and thermosphere.**
D. E. Anderson, Jr., P. D. Feldman, E. P. Gentieu, R. R. Meier.
Geophys. Res. Lett., Vol. 7, 529 - 532 (1980).

082.015 **Measurements of NO, O_3, and temperature at 19.8 km during the total solar eclipse of 26 February 1979.** W. L. Starr, R. A. Craig, M. Loewenstein, M. E. McGhan.
Geophys. Res. Lett., Vol. 7, 553 - 555 (1980).

Local measurements of stratospheric NO and O_3 mixing ratios and air temperature were made during the total solar eclipse of 26 February 1979. The instrumentation was carried aboard a U–2 aircraft flown at an altitude of 19.8 km in the region near 47° N – 112° W. The NO mixing ratio was reduced at least a factor of 25 at the maximum of the eclipse. The decrease and recovery of NO during the passage of the Moon's shadow over the measurement region follows approximately the predictions of two independent models. No change was observed in either the O_3 mixing ratio or the air temperature that could be attributed to the eclipse.

082.016 **Calculation of the earth's and astronomical refraction at large zenith distances.**
V. P. Nelyubina, N. F. Nelyubin.

Astrometr. Astrofiz., Vyp. (No.) 41, p. 82 - 91 (1980). In Russian.

The paper presents the calculation procedures of the earth's and astronomical refraction at the horizon. The methods of calculating the earth's refraction for arbitrary altitude of an object are considered for the case when the elevation profile of the refractive index is known only up to 30 km altitude. An analysis of the optimal distribution of integration points with the numerical solution of the refraction equations is carried out. Recommendations are given on division of the atmosphere into layers where it is necessary to obtain meteorological information to ensure maximum accuracy of calculated refraction values.

082.017 **The source of atmospheric motions.**
A. S. Ginzburg.
Zemlya Vselennaya, 1980, No. 4, p. 28 - 32. In Russian.

082.018 **Global model of density variations of the earth's atmosphere at heights from 0 - 150 km.**
A. A. Ramazov, Yu. G. Sikharulidze.
Kosm. Issled., Tom 18, 527 - 535 (1980). In Russian.

082.019 **Atmospheric aerosol scattering to mass ratio.**
A. W. Harrison, R. Pi.
Canadian J. Phys., Vol. 58, 549 - 553 (1980). – Abstr. in Phys. Abstr., Vol. 83, Abstr. 72628 (1980).

082.020 **A perturbative treatment of aerosol scattering of infrared radiation.** W. R. Yueh, W. L. Chameides.
J. Atmos. Sci., Vol. 36, 1997 - 2005 (1979). – Abstr. in Phys. Abstr., Vol. 83, Abstr. 72630 (1980).

082.021 **Modelling of the composition of the lower thermosphere taking account of the dynamics with applications to tidal variations of the [O I] 5577 Å airglow.**
R. A. Akmaev, G. M. Shved.
J. Atmos. Terr. Phys., Vol. 42, 705 - 716 (1980).

082.022 **On the influence of gravitational waves on the oxygen composition of the lower thermosphere.**
N. M. Gavrilov, A. I. Pogorel'tsev.
Geomagn. Aehron., Tom 20, 664 - 669 (1980). In Russian.

082.023 **Chemical evolution and ammonia in the early earth's atmosphere.** A. Henderson-Sellers, A. W. Schwartz.
Nature, Vol. 287, 526 - 528 (1980).

A constraint on research into the evolution of the earth's atmosphere may have been resolved by recent work on the photochemical reduction of atmospheric nitrogen by a naturally occurring form of TiO_2. The authors show that this source could have provided a small but vital abiological supply of ammonia in the early atmosphere of the earth.

082.024 **Tidal dynamics and composition variations in the thermosphere.** J. M. Forbes, M. E. Hagan.
J. Geophys. Res., Vol. 85, 3401 - 3406 (1980).

The effects of mutual diffusion between O and N_2 are examined for diurnal and semidiurnal tidal oscillations in the thermosphere.

082.025 **Spaced lidar and nightglow observations of an atmospheric sodium enhancement.**
B. R. Clemesha, V. W. J. H. Kirchhoff, D. M. Simonich, H. Takahashi, P. P. Batista.
J. Geophys. Res., Vol. 85, 3480 - 3484 (1980).

082.026 **Seasonal-latitudinal tidal structures of O, N_2, and total mass density in the thermosphere.**
J. M. Forbes, F. A. Marcos.
J. Geophys. Res., Vol. 85, 3489 - 3493 (1980).

The seasonal-latitudinal tidal structures of O, N_2, and total mass density in the thermosphere for minimum and maximum levels of solar activity are investigated by using a theoretical model and accelerometer data from the Atmosphere Explorer C, D, E, and Air Force S3-1 satellites.

082.027 **6300- Å airglow meridional intensity gradients.**
F. A. Herrero, J. W. Meriwether, Jr.
J. Geophys. Res., Vol. 85, 4191 - 4204 (1980).

Airglow meridional intensity gradients (MIG) that moved from east to west with the antisolar meridian have been observed during times of considerable magnetic activity. The magnitude of the gradient is around 0.1 R/km or more. The authors believe the MIG phenomenon develops as a result of a velocity gradient that is produced by the meeting of opposing meridional neutral winds. The conceptual model that evolved from these observations accounts for the overall behavior of the MIG as well as the normal midnight collapse. A simple calculation shows that the MIG can be used to estimate the meridional velocity gradient as well as the magnitude of the winds in the midnight pressure bulge.

082.028 **Atmospheric limitations of narrow-field optical astrometry.** L. Lindegren.
Astron. Astrophys., Vol. 89, 41 - 47 (1980).

The influence of refraction anomalies on optical differential astrometry is analysed. It is found that the mean error of the measured angle θ (rad) between two objects near zenith is $1''\!.3\theta^{0.25}T^{-0.5}$, where $T \gg 300\theta$ is the integration time in seconds. If the position of a star is measured relative to the centroid of several reference stars within a radius R (rad), the mean error in each coordinate is $0''\!.8R^{0.25}T^{-0.5}$, or $0''\!.003$ for $R = 10'$ and $T = 1$ h.

082.029 **On the measurement of atmospheric extinction by Rufener's method.** Z. Kviz.
Proc. Astron. Soc. Australia, Vol. 3, 326 - 327 (1979).

082.030 **Detection of electric quadrupole transitions in the oxygen A band at 7600 Å.** J. W. Brault.
J. Mol. Spectrosc., Vol. 80, 384 - 387 (1980). – Abstr. in Phys. Abstr., Vol. 83, Abstr. 85746 (1980).

082.031 **Phase-correlation scale effects in atmospheric optics.**
A. M. Schneiderman, D. P. Karo.
Opt. Acta., Vol. 27, 661 - 670 (1980). – Abstr. in Phys. Abstr., Vol. 83, Abstr. 94604 (1980).

082.032 **Results of an investigation of the vibration of star images.**
M. V. Bratijchuk, I. I. Motrunich, I. V. Shvalagin.
Problems of cosmic physics. Vyp. 15, (see 003.003), p. 37 - 47 (1980). In Russian.

The results of the reduction of measurements of more than 1000 trails of stars are given. In this paper are presented: the distributions of amplitudes of image tremor, the range of variation of them as function of zenith angle and aperture diameter, the seeing-disk diameter, the fluctuation of the amplitude of the image tremor, the angular radius of correlation region, the frequency spectrum of the image tremor.

082.033 **Ground-based measurements of atmospheric NO_2 by differential optical absorption.**
B. B. McMahon, E. L. Simmons.
Nature, Vol. 287, 710 - 711 (1980).

082.034 **Study of the connection between the zonal circulation of the atmosphere and the sector structure of the interplanetary magnetic field.**
A. A. Dmitriev, V. A. Malinnikov.
Nov. v soln.-zemn. svyazyakh. Mater. soveshch. po soln.-zemn. svyazyam v meteorol. Moskva, 1980, p. 7 - 8. In Russian. Abstr. in Ref. zh., 51. Astron., 8.51.418 (1980).

082.035 **Ion-induced nucleation of atmospheric water vapor at the mesopause.** F. Arnold.
Planet. Space Sci., Vol. 28, 1003 - 1009 (1980).

The possibility of aerosol formation at the mesopause by ion-induced water vapor nucleation is investigated. A simple kinetic model considering new information on ion-growth and -recombination processes as well as on mesospheric water vapor and temperatures is put forward.

082.036 **The behavior of ozone near the stratopause from two years of BUV observations.** R. D. McPeters.
J. Geophys. Res., Vol. 85, 4545 - 4550 (1980).

Two years of Nimbus 4 backscattered ultraviolet data have been recalibrated and reprocessed. A Laplace transform inversion was applied to radiances at 2876 Å and below for 560,000 individual scans for the period April 1970–May 1972. The behavior of ozone near 50 km as a function of time, latitude, and longitude is presented. The high-latitude 1-mbar ozone mixing ratio is maximum at the winter solstice, about 10 μg/g, and is minimum at the summer solstice, about 4 μg/g. Below 30° latitude the ozone is fairly constant at 4 μg/g. Ozone variability is large in winter and spring and small in summer and fall.

082.037 **Comment on 'The problem of nighttime stratospheric NO_3' by J. R. Herman.**
J. F. Noxon, R. B. Norton, W. R. Henderson, with a reply by J. R. Herman.
J. Geophys. Res., Vol. 85, 4556 - 4559 (1980). – See Abstr. 26.082.105.

082.038 **Density and temperature profiles obtained by lidar between 35 and 70 km.**
A. Hauchecorne, M.-L. Chanin.
Geophys. Res. Lett., Vol. 7, 565 - 568 (1980).

082.039 **Investigation of the atmosphere aboard Salyut 4.**
O. A. Avaste, U. Veismann, Ch. Willmann, K. Eerme.
Vestn. AN SSSR, 1980, No. 2, p. 60 - 65. In Russian. – Abstr. in Ref. zh., 62. Issled. kosm. prostranstva, 9.62.273 (1980).

082.040 **Dissipation of the primordial terrestrial atmosphere due to irradiation of the solar EUV.**
M. Sekiya, K. Nakazawa, C. Hayashi.
13th Lunar and Planetary Symposium, (see 012.018), p. 387 - 393 (1980).

A spherical and steady model of the escaping primordial atmosphere of the Earth is studied by integrating the hydrodynamic equations numerically. As heating sources, the authors take into account the solar EUV radiation, which is expected to have been stronger than the present one during T Tauri stage, the solar visible light, and the release of the gravitational energy of accretion planetesimals. The results show that the primordial atmosphere having existed in the early stage of the Earth's history, is dissipated within a period of 5×10^8 yrs.

082.041 **Theoretical infrared stratospheric emission spectra from balloons and satellites.**
B. Rebours, P. Rabache.
J. Quant. Spectrosc. Radiat. Transfer, Vol. 24, 517 - 523 (1980).

A line-by-line and layer-by-layer method is proposed for the analysis and computation of spectra in the far i.r. for a stratospheric medium. Observations of the lower layers of the stratosphere were carried out from a balloon and a satellite. The method is applied to the computation of the spectral luminance of a layer near the maximum ozone concentration. The results are shown as synthetic spectra and the presence of minor gaseous constituents in the stratospheric medium is discussed.

082.042 **Photochemical production of formaldehyde in earth's primitive atmosphere.**
J. P. Pinto, G. R. Gladstone, Y. L. Yung.
Science, Vol. 210, 183 - 185 (1980).

Formaldehyde could have been produced by photochemical reactions in earth's primitive atmosphere, at a time when it consisted mainly of molecular nitrogen, water vapor, carbon dioxide, and trace amounts of molecular hydrogen and carbon monoxide.

082.043 **Influence of the non-stationary motion of the earth's surface on the atmosphere and on the neutral component of the ionosphere.** V. A. Pavlov.
Geomagn. Aehron., Tom 20, 865 - 874 (1980). In Russian.

082.044 **Influence of circulating conditions of the atmosphere on the astronomical refraction.**
A. V. Alekseev, B. D. Belan, N. F. Nelyubin.
Soveshchanie po atmos. opt. Tez. dokl. Ch. 2. Tomsk, 1980, p. 206 - 209. In Russian. – Abstr. in Ref. zh., 51. Astron., 10.51.189 (1980).

082.045 **Optical investigations made by the crew of the second expedition of the orbital scientific station Salyut 6.**
A. I. Lazarev, V. V. Kovalenok, A. S. Ivanchenkov, S. V. Avakyan.
Opt.-mekh. prom-st', 1980, No. 6, p. 43 - 49. In Russian. Abstr. in Ref. zh., 62. Issled. kosm. prostranstva, 10.62.77 (1980).

082.046 **Advances in noctilucent cloud research in the space era.** O. A. Avaste, A. V. Fedynsky (*Fedynskij*), G. M. Grechko, V. I. Sevastyanov (*Sevast'yanov*), Ch. I. Willmann.
Pure Appl. Geophys., Vol. 118, 528 - 580 (1980). – Abstr. in Phys. Abstr., Vol. 83, Abstr. 105095 (1980).

082.047 **Schwächung der solaren UV-Strahlung durch Ozon und andere erdatmosphärische Spurenstoffe.**
K. Dehne.
Evolution der Planetenatmosphären und des Lebens, (see 012.032), p. 140 - 163 (1980).

082.048 **Decrease in CO_2 mixing ratio observed in the stratosphere.** W. Bischof, P. Fabian, R. Borchers.
Nature, Vol. 288, 347 - 348 (1980).

The results reported show that the CO_2 mixing ratio is not constant with altitude but rather decreases in the stratosphere, by about 7 p.p.m.v., between the tropopause and 33 km. One conclusion is that recently increased concentrations of CO_2 in the troposphere have not propagated far into the stratosphere.

082.049 **On a possibility of the measurement of aerosol content in the atmosphere from a geostationary satellite.**
T. Z. Dworak.
Artif. Satell., Vol. 15, No. 1, p. 67 - 71 (1980).

082.050 **Two-dimensional model calculations of stratospheric HCl and ClO.**
C. Miller, J. M. Steed, D. L. Filkin, J. P. Jesson.
Nature, Vol. 288, 461 - 464 (1980).

Most previous attempts to estimate the effects of atmospheric contaminants on stratospheric ozone have involved one-dimensional models of the troposphere and stratosphere. The authors have developed a two-dimensional model which incorporates the refinements of the most advanced one-dimensional models: they report that, contrary to expectation, the calculated profiles of HCl and ClO are insignificantly different from those of comparable one-dimensional models.

082.051 **Investigation of sodium in the upper atmosphere.** I. Yeğingil, H. Ögelman, N. Kiziloglu.
J. Geophys. Res., Vol. 85, 5507 - 5510 (1980).

Upper atmospheric atomic sodium abundances have been measured near Ankara, Turkey, from February 1976 to April 1977. The daytime absorption measurements were made with a Pepsios spectrometer using the solar radiation near the D2 line of sodium at 5890 Å. Seasonal and daily variations of the sodium abundance were examined. A model is proposed in which the atomic sodium has a cloudlike structure, and the parameters obtained in this way suggest that the 'clouds' have a mean column density of 1.5×10^9 atoms/cm^2.

082.052 **Measurement of stratospheric water vapor by cryogenic collection.**
W. Pollock, L. E. Heidt, R. Lueb, D. H. Ehhalt.
J. Geophys. Res., Vol. 85, 5555 - 5568 (1980).

082.053 **Identification of isolated NO lines in balloon-borne infrared solar spectra.**
F. J. Murcray, A. Goldman, D. G. Murcray, G. R. Cook, J. W. Van Allen, R. D. Blatherwick.
Geophys. Res. Lett., Vol. 7, 673 - 676 (1980).

Balloon-borne infrared solar spectra at ~ 0.02 cm^{-1} resolution show a number of atmospheric NO lines isolated from other atmospheric and solar lines in the 1830 - 1930 cm^{-1} region. Typical spectra are presented and NO total column values are derived.

082.054 **Stratospheric NO_2 at night from stellar spectra in the 440 nm region.**
J. P. Naudet, P. Rigaud, D. Huguenin.
Geophys. Res. Lett., Vol. 7, 701 - 703 (1980).

NO_2 was measured in the stratosphere at night by means of a balloon-borne spectrometer, pointing at a bright rising star, in the 427 to 452 nm range. The vertical distribution of NO_2 was deduced using the tangent ray technique between 23 and 38 km. The concentration was $(6 \pm 1.5) \times 10^9$ mol.cm^{-3} at 25 km, passing through a minimum of $(2 \pm 1) \times 10^9$ mol.cm^{-3} at 29 km and rising again up to $(4 \pm 0.8) \times 10^9$ mol.cm^{-3} at 32 km.

082.055 **Far infrared measurement of stratospheric HCl.** K. V. Chance, J. C. Brasunas, W. A. Traub.
Geophys. Res. Lett., Vol. 7, 704 - 706 (1980).

082.056 **An observed annual cycle in tropical upper stratospheric and mesospheric ozone.**
J. E. Frederick, R. B. Abrams, R. Dasgupta, B. Guenther.
Geophys. Res. Lett., Vol. 7, 713 - 716 (1980).

082.057 **The importance of energetic particle precipitation on the chemical composition of the middle atmosphere.** R. M. Thorne.
Pure Appl. Geophys., Vol. 118, 128 - 151 (1980). – Abstr. in Phys. Abstr., Vol. 83, Abstr. 109213 (1980).

082.058 **Solar UV radiation and its absorption in the mesosphere and stratosphere.** M. Nicolet.
Pure Appl. Geophys., Vol. 118, 3 - 19 (1980). – Abstr. in Phys. Abstr., Vol. 83, Abstr. 109248 (1980).

082.059 **Noctilucent clouds over western Europe during 1979.** D. H. McIntosh, M. Hallissey.
Meteorol. Mag., Vol. 109, 182 - 184 (1980). – Abstr. in Phys. Abstr., Vol. 83, Abstr. 109289 (1980).

082.060 **Remote sensing of atmospheric water vapor and cloud liquid.**
G. Elgered, B. O. Rönnäng, J. I. H. Askne.
Res. Lab. Electron. Onsala Space Obs., Res. Rep. No. 141, 27 pp. (1980).

082.061 **Una medicion de lineas de radiacion γ atmosferica a bajas latitudes.**
H. S. Ghielmetti, V. J. Mugherli, I. N. Azcarate.
Inst. Astron. Fis. Espacio, Buenos Aires, Ser. Publ. Cient., No. IAFE-PC-1-79, 28 pp. (1979).

An experiment carried out with a collimated gamma-radiation detector that was transported by a stratospheric balloon is described. As a characteristic feature of the energy loss spectrum, the presence of a peak is observed. That peak corresponds to the 511 KeV line produced by positron annihilation in both the atmosphere and the lead collimator. Upper limits for other atmospheric γ-ray lines are obtained. Finally, the detector sensitivity for the observation of the more intense solar lines is derived from this experiment.

082.062 **Measurements of atmospheric absorption in the range 5 - 17 cm^{-1} and its temperature dependence.**
R. J. Emery, A. M. Zavody, H. A. Gebbie.
J. Atmos. Terr. Phys., Vol. 42, 801 - 807 (1980).

Field measurements of atmospheric absorption made in the wavenumber range 5 - 17 cm^{-1} are found to have values in excess of prediction based on absorption by monomeric water molecules. The measurements also show unexplained variations with time, although on average the absorption in the region of 7.1 cm^{-1} is found to be close to comparison laboratory values. The absorption measured in excess of water monomer prediction is analysed for its dependence on temperature. Results are also given for measurements in fog conditions.

082.063 **Changes in the planetary heat balance with chemical changes in air.** J. W. Chamberlain.
Planet. Space Sci., Vol. 28, 1011 - 1018 (1980).

082.064 **Nonpermanent nighttime H Lyman alpha emissions at low and middle latitudes, detected from the D2A satellite.** S. Cazes, C. Emerich.
J. Geophys. Res., Vol. 85, 6049 - 6054 (1980).

Strong nonpermanent emissions are observed from two H Ly α experiments placed on board the satellite D2A. They are superimposed to the permanent emission due to resonant scattering of the Ly α solar line by telluric atomic hydrogen. These emissions are detected both above and under the altitude of the satellite ($\sim$ 500 km), in a latitude range ±40°. The correlation observed between their occurrence frequency and the *Kp* index strongly suggests that the source of excitation is precipitation of particles.

082.065 **An anomalous low-latitude phenomenon observed by the OGO 6 UV photometer between 400 and 800 km.** G. E. Thomas.
J. Geophys. Res., Vol. 85, 6055 - 6062 (1980).

082.066 **Some results of measurements of atmospheric submillimetre radiation from "SALJUT - 6".**
V. N. Bakun, G. M. Grechko, A. S. Ivanchenkov, V. V. Kovalenok, E. P. Kropotkina, Yu. V. Romanenko, A. E. Salomonovich, S. V. Solomonov.
Space Research, Vol. XX, (see 012.043), 11 - 13 (1980).

The paper gives some new results of measurements of thermal (submillimetre) radiation from the earth's atmosphere in two spectral channels (60-130 μm and 500-800 μm). Measurements were carried out with the telescope BST–IM on the piloted orbital space station "SALJUT - 6" in February, June and September 1978.

082.067 **An empirical temperature model of the middle atmosphere of the southern hemisphere.**
Yu. Koshelkov.
Space Research, Vol. XX, (see 012.043), 41 - 47 (1980).

082.068 **The latitudinal and seasonal variation of atomic oxygen deduced from observations of the E-region O I 557.7 nm airglow.** L. L. Cogger, J. S. Murphree.
Space Research, Vol. XX, (see 012.043), 115 - 120 (1980).
Photometric observations of the airglow limb at 557.7 nm have been made from the ISIS 2 satellite since 1971. Sufficient measurements of peak slant emission rate have been acquired so that prominent characteristics of the seasonal-latitudinal variation have been identified. From these results, the corresponding variation of atomic oxygen concentration can be inferred.

082.069 **Measurements of the spectral profile of Balmer alpha emission from the hydrogen geocorona.**
J. W. Meriwether, Jr., S. K. Atreya, T. M. Donahue, R. G. Burnside.
Geophys. Res. Lett., Vol. 7, 967 - 970 (1980).
Instrumental improvements responsible for a factor of 25 increase in the sensitivity of the Fabry-Perot interferometer enable the authors to observe for the first time the short wavelength depletion of the Balmer α spectral profile due to hydrogen escape. These results are shown to be consistent with the implications of OGO-5 observations by Bertaux.

082.070 **Electromagnetic beam propagation in turbulent media: an update.** R. L. Fante.
Proc. IEEE, Vol. 68, 1424 - 1443 (1980).
In 1975, the author reviewed the recent developments on wave propagation in turbulent media. In this paper, he updates that paper to include recent results on adaptive optics, intensity scintillations, pulse propagation, atmospheric measurements at visible, IR, and millimeter wave frequencies, speckle interferometry, and other related topics.

082.071 **Visibility of halos and rainbows.**
S. D. Gedzelman.
Appl. Opt., Vol. 19, 3068 - 3074 (1980).

082.072 **Atmospheric phenomena before and during sunset.**
M. Menat.
Appl. Opt., Vol. 19, 3458 - 3468 (1980).
The atmospheric transmittance and the astronomical refraction for low-elevation trajectories are discussed and quantitatively developed. The results are used to describe and calculate some of the atmospheric phenomena occurring shortly before and during sunset, such as the diminishing apparent luminance of the sun, its shape during sunset, and the green flash.

082.073 **Statistics of stellar scintillation.**
G. Parry, J. G. Walker.
J. Opt. Soc. America, Vol. 70, 1157 - 1159 (1980).
The authors present the results of measurements of the first five moments of the intensity fluctuations of stellar scintillation. Measurements were made using photon counting techniques. The moments were found to be consistent with a log normal model provided the normalized second moment of the intensity was less than ~1.65. Above this, departure from log normal behavior occurred.

082.074 **An analytical model of the daily variation in air density.** B.-k. Lu, C.-a. Sun.
Acta Astron. Sinica, Vol. 21, 404 - 408 (1980). In Chinese.

082.075 **Taking into account the spectral transmittance of the atmosphere in determining the astronomical refraction and distance of an extraatmospheric object.**
N. F. Nelyubin.
Soveshchanie po atmos. opt. Tez. dokl. Ch. 2. Tomsk, 1980, p. 186 - 189. In Russian. – Abstr. in Ref. zh., 52. Geod. Aehrosemka, 11. 52.70 (1980).

082.076 **Tables of astronomical refraction.**
A. I. Nefed'eva.
Izv. Astron. Ehngel'gardt. Obs., No. 45, p. 3 - 81 (1978). In Russian.

082.077 **Determination of the water vapour content in the earth's atmosphere using stellar spectra.**
V. D. Galkin, A. A. Arkharov.
Astron. Tsirk., No. 1096, p. 5 - 7 (1980). In Russian.

082.078 **Ozone, ultraviolet flux and temperature of the paleoatmosphere.**
J. S. Levine, R. E. Boughner, K. A. Smith.
Limits of life, (see 012.052), p. 105 - 119 (1980).
Surface and atmospheric levels of paleoatmospheric O_3 were calculated using a detailed photochemical model, including the chemistry of the oxygen, nitrogen, and hydrogen species and the effects of vertical transport.

082.079 **Spektrophotometrische Untersuchungen von He-I/II-Resonanzstrahlungen mit der Raketennutzlast ASTRO-HEL.** H. J. Fahr, G. Lay, C. Wulf-Mathies.
Mitt. Astron. Ges., Nr. 50, p. 83 - 89 (1980).

082.080 **Winde und Turbulenzen in der polaren unteren Thermosphäre.** K. Schuchardt, P. Blum.
Mitt. Astron. Ges., Nr. 50, p. 104 - 108 (1980).

082.081 **On the restoration of perturbations of the vertical density profile in the atmosphere from photographs of the sun received on an artificial earth satellite.**
S. V. Sokolovskij.
Soveshchanie po atmos. opt. Tez. dokl. Ch. 2. Tomsk, 1980, p. 16 – 19. In Russian. – Abstr. in Ref. zh., 51. Astron., 12.51.132 (1980).

082.082 **Determination of the turbulence coefficient from results of measurements of meteorological elements and zenith distances.** D. I. Maslich, L. S. Khizhak, I. I. Didukh, N. D. Josipchuk, M. B. Yaskilka.
Soveshchanie po atmos. opt. Tez. dokl. Ch. 2. Tomsk, 1980, p. 179 - 181. In Russian. – Abstr. in Ref. zh., 51. Astron., 12.51.133 (1980).

082.083 **Experimental determination of the astronomical refraction from observations on large zenith distances.** D. I. Maslich, B. A. Kovalenko, M. I. Rusin, A. E. Filippov, P. I. Koval', N. F. Nelyubin.
Soveshchanie po atmos. opt. Tez. dokl. Ch. 2. Tomsk, 1980, p. 182 - 185. In Russian. – Abstr. in Ref. zh., 51. Astron., 12.51.134 (1980).

082.084 **Short-periodic pulsations of a charged particle stream in the stratosphere of the auroral zone.**
Yu. I. Barannikov, O. A. Barsukov, P. F. Gavrilov.
Geomagn. Aehron., Tom 20, 1101 - 1102 (1980). In Russian.

082.085 **Intensity variations of O I 6300 Å emission as evidence of planetary effects in the upper atmosphere.**
E. M. Trunkovskij.
Geomagn. Aehron., Tom 20, 1118 - 1120 (1980). In Russian.

082.086 **Turbopause internal gravity waves, 557.7-nm airglow, and eddy diffusion coefficient.**
E. Battaner, A. Molina.
J. Geophys. Res., Vol. 85, 6803 - 6810 (1980).
A study of the properties of internal gravity waves by means of observations of the 557.7-nm emission line has been carried out. Special emphasis was put on the determination of the kinetic energy density of the waves, which have maxima in winter and in summer, and on the determination of the enhanced diffusion tensor K. The possible relation between

K and *D* region winter anomalous absorption has been also examined.

082.087 **Verification of the dependence of atmospheric extinction coefficients upon the spectral types of stars.** K. Zdanavičius.
Bull. Vilnius Astron. Obs., Nr. 55, p. 3 - 10 (1980). In Russian.

082.088 **On the atmospheric extinction on the Maidanak Mountain.** K. Zdanavičius, D. Macijauskas.
Bull. Vilnius Astron. Obs., Nr. 55, p. 11 - 19 (1980). In Russian.

082.089 **Investigation of ground meteorological fields in the course of the day-time determination of declinations.** V. P. Sibilev.
Astrometr. Astrofiz., Vyp. (No.) 42, p. 77 - 82 (1980). In Russian.

The results of the analysis of the temperature field both in the pavilion and in the ground layer up to the height of 16 m under the ground level are dealt with. The horizontal pressure distribution is also investigated.

082.090 **Investigation of the optical instability of the earth's atmosphere.**
I. I. Motrunich, I. V. Shvalagin.
Soveshchanie po atmos. opt. Tez. dokl. Ch. 2. Tomsk, 1980, p. 120 - 123. In Russian. – From Ref. zh., 51. Astron., 1.51.127 (1981).

082.091 **Analysis of random refraction errors with the help of the collocation method.**
D. I. Maslich, S. D. Volzhanin.
Soveshchanie po atmos. opt. Tez. dokl. Ch. 2. Tomsk, 1980, p. 147 - 150. In Russian. – Abstr. in Ref. zh., 51. Astron., 1.51.128 (1981).

082.092 **Determination of the angles of astronomical refraction from data of observations of reference stars.**
I. F. Kushtin.
Soveshchanie po atmos. opt. Tez. dokl. Ch. 2. Tomsk, 1980, p. 155 - 157. In Russian. – Abstr. in Ref. zh., 51. Astron., 1.51.129; 52. Geod. Aehrosemka, 1.52.102 (1981).

082.093 **Improvement of indirect methods of calculation of refraction of optical rays in the atmosphere.**
I. F. Kushtin.
Soveshchanie po atmos. opt. Tez. dokl. Ch. 2. Tomsk, 1980, p. 158 - 161. In Russian. – Abstr. in Ref. zh., 51. Astron., 1.51.130 (1981).

082.094 **Results of an investigation of astronomical and satellite refraction from aerological data.**
I. I. Motrunich, I. V. Shvalagin.
Vliyanie atmos. na astron. nablyudeniya v optich. i radiodiapazonakh. Tez. dokl. soveshch. probl.-temat. gruppy po teor. astrometr. sekts. astrometr. AS AN SSSR. 1980. Irkutsk, 1980, p. 29 - 31. In Russian. – Abstr. in Ref. zh., 51. Astron., 1.51.132 (1981).

082.095 **Variation of the refraction angles under various climatic conditions.** V. A. Yakovlev.
Vliyanie atmos. na astron. nablyudeniya v optich. i radiodiapazonakh. Tez. dokl. soveshch. probl.-temat. gruppy po teor. astrometr. sekts. astrometr. AS AN SSSR. 1980. Irkutsk, 1980, p. 40 - 42. In Russian. – Abstr. in Ref. zh., 51. Astron., 1.51.133 (1981).

082.096 **On the character of influence and on the method of calculation of the chromatic refraction in determination of declinations of stars with the methods of position astronomy.**
B. K. Bagil'dinskij, E. G. Zhilinskij, V. D. Shkutov.
Vliyanie atmos. na astron. nablyudeniya v optich. i radiodiapazonakh. Tez. dokl. soveshch. probl.-temat. gruppy po teor. astrometr. sekts. astrometr. AS AN SSSR. 1980. Irkutsk, 1980, p. 52 - 54. In Russian. – Abstr. in Ref. zh., 51. Astron., 1.51.135 (1981).

082.097 **Instrumentation, method and results of measurements of the infrared radiation stream of the upper atmosphere on the 25-m artificial earth satellite Meteor.**
Yu. M. Kondrat'ev, Yu. A. Martynov, V. G. Sochnev, S. G. Yakovlev.
Tr. N.-I. tsentra izuch. prirod. resursov, 1980, No. 11, p. 15 - 24. In Russian. – Abstr. in Ref. zh., 62. Issled. kosm. prostranstva, 1.62.118 (1981).

082.098 **Diffusion model of an artificial chemiluminescent cloud in the upper atmosphere.**
N. V. Kulikova, V. N. Balabanova.
Astron. Vestn., Tom 14, 119 - 125 (1980). In Russian.

Data information subsystem – Upper Atmosphere Data Bank. See Abstr. 002.051.

Atmospheric phenomena. See Abstr. 003.072.

Fifth ESA Symposium on European rocket & balloon programmes & related research. Conference held at Bournemouth, England, 14 - 18 April 1980.
See Abstr. 012.058.

The genetic code as fossil vestiges of the primitive atmosphere. See Abstr. 015.013.

A differential correction theory for the determination of atmospheric functions. See Abstr. 021.025.

An accurate representation of the transmission functions of the H_2O and CO_2 infrared bands.
See Abstr. 022.005.

Effective O_2 absorption cross sections for solar absorption data from Lyman α ion chambers.
See Abstr. 022.147.

The O (^1S) quantum yield from O_2^+ dissociative recombination. See Abstr. 022.155.

High resolution spectrum of the N_2^+ Meinel system to 11250 Å. See Abstr. 022.168.

On the accuracy of determination of the angle of astronomical refraction by the visual method with the help of universal field instruments. See Abstr. 031.605.

Taking into account refraction effects in determination of the position of artificial celestial bodies.
See Abstr. 031.624.

Estimate of and allowance for the influence of atmospheric conditions on the accuracy of measurement of displacement of the effective center of solar radio radiation.
See Abstr. 031.626.

An infrared hygrometer for astronomical site testing. See Abstr. 034.051.

Ein Infrarot-Absorptions-Hygrometer.
See Abstr. 034.075.

Atmospheric limitations to narrow-field optical astrometry. See Abstr. 041.018.

Comparative study of some formulae approximating the atmospheric density and their utilization in the theory of artificial satellites motion. See Abstr. 052.041.

Sur la variation du demi-grand axe de l'orbite d'un satellite artificiel, causée par la résistance atmosphérique. See Abstr. 052.046.

Influence of fluctuations of the atmosphere's density on the accuracy of determination and forecast of orbits of artificial earth satellites. See Abstr. 052.056.

Variations in the quasi-nodal period of the satellite Explorer 19 caused by the atmospheric drag. See Abstr. 054.009.

Scattering by nonspherical particles of size comparable to a wavelength: a new semi-empirical theory and its application to tropospheric aerosols. See Abstr. 063.036.

The diffusivity factor re-examined. See Abstr. 063.044.

Seasonal variation of oriental sunspot sightings. See Abstr. 072.016.

Search for optical modulation of the solar corona during the February 16, 1980 total solar eclipse. See Abstr. 074.099.

Observation of Baily's beads during the total solar eclipse of 1980 February 16. See Abstr. 079.107.

Detection of 160-min solar oscillations and atmospheric extinction. See Abstr. 080.005.

Energy transport by gravity waves. See Abstr. 083.040.

Magnetosphere, ionosphere and atmosphere interactions. See Abstr. 084.098.

Die Strahlungssituation der Erde. See Abstr. 085.017.

Ionen- Molekül-Reaktionen in Planetenatmosphären. See Abstr. 091.022.

Die Venus als Glashaus. Die Bedeutung des Kohlendioxid für das Klima von Erde und Venus. See Abstr. 093.008.

Meteors and atmospheres. See Abstr. 104.023.

The influence of the atmosphere on radar meteor rates. See Abstr. 104.024.

Interaction of large bodies with the earth's atmosphere. See Abstr. 104.040.

Optical spectroscopy of interplanetary dust collected in the earth's stratosphere. See Abstr. 106.004.

Dissipation of the rare gases contained in the primordial Earth's atmosphere. See Abstr. 107.025.

The vertical component of 1–20 MeV gamma rays at balloon altitudes. See Abstr. 142.510.

A method to study the atmospheric influence on the isotopic composition of primary cosmic rays applied to the elements carbon and oxygen. See Abstr. 143.044.

083 Ionosphere

083.001 **Ionospheric currents associated with solar flares: a short review.** M. Roy.
Bull. Astron.Soc. India, Vol. 8, 15 - 20 (1980).
The enhanced radiation of the X-rays and the ultra - violets during a chromospheric flare can induce a qualitative change in the ionospheric current system. The observational results of this effect, popularly known as geomagnetic solar flare effect, are briefly reviewed. Various theoretical formulations to explain the morphological results are described.

083.002 **A computer simulation of the midlatitude plasmasphere and ionosphere.**
E. R. Young, D. G. Torr, P. Richards, A. F. Nagy.
Planet. Space Sci., Vol. 28, 881 - 893 (1980).
The main objective of the paper is to report the development of a new computer model to simulate species density, temperature and plasma flow in the ionosphere and plasmasphere. The paper provides a general description of the code for potential users.

083.003 **Midlatitude measurements of the ionospheric electric field during the ALADDIN programme.**
D. Rees, N. C. Maynard.
J. Atmos. Terr. Phys., Vol. 42, 577 - 582 (1980).

083.004 **Global distribution of electric fields and currents in the ionosphere produced by field-aligned currents of magnetospheric origin–I.** K. Maekawa.
J. Atmos. Terr. Phys., Vol. 42, 673 - 682 (1980).

083.005 **Active experiments in space plasmas.**
M. J. Rycroft.
Nature, Vol. 287, 7 - 8 (1980).

083.006 **Quasi-periodic variations of ionospheric absorption with characteristic changes during a solar cycle.**
F. Marcz.
Acta Geod. Geophys. Montan, Vol. 14, No. 1 - 2, p. 227 - 236 (1979). – Abstr. in Phys. Abstr., Vol. 83, Abstr. 77193(1980).

083.007 **The quiet midlatitude *D*-region: a comparison between modeling efforts and experimental measurements.** M. G. Heaps, J. M. Heimerl.
J. Atmos. Terr. Phys., Vol. 42, 733 - 742 (1980).

083.008 **Surplus radiation from data of an experiment at the artificial earth satellite Cosmos 721.**
S. N. Kuznetsov, S. P. Ryumin, G. A. Trebukhovskaya.
Geomagn. Aehron., Tom 20, 595 - 602 (1980). In Russian.

083.009 **Calculation of the relative contribution of high-energy electrons arising at different altitudes in the atmosphere to the albedo stream at a height of ~ 200 km.**
R. N. Basilova, E. I. Kogan-Laskina, G. I. Pugacheva.
Geomagn. Aehron., Tom 20, 603 - 606 (1980). In Russian.

083.010 **Heights of the equatorial F region and their forecast.**
O. V. Chernyshev, B. S. Shapiro.
Geomagn. Aehron., Tom 20, 607 - 610 (1980). In Russian.

083.011 **An approximate solution of the diffusion equation of oxygen ions in the ionospheric F2 region.**
V. M. Polyakov, M. A. Koen, A. D. Kazmirov.
Geomagn. Aehron., Tom 20, 611 - 616 (1980). In Russian.

083.012 **On variations of the electron concentration in the outer ionosphere at midlatitudes.**
M. N. Vlasov, L. G. Tashkinova.
Geomagn. Aehron., Tom 20, 617 - 620 (1980). In Russian.

083.013 **On relaxation of an anisotropic monoenergetic electron stream in the ionosphere.**
N. I. Izhovkina.
Geomagn. Aehron., Tom 20, 621 - 625 (1980). In Russian.

083.014 **Model definition of volume characteristics of ionospheric irregularities.**
V. D. Gusev, N. P. Ovchinnikova.
Geomagn. Aehron., Tom 20, 626 - 631 (1980). In Russian.

083.015 **Boundary conditions for MHD waves in the ionosphere.** P. P. Belyaev, S. V. Polyakov.
Geomagn. Aehron., Tom 20, 637 - 641 (1980). In Russian.

083.016 **Planetary development of the ionospheric substorm of 1974, September 18.**
G. F. Deminova, L. A. Yudovich.
Geomagn. Aehron., Tom 20, 742 - 743 (1980). In Russian.

083.017 **Empirical modelling of variations of the F2 region during positive ionospheric disturbances.**
R. A. Zevakina, M. V. Kiseleva.
Geomagn. Aehron., Tom 20, 746 - 749 (1980). In Russian.

083.018 **Forecasting the development of an ionospheric disturbance.**
V. P. Kuleshova, E. V. Lavrova, L. N. Lyakhova.
Geomagn. Aehron., Tom 20, 749 - 751 (1980). In Russian.

083.019 **Definition of the parameters of the ionosphere by the method of a plasma line.**
V. D. Tereshchenko.
Geomagn. Aehron., Tom 20, 751 - 752 (1980). In Russian.

083.020 **Production of nitrous oxide in the auroral *D* and *E* regions.** E. C. Zipf, S. S. Prasad.
Nature, Vol. 287, 525 - 526 (1980).

083.021 **Investigation of the variations of the O^+ and N^+ ion concentrations, of the dynamics of the ionosphere and of the streams of energetic electrons in the outer ionosphere of the earth aboard the artificial earth satellite Meteor. 2. Measurements in the equatorial ionosphere.**
G. V. Ivanov, I. A. Perkov, L. I. Pogulyaevskij,
Yu. A. Romanovskij, Yu. P. Rylov, A. P. Yaichnikov.
Kosm. Issled., Tom 18, 733 - 739 (1980). In Russian.

083.022 **Investigations of the sporadic radio radiation of the sun and of the parameters of the earth's ionosphere aboard Intercosmos-Copernicus 500. 5. Results of investigations of the electron concentration in the ionosphere.**
V. I. Aksenov, G. M. Artem'eva, G. P. Komrakov,
L. A. Skrebkova, H. Hanasz.
Kosm. Issled., Tom 18, 740 - 747 (1980). In Russian.

083.023 **Photochemistry of the lower F region and structural parameters of the ionosphere from data of complex ground-based and rocket experiments.**
A. V. Biryukov, N. P. Danilkin, P. F. Denisenko,
G. M. Kucherenko, I. A. Knorin, V. T. Rodionova,
V. A. Rudakov, V. V. Sotskij, Yu. N. Faer, L. A. Shnyreva,
N. M. Shyutte.
Kosm. Issled., Tom 18, 748 - 753 (1980). In Russian.

083.024 **Behaviour of the ionospheric E region during aurorae.** V. A. Kurilov, V. P. Nesterov.
Tr. Arkt. i Antarkt. NII, 1980, No. 366, p. 36 - 52. In Russian. Abstr. in Ref. zh., 62. Issled. kosm. prostranstva, 9.62.252 (1980).

083.025 **Effects of non-linearity and the effective recombination coefficient in the E-region of the ionosphere.**
L. A. Antonova, G. S. Ivanov-Kholodnyj.
Geomagn. Aehron., Tom 20, 802 - 808 (1980). In Russian.

083.026 **Investigations of the variability of the ionosphere at 500 km height from data of the artificial earth satellite Cosmos 900 (autumn 1977).**
G. L. Gdalevich, V. D. Ozerov, I. S. Vsekhsvyatskaya, L. N. Novikova, T. N. Soboleva.
Geomagn. Aehron., Tom 20, 809 - 816 (1980). In Russian.

083.027 **The main ionospheric gap in the day-light sector from vertical sounding data.**
N. P. Ben'kova, E. F. Kozlov, A. M. Mozhaev, N. K. Osipov, N. I. Samorokin.
Geomagn. Aehron., Tom 20, 817 - 821 (1980). In Russian.

083.028 **Regular variations of the midlatitude ionospheric structure and systems of convective motions inside the plasmapause.**
Yu. M. Berezin, N. K. Osipov, T. A. Yankovskaya, T. I. Sokolova.
Geomagn. Aehron., Tom 20, 822 - 827 (1980). In Russian.

083.029 **Long-period and short-period variations of the ionospheric parameters from complex observations on Cuba.** B. Laso, L. A. Lobachevskij, N. I. Potapova, I. A. Frejzon, B. S. Shapiro.
Geomagn. Aehron., Tom 20, 828 - 836 (1980). In Russian.

083.030 **Influence of longitudinal electrostatic fields on ionospheric structure.**
M. G. Deminov, V. P. Kim, V. V. Khegaj.
Geomagn. Aehron., Tom 20, 837 - 840 (1980). In Russian.

083.031 **Mechanism of formation of sporadic E-layers in the high-latitude ionosphere.**
M. N. Vlasov, E. V. Mishin, V. A. Telegin.
Geomagn. Aehron., Tom 20, 932 - 933 (1980). In Russian.

083.032 **The role of precipitating hard electrons in the midlatitude night-time lower ionosphere.**
J. Laštovička.
Geomagn. Aehron., Tom 20, 934 - 936 (1980). In Russian.

083.033 **Atmospheric processes and the structure of the lower ionosphere in the winter period.**
V. G. Kidiyarova, Yu. A. Ryabov, A. A. Yastrebov.
Geomagn. Aehron., Tom 20, 937 - 940 (1980). In Russian.

083.034 **On a correlation of variations of the outer ionosphere with the B_z-component of the interplanetary magnetic field.** R. A. Zevakina, M. V. Kiseleva.
Geomagn. Aehron., Tom 20, 940 - 943 (1980). In Russian.

083.035 **On the nature of the night anomaly of the ion composition of the earth's outer ionosphere at solar activity minimum.** A. P. Yaichnikov.
Geomagn. Aehron., Tom 20, 943 - 944 (1980). In Russian.

083.036 **Airborne studies of equatorial *F* layer ionospheric irregularities.** E. J. Weber, J. Buchau, J. G. Moore.
J. Geophys. Res., Vol. 85, 4631 - 4641 (1980).
Radio wave and optical experiments were conducted onboard a U. S. Air Force research aircraft in March 1977 and March 1978 at low magnetic latitudes to investigate the effects of *F* region electron density irregularities on transionospheric communications links. Imaging photometer, ionosonde, and satellite amplitude scintillation measurements were used to monitor the development and motion of *F* region 6300-Å O I airglow depletions, spread *F*, and scintillation producing irregularities that are all associated with low-density bubbles in the postsunset equatorial ionosphere.

083.037 **Determination of model vertical distributions of electron temperature and concentration by satellite and ground measurements.** K. B. Serafimov.
Dokl. Bolg. AN, Vol. 33, 199 - 202 (1980). In Russian. Abstr. in Ref. zh., 62. Issled. kosm. prostranstva, 10.62.131 (1980).

083.038 **News on the luminescence of the ionosphere.**
V. V. Ryumin, G. M. Grechko, G. M. Nikol'skij, S. A. Savchenko, A. I. Simonov.
Zemlya Vselennaya, 1980, No. 6, p. 18. In Russian.

083.039 **Variations of ion composition and electron density accompanying solar flares.** M. Dymek.
Artif. Satell., Vol. 15, No. 1, p. 29 - 51 (1980).
An attempt has been made to investigate changes of ion composition under flare conditions basing on accurate analysis of aeronomic processes and using the Donnelly's model of flare radiation prepared for the ionospheric use. Assuming that the quasistationary condition is fulfilled for related slow component and very slow component of flares, the steady-state positive and negative ions composition in the 70 - 300 km height range has been calculated for several characteristic moments of flare duration.

083.040 **Energy transport by gravity waves.**
R. P. Singh, K. Singh, I. M. L. Das.
Phys. Scr., Vol. 22, 417 - 421 (1980).
Energy transport by gravity waves has been studied. It is shown that the high period waves heat the upper *F*-region whereas low period waves are responsible for lower *F*-region heating. Radio-fading records have been analysed and the rate of heating caused by gravity waves has been computed.

083.041 **A comparison of solar wind and ionospheric ion acoustic waves.** P. M. Kintner, M. C. Kelley.
J. Geophys. Res., Vol. 85, 5162 - 5164 (1980).
Ion acoustic waves produced during the Trigger experiment are compared to ion acoustic waves observed in the solar wind. After normalizing to the Debye length the spectra are nearly identical, although the ionospheric wave relative energy density is 10^2 larger than the solar wind case.

083.042 **On the location of the ionospheric current systems responsible for the lunar and solar magnetic variations.**
E. C. Butcher.
Geophys. J. R. Astron. Soc., Vol. 63, 775 - 782 (1980).
From a study of the sunspot cycle influence on the lunar (*L*) and solar (*S*) daily geomagnetic variations, Malin, Cecere & Palumbo have suggested that the electric currents responsible for these variations do not flow in the *E*-region as has hitherto been accepted. Indeed they suggested that the *L* current may even flow in the *F*2 region. The consequences of such a suggestion are considered and it is shown that an *L* current in the *F*-region is highly unlikely from the dynamical point of view. The evidence for *E*-region currents is presented and it is suggested that the variation of *E*-region electron density used by Malin et al. to indicate conductivity changes in that region may not be a reliable indicator of such changes.

083.043 **Ionospheric refraction in Doppler satellite observation.** K. Sato, M. Aihara.
Publ. Int. Latitude Obs. Mizusawa, Vol. 13, 19 - 32 (1979).

083.044 **Application of a simplified theory of ELF propagation to a simplified worldwide model of the ionosphere.** A. B. Behroozi-Toosi, H. G. Booker.
J. Atmos. Terr. Phys., Vol. 42, 943 - 974 (1980).

083.045 **Photoionization rates in the night-time *E*- and *F*-region ionosphere.**
D. F. Strobel, C. B. Opal, R. R. Meier.
Planet. Space Sci., Vol. 28, 1027 - 1033 (1980).

083.046 **Satellite radio beacon study of the ionospheric variations at Hyderabad during the total solar eclipse of February 16, 1980.** T. R. Tyagi, L. Singh, P. N. Vijayakumar, Y. V. Somayajulu, B. Lokanadham, G. Yellaiah.
Bull. Astron. Soc. India, Vol. 8, 69 - 72 (1980).

Apparently no Travelling Ionospheric Disturbances were generated by the eclipse. The percentage decrease in Faraday rotation showed an excellent correlation with percentage increase in optical obscuration of the Sun with a time lag of about 12 minutes. No amplitude scintillations were observed on the night of the eclipse day, thought there were strong scintillations on adjacent days.

083.047 **Day-time ionospheric distensions of corpuscular origin in the midlatitude D-region.**
J. Laštovička, J. Boška.
Stud. Geophys. Geod., Vol. 24, 191 - 196 (1980).

083.048 **The ionospheric plasma.** D. T. Farley.
Solar system plasma physics, Vol. 3, (see 003.010), 271 - 298 (1979).

083.049 **Understanding plasma instabilities in space: ionospheric research and communications applications.**
F. W. Perkins.
Solar system plasma physics, Vol. 3, (see 003.010), 299 - 313 (1979).

083.050 **Variations in the structure of the F-layer of the polar ionosphere at a change of the sign of the y-component of the interplanetary magnetic field. The Svalgaard-Mansurov effect in the ionosphere.**
Yu. I. Gal'perin, A. G. Zosimova, T. N. Larina, A. M. Mozhaev, N. K. Osipov, Yu. N. Ponomarev.
Kosm. Issled., Tom 18, 877 - 898 (1980). In Russian.

083.051 **On the possible nature of the "hot" zone in the earth's plasmosphere.**
I. A. Krinberg, A. V. Tashchilin, S. V. Fridman.
Geomagn. Aehron., Tom 20, 1028 - 1035 (1980). In Russian.

083.052 **Quasi-stationary electric fields and precipitating particles in the ionosphere.**
O. R. Grigoryan, S. N. Kuznetsov, S. I. Klimov.
Geomagn. Aehron., Tom 20, 1058 - 1066 (1980). In Russian.

083.053 **Influence of an electric field and plasma stream on the behaviour of the night-time midlatitude F2 region during substorms.** V. M. Novikov.
Geomagn. Aehron., Tom 20, 1102 - 1104 (1980). In Russian.

Schumann resonances in the earth-ionosphere cavity. See Abstr. 003.020.

Physical processes in the ionosphere and magnetosphere. See Abstr. 003.078.

Influence of the non-stationary motion of the earth's surface on the atmosphere and on the neutral component of the ionosphere. See Abstr. 082.043.

The latitudinal and seasonal variation of atomic oxygen deduced from observations of the E-region OI 557.7 nm airglow. See Abstr. 082.068.

Large-scale electric fields and currents and related geomagnetic variations in the quiet plasmasphere. See Abstr. 084.008.

Theory of imperfect magnetosphere-ionosphere coupling. See Abstr. 084.063.

Impulsive penetration of solar wind plasma and its effects on the upper atmosphere. See Abstr. 084.089.

Interaction between the magnetospheric boundary layers and the ionosphere. See Abstr. 084.090.

Magnetosphere, ionosphere and atmosphere interactions. See Abstr. 084.098.

Coordinated ionospheric and magnetospheric observations from the ISIS 2 satellite by the ISIS 2 experimenters. Volume 3. High-latitude charged particle, magnetic field, and ionospheric plasma observations during northern summer. See Abstr. 084.109.

Pattern of solar activity influence upon circumpolar baric features. See Abstr. 085.013.

Influence of the parameters of the interplanetary magnetic field on ionization of the F2 layer in the polar cap. See Abstr. 106.009.

The level of cosmic radio noises and integral absorption in the subauroral ionosphere from simultaneous satellite and ground-based measurements at a frequency of 5.5 MHz. See Abstr. 141.111.

084 Aurorae, Geomagnetic Field, Magnetosphere

084.001 **A modeling of magnetic field variations during magnetospheric substorms.**
S.-I. Akasofu, G. K. Corrick.
Planet. Space Sci., Vol. 28, 749 - 755 (1980).

Magnetic field variations in the noon-midnight plane during a magnetospheric substorm are studied in terms of changes of three current systems: the dynamo-driven current on the magnetopause, the cross-tail current and the field-aligned current-auroral electrojet system. The field-aligned current is assumed to be generated as a result of interruption and subsequent diversion of the cross-tail current to the ionosphere. It is concluded that the available observations are consistent with a large increase of the three currents.

084.002 **Effects of the interplanetary magnetic field on the magnetotail structure: large-scale changes of the plasma sheet during magnetospheric substorms.**
S.-I. Akasofu, D. N. Covey.
Planet. Space Sci., Vol. 28, 757 - 762 (1980).

084.003 **The energy coupling function and the power generated by the solar wind–magnetosphere dynamo.**
J. R. Kan, L. C. Lee, S.-I. Akasofu.
Planet. Space Sci., Vol. 28, 823 - 825 (1980).

084.004 **A 2.2-Hz modulation of auroral electrons imposed at the geomagnetic equator.**
D. R. Lepine, D. A. Bryant, D. S. Hall.
Nature, Vol. 286, 469 - 471 (1980).

The authors report 2.2 ± 0.5 Hz oscillations in the intensities of 4–25 keV electrons producing a pulsating aurora.

084.005 **Bunching of 8–10 keV auroral electrons near an artificial electron beam.** M. P. Gough, B. N. Maehlum, G. Martelli, P. N. Smith, G. Ventura.
Nature, Vol. 287, 15 - 17 (1980).

High frequency modulations (~1 MHz) of energetic (keV) electrons in the auroral ionosphere have been detected unambiguously for the first time using a novel technique. These modulations were induced by an artificially injected pulsed electron beam (8 keV, 4% duty cycle), and were accompanied by electrostatic wave emissions at the same frequency.

084.006 **The equatorial latitude of auroral activity during 1972 - 1977.**
N. R. Sheeley, Jr., R. A. Howard.
Sol. Phys., Vol. 67, 189 - 206 (1980).

The equatorial latitude of auroral activity has been derived from both electron and optical observations. All of the observations that were obtained during the 5-year interval June 1972–September 1977 have been used to construct a nearly continuous plot of invariant geomagnetic latitude versus time. The authors conclude that: (a) Modest auroral expansions occur during the main body of high-speed streams from coronal holes; (b) great expansions occur only during intervals of intense interplanetary magnetic fields such as may occur at the leading edge of a high-speed stream or at a flare-produced interplanetary shock.

084.007 **Geomagnetic observation at Kanozan Geodetic Observatory (1978).**
Bull. Geogr. Surv. Inst., Vol. 24, Part 1, 37 - 55 (1980).

084.008 **Large-scale electric fields and currents and related geomagnetic variations in the quiet plasmasphere.**
C.-U. Wagner, D. Möhlmann, K. Schäfer, V. M. Mishin, M. I. Matveev.
Space Sci. Rev., Vol. 26, 391 - 446 (1980).

Based on all morphological and theoretical results a plasmaspheric-ionospheric current system has been constructed and some properties of the plasmaspheric field-aligned current distribution have been derived.

084.009 **Secular variation and excursions of the earth's magnetic field.** C. G. A. Harrison.
J. Geophys. Res., Vol. 85, 3511 - 3522 (1980).

084.010 **Paleomagnetic evidence for the existence of the geomagnetic field 3.5 Ga ago.**
M. W. McElhinny, W. E. Senanayake.
J. Geophys. Res., Vol. 85, 3523 - 3528 (1980).

084.011 **Long-term nondipole components in the geomagnetic field during the last 130 m. y.**
D. H. Coupland, R. Van der Voo.
J. Geophys. Res., Vol. 85, 3529 - 3548 (1980).

084.012 **Investigation of the geomagnetic field polarity during the Jurassic.** M. B. Steiner.
J. Geophys. Res., Vol. 85, 3572 - 3586 (1980).

084.013 **Applicability of Syrovatskii's theory to particle accelerations in the geomagnetic neutral sheet.**
S. Shibuya.
Bull. Yamagata Univ. (Nat. Sci.), Vol. 10, No. 1, p. 35 - 49 (1980). – Abstr. in Phys. Abstr., Vol. 83, Abstr. 72696 (1980).

084.014 **Observations of rapid auroral fluctuations.**
T. Oguti.
J. Geomagn. Geoelectr., Vol. 30, 299 - 314 (1978). – Abstr. in Phys. Abstr., Vol. 83, Abstr. 77186 (1980).

084.015 **Highlights in the studies of the relationship of geomagnetic field changes to auroral luminosity.**
W. H. Campbell.
J. Geomagn. Geoelectr., Vol. 30, 315 - 325 (1978). – Abstr. in Phys. Abstr., Vol. 83, Abstr. 77187 (1980).

084.016 **Observed microstructure of auroral forms.**
T. N. Davis.
J. Geomagn. Geoelectr., Vol. 30, 371 - 380 (1978). – Abstr. in Phys. Abstr., Vol. 83, Abstr. 77188 (1980).

084.017 **Relationships between particle precipitation and auroral forms.** J. L. Burch.
J. Geomagn. Geoelectr., Vol. 30, 395 - 406 (1978). – Abstr. in Phys. Abstr., Vol. 83, Abstr. 77189 (1980).

084.018 **Review of auroral currents and auroral arcs.**
G. Atkinson.
J. Geomagn. Geoelectr., Vol. 30, 435 - 447 (1978). – Abstr. in Phys. Abstr., Vol. 83, Abstr. 77190 (1980).

084.019 **Acceleration mechanisms for auroral electrons.**
D. W. Swift.
J. Geomagn. Geoelectr., Vol. 30, 449 - 459 (1978). – Abstr. in Phys. Abstr., Vol. 83, Abstr. 77210 (1980).

084.020 **Generation mechanisms for magnetic-field-aligned electric fields in the magnetosphere.**
C.-G. Fälthammar.
J. Geomagn. Geoelectr., Vol. 30, 419 - 434 (1978). – Abstr. in Phys. Abstr., Vol. 83, Abstr. 77233 (1980).

084.021 **On resonance properties of force lines of the magnetosphere.**
A. L. Krylov, A. E. Lifshits, E. N. Fedorov.
Geomagn. Aehron., Tom 20, 689 - 692 (1980). In Russian.

084.022 **60-year variation of the geomagnetic field on the territory of Europe.**
N. E. Papitashvili, N. M. Rotanova, A. N. Pushkov.
Geomagn. Aehron., Tom 20, 711 - 717 (1980). In Russian.

084.023 **On an analytical continuation of the main geomagnetic field and its secular variations.**
M. S. Zhdanov, N. M. Rotanova, T. A. Chernova.
Geomagn. Aehron., Tom 20, 718 - 725 (1980). In Russian.

084.024 **Estimate of the precision of various approximations of the geomagnetic field.**
S. M. Zabluda, A. M. Yanshin.
Geomagn. Aehron., Tom 20, 726 - 730 (1980). In Russian.

084.025 **Effects of internal gravitational waves in the period of the magnetospheric substorm of 1978, February 15.** L. Lois, V. M. Shashun'kina, L. A. Yudovich.
Geomagn. Aehron., Tom 20, 744 - 746 (1980). In Russian.

084.026 **Dayside cusp auroral morphology related to nightside magnetic activity.**
P.-E. Sandholt, K. Henriksen, C. S. Deehr, G. G. Sivjee, G. J. Romick, A. Egeland.
J. Geophys. Res., Vol. 85, 4132 - 4138 (1980).
The study is an analysis of meridian-scanning photometer observations of the dayside auroral 6300-Å emission from Svalbard and magnetometer data taken simultaneously on the nightside from the University of Alaska meridian chain of observatories.

084.027 **The chemistry of excited NO^+ in an aurora.**
E. R. Young, D. G. Torr, K. C. Clark.
J. Geophys. Res., Vol. 85, 4223 - 4231 (1980).

084.028 **Effect of a horizontal wind on 1.27–μm auroral emission from $O_2(^1\Delta_g)$ molecules.**
M. H. Rees, R. G. Roble.
J. Geophys. Res., Vol. 85, 4295 - 4298 (1980).

084.029 **On a comparison between inferred and observed electron flux in pulsating auroras.**
H. C. Stenbaek-Nielsen.
J. Geophys. Res., Vol. 85, 4299 - 4303 (1980).

084.030 **Comment on 'Dayside aurora and relevance to substorm current systems and dayside merging' by R. H. Eather, S. B. Mende, and E. J. Weber.**
J. L. Burch, with a reply by R. H. Eather.
J. Geophys. Res., Vol. 85, 4304 - 4305 (1980).

084.031 **Mechanism of magnetospheric substorms and the turbulence of solar wind.** W.-r. Hu.
Acta Geophys. Sinica, Vol. 23, 245 - 253 (1980).

084.032 **Photometric investigation of precipitating particle dynamics.** S. B. Mende.
J. Geomagn. Geoelectr., Vol. 30, 407 - 418 (1978). – Abstr. in Phys. Abstr., Vol. 83, Abstr. 82141 (1980).

084.033 **First evidence and early studies of the Earth's bow shock.** J. W. Dungey.
Nuovo Cimento C, Ser. 1, Vol. 2C, (see 012.013), 655 - 660 (1979). – Abstr. in Phys. Abstr., Vol. 83, Abstr. 85816 (1980).

084.034 **The three-dimensional shape of the bow shock.**
V. Formisano.
Nuovo Cimento C, Ser. 1, Vol. 2C, (see 012.013), 681 - 692 (1979). – Abstr. in Phys. Abstr. Vol. 83, Abstr. 85832 (1980).

084.035 **DC magnetic-field observations at the Earth's bow shock.** U. Villante.
Nuovo Cimento C, Ser. 1, Vol. 2C, (see 012.013), 661 - 680 (1979). – Abstr. in Phys. Abstr., Vol. 83, Abstr. 90306 (1980).

084.036 **Initial ISEE observations of the bow shock.**
C. T. Russell, E. W. Greenstadt.
Nuovo Cimento C, Ser. 1, Vol. 2C, (see 012.013) 737 - 746 (1979). – Abstr. in Phys. Abstr., Vol. 83, Abstr. 90307 (1980).

084.037 **Observations of energetic electrons of magnetospheric origin in the magnetosheath and in the solar wind.**
V. Formisano.
Nuovo Cimento C, Ser. 1, Vol. 2C, (see 012.013) 781 - 788 (1979). – Abstr. in Phys. Abstr., Vol. 83, Abstr. 90310 (1980).

084.038 **Observations of relativistic electrons of the outer radiation belt of the earth aboard the artificial earth satellite Prognoz 3 in 1973 - 1974.** N. N. Volodichev, I. A. Savenko, M. A. Saraeva, G. M. Surova, P. I. Shavrin.
Kosm. Issled., Tom 18, 796 - 801 (1980). In Russian.

084.039 **Relation of the λλ 5577 and 6300 Å [O I] emission from patrol spectral observations of aurorae in the northern hemisphere.** N. N. Bliznyuk, N. I. Dzyubenko.
Problems of cosmic physics. Vyp. 15, (see 003.003), p. 32 - 37 (1980). In Russian.
On the basis of patrol spectrographs worked in the period of IGY the distribution is obtained of λλ 5577 and 6300 Å [O I] emission intensity relations in the northern hemisphere in the coordinates invariant latitude–invariant time. Maximum values of I_{5577}/I_{6300} in the night period are on the low-latitude edge of the auroral oval, and the minimum is noted in the region of the daytime cusp. The asymmetry morning-evening is seen. After midnight and in the morning the relation I_{5577}/I_{6300} is higher, on the average, than in the period before midnight.

084.040 **Theory of magnetic field generation of the earth and celestial bodies.**
A. Z. Dolginov, V. A. Urpin, D. G. Yakovlev.
Fiz. kosm. Yader. fiz. Fiz.-khim. issled. Leningrad, 1979, 46 pp. In Russian. – Abstr. in Ref. zh., 51. Astron., 8.51.178 (1980).

084.041 **Evidence in the auroral record for secular solar variability.** G. L. Siscoe.
Rev. Geophys. Space Phys., Vol. 18, 647 - 658 (1980).
The historical record of aurorae is continuous and usefully dense for at least the last 2000 years. Revival of interest in the secular variability in solar activity motivates a review of the auroral record.

084.042 **The application of artificial electron beams to magnetospheric research.** J. R. Winckler.
Rev. Geophys. Space Phys., Vol. 18, 659 - 682 (1980).
Contents: Generating electron beams in space. The stability and wave emission of electron beams. Optical effects and atmospheric interaction of electron beams. Electron beams as magnetospheric probes. Some directions of future research. Appendix 1: chronology of electron beam rocket experiments with literature summary. Appendix 2: complete bibliography of publications under the University of Minnesota electron Echo program (complete to September 1979).

084.043 **Study of individual geomagnetic storms in terms of the solar wind.** S.-I. Akasofu.
Planet. Space Sci., Vol. 28, 933 - 944 (1980).
The paper summarizes a study of the development of a large number of geomagnetic storms in terms of the solar

wind–magnetosphere energy coupling function ϵ, the AE and *Dst* indices.

084.044 **Coordinated ionospheric and magnetospheric observations from the ISIS 2 satellite by the ISIS 2 experimenters. Volume 1. Optical auroral images and related direct measurements.** J. S. Murphree (Coordinator). National Space Science Data Center/World Data Center A for Rockets and Satellites, National Aeronautics and Space Administration, Goddard Space Flight Center, Greenbelt, Maryland, NSSDC/WDC-A-R & S 80-03, 5 + 206 pp. (1980).

A list of ISIS 2 experimenters, a brief description of the ISIS 2 satellite, followed by more detailed instrument descriptions, format descriptions, data set descriptions, and the data themselves and a bibliography of ISIS 2 published papers are given.

084.045 **Plasma processes in the auroral magnetosphere.** A. A. Galeev, Yu. I. Gal'perin, A. V. Zakharov, V. V. Krasnosel'skikh, V. A. Liperovskij, M. I. Pudovkin. Inst. kosm. issled. AN SSSR. Prepr., 1979, No. 519, 118 pp. In Russian. – Abstr. in Ref. zh., 62. Issled. kosm. prostranstva, 9.62.247 (1980).

084.046 **Structure of the night auroral zone.** K. I. Gorelyj, Eh. I. Nemtsova, O. M. Pirog, V. D. Urbanovich, E. A. Ponomarev. Issled. po geomagn., aehron. i fiz. Solntsa, Moskva, 1980, No. 50, p. 92 - 98. In Russian. – Abstr. in Ref. zh., 62. Issled. kosm. prostranstva, 9.62.248 (1980).

084.047 **Magalert: August 27, 1978.** J. Joselyn, J. F. Bryson, Jr. Solar and interplanetary dynamics, (see 012.020), p. 413 - 420 (1980).

The authors conclude that large disappearing filaments are surely worthy of note as harbingers of significant magnetic activity.

084.048 **Influence of the magnetospheric convection on the distribution of hydrogen ions in the plasmasphere of the earth.** M. A. Koen, Yu. V. Konikov, O. M. Radzhabova. Geomagn. Aehron., Tom 20, 875 - 879 (1980). In Russian.

084.049 **Precipitation of hard electrons (E = 20 - 150 keV) at midlatitudes.** J. Laštovička. Geomagn. Aehron., Tom 20, 880 - 883 (1980). In Russian.

084.050 **Space-time distribution of magnetospheric substorms.** A. P. Nikol'skij. Geomagn. Aehron., Tom 20, 889 - 895 (1980). In Russian.

084.051 **Analysis of anomalies of the variable geomagnetic field on the surface of the spherically-irregular earth.** M. S. Zhdanov. Geomagn. Aehron., Tom 20, 912 - 918 (1980). In Russian.

084.052 **On the morphology of medium-frequency secular variations of the geomagnetic field.** V. P. Golovkov, N. V. Kulanin, T. N. Cherevko. Geomagn. Aehron., Tom 20, 925 - 928 (1980). In Russian.

084.053 **On the role of convection in forming the gap and the plasmapause.** V. V. Klimenko, A. A. Namgaladze. Geomagn. Aehron., Tom 20, 946 - 950 (1980). In Russian.

084.054 **Synchronous television recording of active forms and spectra of aurorae.** A. M. Evtushevskij, G. P. Milinevskij, O. S. Popov, G. A. Gospodinov, V. A. Kravchenko. Geomagn. Aehron., Tom 20, 959 - 961 (1980). In Russian.

084.055 **Dependence of the probability of substorm display on solar wind velocity and on the vertical component of the interplanetary magnetic field.** B. L. Shirman. Geomagn. Aehron., Tom 20, 965 - 968 (1980). In Russian.

084.056 **Modulation instability of magnetoacoustic waves in the radiation belt.** A. V. Gul'el'mi. Geomagn. Aehron., Tom 20, 968 - 969 (1980). In Russian.

084.057 **Satellite measurements and theories of low altitude auroral particle acceleration.** F. S. Mozer, C. A. Cattell, M. K. Hudson, R. L. Lysak, M. Temerin, R. B. Torbert. Space Sci. Rev., Vol. 27, 155 - 213 (1980).

Several previous and new S3–3 satellite results on DC electric fields, field-aligned currents, and waves are described, interpreted theoretically, and applied to the understanding of auroral particle acceleration at altitudes below 8000 km. Further analyses of acceleration mechanisms and energetics are presented.

084.058 **Plasma and field signatures of poleward propagating auroral precipitation observed at the foot of the Geos 2 field line.** G. G. Shepherd, R. Boström, H. Derblom, C.-G. Fälthammar, R. Gendrin, K. Kaila, A. Korth, A. Pedersen, R. Pellinen, G. Wrenn. J. Geophys. Res., Vol. 85, 4587 - 4601 (1980).

The 6300-Å auroral emission intensity at Kilpisjärvi, Finland, was monitored during the 1978 - 1979 winter, and comparisons made with data from the geostationary Geos 2 spacecraft located on the same magnetic field line. Optical 6300-Å events with a rise time of a few minutes and a decay time of about 1 hour were regularly observed, and four events from the night of December 18, 1978, were analyzed in detail.

084.059 **Observation of the plasma boundary layer at lunar distances: direct injection of plasma into the plasma sheet.** G. D. Sanders, L. J. Maher, J. W. Freeman. J. Geophys. Res., Vol. 85, 4607 - 4615 (1980).

Simultaneous data returned by the Apollo Suprathermal Ion Detector Experiments are combined to yield information on the properties of plasmas in the magnetosheath at lunar orbit. An interface between the magnetosheath and the plasma sheet has been identified. This layer is characterized by a systematic change in the magnetosheath flow properties in the vicinity of the magnetopause. The flow is observed to turn inward toward the plasma sheet and become thermalized, indicating that the magnetosheath particles can gain access to the plasma sheet at lunar distances.

084.060 **Rocket-borne auroral EUV measurements.** F. Fischer, G. Schmidtke. J. Geophys. Res., Vol. 85, 4716 - 4720 (1980).

The measurements of two rocket spectrometers on January 22, 1977, and February 8, 1977, over Andoya, Norway are described. While five channels recorded height profiles of selected emissions such as O II 83.4 nm, O I 98.9 nm, N I 120 nm, H I 121.6 nm, and O I 130.4 nm, two channels repeatedly scanned the wavelength range from 50 to 135 nm. Results are discussed and compared with other measurements.

084.061 **Auroral particle acceleration – an example of a universal plasma process.** G. Haerendel. ESA J., Vol. 4, 197 - 210 (1980).

The occurrence of discrete and narrow auroral arcs is attributed to a sudden release of magnetic tensions set up in a magnetospheric-ionospheric current circuit of high strength. At altitudes of several thousand kilometres the condition of frozen-in magnetic fields can be broken temporarily in thin regions corresponding to the observed width of auroral arcs. This implies magnetic-field-aligned potential drops of several kilovolts supported by certain anomalous transport processes which can only be maintained in a quasi-stationary fashion if the current density exceeds a critical limit.

084.062 **Magnetic contour maps at the core-mantle boundary.** E. R. Benton, L. A. Muth, M. Stix.
J. Geomagn.Geoelectr., Vol. 31, 615 - 626 (1979). – Abstr. in Phys. Abstr., Vol. 83, Abstr. 104880 (1980).

084.063 **Theory of imperfect magnetosphere-ionosphere coupling.** J. R. Kan, L. C. Lee.
Geophys. Res. Lett., Vol. 7, 633 - 636 (1980).
A theory of magnetosphere-ionosphere coupling in the presence of field-aligned potential drops is formulated within the framework of magnetohydrodynamic equations.

084.064 **The effects on the earth's magnetotail from shocks in the solar wind.**
J. Lyon, S. H. Brecht, J. A. Fedder, P. Palmadesso.
Geophys. Res. Lett., Vol. 7, 721 - 724 (1980).

084.065 **Dayside and nightside auroral arc systems.** S.-I. Akasofu, J. R. Kan.
Geophys. Res. Lett., Vol. 7, 753 - 756 (1980).
It is shown that the dayside and the nightside arc systems of the auroral oval are topologically distinct, although they constitute an annular belt of auroras, the auroral oval. It is suggested that the dayside arc system is associated with field-aligned currents fed by the low-latitude boundary layer dynamo, while the nightside arc system is associated with field-aligned currents fed by the magnetotail dynamo.

084.066 **Influence of the B_X component of the interplanetary magnetic field on magnetopause reconnection.**
W. D. Gonzalez, A. L. C. Gonzalez.
Geophys. Res. Lett., Vol. 7, 773 - 776 (1980).

084.067 **Initial geomagnetic field model from Magsat vector data.** R. A. Langel, R. H. Estes, G. D. Mead, E. B. Fabiano, E. R. Lancaster.
Geophys. Res. Lett., Vol. 7, 793 - 796 (1980).
Magsat data from the magnetically quiet days of November 5 - 6, 1979, were used to derive a thirteenth degree and order spherical harmonic geomagnetic field model, MGST (6/80).

084.068 **Auroral X ray observations at 60- to 30 km altitudes.** K. K. Vij, J. S. Vogel, D. Venkatesan.
J. Geophys. Res., Vol. 85, 5096 - 5104 (1980).

084.069 **Time-averaged fluxes of heavy ions at the geostationary orbit.** A. Konradi, T. A. Fritz, S.-Y. Su.
J. Geophys. Res., Vol. 85, 5149 - 5152 (1980).

084.070 **Aurora Borealis.**
British Astron. Assoc. Circ., No. 609 (1980).

084.071 **OPEN – a study of the origins of plasma in the Earth's neighborhood.** D. J. Williams.
Space Shuttle: dawn of an era. Part II, (see 012.037), p. 605 - 621 (1980). – Abstr. in Phys. Abstr., Vol. 83, Abstr. 109318 (1980).

084.072 **Quantitative relations between the characteristics of storm-time variations of the geomagnetic field and the changes of interplanetary plasma.** P. Ochabova.
Contrib. Geophys. Inst. Slovak Acad. Sci., Vol. 10, 7 - 34 (1980). – Abstr. in Phys. Abstr., Vol. 84, Abstr. 4703 (1981).

084.073 **Solar wind evolution and recurrent geomagnetic disturbances.** A. Prigancova.
Contrib. Geophys. Inst. Slovak Acad. Sci., Vol. 10, 53 - 61 (1980). – Abstr. in Phys. Abstr., Vol. 84, Abstr. 4704 (1981).

084.074 **Field-aligned currents and large-scale magnetospheric electric fields.** N. D'Angelo.
Ann. Géophys., Vol. 36, 31 - 39 (1980). – Abstr. in Phys. Abstr., Vol. 84, Abstr. 9322 (1981).

084.075 **New advances in thermal plasma research.** C. R. Chappell, C. R. Baugher, J. L. Horwitz.
Rev. Geophys. Space Phys., Vol. 18, 835 - 861 (1980).

084.076 **Dependence of mid-latitude hydromagnetic energy spectra on solar wind speed and interplanetary magnetic field direction.** A. Wolfe.
J. Geophys. Res., Vol. 85, 5977 - 5982 (1980).

084.077 **Observations of the mean ionization states of energetic particles in the vicinity of the earth's magnetosphere.**
L. S. Ma Sung, G. Gloeckler, C. Y. Fan, D. Hovestadt.
J. Geophys. Res., Vol. 85, 5938 - 5991 (1980).

084.078 **Cusp proton signatures and the interplanetary magnetic field.**
P. H. Reiff, J. L. Burch, R. W. Spiro.
J. Geophys. Res., Vol. 85, 5997 - 6005 (1980).

084.079 **Orthohelium emissions at 3889 and 5876 Å in the polar upper atmosphere.**
G. G. Sivjee, K. Henriksen, C. S. Deehr.
J. Geophys. Res., Vol. 85, 6043 - 6046 (1980).

084.080 **Temporal fluctuations in the luminosity of diffuse aurora.** F. T. Berkey.
J. Geophys. Res., Vol. 85, 6075 - 6079 (1980).

084.081 **Global aspects of the earth's magnetopause.** D. H. Fairfield.
Magnetospheric boundary layers, (see 012.041), p. 5 - 13 (1979).
The review outlines the large-scale aspects of the magnetopause; how it responds to changes in solar wind pressure and interplanetary field orientation, what macroscopic instabilities may influence its behavior, and what is the nature of the plasmas that interact at the magnetopause and form the boundary layer plasmas.

084.082 **An electric-current model of the magnetosphere.** H. Alfvén.
Magnetospheric boundary layers, (see 012.041), p. 15 - 21 (1979).
The paper outlines a general approach to producing a current model of the magnetosphere. It is concluded that this approach may facilitate the understanding of several magnetospheric phenomena, and especially the mechanism by which solar-wind energy is transferred to the magnetosphere.

084.083 **Energetic electron bursts in the magnetopause electron layer and in interplanetary space.**
J. W. Bieber, E. C. Stone.
Magnetospheric boundary layers, (see 012.041), p. 131 - 135 (1979).

084.084 **Birkeland currents and the interplanetary magnetic field.** T. A. Potemra, N. A. Saflekos.
Magnetospheric boundary layers, (see 012.041), p. 193 - 198 (1979).

084.085 **Dayside aurora and substorm current systems.** R. H. Eather, S. B. Mende, E. J. Weber.

Magnetospheric boundary layers, (see 012.041), p. 199 - 206 (1979).

084.086 **Solution of the Chapman-Ferraro problem within two dimensions.** H. Rucker, H. Biernat.
Magnetospheric boundary layers, (see 012.041), p. 311 - 314 (1979).

The magnetic field structure of the earth's magnetosphere and its boundary (the magnetopause) is calculated with the method of the conformal mapping. According to this method the solution is restricted to two dimensions, whereby in this investigation the noon-midnight meridian plane is taken into consideration. The configuration of the magnetopause and the magnetospheric field line structure is represented. The shape of the boundary is compared with other magnetopause models indicating a remarkable agreement.

084.087 **Influence of the interplanetary magnetic field on the position of the dayside magnetopause.**
G. H. Voigt.
Magnetospheric boundary layers, (see 012.041), p. 315 - 321 (1979).

084.088 **Rates of mass, momentum, and energy transfer at the magnetopause.** T. W. Hill.
Magnetospheric boundary layers, (see 012.041), p. 325 - 332 (1979).

084.089 **Impulsive penetration of solar wind plasma and its effects on the upper atmosphere.** J. Lemaire.
Magnetospheric boundary layers, (see 012.041), p. 365 - 373 (1979).

Impulsive penetration of solar wind plasma irregularities transfers energy, momentum and particles into the magnetosphere. The effects on the upper atmosphere are examined and discussed. The heat deposition in the E-region (and above) due to Joule dissipation of depolarization currents is evaluated and compared to the energy production due to the absorption of solar radiation. The wind speed imparted to ionospheric plasma in the throat region and over the polar cap is compared to the observed convection velocities.

084.090 **Interaction between the magnetospheric boundary layers and the ionosphere.** V. M. Vasyliunas.
Magnetospheric boundary layers, (see 012.041), p. 387 - 393 (1979).

084.091 **Zur Geschichte der Polarlichtforschung.**
W. Schröder.
Acta Geod. Geophys. Montan. Acad. Sci. Hung., Tomus 15, 207 - 219 (1980).

084.092 **Identification of auroral EUV emissions.**
F. Fischer, G. Stasek, G. Schmidtke.
Geophys. Res. Lett., Vol. 7, 1003 - 1006 (1980).

From measurements of two rocket flights and from laboratory investigations auroral EUV emissions are identified. Contrary to the airglow, emissions from ionized nitrogen are present in the auroral emission spectra below 80 nm.

084.093 **Short-period secular variations (SPSV) of the geomagnetic field recorded in highly scattered palaeomagnetic records of holocene lake sediments from North Poland.** P. Tucholka.
Earth Planet. Sci. Lett., Vol. 48, 379 - 384 (1980).

A simple method of calculating running means over different intervals within palaeomagnetic records has been used to identify short-period secular variations of the geomagnetic field. The records from Polish lake sediments reveal variations in declination with a period between 290 and 372 years. The periodicity of variations in the inclination has been estimated to be between 522 and 670 years.

084.094 **The late 1960's secular variation impulse: further constraints on deep mantle conductivity.**
J. Achache, V. Courtillot, J. Ducruix, J.-L. Le Mouël.
Phys. Earth Planet. Inter., Vol. 23, 72 - 75 (1980) = Contrib. Inst. Phys. Globe, Univ. Paris, NS 394.

084.095 **Spatial power spectra of the crustal geomagnetic field and core geomagnetic field.**
M. G. McLeod, P. J. Coleman, Jr.
Phys. Earth Planet. Inter., Vol. 23, P5 - P19 (1980).

084.096 **Geomagnetic field mapping from a satellite: spatial power spectra of the geomagnetic field at various satellite altitudes relative to natural noise sources and instrument noise.** M. G. McLeod, P. J. Coleman, Jr.
Phys. Earth Planet. Inter., Vol. 23, 222 - 231 (1980) = Publ. Inst. Geophys. Planet Phys., Univ. Calif., No. 1810.

084.097 **Earth's magnetosphere: global problems in magnetospheric plasma physics.** J. G. Roederer.
Solar system plasma physics, Vol. 2, (see 003.010), 1 - 56 (1979).

Contents: Introduction. General characteristics of magnetospheric field configuration. Effect of the interplanetary magnetic field on topography, topology and stability of the magnetospheric boundary. The mechanisms for the entry of solar wind plasma into the magnetosphere. Plasma storage, acceleration and release mechanisms in the magnetospheric tail. Magnetic-field-aligned currents and magnetosphere – ionosphere interactions. Magnetospheric research in coming years.

084.098 **Magnetosphere, ionosphere and atmosphere interactions.** P. M. Banks.
Solar system plasma physics, Vol. 2, (see 003.010), 57 - 103 (1979).

The general character of the earth's space environment is discussed with emphasis on the various physical processes which link the magnetosphere, the ionosphere and the upper atmosphere.

084.099 **Hydromagnetic waves.**
L. J. Lanzerotti, D. J. Southwood.
Solar system plasma physics, Vol. 3, (see 003.010), 109 - 135 (1979).

This paper has discussed some of the areas of current research interest in the physics of low frequency waves in the earth's magnetosphere.

084.100 **Plasma processes in the earth's radiation belts.**
L. R. Lyons.
Solar system plasma physics, Vol. 3, (see 003.010), 137 - 163 (1979).

084.101 **Physics of heavy ions in the magnetosphere.**
J. M. Cornwall, M. Schultz.
Solar system plasma physics, Vol. 3, (see 003.010), 165 - 210 (1979).

Contents: Introduction. Auroral magnetospheric coupling and the O^+ problem. Heavy ions as probes for magnetospheric processes. Multi-ion plasma physics. Active plasma-injection experiments. Heavy-ion physics at Jupiter. The future of magnetospheric heavy-ion physics.

084.102 **Magnetospheric plasma waves.** S. D. Shawhan.
Solar system plasma physics, Vol. 3, (see 003.010), 211 - 270 (1979).

The emphasis of this paper is placed on the review and interpretation of plasma wave observations made in the Earth's magnetosphere. Observations of plasma waves from other planetary magnetospheres and cosmic plasma systems are discussed and interpreted in analogy to the Earth's processes.

084.103 **The character of the geomagnetic secular variation in Europe.** W. Mundt.
Gerlands Beitr. Geophys., Band 89, 451 - 466 (1980).

084.104 **Geomagnetic secular variation on the territory of the GDR.** W. Mundt.
Gerlands Beitr. Geophys., Band 89, 467 - 476 (1980).

084.105 **Secular variations of the earth's magnetic field in Poland.** A. Uhrynowski, A. Żółtowski.
Gerlands Beitr. Geophys., Band 89, 477 - 490 (1980).

084.106 **On the geomagnetic field and its secular variations in Hungary.** E. Aczél, Á. Wallner.
Gerlands Beitr. Geophys., Band 89, 491 - 498 (1980).

084.107 **Secular variation studies in Romania.**
M. Anghel, C. Demetrescu, T. Neştianu.
Gerlands Beitr. Geophys., Band 89, 499 - 510 (1980).

084.108 **Analysis of the secular variations of the earth's magnetic field in Bulgaria for the period 1955 - 1970.**
K. Kostov, C. Georgiev.
Gerlands Beitr. Geophys., Band 89, 511 - 526 (1980).

084.109 **Coordinated ionospheric and magnetospheric observations from the ISIS 2 satellite by the ISIS 2 experimenters. Volume 3. High-latitude charged particle, magnetic field, and ionospheric plasma observations during northern summer.** D. M. Klumpar (Coordinator).
National Space Science Data Center/World Data Center A for Rockets and Satellites, National Aeronautics and Space Administration, Goddard Space Flight Center, Greenbelt, Maryland, NSSDC/WDC-A-R & S 80-05. 5 + 255 pp. (1980).

A list of ISIS 2 experimenters, a brief description of the ISIS 2 satellite, followed by more detailed instrument descriptions, format descriptions, data set descriptions, and the data themselves and a bibliography of ISIS 2 published papers are given.

084.110 **High-energy quasi-captured elements near the equator.** R. N. Basilova, A. A. Gusev, G. I. Pugacheva, E. I. Kogan-Laskina.
Kosm. Issled., Tom 18, 937 - 940 (1980). In Russian.

084.111 **Explosions of electron streams with energies > 200 keV in a geostationary orbit in periods of geomagnetic disturbances.**
I. P. Bezrodnykh, V. V. Klimenko, Yu. G. Shafer.
Kosm. Issled., Tom 18, 941 - 942 (1980). In Russian.

084.112 **Berechnung des Zustands minimaler Freier Energie für ein Modell des Schweifs der Magnetosphäre.**
W. Zwingmann.
Mitt. Astron. Ges., Nr. 50, p. 13 (1980). – Abstract.

084.113 **Numerische Rechnungen zur Entwicklung des Schweifs der Erdmagnetosphäre im Zusammenhang mit magnetischen Teilstürmen.** J. Birn.
Mitt. Astron. Ges., Nr. 50, p. 13 (1980). – Abstract.

084.114 **Zur Stabilität zweidimensionaler Magnetosphärengleichgewichte.** H. Goldstein.
Mitt. Astron. Ges., Nr. 50, p. 108 (1980). – Abstract.

084.115 **Existieren im Schweif der Magnetosphäre magnetische Inseln als Gleichgewichtsstrukturen?**
W. Zwingmann.
Mitt. Astron. Ges., Nr. 50, p. 108 (1980). – Abstract.

084.116 **Hydromagnetic waves of Rossby wave type in the earth's magnetosphere.**
A. G. Khantadze, Z. A. Kereselidze, Ya. M. Gogatishvili.
Geomagn. Aehron., Tom 20, 1047 - 1052 (1980). In Russian.

084.117 **A prognosis of the secular variation of the geomagnetic field with the help of a stochastic model.**
T. N. Bondar', G. I. Kolomijtseva, A. N. Pushkov.
Geomagn. Aehron., Tom 20, 1090 - 1096 (1980). In Russian.

084.118 **Time sequence analysis of flickering auroras 1. Application of Fourier analysis.**
F. T. Berkey, M. B. Silevitch, N. R. Parsons.
J. Geophys. Res., Vol. 85, 6827 - 6843 (1980).

Using a technique that enables one to digitize the brightness of auroral displays from individual fields of a video signal, the authors have analyzed the frequency content of flickering aurora.

084.119 **Geomagnetic and solar data.**
J. V. Lincoln (Editor).
J. Geophys. Res., Vol. 85, 3506, 4306, 4721, 5169, 6081, 6900 (1980).

Nordlyset. See Abstr. 003.025.

Physical processes in the ionosphere and magnetosphere. See Abstr. 003.078.

Geomagnetic field and interior structure of the earth. See Abstr. 003.090.

Annual report for magnetic observatories – 1977. See Abstr. 009.009.

Australian magnetic observatories. See Abstr. 009.010.

A laboratory study of the λ2145 Å auroral mystery feature. See Abstr. 022.144.

Hα and Hβ emission cross sections for low-energy H and H^+ collisions with N_2 and O_2. See Abstr. 022.156.

Spectrometer for measurements of low-energy electrons and ions (E 8). See Abstr. 032.562.

Matched tandem etalon camera–MATEC–and its application to auroral observations. See Abstr. 034.053.

The distant geomagnetic tail observing satellite Geos-3. See Abstr. 051.003.

The International Sun-Earth Explorer mission. See Abstr. 051.008.

Summary of terrestrial and planetary radio and plasma waves observed by the initial phase of JIKIKEN (EXOS-B) observations. See Abstr. 051.010.

MAGSAT – a new satellite to survey the Earth's magnetic field. See Abstr. 051.034.

Investigation of the mutual position of the orbit of a satellite with high apogee and the magnetopause, of the shock wave and the shadow of the earth by means of a machine diagram. See Abstr. 052.024.

Rays and foci in a magneto-ionic medium with linearly varying magnetic field. See Abstr. 062.004.

Slow convection of a magnetized plasma and the earth plasma sheet. See Abstr. 062.013.

Necessary conditions for the magnetohydrodynamic dynamo. See Abstr. 062.041.

The theory of magnetic shocks in collisionless plasma. See Abstr. 062.052.

Adiabatic-drift-loss modification of the electromagnetic loss-cone instability for anisotropic plasma. See Abstr. 062.058.

Very hot plasmas in the solar system. See Abstr. 062.084.

Properties of the longitudinal dielectric function: an application to the auroral plasma. See Abstr. 062.107.

A microscopic description of interpenetrated plasma regions. See Abstr. 062.111.

Shock systems in collisionless space plasmas. See Abstr. 062.126.

Magnetic field reconnection. See Abstr. 062.127.

Freie Energie und Stabilität magnetosphärischer und solarer Plasmastrukturen. See Abstr. 062.130.

Why is geomagnetic activity during the ending years of a solar cycle well-correlated to the maximum of the next cycle? See Abstr. 072.067.

Reviews of solar activities, coronal holes and geomagnetic storms in 1978. See Abstr. 073.049.

Coronal holes, solar wind streams, and geomagnetic disturbances during 1978 and 1979. See Abstr. 074.019.

Energization of solar wind ions by reflection from the earth's bow shock. See Abstr. 074.076.

Deceleration of the solar wind in the earth's foreshock region: ISEE 2 and Imp 8 observations. See Abstr. 074.092.

Optical investigations made by the crew of the second expedition of the orbital scientific station Salyut 6. See Abstr. 082.045.

Measurements of the spectral profile of Balmer alpha emission from the hydrogen geocorona. See Abstr. 082.069.

Active experiments in space plasmas. See Abstr. 083.005.

On the location of the ionospheric current systems responsible for the lunar and solar magnetic variations. See Abstr. 083.042.

Orbits of submicron lunar ejecta in the earth-moon system. See Abstr. 094.027.

Interaction of lunar ejecta and the magnetosphere of the earth. See Abstr. 094.028.

Earth and Jupiter: comparison of magnetopause strûcture. See Abstr. 099.132.

Audible sounds excited by aurorae and meteor fireballs. See Abstr. 104.056.

Australasian, Ivory Coast and North American tektite strewnfields: size, mass and correlation with geomagnetic reversals and other earth events. See Abstr. 105.150.

Observations of backstreaming protons in the solar wind near the Earth's bow shock. See Abstr. 106.015.

Low-frequency waves observed in the vicinity of the Earth's bow shock. See Abstr. 106.016.

Possible generation mechanisms of low-frequency waves ($\lesssim$50 Hz) with application to the bow shock plasma. See Abstr. 106.017.

A review of upstream and bow shock energetic-particle measurements. See Abstr. 106.050.

Abnormal quiet days and the effect of the interplanetary magnetic field on the apparent position of the *Sq* focus. See Abstr. 106.073.

On the role of the B_z component of the interplanetary magnetic field in the force balance at the day-time magnetopause. See Abstr. 106.086.

Connection of the components of the vector of the interplanetary magnetic field with variations of the geomagnetic field at high latitudes of the northern hemisphere. See Abstr. 106.087.

Energy spectra and charge states of low energy cosmic rays in the Skylab experiment. See Abstr. 143.038.

Longitudinal run of cosmic rays under the radiation belts of the earth. See Abstr. 143.047.

085 Solar-terrestrial Relations

085.001 **Effect of solar radiation on climatic variation.** Y. Fu.
Kexue Tongbao, Vol. 25, 695 - 699 (1980).

085.002 **Solar variability and climatic change during the current millennium.** M. Stuiver.
Nature, Vol. 286, 868 - 871 (1980).

The author reports a comparison of climate records with the record of solar change obtained from atmospheric ^{14}C variations. This yielded 'negative' results, that is, a relationship between the climatic time series and the ^{14}C derived record of solar change could not be confirmed.

085.003 **Giant solar flares in Antarctic ice.** R. Stothers.
Nature, Vol. 287, 365 (1980).

085.004 **Solar activity parameters and thermospheric density variations.** L. R. Sharp, E. B. Mayfield.
Bull. American Astron. Soc., Vol. 12, 544 (1980). – Abstract.

085.005 **Geologic support for the astronomic theory of climatic fluctuations.** C. J. Mann.
Mod. Geol., Vol. 7, 25 - 28 (1979). – Abstr. in Phys. Abstr., Vol. 83, Abstr. 72509 (1980).

085.006 **Change of primary ion-electron production rates with solar EUV flux.** G. Schmidtke.
Ann. Géophys., Vol. 35, 141 - 144 (1979). – Abstr. in Phys. Abstr., Vol. 83, Abstr. 72675 (1980).

085.007 **On the mechanism of the influence of solar activity on the velocity of the earth's daily rotation.**
A. I. Laptukhov.
Geomagn. Aehron., Tom 20, 670 - 673 (1980). In Russian.

085.008 **Connection of geomagnetic activity with solar wind velocity and concentration under conditions of "quasi-viscous" interaction.** V. V. Shelomentsev.
Geomagn. Aehron., Tom 20, 680 - 684 (1980). In Russian.

085.009 **Detection of a corona of fast oxygen atoms during solar maximum.**
J. H. Yee, J. W. Meriwether, Jr., P. B. Hays.
J. Geophys. Res., Vol. 85, 3396 - 3400 (1980).

A series of twilight interferometric observations of the near infrared O^+ (^2P) doublets at 7320 and 7330 Å between April 1979 and October 1979 detected excessive amounts of emission at shadow heights above 550 km. The scale height deduced from the vertical brightness profile determined from data taken on September 26, 1979, when the F10.7 was 231, showed a marked increase above 550 km. The equivalent temperature was estimated to be 4000°K or higher. The authors conclude from these results that there exists an atomic oxygen corona overlying the thermosphere during the solar maximum period.

085.010 **Solar cycle variation in geomagnetic external spherical harmonic coefficients.**
L. R. Alldredge, C. O. Stearns, M. Sugiura.
J. Geomagn. Geoelectr., Vol. 31, 495 - 508 (1979). – Abstr. in Phys. Abstr., Vol. 83, Abstr. 85592 (1980).

085.011 **Solar cycle variations of the first-degree spherical harmonic components of the geomagnetic field.**
T. Yukutake, J. C. Cain.
J. Geomagn. Geoelectr., Vol. 31, 509 - 544 (1979). – Abstr. in Phys. Abstr., Vol. 83, Abstr. 85593 (1980).

085.012 **Relations Soleil-Terre. L'activité solaire et l'atmosphère terrestre.** R. Kandel.
Astronomie, Vol. 94, 441 - 452 (1980).

085.013 **Pattern of solar activity influence upon circumpolar baric features.** A. A. Dmitriev.
Nov. v soln.-zemn. svyazyakh. Mater. soveshch. po soln.-zemn. svyazyam v meteorol. Moskva, 1980, p. 4 - 6. In Russian. Abstr. in Ref. zh., 51. Astron., 8.51.417 (1980).

085.014 **Secular variation of the solar activity during the geological age and its associated effect on the planetary environments.** K. Sakurai.
13th Lunar and Planetary Symposium, (see 012.018), p. 403 - 406 (1980).

On the basis of the analyses for various radioactive isotopes in the lunar and meteoritic samples, some evidence on the secular variation of the solar activity has been obtained throughout the geological era since the birth of the solar system. There is a possibility that the large increase of the solar activity during the Cretaceous era and the early Quaternary era produced the terrestrial environment with temperature much warmer than that observed with respect to the mean temperature of the atmosphere for the last 100 years.

085.015 **The OPEN program: an example of the scientific rationale for future solar-terrestrial research programs.** D. J. Williams.
Solar and interplanetary dynamics, (see 012.020), p. 507 - 522 (1980).

The field of solar-terrestrial physics has evolved to a point where quantitative theoretical modeling and cause-effect predictive techniques can be foreseen. Instrumentation now exists to quantitatively test these theories and lead to a significant improvement in our understanding of the solar-terrestrial environment. The proposed Origins of Plasma in the Earth's Neighborhood (OPEN) program will be used to trace this theoretical and experimental evolution and describe a solar-terrestrial research scenario for the 1980's which includes specific problems related to solar and interplanetary dynamics.

085.016 **Radio radiation of the near-earth cosmic space as a result of influence of solar flares on the magnetosphere and ionosphere of the earth.** S. I. Musatenko.
Geomagn. Aehron., Tom 20, 884 - 888 (1980). In Russian.

085.017 **Die Strahlungssituation der Erde.**
P. Blum.
Evolution der Planetenatmosphären und des Lebens, (see 012.032), p. 109 - 139 (1980).

Ziel des Artikels ist eine zusammenfassende Darstellung der Eigenschaften der gesamten auf die Erde auffallenden extraterrestrischen Strahlung und ihrer Modifikation durch die Erdatmosphäre.

085.018 **A study of the relation between the PcI geomagnetic pulsations and the solar wind structure.**
Eh. T. Matveeva, M. N. Gnevyshev.
Soln. Dannye 1980 Byull., No. 4, p. 99 - 103 (1980). In Russian.

It is shown that there is a close relationship between the geomagnetic pulsation PcI and high-speed solar wind streams.

085.019 **The relation between atmospheric trace species variabilities and solar UV variability.**
J. E. Penner, J. S. Chang.
J. Geophys. Res., Vol. 85, 5523 - 5528 (1980).

085.020 **Climate and variability in the solar constant.** D. Gough.
Nature, Vol. 288, 639 - 640 (1980).

085.021 **Implications of solar cycles 19 and 20 geomagnetic activity for magnetospheric processes.**
J. Feynman.
Geophys. Res. Lett., Vol. 7, 971 - 973 (1980).

The solar cycle variations of geomagnetic indices are compared for the last two solar cycles. It is found that the midlatitude aa index and the ring current index, D_{st}, exhibit a markedly different relationship to one another in the two cycles.

085.022 **An ionospheric response to the polarity reversal of the general magnetic field of the sun and of the interplanetary magnetic field.**
N. F. Solonitsyna, M. P. Rudina.
Ionos. i soln.-zemn. svyazi. Alma-Ata, 1980, p. 3 - 13. In Russian. – Abstr. in Ref. zh., 51. Astron., 11.51.479 (1980).

085.023 **Decameter wave propagation during solar flares.** Ya. F. Ashkaliev.
Ionos. i soln.-zemn. svyazi. Alma-Ata, 1980, p. 84 - 91. In Russian. – Abstr. in Ref. zh., 51. Astron., 11.51.480 (1980).

085.024 **On the possible role of geological inhomogeneities in the realization of the corpuscular mechanism of solar-terrestrial relations.**
N. P. Tsimakhovich, N. K. Ozolinya.
Astron. Tsirk., No. 1096, p. 7 - 8 (1980). In Russian.

085.025 **On a connection of geomagnetic activity, velocity of the solar wind and irregularity of the daily rotation of the earth.** Yu. D. Kalinin, V. M. Kiselev.
Geomagn. Aehron., Tom 20, 997 - 1001 (1980). In Russian.

085.026 **Variations of f_0F2 in day-time above the equator in the maximum and minimum of solar activity.**
T. Yu. Leshchinskaya, A. V. Mikhajlov.
Geomagn. Aehron., Tom 20, 1002 - 1008 (1980). In Russian.

085.027 **Investigation of ionospheric effects of solar flares by the Doppler method.** V. A. Vazherkin, B. Laso, L. A. Lobachevskij, V. D. Novikov, I. N. Odintsova.
Geomagn. Aehron., Tom 20, 1009 - 1013 (1980). In Russian.

085.028 **A connection of weak identic disturbances with the solar wind parameters.** V. I. Afanas'eva, A. D. Shevnin, V. M. Litinskij, N. F. Shevnina.
Geomagn. Aehron., Tom 20, 1041 - 1046 (1980). In Russian.

085.029 **On the use of geomagnetic field variations for diagnostics of inhomogeneities of the interplanetary electric field.** P. V. Sumaruk, Ya. I. Fel'dshtejn.
Astron. Vestn., Tom 14, 112 - 118 (1980). In Russian.

It is shown that surface variations of the geomagnetic field may be used for diagnostics of solar wind electric field inhomogeneities with dimensions comparable to those of the earth's magnetosphere.

Sun – climate links. See Abstr. 011.024.

L'activité solaire et les atmosphères planétaires. See Abstr. 072.026.

Comparative magnetospherology, part 9. Solar-terrestrial phenomena as explained by heliomagnetic excursion in 1974. See Abstr. 075.014.

Planetary System

091 Physics of the Planetary System (Dynamics, Figure, Rotation, Interiors, Atmospheres, Magnetic Fields, etc.)

091.001 **Sub-Doppler observations of the failure of local thermodynamic equilibrium in CO_2 planetary atmospheres.**
T. Kostiuk, D. Buhl, G. Chin, D. Deming, F. Espenak, M. J. Mumma, D. Zipoy.
Bull. American Astron. Soc., Vol. 12, 460 (1980). – Abstract.

091.002 **Erdähnliche Himmelskörper in neuer Sicht (I).** M. Reichstein.
Astron. Schule, 17. Jahrg., 79 - 82 (1980).

091.003 **The condensed matter physics of planetary interiors.** D. J. Stevenson.
J. Phys. Colloq., Vol. 41, No. C-2, (see 012.005), p. C2/53 - 59 (1980). – Abstr. in Phys. Abstr., Vol. 83, Abstr. 77307 (1980).

091.004 **A modified Titius-Bode law for planetary orbits.** L. Basano, D. W. Hughes.
Nuovo Cimento C, Ser. 1, Vol. 2C, 505 - 510 (1979). – Abstr. in Phys. Abstr., Vol. 83, Abstr. 82236 (1980).

091.005 **The magnetic field of the planets.** M. Surdin.
Nuovo Cimento C, Ser. 1, Vol. 2C, 527 - 536 (1979). Abstr. in Phys. Abstr., Vol. 83, Abstr. 82249 (1980).

091.006 **Life near the Roche limit: behavior of ejecta from satellites close to planets.**
A. R. Dobrovolskis, J. A. Burns.
Icarus, Vol. 42, 422 - 441 (1980).

The distribution of ejecta from impact craters significantly affects the surface characters of satellites and asteroids. In order to understand better the distinctive features seen on Phobos, Deimos, and Amalthea, the authors study the dynamics of nearby debris but include several factors–planetary tides plus satellite rotation and nonspherical shape–that complicate the problem. The authors have also examined the fate of crater ejecta from the satellites of Mars by numerical integration of trajectories for particles leaving their surfaces in the equatorial plane.

091.007 **The necessity of new dynamical theories of the planets.** J. H. Lieske.
Mitt. Astron. Ges., Nr. 48, (see 012.015), p. 81 - 85 (1980).

091.008 **On the possibility of existence and detection of satellites of planetary satellites.** M. Hoffmann.
Mitt. Astron. Ges., Nr. 48, (see 012.015), p. 92 - 93 (1980).

091.009 **Thermodynamic equations for a black planet with nearest-neighbour surface Carnot interaction.**
L. Sertorio.
Nuovo Cimento C, Ser. 1, Vol. 3C, 37 - 44 (1980). – Abstr. in Phys. Abstr., Vol. 83, Abstr. 85877 (1980).

091.010 **Radio scintillations during occultations by turbulent planetary atmospheres.**
R. Woo, A. Ishimaru, F.-C. Yang.
Radio Sci., Vol. 15, 695 - 703 (1980). – Abstr. in Phys. Abstr., Vol. 83, Abstr. 94734 (1980).

091.011 **Longitudinal instabilities and secondary flows in the planetary boundary layer: a review.** R. A. Brown.
Rev. Geophys. Space Phys., Vol. 18, 683 - 697 (1980).

Within the past decade, satellite pictures have shown persistent cloud patterns which indicate that the flow in the atmospheric planetary boundary layer is often organized into helical secondary circulations aligned parallel to the mean flow. Theory and observation agree that both convection in the presence of shear and the dynamic inflection point instabilities of the Ekman layer lead to these flow patterns. The observations and theory are reviewed with emphasis on the dynamics-dominated flow.

091.012 **Planetary rings.** A. F. Cook.
Solid particles in the solar system, (see 012.019), p. 401 - 416 (1980).

Observations of the rings of Saturn from the Pioneer spacecraft, discovery of the ring of Jupiter, ground based polarimetry of the rings of Saturn and some theoretical studies may be combined to markedly advance the understanding of the rings of Jupiter, Saturn and Uranus. In particular, narrow rings can be self-gravitatingly stable inside Roche's limit and outside another closer limit. They can be created from a sattellite which evolves across its Roche limit either by inward tidal drift or by growth of the planet by accretion.

091.013 **Report of the IAU Working Group on Cartographic Coordinates and Rotational Elements of the Planets and Satellites.** M. E. Davies, V. K. Abalakin, C. A. Cross, R. L. Duncombe, H. Masursky, B. Morando, T. C. Owen, P. K. Seidelmann, A. T. Sinclair, G. A. Wilkins, Y. S. Tjuflin (*Yu. S. Tyuflin*).
Celestial Mech., Vol. 22, 205 - 230 (1980).

This paper is the entire report of the IAU Working Group on Cartographic Coordinates and Rotational Elements of the Planets and Satellites, including three annexes. Tables give the recommended values for the directions of the north poles of rotation and the prime meridians of the planets and satellites. Reference surfaces for mapping these bodies are described. The annexes discuss the guiding principles, given in the body of the report, present explanatory notes, and provide a bibliography of the rotational elements and reference surfaces of the planets and satellites, definitions, and algebraic expressions of relevant parameters.

091.014 **Escape of the atmosphere of a rotating planet.** M. K. M. Ahmed.
J. Astron. Soc. Egypt, Vol. 1, 75 - 83 (1979).

The problem of escape of a planetary atmosphere is reviewed. Formulae are derived for the rates of loss of mass and angular momentum from a unit area and from the whole planetary surface. The resulting formulae are applied to find the residence times of H, O, N_2, O_2, and CO_2 in the Martian atmosphere. All constituents are so stable while hydrogen is never stable and cannot be retained to any extend.

091.015 **Dynamical entropy of the early solar system and origin of commensurability.** K. Ito.
13th Lunar and Planetary Symposium, (see 012.018), p. 149 - 157 (1980).

091.016 **A criterion on convection onset in planetary interiors.** H. Mizutani, S.-i. Kawakami.
13th Lunar and Planetary Symposium, (see 012.018), p. 320 - 329 (1980).

The conventional definition of the Rayleigh number yields a large range of the critical Rayleigh number depending on the viscosity distribution and layer thickness in spherical shells. A new convenient Rayleigh number which gives a fairly constant value for a variety of physical conditions is proposed. The critical Rayleigh number based on the new definition is calculated for various physical conditions, for which previous studies are available and is shown to fall in a significantly narrow range of values compared with those calculated with the conventional definition of Rayleigh number.

091.017 **Infrared spectroscopic observations of the outer planets, their satellites, and the asteroids.**
H. P. Larson.
Annu. Rev. Astron. Astrophys., Vol. 18, (see 003.004), 43 - 75 (1980).

This review delineates the spectroscopic data base for planetary studies. It includes atmospheric studies, which have a long observational history, and studies of surface mineralogies, which are relatively recent applications of IR methods. The atmospheric section includes the planets Jupiter, Saturn, Uranus and Neptune and those satellites with atmospheres (Titan and Triton). Infrared studies of surfaces include the satellites of Jupiter and Saturn, the asteroids, and Pluto.

091.018 **An assessment study on non terrestrial frozen ground.** M. Coradini, E. Flamini.
Mem. Soc. Astron. Italiana, Vol. 51, 329 - 358 (1980).

Accepting the definition of permafrost as planetary material where the moisture is predominantly in the solid state, its presence and effects on the surface of the bodies of the solar system are analyzed. On Mars several morphological features and the presence of both polar deposits and atmospheric moisture suggest that permafrost can be a very effective surface modificatory agent. On Europa, Ganymede and Callisto the presence of a large amount of solid H_2O has been definitely ascertained by Voyager 1 and 2. The possible effects of such an amount of water in the crust of the satellites are reviewed.

091.019 **Disk-satellite interactions.** P. Goldreich, S. Tremaine.
Astrophys. J., Vol. 241, 425 - 441 (1980).

The main purpose of this paper is to evaluate the transfer of angular momentum and energy between a disk and a satellite in order to determine their mutual evolution. The authors present an illustrative application of their results to the interaction between Jupiter and the protoplanetary disk.

091.020 **Far UV photolysis of CH_4–NH_3 mixtures and planetary studies.** A. Bossard, G. Toupance.
Nature, Vol. 288, 243 - 246 (1980).

The authors report new photochemical data on the effect of the mole fraction of NH_3 on the production of N-containing organics. Photolysis of CH_4–NH_3 mixtures at 147 nm leads to the formation of nitriles when the mole fraction of NH_3 is low. Important quantities of nitriles may have been photoproduced on the primitive earth if a low partial pressure of NH_3 remained and nitriles are the main N-containing organics that may be photoproduced in the atmosphere of the giant planets.

091.021 **Klimatische Bedingungen auf den erdnahen Planeten und Indizien für ihre Wandelbarkeit.**
H.-J. Bolle.
Evolution der Planetenatmosphären und des Lebens, (see 012.032), p. 44 - 46 (1980).

091.022 **Ionen- Molekül-Reaktionen in Planetenatmosphären.** W. Lindinger.
Evolution der Planetenatmosphären und des Lebens, (see 012.032), p. 47 - 57 (1980).

091.023 **The negative polarization of light scattered from rough surfaces seems to be largely due to diffraction.**
K. Lumme, E. Bowell, B. Zellner.
Bull. American Astron. Soc., Vol. 12, 663 (1980). – Abstract.

091.024 **Theoretical estimates of C_3 and C_4 hydrocarbon abundances on the outer planets.**
J. C. McConnell.
Bull. American Astron. Soc., Vol. 12, 669 (1980). – Abstract.

091.025 **CH_4 non-LTE in the atmospheres of the outer planets.** J. F. Appleby.
Bull. American Astron. Soc., Vol. 12, 670 (1980). – Abstract.

091.026 **Effects of observational distance on planetary polarization measurements.** K. Kawabata.
Bull. American Astron. Soc., Vol. 12, 671 (1980). – Abstract.

091.027 **Sputtering processes on the icy satellites due to charged particle radiation.** R. E. Johnson, E. Sieveka, J. Boring, L. J. Lanzerotti, W. L. Brown.
Bull. American Astron. Soc., Vol. 12, 673 (1980). – Abstract.

091.028 **Magnetospheric influences on the evolution of planetary ring systems.** L. L. Hood.
Bull. American Astron. Soc., Vol. 12, 676 - 677 (1980). Abstract.

091.029 **Dissipative processes and tidal Q for the major planets.** D. J. Stevenson.
Bull. American Astron. Soc., Vol. 12, 696 (1980). – Abstract.

091.030 **Determination of planetary gravitational fields from sequentially correlated Doppler data.**
J. D. Anderson.
Bull. American Astron. Soc., Vol. 12, 698 (1980). – Abstract.

091.031 **Theoretical predictions of deuterium abundances in the Jovian planets.**
W. B. Hubbard, J. J. MacFarlane.
Bull. American Astron. Soc., Vol. 12, 704 (1980). – Abstract.

091.032 **Turbulence models and the atmosphere of Uranus.** R. G. French, J. L. Elliot, R. V. E. Lovelace.
Bull. American Astron. Soc., Vol. 12, 704 (1980). – Abstract.

091.033 **Equilibrium models of the outer planets.** J. F. Appleby.
Bull. American Astron. Soc., Vol. 12, 711 - 712 (1980). Abstract.

091.034 **CO_2 and H_2O atmospheric and mantle reservoirs on Venus, Mars and Earth.** T. J. Ahrens.
Bull. American Astron. Soc., Vol. 12, 720 (1980). – Abstract.

091.035 **Ar and N systematics in Venus, Earth, and Mars. I.** C. J. Hostetler.
Bull. American Astron. Soc., Vol. 12, 720 - 721 (1980). Abstract.

091.036 **Ar and N systematics in Venus, Earth, and Mars. II.** M. J. Drake, C. J. Hostetler.
Bull. American Astron. Soc., Vol. 12, 721 (1980). – Abstract.

091.037 **Time-evolution model of superrotation with momentum exchange between atmosphere and planet: mechanistic interpretation.**

H. Mayr, I. Harris, B. Conrath.
Bull. American Astron. Soc., Vol. 12, 722 (1980). – Abstract.

091.038 **Integration of the outer planets of a ten-planet solar system.** R. S. Harrington.
Bull. American Astron. Soc., Vol. 12, 740 (1980). – Abstract.

091.039 **Neues über mögliche Ringe im Sonnensystem.** N. Giesinger.
Sternenbote, 23. Jahrg., 114 - 120 (1980).

091.040 **Die Atmosphären der Planeten.** S. Bauer.
Phys. unserer Zeit, 11. Jahrg., 162 - 168 (1980).

091.041 **Two-stream approximations to radiative transfer in planetary atmospheres: a unified description of existing methods and a new improvement.**
W. E. Meador, W. R. Weaver.
J. Atmos. Sci., Vol. 37, 630 - 643 (1980). – Abstr. in Phys. Abstr., Vol. 83, Abstr. 109383 (1980).

091.042 **Formazione ed evoluzione delle superfici planetarie.** M. Coradini.
G. Astron., Vol. 6, 187 - 205 (1980).

091.043 **I satelliti.** S. Catalano.
G. Astron., Vol. 6, 207 - 227 (1980).

091.044 **Problemi di dinamica dei corpi minori del sistema solare.** A. Carusi, G. B. Valsecchi.
G. Astron., Vol. 6, 229 - 239 (1980).

091.045 **Risonanze e anelli nel sistema solare.** P. Farinella, P. Paolicchi, F. Ferrini, A. Milani, A. M. Nobili.
G. Astron., Vol. 6, 253 - 264 (1980).

091.046 **Planetary cratering: statistical recognition of secondary-crater fields.**
M. Fulchignoni, M. Poscolieri.
Nuovo Cimento C, Ser. 1, Vol. 3C, 193 - 199 (1980).
Abstr. in Phys. Abstr., Vol. 84, Abstr. 4799 (1981).

091.047 **Ionospheres of the terrestrial planets.** R. W. Schunk, A. F. Nagy.
Rev. Geophys. Space Phys., Vol. 18, 813 - 852 (1980).

The theory and observations relating to the ionospheres of the terrestrial planets Venus, the earth, and Mars are reviewed. Emphasis is placed on comparing the basic differences and similarities between the planetary ionospheres. The review covers the plasma and electric-magnetic field environments that surround the planets, the theory leading to the creation and transport of ionization in the ionospheres, the relevant observations, and the most recent model calculations. The theory section includes a discussion of ambipolar diffusion in a partially ionized plasma, diffusion in a fully ionized plasma, supersonic plasma flow, photochemistry, and heating and cooling processes. The sections on observations and model calculations cover the neutral atmosphere composition, the ion composition, the electron density, and the electron, ion, and neutral temperatures.

091.048 **A planetological law for orbital proportions of planets and moons.** E. Litzroth.
Gerlands Beitr. Geophys., Band 89, 357 - 359 (1980).

An elementary mathematical scheme for the distribution of the regular velocities of the planets, asteroids, moons and rings in the solar system will be derived. The resulting average relative deviations are smaller than 10^{-2}.

091.049 **One-parametric hetegonic orbit structure of planets and moons.** D. T. F. Möhlmann.
Gerlands Beitr. Geophys., Band 89, 360 - 364 (1980).

The velocities of planets and moons seem to be ordered in dual cascades (Litzroth). This paper gives a mathematical formulation of these cascades, depending on one velocity parameter only. The resulting hetegonic scheme is applied to the planetary system and the satellite systems of Jupiter, Saturn and Uranus.

091.050 **Correlated variations of planetary albedos and coincident solar-interplanetary variations.**
S. T. Suess, G. W. Lockwood.
Sol. Phys., Vol. 68, 393 - 409 (1980).

The apparent brightnesses of Titan and Neptune, near 4718 Å and 5508 Å, have changed by up to 10% and 3% respectively since 1972. Since these changes are larger than any plausible solar variation in visible light, the implied cause is an intrinsic albedo change resulting from an externally driven alteration in atmospheric chemistry. Such an alteration could result from solar emissions in UV/EUV or the infrared – where large changes are suspected to exist, or from energetic particle precipitation modulated by the solar wind. It is found that traditional solar indices (10.7 cm radio flux, sunspot number, etc.) do not directly measure the causative process, as they tend to lag the brightness variations in phase. Conversely, solar EUV and some characteristics of the solar wind correlate well with the brightness variations and lead those variations in phase.

091.051 **Tri-axiality of the Earth, the Moon and Mars.** M. Burša, Z. Šíma.
Stud. Geophys. Geod., Vol. 24, 211 - 217 (1980).

091.052 **Theory of planetary dynamos.** F. H. Busse.
Solar system plasma physics, Vol. 2, (see 003.010), 293 - 317 (1979).

In this article an attempt is made to outline the possibilities of a comparative theory of planetary magnetism and to describe initial steps that have been taken towards the achievement of such a theory.

091.053 **Towards a comparative theory of magnetospheres.** G. L. Siscoe.
Solar system plasma physics, Vol. 2, (see 003.010), 319 - 402 (1979).

The focus of this review is restricted to planetary magnetospheres, where direct comparisons between theories and observations are now possible, and where there is a high probability that opportunities for significant comparisons will continue to occur in the future. Only in this way can a solid, i. e. observationally tested, theory of comparative magnetospheres be achieved.

091.054 **On regularities of the rotation of planets of the solar system around their axes.**
B. I. Ogarkov, A. M. Tikhanov.
Voronezh. s.-kh. inst. Voronezh, 1980. 7 pp. In Russian. Abstr. in ref. zh., 51. Astron., 11.51.175 (1980).

091.055 **Meteorite cratering on terrestrial planets.** V. P. Belov, Ya. G. Kats, V. L. Avdeev.
Kosmogen. struktury Zemli. Mater. semin. Moskva, 1980, p. 11 - 13. In Russian. – Abstr. in Ref. zh., 62. Issled. kosm. prostranstva, 11.62.68 (1980).

091.056 **Morphology of impact craters on the moon and planets.** A. T. Bazilevskij, N. N. Grebennik.
Kosmogen. struktury Zemli. Mater. semin. Moskva, 1980, p. 13 - 17. In Russian. – Abstr. in Ref. zh., 62. Issled. kosm. prostranstva, 11.62.69 (1980).

091.057 **Basaltic volcanism: the importance of planet size.** D. Walker, E. M. Stolper, J. F. Hays.
Proc. Tenth Lunar Planet. Sci. Conf., (see 012.050), p. 1995 - 2015 (1979).

091.058 **Time dependent deformation of a basalt at low differential stress.** N. Brodsky, H. Spetzler.
Proc. Tenth Lunar Planet. Sci. Conf., (see 012.050), p. 2155 - 2163 (1979).

091.059 **Planetary dynamo amplification of ambient magnetic fields.** E. H. Levy.
Proc. Tenth Lunar Planet. Sci. Conf., (see 012.050), p. 2335 - 2342 (1979).

091.060 **Some possible effects of solid-state deformation on the thermal evolution of ice-silicate planetary bodies.**
E. M. Parmentier, J. W. Head.
Proc. Tenth Lunar Planet. Sci. Conf., (see 012.050), p. 2403 - 2419 (1979).

091.061 **The history of an atmosphere of impact origin.** B. M. Jakosky, T. J. Ahrens.
Proc. Tenth Lunar Planet. Sci. Conf., (see 012.050), p. 2727 - 2739 (1979).

091.062 **Monogenetic volcanoes of the terrestrial planets.** C. A. Wood.
Proc. Tenth Lunar Planet. Sci. Conf., (see 012.050), p. 2815 - 2840 (1979).

091.063 **Endogenic craters on basaltic lava flows: size frequency distributions.** R. Greeley, D. E. Gault.
Proc. Tenth Lunar Planet. Sci.Conf., (see 012.050), p. 2919 - 2933 (1979).

091.064 **Tectonic patterns on a tidally distorted planet.** H. J. Melosh.
Icarus, Vol. 43, 334 - 337 (1980).

Tidal deformation of the lithosphere of a synchronously rotating planet or satellite produces stresses that may result in a distinctive tectonic pattern. The lithosphere is treated as a thin elastic shell which maintains the equilibrium shape of a tidally distorted body. Stresses develop as the equilibrium shape changes during orbital evolution. E. M. Anderson's theory of faulting is used to translate this stress pattern into a tectonic pattern of faults on the planet's surface. On a body such as the Moon, which has receded from the Earth, an originally large tidal bulge has collapsed. The predicted tectonic pattern is described. The existence of such a tectonic pattern on the Moon can only be resolved by photogeologic mapping. At present, there is little evidence of this pattern. These tectonic patterns, which could provide geologic evidence for large tidal distortions, may also be present on the Galilean satellites of Jupiter.

091.065 **Determination of the orbits of Venus, the earth and Mars from radar observations of Venus and Mars in 1962 - 1978.**
M. D. Kislik, Yu. F. Kolyuka, V. A. Kotel'nikov, G. M. Petrov, V. F. Tikhonov.
Usp. fiz. nauk, Vol. 131, 511 - 513 (1980). In Russian.
Abstr. in Ref. zh., 51. Astron., 12.51.104 (1980).

091.066 **Formation of primary lithospheres of planets, their possible composition and volcanism on the example of the moon.** Yu. A. Khodak.
Probl. planetol. T. 2. Simpoz. No. 8. Mezhdunar. assots. planetol., Erevan, 1977. Leningrad-Erevan, 1977, p. 171 - 176. In Russian. – Abstr. in Ref. zh., 62. Issled. kosm. prostranstva, 12.62.166 (1980).

091.067 **A new predictive model for determining solar wind-terrestrial planet interactions.**
J. R. Spreiter, S. S. Stahara.
J. Geophys. Res., Vol. 85, 6769 - 6777 (1980).

A computational model has been developed for the determination of the gasdynamic and magnetic field properties of the solar wind flow around a magnetic planet, such as the earth, or a nonmagnetic planet, such as Venus. The procedures are based on an established single-fluid, steady, dissipationless, magnetogasdynamic model and are appropriate for the calculation of axisymmetric, supersonic, super-Alfvénic solar wind flow past a planetary magneto/ionosphere. Sample results are reported for a variety of solar wind and planetary conditions.

091.068 **Planetary exploration.** G. E. Hunt.
The state of the Universe, (see 003.013), p. 144 - 180 (1980).

All the planets from Mercury to Jupiter have now been observed at close range, and, before the end of 1979, we will obtain the first glimpse of Saturn with the Pioneer spacecraft, followed by detailed investigations by Voyager in 1980 and 1981. The author briefly reviews current understanding of these bodies, leaning heavily toward comparative planetary processes, and the challenging problems of the next decade.

091.069 **Some recent results of the exploration of the solar system.** Z. Kneževič.
Vasiona, Année 28, 23 - 30 (1980). In Croatian.

091.070 **Toward the outer planets.** M. Dimitrijević.
Vasiona, Année 28, 31 - 38 (1980). In Serbo-Croatian.

091.071 **What the planets are.**
Yu. V. Aleksandrov, V. A. Zakhozhaj.
Astron. Vestn., Tom 14, 129 - 132 (1980). In Russian.

The development of a general theory of the structure and evolution of the planets needs the definition of the notion "planet". The authors propose such a definition which is based on a certain range of masses.

A bibliography of planetary geology principal investigators and their associates, 1979 - 1980. See Abstr. 002.061.

Atlas of the planets. See Abstr. 002.076.

Interiors of the planets. See Abstr. 003.031.

Parameter disposition in pre-Newtonian planetary theories. See Abstr. 004.050.

Planets and their systems at the XXIIIth plenary meeting of COSPAR. See Abstr. 011.035.

On the possibility of using Laplace series for the gravitational potential of the surface of a planet. I. See Abstr. 021.050.

On the explosive formation of macroscopic hypervelocity projectiles for use in the study of planetary cratering. See Abstr. 022.003.

Large enhancements of electron impact produced OH $A-X$ emission in the presence of certain catalyzer gases. See Abstr. 022.021.

The interplay of molecular spectroscopy and astronomy. See Abstr. 022.106.

Theoretical studies of impact cratering.
See Abstr. 022.128.

Regions of weak methane absorptions in the near infrared: laboratory measurements and planetary applications.
See Abstr. 022.129.

Effects of low temerature and pressure on the reflectance spectra of clay minerals. See Abstr. 022.133.

Spectra of Fe-Ti silicate glasses: implications to remote-sensing of planetary surfaces.
See Abstr. 022.180.

Calculational investigation of impact cratering dynamics: early time material motions.
See Abstr. 022.184.

Correlation spectrometry by pressure modulation: a technique for very high spectral resolution remote sensing of planetary atmospheres from spacecraft.
See Abstr. 031.563.

Observations of the sun, moon, and planets. Six-Inch Transit Circle results. See Abstr. 041.027.

Observations of the moon and outer planets obtained in Nikolaev in 1969 - 1972.
See Abstr. 041.042.

On the 4th order short-period solar perturbations in the motion of the satellites of the major planets.
See Abstr. 042.009.

Modern approaches to research in dynamical astronomy. See Abstr. 042.022.

The preparation of new planetary ephemerides.
See Abstr. 047.023.

Summary of terrestrial and planetary radio and plasma waves observed by the initial phase of JIKIKEN (EXOS-B) observations. See Abstr. 051.010.

Microwave communications from interplanetary space. See Abstr. 051.033.

Planetary exploration by spacecraft.
See Abstr. 053.007.

Study of some processess of nucleosynthesis in the evolving galaxy. See Abstr. 061.034.

The chemical evolution in the Galaxy and isotopic ratios in the solar system. See Abstr. 061.036.

Very hot plasmas in the solar system.
See Abstr. 062.084.

Theory of scan plane flux anisotropies.
See Abstr. 062.106.

Spindown of rotating magnets.
See Abstr. 062.108.

Shock systems in collisionless space plasmas.
See Abstr. 062.126.

Light scattering by an optically thin inhomogeneous, spherically-symmetric planetary atmosphere: brightness at the zenith near the terminator. See Abstr. 063.013.

Radiative transfer through an arbitrarily thick, scattering atmosphere. See Abstr. 063.031.

Generalization of the Curtis-Godson approximation to inhomogeneous scattering atmospheres.
See Abstr. 063.032.

L'activité solaire et les atmosphères planétaires.
See Abstr. 072.026.

The earth and planetary sciences.
See Abstr. 081.002.

Earth's magnetosphere: global problems in magnetospheric plasma physics. See Abstr. 084.097.

Magnetospheric plasma waves.
See Abstr. 084.102.

Parameterized convection within the moon and the terrestial planets. See Abstr. 094.054.

Morphology of impact craters on the moon and other planets. See Abstr. 094.638.

Formation of complex impact craters: evidence from Mars and other planets. See Abstr. 097.018.

Significant achievements in the Planetary Geology Program, 1980. See Abstr. 097.085.

Evolution of eccentricities of the orbits of planets in the process of their accumulation. See Abstr. 107.003.

Electrolysis in space and fate of Phaethon.
See Abstr. 107.012.

Erratum

091.901 **Erratum: 'Prediscovery evidence of planetary rings'** [J. British Interplanet. Soc., Vol. 33, 287 - 294 (1980)]. W. I. McLaughlin.
J. British Interplanet. Soc., Vol. 33, 432 (1980). – See Abstr. 27.091.036.

092 Mercury

092.001 **A note on T. J. J. See's observations of craters on Mercury.** J. Lankford.
J. Hist. Astron., Vol. 11, 129 - 132 (1980).

092.002 **The cartographic aspect of studying the planet Mercury.** K. B. Shingareva, T. A. Romashina.
Izv. vuzov. Geod. i aehrofotosemka, 1980, No. 3, p. 112 - 117. In Russian. – Abstr. in Ref. zh., 52. Geod. Aehrosemka, 10.52.170 (1980).

092.003 **Onset of scarp formation on Mercury.** M. Leake.
Bull. American Astron. Soc., Vol. 12, 677 (1980). – Abstract.

092.004 **Breuken in de korst van Mercurius.** W. van Tend. Zenit, 7e Jaarg., 22 - 25 (1980).

092.005 **The magnetosphere of Mercury.** N. F. Ness. Solar system plasma physics, Vol. 2, (see 003.010), 183 - 206 (1979).
Contents: Introduction. Experiments. Mercury I encounter. Mercury III encounter. Magnetosphere models. Magnetic tail and polar cap. Summary and future problems.

092.006 **Mercurian crater rim heights and some interplanetary comparisons.** M. J. Cintala.
Proc. Tenth Lunar Planet. Sci. Conf., (see 012.050), p. 2635 - 2650 (1979).

092.007 **Large impact basins on Mercury and relative crater production rates.** H. Frey, B. L. Lowry.
Proc. Tenth Lunar Planet. Sci. Conf., (see 012.050), p. 2669 - 2687 (1979).

History of the mass of Mercury.
See Abstr. 004.060.

What is the mass of Mercury? Report of a seminar at Cambridge 1980 May 2. See Abstr. 011.001.

Cometary collisions on the Moon and Mercury.
See Abstr. 094.007.

The interaction of the solar wind with Mars, Venus and Mercury. See Abstr. 097.079.

The circumsolar motion of dust particles at the stage of increasing solar luminosity. See Abstr. 106.034.

093 Venus

093.001 **Ein Pionier enthüllt die Venus.**
J. von Puttkamer.
Umschau, 80. Jahrg., 504 - 505 (1980).

Based on the extensive radar data returned by NASA's Pioneer Venus spacecraft, for the first time scientists have mapped nearly the entire planet. Huge continent-sized features, mountains as high as Everest and deep rift valleys have been identified.

093.002 **A new radar determination of the spin vector of Venus.** S. Zohar, R. M. Goldstein, H. C. Rumsey.
Astron. J., Vol. 85, 1103 - 1111 (1980).

Two radar observations of a set of three relatively small features on the surface of Venus have facilitated a refined determination of the spin vector of Venus. The period is found to be 243.019 ± 0.014 days, while the obliquity is 177°.22 ± 0°.18. The effects of deviations from exact sphericity on the interpretation of the measurements are discussed at length and the question of resonance with earth is reexamined.

093.003 **Theoretical dichotomy of Venus, 1960–2000.**
J. Meeus.
J. British Astron. Assoc., Vol. 90, 442 - 443 (1980).

093.004 **On the contradiction between the radar axial period and the atmospheric circulation of Venus.**
V. A. Firsoff.
J. British Astron. Assoc., Vol. 90, 464 - 466 (1980).

The contradiction between the short atmospheric and the long radar axial period remains unresolved. The Pioneer data add to its difficulty, inasmuch as all the recorded air movements are substantially from east to west. Since such movements arise from Coriolis forces, this would imply that all atmospheric flow is from the equator to the poles, which is absurd and is not borne out by the simultaneous records of meridional air flow. It is suggested that the origin of the scale is wrong, and the true axial period is close to four days, the radar period originating in a counter-rotating belt of plasma.

093.005 **The surface of Venus.**
G. H. Pettengill, D. B. Campbell, H. Masursky.
Sci. American, Vol. 243, No. 2, p. 46 - 57 (1980).

Shrouded by clouds, it has now been mapped by radar from the earth and from a spacecraft in orbit around Venus. The images suggest a geology intermediate between that of the earth and that of Mars.

093.006 **Mariner 10 observations of hydrogen Lyman alpha emission from the Venus exosphere: evidence of complex structure.** P. Z. Takacs, A. L. Broadfoot, G. R. Smith, S. Kumar.
Planet. Space Sci., Vol. 28, 687 - 701 (1980).

The Ultraviolet Spectrometer Experiment on the Mariner 10 spacecraft measured the hydrogen Lyman α emission resonantly scattered in the Venus exosphere at several viewing aspects during the encounter period. Venus encounter occurred at 17:01GMT on 5 February 1974. Exospheric emissions above the planet's limb were measured and were analyzed with a spherically symmetric, single scattering, two-temperature model.

093.007 **Venus revealed.** J. K. Beatty.
Sky Telesc., Vol. 60, 185 - 187 (1980).

093.008 **Die Venus als Glashaus.** Die Bedeutung des Kohlendioxid für das Klima von Erde und Venus.
U. von Zahn.
Umschau, 80. Jahrg., 525 - 526 (1980).

Measurements from the NASA project "Pioneer-Venus" have confirmed that carbon dioxide in an atmosphere contributes significantly to an increase in temperature. This gives even more weight to the demands to reduce anthropogenic CO_2 production on Earth rigorously, so as to prevent a major alteration of the climate.

093.009 **Catalogue of the positions of Venus obtained with the wide-angle astrograph of the Sternberg Astronomical Institute.** B. S. Vozdvizhenskij.
Tr. Gos. Astron. Inst. Shternberga, Tom 49, 213 - 230 (1980). In Russian.

The results of the observations consist of 1041 positions for the period from 1969 to 1972.

093.010 **Surface reflections of Pioneer Venus probe radio signals.** T. A. Croft.
Geophys. Res. Lett., Vol. 7, 521 - 524 (1980).

The author has detected surface reflections from all four Venus probes and shown qualitatively that the ground-scattering mechanism provides an explanation for the structural features of the spectra. Horizontal winds and atmospheric refraction play important roles. The significance of these echoes lies primarily in their relevance to future planetary probes.

093.011 **Venus: discoveries and problems.**
M. Ya. Marov.
Zemlya Vselennaya, 1980, No. 4, p. 13 - 18. In Russian.

093.012 **Gaseous sulfur in the Venus atmosphere.**
N. F. San'ko.
Kosm. Issled., Tom 18, 600 - 608 (1980). In Russian.

093.013 **On the structure of the tail of Venus.**
O. L. Vajsberg, V. N. Smirnov.
Kosm. Issled., Tom 18, 651 - 655 (1980). In Russian.

093.014 **Oxidation state of the atmosphere and crust of Venus from Pioneer Venus results.**
J. S. Lewis, F. A. Kreimendahl.
Icarus, Vol. 42, 330 - 337 (1980).

Pioneer Venus experiments confirm that, as predicted, COS and H_2S are dominant over SO_2 in the lower atmosphere of Venus, and that the equilibrium concentrations of S_2 and S_3 are significant. Many criteria serve to bracket the oxidation state of the crust: it is nearly certain that the S_2^{2-}/SO_4^{2-} buffer regulates the oxygen fugacity, and that FeO is at least as abundant as Fe_2O_3 in crustal silicates. A highly oxidized crust is incompatible with the gas-phase sulfur chemistry. If the Pioneer Venus mass spectrometer estimates of the abundance of sulfur gases are correct, earth-like models for the bulk composition of Venus are seriously in error, and a far lower FeO content is required for Venus.

093.015 **Production of nitrogen and carbon species by thunderstorms on Venus.** A. Bar-Nun.
Icarus, Vol. 42, 338 - 342 (1980).

The effects of the newly discovered thunderstorms on Venus upon the nitrogen and carbon species in its atmosphere were calculated.

093.016 **Transmission of CO_2 in the atmosphere of Venus for the spectral region near ten micrometers.**
L. G. Young.
Icarus, Vol. 43, 102 - 110 (1980).

One hundred forty-five vibration-rotation bands are used to compute the transmission for a clear Venus atmosphere in the region near 10 μm. Synthetic spectra for each layer of a seven-layer model atmosphere show that the atmosphere is effectively black below 40 km above the surface. Comparison of calculated transmission with laboratory measurements shows good agreement up to 759°K.

093.017 **The hidden face of Venus.**
J. C. Gerard, A. I. Stewart.
Recherche, Vol. 11, 614 - 616 (1980). In French. – Abstr. in Phys. Abstr., Vol. 83, Abstr. 94735 (1980).

093.018 **Variations of the nightglow of Venus.**
V. A. Krasnopol'skij, G. V. Tomashova.
Kosm. Issled., Tom 18, 766 - 774 (1980). In Russian.

093.019 **Turbulence of the Venus atmosphere from radiowave fluctuation data emitted by the Venera 9 and Venera 10 stations.**
T. S. Timofeeva, A. I. Efimov, O. I. Yakovlev.
Kosm. Issled., Tom 18, 775 - 782 (1980). In Russian.

093.020 **Vertical temperature gradient of the Venus atmosphere from radio occultation data.**
S. S. Matyugov, E. V. Chub, G. D. Yakovleva, O. I. Yakovlev.
Kosm. Issled., Tom 18, 783 - 786 (1980). In Russian.

093.021 **On the asymmetry of the internal current of the tail of Venus.** O. L. Vajsberg.
Kosm. Issled., Tom 18, 809 - 812 (1980). In Russian.

093.022 **Construction of large-scale maps of regions of the Venus surface.**
Yu. S. Tyuflin, V. I. Kabeshkina, L. M. Kadnichanskaya.
Geod. i kartogr., 1980, No. 4, p. 23 - 27. In Russian. – Abstr. in Ref. zh., 52. Geod. Aehrosemka, 8.52.191 (1980).

093.023 **The location of the dayside ionopause of Venus: Pioneer Venus Orbiter magnetometer observations.**
R. C. Elphic, C. T. Russell, J. A. Slavin, L. H. Brace, A. F. Nagy.
Geophys. Res. Lett., Vol. 7, 561 - 564 (1980).

The location of the dayside Venus ionopause, as observed by the Pioneer Venus Orbiter, is shown to depend on the magnetic pressure in the shocked, highly compressed solar wind plasma just outside the ionopause. Assuming a balance exclusively between this external magnetic pressure and internal ionospheric thermal pressure, invariance of ionospheric conditions, and an isothermal ionosphere, it is possible to determine pressure scale heights for various solar zenith angle intervals. These scale heights yield ionospheric temperatures which agree with direct measurements obtained independently.

093.024 **Radiative heat exchange in the Venus atmosphere.**
K. Ya. Kondrat'ev, N. I. Moskalenko, V. F. Terzi.
Dokl. AN SSSR, Vol. 251, 1345 - 1349 (1980). In Russian. Abstr. in Ref. zh., 51. Astron., 9.51.202 (1980).

093.025 **Interpretation of the photometric observations of Venus in the near infrared.**
S. Mukai, T. Mukai, S. Sato.
13th Lunar and Planetary Symposium, (see 012.018), p. 195 - 200 (1980).

The authors investigate a Venus cloud model, including its vertical structure, to examine the phase angle variation of both intensity and polarization of near infrared radiation. Based on the multiple scattering calculations considering an internal heat source, it is shown that a model composed of two layers gives a fairly good agreement with the observed features.

093.026 **Venera 11 and Venera 12: preliminary evaluations of wind velocity and turbulence in the atmosphere of Venus.** V. V. Kerzhanovich, Yu. F. Mararov, M. Ya. Marov, M. K. Rozhdestvenskiy (*Rozhdestvenskij*), V. P. Sorokin.
Moon Planets, Vol. 23, 261 - 270 (1980).

Data accumulated from Venera 4–10 have revealed unusual properties of the troposphere of Venus. These data have not been sufficient, however, to enable a complete description and account of these properties. Essential supplementary data have now been obtained from experiments carried out on Venera 11 and 12. These results, which are described in this paper, indicate that the atmosphere of Venus encompasses a strong, nonuniform, zonal movement which reveals a new type of circulation of a deep atmosphere.

093.027 **Monografía sobre planetas: Venus.**
A. J. Camponovo.
Rev. Astron., No. 212, p. 3 - 8 (1980).

093.028 **Investigation of the Venus atmosphere and surface by the method of radiosounding using Venera-9 and 10 satellites.** M. A. Kolosov, O. I. Yakovlev, A. I. Efimov, S. S. Matyugov, T. S. Timofeeva, E. V. Chub, A. G. Pavelyev (*Pavel'ev*), A. I. Kucheryavenkov, I. E. I. E. (*I. Eh.*) Kalashnikov, O. E. Milekhin.
Acta Astronaut. Vol. 7, 219 - 234 (1980). – Abstr. in Phys. Abstr., Vol. 83, Abstr. 101543 (1980).

093.029 **A comparative study of geological evolution of Venus.**
R. Bianchi, M. Coradini, M. Fulchignoni, E. Flamini.
Mem. Soc. Astron. Italiana, Vol. 51, 263 - 284 (1980).

Radar analysis carried out by the Pioneer Venus spacecraft during the first period of the mission confirms some hypotheses formulated on the basis of the earth-based radar observations. A geomorphologic analysis of the surface, using Pioneer Venus preliminary data has been carried out; tridimensional projections of the highlands of the planet have been obtained. The hypothesis that Venus is a very flat planet seems to be confirmed.

093.030 **The atmospheres of Venus and Jupiter.**
B. J. Mason.
Contemp. Phys., Vol. 21, 381 - 399 (1980). – Abstr. in Phys. Abstr., Vol. 83, Abstr. 105186 (1980).

093.031 **Observations of CO in the stratosphere of Venus via its $J = 0 \rightarrow 1$ rotational transition.**
F. P. Schloerb, S. E. Robinson, W. M. Irvine.
Icarus, Vol. 43, 121 - 127(1980) = Five Coll. Obs., Contrib. No. 433.

The authors have observed carbon monoxide in the stratosphere of Venus at phase angles of 180 and 120° via the $J = 0 \rightarrow 1$ rotational transition at 115.2712 GHz. The mixing ratio profile of CO has been obtained by fitting the theoretical spectrum produced by a small number of layers with constant CO mixing ratio to the line profile, and the results suggest that the CO mixing ratio below the 1-mb level increased by at least a factor of 10 between 180 and 120° phase angles. The depletion of CO on the nightside of the planet was not anticipated and may require either a new loss mechanism for CO which can operate without sunlight or an increase in the amount of vertical mixing on the nightside.

093.032 **Venus gravity: a high-resolution map from orbiter tracking data.** R. D. Reasenberg, Z. M. Goldberg, P. E. MacNeil, I. I. Shapiro.
Bull. American Astron. Soc., Vol. 12, 689 (1980). – Abstract.

093.033 **Geology of Venus.**
H. Masursky, G. Schaber, M. Strobell, A. Dial.
Bull. American Astron. Soc., Vol. 12, 690 (1980). – Abstract.

093.034 **The radar reflectivity of Venus.**
G. H. Pettengill, S. Nozette, P. G. Ford.
Bull. American Astron. Soc., Vol. 12, 690 (1980). – Abstract.

093.035 **Radar craters on Venus: a steady state population.** J. A. Cutts, T. W. Thompson, B. H. Lewis.
Bull. American Astron. Soc., Vol. 12, 690 (1980). – Abstract.

093.036 **Venus: do sediments cover lowlands?** J. L. Warner.
Bull. American Astron. Soc., Vol. 12, 691 (1980). – Abstract.

093.037 **Measurements of Venusian temperature profiles using fully resolved CO_2 line shapes.**
D. Deming, T. Kostiuk, F. Espenak, D. Jennings, G. Chin, D. Buhl, M. Mumma.
Bull. American Astron. Soc., Vol. 12, 714 (1980). – Abstract.

093.038 **Simultaneous observations of the CO J=1–0 and J=2–1 rotational transitions in the stratosphere of Venus.** F. P. Schloerb, R. W. Wilson, J. C. Good.
Bull. American Astron. Soc., Vol. 12, 714 - 715 (1980). Abstract.

093.039 **Nature of the UV absorber in Venus' atmosphere.** J. B. Pollack, O. B. Toon.
Bull. American Astron. Soc., Vol. 12, 715 (1980). Abstract.

093.040 **Venus upper atmosphere aerosols from polarimetry.** R. Santer, A. Dollfus.
Bull. American Astron. Soc., Vol. 12, 715 (1980). – Abstract.

093.041 **Analysis of aircraft spectra and Pioneer radiometry of Venus near 30 - 100 μm: implications for high altitude aerosol structure.**
G. S. Orton, H. H. Aumann, J. V. Martonchik.
Bull. American Astron. Soc., Vol. 12, 715 - 716 (1980). Abstract.

093.042 **Venus temperature and cloud structure – global retrievals.** J. T. Schofield, F. W. Taylor.
Bull. American Astron. Soc., Vol. 12, 716 (1980). – Abstract.

093.043 **Mesoscale convection in the Venus clouds.** G. Schubert, C. Covey.
Bull. American Astron. Soc., Vol. 12, 716 (1980). – Abstract.

093.044 **Resonant planetary-scale waves in the Venus atmosphere.** C. Covey, G. Schubert.
Bull. American Astron. Soc., Vol. 12, 716 (1980). – Abstract.

093.045 **Progress report on the measurement of Venus winds using a Reticon array.**
E. S. Barker, W. D. Cochran.
Bull. American Astron. Soc., Vol. 12, 717 (1980). – Abstract.

093.046 **Possible explanation of large thermal fluxes measured from Pioneer Venus entry probes.**
A. T. Young.
Bull. American Astron. Soc., Vol. 12, 717 (1980). – Abstract.

093.047 **Cloud structure and the flux of sunlight in the atmosphere of Venus.** L. R. Doose, M. G. Tomasko, P. H. Smith, N. D. Castillo, A. P. Odell.
Bull. American Astron. Soc., Vol. 12, 717 (1980). – Abstract.

093.048 **Individual retrieved temperature profiles in the hot spots and collar regions of the middle atmosphere of Venus.** R. Haskins, M. Chahine, D. McCleese.
Bull. American Astron. Soc., Vol. 12, 717 - 718 (1980). Abstract.

093.049 **A diagnostic model for the solar related circulation above the clouds of Venus from Pioneer VORTEX temperature measurements.** L. Elson.
Bull. American Astron. Soc., Vol. 12, 718 (1980). – Abstract.

093.050 **Preliminary results from Very Large Array (VLA) observations of Venus at 6 and 20 cm wavelengths.**
D. O. Muhleman, G. L. Berge, S. Deguchi.
Bull. American Astron. Soc., Vol. 12, 718 (1980). – Abstract.

093.051 **SO_2 in the Venus atmosphere as constrained by the microwave spectrum.**
M. A. Janssen, M. J. Klein, R. L. Poynter.
Bull. American Astron. Soc., Vol. 12, 718 (1980). – Abstract.

093.052 **Radio occultation evidence for cloud-related gases as the source of the microwave opacity of the middle atmosphere of Venus.** V. R. Eshleman.
Bull. American Astron. Soc., Vol. 12, 718 - 719 (1980). Abstract.

093.053 **Sulfur dioxide and other cloud-related gases as microwave absorbers in the middle atmosphere of Venus.** P. Steffes.
Bull. American Astron. Soc., Vol. 12, 719 (1980). – Abstract.

093.054 **Search for optical pulses from Venus.** W. J. Borucki.
Bull. American Astron. Soc., Vol. 12, 719 (1980). – Abstract.

093.055 **H_2 abundance in the upper atmosphere of Venus.** S. Kumar, D. M. Hunten, H. A. Taylor, Jr.
Bull. American Astron. Soc., Vol. 12, 719 - 720 (1980). Abstract.

093.056 **Structure of the dayside ionosphere of Venus from Pioneer Venus radio occultations.**
A. Kliore, I. R. Patel, T. E. Cravens.
Bull. American Astron. Soc., Vol. 12, 720 (1980). – Abstract.

093.057 **Thermal structure of the mesosphere and lower thermosphere of Venus.**
S. P. Bradley, F. W. Taylor.
Bull. American Astron. Soc., Vol. 12, 720 (1980). – Abstract.

093.058 **Polar cloud morphology of Venus.** J. B. Cimino, D. McCleese, A. Kliore, C. Elachi, I. Patel.
Bull. American Astron. Soc., Vol. 12, 730 (1980). – Abstract.

093.059 **Venus north polar region: coincident infrared and polarimetry data from the Pioneer Orbiter.**
H. Nebel, L. S. Elson, D. J. Diner, L. D. Travis.
Bull. American Astron. Soc., Vol. 12, 730 (1980). – Abstract.

093.060 **Five micron spectroscopy of the middle atmosphere and clouds of Venus.** R. Beer, S. T. Ridgway.
Bull. American Astron. Soc., Vol. 12, 732 (1980). – Abstract.

093.061 **Het oppervlak van de planeet Venus.** C. Titulaer.
Zenit, 7e Jaarg., 366 - 369 (1980).

093.062 **Two-frequency radio occultation measurements with Venera-9 and Venera-10 orbiters.**
M. B. Vasilyev (*Vasil'ev*), A. S. Vyshlov, M. A. Kolosov, A. P. Mesterton (*Mestehrton*), N. A. Savich, V. A. Samovol, L. N. Samoznaev, A. I. Sidorenko.
Acta Astronaut., Vol. 7, 335 - 340 (1980). – Abstr. in Phys. Abstr., Vol. 84, Abstr. 9395 (1981).

093.063 **The solar wind interaction with Venus.** C. T. Russell, R. C. Elphic, J. A. Slavin.
Magnetospheric boundary layers, (see 012.041), p. 231 - 239 (1979).

The Pioneer Venus orbiter reveals that Venus has a well developed bow shock like the earth's but one that is significantly weaker than the earth's shock. The location of the bow shock is highly variable, more so than would have been expected for an obstacle with essentially fixed size. The altitude of the ionopause is also highly variable in response to changes in the solar wind. Venus has a much smaller intrinsic magnetic moment than expected from scaling the terrestrial moment.

093.064 **Radio-images of unexplored regions of Venus from bistatic experiments.** A. G. Pavelyev (*Pavel'ev*), M. A. Kolosov, A. I. Kucherjavenkov (*Kucheryavenkov*), O. E. Milechin (*Milekhin*).
Space Research, Vol. XX, (see 012.043), 193 - 196 (1980).

The results of radio investigation of three regions of the surface of Venus are presented. Maps of the distribution of reflectivity are obtained; the roughness of the small scale relief and the height changes for the large scale relief are measured. In the meridional direction the change of large scale relief height exceeded 3 km. The mean density of the soil was deduced to be 2.4 g/cm^3; this increased in some parts of the third region to a value of 3.1 to 3.2 g/cm^3, typical of basalt rocks. The results of the image analysis suggest a diversity of surface structure and physical properties of the Venusian soil.

093.065 **The geochemical model of the troposphere and lithosphere of Venus based on new data.**
V. L. Barsukov, I. L. Khodakovsky (*Khodakovskij*), V. P. Volkov, K. P. Florensky (*Florenskij*).
Space Research, Vol. XX, (see 012.043), 197 - 207 (1980).

New data on the composition of the Venus dayside troposphere obtained by Venera 11 and 12 are applied to the chemical vertical zoning of the Venus troposphere. They are interpreted as being made in the ascending, reducing part of a tropospheric convection cell, whereas the Pioneer-Venus results are considered to be in the descending, oxidizing part. The possibility of liquid sulphur droplets existing in the Venusian clouds is considered.

093.066 **Venera 11 and 12 lander results on the Venus day-sky spectrum.** V. I. Moroz, B. E. Moshkin, A. P. Ekonomov (*Ehkonomov*), N. F. San'ko, N. A. Parfent'ev, Yu. M. Golovin.
Space Research, Vol. XX, (see 012.043), 209 - 214 (1980).

Day-sky spectra of Venus within the range 4500Å to 12000Å and the angular distribution of the scattered solar radiation were recorded by the Venera 11 and 12 landers during their descent through the atmosphere (from 65 km to the surface). The spectra show absorption band of CO_2 and H_2O, and some absorption in the blue-green part of the spectrum which is probably due to gaseous sulphur. The $[H_2O]$ to $[CO_2]$ ratio is deduced to be 2×10^{-5} in the lowest part of the atmosphere (within one scale of height of the surface). If the identification with sulphur is correct, its abundance is about 10^{-8} in the same region. About 6% of the entire solar flux reaches the planet surface in the subsolar region. The lower boundary of the clouds is at about 48 km; the clouds have a stratified structure.

093.067 **Mass spectrometer measurements of the composition of the lower atmosphere of Venus.**
V. G. Istomin, K. V. Grechnev, V. A. Kotchnev.
Space Research, Vol. XX, (see 012.043), 215 - 218 (1980).

The Venera 11 and 12 landers carried out the mass spectrometer measurements of atmospheric composition. All in all, 22 samples were taken and 176 mass-spectra were transmitted to the earth. The measurements revealed a considerable amount of nitrogen in the Venusian atmosphere (about 4.5%) in addition to the main component, CO_2. The mass-spectrometer also detected (as a small component) water vapour (with mass peaks at 17 and 18 a.m.u.). There are also some indications of the presence of chlorine in comparable amounts (35 a.m.u.) as well as sulphur (32 a.m.u.) and possible chlorine and sulphur-containing compounds.

093.068 **Gas chromatograph analysis of the chemical composition of the Venus atmosphere.**
B. G. Gelman (*Gel'man*), V. G. Zolotukhin, L. M. Mukhin, N. I. Lamonov, B. V. Levchuk, D. F. Nenarokov, B. P. Okhotnikov, V. A. Rotin, A. N. Lipatov.
Space Research, Vol. XX, (see 012.043), 219 - 221 (1980).

The paper deals with the gas-chromatograph experiment aboard Venera 12. Eight samples of the Venus atmosphere were taken from a height of 42 km down to the surface of the planet. The concentration of nitrogen was found to be 2.5 ± 0.5% by volume, that of argon $(4 \pm 2) \times 10^{-3}$% by volume, that of CO $(2.8 \pm 1.4) \times 10^{-3}$% by volume and that of SO_2 $(1.3 \pm 0.6) \times 10^{-2}$% by volume. The upper limits of oxygen and water were estimated to be 2×10^{-3} and 10^{-2}% by volume, respectively.

093.069 **Electrical discharges in the Venusian atmosphere.**
L. V. Ksanfomaliti, N. M. Vasilchikov (*Vasil'chikov*), O. F. Canpantzerova (*Ganpantserova*), E. V. Petrova, A. P. Suvorov, G. F. Filippov, O. V. Yablonskaya, L. V. Yabrova.
Space Research, Vol. XX, (see 012.043), 223 - 226 (1980).

The results of an experiment on electrical activity in the Venusian atmosphere are as follows: (1) electrical discharges occur in the cloud layer, (2) their energy is considered to be approximately the same as that of lightning on the Earth, (3) the mean frequency of discharge is 20 sec^{-1} from one storm centre.

093.070 **Infrared remote sensing of the atmosphere of Venus from the Pioneer 12 orbiter.**
F. W. Taylor, D. J. McCleese, L. S. Elson, J. V. Martonchik, D. J. Diner, J. T. Houghton, J. Delderfield, J. T. Schofield, S. P. Bradley.
Space Research, Vol. XX, (see 012.043), 227 - 230 (1980).

The Pioneer 12 spacecraft, which was placed into orbit about Venus on 4 December 1978, carried a multispectral radiometer designed to investigate the structure and meteorological properties of the middle atmosphere (60 to 140 km). This paper reviews the experiment concept, the design, implementation and calibration of the hardware, and the preliminary scientific results from data obtained through 14 February 1979.

093.071 **The formation of the daytime Venusian ionosphere: the results of dual-frequency occultation experiments.** A. L. Gavrik, G. S. Ivanov-Kholodny (*Ivanov-Kholodnyj*), A. V. Mihailov (*Mikhajlov*), N. A. Savich, L. N. Samoznaev.
Space Research, Vol. XX, (see 012.043), 231 - 235 (1980).

From the dual-frequency radio occultation experiments carried out with the Venera-9 and 10 satellites, 13 electron density profiles were obtained in the day-time ionosphere of Venus. An analysis of these profiles made it possible to study variations of the Venusian ionosphere with solar zenith angle Z_0.

093.072 **Venus nighttime hydrogen bulge.**
H. C. Brinton, H. A. Taylor, Jr., H. B. Niemann, H. G. Mayr, A. F. Nagy, T. E. Cravens, D. F. Strobel.
Geophys. Res. Lett., Vol. 7, 865 - 868 (1980).

The concentration of atomic hydrogen in the Venus thermosphere near 165 km altitude and ~18° north latitude has been derived from Pioneer Venus in situ measurements of $n(H^+)$, $n(O^+)$, $n(O)$, and $n(CO_2)$, under the assumption of chemical equilibrium. Altitude profiles of derived n(H) suggest that chemical equilibrium prevails to an altitude of at least 200 km on the dayside and to 165 km on the nightside. Measurements below these limits were made by the ion and neutral mass spectrometers on the orbiter spacecraft between December 1978 and July 1979, while periapsis traversed a complete diurnal cycle.

093.073 **Observations of large scale steady magnetic fields in the dayside Venus ionosphere.**
J. G. Luhmann, R. C. Elphic, C. T. Russell, J. D. Mihalov, J. H. Wolfe.
Geophys. Res. Lett., Vol. 7, 917 - 920 (1980).

Although the dayside ionosphere of Venus is often "field-free" except for fine scale features, large scale, steady ionospheric magnetic fields with magnitudes sometimes exceeding 100 gammas are occasionally observed by the Pioneer Venus Orbiter magnetometer.

093.074 **The surface of the planet Venus.** Z. Pokorný. Říše hvězd, Vol. 61, 247 - 249 (1980). In Czech.

093.075 **Initial Pioneer Venus magnetometer observations.** C. T. Russell, R. C. Elphic, J. A. Slavin.
Proc. Tenth Lunar Planet. Sci. Conf., (see 012.050), p. 2277 - 2290 (1979).

093.076 **Infrared polarization of Venus.** S. Sato, K. Kawara, Y. Kobayashi, H. Okuda, K. Noguchi, T. Mukai, S. Mukai.
Icarus, Vol. 43, 288 - 292 (1980).

Infrared polarization of Venus was measured at several phase angles during 1977, 1978, and 1979. Observations revealed a significant degree of polarization at wavelengths of both 3.6 and 4.8 μm in contrast with little or no polarization in wavelengths shorter than 3 μm. Although the sulfuric acid cloud model seems a likely explanation of the authors' measurements, they found a small discrepancy between these results and the calculations by S. Mukai and T. Mukai (1979).

093.077 **Venera 9, 10: Spectroscopy of scattered radiation in the atmosphere above clouds.**
V. A. Krasnopol'skij.
Kosm. Issled., Tom 18, 899 - 906 (1980). In Russian.

093.078 **Correlation connections of radiative heat transfer in the Venus atmosphere with its chemical composition and structure.**
K. Ya. Kondrat'ev, N. I. Moskalenko, V. F. Terzi.
Dokl. AN SSSR, Vol. 253, 569 - 572 (1980). In Russian. Abstr. in Ref. zh., 51. Astron., 12.51.192 (1980).

093.079 **Thermal energy transport in the Venus ionosphere: classical and saturated electron temperature profiles.**
D. Merritt, K. Thompson.
J. Geophys. Res., Vol. 85, 6778 - 6782 (1980).

The authors point out that parts of the Venus ionosphere are in a temperature and density regime where flux saturation is important. Simple calculations including this effect can reproduce the observed temperature profiles quite well, without the need for magnetic fields or arbitrary heating mechanisms and with somewhat more modest solar wind energy inputs than earlier calculations (which also neglected the magnetic field) have been forced to assume. The results presented do not, however, rule out the possibility that the heat flux is limited solely by a magnetic field.

093.080 **On the shape of the photographic image of Venus in position observations.** G. K. Gorel'.
Nikolaev. otd. Glav. astron. obs. AN SSSR, Nikolaev, 1979. 16 pp. In Russian. – Abstr. in Ref. zh., 51. Astron., 1.51.99 (1981).

093.081 **Optical characteristics of a sulphuric acid model of Venus clouds.** S. P. Obraztsov, Yu. M. Timofeev.
Probl. fiz. atmos. Leningrad, 1980, No. 16, p. 177 - 192. In Russian. – Abstr. in Ref. zh., 51. Astron., 1.51.153 (1981).

093.082 **Study of the Venus atmosphere.** M. Ya. Marov. Astron. Vestn., Tom 13, 3 - 23 (1979). In Russian.

093.083 **Observations of cloud features on Venus in 1975.** V. V. Prokof'eva, O. M. Starodubtseva.
Astron. Vestn., Tom 14, 65 - 71 (1980). In Russian.

Results of television and photographic observations of Venus UV-cloud features are presented. Contrasts on the disk were recorded and measured in ultraviolet (366, 377 nm), violet (404 nm) and blue (435 nm) spectral regions. In green (530 nm) and red (640 nm) regions constrasts are absent.

093.084 **Relative reflectivity between UV markings on Venus.** O. M. Starodubtseva.
Astron. Vestn., Tom 14, 141 - 147 (1980). In Russian.

The relative spectrophotometry of UV markings on Venus from 3500 to 6500 Å over a phase angle range from 68° to 90° for 1975 is presented. The spectral curves for the relative reflectivity between dark and bright markings are given.

Atlas van Maan, Mars, Venus.
See Abstr. 002.084.

Laboratory simulation of the induced magnetospheres of comets and Venus. See Abstr. 022.090.

The Pioneer-Venus Solar Flux Radiometer.
See Abstr. 032.519.

Absolute brightness temperature measurements at 3.5-mm wavelength. See Abstr. 033.019.

On the nature of obstacles decelerating the solar wind near Venus and Mars and on peculiarities of intetaction between the solar wind and the atmospheres of these planets.
See Abstr. 074.022.

Energetic spectra of electrons in the atmospheres of Mars and Venus. See Abstr. 097.022.

The interaction of the solar wind with Mars, Venus and Mercury. See Abstr. 097.079.

A spectroscopic search for N_2^+ emission in the polar regions of Saturn and on the night side of Venus.
See Abstr. 100.044.

Erratum

093.901 **Erratum: "Venera 9 and 10: thermal radiometry."** [Icarus, Vol. 41, 36 - 64 (1980)].
L. V. Ksanfomality (*Ksanfomaliti*).
Icarus, Vol. 43, 230 (1980). – See Abstr. 27.093.009.

094 Moon (Dynamics, General Aspects, Local Properties)

Moon (Dynamics, General Aspects)

094.001 **The temporal development of microcrater and accretionary grain populations on lunar rocks subjected to meteoroid and solar wind bombardment–I. Theory.** D. G. Ashworth.
Planet. Space Sci., Vol. 28, 617 - 624 (1980).

The development, with time, of microcrater and accretionary particle distributions is investigated for lunar rocks subjected to meteoroid and solar wind bombardment. Equations are derived which enable the evaluation of crater and accreta lifetimes and also allow crater and accreta production, transition and equilibrium populations to be calculated.

094.002 **The temporal development of microcrater and accretionary grain populations on lunar rocks subjected to meteoroid and solar wind bombardment – II. Observations on the lunar surface and Apollo samples.**
J. A. M. McDonnell, D. G. Ashworth, W. C. Carey.
Planet. Space Sci., Vol. 28, 625 - 638 (1980).

The development, with time, of microcrater and accretionary particle distributions is investigated for lunar rocks subjected to meteoroid and solar wind bombardment. Experimental observations of the impact crater size distributions and accretionary particle populations on specially selected areas of Apollo lunar samples are used to derive incident fluxes for the theory of topological development described in paper I.

094.003 **Why is the full moon spotty?**
E. A. Whitaker.
J. British Astron. Assoc., Vol. 90, 411 - 421 (1980).

094.004 **Photometry of the lunar surface during lunar eclipses.** N. Sekiguchi.
Moon Planets, Vol. 23, 99 - 107 (1980).

Photometric observations of the lunar surface during lunar eclipses were carried out on four nights between 1972 to 1978. The photometry was performed in B-, V-, and R-colours, and arranged in accordance with the angular distance from the centre of the Earth's shadow. The results do not show any large systematic differences between the four nights, showing no support for Danjon's proposition.

094.005 **Analyse photométrique de l'éclipse de lune du 16 Septembre 1978.**
J. Dubois, F. Link.
Moon Planets, Vol. 23, 109 - 111 (1980).

A discussion of the eclipse based on the comparison of the photometrical measurements of the shadow with the theory based on pure terrestrial atmosphere, is presented.The differences between the observed and computed density of the shadow are partly due to the absorption in the ozone layer, and the remainder can be attributed to the absorbing medium above the ozone layer.

094.006 **Lunar core detection using two surface magnetometers at very low frequency.** F. Herbert.
Moon Planets, Vol. 23, 127 - 131 (1980).

A method of analysis is developed for combining the measurements of two surface magnetometers so as to deduce empirically the very low frequency electromagnetic transfer function of the Moon. The method is expected to be useful in determining the presence of a lunar metallic core by making usable for this purpose simultaneous magnetic field measurements by the Apollo 15 and 16 surface modules.

094.007 **Cometary collisions on the Moon and Mercury.**
L. L. Hood, with a reply by P. H. Schultz, L. J. Srnka.
Nature, Vol. 287, 86 - 87 (1980).

094.008 **How was the lunar relief formed?**
I. K. Volchanskaya, E. N. Sapozhnikova.
Priroda, 1980, No. 8, p. 76 - 83. In Russian.

094.009 **Formulas and diagrams for the definition of the average lunar photometric function.**
V. V. Shevchenko.
Tr. Gos. Astron. Inst. Shternberga, Tom 50, 105 - 117 (1980). In Russian.

The complete spatial scattering indicatrix of the lunar surface was obtained by photometric treatments of ground and space surveys. The formulas for the calculation of the angular parameters and diagrams of the average photometric function's change according to the angular parameters are given.

094.010 **Photometric investigations of the lunar surface at visible and infrared wavelengths.**
Yu. N. Lipskij, S. G. Pugacheva.
Tr. Gos. Astron. Inst. Shternberga, Tom 50, 118 - 124 (1980). In Russian.

The paper deals with a selective reflectivity of the lunar surface in the far infrared (10 - 12 μm) and visible wavelengths (0.45 μm). A method for investigation of the relation between the reflected solar radiation and the resulting surface temperature of the lunar surface was considered and the correlation coefficient and the linear regression coefficient were obtained for 23 phase angles.

094.011 **Peculiarities of the color of the eastern lunar hemisphere.**
V. V. Novikov, A. P. Popov, V. V. Titov.
Tr. Gos. Astron. Inst. Shternberga, Tom 50, 125 - 134 (1980). In Russian.

Laboratory color investigations of terrestrial igneous rocks were carried out and on this basis a possible interpretation of the color nature of lunar features was given.

094.012 **Polarization as an instrument of remote selenochemistry.** V. V. Novikov.
Tr. Gos. Astron. Inst. Shternberga, Tom 50, 135 - 149 (1980). In Russian.

The results of lunar polarization observations during the last years are generalized. The most important parameters of the polarized light are considered in the connection with a remote mineralogic analysis. A possible chemical interpretation of the maximum degree of lunar features polarization is given.

094.013 **Diagrams of optical characteristics of the lunar surface.** N. N. Evsyukov, D. I. Shestopalov.
Vestn. Khar'k. Univ., No. 190. Fiz. Luny Planet. Fundam. Astrometr., Vyp. 14, p. 38 - 44 (1979). In Russian.

094.014 **On the nature of the interconnection albedo – polarization degree of the lunar surface.**
Yu. G. Shkuratov.
Vestn. Khar'k. Univ., No. 190. Fiz. Luny Planet. Fundam. Astrometr., Vyp. 14, p. 44 - 52 (1979). In Russian.

094.015 **On a method of digital mapping of the normal albedo of the lunar surface from data of a photographic survey from space.** V. A. Psarev.

Vestn. Khar'k. Univ., No. 190. Fiz. Luny Planet. Fundam. Astrometr., Vyp. 14, p. 78 - 90 (1979). In Russian.

094.016 **Lunar asymmetry and palaeomagnetism.** D. J. Stevenson.
Nature, Vol. 287, 520 - 521 (1980).

The compositional asymmetry between the nearside and farside of the moon and the natural remanent magnetism of lunar rocks are poorly understood. The author proposes a model for the early lunar evolution in which the preferred gravitational energy state consisted of an asymmetric accumulation of a liquid iron alloy (Fe–Ni and a small amount of sulphur) which displaces upwards the cold, primordial, undifferentiated core.

094.017 **The main problem of lunar theory solved by the method of Brown.** D. S. Schmidt.
Moon Planets, Vol. 23, 135 - 164 (1980).

Brown's method for solving the main problem of lunar theory has been adapted for the computation by machine with the help of an algebraic processor. Brown's results are first recovered and refined. The solution is then expanded to include most terms of order nine. The terms in the series for the longitude and latitude are listed with an accuracy of 0.000 01″ and of 0.000 001″ for the parallax.

094.018 **The evolution of the lunar orbit revisited, II.** F. Mignard.
Moon Planets, Vol. 23, 185 - 201 (1980).

The author presents a model for the tidal evolution of an isolated two-body system. Equations are derived, including the dissipation in the planet as in the satellite, in a frequency dependent lag model. The set of differential equations obtained is still valid for large eccentricity, as well as for all inclinations. The results obtained show the moon, after having approached the earth with small variations for the inclination and the eccentricity, exhibits strong increase for the two parameters in the vicinity of the closest approach. Some qualitative results are also investigated for the final behaviour of satellites such as Triton and the Galilean satellites.

094.019 **Great volcanic eruptions and the luminosity of the moon during total eclipses.** P. Hédervári.
Strolling Astron., Vol. 28, 158 - 165 (1980).

094.020 **How high the moon: a decade of laser ranging.** J. D. Mulholland.
Sky Telesc., Vol. 60, 274 - 279 (1980).

094.021 **The internal structure and early history of the Moon.** G. H. A. Cole.
J. British Astron. Assoc., Vol. 90, 539 - 559 (1980).

The Apollo missions provided a wealth of detailed data about the Moon's surface conditions to supplement the general data that had resulted from the previous Earth-bound study of the Moon by many observers over more than a hundred years. As a result it has been possible to construct a firm model of the interior structure of the Moon and to deduce something of the early lunar history. This paper draws these various conclusions together setting the arguments within the general context of the Moon as a planetary body. Magnetic effects are briefly touched on and attention is drawn to a number of unsolved problems associated with our present understanding of the Moon.

094.022 **Contributions to the study of Babylonian lunar theory.** A. Aaboe, N. T. Hamilton.
Mat.-Fys. Medd. Danske Vidensk. Selsk., Vol. 40, No. 6, p. 1 - 36 (1979). – Abstr. in Phys. Abstr., Vol. 83, Abstr. 94729 (1980).

094.023 **Some comments on anorthosites and basalts of the moon and the earth in connection with the evolution of these planets.** Eh. M. Spiridonov.
Dokl. AN SSSR, Vol. 251, 951 - 953 (1980). In Russian. From Ref. zh., 51. Astron., 8.51.293 (1980).

094.024 **Selenographic coordinates of objects at the far side of the moon.** S. G. Valeev.
Kemerov. univ., Kemerovo, 1980. 35 pp. In Russian. – Abstr. in Ref. zh., 62. Issled kosm. prostranstva, 8.62.462 (1980).

094.025 **On the internal structure of a moon of fission origin.** A. B. Binder.
J. Geophys. Res., Vol. 85, 4872 - 4880 (1980).

Internal structure models of the moon, based in part on petrological and thermal models of a moon of fission origin, indicate that the moon has an iron or iron-rich core with a radius between 200 and 400 km. The derived STP density of the lower mantle is between 3.35 and 3.51 g/cm^3 and indicates that the Mg′ of the lower mantle is between 73 and 88. This result, together with the Mg′ of the upper mantle ($\simeq$70) and of the crust ($\simeq$70), leads to Mg′ between 72 and 84 for the bulk moon (minus the core). The V_p in the mantle of the moon is found to be between 7.6 and 7.9 km/s; this result is in general agreement with that derived from seismic data.

094.026 **Scientific achievements from ten years of lunar laser ranging.** J. D. Mulholland.
Rev. Geophys. Space Phys., Vol. 18, 549 - 564 (1980).

In the 10 years since lunar laser ranging became a reality the need to analyze the observations has motivated improvements in several aspects of the mathematical model of earth-moon dynamics. Application of the data to improved estimates of the physical parameters of the earth-moon system has yielded significant astronomical, selenophysical, geophysical, and cosmological results. The scientific impact, both in improved theories and in numerical applications, is surveyed. The underlying physics and major difficulties are discussed, as well as the scientific results.

094.027 **Orbits of submicron lunar ejecta in the earth-moon system.** J. D. Chamberlain, W. M. Alexander, J. D. Corbin.
Solid particles in the solar system, (see 012.019), p. 421 - 424 (1980).

Flux measurements of picogram dust particles near the lunar surface and in selenocentric and cis-lunar space made by Lunar Explorer 35, HEOS, and ALSOP dust experiments all indicate, to varying degrees, ejecta from lunar impacts of interplanetary dust particles. The orbits of these submicron particles in the earth-moon system are significantly altered by radiation pressure. Recent orbit calculations show that, in favorable lunar phases, as many as 80% of the ejecta may enter the magnetosphere and 20% may enter the earth's atmosphere. The results of this analysis are presented, and their implications are discussed.

094.028 **Interaction of lunar ejecta and the magnetosphere of the earth.**
W. M. Alexander, J. D. Corbin.
Solid particles in the solar system, (see 012.019), p. 425 - 428 (1980).

A significant flux of ejecta from lunar impacts of interplanetary dust particles leaves selenocentric space and enters the magnetosphere of the earth. During favorable lunar phases, 80% of the ejecta enter the magnetosphere where their orbits are determined by electrodynamic as well as gravitational forces. Initial study of the orbital characteristics and perturbations of these magnetosphere ejecta is presented and its implications are discussed.

094.029 **On the nature of the polarization of the Moon's reflected light.** N. Sekiguchi.
13th Lunar and Planetary Symposium, (see 012.018), p. 291 - 296 (1980).

094.030 **Planetary perturbations of the moon. Comparison of ELP-1900 with Brown's theory.**
M. Chapront-Touzé, J. Chapront.
Astron. Astrophys., Vol. 91, 233 - 246 (1980). – In French.
The paper discusses the construction of the planetary perturbations of the moon, in the frame of the solution ELP-1900, i.e. in connection with the solution for the main problem and its partials, built by the first author, with a system of constants related to 1900. It covers the two types of perturbations on the moon that are commonly distinguished: the direct and indirect actions. The planetary expressions (the developments of the inverses of the distances between earth and planets, and the theory for the earth's motion) are derived from Bretagnon's work.

094.031 **On the estimation of the free libration constants of the moon.** K. Kozieł.
Postępy Astron., Tom 28, 227 - 230 (1980). In Polish.
The results of simultaneous and homogeneous reduction of 8 libration series are presented. On this basis all 6 free libration constants and forced libration constants have been estimated using some new formulae.

094.032 **Interpretation of lunar photometry using a new scattering theory.** K. Lumme, W. M. Irvine.
Bull. American Astron. Soc., Vol. 12, 660 (1980). – Abstract.

094.033 **IUE observations of the lunar surface.**
E. N. Wells, B. W. Hapke.
Bull. American Astron. Soc., Vol. 12, 660 (1980). – Abstract.

094.034 **Regolith vertical structure and the lunar microwave emission.** S. Keihm.
Bull. American Astron. Soc., Vol. 12, 661 (1980). – Abstract.

094.035 **Statistics of ten years of lunar laser ranging data from McDonald Observatory.** P. J. Shelus.
Bull. American Astron. Soc., Vol. 12, 741 - 742 (1980). Abstract.

094.036 **Perturbations due to the shape of the Moon in lunar theory.** J. Henrard.
Celestial Mech., Vol. 22, 335 - 341 (1980).
The author computes the perturbations on the motion of the Moon due to its shape. The accuracy is estimated at $1''10^{-5}$ in longitude and latitude and 5 parts in 10^{11} in distance.

094.037 **Direct perturbations of the planets on the Moon's motion.** D. Standaert.
Celestial Mech., Vol. 22, 357 - 369 (1980).
The aim of the present paper is to present the theoretical background of a method to compute the planetary perturbations on the Moon's motion. The author formulates an algorithm based upon the Lie transform method and well-suited to the particular problem at hand. This algorithm is being implemented using Henrard's Semi-Analytical Lunar Ephemeris (SALE) as solution of the Main Problem and Bretagnon's planetary theory. The accuracy of the solution is intended to be about 0.''001 for terms of period up to 2000 years.

094.038 **The selenographical distribution of deep moonquake epicenters.** A. B. Binder.
Geophys. Res. Lett., Vol. 7, 707 - 708 (1980).
The observed selenographical distribution of the epicenters of the deep moonquakes is explained using a simple lunar model with a low Q deep interior and assuming that the number of epicenters per unit area is proportional to the tidal potential affecting the area.

094.039 **Modification of the lunar surface by solar wind ion bombardment.** A. J. T. Jull, C. T. Pillinger.
Radiat. Eff., Vol. 48, No. 1 - 4, p. 69 - 74 (1980). – Abstr. in Phys. Abstr., Vol. 83, Abstr. 109325 (1980).

094.040 **Hydrostatic tidal model for lunar asymmetry.**
S. Friedlander.
Geophys. Astrophys. Fluid Dyn., Vol. 15, 105 - 122 (1980). Abstr. in Phys. Abstr., Vol. 84, Abstr. 4785 (1981).

094.041 **Analytical lunar libration tables.** A. Migus.
Moon Planets, Vol. 23, 391 - 427 (1980).
The author has developed a theory of the rotation of the Moon, for the purpose of obtaining libration series explicitly dependent upon lunar gravitational field model parameters. A nonlinear model is used in which the rigid Moon, whose motion in space is that of the main problem of lunar theory, and whose gravity potential is represented through its third degree harmonics, is torqued by the Earth and Sun. The analytical series are then obtained as Poisson series. Numerical comparisons with Eckhardt's solution are briefly exposed.

094.042 **Infrared reflection spectra of the moon and lunar soil.** M. N. Markov, V. S. Petrov,
M. V. Akhmanova, B. V. Dementjev (*Dement'ev*).
Space Research, Vol. XX, (see 012.043), 189 - 192 (1980).
Spectra of the moon in the range 2 - 15 μ have been obtained with the help of the infrared telescope-spectrometer aboard Saljut 5. They have been compared with the spectra of regolith returned by Luna 24. The reflected intensity is greatest in the range from 3 to 5 μ (up to 60% for Saljut 5 and 30% for Luna 24), while at 11.5 μ the intensity is reduced. From Luna 24, water absorption bands appear in the spectra at certain depths in the core; the concentration of water is estimated to be 0.1%.

094.043 **Shallow moonquakes: argon release mechanism.**
A. B. Binder.
Geophys. Res. Lett., Vol. 7, 1011 - 1013 (1980).
Apollo 17 surface mass-spectrometer observations have shown that 10^{27} to 10^{28} mol. of ^{40}Ar are released from the moon when shallow moonquakes occur. Model calculations, based on the seismic and strain energies released by the moonquakes, the volumes of the focal zones, and the surface areas of observed shallow thrust faults, indicate that $> 1.5 \times 10^{25}$ to $\lesssim 2 \times 10^{28}$ mol. of ^{40}Ar could be released by each shallow moonquake as a result of mylomization and possible frictional melting of the rocks within a few decimeters of the newly formed parts of the fault surface. The rapid escape of this ^{40}Ar along the fractures of the fault would produce the observed argon release event shortly after each moonquake.

094.044 **On the periodicity of a code of lunar seismograms.**
I. N. Galkin, A. V. Nikolaev, O. B. Khavroshkin,
V. V. Tsyplakov.
Izv. AN SSSR. Fiz. Zemli, 1980, No. 7, p. 82 - 85. In Russian. Abstr. in Ref. zh., 51. Astron., 11.51.356 (1980).

094.045 **Lunar multispectral imaging at 2.26 μm: first results.**
D. W. Davies, T. V. Johnson, D. L. Matson.
Proc. Tenth Lunar Planet. Sci. Conf., (see 012.050), p. 1819 - 1828 (1979).

094.046 **Lunar nearside magnetic anomalies.**
L. L. Hood, P. J. Coleman, Jr., D. E. Wilhelms.
Proc. Tenth Lunar Planet. Sci. Conf., (see 012.050), p. 2235 - 2257 (1979).

094.047 **High spatial resolution measurements of surface magnetic fields of the lunar frontside.**
R. P. Lin.
Proc. Tenth Lunar Planet. Sci. Conf., (see 012.050) p. 2259 - 2264 (1979).

094.048 **Gravity anomaly and structure associated with the Lamont region of the moon.**
J. Dvorak, R. J. Phillips.
Proc. Tenth Lunar Planet. Sci. Conf., (see 012.050) p. 2265 - 2275 (1979).

094.049 **Comparison of theoretical and observed λ 3.55 cm wavelength brightness temperature maps of the full moon.** S. J. Keihm, B. L. Gary.
Proc. Tenth Lunar Planet. Sci. Conf., (see 012.050), p. 2311 - 2319 (1979).

094.050 **On the core of the moon.** B. J. (*B. Yu.*) Levin. Proc. Tenth Lunar Planet. Sci. Conf., (see 012.050), p. 2321 - 2323 (1979).

094.051 **An iron core in the moon generating an early magnetic field?** S. K. Runcorn.
Proc. Tenth Lunar Planet. Sci. Conf., (see 012.050), p. 2325 - 2333 (1979).

094.052 **Models of an early lunar dynamo.**
L. J. Srnka, M. H. Mendenhall.
Proc. Tenth Lunar Planet. Sci. Conf., (see 012.050), p. 2343 - 2355 (1979).

094.053 **Electrical conductivity of the lunar interior: theory, error sources, and estimates.** B. E. Goldstein.
Proc. Tenth Lunar Planet. Sci. Conf., (see 012.050), p. 2357 - 2373 (1979).

094.054 **Parameterized convection within the moon and the terrestrial planets.**
D. L. Turcotte, F. A. Cooke, R. J. Willeman.
Proc. Tenth Lunar Planet. Sci. Conf., (see 012.050), p. 2375 - 2392 (1979).

094.055 **The bounds of the heat production rate within the moon.** A. T. Hsui.
Proc. Tenth Lunar Planet. Sci. Conf., (see 012.050), p. 2393 - 2402 (1979).

094.056 **The lunar interior: a summary report.**
N. R. Goins, M. N. Toksöz, A. M. Dainty.
Proc. Tenth Lunar Planet. Sci. Conf., (see 012.050), p. 2421 - 2439 (1979).

094.057 **Elastic lithosphere thickness on the moon from mare tectonic features: a formal inversion.**
R. P. Comer, S. C. Solomon, J. W. Head.
Proc. Tenth Lunar Planet. Sci. Conf., (see 012.050), p. 2441 - 2463 (1979).

094.058 **Impact melting early in lunar history.**
M. A. Lange, T. J. Ahrens.
Proc. Tenth Lunar Planet. Sci. Conf., (see 012.050), p. 2707 - 2725 (1979).

094.059 **Interconnection of the albedo and polarization characteristics of the moon. I. New optical parameter (preliminary investigations).** Yu. G. Shkuratov, S. P. Red'kin, N. V. Bitanova, A. V. Il'inskij.
Astron. Tsirk., No. 1112, p. 3 - 6 (1980). In Russian.

094.060 **The lunar fundamental photometric constants in the system of true full moon.**
V. V. Shevchenko.
Astron. Zh., Tom 57, 1341 - 1343 (1980). In Russian.
English translation in Soviet Astron., Vol. 24, No. 6.

The values of lunar photometric constants in the system of true full moon have been obtained: the integral stellar magnitude of the moon, the phase integral, the geometrical and the spherical albedo, the light constant of the moon, the stellar magnitude of the moon at unity distance from the earth and from the sun. The empirical expression for the lunar phase curve has been found.

094.061 **Comparative analysis of the gravitational field of the moon.**
G. A. Meshcheryakov, V. V. Kirichuk, P. M. Zazulyak.
Geod., kartogr. i aehrofotosemka, L'vov, 1980, No. 32, p. 101-106. In Russian. – Abstr. in Ref. zh., 52. Geod. Aehrosemka, 12.52.150 (1980).

094.062 **Data on lunar volcanism.** P. Leonardi.
Probl. planetol. T. 2. Simpoz. No. 8. Mezhdunar. assots. planetol., Erevan, 1977. Leningrad-Erevan, 1977, p. 183 - 188. In Russian. – Abstr. in Ref. zh., 62. Issled. kosm. prostranstva, 12.62.168 (1980).

094.063 **Some comparative characteristics of the volcanism of the earth and moon from data of cosmic photographs.** R. A. Arakelyan, K. G. Shirinyan.
Probl. planetol. T. 2. Simpoz. No. 8. Mezhdunar. assots. planetol., Erevan, 1977. Leningrad-Erevan, 1977, p. 43 - 45. In Russian. – Abstr. in Ref. zh., 62. Issled. kosm. prostranstva, 12.62.170 (1980).

094.064 **Construction of a numerical model of the lunar indicatrix.** V. A. Psarev.
Khar'kov. univ. Khar'kov, 1980, 32 pp. In Russian. – Abstr. in Ref. zh., 51. Astron., 1.51.239 (1981).

094.065 **Construction and adjustment of a selenodetic coordinate system from orbital and ground observations.** E. P. Aleksashin, Yu. S. Tyuflin.
Probl. mat. obrab. geod. setej. Mater. Vses. konf., Novosibirsk 1977. Novosibirsk, 1979, p. 85 - 89. In Russian. – Abstr. in Ref. zh., 62. Issled. kosm. prostranstva, 1.62.345 (1981).

094.066 **Observations of instationary phenomena on the moon (list 4).** P. V. Florenskij, V. M. Chernov.
Astron. Vestn., Tom 13, 62 - 63 (1979). In Russian.

094.067 **Infrared brightness temperature of the lunar surface as a function of the angular parameters.**
S. G. Pugacheva.
Astron. Vestn., Tom 14, 133 - 140 (1980). In Russian.

The results of a comparative analysis of the selective reflectivity of the lunar surface in the far infrared (10–12 microns) for 23 phase angles are given. The empirical dependences of the resulting surface temperature of the lunar surface on the incidence angle, reflection angle and phase angle are studied.

A complete lunar map ("Polnaya Karta Lunyi").
See Abstr. 002.080.

Atlas van Maan, Mars, Venus.
See Abstr. 002.084.

Diameter to depth dependence of impact craters.
See Abstr. 022.002.

Collisional processes of iron and steel projectiles on targets of different densities. See Abstr. 022.072.

Observations of the moon and outer planets obtained in Nikolaev in 1969 - 1972.
See Abstr. 041.042.

On the solving of the Hill's differential equation through the method of operational calculus.
See Abstr. 042.003.

Theory of the motions of the Moon and of the satellites with Laplace's variables. See Abstr. 042.065.

Geophysical parameters of the earth-moon system.
See Abstr. 081.007.

Problemi di dinamica dei corpi minori del sistema solare. See Abstr. 091.044.

Tectonic patterns on a tidally distorted planet.
See Abstr. 091.064.

Formation of primary lithospheres of planets, their possible composition and volcanism on the example of the moon. See Abstr. 091.066.

An analysis of total lunar occultations made in the years 1943 to 1974. See Abstr. 096.011.

A comparative analysis of some optical characteristics of asteroids and of the moon. See Abstr. 098.021.

Moon (Local Properties)

094.501 **Morphological characteristics of lunar craters with small depth/diameter ratio. II.**
M. Moutsoulas, P. Preka.
Moon Planets, Vol. 23, 113 - 126 (1980).

The morphological characteristics of craters with relativly small ratio depth/diameter are discussed. It is observed that many morphological similarities exist among craters possessing ratios d/D which do not differ considerably. The distribution of 1933 craters with respect to diameter, depth and the depth/diameter ratio is presented.

094.502 **Solar-flare produced ^{3}He in lunar samples.**
M. N. Rao, T. R. Venkatesan.
Nature, Vol. 286, 788 - 790 (1980).

The authors describe, for the first time, using ^{3}He in samples from three different depths of lunar rocks, a method for deducing the absolute solar cosmic rays (flare) proton fluxes in the past few million years which compare favourably with those derived from ^{26}Al and ^{53}Mn studies of other lunar samples.

094.503 **Late high-titanium basalts of the western maria: geology of the Flamsteed region of Oceanus Procellarum.**
C. M. Pieters, J. W. Head, J. B. Adams, T. B. McCord, S. H. Zisk, J. L. Whitford-Stark.
J. Geophys. Res., Vol. 85, 3913 - 3938 (1980).

The Flamsteed region of Oceanus Procellarum is representative of an unsampled volcanic complex on the lunar nearside. The evolution of this region, as portrayed in the current surface, is investigated in detail.A synthesis of remote sensing data, including multispectral images (digital vidicon images and color composite photographs), spectral reflectance measurements, radar topographic maps, and orbital and earth-based photography, is presented. Additional information is derived from crater degradation studies, radar backscatter, gravity measurements and orbital γ ray data.

094.504 **Unusual property of lunar regolith.**
V. L. Barsukov, O. A. Bogatikov, V. I. Nefedov.
Zemlya Vselennaya, 1980, No. 4, p. 10 - 12. In Russian.

094.505 **Core 14220 and the lateral continuity of soils at Apollo 14 station G.** J. S. Nagle.
Moon Planets, Vol. 23, 165 - 183 (1980).

094.506 **Mesures absolues de la température apparente des mers lunaires en infra-rouge.**
J. Gay, Y. Rabbia, Y. Boudon, J. P. Corteggiani, R. Futaully, N. Gaignebet, P. Granes, J. Kovalevsky, J. L. Sagnier, J. M. Torre, J. P. Villain, J. J. Walch.
Moon Planets, Vol. 23, 203 - 211 (1980).

A series of measurements of infrared apparent temperatures of ten selected sites in lunar maria has been undertaken in CERGA, using an infrared receiver (11.3 μm). The experimental procedures are described. The theory of the receiver is developed and the theory of two methods of calibration of the measurements is given.

094.507 **Modelling of temperatures of selected points in lunar maria.** J. Gay, J.-L. Falin, J. Kovalevsky, J.-P. Corteggiani, M.-T. Dumoulin.
Moon Planets, Vol. 23, 213 - 230 (1980).

The purpose of this paper is to propose a numerical model representing the brightness temperature of ten selected sites of the moon. A theoretical approach is proposed in which the temperature is expressed as a function of the local zenith distance of the sun and a quantity depending upon the local zenith distance of the observer and the difference of their azimuths. The most important coefficients of the formulae have been determined.

094.508 **Small impact craters in the lunar regolith – their morphologies, relative ages, and rates of formation.**
H. J. Moore, J. M. Boyce, D. A. Hahn.
Moon Planets, Vol. 23, 231 - 252 (1980).

Apparently, there are two types of size-frequency distributions of small lunar craters ($\simeq$ 1–100 m across): (1) crater production distributions for which the cumulative frequency of craters is an inverse function of diameter to power near 2.8, and (2) steady-state distributions for which the cumulative frequency of craters is inversely proportional to the square of their diameters. A flux of crater producing objects can be inferred from size-frequency distributions of small craters on the flanks and ejecta of craters of known age. Crater frequency distributions and data on the craters Tycho, North Ray, Cone, and South Ray, when compared with the flux of objects measured by the Apollo Passive Seismometer, suggest that the flux of objects has been relatively constant over the last 100 m.y.

094.509 **Lunar and solar noble gases trapped by volcanic glasses 3.6 AE ago.**
O. Eugster, N. Grögler, P. Eberhardt, J. Geiss.
Meteoritics, Vol. 15, 181 (1980). – Abstract.

094.510 **The separation and distribution of some Luna 24 core materials.** C. T. Pillinger, D. M. Fabian.
Philos. Trans. R. Soc. London, Ser. A, Vol. 297, No. 1428, p. 1 - 6 (1980). – Abstr. in Phys. Abstr., Vol. 83, Abstr. 82241 (1980).

094.511 **Metal and phosphide phases in Luna 24 soil fragments.** H. J. Axon, M. J. Nasir, F. Knowles.
Philos. Trans. R. Soc. London, Ser. A, Vol. 297, No. 1428, p. 7 - 13 (1980). – Abstr. in Phys. Abstr., Vol. 83, Abstr. 82242 (1980).

094.512 **Mineralogy and petrology of fragments from the Luna 24 core.** A. L. Graham, R. Hutchison.
Philos. Trans. R. Soc. London, Ser. A, Vol. 297, No. 1428, p. 15 - 22 (1980). – Abstr. in Phys. Abstr., Vol. 83, Abstr. 82243 (1980).

094.513 **Optical excitation spectroscopy of the Luna 24 sample 24125.** D. J. Telfer, G. Fielder.
Philos. Trans. R. Soc. London, Ser. A, Vol. 297, No. 1428, p. 23 - 25 (1980). – Abstr. in Phys. Abstr., Vol. 83, Abstr. 82244 (1980).

094.514 **^{40}Ar-^{39}Ar ages and irradiation history of Luna 24 basalts.** J. Hennessy, G. Turner.
Philos. Trans. R. Soc. London, Ser A, Vol. 297, No. 1428, p. 27 - 39 (1980). – Abstr. in Phys. Abstr., Vol. 83, Abstr. 82245 (1980).

094.515 **Solar-flare exposure and thermoluminescence of Luna 24 core material.**
S. A. Durrani, R. K. Bull, S. W. S. McKeever.
Philos. Trans. R. Soc. London, Ser. A, Vol. 297, No. 1428, p. 41 - 50 (1980). – Abstr. in Phys. Abstr., Vol. 83, Abstr. 82246 (1980).

094.516 **Determination of the sulphur content of a lunar sample by neutron capture gamma-ray spectrometry.**
M. A. Islam, T. J. Kennett, W. V. Prestwich, C. E. Rees.
J. Radioanal. Chem., Vol. 56, No. 1 - 2, p. 123 - 130 (1980). Abstr. in Phys. Abstr., Vol. 83, Abstr. 94728 (1980).

094.517 **Preliminary results of determining physical properties of lunar microfragments from Luna 24 samples.**
I. A. Turchaninov, R. V. Medvedev, V. V. Chashnikov.
Dokl. AN SSSR, Vol. 252, 358 - 360 (1980). In Russian. Abstr. in Ref. zh., 51. Astron., 9.51.240 (1980).

094.518 **Discovery of anomalously weak oxidability of the elements in lunar regolith.**
Metallofiz., Vol. 2, No. 3, p. 124 - 125 (1980). In Russian. Abstr. in Ref. zh., 62. Issled. kosm. prostranstva, 9.62.128 (1980).

094.519 **A theory of albedo of details of the lunar surface.** N. Kumagai.
13th Lunar and Planetary Symposium, (see 012.018), p. 306 - 311 (1980).

094.520 **A lunar basalt intermediate between mare and highland-type rocks and its survey by future lunar missions.** H. Takeda, T. Ishii, M. Miyamoto.
13th Lunar and Planetary Symposium, (see 012.018), p. 312 - 319 (1980).

Single crystal X-ray diffraction studies and electron microprobe analyses have been made of pyroxenes and plagioclases from feldspathic basalt clasts in lunar breccia 14321,922. The chemical trends and crystallography of pyroxenes in the 14321,973 clast is intermediate between low-K KREEPy basalt and low-Ti aluminous mare basalt.

094.521 **Luna 24 ferrobasalt as a low-Mg primary melt.** M. Norman, G. Ryder.
Moon Planets, Vol. 23, 271 - 292 (1980).

094.522 **Tectonic implications of the mare-ridge pattern of the central parts of Oceanus Procellarum on the Moon.** J. Raitala.
Moon Planets, Vol. 23, 307 - 321 (1980).

A structural analysis is presented of the mare ridge pattern in an area of about 1 000000 km^2 in the central parts of Oceanus Procellarum.

094.523 **On the rate of ancient meteoroidal flux: studies of the pre-mare crater population on the Moon.**
A. T. Basilevsky (*Bazilevskij*), L. B. Ronca, B. A. Ivanov.
Moon Planets, Vol. 23, 355 - 371 (1980).

According to radiometric dating of lunar rocks, meteoroidal bombardment and accompanying cratering on the Moon were intensive in the first 0.7×10^9 y, the so-called terra stage. Recently the hypothesis of a 'terminal cataclysm' has been gaining acceptance, meaning that a sharp increase in the bombardment followed by a steep decay occurred at the end of the terra stage. The purpose of this paper is to investigate possible variations in the intensity of the bombardment during the terra stage by analyzing the population of large (3–1000 km) terra craters and comparing it with results obtained by theoretical models. The proportion of fresh craters is specifically used.

094.524 **Planetary cratering: on the conical shape of lunar fresh craters.**
R. Bianchi, M. Fulchignoni, S. Evangelista.
Mem. Soc. Astron. Italiana, Vol. 51, 285 - 297 (1980).

The results of this work deal with the morphometric characters of craters that allow the definition of the "original" crater shape. Analysis of the morphometric parameters of actual craters allows the description both of the influence of the geology in the first assessment of an impact structure and of the modification processes which have been active during the crater life.

094.525 **Planetary cratering: a model to estimate the morphometric modification of lunar fresh craters.**
R. Bianchi, M. Fulchignoni, S. Evangelista.
Mem. Soc. Astron. Italiana, Vol. 51, 299 - 308 (1980).

A method to evaluate the degree of alteration of the shape of a crater has been described. The method is able to describe the status of modification of crater walls and can be easily applied to various problems related to the morphology of the craters.

094.526 **The Reiner Gamma lunar magnetic anomaly as studied via telescopic reflectance spectra.**
J. F. Bell, B. R. Hawke.
Bull. American Astron. Soc., Vol. 12, 660 - 661 (1980). Abstract.

094.527 **Lunar craters with radar-bright ejecta.** T. Thompson, S. Zisk, R. Shorthill, P. Shultz, J. Cutts.
Bull. American Astron. Soc., Vol. 12, 661 (1980). – Abstract.

094.528 **Small lunar dark-mantle deposits of probable pyroclastic origin.**
B. R. Hawke, T. B. McCord, J. W. Head.
Bull. American Astron. Soc., Vol. 12, 663 - 664 (1980). Abstract.

094.529 **Moon and Earth: compositional differences inferred from siderophiles, volatiles, and alkalis in basalts.**
R. Wolf, E. Anders.
Geochim. Cosmochim. Acta, Vol. 44, 2111 - 2124 (1980).

The authors have compared RNAA analyses of 18 trace elements in 25 low-Ti lunar and 10 terrestrial oceanic basalts. According to Ringwood and Kesson, the abundance ratio in basalts for most of these elements approximates the ratio in the two planets.

094.530 **The lunar highlands crust.**
C. H. Simonds, C. T. Herzberg, J. J. Papike.
EOS Trans. American Geophys. Union, Vol. 61, 473 - 475 (1980). – Abstr. in Phys. Abstr., Vol. 83, Abstr. 109381 (1980).

094.531 **Mare basalt units and the compositions of their magmas.**
A. B. Binder, M. A. Lange, H.-J. Brandt, S. Kähler.
Moon Planets, Vol. 23, 445 - 481 (1980).

A synthesis of the majority of the available mare basalt data shows that basalts and glasses come from 28 different volcanic units. The compositions of the magmas of 12 of these units can be calculated with a high degree of confidence. Reasonable estimates can be made for the compositions of nine of the remaining units. In addition, the compositions of three general magma types can be obtained from data derived from the Luna 16, Luna 24, and Apollo 17 fines. The compositional data presented provide a firm basis for the further study of the characteristics of the mare basalt magma source region.

094.532 **Solar wind nitrogen and indigenous nitrogen in lunar material.** O. Müller.
Origin and distribution of the elements, (see 012.039), p. 47 - 59 (1979).

The author reports on chemically bound nitrogen contents in various lunar samples from the Apollo 11 through Apollo 17 landing sites. He has found by the analysis of grain-size fractions, by leach-experiments, by comparison with trapped noble gases and with agglutinate contents that the main amount of the nitrogen in the fines is derived from the solar wind by ion implantation and is present primarily in a bound state. Bound nitrogen contents have also been determined in several lunar igneous rocks and U.S.G.S. standard rocks using an optimized, program-controlled Kjeldahl method.

094.533 **Solar wind carbon chemistry as revealed by lunar sample analysis.** C. T. Pillinger.
Origin and distribution of the elements, (see 012.039), p. 61 - 72 (1979).

Knowledge of solar wind carbon chemistry is restricted to what has been learned from the analysis of lunar samples. The study of total carbon abundances, the search for the existence of distinct carbon compounds, and the investigation of carbon isotope ratios have demonstrated the behaviour of a reactive element implanted from the solar wind into the surface of an atmosphereless planet, and provided information about the evolution of the lunar regolith.

094.534 **Lunar highland chronology.** T. Kirsten.
Origin and distribution of the elements, (see 012.039), p. 91 - 98 (1979).

The distribution of lunar highland rock ages is of twofold significance. At first, the highest ages provide a lower limit of ~4.4 b.y. for the time of the initial differentiation of the lunar crust. Secondly, the ages reflect the collisional history of the moon in the impact-dominated period between 4.4 and 3.8 b.y. ago in which the large multiring-basins were formed. More than 150 highland rock ages have been dated.

094.535 **The earth-moon system, Chemistry and origin.**
H. Wänke, G. Dreibus.
Origin and distribution of the elements, (see 012.039), p. 99 - 109 (1979).

094.536 **The main features of the geochemistry of lunar rocks.** V. L. Barsukov, L. V. Dmitriev.
Origin and distribution of the elements, (see 012.039), p. 127 - 138 (1979).

Statistical cluster analysis of 635 individual bulk chemical compositions of lunar rocks distinguishes three major chemical groups which correspond respectively to highland rocks, high titanium mare basalts and mare basalts with moderate titanium contents. Petrochemical variations are plotted. The petrogenetic meaning of these characteristic trends are discussed. Also, the similarities with or differences from terrestrial basic magmatism are demonstrated. The evidence for two different (continental and mare) early stages of crustal formation on the moon and terrestrial planets is emphasized and it is considered to be impossible to construct, at the present time, a single universal model for the formation of the earth's crust.

094.537 **The distribution of zirconium and hafnium in terrestrial rocks, meteorites and the moon.**
W. D. Ehman, L. L. Chyi, A. N. Garg, M. Z. Ali.
Origin and distribution of the elements, (see 012.039), p. 247 - 259 (1979).

094.538 **Evidence for a secular variation in the $^{13}C/^{12}C$ ratio of carbon implanted in lunar soils.** R. H. Becker.
Earth Planet. Sci. Lett., Vol. 50, 189 - 196 (1980).

A plot of $\delta\,^{13}C$ against $\delta\,^{15}N$ for all lunar soils and breccias for which both values are available indicates that the 30% change in $\delta\,^{15}N$ previously observed is accompanied by a change in $\delta\,^{13}C$ about one-tenth as large. A rough calculation of the relative production rates of ^{13}C and ^{15}N indicates that spallation reactions in the sun could lead to the observed ratio of the $\delta\,^{13}C$ to $\delta\,^{15}N$ variations.

094.539 **Discovery of anomal low oxidation of elements in lunar regolith recorded.**
Metallofiz., Vol. 2, 124 - 125 (1980). In Russian. – Abstr. in Ref. zh., 51. Astron., 11.51.357 (1980).

094.540 **Classification of the Apollo-11 mare basalts according to $Ar^{39}-Ar^{40}$ ages and petrological properties.**
S. Guggisberg, P. Eberhardt, J. Geiss, N. Grögler, A. Stettler, G. M. Brown, A. Peckett.
Proc. Tenth Lunar Planet. Sci. Conf., (see 012.050), p. 1 - 39 (1979).

094.541 **The petrology and chemistry of basaltic fragments from the Apollo 11 soil, part I.** D. W. Beaty, S. M. R. Hill, A. L. Albee, M.-S. Ma, R. A. Schmitt.
Proc. Tenth Lunar Planet. Sci. Conf., (see 012.050), p. 41 - 75 (1979).

094.542 **The Sr and Nd isotopic record of Apollo 12 basalts: implications for lunar geochemical evolution.**
L. E. Nyquist, C.-Y. Shih, J. L. Wooden, B. M. Bansal, H. Wiesmann.
Proc. Tenth Lunar Planet. Sci. Conf., (see 012.050), p. 77 - 114 (1979).

094.543 **Apollo 12 feldspathic basalts 12031, 12038 and 12072: petrology, comparison and interpretations.**
D. W. Beaty, S. M. R. Hill, A. L. Albee, W. S. Baldridge.
Proc. Tenth Lunar Planet. Sci. Conf., (see 012.050), p. 115 - 139 (1979).

094.544 **The petrology of the Apollo 12 pigeonite basalt suite.**
W. S. Baldridge, D. W. Beaty, S. M. R. Hill, A. L. Albee.
Proc. Tenth Lunar Planet. Sci. Conf., (see 012.050), p. 141 - 179 (1979).

094.545 **Petrology, chemistry, and chronology of Apollo 14 KREEP basalts.**
G. A. McKay, H. Wiesmann, B. M. Bansal, C.-Y. Shih.
Proc. Tenth Lunar Planet. Sci. Conf., (see 012.050), p. 181 - 205 (1979).

094.546 **The unique nature of Apollo 17 VLT mare basalts.** S. Wentworth, G. J. Taylor, R. D. Warner, K. Keil, M.-S. Ma, R. A. Schmitt.
Proc. Tenth Lunar Planet. Sci. Conf., (see 012.050), p. 207 - 223 (1979).

094.547 **Apollo 17 high-Ti mare basalts: new bulk compositional data, magma types, and petrogenesis.**
R. D. Warner, G. J. Taylor, G. H. Conrad, H. R. Northrop, S. Barker, K. Keil, M.-S. Ma, R. Schmitt.
Proc. Tenth Lunar Planet. Sci. Conf., (see 012.050), p. 225 - 247 (1979).

094.548 **Melt inclusions in 75075 and 78505 – the problem of anomalous low-K inclusions in ilmenite revisited.**
E. Roedder.
Proc. Tenth Lunar Planet. Sci. Conf., (see 012.050), p. 249 - 257 (1979).

094.549 **Lead isotope systematics of three Apollo 17 mare basalts.** G. R. Tilton, J. H. Chen.
Proc. Tenth Lunar Planet. Sci. Conf., (see 012.050), p. 259 - 274 (1979).

094.550 **Apollo 15 green glass: chemistry and possible origin.** J. W. Delano.
Proc. Tenth Lunar Planet. Sci. Conf., (see 012.050), p. 275 - 300 (1979).

094.551 **A note on the Apollo 15 green glass vitrophyres.** A. Basu, D. S. McKay, C. H. Moore, N. R. Shaffer.
Proc. Tenth Lunar Planet. Sci. Conf., (see 012.050), p. 301 - 310 (1979).

094.552 **The driving mechanism of lunar pyroclastic eruptions inferred from the oxygen fugacity behavior of Apollo 17 orange glass.** M. Sato.
Proc. Tenth Lunar Planet. Sci. Conf., (see 012.050), p. 311 - 325 (1979).

094.553 **74001 drive tube: siderophile elements match II B iron meteorite pattern.**
J. W. Morgan, G. A. Wandless.
Proc. Tenth Lunar Planet. Sci. Conf., (see 012.050), p. 327 - 340 (1979).

094.554 **Scanning Auger microprobe and atomic absorption studies of lunar volcanic volatiles.**
E. H. Cirlin, R. M. Housley.
Proc. Tenth Lunar Planet. Sci. Conf., (see 012.050), p. 341 - 354 (1979).

094.555 **Armalcolite/ilmenite: mineral chemistry, paragenesis, and origin of textures.**
F. T. Stanin, L. A. Taylor.
Proc. Tenth Lunar Planet Sci. Conf., (see 012.050), p. 383 - 405 (1979).

094.556 **One atmosphere melting experiments on ilmenite basalt 12008.**
J. M. Rhodes, G. E. Lofgren, D. P. Smith.
Proc. Tenth Lunar Planet. Sci. Conf., (see 012.050), p. 407 - 422 (1979).

094.557 **Comparison of dynamic crystallization techniques on Apollo 15 quartz normative basalts.**
G. E. Lofgren, T. L. Grove, R. W. Brown, D. P. Smith.
Proc. Tenth Lunar Planet. Sci. Conf., (see 012.050), p. 423 - 438 (1979).

094.558 **Crystallization kinetics in a multiply saturated basalt magma: an experimental study of Lunar 24 ferrobasalt.** T. L. Grove, A. E. Bence.
Proc. Tenth Lunar Planet. Sci. Conf., (see 012.050), p. 439 - 478 (1979).

094.559 **Partitioning as a cooling rate indicator.** P. I. K. Onorato, H. Yinnon, D. R. Uhlmann, L. A. Taylor.
Proc. Tenth Lunar Planet. Sci. Conf., (see 012.050), p. 479 - 491 (1979).

094.560 **The solubility of sulfur in high-TiO_2 mare basalts.** P. A. Danckwerth, P. C. Hess, M. J. Rutherford.
Proc. Tenth Lunar Planet. Sci. Conf., (see 012.050), p. 517 - 530 (1979).

094.561 **A summary of the petrology and geochemistry of pristine highlands rocks.**
M. D. Norman, G. Ryder.
Proc. Tenth Lunar Planet. Sci. Conf., (see 012.050), p. 531 - 559 (1979).

094.562 **The chemical components of highlands breccias.** G. Ryder.
Proc. Tenth Lunar Planet. Sci. Conf., (see 012.050), p. 561 - 581 (1979).

094.563 **The compositional-petrographic search for pristine nonmare rocks: third foray.**
P. H. Warren, J. T. Wasson.
Proc. Tenth Lunar Planet. Sci. Conf., (see 012.050), p. 583 - 610 (1979).

094.564 **Non-meteoritic siderophile elements in lunar highland rocks: evidence from pristine rocks.**
H. Wänke, G. Dreibus, H. Palme.
Proc. Tenth Lunar Planet. Sci. Conf., (see 012.050), p. 611 - 626 (1979).

094.565 **Lunar highland rocks: element partitioning among minerals I: electron microprobe analyses of Na, Mg, K and Fe in plagioclase; *mg* partitioning with orthopyroxene.**
E. C. Hansen, I. M. Steele, J. V. Smith.
Proc. Tenth Lunar Planet. Sci. Conf., (see 012.050), p. 627 - 638 (1979).

094.566 **Terrestrial and lunar impact breccias and the classification of lunar highland rocks.**
D. Stöffler, H.-D. Knöll, U. Maerz.
Proc. Tenth Lunar Planet. Sci. Conf., (see 012.050), p. 639 - 675 (1979).

094.567 **Ilmenite in the crystallization sequence of lunar rocks.** W. von Engelhardt.
Proc. Tenth Lunar Planet. Sci. Conf., (see 012.050), p. 677 - 694 (1979).

094.568 **Olivine vitrophyres in Apollo 14 breccia 14321: samples of the high-Mg component of the lunar highlands.** F. M. Allen, A. E. Bence, T. L. Grove.
Proc. Tenth Lunar Planet. Sci. Conf., (see 012.050), p. 695 - 712 (1979).

094.569 **Consortium breccia 73255: genesis and history of two coarse-grained "norite" clasts.**
O. B. James, J. J. McGee.

Proc. Tenth Lunar Planet. Sci. Conf., (see 012.050), p. 713 - 743 (1979).

094.570 **^{40}Ar-^{39}Ar age systematics of consortium breccia 73255.**
T. Staudacher, E. K. Jessberger, I. Flohs, T. Kirsten.
Proc. Tenth Lunar Planet. Sci. Conf., (see 012.050), p. 745 - 762 (1979).

094.571 **Consortium breccia 73255: laser ^{39}Ar-^{40}Ar dating of aphanite samples.**
G. Eichhorn, J. J. McGee, O. B. James, O. A. Schaeffer.
Proc. Tenth Lunar Planet. Sci. Conf., (see 012.050), p. 763 - 788 (1979).

094.572 **Breccias 73215 and 73255: siderophile and volatile trace elements.** J. W. Morgan, R. K. Petrie.
Proc. Tenth Lunar Planet. Sci. Conf., (see 012.050), p. 789 - 801 (1979).

094.573 **Remnants from the ancient lunar crust: clasts from consortium breccia 73255.**
D. P. Blanchard, J. R. Budahn.
Proc. Tenth Lunar Planet. Sci. Conf., (see 012.050), p. 803 - 816 (1979).

094.574 **Thermal and mechanical history of granulated norite and pyroxene anorthosite clasts in breccia 73255.**
G. L. Nord, Jr., J. J. McGee.
Proc. Tenth Lunar Planet. Sci. Conf., (see 012.050), p. 817 - 832 (1979).

094.575 **Characterization and depositional and evolutionary history of the Apollo 17 deep drill core.**
R. V. Morris, H. V. Lauer, Jr., W. A. Gose.
Proc. Tenth Lunar Planet. Sci. Conf., (see 012.050), p. 1141 - 1157 (1979).

094.576 **Stratigraphy and depositional history of the Apollo 17 drill core.**
G. J. Taylor, R. D. Warner, K. Keil.
Proc. Tenth Lunar Planet. Sci. Conf., (see 012.050), p. 1159 - 1184 (1979).

094.577 **The Apollo 17 drill core: petrologic systematics and the identification of a possible Tycho component.**
D. T. Vaniman, T. C. Labotka, J. J. Papike, S. B. Simon, J. C. Laul.
Proc. Tenth Lunar Planet. Sci. Conf., (see 012.050), p. 1185 - 1227 (1979).

094.578 **Deposition and irradiation of the Apollo 17 deep drill core.** G. Crozaz, L. M. Ross, Jr.
Proc. Tenth Lunar Planet. Sci. Conf., (see 012.050), p. 1229 - 1241 (1979).

094.579 **History of the Apollo 17 deep drill string during the past few million years.**
J. S. Fruchter, J. H. Reeves, L. A. Rancitelli, R. W. Perkins.
Proc. Tenth Lunar Planet. Sci. Conf., (see 012.050), p. 1243 - 1251 (1979).

094.580 **Depositional history of the Apollo 17 deep drill core based on particle track record.**
J. N. Goswami, D. Lal.
Proc. Tenth Lunar Planet. Sci. Conf., (see 012.050), p. 1253 - 1267 (1979).

094.581 **The Apollo 17 drill core: chemical systematics of grain size fractions.**
J. C. Laul, E. A. Lepel, D. T. Vaniman, J. J. Papike.
Proc. Tenth Lunar Planet. Sci. Conf., (see 012.050), p. 1269 - 1298 (1979).

094.582 **Preliminary description and interpretation of Apollo 14 cores 14211/10.** J. S. Nagle.
Proc. Tenth Lunar Planet. Sci. Conf., (see 012.050), p. 1299 - 1319 (1979).

094.583 **Apollo 15 deep drill core: trace element and metallic iron abundances in size fractions of sample 15002,170.** C.-L. Chou, G. W. Pearce.
Proc. Tenth Lunar Planet. Sci. Conf., (see 012.050), p. 1321 - 1332 (1979).

094.584 **Irradiation stratigraphy and depositional history of the Apollo 16 double drive tube 60009/10.**
G. E. Blanford, J. Blanford, J. A. Hawkins.
Proc. Tenth Lunar Planet. Sci. Conf., (see 012.050), p. 1333 - 1349 (1979).

094.585 **Double drive tube 74001/2: history of the black and orange glass; determination of a pre-exposure 3.7 AE ago by $^{136}Xe/^{235}U$ dating.**
O. Eugster, N. Grögler, P. Eberhardt, J. Geiss.
Proc. Tenth Lunar Planet. Sci. Conf., (see 012.050), p. 1351 - 1379 (1979).

094.586 **Regolith reworking at Shorty Crater.** G. Crozaz.
Proc. Tenth Lunar Planet. Sci. Conf., (see 012.050), p. 1381 - 1384 (1979).

094.587 **Drive tube 76001 – continuous accumulation with complications?** J. S. Nagle.
Proc. Tenth Lunar Planet. Sci. Conf., (see 012.050), p. 1385 - 1399 (1979).

094.588 **The distribution of radionuclides in the Luna 24 lunar soil column.**
Yu. A. Surkov, L. P. Moskalyova (*Moskaleva*), O. S. Manvelyan, V. P. Kharyukova, V. V. Vysochkin, E. L. Kovalchuk, A. A. Pomansky, A. D. Dudin.
Proc. Tenth Lunar Planet. Sci. Conf., (see 012.050), p. 1401 - 1412 (1979).

094.589 **Petrography and provenance of Apollo 15 soils.**
A. Basu, D. S. McKay.
Proc. Tenth Lunar Planet. Sci. Conf., (see 012.050), p. 1413 - 1424 (1979).

094.590 **Regolith layering processes based on studies of low-temperature volatile elements in Apollo core samples.** S. Jovanovic, G. W. Reed, Jr.
Proc. Tenth Lunar Planet. Sci. Conf., (see 012.050), p. 1425 - 1435 (1979).

094.591 **Composition of glasses in Apollo 17 samples and their relation to known lunar rock types.**
R. D. Warner, G. J. Taylor, K. Keil.
Proc. Tenth Lunar Planet. Sci. Conf., (see 012.050), p. 1437 - 1456 (1979).

094.592 **Grain-size dependent distributions of elements and their origin in minerals and agglutinates of soil 75080.** H. R. von Gunten, A. Grütter, D. Jost, U. Krähenbühl, G. Meyer, F. Wegmüller.
Proc. Tenth Lunar Planet. Sci. Conf., (see 012.050), p. 1457 - 1468 (1979).

094.593 **Hydrolysable carbon, magnetic susceptibility and isothermal remanent magnetisation measurements of highland sample 68501: comments on carbon content and size distribution of finely divided lunar iron.**

A. E. Fallick, C. T. Pillinger, A. Stephenson.
Proc. Tenth Lunar Planet. Sci. Conf., (see 012.050), p. 1469 - 1481 (1979).

094.594 **XPS and SAM studies of the surface chemistry of lunar impact glasses including 12054.**
R. M. Housley, R. W. Grant, E. H. Cirlin.
Proc. Tenth Lunar Planet. Sci. Conf., (see 012.050), p. 1483 - 1490 (1979).

094.595 **Chemical peculiarities of particle surface layers of some Apollo 17 regolith samples.**
Yu. P. Dikov, O. A. Bogatikov, V. L. Barsukov, K. P. Florensky (*Florenskij*), A. V. Ivanov, V. V. Nemoshkalenko, V. G. Alyoshin.
Proc. Tenth Lunar Planet. Sci. Conf., (see 012.050), p. 1491 - 1505 (1979).

094.596 **Analytical electron microscopy study of submicroscopic metal particles in glassy constituents of lunar breccias 15015 and 60095.** S. Mehta, J. I. Goldstein.
Proc. Tenth Lunar Planet. Sci. Conf., (see 012.050), p. 1507 - 1521 (1979).

094.597 **Investigations of submicron sized metal particles in glass coatings of lunar breccia 15286.**
S. Mehta, J. I. Goldstein, J. J. Friel.
Proc. Tenth Lunar Planet. Sci. Conf., (see 012.050), p. 1523 - 1530 (1979).

094.598 **Microdistribution patterns of implanted rare gases in a large number of individual lunar soil particles.**
M. Warhaut, J. Kiko, T. Kirsten.
Proc. Tenth Lunar Planet. Sci. Conf., (see 012.050), p. 1531 - 1546 (1979).

094.599 **Noble gas based solar flare exposure history of lunar rocks and soils.** M. N. Rao, T. R. Venkatesan, J. N. Goswami, C. M. Nautiyal, J. T. Padia.
Proc. Tenth Lunar Planet. Sci. Conf., (see 012.050), p. 1547 - 1564 (1979).

094.600 **Predicted versus observed cosmic-ray-produced noble gases in lunar samples: improved Kr production ratios.** S. Regnier, C. M. Hohenberg, K. Marti, R. C. Reedy.
Proc. Tenth Lunar Planet. Sci. Conf., (see 012.050), p. 1565 - 1586 (1979).

094.601 **Xenon component organization in 14301.**
T. J. Bernatowicz, C. M. Hohenberg, F. A. Podosek.
Proc. Tenth Lunar Planet. Sci. Conf., (see 012.050), p. 1587 - 1616 (1979).

094.602 **Some effects of gas adsorption on the high temperature volatile release behavior of a terrestrial basalt, tektite and lunar soil.**
D. G. Graham, D. W. Muenow, E. K. Gibson, Jr.
Proc. Tenth Lunar Planet. Sci. Conf., (see 012.050), p. 1617 - 1627 (1979).

094.603 **Sulphur isotopes in lunar and meteorite samples.**
H. G. Thode, C. E. Rees.
Proc. Tenth Lunar Planet. Sci. Conf., (see 012.050), p. 1629 - 1636 (1979).

094.604 **Near-surface daytime thermal conductivity in the lunar regolith.** G. W. Reed, Jr., S. Jovanovic.
Proc. Tenth Lunar Planet. Sci. Conf., (see 012.050), p. 1637 - 1647 (1979).

094.605 **Properties of microcraters and cosmic dust of less than 1000 Å dimensions.**
D. A. Morrison, U. S. Clanton.
Proc. Tenth Lunar Planet. Sci. Conf., (see 012.050), p. 1649 - 1663 (1979).

094.606 **A model for chemical and isotopic fractionation in the lunar regolith by impact vaporization.**
R. M. Housley.
Proc. Tenth Lunar Planet. Sci. Conf., (see 012.050), p. 1673 - 1683 (1979).

094.607 **Thorium concentrations in the lunar surface: III. Deconvolution of the Apenninus region.**
A. E. Metzger, E. L. Haines, M. I. Etchegaray-Ramirez, B. R. Hawke.
Proc. Tenth Lunar Planet. Sci. Conf., (see 012.050), p. 1701 - 1718 (1979).

094.608 **A comparison of mare surface titanium concentrations obtained by spectral reflectance and gamma-ray spectroscopy: an early assessment.**
A. E. Metzger, T. V. Johnson, D. L. Matson.
Proc. Tenth Lunar Planet. Sci. Conf., (see 012.050), p. 1719 - 1726 (1979).

094.609 **Evidence for early volcanism in Mare Smythii.**
J. Conca, N. Hubbard.
Proc. Tenth Lunar Planet. Sci. Conf., (see 012.050), p. 1727 - 1737 (1979).

094.610 **Are early magnesium-rich basalts widespread on the moon?** C. G. Andre, R. W. Wolfe, I. Adler.
Proc. Tenth Lunar Planet. Sci. Conf., (see 012.050), p. 1739 - 1751 (1979).

094.611 **Regional chemical variations in lunar basaltic lavas.**
N. Hubbard.
Proc. Tenth Lunar Planet. Sci. Conf., (see 012.050), p. 1753 - 1774 (1979).

094.612 **Infrared spectra of lunar soils and related optical constants.**
J. R. Aronson, A. G. Emslie, E. M. Smith, P. F. Strong.
Proc. Tenth Lunar Planet. Sci. Conf., (see 012.050), p. 1787 - 1795 (1979).

094.613 **Lunar and terrestrial potassium and uranium abundances: implications for the fission hypothesis.**
S. R. Taylor.
Proc. Tenth Lunar Planet. Sci. Conf., (see 012.050), p. 2017 - 2030 (1979).

094.614 **Effects of pressure on the crystallization of a "chondritic" magma ocean and implications for the bulk composition of the moon.** P. H. Warren, J. T. Wasson.
Proc. Tenth Lunar Planet. Sci. Conf., (see 012.050), p. 2051 - 2083 (1979).

094.615 **Complex igneous processes and the formation of the primitive lunar crustal rocks.**
J. Longhi, A. E. Boudreau.
Proc. Tenth Lunar Planet. Sci. Conf., (see 012.050), p. 2085 - 2105 (1979).

094.616 **Lunar basalts and pristine highland rocks: comparison of siderophile and volatile elements.**
R. Wolf, A. Woodrow, E. Anders.
Proc. Tenth Lunar Planet. Sci. Conf., (see 012.050), p. 2107 - 2130 (1979).

094.617 **Seismic Q and velocity at depth.**
B. R. Tittmann, H. Nadler, V. Clark, L. Coombe.
Proc. Tenth Lunar Planet. Sci. Conf., (see 012.050), p. 2131 - 2145 (1979).

094.618 **Compressional velocity measurements for a highly fractured lunar anorthosite.**
C. H. Sondergeld, L. A. Granryd, H. A. Spetzler.
Proc. Tenth Lunar Planet. Sci. Conf., (see 012.050), p. 2147 - 2154 (1979).

094.619 **Microwave dielectric measurements of lunar soil with a coaxial line resonator method.**
H. E. Bussey.
Proc. Tenth Lunar Planet. Sci. Conf., (see 012.050), p. 2175 - 2182 (1979).

094.620 **A new magnetic paleointensity value for a "young lunar glass".**
N. Sugiura, Y. M. Wu, D. W. Strangway, G. W. Pearce, L. A. Taylor.
Proc. Tenth Lunar Planet. Sci. Conf., (see 012.050), p. 2189 - 2197 (1979).

094.621 **On the natural remanent magnetism of certain mare basalts.**
M. Fuller, E. Meshkov, S. M. Cisowski, C. J. Hale.
Proc. Tenth Lunar Planet. Sci. Conf., (see 012.050), p. 2211 - 2233 (1979).

094.622 **Electrical conductivity anomalies associated with circular lunar maria.** P. Dyal, W. D. Daily.
Proc. Tenth Lunar Planet. Sci. Conf., (see 012.050), p. 2291 - 2297 (1979).

094.623 **Shallow moonquakes: depth, distribution and implications as to the present state of the lunar interior.**
Y. Nakamura, G. V. Latham, H. J. Dorman, A.-B. K. Ibrahim, J. Koyama, P. Horvath.
Proc. Tenth Lunar Planet. Sci. Conf., (see 012.050), p. 2299 - 2309 (1979).

094.624 **Shock metamorphism of granulated lunar basalt.**
R. B. Schaal, F. Hörz, T. D. Thompson, J. F. Bauer.
Proc. Tenth Lunar Planet. Sci. Conf., (see 012.050), p. 2547 - 2571 (1979).

094.625 **Asymmetric terracing of lunar highland craters: influence of pre-impact topography and structure.**
A. W. Gifford, T. A. Maxwell.
Proc. Tenth Lunar Planet. Sci. Conf., (see 012.050), p. 2597 - 2607 (1979).

094.626 **Smythii basin topography and comparisons with Orientale.** P. L. Strain, F. El-Baz.
Proc. Tenth Lunar Planet. Sci. Conf., (see 012.050), p. 2609 - 2621 (1979).

094.627 **Central peaks in lunar craters: morphology and morphometry.** W. Hale, J. W. Head.
Proc. Tenth Lunar Planet. Sci. Conf., (see 012.050), p. 2623 - 2633 (1979).

094.628 **Monte Carlo simulation of lunar megaregolith and implications.** H. R. Aggarwal, V. R. Oberbeck.
Proc. Tenth Lunar Planet. Sci. Conf., (see 012.050), p. 2689 - 2705 (1979).

094.629 **Alphonsus-type dark-halo craters: morphology morphometry and eruption conditions.**
J. W. Head III, L. Wilson.
Proc. Tenth Lunar Planet. Sci. Conf., (see 012.050), p. 2861 - 2897 (1979).

094.630 **Evidence for ancient mare volcanism.**
P. H. Schultz, P. D. Spudis.
Proc. Tenth Lunar Planet. Sci. Conf., (see 012.050), p. 2899 - 2918 (1979).

094.631 **Thickness of the western mare basalts.**
R. A. De Hon.
Proc. Tenth Lunar Planet. Sci. Conf., (see 012.050), p. 2935 - 2955 (1979).

094.632 **Altitude-age relationships of the lunar maria.**
B. K. Lucchitta, J. M. Boyce.
Proc. Tenth Lunar Planet. Sci. Conf., (see 012.050), p. 2957 - 2966 (1979).

094.633 **Mare Crisium geologic units: implications of additional remote sensing data.**
C. M. Pieters, T. B. McCord, J. W. Head III, J. B. Adams, S. Zisk.
Proc. Tenth Lunar Planet. Sci. Conf., (see 012.050), p. 2967 - 2973 (1979).

094.634 **Charting the southern seas: the evolution of the lunar Mare Australe.** J. L. Whitford-Stark.
Proc. Tenth Lunar Planet. Sci. Conf., (see 012.050), p. 2975 - 2994 (1979).

094.635 **Multispectral mapping of the Apollo 15 – Apennine region: the identification and distribution of regional pyroclastic deposits.**
B. R. Hawke, D. MacLaskey, T. B. McCord, J. B. Adams, J. W. Head, III, C. M. Pieters, S. H. Zisk.
Proc. Tenth Lunar Planet. Sci. Conf., (see 012.050), p. 2995 - 3015 (1979).

094.636 **Belts of lunar maria and their origin as a result of mantle melting.** R. Ivanov.
Probl. planetol. T. 2. Simpoz. No. 8. Mezhdunar. assots. planetol., Erevan, 1977. Leningrad-Erevan, 1977, p. 177 - 182. In Russian. – Abstr. in Ref. zh., 62. Issled. kosm. prostranstva, 12.62.167 (1980).

094.637 **Gravitational anomalies above lunar craters, caldera and meteorite craters: problems and correlations.**
G. S. Shtejnberg.
Probl. planetol. T. 2. Simpoz. No. 8. Mezhdunar. assots. planetol., Erevan, 1977. Leningrad-Erevan, 1977, p. 189 - 194. In Russian. – Abstr. in Ref. zh., 62. Issled. kosm. prostranstva, 12.62.169 (1980).

094.638 **Morphology of impact craters on the moon and other planets.**
K. P. Florenskij, A. T. Bazilevskij, N. N. Grebennik.
Meteoritn. struktury na poverkhnosti planet. Moskva, 1979, p. 192 - 203. In Russian. – Abstr. in Ref. zh., 62. Issled. kosm. prostranstva, 1.62.281 (1981).

094.639 **Preliminary results of determinations of the physical properties of microfragments of lunar rocks from Luna 24 soil.**
I. A. Turchaninov, R. V. Medvedev, V. V. Chashnikov.
Dokl. AN SSSR, Vol. 252, 358 - 360 (1980). In Russian. Abstr. in Ref. zh., 62. Issled. kosm. prostranstva, 1.62.313 (1981).

094.640 **On the role of volatile components in the formation of lunar rocks.**
O. A. Bogatikov, D. I. Frikh-Khar, N. A. Ashikhmina, Yu. P. Dikov, E. E. Laz'ko, E. V. Sveshnikova.

Zap. Vses. mineral. o-va, Vol. 109, 30 - 36 (1980). In Russian. Abstr. in Ref. zh., 62. Issled. kosm. prostranstva, 1.62.320 (1981).

Actinide crystal–liquid partitioning for clinopyroxene and $Ca_3(PO_4)_2$. See Abstr. 022.011.

Reflectance spectrophotometry extended to u.v. for terrestrial, lunar and meteoritic samples. See Abstr. 022.012.

High sensitivity stable carbon isotope analysis for meteorite and lunar studies. See Abstr. 022.037.

Lunar dust grains and the thermal properties of the ancient solar wind. See Abstr. 022.042.

Radiative cooling experiments on lunar glass analogues. See Abstr. 022.176.

A simplified model for glass formation. See Abstr. 022.177.

The olivine-ilmenite thermometer. See Abstr. 022.178.

Experimental studies of nickel and chromium partitioning into olivine from synthetic basaltic melts. See Abstr. 022.179.

Comparison of single crack propagation in lunar analogue glass and the failure strength of rocks. See Abstr. 022.181.

Meteorite cratering on terrestrial planets. See Abstr. 091.055.

Morphology of impact craters on the moon and planets. See Abstr. 091.056.

Mercurian crater rim heights and some interplanetary comparisons. See Abstr. 092.006.

Gardening process of lunar surface layer inferred from the galactic cosmic-ray exposure ages of lunar samples. See Abstr. 094.008.

Infrared reflection spectra of the moon and lunar soil. See Abstr. 094.042.

Ancient crater populations on Mars. See Abstr. 097.034.

Pyroxenes in early crustal cumulates found in achondrites and lunar highland rocks. See Abstr. 105.139.

Magnetic properties and paleointensity of achondrites in comparison with those of lunar rocks. See Abstr. 105.146.

Stochastic model of the dissipation of the regolith layer consisting of material losing its initial properties in the process of cratering. See Abstr. 105.155.

Cosmic ray effects in solar system objects. See Abstr. 143.039.

Erratum

094.901 **Erratum: "On the main parameter of the physical libration of the moon"** [Astron. Tsirk., No. 1059, p. 5 - 6 (1979)]. Yu. A. Chikanov, K. S. Shakirov. Astron. Tsirk., No. 1097, p. 8 (1980). In Russian. – See Abstr. 27.094.024.

095 Lunar Eclipses

095.001 **Total penumbral lunar eclipses.** J. Meeus. J. R. Astron. Soc. Canada, Vol. 74, 291 - 295 (1980).

This paper lists all total penumbral eclipses for the period AD 1600 - 2500. The total penumbral eclipses are unevenly distributed with time; their frequency varies in phase with that of tetrads and that of total eclipses in the umbra.

Photometry of the lunar surface during lunar eclipses. See Abstr. 094.004.

Analyse photométrique de l'éclipse de lune du 16 Septembre 1978. See Abstr. 094.005.

Great volcanic eruptions and the luminosity of the moon during total eclipses. See Abstr. 094.019.

096 Lunar and Planetary Occultations

096.001 **An occultation by Charon.** A. R. Walker.
Mon. Not. R. Astron. Soc., Vol. 192, 47P - 50P (1980).

An occultation of a 13th magnitude star by Charon (1978 P1) gives a minimum diameter of 1200 km for the satellite. The occultation light curve indicates that Charon is a solid body.

096.002 **Illinois occultation summary. I. 1977 - 1978.**
R. Radick, D. Lien.
Astron. J., Vol. 85, 1053 - 1061 (1980).

The authors present results from the first two years of a program undertaken to record lunar occultations at the University of Illinois Prairie Observatory. The 64 events summarized include 30 observations of stars brighter than 7th mag, 40 reappearances, four angular diameter measurements, eight observations of binary stars, and six observations which may indicate multiplicity.

096.003 **Regulus, an important but jinxed star.**
D. W. Dunham.
Occultation Newsl., Vol. 2, 100 - 102 (1980).

096.004 **News of recently-attempted planetary occultations.**
D. W. Dunham.
Occultation Newsl., Vol. 2, 103 - 104 (1980).

096.005 **Lunar occultations of planets.**
Occultation Newsl., Vol. 2, 104 (1980).

096.006 **Plans for upcoming asteroidal occultations and other planetary occultation news.**
D. W. Dunham.
Occultation Newsl., Vol. 2, 104 - 112 (1980).

096.007 **Application of the regularization method to the analysis of lunar occultation observations of close double stars and radio sources.** M. B. Bogdanov.
Astron. Zh., Tom 57, 762 - 766 (1980). In Russian.
English translation in Soviet Astron., Vol. 24, No. 4.

The possibility of application of the regularization method for digital solution of the integral equation which connects the observed light curve with the brightness distribution across the source is examined on the example of the lunar occultation of the close double star 16 Tau.

096.008 **Observations of stellar occultations by asteroids.**
V. G. Tejfel, F. A. Tupieva.
Zemlya Vselennaya, 1980, No. 4, p. 18 - 19. In Russian.

096.009 **Observations of stellar occultations by the moon in the years 1976 - 1977.**
V. M. Kirpatovskij, P. P. Pavlenko, R. M. Shut'eva.
Vestn. Khar'k. Univ., No. 190. Fiz. Luny Planet. Fundam. Astrometr., Vyp. 14, p. 7 - 8 (1979). In Russian.

096.010 **Fading occultations of stars by the Moon: 1943–1977.** G. M. Appleby.
J. British Astron. Assoc., Vol. 90, 572 - 581 (1980).

An investigation into why 423 lunar occultations of stars were observed to occur non-instantaneously reveals that 160 of the observations can be explained by the fact that 140 of the stars are known to be double. A further 19 observations can be attributed to effects due to stellar angular diameter or diffraction.

096.011 **An analysis of total lunar occultations made in the years 1943 to 1974.** L. V. Morrison.
J. British Astron. Assoc., Vol. 91, 14 - 24 (1980).

The times of 62 000 occultations of stars by the Moon observed during 1943 - 74 are analyzed to determine corrections to the Earth's rate of rotation, the ephemeris and figure of the Moon, and the origin of right ascension in the Fourth Fundamental Catalogue of star positions.

096.012 **The observation of occultations.**
G. W. Amery.
J. British Astron. Assoc., Vol. 91, 24 - 25 (1980). – Summary.

096.013 **Occultations rasantes en France en 1981.**
J. Meeus.
Astronomie, Vol. 94, 547 - 550 (1980).

096.014 **Observation de l'occultation rasante d'Aldébaran par la Lune le 11 avril 1978.**
R. Boninsegna, J. Schwaenen.
Ciel Terre, Vol. 96, 387 - 393 (1980).

096.015 **Observations of star occultations by the moon at the Engelhardt Astronomical Observatory in April - November 1975 and January - April 1976.**
I. G. Chugunov, M. I. Shpekin, V. S. Borovskikh.
Izv. Astron. Ehngel'gardt. Obs., No. 45, p. 88 - 90 (1978). In Russian.

096.016 **Occultation of the star AGK3 + 19°599 by the minor planet 65 Cybele.**
S. I. Gerasimenko, N. N. Kiselev, F. N. Masumi, O. Naimov, N. V. Narizhnaya, V. Yu. Rakhimov, F. A. Tupieva, G. P. Chernova.
Astron. Tsirk., No. 1098, p. 6 - 8 (1980). In Russian.

096.017 **Observations of lunar occultations of stars at the Poltava Observatory in 1972 - 1975.**
B. F. Sincheskul.
Vrashchenie i priliv. deformatsii Zemli, Kiev, 1980, No. 12, p. 12 - 33. In Russian. – Abstr. in Ref. zh., 51. Astron., 1.51.116 (1981).

096.018 **Results of observations of occultations of stars by the moon in Ussurijsk in 1974 - 1977.**
V. A. Golubev.
Astron. Vestn., Tom 14, 189 - 190 (1980). In Russian.

A new method of deconvolution and its application to lunar occultations. See Abstr. 021.012.

Restoration of lunar occultation scans.
See Abstr. 021.013.

LODAS – a lunar occultation data acquisition system. See Abstr. 021.019.

An efficient, automated method of finding asteroid occultations of catalog stars. See Abstr. 021.024.

A New Zealander makes WOPAI (*Williams Occultation Position Angle Indicator*). See Abstr. 031.513.

Improving planetary occultation astrometry.
See Abstr. 041.008.

The size and shape of Juno from its occultation of AGK3+0°1022. See Abstr. 098.036.

Occultation of SAO 75392 by (78) Diana on 1980 September 4. See Abstr. 098.063.

Occultation of AGK3 +35°0669 by 187 Lamberta on 1980 November 11. See Abstr. 098.076.

In search of satellites of minor planets. See Abstr. 098.088.

Occultation by Uranus. See Abstr. 101.021.

Occultations by Neptune. See Abstr. 101.023.

Results from the 10 March 1977 occultation by the Uranus system. See Abstr. 101.033.

Angular diameters by the lunar occultation technique. III. See Abstr. 115.008.

Lunar occultation angular diameter measurements. See Abstr. 115.009.

The importance of SAO 93957. See Abstr. 120.022.

Lunar occultation observations of millimeter CO emission in S255. See Abstr. 131.162.

Small scale structure of the CO emission in the biconical nebula LkHα 208 from lunar occultation observations. See Abstr. 134.012.

Ooty lunar occultation survey: list 9. See Abstr. 141.077.

Possible determination of $\bar{q}_0$ using lunar occultations and laser ranging observations. See Abstr. 162.095.

097 Mars, Mars Satellites

097.001 **Rotation period of Mars from transits of albedo stations at central meridian, 1659–1971.**
G. de Vaucouleurs.
Astron. J., Vol. 85, 945 - 960 (1980).

The rotation period of Mars P has been re-derived from an analysis of more than 800 telescopic observations from 1659 to 1971 of transits of albedo markings at the central meridian. All observations were corrected for phase effects and reduced to ephemeris time, and corresponding longitudes λ were calculated in the current American Ephemeris (N.A.3) system, which assumes $P = 24^h 37^m 22^s.6689$. Solutions of the form $\lambda = \lambda_0 + b(T - T_0)$ were made for a multiplicity of stations, time bases, and weighting systems.

097.002 **Phobos and Deimos: the Martian asteroids.**
G. Bowman.
Spaceflight, Vol. 22, 303 - 311 (1980).

097.003 **News about Martian seismology.**
I. N. Galkin.
Priroda, 1980, No. 8, p. 61. In Russian.

097.004 **Variations of the Martian CO_2 band intensities.**
O. G. Taranova, A. M. Romanov.
Astron. Zh., Tom 57, 812 - 815 (1980). In Russian.
English translation in Soviet Astron., Vol. 24, No. 4.

The results of the ground-based observations of the Martian CO_2 bands near λ 1.6 μm during 1971 - 1975 are presented. The variations of the CO_2 bands intensities are generally associated with the seasonal changes of the total pressure in the Martian atmosphere.

097.005 **On the infrared radiative properties of CO_2 ice clouds: application to Mars.** G. E. Hunt.
Geophys. Res. Lett., Vol. 7, 481 - 484 (1980).

The author investigated the effect on the observed 20 μm brightness temperatures measured by the IRTM instrument, of the presence of cloud layers of CO_2 ice particles. The results show that such clouds have a profound effect, and a layer corresponding to visible thickness of ~ 0.5 can create brightness temperatures of 120 - 130 K, even when the surface is at 150 K. This suggests that these clouds may also contribute to the effect of brightness temperatures below the CO_2 frost point, which have been observed over the southern winter polar region.

097.006 **The longitudinal variation of the thermal inertia and of the 2.8 centimeter brightness temperature of Mars.** B. M. Jakosky, D. O. Muhleman.
Astrophys. J., Vol. 239, 403 - 409 (1980).

The spatial variations on Mars of the surface thermal inertia and radiometric albedo are used to predict the variation with sub-earth longitude of the 2.8 cm whole-disk brightness temperature. The maximum variation predicted, about 8 K, agrees well with observations. The sub-earth longitudes at which the temperature maxima and minima are predicted to occur nearly agree with the observations. There are, however, differences in the overall form of the variation with longitude. These discrepancies can be reduced by an ad hoc assumption of spatial variations in either the fraction of the surface covered by rock or the amount of atmospheric dust.

097.007 **Unusual Martian surface features.**
V. Dipietro, G. Molenaar.
Bull. American Astron. Soc., Vol. 12, 450 - 451 (1980). Abstract.

097.008 **Brightness temperature of Mars: 1979 - 1983.**
E. L. Wright, S. Odenwald.
Bull. American Astron. Soc., Vol. 12, 456 (1980). – Abstract.

097.009 **Mapping of non-thermal CO_2 emission features on Mars.**
D. Deming, G. Chin, F. Espenak, T. Kostiuk, M. J. Mumma.
Bull. American Astron. Soc., Vol. 12, 507 (1980). – Abstract.

097.010 **Some characteristics of the soil in the Martian equatorial zone from data of a thermal radiometry aboard Mars 5.** V. V. Vdovin, V. S. Zhegulev, L. V. Ksanfomaliti, V. I. Moroz, E. V. Petrova.
Kosm. Issled., Tom 18, 609 - 622 (1980). In Russian.

097.011 **Analysis of the γ-radiation of Martian rocks from data of the automatic interplanetary station Mars 5.**
Yu. A. Surkov, L. P. Moskaleva, O. S. Manvelyan, V. P. Kharyukova.
Kosm. Issled., Tom 18, 623 - 631 (1980). In Russian.

097.012 **The new Mars.** F. El-Baz.
Interdisciplinary Sci. Rev., Vol. 4, 315 - 317 (1979).
Abstr. in Phys. Abstr., Vol. 83, Abstr. 82255 (1980).

097.013 **Geomorphology of Martian valleys.**
D. C. Pieri.
Advances in planetary geology, (see 003.002), p. 1 - 160 (1980).

The dissertation covers a broad range of topics related to Martian valleys: terminology, classification, distribution, morphology, age, quantitative analysis of pattern, physics of flow, origin, and the relation of Martian valleys to climatic change.

097.014 **Volcano-ice interactions on the Earth and Mars.**
C. C. Allen.
Advances in planetary geology, (see 003.002), p. 161 - 264 (1980).

The interaction of volcanism with ice and water produces a suite of landforms with morphologies highly characteristic of their origin. They include large flood channels, distinctive mountains and ridge systems, and fields of steam explosion craters. Features similar to these in form, size and geologic setting occur on Mars and have proved to be valuable indicators of the history of water on that planet.

097.015 **The stratigraphic sequence of volcanic and sedimentary units in the north polar region of Mars.**
M. E. Botts.
Advances in planetary geology, (see 003.002), p. 265 - 326 (1980).

Based on photogeologic mapping of Viking Orbiter images of Mars, four distinct informal stratigraphic units can be defined for the region north of 70°N latitude. They are: (a) bulbous plains, (b) mantled plains, (c) dune deposits, and (d) layered deposits/perennial ice. These stratigraphic units are discussed in detail.

097.016 **Radar altimetry of South Tharsis, Mars.**
L. E. Roth, G. S. Downs, R. S. Saunders, G. Schubert.
Icarus, Vol. 42, 287 - 316 (1980).

Martian altitudes were measured by radar during the oppositions of 1971 and 1973 using the 64-m antenna at Goldstone (California). The resultant topographic profiles substantiate a zonal classification of the volcanic flows blanketing the south flanks of Arsia Mons, and they confirm the existence of a secondary, parasitic shield attached from the SSW to the main Arsia shield.

097.017 **Modes of sediment transport in channelized water flows with ramifications to the erosion of the Martian outflow channels.** P. D. Komar.
Icarus, Vol. 42, 317 - 329 (1980).

The criteria for quantitatively determining which grain-size ranges are being transported in terrestrial rivers as bed load, suspended load and wash load are applied to an analysis of sediment transport in the large Martian outflow channels, assuming their origin to have been from water flow. Of importance is the balance of the effects of the reduced Martian gravity on the water flow velocity versus the reduction in grain settling velocities. Analyses were performed using grain densities ranging from 2.90 g/cm^3 (basalt) to 1.20 g/cm^3 (volcanic ash).

097.018 **Formation of complex impact craters: evidence from Mars and other planets.** R. J. Pike.
Icarus, Vol. 43, 1 - 19 (1980).

The simple-to-complex transition for impact craters on Mars occurs at diameters between about 3 and 8 km. Ballistically emplaced ejecta surround primarily those craters that have a simple interior morphology, whereas ejecta displaying features attributable to fluid flow are mostly restricted to complex craters. Size-dependent characteristics of 73 relatively fresh Martian craters, emphasizing the new depth/diameter *(d/D)* data of D. W. G. Arthur (1980, to be submitted for publication), test two hypotheses for the mode of formation of central peaks in complex craters.

097.019 **Mars: 1977–1978.** E. H. Collinson.
J. British Astron. Assoc., Vol. 90, 560 - 564 (1980).
A report of the Mars Section.

097.020 **Tectonic nature of the Martian channels.**
G. A. Burba.
Priroda, 1980, No. 9, p. 88 - 89. In Russian.

097.021 **Influence of outer factors on the composition of the Martian atmosphere.** A. K. Pavlov.
XIth seminar on cosmophysics, (see 012.035), p. 203 - 209 (1979). In Russian. – Abstr. in Ref. zh., 51. Astron., 8.51.260 (1980).

097.022 **Energetic spectra of electrons in the atmospheres of Mars and Venus.** L. A. Akatova.
Issled. po geomagn., aehron. i fiz. Solntsa. Moskva, 1980, No. 51, p. 90 - 94. In Russian. – Abstr. in Ref. zh., 62. Issled. kosm. prostranstva, 8.62.126 (1980).

097.023 **A post-Viking view of Martian geologic evolution.**
R. E. Arvidson, K. A. Goettel, C. M. Hohenberg.
Rev. Geophys. Space Phys., Vol. 18, 565 - 603 (1980).

Contents: Mantle and core. Composition and mineralogy of Martian surface materials. Structural geology. Volcanic activity in space time. Thermal evolution. Elemental and isotopic abundances in the Martian atmosphere and the volatile inventory. Geologic constraints on the volatile inventory.

097.024 **Viking orbiters: one last look at Mars.**
Sky Telesc., Vol. 60, 386 - 389 (1980).

097.025 **The mosaics of Mars.** As seen by the Viking Lander Cameras. E. C. Levinthal, K. L. Jones.
NASA Contract. Rep., NASA CR 3326, 8 + 72 pp. (1980).

This publication describes the mosaics and derivative products produced from many individual high resolution images acquired by the Viking Lander Camera Systems. The derived products include special geometric projections of the mosaic data sets, polar stereographic ("donut"), stereoscopic, and orthographic. Additionally, the mosaics were used to produce contour maps and vertical profiles of the topography. For all of these products, the method by which they were created is described.

097.026 **Seasonal change of the Martian north polar cap, 1979–80.**
K. Iwasaki, Y. Saito, T. Akabane.
13th Lunar and Planetary Symposium, (see 012.018), p. 17 - 23 (1980).

The regression curve for the north polar cap of Mars is extracted from photographic observations at the Kwasan Observatory and at the Hida Observatory during the 1979-1980 apparition. The time when the north polar cap started to shrink after the temporary stop appears to be different from year to year.

097.027 **The Martian global dust storms and extraordinary warming of the atmosphere in the polar night regions.**
S. Moriyama.
13th Lunar and Planetary Symposium, (see 012.018), p. 24 - 30 (1980).

The mechanism of the sudden atmospheric warming of the Martian winter hemisphere, which was observed by the Viking IR temperature map measurements at the time of the global dust storms, is discussed. It is also pointed out that dust storms might be a necessary part of the adjustment process in the Martian atmosphere to respond to the strong radiative cooling in polar regions during the very long winter night of Mars.

097.028 **The seasonal variation of atmospheric pressure on Mars.** Y. Narumi.
13th Lunar and Planetary Symposium, (see 012.018), p. 31 - 40 (1980).

The seasonal variation of atmospheric pressure on Mars is calculated using a model of the Martian surface including the heating effect of the polar hood cloud, and the diurnal behavior of surface temperature at various latitudes are predicted with the same model. The model results are in fairly good agreement with the Viking observations.

097.029 **Surface erosion caused on Mars from Viking descent engine plume.** R. E. Hutton, H. J. Moore, R. F. Scott, R. W. Shorthill, C. R. Spitzer.
Moon Planets, Vol. 23, 293 - 305 (1980).

During the Martian landings the descent engine plumes on Viking Lander 1 (VL-1) and Viking Lander 2 (VL-2) eroded the Martian surface materials. This had been anticipated and investigated both analytically and experimentally during the design phase of the Viking spacecraft. This paper presents data on erosion obtained during the tests of the Viking descent engine and the evidence for erosion by the descent engines of VL-1 and VL-2 on Mars. From these and other results, it is concluded that there are four distinct surface materials on Mars: (1) drift material, (2) crusty to cloddy material, (3) blocky material, and (4) rock.

097.030 **Anomalous region on Mars: implications for near-surface liquid water.**
S. H. Zisk, P. J. Mouginis-Mark.
Nature, Vol. 288, 735 - 738 (1980).

An anomalous region has been identified on Mars from the 1971 and 1973 Earth-based Goldstone radar data. This region is characterized by coincident very high radar reflectivities and unusual smoothness. Very few realistic surface morphologies can generate such return-signals. Liquid water within ~50–100 cm of the surface is one possibility, an interpretation strengthened by an apparent seasonal variation in surface reflectivity for the same locality.

097.031 **Feroxyhyte on Mars?**
K. M. Towe, with a reply by R. G. Burns.
Nature, Vol. 288, 196 (1980).

097.032 **Riddles of the Martian soil and origin of life on the earth.** M. D. Nusinov.
Zemlya Vselennaya, 1980, No. 6, p. 57 - 61. In Russian.

097.033 **Phase curves of Phobos and Deimos.**
K. D. Pang, J. W. Rhoads, G. A. Hanover, K. Lumme, E. Bowell.
Bull. American Astron. Soc., Vol.12, 663 (1980). – Abstract.

097.034 **Ancient crater populations on Mars.**
M. Gurnis.
Bull. American Astron. Soc., Vol. 12, 677 (1980). – Abstract.

097.035 **Possible relations between the early Martian geologic record, convective overturn, and core segregation.** R. E. Arvidson, G. F. Davies.
Bull. American Astron. Soc., Vol. 12, 678 (1980). – Abstract.

097.036 **Mars: constraints on a global subpermafrost groundwater system.** S. M. Clifford, R. L. Huguenin.
Bull. American Astron. Soc., Vol. 12, 678 (1980). – Abstract.

097.037 **Chasma Boreale (85°N, 0°W): remnant of a Martian jökulhlaup?** S. M. Clifford.
Bull. American Astron. Soc., Vol. 12, 678 - 679 (1980). Abstract.

097.038 **Processes of Martian valley formation.**
D. Pieri.
Bull. American Astron. Soc., Vol. 12, 679 (1980). – Abstract.

097.039 **Interpretation of topography of Martian central volcanoes.** K. R. Blasius.
Bull. American Astron. Soc., Vol. 12, 679 (1980). – Abstract.

097.040 **Mars: definition and characterization of global surface units, with compositional emphasis.**
T. B. McCord, R. B. Singer, B. R. Hawke, J. B. Adams, J. W. Head, C. M. Pieters, P. J. Mouginis-Mark, S. H. Zisk.
Bull. American Astron. Soc., Vol. 12, 679 - 680 (1980). Abstract.

097.041 **Spectral variety of Martian surface materials: comparison of earthbased and Viking lander data.**
R. B. Singer, E. L. Strickland.
Bull. American Astron. Soc., Vol. 12, 680 (1980). – Abstract.

097.042 **Mars: new regional near-infrared spectrophotometry (0.65 - 2.50 μm) obtained during the 1980 apparition.** R. B. Singer, R. N. Clark, P. D. Owensby.
Bull. American Astron. Soc., Vol. 12, 680 (1980). – Abstract.

097.043 **Mars north pole surface properties: bistatic radar results.** R. A. Simpson, G. L. Tyler.
Bull. American Astron. Soc., Vol. 12, 680 (1980). – Abstract.

097.044 **A preliminary reference map of Martian bolometric albedos from Viking IRTM (*Infrared Thermal Mapper*) data.** L. Pleskot, E. Miner.
Bull. American Astron. Soc., Vol. 12, 680 - 681 (1980). Abstract.

097.045 **Comparison of Viking Orbiter multispectral images and laboratory reflectance spectra of terrestrial samples.** D. L. Evans, J. B. Adams.
Bull. American Astron. Soc., Vol. 12, 681 (1980). – Abstract.

097.046 **Martian surface properties derived from Viking IRTM spectral differences.**
P. R. Christensen, H. H. Kieffer.
Bull. American Astron. Soc., Vol. 12, 688 (1980). – Abstract.

097.047 **The thermal and radar properties of the Martian surface.** B. M. Jakosky, D. O. Muhleman.
Bull. American Astron. Soc., Vol. 12, 688 - 689 (1980). Abstract.

097.048 **Revised location of the Viking 1 Lander on Mars.**
E. Morris.
Bull. American Astron. Soc., Vol. 12, 689 (1980). – Abstract.

097.049 **Are the soils at the Viking landing sites part of a globally homogenized wind deposit?**
E. A. Guinness.
Bull. American Astron. Soc., Vol. 12, 691 (1980). – Abstract.

097.050 **The search for sun dogs on Mars.**
P. Romani, S. Wall, P. Doherty.
Bull. American Astron. Soc., Vol. 12, 702 (1980). – Abstract.

097.051 **Observations of CO in the Mars atmosphere via its J = 1–0 rotational transition.**
J. C. Good, F. P. Schloerb.
Bull. American Astron. Soc., Vol. 12, 702 - 703 (1980). Abstract.

097.052 **Quantitative analysis of CO_2 lines formed in the lower atmosphere of Mars: comparison with Viking and Mariner ground-truth results.**
T. Kostiuk, F. Espenak, D. Buhl, G. Chin, D. Deming, M. Mumma, D. Glenar, D. Zipoy.
Bull. American Astron. Soc., Vol. 12, 703 (1980). – Abstract.

097.053 **Quantitative analysis of CO_2 non-thermal emission lines formed in the upper atmosphere of Mars.**
M. Mumma, T. Kostiuk, D. Deming, F. Espenak, G. Chin, D. Buhl, D. Zipoy.
Bull. American Astron. Soc., Vol. 12, 703 (1980). – Abstract.

097.054 **Application of a radiative-convective model to Mars and its implications for the Martian general circulation.** K. B. Ronnholm, C. B. Leovy.
Bull. American Astron. Soc., Vol. 12, 703 - 704 (1980). Abstract.

097.055 **Mars rotation: detection of seasonal rate variation.**
R. D. Reasenberg, T. M. Eubanks, P. E. MacNeil, I. I. Shapiro.
Bull. American Astron. Soc., Vol. 12, 721 (1980). – Abstract.

097.056 **Martian south polar cap boundary: 1971 and 1973 terrestrial observations.** P. B. James, K. Lumme.
Bull. American Astron. Soc., Vol. 12, 722 (1980). – Abstract.

097.057 **Mars: north polar cap annual radiation budget.**
D. A. Paige.
Bull. American Astron. Soc., Vol. 12, 722 (1980). – Abstract.

097.058 **Digital map of eolian features on Mars.**
A. W. Ward, P. J. Helm, N. Witbeck, M. Weisman.
Bull. American Astron. Soc., Vol. 12, 722 - 723 (1980). Abstract.

097.059 **Mars: comparison of surface wind patterns at high northern and southern latitudes.** P. Thomas.
Bull. American Astron. Soc., Vol. 12, 723 (1980). – Abstract.

097.060 **Field investigations of dark streaks.**
J. Zimbelman, H. Kieffer.
Bull. American Astron. Soc., Vol. 12, 723 (1980). – Abstract.

097.061 **Salt storms on Mars?** R. L. Huguenin.
Bull. American Astron. Soc., Vol. 12, 723 - 724 (1980). – Abstract.

097.062 **Inferring global atmospheric dust opacity from Viking Lander 1 pressure observations.**
R. W. Zurek.
Bull. American Astron. Soc., Vol. 12, 724 (1980). – Abstract.

097.063 **Arsia-Pavonis Montes bore wave: three Mars years of observations.** N. Evans.
Bull. American Astron. Soc., Vol. 12, 724 (1980). – Abstract.

097.064 **A possible triggering mechanism for Solis Lacus dust storms.** L. J. Martin.
Bull. American Astron. Soc., Vol. 12, 724 (1980). – Abstract.

097.065 **Some effects of global dust storms on the atmospheric circulation of Mars.**
R. M. Haberle, C. B. Leovy, J. B. Pollack.
Bull. American Astron. Soc., Vol. 12, 725 (1980). – Abstract.

097.066 **Martian atmospheric lee waves.**
P. Clark, P. Gierasch.
Bull. American Astron. Soc., Vol. 12, 725 (1980). – Abstract.

097.067 **Slope winds in Tharsis – Viking observations over a Martian year.** S. W. Lee, P. C. Thomas, J. Veverka.
Bull. American Astron. Soc., Vol. 12, 725 (1980). – Abstract.

097.068 **Numerical studies of the airflow over the Martian volcanoes.** A. O. Pickersgill, G. E. Hunt.
Bull. American Astron. Soc., Vol. 12, 725 (1980). – Abstract.

097.069 **Baroclinic waves in the Martian atmosphere: two Mars years.** J. R. Barnes.
Bull. American Astron. Soc., Vol. 12, 725 - 726 (1980). Abstract.

097.070 **Mesoscale wind generation on Mars.**
R. Rankin, G. Eckerman, A. Peterfreund.
Bull. American Astron. Soc., Vol. 12, 726 (1980). – Abstract.

097.071 **Near-simultaneous Viking and earth-based observations of Mars in 1978.** L. J. Martin, W. A. Baum.
Bull. American Astron. Soc., Vol. 12, 726 (1980). – Abstract.

097.072 **Return of the Martian Rima Tenuis.** C. Capen.
Bull. American Astron. Soc., Vol. 12, 726 - 727 (1980). – Abstract.

097.073 **Theoretical interpretation of photometric properties of the Martian surface and atmosphere.**
K. Lumme, L. J. Martin, W. A. Baum.
Bull. American Astron. Soc., Vol. 12, 727 (1980). – Abstract.

097.074 **Is er leven op Mars?** F. P. Israel.
Zenit, 7e Jaarg., 174 - 183 (1980).

097.075 **Photographic observations of Mars during the 1977 - 1978 opposition.** T. Akabane, M. Mastui, K. Ishiura, A. Hattori, K. Iwasaki, Y. Saito, T. Asada, S. Saito, Y. Nakai.
Contrib. Kwasan Hida Obs., Univ. Kyoto, No. 242, 27 pp. (1980).

This is the report of photographic observations of Mars at the Kwasan and Hida Observatories during the 1977-1978 apparition. In the observation period, a dust storm occurred at Tempe region. Daedalia and Syrtis exhibited almost the same features of albedo as those in the 1975 apparition. The north polar cap was observed at Ls = 14° for the first time.

097.076 **Martian Valleys: morphology, distribution, age, and origin.** D. C. Pieri.
Science, Vol. 210, 895 - 897 (1980).

Branching valley networks throughout the heavily cratered terrain of Mars exhibit no compelling evidence for formation by rainfall-fed erosion. Rather, the deeply entrenched canyons suggest headward extension by basal sapping. The size-frequency distributions of impact craters in these valleys and in the heavily cratered terrain that surrounds them are statistically indistinguishable, suggesting that valley formation has not occurred on Mars for billions of years.

097.077 **Monografías sobre planetas: Marte.**
A. J. Camponovo.
Rev. Astron., No. 213, p. 2 - 12 (1980).

097.078 **The geological history of Mars.**
Priroda, 1980, No. 12, p. 90 - 91. In Russian.

097.079 **The interaction of the solar wind with Mars, Venus and Mercury.** C. T. Russell.
Solar system plasma physics, Vol. 2, (see 003.010), 207 - 252 (1979).

The author summarizes the history of investigation of the solar wind interaction with the inner planets. He examines Mars, Venus and Mercury. He compares what we know about the interaction of the solar wind with each of these planets and attempts to define a magnetic hierarchy for the terrestrial planets. He concludes with an outline of the remaining problems in this area.

097.080 **Is there and has there been life on Mars?**
M. D. Nusinov.
Zh. Vses. khim. o-va im. D. I. Mendeleeva, Vol. 25, 447 - 451 (1980). In Russian. – Abstr. in Ref. zh., 51. Astron., 11.51.5 (1980).

097.081 **The gravitational field and figure of Mars.**
A. L. Tserklevich, Eh. M. Evseeva.
Izv. AN SSSR. Fiz. Zemli, 1980, No. 7, p. 3 - 15. In Russian. Abstr. in Ref. zh., 51. Astron., 11.51.279; 52. Geod. Aehrosemka, 11.52.163 (1980).

097.082 **Determination of the depth of icy rocks on Mars from the morphology of new craters.**
R. O. Kuz'min.
Dokl. AN SSSR, Vol. 252, 1445 - 1448 (1980). In Russian. Abstr. in Ref. zh., 51. Astron., 11.51.280 (1980).

097.083 **On the classification of ring structures of Mars.**
V. L. Avdeev, Ya. G. Kats.
Kosmogen. struktury Zemli. Mater. semin. Moskva, 1980, p. 17 - 22. In Russian. – Abstr. in Ref. zh., 62. Issled. kosm. prostranstva, 11.62.70 (1980).

097.084 **Variation of the parameters of optical details simulating a Martian dust storm.**
A. I. Akshin, V. K. Baranov, D. A. Karakulev, V. P. Kiryukhin, L. P. Tereshina, G. P. Tikhomirov.
Opt.-mekh. prom-st', 1980, No. 6, p. 10 - 12. In Russian. Abstr. in Ref. zh., 62. Issled. kosm. prostranstva, 11.62.71 (1980).

097.085 **Significant achievements in the Planetary Geology Program, 1980.**
H. E. Holt (Editor), with contributions by G. E. McGill, P. Mouginis-Mark, C. A. Hodges, R. L. Huguenin, D. Nummedal, J. A. Cutts, R. Greeley, K. L. Jones,

L. A. Soderblom, R. G. Strom, M. C. Malin, R. E. D'Alli.
NASA Tech. Memo, NASA TM 82384, 5 + 32 pp. (1980).

Recent developments in planetology research as reported at the 1980 NASA Planetology Program Principal Investigators meeting are summarized. Important developments are summarized in topics ranging from solar system evolution and comparative planetology to geologic processes active on other planetary bodies.

097.086 **Comparison of Viking Lander multispectral images and laboratory reflectance spectra of terrestrial samples.** D. L. Evans, J. B. Adams.
Proc. Tenth Lunar Planet. Sci. Conf., (see 012.050), p. 1829 - 1834 (1979).

097.087 **Mars: large scale mixing of bright and dark surface materials and implications for analysis of spectral reflectance.** R. B. Singer, T. B. McCord.
Proc. Tenth Lunar Planet. Sci. Conf., (see 012.050), p. 1835 - 1848 (1979).

097.088 **Ejecta emplacement of the martian impact crater Bamburg.** P. J. Mouginis-Mark.
Proc. Tenth Lunar Planet. Sci. Conf., (see 012.050), p. 2651 - 2668 (1979).

097.089 **The chronology of the martian volcanoes.**
J. B. Plescia, R. S. Saunders.
Proc. Tenth Lunar Planet. Sci. Conf., (see 012.050), p. 2841 - 2859 (1979).

097.090 **Eolian streaks in southwestern Egypt and similar features in the Cerberus region of Mars.**
F. El-Baz, T. A. Maxwell.
Proc. Tenth Lunar Planet. Sci. Conf., (see 012.050), p. 3017 - 3030 (1979).

097.091 **Origin of Valles Marineris.** E. Schonfeld.
Proc. Tenth Lunar Planet. Sci. Conf., (see 012.050), p. 3031 - 3038 (1979).

097.092 **Geologic problems in the northern plains of Mars.**
D. H. Scott.
Proc. Tenth Lunar Planet. Sci. Conf., (see 012.050), p. 3039 - 3054 (1979).

097.093 **Martian soil stratigraphy and rock coatings observed in color-enhanced Viking Lander images.**
E. L. Strickland III.
Proc. Tenth Lunar Planet. Sci. Conf., (see 012.050), p. 3055 - 3077 (1979).

097.094 **Some physical characteristics of local clouds on Mars deduced from a multicolour TV photometry.**
V. A. Fenchak.
Astron. Tsirk. No. 1098, p. 5 - 6 (1980). In Russian.

097.095 **Statistic characteristics of the radar backscattering for some Martian surface zones.**
V. I. Koptyaev.
Astron. Zh., Tom 57, 1317 - 1319 (1980). In Russian.
English translation in Soviet Astron., Vol. 24, No. 6.

A statistic processing of published radar data for the Martian surface is carried out for five surface zones. The processing results are used for estimating the average backscattering characteristics of these zones.

097.096 **On an investigation of the global density distribution of the Martian interior.**
Yu. P. Dejneka, A. L. Tserklevich.
Geod., kartogr. i aehrofotosemka, L'vov, 1980, No. 32, p. 26 - 33. In Russian. – Abstr. in Ref. zh., 52. Geod. Aehrosemka, 12.52.153 (1980).

097.097 **On a possible interpretation of the Martian gravitational field.**
G. A. Meshcheryakov, Yu. P. Dejneka, A. L. Tserklevich.
Geod., kartogr. i aehrofotosemka, L'vov, 1980, No. 32, p. 96 - 100. In Russian. – Abstr. in Ref. zh., 52. Geod. Aehrosemka, 12.52.154 (1980).

097.098 **Television photometry of cloudy masses on Mars during the period from August 4 to September 5, 1971.** V. A. Fenchak.
Astron. Vestn., Tom 13, 87 - 93 (1979). In Russian.

The results of photometric reduction of 300 television pictures of Mars taken in four spectral regions ($\lambda\lambda_{eff}$ 377, 448, 546, 642 nm) are presented. The relative distribution of energy in the spectrum of the radiation reflected by clouds located over the planet's bright regions relative to the adjacent bright regions free from clouds is obtained. Spectral photometric characteristics of clouds at the limb of the planet's disk and their time variations are obtained.

Viking Orbiter Stereo Imaging Catalog.
See Abstr. 002.003.

Viking Orbiter views of Mars.
See Abstr. 002.071.

Atlas van Maan, Mars, Venus.
See Abstr. 002.084.

The Martian surface. See Abstr. 003.101.

The case against UV photostimulated oxidation of magnetite. See Abstr. 022.076.

Altered basaltic glass – a Martian soil analog.
See Abstr. 022.134.

The formation of the Martian soil by hydrothermal alteration of impact melt sheets. See Abstr. 022.135.

Statistical reconstruction of a Martian scene; G-mode cluster analysis results from multispectral data population.
See Abstr. 031.554.

Determination of the initial entry conditions of a space vehicle into the Martian atmosphere using a detached navigation probe. See Abstr. 052.011.

On the nature of obstacles decelerating the solar wind near Venus and Mars and on peculiarities of interaction between the solar wind and the atmospheres of these planets.
See Abstr. 074.022.

Escape of the atmosphere of a rotating planet.
See Abstr. 091.014.

An assessment study on non terrestrial frozen ground.
See Abstr. 091.018.

Mercurian crater rim heights and some interplanetary comparisons. See Abstr. 092.006.

Erratum

097.901 **Erratum: "The origin of polygonal troughs on the northern plains of Mars."** [Icarus, Vol. 42, 185 - 210 (1980)]. J. C. Pechmann.
Icarus, Vol. 43, 230 (1980). – See Abstr. 27.097.180.

098 Minor Planets

098.001 **Rotation period and photoelectric lightcurves of asteroids 68 Leto and 563 Suleika.**
J. Surdej, H. J. Schober.
Astron. Astrophys., Suppl. Ser., Vol. 41, 335 - 338 (1980).
The rotation lightcurve of 68 Leto displays three identifiable maxima and minima with a total amplitude $\Delta V = 0.19$ mag. and the synodic period of rotation is found to be $14^h 50^m 51^s \pm 52^s$. For 563 Suleika the derived synodic period of rotation is $5^h 41^m 31^s \pm 3^m 36^s$ and its lightcurve shows two nearly symmetric maxima and minima within an amplitude range $\Delta V = 0.21$ mag. The color indices measured for 68 Leto and 563 Suleika are, respectively, $B-V = 0.839 \pm 0.022$ mag., $U-B = 0.472 \pm 0.025$ mag. and $B-V = 0.891 \pm 0.017$ mag., $U-B = 0.459 \pm 0.030$ mag.

098.002 **Observations of minor planets.** W. Landgraf.
Astron. Nachr., Band 301, 195 (1980). In German.
11 positions for minor planets (2), (6), (246), and (386) are given.

098.003 **Observations of minor planets at Dresden Lohrmann Observatory.** E. Asenjo, D. Böhme, K.-G. Steinert.
Astron. Nachr., Band 301, 197 - 199 (1980) = Mitt. Lohrmann Obs., Tech. Univ. Dresden, No. 43.

098.004 **The unusual asteroid 216 Kleopatra.**
D. J. Tholen.
Sky Telesc., Vol. 60, 203 (1980).

098.005 **Positions of the minor planet No. 433 Eros in the years 1974 - 1975.** M. Antal.
Contrib. Astron. Obs. Skalnaté Pleso, Vol. 9, 145 - 162 (1980).
The paper contains results of photographic positional observations of the minor planet No. 433 Eros made from 17 November 1974 to 11 February 1975. Some tens of the observations were obtained and the accurate topocentric positions of Eros were determined from the 22 best of all. The measurements were based on the data about reference stars taken from the Smithsonian Star Catalogue. The Schlesinger method of dependences was used for reductions of the measurements. Except the resulting 22 accurate positions of Eros the paper contains the dependences for the reference stars for each triangle.

098.006 **Speckle interferometric observations of Pallas.**
E. K. Hege, W. J. Cocke, E. Hubbard, M. Gresham, P. A. Strittmatter, S. P. Worden, R. Radick.
Bull. American Astron. Soc., Vol. 12, 509 (1980). – Abstract.

098.007 **Photographic position observations of minor planet Eros in 1975.** P. P. Pavlenko, L. S. Pavlenko.
Vestn. Khar'k. Univ., No. 190. Fiz. Luny Planet. Fundam. Astrometr., Vyp. 14, p. 9 (1979). In Russian.

098.008 **Positions d'astéroïdes au GPO, ESO, La Silla, 1978, 1979.**
H. Debehogne, L. E. Machado, J. F. Caldeira, G. G. Vieira, E. R. Netto.
Astron. Astrophys., Suppl. Ser., Vol. 42, 81 - 83 (1980).
81 precise positions of minor planets are given.

098.009 **Photoelectric lightcurves of the asteroids 139 Juewa and 161 Athor, obtained with the 50 cm photometric telescope at ESO, La Silla.**
H. Debehogne, V. Zappalà.
Astron. Astrophys., Suppl. Ser., Vol. 42, 85 - 89 (1980).
The rotation period of 139 Juewa, if one assumes its lightcurve to have two maxima and two minima per rotation cycle, is twice that reported by Goguen et al. (1976), namely 41.8 hours. The mean $B-V$ and $U-B$ colors are found to be $0^m\!.69 \pm 0^m\!.01$ and $0^m\!.29 \pm 0^m\!.02$, respectively. The lightcurve of 161 Athor is typical of that for a regular figure of revolution. The rotation period is $7^h\!.288 \pm 0^h\!.007$, with a maximum amplitude of 0.27 mag. The authors determined the phase coefficient and mean colors to be $\beta_V = 0.038$ mag/deg, $B-V = 0^m\!.72 \pm 0^m\!.02$ and $U-B = 0^m \pm 0^m\!.03$.

098.010 **Recovery of the long lost minor planet (1370) Hella.** L. D. Schmadel.
Astron. Nachr., Band 301, 251 - 252 (1980).
The orbit of minor planet (1370) Hella, which was supposed to be almost hopelessly lost since the discovery apparition in 1935, is rediscussed. The old plates have been remeasured and reduced anew and the object was recovered in December 1979 by means of these elements. An improved orbit based on 11 positions 1935 - 1979 is given.

098.011 **Positions of minor planets.**
M. Gressmann.
Astron. Nachr., Band 301, 253 - 255 (1980). In German.
A table contains topocentric positions of sixteen numbered and three unnumbered minor planets.

098.012 **Theorie der Bahnbewegung der Planetoiden Ceres, Pallas, Juno und Vesta.** W. Höppner.
Mitt. Astron. Ges., Nr. 48, (see 012.015), p. 87 (1980).

098.013 **Ist das Objekt (2060) Chiron ein kleiner Planet oder ein Komet?** H. Scholl.
Mitt. Astron. Ges., Nr. 48, (see 012.015), p. 91 (1980).

098.014 **Asteroid rotation. III. 1978 observations.**
A. W. Harris, J. W. Young.
Icarus, Vol. 43, 20 - 32 (1980).
Photoelectric observations of 32 asteroids observed from Table Mountain Observatory during the second half of 1978 are reported. Rotation periods were obtained for most objects. Absolute magnitudes and phase functions were not determined for any of these asteroids. The geometric mean rotation period of the 32 asteroids observed is 14.2 ± 1.6 hr, as compared to 9.38 ± 0.35 hr for 182 asteroids analyzed in Paper I.

098.015 **Rotational properties of asteroids: correlations and selection effects.** E. F. Tedesco, V. Zappalà.
Icarus, Vol. 43, 33 - 50 (1980).
The purpose of this paper is twofold: (1) to discuss the selection effects present in asteroid rotational data and (2) to indicate several of the more obvious features displayed by these data.

098.016 **Scoperto il pianetino 1979 SA e ritrovato il pianetino 1975 TN dagli astrofili dell'Osservatorio S. Vittore (Bologna).** E. Colombini, V. Goretti, E. Pancaldi, G. Sassi, G. Sette, C. Vacchi.
Astronomia, N. 3, p. 23 - 29 (1980).

098.017 **Ultraviolet spectra of asteroids.** P. S. Butterworth, A. J. Meadows, G. E. Hunt, V. Moore, D. M. Willis.
Nature, Vol. 287, 701 - 703 (1980).
Spectra over the range 2,100–3,200 Å have been obtained for Ceres, Pallas and Vesta using the International Ultraviolet Explorer satellite. These represent the first detailed UV reflectivity data for asteroids.

098.018 **Posições do asteróide 14 Irene obtidas com a câmara astrográfica Pazos (d = 11.6 cm; f = 1.83 m) do Observatório do Valongo na Estação Astronòmica Municipal do Pico das Cabras, SP, junho 1980.**
L. E. Machado, J. F. Caldeira, E. R. Netto, G. G. Vieira, N. Travnik.
Bol. Astron., Vol. 1, No. 1, p. 5 - 7 (1980).

098.019 **Gli asteroidi che «sfiorano» la Terra.** V. Zappalà.
Orione, Vol. 2, 116 - 121 (1980).

098.020 **A close encounter of two minor planets.** G.-i. Hori.
13th Lunar and Planetary Symposium, (see 012.018), p. 144 - 148 (1980).

098.021 **A comparative analysis of some optical characteristics of asteroids and of the moon.**
L. F. Golubeva, D. I. Shestopalov, Yu. G. Shkuratov.
Astron. Zh., Tom 57, 1047 - 1055 (1980). In Russian.
English translation in Soviet Astron., Vol. 24, No. 5.

Diagrams "albedo $\rho(0.62\mu m)$ vs colour $C(0.62/0.38)$" and "colour $C(0.95/0.62)$ vs colour C (0.62/0.38)" for asteroids with known values of albedo and spectral characteristics have been constructed. A class of asteroids with combinations of optical characteristics specific for the moon's surface has been selected. The low extent of maturation of the regolith of the majority of asteroids has been confirmed.

098.022 **On the asymmetry of distribution of asteroids near exact commensurabilities.** S. G. Zhuravlev.
Astron. Zh., Tom 57, 1056 - 1062 (1980). In Russian.
English translation in Soviet Astron., Vol. 24, No. 5

Results of recent numerical, statistical and qualitative investigations of the motion of asteroids near exact commensurabilities enable one to conclude that an observed asymmetry of gaps relative to the exact commensurabilities is due to a resonant gravitational interaction of asteroids and Jupiter.

098.023 **Peculiar shapes of asteroids: implications for light curves and periods of rotation.** V. Zappalà.
Moon Planets, Vol. 23, 345 - 353 (1980).

This paper presents some simple geometrical models of asteroids with theoretical light curves similar to the observed ones. In some cases the results suggest rotation periods to be double those one can obtain adopting two- or three-axial ellipsoids as models. A possible model, not in terms of a binary system, for asteroids with light curves like those of eclipsing binary stars, is also given.

098.024 **Orbital elements of some unnumbered asteroids.**
J.-x. Yang, S.-l. Wei, Q. Wang, J.-q. Zheng, Y.-l. Ge, C.-l. Yuan, J.-x. Zhang.
Astron. Circ., No. 7, p. 1, 3 - 8, 10 (1980). – In Chinese and English.

098.025 **Spectroscopy of faint asteroids, satellites, and comets.** J. Degewij.
Astron. J., Vol. 85, 1403 - 1412 (1980).

Nineteen asteroids with orbital elements comparable to those of short-period comets and the outer Jovian satellites J6 Himalia and J7 Elara have been observed with the 228-cm telescope and image-tube spectrograph of Steward Observatory. No activity indicating cometary outgassing was detected. If comets are being perturbed into asteroidal orbits, then this lack of activity can be explained by an apparently short transition time between active and extinct phase. In addition, spectra of five faint comets were obtained, showing CN(0,0) and in some cases C_2 (1,0) and C_3 emission.

098.026 **What's in a name? The nomenclature of minor planets.** P. M. Kilmartin.
South. Stars, Vol. 28, 121 - 126 (1980).

098.027 **Photoelectric observations of MP 4 Vesta, 1978 May 19 - 20 U. T.**
B. F. Marino, W. S. G. Walker, G. Herdman.
South. Stars, Vol. 28, 134 - 136 (1980).

098.028 **The remaining large minor planets with unknown rotational properties: 31 Euphrosyne and 65 Cybele.**
H. J. Schober, F. Scaltriti, V. Zappalà, A. W. Harris.
Astron. Astrophys., Vol. 91, 1 - 6 (1980).

The authors present the results of photoelectric photometry on the asteroids 31 Euphrosyne and 65 Cybele, the last two asteroids in the size range of $D \gtrsim 250$ km with unknown periods of rotation. This is an important result in view of statistical studies on the collisional evolution of the Main Belt Objects; some preliminary considerations are made on this topic. The rational properties of 31 Euphrosyne and 65 Cybele were found to be quite similar, both in period P and maximum amplitude Δm. Evidence was found for an opposition effect for 65 Cybele. The $B - V$ and $U - B$ colors were determined. Recent observations of 65 Cybele occulting a star have shown evidence of a possible satellite for this minor planet.

098.029 **Positional observations of minor planets and comets at the Astronomical Observatory of the Silesian Planetarium.** I. Włodarczyk.
Postępy Astron., Tom 28, 231 (1980). In Polish.

098.030 **Conjonctions de petites planètes avec des étoiles brillantes.** J. Meeus.
Astronomie, Vol. 94, 505 - 507 (1980).

098.031 **Radar detection of Vesta.**
S. J. Ostro, D. B. Campbell, G. H. Pettengill, I. I. Shapiro.
Icarus, Vol. 43, 169 - 171 (1980).

Asteroid 4 Vesta was detected on 1979 November 6 with the Arecibo Observatory's S-band (12.6-cm-wavelength) radar. The echo power spectrum, received in the circular polarization opposite to that transmitted, yields a radar cross section $(0.2 \pm 0.1)\ \pi a^2$, for $a = 272$ km. The data are too noisy to permit derivation of Vesta's rotation period.

098.032 **Surface texture of Vesta from optical polarimetry.**
T. Le Bertre, B. Zellner.
Icarus, Vol. 43, 172 - 180 (1980).

Polarimetric, photometric, and reflectance spectroscopic properties of asteroid 4 Vesta are simulated in the laboratory by a preparation of eucrite Bereba consisting of a broad mixture of particle sizes (mainly greater than 50-μm) mixed and partially coated with particles of size 10 μm and less. Photometrically, if coating a sphere, this sample shows a constant brightness on the sunward half of the observed hemisphere, the brightness being given on the other half by the Minnaert reciprocity principle. With such a photometric behavior, the global geometric albedo and the sub-Earth point geometric albedo differ by no more than 5%. The microscopic phase coefficient β is 0.021 magnitude per degree for the sample; the larger value, $\beta = 0.025$, observed telescopically for Vesta indicates that large-scale roughness is present on this asteroid.

098.033 **The lightcurve and phase function of the asteroid 304 Olga.** A. W. Harris, J. W. Young, E. Bowell.
Icarus, Vol. 43, 181 - 183(1980).

Photoelectric lightcurves of 304 Olga were obtained at Table Mountain Observatory in 1978 near opposition. From these observations, and several observations made from Lowell Observatory a month later, the authors obtain a rotation period of 18.36 ± 0.02 hr and lightcurve amplitude of $0^m.20$. The range of solar phase angle covered by the observations is from 2°.0 to 22°. The resulting phase function is well fit by the Bowell and Lumme model (1979), with $Q = 0.02$. This low value of Q is suggestive of a low-albedo object.

098.034 **Asteroids yesterday and today.**
A. N. Simonenko.
Zemlya Vselennaya, 1980, No. 6, p. 10 - 14. In Russian.

098.035 **Revival of the hypothesis on Phaethon?**
V. A. Bronshtehn.
Zemlya Vselennaya, 1980, No. 6, p. 19 - 20. In Russian.

098.036 **The size and shape of Juno from its occultation of AGK3+0°1022.**
R. L. Millis, L. H. Wasserman, E. Bowell, O. G. Franz.
Bull. American Astron. Soc., Vol. 12, 662 (1980). – Abstract.

098.037 **Possible secondaries of asteroids found by speckle interferometry.** E. K. Hege, W. J. Cocke, E. N. Hubbard, J. Christou, R. Radick.
Bull. American Astron. Soc., Vol. 12, 662 (1980). – Abstract.

098.038 **How fast can an asteroid spin?**
S. J. Weidenschilling.
Bull. American Astron. Soc., Vol. 12, 662 (1980). – Abstract.

098.039 **Creation and destruction of multiple asteroids.**
C. R. Chapman, D. R. Davis, S. J. Weidenschilling.
Bull. American Astron. Soc., Vol. 12, 662 - 663 (1980). Abstract.

098.040 **Observations of selected asteroids with the International Ultraviolet Explorer (IUE).**
G. J. Veeder, D. L. Matson, R. M. Nelson, A. L. Lane, T. V. Johnson, T. B. McCord, M. J. Gaffey.
Bull. American Astron. Soc., Vol. 12, 663 (1980). – Abstract.

098.041 **The spectral reflectance (0.3 - 2.5 μm) of 10 S-type asteroids: evidence for an undifferentiated composition?** M. Feierberg, H. Larson, C. Chapman.
Bull. American Astron. Soc., Vol. 12, 664 (1980). – Abstract.

098.042 **Spectroscopic evidence for aqueous alteration products on the surfaces of low-albedo asteroids.**
M. Feierberg, L. Lebofsky, H. Larson.
Bull. American Astron. Soc., Vol. 12, 664 - 665 (1980). Abstract.

098.043 **The near-earth asteroids: their nature as presently understood from reflectance spectroscopy.**
L. A. McFadden.
Bull. American Astron. Soc., Vol. 12, 665 (1980). – Abstract.

098.044 **1979 VA: physical parameters of a possible cometary nucleus.** J. Degewij, J. G. Williams, D. P. Cruikshank, M. J. Gaffey, E. F. Helin, H. Spinrad, J. Stauffer, E. F. Tedesco, D. J. Tholen.
Bull. American Astron. Soc., Vol. 12, 665 (1980). – Abstract.

098.045 **Radar observations of asteroid 1685 Toro.**
S. J. Ostro, I. I. Shapiro.
Bull. American Astron. Soc., Vol. 12, 666 (1980). – Abstract.

098.046 **The Lincoln Laboratory earth-crossing asteroid search.** L. G. Taff.
Bull. American Astron. Soc., Vol. 12, 666 (1980). – Abstract.

098.047 **General report of position observations by the A.L.P.O. Minor Planets Section for the year 1979.**
F. Pilcher.
Minor Planet Bull., Vol. 7, 21 - 26 (1980).

098.048 **657 Gunlöd – a main belt asteroid with a large amplitude lightcurve.** F. Pilcher.
Minor Planet Bull., Vol. 7, 26 - 27 (1980).

The minor planet 657 Gunlöd was found by visual observation in February and March, 1980, to have a large amplitude lightcurve near 1^m.0. The rotation period could not be uniquely determined but may be near 15^h or near 11^h.

098.049 **Minor planet rotations 1978 - 1979.**
A. C. Porter, D. Wallentinsen.
Minor Planet Bull., Vol. 7, 27 - 29 (1980).

098.050 **Special observer's supplement.**
Minor Planet Bull., Vol. 7, 33 - 34 (1980).

098.051 **Observations of minor planets and comets.**
Minor Planet Circ. (M.P.C.), Nos. 5392 - 5394, 5395 - 5409, 5424 - 5439, 5455 - 5515, 5541 - 5596, 5605 - 5637 (1980).

Observations made at the following stations are published: Belgrade, Byurakan, Caussols, Cerro el Roble, Chorzow, Coonabarabran, Crimea (39th and 40th report), El Leoncito, European South. Obs., Falkensee, Geisei, Gissar, Goethe Link Obs., Harvard Agassiz, Haute Provence, Hoher List, JCPM Hamatonbetsu Stat., Kambah (Canberra), Kazan, Kitab, Kleť, La Seyne sur Mer, Leiden South. Stat., Lincoln Lab., Lowell Obs., Lowell Obs. Anderson Mesa Stat., Mauna Kea, Meschede, Mt. Palomar, Mt. Wilson, Peking, Perth, Pino Torinese, Potsdam, Reintal, S. Vittore (Bologna), Sendai, Skalnate Pleso, Sonneberg, Stakenbridge, Tartu, Tautenburg, Tokai, Tokyo Obs. Kiso Stat., Traunstein, Turku, Uccle, Uppsala South. Stat., Woolston Obs., Yatzugatake Obs., Yerkes Obs., Zimmerwald.

098.052 **Ephemerides of minor planets and comets.**
Minor Planet Circ., (M.P.C.), Nos. 5418 - 5422, 5452 - 5454, 5525 - 5540, 5603 - 5604, 5651 - 5658 (1980).

098.053 **Orbital elements of one-opposition minor planets.**
Minor Planet Circ., (M.P.C.), Nos. 5408 - 5409, 5439 - 5440, 5515 - 5517, 5596 - 5597, 5637 - 5638.

098.054 **New names of minor planets.**
Minor Planet Circ., (M.P.C.), Nos. 5449 - 5452, 5523 - 5525 (1980).

098.055 **Identifications and identification changes of minor planets.**
Minor Planet Circ., (M.P.C.), Nos. 5394 - 5395, 5423 - 5424, 5455, 5542, 5605 (1980).

098.056 **Orbital elements of numbered minor planets.**
Minor Planet Circ., (M.P.C.), Nos. 5391 - 5658 (1980).

The minor planets are listed according to their definitive number. Newly numbered objects are indicated by an asterisk. The names of the authors are given behind the respective M.P.C. numbers. (682) 5517 L. K. Kristensen; (937) 5409, (975) 5409 P. Herget; (1037) 5413, (1198) 5640 B. G. Marsden; (1260) 5409, (1275) 5409, (1306) 5410, (1370) 5410, (1480) 5410, (1489) 5410, (1619) 5410, (1626) 5410 P. Herget; (1627) 5640 B. G. Marsden; (1638) 5411, (1642) 5411, (1650) 5411, (1651) 5411 P. Herget; (1685) 5441, (1862) 5641, (1865) 5641 B. G. Marsden; (2035) 5412 W. Landgraf; (2060) 5412 G. Sitarski; (2219) 5415 - 5416 C. M. Bardwell; (2265)* - (2267)* 5413 - 5414 B. G. Marsden; (2268)* - (2270)* 5416 - 5417 C. M. Bardwell; (2271)* 5440 S. Nakano, T. Urata; (2272)* - (2275)* 5441 - 5442 B. G. Marsden; (2276)* - (2289)* 5443 - 5447 C. M. Bardwell; (2290)* - (2291)* 5519 B. G. Marsden; (2292)* - (2293)* 5520 - 5521 C. M. Bardwell; (2294)* - (2295)* 5597 - 5598 S. Nakano, T. Urata; (2296)* 5599 B. G. Marsden; (2297)*

5600 - 5601 C. M. Bardwell; (2298)* - (2305)* 5641 - 5643 B. G. Marsden; (2306)* - (2319)* 5644 - 5648 C. M. Bardwell; (2320)* - (2321)* 5649 - 5650 S. Nakano, T. Urata.

098.057 **Orbital elements of unnumbered minor planets.** Minor Planet Circ., (M.P.C.), Nos. 5391 - 5658 (1980).

The unnumbered minor planets are listed according to their preliminary designation. Objects from the Palomar-Leiden Survey are sorted by number. The names of the authors are given behind the respective M.P.C. numbers. [1928 QB] 5648, [1928 TK] 5417, [1939 PM] 5417 C. M. Bardwell; [1940 GN] 5644 B. G. Marsden; [1941 SS] 5417, [1950 DL] 5447 - 5448 C. M. Bardwell; [1951 RL] 5414 B. G. Marsden; [1953 GM] 5601, [1959 RJ] 5648 C. M. Bardwell; [1961 RA] 5599 B. G. Marsden; [1962 HD] 5417 - 5418, [1964 VY] 5448 C. M. Bardwell; [1964 VM_1] 5599, [1965 LA] 5442 - 5443 B. G. Marsden; [1970 PL] 5521 C. M. Bardwell; [1971 UG_1] 5519, [1971 UM_1] 5599 - 5600 B. G. Marsden; [1972 NC] 5521 C. M. Bardwell; [1972 TF_2] 5650 S. Nakano, T. Urata; [1972 TL_2] 5414 - 5415, [1973 SO_2] 5600 B. G. Marsden; [1974 FG] 5601 C. M. Bardwell; [1974 OS] 5517 S. Nakano, T. Urata; [1974 QE_1] 5601 C. M. Bardwell; [1974 RV_1] 5440, [1974 SJ] 5518 S. Nakano, T. Urata; [1974 SU_4] 5521 - 5522, [1974 TA_1] 5448, [1975 FW] 5649, [1975 RB] 5448 C. M. Bardwell; [1975 VO_2] 5598 S. Nakano, T. Urata; [1975 VD_3] 5418, [1975 WL_1] 5448 - 5449 C. M. Bardwell; [1975 XA_3] 5518 S. Nakano, T. Urata; [1976 AG] 5602, [1976 GQ_1] 5522, [1976 GU_2] 5522 C. M. Bardwell; [1976 GJ_3] 5600 B. G. Marsden; [1976 UH_1] 5418, [1976 UW_{15}] 5602 C. M. Bardwell; [1976 YQ_2] 5520, [1977 ET_1] 5600 B. G. Marsden; [1977 PY_1] 5598, [1977 PZ_1] 5441, [1977 QX_2] 5441, [1977 QM_3] 5518 S. Nakano, T. Urata; [1977 QP_4] 5522 C. M. Bardwell; [1977 RA] 5415, [1977 UP] 5520 B. G. Marsden; [1978 RH] 5602, [1978 SP] 5602, [1978 VG_3] 5418 C. M. Bardwell; [1978 VJ_7] 5518 S. Nakano, T. Urata; [1979 BA] 5443 B. G. Marsden; [1980 KJ] 5650, [1980 LD] 5518 - 5519 S. Nakano, T. Urata; [1980 OB] 5522, [1980 OC] 5523 C. M. Bardwell; [1980 OE] 5651, [1980 OH] 5651 S. Nakano, T. Urata; [1980 PA] 5520 B. G. Marsden; [1980 PN] 5649, [1980 RY] 5649, [2533 P-L] 5523, [7071 P-L] 5603 C. M. Bardwell.

098.058 **(1943) Anteros.**
IAU Circ., No. 3492 (1980).

098.059 **(216) Kleopatra.**
IAU Circ., No. 3495 (1980).

098.060 **1980 LB.**
IAU Circ., No. 3495 (1980).

098.061 **1980 PA.**
IAU Circ., Nos. 3499, 3501, 3505, 3512 (1980).

098.062 **1980 QA.**
IAU Circ., No. 3506 (1980).

098.063 **Occultation of SAO 75392 by (78) Diana on 1980 September 4.**
IAU Circ., No. 3508 (1980).

098.064 **1980 RG_1.**
IAU Circ., Nos. 3522, 3553 (1980).

098.065 **(201) Penelope.**
IAU Circ., Nos. 3523, 3527 (1980).

098.066 **(1862) Apollo.**
IAU Circ., No. 3526 (1980).

098.067 **(1865) Cerberus.**
IAU Circ., No. 3540 (1980).

098.068 **1980 WF.**
IAU Circ., Nos. 3549, 3550 (1980).

098.069 **(1943) Anteros.**
Yamamoto Circ., No. 1939 (1980).

098.070 **1980 PA.**
Yamamoto Circ., Nos. 1940, 1941, 1943 (1980).

098.071 **(216) Kleopatra.**
Yamamoto Circ., No. 1940 (1980).

098.072 **1980 QA.**
Yamamoto Circ., No. 1942 (1980).

098.073 **Planetoïde (2145) Blaauw.**
Zenit, 7e Jaarg., 169 (1980).

098.074 **Merkwaardige Apollo-planetoïde.**
Zenit, 7e Jaarg., 243 (1980).

098.075 **Planetoïde 1862 Apollo dicht langs de Aarde.**
E. P. Bus, G. Comello, H. Feijth.
Zenit, 7e Jaarg., 442 - 445 (1980).

098.076 **Occultation of AGK3 +35°0669 by 187 Lamberta on 1980 November 11.**
British Astron. Assoc. Circ., No. 610 (1980).

098.077 **1980 WF.**
Yamamoto Circ., No. 1949 (1980).

098.078 **1976 WA.**
IAU Circ., No. 3554 (1980).

098.079 **Positionen Kleiner Planeten 1977.**
M. Gressmann.
Astron. Nachr., Band 299, 223 - 224 (1978).

Topocentric positions of six minor planets are given.

098.080 **Return of the feathered serpent.**
E. Goffin, J. Meeus.
Sky Telesc., Vol. 60, 460 - 461 (1980).

Concerning minor planet (1915) Quetzalcoatl.

098.081 **Physical studies of asteroids – an observing programme at ESO.** C.-I. Lagerkvist.
Messenger, No. 22, p. 5 - 7 (1980).

098.082 **Asteroidi.** V. Zappalà.
G. Astron., Vol. 6, 241 - 251 (1980).

098.083 **The asteroids.** J. Gradie.
News Lett. Astron. Soc. N. Y., Vol. 1, No. 7, p. 7 - 8 (1980). – Abstract.

098.084 **Radar detection of Vesta.** S. J. Ostro.
News Lett. Astron. Soc. N. Y., Vol. 1, No. 7, p. 22 (1980). – Abstract.

098.085 **Positions of minor planets obtained at the Chorzów Observatory.** I. Włodarczyk.
Acta Astron., Vol. 30, 469 - 470 (1980).

The paper contains results of photographic observations of the minor planets (2) Pallas, (15) Eunomia, and (324) Bamberga made in the years 1977 and 1978.

098.086 **Observations of minor planets at Abastumani.** R. I. Kiladze.
Byull. Inst. Teor. Astron., Tom 14, 589 (1980). In Russian.

098.087 **Lightcurves and phase relation of asteroid 324 Bamberga.** F. Scaltriti, V. Zappalà, R. Stanzel, C. Blanco, S. Catalano, J. W. Young.
Icarus, Vol. 43, 391 - 398 (1980).

A worldwide photometric investigation of the asteroid 324 Bamberga was conducted during the period September-November 1978. The full-cycle lightcurve shows two maxima and two minima with a maximum amplitude of 0.075 mag; the rotation period was found to be $P_{syn} = 29^h\!.42 \pm 0^h\!.01$. A linear least-squares solution of the phase relation gives $\beta_v = (0.034 \pm 0.001)$mag/degree and $V_0(1,0) = (7.17 \pm 0.01)$ mag. The color indices measured are: $B-V = 0.69$, $U-B = 0.36$, in agreement with the C taxonomic type given for 324 Bamberga. The very long period indicates 324 Bamberga is an unusual object among asteroids with diameters greater than 200 km.

098.088 **In search of satellites of minor planets.** P. D. Maley.
J. R. Astron. Soc. Canada, Vol. 74, 327 - 333 (1980).

Occultations of stars by asteroids have sometimes been observed. Since the early 1970's, reports of anomalous decreases of stellar brightness, both inside and outside the zone of primary occultation, have been recorded for a number of these events. Circumstantial evidence points to the possibility that some of these occurrences are due to a satellite orbiting the minor planet or to the multiple nature of the asteroid. Many more such events need careful study in order to determine whether satellites of minor planets actually exist or whether the source of the observed anomalies lies elsewhere.

098.089 **Apollo asteroids.** C. Cunningham.
J. R. Astron. Soc. Canada, Vol. 74, 367 - 368 (1980). – Abstract.

098.090 **Albedos of asteroids.** Yu. G. Shkuratov.
Astron. Zh., Tom 57, 1320 - 1322 (1980). In Russian. English translation in Soviet Astron., Vol. 24, No. 6.

It is shown that for the low-reflectivity asteroids the method of determination of the albedo based on data of polarization measurements is not correct.

098.091 **Rotational properties of the larger minor planets.** H. J. Schober, F. Scaltriti, V. Zappalà, A. Harris.
Mitt. Astron. Ges., Nr. 50, p. 83 (1980). – Abstract.

098.092 **Minor planet 1550 Tito.** M. Protić.
Vasiona, Année 28, 2 - 3 (1980). In Croatian.

Probable duplicity of Pallas from speckle interferometry and solar shrinkage from eclipse data reported at A.A.S. Meeting. See Abstr. 011.003.

An efficient, automated method of finding asteroid occultations of catalog stars. See Abstr. 021.024.

Real time asteroid identification. See Abstr. 021.029.

Model experiment of collisional breakup of asteroids: on the velocities of fragments. See Abstr. 022.082.

Critique of singular value analysis. See Abstr. 031.518.

A comparison of relative astrometric techniques as applied to minor planets. See Abstr. 031.564.

Method for calculating an object's position using two reference stars. See Abstr. 031.565.

Zur Reduktion photographischer Positionen von kleinen Planeten und Kometen. See Abstr. 031.594.

Equator and equinox of FK4 by 2240 observations of (51) Nemausa. See Abstr. 041.014.

The orbits of five minor planets and corrections to the FK4 equator and equinox. See Abstr. 041.028.

On the character of evolution of orbits in the vicinity of the 1 : 3 Kirkwood gap. See Abstr. 042.029.

Theory of the Trojan asteroids. Part III. See Abstr. 042.040.

On the origin of the Kirkwood Gaps. See Abstr. 042.042.

The calculation of the Trojan period. See Abstr. 042.055.

Der Einfluß des Jupiter im Sonnensystem. See Abstr. 042.067.

On long-periodic perturbations of Trojan asteroids. See Abstr. 042.071.

Infrared spectroscopic observations of the outer planets, their satellites, and the asteroids. See Abstr. 091.017.

Risonanze e anelli nel sistema solare. See Abstr. 091.045.

News of recently-attempted planetary occultations. See Abstr. 096.004.

Observations of stellar occultations by asteroids. See Abstr. 096.008.

Connection of meteor matter with comets and asteroids. See Abstr. 104.047.

The meteorite-asteroid connection: the infrared spectra of eucrites, shergottites, and Vesta. See Abstr. 105.003.

Reflectance spectra of some newly found, unusual meteorites and their bearing on surface mineralogy of asteroids. See Abstr. 105.067.

Mineralogically diagnostic visible and near-IR spectral features of C1 (CI) and C2 (CM) chondritic meteorite mineral assemblages. See Abstr. 105.079.

Origin of differentiated meteorites. See Abstr. 105.093.

The nature of asteroidal differentiation processes: implications for primordial heat sources. See Abstr. 105.145.

Sweeping of the Jovian resonances and the evolution of the asteroids. See Abstr. 107.019.

099 Jupiter, Jupiter Satellites

099.001 **A possible detection of solar variability from photometry of Io, Europa, Callisto, and Rhea, 1976–1979.**
G. W. Lockwood, D. T. Thompson, K. Lumme.
Astron. J., Vol. 85, 961 - 968 (1980).

Observations of Io, Europa, Callisto, and Rhea were made in *b* (472 nm) and *y* (551 nm) from 1976 to 1979 and have been fitted by theoretical solar phase curves which take into account mutual shadowing of the surface particles of the satellites, surface roughness, and multiple scattering. The brightness of Io is intrinsically variable at the 2% level, and the average fluctuations of the other objects are consistent with a reported 0.4% increase of the solar constant between 1976 and 1979.

099.002 **Lyman alpha albedo of Jupiter and solar activity.** A. Vidal-Madjar, C. Emerich, S. Cazes.
Astron. Astrophys., Vol. 87, L12 - L14 (1980).

From a comparison between the evolution over 12 years of both the Lyα Jovian emission and the flux at the center of the solar Lyα emission profile, the Lyα albedo of Jupiter appears not to be always linearly correlated to the solar activity. A spectacular increase of the albedo occurs around solar minimum and is shown to be related to some real change in the upper atmosphere of the planet. One of the potential agents responsible for this phenomenon is suggested to be the solar wind.

099.003 **RAS specialist discussion on results from the *Voyager* encounters with Jupiter and its satellites,** held 1980 January 11.
Observatory, Vol. 100, 106 - 108 (1980).

099.004 **Corotating Birkeland currents in Jupiter's magnetosphere: an Io plasma-torus source.**
A. J. Dessler.
Planet. Space Sci., Vol. 28, 781 - 788 (1980).

The author proposes that a persistent longitudinally asymmetric pattern of Birkeland (magnetically field aligned) currents flow between the Jovian ionosphere and the plasma torus that encircles Jupiter near Io's orbit. A specific longitudinal sector of the torus (i.e., the active sector) contains more plasma and is therefore more massive than the rest of the torus. This asymmetry causes Birkeland currents to flow between Jupiter's ionosphere and the torus. The principal currents are studied in detail.

099.005 **Charged dust in the outer planetary magnetospheres. II. Trajectories and spatial distribution.**
J. R. Hill, D. A. Mendis.
Moon Planets, Vol. 23, 53 - 71 (1980).

The authors consider the dynamics of the electrostatic disruption products of fragile interplanetary dust aggregates which are initially electrically charged on entering the Jovian plasmasphere. The detailed orbits of these charged dust fragments, which are shown to be confined to the equatorial plane, are computed for various launch angles. It is established that the fragments with radii typically around 1μ are magneto-gravitationally trapped within the plasmasphere due to the velocity induced oscillation of their surface potentials. The spatial distribution of these fragments are evaluated and the time evolution of the distributions followed. The observed brightness asymmetries between the leading and trailing sides of the Galilean satellites appears to be a natural consequence of the impact geometries of these charged dust grains with the satellite surfaces.

099.006 **Positions of Jupiter and Galilean satellites in December 1978.**
H. Debehogne, R. R. de Freitas Mourao.
Acta Astron., Vol. 30, 215 - 218 (1980).

35 positions of Jupiter and the Galilean satellites are given.

099.007 **International Information Bureau on Astronomical Ephemerides.**
I. I. B. A. E. (I. A. U. - COSPAR), Information cards Nos. 133 - 135 (1980).
Available from Bureau International d'Information sur les Éphémerides Astronomiques, 77, avenue Denfert-Rochereau, 75014 Paris, France.

099.008 **21 cm maps of Jupiter's radiation belts from all rotational aspects.** I. de Pater.
Astron. Astrophys., Vol. 88, 175 - 183 (1980).

Two-dimensional maps of the radio emission from Jupiter were made in December 1977 at a frequency of 1,412 MHz using the Westerbork telescope in the Netherlands. Pictures in all four Stokes parameters have been obtained every 15° in longitude, each smeared over 15° of the planet's rotation. The maps have an E–W resolution of ~1/3 of the diameter of the disk and a N–S resolution 3 times less. The total intensity and linear polarization maps are accurate to 0.5% and the circular polarized maps to 0.1% of the maximum intensities in *I*. The whole set of maps clearly show the existence of higher order terms in the magnetic field of Jupiter.

099.009 **Multicolor polarimetry of the Galilean satellites of Jupiter.** V. V. Botvinova, V. A. Kucherov.
Astrometr. Astrofiz., Vyp. (No.) 41, p. 59 - 63 (1980). In Russian.

Measurements of the Galilean satellites polarization are given for the wavelengths of 0.390, 0.451, 0.550, 0.619, 0.645, 0.685 μm.

099.010 **Hydrocarbon photochemistry and Lyman alpha albedo of Jupiter.** Y. L. Yung, D. F. Strobel.
Astrophys. J., Vol. 239, 395 - 402 (1980) = Div. Geol. Planet. Sci., Calif. Inst. Technol.,Contrib. No. 3312.

A combined study of hydrocarbon and atomic hydrogen photochemistry is made to calculate self-consistently the Lα albedo of Jupiter. It is shown that the Lα emissions observed by Voyagers I and II can be explained by resonance scattering of sunlight. Precipitation of energetic particles from the magnetosphere can provide the large required source of atomic hydrogen, although the contribution of direct particle excitation to the disk-averaged brightness is insignificant. The variability of the Lα brightness inferred from many observations in recent years is examined. The large difference in the brightness of the He 584 Å resonance line observed by Pioneer and Voyager is briefly discussed.

099.011 **Spectral broadening measurements of the ionospheres of Jupiter and Saturn.**
R. Woo, J. W. Armstrong.
Nature, Vol. 287, 309 - 311 (1980).

The authors present the characteristics of the electron density irregularities for the ionospheres of both Jupiter and Saturn.

099.012 **The dynamics of Jupiter's atmosphere as inferred from Voyager IRIS.**
F. M. Flasar, B. J. Conrath, J. A. Pirraglia.
Bull. American Astron. Soc., Vol. 12, 450 (1980). – Abstract.

099.013 **Observation of X-rays from the Jovian system.**
A. E. Metzger, D. A. Gilman, J. L. Luthey,
K. C. Hurley, J. D. Sullivan, F. D. Seward, H. W. Schnopper.
Bull. American Astron. Soc., Vol. 12, 450 (1980). – Abstract.

099.014 **Geheimnis des Ringes um Jupiter gelüftet?**
E. Hintsches.
Phys. Bl., 36. Jahrg., 246 - 248 (1980).

099.015 **Modeling Jupiter's current disc: Pioneer 10 outbound.**
D. E. Jones, J. G. Melville II, M. L. Blake.
J. Geophys. Res., Vol. 85, 3329 - 3336 (1980).
The magnetic field of the Jovian current disc has been modeled by using Euler functions and the Biot-Savart law applied to a series of concentric, but not necessarily coplanar, current rings.

099.016 **Azimuthal magnetic field at Jupiter.**
J. L. Parish, C. K. Goertz, M. F. Thomsen.
J. Geophys. Res., Vol. 85, 4152 - 4156 (1980).
The authors model the azimuthal component of the magnetic field at Jupiter. They start by writing a current distribution which is the sum of two currents, one flowing along the magnetic field lines and another injected into the current sheet at r_o ($\sim 10\ R_J$). The authors look at two cases, one in which current flows along the field lines into the current sheet at r_o. Each of these two cases results in an azimuthal magnetic field which fits the magnetic field data.

099.017 **The occurrence rate, polarization character, and intensity of broadband Jovian kilometric radiation.**
M. D. Desch, M. L. Kaiser.
J. Geophys. Res., Vol. 85, 4248 - 4256 (1980).
The authors describe the major observational features of one new component of Jupiter's radio emission spectrum, the broadband kilometer wavelength radiation, or bKOM. The study, using the Voyager Planetary Radio Astronomy experiments, reveals that the overall occurrence morphology, dynamic spectra, and polarization character of bKOM are strong functions of the latitude and/or local time geometry of the observations. The strong dependence of the morphology on local time suggests a source whose beam is nearly fixed relative to the Jupiter-sun line.

099.018 **Spatial imaging of hydrogen Lyman α emission from Jupiter.** J. T. Clarke, H. A. Weaver,
P. D. Feldman, H. W. Moos, W. G. Fastie, C. B. Opal.
Astrophys. J., Vol. 240, 696 - 701 (1980).
A sounding rocket measurement of the H I Lα emission from Jupiter made on 1978 December 1 shows limb darkening and an average disk brightness of 13kR. This brightness is significantly higher than in previous measurements, and was confirmed by an *IUE* observation on 1978 December 10. Comparison with a plane-parallel hydrogen layer model indicates that there is enhanced emission from the equatorial regions, reaching a peak near 80° longitude.

099.019 **Jovian weather: like earth's or a star's?**
R. A. Kerr.
Science, Vol. 209, 1219 - 1220 (1980).

099.020 **On the origin of modulation lanes in the dynamic spectra of Jupiter's decametric radiation.**
V. V. Zheleznyakov, V. E. Shaposhnikov.
Proc. Astron. Soc. Australia, Vol. 3, 259 - 262 (1978).

099.021 **Phase effect for the brightness coefficient of the central disk of Saturn and features of Jupiter's disk.**
V. M. Klimenko, A. V. Morozhenko, A. P. Vid'machenko.
Icarus, Vol. 42, 354 - 357 (1980).
Spectral reflectivities $\rho(\lambda)$ of the features on the disk of Jupiter (NTrZ, NEB, EZ, SEB, STrZ) and the center of the disk of Saturn were measured for different phase angles. The phase dependence of $\rho(\lambda)$ is shown to correspond to the shape of the phase function.

099.022 **The effect of dense cores on the structure and evolution of Jupiter and Saturn.**
A. S. Grossman, J. B. Pollack, R. T. Reynolds, A. L. Summers, H. C. Graboske, Jr.
Icarus, Vol. 42, 358 - 379 (1980).
The authors examine the effect on the evolutionary history of Jupiter and Saturn of: the presence of a heavy element core, rotation, improved equation of state, and the uncertainty in the H/He ratio. In addition, static models are computed with the improved equations of state and optimized to agree with the observed gravitational moments. Finally, the two sets of results are compared for consistency.

099.023 **Thermal migration of water on the Galilean satellites.**
N. G. Purves, C.B. Pilcher.
Icarus, Vol. 43, 51 - 55 (1980).
The authors have modeled the thermal migration of water on the Galilean satellites under the assumption of ballistic molecular trajectories. They find that water migrating owing to solar radiation on an icecovered satellite will build up in temperate latitudes, in general not reaching the poles.

099.024 **Io: are vapor explosions responsible for the 5-μm outbursts?** W. M. Sinton.
Icarus, Vol. 43, 56 - 64 (1980).
It is proposed that a vapor explosion of a submerged pool of liquid sulfur will remove the crust overlying an area of ~50-km diameter. Thermal radiation from the exposed liquid sulfur pool with a surface temperature of 600 K is then presumed to be responsible for the 5-μm outbursts that have been observed. The explosive volcanoes are expected to leave black sulfur calderas, which are, indeed, found on the surface. The 5-μm outburst observed by W. M. Sinton (1980), on June 11, 1979 (UT), is identified with a new caldera found on Voyager 2 photographs but which had not been present on Voyager 1 pictures.

099.025 **Lightning synthesis of organic compounds on Jupiter.** J. S. Lewis.
Icarus, Vol. 43, 85 - 95 (1980).
The purpose of this work is to assess the efficiency of electrical discharge and shock-wave energy in the synthesis of disequilibruim species under realistic Jovian conditions, including predictions of the yields of a number of organic species.

099.026 **Ecclissi di Sole su Giove.** P. Senigalliesi.
Astronomia, N. 3, p. 10 - 22 (1980).

099.027 **Charged dust rings in the outer planetary magnetospheres.** J. R. Hill, D. A. Mendis.
Solid particles in the solar system, (see 012.019), p. 417 - 420 (1980).
The authors discuss the motion of charged dust in outer planetary magnetospheres, particularly that of Jupiter.

099.028 **Maximum frequency of the decametric radiation from Jupiter.** C. H. Barrow, J. K. Alexander.
Astron. Astrophys., Vol. 90, L4 - L6 (1980).
The upper frequency limits of Jupiter's decametric radio emission are found to be essentially the same when observed from the Earth or, with considerably higher sensitivity, from the Voyager spacecraft close to Jupiter. This suggests that the maximum frequency is a real cut-off corresponding to a maximum gyrofrequency of about 38-40 MHz at Jupiter. It no longer appears to be necessary to specify different cut-off

frequencies for the Io and Non-Io emission as the maximum frequencies are roughly the same in each case.

099.029 **Volume changes in Ganymede and Callisto and the origin of grooved terrain.** S. W. Squyres.
Geophys. Res. Lett., Vol. 7, 593 - 596 (1980).

Internal melting and differentiation of Ganymede and Callisto may have caused an increase in the surface area of these bodies early in their histories of up to 5-7%. Expansion due to differentiation may have caused formation of grooved terrain on Ganymede. The calculations suggest that grooved terrain formation is essentially a replacement and/or deformation process, with no more than about 15% of grooved terrain actually being new material. The absence of grooved terrain on Callisto may be due to the effects of a thicker crust and a lower expansion rate.

099.030 **Jupiter's magnetosphere.** S. W. H. Cowley.
Nature, Vol. 287, 775 - 776 (1980).

099.031 **The dynamic expansion and contraction of the Jovian plasma sheet.**
J. W. Belcher, R. L. McNutt, Jr.
Nature, Vol. 287, 813 - 815 (1980).

The authors report that near noon in the magnetosphere of Jupiter, plasma is observed to flow away from the equatorial current sheet. The motion reverses near dusk, with a subsequent infall of plasma towards the sheet on the nightside. They associate this motion with the compression of magnetic flux tubes on the dayside and their subsequent expansion on the nightside due to the solar wind interaction with Jupiter.

099.032 **Io control of Jovian radio emission.** M. D. Desch.
Nature, Vol. 287, 815 - 817 (1980).

Recent low-frequency studies have shed some doubt on the conclusion that Io control predominates only at high frequencies. The author demonstrates that observations above 8 MHz require that the emission intensity be taken into account and offers a reinterpretation of these data which is consistent with the recent low-frequency satellite and ground-based results.

099.033 **CH_4 photochemistry in the outer major planets.** O. Ashihara.
13th Lunar and Planetary Symposium, (see 012.018), p. 45 - 51 (1980).

Emissions from hydrocarbons such as C_2H_2, C_2H_6 and C_2H_4 have been detected in the recent infrared observations of some outer planets. Photochemical processes leading to the production of these and higher hydrocarbons have been studied. Self-consistent solution of the molecular species (about 70 neutral and ionic) have been obtained under the assumption of local photochemical equilibrium.

099.034 **On the Planetary Solitary Wave hypothesis of the Jovian Great Red Spot.** Y. Nogami.
13th Lunar and Planetary Symposium, (see 012.018), p. 52 - 61 (1980).

The author has delineated the observational results of the Voyagers pertaining to the Great Red Spot. Based on these results, he has reviewed the GRS models, especially about the Planetary Solitary Wave (PSW) hypothesis.

099.035 **On internal shock in the Jovian magnetodisc.** T. Aoyama, H. Oya.
13th Lunar and Planetary Symposium, (see 012.018), p. 70 - 84 (1980).

The theoretical studies on the internal shock based on the numerical calculation has been carried out for the case where the solar wind interacts with the internal Jovian disc wind with the supersonic flow velocity, sandwitching the Jovian intrinsic magnetic field. The results of this calculation indicate the existence of the internal shock with the clear dawn-dusk asymmetry in both the shape and the strength of the internal shock.

099.036 **The energy sources of Jovian decametric radio waves.** H. Oya, T. Kondo, A. Morioka.
13th Lunar and Planetary Symposium, (see 012.018), p. 85 - 98 (1980).

The contribution of the energy input on the Jovian decametric radio waves from the Io, Io-torus, and the Jovian magnetosphere has been investigated. The results indicate that the main contribution of the energy input for the non Io related A and B sources is mainly in the deep magnetosphere corresponding to the region $L > 20$ while the effect of Io-torus for the non Io related source is very weak. The C source is also related to the energy input from the deep magnetosphere $L > 20$.

099.037 **Joule heating for Io due to electromagnetic induction.** M. Yanagisawa.
13th Lunar and Planetary Symposium, (see 012.018), p. 99 - 106 (1980).

Variation of the Jovian magnetic field at Io generates electromagnetic induction current in Io's interior, and this electric current heats Io's interior. Heating rates have been calculated for two models.

099.038 **Io plasma torus as an origin of Jupiter's non-Io-related decametric radio emissions.** K. Imai.
13th Lunar and Planetary Symposium, (see 012.018), p. 107 - 115 (1980).

Observations of the L-bursts in Jovian decametric radio emissions reveal that the emission time structure of the non-Io-related A source resembles that of the Io-related A source. These results imply that the origins of their sources are within the same L-shell. In the vicinity of this L-shell there exists the equatorial high density plasma torus, which was recently measured with the radio and plasma instrument on board Voyager 1.

099.039 **Significance of mutual phenomena observation in study of Jovian and Saturnian satellites.**
T. Nakamura.
13th Lunar and Planetary Symposium, (see 012.018), p. 136 - 143 (1980).

The author discusses the conditions for the occurrence of mutual phenomena, the observational accuracy, the merit and demerit of these observations, gives a brief review of the analysis of the mutual phenomena of Galilean satellites in 1973 and reports preliminary results of the mutual phenomena observations of Saturnian satellites in 1980.

099.040 **Ganymede and Callisto: thermal and structural evolution model.** S.-i. Kawakami, H. Mizutani.
13th Lunar and Planetary Symposium, (see 012.018), p. 330 - 345 (1980).

Thermal history models of icy Galilean satellites are presented including the effect of the solid-state convective heat transfer in ice crust and silicate core. If the satellites form with a high initial temperature, the ice-silicate differentiation takes place. In such a case, the upward migration of H_2O causes the phase transformation from high pressure ice to low pressure ice with the volume expansion of ice. The fault-like features on Ganymede may be explained by the volume expansion due to the phase transformation of ice.

099.041 **Escape of H_2O from Jovian satellite Io.** H. Mizutani, S.-i. Kawakami.
13th Lunar and Planetary Symposium, (see 012.018), p. 346 - 354 (1980).

There are two hypotheses about the origin of the Galilean satellites. One is the accretion hypothesis, the other is the capture hypothesis. The variation of the H_2O/rock ratio among the Galilean satellites is thought to favor the accretion hypothesis. The authors calculate a thermal history of Io which is assumed to be initially of the same size and mean density as Ganymede, taking into account of the tidal heating in it. Using the thermal model of Io, they also calculate the amount of H_2O which escapes from Ganymede-like Io by the Jeans-Spitzer mechanism. The authors find the amount of H_2O escaped from the satellite is not enough to change the initial Ganymede-like composition to the present composition of Io. Therefore they conclude the capture hypothesis on the origin of the Galilean satellites is unlikely.

099.042 **Ganymede: a relationship between thermal history and crater statistics.**
R. J. Phillips, M. C. Malin.
Science, Vol. 210, 185 - 188 (1980).

An approach for factoring the effects of a planetary thermal history into a predicted set of crater statistics for an icy satellite is developed and forms the basis for subsequent data inversion studies. An example is given for the satellite Ganymede and the effect of the thermal history is easily seen in the resulting predicted crater statistics. A preliminary comparison with the data, subject to the uncertainties in ice rheology and impact flux history, suggests a surface age of 3.8×10^9 years and a radionuclide abundance of 0.3 times the chondritic value.

099.043 **TV pictures from another world (*Jupiter*).**
W. Engelhardt.
Funkschau, Vol. 52, No. 16, p. 53 - 56 (1980). In German.
Abstr. in Phys. Abstr., Vol. 83, Abstr. 101549 (1980).

099.044 **Dark and bright ray systems on Ganymede's surface: preliminary results.**
M. Poscolieri, P. H. Schultz.
Mem. Soc. Astron. Italiana, Vol. 51, 359 - 377 (1980).

Voyager 1 and 2 pictures have shown Ganymede's surface exhibiting many craters with prominent systems of both bright and dark rays. Such evidence is contrasting with that shown by Callisto's surface, where just a few evident rayed craters and many haloed craters are present. The differences in ray systems distribution between the two icy satellites have been studied trying to determine which process was longer active or more important in their ray system evolution.

099.045 **Jupiterreport 1980.**
M. Knülle, C. Schambeck.
Sterne Weltraum, Jahrg. 19, 390 - 391 (1980).

099.046 **Latitudinal beaming of planetary radio emissions.**
D. Jones.
Nature, Vol. 288, 225 - 229 (1980).

The Voyager missions have revealed that Jovian kilometric radiation (emanating from Jupiter's magnetosphere) is beamed away from the zenomagnetic equator. Results from GEOS 1 show that terrestrial non-thermal continuum or myriametric radiation is similarly beamed away from the geomagnetic equator. A mode-coupling mechanism is proposed such that measurements of the direction of propagation of the escaping O-mode radiations may allow the construction of the first high-resolution maps of the shapes of the equatorial plasmapause and Io torus.

099.047 **Airglow from Jupiter's nightside and crescent: ultraviolet spectrometer observations from Voyager 2.** J. C. McConnell, B. R. Sandel, A. L. Broadfoot.
Icarus, Vol. 43, 128 - 142 (1980).

The authors present Lyα measurements of the nightside of Jupiter and crescent and a revised upper limit to H_2 band emissions. They interpret these measurements in terms of reflection of the sky background and particle precipitation. They also briefly discuss day-night and longitudinal asymmetries in the precipitation patterns which appear to be emerging from the data. The implications of these conclusions for the interpretation of dayside intensities are briefly discussed.

099.048 **Low-latitude thermal structure of Jupiter in the region 0.1 - 5 bars.**
D. M. Hunten, M. Tomasko, L. Wallace.
Icarus, Vol. 43, 143 - 152 (1980).

The radiative heat flux from 0.1 to 10 bars is estimated on the basis of a "two-cloud" scattering model that fits available spectral data and Pioneer photometry. Deeper than a few bars, the flux is 4.5 W m^{-2}, compared with the 18.8 W m^{-2} used in an earlier study by Trafton and Stone. A temperature profile is computed, with the H_2 pressure-induced opacity; the temperature at 1 bar is found to be 156°K, rather than the commonly accepted 170°K. An additional optical depth of unity at the 0.67-bar level could restore the conventional value; otherwise a considerably cooler atmosphere is a serious possibility.

099.049 **Electrons from Jupiter.** Yu. V. Mineev.
Zemlya Vselennaya, 1980, No. 6, p. 25 - 27. In Russian.

099.050 **Internal heat and the hydrogen to helium ratio of Jupiter and Saturn.** A. P. Ingersoll, G. S. Orton.
Bull. American Astron. Soc., Vol. 12, 669 - 670 (1980). Abstract.

099.051 **New optical emissions from Jupiter's hot plasma torus.** R. A. Brown.
Bull. American Astron. Soc., Vol. 12, 672 - 673 (1980). Abstract.

099.052 **Observations of potassium emission near Io.**
L. Trafton.
Bull. American Astron. Soc., Vol. 12, 673 (1980). – Abstract.

099.053 **High spatial resolution observations of the Io plasma torus from the Voyager 2 Ultraviolet Spectrometer.**
J. B. Holberg, D. E. Shemansky, B. R. Sandel, A. L. Broadfoot.
Bull. American Astron. Soc., Vol. 12, 673 (1980). – Abstract.

099.054 **Induced electrical current system at Io.**
F. Herbert, B. R. Lichtenstein.
Bull. American Astron. Soc., Vol. 12, 673 - 674 (1980). Abstract.

099.055 **Radial dependence of the Io plasma torus spectrum in the EUV.** D. E. Shemansky, B. R. Sandel.
Bull. American Astron. Soc., Vol. 12, 674 (1980). – Abstract.

099.056 **Ring current impoundment of the Io plasma torus.**
G. L. Siscoe, A. Eviatar, R. M. Thorne, J. D. Richardson, F. Bagenal, J. L. Sullivan.
Bull. American Astron. Soc., Vol. 12, 674 (1980). – Abstract.

099.057 **The interaction of the Jovian magnetosphere with the icy Galilean satellites.**
R. S. Wolff, D. A. Mendis.
Bull. American Astron. Soc., Vol. 12, 674 (1980). – Abstract.

099.058 **Io sputtering hypothesis revisited.**
D. L. Matson, T. V. Johnson, F. P. Fanale.
Bull. American Astron. Soc., Vol. 12, 675 (1980). – Abstract.

099.059 **Transient sodium ejection from Io.**
C. Pilcher.
Bull. American Astron. Soc., Vol. 12, 675 (1980). – Abstract.

099.060 **Sputter-related processes in Io's atmosphere.** C. C. Watson, P. K. Haff, Y. L. Yung.
Bull. American Astron. Soc. Vol. 12, 675 (1980). – Abstract.

099.061 **Voyager 2 observations of Jupiter's south polar plasma environment.** D. P. Hinson, G. L. Tyler.
Bull. American Astron. Soc., Vol. 12, 675 - 676 (1980). Abstract.

099.062 **Limit on rotational energy available to excite Jovian aurora.** A. Eviatar, G. L. Siscoe.
Bull. American Astron. Soc., Vol. 12, 676 (1980). – Abstract.

099.063 **Creation of the Jovian hydrogen bulge by co-rotating magnetospheric convection.**
A. J. Dessler, B. R. Sandel, S. K. Atreya.
Bull. American Astron. Soc., Vol. 12, 676 (1980). – Abstract.

099.064 **The helium abundance of Jupiter from Voyager.** D. Gautier, B. Conrath. M. Flasar, R. Hanel, V. Kunde, A. Chedin, N. Scott.
Bull. American Astron. Soc., Vol. 12, 683 (1980). – Abstract.

099.065 **Albedo, internal heat, and energy balance of Jupiter; preliminary results of the Voyager infrared investigation.** R. Hanel, B. Conrath, L. Herath, V. Kunde, J. Pirraglia.
Bull. American Astron. Soc., Vol. 12, 683 (1980). – Abstract.

099.066 **The atmosphere of Jupiter: an analysis of the Voyager radio occultation measurements.**
G. F. Lindal, V. R. Eshleman.
Bull. American Astron. Soc., Vol. 12, 683 - 684 (1980). Abstract.

099.067 **Winds on Jupiter as inferred from Voyager 1 and 2 images.** A. P. Ingersoll, R. F. Beebe, J. L. Mitchell, G. W. Garneau, G. M. Yagi.
Bull. American Astron. Soc., Vol. 12, 684 - 685 (1980). Abstract.

099.068 **A comparison of recent high-velocity Jovian features at 23°.5 N latitude.** R. Suggs, R. Beebe.
Bull. American Astron. Soc., Vol. 12, 685 (1980). – Abstract.

099.069 **Variations in the structure and motions of Jovian equatorial plumes from Voyager observations.**
G. E. Hunt, J.-P. Muller, P. Gee, R. F. T. Barrey.
Bull. American Astron. Soc., Vol. 12, 685 (1980). – Abstract.

099.070 **Considerations on the energetics of Jupiter's Great Red Spot.** T. Maxworthy, J. L. Mitchell.
Bull. American Astron. Soc., Vol. 12, 685 (1980). – Abstract.

099.071 **Theoretical calculations of the effect of C_2H_2 absorption on the UV albedo of Jupiter.**
G. R. Smith, J. C. McConnell.
Bull. American Astron. Soc., Vol. 12, 685 (1980). – Abstract.

099.072 **Numerical model of long-lived Jovian vortices.** A. P. Ingersoll, P. G. Cuong.
Bull. American Astron. Soc., Vol. 12, 686 (1980). – Abstract.

099.073 **Studies of the colours of the Jovian clouds using multispectral image processing techniques.**
J.-P. Muller, G. E. Hunt.
Bull. American Astron. Soc., Vol. 12, 686 - 687 (1980). Abstract.

099.074 **Lightning on Jupiter: rate, energetics and chemical effects.** J. S. Lewis.
Bull. American Astron. Soc., Vol. 12, 687 (1980). – Abstract.

099.075 **Structure and composition of the upper atmosphere of Jupiter by Voyager 2 UVS stellar occultation experiment.** M. C. Festou, S. K. Atreya, T. M. Donahue, B. R. Sandel, D. E. Shemansky, A. L. Broadfoot.
Bull. American Astron. Soc., Vol. 12, 687 (1980). – Abstract.

099.076 **Results from the Voyager 1 ultraviolet spectrometer solar occultation experiment at Jupiter.**
G. R. Smith, J. C. McConnell, D. E. Shemansky, B. R. Sandel, A. L. Broadfoot.
Bull. American Astron. Soc., Vol. 12, 687 - 688 (1980). Abstract.

099.077 **Voyager UV spectrometer observations of He 584 Å dayglow at Jupiter.**
A. L. Broadfoot, J. C. McConnell, B. R. Sandel.
Bull. American Astron. Soc., Vol. 12, 688 (1980). – Abstract.

099.078 **Jupiter's H Lyman-alpha nightglow.** B. R. Sandel, J. C. McConnell.
Bull. American Astron. Soc., Vol. 12, 692 (1980). – Abstract.

099.079 **Possible infrared aurorae on Jupiter.** J. Caldwell, A. Tokunaga, F. C. Gillett.
Bull. American Astron. Soc., Vol. 12, 692 (1980). – Abstract.

099.080 **A determination of the Jovian PH_3 abundance from Voyager IRIS spectra in the 2200 - 2300 cm^{-1} spectral range.** P. Drossart, T. Encrenaz, V. Kunde, M. Combes.
Bull. American Astron. Soc., Vol. 12, 692 (1980). – Abstract.

099.081 **Vertical distribution of NH_3 in the upper Jovian atmosphere from IUE observations.**
M. Combes, R. Courtin, J. Caldwell, T. Encrenaz, K. H. Fricke, V. Moore, T. Owen, P. S. Butterworth.
Bull. American Astron. Soc., Vol. 12, 692 - 693 (1980). Abstract.

099.082 **The upper atmosphere of Jupiter – modeling the effects of particle precipitation.**
G. R. Gladstone, Y. L. Yung, M. Allen.
Bull. American Astron. Soc., Vol. 12, 693 (1980). – Abstract.

099.083 **Photodissociation in the Jovian atmosphere computed using the δ-Eddington approximation.**
G. Visconti.
Bull. American Astron. Soc., Vol. 12, 693 (1980). – Abstract.

099.084 **Ionization layer formation on Jupiter.** R. H. Chen.
Bull. American Astron. Soc., Vol. 12, 693 - 694 (1980). Abstract.

099.085 **Spatially resolved observations of the H_2S_3 (1) and the NH_3 6457.2 Å and 6454 Å features for Jupiter.**
W. Hayden Smith, K. Baines, W. D. Cochran.
Bull. American Astron. Soc., Vol. 12, 694 (1980). – Abstract.

099.086 **A new observation of PH_3 on Jupiter at 1 mm wavelength.**
T. de Graaw, S. Lindholm, H. van de Stadt, F. van Vliet, C. de Vries, T. Encrenaz, M. Combes, P. Drossart.
Bull. American Astron. Soc., Vol. 12, 694 (1980). – Abstract.

099.087 **Results from the near-infrared Io monitoring program.** W. Sinton, D. Lindwall, F. Cheigh.
Bull. American Astron. Soc., Vol. 12, 694 (1980). – Abstract.

099.088 **Observational constraints on the internal heat source of Io.** D. Morrison, C. M. Telesco.
Bull. American Astron. Soc., Vol. 12, 694 (1980). – Abstract.

099.089 **Dynamics of Io's eruption plumes.**
N. M. Schneider, R. G. Strom, A. F. Cook.
Bull. American Astron. Soc., Vol. 12, 695 (1980). – Abstract.

099.090 **Io post eclipse brightening and the physical chemistry of SO_2.** F. P. Fanale, W. B. Banerdt.
Bull. American Astron. Soc., Vol. 12, 695 (1980). – Abstract.

099.091 **Search for ultraviolet spectral variations of the Galilean satellites.** A. L. Lane, R. M. Nelson, D. L. Matson, F. C. Motteler, M. E. Ockert.
Bull. American Astron. Soc., Vol. 12, 695 (1980). – Abstract.

099.092 **Atmospheric dynamics on Io.**
M. E. Summers, A. P. Ingersoll.
Bull. American Astron. Soc., Vol. 12, 695 - 696 (1980). Abstract.

099.093 **Ground-based observations of the Jovian ring and inner satellites.** D. Jewitt, R. Terrile, E. Danielson.
Bull. American Astron. Soc., Vol. 12, 697 (1980). – Abstract.

099.094 **Bounds on particle size and optical depth of Jupiter's ring from a comparative analysis of radio, infrared, and optical results.** E. A. Marouf, G. L. Tyler.
Bull. American Astron. Soc., Vol. 12, 697 - 698 (1980). Abstract.

099.095 **Source of Jupiter's ring? Unseen satellites by Jove!**
J. A. Burns, M. R. Showalter, J. N. Cuzzi, J. B. Pollack.
Bull. American Astron. Soc., Vol. 12, 698 (1980). – Abstract.

099.096 **Ground-based observations of Io's hot spots and its internal heat flux.** W. M. Sinton.
Bull. American Astron. Soc., Vol. 12, 702 (1980). – Abstract.

099.097 **The effect of NH_3 ice particles in the Jovian atmosphere on outgoing thermal radiation.**
G. S. Orton, J. F. Appleby, J. V. Martonchik.
Bull. American Astron. Soc., Vol. 12, 706 (1980). – Abstract.

099.098 **Preliminary determination of the shape of Io using limb fitting techniques.**
S. P. Synnott, L. A. Morabito.
Bull. American Astron. Soc., Vol. 12, 708 - 709 (1980). Abstract.

099.099 **The surface of Io: geologic units, morphology and tectonics.** G. G. Schaber.
Bull. American Astron. Soc., Vol. 12, 709 (1980). – Abstract.

099.100 **Europa's petrologic thermal history.**
G. A. Ransford, A. A. Finnerty.
Bull. American Astron. Soc., Vol. 12, 709 (1980). – Abstract.

099.101 **Europa surface cracking: a consequence of thermal evolution.**
A. A. Finnerty, G. A. Ransford, D. Pieri.
Bull. American Astron. Soc., Vol. 12, 709 - 710 (1980). Abstract.

099.102 **Global tectonics on Ganymede.**
D. C. Phinney, C. K. Seyfert, J. L. Warner.
Bull. American Astron. Soc., Vol. 12, 710 (1980). – Abstract.

099.103 **Grooved terrain on Ganymede.**
B. K. Lucchitta.
Bull. American Astron. Soc., Vol. 12, 710 (1980). – Abstract.

099.104 **Ice in the interiors of Ganymede and Callisto.**
K. Ellsworth, G. Schubert, D. J. Stevenson.
Bull. American Astron. Soc., Vol. 12, 710 (1980). – Abstract.

099.105 **Ganymede cratering I: the dark terrain.**
J. B. Plescia, J. M. Boyce, E. M. Shoemaker.
Bull. American Astron. Soc., Vol. 12, 710 - 711 (1980). Abstract.

099.106 **Ganymede cratering II: the smooth and grooved terrains.**
J. B. Plescia, J. M. Boyce, E. M. Shoemaker.
Bull. American Astron. Soc., Vol. 12, 711 (1980). – Abstract.

099.107 **Coordinates of features on the Galilean satellites.**
M. E. Davies.
Bull. American Astron. Soc., Vol. 12, 711 (1980). – Abstract.

099.108 **Global distribution of craters and multiring structures on Callisto.**
Q. R. Passey, E. M. Shoemaker.
Bull. American Astron. Soc., Vol. 12, 712 (1980). – Abstract.

099.109 **Global multispectral mosaics of the Galilean satellites.** T. V. Johnson, L. A. Soderblom, J. A. Mosher, G. E. Danielson, A. F. Cook, P. Kupferman.
Bull. American Astron. Soc., Vol. 12, 713 (1980). – Abstract.

099.110 **Ultraviolet albedos of the Galilean satellites from IUE.** R. M. Nelson, D. L. Matson, A. L. Lane, F. C. Motteler, M. E. Ockert.
Bull. American Astron. Soc., Vol. 12, 713 (1980). – Abstract.

099.111 **Galilean satellites: high precision near infrared spectrophotometry (0.65 - 2.5 μm) of the leading and trailing sides.**
R. N. Clark, R. B. Singer, P. D. Owensby, F. P. Fanale.
Bull. American Astron. Soc., Vol. 12, 713 - 714 (1980). Abstract.

099.112 **Voyager photometry of surface features on Ganymede and Callisto.**
S. Squyres, J. Veverka.
Bull. American Astron. Soc., Vol. 12, 714 (1980). – Abstract.

099.113 **Long period effects in the orbits of the Galilean satellites.** B. C. Brown.
Bull. American Astron. Soc., Vol. 12, 744 (1980). – Abstract.

099.114 **Detection of energetic hydrogen molecules in Jupiter's magnetosphere by Voyager 2: evidence for an ionospheric plasma source.**
D. C. Hamilton, G. Gloeckler, S. M. Krimigis, C. O. Bostrom, T. P. Armstrong, W. I. Axford, C. Y. Fan, L. J. Lanzerotti, D. M. Hunten.
Geophys. Res. Lett., Vol. 7, 813 - 816 (1980).

The authors report the discovery of energetic (~1 MeV/nuc) H_3 and H_2 molecules in Jupiter's magnetosphere. The data, obtained with the LECP instrument on Voyager 2, showed these molecules to be present throughout the magnetosphere and as far as 180 R_J from the planet, in the "magnetospheric wind" region. Although the relative abundances of H_3 and H_2 do not show a monotonic trend with distance from Jupiter, the intervals of highest abundance were found in the outer magnetosphere. Since H_3^+ is expected to be an important constituent of Jupiter's ionosphere, the data provide strong evidence that, in addition to Io, the ionosphere may be an important local plasma source for the Jovian energetic particles.

099.115 **Statics of the nightside Jovian plasma sheet.**
L. J. Lanzerotti, C. G. Maclennan, S. M. Krimigis,

T. P. Armstrong, K. Behannon, N. F. Ness.
Geophys. Res. Lett., Vol. 7, 817 - 820 (1980).

The authors present analyses of data from the low energy charged particle (LECP) and magnetic field (MAG) experiments on the Voyager 2 spacecraft that demonstrate that the configuration of the Jovian plasma sheet at distances ~ 80 to ~ 120 R_J is determined by ions (protons and heavier nuclei) of energies ≳ 30 keV. The energy densities of these ions are sufficient to provide the diamagnetic depressions in the magnetic field strengths observed as the spacecraft crossed the Jovian plasma sheet. The particle bulk direction of motion is predominantly across the local magnetic field, consistent with that expected from corotation of the planetary magnetic field.

099.116 **Satellites of Jupiter.**
IAU Circ., No. 3507 (1980).

099.117 **Io: longitudinal distribution of sulfur dioxide frost.**
R. M. Nelson, A. L. Lane, D. L. Matson, F. P. Fanale, D. B. Nash, T. V. Johnson.
Science, Vol. 210, 784 - 786 (1980).

Twenty spectra of Io (0.26 to 0.33 micrometer), acquired with the International Ultraviolet Explorer spacecraft, have been studied. There is a strong ultraviolet absorption shortward of 0.33 micrometer. This spectral feature and its variation are interpreted as indicative of a longitudinal variation in the distribution of sulfur dioxide frost on Io. Variations in spectral reflectivity between 0.4 and 0.5 micrometer correlate inversely with variations in reflectivity between 0.26 and 0.33 micrometer. It is concluded that this is because the Io surface component with the highest visible reflectivity (sulfur dioxide frost) has the lowest ultraviolet reflectivity. At least one other component is present and may be sulfur allotropes or alkali sulfides.

099.118 **1979J2: the discovery of a previously unknown Jovian satellite.** S. P. Synnott.
Science, Vol. 210, 786 - 788 (1980).

The satellite has been observed in transit in both Voyager 1 and Voyager 2 frames; its period is 16 hours 11 minutes 21.25 seconds ± 0.5 second and its semimajor axis is 3.1054 Jupiter radii. The profile observed when the satellite is in transit is roughly circular with a diameter of 80 kilometers. It appears to have an albedo of ~0.05, similar to Amalthea's.

099.119 **Io: ground-based observations of hot spots.**
W. M. Sinton, A. T. Tokunaga, E. E. Becklin, I. Gatley, T. J. Lee, C. J. Lonsdale.
Science, Vol. 210, 1015 - 1017 (1980).

Observations of Io in eclipse demonstrate conclusively that Io emits substantial amounts of radiation at 4.8 and 3.8 micrometers and a measurable amount at 2.2 micrometers. Color temperatures derived from the observations fit blackbody emission at 560 K. The required source area to yield the observed 4.8-micrometer flux is approximately 5×10^{-5} of the disk of Io and is most likely comprised of small hot spots in the vicinity of the volcanoes.

099.120 **Jupiter and Saturn: giant magnetic rotating fluid planets.** R. Hide.
Observatory, Vol. 100, 182 - 193 (1980).

099.121 **Erdähnliche Himmelskörper in neuer Sicht (II).**
M. Reichstein.
Astron. Schule, 17. Jahrg., 99 - 101 (1980).

099.122 **The eruptive evolution of the Galilean satellites: implications for the ancient magnetic field of Jupiter.** E. M. Drobyshevski (*Eh. M. Drobyshevskij*).
Moon Planets, Vol. 23, 483 - 491 (1980).

The hypothesis considering the Jupiter-Sun system as a limiting case of a close binary star implies the initial relative ice abundances in all the Galilean satellites to be essentially equal. The satellites move in the Jovian magnetosphere; thus the unipolar current flowing through their bodies subjected their ices to volumetric electrolysis. Explosions of the electrolysis products resulted in a loss of ices. While Callisto did not explode at all, Ganymede exploded once, Europa twice, and Io two or three times.

099.123 **Sondage à distance de l'atmosphère des planètes géantes par spectroscopie de Fourier.**
M. Combes.
Origin and distribution of the elements, (see 012.039), p. 111 - 119 (1979).

A brief review is given of the definitely or tentatively identified molecular species. The significance of the derived abundance and isotopic ratios is discussed, taking into account the different mechanisms of lines formation involved in remote sensing methods. Possible future studies of the Jovian planets are examined.

099.124 **Colour changes on Jupiter.**
G. C. Browne, A. J. Meadows.
Planet. Space Sci., Vol. 28, 1055 - 1058 (1980).

An analysis of four-colour photographic images of Jupiter taken during 1970 - 1974 is presented. It is shown that major global and hemispheric changes in colour – mainly in the blue – occurred over this period.

099.125 **Dust in Jupiter's magnetosphere: physical processes.**
G. E. Morfill, E. Grün, T. V. Johnson.
Planet. Space Sci, Vol. 28, 1087 - 1100 (1980).

The physical processes acting on charged microscopic dust grains in the Jovian magnetosphere are examined. Such small dust grains are believed to be injected continuously into the magnetosphere via volcanic activity on Io. It is shown that electromagnetic forces dominate the dust particle dynamics, and that these particles behave adiabatically, in the sense that the guiding centre approximation to their motion applies. Based on this fact, the diffusion across field lines, caused by random charge fluctuations of the dust grains, can be determined. Other physical processes (radiation pressure drag, Coulomb drag, sputtering) are also examined regarding their importance for particle transport.

099.126 **Dust in Jupiter's magnetosphere: origin of the ring.**
G. E. Morfill, E. Grün, T. V. Johnson.
Planet. Space Sci, Vol. 28, 1101 - 1110 (1980).

A model for the production of the Jovian ring is proposed. The 'visible' ring particles are micron-sized and produced by erosive collisions between an assumed population of km-sized parent bodies and sub-micron sized magnetospheric dust particles. These small dust particles are ejected by volcanoes from Io. The observed topology of the ring is described quite well with the theory, and properties of the parent bodies are deduced.

099.127 **Dust in Jupiter's magnetosphere: time variations.**
G. E. Morfill, E. Grün, T. V. Johnson.
Planet. Space Sci., Vol. 28, 1111 - 1114 (1980).

The ring topology and intensity of the Jovian dust ring are determined over various relevant time scales, assuming the origin model proposed by Johnson et al. (1980) and Morfill et al. (1980) to apply. It is concluded that the ring is a quasi-permanent and quasi-stable feature of the Jovian system.

099.128 **Dust in Jupiter's magnetosphere: effect on magnetospheric electrons and ions.**
G. E. Morfill, E. Grün, T. V. Johnson.
Planet. Space Sci., Vol. 28, 1115 - 1123 (1980).

The interaction of the Jovian energetic radiation belt electrons, and the Jovian plasma, with an ambient dust population is examined. Firstly the distribution of dust, ejected from

Io, in the inner magnetosphere is calculated. Using the mass loss in submicron particles of ~ 13 g/sec, which is required to model the intensity and shape of the Jovian ring in the model of Morfill et al. (1980), it is possible to quantitatively calculate losses of magnetospheric ions and electrons due to direct collisions with charged dust particles as well as multiple Coulomb scattering with resultant losses in the Jovian atmosphere.

099.129 **Io and its plasma environment.** D. J. Southwood, M. G. Kivelson, R. J. Walker, J. A. Slavin.
J. Geophys. Res., Vol. 85, 5959 - 5968 (1980).

The authors describe and examine the interaction of Io with its plasma torus and the Jovian magnetic field in the context of several currently popular models. They address three specific matters. First they discuss features implied by sub-Alfvenic flow which must be common to all models. Next, they examine the magnetic signature observed near Io by the Goddard Space Flight Center Voyager 1 magnetometer. Lastly, they point out the crucial role of charged particle data for probing the near Io interaction.

099.130 **Jupiter's magnetopause.** E. J. Smith.
Magnetospheric boundary layers, (see 012.041), p. 221 - 224 (1979).

Some of the major conclusions regarding the magnetopause of Jupiter, based on Pioneer and Voyager observations, are discussed. Topics include the overall shape of the magnetosphere, the variability in the location of the magnetopause with changes in the solar wind, intrinsic magnetospheric motions, the character of the field and plasma just inside the magnetopause and evidence of magnetic merging.

099.131 **The Jovian boundary layer as formed by magnetic-anomaly effects.** A. J. Dessler.
Magnetospheric boundary layers, (see 012.041), p. 225 - 226 (1979).

A model is presented in which a plasma boundary layer of Jupiter is formed from plasma of internal origin. It is proposed that Jupiter's boundary layer consists principally of sulphur and oxygen from the Io plasma torus, plus a small component of hydrogen from Jupiter's ionosphere. Fresh plasma is supplied to the boundary layer once each planetary rotation period by a convection pattern that rotates with Jupiter.

099.132 **Earth and Jupiter: comparison of magnetopause structure.** P. H. Reiff, T. W. Hill.
Magnetospheric boundary layers, (see 012.041), p. 227 - 228 (1979).

099.133 **Structure of Jupiter's magnetosphere.** B. T. Tsurutani, E. J. Smith, J. H. Wolfe, B. U. Ö. Sonnerup.
Magnetospheric boundary layers, (see 012.041), p. 229 (1979). Abstract.

099.134 **Voyager exploration of Jupiter: sulfur as a major chromophore.** J. Veverka.
News Lett. Astron. Soc. N. Y., Vol. 1, No. 7, p. 5 - 6 (1980). Abstract.

099.135 **The position angle of Jupiter's linearly polarized synchrotron emission – observations extending over 16 years.** M. M. Komesaroff, P. M. McCulloch, G. L. Berge, M. J. Klein.
Mon. Not. R. Astron. Soc., Vol. 193, 745 - 759 (1980).

Parkes, Owens Valley and Goldstone measurements are presented, showing the variation with central meridian longitude of the position angle of Jupiter's linearly polarized synchrotron emission at wavelengths of 21, 13, 11 and 6 cm. The form of the position angle versus longitude curve shows a slight dependence on wavelength and epoch. The epoch dependence appears to reflect Jupiter's changing aspect.

099.136 **Observations from earth orbit and variability of the polar aurora on Jupiter.**
J. T. Clarke, H. W. Moos, S. K. Atreya, A. L. Lane.
Astrophys. J., Lett., Vol. 241, L179 - L182, plate L3 (1980).

Spatially resolved spectra of Jupiter taken with the International Ultraviolet Explorer satellite show enhanced emission from the polar regions at H Lα (1216 Å) and in the Lyman and Werner bands of H_2 (1175 - 1650 Å). Two types of variability in emission brightness have been observed in these aurorae: an increase in the observed emission as the auroral oval rotates with Jupiter's magnetic pole to face toward the earth and a general variation in brightness of more than an order of magnitude under nearly identical observing conditions. The spectral character of these aurorae appears variable, indicating that the depth of penetration of the auroral particles is not constant.

099.137 **De galileiske måner. Fire forunderlige verdener.** Ø. Hauge.
Astron. Tidsskr., Årg. 13, 166 - 172 (1980).

099.138 **Dynamik der Jupiteratmosphäre.** A. M. Quetsch.
Jupiter, (see 012.045), p. 19 - 43 (1980).

099.139 **Radioemission des Jupiter.** D. Lang.
Jupiter, (see 012.045), p. 45 - 64 (1980).

099.140 **Innerer Aufbau des Jupiter und Entstehung des Magnetfeldes.** H. Gass.
Jupiter, (see 012.045), p. 65 - 72 (1980).

099.141 **Die Galileischen Satelliten.** A. Krabbe.
Jupiter, (see 012.045), p. 73 - 88 (1980).

099.142 **Io.** V. Werle.
Jupiter, (see 012.045), p. 89 - 119 (1980).

099.143 **Wechselwirkungen Io mit der Jupitermagnetosphäre.** G. Thiele.
Jupiter, (see 012.045), p. 121 - 138 (1980).

099.144 **Der Staubring.** G. Morfill.
Jupiter, (see 012.045), p. 139 - 140 (1980).
Summary.

099.145 **Effects of Io's volcanos on the plasma torus and Jupiter's magnetosphere.** A. F. Cheng.
Astrophys. J., Vol. 242, 812 - 827 (1980).

Io's volcanism can have dominant effects on Jupiter's magnetosphere. A model is developed in which a neutral gas torus is formed at Io's orbit by volcanic SO_2 escaping from Io. Ionization and dissociation of volcanic SO_2 is shown to be the dominant source of plasma in Jupiter's magnetosphere. The failure of Voyager observations to confirm predictions of the magnetic anomaly model is naturally explained. Negative ions are predicted in the Io plasma torus.

099.146 **Electromagnetic heating of Io.** D. S. Colburn.
J. Geophys. Res., Vol. 85, 7257 - 7261 (1980).

Electrical-energy deposition within Io is a possible mechanism for heating the interior, the source of the electrical current system being the relative motion of Io with respect to Jupiter's magnetic field. The transverse-electric mode currents circulate entirely within Io, with the frequency of Jupiter's rotation rate as observed in the Io frame. Maximum heating for Io is obtained if the electrical conductivity is 0.02 mho m^{-1}, a value spanned by estimates of rock conductivity for hot interiors. Under present conditions, the maximum heating available from the transverse-electric mode is some 3 orders of magnitude less than would be necessary to maintain Io's estimated heat flow.

099.147 **Spectrophotometry of Io: preliminary Voyager 1 results.** L. Soderblom, T. Johnson, D. Morrison, E. Danielson, B. Smith, J. Veverka, A. Cook, C. Sagan, P. Kupferman, D. Pieri, J. Mosher, C. Avis, J. Gradie, T. Clancy.
Geophys. Res. Lett., Vol. 7, 963 - 966 (1980).

Multispectral images of Io acquired with the Voyager 1 narrow-angle camera agree with Earth-based spectrophotometry to better than 10%. Although the surface materials have general spectral properties similar to various allotropes of sulfur, their ultraviolet (UV) reflectances are much higher. It is likely that varying amounts of SO_2 frost mixed with or absorbed on sulfur-rich materials raises the UV reflectance. The possible association with large amounts of SO_2 with low temperature forms of sulfur in the white patches on Io is consistent with Io surface models in which SO_2 and S exist in thermally stable stratified zones.

099.148 **Tidal dissipation in Europa: a correction.** P. Cassen, S. J. Peale, R. T. Reynolds.
Geophys. Res. Lett., Vol. 7, 987 - 988 (1980). – See Abstr. 26.099.138.

099.149 **Satellites of Jupiter.** J. Bouška.
Říše hvězd, Vol. 61, 203 - 206, 209 - 212 (1980). In Czech.

099.150 **Jupiter's magnetosphere and radiation belts.** C. F. Kennel, F. V. Coroniti.
Solar system plasma physics, Vol. 2, (see 003.010), 105 - 181 (1979).

The authors present a selective review of the Jovian magnetosphere. They briefly review the general features of the Jovian magnetosphere revealed by Pioneers 10 and 11. They discuss Jupiter's radiation belts, concentrating upon the theoretical understanding achieved thus far. They discuss Pioneer 10 measurements of Jupiter's outer magnetosphere, emphasizing their time dependent nature. Various theories of Jupiter's outer magnetosphere are discussed.

099.151 **Ion erosion on the Galilean satellites of Jupiter.** P. K. Haff, C. C. Watson, T. A. Tombrello.
Proc. Tenth Lunar Planet. Sci. Conf., (see 012.050), p. 1685 - 1699 (1979).

099.152 **The surface of Io: geologic units, morphology, and tectonics.** G. G. Schaber.
Icarus, Vol. 43, 302 - 333 (1980).

A preliminary geologic map, representing 26.5% of the surface of Io, has been compiled using best-resolution (0.5 to 5 km/line pair) Voyager 1 images and (as a base) a preliminary pictorial map of Io. Nine volcanic units are identified, including materials of mountains, plains, flows, cones, and crater vents, in addition to seven types of structural features. The mapped area lies within the longitudinal zone (250 to 323°) of least-abundant SO_2 frost, indicating that other sulfurous components dominate the upper surface layers in this area.

099.153 **Observations of radio radiation of Jupiter at low frequencies (May - October 1974).**
V. P. Grigor'eva, E. S. Sakharova.
Astron. Tsirk., No. 1103, p. 1 - 3 (1980). In Russian.

099.154 **Energetic particles in the pre-dawn magnetotail of Jupiter.**
A. W. Schardt, F. B. McDonald, J. H. Trainor.
NASA Tech. Memo., NASA TM 81991, 60 pp. (1980).

A detailed account is given of the energetic electron and proton populations as observed with Voyagers 1 and 2 during their passes through the dawn magnetotail of Jupiter.

099.155 **Superthermal electrons and Bernstein waves in Jupiter's inner magnetosphere.**
D. D. Barbosa, W. S. Kurth.
J. Geophys. Res., Vol. 85, 6729 - 6742 (1980).

A theoretical model for generation of banded electrostatic emissions by low density, superthermal electrons is developed for application to Jupiter's magnetosphere. The model employs a power law form for the energy dependence and a loss cone pitch angle distribution of the superthermals to drive convective instability of Bernstein modes. The authors concentrate on instability in the upper hybrid band and on lower harmonic bands below the upper hybrid frequency.

099.156 **Some spectral characteristics of the hectometric Jovian emission.** A. Lecacheux, B. Møller-Pedersen, A. C. Riddle, J. B. Pearce, A. Boischot, J. W. Warwick.
J. Geophys. Res., Vol. 85, 6877 - 6882 (1980).

The study of the dynamic spectra of hectometric Jovian emission ($f < 1.3$ MHz) for the period January to June 1978 from the planetary radio astronomy experiment on the Voyager 1 and 2 spacecraft shows that its shape is stable with the rotation of the planet. However, there are noticeable differences between the spectra observed at one-month intervals by one spacecraft, or simultaneously by the two spacecraft, which can be clearly related neither to an effect of Io nor to the Jovicentric declination of the observer.

099.157 **Possible fluid dynamical interpretation of some reported features in the Jovian atmosphere.**
T. Maxworthy, L. G. Redekopp.
Science, Vol. 210, 1350 - 1351 (1980).

Flow patterns of some of the features in the Voyager 1 imaging data for Jupiter appear to be consistent with those predicted by a solitary-wave theory.

099.158 **Lightning on Jupiter: rate, energetics, and effects.** J. S. Lewis.
Science, Vol. 210, 1351 - 1352 (1980).

Voyager data on the optical and radio-frequency detection of lightning discharges in the atmosphere of Jupiter suggest a stroke rate significantly lower than on the earth. The efficiency of conversion of atmospheric convective energy flux into lightning is almost certainly less than on the earth, probably near 10^{-7} rather than the terrestrial value of 10^{-4} At this level the rate of production of complex organic molecules by lightning and by thunder shock waves is negligible compared to the rates of known photochemical processes for forming colored inorganic solids.

099.159 **Stars of Medici (4).** T. Kwast.
Urania Kraków, Vol. 51, 299-302 (1980). In Polish.

099.160 **Methane absorption near λ 6800 Å in the spectra of Jupiter and Saturn. I.**
V. V. Avramchuk, A. I. Karmelyuk, Sh. M. Namazov.
Astrometr. Astrofiz., Vyp. (No.) 42, p. 26 - 36 (1980). In Russian.

The intensity and the fine structure of the methane absorption band near 6800 Å in the spectra of Jupiter and Saturn are investigated. On the spectrograms above 30 band components are detected. Most of them are new. For all these components the values of wavelengths are determined. It is found that the components of the fine structure are distributed chaotically.

099.161 **White ovals in the Jovian south tropical belt.** L. Jovanović.
Vasiona, Année 28, 43 - 44 (1980). In Croatian.

099.162 **Mass-loading and diffusion-loss rates of the Io plasma torus.** D. E. Shemansky.
Astrophys. J., Vol. 242, 1266 - 1277 (1980).

Limits to the mass-loading and diffusion-loss rates of ions in the Io plasma torus have been calculated on the assumption that observed optical emissions are controlled by electron-ion collisions. Calculations of the yield of emission from the vicinity of Io limit the mass-loading rate to the order of 10^{27} s^{-1} for S II or O II, on the grounds that electron-excited emissions associated with the location of Io have not been observed in the optical spectrum. This mass-loading limit is dependent on the assumptions that Io is the source of torus particles and that most of the neutral atoms are converted to ions within 1 R_J of Io. The observed partitioning of sulfur ion species in the hot torus at the time of Voyager 1 encounter indicates that the diffusion-loss time of the ions is of the order of $1/D$ = 100 days. The two results suggest that the energy required to maintain the observed radiated power cannot be supplied by acceleration of ions formed at Io in Jupiter's rotating magnetic field.

099.163 **Multicolour photometry of Jovian disk details. I. Relative spectrophotometry in autumn – winter 1977–1978.**
A. P. Vid'machenko, V. M. Klimenko, A. V. Morozhenko.
Astron. Vestn., Tom 14, 80 - 85 (1980). In Russian.

099.164 **Multicolour photometry of Jovian disk details. II. Absolute brightness coefficient.** V. M. Klimenko.
Astron. Vestn., Tom 14, 148 - 153 (1980). In Russian.

A line parameter list for the ν_2 and ν_4 bands of $^{12}CH_4$ and $^{13}CH_4$, extended to J' = 25 and its application to planetary atmospheres. See Abstr. 022.004.

Quantitative laboratory spectra and spectral line parameters for the ν_2 and ν_4 bands of PH_3 applicable to spectral radiative models of the atmosphere of Jupiter.
See Abstr. 022.006.

Coriolis phenomenon in nuclei and planets: rotation alignment. See Abstr. 022.060.

Pressure dependence of the absolute rate constant for the reaction $OH + C_2H_2$ from 228 – 413K: implications for CO modeling of the Jovian atmosphere.
See Abstr. 022.136.

The decomposition of methane by hot hydrogen atoms: a different approach. See Abstr. 022.137.

Laboratory simultations of central pit craters.
See Abstr. 022.141.

SO_2 frost: UV-visible reflectivity and Io surface coverage. See Abstr. 022.143.

Automatic technique for accurately locating planet centers in Voyager images. See Abstr. 031.553.

The search for evolute flash at Jupiter.
See Abstr. 031.561.

Turbulence in deep occultations.
See Abstr. 031.562.

Absolute brightness temperature measurements at 3.5-mm wavelength. See Abstr. 033.019.

A new high-gain, broadband, steerable array to study Jovian decametric emission. See Abstr. 033.024.

Der Einfluß des Jupiter im Sonnensystem.
See Abstr. 042.067.

Satellites galiléens de Jupiter. Phénomènes et configurations pour 1981. See Abstr. 047.044.

Verändertes Jupiter-Programm Galileo.
See Abstr. 051.018.

Die Galileo - Raumflugmission.
See Abstr. 051.035.

Axisymmetric convection driven by latitudinal temperature gradients in rotating spherical shells.
See Abstr. 062.090.

Physics of heavy ions in the magnetosphere.
See Abstr. 084.101.

An assessment study on non terrestrial frozen ground.
See Abstr. 091.018.

Disk-satellite interactions. See Abstr. 091.019.

Magnetospheric influences on the evolution of planetary ring systems. See Abstr. 091.028.

Dissipative processes and tidal Q for the major planets. See Abstr. 091.029.

The atmospheres of Venus and Jupiter.
See Abstr. 093.030.

Spectroscopy of faint asteroids, satellites, and comets. See Abstr. 098.025.

The rings of Saturn, Uranus and Jupiter.
See Abstr. 100.014.

A cosmic dance of moons and rings.
See Abstr. 100.024.

IUE observations of Saturn and Jupiter from 3200 to 1400 Å. See Abstr. 100.034.

Modeling of neutral satellite gas clouds – recent advances. See Abstr. 100.052.

Comets and the Galilean satellites. See Abstr. 102.023

Ions of Jovian origin observed by Voyager 1 and 2 in interplanetary space. See Abstr.106.003.

Errata

099.901 **Correction: "Measurements of wind vectors, eddy momentum transports, and energy conversions in Jupiter's atmosphere from Voyager 1 images "** [Geophys. Res. Lett., Vol. 7, 1 - 4 (1980)]. R. F. Beebe, A. P. Ingersoll, G. E. Hunt, J. L. Mitchell, J.-P. Müller.
Geophys. Res. Lett., Vol. 7, 621 - 622 (1980). – See Abstr. 27.099.036.

099.902 **Erratum: "Non-Io decametric radiation from Jupiter at frequencies above 30 MHz"** [Astron. Astrophys., Vol. 86, 339 - 341 (1980)].
C. H. Barrow, M. D. Desch.
Astron. Astrophys., Vol. 91, 378 (1980). – See Abstr. 27.099.083.

100 Saturn, Saturn Satellites

100.001 **The relative reflectance of Iapetus at 1.6 and 2.2 μm.** G. J. Veeder, D. L. Matson.
Astron. J., Vol. 85, 969 - 972 (1980).

Iapetus was observed at 0.56, 1.6 and 2.2 μm near seven elongations. The infrared reflectance at 2.2 μm relative to 0 0.56 μm varies from 0.6 to 1.6 (±0.1). The low infrared reflectance of the trailing hemisphere of Iapetus is consistent with the previous identification of water ice as the composition of the bright surface. The high infrared reflectance of the leading hemisphere indicates a dark material that is too red to be consistent with a carbonaceous chondritic or "C"-type surface composition.

100.002 **Astrometric observations of the satellites of Saturn during 1975 - 1976.**
J. D. Mulholland, P. J. Shelus.
Astron. J., Vol. 85, 1112 - 1116 (1980).

Absolute astrometric positions of satellites I-IX of Saturn have been obtained from plates taken with the 2.1-m Otto Struve reflector of the McDonald Observatory during 1975 and 1976. The positions are presented both in absolute celestial coordinates and in intersatellitary relative coordinates.

100.003 **Astrometric observations of satellites of the outer planets. IV. The satellites of Saturn during 1977.**
P. Seitzer, P. A. Ianna.
Astron. J., Vol. 85, 1117 - 1121 (1980).

The authors present accurate photographic positions obtained with the McCormick 67-cm refractor of the satellites of Saturn during the opposition of 1977.

100.004 **Saturn 1978–1979.**
A. W. Heath.
J. British Astron. Assoc., Vol. 90, 451 - 463 (1980). – A report of the Saturn section.

100.005 **Saturn's E ring revisited.**
W. A. Feibelman, D. A. Klinglesmith III.
Science, Vol. 209, 277 - 279 (1980).

Saturn's E ring is revealed by image processing of direct photographs of the 1966 edge-on presentation of the planet's ring plane. Two different techniques were used: scanning with an image quantizer operated in the derivative mode and computer-enhanced background subtraction from digitized images.

100.006 **Saturn: tropospheric ammonia and nitrogen.**
S. K. Atreya, W. R. Kuhn, T. M. Donahue.
Geophys. Res. Lett., Vol. 7, 474 - 476 (1980).

Photochemical calculations based on recent data on the Saturn temperature structure and Lyman alpha albedo indicate that detectable amounts of gaseous ammonia may exist between 20 and 35 km above the cloud tops. An instrument that might be able to observe this gas is the spectrometer on board the International Ultraviolet Explorer satellite.

100.007 **Discussion of the Pioneer 11 observations of the F ring of Saturn.** W.-H. Ip.
Nature, Vol. 287, 126 - 128 (1980).

The photopolarimetric and charged particle data from the Pioneer 11 Saturn encounter were used as diagnostic tools in a preliminary analysis of the physical nature of the F ring of Saturn. It is argued that the F ring may have two components, with the outer one composed of solid particles of centimetre-diameter and the inner one of micrometre-sized dust particles. Also, the outer component may have two rings instead of one.

100.008 **The 20-μm brightness temperature of the unilluminated side of Saturn's rings.**
A. T. Tokunaga, J. Caldwell, I. G. Nolt.
Nature, Vol. 287, 212 - 214 (1980).

The 20-μm brightness temperature of the unilluminated (north) side of Saturn's rings is reported to be 56±1 K. The authors discuss the implications of the IR data for several published ring models.

100.009 **Saturn's magnetic field: predictions, measurements and interpretation.**
Sh. Sh. Dolginov, V. A. Sharova.
Pis'ma Astron. Zh., Tom 6, 437 - 441 (1980). In Russian. English translation in Soviet Astron. Lett., Vol. 6.

The magnitude and the unique topology of Saturn's magnetic field is naturally explained in the frame of the dynamo precession model taking into account the influence of the sun and the satellites Mimas and Tethis on Saturn's precession, as well as the difference in the conductivities of Jupiter's and Saturn's liquid cores.

100.010 **Image enhancement of Saturn's E-ring from the 1966 discovery photographs.**
W. A. Feibelman, D. A. Klinglesmith III.
Bull. American Astron. Soc., Vol. 12, 510 (1980). – Abstract.

100.011 **Voyager detection of nonthermal radio emission from Saturn.** M. L. Kaiser, M. D. Desch.
Bull. American Astron. Soc., Vol. 12, 510 (1980). – Abstract.

100.012 **The middle-infrared spectrum of Saturn: evidence for phosphine and upper limits to other trace atmospheric constituents.** H. P. Larson, U. Fink, H. A. Smith, D. S. Davis.
Astrophys. J., Vol. 240, 327 - 337 (1980).

The authors have observed Saturn at high spectral resolution ($\Delta\sigma \gtrsim 0.3\ cm^{-1}$) in the middle-IR spectral region (2.5–5.6 μm) using a Fourier spectrometer at ground-based and airborne observatories. These spectra establish that PH_3 exists on Saturn with an abundance at least equal to the solar P/H value, and probably enhanced by about a factor of 2. The authors find no evidence for gaseous or solid NH_3 on Saturn throughout this spectral region, and they determine upper limits to several other molecules (H_2O, HCN, SiH_4, and GeH_4). By making frequent comparisons of their spectral data for Saturn observations and interpretations of IR studies of Jupiter, the authors find that these two planetary atmospheres are chemically similar, with major observational differences accounted for by reduced H_2O and NH_3 abundances in Saturn's colder atmosphere.

100.013 **Voyager detection of nonthermal radio emission from Saturn.**
M. L. Kaiser, M. D. Desch, J. W. Warwick, J. B. Pearce.
Science, Vol. 209, 1238 - 1240 (1980).

The planetary radio astronomy experiment on board the Voyager spacecraft has detected bursts of nonthermal radio noise from Saturn occurring near 200 kilohertz, with a peak flux density comparable to higher frequency Jovian emissions. The radiation is right-hand polarized and is most likely emitted in the extraordinary magnetoionic mode from Saturn's northern hemisphere. Modulation that is consistent with a planetary rotation period of 10 hours 39.9 minutes is apparent in the data.

100.014 **The rings of Saturn, Uranus and Jupiter.**
A. Brahic.
J. Phys. Colloq., Vol. 41, No. C-3, (see 012.010), p. C3/1 - 16

(1980). In French. – Abstr. in Phys. Abstr., Vol. 83, Abstr. 82281 (1980).

100.015 **The 1978 - 1979 apparition of Saturn.**
J. L. Benton, Jr.
Strolling Astron., Vol. 28, 140 - 150 (1980).

100.016 **On an observation of Saturn: the eye and the astronomical observer.** W. Sheehan.
Strolling Astron., Vol. 28, 150 - 154 (1980).

100.017 **Some recent notes on Saturn.**
A. W. Heath.
Strolling Astron., Vol. 28, 165 - 166 (1980).

100.018 **Tidal dissipation, orbital evolution, and the nature of Saturn's inner satellites.**
S. J. Peale, P. Cassen, R. T. Reynolds.
Icarus, Vol. 43, 65 - 72 (1980).

Estimates of tidal damping times of the orbital eccentricities of Saturn's inner satellites place constraints on some satellite rigidities and dissipation functions Q. These constraints favor rocklike rather than ice-like properties for Mimas and probably Dione. Photometric and other observational data are consistent with relatively higher densities for these two satellites, but require lower densities for Tethys, Enceladus, and Rhea. This leads to a nonmonotonic density distribution for Saturn's inner satellites, apparently determined by different mass fractions of rocky materials. In spite of the consequences of tidal dissipation for the orbital eccentricity decay and implications for satellite compositions, tidal heating is not an important contributor to the thermal history of any Saturnian satellite.

100.019 **A numerical study of aerosol growth in Titan's atmosphere.** M. Podolak, E. Podolak.
Icarus, Vol. 43, 73 - 84 (1980).

The authors present a simple model for the formation and growth of photochemical aerosols in the atmosphere of Titan. They show that, in general, an optically thick layer of particles in the size range required by models of Titan cannot be obtained at pressures less than about 2 mbar. Since the thin model of Titan's atmosphere requires that the inversion not extend below pressures of 0.11 mbar, it seems to be ruled out by the calculations.

100.020 **A new satellite of Saturn: Dione B.**
J. Lecacheux, P. Laques, L. Vapillon, A. Auge, R. Despiau.
Icarus, Vol. 43, 111 - 115 (1980).

On March 1, 1980, observations of Saturn from Pic-du-Midi Observatory using a Lallemand electronographic camera led to the discovery of a new satellite (V magnitude $\simeq +17.5$) whose orbital period is surprisingly similar to that of Dione.

100.021 **A new Saturnian satellite near Dione's L4 Point.**
H. J. Reitsema, B. A. Smith, S. M. Larson.
Icarus, Vol. 43, 116 - 119 (1980).

A new satellite of Saturn, discovered by Laques and Lecacheux, has been observed on a number of occasions and its orbit determined. It has a period of 2.73614 ± 0.00006 days and occupies the leading Lagrangian point (L4) of Dione. A second object has been observed on one night which may be near the L5 point of Dione.

100.022 **Osservazioni sistematiche di Saturno: presentazione 1978 - 1979.** E. Sassone Corsi, P. Sassone Corsi.
Astronomia, N. 3, p. 30 - 36 (1980).

100.023 **Saturn's "new" satellites: a perspective.**
S. Larson, J. W. Fountain.
Sky Telesc., Vol. 60, 356 - 360 (1980).

100.024 **A cosmic dance of moons and rings.**
J. Loudon.
Sky Telesc., Vol. 60, 358 - 359 (1980).

100.025 **Voyager 1 and Saturn.**
Sky Telesc., Vol. 60, 360 - 361 (1980).

100.026 **Strong interaction between the ring system and the ionosphere of Saturn.** M. Shimizu.
13th Lunar and Planetary Symposium, (see 012.018), p. 41 - 44 (1980).

An abnormally low electron density in the Saturnian ionosphere observed by the radio occultation experiment of the Pioneer 11 may be explained in terms of the contamination of water in the Saturnian upper atmosphere from its ring system.

100.027 **The two-components particle model for the Saturn rings.** Y. Kawata.
13th Lunar and Planetary Symposium, (see 012.018), p. 62 - 69 (1980).

The present investigation shows that the two-components particle model can satisfy not only the infrared tilt effect observations but also the observations in the visible region. The scattering characteristics for two kinds of ring particles are obtained. It is found that the Saturn ring B consists of three stratified layers. The ring particles in the middle layer seem to be larger and more absorbing, and their scattering patterns are more backward scattering than those in other layers.

100.028 **The east-west asymmetry of Saturn's rings: a possible indicator of meteoroids?**
J. O. Piironen, J. Lukkari.
Moon Planets, Vol. 23, 373 - 377 (1980).

There seems to be a relationship between the activity of the east-west asymmetry of Saturn's rings and activities of bright meteors and meteorite falls. The relationship supports the suggestion (Hämeen-Antilla and Itävuo, 1976) which explains the asymmetry as a product of impacting bodies on the rings. The correlation indicates that presence of interstellar meteoroids especially during the high activity of meteors and meteorites. Other observed peculiarities of the rings also seem to accumulate around unusually high falling rates of meteoroidal matter.

100.029 **A close look at Saturn.** P. Campbell.
Nature, Vol. 288, 9 - 10 (1980).

100.030 **Determination of the orbital elements of Janus, the tenth satellite of Saturn.** A. Dollfus.
C. R. Acad. Sci. Paris, Tome 291, Sér. B, 177 - 180 (1980). In French.

The tenth satellite of Saturn, Janus was discovered in 1966. Its orbit is equatorial, circular and close to the outer edge of ring A. The new satellite observations of March 1980 permit the reidentification of Janus and the determination of its most probable period, 0.694684 day, but without excluding, however, two other possible values: 0.694783 or 0.694584 day; the orbital semi-axis is 151 500 km. Another object, less bright and provisionally designated 1980 S3, is associated with Janus in a similar orbit. These two objects have to be associated gravitationally.

100.031 **The rings of Saturn: new near-infrared reflectance measurements and a 0.325 - 4.08 μm summary.**
R. N. Clark, T. B. McCord.
Icarus, Vol. 43, 161 - 168 (1980).

A new high photometric precision reflectance spectum of Saturn's rings covering the spectral region 0.65 to 2.5-μm is presented and three previously unreported absorption features at 1.25, 0.85, and probably 1.04 μm are identified. The 1.25- and 1.04-μm absorptions are due to water ice. The 0.85-μm feature may be due to a combination of 0.81- and 0.90-μm ice absorptions. Another possibility is that the 0.85-μm band is due to Fe^{3+} -bearing minerals in an ice-mineral mixture. This explanation could also account for the drop in the visible and ultraviolet reflectance and the rise in reflectance around 3.6μm. A composite spectrum from 0.325 to 4.08μm is presented.

100.032 **Interior structure of Saturn inferred from Pioneer 11 gravity data.**
J. J. MacFarlane, W. B. Hubbard, J. D. Anderson, G. W. Null, E. D. Biller.
Bull. American Astron. Soc., Vol. 12, 666 - 667 (1980). Abstract.

100.033 **IUE observations of Saturn.**
J. T. Clarke, H. W. Moos, S. K. Atreya, A. L. Lane.
Bull. American Astron. Soc., Vol. 12, 667 (1980). – Abstract.

100.034 **IUE observations of Saturn and Jupiter from 3200 to 1400 Å.** V. Moore.
Bull. American Astron. Soc., Vol. 12, 667 (1980). – Abstract.

100.035 **Photometry and polarimetry of Saturn at large phase angles.**
R. McMillan, M. G. Tomasko, L. R. Doose, J. P. Dilley, N. D. Castillo.
Bull. American Astron. Soc., Vol. 12, 667 (1980). – Abstract.

100.036 **Saturn: new UBV photoelectric pinhole scans of the disk.** M. J. Price, O. G. Franz.
Bull. American Astron. Soc., Vol. 12, 667 (1980). – Abstract.

100.037 **Modelling the albedo variation of Saturn from 0.4 to 0.81 μm.** J. T. Bergstralh.
Bull. American Astron. Soc., Vol. 12, 668 (1980). – Abstract.

100.038 **Search for aerosol on Saturn using an eclipse of Titan.** D. W. Smith, P. E. Johnson, R. W. Shorthill, E. Budding, A. S. Asaad.
Bull. American Astron. Soc., Vol. 12, 668 (1980). – Abstract.

100.039 **Interferometric observations of Saturn at 3.4 millimeter wavelength.** J. N. Cuzzi, W. J. Welch.
Bull. American Astron. Soc., Vol. 12, 668 (1980). – Abstract.

100.040 **Longitudinal variability of methane and ammonia on Saturn.** A. L. Cochran, W. D. Cochran.
Bull. American Astron. Soc., Vol. 12, 668 - 669 (1980). Abstract.

100.041 **A study of ethane on Saturn in the 3 μm region.**
G. Bjoraker, H. P. Larson, U. Fink.
Bull. American Astron. Soc., Vol. 12, 669 (1980). – Abstract.

100.042 **The detection of C_2H_2 on Saturn and Titan.**
A. Tokunaga, S. Beck, T. Geballe, J. Lacy.
Bull. American Astron. Soc., Vol. 12, 669 (1980). – Abstract.

100.043 **Observations of eclipses of Titan by Saturn in CH_4 and C_2H_6 spectral bands and at 20μ.**
M. Combes, T. Encrenaz, N. Epchtein, J. Lecacheux, T. Owen, Y. Andrillat.
Bull. American Astron. Soc., Vol. 12, 670 (1980). – Abstract.

100.044 **A spectroscopic search for N_2^+ emission in the polar regions of Saturn and on the night side of Venus.**
E. S. Barker, L. M. Trafton, S. K. Atreya.
Bull. American Astron. Soc., Vol. 12, 670 (1980). – Abstract.

100.045 **The radius of Titan from Pioneer Saturn data.**
P. H. Smith.
Bull. American Astron. Soc., Vol. 12, 670 - 671 (1980). Abstract.

100.046 **Photometry and polarimetry of Titan at large phase angles.** M. G. Tomasko, P. H. Smith.
Bull. American Astron. Soc., Vol. 12, 671 (1980). – Abstract.

100.047 **A physical model of Titan's clouds.**
O. B. Toon, R. P. Turco, J. B. Pollack.
Bull. American Astron. Soc., Vol. 12, 671 (1980). – Abstract.

100.048 **The atmospheric temperature structure of Titan.**
K. A. Rages, J. B. Pollack, O. B. Toon.
Bull. American Astron. Soc., Vol. 12, 672 (1980). – Abstract.

100.049 **Thermal structure of the upper atmosphere of Titan.** A. J. Friedson, Y. L. Yung.
Bull. American Astron. Soc., Vol. 12, 672 (1980). – Abstract.

100.050 **Effective pressure estimates in Titan's atmosphere from modeling observations of the single methane absorption line at 6818.95 Å.**
L. P. Giver, W. H. Smith, K. Baines, K. A. Rages.
Bull. American Astron. Soc., Vol. 12, 672 (1980). – Abstract.

100.051 **Titan: photochemistry of Danielson dust and albedo variations over the sunspot cycle.**
Y. L. Yung, M. Allen, J. P. Pinto.
Bull. American Astron. Soc., Vol. 12, 672 (1980). – Abstract.

100.052 **Modeling of neutral satellite gas clouds – recent advances.** W. H. Smyth.
Bull. American Astron. Soc., Vol. 12, 674 (1980). – Abstract.

100.053 **Saturn: resonance and superposition principle as applied to tidal-elongation waves.** J. Boynton.
Bull. American Astron. Soc., Vol. 12, 696 (1980). – Abstract.

100.054 **Photometry and polarimetry of Saturn's rings from Pioneer/Saturn.**
L. W. Esposito, J. P. Dilley, J. W. Fountain.
Bull. American Astron. Soc., Vol. 12, 698 (1980). – Abstract.

100.055 **Saturn ring C properties.** J. Dilley.
Bull. American Astron. Soc., Vol. 12, 699 (1980). Abstract.

100.056 **Saturn's rings: azimuthal variations, phase curves, and radial profiles observed at three tilt angles.**
W. Thompson, K. Lumme, W. M. Irvine, W. A. Baum, L. W. Esposito.
Bull. American Astron. Soc., Vol. 12, 699 (1980). – Abstract.

100.057 **Saturn's rings in the infrared: Pioneer and earth-based results.** L. Froidevaux, A. P. Ingersoll.
Bull. American Astron. Soc., Vol. 12, 699 (1980). – Abstract.

100.058 **Saturn's rings: 3-mm low-inclination observations and derived properties.**
E. E. Epstein, M. A. Janssen, J. N. Cuzzi.
Bull. American Astron. Soc., Vol. 12, 699 - 700 (1980). Abstract.

100.059 **Saturn's rings: comments on radar observations.**
S. J. Ostro, G. H. Pettengill, D. B. Campbell.
Bull. American Astron. Soc., Vol. 12, 700 (1980). – Abstract.

100.060 **Bimodality of the particles in Saturn's A and B rings.**
T. Gehrels.
Bull. American Astron. Soc., Vol. 12, 700 (1980). – Abstract.

100.061 **Observations of Saturn's edgewise rings in France.** A. Brahic, J. Lecacheux, B. Sicardy.
Bull. American Astron. Soc., Vol. 12, 700 (1980). – Abstract.

100.062 **Profile of Saturn's E ring.** W. A. Baum, T. Kreidl, J. A. Westphal, G. E. Danielson, P. K. Seidelmann, D. Pascu, D. G. Currie.
Bull. American Astron. Soc., Vol. 12, 700 - 701 (1980). Abstract.

100.063 **The E ring of Saturn.** H. J. Reitsema, B. A. Smith, S. M. Larson.
Bull. American Astron. Soc., Vol. 12, 701 (1980). – Abstract.

100.064 **Infrared photometry of Saturn's E ring.** R. J. Terrile, A. Tokunaga.
Bull. American Astron. Soc., Vol. 12, 701 (1980). – Abstract.

100.065 **Sizes and densities of the Saturn satellites: a pre-Voyager analysis.** D. Morrison.
Bull. American Astron. Soc., Vol. 12, 727 (1980). – Abstract.

100.066 **Saturn's syzygistic co-orbital satellites.** B. A. Smith, H. J. Reitsema, J. W. Fountain, S. M. Larson.
Bull. American Astron. Soc., Vol. 12, 727 - 728 (1980). Abstract.

100.067 **Searches for Saturn outer ring and inner satellites with a focal coronograph.**
A. Dollfus, S. Brunier.
Bull. American Astron. Soc., Vol. 12, 728 (1980). – Abstract.

100.068 **The 1966 observations of Saturn's faint inner satellites.**
S. M. Larson, J. W. Fountain, B. A. Smith, H. J. Reitsema.
Bull. American Astron. Soc., Vol. 12, 728 (1980). – Abstract.

100.069 **Observations of the new Saturnian satellite Dione B.** H. J. Reitsema, B. A. Smith, J. W. Fountain, S. M. Larson.
Bull. American Astron. Soc., Vol. 12, 728 (1980). – Abstract.

100.070 **The new satellite Dione B and outer ring of Saturn.** P. L. Lamy, N. Mauron.
Bull. American Astron. Soc., Vol. 12, 728 - 729 (1980). Abstract.

100.071 **Saturn's inner moons and rings: their origin and composition.** A. J. R. Prentice.
Bull. American Astron. Soc., Vol. 12, 729 (1980). – Abstract.

100.072 **Astrometric observations of Saturn.** D. Pascu, R. E. Schmidt.
Bull. American Astron. Soc., Vol. 12, 740 - 741 (1980). Abstract.

100.073 **The ephemeris of Saturn for the Voyager mission.** E. M. Standish.
Bull. American Astron. Soc., Vol. 12, 742 (1980). – Abstract.

100.074 **Saturn gravity results obtained from Pioneer Saturn spacecraft tracking data and earth-based Saturn satellite data.**
G. W. Null, J. D. Anderson, E. D. Biller, E. L. Lau.
Bull. American Astron. Soc., Vol. 12, 742 (1980). – Abstract.

100.075 **First Voyager view of the rings of Saturn.** S. A. Collins, A. F. Cook II, J. N. Cuzzi, G. E. Danielson, G. E. Hunt, T. V. Johnson, D. Morrison, T. Owen, J. B. Pollack, B. A. Smith, R. J. Terrile.
Nature, Vol. 288, 439 - 442 (1980).

Pictures taken from Voyager 1 between 3 September and 13 October 1980 at a resolution of ~1,000 km/lp reveal new data on Saturn rings.

100.076 **On the relationship between secular brightness changes of Titan and solar variability.**
J. B. Pollack, K. Rages, O. B. Toon, Y. L. Yung.
Geophys. Res. Lett., Vol. 7, 829 - 832 (1980).

Titan's geometric albedo varied noticeable from 1972 to 1978, in phase with variations in solar activity. The authors carry out a series of radiative transfer and aerosol formation calculations in order to demonstrate the feasibility of the following scenario for these secular brightness changes: solar activity changes, especially in the UV output of the sun, result in alterations to the mass production rate of aerosols in Titan's atmosphere, which lead to modifications of their microphysical properties. The latter, in turn, cause the albedo to vary.

100.077 **Satellites of Saturn.** IAU Circ., Nos. 3491, 3495 - 3497, 3532, 3534, 3539, 3545, 3549 (1980).

100.078 **Saturn.** IAU Circ., No. 3507 (1980).

100.079 **ISVTOP (*International Saturn Voyager Telescope Observations*).**
Yamamoto Circ., No. 1946 (1980).

100.080. **Het Pioneer-onderzoek van Saturnus.** T. Gehrels, O. Namba.
Zenit, 7e Jaarg., 322 - 330 (1980).

100.081 **Saturnerkundung durch Voyager 1 beginnt.** H. Mucke.
Sternenbote, 23. Jahrg., 166 - 171 (1980).

100.082 **Voyager 1 beim Saturn: erste Ergebnisse.** H. Mucke.
Sternenbote, 23. Jahrg., 186 - 194 (1980).

100.083 **Brightness temperatures of Saturn's disk and rings at 400 and 700 micrometers.**
S. E. Whitcomb, R. H. Hildebrand, J. Keene.
Science, Vol. 210, 788 - 789 (1980).

Saturn was observed in two broad submillimeter photometric bands with the rings nearly edge-on. The observed brightness temperatures fall below the prediction of atmospheric models constructed from data at shorter wavelengths, indicating the presence of an opacity source besides pressure-broadened hydrogen lines in the submillimeter region. In combination with earlier measurements at larger inclination angles, these results yield a 400-micrometer brightness temperature for the rings of ~75 K.

100.084 **On the thickness of Saturn rings.** B. Sicardy, G. Wlérick, A. Brahic, P. Laques, J. Lecacheux, B. Servan, R. Despiau, D. Michet, L. Renard.
C. R. Acad. Sci. Paris, Tome 291, Sér. B, 209 - 214 (1980). In French.

Electronographic plates of Saturn were taken during the transit of the Earth through the ring plane. The authors deduce an apparent photometric ring thickness equal to 1.5 ± 0.3 km. For a homogeneous layer of small particles colliding inelastically, theory predicts a thickness of the order of a few particles radii, i. e. a few tens of meters. The observed brightness could be explained by the E ring, the brightness of large chunks, condensations and warping of the ring.

100.085 **Voyager 1 at Saturn.** M. M. Waldrop.
Science, Vol. 210, 1107 - 1111 (1980).

100.086 **Rings within rings within rings within ...**
R. A. Kerr.
Science, Vol. 210, 1111 - 1113 (1980).

Theories already proposed for the Uranian rings could explain much of the odd behavior of rings and satellites around Saturn.

100.087 **Saturn – the latitudes of the belts 1946 - 1976.**
A. J. Hollis.
J. British Astron. Assoc., Vol. 91, 41 - 53 (1980).

Variations in the apparent visibility of the planetary features are noted as is the drift of some belts in latitude. The interaction of belts is noted.

100.088 **If you've seen one magnetosphere, you haven't seen them all: energetic particle observations in the Saturn magnetosphere.**
F. B. McDonald, A. W. Schardt, J. H. Trainor.
NASA Tech. Memo., NASA TM 81980, 68 pp. (1980).
See Abstr. 100.103.

100.089 **Closing in on Saturn.** L. J. Robinson.
Sky Telesc., Vol. 60, 481 (1980).

100.090 **Quer durch das Saturnsystem. Phantastische Ergebnisse der Mission von Voyager 1 beim Ringplaneten.**
W. Engelhardt.
Umschau, 80. Jahrg., 743 - 744, 746 - 747 (1980).

The fly-by of the NASA space probe "Voyager 1" past Saturn in mid-November, 1980, has provided profuse information on the planet, its moons, and its rings. An initial evaluation of the television pictures has shown, in particular, that the system of rings is much more complicated than was believed to date.

100.091 **Saturn's magnetosphere and its interaction with the solar wind.**
E. J. Smith, L. Davis, Jr., D. E. Jones, P. J. Coleman, Jr., D. S. Colburn, P. Dyal, C. P. Sonett.
J. Geophys. Res., Vol. 85, 5655 - 5674 (1980).

The paper is a comprehensive report of magnetic field observations made in the vicinity of Saturn by the Pioneer 11 vector helium magnetometer. The planetary magnetic field of Saturn, the properties of the magnetosphere near the noon and dawn meridians, and the characteristics of the Saturnian magnetopause and bow shock are described.

100.092 **The magnetic field of Saturn: further studies of the Pioneer 11 observations.**
M. H. Acuña, N. F. Ness, J. E. P. Connerney.
J. Geophys. Res., Vol. 85, 5675 - 5678 (1980).

It is important to emphasize that the Pioneer 11 encounter trajectory provided a very limited sampling of the planetary field in terms of latitude and longitude coverage of the planet at close range ($r<5R_s$). It is the purpose of this paper to explore some of the possible consequences of this trajectory upon the analysis of the magnetic field data obtained by the Goddard Space Flight Center high-field flux gate magnetometer experiment.

100.093 **Sources and sinks of energetic electrons and protons in Saturn's magnetosphere.**
J. A. Van Allen, B. A. Randall, M. F. Thomsen.
J. Geophys. Res., Vol. 85, 5679 - 5694 (1980).

This paper reports the results of continuing analysis and interpretation of energetic particle observations obtained by the University of Iowa instrument on Pioneer 11 during traversal of Saturn's magnetosphere in August - September 1979. On the basis of the radial dependence of the phase space density of very energetic protons ($E_p > 80$ MeV) and estimates of the necessary source strength, it is reasonably certain that cosmic ray neutron albedo from the planet's atmosphere and Rings A and B is the source of such particles. The spectrum of electrons $0.040 < E_e <$ (few MeV) and the radial dependence of the phase space density of such electrons are derived. The source of these electrons is at the magnetosheath. The source of protons $E_p \sim 1$ MeV is also at the magnetosheath and may be either thermalized solar wind or energetic interplanetary protons which were unusually abundant at the time of Pioneer 11's encounter.

100.094 **Plasmas in Saturn's magnetosphere,**
L. A. Frank, B. G. Burek, K. L. Ackerson, J. H. Wolfe, J. D. Mihalov.
J. Geophys. Res., Vol. 85, 5695 - 5708 (1980).

The solar wind plasma analyzer on board Pioneer 11 provides first observations of low-energy positive ions in the magnetosphere of Saturn. Measurable intensities of ions within the energy per unit charge range 100 eV to 8 keV are present over the planetocentric radial distance range ~4 - 16 R_s in the dayside magnetosphere. The properties of the plasmas are discussed in detail.

100.095 **The energetic charged particle absorption signature of Mimas.**
J. A. Van Allen, M. F. Thomsen, B. A. Randall.
J. Geophys. Res., Vol. 85, 5709 - 5718 (1980).

Data are presented for a 1-min dip in electron intensity that was observed as Pioneer 11 crossed the orbit of Mimas inbound during its encounter with the Saturn system. The authors develop the hypothesis that this microsignature is the shadow of Mimas in the sense that the observations reveal the effect of the satellite on a distribution of particles which interacted with it at some time in the recent past and then drifted in longitude (by means of corotation, gradient, and curvature drifts) to the observational location. Such a point of view leads to a characterization of the electron energy spectrum in Saturn's inner magnetosphere and to an estimate of the radial diffusion coefficient for such electrons.

100.096 **Corotation of Saturn's magnetosphere: evidence from energetic proton anisotropies.**
M. F. Thomsen, T. G. Northrop, A. W. Schardt, J. A. Van Allen,
J. Geophys. Res., Vol. 85, 5725 - 5730 (1980).

The theory and technique of Northrop and Thomsen (1980) are applied to observations of energy spectra and directional anisotropies of 0.61- to 3.41-MeV protons in Saturn's magnetosphere. Fourier fits to 15-min intervals of data are combined with spectral indices to yield information about the $\mathbf{E} \times \mathbf{B}$ convection velocity and temporal changes in the particle population. It is found that although these data do not by themselves allow an unambiguous determination of the extent of corotation in Saturn's outer magnetosphere, they are consistent with exact corotation at the nominal rotation period in the presence of significant but not unreasonable temporal variations in the energetic proton population.

100.097 **The trapped radiations of Saturn and their absorption by satellites and rings.**
J. A. Simpson, T. S. Bastian, D. L. Chenette, R. B. McKibben, K. R. Pyle.
J. Geophys. Res., Vol. 85, 5731 - 5762 (1980).

The paper contains in detail the final energetic charged-particle measurements and new observations obtained from the University of Chicago instrumentation on Pioneer 11, including the overall characteristics of the trapped electron, proton, and helium radiation, which was found to lie inside ~20 Saturn radii from the planet, and the regions extending outward to beyond the planetary bow shocks and into the interplanetary medium. The authors also treat the absorption of trapped radiation by the known satellites and rings and show that additional satellites and/or rings of matter have been discovered by using the charged-particle absorption technique.

The authors also include information such as the trajectory, instrumentation, and absorption due to satellites and rings.

100.098 **Charged particle anisotropies in Saturn's magnetosphere.** T. S. Bastian, D. L. Chenette, J. A. Simpson.
J. Geophys. Res., Vol. 85, 5763 - 5771 (1980).

The authors report observations of anisotropies and pitch angle distributions for 0.5 - 1.8 MeV protons, 7 - 17 MeV electrons and $\gtrsim$3.4 MeV electrons in Saturn's magnetosphere made with the University of Chicago experiments on Pioneer 11. Pitch angle distributions of charged particles throughout the magnetosphere are vital for understanding the physical processes governing the trapped radiation, and the authors have therefore concentrated upon these measurements in this paper.

100.099 **Charged particle diffusion and acceleration in Saturn's radiation belts.**
R. B. McKibben, J. A. Simpson.
J. Geophys. Res., Vol. 85, 5773 - 5783 (1980).

The authors use observations of high-energy ($\gtrsim$1 MeV) protons and electrons in the core of Saturn's magnetosphere ($R \lesssim 4\ R_s$) to investigate the hypothesis that the trapped radiation is accelerated and maintained by inward diffusion in violation of the third adiabatic invariant.

100.100 **High-energy trapped radiation penetrating the rings of Saturn.**
D. L. Chenette, J. F. Cooper, J. H. Eraker, K. R. Pyle, J. A. Simpson.
J. Geophys. Res., Vol. 85, 5785 - 5792 (1980).

Electrons and protons in the energy ranges 2 - 25 MeV and >67 MeV, respectively, have been discovered throughout the entire equatorial region inward from the outer edge of the A ring at $L = 2.3$ to the periapsis of the Pioneer trajectory at $L \sim 1.3$. The trapped radiation which populates Saturn's magnetosphere beyond $L = 2.3$ is totally absent in this region.

100.101 **Sources of high-energy protons in Saturn's magnetosphere.** J. F. Cooper, J. A. Simpson.
J. Geophys. Res., Vol. 85, 5793 - 5802 (1980).

The Pioneer 11 passage through the magnetosphere and trapped radiation of Saturn revealed an especially intense region of high-energy particle fluxes that places unique constraints on models for sources of high-energy protons in the innermost radiation zones. Of special interest is the high-intensity flux of protons with energies >35 MeV in the inner magnetosphere between the *A*-Ring and the orbit of Mimas. The authors have examined the possibility that this component of high-energy protons could result from direct injection in situ by the decay of secondary neutrons produced by high-energy cosmic ray impacts on Saturn's atmosphere or rings.

100.102 **Very energetic protons in Saturn's radiation belt.** W. Fillius, C. McIlwain.
J. Geophys. Res., Vol. 85, 5803 - 5811 (1980).

Very energetic protons are trapped in the inner Saturnian radiation belt. The authors have definitely identified protons of energy >80 MeV and have tentatively detected protons of energy >600 MeV. The spatial distribution of the protons is distinct from that of the trapped electrons, the main difference being that the protons are strongly absorbed by the innermost moons and that the electrons are not. The authors have estimated the source strength for injecting protons by the decay of cosmic ray albedo neutrons generated in the rings of Saturn. The required proton lifetime is ~20 years.

100.103 **If you've seen one magnetosphere, you haven't seen them all: energetic particle observations in the Saturn magnetosphere.**
F. B. McDonald, A. W. Schardt, J. H. Trainor.
J. Geophys. Res., Vol. 85, 5813 - 5830 (1980).

The Goddard Space Flight Center/University of New Hampshire Pioneer 11 cosmic ray experiment was especially well suited for studying Saturn's magnetosphere. Energetic electrons (>0.16 MeV) and protons (E>0.2 MeV) were identified as major magnetospheric constituents, along with trace amounts of alpha particles (>0.65 MeV/nuc) and possible heavier nuclei. A systematic study of the temporal and spatial variations of the angular distributions and energy spectra of the protons and electrons, as well as of the proton phase space distribution, is presented. The entry of solar cosmic rays into the Saturn magnetosphere and possible tail-field configurations are also discussed.

100.104 **Motion of trapped electrons and protons in Saturn's inner magnetosphere.**
M. F. Thomsen, J. A. Van Allen.
J. Geophys. Res., Vol. 85, 5831 - 5834 (1980).

A summary is given of basic formulas for the guiding center motion of energetic charged particles trapped in a dipolar magnetic field. These formulas for longitudinal drift rates, latitudinal bounce periods, equatorial gyroradii, and equatorial gyroperiods are then stated in convenient numerical form for electrons and protons as functions of kinetic energy E, magnetic shell parameter L, and equatorial pitch angle α_0 for a slightly simplified model of the observed magnetic field of Saturn. To aid in the study of the interaction of charged particles with the rings and inner satellites of Saturn, additional formulas are given.

100.105 **A possible magnetic wake of Titan: Pioneer 11 observations.**
D. E. Jones, B. T. Tsurutani, E. J. Smith, R. J. Walker, C. P. Sonett.
J. Geophys. Res., Vol. 85, 5835 - 5840 (1980).

The characteristics of the magnetic signature measured by Pioneer 11 during the several hour interval around the crossing of Titan's L shell have been found to be consistent with plausible models of the interaction of a 200 km/s corotating magnetized plasma with a conducting or magnetized object. The observed manner in which the magnetic fluctuations varied throughout the interval, the average magnetosheath/ambient field amplitude ratio, and the occurrence of a field minimum at the time of closest approach to the axis of the extended Titan tail are all consistent with the detection of a magnetic wake roughly 145 R_T downstream from Titan.

100.106 **Observations of extreme ultraviolet emissions from the Saturnian plasmasphere.**
F. M. Wu, D. L. Judge, R. W. Carlson.
J. Geophys. Res., Vol. 85, 5853 - 5856 (1980).

Emission signals from the Saturnian plasmasphere at wavelengths shortward of 800 Å have been detected by the Pioneer ultraviolet photometer. The surface brightness of the emissions is about 0.3 ± 0.2 R. These short-wavelength emissions are interpreted as arising primarily from the radiative decay of electron excited atomic oxygen ions (O^{++}), in the region between 5 and 7 R_s from Saturn.

100.107 **Structure of the ionosphere and atmosphere of Saturn from Pioneer 11 Saturn radio occultation.**
A. J. Kliore, I. R. Patel, G. F. Lindal, D. N. Sweetnam, H. B. Hotz, J. H. Waite, Jr., T. R. McDonough.
J. Geophys. Res., Vol. 85, 5857 - 5870 (1980).

During the flyby of Saturn by Pioneer 11 on September 1, 1979, radio occultation measurements of the ionosphere and upper neutral atmosphere were made near the terminatior at latitudes 9.7° south and 11.6° south. The principal electron density peak of the ionosphere occurs at an altitude of about 1,800 km and has a magnitude of 11,400 cm^{-3}, with a sharp lower peak of about 9,000 cm^{-3} at about 1,200 km. Ionization appears to extend to 30,000 km, with a broad peak of magnitude 7,000 cm^{-3} at an altitude of about 14,500 km, corresponding to the inner edge of the C ring. In the neutral atmosphere, measurements were made to a pressure level of about 180 mbar, showing a temperature inversion region with a triple minimum. The temperature at the principal minimum at 60,344 km is 88 ± 4 K at a pressure of 74 mbar. Comparisons with the temperature structure derived from the infrared radiometer data suggest that the helium fraction in the Saturnian atmosphere is 10 ± 4%.

100.108 **Saturn's atmospheric temperature structure and heat budget.** G. S. Orton, A. P. Ingersoll.
J. Geophys. Res., Vol. 85, 5871 - 5881 (1980).

The effective temperature of Saturn from 30°S to 10°N is 96.5 ± 2.5 K. The atmospheric mole fraction of H_2 + He is 90 ± 3%. The high value of the effective temperature suggests that Saturn has an additional energy source besides cooling and contraction. The high mole fraction of H_2 suggests that separation of heavier He toward the core may be supplying the additional energy. Atmospheric temperatures in the 60- to 600-mbar range are 2.5 K lower within 7° of the equator than at higher latitudes. An almost isothermal layer exists between 60 and 160 mbar at all latitudes.

100.109 **Cloud forms on Saturn.**
J. J. Burke, T. Gehrels, R. N. Strickland.
J. Geophys. Res., Vol. 85, 5883 - 5890 (1980).

The imaging photopolarimeter (IPP) on Pioneer Saturn provided spin-scan images of Saturn's cloudtops. Only subtle departures from a uniform brightness distribution were apparent, except in the polar regions. At other latitudes the images show only a few features; they primarily support the conclusion that the visible atmosphere is a deep haze. Belts and zones are seen, and some detail in a zone near ± 60° latitude. The North Equatorial Belt consists of two dark belts separated by a brighter zone exhibiting longitudinal structure.

100.110 **Photometry of Saturn at large phase angles.**
M. G. Tomasko, R. S. McMillan, L. R. Doose, N. D. Castillo, J. P. Dilley.
J. Geophys. Res., Vol. 85, 5891 - 5903 (1980).

Limb-darkening curves are derived from Pioneer 11's imaging photometry of a bright zone ($-11° \leqslant$ latitude $\leqslant -6°$) and a dark belt ($-17° \leqslant$ latitude $\leqslant -15°$) on Saturn in red and blue light at 12 phase angles between 13° and 151°. The atmosphere of Saturn scatters light only 1/3 as efficiently in the forward direction as does the atmosphere of Jupiter, and the belt on Saturn is only ~5% fainter than the zone instead of the factor of 2 contrast seen on Jupiter in blue light. Vertically inhomogeneous models of the scattering of light in the atmosphere of Saturn were computed to fit these photometric data by using the cloud structure derived by Gehrels et al. (1980).

100.111 **Graphical measurement of Saturn's oblateness and the radius of the Encke gap.** J. J. Burke.
J. Geophys. Res., Vol. 85, 5904 - 5908 (1980).

The geometric distortions attendant to spin-scan imaging were extraordinarily manifest in the Pioneer encounter with Saturn. By overlapping graphical predictions of these distortions with raster scans of the raw intensity data, the authors are able to determine the precise geometry of a data-taking sequence. The predictions are quite sensitive to the assumed dimensions of Saturn and its rings. The authors accordingly conclude that the ratio of polar to equatorial radius is 0.91 and that the Encke gap is 133,500 km from Saturn's center.

100.112 **Interior structure of Saturn from Pioneer 11 gravity data.**
W. B. Hubbard, J. J. MacFarlane, J. D. Anderson, G. W. Null, E. D. Biller.
J. Geophys. Res., Vol. 85, 5909 - 5916 (1980).

The structure of Saturn is studied via a fourth-order theory for rotating planets and equations of state for the envelope which depend parametrically on the helium abundance, on the starting temperature for the adiabat, and on adopted forms of the pressure-density curve in the region of transition from molecular to metallic hydrogen. Models are constrained by the values of J_2 and J_4 obtained from the Pioneer-Saturn celestial mechanics experiment.

100.113 **Bimodality and the formation of Saturn's ring particles.** T. Gehrels.
J. Geophys. Res., Vol. 85, 5917 - 5924 (1980).

The Pioneer and earth-based observations of Saturn's rings are described and explained with a model for the *B* and *A* rings to some extent of a bimodal size distribution of particles; the larger ones may be original accretions, while small debris diffuses inward through the Cassini Division and the *C* ring. During the formation of the ring system, differential gravitation allowed only silicaceous grains of higher density ($\rho \gtrsim 3$ g cm^{-3}) to coagulate. These serve as interstitial cores for snowy carbonaceous grains, between the times of accretion from interplanetary cometary grains and liberation by collision followed by diffusion inward to Saturn and final evaporation.

100.114 **An extraordinary view of Saturn's rings.**
J. J. Burke, C. E. KenKnight.
J. Geophys. Res., Vol. 85, 5925 - 5928 (1980).

The authors report a unique set of photometric and polarimetric observations made by the imaging photopolarimeter aboard the Pioneer spacecraft shortly after it penetrated the ring plane from above on its way to closest approach. The bright planet limb is observed through the rings, which themselves weakly reflect direct sunlight. Polarizations range from 41% (16%) at the extreme planet limb to 2% (1.5%) at the darkest part of the B ring in the blue (red) channel. Detailed structure of the A ring is evident, most notably the Encke gap.

100.115 **Temperatures and optical depths of Saturn's rings and a brightness temperature for Titan.**
L. Froidevaux, A. P. Ingersoll.
J. Geophys. Res., Vol. 85, 5929 - 5936 (1980).

The Pioneer Saturn infrared radiometer viewed Saturn's rings at 20- and 45-μm wavelength under several conditions of illumination. The data are analyzed to infer radial locations of major ring boundaries, temperatures and temperature gradients, and normal optical depths. Titan's 45-μm brightness temperature is 75 ± 5 K, in good agreement with earth-based observations.

100.116 **Preliminary results of polarimetry and photometry of Titan at large phase angles from Pioneer 11.**
M. G. Tomasko.
J. Geophys. Res., Vol. 85, 5937 - 5942 (1980).

The imaging photopolarimeter (IPP) aboard the Pioneer 11 spacecraft measured the linear poarization of the integrated disk of Titan in red and blue light at a variety of phase angles from 15° to 97°. The large polarization (54%) measured in blue light at 90° phase constrains the size of the aerosols near the top of Titan's atmosphere to have radii smaller than about 0.09 μm if they have a refractive index of 2.0. The polarization at 90° phase in red light is smaller (41%) and implies that the optical thickness of the layer of small aerosols is about 0.6 above and effectively depolarizing surface. The shape of the polarization versus phase curve in blue light suggests increasing particle size with increasing depth into the atmosphere. The limb darkening of Titan was measured at 28°. It is reasonably consistent with that given by the scattering models derived from the polarization observations.

100.117 **The radius of Titan from Pioneer Saturn data.**
P. H. Smith.
J. Geophys. Res., Vol. 85, 5943 - 5947 (1980).

The radius of Titan has been determined from Pioneer Saturn observations. The data sets, one in the red (0.64 μm), and one in the blue (0.44 μm), yield radii of 2840 ± 25 km and 2880 ± 22 km, respectively. The discrepancy between the radius values in the two colors is felt to be the consequence of an optically thin submicron haze above the nominal haze layer. Using a cloud model derived from the Pioneer Saturn data, the lunar occultation radius (Elliot et al., 1975) has been revised to 2845 ± 40 km, in good agreement with the present result. A lower limit of 1.37 g cm^{-3} can now be set on Titan's bulk density, using the red radius at the haze top. The actual bulk density probably falls in the range 1.65 - 1.85 g cm^{-3} for an atmosphere of 150- to 250-km thickness below the red limb.

100.118 **Photometry and polarimetry of Saturn's rings from Pioneer Saturn.**
L. W. Esposito, J. P. Dilley, J. W. Fountain.
J. Geophys. Res., Vol. 85, 5948 - 5956 (1980).

Images of Saturn's rings from Pioneer 11 provide the first close views of the unilluminated side of Saturn's rings, the rings' shadows on the planet, and the directly transmitted light from the planet which shines through the rings. These data are analyzed to determine the optical depth of the rings as a function of radial distance from the planet and to constrain the individual and bulk properties of the scattering particles in the rings. Polarimetry of the illuminated side of the rings shows the rings to be unlike other solar system objects in its bulk polarizing properties.

100.119 **Saturn's satellite situation.** B. G. Marsden.
J. Geophys. Res., Vol. 85, 5957 - 5958 (1980).

The situation with regard to the recent claims of discoveries of Saturnian satellites is reviewed. It appears that Saturn now has twelve definite satellites.

100.120 **Lyman-alpha observations in the vicinity of Saturn with Copernicus.**
E. Barker, S. Cazes, C. Emerich, A. Vidal-Madjar, T. Owen.
Astrophys. J., Vol. 242, 383 - 394 (1980).

High-resolution $L\alpha$ observations of the Saturn vicinity showed that near a minimum of solar activity the emissions related to several sources are 250 ± 50 rayleighs for the interplanetary medium in a near-downwind direction, less than 100 rayleighs for the rings, 200 ± 100 rayleighs for a torus linked to the Titan orbit, and 1400 ± 450 rayleighs for the disk of Saturn. These results induce some constraints through the corresponding theoretical evaluations: the B rings as the primary source of the atoms for the ring emissions; an efficient production mechanism for hydrogen atoms in the Titan torus; and a slightly larger eddy diffusion coefficient in the Saturn atmosphere than in the Jupiter atmosphere near solar minimum.

100.121 **Radius and brightness temperature observations of Titan at centimeter wavelengths by the Very Large Array.** W. Jaffe, J. Caldwell, T. Owen.
Astrophys. J., Vol. 242, 806 - 811 (1980).

The authors present brightness and radius measurements of the surface of Titan at 6, 2, and 1.3 cm wavelengths obtained with the Very Large Array radio interferometer. The new data do not permit a choice between an inversion model for the atmosphere of Titan that predicts a surface temperature of 78 K (value from J. Caldwell) and a model with both a stratospheric temperature inversion and a modest greenhouse effect that would increase the surface temperature by 10–40 K

100.122 **Titan: aerosol photochemistry and variations related to the sunspot cycle.**
M. Allen, J. P. Pinto, Y. L. Yung.
Astrophys. J., Lett., Vol. 242, L125 - L128 (1980) = Contrib. No. 3436, Div. Geol. Planet. Sci., Calif. Inst. Technol.

A photochemical theory is proposed for producing complex polymers in a methane atmosphere. It is argued that the polyacetylenes are the most likely precursor molecules for the formation of the stratospheric haze layer on Titan. The production of polyacetylenes involves a strong positive feedback, leading to more production of polyacetylenes. The thermosphere of Titan may undergo substantial expansion and contraction over a solar cycle, with important consequences for the chemistry of the upper atmosphere.

100.123 **How many satellites does Saturn have?**
V. Pohánka.
Kozmos, Vol. 11, 170 - 172 (1980). In Slovak.

100.124 **Satellites of Saturn.** J. Bouška.
Říše hvězd, Vol. 61, 183 - 187, 250 - 251 (1980). In Czech.

100.125 **Observations of Saturn on the 16" astrograph of the Engelhardt Astronomical Observatory in 1973 - 1975.** I. G. Chugunov, N. A. Smirnova.
Izv. Astron. Ehngel'gardt. Obs., No. 45, p. 82 - 87 (1978). In Russian.

100.126 **A physical model of Titan's clouds.**
O. B. Toon, R. P. Turco, J. B. Pollack.
Icarus, Vol. 43, 260 - 282 (1980).

The authors have constructed a model of the physical processes controlling Titan's clouds. Their model produces clouds that qualitatively match the present observational constraints in a wide variety of model atmospheres. They find the following: (1) high atmospheric temperatures (160°K) are important so that there is a large scale height in the first few optical depths of cloud; (2) the aerosol mass production occurs at very low aerosol optical depth so that the cloud particles do not directly affect the photochemistry producing them; (3) the production rate of aerosol mass by chemical processes is probably greater than 3.5×10^{-14} g cm^{-2} sec^{-1}; (4) and the eddy diffusion coefficient is less than 5×10^6 cm^2 sec^{-1} except perhaps in the top optical depth of the cloud.

100.127 **Far-infrared photometry of Titan and Iapetus.**
R. F. Loewenstein, D. A. Harper, R. H. Hildebrand, H. Moseley, E. Shaya, J. Smith.
Icarus, Vol. 43, 283 - 287 (1980).

Titan was observed in four broad passbands between 35 and 150 μm. The brightness temperature in this interval is roughly constant at 76 ± 3° K. Integrating Titan's spectrum from 5 to 150 μm yields an effective temperature of 86 ± 3° K. Both the bright and dark hemispheres of Iapetus were observed in one broadband filter with $\lambda_e \sim 66$ μm; The brightness temperatures for these two sides of Iapetus are 96 ± 9° K and 114 ± 10° K, respectively.

100.128 **Saturn satellite observations with reflecting coronagraph optics.** A. W. Harris, J. Gibson.
Publ. Astron. Soc. Pacific, Vol. 92, 547 (1980). – Abstract.

100.129 **How Titan will appear to us?** V. G. Surdin.
Astron. Tsirk., No. 1101, p. 6 - 8 (1980). In Russian.

100.130 **Photographic observations of Saturn satellites.**
I. G. Chugunov, Yu. A. Nefed'ev.
Astron. Tsirk., No. 1114, p. 7 - 8 (1980). In Russian.

100.131 **Spectral albedo of Titan.**
L. A. Bugaenko, L. R. Lisina.
Astrometr. Astrofiz., Vyp. (No.) 42, p. 36 - 40 (1980). In Russian.

Moderate resolution photoelectric spectral scans of Titan were used for geometric albedo calculations. The geometric albedo is given as a function of wavelength in the spectral region from 0.45 to 0.78 μm. The maximum value of P(0) equals 0.307 at $\lambda = 0.63$ μm, the minimum value occurs at the centre of the strong methane absorption band at $\lambda = 0.725\ \mu m$, and equals 0.14. The equivalent widths for methane absorption bands at $\lambda\lambda\, 0.619, 0.668, 0.685, 0.702$ and $0.725\ \mu m$ are derived.

100.132 **Perturbations in the motion of Saturn satellites. Part 2.** I. G. Chugunov.
Mordovsk. univ., Saransk, 1980. 33 pp. In Russian. – Abstr. in Ref. zh., 51. Astron., 1.51.77 (1981).

The planet Saturn. See Abstr. 003.015.

Fishing on Saturn with the KAO (*Kuiper Airborne Observatory*). See Abstr. 032.533.

Absolute brightness temperature measurements at 3.5-mm wavelength. See Abstr. 033.019.

Voyager 1 bei Saturn angelangt. See Abstr. 053.006.

L'exploration de Saturne par la sonde spatiale Pioneer. See Abstr. 053.008.

Voyager 1 bereikt Saturnus. See Abstr. 053.009.

Pioneer 11 Saturn encounter. See Abstr. 053.010.

Axisymmetric convection driven by latitudinal temperature gradients in rotating spherical shells. See Abstr. 062.090.

Magnetospheric influences on the evolution of planetary ring systems. See Abstr. 091.028.

A possible detection of solar variability from photometry of Io, Europa, Callisto, and Rhea, 1976–1979. See Abstr. 099.001.

International Information Bureau on Astronomical Ephemerides. See Abstr. 099.007.

Spectral broadening measurements of the ionospheres of Jupiter and Saturn. See Abstr. 099.011.

Phase effect for the brightness coefficient of the central disk of Saturn and features of Jupiter's disk. See Abstr. 099.021.

The effect of dense cores on the structure and evolution of Jupiter and Saturn. See Abstr. 099.022.

CH_4 photochemistry in the outer major planets. See Abstr. 099.033.

Significance of mutual phenomena observation in study of Jovian and Saturnian satellites. See Abstr. 099.039.

Internal heat and the hydrogen to helium ratio of Jupiter and Saturn. See Abstr. 099.050.

Jupiter and Saturn: giant magnetic rotating fluid planets. See Abstr. 099.120.

Sondage à distance de l'atmosphère des planètes géantes par spectroscopie de Fourier. See Abstr. 099.123.

Methane absorption near λ 6800 Å in the spectra of Jupiter and Saturn. I. See Abstr. 099.160.

Results of Pioneer 10 and 11 meteoroid experiments: interplanetary and near-Saturn. See Abstr. 106.079.

101 Uranus, Neptune, Pluto, Transplutonian Planets

101.001 **The orbit of the satellite of Pluto.** R. S. Harrington, J. W. Christy.
Bull. American Astron. Soc., Vol. 12, 509 (1980). – Abstract.

101.002 **The origin of Pluto.** J. R. Dormand, M. M. Woolfson.
Mon. Not. R. Astron. Soc., Vol. 193, 171 - 174 (1980).

The origin of the present orbit of Pluto, with regard to Neptune and Triton, is considered in the light of the recent discovery of the Plutonian satellite and consequent mass revision. It is shown that if Pluto originated as a satellite of Neptune an encounter with Triton could perturb it sufficiently to eject it from the Neptunian system to pursue a heliocentric orbit similar to that observed at the present day.

101.003 **Astrometric study of the Uranus satellite Miranda.** C. Veillet, G. Ratier.
Astron. Astrophys., Vol. 89, 342 - 344 (1980).

Positions of Miranda relative to Uranus are given for 1977, 1978, and 1979 oppositions from plates taken at the Pic-du-Midi with an innovative photographic method. The comparison of these new observations and recently published ones with the calculated positions derived from the nominal orbital elements shows a longitude variation with a period of 12.2 yr, very close to the circulation period of the near-commensurability between Miranda, Ariel, and Umbriel.

101.004 **Application of methane band-model parameters to the visible and near-infrared spectrum of Uranus.**
D. C. Benner, U. Fink.
Icarus, Vol. 42, 343 - 353 (1980).

Laboratory band-model absorption coefficients of CH_4 are used to calculate the Uranus spectrum from 5400 to 10,400 Å. A good fit of both strong and weak bands for the Uranus spectrum over the entire wavelength interval is achieved for the first time. Three different atmospheric models are employed: a reflecting layer model, a homogeneous scattering layer model, and a clear atmosphere sandwiched between two scattering layers.

101.005 **Observations of Uranus made with the Danjon astrolabe of Santiago, Chile, during 1978.**
F. Noël, K. Contreras, H. Repetur.
Astron. Astrophys., Suppl. Ser., Vol. 42, 193 - 194 (1980).

This paper gives the results of the observations of Uranus. The residuals in zenith distance of the planet, the number of the fundamental star group to which the residuals are referred as well as its weight are given among other data.

101.006 **Charon's diameter.** D. W. Hughes.
Nature, Vol. 287, 677 - 678 (1980).

101.007 **Correlated variations of planetary albedos and solar-interplanetary parameters.**
G. W. Lockwood, S. T. Suess, D. T. Thompson.
Solar and interplanetary dynamics, (see 012.020), p. 163 - 166 (1980).

The brightnesses of Titan and Neptune have been monitored photoelectrically at 472 and 551 nm since 1972 at the Lowell Observatory. Both objects increased steadily in brightness until 1976 and declined thereafter (Lockwood 1977, Lockwood and Thompson 1979). The authors conclude that the solar EUV output and the solar wind are the most viable candidates for the cause of the observed planetary variations.

101.008 **Measurements of the H_2 4 - 0 quadrupole bands of Uranus and Neptune.**
W. H. Smith, W. Macy, Jr., C. B. Pilcher.
Icarus, Vol. 43, 153 - 160 (1980).

The authors find that the equivalent widths of the lines of the 4 - 0 H_2 quadrupole band on Uranus and Neptune are substantially smaller than the values found by some previous observers. An analysis of the authors' results based on a range of atmospheric models yields H_2 abundances of 240 ± 60 km-amagats for Uranus and $\gtrsim$ 200 km-amagats for Neptune.

101.009 **Where are the rings of Neptune?**
A. R. Dobrovolskis.
Icarus, Vol. 43, 222 - 226 (1980).

Stable rings can exist at inclinations of 0 - 15°, 165°- 180°, or ~90° to Neptune's equator, but perturbations due to the massive satellite Triton would produce a severe "warping" of the ring plane. If Neptune possesses rings, they may not lie in the plane of its equator.

101.010 **Detection of a CH_4 atmosphere on Pluto.**
U. Fink, B. A. Smith, D. C. Benner, J. R. Johnson, H. J. Reitsema, J. A. Westphal.
Bull. American Astron. Soc., Vol. 12, 696 - 697 (1980). Abstract.

101.011 **Upper limit of gaseous CH_4 on Triton.**
J. R. Johnson, U. Fink, B. A. Smith, H. J. Reitsema.
Bull. American Astron. Soc., Vol. 12, 697 (1980). – Abstract.

101.012 **Uranus ring orbits.** J. L. Elliot, R. G. French.
Bull. American Astron. Soc., Vol. 12, 701 (1980). Abstract.

101.013 **The rotational period of Uranus.**
D. B. Slavsky, H. J. Smith.
Bull. American Astron. Soc., Vol. 12, 704 (1980). – Abstract.

101.014 **Further evidence for the longer rotation period of Neptune.** H. J. Smith, D. B. Slavsky.
Bull. American Astron. Soc., Vol. 12, 704 (1980). – Abstract.

101.015 **The rotation period of Neptune.**
R. H. Brown, D. P. Cruikshank, A. T. Tokunaga.
Bull. American Astron. Soc., Vol. 12, 704 - 705 (1980). Abstract.

101.016 **Where is Neptune's pole?** A. W. Harris.
Bull. American Astron. Soc., Vol. 12, 705 (1980). Abstract.

101.017 **Photometric spectra of the elusive Neptune haze.**
J. Apt, R. N. Clark, R. B. Singer.
Bull. American Astron. Soc., Vol. 12, 705 (1980). – Abstract.

101.018 **Uranian satellites: water ice on Ariel and Umbriel.**
D. P. Cruikshank, R. H. Brown.
Bull. American Astron. Soc., Vol. 12, 729 (1980). – Abstract.

101.019 **Photometric observations of Pluto in 1980.**
E. F. Tedesco, D. J. Tholen.
Bull. American Astron. Soc., Vol. 12, 729 (1980). – Abstract.

101.020 **Origin of the eccentricity gradient and the apse alignment of the ϵ ring of Uranus.**
S. F. Dermott, C. D. Murray.
Bull. American Astron. Soc., Vol. 12, 745 (1980). – Abstract.

101.021 **Occultation by Uranus.**
IAU Circ., Nos. 3503, 3515 (1980).

101.022 **1978 P 1.**
IAU Circ., Nos. 3509, 3515, 3522 (1980).

101.023 **Occultations by Neptune.**
IAU Circ., No. 3515 (1980).

101.024 **Op zoek naar Pluto.** C. W. Tombaugh.
Zenit, 7e Jaarg., 46 - 49 (1980).

101.025 **Pluto tussen 1930 en 1978: de grenzen van het meetbare.** G. W. E. Beekman.
Zenit, 7e Jaarg., 50 - 51 (1980).

101.026 **Pluto en Charon nu.** O. Namba.
Zenit, 7e Jaarg., 52 - 55 (1980).

101.027 **De zichtbaarheid van Pluto.** G. Comello.
Zenit, 7e Jaarg., 56 - 57 (1980).

101.028 **De absolute helderheid van Pluto.** H. Feijth.
Zenit, 7e Jaarg., 503 (1980).

101.029 **Conjonction de Pluton et d'une étoile brillante.**
R. Néel.
Astronomie, Vol. 94, 535 - 536 (1980).

101.030 **Planet Pluto.** D. B. Herrmann.
Astron. Schule, 17 Jahrg., 101 - 104 (1980).

101.031 **Pluton peut-il encore faire partie des grosses planètes?** H. Debehogne, R. R. de Freitas Mourao.
Ciel Terre, Vol. 96, 365 - 369 (1980).

101.032 **Speckle interferometry with the CFHT 3.60 m. I. Resolution of the system Pluto-Charon.**
D. Bonneau, R. Foy.
Astron. Astrophys., Vol. 92, L1 - L4 (1980). In French.

The authors resolved the system Pluto-Charon, using the Labeyrie's speckle interferometer and a photon counting TV camera. They derived the magnitude difference between Pluto and Charon $\Delta m_v = 1.6 \pm 0.2$ mag. From the radial profiles of the autocorrelations of speckles, the authors derived the diameter of Pluto and Charon, respectively $\Theta_P = 0''18 \pm 0''02$ and $\Theta_C = 0''09 \pm 0''01$, or $D_P = 4000 \pm 400$ km and $D_C = 2000 \pm 200$ km. The densities are $\rho_C \cong \rho_P = 0.5$ g/cm³. This confirms the hypothesis of the common origin of Pluto and Charon, which could have been a satellite of Neptune.

101.033 **Results from the 10 March 1977 occultation by the Uranus system.** W. B. Hubbard, B. H. Zellner.
Astron. J., Vol. 85, 1663 - 1669 (1980).

The authors present timings of occultation events observed at Perth, Australia during the appulse of Uranus to SAO 158687. In addition to pre- and post-appulse observations of the α ring, a search of the data confirms events by the "4", "5", and "6" rings on the pre-appulse side of Uranus, and the

"5" and β rings on the post-appulse side. The authors present a table of other suspected events. A fairly strong "event," so far unconfirmed, is noted near the 5:1 Miranda resonance orbit. Examination of the light curves at high time resolution indicates that the α ring was, during this aspect, about 10 km in radial extent and about 50% transparent.

101.034 **The structure of the Uranus atmosphere.**
L. Wallace.
Icarus, Vol. 43, 231 - 259 (1980).

A series of radiative/convective models is presented for the Uranus atmosphere for various methane-to-hydrogen mixing ratios and internal heat fluxes. From model spectra calculated for the visible, thermal infrared, and microwave regions, it is concluded that the methane-to-hydrogen mixing ratio is greater than 0.01 and probably less than 0.10. The lower limit to the internal heat flux is nonzero but less than ~ 1/2000 th of the total flux. In addition, the specific heat of the molecular hydrogen is found to be very close to that for normal hydrogen. Peculiarities in thermal structure are found to be of no help in understanding the microwave spectrum, but H_2S-to-NH_3 mixing ratios somewhat greater than unity are almost as good in explaining the spectrum.

101.035 **Origin of the eccentricity gradient and the apse alignment of the ϵ ring of Uranus.**
S. F. Dermott, C. D. Murray.
Icarus, Vol. 43, 338 - 349 (1980).

The suggestion that the apse alignment of the eccentric Uranian ϵ ring is maintained by self-gravitation alone is criticized. If self-gravitation were the only factor involved, then the extreme variation in width of the ring would be a chance product. The authors consider that the ring particles move in a near monolayer which is close packed at pericenter. They suggest that it is the close packing of the particles which prevents differential precession. They describe how differential precession, particle collisions, and self-gravitation, acting together, can transform a narrow eccentric ring of uniform width into a ring with a large, positive eccentricity gradient. The model they present is supported by the observed occultation profiles of the ϵ ring. The F ring of Saturn should provide further tests of their model.

101.036 **Brightness decrease of Pluto.**
A. N. Abramenko, V. V. Avramchuk,
V. A. Kucherov, L. R. Lisina, V. V. Prokof'eva.
Astron. Tsirk., No. 1100, p. 1 - 3 (1980). In Russian.

101.037 **Dynamic peculiarities of particles ejected from the Uranus satellites.** A. S. Guliev.
Komet. Tsirk., Kiev, No. 265 (1980). In Russian.

101.038 **Uranus and Neptune: spectrophotometry in the region λλ 4300–7000 Å.** A. A. Atai.
Astron. Vestn., Tom 14, 154 - 161 (1980). In Russian.

Spectra of Uranus and Neptune were examined to obtain the detail structure of molecular absorption bands.

Out of the darkness: the planet Pluto.
See Abstr. 003.103.

The planet Pluto. See Abstr. 003.112.

Galileo saw Neptune in 1612.
See Abstr. 004.014.

Galileo's observations of Neptune.
See Abstr. 004.015.

Galileo's sighting of Neptune.
See Abstr. 004.052.

Dynamics of particle orbits for narrow eccentric rings maintained by small satellites. See Abstr. 042.053.

Turbulence models and the atmosphere of Uranus.
See Abstr. 091.032.

An occultation by Charon. See Abstr. 096.001.

News of recently-attempted planetary occultations.
See Abstr. 096.004.

CH_4 photochemistry in the outer major planets.
See Abstr. 099.033.

Sondage à distance de l'atmosphère des planètes géantes par spectroscopie de Fourier. See Abstr. 099.123.

The rings of Saturn, Uranus and Jupiter.
See Abstr. 100.014.

On the existence of a comet belt beyond Neptune.
See Abstr. 102.001.

On an approximate method for investigation of the mutual gravitational influence of bodies of the protoplanetary cloud. On the evolution of Pluto's orbit.
See Abstr. 107.008.

102 Comets (Origin, Structure, Atmospheres, Dynamics)

102.001 **On the existence of a comet belt beyond Neptune.** J. A. Fernández.
Mon. Not. R. Astron. Soc., Vol. 192, 481 - 491 (1980).
The possible existence of a comet belt in connection with the origin of the short-period comets is analysed. It is noted that the current theory – that these comets originate as near-parabolic comets captured by Jupiter and the other giant planets – implies an excessive wastage of comets lost in hyperbolic orbits, which is avoided in the present model.

102.002 **The perihelion longitude distribution of Jupiter family comets.** R. A. Beck.
Irish Astron. J., Vol. 14, 14 - 20 (1979).

102.003 **Influence of a phase transition of ice on the heat and mass balance of comets.** J. Klinger.
Science, Vol. 209, 271 - 272 (1980).
Differences in gas production rates of comets may be explained in part by the phase transition of ice in the comet nuclei.

102.004 **On the capture of comets by Jupiter.** V. P. Tomanov.
Astron. Zh., Tom 57, 816 - 823 (1980). In Russian.
English translation in Soviet Astron., Vol. 24, No. 4.
The planar problem of gravitational interaction of parabolic comets with Jupiter is considered. The main objections against the hypothesis of capture of short-period comets are removed. It is shown that short-period comets must have only direct motions. Hyperbolic comets are not observable due to their large perihelion distance. The deficiency of comets with the orbit major semiaxis of 20 - 30 a. u. is explained by the small probability of their capture.

102. 005 **Fluorescence of OH in comets.** D. G. Schleicher, M. F. A'Hearn.
Bull. American Astron. Soc., Vol. 12, 462 (1980). – Abstract.

102.006 **Millimeter-wave radiometry of the cometary nucleus.** R. W. Hobbs, J. C. Brandt, S. P. Maran, D. Buhl, P. Gloersen, A. H. Delsemme, P. D. Feldman, M. J. Klein, P. Swanson, L. E. Snyder.
Bull. American Astron. Soc., Vol. 12, 510 - 511 (1980). Abstract.

102.007 **Physical similarities between dissipating comets and short-lived fragments of the split comets.**
Z. Sekanina.
Bull. American Astron. Soc., Vol. 12, 511 (1980). – Abstract.

102.008 **Comets and the interplanetary magnetic field.** E. V. Ivanov.
Zemlya Vselennaya, 1980, No. 4, p. 37. In Russian.

102.009 **Physicochemical and dynamical processes in cometary ionospheres. I. The basic flow profile.**
H. L. F. Houpis, D. A. Mendis.
Astrophys. J., Vol. 239, 1107 - 1118 (1980).
The Newtonian thin layer approximation for the interaction between two hypersonic streams is used to derive the three-dimensional axisymmetric flow structure of the cometary ionosphere. The strong collisional coupling of the cometary ions to the supersonically expanding neutral species in the inner coma causes them to expand supersonically too. This supersonic outflow of the cometary ions is then braked and diverted to the tail by an "inner shock" whose size and shape have been determined.

102.010 **Evolution of comet orbits under the perturbing influence of the giant planets and nearby stars.**
J. A. Fernández.
Icarus, Vol. 42, 406 - 421 (1980).
The orbital evolution of 500 hypothetical comets during 10^9 years is studied numerically. It is assumed that the birthplace of such comets was the region of Uranus and Neptune from where they were deflected into very elongated orbits by perturbations of these planets. Then, the author adopted the following initial orbital elements: perihelion distances between 20 and 30 AU, inclinations to the ecliptic plane smaller than 20°, and semimajor axes from 5×10^3 to 5×10^4 AU. Gravitational perturbations by the four giant planets and by hypothetical stars passing at distances from the Sun smaller than 5×10^5 AU are considered.

102.011 **Brightness variations of comets and solar corpuscular activity.** D. A. Andrienko, A. V. Karpenko.
Problems of cosmic physics. Vyp. 15, (see 003.003), p. 103 - 108 (1980). In Russian.
On the basis of investigations of brightness and outburst variations of 15 comets whose orbits are near the ecliptic the possibility is shown of definition of the solar interplanetary plasma flux velocities. It has been found that a comet's brightness and outburst variations are due to the influence of high-velocity fluxes of 500 - 750 km/sec.

102.012 **Disintegration of comets and solar activity.** D. A. Andrienko, V. N. Vashchenko, L. M. Kostenko.
Problems of cosmic physics. Vyp. 15, (see 003.003), p. 108 - 114 (1980). In Russian.
The disintegration process of a comet is studied depending on orbital inclinations, perihelion distance and period of revolution around the sun.

102.013 **Study of the orbits of the cometary families of Saturn, Uranus and Neptune as compared with the conclusions of the eruption theory.**
S. K. Vsekhsvyatskij, A. A. Demenko.
Problems of cosmic physics. Vyp. 15, (see 003.003), p. 114 - 127 (1980). In Russian.
Observational and theoretical distributions of elements of orbits for Saturn's, Uranus' and Neptune's families comets are compared. It is shown that for the period of the existence of the mentioned cometary families (150 years) 670 comets should come off the sphere of influence of Saturn, 103 comets from Uranus' sphere, 273 from Neptune's and 637 comets from Jupiter's sphere of influence.

102.014 **On the appearance of metal atoms in the heads of comets as a result of molecular destruction.**
S. Ibadov.
Dokl. AN TadzhSSR, Vol. 23, 76 - 79 (1980). In Russian.
Abstr. in Ref. zh., 51. Astron., 8.51.310 (1980).

102.015 **Physical characteristics of cometary dust from optical studies.** M. S. Hanner.
Solid particles in the solar system, (see 012.019), p. 223 - 236 (1980).
Observations of the scattered sunlight and thermal emission from cometary dust provide information on the composition and dominant size of the dust. The observations are summarized and compared to theoretical models for dielectric and absorbing materials, with emphasis on the thermal emission. The compatibility of the optical data with the size distribution derived from dynamical studies is discussed.

102.016 **Physical characteristics of cometary dust from dynamical studies: a review.** Z. Sekanina.
Solid particles in the solar system, (see 012.019), p. 237 - 250 (1980).

Attention is focused on those areas of the dynamical investigation of cometary dust in which significant progress has been achieved since previous review papers (Sekanina 1976, 1977). Much of the progress stems from work based on the model of dust comets formulated by Finson and Probstein (1968). Their introduction of a new technique for dust-tail studies has made it possible for the first time to gain insight into such properties of cometary dust as the size distribution function of particles shortly after emission from the comet nucleus, the distribution of particle ejection velocities, and the production of dust versus time.

102.017 **On the particle-size distribution function of cometary dust.** Z. Sekanina.
Solid particles in the solar system, (see 012.019), p. 251 - 254 (1980).

102.018 **On the dust production rates of comets.**
R. Hellmich, H. U. Keller.
Solid particles in the solar system, (see 012.019), p. 255 - 258 (1980).

The dust production rates of comets on various orbits with perihelia between 0.1 and 0.5 a. u. and semimajor axes between 5 and 40 a.u. are calculated. A new model is used taking into account the screening of solar radiation by the dust and the multiple scattering of the photons. Screening and multiple scattering effects tend to compensate each other over a range of moderate optical thickness.

102.019 **On the photochemical radicals in cometary comae.**
T. Yamamoto.
13th Lunar and Planetary Symposium, (see 012.018), p. 158 - 165 (1980).

The formation of CN, C_2 and C_3 radicals in cometary comae is investigated. It is shown that CN and C_2 are the primary products in the photolysis of some parent molecules sublimating from cometary nucleus, whereas C_3 is a parent molecule which does not suffer photodissociation. The chemical composition of cometary nuclei is discussed in relation to these molecules.

102.020 **On "seasonal" variations of cometary flare activity.**
D. A. Andrienko, V. N. Vashchenko.
Astron. Zh., Tom 57, 1094 - 1098 (1980). In Russian.
English translation in Soviet Astron., Vol. 24, No. 5.

The paper deals with the character of variations of the annual spectrum of cometary flare activity from observational data for the last 130 years. The spectrum obtained is in a good agreement with "corpuscular seasons" noted by Mustel in solar-terrestrial relations. The variation of frequency and of mean amplitudes of cometary flares is similar to the intensity seasonal variation of different geophysical events.

102.021 **Stellar perturbations of the cometary cloud.**
P. R. Weissman.
Nature, Vol. 288, 242 - 243 (1980).

The existence of a cloud of $\sim 2 \times 10^{11}$ comets surrounding the solar system and extending out to interstellar distances has been suggested by Oort (1950) based on the observed distribution of inverse semi-major axes of the long-period comets. Simple statistical arguments can be used to update the current understanding of how stellar perturbations act on the Oort cloud, and to infer some of the cloud properties.

102.022 **Moleküle in Kometen.** R. Lüst.
Evolution der Planetenatmosphären und des Lebens, (see 012.032), p. 190 - 197 (1980).

102.023 **Comets and the Galilean satellites.**
E. M. Shoemaker, R. F. Wolfe.
Bull. American Astron. Soc., Vol. 12, 712 - 713 (1980).
Abstract.

102.024 **The forbidden oxygen lines in comets.**
P. D. Feldman, M. C. Festou.
Bull. American Astron. Soc., Vol. 12, 731 (1980). – Abstract.

102.025 **Amorphous ice and the behavior of cometary nuclei.**
R. Smoluchowski.
Bull. American Astron. Soc., Vol. 12, 731 (1980). – Abstract.

102.026 **Chemical model calculations of C_2, C_3, CH, CN, OH and NH_2 abundances in cometary comae.**
G. F. Mitchell, S. S. Prasad, W. T. Huntress, Jr.
Bull. American Astron. Soc., Vol. 12, 731 - 732 (1980).
Abstract.

102.027 **Cometary mass determination.** D. K. Yeomans.
Bull. American Astron. Soc., Vol. 12, 743 (1980).
Abstract.

102.028 **Dynamical evolution of Oort cloud comets.**
P. R. Weissman.
Bull. American Astron. Soc., Vol. 12, 744 (1980). – Abstract.

102.029 **Interplanetary gas. XXV. A solar wind and interplanetary magnetic field interpretation of cometary light outbursts.** M. B. Niedner, Jr.
Astrophys. J., Vol. 241, 820 - 829 (1980).

Cometary brightness outbursts have been examined for possible relationships with, and causes in, the solar wind and interplanetary magnetic field (IMF). Two classes of outburst have been defined, both of which have plausible explanations in terms of a solar wind/IMF model. Class I flares are characterized by an increase of ionization in the head which strengthens the plasma tail shortly before a disconnection event (DE) occurs. The ionization surge is caused by electron jetting in the reconnecting current sheet which is produced by the passage of an interplanetary sector boundary. The time scale of ionization of CO is found to be $\sim 10^4$ s. Class II outbursts involve the explosive release of gas and dust from the nucleus (in contrast to Class I, in which the brightening is atmospheric in nature); and in the cases of three flares in comet P/Tuttle-Giacobini-Kresak 1973b and comet P/Pons-Brooks 1883b, correlations with corotated high-speed streams and sector boundaries were found. The interpretation of these events is that the bombardment of the nucleus by disturbed solar wind triggers highly exothermic chemical reactions, as proposed by some previous workers.

102.030 **Neutral cometary atmospheres. III. Acceleration of cometary CN by solar radiation pressure.**
M. R. Combi.
Astrophys. J., Vol. 241, 830 - 837 (1980).

The acceleration of cometary CN radicals due to solar radiation pressure has been determined by fitting Monte Carlo models to nine observed sunward-tailward pairs of brightness profiles of the (0–0) band of CN at 3883 Å. The profiles were determined from spectrograms of comets Bennett 1970 II and West 1976 VI. The values of the observed acceleration agree with those computed from resonance fluorescence calculations to within the expected uncertainties. This provides an independent confirmation of the identification of the observed scale lengths with the photochemical lifetimes and velocities associated with the production of observed cometary CN by the photodissociation of HCN. The ratio of the intensity of the (0–1) band of CN at 4216 Å to the (0–0) band at 3883 Å has been determined from spectrograms of comet West, and is compared with theoretical values.

102.031 **The effect of solar-cycle ultraviolet flux variations on cometary gas.** M. Oppenheimer, C. J. Downey.
Astrophys. J., Lett., Vol. 241, L123 - L127 (1980).

The effects on comet gas of variations in the solar ultraviolet flux during solar cycle 21 are discussed. The photoionization, photodissociation, and resonance fluorescence and scattering rates of individual atoms and molecules increase by factors ranging up to 4, leading to potential order of magnitude variations in emission-line fluxes between minimum and maximum solar activity. These effects are illustrated for H_2O and CO. Recent observations of comet Bradfield (1979*l*) are discussed, and it is suggested that comets appearing near solar maximum be extensively observed to provide information on the response of cometary gas to solar flux variations, which can be used to discriminate between cometary models.

102.032 **Abundance correlations among comets.** M. F. A'Hearn, R. L. Millis.
Astron. J., Vol. 85, 1528 - 1537 (1980).

The authors present the results of narrowband filter photometry for comets 1977g, 1978b, 1978f, 1978j, 1979c, and 1979i. The ratio of CN to C_2 production is much larger at heliocentric distances > 2.0 AU than at distances < 1.5 AU. Except for this change, the ratio is remarkably constant (±0.1 in the log) from comet to comet. Correlations of CN and C_2 with C_3 and with published data for OH show greater scatter but there is still no evidence for any systematic variation. There is no correlation with gas-to-dust ratio or with the dynamical age of the comet. The correlation with visual magnitude is surprising both in that it is better for CN than for C_2 and in the slope is significantly different from 0.4. These relations will make useful tools for predicting observational circumstances for newly discovered comets.

102.033 **Comets and cometary missions.** D. A. Mendis.
Space Research, Vol. XX, (see 012.043), 177 - 188 (1980).

The proposed joint mission to Comet P/Halley and Comet P/Temple 2 in the latter half of the next decade is presented. The present understanding of the physical structure and the chemical composition of the cometary nucleus as well as the nature of the comet-solar wind interaction is discussed, with emphasis on the more recent findings.

102.034 **Cometary atmospheres. I. Solar wind modification of the outer ion coma.** W.-H. Ip.
Astron. Astrophys., Vol. 92, 95 - 100 (1980).

Using a quasi-one-dimensional fluid approximation for the solar wind interaction with a comet, velocity profiles for the inward streaming plasma in a cometary atmosphere are generated and, in turn, adopted to compute the number density distributions of cometary ions produced upstream of the comet head. The external solutions so obtained are interpolated to the inner ion coma models. The dynamical behaviour of the cometary ionospheric plasma of an Halley-type comet is approximated as a combination of subsonic inward flow and outflow from the central source.

102.035 **On the origin of striae in cometary dust tails.** J. R. Hill, D. A. Mendis.
Astrophys. J., Vol. 242, 395 - 401 (1980).

It has been argued by Sekanina that the so-called striae or pseudo-synchronic bands observed in the dust tails of several comets are a consequence of the sudden disruption of elongated or chainlike parent grains in the tail. The authors offer a specific mechanism for this disruption process, that is, electrostatic disruption following sudden charging to high electrostatic potentials.

102.036 **Cometary dust on the Earth.** A. Vítek.
Vesmír, Vol. 59, 306 - 307 (1980). In Czech.

102.037 **Pristine nature of comets as revealed by their UV spectrum.** A. H. Delsemme.
Appl. Opt., Vol. 19, 4007 - 4014 (1980).

The observation of the VUV spectrum of comets from rockets and satellites has brought to light some new clues on those primitive bodies that could be the link between interstellar molecules and early planetary atmospheres as well as the bridge between stellar and planetary astrophysics.

102.038 **The interaction of the solar wind with comets.** J. C. Brandt, D. A. Mendis.
Solar system plasma physics, Vol. 2, (see 003.010), 253 - 292 (1979).

Contents: Overview. Continuum description of the interaction. Cometary plasma. Generation of tail streamers. Gross tail structure. Fine structure in the plasma tail. The problem of anomalous comets. Epilogue.

102.039 **Determination of magnetic field direction in a comet's head.** D. A. Varshalovich, G. F. Chörny.
Icarus, Vol. 43, 385 - 390 (1980).

The Stokes parameters of resonance radiation scattered by a Na atom with an angular momentum **F** aligned by directed unpolarized radiation in a magnetic field $H \sim 10^{-5} - 10^{-1}$ Oe are presented. The influence of the orientation of the magnetic field on these parameters is studied; the intensity ratio $I(D_2)/I(D_1)$ changes within ±5%, and the polarization degree $P(D_2)$ within ±25%. Measurements of $I(D_2)/I(D_1)$ and $P(D_2)$, if the geometry of scattering is known, may give information on the direction of the magnetic field in the sodium atmospheres of comets, as well as Io's sodium cloud or man-made cosmic clouds.

102.040 **Feedback of the dust coma on the evaporation process.** H. U. Keller.
Comet Halley micrometeoroid hazard workshop, (see 012.049), p. 57 - 60 (1979).

Some results are presented for a new model taking into account the multiple scattering of the solar radiation in the dust coma. The feedback of the dust coma on the sublimation of the surface of the cometary nucleus leads to a slight increase of gas and dust production. For the situation of the comet Halley flyby, the terminal velocities of the dust grains resulting from the hydrodynamic interaction with the gas can be approximated in a simple form. The apex distance and maximum grain size are calculated.

102.041 **Summary of discussions in Group A (Models for the production of cometary dust particles).**
Z. Sekanina.
Comet Halley micrometeoroid hazard workshop, (see 012.049), p. 139 (1979).

102.042 **Models of cometary comae.** G. F. Mitchell, M. B. Swift.
J. R. Astron. Soc. Canada, Vol. 74, 356 (1980). – Abstract.

102.043 **The latitude selection effect in the distribution of cometary perihelia.**
I. N. Potapov, V. P. Tomanov.
Astron. Tsirk., No. 1106, p. 6 - 8 (1980). In Russian.

102.044 **Dependence of flare activity of comets on heliographic latitude.**
D. A. Andrienko, V. N. Vashchenko, L. M. Kostenko.
Astron. Tsirk., No. 1116, p. 5 - 7 (1980). In Russian.

102.045 **Cometary brightness outbursts and solar corpuscular activity.**
D. A. Andrienko, V. N. Vashchenko.
Astron. Zh., Tom 57, 1310 - 1316 (1980). In Russian.
English translation in Soviet Astron., Vol. 24, No. 6.

On the basis of an analysis of the cometary brightness outbursts catalogue compiled by the authors the evidence of their relation to corpuscular solar radiation is given. It is shown that flare cometary activity depends on the phase of the 11-year cycle of solar activity. Active events in comets correlate not only with the appearance of high-velocity solar plasma streams but also with numerous geophysical events appearing as the result of the solar plasma influence on the earth's magnetosphere and atmosphere.

102.046 **Cyanide – possible parent molecule of cometary nuclei.** K. I. Churyumov.
Komet. Tsirk., Kiev, No. 265 (1980). In Russian.

102.047 **Possibility of origin of X-rays of comets.** S. Ibadov.
Komet. Tsirk., Kiev, No. 266 (1980). In Russian.

102.048 **Dependence of the angles of inclination of planes on the elliptic length of aphelia of the orbits of periodic comets.** P. T. Veleshchuk.
Komet. Tsirk., Kiev, No. 268 (1980). In Russian.

102.049 **Peculiarities of the Saturn cometary family.** A. S. Guliev.
Komet. Tsirk., Kiev, No. 268 (1980). In Russian.

102.050 **Distribution of cometary outbursts according to heliocentric distances.**
D. A. Andrienko, V. N. Vashchenko.
Komet. Tsirk., Kiev, No. 269 (1980). In Russian.

102.051 **Celestial-mechanical aspects of the ejection hypothesis.** V. V. Radzievskij.
Astron. Vestn., Tom 13, 32 - 41 (1979). In Russian.

New data in favour of the concentration reality of cometary perihelia near celestial longitude $\lambda = 270°$ have been obtained. The destiny of matter ejected by a planet or its satellite in the case of normal distribution of ejection velocities has been investigated. The first result stated is that a great majority of comets ejected by a satellite must revolve around the planet.

102.052 **On a dynamic characteristic of cometary orbits.** E. N. Kramer, V. I. Musij, I. S. Shestaka.
Astron. Vestn., Tom 13, 42 - 49 (1979). In Russian.

In order to study perturbations of osculating elements of periodical comets, to detect new cometary families and verify the known ones, a method of estimation of the phase distance between orbital elements has been used. The phase distance formally depends on perturbations transforming one orbit into another. If this transformation is physically justified, the objects moving along corresponding orbits are supposed to be related.

102.053 **Statistics of short-period comets.** V. P. Tomanov.
Astron. Vestn., Tom 14, 162 - 167 (1980). In Russian.

The distributions of the elements of the orbits for the system of comets with period of revolution < 165 years are presented. The statistical analysis of the Jupiter family of short-period comets gives new arguments for their capture from the system of long-period comets.

MHD-calculations for cometary plasmas. See Abstr. 021.011.

Automatic calculation of the moonless period in observations of comets. See Abstr. 021.053.

Rotational analysis of the (2,0) band of $^{13}C^{16}O^{+}$ molecule of comet-tail system. See Abstr. 022.030.

Laboratory simulation of the induced magnetospheres of comets and Venus. See Abstr. 022.090.

The interplay of molecular spectroscopy and astronomy. See Abstr. 022.106.

Similarity solutions of a strong shock wave propagation in a mixture of a gas and dusty particles. See Abstr. 022.151.

Dust experiment for a rendezvous cometary mission. See Abstr. 032.530.

A new instrument for cometary dust studies. See Abstr. 032.545.

Characteristics of trajectories to the comets of the Jovian group. See Abstr. 052.029.

Plasma acceleration by ion-acoustic turbulence. See Abstr. 062.120.

The variable nature of the solar wind interaction with cometary atmospheres. See Abstr. 074.083.

Determination of the kinematic characteristics of solar corpuscular streams from outbursts of comets. See Abstr. 078.007.

Problemi di dinamica dei corpi minori del sistema solare. See Abstr. 091.044.

Ist das Objekt (2060) Chiron ein kleiner Planet oder ein Komet? See Abstr. 098.013.

Emission band and continuum photometry of Comet West (1975n) – II. Emission profiles of the neutral coma, lifetimes of molecules and distribution of the molecules and dust within the coma. See Abstr. 103.102.

Effect of solar radiation on a swarm of meteoric particles. See Abstr. 104.004.

Connection of meteor matter with comets and asteroids. See Abstr. 104.047.

On the reliability of identification of meteor streams with comets. See Abstr. 104.064.

Optical investigation of dust in the solar system. See Abstr. 106.021.

From interstellar dust to comets to the zodiacal light. See Abstr. 106.045.

103 Comets (Individual Objects)

103.001 **Magnitude analyses of four comets.**
D. W. E. Green.
Strolling Astron., Vol. 28, 134 - 140 (1980).

103.002 **Extra-atmospheric cometary observations. I. Instruments.** V. P. Tarashchuk.
Problems of cosmic physics. Vyp. 15, (see 003.003), p. 79 - 92 (1980). In Russian.
The work is a review of investigations of extra-atmospheric cometary observations in 1970 - 1977. The first part contains the description and basic parameters of all the instruments available on board of spacecraft and rockets used for observations of comets in the ultraviolet wavelength range.

103.003 **Observations of minor planets and comets.**
Minor Planet Circ., (M.P.C.), Nos. 5392 - 5394, 5395 - 5409, 5424 - 5439, 5455 - 5515, 5541 - 5596, 5605 - 5637 (1980).
Concerning observations of the following comets: 1949 I Wirtanen, 1949 IV Bappu-Bok-Newkirk, 1951 VIII Tempel 2, 1954 IX Encke, 1965 I Tsuchinshan 1, 1965 II Tsuchinshan 2, 1977 XI Encke, 1977 XIV Kohler, 1978 III Arend-Rigaux, 1978 XIV Ashbrook-Jackson, 1978 XV Seargent, 1978 XVII Comas Solá, 1978 XX Haneda-Campos, 1978 XXI Meier, 1978h Giacobini-Zinner, 1979d Russell 1, 1979e Torres, 1979*l* Bradfield, 1980a Forbes, 1980b Bowell, 1980c Honda-Mrkos-Pajdušáková, 1980d Wild 3, 1980e Torres, 1980f Brooks 2, 1980g Stephan-Oterma, 1980h Tuttle, 1980i Borrelly, 1980j Kohoutek, 1980k Černis-Petrauskas, 1980*l* Russell, 1980m Harrington, 1980n Reinmuth 2, 1980o Russell 2, 1980q Meier.

103.004 **Ephemerides of minor planets and comets.**
Minor Planet Circ., (M.P.C.), Nos. 5418 - 5422, 5452 - 5454, 5525 - 5540, 5603 - 5604, 5651 - 5658 (1980).
Concerning ephemerides of the following comets: 1974 X Finlay, 1974 XIV Longmore, 1980b Bowell, 1980d Wild 3, 1980e Torres, 1980*l* Russell, 1980o Russell 2, 1980q Meier.

103.005 **Orbital elements of comets.**
Minor Planet Circ., (M.P.C.), Nos. 5391 - 5658 (1980).
The comets are listed according to their Roman numeral designation or preliminary designation. The names of the authors are given behind the respective M.P.C. numbers. 1937 II Wilk 5411 W. Landgraf; 1972 VII Swift-Gehrels 5638 G. Sitarski; 1973 XI Gehrels 2 5640 B. G. Marsden; 1974 X Finlay 5639 D. K. Yeomans; 1979d Russell 1 5639, 1979e Torres 5639 B. G. Marsden; 1979*l* Bradfield 5411 W. Landgraf; 1980b Bowell 5413, 1980d Wild 3 5413, 1980e Torres 5412, 5639, 1980k Černis-Petrauskas 5640, 1980*l* Russell 5640, 1980o Russell 2 5639, 1980q Meier 5640 B. G. Marsden.

103.006 **Comet on Palomar Sky Survey.**
IAU Circ., No. 3540 (1980).

103.007 **De kometen van 1979.** R. J. Bouma.
Zenit, 7e Jaarg., 233 (1980).

103.008 **Observations of comets.**
British Astron. Assoc. Circ., No. 607 (1980).
Concerning comets Bradfield 1979c, Bradfield 1979*l*, and Meier 1979i.

103.009 **Comet on Palomar Sky Survey.**
Yamamoto Circ., No. 1949 (1980).

103.010 **Comètes: l'année record 1978.**
J. Meeus.
Astronomie, Vol. 94, 539 - 541 (1980).

103.011 **Les comètes de 1979.** J. Sauval.
Ciel Terre, Vol. 96, 382 - 384 (1980).

103.012 **Recent news of comets.**
Int. Comet Q., Vol. 2, 42, 61, 67 - 68, 86 (1980).
Concerning comets: 1977 XI Encke, 1980c Honda-Mrkos-Pajdušáková, 1980d Wild 3, 1980e Torres, 1980f Brooks 2, 1980g Stephan-Oterma, 1980h Tuttle, 1980i Borrelly, 1980j Kohoutek, 1980k Černis-Petrauskas, 1980*l* Russell, 1980m Harrington, 1980n Reinmuth 2, 1980o Russell 2, 1980p Helin-Dunbar, 1980r West-Kohoutek-Ikemura.

103.013 **Observations of comets.**
Int. Comet Q., Vol. 2, 47 - 59, 75 - 85 (1980).
Concerning comets: 1974 II Schwassmann-Wachmann 1, 1974 III Bradfield, 1974 XVI Honda-Mrkos-Pajdušáková, 1975 IX Kobayashi-Berger-Milon, 1975 XII Mori-Sato-Fujikawa, 1977 VI Grigg-Skjellerup, 1977 XI Encke, 1977 XIV Kohler, 1978 XXI Meier, 1979c Bradfield, 1979i Meier, 1979*l* Bradfield, 1980g Stephan-Oterma, 1980h Tuttle, 1980k Černis-Petrauskas, 1980q Meier.

103.014 **Comets in the year 1979.** J. Bouška.
Vesmír, Vol. 59, 316 (1980). In Czech.

103.015 **Observations of comets at the station of the Harvard College.**
Komet. Tsirk., Kiev, No. 269 (1980). In Russian.

103.016 **Comet digest.** J. E. Bortle.
Sky Telesc., Vol. 60, 198, 290, 378, 474 (1980).

Spectroscopy of faint asteroids, satellites, and comets. See Abstr. 098.025.

Positional observations of minor planets and comets at the Astronomical Observatory of the Silesian Planetarium. See Abstr. 098.029.

On a possible relation between the Quadrantid meteor stream and comets. See Abstr. 104.068.

Comet 1976 VI West

103.101 **Vaporization in comets: the icy grain halo of Comet West.** M. F. A'Hearn, J. J. Cowan.
Moon Planets, Vol. 23, 41 - 52 (1980).
The variation with heliocentric distance of the production rates of various species in Comet West (1975*n* = 1976 VI) is explained with a cometary model consisting of a CO_2-dominated nucleus plus a halo of icy grains of H_2O or clathrate hydrate. The authors conclude that the parents of CN and C_3 are released primarily from the nucleus but that the parent of C_2 is released primarily from the halo of icy grains.

103.102 **Emission band and continuum photometry of Comet West (1975n) – II. Emission profiles of the neutral coma, lifetimes of molecules and distribution of the molecules and dust within the coma.**
M. K. V. Bappu, M. Parthasarathy, K. R. Sivaraman, G. S. D. Babu.

Mon. Not. R. Astron. Soc., Vol. 192, 641 - 650 (1980).

The scale lengths of the CN and C_2 molecules and their parents are derived based on Haser's model from drift scans across the coma of Comet West in the light on CN (0,0), C_2 (0,0) and the continuum around 5000 Å. The CN molecules are produced by the dissociation of two species of parent molecules having entirely different lifetimes. A similar result is also obtained for the C_2 molecules. There is an increase of reddening of the continuum, from the centre of the nucleus outwards.

103.103 **Sodium D-line emission in comet West (1975n) and the sodium source in comets.** M. Oppenheimer.
Astrophys. J., Vol. 240, 923 - 928 (1980).

A spectrum of the sodium D-line emission from comet West (1975n) taken at heliocentric distance 1.4 AU is analyzed by comparing the D-line brightness to that in the underlying continuum. Using the observations of this comet by Ney and Merrill, the author finds that the dust grains which dominate the visible and infrared continua are too cool to provide the observed sodium atoms through evaporation of sodium metal or a sodium compound from grain surfaces. Though sodium metal may evaporate from a small-grain component, the author suggests that molecules embedded in the volatile nuclear matrix are a more plausible source of sodium. The relationship between this source and the interstellar sodium abundance is discussed.

103.104 **Photoelectric observations of comets West 1976 VI and Kohler 1977 XIV.** V. P. Tarashchuk.
Problems of cosmic physics. Vyp. 15, (see 003.003), p. 92-98 (1980). In Russian.

The possibilities are discussed of photoelectric observations of comets in the system *U, B, V.* The observational results of comets West 1976 VI and Kohler 1977 XIV are presented. It has been found that the colour index of the comet West A, B, C nuclei are different. The most blue proved nucleus C. The colour characteristics of comet Kohler were found to be typical for comets which lack a dust component.

103.105 **Ultraviolet albedo of comet West (1976 VI).**
P. D. Feldman.
Solid particles in the solar system, (see 012.019), p. 263 - 266 (1980).

The ultraviolet spectrum of comet West (1976 VI) in the range 1200–3200 Å was recorded by rocket-borne instruments on March 5.5, 1976. Longward of 2100 Å the continuum of solar radiation scattered by cometary dust is detected and is found to closely follow the solar spectrum. Since the dust coma is completely included in the spectrometer slit, the ultraviolet albedo can be determined relative to the visible and this ratio is found to be ≈ 0.3 at 2700 Å. There is evidence for a further decrease in albedo near 2200 Å. Using a visible albedo of 0.2 gives a value of 0.06 for the cometary albedo at 2700 Å, a value similar to that found for the moon and lunar dust in this spectral region.

103.106 **Evidence for fragmentation of strongly nonspherical dust particles in the tail of comet West (1976 VI).**
Z. Sekanina, J. A. Farrell.
Solid particles in the solar system, (see 012.019), p. 267 - 270 (1980).

Following Sekanina's suggestion that striae observed in the dust tails of several comets could be products of fragmentation of friable dust particles ejected from the nucleus, the authors have completed an investigation of these structures in comet West (1976 VI), which shows that the dynamical solutions, provided by the fragmentation model on the assumption that solar attraction and radiation pressure are the only forces involved, are indeed in agreement with the observed motions of 16 striae through the tail over a time interval of more than three days.

103.107 **The second tail of Comet West 1976 VI.**
T. Akabane.
13th Lunar and Planetary Symposium, (see 012.018), p. 166 - 174 (1980).

103.108 **Matters of the dust tails of the comets West (1976 VI) and Ikeya-Seki (1965 VIII).**
K. Saito, S. Isobe, K. Nishioka, T. Ishii.
13th Lunar and Planetary Symposium, (see 012.018), p. 175 - 182 (1980).

Dust tails of the comets West (1976 VI) and Ikeya-Seki (1965 VIII) are analysed by the mechanical theory. Two results are shown: (1) Silicates grains and iron compounds grains are suitable to explain the repulsive forces acting on the particles, while matters like graphite grains, as suffered big forces, are not. (2) The far ending part of the tail of Comet Ikeya-Seki must have been composed of silicates alone, as iron compounds should have been sublimated and lost.

103.109 **Interpretation of the polarization distribution of Comet West.** S. Isobe.
13th Lunar and Planetary Symposium, (see 012.018), p. 183 - 192 (1980).

It is found that the observation of Comet West resulting in high and low polarizations at the region to the anti-solar and solar directions from the cometary core, respectively, can not be interpreted by any simple grain model. The heterogeneous grain model is found to give a fairly good fit to the polarization observation.

103.110 **The striated dust tail of Comet West 1976 VI as a particle fragmentation phenomenon.**
Z. Sekanina, J. A. Farrell.
Astron. J., Vol. 85, 1538 - 1554 (1980).

The motions of 16 striae in the dust tail of Comet West between 4 and 7 March 1976 have been successfully fitted on four small-scale photographs. The model assumes that the striae are the result of the ejection of dust particles that subsequently fragment in the tail. The particles responsible for the formation of a discrete stria must be emitted simultaneously, be subjected to the same repulsive acceleration in the tail, and break up simultaneously. The results of the analysis indicate a strong correlation between the ejection times and the times of known explosive events. The mass of dust in an average stria is estimated to be about 10^9 g. The authors consider rotational bursting caused by a "windmill' effect of radiation pressure to be a possible fragmentation mechanism.

103.111 **Production of carbon, sulfur and CS in comet West.**
A. M. Smith, T. P. Stecher, L. Casswell.
Astrophys. J., Vol. 242, 402 - 410, plate 8 (1980).

Results from ultraviolet spectrographic observations of comet West (1975n) in the $\lambda\lambda$ 1620 - 3960 wavelength range are presented. Objective grating imagery of the comet reveals for the first time emission from sulfur atoms and CS molecules as well as previously observed lines or bands in carbon atoms and CO^+, CO_2^+, and OH molecules. Some evidence exists for emission in the C_2 Mulliken bands and in the CO cameron bands. On the basis of the observations, producing rates of C (ground state), C (^{1}D), S, and CS are inferred.

103.112 **Parameters of the envelopes in the head of comet West (1976 VI).**
R. A. Zayats, K. I. Churyumov.
Komet. Tsirk., Kiev, No. 266 (1980). In Russian.

103.113 **Brightness of the head and nuclei of comet West (1976 VI).**
Komet. Tsirk., Kiev, No. 270 (1980). In Russian.

Periodic comet Brorsen-Metcalf

103.121 **Influence of non-gravitational forces on the motion of comet Brorsen-Metcalf 1847 V (1919 III).**
L. M. Belous.
Problems of cosmic physics. Vyp. 15, (see 003.003), p. 128 - 132 (1980). In Russian.

The orbital elements of the comet 1847 V are improved. The tendency of changes of the elements due to the effect of non-gravitational forces is shown. New data confirm the author's calculation on the rotational axis position of the comet's nucleus.

Periodic comet Stephan-Oterma

103.141 **On the motion of comet Coggia-Stephan-Oterma in 1867 - 1943.** L. M. Belous.
Problems of cosmic physics. Vyp. 15, (see 003. 003), p. 132 - 136 (1980). In Russian.

Orbital elements of the comet are improved from two apparitions by the author's technique. Approximate values of the changes of elements due to the effect of non-gravitational forces are found.

103.142 **Periodic comet Stephan-Oterma (1980g).**
IAU Circ., Nos. 3515, 3528, 3530, 3531, 3543, 3557 (1980)

103.143 **Periodic comet Stephan-Oterma (1980g).**
Yamamoto Circ., Nos. 1937, 1938, 1944, 1946, 1947, 1949 (1980).

103.144 **Comet P/Stephan-Oterma 1980g.**
British Astron. Assoc. Circ., Nos. 607, 609 (1980).

103.145 **Favorable apparition predicted for P/Comet Stephan-Oterma 1980g.**
C. S. Morris, D. W. E. Green.
Int. Comet Q., Vol. 2, 43 - 45, 60 - 61 (1980).

103.146 **Observation of short-period comet Stephan-Oterma (1980g).**
Komet. Tsirk., Kiev, Nos. 268, 269 (1980). In Russian.

Drie heldere periodieke kometen in 1980 (2).
See Abstr. 103.743.

Periodic comet Ashbrook-Jackson

103.161 **Interference filter photometry of P/Ashbrook-Jackson.**
R. L. Newburn, Jr., T. McCord, J. Bell.
Bull. American Astron. Soc., Vol. 12, 730 (1980). – Abstract.

103.162 **Nature of dust grains in the atmosphere of comet Ashbrook.** O. V. Dobrovolsky (*Dobrovol'skij*), N. N. Kiselev, G. P. Chernova, F. A. Tupieva, N. V. Narizhnaja (*Narizhnaya*).
Solid particles in the solar system, (see 012.019), p. 259 - 262 (1980).

Anomalous polarization at small phase angles is confirmed as a common feature of dusty cometary atmospheres. The opposition effect is detected and interpreted as evidence of similarity between grains covering interplanetary, cometary and asteroidal surfaces. The prevailing radii of dust grains in the comet's atmosphere are estimated to be 0.15–0.19 μm.

Comet 1975 XI Bradfield

103.171 **Brightness of comet Bradfield (1975 XI).**
Komet. Tsirk., Kiev, No. 270 (1980). In Russian.

Comet 1979 X Bradfield

103.201 **Observations of comet Bradfield 1979*l*.**
J. C. Bennett.
Mon. Notes Astron. Soc. South. Africa, Vol. 39, 18 - 21 (1980).

103.202 **Spectroscopic and photographic observations of Comet Bradfield (1979*l*).**
C. B. Cosmovici, S. Ortolani.
Astron. Astrophys., Vol. 88, L16 - L20 (1980).

Spectra of Comet Bradfield (1979*l*) obtained in the period 4 - 12 February 1980 with the 122 and 182 cm Asiago telescope show, besides the usual CN, C_2, C_3, NH_2, and CH bands in the visual part of the spectrum, the near infrared (3 - 0) Philipps band system of C_2 and the (8 - 0), (7 - 0), (6 - 0) bands of H_2O^+. Photographic plates taken with the 90 cm Schmidt telescope show a dust poor comet with a very narrow and faint plasma tail containing several filaments.

103.203 **Investigation of water production in comet Bradfield (1979*l*).**
H. A. Weaver, P. D. Feldman, M. C. Festou.
Bull. American Astron. Soc., Vol. 12, 511 (1980). – Abstract.

103.204 **A very rapid rotation of the plasma tail axis of comet Bradfield 1979*l* on 1980 February 6.**
J. C. Brandt, J. Hawley, M. B. Niedner.
Bull. American Astron. Soc., Vol. 12, 511 - 512 (1980). Abstract.

103.205 **Small band width imagery of comet Bradfield.**
C. A. Harvel, T. R. Gull.
Bull. American Astron. Soc., Vol. 12, 530 (1980). – Abstract.

103.206 **A very rapid turning of the plasma-tail axis of comet Bradfield 1979*l* on 1980 February 6.**
J. C. Brandt, J. D. Hawley, M. B. Niedner, Jr.
Astrophys. J., Lett., Vol. 241, L51 - L54, plate L2 (1980).

Schmidt camera photographs of comet Bradfield 1979*l* indicate that a rapid change took place in the comet's plasma tail. A sequence of photographs spanning 27.5 minutes shows a 10° shift occuring in the plasma-tail axis between the first and last exposures. An interpretation based on the windsock theory of plasma tails is that the comet entered a region of rapidly changing solar-wind flow direction.

103.207 **Photometry of comet Bradfield 1979*l*.**
M. F. A'Hearn, R. L. Millis, P. V. Birch.
Bull. American Astron. Soc., Vol. 12, 730 (1980). – Abstract.

103.208 **Water production models of comet Bradfield (1979*l*).**
H. A. Weaver, P. D. Feldman, M. C. Festou, M. F. A'Hearn.
Bull. American Astron. Soc., Vol. 12, 731 (1980). – Abstract.

103.209 **Komeet Bradfield 1979*l*.** H. Feijth.
Zenit, 7e Jaarg., 344 - 345 (1980).

103.210 **Near-infrared spectroscopy of Comet Bradfield (1979*l*).** A. C. Danks, M. Dennefeld.
ESO Sci. Prepr. No. 126, 11 pp. (1980). – Submitted to Astron. J.

103.211 **IUE observations of the UV spectrum of Comet Bradfield.** P. D. Feldman, H. A. Weaver, M. C. Festou, M. F. A'Hearn, W. M. Jackson, B. Donn, J. Rahe, A. M. Smith, P. Benvenuti.
Nature, Vol. 286, 132 - 135 (1980).

The orbit of Comet Bradfield (1979*l*) was very favourable for observation by the IUE satellite, and the first observations, the preliminary results of which are reported here, were made on 10 and 11 January 1980. Previously, comprehensive vacuum UV and cometary spectra were available only for Comet West (1976 VI) from rocket observations and Comet Seargent (1978m) obtained with IUE, hence three comets have now been observed in the wavelength region where most of the major constitutents are detectable spectroscopically. The UV spectra of these comets are remarkably similar, and although the statistical basis is still very small, the implication for cosmogony is that nearly all comets have the same primary composition and origin.

103.212 **Comet Bradfield (1979*l*).**
Komet. Tsirk., Kiev, No. 265 (1980). In Russian.

103.213 **Carbon in comet Bradfield 1979*l*.**
M. F. A'Hearn, P. D. Feldman.
Astrophys. J., Lett., Vol. 242, L187 - L190 (1980).

Ultraviolet spectra of comet Bradfield obtained with *IUE* show a surprising absence of CO^+ despite the presence of CO_2^+, and they also show, for the first time, the unambiguous presence of the Mulliken bands of C_2. The anomalous CO^+/CO_2^+ ratio can be accounted for by the different scale lengths of the species rather than by an abundance difference. Using the model of Feldman, the upper limit for CO^+ is found to be consistent with the observed abundance of CO. On that same model, however, CO cannot produce nearly as much atomic carbon as is observed.

Orbital elements of comets. See Abstr. 103.005.

On the absence of CO^+ in comets Seargent and Bradfield. See Abstr. 103.231.

Comet 1980b Bowell

103.221 **Comet Bowell (1980b): a case of three-body preliminary-orbit determination?** B. G. Marsden.
Bull. American Astron. Soc., Vol. 12, 744 (1980). – Abstract.

103.222 **Comet Bowell (1980b).**
Yamamoto Circ., No. 1945 (1980).

103.223 **Comet Bowell (1980b).**
Komet. Tsirk., Kiev. Nos. 266, 270 (1980). In Russian.

Orbital elements of comets. See Abstr. 103.005.

Comet 1978 XV Seargent

103.231 **On the absence of CO^+ in comets Seargent and Bradfield.** M. F. A'Hearn, P. D. Feldman.
Bull. American Astron. Soc., Vol. 12, 751 (1980). – Abstract.

Periodic comet Schaumasse

103.241 **Results of an investigation of the motion of comet Schaumasse.** K. P. Matsukov.
Komet. Tsirk., Kiev. No. 264 (1980). In Russian.

Periodic comet Schwassmann-Wachmann 1

103.301 **Are outbursts in P/Schwassmann-Wachmann 1 explosive?** J. J. Cowan, M. F. A'Hearn.
Bull. American Astron. Soc., Vol. 12, 511 (1980). – Abstract.

103.302 **The search for a mechanism of the Comet 1925 II (Schwassmann-Wachmann 1) outbursts.**
S. Grudzińska.
Acta Astron., Vol. 30, 367 - 371 (1980).

The observations of the sudden brightness variations of the Comet Schwassmann-Wachmann 1 (1925 II) are discussed. They are not in contradiction with the supposition that the outbursts are caused by the collisions with the meteor streams.

103.303 **Short-period comet Schwassmann-Wachmann 1 (1925 II).**
Komet. Tsirk., Kiev, Nos. 266, 269 (1980). In Russian.

Comet 1965 VII Ikeya-Seki

Matters of the dust tails of the comets West (1976 VI) and Ikeya-Seki (1965 VIII).
See Abstr. 103.108.

Comet 1937 II Wilk

Orbital elements of comets. See Abstr. 103.005.

Periodic comet Swift-Gehrels

Orbital elements of comets. See Abstr. 103.005.

Periodic comet Gehrels 2

Orbital elements of comets. See Abstr. 103.005.

Periodic comet Lexell

The summer ecliptical meteor showers and Comet Lexell. See Abstr. 104.013.

Periodic comet Finlay

Orbital elements of comets. See Abstr. 103.005.

Periodic comet Russell 1

Orbital elements of comets. See Abstr. 103.005.

Periodic comet Brooks 2

103.431 **Periodic comet Brooks 2 (1980f).**
Yamamoto Circ., No. 1937 (1980).

103.432 **Comet P/Brooks (2) 1980f.**
British Astron. Assoc. Circ., No. 607 (1980).

103.433 **Rediscovery of the short-period comet Brooks 2 (1980f).**
Komet. Tsirk., Kiev, No. 264 (1980). In Russian.

Comet 1979 VI Torres

Orbital elements of comets. See Abstr. 103.005.

Comet 1980e Torres

103.461 **Comet Torres (1980e).**
Yamamoto Circ., Nos. 1937, 1938, 1945 (1980).

103.462 **New comet Torres 1980e.**
British Astron. Assoc. Circ., No. 607 (1980).

103.463 **New comet Torres (1980e).**
Komet. Tsirk., Kiev, No. 264 (1980). In Russian.

Orbital elements of comets. See Abstr. 103.005.

Periodic comet Halley

103.501 **Thermal models of comets Halley and Tempel 2.** P. R. Weissman, H. H. Kieffer.
Bull. American Astron. Soc., Vol. 12, 731 (1980). – Abstract.

103.502 **Der Komet P/Halley.** Seine Erscheinungen in der Vergangenheit sowie besonders jene von +1986 und +2061. H. Mucke.
Sternenbote, 23. Jahrg., 126 - 138 (1980).

103.503 **Expected characteristics of large dust particles in periodic comet Halley.** Z. Sekanina.
Comet Halley micrometeoroid hazard workshop, (see 012.049), p. 25 - 34 (1979).

103.504 **Physical models of comet Halley based upon qualitative data from the 1910 apparition.**
R. L. Newburn, Jr.
Comet Halley micrometeoroid hazard workshop, (see 012.049), p. 35 - 50 (1979).

103.505 **Magnetic tail and electrodynamical forces in comet Halley (analysis of laboratory and observational data).** Eh. M. Dubinin, P. L. Izrajlevich, I. M. Podgornyj, S. I. Shkol'nikova.
Kosm. Issled., Tom 18, 907 - 913 (1980). In Russian.

An impact mass-spectrometer for the Halley-probe. See Abstr. 032.531.

The International Comet Mission (ICM). See Abstr. 051.006.

The joint NASA/ESA cometary mission to comets Halley and Tempel 2. See Abstr. 051.040.

The micrometeoroid hazard to a space probe in the vicinity of the nucleus of Halley's comet. See Abstr. 051.041.

Dust hazard of the cometary probe. See Abstr. 051.042.

The treatment of the problems due to hypervelocity impact during a fast fly-by (57 km/s) of Halley's comet. See Abstr. 051.043.

Preliminary design concepts for the Halley probe shield. See Abstr. 053.011.

A particle impact detector for the Halley probe. See Abstr. 053.012.

The core of the meteor stream associated with comet Halley. See Abstr. 104.034.

Periodic comet Wild 3

103.521 **New comet P/Wild (3) 1980d.**
British Astron. Assoc. Circ., No. 607 (1980).

103.522 **Short-period comet Wild 3 (1980d).**
Komet Tsirk., Kiev, No. 266 (1980). In Russian.

Orbital elements of comets. See Abstr. 103.005.

Periodic comet Tempel 2

Experiments on dust collection for a cometary mission. See Abstr. 022.069.

The International Comet Mission (ICM). See Abstr. 051.006.

The joint NASA/ESA cometary mission to comets Halley and Tempel 2. See Abstr. 051.040.

Thermal models of comets Halley and Tempel 2. See Abstr. 103.501.

Comet 1980k Černis-Petrauskas

103.621 **Comet Černis-Petrauskas (1980k).**
IAU Circ., Nos. 3498, 3499, 3504, 3508, 3514, 3516, 3542 (1980).

103.622 **Comet Černis-Petrauskas (1980k).**
Yamamoto Circ., Nos. 1939 - 1942, 1944 (1980).

103.623 **New comet Černis-Petrauskas 1980k.**
British Astron. Assoc. Circ., No. 608 (1980).

103.624 **New comet Černis-Petrauskas (1980k).**
Komet Tsirk., Kiev, Nos. 266, 268, 269, 270 (1980). In Russian.

Orbital elements of comets. See Abstr. 103.005.

Comet 1980*l* Russell

103.641 **Comet Russell (1980*l*).**
IAU Circ., Nos. 3510, 3513, 3519, 3524 (1980).

103.642 **Comet Russell (1980*l*).**
Yamamoto Circ., Nos. 1942, 1943 (1980).

103.643 **New comet Russell 1980*l*.**
British Astron. Assoc. Circ., No. 609 (1980).

Orbital elements of comets. See Abstr. 103.005.

Comet 1977 XIV Kohler

103.651 **Observations of comet Kohler (1977 XIV).**
Komet Tsirk., Kiev, No. 265 (1980). In Russian.

Photoelectric observations of comets West 1976 VI and Kohler 1977 XIV. See Abstr. 103.104.

Periodic comet Russell 2

103.661 **Comet Russell 2 (1980o).**
IAU Circ., Nos. 3522, 3523, 3525 (1980).

103.662 **Comet Russell 2 (1980o).**
Yamamoto Circ., Nos. 1944 - 1946 (1980).

103.663 **New comet Russell 1980o.**
British Astron. Assoc. Circ., No. 611 (1980).

103.664 **New short-period comet Russell 2 (1980o).**
Komet. Tsirk., Kiev, No. 268 (1980). In Russian.

Orbital elements of comets. See Abstr. 103.005.

Comet 1980q Meier

103.681 **Comet Meier (1980q).**
IAU Circ., Nos. 3535, 3536, 3539, 3546, 3551 (1980).

103.682 **Comet Meier (1980q).**
Yamamoto Circ., Nos. 1946 - 1949 (1980).

103.683 **New comet Meier 1980q.**
British Astron. Assoc. Circ., Nos. 611, 612 (1980).

103.684 **New comet Meier (1980q).**
Komet. Tsirk., Kiev. Nos. 268, 270 (1980). In Russian.

Orbital elements of comets. See Abstr. 103.005.

Comet 1979 IX Meier

103.691 **Observations of comet Meier (1979i).**
Komet. Tsirk., Kiev, No. 264 (1980). In Russian.

Comet 1978 XXI Meier

103.701 **Radio observation of comet Meier (1978f) in 18-cm OH lines.**
P. T. Giguere, W. F. Huebner, T. M. Bania.
Astron. J., Vol. 85, 1276 - 1280 (1980).
Observations of 18-cm OH spectral lines in comet Meier (1978f) with the 1000-ft. Arecibo telescope show spatial resolution of the OH coma by the 2!9 beam (= 3.7 × 10^5 km). The data agree with predictions of the solar Fraunhofer spectrum-pumping theory of comet OH excitation. On the assumption that the OH parent molecule (e.g., H_2O) has a Haser-model scale length ≈ 1.0 × 10^5 km at heliocentric distance 1 AU, the authors derive an OH scale length < 10^6 km, and probably near 1.0 × 10^5 km. Assuming a recently calculated value of the OH lifetime and the solar radiative pumping model, the results indicate an OH production rate ~ 10^{29} s^{-1} at heliocentric distance 2 AU.

Periodic comet Tuttle

103.741 **Periodic comet Tuttle (1980h).**
IAU Circ., Nos. 3493, 3520, 3528, 3531, 3540, 3552 (1980).

103.742 **Periodic comet Tuttle (1980h).**
Yamamoto Circ., Nos. 1939, 1946, 1947, 1949 (1980).

103.743 **Drie heldere periodieke kometen in 1980 (2).**
R. J. Bouma.
Zenit, 7e Jaarg., 156 - 159 (1980). – Concerning periodic comets Tuttle and Stephan-Oterma.

103.744 **Comet P/Tuttle 1980h.**
British Astron. Assoc. Circ., Nos. 607, 609 (1980).

103.745 **An ephemeris and magnitude analysis of P/Comet Tuttle for 1980/1.** D. W. E. Green.
Int. Comet Q., Vol. 2, 62 - 64 (1980).

103.746 **Rediscovery of short-period comet Tuttle (1980h).**
Komet. Tsirk., Kiev, Nos. 265, 268 (1980). In Russian.

Comet 1980u Panther

103.751 **Comet Panther (1980u).**
IAU Circ., Nos. 3556, 3557 (1980).

Periodic comet Borrelly

103.761 **Periodic comet Borrelly (1980i).**
IAU Circ., No. 3494 (1980).

103.762 **Periodic comet Borrelly (1980i).**
Yamamoto Circ., No. 1940 (1980).

103.763 **Comet P/Borrelly 1980i.**
British Astron. Assoc. Circ., No. 608 (1980).

103.764 **Rediscovery of short-period comet Borrelly (1980i).**
Komet. Tsirk., Kiev, Nos. 265, 269 (1980). In Russian.

Periodic comet Kohoutek

103.781 **Periodic comet Kohoutek (1980j).**
IAU Circ., No. 3499 (1980).

103.782 **Periodic comet Kohoutek (1980j).**
Yamamoto Circ., No. 1940 (1980).

103.783 **Comet P/Kohoutek 1980j.**
British Astron. Assoc. Circ., No. 608 (1980).

103.784 **Rediscovery of short-period comet Kohoutek (1980j).**
Komet. Tsirk., Kiev, No. 265 (1980). In Russian.

Periodic comet Harrington

103.791 **Periodic comet Harrington (1980m).**
IAU Circ., No. 3513 (1980).

103.792 **Periodic comet Harrington (1980m).**
Yamamoto Circ., No. 1943 (1980).

103.793 **Recoveries of periodic comets.**
British Astron. Assoc. Circ., No. 609 (1980).
Concerning periodic comets Harrington 1980m and Reinmuth 2 1980n.

103.794 **Rediscovery of short-period comet Harrington (1980m).**
Komet. Tsirk., Kiev, No. 268 (1980). In Russian.

Periodic comet Encke

103.801 **The 1980 return of Encke's comet.** D. Milon.
Strolling Astron., Vol. 28, 170 - 172 (1980).

103.802 **Periodic comet Encke.**
IAU Circ., Nos 3501, 3521, 3526, 3528, 3531, 3533, 3545 (1980).

103.803 **Periodic comet Encke.**
Yamamoto Circ., Nos. 1941, 1943, 1944, 1946, 1947, 1949 (1980).

103.804 **Drie heldere periodieke kometen in 1980 (1). Komeet van Encke.** R. J. Bouma.
Zenit, 7e Jaarg., 112 - 116 (1980).

103.805 **Comet P/Encke.**
British Astron. Assoc. Circ., No. 609 (1980).

103.806 **A possible explanation of the light curve of comet Encke.** I. Ferrin, O. Naranjo.
Mon. Not. R. Astron. Soc., Vol. 193, 667 - 681 (1980).
The authors have constructed a theoretical model for the brightness of comet Encke as a function of heliocentric distance, R, using conservation of solar energy, and assuming that C_2 molecules are evaporated at the same rate as water molecules. An equation relating the total number of C_2 molecules in the cometary coma, and the total visual magnitude was used to produce a graph of m_V versus log R, as a function of surface albedo, A. The derived values were: for the nuclear radius, $r_N = 0.80 \pm 0.10$ km; for the Bond albedo, $A_{bond} = 0.77 \pm 0.02$; for the ratio of C_2 molecules to H_2O molecules, $\alpha = (8.2 \pm 2.2) \times 10^{-3}$.

103.807 **Short-period comet Encke.**
Komet. Tsirk., Kiev, Nos. 264, 268, 269 (1980). In Russian.

Periodic comet Reinmuth 2

103.821 **Periodic comet Reinmuth 2 (1980n).**
IAU Circ., No. 3514 (1980).

103.822 **P/Reinmuth 2 (1980n).**
Yamamoto Circ., No. 1944 (1980).

103.823 **Rediscovery of short-period comet Reinmuth 2 (1980n).**
Komet. Tsirk., Kiev, No. 268 (1980). In Russian.

Recoveries of periodic comets.
See Abstr. 103.793.

Comet 1980p Helin-Dunbar

103.841 **Comet Helin-Dunbar (1980p).**
IAU Circ., Nos. 3528, 3530 (1980).

103.842 **Comet Helin-Dunbar (1980p).**
Yamamoto Circ., Nos. 1945, 1947 (1980).

103.843 **New comet Helin-Dunbar (1980p).**
Komet. Tsirk., Kiev, No. 268 (1980). In Russian.

Periodic comet West-Kohoutek-Ikemura

103.861 **Periodic comet West-Kohoutek-Ikemura (1980r).**
IAU Circ., No. 3538 (1980).

103.862 **Periodic comet West-Kohoutek-Ikemura (1980r).**
Yamamoto Circ., No. 1947 (1980).

103.863 **Comet P/West-Kohoutek-Ikemura 1980r.**
British Astron. Assoc. Circ., No. 612 (1980).

103.864 **Rediscovery of comet West-Kohoutek-Ikemura (1980r).**
Komet. Tsirk., Kiev, No. 268 (1980). In Russian.

Comet 1980s Lovas

103.881 **Comet Lovas (1980s).**
IAU Circ., Nos. 3547, 3551, 3553 - 3555 (1980).

103.882 **Comet Lovas 1980s.**
Yamamoto Circ., Nos. 1948 - 1950 (1980).

Comet 1980t Bradfield

103.891 **Comet Bradfield (1980t).**
Yamamoto Circ., Nos. 1949, 1950 (1980).

103.892 **Comet Bradfield (1980t).**
IAU Circ., Nos. 3554, 3555 (1980).

103.893 **New comet Bradfield (1980t).**
Komet. Tsirk., Kiev, No. 270 (1980). In Russian.

104 Meteors, Meteor Streams

104.001 **A meteoric nightglow?** M. Gadsden.
Mon. Not. R. Astron. Soc., Vol. 192, 581 - 594 (1980).

The original reports of sky glows during spectacular meteor showers in the nineteenth century leave some doubt whether the observers might not simply have been over-enthusiastic. The nightglow photometric data available for the sole spectacular shower so far in the twentieth century are against their being any noticeable atmospheric effects resulting from the deposition of the meteoric material; calculations confirm the conclusions of Baggaley that sunlight scattered from the meteor stream might be visible near the radiant or the anti-radiant but, with the sole exception of one observer of the 1833 Leonids, no reports have been made of such an area of light.

104.002 **Measurements of the initial radii of the ionization columns of bright meteors.**
W. J. Baggaley, G. W. Fisher.
Planet. Space Sci.,Vol. 28, 575 - 580 (1980).

A multi-wavelength radar backscatter study of the echo characteristics of radio-meteors has yielded measurements of the height dependence of the radii, r_i, of overdense plasma meteor columns. For electron line densities $\alpha \sim 10^{15} m^{-1}$ it is found that $r_i \propto p^{-0.63}$ (p atmospheric density) with r_i=5 m at a height of 100 km.

104.003 **Meteor activity: 1979 July to 1980 April.**
G. H. Spalding.
J. British Astron. Assoc., Vol. 90, 467 - 468 (1980).

104.004 **Effect of solar radiation on a swarm of meteoric particles.**
R. A. Lyttleton.
Moon Planets, Vol. 23, 27 - 39 (1980).

The theory of the Poynting–Robertson effect is applied to the motion of meteors relative to a parent-comet describing an undisturbed elliptic orbit. It is shown that initially any emitted particle proceeds to move retrogressively away from the comet to a certain maximum angular distance (as seen from the Sun) depending on its σs-value, and thereafter undergoes relative motion in the opposite forward dirction. For comet Encke the time for the elongation to return to zero is about 6600 y, for Halley it is about 2×10^5 y, and for Tempel–Tuttle (1965 IV) just over 10^5 y.

104.005 **On some aims of observations of fireballs from space.**
R. L. Khotinok.
Meteoritika, Vyp. 38, p. 104 - 105 (1979). In Russian.

104.006 **Prairie network fireball data. II. Trajectories and light curves.**
R. E. McCrosky, S.-Y. Shao, A. Posen.
Meteoritika, Vyp. 38, p. 106 - 156 (1979). In Russian.

104.007 **IR observation of a persistent meteor train.**
M. A. Hapgood.
Nature, Vol. 286, 582 - 583 (1980).

A long-lived (>32 min) persistent meteor train has been observed with a low light television camera which was sensitive only to radiation between wavelengths of 700 and 900 nm. It is suggested that the IR emission arises from the excitation of the atmospheric band of molecular oxygen ($b^1\Sigma_g \rightarrow X^3\Sigma_g$) by the sodium catalytic process.

104.008 **Results of photographic observations of meteors in Kiev for 1957 - 1966. I. Main equations.**
V. V. Benyukh, V. G. Kruchinenko, L. M. Sherbaum.
Astrometr. Astrofiz., Vyp. (No.) 41, p. 68 - 81 (1980). In Russian.

The method of kinematic reduction of photographed meteors applied at the Kiev Astronomical Observatory is described. The detailed results of the reduction are given for 100 bright meteors: heights, velocities, decelerations, heliocentric orbits and other data.

104.009 **Changes in radar meteor echo rates at sunrise.**
A. Hajduk, E. M. Pittich, B. A. McIntosh.
Contrib. Astron. Obs. Skalnaté Pleso, Vol. 9, 115 - 120 (1980).

The authors present the change in the proportion of meteor echoes in all duration classes recorded. The magnitude of the relative increases is found to be approximately the same for all duration classes, but the rate of increase after sunrise is slower for counts down to shorter duration, and the maximum of the increase occurs later in time.

104.010 **Meteor echo drifts and echo duration.**
A. Hajduk, P. Prikryl.
Contrib. Astron. Obs. Skalnaté Pleso, Vol. 9, 121 - 124 (1980).

Analysis of the duration and radial components of the velocity of 215 echoes showing drift on the range-time radar record confirms the interpretation of echo drifts in terms of the motion of the effective reflexion point along the meteor trajectory.

104.011 **Geminid meteor shower: activity and magnitude distribution.**
V. Porubčan, M. Kresáková, J. Štohl.
Contrib. Astron. Obs. Skalnaté Pleso, Vol. 9, 125 - 144 (1980).

A series of visual observations of the Geminid meteor shower, comprising over 8000 records of meteors during nine revolutions of the shower in 1944 - 1974 is analysed. A mean curve of activity is constructed, and the individual deviations are discussed. The magnitude distribution exhibits a definite decrease of its index around the peak of the shower activity. This is in qualitative agreement with the predicted mass separation by the Poynting–Robertson effect, as well as with the radar observations of fainter meteors.

104.012 **Radar meteor rates and atmospheric density changes.**
C. D. Ellyett, J. A. Kennewell.
Nature, Vol. 287, 521 - 522 (1980).

As a way of explaining the evidence linking recorded radar meteor rates with solar activity, it has been suggested that variations of the atmospheric scale height at meteor ablation altitudes, themselves controlled by the sun, may be responsible. A quantitative examination of a theoretical treatment now shows that the above suggestion is certainly plausible, and details the necessary scale height changes as a function of the meteor mass exponent.

104.013 **The summer ecliptical meteor showers and Comet Lexell.** M. Kresáková.
Bull. Astron. Inst. Czechoslovakia, Vol. 31, 193 - 206 (1980).

The existence of meteors produced by Comet Lexell is of particular interest, because this is the only comet for which the period of injection of meteoroids into Earth-crossing orbits can be exactly specified. A number of authors have claimed an association with the comet for minor meteor showers observed in June - August and December. The reality of the relationship is examined. A comparison of the distributions in radiant positions, velocities, orbital elements and Tisserand invariant with the sporadic background does not reveal any of the expected anomalies above the threshold of random fluctuations, lending no support to the association.

104.014 **Comparison of the theory of a dynamically significant coma with PN-fireballs.** V. Padevět.
Bull. Astron. Inst. Czechoslovakia, Vol. 31, 206 - 223 (1980).

Some of the substantial observed properties of fireballs (the discrepancy between the dynamic and photometric masses, fragmentation into large fragments, different heights of the ends of otherwise identical fireballs and the corresponding types of light curves, etc.) can be explained by the hypothetical assumption of an unclassically high value of the aerodynamic head pressure on the solid component of the fireball. The theory of a dynamically significant coma provides one of the possible explanations for the generation of the high value of the aerodynamic pressure. The comparison of this theory with 116 fireballs selected from the PN-data (and with the Innisfree meteoritic fall) also provides the possibility of coordinating the data of known meteorites with all the fireball groups.

104.015 **Multi-frequency studies of radio-meteor train diffusion.** W. J. Baggaley,
Bull. Astron. Inst. Czechoslovakia, Vol. 31, 305 - 308 (1980).

Radio-meteor heights are often estimated from measurements of echo decay time-constants. However, it is known that even for a fixed height these time constants show a large dispersion. Dual wavelength measurements of echo decay times are reported and the results compared with those of two previous multi-frequency studies in order to provide information about the origin of the dispersion and hence of the uncertainties involved in measurements of heights.

104.016 **Measurements of the velocity dependence of the initial radii of meteor trains.** W. J. Baggaley.
Bull. Astron. Inst. Czechoslovakia, Vol. 31, 308 - 311 (1980).

Dual frequency radio-meteor measurements have been made of the initial radii of the plasma columns of both Geminid and sporadic meteors. The significantly lower Geminid radii suggest that train initial radius is nearly directly proportional to meteoroid velocity.

104.017 **Diurnal and seasonal variations of meteor rates from LF observation.** B. Saha, A. K. Sen.
Ann. Géophys., Vol. 35, 203 - 206 (1979). – Abstr. in Phys. Abstr., Vol. 83, Abstr. 82309 (1980).

104.018 **Some observed visual telescopic meteor rates.** D. Machholz.
Strolling Astron., Vol. 28, 166 - 167 (1980).

104.019 **Orbital elements of meteors from photographic observations during 1961 - 1965.**
E. N. Kramer, A. K. Markina.
Problems of cosmic physics. Vyp. 15, (see 003.003), p. 53 - 63 (1980). In Russian.

104.020 **Densities of meteor bodies. Photometric method.** V. G. Kruchinenko, A. V. Kuznetsova.
Problems of cosmic physics. Vyp. 15, (see 003.003), p. 63 - 68 (1980). In Russian.

The group of parameters containing specific energy of destruction, heat tranfer coefficient and meteor body density is determined on the basis of the equation of destruction and from observations of 102 bright meteors. The heat transfer coefficient is determined independently.

104.021 **Method and some results of an interpretation of the radio echoes of intermediate meteor trails.**
R. I. Mojsya, Yu. V. Chumak.
Problems of cosmic physics. Vyp. 15, (see 003.003), p. 68 - 79 (1980). In Russian.

A method of interpretation of intermediate meteor trails radio echoes is described. It is based on the comparison on the experimental amplitude-time and phase-time characteristics of meteor radio echoes with the standards. The latter were obtained from the solution of the problem of radio-wave scattering by a plasma cylinder with Gaussian profile.

104.022 **Bolides photographed in Dushanbe.** P. B. Babadzhanov, T. I. Getman.
Meteoritika, Vyp. 39, p. 15 - 18 (1980). In Russian.

104.023 **Meteors and atmospheres.** W. J. Baggaley.
Solid particles in the solar system, (see 012.019), p. 85 - 100 (1980).

Meteoroid ablation in an atmosphere produces a weak plasma column. The processes important in governing the state of the meteoritic species subsequent to column formation are reviewed. The actions of transport and of chemical reactions in controlling the life of the various species in the column are described.

104.024 **The influence of the atmosphere on radar meteor rates.** W. G. Elford.
Solid particles in the solar system, (see 012.019), p. 101 - 104 (1980).

104.025 **Serial correlation of meteor radar rates.** B. A. Lindblad.
Solid particles in the solar system, (see 012.019), p. 105 - 108 (1980).

104.026 **Orbit, chemical composition and atmospheric fragmentation of a meteoroid from instantaneous photographs.** P. B. Babadzhanov, V. S. Getman.
Solid particles in the solar system, (see 012.019), p. 111 - 115 (1980).

104.027 **Two station television meteor studies.** R. L. Hawkes, J. Jones.
Solid particles in the solar system, (see 012.019), p. 117 - 120 (1980).

The paper deals with some of the results of the first two station television intensifier study of faint meteors, and the implications of these results concerning the validity of the current theories of the structure and ablation of dustball meteors.

104.028 **One hundred and fifteen years of meteor spectroscopy.** P. M. Millman.
Solid particles in the solar system, (see 012.019), p. 121 - 128 (1980).

104.029 **Correlation of height and forbidden oxygen line strength for Perseid meteors.** J. A. Russell.
Solid particles in the solar system, (see 012.019), p. 129 - 132 (1980).

During the 1977 and 1978 Perseid showers, seventeen meteors were simultaneously photographed with a spectrograph and with a camera equipped with a rotating shutter. Of the eight best examples, the half in whose spectra the forbidden line of neutral oxygen at λ5577 was relatively strongest appeared and disappeared at heights about 9 km greater than did the half in whose spectra the line at λ5577 was relatively weak. Structural or compositional differences in the meteroroidal particles appear to be the most likely explanation for these height variations. A by-product of the investigation was an average value of -8.8 km/sec^2 for the deceleration of the meteors over the observed trajectories.

104.030 **Experimental and theoretical study of radiometeors.** J. Delcourt.
Solid particles in the solar system, (see 012.019), p. 133 - 136 (1980).

The author studies the long-term evolution of the orbits

of meteoric particles subjected to planetary perturbations and to Poynting-Robertson drag. Solar wind erosion of the particle is considered. The author computes the long-period elliptic elements of the meteoric orbit when it intersects the orbit of the earth. He compares experimental results with those of the dynamical study.

104.031 **Ground radar detection of meteoroids in space.** D. J. Kessler, P. M. Landry, J. R. Gabbard, J. L. T. Moran.
Solid particles in the solar system, (see 012.019), p. 137 - 139 (1980).

During a special test conducted by NORAD (North American Air Defense Command) for NASA to lower the detection threshold for small satellites, part of the radar software was modified, resulting in acquisition of what appears to be important new meteoroid data. The purpose of the paper is to describe the special test and give a preliminary analysis of the meteoroid data.

104.032 **On time-dependent models of the meteoric background complex.** J. Stohl.
Solid particles in the solar system, (see 012.019), p. 141 - 144 (1980).

Continuous models of the meteoric background complex are discussed and analysed on the basis of observational data on radiants and hourly rates of sporadic meteors. A time-dependent, bi-elliptical model of the true radiant distribution fits the observational results.

104.033 **Evolution of meteoroid orbits over millenia.** I. V. Galibina, A. K. Terentjeva (*Terent'eva*).
Solid particles in the solar system, (see 012.019), p. 145 - 148 (1980).

The evolution of meteoroid orbits over long time periods has been studied using the Gauss-Halphen-Gorjatschew method taking into account perturbations from the outer four planets (Galibina 1970). As initial elements, 15 different meteor orbits were taken from photographic observations (McCrosky 1968). Calculations were carried out from the modern epoch back in time for intervals of 10000 to 100000 years.

104.034 **The core of the meteor stream associated with comet Halley.** A. Hajduk.
Solid particles in the solar system, (see 012.019), p. 149 - 152 (1980).

Data on 240,000 meteor echoes recorded in 1958 - 1967 with the Springhill Meteor Radar, have been used for determining the structural features of the Eta Aquarid stream. A zone with a distinct enhancement in meteor rates was located at solar longitudes of 43° to 47°. The core of the stream, corresponding to the four peak days, extends from the orbit of the comet Halley to a distance of 0.08 AU. Other observations, from both hemispheres, support these results.

104.035 **The quadrantid meteor stream: past, present and future.**
D. W. Hughes, I. P. Williams, C. D. Murray.
Solid particles in the solar system, (see 012.019), p. 153 - 156 (1980).

104.036 **Evolution of orbits and intersection conditions with the earth of the Geminid and Quadrantid meteor streams.** P. B. Babadzhanov, Yu. V. Obrubov.
Solid particles in the solar system, (see 012.019), p. 157 - 162 (1980).

The paper presents the results of a study of the orbital evolution of the Geminids and Quadrantids and the intersection conditions of the streams with the earth taking into account nongravitational effects, as well as the perturbations from six planets.

104.037 **The structure of the Taurid, Geminid and Quadrantid meteor streams.** B. Lokanadham.
Solid particles in the solar system, (see 012.019), p. 163 - 166 (1980).

The Taurids, Geminids and Quadrantids represent meteor streams existing at different stages of stream evolution. The structure of these streams can be deduced from a study of the variations in incident flux of shower meteors and their mass exponents as a function of solar longitude. The paper is an attempt to deduce such structure from recent observations in India combined with data available for these showers from the early years of this century.

104.038 **Television observations of the Delta-Aquarid shower.** T. Sarma, J. Jones.
Solid particles in the solar system, (see 012.019), p. 167 - 170 (1980).

104.039 **Observational and theoretical aspects of fireballs.** Z. Ceplecha.
Solid particles in the solar system, (see 012.019), p. 171 - 183 (1980).

Recent theoretical concepts of large meteoroid entry into the atmosphere are compared and an evident lack of a theory applicable to all observed fireballs is found. A new empirical criterion separating different fireball groups according to structure and composition of their bodies and containing only values directly obtained by photographic observations is proposed.

104.040 **Interaction of large bodies with the earth's atmosphere.** D. O. ReVelle.
Solid particles in the solar system, (see 012.019), p. 185 - 198 (1980).

During the last decade much work has proceeded on the problem of understanding the complex physicochemical processes associated with the entry of large meteoroids into the atmosphere. In this paper the respective areas of ablation, luminosity and infrasonic wave generation are surveyed from the viewpoint of the physics of fluids. Related work on meteorite cratering processes is not considered.

104.041 **On the probability of collision with earth, and orbital life-time of bodies of asteroidal and cometary origin.** E. N. Kramer, V. I. Musiy (*Musij*), E. A. Timchenko-Ostroverkhova, I. S. Shestaka.
Solid particles in the solar system, (see 012.019), p. 199 - 204 (1980).

104.042 **Submicron particles in meteor streams.** S. F. Singer, J. E. Stanley.
Solid particles in the solar system, (see 012.019), p. 329 - 332 (1980).

Submicron particles (mass$<10^{-15}$ g) were observed in certain meteor streams by impact detectors on satellite Explorer 46. This unexpected finding leads to conclusions about the nature and origin of the particles and about the importance of nongravitational forces.

104.043 **On some physical phenomena in the plasma of meteor trains.** N. Abdrakhmanov.
Spektrosk. i prikl. fiz., Karaganda, 1979, p. 82 - 88. In Russian. Abstr. in Ref. zh., 51. Astron., 9.51.270 (1980).

104.044 **Anomalous sounds from the entry of meteor fireballs.** C. S. L. Keay.
Science, Vol. 210, 11 - 15 (1980).

A very bright fireball observed over New South Wales in 1978 produced anomalous sounds clearly audible to some of the observers. An investigation of the phenomenon indicates that bright fireballs radiate considerable electromagnetic energy in the very-low-frequency (VLF) region of the spectrum. A mechanism for the production of VLF emissions from the highly energetic wake turbulence of the fireball is proposed.

104.045 **The search for the Zvolen fireball.**
D. W. Hughes.
Nature, Vol. 288, 118 (1980).

104.046 **Account for the sphericity of the earth's surface in the statistical theory of meteor radio signals.**
A. A. Gajdaev, A. M. Zulliev.
Geomagn. Aehron., Tom 20, 957 - 958 (1980). In Russian.

104.047 **Connection of meteor matter with comets and asteroids.** B. Yu. Levin.
Zemlya Vselennaya, 1980, No. 6, p. 5 - 9. In Russian.

104.048 **New meteor shower.**
IAU Circ., Nos. 3528, 3542, 3545 (1980).

104.049 **Results of the examination of the Skylab/Apollo windows for micrometeoroid impacts.**
B. G. Cour-Palais.
Proc. Tenth Lunar Planet. Sci. Conf., (see 012.050), p. 1665 - 1672 (1979).

104.050 **De vuurbol van 21 augustus 1979.**
B. Apeldoorn.
Zenit, 7e Jaarg., 118 - 120 (1980).

104.051 **De vuurbol van 21 augustus 1979.** P. Roggemans.
Zenit, 7e Jaarg., 225 (1980).

104.052 **Werkgroep Meteoren fotografeert bijzondere meteoor.** N. de Kort.
Zenit, 7e Jaarg., 394 - 395 (1980).

104.053 **1980 Perseid meteor shower.**
British Astron. Assoc. Circ., No. 609 (1980).

104.054 **The Geminids and Quadrantids.**
British Astron. Assoc. Circ., No. 612 (1980).

104.055 **Some Ursids on 1980?**
British Astron. Assoc. Circ., No. 612 (1980).

104.056 **Audible sounds excited by aurorae and meteor fireballs.** C. S. L. Keay.
J. R. Astron. Soc. Canada, Vol. 74, 253 - 260 (1980).

104.057 **Flux of optical meteors down to $M_{pg} = +12$.**
A. F. Cook, T. C. Weekes, J. T. Williams, E. O'Mongain.
Mon. Not. R. Astron. Soc., Vol. 193, 645 - 666 (1980).

The authors derive a cumulative flux-density relationship of sporadic meteors: $\log \phi = -17.88 + 0.533\, M_{pg}$.

104.058 **The luminosity functions of the 1969 Perseid and Orionid meteor showers.** K. Krisciunas.
Icarus, Vol. 43, 381 - 384 (1980).

Visual counts of the 1969 Perseid and Orionid meteor showers are presented. On the basis of the maximum-likelihood method of determining the power law luminosity function index, the author derives $s = 1.56 \pm 0.06$ for the Perseids with $m_v = +1$ to -5, and $s \approx 1.85 \pm 0.1$ for the Orionids with $m_v = +2$ to -3. These values are somewhat lower than those found by other observers. Under the assumption that the masses of visual meteors are proportional to a power law function of the luminosities, this implies power law mass functions.

104.059 **On deionization of meteor tracks in the earth's atmosphere.**
R. Sh. Bibarsov, A. V. Blokhin, G. G. Novikov.
Geomagn. Aehron., Tom 20, 1116 - 1118 (1980). In Russian.

104.060 **Results of photographic observations of meteors in Kiev during 1957 - 1966. II. Photometry of meteors.**
V. V. Benyukh, V. G. Kruchinenko, L. M. Sherbaum.
Astrometr. Astrofiz., Vyp. (No.) 42, p. 41 - 54 (1980). In Russian.

The method of photographic photometry of meteors elaborated at the Kiev State University Astronomical Observatory is described. Photometric processing of 100 bright basic meteors results in the curves of brightness and in the masses of meteor bodies.

104.061 **On the formation of the Geminid meteor stream.**
Yu. V. Obrubov.
Dokl. AN TadzhSSR, Vol. 23, 175 - 179 (1980). In Russian.
Abstr. in Ref. zh., 51. Astron., 1.51.266 (1981).

104.062 **Motion and evaporation of meteoric particles in the upper layers of the atmosphere.**
Eh. Z. Apshtejn, N. N. Pilyugin.
Neravnovesn. techeniya gaza s fiz.-khim. prevrashcheniyami, Moskva, 1980, p. 76 - 91. In Russian. – Abstr. in Ref. zh., 51. Astron., 1.51.269 (1981).

104.063 **Evolution of the radiant of the 13-Lyrid meteor stream.** A. K. Terent'eva.
Komet. Tsirk., Kiev, No. 270 (1980). In Russian.

104.064 **On the reliability of identification of meteor streams with comets.** B. A. Vorontsov-Vel'yaminov.
Komet. Tsirk., Kiev, No. 269 (1980). In Russian.

104.065 **Thermal fracture of meteor bodies.**
O. N. Salimov, I. E. Konstantinov, S. G. Mikheenko.
Astron. Vestn., Tom 13, 50 - 55 (1979). In Russian.

The average diameter of particles separated from a meteor body by thermal stresses is calculated. It is assumed that the fracture is in the surface layer.

104.066 **Relation between the minimal value of the coefficient of heat transfer to a meteoric body and the depth of its penetration into the atmosphere.**
V. V. Kalenichenko.
Astron. Vestn., Tom 14, 86 - 88 (1980). In Russian.

Numerical solution of the equation of meteoric body evaporation permitted to obtain the relation between the minimal value of the coefficient of heat transfer to the meteoric body and the depth of its penetration into the atmosphere.

104.067 **Possibility of using optimization methods for determination of the parameters of meteors. I.**
K. V. Kostylev, K. K. Kostylev.
Astron. Vestn., Tom 14, 89 - 96 (1980). In Russian.

The possibility is considered of using non-linear optimization methods for determining the main parameters of a meteoric particle by analysis of amplitude-time characteristics obtained from uncompressed meteor trails. The following parameters are deduced: reflection-point height; meteor velocity and zenith angle of its path; the position of the reflection-point on the ionization curve; the parameter which describes meteor fragmentation.

104.068 **On a possible relation between the Quadrantid meteor stream and comets.** E. D. Kondrat'eva.
Astron. Vestn., Tom 14, 97 - 101 (1980). In Russian.

The velocity values necessary for particle transfer from a cometary orbit to the orbit of the Quadrantids are presented. The possibility of connection of the shower with the comets Tuttle, Pons–Brooks, Stephan–Oterma, Kozik–Peltier, 1860 I is considered.

104.069 **Physically-mathematical modelling of formation and evolution of meteor streams. I.**
L. A. Katasev, N. V. Kulikova.
Astron. Vestn., Tom 14, 168 - 175 (1980). In Russian.

Assuming that meteor streams are formed by cometary nuclei disintegration the ejection rate of meteoric particles from the nuclei of comets is estimated by a statistical test method and the meteor stream formation process is simulated. The theoretical calculations are compared with the results of observations of Draconids, Perseids, Taurids and Leonids. Similar calculations have been made for the streams of Lyrids, Ursids, α-Capricornids, Orionids, η-Aquarids, Andromedids in order to ascertain their possible relation with the comets 1861 I, 1939 X, 1954 III, 1910 II, 1852 III. The initial orbit elements of streams and comets as well as simulation results of matter ejection from the nuclei of corresponding comets in the perihelia of cometary orbits with the rate up to 100 m/s –comets 1861 I, 1852 III, 1910 II, and up to 350 m/s – comets 1910 II, 1939 X, 1954 III are presented.

104.070 **Perturbed motion of meteoric swarms with various dimensions and orbital inclinations.**
V. G. Kruchinenko, L. M. Sherbaum.
Astron. Vestn., Tom 14, 176 - 181 (1980). In Russian.

The authors have carried out model calculations for the motion of nine meteoric swarms with various orbital dimensions (a = 2.5, 4.5, 6.5 a.u.) and inclinations (i = 2, 30, 60°) taking into account the perturbing action of Jupiter and Saturn in the interval of 250 years. The results show that for short-period swarms the inference of perturbations on the variations of orbital elements of individual particles is small (except the case of commensurability of orbital periods of the swarm and Jupiter). The swarms approaching the planets deform and expand themselves significantly during 1–2 years. If the approach does not take place (i = 30, 60°), the deformations are small and develop slowly.

Results of radar observations of faint meteors. Catalogue of orbits to + 12^m. See Abstr. 002.034.

The use of a large-aperture radio system for meteor studies. See Abstr. 031.536.

The analysis of meteor data. See Abstr. 031.544.

Problemi di dinamica dei corpi minori del sistema solare. See Abstr. 091.044.

105 Meteorites, Meteorite Craters

105.001 **A study of chondrule rims and chondrule irradiation records in unequilibrated ordinary chondrites.**
J. S. Allen, S. Nozette, L. L. Wilkening.
Geochim. Cosmochim. Acta, Vol. 44, 1161 - 1175 (1980).

In order to investigate the possibility that chondrules may have had an independent existence in space, the authors have searched for unusual nuclear track densities in chondrules and studied the compositions of chondrule rims on chondrules from thirteen unequilibrated ordinary chondrites.

105.002 **Il meteorite del 18 aprile 1977.**
S. Ghedini, L. Baldinelli.
Coelum, Vol. 49, 137 - 145 (1980).

105.003 **The meteorite-asteroid connection: the infrared spectra of eucrites, shergottites, and Vesta.**
M. A. Feierberg, M. J. Drake.
Science, Vol. 209, 805 - 807 (1980).

Infrared reflectance spectra have been obtained for the meteorites Shergotty and Allan Hills (ALHA) 77005, a unique achondrite apparently related to the shergottites. Comparisons with the reflectance spectra of eucrites and asteroid 4 Vesta indicate that the surface of Vesta is covered with eucrite-like basalts and that, if shergottite-like basalts are present on the surface of Vesta, they must be a minor rock type. The paradox that both the eucrite and shergottite parent bodies should presently exist is examined. The preferred solution is that both eucrites and shergottites are derived from Vesta, and that this asteroid is compositionally and isotopically heterogeneous; however, other possible solutions cannot be ruled out.

105.004 **The origin of the 'FUN' anomalies and the high temperature inclusions in the Allende meteorite.**
G. J. Consolmagno, A. G. W. Cameron.
Moon Planets, Vol. 23, 3 - 25 (1980).

The discovery of isotopic anomalies in white inclusions of the meteorite Allende has led to fundamental questions concerning the origin of these anomalies and of the white inclusions themselves. An analysis of the 'FUN' anomalies in the inclusions C1 and EK1-4-1 demonstrates that these isotopic anomalies may be decomposed into individual nucleosynthetic components, which have been subjected to separate mass and component fractionations. Giant gaseous protoplanets, as described for the early solar nebula by Cameron (1978), are a potential site for achieving both mass and component fractionations, and for producing white inclusions in general.

105.005 **Origin of iron meteorite groups IAB and IIICD.**
J. T. Wasson, J. Willis, C. M. Wai, A. Kracher.
Z. Naturforsch., Band 35a, 781 - 795 (1980).

Several low-Ni iron meteorites previously assigned to group IAB are reclassified IIICD on the basis of lower Ge, Ga, W and Ir concentrations and higher As concentrations; the low-Ni extreme of IIICD is now 62 mg/g, that of IAB is 64 mg/g. The resulting fractionation patterns in the two groups are quite similar. Models have been proposed, but all have serious flaws. A new model is proposed involving the formation of each iron in small pools of impact melt on a parent body consisting of material similar to the chondritic inclusions found in some IAB and IIICD irons, but initially unequilibrated.

105.006 **The Sikhote-Alin meteorite shower as a classical meteorite fall.** E. L. Krinov, V. I. Tsvetkov.
Meteoritika, Vyp. 38, p. 19 - 26 (1979). In Russian.

105.007 **Rare gases in carbonaceous chondrites.**
L. K. Levskij.
Meteoritika, Vyp. 38, p. 27 - 36 (1979). In Russian.

105.008 **Investigation of the Elenovka chondrite (petrography and mineralogy).**
G. V. Baryshnikova, A. K. Lavrukhina.
Meteoritika, Vyp. 38, p. 37 - 44, 159, 161 - 163 (1979). In Russian.

105.009 **Composition and structure of nickel iron of some chondrites.**
V. P. Semenenko, A. F. Nikityuk, N. S. Stetsenko.
Meteoritika, Vyp. 38, p. 45 - 49, 159, 164 - 165 (1979). In Russian.

105.010 **Investigation of the Baku iron meteorite from the collection of the Committee on meteorites of the USSR Academy of Sciences.**
N. I. Zaslavskaya, L. D. Barsukova, V. Ya. Kharitonova, G. M. Kolesov.
Meteoritika, Vyp. 38, p. 50 - 54, 160, 166 - 167 (1979). In Russian.

105.011 **A study of the thermal history of the pallasites Marjalahti, Lipovsky Khutor and Eagle Station using track and thermoluminescence analysis.**
V. P. Perelygin, V. G. Kashkarova.
Meteoritika, Vyp. 38, p. 55 - 58 (1979). In Russian.

105.012 **Radiation age and preatmospheric mass of the Boguslavka meteorite.**
A. V. Fisenko, V. A. Alekseev, L. K. Levskij.
Meteoritika, Vyp. 38, p. 59 - 61 (1979). In Russian.

105.013 **Some results of experiments on vaporization of the metallic phase of meteorites in a vacuum.**
A. K. Lavrukhina, A. V. Fisenko, N. K. Sazhina, S. A. Stakheeva.
Meteoritika, Vyp. 38, p. 62 - 64 (1979). In Russian.

105.014 **On losses when collecting iron meteorites by a metal detector.** V. I. Tsvetkov.
Meteoritika, Vyp. 38, p. 65 - 67 (1979). In Russian.

105.015 **A simple model of crater formation.**
B. A. Ivanov.
Meteoritika, Vyp. 38, p. 68 - 85 (1979). In Russian.

105.016 **Geologically-geophysical characteristic of the meteoritic crater Zhamanshin (on the results of the expedition of 1977).**
P. V. Florenskij, A. I. Dabizha, A. O. Aaloe, Eh. S. Gorshkov, V. I. Miklyaev.
Meteoritika, Vyp. 38, p. 86 - 98, 160, 168 (1979). In Russian.

105.017 **Meteoritic crater Shunak.**
V. I. Fel'dman, A. I. Dabizha, L. B. Granovskij.
Meteoritika, Vyp. 38, p. 99 - 103 (1979). In Russian.

105.018 **Ordering of FeNi in clear taenite from meteorites.**
E. R. D. Scott, R. S. Clarke, Jr.
Nature, Vol. 287, 255 (1980).

105.019 **Oxygen isotopic anomalies in Allende inclusion HAL.** T. Lee, T. K. Mayeda, R. N. Clayton.
Geophys. Res. Lett., Vol. 7, 493 - 496 (1980).

105.020 **Chemical energy in cold-cloud aggregates: the origin of meteoritic chondrules.** D. D. Clayton.
Astrophys. J., Lett., Vol. 239, L37 - L41 (1980).

If interstellar particles and molecules accumulate into larger particles during the collapse of a cold cloud, the resulting aggregates contain a large store of internal chemical energy. It is proposed that subsequent warming of these accumulates leads to a thermal runaway when exothermic chemical reactions begin within the aggregate. These, after cooling, are the crystalline chondrules found so abundantly within chondritic meteorites. Chemical energy can also heat meteoritic parent bodies of any size, and both thermal metamorphism and certain molten meteorites are proposed to have occurred in this way. If this new theory is correct, (1) the model of chemical condensation in a hot gaseous solar system is eliminated, and (2) a new way of studying the chemical evolution of the interstellar medium has been found. A simple dust experiment on a comet flyby is proposed to test some features of this controversy.

105.021 **Cosmic radiation effects in Dhajala meteorite.** H. S. Virk.
Curr. Sci., Vol. 48, 1067 - 1068 (1979). – Abstr. in Phys. Abstr., Vol. 83, Abstr. 72733 (1980).

105.022 **Endemic isotopic anomalies in titanium.** F. R. Niederer, D. A. Papanastassiou, G. J. Wasserburg.
Astrophys. J., Lett., Vol. 240, L73 - L77 (1980).

Abundances of the titanium isotopes were determined using a new high-precision technique that shows terrestrial, lunar, and bulk meteorite samples to be distinguishable. Ca-Al-Ti–rich inclusions in the Allende meteorite are found to contain Ti of widely varying isotopic composition reflecting the presence of at least three nucleosynthetic components. The anomalies in Ti appear to be relatively widespread and, when correlated with Ca data, provide a clue to nucleosynthesis in the neighborhood of the iron peak and to the late-stage nucleosynthetic process which immediately preceded formation of the solar nebula.

105.023 **Superheavy-element fission tracks in iron meteorites.** S. K. Runcorn, W. F. Libby, L. M. Libby, with a reply by R. K. Bull.
Nature, Vol. 287, 565 (1980).

105.024 **Transmission electron microscopy of the Jilin (Kirin) meteorite.**
P. Zhang, K. Tao, D. Yang, X. Chen.
Kexue Tongbao, Vol. 25, 767 - 769 (1980).

105.025 **A major meteorite impact on the earth 65 million years ago: evidence from the Cretaceous-Tertiary boundary clay.** R. Ganapathy.
Science, Vol. 209, 921 - 923 (1980).

Evidence for a major meteorite impact on the earth 65 million years ago is shown by the presence of meteoritic debris in the "fish clay" from Denmark representing the Cretaceous-Tertiary boundary. With the exception of rhenium, all the enriched noble metals in the clay are present in cosmic proportions, indicating that the impacting celestial body had not undergone gross chemical differentiation.

105.026 **Deformation mechanisms in mildly shocked chondritic diopside.** J. R. Ashworth.
Meteoritics, Vol. 15, 105 - 115 (1980).

105.027 **Nepheline and sodalite in a barred olivine chondrule from the Allende meteorite.** G. R. Lumpkin.
Meteoritics, Vol. 15, 139 - 147 (1980).

105.028 **A new specimen of the Mount Dooling iron meteorite from Mount Manning, Western Australia.**
J. R. De Laeter.
Meteoritics, Vol. 15, 149 - 155 (1980).

A 701 kg iron meteorite has recently been discovered near the Mount Manning Range in Western Australia. The meteorite is a member of Chemical Group IC. A comparison of the chemical composition, surface features, microstructure and location of this meteorite with the Mount Dooling meteorite confirms that the find is a larger specimen of Mount Dooling.

105.029 **Impact and impact-like structures in Algeria. Part I. Four bowl-shaped depressions.**
P. Lambert, J. F. McHone, Jr., R. S. Dietz, M. Houfani.
Meteoritics, Vol. 15, 157 - 179 (1980).

105.030 **Neon-E in a Murchison residue.** F. O. Meier, P. Eberhardt, L. Alaerts, R. Lewis.
Meteoritics, Vol. 15, 181 - 182 (1980). – Abstract.

105.031 **Chemistry, mineralogy, and irradiation records of the Acapulco meteorite.**
M. Christophe-Michel-Lévy, J. C. Lorin, H. Palme, L. Schultz, B. Spettel, H. Wänke, H. W. Weber.
Meteoritics, Vol. 15, 182 - 183 (1980). – Abstract.

105.032 **$^{40}Ar-^{39}Ar$ ages of Allende.** E. K. Jessberger, B. Dominik, T. Staudacher, G. F. Herzog.
Icarus, Vol. 42, 380 - 405 (1980).

105.033 **Hg and Pt-metals in meteorite carbon-rich residues: suggestions for possible host phase for Hg.**
S. Jovanovic, G. W. Reed, Jr.
Geochim. Cosmochim. Acta, Vol. 44, 1399 - 1407 (1980).

Carbon-rich and oxide residual phases have been isolated from Allende and Murchison by acid demineralization for the determination of their Hg, Pt-metal, Cr, Sc, Co and Fe contents. Large enrichments of Hg and Pt-metals were found in Allende but not in Murchison residues. Some implications of how Hg is associated with residual phase(s) based on the stepwise heating release pattern and the activation energy to mobilize it are discussed.

105.034 **Carbonaceous chondrites – II. Carbonaceous chondrite phyllosilicates and light element geochemistry as indicators of parent body processes and surface conditions.** T. E. Bunch, S. Chang.
Geochim. Cosmochim. Acta, Vol. 44, 1543 - 1577 (1980).

The authors report detailed work on the matrices of Jodzie, Murray, Murchison, and Nogoya, in which they find ample evidence to confirm the earlier contention that the 'low-temperature' matrices of carbonaceous chondrites formed in aqueous media under low temperature (<400 K) on or in the near surface of a planetary body. The authors' observations of some C3(V) and C3(O) matrices also suggest they were partially altered and that in situ alteration of carbonaceous chondrites was more prevalent that previously recognized. How and where the low-temperature fraction (matrices) formed is important to understanding the evolution of organic compounds, entrapment and retention of noble gases (the carrier is a matrix phase) and environmental constraints on parent body surfaces.

105.035 **The nature and origin of ureilites.** J. L. Berkley, G. J. Taylor, K. Keil, G. E. Harlow, M. Prinz.
Geochim. Cosmochim. Acta, Vol. 44, 1579 - 1597 (1980).

The authors performed a comprehensive mineralogical and petrological study on eight ureilites the results of which were used to construct a hypothesis for the origin of ureilites.

105.036 **Carbynes: carriers of primordial noble gases in meteorites.**
A. G. Whittaker, E. J. Watts, R. S. Lewis, E. Anders.
Science, Vol. 209, 1512 - 1514 (1980).
Five carbynes have been found in the Allende and Murchison carbonaceous chondrites: carbon VI, VIII, X, XI, and (tentatively) XII. From the isotopic composition of the associated noble-gas components, it appears that the carbynes in Allende are local condensates from the solar nebula, whereas at least two carbynes in Murchison are of exotic, presolar origin.

105.037 **Carbynes in meteorites: detection, low-temperature origin, and implications for interstellar molecules.**
R. Hayatsu, R. G. Scott, M. H. Studier, R. S. Lewis, E. Anders.
Science, Vol. 209, 1515 - 1518 (1980).
Low-temperature formation by surface catalysis may be the dominant source of carbynes on the earth and in meteorites, and a major source of interstellar carbynes and cyanopoly-acetylenes.

105.038 **Morphology of meteorites as an investigation method of their fall on the earth.** E. L. Krinov.
Meteoritika, Vyp. 39, p. 3 - 14 (1980). In Russian.

105.039 **On the classification of stony and stony-iron meteorites.** A. A. Yavnel'.
Meteoritika, Vyp. 39, p. 19 - 27 (1980). In Russian.

105.040 **Distribution of some elements between the phases of iron meteorites.** A. A. Yavnel'.
Meteoritika, Vyp. 39, p. 28 - 37 (1980). In Russian.

105.041 **Determination of the petrological types of chondrites.** A. Ya. Skripnik, O. A. Kirova.
Meteoritika, Vyp. 39, p. 38 - 42 (1980). In Russian.

105.042 **Mineralogically-petrographic investigation of the Vetluga meteorite.** A. Ya. Skripnik.
Meteoritika, Vyp. 39, p. 43 - 48, 144 - 145, 161 (1980). In Russian.

105.043 **Investigation of the Genichesk meteorite.**
O. A. Kirova, L. D. Barsukova, V. Ya. Kharitonova, A. Ya. Skripnik.
Meteoritika, Vyp. 39, p. 49 - 53, 146 - 147, 161 (1980). In Russian.

105.044 **The Aliskerovo iron meteorite.**
S. G. Zhelnin, A. A. Plyashkevich, L. N. Plyashkevich, S. M. Sandomirskaya.
Meteoritika, Vyp. 39, p. 54 - 63, 148 - 152, 161 - 162 (1980). In Russian.

105.045 **The Lazarev meteorite.**
N. I. Zaslavskaya, G. M. Kolesov.
Meteoritika, Vyp. 39, p. 64 - 69, 153 - 154, 162 (1980). In Russian.

105.046 **Mineralogy and structure of the Egvekinot meteorite.** L. N. Plyashkevich, S. M. Sandomirskaya, N. I. Zaslavskaya.
Meteoritika, Vyp. 39, p. 70 - 78, 155 - 158, 163 (1980). In Russian.

105.047 **X-ray study of the meteoritic magnesioferrite.**
V. P. Semenenko, V. S. Mel'nikov.
Meteoritika, Vyp. 39, p. 79 - 80 (1980). In Russian.

105.048 **Investigation of specular material on the walls of crystal olivine imprints in nickel iron of the Marjalahti pallasite.** V. D. Kolomenskij, I. A. Yudin.
Meteoritika, Vyp. 39, p. 81 - 82, 159, 163 (1980). In Russian.

105.049 **Investigation of the fused crust of the Marjalahti pallasite.** V. D. Kolomenskij, I. A. Yudin.
Meteoritika, Vyp. 39, p. 83 - 84, 160, 163 (1980). In Russian.

105.050 **On preatmospheric sizes of the Lipovsky Khutor meteorite.** T. P. Zholud', L. K. Levskij, D. Lkhagvasurehn, O. Otgonsurehn, V. P. Perelygin, S. G. Stetsenko.
Meteoritika, Vyp. 39, p. 85 - 89 (1980). In Russian.

105.051 **Magnetic properties of some tektites.**
E. G. Gus'kova.
Meteoritika, Vyp. 39, p. 90 - 94 (1980). In Russian.

105.052 **Chinge: new discovery of meteorites and a possible crater.** V. I. Tsvetkov.
Meteoritika, Vyp. 39, p. 95 - 101 (1980). In Russian.

105.053 **Shock metamorphosed rocks of the Elgygytgyn meteorite crater on Chukotka.**
E. P. Gurov, E. P. Gurova.
Meteoritika, Vyp. 39, p. 102 - 109 (1980). In Russian.

105.054 **Some peculiarities of geochemistry of Elgygytgyn impactites, Chukotka, USSR.**
V. I. Fel'dman, L. B. Granovskij, I. G. Naumova, N. N. Nikishina.
Meteoritika, Vyp. 39, p. 110 - 113 (1980). In Russian.

105.055 **Search for the last parent body of the Farmington meteorite.**
I. V. Galibina, A. N. Simonenko, B. Yu. Levin.
Meteoritika, Vyp. 39, p. 114 - 120 (1980). In Russian.

105.056 **Definition of the finding place of the Pallas Iron meteorite.** A. I. Eremeeva.
Meteoritika, Vyp. 39, p. 134 - 143 (1980). In Russian.

105.057 **The stream of crater forming meteorites on the earth (within the last two aeons).**
V. V. Fedynsky (*Fedynskij*), A. I. Dabizha, I. T. Zotkin.
Solid particles in the solar system, (see 012.019), p. 205 (1980). – Abstract.

105.058 **On the mass distribution of meteorites and their influx rate.** D. W. Hughes.
Solid particles in the solar system, (see 012.019), p. 207 - 210 (1980).

105.059 **Thermomagnetic study of chondrules.**
B. Lang, M. Pekala, E. Król, A. Nowakowski, P. Martin, Yu. Stakheev, G. Baryshikova (*Baryshnikova*).
Solid particles in the solar system, (see 012.019), p. 371 - 374 (1980).
Thermomagnetic and saturation magnetization curves were obtained for separate chondrules and fine matrix grains from one H- (Allegan) and four L-type (Bjurböle, Elenovka, Saratov, Nikolskoe) chondrites. Mineralogical and petrological data were recovered from chondrule thin sections. The differences observed in the obtained curves seem to suggest episodes in the story of chondrules predating their incorporation into matrices.

105.060 **Thermal history of the H-group of chondritic meteorites.**
R. Hutchison, A. W. R. Bevan, S. O. Agrell, J. R. Ashworth.
Nature, Vol. 287, 787 - 790 (1980).
The Tieschitz unequilibrated H-group chondrite accreted at (800 ± 100)°C during rapid cooling. From their mineral

chemistry, nine other H-group chondrites, exhibiting greater degrees of equilibration, also formed hot. Degree of equilibration is equated with slowness of cooling, especially below about 700°C. If metamorphism is defined as change brought about by an increase in temperature, in the H-group chondrites it is recognized only in the localized effects of transient reheating, probably induced by shock.

105.061 **Measuring metamorphic history of unequilibrated ordinary chondrites.** D. W. Sears, J. N. Grossman, C. L. Melcher, L. M. Ross, A. A. Mills.
Nature, Vol. 287, 791 - 795 (1980).
A thermoluminescence sensitivity technique is used to give a new measurement of the degree of metamorphism of unequilibrated ordinary chondrites. Consequently the petrological assignment of these meteorites is modified.

105.062 **Si-rich Fe–Ni grains in highly unequilibrated chondrites.** E. R. Rambaldi, D. W. Sears, J. T. Wasson.
Nature, Vol. 287, 817 - 820 (1980).
The authors show that improvements in the Si activity coefficients and plausible constraints on the nucleation of mafic silicate condensates allow Si mole fractions of 0.003 at nebular pressures of $<10^{-6}$ atm. Thus such chondrites could have formed over a wide range of distances from the sun.

105.063 **On the effect of emission upon the earth during the flight of large meteorite bodies.** B. V. Putyatin.
Dokl. AN SSSR, Vol. 252, 318 - 320 (1980). In Russian.
Abstr. in Ref. zh., 51. Astron., 9.51.345 (1980).

105.064 **Thermal histories of pallasitic parent bodies.** T. Matsui, S.-i. Kawakami, H. Mizutani.
13th Lunar and Planetary Symposium, (see 012.018), p. 247 - 255 (1980).
Thermal histories of pallasitic parent bodies are calculated. New constraints on thermal history calculation, which were derived from observations of dislocation structures of pallasitic olivines, are used in addition to the cooling rate data. A reasonable size of pallasitic parent bodies is estimated to be 500 to 800 km.

105.065 **Cosmogenic nuclides in the Kirin meteorite.** M. Honda, K. Horie, O. Nitoh, M. Imamura, K. Nishiizumi, N. Takaoka, K. Komura.
13th Lunar and Planetary Symposium, (see 012.018), p. 256 - 266 (1980).

105.066 **Distributions of Ni, Co, Ga and Cu in iron meteorites.** J. Okano, H. Nishimura.
13th Lunar and Planetary Symposium, (see 012.018), p. 267 - 272 (1980).

105.067 **Reflectance spectra of some newly found, unusual meteorites and their bearing on surface mineralogy of asteroids.** L. A. McFadden, M. J. Gaffey, H. Takeda.
13th Lunar and Planetary Symposium, (see 012.018), p. 273 - 280 (1980).
The spectral properties of seven meteorites from the Antarctic collection are studied and related to the spectral properties of asteroid surfaces. Many of them show some unusual mineralogical characteristics. Preliminary examination of their reflectance spectra also reveals interesting aspects of the mineralogy of these samples. From studying more unusual meteorites it is hoped to discover spectral characteristics which define new classes of different meteorite types.

105.068 **Origin and evolution of the unique meteorites: Nakhlites and Chassignites.** N. Nakamura.
13th Lunar and Planetary Symposium, (see 012.018), p. 281 - 289 (1980).

105.069 **Dislocation loop in olivine of Allende chondrite and heating event on chondrules and grains in the protosolar nebula.** M. Toriumi.
13th Lunar and Planetary Symposium, (see 012.018), p. 363 - 370 (1980).

105.070 **Thermal history of chondrules in the primitive solar nebula.** M. Kimura.
13th Lunar and Planetary Symposium, (see 012.018), p. 371 - 378 (1980).

105.071 **Schwierigkeiten beim Nachweis von Aminosäuren in kohlenstoffhaltigen Chondriten.**
A. Liebold.
Sterne Weltraum, Jahrg. 19, 383 - 384 (1980).

105.072 **Main regularities of magnetization of stony meteorites as compared to terrestrial and lunar rocks.**
E. G. Gus'kova.
Geomagn. pole i vnutr. stroenie Zemli. Moskva, 1980, p. 164 - 176. In Russian. – Abstr. in Ref. zh., 51. Astron., 10.51.449 (1980).

105.073 **$^{40}Ar/^{39}Ar$ dating, Ar diffusion properties, and cooling rate determinations of severely shocked chondrites.** D. D. Bogard, W. C. Hirsch.
Geochim. Cosmochim. Acta, Vol. 44, 1667 - 1682 (1980).
The purpose of this investigation is to determine the times of the shock events and to investigate the effects of high post-shock temperatures on argon loss from seven meteorites. In principle, the high reheating temperatures and relatively long cooling times determined for several of these chondrites should have completely degassed them of radiogenic argon shortly after the shock event. However, the ^{40}Ar-^{39}Ar release curves of severely shocked chondrites are more complex than those of moderately shocked chondrites.

105.074 **Isotopic anomalies of noble gases in meteorites and their origins – VII. C3V carbonaceous chondrites.**
J.-i. Matsuda, R. S. Lewis, H. Takahashi, E. Anders.
Geochim. Cosmochim. Acta, Vol. 44, 1861 - 1874 (1980).
The authors have measured noble gases in three C3V chondrites: Grosnaja, Vigarano, and Leoville. Together with Allende, they cover at least two of the major subdivisions of this motley group (McSween, 1977): Leoville and Vigarano are reduced, whereas Allende and Grosnaja are oxidized; all but Grosnaja have low matrix/chondrule ratios (0.6–0.7 vs 1.15).

105.075 **Formation of *E* chondrites and aubrites – a thermodynamic model.** D. W. Sears.
Icarus, Vol. 43, 184 - 202 (1980).
Condensation and accretion models for the formation of *E* chondrites have been examined. It is concluded that there is no simple equilibrium process which can explain all their fundamental properties. The nearest would seem to involve a complex accretion history, whereby metal and silicates which ceased to equilibrate at high temperatures and pressures (say about 1200 - 1600°K and about 1 atm) were mixed with material which ceased to equilibrate at the same pressures but over the temperature range 600 - 700°K. In this way the level of reduction displayed by this class, and the fractionation of several major, minor, and trace elements, may be explained.

105.076 **Organische Moleküle in Meteoriten.** H. Wänke.
Evolution der Planetenatmosphären und des Lebens, (see 012.032), p. 198 - 203 (1980).

105.077 **About what iron meteorites are telling.** A. A. Yavnel'.
Zemlya Vselennaya, 1980, No. 6, p. 14 - 18. In Russian.

105.078 **Spurs of Pallas Iron.** A. I. Eremeeva.
Zemlya Vselennaya, 1980, No. 6, p. 50 - 52. In Russian.

105.079 **Mineralogically diagnostic visible and near-IR spectral features of C1 (CI) and C2 (CM) chondritic meteorite mineral assemblages.** M. J. Gaffey.
Bull. American Astron. Soc., Vol. 12, 665 (1980). – Abstract.

105.080 **The Abee meteorite : a possible asteroidal regolith probe.** J. F. Wacker, K. Marti.
Bull. American Astron. Soc., Vol. 12, 665 - 666 (1980). Abstract.

105.081 **A revision of the meteorite based cosmic abundance of boron.** D. Curtis, E. Gladney, E. Jurney.
Geochim. Cosmochim. Acta, Vol. 44, 1945 - 1953 (1980).
The authors initiated an experiment to measure boron abundances in a large number of meteorites and meteoritic material. The result is the recognition of an anomalous enhancement of boron in many chondritic meteorites with unknown terrestrial histories. Presumably this enhancement is due to terrestrial contamination. Analyses of freshly prepared interior pieces of chondritic meteorites produce an estimate of the cosmic abundance of boron that is significantly less than that based upon previous results.

105.082 **Thermal metamorphism of primitive meteorites – XI. The enstatite meteorites: origin and evolution of a parent body.**
S. Biswas, T. Walsh, G. Bart, M. E. Lipschutz.
Geochim. Cosmochim. Acta, Vol. 44, 2097 - 2110 (1980).
The authors study the trace element contents of a large number of aubrites and similar material from Mt. Egerton – a unique, possibly-related, stony-iron meteorite – to shed further light on their genesis and relationship to E4–6 chondrites. The authors extended the heating experiments of Ikra-Muddin et al. (1976) to temperatures >1000°C, i.e. to study volatilization in a regime where partial melting of a parent body would become important. The trace elements chosen for study are chalcophile, lithophile or siderophile in E4–6 chondrites and represent a range of thermal mobilities.

105.083 **Comments on "Chemical relationships among irghizites, zhamanshinites, Australasian tektites and Henbury impact glass".**
J. A. O'Keefe, with a reply by S. R. Taylor, S. M. McLennan.
Geochim. Cosmochim. Acta, Vol. 44, 2151 - 2157 (1980).

105.084 **Papuanites: pseudo-tektites from New Guinea.**
D. A. Visker, W. D. Ehmann, R. C. Young III.
Meteoritics, Vol. 15, 189 - 192 (1980).

105.085 **A note on the Allan Hills A77278 unequilibrated ordinary chondrite.**
H. Y. McSween, Jr., L. L. Wilkening.
Meteoritics, Vol. 15, 193 - 199 (1980).
Petrographic measures of disequilibrium in the ALHA 77278 chondrite indicate that this meteorite is more equilibrated than its exceptionally high volatile element contents suggest. Based on its metal compositions, this meteorite should be classified as an LL3 rather than an L3 chondrite.

105.086 **Gomez, Terry County, Texas: a new meteorite find.**
P. P. Sipiera, J. Tarter, C. B. Moore, B. D. Dod, R. A. Johnston.
Meteoritics,Vol. 15, 201 - 210 (1980).

105.087 **An H4 - 6 chondrite: Motta di Conti.**
G. R. Levi-Donati, A. Maras, G. P. Sighinolfi.
Meteoritics, Vol. 15, 211 - 223 (1980).
The mineralogical and chemical compositions of meteorites from the Motta di Conti, Vercelli, Italy, shower (February 29, 1868) have been determined.

105.088 **The Parsa enstatite chondrite.**
N. Bhandari, V. B. Shah, J. T. Wasson.
Meteoritics, Vol. 15, 225 - 233 (1980).

105.089 **The Meteoritical Bulletin, No. 58.** Sponsored by The Meteoritical Society. A. L. Graham (Editor).
Meteoritics, Vol. 15, 235 - 240 (1980).
Place of fall/find, class and type, number of individual specimens, total weight and circumstances of fall/find of some meteorites are given.

105.090 **Lu-Hf total-rock isochron for the eucrite meteorites.**
P. J. Patchett, M. Tatsumoto.
Nature, Vol. 288, 571 - 574 (1980).

105.091 **A Late Cretaceous ^{40}Ar-^{39}Ar range for the Lappajarvi impact crater, Finland.**
E. K. Jessberger, W. U. Reimold.
J. Geophys., Vol. 48, No. 2, p. 57 - 59 (1980). – Abstr. in Phys. Abstr., Vol. 84, Abstr. 4498 (1981).

105.092 **Morphology of Lonar Crater, India: comparisons and implications.**
R. F. Fudali, D. J. Milton, K. Fredriksson, A. Dube.
Moon Planets, Vol. 23, 493 - 515 (1980).

105.093 **Origin of differentiated meteorites.**
G. W. Wetherill, J. G. Williams.
Origin and distribution of the elements, (see 012.039), p. 19 - 31 (1979).
The larger asteroids with low inclination in the inner portion of the asteroid belt are evaluated as sources of differentiated meteorites (irons, stony irons and achondrites). It is found that the calculated earth impact rate is in agreement with the observed flux of these bodies. Although contributions from other sources are possible, these objects appear to be the most plausible sources of differentiated meteorites. Chemical and isotopic data on these bodies suggest a relationship to the earth and moon, and it may be that they represent residual material from earth's accretional zone.

105.094 **Halogens in meteorites and their primordial abundances.** G. Dreibus, B. Spettel, H. Wänke.
Origin and distribution of the elements, (see 012.039), p. 33 - 38 (1979).
The authors have developed a procedure which allows to measure all the halogen elements F, Cl, Br and I on the same sub-sample. With this technique they have determined the halogen elements in both undifferentiated and differentiated meteorites.

105.095 **Noble gases in chondritic polymict breccias: clues to their origin.** L. Schultz.
Origin and distribution of the elements, (see 012.039), p. 39 - 45 (1979).

105.096 **On the anomalous fading of thermoluminescence in meteorites.** S. W. S. McKeever, S. A. Durrani, M. J. Aitken.
Mod. Geol., Vol. 7, 75 - 79 (1980). – Abstr. in Phys. Abstr., Vol. 84, Abstr. 9434 (1981).

105.097 **The analysis of thermoluminescence glow-curves from meteorites.** S. W. S. McKeever.
Mod. Geol., Vol. 7, 105 - 114 (1980). – Abstr. in Phys. Abstr., Vol. 84, Abstr. 9435 (1981).

105.098 **Experiments on the chemical concentration of a new spontaneously fissioning nuclide from Allende**

meteorite. II. A repeated treatment of sublimated products.
B. L. Zhuikov (*Zhujkov*), I. Zvara.
Radiochem. Radioanal. Lett., Vol. 44, No. 1, p. 47 - 60 (1980). – Abstr. in Phys. Abstr., Vol. 84, Abstr. 9442 (1981).

105.099 **Dislocation structures of olivine from pallasite meteorites.** T. Matsui, S.-i. Karato, T. Yokohura.
Geophys. Res. Lett., Vol. 7, 1007 - 1010 (1980).

105.100 **Evidence for live ^{247}Cm in the early solar system.**
M. Tatsumoto, T. Shimamura.
Nature, Vol. 286, 118 - 122 (1980).

Variations of the $^{238}U/^{235}U$ ratio in the Allende meteorite, ranging from −35% to +19%, are interpreted as evidence of live ^{247}Cm in the early solar system. The amounts of these and other r-products in the solar system indicate values of (9,000 ± 3,000) Myr for the age of the Galaxy and ~8 Myr for the time between the end of nucleosynthesis and the formation of meteoritic grains. Three possible explanations are presented for the different values of the latter time period which are indicated by the decay products of ^{247}Cm, ^{26}Al, ^{244}Pu and ^{129}I.

105.101 **Uranium-lead age of the Bruderheim L6 chondrite and the 500-Ma shock event in the L-group parent body.** N. H. Gale, J. W. Arden, M. C. B. Abranches.
Earth Planet. Sci. Lett., Vol. 48, 311 - 324 (1980).

105.102 **Siderophile element fractionation in enstatite chondrites.** E. R. Rambaldi, M. Cendales.
Earth Planet. Sci. Lett., Vol. 48, 325 - 334 (1980).

105.103 **I-Xe age and trapped Xe components of the Murray (C-2) chondrite.** S. Niemeyer, A. Zaikowski.
Earth Planet. Sci. Lett., Vol. 48, 335 - 347 (1980).

105.104 **$^{40}Ar/^{39}Ar$ age and thermal history of the Kirin chondrite.**
S. Wang, I. McDougall, N. Tetley, T. M. Harrison.
Earth Planet. Sci. Lett., Vol. 49, 117 - 131 (1980).

The Kirin meteorite, a large (>2800 kg) H5 chondrite, fell in Kirin Province, China in 1976. A sample from each of the two largest fragments (K-1, K-2) yield $^{40}Ar/^{39}Ar$ total fusion ages of 3.63 ± 0.02 b.y. and 2.78 ± 0.02 b.y. respectively. Modelling of possible thermal events in the parent body indicates that samples K-1 and K-2 were at a depth of less than 3 m from the base of an impact melt of a thickness less than 7 m and separated by no more than ~2 m from one another at the time of the heating event about 0.5 b.y. ago. Further, the duration of heating was probably less than a few years. Calculations from ^{38}Ar data yield exposure ages for samples K-1 and K-2 of about 5 m.y., similar to that found for many other H chondrites.

105.105 **Meteoritic chondrules and the Weibull function.**
P. M. Martin, D. W. Hughes.
Earth Planet. Sci. Lett., Vol. 49, 175 - 180 (1980).

Chondrule mass frequency distributions determined from the Bjurböle, Chainpur, Allegan, Saratov, Elenovka and Nikolskoe meteorites have been tested to see if they could be fitted to either the Rosin or Weibull statistical functions. Whereas none of the distributions gave a fit to Rosin's law, they could all (with the exception of Nikolskoe) be fitted to the Weibull function suggesting similar origins and/or histories.

105.106 **Fission-track retention age of the Bondoc mesosiderite.** R. K. Bull, S. A. Durrani.
Earth Planet. Sci. Lett., Vol. 49, 181 - 187 (1980).

105.107 **Sm-Nd isotopic evolution of chondrites.**
S. B. Jacobsen, G. J. Wasserburg.
Earth Planet. Sci. Lett., Vol. 50, 139 - 155 (1980) = Div. Geol. Planet. Sci., Calif. Inst. Technol., Contrib. No. 3371 (336).

105.108 **Cosmic ray exposure ages of chondrites, pre-irradiation and constancy of cosmic ray flux in the past.**
K. Nishiizumi, S. Regnier, K. Marti.
Earth Planet. Sci. Lett., Vol. 50, 156 - 170 (1980).

A systematic calibration of the production rate of one specific cosmic-ray-produced nuclide in chondrites, that of ^{21}Ne, was achieved by using four independent methods. Based on these results, the authors recommend a value $P_{21}(1.11)$ = 0.31 (in units of 10^{-8} cm^3 STP/g My).

105.109 **Elemental abundances in chondrules from unequilibrated chondrites: evidence for chondrule origin by melting of pre-existing materials.**
J. L. Gooding, K. Keil, T. Fukuoka, R. A. Schmitt.
Earth Planet. Sci. Lett., Vol. 50, 171 - 180 (1980).

105.110 **Sr isotopic fractionation in Allende chondrules: a reflection of solar nebular processes.**
P. J. Patchett.
Earth Planet. Sci. Lett., Vol. 50, 181 - 188 (1980).

True relative Sr isotopic compositions, determined by the double-spike technique, are reported for 8 olivine chondrules from Allende and a single chondrule from Richardton. The Richardton chondrule has an Sr composition identical with the whole meteorite, but the Allende chondrules are up to 1.4% per mass unit light-isotope enriched. The lack of any detectable non-linear Sr isotopic anomaly in the objects suggests that their Sr compositions did not have some exotic or extrasolar origin, but were derived from normal solar system Sr by mass fractionation. The consistent light-Sr enrichment of Allende objects may be explained by several schemes, and all are heavily model-dependent.

105.111 **Preferred chondrule orientations in meteorites.**
P. M. Martin, A. A. Mills.
Earth Planet. Sci. Lett., Vol. 51, 18 - 25 (1980).

The orientation of chondrule long-axes has been examined in relatively friable (Allegan, Bjurböle, Chainpur) and more compacted (Allende, Kediri, Knyahinya, Parnallee) meteorites. Preferential chondrule orientations were detected in the more compacted specimens but were found to be much less well developed in the friable meteorites. The authors conclude therefore that the chondrule petrofabric was imposed during compaction in the meteorite parent bodies.

105.112 **The dependence of chondrule density on chondrule size.** D. W. Hughes.
Earth Planet. Sci. Lett., Vol. 51, 26 - 28 (1980).

Spherical chondrules disaggregated from the meteorite Bjurböle have been found to have densities which vary as a function of size. The relationship is: $\rho = (3.78 \pm 0.06) - (1.25 \pm 0.37) d$ where ρ is the chondrule density in g cm^{-3} and d is the diameter in cm. The chondrule size range was $0.034 < d < 0.32$ cm.

105.113 **The composition of mesosiderite olivine clasts and implications for the origin of pallasites.**
D. W. Mittlefehldt.
Earth Planet. Sci. Lett., Vol. 51, 29 - 40 (1980).

105.114 **Lithium isotopic composition in some stone meteorites.**
R. S. Rajan, L. Brown, F. Tera, D. J. Whitford.
Earth Planet. Sci. Lett., Vol. 51, 41 - 44 (1980).

105.115 **Depth and size dependence of ^{53}Mn activity in chondrites.**
S. K. Bhattacharya, M. Imamura, N. Sinha, N. Bhandari.
Earth Planet. Sci. Lett., Vol. 51, 45 - 57 (1980).

105.116 **Comments on: "Solubility of noble gases in serpentine: implications for meteoritic noble gas abundances" by A. Zaikowski and O. A. Schaeffer.**
O. K. Manuel, D. D. Sabu.
Earth Planet. Sci. Lett., Vol. 51, 233 - 234 (1980). – See also Abstract 26.105.114.

105.117 **A 3.6-b.y.-old impact-melt rock fragment in the Plainview chondrite: implications for the age of the H-group chondrite parent body regolith formation.**
K. Keil, R. V. Fodor, P. M. Starzyk, R. A. Schmitt, D. D. Bogard, L. Husain.
Earth Planet. Sci. Lett., Vol. 51, 235 - 247 (1980).

105.118 **Some mechanical aspects of cratering.**
B. A. Ivanov.
Kosmogen. struktury Zemli. Mater. semin., Moskva, 1980, p. 39 - 45. In Russian. – Abstr. in Ref. zh., 51. Astron., 11.51.412 (1980).

105.119 **Reflection of meteorite craters in geophysical fields.**
A. I. Dabizha, I. T. Zotkin.
Kosmogen. struktury Zemli. Mater. semin., Moskva, 1980, p. 45 - 49. In Russian. – Abstr. in Ref. zh., 51. Astron., 11.51.413 (1980).

105.120 **Typomorphous features of solution and melting inclusions in minerals and rocks of meteorite craters.**
F. P. Mel'nikov, L. M. Valyashko, V. P. Belov, V. I. Fel'dman.
Kosmogen. struktury Zemli. Mater. semin., Moskva, 1980, p. 77 - 79. In Russian. – Abstr. in Ref. zh., 51. Astron., 11.51.414 (1980).

105.121 **Complex study of tektites and impactites from the Zhamashin meteorite crater.**
P. V. Florenskij, T. S. Gendler, Eh. S. Gorshkov, Yu. P. Dikov, T. A. Eremenko, T. Kh. Pencheva, V. P. Perelygin.
Kosmogen. struktury Zemli. Mater. semin., Moskva, 1980, p. 74 - 77. In Russian. – Abstr. in Ref. zh., 51. Astron., 11.51.415 (1980).

105.122 **Spinel framboids and fremdlinge in Allende inclusions: possible sequential markers in the early history of the solar system.**
A. El Goresy, K. Nagel, P. Ramdohr.
Proc. Tenth Lunar Planet. Sci. Conf., (see 012.050), p. 833 - 850 (1979).

105.123 **Magnetite-sulfide-metal complexes in the Allende meteorite.** S. E. Haggerty, B. M. McMahon.
Proc. Tenth Lunar Planet. Sci. Conf., (see 012.050), p. 851 - 870 (1979).

105.124 **Petrography and olivine mineral chemistry of chondrules and inclusions in the Allende meteorite.**
S. B. Simon, S. E. Haggerty.
Proc. Tenth Lunar Planet. Sci. Conf., (see 012.050), p. 871 - 883 (1979).

105.125 **Li, Be, and B in minerals of a refractory -rich Allende inclusion.**
D. Phinney, B. Whitehead, D. Anderson.
Proc. Tenth Lunar Planet. Sci. Conf., (see 012.050), p. 885 - 905 (1979).

105.126 **Iron group cosmic-ray tracks in the Allende olivine: evidence of complex exposure history.**
N. N. Korotkova, A. K. Lavrukhina, V. P. Perelygin.
Proc. Tenth Lunar Planet. Sci. Conf., (see 012.050), p. 907 - 920 (1979).

105.127 **Characterization of submicron matrix phyllosilicates from Murray and Nogoya carbonaceous chondrites.**
T. R. McKee, C. B. Moore.
Proc. Tenth Lunar Planet. Sci. Conf., (see 012.050), p. 921 - 935 (1979).

105.128 **High resolution transmission electron microscopy of two stony meteorites: Murchison and Kenna.**
I. D. R. Mackinnon, P. R. Buseck.
Proc. Tenth Lunar Planet. Sci. Conf., (see 012.050), p. 937 - 949 (1979).

105.129 **Microcharacterization of "Brownlee" particles: features which distinguish interplanetary dust from meteorites?** P. Fraundorf, J. Shirck.
Proc. Tenth Lunar Planet. Sci. Conf., (see 012.050), p. 951 - 976 (1979).

105.130 **On the problem of the origin of enstatite chondrites.**
T. V. Malysheva, K. I. Tobelko, D. A. Khramov, O. A. Matveeva.
Proc. Tenth Lunar Planet. Sci. Conf., (see 012.050), p. 977 - 988 (1979).

105.131 **Fractionation of refractory lithophile elements among chondritic meteorites.** J. F. Kerridge.
Proc. Tenth Lunar Planet. Sci. Conf., (see 012.050), p. 989 - 996 (1979).

105.132 **Interelement refractory siderophile fractionation in ordinary chondrites.**
E. R. Rambaldi, H. Wänke, J. W. Larimer.
Proc. Tenth Lunar Planet. Sci. Conf., (see 012.050), p. 997 - 1010 (1979).

105.133 **U-Th-Pb age of the Barwell chondrite: anatomy of a "discordant" meteorite.**
D. M. Unruh, R. Hutchison, M. Tatsumoto.
Proc. Tenth Lunar Planet. Sci. Conf., (see 012.050), p. 1011 - 1030 (1979).

105.134 **Thermal history of the Shaw chondrite.**
E. R. D. Scott, R. S. Rajan.
Proc. Tenth Lunar Planet. Sci. Conf., (see 012.050), p. 1031 - 1043 (1979).

105.135 **Total carbon and sulfur abundances in Antarctic meteorites.** E. K. Gibson, Jr., K. Yanai.
Proc. Tenth Lunar Planet. Sci. Conf., (see 012.050), p. 1045 - 1051 (1979).

105.136 **^{14}C and ^{39}Ar abundances in Allan Hills meteorites.**
E. L. Fireman.
Proc. Tenth Lunar Planet. Sci. Conf., (see 012.050), p. 1053 - 1060 (1979).

105.137 **^{26}Al content of Antarctic meteorites: implications for terrestrial ages and bombardment history.**
J. C. Evans, L. A. Rancitelli, J. H. Reeves.
Proc. Tenth Lunar Planet. Sci. Conf., (see 012.050), p. 1061 - 1072 (1979).

105.138 **Aubrites: their origin and relationship to enstatite chondrites.** T. R. Watters, M. Prinz.
Proc. Tenth Lunar Planet. Sci. Conf., (see 012.050), p. 1073 - 1093 (1979).

105.139 **Pyroxenes in early crustal cumulates found in achondrites and lunar highland rocks.**
H. Takeda, M. Miyamoto, T. Ishii.
Proc. Tenth Lunar Planet. Sci. Conf., (see 012.050), p. 1095 - 1107 (1979).

105.140 **The pyroxene chemistry of four mesosiderites.** R. H. Hewins.
Proc. Tenth Lunar Planet. Sci. Conf., (see 012.050), p. 1109 - 1125 (1979).

105.141 **Origin of impact melt rocks in the Bununu howardite.** L. C. Klein, R. H. Hewins.
Proc. Tenth Lunar Planet. Sci. Conf., (see 012.050), p. 1127 - 1140 (1979).

105.142 **Vacuum ultraviolet reflectance spectra of group H chondrites.**
J. K. Wagner, A. J. Cohen, B. W. Hapke, W. D. Partlow.
Proc. Tenth Lunar Planet. Sci. Conf., (see 012.050), p. 1797 - 1818 (1979).

105.143 **A Monte Carlo fragmentation model for the production of meteorites: implications for gas retention ages.** G. Turner.
Proc. Tenth Lunar Planet. Sci. Conf., (see 012.050), p. 1917 - 1941 (1979).

105.144 **Noble gas trapping and fractionation during synthesis of carbonaceous matter.**
U. Frick, R. Mack, S. Chang.
Proc. Tenth Lunar Planet. Sci. Conf., (see 012.050), p. 1961 - 1973 (1979).

105.145 **The nature of asteroidal differentiation processes: implications for primordial heat sources.**
D. W. Mittlefehldt.
Proc. Tenth Lunar Planet. Sci. Conf., (see 012.050), p. 1975 - 1993 (1979).

105.146 **Magnetic properties and paleointensity of achondrites in comparison with those of lunar rocks.**
T. Nagata.
Proc. Tenth Lunar Planet. Sci. Conf., (see 012.050), p. 2199 - 2210 (1979).

105.147 **The distribution of volatile and siderophile elements in the impact melt of East Clearwater (Quebec).**
H. Palme, E. Göbel, R. A. F. Grieve.
Proc. Tenth Lunar Planet. Sci. Conf., (see 012.050), p. 2465 - 2492 (1979).

105.148 **Petrology of impactites from Lake St. Martin structure, Manitoba.** C. H. Simonds, P. E. McGee.
Proc. Tenth Lunar Planet. Sci. Conf., (see 012.050), p. 2493 - 2518 (1979).

105.149 **Structural deformation at the Flynn Creek impact crater, Tennessee: a preliminary report on deep drilling.** D. J. Roddy.
Proc. Tenth Lunar Planet. Sci. Conf., (see 012.050), p. 2519 - 2534 (1979).

105.150 **Australasian, Ivory Coast and North American tektite strewnfields: size, mass and correlation with geomagnetic reversals and other earth events.**
B. P. Glass, M. B. Swincki, P. A. Zwart.
Proc. Tenth Lunar Planet. Sci. Conf., (see 012.050), p. 2535 - 2545 (1979).

105.151 **Meteorite impact in the ocean.** R. Strelitz.
Proc. Tenth Lunar Planet. Sci. Conf., (see 012.050), p. 2799 - 2813 (1979).

105.152 **Volatile matter in meteorites, in the protoplanetary nebula, and formation of planetary atmospheres.**
M. N. Izakov.
Kosm. Issled., Tom 18, 918 - 932 (1980). In Russian.

105.153 **Die Genese der Impaktschmelze des finnischen Meteoritenkraters Lappajärvi aus geochemischer und petrographischer Sicht.** W. U. Reimold.
Mitt. Astron. Ges., Nr. 50, p. 90 - 91 (1980).

105.154 **Meteorite craters at Morasko.** J. Classen.
Urania Kraków, Vol. 51, 361 - 372 (1980). In Polish.

105.155 **Stochastic model of the dissipation of the regolith layer consisting of material losing its initial properties in the process of cratering.** G. A. Lejkin, E. V. Zabalueva.
Astron. Vestn., Tom 13, 24 - 31 (1979). In Russian.

The depth distribution of material having some definite properties is presented as a function of time, crater shape, radius distribution of craters and initial depth of the layer with day surface as datum level of depth.

105.156 **The regmaglyptic relief of meteorites.**
E. L. Krinov.
Astron. Vestn., Tom 13, 56 - 61 (1979). In Russian.

As a result of the study of the regmaglyptic relief of 2657 meteorites the author determined the relief factor k being the ratio of the mean diameter of regmaglypts to the mean diameter of meteorites. It is found that k changes from 0.23 to 0.05 when the mean diameter of meteorites changes from 1.2 to 54.2 cm or the meteorite mass changes from 4 g to 850 kg.

105.157 **On fine-dispersed matter of the meteorite Kaalijarv.**
V. I. Koval'.
Astron. Vestn., Tom 14, 102 - 111 (1980). In Russian.

105.158 **Evolution of a terrestrial meteorite crater system.**
A. I. Dabizha, I. T. Zotkin.
Astron. Vestn., Tom 14, 182 - 188 (1980). In Russian.

Starting from the known values of the age T and diameter D of 114 earth meteorite craters, it is concluded that the mean life-time of an earth crater is $T_e = D^2\kappa$ where $\kappa = 1/30\ m^2 year^{-1}$ is the coefficient of macrodiffusion. The dissipation of a crater is described by the process of random walk. The diagnostic signs of craters are proportional to the initial area and inversely proportional to the age. Craters exist at the background of geological noises with normal (Gaussian) spectrum.

The meteorite crater Zhamanshin.
See Abstr. 003.047.

Heralds of the universe. See Abstr. 003.069.

Ways of development and progress of Soviet meteoritics. See Abstr. 013.001.

Progress in investigation of meteoritic matter in the USSR. See Abstr. 013.002.

Actinide crystal–liquid partitioning for clinopyroxene and $Ca_3(PO_4)_2$. See Abstr. 022.011.

Reflectance spectrophotometry extended to u.v. for terrestrial, lunar and meteoritic samples.
See Abstr. 022.012.

Carbon in solid solution in forsterite – a key to the untractable nature of reduced carbon in terrestial and cosmogenic rocks. See Abstr. 022.013.

Mid and far infrared spectra of hydrous silicates.
See Abstr. 022.078.

Evaporation experiment: implication to the origin of cosmic materials. See Abstr. 022.079.

Low temperature sintering of fine powders – an application to the consolidation process of meteorites.
See Abstr. 022.080.

Abiotic organic synthesis in space.
See Abstr. 022.105.

Xenon from intermediate zones of supernovae.
See Abstr. 061.044.

Solar and meteoritic abundance of silicon.
See Abstr. 071.040.

On the classification of annular structures of the earth. See Abstr. 081.031.

Meteorite cratering on terrestrial planets.
See Abstr. 091.055.

Morphology of impact craters on the moon and planets. See Abstr. 091.056.

The distribution of zirconium and hafnium in terrestrial rocks, meteorites and the moon.
See Abstr. 094.537.

74001 drive tube: siderophile elements match II B iron meteorite pattern. See Abstr. 094.553.

Sulphur isotopes in lunar and meteorite samples.
See Abstr. 094.603.

The near-earth asteroids: their nature as presently understood from reflectance spectroscopy.
See Abstr. 098.043.

1979 VA: physical parameters of a possible cometary nucleus. See Abstr. 098.044.

The asteroids. See Abstr. 098.083.

Comparison of the theory of a dynamically significant coma with PN-fireballs. See Abstr. 104.014.

Noble gas anomalies and synthesis of the chemical elements. See Abstr. 106.012.

Isotopic anomalies in the early solar system.
See Abstr. 107.022.

Erratum

105.901 **Erratum: "Chemical variations among L-group chondrites 1. The Air, Apt and Tourinnes-la-Grosse (L6) chondrites".** [Meteoritics, Vol. 15, 69 - 83 (1980)].
R. T. Dodd, E. Jarosewich.
Meteoritics, Vol. 15, 249 (1980). – See Abstr. 27.105.203.

106 Interplanetary Matter, Interplanetary Magnetic Field, Zodiacal Light

106.001 **A model of the heliospheric magnetic field configuration.** S.-I. Akasofu, P. C. Gray, L. C. Lee.
Planet. Space Sci., Vol. 28, 609 - 615 (1980).

A three-dimensional model of the magnetic field configuration in the heliosphere is constructed by assuming that the interplanetary magnetic field consists of four components, (1) the solar dipole, (2) a large number of small spherical dipoles located along an equatorial circle just inside the sun (representing the magnetic field line arcade), (3) the field of the poloidal current system generated by the solar unipolar induction and (4) the field of an extensive current disk around the sun lying in the ecliptic plane. The magnetic field intensity at a distance of 1 A. U. (about 20 $R_{\odot}$ above the ecliptic plane) is normalized to fit the observed spiral configuration.

106.002 **Method for the determination of density and phase functions of interplanetary dust.**
A. Mujica, G. López, F. Sánchez.
Planet. Space Sci., Vol. 28, 657 - 660 (1980).

A new method of determination of the scattered light intensity, $I(r, \epsilon)$, by a unit-volume of interplanetary space is presented. From ground base Zodiacal Light measurements and the experimental results of Pioneer X the density, $\rho(r)$, and phase functions, $\sigma(\epsilon)$, are obtained without any previous assumptions about them.

106.003 **Ions of Jovian origin observed by Voyager 1 and 2 in interplanetary space.**
R. D. Zwickl, S. M. Krimigis, T. P. Armstrong, L.J. Lanzerotti.
Geophys. Res. Lett., Vol. 7, 453 - 456 (1980).

The authors report the observation of Jovian ions $\lesssim$0.6 AU from the planet and discuss the general morphology of Jovian ion events measured with the low energy charged particle (LECP) detectors onboard Voyager 1 and 2.

106.004 **Optical spectroscopy of interplanetary dust collected in the earth's stratosphere.** P. Fraundorf, R. I. Patel, J. Shirck, R. M. Walker, J. J. Freeman.
Nature, Vol. 286, 866 - 868 (1980).

A new class of extraterrestrial material has recently become available for study: interplanetary dust particles (2–30 μm in size) collected in the atmosphere at 20-km altitude. The authors report the first optical absorption spectra of these interplanetary dust particles. They show that absorption features in the visible at ~9,800 cm^{-1} and in the IR at ~1,000 cm^{-1} are present in spectra of these particles.

106.005 **On the origin of the B_x and B_z components of the interplanetary magnetic field.** D. I. Ponyavin.
Pis'ma Astron. Zh., Tom 6, 432 - 436 (1980). In Russian.
English translation in Soviet Astron. Lett., Vol. 6.

A method to calculate the total vector of the super-large-scale solar magnetic field from synoptic maps in the potential approximation accounting for its temporal evolution is proposed. It is found that during the first half of 1974 the first terms of the field expansion in spherical harmonics are responsible for the formation of the super-large-scale and interplanetary magnetic fields.

106.006 **A reanalysis of the observed interplanetary hydrogen Lα emission profiles and the derived local interstellar gas temperature and velocity.**
F. M. Wu, D. L. Judge.
Astrophys. J., Vol. 239, 389 - 394 (1980).

An improved model of the interplanetary emission profile is presented. In this model the authors utilize a new version of the modified Danby-Camm velocity distribution function to calculate emission profiles, and it appears to be very efficient. Using this model, the hydrogen Lα profiles are determined for various viewing directions, and at a solar distance of 1 AU. It is found that the linewidth of the emission profiles and their Doppler shifts depend strongly on the viewing direction, as well as the assumed interstellar temperature. Previous determinations of the interstellar temperature T_0 and the inflow velocity V_b, using approximate theoretical hydrogen profiles, are found to be inadequate. These values are recalculated and are determined to be $T_0 = 7000 \pm 1200$ K and $V_b = 19 \pm 3$ km s^{-1}.

106.007 **Interplanetary radio manifestations of large flare events.**
H. V. Cane, R. G. Stone, J. Fainberg, J. L. Steinberg.
Bull. American Astron. Soc., Vol. 12, 546 (1980). – Abstract.

106.008 **Observation of oxygen-enriched plasma in the driving piston of a flare-generated shock wave.**
C. Bonifazi, A. Egidi, G. Moreno.
Nuovo Cimento, Lett., Ser. 2, Vol. 28, 39 - 43 (1980).
Abstr. in Phys. Abstr., Vol. 83, Abstr. 72740 (1980).

106.009 **Influence of the parameters of the interplanetary magnetic field on ionization of the F2 layer in the polar cap.** A. S. Besprozvannaya, L. N. Makarova.
Geomagn. Aehron., Tom 20, 737 - 739 (1980). In Russian.

106.010 **Microstructure of the interplanetary magnetic field near 4 and 5 AU.** G. D. Parker.
J. Geophys. Res., Vol. 85, 4275 - 4282 (1980).

Seventy-two days of vector magnetic field measurements from Pioneer 10 and 11 are analyzed for information about magnetic field fluctuations in the quiet solar wind near 4 and 5 AU. Calculated as functions of frequency over the range 4×10^{-5} to 9×10^{-3} Hz, directional properties of magnetic field fluctuations are presented and are discussed with reference to theoretical predictions for MHD plane waves.

106.011 **Propagation directions of hydromagnetic waves in interplanetary space: Pioneer 10 and 11.**
G. D. Parker.
J. Geophys. Res., Vol. 85, 4283 - 4287 (1980).

The real and imaginary parts of spectral matrices of magnetic field fluctuations near 4 and 5 AU are used separately to generate estimates of hydromagnetic wave normal directions over the frequency range 10^{-5} to 10^{-2} Hz in the spacecraft frame. Both procedures suggest the predominance of the Alfven mode over the fast and slow MHD modes among transverse magnetic field perturbations. Both procedures also show that the Alfvén mode is more common among transverse perturbations at 4 AU than at 5AU. The author develops statistical techniques for estimating uncertainties in computed wave normal directions and concludes that the real parts of the spectral matrices are more useful than the imaginary parts in specifying wave normals at times of magnetic quiet in the solar wind near 4 and 5 AU.

106.012 **Noble gas anomalies and synthesis of the chemical elements.** D. D. Sabu, O. K. Manuel.
Meteoritics, Vol. 15, 117 - 138, with a correction p. 249 (1980).

Mixing of nucleosynthesis products from different regions of a supernova is responsible for the observed correlations between elemental and isotopic ratios of planetary noble gases in different classes of meteorites. The solar system condensed directly from the chemically and isotopically heterogeneous debris of a single supernova. Variations in the abundance pattern of planetary noble gases are primarily the result of stellar fusion reactions and physical adsorption, rather than gas solubility.

106.013 **A search for natural or artificial objects located at the earth-moon libration points.**
R. A. Freitas, Jr., F. Valdes.
Icarus, Vol. 42, 442 - 447 (1980).
Photographs in the vicinty of the earth–moon triangular libration points L4 and L5, and of the solar-synchronized positions in the associated halo orbits were made during August - September 1979, using the 30-in Cassegrain telescope at Leuschner Observatory, Lafayette, California. An effective 2° square field was covered at each position. No discrete objects, either natural or artificial, were found. The detection limit was about 14th magnitude.

106.014 **Observations of nonlinear turbulence in the upstream solar wind.** M. Dobrowolny, S. Orsini, C. C. Harvey, A. Mangeney, J. Etcheto, C. T. Russell.
Nuovo Cimento C, Ser. 1, Vol. 3C, 17 - 36 (1980). – Abstr. in Phys. Abstr., Vol. 83, Abstr. 85833 (1980).

106.015 **Observations of backstreaming protons in the solar wind near the Earth's bow shock.**
C. Bonifazi, P. Cerulli-Irelli, M. Dobrowolny, A. Egidi, G. Moreno, S. Orsini.
Nuovo Cimento C, Ser. 1, Vol. 2C, (see 012.013), 772 - 780 (1979). – Abstr. in Phys. Abstr., Vol. 83, Abstr. 90309 (1980).

106.016 **Low-frequency waves observed in the vicinity of the Earth's bow shock.** V. Formisano.
Nuovo Cimento C, Ser. 1, Vol. 2C, (see 012.013), 789 - 794 (1979). – Abstr. in Phys. Abstr., Vol. 83, Abstr. 90311 (1980).

106.017 **Possible generation mechanisms of low-frequency waves ($\lesssim$50 Hz) with application to the bow shock plasma.** N. D'Angelo.
Nuovo Cimento C, Ser. 1, Vol. 2C, (see 012.013), 815 - 833 (1979). – Abstr. in Phys. Abstr., Vol. 83, Abstr. 90312 (1980).

106.018 **The thickness of interplanetary collisionless shock waves.** S. Pinter.
Report KFKI-1980-27, Hungarian Acad. Sci., Budapest. 26pp. (1980). – Abstr. in Phys. Abstr., Vol. 83, Abstr. 90314 (1980).

106.019 **Energetic particle, solar wind plasma and magnetic field measurements on board Prognoz-6 during the large scale interplanetary disturbance of Jan. 3 - 4, 1978.**
V. G. Kurt, V. G. Stolpovskii (*Stolpovskij*), T. I. Gombosi, K. Kecskemety, A. J. Somogyi, K. I. Gringauz, G. A. Kotova, M. I. Verigin, V. A. Styazhkin.
Report KFKI-1980-32, Hungarian Acad. Sci., Budapest. 17 pp. (1980). – Abstr. in Phys. Abstr., Vol. 83, Abstr. 90315 (1980).

106.020 **Background interplanetary magnetic fields in the earth's orbit and anisotropy of galactic cosmic rays.**
G. Ya. Vasil'eva, M. A. Kuznetsova, L. M. Kotlyar.
XIth seminar on cosmophysics, (see 012.035), p. 295 - 304 (1979). In Russian. – Abstr. in Ref. zh., 51. Astron., 8.51.391; 62. Issled. kosm. prostranstva, 8.62.133 (1980).

106.021 **Optical investigation of dust in the solar system.** R. H. Giese.
Solid particles in the solar system, (see 012.019), p. 1 - 13 (1980).
An introduction to definitions and methods used for interpretation of zodiacal light is given together with a short summary of observational results. After this, predictions for probing the zodiacal light by an out of ecliptic space probe and possibilities of explaining the scattering of interplanetary and cometary dust by irregular and fluffy particles are presented.

106.022 **Four years of zodiacal light observations from the Helios space probes: evidence for a smooth distribution of interplanetary dust.**
C. Leinert, I. Richter, E. Pitz, M. Hanner.
Solid particles in the solar system, (see 012.019), p. 15 - 18 (1980).
The zodiacal light experiments on Helios 1 and Helios 2 have been continually operating since launch in December 1974 and January 1976, respectively. The data support the view that the distribution of interplanetary dust is rather simple in space and quite constant in time. The authors present Helios 2 data only.

106.023 **Brightness and polarization of the zodiacal light: results of fixed–position observations from Skylab.**
J. L. Weinberg, R. C. Hahn.
Solid particles in the solar system, (see 012.019), p. 19 - 22 (1980).
The brightness and degree of polarization of zodiacal light at north celestial pole, south ecliptic pole, vernal equinox, and two places near the north galactic pole are derived using Pioneer 10 observations of background starlight from beyond the asteroid belt and the assumption that the zodiacal light is solar color in total light.

106.024 **Change in the zodiacal light with solar activity.** R. Robley.
Solid particles in the solar system, (see 012.019), p. 33 - 36 (1980).
The Gegenschein observed at the Pic du Midi Observatory during a solar cycle has a behaviour similar to that of the brightness of the eclipsed Moon. It is assumed that in both cases the solar wind ejected through the polar coronal holes is responsible for these variations.

106.025 **Two-dimensional photographic photometry of the zodiacal light from spatial observations.**
P. L. Lamy, A. Llebaria, S. Koutchmy.
Solid particles in the solar system, (see 012.019), p. 37 - 39 (1980).
Preliminary results for the photometry of a photograph of the inner zodiacal light taken aboard Salyut 6 are reported. The Very Wide Field Camera to be flown aboard Spacelab 1 is described and the program of observations of the zodiacal light is presented.

106.026 **Far-ultraviolet studies. VIII. Apollo 17 search for zodiacal light.**
R. C. Henry, R. C. Anderson, W. G. Fastie.
Solid particles in the solar system, (see 012.019), p. 41 - 44 (1980).
In paper VI of this series the authors reported analysis of a large quantity of far-ultraviolet spectrometer data from the Apollo 17 mission. The discussion centered on a search for diffuse galactic light. In the present paper the authors re-examine these data, searching for zodiacal light. They set an upper limit similar to existing upper limits at somewhat longer wavelengths.

106.027 **Photometric axis measurements of the zodiacal light at large elongations.**
H. Tanabe, A. Takechi, A. Miyashita.
Solid particles in the solar system, (see 012.019), p. 45 - 48 (1980).
The paper is a report of photoelectric measurements of the photometric axis of the zodiacal light at $90° < \lambda - \lambda_{\odot} < 270°$ made at the Kiso Station of the Tokyo Astronomical Observatory on nine clear nights in 1978.

106.028 **The symmetry plane of the zodiacal cloud near 1 AU.** N. Y. Misconi.

Solid particles in the solar system, (see 012.019), p. 49 - 53 (1980).

Analysis of zodiacal light observations from Mt. Haleakala, Hawaii show that the symmetry plane of the zodiacal cloud near 1 AU is close to the invariable plane of the solar system. Since the symmetry plane of the inner zodiacal cloud is close to the orbital plane of Venus the author suggests that the symmetry plane changes inclination with heliocentric distance.

106.029 **Evidence that the properties of interplanetary dust beyond 1 AU are not homogeneous.**
D. W. Schuerman.
Solid particles in the solar system, (see 012.019), p. 71 - 74 (1980).

Traditionally, earth-based observations of the zodiacal light require two assumptions for further analysis: (1) the dust density is a power of heliocentric distance, $n \propto R^{-\nu}$; (2) the nature of the dust is independent of location. Observations from Pioneer 10 do not verify these assumptions.

106.030 **Zodiacal light models with a bimodal population.**
P. L. Lamy, J. M. Perrin.
Solid particles in the solar system, (see 012.019), p. 75 - 80 (1980).

The compatibility of the observed properties of the zodiacal light with a model made of two populations is investigated. It is shown that a population of submicron grains may play a non-negligible role.

106.031 **Wavelength dependent models of the zodiacal light.** H. J. Staude, S. Röser.
Solid particles in the solar system, (see 012.019), p. 81 - 84 (1980).

The authors discuss the significance of their zodiacal light models, which are based on Mie theory and refractive indices measured in the range $0.15\mu m \leqslant \lambda \leqslant 100$ μm. The models include scattering and thermal emission. They are absolute, since particle number density at 1 AU, size distribution and radial density gradient in the ecliptic are fixed by lunar microcrater counts and by Helios and Pioneer results. The only free parameter is the chemical composition of the dust.

106.032 **Sources of interplanetary dust.**
Ľ. Kresák.
Solid particles in the solar system, (see 012.019), p. 211 - 222 (1980).

It is generally assumed that ejections from active cometary nuclei are the major source of replenishment of the interplanetary dust complex. Conjectures against this concept are usually based on comparison of the quantitative efficiency of the dust production by comets with the efficiency of all the dissipative processes involved. The paper discusses this problem from the dynamical point of view, tracing the evolution of swarms of cometary ejecta as they pass through different evolutionary stages. It is concluded that the contribution of the present population of active comets, of all revolution periods, is not only inadequate to explain the abundance of interplanetary particles, but is also inconsistent with the distribution of their orbits. Other potential sources and their implications for the equilibrium problem are reviewed.

106.033 **In situ measurements of interplanetary dust in the inner solar system.** E. Grün.
Solid particles in the solar system, (see 012.019), p. 277 - 278 (1980).

106.034 **The circumsolar motion of dust particles at the stage of increasing solar luminosity.**
A. N. Simonenko, B. J. (*B. Yu.*) Levin.
Solid particles in the solar system, (see 012.019), p. 279 (1980). – Abstract.

106.035 **Collisions among interplanetary dust grains.**
L. B. Le Sergeant, P. L. Lamy.
Solid particles in the solar system, (see 012.019), p. 289 - 292 (1980).

Assuming that the zodiacal cloud is composed of two populations, the effect of collisions among the largest grains is investigated. It is found that the population of fragments is unable to explain the observed flux of submicron grains.

106.036 **Dynamics of micrometeoroids.**
E. Grün, H. A. Zook.
Solid particles in the solar system, (see 012.019), p. 293 - 298 (1980).

106.037 **Poynting-Robertson effect and collisions in the interplanetary dust cloud.**
J. Trulsen, A. Wikan.
Solid particles in the solar system, (see 012.019), p. 299 - 302 (1980).

A simulation model has been developed to study the results of the Poynting-Robertson (PR) effect and collisions on the dynamical evolution of an interplanetary dust cloud. Fragmentational and accretional effects are neglected. With a mean free collision time of the order of the PR lifetime, collisional effects become important. As the individual grains still spiral inwards, collisions act to make the mean eccentricity and inclination of the grain orbits both decrease at comparable rates, giving rise to an expanding fan shaped dust cloud.

106.038 **The electrostatic potential of interplanetary grains.**
J. P. J. Lafon, P. L. Lamy, J. M. Millet.
Solid particles in the solar system, (see 012.019), p. 303 - 308 (1980).

A new method is presented for calculating the electrostatic potential of cosmic grains taking into account the plasma sheath and its bulk velocity. Results for various solar wind situations are given.

106.039 **The motion of charged dust particles in interplanetary space.** G. E. Morfill, E. Grün.
Solid particles in the solar system, (see 012.019), p. 309 - 310 (1980). – Abstract.

106.040 **Electromagnetic effects on the zodiacal dust cloud.**
E. Grün, G. E. Morfill.
Solid particles in the solar system, (see 012.019), p. 311 (1980). – Abstract.

106.041 **A two-stream instability in streams of charged grains.** O. Havnes.
Solid particles in the solar system, (see 012.019), p. 315 - 318 (1980).

The author studies the conditions for the onset of two-stream instabilities in interacting streams of charged grains. These conditions are such that instabilities cannot be important in the solar system of today and have probably been of very limited importance in the protoplanetary cloud.

106.042 **Trajectories of sublimating interplanetary dust grains.** G. H. Schwehm.
Solid particles in the solar system, (see 012.019), p. 319 - 320 (1980). – Abstract.

106.043 **Orbital elements of micrometeoroids detected by the Helios 1 space probe in the inner solar system.**
K. D. Schmidt, E. Grün.
Solid particles in the solar system, (see 012.019), p. 321 - 324 (1980).

The Helios spacecraft is able to detect micrometeoroids along its orbit in the inner solar system between 0.3 AU and 1 AU distance from the sun. Data of the first 6 orbits are presented here. 168 particles have been identified in such a way that their osculating orbital elements (here especially semi-major axis a, eccentricity e and inclination i) can be calculated.

106.044 **Analysis of interplanetary dust collections.** D. E. Brownlee, L. Pilachowski, E. Olszewski, P. W. Hodge.
Solid particles in the solar system, (see 012.019), p. 333 - 342 (1980).

Interplanetary dust particles collected in the form of micrometeorites in the stratosphere and meteor ablation spherules in deep sea sediments are possibly a relatively unbiased sample of the micrometeoroid complex near 1 AU. Detailed laboratory analysis of the particles has provided information on physical properties which may be useful in modeling a variety of aspects of interplanetary dust.

106.045 **From interstellar dust to comets to the zodiacal light.** J. M. Greenberg.
Solid particles in the solar system, (see 012.019), p. 343 - 350 (1980).

The author considers the consequences of the assumption that the interplanetary particles which produce the zodiacal light have evolved from interstellar dust via comets. The chemical evolution of interstellar dust followed by the process of aggregation into the cometary nucleus and the subsequent ejection of cometary debris provide the basis for a model for the interplanetary particles. The scattering properties of these particles are reasonably consistent with current observations of the variation with elongation angle of the brightness and polarization of the zodiacal light. The major chemical constituents of the model are in the form of a matrix of volatile ices and complex nonvolatile molecules containing C, N and O in which are imbedded silicate and metallic inclusions.

106.046 **Evidence for ice meteoroids beyond 2 AU.** H. Zook.
Solid particles in the solar system, (see 012.019), p. 375 - 380 (1980).

106.047 **Existence and role of amorphous grains in the solar system.** R. Smoluchowski.
Solid particles in the solar system, (see 012.019), p. 381 - 384 (1980).

106.048 **Grain disruption by collisions with solar energetic particles.** T. Mukai.
Solid particles in the solar system, (see 012.019), p. 385 - 389 (1980).

The catastrophic disruption of interplanetary dust grains, including water-ice, obsidian and magnetite, by impinging solar cosmic rays is investigated. The disruption is caused by the stress wave emanating from the heated lattice atoms along the path of an impinging particle. The author finds that the disruption plays an important role in the mass loss rate of grains compared with that due to sublimation and sputtering by solar particles.

106.049 **Radiation induced rotation of interplanetary dust particles; a feasibility study for a space experiment.**
K. F. Ratcliff, N. Y. Misconi, S. J. Paddack.
Solid particles in the solar system, (see 012.019), p. 391 - 394 (1980).

Irregular interplanetary dust particles may acquire a considerable spin rate due to non-statistical dynamical mechanisms induced by solar radiation. These arise from variations in surface albedo and from irregularities in surface geometry. The authors report on an experiment which will lead to an evaluation in space of the effectiveness of these two spin mechanisms. They utilize the technique of optical levitation in an argon laser beam to provide a stable trap for particles (10–60 microns in diameter). The spin rate and direction of the spin axis are measured in a straightforward manner. The objective is to design an optical trap for dielectric particles in vacuum which can be used to study these rotation mechanisms in the gravity-free environment of a Spacelab experiment.

106.050 **A review of upstream and bow shock energetic-particle measurements.** K. A. Anderson.
Nuovo Cimento C, Ser. 1, Vol. 2C, (see 012.013), 747 - 771 (1979). – Abstr. in Phys. Abstr.,Vol. 83, Abstr. 90308 (1980).

106.051 **Photographs of the zodiacal light.**
Sky Telesc., Vol. 60, 444 - 445 (1980).

106.052 **Cosmic spherules.** D. W. Hughes.
Nature, Vol. 287, 778 - 779 (1980).

106.053 **Energetic particles in space.** R. P. Lin.
Sol. Phys., Vol. 67, 393 - 399 (1980). Review talk given at the meeting of Commission 10 of the International Astronomical Union General Assembly, Montreal, Canada, August 15, 1979.

Particles ranging in energy from just above solar wind, $\gtrsim 1$ keV, to galactic cosmic rays of many GeV or greater are observed to be always present in the interplanetary medium. These suprathermal particles appear to come from many different sources: among them the galaxy and nearby interstellar medium, the Sun, planetary magnetospheres and bow shock waves. Recent studies have shown that the interplanetary medium itself is a major source of low energy, $\lesssim 10^2$ MeV ions. It appears that collisionless shock waves are often involved in the acceleration of these particles. The author reviews previous observations of these suprathermal particles and presents some preliminary new observations of low energy, $< 10^2$ keV particles from experiments aboard the ISEE-1, 2, and 3 spacecraft.

106.054 **Influence of anisotropic shock waves on the parameters of the interplanetary plasma.** S. A. Grib.
Fiz. protsessy v ionos. i magnitosfere, Moskva, 1979, p. 23 - 28. In Russian. – Abstr. in Ref. zh., 62. Issled. kosm. prostranstva, 9.62.203 (1980).

106.055 **Possible explanation for the variation of zodiacal light brightness, as observed from ground base and outer space.** A. S. Asaad.
J. Astron. Soc. Egypt, Vol. 1, 84 - 95 (1979).

Past and new observations of the zodiacal light are reviewed. Variations noted in its brightness and polarization from ground base and outer space are given. Possible explanations for these variations are mentioned.

106.056 **Disruption of a small interplanetary grain.** T. Mukai.
13th Lunar and Planetary Symposium, (see 012.018), p. 11 - 16 (1980).

It is shown that in interplanetary space electrostatic disruption occurs in silicate grains when the radius S is less than 0.01 μm. The lifetime of silicate grains with $0.01\ \mu m \lesssim S \lesssim 1 \mu m$ is mainly affected by mass loss rate due to impact of high energy particles on the grain, and consequently such a grain will disappear before it falls into the sun by the Poynting-Robertson effect. For a silicate grain with $S \gtrsim 1 \mu m$, however, the Poynting-Robertson effect governs its lifetime.

106.057 **Photometric axis measurements of the zodiacal light at large elongations.**
H. Tanabe, A. Takechi, A. Miyashita.
13th Lunar and Planetary Symposium, (see 012.018), p. 193 - 194 (1980). – Abstract.

106.058 **Estimate of typical cross dimensions of interplanetary shock waves from powerful isolated solar flares.**
K. G. Ivanov, L. V. Evdokimova, N. V. Mikerina.

Pis'ma Astron. Zh., Tom 6, 645 - 647 (1980). In Russian. English translation in Soviet Astron. Lett., Vol. 6.

Typical cross dimensions near the earth of interplanetary shock waves from powerful solar flares are estimated by considering the occurrence of geomagnetic disturbances.

106.059 **The origin of interplanetary sectors.**
K. H. Schatten.
Solar and interplanetary dynamics, (see 012.020), p. 67 - 72 (1980).

106.060 **Interplanetary response to solar long time-scale phenomena.** C. D'Uston, J. M. Bosqued.
Solar and interplanetary dynamics, (see 012.020), p. 105 - 125 (1980).

In this paper the authors review the experimental knowledge gained in the recent years on the interplanetary response to solar long-time scale phenomena such as the coronal magnetic structure and its evolution. Observational evidence that solar wind flow in the outer corona comes from the unipolar diverging magnetic regions of the photosphere is discussed along with relations to coronal holes.

106.061 **Large-scale magnetic field structure at the earth's orbit, its correlation with solar activity and orientation and motion of the solar system in the Galaxy.**
G. J. Vassilyeva (*G. Ya. Vasil'eva*), M. A. Kuznetsova, L. M. Kotlyar.
Solar and interplanetary dynamics, (see 012.020), p. 167 - 172 (1980).

Interplanetary magnetic field data from the different satellites obtained during the period 1963–1973 at 1 A.U. and compiled by J. King have been analysed in heliocentric ecliptic coordinates. The peculiarities of the background interplanetary magnetic field are discussed in relation to the orientation of the solar system in the Galaxy and the variable helioefficiency of the planets. The results of the direct cosmic experiments are evidence of the solar activity being a complex phenomenon of the solar system as a whole.

106.062 **Transient phenomena originating at the sun – an interplanetary view.** D. S. Intriligator.
Solar and interplanetary dynamics, (see 012.020), p. 357 - 374 (1980).

An overview is given of various phenomena observed by numerous spacecraft in the interplanetary medium. These phenomena are related to transient solar events such as flares and coronal holes. The effects of such transient solar events are extensive. The observations show that phenomena originating at the sun affect the structure of the particles and fields regimes throughout the explored solar system, extending out to at least 17-20 AU in the vicinity of the ecliptic plane. Transient phenomena are significantly modified by their passage through the interplanetary medium.

106.063 **Variations of interplanetary parameters and cosmic-ray intensities.** A. Geranios.
Solar and interplanetary dynamics, (see 012.020), p. 393 - 398 (1980).

Cosmic ray intensity depressions correlate with the average propagation speed of interplanetary shocks as well as with the amplitude of the interplanetary magnetic field after the eruption of a solar flare. About one fourth of the observed events correlates with corotating fast solar wind streams.

106.064 **IPS observations of flare-generated disturbances.**
T. Watanabe.
Solar and interplanetary dynamics, (see 012.020), p. 409 - 412 (1980).

106.065 **Theoretical interpretation of traveling interplanetary phenomena and their solar origins.** S. T. Wu.
Solar and interplanetary dynamics, (see 012.020), p. 443 - 458 (1980).

Recent theoretical studies on traveling interplanetary phenomena and their relation or presumed relation to their solar origins will be reviewed. An attempt is made to outline the theoretical studies in the context of mathematical methods and physical processes.

106.066 **Physical processes and models of interplanetary responses: suggested theoretical studies.**
S. Cuperman.
Solar and interplanetary dynamics, (see 012.020), p. 459 - 474 (1980).

Three ways to improve the theory and therefore the understanding of the physical processes in the interplanetary medium (during both quiet and disturbed periods of solar activity) are suggested.

106.067 **Global modeling of disturbances in the corona–interplanetary space.** T. Yeh.
Solar and interplanetary dynamics, (see 012.020), p. 503 - 506 (1980).

In order to fully account for the global features of the corona–interplanetary phenomenon, it is necessary to do three-dimensional calculations. The author discusses some groundwork for the formulation of a time-dependent three-dimensional code.

106.068 **On the deformation of the plane of the interplanetary current layer by solar wind streams.**
M. S. Bobrov.
Geomagn. Aehron., Tom 20, 929 - 932 (1980). In Russian.

106.069 **Numerical simulation of Poynting-Robertson and collisional effects in the interplanetary dust cloud.**
J. Trulsen, A. Wikan.
Astron. Astrophys., Vol. 91, 155 - 160 (1980).

Due to the solar radiation pressure individual grains of the interplanetary dust cloud are subject to the Poynting-Robertson effect. The grains are spiralling inwards, the orbits becoming more circular while the orbital planes remain unchanged. At the same time the grains suffer collisions among themselves. Such collisions act to establish a dynamical equilibrium between the distributions of individual eccentricities and inclinations in the dust cloud with approximately equal mean values of these quantities. A numerical simulation model is developed to study the combined result of these two effects on the dynamical evolution of a dust population.

106.070 **Influence of the interplanetary magnetic field on cometary and primordial dust orbits: applications of Lorentz scattering.** G. J. Consolmagno
Icarus, Vol. 43, 203 - 214 (1980).

The equations describing the change in orbital elements of interplanetary dust due to Lorentz-force accelerations are presented in a simplified form. Such accelerations depend on the charge state of the dust. The scattering of dust by a randomly changing magnetic field can be viewed analogously to the dust diffusing in space; the equations presented thus can be used to interpret observations of the present distribution of dust in terms of its possible sources and sinks. The stronger magnetic fields of the early solar system would have led to more vigorous scattering of the dust; particles as large as 1 mm could have been significantly transported by Lorentz scattering during this time.

106.071 **The distribution of temperature maxima for micrometeorites decelerated in the earth's atmosphere without melting.** P. Fraundorf.
Geophys. Res. Lett., Vol. 7, 765 - 768 (1980).

A simple model for the statistics of heating of micrometeorites decelerated in the earth's atmosphere without melting predicts that for 10 μm particles with thermal emissivity near 1, roughly half of those with density 1 g/cc are heated above 550°C, while half of those with density 3 g/cc are heated above 800°C.

106.072 **Het zodiakale licht en interplanetaire deeltjes.** J. W. Hovenier.
Zenit, 7e Jaarg., 90 - 95 (1980).

106.073 **Abnormal quiet days and the effect of the interplanetary magnetic field on the apparent position of the *Sq* focus.** E. C. Butcher, G. M. Brown.
Geophys. J. R. Astron. Soc., Vol. 63, 783 - 789 (1980).

106.074 **A simplified method of determination of the geocentric ephemeris of the two dust moons near the libration points in the earth-moon-system and the ephemeris for the year 1982.** K. Kordylewski.
Cracow Obs. Repr. No. 125, 4 pp. (1980).

106.075 **The propagation of cosmic-rays in the interplanetary region (the theory).**
L. J. Gleeson, G. M. Webb.
Fundam. Cosmic Phys., Vol. 6, 187 - 312 (1980).

106.076 **Further studies on cosmic spherules from deep-sea sediments.** D. W. Parkin, R. A. L. Sullivan, J. N. Andrews.
Philos. Trans. R. Soc. London, Ser. A, Vol. 297, 495 - 518 (1980). – Abstr. in Phys. Abstr., Vol. 84, Abstr. 4822 (1981).

106.077 **The thickness of interplanetary collisionless shock waves.** S. Pintér.
Bull. Astron. Inst. Czechoslovakia, Vol. 31, 368 - 379 (1980).

The thickness of magnetic structures of interplanetary shock waves related to the upstream solar wind plasma parameters have been studied. The following results have been obtained: The measured shock thickness increases for a decreasing upstream proton number density and decreases for an increasing flux energy. The shock thickness strongly depends on the ion pressure β, i. e. for higher values of β the thickness decreases. The shock wave becomes thicker with increasing M_{An} (Alfvén Mach number taken in the direction of the normal).

106.078 **On the equivalence of expressions for inverting the 3-dimensional zodiacal light brightness integral.**
D. W. Schuerman.
Planet. Space Sci., Vol. 28, 1077 - 1079 (1980).

Both Schuerman (1979) and Buitrago (1979) independently derived expressions for the general, mathematical inversion of the zodiacal light brightness integral. It is shown that the expressions are equivalent, differing only in the choice of reference systems.

106.079 **Results of Pioneer 10 and 11 meteoroid experiments: interplanetary and near-Saturn.** D. H. Humes.
J. Geophys. Res., Vol. 85, 5841 - 5852 (1980).

The concentration of meteoroids in interplanetary space between 1 and 18 AU and near the planets Jupiter and Saturn has been measured with the pressurized cell meteoroid detectors on Pioneer 10 and Pioneer 11. It is found that the meteoroids between 4 and 5 AU are not in direct circular or near-circular orbits near the ecliptic plane. The Pioneer 11 data obtained between 4 and 5 AU are best explained by the meteoroids being in randomly inclined orbits of high eccentricity. If meteoroids are in these cometlike orbits, the great increase in penetration flux previously measured near Jupiter with the Pioneer 10 experiment cannot be attributed to gravitational focusing unless the size distribution of meteoroids changes substantially between 1 and 5 AU. At Saturn encounter, the penetration flux increased by about three orders of magnitude, probably as the result of impacts from ring particles.

106.080 **On the size distribution and physical properties of interplanetary dust grains.**
L. B. Le Sergeant D'Hendecourt, P. L. Lamy.
Icarus, Vol. 43, 350 - 372 (1980).

This paper synthesizes information on the size distribution and physical properties of interplanetary dust grains obtained from analyses of lunar microcraters performed until 1979. The different aspects of these analyses (counting methods, simulation, calibrations) are summarized and a large amount of data is collected and discussed in order to clarify past contradictions. All results converge to a two-component dust population: Population I consists principally of large grains ($d > 2$ μm) with density typical of silicates while Population 2 consists of small grains ($d < 2$ μm) with higher density typical of iron, with a minor component of silicates.

106.081 **On the albedo of the interplanetary dust.** M. S. Hanner.
Icarus, Vol. 43, 373 - 380 (1980).

The zodiacal light brightness and measured spatial density of the interplanetary dust lead to a mean geometric albedo of 0.24 for the dust particles near 1 AU; whereas the composition of collected micrometeoroids suggests a geometric albedo $\lesssim 0.1$. The evidence is against a change in the mean particle albedo between 0.1 and 2 AU. Beyond 2 AU the data are unclear and a change in albedo is not ruled out.

106.082 **Behaviour of the interplanetary magnetic field near the warped current sheet.** M. A. Livshits.
Astron. Tsirk., No. 1116, p. 1 - 5 (1980). In Russian.

106.083 **Staubgenerierte Neutralgase in Sonnennähe.** H. W. Ripken, H. J. Fahr.
Mitt. Astron. Ges., Nr. 50, p. 46 - 50 (1980).

106.084 **Wechselwirkungsprozesse zwischen Sonnenwind und interplanetarem Staub.**
H. W. Ripken, H. J. Fahr, G. Lay.
Mitt. Astron. Ges., Nr. 50, p. 133 - 137 (1980).

106.085 **On a connection of the configuration of interplanetary shock waves from powerful isolated flares with proton events.**
K. G. Ivanov, L. V. Evdokimova, N. V. Mikerina, A. F. Kharshiladze.
Geomagn. Aehron., Tom 20, 982 - 989 (1980). In Russian.

106.086 **On the role of the B_z component of the interplanetary magnetic field in the force balance at the day-time magnetopause.** T. V. Kuznetsova.
Geomagn. Aehron., Tom 20, 1036 - 1040 (1980). In Russian.

106.087 **Connection of the components of the vector of the interplanetary magnetic field with variations of the geomagnetic field at high latitudes of the northern hemisphere.**
R. G. Afonina, B. A. Belov, A. E. Levitin, M. Yu. Markova, Ya. I. Fel'dshtejn, M. V. Fiskina.
Geomagn. Aehron., Tom 20, 1073 - 1083 (1980). In Russian.

106.088 **The Parker spiral configuration of the interplanetary magnetic field between 1 and 8.5 AU.**
B. T. Thomas, E. J. Smith.
J. Geophys. Res., Vol. 85, 6861 - 6867 (1980).

An analysis is presented of the magnetic field data obtained by the Pioneer 10 and 11 spacecraft. The purpose is to provide a quantitative picture of the overall configuration of the interplanetary magnetic field in the outer heliosphere.

106.089 **On the electrostatic potential of interplanetary grains: influence of the thermoionic effect.**
J. Millet, J. P. L. Lafon, P. L. Lamy.
Astron. Astrophys., Vol. 92, 6 - 12 (1980).

The authors investigate the importance of the thermoionic emission for interplanetary iron and carbon grains in the vicinity of the Sun. This emission prevents the potential of carbon grains from becoming strongly negative and allows small grains to exist in regions where they would otherwise be destroyed.

Collisionless shocks: simulation and laboratory experiments. See Abstr. 022.059.

Determination of particle densities by penetration studies. See Abstr. 022.073.

Method for the determination of density and phase functions of interplanetary dust. See Abstr. 031.541.

Method of scattering plane scanning. See Abstr. 031.542.

Inversion of the zodiacal brightness integral: a new geometric approach suitable for the out-of-ecliptic zodiacal light programme. See Abstr. 031.543.

Planned observations of the diffuse sky radiation during Shuttle mission STS–4. See Abstr. 032.528.

An attempt to observe zodiacal light at 5μ with a balloon experiment. See Abstr. 032.529.

Micrometeoroid multiple foil penetration and particle recovery experiments on board Space Shuttle's Long Duration Exposure Facility (LDEF). See Abstr. 032.532.

The plasma experiment on Helios (E 1). See Abstr. 032.555.

Scientific results obtained by the Helios Technical University of Braunschweig flux-gate (E 2) and search-coil (E 4) magnetometer experiments. See Abstr. 032.556.

Rome-GSFC magnetic field experiment (E 3). See Abstr. 032.557.

Spectrometer for measurements of low-energy electrons and ions (E 8). See Abstr. 032.562.

Zodiacal light experiment (E 9). See Abstr. 032.563.

The micrometeoroid analyzer (E 10). Micrometeoroid experiment – data analysis. See Abstr. 032.564.

Helios occultation experiment – time delay measurements (E OC). See Abstr. 032.566.

ASTRO-HEL, ein Raketenexperiment zur Beobachtung der Heliumhintergrundstrahlung bei 30,4 nm und 58,4 nm. See Abstr. 032.578.

Interplanetary scintillation – preliminary observations at 103 MHz. See Abstr. 033.012.

The effect of radiation pressure on the restricted three-body problem. See Abstr. 042.031.

Planet-A project on a magnetic measurement in the interplanetary space during the Halley-Venus mission. See Abstr. 051.009.

The International Solar Polar Mission: zodiacal light/background starlight experiment (ISPM–ZLE). See Abstr. 051.012.

Les satellites et les manifestations de l'activité solaire dans l'espace interplanétaire. See Abstr. 054.004.

Very hot plasmas in the solar system. See Abstr. 062.084.

A simple derivation of the radiation forces felt by scattering particles. See Abstr. 063.041.

Scattering by ensembles of small particles. Experiment, theory and application. See Abstr. 063.052.

Classification and investigation of solar flare situations conformably to interplanetary and magnetospheric disturbances. See Abstr. 073.088.

Distribution of chromospheric flares relative to the sector boundaries of the interplanetary magnetic field extrapolated to the sun. See Abstr. 073.121.

Observations of large fluxes of He^+ in the solar wind following an interplanetary shock. See Abstr. 074.028.

An observational picture of solar-wind MHD turbulence. See Abstr. 074.035.

Observation of dust generated hydrogen in the solar vicinity. See Abstr. 074.055.

A large decametric wavelength antenna array for IPS (*interplanetary scintillation*) observations of radio sources. See Abstr. 074.069.

Observations of interplanetary scintillation and a theory of high-speed solar wind. See Abstr. 074.072.

Hydromagnetic waves and turbulence in the solar wind. See Abstr. 074.107.

Effect of temporal variations of solar wind flux velocity on the structure of the interplanetary magnetic field. See Abstr. 074.109.

On the role of Alfvénic fluctuations in the inner solar system. See Abstr. 074.116.

Comparative magnetospherology, part 9. Solar-terrestrial phenomena as explained by heliomagnetic excursion in 1974. See Abstr. 075.014.

Evolution of coronal and interplanetary magnetic fields. See Abstr. 075.017.

Connection of characteristics of the interplanetary medium with the intensity variations of charged particles from the flare of 1973, September 7. See Abstr. 078.003.

Study of the connection between the zonal circulation of the atmosphere and the sector structure of the interplanetary magnetic field. See Abstr. 082.034.

Spektrophotometrische Untersuchungen von He-I/II-Resonanzstrahlungen mit der Raketennutzlast ASTRO-HEL. See Abstr. 082.079.

On a correlation of variations of the outer ionosphere with the B_z-component of the interplanetary magnetic field. See Abstr. 083.034.

Variations in the structure of the F-layer of the polar ionosphere at a change of the sign of the y-component of the interplanetary magnetic field. The Svalgaard-Mansurov effect in the ionosphere. See Abstr. 083.050.

The three-dimensional shape of the bow shock. See Abstr. 084.034.

Dependence of the probability of substorm display on solar wind velocity and on the vertical component of the interplanetary magnetic field. See Abstr. 084.055.

Influence of the B_X component of the interplanetary magnetic field on magnetopause reconnection. See Abstr. 084.066.

Quantitative relations between the characteristics of storm-time variations of the geomagnetic field and the changes of interplanetary plasma. See Abstr. 084.072.

Dependence of mid-latitude hydromagnetic energy spectra on solar wind speed and interplanetary magnetic field direction. See Abstr. 084.076.

Observations of the mean ionization states of energetic particles in the vicinity of the earth's magnetosphere. See Abstr. 084.077.

Cusp proton signatures and the interplanetary magnetic field. See Abstr. 084.078.

Energetic electron bursts in the magnetopause electron layer and in interplanetary space. See Abstr. 084.083.

Birkeland currents and the interplanetary magnetic field. See Abstr. 084.084.

Influence of the interplanetary magnetic field on the position of the dayside magnetopause. See Abstr. 084.087.

Pattern of solar activity influence upon circumpolar baric features. See Abstr. 085.013.

An ionospheric response to the polarity reversal of the general magnetic field of the sun and of the interplanetary magnetic field. See Abstr. 085.022.

On the use of geomagnetic field variations for diagnostics of inhomogeneities of the interplanetary electric field. See Abstr. 085.029.

Correlated variations of planetary albedos and coincident solar-interplanetary variations. See Abstr. 091.050.

Orbits of submicron lunar ejecta in the earth-moon system. See Abstr. 094.027.

Comets and the interplanetary magnetic field. See Abstr. 102.008.

Interplanetary gas. XXV. A solar wind and interplanetary magnetic field interpretation of cometary light outbursts. See Abstr. 102.029.

Microcharacterization of "Brownlee" particles: features which distinguish interplanetary dust from meteorites? See Abstr. 105.129.

Solar wind interaction with interstellar helium. See Abstr. 131.004.

Survival probabilities for interstellar hydrogen flowing into the interplanetary system from far regions of the heliosphere. See Abstr. 131.005.

Electromagnetic effects on hyperbolic cosmic dust particles. See Abstr. 131.116.

A Monte Carlo simulation of the mass distribution in an accreting system of dust particles. See Abstr. 131.117.

Formation of amorphous grains in cosmic environments. See Abstr. 131.126.

Comparison of solar backscatter and interstellar absorption measurements of the ISM. See Abstr. 131.148.

Particle trapping and acceleration during the August 1972 event. See Abstr. 143.002.

Kinetic theory of modulation of galactic cosmic rays by an interplanetary magnetic piston. See Abstr. 143.010.

Rigidity dependence of cosmic ray scintillations in the 50- to 300-GV range. See Abstr. 143.011.

On proton energy variations in their motion towards the sun. See Abstr. 143.031.

Erratum

106.901 **Erratum: "IMF sector behavior deduced from geomagnetic data"** [J. Geophys. Res., Vol. 85, 2357 - 2365 (1980)]. S. Matsushita, D. E. Trotter. J. Geophys. Res., Vol. 85, 5165 - 5168 (1980). – See Abstr. 27.106.034.

107 Cosmogony

107.001 **On the origin of the solar system.**
W. Dai, Z. Hu.
Sci. Sinica, Vol. 23, 862 - 879 (1980).

After a critical review of a variety of theories on the origin of the solar system the authors made a laborious research on this problem, the result of which is presented. The origin of main features and different kinds of celestial bodies in the solar system are discussed.

107.002 **Vers une théorie moderne de la formation des planètes.** W. K. Hartmann.
Astronomie, Vol. 94, 385 - 401 (1980).

107.003 **Evolution of eccentricities of the orbits of planets in the process of their accumulation.**
G. V. Pechernikova, A. V. Vityazev.
Astron. Zh., Tom 57, 799 - 811 (1980). In Russian.
English translation in Soviet Astron., Vol. 24, No. 4.

In the theory of accumulation of planets, it is shown that mutual gravitational perturbations result in the increase of the orbits' eccentricities and inclinations in the planets' feed zones with the growth of masses of the greatest bodies. The equations connecting the eccentricity of the planetary orbit with the average eccentricity of the orbits of preplanetary bodies and with the mass distribution function of the encountering and falling onto the planetary bodies are deduced.

107.004 **'Dual crystallization' in the solar system.**
S. Berczi.
Fiz. Sz., Vol. 29, 412 - 417 (1979). In Hungarian. – Abstr. in Phys. Abstr., Vol. 83, Abstr. 72795 (1980).

107.005 **Formation of the solar system.**
A. J. R. Prentice.
Proc. Astron. Soc. Australia, Vol. 3, 300 - 308 (1979).

107.006 **On the formation of planetesimals.**
K. Hourigan, A. J. R. Prentice.
Proc. Astron. Soc. Australia, Vol. 3, 389 - 390 (1979).

107.007 **Numerische Untersuchungen zur Entwicklung des Planetensystems.** R. Dvorak.
Mitt. Astron. Ges., Nr. 48, (see 012.015), p. 87 - 89 (1980).

107.008 **On an approximate method for investigation of the mutual gravitational influence of bodies of the protoplanetary cloud. On the evolution of Pluto's orbit.**
S. I. Ipatov.
Inst. prikl. mat. AN SSSR. Prepr., 1980, No. 43. 33 pp. In Russian. – Abstr. in Ref. zh., 51. Astron., 8.51.236 (1980).

107.009 **On the frame of the solar system – the importance of giant planets.** H. Oya.
13th Lunar and Planetary Symposium, (see 012.018), p. 355 - 362 (1980).

107.010 **Growth and sedimentation of dust grains in the primordial solar nebula.**
Y. Nakagawa, K. Nakazawa, C. Hayashi.
13th Lunar and Planetary Symposium, (see 012.018), p. 379 (1980). – Abstract.

107.011 **Capture process of planetesimals by proto-Earth.**
S. Nishida.
13th Lunar and Planetary Symposium, (see 012.018), p. 380 - 386 (1980).

Simulational analyses are made for the motions of planetesimals in the Hill sphere of the protoplanet on the orbit of the present Earth. Planetesimals whose mass is as large as 10^{20}g, lose energy due to the gas drag effect of the dense proto-atmosphere during circulating round the planet, and are closed in the Hill sphere. The author calculated for three cases of the proto-Earth's mass M, that is, for the cases of $M = 0.01\,M_E$, $0.1\,M_E$ and M_E where M_E is the present Earth's mass. The time needed to be closed in the sphere is less than 10^2 yrs for the case of $M = 0.01\,M_E$ and is about a year for the case of $M = M_E$. Planetesimals are really captured in the sphere with the probability of 35% to 60%, dependent upon the protoplanetary mass.

107.012 **Electrolysis in space and fate of Phaethon.**
E. M. Drobyshevski (*Eh. M. Drobyshevskij*).
Moon Planets, Vol. 23, 339 - 344 (1980).

107.013 **Formation of the terrestrial planets.**
G. W. Wetherill.
Annu. Rev. Astron. Astrophys., Vol. 18, (see 003.004), 77 - 113 (1980).

In this review emphasis will be placed on the present state of our understanding of the ways in which the processes of (1) gravitational collapses and (2) accumulation operate, with special attention given to their relevance to the formation of the terrestrial planets.

107.014 **Accretional heating as the major cause of compositional differences among meteorite parent bodies, the Moon, and Earth.** G. P. Horedt.
Icarus, Vol. 43, 215 - 221 (1980).

Heating of meteorite parent bodies occurs mainly after their accretion, by destructive collisions. The heating was generally not sufficient to differentiate the parent bodies completely so that iron meteorites would originate from the mantle, rather than from the core of a meteorite parent body. Assuming that the Earth and Moon accreted from material of similar chemical composition, the author suggets that only from the outer lunar shell is there a loss of gases and volatiles due to accretional melting. The Earth melted completely and degassing was efficient for the whole mass of the Earth leading to its ≈20% higher uncompressed mean density in comparison to the Moon. Because of its lower gravitational field, gases and volatiles escaped much more easily from the lunar atmosphere than from the terrestrial one, leading to the observed depletion in volatiles of the outer parts of the Moon.

107.015 **On the Laplacian plane of a self-gravitating circum-planetary disc.** W. R. Ward.
Bull. American Astron. Soc., Vol. 12, 707 (1980). – Abstract.

107.016 **Iron-60 as a heat source and chronometer in the early solar system.** T. P. Kohman, M. S. Robison.
Bull. American Astron. Soc., Vol. 12, 707 (1980). – Abstract.

107.017 **Planetesimal growth: comparison of numerical and analytical models.**
R. Greenberg, S. J. Weidenschilling.
Bull. American Astron. Soc., Vol. 12, 707 (1980). – Abstract.

107.018 **Thermal and mineralogic evolution of C2-type planetesimals.** M. J. Gaffey.
Bull. American Astron. Soc., Vol. 12, 708 (1980). – Abstract.

107.019 **Sweeping of the Jovian resonances and the evolution of the asteroids.** M. Torbett, R. Smoluchowski.
Bull. American Astron. Soc., Vol. 12, 708 (1980). – Abstract.

107.020 **Origine dei pianeti.**
A. Coradini, C. Federico, G. Magni.
G. Astron., Vol. 6, 175 - 186 (1980).

107.021 **Protoplanetary core formation by rain-out of minerals.**
W. L. Slattery, W. M. DeCampli, A. G. W. Cameron.
Moon Planets, Vol. 23, 381 - 390 (1980).

Models of giant gaseous protoplanets indicate that iron and probably other minerals in the interior of a planet would be in the liquid state during part of the protoplanet evolution. The authors have modeled this process by using the 'stochastic collection' equation (Slattery, 1978) for various initial conditions. In all of the cases considered, the growth time (to centimeter-size droplets) is much shorter than the time that the drops are in the liquid state. Since the particles are rapidly swept from interstellar grain sizes to much larger sizes, the opacity in the cloud layer is expected to drop sharply following melting of the grains.

107.022 **Isotopic anomalies in the early solar system.**
R. N. Clayton.
Origin and distribution of the elements, (see 012.039), p. 121 - 125 (1979).

The primitive meteorites known as "carbonaceous chondrites" provide the best available information concerning the nuclear composition of the matter in the solar system at the time of formation of the sun and planets. Several light elements show "isotopic anomalies" which appear to result from major nucleosynthetic events occurring just before collapse of the solar nebula. The isotopic anomalies have been interpreted as evidence for a supernova "trigger" for formation of the solar system. The ^{16}O "anomaly" is not restricted to the primitive meteorites, but is found to varying degrees in all meteorites, reflecting the inhomogeneous distribution of the ^{16}O-rich component in the early solar system.

107.023 **Formation of the giant planets.** H. Mizuno.
Prog. Theor. Phys., Vol. 64, 544 - 557 (1980).
Abstr. in Phys. Abstr., Vol. 84, Abstr. 9414 (1981).

107.024 **Mass loss from planetary protoatmospheres and from the protoplanetary nebula.** G. P. Horedt.
Astron. Astrophys., Vol. 92, 267 - 272 (1980).

The maximum mass loss rates for a planetary atmosphere due to heat input from solar extreme ultraviolet radiation and from a T Tauri like solar wind are estimated. The maximum mass loss rates for hydrogen are found to be 10^{-3}–10^{-4} parts of the planetary mass for the terrestrial planets and 10^{-7}–10^{-8} parts for the outer planets. The mass loss rates due to a planetary wind are generally about 2–4 times larger. The present-day extreme UV radiation of the Sun or heating from the T Tauri like solar wind is not sufficient to dissipate the protoplanetary nebula in the terrestrial region.

107.025 **Dissipation of the rare gases contained in the primordial Earth's atmosphere.**
M. Sekiya, K. Nakazawa, C. Hayashi.
Earth Planet. Sci. Lett., Vol. 50, 197 - 201 (1980).

If the Earth was formed by accumulation of rocky bodies in the presence of the gases of the primordial solar nebula, the Earth at this formation stage was surrounded by a massive primordial atmosphere (of about 1×10^{26} g) composed mainly of H_2 and He. The authors suppose that the H_2 and He escaped from the Earth, owing to the effects of strong solar wind and EUV radiation, in stages after the solar nebula itself dissipated into the outer space. The primordial atmosphere also contained the rare gases Ne, Ar, Kr and Xe whose amounts were much greater than those contained in the present Earth's atmosphere. The authors have studied the dissipation of these rare gases due to the drag effect of outflowing hydrogen molecules.

107.026 **Dissolution of the primordial rare gases into the molten Earth's material.**
H. Mizuno, K. Nakazawa, C. Hayashi.
Earth Planet. Sci. Lett., Vol. 50, 202 - 210 (1980).

Using the results of their previous and new calculations (see 107.025) on the structure of the primordial atmosphere, the authors have investigated the amount of dissolution of the rare gases, which were contained in the primordial atmosphere, into the molten Earth's material. If a considerable amount of neon with nearly the solar isotopic ratio is discovered in present mantle material, this offers direct evidence for the proposition that the proto-Earth was once surrounded by the primordial atmosphere.

107.027 **On the equation of the accumulation process in the formation of planetary systems.**
T. M. Ehneev.
Dokl. AN SSSR, Vol. 253, 69 - 73 (1980). In Russian.
Abstr. in Ref. zh., 51. Astron., 11.51.266 (1980).

107.028 **Star and planetary system formation in collapsing, viscous, rotating clouds.**
P. J. Wiita, D. N. Schramm, E. M. D. Symbalisty.
Proc. Tenth Lunar Planet. Sci. Conf., (see 012.050), p. 1849 - 1865 (1979).

107.029 **A comparison of two accretional heating models.**
G. A. Ransford.
Proc. Tenth Lunar Planet. Sci. Conf., (see 012.050), p. 1867 - 1879 (1979).

107.030 **Collisional evolution of the mass-distribution spectrum of planetesimals. II.** T. Matsui.
Proc. Tenth Lunar Planet. Sci. Conf., (see 012.050), p. 1881 - 1895 (1979).

107.031 **A special class of planetary collisions: theory and evidence.** W. K. Hartmann.
Proc. Tenth Lunar Planet. Sci. Conf., (see 012.050), p. 1897 - 1916 (1979).

107.032 **Influence of the greenhouse effect in primordial atmospheres on the formation of the earth and Venus.** D. D. Kvasov, A. P. Gal'tsev, A. S. Safraj.
Astron. Vestn., Tom 14, 72 - 79 (1980). In Russian.

The greenhouse effect has been calculated in the primordial atmospheres which could form in the case of instantaneous outgassing of part of the volatiles present in the interiors of the earth and Venus. The result creates insurmountable obstacles for the theories of condensation of the terrestrial planets from a gas solar nebula, rapid accretion of meteoric matter and capture of the moon by the earth. It is only the theory of slow accretion of the earth and the theory of the formation of the moon at the terrestrial orbit that allow the existence of primordial atmosphere of the earth at low pressure.

Cosmogonia del sistema solare.
See Abstr. 003.058.

The significance of low melting-temperature materials in consolidation of planetesimals.
See Abstr. 022.081.

On the origin of the Kirkwood Gaps.
See Abstr. 042.042.

Dissipation of the primordial terrestrial atmosphere due to irradiation of the solar EUV. See Abstr. 082.040.

Revival of the hypothesis on Phaethon?
See Abstr. 098.035.

Chemical energy in cold-cloud aggregates: the origin of meteoritic chondrules. See Abstr. 105.020.

Endemic isotopic anomalies in titanium. See Abstr. 105.022.

Dislocation loop in olivine of Allende chondrite and heating event on chondrules and grains in the protosolar nebula. See Abstr. 105.069.

Thermal history of chondrules in the primitive solar nebula. See Abstr. 105.070.

Evidence for live ^{247}Cm in the early solar system. See Abstr. 105.100.

Sr isotopic fractionation in Allende chondrules: a reflection of solar nebular processes. See Abstr. 105.110.

Volatile matter in meteorites, in the protoplanetary nebula, and formation of planetary atmospheres. See Abstr. 105.152.

The circumsolar motion of dust particles at the stage of increasing solar luminosity. See Abstr. 106.034.

A two-stream instability in streams of charged grains. See Abstr. 106.041.

Hydrodynamics of stellar encounters. See Abstr. 117.020.

Computer simulation of planet formation in a binary star system: terrestrial planets. See Abstr. 117.058.

Stars

111 Parallaxes, Proper Motions, Radial Velocities, Space Motions, Distances

111.001 **Rubin-152: a massive star on the galactic fringe?** F. R. Chromey.
Astron. J., Vol. 85, 853 - 857 (1980).

Rubin-152 is spectroscopically identical with a type O4–5 main-sequence star, and the standard luminosity calibration places it at the unexpected distance of 50 kpc from the galactic center. Although its radial velocity is inconsistent with a circular orbit, a highly eccentric one cannot be excluded. On the other hand, the data reported here are also consistent with the hypothesis that R-152 is in an advanced stage of evolution, and that it is a nearby, low-mass, low-luminosity object which by perverse coincidence exhibits the effective temperature and surface gravity of a very massive star.

111.002 **Radial velocity study of four southern RS CVn candidates and related field stars.**
J. G. Stacy, R. E. Stencel, E. J. Weiler.
Astron. J., Vol. 85, 858 - 866 (1980).

Radial velocity variations are demonstrated for four southern RS CVn candidates: HD 39937, 101379, 155555, and 174429. The period of HD 155555 appears to have decreased by about one part in 100 000 over the past 18 yr. In addition, high-resolution observations of the Ca II H and K profiles and the Mg II 2800-Å emission doublet are presented. These enhance the likelihood of these being RS CVn objects. Radial velocity data for 15 other late-type bright field stars are presented. Some of these were observed by Copernicus and may exhibit unusual chromospheres, while the rest lack radial velocity information as tabulated in the Bright Star Catalog. All velocity measures have been corrected to the IAU heliocentric system.

111.003 **Systematic effects in trigonometric parallaxes – I. The distribution of observed parallaxes.**
R. B. Hanson.
Mon. Not. Astron. Soc., Vol. 192, 347 - 357 (1980).

The frequency distributions of trigonometric parallaxes determined by the four major General Catalogue observatories (Allegheny, McCormick, Cape and Yale) are used to study their absolute zero points, systematic differences and external errors. The method is to fit the observed distributions by convolving power-law space distributions with Gaussian observational errors. Given a nominal correction (+0.003 arcsec) to absolute, the zero point of the General Catalogue parallaxes is confirmed to within ± 0.001 arcsec, without systematic observatory differences generally found from comparisons using multiple observations. The external error estimates (0.012, 0.019, 0.017, 0.015 arcsec respectively) agree with the General Catalogue precepts and other recent determinations. These results indicate possible precepts for the new Yale Parallax Catalogue. The space distributions of stars selected for parallax observations give useful information to aid luminosity calibrations.

111.004 **Proper motion of star No. 154 on Plate 1329 of the AC San Fernando Zone (=3A2254-033).**
L. Quijano.
Observatory, Vol. 100, 119 (1980).

111.005 **Dispersion inhomogeneities of nearby stars velocities along the direction $l = 330°$, $b = 0°$.**
S. Menge de Freitas.
Astron. Astrophys., Suppl. Ser., Vol. 41, 433 - 436 (1980).

The author has taken a population of 726 nearby stars and computed their velocity components along a direction expected to be approximately perpendicular to the local spiral arm or at least to its Cygnus branch. Averages and standard deviations of such velocities are calculated for the whole population and for some samples, the stratification being made by grouping stars according to their galactic longitudes. Variances among groups (longitude sectors) are found to be inhomogeneous in several partitions.

111.006 **Absolute parallaxes to ± 1 milliarc-second.** J. W. Stein, G. D. Gatewood.
Bull. American Astron. Soc., Vol. 12, 454 - 455 (1980). Abstract.

111.007 **A study of Barnard's star.** L. W. Fredrick, P. A. Ianna.
Bull. Americam Astron. Soc., Vol. 12, 455 (1980). – Abstract.

111.008 **Stars nearer than 22 parsecs; changes made to Gliese's catalogue by means of photometric parallaxes.** A. G. D. Philip, B. Hauck.
Bull. American Astron. Soc., Vol. 12, 523 - 524 (1980). Abstract.

111.009 **An analysis of parallaxes determined in two coordinates.** T. E. Lutz, A. R. Upgren.
Bull. American Astron. Soc., Vol. 12, 524 (1980). – Abstract.

111.010 **The radial velocity of the Am star 68 Tau.** W. Häupl.
Astron. Nachr., Band 301, 207 - 208 (1980). In German.

The radial velocity of the star 68 Tau is searched in dependence on the light variation period. No variations have been found. The mean radial velocity of 22 spectra is 40.4 km/sec ± 0.7 km/s.

111.011 **The proper motions of the Beta-Cephei stars of the FK4.** K. Ferrari d'Occhieppo.
Mitt. Astron. Ges., Nr. 48, (see 012.015), p. 56 (1980).

111.012 **Die systematischen Fehler der trigonometrischen Parallaxen.** M. Buchholz, T. Schmidt-Kaler.
Mitt. Astron. Ges., Nr. 48, (see 012.015), p. 149 (1980).

111.013 **Untersuchungen von 9000 Raumgeschwindigkeiten.** W. Dieckvoss.
Mitt. Astron. Ges., Nr. 48, (see 012.015), p. 176 (1980).

111.014 **An analysis of parallaxes determined in two coordinates.** T. E. Lutz, A. R. Upgren.
Astron. J., Vol. 85, 1390 - 1398 (1980).

About ten percent of all 12 000 published trigonometric parallaxes list the parallax data in both x and y coordinates. Almost all of these were published in recent years at only four observatories: United States Naval Observatory (484 stars), Van Vleck (248 stars), Sproul (199 stars), and McCormick (75 stars). An analysis of these four sets of data shows that π_y measures the same quantity as π_x, although, of course, with lower precision. Probability plots have been analyzed and show that the precision in the difference $\pi_x - \pi_y$ has been overestimated at all four observatories by different amounts.

111.015 **The impact of star parallaxes and very accurate proper motions on galactic structure and dynamics studies.** P. O. Lindblad.
Highlights of Astronomy, Vol. 5, (see 012.029), 799 - 804 (1980).

111.016 **The needs in the radial velocity area in view of impending space astrometry projects.**
J. Andersen.
Highlights of Astronomy, Vol. 5, (see 012.029), 805 - 807 (1980).

111.017 **Research priorities for nearby stars – I.**
M. J. Halliwell.
Vistas Astron., Vol. 24, 259 - 272 (1980).

A compilation of stars brighter than 10th magnitude, probably within 25 parsecs, and needing photometry, spectral type or parallax as of January 1980 is given.

111.018 **Research priorities for nearby stars – II.**
M. J. Halliwell.
Vistas Astron., Vol. 24, 273 - 299 (1980).

The lists of photometry, spectral type and parallax candidates given in paper I (see 28.111.017) is completed to 40 pc.

111.019 **Classification of stellar populations and luminosity classes from accurate proper motions.**
L.-T. G. Chiu.
Astrophys. J., Suppl. Ser., Vol. 44, 31 - 71 (1980).

Proper motions of stars with $V \leq 20.5$ in SA 51, SA 57, and SA 68 were derived from a set of Lick 3 m, Hale 5 m, and KPNO 4 m prime-focus plates. Proper motions and photographically determined V and $B - V$ were used to construct reduced proper motion diagrams. A theory is presented for classifying the stars into different populations and luminosity classes with a probabilistic approach based on the distribution function of the reduced proper motion as a function of $B - V$. It is concluded that the Population I main sequence in the sample has both a higher velocity dispersion and a lower metallicity than those in the solar neighborhood.

111.020 **New parallaxes for old: a coming improvement in the distance scale of the universe.**
A. R. Upgren.
Mercury, Vol. 9, 143 - 148 (1980).

111.021 **Radial velocities of southern HR stars.**
W. I. Beavers, J. J. Eitter.
Publ. Astron. Soc. Pacific, Vol. 92, 713 - 716 (1980).

Photoelectric measurements of the radial velocities of 105 southern HR stars are reported. This study reveals 11 possible variable velocity stars and one double-line spectroscopic binary within this sample.

111.022 **Motion of high-velocity stars.** L. P. Osipkov.
Astron. Tsirk., No. 1105, p. 5 - 7 (1980). In Russian.

111.023 **Comparative analysis of a determination of absolute proper motions of stars in the meridional cross section of the Galaxy.** N. V. Kharchenko.
Astrometr. Astrofiz., Vyp. (No.) 42, p. 70 - 76 (1980). In Russian.

A program of observations of 21 selected areas of the sky was compiled to determine absolute proper motions of stars with respect to galaxies in the main meridional cross section of the Galaxy. Comparison of the absolute proper motions in five areas with Fatchikhin and Rakhimov's catalogues shows that the author's catalogue exceeds the second one in accidental accuracy and is identical with the first one.

NLTT Catalogue. Volume IV. –30° to –90°.
See Abstr. 002.021.

The nearest stars. See Abstr. 002.022.

Catalogs of proper motion stars. II. Stars brighter than visual magnitude 15 and south of declination +30° with annual proper motion between 0.″7 and 1.″0.
See Abstr. 002.031.

New USNO Zodiacal Star Catalog.
See Abstr. 002.050.

Lowell proper motions XIX.
See Abstr. 002.055.

Lowell Proper Motion Survey.
See Abstr. 002.056.

Catalogue of proper motions of stars in selected regions of the sky with galaxies. II.
See Abstr. 002.065.

Tests of two different Maximum Likelihood algorithms for determining statistical parallaxes.
See Abstr. 021.014.

Radial velocities of faint stars from objective prism plates. See Abstr. 031.548.

Astrometric analyses of 14 Sproul plate series.
See Abstr. 031.549.

On the construction of a comprehensive general catalogue of star positions. See Abstr. 041.001.

Systematische Bewegungen von Fundamental-Sternen. See Abstr. 041.015.

Recent results of BVRI photometry of late-type dwarf stars in nearby clusters and in the solar neighborhood.
See Abstr. 113.008.

Study of the variable F-type supergiants HD 161796 and HD 163506 in radial velocity and photometry.
See Abstr. 113.023.

The distant companion of θ Cygni.
See Abstr. 118.001.

FG Sagittae: a binary? See Abstr. 122.001.

Distances and radii of classical cepheids.
See Abstr. 122.093.

The possible nature of the high-velocity OB stars: hot UV-bright stars in the galactic disk. See Abstr. 155.044.

112 Circumstellar Matter (Shells, Dust, Masers, Stellar Winds, etc.)

112.001 **Molecules in celestial objects – I. Circumstellar CO in 9 Cephei (B2Ib).**
S. P. Tarafdar, K. S. Krishna Swamy, M. S. Vardya.
Mon. Not. R. Astron. Soc., Vol. 192, 417 - 426 (1980).
The column density of interstellar CO towards 9 Cep has been determined and a tentative identification of circumstellar CO around the star made. This is the first identification of a molecule around a B star. The circumstellar column density of CO and rate of mass loss have been estimated.

112.002 **An atlas of the shell spectrum of Pleione between 3 167 Å and 4 924 Å.** D. Ballereau.
Astron. Astrophys., Suppl. Ser., Vol. 41, 305 - 318 (1980).
Since 1972, Pleione has entered a new shell-star-phase. In 1977 a greater number of lines appeared in the near UV spectrum. The radial velocities measured since 1976 show a progressive infall of matter onto the star. The author gives a tracing of the spectrum of Pleione, converted to intensity between 3 167 Å and 4 924 Å and an identification of the shell lines between these two limits.

112.003 **Circumstellar matter in young clusters. III. Diffuse lines in optical region.**
J. Krełowski, A. Strobel.
Acta Astron., Vol. 30, 183 - 189 (1980).
Observational data concerning the diffuse lines indicate that the relation line intensity *vs.* E_{B-V} is similar to that found for the 2200 hump and can be interpreted as the result of strong circumstellar contribution to this line. Another effect, connected probably with interstellar clouds, contributes considerably to E_{B-V} and very weakly to diffuse bands intensities. Similar behaviour of all diffuse lines together with the well known correlations of their intensities makes possible the conclusion of the common origin (mainly circumstellar) of all diffuse features.

112.004 **SiO maser emission from VY CMa.**
G. C. McIntosh, A. P Lane, D. P. Clemens.
Bull. American Astron. Soc., Vol. 12, 456 (1980). – Abstract.

112.005 **Shell nebulae associable with Wolf-Rayet stars.**
J. N. Heckathorn, T. R. Gull.
Bull. American Astron. Soc., Vol. 12, 458 (1980). – Abstract.

112.006 **Weak shells around rapidly rotating early type stars.**
W. Oegerle, R. Polidan.
Bull. American Astron. Soc., Vol. 12, 463 (1980). – Abstract.

112.007 **An investigation of the nature of dust in circumstellar shells.** W. M. Rumpl.
Bull. American Astron. Soc., Vol. 12, 485 - 486 (1980). Abstract.

112.008 **SiO maser emission from late-type stars and Orion.**
A. P. Lane, C. R. Predmore, R. Genzel, J. M. Moran.
Bull. American Astron. Soc., Vol. 12, 499 (1980). – Abstract.

112.009 **A far-infrared emission feature in IRC+10216, IC 418, and NGC 6572.**
W. J. Forrest, J. R. Houck, J. F. McCarthy.
Bull. American Astron. Soc., Vol. 12, 505 (1980). – Abstract.

112.010 **OH maser pumping by line overlap at 2.8 microns.**
M. Cimerman, N. Scoville.
Astrophys. J., Vol. 239, 526 - 532 (1980).
The pumping of OH masers in late-type variable stars by direct stellar radiation of 2.8 μm is studied. Two pumping schemes are suggested based on line coincidences between OH $V = 0-1$ transitions and strong H_2O ν_3 transitions. The water molecules perturb the stellar radiation by absorption and thus cool the coinciding OH transitions. It is demonstrated that this cooling is capable of producing strong OH masers in the circumstellar envelopes of late-type variable stars. This model can account for the existence of main-line ($\Delta F = 0$) OH masers, which is hard to explain otherwise.

112.011 **The size and surface brightness of the circumstellar gas shell surrounding Betelgeuse.**
R. K. Honeycutt, A. P. Bernat, J. E. Kephart, C. E. Gow, M. T. Sandford II, D. L. Lambert.
Astrophys. J., Vol. 239, 565 - 569 (1980).
The authors have obtained direct images of the K I gas shell surrounding the M supergiant Betelgeuse using a two-dimensional television system and a 2 Å bandpass filter. The emission extends to at least 50″.

112.012 **Mass loss and rotation in early-main-sequence B stars.**
I. Furenlid, A. Young.
Astrophys. J., Lett., Vol. 240, L59 - L61 (1980).
Using a CCD detector, the authors have made high signal-to-noise ratio, intermediate-resolution spectroscopy of Hα in 60 normal main-sequence stars of spectral type B0–B3. A majority of the stars show asymmetry in Hα, which the authors interpret as mass loss. The mass loss is coupled to the rotational velocity. A minimum velocity of around 200 km s^{-1} is required to trigger significant mass loss in these stars. The mass loss occurs only parallel to the equatorial plane; no mass loss is observed over the poles.

112.013 **IUE observations of circumstellar lines and mass loss from B-stars.**
S. P. Tarafdar, K. S. Krishna Swamy, M. S. Vardya.
Stellar turbulence, (see 012.008), p. 295 (1980). – Abstract.

112.014 **Circumstellar chlorine chemistry and a search for AlCl.** R. E. S. Clegg, H. A. Wootten.
Astrophys. J., Vol. 240, 828 - 833 (1980).
The authors have searched for the $J = 7-6$ transition of AlCl in carbon-rich circumstellar shells, Orion A, and Sgr B-2. The upper limit for IRC + 10216, $T_A^* < 0.02$ K, is 20 times less than the predicted value for fully associated AlCl. The authors conclude that either the initial density in the circumstellar gas flow $n_o \lesssim 10^{10}$ cm^{-3} or the gaseous Al and Cl are depleted by grains in this object.

112.015 **Infrared observations of circumstellar ammonia in OH/IR supergiants.** R. A. McLaren, A. L. Betz.
Astrophys. J., Lett., Vol. 240, L159 - L163 (1980).
Ammonia has been detected in the circumstellar envelopes of VY Canis Majoris, VX Sagittarii, and IRC +10420 by means of several absorption lines in the ν_2 vibration-rotation band near 950 cm^{-1}. The line profiles are well resolved and show the gas being accelerated to terminal expansion velocities near 30 km s^{-1}. The observations reveal a method for determining the position of the central star on VLBI maps of OH maser emission to an accuracy of ~0.″2. A firm lower limit of 2×10^{15} cm^{-2} is obtained for the NH_3 column density in VY Canis Majoris.

112.016 **An anonymous ring nebula around a WC6 star in Carina.** M. C. Lortet, V. S. Niemela, R. Tarsia.
Astron. Astrophys., Vol. 90, 210 - 213 (1980).
The authors draw attention to a ring-shaped filamentary nebula around the WC6 star in HD92809. The optical spectrum of the star is briefly described and with the wind

velocity of 2100 km s^{-1} determined from this spectrum they calculate some parameters of the nebula.

112.017 **Envelopes around late-type giant stars.**
B. Zuckerman.
Annu. Rev. Astron. Astrophys., Vol. 18, (see 003.004), 263 - 288 (1980).

Giant stars have long been a favorite of optical astronomers. Now the cool circumstellar matter that surrounds most giants can also be studied in the infrared and microwave domains. The following topics are considered: chemical composition, physical properties, masers, mass loss, and late stellar evolution.

112.018 **An investigation of the maser radiations from NML Cyg at OH 1612 MHz.** Z.-p. Zhou.
Acta Astron. Sinica, Vol. 21, 211 - 218 (1980). In Chinese.

In the analysis of LBI observations of maser radiations at OH 1612 MHz made by Masheder et al. in the direction of NML Cyg, a new idea has been introduced that the red-shift components and the blue-shift components separately come from different shells of the circumstellar envelope. A spherical symmetry model of the air-flow velocity field in the circumstellar envelope of NML Cyg has been obtained. Mass loss rate, coherent length and pumping mechanism have been discussed and the results agree with the observations very well.

112.019 **Circumstellar absorption and intrinsic colours of massive stars.** A. Ardeberg, E. Maurice.
Astron. Astrophys., Vol. 91, 53 - 58 (1980).

The reality of significant amounts of circumstellar obscuration for massive stars is discussed. At the same time, the authors examine the question of possible incorrectness of currently adopted data for intrinsic colour indices of O stars. The work is based primarily on recent data for stars in the Large Magellanic Cloud and on new observational results for the galactic stellar aggregate surrounding HD 101205. For the high-luminosity stars in the Large Magellanic Cloud only marginal effects of possible circumstellar obscuration may be traced, whereas for the galactic stellar aggregate no such effects are noted. There is no convincing overall indication that currently adopted intrinsic colours for O stars should be significantly in error.

112.020 **Behavior and significance of circumstellar clouds.**
B. Zuckerman.
Interstellar molecules, (see 012.033), p. 479 - 486 (1980).

The author discusses molecular envelopes around post main sequence stars. Topics that are covered include chemical composition, physical properties, mass loss and evolution.

112.021 **Observations of circumstellar clouds.**
P. G. Wannier, R. O. Redman, T. G. Phillips, R. B. Leighton, G. R. Knapp, P. J. Huggins.
Interstellar molecules, (see 012.033), p. 487 - 493 (1980).

Observations have been made of J = 2 – 1 CO in eleven circumstellar clouds including seven carbon stars and four oxygen-rich stars. Several results are discussed with special emphasis on the implications for two sources, namely IRC+10216 and Mira. The observations of IRC+10216 show CO emission over a diameter of 6 arcmin, a result suggesting a very large mass-loss rate. Mira is unique among the objects studied in displaying a small CO opacity and a high CO excitation temperature.

112.022 **Radio detection of ammonia in IRC+10216.**
M. B. Bell, S. Kwok, P. A. Feldman.
Interstellar molecules, (see 012.033), p. 495 - 496 (1980).

IRC+10216 (CW Leo) is a carbon star surrounded by an expanding circumstellar envelope which is rich in molecules. The authors report the detection of the (1,1) and possibly the (2,2) inversion transitions of (para) NH_3, the first radio detection of ammonia in a star. This information can be used to determine the thermal structure of the envelope of IRC+10216.

112.023 **Molecular abundances in IRC+10216.**
E. M. McCabe, R. C. Smith, R. E. S. Clegg.
Interstellar molecules, (see 012.033), p. 497 - 502 (1980).

The observed molecular column-densities in IRC+10216 can be matched by chemical-equilibrium calculations for T ~ 1250° K, P ~ 100 dyn cm^{-2} and no graphite grain formation. Condensation of silicon carbide into grains may explain the low observed abundances of SiO and SiS.

112.024 **Infrared heterodyne spectroscopy of circumstellar molecules.** A. L. Betz, R. A. McLaren.
Interstellar molecules, (see 012.033), p. 503 - 508 (1980).

Ammonia has been detected in the circumstellar envelopes of IRC+10216, VY CMa, VX Sgr, and IRC+10420. A number of absorption lines of $^{14}NH_3$ in the ν_2 vibration-rotation band around 28 THz (950 cm^{-1}) have been observed at a velocity resolution of 0.2 km/s. Typical linewidths are 1 to 4 km/s, and the details of the line profiles provide additional insights on the process of mass loss in these stars.

112.025 **Spectroscopic studies of IRC+10216 and similar objects.** S. T. Ridgway, D. N. B. Hall.
Interstellar molecules, (see 012.033), p. 509 - 514 (1980).

Spectroscopy of circumstellar molecular species in the 2-14μm range provides evidence for a range of shell optical depths in the +10216 class of stars. In some cases a photospheric spectrum is present. The spectrum of +10216 shows absorption over a range of velocities including two distinct velocities of different excitation temperature. The occurrence of multiple velocities may be common in similar objects.

112.026 **Infrared spectroscopy of molecules in circumstellar material.** D. N. B. Hall.
Interstellar molecules, (see 012.033), p. 515 - 524 (1980).

High resolution spectra of red giants and long periods variables exhibit lines of infrared CO vibration-rotation bands arising in circumstellar material. In the few such so far observed at very high resolution ($\lesssim$ 1 km/s) the circumstellar material appears localized in 3 distinct regimes with temperatures of 800K, 200K and 75K and expansion velocities of 0, 10 and 16 km/s rather than being uniformly distributed.

112.027 **Observational characteristics of masers associated with stars.** L. E. Snyder.
Interstellar molecules, (see 012.033), p. 525 - 533 (1980).

OH, H_2O and SiO are the 3 known molecular masers associated with stars. The known transitions, their line profiles, radial velocities, temporal intensity variations and polarization properties are discussed. Current work on determination of circumstellar shell sizes is reviewed. Possible future research topics are outlined.

112.028 **VLBI observations of the V = 1 and V = 2 SiO masers in W Hydra and VX Sagittarius.**
A. P. Lane, P. T. P. Ho, C. R. Predmore, J. M. Moran, R. Genzel, S. S. Hansen, M. J. Reid.
Interstellar molecules, (see 012.033), p. 535 - 536 (1980).

112.029 **Vibrationally excited silicon monoxide masers.**
D. Buhl, F. O. Clark, G. Chin, D. Glenar, T. Kostiuk, M. J. Mumma, F. J. Lovas.
Interstellar molecules, (see 012.033), p. 537 - 538 (1980).

Published data on a select group of SiO maser sources have been analyzed for velocity variations as a function of phase. No apparent correlation was found to a level of about 2 km/s. This places constraints on the location of the maser molecules.

112.030 **Time variation of SiO maser emissions.**
N. Ukita, N. Kaifu.
Interstellar molecules, (see 012.033), p. 539 - 540 (1980).

The authors report a series of observations of the v = 1, J = 2 – 1 transition of SiO (86243.28 MHz) toward four Mira variables (o Cet, R Leo, W Hya, and R Cas) and Orion A over the period September 1976 - February 1978.

112.031 **Time variability of the Orion A, R Leo and o Ceti SiO (v = 1, J = 2 – 1) masers.**
Å. Hjalmarson, H. Olofsson.
Interstellar molecules, (see 012.033), p. 541 - 542 (1980).

112.032 **Polarized emission in the broad SiO feature from R Leo.** F. O. Clark, D. R. Johnson, T. H. Troland, C. E. Heiles.
Interstellar molecules, (see 012.033), p. 543 - 544 (1980).

Linearly polarized SiO emission spread over 12 km/s has been detected from the star R Leo. The position angle of polarized emission varies systematically with respect to the spectral line center. Interpreted in terms of radiative transfer theory, this change in position angle may be due to magneto-rotation, which allows the determination of the magnetic field and the SiO systemic velocity.

112.033 **The OH circumstellar maser in late-type stars.**
Nguyen-Q-Rieu, V. Bujarrabal, J. Guibert, A. Omont.
Interstellar molecules, (see 012.033), p. 549 - 550 (1980).

112.034 **Interpretation of circumstellar masers.**
P. Goldreich.
Interstellar molecules, (see 012.033), p. 551 - 558 (1980).

The author presents the interpretation of masers which operate in the circumstellar envelopes about oxygen-rich Mira variables. Most of them also apply to the masers associated with oxygen-rich supergiants.

112.035 **Pumping mechanisms of OH masers.**
A. Omont, J. Guibert, S. Guilloteau, V. Bujarrabal, Nguyen-Q-Rieu.
Interstellar molecules, (see 012.033), p. 559 - 564 (1980).

The authors analyze in detail the pumping of 18 cm OH masers by the overlap of far infrared lines in H II/OH regions and in circumstellar shells. The authors present some results of a model of pumping of H II/OH masers by thermal overlap. Other types of pumping, mainly by collisions with H or by near infrared radiation, are possible if certain conditions are met for the physical and astrophysical parameters.

112.036 **Circular polarization from scattering by circumstellar grains.** A. Shafter, M. Jura.
Astron. J., Vol. 85, 1513 - 1519 (1980).

The authors suggest that observations of circular polarization can be a valuable tool for the study of circumstellar nebulae. Here they present a simple model to predict the percentage of circular polarization produced by scattering from the dust contained within such nebulae, and they find that a significant component of circular polarization can be produced in nebulae which have a net linear polarization. As examples, the authors discuss possible observations of the red supergiants μ Cep and α Ori.

112.037 **o Andromedae.**
IAU Circ., No. 3493 (1980).

112.038 **μ Centauri.**
IAU Circ., No. 3495 (1980).

112.039 **W Hydrae.**
IAU Circ., No. 3532 (1980).

112.040 **R Aquarii.**
IAU Circ., No. 3535 (1980).

112.041 **Expandierende Hüllen in Doppelsternsystemen.**
W. Neutsch.
Veröff. Astron. Inst. Bonn, Nr. 93, 52 pp. (1979).

A physical model for the structure of the envelope of the binary HD 152270 (consisting of a main sequence star and a Wolf Rayet companion) is constructed. The equations of motion of atoms in the neighbourhood of two hot stars are discussed in detail. It is found that in the limit of very high temperature of at least one of the components Euler's solution of a particle's motion under the influence of two fixed centers (neglecting Coriolis and centrifugal forces) is – to a certain degree – quite adequate to understand the gross features of such an envelope. Results of a detailed and more exact physical model which takes into account gravitation, radiation pressure, line absorption and the local variations of the radiation field due to differences in the configuration of the stars as seen from an emitting atom are given and compared with the measurements for a number of phases.

112.042 **The size of a Wolf-Rayet star's dust shell measured by speckle interferometry.**
D. A. Allen, J. R. Barton, P. T. Wallace.
Anglo-Australian Obs. Prepr. No. 136, 8 pp. (1980).

The authors have resolved the WC 9 star Ve2-45 at a wavelength of 2.2 μm. The shell of circumstellar dust is optically thin, and has a characteristic radius of 0.04 arcsec. This is in agreement with simple models of the dust shell, and will help to constrain more elaborate models, when they become available.

112.043 **Observations of stellar winds from cool stars.**
W. Hagen.
Smithsonian Astrophys. Obs., Spec. Rep. 389, (see 012.038), p. 143 - 152 (1980).

112.044 **A search for atomic hydrogen from evolved stars and planetary nebulae.**
B. Zuckerman, Y. Terzian, P. Silverglate.
Astrophys. J., Vol. 241, 1014 - 1020 (1980).

With the 305 m Arecibo antenna, the authors searched for 21 cm wavelength radiation from atomic hydrogen in expanding envelopes around a total of 13 red giants and planetary nebulae. For some stars upper limits on the mass of atomic hydrogen in the circumstellar envelopes are a few percent of the total envelope mass estimated from other considerations. No 21 cm radiation was detected that is directly associated with the planetary nebulae NGC 6210, NGC 6572, NGC 6720, and NGC 2346. Possible 21 cm absorption of the continuum radiation from NGC 6572 by foreground hydrogen would suggest a distant location for this nebula.

112.045 **Line formation in winds with enhanced equatorial mass-loss rates and its application to the Wolf-Rayet star HD 50896.** W. M. Rumpl.
Astrophys. J., Vol. 241, 1055 - 1069 (1980).

The author proposed a simple model for a non-spherically symmetric wind. In particular, he assumed that, owing to rotation, a star's mass loss is greater at its equator than at its poles. Using this model, he demonstrated the effects on line formation of several parameters: the velocity distribution of the wind, the amount of enhancement of the equatorial mass-loss rate relative to the polar one, the orientation of the star-wind system with respect to a distant observer.

112.046 **Recent results on red giant circumstellar shells.**
R. E. S. Clegg.
Variability in stars and galaxies, (see 012.044), p. E.2.1 - 2.8 (1980).

Results from the last 1-2 years on the shells of gas and

dust surrounding red giant stars are reviewed. The subject areas covered include the circumstellar SiO masers, microwave spectral lines and their excitation, infrared spectroscopy and circumstellar chemistry.

112.047 **Detection of CO emission at 1.3 millimeters from the Betelgeuse circumstellar shell.**
G. R. Knapp, T. G. Phillips P. J. Huggins.
Astrophys. J., Lett., Vol. 242, L25 - L28 (1980).

The paper reports the detection of millimeter-wave molecular emission from Betelgeuse in the CO $J = 2 - 1$ line at 230 GHz. The observations suggest that the mass-loss rate for this star is low and that the CO emission is optically thin, and provide a measurement of the stellar systemic velocity and the envelope outflow velocity.

112.048 **The linear shell diameter of IRC + 10011.**
P. R. Jewell, J. C. Webber, L. E. Snyder.
Astrophys. J., Lett., Vol. 242, L29 - L31 (1980).

The line-of-sight diameter of the circumstellar OH shell about IRC + 10011 has been measured at $(6.6 \pm 1.4) \times 10^{16}$ cm. This size was determined from the phase relation between variations in the 1612 MHz OH maser emission originating from opposite sides of the shell. The measurement is hence independent of any assumptions about the distance to the source and constitutes the first direct evaluation of the size of such an object. The linear extent of the maser region extrapolated from this measurement provides an independent check to VLBI results.

112.049 **Empirical ionization fractions in the winds and the determination of mass-loss rates for early-type stars.**
H. J. G. L. M. Lamers, R. Gathier, T. P. Snow.
Astrophys. J., Lett., Vol. 242, L33 - L36 (1980).

From a study of the UV lines in the spectra of 25 stars from O4 to B1 the authors derived the empirical relations between the mean density in the wind and the ionization fractions of O VI, N V, Si IV, and C III. They derived a simple relation between the mass-loss rate and the column density of any of these four ions. This relation can be used for a simple determination of the mass-loss rate from O4 to B1 stars;

112.050 **Infrared photometry of HDE 226868 (Cyg X-1) from 2.3 to 10μ: mass loss rate.**
P. Persi, M. Ferrari-Toniolo, G. L. Grasdalen, G. Spada.
Astron. Astrophys., Vol. 92, 238 - 241 (1980).

Infrared photometry between 2.3 and 10μ, shows an ionized gaseous envelope surrounding the supergiant OB star HDE 226868, the optical counterpart of the X-ray binary system Cyg X-1. The mean extent of the envelope at 10μ is ~ 1.3 R*. The mass loss rate of HDE 226868 is 3.5×10^{-6} $M_\odot$ y^{-1}. This value of the mass loss rate is sufficient to account for the X-ray emission of Cyg X-1, under the assumption that the X-ray source is powered by accretion from a stellar wind.

112.051 **Statistics of correlation between the velocity separation of double peaks of maser sources and associated variable's period, and mass loss from variables.**
Y. Fan, J. Sun.
Acta Astron. Sinica, Vol. 21, 379 - 388 (1980). In Chinese.

A study of the relation between velocity separation of double peaks ΔV of maser sources and associated variable's period Π is presented. A linear correlation between ΔV_{OH} (or ΔV_{SiO}) and Π has been obtained, the correlation coefficient r is greater than 0.9. It indicates that the model of expanding shell of a maser star is correct. The distribution of velocity of matter flow is used to calculate the position of the SiO shell.

112.052 **Stellar winds in close binary systems.**
R. S. Polidan, G. J. Peters.
Publ. Astron. Soc. Pacific, Vol. 92, 549 (1980). – Abstract.

112.053 **The rate of mass loss and variations in the wind from the Be star δ Centauri.**
T. P. Snow, Jr., W. R. Oegerle, R. S. Polidan.
Astrophys. J., Vol. 242, 1077 - 1082 (1980).

The paper reports a change found in the wind of the Be star δ Cen. This change, resulting in the disappearance of any evidence of what had been a substantial wind, took place in a time of 3 yr. Optical and previous ultraviolet data on δ Cen are summarized, and the Copernicus observations are described. A quantitative assessment of the rate of mass loss before and after the change is made and a discussion of the photospheric line profiles is presented.

112.054 **The structure of OH masers around late-type stars.**
P. F. Bowers, M. J. Reid, K. J. Johnston, J. H. Spencer, J. M. Moran.
Astrophys. J., Vol. 242, 1088 - 1101 (1980).

Results of very long baseline interferometric observations at 1612 MHz are presented for 17 distant ($\gtrsim 2$ kpc) OH/IR stars of the type commonly associated with cool Mira variables and supergiant stars. From comparison of the cirumstellar shell structures of nearby and distant sources, the authors show that the extent of the 1612 MHz masing regions in OH/IR stars is always $\gtrsim 100$ AU. They suggest that the mass-loss rate plays an important role in determining the total size of the masing region. They also suggest that the formation of OH in the outer envelopes by ultraviolet dissociation of H_2O is unlikely to be important for OH/IR stars with low mass-loss rates. In addition, the percentage of sources with evidence for unresolved ($\lesssim 0\farcs05$) structure is determined.

On the structure and composition of the Wolf-Rayet atmospheres. See Abstr. 064.001.

Self accreting stellar winds.
See Abstr. 064.003.

An envelope model of Phi Persei.
See Abstr. 064.006.

Chemical structure of circumstellar shells.
See Abstr. 064.012.

Atmospheric structure of a carbon star with thin circumstellar shell. See Abstr. 064.042.

X-ray emission from the winds of hot stars.
See Abstr. 064.048.

Chromospheres, coronae, and mass loss in stars hotter than the sun. See Abstr. 064.054.

Winds from supergiants. See Abstr. 064.062.

Mass loss from red giants: some effects on the interstellar medium. See Abstr. 064.082.

On the nature of MWC 349. See Abstr. 114.039.

8–13 μm spectrophotometry of V1016 Cyg and the shape of the 'silicate' feature. See Abstr. 114.049.

Variability in early-type stars.
See Abstr. 114.130.

Scanner observations of Gamma Cassiopeiae.
See Abstr. 114.137.

Polarimetry of emission lines as a probe of stellar winds. See Abstr. 116.016.

Circular and linear polarization in A-type shell stars. See Abstr. 116.025.

A search for linear polarization variability in pole-on Be stars. See Abstr. 116.029.

Observational determination of the gravity darkening exponent and bolometric albedo for close binary star systems. See Abstr. 117.015.

SS 433 as a prototype of astrophysical jets. See Abstr. 117.066.

Die Hülle der Wolf-Rayet-Komponente des Doppelsterns HD 152270. See Abstr. 120.019.

The spectrum of the nebulosity around the symbiotic long-period variable, R Aquarii. See Abstr. 122.009.

Stellar absorption lines of NH_3 in variable stars. See Abstr. 122.040.

Photoelectric activity of the shell star *o* And: new observations and discussion. See Abstr. 122.082.

Red variables: some current topics. See Abstr. 122.156.

Infrared emission by dust grains near variable primary sources. II. A model for infrared novae. See Abstr. 124.007.

Theoretical investigation of variable infrared emission by cosmic dust sources. See Abstr. 131.151.

The interaction of T-Tauri stars with molecular clouds. See Abstr. 131.193.

The extinction of HD 200775 by dust in NGC 7023. See Abstr. 131.260.

Sternenstaub. See Abstr. 131.273.

An infrared candidate for OH 205.1 – 14.1. See Abstr. 133.004.

Simultaneous far-infrared, near-infrared, and radio observations of OH/IR stars. See Abstr. 133.007.

Maser emission from infrared stars. I. New OH and H_2O observations. See Abstr. 133.008.

An upper limit to the mass loss rate from the nuclei of planetary nebulae. See Abstr. 135.019.

Mass loss and stellar wind in massive X-ray binaries. See Abstr. 142.081.

X-ray observations of stellar coronae and winds. See Abstr. 142.083.

A search for X-ray binary stars in their quiescent phase. See Abstr. 142.126.

Erratum

112.901 **Addendum: "Detection of mass loss in stellar chromospheres"**
[Astrophys. J., Vol. 238, 221 - 228 (1980)].
R. E. Stencel, D. J. Mullan.
Astrophys. J., Vol. 240, 718 (1980). – See Abstr. 27.112.018.

113 Photometric Properties

113.001 **Infrared photometry of southern early-type stars.** D. C. B. Whittet, I. G. van Breda.
Mon. Not. R. Astron. Soc., Vol. 192, 467 - 480 (1980).

Infrared photometry tied to the *JHKL* (1.2–3.5 μm) system is presented for 229 southern early-type stars. Data for stars of low reddening are used to determine intrinsic visual – IR colour indices. The E_{V-K}/E_{B-V} diagram is used to evaluate the ratio of total selective extinction ($R = A_V/E_{B-V}$). A mean value of $R = 3.12 \pm 0.05$ is found for stars close to the galactic plane, but a higher value ($R \sim 4.0$) applies to the Orion and Sco–Oph regions. Infrared two-colour diagrams are used to investigate the occurrence of infrared excess emission in different classes of shell stars.

113.002 **HD 219150: a star with a remarkable ultraviolet excess.** J. D. Fernie, C. T. Bolton.
Publ. Astron. Soc. Pacific, Vol. 92, 328 - 332 (1980).

Attention is drawn to HD 219150, an F0 V star with a one-quarter magnitude excess in $(U-B)$. Both *UBVRI* and four-color-Hβ observations are consistent in showing the reality of this excess. Moreover, independent observations down to λ 1565 show strong and complicated excesses at more extreme ultraviolet wavelengths. Spectrograms at both classification and higher dispersions show conclusively that the star is not metal-weak and that it appears entirely normal in the blue-green region of the spectrum. However, the measured β-index appears slightly weak, but this is not supported by the profile of Hβ on the 12 Å mm^{-1} plates.

113.003 **UBV observations of two helium weak stars.** J. F. Dean.
Mon. Notes Astron. Soc. South. Africa, Vol. 39, 13 - 18 (1980).

UBV observations are presented for the He weak variable HR 1441 and the He weak candidate HD 70084.

113.004 **Fainter standards for VRI photometry in the E regions.** A. W. J. Cousins.
Mon. Notes Astron. Soc. South. Africa, Vol. 39, 22 - 32 (1980).

New measures of V, $R-I_{KC}$ and $V-I_{KC}$ are presented for 311 E region stars, three-quarters of them between V magnitudes 7.0 and 9.0.

113.005 **Mehrfarben-Photometrie: Werkzeug der Stellarstatistik.**
H.-A. Ott.
Sterne Weltraum, Jahrg. 19, 295 - 301 (1980).

113.006 **The IR variability of SS433.** A. B. Giles, A. R. King, R. F. Jameson, M. R. Sherrington, J. H. Hough, J. A. Bailey, E. C. Cunningham.
Nature, Vol. 286, 689 - 691 (1980).

The authors report the results of a one-year monitoring programme of SS433 at IR and visual wavelengths. The most striking feature of these observations is the near constancy of the IR colours while the source's flux varies over a range of ~1.2 mag. There is no evidence for the 13.1-day period.

113.007 **The optical light curve of the AM Herculis system 2A 0311–227.**
M. G. Watson, S. K. Mayo, A. R. King.
Mon. Not. R. Astron. Soc., Vol. 192, 689 - 696 (1980).

Optical *V*-band light curves of this AM Herculis binary are presented. These show two equal minima and two unequal maxima in each binary cycle as well as strong flickering on a short time-scale. The light curves are interpreted in terms of a model involving a phase-locked white dwarf with two active magnetic poles; unusually, the system appears to undergo star-by-star eclipses.

113.008 **Recent results of BVRI photometry of late-type dwarf stars in nearby clusters and in the solar neighborhood.** E. W. Weis, A. R. Upgren.
Bull. American Astron. Soc., Vol. 12, 524 (1980). – Abstract.

113.009 **Measurement of the absolute flux from Vega in the *K* band (2.2 μm).**
M. J. Selby, D. E. Blackwell, A. D. Petford, M. J. Shallis.
Mon. Not. R. Astron. Soc., Vol. 193, 111 - 114 (1980).

A method is described for determining absolute stellar flux in the infrared, based on a direct comparison between a star and a furnace at a known temperature. Observations have been made of Vega in the *K* band (2.2 μm) using the Tenerife flux collector. The absolute flux from this star at the Earth, reduced to the wavelength 2.20 μm, is found to be 0.375×10^{-9} W m^{-2} μm^{-1} with an uncertainty of 8 per cent.

113.010 ***uvbyg$_1$g$_2$*-photometry of some fainter Ap-stars.**
H. M. Maitzen.
Astron. Astrophys., Vol. 89, 230 - 233 (1980).

$uvbyg_1g_2$ photometry (Maitzen, 1976) in a slightly modified system is presented and its performance compared with previous photometry. The resulting peculiarity parameter values Δa are compared with published Geneva system $\Delta(V1-G)$ values. As two stars of the sample are markedly reddened the problem of dereddening $(b-y)$ for Ap stars had to be discussed. Reddening corrected Δa-values were obtained.

113.011 **On zero-point drift in astronomical photoelectric photometry.** Z. Kviz.
Proc. Astron. Soc. Australia, Vol. 3, 275 - 278 (1978).

113.012 **A search for Hα-emission objects in a region in Ara.**
E. I. Vega, M. Rabolli, J. C. Muzzio, A. Feinstein.
Astron. J., Vol. 85, 1207 - 1217 (1980).

A search for Hα-emission objects covering an area of 25 deg^2 in the Ara region of the Milky Way resulted in the discovery of 124 objects. The limiting magnitude of the survey is about $R=14.5$. More than two-thirds of the objects were not previously known. *UBV* photographic photometry was also obtained for objects brighter than $V=15.4$, $B=16.3$, and $U=16.4$; the photometric data allow to make a preliminary discussion of the early-type objects which suggests that most of them lie beyond the Sagittarius arm.

113.013 **Intrinsic colours of MK types in the Geneva photometric system.** G. Meylan, M. Python, B. Hauck.
Astron. Astrophys., Vol. 90, 83 - 87 (1980).

The authors have used the catalogue of selected MK types established by Jaschek (1978) in order to determine intrinsic colours in the Geneva photometric system. Their results are given for O to M stars of luminosity classes V to Ia. In addition, the authors give relations between their indicator of spectral type, (B_2-V_1), and that of the *UBV* photometry, $(B-V)$. These relations were obtained for the various luminosity classes.

113.014 **Ultraviolet photometry with the Astronomical Netherlands Satellite (ANS): faint blue stars in the halo.** K. S. de Boer, P. R. Wesselius.
Astron. J., Vol. 85, 1354 - 1360 (1980).

Blue stars at high galactic latitudes have been observed with the UV telescope on board ANS. In this paper the authors discuss a subset of the collected data pertaining to the cooler stars. Most of them have energy distributions in general agreement with the visual spectral type. One star is exceptionally blue, and of seven possible horizontal-branch stars, two have UV energy distributions distinct from main-sequence stars in the sense that they have an excess at 1550 Å and a large Balmer jump.

113.015 **Spectrophotometry of B, A, and F stars. II.** S. J. Adelman, D. M. Pyper, R. E. White.
Astrophys. J., Suppl. Ser., Vol. 43, 491 - 499 (1980).
Optical region energy distributions of 35 normal B, A, and F stars are presented consistent with the Hayes-Latham calibration of Vega. Upper limits on the photometric accuracy are derived by a comparison of observed and synthesized $u - b$ and $b - y$ indices for each observer. Effective temperatures are found using the empirical calibration of Code and his colleagues and the fully line blanketed solar composition model atmospheres of Kuruzc.

113.016 **Ultraviolet photometry from the Orbiting Astronomical Observatory. XXXIV. Filter photometry of 531 stars of diverse types.**
A. D. Code, A. V. Holm, R. L. Bottemiller.
Astrophys. J., Suppl. Ser. Vol. 43, 501 - 545 (1980).
This paper tabulates ultraviolet magnitudes for 531 stars observed with the Wisconsin Experiment Package on OAO 2. The tabulation includes previously published data all reduced to a uniform magnitude system. Eleven magnitudes are presented designated by their centroid wavelengths. The tabulated magnitudes are on an absolute energy basis.

113.017 **The visibility of two faint star sequences on the Palomar and ESO sky surveys.** G. L. White.
Mon. Not. R. Astron. Soc., Vol. 193, 329 - 332 (1980).
Stars with published electronographic *B* magnitudes fainter than the accepted limiting magnitudes of both the Palomar and ESO B sky surveys are clearly visible on both surveys. An independent estimate of the magnitudes of these stars based on another deep sequence indicates a possible inconsistency in the electronographic magnitudes.

113.018 **Photoelectric observations of a magnitude and color variable object in Perseus.**
Y.-h. Chu, C.-j. Wang, X.-y. Yang, C.-l. Xia, X.-z. Shi.
Acta Astron. Sinica, Vol. 21, 252 - 256 (1980). In Chinese.
After seven nights of $U' B' V'$ photoelectric observations this object looks like an object with shell activity of an early-type star in its center. It varies with a more violent U' band than B' and V'.

113.019 **Changement de systѐme photométrique dans l'Annuaire du Bureau des Longitudes.**
F. Ochsenbein.
Astronomie, Vol. 94, 501 - 505 (1980).

113.020 **Astrophysical parameters from four-color photometry.** A. G. D. Philip.
Bull. Inf. Cent. Données Stellaires, No. 19, p. 2 - 10 (1980).

113.021 ***uvbyβ* photometry of Equatorial and Southern bright stars. II.** A. Heck, J. Manfroid.
Astron. Astrophys., Suppl. Ser., Vol. 42, 311 - 318 (1980).
uvbyβ photometric observations are reported for 330 Equatorial and Southern bright stars. Complementary observations of stars suspected of variability from previous runs are also discussed.

113.022 ***UBV* sequences in three northern Milky Way regions and a comment on the interstellar extinction around $l = 90°$.** H. Lindgren, K. Bern.
Astron. Astrophys., Suppl. Ser., Vol. 42, 335 - 346 (1980).
Photoelectric *UBV*-data are given for 317 stars in three different Milky Way regions. The interstellar extinction for a Cygnus field is investigated by a study of colour excesses and by a search for faint galaxies.

113.023 **Study of the variable F-type supergiants HD 161796 and HD 163506 in radial velocity and photometry.**
G. Burki, M. Mayor, F. Rufener.
Astron. Astrophys., Suppl. Ser., Vol. 42, 383 - 389 (1980).
HD 161796 is discovered to be variable in light and radial velocity. The variations are quite normal for a supergiant star, i.e. they can be described as a pseudo-periodic phenomenon, with a characteristic time of 54 d. The variability of HD 163506 is more complex. On the one hand relatively regular cycles of variation of period between 50 and 75 d were observed in light and radial velocity. But quasi-erratic variations are also observed in radial velocity when simultaneous light variations are quite smooth. The correlation function given by the spectrophotometer CORAVEL reveals that in both supergiants a large number of photospheric lines from neutral metals have a P Cygni profile. This new fact gives photospheric evidence of mass loss from these two supergiants. One of these stars (HD 163506) is known for exhibiting variable P Cygni profiles of the circumstellar lines, with a terminal velocity of 150-200 km/s (Sargent and Osmer, 1969). A discussion of the luminosity of these two supergiants is presented. It is concluded that these FIa-type stars are probably not "normal", massive, supergiants of population I.

113.024 **HD 44179.**
IAU Circ., No. 3532 (1980).

113.025 **AM Herculis.**
Yamamoto Circ., No. 1938 (1980).

113.026 **Wide- and narrow-band photometry of stars in a field around Collinder 135.**
J. J. Clariá, S. O. Kepler.
Publ. Astron. Soc. Pacific, Vol. 92, 501 - 513 (1980).
A field of about 10^4 square arc minutes around the cluster Cr 135 is studied using *UBV*, Hα, and Hβ photometries. The photometric data do not favor the existence of Cr 135 as a real open cluster. Only eight B-type stars lie on a very well-defined main sequence in the *UBV* photometric diagrams. These eight stars – which are shown not to be gravitationally bound – are located at a mean distance of 630 pc with a foreground color excess amounting to $E(B - V) = 0^m.06$. These stars, together with most of the B stars in the region, can be considered probable members of the large, old, Pup-Can association.

113.027 **Photometry and polarimetry of the extreme blue straggler K1211 in NGC 7789.**
M. Breger, J. C. Wheeler.
Publ. Astron. Soc. Pacific, Vol. 92, 514 - 517 (1980).
A discrepancy in the literature has led the authors to redetermine the $(B - V)$ color index of the blue straggler K1211 in the old open cluster NGC 7789. A reanalysis of the cluster reddening leads them to adopt $E(B - V) = 0^m.27$, and $(B - v)_O = -0.08$ for K1211. If this star is evolving normally off the main sequence it has a mass $M \sim 3\,M_\odot$ in a cluster with turnoff mass $\sim 1.5_\odot$. Thus K1211 is just at or slightly beyond the limiting mass allowed in the simplest scenario in which blue stragglers are produced by binary mass transfer. Several cluster stars and K1211 were checked for polarization. The polarization is due to a uniform cloud along the line of sight which spans the cluster.

113.028 **On the variability of central stars of two planetary nebulae.**
B. E. Zhiljaev (*Zhilyaev*), A. G. Totochava.
Inf. Bull. Variable Stars, No. 1848, 3 pp. (1980).

113.029 **14 Lac: a new light variable Be star.** L. Mantegazza.
Inf. Bull. Variable Stars, No. 1886, 2 pp. (1980).

113.030 **Photographic observations of SS 433.** W. Wenzel.
Mitt. Veränderl. Sterne (MVS), Band 8, 141 - 142 (1980).

Sections of the lightcurve, observed on Sonneberg plates, are given. No persistent periodicity could be found.

113.031 **A search for ultraviolet-excess objects.** T. Noguchi, H. Maehara, M. Kondo.
Ann. Tokyo Astron. Obs., Second Ser., Vol. 18, 55 - 70 (1980).

A search for UV-excess objects has been undertaken with the use of the 105-cm Schmidt telescope at the Kiso Observatory by means of *UGR* three-image method. A general description of the method is given, and positions, magnitudes, and color indices of detected 588 objects covering 22 fields (600 square degrees) are listed. The brightness of the objects ranges from 12.5 to 18.5 magnitudes. About 13 percent of all are identified with white dwarfs or quasars.

113.032 **Photometric standard stars for the UBV and $(RI)_{KC}$ systems.**
J. W. Menzies, R. M. Banfield, J. D. Laing.
South African Astron. Obs. Circ., Vol. 1, 149 - 162 (1980).

The authors present a table of magnitudes and colours for standard stars in the E- and F-regions and in the SMC and LMC. Data have been combined from several sources and the list is currently used at SAAO, Sutherland for UBV $(RI)_{KC}$ photometry.

113.033 **VRI photometry of nearby stars.** A. W. J. Cousins.
South African Astron. Obs. Circ., Vol. 1, 166 - 171 (1980).

Observations have been made in the VRI_{KC} system of nearly all the stars in Gliese's "Catalogue of Nearby Stars" that are brighter than $9^m.0$ and south of $+10°$ declination.

113.034 **Photoelectric standards of intermediate brightness in the E-regions. I. UBV observations.**
J. W. Menzies, J. D. Laing.
South African Astron. Obs. Circ., Vol. 1, 175 - 182 (1980).

The authors present the results of photoelectric UBV photometry of 175 stars selected from the E-region lists of Cousins and Stoy (1962). The stars in each E-region were chosen to cover a wide range in colour, and most of them have $8^m.5 \leqslant V \leqslant 11^m.5$. They are suitable for use as standard stars with pulse counting photometers on large telescopes.

113.035 **VRI photometry of southern stars.** A. W. J. Cousins.
South African Astron. Obs. Circ., Vol. 1, 234 - 256 (1980).

Mean values of $V\text{-}R_{KC}$ and $V\text{-}I_{KC}$, measured with a Quantacon photomultiplier, are given for 1425 stars south of $+10°$ declination. These means have been compared with published photometry in other similar systems to obtain colour transformations.

113.036 **Rotational variations in the nonflaring optical and X-ray fluxes of YZ CMi.**
B. R. Pettersen, S. Kahler, L. Golub, G. S. Vaiana.
Smithsonian Astrophys. Obs., Spec. Rep. 389, (see 012.038), p. 113 - 117 (1980).

Rotational variations in V band fluxes from the dMe star YZ CMi are compared with variations in the quiescent X-ray fluxes observed with Einstein. The V band modulation due to presumed starspots is 12%, but the soft X-ray fluxes vary by no more than 50%, substantially less than expected from a simple extrapolation of the solar case.

113.037 **Infrared observations of the WC 5 Wolf-Rayet star HD 115473.** P. M. Williams, D. A. Allen.
Observatory, Vol. 100, 202 - 206 (1980).

Near-infrared photometry shows no evidence of the circumstellar dust emission which previously-reported infrared and ultraviolet observations of HD 115473 might lead one to expect. The two-micron spectrum is dominated by the 2.06 μm He I emission line which carries about one-third of the K-filter flux, while about 13 per cent of the H-filter flux is carried in strong lines at 1.74 and 1.80 μm identified with C IV.

113.038 **UV photometry of Am stars.**
C. van't Veer-Menneret, R. Faraggiana, C. Burkhart.
Astron. Astrophys., Vol. 92, 13 - 21 (1980).

UV properties of Am stars are investigated on the basis of S2/68 UV fluxes and compared with the behaviour of a selection of standard stars and with theoretical data. A very metallicity sensitive UV index, $\delta(m_{23} - V)_0$, independent of luminosity has been introduced. The important fact that Am behaviour in the $(m_{i,UV} - m_{j,UV})_0$ colours diagrams cannot be confused with a luminosity effect is brought out and is well reproduced by the computed colours.

113.039 **Line blanketed model atmospheres of Ap stars. IV. *UBV* colors and effective temperatures of Ap stars.** K. Stępień, H. Muthsam.
Astron. Astrophys., Vol. 92, 171 - 174 (1980).

UBV data for Ap stars at the distance less than 100 pc show that hotter stars lie on or close to the main sequence in the two-color diagram but stars with $B-V \gtrsim +0.05$ lie below it. Model atmospheres of "standard peculiar" and individual peculiar stars confirm this behaviour. It is shown that unreddened $B-V$ of an Ap star can be used to find T_e from normal relation, if it is within the range $+0.02 \gtrsim B-V \gtrsim -0.08$.

113.040 **Properties of Am stars in the Geneva photometric system.** B. Hauck, A. Curchod.
Astron. Astrophys., Vol. 92, 289 - 295 (1980).

The authors have investigated the properties of Am stars in the Geneva system. They give a table containing the observed photometric data and the data M_v and Δm_2, the metallicity parameter. The metallicity is more important for the cooler Am stars than for the hotter, while – as already remarked by other authors – the metallicity decreases when rotational velocity increases. Other stars for which Δm_2 can also be important are reviewed.

113.041 **Infrared observations of southern bright stars.** D. Engels, W. A. Sherwood, W. Wamsteker, G. V. Schultz.
ESO Sci. Prepr. No. 130, 13 pp. (1980). – Submitted to Astron. Astrophys., Suppl. Ser.

113.042 **A search for variability of 8 stars with anomalous helium content.**
B. Vetö, W. Schöneich, Yu. S. Rustamov.
Astron. Nachr., Band 301, 317 - 326 (1980). In German.

For eight stars with anomalous helium content a search for periodic lightvariability was carried out. Three of them HD 178993, HD 207538, and HD 209339 did not show any detectable variability during the period of observations. For HD 184927, the period, obtained by Bond, Levato, was confirmed. For HD 191980, HD 208266, and HD 217833 photometric variability was detected and the periods were obtained.

113.043 **Vector space methods of photometric analysis. II. Refinement of the MK grid for B stars.**
D. Massa.
Astron. J., Vol. 85, 1644 - 1650 (1980).

This paper discusses systematic errors which can arise from exclusive use of the MK system (especially when supergiant stars are included) to determine reddening. It is shown that implementation of *uvby*, β photometry to refine the qualitative MK grid substantially reduces stellar mismatch error. A working definition of "identical" *uvby*, β types is introduced and the relationship of *uvby* to $B - V$ color excesses is determined. Next, the MK types are assigned numerical values and employed to check for possible systematic errors in the photoelectric types. Finally, a comparison is made of the hydrogen-based *uvby*, β types to the MK system which is based on He and metal lines. This analysis reveals a small correlated effective temperature-luminosity error in the MK system for the early B stars and a breakdown of the MK-luminosity criteria for the late B stars.

113.044 **Vector space methods of photometric analysis. III. The two components of ultraviolet reddening.**
D. Massa.
Astron. J., Vol. 85, 1651 - 1662 (1980).

This paper investigates the wavelength dependence of interstellar extinction in the ultraviolet-wavelength range observed by the TD-1 satellite. The sets of identical stars employed to find reddening are determined more precisely than before and consist only of normal, nonsupergiant stars. A multivariate analysis of variance technique in an unbiased coordinate system is used to obtain the wavelength dependence of reddening. This is an optimal technique and reveals two distinct variances in the UV spectra of "identical" stars well beyond the expected stellar mismatch error. The author proceeds on the assumption that both of the observed variances are of interstellar origin. By imposing very general observational constraints on which linear combinations of the two reddening curves are possible, the following conclusions are drawn. First, no acceptable combination is completely gray in the visible. Second, no allowed combination is compatible with the currently available theoretical forms of graphite extinction. Third, a relatively featureless combination is possible which, for fixed $B - V$, accounts for most of the extinction in the range $5.5\,\mu^{-1} < \lambda^{-1} < 7.0\,\mu^{-1}$. Fourth, there is no a priori reason for hypothesizing more than two species of grains to explain ultraviolet extinction.

113.045 **Calibration of photometric abundance determinations.** G. Cayrel de Strobel.
ESO workshop on methods of abundance determination for stars, (see 012.021), p. 49 - 56 (1980).

113.046 **A photometric system for limiting *JF* photography I: the primary standards.**
W. J. Couch, E. B. Newell.
Publ. Astron. Soc. Pacific, Vol. 92, 746 - 751 (1980).

The authors present photoelectric observations of 90 standard stars in the southern E-regions ($\delta = -45°$). These standards define a *JF*-photometric system that is intended to match the photographic system based on the following combinations: *J* (IIIa-J + GG385), *F* (IIIa-F + RG630). The authors use their observations to derive color-dependent transformations that can be used to estimate *JF* values for stars observed in the Cape-Cousins *B* and *R* bands.

113.047 **On the reduction of photometric systems.**
A. P. Varakin.
Astron. Tsirk., No. 1108, p. 5 - 7 (1980). In Russian.

113.048 **Transformations between magnitudes of the photometric systems RGU and VILGEN.**
V. Straižys, A. Bogdanovič.
Bull. Vilnius Astron. Obs., Nr. 55, p. 20 - 26 (1980). In Russian.

Diagrams and formulas relating magnitudes of the photometric systems RGU and VILGEN are computed. It is shown that the system U, Y, S consisting of three magnitudes of the VILGEN system has some advantages in comparison with the RGU system.

Vilnius photometric catalogue (magnetic tape). See Abstr. 002.002.

Photometric catalogues:present and future. See Abstr. 002.015.

Catalogue général d'étoiles de type O. Données spectroscopiques et photométriques. Bande magnétique et listage. See Abstr. 002.023.

Catalogs of proper motion stars. II. Stars brighter than visual magnitude 15 and south of declination +30° with annual proper motion between 0.″7 and 1.″0. See Abstr. 002.031.

Catalogue of extinction data of 12547 O . . . F stars, galactic clusters and δ-Cephei stars. See Abstr. 002.039.

UBV data 1976 - 1979. See Abstr. 002.040.

A compilation of Balmer lines photometric data. See Abstr. 002.041.

Fortran programmes for reducing uvby-Hβ photometry and for deriving physical properties of stars. See Abstr. 021.032.

On an approximation of photoelectric profiles of star images. See Abstr. 031.526.

The determination of the own photometric system and the passing at the international system. See Abstr. 031.530.

Fast photometry – new facilities at La Silla. See Abstr. 031.579.

Atmospheres for hot, high-gravity stars. I. Pure hydrogen models. See Abstr. 064.011.

Circumstellar absorption and intrinsic colours of massive stars. See Abstr. 112.019.

Spectroscopic and photometric observations of the object He2-442. See Abstr. 114.018.

Infrared spectrophotometry of SS 433. See Abstr. 114.064.

Correlations between line-profile and photometric variations in the B2 IV [*e*] star HD 45677. See Abstr. 114.072.

The sun among the stars. III. Energy distributions of 16 northern G-type stars and the solar flux calibration. See Abstr. 114.088.

Spectrophotometry of peculiar B and A stars. VI. HD 32633, HD 34452 and HD 133029. See Abstr. 114.094.

Spectrophotometry of peculiar B and A stars. VII. HD 6164, HD 8855, HD 11187, HD 171782, HD 190068, HD 200311, and HD 220147. See Abstr. 114.095.

Studies of the Carina Nebula. III. The spectral energy distribution of the very hot and massive star HD 93250. See Abstr. 114.100.

Analyse d'une des caractéristiques des étoiles Ap: la dépression à 5200 Å. See Abstr. 114.117.

The absolute calibration of stellar spectrophotometry. See Abstr. 114.119.

The distribution of MK spectral types in photometric boxes of the four-color system. See Abstr. 114.122.

Scanner observations of Gamma Cassiopeiae. See Abstr. 114.137.

A plot of *UBV* diagram. See Abstr. 115.007.

Relation between surface magnetic field intensities and Geneva photometry. See Abstr. 116.001.

The binary nature of the single-line Wolf-Rayet star EZ Canis Majoris = HD 50896. See Abstr. 117.010.

Long-term period changes and photometric orbital solution of RZ Draconis. See Abstr. 119.003.

Ultraviolet colors of W Ursae Majoris: gravity darkening, temperature differences, and the cause of W-type light curves. See Abstr. 119.049.

A search for northern hemisphere double mode cepheids – II. New *UBV* cepheid photometry. See Abstr. 122.006.

A discussion on three yellow variable supergiants in and near the Cepheid instability strip: V 810 Cen (=HD 101947), Tr. 27–102 (=HD 159378) and BL Tel (F), based on *VBLUW* photometry and the long-period Cepheids absence in the Galaxy. See Abstr. 122.026.

Outburst photometry of the dwarf nova SS Cygni. See Abstr. 122.058.

***ubv* observations of the δ Scuti star HD 181333.** See Abstr. 122.060.

Observational stability in Am stars. See Abstr. 122.069.

The variable, single-line Wolf-Rayet star HD 96548 with a low-mass companion. See Abstr. 122.085.

Color excesses of classical cepheids. II. See Abstr. 122.206.

UBV observations of FG Sagittae in 1979. See Abstr. 122.217.

A hot blue star near the center of the remnant of supernova A. D. 1006. See Abstr. 125.058.

Interstellar reddening towards the south galactic pole. See Abstr. 131.081.

The interstellar reddening law in the visible. See Abstr. 131.132.

On the variation of the colour excess in the Carina-Crux-Centaurus-Norma region of the Milky Way. See Abstr. 131.147.

Interstellare Verfärbung und Dreifarben-Photometrie. See Abstr. 131.272.

UBVRI photometry of stars in the field of dark clouds of Taurus. See Abstr. 131.311.

The galactic foreground reddening in the direction of the Magellanic Clouds. See Abstr. 131.317.

Results of synchronous observations of the peculiar object SS 433 in radio and optical ranges. See Abstr. 141.121.

Astrometry and high-speed photometry of an optical candidate for PSR 1913+16. See Abstr. 141.531.

Luminous stars beyond the solar circle: investigation of a galactic field at $l = 231°$. See Abstr. 152.006.

***uvbyβ* photometry of the young southern cluster NGC 3293 and comparison with other young clusters.** See Abstr. 153.003.

A *UBV* photometric study of the open cluster NGC 654. See Abstr. 153.019.

Five open clusters with suspected Wolf-Rayet type members. See Abstr. 153.021.

Infrared observations of red giants in globular clusters. See Abstr. 154.034.

The chemical evolution of the solar neighborhood. I. A bias-free reduction technique and data sample. See Abstr. 155.037.

Studies of luminous stars in nearby galaxies. VI. The brightest supergiants and the distance to M33. See Abstr. 158.168.

Studies of luminous stars in nearby galaxies. VII. The brightest blue stars in the spiral galaxies M101 and NGC 2403. See Abstr. 158.169.

Luminosities and temperatures of the reddest stars in three LMC clusters. See Abstr. 159.006.

Photoelectric photometry of stars in the Small Magellanic Cloud. See Abstr. 159.008.

Five-colour photometry of blue stars in the Magellanic-Cloud region. See Abstr. 159.015.

Erratum

113.901 **Erratum: 'Final catalogue of 229 photometric standards in UBV system near the Selected Areas 1-115'** [Acta Astron., Vol. 29, 177 - 186 (1979)]. E. Rybka. Acta Astron., Vol. 30, 218 (1980). – See Abstr. 25.113.070.

114 Spectra, Temperatures, Chemical Composition, etc.

114.001 **Spectrophotometry of peculiar B and A stars. IV. Secondary standards and the Ap stars HD 10783 and CU Virginis.** R. E. White, D. M. Pyper, S. J. Adelman. Astron. J., Vol. 85, 836 - 847 (1980) = Contrib. No. 328 Five College Obs.

Spectrophotometric measurements of two magnetic Ap stars, HD 10783 and CU Vir, are presented for λλ3300-7850. The use of the secondary standards in the data reductions is discussed. The λ4200 and λ5200 broad continuum features are present in HD 10783 but show no definite variation. The energy distribution of this star does not fit the predictions of any solar composition model with a main-sequence-type value of the gravity over the effective temperature range expected from its $u - b$ and $b - y$ colors. The Balmer jump fits a model of higher effective temperature than does the Paschen continuum, a circumstance which probably results from line blocking and flux redistribution from the ultraviolet.

114.002 **Near ultraviolet spectra of a group of early-type stars with Balmer emission.** A. E. Ringuelet. Mon. Not. R. Astron. Soc., Vol. 192, 399 - 407 (1980).

In the near ultraviolet region some absorption features in B stars are easily recognizable as criteria for assigning luminosity class. In the same region the characteristics displayed by Be stars suggest higher luminosities than in the photographic region. It appears that Fe II 2538–2548 Å should be useful for studying the outer layers of stellar atmospheres.

114.003 **Near infrared spectrometry of WC stars.** P. M. Williams, D. J. Adams, S. Arakaki, D. H. Beattie, J. Born, T. J. Lee, D. J. Robertson, J. M. Stewart. Mon. Not. R. Astron. Soc., Vol. 192, 25P - 30P (1980).

Low resolution spectra of Wolf–Rayet stars in the 1.4 to 2.5 μm region are shown to be dominated by He I and He II blends at 1.87 μm and, in the WC stars only, by a He I line at 2.058 μm. The WC stars also show strong unidentified features at 1.72 and 2.41 μm. The (H–K) differences between WN and WC stars are attributed to the influence of the emission lines.

114.004 **The identification of rare earths in the silicon star HD 187473.** H. Hensberge, C. R. Cowley, W. van Rensbergen, G. C. L. Aikman. Astron. Astrophys., Vol. 87, 369 - 372 (1980).

A twelve Å/mm blue spectrum of HD 187473, taken near phase of minimum light in uvby, has been analysed for element identifications. The silicon type classification (Bidelman, Mc Connell, 1973) is confirmed. The iron group spectra are weak, except for iron itself. Although blending is a serious problem, since rotational line broadening is important, some light and intermediate lanthanides have been readily identified. Strong lines of ionized europium have been found, while weaker stellar features have been attributed to La II and Pr II. Furthermore, there is statistical evidence for the presence of Ce II, Nd II, and Sm II. HD 187473 is the hottest peculiar star with a convincing evidence for ionized lanthanum.

114.005 **A radial-velocity study of the nitrogen O supergiant HD 105056.** N. R. Walborn, P. S. Conti, J.-M. Vreux. Publ. Astron. Soc. Pacific, Vol. 92, 284 - 287 (1980).

An investigation of the radial velocity of the ON supergiant HD 105056 has shown it to be variable with a small range (30–40 km sec^{-1}) and short time scale (a few days), but no satisfactory periodicity has been found in the variations. Evidence is found for a stellar wind of variable magnitude, and there are small absorption-line intensity changes in the spectrum possibly with a very short time scale ($\lesssim 1$ hour). Either the star is not a binary and the variations arise entirely in the stellar wind; or, if it is a binary, it must have a low-mass companion and/or low inclination, and the small-amplitude velocity period is obscured by the combination of measurement errors and stellar wind effects.

114.006 **Spectroscopic observations of the surface-active binary II Pegasi (HD 224085).** B. W. Bopp, P. V. Noah. Publ. Astron. Soc. Pacific, Vol. 92, 333 - 337 (1980).

Spectroscopy of the surface-active binary II Peg (HD 224085) shows the Hα emission feature to be variable in equivalent width (EW) by a factor of ten on time scales of several days. Most of the EW variability is in phase with the stellar rotation period, suggesting strong localization of Hα emitting regions. While it is a single-lined system, with an active primary component, II Peg has all the spectroscopic characteristics of a true RS CVn. Apparently the RS CVn-type of stellar surface activity can appear in systems of widely differing make-up.

114.007 **Abundance analyses of Theta Aquilae (B9.5 III), Nu Capricorni (B9.5 V), and Sigma Aquarii (A0 IVs).** S. J. Adelman, M. A. Nasson. Publ. Astron. Soc. Pacific, Vol. 92, 346 - 356 (1980).

Abundance analyses for three moderately sharp-lined stars with spectral types near A0, θ Aql, υ Cap, and σ Aqr have been performed via model atmospheres techniques. The derived abundances of the first two stars are quite similar to those of the sun and other normal main-sequence stars. Those of σ Aqr suggest that this star is a hot metallic-line star.

114.008 **An automatic procedure for a determination of the effective temperature and gravity. Application to 100 O-type stars.** C. Morossi, L. Crivellari. Astron. Astrophys., Suppl. Ser., Vol. 41, 299 - 304 (1980).

The authors have developed an automatic procedure to determine T_{eff} and $\log g$ from the comparison of observational data with their theoretical values. After a description of the method they compare their results for a sample of 100 O-type stars with the classifications of Conti and Walborn and with the determinations of Morrison by photometric indices. The agreement is fair and the authors present a HR diagram for O type stars based on their data.

114.009 **Spectroscopic data for bright southern type F supergiants.** J. C. Castley, R. D. Watson. Astron. Astrophys., Suppl. Ser., Vol. 41, 397 - 404 (1980).

Spectroscopic data are presented for eleven bright southern stars for which high dispersion spectra were obtained during a study of early F type supergiants. These data include equivalent widths, line profiles of H_γ, H_δ, H8 and calcium K, together with line blocking measurements.

114.010 **A catalogue of [Fe/H] determinations.** G. Cayrel de Strobel, C. Bentolila, B. Hauck, A. Curchod. Astron. Astrophys., Suppl. Ser., Vol. 41, 405 - 419 (1980).

A catalogue resulting from the compilation of published values of iron/hydrogen abundances is given for 628 stars.

114.011 **8 - 13 μm spectra of very late type Wolf-Rayet stars.** D. K. Aitken, M. J. Barlow, P. F. Roche, P. M. Spenser. Mon. Not. R. Astron. Soc., Vol. 192, 679 - 687 (1980).

8 - 13 μm spectra are presented of the late Wolf-Rayet stars, Ve 2 - 45 (WC9), CRL 2104 (WC8), He 2 - 113 (WC10) and CPD - 56° 8032 (WC10). Both WC10 stars show the un-

identified feature at 11.25 μm and one of them that at 8.6 μm; their spectra resemble those of some planetary nebulae. These features are absent in the WC8/9 stars. The WC 10 objects are characterized by much lower dust temperatures and their evolutionary status appears to be very different from that of the WC8/9 stars.

114.012 **Empirical effective temperatures of B and early A stars.** E. Kontizas, E. Theodossiou.
Mon. Not. R. Astron. Soc., Vol. 192, 745 - 753 (1980).

Empirical effective temperatures of 112 B and early A type stars have been derived by comparing observed and computed fluxes in a wide and appropriate wavelength range.

114.013 **Simultaneous spectroscopic (UV Mg II and Al II lines) and photometric variations of ζ^1 Sco (HD 152236).**
A. Heck, G. Burki, L. Bianchi, A. Cassatella, J. Clavel.
Mon. Not. R. Astron. Soc., Vol. 192, 59P - 65P (1980).

Repeated IUE observations of ζ^1 Sco show variations in the line profiles of the two Mg II doublets at about 2800 Å and of the Al II λ 1670 line. A series of puffs in the expanding envelope have been identified, implying a non-stationary mass loss process.

114.014 **Some spectroscopic investigations of twenty-six K stars.** J. N. Rai, J. P. Chaturvedi.
Astrophys. Space Sci., Vol. 71, 333 - 336 (1980).

Some spectroscopic characteristics of twenty-six K stars, in the region 4500–7500 Å, have been studied. The observations were recorded on 52 cm and 104 cm reflecting telescopes at the State Observatory, Naini Tal, India. The telescopes were coupled with a spectrum scanner which consisted of a plane diffraction grating. The frequencies have been noted and their probable assignments have been given. From these assignments, the probable model of K stars has been concluded.

114.015 **Calibration of the spectra of selected stars. II. Spectra of eight standard stars.** A. V. Kharitonov, V. M. Tereshchenko, L. N. Knyazeva, P. N. Bojko.
Astron. Zh., Tom 57, 725 - 730 (1980). In Russian.
English translation in Soviet Astron., Vol. 24, No. 4.

The energy distribution of the spectral density of the energetic illumination for 8 stars by means of fastening to the spectra of two ribbon lamps is obtained. The suspicion that α Lyr and η UMa are slightly variable ($0^m.02 - 0^m.05$) is substantiated.

114.016 **Energy distribution in the spectra of main-sequence B – F stars. Comparison with theoretical models.**
I. N. Glushneva, V. T. Doroshenko.
Astron. Zh., Tom 57, 731 - 737 (1980). In Russian.
English translation in Soviet Astron., Vol. 24, No. 4.

The values of four spectrophotometric gradients and of the Balmer jumps are obtained on the basis of stellar energy distribution in the region λλ 3200 - 7600 Å. The spectra of 152 main sequence B1 - G0 stars observed photoelectrically were investigated. A comparison with the theoretical values calculated using model stellar atmospheres was made. It is shown that theoretical and observational data are in good agreement.

114.017 **The effects of chemical composition in the atmospheres of M, S, and C stars.**
V. V. Tsymbal, V. E. Panchuk.
Astron. Zh., Tom 57, 881 - 883 (1980). In Russian.
English translation in Soviet Astron., Vol. 24, No. 4.

On the basis of the solution of the system of equations of thermal-chemical equilibrium, it is shown that a portion of S stars with an excess of Y, Zr, La may have spectra undistinguishable from those of M stars.

114.018 **Spectroscopic and photometric observations of the object He2-442.**
V. F. Esipov, O. G. Taranova, B. F. Yudin.
Pis'ma Astron. Zh., Tom 6, 418 - 420 (1980). In Russian.
English translation in Soviet Astron. Lett., Vol. 6.

Photoelectric and spectroscopic observations of the object He2-442 in the *UVRK* system and in the Hα line region are presented. Molecular bands of TiO λλ 7054 and 7198 are detected.

114.019 **The nature of the optical continuum of flares on red dwarfs.**
M. M. Katsova, A. G. Kosovichev, M. A. Livshits.
Pis'ma Astron. Zh., Tom 6, 498 - 504 (1980). In Russian.
English translation in Soviet Astron. Lett., Vol. 6.

Gas-dynamical processes caused by heating of accelerated electrons of a partially ionized hydrogen stellar chromosphere are investigated numerically. Radiative energy losses in optical thick layers are estimated. It is obtained that a condensation is moving down into the chromosphere. The luminosity, color-index and Balmer jump of the optical continuum of this condensation are consistent with observations of flares on red dwarfs.

114.020 **The energy distribution for 15 stars in the spectral region λλ3200 - 10800 Å.**
E. A. Kolotilov, I. N. Glushneva, I. B. Voloshina, M. F. Novikova, T. S. Fetisova, V. I. Shenavrin, L. V. Mossakovskaya, V. T. Doroshenko.
Soobshch. Gos. Astron. Inst. Shternberga, No. 219, 29pp. (1980). In Russian.

The values of stellar energy distribution for 15 stars in the wide spectral region λλ3200 - 10800 Å are presented. A comparison is presented between the data obtained and results of Johnson and Mitchell's photometric investigations and spectrophotometric investigations being done at the Astrophysics Institute of Kazakh Academy of Sciences.

114.021 **Mercury in a thin layer in HgMn stars: a test of a diffusion model.**
C. Mégessier, G. Michaud, E. J. Weiler.
Astrophys. J., Vol. 239, 237 - 247 (1980).

Lines of the first three states of ionization of mercury have been observed in μ Leporis and χ Lupi using the Copernicus satellite. Lines of Hg II and Hg III have been observed in α Andromedae. There appears to be an absorption feature at every wavelength where there is expected to be a mercury line. The presence of all three states of ionization is likely in μ Lep and χ Lup. The relative equivalent widths of the lines of the various states of ionization do not depend on the effective temperature of the stars, in contradiction to what is expected if mercury were uniformly distributed in the atmosphere. It is, however, expected if mercury has been concentrated, by diffusion, in a thin layer, where the radiative forces just equal the gravitational forces on mercury.

114.022 **Spectroscopic observations of Near Infrared Photographic Sky Survey objects.**
D. E. Turnshek, E. R. Craine, P. C. Boeshaar, D. A. Turnshek.
Bull. American Astron. Soc., Vol. 12, 463 (1980). – Abstract.

114.023 **Far ultraviolet flux distributions of Cygnus stars.** G. R. Carruthers, H. M. Heckathorn, C. B. Opal.
Bull. American Astron. Soc., Vol. 12, 521 (1980). – Abstract.

114.024 **The relation between color and effective temperatures for K and M stars.**
J. Piccirillo, A. P. Bernat, H. R. Johnson.
Bull. American Astron. Soc., Vol. 12, 522 (1980). – Abstract.

114.025 **Oxygen abundances in metal poor field stars.**
E. M. Leep, G. Wallerstein.
Bull. American Astron. Soc., Vol. 12, 522 (1980). – Abstract.

114.026 **The chemical composition, gravity and temperature of Vega.** L. A. Dreiling, R. A. Bell.
Bull. American Astron. Soc., Vol. 12, 522 (1980). – Abstract.

114.027 **Rubin-152: a massive star on the galactic fringe?**
F. R. Chromey.
Bull. American Astron. Soc., Vol. 12, 523 (1980). – Abstract.

114.028 **A high dispersion temperature index for cool stars.**
L. W. Ramsey.
Bull. American Astron. Soc., Vol. 12, 527 (1980). – Abstract.

114.029 **Extension of line identifications in Arcturus shortward to 2250 Å.**
K. G. Carpenter, R. E. Stencel, R. F. Wing.
Bull. American Astron. Soc., Vol. 12, 529 (1980). – Abstract.

114.030 **Hot companions of G supergiants observed with Skylab and IUE.** S. B. Parsons.
Bull. American Astron. Soc., Vol. 12, 529 (1980). – Abstract.

114.031 **An IUE survey of the ultraviolet emission line spectra of dMe and dM stars.**
K. G. Carpenter, R. F. Wing, P. L. Bornmann, J. L. Linsky, M. S. Giampapa, S. P. Worden.
Bull. American Astron. Soc., Vol. 12, 538 (1980). – Abstract.

114.032 **IUE observations of UV continuum emission from TV Gem (M1 Iab).**
A. G. Michalitsianos, R. W. Hobbs, M. Kafatos.
Bull. American Astron. Soc., Vol. 12, 539 (1980). – Abstract.

114.033 **Horizontal branch stars and globular clusters with IUE.** K. S. de Boer, A. D. Code.
Bull. American Astron. Soc., Vol. 12, 539 (1980). – Abstract.

114.034 **Effective temperatures and luminosities for B1 Ia+, O4 and WN7 stars.** A. B. Underhill.
Bull. American Astron. Soc., Vol. 12, 539 - 540 (1980).
Abstract.

114.035 **Six day periodicity in wavelengths of "moving" lines in SS433.** G. H. Newsom, G. W. Collins, II.
Bull. American Astron. Soc., Vol. 12, 540 (1980). – Abstract.

114.036 **The riddle of SS433.**
V. M. Lipunov, V. G. Surdin.
Zemlya Vselennaya, 1980, No. 4, p. 20 - 25. In Russian.

114.037 **Ca II H and K lines in A type stellar spectra.**
R. J. Panek, D. R. Stutman.
Astrophys. J., Vol. 239, 598 - 606 (1980).

High-quality, high-resolution photoelectric and photographic spectra of α CMa and γ Gem are used to obtain the profiles of the Ca II H and K lines. These profiles are analyzed using theoretical spectra of model atmospheres to determine the calcium abundance and to study the structure of the stellar atmosphere. The authors find calcium underabundant in Sirius.

114.038 **The white dwarf companion of the barium star ζ Capricorni.** E. Böhm-Vitense.
Astrophys. J., Lett., Vol. 239, L79 - L83 (1980).

The author shows that the barium star ζ Cap has a white dwarf companion with $M \sim 1.0\,M_\odot$ and $T_{eff} \sim 23,000$ K. Strong emission lines and very deep reversals in the Mg II emission line cores indicate circumstellar gas which may be due to mass transfer in the system. The Ba star HD 65699 also shows the very deep central reversals in very strong Mg II emission cores. The weak spectra of the barium stars ζ Cyg, HD 65699, and HD 116713 probably show excessive UV radiation also, but further studies are necessary.

114.039 **On the nature of MWC 349.**
L. Hartmann, D. Jaffe, J. P. Huchra.
Astrophys. J., Vol. 239, 905 - 910 (1980).

Line profile observations of MWC 349 confirm the suggestion of Olnon that the compact H II region surrounding the star is a stellar wind. The broad wings on Hα are caused by electron-scattering redistribution; the appreciable Thomson optical depth in the wind indicates that the photosphere of the star is extended. The terminal velocity of the wind is ~ 50 km s^{-1}, extremely slow for mass ejection from an early-type star. It is suggested that the slow wind and extended photosphere can be reconciled with the radiatively driven wind theory of Castor, Abbott, and Klein if MWC 349 has a low effective surface gravity, possibly because the star is overluminous for its mass.

114.040 ***IUE* spectra of a flare in the RS Canum Venaticorum–type system UX Arietis.**
T. Simon, J. L. Linsky, F. H. Schiffer III.
Astrophys. J., Vol. 239, 911 - 918 (1980).

IUE spectra of UX Ari obtained during the large flare of 1979 January 1 exhibit chromospheric and transition-region emission-line fluxes about 2.5 and 5.5 times brighter than quiescent fluxes, respectively, and up to 1400 times brighter than the quiet Sun. A high-dispersion spectrum of the 2000–3000 Å region exhibits enhanced Fe II emission, which is probably associated mainly with the K0 IV star, and enhanced Mg II emission with asymmetric wings extending to +475 km s^{-1}. The authors interpret these line wings as evidence for mass flow from the K0 IV star to the G5 V star.

114.041 ***IUE* observations of RW Hydrae (gM2 + pec).**
M. Kafatos, A. G. Michalitsianos, R. W. Hobbs.
Astrophys. J., Vol. 240, 114 - 124 (1980).

IUE observations of the late type star RW Hya (gM2 + pec) have been obtained. Analysis of the intense UV continuum observed between 1100 and 2000 Å suggests that this object is a binary system in which the secondary is identified as the central star of a planetary nebula with $T_{eff} \sim 10^5$ K. The ultraviolet spectrum is characterized by semiforbidden and allowed transition lines, of which the C IV (1548 Å, 1550 Å) doublet is particularly strong. A general absence of strong forbidden line emission suggests that the compact nebula in which both primary and secondary stars are embedded has particularly high densities of $\sim 10^8 - 10^9$ cm^{-3}. Tidal interaction rather than steady state mass flow from the M giant is suggested as a means to form a nebula with the characteristic densities inferred from the UV line analysis. A general discussion concerning the ionic abundances of various atomic ions present is given.

114.042 **Variations in the K-line emission of Arcturus.**
D. F. Gray.
Astrophys. J., Vol. 240, 125 - 129 (1980).

High-resolution photoelectric measurements of the Ca II K line of Arcturus show the K_{2V} emission peak to vary 1.7 times as much and in the opposite sense as the K_{2R} emission. During the span of observations the ratio of violet to red emission peaks (V/R) varied from 0.80 to 1.05. The wavelength position of K_3 is found to change with V/R. The rise from K_{1V} to K_{2V} is more abrupt than the corresponding rise on the red side. Solar analog models and mass-loss models are considered in the light of these data.

114.043 **The spectrum variations of HD 153919.**
G. G. Fahlman, G. A. H. Walker.
Astrophys. J., Vol. 240, 169 - 179 (1980).

New spectroscopic observations of the Of star

HD 153919, the primary in the 4U 1700–37 X-ray binary system, in the wavelength range 5600–6800 Å, are discussed. The phase-dependent absorption components present in the Hα and the He I λ5876 lines indicate the presence of a gas stream in the stellar wind of the primary. The published X-ray photometry is used in conjunction with the authors' spectroscopy to construct a geometrical model of the system which includes a radial symmetric wind from the primary, a heated zone around the secondary, a small coronal region around the primary, and the gas stream. A series of 27 successive spectra obtained over a 95 minute time interval at a phase during X-ray source eclipse is also discussed.

114.044 **Hα line in the spectrum of the star HDE 245770.**
N. F. Vojkhanskaya.
Pis'ma Astron. Zh., Tom 6, 582 - 586 (1980). In Russian. English translation in Soviet Astron. Lett., Vol. 6.

The emission line Hα in the spectrum of HDE 245770 is divided in constant and variable components. The latter is shown to be a line arising in the vicinity of the degenerated components of the system. The radial velocity of the constant part of the line does not vary and equals to +10 km/s. For the variable line component a curve of radial velocities that varies with a period of 104 s is obtained.

114.045 **Quasi-periodic variations of the Hα emission in the β Lyrae system.** V. I. Burnashev, M. Yu. Skul'skij.
Pis'ma Astron. Zh., Tom 6, 587 - 591 (1980). In Russian. English translation in Soviet Astron. Lett., Vol. 6.

Photometry of the close binary system β Lyrae was carried out, an absolute calibration of the data being performed. Quasi-periodic variations of the radiation of circumstellar gaseous structures are derived. The effective radius of the gaseous ring surrounding the large component of the system is estimated to be 1/3 of the distance between the centers of the components.

114.046 **The ultraviolet spectrum of P Cygni.**
L. Luud, A. Sapar.
Tartu Astrofüüs. Obs., Teated Nr. 60, 22 pp. = Tartu Astrofiz. Obs., Prepr. A-1 (1980).

Using IUE observations the ultraviolet spectral lines of P Cygni in the region from 1230.3 to 3096.0 Å have been identified and the line displacements have been determined. It has been found that the relationship between the absorption line displacement velocity and the excitation energy is the same as for the lines observed in the visual spectral region, except the resonance lines that have almost equal absorption velocities. It has been found that sharp lines with approximately zero displacement and zero excitation are probably of interstellar (but not of circumstellar) origin.

114.047 **Spectral classification of carbon stars in Magellanic Cloud clusters.** T. Lloyd Evans.
Mon. Not. R. Astron. Soc., Vol. 193, 97 - 105 (1980).

Image tube spectra of 33 stars in 17 Magellanic Cloud clusters and six stars in the field of the SMC are discussed. Nineteen of the 21 carbon stars have been classified. None of these stars, which have $M_V \gtrsim -3$, has strong ^{13}C features. The weakness of CN relative to C_2 suggests an affinity to CH stars in the Galaxy, in accordance with the idea that intermediate age clusters in the Magellanic Clouds are metal deficient.

114.048 **The optical spectrum of SS 433.**
P. Murdin, D. H. Clark, P. G. Martin.
Mon. Not. R. Astron. Soc,, Vol. 193, 135 - 151 (1980).

SS 433 has a continuum spectrum, luminosity and surface gravity suggesting that it is an early O-type star suffering interstellar absorption of $A_V \sim 8$ mag. The rest emission lines appear to be a complex arising from a narrow system which is fixed to the primary object (type O5f) in a binary star and a broad system presumably representing mass loss. The moving features seem to arise from discrete 'bullets' symmetrically shot out in two opposite jets at 0.25 c, in accord with earlier models.

114.049 **8–13 μm spectrophotometry of V1016 Cyg and the shape of the 'silicate' feature.**
D. K. Aitken, P. F. Roche, P. M. Spenser.
Mon. Not. R. Astron. Soc., Vol. 193, 207 - 212 (1980).

8–13 μm spectrophotometry of V1016 Cyg shows a broad emission feature attributed to radiation from silicate grains. This emission feature more closely resembles that of the circumstellar shells of oxygen-rich supergiants than the more dilute feature, typical of the interstellar medium, which is observed from the Trapezium source in the Orion nebula. It appears to be possible to distinguish the evolutionary status of an object from the form of its silicate excess.

114.050 **A classification of Be stars.**
M. Jaschek, A.-M. Hubert-Delplace, H. Hubert, C. Jaschek.
Astron. Astrophys., Suppl. Ser., Vol. 42, 103 - 114 (1980).

Based upon a sample of 140 stars observed over 20 years for which about 5 000 spectrograms are available, a classification scheme of Be stars is presented. This is the first attempt to subdivide the Be star group into physically significant subgroups, from which typical objects can be selected for further study. The four groups proposed are based upon a discussion of spectrum characteristics, multicolor photometry, polarization, rotational velocities, UV spectral types and time variability. Starting with the group membership of a Be star, predictions can be made of the future behaviour of it.

114.051 **Ground-based observations of some stars classified in the satellite ultraviolet with spectral particularities.** M. Jaschek, C. Jaschek.
Astron. Astrophys., Suppl. Ser., Vol. 42, 115 - 118 (1980).

Spectral classifications based on 80 Å/mm plates were obtained for 48 stars classified in the UV order to examine existing discrepancies. Although reclassification generally improves the agreement, there remain several cases of significant disagreements.

114.052 **Copernicus observations of β Cephei stars.**
J. B. Hutchings, G. Hill.
Astron. Astrophys., Suppl. Ser., Vol. 42, 135 - 140 (1980).

Repeated scans at ~ 0.2 Å resolution have been obtained of the λ1100 spectral region of 8 bright β Cep stars. The data reveal simultaneous light and velocity changes which, under the assumption of radial pulsation, can be related to surface area and temperature change. The star α Lup shows different behaviour for different lines, and line profile variations. New orbital parameters are derived for α Vir from its UV spectrum, and large line profile changes are seen in high resolution data.

114.053 **The ultraviolet absorption by the photoionisation spectrum of Fe I in Ap stars.** C. Jamar.
Astron. Astrophys., Vol. 89, 22 - 25 (1980).

The autoionisation spectrum of Fe I is shown to be a major source of opacity between 1300 and 1450 Å in stellar spectra. The strength of the absorption is prominent in cool stars and naturally in cool Ap stars where Fe is overabundant. When the Fe abundance is increased by factor 100, the feature could appear for temperatures as large as 11000 K.

114.054 **I.U.E. observations of HD 102567, the proposed optical counterpart of 4U 1145–61.**
L. Bianchi, P. L. Bernacca.
Astron. Astrophys., Vol. 89, 214 - 219 (1980).

The B1V star HD 102567, proposed optical counterpart of the X-ray source 4U 1145–61, has been observed with the two SWP and LWR cameras on board the IUE satellite in the low resolution mode. The ultraviolet continuum flux can adequately be fitted to the emergent flux from a model atmosphere having $T_{eff} \sim 22,000$ K and $\log g = 4.5$ in agreement with the spectral type of the star. The line spectrum shows the presence of a low density envelope around the star expanding with a terminal velocity of about 1800 km s^{-1}.

114.055 **Visible spectrum of π^1 Gruis: identification of new $e\,^1\Pi - X\,^1\Sigma^+$ and $^1\Sigma^+ - X\,^1\Sigma^+$ bands of ZrO.**
P. S. Murty.
Astrophys. J., Vol. 240, 585 - 587, plate 5 (1980).

New molecular features identified with the $e\,^1\Pi - X\,^1\Sigma^+$ and $^1\Sigma^+ - X\,^1\Sigma^+$ transitions of ZrO, in the 5840–7210 Å spectral region of the S star π^1 Gru, are presented. The measured band positions are in accord with the laboratory and predicted spectra of the molecule. The proposed new identifications suggest that molecular features due to the oxides of *s*-process elements are stronger and more extensive in stellar spectra than in the laboratory.

114.056 **A lower limit on the magnitude of the companion to HDE 226868 (Cygnus X-1).**
A. W. Shafter, R. J. Harms, B. Margon, J. L. Katz.
Astrophys. J., Vol. 240, 612 - 618 (1980).

High-precision digital spectra of the HDE 226868/Cygnus X-1 system have been obtained at a variety of orbital phases, using a Digicon photon counting array at the coude focus of the Lick Shane reflector. The spectra have been cross-correlated to search for evidence for the existence of a luminous stellar companion to HDE 226868, as has been suggested by several models where the secondary star is not a black hole. The authors find no evidence for such a star and place a lower limit on the difference between HDE 226868 and its companion of 4 magnitudes. This limit rules out all of the models for the Cygnus X-1 system calculated by Avni and Bahcall which include main-sequence stars as secondaries rather than black holes. A merit function generally useful for cross-correlation of digital spectra is described.

114.057 **Differential line shifts in late type stars.**
A. Stawikowski.
Stellar turbulence, (see 012.008), p. 52 (1980). – Abstract.

114.058 **Differential line-shifts.** W. Buscombe.
Stellar turbulence, (see 012.008), p. 157 - 162 (1980).

114.059 **Time dependence of Balmer progression in the spectrum of HD 92207.** H. G. Groth.
Stellar turbulence, (see 012.008), p. 163 (1980). – Abstract.

114.060 **High luminosity F-K stars motions and Hα emissions.**
J. Smolinski, J. L. Climenhaga, B. L. Harris.
Stellar turbulence, (see 012.008), p. 170 (1980). – Abstract.

114.061 **Changes of photospheric line asymmetries with effective temperature.** D. F. Gray.
Stellar turbulence, (see 012.008), p. 297 - 298 (1980). Abstract.

114.062 **The chemically peculiar star HR 6127. 2. Detailed analysis.** J. Žižňovský.
Bull. Astron. Inst. Czechoslovakia, Vol. 31, 300 - 302 (1980).

The value $v \sin i \leq 10$ km s^{-1} was determined from the halfwidth of the spectral line Mg II λ 448.13 nm. The following atmospheric parameters were determined from the observed profile of Hγ : $\Theta_e = 0.55$ and $\log g = 3.0$. The microturbulent velocity is 7 km s^{-1}. As a result of the abundances of the individual elements HR 6127 was classified as a cool Ap star.

114.063 **Carbon stars in the Fornax dwarf spheroidal galaxy.**
M. Aaronson, J. Mould.
Astrophys. J., Vol. 240, 804 - 810 (1980).

A number of the very red stars in the Fornax dwarf galaxy recently discovered by Demers and Kunkel have been observed spectroscopically and in the infrared. For $B - V$ greater than ~2.1, the giant branch of Fornax may consist entirely of carbon stars. The luminosities of these stars range from −4.5 to −5.5 in M_{bol}, placing them on the upper asymptotic giant branch, in analogy with the carbon stars found in Magellanic Cloud globular clusters. The authors argue that such stars can be produced only by a population of intermediate age.

114.064 **Infrared spectrophotometry of SS 433.**
C. W. McAlary, R. A. McLaren.
Astrophys. J., Vol. 240, 853 - 858 (1980).

Infrared spectrophotometry of the peculiar object SS 433 from 2.0 to 2.4 μm has revealed the presence of strong emission lines of the Brackett and Paschen series of hydrogen. The authors have detected Bγ at its rest wavelength, as well as redshifted components of Bγ, Bδ, and Pα, whose velocities correspond closely to those predicted for the Balmer emission lines at the epoch of observation. The strengths of the Bγ and Hα lines are used to estimate a value of A_v= 7.0–7.5 mag for the reddening toward SS 433.When corrected for reddening and free-free emission, the observed 0.55–3.5 μm continuum of SS 433 can be fitted to a single 15,700 K blackbody. The implications of this model are discussed briefly.

114.065 **Four probable late subdwarfs with extreme metal deficiency.** T. B. Ake, J. L. Greenstein.
Astrophys. J., Vol. 240, 859 - 864 (1980).

The authors present multichannel spectrophotometer observations of four extreme subdwarf M stars: LHS 453, LHS 3382, G7-17, and LHS 489. They are characterized as having all molecular features weak, but Ca I λ4227 anomalously strong. A classification-dispersion SIT spectrogram of LHS 3382 exhibits the Na D lines in normal strength. The behavior of the spectral features can be understood from the interplay of ionization and molecular dissociation equilibria with the greater gas and electron pressures in abnormally transparent atmospheres. In comparison with other low-metal-abundance sdM stars, these few objects have extreme metal deficiency, such as are found in extreme sdG stars, and are subluminous for their color.

114.066 **Spectral scans of SS 433.**
G. H. Newsom, H. Jenkner, R. M. Wagner.
Astron. J., Vol. 85, 1229 - 1231 (1980).

The far-red spectrum of SS 433 was scanned on seven nights in June 1979 and compared to spectra of standard stars to provide absolute fluxes. Integrated fluxes of the unshifted and shifted Hα lines are tabulated, along with redshifts of the moving lines.

114.067 **The ultraviolet flux of HD 122563.**
B. Gustafsson, R. A. Bell, K. Fredga, G. F. Gahm.
Astron. Astrophys., Vol. 89, 255 - 258 (1980).

Previous work has suggested that the very metal deficient giant stars are brighter in the near ultraviolet than are the corresponding model atmospheres. Observations of the star HD 122563 made with the IUE satellite suggest that this difference is less significant and may well be nonexistent at wavelengths less than 2700 Å.

114.068 **On the UV classification of the Am stars.**
M. Jaschek, C. Jaschek, A. Cucchiaro.
Astron. Astrophys., Vol. 89, 380 - 381 (1980).

The ultraviolet spectra, observed by TD1, of 33 Am stars and of 25 normal stars of luminosity class IV and V were analyzed. If two different temperature indicators are used, normal stars fall, as expected, on a narrow sequence; in the Am stars on the other hand the two indicators disagree. This is attributed to increased blanketing caused by the enhancement of metals, but no detailed atmospheric models are available to check the results.

114.069 **Low-dispersion spectral classification and *UBV* photographic photometry of Hα-emission objects in the Coalsack region.** R. E. Martinez, J. C. Muzzio, S. Waldhausen.
Astron. Astrophys., Suppl. Ser., Vol. 42, 179 - 184 (1980).

The authors present approximate spectral types, obtained from objective-prism plates of 1 360 Å/mm at Hγ, and *UBV* photographic photometry of Hα-emission objects in the Coalsack region. Nearly all the objects were selected from published lists, but half a dozen were found by themselves. Their results enable them to recognize different kinds of objects from those lists. For the early OB stars the authors obtain approximate color excesses and distance moduli that suggest that the bulk of them lie at the same distance as the Cen OB1 association; quite a few of the OB stars lie even farther away, beyond 5 kpc from the Sun.

114.070 **Thorium in Arcturus, Pollux, Procyon and the sun.** H. Holweger.
Observatory, Vol. 100, 155 - 160 (1980).

114.071 **On short time changes of the emission lines in the spectra of some WR-type stars.**
M. B. Babaev, A. A. Atai, L. Kh. Gasanalizade.
Izv. AN AzSSR. Ser. fiz.-tekh. i mat. nauk, 1979, No. 4, p. 3 - 10. In Russian. – Abstr. in Ref. zh., 51. Astron., 8.51.487 (1980).

114.072 **Correlations between line-profile and photometric variations in the B2 IV [*e*] star HD 45677.**
J. P. Swings, R. Barbier, M. Klutz, A. Surdej, J. Surdej.
Astron. Astrophys., Vol. 90, 116 - 122 (1980).

The results deduced from two sets of homogeneous simultaneous photometric and spectrographic observations of HD 45677 in 1977 (6 nights of *UBV* data; 38 spectra) and in 1979 (7 nights of *uvby* data; 30 spectra) in a search for correlated variations are described. It appears that photometric data and some spectral features are well correlated (or anticorrelated) at certain epochs, but not at all at others. Because of these conflicting results, no explanation of the behaviour of the extended complex atmosphere surrounding this peculiar object can be proposed.

114.073 **Ultraviolet observations of the blue halo star: HD 93521.** M. Ramella, C. Morossi, P. Santin.
Astron. Astrophys., Vol. 90, 146 - 150 (1980).

The ultraviolet spectrum of the high galactic latitude O-type star HD 93521 has been studied. The authors furnish a list of interstellar and stellar features, identified from IUE high resolution spectrograms. A comparison between the observed UV flux distribution obtained with the S 2/68 satellite and that derived from theoretical models by Kurucz, is given. The authors also present an alternative procedure to subtract a background from the IUE "gross" flux. The evolutionary stage of HD 93521 is briefly discussed.

114.074 **The ultraviolet spectrum of β Canis Majoris stars.**
M. Burger, C. de Jager, T. M. Kamperman, L. Neven.
Astron. Astrophys., Vol. 90, 170 - 175 (1980).

Ultraviolet spectra of 15 β Canis Majoris stars in wavelength bands of ~100 Å around 2100, 2500, and 2800 Å (resolution 1.8 Å), obtained with the Ultraviolet Stellar Spectrophotometer S59 on board the ESRO TD-1A satellite are discussed. In general the spectra are similar to those of "normal" stars, only the star α Vir has He I, C II, and Mg II lines slightly weaker than normal. Comparison with theoretical computations shows that the Fe abundance in the β CMa stars is solar and that the average microturbulent velocity is about 4 km s^{-1}. The UV spectral lines of β Cep do not show any significant variations in equivalent width with phase.

114.075 **Spectral types of S and SC stars on the revised MK system.** P. C. Keenan, P. C. Boeshaar.
Astrophys. J., Suppl. Ser., Vol. 43, 379 - 391 (1980).

The catalog of 101 stars provides standards of spectral type in the sequence from MS through S and SC to the carbon stars with weak C_2 bands. In addition to temperature types and C/O indices, intensities of ZrO, TiO, Na D lines, YO, and Li 6708 are given whenever possible. For variables of large amplitude the observations at different dates are tabulated.

114.076 **Energy distribution in the spectra of stars used as spectrophotometric standards in the near infrared.**
I. B. Voloshina, I. N. Glushneva, V. I. Shenavrin.
Astron. Zh., Tom 57, 1003 - 1009 (1980). In Russian.
English translation in Soviet Astron., Vol. 24, No. 5.

Data on the energy distribution in the spectra of 7 stars, used as spectrophotometric standards, are obtained by means of comparison with α Lyr in the near infrared (λλ 6300 - 10800 Å).

114.077 **The bizarre spectrum of SS 433.** B. Margon.
Sci. American, Vol. 243, No. 4, p. 44 - 55 (1980).

Displaced emission lines found in the spectrum of this extraordinary stellar object suggest that it is spewing two narrow high-speed jets of matter in opposite directions. How can such behavior be explained?

114.078 **Spectroscopic observations of SS 433 in 1979.** K. K. Chuvaev.
Pis'ma Astron. Zh., Tom 6, 623 - 627 (1980). In Russian.
English translation in Soviet Astron. Lett., Vol. 6.

For 17 nights in the second half of 1979 radial velocities in the spectrum of SS 433 were found from the shifted Hα line. Variations of position, intensity and structure of the shifted Hα line for the time intervals of a day and of several hours are shown. Comparison between profiles of shifted Hβ and Hα lines allowed to conclude that there is a significant difference in Balmer decrement of particular gas clouds forming jets. Spectra with strong absorptions in short-wave wings of "stationary" H and He I lines are described.

114.079 **The metallicity of F and G dwarfs and the problem of star formation.** A. A. Suchkov.
Pis'ma Astron. Zh., Tom 6, 632 - 636 (1980). In Russian.
English translation in Soviet Astron. Lett., Vol. 6.

The metallicity function of disk population F and G dwarfs is found to have a central gap at [Fe/H] = –0.1 at a high level of statistical significance. The metallicity functions of both types of stars are very similar, dropping to [Fe/H] = –0.5 and +0.3, having almost the same average values and dispersions, and a gap at the center which subdivides the stars into two metallicity groups. However, comparison of the kinematics of these groups reveals a paradox: the metal-rich group of G dwarfs turns out to be kinematically older than the group of F dwarfs with two-times smaller metal abundance. The implication of this result to the star formation problem is discussed.

114.080 **Abundance of chemical elements in the sun's neighbourhood.**
N. S. Komarov, A. N. Shcherbak.
Pis'ma Astron. Zh., Tom 6, 637 - 640 (1980). In Russian.
English translation in Soviet Astron. Lett., Vol. 6.

Abundances of chemical elements in the atmospheres of 21 cool giant stars have been determined by means of a modified differential curve-of-growth method. These stars belong to open clusters and moving groups. The iron abundance decreases in the sun's neighbourhood towards the anticenter of the Galaxy by ~0.8 dex.

114.081 **Methods for the analysis of stellar spectra veiled by lines (III).** A. Greve, C. Zwaan.
Astron. Astrophys., Vol. 90, 239 - 245 (1980).

Ultraviolet spectrograms of the Sun and of αCMi are analyzed to investigate effects of finite spectral resolution in the interpretation of stellar spectra veiled by lines. The recorded completeness of intrinsic spectral detail is discussed. Solar spectrograms of approximately 0.03 Å resolution, i.e. of the order of the Doppler widths of spectral lines, are a good representation of the true spectral distribution. Even in case UV spectra are completely resolved, the unresolvable lines produce a line haze which should be included in the total opacity. The empirical total opacity derived from the centre to limb variation of the solar UV intensity is quantitatively explained by the known continuous opacities plus the opacity of unresolvable lines. The comparison of calculated spectra and spectrograms observed with incomplete resolution is discussed.

114.082 **Ultraviolet, optical, and infrared observations of the Herbig Be star HD 200775.** A. Altamore, G. B. Baratta, A. Cassatella, G. L. Grasdalen, P. Persi, R. Viotti.
Astron. Astrophys., Vol. 90, 290 - 296 (1980).

The ultraviolet low-dispersion spectrum shows several strong absorption resonance lines probably formed in an expanding envelope. From the IR observations up to 12.6 μ the authors derive a value of mass loss rate $M/v_{exp} = 2.3 \times 10^{-9} M_\odot \, yr^{-1}/km \, s^{-1}$. The infrared energy distribution shows also the presence of a dust shell.

114.083 **The chemical compositions of 26 distant late-type supergiants and the metallicity gradient in the galactic disk.** R. E. Luck, H. E. Bond.
Astrophys. J., Vol. 241, 218 - 228 (1980).

From an analysis of high-dispersion Mount Wilson spectroscopic data, the authors have obtained atmospheric parameters and chemical abundances for 26 distant supergiants of spectral types G through M. Iron-to-hydrogen ratios with respect to the Sun, [Fe/H], can be determined from this material with an uncertainty of ± 0.2 dex. It is found that supergiants more than about 0.5 kpc from the Sun in the direction of the galactic anticenter are significantly metal-deficient relative to those in the solar neighborhood. The authors derive a radial metallicity gradient $d[Fe/H]/dR = -0.24 \pm 0.04 \, kpc^{-1}$ for the region of the galactic disk between 7.7 and 10.2 kpc from the galactic center.

114.084 **Outer atmospheres of cool stars. V. IUE observations of Capella: the rotation-activity connection.** T. R. Ayres, J. L. Linsky.
Astrophys. J., Vol. 241, 279 - 299 (1980).

The authors summarize stellar parameters for Capella Aa and Ab, and an evolutionary scenario for the system. They describe their ultraviolet spectra and previous observations of Capella, including line identifications, line fluxes, and estimates of the relative contribution of each star to the composite line profiles. The authors discuss the importance of rotation and consider possible transition-region models for the Capella secondary.

114.085 **The 164 and 13 day periods of SS 433: confirmation of the kinematic model.** B. Margon, S. A. Grandi, R. A. Downes.
Astrophys. J., Vol. 241, 306 - 315 (1980).

Spectroscopy of the unique emission line star SS 433 is used to examine the kinematics of the "moving" emission lines discovered by Margon et al. (1979). In particular, coverage of the previously unobserved 30% of the 164 day period has been obtained, and verifies the model. The authors find in addition a 13 day periodic variation of emission line intensities, which provides independent verification of the low-amplitude radial velocity period discovered by Crampton, Cowley, and Hutchings (1980), and strongly suggests that SS 433 is a binary system related to the better-studied galactic X-ray sources.

114.086 **Optical observations of the ultrahigh-excitation Wolf-Rayet star Sanduleak 3.** M. J. Barlow, J. C. Blades, D. G. Hummer.
Astrophys. J., Lett., Vol. 241, L27 - L31 (1980).

The authors have identified recombination lines of O VII, O VIII, and C V in the optical spectrum of an O VI Wolf-Rayet star, representing the first non-X-ray detection of these ions in astronomical spectra and implying excitation energies in excess of 800 eV. Rapid variations on a time scale of about 150 s have been observed in the profile of one of the O VII lines.

114.087 **Periodic variations in the spectrum of X Persei identified with the X-ray source 3U 0352+30.** T. S. Galkina.
Izv. Krymskoj Astrofiz. Obs., Tom 61, 77 - 89 (1980). In Russian.

On the basis of an analysis of the emission spectrum of X Persei there have been discovered two kinds of periodicities. One of them is a 22-hr periodicity, analogous to the 22-hr periodicity of the X-ray variations of the source 3U 0352+30. The other is a 581-day one. All this led to the evident conclusion of the relation between X Persei and the X-ray source 3U 0352+30. The X Per 3U 0352+30 system may be considered as a triple system: a neutron star moving around a B0e primary, and this double system moving around a far third companion with a 581-day period.

114.088 **The sun among the stars. III. Energy distributions of 16 northern G-type stars and the solar flux calibraton.** J. Hardorp.
Astron. Astrophys., Vol. 91, 221 - 232 (1980).

Energy distributions from 3308 to 8390 Å of two candidates for a solar spectral analog and of 14 other northern G-type dwarfs are compared to the solar energy distribution via stellar spectrophotometric standards. The reliability of the stellar and solar flux-calibrations is evaluated. While the stellar calibration seems to be in good shape, solar calibrations differ widely.

114.089 **Ultraviolet spectroscopic measurements of cool stars.** A. K. Dupree.
Highlights of Astronomy, Vol. 5, (see 012.024), 263 - 276 (1980).

Drawing on the most recent material from IUE, the author discusses the presence and structure of chromospheres and coronae in single stars of varying gravities, surface temperatures, and activity. Evidence of mass loss and the concurrent presence of a corona are also noted. Binary systems of late-type stars (the RS CVn and W UMa type) are briefly discussed.

114.090 **Ultraviolet spectroscopy of chemically peculiar stars.** D. S. Leckrone.
Highlights of Astronomy, Vol. 5, (see 012.024), 277 - 284 (1980).

This discussion focuses on high dispersion spectroscopic observations of chemically peculiar stars of the upper main sequence, classical HgMn, Si and SrCrEu stars, obtained with the IUE. Two subjects are emphasized – the confirmation of composition anomalies previously identified in ground-based spectra and evidence relating to the validity of the diffusion theory for the production of chemical peculiarities.

114.091 **Stellar atmospheres and chemical compositions of galaxies.** L. H. Aller.
Highlights of Astronomy, Vol. 5, (see 012.030), 811 - 815 (1980).

114.092 **He I 10830 Å observations of chromospheres in late type stars.** G. O'Brien.
Bull. American Astron. Soc., Vol. 12, 747 (1980). – Abstract.

114.093 **He I λ 10830 in some O6 - B3 III, IV and V stars.** Z. Frank, D. D. Meisel, B. Saunders.
Bull. American Astron. Soc., Vol. 12, 751 (1980). – Abstract.

114.094 **Spectrophotometry of peculiar B and A stars. VI. HD 32633, HD 34452 and HD 133029.**
S. J. Adelman, R. E. White.
Astron. Astrophys., Suppl. Ser., Vol. 42, 289 - 298 (1980) = Contrib. Five Coll. Obs., Amherst, Mass, No. 330.

Optical region spectrophotometry of λλ3300-7530 is presented for three magnetic peculiar A stars, which exhibit definite λ4200 and λ5200 broad, continuum features, HD 32633, HD 34452, and HD 133029. The synthesized *u-b* and *b-y* colors vary in good qualitative agreement with published photometry and are consistent with the period of Preston and Stepien for HD 32633, of Deutsch for HD 34452, and of Bonsack for HD 133029. HD 32633 is reddened, HD 34452 shows a very sharp core to its λ5200 feature, and HD 133029 exhibits complicated changes in its energy distribution.

114.095 **Spectrophotometry of peculiar B and A stars. VII. HD 6164, HD 8855, HD 11187, HD 171782, HD 190068, HD 200311, and HD 220147.** S. J. Adelman.
Astron. Astrophys., Suppl. Ser., Vol. 42, 375 - 382 (1980).

Spectrophotometry of the optical region λλ3300-7100 is presented for seven moderately sharp-lined silicon stars: HD 6164, HD 8855, HD 11187, HD 171782, HD 190068, HD 200311, and HD 220147. The energy distributions show definite λ4200 and λ5200 broad, continuum features as well as other deviations from the predictions of normal stellar model atmospheres. Evidence is presented for photometric variability in these 7th and 8th magnitude stars. Problems in correcting for possible reddening and in matching the data with the predictions of normal stellar model atmospheres are discussed.

114.096 **A study of H_α profile variations in κ Orionis, B0.5 Ia.**
L. Rusconi, G. Sedmak, R. Stalio, C. Arpigny.
Astron. Astrophys., Suppl. Ser., Vol. 42, 347 - 356 (1980).

Variations in the H_α profile in the spectrum of κ Orionis, B0.5 Ia, are determined from coudé plates at 12.7 Å. mm^{-1} reciprocal dispersion. The profile consists of a double-peaked-absorption-plus-central-emission core over broad absorption wings, more extended in the negative velocity part. Changes are observed typically over a long time scale, from night to night, in the profile shape. There are also evidences of short time scale variations, of the order of several tens of minutes, in the central reversal. The self reversal may provide evidence for the presence of hot regions surrounding the star's photosphere. Their limited extension can be inferred from the time scale of the emission variability. Changes in the overall profile could be associated with variations in ionization structure of the wind.

114.097 **Line blocking and equivalent widths in the spectrum of Pollux.** F. Ruland, R. Griffin, R. Griffin, D. Biehl, H. Holweger.
Astron. Astrophys., Suppl. Ser., Vol. 42, 391 - 401 (1980).

High-dispersion spectrograms of Pollux (β Gem, K0 III) have been used to derive line blocking coefficients for the spectral range λλ4900-9300 Å, and equivalent widths of lines of 43 atoms, ions, and molecules.

114.098 **The chemical composition, gravity, and temperature of Vega.** L. A. Dreiling, R. A. Bell.
Astrophys. J., Vol. 241, 736 - 758 (1980).

Several flux-constant, line-blanketed model stellar atmospheres have been computed for the bright star Vega. The stellar effective temperature has been found to be 9650 K by comparisons of observed and computed absolute fluxes at 5556 Å and of relative absolute fluxes between 3300 and 10800 Å. The Balmer line profiles give log g=3.9±0.2, consistent with the result from the Balmer jump and with that found from the radius and (estimated) mass. The Ti II and Fe II curves of growth give abundances of log N(Ti)=4.7±0.3 and log N(Fe)=7.1±0.3 on the scale log N(H)=12.0. The local thermodynamic equilibrium (LTE) Fe I curve of growth gives log N(Fe)=6.9±0.3, whereas correction for non-LTE effects suggests log N(Fe)=7.5±0.3. The model fluxes between 1200 and 3300 Å generally agree with observations of Vega from satellites. Infrared model fluxes have been computed between 1 and 30 μm.

114.099 ***IUE* observations of a luminous M supergiant that exhibits emission continuum in the far ultraviolet.**
A. G. Michalitsianos, M. Kafatos, R. W. Hobbs.
Astrophys. J., Vol. 241, 774 - 778 (1980).

IUE observations of the late-type M supergiant TV Gem (M1 Iab) have been obtained that reveal strong UV continuum between 1200 and 3200 Å. The continuum is essentially featureless with the exception of a number of broad absorption features in the short wavelength spectral range. UV emission from this star is unexpected because earlier ground-based observations give no indication of a possible association with an early companion or circumstellar ionized nebulosity.

114.100 **Studies of the Carina Nebula. III. The spectral energy distribution of the very hot and massive star HD 93250.**
P. S. Thé, H. R. E. Tjin A Djie, R. P. Kudritzki, P. R. Wesselius.
Astron. Astrophys., Vol. 91, 360 - 364 (1980).

Photometric observations in the spectral region between 0.33 - 5 μm are used to determine the spectral energy distribution of the very massive O 3-type star HD 93250 (located in the Carina Nebula). From the agreement of the observed extinction-free and the calculated theoretical energy distribution of HD 93250 it is concluded that the photosphere of the star is well described by a plane-parallel, hydrostatic non-LTE model. This confirms the result of a non-LTE analysis of the stellar line spectrum, which has been carried out recently by Kudritzki (1980). A study of ANS ultraviolet observations of HD 93250 shows that the UV part of the extinction law is probably also anomalous.

114.101 **Rare-earth systematics in Ap and Am stellar spectra: are new theoretical developments required?**
C. R. Cowley.
Vistas Astron., Vol. 24, 245 - 257 (1980).

Attention is called to the observational fact that the cool Am stars have deeper line cores than chemically normal main sequence stars or cool Ap stars. These line cores resemble those of more luminous stars, a fact which comports well with some low dispersion similarities between Am stars and evolved objects. Some modification of the traditional models and/or line formation theory is indicated in order to explain the observations.

114.102 **SS 433.**
IAU Circ., Nos. 3494, 3517, 3521, 3522, 3527, 3547 (1980).

114.103 **AM Herculis.**
IAU Circ., Nos. 3496, 3500 (1980).

114.104 **59 Cygni.**
IAU Circ., No. 3498 (1980).

114.105 **Near-infrared observations of emission objects.**
IAU Circ., No. 3498 (1980).

114.106 **The lithium in weak G-band stars.**
M. Parthasarathy, N. Kameswara Rao.
Astrophys. Space Sci., Vol. 73, 251 - 257 (1980).

The lithum abundance in weak G-band stars is found to be linearly correlated with CH deficiency. Weak G-band stars with a high Li abundance are found to show relatively least carbon deficiency, while the stars with a relatively low Li abundance show a high under-abundance of carbon.

114.107 **A survey of chromospheric Ca II H and K emission in field stars of the solar neighborhood.**
A. H. Vaughan, G. W. Preston.
Publ. Astron. Soc. Pacific, Vol. 92, 385 - 391 (1980).

Fluxes in 1 Å bands at the centers of the H and K lines are being measured in main-sequence F-G-K-M stars in the northern half of the Woolley et al. (1970) Catalog of stars within twenty-five parsecs of the Sun, in a survey not yet completed. Results for 486 stars are presented in the form of flux-color diagrams and discussed in light of evidence that chromospheric activity declines with age in main-sequence stars. Support is noted for the reality of the Sirius moving group. The relative numbers of more-active (Hyades-like) and less- active (solar like) F-G stars are tolerably in agreement with a nearly constant rate of formation, but there exists an apparent deficiency in the number of F-G stars exhibiting intermediate activity.

114.108 **Comparison of activity cycles in old and young main-sequence stars.** A. H. Vaughan.
Publ. Astron. Soc. Pacific, Vol. 92, 392 - 396 (1980).

Evidence cited indicates that solar-like activity cycles (smooth, undulating variations in chromospheric H-K emission flux, with a rapidly varying component that increases in amplitude in the maximum of the cycle) are characteristically found only in old stars, among the 91 main-sequence stars studied by Wilson (1978).

114.109 **A search for short-time-scale variations in Hα emission in the spectra of four bright Be stars.**
R. C. Reynolds, A. Slettebak.
Publ. Astron. Soc. Pacific, Vol. 92, 472 - 474 (1980).

Simultaneous Hα observations of four bright Be stars with a photoelectric scanner and an interference filter photometer during seven consecutive nights in May 1978 show (1) no evidence for real changes larger than about 5% in Hα emission strengths during any given night; (2) possible night-to-night changes for two of the stars, though the evidence is uncertain; and (3) definite changes on time scales of months for all four stars.

114.110 **The spectrum of R Sextantis.** W. P. Bidelman.
Inf. Bull. Variable Stars, No. 1858 (1980).

114.111 **Further infrared studies of the pre-main-sequence object HD 97048.** D. K. Aitken, P. F. Roche.
Anglo-Australian Obs., Prepr. No. 141, 14 pp. (1980).
Submitted to Mon. Not. R. Astron. Soc.

114.112 **Spectroscopy of the Small Magellanic Cloud emission line star Hen S 18.**
M. Azzopardi, J. Breysacher, G. Muratorio.
ESO Sci. Prepr. No. 118, 10 pp. (1980). – Submitted to Astron. Astrophys.

114.113 **The spectrum of the SC star UY Cen 5413 Å to 8839 Å.** R. M. Catchpole.
South African Astron. Obs. Circ., Vol. 1, 183 - 233 (1980).

A line list and atlas is presented for the SC star UY Cen. The wavelength coverage is from 5413 to 8840 Å on 13 Å mm^{-1} spectra. The majority of the 2484 lines measured have been identified.

114.114 **Solar-type phenomena on late-type stars.**
L. Hartmann.
Smithsonian Astrophys. Obs., Spec. Rep. 389, (see 012.038), p. 15 - 34 (1980).

114.115 **An estimate of active region filling factors for dMe and dM stars.** M. S. Giampapa.
Smithsonian Astrophys. Obs., Spec. Rep. 389, (see 012.038), p. 119 - 122 (1980).

114.116 **Chromospheric and coronal activity in F, G and K type stars.** R. Mewe, C. Zwaan.
Smithsonian Astrophys. Obs., Spec. Rep. 389, (see 012.038), p. 123 - 126 (1980).

114.117 **Analyse d'une des caractéristiques des étoiles Ap: la dépression à 5200 Å.** G. Joncas.
J. R. Astron. Soc. Canada, Vol. 74, 261 - 270 (1980).

The structure of the broad-band flux depression at 5200 Å, characteristic of Ap stars, is studied. The author compares a parameter, Δa, quantifying this depression with two other Ap parameters: $\Delta(V_1 - G)$ from the Geneva Photometric System (Hauck 1978) and the blue excesses $\Delta(b-y)$ obtained from a relation between $(b-y)$ and $(B-V)$ for normal stars. The results are then compared with those obtained by Maitzen and Seggewiss (1980). They arrived at the conclusion that the λ5200 depression is composed of two main components having different temperature behaviour. The author is unable to reproduce the results of Maitzen and Seggewiss.

114.118 **SS433 – an October 1979 view.**
J. Shaham.
Comments Astrophys., Vol. 9, 1 - 11 (1980).

114.119 **The absolute calibration of stellar spectrophotometry.**
H. L. Johnson.
Rev. Mexicana Astron. Astrofis., Vol. 5, 25 - 30 (1980).

The combination of 13-color photometry with recent Michelson spectrophotometry, covering the entire spectral range from 4 000 to 10 300 Å, has made possible new absolute calibrations for 16 stars. The new calibrations are remarkably similar to the recent one by Tüg, White and Lockwood. The absolute calibration of stellar spectrophotometry appears now to be quite firm over the spectral range from 4 000 to 10 300 Å.

114.120 **Spectral variations of HD 192163.**
G. Koenigsberger, C. Firmani, G. F. Bisiacchi.
Rev. Mexicana Astron. Astrofis., Vol. 5, 45 - 49 (1980).

Medium dispersion (0.7 Å $channel^{-1}$) spectra of HD 192163 in the 4470-4800 Å region are examined for profile and radial velocity variations. The star is variable in both senses. This analysis supports the idea that HD 192163 may be a binary system. The presence of a collapsed companion is suggested.

114.121 **The symbiotic star near the globular cluster NGC 6401.**
S. Torres-Peimbert, E. Recillas-Cruz, M. Peimbert.
Rev. Mexicana Astron. Astrofis., Vol. 5, 51 - 58 (1980).

Spectrophotometric observations of the emission-line object near the globular cluster NGC 6401 are presented. It is found that it is a symbiotic star and not a planetary nebula. Its spectra show: TiO bands in absorption, very strong H and He lines in emission; there are no forbidden lines present.

114.122 **The distribution of MK spectral types in photometric boxes of the four-color system.**
A. G. D. Philip.

News Lett. Astron. Soc. N. Y., Vol. 1, No. 7, p. 10 - 12 (1980). Abstract.

114.123 **A far-infrared emission feature in the carbon star IRC + 10216 and the planetary nebulae IC 418 and NGC 6572.** W. J. Forrest, J. R. Houck.
News Lett. Astron. Soc. N. Y., Vol. 1, No. 7, p. 17 (1980). Abstract.

114.124 **A large Magellanic Cloud member intermediate between Of and WN7.** K. Nandy, D. H. Morgan, A. J. Willis, P. M. Gondhalekar.
Mon. Not. R. Astron. Soc., Vol. 193, 43P - 47P (1980).

IUE observations of SK-65-22 (O6Iaf+) in the Large Magellanic Cloud show N IV]λ1486 and He IIλ1640 emission lines, characteristics of 'transition WR' stars, as well as the normal Of P Cygni lines. The spectrum of this star is compared with those of WN7 and normal Of LMC members and seen to be intermediate in character. The absolute magnitude is obtained and found to be consistent with the theory that this type of object represents an evolutionary link between Of and WN7 stars.

114.125 **Analysis of shapes of CaII and MgII emission cores in late-type stars.** R. Głębocki, A. Stawikowski.
Acta Astron., Vol. 30, 285 - 298 (1980).

An analysis of well calibrated profiles of CaII H and K emissions for about 70 F, G, K, M type stars indicates that Wilson's parameters W_0 and I_K are good quantitative measures of full width at half maximum and of emission core intensity. It is also shown that the shape of emission profiles does not depend on T_{eff} and M_v but the peak separation (W_2) decreases while the width at the base of emission (W_1) increases with increasing strength of emission. Similar dependence is obtained for MgII h and k lines from well calibrated data. A new $M_v - \log W$ relation is derived for the MgII lines.

114.126 **Spectroscopic analysis of Pollux relative to the Sun with special reference to Arcturus.**
F. Ruland, H. Holweger, R. Griffin, R. Griffin, D. Biehl.
Astron. Astrophys., Vol. 92, 70 - 85 (1980).

A model-atmosphere analysis of the MK K 0 III standard Pollux based on high-resolution photographic and FTS spectra is given. Stellar surface parameters are found to be $T_{eff} = 4840 \pm 50°K$, $\log g = 2.24 \pm 0.35$. In conjunction with the radius, $R = 9.1\ R_\odot$, this value of $\log g$ leads to a mass for Pollux 2.3 times that of Arcturus. Evidence is found for chromospheric inhomogeneities, large-scale photospheric flow patterns (giant granules), and severe departures from LTE that vitiate abundances derived from low-excitation lines of neutral atoms. The chemical composition determined for 30 elements, lithium to thorium, characterizes Pollux as a normal population I object.

114.127 **The atmospheric abundances of the giant Am star 22 Bootis.**
C. Burkhart, C. Van't Veer, M. F. Coupry.
Astron. Astrophys., Vol. 92, 132 - 138 (1980).

The fine abundance analysis of 22 Bootis relative to 15 Vulpeculae shows that an evolved Am star has the characteristic abundance anomalies of the Am stars: Sc and Ca are underabundant, the group (Ti, V, Cr, Mn, Fe) is almost normal, the elements heavier than Fe are overabundant. The evolved Am star 22 Boo does not have the same abundances (referred to Fe) as the five anomalous-abundance δ Del stars (Kurtz, 1976).

114.128 **HR 4453: an anomalously bright UV source?**
R. S. Polidan, W. R. Oegerle, B. Margon.
Astron. Astrophys., Vol. 92, 212 - 213 (1980).

Attention was drawn to HR 4453 recently, when Crawford et al. (1979) reported that this visual binary system has an anomalously large UV flux in the λλ 1350 - 1600 Å band. The authors have investigated this system further and report their results of ultraviolet spectrophotometry of HR 4453 obtained with the Copernicus satellite. Portions of the spectrum from 1120 Å to 2660 Å have been scanned. No stellar signal was detected in any wavelength interval. This result is consistent with both components of the binary being normal A2V stars.

114.129 **A comparison of the Mg resonance lines in Am and non Am stars of similar temperatures.**
E. Böhm-Vitense.
Astron. Astrophys., Vol. 92, 219 - 221 (1980).

A comparison of the Mg II resonance lines in Am and non Am stars of similar effective temperatures does not show any measurable differences. The Mg I resonance lines may be weaker in the Am stars. The mechanism reducing the Ca II K line intensities in Am stars does not work for the Mg II k and h lines.

114.130 **Variability in early-type stars.**
R. Stalio.
Variability in stars and galaxies, (see 012.044), p. E.5.1 - 5.17 (1980).

Variability in light and in the profile shape of Hα and of the ultraviolet resonance lines gives informations of motions and temperature/density structures of the atmospheric layers of early-type stars.

114.131 **Effects of stellar rotation on spectral classification.**
A. Slettebak, T. J. Kuzma, G. W. Collins II.
Astrophys. J., Vol. 242, 171 - 187 (1980).

Line profiles and equivalent widths of spectrum lines used in the classification of early-type stars have been calculated for rigidly rotating B0 - F8 models, taking shape distortion and gravity darkening into account. Changes in line strengths of He I 4144 and 4471, Si II 4128 - 4131, Hγ, Fe I 4476, and Mg II 4481 are shown as functions of the fractional angular velocity and inclination of the rotation axis of the models.

114.132 **Observational studies of the symbiotic stars. I. Hα profile variations in CH Cygni.**
C. M. Anderson, N. A. Oliversen, K. H. Nordsieck.
Astrophys. J., Vol. 242, 188 - 194 (1980).

Moderately high resolution Hα profiles of the symbiotic star CH Cygni obtained with the Washburn Observatory echelle spectrograph and Reticon detector system over the course of 13 months are presented. Several episodes of profile variability were observed. Various kinematic models for these variations are presented and compared.

114.133 **Discovery of the first SC star in the Magellanic Clouds.** H. B. Richer, J. A. Frogel.
Astrophys. J., Lett., Vol. 242, L9 - L12 (1980).

This paper presents photometric and spectroscopic observations of the first SC-type star found in the Magellanic Clouds. The authors point out that in most cases, these stars should be readily distinguishable from C and M stars by infrared photometry alone.

114.134 **HV 11417: a peculiar M supergiant in the Small Magellanic Cloud.**
J. H. Elias, J. A. Frogel, R. M. Humphreys.
Astrophys. J., Lett., Vol. 242, L13 - L17 (1980).

The semiregular Small Magellanic Cloud variable HV 11417 is a cool, luminous star of spectral type M5e I near maximum light. Its spectroscopic and photometric properties resemble those of VX Sgr and similar peculiar galactic M supergiant-like objects. HV 11417 is the first identified extragalactic member of this exotic class of evolved objects. Evidence is presented which suggests that HV 11417 and related galactic stars may be luminous long-period variables rather than true supergiants.

114.135 **Three new hydrogen-deficient B-type stars.** J. S. Drilling.
Astrophys. J., Lett., Vol. 242, L43 - L44,plate L1 (1980) = Contrib. Louisiana State Univ. Obs., No. 159.

The stars designated as LS IV $-1°2$, LSS 1922, and LSS 4300 in the Case-Hamburg Luminous Stars surveys have been found to be extremely hydrogen deficient. The absorption spectra of these stars appear to form a continuous sequence linking the extreme helium stars with the close binary system ν Sagittarii.

114.136 **IUE observations of two late-type stars: R Aql and W Hya.** M. Kafatos, A. G. Michalitsianos, R. W. Hobbs.
Astron. Astrophys., Vol. 92, 320 - 322 (1980).

Ultraviolet spectra of two M stars R Aql and W Hya were obtained. Spectra were obtained of R Aql near maximum ($\Phi = 0.21$) and minimum ($\Phi = 0.65$) of the visible light curve. The authors find that the absolute flux intensity of the Mg II resonance doublet is essentially the same at these phases in the visible light curve. A nebular emission feature at 3133 Å is detected at minimum light in R Aql that is possibly due to O III. Mg II emission is totally absent in W Hya, which contradicts earlier predictions that this star has an 8000 K permanent chromosphere. These results are discussed as they pertain to the formation of silicate grains in cool M giant atmospheres.

114.137 **Scanner observations of Gamma Cassiopeiae.** P. S. Goraya.
Astrophys. Space Sci., Vol. 73, 319 - 326 (1980).

The photoelectric spectrophotometric scans of γ Cas have been analysed to find the stellar and envelope parameters. The absolute energy distribution of γ Cas covering the wavelength interval λ350–700 nm have been given. Its effective temperature and gravity have been estimated by a comparison of the observed energy distribution curves with appropriate model atmospheres. The temperature has also been determined by Zanstra's method from the total energy emitted in the Hα-line. The mass, radius, luminosity, photospheric electron density and mass-ejection rate for γ Cas have been derived. An estimate of the extension of the stellar envelope has been made.

114.138 **Spectroscopic studies of Wolf-Rayet stars with absorption lines. II. The WN8 standard star HD 177230.** P. Massey, P. Conti.
Astrophys. J., Vol. 242, 638 - 645 (1980).

Observations and the reductions of HD 177230 are presented. The absence of variations in radial velocities is discussed. The spectrum is shown and described.

114.139 **Stellar abundances from line statistics.** C. R. Cowley, G. C. L. Aikman.
Astrophys. J., Vol. 242, 684 - 698 (1980).

The method of linear statistical modeling has been combined with a rudimentary model of stellar photospheres to obtain abundance estimates for a large member of "normal" and peculiar, upper main sequence stars. The adopted standard abundances, obtained from published fine analyses, can be fitted by the authors' algorithms with about 0.5 dex as the standard deviation. Abundances, rounded to the nearest 0.5 dex, are reported for Cr, Mn, Fe, and Y. The largest ranges of abundances are found for chromium and yttrium. A remarkable constancy has been found for iron.

114.140 **Spectroscopic and polarimetric observations of NGC 1333 and the surrounding dark cloud complex.** D. A. Turnshek, D. E. Turnshek, E. R. Craine.
Astron. J., Vol. 85, 1638 - 1643 (1980).

Spectroscopy and polarimetry of several very red stars, noted on Near Infrared Photographic Sky Survey photographs in the region of NGC 1333, have yielded a "map" suggesting the extent of polarization arising from a two-cloud structure in the region. Spectral types and color indices of the stars allow to infer that grain radii in the clouds exceed typical interstellar medium values and, further, that they increase with optical depth into the cloud. The authors' observations indicate that the cloud structure is far more extensive than previously realized.

114.141 **Why abundances?** B. E. J. Pagel.
ESO workshop on methods of abundance determination for stars, (see 012.021), p. 1 - 10 (1980).

Motivation for the study of abundances in astronomical objects comes both from a desire to make sure that one understands the physical processes leading to absorption and emission features and from the role played by abundance studies in helping to understand the origin of elements and the evolution of stars, galaxies and the universe. Some comments are made on these various aspects.

114.142 **Spectral analysis of early-type stars (O - F5).** K. Hunger.
ESO workshop on methods of abundance determination for stars, (see 012.021), p. 29 - 30 (1980).

114.143 **How could we improve abundance analyses of late-type stars – and should we?** B. Gustafsson.
ESO workshop on methods of abundance determination for stars, (see 012.021), p. 31 - 44 (1980).

The possibilities for improving abundance analyses of late-type stars are discussed. It is concluded that the most important obstacle for a considerable increase in the accuracy of detailed analyses is our insufficient knowledge of many aspects of the physics of stellar atmospheres. Strategies to improve this situation are commented on.

114.144 **Spectral analysis of special stars.** V. Weidemann.
ESO workshop on methods of abundance determination for stars, (see 012.021), p. 45 - 47 (1980).

114.145 **Absolute spectrophotometry of four supergiants.** V. I. Burnashev.
Izv. Krymskoj Astrofiz. Obs., Tom 62, 3 - 16 (1980). In Russian.

The absolute energy distribution in the supergiant spectra of α Cyg, α Per, γ Cyg, ρ Cas within the spectral range 3300 to 7550 Å has been determined. These absolute spectrophotometric data, corrected for the interstellar absorption and blanketing-effect, are compared to the energy distribution calculated by Parsons and Michalas for supergiant models.

114.146 **Catalogue of the profiles and equivalent widths of the K-line of ionized calcium in metallic-line stars.** V. M. Dobrichev.
Izv. Krymskoj Astrofiz. Obs., Tom 62, 34 - 38 (1980). In Russian.

The K-line profiles of 87 bright Am, A- and F-stars have been measured from spectra with dispersion 15 Å/mm. The halfwidths of the profiles for fixed values of the line depth are presented. The central depths and equivalent widths are also given. In contrast to peculiar stars, the K-line profiles in the metallic-line stars do not show any noticeable peculiar structure.

114.147 **On the origin of anomalous Balmer decrements in the spectra of eruptive stars.** V. P. Grinin.
Izv. Krymskoj Astrofiz. Obs., Tom 62, 54 - 58 (1980). In Russian.

Balmer decrements which are characterized by anomalous

high inverse relation $I_{H\gamma}/I_{H\beta}$ have been observed in the spectra of some eruptive stars. It is shown that the appearance of such Balmer decrements may be considered as evidence of the existence of a large gradient of the physical conditions in the emission region in combination with radiative coupling in the spectral lines between regions with high and low density of the gas.

114.148 **The Balmer decrement in the spectra of moving envelopes of stars.** V. P. Grinin, N. A. Katysheva.
Izv. Krymskoj Astrofiz. Obs., Tom 62, 59 - 65 (1980). In Russian.

The multilevel problem for the hydrogen atom with nonzero boundary conditions is considered. On the basis of the escape-probability method by V. V. Sobolev the state of non-LTE gas lightened by a star with $T_* = 5000$ K has been calculated for the following parameters: $T_e = 10\,000, 15\,000, 20\,000$ K, electron density $n_e = 10^8 - 10^{12}$ cm^{-3}, dilution coefficient $W = 10^{-1}, 10^{-2}, 10^{-3}$, L$\alpha$-quantum escape probability $\beta_{12} = 10^{-3} - 10^{-8}$. The characteristics as relative intensities of Hα, Hβ and Hγ, intensity Hα in units of neighbouring continuum and degree of ionization of the gas are given.

114.149 **The composite spectrum of BY Draconis.**
P. C. Keenan.
Publ. Astron. Soc. Pacific, Vol. 92, 548 (1980). – Abstract.

114.150 **Transition region and chromospheric models of 24 UMa based on IUE ultraviolet spectrograms.**
T. Simon, R. E. Stencel, B. W. Lites.
Publ. Astron. Soc. Pacific, Vol. 92, 550 (1980). – Abstract.

114.151 **A new weak-banded, probable carbon star, having bright Hα.** C. B. Stephenson.
Publ. Astron. Soc. Pacific, Vol. 92, 653 (1980).

114.152 **Spectrophotometry of G-type stars.**
A. Gutiérrez-Moreno, H. Moreno, G. Cortés.
Publ. Astron. Soc. Pacific, Vol. 92, 666 - 674 (1980).

Photoelectric spectrophotometry of 19 G-type stars is presented. The outside-of-the-atmosphere energy distributions, given in a relative system, are included. Both the scans and the energy distributions are used for obtaining spectral classification criteria.

114.153 **Abundances in the atmospheres of cool stars of different age.** N. S. Komarov, A. N. Shcherbak.
Astron. Tsirk., No. 1095, p. 4 - 7 (1980). In Russian.

114.154 **Synthetic spectra of late-type stars.**
V. V. Tsymbal.
Astron. Tsirk., No. 1103, p. 5 - 6 (1980). In Russian.

114.155 **On rapid variations of radial velocities of "relativistic" components of the Hα line in the spectrum of SS433.** V. P. Arkhipova, V. F. Esipov, N. E. Kurochkin.
Astron. Tsirk., No. 1104, p. 1 - 5 (1980). In Russian.

114.156 **A study of stellar spectra with IUE.**
P. Delaney.
J. R. Astron. Soc. Canada, Vol. 74, 355 - 356 (1980). Abstract.

114.157 **Determination of the abundances of chemical elements in the atmospheres of cool giant stars.**
N. S. Komarov, A. N. Shcherbak.
Astron. Tsirk., No. 1106, p. 4 - 6 (1980). In Russian.

114.158 **New carbon stars in selected regions of the Milky Way.** O. M. Kurtanidze.
Astron. Tsirk., No. 1109, p. 2 - 6 (1980). In Russian.

114.159 **Hα line profile in the spectrum of the shell star V923 Aquilae.** V. Ya. Alduseva, E. A. Kolotilov.
Astron. Tsirk., No. 1111, p. 2 - 4 (1980). In Russian.

114.160 **Comparison between three independent spectrophotometric investigations of stars.**
I. N. Glushneva.
Astron. Zh., Tom 57, 1215 - 1217 (1980). In Russian.
English translation in Soviet Astron., Vol. 24, No. 6.

A comparison of stellar spectrophotometric data obtained at the Sternberg Institute with the results of the astrophysical expedition of the USSR Academy of Sciences to Chile and the observations at the Crimean Observatory is performed. It is shown that these data are in satisfactory agreement.

114.161 **Spectrophotometric study of AM Herculis. II.**
N. F. Vojkhanskaya.
Astron. Zh., Tom 57, 1218 - 1226 (1980). In Russian.
English translation in Soviet Astron., Vol. 24, No. 6.

The results of a study of the spectrum at maximum brightness are reported and spectral characteristics at different brightness of the system are compared. Rapid variations of equivalent widths of spectral lines on a time-scale ~1 minute are detected; they are independent on the phase of the orbital period. It is shown that at minimum brightness the degree of excitation of the spectrum goes down, the contribution of radiation of the hot source decreases and the cooler and more rarefied parts of the emitting region become distinguishable.

114.162 **Die Metallgehalte von F- und G-Sternen in den Basler Halofeldern.** C. Trefzger.
Mitt. Astron. Ges., Nr. 50, p. 33 (1980). – Abstract.

114.163 **Das Ultraviolett-Spektrum des Sterns HD 269546.**
M. Grewing.
Mitt. Astron. Ges., Nr. 50, p. 164 - 168 (1980).

114.164 **Supernovae and the Ap phenomenon.**
R. Rajamohan, A. K. Pati.
J. Astrophys. Astron., Vol. 1, 155 - 164 (1980).

The authors suggest that supernova explosions in clusters and associations together with the magnetic accretion hypothesis of Havnes and Conti (1971) and Havnes (1974, 1975) provides the most natural explanation for the wide range of phenomena associated with the chemically peculiar stars.

114.165 **The outer atmospheres of cool stars. VII. High resolution, absolute flux profiles of the Mg II *h* and *k* lines in stars of spectral types F8 to M5.** R. E. Stencel, D. J. Mullan, J. L. Linsky, G. S. Basri, S. P. Worden.
Astrophys. J., Suppl. Ser., Vol. 44, 383 - 402 (1980).

The authors present high-resolution *IUE* spectra of the emission cores of the Mg II resonance doublet at 280 nm in a selection of 54 stars covering a range of spectral type from F8 to M5 and of luminosity class from supergiant (Ia) to subgiant (IV). The authors discuss the qualitative line profile groupings and derive chromospheric radiative losses in the *h* and *k* lines; they discuss these loss rates as functions of effective temperature and luminosity class. The authors make further comparisons of these rates with rates derived for the Ca II H and K lines. Chromospheric velocity fields and indicators of circumstellar envelopes are discussed in terms of profile asymmetries and other diagnostics. Line width measures and velocity shifts of the central reversals are tabulated, among other quantities, and several correlations noted. Finally, the authors discuss the relation of the Wilson K index and stellar coronae to Mg II emission, and note the occurrence of Fe II emission lines in the middle range of the UV of late-type stars.

114.166 **SS 433 – the object nothing like as everything known.** M. Kozłowski.
Urania Kraków, Vol. 51, 354 - 360 (1980). In Polish.

114.167 **Energy distribution in stellar spectra of different spectral types and luminosities. III.**
Z. Sviderskienė.
Bull. Vilnius Astron. Obs., Nr. 55, p. 27 - 47 (1980). In Russian.

114.168 **The stars as suns.** D. E. Blackwell.
The state of the Universe, (see 003.013), p. 68 - 92 (1980).

The author is concerned with the immediate neighborhood of the Sun, out to a distance of a few hundred light-years. He discusses some of the properties of ordinary undistinguished stars, and relates these to current ideas about the way in which stars form and the earlier phases of their evolution.

114.169 **Spectroscopic studies of O type stars. IX. Binary frequency.**
C. D. Garmany, P. S. Conti, P. Massey.
Astrophys. J., Vol. 242, 1063 - 1076 (1980).

The authors have studied a sample of 67 O type stars to determine the frequency of binaries. Binaries are detected in 36% (±7%) of the sample, and of these, 85% have mass ratios less than 2.5. The binary frequency is lower than has been found in earlier studies because a number of stars, previously identified as variable, in fact show only random photospheric variations up to 30 km s^{-1}. It appears that mass ratios greater than 3 may not be present in unevolved O type systems. The authors also discuss the relationship of the highly evolved X-ray binary systems to the unevolved O types. They find that a few of the 67 O stars in the sample could contain neutron star companions and they list several possible candidates.

114.170 **Ultraviolet, X-ray, and infrared observations of HDE 226868 = Cygnus X-1.**
A. Treves, L. Chiappetti, E. G. Tanzi, M. Tarenghi, H. Gursky, A. K. Dupree, L. W. Hartmann, J. Raymond, R. J. Davis, J. Black, T. A. Matilsky, P. Vanden Bout, F. Sanner, G. Pollard, P. W. Sanford, R. D. Joseph, W. P. S. Meikle.
Astrophys. J., Vol. 242, 1114 - 1123 (1980).

During 1978 April, May, and July, HDE 226868, the optical counterpart of Cygnus X-1, was repeatedly observed in the ultraviolet with the IUE satellite. Some X-ray and infrared observations have been made during the same period. The general shape of the spectrum is that expected from a late O supergiant. Strong absorption features are apparent in the ultraviolet, some of which have been identified. The equivalent widths of the most prominent lines appear to be modulated with the orbital phase. This modulation is discussed in terms of the ionization contours calculated by Hatchett and McCray, for a binary X-ray source in the stellar wind of the companion.

Catalogue général d'étoiles de type O. Données spectroscopiques et photométriques. Bande magnétique et listage. See Abstr. 002.023.

The results of a treatment of the tape version of the [Fe/H] catalogue by Morel et al. See Abstr. 002.036.

Catalogue of extinction data of 12547 O . . . F stars, galactic clusters and δ-Cephei stars.
See Abstr. 002.039.

Spectroscopy – what are the needs once space astrometry has given us a new data base.
See Abstr. 002.048.

Spectrophotometry with a self-scanned silicon diode array. III. A catalogue of energy distributions.
See Abstr. 002.053.

Sample spectral atlas for Sirius.
See Abstr. 002.054.

The brightest stars. See Abstr. 003.060.

Proceedings of the 2nd European IUE Conference. Tübingen, Germany, 26 - 28 March 1980.
See Abstr. 012.057.

Ultraviolet astronomy enters the eighties.
See Abstr. 013.007.

On the $e\ ^1\Pi - X\ ^1\Sigma^+$ and $^1\Sigma^+ - X\ ^1\Sigma^+$ transitions of ZrO related to S star spectra. See Abstr. 022.036.

Theoretical profiles for the 1–0 $S(1)$ H_2 line in carbon stars. See Abstr. 022.042.

Can lines of C_2^- be observed in carbon stars?
See Abstr. 022.088.

Relative intensities and predicted new bands of the A (E – X) and B (D – A) systems of ZrO.
See Abstr. 022.153.

Analysis of high resolution stellar line profiles.
See Abstr. 031.529.

The application of a coherent optical data processing system to photographically recorded astronomical spectra.
See Abstr. 031.550.

Measuring stellar temperatures. (Part I).
See Abstr. 031.551.

The two new ESO tools for stellar abundance work.
See Abstr. 034.067.

Synthesis of light metals in the Galaxy. Aluminium abundances in cool halo stars. See Abstr. 061.012.

Models of the P Cygni envelope. An analysis of the formation of hydrogen spectral lines. See Abstr. 064.015.

Observations of the outer atmospheric regions of α Orionis. See Abstr. 064.032.

A comparison between the observed and predicted UV line blocking for blanketed model atmospheres of early type stars. See Abstr. 064.051.

Chromospheres, coronae, and mass loss in stars hotter than the sun. See Abstr. 064.054.

Chromospheres, coronae and mass loss in solar and late-type stars. See Abstr. 064.055.

On the possible cause of observable underabundance of He in the atmospheres of Bp stars.
See Abstr. 064.061.

Coronal dichotomies. See Abstr. 064.070.

Theoretical surface abundances of Population I giants. See Abstr. 065.002.

History of the stellar birthrate from lithium abundances in red giants. See Abstr. 065.031.

Solar plages and the interpretation of stellar Ca II H and K line variations in late type dwarfs.
See Abstr. 071.011.

Rubin-152: a massive star on the galactic fringe?
See Abstr. 111.001.

Radial velocity study of four southern RS CVn candidates and related field stars. See Abstr. 111.002.

An atlas of the shell spectrum of Pleione between 3 167 Å and 4 924 Å. See Abstr. 112.002.

Mass loss and rotation in early-main-sequence B stars.
See Abstr. 112.012.

Molecular abundances in IRC+10216.
See Abstr. 112.023.

Spectroscopic studies of IRC+10216 and similar objects. See Abstr. 112.025.

Infrared spectroscopy of molecules in circumstellar material. See Abstr. 112.026.

Line formation in winds with enhanced equatorial mass-loss rates and its application to the Wolf-Rayet star HD 50896. See Abstr. 112.045.

HD 219150: a star with a remarkable ultraviolet excess. See Abstr. 113.002.

Spectrophotometry of B, A, and F stars. II.
See Abstr. 113.015.

Infrared observations of the WC 5 Wolf-Rayet star HD 115473. See Abstr. 113.037.

Vector space methods of photometric analysis. II. Refinement of the MK grid for B stars.
See Abstr. 113.043.

Vector space methods of photometric analysis. III. The two components of ultraviolet reddening.
See Abstr. 113.044.

Calibration of photometric abundance determinations. See Abstr. 113.045.

Angular diameters by the lunar occultation technique. III. See Abstr. 115.008.

The occultation of 119 Tauri and the effective temperatures of three M supergiants.
See Abstr. 115.010.

Variation of radial velocity of the Ap star ε Ursae Majoris. See Abstr. 116.022.

The binary nature of the single-line Wolf-Rayet star EZ Canis Majoris = HD 50896. See Abstr. 117.010.

The ultraviolet spectrum and flux of HD 152667 (= X-ray source 1653−40?). See Abstr. 117.013.

***IUE* ultraviolet spectra and chromospheric models of HR1099 and UX Arietis.** See Abstr. 117.036.

New spectral classifications on the MK system for visual double stars. See Abstr. 118.023.

Short-time fluctuations of hydrogen and helium emissions in the spectrum of β Lyrae.
See Abstr. 119.038.

The ultraviolet spectrum of the T Tauri star RW Aurigae. See Abstr. 121.006.

UV spectrograms of T Tauri stars.
See Abstr. 121.007.

Electron scattering in the infalling envelope of the protostar S CrA. See Abstr. 121.008.

Multichannel spectrophotometry of stellar flares.
See Abstr. 122.029.

Ultraviolet spectroscopy of F and G supergiants with *IUE*. I. First results on Cepheid variables.
See Abstr. 122.047.

Periodic spectral variability of the Ap star HR 234.
See Abstr. 122.070.

BI Crucis: a new symbiotic star. See Abstr. 122.109.

Spectrum variability in HR 8752.
See Abstr. 122.111.

Fast time-resolution spectroscopy of BW Vulpeculae. See Abstr. 122.174.

A hot blue star near the center of the remnant of supernova A. D. 1006. See Abstr. 125.058.

Spectrophotometry of W50: the supernova remnant around SS 433. See Abstr. 125.070.

A proposed new white dwarf spectral classification.
See Abstr. 126.005.

IUE observations of the early-type subdwarf HD 49798. See Abstr. 126.017.

Hot subluminous stars. See Abstr. 126.027.

Studies of ultraviolet interstellar extinction with the Sky-Survey telescope of the TD-1 satellite. Results for different OB-associations. See Abstr. 131.110.

Interstellar extinction in the direction of CI Cyg.
See Abstr. 131.279.

The calculation of the optical spectra of NGC 6888.
See Abstr. 134.004.

IUE observations of two planetary nuclei with temperatures of 150,000 K. See Abstr. 135.011.

Ultraviolet spectra of planetary nebulae and their central stars. See 135.027.

Optical spectroscopy of Centaurus X-3.
See Abstr. 142.067.

Optical and X-ray studies of 2A 1822 - 371.
See Abstr. 142.111.

Spectral types in the open cluster NGC 6231.
See Abstr. 153.002.

The metallicity of M67. See Abstr. 153.006.

Analysis of low dispersion spectra of globular cluster stars. See Abstr. 154.017.

Abundances in globular cluster red giants. III. M71, M67, and NGC 2420. See Abstr. 154.025.

The chemical evolution of the solar neighborhood. I. A bias-free reduction technique and data sample. See Abstr. 155.037.

Iron-peak abundances in the nuclear regions of M31, M81, and M94. See Abstr. 158.100.

Studies of luminous stars in nearby galaxies. VI. The brightest supergiants and the distance to M33. See Abstr. 158.168.

Studies of luminous stars in nearby galaxies. VII. The brightest blue stars in the spiral galaxies M101 and NGC 2403. See Abstr. 158.169.

IUE observations of Large Magellanic Cloud members. See Abstr. 159.003.

Highly ionized species in the spectra of Small Magellanic Cloud stars. See Abstr. 159.012.

Ultraviolet studies of the Magellanic Clouds – I. Interstellar lines in the spectra of FD 70 and SK-71-45. See Abstr. 159.019.

Errata

114.901 **Erratum:"A distant supergiant in the Norma section of the Milky Way"** [Publ. Astron. Soc. Pacific, Vol. 92, 36 - 37 (1980)]. J. C. Muzzio, H. Levato.
Publ. Astron. Soc. Pacific, Vol. 92, 384 (1980). – See Abstr. 27.114.143

114.902 **Erratum; 'The effective temperature, radius, rate of mass loss, and luminosity of P Cygni, HD 190603, κ Cassiopeiae, and ρ Leonis'** [Astrophys. J., Vol. 234, 528 - 537 (1979)]. A. B. Underhill.
Astrophys. J., Vol. 239, 414 (1980). – See Abstr. 26.114.088.

114.903 **Erratum: "Spectral atlas of helium-rich stars."** [Astron. Astrophys., Suppl. Ser., Vol. 41, 271 - 294 (1980)]. J. P. Kaufmann, U. Theil.
Astron. Astrophys., Suppl. Ser., Vol. 42, 282 (1980).
See Abstr. 27.114.190.

115 Luminosities, Masses, Diameters, HR and other Diagrams

115.001 **Luminosity functions of disk and halo populations in SA 51, SA 57, and SA 68.** L.-T. G. Chiu.
Astron. J., Vol. 85, 812 - 823 (1980).

An investigation of the luminosity functions of Population I main-sequence stars, Population I white dwarfs, and Population II subdwarfs in SA 51, SA 57, and SA 68 is made using the V'/V'_m method. The luminosity function of Population I main-sequence stars in the author's sample depends on the scale height z_0 adopted. The permissible range of z_0, using the V'/V'_m test, is 300-500 pc. From kinematic considerations, it is argued that the appropriate z_0 for the present sample is 500 pc. The space density of Population I white dwarfs is estimated to be 2×10^{-2} pc^{-3}. The scale height for Population I white dwarfs is 400-500 pc. The local mass density of Population II objects is estimated to be 3×10^{-4} $M_\odot$ pc^{-3}.

115.002 **An observational H-R diagram for early-type stars.** A. B. Underhill.
Astrophys. J., Vol. 239, 220 - 236 (1980).

The positions of 164 O and B stars, five Wolf-Rayet stars, and P Cyg in the (log $L/L_\odot$, log T_{eff})-plane are compared with the theoretical position of the core-hydrogen–burning band. It is concluded that the observations can be construed as providing support for the choice of Carson opacities and that at type B3 and later there is a significant discrepancy between the position of the observed zero-age main sequence (ZAMS) demonstrated by nearby stars, and the predicted ZAMS, no matter what choice is made for the opacity tables. The star P Cyg falls in the region of the late B-type supergiants. Evolved massive stars cannot be identified with models of a particular mass and age unless an algorithm which accurately represents the variation of the rate of mass loss as a function of the instantaneous properties of the star is used. The Wolf-Rayet stars are shown to have effective temperatures near 30,000 K and to fall in the core-hydrogen–burning band just above the position of the B0 III stars. They are discussed in detail.

115.003 **What size is Aldebaran?**
D. S. Evans, D. A. Edwards, B. R. Pettersen, E. L. Robinson, W. H. Sandmann, J. R. Wiant.
Astron. J., Vol. 85, 1262 - 1264 (1980).

In the visible region the best value for the angular diameter of Aldebaran, assumed to have a fully darkened disk, is 19.9 ± 0.3 arc ms. None of the Aldebaran observations show the peculiarities found in the case of Antares. There is no indication of variation with time. The linear diameter is near 45 solar diameters.

115.004 **Stellar masses.** D. M. Popper.
Annu. Rev. Astron. Astrophys., Vol. 18, (see 003.004), 115 - 164 (1980).

In an earlier review (Popper 1967), the problems of determining masses from data for eclipsing binaries were discussed and some of the available results tabulated. The present review may be considered a sequel, although the most definitive visual binaries are also included. The same viewpoint is adopted, namely that only individual masses of considerable accuracy, determined directly from the observational data, are treated.

115.005 **Supergiant variability: amplitudes and pulsation constants in relation with mass loss and convection.**
A. Maeder.
Astron. Astrophys., Vol. 90, 311 - 317 (1980).

The distribution in the HR diagram of the amplitudes of light variations for supergiants is established on the basis of 2420 observations of supergiants made over the last 20 yr in the Geneva photometry (Rufener, 1979). It is shown that (1) for any spectral type the amplitudes increase with the luminosity, (2) for Ia supergiants there is a small local maximum of amplitudes in B-type stars, (3) for G–M supergiants the higher is the luminosity, the earlier is the spectral type at which important amplitudes (> $0^m.1$) appear. A relation is found between the amplitudes of light variations and the rate of mass loss $\dot{M}$ for B and A supergiants; it indicates an exponential increase of $\dot{M}$ with the amplitudes. Empirical pulsation constants Q for B-G supergiants are determined and they are found to be systematically larger than the theoretical Q for the fundamental mode of radial oscillation.

115.006 **A period–luminosity relation for supergiant red variables in the Large Magellanic Cloud.**
M. W. Feast, R. M. Catchpole, B. S. Carter, G. Roberts.
Mon. Not. R. Astron. Soc., Vol. 193, 377 - 380 (1980).

Infrared photometry for 24 red supergiant variables in the LMC is used to derive bolometric magnitudes. The existence of a period–luminosity relation for these stars is demonstrated and compared with theory.

115.007 **A plot of *UBV* diagram.** B. Nicolet.
Astron. Astrophys., Suppl. Ser., Vol. 42, 283 - 284 (1980).

A plot containing more than 46000 stars having both *U-B* and *B-V* colour indices in the catalogue of Nicolet (1978) is presented. The plot has been drawn by a BENSON tracer.

115.008 **Angular diameters by the lunar occultation technique. III.**
S. T. Ridgway, G. H. Jacoby, R. R. Joyce, D. C. Wells.
Astron. J., Vol. 85, 1496 - 1504 (1980).

Diameters for 13 stars are reported. Two of these involve simultaneous observations at two telescopes, and repeat measurements are presented for three additional stars. A highly reliable diameter of the carbon star SZ Sgr is reported, and included in an analysis of the relationship between the effective and color temperatures for carbon stars. Formal error estimates for the diameters from least-squares solutions are discussed based on accumulated multiple measurements for 24 stars.

115.009 **Lunar occultation angular diameter measurements.**
W. I. Beavers, J. J. Eitter, D. W. Dunham, W. L. Stein.
Astron. J., Vol. 85, 1505 - 1508 (1980).

The analyses of one dozen lunar occultation diameter candidate observations are reported. Within this set of occultation measurements at Fick Observatory, six of the stars provide sensible angular diameters, and the remainder appear as virtual point sources. Angular diameter measurements are reported for ϵ Gem, BD + 24°0571, BD + 01°2519, υ Cap, R Gem, and BD + 23°1518.

115.010 **The occultation of 119 Tauri and the effective temperatures of three M supergiants.**
N. M. White.
Astrophys. J., Vol. 242, 646 - 656 (1980).

The angular diameter of 119 Tauri has been measured in three colors by lunar occultation, and it is used to derive the effective temperature and linear radius of this M2.2 Iab$^-$-type star. These values are compared with those calculated for two stars of similar spectral type, α Orionis and α Scorpii. Consistent results are obtained for the three stars and with recent model-atmosphere effective temperatures if α Orionis is assumed to have a variable photospheric diameter and if an empirical correction for the possible effects of circumstellar clouds is applied to the three stars. New mean spectral types and near-infrared photometry are given.

115.011 **The angular diameters of supergiant stars from speckle interferometry.**
G. L. Welter, S. P. Worden.
Astrophys. J., Vol. 242, 673 - 683, plate 18 (1980).

The Lynds et al. image reconstruction algorithm is applied to speckle interferometric data for the supergiant stars α Ori, o Cet, α Tau, ρ Per, and α Her. Further restoration is applied to the images of α Ori and o Cet. Estimations of angular diameter as a function of wavelength and limb darkening are made, confirming the observation by Bonneau and Labeyrie that such stars appear smaller at long wavelengths. An upper limit is placed on the degree of large-scale surface structure on α Ori.

A search for carbon stars in the AFGL Catalogue.
See Abstr. 002.027.

Catalogue of masses and ages of stars in 68 open clusters. See Abstr. 002.042.

A catalogue of masses, radii, and luminosities of 71 double-lined spectrum eclipsing binaries.
See Abstr. 002.063.

The brightest stars. See Abstr. 003.060.

Het Hertzsprung-Russell-Diagram.
See Abstr. 014.010.

Estimate of the spectral density of images with the help of a coherent-optical spectrum analyser.
See Abstr. 031.525.

Stellar age and mass determinations from kinematic data. See Abstr. 065.073.

Illinois occultation summary. I. 1977 - 1978.
See Abstr. 096.002.

Fading occultations of stars by the Moon: 1943–1977. See Abstr. 096.010.

Classification of stellar populations and luminosity classes from accurate proper motions. See Abstr. 111.019.

Measurement of the absolute flux from Vega in the *K* band (2.2 μm). See Abstr. 113.009.

Vector space methods of photometric analysis. II. Refinement of the MK grid for B stars.
See Abstr. 113.043.

Effective temperatures and luminosities for B1 Ia+, O4 and WN7 stars. See Abstr. 114.034.

A survey of chromospheric Ca II H and K emission in field stars of the solar neighborhood. See Abstr. 114.107.

Comparison of activity cycles in old and young main-sequence stars. See Abstr. 114.108.

Scanner observations of Gamma Cassiopeiae.
See Abstr. 114.137.

Slightly detached binaries as calibrators of the main-sequence. See Abstr. 117.023.

Luminosity of the Mira variables.
See Abstr. 122.056.

Distances and radii of classical cepheids.
See Abstr. 122.093.

The absolute magnitudes of RR Lyrae stars.
See Abstr. 122.194.

On solar type protostars. See Abstr. 131.308.

The extended giant branches of intermediate age globular clusters in the Magellanic Clouds.
See Abstr. 154.009.

The stellar content of dwarf spheroidal galaxies.
See Abstr. 158.124.

Studies of luminous stars in nearby galaxies. VI. The brightest supergiants and the distance to M33.
See Abstr. 158.168.

On the stellar content and structure of the spiral galaxy M33. See Abstr. 158.318.

Luminosities and temperatures of the reddest stars in three LMC clusters. See Abstr. 159.006.

116 Magnetic Fields, Polarization, Figure, Rotation, Radio Radiation

116.001 **Relation between surface magnetic field intensities and Geneva photometry.**
N. Cramer, A. Maeder.
Astron. Astrophys., Vol. 88, 135 - 140 (1980).

B-type stars may be classified in a reddening free 3-dimensional orthogonal space based on Geneva photometry, such that the X-axis corresponds to the direction of T_{eff} effects and the Y-axis to luminosity effects. The third dimension Z is shown to be related in a very useful way to the intensity of the surface magnetic field for Ap and Bp stars with T_{eff} between 10,000° and 20,000° K; this clear relation, which is valid up to 5 kG, is attributed to the 5200 Å depletion present in Ap and Bp stars. The observed relation does not correspond to the prediction of models of magnetic stars; this may be due to the fact that the autoionization of spectral lines is not incorporated in the existing models.

116.002 **Radio studies of stellar activity with the Arecibo interferometer.**
K. C. Turner, M. M. Davis, Y. Terzian.
Bull. American Astron. Soc., Vol. 12, 499 - 500 (1980). Abstract.

116.003 **Variations in the radio flux of SS433 at 11.1 and 3.75 cm.**
K. Johnston, N. Santini, J. Spencer, G. Kaplan, W. Klepczynski, J. Josties, D. Matsakis, P. Angerhofer.
Bull. American Astron. Soc., Vol. 12, 540 (1980). – Abstract.

116.004 **A search for radio spectral lines from SS 433.**
N. L. Cohen, F. D. Drake.
Astron. Astrophys., Vol. 89, L6 - L7 (1980).

A search has been made for radio spectral lines associated with SS433, from 1345 to 1412.5 MHz. No evidence for lines with velocity widths from 9 to 7000 km s^{-1} was found to a best limit of 40 mJy. This lack of lines may be explained by a variety of line broadening or suppression mechanisms.

116.005 **Differential rotation and magnetic activity of the lower main sequence stars.**
G. Belvedere, L. Paterno, M. Stix.
Stellar turbulence, (see 012.008), p. 296 - 297 (1980). Abstract.

116.006 **Radial velocity and magnetic field measurements of the A-type supergiant ν Cep (HD 207260).**
G. Scholz, E. Gerth.
Astron. Nachr., Band 301, 211 - 216 (1980).

On the basis of 55 Zeeman spectrograms obtained at Tautenburg in the years 1975 - 1979 the effective magnetic fields and the radial velocities of the supergiant ν Cep (HD 207260) were determined. For the effective magnetic field a slow variation occurring on a time scale of years was found. In 1978 values of +2000 Gauss were detected. The measured radial velocities show a longtime variability similar to that of the magnetic field as well as more rapid changes. A periodical variation of the radial velocity with a period of ~ 39.9 days, perhaps produced by the rotation of the star, is indicated.

116.007 **Intrinsic polarization variations of Mira late-type stars due to optical changes of irradiated grains.**
J. Svatoš.
Bull. Astron. Inst. Czechoslovakia, Vol. 31, 302 - 305 (1980).

Intrinsic time-dependent polarization of some individual late-type stars (e. g., Mira) is explained due to optical changes of UV and X-ray irradiated amorphous silicate grains.

116.008 **18-cm observations of an outburst in V711 Tau.**
J. D. Fix, M. J. Claussen, D. J. Doiron.
Astron. J., Vol. 85, 1238 - 1239 (1980) = Iowa Radio Astron. Group, Contrib. No. 79.

The authors observed V711 Tau at 18 cm using the Arecibo telescope during a modest radio outburst on 17 December 1979. The authors found that the emission from V711 Tau was more highly circularly polarized (predominantly left circular) than ever previously reported. The emission, especially in left circular polarization, was highly variable on a time scale of a few minutes.

116.009 **Optical polarization and infrared spectrum of a possible protostar in a reflection nebula.**
E. P. Ney, B. F. Hatfield, R. D. Gehrz.
Proc. Natl. Acad. Sci. USA, Vol. 77, 14 - 17 (1980). – Abstr. in Phys. Abstr., Vol. 83, Abstr. 85944 (1980).

116.010 **Near infrared polarimetry of cool stars.**
A. McCall, J. H. Hough.
Astron. Astrophys., Suppl. Ser., Vol. 42, 141 - 154 (1980).

Broad band near infrared polarimetry of 29 cool stars is presented. Measurements were made at the effective wavelengths 0.91 μm (I), 1.21 μm (J), 1.65 μm (H) and 2.18 μm(K). The program stars include M giants and supergiants, M and S type Mira variables and carbon stars. Theoretical calculations, interpreting the polarization in terms of the scattering properties of circumstellar dust grains, are also presented.

116.011 **Coherent properties of the radiation of a rotating star.**
S. G. Kravchuk, A. V. Mandzhos, F. E. Khlystun.
Problems of cosmic physics. Vyp. 15, (see 003.003), p. 147 - 152 (1980). In Russian.

The interference properties of the radiation of rotating stars have been investigated. It has been shown that taking into account the rotation effect one may determine the value and orientation of the angular velocity projection on the plane perpendicular to the line of sight.

116.012 **Discovery of intrinsic linear polarization in SS433.**
I. S. McLean, S. Tapia.
Nature, Vol. 287, 703 - 705 (1980).

The authors report new observations which reveal that the linear polarization of the optical radiation from SS433 exhibits a long-term variation as well as large night-to-night changes. This establishes that a substantial fraction of the observed polarization is intrinsic to SS433 whereas the remainder may be produced in the interstellar medium.

116.013 **Polarimetric observations of MV Lyr.**
Yu. S. Efimov, N. M. Shakhovskoj.
Izv. Krymskoj Astrofiz. Obs., Tom 61, 120 - 123 (1980). In Russian.

MV Lyr was observed for linear polarization. There were detected irregular variations of the light and linear polarization. No correlation was found between the light and polarization fluctuations.

116.014 **Rotational studies of lower main sequence stars.**
M. A. Smith.
Highlights of Astronomy, Vol. 5, (see 012.030), 827 - 830 (1980).

This review summarizes the techniques and limitations involved in determining small rotational velocities in late-type stars. Recent results from photoelectric line profiles of field main sequence G stars are presented.

116.015 **The rotation of pre-main sequence stars.**
L. V. Kuhi, S. Vogel.
Highlights of Astronomy, Vol. 5, (see 012.030), 835 - 837 (1980).

116.016 **Polarimetry of emission lines as a probe of stellar winds.** I. S. McLean.
Highlights of Astronomy, Vol. 5, (see 012.031), 859 (1980).

116.017 **The quiescent radio spectrum of SS 433.**
E. R. Seaquist, W. Gilmore, G. J. Nelson, W. J. Payten, O. B. Slee.
Astrophys. J., Lett., Vol. 241, L77 - L81 (1980).

Nearly simultaneous low- and high-radiofrequency observations of SS 433 on two epochs for which the source was relatively quiescent show that its nonthermal spectrum flattens and probably turns over near 0.3 GHz. Several physical mechanisms are examined as possible causes of the observed spectrum. The analysis permits to rule out foreground absorption by the interstellar medium or an intervening H II region, but several other mechanisms intrinsic to the source are plausible. A map made at 160 MHz shows the source is essentially unresolved, indicating an angular diameter less than about 1′ at that frequency.

116.018 **Aperture synthesis of the radio structure of SS433.**
W. Gilmore, E. R. Seaquist.
Astron. J., Vol. 85, 1486 - 1495 (1980).

On the epoch 22 July 1979 VLA radio synthesis maps of SS433 showed the existence of radio structure at both 1.5 and 4.9 GHz. The structure has size of a few arcseconds and is closely aligned with the bulges of the SNR W50 lying nearly a degree away to the east and west. Thus the observed structure may be due either to the hypothesized precessing beams or to flaring activity from the central source, or both. Linear polarization was detected within $\lesssim 0''.5$ of the unresolved core at a level ~ 1% of the core total intensity.

116.019 **X-ray emission from LSI + 61°303.**
IAU Circ., No. 3518 (1980).

116.020 **The millimeter wavelength spectrum of MWC 349.**
P. R. Schwartz,
Publ. Astron. Soc. Pacific, Vol. 92, 534 - 535 (1980).

The millimeter wavelength spectrum of MWC 349 has been measured and fit to an empirical spectrum. It is suggested that objects like MWC 349 are attractive calibrators at millimeter, submillimeter, and far-infrared wavelengths.

116.021 **A VLA sequel to HEAO-1 detections of main sequence stars.** H. M. Johnson, W. C. Cash, Jr.
Smithsonian Astrophys. Obs.,Spec. Rep. 389,(see 012.038), p. 137 - 139 (1980).

VLA observations establish upper limits of about 1 mJy on the 4885-MHz continuum of a sample of stars of known X-ray luminosity, L_x. The combined data are presented in a diagram which shows that, scaling from the Sun, a direct increase of radio flux density S_ν with L_x is permissible, but the radio emission from RS CVn stars increases more rapidly than this, and their scaling law ($S_\nu \propto L_x^{3/2}$) can be ruled out for the isolated main sequence stars.

116.022 **Variation of radial velocity of the Ap star ϵ Ursae Majoris.** A. Woszczyk, M. Jasińki.
Acta Astron., Vol. 30, 331 - 345 (1980).

Radial velocities at several phases from spectral lines of different elements were determined. The variations are sinusoidal for Fe, Cr and Ti. The results are discussed on the basis of an oblique rotator model. The conclusion is drawn that on the surface of the star exists a spot of increased abundance of some elements.

116.023 **Ordered linear polarization changes in Betelgeuse.**
D. P. Hayes.
Astrophys. J., Lett., Vol. 241, L165 - L168 (1980).

Betelgeuse's linear polarization was intensively monitored over the 1979 - 1980 observing season, an interval of 7 months. Significant ordered (as opposed to stochastic) changes were detected. The polarization variations arise from changes in the lower atmosphere – either directly through changes in the photospheric scattering regime, or indirectly through changes in the illumination of circumstellar material. Mechanisms to account for these lower atmospheric variations are indicated.

116.024 **Radio emission from Cyg OB 2 No. 12.**
H. J. Wendker, W. J. Altenhoff.
Astron. Astrophys., Vol. 92, L5 - L6 (1980).

Radio emission of the star No. 12 of the Cygnus OB 2 association has been detected. It is discussed in terms of the rate of mass loss. The rate is still compatible with current understanding of a fast line-driven stellar wind of early type stars, even at this rather late spectral type and very high luminosity. It seems to indicate that for evolved stars the strength of this wind may be independent of spectral type.

116.025 **Circular and linear polarization in A-type shell stars.**
G. C. Clayton, J. M. Marlborough.
Astrophys. J., Vol. 242, 165 - 170 (1980).

An attempt was made to determine longitudinal surface magnetic fields in eight A-type shell stars from observations of circularly polarized light in the wings of Hα and Hβ produced by the Zeeman effect. No fields were detected above the 2 σ level of significance. An upper limit of approximately 300 gauss can be placed on any fields which may exist. The same sample of stars were also observed for continuum linear polarization. No intrinsic polarization was found. The lack of intrinsic linear polarization supports the conclusions that the A-type shells are predominantly neutral.

116.026 **Rotational modulation of chromospheric variations of main-sequence stars.** R. W. Stimets, R. H. Giles.
Astrophys. J., Lett., Vol. 242, L37 - L41 (1980).

The data of O.C. Wilson on the Ca H-K flux variations of main-sequence stars have been analyzed for periodicities by an autocorrelation technique which can be successfully employed with the sparse and quasi-periodic sampling schedule used to take the data. Rotational periods are determined for 10 stars. When the rotational frequencies are plotted versus spectral type for these stars and the Sun, the results are generally consistent with a rule relating degree and type of chromospheric activity to rotation, postulated on the basis of previous data.

116.027 **The magnetic Ap stars.** J. Zverko.
Kozmos, Vol. 11, 135 - 138 (1980). In Slovak.

116.028 **Broad-band polarization observations of SS 433.**
J. J. Michalsky, G. M. Stokes, P. Szkody, N. R. Larson.
Publ. Astron. Soc. Pacific, Vol. 92, 654 - 656 (1980).

The authors report unfiltered visible linear and circular polarization observations of SS 433. The source does not exhibit circular polarization, but it does have a somewhat unusual interstellar linear polarization.

116.029 **A search for linear polarization variability in pole-on Be stars.** D. P. Hayes.
Publ. Astron. Soc. Pacific, Vol. 92, 661 - 665 (1980).

Five Be stars classified as pole-on were monitored for linear polarization variability. Only one star (ω Ori) showed polarization changes. These changes are attributed to variable mass loss. Observations of this star indicate an axially symmetric distribution of envelope scattering material.

116.030 **Periodic radio emission from LSI + 61°303.**
A. R. Taylor, P. C. Gregory.
J. R. Astron. Soc. Canada, Vol. 74, 358 (1980). – Abstract.

116.031 **Radio superflares in RS Canum Venaticorum binaries.** P. A. Feldman.
J. R. Astron. Soc. Canada, Vol. 74, 358 - 359 (1980). Abstract.

Circular polarization of molecular spectra. See Abstr. 022.025.

Magnetic field measurements on stellar sources: a new method. See Abstr. 031.527.

Rapidly rotating α^2-dynamo models. See Abstr. 062.006.

Magneetvelden in de astrofysika. See Abstr. 062.092.

The magnetospheres of magnetic A stars and of pulsars. See Abstr. 062.109.

Cyclotron absorption in accreting magnetic white dwarfs. See Abstr. 062.113.

Stellar atmospheres with magnetic fields. See Abstr. 064.081.

Stellar winds, fast rotators, and magnetic acceleration. See Abstr. 064.085.

Turbulent surface layer of rotating B-type stars. See Abstr. 065.048.

On dynamos in the cores of magnetic stars. See Abstr. 065.059.

Magnetic cycles of lower main sequence stars. See Abstr. 065.060.

On unstably stratified toroidal magnetic fields in stars. See Abstr. 065.067.

Global circulation of the sun: where are we and where are we going? See Abstr. 080.045.

Theory of magnetic field generation of the earth and celestial bodies. See Abstr. 084.040.

The radial velocity of the Am star 68 Tau. See Abstr. 111.010.

Weak shells around rapidly rotating early type stars. See Abstr. 112.006.

Mass loss and rotation in early-main-sequence B stars. See Abstr. 112.012.

Radio detection of ammonia in IRC+10216. See Abstr. 112.022.

Polarized emission in the broad SiO feature from R Leo. See Abstr. 112.032.

Photometry and polarimetry of the extreme blue straggler K1211 in NGC 7789. See Abstr. 113.027.

Outer atmospheres of cool stars. V. IUE observations of Capella: the rotation-activity connection. See Abstr. 114.084.

AM Herculis. See Abstr. 114.103.

Effects of stellar rotation on spectral classification. See Abstr. 114.131.

Spectroscopic and polarimetric observations of NGC 1333 and the surrounding dark cloud complex. See Abstr. 114.140.

Supernovae and the Ap phenomenon. See Abstr. 114.164.

The variability of the polarization of VV Cep. See Abstr. 117.007.

Rotation in close binary stars. See Abstr. 117.028.

Radio aspects of stellar activity in close binaries. See Abstr. 117.031.

The May-June 1979 radio outburst of HR 5110. See Abstr. 119.021.

A polarimetric study of U Cephei. Part I. See Abstr. 119.034.

Variability in T Tauri stars: some new aspects. See Abstr. 121.011.

U Orionis: corrélation entre l'émission sur 1612 MHz et la courbe de lumière. See Abstr. 122.020.

Linear polarization of three α^2 CVn stars. See Abstr. 122.038.

VLA observations of M-dwarf flare stars. See Abstr. 122.039.

Coordinated meter-wavelength observations of the X-ray flare from YZ Canis Minoris. See Abstr. 122.043.

A magnetic study of spotted UV Ceti flare stars and related late-type dwarfs. See Abstr. 122.059.

The periods of 21 Com. See Abstr. 122.067.

Rapid oscillations in cataclysmic variables. IV. WZ Sagittae. See Abstr. 122.073.

Polarimetric observations of R CrB. See Abstr. 122.078.

Photometry and polarimetry of VW Hydri during the October 1978 supermaximum. See Abstr. 122.083.

VLBI observations of main-line OH emission from U Orionis. See Abstr. 122.099.

Variables spectrales et magnétiques. See Abstr. 122.160.

Is there a magnetic field-period relation for the hotter Ap-stars? See Abstr. 123.010.

Hot subluminous stars. See Abstr. 126.027.

Circular polarization measurements of selected white dwarfs. See Abstr. 126.037.

VLA maps of V1016 Cyg, Hb12, CRL2591, HR8752, and P Cyg. See Abstr. 141.030.

Results of synchronous observations of the peculiar object SS 433 in radio and optical ranges. See Abstr. 141.121.

A critique of the polarimetric evidence on the nature of Cygnus X-1. See Abstr. 142.078.

On the phase-locked polarization variations in Cygnus X-1. See Abstr. 142.079.

117 Close Binaries (Observations, Theory)

117.001 **Evolution of a blue supergiant with a neutron star companion immersed in its envelope.**
A. J. Delgado.
Astron. Astrophys., Vol. 87, 343 - 348 (1980).

The evolution of a binary system consisting of 1 $M_\odot$ neutron star and a 25 $M_\odot$ blue supergiant through a phase of common envelope is investigated. The author includes the effects of an additional energy source on the supergiant's envelope, due to the presence of the neutron star, and variable mass loss from the system, taken as proportional to the total luminosity. The results indicate that, independently of the initial period, the system loses its whole envelope as a consequence of the common envelope phase, the final product of this being a detached system, consisting of a neutron star and a helium star.

117.002 **Fundamental photometric data for two contact binaries: MW Pavonis and TY Mensae.**
E. Lapasset.
Astron. J., Vol. 85, 1098 - 1102 (1980).

The *UBV* light curves of MW Pav and TY Men were reanalyzed by means of the Wilson and Devinney method. They were found to have contact configurations. The asymmetrics in the light curves were handled as if they were caused by a hot spot. The evolutionary stage was estimated by two different methods, and in both cases the systems were found to be evolved. It is concluded that MW Pav is a classic A-type W UMa system, while TY Men is an early-type contact binary with marginal physical and thermal contact.

117.003 **An analysis of the radial velocity curves of close binary systems, II.** Z. Kopal.
Astrophys. Space Sci., Vol. 71, 65 - 74 (1980).

The aim of the paper has been to investigate the effects, on radial velocities of the components of close binary systems, of atmospheric gas motions caused by mutual irradiation of the two stars. Quantitative determination of this effect is given for the simplified case in which the symmetry of gas motion can be described in terms of zonal harmonics of arbitrary degree, and a brief comparison is made with the observed radial velocities of the B9-component of the eclipsing system U Cephei, which is known to move in a circular orbit, but exhibits a radial-velocity curve of marked skew-symmetry.

117.004 **On the influence of radiation pressure on the light curve of HZ Herculis.** J. Krebs.
Astron. Astrophys., Vol. 88, 363 - 364 (1980).

The radiation pressure exerted on HZ Her by its X-ray companion is estimated by means of a perturbation method.

117.005 **An analysis of the light curves of V711 Tau using the starspot model.** J. D. Dorren, E. F. Guinan.
Bull. American Astron. Soc., Vol. 12, 452 (1980). – Abstract.

117.006 **New photoelectric light curves of AW UMa.**
B. J. Hrivnak.
Bull. American Astron. Soc., Vol. 12, 452 (1980). – Abstract.

117.007 **The variability of the polarization of VV Cep.**
R. J. Pfeiffer, R. H. Koch.
Bull. American Astron. Soc., Vol. 12, 499 (1980). – Abstract.

117.008 **Absorption line asymmetries in HR 1099.**
F. C. Fekel.
Bull. American Astron. Soc., Vol. 12, 500 - 501 (1980). Abstract.

117.009 **New BVRI light curves of HD 224085.**
H. L. Nations, L. W. Ramsey.
Bull. American Astron. Soc., Vol. 12, 530 (1980). – Abstract.

117.010 **The binary nature of the single-line Wolf-Rayet star EZ Canis Majoris = HD 50896.**
C. Firmani, G. Koenigsberger, G. F. Bisiacchi, A. F. J. Moffat, J. Isserstedt.
Astrophys. J., Vol. 239, 607 - 621 (1980).

Results of spectral and photoelectric observations of HD 50896 obtained in the period 1975 February–1978 March are presented. The variations in emission-line profiles, radial velocities, and light are found to be consistent with the 3.76 day period. It is concluded that the binary hypothesis is the most likely one to explain the periodic variations, and that the companion is probably a neutron star of mass $m_2 = 1.3 \pm 0.4\, M_\odot$. In addition, it is found that the eccentricity differs considerably from zero. It is suggested that this eccentricity is responsible for some of the more extraordinary variations observed in HD 50896.

117.011 **On various criticisms of the contact discontinuity model.** F. H. Shu, S. H. Lubow, L. Anderson.
Astrophys. J., Vol. 239, 937 - 940 (1980).

The authors discuss various criticisms raised recently by Lucy and Wilson, by Papaloizou and Pringle, and by Smith, Robertson, and Smith against the contact discontinuity model for contact binaries. In the process they discover a promising means by which the filled fraction occupied by the common envelope could, in principle, be determined mechanistically for a contact binary of given total mass, angular momentum, initial chemical composition, and age.

117.012 **Gravitational radiation and the evolution of cataclysmic binaries.**
R. E. Taam, B. P. Flannery, J. Faulkner.
Astrophys. J., Vol. 239, 1017 - 1023 (1980).

Results are presented for several evolutionary sequences of close binary systems which resemble dwarf novae. The systems, containing white dwarfs and main-sequence companions, have masses less than 3.4 $M_\odot$ and periods less than 17 hours. The loss of angular momentum by gravitational radiation is included in the calculations. It is found that a system can be captured as a result of evolution dominated by gravitational radiation losses when the primary encounters its Roche lobe while near the main sequence. In such a case, the system evolves to even shorter periods, with the nuclear time scale of the primary increasing so that the latter is unable to exhaust hydrogen.

117.013 **The ultraviolet spectrum and flux of HD 152667 (=X-ray source 1653–40?).**
J. B. Hutchings, A. K. Dupree.
Astrophys. J., Vol. 240, 161 - 168 (1980).

IUE data are presented on the massive binary system HD 152667 (spectral type B0 I), covering many phases of the 8 day orbit. Low-dispersion data are used to derive T_{eff}, E_{B-V}, and the UV light curve. The secondary object is much fainter than the primary at all wavelengths. The stellar wind shows complex multiple components and is unique in the authors' experience. Weak lines show the primary orbital motion, with minor distortions. The interstellar absorption spectrum shows high temperature species most probably arising in the H II region produced by the star. No clear evidence is found regarding the nature of the secondary object. Discrepancies in published ephemerides of the system are clarified.

117.014 **Transient mass transfer caused by local surface heating in close binaries.**
J. J. Modisette, Y. Kondo.
Astrophys. J., Vol. 240, 180 - 184 (1980).

The surge of mass from one component of a binary system resulting from local surface heating is analyzed. The impact of such surges on the companion can produce transient phenomena such as those seen in X-ray binaries, RS CVn objects, and cataclysmic variables. The heating may be caused by nonlinear *g*-mode oscillations or by X-ray heating by the companion in X-ray binaries, among other possible mechanisms. As an example, the authors have performed model calculations for a surge, triggered by a relatively moderate local heating, in a hypothetical X-ray binary; the results show that such a surge can account for X-ray turn-ons.

117.015 **Observational determination of the gravity darkening exponent and bolometric albedo for close binary star systems.** J. B. Rafert, L. W. Twigg.
Mon. Not. R. Astron. Soc., Vol. 193, 79 - 86 (1980).

Stars with radiative envelopes were found to have an average value of $g_{rad} = 0.96$ and an average value of $A_{rad} = 1.02$, while stars with convective envelopes gave an average value of $g_{conv} = 0.31$ and an average value of $A_{conv} = 0.56$. For a subgroup composed exclusively of A and W-type W UMa systems, the A-types were found to have a slightly lower average value of $g_A = 0.28$, while the W-type binaries were found to have both higher average values of $g_W = 0.37$, and a temperature differential of several hundred degrees, with the secondary being the hotter component.

117.016 **Time resolved spectroscopy of cataclysmic variables: SS Cygni.**
R. J. Stover, E. L. Robinson, R. E. Nather, T. J. Montemayor.
Astrophys. J., Vol. 240, 597 - 607 (1980).

Spectroscopic observations of radial velocity variations in the dwarf nova SS Cygni are presented, and the observing and analysis techniques are discussed in detail. The authors' analysis leads to the following orbital elements: $K_R = 153$ km s^{-1}, $K_B = 90$ km s^{-1}, $P_{ORB} = 0.275129$ days. Both K_R and P_{ORB} are significantly different from previously published measurements. They conclude that the white dwarf in SS Cygni may have a mass very near the Chandrasekhar limit.

117.017 **Close binary systems in the phase of mass loss. VW Cygni and SZ Herculis.**
G. Chiş, G. D. Chiş.
Proceedings of the colloquium of astronomy, Cluj-Napoca, (see 012.009), p. 1 - 17 (1978).

117.018 **Ellipsoid-ellipsoid approximation for close binary systems with elliptical orbits.** V. Ureche.
Proceedings of the colloquium of astronomy, Cluj-Napoca, (see 012.009), p. 18 - 22 (1978). – Summary.

117.019 **On the spectrum of VV Puppis.**
D. T. Wickramasinghe, N. Visvanathan.
Proc. Astron. Soc. Australia, Vol. 3, 311 - 312 (1979).

117.020 **Hydrodynamics of stellar encounters.**
R. A. Gingold, J. J. Monaghan.
Proc. Astron. Soc. Australia, Vol. 3, 364 - 366 (1979).

The authors examine some aspects of two tidal problems of current astrophysical interest; a proposed tidal origin of the solar system and the evolution of orbital elements in a binary similar to that proposed for Cir X-1.

117.021 **VV Puppis and AN Ursae Majoris: a radial velocity study.** D. P. Schneider, P. Young.
Astrophys. J., Vol. 240, 871 - 884 (1980).

The emission lines of VV Puppis and AN Ursae Majoris have been observed through complete orbits at 2 Å spectral and 5 minutes temporal resolution. VV Puppis exhibits spectacular emission line variations. The lines have a mean radial velocity amplitude $K = 376$ km s^{-1} and have clearly defined broad and sharp components. The sharp component becomes very strong when maximally redshifted (+450 km s^{-1}) and remains strong while moving to its maximum blueshift (−450 km s^{-1}), then suddenly becomes weak. Its movement lags the broad pedestal of emission on which it is superposed by 55° in phase. AN UMa exhibits similar but less spectacular variations. The mean radial velocity amplitude $K = 256$ km s^{-1} and the base and peak of the emission lines are 49° out of phase. A milder version of the sharp component effect is seen in this system. Attempts are made to integrate particle trajectories in the Roche geometry where a dipole magnetic field dominates the gas flow.

117.022 **Cataclysmic binaries – from the point of view of stellar evolution.** H. Ritter.
Messenger, No. 21, p. 16 - 18 (1980).

117.023 **Slightly detached binaries as calibrators of the main-sequence.** R. E. Wilson, J. B. Rafert.
Astron. Astrophys., Suppl. Ser., Vol. 42, 195 - 207 (1980).

The authors have analysed the light curves of six close binaries (RX Ari, RW CrB, BH Vir, TZ CrA, DI Peg and KR Cyg) by the method of differential corrections. All of these are probably detached, although DI Peg may be semi-detached. They use the results to insert observational data points into the mass-luminosity and mass-radius diagrams in the low mass regime where only visual binary data are normally available. One is able to do this by supplementing single-lined radial velocity information (mass functions) with photometric mass ratios, thus finding individual masses, radii, and luminosities (although the luminosities are sensitive to the assumed primary spectral types).

117.024 **On the motion of the apsidal line in interacting binary systems.** J. Papaloizou, J. E. Pringle.
Mon. Not. R. Astron. Soc., Vol. 193, 603 - 615 (1980).

The authors investigate the effect of resonance between stellar oscillations and the orbital motion on the standard calculation of the rate of advance of the line of apsides in an eccentric binary system. As an illustration the authors apply their results to the X-ray binary system 4U 0900–40 for which the pulsing X-ray star enables accurate measurement of orbital parameters. It is shown that the standard formula for apsidal advance can be significantly in error, even to the extent of predicting the wrong direction of the apsidal precession.

117.025 **Close binary systems before and after mass transfer. II. Semi-detached systems.**
J. P. de Grève.
Astrophys. Space Sci., Vol. 72, 411 - 432 (1980).

Recent data on semi-detached systems are compared with general results of close binary computations to determine the origin of Algol-type systems (classical case A and B results are considered). Both the conservative and the non-conservative mode of mass exchange are investigated. It is found that the assumption of considerable mass loss from the system gives satisfying agreement with the observations. It is also concluded that the calculation of an extensive homogeneous network of case A computations is needed.

117.026 **On the size of accretion disks in cataclysmic binaries.**
H. Ritter.
Astron. Astrophys., Vol. 91, 161 - 164 (1980).

It is shown that the analysis of light curves of cataclysmic binaries which undergo double total eclipses yields a relation for the size of the accretion disk which can be expressed as a slowly varying function of the binary's mass ratio. Therefore, the light curve analysis provides a reliable estimate of the disk's size even in cases where the mass ratio is not exactly known. Results of applying the method to recent observations of the dwarf novae Z Cha and OY Car are briefly discussed.

117.027 **Stellar evolution and close binaries.**
B. Paczyński.
Highlights of Astronomy, Vol. 5, (see 003.006), 27 - 44 (1980).
A general scheme of evolution of close binaries is outlined. Many types of observed systems are classified according to their evolutionary status. Present theory can account reasonably well for the mass transfer between the two components. Most close binaries sooner or later should develop extended common envelopes. A loss of a common envelope may remove a large fraction of mass and most of angular momentum from a binary, and leave as a remnant a very short period and highly evolved system. Binary nuclei of planetary nebulae, cataclysmic variables and some X-ray binaries are produced this way.

117.028 **Rotation in close binary stars.** M. J. Plavec.
Highlights of Astronomy, Vol. 5, (see 012.030), 831 - 834 (1980).
In binary stars of short period, axial rotation of the components tends to be synchronized with the orbital revolution. Rotation of B and A stars is therefore slowed down, while for the later-type stars, it is accelerated. This latter fact probably contributes to the phenomenon of the RS CVn stars. Notable deviations from synchronism among short-period systems are probably related to mass transfer between the components.

117.029 **A baker's dozen: unsolved problems with the RS Canum Venaticorum and BY Draconis stars.**
D. S. Hall.
Highlights of Astronomy, Vol. 5, (see 012.031), 841 - 845 (1980).

117.030 **A review of recent ultraviolet observations of close binaries from space.** Y. Kondo.
Highlights of Astronomy, Vol. 5, (see 012.031), 849 - 851 (1980).

117.031 **Radio aspects of stellar activity in close binaries.**
R. M. Hjellming.
Highlights of Astronomy, Vol. 5, (see 012.031), 857 (1980).

117.032 **On the light curve of HZ Her in the minimum.**
N. N. Kilyachkov, V. S. Shevchenko.
Pis'ma Astron. Zh., Tom 6, 717 - 722 (1980). In Russian. English translation in Soviet Astron. Lett., Vol. 6.
56 photoelectric observations of HZ Her = Her X-1 in the *BV*-system near the primary minimum are obtained. Considerable asymmetry of the light curves has been observed again. A model for the system HZ Her = Her X-1 with marked asymmetry of the X-ray source is discussed.

117.033 **RT Scl: a latent contact binary?** J. B. Rafert.
Bull. American Astron. Soc., Vol. 12, 748 (1980). Abstract.

117.034 **Evolution of low mass, very close binary systems including rotational and tidal distortions and gravitational radiational losses.**
W. Y. Chau, C. Chiosi, K. L. Chan, L. Nelson.
Bull. American Astron. Soc., Vol. 12, 748 (1980). – Abstract.

117.035 **Accretion disk radii in cataclysmic variables.**
M. E. Sulkanen, L. W. Brasure, J. Patterson.
Bull. American Astron. Soc., Vol. 12, 749 (1980). – Abstract.

117.036 ***IUE* ultraviolet spectra and chromospheric models of HR 1099 and UX Arietis.**
T. Simon, J. L. Linsky.
Astrophys. J., Vol. 241, 759 - 773 (1980).
The authors present and analyze *IUE* spectra (1150–3200 Å) of the RS CVn–type systems HR 1099 and UX Ari obtained at three quiescent phases each. High-resolution Mg II spectra suggest, at least for UX Ari, that the strong emission in these lines originates with the K star rather than the G star in these systems. The authors have computed plane-parallel chromospheric models, extending from the temperature minimum to 40,000 K, and synthesized line fluxes for Mg II, Si II, Si III, and C II for comparison with the observations. Using four atmospheric temperature structures and several different transition-region pressures, P_{TOP}, the authors conclude that P_{TOP} probably lies in the range 0.18–1.0 dynes cm^{-2}.

117.037 **De RS Canum Venaticorumsterren.**
J. H. H. M. Potters.
Zenit, 7e Jaarg., 186 - 189 (1980).

117.038 **New BV light curves and minima of the eclipsing binaries: W UMa, AW UMa and 44i Boo.**
J. Mikolajewska, M. Mikolajewski.
Inf. Bull. Variable Stars, No. 1812, 4 pp. (1980).

117.039 **Photoelectric minima of AB Andromedae.**
P. Rovithis, H. Rovithis-Livaniou.
Inf. Bull. Variable Stars, No. 1823, 2 pp. (1980).

117.040 **Light elements of AG Phoenicis.** M. A. Cerruti.
Inf. Bull. Variable Stars, No. 1830, 4 pp. (1980).

117.041 **Photoelectric light curves of RZ Columbae.**
B. B. Bookmyer, D. R. Faulkner.
Inf. Bull. Variable Stars, No. 1837, 2 pp. (1980).

117.042 **Photoelectric observations of XY Leonis.**
J. S. Hebron, M. R. A. Shegelski.
Inf. Bull. Variable Stars, No. 1838, 3 pp. (1980) = Rothney Astrophys. Obs., Calgary, Canada, Ser. B, Publ. No. 5.

117.043 **Eruptive mass-transfer events in RW Tauri.**
E. C. Olson.
Inf. Bull. Variable Stars, No. 1839, 3 pp. (1980).

117.044 **Photoelectric minima observations of the eclipsing binary SV Centauri.** H. Drechsel,
H.-D. Radecke, J. Rahe, G. Rupprecht, B. Wolf.
Inf. Bull. Variable Stars, No. 1842, 2 pp. (1980).

117.045 **Period change of AW Ursae Majoris.**
M. Kurpinska-Winiarska.
Inf. Bull. Variable Stars, No. 1843, 2 pp. (1980).

117.046 **A light curve and time of minimum for W UMa obtained with the IUE satellite.** S. M. Rucinski,
P. Gondhalekar, J. E. Pringle, J. A. J. Whelan.
Inf. Bull. Variable Stars, No. 1844, 4 pp. (1980).

117.047 **Optical variability of the RS CVn candidate HD 174429.**
D. W. Coates, L. Halprin, P. Sartori, K. Thompson.
Inf. Bull. Variable Stars, No. 1849, 2 pp. (1980).

117.048 **Photoelectric y and λ6585 observations of VW Cephei.**
E. F. Guinan, S.-I. Najafi, F. Zamani-Noor.
Inf. Bull. Variable Stars, No. 1850, 2 pp. (1980).

117.049 **Narrow- and intermediate-band Hα photoelectric photometry of VW Cephei.**
E. F. Guinan, A. G. Weisenberger.
Inf. Bull. Variable Stars, No. 1856, 5 pp. (1980).

117.050 **TZ Bootis' 1980 light curve.** M. Hoffmann.
Inf. Bull. Variable Stars, No. 1877, 3 pp. (1980).

117.051 **Observations of the spot stars EQ Vir and UZ Lib in 1977.** M. Hoffmann.
Inf. Bull. Variable Stars, No. 1878, 3 pp. (1980).

117.052 **On the suspected ultrashort period W UMa type binaries BG CrA and AB Tel.** M. Hoffmann.
Inf. Bull. Variable Stars, No. 1879, 4 pp. (1980).

117.053 **X-ray emission from RS CVn systems.**
F. Walter, P. Charles, S. Bowyer.
Smithsonian Astrophys. Obs., Spec. Rep. 389, (see 012.038), p. 35 - 45 (1980).

The authors review the state of knowledge of the coronae of RS CVn binaries prior to HEAO-2. They present preliminary Einstein Observatory results from the RS CVn survey, and consider their significance. The authors find that (1) L_x/L_{bol} is inversely related to orbital period, (2) an apparent bimodality in the RS CVn luminosity function may be due to two types of coronal activity in these systems, one perhaps similar to solar large scale structures and the other similar to solar active regions.

117.054 **Spectroscopic evidence for starspots on HR1099.**
L. W. Ramsey, H. L. Nations.
Smithsonian Astrophys. Obs., Spec. Rep. 389, (see 012.038), p. 97 - 99 (1980).

117.055 **Ultraviolet and optical chromospheric activity in λ Andromedae: evidence for starspots and active regions.** S. L. Baliunas, A. K. Dupree.
Smithsonian Astrophys. Obs., Spec. Rep. 389, (see 012.038), p. 101 - 105 (1980).

117.056 **A spectroscopic search for new RS CVn binaries.**
F. C. Fekel, Jr.
Smithsonian Astrophys. Obs., Spec. Rep. 389, (see 012.038), p. 133 - 136 (1980).

117.057 **The close binary SZ Camelopardalis – a semi-detached system.** D. Chochol.
Bull. Astron. Inst. Czechoslovakia, Vol. 31, 321 - 343 (1980).

Drawing on a comprehensive study of photometric and spectroscopic material, the author presents a new model of the close binary SZ Camelopardalis as a semi-detached system, valid after the year 1970. The secondary component has the spectral type B0 and is overluminous for its mass. An increase of period was observed in the system. In the simplest conservative case the amount of mass transferred from the secondary component to the primary is $\Delta m = 1.51 \times 10^{-5}\ M_\odot$ per year. The gaseous streams in the system are indicated photometrically and photoscopically and their model was constructed. The analysis of light curves has shown that the eclipse in the system is total and annular. Arguments in favour of an increase in the radius of the secondary component during the last 40 years are given.

117.058 **Computer simulation of planet formation in a binary star system: terrestrial planets.**
B. B. Djakov (*D'yakov*), B. I. Reznikov.
Moon Planets, Vol. 23, 429 - 443 (1980).

A model of planetary formation in a binary system with a small relative mass of the primary is computed on the assumption of a mass transfer from the less massive component to the more massive one with no mass and angular momentum carried away from the system under consideration. At the last stage of mass transfer the condensed Moon-like objects (planetoids) are ejected through the inner Lagrange point of the primary Roche lobe with the outflow of gaseous matter. The mass transfer ceases at the primary relative mass 10^{-3} which corresponds to the present Sun-Jupiter system. The total mass of planetoids approximates that of the terrestrial planets. Numerical computations reveal the division of the planetoids into 4 groups along their distances from the Sun. Further, each group forms a single planet and a less massive body at the nearest orbits. The parameters of simulated planet orbits are close to the present ones and the interplanetary spacings are in accord with the Titius-Bode law.

117.059 **The slaved disc model for SS 433.**
D. P. Whitmire, J. J. Matese.
Mon. Not. R. Astron. Soc., Vol. 193, 707 - 712 (1980).

The authors investigate a slaved disc model for SS 433 in which the beams originate normal to the surface of an accretion disc around a compact object in a binary system. The 13-day period in the 'stationary' system of lines is assumed to be associated with binary orbital motion and the 164-day periodicity in the moving line system is identified with disc precession. The authors show that the viability of the basic model requires that the mass of the compact object be $\gtrsim 10$ times the mass of the secondary. Thus, if the slaved disc model is applicable to SS 433 and if the mass of the secondary is $\gtrsim 1 M_\odot$ it follows that the compact object is a massive black hole.

117.060 **Anomalous limb darkening coefficients of eclipsing binary systems.** L. W. Twigg, J. B. Rafert.
Mon. Not. R. Astron. Soc., Vol. 193, 775 - 779 (1980).

Values of the linear limb darkening coefficients in two passbands (V and B) derived by several authors using the Wilson–Devinney Computer Code are compared with theoretical values derived from currently accepted model atmospheres. The results show that (1) contact and semi-detached binary systems yield linear limb darkening coefficients that systematically disagree with theoretical values, and (2) the few detached systems analysed show much better agreement with theory than the contact or semi-detached systems.

117.061 **Inner accretion disk in a close binary.**
B. Paczyński, B. Rudak.
Acta Astron., Vol. 30, 237 - 247 (1980).

The authors approximate stationary streamlines in a disk in a close binary with simple periodic particle orbits. They find that the disk may extend beyond 1:2 resonance provided the mass ratio is very extreme ($\mu \leqslant 0.00618$). There is a wide gap in the disk at the 1:2 resonance. The pressure must be very high and the disk must be very thick if the gap is to be closed. Therefore, the authors do not expect the 1:2 resonance to be important in known close binaries.

117.062 **SW Lacertae and the spot activity on W UMa stars.**
K. Stępień.
Acta Astron., Vol. 30, 315 - 321 (1980).

UBV photometry of SW Lac obtained in 1967 is given. The discussion of all available *UBV* data for this star shows that color index curves are virtually insensitive to the temporal variations of relative depths of brightness minima, contrary to the model with variable temperature excess of the secondary. Spot activity is suggested as a possible explanation of abnormal and variable relative depths of minima.

117.063 **A note on the existence of W UMa-type systems in the cluster Cr 359.** S. M. Ruciński.
Acta Astron., Vol. 30, 373 - 379 (1980).

Arguments are presented that numerous W UMa-type systems in Cr 359 might in fact be field objects with the expected frequency of occurrence; in addition (at least) four out of the listed nine objects might be either non-variable or of the RR Lyr-type. The physical association of stars forming Cr 359 is also questioned but the present data do not allow to disprove it entirely. The far reaching conclusion of van't Veer regarding the presence of W UMa systems in very young clusters finds no support from these results.

117.064 **Reply to the note of Rucinski on the existence of W UMa-type systems in the cluster Cr 359.**
F. van't Veer.
Acta Astron., Vol. 30, 381 - 385 (1980).

The following four questions are commented by the author: (1) Is Cr 359 really a young physical association? (2) If so, are the observed W UMa binaries members of this association? (3) If not, can one explain them from the generally adopted estimates of their space density? (4) What are the implications of the answer on the preceding questions on our understanding of the evolutionary history of W UMa stars?

117.065 **On the rapid variability of AM Herculis.**
R. J. Panek.
Astrophys. J., Vol. 241, 1077 - 1081 (1980).

New observations of the rapid brightness flickering of AM Her are presented, with 10 hours of data from three nights and with a time resolution of 1.6 s, and the statistical properties of the flickering are described. The flickering is assumed to be the superposition of bursts produced as individual clumps in the accretion flow encounter the region of strong emission just above the surface of the white dwarf at the magnetic poles. A statistical shot-noise analysis is then appropriate, and a pulse duration of 70 - 90 s is determined, with many pulses overlapping at any instant. The clumps must then have the form of long, thin strings as they reach the emission region.

117.066 **SS 433 as a prototype of astrophysical jets.**
K. Davidson, R. McCray.
Astrophys. J., Vol. 241, 1082 - 1089 (1980).

SS 433, a "small-scale" object within our Galaxy having cool gas within its jets, offers unprecedented clues to the nature of astrophysical jets in general. The jets are probably accelerated by radiation pressure and partially collimated by nozzles, formed where a dense accretion disk encounters heating and high pressures close to the central source. Interaction with an ambient medium, perhaps a stellar wind, is responsible for the line emission.

117.067 **Comments on gas stream effects in typical Algol systems.** K. Walter.
Astron. Astrophys., Vol. 92, 86 - 94 (1980).

The presence of circumstellar matter in typical Algol systems raises problems in obtaining photometric solutions which approximate as near as possible the true geometric parameters of the systems. These problems are discussed and a modified way to solutions is described. Results about photometric gas stream effects, obtained with this method from the investigation of nine systems with total primary eclipses, are compiled. At almost all epochs of observation hot spots near the poles were found. For epochs of low mass transfer a model of typical Algol systems is given.

117.068 **Period changes in close binaries caused by the presence of a third companion.** O. Havnes.
Astron. Astrophys., Vol. 92, 151 - 155 (1980).

The author examines under what conditions orbital changes of W UMa stars may be produced during periastron passages of a third body moving in an orbit of high eccentricity around the contact binary. The author suggests that period jumps may result as secondary effects from close passages which directly produce smaller period jumps which in turn produce, or trigger, mass transfer or ejection events. The third body mass has an upper limit of about 0.3 - 0.4 $M_{\odot}$.

117.069 **Stability of tidal equilibrium.** P. Hut.
Astron. Astrophys., Vol. 92, 167 - 170 (1980).

It is proved rigorously that a binary system can be in tidal equilibrium only if coplanarity, circularity and corotation have been established. A complete analysis is given of the stability of this equilibrium against general perturbations. Stability occurs only if the orbital angular momentum exceeds the sum of the spin angular momenta of the stars by more than a factor of three. This condition is identical to the criterium for stability against perturbations of corotation only, as given by Counselman (1973).

117.070 **Binary X-ray sources and mass exchange in binaries.**
C. de Loore.
Variability in stars and galaxies, (see 012.044), p. D.1.1 - 1.23 (1980).

117.071 **Infrared observations of AE Aqr.**
E. G. Tanzi, G. Chincarini, M. Tarenghi.
ESO Sci. Prepr. No. 128, 8 pp. (1980). – Submitted to Publ. Astron. Soc. Pacific.

117.072 **Evolution in semi-detached binary systems.**
U. S. Chaubey.
Astrophys. Space Sci., Vol. 73, 503 - 506 (1980).

The evolutionary process of semi-detached binary systems is examined on the basis of non-conservation of orbital angular momentum. The author concludes that the semi-detached binary systems follow type B evolution.

117.073 **On the interconnection of the pulsational and orbital period values for binary systems.** M. S. Frolov, E. N. Pastukhova, A. V. Mironov, V. G. Moshkalev.
Inf. Bull. Variable Stars, No. 1894, 4 pp. (1980).

117.074 **A 5.57 hr modulation in the optical counterpart of 2S 1822–371.** K. O. Mason, J. Middleditch, J. E. Nelson, N. E. White, P. Seitzer, I. R. Tuohy, L. K. Hunt.
Astrophys. J., Lett., Vol. 242, L109 - L113 (1980).

A periodic 5.57 hr modulation has been observed in the optical counterpart of the X-ray source 2S 1822–371. Two alternative periods are consistent with the data. The light curve of the star can be understood in terms of a highly inclined close binary system which contains a relatively large, luminous accretion disk that is periodically occulted by a companion star and an associated gas stream. On the assumption that the companion is a main-sequence star, the distance of the system is found to be more than 600 pc.

117.075 **Double stars.** S. Kříž.
Kozmos, Vol. 11, 133 - 134 (1980). In Czech.

117.076 **IUE observations of the ultraviolet spectrum of the close binary δ Pictoris.**
Y. Kondo, G. E. McCluskey, W. A. Feibelman.
Publ. Astron. Soc. Pacific, Vol. 92, 688 - 690 (1980).

The eclipsing binary δ Pic has been observed with the International Ultraviolet Explorer (IUE) satellite. Two high-resolution spectrograms in the far-ultraviolet (λλ1150-1900) and one high-resolution spectrogram in the mid-ultraviolet of (λλ1900-3200) were obtained. The far-ultraviolet resonance lines of C III (λ1175) and Si IV (λλ1393 and 1402) show the presence of excess absorption due to gas leaving the primary star with velocities as high as 600-700 km s^{-1}. Many of the interstellar lines are doubled with the two components having mean velocities of –40.6 and + 9.0 km s^{-1}. It is suggested that the negative velocity component may be due to an expanding circumstellar shell surrounding the binary system.

117.077 **Evolution of low-mass, distorted stars in a very close binary system with the inclusion of gravitational-radiation losses.** W. Y. Chau, C. Chiosi, K. L. Chan, L. A. Nelson.
J. R. Astron. Soc. Canada, Vol. 74, 359 (1980). – Abstract.

117.078 **Perturbation of a binary system with the passage of gravitational waves.** L. A. Nelson, W. Y. Chau. J. R. Astron. Soc. Canada, Vol. 74, 359 (1980). – Abstract.

117.079 **Phasenkorrelierte Variation der P Cygni Profilstruktur von C III (λ 1175) im Spektrum des engen Doppelsternsystems UW CMa.**
H. Drechsel, J. Rahe, Y. Kondo, G. E. McCluskey, Jr.
Mitt. Astron. Ges., Nr. 50, p. 137 - 140 (1980).

International Astronomical Union, Commission 42. Bibliography and Program Notes on Close Binaries, Nos. 33, 34. See Abstr. 002.028.

First supplement to the catalogue of modern light curve synthesis photometric solutions of close binary systems. See Abstr. 002.032.

An envelope model of Phi Persei. See Abstr. 064.006.

On the differences at chromospheric levels between RS CVn-type binaries, active and quiet chromosphere single stars, and active and quiet regions in the sun. See Abstr. 064.059.

Mass transfer between binary stars. See Abstr. 064.060.

Nonradial accretion onto magnetized neutron stars. See Abstr. 064.096.

On the binary frequency distribution and evolution of Wolf-Rayet stars. See Abstr. 065.007.

Relativistic corrections to the gravitational radiation of a binary system and the fine structure of the spectrum. See Abstr. 066.090.

Gravitationswellen von Doppelsternsystem? See Abstr. 066.133.

Stellar winds in close binary systems. See Abstr. 112.052.

The white dwarf companion of the barium star ζ Capricorni. See Abstr. 114.038.

***IUE* spectra of a flare in the RS Canum Venaticorum-type system UX Arietis.** See Abstr. 114.040.

***IUE* observations of RW Hydrae (gM2 + pec).** See Abstr. 114.041.

The 164 and 13 day periods of SS 433: confirmation of the kinematic model. See Abstr. 114.085.

Ultraviolet spectroscopic measurements of cool stars. See Abstr. 114.089.

Spectroscopic studies of O type stars. IX. Binary frequency. See Abstr. 114.169.

Ultraviolet, X-ray, and infrared observations of HDE 226868 = Cygnus X-1. See Abstr. 114.170.

Stellar masses. See Abstr. 115.004.

18-cm observations of an outburst in V711 Tau. See Abstr. 116.008.

Radio superflares in RS Canum Venaticorum binaries. See Abstr. 116.031.

Is SS433 a trinary? See Abstr. 118.021.

A model of the subdwarf binary system LB 3459. See Abstr. 119.002.

Geometry and dynamics of the Algol system. See Abstr. 119.024.

Some aspects related to the reflection effect in the eclipsing binary system AB Andromedae. See Abstr. 119.029.

A polarimetric study of U Cephei. Part I. See Abstr. 119.034.

AM Eridani: the definite period. See Abstr. 119.078.

Verbesserte Elemente des β-Lyrae-Sterns NN Cephei. See Abstr. 119.085.

Analysis of the change of period and the photometry of the minima of the eclipsing binary system TX Ursae Maioris. See Abstr. 119.087.

HR 5110 – an Algol-type system with RS CVn properties. See Abstr. 120.002.

Infrared observations of polars: AM Her, VV Pup, and AN UMa. See Abstr. 122.004.

Spectroscopy of EX Hydrae. See Abstr. 122.008.

A bulge model of a dwarf nova. See Abstr. 122.036.

HR 1099 and the starspot hypothesis for RS Canum Venaticorum binaries. See Abstr. 122.050.

The variable, single-line Wolf-Rayet star HD 96548 with a low-mass companion. See Abstr. 122.085.

BY Draconis and RS Canum Venaticorum stars: the discoveries of classical photometry and spectroscopy. See Abstr. 122.098.

Cataclysmic variables. See Abstr. 122.162.

A wave model for dwarf novae. See Abstr. 122.201.

The puzzle of the symbiotic stars: new results from the International Ultraviolet Explorer satellite. See Abstr. 122.205.

The detection of a companion star to the Cepheid variable T Monocerotis. See Abstr. 122.223.

Eruptive binaries. X. DQ Herculis. See Abstr. 124.303.

Simultaneous X-ray, UV, and optical observations of the recent nova HR Delphini. See Abstr. 124.701.

A search for OB stars in supernova remnants. See Abstr. 125.026.

Discovery of a third radio pulsar in a binary system. See Abstr. 141.511.

Binary model of Circinus X-1. I. Eccentricity from combined X-ray and radio observations.
See Abstr. 142.001.

Binary model of Circinus X-1. II. Radio emission.
See Abstr. 142.002.

The system AM Her = 4U 1814 + 50.
See Abstr. 142.015.

Precise location of 3A 2352+28: the high galactic latitude transient A0000+28. See Abstr. 142.029.

Determination of a binary period for the variable X-ray source A1907+09. See Abstr. 142.051.

***UBV* photometry of HZ Herculis: the shape of the primary minimum.** See Abstr. 142.066.

The optical identification of H2252–035 with a cataclysmic variable. See Abstr. 142.073.

Ultraviolet emission from strong X-ray sources as observed with IUE. See Abstr. 142.080.

X-ray observations of stellar coronae and winds.
See Abstr. 142.083.

AM Herculis: simultaneous X-ray, optical, and near-IR coverage. See Abstr. 142.110.

Discovery of soft X-ray emission from V471 Tauri and UU Sagittae: two highly evolved, low-mass binaries.
See Abstr. 142.120.

X-ray emission from spherical accretion onto white dwarfs in binary systems. See Abstr. 142.139.

A search for pulsations and eclipses from X-ray burst sources. See Abstr. 142.142.

Dynamical theory of binaries in clusters.
See Abstr. 151.086.

The blue stragglers in NGC 7789 and the binary post-main-sequence mass-exchange hypothesis.
See Abstr. 153.025.

Random gravitational encounters and the evolution of spherical systems. VIII. Clusters with an initial distribution of binaries. See Abstr. 154.022.

118 Visual Binaries, Multiple Stars, Astrometric Binaries

118.001 **The distant companion of θ Cygni.**
W. P. Bidelman.
Publ. Astron. Soc. Pacific, Vol. 92, 345 (1980).
Confirmatory evidence is reported for a distant faint companion to the F dwarf θ Cyg.

118.002 **Redressement des orbites des couples visuels ADS 1833 et 6871 AB.** A. Valbousquet.
Astron. Astrophys., Suppl. Ser., Vol. 41, 295 - 298 (1980).
The orbits of the visual double stars ADS 1833 and 6871 AB are recomputed. The method used is the analytical method of the author (Valbousquet, 1979). The weights used for the observations are proportional to the numbers of nights.

118.003 **Photographic measures of double stars.**
R. Pannunzio, F. Siciliano.
Astron. Astrophys., Suppl. Ser., Vol. 41, 319 - 322 (1980).
132 photographic measures of visual double stars of particular interest, made with the astrometric reflector of the Astronomical Observatory of Torino are given.

118.004 **New double stars.** D. W. Dunham.
Occultation Newsl., Vol. 2, 98 - 100 (1980).

118.005 **The astrometric binary Mu Cas.**
S. L. Lippincott.
Bull. American Astron. Soc., Vol. 12, 453 (1980). – Abstract.

118.006 **The nearby multiple system G107 - 69/70.**
J. W. Christy, R. S. Harrington, C. C. Dahn, K. A. Strand.
Bull. American Astron. Soc., Vol. 12, 454 (1980). – Abstract.

118.007 **On the kinematics and ages of wide binaries containing a white dwarf.** G. Wegner.
Bull. American Astron. Soc., Vol. 12, 464 (1980). – Abstract.

118.008 **Speckle interferometric measurements of binary stars. V.** H. A. McAlister, F. C. Fekel.
Astrophys. J., Suppl. Ser., Vol. 43, 327 - 337 (1980).
Three hundred and twenty-seven measurements of 217 binary stars observed by means of speckle interferometry with the Mayall 4 m telescope at KPNO are presented. Measured separations range from 0".030 to 2".292. New directly resolved systems include 51 Psc, HR 233, 64 Ori, 65 Gem, 9 Cyg, ϕ Cyg, λ Cyg, μ Aqr, 12 Aqr, and 5 Lac. Continued observations of the two newly resolved double-lined spectroscopic systems with giant components, 65 Gem (K III, K III) and ϕ Cyg (G8 III–IV, G8 III–IV) offer the possibility of particularly important mass/luminosity determinations.

118.009 **Three 40-year intensive Sproul plate series and planetary detection capability and probability.**
J. L. Hershey, E. R. Borgman, M. D. Worth.
Astrophys. J., Vol. 240, 130 - 136 (1980).
Forty-year interval plate series with intensive coverage on the nearby stars Wolf 294, Groombridge 1618, and Ross 128 were measured and analyzed for long-term telescope stability and upper limits on the masses of dark companions. The limits are found to be at the level of one to several Jupiter masses for periods ranging from four decades to one decade, respectively. Discussion is given of the statistics of Sproul astrometric binaries and expected discovery statistics for substellar mass companions of the nearby stars. A distant triple system was discovered in the field of Groombridge 1618.

118.010 **Orbites provisoires des étoiles doubles visuelles ADS 102 et ADS 281.** M. Scardia.
Astron. Nachr., Band 301, 233 - 240 (1980).
The orbital elements of the visual binary stars ADS 102 and ADS 281 are given. The provisional elements obtained by the Kowalsky method have been corrected with the differential method of Hellerich. Finally, the dynamical parallaxes and total masses of the systems have been calculated.

118.011 **Orbites provisoires des étoiles doubles visuelles ADS 161, ADS 2236 et ADS 8739.**
M. Scardia.
Astron. Nachr., Band 301, 241 - 250 (1980).
The orbital elements of the visual binary stars ADS 161, ADS 2236 and ADS 8739 are given. The provisional elements obtained by the Kowalsky method have been corrected with the differential method of Hellerich. Finally, the dynamical parallaxes and total masses of the systems have been calculated.

118.012 **Transformation de la trajectoire rectiligne de l'étoile double ADS 8486 = Σ 1621 dans une orbite elliptique.** S. Vlaicu.
Proceedings of the colloquium of astronomy, Cluj-Napoca, (see 012.009), p. 37 - 40 (1978).

118.013 **Méthode graphique de détermination des éléments de l'orbite des étoiles triples visuelles.**
S. Vlaicu.
Proceedings of the colloquium of astronomy, Cluj-Napoca, (see 012.009), p. 41 - 45 (1978).

118.014 **A "Narrabri" binary star resolved by speckle interferometry.**
W. J. Tango, J. Davis, R. J. Thompson, R. Hanbury Brown.
Proc. Astron. Soc. Australia, Vol. 3, 323 - 324 (1979).

118.015 **The triple system Gliese 815 (Furuhjelm 54): an astrometric study.** J. Russell, G. Gatewood.
Astron. J., Vol. 85, 1270 - 1273 (1980).
A total of 332 exposures of Gliese 815 were measured and analyzed for its position, proper motion, parallax, and orbit. Its absolute parallax is 0.0692 ± 0.0029 arcsec. The astrometric orbit of the wide pair, the visual observations, and the spectroscopic orbit of the close pair yield masses $M_{Aa} = 0.9\,M_\odot$, $M_{Ab} = 0.6\,M_\odot$, and $M_B = 0.5\,M_\odot$. Astrometric constants were determined for 32 other stars in the region, including BD + 39°4400. The absolute parallax of the latter is 0.0200 ± 0.0052 arcsec. Two other reference stars, BD + 39°4403 and AO 584, were discovered to be a common proper motion pair.

118.016 **Speckle interferometric measurements of binary stars.** D. Bonneau, A. Blazit, R. Foy, A. Labeyrie.
Astron. Astrophys., Suppl. Ser., Vol. 42, 185 - 188 (1980).
The authors report speckle interferometric measurements of 104 binaries or suspected binaries with the 3.60 m ESO and the 1.93 OHP telescopes. Five binaries are first resolved.

118.017 **Visual double stars for the amateur.**
J. Ashbrook.
Sky Telesc., Vol. 60, 379 - 381 (1980).

118.018 **Visual multiples. IV. Radial velocities of 13 systems.**
E. Roemer, N. B. Sanwal.
Astrophys. J., Suppl. Ser., Vol. 43, 547 - 548 (1980).
For 13 visual systems with relatively short periods (2.6 to 26.4 yr), a total of 120 radial velocities are given from spectra obtained with the Lick Observatory Mills spectrograph, mostly in 1953 - 1958.

118.019 **Visual multiples. V. Radial velocities of 160 systems.**
H. A. Abt, N. B. Sanwal, S. G. Levy.
Astrophys. J., Suppl. Ser., Vol. 43, 549 - 575 (1980).
The authors list 937 radial velocities from coudé spectra of 160 visual multiples with known visual orbital elements; these, plus the velocities in paper IV (see abstract 28.118.018), are discussed.

118.020 **Motions of 4 southern double stars.**
R. H. Wilson, Jr.
Bull. American Astron. Soc., Vol. 12, 738 (1980). – Abstract.

118.021 **Is SS433 a trinary?**
B. M. Barker, G. G. Byrd, R. F. O'Connell.
Bull. American Astron. Soc., Vol. 12, 740 (1980). – Abstract.

118.022 **Vijfvoudig stersysteem Sigma Orionis.**
M. Drummen.
Zenit, 7e Jaarg., 228 - 232 (1980).

118.023 **New spectral classifications on the MK system for visual double stars.** C. J. Corbally, R. F. Garrison.
Publ. Astron. Soc. Pacific, Vol. 92, 493 - 496 (1980).
MK classifications are presented for the components of close visual double stars with separations mostly between one and three arc seconds. Twenty-one pairs have individual components classified, four pairs have just the primary presented, and the composite classification is given for a further five pairs. Good agreement is found between the direct classification of individual components and an indirect classification method of Edwards (1976). Six evolved pairs are related to the theoretical isochrones of Ciardullo and Demarque (1977).

118.024 **Orbital analysis and parallax of the unresolved astrometric binary CC 1299 from plates taken with the 61-cm Sproul refractor.**
S. L. Lippincott, A. Turner.
Publ. Astron. Soc. Pacific, Vol. 92, 531 - 533 (1980).
The unseen companion of BD + 27° 4120, CC 1299, was discovered in 1964 from plates taken with the Sproul refractor from 1939 - 63. The observations extending to 1979 are satisfied by a period of 85 years. With an adopted mass range for CC 1299, 0.44 - 0.55 $M_{\odot}$, the most likely mass for the unseen companion ranges between 0.23 and 0.29 $M_{\odot}$ with $M_v \sim 13$. This interpretation leads to a somewhat underluminous M dwarf companion. The semi-major axis of the photocentric orbit, $\alpha = 0''.436 \pm 0''.004$ (p.e.) and the Sproul absolute parallax, $\pi = +0''.081 \pm 0''.002$ (p.e.).

118.025 **Orbites nouvelles.**
Circ. Inf., No. 82 (1980).

118.026 **Etoiles doubles nouvelles.** P. Couteau, P. Muller.
Circ. Inf., No. 82 (1980).

118.027 **Micrometer measures of 1980 double stars.**
C. E. Worley.
Publ. United States Naval Obs., Washington, Second Ser., Vol. 24, Part VI, 186 pp. (1978).
This paper lists 7359 measures of 1980 double stars, principally made with the 26-inch refractor in Washington and the 36 and 60-inch reflectors at Cerro Tololo, Chile. A few miscellaneous measures made with the 12-inch refractor in Washington, and the 40 and 61-inch reflectors at Flagstaff, Arizona, are also included.

118.028 **The detection of other planetary systems.**
D. C. Black.
Mercury, Vol. 9, 105 - 111 (1980).

118.029 **Variability of ξ Boo A + B in 1955 to 1966.**
S. M. Ruciński.
Acta Astron., Vol. 30, 323 - 330 (1980).
Photometric observations obtained during the Solar Variability Programme at the Lowell Observatory have been analysed. The small amplitude variability of ξ Boo is consistent with the hypothesis of spots on the component A which rotates with the period of about 10.15 days.

118.030 **Micrometer observations of double stars and new pairs. X.** W. D. Heintz.
Astrophys. J., Suppl. Ser., Vol. 44, 111 - 136 (1980).
Measurements of 1907 pairs obtained in Swarthmore and Cerro Tololo in the time 1977.87 to 1979.99 and a list of 91 new objects are given.

118.031 **On the detection of other planetary systems: detection of intrinsic thermal radiation.**
D. C. Black.
Icarus, Vol. 43, 293 - 301 (1980).
One possible way to detect and study other planetary systems is infrared observations of continuum blackbody radiation from planets revolving around other stars. It is shown that the effective temperature of large planets revolving around mid- to late-spectral-type main-sequence stars is set by energy sources internal to the planet rather than by equilibrium with the radiation field of the central star, making them easier to detect than had been previously thought. Consideration is given to the two major observational constraints on detecting planetary companions to nearby stars, namely, angular resolution and sensitivity. A comparison is made between the performance of an ambient ($T \sim 200°$K), single-aperture telescope and a cooled interferometer.

118.032 **Interferometric observations of double stars.**
A. A. Tokovinin.
Astron. Tsirk., No. 1097, p. 3 - 5 (1980). In Russian.

118.033 **The orbits of five visual binaries.**
G. A. Starikova.
Astron. Tsirk., No. 1097, p. 5 - 6 (1980). In Russian.

118.034 **ADS 7114 AB.** I. N. Latyshev.
Astron. Tsirk., No. 1105, p. 7 - 8 (1980). In Russian.

118.035 **Determination of the orbit of a visual double star with the method of apparent motion parameters from short arc observations.**
A. A. Kiselev, O. V. Kiyaeva.
Astron. Zh., Tom 57, 1227 - 1241 (1980). In Russian.
English translation in Soviet Astron., Vol. 24, No. 6.
The problem of determination of the elements of a visual double star on the basis of high-accuracy short-arc observations is considered. It is shown that the problem may be solved by a differential dynamic method using the vectors of the relative position and relative velocity deduced from the observations. The solution of the problem by the method of apparent motion parameters is described.

118.036 **The astrometric search for neighboring planetary systems.** G. Gatewood, J. Stein, L. Breakiron, R. Goebel, S. Kipp, J. Russell.
Strategies for the search for life in the universe, (see 012.053), p. 111 - 154 (1980).
Extensive testing suggests that astrometric techniques can be used to detect and study virtually any planetary system that may exist within 40 light years of the sun. The astrometric group at the Allegheny Observatory began an intensive survey of 20 nearby stars to detect the nonlinear variations in their motion that planetary systems would induce. A new photoelectric detector has been designed and a simplified prototype of it successfully tested. This is sufficient to determine relative positions to within an accuracy of 2 mas/hr.

Project Orion. See Abstr. 003.019.

Laboratory exercises in astronomy – the orbit of a visual binary. See Abstr. 014.001.

Search for planets and early life in other solar systems. An introduction. See Abstr. 015.046.

The search for early forms of life in other planetary systems: future possibilities afforded by spectroscopic techniques. See Abstr. 015.047.

Stellar velocity fields and the spectroscopic detectability of planetary systems. See Abstr. 031.523.

Search for planets by spectroscopic methods. See Abstr. 031.617.

The search for planets in other solar systems through use of the Space Telescope. See Abstr. 031.618.

A comparison of alternative methods for detecting other planetary systems. See Abstr. 031.619.

The ability of the Space Telescope to detect extrasolar planetary systems. See Abstr. 032.505.

High-precision astrometry. See Abstr. 041.007.

A statistical theory for the disruption of four-body systems. See Abstr. 042.006.

Illinois occultation summary. I. 1977 - 1978. See Abstr. 096.002.

Periodic variations in the spectrum of X Persei identified with the X-ray source 3U 0352+30. See Abstr. 114.087.

Simulated analysis of unseen companions. See Abstr. 120.005.

Speckle interferometry of the spectroscopic/astrometric binary Chi Draconis. See Abstr. 120.007

A spectroscopic orbit for HR 6626. See Abstr. 120.011.

Intrinsic variability of η Ori. See Abstr. 122.030.

Spectroscopic evidence for pulsation in the system η Orionis. See Abstr. 122.031.

Continued astrometric study of BD + 43°4305. See Abstr. 122.033.

Gliese 867 – further observations of a multiple flare star system. See Abstr. 122.076.

HD 113001: photometric separation. See Abstr. 126.004.

Are binaries concentrated toward the centers of open clusters? See Abstr. 153.012.

Etoiles doubles dans les amas proches. See Abstr. 153.016.

Erratum

118.901 **Erratum: " Low-mass unseen companions to two nearby red dwarfs, CC 1228 and Wolf 922"**
[Publ. Astron. Soc. Pacific, Vol. 91, 784 - 788 (1979)].
S. L. Lippincott.
Publ. Astron. Soc. Pacific, Vol. 92, 384 (1980). – See Abstr. 27.118.014.

119 Eclipsing Binaries

119.001 **RY Cancri is not a member of Praesepe.**
N. S. Awadalla, E. Budding.
Observatory, Vol. 100, 108 - 112 (1980).
BV photometry shows the eclipsing binary system RY Cnc, located in the Praesepe (NGC 2632) field, to be a classical Algol-type system four times as far away as the cluster.

119.002 **A model of the subdwarf binary system LB 3459.**
B. Paczyński.
Acta Astron., Vol. 30, 113 - 125 (1980).
A model is presented for the short-period eclipsing binary LB 3459 (= CPD – 60°389 = HDE 269696). The primary of 0.36 $M_\odot$ and effective temperature of 64000 K burns hydrogen in a shell source surrounding a degenerate helium core. The secondary of 0.054 $M_\odot$ is nearly degenerate, and probably a hydrogen-rich star. The hemisphere facing the primary is heated to 200000 K. After 5×10^{10} years the degenerate, hydrogen-rich secondary will overflow its Roche lobe and LB 3459 will become a catalysmic variable.

119.003 **Long-term period changes and photometric orbital solution of RZ Draconis.** D.-s. Zhai, Y.-l. Huang.
Acta Astron. Sinica, Vol. 21, 158 - 171 (1980). In Chinese.
The short period eclipsing system RZ Dra was observed photo-electrically. Four epochs of primary minima were obtained and from these, together with 99 previously observed minima, an improved light curve was derived. Two alternatively possible figures of RZ Dra are discussed so as to explain the secular period changes and to estimate the masses of the two components.

119.004 **IUE observations of the atmospheric eclipsing binary system ζ Aurigae.** R. D. Chapman.
Nature, Vol. 286, 580 - 582 (1980).
The author reports IUE observations of the ζ Aur stellar system carried out between September 1979 and January 1980. There is strong evidence for cool, outflowing material near the primary star and hotter material farther from the primary perhaps in a shockwave. The onset of the atmospheric phase of the eclipse of the system began between 15 September and 1 November 1979.

119.005 **V368 Cas - Ein interessantes Beobachtungsergebnis liegt vor.** D. Lichtenknecker.
BAV Rundbrief, 29. Jahrg., 33 (1980).

119.006 **YY CMi - Elementenverbesserung durch BAV-Beobachtungen.** D. Lichtenknecker.
BAV Rundbrief, 29. Jahrg., 34 (1980).

119.007 **DM Del - Ein Mysterium ?**
D. Lichtenknecker, M. Fernandes.
BAV Rundbrief, 29. Jahrg., 35 - 37 (1980).

119.008 **Eine Liste von Bedeckungsveränderlichen, die kontinuierlich überwacht werden sollten.**
M. Fernandes.
BAV Rundbrief, 29. Jahrg., 43 - 44 (1980).

119.009 **On the relativistic motion of the periastron in the eclipsing binary system DI Herculis.**
D. Ya. Martynov, Kh. F. Khaliullin.
Astrophys. Space Sci., Vol. 71, 147 - 170 (1980).
An analysis of fifteen-year photoelectric observations of DI Herculis shows that the upper estimate for the relativistic part of apsidal motion in this binary system is almost three times lower than the theoretical value. The total observed angular velocity of apsidal rotation is 0°.0124 yr^{-1}. The relativistic part is less than 0°.008 yr^{-1} instead of 0°.023 yr^{-1} required by the theory.

119.010 **Preliminary study of light variations of the eclipsing binary AB Cassiopeiae.** H. Ando.
Astrophys. Space Sci., Vol. 71, 249 - 256 (1980).
Preliminary study of the eclipsing binary AB Cas is presented by using photometric observational data. The primary component is a δ Sct variable with a period of 0ᵈ.054, and whether the oscillation is of a radial mode or of a non-radial one is discussed. Two color indices ($B-V$ and $U-B$) data and the light curve analysis suggest that this binary system is a typical Algol system, in which the primary component is near the ZAMS with about 2.3 $M_\odot$ and the secondary one is a subgiant star with about 0.5 $M_\odot$.

119.011 **Revised photometric elements of the eclipsing binaries XY Cet and V380 Cyg.**
M. Ramella, G. Giuricin, F. Mardirossian, M. Mezzetti.
Astrophys. Space Sci., Vol. 71, 385 - 392 (1980).
Photoelectric light curves of the double-lined binaries XY Cet and V 380 Cyg, known to be detached systems, have been reanalyzed by means of Wood's program in order to obtain homogeneous photometric elements. Masses, radii and luminosities are given.

119.012 **Revised photometric elements of the eclipsing binary AI Cru.**
G. Giuricin, F. Mardirossian, M. Mezzetti, F. Predolin.
Astrophys. Space Sci., Vol. 71, 411 - 413 (1980).
Using Wood's (1972) model the authors have re-analyzed the photoelectric light curve of the eclipsing binary AI Cru in order to obtain reliable photometric elements.For the authors' photometric mass ratio $q = 0.8$, both components, though detached from their Roche lobes, are fairly close to a contact configuration. AI Cru, composed of a brighter and larger B5 primary and a fainter and smaller (around B9.5) secondary, is probably not a normal main sequence system.

119.013 **Photometric elements of the system DQ Her and the structure of the disk-like envelope.**
E. S. Dmitrienko, A. M. Cherepashchuk.
Astron. Zh., Tom 57, 749 - 761 (1980). In Russian.
English translation in Soviet Astron., Vol. 24, No. 4.
The analysis of the eclipses in the system DQ Her based on photometric observations of different authors during the years 1954, 1956, and 1975 allowed to determine its photometric elements as well as the structure of the disk-like envelope localized around the white dwarf.

119.014 ***J, K, L,* infrared observations of RZ Scutum.**
R. Akinci, R. F. Jameson.
Astron. Astrophys., Vol. 88, 320 - 322 (1980).
Light curves of the eclipsing binary RZ Scutum are given for 1.2, 2.2, and 3.8 μm. These observations are interpreted in terms of gas disk, $N_e \sim 10^{12}$ cm^{-3} about the primary component together with a gas stream between the components.

119.015 **The system of β Lyrae.** J. Sahade.
Space Sci. Rev., Vol. 26, 349 - 389 (1980).
The observational information that is available and the present knowledge on the system of β Lyrae are presented with emphasis on the problems posed by the object.

119.016 **Spectroscopic studies of O-type binaries. VI. The quadruple system QZ Carinae (HD 93206).**
N. D. Morrison, P. S. Conti.
Astrophys. J., Vol. 239, 212 - 219 (1980).

QZ Carinae is an O-type, double-lined, eclipsing binary in the Carina Nebula. The authors' radial velocity study shows that this system actually consists of two single-lined spectroscopic binaries in the same line of sight, probably in wide orbit about each other. In one of these binaries, which undergoes eclipses with period 5.9980 days, the primary has spectral type roughly O9 III and produces the weaker of the two observed sets of spectral lines. The star responsible for the stronger set is 0.6 mag brighter, has spectral type O9.5 I, and moves in an eccentric orbit about an unseen companion with period 20.73 days. The authors present radial velocities and determine orbits for both systems.

119.017 **Analysis of the light curve, orbital elements and the period changes of U Cephei.**
J. Tremko, G. A. Bakos.
Contrib. Astron. Obs. Skalnaté Pleso, Vol. 9, 163 - 190 (1980).

From photoelectric observations from 1967 to 1976 the authors derive the completelight curve of U Cephei. The orbital elements are essentially the same as derived by Batten. Combined with other photoelectric timings of middle-eclipse the variations of the period of U Cephei are investigated. It is shown that a departure of the residuals from a parabolic distribution is caused by two factors: One is physical resulting from the mass transfer process and the other is observational caused by errors in deriving the epochs of middle-eclipse from the observed light curve. The authors study the long-term increase of the period. They analyse the light curve of U Cephei. They find that the overall depth of the primary eclipse changes by $0^{m}.1$ between times of active and quiet stages.

119.018 **W Ursae Minoris, a non-classical Algol system.**
E. J. Devinney, Jr., C. S. Sutton.
Bull. American Astron. Soc., Vol. 12, 452 - 453 (1980). Abstract.

119.019 **On the existence of "hot spots" in Algol-type binaries.** L. W. Twigg.
Bull. American Astron. Soc., Vol. 12, 453 (1980). – Abstract.

119.020 **The period of RZ Draconis.**
A. D. Mallama.
Bull. American Astron. Soc., Vol. 12, 453 (1980). – Abstract.

119.021 **The May-June 1979 radio outburst of HR 5110.**
P. A. Feldman, M. R. Viner.
Bull. American Astron. Soc., Vol. 12, 508 (1980). – Abstract.

119.022 **RW Comae Berenices. I. Early photometry and *UBV* light curves.**
E. F. Milone, T. T. Chia, K. G. Castle, R. M. Robb, J. E. Merrill.
Astrophys. J., Suppl. Ser., Vol. 43, 339 - 364 (1980) = Publ. Rothney Astrophys. Obs., No. 12.

Photometric observations by the authors are presented and these and all published light curves of RW Com have been analyzed, revealing a variable asymmetric light curve and a variable period. The strongest complication in the light curve is the sin ϕ term, which is variable, and currently has a large, negative value. The O-C analysis is carried out for each extremum yielding a weighted mean of $\dot{P} = -0.40$ (±0.03) dex (−10) day per day, assuming a parabolic fit for the O-C data and therefore a continuous period variation. Using Rucinski's A_4/A_2 relation, the fill-out parameter is $f \approx 1.0$. The system is redder at all phases than is expected of early-to mid-G components, but there is a tendency for the brighter maximum to be slightly bluer.

119.023 **Revised photometric elements of eight eclipsing binaries.**
M. Mezzetti, F. Predolin, G. Giuricin, F. Mardirossian.
Astron. Astrophys., Suppl. Ser., Vol. 42, 15 - 22 (1980).

Photoelectric lightcurves of eight eclipsing binaries, known as detached systems, have been reanalysed by means of Wood's model in order to obtain homogeneous photometric elements. All binaries are confirmed to be detached. TU Cam, CW CMa, YZ Cas, CW Eri, CO Lac and EE Peg appear to be normal main-sequence (or near main-sequence) detached systems, but only the absolute elements of CO Lac are well-known. The detached binaries EK Cep and IQ Per are shown to be anomalous.

119.024 **Geometry and dynamics of the Algol system.**
S. Söderhjelm.
Astron. Astrophys., Vol. 89, 100 - 112 (1980).

A comprehensive analysis of previously published observations of Algol is presented. The two orbits are found to be co-planar, but with opposite senses of revolution. The close orbit is strictly circular, and thus apsidal motion cannot explain the 32^{y} period in the times of minima. The secondary component fills its Roche lobe, and its temperature is near 5000 K. The period-changes indicate that mass and angular momentum are lost from the system, and again magnetic interactions could be involved. Speculatively, the 32^{y} periodicity (which is not strict) is interpreted as some kind of magnetic cycle.

119.025 **Turbulence in the atmospheres of eclipsing binary stars.** K. O. Wright.
Stellar turbulence, (see 012.008), p. 144 - 156 (1980).

A review of the orbits and dimensions of the ζ Aurigae systems is given, based on photometric and spectrographic observations. The Ca II K-line has been studied intensively to determine the extent and uneven structure of the chromospheres of these stars; the multiple structure of this line observed at several eclipses confirms the presence of large-scale clouds in the atmospheres.

119.026 **Strongly interacting binary RX Cas.**
S. Kříž, J. Arsenijević, J. Grygar, P. Harmanec, J. Horn, P. Koubský, K. Pavlovski, J. Zverko, F. Ždárský.
Bull. Astron. Inst. Czechoslovakia, Vol. 31, 284 - 292 (1980).

UBV photoelectric observations of RX Cas made in the years 1975 to 1977 at Hvar and Skalnaté Pleso Observatories are presented. The asymmetry of the light curve and the long-term light variations with the period of 516 days are most conspicuous in the *U* colour. The high increase of the orbital period ($\Delta P/P = 6.3 \times 10^{-7}$ per cycle) leads to an estimated rate of mass exchange greater than $10^{-6} M_{\odot}$/year. The spectroscopic elements by Struve (1944) are re-computed. The case *B* of mass transfer or the second reverse transfer seems to be appropriate for RX Cas. Possible explanations of light anomalies are discussed.

119.027 **Increasing periods of AH Cep and V 382 Cyg.**
P. Mayer.
Bull. Astron. Inst. Czechoslovakia, Vol. 31, 292 - 297 (1980).

New light elements are derived for the early-type eclipsing variables AH Cep and V 382 Cyg. For both binaries, the available data may be explained by a prolongation of the periods. The prolongation is estimated as $\Delta P/P = 3.5^{d} \times 10^{-9}$ and $1.0^{d} \times 10^{-9}$ per cycle for AH Cep and V 382 Cyg, respectively. In the case of AH Cep the minima were deeper in recent years than before.

119.028 **Orbital elements of the eclipsing binaries RW CrA and HO Tel from multicolour light curves.**
J. Grygar, T. B. Horák.
Bull. Astron. Inst. Czechoslovakia, Vol. 31, 297 - 300 (1980).

Five-colour *WULBV* photometry of the eclipsing binaries RW CrA and HO Tel was employed for the orbital element analysis using the limb-darkening scan technique (Horák, 1981). The light curves of both the systems exhibit only slight proximity effects and the mutual agreement of the elements in all colours is fairly good. The presence of the extremely high "third light" in the system RW CrA is puzzling and no plausible explanation of its existence was found.

119.029 **Some aspects related to the reflection effect in the eclipsing binary system AB Andromedae.**
H. Minti.
Proceedings of the colloquium of astronomy, Cluj-Napoca, (see 012.009), p. 23 - 36 (1978).

A reflection effect dependent on the wavelength as well as the phase is obtained. The highest effect on the synthetic light curve is about $0^m_.02$ at phase 40° and 140°, respectively.

119.030 **1977-78 and 1978-79 photoelectric light curves of the RS CVn-type binaries VV Mon, RU Cnc and CQ Aur.** M. Cerruti-Sola, F. Scaltriti, C. Blanco, S. Catalano, E. Marilli, M. Rodonò, G. Strazzulla, C. R. Chambliss.
Astron. Astrophys., Suppl. Ser., Vol. 42, 245 - 250 (1980).

The authors present *UBV* observations of RS CVn-type eclipsing binaries carried out at Torino Astronomical Observatory, at Catania Astrophysical Observatory, and at Kitt Peak National Observatory in the years 1977-78 and 1978-79. These two sets of observations show that remarkable changes have occurred in the outside eclipse lightcurve of the binaries. VV Mon shows a significant increase in brightness around phase 0.25 with respect to the level observed in 1977-78. RU Cnc shows a bump around phase 0.75 completely absent during 1977-78. A retrograde shift of the maximum after the secondary minimum is apparent in CQ Aur.

119.031 **Campagna osservativa EE Cep 1980.** A. Maitan.
G. A. A. B., No. 59-60, p. 3 - 7 (1980).

119.032 **Fourier analysis of the light curves of eclipsing variables. XXVI: Comparison with Kitamura's method.** Z. Kopal, A. Yamasaki.
Astrophys. Space Sci., Vol. 72, 3 - 14 (1980).

The aim of the present paper has been to construct analytic expressions for incomplete Fourier transforms underlying Kitamura's method for an analysis of the light curves of eclipsing binary systems (Kitamura, 1965). The expansions established for Kitamura's coefficients c_n and s_n have been used as a basis for checking numerical accuracy of these coefficients tabulated by Kitamura in 1967; and the outcome bespeaks a high quality of his tables constructed by numerical quadratures.

119.033 **Light curve variations of RT Lacertae.**
C. Ibanoğlu, M. Kurutaç, O. Tümer, S. Evren, Z. Tunca, A. Y. Ertan.
Astrophys. Space Sci., Vol. 72, 61 - 70 (1980).

The eclipsing variable RT Lac, which is classified as an RS CVn-type binary by Hall, was observed in two colours, *B* and *V*, during the summer seasons of 1978 and 1979. The observations made during two successive seasons indicate that, outside eclipses, the system has brightened by about 0.12 and 0.15 magnitudes on average in *B* and *V*, respectively. Moreover, RT Lac was even fainter when observed by Milone in 1965. While the total brightness of the system remained unchanged at mid-secondary, it increased at mid-primary and outside eclipses. This may be interpreted as indicating that the cooler component is responsible for the peculiar light variations.

119.034 **A polarimetric study of U Cephei. Part I.**
V. Piirola.
Astron. Astrophys., Vol. 90, 48 - 53 (1980).

Polarimetric observations of U Cephei in 1972-78 are presented. Time and phase dependent variations in polarization and changes in the shape of the light are discussed.

119.035 **New photoelectric observations of the Wolf-Rayet star HD 5980 in the Small Magellanic Cloud.**
J. Breysacher, C. Perrier.
Astron. Astrophys., Vol. 90, 207 - 209 (1980).

New photoelectric observations of the eclipsing Wolf-Rayet binary HD 5980 in the Small Magellanic Cloud are presented. The phase ϕ =0.36 of the secondary minimum indicates a strong eccentric orbit (e =0.47 for i =80°). The period of 25.56d determined by Hoffmann et al. (1978) is not confirmed, P = 19.266±0.003d is the new value found. A qualitative interpretation of the light curve is proposed in prelude to a further detailed analysis of the system.

119.036 **Observações fotoelétricas do sistema binário eclipsante V505 Sgr na Estação Astronômica Municipal do Pico das Cabras, Campinas, SP, junho/julho 1980.**
G. J. Vilar, J. A. Vieira, J. A. S. Campos, J. A. B. Nazareth, T. J. Vives, R. C. Andrade, E. P. Andrade.
Bol. Astron., Vol. 1, No. 1, p. 8 - 12 (1980).

119.037 **Interpretation of the λ 4500 broad-band light curve of V444 Cyg.**
N. A. Lipunova, A. M. Cherepashchuk.
Astron. Zh., Tom 57, 1033 - 1037 (1980). In Russian.
English translation in Soviet Astron., Vol. 24, No. 5.

The results of the direct-method solution for the λ 4500 broad-band light curve obtained by Kron and Gordon for the system V444 Cyg (WN5 + O6) are given.

119.038 **Short-time fluctuations of hydrogen and helium emissions in the spectrum of β Lyrae.**
M. Yu. Skul'skij.
Pis'ma Astron. Zh., Tom 6, 628 - 631 (1980). In Russian.
English translation in Soviet Astron. Lett., Vol. 6.

From spectral observations variations in equivalent widths and line profiles of strong hydrogen and helium emissions are analysed. Arguments in favour of reality for rare sudden fluctuations of radiation flux with time-scale of ~5 - 15 min and amplitudes of equivalent widths up to 40% for Hα emission and up to 100% for the He I λ 5875 line are obtained.

119.039 **Accretion disks in cataclysmic variables. I. The eclipse-related phase shifts in DQ Herculis and UX Ursae Majoris.** J. A. Petterson.
Astrophys. J., Vol. 241, 247 - 256 (1980).

A model is described for a cataclysmic-variable binary system in which a large accretion disk reflects radiation from a progradely rotating source on or near the white dwarf surface, thus producing rapid oscillations in the optical light. Numerical calculations of the reflection process are presented which simulate the behavior of the oscillations as the system goes through eclipse. The model reproduces the phase shift and amplitude variations of the DQ Herculis oscillations during eclipse very well. The same model, with a progradely rotating source, also reproduces the completely different eclipse-related phase shift behavior of UX Ursae Majoris. Only one new element is added: a small fraction (~0.1%) of the oscillating light near the white dwarf is observed directly, without reflection by the disk.

119.040 **U Cephei: effects of disk, hot spot, and stream on primary eclipse light curves.** E. C. Olson.
Astrophys. J., Vol. 241, 257 - 268 (1980).

The purpose of the present paper is to complete a discussion of photometric perturbations in ingress and egress of primary eclipses of U Cep during eruptive transfer events, to present a brief summary of photometric activity since late 1974 and to propose a simple process to account for the photometric activity.

119.041 **Ephemerides of eclipsing binaries for the year 1981.**
P. Flin.
Rocznik Astronomiczny Obserwatorium Krakowskiego 1981, (see 047.019), p. 1 - 105 (1980).

119.042 **Ephemerides of eclipsing binaries among cataclysmic variables for the year 1981.** J. M. Kreiner.
Rocznik Astronomiczny Obserwatorium Krakowskiego 1981, (see 047.019), p. 107 - 111 (1980).

119.043 **The primary component of the close binary system UU Ophiuchi.** A. Okazaki.
Publ. Astron. Soc. Japan, Vol. 32, 445 - 450 (1980).

Spectroscopic observations of UU Oph, known as a member of the "R CMa type," were made. The observed Balmer-line profiles give log $g \gtrsim 4.4 \pm 0.3$ for the primary component, which consists of normal main-sequence stars. This suggests that UU Oph is not an R CMa-type but an ordinary Algol-type system. The rotational velocity of the primary component V_{rot} sin i is found to be ~ 250 km s^{-1}.

119.044 **Fourier techniques for an analysis of eclipsing binary light curves, VIb.** O. Demircan.
Astrophys. Space Sci., Vol. 72, 281 - 286 (1980).

The paper is a continuation of a previous paper (Demircan, 1980) and aims at ascertaining some other relations between the integral transforms of the light curves of eclipsing binary systems. The appropriate use of these relations should facilitate the numerical computations for an analysis of eclipsing binary light curves by different Fourier techniques.

119.045 **Fourier techniques for an analysis of eclipsing binary light curves, VII.** O. Demircan.
Astrophys. Space Sci., Vol. 72, 287 - 291 (1980).

A Laplace and a Hankel transform of the light curves of eclipsing binary star systems have been expressed in terms of the eclipse elements (L_1, $r_{1,2}$ and i) of the spherical basic model. These general expressions can be used for the solution of the eclipse elements of the systems in terms of the observed quantities.

119.046 **The period distribution of eclipsing and spectroscopic binary systems, II.**
E. Antonello, P. Farinella, G. Guerrero, L. Mantegazza, P. Paolicchi.
Astrophys. Space Sci., Vol. 72, 359 - 367 (1980).

A statistical analysis of period distribution for eclipsing and spectroscopic binary systems, based on the spectral types of the components, shows several common features between the two independent samples. The similarity is increased if one eliminates the geometrical selection effect on the eclipsing binaries sample by means of the method described in previous papers. The period distribution becomes broader (and probably non-unimodal) for advanced spectral types. The authors also performed an analysis of the mean separation of systems as a function of the spectral type.

119.047 **The system VV Ori and the consistency of photometric analysis of eclipsing binary light curves.**
E. Budding, N. N. Najim.
Astrophys. Space Sci., Vol. 72, 369 - 396 (1980).

The extensive, homogeneous and automatically obtained OAO-2 *UV* data on VV Ori at six different wavelengths, as presented by Eaton (1975), are subjected to detailed numerical analysis, and the extent of derivable information on this star from such photometry is critically assessed. An economic (analytical integration) fitting function which can depend on up to 16 independently specifiable parameters is introduced. The use of frequency domain techniques along previously described lines is also discussed. The determination of sought parameter values is examined in some generality, and it is shown that a simultaneous unique optimal specification of more than six or seven such parameters is highly unlikely. By adopting means from all the solution sets of a five-element specification for basic parameters and carrying out statistical tests for self consistency it is possible to investigate gravity darkening and reflection coefficients. The authors find that there is no great departure in the corresponding values from simple standard theory.

119.048 **Photometry of HU Tauri.** M. Parthasarathy, M. B. K. Sarma.
Astrophys. Space Sci., Vol. 72, 477 - 496 (1980).

Photoelectric observations of the bright eclipsing binary system HU Tauri are presented. The observations were made with standard B and V filters. Observations made during the eclipses yielded six epochs of minimum light. A study of all the available times of minimum light has been made. Analysis of all the photoelectric times of minimum light yielded the following new ephemeris: $JD_{\odot}$ 2441275.3166 + $2^{d}0563107E$. Asymmetries in the light curves of HU Tauri were noticed.

119.049 **Ultraviolet colors of W Ursae Majoris: gravity darkening, temperature differences, and the cause of W-type light curves.** J. A. Eaton, C.-C. Wu, S. M. Rucinski.
Astrophys. J., Vol. 239, 919 - 927 (1980).

The authors report three-color satellite ultraviolet photometry for the W-type system W UMa. This system has been well studied from the ground and has been observed extensively enough by the authors to obtain a useful color curve in *(2200 – 3300)*. The observations are presented and the data are analyzed obtaining limits on the parameters which describe the surface brightness. The implications of those results are discussed.

119.050 **Revised photometric elements of the eclipsing binary DI Peg.**
F. Mardirossian, F. Predolin, G. Giuricin.
Astron. Astrophys., Vol. 91, 254 - 256 (1980).

Using Wood's (1972) model the authors have analyzed Binnendijk's (1973) two-colour photoelectric observations of the eclipsing binary DI Peg and the photoelectric light curves published by Ruciński (1967). The photoelectric elements, though still in favour of a semidetached configuration, considerably differ from Ruciński's previous solution. The F4 primary is accompanied by a fainter and smaller cooler star, which fills its Roche lobe for the authors' photometric mass ratio $q = 0.3$. The absolute elements of DI Peg reveal that the secondary appears to be clearly undermassive for its temperature, size and luminosity, like common mass-exchange cooler remnants of Algol-type binaries.

119.051 **A photometric study of the eclipsing binary RX Herculis.** K. W. Jeffreys.
Astron. Astrophys., Suppl. Ser., Vol. 42, 285 - 288 (1980).

A new photoelectric light curve of RX Herculis, a binary system with similar components, has been analysed using Wood's computer model. RX Her, using Popper's spectroscopic mass ratio of $q = 0.8472$, turned out to be composed of a dimmer A0 component and a larger B9.5 component. This detached system, upon analysis of the residuals in secondary minimum, shows some asymmetry during ingress which then disappears just before secondary minimum. The eccentricity $e = 0.022$ determined in this study is a little larger than previously published values of $e = 0.018$. In combination with the spectroscopic analysis of Popper, and *ubvy* data of Olson and Hill and Hilditch new photometric elements for RX Her were found.

119.052 **Abell 46.**
IAU Circ., No. 3503 (1980).

119.053 **CI Cygni.**
IAU Circ., No. 3518 (1980).

119.054 **New photoelectric observations of VW Cephei.**
P. Rovithis, H. Rovithis-Livaniou.
Astrophys. Space Sci., Vol. 73, 27 - 32 (1980).

The aim of the present paper is to present the new B and V light curves of the eclipsing binary VW Cep. Some observational and reductional details are given and the obtained light curves are presented. The period of the system was found to continue its shortening. Finally, a general discussion concerning the light curves is given.

119.055 **Photovisual light curve and photometric elements of the eclipsing nucleus of the planetary nebula Abell 63.** E. Budding, Z. Kopal.
Astrophys. Space Sci., Vol. 73, 83 - 100 (1980).

The first V-photoelectric light curve (on the UBV system) of the eclipsing variable UU Sagittae (constituting the central star of a faint planetary nebula Abell 63) was obtained. Some of the geometrical elements obtained differ significantly from those previously deduced. The most significant feature of the light curve of UU Sge (in both colours) is the large amplitude of the reflection effect exhibited between minima, as well as the fact that the secondary minimum appears to be almost wholly due to an eclipse of reflected light. This, combined with the depths of the alternate minima observed in both colours, leads the authors to conclude that the effective temperature of the O-type component is probably not much higher than 30 000 K, while that of the secondary component is not less than 6000 K (corresponding to a subgiant of spectral class close to G0).

119.056 **The period of EM Cephei: constant or variable?**
R. A. Breinhorst, M. T. Karimie.
Publ. Astron. Soc. Pacific, Vol. 92, 432 - 457 (1980).

Details of a recent photoelectric three-color photometry and an unpublished series of earlier UBV observations of the system EM Cep are presented. It is suggested that the previously reported light-curve anomalies result from abrupt transitions between two distinct modes of light variation with different amplitudes rather than from continuous changes of the light curve. An extension of the $(O - C)$ diagram until 1978 is interpreted in terms of only a marginal secular decrease of the period.

119.057 **Photoelectric light curve and period study for RZ Draconis.** A. D. Mallama.
Publ. Astron. Soc. Pacific, Vol. 92, 463 - 467 (1980).

Photoelectric observations in V and B are presented. The magnitudes and color indices at the four phases of principal interest are $V = 11^m.01$, $(B - V) = 0^m.31$ at phase 0.0; $V = 10^m.11$, $(B - V) = 0^m.25$ at phase 0.25; $V = 10^m.38$, $(B - V) = 0^m.25$ at phase 0.5; and $V = 10^m.14$, $(B - V) = 0^m.24$ at phase 0.75. The $(O - C)$ diagram indicates that the orbital period was constant from the time of the star's discovery as a variable in 1907 until about 1968. At that time the period suddenly decreased by about 2.9×10^{-6} days. The current period is $0^d.5508738$, and the time of minimum derived from the observations in this paper is JD (hel.) 2444177.5555.

119.058 **Photometry and period study of YY Geminorum.**
A. D. Mallama.
Publ. Astron. Soc. Pacific, Vol. 92, 468 - 471 (1980).

Photoelectric observations in 1979 show a second-order light variation superposed on the eclipse light curve, such that the maximum at phase 0.75 is brighter than that at phase 0.25 by about 0.06 in V light. The $(B - V)$ color curve is about 0.08 redder at phase 0.0 than at phase 0.5. Two times of secondary minimum light and one time of primary were observed at the following heliocentric Julian Dates: 2443960.6731, 2443969.6277, and 2443949.679, respectively. An analysis of these and other recent timings indicate that the orbital period has remained constant between van Ghent's photographic observations in the 1920s and the present time.

119.059 **The light curve of Stepanian's star.**
P. Rovithis.
Inf. Bull. Variable Stars, No. 1814, 2 pp. (1980).

119.060 **UBV photometry of DN Cassiopeiae.**
T. J. Davidge.
Inf. Bull. Variable Stars, No. 1817, 2 pp. (1980) = Rothney Astrophys. Obs., Calgary, Canada, Ser. B, Publ. No. 4.

119.061 **HD39917: a new, probably non-eclipsing RS CVn-type variable.**
J. Andersen, B. Nordström, E. H. Olsen.
Inf. Bull. Variable Stars, No. 1821, 3 pp. (1980).

119.062 **Photoelectric photometry of the eclipsing binary DM Del.** R. Diethelm.
Inf. Bull. Variable Stars, No. 1822, 3 pp. (1980).

119.063 **Photoelectric photometry of the bright metallic-line eclipsing binary δ Capricorni.**
J. D. Dorren, M. J. Siah, E. F. Guinan, G. P. McCook, M. J. Acierno, R. R. Del Conte, O. L. Lupie, K. N. Young.
Inf. Bull. Variable Stars, No. 1826, 4 pp. (1980).

119.064 **Revised photometric elements of XZ Sgr.**
G. Russo, C. Sollazzo.
Inf. Bull. Variable Stars, No. 1827, 3 pp. (1980).

119.065 **93 Leonis.** A. H. Batten.
Inf. Bull. Variable Stars, No. 1833 (1980).

119.066 **Provisional elements for an eclipsing variable in Sagittarius.** A. Schwarzmann.
Inf. Bull. Variable Stars, No. 1836, 2 pp. (1980).

119.067 **Times of minima of six Algol-like eclipsing binaries.**
E. C. Olson.
Inf. Bull. Variable Stars, No. 1840, 2 pp. (1980).

119.068 **On the variability of V1068 Cygni.**
R. I. Chuprina, V. G. Derevyagin.
Inf. Bull. Variable Stars, No. 1852, 2 pp. (1980).

119.069 **H and K emissions in V 471 Tauri (BD +16°516).**
E. Hamzaoglu.
Inf. Bull. Variable Stars, No. 1860, 4 pp. (1980).

119.070 **A sudden acceleration in the migration rate of RS CVn.**
J. A. Eaton, D. S. Hall, G. W. Henry.
Inf. Bull. Variable Stars, No. 1862, 4 pp. (1980).

119.071 **Photoelectric minima of DO Cassiopeiae.**
O. Tümer, S. Evren.
Inf. Bull. Variable Stars, No. 1863, 2 pp. (1980).

119.072 **Period of SV Centauri stopped decreasing.**
Z. Kviz, K. J. Murray.
Inf. Bull. Variable Stars, No. 1864 (1980).

119.073 **Light curve of RS Scuti.** D. A. H. Buckley.
Inf. Bull. Variable Stars, No. 1867, 3 pp. (1980).

119.074 **The eclipse of CI Cyg in 1980.**
R. Burchi, A. di Paolantonio, S. Mancuso, L. Milano, A. Vittone.
Inf. Bull. Variable Stars, No. 1871, 2 pp. (1980).

119.075 **On the photometric elements of RX Her.**
G. Giuricin, F. Mardirossian.
Inf. Bull. Variable Stars, No. 1872, 2 pp. (1980).

119.076 **82nd - 84th list of minima of eclipsing binaries.** Compiled by: M. Andrakakou, R. Diethelm, D. P. Elias, R. Germann, R. Leyman, K. Locher, G. Mavrofridis, D. Mourikis, A. Parris, H. Peter, N. Stoikidis, M. Franchini, I. Nikolaou, C. Pampaloni, G. Boistel, A. Buzzoni, D. Leyman, E. Nezry, P. Ralincourt, G. Troispoux.
BBSAG Bull., No. 49, p. 1 - 6, No. 50, p. 1 - 5, No. 51, p. 1 - 5 (1980).

119.077 **DP Cephei, a reinterpretation.** K. Locher.
BBSAG Bull., No. 49, p. 6 (1980).

119.078 **AM Eridani: the definite period.** K. Locher.
BBSAG Bull., No. 50, p. 5 (1980).

119.079 **Index of star names** (BBSAG Bulletins 41 to 50).
BBSAG Bull., No. 50, p. 6 - 8 (1980).

119.080 **GO Monocerotis: the minimum brightness.** K. Locher.
BBSAG Bull., No. 51, p. 5 (1980).

119.081 **Photoelectric times of minima of TT Her.** A. D'Orsi, S. Marcozzi, L. Milano.
Inf. Bull. Variable Stars, No. 1875 (1980).

119.082 **Supplementary observations of AR Lac in 1974.** M. Hoffmann.
Inf. Bull. Variable Stars, No. 1880, 4 pp. (1980).

119.083 **The new photoelectric ephemeris and light curves of NN Cephei.** N. Güdür, Ö. Gülmen.
Inf. Bull. Variable Stars, No. 1881, 2 pp. (1980).

119.084 **Request for confirmation of a new discovered eclipsing binary (SAO 072799).** P. Frank.
Inf. Bull. Variable Stars, No. 1885 (1980).

119.085 **Verbesserte Elemente des β-Lyrae-Sterns NN Cephei.** T. Berthold.
Mitt. Veränderl. Sterne (MVS), Band 8, 162 - 163 (1980).

119.086 **Beobachtung eines Bedeckungsminimums bei η Geminorum.** D. Böhme.
Mitt. Veränderl. Sterne (MVS), Band 8, 163 - 164 (1980).

119.087 **Analysis of the change of period and the photometry of the minima of the eclipsing binary system TX Ursae Maioris.** J. M. Kreiner, J. Tremko.
Bull. Astron. Inst. Czechoslovakia, Vol. 31, 343 - 351 (1980).

The authors investigated the changes in the shape of the light curve of the close binary TX Ursae Maioris in the vicinity of the primary minimum and analysed the changes of period. It was proved that the change of shape of the light curve was asymmetric and the effect on determining the minimum epoch was established. The hypothesis of the existence of the effect of the rotation of the line of apsides was disproved. It was found that processes, leading to a change in period, occurred at least three times. The light changes during the recent period can be expressed by a linear ephemeris, derived from the photoelectric epochs of the minimum.

119.088 **Two eclipses of AL Velorum.** F. B. Wood, R. R. D. Austin.
Mon. Not. R. Astron. Soc., Vol. 193, 867 - 873 (1980).

Photoelectric observations in two eclipses confirm earlier work indicating that light loss in primary shows effects similar to those caused by atmospheric extinction in systems such as ζ Aurigae.

119.089 **Period changes and mass loss rates in 34 RS CVn binaries.** D. S. Hall, J. M. Kreiner.
Acta Astron., Vol. 30, 387 - 451 (1980).

In this paper the authors collect all times of minimum available to them, analyze the period variations, and compute mass loss rates by assuming the active (probably spotted) star has a convectively driven wind which may be anisotropic. They include not only the best known 23 eclipsing RS CVn binaries but also 7 short-period and 4 long-period RS CVn binaries.

119.090 **Revised photometric elements of the eclipsing binary TW Dra.**
G. Giuricin, F. Mardirossian, F. Predolin.
Astrophys. Space Sci., Vol. 73, 389 - 394 (1980).

The photoelectric light curves of TW Dra have been reanalysed in order to obtain accurate photometric elements. TW Dra is confirmed to be a typical sd-system, having a distinctly oversized and overluminous secondary. The primary appears to be slightly more luminous than expected for a main-sequence star, in agreement with theoretical predictions for present primaries of mass-exchange binary systems.

119.091 **New ephemerides for 120 eclipsing binary stars.** A. D. Mallama.
Astrophys. J., Suppl. Ser., Vol. 44, 241 - 272 (1980).

New ephemerides based on data from 1968 to 1978 have been derived for 120 eclipsing binary stars with Algol-like light curves. The original visual timings are grouped into yearly normals of rather high accuracy.

119.092 **Photographic observations of 1980 eclipse of CI Cygni.** M. Huruhata.
Inf. Bull. Variable Stars, No. 1896, 2 pp. (1980).

119.093 **Photoelectric minima of the eclipsing binary DM Persei.** C. Sezer.
Inf. Bull. Variable Stars, No. 1899 (1980).

119.094 **Determinations of six times of minima, and a new ephemeris for BS Dra.** P. B. Etzel.
Inf. Bull. Variable Stars, No. 1900, 4 pp. (1980).

119.095 **The 1981 eclipse of RZ Oph.** J. Smak.
Inf. Bull. Variable Stars, No. 1901, 2 pp. (1980).

119.096 **12 Camelopardalis: a new variable star.** J. A. Eaton, D. S. Hall, G. W. Henry, H. J. Landis, T. G. McFaul, T. R. Renner.
Inf. Bull. Variable Stars, No. 1902, 5 pp. (1980)

119.097 **Visual minima of eclipsing variables.** J. M. Kreiner, A. Mistecka, M. Winiarski.
Astron. Nachr., Band 301, 327 - 328 (1980).

This paper contains 57 heliocentric minima for 37 eclipsing variables.

119.098 **Photoelectric observations of minima of TX Herculis.** R. A. Botsula.
Izv. Astron. Ehngel'gardt. Obs., No. 45, p. 91 - 93 (1978). In Russian.

119.099 **Qualitative schemata of light curves of eclipsing binary systems in the presence of diffuse matter in the system and spots on the surface of the components.**
R. A. Botsula.
Izv. Astron. Ehngel'gardt. Obs., No. 45, p. 94 - 103 (1978). In Russian.

119.100 **The spectrum of RS Canum Venaticorum.** S. A. Naftilan, S. A. Drake.
Publ. Astron. Soc. Pacific, Vol. 92, 675 - 681 (1980).

Two dozen spectrograms of the eclipsing binary RS CVn, taken over the past ten years, are examined. Variable Ca II H and K emission and Balmer-line emission are seen. The variability is examined for correlations with both phase and position of the migrating photometric wave, but the limited data fail to reveal any clear correlation. There is an indication that at least Hα is strongest at primary minimum. The Ca II emission is consistent with a chromospheric origin, but the Hα emission is very broad, and may arise from circumstellar material. The secondary star appears to be metal deficient.

119.101 **Variations of the light curve of the eclipsing binary EM Cephei.** J. Tremko, G. A. Bakos.
J. R. Astron. Soc. Canada, Vol. 74, 321 - 326 (1980).

New photoelectric observations of EM Cep have been compared with those obtained earlier. The authors find that the light curve exhibits large variations in brightness, especially near the primary minimum and the following maximum. In order to explain the variations, a shell surrounding the contact system is adopted as a model.

119.102 **AC Cnc – unique eclipsing system.**
N. E. Kurochkin, S. Yu. Shugarov.
Astron. Tsirk., No. 1114, p. 1 - 3 (1980). In Russian.

119.103 **P Cygni-Profile im UV Spektrum von UW Canis Majoris.** H. Drechsel.
Mitt. Astron. Ges., Nr. 50, p. 21 - 24 (1980).

119.104 **Neue photoelektrische Beobachtungen von Bedekkungsveränderlichen: UZ Octantis.**
G. Wolfschmidt.
Mitt. Astron. Ges., Nr. 50, p. 30 - 32 (1980).

A catalogue of masses, radii, and luminosities of 71 double-lined spectrum eclipsing binaries.
See Abstr. 002.063.

The spectrum variations of HD 153919.
See Abstr. 114.043.

Quasi-periodic variations of the Hα emission in the β Lyrae system. See Abstr. 114.045.

New BV light curves and minima of the eclipsing binaries: W UMa, AW UMa and 44i Boo.
See Abstr. 117.038.

Photoelectric minima of AB Andromedae.
See Abstr. 117.039.

Light elements of AG Phoenicis.
See Abstr. 117.040.

Photoelectric light curves of RZ Columbae.
See Abstr. 117.041.

Photoelectric observations of XY Leonis.
See Abstr. 117.042.

Photoelectric minima observations of the eclipsing binary SV Centauri. See Abstr. 117.044.

A light curve and time of minimum for W UMa obtained with the IUE satellite. See Abstr. 117.046.

Optical variability of the RS CVn candidate HD 174429. See Abstr. 117.047.

Photoelectric y and λ6585 observations of VW Cephei. See Abstr. 117.048.

Narrow- and intermediate-band Hα photoelectric photometry of VW Cephei. See Abstr. 117.049.

TZ Bootis' 1980 light curve.
See Abstr. 117.050.

On the suspected ultrashort period W UMa type binaries BG CrA and AB Tel. See Abstr. 117.052.

The close binary SZ Camelopardalis – a semi-detached system. See Abstr. 117.057.

Anomalous limb darkening coefficients of eclipsing binary systems. See Abstr. 117.060.

SW Lacertae and the spot activity on W UMa stars.
See Abstr. 117.062.

Comments on gas stream effects in typical Algol systems. See Abstr. 117.067.

IUE observations of the ultraviolet spectrum of the close binary δ Pictoris. See Abstr. 117.076.

Variation of the Hβ emission lines of YY Geminorum. See Abstr. 120.010.

A discussion on three yellow variable supergiants in and near the Cepheid instability strip: V 810 Cen (=HD101947), Tr. 27–102 (=HD 159378) and BL Tel (F), based on *VBLUW* photometry and the long-period Cepheids absence in the Galaxy. See Abstr. 122.026.

Intrinsic variability of η Ori.
See Abstr. 122.030.

Rapid oscillations in cataclysmic variables. IV. WZ Sagittae. See Abstr. 122.073.

Stepanyan's star: a new eclipsing cataclysmic variable. See Abstr. 122.224.

A list of eclipsing binaries to be continuously monitored. See Abstr. 123.006.

Revised Blue Mountain Observatory timings.
See Abstr. 123.012.

Detection of periodic light variations in the old nova V 603 Aquilae (1918). See Abstr. 124.201.

M1–2, a possible eclipsing binary planetary nebula central star. See Abstr. 135.006.

X-ray observations of Algol.
See Abstr. 142.041.

Hard X-ray pulses from 4U 0900–40.
See Abstr. 142.043.

120 Spectroscopic Binaries

120.001 **Spectroscopic binary orbits from photoelectric radial velocities. Paper 33: HR 913.**
R. F. Griffin
Observatory, Vol. 100, 113 - 118 (1980).

120.002 **HR 5110 – an Algol-type system with RS CVn properties.** J. D. Dorren, E. F. Guinan.
Astron. J., Vol. 85, 1082 - 1085 (1980).

Intermediate- and narrowband Hα and Strömgren *y* and *u* photometry of the RS CVn-type variable HR 5110 is presented. The Wilson-Devinney program is employed to fit existing data, and the results confirm Hall's suggestion that the low-amplitude (~0ᵐ010) light variation is attributable to the reflection effect. HR 5110 has also been classified as an Algol-type system of low inclination (i~13°) and it is suggested that Algol-type systems might often contain secondaries with highly active chromospheres but have not been classified as RS CVn-type variables because the light from the system is dominated by the primary component.

120.003 **Hα variability in HR 1099 and other RS CVn stars.** H. L. Nations, L. W. Ramsey.
Astron. J., Vol. 85, 1086 - 1091 (1980).

Observations of the RS CVn stars HR 1099, UX Ari, HD 224085, AR Lac, and SZ Psc have been carried out with a CCD detector at KPNO to study the variability of the Hα line. The Hα equivalent width in HR 1099 is shown to undergo a modulation over the orbital period of this spectroscopic binary with the modulation being in antiphase with the photometric distortion wave of the system. Profile and/or intensity variations of Hα with time scales of a few days are also demonstrated for the other systems studied here.

120.004 **A study of the cataclysmic variable TX CVn.** J. W. Fried.
Astron. Astrophys., Vol. 88, 141 - 144 (1980).

TX CVn is a spectroscopic binary. The components are a B9V and a KOIII star, the orbital period is about 71 d. The distance of the star is about 1300 pc. The spectrum of the star shows emission- and absorption features due to a shell. The flux in the Hα emission cannot be explained if the B9V star is the only source of ionizing radiation. The IR measurements of Swings and Allen (1972) show that TX CVn has a strong IR excess which cannot be explained by thermal bremsstrahlung and may be due to dust emission.

120.005 **Simulated analysis of unseen companions.** D. G. Monet.
Bull. American Astron. Soc., Vol. 12, 455 (1980). – Abstract.

120.006 **BD –0°4234: a high-velocity, metal-poor, double-lined spectroscopic binary.** R. C. Peterson, D. W. Willmarth, B. W. Carney, F. H. Chaffee, Jr.
Astrophys. J., Vol. 239, 928 - 936 (1980).

The high-velocity, moderately metal-poor star BD – 0°4234 is shown to be a double-lined spectroscopic binary. The Ca II H and K lines and the Balmer lines Hα and Hε are always seen in emission. The authors observationally derive the primary's temperature and the system orbital elements, then with the aid of model isochrones they estimate the secondary's temperature, the component masses and radii, and the orbital inclination. Both stars are shown to be dwarfs. The authors estimate the rotational velocity assuming synchronous rotation and measure it by using an unusual technique, with results that v_{rot} = 7 ± 2 km s⁻¹. The system's similarities to RS CVn and BY Dra variables are discussed, and one concludes that BD –0°4234 represents an intermediate case.

120.007 **Speckle interferometry of the spectroscopic/astrometric binary Chi Draconis.** H. A. McAlister.
Astron. J., Vol. 85, 1265 - 1269 (1980).

Sixteen speckle interferometric observations of the 281-day spectroscopic and astrometric binary star χ Dra are used to determine the apparent orbit, the masses, and the luminosities of the system. It is shown that, although the discrepancy between the astrometric and interferometric orbits cannot be eliminated, the independent speckle observations from different observers lead to a well determined, highly inclined orbit for χ Dra.

120.008 **Spectroscopic binary orbits from photoelectric radial velocities. Paper 34: HR 4249 A.**
R. F. Griffin.
Observatory, Vol. 100, 161 - 164 (1980).

120.009 **Stellar surface phenomena: stellar rotation and the BY Draconis syndrome in the high-eccentricity binary BD + 24°692.** B. W. Bopp, P. V. Noah, A. Klimke.
Astron. J., Vol. 85, 1386 - 1389 (1980).

The double-line spectroscopic binary (SB2) BD+24°692 (sp. dK3) is found to be a BY Draconis variable, showing a light variation of ~0.1 mag owing to spots on the rotating stellar surface. The rotation period, 6.82 days, differs markedly from the orbital period of 11.93 days.

120.010 **Variation of the Hβ emission lines of YY Geminorum.** K. Kodaira, K. Ichimura.
Publ. Astron. Soc. Japan, Vol. 32, 451 - 462 (1980).

Forty-four image-tube spectra of YY Gem (4 Å mm⁻¹, λλ 4820 - 4900 Å) are analyzed to yield the radial velocity curves and the variations in the intensity and the width of Hβ emission lines during the quiescent phase at epoch 1978 January 4 - 8. The intensity weakly varies in a single-wave mode over an orbital period for the primary, while it strongly varies in a double-wave mode for the secondary with the maximum around phase 0.7 corresponding to the phase of the "star spot" phenomenon. This anomaly is interpreted in terms of non-axisymmetric sectorial distributions of the emission intensity on the stellar surfaces subject to synchronous rotation.

120.011 **A spectroscopic orbit for HR 6626.** M. Mayor, R. F. Griffin.
Astron. Astrophys., Vol. 91, 112 - 114 (1980).

Photoelectric radial-velocity measurements are used to determine the orbit of the late-type giant star HR 6626, and to provide some information (albeit of a negative character) concerning its tidal distortion. The visual companion, 8″away, is confirmed to be physically related to HR 6626.

120.012 **Binarité des étoiles supergéantes.** A. Pedoussaut, N. Ginestet, J. M. Carquillat.
Bull. Inf. Cent. Données Stellaires, No. 19, p. 35 - 38 (1980).

120.013 **Orbites spectroscopiques d'étoiles doubles déterminées avec le spectromètre photoélectrique à vitesses radiales Coravel. II-HD 195987.** M. Imbert.
Astron. Astrophys., Suppl. Ser., Vol. 42, 331 - 334 (1980).

A spectroscopic orbit has been determined for the star HD 195987. The radial velocities were obtained with the Coravel radial velocity spectrometer. Calculations, taking into account previous spectroscopic radial velocities, made with several series of observations at different epochs, give a period with good accuracy: P = 57.3210 ± 0.0014 days. The other orbital elements were also determined.

120.014 **The disappearance of V-V 1-7 and the nature of its central star.**
R. H. Méndez, P. Lee, A. O'Brien, W. Liller.
Astron. Astrophys., Vol. 91, 331 - 334 (1980).

New observations of HD 62001, the central star of V-V 1-7, suggest that it is a normal A-type star which has started to depart from the main sequence; probably it is a spectroscopic binary. The authors show that V-V 1-7 is no longer visible and discuss its nature.

120.015 **Period variations in spectroscopic binaries: SS Cygni and AE Aquarii.** A. N. Feldt, G. Chincarini.
Publ. Astron. Soc. Pacific, Vol. 92, 528 - 530 (1980).

New orbital elements are computed for the variables SS Cyg and AE Aqr. Both systems are X-ray sources.

120.016 **Infrared photometry of HR 1099 during late 1979.**
E. Antonopoulou.
Inf. Bull. Variable Stars, No. 1816, 3 pp. (1980).

120.017 **Results of the photoelectric observations of 88 Her.**
N. L. Magalashvili, J. I. (*Ya. I.*) Kumsishvili.
Inf. Bull. Variable Stars, No. 1868, 2 pp. (1980).

120.018 **Further on the period of NY Cephei.**
D. J. Barlow, D. W. Forbes.
Inf. Bull. Variable Stars, No. 1882, 2 pp. (1980).

120.019 **Die Hülle der Wolf-Rayet-Komponente des Doppelsterns HD 152270.**
W. Neutsch, H. Schmidt, W. Seggewiss.
Veröff. Astron. Inst. Bonn, Nr. 92, 15 pp. (1979).

The spectroscopic binary HD 152270, a member of the open cluster NGC 6231, is composed of an O5 star and a Wolf-Rayet star (WC6-7) which is surrounded by an expanding envelope. The period of revolution is found to be $P = 8\overset{d}{.}893$. The flat, unblended CIII emission line $\lambda = 5696$ Å shows variations of its profile with the phase. Two "peaks" are superimposed on top of this line and are shifted synchronously with the period P of the orbital revolution with an amplitude of about 660 km/s. The separation of the "peaks" is of the order of 1700 km/s but varies around the mean position with half the period P and an amplitude of around 170 km/s. For the interpretation of these phenomena a working model has been developed in which an extended thin CIII shell is disturbed by the hot O-star.

120.020 **Preliminary results of OGS X-ray observations of Capella.** R. Mewe, E. H. B. M. Gronenschild, A. C. Brinkman, J. H. Dijkstra, J. Schrijver, J. Heise, F. D. Seward, H. W. Schnopper, J. P. Delvaille.
Smithsonian Astrophys. Obs., Spec. Rep. 389, (see 012.038), p. 107 - 111 (1980).

A soft X-ray spectrum of Capella has been obtained with the Objective Grating Spectrometer (OGS) onboard the Einstein Observatory in the wavelength range 10–50 Å with a resolution of 0.4 to 1 Å. Emission features have been detected at ≈ 15 and 19 Å, which can adequately be explained as emission from a blend of Fe XVII–XIX lines and from the O VIII Lyman α line, respectively. The spectrum is consistent with emission from an optically thin and isothermal plasma.

120.021 **Spectroscopic binary orbits from photoelectric radial velocities. Paper 35: HD 170737.**
R. F. Griffin.
Observatory, Vol. 100, 193 - 197 (1980).

HD 170737 is shown to be a spectroscopic binary with a γ-velocity of -141 km/s. This is the highest $|\gamma|$ ever found for a galactic binary system.

120.022 **The importance of SAO 93957.**
D. S. Evans, D. A. Edwards.
Observatory, Vol. 100, 206 - 207 (1980).

120.023 **Photoelectric radial velocities. Paper VIII. The orbit of the barium star HD 101013.** R. & R. Griffin.
Mon. Not. R. Astron. Soc., Vol. 193, 957 - 960 (1980).

The barium star HD 101013 (HR4474) is shown to be a spectroscopic binary with a period of about five years, and its orbit – the first ever published for a barium star – is determined.

120.024 **HD 137569: a Population II ultraviolet-bright spectroscopic binary.**
C. T. Bolton, J. R. Thomson.
Astrophys. J., Vol. 241, 1045 - 1049 (1980).

The authors find that HD 137569 is a single-line spectroscopic binary with period 529.8 days. The mass function combined with limits on the mass of the visible star, indicates that the invisible star is the more massive. The present separation of the binary pair is such that the visible star could have evolved without filling its Roche lobe only if its present mass does not exceed $0.6\,M_\odot$ and the available evolutionary models do not underestimate the radius attained near the peak of the asymptotic giant branch.

120.025 **HDE 311884: a massive Wolf-Rayet binary.**
V. S. Niemela, P. S. Conti, P. Massey.
Astrophys. J., Vol. 241, 1050 - 1054 (1980).

A preliminary orbit of this WN6 + O Wolf-Rayet binary reveals that the mass of the WN6 star must be greater than $40\,M_\odot$, contrary to the widely-held belief that all W-R stars have masses $\sim 10\,M_\odot$. The slightly more massive O-type companion is a fairly early (O5?) main-sequence star, and about a magnitude fainter than the WN6 star.

120.026 **Duplicity in the solar neighborhood. I. A new spectroscopic orbit for BY Draconis.**
P. B. Lucke, M. Mayor.
Astron. Astrophys., Vol. 92, 182 - 185 (1980).

A new orbit has been determined for the double lined spectroscopic binary BY Draconis (HDE 234677).

120.027 **Search for an eclipse of HR 913.** G. W. Henry.
Inf. Bull. Variable Stars, No. 1893, 2 pp. (1980).

120.028 **A redetermination of the orbit of HD 192276.**
D. P. Hube, A. Lowe.
J. R. Astron. Soc. Canada, Vol. 74, 342 - 347 (1980).

The previously published orbital solution for HD 192276 is shown to be in error due to an incorrect determination of the period. A period of 7.18584 days is found to satisfy the sixty radial velocities accumulated in 1962 - 79. There is marginal evidence for non-orbital velocity variations on a time scale of a few hours.

120.029 **The spectroscopic orbit of HD 7308.**
R. F. Griffin.
J. R. Astron. Soc. Canada, Vol. 74, 348 - 352 (1980).

HD 7308 is shown to be a spectroscopic binary with a highly eccentric orbit and a period of 660 days.

120.030 **33 Tauri – a new ellipsoidal variable.**
D. P. Hube.
J. R. Astron. Soc. Canada, Vol. 74, 365 - 366 (1980).
Abstract.

Catalogue of physical parameters of spectroscopic binary stars. See Abstr. 002.043.

A catalogue of masses, radii, and luminosities of 71 double-lined spectrum eclipsing binaries.
See Abstr. 002.063.

A radial-velocity study of the nitrogen O supergiant HD 105056. See Abstr. 114.005.

Outer atmospheres of cool stars. V. IUE observations of Capella: the rotation-activity connection.
See Abstr. 114.084.

The composite spectrum of BY Draconis.
See Abstr. 114.149.

Speckle interferometric measurements of binary stars. See Abstr. 118.016.

Spectroscopic studies of O-type binaries. VI. The quadruple system QZ Carinae (HD 93206).
See Abstr. 119.016.

The period distribution of eclipsing and spectroscopic binary systems, II. See Abstr. 119.046.

Analysis of the change of period and the photometry of the minima of the eclipsing binary system TX Ursae Maioris. See Abstr. 119.087.

Spectroscopic evidence for pulsation in the system η Orionis. See Abstr. 122.031.

SS Cygni: a new period and discussion of orbital parameters. See Abstr. 122.074.

AH Velorum – another classical cepheid with a blue companion. See Abstr. 122.110.

Infrared photometry of classical Cepheids.
See Abstr. 122.143.

β Cephei stars as non-radial oscillators. II. Multiperiodic variables σ Sco and ν Eri.
See Abstr. 122.153.

Stellar surface phenomena: asymmetric light curves of the RS Canum Venaticorum binaries λ Andromedae and II Pegasi. See Abstr. 122.208.

Eruptive binaries. X. DQ Herculis.
See Abstr. 124.303.

Detection of X-ray emission from the vicinity of two short-period RS CVn-like binaries.
See Abstr. 142.011.

HR976 and 4C34.13: an X-ray odd couple.
See Abstr. 142.089.

Are binaries concentrated toward the centers of open clusters? See Abstr. 153.012.

121 Early-stage Stars (T Tauri Stars, Herbig-Haro Objects, etc.)

121.001 **The red nebulosity associated with Allen's infrared object in NGC 2264.**
J. R. Walsh, N. J. White.
Observatory, Vol. 100, 119 - 121 (1980).

Relative-intensity contour maps at two wavelengths of the reflection nebulosity associated with the infrared source in NGC 2264 are presented and compared. The characteristics of the nebulosity suggest that it is the remnant of a Herbig–Haro object.

121.002 **Surprising DR Tauri.**
J. Krautter, U. Bastian.
Astron. Astrophys., Vol. 88, L6– L8 (1980).

New spectroscopic observations of the peculiar T Tauri star DR Tauri show that strong qualitative changes in the Balmer line profiles can occur within 24 hours. Furthermore, on one plate ordinary P Cygni profiles (suggesting mass-outflow) and inverse P Cygni profiles (suggesting mass-infall) have been observed to occur in the same spectrogram at different Balmer lines.

121.003 **Simultaneous UBVRI and JHKL photometry of T Tauri stars.** A. E. Rydgren, F. J. Vrba.
Bull. American Astron. Soc., Vol. 12, 453 (1980). – Abstract.

121.004 **The chromospheric explanation of T Tauri spectra.**
G. Basri, N. Calvet, L. V. Kuhi.
Bull. American Astron. Soc., Vol. 12, 520 (1980). – Abstract.

121.005 **Upper limits to the coronal line emission from the T Tauri star RU Lupi.**
G. F. Gahm, R. Liseau, K. Fredga.
Stockholms Obs. Rep. No. 16, 7 pp. (1979).

Three spectrograms of the T Tauri star RU Lupi obtained with the IUE satellite and covering the spectral region λλ 1150 to 2000 Å have been searched for coronal lines. An upper limit of 3×10^6 times the solar values of coronal line fluxes can be derived. The N V lines at λλ1239/43 Å ($T \sim 2 \times 10^5$ K) are present, the absolute flux being 5×10^4 times the corresponding solar values ($A_V = 0.6$). Some implications on predictions of soft X-ray fluxes from T Tauri stars are discussed.

121.006 **The ultraviolet spectrum of the T Tauri star RW Aurigae.** C. L. Imhoff, M. S. Giampapa.
Astrophys. J., Lett., Vol. 239, L115 - L119 (1980).

Ultraviolet spectra of the T Tauri star RW Aurigae have been obtained with the International Ultraviolet Explorer. Emission lines include Mg II, Fe II, Cr II, Si I, Si II, Si III, Si IV, C II, C IV, and Lα. The spectrum roughly resembles the solar chromospheric spectrum, but the surface fluxes are 200 times greater. N V and He II appear to be absent, a possible indication of either an extended chromosphere cooled by mass loss or an optically thick, hot shocked region. The ultraviolet excess in RW Aur is strong, consistent with Balmer continuum emission. The observed continuum may be used to set limits on the type and amount of extinction affecting the star.

121.007 **UV spectrograms of T Tauri stars.**
I. Appenzeller, C. Chavarria, J. Krautter, R. Mundt, B. Wolf.
Astron. Astrophys., Vol. 90, 184 - 191 (1980).

The IUE satellite has been used to observe the ultraviolet spectra ($1200 \leqq \lambda \leqq 3200$ Å, resolution ~6 Å) of the T Tauri stars DR Tauri, CoD −35° 10525, and AS205. In addition, earlier IUE spectrograms of the T Tauri star S CrA were rereduced using a more effective procedure. The new IUE observations were supplemented by simultaneous (or nearly simultaneous) ground-based observations. All observed UV spectra show emission lines of a great variety of different atoms and ions. The ion of highest ionization stage observed is N V. Two of the observed T Tauri stars show a strong UV continuum and UV envelope absorption lines. The occurrence of such absorption lines and of a strong UV continuum seems to be related to the envelope density of these objects.

121.008 **Electron scattering in the infalling envelope of the protostar S CrA.** O. Stahl, B. Wolf.
Astron. Astrophys., Vol. 90, 338 - 340 (1980).
Very broad shallow wings (total width ≈40 Å) of the Balmer lines H_α and H_β of the YY Orionis star S CrA were measured. The observed broad wings are interpreted in terms of noncoherent electron scattering of line photons in the infalling envelope of the low mass protostar S CrA. A crude model was adopted to fit the line profiles. From the fitting procedure the authors estimated the mean electron temperature of the shell to $T_e \approx 7000–8000$ K. The density at the inner boundary of the infalling envelope was estimated to $q_s \approx 0.9 \times 10^{-12}$ g cm^{-3}.

121.009 **The peculiar, galactic object ESO 313-N*10.**
R. M. West.
Astron. Astrophys., Vol. 90, 366 - 371 (1980).
Deep *B*, *J*, and *R* Schmidt photographs show that the peculiar 13^m object ESO 313-N*10 ($\alpha = 8^h40^m27^s$; $\delta = -40°33\!'\!.4$; 1950.0) consists of at least nine condensations, embedded in a nebulosity (diameter 35"). The condensations are red and have strong Hα-emission. The nebulosity is blue and has a composite, B0-B2 absorption spectrum.
ESO 313-N*10 is probably associated with a large cloud complex at a distance of ~1 kpc. It is provisionally interpreted as intermediate between a typical Herbig-Haro (H–H) object with multiple condensations and a group of Herbig Be-stars with nebulosity and may represent the late, pre-main sequence phase of a compact group of early-type stars.

121.010 **H_2 emission from Herbig-Haro objects.**
J. H. Elias.
Astrophys. J., Vol. 241, 728 - 735 (1980).
Molecular hydrogen emission lines have been detected in six Herbig-Haro objects. The line intensities suggest that the H_2 emission arises in a moderate-density, shock heated gas, consistent with evidence for a similar origin of the visible emission-line spectra in Herbig-Haro objects. Indirect arguments indicate that the typical H_2 line widths are less than 70 km s^{-1} and that typical heliocentric radial velocities are no more than 30 km s^{-1}in magnitude.

121.011 **Variability in T Tauri stars: some new aspects.**
C. Bertout.
Variability in stars and galaxies, (see 012.044), p. E.1.1 - 1.17 (1980).
After a short discussion of the general characteristics of variations in T Tauri stars, the variability of DR Tau in the optical and UV ranges is discussed. The radio data available for pre-main-sequence stars are then reviewed, and the star T Tau is used to show how radio and UV data can be combined to give a picture of the emitting layers. Finally, the possibility of variations in the radio spectrum is discussed and the suggestion made that widely different time-scales of variations could result from differing velocity fields in the emitting envelopes.

121.012 **Observations and interpretation of the near-infrared line spectra of T Tauri stars.**
G. H. Herbig, D. R. Soderblom.
Astrophys. J., Vol. 242, 628 - 637, plates 16 - 17 (1980) = Lick Obs. Bull., No. 875.
This paper is essentially a description, with some tentative interpretation, of the spectra of T Tauri stars in the 7500–8700 Å region, as observed with a resolution of about 1 Å.

121.013 **IUE observations of the Mg II emission lines in T Tauri stars.**
C. L. Imhoff, M. S. Giampapa.
Publ. Astron. Soc. Pacific, Vol. 92, 548 (1980). – Abstract.

121.014 **Line profiles in YY Orionis.**
M. F. Walker, D. Burstein.
Publ. Astron. Soc. Pacific, Vol. 92, 648 - 652 (1980) = Lick Obs. Bull. No. 873.
Scanner observations of line profiles in YY Ori on two nights are presented in digital form. In addition to the inverse P Cyg structure, the Balmer lines show multiple emission components that vary in intensity with quantum number and time.

121.015 **On the inapplicability of Ulrich's accretion model to T Tauri-type stars.** S. A. Lamzin.
Astron. Tsirk., No. 1101, p. 1 - 2 (1980). In Russian.

121.016 **Deceleration of rotation and energy of flares of T Tau-type stars.** U. A. Nurmanova.
Astron. Tsirk., No. 1114, p. 4 - 7 (1980). In Russian.

121.017 **UV-spectroscopy of T Tauri stars.**
I. Appenzeller, C. Chavarria, J. Krautter, R. Mundt, B. Wolf.
Mitt. Astron. Ges., Nr. 50, p. 141 (1980). – Abstract.

121.018 **X-ray observations of T Tauri stars.** G. F. Gahm.
Astrophys. J., Lett., Vol. 242, L163 - L166 (1980).
Observations of seven T Tauri stars with the Einstein X-ray observatory are presented. One star, Th 12, was detected in the hard-energy band 0.5–4.0 keV. For this star the authors derive an intrinsic X-ray luminosity of 6×10^{30}erg s^{-1}, corresponding to $\sim 10^5$ times that of the quiet Sun. The authors find that the majority, and possibly all, of the T Tau stars detected up to now at X-ray energies belong to a subclass of T Tau stars which are most frequently bright or show extremely rapid light variations. They suggest that the X-ray emission observed originates in flarelike events on the stars. One Herbig-Haro object, HH 55, was observed but not detected in the present investigation.

121.019 **Elektronenstreuung in der einfallenden Hülle von S CrA.** O. Stahl, B. Wolf.
Mitt. Astron. Ges., Nr. 50, p. 122 - 123 (1980).

Balmerlinienprofile von Protosternhüllen.
See Abstr. 064.097.

Modellrechnungen zur Struktur der Schockkühlschicht von YY-Orionis-Sternen. See Abstr. 064.100.

Optical polarization and infrared spectrum of a possible protostar in a reflection nebula. See Abstr. 116.009.

Protostars and pre-main sequence objects.
See Abstr. 131.021.

A stellar-wind focusing mechanism as an explanation for Herbig-Haro objects. See Abstr. 131.070.

The nature of the Kleinmann–Low nebula. See Abstr. 131.080.

The OH maser near the Herbig–Haro object GGD37. See Abstr. 131.141.

The interaction of T-Tauri stars with molecular clouds. See Abstr. 131.193.

Observations of the J = 1 – 0 and J = 2 – 1 lines of ^{12}CO in L1551: evidence for anisotropic mass loss. See Abstr. 131.194.

Ammonia observations of dark clouds containing Herbig-Haro objects. See Abstr. 131.195.

Observations of radio emission in the 18-cm hydroxyl lines in the direction of Herbig-Haro objects and reflection nebulae. See Abstr. 131.321.

The structure of NGC 1999. See Abstr. 134.001.

A survey of emission-line stars in the Per OB2 dark cloud. See Abstr. 152.008.

122 Intrinsic Variables (Pulsating Variables, Spectrum Variables, etc.)

122.001 **FG Sagittae: a binary?**
R. D. Cohen, G. W. Marcy, E. A. Harlan.
Astron. J., Vol. 85, 867 - 870 (1980) = Lick Obs. Bull., No. 867.

Thirty-four high-dispersion spectrograms of the variable star FG Sge have been used to look for periodicity in its radial velocity. The search has yielded no evidence of any periodicity with an amplitude greater than the expected measurement error of 2.5 km s^{-1}, on timescales of 15 to 200 days. This result strongly indicates that FG Sge, the central star of the planetary nebula 60 – 7° 1, is not a binary star system as had been earlier suggested. Rather, the interpretation of the secular changes in the spectrum of this object must be made in terms of the evolution of a single star.

122.002 **Starspots and the rotation of the flare star EV Lac.**
B. R. Pettersen.
Astron. J., Vol. 85, 871 - 874 (1980).

Sinusoidal variations in the magnitude of EV Lac outside of flares are interpreted as due to intensity modulations from a photospheric spot group. No variations are detected in the color indices. An upper limit set by observational uncertainties indicates that the temperature difference between the photosphere and the spot region is only a few hundred degrees. The period of the cyclic magnitude variations is determined for the first time for EV Lac, and the best estimate is P = 4.378 days. This implies an equatorial rotational velocity of 4.2±0.5 kms^{-1}.

122.003 **Narrowband polarimetry of southern red long-period variables.**
S. J. Codina-Landaberry, A. M. Magalhães.
Astron. J., Vol. 85, 875 - 881 (1980).

Narrowband linear polarization observations are reported for four Mira variables and the semiregular variable L_2 Pup. The polarization has been observed to change, across at least one of the two TiO bands studied, in R Car, R Cen, S Lep, and L_2 Pup. This is most likely due to the variation of the source function and the ratio of absorption to scattering as a function of optical depth in the stellar photosphere. In L_2 Pup the polarization also varies across the Ca I 4226-Å line, indicating that Ca I is unevenly distributed across the stellar surface.

122.004 **Infrared observations of polars: AM Her, VV Pup, and AN UMa.** P. Szkody, R. W. Capps.
Astron. J., Vol. 85, 882 - 885 (1980).

Infrared observations of the three polars AM Her, VV Pup, and AN UMa are presented. The 1.2–2.2 μm energy distribution of VV Pup during the low state is consistent with that of an M4–M5 star and a light curve obtained with the K filter shows small variability over the 100-min orbital period. 1.2–20-μm observations of AM Her during the high state show flux maxima near 1 and 10 μm. Large variations at all wavelengths were also observed from AM Her during the course of one year. No linear polarization was detected at filter K. J, H, and K measurements of AN UMa, made during a low state, reveal a small infrared excess which is probably not due to a late-type companion.

122.005 **On the determination of the colour term in the P–L–C relation for cepheids.**
M. W. Feast, L. A. Balona.
Mon. Not. R. Astron. Soc., Vol. 192, 439 - 443 (1980).

The recent speculations by Brodie & Madore and by Clube & Dawe that the coefficient of the colour term in the P–L–C relation for cepheids may be considerably different from 2.7 are inconsistent with existing data on LMC cepheids. The colour coefficient recently derived by Martin, Warren & Feast is not significantly affected by taking into account an absorption induced correlation between the errors in $\langle V_0\rangle$ and $(\langle B_0\rangle - \langle V_0\rangle)$. The computer simulation of Brodie & Madore which shows the bias introduced in the colour coefficient when this is derived by an inappropriate method (simple least squares) is consistent with theory and in accord with actual results in the LMC. This gives added confidence in the unbiased nature of the results of Martin, Warren & Feast derived by a maximum likelihood method which takes into account errors in each variable.

122.006 **A search for northern hemisphere double mode cepheids – II. New *UBV* cepheid photometry.**
A. A. Henden.
Mon. Not. R. Astron. Soc., Vol. 192, 621 - 623, Microfiche MN 192/1 (1980).

Approximately 500 *UBV* observations of 32 short period cepheids have been made. The observations reported here complete photoelectric light curves for all known cepheids north of –25° declination, brighter than B = 12.5 mag at minimum, and with periods shorter than 5 day. These observations were made with respect to standardized comparison stars and have a median of about 15 observations per variable. Newly derived periods for some of the variables were found using an autocorrelation program. A list of the observations is presented, along with light curves and finding charts for some of the lesser known cepheids.

122.007 **On the nature of AI Velorum.**
L. A. Balona, R. S. Stobie.
Mon. Not. R. Astron. Soc., Vol. 192, 625 - 633, Microfiche MN 192/1 (1980).

Simultaneous BVRI photometry and photoelectric radial velocities are presented for the dwarf cepheid AI Velorum. The Wesselink method of radius estimation yields a radius of 3.1 solar units. Pulsation properties deduced from the observations are compared with published models and found to be in good agreement with Population I masses.

122.008 **Spectroscopy of EX Hydrae.**
J. Breysacher, N. Vogt.
Astron. Astrophys., Vol. 87, 349 - 353 (1980).

Spectroscopic observations of the dwarf nova and X-ray binary EX Hydrae were obtained with a dispersion of 124 Å/mm and high time resolution (15 min). The spectrum is characterised by the emission lines of the Balmer Series, of He I and He II. Consequences for possible models of EX Hya are briefly discussed. In particular, the present data give evidence for a continuous accretion of gas onto the white dwarf.

122.009 **The spectrum of the nebulosity around the symbiotic long-period variable, R Aquarii.**
G. Wallerstein, J. L. Greenstein.
Publ. Astron. Soc. Pacific, Vol. 92, 275 - 283 (1980).

Multichannel spectrophotometric and SIT spectrograph data are reported for R Aqr during its deep minimum of September 1977. Line identifications and fluxes are presented. From the ratio of Balmer continuum emission to Hβ an electron temperature near 11,000 K is derived. A model of R Aqr with a white dwarf or O subdwarf accreting material that is lost by the long-period variable is discussed but to date there is no direct evidence that two separate stars are present. The evidence that the maxima were suppressed when the hot source was bright is used to question the binary hypothesis and to suggest that R Aqr may be a single star with a magnetically active region.

122.010 **HD 185332 a new Delta Scuti star.**
R. Peniche, J. H. Peña. G. Sánchez, J. Warman.
Publ. Astron. Soc. Pacific, Vol. 92, 300 - 302 (1980).

Recent photometric observations show that HD 185332 is a new δ Scuti variable star. Spectrograms of this star fix its spectral type as A3V. Indirect determination of M_v and $(b-y)$ plus the deduced period place this star in agreement with the PLCR and establish it as a first overtone pulsator.

122.011 **A photoelectric study of three southern δ Scuti stars.** S. Morris, D. L. DuPuy.
Publ. Astron. Soc. Pacific, Vol. 92, 303 - 314 (1980).

Differential photoelectric observations of the three δ Scuti stars AI Scl, WZ Scl, and XX Scl were obtained on ten nights during October 1978. AI Scl did not exhibit strictly periodic behavior, but showed a tendency to pulsate at 134 minutes. WZ Scl and XX Scl were both found to be multiperiodic, with period ratios of 0.6853 and 0.9522, respectively. As the period ratio for radial fundamental and first overtone pulsation is 0.76, the observed period ratios are interpreted as caused by nonradial pulsation. A slightly modified Jurkevich period search method, used to obtain these results, is described.

122.012 **Concerning the incidence of duplicity among cepheid variables.** B. F. Madore, J. D. Fernie.
Publ. Astron. Soc. Pacific, Vol. 92, 315 - 318 (1980).

A "phase-shift" technique for detecting photometric companions to cepheids is applied to 202 variables. The inferred incidence of duplicity is 35 ± 5%. An earlier "color-color loop" method suggested 20%–27% duplicity. However, when positive results are required of both methods for a given star, the incidence falls to 15%. This lack of detailed agreement, as well as an apparent correlation of duplicity with period, suggests neither method is infallible.

122.013 **A photometric history of KR Aurigae.**
M. H. Liller.
Astron. J., Vol. 85, 1092 - 1097 (1980).

A search of Harvard plates has yielded 600 magnitudes and upper limits to the brightness of KR Aurigae. The star shows periods of relative quiescence near its brighter limit, times of brightening and fading by several magnitudes, and a probable extended period of faintness.

122.014 **Disk accretion in U Geminorum.**
B. Paczyński, A. Schwarzenberg-Czerny.
Acta Astron., Vol. 30, 127 - 141 (1980).

The analysis of optical and near infrared observations demonstrates that the rate of disk accretion at minimum light of U Geminorum must be much lower than the rate of mass transfer from the red star onto the disk. The optical luminosity of outbursts is consistent with disk accretion of matter stored at the outer disk rim between the outbursts.

122.015 **Identification of Paschen series in emission in the spectrum of the long-period variable star R Cas.**
E. Duch.
Acta Astron., Vol. 30, 191 - 192 (1980).

21 emission lines from 30 which could be observed were found in the spectrum of R Cas. The absence of the remaining 9 lines is explained by strong line or band absorption. The enhancement of continuous spectrum outside the Paschen series limit was also observed.

122.016 **Polarization models for hot nonradial pulsators.**
P. A. Stamford, R. D. Watson.
Acta Astron., Vol. 30, 193 - 214 (1980).

Polarization calculations have been made for some accurate model atmospheres with parameters applicable to β Cephei and 53 Persei variables. Variable polarization effects expected from nonradial pulsations of such stars have been modelled. It is shown that such effects are too small to be detected in the visual but should be large enough at satellite ultraviolet wavelengths to provide a further mode discrimination test in these hot pulsators.

122.017 **Variable stars and the AAVSO.**
J. A. Mattei, E. H. Mayer, M. E. Baldwin.
Sky Telesc., Vol. 60, 180 - 184 (1980).

122.018 **A soft X-ray halo around SU UMa.**
F. A. Cordova, K. O. Mason.
Nature, Vol. 287, 25 - 27 (1980).

The dwarf nova SU UMa was observed with the imaging proportional counter (IPC) detector on the Einstein X-ray Observatory on 4 October 1979 at 4 h UT for ~3,000 s. An X-ray source was detected at the position of the star. It had a flux of 1.3×10^{-11} erg cm^{-2} s^{-1} in the 0.1–4.5 keV energy range of the IPC. In addition, several areas of weaker emission were observed which form a ring that is symmetrically disposed about the position of the dwarf nova. The authors discuss the possibility that it arises in material that has been ejected from the star in a nova-like event.

122.019 **Three-mode resonances in double-mode cepheids.**
M. Takeuti, T. Aikawa.
Mon. Not. R. Astron. Soc., Vol. 192, 697 - 707 (1980).

The three-wave resonances are studied for models with helium enriched inhomogeneous outer layers. The resonance among the fundamental mode, the first overtone, and the third overtone is studied for several models with $Y_S = 0.5$ and $Z_S/X_S = 0.02/0.70$. The model having the evolutionary mass cannot give both the observed period ratio and a sufficiently strong resonance. Masses less than the evolutionary mass or with more enriched helium abundance for the outermost layers are needed to explain the observation.

122.020 **U Orionis: corrélation entre l'émission sur 1612 MHz et la courbe de lumière.**
M. O. Mennessier, J. P. Garrigue.
Bull. AFOEV, Tome 14, 38 - 39 (1980).

122.021 **Le point sur les Mira: quoi de neuf?**
D. Proust.
Bull. AFOEV, Tome 14, 40 - 45 (1980).

122.022 **Beta - Cephei - Sterne.**
M. Fernandes.
BAV Rundbrief, 29. Jahrg., 38 - 43 (1980).

122.023 **RR Leonis.**
W. Wenske.
BAV Rundbrief, 29. Jahrg., 45 - 49 (1980).

122.024 **IUE and ground based observations of the LMC star S Doradus.**
B. Wolf, I. Appenzeller, A. Cassatella.
Astron. Astrophys., Vol. 88, 15 - 22 (1980).

The IUE satellite was used to obtain two low resolution (~6 Å) spectrograms (effective wavelength ranges $1150 \lesssim \lambda \lesssim 1950$ and $1900 \lesssim \lambda \lesssim 3100$ Å) of the luminous Hubble-Sandage variable S Dor of the LMC. In addition ground based high resolution coudé spectrograms were obtained using the ESO 1.5 m telescope. The UV spectrograms are dominated by strong absorption features of singly ionized metallic lines. In addition the resonance lines of C IV (λλ 1548, 1551) and Si IV (λλ 1394, 1402) are observed in absorption. The observed continuum energy distribution of S Dor is explained by a superposition of the radiation of an early A type supergiant photosphere and a surrounding envelope which produces Balmer continuum radiation. The line profiles observed in the visual region indicate a very strong timedependent mass loss of $\dot{M} \approx 7 \times 10^{-4} M_\odot$ yr^{-1}. Possible implications of this extremely strong mass loss on the evolution of S Dor are discussed.

122.025 **The SU UMa stars, an important sub-group of dwarf novae.** N. Vogt.
Astron. Astrophys., Vol. 88, 66 - 76 (1980).

The aim of this paper is: (1) to collect the basic properties of SU UMa stars as known at present, (2) to analyse periodicities in the occurrence of supermaxima, with some tentative conclusions about their origin, (3) to propose a new definition of the SU UMa sub-group, and (4) to point out some consequences for the nature of ultrashort period cataclysmic binaries.

122.026 **A discussion on three yellow variable supergiants in and near the Cepheid instability strip: V 810 Cen (=HD 101947), Tr. 27–102 (=HD 159378) and BL Tel (F), based on *VBLUW* photometry and the long-period Cepheids absence in the Galaxy.** A. M. van Genderen.
Astron. Astrophys., Vol. 88, 77 - 83 (1980).

The present paper deals with recent *VBLUW* photometry of V 810 Cen. A comparison is made with Tr. 27 - 102 and the F type supergiant of the eclipsing binary BL Tel (also near the Cepheid strip) with a characteristic time of 65 d (van Genderen, 1977) of which also *VBLUW* photometry is available (Walraven and Walraven, 1970). Their presence in and near the long-period part of the Cepheid strip is discussed in connection with the absence of long-period Cepheids in the Galaxy.

122.027 **Photoelectric study of the δ Scuti star HR 1225.** S. K. Gupta.
Astrophys. Space Sci., Vol. 71, 377 - 384 (1980).

The periodogram analysis of the V observations of the δ Scuti star HR 1225 has been carried out. Two frequencies of 6.415 cd ($P_0 = 0\overset{d}{.}1558$) and 8.418 cd ($P_1 = 0\overset{d}{.}1188$) have been determined. The period ratio of $P_1/P_0 = 0.762$ indicates radial pulsation. The absolute magnitude, effective temperature and mass of the star are derived to be $1\overset{m}{.}05$, 7600 K and 1.9 $M_\odot$, respectively.

122.028 **Infrared variability of the stars HM Sge and V1016 Cyg.** O. G. Taranova, B. F. Yudin.
Pis'ma Astron. Zh., Tom 6, 495 - 497 (1980). In Russian. English translation in Soviet Astron. Lett., Vol. 6.

Observations in the VJHKL system of HM Sge and V1016 Cyg have been carried out during 1978 - 1979. These observations bear evidence to the existence of late-type stars in these systems.

122.029 **Multichannel spectrophotometry of stellar flares.** S. W. Mochnacki, H. Zirin.
Astrophys. J., Lett., Vol. 239, L27 - L31 (1980).

The authors have observed stellar flares using the 32 channel spectrophotometer on the 5 m telescope. Net flare fluxes in the region 3200–7000 Å are presented. A simple model of blackbody radiation and hydrogen recombination emission appears to fit the continuum points well. The region between 4200 and 7000 Å was used for a detailed fit to the Planck function to obtain apparent temperatures and effective areas. The rise of each flare was associated with an increase of the area, while the initial steep decline of the light was associated with a similar decrease of the blackbody temperature. The maximum temperatures, coincident with maximum light, were 7500–9500 K, similar to values for solar flares. The hydrogen line emission rose simultaneously with the continuum but declined more slowly. The ratio of Hγ to Hα was about 1.5 at the peak, declining to about 1.0 after the peak.

122.030 **Intrinsic variability of η Ori.** R. H. Koch, B. J. Hrivnak, D. H. Bradstreet.
Bull. American Astron. Soc., Vol. 12, 452 (1980). – Abstract.

122.031 **Spectroscopic evidence for pulsation in the system η Orionis.** W. R. Beardsley, E. R. Zizka.
Bull. American Astron. Soc., Vol. 12, 452 (1980). – Abstract.

122.032 **IUE spectrophotometry of R Coronae Borealis variables.** A. V. Holm, C.-C. Wu.
Bull. American Astron. Soc., Vol. 12, 453 (1980). – Abstract.

122.033 **Continued astrometric study of BD + 43°4305.** P. van de Kamp, S. L. Lippincott.
Bull. American Astron. Soc., Vol. 12, 455 (1980). – Abstract.

122.034 **The cepheid luminosity scale.** E. G. Schmidt.
Bull. American Astron. Soc., Vol. 12, 457 - 458 (1980). Abstract.

122.035 **The detection of an early type companion star to the classical cepheid T Mon.** J. T. Mariska, G. A. Doschek, U. Feldman.
Bull. American Astron. Soc., Vol. 12, 462 (1980). – Abstract.

122.036 **A bulge model of a dwarf nova.** W. M. Sparks, G. S. Kutter.
Bull. American Astron. Soc., Vol. 12, 466 - 467 (1980). Abstract.

122.037 **Nuclear burning and dwarf novae.** J. E. Steiner.
Bull. American Astron. Soc., Vol. 12, 467 (1980). – Abstract.

122.038 **Linear polarization of three α^2 CVn stars.** C. A. Koegler.
Bull. American Astron. Soc., Vol. 12, 499 (1980). – Abstract.

122.039 **VLA observations of M-dwarf flare stars.** D. M. Gibson, P. L. Fisher.
Bull. American Astron. Soc., Vol. 12, 500 (1980). – Abstract.

122.040 **Stellar absorption lines of NH_3 in variable stars.** D. Buhl, G. Chin, D. Deming, T. Kostiuk, M. J. Mumma, D. Zipoy.
Bull. American Astron. Soc., Vol. 12, 520 (1980). – Abstract.

122.041 **The IR derived effective temperature of Mira variables.** L. L. Smith.
Bull. American Astron. Soc., Vol. 12, 521 - 522 (1980). Abstract.

122.042 **Coordinated X-ray, optical, and radio observations of flares from the dMe star YZ Canis Minoris.** S. W. Kahler, L. Golub, F. R. Harnden, F. D. Seward, G. S. Vaiana.
Bull. American Astron. Soc., Vol. 12, 526 - 527 (1980). Abstract.

122.043 **Coordinated meter-wavelength observations of the X-ray flare from YZ Canis Minoris.** C. J. Crannell, R. W. Hobbs.
Bull. American Astron. Soc., Vol. 12, 527 (1980). – Abstract.

122.044 **Metal abundance of Magellanic Cloud variable stars.** D. Butler, P. Demarque, H. A. Smith.
Bull. American Astron. Soc., Vol. 12, 538 (1980). – Abstract.

122.045 **A theoretical calibration of the ΔS system.** A. Manduca, R. A. Bell.
Bull. American Astron. Soc., Vol. 12, 538 (1980). – Abstract.

122.046 **XX Cam – the inactive R CrB star.** N. Kameswara Rao, N. M. Ashok, P. V. Kulkarni.
J. Astrophys. Astron., Vol. 1, 71 - 78 (1980).

Infrared observations obtained six years apart of the R CrB type star XX Cam do not show any infrared excess, unlike all the other members of the class. The observed colours match a 7000 K black body energy distribution quite well. From the year 1898 till to date, apparently XX Cam has undergone only one visual light minimum in 1940. The lack of infrared excess, the abundance peculiarities and further lack of small amplitude light variations with periods of few tens of days, which are characteristic of R CrB type stars, are discussed in terms of theoretical pulsation models of helium stars.

122.047 **Ultraviolet spectroscopy of F and G supergiants with *IUE*. I. First results on Cepheid variables.**
S. B. Parsons.
Astrophys. J., Vol. 239, 555 - 564 (1980).

The classical Cepheid variables δ Cep and β Dor were observed with the *IUE* satellite in 1978 December at several phases each, especially in the interval 0.7–1.0 *P*. A slowly decreasing period for δ Cep is confirmed and an improved ephemeris is derived. Similar spectra were obtained of non-pulsating supergiants over the spectral type range F0 Ib–G2 Ib. It is speculated that the presence or absence of a substantial chromosphere is related to the energetics of the pulsation or to the small difference in mean temperature of the Cepheids rather than to the actual values of the temperature during their pulsations.

122.048 **GD 385: a new ZZ Ceti variable.**
G. Fontaine, J. T. McGraw, L. Coleman, P. Lacombe, J. Patterson, G. Vauclair.
Astrophys. J., Vol. 239, 898 - 904 (1980).

The authors report the discovery of a new ZZ Ceti variable, GD 385, thus increasing the number of stars belonging to this class to a total of 13. On time scales of a month the luminosity variations of this star are dominated by a single period of about 256 s; however, the light curve may abruptly change its character on time scales on the order of days. Variations have, on several occasions, been virtually undetectable during an entire night's observing run. This behavior cannot be explained in terms of simple beating phenomena. The authors suggest that nonlinear coupling of pulsational energy among a large number of nonradial *g*-modes may account for the photometric behavior of GD 385.

122.049 **The structure of the Cepheid instability strip.**
B. C. Cogan.
Astrophys. J., Vol. 239, 941 - 952 (1980).

The properties of Cepheids in the Galaxy, LMC, and SMC are analyzed to compare them with theoretical models. The PLC relationships of the LMC and SMC agree with theoretically derived PLC relationships when fitting of data is done by the maximum likelihood method. Differences in the two relationships can be ascribed to abundance differences.

122.050 **HR 1099 and the starspot hypothesis for RS Canum Venaticorum binaries.**
L. W. Ramsey, H. L. Nations.
Astrophys. J., Lett., Vol. 239, L121 - L124 (1980).

Observations of the RS CVn binary HR 1099 (V711 Tau) have been made using the CCD camera at KPNO to test the starspot hypothesis spectroscopically. A TiO band system near 8860 Å is shown to strengthen greatly at phases when starspots have been predicted to be present on the visible hemisphere. Data from other spectral regions support this result. A minimum temperature difference of ~1000 K between the spot and photosphere is indicated.

122.051 **Pulsational mode-typing in line profile variables. III. Multimode behavior in 12 Lacertae.**
M. A. Smith.
Astrophys. J., Vol. 240, 149 - 160 (1980).

Line profile observations of the Si III λ4567 line in the β Cephei star 12 Lacertae have been successfully fitted over three observing runs with a four-mode solution consistent with periods determined earlier by Jerzykiewicz. It is found that the variations in line shape can be fitted only with a radial mode of amplitude 13 km s^{-1}and with three nonradial modes of amplitude 40 km s^{-1} whose states are described by $l = 2$, $m = 0, -1$, and -2. The latter three modes are evidently equipartitioned in energy. The $l = 2$ identification is also compatible with the observed light and color amplitudes for these modes, but $l \geq 3$ fails to meet these tests. Two independent types of analysis show that 12 Lac is observed from a nearly pole-on orientation ($i \approx 15°$).

122.052 **Radial and non-radial oscillations in the very high metallicity δ Scuti star HD 188136.** D. W. Kurtz.
Mon. Not. R. Astron. Soc., Vol. 193, 29 - 49 (1980).

Observations which show HD 188136 to be a δ Scuti star ($\Delta V \leqslant 0.10$ mag) are presented. The metallicity index $\delta m_1 = -0.181$ and the Sr 4077 line which is almost as strong as Hδ indicate that this star has a very high metallicity. Two frequencies, $f_1 = 8.0062$ day^{-1} and $f_2 = 12.5763$ day^{-1}, are shown to be present. A third frequency, $f_3 = 8.11$ day^{-1}, is also shown to be present throughout the data set although its exact value is less certain due to aliasing problems. The proximity of f_3 to the radial mode frequency f_1 suggests that f_3 is due to a non-radial mode. Some comparisons with β Cep stars are made.

122.053 **The frequency analysis of low amplitude δ Scuti stars – V. HD 188520.** D. W. Kurtz.
Mon. Not. R. Astron. Soc., Vol. 193, 51 - 59 (1980).

HD 188520 is an A7IV/V star which is announced here to be a new δ Scuti variable. *uvby* β photometry is presented for this star which is consistent with the A7IV/V spectral classification. A frequency analysis is presented. Two frequencies, $f_1 = 4.762$ day^{-1} and $f_2 = 18.345$ day^{-1} are shown to be present. It is suggested that the frequency at 4.762 day^{-1} may be due to either *g*-mode pulsation or ellipsoidal variability.

122.054 **On the stability of observed frequencies in δ Scuti stars: a reanalysis of θ Tuc.** D. W. Kurtz.
Mon. Not. R. Astron. Soc., Vol. 193, 61 - 77 (1980).

The claim that some δ Scuti stars change their frequencies on time-scales as short as 24 hr is examined. It is suggested that the hypothesis that these δ Scuti stars have stable frequencies is viable. The frequency of highest amplitude is shown to be present at a constant amplitude over the 7 yr time span of the entire data set. A possible set of frequencies is fitted to the θ Tuc data. Other δ Scuti stars which have been claimed to have variable frequencies are discussed.

122.055 **Photometric features near the initial phase of flares on UV Cet-type stars.**
S. Cristaldi, R. E. Gershberg, M. Rodonò.
Astron. Astrophys., Vol. 89, 123 - 125 (1980).

The characteristics of wide-band light variations near the initial phase of flares on UV Cet-type stars, as light decrease (*"dip"*) or increase relative to the quiescent star luminosity, are found to be correlated with the flare amplitude, equivalent duration and rise-time to light maximum. *U*, *B* and *V* observations of a few EQ Peg flares show that the star becomes bluer at pre-flare increase and redder at *dip* minimum.

122.056 **Luminosity of the Mira variables.** L. Celis S.
Astron. Astrophys., Vol. 89, 145 - 149 (1980).

With the elements of the light curve it is possible to extract the luminosity properties of *Me* giants and supergiants, particularly of the Mira variables. In the present work, a relation between the period and the partial periods is found. The most interesting result is a proposed relation between period, spectral type and luminosity (*P–Sp–L*) at maximum light: $M_V = -2.25 \log P + n_o Sp + m_o$, where $n_o = 0.54$ and

$m_o = 1^m.37$, for Miras with $Sp \geqslant$ M 3 and where $n_o = 0.20$ and $m_o = 2^m.38$ for Miras with $Sp <$ M 3. The P–Sp–L relation describes the visual luminosity of the large amplitude red variables and all the behaviour observed with statistical parallaxes.

122.057 **EK Trianguli Australis, a new SU UMa type dwarf nova.** N. Vogt, I. Semeniuk.
Astron. Astrophys., Vol. 89, 223 - 224 (1980).

Photoelectric *UBV* photometry of EK TrA during both types of outbursts (short eruption and supermaximum) is presented. The light curve of the short eruption is smooth without periodic features. During supermaximum, pronounced superhumps with an amplitude of 0.24 mag and a period of 0.0649 d were observed. The superhump period is confirmed by visual observations. A short comparison with other SU UMa stars is given.

122.058 **Outburst photometry of the dwarf nova SS Cygni.** U. Hopp, S. Witzigmann.
Astron. Astrophys., Vol. 89, 227 - 229 (1980).

Photoelectric *B* and *V* observations on 35 nights during three outbursts and the quiet state of the dwarf nova SS Cyg are presented and discussed. It was found that the change of luminosity and the change of temperature during the outbursts are in all observed cases similar. The energy of the outbursts varies from outburst to outburst owing to their variable durations.

122.059 **A magnetic study of spotted UV Ceti flare stars and related late-type dwarfs.** S. S. Vogt.
Astrophys. J., Vol. 240, 567 - 584 (1980).

A multichannel photoelectric Zeeman analyzer has been used to investigate the magnetic nature of the spotted UV Ceti flare stars. Magnetic observations were obtained on a sample of 19 program objects, of which 5 were currently spotted dKe–dMe stars, 7 were normal dK–dM stars, 7 were UV Ceti flare stars, and 1 was a possible post–T Tauri star. Contrary to most previously published observations and theoretical expectations, no magnetic fields were detected on any of these objects from either the absorption lines or the Hα emission line down to an observational uncertainty level of 100–160 gauss (standard deviation).

122.060 ***ubv* observations of the δ Scuti star HD 181333.** G. Hildebrandt, D. Lange.
Astron. Nachr., Band 301, 209 - 210 (1980). In German.

The δ Scuti star HD 181333 was observed in summer 1979 during three nights with the twin telescope in the *ubv*-system. A period of 0.1277 days was found.

122.061 **Photoelectric observations of XZ Cygni.** V. Pop. Proceedings of the colloquium of astronomy, Cluj-Napoca, (see 012.009), p. 52 - 72 (1978).

122.062 **The Beta Cephei variables: current problems.** R. R. Shobbrook.
Proc. Astron. Soc. Australia, Vol. 3, 296 - 299 (1979).

122.063 **The boundary layer method for pulsating stars.** C. A. Jones, P. H. Roberts.
Geophys. Astrophys. Fluid Dyn., Vol. 14, 61 - 101 (1979).
Abstr. in Phys. Abstr., Vol. 83, Abstr. 82389 (1980).

122.064 **Period stability of the pulsating white dwarf R548 (=ZZ Ceti).**
R. J. Stover, J. E. Hesser, B. M. Lasker, R. E. Nather, E. L. Robinson.
Astrophys. J., Vol. 240, 865 - 870 (1980).

R548 is a pulsating DA white dwarf and the prototype of the ZZ Ceti class of variable stars. R548 is pulsating in four pulsation modes simultaneously, the pulsations having periods of 212.77, 213.13, 274.25, and 274.77 s. The authors have combined all of the high-speed photometry of R548 acquired at McDonald and Cerro Tololo Inter-American Observatories. The data, which now cover the years from 1970 to 1978, demonstrate that the properties of the pulsations have remained remarkably constant during this time.

122.065 **Period changes in RR Lyrae stars in the globular cluster NGC 6934.** C. Stagg, A. Wehlau.
Astron. J., Vol. 85, 1182 - 1192 (1980).

An investigation has been made of the period changes in RR Lyrae variables in the globular cluster NGC 6934. The study was based on 211 plates taken between 1911 and 1978. A total of 50 RR Lyrae variables were studied. Of these, two have been changing period irregularly, six have been increasing in period, 11 have been decreasing, and 31 have been constant. The median rate of period increase is +0.03 day per million years and the median rate of period decrease is −0.08 day per million years. One star exhibits the Blazhko effect, and another star had an abrupt change of period after 1938. The theoretical implications of the pattern of period changes for both this cluster and others are discussed.

122.066 **Infrared observations of UV Cas.**
N. Kameswara Rao.
Observatory, Vol. 100, 164 - 165 (1980).

122.067 **The periods of 21 Com.** W. W. Weiss, M. Breger, K. D. Rakosch.
Astron. Astrophys., Vol. 90, 18 - 25 (1980).

There seems to be at least one additional period ($0^d.9178$) possible for 21 Com (HD 108945) which fits the data at least as good as the published long (=rotational) periods. All available photometric and spectroscopic data are reanalyzed in an attempt to improve the long and short (= pulsational) periods. However, no definite conclusion can be drawn for the long period since the authors could not find a period which fits all the data simultaneously in a sufficient way.

122.068 **ZZ Ceti stjerner. Pulserende hvite dverger.**
B. R. Pettersen.
Astron. Tidsskr., Årg. 13, 173 - 180 (1980).

122.069 **Observational stability in Am stars.**
S. F. González B., J. Warman, J. H. Peña.
Astron. J., Vol. 85, 1361 - 1365 (1980).

Differential photoelectric photometry is carried out in metallic-line stars suspected of variability and in the pulsating Am: star HR 8210. The pulsational stability of classical Am stars is confirmed and the hypothesis that the Am phenomenon is the result of diffuse element separation is still consistent with the observations. The variability of HR 8210 is confirmed too. The behaviour of this star should be interpreted as evidence of a multiple-period structure.

122.070 **Periodic spectral variability of the Ap star HR 234.**
R. J. Panek.
Astron. Astrophys., Vol. 90, 341 - 343 (1980).

Spectra of the Ap CrSrEu star HR 234 (HD 4778) at 27 Å/mm were obtained on 13 nights. Contrary to earlier reports, strontium is found to have the largest variation in line strengths, with three maxima occurring with a $2^d.156$ period. A broad feature in the Ca II K line is also variable. Representative spectra are illustrated. Radial velocities show evidence for variability associated with spots but no spectroscopic binary motion.

122.071 **On the space distribution of semi-regular variables.**
Z. Aslan.
Astron. Astrophys., Vol. 90, 355 - 358 (1980).

M type SRa variables within two kiloparsecs of the Sun

are less uniformly distributed, and more concentrated to the plane, than SRb variables. Mean distances from the galactic plane and median latitudes decrease as the period increases. This and the mean motions suggest the shorter period stars may be older than the longer period, and SRa variables may be younger than SRb.

122.072 **A different view of variable stars.**
W. S. G. Walker.
South. Stars, Vol. 28, 126 - 133 (1980).

122.073 **Rapid oscillations in cataclysmic variables. IV. WZ Sagittae.** J. Patterson.
Astrophys. J., Vol. 241, 235 - 246 (1980).

An extensive study of the rapid coherent oscillations in the light curve of the dwarf nova WZ Sagittae is presented. A model is presented in which the oscillation is attributed to the rotation of an accreting, magnetized white dwarf. Most of the observed phenomena can be understood in this model, but the large period increase during 1976 - 1978 remains a problem.

122.074 **SS Cygni: a new period and discussion of orbital parameters.** A. P. Cowley, D. Crampton, J. B. Hutchings.
Astrophys. J., Vol. 241, 269 - 274 (1980).

New spectroscopic observations of the dwarf nova SS Cygni show the accepted orbital period to be substantially in error so that the phases computed from the published ephemeris cannot be used. A revised period and velocity amplitude for each star lead to new estimates of the component masses. However, systematic differences exist in velocities from different data sets, which appear to be real. The new ephemeris is essential for the understanding of the phase correlation of recent X-ray and UV data. Through examination of the emission-line profiles and other quantities, limits are set on the orbital inclination.

122.075 **Spectra of red supergiant variables in the SMC.**
T. Lloyd Evans.
Mon. Not. R. Astron. Soc., Vol. 193, 333 - 336 (1980).

Spectra of eight bright red variables in the SMC show that five are M stars, one shows no strong molecular bands, and two are carbon stars. The latter are variables of large amplitude, and it appears that there are both M stars and carbon stars among those SMC stars which appear to be supergiant analogues of Mira variables. Current theory predicts the weak C_2 bands of these luminous carbon stars, but also requires strong ^{13}C features which are not observed.

122.076 **Gliese 867 – further observations of a multiple flare star system.** P. B. Byrne, J. McFarland.
Mon. Not. R. Astron. Soc., Vol. 193, 525 - 532 (1980).

Light curves of flares on both G867A and G867B are presented. It is shown that activity levels determined during 1977 are maintained. Differential magnitudes between G867A and a number of nearby comparison stars are discussed and used to place upper limits to BY Draconis variations during 1978 and 1979.

122.077 **Synchronous photoelectric observations of flare stars in the visible and near infrared regions.**
V. V. Bruevich, V. I. Burnashev, V. P. Grinin, N. N. Kilyachkov, V. V. Kotyshev, N. I. Shakhovskaya, V. S. Shevchenko.
Izv. Krymskoj Astrofiz. Obs., Tom 61, 90 - 109 (1980). In Russian.

The results of synchronous photoelectric observations of the flare stars AD Leo and EV Lac made in 1975 with a *B*-filter and in the near infrared and of observations of UV Cet and EV Lac made in 1976 are given.

122.078 **Polarimetric observations of R CrB.**
Yu. S. Efimov.
Izv. Krymskoj Astrofiz. Obs., Tom 61, 110 - 119 (1980). In Russian.

Linear polarization of R CrB has been measured since 1972 to 1976. The results are given. The hypothesis of Faraday rotation in a magnetoactive medium is assumed to explain the wavelength dependence on the position angle of polarization observed during the minimum of R CrB in 1972.

122.079 **New variable star of BY Dra-type: HD 1835.**
P. F. Chugajnov.
Izv. Krymskoj Astrofiz. Obs., Tom 61, 124 - 126 (1980). In Russian.

Photoelectric observations of the G2-star HD 1835 in the *B, V* system are presented. It is shown that the brightness of the star varies with period $7^{d}.655$, the amplitudes in V and B being $0^{m}.032$ and $0^{m}.037$ respectively. The light variations are probably due to the presence of dark spots on the star's surface.

122.080 **Light variations of the late spectral-type star (HD 34454) embedded in a nebula.**
P. F. Chugajnov.
Izv. Krymskoj Astrofiz. Obs., Tom 61, 127 - 130 (1980). In Russian.

Light variations of the M5 III star HD 34454 associated with a small reflective nebula are discovered.The star probably belongs to the group of small amplitude red variables found by Eggen. It is possible that the nebula belongs to the Orion OB-association.

122.081 **Ephemerides of RR Lyrae-type variables for the year 1981.** W. P. Zessewitsch (*V. P. Tsesevich*), B. N. Firmaniuk, J. M. Kreiner.
Rocznik Astronomiczny Obserwatorium Krakowskiego 1981, (see 047.019), p. 113 - 130 (1980).

122.082 **Photoelectric activity of the shell star o And: new observations and discussion.**
M. Bossi, G. Guerrero, L. Mantegazza, M. Scardia.
Astrophys. Space Sci., Vol. 72, 433 - 437 (1980).

Photoelectric observations of the shell star *o* And, obtained in 1979, are presented. The star shows variations of some hundredth of magnitudes during a few hours. The trend to retake the values of the luminosity and colour indices prior to the reduction happened between JD 42,714 - 27, seems to continue. The authors also give a qualitative model which explains satisfactorily the main features observed after the 1975 shell episode.

122.083 **Photometry and polarimetry of VW Hydri during the October 1978 supermaximum.**
R. Schoembs, N. Vogt.
Astron. Astrophys., Vol. 91, 25 - 31 (1980).

Photoelectric observations of VW Hyi at supermaximum are presented. The well-known superhump phenomenon was confirmed. During decline the superhumps weakened; several additional peaks appeared in the light curve and repeated with the superhump period. Two ultrashort period oscilations, with periods 33 s and 74 s, were found near the detection limit. Periodic colour variations within the superhump cycle resemble those of a pulsating variable, indicating an intrinsic variability of the relevant light source. The nightly mean linear polarization ranges between 0.02% and 0.1% and was found to increase with decreasing brightness. The authors interpret the polarization as being caused by single particle scattering in the disc. The colour curves, as well as the long-term variability of the polarization, suggest that the disc has a more spherical shape at early outburst state than in quiescence. During decline it gradually flattens.

122.084 **HR 7308, a new cepheid with variable amplitude and very-short period (1.5 d).**
G. Burki, M. Mayor.
Astron. Astrophys., Vol. 91, 115 - 121 (1980).

HR 7308 (HD 180583) was observed 132 times with the spectrophotometer CORAVEL (radial velocity) and 10 times in the Geneva photometric system. These data reveal that this bright star is a new cepheid. It is shown that it probably belongs to population I. If this is true, the period (1.49107 d) is the shortest actually known for classical cepheids. The peak-to-peak amplitude of variation in radial velocity has decreased during the survey period from about 18 to 4 km s^{-1}. Thus, HR 7308 is the first case of a single-mode classical cepheid exhibiting strong amplitude variations. If this variation is periodic (period of at least 600 d), HR 7308 would be the first classical cepheid exhibiting the Blazhko effect.

122.085 **The variable, single-line Wolf-Rayet star HD 96548 with a low-mass companion.**
A. F. J. Moffat, J. Isserstedt.
Astron. Astrophys., Vol. 91, 147 - 154 (1980).

Extensive narrow-band photoelectric photometry and high dispersion optical photographic spectroscopy indicate in the WN8 star HD 96548 the presence of two variable components: (1) random noise with ~0.02 mag rms both in the continuum and, uncorrelated, in emission lines, on a time scale of days; (2) a possible periodic modulation (P = 4.762 d) with full amplitude 0.04 mag in the light curve with double minimum and semiamplitude 8 - 10 kms^{-1} in the radial velocity curve. These are tentatively interpreted in terms of a WR + compact binary.

122.086 **Variability in β Cephei and B stars.**
M. A. Smith.
Highlights of Astronomy, Vol. 5, (see 012.026), 457 - 460 (1980).

122.087 **Theoretical nonradial pulsation of β Cephei models.**
H. Sato.
Highlights of Astronomy, Vol. 5, (see 012.026), 461 - 462 (1980).

The author concludes that the opacity bump near the He^+ ionization zone as given by current opacities is not enough to excite any pulsations (radial or nonradial) in the β Cephei models.

122.088 **Amplitudes and frequencies of δ Scuti stars: systematics.** M. Breger.
Highlights of Astronomy, Vol. 5, (see 012.026), 463 - 466 (1980).

122.089 **Recent theoretical results for δ Scuti and dwarf cepheid variables.** J. R. Percy.
Highlights of Astronomy, Vol. 5, (see 012.026), 467 - 468 (1980).

122.090 **White dwarf pulsations.** W. Dziembowski,
Highlights of Astronomy, Vol. 5, (see 012.026), 469 - 472 (1980).

122.091 **Supergiant pulsations.** A. Maeder.
Highlights of Astronomy, Vol. 5, (see 012.026), 473 - 476 (1980).

The main data concerning the amplitudes and periods of supergiant variations are reviewed. The period-luminosity-colour relation of intermediate-term variations is compatible with pulsation motions. The Q-values seem to be larger than those corresponding to the fundamental mode and there are arguments favouring non-radial oscillations.

122.092 **Observations of classical cepheids.**
B. F. Madore.
Highlights of Astronomy, Vol. 5, (see 012.026), 477 - 478 (1980).

122.093 **Distances and radii of classical cepheids.**
T. G. Barnes III.
Highlights of Astronomy, Vol. 5, (see 012.026), 479 - 482 (1980).

The author shows that the surface brightnesses of cepheid variables may be determined from the cepheids themselves, that for short-period cepheids the resultant values are in good agreement with non-variables of the same color, and that the preliminary distance scale to which these values lead supports other recent suggestions for an enlarged cepheid distance scale.

122.094 **Recent work on beat cepheids at Mount Stromlo Observatory.** S. L. Barrell, B. C. Cogan, D. J. Faulkner, R. R. Shobbrook.
Highlights of Astronomy, Vol. 5, (see 012.026), 483 - 485 (1980).

122.095 **Recent theoretical results for cepheid pulsation.**
A. N. Cox.
Highlights of Astronomy, Vol. 5, (see 012.026), 487 - 488 (1980).

122.096 **RR Lyrae variable stars.** Yu. S. Romanov, V. P. Tsesevich, L. P. Zaikova (*Zajkova*), B. N. Firmanyuk, S. N. Udovichenko, Z. N. Fenina, L. I. Shakun.
Highlights of Astronomy, Vol. 5, (see 012.026), 489 - 492 (1980).

122.097 **The pulsation properties of red variables.**
M. W. Feast.
Highlights of Astronomy, Vol. 5, (see 012.026), 493 - 496 (1980).

Recently derived bolometric absolute magnitudes are used to investigate the pulsational properties of some classes of red variables. The Mira variables with periods over 200 days have pulsation constants appropriate to overtone pulsators but the shorter period 135 day Miras seem to be hotter and less luminous and may be fundamental pulsators. The large amplitude Shapley-Nail variables in the SMC are probably fundamental pulsators with masses in the 5-10 solar range. Bright supergiant red variables in the LMC show evidence of the period-luminosity relation expected if they are in the early stages of core helium burning.

122.098 **BY Draconis and RS Canum Venaticorum stars: the discoveries of classical photometry and spectroscopy.**
B. W. Bopp.
Highlights of Astronomy, Vol. 5, (see 012.031), 847 - 848 (1980).

122.099 **VLBI observations of main-line OH emission from U Orionis.** J. D. Fix, R. L. Mutel, J. M. Benson, M. L. Claussen.
Astrophys. J., Lett., Vol. 241, L95 - L98 (1980).

Compact structure in the 1665 MHz OH emission from the long-period variable star U Orionis was detected using a five-telescope VLBI array. The U Orionis spectrum contains many compact, highly circularly polarized features which are spread over an area about 0.″5 in size. The polarization of the 1665 MHz emission from U Orionis cannot be understood by a simple Zeeman model. The extent of the main-line emission from U Orionis is significantly larger than that of previously measured 1612 MHz emission.

122.100 **HD 9250 and HD 14662 (HR 690), two new bright cepheids with very small amplitude.**
G. Burki, M. Mayor, C. Waelkens.
Astron.Astrophys., Vol. 91, 276 - 278 (1980).

From radial velocity and photometric surveys, these two new bright cepheids have been discovered. Radial velocity curves are presented exhibiting a residual dispersion of only about 0.4 km/sec. These two stars are among the cepheids of smallest known amplitude.

122.101 **HR 7308: a unique Cepheid.**
J. R. Percy, N. R. Evans.
Astron. J., Vol. 85, 1509 - 1512 (1980).

The authors have obtained 44 photometric observations and nine radial velocities of the F6I-IIb star HR 7308. The spectral type, the period ($1^{d}.49$), and the relative range and phase of light, color, and velocity variations all suggest that the star is a small-amplitude Cepheid. The metal abundance appears to be normal. However, the light amplitude decreased slowly during 1978 from $0^{m}.3$ to $0^{m}.1$, and increased slightly during 1979 from $0^{m}.05$ to $0^{m}.07$. Such behavior is unprecedented in a Population I Cepheid.

122.102 **The masses of cepheids.** A. N. Cox.
Annu. Rev. Astron. Astrophys., Vol. 18, (see 003.004), 15 - 41 (1980).

The author describes six ways of deriving cepheid masses. Only anomalies involving the period ratios Π_1/Π_0 for beat cepheids and Π_2/Π_0 for bump cepheids seem to remain with the others largely resolved. The hypotheses of surface helium enhancement, tangled surface magnetic fields, and possibly nonradial effects seem to be the only viable ones to resolve mass anomalies for the bump and beat cepheids.

122.103 **SU Tauri.**
IAU Circ., Nos. 3524, 3546 (1980).

122.104 **U Geminorum.**
IAU Circ., No. 3526 (1980).

122.105 **UX Ursae Majoris.**
IAU Circ., Nos. 3538, 3546 (1980).

122.106 **CH Cygni.**
IAU Circ., No. 3549 (1980).

122.107 **SU Tauri.**
Yamamoto Circ., No. 1945 (1980).

122.108 **Hydromagnetics and period changes in RR Lyrae stars.** R. Stothers.
Publ. Astron. Soc. Pacific, Vol. 92, 475 - 478 (1980).

Balázs-Detre and Detre have drawn an interesting parallel between the observed time scales of variability in the sun and in RR Lyrae stars. Additional information is presented to support their conjecture that an analog of the solar magnetic cycle is operating in RR Lyrae stars. Rough considerations of the expected changes of photospheric radius and of magnetic energy content during a magnetic cycle suggest that the pulsation periods of these stars should also change in time. Within the large observational and theoretical uncertainties, the predicted period changes are compatible with those observed.

122.109 **BI Crucis: a new symbiotic star.**
K. G. Henize, E. D. Carlson.
Publ. Astron. Soc. Pacific, Vol. 92, 479 - 483 (1980).

A Mount Stromlo spectrogram of BI Cru taken in 1962 shows emission lines of H I, He I, He II, Fe II, N III, [O III], [Ne III], and [S II] superposed on a weak bluish continuum. A spectrogram by Allen in 1974 shows emission lines of H I and Fe II and possibly weak He I, [Fe II], and [O I] superposed on an M-star absorption spectrum. The object is evidently a symbiotic star showing large variations in its spectral character. Significant differences exist in the mean ion velocities and appear to be correlated with ionization potential.

122.110 **AH Velorum – another classical cepheid with a blue companion.** W. Gieren.
Publ. Astron. Soc. Pacific, Vol. 92, 484 - 488 (1980).

The binary nature of AH Vel previously suspected on the basis of radial-velocity measurements is confirmed by three independent photometric methods, using extensive *UBV* photometry of the star. The open loop shown by AH Vel in the $(U-B)$, $(B-V)$ diagram gives a B-type main-sequence companion about three magnitudes fainter than the cepheid, using the calibration of Madore (1977). The shift of the $(U-V)$ minimum with respect to the minimum of the V light curve (Fernie 1979) suggests a companion that is about $2^{m}.8$ fainter in visual magnitude than AH Vel, in agreement with the result from the color-color diagram. Adopting $\langle M_V \rangle = -4^{m}.58$ for AH Vel (Gieren 1980) the companion is a B3 V star.

122.111 **Spectrum variability in HR 8752.**
S. C. Barden, L. W. Ramsey.
Publ. Astron. Soc. Pacific, Vol. 92, 497 - 500 (1980).

Observations in the region of Hα are presented for the luminous supergiant HR 8752. The authors detect two time scales for the variation of emission features. Substantial changes in Hα appear to take months or years. More subtle changes can occur in a matter of days. The photospheric spectrum also shows evolution on both long and short time scales as has been noted by previous observers.

122.112 **On the large-scale variations of MV Lyrae.**
W. Wenzel.
Inf. Bull. Variable Stars, No. 1810, 2 pp. (1980).

122.113 **Photoelectric observations of symbiotic stars.**
R. Burchi.
Inf. Bull. Variable Stars, No. 1813, 3 pp. (1980).

122.114 **V and B-V magnitudes of bright Cepheids from electrospectrophotometric observations.**
E. A. Depenchuk.
Inf. Bull. Variable Stars, No. 1819, 3 pp. (1980).

122.115 **HR 3593, a new bright β Cephei star.**
G. Burki, M. Burnet, F. Rufener.
Inf. Bull. Variable Stars, No. 1820, 3 pp. (1980).

122.116 **Variability of HD 137147 (comparison star of UCrB) and "a flare" of HD 137050.** E. C. Olson.
Inf. Bull. Variable Stars, No. 1825, 5 pp. (1980).

122.117 **The variability of HR 7442.**
D. R. Skillman, T. G. McFaul.
Inf. Bull. Variable Stars, No. 1829, 2 pp. (1980).

122.118 **Shorter secondary variation of RS Bootis.**
S. Kanyó.
Inf. Bull. Variable Stars, No. 1832, 4 pp. (1980).

122.119 **DY Pegasi – a double mode dwarf Cepheid?**
T. Kozar.
Inf. Bull. Variable Stars, No. 1834, 5 pp. (1980).

122.120 **Spectroscopy of the nova-like object Kuwano (Nova Vulpeculae 1979) in the year 1979.**
L. Hric, D. Chochol, J. Grygar.
Inf. Bull. Variable Stars, No. 1835, 3 pp. (1980).

122.121 **Rapid variability of symbiotic stars: CH Cygni and EG Andromedae.**
J. Mikołajewska, M. Mikołajewski.
Inf. Bull. Variable Stars, No. 1846, 2 pp. (1980).

122.122 **Photoelectric behaviour of V 818 Sco and X Per.** B. B. Sanwal.
Inf. Bull. Variable Stars, No. 1847, 3 pp. (1980).

122.123 **HD 37819, a new δ Scuti star.** G. Burki, M. Mayor.
Inf. Bull. Variable Stars, No. 1851, 2 pp. (1980).

122.124 **A cepheid variable in NGC 6067.** O. J. Eggen.
Inf. Bull. Variable Stars, No. 1853, 4 pp. (1980).

122.125 **UBV photometry of SZ Lyn.** G. A. Garbusov.
Inf. Bull. Variable Stars, No. 1854, 3 pp. (1980).

122.126 **The maximum times and new light elements of 28 Aquilae.** A. Y. Ertan, Z. Tunca, O. Tümer, S. Evren, M. Kurutac, C. Ibanoglu.
Inf. Bull. Variable Stars, No. 1855, 2 pp. (1980).

122.127 **The period of V154 in NGC 5272 (M3).** U. Hopp.
Inf. Bull. Variable Stars, No. 1857, 2 pp. (1980).

122.128 **Surprisingly high optical polarization of μ Cephei.** J. Arsenijevic, A. Kubicela, I. Vince.
Inf. Bull. Variable Stars, No. 1859 (1980).

122.129 **Present activity of CH Cygni.** R. Faraggiana, M. Hack.
Inf. Bull. Variable Stars, No. 1861, 3 pp. (1980).

122.130 **An interesting phenomenon of the flare star BD +22°3406.**
F. M. Mahmoud, M. A. Soliman.
Inf. Bull. Variable Stars, No. 1866, 3 pp. (1980).

122.131 **HD 200925, a pulsating variable.** S. K. Gupta, T. D. Padalia.
Inf. Bull. Variable Stars, No. 1870, 2 pp. (1980).

122.132 **CW UMa – a new flare star?** W. P. Bidelman.
Inf. Bull. Variable Stars, No. 1873 (1980).

122.133 **Spectroscopic measurements of the yellow supergiant CO Aur.** M. Hoffmann.
Inf. Bull. Variable Stars, No. 1876, 2 pp. (1980).

122.134 **Tau Cygni.** C. Bartolini, A. Dapergolas.
Inf. Bull. Variable Stars, No. 1884, 2 pp. (1980).

122.135 **Period changes of dwarf cepheids, I. CY Aquarii, EH Librae, DY Pegasi.**
H. A. Mahdy, B. Szeidl.
Mitt. Sternw. Ungarische Akad. Wiss., No. 74, 46 pp. (1980).

Period changes of the dwarf cepheids CY Aqr, EH Lib and DY Peg are discussed and O-C diagrams of these stars are constructed. EH Librae has a highly stable, constant period. The period of CY Aquarii changed suddenly in 1952 ($\Delta P = -1.81 \times 10^{-7}$ day = 0.016 sec). Before the abrupt change its period showed small fluctuations, such fluctuations have also been shown since that time. The period of DY Pegasi has become shorter during the past 45 years.

122.136 **Photoelectric UBV photometry of northern Cepheids, II.** L. Szabados.
Mitt. Sternw. Ungarische Akad. Wiss., No. 76, 134 pp. (1980).

New UBV photoelectric observational data on 42 northern Cepheids with periods of 5 - 10 days are presented. The period changes and the variations in the light curve of the observed Cepheids are investigated. No secular light curve variation has been discovered.

122.137 **The pulsating helium star BD +13°3224.**
P. W. Hill, D. Kilkenny, D. Schönberner, H. J. Walker.
Anglo-Australian Obs. Prepr. No. 138 (1980).

The period of the small-amplitude photometric variable extreme hydrogen-deficient star BD +13°3224 is improved. Measurements of radial velocity show it to vary with the same period as the light and provide for the first time clear evidence for pulsation of a hot hydrogen-deficient star. The observed variation is compared with theoretical models of pulsation in helium stars and the connection with the R CrB variables examined.

122.138 **High-time resolution spectroscopy of VW Hydri and WX Hydri.** R. Schoembs, N. Vogt.
ESO Sci. Prepr. No. 124, 28 pp. (1980). – Submitted to Astron. Astrophys.

122.139 **On the nature of the 125-day Cepheid V 810 Cen (= HR 4511): IUE spectra.**
W. Eichendorf, A. Heck, J. Isserstedt, J. Lub, M. Pakull, B. Reipurth, A. M. van Genderen.
ESO Sci. Prepr. No. 125, 8 pp. (1980). – Submitted to Astron. Astrophys.

122.140 **The convection and stellar activity in Cepheid variables.** M. Takeuti.
Sci. Rep. Tôhoku Univ., Ser. 1, Vol. 62, 115 - 120 (1980).

The kinetic energy included in the surface convection layers is roughly estimated for Cepheid variables. For homogeneous envelopes with the normal Population I composition, the total kinetic energy of convection in beat Cepheids is nearly equal to that in Cepheids without beat. This energy becomes larger for beat Cepheids than for normal Population I Cepheids, as the helium is enriched in their envelopes. The suggested helium enrichment in the outer envelopes of Cepheids may induce the spontaneously enhanced Hα emission in beat Cepheids, while the emission is not found in normal Cepheids but redder ones. Distortions or emission of Hα line in Cepheids could be an evidence to support the inhomogeneous envelope theory.

122.141 **Zum Helligkeitsanstieg von DR Tauri.** W. Götz.
Mitt. Veränderl. Sterne (MVS), Band 8, 143 - 150 (1980).

The star was observed on 693 blue-sensitive and on 370 photovisual sky patrol plates of Sonneberg Observatory covering the years from 1931 respectively 1962 to February 1980, in order to complete the light curve of the object given by C. Chavarria-K. The results and a discussion of the light curve are given in detail.

122.142 **Beobachtungen von V 154 im Kugelhaufen M3.** I. Meinunger.
Mitt. Veränderl. Sterne (MVS), Band 8, 161 - 162 (1980).

122.143 **Infrared photometry of classical Cepheids.** T. Lloyd Evans.
South African Astron. Obs. Circ., Vol. 1, 163 - 165 (1980).

A total of 59 sets of JHKL measurements are presented for eight classical Cepheids, most of which are known or suspected spectroscopic binaries.

122.144 **Multicolour photoelectric photometry of Magellanic Cloud Cepheids. III: BVI observations of ten LMC Cepheids.** W. L. Martin.
South African Astron. Obs. Circ., Vol. 1, 172 - 174 (1980).

BVI observations of ten LMC Cepheids in the period range 7 - 13 days are presented and relevant light curve parameters derived.

122.145 **Radial velocities of southern Cepheids.** T. Lloyd Evans.

South African Astron. Obs. Circ., Vol. 1, 257 - 270 (1980).
Radial velocities are presented for 50 classical Cepheid variables in the southern sky.

122.146 **Non-emission-line flare stars.**
B. R. Pettersen, R. F. Griffin.
Observatory, Vol. 100, 198 - 202 (1980).
The data for the three flare stars discussed lend support to the impression that non-emission-line flare stars show abnormally low activity levels. But the authors have also demonstrated that non-emission-line stars do flare.

122.147 **Optical behaviour of SS Cygni in 1978 and 1979.**
J. A. Mattei.
J. R. Astron. Soc. Canada, Vol. 74, 317 - 320 (1980).

122.148 **Objets célestes remarquables. FG Sagittae, étoile variable exceptionnelle.** J. Sauval.
Ciel Terre, Vol. 96, 371 - 381 (1980).
The author presents a review of the main observations of the variable star FG Sagittae and of its surrounding planetary nebula. The exceptional properties and the strange behavior of this object are particularly described. The observations are rather well interpreted in terms of "helium-shell flashes" which occur at advanced stages of the evolution during the red-giant phase. FG Sagittae remains as yet the only star which proves that s-process nucleosynthesis is really going on in stars.

122.149 **On the periods of the Delta Scuti star HR 1170.**
J. H. Peña, J. Warman.
Rev. Mexicana Astron. Astrofis., Vol. 5, 5 - 7 (1980).
Analysis of the periodic content of the δ Scuti star HR 1170 is carried out. From the periods found for this and other available data, $0^{d}.0994$ and $0^{d}.0913$, it is concluded that this star pulsates in non-radial modes.

122.150 **Photoelectric photometry of Delta Scuti stars: HR 4715, HR 5329 and HR 7331.**
M. A. Moreno.
Rev. Mexicana Astron. Astrofis., Vol. 5, 19 - 24 (1980).
Differential photometry in Johnson's *V* filter is presented.

122.151 **Additional photoelectric observations and analysis of the variability of the β Cephei stars 12 and 16 Lacertae.** T. Jarzebowski, M. Jerzykiewicz, M. Ríos Herrera, M. Ríos Berumen.
Rev. Mexicana Astron. Astrofis., Vol. 5, 31 - 37 (1980).
The results of a frequency analysis are given.

122.152 **An interactive procedure to determine the mean radius of Cepheids.**
B. Caccin, A. Onnembo, G. Russo, C. Sollazzo.
Mem. Soc. Astron. Italiana, Vol. 51, (see 012.040), 505 - 515 (1980).
The authors present a procedure to compute radii for classical Cepheids, using light, colour and velocity curves. This procedure is the application of a new global method proposed by Caccin et al. (1980); its speed and interactivity (by means of a videographic device) are very suitable for an extensive application to recalibrate the log R - log P relation.

122.153 **β Cephei stars as non-radial oscillators. II. Multiperiodic variables σ Sco and ν Eri.**
M. Kubiak.
Acta Astron., Vol. 30, 219 - 235 (1980).
Observed radial velocity and line profile variations for multiperiodic variables σ Sco and ν Eri are interpreted in terms of non-radial oscillations. In σ Sco excited are modes (2, −1) and (3,0); the star is seen at inclination angle about 50°. In ν Eri excited are modes (2,0), (1, +1), (1, 0), (1, −1); the star is a slow rotator seen nearly equator-on. Importance of mode identification for proper interpretation of observed $2K/\Delta m$ ratio is discussed. Identification of modes excited in β Cephei stars seems to indicate that there is no preference for any particular *l*-values.

122.154 **New photometric observations of the short period variable CY Aquarii.** E. Bohusz, A. Udalski.
Acta Astron., Vol. 30, 359 - 366 (1980).
New observations of CY Aquarii are presented. They are used together with the earlier ones to study the period variations. During the years 1966 - 1978 the period seems to be nearly constant and close to the previous value. Small period changes are possible.

122.155 **Radial velocity curve, and radius of the pulsating star FG Sge.** M. Mayor, A. Acker.
Astron. Astrophys., Vol. 92, 1 - 5 (1980).
The radial velocity curve of FG Sge is established with 80 radial velocities measured from 1977 to 1979. The mean period is 120 ± 10 days, the mean amplitude and velocity being $K = 3.0 \pm 0.2$ km/s and $V_0 = 37.5 \pm 0.1$ km/s. The value of the radius is estimated to about 140 $R_\odot$, which leads to a pulsation mass of about 1 $M_\odot$ and a distance of about 2.5 kpc.

122.156 **Red variables: some current topics.** M. W. Feast.
Variability in stars and galaxies, (see 012.044), p. B.1.1 - 1.4 (1980).
The following topics are discussed: (1) A period-luminosity relation for red supergiant variables in the LMC and a comparison with theory. (2) Possibilities of confusion between red supergiant variables and Mira variables. (3) The effective temperatures of Miras in so far as this is relevant to their pulsational state. (4) The relation between optical/infrared and OH/H_2O masering properties of normal Miras. (5) The effect of binary system membership on the dust shells around Miras and their OH/H_2O masering properties.

122.157 **Cepheids and RR Lyrae variables.** R. S. Stobie.
Variability in stars and galaxies, (see 012.044), p. B.2.1 - 2.10 (1980).
Contents: Multicolour photometry of galactic population I cepheids. Extragalactic cepheids and the PLC relationship. Mass (or radius) anomalies for population I cepheids. Double-mode cepheids. BL Herculis variables. Anomalous cepheids in dwarf elliptical galaxies. Observed physical parameters of RR Lyrae variables.

122.158 **Delta Scuti stars: observational and theoretical aspects.** A. Baglin, M. Auvergne, J.-C. Valtier, M. Saez.
Variability in stars and galaxies, (see 012.044), p. B.3.1 - 3.24 (1980).
The authors recall rapidly the main well established properties of Delta Scuti stars and emphasize on the more recent results.

122.159 **β Cephei stars – a review of recent observational data.** M. Jerzykiewicz, C. Sterken.
Variability in stars and galaxies, (see 012.044), p. B.4.1 - 4.19 (1980).
Results of frequency analyses of the light and radial-velocity variations of multiperiod β Cephei stars are discussed. Updated versions of the H–R and P–L diagrams for the β Cephei stars are presented. The P–L–T_{eff} relation is re-examined and shown to be indeterminate. A $0^{m}.5$ downward revision of the Hβ line absolute magnitudes is proposed. Consequently, the current opinion according to which the β Cephei variables are predominantly first overtone pulsators turns out to be unfounded. Implications of these observational results for the β Cephei pulsation mechanism, in particular the Stellingwerf's opacity bump driving, are briefly considered.

122.160 **Variables spectrales et magnétiques.** P. Renson. Variability in stars and galaxies, (see 012.044), p. B.5.1 - 5.5 (1980).

122.161 **Supergiant variability.** A. Maeder. Variability in stars and galaxies, (see 012.044), p. B.6.1. - 6.16 (1980).

122.162 **Cataclysmic variables.** G. T. Bath, J. E. Pringle. Variability in stars and galaxies, (see 012.044), p. D.2.1 - 2.11 (1980).

Models of time-dependent evolving accretion discs are described and applied to dwarf nova eruptions.

122.163 **Flare stars in the γ Cygni region.** K. P. Tsvetkova. Inf. Bull. Variable Stars, No. 1887, 2 pp. (1980).

122.164 **Flare stars in the Pleiades.** M. K. Tsvetkov, A. G. Tsvetkova, S. A. Tsvetkov. Inf. Bull. Variable Stars, No. 1888 (1980).

122.165 **Flare stars in Orion.** M. K. Tsvetkov, S. A. Tsvetkov, A. G. Tsvetkova. Inf. Bull. Variable Stars, No. 1889 (1980).

122.166 **Photoelectric observations of the flare star BY Dra in 1975.** L. N. Mavridis, P. Varvoglis. Inf. Bull. Variable Stars, No. 1891, 3 pp. (1980).

122.167 **Photometry of two suspected long period Cepheids.** J. F. Dean. Inf. Bull. Variable Stars, No. 1892, 3 pp. (1980).

122.168 **New period increase in the Cepheid S Vul.** F. Mahmoud, L. Szabados. Inf. Bull. Variable Stars, No. 1895, 7 pp. (1980).

122.169 **Flare activity of V914 Sco.** I. C. Busko, F. J. Jablonski, G. R. Quast, C. A. O. Torres. Inf. Bull. Variable Stars, No. 1897, 3 pp. (1980).

122.170 **Vyss 124 as a BY Dra variable.** I. C. Busko, G. R. Quast, C. A. O. Torres. Inf. Bull. Variable Stars, No. 1898, 2 pp. (1980).

122.171 **The symbiotic stars.** L. Hric. Kozmos, Vol. 11, 139 - 140 (1980). In Slovak.

122.172 **Beta Cephei and related stars.** J. R. Percy. Space Sci. Rev., Vol. 27, (see 012.046), 313 - 322 (1980).

This review examines some recent observational and theoretical results on β Cep stars and their relatives.

122.173 **Linear and nonlinear pulsations of β Cephei stars.** A. N. Cox, S. W. Hodson. Space Sci. Rev., Vol. 27, (see 012.046), 323 - 327 (1980).

122.174 **Fast time-resolution spectroscopy of BW Vulpeculae.** A. Young, I. Furenlid. Space Sci. Rev., Vol. 27, (see 012.046), 329 - 335 (1980).

An experimental Fairchild CCD-211 was placed in the 2.1-m coudé spectrograph at the Kitt Peak National Observatory and used to record an 87 Å spectral band centered on Hα. This system was used to observe continuously the extreme β Cephei variable BW Vulpeculae throughout one of its 4^h49^m cycles. The results show dramatic profile variations of Hα, including features not previously reported, and extreme variations of the C II $\lambda\lambda$6578, 6582 lines, including variations of equivalent width. The observations have led directly to important new interpretations of the complex atmospheric pulsations in this star, including effects of altered opacity on the formation of spectral lines and the suggestion of a helium-ionization heat engine as a mechanism for driving atmospheric pulsations.

122.175 **The light and velocity curve bumps for BW Vulpeculae.** W. D. Pesnell, A. N. Cox. Space Sci. Rev., Vol. 27, (see 012.046), 337 - 343 (1980).

The most outstanding feature of the observations of BW Vul is a standstill or bump, on the light curve, accompanied by doubled lines in the radial velocity observations. In this papers, the authors have attempted to explain these phenomena with two different models. Firstly, the resonance mechanism, from Simon and Schmidt, and secondly, a non-linear calculation.

122.176 **Non-radial pulsation models of the H-alpha profile in BW Vulpeculae.** A. P. Odell. Space Sci. Rev., Vol. 27, (see 012.046), 345 - 350 (1980).

The profile of the hydrogen alpha spectral line in the large-amplitude Beta Cephei star BW Vulpeculae was modeled in terms of a linear, non-radially moving, and rotating stellar atmosphere. The pulsation and rotation parameters were derived from fitting the radial velocity amplitude. When the line profiles are compared to this model, the fit is quite encouraging for most phases, but several discrepancies remain.

122.177 **Pulsation modes and luminosities of the β Cephei stars.** M. Jerzykiewicz, C. Sterken. Space Sci. Rev., Vol. 27, (see 012.046), 351 - 352 (1980). Abstract.

122.178 **Are there different types of variability among B stars?** J.-P. Sareyan, J.-M. Le Contel, J.-C. Valtier, D. Ducatel. Space Sci. Rev., Vol. 27, (see 012.046), 353 - 359 (1980).

122.179 **Delta Scuti stars and dwarf Cepheids: review and pulsation modes.** M. Breger. Space Sci. Rev., Vol. 27, (see 012.046), 361 - 370 (1980).

Recent developments in the field of Delta Scuti stars are discussed. Considerable progress has been made in determining the multiple-period structure of individual variable stars and in identifying the excited radial and nonradial modes through a variety of methods. This allows an analysis of the change of excited modes as a function of the position in the instability strip.

122.180 **Radial and nonradial oscillations in Delta Scuti stars.** D. W. Kurtz. Space Sci. Rev., Vol. 27, (see 012.046), 377 - 380 (1980).

122.181 **Theoretical evolution sequences for homogeneous and two-zone models of Delta Scuti variables.** G. K. Andreasen, P. M. Hejlesen, J. O. Petersen. Space Sci. Rev., Vol. 27, (see 012.046), 381 - 387 (1980).

Radial pulsation periods are calculated for models of variables in the lower Cepheid instability strip. Evolution tracks in several diagrams are constructed and used for comparisons with observed periods, period ratios, and effective temperatures. The comparisons are simpler and more direct than in earlier studies with separate evolution and pulsation calculations.

122.182 **Recent work on Cepheid variables.** J. P. Cox. Space Sci. Rev., Vol. 27, (see 012.046), 389 - 399 (1980).

Recent work on Cepheids is reviewed in the areas of the large-amplitude mode behavior, convection, and Cepheid masses.

122.183 **Double-mode Cepheids.** R. S. Stobie.
Space Sci. Rev., Vol. 27, (see 012.046), 401 - 412 (1980).

122.184 **Full amplitude models of 15 day Cepheids.** B. C. Cogan, A. N. Cox, D. S. King.
Space Sci. Rev., Vol. 27, (see 012.046), 419 - 424 (1980).

Numerical models of Cepheids have been computed with a range of effective temperatures and compositions. The amplitudes increase if the helium abundance increases or if the effective temperature decreases. The latter effect is contrary to observational data. The models also exhibit velocity amplitudes which are much lower than those observed.

122.185 **HR 7308: a unique Cepheid.** J. R. Percy, N. R. Evans.
Space Sci. Rev., Vol. 27, (see 012.046), 425 - 428 (1980).

The authors have obtained photometric observations and radial velocities of the F6I–IIb star HR 7308. The spectral type, the period, the relative range and phase of light, colour and velocity variations all suggest that the star is a small-amplitude Cepheid. The metal abundance appears to be normal. However, the light amplitude decreased slowly during 1978 from $0\overset{m}{.}3$ to $0\overset{m}{.}1$, and increased during 1979 and 1980 from $0\overset{m}{.}05$ to $0\overset{m}{.}15$. Such behaviour is unprecedented in a population I Cepheid.

122.186 **HR 7308, a short period Cepheid with variable amplitude.** G. Burki, M. Mayor.
Space Sci. Rev., Vol. 27 (see 012.046), 429 - 430 (1980).

122.187 **The unusual classical Cepheid HR 7308: 1966-1969.** M. Breger.
Space Sci. Rev., Vol. 27, (see 012.046), 431 - 436 (1980).

The small-amplitude classical Cepheid HR 7308 has the shortest period known in our Galaxy for a classical Cepheid. 1966-1969 photoelectric observations together with more recent data show a period of $1\overset{d}{.}49078$ with sinusoidal light variations. The amplitude of HR 7308 varies on a timescale near 970 days. The variations may be periodic. The available data do not at present allow to distinguish between the Blazhko Effect and a model with two very close interfering periods in HR 7308.

122.188 **Ultra-short period Cepheids in the LMC.** L. Connolly.
Space Sci. Rev., Vol. 27, (see 012.046), 443 - 448 (1980).

Six Cepheids in the Large Magellanic Cloud have been discovered with periods less than one day. They lie at the faint, blue end of the instability strip as would be expected but form a separate period-luminosity relation from that for normal Cepheids. Membership in the LMC is based upon the existence of the P–L relationship. Several of the variables have uncertain periods although all apparently have periods under one day. One additional variable has tentatively been identified as a bright anomalous Cepheid.

122.189 **The Cepheid luminosity scale.** E. G. Schmidt.
Space Sci. Rev., Vol. 27, (see 012.046), 449 - 455 (1980).

The distance moduli of galactic clusters containing classical Cepheids are being redetermined using the four-color and $H\beta$ photometric system. The results for four clusters are presented and it is found that the distance moduli are smaller than previous values. Possible reasons for this discrepancy are discussed.

122.190 **Pulsation properties of Mira long period variables.** J. H. Cahn.
Space Sci. Rev., Vol. 27, (see 012.046), 457 - 466 (1980).

The author first gives observational evidence for the pulsation constant $Q = P(M/R^3)^{1/2}$ for Mira long period variables, where P is the period and M and R are the mass and radius in solar units. He then compares the observations with calculations. Next, the author reviews two interesting groups of papers dealing with hydrodynamic properties of long period variables. Finally, he reviews the paper by Wood-Zarro (1980) dealing with the pulsation constant Q.

122.191 **RR Lyrae and BL Herculis variables.** A. N. Cox.
Space Sci. Rev., Vol. 27, (see 012.046), 475 - 481 (1980).

122.192 **Bump masses and radii of BL Herculis variables.** J. O. Petersen.
Space Sci. Rev., Vol. 27, (see 012.046), 495 - 501 (1980).

Bump masses and radii are derived for 18 BL Her stars from the observed bump phase and the accurately known fundamental period. The mean mass $M/M_\odot = 0.60 \pm 0.09$ agrees precisely with predictions from standard stellar evolution theory and gives a new test of the theoretical models. The derived radius of V553 Centauri is in good agreement with the radius recently determined by an independent modified Baade-Wesselink method by Balona. Finally, a preliminary discussion of possible continuations of the BL Her bump progression is given.

122.193 **Nonlinear calculations for BL Her stars.** S. W. Hodson, A. N. Cox, D. S. King.
Space Sci. Rev., Vol. 27, (see 012.046), 503 - 508 (1980).

122.194 **The absolute magnitudes of RR Lyrae stars.** A. Manduca, R. A. Bell.
Space Sci. Rev., Vol. 27, (see 012.046), 509 (1980). Abstract.

122.195 **Blazhko effect in the RR Lyrae variables as a result of double mode pulsation.** K. J. Borkowski.
Space Sci. Rev., Vol. 27, (see 012.046), 511 - 517 (1980).

122.196 **R Coronae Borealis pulsations.** D. S. King.
Space Sci. Rev., Vol. 27, (see 012.046), 519 - 527 (1980).

The author briefly reviews the evidence for the Cepheid-like pulsations of some of the R CrB stars and discusses the problem of how these low-mass, hydrogen-deficient carbon stars could have evolved to their present position in the H-R diagram. Linear and nonlinear pulsation calculations are reviewed.

122.197 **Pulsations of the R Coronae Borealis stars.** J. P. Cox, D. S. King, A. N. Cox, J. C. Wheeler, C. J. Hansen, S. W. Hodson.
Space Sci. Rev., Vol. 27, (see 012.046), 529 - 535 (1980).

The authors have examined the radial oscillations of models of essentially helium stars of large luminosity and small mass, i. e., $L/M \sim 10^4$ (solar units). Such oscillations may be represented in nature by the pulsations of certain R CrB stars.

122.198 **White dwarf pulsations: a review.** J. T. McGraw.
Space Sci. Rev., Vol. 27, (see 012.046), 601 - 611 (1980).

A subset of the DA dwarfs, the ZZ Ceti stars, form a highly homogeneous class of nonradially pulsating variable stars. The author reviews the observations from which both the physical properties of the stars and the characteristics of the pulsations have been derived. Data obtained since the last review of these variables (Robinson 1979) is stressed, as these data are forcing a somewhat revised understanding of the ZZ Ceti stars and their relationship to investigations of white dwarfs and to pulsating variable stars, in general.

122.199 **The ZZ Ceti stars and the rate of evolution of white dwarfs.** E. L. Robinson, S. O. Kepler.
Space Sci. Rev., Vol. 27, (see 012.046), 613 - 620 (1980).

122.200 **A new instability strip for hot degenerates.** S. G. Starrfield, A. N. Cox, S. W. Hodson.
Space Sci. Rev., Vol. 27, (see 012.046), 621 - 626 (1980).

122.201 **A wave model for dwarf novae.** W. M. Sparks, G. S. Kutter.
Space Sci. Rev., Vol. 27, (see 012.046), 643 - 648 (1980).

The rapid coherent oscillation during a dwarf nova outburst is attributed to an accretion-driven wave going around the white dwarf component of the binary system. The increase and decrease in the period of this oscillation is due to the change in the velocity of the wave as it is first being driven and then damped. Qualitatively, a large number of observations can be explained with such a model. The beginnings of a mathematical representation of this model are developed.

122.202 **On the period-age relation for cepheids.** Ts. G. Tsvetkov.
Pis'ma Astron. Zh., Tom 6, 756 - 758 (1980). In Russian. English translation in Soviet Astron. Lett., Vol. 6.

Ages of 156 classical cepheids in the Galaxy are calculated. A semiempirical period-age relation for cepheids with small ($P \lesssim 10$ days) and large periods is obtained. This relation is consistent with that derived theoretically and verified by means of cluster cepheids.

122.203 **Variable stars in the globular cluster NGC 6284.** C. M. Clement, H. Sawyer Hogg, T. R. Wells.
Astron. J., Vol. 85, 1604 - 1611 (1980).

Five new variables have been found in NGC 6284. Of 15 variables in and around the cluster, periods have been determined for 13. Nine of these are RR Lyrae variables, two are population II cepheids, and two are Mira variables. Four of the RR Lyrae stars and both cepheids may be members. This cluster has the highest metallicity for a cluster with population II cepheids. A distance of 12.9 kpc has been determined for NGC 6284.

122.204 **Polarimetric observations of R Coronae Borealis during its minimum in 1977.** Yu. S. Efimov.
Izv. Krymskoj Astrofiz. Obs., Tom 62, 17 - 26 (1980). In Russian.

Photometry and polarimetry of R CrB in five spectral bands during its light minimum in 1977 was made. An extremely large (up to 14%) and variable linear polarization with practically constant position angle was found. The variations of the wavelength dependence of polarization were studied.

122.205 **The puzzle of the symbiotic stars: new results from the International Ultraviolet Explorer satellite.**
M. H. Slovak.
Publ. Astron. Soc. Pacific, Vol. 92, 550 (1980). – Abstract.

122.206 **Color excesses of classical cepheids. II.** K. A. Feltz Jr., D. H. McNamara.
Publ. Astron. Soc. Pacific, Vol. 92, 609 - 647 (1980).

New photometric data in the $uvby\beta$ and (khg_2) systems are presented for 41 cepheids. Intrinsic relations are established, using the color indices determined from field stars of five cepheids, which allow color excesses to be obtained for the other 36 variables by means of curve-matching techniques. It appears that serious systematic errors are present in previously published color excesses.

122.207 **Nonradial period structure of the δ Scuti star HR 1170.**
J. Warman, J. H. Peña, T. E. Margrave.
Publ. Astron. Soc. Pacific, Vol. 92, 696 - 699 (1980).

Recent differential photometry of the δ Scuti star HR 1170 has been carried out. Analysis of these data and some earlier data for this star strongly supports the derived periods of $0\overset{d}{.}3307$, $0\overset{d}{.}0994$, $0\overset{d}{.}0915$, and $0\overset{d}{.}0789$. If one considers the period ratios, the result indicates that this star is pulsating in a nonradial mode.

122.208 **Stellar surface phenomena: asymmetric light curves of the RS Canum Venaticorum binaries λ Andromedae and II Pegasi.** B. W. Bopp, P. V. Noah.
Publ. Astron. Soc. Pacific, Vol. 92, 717 - 724 (1980).

The authors examine the feasibility of applying the starspot model to explain the unusual asymmetric light curves exhibited by the noneclipsing RS CVn-like binaries λ And and II Peg. A simple spot model incorporating two cool dark regions, separated by 40°–100° in longitude and with size ratio ~2:1 can reproduce the observed light variations to a precision of $\sim 0\overset{m}{.}01$. Considering all the existing light curves of noneclipsing stars of high surface activity (RS CVn and dMe) the authors note that: (1) No single temperature difference ΔT between photosphere and spot can be applied to all stars, and some objects may exhibit changes in ΔT from season to season. (2) There is no preference for the larger spot to lead or trail the rotation. (3) The distribution of starspots on stellar surfaces is characterized by very large extents (~100°–180°) in longitude.

122.209 **On brightness variations of XX Persei.** V. P. Smykov.
Astron. Tsirk., No. 1095, p. 7 - 8 (1980). In Russian.

122.210 **The Blazhko effect of RR Lyr-type stars in globular clusters. I. V14 in M5 (NGC 5904).**
V. P. Goranskij.
Astron. Tsirk., No. 1096, p. 3 - 5 (1980). In Russian.

122.211 **Star activity cycles of R Coronae Borealis-type variables.** T. S. Khruzina.
Astron. Tsirk., No. 1098, p. 1 - 3 (1980). In Russian.

122.212 **HR 7308: a unique cepheid.** J. R. Percy, N. R. Evans.
J. R. Astron. Soc. Canada, Vol. 74, 360 (1980). – Abstract.

122.213 **Ninety years of variable stars in globular clusters.** H. Sawyer Hogg.
J. R. Astron. Soc. Canada, Vol. 74, 363 (1980). – Abstract.

122.214 **The variable stars of the globular cluster NGC 6284.** C. M. Clement, T. R. Wells.
J. R. Astron. Soc. Canada, Vol. 74, 363 (1980). – Abstract.

122.215 **A slow flare of the emission-line variable V630 Ori in the association near FU Ori.** G. A. Ponomareva.
Astron. Tsirk., No. 1106, p. 2 - 4 (1980). In Russian.

122.216 **Electrospectrophotometry of the bright cepheids η Aql and ζ Gem.** E. A. Depenchuk.
Astron. Tsirk., No. 1107, p. 6 - 8 (1980). In Russian.

122.217 **UBV observations of FG Sagittae in 1979.** V. P. Arkhipova, G. V. Zajtseva, R. I. Noskova.
Astron. Tsirk., No. 1111, p. 1 - 2 (1980). In Russian.

122.218 **The Blazhko effect of RR Lyrae-type stars in globular clusters. II. V 63 in M5 (NGC 5904).**
V. P. Goranskij.
Astron. Tsirk., No. 1111, p. 4 - 5 (1980). In Russian.

122.219 **The Blazhko effect of RR Lyrae-type stars in globular clusters. III. Slow irregular amplitude vari-**

ations of V 79 in M3 (NGC 5272). V. P. Goranskij.
Astron. Tsirk., No. 1111, p. 6 - 8 (1980). In Russian.

122.220 **Observations of V426 Oph.**
S. Yu. Shugarov.
Astron. Tsirk., No. 1113, p. 7 - 8 (1980). In Russian.

122.221 **Spektroskopische Beobachtungen der Zwergnova V 436 Cen.** W. Wargau, J. Rahe, N. Vogt.
Mitt. Astron. Ges., Nr. 50, p. 28 - 30 (1980).

122.222 **Observational studies of Cepheids. I. *BVRI* photometry of bright Cepheids.**
T. J. Moffett, T. G. Barnes III.
Astrophys. J., Suppl. Ser., Vol. 44, 427 - 450 (1980).
Over 1,000 differentially determined photoelectric *BVRI* observations and the resulting light curves are presented for 24 bright Cepheids accessible from northern hemisphere observatories. The internal precision of these data is shown to be better than ±0.01 mag, and the accuracy of transformation to the Johnson *BVRI* system is nearly as good.

122.223 **The detection of a companion star to the Cepheid variable T Monocerotis.**
J. T. Mariska, G. A. Doschek, U. Feldman.
Astrophys. J., Vol. 242, 1083 - 1087 (1980).
The authors have obtained ultraviolet spectra with the International Ultraviolet Explorer spacecraft of the classical Cepheid T Mon at several phases in the 27 day period. Significant ultraviolet emission is detected at wavelengths less than 1600 Å, where little flux is expected from classical Cepheids. Furthermore, the emission at wavelengths less than about 1900 Å does not vary with phase. Comparison with model atmosphere flux distributions shows that the emission is consistent with the flux expected from a companion star with an effective temperature of about 10,000 K (approximately A0) near the main sequence.

122.224 **Stepanyan's star: a new eclipsing cataclysmic variable.** K. Horne.
Astrophys. J., Lett., Vol. 242, L167 - L171 (1980).
High-speed photometry of the nova-like variable Stepanyan's star shows a 3^h48^m orbital period marked by a 2.2 mag eclipse which spans orbital phases 0.91–0.09. The uneclipsed light is variable by typically 0.15 mag on time scales of minutes to seconds and shows no obvious modulation with orbital phase. Intermediate-band colors show an enhancement in Balmer continuum emission and in red light during the eclipse.

UBV data 1976 - 1979. See Abstr. 002.040.

The AAVSO Variable Star Atlas.
See Abstr. 002.072.

Theory of stellar pulsation.
See Abstr. 003.033.

Flare stars. See Abstr. 003.053.

A least-squares determination of cepheid maxima.
See Abstr. 031.616.

The critical frequency in the stellar pulsation theory. See Abstr. 064.089.

Hydrodynamic modeling of mass loss from cataclysmic variable secondaries. See Abstr. 064.090.

Asymptotic approximations for stellar nonradial pulsations. See Abstr. 065.044.

Turbulent convection and pulsational stability of variable stars. See Abstr. 065.045.

Types of stellar instabilities.
See Abstr. 065.053.

Non-radial oscillations in red giants.
See Abstr. 065.056.

Current problems on horizontal-branch (HB) stars. IV. The influence of rotation on the expected properties of RR Lyrae pulsators. See Abstr. 065.061.

A new method for determining the internal rotational angular velocity of the stars. See Abstr. 065.062.

Mode discrimination in nonradially oscillating stars from light, colour and velocity observations.
See Abstr. 065.078.

An extended work integral for pulsating stars.
See Abstr. 065.079.

Spherical oscillation patterns.
See Abstr. 065.080.

Low-mass evolution from He ignition to beyond the horizontal branch. See Abstr. 065.081.

The effect of a magnetic field on stellar pulsations as a singular perturbation problem. See Abstr. 065.086.

Qualitative analysis of star flare models.
See Abstr. 065.090.

Continuous emission of solar and stellar flares.
See Abstr. 073.061.

The proper motions of the Beta-Cephei stars of the FK4. See Abstr. 111.011.

Observations of circumstellar clouds.
See Abstr. 112.021.

Infrared spectroscopy of molecules in circumstellar material. See Abstr. 112.026.

VLBI observations of the V = 1 and the V = 2 SiO masers in W Hydra and VX Sagittarius. See Abstr. 112.028.

Time variation of SiO maser emissions.
See Abstr. 112.030.

Time variability of the Orion A, R Leo and o Ceti SiO (v = 1, J = 2 – 1) masers. See Abstr. 112.031.

Polarized emission in the broad SiO feature from R Leo. See Abstr. 112.032.

The OH circumstellar maser in late-type stars.
See Abstr. 112.033.

Interpretation of circumstellar masers.
See Abstr. 112.034.

W Hydrae. See Abstr. 112.039.

R Aquarii. See Abstr. 112.040.

The structure of OH masers around late-type stars.
See Abstr. 112.054.

The IR variability of SS433.
See Abstr. 113.006.

Spectroscopic observations of the surface-active binary II Pegasi (HD 224085). See Abstr. 114.006.

***IUE* spectra of a flare in the RS Canum Venaticorum–type system UX Arietis.** See Abstr. 114.040.

Copernicus observations of β Cephei stars.
See Abstr. 114.052.

On short time changes of the emission lines in the spectra of some WR-type stars. See Abstr. 114.071.

The ultraviolet spectrum of β Canis Majoris stars.
See Abstr. 114.074.

The spectrum of R Sextantis.
See Abstr. 114.110.

Spectral variations of HD 192163.
See Abstr. 114.120.

Observational studies of the symbiotic stars. I. Hα profile variations in CH Cygni. See Abstr. 114.132.

On the origin of anomalous Balmer decrements in the spectra of eruptive stars. See Abstr. 114.147.

The composite spectrum of BY Draconis.
See Abstr. 114.149.

Hα line profile in the spectrum of the shell star V923 Aquilae. See Abstr. 114.159.

A period–luminosity relation for supergiant red variables in the Large Magellanic Cloud. See Abstr. 115.006.

Intrinsic polarization variations of Mira late-type stars due to optical changes of irradiated grains.
See Abstr. 116.007.

A baker's dozen: unsolved problems with the RS Canum Venaticorum and BY Draconis stars.
See Abstr. 117.029.

A note on the existence of W UMa-type systems in the cluster Cr 359. See Abstr. 117.063.

On the interconnection of the pulsational and orbital period values for binary systems. See Abstr. 117.073.

Accretion disks in cataclysmic variables. I. The eclipse-related phase shifts in DQ Herculis and UX Ursae Majoris. See Abstr. 119.039.

The eclipse of CI Cyg in 1980.
See Abstr. 119.074.

Stellar surface phenomena: stellar rotation and the BY Draconis syndrome in the high-eccentricity binary BD + 24°692. See Abstr. 120.009.

Variation of the Hβ emission lines of YY Geminorum. See Abstr. 120.010.

Periods and photographic mean light curves of 17 long period variables in a field around $\alpha = 17^h$, $\delta = -70°$.
See Abstr. 123.008.

The symbiotic-nova system AS 239.
See Abstr. 124.023.

A photographic light curve of Nova V 1668 Cygni and some observations of SS Cygni. See Abstr. 124.105.

Constraints on possible long-term variability of ZZ Ceti stars. See Abstr. 126.002.

Non-radial oscillations of white dwarfs: the effect of compositional stratification on the g-mode periods and eigenfunctions. See Abstr. 126.008.

33 second X-ray pulsations in AE Aquarii.
See Abstr. 142.060.

Einstein X-ray observations of Proxima Centauri and the surrounding region. See Abstr. 142.119.

Association membership for the 20 day Cepheid RU Scuti. See Abstr. 152.003.

A detailed investigation of the R association containing the 1ᵈ95 cepheid SU Cassiopeiae.
See Abstr. 152.010.

The beat Cepheid V367 Scuti and NGC 6649.
See Abstr. 153.007.

Observations of pre-main-sequence stars in the Pleiades. See Abstr. 153.011.

The stellar content of dwarf spheroidal galaxies.
See Abstr. 158.124.

Erratum

122.901 **Erratum: "The frequency analysis of low-amplitude δScuti stars – II. HD 52788: a new variable δ Delphini star. III. HD 8781. IV. HD 31908."**
[Mon. Not. R. Astron. Soc., Vol. 186, 575 - 582 (1979), Mon. Not. R. Astron. Soc., Vol. 190, 103 - 110 (1980), Mon. Not. R. Astron. Soc., Vol. 190, 479 - 486 (1980)]. D. W. Kurtz. Mon. Not. R. Astron. Soc., Vol. 192, 959 (1980). – See abstracts 25.122.009, 27.122.005, 27.122.026.

123 Variable Stars (Surveys, Lists of Observations, Charts, etc.)

123.001 **Maxima et minima de variables à longue période observées par l'AFOEV en 1979.**
E. Schweitzer.
Bull. AFOEV, Tome 14, 45 - 49 (1980).

123.002 **Tableaux des observations faites par les sociétaires de l'AFOEV en mai et juin 1980.**
Bull. AFOEV, Tome 14, 51 - 63 (1980).

123.003 **Observing variable stars.**
J. A. Mattei, E. H. Mayer, M. E. Baldwin.
Sky Telesc., Vol. 60, 285 - 289 (1980).

123.004 **AG Draconis.**
IAU Circ., Nos. 3540, 3554 (1980).

123.005 **BD + 15°4264.**
IAU Circ., No. 3548 (1980).

123.006 **A list of eclipsing binaries to be continuously monitored.** A. Gimenez, A. J. Delgado.
Inf. Bull. Variable Stars, No. 1815, 3 pp. (1980).

123.007 **Nouvelle recherche de périodes d'étoiles Ap observées a l'ESO. V.**
J. Manfroid, P. Renson.
Inf. Bull. Variable Stars, No. 1824, 3 pp. (1980).

123.008 **Periods and photographic mean light curves of 17 long period variables in a field around $\alpha = 17^h$, $\delta = -70°$.** M. Goossens, C. Waelkens.
Inf. Bull. Variable Stars, No. 1828, 8 pp. (1980).

123.009 **Variability of cool carbon stars situated near or in intermediate-age open clusters.**
Z. Alksne, A. Alksnis.
Inf. Bull. Variable Stars, No. 1831, 4 pp. (1980).

123.010 **Is there a magnetic field-period relation for the hotter Ap-stars?** W. W. Weiss.
Inf. Bull. Variable Stars, No. 1841, 6 pp. (1980).

123.011 **New variable stars in the field of 9 Ursae Majoris.**
G. Romano.
Inf. Bull. Variable Stars, No. 1865, 2 pp. (1980).

123.012 **Revised Blue Mountain Observatory timings.**
T. E. Margrave.
Inf. Bull. Variable Stars, No. 1869, 4 pp. (1980).

123.013 **Variable star in the galactic cluster NGC 7654.**
W. Pfau.
Inf. Bull. Variable Stars, No. 1874, 3 pp. (1980).

123.014 **Observations of variable stars.** January - December 1979. H. Feijth, L. Plaut.
Nederlandse Ver. Weer Sterrenkunde. Kapteyn Astron. Lab., Groningen, Netherlands, Rep. No. 34 (1980).

123.015 **Photographische Beobachtungen an veränderlichen Sternen des Praesepe-Haufens.** W. Götz.
Mitt. Veränderl. Sterne (MVS), Band 8, 150 - 157 (1980).
Results from the investigation of 17 variable stars on 46 blue plates of the Schmidt telescope 50/70/172 cm at Sonneberg Observatory from 1977 March 8 to 1979 March 23 are given. 16 of these stars are members of the Praesepe cluster and belong to the eclipsing and flare stars. Dates of minima and flares and the brightness of the objects in B are summarized.

123.016 **6 neue Veränderliche im Feld β Aurigae.**
H. Geßner.
Mitt. Veränderl. Sterne (MVS), Band 8, 157 (1980).

123.017 **Mehrfarben-Beobachtungen von V 95 im Kugelhaufen M3.** I. Meinunger.
Mitt. Veränderl. Sterne (MVS), Band 8, 158 - 159 (1980).

123.018 **Neue veränderliche Sterne im Kugelhaufen M3.**
I. Meinunger.
Mitt. Veränderl. Sterne (MVS), Band 8, 159 - 161 (1980).

123.019 **Informations.** E. Schweitzer.
Bull. AFOEV, Tome 14, 70 - 74 (1980).

123.020 **Tableaux des observations faites par les sociétaires de l'AFOEV en juillet, août et septembre 1980.**
Bull. AFOEV, Tome 14, 75 - 113 (1980).

123.021 **On the period determination of variable stars.**
P. Renson.
Astron. Astrophys., Vol. 92, 30 - 32 (1980). In French.
The periods announced for 11 Ori and 137 Tau by another author are in fact not the true ones, but related periods. This seems to be also true in other cases.

123.022 **Two new variable stars in the γ Cygni region.**
K. P. Tsvetkova.
Inf. Bull. Variable Stars, No. 1890, 2 pp. (1980).

123.023 **New variable stars in Virgo.**
V. P. Goranskij.
Astron. Tsirk., No. 1103, p. 4 - 5 (1980). In Russian.

123.024 **Information on photoelectric observations of variable stars deposited at the Odessa Astronomical Observatory.** V. P. Tsesevich, E. N. Makarenko.
Astron. Tsirk., No. 1112, p. 7 - 8 (1980). In Russian.

123.025 **Variable star notes.** J. A. Mattei.
J. R. Astron. Soc. Canada, Vol. 74, 317 - 320 (1980).

123.026 **La page de l'observateur.** M. Duruy.
Bull. AFOEV, Tome 14, 50 (1980).

Charts for southern variables. Series No. 12.
See Abstr. 002.030.

82nd - 84th list of minima of eclipsing binaries.
See Abstr. 119.076.

Erratum

123.901 **Corrections to Bull. AFOEV, Tomes 13/1, 13/2, 13/3.**
Bull. AFOEV, Tome 14, 54 (1980).

124 Novae

124.001 **On the late-type components of slow novae and symbiotic stars.** D. A. Allen.
Mon. Not. R. Astron. Soc., Vol. 192, 521 - 530 (1980).

It is argued that the various types of symbiotic stars and the slow novae are the same phenomena exhibiting a range of associated time-scales, the slow novae being of intermediate speed. Evidence is summarized showing that both types of object contain normal M giants or mira variables. This fact is at odds with currently fashionable single-star models for slow novae, according to which the M star is totally disrupted before the outburst. Spectral types of the late-type components are presented for nearly 80 symbiotic stars and slow novae, derived from 2 μm spectroscopy. It is found that both the intensity of the emission spectrum and the electron density of the gas are functions of the spectral type of the late-type star. Explanations for these correlations are given. On the assumption that the late-type components are normal giants, spectroscopic parallaxes are determined; credible distances are derived which indicate that the known symbiotic stars have been sampled as far afield as the galactic centre.

124.002 **White dwarf rotation and the shapes of nova remnants.** R. L. Fiedler, T. W. Jones.
Astrophys. J., Vol. 239, 253 - 256 (1980).

The authors explain the elongated appearance of many nova remnants as due to rapid rotation of the parent white dwarf. The principal influence of this rotation is seen to be the spin angular momentum given to material ejected in the equatorial directions. Resulting shells should be oblate. Implied white dwarf rotation periods are of the order of minutes, consistent with that apparently observed for DQ Herculis (N Her 1934) and V533 Herculis (N Her 1963).

124.003 **Theoretical models of nova outbursts.**
G. S. Kutter, W. M. Sparks.
Bull. American Astron. Soc., Vol. 12, 467 (1980). – Abstract.

124.004 **On X-ray survey of nine nearby historical novae.**
R. H. Becker.
Bull. American Astron. Soc., Vol. 12, 500 (1980). – Abstract.

124.005 **The effects of ^{3}He on nova outbursts–a possible recurrent nova trigger.** M. M. Shara.
Astrophys. J., Vol. 239, 581 - 585 (1980).

The author discusses the constraints and difficulties imposed on thermonuclear runaway models of recurrent novae by observations and theory. He shows that giant secondaries in RN should transfer significant quantities of ^{3}He to their primaries, easing the difficulties and motivating this study of the effects of ^{3}He on novae. The initial conditions and computational method for ^{3}He-rich ($Y_3 = 2 \times 10^{-3}$) and ^{3}He-poor ($Y = 0$) models of novae are given. The simulation results are presented and discussed. The author discusses the significance of the results to recurrent novae.

124.006 **What determines the speed class of novae?**
M. M. Shara, D. Prialnik, G. Shaviv.
Astrophys. J., Vol. 239, 586 - 591 (1980).

The authors propose a unified picture for novae of different speed classes as a combination of CNO enrichment and envelope mass. Several consequences of their scheme are discussed. The observational and theoretical results are explained in the context of the unified picture.

124.007 **Infrared emission by dust grains near variable primary sources. II. A model for infrared novae.**
M. F. Bode, A. Evans.
Astron. Astrophys., Vol. 89, 158 - 168 (1980).

The authors review the observational data on the visual, infrared and ultraviolet behaviour of the Novae Serpentis 1970 and Aquilae 1975 after outburst. They suggest that the grains giving rise to the well-known infrared development of novae already exist near nova progenitors prior to outburst, and that the evidence for the rapid formation of grains after outburst is not convincing. Their model enables the authors to produce detailed infrared light curves, at all relevant wavelengths, that closely match the infrared development of Novae Serpentis and Aquilae.

124.008 **Are novae shells optically thick in the infrared?**
E. P. Ney, B. F. Hatfield.
Astron. J., Vol. 85, 1292 - 1293 (1980).

The optical depth of the dust shells of novae in the infrared is an important parameter in discussing the energy balance and the morphology of the nova ejecta. A recent publication has questioned the authors' previous interpretation of infrared data for NQ Vul. The authors present their argument concerning the infrared optical depths of this and other novae.

124.009 **Origine des raies coronales des novae.**
I. Malakpour.
Astrophys. Space Sci., Vol. 72, 143 - 157 (1980).

First, the lines identified on the spectrum region 3060–6100 Å of Nova Delphini 1967, taken in November 1972, are given. The coronal lines of novae Herculis 1960, Herculis 1963, Delphini 1967, Serpentis 1970, Scuti 1975, Cygni 1975 have been studied and the observational results of these as well as those obtained for older novae and the recurrent nova RS Ophiuchi are given. By studying all the proposed mechanisms for the formation of the coronal lines and the observational results mentioned above, it is found that the shock waves, produced by the collision of the principal envelope with the dusty circumstellar envelope, heat the circumstellar envelope and the heated material then emits the coronal lines.

124.010 **Radial accretion of H-rich material onto a He white dwarf.** G. S. Kutter, W. M. Sparks.
Astrophys. J., Vol. 239, 988 - 998 (1980).

The authors are attempting to model the nova outburst by spherically accreting H-rich material onto a 1 $M_\odot$ He white dwarf at a rate of $10^{-8} M_\odot$ yr^{-1}. The star accretes for 5848 years, when the nuclear reactions run away near the base of the accreted envelope. The nuclear-energy generation rate rises to 4.6 (8) $L_\odot$, and the envelope expands in response to it. However, nova-like mass ejection does not occur because the envelope is of insufficient mass, the base of the envelope is only mildly electron degenerate, and there is no enrichment of the CNO abundance. To overcome these limiting conditions, the authors suggest that the H-rich material be accreted either more slowly than $10^{-10} M_\odot$ yr^{-1} or with angular momentum.

124.011 **IUE observations of novae.** W. M. Sparks, C.-C. Wu, A. V. Holm, F. H. Schiffer, III.
Highlights of Astronomy, Vol. 5, (see 012.024), 285 - 291 (1980).

The authors present IUE observations of Nova Cygni 1978, WZ Sagittae, and U Scorpii.

124.012 **Nova modeling.** W. M. Sparks, G. S. Kutter, S. Starrfield, J. W. Truran.
Highlights of Astronomy, Vol. 5, (see 012.026), 505 - 508 (1980).

124.013 **Novalike object in Vulpecula.**
IAU Circ., No. 3494 (1980).

124.014 **VY Sculptoris.**
IAU Circ., Nos 3502, 3508 (1980).

124.015 **Novalike object in Vulpecula.**
Yamamoto Circ., Nos. 1937, 1940 (1980).

124.016 **TT Arietis.**
IAU Circ., Nos. 3541, 3552 (1980).

124.017 **Probable nova in Cygnus.**
IAU Circ., Nos. 3546, 3548 (1980).

124.018 **Probable nova in Cygnus.**
Yamamoto Circ., No. 1948 (1980).

124.019 **Nevels rond novae.**
Zenit, 7e Jaarg., 423 (1980).

124.020 **Possible nova in Cygnus.**
British Astron. Assoc. Circ., No. 612 (1980).

124.021 **The 1979 outburst of U Scorpii.** M. J. Barlow, J. P. Brodie, C. C. Brunt, D. A. Hanes, P. W. Hill, S. K. Mayo, J. E. Pringle, M. J. Ward, M. G. Watson, J. A. J. Whelan, A. J. Willis.
Anglo-Australian Obs. Prepr. No. 137 (1980).

Optical and ultraviolet observations are presented of the 1979 outburst of the recurrent nova U Sco. For the first time the evolution through outburst is documented photometrically and spectroscopically. Lines of the following ions are identified: H I, He II, C IV, N III, N IV, N V, O IV, O VI, Si IV. No forbidden lines were observed. Mg I was seen in absorption at a late stage in the decline. The Balmer lines have broad and narrow components which change with time. There is evidence that nitrogen is overabundant with respect to carbon and the helium to hydrogen number ratio is about 2.

124.022 **Photometric observations of the Honda-Kuwano object in Vulpecula.** M. Nakagiri, Y. Yamashita.
Tokyo Astron. Bull., Second Ser., No. 263, p. 2993 - 2998 (1980).

UBV observations were made from April 17, 1979 to May 29, 1980 for the nova-like object in Vulpecula found by Honda and Kuwano. In addition, *V–R* and *V–I* colours were obtained on eight nights during the period from February through May 1980.

124.023 **The symbiotic-nova system AS 239.**
M. W. Feast, I. S. Glass.
Observatory, Vol. 100, 208 (1980).

124.024 **Hydrodynamic studies of the nova outburst.**
S. Starrfield.
Space Sci. Rev., Vol. 27, (see 012.046), 635 - 640 (1980).

124.025 **Zur Struktur von Novahüllen.**
W. C. Seitter, H. W. Duerbeck.
Mitt. Astron. Ges., Nr. 50, p. 70 - 75 (1980).

124.026 **Recent outburst of U Scorpii.**
B. S. Shylaja, T. P. Prabhu.
Kodaikanal Obs. Bull., Ser. A, Vol. 2, 213 - 216 (1979).

A photoelectric scan of Hβ and a photographic spectrum between λλ4500 - 7000 obtained during the 1979 outburst of the recurrent nova U Sco have been studied. The electron density and the mass of ionized hydrogen in the nova shell on June 28.77 U.T. are given.

The Japanese record of the guest-star of 1408.
See Abstr. 004.001.

UK nova/supernova search programme.
See Abstr. 013.035.

The SU UMa stars, an important sub-group of dwarf novae. See Abstr. 122.025.

Nucleosynthesis during nova explosions and grain formation in ejected envelopes. See Abstr. 126.025.

Nova explosion of mass-accreting white dwarfs.
See Abstr. 126.026.

Discovery of a large X-ray burst from an X-ray nova, Centaurus X-4. See Abstr. 142.061.

Nova Cygni 1978 = V1668 Cygni

124.101 **Multifilter photometry and polarimetry of Nova Cygni 1978 (V1668 Cygni).**
W. Blitzstein, D. H. Bradstreet, B. J. Hrivnak, A. B. Hull, R. H. Koch, R. J. Pfeiffer, A. P. Galatola.
Publ. Astron. Soc. Pacific, Vol. 92, 338 - 344 (1980).

The results from more than 2700 filtered photoelectric observations of Nova Cyg 1978 are summarized. The nova's decline through about 5 magnitudes is documented. Variability on time scales up to 0.08 day and with peak-to-peak amplitudes up to $0^m.13$ were common. No short-term periodicity was found. A distance of 3 kpc is suggested. A few linear polarization measures, taken before dust formation, are listed. From the large interstellar component, a net intrinsic polarization is derived for the post-dust stages.

124.102 **Spectral and photometric observations of Nova Cygni 1978 (V1668 Cygni).** E. A. Kolotilov.
Pis'ma Astron. Zh., Tom 6, 486 - 490 (1980). In Russian.
English translation in Soviet Astron. Lett., Vol. 6.

Spectral and photometric observations of Nova V1668 Cyg were carried out during 29 nights in 1978 - 1979. Continuous observations of the nova during 5 - 6 hours with the purpose to search shorttime light variations carried out during 3 nights were unsuccesful. The observed Hα emission line profiles in the spectrum of V1668 Cyg were compared with the theoretical calculations by Boyarchuk and Gershberg (1977).

124.103 **The optically thin dust shell of Nova Cygni 1978.**
R. D. Gehrz, J. A. Hackwell, G. L. Grasdalen, E. P. Ney, G. Neugebauer, K. Sellgren.
Astrophys. J., Vol. 239, 570 - 580 (1980).

Nova Cygni 1978 was monitored photometrically from *V* to 19.5 μm for 120 days after the eruption. Following the initial expansion of the hot gas shell, an optically thin dust shell formed and reached a maximum visual optical depth of about 0.1 by day 60. No visible transition phase of the type observed in the very dusty DQ Herculis novae occurred in Nova Cygni 1978. The authors argue that dust grain growth was inhibited because of the low mass of condensable atoms in the shell.

124.104 **Spectral evolution of Nova Cygni 1978.**
G. Klare, B. Wolf, J. Krautter.
Astron. Astrophys., Vol. 89, 282 - 290 (1980).

69 spectrograms (dispersion 22 and 44 Å mm^{-1}) of Nova Cygni 1978 were obtained during the time interval September 13 to December 18, 1978, covering the evolutionary stages from the principal to the nebular stage. The spectral evolution of Nova Cyg 1978 was found to be very fast. The radial velocities of the absorption systems ranged from –661 km s^{-1} (principal spectrum) to –2115 km s^{-1} (Orion spectrum). Four rather long-lived components of the emission lines were ob-

served. An additional highly blue shifted (≈ −1060 km s^{-1}) but short-lived emission component was seen at H_β, H_γ, and the Fe II lines in the spectrograms from September 16 to about September 19. The principal absorption spectrum was found to be rather pronounced. The absolute visual magnitude at maximum was found to be M_V (max) = −8.0.

124.105 **A photographic light curve of Nova V1668 Cygni and some observations of SS Cygni.**
H. W. Duerbeck, H. Pollok.
Inf. Bull. Variable Stars, No. 1845, 7 pp. (1980).

124.106 **TV observations of nova Cygni 1978 = V1668 Cyg.**
E. P. Pavlenko.
Astron. Tsirk., No. 1097, p. 1 - 3 (1980). In Russian.

124.107 **Spektrale Entwicklung von Nova Cygni 1978.**
G. Klare, B. Wolf, J. Krautter.
Mitt. Astron. Ges., Nr. 50, p. 125 - 127 (1980).

Nova Aquilae 1918 = V603 Aquilae

124.201 **Detection of periodic light variations in the old nova V 603 Aquilae (1918).**
J. Rahe, A. Boggess, H. Drechsel, A. Holm, J. Krautter.
Astron. Astrophys., Vol. 88, L9 - L10 (1980).

Periodic visible light variations of the old nova V 603 Aql (1918) were detected with the Fine Error Sensor Instrument (FES) aboard the IUE satellite. Continuous observations were carried out during more than two complete cycles. The light curve shows a broad maximum and a pronounced minimum which was covered three times, and is tentatively interpreted as an eclipse of the accretion disk around the white dwarf by the late main sequence component.

124.202 **V603 Aquilae.**
IAU Circ., No. 3493 (1980).

124.203 **Simultaneous optical and satellite observations provide new understanding of a famous nova.**
H. Drechsel, J. Rahe, A. Holm, J. Krautter.
Messenger, No. 22, p. 10 - 12 (1980).

Nova Herculis 1934 = DQ Herculis

124.301 **On the carbon composition of the ejecta of DQ Herculis.** G. J. Ferland, J. W. Truran.
Astrophys. J., Vol. 240, 608 - 611 (1980).

The authors call attention to the fact that existing spectrophotometric data for Nova DQ Herculis 1934, spanning virtually its entire postoutburst history, reveal a time dependence of the inferred nebular composition. The observed trends may suggest that the intrinsic carbon abundance is log C/H ≈ − 1.7, and that the nebula has grown progressively cooler over the past 45 yr. The presence of a source of hard radiation in the DQ Herculis system is suggested. They stress the importance of similar long-term studies of recent novae.

124.302 **On the distance and luminosity of nova DQ Herculis.**
G. J. Ferland.
Observatory, Vol. 100, 166 - 167 (1980).

124.303 **Eruptive binaries. X. DQ Herculis.** J. Smak.
Acta Astron., Vol. 30, 267 - 283 (1980).

Radial velocities have been measured from peaks of double emission lines on 27 spectrograms obtained in 1978. Radial velocities from double emission lines are analyzed consistently within a three-body approximation for the rotation of the outer parts of the disk. The mass of the primary, white dwarf component is $M_1 = 0.5\pm0.2\ M_\odot$. The secondary component, with $M_2 = 0.35\pm0.1\ M_\odot$, is slightly oversized and probably underluminous, as compared with main sequence stars. The luminosity of the disk requires an accretion rate of about 3×10^{17} g/s.

Photometric elements of the system DQ Her and the structure of the disk-like envelope. See Abstr. 119.013.

Accretion disks in cataclysmic variables. I. The eclipse-related phase shifts in DQ Herculis and UX Ursae Majoris. See Abstr. 119.039.

Nova Vulpeculae 1976 = NQ Vulpeculae

124.401 **Spectroscopic observations of Nova Vulpeculae 1976.** J. W. Younger.
Astron. J., Vol. 85, 1232 - 1237 (1980).

SIT Vidicon spectra with a resolution of 2.5 Å were obtained over an interval of approximately one month beginning near nova maximum. Particular attention was focused on the evolution of the Hα, Hβ, and Fe II (RMT 42) line profiles and their radial velocities. The radial velocities ranged from −550 to −1600 km/s for Hα and −300 to −1180 km/s for the Hβ and Fe II lines. An interstellar band located at 6614 Å is used to derive a color excess of 0.7 ± 0.2, a distance of 1200 ± 250 pc, and an absolute visual magnitude of −6.0 ± 1.0. Significant time variations on the order of minutes were detected in the central portion of the Hα-emission profile on the night of 1 November 1976. The simplest interpretation of these variations requires material with high velocity near the surface of the stellar remnant.

Nova Sagittarii 1980

124.501 **Nova Sagittarii 1980.**
IAU Circ., Nos. 3533, 3534, 3548 (1980).

124.502 **Nova Sagittarii 1980.**
Yamamoto Circ., Nos. 1946, 1947 (1980).

Nova Cygni 1980

124.601 **Nova Cygni 1980.**
IAU Circ., Nos. 3551, 3553 (1980).

Nova Delphini 1967 = HR Delphini

124.701 **Simultaneous X-ray, UV, and optical observations of the recent nova HR Delphini.** J. B. Hutchings.
Publ. Astron. Soc. Pacific, Vol. 92, 458 - 462 (1980).

HR Del was observed simultaneously with the IUE satellite, the DAO 1.8-m telescope, and the Einstein observatory over part or all of one 4-hour orbital cycle. The orbital ephemeris is refined and accurate phases computed for the data. The UV spectra show changes in the C IV P Cygni profile which suggest a spiralling stellar wing. The UV data are also used to determine the continuum flux and temperature, and its variability. Optical spectroscopy shows some small changes since 1978 and confirms the strength of C features in the accretion disk. The X-ray observations yield a marginal detection, indicating a luminosity of 2×10^{31} erg s^{-1} in the 0.5–3 keV flux. New estimates of the reddening and luminosity are derived.

124.702 **On the variability of the old nova HR Dephini 1967.** H. Drechsel, J. Rahe.
Inf. Bull. Variable Stars, No. 1811, 3 pp. (1980).

124.703 **Light variability of Nova Delphini 1967 in 1977 and 1979.** L. Kohoutek, R. Pauls.
Astron. Astrophys., Vol. 92, 200 - 203 (1980).

Photoelectric *UBV* photometry of N Del 1967 indicates a short-period, sinusoidal light variability with $P = 0\overset{d}{.}1775$ and with the *V*-amplitude 0.16 (in 1977) and 0.10 mag (in 1979), respectively. The photometric period seems to be identical with the spectroscopic one which very likely reflects the orbital motion of the binary.

Nova Cygni 1975 = V1500 Cygni

124.801 **UBV photometry of the Nova Cygni 1975 on September 1975. A photoelectric lightcurve.**
M. E. Contadakis.
Inf. Bull. Variable Stars, No. 1818, 4 pp. (1980).

Nova Vulpeculae 1979

Spectroscopy of the nova-like object Kuwano (Nova Vulpeculae 1979) in the year 1979.
See Abstr. 122.120.

125 Supernovae, Supernova Remnants

125.001 **Further high resolution radio observations of the supernova remnant G84.2–0.8.**
H. E. Matthews, P. A. Shaver.
Astron. Astrophys., Vol. 87, 255 - 260 (1980).

The supernova remnant G84.2–0.8 has been mapped with the Westerbork Synthesis Radio Telescope at 1,415 MHz with a resolution of ~ 27″, and with the 100-m Effelsberg telescope at 2,695 MHz with a resolution of 4.′4. Striking filamentary structure is observed at 1,415 MHz, which appears to be aligned parallel to the major axis of the remnant and to the galactic plane. A strong asymmetry in the observed polarized flux density distribution may be related to this filamentary structure. The possibility of a physical association between the SNR and a relatively strong point source located within its boundary is discussed.

125.002 **The optical spectrum of W50.**
J. M. Shuder, B. F. Hatfield, R. D. Cohen.
Publ. Astron. Soc. Pacific, Vol. 92, 259 - 261 (1980) = Lick Obs. Bull. No. 868.

Spectrophotometry of a faint filament near the position of the radio source W50 is presented. Emission lines that are characteristic of supernova remnants are seen with a radial velocity of about + 100 km s^{-1} relative to the local standard of rest. The line-intensity measurements indicate that W50 is an old supernova remnant, perhaps having a nitrogen overabundance.

125.003 **The influence of the reverse shock wave on the evolution of young supernova remnants.**
Z.- x. Cui.
Acta Astron. Sinica, Vol. 21, 184 - 188 (1980). In Chinese.

The author discusses the reverse shock wave model that was suggested by McKee. McKee's formula of the propagation of the reverse shock wave is improved. Finally, the author discusses the influence of the reverse shock wave on the evolution of young supernova remnants.

125.004 **Discovery of a very distant supernova.**
M. R. S. Hawkins.
Nature, Vol. 286, 467 - 469, with a corrigendum Vol. 287, 466 (1980).

A very distant ($z \simeq 0.5$) supernova is reported together with photometric observations of its light curve.

125.005 **The interaction between the relativistic jets of SS433 and the interstellar medium.**
W. J. Zealey, M. A. Dopita, D. F. Malin.
Mon. Not. R. Astron. Soc., Vol. 192, 731 - 743 (1980).

A sky limited, red sensitive IIIaF plate taken of the field of the W50 supernova remnant reveals two areas of nebulosity, 20 arcmin in extent, centred on SS433 (1909+048) and separated from each other by 70 arcmin. The authors conclude that these optical filaments are excited by the relativistic beams of SS433 and they derive, by two independent techniques, a mass loss rate in the beams of $\sim 3\times 10^{-6} M_{\odot} yr^{-1}$. The age of the system is about 10^5 yr. The data indicate supercritical accretion on to a compact object as the power source of the beams. If this compact object is a black hole it has mass in the range 2 - 40 $M_{\odot}$ with a probable value of order 12 $M_{\odot}$.

125.006 **The distance to Tycho's supernova remnant (3C10) – a rediscussion.**
U. J. Schwarz, E. M. Arnal, W. M. Goss.
Mon. Not. R. Astron. Soc., Vol. 192, 67P - 71P (1980).

New H I observations of 3C10 at high velocity-resolution (1 km s^{-1}) have been made. The resulting lower limit for the distance is 4 kpc. A weaker upper limit of 5 kpc has also been determined by a comparison with 3C11.1, a nearby extragalactic source.

125.007 **Ultraviolet spectroscopy of the Vela supernova remnant.** I. J. Danziger, R. Wood, D. H. Clark.
Mon. Not. R. Astron. Soc., Vol. 192, 83P - 86P (1980).

A preliminary ultraviolet spectrum secured from the IUE satellite, of a bright filament in the Vela supernova remnant, displays both emission-line and continuum components. The emission-line spectrum shows the same anomalously low carbon-line strength reported previously for the Cygnus Loop. The continuum emission may be explained in terms of normal recombination processes in hydrogen and helium in a dense hot knot which dominates the field. The interpretation of the UV spectrum is supported by optical data for this region of Vela.

125.008 **Supernovae gamma astronomy?**
Priroda, 1980, No. 7, p. 101 - 102. In Russian.

125.009 **Supernovae in multiple systems.**
I. S. Shklovskij.
Astron. Zh., Tom 57, 673 - 676 (1980). In Russian.
English translation in Soviet Astron., Vol. 24, No. 4.

It follows from the statistics of X-ray sources that while in massive multiple systems a more rapidly evolving component after the Wolf-Rayet stage almost always explodes as a supernova and turns into a neutron star this conversion has a probability of about 10^{-2} for small-mass multiple systems. Arguments are given that SS433 is an early evolution phase of Hercules X-1 type X-ray sources forming in small-mass binaries.

125.010 **To what type of supernovae old galactic remnants belong; evolution of envelopes in the interstellar medium.** T. A. Lozinskaya.
Astron. Zh., Tom 57, 707 - 715 (1980). In Russian.
English translation in Soviet Astron., Vol. 24, No. 4.

The investigation of the kinematics of old optical supernova remnants allows to estimate the mean initial energy of the envelopes, this gives the mass ejected $\cong 0.3 M_{\odot}$ for the type I, and 1 - 2 $M_{\odot}$ for the type II supernovae. For determining the supernova type the physical association of the remnants with gas-and-dust complexes and OB-associations is considered. It is shown that the optical remnants IC443, the Monoceros Loop, the weak envelope of W1, W28, Vela X, and possibly HB3, may have been formed in a type II supernova explosion.

125.011 **Intensity and spectrum of the continuum gamma ray emission from supernovae.**
G. Cavallo, F. Pacini.
Astron. Astrophys., Vol. 88, 367 - 369 (1980).

The authors discuss the emission of gamma rays from π^0 decay in the interaction between the cosmic rays produced by a young pulsar and the surrounding supernova shell. The mechanism of photon degradation via pair production in the field of charged particles allows the authors to draw qualitative conclusions about the time evolution of the spectrum of the emitted X- and γ-rays from 10 keV upwards for this specific process. In the case of the recent supernova in NGC 4321 the resulting flux may well have reached values of the order of 10^{-6} photons $cm^{-2} s^{-1}$ during a few months, and should have been observable with the available techniques.

125.012 **The nature of X- and gamma-ray emission from the Vela supernova remnant.**
Yu. P. Ochelkov, V. V. Usov.
Pis'ma Astron. Zh., Tom 6, 414 - 417 (1980). In Russian.

English translation in Soviet Astron. Lett., Vol. 6.

The nonthermal X-ray emission from the extended source 2U 0832−45 surrounding PSR 0833−45 can be explained by synchrotron radiation of particles ejected from the pulsar. During the ejection the particles can generate the γ-radiation of PSR 0833−45 via the curvature mechanism.

125.013 **Effects of line overlapping in spectra of supernovae in the case of conservative scattering.** N. N. Chugaj.
Pis'ma Astron. Zh., Tom 6, 481 - 485 (1980). In Russian. English translation in Soviet Astron. Lett., Vol. 6.

The scattering of radiation in the lines of atoms and ions of metals in spectra of type I supernovae seems to be approximately conservative. Effects of overlapping of a great number of scattering lines are considered in the case of a freely expanding atmosphere of supernovae. It is shown that under certain conditions the overlapping may result in appearence of some features in spectra of type I supernovae which would be complicate to identify.

125.014 **A radioactive excitation source model for the late time spectra of type I supernovae.** R. E. Meyerott.
Astrophys. J., Vol. 239, 257 - 270 (1980).

The synthetic spectrum of Fe^{+} and Fe^{++} is calculated as a function of electron temperature, assuming a Boltzmann population of the metastable levels of Fe^{+} and Fe^{++}. This synthetic spectrum is compared with the spectrum of the type I supernova 1972e in NGC 5253, taken 245 days after maximum. It is shown that good quantitative agreement can be obtained between the observed spectra and the synthetic spectra at an electron temperature of ~4000 K with a ratio of Fe^{++} to Fe^{+} of ~4 to 1. The radioactive excitation source model of late time spectra is used to interpret this result. The results are consistent with an interpretation that the supernova envelope at late time consists mostly of Fe. The evidence for an exterior He shell is discussed.

125.015 **X-ray observations of the North Polar Spur.** DeA. Iwan.
Astrophys. J., Vol. 239, 316 - 327 (1980).

The X-ray enhancement associated with the North Polar Spur (Radio Loop I) was observed by the University of Wisconsin soft X-ray instrument on OSO 8 during August 1975. The paper presents image-processed maps of the region scanned, a detailed analysis of the X-ray data, and interpretations of the data in terms of supernova remnant models. If the neutral hydrogen ridge, radio continuum ridge, and X-ray emission ridge are all associated, then they cannot all be fitted by a single supernova remnant model because they require different energies of explosion. A reheated supernova remnant model does appear compatible with all the available observations.

125.016 **Spectrophotometry of the supernova remnant IC 443.** R. A. Fesen, R. P. Kirshner.
Bull. American Astron. Soc., Vol. 12, 446 (1980). – Abstract.

125.017 **Spectrophotometry of supernova remnants in M31.** W. P. Blair, R. P. Kirshner, R. A. Chevalier.
Bull. American Astron. Soc., Vol. 12, 446 (1980). – Abstract.

125.018 **Further measurements of the decay rate of Cassiopeia A at 8 GHz.**
T. H. Williams, H. D. Aller, D. G. Schleicher.
Bull. American Astron. Soc., Vol. 12, 498 - 499 (1980). Abstract.

125.019 **Spectroscopic observations of the Cygnus Loop.** R. P. Kirshner, R. A. Fesen, W. P. Blair.
Bull. American Astron. Soc., Vol. 12, 506 (1980). – Abstract.

125.020 **X-ray line emission from LMC supernova remnants.** P. F. Winkler, C. R. Canizares, T. H. Markert, G. W. Clark.
Bull. American Astron. Soc., Vol. 12, 532 (1980). – Abstract.

125.021 **X-rays from G74.9 + 1.2, an elderly Crab Nebula.** A. S. Wilson.
Bull. American Astron. Soc., Vol. 12, 541 (1980). – Abstract.

125.022 **Neutrino mediated shocks in supernovae.** D. Kazanas.
Bull. American Astron. Soc., Vol. 12, 542 (1980). – Abstract.

125.023 **Measurements of the extent of ionization disequilibrium in supernova remnants.**
T. H. Markert, P. F. Winkler, C. R. Canizares, J. G. Jernigan.
Bull. American Astron. Soc., Vol. 12, 542 (1980). – Abstract.

125.024 **A high-resolution X-ray image of the Puppis A supernova remnant.**
R. Petre, P. F. Winkler, G. A. Kriss.
Bull. American Astron. Soc., Vol. 12, 542 - 543 (1980). Abstract.

125.025 **Pulsar activity and the morphology of supernova remnants.** V. Radhakrishnan, G. Srinivasan.
J. Astrophys. Astron., Vol. 1, 25 - 32 (1980).

The authors use the recently introduced concept of a 'window' of magnetic field strengths in which pulsars can be active to explain the variation in morphology of supernova remnants. The striking difference between shell-type and filled-typed remnants is attributed to differences in the magnetic field strengths of the neutron stars left by the respective supernovae. Field strengths of a value permitting pulsar activity result in particle production and Crab-like centrally concentrated remnants. Other field values lead to strong magnetic dipole radiation and consequent shell formation (e.g. Cas A).

125.026 **A search for OB stars in supernova remnants.** S. van den Bergh.
J. Astrophys. Astron., Vol. 1, 67 - 70 (1980).

A massive binary, in which the primary becomes a supernova, should leave a luminous secondary near the centre of its remnant. Contrary to expectation no statistically significant excess of OB stars is, however, found near the centres of optically visible galactic supernova remnants.

125.027 **Rayleigh-Taylor convective overturn as a possible solution to the supernova puzzle.**
M. Livio, J. R. Buchler.
J. Phys. Colloq., Vol. 41, No. C-2, (see 012.005), p. C2/155 157 (1980). – Abstr. in Phys. Abstr., Vol. 83, Abstr. 77429 (1980).

125.028 **Supernova radio remnants and the filling factor of the hot interstellar medium.**
J. C. Higdon, R. E. Lingenfelter.
Astrophys. J., Vol. 239, 867 - 872 (1980).

If a hot (~ 10^6 K), tenuous gas fills ~90% of interstellar space, then the observed number and surface brightness distribution of galactic radio remnants implies a galactic supernova rate of once every ~30 years, which is consistent with estimates based on observations of historical supernovae, pulsars, and supernovae in other spiral galaxies.

125.029 **Nonuniform abundances in young supernova remnants.** W. D. Arnett.
Astrophys. J., Vol. 240, 105 - 108 (1980).

In a young SNR, the possible variation in composition should not be ignored in interpreting observational data. As an example, it is explicitly shown that the Becker et al. observations of Tycho's supernova remnant with HEAO 2 (Einstein) Observatory are consistent with a previously calculated numerical model of a Type I supernova explosion incorporating decay of ^{56}Ni. In young SNRs it may be possible to detect directly the compositions characteristic of the layered structure of the presupernova, not merely the average abundance.

125.030 **Is the remnant of SN 1006 Crablike?** R. H. Becker, A. E. Szymkowiak, E. A. Boldt, S. S. Holt, P. J. Serlemitsos.
Astrophys. J., Lett., Vol. 240, L33 - L35 (1980).

The solid-state spectrometer on the Einstein Observatory and the GSFC cosmic X-ray spectrometer on OSO 8 have observed the X-ray spectrum of SN 1006. The data can be well represented by a power-law model with $\alpha = 1.2$, similar to the spectrum of the Crab Nebula. This is in contrast to the radio and X-ray maps of SN 1006 which show a shell structure more typical of SNR with thermal X-ray emission. The X-ray spectrum is suggestive of nonthermal synchrotron emission, raising the possibility that the remnant of SN 1006 contains a source of relativistic electrons.

125.031 **X-ray line emission from the supernova remnant G292.0+1.8.**
D. H. Clark, I. R. Tuohy, R. H. Becker.
Mon. Not. R. Astron. Soc., Vol. 193, 129 - 133 (1980).

The solid state spectrometer on the Einstein Observatory has detected line emission from Mg, Si and S in the thermal X-ray spectrum of the supernova remnant G292.0+1.8. An over-abundance of sulphur and a probable under-abundance of iron relative to their solar values is indicated by the data. The X-ray and optical data for this object support an interpretation of emission originating in ejecta from the comparatively recent explosion of a massive ($\sim 20 M_{\odot}$) progenitor star.

125.032 **The X-ray structure and mass of the Cassiopeia A supernova remnant.** A. C. Fabian, R. Willingale, J. P. Pye, S. S. Murray, G. Fabbiano.
Mon. Not. R. Astron. Soc., Vol. 193, 175 - 188 (1980).

The X-ray images of the Cassiopeia A supernova remnant from the Einstein Observatory have been processed by a maximum-entropy algorithm. The emission appears to originate in two concentric thin shells. The mass of X-ray emitting gas is found to be at least $15 M_{\odot}$, which is considerably more than observed directly at other wavelengths. The X-ray structure and dynamics of Cas A are consistent with it being in a 'free-expansion' phase of evolution, with the bulk of the emission from a reverse shock in the ejecta. The progenitor star is likely to have been massive, as seems to be required by the element abundances of the optical knots.

125.033 **Dynamical determination of the mass ejected by the Cassiopeia A supernova.**
K. Brecher, I. Wasserman.
Astrophys. J., Lett., Vol. 240, L105 - L106 (1980).

The authors estimate the total mass, M_e, for the Cas A supernova remnant by combining the present-day astronomical observations of its current expansion rate with a recent historical determination of the date of the Cas A supernova explosion. They find $M_e \gtrsim 10-12 M_{\odot}$.

125.034 **Fleurs synthesis telescope observations of galactic SNRs.** D. K. Milne.
Proc. Astron. Soc. Australia, Vol. 3, 341 - 343 (1979).

125.035 **Supernova remnants: a generalized theoretical approach to the radio evolution.**
J. L. Caswell, I. Lerche.
Proc. Astron. Soc. Australia, Vol. 3, 343 - 347 (1979).

125.036 **On interpretation of observed radio properties of Cas A.** I. Lerche, J. L. Caswell.
Proc. Astron. Soc. Australia, Vol. 3, 347 - 348 (1979).

125.037 **Search for far-infrared emission from young supernova remnants.** E. L. Wright, D. A. Harper, R. F. Loewenstein, J. Keene, S. E. Whitcomb.
Astrophys. J., Lett., Vol. 240, L157 - L158 (1980).

The authors have searched for 80–350 μm emission from the Tycho, Kepler, and Cas A supernova remnants. The bolometric fluxes are less than 5×10^{-11}, 4×10^{-12}, and 9×10^{-12} W m^{-2}, respectively, for the entire radio shell, corresponding to luminosities less than 2.5×10^4, 1.1×10^3, and 1.5×10^3 $L_{\odot}$. The ratio of dust to gas in Cas A is at least 10 times less than the "normal" ratio.

125.038 **Optical and radio studies of supernova remnants in the Local Group galaxy M33.**
I. J. Danziger, S. D'Odorico, W. M. Goss.
Messenger, No. 21, p. 7 - 11 (1980).

125.039 **An investigation of a faint nebula – the supernova remnant HB3.** T. A. Lozinskaya, T. G. Sitnik.
Astron. Zh., Tom 57, 997 - 1002 (1980). In Russian. – English translation in Soviet Astron., Vol. 24, No. 5.

Hα interferometric observations of the supernova remnant HB3 have been carried out. The gas motions in the remnant have been studied. The energetics of the remnant and its association with the W3 - W4 - W5 complex are discussed.

125.040 **Secular decrease of radio flux density of Cassiopeia A at 2924 MHz.**
E. N. Vinyajkin, V. A. Razin, V. V. Khrulev.
Pis'ma Astron. Zh., Tom 6, 620 - 622 (1980). In Russian. English translation in Soviet Astron. Lett., Vol. 6,

From measurements of ratios of radio flux densities of Cassiopeia A and Cygnus A at 2924 MHz carried out in September 1962 and December 1979 the mean annual decrease of the radio flux density of Cassiopeia A is found to be $(0.80 \pm 0.07)\%$. A comparison with measurements at lower frequencies confirms the conclusion that the spectrum of radio emission of Cassiopeia A in the centimeter-meter wavelength region flattens as time changes.

125.041 **Vela X and the evolution of plerions.** K. W. Weiler, N. Panagia.
Astron. Astrophys., Vol. 90, 269 - 282 (1980).

Based on the radio, optical, and X-ray observations available for the supernova remnant consisting of Vela X, Y, and Z, the suggestion is made that Vela X in fact resembles the Crab Nebula more closely than it resembles the vast majority of shell-type remnants present in radio supernova remnant surveys. Vela X is dependent for its energy on the rotational energy losses of the pulsar rather than on the energy release in the original supernova explosion. It is thus included in the class of "plerions" or "filled center" supernova remnants containing, at present, about a dozen established or suggested members. An evolutionary model for the plerions is developed and used to describe the size, expansion velocity, magnetic field, flux density, and electromagnetic spectrum and their changes with time.

125.042 **X-rays from G74.9+1.2: an elderly Crab Nebula?** A. S. Wilson.
Astrophys. J., Lett., Vol. 241, L19 - L22 (1980).

The supernova remnant G74.9+1.2, which resembles the Crab Nebula in many of its radio properties, has been detected in 0.15–3 keV X-rays. The intrinsic luminosity of the object is about 100 times weaker than the Crab Nebula in this band. The nearby, intense, compact radio source 2013+370 is also probably detected as an extended X-ray source. It is argued that G74.9+1.2 represents an object similar to the Crab Nebula but of age 4000–7000 years.

125.043 **Hot plasmas in supernova remnants.**
F. D. Kahn.
Highlights of Astronomy, Vol. 5, (see 012.025), 365 - 374 (1980).

Contents: The explosion. Interaction with the interstellar medium. Production of hot gas and its consequences. Limits on the effects of thermal conduction. The influence of inhomogeneities in the interstellar medium.

125.044 **Supernovae: the ultimate instability.**
R. P. Kirshner.
Highlights of Astronomy, Vol. 5, (see 012.026), 513 - 515 (1980).

All the available lines of evidence from statistics, from the state of the star at maximum light, and from the chemistry of the ejecta point to the origin of supernovae in massive and short-lived stars. The most controversial point is the origin of SN I in elliptical galaxies, since the presence of supernovae that come from massive stars requires substantial present-day star formation in these systems.

125.045 **The Rayleigh-Taylor instability overturn of supernova cores during bouncing and resulting neutrino energy release.** S. A. Colgate, A. G. Petschek.
Highlights of Astronomy, Vol. 5, (see 012.026), 517 - 519 (1980).

125.046 **Supernovae.**
IAU Circ., No. 3491 (1980).

125.047 **Possible supernova in Virgo cluster.**
IAU Circ., Nos. 3492, 3500 (1980).

125.048 **Supernova in anonymous galaxy.**
IAU Circ., No. 3494 (1980).

125.049 **Possible radio identification of Tycho's supernova.**
IAU Circ., Nos. 3502, 3511 (1980).

125.050 **Supernova in anonymous galaxy.**
IAU Circ., No. 3550 (1980).

125.051 **Possible supernova in Virgo cluster.**
Yamamoto Circ., No. 1939 (1980).

125.052 **Supernova met satelliet waargenomen.**
C. Titulaer.
Zenit, 7e Jaarg., 16 (1980).

125.053 **Ultraviolet spectroscopy of the Vela supernova remnant.** I. J. Danziger, R. Wood, D. H. Clark.
ESO Sci. Prepr. No. 104, 7 pp. (1980). – Submitted to Mon. Not. R. Astron. Soc.

125.054 **Astrophysical interpretation of the λλ 1200 - 7300 Å emission line spectrum of a filament in the Cygnus Loop supernova remnant.**
S. D'Odorico, P. Benvenuti, M. Dennefeld, M. A. Dopita, A. Greve.
ESO Sci. Prepr. No. 121, 13 pp. (1980). – Submitted to Astron. Astrophys.

125.055 **Effects of heavy element abundance on evolution of supernova remnants.**
M. Fukunaga, Y. Sabano.
Sci. Rep. Tôhoku Univ., Ser. 8, Vol. 1, No. 1, p. 30 - 41 (1980).

It is shown that the time-evolution of a supernova remnant is delayed with the decrease in the heavy element abundance because of the suppressed cooling efficiency in the gas, and that the dynamical behavior of a remnant in the regime of z (the ratio of heavy element abundance to cosmic one) less than 10^{-2} is essentially the same as that in the case of no heavy element. The heavy element cooling also affects properties of a dense shell behind the shock front, and the kinetic energy and the radius of a remnant. A discussion is given that the pressure in the hot gas, produced by supernova explosions, could be sufficiently high to support against the gravitational contraction in the initial phase of the Galaxy.

125.056 **Searching for supernovae.** C. R. Martys.
J. British Astron. Assoc., Vol. 91, 54 - 62 (1980).

This paper examines the role of supernovae in modern observational astronomy, and recommends patrolling for supernovae in external galaxies as a useful activity for properly equipped amateur astronomers. The types of galaxies most suitable for monitoring and the various methods used and equipment needed are described, with particular emphasis on photographic search methods.

125.057 **The X-ray structure of Kepler's supernova remnant.**
R. Pisarski.
News Lett. Astron. Soc. N. Y., Vol. 1, No. 7, p. 20 (1980). Abstract.

125.058 **A hot blue star near the center of the remnant of supernova A. D. 1006.**
F. Schweizer, J. Middleditch.
Astrophys. J., Vol. 241, 1039 - 1044 (1980).

The authors have searched for the remnant star of SN 1006 with the CTIO 4 m telescope using two techniques: trailed prime-focus plates to see whether, against expectation, optical pulses from a bright object could be detected, and two-color plates to look for an unusually blue object. They describe this search and the spectroscopic and photometric observations of a very blue star of B = 16.6 discovered in the process.

125.059 **Astrophysical interpretation of the λλ1200 - 7300 Å emission line spectrum of a filament in the Cygnus Loop supernova remnant.** S. D'Odorico, P. Benvenuti, M. Dennefeld, M. A. Dopita, A. Greve.
Astron. Astrophys., Vol. 92, 22 - 25 (1980).

An IUE spectrum of a filament in the Cygnus Loop supernova remnant has been combined with earlier far UV and optical data to obtain a complete spectral coverage between 1200 and 7350 Å. The observed continuum is fully understood as a combination of free-free emission, recombination and hydrogen two photon emission. The emission line spectrum is well interpreted in the framework of a shockwave model of moderate velocity (v_s = 90 km/s) and depleted abundances.

125.060 **G 40.5 - 0.5: a previously unrecognised supernova remnant in Aquila.**
A. J. B. Downes, T. Pauls, C. J. Salter.
Astron. Astrophys., Vol. 92, 47 - 50 (1980).

Radio observations at frequencies of 1720 and 2700 MHz of the previously unrecognized supernova remnant G 40.5 - 0.5 are reported. The radio spectral index of the remnant is 0.41 ± 0.05 over the range 750 - 2700 MHz. The distance to the remnant is 5.5 - 8.5 kpc.

125.061 **G93.3 + 6.9: a highly polarized supernova remnant.**
C. G. T. Haslam, T. Pauls, C. J. Salter.
Astron. Astrophys., Vol. 92, 57 - 62 (1980).

The authors present total power and linear polarization maps of the SNR G93.3 + 6.9 at 1720 MHz and 2700 MHz. The source appears to show certain characteristics of both shell-type and Crab-like remnants. Extremely high polarization percentages are found on the object, reaching beyond 50% at 2700 MHz.

125.062 **Observations of the old supernova remnant S 147 at 11.1 and 18.2 cm wavelengths.**
M. R. Kundu, P. E. Angerhofer, E. Fürst, W. Hirth.
Astron. Astrophys., Vol. 92, 225 - 229 (1980).

The authors report new high resolution and high sensitivity radio observations of the old galactic supernova remnant S 147, made at wavelengths of 11.1 and 18.2 cm. Their data reveal the presence of a well-developed radio shell. The integrated flux density spectrum shows a distinct break at 1 GHz, which the authors attribute to enhanced volume emissivity of the swept up interstellar material. Derived physical parameters are reported.

125.063 **Supernovae.** F. Ciatti.
Variability in stars and galaxies, (see 012.044), p. F.1.1 - 1.8 (1980).

125.064 **The evolution of supernova remnants in different galactic environments, and its effects on supernova statistics.** M. Kafatos, S. Sofia, F. Bruhweiler, T. Gull.
Astrophys. J., Vol. 242, 294 - 305 (1980).

By examining the interaction between supernova (SN) ejecta and the various environments in which the explosive event might occur, the authors conclude that only a small fraction of the many SNs produce observable supernova remnants (SNRs). This fraction, which is found to depend weakly upon the lower mass limit of the SN progenitors, and more strongly on the specific characteristics of the associated interstellar medium, decreases from approximately 15% near the galactic center to 10% at $R_{gal} \sim 10$ kpc and drops nearly to zero for $R_{gal} > 15$ kpc. The results explain the differences between the substantially larger SN rates in the Galaxy derived both from pulsar statistics and from observations of SN events in external galaxies, when compared to the substantially smaller SN rates derived from galactic SNR statistics. The resuts also explain the very large number of SNRs observed toward the galactic center in comparison to few SNRs found in the anticenter direction.

125.065 **X-ray line emission from the Cygnus Loop.**
S. M. Kahn, P. A. Charles, S. Bowyer, R. J. Blissett.
Astrophys. J., Lett., Vol. 242, L19 - L22 (1980).

The HEAO 1 satellite has observed the X-ray spectrum of the Cygnus Loop in the 0.1 - 2.0 keV band. The observed count-rate spectrum is deconvolved by the Kahn and Blissett technique to demonstrate conclusively the presence of oxygen- and iron-line emission between 0.6 and 0.9 keV. The authors have fitted the spectrum with single- and two-component Raymond-Smith plasma emission models.

125.066 **The infrared echo of a supernova in a dust cloud.**
E. L. Wright.
Astrophys. J., Lett., Vol. 242, L23 - L24 (1980).

A simple model is given for the infrared light-echo from a supernova explosion in a molecular cloud. The color temperature is determined by the vaporization temperature of the dust grains, and the time scale is given by the light travel time across the distance to which grains are vaporized.

125.067 **The calculation of the optical spectra of the Cygnus Loop.** M. Contini, B. Z. Kozlovsky, G. Shaviv.
Astron. Astrophys., Vol. 92, 273 - 280 (1980).

The optical spectra of the Cygnus Loop supernova remnant were calculated using a shock wave propagation model. The authors fit the observed line ratios of the single filaments of the Cygnus Loop and investigate their structure. The best fitting parameters were obtained and discussed.

125.068 **The discovery of optical emission from the SNR G126.2 + 1.6.** W. P. Blair, R. P. Kirshner, T. R. Gull, D. L. Sawyer, R. A. R. Parker.
Astrophys. J., Vol. 242, 592 - 595, plates 14 - 15 (1980).

The authors have used interference filter photographs to identify an arc of nebulosity that is coincident with the radio contours of the galactic supernova remnant G126.2 + 1.6. Spectrophotometry of the filament shows that the emission-line spectrum matches the spectra of other galactic supernova remnants. The spectrum can be adequately matched by a shock of velocity near 100 km s^{-1} in an interstellar cloud of density 3.

125.069 **A new oxygen-rich supernova remnant in the Large Magellanic Cloud.**
D. S. Mathewson, M. A. Dopita, I. R. Tuohy, V. L. Ford.
Astrophys. J., Lett., Vol. 242, L73 - L76 (1980).

Narrow-band imaging observations with the Anglo-Australian Telescope have been made of the supernova remnant 0540–69.3 in the Large Magellanic Cloud. The authors report the optical identification of the radio/X-ray object with a new oxygen-rich remnant. Spectroscopic observations of the intense annulus in the [O III] lines show it to have a high velocity dispersion (~3000 km s^{-1}), indicative of a very young remnant.

125.070 **Spectrophotometry of W50: the supernova remnant around SS 433.** R. P. Kirshner, R. A. Chevalier.
Astrophys. J., Lett., Vol. 242, L77 - L81 (1980).

The authors have obtained spectra of two regions of nebulosity associated with the radio supernova remnant W50 that surrounds SS433. In general, the interpretation of the spectra supports the idea that the nebulosity is associated with the supernova remnant and is consistent with the association of W50 with SS 433.

125.071 **Core collapse, bounce and shock propagation.**
R. L. Bowers, J. R. Wilson.
Space Sci. Rev., Vol. 27, (see 012.046), 537 - 544 (1980).

Recent developments in the physics input for iron core collapse models of type II supernovae are reviewed. The effect of these developments on collapse calculations is also discussed.

125.072 **The light curve epochs of supernovae explosions.**
S. W. Falk.
Space Sci. Rev., Vol. 27, (see 012.046), 545 - 553 (1980).

A brief description of the current state of knowledge for Type II light curves is given, including discussion of some of the remaining problems. The use of supernovae of both types as independent distance indicators is outlined and difficulties are pointed out. A summary of new ideas regarding the origin of Type I light curves is presented.

125.073 **White dwarf models for type I supernovae and quiet supernovae, and presupernova evolution.**
K. Nomoto.
Space Sci. Rev., Vol. 27, (see 012.046), 563 - 570 (1980).

Supernova mechanisms in accreting white dwarfs are presented, i. e., the carbon deflagration as a plausible mechanism for producing type I supernovae and electron captures to form quiet supernovae leaving neutron stars. These outcomes depend on accretion rate of helium, initial mass and composition of the white dwarf. The various types of hydrogen shell-burning in the presupernova stage are also discussed.

125.074 **Supernova explosions – the role of a Rayleigh-Taylor instability.** J. R. Buchler, M. Livio, S. A. Colgate.
Space Sci. Rev., Vol. 27, (see 012.046), 571 - 577 (1980).

A two dimensional hydrodynamic study indicates that convectively unstable gradients which develop during core collapse and bounce give rise to large scale core overturn. It is also shown that the concomitant release of neutrinos can deposit large amounts of energy and momentum in the infalling envelope and give rise to a powerful supernova explosion.

125.075 **The effect of stellar structure on supernova remnant evolution.** E. M. Jones, B. W. Smith, W. C. Straka.
Space Sci. Rev., Vol. 27, (see 012.046), 579 - 584 (1980).
One-dimensional hydrodynamic calculations have been done of 1E51 erg explosions in $15M_\odot$ stars. The authors have appended a steep external density gradient to the pre-supernova model of Weaver et al. and find: (1) the outer shock wave decelerates throughout the pre-Sedov phase, (2) the expanding stellar envelope and the shocked interstellar material are Rayleigh-Taylor stable until the Sedov phase, and (3) steep internal density gradients are R–T unstable during the early expansion and may be the source of high velocity knots seen in Cas A.

125.076 **Peculiarities of supernova remnant and pulsar distribution relative to the local spiral arm of the Galaxy.** G. N. Kimeridze.
Pis'ma Astron. Zh., Tom 6, 742 - 744 (1980). In Russian.
English translation in Soviet Astron. Lett., Vol. 6.
The selectivity in the distribution of supernova remnants and pulsars relative to separate branches and their offshoots in the solar neighbourhood is revealed; their distribution along the separate branches of the local spiral arm is different.

125.077 **Gamma ray emission from supernova remnants.** R. Cowsik, S. Sarkar.
Non-solar gamma-rays, (see 012.047), p. 57 - 60 (1980).
In this work the authors point out that gamma-ray astronomy can be an additional powerful tool in clarifying some important features of supernova remnants. To illustrate this claim, they obtain a lower limit to the magnetic field in the radio-emitting shell of the young remnant Cassiopeia-A.

125.078 **A search for radio remnants of ancient oriental "guest stars" with the Westerbork Telescope.**
G. G. C. Palumbo, P. Schiavo Campo, G. K. Miley.
Acta Astron. Sinica, Vol. 21, 334 - 339 (1980). In Chinese.
Seven $1.5° \times 1.5°$ fields in which " guest stars" had been reported by ancient oriental astronomers have been surveyed at 610 MHz for radio remnants with the Westerbork Telescope. No diffuse radio emission was detected.

125.079 **The nature of high-frequency pulsar radiation and activity of supernova remnants.**
D. G. Lominadze, G. Z. Machabeli, A. B. Mikhajlovskij, Yu. P. Ochelkov, V. V. Usov.
Usp. fiz. nauk, Vol. 131, 516 - 518 (1980). In Russian.
Abstr. in Ref. zh., 51. Astron., 11.51.584 (1980).

125.080 **The spectrum of the supernova remnant MSH 15-5$_6$ (G326.3-1.8).** M. Dennefeld.
Publ. Astron. Soc. Pacific, Vol. 92, 603 - 605 (1980).
Spectroscopy confirms that the optical filaments found by van den Bergh within the nonthermal radio source MSH 15-5$_6$ belong to the supernova remnant. The spectral characteristics are typical of other shell-type galactic SNR and indicate a diameter of 30 to 40 pc, consistent with the estimated radio distance of 3.2 kpc. A lower limit to the visual absorption is found to be $A_v = 5.1$ magnitudes. The data do not allow any conclusion about the peculiarity of this source, as seen in radio, but support the morphological similarity with other objects like HB3, HB9, or VRO 42.05.01.

125.081 **Anomalous beta decay in type-I supernovae.** T. Pankey, Jr.
Publ. Astron. Soc. Pacific, Vol. 92, 606 - 608 (1980).
The maximal effects of the absence of near neighbors on the beta decay of Co^{56} were calculated. An increase in the number of low-energy positrons was found, but the effects on K-capture were infinitestimally small. The corrections were too small to explain the accelerated decay rate of the luminosity of type-I supernovae.

125.082 **On the possible presence of emission lines of Fe II and Mg I 4571 Å in the spectra of type II supernovae at the late stage after light maximum.**
N. N. Chugaj.
Astron. Tsirk., No. 1105, p. 1 - 3 (1980). In Russian.

125.083 **Radial distribution of supernovae in elliptical galaxies.** D. Yu. Tsvetkov.
Astron. Tsirk., No. 1113, p. 5 - 7 (1980). In Russian.

125.084 **Filamentary structure of old optical supernova remnants.** T. A. Lozinskaya.
Astron. Zh., Tom 57, 1197 - 1203 (1980). In Russian.
English translation in Soviet Astron., Vol. 24, No. 6.
The results of interferometric and spectral observations are presented which approach to the problem of thin filaments formation. A morphological classification of the nebulae investigated is discussed in connection with physical parameters of the SNRs and of the interstellar environment. An analysis of the spatial geometry and optical emission of thin filaments is given.

125.085 **Observations of supernova remnants with the Einstein Observatory.** G. Fabbiano.
X-ray astronomy, (see 012.055), p. 15 - 34 (1980).
Some of the SNR's observed with the Einstein Observatory are discussed, namely the Crab Nebula, the Vela Nebula and pulsar, the Tycho SNR, SN 1006 and Cas A.

125.086 **X-ray spectra of supernova remnants.** S. S. Holt.
X-ray astronomy, (see 012.055), p. 35 - 45 (1980).

125.087 **Observations of supernova remnants in the Large Magellanic Cloud with the Einstein Observatory.**
D. J. Helfand, K. S. Long.
X-ray astronomy, (see 012.055), p. 47 - 59 (1980).
The authors have detected 13 SNR in the course of their continuing X-ray survey of the LMC with the Einstein Observatory. A few other newly discovered sources with similarly soft spectra may also turn out to be remnants when their positions are compared with optical and radio maps. Several of the sources are bright enough for detailed study of their X-ray brightness distribution and its correlation with radio and optical properties. These data offer an excellent opportunity to compare a uniform, complete, luminosity-limited sample of SNR with theoretical models of SNR evolution.

125.088 **High resolution X-ray spectroscopy from the Einstein observatory.** P. F. Winkler, C. R. Canizares, G. W. Clark, T. H. Markert, C. Berg, J. G. Jernigan, M. L. Schattenburg.
X-ray astronomy, (see 012.055), p. 61 - 71 (1980).
The authors begin with a few general remarks about X-ray spectroscopy, followed by a brief description of the M. I. T.'s Focal Plane Crystal Spectrometer instrument. The results presented deal primarily with supernova remnants: Puppis A and Cas A in the Galaxy, and N132D and N63A in the Large Magellanic Cloud. In addition the authors report their detection of oxygen emission from the vicinity of M87 in the Virgo Cluster.

125.089 **Spectrophotometry of the supernova remnant IC 443.** R. A. Fesen, R. P. Kirshner.
Astrophys. J., Vol. 242, 1023 - 1040, plates 21 - 24 (1980).
Spectrophotometry of the line emission at six locations in IC 443 is presented and analyzed. Comparison of the observations with published shock-wave model calculations

shows excellent agreement and indicates shock velocities for the filaments between 65 and 100 km s^{-1}, with a preshock number density of 10–20 cm^{-3} for the denser filaments. The observations suggest that the remnant's shock front is encountering an inhomogeneous interstellar medium whose elemental abundances are depleted with respect to cosmic values. A comparison of X-ray and optical data indicates that the gas in the optical filaments has a lower pressure than the X-ray–emitting gas.

Tycho's supernova. See Abstr. 004.022.

UK nova/supernova search programme. See Abstr. 013.035.

X-ray astronomical spectroscopy. See Abstr. 031.567.

A new equation of state of supernova matter. See Abstr. 061.015.

Neutrino emission from a supernova shock. See Abstr. 061.028.

Seed abundances for *r*-processing in the helium shells of supernovae. See Abstr. 061.039.

Xenon from intermediate zones of supernovae. See Abstr. 061.044.

The chemical abundances of the Cassiopeia A fast-moving knots: explosive nucleosynthesis on a minicomputer. See Abstr. 061.047.

Stability of the Primakoff-Sedov blast wave and its generalizations. See Abstr. 062.032.

On equilibrium states of heated self-gravitating gas clouds cooling by conduction in an external gravitational field. See Abstr. 062.134.

Supernova triggered by electron captures. See Abstr. 065.012.

Final stages of stellar evolution and supernova outbursts. See Abstr. 065.091.

Final stages of stellar evolution and the hydrodynamic theory of supernova outbursts. Detailed review. See Abstr. 065.092.

Supernovae and the Ap phenomenon. See Abstr. 114.164.

Off-center detonation supernovae in accreting white dwarfs. See Abstr. 126.009.

Carbon monoxide observations of the W44 region. See Abstr. 131.060.

Identification of the anomalous 26.131-MHz nitrogen line observed towards Cas A. See Abstr. 131.113.

Molecular clouds near supernova remnants. See Abstr. 131.235.

Evaporation of H I clouds in supernova remnants. See Abstr. 132.007.

Upper limits on X-ray line emission from the Crab Nebula. See Abstr. 134.036.

No detectable supernova remnant near the pulsar PSR 1930 + 22. See Abstr. 141.541.

Discovery of a compact X-ray source at the center of the supernova remnant RCW 103. See Abstr. 142.044.

An extraordinary new celestial X-ray source. See Abstr. 142.068.

A search for prompt transient emission of high energy γ-rays ($\geq 10^{12}$ eV) from supernovae. See Abstr. 142.509.

Discrete sources of gamma-radiation, historic supernovae and pulsars. See Abstr. 142.513.

New evidence for links between SNOBs and galactic γ-ray sources. See Abstr. 142.525.

A search for gamma ray spectral features from the galactic plane and Cas-A. See Abstr. 157.005.

X-ray characteristics of Loop I and the local interstellar medium. See Abstr. 157.010.

Radio observations of three supernova remnants in M33. See Abstr. 158.203.

Supernova in NGC 4321 (1979c)

125.101 **Coordinated optical, ultraviolet, radio and X-ray observations of supernova 1979c in M 100.**
N. Panagia, G. Vettolani, A. Boksenberg, F. Ciatti, S. Ortolani, P. Rafanelli, L. Rosino, C. Gordon, D. Reimers, K. Hempe, P. Benvenuti, J. Clavel, A. Heck, M. V. Penston, F. Macchetto, D. J. Stickland, J. Bergeron, M. Tarenghi, B. Marano, G. G. C. Palumbo, A. N. Parmar, G. S. W. Pollard, P. W. Sanford, W. L. W. Sargent, R. A. Sramek, K. W. Weiler, P. Matzik.
Mon. Not. R. Astron. Soc., Vol. 192, 861 - 879 (1980).

The authors present and discuss observations made in the first six weeks after discovery (1979 April 20 - June 5). On the basis of the optical data, the SN is classified as peculiar type II. The main results form the ultraviolet line spectrum are that: (1) Most absorption features are produced either in the disc or the halo of our Galaxy and M 100. (2) The emission lines indicate the presence of a shell where the gas is highly ionized. Radio and X-ray observations have provided only upper limits to the SN emission. The inferences from these limits confirm the ultraviolet results.

125.102 **Infrared emission by dust grains near variable primary sources – III. Type II supernovae.**
M. F. Bode, A. Evans.
Mon. Not. R. Astron. Soc., Vol. 193, 21P - 24P (1980).

The authors consider the radiative heating of an extensive pre-existing circumstellar dust shell surrounding the recent supernova in M100. The results are consistent with observations of a thermal infrared excess in this object after ~260 days. The authors also examine emission by interstellar dust grains in the galactic plane of M100. Although the derived fluxes are too low by $\sim 10^3$, re-emission by interstellar dust heated by supernovae in Local Group galaxies should be detectable for a long time after outburst.

Supernovae. See Abstr. 125.046.

A radio supernova in M100.
See Abstr. 141.127.

Supernova in NGC 6946

125.201 **Supernova in NGC 6946.**
IAU Circ., Nos. 3532, 3534, 3537, 3542, 3544, 3548, 3557 (1980).

Supernova in NGC 7448

125.301 **Supernova in NGC 7448.**
IAU Circ., No. 3547 (1980).

Supernova in NGC 1316

125.401 **Supernova in NGC 1316.**
IAU Circ., Nos. 3548, 3556 (1980).

Supernova in NGC 1058 (1969*l*)

125.501 **Line-formation – Rechnungen in der Supernovahülle 1969*l*.** K. Hempe.
Mitt. Astron. Ges., Nr. 50, p. 69 - 70 (1980).

126 Low-luminosity Stars, Subdwarfs, White Dwarfs, Degenerate Stars

126.001 **The basic structure of hot white dwarfs atmospheres as a function of composition.**
K. H. Böhm, S. Kapranidis.
Astron. Astrophys., Vol. 87, 307 - 314 (1980).

The purpose of the paper is to obtain some additional general insight into the properties of hot white dwarfs atmospheres and the resulting surface fluxes. The authors shall be mostly concerned with the typical qualitative features of hot white dwarf model atmospheres (in contradistinction to model atmospheres of, say, main sequence stars) and with the dependence of these properties on chemical composition. For this purpose the authors study model atmospheres with a wide variety of chemical compositions, including typical DA, non-DA, normal population I and a number of other abundance distributions.

126.002 **Constraints on possible long-term variability of ZZ Ceti stars.** M. H. Liller, J. E. Hesser.
Publ. Astron. Soc. Pacific, Vol. 92, 319 - 322 (1980).

Magnitude estimates for 15 ZZ Ceti (white-dwarf) variable stars determined from plates in the Harvard collection reveal no evidence for variability in excess of measurement errors (~$0^m.15$) since the 1890s. This suggests that these DA variables are not distinguished from their nonvariable counterparts by any large-amplitude, low-frequency, or secular variations.

126.003 **SB 21, an extremely helium-rich subdwarf O-star.**
K. Hunger, R. P. Kudritzki.
Astron. Astrophys., Vol. 88, L4 - L5 (1980).

The subdwarf O-star SB 21 is found to be extremely helium-rich from 29 Å/mm spectrograms taken with ESO's 3.6 m telescope. H_γ is absent. The spectrum closely resembles that of CPD −31° 1701.

126.004 **HD 113001: photometric separation.** G. Goy.
Astron. Astrophys., Vol. 88, 370 (1980). In French.

The distance and absolute magnitude of SdO-type stars are very badly known. The existence of a SdO + F binary may therefore bring a solution to the problem. By using a method already applied in a previous paper (Goy, 1977) the author shows that the spectral distribution of the F component is exactly similar to that of a normal F 2 V star. The absolute magnitude of the hot star can thus be deduced with good accuracy.

126.005 **A proposed new white dwarf spectral classification.**
E. M. Sion.
Bull. American Astron. Soc., Vol. 12, 456 (1980). – Abstract.

126.006 **The white dwarf stars.**
H. M. van Horn.
Bull. American Astron. Soc., Vol. 12, 463 - 464 (1980). Abstract.

126.007 **A photometric search for white dwarfs in IC 2602.**
B. J. Anthony-Twarog.
Bull. American Astron. Soc., Vol. 12, 464 (1980). – Abstract.

126.008 **Non-radial oscillations of white dwarfs: the effect of compositional stratification on the g-mode periods and eigenfunctions.** D. E. Winget, H. M. Van Horn, C. J. Hansen.
Bull. American Astron. Soc., Vol. 12, 464 - 465 (1980). Abstract.

126.009 **Off-center detonation supernovae in accreting white dwarfs.** K. Nomoto, M. Y. Fujimoto.
Bull. American Astron. Soc., Vol. 12, 465 (1980). – Abstract.

126.010 **The evolution of accreting carbon white dwarfs.**
R. E. Taam.
Bull. American Astron. Soc., Vol. 12, 465 (1980). – Abstract.

126.011 **A spectrophotometric analysis of the hot white dwarf HD 149499B.**
E. F. Guinan, E. M. Sion, F. Wesemael.
Bull. American Astron. Soc., Vol. 12, 465 - 466 (1980). Abstract.

126.012 **A theoretical interpretation of the hot subdwarfs.**
W. Cabot, H. M. van Horn, D. E. Winget, F. Wesemael.
Bull. American Astron. Soc., Vol. 12, 466 (1980). – Abstract.

126.013 **Infrared photometry of white dwarf-red dwarf composites.** R. G. Probst.
Bull. American Astron. Soc., Vol. 12, 507 (1980). – Abstract.

126.014 **Ultraviolet line spectra of white-dwarf stars.**
H. L. Shipman.
Bull. American Astron. Soc., Vol. 12, 508 (1980). – Abstract.

126.015 **Discovery of a CH subdwarf.**
L.-T. G. Chiu, R. G. Kron.
Bull. American Astron. Soc., Vol. 12, 521 (1980). – Abstract.

126.016 **Subdwarf M stars of extreme metal-deficiency.**
T. B. Ake, J. L. Greenstein.
Bull. American Astron. Soc., Vol. 12, 522 (1980). – Abstract.

126.017 **IUE observations of the early-type subdwarf HD 49798.**
F. C. Bruhweiler, Y. Kondo, G. E. McCluskey.
Bull. American Astron. Soc., Vol. 12, 539 (1980). – Abstract.

126.018 **An absorption line in the ultraviolet spectrum of 40 Eridani B.** J. L. Greenstein.
Astrophys. J., Lett., Vol. 241, L89 - L93 (1980).

Two excellent low-resolution spectra show an absorption line of equivalent width 3 Å, near λ1391, in the typical DA (hydrogen atmosphere) white dwarf 40 Eri B. The line is confirmed by a high-resolution spectrum and is the first seen in any DA star.

126.019 **The discovery of an O subdwarf in the globular cluster NGC 6712.** R. A. Remillard, C. R. Canizares, J. E. McClintock.
Astrophys. J., Vol. 240, 109 - 113 (1980).

Photographic plates of the globular cluster NGC 6712 revealed a star, C49 (Sandage and Smith) with a large UV excess. Spectrophotometry using the 4 m CTIO telescope showed the star to have an O subdwarf spectrum. Cluster membership is favored. Analysis of the spectral data using nonlocal thermodynamic equilibrium (NLTE) models gives surface parameters $T \gtrsim 50{,}000$ K, $\log g = 5.5 \pm 0.5$, and $N(\mathrm{He})/N(\mathrm{H}) \approx 0.1$. The results are consistent with the evolutionary models for low-mass stars that are contracting into white dwarfs. Globular cluster membership provides advantages for establishing this phase of stellar evolution as one explanation of sdO stars.

126.020 **The effect of photoelectric absorption on the Comptonized X-ray spectrum of an accreting white dwarf.** R. R. Ross, A. C. Fabian.
Mon. Not. R. Astron. Soc., Vol. 193, 1P - 5P (1980).

The authors have computed self-consistent models for spherical accretion on to a degenerate dwarf when the infalling gas is optically thick to Compton scattering. They find that photoelectric absorption by the heavy elements has an important effect on the emergent X-ray spectrum. This indicates that the compact object in Cygnus X-2 is most unlikely to be a degenerate dwarf.

126.021 **Cyclotron absorption in GD229?** D. T. Wickramasinghe, B. Martin.
Proc. Astron. Soc. Australia, Vol. 3, 269 - 271 (1978).

126.022 **White dwarfs.** H. M. Van Horn.
J. Phys. Colloq., Vol. 41, No. C-2, (see 012.005), p. C2/97 - 104 (1980). – Abstr. in Phys. Abstr., Vol. 83, Abstr. 82356 (1980).

126.023 **A new gravitational redshift for the white dwarf o^2 Eri B.** G. Wegner.
Astron. J., Vol. 85, 1255 - 1261 (1980).

Differential measurements of the radial velocities of the DA white dwarf o^2 Eri B and its M4Ve companion, o^2 Eri C, using the coudé spectrograph of the 100-in. Mount Wilson telescope, yield a new gravitational redshift of $V_{RS} = +23.9 \pm 1.3$ (s.d.) km s^{-1}. Corrections for the binary's orbital motion, the small gravitational redshift of the non-degenerate companion, and atmospheric refraction have been included in the reduction.

126.024 **White dwarf stars.** J. Liebert.
Annu. Rev. Astron. Astrophys., Vol. 18, (see 003.004), 363 - 398 (1980).

Contents: Observational constraints, stellar and atmospheric parameters, the origin and evolution of degenerate dwarfs.

126.025 **Nucleosynthesis during nova explosions and grain formation in ejected envelopes.**
M. Y. Fujimoto.
Publ. Astron. Soc. Japan, Vol. 32, 463 - 471 (1980).

Evolution of CNO elements during the hydrogen shell flash on the surface of white dwarfs is discussed in relation to nova explosions. Two typical groups of the C/O ratio result from the investigation: oxygen-rich in the case of small ΔM_H (mass of the accreted hydrogen-rich envelope) and small M_{WD} (mass of the white dwarf), and carbon-rich in the case of large ΔM_H or large M_{WD}. Under these predicted chemical compositions, the growth of grains in the expanding shell is investigated. There exists a lower mass limit of the expanding shell if infrared brightening is to be observed; the IR brightening is due to thermal emission from grains. Grains consist of graphites when M_{WD} is greater than 0.8 $M_\odot$ and they are composed mainly of silicates when M_{WD} is smaller than 0.6 $M_\odot$.

126.026 **Nova explosion of mass-accreting white dwarfs.** K. Nariai, K. Nomoto, D. Sugimoto.
Publ. Astron. Soc. Japan, Vol. 32, 473 - 494 (1980).

The evolution of mass-accreting white dwarfs has been computed from the onset of accretion through nova explosion. Two cases have been considered: a carbon-oxygen white dwarf of 1.3 $M_\odot$ with the accretion rate of $1.0 \times 10^{-10} M_\odot$ yr^{-1} and a helium white dwarf of 0.4 $M_\odot$ with $1.0 \times 10^{-8} M_\odot$ yr^{-1}. Because the thermal evolution during the accretion phase has been fully taken into consideration in computation, the mass of the accreted hydrogen-rich envelope and the corresponding temperature distribution in the envelope have been determined. When a certain amount of the hydrogen-rich envelope has been formed, a hydrogen shell flash commences. The flashing shell lies in the midway between the bottom of the envelope and the stellar surface. In the case of the 1.3 $M_\odot$ white dwarf, the flash has been found to grow strong enough to lead to a nova-like explosion, even for the normal abundance of CNO elements.

126.027 **Hot subluminous stars.** J. L. Greenstein.
Highlights of Astronomy, Vol. 5, (see 012.024), 255 - 262 (1980).

Extensive mass loss is observed for hot subluminous stars, through P Cygni lines in the ultraviolet. The ultraviolet provides determination of effective temperatures. Among nine sdO's, the maximum temperature reported is definitely below 60,000 K. The sdO's show a wide variety of line strengths, notably in N V, C IV and Si IV, as well as He II. One halo sdB is reported as rich in peculiar elements. The comparison of effective temperatures of white dwarfs observed from space and from the ground gives excellent agreement. Another helium-rich close binary probably has a accretion disk; it is the only white dwarf to show the expanding shell of N V, C IV, Si IV characteristic of some subdwarfs. Two magnetic white dwarfs have been observed; one has strong unidentifiable features and the smallest known radius.

126.028 **Diffusion and the chemical abundances in white dwarfs.** D. O. Muchmore.
Bull. American Astron. Soc., Vol. 12, 747 - 748 (1980). Abstract.

126.029 **The evolutionary status of the hot subdwarfs in the extended horizontal branch.** D. Winget.
News Lett. Astron. Soc. N. Y., Vol. 1, No. 7, p. 13 (1980). Abstract.

126.030 **Theory and observations of the optical continuum and line spectra of accretion discs around white dwarfs.** S. K. Mayo, D. T. Wickramasinghe, J. A. J. Whelan.
Mon. Not. R. Astron. Soc., Vol. 193, 793 - 824 (1980).

Theoretical calculations of the optical continuum ($\lambda \gtrsim 3000$ Å), *UBV* colours and Hα, Hβ and Hγ absorption line profiles are presented for steady-state accretion discs around white dwarfs. The theoretical results are compared with observations of *UBV* colours obtained from the literature and with line profiles reported here of dwarf-novae in outburst and UX UMa 'disc' stars. In general the overall agreement between theory and observations is good for the *UBV* colours and reasonable for the line profiles.

126.031 **A search for variability in white dwarfs in the region of the Hyades.** D. W. Peterson, W. I. Beavers.
Astron. Astrophys., Vol. 92, 214 - 218 (1980).

Three white dwarfs, G8 - 7, G7 - 41, and G7 - 255, were examined for variability. Data analysis by the maximum entropy method shows all three to be nonvariable in the period range 30 s to 1200 s. Upper limits on variability are placed at 0.006 to 0.02 mag. No evidence was found for variability at longer periods with higher amplitudes.

126.032 **Calcium and magnesium (?) in the helium white dwarf GD 401.** P. L. Cottrell, J. L. Greenstein.
Astrophys. J., Vol. 242, 195 - 198 (1980).

A calcium abundance (Ca/He ≈ 6 - 9×10^{-9}) has been derived for the DB degenerate object GD 401. The authors have derived upper limits for the Mg/He and Fe/He ratios of $\sim 1 \times 10^{-8}$ and $\sim 6 \times 10^{-9}$, respectively. The errors on these abundances are relatively high. Two conclusions can be made from this analysis: (1) calcium is deficient relative to the Sun, and (2) the Mg/Ca ratio limit is much smaller than the solar value, in contrast to the results of the two other well-studied metal-line white dwarfs (Ross 640, van Maanen 2). The second result for GD 401 is at variance with recent diffusion theories for white dwarfs.

126.033 **Degenerate stars. XII. Recognition of hot non-degenerates.** J. L. Greenstein.
Astrophys. J., Vol. 242, 738 - 748 (1980).

A table lists 51 newly observed degenerates, together with further data on 16 others. Special attention was paid in this list to objects of relatively small proper motion; thus, an additional 14 nondegenerates are included. Equivalent widths of Hγ are given from the SIT. There are 13 faint red degenerates; one previously suspected red degenerate (Gr 422) proves to be a sdM. Most of these red degenerates do not show any lines except for a DF, Gr 554. Hot subdwarfs and an X-ray source are included. The problem of low-resolution spectroscopic classification of dense hot stars is extensively discussed. The multichannel spectrum of the carbon-rich magnetic star LP 790-29 (Gr 535) is analyzed by fitting the undisturbed portions of the spectrum to a blackbody of 7625 K by least squares.

126.034 **The long-term evolution of accreting carbon white dwarfs.** R. E. Taam.
Astrophys. J., Vol. 242, 749 - 755 (1980).

The long-term evolution of an accreting carbon white dwarf of 1.2 $M_\odot$ is investigated. It is found that for a mass accretion rate of $10^{-9} M_\odot$ yr^{-1} a degenerate-helium-shell flash develops, whereas for an accretion rate of $5 \times 10^{-10} M_\odot$ yr^{-1} a central carbon flash develops. The results are discussed in the context of the nova and Type I supernova phenomena.

126.035 **Extreme-UV and far-UV observations of the white dwarf HZ 43 from Voyager 2.**
J. B. Holberg, B. R. Sandel, W. T. Forrester, A. L. Broadfoot, H. L. Shipman, D. C. Barry.
Astrophys. J., Lett., Vol. 242, L119 - L123 (1980).

Using the ultraviolet spectrometer on the Voyager 2 spacecraft, the authors have obtained the first spectrum of the white dwarf HZ 43 in the 500–900 Å and the 900–1200 Å regions. The 500–900 Å data are used to determine an effective temperature of 55,000 K for HZ 43 and an average interstellar H I density of 0.002 ± 0.001 atoms cm^{-3} in the direction of HZ 43.

126.036 **Mass defect of white dwarfs with low masses.**
V. F. Balek, Eh. V. Chubaryan.
Uch. zap. Erevan. univ. Estestv. nauk, 1980, No. 1, p. 57 - 62.
In Russian. – Abstr. in Ref. zh., 51. Astron., 11.51.494 (1980).

126.037 **Circular polarization measurements of selected white dwarfs.** J. Liebert, H. S. Stockman.
Publ. Astron. Soc. Pacific, Vol. 92, 657 - 660 (1980).

Broad-band circular polarization measurements for 32 white dwarfs are presented. None of the stars discussed here represent statistically significant detections, though a few are marginal, and one in particular (LP777-1) deserves closer study. The null results for cool degenerates include the very low luminosity stars LP701-29, LP131-66, and Ross 193B (VB11), and support the hypothesis that the temperature distribution of magnetic stars is not significantly different from the distribution of all white dwarfs. G35-26, a likely low-field case with Zeeman-split lines, does not show significant continuum circular polarization.

126.038 **Linienverbreiterung von Ca II- und C II-Linien unter Druck- und Temperaturbedingungen in heliumreichen Weißen Zwergsternen.** I. Bues.
Mitt. Astron. Ges., Nr. 50, p. 24 - 28 (1980).

126.039 **Acoustic fluxes in white dwarfs.**
J.-P. Arcoragi, G. Fontaine.
Astrophys. J., Vol. 242, 1208 - 1225 (1980).

Estimates of the acoustic flux generated in white dwarf convection zones are given. Wide ranges of mass, effective temperature, and convective efficiency have been considered. Homogeneous hydrogen- and helium-rich convective envelope models were computed as well as inhomogeneous models representing DA white dwarfs with thin layers of hydrogen surrounding helium-dominated regions. The calculation of the transition zone that separates the hydrogen- and helium-rich layers of these models is extended to take into account the fact that both hydrogen and helium are non-trace species. Only very thin layers of hydrogen are allowed if an underlying helium convection zone is to produce acoustic noise in these inhomogeneous models. It is suggested that such layers are unstable against mixing. The soft X-ray luminosities of acoustically heated white dwarf coronae are estimated according to Hearn's coronal model.

A method of numerical integration of differential equations of order II and its application to the determination of a white dwarf model. See Abstr. 021.015.

Methods for calculating circular polarisation in magnetic white dwarfs. See Abstr. 031.532.

Cyclotron absorption in accreting magnetic white dwarfs. See Abstr. 062.113.

Atmospheres for hot, high-gravity stars. I. Pure hydrogen models. See Abstr. 064.011.

Improved secular stability limits for differentially rotating polytropes and degenerate dwarfs.
See Abstr. 065.021.

A eutectic in carbon-oxygen white dwarfs?
See Abstr. 065.029.

The collapse of carbon-oxygen white dwarfs.
See Abstr. 065.046.

Secular instabilities. See Abstr. 065.054.

Shell flashes. See Abstr. 065.055.

The surface chemistry of stars. III. The electric field of a chemically inhomogeneous star. See Abstr. 065.074.

Convective mixing in extended horizontal branch envelope models: the sdB/sdO transition.
See Abstr. 065.093.

Thermogalvanomagnetic phenomena in neutron stars and white dwarfs. See Abstr. 066.505.

Neutron star formation by collapse of white dwarfs.
See Abstr. 066.532.

The white dwarf companion of the barium star ζ Capricorni. See Abstr. 114.038.

Spectral analysis of special stars.
See Abstr. 114.144.

Luminosity functions of disk and halo populations in SA 51, SA 57, and SA 68. See Abstr. 115.001

On the kinematics and ages of wide binaries containing a white dwarf. See Abstr. 118.007.

H and K emissions in V 471 Tauri (BD +16°516).
See Abstr. 119.069.

A bulge model of a dwarf nova.
See Abstr. 122.036.

Period stability of the pulsating white dwarf R548 (=ZZ Ceti). See Abstr. 122.064.

ZZ Ceti stjerner. Pulserende hvite dverger.
See Abstr. 122.068.

White dwarf pulsations. See Abstr. 122.090.

White dwarf pulsations: a review. See Abstr. 122.198.

The ZZ Ceti stars and the rate of evolution of white dwarfs. See Abstr. 122.199.

On the late-type components of slow novae and symbiotic stars. See Abstr. 124.001.

White dwarf rotation and the shapes of nova remnants. See Abstr. 124.002.

Theoretical models of nova outbursts. See Abstr. 124.003.

Radial accretion of H-rich material onto a He white dwarf. See Abstr. 124.010.

White dwarf models for type I supernovae and quiet supernovae, and presupernova evolution. See Abstr. 125.073.

The system AM Her = 4U 1814 + 50. See Abstr. 142.015.

X-ray emission from spherical accretion onto white dwarfs in binary systems. See Abstr. 142.139.

Interstellar Matter, Nebulae

131 Interstellar Matter, Star Formation

131.001 **New results on interstellar reddening in the near infrared.** T. J. Jones, A. R. Hyland.
Mon. Not. R. Astron. Soc., Vol. 192, 359 - 364 (1980).

New *JHK* photometry of highly reddened stars yields a value for the colour excess ratio $E(J-H)/E(H-K)$ of 2.09±0.10, which is larger than previously reported values. There is no evidence for variation in this ratio over the sky.

131.002 **Iron hydrides formation in interstellar clouds.** A. Bar-Nun, M. Pasternak, P. H. Barrett.
Astron. Astrophys., Vol. 87, 328 - 329 (1980).

A recent Mössbauer study with ^{57}Fe in a solid hydrogen or hydrogen-argon matrix demonstrated the formation of an iron hydride molecule (FeH_2) at 2.5–5 K. Following this and other studies, the authors propose the possible existence of iron hydride molecules in interstellar clouds. In clouds, the iron hydrides FeH and FeH_2 would be formed only on grains, by encounters of H-atoms or H_2 molecules with Fe-atoms which are adsorbed on the grains. The other transition metals; Sc, Ti, V, Cr, Mn, Co, N; Cd and also Cu and Ca form hydrides of the type M–H, which could be responsible, at least in part, for the depletion of these metals in clouds.

131.003 **The first stars.** W. H. McCrea.
Irish Astron. J., Vol. 14, 41 - 49 (1979).

131.004 **Solar wind interaction with interstellar helium.** S. Grzedzielski.
Planet. Space Sci., Vol. 28, 799 - 802 (1980).

Solar wind interaction with neutral interstellar helium focused by the sun's gravity in the downwind solar cavity is discussed in a hydrodynamical approach.

131.005 **Survival probabilities for interstellar hydrogen flowing into the interplanetary system from far regions of the heliosphere.** J. A. Kunc.
Planet. Space Sci., Vol. 28, 815 - 821 (1980).

The expressions for "survival" probabilities are presented for an atomic hydrogen particle moving on a trajectory from far regions of the heliosphere to the vicinity of the sun. Three "destroying" processes have been considered; photoionization, charge transfer and electron ionization. The solar wind has been assumed to be a two-flux steady stream radially expanding with constant flow velocity. Recent profiles of solar-wind electron temperature have been used. The results can be useful for theoretical analyses as well as for analysis of spaceflight observations.

131.006 **Observations of HC_5N and HC_7N in Sgr B2 and Cloud 2.** L. F. Rodriguez, E. J. Chaisson.
Mon. Not. R. Astron. Soc., Vol. 192, 651 - 658 (1980).

The authors present observations of the $J = 3 \rightarrow 2$ transition of HC_5N and of the $J = 7 \rightarrow 6$ transition of HC_7N in Sgr B2 and Cloud 2. In Sgr B2 several of the low J lines of HC_5N are weak masers. The available data were satisfactorily fitted to a simple local thermodynamic equilibrium model with a column density of 1.6×10^{14} cm^{-2} and an excitation temperature of 15 K. HC_7N was not detected in Sgr B2. The $J = 7 \rightarrow 6$ transition of HC_7N measured in Cloud 2 gives a [HC_5N/HC_7N] abundance ratio of 7 ± 3.

131.007 **The collapse of interstellar gas clouds – V. On the stability of non-uniform collapse.**
D. McNally, J. J. Settle.
Mon. Not. R. Astron. Soc., Vol. 192, 917 - 930 (1980).

The stability of spherically symmetric free-fall collapse to small radial perturbations is examined for non-uniform clouds. It is concluded that fragmentation of the central region of a collapsing gas cloud is possible if: (1) the density distribution is sufficiently smooth; and (2) the collapse is nearly free fall.

131.008 **The collapse of interstellar gas clouds – VI. Formation of molecular hydrogen and its effect on the collapse of spherically symmetrical interstellar gas clouds.**
P. Barr, D. McNally.
Mon. Not. R. Astron. Soc., Vol. 192, 669 - 677 (1980).

The effect of formation of molecular hydrogen on the collapse of spherically symmetrical interstellar gas cloud is computed. It is shown that: (1) the volume dominated by molecular hydrogen accounts for 0.1 per cent of the total mass of the cloud; (2) the volume shielded from (but not dominated by molecular hydrogen) the background radiation field responsible for the photo-dissociation of molecular hydrogen accounts for $\gtrsim$ 10 per cent of the total cloud mass; (3) the formation of molecular hydrogen produces only marginal dynamical effects; and (4) the presence of molecular hydrogen permits collapse to be maintained under conditions of adequate cooling for a longer period than if it were totally absent.

131.009 **The interstellar spectrum of the central object of NGC 3603 (HD 97950).**
W. B. Somerville, J. C. Blades.
Mon. Not. R. Astron. Soc., Vol. 192, 719 - 723 (1980).

New observations are presented of interstellar features in the spectrum of HD 97950, the central object of NGC 3603. The authors confirm the detection by Walborn of very strong λ4430 but with central depth 17.5 ± 5 per cent rather than 29.6 ± 1.5 per cent.

131.010 **Interstellar chemistry of sulphur.** W. W. Duley, T. J. Millar, D. A. Williams.
Mon. Not. R. Astron. Soc., Vol. 192, 945 - 957 (1980).

To the present date, 10 molecules containing sulphur have been detected within dense clouds in the interstellar medium. In this paper, the authors show that, in the light of recent laboratory determinations of reaction rate coefficients, and observational and theoretical evidence on the depletion of heavy elements, the proposed ion–molecule schemes for the formation of these molecules are extremely inefficient in producing the observed species.

131.011 **IUE observations of N V, a diagnostic of hot interstellar gas.** L. J. Smith, T. W. Hartquist.
Mon. Not. R. Astron. Soc., Vol. 192, 73P - 77P (1980).

The authors report IUE observations of interstellar N V toward a number of Wolf-Rayet stars and compare the derived number density with that predicted by one theory of the interstellar medium. The observed N V/O VI ratio implies that most of the hot gas in the interstellar medium is at $T \cong 2.3 \times 10^5$ K.

131.012 **Detection of the $J=4\to 3$ transition of HCN at 354 GHz in the Orion Molecular Cloud.**
R. Padman, R. E. Hills, N. J. Cronin, W. B. Rose.
Mon. Not. R. Astron. Soc., Vol. 192, 87P - 91P (1980).

The $J=4\to 3$ transition of HCN at 354 GHz has been detected in the Orion Molecular Cloud. Both the "spike" and "plateau" sources are seen, and the physical parameters of these are derived from a model of the cloud.

131.013 **HCN absorption towards Cassiopeia A.**
P. J. Encrenaz, A. A. Stark, F. Combes, R. A. Linke, R. Lucas, R. W. Wilson.
Astron. Astrophys., Vol. 88, L1 - L3 (1980).

The HCN $J = 1\to 0$ lines have been observed in absorption in front of the continuum source Cas A. The spectrum exhibits several features corresponding to H I clouds in the Orion and Perseus arm. The lines are found to be optically thin. For an Orion arm cloud with $A_v \sim 1$ magnitude, the authors find a column density $N_{HCN} = 4\times 10^{11}\,cm^{-2}$.

131.014 **A unified model of interstellar grains: a connection between alignment efficiency, grain model size, and cosmic abundance.** S. S. Hong, J. M. Greenberg.
Astron. Astrophys., Vol. 88, 194 - 202 (1980).

It is shown, on the basis of a core-mantle model of interstellar grains, that the cosmic abundance constraints, the polarization to extinction ratio, and the wavelength dependences of extinction, linear polarization and circular polarization are all linked by the question of the strength of the alignment mechanism. One of the implications of the model is that, in order to obtain the best agreement with the principal observational features, the authors must invoke some enhancement of the Davis-Greenstein mechanism.

131.015 **Ammonia and cyanoacetylene observations of the high density core of L 183 (L 134 N).**
H. Ungerechts, C. M. Walmsley, G. Winnewisser.
Astron. Astrophys., Vol. 88, 259 - 266 (1980).

The NH_3 (1,1) transition has been mapped in the core of the nearby dark dust cloud L 183 (often called L 134 N) using the Bonn 100 m telescope. The authors also present observations at selected positions of the NH_3 (2,2) line and of the $J=1\to 0$, $F=2\to 1$ transition of HC_3N. The ammonia observations are used to derive a rotation temperature of 9 ± 1 K which appears to be uniform throughout the high density region where NH_3 (2,2) was observed. There is no evidence, either from the temperature estimates or from the line widths, for the presence of protostellar objects embedded within the cloud. The NH_3 (1,1) results, however, suggest that the core of L 183 consists of two condensations, which are in gravitational equilibrium and are presently quiescent. The authors discuss how such condensations might evolve.

131.016 **Grain growth by radiation pressure induced coagulation.** I. C. Simpson, S. Simons, I. P. Williams.
Astrophys. Space Sci., Vol. 71, 3 - 24 (1980).

A theoretical treatment is given of the growth of grains as a consequence of their mutual coagulation brought about by relative motions induced by radiation pressure. Analytical and numerical techniques are employed to tackle the relevant coagulation equation. It was found that in interstellar clouds, grains composed of iron, graphite and glassy carbon, being typical examples of three basic types of material, could grow to a size where condensation of volatiles was possible. On the other hand, olivine, a typical silicate, could not. If a source of radiation existed at the centre of the cloud, then growth could occur if the cloud was turbulent or if the density was high enough; otherwise the grains were driven out of the regions of interest at high velocity. In the latter case, with a high cloud density, re-radiation has to be taken into account.

131.017 **Comments on the existence of circumstellar clouds derived from interstellar Mg observations.**
D. L. Giaretta, B. Bates.
Astrophys. Space Sci., Vol. 71, 211 - 217 (1980).

From observations of interstellar Mg I, II resonance lines, Gurzadyan has proposed recently that for the majority of hot stars most of the line-of-sight gas originates in dense, circumstellar clouds. The authors have now considered the same data and have included some additional observations of interstellar Mg lines. They suggest that an empirical relationship between Mg II equivalent width and stellar effective temperature, which is central to the model proposed by Gurzadyan, may be explained by an observational selection effect. Further, they suggest that while circumstellar material may well contribute in part to observed column densities, there is no firm evidence that most of the gas is located in circumstellar clouds.

131.018 **Collisional excitation of interstellar molecules due to H_2: linear molecules CO, OCS, SiO, HCN, and HC_3N in $^1\Sigma$ state.** S. C. Mehrotra.
Astrophys. Space Sci., Vol. 71, 507 - 514 (1980).

Using a normalized perturbative, semi-classical approach, collision-induced rotational excitation rates of CO, OCS, SiO, HCN, HC_3N due to H_2 are computed. The calculated excitation rates for CO$-H_2$ and OCS$-H_2$ systems at 100 K are in good agreement with the results of close coupling approximation at low values of J, where J is the rotational quantum number. The rates are found to be very sensitive with respect to ortho and para states of H_2.

131.019 **An exceptional cold diffuse cloud.**
J. Crovisier, I. Kazès.
Astron. Astrophys., Vol. 88, 329 - 333 (1980).

The authors report a new determination of the H I spin temperature of a cloud at $l\sim 226°$, $b\sim 44°$, formerly studied by Knapp and Verschuur (1972). This determination is based on the observation of 21-cm emission and absorption profiles towards the background source 3 C 225.0 ($l\sim 220°$, $b\sim 44°$). The authors confirmed the cold spin temperature ($T_S\sim 20$ K) and failed to detect CO and OH lines in this cloud.

131.020 **Star formation and activity in the nuclei of barred galaxies.** T. M. Heckman.
Astron. Astrophys., Vol. 88, 365 - 366 (1980).

The data published by Heckman et al. (1980)on a large sample of normal galactic nuclei is used to show that recent detectable star formation has occurred more frequently in the nuclei of barred compared to non-barred disk galaxies. However, the presence or absence of a bar does not enhance the detectability of low-level nuclear activity (anomalous emission lines and/or flat spectrum radio sources).

131.021 **Protostars and pre-main sequence objects.**
I. Appenzeller.
Star formation, (see 012.001), p. 1 - 73 (1980).

The author shows that the basic facts of the final stages of the star formation process seem now to be - at least qualitatively - well understood on the basis of the hydrodynamic evolution of protostars. However, there are still many details which remain unsolved and the quantitative results are very uncertain. The agreement between the theoretical predictions and the observations is surprisingly good in view of the many simplifications of the theory. On the other hand, detailed comparisons are at present possible only in the case of the low mass protostars. Even there most of our "hard" knowledge is based on a very small sample of high resolution spectroscopic observations of T Tauri stars.

131.022 **Empirical information on star formation in galaxies.**
J. Lequeux.
Star formation, (see 012.001), p. 75 - 130 (1980).

The author discusses the initial mass function and the

global rate of star formation in galaxies and large regions inside a galaxy. A section examines in detail the empirical evidence on more detailed aspects of star formation (contagious star formation etc.) and another one deals with the consequences of the initial mass function and the rate of star formation on the evolution of galaxies.

131.023 **Molecular clouds and star formation.** J. Silk.
Star formation, (see 012.001), p. 131 - 222 (1980).

The author summarizes the key characteristics of molecular clouds and of star formation theories that are determined by observational considerations. He describes evolutionary models for the two classes, cold and warm, of molecular clouds. Ongoing star formation plays an essential role in these models, and successive sections are devoted to the theory of the fragmentation of molecular clouds and to the development of the initial stellar mass function. It is of interest to also consider the cosmological aspects of star formation. What can theory tell us about primordial star formation? And how does the rate of star formation vary with time? The principal uncertainty in developing a theory of galaxy formation centers on the inadequacies of our knowledge of star formation, and the final section is devoted to exploring how star formation theory may yield insight into galactic evolution.

131.024 **A model for gas phase chemistry in interstellar clouds: I. The basic model, library of chemical reactions, and chemistry among C, N, and O compounds.**
S. S. Prasad, W. T. Huntress, Jr.
Astrophys. J., Suppl. Ser., Vol. 43, 1 - 35 (1980).

A time-dependent gas phase chemical model of the chemistry in interstellar clouds is presented which uses a comprehensive library consisting of over 1400 reactions for 137 species. Specified constant density, temperature, and radiation field intensity are used to generate the evolution with time of these species for 10^7 years in four model clouds to simulate the outer and inner components of ζ Oph and of Orion. The C, N, and O families of chemical species are discussed in detail with particular emphasis on production and loss processes, and on the interrelationships among the chemistries of these families.

131.025 **Fragmentation of cosmic gas clouds due to thermal instabilities.** Y. Yoshii, Y. Sabano.
Publ. Astron. Soc. Japan, Vol. 32, 229 - 245 (1980).

Thermal evolution of gravitationally contracting gas clouds, which have various abundances of heavy elements, is investigated with regard to thermal stability of the condensation mode. It is shown that the thermal behaviour of a gas cloud in the regime of z (the ratio of the heavy element abundance to the solar one) less than 10^{-4} is essentially similar to that in the case of no heavy element, and that the heavy-element cooling brings about thermal instability in a wide range of parameters in the regime of z greater than 10^{-3}. The growth of the instability is followed by numerical computation and the results in the case of z=10^{-2} show that a density perturbation with a rather small mass of about 10 $M_\odot$ can grow rapidly to be bound by self-gravitation in a contracting gas cloud. A discussion is given that formation of population II objects is much activated according to the significant change of thermal conditions at z=10^{-3}.

131.026 **Molecule formation in shock waves from the galactic nucleus.** M. Saitō, S. Deguchi.
Publ. Astron. Soc. Japan, Vol. 32, 257 - 290 (1980).

The thermal and chemical history of a shock wave propagating from the galactic nucleus is calculated between 50 and 350 pc in the galactic plane. Molecules are formed in the cold shell of the wave, while H I gas exists mainly near the boundaries of the shell. The resulting longitude–velocity maps of H I, CO, OH, and H_2CO are compared with the expanding molecular ring observed at $|l| \lesssim 1^\circ.5$. The characteristic features of the molecular ring can be interpreted in terms of the shock-wave model.

131.027 **On the distribution of molecules in the vicinity of young B stars.** A. A. Rejtblat.
Pis'ma Astron. Zh., Tom 6, 509 - 512 (1980). In Russian. English translation in Soviet Astron. Lett., Vol. 6.

The distributions of the most abundant molecules are calculated depending on the distance from the star embedded in a dense cloud. C_2H and HCN molecules can be indicators of young B stars hidden from the observer by a thick layer of dust.

131.028 **Heating and cooling of dense interstellar clouds.** L. N. Arshutkin.
Astrometr. Astrofiz., Vyp. (No.) 41, p. 29 - 39 (1980). In Russian.

The mechanisms of dense interstellar clouds heating and cooling are discussed. The most important heating mechanisms are as follows: photoelectric effect on dust grains, cosmic ray ionization, H_2 formation on the surface of dust grains, chemical heating and dissipation of turbulence. Cooling by C^+ is dominating up to $N_p \sim 1.5 \times 10^{21}$ cm^{-2} in the outer layers of the cloud. In deeper layers the cloud cooling by interstellar grains and carbon monoxide molecules appears to be the most effective.

131.029 **The structure of interstellar hydroxyl masers: VLBI synthesis observations of W3(OH).**
M. J. Reid, A. D. Haschick, B. F. Burke, J. M. Moran, K. J. Johnston, G. W. Swenson, Jr.
Astrophys. J., Vol. 239, 89 - 111 (1980).

The authors analyzed data from 13 baselines and mapped W3(OH) in the $^2\Pi_{3/2}, J = {}^3/_2, F = 1{-}1$ transition of OH having a rest frequency of 1665.4018 MHz with a synthesized beam of 0″.01 and a spectral resolution of 0.14 km s^{-1}. The map shows 70 components, and has greatly increased sensitivity, dynamic range, and angular resolution compared to previous studies of this source. The study suggests a simple dynamical model for the masing region. It also offers evidence for the Zeeman effect in OH maser emission and provides new insights into the maser process.

131.030 **Structure and evolution of molecular clouds near H II regions. I. CO observations of an expanding molecular shell surrounding the Pelican Nebula.**
J. Bally, N. Z. Scoville.
Astrophys. J., Vol. 239, 121 - 136 (1980) = Five College Astron. Dep., Contrib. No. 314.

The paper reports ^{12}CO observations of a fragmented, expanding cloud network surrounding an old H II region, W80. Systematic velocity gradients and line splitting indicate the presence of a 3–6 $\times$ $10^4 M_\odot$ molecular shell expanding with a velocity V = 5 km s^{-1} from a point centered near the peak of the radio *f-f* emission. The authors interpret this feature as the remnant of the ionization shock front system from the evolving H II region which was formed off-center in the original molecular cloud. A simple model is constructed to follow the evolution of the shocked gas layer.

131.031 **Wave-wave interactions in a rotating gravitating gas cloud.** V. B. Bhatia, G. L. Kalra.
Astrophys. J., Vol. 239, 146 - 150 (1980).

Wave-wave interactions in a rotating gravitating gas cloud have been analyzed between the same or different wave modes. It is found that when the interacting waves resonate with the wave frequencies given by the linear theory the amplitude grows aperiodically. The growth of amplitude is fastest when both the wave modes given by the linear theory have identical wave frequencies and wave vectors so that resonance takes place between both these modes and the interacting waves.

131.032 **A model for gas phase chemistry in interstellar clouds. II. Nonequilibrium effects and effects of temperature and activation energies.**
S. S. Prasad, W. T. Huntress, Jr.
Astrophys. J., Vol. 239, 151 - 165 (1980).

The chemical evolution of diffuse and dense interstellar clouds is examined via the time-dependent model outlined in Paper I. The chemistry and the observational data pertaining to CH, CO, O_2, CH_4, CH_2O, CN, C_2, C_2H, HC_3N, and NH_3 are discussed and are shown to have significant implications regarding radiative association reactions and their temperature dependence, branching ratios in dissociative recombinations, and activation energies in neutral radical-radical reactions. The approach adopted in the present study illustrates the potential role of interstellar clouds as unique natural laboratories for addressing some fundamental aspects of chemical kinetics, in particular those which are not easily studied in terrestrial laboratories.

131.033 **Numerical calculations of the collapse of non-rotating, magnetic gas clouds.**
E. H. Scott, D. C. Black.
Astrophys. J., Vol. 239, 166 - 172 (1980).

The authors present results of the first self-consistent numerical calculations of the dynamic collapse of a magnetized protostellar gas cloud. Several simplifications have been incorporated into these calculations. The authors have calculated the time evolution of the clouds for roughly two initial free-fall times, at which point the central density has increased by a factor of $\sim 10^4$ to $\sim 10^6$. Several such calculations have been performed for different values of the cloud's initial thermal, magnetic, and gravitational energies.

131.034 **OB stars and the structure of the interstellar medium: cloud formation and effects of different equations of state.** T. M. Bania, J. G. Lyon.
Astrophys. J., Vol. 239, 173 - 192 (1980).

The effects of OB stars on the physical state of the interstellar medium have been investigated using a time-dependent numerical simulation which includes explicit two-dimensional hydrodynamics and implicit, coupled radiative transfer and thermodynamics. Using the observed birthrates, OB stars are placed randomly in space and time on an 180 pc square numerical grid where they ionize and push ambient gas for the duration of their main sequence lifetimes. Though they contribute a minuscule amount of ionization, the B stars are efficient excavators of ambient matter because their initial Strömgren radii are much smaller than the distance that the shock front around the H II region can travel in the stellar lifetime. In 10^7 years the superposition of the radially expanding H II region shock fronts about OB stars produces a clumped distribution of neutral clouds from an initially homogeneous and isothermal medium. These clouds are similar in many respects to the observed diffuse clouds.

131.035 **The evolution of refractory interstellar grains in the solar neighborhood.** E. Dwek, J. M. Scalo.
Astrophys. J., Vol. 239, 193 - 211 (1980).

The abundance of thermally condensed refractory grains and corresponding interstellar metal depletions are investigated using models for the evolution of the solar neighborhood. The calculations include a self-consistent treatment of red giant winds; planetary nebulae; protostellar nebulae and supernovae as sources of grains; star formation and encounters with supernova blast waves as sinks, for a variety of birthrate histories; chemical evolution models; and mass-loss parameters.

131.036 **Observations of CO in L1551: evidence for stellar wind driven shocks.**
R. L. Snell, R. B. Loren, R. L. Plambeck.
Astrophys. J., Lett., Vol. 239, L17 - L22 (1980).

CO observations reveal the presence of a remarkable, double-lobed structure in the molecular cloud L1551. The two lobes extend for ~ 0.5 pc in opposite directions from an infrared source buried within the cloud; one lobe is associated with the Herbig-Haro objects HH28, HH29, and HH102. The authors suggest that the CO emission in the double-lobed structure arises from a dense shell of material which has been swept up by a strong stellar wind from the infrared source.

131.037 **Evidence for a second class of interstellar extinction curves in the ultraviolet.**
A. N. Witt, R. C. Bohlin, T. P. Stecher.
Bull. American Astron. Soc., Vol. 12, 446 (1980). – Abstract.

131.038 **Physical properties of the dark cloud B5.**
J. S. Young, P. F. Goldsmith, W. D. Langer, E. Carlson, R. W. Wilson.
Bull. American Astron. Soc., Vol. 12, 446 - 447 (1980). Abstract.

131.039 **Turbulence and star formation in molecular clouds.**
R. B. Larson.
Bull. American Astron. Soc., Vol. 12, 447 (1980). – Abstract.

131.040 **Numerical simulations of the interstellar medium: OB stars and the formation and destruction of diffuse clouds.** T. M. Bania, J. G. Lyon,
Bull. American Astron. Soc., Vol. 12, 460 (1980). – Abstract.

131.041 **The spectrum and latitude dependence of the local interstellar radiation field.**
R. C. Henry, R. C. Anderson, W. G. Fastie.
Bull. American Astron. Soc., Vol. 12, 467 - 468 (1980). Abstract.

131.042 **Interstellar atomic and molecular abundances along the line of sight towards X Perseus.**
D. Lien, D. Buhl, R. M. Crutcher, B. Donn, A. M. Smith, L. E. Snyder, L. J. Stief.
Bull. American Astron. Soc., Vol. 12, 469 (1980). – Abstract.

131.043 **On the Zeta Ophiuchi diffuse interstellar cloud.**
R. M. Crutcher, W. D. Watson.
Bull. American Astron. Soc., Vol. 12, 469 (1980). – Abstract.

131.044 **Extreme ultraviolet/soft X-ray background: the temperature distribution of the emitting gas.**
R. Stern, F. Paresce.
Bull. American Astron. Soc., Vol. 12, 469 (1980). – Abstract.

131.045 **Evidence for shocked interstellar gas toward the Perseus OB2 association.** S. R. Federman.
Bull. American Astron. Soc., Vol. 12, 469 (1980). – Abstract.

131.046 **Carbon monoxide observations of molecular clouds associated with H II regions beyond the solar circle.**
J. R. Sewall, P. D. Jackson.
Bull. American Astron. Soc., Vol. 12, 482 (1980). – Abstract.

131.047 **Two CO surveys of the first galactic quadrant.**
R. S. Cohen, T. M. Dame, P. Thaddeus.
Bull. American Astron. Soc., Vol. 12, 482 (1980). – Abstract.

131.048 **The age of molecular clouds.**
T. M. Dame, R. S. Cohen, P. Thaddeus.
Bull. American Astron. Soc., Vol. 12, 483 (1980). – Abstract.

131.049 **The position of the hot molecular source in Orion.**
B. Zuckerman, P. Palmer, M. Morris.
Bull. American Astron. Soc., Vol. 12, 483 (1980). – Abstract.

131.050 **Methane emission at 1.6 cm in Orion A.**
D. E. Jennings, K. Fox.
Bull. American Astron. Soc., Vol. 12, 483 (1980). – Abstract.

131.051 **Quiescent molecular clouds near KL/BN.**
T. L. Wilson, C. M. Walmsley, G. Winnewisser, A. G. Kislyakov, P. Bastien.
Bull. American Astron. Soc., Vol. 12, 483 - 484 (1980). Abstract.

131.052 **Proper motion of H_2O maser sources: the outflow in Orion-KL and W51 Main.**
R. Genzel, M. J. Reid, J. M. Moran, M. H. Schneps, D. Downes, V. I. Kostenko, L. I. Matveyenko (*Matveenko*), B. Rönnäng.
Bull. American Astron. Soc., Vol. 12, 484 (1980). – Abstract.

131.053 **VLBI aperture synthesis observations of the OH maser source W75 N.**
A. D. Haschick, M. J. Reid, J. M. Moran, G. Miller.
Bull. American Astron. Soc., Vol. 12, 484 (1980). – Abstract.

131.054 **Observations of a second rotating interstellar cloud with a retrograding core.**
F. O. Clark, D. R. Johnson.
Bull. American Astron. Soc., Vol. 12, 484 (1980). – Abstract.

131.055 **Observations of the J = 17 - 16 transition of HC_5N.**
R. L. Snell, F. P. Schloerb, J. S. Young.
Bull. American Astron. Soc., Vol. 12, 484 - 485 (1980). Abstract.

131.056 **Detection of DC_3N in TMC 1.** W. D. Langer, F. P. Schloerb, R. L. Snell, J. S. Young.
Bull. American Astron. Soc., Vol. 12, 485 (1980). – Abstract.

131.057 **Observations of HC_5N at 18.6, 21.3, 24.0, and 26.6 GHz in Sgr B2, TMC-1, and IRC+10216.**
K. Fox, D. E. Jennings.
Bull. American Astron. Soc., Vol. 12, 485 (1980). – Abstract.

131.058 **New evidence for contraction in dark clouds.**
P. C. Myers.
Bull. American Astron. Soc., Vol. 12, 485 (1980). – Abstract.

131.059 **Dark cloud detections of HC_5N.**
P. J. Benson, P. C. Myers.
Bull. American Astron. Soc., Vol. 12, 485 (1980). – Abstract.

131.060 **Carbon monoxide observations of the W44 region.**
G. Chin, R. H. Cornett.
Bull. American Astron. Soc., Vol. 12, 506 (1980). – Abstract.

131.061 **Orion-KL water maser superflare.**
N. Cohen, S. Zisk.
Bull. American Astron. Soc., Vol. 12, 507 - 508 (1980). Abstract.

131.062 **The galactic foreground reddening in the direction of the Magellanic Clouds.**
D. H. McNamara, K. A. Feltz, Jr.
Bull. American Astron. Soc., Vol. 12, 538 (1980). – Abstract.

131.063 **Evidence for a large population of shocked interstellar clouds.** V. Radhakrishnan, G. Srinivasan.
J. Astrophys. Astron., Vol. 1, 47 - 66 (1980).

A 21 cm absorption measurement over a long path length free of the effects of differential galactic rotation indicates the existence of two distinct cloud populations in the plane. One of them consisting of cold, dense clouds has been well studied before. The newly found hot clouds appear to be at least five times more numerous. The authors propose that they are shocked clouds found only within supernova bubbles and that the cold clouds are found in the regions in-between old remnants, immersed in an intercloud medium. It is concluded that the solar neighbourhood must be located between old supernova remnants rather than within one.

131.064 **Highly-ionized species in the interstellar medium.**
J. H. Black, A. K. Dupree, L. W. Hartmann, J. C. Raymond.
Astrophys. J., Vol. 239, 502 - 514 (1980).

Interstellar absorption lines have been observed toward 25 stars with the International Ultraviolet Explorer satellite. Results are presented for observations of interstellar lines of C III, C IV, N V, Si III, and Si IV. Comparison with simple models of the ionization structures of H II regions around hot stars suggests that the observed column densities of Si III, Si IV, and C IV are in harmony with those expected in normal photoionized nebulae. Stellar wind velocities are presented for 21 of the stars observed in this survey of interstellar absorption lines.

131.065 **SiO emission in Orion–KL: an evolved star in a region of star formation or a unique object in the Galaxy?** R. Genzel, D. Downes, P. R. Schwartz, J. H. Spencer, V. Pankonin, J. W. M. Baars.
Astrophys. J., Vol. 239, 519 - 525 (1980).

The authors report observations of the $J = 1–0$ transition at 43 GHz in the $v = 0$, 1, and 2 states toward Orion–KL. They have also searched for SiO emission toward 27 regions of star formation.

131.066 **On the Taurus dust cloud toward 3C 123.**
R. M. Crutcher.
Astrophys. J., Vol. 239, 549 - 554 (1980).

The author has carried out observations of the interstellar D lines of Na I and the λ7699 line of K I in the spectra of stars behind the Taurus dark clouds. Ionization equilibrium calculations yield $n_e \gtrsim 10^{-2}\,cm^{-3}$; this value is so high that charged particle collisions will nearly thermalize the 1665 and 1667 MHz OH lines. He has made detailed calculations of the excitation of OH and CO toward 3C 123. A two-component model is required.

131.067 **A search for ultracold molecular gas in our Galaxy.**
N. J. Evans II, R. H. Rubin, B. Zuckerman.
Astrophys. J., Vol. 239, 839 - 843 (1980).

Existing galactic plane CO surveys are able to detect molecular clouds which contain substantial CO with an excitation temperature $\gtrsim 5$ K. The authors have tested the possibility that substantial numbers of molecular clouds exist where $T_{ex} \lesssim 5$ K. CO observations were obtained in the direction of 17 continuum sources which showed multiple H_2CO absorption features at a wavelength of 6 cm. The H_2CO features were taken as evidence of a molecular cloud, and CO observations were used to determine T_{ex} of that cloud. Although a few clouds may have T_{ex} slightly less than 5 K, a large amount of mass does not appear to be hidden in such ultracold molecular clouds with detectable H_2CO absorption.

131.068 **Detection of C_2H in cold dark clouds.**
A. Wootten, E. P. Bozyan, D. B. Garrett, R. B. Loren, R. L. Snell.
Astrophys. J., Vol. 239, 844 - 854 (1980).

The authors have surveyed a wide variety of interstellar clouds for the radical C_2H including a number of cold dust clouds in which star formation is not known to be occurring (yet). They have also examined small dust clouds with single infrared sources or Herbig-Haro objects. Definite detections of C_2H were made in 19 of the 30 regions examined with a further 3 possible detections. C_2H was detected in 8 cold dust clouds. The authors have made extensive maps of the unusual dust cloud L1534 (TMC 1) and of the M17SW star-formation region. Smaller maps of B227, L43, and L134N were also made. The L1534 and M17SW maps show that the C_2H distribution is well correlated with that of other molecular species in these clouds. The fractional abundance of C_2H in cold clouds is typically found to be $X(C_2H) \approx 6 \times 10^{-9}$.

131.069 **Photodissociation cross sections and rates for CH^+ in interstellar clouds.**
K. Kirby, W. G. Roberge, R. P. Saxon, B. Liu.
Astrophys. J., Vol. 239, 855 - 858 (1980).

Photodissociation cross sections have been calculated for transitions arising from the ground state of CH^+ to three excited states of ${}^1\Sigma^+$ and ${}^1\Pi$ symmetry accessible with photon energies $\lesssim 13.6$ eV. Two of these dissociation channels have large cross sections and are therefore particularly relevant to the destruction of CH^+ in the interstellar medium. Photodissociation rates in the interstellar radiation field for transitions into these three states are presented as a function of optical depth. In the unattenuated interstellar field these rates are several orders of magnitude larger than the photodissociation rate through the $A\,{}^1\Pi$ state.

131.070 **A stellar-wind focusing mechanism as an explanation for Herbig-Haro objects.**
J. Canto, L. F. Rodríguez.
Astrophys. J., Vol. 239, 982 - 987 (1980).

The steady state solution for the interaction of an isotropic stellar wind with a homogeneous, isothermal nebular environment is a spherical cavity, filled by unshocked stellar wind, and bounded by a shock. The radius of the shock is determined by the condition of balance between the ram pressure of the wind and the ambient gas pressure. In this paper the authors consider the interaction of an isotropic stellar wind with a nebular environment having a pressure gradient. This situation is expected to apply for recently formed stars in the edge of a dark cloud.

131.071 **Detection of deuterated cyanoacetylene in the interstellar cloud TMC 1.**
W. D. Langer, F. P. Schloerb, R. L. Snell, J. S. Young.
Astrophys. J., Lett., Vol. 239, L125 - L128 (1980) = Five College Astron. Dep., No. 432.

Deuterated cyanoacetylene has been observed for the first time in an interstellar cloud, TMC 1, by the detection of emission from its $J=5\rightarrow4$ transition at 42.2 GHz. Comparison of DC_3N with HC_3N implies that the abundance ratio DC_3N/HC_3N is larger than that observed for DCO^+/HCO^+. This larger enhancement of deuterium indicates that HC_3N in this cloud is formed by gas-phase ion-molecule reactions rather than by catalysis on grain surfaces or the dissociation of even longer molecules formed in stellar atmospheres.

131.072 **Detection of CO $J = 21 \rightarrow 20$ (124.2 μm) and $J = 22 \rightarrow 21$ (118.6 μm) emission from the Orion Nebula.** D. M. Watson, J. W. V. Storey, C. H. Townes, E. E. Haller, W. L. Hansen.
Astrophys. J., Lett., Vol. 239, L129 - L132 (1980).

The authors report the first observation of far-infrared molecular line emission from the interstellar medium. Strong emission from the $J = 21 \rightarrow 20$ and $J = 22 \rightarrow 21$ rotational transitions of carbon monoxide was detected in the Kleinmann-Low/shocked H_2 region of the Orion Molecular Cloud. The results imply that the region is optically thin in these lines and that much of the carbon is in the form of CO. This work also represents the first use of an antimony-doped germanium photoconductor in an astronomical application.

131.073 **Isotopic abundance ratios in interstellar carbon monosulfide.** M. A. Frerking, R. W. Wilson, R. A. Linke, P. G. Wannier.
Astrophys. J., Vol. 240, 65 - 73 (1980).

A survey of the relative abundances of the isotopic species, ${}^{12}C^{32}S$, ${}^{12}C^{34}S$, and ${}^{13}C^{32}S$ has been made in 14 dense molecular clouds through observations of the $J = 2-1$ rotational transitions at 3 mm.

131.074 **The formation of elephant-trunk globules in the Rosette nebula: CO observations.**
M. H. Schneps, P. T. P. Ho, A. H. Barrett.
Astrophys. J., Vol. 240, 84 - 98, plates 2 - 3 (1980).

The prominent elephant-trunk globules in the northwest quadrant of the Rosette nebula have been observed in the microwave lines of CO and ${}^{13}CO$ ($J = 1\rightarrow 0$). The CO emission closely follows the optical outline of the obscuring material and leaves little doubt that the emission is associated with the globules. The physical characteristics derived are typical of those observed in other dust globules which are not necessarily associated with H II regions. The globules are observed to be blueshifted by $\sim$ 17 km s^{-1} with respect to the mean rest velocity of the H II region. The globules also show a systematic radial velocity gradient of $\sim$ 1.1 km s^{-1} arcmin^{-1} along the lengths of the elephant-trunks. The authors suggest that the formation of the elephant-trunk globules is the result of stellar-wind action in the H II region. The dense shell swept up by these winds can lead to globule formation either through a Raleigh-Taylor instability or through clumping in the ambient medium. Models show that either mechanism can account for the observed velocity gradients with reasonable values for the parameters.

131.075 **Far infrared observations of the globule B335.**
J. Keene, D. A. Harper, R. H. Hildebrand, S. E. Whitcomb.
Astrophys. J., Lett., Vol. 240, L43 - L46 (1980).

The authors have observed far-infrared and submillimeter continuum emission from the dark globule B335. The spectrum is that of optically thin thermal emission from dust with temperature 13–16 K. The estimated dust mass is 0.07–0.17 $M_\odot$. The observed luminosity, 5 $L_\odot$, can be supplied by the interstellar radiation field.

131.076 **Evidence for Mg and (Fe, Si) phase separation in interstellar dust.** W. W. Duley.
Astrophys. J., Lett., Vol.240, L47 - L51 (1980).

Analysis of the depletions of Mg, Si, and Fe in diffuse clouds shows that (1) distinct Mg-rich and (Fe, Si)-rich dust components exist in diffuse clouds; (2) it is unlikely that more than 65% of interstellar Mg is ever contained in silicates; and (3) depletion in diffuse clouds may occur by selective accretion on (Fe, Si)-oxide cores.

131.077 **On the nature of inhomogeneities of the interstellar medium.** I. V. Chashej, V. I. Shishov.
Pis'ma Astron. Zh., Tom 6, 574 - 578 (1980). In Russian. English translation in Soviet Astron. Lett., Vol. 6.

A qualitative interpretation of the interstellar medium inhomogeneities spectrum following from the observations of pulsar scintillations is carried out. It is shown that the large-scale ($\sim 10^{13}$ cm) inhomogeneities cannot be connected with waves in the interstellar plasma. On the contrary, one can explain the small-scale ($\sim 10^{10}$ cm) inhomogeneities by Alfvén and magnetosonic waves.

131.078 **The 3.4-μm interstellar absorption feature.**
D. T. Wickramasinghe, D. A. Allen.
Nature, Vol. 287, 518 - 519 (1980).

The authors report the detection of a 3.4-μm absorption feature in three IR sources and hence argue that an appreciable mass fraction of interstellar grains are predominantly organic in character.

131.079 **Ammonia observations of the molecular clouds near S68, S140, OMC2 and S106.** L. T. Little, A. T. Brown, G. H. Macdonald, P. W. Riley, D. N. Matheson.
Mon. Not. R. Astron. Soc., Vol. 193, 115 - 128 (1980).

The $J = 1, K = 1$ and $J = 2, K = 2$ transitions of interstellar ammonia have been observed in the molecular clouds near S68, S140, OMC2 and S106. Maps of the ammonia emission obtained with a 2.2-arcmin beam are presented and compared with observations of other interstellar molecules, in particular carbon monoxide and formaldehyde. The densities of hydrogen molecules derived from the ammonia observations on the basis of a simple uniform-density model for the source are much lower than those obtained from 2-mm formaldehyde observations. This discrepancy may be resolved either by assuming a 'core–halo or a 'clumped' structure for the source.

131.080 **The nature of the Kleinmann–Low nebula.**
J. P. Phillips, J. E. Beckman.
Mon. Not. R. Astron. Soc., Vol. 193, 245 - 260 (1980).

Recent data on, and interpretations of the Kleinmann–Low nebula are examined. It is suggested that associated high velocity gas is radiatively accelerated by a protostar in an early stage of pre-main-sequence contraction.

131.081 **Interstellar reddening towards the south galactic pole.**
R. Albrecht, H. M. Maitzen.
Astron. Astrophys., Suppl. Ser., Vol. 42, 9 - 13 (1980).

90 B, A and F type stars were observed in an area of about 400 square degrees around the galactic south pole using the Strömgren-Crawford *uvbyβ*-system. The authors derived intrinsic $(b\text{-}y)_0$ colors. The most plausible way of determining the galactic extinction perpendicular to the galactic plane seems at present to average the reddening values of distant stars in order to consider both the reddened and unreddened areas at the SGP. This way, a mean value $E(b\text{-}y)_{SGP} = 0\overset{m}{.}019$ is derived.

131.082 **Structure of molecular clouds: I. Observational constraints and CO line formation.**
L. G. Stenholm.
Astron. Astrophys., Suppl. Ser., Vol. 42, 23 - 41 (1980).

The present status of molecular line observations is discussed. A classification of molecular clouds is suggested: type I shows little or no indication of star formation and type II has an internal H II-region. Theoretical calculations of ^{12}CO and ^{13}CO are made to show how variations in the cloud structure influence the observed quantities.

131.083 **The effect of turbulent viscosity on stability and collapse of a rotating protostellar cloud.**
O. Regev, G. Shaviv.
Astron. Astrophys., Vol. 89, 61 - 66 (1980).

The authors investigate the effect of turbulent viscosity in two phases of the generally accepted picture of star formation: the onset of instability in an interstellar cloud, and the formation of toroidal structures during the collapse. They present a linear stability analysis for an idealized, uniform, infinite, isothermal medium with uniform rotation and include in it the effects of turbulent viscosity on the perturbation growth scale and rate. They find that the turbulent viscosity introduces a characteristic scale length for the gravitationally unstable subregions. The authors find also that the turbulent transport of angular momentum during the collapse of a rotating cloud is likely to reduce the probability of ring formation.

131.084 **Contributions to the theory of spiral structure. IV. The propagation of sound waves in an inhomogeneous interstellar medium.** T. Schmidt-Kaler, R. Wiegandt.
Astron. Astrophys., Vol. 89, 67 - 69 (1980).

Taking into account the presence of a very hot gas component ($T \cong 7 \times 10^5$ K) in the complex interstellar medium, the authors calculate the effective speed of sound in a binary gas-mixture. It is found that in the long wave approximation, valid for spiral density waves, the sound velocity increases to a mean value of about 16 km s^{-1}. The resulting effective speed of sound leads to a kinematical lifetime of spiral waves of 6×10^8 yr and provides the possibility of strong shocks even in the presence of an extended very hot gas component.

131.085 **The distribution of the interstellar dust in the galactic plane within 3 kpc.** J. Krautter.
Astron. Astrophys., Vol. 89, 74 - 79 (1980).

Interstellar polarization data have been used to derive the spatial distribution of the interstellar dust within the galactic plane. The observed distribution shows a strongly irregular and cloudy structure. The correlation between the observed dust distribution and the spiral arm indicators (young open star clusters and *R*-associations) is found to be relatively poor.

131.086 **Modeling of diffuse interstellar clouds: the case of Gamma Arae.** S. R. Federman, A. E. Glassgold.
Astron. Astrophys., Vol. 89, 113 - 117 (1980).

Observations of the interstellar cloud toward γ Arae are analyzed with an improved version of the isobaric, steady-state model of Glassgold and Langer (1974). Good agreement is obtained using parameters which are close to those which represent the average properties of many diffuse interstellar clouds. The successful correlation of the data for H, H_2, C II, C I, O I, CO and OH supports current concepts on interstellar thermodynamics and chemistry.

131.087 **Formaldehyde as a probe of dark clouds.**
Aa. Sandqvist, C. Bernes.
Astron. Astrophys., Vol. 89, 187 - 197 (1980).

The authors have mapped the 6-cm, 2-cm, and 2-mm H_2CO lines in three dark clouds (L1551, L1317, and the cloud complex near NGC 7023) in order to investigate the physical conditions in these regions. The kinetic temperature and the density increase from the periphery to the central regions of L1551. Emission in the 2-mm H_2CO line has been observed from a region near an embedded 2 μm source. The authors find a maximum H_2 density of 4×10^4 cm^{-3} for this region. The observations of L1317 show this cloud to be divided into two components with different radial velocities. Hyperfine analysis has been performed on a high-quality 2-cm H_2CO profile obtained towards the dark cloud L134.

131.088 **Studies of the Carina Nebula. II. The extinction law in the direction of 14 O-type stars.**
P. S. Thé, R. Bakker, H. R. E. Tjin A Djie.
Astron. Astrophys., Vol. 89, 209 - 213 (1980).

UBV, RI, JHKLM photometry of 14 O-type stars located in the direction of the Carina Nebula have been analysed in connection with the anomalously high values of the ratio of total to selective extinction (*R*). The authors' analysis shows that the extinction laws in the directions of the 14 stars are anomalous and are characterized by a mean value of $\bar{R} = 3.9 \pm 0.1$.

131.089 **Nonlinear hydrodynamics of acoustic instabilities in diffuse clouds.** S. L. W. McMillan, B. P. Flannery, W. H. Press.
Astrophys. J., Vol. 240, 488 - 498 (1980).

Using a linearized analysis, Flannery and Press demonstrated the existence of an ionization-coupled acoustic instability in the thermally stable phase of diffuse cold clouds in the ISM (interstellar medium); in this paper the authors describe a series of one-dimensional, time-dependent, hydrodynamic calculations of the nonlinear development of that instability. They find two qualitatively different types of behavior, depending on the initial temperature of the slightly perturbed (instable) equilibrium model.

131.090 **Neutral hydrogen emission-absorption observations of high-velocity clouds.**
H. E. Payne, E. E. Salpeter, Y. Terzian.
Astrophys. J., Vol. 240, 499 - 513 (1980).

Neutral hydrogen emission-absorption studies were carried out at Arecibo Observatory for some "high-velocity clouds" with velocity $|\nu| > 50$ km s^{-1}. Conspicuous, narrow absorption features are rare but were found in a few cases. In addition to these few confirmed features, statistical tests indicate the presence of more widespread weak absorption. Harmonic mean spin temperatures of a few hundred K are obtained from averages over velocity and sky position for the high-velocity material, comparable to spin temperatures for low-velocity local clouds of equally small column density.

131.091 **High angular resolution observations of CS in the Orion Nebula.** P. F. Goldsmith, W. D. Langer, F. P. Schloerb, N. Z. Scoville.
Astrophys. J., Vol. 240, 524 - 531 (1980).

The authors report on high spatial resolution (33″) observations of the CS $J = 3 \rightarrow 2$ transition in the Orion molecular cloud. These observations are well suited to distinguish among the possible density models since the CS $J = 3 \rightarrow 2$ transition is an excellent tracer of density and the resolution obtained in this transition by the Five College Astronomy Observatory 14 m antenna is fine enough to resolve clumping at the scale suggested by the analysis of some molecules (H_2CO, NH_3). During the course of the Orion mapping, the authors discovered high-velocity CS emission from the plateau source: these results are discussed.

131.092 **Models of molecular clouds and the abundances of H_2CO and HCO^+.**
A. Wootten, R. Snell, N. J. Evans II.
Astrophys. J., Vol. 240, 532 - 546 (1980).

Observations of HCO^+ and H_2CO in a sample of 13 molecular clouds have been analyzed by construction of uniform, spherical cloud models. The total densities and the abundance of HCO^+ and H_2CO relative to H_2 which result from these models fall into two domains: one group of clouds has a low temperature, moderate density, and high abundances; the other group has higher temperature and density, but lower abundances. The factor distinguishing these groups may be depletion onto grains in the denser sources.

131.093 **Observations of H_2 emission from NGC 7538.**
J. Fischer, G. Righini-Cohen, M. Simon, R. R. Joyce, T. Simon.
Astrophys. J., Lett., Vol. 240, L95 - L98 (1980).

Observations of the molecular hydrogen emission at 2 μm in NGC 7538 are presented. A map of the $\nu = 1-0\ S(1)$ line obtained at 34″ resolution indicates that the peak of the molecular line emission occurs between the interface of the visible H II region and the dense molecular cloud and the infrared cluster within the molecular cloud. The apparent luminosity in the $S(1)$ line of $\sim 1.2\ L_\odot$ is similar to that measured for Orion, $\sim 2.5 L_\odot$, before correction for extinction by overlying dust.

131.094 **Detection of the 157 micron (1910 GHz) [C II] emission line from the interstellar gas complexes NGC 2024 and M42.**
R. W. Russell, G. Melnick, G. E. Gull, M. Harwit.
Astrophys. J,, Lett., Vol. 240, L99 - L103 (1980).

The authors present the first detection of the [C II] fine-structure emission line at a wavelength of 157 μm. The [C II] line strengths are 7.1×10^{-16} and 1.0×10^{-15} W cm^{-2}, respectively, in NGC 2024 and M42. The line-to-continuum ratio is higher in NGC 2024 where the continuum is 7.0×10^{-16} W cm^{-2} μm^{-1}, in contrast to M42 where it assumes a value of 2.6×10^{-15} W cm^{-2} μm^{-1}. The respective luminosities in the line are $\sim$50 and 80 $L_\odot$. The observations were obtained with a stressed Ge:Ga photoconductor.

131.095 **Star formation in a galactic wind.**
A. C. Fabian, P. E. J. Nulsen, G. C. Stewart.
Nature, Vol. 287, 613 - 614 (1980).

Malin and Carter (1980) have discovered giant shells of faint optical emission around several normal elliptical galaxies. The authors suggest here that these shells are composed of stars formed in shocked galactic wind material.

131.096 **Evidence for the existence of a low-velocity molecular cloud near Sgr A.**
J. B. Whiteoak, F. F. Gardner.
Proc. Astron. Soc. Australia, Vol. 3, 266 - 269 (1978).

131.097 **Dust and gas correlations in the region of the South Celestial Pole.** D. J. King, K. N. R. Taylor.
Proc. Astron. Soc. Australia, Vol. 3, 315 - 316 (1979).

131.098 **Energy shifts in molecular hydrogen on grain surfaces.** M. I. Darby, S. Papaconstantinopoulou, K. N. R. Taylor.
Proc. Astron. Soc. Australia, Vol. 3, 317 - 319 (1979).

The spectrophotometric results from the Copernicus satellite have stimulated considerable interest in interstellar molecular hydrogen. Calculations are presented which suggest that it may be possible for hydrogen molecules on grains to give rise to diffuse interstellar lines in the ultraviolet.

131.099 **The chemical role of grains in the interstellar medium and related physical problems.**
D. A. Williams.
J. Phys. Colloq., Vol. 41, No. C-3, (see 012.010), p. C3/225 - 232 (1980). – Abstr. in Phys. Abstr., Vol. 83, Abstr. 82470 (1980).

131.100 **Observations and analyses of interstellar cyanogen.**
E. Churchwell.
Astrophys. J., Vol. 240, 811 - 827 (1980).

New observations of CN (113.5 GHz) are reported for eight locations in the Taurus dark cloud complex and 11 molecular clouds associated with H II regions including partial maps of CN toward DR 21 and Sgr A. CN column densities are estimated for 13 clouds. Using independent estimates of the total gas density and kinetic temperature, the fractional abundance of CN versus gas density is considered in these clouds. The correlations of CN with NH_3 and HC_5N in the Taurus clouds are discussed. Finally, the structure and kinematics of the DR 21 molecular cloud are discussed on the basis of the CN observations.

131.101 **Interstellar gas in the Gum Nebula.**
G. Wallerstein, J. Silk, E. B. Jenkins.
Astrophys. J., Vol. 240, 834 - 845 (1980).

The authors have surveyed the interstellar gas in and around the Gum Nebula by optically observing 67 stars at Ca II, 42 stars at Na I, and 14 stars in the ultraviolet with the Copernicus satellite. Radial velocities and column densities for all resolved absorption components are given. Velocity dispersions for gas in the Gum Nebula, excluding the region of Vela remnant filaments, are not significantly larger than in the general interstellar medium. The ionization structure is predominantly that of an H II region with moderately high ionization. Clouds in the Gum Nebula do not show the anomalously high ionization seen in the Vela remnant clouds. The observational data are generally consistent with a model of the Gum Nebula as an H II region ionized by OB stars and stirred up by multiple stellar winds.

131.102 **Star formation in W3 and W4: discovery of 135 possibly embedded near-infrared stars.**
D. M. Elmegreen.
Astrophys. J., Vol. 240, 846 - 852, plates 9 - 10 (1980).

The W3-W4 region (IC 1795 and IC 1805) is an area of extensive star formation in our Galaxy. A 2° X 3° field including W3 and most of W4 has been surveyed in the near-infrared;

137 stars were found to be much brighter in the near-infrared than at red wavelengths. The locations of these stars are compared with CO and far-IR peaks and to optically prominent areas such as bright rims and emission regions. The most probable spectral types and the extinctions are given.

131.103 **Redox reactions and the optical properties of interstellar grains.** W. W. Duley.
Astrophys. J., Vol. 240, 950 - 955 (1980).

Some basic redox reactions leading to chemical changes in grain material are investigated. The role of both chemical reactions on grain surfaces and photochemical processes within grains on the oxidation state of grain material is examined. To illustrate these processes, the relationship between Fe, FeO, Fe_3O_4, and Fe_2O_3 in the interstellar medium is discussed. Finally, it is shown that such diverse effects as irregularities in visual extinction in H II regions, the variability of λ_{max}, the wavelength of maximum linear polarization, and the ultraviolet properties of dust in regions of nebulosity may be understood in terms of chemical changes in grain material.

131.104 **Anisotropic mass outflow in Cepheus A.** L. F. Rodríguez, P. T. P. Ho, J. M. Moran.
Astrophys. J., Lett., Vol. 240, L149 - L152 (1980).

The authors report the detection of high-velocity wings on the CO profile observed toward Cepheus A, a region of star formation in the Cepheus OB3 association. The velocity extent of the emission is ~ 50 km s^{-1}. The blueshifted and redshifted wings have peaks of emission which are spatially separated by $\sim 90''$ (0.3 pc), suggesting an anisotropic outflow in two opposite directions.

131.105 **2 centimeter H_2CO emission in the ρ Ophiuchi cloud.** R. B. Loren, A. Wootten, Aa. Sandqvist, C. Bernes.
Astrophys. J., Lett., Vol. 240, L165 - L169 (1980).

Observations of a 2 cm H_2CO emission line in the ρ Oph cloud have revealed a very dense cold core not associated with known signposts of star formation. The density of this region is $n \gtrsim 10^6$ cm^{-3}, and it has $A_V \approx 200$ mag and a total mass of $100 M_\odot$ within a radius of 0.08 pc. The temperature of the region is low, probably $T_K \lesssim 20$ K. Higher temperatures deduced from the CO line probably originate in a less dense, warmer region on the far side of the cloud. The high core mass and the velocity structure in the CO line suggest that this region is a propitious site for future star formation, while its low temperature suggests a current lack of high-luminosity internal heating sources.

131.106 **Studies of high-velocity clouds. I. A high-sensitivity survey.** R. Giovanelli.
Astron. J., Vol. 85, 1155 - 1181 (1980).

The results of the first part of a high-sensitivity, high-velocity survey in the 21-cm transition of H I are presented. Over 6000 spectra have been collected, with an rms noise of less than 0.01 K and usable velocity coverage between about -900 and $+900$ km s^{-1}. Many previously unreported features have been found, including three objects believed to be nearby galaxies. On the basis of the findings of the survey, the large-scale properties of the high-velocity gas are analyzed.

131.107 **On the accuracy and precision of extinctions derived from general star counts.** G. S. Rossano.
Astron. J., Vol. 85, 1218 - 1228 (1980).

The accuracy and the applicability of the method of general star counts are discussed. The effects of statistical errors, pseudorandom errors, systematic errors, and sampling errors are considered. The conditions under which quantitatively useful results can be derived are identified. It is found that for plates reaching 17th to 20th mag, general star counts can be a useful technique for investigating the structure of dark cloud complexes located within 500 to 1000 pc of the sun.

131.108 **Structure of molecular clouds. II. Clouds without prominent star formation.** L. G. Stenholm.
Astron. Astrophys., Vol. 89, 264 - 271 (1980).

The available observations (mainly CO line observations) are used to derive the probable structure of molecular clouds which do not have prominent star formation. The probabilities for turbulence, collapse and random motion models are discussed. It is concluded that no definite choice can be made, but it is possible to strongly limit the structural freedom in each model.

131.109 **CO observations in galactic clouds.** A. R. Gillespie.
Messenger, No. 21, p. 20 - 22 (1980).

131.110 **Studies of ultraviolet interstellar extinction with the Sky-Survey telescope of the TD-1 satellite. Results for different OB-associations.** C. Morales, F. Llorente de Andrés, J. A. Ruiz del Arbol, J. Pérez Molla.
Astron. Astrophys., Suppl. Ser., Vol. 42, 155 - 161 (1980).

From a sample of about 100 reddened early type stars, a mean interstellar extinction law in the wavelength range from 2740 to 1380 Å has been derived by means of a grid of ultraviolet intrinsic colours. The ultraviolet spectra were obtained with the Sky-Survey telescope (S2/68) in the TD-1 satellite.

131.111 **Formaldehyde in the galactic center region: observations.** J. Bieging, D. Downes, T. L. Wilson, A. H. M. Martin, R. Güsten.
Astron. Astrophys., Suppl. Ser., Vol. 42, 163 - 176 (1980).

The galactic center region has been mapped in formaldehyde absorption at 4.8 GHz with the Effelsberg 100-m telescope. The data are presented in the form of longitude-velocity diagrams and as spatial maps of apparent optical depth within selected velocity intervals. Detailed maps are given of the molecular clouds near Sgr A, Sgr B2, and the continuum *Arc* near $l = 0.2°$. A map of the $H_2{}^{13}CO$ isotope over Sgr B2 shows that the apparent $^{12}C/^{13}C$ ratio is roughly constant at all positions over the source.

131.112 **The spatial distribution of the interstellar extinction.** T. Neckel, G. Klare.
Astron. Astrophys., Suppl. Ser., Vol. 42, 251 - 281 (1980).

Extinction values and distances have been computed from *UBV*, MK and β data for more than 11 000 O to F stars, including 7 565 O and B stars. For 1 020 stars two independent distance moduli were derived using M_v (MK) as well as M_v (β). The mean value of their differences is less than 0^m01. With the aid of photographs of the Milky Way 325 fields in the galactic belt $|b| \leq 7.^\circ6$ were demarcated in which the extinction and the star density is rather homogeneous. The $A_v(r)$ diagrams of these fields are discussed. From the fields with the most reliable $A_v(r)$-relations the galactic distribution of the dust up to 3 kpc has been derived.

131.113 **Identification of the anomalous 26.131-MHz nitrogen line observed towards Cas A.** D. H. Blake, R. M. Crutcher, W. D. Watson.
Nature, Vol. 287, 707 - 708 (1980).

Konovalenko and Sodin (1979) reported the detection of an absorption line at a frequency of 26.131 MHz towards Cassiopeia A and attributed the line to the 26.127 MHz $F = 5/2 \rightarrow 3/2$ hyperfine transition of ^{14}N. The authors suggest here that the line is a 631 α recombination line due to neutral atoms of heavy elements – either carbon atoms in the H I clouds or possibly heavier atoms (Mg, Ca, Fe) in hot ($\sim 10^4$ K) ionized gas along the line of sight to Cas A.

131.114 **Organic material and the 2.5–4 μm spectra of galactic sources.** F. Hoyle, N. C. Wickramasinghe.
Astrophys. Space Sci., Vol. 72, 183 - 188 (1980).

The spectra of galactic infrared sources over the

waveband 2.5–4 μm provide clear evidence for an organic composition of interstellar grains.

131.115 **Dry polysaccharides and the infrared spectrum of OH 26.5 + 0.6.** F. Hoyle, N. C. Wickramasinghe.
Astrophys. Space Sci., Vol. 72, 247 - 249 (1980).
The spectrum of the infrared source OH 26.5 + 0.6 over the waveband 2–40 μm is interpreted on the basis of a model involving emission and absorption of radiation by anhydrous cellulose grains.

131.116 **Electromagnetic effects on hyperbolic cosmic dust particles.** G. E. Morfill, E. Grün.
Solid particles in the solar system, (see 012.019), p. 313 - 314 (1980). – Abstract.

131.117 **A Monte Carlo simulation of the mass distribution in an accreting system of dust particles.**
P. A. Daniels, D. W. Hughes.
Solid particles in the solar system, (see 012.019), p. 325 - 328 (1980).
The paper investigates the way in which the mass distribution index, s, of the particles in a dust cloud varies as a function of time and particle mass, assuming that accretion, and only accretion, takes place after interparticle collisions.

131.118 **H II bubbles and disruption of molecular clouds.** T. J. Mazurek.
Astron. Astrophys., Vol. 90, 65 - 69 (1980).
The possibility that molecular clouds are disrupted by expanding H II regions within them is examined. The results show that the self-gravity of the dense shell that is swept up can halt the expansion if the cloud is sufficiently massive. For hydrostatic clouds the mass required to prevent disruption by the expansion of the H II region of a typical star is $\sim 7 \cdot 10^4$ $M_\odot$. This upper mass limit for disruption drops dramatically if the molecular clouds are in free-fall. However, indications exist that the clouds are not in free-fall, and this mode of cloud disruption could account for the absence of CO emission in older open clusters, the configuration of the region around λOri, and observed H I shells.

131.119 **CO observations of interstellar clouds: isotopic ratios.** F. Combes, E. Falgarone, J. Guibert, Nguyen-Q-Rieu.
Astron. Astrophys., Vol. 90, 88 - 96 (1980).
The authors have observed the $J = 1 \rightarrow 0$ transition of carbon monoxide and its isotopic species in a variety of interstellar clouds, in order to study the correlation of the ratio $^{12}CO/^{13}CO$ with the physical conditions. Kinetic temperatures and hydrogen densities, required to derive the CO column densities, are provided to some extend by available H I, OH, and H_2CO observations of the same clouds. The large uncertainties inherent in the interpretation of the CO lines are therefore reduced.

131.120 **On the motion and destruction of grains in interstellar clouds.** O. Havnes.
Astron. Astrophys., Vol. 90, 106 - 112 (1980).
The author has studied the possibilities of a two stream instability when charged grains stream through the post shock gas of a shock moving through interstellar gas. The condition for the onset of such an instability is found. The results indicate that small grains of radius less than a few times 10^{-6} cm are brought to rest with respect to the gas in a very short distance while larger grains are practically unaffected by this and other braking mechanisms.

131.121 **Structure of molecular clouds. III. Effects of MHD waves in collapsing fragments.**
G. E. Morfill, L. G. Stenholm.
Astron. Astrophys., Vol. 90, 134 - 139 (1980).
The authors have calculated the signature of early protostars in CO line emission, taking into account the possibility that MHD waves may be generated in the differentially rotating core. They obtain the result that the CO line shape of unresolved protostars is broadened, corresponding to supersonic but subalfvenic turbulent velocities, and that it has extended low intensity wings.

131.122 **A study of cold hydrogen in the dark cloud Lynds 134.**
A. Winnberg, M. Grasshoff, W. M. Goss, R. Sancisi.
Astron. Astrophys., Vol. 90, 176 - 183 (1980).
The dark cloud Lynds 134 has been mapped in the 21-cm H I line using the 100-m telescope at Effelsberg and the 64-m telescope at Parkes. Two narrow self-absorption lines have been detected: Comp. *A* at 2.7 km s^{-1} and Comp. *B* at 0.7 km s^{-1}.

131.123 **Ultraviolet, visible and infrared optical depths inside an interstellar dust cloud.** A. Aiad.
Proc. Math. Phys. Soc. Egypt, No. 43, p. 185 - 191 (1977).
Mean absorption values are used with a standard interstellar cloud model to estimate the ultraviolet and visual optical depths inside an interstellar dust cloud. A method is described for the computation of the infrared optical depth inside the cloud for temperatures not much larger than 100°K.

131.124 **Interstellar helium in the earth's orbit.** M. A. Gruntman.
Inst. kosm. issled. AN SSSR. Prepr., 1980, No. 543, 22 pp. In Russian. – Abstr. in Ref. zh., 51. Astron., 9.51.620 (1980).

131.125 **Contraction of massive interstellar clouds and the formation of stars.**
A. Aiad, M. El-Shalaby, I. Ahmed.
J. Astron. Soc. Egypt, Vol. 1, 7 - 40 (1979).
The contraction of a gas cloud of about 600 $M_\odot$ was numerically studied. After a contraction time of 5.3×10^{13} sec the temperature starts to rise from a minimum value of 4 K in the center. The importance of both molecular and atomic cooling as well as formation of molecules is demonstrated.

131.126 **Formation of amorphous grains in cosmic environments.** J. Seki, H. Hasegawa.
13th Lunar and Planetary Symposium, (see 012.018), p. 201 - 209 (1980).
The authors investigate formation processes of amorphous dust grains through their condensation theoretically. The smaller the grain size is and the lower the pressure is and the shorter the cooling time is, the higher the probability of grains forming as amorphous is. Iron condenses as crystalline grains in astrophysical circumstances. As for silicates, grains can be either amorphous or crystal according to materials, grain sizes and circumstances of grain formation sites.

131.127 **Statistical evidence of absorption at high latitudes.** B. I. Fesenko.
Astron. Zh., Tom 57, 953 - 958 (1980). In Russian. – English translation in Soviet Astron., Vol. 24, No. 5.
Evidence is considered which indicates the significant effect of irregular interstellar absorption at high latitudes *b*. The number density of faint galaxies grows with increasing $|b|$ even at values of $|b|$ exceeding 50°. The effects of the interstellar medium are traced even in the directions of stars and globular clusters with very low values of the colour excess. The coefficient of absorption, 0.29 ± 0.05, was estimated from the colours of bright E-galaxies.

131.128 **Interstellar shock waves.** C. F. McKee, D. J. Hollenbach.
Annu. Rev. Astron. Astrophys., Vol. 18, (see 003.004), 219 - 262 (1980).

The authors focus on the structure of interstellar shocks, not on the propagation. Contents: Jump conditions, structure of shock fronts, nonradiative shocks, radiative atomic shocks, molecules in shock waves, current problems and applications.

131.129 **Nuclear abundances and evolution of the interstellar medium.** P. G. Wannier.
Annu. Rev. Astron. Astrophys., Vol. 18, (see 003.004), 399 - 437 (1980).
Contents: Molecular isotope abundances and chemical fractionation, giant clouds: millimeter-wave and centimeter-wave spectra, current enrichment of the interstellar matter: mass loss from evolved stars, chemical abundance gradients: H II regions and planetary nebulae, cosmic rays.

131.130 **Interstellar titanium monoxide: limits and implications.**
E. Churchwell, W. H. Hocking, A. J. Merer, M. C. L. Gerry.
Astron. J., Vol. 85, 1382 - 1385 (1980).
A search for TiO in two mm-wave transitions toward several molecular clouds implies that the total column density of TiO is typically $< 1 \times 10^{12}$ cm^{-3}. This is nominally within the range allowed by present ideas of TiO chemistry in molecular clouds if gas phase Ti and Ca are at least as depleted as they are in diffuse clouds. The authors also did not detect TiO toward several oxygen-rich late-type stars. Neither of the mm-wave transitions of TiO is strongly mased in stars or molecular clouds. The depletion of interstellar atomic gas phase Ti cannot be due to the presence of interstellar TiO.

131.131 **OH observations of molecular complexes in Orion and Taurus.** B. Baud, J. G. A. Wouterloot.
Astron. Astrophys., Vol. 90, 297 - 303 (1980).
The molecular complexes in Orion and Taurus have been mapped in both OH main lines over a large (20° × 20°) area of sky in order to trace out their full extent. The derived column density and total mass of the complexes are in good agreement with CO results.

131.132 **The interstellar reddening law in the visible.** P. B. Lucke.
Astron. Astrophys., Vol. 90, 350 - 354 (1980).
The interstellar reddening law in the visible has been determined by multivariate analysis of the Geneva seven color photometry for 239 O stars. Very little variation is found in the reddening law with galactic longitude with the exception that the law for the Cygnus region (60° - 90°) shows a significantly higher slope at the blue end. Reddening laws are presented for eleven longitude groupings of the sample of O stars.

131.133 **HCO^+ emission in the galactic center region. I. Observations.** Y. Fukui, N. Kaifu, M. Morimoto, T. Miyaji.
Astrophys. J., Vol. 241, 147 - 154 (1980).
HCO^+ emission (3.4 mm, $J = 1-0$) in the galactic center region has been observed toward the dense molecular clouds near Sgr A and Sgr B2 at over 20 points. A position-velocity map along the galactic plane including the galactic nucleus was obtained. Comparison with the HCN emission leads to the conclusion that the number density ratio $[HCO^+]/[HCN]$ is higher by an order of magnitude in the diffuse region. A dip at 0 km s^{-1}, which is probably due to self-absorption by foreground cold molecular gas, was found in the HCO^+ profiles as well as in spectra of HCN. The abundance of HCO^+ relative to H_2 in the cold cloud is estimated.

131.134 **A search for interstellar pyrrole: evidence that rings are less abundant than chains.**
P. C. Myers, P. Thaddeus, R. A. Linke.
Astrophys. J., Vol. 241, 155 - 157 (1980).
Searches for three transitions of pyrrole (C_4H_5N) give maximum column density $N_{max} = 3-10 \times 10^{13}$ cm^{-2} in Sgr B2. This limit is more than 10 times lower than previous ring molecule limits, and is slightly lower than column densities of known interstellar molecules with from four to six heavy atoms.

131.135 **A radio search for interstellar phosphorus compounds.** J. M. Hollis, L. E. Snyder, F. J. Lovas, B. L. Ulich.
Astrophys. J., Vol. 241, 158 - 160 (1980).
The $J = 1-0$ and $3-2$ transitions of phosphorus nitride and transitions of phosphine have been searched for but not found in interstellar molecular clouds. The $J = 3/2-1/2$, $F = 3/2-3/2$ transition of nitric oxide and the $J_{K-K+} = 16_{4,12}-15_{5,11}$ transition of sulfur dioxide have been detected in Orion and Sagittarius B2. An unidentified emission line, U140921.8 MHz, has been observed in IRC + 10216.

131.136 **Interstellar depletions and far-ultraviolet extinction in the Rho Ophiuchi Cloud.**
T. P. Snow, Jr., E. B. Jenkins.
Astrophys. J., Vol. 241, 161 - 172 (1980).
Copernicus ultraviolet data on six stars embedded in the well-studied ρ Ophiuchi Cloud complex have been analyzed for interstellar abundances, densities, and far-ultraviolet extinction parameters, in an effort to determine how interstellar depletions depend on physical conditions. A picture is suggested in which the cloud core has a small velocity dispersion or is either so strongly depleted or so dense (hence not ionized) that the lines of first ions of heavy elements do not arise there, but rather form in a less-dense or less-depleted outer region. The cloud as a whole has substantial depletions, average to low far-ultraviolet extinction, and a low ratio of molecular to atomic hydrogen.

131.137 **An ultraviolet extinction study of the Rho Ophiuchi dark cloud.** C.-C. Wu, D. P. Gilra, R. J. van Duinen.
Astrophys. J., Vol. 241, 173 - 182 (1980).
Five-band ultraviolet spectrophotometry has been carried out for 54 stars in the upper Scorpius complex. Of these, 14 stars in the ρ Ophiuchi dark cloud and 13 in the surrounding area are found to be free of spectral peculiarities and contamination by scattered light from nearby bright stars. The extinction curves are significantly lower than the average extinction curves derived by other investigators and also those observed for localized regions. This implies that the dust in the upper Scorpius region, expecially that in the ρ Ophiuchi dark cloud, has larger average size than the normal interstellar medium. The ratio of total-to-selective extinction, is found to be 4.2. The effects of a lower extinction curve on the chemistry of a dark cloud is also briefly discussed.

131.138 **H I self-absorption in the Southern Coalsack dust complex.** P. F. Bowers, F. J. Kerr, T. G. Hawarden.
Astrophys. J., Vol. 241, 183 - 196, plate 1, (1980).
The authors report the results of an extensive search for H I self-absorption toward globules and dense patches of obscuration in the Coalsack. Cold H I gas was detected and mapped in the region containing globules. The authors discuss the relation of the H I to the molecules and dust in this region and also present a catalog of other globules and dark regions for which no H I self-absorption was detected.

131.139 **Star formation and ionization in the 3 kiloparsec arm.** F. J. Lockman.
Astrophys. J., Vol. 241, 200 - 207 (1980).
This paper presents new observations of H166α recombination lines that are very sensitive to large optically thin H II regions, and uses existing surveys of dense H II regions and type I OH masers to determine the amount of star formation,

the maximum mass of ionized gas, and the total hydrogen ionization rate in the 3 kpc arm.

131.140 **Molecule formation and infrared emission in fast interstellar shocks. II. Dissociation speeds for interstellar shock waves.** D. Hollenbach, C. F. McKee.
Astrophys. J., Lett., Vol. 241, L47 - L50 (1980).

The postshock destruction of molecules is examined, including the processes of (1) collisions with neutral hydrogen atoms and molecules, (2) electronic collisions, and (3) neutral chemical reactions with atoms, particularly atomic hydrogen. The results demonstrate that many of the observed high-speed interstellar molecules, if shock accelerated, must have dissociated and reformed in the postshock gas.

131.141 **The OH maser near the Herbig–Haro object GGD37.** R. P. Norris.
Mon. Not. R. Astron. Soc., Vol. 193, 39P - 41P (1980).

H_2O masers are often associated with Herbig–Haro (HH) objects, but OH masers have been found near only one HH object, GGD37. The author has measured the position of this maser and finds it to be coincident with the compact H II region from which the HH object was probably ejected.

131.142 **On the production of suprathermal grains.** A. K. Dasgupta.
Astrophys. Space Sci., Vol. 72, 271 - 280 (1980).

The physical conditions under which suprathermal grains can be produced when they are accelerated by radiation pressure against the drag of ambient gas are investigated. It is found that dust grains may attain a terminal velocity $U(= 10^5\ \mathrm{cm\ s^{-1}})$ in most regions and move out of the midplane of the source region about a distance $|z| \leqslant 100$ pc. Once clear of the main gas/dust layer the dust grains ($a \simeq 3 \times 10^{-6}$ cm) may then attain 'suprathermal energy' ($V_g \simeq 3 \times 10^8\ \mathrm{cm\ s^{-1}}$) by the Fermi process.

131.143 **The predicted $1s^2$ - $1s\,6p$ H^- auto-ionization resonance observed as a diffuse interstellar line at 7581 Å.**
P. Gammelgaard, M. Rudkjøbing.
Astrophys. Space Sci., Vol. 72, 319 - 322 (1980).

The $1s^2$ - $1s\,6p$ H^- auto-ionization resonance has been observed as a diffuse interstellar line in several highly reddened O and B stars. The identification based on its wavelength has been confirmed by its equivalent width in HD 183143. The addition of the $1s^2$ - $1snp$ diffuse-line series raises the total H^- oscillator strength of absorption to a value close to 2.

131.144 **Silicon chemistry in dense clouds.** T. L. Millar.
Astrophys. Space Sci., Vol. 72, 509 - 517 (1980).

The equilibrium chemistry of silicon in dense interstellar clouds is discussed in terms of both gas phase and grain surface reactions. Unless the metal depletion is very large, the gas phase scheme tends to over-produce SiO and/or SiS when compared to the observations of Sgr B2. The scheme also predicts SiC to be an abundant form of silicon. Reactions between positively charged gas phase ions and small grains can lead to the formation of SiO and SiS. This type of reaction seems to offer a simple explanation for the observed differences between sulphur and silicon chemistry in dense clouds.

131.145 **COS-B observation of high-energy gamma-ray emission from the Orion cloud complex.**
P. A. Caraveo, K. Bennett, G. F. Bignami, W. Hermsen, G. Kanbach, F. Lebrun, J. L. Masnou, H. A. Mayer-Hasselwander, J. A. Paul, B. Sacco, L. Scarsi, A. W. Strong, B. N. Swanenburg, R. D. Wills.
Astron. Astrophys., Vol. 91, L3 - L5 (1980).

Emission of high-energy gamma rays ($70 < E < 5000$ MeV) has been observed from the direction of the Orion cloud complex by the COS-B satellite. The emission is resolved and the existence of structure is apparent within the excess with two maxima coinciding with L1630 andL1641, i.e. the northern and the southern major clouds in the complex. The flux is evaluated and, coupled with an emissivity value deduced for the same general region of the sky, yields a total mass for the complex well in agreement with independent radio astronomical evaluations.

131.146 **Observations of the 3_{12}–3_{13} line of H_2CO.**
T. L. Wilson, C. M. Walmsley, C. Henkel, T. Pauls, H. Mattes.
Astron. Astrophys., Vol.91, 36 - 40 (1980).

Measurements of the 3_{12}–3_{13} line of H_2CO at 28.974 GHz have been made with an angular resolution of 35″. The line was observed in absorption toward Sgr B2 and W 33, and mapped in emission toward the Orion Molecular Cloud. The authors also obtained upper limits for the optical depth toward W 3 A, W 51 E, and DR 21. The measurement for W 33 and the upper limits are consistent with the H_2 densities, $\sim 10^5\ \mathrm{cm^{-3}}$, obtained by Henkel et al. (1980). For Sgr B 2, from model fits to the new data and previous H_2CO results, the H_2 density is $10^{4.2}\ \mathrm{cm^{-3}}$, and the $(H_2{}^{12}CO/H_2{}^{13}CO)$ ratio corrected for photon trapping is ~ 25. The OMC-1 line emission is formed in a $\sim 40'' \times \sim 20''$ region; the model of Evans et al. (1979) is consistent with the results.

131.147 **On the variation of the colour excess in the Carina-Crux-Centaurus-Norma region of the Milky Way.**
L. O. Lodén, A. Sundman.
Astron. Astrophys., Vol. 91, 59 - 61 (1980).

For a material consisting of 2263 stars distributed over the Milky Way between $l = 280°$ and $l = 332°$ the average $B - V$ colour excess has been studied as a function of (uncorrected) distance modulus. The general impression is that although the colour excess may vary considerably from one direction to another – even when the angular separation between those directions is very small – the average value for large regions seems to repeat fairly well from one region to another up to at least 1 - 1.5 kpc. The authors conclude that average values of the colour excess over regions of several square degrees should not be used for mapping the structure of interstellar dust clouds or for correcting stellar density functions.

131.148 **Comparison of solar backscatter and interstellar absorption measurements of the ISM.** R. R. Meier.
Astron. Astrophys., Vol. 91, 62 - 67 (1980).

Measurements of density, temperature, and velocity of the interstellar wind in the immediate vicinity of the solar system are reviewed and compared with observations from Copernicus. The dynamic properties of the nearby interstellar medium appear to extend to at least a few tens of pcs from the sun but do not appear to be associated with any well-known phenomenon. The available evidence suggests that the H density is lower at the solar system, and in the directions of α Aur and HR 1099 than elsewhere in the nearby interstellar medium. The fractional ionization of H deduced locally is compatible with that derived for other species, assuming photoionization equilibrium. These results show that the region near the solar system is part of a warm, ionized, low density phase of the interstellar medium.

131.149 **Hydrostatic models of molecular clouds. I. Steady state models.** T. de Jong, A. Dalgarno, W. Boland.
Astron. Astrophys., Vol. 91, 68 - 84 (1980).

The authors have calculated the gas temperature in molecular clouds by solving simultaneously the equations of chemical equilibrium and of thermal balance. The gas is heated predominantly by photoelectrons from dust grains in the outer parts of the clouds and by cosmic ray ionizations of molecular hydrogen in the inner parts. It is cooled by fine-structure excitation followed by radiative decay of C^+ ions and C atoms in the outer parts of the clouds and by rotational excitation of

CO molecules followed by radiative decay in the inner parts. The effect of radiation trapping is included in the calculation of the escape of the cooling photons from the clouds. Equations of state of the gas have been constructed from these results. The authors also constructed plane-parallel hydrostatic models of molecular clouds which are supported by turbulent pressure and have studied the effects on the cloud models of the depletion of molecules by collisions with dust grains.

131.150 **The area around the Orion Nebula observed in the CO ($J = 1 - 0$) transition.**
A. R. Gillespie, G. J. White.
Astron. Astrophys., Vol. 91, 257 - 258 (1980).

An area 1 deg^2 around the Orion Nebula has been surveyed with a 2.6 km s^{-1} wide single-channel receiver. The map shows considerable structure in the east including a bright spot 24′(2.8 pc) from the main CO peak.

131.151 **Theoretical investigation of variable infrared emission by cosmic dust sources.** M. F. Bode.
Thesis, Univ. Keele, Staffs., England (1979). – Abstr. in Phys. Abstr., Vol. 83, Abstr. 105324 (1980).

131.152 **Abundances in diffuse interstellar clouds.**
M. Jura.
Highlights of Astronomy, Vol. 5, (see 012.024), 293 - 300 (1980).

131.153 **Models of the interstellar medium.**
T. de Jong.
Highlights of Astronomy, Vol. 5, (see 012.024), 301 - 310 (1980).

Up-to-date models of the interstellar medium should account for the existence of at least three phases of the interstellar gas, a hot phase at $\sim 10^6$ K, a warm phase at $\sim 10^4$ K and a cold phase at $< 10^2$ K. A model of a 3-phase ISM was developed by McKee and Ostriker (1977). The author concentrates on several areas where the model of McKee and Ostriker can be improved and refined by making use of recent observations of interstellar absorption lines.

131.154 **The hot component of the interstellar medium.**
E. B. Jenkins.
Highlights of Astronomy, Vol. 5, (see 012.025), 361 - 363 (1980).

131.155 **Moleküle im interstellaren Raum.**
G. Winnewisser.
Evolution der Planetenatmosphären und des Lebens, (see 012.032), p. 184 - 189 (1980).

131.156 **Physical and dynamical conditions in interstellar clouds.** N. J. Evans II.
Interstellar molecules, (see 012.033), p. 1 - 19 (1980).

131.157 **Detection of submillimeter lines of CO (0.65 mm) and H_2O (0.79 mm).**
T. G. Phillips, J. Kwan, P. J. Huggins.
Interstellar molecules, (see 012.033), p. 21 - 24 (1980).

The 91.5 cm telescope of NASA's Kuiper Airborne Observatory has been used, in conjunction with an InSb heterodyne receiver, to detect the $J = 4 - 3$ submillimeter transition of CO (461 GHz) and the $4_{14} - 3_{21}$ transition of H_2O (380 GHz). The water emission was detected from the Orion "plateau" region.

131.158 **New detections of spectral lines in the frequency range 260 - 285 GHz.** N. Erickson, J. H. Davis, N. J. Evans II, R. B. Loren, L. Mundy, W. L. Peters III, M. Scholtes, P. A. Vanden Bout.
Interstellar molecules, (see 012.033), p. 25 - 30 (1980).

131.159 **Further observational studies of the high velocity molecular source ("plateau") in Orion.**
T. B. H. Kuiper, E. N. Rodriguez Kuiper, B. Zuckerman.
Interstellar molecules, (see 012.033), p. 31 - 32 (1980).

Sensitive measurements have been made of the profiles of various molecular lines in the direction of the BN/KL infrared complex in Orion. Similar line shapes are found for the high velocity wings ("plateau") of most lines observed. The position of the high velocity source (HVS) was found in SO and HCO^+ relative to the SiO maser. A search for HV emission in other sources was unsuccessful.

131.160 **High velocity gas in the Orion nebula.**
N. Z. Scoville.
Interstellar molecules, (see 012.033), p. 33 - 38 (1980).

131.161 **High spatial resolution studies of $H^{13}CN$, $H^{12}CN$ and HCO^+ $J = 1-0$ emissions in Orion A.**
O. E. H. Rydbeck, Å. Hjalmarson, G. Rydbeck, J. Ellděr, A. Sume, S. Lidholm.
Interstellar molecules, (see 012.033), p. 39 - 40 (1980).

131.162 **Lunar occultation observations of millimeter CO emission in S255.** F. P. Schloerb, N. Z. Scoville.
Interstellar molecules, (see 012.033), p. 41 - 42 (1980).

The authors have used lunar occultations of the region around two infrared sources embedded in the S255 molecular cloud to determine the angular structure of the CO emission of the surrounding gas, and they have found that this region closely resembles the core of the Orion molecular cloud.

131.163 **Radio-frequency observations of interstellar CH_4 and HC_5N.** K. Fox, D. E. Jennings.
Interstellar molecules, (see 012.033), p. 43 - 44 (1980).

The authors detected new $^{12}CH_4$ transitions at 21,303 and 94,089 MHz in emission from Orion A, and observed temporal variations of intensity among several velocity components at 76,700 MHz. The authors also detected an HC_5N transition at 21,301 MHz in emission from IRC+10216.

131.164 **Long chain carbon molecules in the interstellar medium.** L. W. Avery.
Interstellar molecules, (see 012.033), p. 47 - 58 (1980).

The long chain carbon molecules known as the cyanopolyynes (HC_nN, n = 3,5,7,9) are becoming increasingly more important in astrophysics. As a family, these molecules share a number of properties that are unique or, at least, unusual among interstellar species. The author reviews briefly some of these properties and how they contribute to the usefulness of the cyanopolyynes as astrophysical probes. He presents some specific studies in which these molecules have been utilized.

131.165 **The relative distribution of ammonia and cyanobutadiyne emission in Heiles 2 dust cloud.**
L. T. Little, G. H. Macdonald, P. W. Riley, D. N. Matheson.
Interstellar molecules, (see 012.033), p. 67 - 68 (1980).

131.166 **Observations of HC_5N (J=12–11 and 13–12) and HC_7N (J=28–27), including a detection of B335.** Å. Hjalmarson, P. Friberg.
Interstellar molecules, (see 012.033), p. 69 - 70 (1980).

131.167 **Spectra of the 1_0-0_1 transition of sulfur monoxide in interstellar clouds.** O. E. H. Rydbeck, Å. Hjalmarson, G. Rydbeck, J. Ellděr, E. Kollberg, W. M. Irvine.
Interstellar molecules, (see 012.033), p. 71 - 76 (1980).

131.168 **CN observations in Taurus dark cloudlets.**
E. Churchwell.
Interstellar molecules, (see 012.033), p. 77 - 80 (1980).

CN has been searched for toward 8 locations in the

Taurus dark cloud complex. CN appears to be correlated with NH_3 and anti-correlated with HC_5N and HC_3N.

131.169 **C_2H and HC_3N in interstellar clouds.** A. Wootten, G. P. Bozyan, D. B. Garrett, R. B. Loren, R. L. Snell, P. Vanden Bout.
Interstellar molecules, (see 012.033), p. 81 - 82 (1980).

131.170 **Detection of new ammonia sources.** G. H. Macdonald, A. T. Brown, L. T. Little, D. N. Matheson, M. Felli.
Interstellar molecules, (see 012.033), p. 83 - 84 (1980).
The authors have undertaken a systematic survey of compact H II regions, H_2O masers, Herbig-Haro objects and other protostellar indicators for emission in the (1,1) and (2,2) transitions of NH_3. This program has yielded 33 detections which the authors report.

131.171 **Ammonia observations of the molecular clouds near S68, S140 and OMC2.** L. T. Little, A. T. Brown, G. H. Macdonald, P. W. Riley, D. N. Matheson.
Interstellar molecules, (see 012.033), p. 85 - 87 (1980).
Maps of the J = 1, K = 1 inversion transition of interstellar ammonia are presented and compared with observations of carbon monoxide and formaldehyde.

131.172 **Ammonia observations of the molecular cloud near S106.** L. T. Little, G. H. Macdonald, P. W. Riley, D. N. Matheson.
Interstellar molecules, (see 012.033), p. 89 - 90 (1980).
The region surrounding the optical nebula S106 has been mapped in the J = 1, K = 1 (1,1) transition of ammonia. The source is observed as two components with the H II region sandwiched between them. Observations of the (1,1) and (2,2) transitions yield a kinetic temperature for the NH_3 molecules in the range 18 - 22K. There is evidence for "clumpiness" in the distribution of matter.

131.173 **The core of a quiescent cloud, L183.** C. M. Walmsley, H. Ungerechts, G. Winnewisser.
Interstellar molecules, (see 012.033), p. 91 - 92 (1980).

131.174 **Observations of NH_3 toward DR21.** T. Pauls, T. L. Wilson.
Interstellar molecules, (see 012.033), p. 93 - 94 (1980).

131.175 **High resolution 4.8 GHz mapping of H_2CO using the Westerbork synthesis radio telescope.** J. R. Forster, W. M. Goss, T. de Jong, C. A. Norman, H. J. Habing, H. R. Dickel.
Interstellar molecules, (see 012.033), p. 95 - 98 (1980).
The goal of this work is to measure scale sizes of H_2CO in molecular clouds near H II regions and to study the kinematics of the clouds in the molecular line.

131.176 **Surveys of the 4.8 GHz formaldehyde absorption line in dark clouds in M17 and NGC 2024.** Y. K. Minn.
Interstellar molecules, (see 012.033), p. 99 - 100 (1980).

131.177 **Cloud-to-cloud variations in H_2CO-to-H_2 ratios.** W. A. Sherwood.
Interstellar molecules, (see 012.033), p. 101 - 102 (1980).

131.178 **Formaldehyde in L1551, L134 and the galactic centre.** Aa. Sandqvist, C. Bernes.
Interstellar molecules, (see 012.033), p. 103 - 107 (1980).
The formaldehyde molecule is an excellent probe of physical conditions inside interstellar clouds. The authors illustrate this by presenting results of their recent series of observations using the MPIfR Effelsberg 100-m radio telescope for the 6-cm transition, the NRAO Green Bank 43-m radio telescope for the 2-cm transition and the NRAO Kitt Peak 11-m mm-wave telescope for the 2-mm transition.

131.179 **Observations of the 3.4-mm HCO^+ line toward the galactic center.** Y. Fukui, N. Kaifu, M. Morimoto, T. Miyaji.
Interstellar molecules, (see 012.033), p. 109 - 110 (1980).

131.180 **CO and OH in the galactic center region.** J. Inatani, N. Ukita, N. Kaifu, S. Kodaira, K. Ishii.
Interstellar molecules, (see 012.033), p. 111 - 112 (1980).

131.181 **In search of $\lesssim$ 5 K galactic molecular gas.** R. H. Rubin, N. J. Evans II, B. Zuckerman.
Interstellar molecules, (see 012.033), p. 113 - 114 (1980).
Existing surveys of our galaxy are able to detect CO clouds with an excitation temperature $T_{ex} \gtrsim 5$ K. The authors have made observations to determine if a substantial molecular component is colder than ~ 5 K. The results do not suggest the existence of a large amount of mass in ultra-cold molecular clouds.

131.182 **CO (J = 2 – 1) observations of several galactic H II regions.** T. de Graauw, S. Lidholm, B. Fitton, F. P. Israel, A. Sargent, T. B. H. Kuiper, H. Nieuwenhuyzen.
Interstellar molecules, (see 012.033), p. 115 - 116 (1980).

131.183 **Molecular cloud densities from observations of carbon monosulfide.** R. A. Linke, P. F. Goldsmith.
Interstellar molecules, (see 012.033), p. 117 - 121 (1980).
The authors have made carefully calibrated measurements of the J = 1 → 0 and J = 2 → 1 transitions of CS in 32 molecular sources in order to obtain density and fractional abundance information from the excitation of this molecule. The authors' observations are seen to be in good agreement with the results of excitation calculations involving either a velocity gradient model or a purely micro-turbulent model.

131.184 **CO observations in the southern hemisphere.** A. R. Gillespie.
Interstellar molecules, (see 012.033), p. 123 - 124 (1980).
CO maps of areas around H II regions show many secondary hotspots. J = 2 – 1 observations of 2 dark clouds are discussed.

131.185 **1.0 mm continuum observations of cool southern clouds.** D. Y. Gezari, L. Cheung, M. G. Hauser, J. A. Frogel.
Interstellar molecules, (see 012.033), p. 129 - 132 (1980).
High surface brightness 1.0 mm continuum emission has been mapped in nine southern hemisphere H II/molecular cloud complexes. This paper presents new 1.0 mm continuum mapping results with 65 arc sec resolution.

131.186 **The extinction efficiency of dust grains at 1 mm.** W. A. Sherwood, E. M. Arnold, G. V. Schultz.
Interstellar molecules, (see 012.033), p. 133 - 134 (1980).

131.187 **Comparison of submillimeter and CO brightness in Orion and Mon R2.** D. Cudaback, L. Anderson, D. Lynch, J. Smith.
Interstellar molecules, (see 012.033), p. 135 - 136 (1980).

131.188 **The evolution of giant molecular clouds.** C. Norman, J. Silk.
Interstellar molecules, (see 012.033), p. 137 - 149 (1980).
The authors discuss the origin, lifetime, destruction, spatial distribution and relation to star formation of giant molecular clouds. A coagulation model including the effects of spiral density wave shocks is described. The authors explore implications for CO observations of external galaxies.

The collective effects of OB star winds and supernova remnants in disrupting clouds are considered.

131.189 **The disruption of the molecular cloud associated with the North America and Pelican Nebulae.**
J. Bally.
Interstellar molecules, (see 012.033), p. 151 - 156 (1980).

131.190 **The structure and evolution of the W3 molecular cloud.** H. R. Dickel.
Interstellar molecules, (see 012.033), p. 157 - 158 (1980).

A model of the W3 molecular cloud derived from molecular observations is presented and the evolution of the cloud is discussed.

131.191 **Atomic hydrogen in and around the giant molecular cloud near W3 and W4.**
T. Hasegawa, F. Sato, Y. Fukui.
Interstellar molecules, (see 012.033), p. 159 - 162 (1980).

Cold H I gas appears as self-absorption dips in the 21-cm line profiles in and around the giant molecular cloud near W3 and W4. The cold H I cloud is ~ 150 pc long and extends along the galactic plane. It consists of several fragments. The $[H_2]/[H\ I]$ ratio is estimated to be 15 - 50. The mass of the entire H I cloud amounts to $\sim 10^5\ M_\odot$, which is comparable to that observed in CO emission.

131.192 **High rate of destruction of molecular clouds by hot stars.** M. Heydari-Malayeri, M. C. Lortet, L. Deharveng.
Interstellar molecules, (see 012.033), p. 163 - 164 (1980).

131.193 **The interaction of T-Tauri stars with molecular clouds.** J. Silk, C. Norman.
Interstellar molecules, (see 012.033), p. 165 - 172 (1980).

It is demonstrated that T-Tauri stars could provide significant dynamical input into molecular clouds. The authors consider in detail the nature of the interaction between T-Tauri stellar winds and the ambient molecular cloud.

131.194 **Observations of the J = 1 – 0 and J = 2 – 1 lines of ^{12}CO in L1551: evidence for anisotropic mass loss.** R. L. Snell, R. B. Loren, R. L. Plambeck.
Interstellar molecules, (see 012.033), p. 173 - 174 (1980).

131.195 **Ammonia observations of dark clouds containing Herbig-Haro objects.** P. T. P. Ho, A. H. Barrett.
Interstellar molecules, (see 012.033), p. 175 - 176 (1980).

A study in NH_3 was conducted towards H-H objects. Cloud fragmentation appears to have occurred in the regions mapped. Rotation is present with velocity gradients of 1 - 2 km s^{-1} pc^{-1}. The authors suggest that in the regions containing H-H objects, formation of stars of different spectral types may be taking place.

131.196 **Hydrostatic models of molecular clouds.**
T. de Jong, A. Dalgarno, W. Boland.
Interstellar molecules, (see 012.033), p. 177 - 181 (1980).

The authors have constructed a plane-parallel model of the well-studied molecular cloud L134 assuming that it is in hydrostatic equilibrium supported by turbulent pressure. This cloud model is quite centrally condensed. Molecular abundances are calculated by solving the coupled set of equations of chemical equilibrium and thermal balance as a function of depth in the cloud. Depletion of atoms and molecules onto dust grains is taken into account. Column densities of several molecules are predicted and compared with the observations.

131.197 **Contagious B star formation in the Rho Ophiuchi dark cloud.** E. Falgarone.
Interstellar molecules, (see 012.033), p. 183 - 184 (1980).

131.198 **A nearby example of a giant molecular cloud.**
J. W. Barrett, R. L. deZafra, D. B. Sanders, P. M. Solomon.
Interstellar molecules, (see 012.033), p. 185 - 186 (1980).

131.199 **Cold H I gas in the region of the giant molecular cloud near M17.** F. Sato, Y. Fukui, T. Hasegawa.
Interstellar molecules, (see 012.033), p. 187 - 188 (1980).

131.200 **Search for CO in atomic hydrogen clouds.**
I. Kazès, J. Crovisier.
Interstellar molecules, (see 012.033), p. 189 - 190 (1980).

131.201 **Molecular clouds in Orion and Monoceros.**
M. Morris, J. Montani, P. Thaddeus.
Interstellar molecules, (see 012.033), p. 197 - 203 (1980).

The authors report the results of a survey of CO at negative galactic latitudes mainly in Orion and Monoceros. Several new features emerge, the most striking of which are very long (10 - 20°) molecular filaments which appear to connect molecular clouds lying far below the galactic plane to molecular features in the plane itself. Many new clouds covering a large range of sizes have been found.

131.202 **Columbia CO survey: molecular clouds and spiral structure.** R. S. Cohen, T. M. Dame, P. Thaddeus.
Interstellar molecules, (see 012.033), p. 205 - 207 (1980).

131.203 **Molecular fan of 360-pc radius in the galactic center region.** Y. Fukui.
Interstellar molecules, (see 012.033), p. 209 - 211 (1980).

131.204 **Optical observations of interstellar molecules.**
T. P. Snow, Jr.
Interstellar molecules, (see 012.033), p. 247 - 256 (1980).

Recent observational data are summarized on molecular species in diffuse interstellar clouds, and a comprehensive list of all species detected or sought is included. The discussion is focussed on the use of molecular observations to determine physical conditions in clouds, and on important species which provide constraints on cloud chemistry models.

131.205 **Observations of interstellar molecules with the International Ultraviolet Explorer.**
J. H. Black.
Interstellar molecules, (see 012.033), p. 257 - 260 (1980).

The ultraviolet spectra of 25 early-type stars have been obtained with the International Ultraviolet Explorer observatory. Bands of the 4th-positive system of interstellar CO are seen towards 12 of these stars. Spectra of HD46223 have been examined for interstellar lines of CH, C_2, CH_2, OH, HCl, and H_2O.

131.206 **Interstellar line spectra of a dense cloud: the VI Cygni association.** S. P. Souza, B. L. Lutz.
Interstellar molecules, (see 012.033), p. 261 - 262 (1980).

Spectroscopic observations of the dense interstellar cloud which obscures the VI Cygni association have yielded an optical map of the atomic and molecular constituents in a region which may represent a transition between the diffuse and radio clouds. The velocity of the optical components found is consistent with that of a CO cloud seen in front of No. 12.

131.207 **Rotational fine structure lines of interstellar C_2 toward ζ Persei.** F. H. Chaffee, Jr., B. L. Lutz, J. H. Black, P. A. Vanden Bout, R. L. Snell.

Interstellar molecules, (see 012.033), p. 263 - 267 (1980).

The authors have detected 9 of the rotational fine structure lines of the 2-0 Phillips band of interstellar C_2 toward ζ Persei. These data yield a total C_2 column density of 1.2×10^{13} cm^{-2} and a rotational temperature of 97 K compared to 1.4×10^{13} cm^{-2} and 45 K predicted by Black, Hartquist and Dalgarno. The authors suggest that radiative pumping through the Mulliken and Phillips systems has modified the C_2 level populations in such a way as to produce an observed rotational temperature which exceeds that arising in pure thermal equilibrium.

131.208 **Chlorine chemistry in diffuse interstellar clouds.** J. H. Black, P. L. Smith.
Interstellar molecules, (see 012.033), p. 271 - 272 (1980).

131.209 **Molecular formation in hot diffuse clouds.** A. Dalgarno.
Interstellar molecules, (see 012.033), p. 273 - 280 (1980).

A description is given of the processes of molecular formation and destruction in diffuse interstellar clouds and detailed models of the clouds lying towards ζ Ophiuchi, ζ Persei and o Persei are used to assess the validity of gas phase chemistry. Modifications that may arise from shock-heated regions are discussed.

131.210 **Formation of simple molecules by C^+ reactions on oxide grains in diffuse clouds.** W. W. Duley.
Interstellar molecules, (see 012.033), p. 281 - 282 (1980).

The reaction between C^+ ions and OH^- ions on the surface of oxide or silicate grains in diffuse clouds is shown to be a source of CH, OH, CO, HCO and H_2CO molecules.

131.211 **The photodissociation of interstellar CH^+.** K. Kirby.
Interstellar molecules, (see 012.033), p. 283 - 286 (1980).

Photodissociation cross sections have been calculated, and the CH^+ photodissociation rates in the interstellar radiation field as a function of optical depth have been determined. The author has compared these new photodissociation rates with the rates of other CH^+ destruction mechanisms in three different models of the interstellar medium. Photodissociation appears to be a significant destruction mechanism in any model requiring large interstellar radiation fields.

131.212 **Molecular synthesis in interstellar clouds: the radiative association reaction $H + OH \rightarrow H_2O + h\nu$.**
D. Field, N. G. Adams, D. Smith.
Interstellar molecules, (see 012.033), p. 289 - 290 (1980).

The authors tentatively suggest that the radiative association reaction of H and OH can be a significant mechanism for the production of H_2O in interstellar clouds.

131.213 **Interstellar sulfur chemistry.** S. S. Prasad, W. T. Huntress, Jr.
Interstellar molecules, (see 012.033), p. 297 - 298 (1980).

This paper summarizes results of a chemical model of SO, CS, and OCS chemistry in dense clouds.

131.214 **Gas phase synthesis of amino-, cyano- and nitroso-compounds in interstellar clouds.**
N. G. Adams, D. Smith.
Interstellar molecules, (see 012.033), p. 311 - 315 (1980).

From laboratory data relating to several hundreds of ion-atom and ion-molecule reactions at thermal energies the authors qualitatively describe probable chemical paths to the synthesis of amino-, cyano- and nitroso-compounds in interstellar clouds.

131.215 **Molecular evolution in dense clouds.** H. Suzuki.
Interstellar molecules, (see 012.033), p. 337 - 338 (1980).

131.216 **The determination of electron abundances in interstellar clouds.** A. Wootten, R. Snell,
A. E. Glassgold.
Interstellar molecules, (see 012.033), p. 339 - 340 (1980).

131.217 **Molecule formation in cool, dense interstellar clouds.** W. D. Watson.
Interstellar molecules, (see 012.033), p. 341 - 353 (1980).

A discussion is given of the general processes and considerations that arise in attempting to understand molecular reactions in cool, dense interstellar clouds.

131.218 **Interstellar molecules on dust mantles.** N. Nakagawa.
Interstellar molecules, (see 012.033), p. 365 - 366 (1980).

Condensation temperatures of various interstellar molecules in dark clouds are calculated. Chemical reactions in dust mantles are discussed.

131.219 **The chemical identification of grain mantles by infrared spectroscopy.** L. J. Allamandola,
J. M. Greenberg, C. A. Norman, W. Hagen.
Interstellar molecules, (see 012.033), p. 373 - 380 (1980).

The authors discuss the infrared spectra associated with astrophysical objects which show spectral characteristics indicative of molecular mantles.

131.220 **Infrared molecular absorption features.** S. P. Willner, R. C. Puetter, R. W. Russell,
B. T. Soifer.
Interstellar molecules, (see 012.033), p. 381 - 386 (1980).

Spectra of infrared sources associated with molecular clouds have shown absorption features at wavelengths of 6.0 and 6.8 μm. The authors suggest that the 6.0 μm feature can be identified with the stretching vibration of C=O and the 6.8 μm feature with the bending vibrations of CH_2 and CH_3. The amount of carbon in the form of hydrocarbon molecules may be comparable to the amount in CO. This abundance of hydrocarbons is probably too large to be consistent with radio observations if the molecules are gaseous, but large abundances of hydrocarbons on the surfaces of grains may explain the infrared features, yet be unobservable in the radio.

131.221 **Correlations between the λ2200 feature, the diffuse λ4430 band and E_{B-V}.** A. C. Danks.
Interstellar molecules, (see 012.033), p. 389 - 394 (1980).

The λ2200 extinction feature has been measured from ANS observations of 30 stars. For each star, the depth of the λ2200 feature is compared to E_{B-V} and the equivalent width of the diffuse band λ4430. A good correlation appears out to $E_{B-V} = 1.13$. The various mechanisms for producing the diffuse and λ2200 features are discussed, and a discriminatory test is put forward based on observations of the Magellanic Clouds.

131.222 **Correlations for interstellar molecules and diffuse bands.** W. B. Somerville.
Interstellar molecules, (see 012.033), p. 395 - 396 (1980).

Results are presented from a programme of optical spectroscopy and related studies which has the double purpose of investigating the structure of diffuse molecular clouds and of establishing tighter correlations for the unidentified interstellar diffuse bands.

131.223 **Measurements of isotopic abundances in interstellar clouds.** A. A. Penzias.
Interstellar molecules, (see 012.033), p. 397 - 404 (1980).

131.224 **Isotopic abundance ratios from microwave observations of formaldehyde.** T. L. Wilson, C. Henkel,

C. M. Walmsley, T. Pauls.
Interstellar molecules, (see 012.033), p. 405 - 408 (1980).

131.225 **Isotope ratios in interstellar formaldehyde.** M. L. Kutner, D. E. Machnik, K. D. Tucker, W. Massano.
Interstellar molecules, (see 012.033), p. 409 - 410 (1980).

131.226 **The $^{12}C/^{13}C$ ratio in interstellar dark clouds.** W. H. McCutcheon, R. L. Dickman, W. L. H. Shuter, R. S. Roger.
Interstellar molecules, (see 012.033), p. 411 - 416 (1980).

131.227 **CO abundance and isotopic fractionation in dark clouds.** P. F. Goldsmith, W. D. Langer, E. R. Carlson, R. W. Wilson.
Interstellar molecules, (see 012.033), p. 417 - 420 (1980).

131.228 **CO isotope line shapes in dark clouds.** P. C. Myers, R. B. Buxton, P. T. P. Ho.
Interstellar molecules, (see 012.033), p. 421 - 422 (1980).

131.229 **Isotopic fractionation in interstellar carbon-bearing molecules unrelated to carbon monoxide.** V. Vanýsek.
Interstellar molecules, (see 012.033), p. 423 - 426 (1980).

The isotopic abundance ratio $^{12}C/^{13}C$ in some carbon-bearing molecules is discussed in the context of chemical fractionation via ion-molecule and exchange reactions in dense interstellar clouds. These processes can lead to enhancement of ^{12}C in molecules not related to carbon monoxide. The effect is transient and takes place preferentially outside the cores of the interstellar clouds.

131.230 **Detection of deuterated formaldehyde in interstellar clouds.** W. D. Langer, M. A. Frerking, R. A. Linke, R. W. Wilson.
Interstellar molecules, (see 012.033), p. 439 - 443 (1980).

Deuterated formaldehyde has been detected for the first time in interstellar clouds; the observed ratio $HDCO/H_2CO$ implies formation by gas phase chemistry.

131.231 **Theoretical considerations of shock wave behavior.** D. Hollenbach.
Interstellar molecules, (see 012.033), p. 445 - 453 (1980).

The purpose of this paper is to review the overall structure near interstellar shock waves and to focus on the interaction of shocks with interstellar molecules.

131.232 **Observations of shock waves in interstellar clouds.** S. Beckwith.
Interstellar molecules, (see 012.033), p. 455 - 463 (1980).

Some techniques for observing shock waves in interstellar clouds are discussed. It is concluded that recent measurements of molecular hydrogen emission provide the best currently available technique for studying shocks. The results of measurements toward the Orion nebula are discussed, and a discussion and summary of the currently known H_2 sources is given.

131.233 **Observations of the V = 0 S(2) line of molecular hydrogen at 12.28 μm in the Orion molecular cloud.** T. R. Geballe, S. C. Beck, J. H. Lacy.
Interstellar molecules, (see 012.033), p. 465 - 468 (1980).

The 12.28μm pure rotational line of molecular hydrogen has been detected in emission from the region of vibration-rotation line emission in Orion. The line shapes, widths, and velocities are similar to those observed in the $V = 1 \rightarrow 0$ transition at 2.12μm. Constraints imposed by these new results on models of the emitting region are discussed.

131.234 **Spectra of the 2.12μm quadrupole line of H_2 in the Orion molecular cloud.** D. Nadeau, G. Neugebauer, T. R. Geballe.
Interstellar molecules, (see 012.033), p. 469 - 470 (1980).

131.235 **Molecular clouds near supernova remnants.** V. I. Slysh, T. L. Wilson, T. Pauls, C. Henkel.
Interstellar molecules, (see 012.033), p. 473 - 478 (1980).

A survey of 14 SNR's in the 4.8 GHz absorption line of H_2CO shows that two of them, W28 and W44, possibly interact with molecular clouds. The interaction leads to acceleration of a part of the molecular cloud to a velocity of ~ 5 km s^{-1} without a significant increase in the kinetic temperature or turbulence.

131.236 **SiO emission from the Orion Nebula.** B. Baud, J. H. Bieging, R. L. Plambeck, D. D. Thornton, W. J. Welch, M. C. H. Wright.
Interstellar molecules, (see 012.033), p. 545 - 548 (1980).

SiO emission in both a strong maser transition in the first excited vibrational state and in a weaker transition in the ground vibrational state both arise from the same small region in the Kleinmann-Low Nebula in Orion. Within the errors, the source position coincides with that of the infrared source IRC2.

131.237 **Observations of masers in regions of star formation.** D. Downes, R. Genzel.
Interstellar molecules, (see 012.033), p. 565 - 577 (1980).

This review covers recent progress in observations of masers in regions of star formation. Four regions of special interest are described in more detail: Orion, W51, W49 and W3.

131.238 **Absolute positions of OH masers associated with H II regions.** R. S. Booth, R. P. Norris.
Interstellar molecules, (see 012.033), p. 579 - 580 (1980).

131.239 **The pumping of interstellar OH main line masers: an efficient mechanism.** R. Lucas.
Interstellar molecules, (see 012.033), p. 581 - 582 (1980).

131.240 **Pumping of strong H_2O cosmic masers.** V. S. Strelnitsky (*Strel'nitskij*).
Interstellar molecules, (see 012.033), p. 591 - 592 (1980).

131.241 **Time variations of interstellar water masers in H II regions.** G. J. White, G. H. Macdonald.
Interstellar molecules, (see 012.033), p. 593 - 598 (1980).

131.242 **The interpretation of high velocity H_2O masers.** V. V. Burdyuzha.
Interstellar molecules, (see 012.033), p. 603 - 609 (1980).

Kinematic and physical models of high velocity H_2O masers are briefly discussed.

131.243 **Far ultraviolet objective spectrographic surveys for mapping of interstellar H_2, H, and CO.** G. R. Carruthers.
Interstellar molecules, (see 012.033), p. 611 - 612 (1980).

131.244 **OH 205.1 - 14.1.**
IAU Circ., No. 3502 (1980).

131.245 **Far-infrared interstellar carbon.**
Nature, Vol. 288, 321 (1980).

131.246 **Submillimeter heterodyne observations of CO at 433 microns.** H. Fetterman, G. Koepf, P. Goldsmith, B. Clifton, D. Buhl, N. Erickson, D. Peck, N. McAvoy, P. Tannenwald.
Bull. American Astron. Soc., Vol. 12, 751 (1980). – Abstract.

131.247 **Star formation in spiral galaxies.** A. V. Zasov.
Spiral structure of the Galaxy (see 012.034), Abastumansk. Astrofiz. Obs. Byull., No. 52, p. 55 - 58 (1980). In Russian.

The dependence of rate and efficiency of star formation on the gas density and the distance from the centre of a galaxy are considered.

131.248 **Observable traces of the star formation process.** R. B. Shatsova.
Spiral structure of the Galaxy (see 012.034), Abastumansk. Astrofiz. Obs. Byull., No. 52, p. 65 - 92 (1980). In Russian.

The stellar arrangement and motion in the Hercules region was investigated on the basis of data from Wilson's catalogue. The files arrangement fairly well follows the arrangement of gas-dust filaments. Apparently, the files have remained the traces of a star formation process.

131.249 **Distribution of the sources of maser radio emission in the spiral arms of the Galaxy.** G. M. Rudnitskij.
Spiral structure of the Galaxy (see 012.034), Abastumansk. Astrofiz. Obs. Byull., No. 52, p. 109 - 116 (1980). In Russian.

To check the density-wave theory of the spiral structure of the Galaxy, systematic differences in the kinematics of neutral hydrogen and of 18-cm OH line maser sources associated to the regions of star formation are investigated. The data on the radial velocities of OH masers and of 21-cm lines are interpreted in the framework of two versions of the wave theory – the linear density-wave theory and the shock-front wave theory. The result better fits the linear theory.

131.250 **Ammonia observations of DR 21, W51, NGC 1333, and other sources.** D. N. Matsakis, J. M. Bologna, P. R. Schwartz, A. C. Cheung, C. H. Townes.
Astrophys. J., Vol. 241, 655 - 675 (1980).

The ammonia clouds in DR 21, W51, NGC 1333, NGC 2264, L 134 N, and S140 have been mapped in several transitions. Ammonia has also been detected in several other sources as well. The DR 21 data reveal a long, narrow emission ridge and suggest a slight anomaly in the NH_3 (1,1) satellite emission, as was previously predicted. The validity of the single-temperature local thermodynamic equilibrium (LTE) model is discussed, and curves describing the effect of a gradient in excitation temperature are presented.

131.251 **Giant molecular complexes and OB associations. I. The Rosette molecular complex.**
L. Blitz, P. Thaddeus.
Astrophys. J., Vol. 241, 676 - 696, plates 9 - 15 (1980).

The authors present CO observations of the Rosette molecular complex which is associated with Mon OB2. The complex is extended along the galactic plane, has a maximum extent of ~100 pc, and shows a striking interaction with the Rosette Nebula. The complex appears to have a well defined boundary, and it is likely that the complex is no more than ~20% larger in any one dimension than the extent defined by our lowest contour. The molecular gas appears to be embedded in an H I cloud with a mass comparable to the molecular mass. The authors estimate the molecular mass to be $1.3 \times 10^5 M_\odot$, and the inferred volume density to be ~ 30 cm^{-3}. The complex possesses an overall velocity gradient of 0.20 km s^{-1} pc^{-1}. At high resolution the CO appears to be very clumpy. Evidence for very recent star formation appears at the site of unusual CO self-reversal, but not at the position of maximum CO emission.

131.252 **The interstellar material in front of χ Ophiuchus. II. Ultraviolet observations.** P. C. Frisch.
Astrophys. J., Vol. 241, 697 - 708 (1980).

Ultraviolet observations of the interstellar material in front of the star χ Oph are reported and analyzed: The observations made with the Copernicus satellite are described; the column densities of over 20 ions and molecules are found from these data; from these column densities the space density of the neutral regions is estimated using several different approaches; the ionized material in the line of sight is discussed; the results of this analysis are discussed. The conclusions drawn from these data support the general picture developed in Paper I for the interstellar gas in the direction of χ Oph.

131.253 **Interstellar C_3N: detection in Taurus dark clouds.** P. Friberg, Å. Hjalmarson, W. M. Irvine, M. Guélin.
Astrophys. J., Lett., Vol. 241, L99 - L103 (1980).

The authors report the first detection of the radical C_3N in interstellar sources, at the centers of the two dark clouds TMC 1 and TMC 2. The very small line widths observed for this low-order transition ($N = 3 \rightarrow 2$) allow a partial resolution of the hyperfine structure and a tentative estimate of this radical's hyperfine constants. The C_3N/HC_3N abundance ratio observed in the two clouds is similar to that in IRC + 10216, but much larger than the upper limit derived for Orion A and Sagittarius B2.

131.254 **On the origin of the grain-size spectrum of interstellar dust.** P. Biermann, M. Harwit.
Astrophys. J., Lett., Vol. 241, L105 - L107 (1980).

The authors show that grain-grain collisions are frequent in the atmospheres of red giants, where interstellar grains currently are believed to originate. Theoretical, as well as observational, arguments therefore suggest that grains should be injected into interstellar space with radii, a, that obey a power-law spectrum $a^{-3.5}$. This is in accord with models constructed by Mathis, Rumpl, and Nordsieck, and Mathis, who sought to fit interstellar extinction characteristics with simple grain-size distributions. The authors suggest that grain-grain collisions are also relevant in shells around novae and planetary nebulae and, thus, give the same grain-size spectrum from such sources of grains.

131.255 **Evidence for shocked interstellar gas toward the Perseus OB2 association.** S. R. Federman.
Astrophys. J., Lett., Vol. 241, L109 - L112 (1980).

Optical measurements of absorption lines of CH^+ and CH toward ζ, o, and ξ Persei reveal a shift in velocity between the two molecular species. The equivalent width of the CH^+ varies directly with both the amount of shift seen and the line width. The present observations confirm predictions of the chemical model formulated by Elitzur and Watson. The model requires shocks for producing sufficient amounts of CH^+; therefore, strong evidence for shocked interstellar gas is now available. Reexamination of the optical results of OH toward ζ and o Persei indicates that a substantial fraction of the OH also was formed in the high-temperature shocked gas.

131.256 **Ammonia excitation: the absorbing cloud toward W3 (OH).** T. Pauls, T. L. Wilson.
Astron. Astrophys., Vol. 91, L11 - L14 (1980).

The authors present new measurements of four NH_3 metastable and non-metastable absorption lines toward W3 (OH), and they have combined these data with previous results. The authors confirm that the size of the absorption region is 1″. The rotational temperatures between various combinations of these lines are all equal to ~ 50 K, which suggests that collisions are the dominant source of excitation. The cloud may be associated with some or all of the OH maser sources. The concentration of ammonia relative to molecular hydrogen is $\lesssim 6 \times 10^{-8}$.

131.257 **Interpretation of OH main line anomalies in interstellar clouds.** V. Bujarrabal, Nguyen-Q-Rieu.
Astron. Astrophys., Vol. 91, 283 - 289 (1980).

An excitation model has been developed to interpret the non-thermal OH 18-cm main line emission observed in interstellar clouds. A large scale velocity field produces the infrared line overlap which can be responsible for the difference of about 1-3 K between the main line excitation temperatures. The 1667 MHz anomaly can be achieved for small collapse velocities ($\lesssim 1$ km s^{-1}), whereas the 1665 MHz anomaly appears for higher velocities. The situation is reversed when the cloud is expanding. The effect of various collision laws on main line anomalies is discussed.

131.258 **A model for the H I cloud spectrum in the solar neighbourhood.** J. P. Chièze, B. Lazareff.
Astron. Astrophys., Vol. 91, 290 - 301 (1980).

The authors present a model for the evolution of the H I cloud spectrum consistent with a picture of the interstellar medium regulated by supernova explosions. They consider the net effect of evaporation of the clouds in the hot coronal gas, and of collisions with supernova remnants before and after the radiative phase. Besides these interactive effects with the intercloud medium, the authors treat collisions among clouds paying special attention to ensuing possible fragmentation. The model predicts a noticeable depletion of low mass clouds, mainly due to evaporation.

131.259 **OH 351.78-0.54.**
IAU Circ., No. 3509 (1980).

131.260 **The extinction of HD 200775 by dust in NGC 7023.**
G. A. H. Walker, S. Yang, G. G. Fahlman, A. N. Witt.
Publ. Astron. Soc. Pacific, Vol. 92, 411 - 417 (1980).

HD 200775 is obscured by the dust of the bright reflection nebula NGC 7023. In ground-based Reticon spectra between λλ5500 and 6800 the diffuse interstellar features are extremely weak and curvature, or very broad-band structure, in the extinction curve is significantly less than for other stars. The UV extinction curve has been generated from IUE spectra between λλ1250 and 3200. The λ2175 absorption is weak and the λ1540 feature found by Nandy et al. (1976) is absent, which resembles the case of θ^1 Ori, but the marked rise in extinction for $\lambda^{-1} > 6\mu^{-1}$ is typical of the 'average' extinction curve. Differences between the extinction curve for HD 200775 and the 'average' curve and that for θ^1 Ori give profiles of the λ2175 + λ1540 features, and the rapid increase in UV extinction, respectively. It is suggested that a three-component dust model is necessary to explain these observations.

131.261 **Onsala high spatial resolution observations of HCN, HCO$^+$, and their isotopes in Orion A and S140.**
O. E. H. Rydbeck, Å. Hjalmarson, G. Rydbeck, J. Elldér, H. Olofsson, A. Sume.
Res. Lab. Electron. Onsala Space Obs., Res. Rep. No. 140, 29 pp. (1980).

131.262 **The collision of clouds with a galactic disk.**
G. Tenorio-Tagle.
ESO Sci. Prepr. No. 102, 19 pp. (1980). – Submitted to Astron. Astrophys.

131.263 **The interstellar medium on the Gamma Cas line of sight.** R. Ferlet, A. Vidal-Madjar, C. Laurent, D. G. York.
ESO Sci. Prepr. No. 106, 36 pp. (1980). – Submitted to Astrophys. J.

131.264 **Hot-centred and cold molecular clouds.**
M. Rowan-Robinson.
Observatory, Vol. 100, 177 - 178 (1980). – Abstract.

131.265 **A random view of Cygnus X.**
S. Harris.
Observatory, Vol. 100, 178 (1980). – Abstract.

131.266 **Amplification of protostellar magnetic fields.**
A. P. Whitworth.
Observatory, Vol. 100, 179 (1980). – Abstract.

131.267 **Fragmentation of isothermal collapsing clouds.**
D. Wood.
Observatory, Vol. 100, 179 - 180 (1980). – Abstract.

131.268 **Star-formation and star-rotation.**
W. H. McCrea.
Observatory, Vol. 100, 180 (1980). – Abstract.

131.269 **Fünfzig Jahre interstellare Extinktion.**
J. Dorschner, J. Gürtler.
Sterne, 56. Band, 267 (1980).

131.270 **Staub und interstellare Extinktion.** E. J. Öpik.
Sterne, 56. Band, 268 - 270 (1980).

131.271 **Die interstellare Extinktionskurve.** C. Schalén.
Sterne, 56. Band, 271 - 281 (1980).

131.272 **Interstellare Verfärbung und Dreifarben-Photometrie.** W. Becker.
Sterne, 56. Band, 282 - 287 (1980).

131.273 **Sternenstaub.** E. P. Ney.
Sterne, 56. Band, 288 - 299 (1980). – Translation of a paper which appeared in Science, Vol. 195, 541 - 546 (1977). See Abstr. 19.131.059.

131.274 **Behavior of grains in the drift of plasma and magnetic field in dense interstellar clouds.**
T. Nakano, T. Umebayashi.
Publ. Astron. Soc. Japan, Vol. 32, 613 - 621 (1980).

Because most grains are negatively charged in dense interstellar clouds, they retard the drift of plasma and magnetic field and consequently the magnetic flux leakage from clouds. The authors investigate this effect for different situations.

131.275 **Chain-reacting thermal instability in interstellar CO clouds.** Y. Sofue, Y. Sabano.
Publ. Astron. Soc. Japan, Vol. 32, 623 - 627 (1980).

A new type of nonlinear, chain-reacting instability (CRI) is presented in which a sequential condensation occurs in a thermally unstable interstellar CO cloud, triggered by a local density perturbation. The CRI mechanism could be related to the sequential star formation, unless other stabilizing effects, e.g., magnetic pressure, turbulence, or rotation are present.

131.276 **Possible new interstellar masers.**
E. Gold, R. E. Hammersley, W. G. Richards.
Proc. R. Soc. London, Ser. A, Vol. 373, 269 - 284 (1980). Abstr. in Phys. Abstr., Vol. 84, Abstr. 9640 (1981).

131.277 **Observations of the ^{12}C and ^{13}C isotopes of formamide at 19 cm.**
F. F. Gardner, P. D. Godfrey, D. R. Williams.
Mon. Not. R. Astron. Soc., Vol. 193, 713 - 721 (1980).

Observations towards the Galactic Centre of the $1_{10}-1_{11}$ transition of the ^{13}C and ^{12}C isotopes of formamide (NH_2CHO) have been made. Spectra of $NH_2{}^{12}CHO$ showing all hyperfine components in emission were obtained for Sgr A and Sgr B2. No evidence for non-LTE behaviour was obtained.

131.278 **Very-high-resolution spectroscopy of interstellar Na I.** J. C. Blades, I. Wynne-Jones, R. C. Wayte.
Mon. Not. R. Astron. Soc., Vol. 193, 849 - 866 (1980).

A Michelson interferometer, operating at a spectroscopic resolution of $\sim 6 \times 10^5$, has been used to study interstellar Na I in the lines-of-sight to α, δ Cyg and η Tau. Theoretical line profiles have been fitted to the observations to determine the column density and velocity width of Na I. The major new result of the work is the discovery that Na I line widths are significantly smaller than previously supposed.

131.279 **Interstellar extinction in the direction of CI Cyg.** J. Mikołajewska, M. Mikołajewski.
Acta Astron., Vol. 30, 347 - 358 (1980).

B and *V* magnitudes of 190 stars in the vicinity of CI Cyg have been determined. The two-parameter spectral classification has been made. These data were used to estimate the interstellar extinction in the direction of CI Cyg ($l = 70°$, $b = +6°$).

131.280 **A search for the lowest-energy conformer of interstellar glycine.**
J. M. Hollis, L. E. Snyder, R. D. Suenram, F. J. Lovas.
Astrophys. J., Vol. 241, 1001 - 1006 (1980).

The authors have conducted the first search for the lowest-energy conformation of interstellar glycine. They have detected an emission line in Sgr B2 which is coincident in frequency with the $J_{K-K+} = 14_{1,14} - 13_{1,13}$ transition of conformer I glycine, but other transitions have not yet been found. Five unidentified interstellar lines were found.

131.281 **VLBI observations of the H_2O masers in Sagittarius B2.** B. G. Elmegreen, R. Genzel, J. M. Moran, M. J. Reid, R. C. Walker.
Astrophys. J., Vol. 241, 1007 - 1013 (1980).

The authors have mapped 33 velocity features of the H_2O maser emission at 1.35 cm wavelength from Sagittarius B2 to a relative accuracy of 4 milli-arcsec with a two-station VLBI experiment. Absolute positions were obtained by comparison with the Hat Creek interferometer observations (Forster et al. 1978) and are estimated to be accurate to 1″. The masers originate in three main groups spaced by 44″ in the north-south direction.

131.282 **Abundances and excitation of interstellar methyl formate.** E. Churchwell, A. Nash, J. Rahe, C. M. Walmsley, O. Lochner, G. Winnewisser.
Astrophys. J., Lett., Vol. 241, L169 - L174 (1980).

The authors report the detection of methyl formate ($HCOOCH_3$) in four transitions toward Orion KL, in two transitions toward Sagittarius B2, and in one transition toward W51. The relatively large column densities of $HCOOCH_3$ in clouds outside the galactic-center region probably imply that this molecule is widely distributed in the Galaxy.

131.283 **Detection of the $S(9)$, $\nu = 0 \rightarrow 0$ rotation line of the hydrogen molecule in Orion.**
R. F. Knacke, E. T. Young.
Astrophys. J., Lett., Vol. 242, L183 - L186 (1980).

The authors have detected the $S(9)$, $\nu = 0 \rightarrow 0$, $J = 11 \rightarrow 9$ line of H_2 in the molecular-hydrogen emission region in Orion. The line is much weaker (X 6) than expected from level populations at the previously determined excitation temperature of 2000 K in the shock region.

131.284 **Ionization front structure dependence on boundary conditions.** D. J. Mason.
Astron. Astrophys., Vol. 92, 117 - 127 (1980).

The dependence of ionization front structure on local interstellar conditions is investigated. Variations in physical and emission properties of *I*-fronts are evaluated for upstream temperatures from 10 to 5×10^3 °K, front velocities from 0.1 cm s^{-1} to 1000 km s^{-1}, stellar temperatures from 10^4 to 10^5 °K, upstream densities from 1 to 10^8 cm^{-3}, and for a range of helium, trace element and dust particle abundances. Empirical relationships between upstream and downstream conditions across an *I*-front are discussed. Departures from constant pressure and constant density flows for *D* and *R* type fronts respectively are found to be small, except near strong *D* or strong *R* conditions.

131.285 **Structure of molecular clouds. IV. Clouds with prominent star formation.** L. G. Stenholm.
Astron. Astrophys., Vol. 92, 142 - 150 (1980).

The available observations (mainly lines of CO) are used to derive the structure of molecular clouds with prominent star formation. Turbulent collapse, random motion and mixed models are discussed. It is concluded that only the mixed model agrees with the observations. The qualitative structure can be determined but the absolute values of H_2 density is very uncertain. This leads to a large uncertainty in the total cloud mass, by a factor of at least 10 - 100.

131.286 **High energy phenomena in molecular clouds.** A. Natta.
Variability in stars and galaxies, (see 012.044), p. E.4.1 - 4.11 (1980).

Observations at IR and mm wavelengths in the vicinity of the BN-KL cluster in Orion have revealed the presence of a violent energetic activity. Observations of collisionally excited H_2 emission at 2.2 μm from other molecular clouds suggest that shocks occur frequently in the life of molecular clouds. Stellar winds and supernovae explosions are discussed as possible sources of the observed shocks. The lifetime of the observed activity is in both cases shorter than 10^3 yrs.

131.287 **The W3 molecular cloud core: kinematics and ^{12}CO line inversion reexamined.**
E. Brackmann, N. Scoville.
Astrophys. J., Vol. 242, 112 - 120 (1980).

High angular resolution CO and ^{13}CO observations of the W3 molecular cloud are presented. Within the core at a radius of 0.6 pc high-velocity line wings are seen covering a full width of 20 km s^{-1}. The authors suggest that the central region is undergoing either a rapid, differential rotating or an asymmetric radial motion. If it is rotation, the angular momentum vector is closely aligned with that of the Milky Way. If it is radial motion, the flow could be an expanding envelope similar to what seems to occur in the Orion high-velocity core. The "absorption" dips previously noted in the ^{12}CO emission from the core source are probably due to the much larger W3 molecular cloud rather than radiative transfer effects within the core itself.

131.288 **Velocity structure in the Canis Major R1 molecular clouds.** D. E. Machnik, M. C. Hettrick, M. L. Kutner, R. L. Dickman, K. D. Tucker.
Astrophys. J., Vol. 242, 121 - 131 (1980).

The authors have mapped CO emission in the Canis Major OB1/R1 region. While most of the emission falls in the LSR velocity range 10 - 20 km s^{-1}, they find some material over the full velocity range covered (−30 to +45 km s^{-1}). The observations are consistent with the idea that some energetic process, which took place in an initially inhomogeneous cloudy medium, was responsible for the observed morphology of the region. Simple arguments suggest that a supernova explosion is the most likely candidate for this energetic process. The relationship between the process that shaped the clouds and star formation in the region is discussed.

131.289 **Infrared observations of a Bok globule in the Southern Coalsack.**
T. J. Jones, A. R. Hyland, G. Robinson, R. Smith, J. Thomas.
Astrophys. J., Vol. 242, 132 - 140 (1980).

The authors present near-infrared *JHK* photometry of 75 stars detected in a 2.2 μm survey of a 200 $arcmin^2$ area centered on Bok globule 2 in the Southern Coalsack. There is no evidence for a young cluster associated with the dark cloud; rather, the sample consists entirely of field stars. The color excess $E(J-K)$ is used to establish the density distribution in the globule. From these observations the mass of globule 2 is estimated to be a relatively low 11 $M_\odot$. There is inconclusive evidence that the gas-to-dust ratio in the Coalsack is abnormally low.

131.290 **Criteria for fragmentation in a collapsing rotating cloud.** P. Bodenheimer, J. E. Tohline, D. C. Black. Astrophys. J., Vol. 242, 209 - 218 (1980) = Lick Obs. Bull., No. 856.

The isothermal collapse of a rotating protostar with pressure and self-gravity is calculated with a hydrodynamic computer code that treats three space dimensions. A variety of initial conditions are tested to determine the conditions under which a cloud is unstable to fragmentation. The results show that fragmentation does not proceed to a significant extent during the first free-fall time, but that, in most cases, fragmentation has occurred by 1.5 to 2 times the initial free-fall time. The fragments are unstable to further collapse. The dependence of the results on spatial resolution is examined.

131.291 **Instabilities and star formation within ionization-shock fronts.** J. L. Giuliani, Jr. Astrophys. J., Vol. 242, 219 - 225 (1980).

The maximal growth rate of the fastest growing wavelength for the photon-driven-slab (PDS) instability of ionization-shock (I-S) fronts is compared with that one for the gravitational instability of shock-compressed slabs. For both instabilities the maximal growth rate changes in time due to the increasing column density of the slab. The analysis shows that the PDS instability can dominate the gravitational one until large column densities are attained within the I-S front. The equivalent time span is extremely long in low density H II regions with high velocity fronts, but it can also be significant in other regimes. The results are presented in a manner such that a sample of observed bright-rimmed clouds could test the conjecture.

131.292 **Infrared spectra of hydrated silicates, carbonaceous chondrites, and amorphous carbonates compared with interstellar dust absorptions.** R. F. Knacke, W. Krätschmer. Astron. Astrophys., Vol. 92, 281 - 288 (1980).

Infrared (2.5–30 μm) spectra of C1 and C2 carbonaceous chondrites have been obtained for mineral identification and comparison with interstellar dust. The authors discuss interstellar infrared absorption spectra using carbonaceous chondrites as a likely analogue of the interstellar material. Absorption bands at 2.7 and near 2.9 μm could be used to test for a hydrated silicate component in the interstellar dust. The interstellar 3.07 μm band is shown not to be absorption by bound water or hydroxyls in a hydrated silicate, but a broad feature between 3.1 and 4 μm in hydrated silicates resembles a feature observed in grain spectra. A band at 7 μm in carbonaceous chondrites is probably a carbonate absorption. The authors discuss the possibility that carbonates contribute to an interstellar absorption near 6.9 μm.

131.293 **Intermediate velocity clouds in the region of the south celestial pole.** R. Morras. Astron. Astrophys., Vol. 92, 315 - 319 (1980).

A correlation is shown between intermediate velocity clouds and a relative lack of material at low positive velocities in the south celestial pole region. This correlation can be explained by a supernova shell sweeping the interstellar medium. The counterpart at positive velocities would expand in the halo of the Galaxy. The IVC's gas is also correlated in position with high velocity clouds and such correlation would be explained by explosive events with energies higher than 10^{52} erg.

131.294 **On the density of star formation in the universe.** B. M. Tinsley, L. Danly. Astrophys. J., Vol. 242, 435 - 442 (1980).

The mean star formation rate per unit volume in the universe is estimated, from the distribution of colors of spiral galaxies, to be 0.05 $(H_0/100)^2\ M_\odot\ \mathrm{yr}^{-1}\ \mathrm{Mpc}^{-3}$. Implications of this star formation density are that the gas within spiral galaxies is being consumed on a time scale of about 0.25 of the Hubble time; stars in the disks of spiral galaxies form on about the Hubble time scale; and the amount of gas associated with galaxies was significantly greater in the recent past.

131.295 **Search for interstellar pyrrole and furan.** M. L. Kutner, D. E. Machnik, K. D. Tucker, R. L. Dickman. Astrophys. J., Vol. 242, 541 - 544 (1980).

The authors report unsuccessful searches for the organic ring molecules pyrrole (C_4H_4N) and furan (C_4H_4O). During these searches, a number of unidentified millimeter wavelength lines were also found; their frequencies are reported.

131.296 **The abundance of CO in diffuse interstellar clouds – an ultraviolet survey.** S. R. Federman, A. E. Glassgold, E. B. Jenkins, E. J. Shaya. Astrophys. J., Vol. 242, 545 - 559 (1980).

An ultraviolet survey of interstellar CO has been made with Copernicus in the $C–X$ 1088 Å and $E–X$ 1076 Å lines toward 48 bright stars. CO was detected in 17 directions, and upper limits were estimated for 21 others. Nine of the detections involve unsaturated features from which column densities are deduced. An overall picture of CO column densities toward bright stars is obtained by combining these results with those of other Copernicus investigations made with weaker transitions. Theoretical considerations and calculations are also given which suggest that the overall trends of the UV observations are in accord with current concepts of gas phase chemistry.

131.297 **Comparisons of interstellar CH^+ and H_2.** P. C. Frisch, M. Jura. Astrophys. J., Vol. 242, 560 - 567 (1980).

The authors have obtained Copernicus observations of H_2 toward stars in the Pleiades and 23 Orionis because of their importance in understanding the formation of interstellar CH^+. No model of CH^+ equilibrium seems to agree very well with these observations. These data do suggest that there is a correlation between the amounts of rotationally excited H_2 and CH^+ which are present.

131.298 **Interstellar HCO^+ self-reversals.** R. B. Loren, A. Wootten. Astrophys. J., Vol. 242, 568 - 575 (1980).

The authors have obtained high-resolution observations of HCO^+ toward the cores of several star-forming clouds which have self-reversals or enhanced wings in the CO profiles. The authors compare these observations with the corresponding CO profiles and demonstrate their usefulness for probing the HCO^+ abundance and excitation in intermediate-density regions of molecular clouds.

131.299 **The interstellar medium on the Gamma Cassiopeiae line of sight.** R. Ferlet, A. Vidal-Madjar, C. Laurent, D. G. York. Astrophys. J., Vol. 242, 576 - 583 (1980).

The interstellar medium on the γ Cas line of sight is studied through the observations of O I, H_2, H I, D I, and Ar I absorption features with the Copernicus satellite. By using the velocity structure of the line of sight previously determined through atomic nitrogen, the authors demonstrate that these neutrals are located in the same physical regions and that the O I and Ar I abundances could be solar, on the average, relative to N I in all components. The authors also reevaluated the D/H ratio. Furthermore they report a peculiar behavior of the Ar I abundance in the different components, possibly associated with high-velocity interstellar gas present on that line of sight. They conclude that the γ Cas line of sight could intercept a dense cloud which has been engulfed and disrupted by a shock.

131.300 **Star formation in IC 1848 A.**
H. A. Thronson, Jr., R. I. Thompson, P. M. Harvey, L. J. Rickard, A. T. Tokunaga.
Astrophys. J., Vol. 242, 609 - 614 (1980).

Recent far-infrared photometric and near-infrared spectroscopic observations of IC 1848 A/W5 East are reported. The source appears to be excited by a star of spectral type near B0, which can explain the far-infrared luminosity, 2 μm spectrum, radio continuum, and radio molecular observations. Although this star appears surrounded by a local very compact H II region, the density over a larger scale is relatively low. This puts constraints on models for the formation of this star.

131.301 **An anomalous ultraviolet extinction curve in the Taurus dark cloud.** T. P. Snow, Jr., C. G. Seab.
Astrophys. J., Lett., Vol. 242, L83 - L86 (1980).

Data from *IUE* have been used to derive an ultraviolet extinction curve for HD 29647, a B star embedded in the Taurus dark cloud complex. The curve appears normal, except that the 2200 Å extinction bump is absent. Arguments are developed to show that this is most likely due to a modification of the grain optical properties through the accretion of mantles in this dense cloud. If this conclusion is correct, it will have some impact on models for the formation of the 2200 Å bump.

131.302 **Detection of HC_5N in four dark clouds.**
P. J. Benson, P. C. Myers.
Astrophys. J., Lett., Vol. 242, L87 - L91 (1980).

The authors report new detections of the $J = 9 \rightarrow 8$ rotational transition of HC_5N in two dark clouds outside the Taurus complex, L134N and L778, and in two clouds in the Taurus complex, L1489 and L1536. Observations of the $(J, K) = (1, 1)$ and $(2, 2)$ rotation inversion transitions of NH_3 indicate line widths of $\sim$0.2 km s^{-1}, kinetic temperatures of $\sim$10 K, and number densities of $\sim 10^4$ cm^{-3}, properties typical of dark cloud cores. However, the abundance of HC_5N in these new sources is $\sim 10^{-9}$, $\sim$10 times smaller than in TMC-1.

131.303 **On methyl formate, methane, and deuterated ammonia in Orion A.** J. Elldér, P. Friberg, Å. Hjalmarson, B. Höglund, W. M. Irvine, L. E. B. Johansson, H. Olofsson, G. Rydbeck, O. E. H. Rydbeck, M. Guélin.
Astrophys. J., Lett., Vol. 242, L93 - L97 (1980).

Some 25 previously unreported interstellar lines have been detected in the 3–4 mm spectrum of Orion A and are attributed to methyl formate. A rotational temperature $T \gtrsim 50$ K and a beam-averaged column density $N \approx 2(10)^{14}$ cm^{-2} are estimated. A new limit on the abundance of formic acid suggests that $[HCOOCH_3]/[HCOOH] \gtrsim 3$. Methyl formate transitions at 76.702 and 76.711 GHz, at 85.927 GHz and at 86.224 GHz clearly correspond to previously reported spectral lines attributed to CH_4, NH_2D, and $(CH_3)_2O$, respectively. These results suggest that the detections of methane and deuterated ammonia in Orion A should be regarded as tentative.

131.304 **Interpretation of type I OH maser sources.**
W. H. Kegel, D. A. Varshalovich.
Nature, Vol. 286, 136 - 138 (1980).

A model for the interpretation of type I OH maser sources in suggested. This model combines the effect of velocity and magnetic field gradients with radiative pumping by means of the $(^2\Pi_{3/2}, J = 5/2 \rightarrow {}^2\Pi_{1/2}, J = 5/2)$ transition. Due to an overlap of particular hyperfine components, inversion of the 1,665 MHz line is favoured.

131.305 **1D, 2D and 3D collapse of interstellar clouds.**
W. M. Tscharnuter.
Space Sci. Rev., Vol. 27, (see 012.046), 235 - 246 (1980).

The review is concerned with recent theoretical investigations and numerical models of star formation with various symmetries.

131.306 **Fragmentation in rotating isothermal protostellar clouds.** P. Bodenheimer, J. E. Tohline, D. C. Black.
Space Sci. Rev., Vol. 27, (see 012.046), 247 - 253 (1980).

The authors report briefly the results of an extensive set of 3-D hydrodynamic calculations that have been performed during the past two and one-half years to investigate the susceptibility of rotating clouds to gravitational fragmentation. Because of the immensity of parameter space and the expense of computations, the authors have chosen to restrict this investigation to strictly isothermal collapse sequences.

131.307 **Collapse, equilibrium, and fragmentation of rotating, adiabatic clouds.** A. P. Boss.
Space Sci. Rev., Vol. 27, (see 012.046), 255 - 259 (1980).

Numerical calculations of the collapse of adiabatic clouds from uniform density and rotation initial conditions show that when restricted to axisymmetry, the clouds form either near-equilibrium spheroids or rings.

131.308 **On solar type protostars.**
K.-H. A. Winkler, M. J. Newman.
Space Sci. Rev., Vol. 27, (see 012.046), 261 - 266 (1980).

The formation of a 1 $M_\odot$ protostar in spherical symmetry has been followed in time dependent hydrodynamics with a detailed description of the equation of state and a careful treatment of radiative transport. The comparison of the dynamic evolution with observation is made in terms of the Hertzsprung-Russell diagram.

131.309 **Two-dimensional radiation-hydrodynamics calculations of the formation of O-B associations in dense molecular clouds.** R. I. Klein, M. T. Sandford II, R. W. Whitaker.
Space Sci. Rev., Vol. 27, (see 012.046), 275 - 282 (1980).

Two-dimensional calculations of ionization-shockwave propagation into a curved molecular cloud are presented. Density enhancement occurs due to the combined effects of cloud curvature and radiation flow. The star formation process is expected to be enhanced near the edges of irregularly shaped molecular clouds.

131.310 **Molecular spectroscopy of the interstellar medium.**
D. A. Varshalovich, V. K. Khersonskij.
Priroda, 1980, No. 12, p. 44 - 52. In Russian.

131.311 **UBVRI photometry of stars in the field of dark clouds of Taurus.**
V. E. Slutskij, O. I. Stal'bovskij, V. S. Shevchenko.
Pis'ma Astron. Zh., Tom 6, 750 - 755 (1980). In Russian.
English translation in Soviet Astron. Lett., Vol.6.

Results of UBVRI observations of 127 stars in the field of Taurus dark clouds are presented. The distance of these clouds is 132 ± 10 pc. The reddening law in direction of the clouds is close to the normal one. The stars associated with the clouds have been picked out. The space density of these stars is several times higher than that in the solar neighbourhood.

131.312 **On the distribution of molecules in the neighbourhood of young B stars.** A. A. Rejtblat.
Inst. kosm. issled. AN SSSR. 1980. 59 pp. In Russian.
Abstr. in Ref. zh., 51. Astron., 11.51.628 (1980).

131.313 **On the high velocity gas motions in the Orion molecular cloud.** R. A. Chevalier.
Astrophys. Lett., Vol. 21, 57 - 61 (1980).

The observed high velocity motions in the Orion molecular cloud may be the motion of many knots of gas. Shock waves which emit H_2 line radiation are driven into the knots when the knots interact with dense regions of the molecular cloud. The shock waves in the molecular cloud may be observable in Brackett α emission. Assuming $CO/H_2 \approx 3 \times 10^{-5}$, the momentum in the motions is too low by a factor 10 - 100 to be explained by a supernova. The knots are probably similar to the clumps of gas which give H_2O masers and Herbig-Haro objects.

131.314 **Collapse of ionized gas in galactic nuclei.** H.-H. Loose, K. J. Fricke.
Astrophys. Lett., Vol. 21, 65 - 70 (1980).

A simplified version of the virial equation is discussed for a spherical gas cloud centered at a spherical star cluster or a galactic nucleus having a stellar density distribution $\rho_* \propto r^{-\nu}$. It is shown, how the ordinary Jeans criterion for gravitational instability has to be modified for the gas in the cluster. For values $\nu < 2$ a gas cloud with a mass greater than a critical mass M_{crit} – different from the Jeans mass – will be gravitational unstable. In particular, for ionized gas in the core of an elliptical galaxy, M_{crit} is of the order of 10^5 solar masses which is in good correspondence to spectroscopic observations. Numerical calculations for the dynamical evaluation of the gas cloud confirm this analytical result.

131.315 **Linear polarization and grain growth.** P. A. Aannestad.
Publ. Astron. Soc. Pacific, Vol. 92, 546 (1980). – Abstract.

131.316 **Infrared and radio observations of the dark cloud L1084.** H. A. Thronson, Jr., C. J. Lada, M. F. Campbell, W. F. Hoffmann, B. Wilking.
Publ. Astron. Soc. Pacific, Vol. 92, 551 (1980). – Abstract.

131.317 **The galactic foreground reddening in the direction of the Magellanic Clouds.**
D. H. McNamara, K. A. Feltz, Jr.
Publ. Astron. Soc. Pacific, Vol. 92, 587 - 591 (1980).

Intermediate-band ($uvby\beta$) photometry has been secured of galactic foreground A and F stars in the direction of the SMC and LMC. The photometry yields mean color-excess values of $\langle E(b-y)\rangle = 0^m013$ for the SMC and $\langle E(b-y)\rangle = 0^m024$ for the LMC. By adopting de Vaucouleurs' apparent distance moduli of the Clouds and absorptions of $A_B = 0^m08$ (SMC) and $A_B = 0^m015$ (LMC) the authors find true distance moduli of $\mu_0 = 18.85$ (SMC) and $\mu_0 = 18.59$ (LMC).

131.318 **Spectral-line observations of the IC 5146 region.** W. H. McCutcheon, R. S. Roger, R. L. Dickman.
J. R. Astron. Soc. Canada, Vol. 74, 362 (1980). – Abstract.

131.319 **On the dynamics of giant interstellar gas clouds.** V. G. Surdin.
Astron. Tsirk., No. 1113, p. 3 - 4 (1980). In Russian.

131.320 **Scintillations of pulsars and the parameters of inhomogeneities of the interstellar plasma.**
A. V. Pynzar', V. I. Shishov.
Astron. Zh., Tom 57, 1187 - 1196 (1980). In Russian.
English translation in Soviet Astron., Vol. 24, No. 6.

It is shown that the interstellar plasma inhomogeneities have two characteristic scales: $a_0 = 3 \times 10^{10}$ cm and $a_1 = 3 \times 10^{13}$ cm. The inhomogeneities with the scale a_0 are responsible for the main effect of the pulsar scintillations in the decimetric waveband. The inhomogeneities with the scale a_1 are responsible for the long-time variations of the flux of the pulsars and quasars in the waveband near $\lambda = 1$m.

131.321 **Observations of radio emission in the 18-cm hydroxyl lines in the direction of Herbig-Haro objects and reflection nebulae.**
M. I. Pashchenko, G. M. Rudnitskij.
Astron. Zh., Tom 57, 1204 - 1214 (1980). In Russian.
English translation in Soviet Astron., Vol. 24, No. 6.

In 1978 observations of Herbig-Haro objects and R-associations in the main lines 1665 and 1667 MHz of the ground state of the OH molecule have been carried out. In most cases the observed line profiles have a simple single-peak structure with linewidths of 1 - 3 km/sec. This emission most probably originates in interstellar dust clouds surrounding the Herbig-Haro objects and R-associations. Some implications of the results obtained are briefly discussed.

131.322 **The thermal evolution of gas with primeval chemical composition contracting under the influence of gravitation.** Yu. A. Shchekinov, M. A. Ehdel'man.
Astron. Zh., Tom 57, 1287 - 1294 (1980). In Russian.
English translation in Soviet Astron., Vol. 24, No. 6.

The thermal evolution of spherical gas clouds with primeval chemical composition, contracting in a free-fall regime is calculated. The initial densities and temperatures vary from 0.1 cm^{-3} and 3000 K to 10 cm^{-3} and 10^4 K respectively. The cooling of protogalactic gas to low temperatures ($T \sim 10^2$ K) and a successful star formation are provided by H_2 molecules. The conditions needed for the formation of H_2 molecules are investigated. The influence of radiation from protoclusters of galaxies on the H_2 abundance is considered.

131.323 **A Fokker-Planck model for the stellar initial mass function.** H. Zinnecker, S. Drapatz, R. Cowsik.
Mitt. Astron. Ges., Nr. 50, p. 64 - 68 (1980).

Though star formation has become a very active field of astrophysical research, the distribution function for the masses of stars at birth, i. e. the initial mass function, is only poorly understood. The authors present a new statistical model for this distribution function.

131.324 **Untersuchung des (H_2 ^{12}CO)/(H_2 ^{13}CO) – Verhältnisses in interstellaren Molekülwolken.**
C. Henkel, C. M. Walmsley, T. L. Wilson.
Mitt. Astron. Ges., Nr. 50, p. 69 (1980).

131.325 **Zum Ionisationszustand des lokalen interstellaren Mediums.** N. Witt, P. W. Blum, S. Grzedzielski.
Mitt. Astron. Ges., Nr. 50, p. 115 - 118 (1980).

131.326 **H I und Staub in Kutners Wolke.** W. Batrla, T. L. Wilson.
Mitt. Astron. Ges., Nr. 50, p. 118 - 121 (1980).

131.327 **The hydrostatic equilibrium of interstellar gas and magnetic fields in the 6 kpc region of the Galaxy.**
B. Fuchs, H. Spreckels, K. O. Thielheim.
Mitt. Astron. Ges., Nr. 50, p. 156 - 160 (1980).

The authors apply a two-component gas model to the vertical hydrogen distribution in the 6 kpc region of the Galaxy as evaluated by Celnik, Rohlfs and Braunsfurth (1979) from 21-cm measurements of Westerhout et al. (1972) and Braunsfurth and Rohlfs (1978).

131.328 **Formaldehyd (H_2CO), Kohlenmonoxid (CO) und Kohlenstoffmonosulfat (CS) in TMC1.**
C. Henkel, T. L. Wilson, V. Pankonin.
Mitt. Astron. Ges., Nr. 50, p. 168 (1980). – Abstract.

131.329 **Interstellar medium.** W. L. Kraushaar.
X-ray and gamma-ray astronomy in the 1980's, (see 012.054), p. 24 - 26 (1979).

131.330 **A high-resolution optical survey of interstellar absorption lines toward globular clusters and extragalactic objects. I. Basic data.** A. Songaila, D. G. York. Astrophys. J., Vol. 242, 976 - 986 (1980).

The authors present the results of an investigation into the existence of gas in the halo of our own and neighboring galaxies, by means of a study at high resolution of interstellar sodium and calcium lines in the lines of sight to globular cluster cores, individual Magellanic Cloud supergiants, and the nuclei of Seyfert galaxies. A preliminary discussion of the number density of halo components gives a number of 0.5 per 100 kpc above the sensitivity limit, comparable to, but somewhat higher than, is indicated by 21 cm studies of QSO-galaxy pairs.

131.331 **The relative abundances of ^{28}Si, ^{29}Si, and ^{30}Si in the interstellar medium.** R. S. Wolff. Astrophys. J., Vol. 242, 1005 - 1012 (1980).

Millimeter-wave observations of the $J = 2 \rightarrow 1$ ($v = 0$) rotational transition of ^{28}SiO and the isotopically substituted species ^{29}SiO and ^{30}SiO have been made toward four dense molecular clouds. Additional observations of the $J = 3 \rightarrow 2$ ($v = 0$) transition of ^{28}SiO toward the same clouds were also made. The results are utilized to derive relative isotopic abundances in the interstellar medium. No evidence for variation of isotope abundance with galactic radius is found.

131.332 **Asymmetrical ^{13}CO lines in dark clouds: evidence for contraction.** P. C. Myers. Astrophys. J., Vol. 242, 1013 - 1018 (1980).

Four ^{13}CO spectra having similar and asymmetrical shapes (blueshifted peaks) were observed with 30 kHz resolution toward TMC2 and L134N. A radiative transfer model of a symmetric cloud indicates that each cloud is contracting along the two observed lines of sight. Contraction onto a static core fits the observed ^{13}CO and C^{18}O lines slightly better than contraction with $v \propto r$. The contraction speed ~0.7 km s^{-1} and velocity dispersion ~0.3 km s^{-1} are consistent with a 15 K cloud having a free-fall envelope and sonic turbulence.

131.333 **Turbulence and the stability of molecular clouds.** R. C. Fleck, Jr. Astrophys. J., Vol. 242, 1019 - 1022 (1980).

Molecular clouds may be stabilized against gravitational collapse by the turbulent velocity field within them. It is suggested that the energy derived from differential galactic rotation can maintain the turbulent flow in the interstellar medium. The characteristic decay time for interstellar turbulence is found to be ~10^{10} years. The rate and efficiency of star formation in giant molecular clouds reflect the stochastic nature of turbulence.

131.334 **X-ray photoionized nebulae.** J. P. Halpern, J. E. Grindlay. Astrophys. J., Vol. 242, 1041 - 1055 (1980).

The authors explicitly calculate the integrated optical emission from a spherical cloud in an intermediate case, $n = 10^4$ cm^{-3}, where collisional de-excitation of forbidden lines is density dependent. They make use of the recent charge-transfer calculations by Butler, Heil, and Dalgarno (1980) and Butler and Dalgarno (1979, 1980), and photoelectron interactions as calculated by Shull (1979). Approximate methods are developed for treating ionization by incoherent Compton scattering of X-rays and ionization by the diffuse ultraviolet flux.

131.335 **Infrared observations of Barnard 35: heat sources for bright-rimmed molecular clouds.** C. J. Lada, B. A. Wilking. Astrophys. J., Vol. 242, 1056 - 1062 (1980).

The authors have conducted a deep 2 μm wavelength infrared survey of an extended region of enhanced ^{12}CO emission in the bright-rimmed molecular cloud Barnard 35 in order to search for embedded heat sources. Twelve infrared sources were detected, all of which have been identified with visible, apparently foreground stars. Optical spectra were obtained for most of these sources, and the authors found them to be relatively late spectral type (i.e., F8 and later). They found no evidence for an embedded population of either early type (i.e., B and earlier) or intermediate type (F and earlier) stars in the cloud. The observations strengthen the suggestions that the energy sources for heating of certain bright-rimmed clouds are not embedded neo- or proto-stellar objects.

131.336 **H_2 fluorescence and the diffuse galactic light in the vacuum ultraviolet.** W. W. Duley, D. A. Williams. Astrophys. J., Lett., Vol. 242, L179 - L182 (1980).

Fluorescence by molecular H_2 within the Lyman and Werner bands, as well as to continuum levels of the $X^1\Sigma_g^+$ state, occurs at wavelengths $\lambda \leqslant 0.18$ μm. This emission is shown to be of intensity comparable to that scattered by dust over the same wavelength range in diffuse clouds. The implied increase in dust albedo in the vacuum ultraviolet (VUV) is likely due to this effect. Thus the small particles responsible for VUV extinction are of conventional chemical composition, i.e., absorbing oxides, silicates, or carbons.

A catalog of nonstellar water maser sources. See Abstr. 002.025.

Catalogue of extinction data of 12547 O . . . F stars, galactic clusters and δ-Cephei stars. See Abstr. 002.039.

Physics of the interstellar medium. See Abstr. 003.041.

Hundert Milliarden Sonnen. See Abstr. 003.066.

Infrared astronomy comes of age. See Abstr. 011.009.

Giant molecular clouds in the Galaxy. See Abstr. 012.064.

Detection of interstellar BS in the Cirrus dark cloud of the Numbbum association. I. An intuitive model and its subsequent observation. See Abstr. 015.027.

Large enhancements of electron impact produced OH $A-X$ emission in the presence of certain catalyzer gases. See Abstr. 022.021.

A kinetic theory of grain formation. See Abstr. 022.023.

Vibrational disequilibrium in astrophysical systems. See Abstr. 022.026.

Population inversion and suprathermal excitation in carbon monoxide. See Abstr. 022.039.

An analysis of collisional vibrational excitations and the astrophysical SiO maser phenomenon. See Abstr. 022.041.

A new, dielectronic-like recombination process for low temperatures and the radio recombination lines of carbon. See Abstr. 022.049.

Theoretical microwave spectral constants for C_2N, C_2N^+, and C_3H. See Abstr. 022.050.

Theoretical microwave spectral constants for C_3H^+ and C_4H^+. See Abstr. 022.051.

An ab initio prediction of the $J = 1 \leftarrow 0$ transition frequency of $HNCH^+$. See Abstr. 022.053.

Microwave spectra of molecules of astrophysical interest. XVIII. Formic acid. See Abstr. 022.063.

Amorphous interstellar grains: wavelength dependence of far-infrared emission efficiency. See Abstr. 022.068.

Laboratory measurements on the infrared features of interstellar silicate grains. See Abstr. 022.070.

Cosmic dust synthesized in reducing environments. See Abstr. 022.071.

The microwave spectrum of CH_2DOH. See Abstr. 022.095.

An additional uncertainty in calculated radiative association rates of molecular formation at low temperatures. See Abstr. 022.096.

Laboratory production of candidates for the diffuse interstellar bands. See Abstr. 022.098.

Recombination of ions and electrons on grains and the ionization degree in dense interstellar clouds. See Abstr. 022.100.

The interplay of molecular spectroscopy and astronomy. See Abstr. 022.106.

On the identification of mm-wavelength U-lines. See Abstr. 022.107.

Long carbon chain molecules in the laboratory and in space. See Abstr. 022.108.

Population inversion and suprathermal excitation in carbon monoxide. See Abstr. 022.109.

The hydrogen molecule as a collision partner. See Abstr. 022.110.

Recent laboratory work on molecules of possible importance for interstellar studies. See Abstr. 022.111.

Far infrared laser magnetic resonance spectroscopy. See Abstr. 022.112.

Laboratory measurements of oscillator strengths of ultraviolet molecular lines of HCl and H_2O and column densities of these molecules in the Zeta Ophiuchi cloud. See Abstr. 022.113.

Experimental measurements of ion-molecule reactions. See Abstr. 022.115.

An ICR study of ion-molecule reactions in the C_2H_2/HCN system. See Abstr. 022.116.

An ICR study of an association reaction at low pressure. See Abstr. 022.117.

Laboratory studies of interstellar carbon/nitrogen ion chemistry. See Abstr. 022.118.

The formation of complex interstellar molecules by radiative association. See Abstr. 022.119.

The formation of interstellar molecules via radiative association reactions. See Abstr. 022.120.

On the formation of interstellar linear molecules. See Abstr. 022.121.

Laboratory and modeling studies of chemistry in dense molecular clouds. See Abstr. 022.122.

Laboratory and theoretical results on interstellar molecule production by grains in molecular clouds. See Abstr. 022.123.

The formation of hydrocarbons and iron-hydrides on cold interstellar grains - experimental studies. See Abstr. 022.124.

Reproduction of the interstellar ice band by grain mantle analogs. See Abstr. 022.125.

Resonant interaction between an active molecular medium and a free electron laser. See Abstr. 022.146.

Ground state spectroscopic constants of $H^{15}NCS$, $HN^{13}CS$, and $HNC^{34}S$, and the molecular structure of isothiocyanic acid. See Abstr. 022.148.

Die Rolle des Laboratoriumsexperiments bei der Erforschung des kosmischen Staubes. See Abstr. 022.149.

A possible role for triplet H_2CN^+ isomers in the formation of HCN and HNC in interstellar clouds. See Abstr. 022.152.

Behavior of molecules on interstellar grains: application of the Langevin equation and iterative extended Hückel. See Abstr. 022.159.

Laboratory studies of isotope fractionation in the reactions of C^+ and HCO^+ with CO: interstellar implications. See Abstr. 022.160.

Sind Laserübergänge in chemisch erzeugtem OH möglich? See Abstr. 022.190.

New experimental possibilities and the future at far IR wavelengths. See Abstr. 031.560.

Future possibilities for ultraviolet observations of interstellar molecules. See Abstr. 032.542.

Ultraviolet spectroscopy of interstellar and intergalactic matter. See Abstr. 032.571.

ASTRO-HEL, ein Raketenexperiment zur Beobachtung der Heliumhintergrundstrahlung bei 30,4 nm und 58,4 nm. See Abstr. 032.578.

Molecular line observations at Parkes with a 3 GHz maser. See Abstr. 033.004.

Future spectral line research with the VLA. See Abstr. 033.017.

New experimental possibilities and future prospects for $1-5\mu m$ infrared spectroscopy of interstellar molecules. See Abstr. 034.028.

Interpretation of isotopic abundances in interstellar clouds. See Abstr. 061.030.

On absolute and convective instabilities. See Abstr. 062.008.

A new calculation on rotating protostar collapse. See Abstr. 062.029.

On exact equilibrium states in external gravitational fields of heated, self-gravitating gas clouds cooling by conduction and radiation. See Abstr. 062.054.

Expansions into magnetized media. See Abstr. 062.099.

Interstellar shock waves with magnetic precursors. See Abstr. 062.112.

Collapse and equilibrium of rotating, adiabatic clouds. See Abstr. 062.121.

Scattering by ensembles of small particles. Experiment, theory and application. See Abstr. 063.052.

Light scattering by spheroidal grains. See Abstr. 063.053.

Radiative transfer in dust clouds. I. Hot-centered clouds associated with regions of massive star formation. See Abstr. 063.068.

Star dust, mass loss and the late stages of stellar evolution. See Abstr. 064.016.

Mass loss from red giants: some effects on the interstellar medium. See Abstr. 064.082.

The evolution of protostars. II. The hydrostatic core. See Abstr. 065.070.

Model calculations of solar wind expansion including an enhanced fraction of ionizing electrons. See Abstr. 074.056.

Einflüsse hoher interstellarer Gasdichten auf die Sonnenwind-Expansion. See Abstr. 074.113.

Chemical energy in cold-cloud aggregates: the origin of meteoritic chondrules. See Abstr. 105.020.

Carbynes in meteorites: detection, low-temperature origin, and implications for interstellar molecules. See Abstr. 105.037.

A reanalysis of the observed interplanetary hydrogen Lα emission profiles and the derived local interstellar gas temperature and velocity. See Abstr. 106.006.

Star and planetary system formation in collapsing, viscous, rotating clouds. See Abstr. 107.028.

Molecules in celestial objects – I. Circumstellar CO in 9 Cephei (B2Ib). See Abstr. 112.001.

Circumstellar matter in young clusters. III. Diffuse lines in optical region. See Abstr. 112.003.

SiO maser emission from late-type stars and Orion. See Abstr. 112.008.

Time variation of SiO maser emissions. See Abstr. 112.030.

Time variability of the Orion A, R Leo and o Ceti SiO (v = 1, J = 2 – 1) masers. See Abstr. 112.031.

Pumping mechanisms of OH masers. See Abstr. 112.035.

Circular polarization from scattering by circumstellar grains. See Abstr. 112.036.

***UBV* sequences in three northern Milky Way regions and a comment on the interstellar extinction around $l = 90°$.** See Abstr. 113.022.

Vector space methods of photometric analysis. III. The two components of ultraviolet reddening. See Abstr. 113.044.

The metallicity of F and G dwarfs and the problem of star formation. See Abstr. 114.079.

Spectroscopic and polarimetric observations of NGC 1333 and the surrounding dark cloud complex. See Abstr. 114.140.

Das Ultraviolett-Spektrum des Sterns HD 269546. See Abstr. 114.163.

The peculiar, galactic object ESO 313-N*10. See Abstr. 121.009.

The interaction between the relativistic jets of SS433 and the interstellar medium. See Abstr. 125.005.

To what type of supernovae old galactic remnants belong; evolution of envelopes in the interstellar medium. See Abstr. 125.010.

Supernova radio remnants and the filling factor of the hot interstellar medium. See Abstr. 125.028.

The infrared echo of a supernova in a dust cloud. See Abstr. 125.066.

Infrared emission by dust grains near variable primary sources – III. Type II supernovae. See Abstr. 125.102.

Radio continuum interferometry of dark clouds. I. A search for newly formed H II regions. See Abstr. 132.001.

Radio continuum interferometry of dark clouds. II. A study of the physical properties of local newly formed H II regions. See Abstr. 132.002.

The large-scale far-infrared structure of W3 and W4. See Abstr. 132.012.

1.0 millimeter maps and radial density distributions of southern H II/molecular cloud complexes. See Abstr. 132.013.

Radio observations of H II regions in external galaxies. III Thermal emission, H II regions and star formation in 14 late-type galaxies. See Abstr. 132.025.

Extinction of H II regions. See Abstr. 132.032.

The spectrum and the structure of the bipolar nebula S 106. See Abstr. 132.036.

Analysis and interpretation of H I self-absorption lines. I. See Abstr. 132.037.

CO observations of several galactic H II regions. See Abstr. 132.041.

On the nature of the peculiar infrared source AFGL 2636. See Abstr. 133.002.

Ice-band polarimetry of GL 2591. See Abstr. 133.003.

An infrared candidate for OH 205.1 – 14.1. See Abstr. 133.004.

Maser emission from infrared stars. I. New OH and H_2O observations. See Abstr. 133.008.

Carbon monoxide observations of IRC 10442. See Abstr. 133.009.

A low-luminosity far infrared source in the L1551 molecular cloud. See Abstr. 133.010.

Infrared objects near to H_2O masers in regions of active star formation. See Abstr. 133.012.

Observations of H_2CO in the Orion Nebula at 1 centimeter wavelength. See Abstr. 134.018.

Discovery of optical molecular emission from the bipolar nebula surrounding HD 44179. See Abstr. 134.020.

Approximation of non-LTE coefficients. See Abstr. 134.023.

Globules in the Orion Nebula. I. Observations of interstellar sodium D 1 and D 2 lines. See Abstr. 134.030.

CO (J = 2 – 1) observations of the Carina nebula and G333.6–0.2 and a search for CO in LMC and SMC. See Abstr. 134.031.

On the helium and nitrogen enrichment of the interstellar medium by planetary nebulae. See Abstr. 135.034.

Ring nebulae associated with Wolf-Rayet stars in the Large Magellanic Cloud. See Abstr. 135.041.

Radio observations of W3 at 2.7 and 15.4 GHz. See Abstr. 141.109.

Spatial correlation of variables in the decimeter range of extragalactic sources with interstellar structures: loops, spurs, ridges. See Abstr. 141.175.

An unusual radio point source in M17. See Abstr. 141.183.

Revised stellar birthrates and the genesis of pulsars. See Abstr. 141.517.

The soft X-ray diffuse background and the structure of the local interstellar medium. See Abstr. 142.141.

Gamma rays and the local origin of cosmic rays. See Abstr. 142.532.

A cosmic-ray age based on the abundance of ^{10}Be. See Abstr. 143.008.

The acceleration of interstellar grains and the composition of the cosmic rays. See Abstr. 143.037.

Comments on stochastic acceleration of cosmic rays. See Abstr. 143.040.

On the sequential formation of subgroups in OB associations. See Abstr. 152.002.

Multicolour *UBVRI* photometry of stars in M 17. See Abstr. 152.005.

21-cm observations of the Cep IV star-formation region. See Abstr. 152.009.

The young open cluster NGC 3293 and its relation to Car OB1 and the Carina Nebula complex. See Abstr. 153.008.

The mass function for stars in a cluster: a theoretical derivation. See Abstr. 153.015.

O stars and the interstellar medium. See Abstr. 155.001.

The local density enhancement or local system: a new approach to determining the density function in the galactic plane. See Abstr. 155.003.

Surface density, warping and thickness of the galactic H I layer beyond the solar circle. See Abstr. 155.005.

Further high resolution H I observations of the Southern Milky Way and presentation of a complete longitude – velocity map for the galactic equator. See Abstr. 155.006.

The Parkes completely sampled survey of neutral hydrogen in the Southern Milky Way. See Abstr. 155.007.

H I observations towards the Puppis Window of the Galaxy. See Abstr. 155.008.

Galactic spiral structure in a cloudy, supernova-dominated interstellar medium. See Abstr. 155.009.

Molecular clouds and galactic spiral structure. See Abstr. 155.014.

The galactic rotation curve to R = 18 kpc. See Abstr. 155.027.

Carbon monoxide in the inner Galaxy: the 3 kiloparsec arm and other expanding features. See Abstr. 155.042.

The chemical evolution of the solar neighborhood. II. The age-metallicity relation and the history of star formation in the galactic disk. See Abstr. 155.043.

The dynamics of the molecular gas in the galactic center. See Abstr. 155.049.

A recombination line survey of the Milky Way. See Abstr. 156.004.

2.4 μm emission from the galactic centre region. See Abstr. 156.007.

Further observations of radio recombination lines from the direction of the Galactic Centre. See Abstr. 156.010.

X-ray characteristics of Loop I and the local interstellar medium. See Abstr. 157.010.

Neutral hydrogen study of 40 Sa spiral galaxies. See Abstr. 158.023.

An optical analysis of dust complexes in spiral galaxies. See Abstr. 158.028.

The galactic distribution of angular momentum density and star formation. See Abstr. 158.036.

A study of star formation and spiral structure in the Sc galaxy M33. See Abstr. 158.037.

Detection of the CO $J = 2 \rightarrow 1$ line in M82 and IC 342. See Abstr. 158.080.

On the abundance of carbon monoxide in galaxies: a compairson of spiral and Magellanic irregular galaxies. See Abstr. 158.094.

Neutral hydrogen in isolated galaxies: first results for five early-type systems. See Abstr. 158.096.

The size distribution of dark clouds as a new method to determine the distances of galaxies and the amount of dust. See Abstr. 158.116.

Bursts of star formation in the central region of the hot-spot nucleus galaxy NGC 4314. See Abstr. 158.131.

Star formation in the nuclei of normal galaxies. See Abstr. 158.142.

Centers of star formation in the nuclei of galaxies. See Abstr. 158.143.

Optical and theoretical studies of giant clouds in spiral galaxies. See Abstr. 158.158.

Further observations of the H_2O emission from NGC 4945. See Abstr. 158.160.

Centres of star formation in galaxies having no spiral structure. See Abstr. 158.163.

H I observations and star formation in the blue compact galaxy IZw 18. See Abstr. 158.172.

The ionised gas in NGC 5128: evidence for a shock-heated component. See Abstr. 158.183.

Dust-to-gas ratio in the outer regions of spiral galaxies. See Abstr. 158.206.

Are "anemic" spirals deficient in neutral hydrogen? See Abstr. 158.321.

Neutral hydrogen in elliptical galaxies: a bimodal distribution. See Abstr. 158.323.

The giant and supergiant shells of the Magellanic Clouds. See Abstr. 159.002.

Ultraviolet studies of the Magellanic Clouds. II. Internal extinction, formation of massive stars, comparison with other galaxies. See Abstr. 159.014.

Ultraviolet studies of the Magellanic Clouds – I. Interstellar lines in the spectra of FD 70 and SK-71-45. See Abstr. 159.019.

A note on interstellar absorption in the Magellanic Clouds. See Abstr. 159.022.

The gravitational fragmentation of primordial gas clouds. See Abstr. 162.056.

132 H I, H II Regions

132.001 **Radio continuum interferometry of dark clouds. I. A search for newly formed H II regions.**
W. Gilmore.
Astron. J., Vol. 85, 894 - 911 (1980).

A search for compact H II regions embedded in dark clouds has been carried out in an effort to study local massive star formation. Approximately 20% of the total area of opaque dark cloud material in the sky with $A_v \geqslant 6$ mag was surveyed with the NRAO three-element interferometer at 2695 MHz, and at least 5% more was surveyed with the NRAO 300-ft. telescope at 4750 MHz. The regions surveyed include the dark cloud complexes in Perseus, Taurus, Orion, and Ophiuchus, as well as several smaller cloud complexes and individual clouds. No hidden compact H II regions embedded inside dark clouds were detected with certainty in the radio continuum. However, 11 H II regions with associated visible emission and 18 other sources identified as possible H II regions were detected. Comparison of the single dish and interferometer results indicates that selection effects, which may be the result of the process of local massive star formation, were present in the survey. This paper reports the general results of the observations.

132.002 **Radio continuum interferometry of dark clouds. II. A study of the physical properties of local newly formed H II regions.** W. Gilmore.
Astron. J., Vol. 85, 912 - 944 (1980).

In this paper the individual regions of star formation observed in Paper I are studied in detail. By learning more about several individual regions, one may investigate correlations of the presence or absence of various signposts for star formation, and surmise general physical properties of local newly formed massive stars. Furthermore, one may use these properties to deduce observational selection effects that might exist in the specific techniques used in this survey.

132.003 **Radio observations of DR15 at 408, 1407 and 2695 MHz.** D. Colley.
Mon. Not. R. Astron. Soc., Vol. 192, 377 - 388 (1980).

Observations of the galactic H II region DR15 have been made at frequencies of 408 and 1407 MHz with the Cambridge One-Mile telescope. Adjacent to the main source DR15, several previously unknown weak radio sources were found, some of them being correlated with optical emission present on the PSS E print. The source has also been observed at 2695 MHz using the Cambridge 5-km telescope with a resolution of 3.7 × 5.7 arcsec. The brightest source shows considerable internal detail, comprising ionization ridges and individual H II regions. One radio source, component B, is probably excited by two stars visible on the PSS prints and of spectral type O9.5–B0. This source appears to be well described by the Blister model and has an estimated extinction in the visual of 6.4 mag.

132.004 **The helium abundance of galactic H II regions.**
C. Thum, P. G. Mezger, V. Pankonin.
Astron. Astrophys., Vol. 87, 269 - 275 (1980).

Hydrogen and helium recombination lines have been observed at 15 and 22 GHz from 13 galactic H II regions. These data support earlier evidence for a decrease in the total helium abundance from 7 to 15 kpc galactic radius and an increase in the electron temperature of H II regions from 5 to 15 kpc. The present high angular resolution data show that the low values of the ionized helium abundance observed earlier for some H II regions are due to the presence of neutral helium in the H II regions. Inside 7 kpc this neutral fraction confuses the evaluation of the total helium abundances.

132.005 **8–13 μm spectrophotometry of S 106.**
H. Hefele, E. Hölzle.
Astron. Astrophys., Vol. 88, 145 - 148 (1980).

8–13 μm spectrophotometry with mean resolution $\Delta\lambda/\lambda$=0.02 of three compact infrared condensations associated with the galactic H II region S 106 are presented. Strong emission in the [Ne II] λ 12.81 μm fine-structure line has been detected. The line seems to be emitted in a moderate excitation nebula compatible with a late O-type ZAMS star as central source. The unidentified 11.3 μm emission feature appears rather conspicuously in the spectra. Silicate absorption at 9.7 μm is only marginally present.

132.006 **A continuum map of IC 1396 at 2695 MHz.**
H. E. Matthews, C. G. T. Haslam, D. L. Hills, C. J. Salter.
Astron. Astrophys., Vol. 88, 285 - 288 (1980).

The H II region IC 1396 has been mapped in the radio continuum at 2695 MHz with a half power beamwidth of 4.'4. The sensitivity of the present observations greatly exceeds that of previous work. Both the angular resolution and the observing frequency are the highest yet employed in mapping the complete nebula.

132.007 **Evaporation of H I clouds in supernova remnants.**
H. Tsunemi, H. Inoue.
Publ. Astron. Soc. Japan, Vol. 32, 247 - 256 (1980).

Evaporation of H I clouds in the supernova remnants is studied. The authors suppose an interstellar environment where H I clouds of a density of the order 10^2 H atoms cm^{-3} are interspersed in a tenuous plasma of a density 10^{-2} H atoms cm^{-3}. If it is assumed that a shock wave with a pressure about H^6 K cm^{-3} propagates within such an interstellar medium, the tenuous medium will be heated as hot as 10^8 K. The evaporation of the H I clouds in these circumstances is numerically studied. The results show that the hot plasma produced by the evaporation of the clouds well accounts for the observed spectrum of the Cygnus Loop, and that the predicted spectrum is qualitatively different from that a simple blast-wave theory predicts. This model also explains the association of the X-ray emission with the optical fragments.

132.008 **The motions in N51D, a bubble-like nebula in the Large Magellanic Cloud.** B. M. Lasker.
Astrophys. J., Vol. 239, 65 - 73 (1980).

A large bubble-like H II region in the Large Magellanic Cloud, N51D, is shown to have internal motions of about 30 km s^{-1} but structure that is more complex than that expected for an expanding spherical shell. It is proposed that the motions are mainly driven by supersonic stellar winds from early-type stars in the central cluster, NGC 1955, and that the departures from spherical symmetry arise from interactions with a nonuniform component of the interstellar medium.

132.009 **Long-slit spectroscopy in the rocket ultraviolet of the Orion Nebula.**
R. C. Bohlin, J. K. Hill, T. P. Stecher, A. N. Witt.
Astrophys. J., Vol. 239, 137 - 145, plate 5 (1980).

Ultraviolet spectra of the Orion Nebula were obtained with a rocket-borne telescope in three 1 minute exposures by using a microchannel plate detector and film. The slit, 17′ long and 22″ wide, was centered on a line passing near θ^2 Orionis B and through a point 38″ northeast of the Trapezium star θ^1 Orionis C. In addition to the spectrum of θ^2 Ori B, the rocket data consist of the spectrum of the nebula.Three emission lines are present in the nebula, the C III] intercombination doublet at 1908 Å, the C II] intercombination line at 2326 Å and the [O II] forbidden doublet at 2470 Å. The absolute intensities of the lines are tabulated and a simple determination gives a C/O abundance ratio of unity, compared to 0.5 in the sun. The continuum is produced primarily by the scattering of starlight from θ^1 Ori C on the dust particles in the nebula. The authors discuss the implications of their observations in terms of the grain albedo and scattering properties.

132.010 **H I shells in the Galaxy.** E. M. Hu.
Bull. American Astron. Soc., Vol. 12, 468 (1980). Abstract.

132.011 **The largest H II regions in M101.**
L. Blitz, G. Neugebauer, M. Matthews, F. H. Israel.
Bull. American Astron. Soc., Vol. 12, 493 (1980). – Abstract.

132.012 **The large-scale far-infrared structure of W3 and W4.**
H. A. Thronson, Jr., M. F. Campbell, W. F. Hoffmann.
Astrophys. J., Vol. 239, 533 - 539 (1980).

The large-scale structure of the giant H II region complex W3/W4 has been mapped in two wavelength bands covering 60 μm to 120 μm at an angular resolution of 12′. The authors present a simple model relating the far-infrared and CO emission.

132.013 **1.0 millimeter maps and radial density distributions of southern H II/molecular cloud complexes.**
L. H. Cheung, J. A. Frogel, D. Y. Gezari, M. G. Hauser.
Astrophys. J., Vol. 240, 74 - 83 (1980).

1.0 mm continuum mapping observations have been made of seven southern-hemisphere H II/molecular cloud complexes with 65″ resolution. The radial density distribution of the clouds with central luminosity sources was determined. Strong similarities in morphology and general physical conditions were found to exist among all of the southern clouds in the sample.

132.014 **Neutral hydrogen to the west of λ Orionis.**
N. V. Bystrova.
Pis'ma Astron. Zh., Tom 6, 579 - 581 (1980). In Russian.
English translation in Soviet Astron. Lett., Vol. 6.

Results of investigation of the neutral hydrogen within the region of the Hα emission from two filaments and a loop near λ Ori are discussed.

132.015 **On the composition of H II regions in southern galaxies – II. NGC 6822 and 1313.**
B. E. J. Pagel, M. G. Edmunds, G. Smith.
Mon. Not. R. Astron. Soc., Vol. 193, 219 - 230 (1980).

AAT/IPCS spectra have been obtained from seven outlying H II regions in the Local Group Im galaxy NGC 6822 and from the nucleus and six H II regions at various radial distances in the isolated SBd galaxy NGC 1313. Abundances are deduced from emission-line intensities. Both galaxies have quite uniform abundances all over, in agreement with previous findings for the arm regions of a barred spiral and for Magellanic Irregulars.

132.016 **High-resolution radio observations of three southern H II regions.** D. S. Retallack, W. M. Goss.
Mon. Not. R. Astron. Soc., Vol. 193, 261 - 267 (1980).

The authors present high-resolution radio observations of the sources G333.6–0.2, G291.3–0.7 (NGC 3576) and G291.6–0.5 (NGC 3603) at 1415 MHz. These sources are among the most luminous radio H II regions in the Galaxy and are associated with bright infrared sources. Improved positions and parameters are presented, and the relation between the radio-emitting regions and the infrared sources is discussed.

132.017 **The evolved H II region S 115.**
R. Harten, M. Felli.
Astron. Astrophys., Vol. 89, 140 - 144 (1980).

Aperture synthesis radio observations at 0.610 GHz of the extended H II region S 115 are presented. A shell of ionized gas containing about 650 $M_\odot$ and a less regular blob of ionized gas situated on opposite sides of the open cluster Berkeley 90 were detected. From the ionization balance it appears that the nebula is excited by the stars in the open cluster, in particular the O6 star LS III 4612. The region seems to be in an advanced stage of its evolutionary development. The observed structures can be successfully interpreted using the recent champagne flow model.

132.018 **Aperture synthesis observations of recombination lines from compact H II regions. I. W49A and W51A.** J. H. van Gorkom, W. M. Goss, P. A. Shaver, U. J. Schwarz, R. H. Harten.
Astron. Astrophys., Vol. 89, 150 - 157 (1980).

The authors have observed the H109α line in the sources W49A and W51A. In one component in W51A the observed ratio is significantly greater than the expected LTE value; the most likely explanation is stimulated emission. In both sources the more compact components have low line to continuum ratios. This effect can be interpreted in terms of pressure broadening at electron densities in excess of 10^5 cm^{-3}.

132.019 **A three-dimensional model of IC 1396.**
H. J. Wendker, J. W. M. Baars.
Astron. Astrophys., Vol. 89, 180 - 186 (1980).

The authors present a high resolution map of the H II of IC 1396 at 2695 MHz. They propose the exciting star, HD 206267, is approaching a group of denser clouds and has driven a horseshoe type ionization front partly into them. The authors speculate that the early type star possesses an extended region of unimpeded stellar wind flow and that a possible faint central source is the beginning of a nebular shell resembling shells around a few Wolf-Rayet stars.

132.020 **The Cygnus X region. XII. On the excitation and distance of the Gamma Cygni H II region.**
I. Appenzeller, H. J. Wendker.
Astron. Astrophys., Vol. 89, 239 - 240 (1980).

The star suspected of exciting the γ Cyg H II region is classified as O 9 V from new spectrograms. The H II region, at a newly derived distance of 1.5 kpc, is in the transition stage from compact to dense. It most probably belongs to the H II region complex of IC1318 b and c instead of to the γ Cyg supernova remnant.

132.021 **The structure and dynamics of H II regions.**
G. L. Welter.
Astrophys. J., Vol. 240, 514 - 523 (1980).

The early evolution of H II regions, through the first few times 10^5 years, is investigated under conditions of enhanced central density and increasing cluster luminosity. It is found that a D-type ionization front can overtake the precursor shock under conditions of a fairly steep density gradient such as may exist just beyond a stellar dust cocoon, or as new stars turn on within the H II region. The shock is converted into a strong, outward moving pressure pulse as it is ionized.

132.022 **Observations of galactic H II regions with the Fleurs Synthesis Telescope.** D. S. Retallack.
Proc. Astron. Soc. Australia, Vol. 3, 278 - 279 (1978).

132.023 **An H 167α line map of the extended H II region S 252 (NGC 2175).**
A. Donati Falchi, M. Felli, G. Tofani.
Astron. Astrophys., Vol. 89, 363 - 369 (1980).

The recombination line H 167α has been observed in 21 selected positions of the H II region S 252. The T_e^* and V_{LSR} dependence on the position has been studied and related to the morphology of the region where the exciting star is at the center of the ionized gas and a molecular cloud is located on the west border. There is an increase of T_e^* from the west to the east side which can be only partially explained by a density model. The V_{LSR} shows a decrease from the molecular cloud to the outer ionized gas. This trend qualitatively agrees with a champagne model of a gas flow from the surface of a molecular cloud.

132.024 **Why is observable radio recombination line emission from galactic H II regions always close to LTE?**
P. A. Shaver.
Astron. Astrophys., Vol. 90, 34 - 43 (1980).

There is no evidence for significant deviations from LTE in single-dish observations of radio recombination line emission from galactic H II regions. This is in agreement with the known properties of H II regions, particularly their density variations and limited range of excitation parameters; the optimum configuration for strong observable non-LTE effects, low electron density and high emission measure, simply does not exist in galactic H II regions, and the observed lines are emitted under near-LTE conditions. Models of the Orion Nebula and NGC 6604 are presented which fit all available data and show only weak stimulated emission. It is concluded that reliable electron temperatures can indeed be obtained from straightforward analysis of appropriate radio recombination lines.

132.025 **Radio observations of H II regions in external galaxies. III. Thermal emission, H II regions and star formation in 14 late-type galaxies.** F. P. Israel.
Astron. Astrophys., Vol. 90, 246 - 268 (1980).

High-resolution radio maps of 14 late-type galaxies are analysed to yield the properties of bright H II regions. The total sample contains 252 H II regions in 14 galaxies. The brightest H II regions tend to be found at 40% of the de Vaucouleurs radius; they are responsible for a considerable part of the total thermal emission of each galaxy. Estimates are given for the total amount of thermal emission and for the present rate of star formation in each galaxy; an estimate is also given for the number of supernova remnants expected to be present in the radio maps.

132.026 **Infrared spectroscopy with a balloon borne Michelson interferometer. II. Observation of O III, O I, and N III fine structure lines in H II regions.**
A. F. M. Moorwood, P. Salinari, I. Furniss, R. E. Jennings, K. J. King.
Astron. Astrophys., Vol. 90, 304 - 310 (1980).

Observations of the [O III] ionic fine structure lines at 52 μm and 88 μm, made at a resolution of 0.05 cm^{-1} with a balloon borne telescope and Michelson interferometer, are presented for the H II regions W 51, G 333.6–0.2, M 17 S, M 17 N, NGC 6357, and NGC 6334. Values for the electron density deduced from the line ratios are found to agree with the radio data, while the O^{++} abundances indicate a lower excitation than expected in many cases. The authors report the first detection of the [N III] line at 57 μm which was observed from both sources in M 17 and gives the abundance ratio N/O = 0.13. This line was also marginally detected on W 51.

132.027 **Two large H I expanding shells in the field of NGC 7538.** P. L. Read.
Mon. Not. R. Astron. Soc., Vol. 193, 487 - 494 (1980).

Observations are presented of two large 'filamentary' H I emission features which appeared, by chance, in an aperture synthesis survey of 21-cm emission towards NGC 7538. One feature is part of the large, roughly spherical expanding shell of gas related to the Cep OB3 association. The feature is consistent with a type II supernova remnant as suggested by Assousa, Herbst & Turner but could also represent a 'bubble' produced by stellar winds from the stars in Cep OB3. The other feature also appears to be an expanding spherical shell. This object is more distant and larger in extent than the Cep OB3 shell. It may be an old supernova remnant. A stellar wind origin is also plausible energetically, but no correlation is observed between the shell and any known OB association.

132.028 **Fine structure in positive-velocity clouds.**
R. J. Cohen, R. A. Ruelas Mayorga.
Mon. Not. R. Astron. Soc., Vol. 193, 583 - 591 (1980).

The authors present a detailed study of one positive-velocity cloud complex made with angular resolutions 0.°5 and 0.°2 and velocity resolutions 7.3 and 3.7 km s^{-1}, and compare the data with those obtained from negative-velocity clouds at similar resolution.

132.029 **The infrared emission of G333.6–0.2, an extremely nonspherical H II region.** A. R. Hyland, P. J. McGregor, G. Robinson, J. A. Thomas, E. E. Becklin, I. Gatley, M. W. Werner.
Astrophys. J., Vol. 241, 709 - 718 (1980).

The southern H II region G333.6–0.2, which has a total luminosity of $3.3 \times 10^6 L_\odot$ (for an assumed distance of 4 kpc), has been mapped at 2.2, 10, 30, 50, and 100 μm. At all wavelengths, the surface brightness of the infrared radiation is unusually high and the structure of the source is compact and symmetrical. The observations at 10 μm with a beam size of 1.″1 show a bright central core, $\lesssim$0.07 pc in extent, coincident with the center of the ionized region. Most of the luminosity of the source is emitted at far-infrared wavelengths ($\lambda \gtrsim 30\ \mu$m); the far-infrared optical depth is $\gtrsim$1. The present observations, together with the previous data on this subject, suggest strongly that G333.6–0.2 is excited by a single luminous object, or a very compact cluster, which has formed on the front surface of a dense molecular cloud as seen from Earth. It is shown that the spectral and spatial characteristics of the infrared radiation can be understood in terms of this "blister" model.

132.030 **Accurate electron temperatures from radio recombination lines.** P. A. Shaver.
Astron. Astrophys., Vol. 91, 279 - 282 (1980).

There is a unique frequency at which $T_e^* = T_e$, for any given emission measure. It is given by $\nu \sim 0.081\ EM^{0.36}$, a relation fixed only by atomic parameters and virtually independent of the density, temperature, or structure of the H II region. This remarkable property makes it possible to minimize non-LTE corrections, and thus to obtain the most reliable and accurate electron temperature.

132.031 **The "helium problem" in the source DR 21.**
A. Pitault.
Astron. Astrophys., Vol. 91, 374 - 375 (1980).

Following the paper of Pitault and Cesarsky (1980) concerning Sgr B2, the author tries to give a similar interpretation of the radio recombination line observations of the source DR 21. Particularly, he determines the influence of pressure broadening on the [He^+/H^+] ratio, and shows that it contributes greatly to the failure to detect the He 109α line.

132.032 **Extinction of H II regions.**
F. P. Israel, R. C. Kennicutt.
Astrophys. Lett., Vol. 21, 1 - 9, with a correction p. 81 (1980).

Visual extinction of H II regions in nine nearby galaxies as derived from the ratio of radio continuum emission to Hα emission is systematically larger than visual extinction deduced from the Balmer lines alone, if one assumes a value of 3. An optically-limited sample of about 30 extragalactic H II regions has a mean extinction of 1.^{m}7 in the visual; about 1.^{m}2 is not seen in the reddening of the Balmer lines. Both reddening and extinction decreases with increasing galactic radius, at least for M33 and M101.

132.033 **Internal motions in H II regions. VIII. The nebular complex S152 - S153.**

P. Pişmiş, I. Hasse.
Rev. Mexicana Astron. Astrofis., Vol. 5, 39 - 44 (1980).

Radial velocities at 170 points in the nebular complex S152 and S153 are determined by Fabry-Perot interferometry. It is shown that the nebulae are at the same distance. The kinematic distance of the complex is found to be larger than the distance of the exciting stars. Evidence is given for the younger age of S152 compared to S153.

132.034 **Abundances of argon, sulphur and neon in six galactic H II regions from infrared forbidden lines.**
T. Herter, H. L. Helfer, J. L. Pipher, W. J. Forrest, J. McCarthy, J. R. Houck, R. C. Puetter, R. S. Rudy, S. P. Willner, B. T. Soifer, F. C. Gillett.
News Lett. Astron. Soc. N. Y., Vol. 1, No. 7, p. 18 (1980). Abstract.

132.035 **H66α radio recombination line observations of southern H II regions.**
Z. Abraham, J. R. D. Lépine, M. A. Braz.
Mon. Not. R. Astron. Soc., Vol. 193, 737 - 743 (1980).

The H66α recombination line has been observed in Orion A and in four prominent southern H II regions: RCW 38, Car I, RCW 57 and Norma III. The physical parameters line-to-continuum temperature ratio, line halfwidth, and LTE electron temperature, are given.

132.036 **The spectrum and the structure of the bipolar nebula S 106.** J. Solf.
Astron. Astrophys., Vol. 92, 51 - 56 (1980).

Optically the compact H II region S 106 appears as a bipolar nebula with the exciting stellar source located between the lobes and embedded in a flat disk of material of high visual extinction. Associated with the nebula is a massive molecular cloud exhibiting a rotating disk-like structure. The emission line spectrum of both lobes resembles that of the Orion nebula and indicates high electron density throughout. The nebular continuum discovered in both lobes is interpreted as originating from an early-type stellar source between the lobes, and scattered by dust particles coexisting with the ionized gas within the lobes.

132.037 **Analysis and interpretation of H I self-absorption lines. I.** F. H. Levinson, R. L. Brown.
Astrophys. J., Vol. 242, 416 - 423 (1980).

The authors discuss the potential and actual problems associated with observations of hydrogen 21 cm self-absorption lines; they refer this discussion to specific observations of H I self-absorption seen toward nearby identifiable dark clouds.

132.038 **Radial velocity observations of IC 410.**
P. G. Johnson, N. J. White.
Astrophys. Space Sci., Vol. 73, 411 - 415 (1980).

The velocity of the ionized gas has been obtained at many positions over the H II region IC 410 (S 236), using a pressure scanned Fabry-Pérot polychromator/imager. The structure of the object is discussed and its possible association with a nearby supernova remnant discounted.

132.039 **The giant galactic H II region NGC 3603: optical studies of its structure and kinematics.**
B. Balick, G. O. Boeshaar, T. R. Gull.
Astrophys. J., Vol. 242, 584 - 591 (1980).

The authors report the results of the photographic and kinematic portions of a survey of NGC 3603 and some preliminary results of the moderate-dispersion photographic spectroscopy. The spatial and kinematic structure of NGC 3603 and the region near the central stars are discussed. The authors compare NGC 3603 to 30 Dor and other extragalactic H II regions. Finally, they discuss some problems posed by NGC 3603 and other giant H II regions.

132.040 **Neutral hydrogen in the vicinity of the Sagittarius B2 radio source.**
Z. A. Alferova, A. P. Venger, I. V. Gosachinskij, V. G. Grachev, V. G. Mogileva, N. F. Ryzhkov, N. A. Yudaeva.
Pis'ma Astron. Zh., Tom 6, 759 - 762 (1980). In Russian. English translation in Soviet Astron. Lett., Vol. 6.

Results of 21-cm observations of the H I cloud at the velocity of +60 km/s around the Sagittarius B2 gas-dust complex are presented. This cloud has a diameter ≈40 pc, neutral hydrogen mass $\approx 1.5 \times 10^3 M_\odot$ and a rather low kinetic temperature (< 20 K). The cloud rotates with a velocity ≈20 km/s at its edge. Any radial motions with velocities larger than 5 km/s are not found.

132.041 **CO observations of several galactic H II regions.**
F. P. Israel.
Astron. J., Vol. 85, 1612 - 1630 (1980).

Maps of the CO emission at a resolution of 8 arcmin are presented for seven H II regions or H II region groups; the results of a search for CO emission near seven more are also given. The results show a complicated molecular cloud structure near most observed H II regions. There is a tendency for decreasing CO cloud mass when H II regions in more evolved stages are observed. The relevance of these results for the understanding of individual H II region/molecular cloud complexes, and for general aspects of the star-formation process, is briefly discussed.

132.042 **Spectrophotometry and abundances in H II regions in external galaxies.** G. A. Shields.
Photometry, kinematics and dynamics of galaxies, (see 012.051), p. 329 - 331 (1979).

The author reports some initial results of a program of observations of extragalactic H II regions. The procedure has been to compare the H II region spectra of a variety of galaxies with those of the well-studied Scd galaxy M101. The goal is to isolate how differences in morphology, size, and environment correlate with abundances and abundance gradients.

132.043 **Neutral hydrogen in the radiogalaxy Centaurus A.**
I. V. Gosachinskij, V. G. Grachev, N. F. Ryzhkov.
Astron. Zh., Tom 57, 1121 - 1128 (1980). In Russian. English translation in Soviet Astron., Vol. 24, No. 6.

The observational results of the 21-cm H I line in emission and absorption from the radiogalaxy Centaurus A are reported. Two new H I details in absorption and four in emission are found.

Interstellar shock waves with magnetic precursors.
See Abstr. 062.112.

Extreme-UV and far-UV observations of the white dwarf HZ 43 from Voyager 2. See Abstr. 126.035.

Carbon monoxide observations of molecular clouds associated with H II regions beyond the solar circle.
See Abstr. 131.046.

Highly-ionized species in the interstellar medium.
See Abstr. 131.064.

Detection of C_2H in cold dark clouds.
See Abstr. 131.068.

The formation of elephant-trunk globules in the Rosette nebula: CO observations. See Abstr. 131.074.

Detection of the 157 micron (1910 GHz) [C II] emission line from the interstellar gas complexes NGC 2024 and M42. See Abstr. 131.094.

H II bubbles and disruption of molecular clouds. See Abstr. 131.118.

H I self-absorption in the Southern Coalsack dust complex. See Abstr. 131.138.

Star formation and ionization in the 3 kiloparsec arm. See Abstr. 131.139.

Ammonia observations of the molecular cloud near S106. See Abstr. 131.172.

Observations of NH_3 toward DR21. See Abstr. 131.174.

High resolution 4.8 GHz mapping of H_2CO using the Westerbork synthesis radio telescope. See Abstr. 131.175.

CO (J = 2 – 1) observations of several galactic H II regions. See Abstr. 131.182.

CO observations in the southern hemisphere. See Abstr. 131.184.

1.0 mm continuum observations of cool southern clouds. See Abstr. 131.185.

The disruption of the molecular cloud associated with the North America and Pelican Nebulae. See Abstr. 131.189.

Atomic hydrogen in and around the giant molecular cloud near W3 and W4. See Abstr. 131.191.

High rate of destruction of molecular clouds by hot stars. See Abstr. 131.192.

Cold H I gas in the region of the giant molecular cloud near M17. See Abstr. 131.199.

Search for CO in atomic hydrogen clouds. See Abstr. 131.200.

Absolute positions of OH masers associated with H II regions. See Abstr. 131.238.

Time variations of interstellar water masers in H II regions. See Abstr. 131.241.

VLBI observations of the H_2O masers in Sagittarius B2. See Abstr. 131.281.

Ionization front structure dependence on boundary conditions. See Abstr. 131.284.

Instabilities and star formation within ionization-shock fronts. See Abstr. 131.291.

Spectral-line observations of the IC 5146 region. See Abstr. 131.318.

X-ray photoionized nebulae. See Abstr. 131.334.

High angular resolution far-infrared observations of the W3 region. See Abstr. 133.013.

Aperture synthesis observations of the NGC 7000/IC 5070 complex at 610 MHz. See Abstr. 134.005.

The helium ionization structure of the Orion Nebula. See Abstr. 134.021.

Approximation of non-LTE coefficients. See Abstr. 134.023.

Discovery of the exciting star in the North America – Pelican Nebula complex? See Abstr. 134.037.

The problem of sulfur abundance determination in gaseous nebulae. See Abstr. 134.038.

H I in the region around the reflection nebula NGC 1579. See Abstr. 134.042.

A new faint planetary nebula behind the H II region S 232 and close to the galactic anticenter. See Abstr. 135.004.

Radio observations of W3 at 2.7 and 15.4 GHz. See Abstr. 141.109.

On the sequential formation of subgroups in OB associations. See Abstr. 152.002.

Multicolour *UBVRI* photometry of stars in M 17. See Abstr. 152.005.

A 2.3 GHz radio continuum map of the upper Scorpio region. See Abstr. 152.007.

O stars and the interstellar medium. See Abstr. 155.001.

The galactic rotation curve to R = 18 kpc. See Abstr. 155.027.

Structure of the local spiral arm from the 21-cm hydrogen line profiles. See Abstr. 155.040.

A new look at the North Polar Spur. See Abstr. 155.045.

Further observations of radio recombination lines from the direction of the Galactic Centre. See Abstr. 156.010.

Galaxies with the spectra of giant H II regions. See Abstr. 158.079.

Search for H_2O maser emission in nearby galaxies. See Abstr. 158.134.

H II regions in NGC 628. III. Hα luminosities and the luminosity function. See Abstr. 158.167.

H I observations and star formation in the blue compact galaxy IZw18. See Abstr. 158.172.

Late-type galaxies with extended envelopes of neutral hydrogen. See Abstr. 158.175.

Velocity fields in late-type galaxies from Hα Fabry-Perot interferometry. II. Kinematics and dynamics of the Sd spiral NGC 7793. See Abstr. 158.216.

NGC 7385: generation of an H II region by thermal instability associated with the creation of a radio source. See Abstr. 158.227.

Multicolour photometry of bright patches in the galaxy NGC 4303. See Abstr. 158.237.

133 Infrared Sources

133.001 **Near-infrared slit scans of molecular cloud sources.**
H. M. Dyck.
Astron. J., Vol. 85, 891 - 893 (1980).

High-resolution slit scans (~ 1 arcsec) of three compact infrared sources in molecular clouds are described. No structure and an upper limit to the diameter of 0.5 arcsec have been found for the Kleinmann-Wright source in M17. Unresolved double structure with separations on an arcsec scale has been found for S140/IRS 1 and W33A.

133.002 **On the nature of the peculiar infrared source AFGL 2636.**
R. D. Gehrz, J. A. Hackwell, G. L. Grasdalen, K. M. Merrill, R. M. Humphreys, F. O. Williamson, R. C. Puetter, R. W. Russell, S. P. Willner.
Astron. J., Vol. 85, 1071 - 1076, with a correction p. 1676 (1980).

Infrared and visual observations suggest that AFGL 2636 is a region of star formation that is similar to the θ^1-BNKL complex in the Orion Nebula. The geometrical position of AFGL 2636 within the Cygnus-X region may be an important clue to the processes which precipitated its formation.

133.003 **Ice-band polarimetry of GL 2591.**
H. M. Dyck, C. J. Lonsdale.
Astron. J., Vol. 85, 1077 - 1081 (1980).

The authors have measured the linear polarization across the 3-μ ice absorption band in the protostellar source GL 2591. Assuming that the polarization is caused by aligned grains, it seems likely that the grains near GL 2591 are larger than near BN, and have significant albedos at 1 to 4 μm.

133.004 **An infrared candidate for OH 205.1 – 14.1.**
D. A. Allen, J. R. Barton, P. R. Gillingham.
Mon. Not. R. Astron. Soc., Vol. 192, 805 - 807 (1980).

The authors have located an infrared source coincident with the OH and H_2O complex OH 205.1 – 14.1 in the Lynds 1630 dark cloud. The infrared source is probably a late B dwarf embedded in a circumstellar dust cloud.

133.005 **Narrow-band polarimetry of the BN object and AFGL 2591 between 2 and 4 μm.**
Y. Kobayashi, K. Kawara, S. Sato, H. Okuda.
Publ. Astron. Soc. Japan, Vol. 32, 295 - 302 (1980).

Narrow-band polarimetry of the BN object and AFGL 2591 has been made between 2 and 4 μm in conjunction with a CVF spectrophotometer. The results show an ice-band excess in the BN object, but no trace of an excess larger than 1.6% in AFGL 2591. In the case of the BN object the polarization maximum shifts to a longer wavelength compared with that of the extinction peak. Polarization and extinction profiles of elongated ice particles have been calculated using the perfect Davis-Greenstein alignment model. They can well explain the observed shift of the polarization maximum.

133.006 **2 μ spectra of sources in the galactic center.**
E. R. Wollman, H. A. Smith, H. P. Larson, G. C. Augason.
Bull. American Astron. Soc., Vol. 12, 443 (1980). – Abstract.

133.007 **Simultaneous far-infrared, near-infrared, and radio observations of OH/IR stars.**
M. W. Werner, S. Beckwith, I. Gatley, K. Sellgren, G. Berriman, D. L. Whiting.
Astrophys. J., Vol. 239, 540 - 548 (1980).

Simultaneous far-infrared, near-infrared, and radio observations have been made of five infrared stars which show OH maser emission at 1612 MHz. The data permit a direct comparison of the far-infrared and maser emission from these sources, which strongly supports the hypothesis that the maser emission is pumped by 35 μm photons. A comparison with data obtained at earlier epochs suggests that the maser emission is saturated. The infrared and radio data are used together with estimates of the source distances to determine the luminosities and mass loss rates for these objects.

133.008 **Maser emission from infrared stars. I. New OH and H_2O observations.**
F. M. Olnon, A. Winnberg, H. E. Matthews, G. V. Schultz.
Astron. Astrophys., Suppl. Ser., Vol. 42, 119 - 133 (1980).

New sensitive observational data are presented of the λ 18-cm OH and λ 1.35 cm H_2O line profiles of about 70 stellar sources. A considerable number of new features are found. There is a striking correspondence between the OH main line profiles for each source: the 1612 MHz OH profiles are however usually quite different. This result is in agreement with current models of OH excitation. The authors find that the correlation between H_2O sources and Type 1 OH/IR objects reported previously was due to a selection effect. Attention is drawn to a correlation between the H_2O and OH line profiles.

133.009 **Carbon monoxide observations of IRC 10442.**
J. Bally, T. J. Balonek, N. L. Cohen, M. Zeilik, II.
Astron. J., Vol. 85, 1242 - 1246 (1980).

Three molecular clouds have been detected in the directions of IRC 10442 (GL 5268-S). The feature at V_{LSR} = 57 km s^{-1} may be a typical giant molecular cloud associated with the giant H II region G 25.4 –0.2, located at a distance of 13.5 kpc. It is difficult to associate the infrared source with any particular CO feature seen in this direction owing to the absence of an obvious hot spot at the IR position. Three possible models for the infrared peak are considered.

133.010 **A low-luminosity far infrared source in the L1551 molecular cloud.**
C. V. M. Fridlund, H. L. Nordh, R. J. van Duinen, J. W. G. Aalders, A. I. Sargent.
Astron. Astrophys., Vol. 91, L1 - L2 (1980).

Searches at 85 μm and 150 μm of the molecular cloud L1551 have resulted in the discovery of an embedded source of total luminosity ~ 25 $L_\odot$. Its position coincides with an obscured, near-infrared, point source and with a gas density peak. The authors suggest that the embedded object is a low-mass pre-main-sequence star.

133.011 **IRC + 30219.**
IAU Circ., No. 3552 (1980).

133.012 **Infrared objects near to H_2O masers in regions of active star formation.**
A. F. M. Moorwood, P. Salinari.
ESO Sci. Prepr. No. 103, 23 pp. (1980). – Submitted to Astron. Astrophys.

133.013 **High angular resolution far-infrared observations of the W3 region.** M. W. Werner, E. E. Becklin, I. Gatley, G. Neugebauer, K. Sellgren, H. A. Thronson, Jr., D. A. Harper, R. Loewenstein, S. H. Moseley.
Astrophys. J., Vol. 242, 601 - 608 (1980).

A 2′ × 3′ region surrounding the W3 cluster of near-infrared sources and compact H II regions has been mapped at 30, 50, and 100 μm with an angular resolution of ~30″. The data have been used to produce maps of the distribution of luminosity, color temperature, and opacity in the far-infrared which are used to analyze the properties and evolutionary states of the individual compact sources in the cluster and of the molecular cloud in which they are embedded. The total luminosity of the near-infrared source W3-IRS 5 is estimated and it is identified as a forming O star.

Infrared astronomical data base and Master Catalogue. See Abstr. 002.018.

Distribution of astronomical sources in the second Equatorial Infrared Catalogue. See Abstr. 002.026.

A search for carbon stars in the AFGL Catalogue. See Abstr. 002.027.

Near infrared photographic sky survey: a field index. See Abstr. 002.083.

Infrared astronomy comes of age. See Abstr. 011.009.

Infrarotastronomie. See Abstr. 061.018.

A far-infrared emission feature in IRC+10216, IC 418, and NGC 6572. See Abstr. 112.009.

Infrared observations of circumstellar ammonia in OH/IR supergiants. See Abstr. 112.015.

Spectroscopic observations of Near Infrared Photographic Sky Survey objects. See Abstr. 114.022.

A far-infrared emission feature in the carbon star IRC+10216 and the planetary nebulae IC 418 and NGC 6572. See Abstr. 114.123.

The infrared echo of a supernova in a dust cloud. See Abstr. 125.066.

Star formation in W3 and W4: discovery of 135 possibly embedded near-infrared stars. See Abstr. 131.102.

Organic material and the 2.5–4 μm spectra of galactic sources. See Abstr. 131.114.

Dry polysaccharides and the infrared spectrum of OH 26.5 + 0.6. See Abstr. 131.115.

The chemical identification of grain mantles by infrared spectroscopy. See Abstr. 131.219.

Infrared molecular absorption features. See Abstr. 131.220.

Infrared observations of a Bok globule in the Southern Coalsack. See Abstr. 131.289.

Star formation in IC 1848 A. See Abstr. 131.300.

Infrared observations of Barnard 35: heat sources for bright-rimmed molecular clouds. See Abstr. 131.335.

The large-scale far-infrared structure of W3 and W4. See Abstr. 132.012.

1.0 millimeter maps and radial density distributions of southern H II/molecular cloud complexes. See Abstr. 132.013.

High-resolution radio observations of three southern H II regions. See Abstr. 132.016.

The infrared emission of G333.6–0.2, an extremely nonspherical H II region. See Abstr. 132.029.

Discovery of the exciting star in the North America – Pelican Nebula complex? See Abstr. 134.037.

Infrared bursts from Liller I/MXB 1730-333. See Abstr. 154.014.

Observations of the motion and distribution of the ionized gas in the central parsec of the Galaxy. II. See Abstr. 155.021.

Infrared observations of the galactic nucleus. See Abstr. 155.025.

Medium resolution IR-maps of the Cygnus X region. See Abstr. 156.002.

134 Emission Nebulae, Reflection Nebulae

134.001 **The structure of NGC 1999.**
R. F. Warren-Smith, S. M. Scarrott, D. J. King, K. N. R. Taylor, R. G. Bingham, P. Murdin.
Mon. Not. R. Astron. Soc., Vol. 192, 339 - 345 (1980).

An optical polarization map is presented for NGC 1999, the reflection nebulosity surrounding V380 Orionis. This star is slightly embedded on the near side of an extensive dust cloud which has an oblique boundary in front of the star. The dark globule silhouetted against NGC 1999 is physically associated with the nebula.

134.002 **Roberts 22: a bipolar nebula with OH emission.**
D. A. Allen, A. R. Hyland, J. L. Caswell.
Mon. Not. R. Astron. Soc., Vol. 192, 505 - 519 (1980).

Roberts 22 is a bipolar reflection nebula illuminated by a hidden A2 Ie star. Most of its energy is radiated at infrared wavelengths. It also shows strong OH maser emission (OH284.18 – 0.79) on the 1612 and 1665 MHz transitions, generally similar to the masers associated with M stars having infrared excesses. But the system contains no late-type star. From their analysis of the properties of Roberts 22 the authors question some published interpretations of other bipolar nebulae, in particular the derivation of spectral types for their underlying stars by the assumption of photoionization of the gas, and their evolutionary description as protoplanetary nebulae.

134.003 **Spectra of gaseous nebulae.** M. J. Seaton.
Q. J. R. Astron. Soc., Vol. 21, 229 - 244 (1980).

134.004 **The calculation of the optical spectra of NGC 6888.**
M. Contini, G. Shaviv.
Astron. Astrophys., Vol. 88, 117 - 126 (1980).

Theoretical calculations of the spectra of NGC 6888 are compared with Parker's (1978) observations of the single filaments. The model accounts for the shock conditions connected with the propagation of the filament, for the thermal radiation from the "central" WR star HD 192163 and for the secondary diffused radiation, simultaneously. The values of the most indicative parameters were obtained by fitting the calculated to the observed spectra. The most important results are: a) The line ratios (particularly [O III]/[O II]) are correlated with the distance from the star. b) The stellar temperature is $(5–6) \times 10^4$ °K. c) The shock velocities range between 50 and 100 km s^{-1}. d) The N/O calculated relative abundances (0.8–2) are of the order of the average of the observed [N II]/[O II] line ratios. e) N/O is accompanied by He/H (~0.18). f) The unobservable neutral gas – inside the core of many filaments – indicates that previous nebula's mass estimates were lower limits.

134.005 **Aperture synthesis observations of the NGC 7000/IC 5070 complex at 610 MHz.**
H. E. Matthews, W. M. Goss.
Astron. Astrophys., Vol. 88, 267 - 272 (1980).

The central part of the North America and Pelican nebulae (NGC 7000/IC 5070) has been observed at 610 MHz using the Westerbork Synthesis Radio Telescope. The angular resolution is 1′, which corresponds to a relative resolution of less than 1%. Due to the very smooth nature of the radio brightness distribution, only a small fraction of total flux density has been detected. Fifty-six background sources appear in the map, along with 13 bright-rim structures. The latter have typical emission measures of 1600 cm^{-6} pc and r.m.s. electron densities of 25 cm^{-3}. From these bright-rim structures the location of the exciting star or stars can be inferred.

134.006 **Dust grains in reflection nebulae. II. Infinite circular cylinders.** N. V. Voshchinnikov.
Astron. Zh., Tom 57, 716 - 724 (1980). In Russian.
English translation in Soviet Astron., Vol. 24, No. 4.

The *(U–B), (B–V), (V–R)* colors and the *U, B, V, R* polarization in single scattering have been calculated for a homogeneous spherical nebula. Lind and Greenberg's theory has been used to calculate the scattering functions for infinite circular cylinders consisting of dirty ice or silicate material. A class of cylindrical grains randomly oriented in space or in the plane transversal to the direction of the starlight has been considered.

134.007 **On the variability of CRL 2688 (the "Egg" Nebula).**
O. G. Franz.
Bull. American Astron. Soc., Vol. 12, 445 (1980). – Abstract.

134.008 **What is Sharpless 216?**
T. R. Gull, R. A. Fesen, W. P. Blair.
Bull. American Astron. Soc., Vol. 12, 445 (1980). – Abstract.

134.009 **Photoionization models of the Crab Nebula filaments.**
R. Henry, G. M. MacAlpine, R. A. Fesen.
Bull. American Astron. Soc., Vol. 12, 446 (1980). – Abstract.

134.010 **Preliminary analysis of the ionization structure in NGC 7000.** D. A. Klinglesmith, T. R. Gull.
Bull. American Astron. Soc., Vol. 12, 468 (1980). – Abstract.

134.011 **High resolution images of the Orion Nebula at four wavelengths in the rocket ultraviolet.**
J. K. Hill, R. Bohlin, T. P. Stecher.
Bull. American Astron. Soc., Vol. 12, 483 (1980). – Abstract.

134.012 **Small scale structure of the CO emission in the biconical nebula LkHα 208 from lunar occultation observations.** J. Good, J. Bally, F. P. Schloerb, N. Z. Scoville.
Bull. American Astron. Soc., Vol. 12, 486 (1980). – Abstract.

134.013 **UV surface mapping of regions in the Orion Nebula.**
P. M. Perry, B. E. Turnrose, C. A. Harvel, R. W. Thompson, A. D. Mallama.
Bull. American Astron. Soc., Vol. 12, 505 (1980). – Abstract.

134.014 **Preliminary ultraviolet observations of filaments in the Crab Nebula.** K. Davidson, T. R. Gull, S. P. Maran, T. P. Stecher, M. Kafatos, V. Trimble.
Bull. American Astron. Soc., Vol. 12, 541 (1980). – Abstract.

134.015 **Hard X-ray spectrum of the Crab Nebula from HEAO-1.**
D. E. Gruber, J. L. Matteson, F. K. Knight, R. E. Rothschild, P. L. Nolan, S. Howe, A. M. Levine, F. A. Primini.
Bull. American Astron. Soc., Vol. 12, 541 (1980). – Abstract.

134.016 **Upper limits on X-ray line emission from the Crab Nebula.** M. L. Schattenburg, C. R. Canizares, G. W. Clark, T. H. Markert, P. F. Winkler.
Bull. American Astron. Soc., Vol. 12, 541 - 542 (1980). Abstract.

134.017 **High resolution observation of the Crab Nebula.**
R. Schwartz, R. P. Lin, R. Pelling, K. Hurley.
Bull. American Astron. Soc., Vol. 12, 542 (1980). – Abstract.

134.018 **Observations of H_2CO in the Orion Nebula at 1 centimeter wavelength.**
P. C. Myers, R. B. Buxton.
Astrophys. J., Vol. 239, 515 - 518 (1980).

The $J_{K_{-1}K_{+1}} = 3_{12} \rightarrow 3_{13}$ line of formaldehyde (H_2CO) at 29 GHz has been detected in emission from OMC-1. The emission has spatial extent ~2′, similar to that of other H_2CO lines in Orion. Upper limits are placed on the degree of clumping which may be expected.

134.019 **An aperture synthesis map of the Crab Nebula at 23 gigahertz.** M. C. H. Wright, J. R. Forster.
Astrophys. J., Vol. 239, 873 - 879, plate 12 (1980).

The authors present a map of the polarization and total brightness distribution of the Crab Nebula at 23 GHz with a resolution of 7″ × 8″. The total intensity distribution is quite similar to that at 5 GHz and shows radio ridges which have a similar spectral index to the general background, consistent with a model in which the enhanced radio emission of the ridges is due to an increased magnetic field near the optical filaments. The polarized brightness distribution at 23 GHz lacks the detailed structure visible at 5 GHz which is attributed to Faraday rotation at frequencies lower than 23 GHz. In the north of the nebula and in the region of the east optical bay, there appears to be substantial depolarization between the optical and 23 GHz emission, which implies somewhat higher rotation measures than have been previously suggested.

134.020 **Discovery of optical molecular emission from the bipolar nebula surrounding HD 44179.**
G. D. Schmidt, M. Cohen, B. Margon.
Astrophys. J., Lett., Vol. 239, L133 - L138 (1980) = Lick Obs. Bull., No. 869.

The authors present spectrophotometry and spectropolarimetry with HD 44179. These measurements reveal that the very broad bump evident in previous low-resolution spectra possesses a large amount of structure, including groups of narrow emission lines and several diffuse features. A reduction in polarization, but constant position angle, through the bump indicates that this emission originates within the nebula itself and merely dilutes the polarized scattered starlight. A few very weak atomic emission lines are detected, but the overall feature, which strongly resembles the emission spectra of some molecules, remains unidentified. Constraints on the excitation mechanism are discussed.

134.021 **The helium ionization structure of the Orion Nebula.**
V. Pankonin, C. M. Walmsley, C. Thum.
Astron. Astrophys., Vol. 89, 173 - 179 (1980).

The ionized helium volume is found to be smaller than the volume of ionized hydrogen; the ionized helium abundance decreases with increasing distance from the principal exciting star. However, the gradient is much steeper to the east than to the west of the star. These data are consistent with new calculations of stellar model atmospheres (Kurucz, 1979) which show that line blanketing considerably reduces the ratio of helium to hydrogen ionizing photons. The nebular electron temperatures which are derived from hydrogen line-to-continuum ratios show no systematic variations across the Nebula.

134.022 **A Monte Carlo model for light scattering by dark nebulae.**
M. I. Darby, D. J. King, K. N. R. Taylor.
Proc. Astron. Soc. Australia, Vol. 3, 312 - 315 (1979).

134.023 **Approximation of non-LTE coefficients.**
S. A. Gulyaev.
Astron. Zh., Tom 57, 1108 - 1110 (1980). In Russian.
English translation in Soviet Astron., Vol. 24, No. 5.

Simple formulae which approximate the dependence of b_n and $c_n = d(\ln b_n)/dn$ (coefficients which characterize the departure of the populations of the excited states of hydrogen from LTE in nebulae and interstellar plasmas) on temperature, density and quantum number are suggested. Calculations of H II region models can be significantly simplified by these formulae.

134.024 **Far-infrared polarization of the Kleinmann-Low Nebula.**
G. E. Gull, R. W. Russell, G. Melnick, M. Harwit.
Astron. J., Vol. 85, 1379 - 1381 (1980).

Further data on the polarization of the far-infrared ($\lambda > 40\ \mu m$) emission from the Orion Nebula shows no evidence for polarized emission by aligned grains. Results at $\lambda \cong 71 - 115\ \mu m$ are consistent with a small (~ 1.5%) absorption-induced polarization with about the same position angle on the sky as the polarization measured at shorter wavelengths.

134.025 **Polarization in reflection nebulae. III. How good a grain diagnostic?** R. L. White, F. H. Schiffer III, J. S. Mathis.
Astrophys. J., Vol. 241, 208 - 217 (1980).

The practicality of determining single-scattering polarization properties of dust grains from observations of the polarization is investigated. A simple model resembling the Orion Nebula, consisting of a hemispherical shell of dust in front of an optically thick plane slab, is considered.

134.026 **The Boomerang Nebula: a highly polarized bipolar.**
K. N. R. Taylor, S. M. Scarrott.
Mon. Not. R. Astron. Soc., Vol. 193, 321 - 327 (1980).

An optical linear polarization map of a bipolar nebula is presented. Polarizations of ~ 60 per cent are observed in the optically thin lobes. The map leads to a geometry of the object consisting of a central star with an equatorial disc of dust and optically thin lobes illuminated by the central star. The grains in the disc are aligned. The object is a protoplanetary nebula.

134.027 **A study of the Crab Nebula – II. Radio emission from the filaments.** E. Swinbank.
Mon. Not. R. Astron. Soc., Vol. 193, 451 - 468 (1980).

Maps of the Crab Nebula at 2.7, 5 and 23 GHz, with resolution of a few arcseconds, are compared in detail. The synchrotron spectrum of the filaments does not differ systematically from that of the rest of the source, indicating that the relativistic electrons in the filaments and the surrounding material have the same origin – the pulsar. Polarization maps at 2.7 and 5 GHz are compared with new optical radial velocity data on the filaments, confirming that depolarization between these frequencies is caused chiefly by bright filaments on the near side of the nebula. A new model for the magnetic fields in the filaments, based on the properties of a thin slab of randomly-oriented field, is used to explain the observations.

134.028 **Complex flows in the outer regions of NGC 6302 – II.** J. Meaburn, J. R. Walsh.
Mon. Not. R. Astron. Soc., Vol. 193, 631 - 640 (1980).

The motions and densities of the outer filamentary regions of the nebula are considered. Many separate volumes of ionized material with heliocentric radial velocities between – 186 and + 140 km s^{-1} have been discovered. Wind driven flows of radiatively ionized material are proposed to explain these phenomena.

134.029 **The evolution of ideas on the Crab Nebula.**
L. Woltjer.
Oort and the universe, (see 003.005), p. 117 - 122 (1980).

134.030 **Globules in the Orion Nebula. I. Observations of interstellar sodium D 1 and D 2 lines.**
S. Isobe.
Publ. Astron. Soc. Japan, Vol. 32, 423 - 433 (1980).

Observations of interstellar sodium D 1 and D 2 lines

have been carried out for the four Trapezium stars in the Orion Nebula; they show that column densities toward the stars are $(1-6) \times 10^{12}$ cm^{-2} and seem to vary from one star to another, although the observational error is of the order of 10^{12} cm^{-2}. It is shown that most of the neutral sodium atoms are in normal H I regions.

134.031 **CO (J = 2 – 1) observations of the Carina nebula and G333.6–0.2 and a search for CO in LMC and SMC.** T. de Graauw, S. Lidholm, B. Fitton, F. P. Israel, J. Beckman, H. Nieuwenhuyzen, J. Vermue.
Interstellar molecules, (see 012.033), p. 125 - 126 (1980).

134.032 **Zes jaar puzzelen op Eta Carinae.** C. D. Andriesse.
Zenit, 7e Jaarg., 130 - 136 (1980).

134.033 **Relative O II intensities in gaseous nebulae: inadequacy of conventional radiation theory.**
C. J. Zeippen.
J. Phys. B, Vol. 13, L485 - L490 (1980). – Abstr. in Phys. Abstr., Vol. 83, Abstr. 109513 (1980).

134.034 **Discovery of a peculiar stellar object with surrounding nebulosity.**
N. Vogt, W. Wamsteker, J. Breysacher, H. E. Schuster.
ESO Sci. Prepr. No. 111, 14 pp. (1980). – Submitted to Astron. Astrophys.

134.035 **Radiative transfer in dusty nebulae. III. The effects of dust albedo.** V. Petrosian, R. A. Dana.
Astrophys. J., Vol. 241, 1094 - 1106 (1980).

The authors have used the quasi-diffusion method and investigated the effects of an albedo of dust internal to nebulae on the observable parameters of the nebulae, such as ionization structure and temperature of dust grain. Using the Eddington approximation, they find that the general effect of the albedo is to increase the average intensity of the diffuse photons, which increases the ionization level of the hydrogen and helium and heavy elements and reduces the dust temperature gradient so that the infrared spectrum approaches a single-temperature blackbody spectrum as the albedo increases.

134.036 **Upper limits on X-ray line emission from the Crab Nebula.** M. L. Schattenburg, C. R. Canizares, C. J. Berg, G. W. Clark, T. H. Markert, P. F. Winkler.
Astrophys. J., Lett., Vol. 241, L151 - L154 (1980).

The authors report the results of a search for line emission from the Crab carried out with the Focal Plane Crystal Spectrometer on the Einstein Observatory. Their nondetection of line emission allows them to place stringent limits on the amount and temperature of any thermal material in the Crab.

134.037 **Discovery of the exciting star in the North America – Pelican Nebula complex?**
T. Neckel, A. W. Harris, C. Eiroa.
Astron. Astrophys., Vol. 92, L9 - L10 (1980).

In the centre of the dark lane which divides the W80 complex into NGC 7000 (North America Nebula) and IC 5070 (Pelican Nebula), the authors have discovered a very highly reddened star ($A_V \sim 20$ mag) which is very bright in the near infrared (L = 2.1 mag). Its location suggests that it might be a major source of excitation for both nebulae. If this is so, then the assumed distance of the W80 complex must be reduced from the normally quoted value of 1 kpc to 200 pc or less.

134.038 **The problem of sulfur abundance determination in gaseous nebulae.**
A. Natta, N. Panagia, A. Preite-Martinez.
Astrophys. J., Vol. 242, 596 - 600 (1980).

The problem of the ionization equilibrium and the abundance determination of sulfur is discussed. It is found that the commonly used ionization-correction-factor procedure can overestimate the S^{+3} abundance by a factor of about 10 for most planetary nebulae and usually more than 3 for H II regions. A method for deriving S^{+3} abundance in medium- and high-excitation objects is presented and applied to 41 planetary nebulae in order to get total sulfur abundances.

134.039 **Model of the Crab nebula. Emission spectrum and brightness distribution.** A. A. Stepanyan.
Izv. Krymskoj Astrofiz. Obs., Tom 62, 79 - 97 (1980). In Russian.

The synchroton emission of high-energy particles contained in a time-variable magnetic field is considered. It is shown that the spectrum is complex. The emission flux does not depend on the magnetic field strength for the high frequency region. Application of these results to a modified model of Rees and Gunn for the Crab nebula has allowed to find some important features of the nebula and the pulsar NP 0532. A model of the external part of the Crab nebula is suggested. It contributes to the explanation of the spectrum in the radio frequency region.

134.040 **Identification of a nebulous blue object near Maffei 1.** R. J. Buta, M. L. McCall, A. K. Uomoto.
Publ. Astron. Soc. Pacific, Vol. 92, 725 - 729 (1980).

Photometric and spectroscopic data have been acquired for a small nebulous blue object near the heavily obscured galaxy Maffei 1. The observations show that the object is a galactic reflection nebula illuminated by an early, main-sequence B star at a distance of about 3 kpc. Application of the Q-method to the photometry of the star indicates that Maffei 1 suffers at least 3.3 magnitudes of visual extinction.

134.041 **Studies of bipolar nebulae. VI. Optical spectrophotometric mapping of GL 2688 (the Cygnus egg nebula).** M. Cohen, L. V. Kuhi.
Publ. Astron. Soc. Pacific, Vol. 92, 736 - 745 (1980).

Optical scanner spectra are presented for ten positions in the lobes of GL 2688. Color gradients exist across the nebulae, probably due to systematic variations in the sizes of typical scattering grains. Molecular emissions of C_2, C_3, and SiC_2 are found, highly reminiscent of the spectra of comets. Resonance fluorescence seems to be indicated.

134.042 **H I in the region around the reflection nebula NGC 1579.** P. E. Dewdney, R. S. Roger.
J. R. Astron. Soc. Canada, Vol. 74, 361 (1980). – Abstract.

134.043 **Observations of Hα-emission regions with a Fabry-Perot-field spectrograph.** E. H. Geyer, A. Hänel.
Mitt. Astron. Ges., Nr. 50, p. 101 (1980).

134.044 **Reflection nebula in the far-ultraviolet. 2. Silicate and graphite grains.** G. A. Shah, K. S. Krishna Swamy.
Kodaikanal Obs. Bull., Ser. A, Vol. 2, 195 - 202 (1979).

Simple models of a reflection nebula in the form of a homogeneous plane-parallel slab containing single scattering silicate and graphite grains have been considered in view of the far-ultraviolet surface brightness observations obtained from the ANS satellite. The colour difference between the star and the nebula as well as the polarization of the nebular light in the far-ultraviolet have also been given.

Collisional and infrared radiative pumping of molecular vibrational states: the carbon monoxide infrared bands. See Abstr. 022.047.

Ultraviolet pumping of molecular vibrational states: the CO infrared bands. See Abstr. 022.048.

Hydrogenic emission and recombination coefficients for a wide range of temperature and wavelength.
See Abstr. 022.187.

Ultraviolet, optical, and infrared observations of the Herbig Be star HD 200775. See Abstr. 114.082.

Studies of the Carina Nebula. III. The spectral energy distribution of the very hot and massive star HD 93250.
See Abstr. 114.100.

Optical polarization and infrared spectrum of a possible protostar in a reflection nebula. See Abstr. 116.009.

The peculiar, galactic object ESO 313-N*10.
See Abstr. 121.009.

SiO emission in Orion–KL: an evolved star in a region of star formation or a unique object in the Galaxy?
See Abstr. 131.065.

Detection of CO $J = 21 \rightarrow 20$ (124.2 μm) and $J = 22 \rightarrow 21$ (118.6 μm) emission from the Orion Nebula.
See Abstr. 131.072.

Studies of the Carina Nebula. II. The extinction law in the direction of 14 O-type stars.
See Abstr. 131.088.

High angular resolution observations of CS in the Orion Nebula. See Abstr. 131.091.

Interstellar gas in the Gum Nebula.
See Abstr. 131.101.

The area around the Orion Nebula observed in the CO ($J = 1 - 0$) transition. See Abstr. 131.150.

High velocity gas in the Orion nebula.
See Abstr. 131.160.

SiO emission from the Orion Nebula.
See Abstr. 131.236.

The extinction of HD 200775 by dust in NGC 7023.
See Abstr. 131.260.

Detection of the $S(9)$, $\nu = 0 \rightarrow 0$ rotation line of the hydrogen molecule in Orion. See Abstr. 131.283.

Long-slit spectroscopy in the rocket ultraviolet of the Orion Nebula. See Abstr. 132.009.

Chemical compositions of planetary and diffuse nebulae. See Abstr. 135.018.

Spectroscopic observations of galactic nebulae and galaxies with the Imaging Photon Counting System (IPCS).
See Abstr. 135.024.

A model for the magnetic-field structure in extended radio sources. See Abstr. 141.108.

An unusual radio point source in M17.
See Abstr. 141.183.

Observation of microbursts of hard X radiation by SIGNE-2MP aboard the Prognoz 6 station.
See Abstr. 142.048.

135 Planetary Nebulae

135.001 **Molecular hydrogen emission in NGC 7027.** S. Beckwith, G. Neugebauer, E. E. Becklin, K. Matthews, S. E. Persson.
Astron. J., Vol. 85, 886 - 890 (1980).

In this paper the authors present spatial data on the emission in the $H_2 \nu = 1 \rightarrow 0\, S\,(1)$ and H I B γ lines together with measurements of the strengths of several other H_2 lines. The data are discussed in terms of a simple shock-wave excitation model, and some estimates for the mass of the NGC 7027 molecular cloud are made.

135.002 **A huge new nearby planetary nebula.** A. Purgathofer, R. Weinberger.
Astron. Astrophys., Vol. 87, L5 - L6 (1980).

A giant, circular (20′ × 20′) new planetary nebula of low surface brightness has been discovered on a Palomar Sky Survey print. As to the angular diameter, it is the second largest ever found and competes with the Helix nebula for being the closest planetary to the solar system. The object is of low excitation and shows a pronounced ionization stratification; the central star exhibits an sdO spectrum.

135.003 **Distances of planetary nebulae.** W. J. Maciel, S. R. Pottasch.
Astron. Astrophys., Vol. 88, 1 - 7 (1980).

The distances to 121 planetary nebulae are calculated through the use of an empirical relationship between the nebular ionized mass and radius. Such relation is established on the basis of recently determined electron densities and selected distances. The new distances are compared with values previously published, and reasons are given for their adoption, based on observed radio and H β fluxes. Finally, the adopted model is compared with the results produced by simple theoretical models for the central stars.

135.004 **A new faint planetary nebula behind the H II region S 232 and close to the galactic anticenter.**
A. Purgathofer.
Astron. Astrophys., Vol. 88, 275 - 276 (1980).

A new weak planetary nebula near the galactic anticenter and located behind the H II region S 232 has been detected. The galactic coordinates and the measured radial velocity of +45 km s^{-1} suggest a noticeable excentric galactic orbit.

135.005 **Photoionization models and chemical abundances of three halo planetary nebulae.**
S. M. V. Aldrovandi.
Astrophys. Space Sci., Vol. 71, 393 - 404 (1980).

Photoionization models have been constructed to explain the line emission of the three halo planetary nebulae: K 648, 49 + 88°.1 and 108 – 76°.1. The charge transfer reactions between neutral hydrogen and different ions were taken into account in ionization equations. Due to lack of observational information, radiation-bounded as well as matter-bounded models have been considered. Both these models indicate that He and C abundances are close to solar values whereas the heavier elements are depleted.

135.006 **M1–2, a possible eclipsing binary planetary nebula central star.** J. D. Drummond.
Astron. Astrophys., Vol. 88, L11 (1980).

The planetary nebula M1–2 (133 –8°1) has been found to exhibit light variability, suggesting that it may contain the third eclipsing binary central star discovered among planetary nebulae to date. Its 4.0002 ± 0003 hour period is in conflict with previous spectroscopic classifications of the central star as a G supergiant. Further photometric and spectroscopic observations are called for.

135.007 **Two southern planetary nebulae: ESO 263-PN 02 and SchuWe-3.** R. M West, H.-E. Schuster.
Astron. Astrophys., Vol. 88, 350 - 353 (1980).

Spectroscopic observations show that two newly discovered southern objects, ESO 263-PN 02 and SchuWe-3, are planetary nebulae. ESO 263-PN 02 has a high excitation at the center and the 15^m central star is of very early type. SchuWe-3 has ring-shape, is reddened ($A_V \sim 1^m$) and has a low excitation and electron density in the ring. The distance is estimated between 0.6 and 5 kpc, but is probably closer to the lower value. The measured velocities are +33 ± 20 and –194 ± 15 km s^{-1}, respectively.

135.008 **The oxygen enrichment of the Galaxy.** J. B. Kaler.
Astrophys. J., Vol. 239, 78 - 88 (1980).

The purpose of the paper is to make as complete an assessment of O/H in planetaries as is possible, by making use of all the available reliable data. The calculation of the abundances and observational data are treated and correlations with regard to various galactic parameters are considered. The influence of the progenitor star on the nebular O/H and the variation of O/H in the Galaxy with time are discussed.

135.009 **Radio observations of compact planetary nebulae.** S. Kwok, C. R. Purton.
Bull. American Astron. Soc., Vol. 12, 470 (1980). – Abstract.

135.010 **The role of ionization fronts in the colliding wind model of planetary nebulae.** J. L. Giuliani.
Bull. American Astron. Soc., Vol. 12, 470 (1980). – Abstract.

135.011 **IUE observations of two planetary nuclei with temperatures of 150,000 K.**
R. C. Bohlin, J. P. Harrington, T. P. Stecher.
Bull. American Astron. Soc., Vol. 12, 470 (1980). – Abstract.

135.012 **An atlas of emission line fluxes for twenty-six planetary nebulae in the 1150 - 3200 Å region.**
A. Boggess, W. A. Feibelman, C. W. McCracken.
Bull. American Astron. Soc., Vol. 12, 528 (1980). – Abstract.

135.013 **Neutral oxygen and nitrogen lines in planetary nebulae.** J. B. Kaler.
Astrophys. J., Vol. 239, 592 - 597 (1980).

The strength of the λ6300 [O I] line in optically thin planetary nebulae is linearly dependent on that of λ3727 [O II] over a range of more than two orders of magnitude, and averages 6.7% of $I(\lambda 3727)$. In optically thick planetaries with cool central stars ($\log T_* < 4.65$), which contain significant neutral helium, the strengths of [O I] and [O II] are also correlated, but for these the strength of $I(\lambda 6300)$ is suppressed and averages only 1.9% of $I(\lambda 3727)$. The λ5199 [N I] and λ6584 [N II] lines show an analogous but not quite so linear behavior; $I(\lambda 5199)$ averages 1.3% of $I(\lambda 6584)$ for thin nebulae, and only 0.3% for above thick nebulae.

135.014 **The ionization structure of the Ring Nebula. I. Sulfur and argon.** T. Barker.
Astrophys. J., Vol. 240, 99 - 104 (1980).

Measurements of line intensities in the Ring Nebula have been made in four of the six positions observed by Hawley and Miller over a wider wavelength range. The conclusions of Hawley and Miller about the physical conditions and ionization of He, O, Ne, and N are generally confirmed. New observations of S and Ar line intensities have been made, and new ionization correction formulae for these elements are examined.

135.015 **Detection of [O IV] and [Ne V] infrared emission lines from NGC 7027.**
W. J. Forrest, J. F. McCarthy, J. R. Houck.
Astrophys. J., Lett., Vol. 240, L37 - L41 (1980).

Medium-resolution ($\Delta\lambda \approx 0.16\ \mu m$) spectroscopy of NGC 7027 in the wavelength range 16–30μm using the NASA C-141 Kuiper Airborne Observatory has revealed the presence of two emission lines due to fine-structure transitions in Ne V and O IV superposed on a smooth continuum. The [Ne V] line is at a wavelength of 24.28 μm with a flux of 3.0×10^{-10} ergs cm^{-2} s^{-1}. The [O IV] line is at a wavelength of 25.87 μm with a flux of 5.8×10^{-10} ergs cm^{-2} s^{-1}. The line strengths are in good agreement with the model predictions of Péquignot, Aldrovani, and Stasinska. The abundance of these high-excitation ions and the continuum are discussed in detail.

135.016 **Monochromatic isophotometry of NGC 2440 and 7009.** J. P. Phillips, N. K. Reay, S. P. Worswick.
Mon. Not. R. Astron. Soc., Vol. 193, 231 - 243 (1980).

New monochromatic isophotal contour maps are presented for the planetary nebulae NGC 7009 and 2440. The structure of both nebulae is shown to be characterized by bright unresolved condensations which exhibit enhanced low excitation emission. These are interpreted as being high density neutral mass components, and are thought to exist throughout the nebular shells on scales very much less than the 'seeing' limited resolution.

135.017 **Relations between chemical, spatial, and kinematic properties of planetary nebulae.** A. Acker.
Astron. Astrophys., Vol. 89, 33 - 40 (1980).

The author classified 97 PN into four groups, ranging from class-1 PN with high He/H and N/O enrichment values, to class-2 PN with small ratios. Most class-1 PN are of morphological type "B"; the spectra of their nuclei are continuous or of type O, and their mean excitation class is high. The distance from the galactic plane of the class-1 PN is low and their velocity ellipsoid is very flat; their probable age must be in the range 5×10^8–10^9 yr. The spatial and kinematic parameters of the class-2 PN are those of a mixed "old disk-halo" population; these PN are more than 10 times as old than those of class 1.

135.018 **Chemical compositions of planetary and diffuse nebulae.** L. H. Aller.
Proc. Astron. Soc. Australia, Vol. 3, 213 - 219 (1978).

135.019 **An upper limit to the mass loss rate from the nuclei of planetary nebulae.**
A. R. Thompson, R. P. Sinha.
Astron. J., Vol. 85, 1240 - 1241 (1980).

Recent theoretical results and ultraviolet observations both suggest the existence of mass outflow from the central stars of planetary nebulae. Observations of the central stars of five nebulae with the VLA at 6-cm wavelength are described, in order to look for a resulting enhancement of radio emission at the position of the star. Upper limits derived indicate that for an assumed terminal velocity of 2000 km s^{-1} the mass loss rate does not exceed $2 \times 10^{-6}\ M_\odot$ yr^{-1}.

135.020 **Masses of planetary nebulae.** S. R. Pottasch.
Astron. Astrophys., Vol. 89, 336 - 341 (1980).

A new determination of the ionized mass of planetary nebulae is given. The total mass is also discussed. Aside from the mass determination, the most important conclusions are (1) the ionized mass usually is only a small fraction of the total mass, i.e. the nebulae are usually optically thick to ionizing radiation, and (2) the so-called Shklovsky method of determining the distance of nebulae usually gives a substantial overestimate of the distance.

135.021 **Charge transfer of Ne^{2+} with helium.**
A. Dalgarno, S. E. Butler, T. G. Heil.
Astron. Astrophys., Vol. 89, 379 (1980).

From an analysis of the emission spectrum of NGC 7027, it has been argued that charge transfer of Ne^{2+} proceeds rapidly in collision with helium with a rate coefficient exceeding 10^{-10} cm^3 s^{-1}. The authors put forward theoretical considerations which suggest that it is very slow and that the rate coefficient is unlikely to much exceed 10^{-14} cm^3 s^{-1}. The rate coefficient has been measured at 300 K where it has a value of 5×10^{-15} cm^3 s^{-1}. The authors attribute the measurement to a radiative charge transfer process whose rate will change only slowly with increasing temperature.

135.022 **International Ultraviolet Explorer satellite observations of seven high-excitation planetary nebulae.**
L. H. Aller, C. D. Keyes.
Proc. Natl. Acad. Sci. USA, Vol. 77, 1231 - 1234 (1980).
Abstr. in Phys. Abstr., Vol. 83, Abstr. 94860 (1980).

135.023 **Theoretical models of planetary nebulae. IV. Influence of charge exchange between neutral hydrogen and doubly ionized oxygen.** L. H. Aller, C. D. Keyes.
Astrophys. Space Sci., Vol. 72, 203 - 210 (1980).

Introduction of the $O^{++} + H^0 \rightarrow O^+ + H^+$ charge exchange rate suggested by Butler, Bender and Dalgarno leads to theoretical nebular models that differ from those previously computed in several important respects. Different elemental abundances, 'bluer' central star energy distributions, and truncated (material limited) models are required. If the argon abundance is fixed from the λ 7135 [Ar III] line intensity, λ 4740 [Ar IV] is invariably predicted to be much too strong.

135.024 **Spectroscopic observations of galactic nebulae and galaxies with the Imaging Photon Counting System (IPCS).** C. T. Hua, J. Donas, N. H. Doan.
Astron. Astrophys., Vol. 90, 8 - 13 (1980).

The authors present a recent spectrophotometric study of some galactic gaseous nebulae and galaxies made using the image photon counting techniques (IPCS) developed at the Laboratoire d'Astronomie Spatiale (L.A.S.), coupled to the Pellet-Deharveng spectrograph attached at the Cassegrain focus of the 193 cm telescope of the Haute Provence Observatory. From the examination of the spectra obtained it turns out that the IPCS performance is very useful in the observation of low surface brightness nebulae and in the detection of faint emission lines.

135.025 **Ultraviolet spectra of planetary nebulae – II. The young planetary nebula IC 4997.** D. R. Flower.
Mon. Not. R. Astron. Soc., Vol. 193, 511 - 520 (1980).

Observations of the planetary nebula IC 4997 made with the International Ultraviolet Explorer satellite are analysed in conjunction with visual photoelectric observations. The extinction to the nebula and the electron temperature and density are deduced from the relative intensities of lines emitted by O^+ and O^{2+} ions. The relative abundance of C and O is found to be similar to the solar value, whereas the abundances of these elements relative to H are substantially less than solar. The effects of the variability of the nebular emission are considered and found to be only marginally significant.

135.026 **A second list of new planetary nebulae found on United Kingdom 1.2-m Schmidt telescope plates.**
A. J. Longmore, S. B. Tritton.
Mon. Not. R. Astron. Soc., Vol. 193, 521 - 524 (1980).

Positions, photographs and descriptions are given for 11 new planetary nebulae discovered on United Kingdom Schmidt plates. One of the planetary nebulae has the highest galactic latitude of any known planetary, and may be associated with a magnitude 9 G5 star. Near-infrared (*J, H, K*) magnitudes are given for the star.

135.027 **Ultraviolet spectra of planetary nebulae and their central stars.** M. J. Seaton.
Highlights of Astronomy, Vol. 5, (see 012.024), 247 - 253 (1980).

IUE observations have been made of 54 planetaries, and work continues on making observations and reducing and interpreting the data. The present report describes some preliminary results.

135.028 **Planetary nebulae formation.**
Y. Tuchman, N. Sack, Z. Barkat.
Highlights of Astronomy, Vol. 5, (see 012.026), 509 - 512 (1980).

It is a well accepted idea that planetary nebulae formation is due to mass ejection from red giant envelopes. The extreme non-adiabatic behaviour of red giant envelopes which prevents their ejection in "one shot" mechanisms, turns out to be the main cause for multiple ejection.

135.029 **A new explanation for the elongation of planetary nebulae.** A. F. Bentley.
Bull. American Astron. Soc., Vol. 12, 749 (1980). – Abstract.

135.030 **Nebular and auroral transitions of [Ar IV] in some planetary nebulae.** S. J. Czyzak, G. Sonneborn, L. H. Aller, S. A. Shectman.
Astrophys. J., Vol. 241, 719 - 724 (1980).

Measurements of auroral and nebular type transitions in several planetary nebulae of high surface brightness lead to the conclusion that presently available collisional cross sections and transition probabilities for $3p^3$ configurations in Ar^{+3} may be in error. The observed auroral/nebular line ratio is always larger than the predicted value, and the disagreement is further aggravated if auroral lines are weakened by telluric line absorption.

135.031 **Electron densities for six planetary nebulae and HM Sagittae derived from the C III]λ1907/1909 ratio.** W. A. Feibelman, A. Boggess, R. W. Hobbs, C. W. McCracken.
Astrophys. J., Vol. 241, 725 - 727 (1980).

Electron densities for IC 418, NGC 6572, IC 1297, NGC 3242, NGC 6818, NGC 3211, and HM Sagittae derived from high-dispersion *IUE* C III] spectrograms are consistently higher than those derived from either surface brightness measurements or forbidden line intensity ratios in the visible. The nebulae were selected for a range of excitation classes from 3 to 9. Line splitting due to expansion velocities is observed for three objects. The great width of the λ1909 C III] line in HM Sagittae suggests large expansion velocities.

135.032 **Gas-phase abundances of refractory elements in planetary nebulae: a hot-wind model.**
G. A. Shields.
Publ. Astron. Soc. Pacific, Vol. 92, 418 - 421 (1980).

Planetary nebulae (PN) characteristically show large gas-phase depletions of some refractory elements, $[Fe/H] \simeq [Ca/H] \simeq -1.5$. In contrast, the gas-phase abundance of carbon is large, $[C/H] \geq +0.3$. This pattern is difficult to understand in terms of grain formation and destruction during PN formation. However, these abundances are consistent with a model (Kwok, Purton, and FitzGerald 1978) in which the PN shell consists of material expelled as a wind during the red-giant phase and subsequently compressed and accelerated by the impact of a hot stellar wind from the central star.

135.033 **Where exactly is the planetary nebula in M 15?**
S. Adams, C. J. Penn, M. J. Seaton.
Observatory, Vol. 100, 209 (1980).

135.034 **On the helium and nitrogen enrichment of the interstellar medium by planetary nebulae.**
M. Peimbert, A. Serrano.
Rev. Mexicana Astron. Astrofis., Vol. 5, 9 - 18 (1980).

A discussion of the various types of planetary nebulae (PN) is made. From observations of PN in the Galaxy and in the Magellanic Clouds it is found that PN of Type I comprise about 20% of the total number of PN. Type II PN delineate well defined abundance gradients across the galactic disk. Type I PN enrich significantly the interstellar medium with freshly made helium while Type II PN present the He/H abundance ratios with which they were formed. Evidence is presented that suggests that Type I PN are more massive than Type II PN and that the dividing line between them corresponds to objects of $\sim 2.4\ M_\odot$.

135.035 **The dependence of the 8 - 13 micron spectrum of NGC 7027 on position in the nebula.**
B. Jones, K. M. Merrill, W. Stein, S. P. Willner.
Astrophys. J., Vol. 242, 141 - 148 (1980).

Infrared spectra of NGC 7027 at four positions across the nebula show a variation in spectral shape. The spectra are similar to those observed from the Orion ionization-front region, and it is suggested that interpretation of this type of spectrum in terms of silicate absorption may not be valid for NGC 7027 and hence may not be valid for some other objects. The surface brightness in the [S IV] 10.52 μm emission line observed with a 3″ beam is higher than that found in previous work for the nebula as a whole. The surface brightness of emission in the 11.3 μm broad feature observed with a 3″ beam is the same as previously observed for the entire nebula.

135.036 **On multiple-shell planetary nebula formation.**
Y. Tuchman, Z. Barkat.
Astrophys. J., Vol. 242, 199 - 208 (1980).

Predictions of the combination of the authors' recent model for the formation of planetary nebulae (due to a diverging pulsational instability) together with nuclear-burning shell flashing, with regard to formation of multiple shells, are explored and discussed.

135.037 **Efficiency of the Bowen fluorescence mechanism in static nebulae.** T. Kallman, R. McCray.
Astrophys. J., Vol. 242, 615 - 627 (1980).

A simplified theory of the Bowen resonance fluorescence mechanism is presented for both the pumping of the O III Bowen lines by He II λ304 and for the pumping of the N III Bowen lines by O III λ374. The O III Bowen fluorescence yields calculated for a model planetary nebula agree well with those obtained in more rigorous calculations. Intensities of O III and N III Bowen lines observed in the spectra of planetary nebulae and compact X-ray sources are discussed.

135.038 **The evidence for planetary progenitor rotation and for the long term acceleration of the ejected nebular shell.** R. C. Kirkpatrick.
Space Sci. Rev., Vol. 27, (see 012.046), 467 - 472 (1980).

135.039 **The low excitation structures of NGC 6720 and NGC 2474 - 75.** J. P. Phillips, N. K. Reay.
Astrophys. Lett., Vol. 21, 47 - 52 (1980).

135.040 **A high velocity knot in the Helix nebula.**
J. Meaburn, J. R. Walsh.
Astrophys. Lett., Vol. 21, 53 - 56 (1980).

A high velocity ($\sim 66\ km\ s^{-1}$) split feature ~15 arcseconds in extent has been detected in [O II] emission over a dark knot in the loop of the Helix nebula. This velocity splitting is much greater than the $\sim 20\ km\ s^{-1}$ large splitting observed previously, and several mechanisms are proposed to account for this feature.

135.041 **Ring nebulae associated with Wolf-Rayet stars in the Large Magellanic Cloud.**

Y.-H. Chu, B. M. Lasker.
Publ. Astron. Soc. Pacific, Vol. 92, 730 - 735 (1980).
Eight ring-shaped nebulae associated with Wolf-Rayet stars are identified in the Large Magellanic Cloud. These nebulae are typically ten times larger than morphologically similar ones associated with galactic Wolf-Rayet stars.

135.042 **He λ10830 in planetary nebulae.**
J. N. Scrimger.
J. R. Astron. Soc. Canada, Vol. 74, 356 (1980). – Abstract.

135.043 **Planetary nebula K3 - 61.**
L. N. Kondrat'eva, V. I. Afanas'ev, V. A. Lipovetskij, A. I. Shapovalova.
Astron. Tsirk., No. 1107, p. 1 - 3 (1980). In Russian.

The ESO/Uppsala survey fo the ESO (B) atlas of the southern sky – VIII. See Abstr. 002.057.

On the derivation of electron density and temperature from [S II] and [O II] line intensity ratios. See Abstr. 022.158.

Mass-loss from the central star of NGC 6543. See Abstr. 064.010.

Mass loss from red giants: some effects on the interstellar medium. See Abstr. 064.082.

A far-infrared emission feature in IRC+10216, IC 418, and NGC 6572. See Abstr. 112.009.

A search for atomic hydrogen from evolved stars and planetary nebulae. See Abstr. 112.044.

On the variability of central stars of two planetary nebulae. See Abstr. 113.028.

***IUE* observations of RW Hydrae (gM2 + pec).** See Abstr. 114.041.

The symbiotic star near the globular cluster NGC 6401. See Abstr. 114.121.

A far-infrared emission feature in the carbon star IRC+10216 and the planetary nebulae IC 418 and NGC 6572. See Abstr. 114.123.

Abell 46. See Abstr. 119.052.

Photovisual light curve and photometric elements of the eclipsing nucleus of the planetary nebula Abell 63. See Abstr. 119.055.

The disappearance of V-V 1-7 and the nature of its central star. See Abstr. 120.014.

FG Sagittae: a binary? See Abstr. 122.001.

Objets célestes remarquables. FG Sagittae, étoile variable exceptionnelle. See Abstr. 122.148.

Spectra of gaseous nebulae. See Abstr. 134.003.

The Boomerang Nebula: a highly polarized bipolar. See Abstr. 134.026.

The problem of sulfur abundance determination in gaseous nebulae. See Abstr. 134.038.

Erratum

135.901 **Erratum: "The far ultraviolet emission of the central stars of planetary nebulae"** [Astron. Astrophys., Vol. 84, 284 - 296 (1980)].
A. Natta, S. R. Pottasch, A. Preite-Martinez.
Astron. Astrophys., Vol. 91, 378 (1980). – See Abstr. 27.135.028.

Radio Sources, X-ray Sources, Cosmic Radiation

141 Radio Sources, Quasars, Pulsars

Radio Sources, Quasars

141.001 **A confusion-limited extragalactic source survey at 4.755 GHz. I. Source list and areal distributions.**
J. E. Ledden, J.J. Broderick, J. J. Condon, R. L. Brown.
Astron. J., Vol. 85, 780 - 788 (1980).

A confusion-limited 4.755-GHz survey covering 0.00 956 sr between right ascensions 07^h05^m and 18^h near declination +35° has been made with the NRAO 91-m telescope. The survey found 237 sources and is complete above 15 mJy. Source counts between 15 and 100 mJy were obtained directly. The *P(D)* distribution was used to determine the number counts between 0.5 and 13.2 mJy, to search for anisotropy in the density of faint extragalactic sources, and to set a 99%-confidence upper limit of 1.83 mK to the rms temperature fluctuation of the 2.7-K cosmic microwave background on angular scales smaller than 7.3 arcmin. The discrete-source density, normalized to the static Euclidean slope, falls off sufficiently rapidly below 100 mJy that no new population of faint flat-spectrum sources is required to explain the 4.755-GHz source counts.

141.002 **The behaviour of hot spots in classical double radio sources.** S. G. Neff, L. Rudnick.
Mon. Not. R. Astron. Soc., Vol. 192, 531 - 544 (1980).

The relationship between the structure and luminosity of strong extragalactic radio sources (reported by Jenkins & McEllin) and between the source structure and redshift is shown to be largely an effect of observing the sources with unequal linear resolutions. Weak trends still exist after statistical correction for resolution effects, but they are dominated by inter- and intrasource scatter. At 5-km resolution, the fraction of total flux which comes from 'hot spots' of high surface brightness could be dependent on some as-yet-undetermined source characteristics, or it could be caused by some short-lived stochastic process that is unrelated in detail to general source properties.

141.003 **Optical spectra of QSOs from the Molonglo Deep Survey – I.** G. L. White, H. S. Murdoch, R. W. Hunstead.
Mon. Not. R. Astron. Soc., Vol. 192, 545 - 554 (1980).

Details of the optical spectra are given for 14 QSOs identified with radio sources from the Molonglo Deep Survey (MDS) which consists of a narrow strip at $\delta = -20°$ and includes sources with 408 MHz flux densities between 88 mJy and 1 Jy. The spectra have been obtained with the Image Dissector Scanner on the 3.9-m Anglo-Australian Telescope. Analysis of the spectra has been aimed at extracting as much quantitative information as possible with special attention being given to the estimation of magnitudes from both IDS data and the Palomar Sky Survey. Several interesting individual objects are dicsussed and attention is drawn to the wide range of equivalent width for Ly α. There is evidence for a trend towards fainter optical magnitudes and higher redshifts for the weak radio sources of the MDS.

141.004 ***VLBI* observations of 3C 345 at 1.67 GHz.**
D. Stannard, R. S. Booth, R. E. Spencer, L. B. Baath.
Mon. Not. R. Astron. Soc., Vol. 192, 555 - 560 (1980).

VLBI observations of the quasar 3C 345 suggest that the milli-arcsec subcomponents in the core have approximately equal intensities at 1.67 GHz. A new compact feature is detected in the inner jet structure at a distance of ~ 12 m.arcsec from the core.

141.005 **Detailed ultraviolet observations of the quasar 3C 273 with *IUE*.** M. H. Ulrich, A. Boksenberg, G. Bromage, R. Carswell, A. Elvius, A. Gabriel, P. M. Gondhalekar, J. Lind, L. Lindegren, M. S. Longair, M. V. Penston, M. A. C. Perryman, M. Pettini, G. C. Perola, M. Rees, D. Sciama, M. A. J. Snijders, E. Tanzi, M. Tarenghi, R. Wilson.
Mon. Not. R. Astron. Soc., Vol. 192, 561 - 580 (1980).

The authors have obtained nine spectra of the quasar 3C 273 with the *IUE* with the aim of combining them and obtaining a high signal-to-noise spectrum between 1100 and 3300 Å. This permits them to make a detailed analysis of the emission line intensity ratios and of the continuous energy distribution. The authors find evidence for a thermal contribution to the ultraviolet continuum. The ultraviolet spectrum of 3C 273, which they obtained, shows eight absorption lines at zero redshift due to the intervening material in our Galaxy. The C IV λ 1550 absorption is much stronger in the spectrum of 3C 273 than in the spectra of halo stars. This is taken as evidence for the existence of a hot gaseous component in the outer regions of the halo of the Galaxy.

141.006 **1610 – 771: a QSO with a steep optical spectrum.**
R. W. Hunstead, H. S. Murdoch.
Mon. Not. R. Astron. Soc., Vol. 192, 31P - 35P (1980).

1610 – 771 is a strong flat-spectrum radio source identified with a QSO. The optical spectrum is found to have a steep power-law continuum with an observed spectral index α of 3.8 ($f_\nu \propto \nu^{-\alpha}$). Corrections due to atmospheric dispersion and to reddening within the Galaxy lead to a best estimate for α of 2.7. The redshift (1.710) is the highest yet found for a QSO with a steep optical continuum bringing the red-shifted Lyman α line within the atmospheric window.

141.007 ***JHK* observations of two $z = 3$ QSOs.**
I. S. Glass.
Mon. Not. R. Astron. Soc., Vol. 192, 37P - 39P (1980).

Infrared observations of the QSOs PKS 2126–15 (z=3.27) and Q0420–388 (z=3.13) taken with data from other spectral regions show that they have bolometric luminosities in excess of 3×10^{41} W ($q_0 = 0$).

141.008 **Quasare, Blasare, Schwerkraftstrudel.**
W. Priester.
Phys. Bl., 36. Jahrg., 241 - 245 (1980).

141.009 **A search for rapid variability in eight quasars and BL Lacertae objects.** B. E. Schaefer.
Publ. Astron. Soc. Pacific, Vol. 92, 255 - 258 (1980).

Data are presented on the results of a photographic patrol of eight quasars between November 1977 and April 1978. Four of the objects were observed ten to 20 times within a week. PKS 0735+178 was found to vary significantly in brightness on a time scale of weeks.

141.010 **A model for the complex extragalactic radio source 3C 288.** G. G. Byrd, M. J. Valtonen.
Publ. Astron. Soc. Pacific, Vol. 92, 262 - 265 (1980).

The authors describe a method of modeling complex double radio sources and head-tail radio sources. The method is most easily applied to the slingshot theory of radio sources but may be used also in other theories. As an example, they present a model of 3C 288 based on the slingshot theory and calculate the brightness distribution at two different observing frequencies and spatial resolutions.

141.011 **The quasar 3C351: VLA maps and a deep search for optical emission in the outer lobes.**
P. P. Kronberg, J. N. Clarke, S. van den Bergh.
Astron. J., Vol. 85, 973 - 980 (1980).

VLA radio maps of the quasar 3C351 ($z = 0.371$) at $\sim 2''$ and $0\overset{''}{.}4$ resolution (1) show interaction with a relatively dense intergalactic medium, (2) show that there is electron acceleration within at least one of the radio lobes, and (3) imply that the intergalactic gas density is different on one side of the source than on the other. Striking similarities are found between the northern radio lobe of 3C351 and one of the outer hotspots of Cygnus A. A search for optical emission in the lobes shows no emission stronger than 22^m in the J band and $\sim 21^m$ in the F band.

141.012 **An analysis of the cosmological evolution of radio sources. I. Spectral-index dependent counts of sources and spectral index distributions at 1400 MHz.**
J. Machalski.
Astron. Astrophys., Suppl. Ser., Vol. 41, 323 - 327 (1980).

A low intensity extension of the spectral-index dependent counts of GB2 radio sources has been attempted on the basis of the Westerbork surveys at 1415 MHz. Also Westerbork data at 610 MHz have been used to combine the low frequency spectral index distributions for the samples of strong and weak sources which come from the Green Bank and Westerbork surveys.

141.013 **A radio continuum survey at 1.4 GHz of the galaxies in the Virgo region.** C. G. Kotanyi.
Astron. Astrophys., Suppl. Ser., Vol. 41, 421 - 432 (1980).

Radio continuum observations made with the Westerbork Synthesis Radio Telescope at 1.4 GHz are presented for 274 galaxies in the Virgo Cluster region. The observations are partly full syntheses and partly East-West strip distributions. For the 55 detected galaxies the emission is separated into central and extended components. A brief summary of the analysis is also given.

141.014 **The relation between redshift and apparent magnitude of QSOs with light variation not larger than $0\overset{m}{.}5$.**
L.-t. Yang, X.-b. Ai, X.-h. Xiao, C.-l. Cao, Y.-l. Bian, X.-y. Tang.
Acta Astron. Sinica, Vol. 21, 208 - 210 (1980). In Chinese.

141.015 **The double quasar 0957+561 as a gravitational lens: further VLA observations.**
P. E. Greenfield, B. F. Burke, D. H. Roberts.
Nature, Vol. 286, 865 - 866 (1980).

The authors report a full 12-h synthesis made at the Very Large Array at 6-cm wavelength which confirmed the earlier maps and showed additional features at lower flux levels. In particular, this map revealed two small radio 'jets', one close to each of the quasars. The absence of corresponding second images of either of these jets near the other quasar was used to rule out models in which the gravitator was located near the midpoint of the line between A and B.

141.016 **Optical identifications of reference frame benchmark radio sources.** A. N. Argue, C. Sullivan.
Mon. Not. R. Astron. Soc., Vol. 192, 779 - 786 (1980).

Optical positions have been measured and finding charts verified for 84 of the sources that have been provisionally selected for establishing a positional reference frame based on extragalactic radio sources. The accuracy is ± 0.7 arcsec in each coordinate.

141.017 **The structure of Cygnus A at 150 MHz.**
A. J. B. Winter, D. M. A. Wilson, P. J. Warner, E. M. Waldram, D. Routledge, A. T. Nicol, R. C. Boysen, D. W. J. Bly, J. E. Baldwin.
Mon. Not. R. Astron. Soc., Vol. 192, 931 - 944 (1980).

Observations of the structure of Cyg A at 150 MHz with a resolution of 10.5×16 arcsec show a strong bridge of emission linking the two components. Radio spectra between 150 MHz and 5 GHz at points along the bridge are consistent with a model in which the hot spots move outwards at a roughly uniform velocity and leave behind a population of electrons which subsequently ages by synchrotron losses in a magnetic field.

141.018 **Radio jets and bridges in the classical double sources 3C388 and 0816+526.**
J. O. Burns, W. A. Christiansen.
Nature, Vol. 287, 208 - 210 (1980).

The authors report on 4885-MHz full synthesis high-resolution observations of the 'classical'double radio sources 0816+526 and 3C388. These observations clearly illustrate that a variety of source substructures, which must be dealt with in theoretical models, exist within radio galaxies which were previously classified simply as classical doubles.

141.019 **Multifrequency observations of extended radio galaxies III: 3C 465.** W. J. M. van Breugel.
Astron. Astrophys., Vol. 88, 248 - 258 (1980).

Total intensity and polarization maps made at 610 MHz and 4995 MHz using the Westerbork Synthesis Radio Telescope are presented for the wide tail radio galaxy 3C 465. A multifrequency comparison has been made using these and previously published observations at 1.4 GHz. Polarization data are available at the three highest frequencies for three different regions along the tails. An analysis of these data was made using simple Faraday dispersion models (Burn, 1966). For each region the depolarization and the rotation of the polarization position angles at the various frequencies were constrained to fit simultaneously. In this way the foreground rotation measure (RM_F), the rotation measure intrinsic to the source (RM_S) and a parameter (ξ) related to the importance of the random component of the magnetic field are determined. After correction for the foreground rotation the projected magnetic field along the tails of 3C 465 has been derived. The field is directed along these tails and is indicative of a helical symmetry.

141.020 **On the variability of extragalactic radio sources under the long-term injection of relativistic electrons.** V. T. Pechorin.
Astron Zh., Tom 57, 687 - 695 (1980). In Russian.
English translation in Soviet Astron., Vol. 24, No. 4.

On the basis of the B-type model (the energy density of the magnetic field is dominating), which has been discussed in previous papers, the theoretical relations describing the behaviour of the spectral flux density variations of radio emission under the condition of long-term injection of relativistic particles are given. A preliminary comparison shows that the theory is in satisfactory agreement with the observational data on the centimetric radio emission variability of the quasar 3C345.

141.021 **Evolution of radio emission in the "hedgehog" model for instationary sources.**
Yu. A. Kovalev, V. P. Mikhajlutsa.
Astron. Zh., Tom 57, 696 - 706 (1980). In Russian.

English translation in Soviet Astron., Vol. 24, No. 4.

Relations for the numerical calculation of the flux density and polarized radiation intensity of an arbitrary geometry cloud consisting of ultrarelativistic electrons which dissipate in the radial magnetic field of the nucleus under a random angle to the observer are obtained. It is shown that there is a maximum observed distance of the moving cloud from the nucleus centre in the picture plane and the observed cloud lifetime is determined by the radiation cut-off due to the pitch-angles evolution or the absorption of electrons by the nucleus. In the case of a compact cloud analytical and quantitative results are obtained.

141.022 **Radio observations of optically selected quasars.** P. A. Strittmatter, P. Hill, I. I. K. Pauliny-Toth, H. Steppe, A. Witzel.
Astron. Astrophys., Vol. 88, L12 - L15 (1980).

A sample of 70 optically selected QSOs from the Michigan Curtis Schmidt survey has been observed at wavelengths of 6 cm and 2.8 cm using the 100-m Effelsberg telescope. Ten sources stronger than 10 mJy have been detected and exhibit a flux density distribution strongly weighted towards higher values (> 200 mJy). The results are apparently inconsistent with the hypothesis that the radio/optical flux density ratio in QSOs is dominated by the effects of relativistic jets. They do, however, support the view that the ratio is bimodal and that among the detected (i.e. strong) radio sources the optical and radio luminosities are correlated.

141.023 **The Broad-Line Region in active nuclei and quasars: correlations with luminosity and radio emission.** T. M. Heckman.
Astron. Astrophys., Vol. 88, 311 - 316 (1980).

Using recently published infrared and optical continuum fluxes and emission line parameters the author has explored the luminosity dependence of the properties of the Broad-Line Region (BLR) in active nuclei and quasars and looked for systematic differences between optical and radio-selected BLR's in terms of these luminosity dependences. He finds that most properties of the BLR are not strongly luminosity dependent, with the He II $\lambda 4686/H\beta$, and Fe II/$H\beta$ line ratios being two exceptions. Interpretation in terms of a "softening" of the ionizing continuum with increasing luminosity is discussed. The author finds that BLR's occurring in conjunction with extended and strong radio emission are peculiar in many ways.

141.024 **Radio observations of globular clusters and galactic bulge X-ray sources.** Gopal-Krishna, H. Steppe.
Astron. Astrophys., Vol. 88, 354 - 359 (1980).

Radio continuum observations at 2.7, 4.8, and 10.7 GHz are reported for (a) 10 radio sources known to lie in the fields of 6 globular clusters, (b) for the cores of 2 globular clusters and (c) for 5 galactic bulge X-ray sources. All 10 sources in the first group were detected and some of them show a flat radio spectrum. Information on angular structure is obtained by comparing the published interferometric data with pencil-beam measurements and cases of extended radio emission are found. In the second group, a radio source is seen within ~1′ of the centre of NGC 6440. In the third group, GX 17 + 2 is found to coincide with a radio source showing both compact and extended radio emission.

141.025 **Observations of scintillations of the quasars 3C 345 and 3C 454.3.** V. S. Artyukh, Yu. N. Vetukhnovskaya.
Pis'ma Astron. Zh., Tom 6, 402 - 404 (1980). In Russian.
English translation in Soviet Astron. Lett., Vol. 6.

Interplanetary scintillations of the quasars 3C 345 and 3C 454.3 have been observed at a frequency of 102 MHz. Angular diameters and flux densities of the scintillating components of the sources have been estimated.

141.026 **Observations of three bright extragalactic radio sources at 1.38 cm wavelength with a resolution up to 8″.** A. B. Berlin, Yu. V. Korenev, V. Yu. Lesovoj, Yu. N. Parijskij, V. I. Smirnov, N. S. Soboleva.
Pis'ma Astron. Zh., Tom 6, 470 - 475 (1980). In Russian.
English translation in Soviet Astron. Lett., Vol. 6.

New observations of radio galaxies in the shortest wavelength region of the radiotelescope RATAN-600 were performed using the 1.38-cm radiometer. One-dimensional radio brightness distribution of 3C 405 (Cyg A) and Cen A as well as instantaneous spectra of the nuclear sources in 3C 111, 3C 405 and Cen A are presented.

141.027 **A burst of the H_2O maser source in Orion.** L. I. Matveenko, L. R. Kogan, V. I. Kostenko.
Pis'ma Astron. Zh., Tom 6, 505 - 508 (1980). In Russian.
English translation in Soviet Astron. Lett., Vol. 6.

A burst of H_2O maser emission in the radio source Orion A has been investigated in September - November 1979. The radio flux density, the width of spectral line and the size of the burst region were measured. It is shown that the burst emission is highly directional and consists of fine spectral lines corresponding to the emission of individual zones.

141.028 **4C 32.69: a quasar with a radio jet.** R. I. Potash, J. F. C. Wardle.
Astrophys. J., Vol. 239, 42 - 49 (1980).

The authors present high resolution radio observations of the quasar 4C 32.69 ($z = 0.67$). This source exhibits a narrow radio jet, some 200 kpc in length, that reaches from the quasar nucleus to one of the outer radio lobes. This is the first such jet found in a quasar and is two to three orders of magnitude more luminous than similar jets found in relatively nearby radio galaxies. The jet contains a highly ordered magnetic field which is aligned parallel to the jet over its entire length. The physical properties of the jet are discussed in some detail.

141.029 **VLA observations of the M87 jet at 6 and 2 centimeters.** F. N. Owen, P. E. Hardee, R. C. Bignell.
Astrophys. J., Lett., Vol. 239, L11 - L15 (1980).

Observations of M87 made with the VLA in 1979 March at 2 and 16 cm are presented. Both total and polarized intensity maps are shown at both frequencies with 1″ (6 cm) and 0″.6 (2 cm) resolution. The maps show that the brightness distribution of the jet is very similar in the radio and optical regions of the spectrum. The brighter knots are resolved transverse to the jet. Minimum pressure calculations show that, although the knots are slightly larger than previously thought, the regions of emission are apparently out of pressure equilibrium with a surrounding medium unless the pressure in the ISM is much higher than previously thought. The jet is also not strictly linear, but seems to wiggle with sinusoidal oscillation and amplitude increasing as angular separation from the nucleus increases. Various ways of explaining the last two results are discussed briefly.

141.030 **VLA maps of V1016 Cyg, Hb12, CRL2591, HR8752, and P Cyg.**
R. T. Newell, R. M. Hjellming.
Bull. American Astron. Soc., Vol. 12, 458 - 459 (1980). Abstract.

141.031 **The transport of energetic electrons in radio sources.** S. R. Spangler, J. P. Basart.
Bull. American Astron. Soc., Vol. 12, 461 - 462 (1980). Abstract.

141.032 **The polarization spectra of radio outbursts in extragalactic variable sources.**
H. D. Aller, M. F. Aller, P. E. Hodge.
Bull. American Astron. Soc., Vol. 12, 462 (1980). – Abstract.

141.033 **Clustering of radio sources in the 4C catalog.**
M. Seldner, P. J. E. Peebles.
Bull. American Astron. Soc., Vol. 12, 471 (1980). – Abstract.

141.034 **Einstein X-ray and VLBI radio observations of the variable quasars NRAO 140 and NRAO 530.**
A. P. Marscher, J. J. Broderick.
Bull. American Astron. Soc., Vol. 12, 487 (1980). – Abstract.

141.035 **Extragalactic radio source positions determined with VLBI.** C. Ma.
Bull. American Astron. Soc., Vol. 12, 489 (1980). – Abstract.

141.036 **Long term optical behavior of 114 extragalactic sources.** A. J. Pica, J. T. Pollock, A. G. Smith, R. J. Leacock, P. L. Edwards, R. L. Scott.
Bull. American Astron. Soc., Vol. 12, 493 (1980). – Abstract.

141.037 **Jet formation in asymmetric radio galaxies.**
M. J. Siah, P. J. Wiita.
Bull. American Astron. Soc., Vol. 12, 493 (1980). – Abstract.

141.038 **High-resolution radio observations at 6 and 20 cm of the jet in NGC 6251.** R. A. Perley, A. G. Willis.
Bull. American Astron. Soc., Vol. 12, 494 (1980). – Abstract.

141.039 **Broadband spectra and variability of compact non-thermal sources.** L. Rudnick, T. W. Jones, F. N. Owen, J. J. Puschell, D. J. Ennis, M. W. Werner.
Bull. American Astron. Soc., Vol. 12, 495 (1980). – Abstract.

141.040 **Quasar proper motions study of 1038+528.**
M. J. Reid, F. N. Owen, D. B. Shaffer, K. I. Kellermann, A. Witzel.
Bull. American Astron. Soc., Vol. 12, 497 (1980). – Abstract.

141.041 **The radio/optical structure of the QSO 0812+020.**
S. Wyckoff, K. Johnston, F. Ghigo, L. Rudnick, P. Wehinger, T. Gehren, A. Boksenberg.
Bull. American Astron. Soc., Vol. 12, 497 (1980). – Abstract.

141.042 **A model for 3C84.**
C. O'Dea, W. Dent, E. Tademaru, T. Balonek.
Bull. American Astron. Soc., Vol. 12, 497 - 498 (1980).
Abstract.

141.043 **VLA observations of 3C227.**
G. Swarup, R. P. Sinha.
Bull. American Astron. Soc., Vol. 12, 498 (1980). – Abstract.

141.044 **VLBI observations of the "twin quasar" 0957+56A, B.**
M. V. Gorenstein, I. I. Shapiro, N. L. Cohen, E. Falco, N. Kassim, A. E. E. Rogers, A. R. Whitney, R. A. Preston, A. Rius.
Bull. American Astron. Soc., Vol. 12, 498 (1980). – Abstract.

141.045 **CNO in Quasars.** A. K. Uomoto.
Bull. American Astron. Soc., Vol. 12, 498 (1980).
Abstract.

141.046 **Measurements of extragalactic variable sources at five radio frequencies.**
T. J. Balonek, W. A. Dent, C. P. O'Dea.
Bull. American Astron. Soc., Vol. 12, 507 (1980). – Abstract.

141.047 **Line and continuum cooling in QSO emission line regions.**
R. C. Canfield, R. C. Puetter, P. J. Ricchiazzi.
Bull. American Astron. Soc., Vol. 12, 536 (1980). – Abstract.

141.048 **Fe II lines in X-ray selected QSO.**
J. E. Steiner, J. E. Grindlay.
Bull. American Astron. Soc., Vol. 12, 536 (1980). – Abstract.

141.049 **The properties of the gas producing the broad absorption features in QSO spectra.**
D. A. Turnshek, R. J. Weymann.
Bull. American Astron. Soc., Vol. 12, 536 (1980). – Abstract.

141.050 **Spectrophotometry of broad absorption line QSOs.**
V. T. Junkkarinen, E. M. Burbidge, H. E. Smith.
Bull. American Astron. Soc., Vol. 12, 537 (1980). – Abstract.

141.051 **The relative luminosity calibration of quasars.**
J. A. Baldwin, W. L. Burke, C. M. Gaskell, E. J. Wampler.
Bull. American Astron. Soc., Vol. 12, 537 (1980). – Abstract.

141.052 **On observations of cosmic radio sources with the method of very long baseline radiointerferometry.**
V. S. Gubanov, N. D. Umarbaeva, P. A. Fridman, L. I. Yagudin.
Kosm. Issled., Tom 18, 632 - 642 (1980). In Russian.

141.053 **Fluid jets in radio sources.** W. A. Baan.
Astrophys. J., Vol. 239, 433 - 444 (1980).

An axisymmetric beam of viscous fluid moving with relativistic bulk motion is considered to account for the observed radio jets in double radio sources. The confinement of the beam is due to its viscous interaction with the ambient medium. A fluid with a magnetic viscosity, based on the finite size of the Larmor radius, gives a rapid widening of the beam at the nozzle and a steady opening angle further along the beam; the same are observed properties of radio jets. Other sources of viscosity are investigated. The viscous beam shapes can be fitted well to the contours of the observed jets in NGC 315 and NGC 6251. A brightness map based on a simple temperature dependence of the volume emissivity agrees well with the jet of NGC 315. Radio properties are interpreted in the context of Maxwellian electron distributions. The energy transport efficiency of the beam is discussed and also the general stability of the flow pattern.

141.054 **Analysis of quasars found in the CTIO Curtis Schmidt survey in the −40° zone.** H. Arp.
Astrophys. J., Vol. 239, 463 - 468 (1980).

The Curtis Schmidt telescope at CTIO has been used with an objective prism to discover quasars in the south galactic polar cap region (Smith 1976; Osmer and Smith 1976; Osmer and Smith 1977). The zone most completely surveyed was 5° between decl.= −37°.5 to −42°.5. Since this is the first complete survey of radio-quiet quasars over a large area of the sky, it is important to examine their distribution on the sky in an attempt to decide between the two long-conflicting views of quasars: (1) that they are all very distant objects distributed randomly and homogeneously over the sky or (2) that some are at the distances of the nearer galaxies and show groupings and associations on the scale of the nearby galaxies. An examination is carried out in the following paper from the standpoint of spatial correlation on the sky, as well as the properties of the quasars themselves, such as redshift and apparent magnitude.

141.055 **Observations of quasars with the International Ultraviolet Explorer satellite.**
R. F. Green, J. R. Pier, M. Schmidt, F. B. Estabrook, A. L. Lane, H. D. Wahlquist.
Astrophys. J., Vol. 239, 483 - 494 (1980).

The authors report here on observations of six quasars: PG 0953 + 415, PKS 1302−102, 3C 351, PKS 0405−123, PG 1115 + 080, and PG 1247 + 268. An observing log for these objects is presented and finding charts for the newly reported

Palomar-Green quasars. The authors discuss briefly the observations and reduction techniques with *IUE*; they present the observational results for each object individually, and a general discussion follows.

141.056 **Discovery of low-redshift X-ray selected quasars: new clues to the QSO phenomenon.**
J. E. Grindlay, J. E. Steiner, W. R. Forman, C. R. Canizares, J. E. McClintock.
Astrophys. J., Lett., Vol. 239, L43 - L48, plate L1 (1980).
Using the CTIO 4 m and 1.5 m telescopes to identify serendipitous X-ray sources discovered by the Einstein Observatory, the authors have identified as quasars all six sources in four high-latitude IPC fields (~4 deg²). These X-ray selected QSOs have low redshifts ($z \sim 0.08-0.23$ for five objects and $z = 0.73$ for the sixth) and relatively faint optical magnitudes ($\langle m_B \rangle \approx 18$) and are radio quiet.

141.057 **Distorted radio sources in Abell 2255: evidence of intergalactic gas 2.5 to 5 megaparsecs from the cluster center.** P. Hintzen, J. S. Scott.
Astrophys. J., Vol. 239, 765 - 768, plate 9 (1980).
The authors present spectroscopic data for radio galaxies in the A2255 region. Their results indicate that the galaxies associated with distorted radio sources 2.5–5 Mpc projected radius from the cluster center are indeed cluster members. This implies the presence of substantial intergalactic gas 10–20 core radii from the cluster center. Reasonable assumptions concerning the gas distribution result in an estimated minimum mass for the A2255 intracluster gas of $4 \times 10^{14} M_{\odot}$.

141.058 **High-resolution observations of the neutral hydrogen absorption and radio continuum emission of the radio source 3C 178.** A. D. Haschick, P. C. Crane, P. E. Greenfield, B. F. Burke, W. A. Baan.
Astrophys. J., Vol. 239, 774 - 782 (1980).
The radio source 3C 178, which has been identified with the spiral galaxy NGC 2377, has been observed with the Very Large Array. These observations show that the source has a double radio structure with roughly equal unresolved components separated by 24″. The radio lobes are offset by 25″ and 45″, respectively, from the bright nucleus of the galaxy. It is concluded that the radio source is not associated with the galaxy and is possibly a background quasar or radio galaxy.

141.059 **The compact radio source at the galactic center.**
S. P. Reynolds, C. F. McKee.
Astrophys. J., Vol. 239, 893 - 897 (1980).
The authors show that it is unlikely that the compact nonthermal radio source at the galactic center is confined, and consider models involving relativistic outflow, spherically or in jets. They show that several dynamically self-consistent models can reproduce the observations, requiring no more total power than the Crab pulsar. The authors discuss observable characteristics of these models and how future observations may distinguish among them.

141.060 **Carbon emission line profiles of high redshift quasars.** D. O. Richstone, K. Ratnatunga, J. Schaeffer.
Astrophys. J., Vol. 240, 1 - 9 (1980).
Emission line profiles of C IV λ1550 lines in 22 quasars are obtained from spectroscopy with a cooled SIT Vidicon. In 10 of the objects the C III] lines are found to have similar profiles. The observed line profiles are well approximated by the Blumenthal Mathews logarithmic model. The distribution of line breadths – augmented by those in the literature – is inconsistent with a randomly oriented collection of disks, all spinning at the same rate. The Baldwin C IV luminosity effect is not seen in these data. The similarity of line profiles of species of differing ionization potentials suggests that the lines are formed in a number of clouds of different velocities. The line profiles and distribution of line breadths suggest that the cloud formation is three-dimensional rather than planar.

141.061 **The redshift and magnitude distribution of faint quasars.** B. G. Vaucher, D. W. Weedman.
Astrophys. J., Vol. 240, 10 - 19 (1980).
Microdensitometry of survey plates obtained with a transmission grating on the KPNO 4 m telescope has produced magnitudes for 53 faint quasars. The results supplement and confirm 4 existing samples obtained with similar techniques. All such samples are summarized, giving a total of 467 quasars with redshifts. From this summary, the differential number counts of quasars do not increase by more than about 3.5 times per magnitude interval, fainter than 18th mag. Such an increase reflects primarily the luminosity function of quasars with $1.8 < z \leq 2.4$.

141.062 **3C 206: a resolved quasar in a cluster of galaxies.**
S. Wyckoff, P. A. Wehinger, H. Spinrad, A. Boksenberg.
Astrophys. J., Vol. 240, 25 - 31, plate 1 (1980).
Spectroscopic and photometric observations obtained of the quasar 3C 206 and its resolved optical structure show that 3C 206 is a normal (optically variable) QSO, which is seated in a luminous elliptical galaxy. The resolved underlying galactic disk has the color and intrinsic luminosity considered normal for an elliptical galaxy with $z = 0.2$; the spectrum displays low-excitation emission lines and Ca II (H and K) absorption lines with the QSO redshift ($z = 0.2$). Moreover, 3C 206 is found to be centered in a rather compact cluster of fainter galaxies (Bautz-Morgan type I), at least one of which also has the QSO redshift.

141.063 **A second correlated radio-optical outburst in the BL Lacertae-type quasi-stellar object 0235+164.**
T. J. Balonek, W. A. Dent.
Astrophys. J., Lett., Vol. 240, L3 - L5 (1980) = Five College Astron. Dep., Contrib. No. 316.
The authors present evidence for a correlation between a recent radio outburst in 0235+164 and the optical outburst of early 1979. The simultaneous peaks of the optical and radio emission have the same pattern as the 1975 outburst, implying a common energy source for both the radio and optical radiation. Measurements at five radio frequencies show that the radio spectrum has remained very flat at all epochs.

141.064 **High-resolution observations of the nucleus of 3C 390.3.** E. Preuss, K. I. Kellermann, I. I. K. Pauliny-Toth, D. B. Shaffer.
Astrophys. J., Lett., Vol. 240, L7 - L10 (1980).
The nucleus of the radio galaxy 3C 390.3 ($z = 0.0569$) has been observed with a three-element VLB interferometer system at 5 GHz (6 cm). The radio core appears to be an asymmetric core-jet oriented in position angle 143° ± 5°, which is the same as the axis of the 300 kpc double source. About 40% of the flux of the core at 6 cm comes from an unresolved component with diameter ≲ 0.4 milli-arcsec, or 0.6 pc. Comparison of the data from two epochs indicates that there is structural variability on the pc scale, but any relative apparent motion is with a velocity of less than $0.5c$.

141.065 **Millimeter continuum observations of flat spectra radio sources.** E. Kreysa, I. I. K. Pauliny-Toth, G. V. Schultz, W. A. Sherwood, A. Witzel.
Astrophys. J., Lett., Vol. 240, L17 - L19 (1980).
Photometric observations of 14 radio sources with flat or inverted spectra at short cm- and mm- wavelengths have been made near a wavelength of 1 mm. None of the sources shows an excess, thereby ruling out dust emission as a significant contributor to the energy distribution. A comparison of the radio source diameters derived from synchrotron theory with those measured directly with VLBI shows good agreement be-

tween the two. The energy distribution of flat spectra radio sources may be explained by a superposition of synchrotron sources of different sizes.

141.066 **Polarization variability of the optical radiation of the compact extragalactic object B2 1418+54.**
S. G. Marchenko.
Pis'ma Astron. Zh., Tom 6, 564 - 566 (1980). In Russian. English translation in Soviet Astron. Lett., Vol. 6.

Results are given of polarimetric and photometric observations of the object B2 1418+54 made in 1979. Rapid polarization variability is found. Some connection probably exists between variations of brightness and polarization.

141.067 **Morphology of the triple QSO PG1115+08.**
E. K. Hege, J. R. P. Angel, R. J. Weymann, E. N. Hubbard.
Nature, Vol. 287, 416 - 417 (1980).

Observations were undertaken to search for evidence of an intervening galaxy in PG1115+08, appearing either directly in a deep image, or as a difference in colour between components, and also to look for any structure in the three images. A colour difference could arise if a galaxy is nearly coincident with one of the QSO images.

141.068 **Where does particle acceleration occur in extended extragalactic radio sources?** P. A. Hughes.
Mon. Not. R. Astron. Soc., Vol. 193, 277 - 283 (1980).

The author suggests that particle acceleration does not occur in the extended lobes of extragalactic radio sources, but only in the compact heads. Away from these, waves capable of accelerating particles may not propagate. Although wave generation within the lobes would allow acceleration there, it is not obvious that the plasma is sufficiently disturbed for this to occur.

141.069 **Radio variability in the nuclei of double radio galaxies and quasars.** R. G. Hine, P. A. G. Scheuer.
Mon. Not. R. Astron. Soc., Vol. 193, 285 - 293 (1980).

A search for radio variability in the central components of eight extended double radio sources has been made. The three radio galaxies observed all showed variability, whereas the central components of the five quasars remained constant. These variations violate the conventional brightness temperature limit if the source is static, but they are consistent with a relativistic jet with $\gamma \cong 2$, a value which is also admitted by the statistical properties of central components of double sources.

141.070 **Spectral properties of Ooty occultation radio sources.**
Gopal-Krishna, H. Steppe, A. Witzel.
Astron. Astrophys., Vol. 89, 169 - 172 (1980).

Precise measurements of radio spectra are reported for a sample of 50 extragalactic radio sources whose metre-wavelength flux-densities, structures and optical identifications have been determined in the Ooty lunar occultation survey. The interrelationship between these parameters and the radio spectrum is examined.

141.071 **A study of the 5′ halo of 3C84.**
W. Reich, U. Stute, R. Wielebinski.
Astron. Astrophys., Vol. 89, 204 - 206 (1980).

High dynamic range observations were made at $\lambda\lambda$6.2 and 2.8 cm wavelength of 3C84 (NGC 1275, Perseus A) with the 100-m Effelsberg radio telescope. The λ 6.2 cm map shows the asymmetric 5′ halo previously detected only at longer wavelengths. The steepening of the spectral index of this halo from north to south across its western extension is confirmed. In addition, a low upper limit can be set for the halo emission at λ 2.8 cm wavelength. Some consequences of these observations are discussed.

141.072 **Anomalous redshifts of quasi-stellar objects.**
J. V. Narlikar, P. K. Das.
Astrophys. J., Vol. 240, 401 - 414 (1980).

This paper is based on the assumption that the observational evidence to date does point to the possibility that high-redshift quasars are physically associated with low redshift galaxies. It is first argued that the excess (or anomalous) redshifts of the quasars in such associations are unlikely to be either of Doppler or of gravitational origin. A new source for this excess redshift was suggested by Narlikar on the basis of the Hoyle-Narlikar theory of gravitation which is based on Mach's principle. This idea is applied to the hypothesis that quasars may have been ejected from galactic nuclei. The dynamics of such an ejection and its observable consequences are discussed. In particular, it is shown that quasar alignments and redshift bunching which have been observed recently can be understood within the framework of this theory. Further tests of this hypothesis are discussed.

141.073 **The X-ray spectrum of QSO 0241 +622.**
D. M. Worrall, E. A. Boldt, S. S. Holt, P. J. Serlemitsos.
Astrophys. J., Vol. 240, 421 - 428 (1980).

The authors present the X-ray spectrum of QSO 0241 +622 in the range 2–50 keV measured with the Goddard Space Flight Center proportional counters on *OSO 8*. The best power-law fit has a photon spectral index and 90% errors $\Gamma = 1.93\ (+0.5\ -0.3)$ and low-energy absorption consistent with reported gas column densities, but a thermal bremsstrahlung form with temperature 13.1 keV cannot be excluded. No indication of spectral variability is found in three observations of the source with *HEAO* A-2, although they observe a possible 15–30% intensity change over a period of 6 months.

141.074 **Orbital dynamics of the radio galaxy 3C 129. II. Internal tail structure.**
M. J. Valtonen, G. G. Byrd.
Astrophys. J., Vol. 240, 442 - 446 (1980).

Detailed calculations have been carried out on the tail structure of 3C 129 assuming that it results from trails of two supermassive particle accelerators. Special attention has been paid in explaining the high-brightness jetlike feature near the galaxy, which has been previously observed with the VLA. The calculations include the effect of the flow of an intergalactic medium around a circumgalactic medium which travels with the galaxy. Models with various supermassive object orbits have been tried. A contour map of the best model is presented for supersonic motion of the galaxy. This model reproduces the brightness structure of the near tail (within 3′ of the galaxy) fairly well.

141.075 **Interferometric limits on very small-scale fluctuations in the cosmic microwave background.**
H. M. Martin, R. B. Partridge, R. T. Rood.
Astrophys. J., Lett., Vol. 240, L79 - L82 (1980).

Using the three-element interferometer at NRAO Green Bank, the authors have searched for fluctuations in the cosmic microwave background at 2695 and 8085 MHz.

141.076 **Is PKS1921–29 a quasar with correlated radio and optical variations?** G. Gilmore.
Nature, Vol. 287, 612 - 613 (1980).

Optical observations of PKS1921–29 which may show correlated radio and optical variability are reported.

141.077 **Ooty lunar occultation survey: list 9.**
M. N. Joshi, A. K. Singal.
Mem. Astron. Soc. India, Vol. 1, 49 - 72 (1980).

The authors present position and structure information at 327 MHz for 240 radio sources derived from their lunar occultation observations made with the Ooty radio telescope. The flux densities of the sources lie in the range 0.2 to 4.5 Jy with a median value of 0.7 Jy. Resolutions of 6 arcsec or better have been achieved for more than 50% of the sources. A search for optical identifications using the Palomar Sky Survey Prints was made for all the sources and the results are presented.

141.078 **Multiple quasar may indicate another gravitational lens.** G. B. Lubkin.
Phys. Today, Vol. 33, No. 9, p. 17 - 19 (1980).

141.079 **Circinus X-1 at 408 MHz.** A. G. Little.
Proc. Astron. Soc. Australia, Vol. 3, 279 - 282 (1978).

141.080 **Possible anisotropy of the spatial distribution of QSOs from the MC2 and MC3 catalogues.**
H. S. Murdoch.
Proc. Astron. Soc. Australia, Vol. 3, 282 - 283 (1978).

141.081 **The time scale of decimetre flux density variations.** W. B. McAdam.
Proc. Astron. Soc. Australia, Vol. 3, 283 - 285 (1978).

141.082 **80 MHz survey of extra-galactic X-ray sources.** O. B. Slee, P. J. Quinn.
Proc. Astron. Soc. Australia, Vol. 3, 332 - 341 (1979).

141.083 **Accurate radio positions with the Tidbinbilla interferometer.**
M. J. Batty, D. L. Jauncey, P. T. Rayner, S. Gulkis.
Proc. Astron. Soc. Australia, Vol. 3, 400 - 402 (1979).

141.084 **The observed wavelengths of C IV and C III] emission in QSOs.** D. Wills.
Astrophys. J., Vol. 240, 721 - 725 (1980).

Following the suggestion by Walsh, Carswell, and Weymann that the ratio of the observed wavelengths of the C IV and C III] (λλ 1549, 1909) emission lines in the double QSO 0957 + 561 differs significantly from the expected value, the distribution of this ratio in a sample of 189 other QSOs is examined. The mean value of the ratio is consistent with the expected one, and the dispersion is consistent with typical measuring uncertainties. An unexpected finding, with no ready explanation, is that the ratio of the C IV and C III] wavelengths is correlated with emission-line redshift, at the 4 σ level.

141.085 **Peculiar configurations of quasars in two adjacent areas of the sky.** H. Arp, C. Hazard.
Astrophys. J., Vol. 240, 726 - 736 (1980).

On a U.K. Schmidt objective prism plate, C. Hazard has found two areas which contain unusual groupings of quasars. One region at $\alpha = 11^h 46^m 14^s$ and $\delta = 11° 11'42''$(1950) contains five quasars brighter than $\nu = 19.5$ mag within an 8′ diameter circle. The closeness of the groupings and the closeness of four of the five redshifts indicate these quasars are physically associated. The dispersion in redshift, however, is too great to arise from velocity dispersion in a conventional cluster of galaxies. The other area is at $11^h 30^m 24^s$ and $10° 40'17''$(1950) and contains two separate triplets of quasars. In each triplet a very bright quasar is at the center and two fainter quasars are aligned exactly on a straight line on either side. The redshift of the central quasar is $z = 0.54$ in the first case and $z = 0.51$ in the second case. The flanking quasars are $z = 1.61$ and 2.12 in the first case and $z = 1.72$ and 2.15 in the second case. The chances of any of these configurations being accidental is extremely small.

141.086 **QSO evolution and the intergalactic medium.** R. D. Sherman.
Astrophys. J., Vol. 240, 737 - 743 (1980).

X-ray flux data acquired in the last year by the HEAO 2 Einstein Observatory strongly suggest an extragalactic hard X-ray background substantially due to evolving discrete sources. The latest IUE data show that optically selected QSOs have a steep spectral index ($\alpha \approx -2.3$) in the far-UV and recent results of the Palomar Bright Quasar Survey indicate much stronger evolution than previously supposed ($10 \lesssim \kappa \lesssim 24$). These data have been combined with the Braccesi Quasar Survey to model a uniformly distributed intergalactic medium (IGM) in the cosmos.

141.087 **The production of flat radio spectra by superposition of source subcomponents.**
D. B. Cook, S. R. Spangler.
Astrophys. J., Vol. 240, 751 - 758 (1980).

The authors have investigated the suggestion that the broad radio spectra of compact extragalactic radio sources may be produced by superposition of a number of subcomponents. If the subcomponents have radio spectra indicative of a homogeneous synchrotron source, then typically three to four subcomponents are needed, and their frequencies of maximum must be arranged in a highly nonrandom and apparently artificial fashion. If the subcomponents have inhomogeneous synchrotron source spectra, fewer subcomponents are required, and the nonrandomness of the turnover frequency distribution is weakened.

141.088 **Quasars, isotropy of H_0 and the Local Supercluster of galaxies.** H. J. Reboul.
Astron. Astrophys., Vol. 89, 272 - 281 (1980).

A method is described and applied to test the isotropy of H_0 on a sample of quasars ($0.2 \leqslant z \leqslant 3.5$). Quasars are selected by their radio index ($-\alpha \geqslant 0.7$ or $|\alpha| \leqslant 0.3$). Generalized Hubble moduli HM* are computed for each object taking into account q_0, colour, galactic extinction and K-correction. HM* is then an individual measure of H_0. Further study seems to show that HM* is minimum through the disk and in the general direction of the centre of the Local Supercluster. This could be a sign of supergalactic extinction.

141.089 **Two-photon continuum emission in quasar spectra.** C. M. Gaskell.
Observatory, Vol. 100, 148 - 151 (1980).

The peak emissivity (ergs/s/Hz) of the two-photon transition from $2s$ to $1s$ in the hydrogen atom occurs at 1602.9 Å (not 2431.4 Å as is sometimes believed). This has some consequences for our understanding of quasar spectra. A new tabulation of the two-photon emission coefficients is given.

141.090 **The Doppler effect: a consideration of quasar redshifts.** K. J. Gordon.
American J. Phys., Vol. 48, 514 - 517 (1980). – Abstr. in Phys. Abstr., Vol. 83, Abstr. 90621 (1980).

141.091 **Correlation between colors and redshift of optically violent QSOs.** D. Basu.
Astrophys. Space Sci., Vol. 72, 241 - 245 (1980).

The author reports the findings of a statistically significant relation between the continuum color indices and redshifts of optically violent variable QSOs. No such correlation is found to exist for non-variable and moderately variable QSOs.

141.092 **VLA observations of the absorption-line quasars PHL 938 and PHL 5200.** Gopal-Krishna,
R. A. Sramek.
Astron. Astrophys., Vol. 90, L1 - L3 (1980).

Sensitive VLA observations at 1.4 GHz are reported for two fields containing the absorption-line quasars PHL938 and PHL5200. At this frequency, both quasars remained undetected at a 3 rms limit of 0.5 mJy. In additional observations, upper limits of 10 mJy were established at 10.7 GHz for both quasars. The VLA maps show interesting radio sources within ~1′ of these quasars, which may be relevant to the 'intervening galaxy' hypothesis for the origin of absorption lines in quasar spectra.

141.093 **One-sided jets in extragalactic radio sources.** E. van Groningen, G. K. Miley, C. A. Norman.
Astron. Astrophys., Vol. 90, L7 - L9 (1980).

Some explanations for the existence of one-sided jets in symmetrical extended radio sources are discussed. It is shown that in the case of the quasar 4C32.69, a relativistic Doppler interpretation is improbable. Observational constraints on a model involving anisotropic radiation are also examined. This model cannot at present be ruled out, although non-relativistic interpretations of one-sided jets are considered to be most likely.

141.094 **Physical processes in the envelopes of quasars and nuclei of Seyfert galaxies.** A. S. Zentsova.
Astron. Zh., Tom 57, 936 - 941 (1980). In Russian. – English translation in Soviet Astron., Vol. 24, No. 5.

The characteristic properties of quasars and Seyfert galaxies can be explained by the interaction of the central X-ray source radiation with the atmospheres of the stars in Seyfert galaxies and nuclei of quasars.

141.095 **Flat-spectrum radio sources: victims of a conspiracy?** A. P. Marscher.
Nature, Vol. 288, 12 - 13 (1980).

141.096 **Rotationally symmetric structure in two extragalactic radio sources.** C. J. Lonsdale, I. Morison.
Nature, Vol. 288, 66 - 69 (1980).

The multi-telescope radio-linked interferometer (MTRLI) at Jodrell Bank was used during January and February 1980 at a frequency of 408 MHz to map the extragalactic radio sources 3C196 and 3C305 with a resolution of ~1 arc s. The authors show that both the markedly symmetric structures observed and the spectral index distributions inferred from comparison with previously published 5-GHz maps provide evidence for the source axes having rotated during the lifetime of the emitting regions.

141.097 **MTRLI observations of the double QSO at 408 MHz.** R. G. Noble, D. Walsh.
Nature, Vol. 288, 69 - 70 (1980).

The suggestion that the twin QSOs 0957+561 A, B are images of a single QSO formed by a gravitational lens now seems to be generally accepted. The authors present a 408-MHz map which provides spectral information on all components of the radio source, removes the doubt raised by the earlier 408 MHz observations and sheds light on the nature of the source coincident with G1.

141.098 **The structure of extended extragalactic radio sources.** G. Miley.
Annu. Rev. Astron. Astrophys., Vol. 18, (see 003.004), 165 - 218 (1980).

This article deals with the appearances of strong extended extragalactic radio sources. Included as "extended" are sources whose sizes are comparable with or larger than a typical galactic diameter, and as "strong", those with intrinsic luminosities at 408 MHz greater than $\sim 10^{23}$ W Hz^{-1}. Extended radio emission from spiral galaxies will not be discussed here.

141.099 **Optical and infrared polarization of active extragalactic objects.**
J. R. P. Angel, H. S. Stockman.
Annu. Rev. Astron. Astrophys., Vol. 18, (see 003.004), 321 - 361 (1980).

Contents: Strongly polarized objects, properties of blazars, active objects with low polarization, theoretical models and the origin of polarization, relativistic jets and optical polarization.

141.100 **Instantaneous spectrum of 3C 84 from observations at 16 frequencies with RATAN-600.**
A. B. Berlin, V. Ya. Gol'nev, D. V. Korol'kov, I. M. Lovkova, N. A. Nizhel'skij, E. E. Spangenberg, G. M. Timofeeva, V. M. Bogod, S. I. Boldyrev, I. A. Ipatova, N. A. Yudaeva, A. F. Smirnov, V. R. Amirkhanyan, A. A. Kapustkin, V. K. Konnikova, A. N. Lazutkin, M. G. Larionov, O. I. Khromov, N. I. Arzamasova, A. P. Venger, G. N. Il'in, V. A. Prozorov, N. F. Ryzhkov, V. I. Dokuchaev, M. G. Mingaliev, M. N. Naugol'naya, Yu. N. Parijskij, N. S. Soboleva, S. A. Trushkin, L. M. Sharipova, S. N. Yusupova, I. V. Gosachinskij.
Pis'ma Astron. Zh., Tom 6, 617 - 619 (1980). In Russian. English translation in Soviet Astron. Lett., Vol. 6.

The instantaneous multifrequency spectrum of the radio source 3C 84 at two epochs 1979.5 and 1980.0 obtained with the RATAN-600 radiotelescope is presented. The conclusion on the gradual increase of particles and field energy at a rate of about 15% per year is confirmed.

141.101 **Are QSO absorption lines extrinsic?** D. Wills.
Nature, Vol. 288, 114 - 115 (1980).

141.102 **Extremely relativistic electron-positron twin-jets form extragalactic radio sources.**
W. Kundt, Gopal-Krishna.
Nature, Vol. 288, 149 - 150 (1980).

The authors suggest that the beams observed in extended extragalactic double radio sources consist of extremely relativistic electrons and positrons, of typical Lorentz factor $\gamma \gtrsim 10^2$.

141.103 **Turbulent generation of magnetic fields in extended extragalactic radio sources.** D. S. De Young.
Astrophys. J., Vol. 241, 81 - 97 (1980).

The need for in situ generation of magnetic energy in extended radio sources is shown to be present in all models. The question of whether such energy can be supplied by fully developed hydrodynamic turbulence is investigated through use of the nonlinear time-dependent equations of three-dimensional MHD turbulence. It is found that seed fields of 10^{-9} gauss can be amplified to equipartition values in $\sim 10^8$ yr. Dynamical interactions with an external medium are considered to be a source of large-scale field structure.

141.104 **Turbulence-related morphology in extragalactic radio sources.** G. Benford, A. Ferrari, E. Trussoni.
Astrophys. J., Vol. 241, 98 - 110 (1980).

As particle beams propagate through the intergalactic medium, unavoidable instabilities from shear flows produce turbulent magnetic waves. This wave energy may enhance luminosity and alter morphology. For reasonable parameters the dominant nonlinear process is an energy cascade from long wavelengths to short wavelengths where particles are reaccelerated in quasi-linear fashion. The authors construct a phenomenological turbulence theory to describe this. In an ambient magnetic field, wave-particle scatterings which cause reacceleration can also lead to spatial cross-field diffusion, broadening the beam. Thus beams can flare rapidly as they propagate. This relates luminosity to morphology in a new way. A variety of radio source types may be related to this effect.

141.105 **On collimation of relativistic jets from quasars.** M. A. Abramowicz, T. Piran.
Astrophys. J., Lett., Vol. 241, L7 - L11 (1980).

Thick accretion disks orbiting supermassive black holes can explain high luminosities of quasars and other active galactic nuclei as well as fundamental properties of jets and beams emerging from them.

141.106 **Absorption lines and ion abundances in the QSO PKS 0528 – 250.**
D. C. Morton. J.-s. Chen, A. E. Wright, B. A. Peterson, D. L. Jauncey.
Mon. Not. R. Astron. Soc., Vol. 193, 399 - 413 (1980).

Spectra of the QSO PKS 0528 – 250 ($z_e = 2.765$) have been obtained with the AAT at 2 Å resolution from 3100 to 7180 Å. Absorption line systems have been identified at $z_{A1} = 2.81322$, $z_{A2} = 2.81100$, $z_B = 2.53758$ and $z_C = 2.14077$. The ionization ranges from H I, Al II and Fe II to N V or O VI in system A, from H I and possibly Si II to C IV in B and from H I, Al II and Si II to C IV in C.

141.107 **Observations of M87 at 15.4 GHz with the 5-km telescope.** R. A. Laing.
Mon. Not. R. Astron. Soc., Vol. 193, 427 - 437 (1980).

The radio galaxy M87 has been mapped with a resolution of 0.67 × 2.1 arcsec² at a frequency of 15.4 GHz. The radio jet consists of discrete knots separated by regions of low surface brightness within ~ 20 arcsec of the nucleus, but merges into a diffuse ridge of emission at larger distances. The jet appears to oscillate from side to side with an amplitude which increases with increasing distance from the nucleus; observations of the linearly-polarized emission show that the magnetic field within the jet is well-ordered and aligned with its axis.

141.108 **A model for the magnetic-field structure in extended radio sources.** R. A. Laing.
Mon. Not. R. Astron. Soc., Vol. 193, 439 - 449 (1980).

This paper describes a model for the magnetic-field configuration in extended radio sources in which an initially random field has been compressed or sheared to lie entirely in one plane. The polarization properties of radiation from electrons in such a field are evaluated and the expected degree and direction for some simple geometries are derived. The predictions of the model are in good agreement with observations of extragalactic radio sources and of the filaments in the Crab Nebula.

141.109 **Radio observations of W3 at 2.7 and 15.4 GHz.** D. Colley.
Mon. Not. R. Astron. Soc., Vol. 193, 495 - 509 (1980).

The continuum emission from several compact H II regions in W3 has been mapped at 2.7 and 15.4 GHz with a maximum resolution of 0.65 arcsec. Three of the radio components provide evidence to support a model in which star formation has occurred within a molecular cloud and has produced dense neutral shells around H II regions. These neutral shells have been ruptured or dispersed at certain points, allowing ionizing radiation into a region having a lower neutral density.

141.110 **Is C IV λ 1549 a standard cosmic candle?** B. J. T. Jones, J. E. Jones.
Mon. Not. R. Astron. Soc., Vol. 193, 537 - 548 (1980).

It is difficult to observe emission lines of small equivalent width in intrinsically faint QSOs. This selection effect biasses the slope of the log (intrinsic continuum luminosity): log (C IV equivalent width) diagram for samples of QSOs such as those of Baldwin and those taken from slitless spectrum surveys. There is no such bias for QSO samples drawn from radio surveys, and such data suggest that the total flux of the C IV line may itself be a good standard candle. The best way to determine q_0 from QSO samples may be to look at the upper envelope of the $m_{CIV} - z$ relationship.

141.111 **The level of cosmic radio noises and integral absorption in the subauroral ionosphere from simultaneous satellite and ground-based measurements at a frequency of 5.5 MHz.** V. A. Boldyrev, S. Z. Kershengol'ts, Yu. M. Knyaz'kin, A. P. Mamrukov, A. A. Polyntsev, V. M. Filippov, L. D. Filippov.
Geomagn. Aehron., Tom 20, 951 - 953 (1980). In Russian.

141.112 **Investigations of rapid radio emission variations of some extragalactic objects.** V. A. Efanov, I. G. Moiseev, N. S. Nesterov, A. Wright.
Izv. Krymskoj Astrofiz. Obs., Tom 61, 70 - 76 (1980). In Russian.

The results of observations at wavelengths 0.82, 1.35, 2 and 6 cm of some extragalactic radio sources are discussed. It is shown that emission variations in the course of several days are intrinsic to Lacertids and quasars mainly taking place throughout some time intervals only. The existence of radio emission fluctuations during several hours is confirmed not only in BL Lac objects but in the quasar 3C 273 too.

141.113 **On the role of relativistic effects in the ejection and expansion of components of extragalactic double radio sources.** B.-l. Liang.
Acta Astron. Sinica, Vol. 21, 219 - 227 (1980). In Chinese.

Using a sample of 81 typical double radio sources, the author compares the diameter ratio and the flux ratio of the two components with their separations from the optical object respectively. It was found that it seems hardly possible to analyze the observational characteristics of the extended double radio sources based on the relativistic effects alone. There seems to be a correlation between the "cones of ejection" and linear separations for the components, and the expansion of the components practically ceased in the stage of extended double sources.

141.114 **Statistical analysis of the optical variability of QSOs.** X.-h. Xiao, X.-y. Tang, F.-z. Cheng.
Acta Astron. Sinica, Vol. 21, 228 - 236 (1980). In Chinese.

It is shown that the mean rate of optical variation of QSOs is decreasing with increasing redshift, whereas the mean rate of optical variation of active galaxies is higher than those of QSOs. From the viewpoint of cosmological evolution it seems that the optical variation of QSOs becomes more violent with cosmological time, and finally it connects itself with that of active galaxies.

141.115 **Circular polarization as a probe of radio sources (high accuracy polarization measurements at λ 49 cm).** K. W. Weiler, I. de Pater.
Astron. Astrophys., Vol. 91, 41 - 48 (1980).

High accuracy measurements of integrated total intensity and linear and circular polarization have been obtained for 28 relatively compact (size < 1') radio sources. Combining the present results with earlier results, a subset of sources complete over an area of the sky to specified limits can be formed. Its analysis indicates that ~ 1/4 of all radio sources have a detectable circularly polarized component in their synchrotron continuum radiation at λ 49 cm. For one source 0316 + 162 (CTA 21), both the total intensity radio spectrum and circular polarization spectrum are sufficiently well known to establish for the first time, the change in sign expected for circular polarization as the source changes from optically thick to optically thin.

141.116 **Alfvén-driven cyclotron corona as a model for quasar infrared.** W. H. Zurek.
Astron. Astrophys., Vol. 91, 90 - 96 (1980).

The author analyzes the idea that Alfvén waves transport the needed energy from a powerhouse of unknown character in the core of the quasar nucleus to the outer radiating shell or "corona". He finds that this "Alfvén-drive mechanism" can only then deliver adequate power to the corona when the magnetic field in the vicinity of the core, where the Alfvén wave is generated by movements of magnetized matter, has a strength of the order of megagauss. Strongly self-absorbed cyclotron radiation emitted by the plasma under these conditions is able to account for the observed predominance of nonthermal infrared in the spectrum of the typical quasar.

141.117 **Energy transport in radio sources.**
P. A. G. Scheuer.
Highlights of Astronomy, Vol. 5, (see 012.028), 667 - 669 (1980).

141.118 **Optical observations of radio jets.**
W. van Breugel, G. Miley, H. R. Butcher.
Highlights of Astronomy, Vol. 5, (see 012.028), 671 - 675 (1980).

141.119 **X-ray studies of quasars and active galaxies with the Einstein Observatory.** W. H.-M. Ku.
Highlights of Astronomy, Vol. 5, (see 012.028), 677 - 687 (1980).

Preliminary results from the first six months of the Columbia Astrophysics Laboratory survey of active galaxies are presented. Simple statistical tests are applied to determine whether X-ray properties can be used to understand the differences and similarities between the various classes of active galaxies. Particular emphasis is placed on the quasars in the sample.

141.120 **Constraint on quasar number counts from their contribution to the X-ray background.**
J. P. Henry, A. Soltan, H. Tananbaum, G. Zamorani.
Highlights of Astronomy, Vol. 5, (see 012.028), 771 - 774 (1980).

Observations with the Einstein X-ray telescope show that quasars, as a class, are luminous X-ray emitters. By coupling X-ray observations with quasar optical number counts, the authors show that the quasars contribute significantly to the diffuse X-ray background. In fact, the X-ray data strongly suggest that somewhat above 20^m the slope of the optical counts must flatten.

141.121 **Results of synchronous observations of the peculiar object SS 433 in radio and optical ranges.**
S. I. Neizvestnyj, S. A. Pustil'nik, V. G. Efremov.
Pis'ma Astron. Zh., Tom 6, 700 - 705 (1980). In Russian.
English translation in Soviet Astron. Lett., Vol. 6.

For SS 433 the data on *V*, *R*-photometry and radio flux density measurements at wavelengths 2.08, 3.9, 8.2, 13 and 31 cm are presented. During the period of observations (late May - June 1979) the object has shown strong activity in both ranges. Radio flares are correlated with optical ones.

141.122 **The discovery of a gravitational lens.**
F. H. Chaffee, Jr.
Sci. American, Vol. 243, No. 5, p. 60 - 68 (1980).

A recently discovered pair of quasars turns out to be not a pair at all but two images of a single quasar formed by a gravitational lens: an elliptical galaxy halfway between the quasar and our own galaxy.

141.123 **X-ray properties of quasars.**
W. H.-M. Ku, D. J. Helfand, L. B. Lucy.
Nature, Vol. 288, 323 - 328 (1980).

The X-ray properties of 111 catalogued quasars have been examined with the imaging proportional counter on board the Einstein Observatory. Thirty-five of the objects, of redshift between 0.064 and 3.53, were detected as X-ray sources. The 0.5–4.5 keV X-ray properties of these quasars are correlated with their optical and radio continuum properties and with their redshifts and variability characteristics. The X-ray luminosity of quasars tends to be highest for those objects which are bright in both the optical and radio regimes and which exhibit optically violent variability. These observations suggest that quasars should be divided into two classes on the basis of radio luminosities, spectra, evolution and underlying morphology.

141.124 **Prominent VLBI cores in powerful radio sources with arc second structure.**
Gopal-Krishna, E. Preuss, R. T. Schilizzi.
Nature, Vol. 288, 344 - 347 (1980).

The authors have selected all sources lying north of declination $-25°$ for which a flux density $\geqslant 1$ Jy and an overall size between 1 and 4 arc s have been estimated at 327 MHz in the occultation observations. They report a search for compact cores among these sources by VLBI at 5 GHz using the large antennas at Effelsberg and Westerbork. Radio cores were found to be much more prominent in this representative sample of 30 few arc-second sources, as compared with the cores found typically in extended double sources.

141.125 **The double quasar 0957 + 561 A, B.**
J. M. Barnothy.
Bull. American Astron. Soc., Vol. 12, 739 (1980). – Abstract.

141.126 **Photometry of 3C 446 during 1978 and 1979.**
H. R. Miller.
Bull. American Astron. Soc., Vol. 12, 752 (1980). – Abstract.

141.127 **A radio supernova in M 100.** K. W. Weiler,
J. M. van der Hulst, R. A. Sramek, N. Panagia.
Bull. American Astron. Soc., Vol. 12, 752 (1980). – Abstract.

141.128 **High dynamic range observations in the fields of strong extragalactic radio sources.**
U. Stute, W. Reich, P. M. W. Kalberla.
Astron. Astrophys., Suppl. Ser., Vol. 42, 299 - 310 (1980).

23 strong extragalactic radio sources have been surveyed for weak emission in their vicinity at 11 cm wavelength with the Effelsberg 100 m telescope. A catalogue of 257 sources is given which is thought to be complete down to 15 mJy. The results of a statistical analysis are: 1. Source counts near the 23 strong radio galaxies and quasars are not significantly different from source counts in empty fields. Hence the majority of the 257 sources probably are unrelated background sources. 2. Testing whether there is an excess of components aligned with the axes of the strong central sources, the authors found a significant detection rate only for core-jet (D2-type) sources.

141.129 **The double quasar Q0957 + 561 A, B: a gravitational lens image formed by a galaxy at $z = 0.39$.**
P. Young, J. E. Gunn, J. Kristian, J. B. Oke, J. A. Westphal.
Astrophys. J., Vol. 241, 507 - 520 (1980).

The authors believe that they have observed the gravitational lens that is responsible for producing the double quasar. Extremely deep CCD pictures of the region show that the QSOs are behind a rich cluster of galaxies. The CCD data and spectrophotometry of the QSOs indicate that the southern QSO image is seen through the brightest cluster galaxy, whose redshift is 0.39. Calculations of gravitational imaging by King model mass distributions show that the cluster and the brightest galaxy together, acting as a gravitational lens on the light from a single, more distant QSO, can easily reproduce all of the present observations. The authors conclude that the double quasar is almost certainly the multiple image of a single object produced by a gravitational lens.

141.130 **Structure of the compact nuclear radio source in M82.** R. L. Brown, S. G. Neff.
Astrophys. J., Vol. 241, 561 - 566 (1980).

The authors review the observations of the compact radio source 41.9+58 and demonstrate that the simplest interpretation of the radio source as a single Gaussian component is inconsistent with the X-ray and γ-ray observations of M82 and hence that a different radio brightness distribution is required.

As alternatives they consider both double and disk-like structures and conclude that the latter provide for a more plausible model of the observational properties of 41.9+58. They assess the ramifications of this conclusion.

141.131 **Temporal variation in the compact radio structure of NRAO 150 at 1671 MHz.**
R. L. Mutel, R. B. Phillips.
Astrophys. J., Lett., Vol. 241, L73 - L76 (1980).

The milli-arcsec structure of the unidentified radio source NRAO 150 has been examined at epochs 1978.5 and 1979.5 using a multibaseline VLBI array at 1671 MHz. The structure consists of two bright, barely resolved components ($\lesssim 5$ milli-arcsec) and a weak extended component displaced by ~ 25 milli-arcsec. Comparison with published visibility data on a single baseline from epoch 1974.4 shows significant differences. The observed changes are consistent with a 1 Jy increase in the flux of a stationary bright component.

141.132 **'B2 1141+37: a giant radio galaxy with remarkable radio and optical properties.**
M. H. Ulrich, H. Butcher, D. L. Meier.
Nature, Vol. 288, 459 - 461 (1980).

The radio galaxy B2 1141+37 is remarkable for both its radio and optical properties. The radio source has a double structure for which the ratio of the separation between the components to their dimension is exceptionally large. The authors report on the redshift measurement and on some VLA observations at 4.9 GHz.

141.133 **Extragalactic radio sources: rapid variability at 90 GHz.**
E. E. Epstein, R. Landau, J. D. G. Rather.
Astron. J., Vol. 85, 1427 - 1433 (1980).

Thirty-three extragalactic variable radio sources have been observed at 90 GHz (3.3 mm) over two several-day-long periods in a search for daily and hourly variations. OV -236 (1921 - 29) showed a decrease in flux density by a factor of 2.7 over a three-day interval, recovering its previous flux density two days later. (OV -236 began a dramatic outburst a few months later.) The other sources exhibited no significant variations $\gtrsim 20\%$. A summary of previous radio observations of rapid variability is contained in the appendix.

141.134 **Observations of variable radio sources at 18-cm wavelength. III. 1973 - 1979.**
J. C. Webber, K. S. Yang, G. W. Swenson, Jr.
Astron. J., Vol. 85, 1434 - 1441 (1980).

A sample of 58 northern-sky radio sources has been monitored for total flux density variability between 1973 and 1979. In some cases of unquestionable variability, there is correlation with activity at shorter wavelengths. In a few cases, variation of a component physically distinct from that observed at shorter wavelengths is indicated. The observations rule out a simple expanding synchrotron model.

141.135 **Long-term optical behaviour of 114 extragalactic sources.** A. J. Pica, J. T. Pollock, A. G. Smith, R. J. Leacock, P. L. Edwards, R. L. Scott.
Astron. J., Vol. 85, 1442 - 1461 (1980).

Photometric data for over 200 extragalactic sources have been obtained during an 11-yr monitoring program. Twenty that are optically violent variables were reported on by Pollock et al. (1979). The present paper provides data for 114 less active sources, 58 of which exhibit optical variations at a confidence level of 95% or greater. Light curves are given for the 26 most active sources. In addition, the overall monitoring program at the Rosemary Hill Observatory is reviewed, providing information on the status of 206 objects in all.

141.136 **Q0957+561.**
IAU Circ., Nos. 3533, 3552 (1980).

141.137 **Quasars.** T. Dethier.
Zenit, 7e Jaarg., 296 - 302 (1980).

141.138 **Nebulosities around QSOs.**
R. K. Thakur, R. K. Sood.
Astrophys. Space Sci.,Vol. 73, 241 - 249 (1980).

Independent evidence in favour of the hypothesis proposed by Thakur and Sapre (1979) that a QSO consists of a bright central object embedded in an extended nebulosity has been presented. α_{U-B} and α_{B-V}, the spectral indices in $(U-B)$ and $(B-V)$ colours, have been calculated for a sample of 80 QSOs with redshift $z \leqslant 0.76$. $\alpha_{B-V} - \alpha_{U-B}$ has been plotted against $\langle\alpha\rangle = (\alpha_{B-V} + \alpha_{U-B})/2$. In this figure the QSOs in which detectability of nebulosities has been predicted by Thakur and Sapre (1979) occupy a separate but adjacent part of the diagram as compared to those for which such a prediction has not been made.

141.139 **Radio galaxies and quasars.** K. I. Kellermann.
Ann. New York Acad. Sci., Vol. 336, 1 - 11 (1980) = Natl. Radio Astron. Obs., Green Bank, Repr., Ser. B, No. 509.

Observations show that the radio emission from galaxies and quasars originates from a wide range of dimensions, extending from a small fraction of a parsec to more than 3 million parsecs. The smallest sources found in the nuclei of nearby galaxies such as M81, M82, and M104 are only about 0.01 pc in extent, and the weak radio source at the nucleus of the Galaxy is a factor of 10^3 smaller still and comparable in size and luminosity with a variety of stellar radio sources. At the other extreme, the giant radio galaxies such as 3C 236, DA 240, and NGC 315 are the largest objects known in the universe, and observations of the polarized radio emission indicate large scale ordered magnetic fields extending over millions of parsecs.

141.140 **Ca II absorption lines in the spectrum of the quasar PKS 2020-370 due to galactic material in the group Klemola 31.**
A. Boksenberg, I. J. Danziger, R. A. E. Fosbury, W. M. Goss.
ESO Sci. Prepr. No. 115, 20 pp. (1980). – Submitted to Astrophys. J., Lett.

141.141 **3C 273: a review of recent results.**
M.-H. Ulrich.
ESO Sci. Prepr. No. 122, 34 pp. (1980). – Submitted to Space Sci. Rev.

141.142 **The spectrum of the QSO 0805+046 (4C 05.34) at intermediate dispersion.** J.-s. Chen, D. C. Morton, B. A. Peterson, A. E. Wright, D. L. Jauncey.
Anglo-Australian Obs., Prepr. No. 142, 31 pp. (1980).

Spectra of the radio QSO PKS 0805+046 ($z_{em} = 2.8772$) have been obtained with the AAT at 2 Å resolution from 3300 Å to 6100 Å. Two absorption line systems at $z_{abs} =$ 2.87717 and 2.47568 are certain and three more at 1.01422, 0.95915 and 0.70280 are possible. These systems leave unidentified about 80% of the absorption lines shortward of $L\alpha$ emission and 26% longward. A strong peak in the cross-correlation corresponding to $L\alpha$-$L\beta$ pairs and the explicit identification of many of these pairs supports the hypothesis that these lines originate in neutral hydrogen clouds.

141.143 **Multiple quasars and gravitational lenses.**
W. E. Shawcross.
Sky Telesc., Vol. 60, 486 - 487 (1980).

141.144 **Recent progress and problems in the theory of extragalactic radio sources.** D. S. De Young.
News Lett. Astron. Soc. N. Y., Vol. 1, No. 7, p. 14 (1980). Abstract.

141.145 **Models of the cosmological evolution of extragalactic radio sources – I. The 408-MHz source count.**
J. V. Wall, T. J. Pearson, M. S. Longair.
Mon. Not. R. Astron. Soc., Vol. 193, 683 - 706 (1980).

A simple numerical procedure for analysing the counts of extragalactic radio sources is described. The authors show that the technique provides strong support for the qualitative conclusions of previous analyses: the comoving density of the more powerful radio sources with extended structure is now less than at earlier epochs by a factor of $\gtrsim 10^3$. New models are derived which adequately describe the current data. These models make contrasting predictions about redshift distributions amongst the faint radio sources.

141.146 **Quasar Lα absorbers: are precise conclusions possible?** A. L. Melott.
Astrophys. J., Vol. 241, 889 - 893 (1980).

The conclusions of a massive study of QSO absorption lines are critically examined. It is found that tight constraints may not be placed on the nature of pressure-confined clouds nor on the intergalactic medium presumably confining them. An alternative model, of gravitationally bound clouds, is found to be equally viable.

141.147 **Optical continuum and emission-line luminosity of active galactic nuclei and quasars.**
H. K. C. Yee.
Astrophys. J., Vol. 241, 894 - 902 (1980).

Published multichannel spectrophotometry data of 105 quasars, radio galaxies, and Seyfert galaxies are used to study properties of optical continua and emission lines of these objects, and continuity between active galactic nuclei and quasars.

141.148 **A range of time delays for the double quasar 0957 + 561 A, B.** C. C. Dyer, R. C. Roeder.
Astrophys. J., Lett., Vol. 241, L133 - L136 (1980).

The authors have extended the complex formalism of Bourassa and Kantowski for a transparent gravitational lens to a multilens situation. By using a grid of assumed positions for the cluster center within its measured error box, they have then calculated a range of possible time delays between the two images, A and B. The results indicate a range from 0.03 yr to 1.7 yr, which is somewhat smaller than might have been anticipated.

141.149 **Self-absorption in the Balmer line profiles of the QSO 2141 + 174 (=OX 169).** H. E. Smith.
Astrophys. J., Lett., Vol. 241, L137 - L140 (1980).

The QSO 2141 + 174 shows apparent self-absorption in the profiles of the emission lines Hα, Hβ, and possibly Hγ. It is suggested that this self-absorption is most readily interpreted as due to a "broad-line emission cloud" viewed along the line of sight to the continuum source. Under this assumption, the minimum optical depth in Hα, $\tau_0 \approx 40$, confirms previous suggestions that QSO broad-line clouds must be optically thick in the Balmer lines and very optically thick in the Lyman lines. It is possible that variations in the column density of absorbing material along the line of sight may account for the X-ray variability of 2141 + 174.

141.150 **Collimation of the radio jets in 3C 31.**
A. H. Bridle, R. N. Henriksen, K. L. Chan, E. B. Fomalont, A. G. Willis, R. A. Perley.
Astrophys. J., Lett., Vol. 241, L145 - L149 (1980).

The authors show how the rates of expansion of the jets in 3C 31 vary with distance from the galactic nucleus; they also use the beam dynamics given by Chan and Henriksen to examine how the observed variations in expansion rate could result from pinching of the beams by the helical magnetic field and/or from variations in an external confining pressure.

141.151 **Variability in QSOs and active galactic nuclei.**
C. Hazard.
Variability in stars and galaxies, (see 012.044), p. GL.2.1 - 2.11 (1980).

141.152 **Variabilité du rayonnement continu dans les domaines optique et ultraviolet.**
G. Wlérick.
Variability in stars and galaxies, (see 012.044), p. C.2.1. - 2.19 (1980).

141.153 **Flux variations in radio sources.**
R. Fanti, M. Salvati.
Variability in stars and galaxies, (see 012.044), p. C.3.1 - 3.18 (1980).

Flux variations at cm wavelengths are a common feature of flat spectrum, compact, extragalactic sources. The variations propagate too fast and with an amplitude too large toward lower frequencies; the authors show how this behaviour is indicative of a continuous energy supply and a consequent accelerated expansion. As for flux variations at meter wavelengths, the main issue so far has been about their reality. The authors present data from a Bologna monitoring program, which, together with similar work at other observatories, gives a final positive answer. It is not clear whether there is any correlation between high and low frequency variability. A crucial question is the source size; should it turn out larger than that implied by the causality argument, then models with built-in superluminal properties would be favored. A final point is whether the variations in the meter range occur simultaneously and have comparable magnitudes.

141.154 **Structure variations in radio sources.**
I. I. K. Pauliny-Toth.
Variability in stars and galaxies, (see 012.044), p. C.4.1. (1980). – Abstract.

141.155 **A new investigation of the redshift – angular-diameter relation for quasars.** C. R. Masson.
Astrophys. J., Vol. 242, 8 - 13 (1980).

A new test of the redshift–angular-diameter relation is described and applied to a sample of 3CR and 4C quasars. The results show no evidence for size evolution of quasars. The apparent deficit of large quasars at high redshifts is most likely due to an inverse correlation between size and radio power.

141.156 **New identifications of weak emission lines in the spectra of quasars.**
B. J. Wills, H. Netzer, D. Wills.
Astrophys. J., Lett., Vol. 242, L1 - L4 (1980).

Following the discovery of strong emission features between 2300 and 3000 Å in the spectra of quasars and their identification with blended Fe II resonance lines from the 5 eV level of Fe^+ the authors predict the presence of weak emission from more highly excited levels. They present new observations near 2000 Å that support this prediction and suggest that weak emission features previously noticed near 1610 Å, 1670 Å, and 1860 Å are also due to Fe II. This Fe II emission could also account for some weak features seen between 3000 and 4000 Å. Weak Fe II emission will probably set an ultimate limit to studies of line strengths and profiles in quasar spectra.

141.157 **Collimation of electromagnetic beams in extragalactic radio sources.**
A. Ferrari, S. Massaglia, M. Dobrowolny.
Astron. Astrophys., Vol. 92, 246 - 252 (1980).

The energy transfer in extragalactic radio sources is studied in terms of strong electromagnetic beams generated inside parent galactic nuclei by a many-pulsar core or a compact rotating, magnetized object. Self-trapping by nonlinear refractive index effects is shown to be very efficient in collimating jets very close to the central core. Formation of radio lobes can be attributed to parametric instabilities of the electromagnetic wave causing the sudden termination of beams in regions of increasing density.

141.158 **Evidence for the location of quasars in superclusters.**
J. H. Oort, H. Arp, H. de Ruiter.
ESO Sci. Prepr. No. 129, 19 pp. (1980). – Submitted to Astron. Astrophys.

141.159 **Radio observations of a new class of optically selected quasi-stellar objects.**
J. J. Condon, M. A. Condon, K. J. Mitchell, P. D. Usher.
Astrophys. J., Vol. 242, 486 - 491 (1980).

Most QSOs found by low-resolution spectroscopy are known to be radio-quiet, but QSOs selected by polarization, power-law continuum spectra, or optical variability might have different radio characteristics. The authors have observed a sample of 96 QSOs and other blue objects, chosen on the basis of optical variability, with the VLA at 4885 MHz. Only three were detected (LB 8755, LB 8956, and LB 9013), and two of these (LB 8755 = OJ 130 and LB 9013 = 4C 17.46) had previously been found in radio-selected samples. For the rest, 5 σ upper limits of 2.0 mJy can be placed on their 4885 MHz flux densities.

141.160 **A "cluster" of quasi-stellar objects near M82.**
E. M. Burbidge, V. T. Junkkarinen, A. T. Koski, H. E. Smith, A. A. Hoag.
Astrophys. J., Lett., Vol. 242, L55 - L57, plates L2 - L3 (1980).

Three QSOs discovered by the grism (grating prism) technique lie in the field of M82. Their similar redshifts ($z \approx 2$) and small angular separations ($\lesssim 3\overset{'}{.}6$) suggest a physical association. A cosmological interpretation of the redshifts yields a "cluster of QSOs" of reasonable physical dimensions ($d \approx 3$ Mpc, $\sigma_v \approx 600$ km s^{-1}). However, the nearness ($\sim 8'$) of the disturbed galaxy M82 may suggest a noncosmological origin for the redshifts.

141.161 **Discovery of nebulosity associated with the quasar 3C 273.** S. Wyckoff, P. A. Wehinger, T. Gehren, D. C. Morton, A. Boksenberg, R. Albrecht.
Astrophys. J., Lett., Vol. 242, L59 - L63 (1980).

In this Letter the authors report the discovery of diffuse nebulosity physically associated with the quasar 3C 273. Comparison of photometric and spectroscopic data indicate that the dominant emission in the nebulosity to the north of the quasar is continuum radiation. A plausible source of the continuous radiation is integrated starlight from an underlying galaxy associated with 3C 273.

141.162 **Interstellar formaldehyde in the Sagittarius B2 radio source.** I. V. Gosachinskij, V. G. Grachev, T. M. Egorova, S. R. Zhelenkov, G. N. Il'in, N. P. Komar, E. N. Kurochkina, V. A. Prozorov, N. F. Ryzhkov.
Pis'ma Astron. Zh., Tom 6, 763 - 767 (1980). In Russian. English translation in Soviet Astron. Lett., Vol. 6.

Results of observations of the 6.2-cm absorption line of the H_2CO molecule are presented. It is shown from the H_2CO optical depth distribution over the Sagittarius B2 source that there is a gas shell of 11 pc in diameter, its mean radial velocity being close to that of recombination lines. The shell contracts with a velocity of 20 - 25 km/s and rotates with a velocity of about 15 km/s at its edge.

141.163 **The statistical character of optical continuous spectra of QSOs and *K*-correction.**
X.-h. Xiao, S.-l. Cao, F.-z. Cheng, X.-y. Tang, L.-t. Yang.
Acta Astron. Sinica, Vol. 21, 368 - 378 (1980). In Chinese.

The authors researched the statistical character of optical continuous spectra of QSOs, by the use of data of 355 QSOs with colour indices and redshifts. Then they obtained a *K*-correction which is more effective than before.

141.164 **Optical variability and redshift of QSOs.**
D. Basu.
Astrophys. Lett., Vol. 21, 63 - 64 (1980).

Analysis of a homogeneous sample of optically variable QSOs shows a statistically significant dependency of the degree of variability on redshift.

141.165 **Optical polarimetry of quasi-stellar and BL Lac objects.**
D. Wills, B. J. Wills, M. Breger, J.-C. Hsu.
Astron. J., Vol. 85, 1555 - 1558 (1980).

The authors report observations of 39 extragalactic objects, using a new polarimeter at McDonald Observatory. Most of the observations were made in January and June 1980. Some new BL Lac objects were found, and some of the previously known members of this class were also observed. Polarization exceeding 3 - 4 times the formal rms uncertainties was found in about ten QSOs. Rapid changes of polarization and/or position angle occurred in some of the BL Lacertae objects, usually with < 0.1-mag change in the total light.

141.166 **Evidence for rotation in compact extragalactic radio sources.** D. R. Altschuler.
Astron. J., Vol. 85, 1559 - 1564 (1980).

Linear polarization measurements for 0048 –097, 0235 +164, 0133 +476, 1749 +096, and 1510 –089 are discussed. For all these sources a linear variation of the polarization position angle with time is observed during a flux density outburst. The data suggest that this rotation might be a general feature of variable sources, only seen when it is not masked by the juxtaposition of polarized components.

141.167 **High-frequency structure of Ooty occultation sources. II.** T. K. Menon.
Astron. J., Vol. 85, 1577 - 1581 (1980).

A sample of 57 sources from the Ooty Lunar Occultation survey at 327 MHz has been studied using the 3-element interferometer system at Green Bank at frequencies 2.7 and 8.1 GHz. A comparison of the angular separation at low and high frequencies for well resolved double sources does not show any frequency-dependent effects. The mean spectral index of the 42 doubles in the sample is 0.92, considerably steeper than the usual mixed samples. The source 1417-19 is found to be a triple source, with a strong central component coinciding with the N galaxy suggested as the identification.

141.168 **Some results of observations of extragalactic radio sources in the microwave range.**
V. A. Efanov, I. G. Moiseev, N. S. Nesterov.
Izv. Krymskoj Astrofiz. Obs., Tom 62, 108 - 116 (1980). In Russian.

Flux densities of 104 extragalactic radio sources measured at wavelength 1.35 cm in the course of 4 observational periods in 1977 - 1978 are presented.

141.169 **Comparisons of the orientations of double-lobed radio sources and their associated elliptical galaxies.**
M. S. Wilkerson, W. Romanishin.
Publ. Astron. Soc. Pacific, Vol. 92, 551 (1980). – Abstract.

141.170 **Comments on the relativistic ejection of QSOs from galaxies.** R. D. Schwartz.
Publ. Astron. Soc. Pacific, Vol. 92, 570 - 572 (1980).

The hypothesis that some QSOs are ejected from galaxies is explored with respect to the well-known relativistic Doppler effect. It is shown that the compact QSO-like object which Arp finds silhouetted on NGC 1199 could, in principle, have a redshift dominated by a transverse velocity component.

Implications of this hypothesis for other possible galaxy-QSO associations are discussed.

141.171 **Faint QSO candidates in the field of the globular cluster NGC 6752.**
H. B. Richer, B. I. Olson.
Publ. Astron. Soc. Pacific, Vol. 92, 573 - 575 (1980).

Based on their location in the $(U-B)$, $(B-V)$ color-color diagram, 25 faint QSO candidates in the direction of the globular cluster NGC 6752 were found. One of these was confirmed spectroscopically as a QSO and has a redshift of $Z = 1.918 \pm 0.007$ from three lines. These objects may prove useful for ground-based or Space Telescope observers desiring a zero proper-motion reference system in order to carry out an astrometric study of this cluster.

141.172 **A survey of the sky between 0° to +2° declination.** V. R. Amirkhanyan, A. G. Gorshkov, A. A. Kapustkin, V. K. Konnikova, A. N. Lazutkin, M. G. Larionov, A. S. Nikanorov, V. N. Sidorenkov, L. S. Ugol'kova, O. I. Khromov.
Astron. Tsirk., No. 1099, p. 2 - 6 (1980). In Russian.

141.173 **Distance of the double quasar 0957+561.**
Y. P. Varshni.
J. R. Astron. Soc. Canada, Vol. 74, 357 (1980). – Abstract.

141.174 **On the rotation measures (RM) of the radio components of Cyg A.** V. N. Kuril'chik.
Astron. Tsirk., No. 1106, p. 1 - 2 (1980). In Russian.

141.175 **Spatial correlation of variables in the decimeter range of extragalactic sources with interstellar structures: loops, spurs, ridges.**
S. V. Pogrebenko, N. Ya. Shapirovskaya.
Inst. kosm. issled. AN SSSR. Prepr., 1980, No. 568, 16 pp. In Russian. – Abstr. in Ref. zh., 51. Astron., 12.51.654 (1980).

141.176 **Spatial distribution of quasars and cosmology.**
P. V. Vorob'ev.
Inst. yader. fiz. SO AN SSSR. Prepr., 1980, No. 62, 10 pp. In Russian. – Abstr. in Ref. zh., 51. Astron., 12.51.673 (1980).

141.177 **Einstein observations of quasars.**
H. Tananbaum.
X-ray astronomy, (see 012.055), p. 311 - 325 (1980).

The author has found that the quasars are luminous X-ray emitters, not only as a few individual objects, but as an entire class. These objects are a major contributor to the 2 keV X-ray background. The X-ray observations of quasars can be combined with the X-ray background to set limits on the number of faint quasars.

141.178 **A search for neutral hydrogen absorption in the spectra of quasi-stellar objects.**
B. M. Peterson, C. B. Foltz.
Astrophys. J., Vol. 242, 879 - 883 (1980).

The radio spectra of eight QSOs which have redshifted Mg II $\lambda\lambda$2795, 2802 absorption features in their spectra have been examined in a search for 21 cm absorption at the same redshift as the optical lines. To a 3 σ limiting optical depth typically ~0.1, no H I absorption features of width ~10–100 km s^{-1} have been identified, which implies that the maximum neutral-hydrogen column densities through the absorbing regions are generally less than $\sim 5 \times 10^{18} T_s$ cm^{-2}, where T_s is the spin temperature. Conditions under which H I absorption should be detectable are discussed. It is shown that these observations are not inconsistent with simple H I region models of the absorbing clouds.

141.179 **Observations of variable radio sources in the 300 to 1000 MHz range.** J. R. Fisher, W. C. Erickson.
Astrophys. J., Vol. 242, 884 - 893 (1980).

Of 52 radio sources observed at six frequencies between 321 and 920 MHz, the flux densities of seven were seen to be definitely variable, and five other sources were probably variable. The variable portion of the flux densities were nearly constant over the frequency range, and source strengths appeared to increase simultaneously with sample intervals of 2–6 months. Using source sizes observed with VLBI techniques, the brightness temperatures of the variable components of the radiation are less than $\sim 10^{12}$ K, but the time scales of the variations imply much smaller source sizes and, hence, much higher brightness temperatures.

141.180 **The effect of undetected gravitational lenses on statistical measures of quasar evolution.**
E. L. Turner.
Astrophys. J., Lett., Vol. 242, L135 - L139 (1980).

Brightness amplifications by undetected gravitational lenses could be responsible in part for the apparent evolution of quasars, particularly for those which appear to be of high luminosity. It is shown that values of $\overline{V/V_M} \geqslant 0.6$ and number-magnitude slopes $\geqslant 0.9$ need not necessarily imply source density evolution if lensing events are common. Quasar samples which are defined by flux limits and minimum luminosities will preferentially include gravitational lens systems. Even if lensing events are quite rare, a large fraction of the lensed quasars will appear more luminous than the most luminous unlensed quasar.

141.181 **The lens galaxy of the twin QSO 0957+561.**
A. Stockton.
Astrophys. J., Lett., Vol. 242, L141 - L144, plates L9 - L10 (1980).

Direct imaging observations have been obtained under good seeing conditions for the twin QSO 0957+561 and the galaxy responsible for producing the gravitational imaging. The center of the galaxy is found to be $1''.00 \pm 0''.03$ north and $0''.19 \pm 0''.03$ east of the southern QSO image. The apparent core radius of the galaxy falls within the range $0''.24 \pm 0''.09$; however, if the expected third QSO image is nearly coincident with the galaxy and contributes a significant amount of the total light, the actual core radius would be larger. The shape of the galaxy profile cannot discriminate between the case of a small core radius with a faint third image and that of a larger core radius with a bright third image.

141.182 **Ca II absorption lines in the spectrum of the quasar PKS 2020–370 due to galactic material in the group Klemola 31.**
A. Boksenberg, I. J. Danziger, R. A. E. Fosbury, W. M. Goss.
Astrophys. J., Lett., Vol. 242, L145 - L148, plate L11 (1980).

In the spectrum of the quasar ($z_{em} = 1.050$), the authors find Ca II absorption lines both at near zero redshift and at $z_{abs} = 0.02865 \pm 0.00007$. The latter is closely similar to the redshifts of two galaxies in the group Klemola 31, which are nearby on the plane of the sky. This observation adds to the evidence that the narrow-lined heavy-element absorption systems in quasar spectra in general arise in the extended halos of intervening galaxies.

141.183 **An unusual radio point source in M17.**
M. Felli, K. J. Johnston, E. Churchwell.
Astrophys. J., Lett., Vol. 242, L157 - L161 (1980).

An unusual radio point source has been found near the southwestern boundary of M17 with the VLA. It has a diameter less than $0''.3$ and flux densities of 118, 28, and <7 mJy at 2, 6, and 21 cm, respectively. The authors infer an electron temperature >25,000 K, and emission measure $> 3 \times 10^9$ pc cm^{-6}, and a mean electron density $> 10^6$ cm^{-3}. The morphology suggests an intimate connection between the point source and an arc-shaped ionization front. The authors conclude that this may be an example of shock-induced star formation in the cool, compressed gas between the I-front and the shock front which is propagating into the molecular cloud. The geometry suggests that the shock waves were or are now focused roughly at the position of the radio point source.

A revised optical catalog of quasi-stellar objects. See Abstr. 002.013.

On the derivation of a catalogue of radio source positions from interferometric observations. See Abstr. 002.024.

An update of the status of the Revised 3C Catalog of Radio Sources; 22 new galaxy redshifts. See Abstr. 002.066.

Oort and extragalactic astronomy. See Abstr. 005.018.

Radio astronomy at the 1979 WARC (World Administrative Radio Conference). See Abstr. 011.007.

Ten years of discovery with Oort's synthesis radio telescope. See Abstr. 013.015.

An astrometric test of the doublet prime-focus corrector on the Anglo-Australian Telescope, and the optical position of OR 103. See Abstr. 031.540.

The use of objective prism plates from the UK Schmidt telescope for low resolution spectrophotometry of quasars. See Abstr. 031.555.

VLB interferometer sensitivity and the number of compact radio sources. See Abstr. 033.002.

Initial observations with the Tidbinbilla 64-m telescope at $\lambda = 13.5$ mm. See Abstr. 033.005.

The Jodrell Bank radio-linked interferometer network. See Abstr. 033.010.

Interplanetary scintillation – preliminary observations at 103 MHz. See Abstr. 033.012.

Astrometric radio source catalogs. See Abstr. 041.003.

Optical and radio positions of 3C 273B in the FK4-system. See Abstr. 041.019.

On the connection of the radio and optical systems of positions and proper motions. See Abstr. 041.036.

Spontaneous formation of knots in relativistic flows: a model for variability in compact synchrotron sources. See Abstr. 062.016.

Magnetohydrodynamic Kelvin-Helmholtz instabilities in astrophysics – I. Relativistic flows – plane boundary layer in vortex sheet approximation. See Abstr. 062.072.

Relativistic hydrodynamics of a free expansion and a shock wave in one-dimension. Super-light expansion of extragalactic radio sources. See Abstr. 062.076.

On the supersonic dynamics of magnetized jets of thermal gas in radio galaxies. See Abstr. 062.091.

Absorption of intense electromagnetic beams in a magnetoplasma. See Abstr. 062.095.

An explosion model in magnetogasdynamics. See Abstr. 062.119.

Comoving frame calculations of spectral lines formed in rapidly expanding media with the partial frequency redistribution function for zero natural line width. See Abstr. 063.019.

Monte Carlo simulation of relativistic Comptonization. See Abstr. 063.037.

The Cerenkov line radiation and the emission-line spectra of QSOs. See Abstr. 063.050.

Dipole anisotropy in the 2.7 K background. See Abstr. 066.020.

The material vacuum. See Abstr. 066.111.

Kosmische lenzen en quasars. See Abstr. 066.130.

Stellar tidal disruption by a massive binary black hole. See Abstr. 066.172.

$L\alpha/H\alpha$ in solar flares and QSOs. See Abstr. 073.027.

Aperture synthesis of the radio structure of SS433. See Abstr. 116.018.

Further high resolution radio observations of the supernova remnant G84.2–0.8. See Abstr. 125.001.

The optical spectrum of W50. See Abstr. 125.002.

Further measurements of the decay rate of Cassiopeia A at 8 GHz. See Abstr. 125.018.

Supernova radio remnants and the filling factor of the hot interstellar medium. See Abstr. 125.028.

Fleurs synthesis telescope observations of galactic SNRs. See Abstr. 125.034.

Supernova remnants: a generalized theoretical approach to the radio evolution. See Abstr. 125.035.

On interpretation of observed radio properties of Cas A. See Abstr. 125.036.

G 40.5 - 0.5: a previously unrecognized supernova remnant in Aquila. See Abstr. 125.060.

Observations of the old supernova remnant S 147 at 11.1 and 18.2 cm wavelengths. See Abstr. 125.062.

A new oxygen-rich supernova remnant in the Large Magellanic Cloud. See Abstr. 125.069.

Neutral hydrogen emission-absorption observations of high-velocity clouds. See Abstr. 131.090.

A search for the lowest-energy conformer of interstellar glycine. See Abstr. 131.280.

A high-resolution optical survey of interstellar absorption lines toward globular clusters and extragalactic objects. I. Basic data. See Abstr. 131.330.

Radio observations of DR15 at 408, 1407 and 2695 MHz. See Abstr. 132.003.

An aperture synthesis map of the Crab Nebula at 23 gigahertz. See Abstr. 134.019.

A study of the Crab Nebula – II. Radio emission from the filaments. See Abstr. 134.027.

On the nature of the faint ($B \simeq 20$) ultraviolet excess objects and the problem of the X-ray background. See Abstr. 142.063.

HR976 and 4C34.13: an X-ray odd couple. See Abstr. 142.089.

An upper limit on the flux density of radio bursts at 327 MHz from MXB 1730 - 335. See Abstr. 142.116.

Radio properties of extragalactic X-ray sources. See Abstr. 142.135.

Low-energy gamma-ray emission close to CG 135+1. See Abstr. 142.508.

The effect of convection on the propagation of relativistic galactic electrons. See Abstr. 143.007.

Radio continuum observations of the nucleus of our Galaxy and other normal galaxies. See Abstr. 155.023.

A 21 cm radio continuum survey of the galactic plane between $l = 93°$ and $l = 162°$. See Abstr. 156.006.

The dynamics of three radio galaxies – NGC 741, 1316 and 7626. See Abstr. 158.005.

Radio continuum observations of galaxies at 11 cm. See Abstr. 158.006.

Extended radio sources and elliptical galaxies. IV. Structures of 40 resolved sources. See Abstr. 158.008.

Extended radio sources and elliptical galaxies. V. Optical positions for 40 identified sources. See Abstr. 158.009.

Radio observations of a complete sample of spiral galaxies at 408 MHz. See Abstr. 158.013.

The complex radio galaxy 4C47.51 (1919+479). See Abstr. 158.019.

The dynamical evolution of NGC 5128. See Abstr. 158.054.

An observed change in the radio absorption spectrum of the BL Lac object AO 0235 + 164. See Abstr. 158.055.

Radio structures of Seyfert galaxies. See Abstr. 158.059.

Internal tail structure and orbit dynamics of the radio galaxy 3C129. See Abstr. 158.063.

Galaxy-QSO associations: a gravitational lens effect of globular clusters. See Abstr. 158.068.

Neutrino emission from galaxies, and mechanisms for producing radio lobes. See Abstr. 158.069.

Large- and small-scale structure in the continuum energy distributions of quasi-stellar objects and Seyfert 1 galaxies. See Abstr. 158.075.

The radio continuum emission from spiral galaxies in double systems. See Abstr. 158.089.

High-redshift objects near the companion galaxies to NGC 2859. See Abstr. 158.091.

Radio structures of Seyfert galaxies. I. See Abstr. 158.092.

Observations of radio galaxies. See Abstr. 158.107.

On the kinematics of high-velocity gas in the envelopes of the nuclei of type 1 Seyfert galaxies and quasars. See Abstr. 158.112.

Ion abundance and chemical composition in the nuclei of type 2 Seyfert galaxies and radio galaxies with narrow lines. See Abstr. 158.113.

Radio continuum emission from the nuclei of normal galaxies. See Abstr. 158.137.

Extragalactic astronomy and IUE. See Abstr. 158.146.

X-ray variability in active galaxy nuclei and quasars in less than one day. See Abstr. 158.151.

Observations of three radio galaxies with the Einstein X-ray Observatory. See Abstr. 158.154.

Why extended radio doubles are found in elliptical galaxies. See Abstr. 158.170.

Radio survey of Markarian galaxies at 6 and 11 cm. See Abstr. 158.176.

The ionised gas in NGC 5128: evidence for a shock-heated component. See Abstr. 158.183.

A spectroscopic survey of emission-line objects in two fields. See Abstr. 158.185.

The dynamical importance of magnetic fields in beam models of radio galaxies. See Abstr. 158.200.

VLBI observations of galactic nuclei. See Abstr. 158.202.

The dynamics of the broad-line-emitting regions of active galactic nuclei and quasars. I. Broad-line profiles. See Abstr. 158.205.

Lyman alpha fluxes of Seyfert galaxies and low-redshift quasars. See Abstr. 158.215.

Radio observations at 408 MHz of E and S0 nearby galaxies. See Abstr. 158.219.

NGC 7385: generation of an H II region by thermal instability associated with the creation of a radio source. See Abstr. 158.227.

On the origin of the radio, optical emission-line, and X-ray structure of M87. See Abstr. 158.228.

VLA observations of the 6 cm radio continuum emission of galactic nuclei. See Abstr. 158.293.

Strong radio sources in bright spiral galaxies. See Abstr. 158.320.

The 21 centimeter line width as an extragalactic distance indicator. See Abstr. 158.325.

On the distribution of radio emission in the X-ray cluster of galaxies Abell 401. See Abstr. 160.001.

A radio survey of clusters of galaxies. III. 6.2 cm observations, radio spectra and optical identifications of sources in 29 Abell clusters. See Abstr. 160.003.

Observations of clusters of galaxies at meter wavelengths. See Abstr. 160.011.

Diffuse radio emission in the Coma cluster and Abell 1367 – observations at 430 and 1400 MHz. See Abstr. 160.023.

Formation of radio halos in clusters of galaxies from cosmic-ray protons. See Abstr. 160.025.

Radio emission of near clusters of galaxies at 102.5 MHz. See Abstr. 160.026.

A Westerbork survey of clusters of galaxies. XIII. Deep 610 MHz source counts from the Cancer cluster field. See Abstr. 160.028.

Radio and X-ray observations of Abell 754. See Abstr. 160.034.

Stephan's Quintet: H I distribution at high redshifts. See Abstr. 160.036.

The Westerbork survey of rich clusters of galaxies. See Abstr. 160.039.

Clustering of galaxies about extragalactic radio sources – implications for observations with the Einstein X-ray Observatory. See Abstr. 160.040.

Westerbork observations of B2 radio sources in Abell clusters of galaxies. See Abstr. 160.046.

On the morphology of the magnetic field in galaxy clusters. See Abstr. 160.063.

Diffuse radio emission in the Coma cluster and Abell 1367: observations at 430 and 1400 MHz. See Abstr. 160.069.

The effects of X-ray absorption on the spectra of distant objects. See Abstr. 161.002.

Uniformity of the radial distribution of quasars in the chronometric cosmology and the X-ray background. See Abstr. 162.015.

The Hubble diagram of QSOs by statistical method. See Abstr. 162.020.

The effect of variability on the V/V_{MAX}-test. See Abstr. 162.024.

High energy astrophysics and cosmology. See Abstr. 162.069.

Pulsars

141.501 **On the self-consistent description of axisymmetric pulsar magnetospheres – II. A method of solution.**
R. Schmalz, H. Ruder, H. Herold, C. Rossmanith.
Mon. Not. R. Astron. Soc., Vol. 192, 409 - 416 (1980).

The authors investigate the general structure of the system of pulsar equations derived in an earlier paper for the magnetosphere of a parallel rotator including particle inertia. The complete system reduces to one parabolic and two elliptic differential equations. The characteristics are found to be the streamlines. They specify appropriate variational principles and propose a method for solving the pulsar equations numerically. Complete boundary conditions are given. First results for the solution within the corotating zone are presented.

141.502 **The millisecond intensity variation in the emission of radio pulsars.** N. Bartel, W. Sieber, D. A. Graham.
Astron. Astrophys., Vol. 87, 282 - 291 (1980).

A comprehensive study of single pulse intensities on the time scale of milliseconds was made for 36 radio pulsars at several frequencies. In spite of a large subpulse width variation of $\sim 50\%$ (1σ) around the mean for individual single pulses the authors find a strong relation between mean properties of subpulses and integrated pulse profiles so that various functional dependences for subpulses and integrated pulses are similar.

141.503 **Low energy γ rays from PSR 1822–09.**
P. Mandrou, G. Vedrenne, J. L. Masnou.
Nature, Vol. 287, 124 - 126 (1980).

The authors report observations of PSR1822–09 between 80 keV and 6 MeV at balloon altitudes.

141.504 **Frequency – time structure of pulsar scintillation.**
A. Hewish.
Mon. Not. R. Astron. Soc., Vol. 192, 799 - 804 (1980).

Observations of four pulsars are used to compare the density variations in the large- and small-scale irregularities. The values obtained are compatible with a Kolmogorov spectrum for scales in the range $10^{10} - 10^{13}$ cm, but suggest that the power-law index may exceed 11/3.

141.505 **Astrometry and high-speed photometry of an optical candidate for PSR 1913 + 16.**
K. H. Elliott, B. A. Peterson, P. T. Wallace, D. H. P. Jones, E. D. Clements, K. F. Hartley, R. N. Manchester.
Mon. Not. R. Astron. Soc., Vol. 192, 51P - 58P (1980).

A deep IIIaF plate has been obtained at the prime focus of the AAT which shows the candidate star described by Kristian & Westphal and Crane et al. near the timing position of the binary pulsar PSR 1913 + 16. The authors' astrometry shows that this star does not coincide with the timing position.

141.506 **H I absorption measurements of seven low-latitude pulsars.** J. M. Weisberg, J. Rankin, V. Boriakoff.
Astron. Astrophys., Vol. 88, 84 - 93 (1980).

H I absorption spectra have been measured for seven low-latitude pulsars with the Arecibo 305-m telescope. Distance limits based on the Schmidt (1965) galactic rotation model are: pulsars 1919 + 21 and 1929 + 10, $d \lesssim 1.5$ kpc; and pulsars 2016 + 28 and 2020 + 28, $d \gtrsim 1.3$ kpc. An attempt is made to apply the TASS galactic rotation model (Roberts, 1972) to absorption spectra of pulsars 0525 + 21 and 0540 + 23, but no evidence is found for the predicted Perseus arm shock. Similarities of absorption spectra, dispersion and rotation measures, and Strömgren spheres along their lines of sight indicate that PSR 0525 + 21 is approximately as distant as the Crab Nebula pulsar (~2 kpc), in accord with the findings of Gómez-González and Guélin (1974). Similar arguments further suggest strongly that PSR 0540 + 23 is more distant than the Crab Nebula pulsar since its dispersion measure is ~30% greater. Finally, absorption in PSR 0611 + 22 is similar to possible absorption seen in SNR IC 443 (DeNoyer, 1977), but the resemblance may follow merely from the small value of $|dv/dr|$ in the anticenter direction.

141.507 **A radiation model for the Vela pulsar.**
J. Davila, C. Wright, G. Benford.
Astrophys. Space Sci., Vol. 71, 51 - 63 (1980).

The authors model the Vela emission in radio, optical and gamma wavelengths. They assume that radio emission occurs near the polar axis deep in the magnetosphere. Optical and gamma radiation arises through ordinary synchrotron mechanisms in a wide hollow cone, near the light cylinder. Fitting of observed frequencies and luminosities in the gamma and optical give reasonable plasma parameters, and indicate that field-plasma pressure balance breaks down before the light cylinder, as required by the picture. The authors suggest a single-pole orthogonal rotor picture which might account for the Vela pulse phases.

141.508 **New upper limits for pulsed soft X-rays from the Vela pulsar PSR 0833 - 45.** H. U. Zimmermann.
Astron. Astrophys., Vol. 88, 309 - 310 (1980).

In January 1977 the pulsar PSR 0833 - 45 was observed by a rocket experiment consisting of 12 X-ray paraboloidal collectors. No indication for pulsed flux could be detected. The new upper limits are considerably lower than former measurements in the energy band 0.09 to 2 keV.

141.509 **Short time scale integrated pulse shape variations in PSR 0611 + 22.** D. C. Ferguson, V. Boriakoff.
Astrophys. J., Vol. 239, 310 - 315 (1980).

Variations in integrated pulse shape and phase on time scales of minutes are reported for pulsar PSR 0611 + 22. The pulsar, despite its youth, shows sporadic drifting subpulse behavior interspersed with periods of "stable" pulse arrival time, occurring at more than one longitude. It is hypothesized that the proximity of PSR 0611 + 22 to an apparent gap in the distribution of period derivatives for pulsars is related to the instability of individual pulse phases.

141.510 **Timing observations of pulsars.**
P. R. Backus, J. H. Taylor, M. Damashek.
Bull. American Astron. Soc., Vol. 12, 502 (1980). – Abstract.

141.511 **Discovery of a third radio pulsar in a binary system.**
M. Damashek, P. R. Backus, J. H. Taylor.
Bull. American Astron. Soc., Vol. 12, 512 (1980). – Abstract.

141.512 **Asymmetries in the galactic distribution of pulsars.**
A. K. Harding.
Bull. American Astron. Soc., Vol. 12, 513 - 514 (1980). Abstract.

141.513 **X-rays from radio pulsars.**
D. J. Helfand, G. A. Chanan, R. Novick.
Bull. American Astron. Soc., Vol. 12, 514 (1980). – Abstract.

141.514 **Hard X-ray and gamma-ray pulsed emission from the Crab pulsar.** F. K. Knight, W. A. Baity, D. E. Gruber, L. E. Peterson, M. Bautz, F. Lang, W. H. G. Lewin.
Bull. American Astron. Soc., Vol. 12, 541 (1980). – Abstract.

141.515 **Relativistic observable effects in the binary pulsar PSR 1913 + 16.** E. Rudolph.
Isolating gravitating systems in general relativity, (see 012.004), p. 318 - 321 (1979). – Abstr. in Phys. Abstr., Vol. 83, Abstr. 77438 (1980).

141.516 **Pulsar timing. III. Timing noise of 50 pulsars.**
J. M. Cordes, D. J. Helfand.
Astrophys. J., Vol. 239, 640 - 650 (1980).

From an examination of the timing behavior of 50 pulsars, the authors have found that timing activity is $\gtrsim 50\%$ correlated with pulsar period derivative, weakly correlated with period, and uncorrelated with radio luminosity, galactic altitude, and other source parameters. A detailed analysis of 11 pulsars indicates that random walks in the rotational phase (2 objects), frequency (4–7 objects), and frequency derivative (2 objects) are consistent with the data.

141.517 **Revised stellar birthrates and the genesis of pulsars.**
H. L. Shipman, R. F. Green.
Astrophys. J., Lett., Vol. 239, L111 - L113 (1980).

Recent revisions in the local stellar birthrate function significantly affect conclusions regarding the determination of the masses of pulsar progenitors from pulsar statistics. Use of the most recent birthrates, along with recently revised pulsar birthrates, shows that stars in astrophysically reasonable mass ranges (e.g., between 5 and 8 $M_\odot$ or above 7 $M_\odot$) can produce an adequate number of pulsars. Use of the correct stellar birthrates changes the allowable mass range by amounts exceeding the changes produced by recently proposed alterations in the pulsar birthrate. If pulsar birthrates are as low as current observations allow, the mass range can be extremely narrow (e.g., 5.0–5.6 $M_\odot$), or stars with masses exceeding 15 $M_\odot$ can produce the required number of pulsars.

141.518 **The pulsar magnetosphere as a giant radio aerial.**
V. G. Endean.
Mon. Not. R. Astron. Soc., Vol. 193, 213 - 218 (1980).

New formulae for the radiation field of a synchronously rotating charge distribution are derived. The field is stationary in the rotating frame but the wave shape is non-sinusoidal. The synchrotron radiation from a point charge may be deduced as a special case. The formulae are applied to a rotating charged disc to show how the main features of observed pulsar radiation may be generated.

141.519 **The magnetic pole model for pulsar emission.**
R. N. Manchester.
Proc. Astron. Soc. Australia, Vol. 3, 200 - 205 (1978).

141.520 **A survey for sharply pulsed emissions.**
T. W. Cole, R. D. Ekers.
Proc. Astron. Soc. Australia, Vol. 3, 328 - 330 (1979).

141.521 **Radio observations with a wide fractional bandwidth.** T. W. Cole.
Proc. Astron. Soc. Australia, Vol. 3, 330 - 332 (1979).

141.522 **The magnetosphere of X-ray pulsars.**
V. M. Lipunov.
Priroda, 1980, No. 10, p. 52 - 61. In Russian.

141.523 **On inertial drifts in pulsar magnetospheres.**
R. Burman.
Astrophys. Space Sci., Vol. 72, 251 - 253 (1980).

It is pointed out that, in a number of papers, Ardavan has obtained incorrect results because of his invalid neglect of inertial drifts in attempting to obtain an integral of the motion for steadily rotating pulsar magnetospheres.

141.524 **Pulse to pulse intensity modulation from radio pulsars with particular reference to frequency dependence.** N. Bartel, W. Sieber, A. Wolszczan.
Astron. Astrophys., Vol. 90, 58 - 64 (1980).

The pulse to pulse mean intensity modulation was analysed for 36 radio pulsars at various frequencies. The modulation index decreases in general according to $m \propto f^{-0.4}$ up to a critical frequency $f_{cm} \sim 1$ GHz and was found to increase again from there on according to $m \propto f^{0.6}$ where high frequency data were available. f_{cm} coincides with the critical frequencies in the flux density spectrum and/or in the component width-frequency diagram. f_{cm} may be the frequency at which the coherent amplification mechanism breaks down.

141.525 **Detailed characteristics of the high-energy gamma radiation from PSR 0833–45 measured by COS-B.**
G. Kanbach, K. Bennett, G. F. Bignami, R. Buccheri, P. Caraveo, N. D'Amico, W. Hermsen, G. G. Lichti, J. L. Masnou, H. A. Mayer-Hasselwander, J. A. Paul, B. Sacco, B. N. Swanenburg, R. D. Wills.
Astron. Astrophys., Vol. 90, 163 - 169 (1980).

A series of COS-B observations of the high-energy gamma-ray emission from PSR 0833–45 has been combined to derive detailed temporal and spectral characteristics of this source. The gamma-ray period of the pulsar was determined for the observations spanning the time interval October 1975 to June 1977.

141.526 **Properties of slowly rotating helium II and the superfluidity of pulsars.**
J. S. Tsakadze, S. J. Tsakadze.
J. Low Temp. Phys., Vol. 39, 649 - 688 (1980). – Abstr. in Phys. Abstr., Vol. 83, Abstr. 98683 (1980).

141.527 **Pulsar PSR 0820+02 – stellar remnant of a silent collapse?** I. S. Shklovskij.
Astron. Zh., Tom 57, 897 - 898 (1980). In Russian. – English translation in Soviet Astron., Vol. 24, No. 5.

The hypothesis is substantiated that the pulsar PSR 0820+02 located in a wide binary system could be a neutron star formed in a gravitational collapse which was not followed by a supernova explosion phenomenon.

141.528 **Observation of the flare X-ray pulsar FXP 0520–66 aboard the Cosmos 856 satellite.**
V. M. Loznikov, A. S. Melioranskij, M. I. Kudryavtsev, I. A. Savenko, V. M. Shamolin.
Pis'ma Astron. Zh., Tom 6, 614 - 616 (1980). In Russian. English translation in Soviet Astron. Lett., Vol. 6.

X-ray pulsations in the range 32 - 124 keV with the period 8.6 ± 0.6 s and the average flux $\sim 1.4 \times 10^{-3}$ phot/cm^2 s keV were detected on 1976 September 26 during 90 s by the Cosmos 856 satellite from a region of the sky in the direction of the X-ray flare pulsar FXP 0520–66.

141.529 **Space distribution and birthrate of pulsars.**
K.-l. Huang, Q.-h. Peng, X.-t. He, Y. Tong.
Acta Astron. Sinica, Vol. 21, 237 - 242 (1980).

With the data of 321 pulsars the galactic distribution and the luminosity function of pulsars have been investigated. The total number of pulsars in the Galaxy: $N_G = 9 \times 10^4$ is obtained. If the mean age of pulsars is 1.8×10^6 years, the birthrate of pulsars will be one every twenty years in the Galaxy.

141.530 **The superfluid phase transition in pulsars.**
R. J. Green, A. Love.
J. Phys. A, Vol. 13, 2853 - 2856 (1980). – Abstr. in Phys. Abstr., Vol. 83, Abstr. 101607 (1980).

141.531 **Astrometry and high-speed photometry of an optical candidate for PSR 1913+16.**
K. H. Elliott, B. A. Peterson, P. T. Wallace, D. H. P. Jones, E. D. Clements, K. F. Hartley, R. N. Manchester.
Anglo-Australian Obs., Prepr. No. 134, 20 pp. (1980). Submitted to Mon. Not. R. Astron. Soc.

141.532 **A study of PSR 1237 + 25 at 430 MHz.**
T. H. Hankins, G. A. E. Wright.
Nature, Vol. 288, 681 - 684 (1980).

The authors have analysed single-pulse data from this pulsar at 430 MHz to throw light on the still obscure mechanism of pulsar emission. They conclude that the behaviour of PSR 1237 + 25 can be explained on the polar-cap model if regular short bursts spiralling in towards the magnetic pole are occasionally interrupted by long bursts along the magnetic axis.

141.533 **Pulsar extinction: a plasma inertia effect?**
T. W. Hill.
Astrophys. Lett., Vol. 21, 11 - 12 (1980).

If one assumes that corotation near the light cylinder is a necessary condition for pulsar action, then an upper limit is implied for the inertia of the corotating plasma. The Goldreich-Julian charge density implies a minimum plasma mass density which then implies an upper limit for the pulsar spin period T. For typically-assumed pulsar parameters the observed cut-off at T ~ 4 sec is obtained if one further assumes that the plasma is $\gtrsim 1\%$ charge-separated.

141.534 **The boundary of a pulsar magnetosphere. Gamma-ray pulsars.** N. J. Holloway.
Magnetospheric boundary layers, (see 012.041), p. 281 - 284 (1979).

The paper considers the energetics of an aligned or nearly aligned pulsar magnetosphere, and interprets the "boundary" as that region in which the principal energetic processes take place. It is argued that these processes accur at or near the light cylinder, and that they consist primarily of gamma ray emission and classical magnetic dipole radiation. The gamma ray emission becomes dominant at very small angles of non-alignment, as the dipole radiation is reduced by alignment. This dominance of the gamma ray emission represents the late stages of pulsar evolution.

141.535 **Pulsars.** A. G. Lyne.
Variability in stars and galaxies, (see 012.044), p. F.2.1 - 2.10 (1980).

The observations and the advances made in the understanding of the galactic distribution of pulsars and their origin are discussed and the evidence for their origin in supernova events is considered.

141.536 **Test observations of pulsar PSR 0329 + 54 on the RATAN-600 radio telescope.**
V. I. Dokuchaev, A. V. Ipatov, Z. E. Petrov.
Astron. Tsirk., No. 1099, p. 1 - 2 (1980). In Russian.

141.537 **On the relative position of various axes in the pulsars PSR 0525 and PSR 1133.** I. F. Malov.
Astron. Tsirk., No. 1110, p. 1 - 2 (1980). In Russian.

141.538 **Die axialsymmetrische Pulsarmagnetosphäre.**
R. Schmalz, H. Herold, C. Herold, H. Ruder.
Mitt. Astron. Ges., Nr. 50, p. 127 - 128 (1980).

141.539 **Pulsars and compact X-ray sources – cosmic laboratories for the study of neutron stars and hadron matter.** D. Pines.
Usp. fiz. nauk, Vol. 131, 479 - 494 (1980). – Abstr. in Ref. zh., 51. Astron., 12.51.452 (1980).

141.540 **The structure of integrated pulse profiles.**
M. Vivekanand, V. Radhakrishnan.
J. Astrophys. Astron., Vol. 1, 119 - 128 (1980).

The authors offer two possible explanations to account for the characteristics of integrated pulse profiles, in particular their degree of complexity, their variation from pulsar to pulsar, their stability, and the tendency of complex profiles to be associated with older pulsars.

141.541 **No detectable supernova remnant near the pulsar PSR 1930 + 22.** W. M. Goss, D. Morris.
J. Astrophys. Astron., Vol. 1, 189 - 192 (1980).

No supernova remnant has been found near the third youngest pulsar PSR 1930 +22 down to a limiting brightness temperature of 1.4 K at 610 MHz. This is 6 - 8 times less than expected of a typical remnant whose age is that of the pulsar (3.6×10^4 years).

141.542 **Application of a two-dimensional correlation analysis to investigations of pulsar radiation.**
A. A. Bocharov, S. V. Pogrebenko.
Inst. kosm. issled. AN SSSR. Prepr., 1980, No. 583, 20 pp. In Russian. – Abstr. in Ref. zh., 51. Astron., 1.51.514 (1981).

Infinitely magnetized hydrogen atom and pulsar crust. See Abstr. 062.026.

Pulsar radio emission from beam plasma instability. See Abstr. 062.027.

The quiet aligned rotator. See Abstr. 062.060.

Magneetvelden in de astrofysika. See Abstr. 062.092.

The magnetospheres of magnetic A stars and of pulsars. See Abstr. 062.109.

A new approximation for the high magnetic field Compton cross-section. See Abstr. 063.059.

Indirekter Nachweis von Gravitationswellen gelungen. See Abstr. 066.003.

Gravitationswellen von Doppelsternsystem? See Abstr. 066.133.

Introduction to black holes and pulsars. See Abstr. 066.154.

Pair formation and electric field boundary conditions at neutron star magnetic poles. See Abstr. 066.503.

Pulsars and compact X-ray sources: cosmic laboratories for the study of neutron stars and hadron matter. See Abstr. 066.510.

Simultaneous measurement of coronal Faraday rotation and total electron content during solar occultation of PSR 0525 + 21. See Abstr. 074.021.

Intensity and spectrum of the continuum gamma ray emission from supernovae. See Abstr. 125.011.

Pulsar activity and the morphology of supernova remnants. See Abstr. 125.025.

Vela X and the evolution of plerions. See Abstr. 125.041.

Peculiarities of supernova remnant and pulsar distribution relative to the local spiral arm of the Galaxy. See Abstr. 125.076.

Scintillations of pulsars and the parameters of inhomogeneities of the interstellar plasma. See Abstr. 131.320.

Discrete sources of gamma-radiation, historic supernovae and pulsars. See Abstr. 142.513.

The discrete sources of high energy gamma-ray radiation. See Abstr. 142.521.

Discrete gamma-ray sources: theory. See Abstr. 142.522.

The effect of the diffuse gamma radiation on pulsars. See Abstr. 142.531.

Erratum

141.901 **Erratum: 'The physics of QSO absorption line regions'** [Phys. Scr., Vol. 17, 205 - 214 (1978)].
A. Boksenberg.
Phys. Scr., Vol. 22, 666 - 668 (1980). – See Abstr. 21.141.125.

142 UV Sources, X-ray Sources, X-ray Background, Gamma-ray Sources, Gamma-ray Background

UV Sources, X-ray Sources, X-ray Background

142.001 **Binary model of Circinus X-1. I. Eccentricity from combined X-ray and radio observations.**
P. Murdin, D. L. Jauncey, R. F. Haynes, I. Lerche, G. D. Nicolson, S. S. Holt, L. J. Kaluzienski.
Astron. Astrophys. Vol. 87, 292 - 298 (1980).

A binary star model is used to account for the 16.59d flaring behaviour of the X-ray emission from Circinus X-1. The orbital eccentricity of 0.8±0.1 is derived from the X-ray light curve by assuming that the sharp X-ray cut-off every 16.59d is a result of bound-free absorption in the primary star's stellar wind. The shape of the light curve has changed over the last eight years, and the authors interpret this as due to orbital precession of the binary system. Simultaneous radio and X-ray observations of the flare from Circinus X-1 on 1978 February 1–5 are reported. These are accounted for within the framework of their model. The radio observations at 5GHz are used independently to derive a high value of the orbital eccentricity (e=0.7).

142.002 **Binary model of Circinus X-1. II. Radio emission.**
R. F. Haynes, I. Lerche, P. Murdin.
Astron. Astrophys., Vol. 87, 299 - 302 (1980).

A binary star model is used to account for the 16.59d flaring behaviour of the radio emission from Circinus X-1. Mass transfer between the primary star and the compact companion star triggers one or more expanding Eddington luminosity-driven shocks in the vicinity of the compact star which, in turn, produce radio emission.

142.003 **Iron line emission from the Alfvén shell in X-ray binaries.** M. M. Basko.
Astron. Astrophys., Vol. 87, 330 - 338 (1980).

The paper presents a model in which the iron line emission ϵ=6–7 keV, discovered in spectra of some X-ray binaries, is interpreted as a reradiation of the X-ray continuum in the gaseous shell at the Alfvén surface. The solid angle Ω subtended by the shell, its radius, the Thomson optical thickness and the electron density are treated as free parameters. The temperature and the ionization equilibrium are determined by the interaction with the external X-ray flux. The allowed range of variation of the shell parameters is evaluated from the requirement that the values of the iron line equivalent width predicted by the model agree with the observed ones. The restrictions imposed by such an agreement on the Alfvén shell in Her X-1 are shown to be compatible with the limitations deduced from the soft X-ray data. The model predicts a considerable width of the iron line profile, $\Delta\epsilon$>200 eV. The problem of escape of the line photons from the shell in the course of multiple resonance scattering is discussed in detail. A number of observational consequences and tests are proposed for the model under discussion.

142.004 **Unidentified galactic X-ray sources positioned with the HEAO-1 scanning modulation collimator.**
C. A. Reid, M. D. Johnston, H. V. Bradt, R. E. Doxsey, R. E. Griffiths, D. A. Schwartz.
Astron. J., Vol. 85, 1062 - 1070 (1980).

Data obtained with the modulation collimator on the HEAO-1 satellite during its first year of operation yield 16 precise (10"- 30") positions of unidentified X-ray sources at low galactic latitude ($|b|$< 15°).

142.005 **Geometrical model of Cygnus X-2.**
J. Ziołkowski, B. Paczyński.
Acta Astron., Vol. 30, 143 - 156 (1980).

The authors obtain the following limits on the distance to Cyg X-2 and masses of its components: $6 \lesssim d$ [kpc] $\lesssim 17$; $0.3 \lesssim M_{opt}/M_{\odot} \lesssim 5, M_x/M_{\odot} \lesssim 1.2$. They also discuss the geometry of the X-ray emission. The results are in conflict with those derived by Branduardi et al. (1979) from spectral X-ray data.

142.006 **The hard X-ray emission lines of Her X-1.**
Q.-y. Qu, Z.-r. Wang, H.-q. Zhang.
Acta Astron. Sinica, Vol. 21, 180 - 183 (1980). In Chinese.

Using the observational data of hard X-ray emission lines of Her X-1, the authors study the physical situation of the emission region. It is found that the emission region is opaque to X-ray lines radiation.

142.007 **No IR burst from the X-ray rapid burster MXB1730–335.** S. Sato, K. Kawara, Y. Kobayashi, T. Maihara, H. Okuda, J. Jugaku.
Nature, Vol. 286, 688 - 689 (1980).

The authors monitored MXB1730–335 using a multiband IR photometer for about 10 h during April and August 1979. In the latter period, although the X-ray bursts were very active, no coincident IR and X-ray burst could be detected.

142.008 **A high sensitivity determination of the hard X-ray spectrum of Sco X-1.** R. E. Rothschild, D. E. Gruber, F. K. Knight, P. L. Nolan, Y. Soong, A. M. Levine, F. A. Primini, W. A. Wheaton, W. H. G. Lewin.
Nature, Vol. 286, 786 - 788 (1980).

The results of hard X-ray observations of Sco X-1 by HEAO 1 and OSO 7 are reported.

142.009 **Simultaneous IR and X-ray burst observation of Ser X-1.** W. H. G. Lewin, L. R. Cominsky, A. R. Walker, B. S. C. Robertson.
Nature, Vol. 287, 27 - 28 (1980).

During the 1977 coordinated worldwide burst watch, simultaneous X-ray and IR observations were made. On UT June 17,00h 00min 25s, a Type I X-ray burst was observed, which was almost certainly emitted by Ser X-1 (MXB 1837+05, 4U1837+04). Simultaneous IR observations failed to detect any bursts. The authors report on these observations.

142.010 **The K dwarfs associated with the X-ray transients A0620–00 and A1742–28.**
P. Murdin, D. A. Allen, D. C. Morton, J. A. J. Whelan, R. M. Thomas.
Mon. Not. R. Astron. Soc., Vol. 192, 709 - 717 (1980).

Spectra of A0620–00 taken in 1978, two and a half years after its outburst, show a stellar continuum of type K4 - 5V and emission lines characteristic of an accretion disc. Spectra of a star positionally coincident with the radio transient associated with A1742–28 show it to be K3V. The authors suggest there is a class of X-ray transients consisting of a cool dwarf star and an accretion disc around an unseen companion, similar to the common model for dwarf novae.

142.011 **Detection of X-ray emission from the vicinity of two short-period RS CVn-like binaries.**
P. C. Agrawal, G. R. Riegler, G. P. Garmire.
Mon. Not. R. Astron. Soc., Vol. 192, 725 - 730 (1980).

Two new soft X-ray sources H0118 + 067 and H0630 +

82 have been detected. Short-period RS CVn-like binaries UV Psc and SV Cam lie within the error boxes of these sources and are proposed as the optical identifications for the two X-ray sources. The observed X-ray luminosity of the sources are in the same range as observed for other RS CVn binaries. The X-ray emission is probably of coronal origin.

142.012 **Short-term variability of Cyg X-1 and the accretion disk temperature fluctuation.** K. Doi.
Nature, Vol. 287, 210 - 212 (1980).

A balloon observation was carried out to study the variability in the hitherto unknown hard X-ray range ($\gtrsim$ 20 keV). The results of the analysis, together with those of the former rocket observation, suggest that the variation is essentially spectral, implicating that it originates from temperature fluctuation in an accretion disk. Such a model is discussed in detail.

142.013 **Characteristics of the Cen X-3 neutron star from correlated spin-up and X-ray luminosity measurements.** M. van der Klis, J. M. Bonnet-Bidaud, N. R. Robba.
Astron. Astrophys., Vol. 88, 8 - 14 (1980).

The authors present data obtained with the high time resolution mode of the ESA COS-B satellite during a one-month continuous observation of Cen X-3 from December 24, 1975 to January 23, 1976. From these high time resolution data, they were able to study the pulsation period changes from one binary cycle to the next one. These results together with the overall monitoring of the source flux during the same period (Bonnet-Bidaud, van der Klis 1979), allow an interesting study of the correlation between pulsation period changes and the luminosity.

142.014 **Pulsating X-ray sources: the oblique dipole configuration.** H. Riffert.
Astrophys. Space Sci., Vol. 71, 195 - 201 (1980).

The author studies the problem of magnetic field computation for a thin diamagnetic accretion disk around the oblique dipole of a magnetic neutron star. An analytic expression of the general solution is given, containing a free constant which describes the amount of magnetic flux through the hole in the middle of the disk.

142.015 **The system AM Her = 4U 1814 + 50.**
L. Chiappetti, E. G. Tanzi, A. Treves.
Space Sci. Rev., Vol. 27, 3 - 33 (1980).

The binary system AM Herculis gives the first well ascertained example of an X-ray emitting magnetic white dwarf. The orbital period (3.1^h) is apparent from X-ray to IR frequencies and in linear and circular polarization. Since the time of the identification of the X-ray source the system has been extensively studied. The observations (which range from 1MeV to 20 μm) are reviewed and compared with the present theory of X-ray emitting white dwarfs.

142.016 **Prognoz 6 observations of X-ray sources. II. Observations of Sco X-1, X Per and the galactic center.**
Sh. Yu. Rakhamimov, I. V. Ehstulin, G. Vedrenne, M. Niel.
Pis'ma Astron. Zh., Tom 6, 476 - 480 (1980). In Russian.
English translation in Soviet Astron. Lett., Vol. 6.

Results of observations of the X-ray sources Sco X-1, X Per and the galactic center by the SIGNE-2MP instrument on board the Prognoz 6 station are given. The observed energy spectra as well as the fluxes are discussed. Parameters of the stellar wind are estimated for the X Per system.

142.017 **On the X-ray continuum and iron-line emission of the low state of Hercules X-1.** T. Bai.
Astrophys. J., Vol. 239, 328 - 334 (1980).

The hard X-ray continuum observed during low-state periods of Hercules X-1 is interpreted as being due to Compton scattering, by the coronal gases above and below the accretion disk, of the primary hard X-rays from the neutron star, which is hidden from our view during these periods. A similar interpretation holds for the low-state soft X-rays. The low-state iron line at 6.4 keV cannot be due to the coronal gases. The main production mechanism of this line is most likely fluorescence from the accretion disk. The X-ray continuum reflected by HT Herculis, which is expected to show both a strong binary-phase dependence and a peak near 15 keV, and the fluorescent iron K line from HZ Herculis, whose strength also is expected to show a strong binary-phase dependence, are much less intense than theoretical predictions. A likely explanation of this is that a large shadow is cast on HZ Herculis by the accretion disk. The assumption that the abundances (especially the iron abundance) of HZ Herculis are similar to the solar abundances is compatible with the existing data.

142.018 **Observations of Cygnus X-3 with the Einstein (HEAO 2) X-ray Observatory: the period derivative and the asymmetric X-ray light curve.** R. F. Elsner, P. Ghosh, W. Darbro, M. C. Weisskopf, P. G. Sutherland, J. E. Grindlay.
Astrophys. J., Vol. 239, 335 - 344 (1980).

The authors present the results of observations of Cygnus X-3 in late 1978 by the Time Interval Processor (TIP) circuitry of the Monitor Proportional Counter aboard the Einstein (HEAO 2) X-ray Observatory. The authors use these observations together with a re-examination of previous work to establish the values for the period and period derivative of the 4.8 hour variation in the X-ray intensity of this source. In addition, they find that, in agreement with previous investigations, the average 4.8 hour X-ray light curve is asymmetric with a sharp fall to X-ray minimum followed by a gradual rise. The authors discuss the possibility that this asymmetry results from an elliptic binary orbit.

142.019 **The coronae of 40 Eridani.**
W. Cash, P. Charles, H. M. Johnson.
Astrophys. J., Lett., Vol. 239, L23 - L26 (1980).

The authors have obtained high-resolution X-ray and radio images of the 40 Eridani triple star system. The X-ray image, obtained with the Einstein Observatory, shows a corona of approximately solar intensity at the position of the K1V star; the dM4e flare star is a strong X-ray source estimated at 2×10^{28} ergs s^{-1}. The hot white dwarf 40 Eridani B shows no X-ray signal down to a level of 8×10^{26} ergs s^{-1}. A VLA 4885 MHz radio image shows no evidence of signal from any of the three stars, down to the 0.95 mJy level.

142.020 **Implications of UHE γ-radiation from Cyg X-3.**
W. T. Vestrand, D. S. Eichler.
Bull. American Astron. Soc., Vol. 12, 449 (1980). – Abstract.

142.021 **Results from the HEAO-1 high energy X-ray sky survey.** A. Levine, M. Bautz, S. Howe, F. Lang, F. Primini, W. H. G. Lewin, W. Baity, D. Gruber, F. Knight, L. Peterson, R. Rothschild, J. Matteson, P. Nolan.
Bull. American Astron. Soc., Vol. 12, 463 (1980). – Abstract.

142.022 **The Ariel 5 Sky Survey has ended.** M. G. Watson.
Bull. American Astron. Soc., Vol. 12, 486 (1980).
Abstract.

142.023 **A cosmic X-ray background constraint on the spectral evolution of compact sources.** E. A. Boldt.
Bull. American Astron. Soc., Vol. 12, 495 (1980). – Abstract.

142.024 **The rapid Hα variability of the X-ray binary X Persei.**
S. L. Mufson, R. H. Kaitchuck, D. J. Hutter, K. S. Wood.
Bull. American Astron. Soc., Vol. 12, 500 (1980). – Abstract.

142.025 **Discovery of a 32 day periodicity in the X-ray intensity of LMC X-4.**
F. Lang, A. Levine, M. Bautz, S. Howe, F. Primini, W. H. G. Lewin, W. Baity, D. Gruber, R. Rothschild.
Bull. American Astron. Soc., Vol. 12, 512 (1980). – Abstract.

142.026 **Discovery of pulsed X-ray emission from 2S1417-624.** R. Kelley, K. Apparao, H. Bradt, J. G. Jernigan, S. Naranan, S. Rappaport.
Bull. American Astron. Soc., Vol. 12, 512 (1980). – Abstract.

142.027 **Implications of the absence of narrow X-ray line emission from Scorpius X-1.**
C. Berg, C. R. Canizares, T. H. Markert, G. W. Clark.
Bull. American Astron. Soc., Vol. 12, 513 (1980). – Abstract.

142.028 **A 60-day survey of the period variability of the 4.8 second pulsations of Cen X-3.**
D. F. Johnston, J. Kulik, J. G. Jernigan.
Bull. American Astron. Soc., Vol. 12, 513 (1980). – Abstract.

142.029 **Precise location of 3A 2352+28: the high galactic latitude transient A0000+28.**
D. A. Schwartz, M. Garcia, M. Conroy, E. Ralph, W. Roberts, R. E. Griffiths, M. Johnston, I. McHardy, A. Lawrence, J. Pye.
Bull. American Astron. Soc., Vol. 12, 513 (1980). – Abstract.

142.030 **Balloon observations of GX 301-2.** G. S. Maurer, W. N. Johnson, J. D. Kurfess, M. S. Strickman.
Bull. American Astron. Soc., Vol. 12, 513 (1980). – Abstract.

142.031 **On the X-ray periodicity of GX 1+4.**
B. R. Dennis, E. P. Cutler, R. H. Becker, J. F. Dolan, C. J. Crannell, K. J. Frost, L. E. Orwig, N. E. White.
Bull. American Astron. Soc., Vol. 12, 513 (1980). – Abstract.

142.032 **Constraints on the properties of the emission regions of X-ray bursters.** H. L. Marshall.
Bull. American Astron. Soc., Vol. 12, 513 (1980). – Abstract.

142.033 **Locations of unidentified X-ray sources with the HEAO-1 scanning modulation collimator.**
M. Garcia, M. Conroy, R. Doxsey, R. E. Griffiths, M. Johnston, E. Ralph, W. Roberts, D. A. Schwartz.
Bull. American Astron. Soc., Vol. 12, 527 (1980). – Abstract.

142.034 **The diffuse 2-60 keV background.**
D. C. Iwan, E. A. Boldt.
Bull. American Astron. Soc., Vol. 12, 543 (1980). – Abstract.

142.035 **Log N-Log S of extragalactic X-ray sources from the HEAO 1 A2 All Sky Survey.**
G. Piccinotti, R. F. Mushotzky, E. A. Boldt.
Bull. American Astron. Soc., Vol. 12, 543 (1980). – Abstract.

142.036 **Fluctuations in the diffuse X-ray background with HEAO-1: simple models and their comparison with resolved sources.** R. A. Shafer, E. A. Boldt, G. Piccinotti.
Bull. American Astron. Soc., Vol. 12, 543 - 544 (1980). Abstract.

142.037 **Two X-ray pulsars: 2S 1145–619 and 1E 1145.1–6141.** R. C. Lamb, T. H. Markert, R. C. Hartman, D. J. Thompson, G. F. Bignami.
Astrophys. J., Vol. 239, 651 - 654 (1980).

Observations from the Einstein Observatory reveal a previously unreported source, 1E 1145.1–6141, within 15′ of 2S 1145–619 and of comparable intensity during 1979 July. Periodicity analysis of the data shows a 290 ± 2 s period for the 2S source and a 298 ± 4 s period for the 1E source, confirming the previous Ariel 5 report of two periods in this range from this region of the sky.

142.038 **The X-ray pulsars 4U 1145–61 and 1E 1145.1–6141.**
N. E. White, S. H. Pravdo, R. H. Becker, E. A. Boldt, S. S. Holt, P. J. Serlemitsos.
Astrophys. J., Vol. 239, 655 - 660 (1980).

Observations of the X-ray sources near 4U 1145–61, with three separate experiments spanning 4 years, are presented. The authors confirm the presence of two periodicities at 291 and 297 s. The spectrum of 4U 1145–61 is a power law out to at least 60 keV with a 500 eV EW iron line at 6.7 keV. The inferred spectrum of the other is absorbed. The authors find no evidence for a monotonic variation in either pulse period ($\dot{P}/P < 10^{-4}$ yr^{-1}) over 4 years.

142.039 **The hard X-ray periodicity of GX 1+4.**
J-W. C. Koo, R. C. Haymes.
Astrophys. J., Lett., Vol. 239, L57 - L60 (1980).

Pulses that recur with a 4.27 ± 0.01 minute period were detected on 1974 April 2 in the 20–64 keV X-radiation from the galactic center region. Apparently emanating from GX 1+4, a major contributor to the high-energy radiation from that region, the pulses were 1.25 ± 0.42 minutes in duration. If it is in fact the same source that has been observed on all occasions, the period is twice that previously believed.

142.040 **A survey of soft X-ray emission from hot stars.**
K. S. Long, R. L. White.
Astrophys. J., Lett., Vol. 239, L65 - L68 (1980).

Sixteen O and B stars were observed with the Einstein X-ray observatory; X-ray emission from all seven O stars and four of the B stars was discovered. The 0.15–4.5 keV luminosities range from $\sim 10^{28}$ to $\sim 2 \times 10^{32}$ ergs s^{-1} and constitute $10^{-6} - 10^{-8}$ of the total luminosities of the stars detected. Most of the photons emerge at energies less than 1 keV, in contradiction to the prediction made by Cassinelli and Olson based on a hot corona-cool wind model.

142.041 **X-ray observations of Algol.**
N. E. White, S. S. Holt, R. H. Becker, E. A. Boldt, P. J. Serlemitsos.
Astrophys. J., Lett., Vol. 239, L69 - L71 (1980).

The authors have used the Solid State Spectrometer on Einstein to study Algol. Two observations six months apart were made, both including a primary optical eclipse. No corresponding X-ray eclipses were seen. During the second observation the source was flaring and was on the average a factor of 3 brighter. The spectrum on both occasions was consistent with a two-component thermal equilibrium model with temperatures of ~7.5 and 40 million degrees. Attempts to insert a third component indicate the temperature distribution to be bimodal. They discuss models for the X-ray emission and suggest that it most likely originates from an active corona surrounding the K star.

142.042 **Implications of the high-state iron-line feature and soft X-ray emission of Hercules X-1.** T. Bai.
Astrophys. J., Vol. 239, 999 - 1009 (1980).

The interpretation that the soft X-ray emission of Her X-1 is due to reprocessing of the primary hard X-rays intercepted by a plasma layer in the Alfven surface imposes quite a stringent – but not an unreasonable – constraint on its parameters. By using the Alfven shell parameters constrained by this interpretation, the degree of ionization of the Alfvén shell is calculated. The author presents Monte Carlo calculations of the iron K fluorescence efficiency for various models. From the spectral shape of the hard X-rays he discusses the matter distribution in the Alfven surface and the radiation pattern of the hard X-ray emission. Using his results and the observed equivalent width of the iron line he derives information on the iron abundance. Related works are discussed and the summary and conclusions are provided.

142.043 **Hard X-ray pulses from 4U 0900−40.**
R. Staubert, E. Kendziorra, W. Pietsch, C. Reppin, J. Trümper. W. Voges.
Astrophys. J., Vol. 239, 1010 - 1016 (1980).

The 283 s pulsation of 4U 0900−40 (Vela X-1) has been observed at high X-ray energies ($\gtrsim$ 18 keV). Time-averaged light curves with simple double peak structures are detected up to about 70 keV. Individual pulse trains could be observed in the energy range 18−50 keV. Significant pulse-to-pulse variations have been found with respect to pulse intensity. The photon number spectrum can be described by a thermal bremsstrahlung law which has $kT \approx 10$ keV for energies above 30 keV but shows a cutoff below this energy. The spectral shape is stable and independent of source intensity during the authors' observations.

142.044 **Discovery of a compact X-ray source at the center of the supernova remnant RCW 103.**
I. Tuohy, G. Garmire.
Astrophys. J., Lett., Vol. 239, L107 - L110, plate L6 - L7 (1980).

A point source of X-ray emission has been detected at the center of the supernova remnant RCW 103 using the *Einstein* Observatory. The 10″ radius error circle is centered at $\alpha = 16^h 13^m 47^s.8$, $\delta = -50°55'05''$ (1950) and contains two candidate stars of approximately 20 th magnitude. No coincident point source of radio emission is known, but the authors point out that the X-ray object lies 1°.2 away from the centroid position of a γ-ray source, CG 333 + 0. The observed flux from the object at the Earth is $\sim 7 \times 10^{-13}$ ergs cm^{-2} s^{-1} in the 0.6−2 keV band.

142.045 **On the diffuse cosmic ultraviolet background measured from Aries A-8.**
S. D. Price, T. L. Murdock, A. McIntyre, R. E. Huffman, D. E. Paulsen.
Astrophys. J., Lett., Vol. 240, L1 - L2 (1980).

The time- and altitude-dependent UV photometry measured by the diffuse UV experiment of Anderson et al. is attributed to the emission from the rocket exhaust due to combustion of residual fuel after burnout of the motor.

142.046 **The hard X-ray pulse profile of GX 1+4.**
M. S. Strickman, W. N. Johnson, J. D. Kurfess.
Astrophys. J., Lett., Vol. 240, L21 - L25 (1980).

The authors report results from 1976 and 1977 observations of the galactic center X-ray source GX 1+4. On these two occasions, epoch folding at twice the previously reported period of ~2 minutes has resulted in an asymmetry between the shapes of the two resultant peaks in the light curve, indicating that the true period may be ~4 minutes. Mean intensity is shown to vary from pulse to pulse.

142.047 **Repeatable, multiple-peaked structure in Type I X-ray bursts.** J. A. Hoffman, L. Cominsky, W. H. G. Lewin.
Astrophys. J., Lett., Vol. 240, L27 - L31 (1980).

A distinct, identifiable, multiple-peaked structure has been observed in Type I X-ray bursts from three sources. At energies below 6 keV, the light curves look like typical Type I bursts. At higher energies, the burst is double-peaked, with both the depth of the dip and the separation between the peaks increasing with energy. A light curve of the energy-integrated intensity shows no distinct double peak, suggesting that only a single energy release occurs. Blackbody fits to the evolving bursts spectra yield changing radii and temperatures, inversely correlated, during the early part of the burst. The physical interpretation of these changes is uncertain. Burst decay spectra yield relatively constant radii with decreasing temperatures. The authors propose that Compton scattering may be responsible for the dips in the higher-energy light curves by shifting photons to lower energies.

142.048 **Observation of microbursts of hard X radiation by SIGNE-2MP aboard the Prognoz 6 station.**
I. V. Ehstulin.
Pis'ma Astron. Zh., Tom 6, 567 - 570 (1980). In Russian. English translation in Soviet Astron. Lett., Vol. 6.

Pulses of hard X radiation with energy release at the earth of 4×10^{-7} erg/cm^2 were discovered from the Crab nebula.

142.049 **An optical burst from the star identified with the X-ray source 2S1254−690.**
K. O. Mason, J. Middleditch, J. E. Nelson, N. E. White.
Nature, Vol. 287, 516 - 518 (1980).

High time-resolution photometry of the proposed optical counterpart of the X-ray source 2S1254−690 is reported. An event of high statistical significance was observed from the star that has all the temporal characteristics of an X-ray burst. The authors also derive the X-ray spectrum of 2S1254−690 using data obtained with the HEAO 1 satellite.

142.050 **X-ray observations of 4U/MXB 1735−44.**
W. H. G. Lewin, J. van Paradijs, L. Cominsky, S. Holzner.
Mon. Not. R. Astron. Soc., Vol. 193, 15 - 28 (1980).

MXB 1735−44 was observed with the *SAS-3* X-ray observatory on five occasions in 1977 and 1978. It was always 'burst active'. Fifty-three bursts were observed in a total of about 20 days. The intervals between bursts were highly irregular, varying from ~2 to more than 50 hr. The distributions of the burst sizes in terms of their integrated burst flux E_b and maximum burst flux F_{max} both have standard deviations (normalized to the mean value) of ~37 per cent. Taking the smallest and largest burst, the total range in both E_b and F_{max} is about a factor of 7. However, the ratio E_b/F_{max} is approximately constant.

142.051 **Determination of a binary period for the variable X-ray source A1907+09.**
N. Marshall, M. J. Ricketts.
Mon. Not. R. Astron. Soc., Vol. 193, 7P - 13P (1980).

An analysis of observations of A1907+09 made between 1974 November and 1980 February has revealed a binary period of 8.380±0.002 day. The light curve supports the proposition that A1907+09 is a member of the class of X-ray binaries containing a massive optical component.

142.052 **2S1417-624: a variable galactic X-ray source near CG312-1.**
K. M. V. Apparao, S. Naranan, R. L. Kelley, H. V. Bradt.
Astron. Astrophys., Vol. 89, 249 - 250 (1980).

The authors report the precise (30″) position of an X-ray source, probably 4U1416-62, as determined from an observation with the rotation modulation collimators of SAS-3. The data show variations of ~30% on time scales of a few days and possibly 1−2 h. An estimate of the energy spectrum shows it to be rather flat in the 2−11 keV range. The source lies in the region of the gamma-ray source CG312-1.

142.053 **Galactic and extragalactic contributions to the far-ultraviolet background.**
F. Paresce, C. F. McKee, S. Bowyer.
Astrophys. J., Vol. 240, 387 - 400 (1980).

The authors have analyzed far-UV data by Paresce et al. (1979) to extract possible stellar and airglow contamination that could have influenced earlier conclusions. In this paper they present the results of this analysis and discuss their impact on the physics of the interstellar and intergalactic medium and the prospects for a measurement of the extragalactic background.

142.054 **Observations of outbursts from the recurrent X-ray transient A0538 − 66 and LMC X-4.** G. K. Skinner, S. Shulman, G. Share, W. D. Evans, D. McNutt, J. Meekins,

H. Smathers, K. Wood, D. Yentis, E. T. Byram, T. A. Chubb, H. Friedman.
Astrophys. J., Vol. 240, 619 - 627 (1980).

The authors report observations made with the NRL Large-Area Sky Survey (LASS) instrument on HEAO 1 of a number of further flares and outbursts from the region around A0536 – 66 and LMC X-4. From A0538 – 66 they have observed four further outbursts conforming to the 16.66 day periodicity. Two of these differ significantly in character from those reported previously. Other flares are shown to originate not in A0538 – 66 but from the vicinity of LMC X-4.

142.055 **4U 1626–67 and the character of highly compact binary X-ray sources.** F. K. Li, P. C. Joss, J. E. McClintock, S. Rappaport, E. L. Wright.
Astrophys. J., Vol. 240, 628 - 635 (1980).

The authors present the results of new observations of the 7.7 s X-ray pulsar 4U 1626–67 with *SAS 3*. They confirm the presence of quasi-periodic oscillations in the X-ray intensity and demonstrate that these oscillations have a preferred time scale of ~ 1×10^3 s but are not strictly periodic. The authors also place stringent new upper limits on orbital motion of the X-ray star for orbital periods between 10 s and 7 hr.

142.056 **Optical identification of H0123+075 and 4U 1137–65: hard X-ray emission from RS Canum Venaticorum systems.** M. Garcia, S. L. Baliunas, M. Conroy, M. D. Johnston, E. Ralph, W. Roberts, D. A. Schwartz, J. Tonry.
Astrophys. J., Lett., Vol. 240, L107 - L110, plates L4, L5 (1980).

The authors identify the X-ray transient H0123+075 with HD 8357, and the *Uhuru* source 4U 1137 – 65 with HD 101379. The identifications are based on precise positions obtained with the *HEAO 1* scanning modulation collimator. Optical studies of both stars indicate they are members of the RS CVn class.

142.057 **Balloon observations of several southern X-ray sources.** J. G. Greenhill, M. L. Duldig, M. W. Emery, A. G. Fenton, K. B. Fenton, R. M. Thomas, D. J. Watts.
Proc. Astron. Soc. Australia, Vol. 3, 349 - 350 (1979).

142.058 **On the binary nature of X-ray bursters.** R. Hoshi.
Prog. Theor. Phys., Vol. 63, 1442 - 1443 (1980)
Abstr. in Phys. Abstr., Vol. 83, Abstr. 82548 (1980).

142.059 **Hercules X-1 hard X-ray pulsations observed from *HEAO 1*.** D. E. Gruber, J. L. Matteson, P. L. Nolan, F. K. Knight, W. A. Baity, R. E. Rothschild, L. E. Peterson, J. A. Hoffman, A. Scheepmaker, W. A. Wheaton, F. A. Primini, A. M. Levine, W. H. G. Lewin.
Astrophys. J., Lett., Vol. 240, L127 - L131 (1980).

The 1.24 s pulsations of Her X-1in the energy range 13–75 keV have been analyzed in data obtained on three dates in 1978 February and three in 1978 August. Observational results are (1) the main pulse broadens somewhat with increasing energy; (2) the pulsation light curve undergoes pronounced changes at the leading edge of the main pulse from day to day; (3) spectral hardening within the main pulse is confirmed; (4) a 40–60 keV spectral feature in the spectrum is confirmed; (5) this feature is resolved and its centroid varies with pulsation phase; and (6) the 13–75 keV spectrum does not noticeably vary from day to day, except for an overall intensity factor. Some implications of these results for the prevailing models of Her X-1 and the HZ Her-Her X-1system are briefly discussed.

142.060 **33 second X-ray pulsations in AE Aquarii.** J. Patterson, D. Branch, G. Chincarini, E. L. Robinson.
Astrophys. J., Lett., Vol. 240, L133 - L136 (1980).

The authors report the discovery of 33 s pulsations in the 0.1–4.0 keV light curve of the nova-like variable AE Aquarii. These pulsations agree in period and phase with the optical pulsations. The periodicity probably originated from an accretion-induced hot spot on a rapidly rotating, magnetized white dwarf. It is possible that transient pulsations at nearby periods, similar to those seen in the optical light curve, are also present in the X-ray light curve.

142.061 **Discovery of a large X-ray burst from an X-ray nova, Centaurus X-4.** M. Matsuoka, H. Inoue, K. Koyama, K. Makishima, T. Murakami, M. Oda, Y. Ogawara, T. Ohashi, N. Shibazaki, Y. Tanaka, I. Kondo, S. Hayakawa, H. Kunieda, F. Makino, K. Masai, F. Nagase, Y. Tawara, S. Miyamoto, H. Tsunemi, K. Yamashita.
Astrophys. J., Lett., Vol. 240, L137 - L141 (1980).

The X-ray satellite Hakucho detected an X-ray burst from the nova-like transient Cen X-4 on 1979 May 31. The time profile and the spectral softening of the burst show the same properties as those of a type I X-ray burst. The peak intensity of ~ 25 times the Crab in the energy range 1.5–12 keV is the strongest X-ray burst observed so far. The burst occured during a nova-like outburst which is similar to one observed in 1969. The present discovery suggests that a soft transient source such as Cen X-4 comprises a neutron star and a late dwarf and can be an X-ray burster.

142.062 **Properties of X-ray bursts from the X-ray transient 1608–522.** T. Murakami, H. Inoue, K. Koyama, K. Makishima, M. Matsuoka, M. Oda, Y. Ogawara, T. Ohashi, N. Shibazaki, Y. Tanaka, Y. Tawara, S. Hayakawa, H. Kunieda, F. Makino, K. Masai, F. Nagase, S. Miyamoto, H. Tsunemi, K. Yamashita, I. Kondo.
Astrophys. J., Lett., Vol. 240, L143 - L147 (1980).

The recurrent X-ray transient 1608–522 was observed from the X-ray astronomy satellite Hakucho through the high- and low-luminosity states of its persistent flux. 1608–522 is identified to be a burst source from which 22 X-ray bursts were recorded. The burst peak intensity is found to fluctuate by a factor as large as 7. 1608–522 exhibited two distinctly different burst modes with respect to the burst profile and the peak luminosity distribution. The burst mode seems to have changed in correlation with the persistent flux, whereas the burst frequency as well as the time-averaged burst luminosity were essentially constant despite a large change in the persistent flux.

142.063 **On the nature of the faint ($B \simeq 20$) ultraviolet excess objects and the problem of the X-ray background.** F. Bònoli, A. Braccesi, B. Marano, R. Merighi, V. Zitelli.
Astron. Astrophys., Vol. 90, L10 - L12 (1980).

A considerable fraction of the ultraviolet excess objects of the Braccesi et al. (1970) list and of the $13^h+36°$ very faint sample have shown extended images on an excellent IIIa–J, Palomar 48″ plate. The authors show that, at the fainter magnitudes ($B \simeq 20$), the number of resolved objects conflicts with the *z* distribution of QSOs derived from the assumption of simple density evolution. By dismissing the extended objects, the surface density fits the predictions of a luminosity evolution model. On this basis the authors outline a picture of the QSO contribution to the X-ray background in fair agreement with observations.

142.064 **Spectra and pulse formation mechanism in X-ray pulsars: application to Her X-1.** R. Z. Yahel.
Astron. Astrophys., Vol. 90, 26 - 33 (1980).

Results of detailed radiative transport calculations of X-ray pulsars models have been compared to the spectra and light curves of Her X-1. Attempts to reconcile model calculations and observations lead to several conclusions.

142.065 **A discussion of the eccentric binary hypothesis for transient X-ray sources. II. Gradual acceleration stellar wind model.** Y. Avni, I. Goldman.
Astron. Astrophys., Vol. 90, 44 - 47 (1980).

The authors examine the eccentric binary hypothesis for transient X-ray sources in the framework of the gradual acceleration stellar wind model proposed by Barlow and Cohen. They find that a consideration of the ratio of maximum to minimum luminosities and of the ratio of the durations of the high and low states, for a typical transient X-ray source, yields a rather high eccentricity, despite the gradual acceleration of the wind. When typical physical parameters for the binary members are taken into account, they find that a consistent description is possible only for very eccentric orbits ($e \geqslant 0.9$), thus the model is inadequate as a general explanation of the X-ray transient phenomenon. We study the recurrent transient X-ray source 4U 1630–47, which was considered in the past to be a realization of the eccentric binary model, and demonstrate that it cannot be described consistently within the framework of the model, unless the optical primary is very peculiar.

142.066 ***UBV* photometry of HZ Herculis: the shape of the primary minimum.**
R. Kippenhahn, H. U. Schmidt, H.-C. Thomas.
Astron. Astrophys., Vol. 90, 54 - 57 (1980).

Five nights of photoelectric observations in *UBV* are discussed and compared with model predictions. It is shown that a total eclipse phase exists, which is shorter than the X-ray eclipse and that a marked asymmetry is present near primary minimum connected with a rather hot source of radiation.

142.067 **Optical spectroscopy of Centaurus X-3.**
M. Mouchet, S. A. Ilovaisky, C. Chevalier.
Astron. Astrophys., Vol. 90, 113 - 115 (1980).

Analysis of ten 62 Å/mm electronographic spectra of Cen X-3 (4000-5000 Å) obtained over a five-day interval in March 1976 reveals periodic radial velocity variations in the He II λ 4686 emission line, with a semi-amplitude of 400 km s^{-1}, in anti-correlation with the lower-amplitude radial velocity variations of the Balmer, He I and He II absorption lines. This suggests that during at least two orbital cycles the He II emission line was produced near the X-ray source, probably in an accretion disk.

142.068 **An extraordinary new celestial X-ray source.**
P. C. Gregory, G. G. Fahlman.
Nature, Vol. 287, 805 - 806 (1980).

The discovery of an extraordinary new X-ray source detected with the Einstein X-ray Observatory is reported. The object, designated G109.1–1.0, was found using the imaging proportional counter detector on 17 December 1979. A picture shows a clear semicircular arc with an angular diameter of 36 arc min which appears to be the outer shell of the SNR. At the exact centre of the curvature of the shell is a strong compact source designated GF2259+586.

142.069 **Diffuse X-ray emission from the jets of SS433.**
F. Seward, J. Grindlay, E. Seaquist, W. Gilmore.
Nature, Vol. 287, 806 - 808 (1980).

The authors report the detection of large-scale X-ray 'jets' from SS433. The X-ray emission is diffuse, extending at least 30 arc min from SS433, and is exactly aligned with both SS433 and the bulges of the shell of the huge supernova remnant (SNR) W50. This detection now (1) directly confirms the existence of jets related to SS433, (2) shows the link between SS433 and W50, proving that SS433 is galactic, (3) establishes a minimum age of the jet phenomenon of $\sim 10^3$ yr, and (4) offers an explanation of why W50 is so much larger than any other known SNR in at least one dimension.

142.070 **The diffuse UV background.**
F. Paresce, P. Jakobsen.
Nature, Vol. 288, 119 - 126 (1980).

The diffuse radiation field in the UV (900–3,000 Å) affects the structure of galactic molecular clouds and conveys important information concerning the physical characteristics and spatial distribution of gas and dust in the Universe. Continuum emission in this range is probably dominated by interstellar dust scattering in our Galaxy.

142.071 **Observations of optical flares in the recurrent X-ray transient A0538–66.** G. K. Skinner.
Nature, Vol. 288, 141 - 143 (1980).

Observation of the n = 3, 4, 11 outbursts recently led to an improved error box and the proposal of a $B \sim 15$ optical counterpart shown on archival plates to be variable. The author reports an examination of 85 plates between 1974 and 1979. Dramatic flares observed on three of the plates, separated by intervals of 16 days and synchronized with the X-ray outbursts, confirm the proposed identification.

142.072 **Far-ultraviolet observations of Cygnus X-2.**
L. Maraschi, E. G. Tanzi, A. Treves.
Astrophys. J., Lett., Vol. 241, L23 - L26 (1980).

Two far-ultraviolet (1150-1950 Å) spectra of V 1341 Cyg ≡ Cyg X-2, taken with the International Ultraviolet Explorer, at orbital phase 0.3 and 0.6 show an intensity variation of a factor of 2. Strong emission lines (NV, Si IV, He II) are apparent on a poorly defined continuum. The observations can be interpreted in terms of UV emission by the X-ray heated atmosphere of the non-collapsed star in the binary system. A neutron star rather than a white dwarf is indicated as the compact companion.

142.073 **The optical identification of H2252–035 with a cataclysmic variable.** R. E. Griffiths, D. Q. Lamb, M. J. Ward, A. S. Wilson, P. A. Charles, J. Thorstensen, I. M. McHardy, A. Lawrence.
Mon. Not. R. Astron. Soc., Vol. 193, 25P - 33P (1980).

The authors present an improved position for the X-ray source H2252 – 033 using the Ariel V Sky Survey Instrument, and precise positioning using the Scanning Modulation Collimator on HEAO-1. They identify the source optically with a 13th magnitude emission-line object, which is probably a white dwarf binary with a well-formed accretion disc, with similarities to the dwarf novae (e.g. SS Cyg) at minimum.

142.074 **A long-term radio study of Hercules X-1.**
M. J. Coe, P. C. Crane.
Mon. Not. R. Astron. Soc., Vol. 193, 35P - 37P (1980).

The pulsating X-ray binary system Hercules X-1 has been studied at 4.76 GHz for 35 days. At no time was any emission detected and an upper limit of 2mJy (5σ) was found for any emission in phase with the 35-day X-ray cycle. This result, combined with other observations, suggests that pulsating X-ray binary systems, unlike non-pulsating systems, are probably not radio sources.

142.075 **The periodic component of high-energy gamma-radiation of the source Cyg X-3.**
Yu. I. Neshpor, Yu. L. Zyskin, B. M. Vladimirskij, S. A. Gerasimov, A. A. Stepanyan, V. P. Fomin.
Izv. Krymskoj Astrofiz. Obs., Tom 61, 61 - 66 (1980). In Russian.

The period of high-energy γ-ray emission of Cyg X-3 was found on observational data during 6 years (1972 - 1977) to be $0^d.199683 \pm 0^d.000001$. The light curve of the source has two narrow peaks. The flux value averaged over all years of the measurements is equal to 1.8×10^{-10} quanta $cm^{-2}\,s^{-1}$ (for the peak).

142.076 **On the radio emission of the X-ray source Cyg X-3.** Yu. I. Neshpor.
Izv. Krymskoj Astrofiz. Obs., Tom 61, 67 - 69 (1980). In Russian.

It is evident that the absence of the periodic component of the radio emission $T = 4^h\!.8$ at the frequency 8.1 GHz may be caused by scattering of radio waves due to the electron density inhomogeneities in the plasma cloud around the source. If one assumes this explanation it is possible to observe this periodic component of the radio emission at the frequency of more than 16 GHz.

142.077 **Two successive X-ray bursts at ten-minute interval from 1608–522.** T. Murakami, H. Inoue, K. Koyama, K. Makishima, M. Matsuoka, M. Oda, Y. Ogawara, T. Ohashi, N. Shibazaki, Y. Tanaka, S. Hayakawa, H. Kunieda, F. Makino, K. Masai, F. Nagase, Y. Tawara, S. Miyamoto, H. Tsunemi, K. Yamashita, I. Kondo.
Publ. Astron. Soc. Japan, Vol. 32, 543 - 546 (1980).

Two successive X-ray bursts from 1608–522 at an interval of ten minutes were observed by the Japanese X-ray satellite Hakucho. According to the current model of nuclear flash for the bursts, this interval is much shorter than expected from the observed upper limit of the fuel supply rate, if one assumes that the nuclear fuel available is consumed totally in every burst.

142.078 **A critique of the polarimetric evidence on the nature of Cygnus X-1.**
J. F. L. Simmons, C. Aspin, J. C. Brown.
Astron. Astrophys., Vol. 91, 97 - 107 (1980).

The effect of data noise on the evaluation of parameters and on the testing of models for binary star systems from their linear polarimetric variability is investigated and applied to Cyg X-1. In particular an analytic method is devised for finding the constrained optimum fit to the data, with a χ^2 measure of acceptability, over the parameters characterising the canonical model for polarimetric binaries (i.e. a corotating optically thin Thomson scattering envelope with no variable obscuration effects and with an arbitrary density distribution).

142.079 **On the phase-locked polarization variations in Cygnus X-1.** J. C. Kemp.
Astron. Astrophys., Vol. 91, 108 - 111 (1980).

The present situation with the accumulated *V*-band orbital phase curves of the polarization is reviewed, especially with respect to an analysis of some of those data by Simmons, Aspin, and Brown. Contrary to them, the author does not find the curves to conform to the so-called canonical model, which allows no obscuration or shadowing effects.

142.080 **Ultraviolet emission from strong X-ray sources as observed with IUE.** H. Gursky, R. Davis.
Highlights of Astronomy, Vol. 5, (see 012.024), 231 - 245 (1980).

Observations in the ultraviolet of the strong X-ray sources have provided some valuable insights into the nature of these systems. The authors discuss the low mass binaries (e.g. HZ Her), the high mass binaries (e.g. Cyg X-1) and the globular clusters.

142.081 **Mass loss and stellar wind in massive X-ray binaries.** H. F. Henrichs.
Highlights of Astronomy, Vol. 5, (see 012.027), 541 - 547 (1980).

The author derives an upper limit to mass loss rates of unevolved early type stars by studying X-ray pulsars. He considers theoretical predictions concerning the influence of X-rays on the stellar wind and compares these with the observations. Using new data from IUE, he draws some conclusions about mass loss rates and velocity laws as derived from X-ray binaries.

142.082 **Implications of results from the Einstein Observatory for the X-ray background.**
P. Gorenstein, D. A. Schwartz.
Highlights of Astronomy, Vol. 5, (see 012.028), 763 - 770 (1980).

The final result is that the Einstein Deep Survey and quasar investigations converge to the conclusion that the X-ray background is composed principally of unresolved sources and that they are probably QSO's.

142.083 **X-ray observations of stellar coronae and winds.** J. H. Swank.
Highlights of Astronomy, Vol. 5, (see 012.031), 853 - 856 (1980).

142.084 **Kα-lines in the X-ray background spectrum and interstellar gas in galaxies.**
L. A. Vajnshtejn, R. A. Syunyaev.
Pis'ma Astron. Zh., Tom 6, 673 - 679 (1980). In Russian. English translation in Soviet Astron. Lett., Vol. 6.

Atoms and ions of heavy elements in the interstellar space absorb X-ray background radiation and radiate Kα-line photons. Therefore Kα-lines of iron, sulphur and silicon must be observed in the X-ray background spectrum in the direction to the plane of the Galaxy. Their intensities are estimated.

142.085 **The X-ray light curve of 2A0311 - 227.** G. Williams, J. Patterson, W. A. Hiltner.
Bull. American Astron. Soc., Vol. 12, 748 - 749 (1980). Abstract.

142.086 **On the origin and persistence of long-period pulsating X-ray sources.**
R. F. Elsner, P. Ghosh, F. K. Lamb.
Bull. American Astron. Soc., Vol. 12, 749 (1980). – Abstract.

142.087 **The 1979 X-ray outburst of Centaurus X-4.** L. J. Kaluzienski, S. S. Holt, J. H. Swank.
Astrophys. J., Vol. 241, 779 - 786 (1980).

In late spring 1979 the classical transient X-ray source Cen X-4 underwent its first major outburst since its initial discovery in 1969. The flare light curve exhibits a double-peaked maximum at a level of ~4 times the Crab nebula, and its duration and characteristic decay time scale are the shortest yet observed from the class of soft X-ray transients. The authors estimate a total X-ray output of $\sim 3 \times 10^{43}$ ergs, a factor of ~20 less than that of the 1969 outburst. In addition, evidence is found for a regular modulation of the flux during the decline phase at a period of 8.2±0.2 hours. The existing data are consistent with a source model involving episodic mass exchange from a late-type dwarf onto a neutron star companion in a relatively close binary system.

142.088 **The very long type II X-ray bursts from the Rapid Burster.** E. M. Basinska, W. H. G. Lewin, L. Cominsky, J. van Paradijs, F. J. Marshall.
Astrophys. J., Vol. 241, 787 - 792 (1980).

The authors have observed two very long type II bursts (>200 s) from the Rapid Burster (MXB 1730–335) on 1979 March 3 UT. They present the results of the analysis of the bursts.

142.089 **HR 976 and 4C 34.13: an X-ray odd couple.** W. Cash, T. P. Snow, Jr.
Astron. Astrophys., Vol. 91, L7 - L8 (1980).

The single-lined spectroscopic binary Am star (HR 976), reported to be an X-ray source, was observed by the Imaging Proportional Counter on the Einstein Observatory. Two sources with similar X-ray flux were detected in the IPC field: HR 976 at 2.6×10^{-13} erg cm^{-2} s^{-1} and 4C34.13, a quasar, at 2.4×10^{-13} erg cm^{-2} s^{-1}. The soft X-ray flux from HR 976 is a factor of one hundred less than the previous report, and its total X-ray luminosity is about 8×10^{28} erg s^{-1}. The quasar, which is a much harder X-ray source than the A star, has a total luminosity estimated at 8×10^{44} erg s^{-1}.

142.090 **Variable linear polarization in the X-ray binary HD 77581.** T. Korhonen, V. Piirola.
Astron. Astrophys., Vol. 91, 372 - 373 (1980).

Observations are presented on linear polarization of the X-ray binary HD 77581 in the *UBV* passbands made during a period from October 31 to November 14, 1979. The degree of polarization shows a variation of about 0.2% and the position angle 1°.5. The variations are similar in each colour. No clear periodicity can be found. Possible explanations of the polarization variations are discussed.

142.091 **Cygnus X-1.**
IAU Circ., Nos. 3491, 3492, 3502, 3517, 3541 (1980).

142.092 **2A 0526-328.**
IAU Circ., No. 3492 (1980).

142.093 **X-ray bursts.**
IAU Circ., No. 3506 (1980).

142.094 **4U 0115+634.**
IAU Circ., Nos. 3510, 3543 (1980).

142.095 **H 2252-035.**
IAU Circ., Nos. 3511, 3514, 3519, 3525 (1980).

142.096 **MXB 1730-333.**
IAU Circ., No. 3520 (1980).

142.097 **Cygnus X-3.**
IAU Circ., Nos. 3522, 3543 (1980).

142.098 **X-ray nova (A0535+262=HDE 245770).**
IAU Circ., Nos. 3525, 3527, 3544 (1980).

142.099 **4U 1849-31 = V1223 Sagittarii.**
IAU Circ., No. 3529 (1980).

142.100 **1E 0643.0-1648**
IAU Circ., No. 3529 (1980).

142.101 **Optical candidate for LMC X-1.**
IAU Circ., No. 3543 (1980).

142.102 **1E 1145.1 - 6141.**
IAU Circ., No. 3544 (1980).

142.103 **Scorpius X-1.**
British Astron. Assoc. Circ., No. 609 (1980).

142.104 **Photographic observations of the possible counterpart of the X-ray source 1 E 0643.0-1648.**
B. Fuhrmann.
Inf. Bull. Variable Stars, No. 1883 (1980).

142.105 **X-ray spectroscopy of late-type stars.**
J. H. Swank, N. E. White.
Smithsonian Astrophys. Obs., Spec. Rep. 389, (see 012.038), p. 47 - 64 (1980).

142.106 **Super-Röntgenstrahlenhalo entdeckt.**
H. W. Köhler.
Phys. Bl., 36. Jahrg., 360 - 362 (1980).

142.107 **The discovery of 'Sco X-1 type' behaviour from the X-ray burster 4U 1735 – 44.**
N. E. White, P. A. Charles, J. R. Thorstensen.
Mon. Not. R. Astron. Soc., Vol. 193, 731 - 736 (1980).

A series of Copernicus, Ariel V and optical observations of the X-ray burst source 4U 1735 –44 (=MXB 1735 – 44) revealed properties very similar to those of the 'Sco X-1 like' sources. The non-burst X-ray flux varied by a factor of 2 on a time-scale of hours and this was associated with a general hardening of the spectrum. An Ariel V spectrum obtained during a quiescent interval deviated from a simple thermal model in that it showed both a high energy deficiency and a low energy excess.

142.108 **X-ray observations of the OAO1653-40 field.**
A. N. Parmar, G. Branduardi-Raymont, G. S. G. Pollard, P. W. Sanford, A. C. Fabian, G. C. Stewart, E. J. Schreier, R. S. Polidan, W. R. Oegerle, M. Locke.
Mon. Not. R. Astron. Soc., Vol. 193, 49P - 53P (1980).

The authors have observed the region of sky containing OAO1653-40 simultaneously with the Einstein and Copernicus satellites in 1979 August. They obtain a position and a period of 38.019±0.009 s for the X-ray pulsar discovered by White & Pravdo. The X-ray spectrum shows considerable photoelectric absorption with an equivalent hydrogen column density of 2.3×10^{23} atom cm^{-2}. The authors also obtain an upper limit to any 0.5–3.0 keV flux from V861 Sco of $(4 \pm 1) \times 10^{-13}$ erg cm^{-2} s^{-1}.

142.109 **On observations of the cosmic radiation background.**
R. A. Matzner.
Astrophys. J., Vol. 241, 851 - 857 (1980).

The author presents here models in which the anisotropies are true relics of the very early history of the universe, and in which the microwave dipole observations in particular are the result of amplification mechanisms in essentially homogeneous models. One of the models considered is an open, slightly anisotropic universe with very small density. It predicts – apparently consistently with observations – quadrupole X-ray anisotropy together with dipole microwave anisotropy.

142.110 **AM Herculis: simultaneous X-ray, optical, and near-IR coverage.** P. Szkody, F. A. Córdova, I. R. Tuohy, H. S. Stockman, J. R. P. Angel, W. Wisniewski.
Astrophys. J., Vol. 241, 1070 - 1076 (1980).

A 6 hour X-ray pointing at AM Her using the *HEAO 1* satellite is correlated with simultaneous broad-band *V* and *I* photometry and visual circular polarimetry. The absence of correlations on either a flickering or an orbital time scale implies distinct regions for the visual and X-ray emission. Significant changes in the light curves are observed from one binary cycle to the next.

142.111 **Optical and X-ray studies of 2A 1822 - 371.**
P. Charles, J. R. Thorstensen, P. Barr.
Astrophys. J., Vol. 241, 1148 - 1152 (1980).

Optical spectroscopy in the λλ 3400 - 5000 band of the counterpart of 2A 1822 - 371 obtained at CTIO is presented. This very-ultraviolet object displays the C III/N III λλ 4640 - 4650 complex and He II λ 4686, together with two broad emission features in the UV at λλ 3765 and 3815. The authors discuss the identification of these features and suggest they are due to O III and He II. Their photographic plate material enables a refinement of the Seitzer et al. binary period to 5.5704 ± 0.0003 hr. The X-ray spectrum of this object obtained by Ariel 5 is best fitted by a 2.4 keV blackbody spectrum. No evidence for binary modulation, pulsations, or bursts was found in the X-ray data.

142.112 **On the origin and persistence of long-period pulsating X-ray sources.**
R. F. Elsner, P. Ghosh, F. K. Lamb.
Astrophys. J., Lett., Vol. 241, L155 - L159 (1980).

The authors argue that the available evidence strongly suggests that most of the bright neutron star X-ray sources, including those with massive companions, are disk fed and show that this hypothesis provides natural answers to these questions.

142.113 **Spectroscopic observations of the optical counterpart of Centaurus X-4.**
J. van Paradijs, F. Verbunt, T. van der Linden, H. Pedersen, W. Wamsteker.
Astrophys. J., Lett., Vol. 241, L161 - L164 (1980).
The authors present the results of spectroscopic observations of Cen X-4 object made 5 weeks after the source reached its maximum.

142.114 **X-ray bursters.** J. van Paradijs.
Variability in stars and galaxies, (see 012.044), p. GL.3.1.- 3.14 (1980).
The author presents a brief description of the main properties of X-ray bursts, and of the role they have played in understanding low-mass X-ray binaries. He gives an example which illustrates how observations of X-ray bursts can be used to obtain information about properties of neutron stars.

142.115 **X-ray observations of six BL Lacertae fields.**
D. Maccagni, M. Tarenghi.
ESO Sci. Prepr. No. 131, 17 pp. (1980). – Submitted to Astrophys. J.

142.116 **An upper limit on the flux density of radio bursts at 327 MHz from MXB 1730 - 335.**
A. P. Rao, V. R. Venugopal.
Bull. Astron. Soc. India, Vol. 8, 41 - 42 (1980).
The rapid X-ray burster MXB 1730 - 335 was observed on 13 and 14 August 1979. No radio bursts were recorded and the upper limit on the radio flux density from the burster at 327 MHz is estimated to be about 1 Jansky.

142.117 **X-ray burst sources in globular clusters and the galactic bulge.** W. H. G. Lewin.
Globular clusters, (see 003.008), p. 315 - 350 (1980).

142.118 **X-rays from globular clusters: steady emission.**
J. G. Jernigan.
Globular clusters, (see 003.008), p. 351 - 360 (1980).

142.119 **Einstein X-ray observations of Proxima Centauri and the surrounding region.**
B. M. Haisch, J. L. Linsky, F. R. Harnden, Jr., R. Rosner, F. D. Seward, G. S. Vaiana.
Astrophys. J., Lett., Vol. 242, L99 - L103, plate L8 (1980).
The authors report the first detection of both quiescent and flaring soft X-ray emission from a dMe flare star, Proxima Centauri (dM5e). The quiescent state is characterized by a mean X-ray luminosity of $\sim 1.5 \times 10^{27}$ ergs s^{-1}, corresponding to a mean surface flux of $\sim 7 \times 10^5$ ergs cm^{-2} s^{-1}, and an inferred temperature of $\sim 4 \times 10^6$ K. The flare the authors have detected has a peak flux of $\sim 7.4 \times 10^{27}$ ergs s^{-1} and a peak temperature of $\sim 17 \times 10^6$ K. The authors discuss implications of these data for models of the quiescent and flare coronae of dMe stars.

142.120 **Discovery of soft X-ray emission from V471 Tauri and UU Sagittae: two highly evolved, low-mass binaries.** D. Van Buren, P. A. Charles, K. O. Mason.
Astrophys. J., Lett., Vol. 242, L105 - L108 (1980).
The authors report here the discovery of soft X-ray emission from two highly evolved low-mass binaries using the HEAO 1 A-2 experiment. V471 Tau is a 5×10^{30} ergs s^{-1} source in the Hyades with a white dwarf and K V star as components. UU Sge is a planetary nebula nucleus containg an sd0-B star and a K V-M V companion with a soft X-ray luminosity of $\sim 10^{32}$ ergs s^{-1}.

142.121 **Discovery of soft X-ray emission from a BL Lacertae object.** Z. Urban.
Říše hvězd, Vol. 61, 206 - 207 (1980). In Czech.

142.122 **Radiological astronomy.** J. Svatoš.
Říše hvězd, Vol. 61, 245 - 246 (1980). In Czech.

142.123 **X-ray burst sources – a simple two zone model.**
J. R. Buchler, M. Barranco, M. Livio.
Space Sci. Rev., Vol. 27, (see 012.046), 585 - 590 (1980).
With the help of a very simple two zone model, the authors demonstrate the possibility of periodic thermal relaxation (limit cycle) oscillations in the helium burning envelope of accreting neutron stars. Physically reasonable model parameters can be chosen which yield agreement with the observed features of X-ray bursts and the authors suggest that this limit cycle is operative in neutron stars which have an accretion rate in a specific range.

142.124 **Results of observations of the X-ray source Cyg X-3 for the energy range > 10^{12} eV by the Tien-Shan installation for detection of Čerenkov flashes from extensive air showers.**
D. B. Mukanov, N. M. Nesterova, A. A. Stepanyan, V. P. Fomin.
Izv. Krymskoj Astrofiz. Obs., Tom 62, 98 - 102 (1980). In Russian.
The Tien-Shan installation for detection of Čerenkov radiation from cosmic-ray air showers has been described. The observational results of the X-ray source Cyg X-3 for energy more than 3×10^{12} eV which have been obtained during July - November 1977 are presented. The periodic component of γ-ray flux is explored.

142.125 **Activity of the X-ray source Cygnus X-3 at λ = 1.35 cm in 1975 - 1978.**
S. L. Domnin, V. A. Efanov, I. G. Moiseev, N. S. Nesterov.
Izv. Krymskoj Astrofiz. Obs., Tom 62, 103 - 107 (1980). In Russian.
The results of Cyg X-3 patrol observations at wavelength 1.35 cm obtained during 1975 - 1978 are discussed. Despite the presence of radio bursts with several minutes and days duration period the source activity has evidently diminished last years.

142.126 **A search for X-ray binary stars in their quiescent phase.** D. J. Helfand.
Publ. Astron. Soc. Pacific, Vol. 92, 691 - 695 (1980) = Columbia Astrophys. Lab. Contrib. No. 187.
Fourteen early-type stars representative of systems which may be harboring a neutron star companion and are thus potential progenitors of massive X-ray binaries have been examined for X-ray emission with the HEAO A-1 experiment. Limits on the 0.5–20 keV luminosity for these objects lie in the range 10^{31} to 10^{33} erg s^{-1}. In several cases, the hypothesis of a collapsed companion, in combination with the X-ray limit, places a serious constraint on the mass-loss rate of the primary star. In one instance, an X-ray source was discovered coincident with a candidate star, although the luminosity of 5×10^{31} erg s^{-1} is consistent with that expected from a single star of the same spectral type. The prospects for directly observing the quiescent phase of a binary X-ray source with the Einstein Observatory are discussed in the context of these results.

142.127 **Interpretation of the optical UBV and ultraviolet λ 1600 light curves of the Cyg X-1 system.**
N. I. Balog, A. V. Goncharskij, A. M. Cherepashchuk.
Astron. Tsirk., No. 1095, p. 1 - 4 (1980). In Russian.

142.128 **Possible identification of the X-ray source H 1926 + 503 with the variable star CH Cygni.**

L. Luud.
Astron. Tsirk., No. 1098, p. 3 - 4 (1980). In Russian.

142.129 **Compact X-ray sources.** D. Pines.
X-ray and gamma-ray astronomy in the 1980's, (see 012.054), p. 20 - 23 (1979).

142.130 **X-ray astronomy and cosmology.** M. Rees.
X-ray and gamma-ray astronomy in the 1980's, (see 012.054), p. 35 - 39 (1979).

142.131 **Description of ten investigations.**
X-ray and gamma-ray astronomy in the 1980's, (see 012.054), p. 42 - 54 (1979).

Contents: All-sky X-ray survey. LAMAR. A compact object observatory (COO). The investigation of transient X-ray phenomena. X-ray spectroscopy. X-ray polarimetry. Hard X-ray imaging. Hard X-ray spectroscopy. Gamma ray observatory IIa: spectroscopy. Gamma-ray observatory IIb: diffuse and point source studies.

142.132 **X-ray spectra of galactic X-ray sources.**
S. S. Holt.
X-ray astronomy, (see 012.055), p. 89 - 102 (1980).

Contents: Binary X-ray sources. Cataclysmic variables. Bulge sources. X-ray pulsars. Stars.

142.133 **Study of high luminosity X-ray sources in external galaxies (M 31).** R. Giacconi.
X-ray astronomy, (see 012.055), p. 103 - 113 (1980).

142.134 **X-ray sources and stellar evolution.**
E. P. J. van den Heuvel.
X-ray astronomy, (see 012.055), p. 115 - 127 (1980).

Contents: Types of strong galactic X-ray sources. The nature and origin of group I sources. Nature of the group II sources: the low-mass binary model. The rate of mass transfer in group II sources. The formation of the X-ray sources in globular clusters. The formation of the sources in the bulge.

142.135 **Radio properties of extragalactic X-ray sources.**
P. Katgert.
X-ray astronomy, (see 012.055), p. 253 - 272 (1980).

Contents: Relationships between X-ray and radio emission. Radio emission from clusters of galaxies. Radio emission from Seyfert galaxies. Radio emission from quasi-stellar objects. Radio observations of Einstein observatory deep survey areas. The time scale for radio source evolution. The redshift dependence of the radio luminosity function.

142.136 **Categories of extragalactic X-ray sources.**
M. J. Rees.
X-ray astronomy, (see 012.055), p. 363 - 375 (1980).

The paper is concerned with the various classes of extragalactic sources that may appear in X-ray surveys, and the implications for the X-ray background.

142.137 **The X-ray background as a probe of the matter distribution on very large scales.** M. J. Rees.
X-ray astronomy, (see 012.055), p. 377 - 384 (1980).

The X-ray background obviously holds important clues to the evolutionary history and spatial distribution of its sources, be they quasars, hot gas, or some as yet unknown population. The paper deals with just one aspect of this subject: the sensitivity of the X-ray background as a probe for density irregularities on scales larger than superclusters.

142.138 **Deep surveys with the Einstein observatory.**
R. Giacconi.
X-ray astronomy, (see 012.055), p. 385 - 398 (1980).

The detailed study of X-ray luminosities for stars spanning the entire range of stellar spectral types which is being carried out by Einstein, combined with the known distribution of stars in the galaxy will ultimately provide a great deal of information on the discrete source contributions to the soft background. Preliminary results of the surveys of Einstein are summarized in the paper.

142.139 **X-ray emission from spherical accretion onto white dwarfs in binary systems.**
V. A. Krol', P. I. Fomin.
Astrometr. Astrofiz., Vyp. (No.) 42, p. 55 - 63 (1980). In Russian.

The paper is concerned with studying X-ray emission from white dwarfs in close binary systems under accretion of matter from its ordinary star companion. An analytic expression for the temperature distribution in the accreting shell is obtained. The spectrum of thermal X-rays is obtained and studied. It is shown that this spectrum is in good agreement with the X-ray spectrum from discrete sources of the Sco X-1 type.

142.140 **The X-ray Universe.** K. A. Pounds.
The state of the Universe, (see 003.013), p. 93 - 120 (1980).

The nature of both the galactic and extragalactic X-ray sources are described.

142.141 **The soft X-ray diffuse background and the structure of the local interstellar medium.**
P. M. Fried, J. A. Nousek, W. T. Sanders, W. L. Kraushaar.
Astrophys. J., Vol. 242, 987 - 1004 (1980).

Data from five rocket flights of the Wisconsin soft X-ray sky survey, covering the southern galactic hemisphere with $\sim 6\overset{\circ}{.}5$ spatial resolution between 0.1 and 2.0 keV, are presented as intensity maps. Analysis of the data in two energy bands below 0.3 keV is presented here. No evidence is found for the presence of nearby small (radius ~0.6 pc) clouds of cool neutral gas, as has been proposed by McKee and Ostriker. The X-ray measurements are consistent with the small amounts of local neutral material inferred from $L\alpha$ absorption measurements.

142.142 **A search for pulsations and eclipses from X-ray burst sources.** L. Cominsky, J. G. Jernigan, W. Ossmann, J. Doty, J. van Paradijs, W. H. G. Lewin.
Astrophys. J., Vol. 242, 1102 - 1106 (1980).

SAS 3 data on the persistent emission from five X-ray burst sources has been analyzed for the presence of pulsations and eclipses. None were found; 90% confidence upper limits to the pulsed fraction in the persistent X-ray emission are given, and are typically a few percent for periods in the range 1.6–66 s and 5–10% for periods in the range 66–1000 s. The implications of these results for the standard model of a compact object in a binary system are discussed.

142.143 **Two fast soft X-ray transients at high galactic latitude.**
J. A. Nousek, F. A. Córdova, G. P. Garmire.
Astrophys. J., Vol. 242, 1107 - 1113 (1980).

Two soft X-ray sources appeared at a level 10 times the HEAO 1 A-2 experiment's scanning sensitivity for $\sim 10^4 - 10^5$ s, and then dropped below detectability. Both sources were discovered at high galactic latitude. The transient designated H0850+13 appeared on 1977 November 4. In three consecutive observations made within 6 hours a softening of the source spectrum was observed, consistent with decline of temperature of $\sim 2 \times 10^6$ K to $\sim 1 \times 10^6$ K. The transient designated H0336–26 appeared on 1978 February 3 and had a temperature no higher than 10^6 K. The 90% confidence error boxes for these sources contain no convincing optical counterparts. Assuming a hypothetical distance to the sources of 10 pc, the observed intensity implies their luminosities are

$\sim 10^{30}$ ergs s^{-1}. The authors estimate there may be ~ 100 such transients per year.

Cosmic X-ray astronomy.
See Abstr. 003.014.

Röntgenastronomie. See Abstr. 003.039.

Some new results in X-ray astronomy presented at the annual meeting of the High Energy Astrophysics Division of AAS. See Abstr. 011.002.

X-ray astronomy. See Abstr. 013.041.

A new statistical test with application to globular cluster X-ray source masses. See Abstr. 021.020.

X-ray astronomical spectroscopy.
See Abstr. 031.567.

The IHXR80 hard X-ray experiment.
See Abstr. 032.572.

Description of the EXUV, GRO and AXAF satellites.
See Abstr. 032.579.

The Einstein Observatory: New perspectives in astronomy. See Abstr. 051.002.

The extreme-ultraviolet and soft X-ray sky-survey project EXUV. See Abstr. 051.005.

Ruimte-onderzoek en sterrekunde (1). (UV-astronomie). See Abstr. 051.023.

Ruimte-onderzoek en sterrekunde (2).
See Abstr. 051.026.

Quantum effects in cyclotron plasma absorption.
See Abstr. 062.093.

The radiation of a hot magnetized plasma accreted by degenerate stars with a strong magnetic field.
See Abstr. 062.094.

Thermal and vacuum effects on the cyclotron process in X-ray pulsars. See Abstr. 063.018

Resonance radiative transfer for cyclotron line emission with recoil. See Abstr. 063.056.

Thermal effects on the cyclotron line formation process in X-ray pulsars. See Abstr. 063.057.

The structure of X-ray illuminated stellar atmospheres. See Abstr. 064.009.

Stability of accretion column flows.
See Abstr. 064.046.

X-ray emission from the winds of hot stars.
See Abstr. 064.048.

Dipole confined by a disk.
See Abstr. 064.064.

Stellar coronae: overview of the Einstein/CFA (*Harvard-Smithsonian Center for Astrophysics*) stellar survey.
See Abstr. 064.075.

Theory of stellar coronae: an interpretation of X-ray emission from non-degenerate stellar sources.
See Abstr. 064.076.

Stars and their coronae. See Abstr. 064.101.

Stellar coronae from Einstein: observations and theory. See Abstr. 064.104.

Red and blue shifts near compact objects.
See Abstr. 066.174.

Pulsars and compact X-ray sources: cosmic laboratories for the study of neutron stars and hadron matter.
See Abstr. 066.510.

On the mass-radius relationship of neutron stars.
See Abstr. 066.512.

Iron K photons from weakly magnetized neutron stars in X-ray binaries. See Abstr. 066.515.

X-ray bursts from thermonuclear runaways on accreting neutron stars. See Abstr. 066.520.

Instabilities of accretion disk – magnetosphere interfaces. See Abstr. 066.525.

Interaction of accretion disk and rotating magnetic field of a neutron star. See Abstr. 066.526.

X-rays from the surface of neutron stars.
See Abstr. 066.533.

Thermal limit cycle oscillations on the surface of accreting neutron stars – X-ray bursters.
See Abstr. 066.535.

Infrared photometry of HDE 226868 (Cyg X-1) from 2.3 to 10μ: mass loss rate. See Abstr. 112.050.

The optical light curve of the AM Herculis system 2A 0311–227. See Abstr. 113.007.

Rotational variations in the nonflaring optical and X-ray fluxes of YZ CMi. See Abstr. 113.036.

The spectrum variations of HD 153919.
See Abstr. 114.043.

I.U.E. observations of HD 102567, the proposed optical counterpart of 4U 1145–61. See Abstr. 114.054.

A lower limit on the magnitude of the companion to HDE 226868 (Cygnus X-1). See Abstr. 114.056.

Periodic variations in the spectrum of X Persei identified with the X-ray source 3U 0352+30.
See Abstr. 114.087.

Ultraviolet, X-ray, and infrared observations of HDE 226868 = Cygnus X-1. See Abstr. 114.170.

X-ray emission from LSI +61°303.
See Abstr. 116.019.

On the influence of radiation pressure on the light curve of HZ Herculis. See Abstr. 117.004.

Transient mass transfer caused by local surface heating in close binaries. See Abstr. 117.014.

On the motion of the apsidal line in interacting binary systems. See Abstr. 117.024.

On the light curve of HZ Her in the minimum. See Abstr. 117.032.

X-ray emission from RS CVn systems. See Abstr. 117.053.

On the rapid variability of AM Herculis. See Abstr. 117.065.

Binary X-ray sources and mass exchange in binaries. See Abstr. 117.070.

A 5.57 hr modulation in the optical counterpart of 2S 1822–371. See Abstr. 117.074.

Preliminary results of OGS X-ray observations of Capella. See Abstr. 120.020.

X-ray observations of T Tauri stars. See Abstr. 121.018.

Spectroscopy of EX Hydrae. See Abstr. 122.008.

A photometric history of KR Aurigae. See Abstr. 122.013.

A soft X-ray halo around SU UMa. See Abstr. 122.018.

Photoelectric behaviour of V 818 Sco and X Per. See Abstr. 122.122.

On X-ray survey of nine nearby historical novae. See Abstr. 124.004.

Simultaneous X-ray, UV, and optical observations of the recent nova HR Delphini. See Abstr. 124.701.

Supernovae in multiple systems. See Abstr. 125.009.

X-ray observations of the North Polar Spur. See Abstr. 125.015.

X-ray line emission form LMC supernova remnants. See Abstr. 125.020.

X-rays from G74.9 + 1.2, an elderly Crab Nebula. See Abstr. 125.021.

A high resolution X-ray image of the Puppis A supernova remnant. See Abstr. 125.024.

Is the remnant of SN 1006 Crablike? See Abstr. 125.030.

X-ray line emission from the supernova remnant G292.0+1.8. See Abstr. 125.031.

The X-ray structure and mass of the Cassiopeia A supernova remnant. See Abstr. 125.032.

X-rays from G74.9+1.2: an elderly Crab Nebula? See Abstr. 125.042.

X-ray line emission from the Cygnus Loop. See Abstr. 125.065.

A new oxygen-rich supernova remnant in the Large Magellanic Cloud. See Abstr. 125.069.

Observations of supernova remnants with the Einstein Observatory. See Abstr. 125.085.

X-ray spectra of supernova remnants. See Abstr. 125.086.

Observations of supernova remnants in the Large Magellanic Cloud with the Einstein Observatory. See Abstr. 125.087.

High resolution X-ray spectroscopy from the Einstein observatory. See Abstr. 125.088.

The effect of photoelectric absorption on the Comptonized X-ray spectrum of an accreting white dwarf. See Abstr. 126.020.

Extreme ultraviolet/soft X-ray background: the temperature distribution of the emitting gas. See Abstr. 131.044.

X-ray photoionized nebulae. See Abstr. 131.334.

Hard X-ray spectrum of the Crab Nebula from HEAO-1. See Abstr. 134.015.

Upper limits on X-ray line emission from the Crab Nebula. See Abstr. 134.016.

High resolution observation of the Crab Nebula. See Abstr. 134.017.

Upper limits on X-ray line emission from the Crab Nebula. See Abstr. 134.036.

Radio observations of globular clusters and galactic bulge X-ray sources. See Abstr. 141.024.

Einstein X-ray and VLBI radio observations of the variable quasars NRAO 140 and NRAO 530. See Abstr. 141.034.

Fe II lines in X-ray selected QSO. See Abstr. 141.048.

Discovery of low-redshift X-ray selected quasars: new clues to the QSO phenomenon. See Abstr. 141.056.

The X-ray spectrum of QSO 0241 +622. See Abstr. 141.073.

80 MHz survey of extra-galactic X-ray sources. See Abstr. 141.082.

X-ray studies of quasars and active galaxies with the Einstein Observatory. See Abstr. 141.119.

Constraint on quasar number counts from their contribution to the X-ray background. See Abstr. 141.120.

X-ray properties of quasars. See Abstr. 141.123.

Self-absorption in the Balmer line profiles of the QSO 2141 + 174 (=OX 169). See Abstr. 141.149.

Einstein observations of quasars. See Abstr. 141.177.

New upper limits for pulsed soft X-rays from the Vela pulsar PSR 0833 - 45. See Abstr. 141.508.

X-rays from radio pulsars. See Abstr. 141.513.

Hard X-ray and gamma-ray pulsed emission from the Crab pulsar. See Abstr. 141.514.

Pulsars and compact X-ray sources – cosmic laboratories for the study of neutron stars and hadron matter. See Abstr. 141.539.

The unique cosmic event of 1979 March 5. See Abstr. 142.519.

Galactic X-ray and gamma-ray emission and the nature of the interstellar electron spectrum. See Abstr. 143.041.

The Einstein central Hyades survey: a progress report. See Abstr. 153.022.

Accurate equatorial positions for the centers of six X-ray globular clusters. See Abstr. 154.004.

Time-resolved imaging and spectral studies of an X-ray burst from the globular cluster Terzan 2. See Abstr. 154.012.

Infrared bursts from Liller I/MXB 1730-333. See Abstr. 154.014.

X-ray characteristics of Loop I and the local interstellar medium. See Abstr. 157.010.

The UV spectrum of the narrow emission line X-ray emitting nucleus of the galaxy NGC 7582. See Abstr. 158.021.

X-ray observations of the Seyfert 1 galaxies AKN 120 and MCG 8-11-11. See Abstr. 158.034.

Detection of X-ray emission from Markarian 180. See Abstr. 158.039.

X-ray variability of active galactic nuclei. See Abstr. 158.041.

HEAO-1 X-ray observations of time variability in active galactic nuclei. See Abstr. 158.070.

The optical spectra of narrow-line X-ray galaxies. See Abstr. 158.078.

Iron line emission and variability in the X-ray spectrum of the Seyfert galaxy NGC 5548. See Abstr. 158.087.

The continuous spectrum of Markarian 421 during periods of X-ray satellite observations. See Abstr. 158.123.

X-ray spectral constraints on the broad-line cloud geometry of NGC 4151. See Abstr. 158.126.

New optical and radio observations of the X-ray galaxies NGC 7582 and NGC 2992. See Abstr. 158.128.

X-ray properties of galactic nuclei. See Abstr. 158.148.

Recent X-ray observations of Seyferts and emission-line galaxies. See Abstr. 158.150.

X-ray variability in active galaxy nuclei and quasars in less than one day. See Abstr. 158.151.

X-ray observations of M 87's halo. See Abstr. 158.153.

Observations of three radio galaxies with the Einstein X-ray Observatory. See Abstr. 158.154.

X-ray measurements of the mass of M87. See Abstr. 158.165.

High-resolution spectrophotometry of the "low-excitation" X-ray galaxies NGC 1672 and NGC 6221. See Abstr. 158.187.

Short term variability (and a characteristic time-scale?) in the X-ray emission of active galaxies. See Abstr. 158.213.

NGC 4507: A weak Seyfert 1 and X-ray galaxy. See Abstr. 158.220.

X-ray observations of Seyfert galaxies with the Einstein Observatory. See Abstr. 158.226.

On the origin of the radio, optical emission-line, and X-ray structure of M87. See Abstr. 158.228.

Balloon-borne observations of NGC 4151 using the MISO telescope. See Abstr. 158.233.

Active galaxies. See Abstr. 158.309.

X-ray emission from active galaxies. See Abstr. 158.311.

Einstein observations of active galaxies. See Abstr. 158.312.

X-ray spectra of active galactic nuclei. See Abstr. 158.313.

X-ray emission from galactic nuclei. See Abstr. 158.314.

UV and optical observations of X-ray sources in the Magellanic Clouds. See Abstr. 159.016.

HEAO-1 observations of the Perseus cluster above 15 keV. See Abstr. 160.015.

The HEAO A-2 survey of Abell clusters and the X-ray luminosity function. See Abstr. 160.018.

X-ray properties of poor clusters with dominant galaxies. See Abstr. 160.019.

Discovery of a bright X-ray cluster of galaxies at low galactic latitude. See Abstr. 160.020.

X-ray properties of clusters of galaxies. See Abstr. 160.021.

The heating of gas in clusters of galaxies by relativistic electrons: collective effects. See Abstr. 160.024.

Review of recent observations of cluster X-ray sources using the Einstein Observatory.
See Abstr. 160.041.

X-ray spectra of clusters of galaxies.
See Abstr. 160.042.

X-ray imaging studies of NGC 1275 and the core of the Perseus cluster. See Abstr. 160.043.

X-ray studies of clusters of galaxies with the Einstein Observatory. See Abstr. 160.044.

X-ray observations of clusters of galaxies.
See Abstr. 160.045.

The X-ray luminosity function of galaxy clusters.
See Abstr. 160.055.

On the observability of young X-ray clusters of galaxies. See Abstr. 160.057.

The X-ray structure of two rich galaxy clusters and its implications for the detectability of microwave diminutions. See Abstr. 160.061.

Clusters of galaxies. See Abstr. 160.075.

The structure and evolution of X-ray clusters of galaxies. See Abstr. 160.077.

The X-ray spectra of clusters of galaxies.
See Abstr. 160.078.

Einstein observations of the Virgo cluster.
See Abstr. 160.079.

Models of X-ray emission from clusters of galaxies.
See Abstr. 160.081.

X-ray structure of the Coma cluster from optical data. See Abstr. 160.082.

The HEAO A-2 survey of Abell clusters and the X-ray luminosity function. See Abstr. 160.084.

The revised HEAO A-2 X-ray luminosity function for Abell clusters. See Abstr. 160.085.

The effects of X-ray absorption on the spectra of distant objects. See Abstr. 161.002.

High resolution X-ray spectroscopy of the gas surrounding M 87 and NGC 1275: emission line detection and evidence for radiatively regulated accretion.
See Abstr. 161.005.

High energy astrophysics and cosmology.
See Abstr. 162.069.

Gamma-ray Sources, Gamma-ray Background

142.501 **γ-ray burst observations from the UCB/LASL experiment on ISEE-3.** W. D. Evans, R. W. Klebesadel, J. G. Laros, J. Terrell, S. R. Kane.
Nature, Vol. 286, 784 - 786 (1980).

142.502 **Origin of the 5 March 1979 γ-ray transient: a vibrating neutron star.** R. Ramaty, S. Bonazzola, T. L. Cline, D. Kazanas, P. Mészáros, R. E. Lingenfelter.
Nature, Vol. 287, 122 - 124 (1980).

An unusual γ-ray transient was observed on 5 March 1979, with 12 different instruments on 9 different spacecraft. The authors propose that a vibrating neutron star in the LMC is the source of the 5 March transient. This may be both the firs detection of a vibrating neutron star and indirect evidence for gravitation radiation.

142.503 **Periodicities in gamma ray bursts.** G. Pizzichini.
Bull. American Astron. Soc., Vol. 12, 448 (1980). – Abstract.

142.504 **The 1979 March 5 gamma ray transient reviewed: its source location in N49 within the LMC and its characteristics as evidence for a vibrating neutron star.**
T. L Cline, U. Desai, R. Mushotzky, R. Ramaty, B. J. Teegarden, W. D. Evans, J. Laros, R. Klebesadel, K. Hurley, M. Niel, G. Vedrenne, I. V. Estuline, A. Kuznetsov, V. Zenchenko.
Bull. American Astron. Soc., Vol. 12, 448 (1980). – Abstract.

142.505 **The gamma-ray burst of 19 November 1978: evidence for lines.** B. J. Teegarden, T. L. Cline.
Bull. American Astron. Soc., Vol. 12, 448 (1980). – Abstract.

142.506 **Synchrotron and annihilation radiations of e^+-e^- pairs in the magnetosphere of a vibrating neutron star: the radiation mechanism for the March 5, 1979 gamma ray transient.** R. Ramaty.
Bull. American Astron. Soc., Vol. 12, 489 - 490 (1980). Abstract.

142.507 **On the origin of the March 5, 1979 gamma ray transient: a vibrating neutron star in the Large Magellanic Cloud.** R. Ramaty, S. Bonazzola, T. L. Cline, D. Kazanas, P. Mészáros, R. E. Lingenfelter.
NASA Tech. Memo., NASA TM 80738, 13 pp. (1980).

The authors propose that a vibrating neutron star in the LMC is the source of the March 5 transient. Neutron star vibrations transport energy rapidly to the surface, heat the atmosphere by wave dissipation, and decay by gravitational radiation reaction. The redshift implies a gravitational radiation damping time which agrees with the 0.15 second duration of the impulsive phase of the event. Thus, the March 5 transient may be both the first detection of a vibrating neutron star and indirect evidence for gravitational radiation.

142.508 **Low-energy gamma-ray emission close to CG 135+1.** F. Perotti, A. Della Ventura, G. Villa, G. di Cocco, R. C. Butler, A. J. Dean, R. I. Hayles.
Astrophys. J., Lett., Vol. 239, L49 - L52 (1980).

A region of the sky close to the *COS B* source CG 135+1 was studied over the photon energy range 20 keV-25MeV from balloon altitudes on 1978 October 8. A 5 σ excess was measured in the counting rate of the telescope above 120 keV, and the spectrum of the source was evaluated.

142.509 **A search for prompt transient emission of high energy γ-rays (≥10^{12} eV) from supernovae.**
D. J. Fegan, D. O'Brien, C. T. O'Sullivan, B. McBreen.
Astrophys. J., Vol. 240, 344 - 349 (1980).

A cosmic-ray experiment is described which employs ground-based detection techniques at two independent stations, separated by 250 km. The objective is to search for coincident or correlated bursts of high energy γ-rays (≥10^{12} eV) over characteristic emission time scales of between 10μs and 1 s. The combined results of three years of observations are presented. The search proved negative over the time scales in question and for system sensitivities ~5×10^{-5} ergs cm^{-2}. The results are discussed in relation to the predictions by Colgate of prompt transient pulses of electromagnetic radiation from supernovae explosions together with some published results at X-ray and microwave frequencies.

142.510 **The vertical component of 1–20 MeV gamma rays at balloon altitudes.**
V. Schönfelder, F. Graml, F.-P. Penningsfeld.
Astrophys. J., Vol. 240, 350 - 362 (1980).

From the gamma-ray data obtained during a balloon flight of the large-area Compton telescope of the Max-Planck-Institut, the diffuse cosmic and the vertical atmospheric component of 1–20 MeV gamma rays were determined. The resulting energy spectrum of cosmic gamma rays from high galactic latitudes supports the existence of a bump in the spectrum at MeV energies above the extrapolation from hard X-ray energies. Above about 5 MeV the spectrum becomes very steep and nicely connects to the SAS 2 measurements between 35 and 200 MeV. The vertical component of downward-moving atmospheric gamma rays between 1 and 20 MeV can be fitted by one single power law. The question about the origin of the diffuse cosmic gamma-ray component is discussed, and the results about the atmospheric gamma-ray component are compared with different model calculations.

142.511 **Ungewöhnlicher Ausbruch kosmischer Gammastrahlung.** H. Scheffler.
Naturwissenschaften, 67. Jahrg., 508 (1980).

142.512 **Galactic origin of cosmic gamma-ray bursts.**
E. P. Mazets, S. V. Golenetskij, R. L. Aptekar', Yu. A. Gur'yan, V. N. Il'inskij.
Pis'ma Astron. Zh., Tom 6, 609 - 613 (1980). In Russian.
English translation in Soviet Astron. Lett., Vol. 6.

Results of a statistical analysis of γ-burst data agree with the assumption of burst sources being neutron stars in our Galaxy.

142.513 **Discrete sources of gamma-radiation, historic supernovae and pulsars.** B. M. Vladimirskij.
Izv. Krymskoj Astrofiz. Obs., Tom 61, 55 - 60 (1980). In Russian.

The gamma-ray sources which were revealed by the SAS-2 and COS B satellites may be identified with supernova remnants (SNR). As an analysis shows, the SNR associated with γ-sources have spectral indexes $\alpha \lesssim 0.3$. Such SNR probably contains a pulsar which has a magnetic field of large intensity. There are more than 150 γ-sources in the Galaxy. These sources may essentially contribute to diffuse γ-radiation of the galactic disk.

142.514 **Lines in gamma-burst energy spectra.**
E. P. Mazets, S. V. Golenetskij, R. L. Aptekar', Yu. A. Gur'yan, V. N. Il'inskij.
Pis'ma Astron. Zh., Tom 6, 706 - 711 (1980). In Russian.
English translation in Soviet Astron. Lett., Vol. 6.

Lines in γ-burst spectra observed by Venera 11 and Venera 12 imply the burst sources to contain neutron stars with strong magnetic field. The 511 keV annihilation line suffers gravitational redshift becoming pronounced at 400 - 450 keV. Absorption of radiation in the sources at cyclotron frequency under fields of ~5×10^{12} Oe results in the appearance of lines in the 30 - 60 keV range.

142.515 **Statistical evaluation of gamma-ray line observations.**
M. L. Cherry, E. L. Chupp, P. P. Dunphy, D. J. Forrest, J. M. Ryan.
Astrophys. J., Vol. 242, 1257 - 1265 (1980).

The statistical reliability of reported positive observations of solar and cosmic γ-ray lines has been evaluated. The authors have determined the relative probability that each measurement is due to a real source rather than to an accidental fluctuation in the background, and find that the results are statistically compelling in only a small fraction of the reported observations.

142.516 **The log *N*-log *S* graph for the *COS*-B galactic gamma-ray sources.** G. F. Bignami, P. A. Caraveo.
Astrophys. J., Vol. 241, 1161 - 1165 (1980).

As a tool to the study of the collective properties of the 26 low-galactic-latitude γ-ray sources in the second *COS*-B catalog their integral number-intensity relation, log *N* - log *S*, is investigated. The difficulty of accounting for the large flux errors is overcome by a simple Monte Carlo method. Especially when experimental selection effects are removed, significant differences are evident in the graph referring to the "center" and "anticenter" regions of the Galaxy. Such differences may be explained in terms of two populations of γ-ray sources.

142.517 **The diffuse gamma-ray background in the Hoyle-Narlikar cosmology.**
V. M. Canuto, J. R. Owen, J. V. Narlikar.
Astron. Astrophys., Vol. 92, 26 - 29 (1980).

The authors analyze the contributions of QSO's, BL-Lac's and Seyfert galaxies to the diffuse gamma-ray background within the framework of the Hoyle-Narlikar theory. It is shown that the inconsistency reported in standard theory, namely that the evolutionary function needed to explain the gamma-ray data is very different from the one derived from the optical part of the spectrum, is no longer present. It is also shown that the contribution of a variable gravitational "constant" to the expression for the diffuse background is the same as that of a density evolution function.

142.518 **Collisions of asteroids on neutron stars as a cause of cosmic gamma-ray bursts.**
M. J. Newman, A. N. Cox.
Astrophys. J., Vol. 242, 319 - 325 (1980).

The direct collision of asteroid-size bodies with a neutron star has been studied by one-dimensional hydrodynamic calculations with radiation diffusion. Gravitational energy is efficiently converted to thermal energy; very high temperatures at the contact point are produced on short time scales. Gamma-ray bursts of observed durations may result from the superposition of many short events from a tidally disrupted parent body. The probability of occurrence of such collisions is discussed.

142.519 **The unique cosmic event of 1979 March 5.**
T. L. Cline.
Comments Astrophys., Vol. 9, 13 - 21 (1980).

A spectral transient that appears to be neither a typical γ-ray burst nor an X-ray burster has been found to possess a variety of unusual properties that would seem to be mutually inconsistent. These observed parameters include an onset time of less than 200 μs, a subsequent temporal intensity oscillation with an 8-s period, a spectral feature consistent with a moderately red-shifted positron annihilation line, a maximum photon flux greater than any known γ-ray of X-ray transient, and a very accurate source-location measurement consistent with that of the N49 supernova remnant associated with the Large Magellanic Cloud at 55-kpc distance.

142.520 **Discrete gamma-ray sources (E < 30 MeV).** V. Schönfelder.
Non-solar gamma-rays, (see 012.047), p. 3 - 15 (1980).

At present there exist only a few results on discrete gamma-ray sources in the energy range between 0.5 and 30 MeV and even these results are sometimes conflicting. In this review observations of galactic and extragalactic sources are summarised and future aspects are outlined. It seems that the development of extragalactic gamma-ray astronomy may be especially promising at these low gamma-ray energies.

142.521 **The discrete sources of high energy gamma-ray radiation.** R. Buccheri.
Non-solar gamma-rays, (see 012.047), p. 17 - 27 (1980).

The observational material concerning gamma-ray sources at energies above ~ 50 MeV is reviewed and then discussed in order to gain insight about their physical nature. Emphasis is given to pulsars for which experimental and theoretical arguments prove that gamma-ray emission plays a major role in their evolution. The contribution of pulsars to the observed galactic gamma-ray emission is finally estimated.

142.522 **Discrete gamma-ray sources: theory.** M. Salvati.
Non-solar gamma-rays, (see 012.047), p. 29 - 42 (1980).

The author presents a brief account of the various emission models and pattern geometries: the so-called curvature-radiation mechanism and the wide-beam option appear to be favored. He also elaborates upon the possibility that pulsars (or a subclass thereof) are a sizeable fraction of the galactic discrete sources. In the same spirit alternative models are examined: these include accreting black holes and dense clouds irradiated by cosmic-ray protons.

142.523 **High-energy gamma-ray sources observed by COS-B.** R. D. Wills, K. Bennett, G. F. Bignami, R. Buccheri, P. Caraveo, N. D'Amico, W. Hermsen, G. Kanbach, G. G. Lichti, J. L. Masnou, H. A. Mayer-Hasselwander, J. A. Paul, B. Sacco, B. N. Swanenburg.
Non-solar gamma-rays, (see 012.047), p. 43 - 47 (1980).

COS-B has now detected 29 localised sources of high-energy (> 100 MeV) gamma-rays. The 18 sources left show, in addition to their concentration near the Galactic equator, a marked concentration in the first and fourth quadrants of longitude. The energy spectra of four of the sources, in the range 100 MeV to 3 GeV, have been derived.

142.524 **COS-B measurements of the spectra of the Crab and Vela gamma-ray sources.** G. G. Lichti, R. Buccheri, P. Caraveo, G. Gerardi, W. Hermsen, G. Kanbach, J. L. Masnou, H. A. Mayer-Hasselwander, J. A. Paul, B. N. Swanenburg, R. D. Wills.
Non-solar gamma-rays, (see 012.047), p. 49 - 53 (1980).

COS-B measurements above 50 MeV of the pulsed and total energy spectra of the Crab and Vela gamma-ray sources are reported here. It is found that the pulsed fraction of the Crab is consistent with 100% above 400 MeV while below this energy the pulsed fraction lies between 45% and 65%. The emission from Vela is pulsed at least 90% above 50 MeV.

142.525 **New evidence for links between SNOBs and galactic γ-ray sources.** T. Montmerle, C. J. Cesarsky.
Non-solar gamma-rays, (see 012.047), p. 61 - 66 (1980).

On the basis of the most recent COS-B data, it is shown that supernova remnants physically linked with OB associations are likely to be a class of γ-ray sources.

142.526 **Ultra high energy gamma rays.** P. V. Ramana Murthy.
Non-solar gamma-rays, (see 012.047), p. 71 - 84 (1980).

Experiments carried out over the years at many laboratories to detect high energy gamma-rays are summarised and the implications of the results discussed. Besides the traditional celestial sources (e.g. SN remnants and pulsars) one has to consider also the possibility of Primordial Black Holes being sources of high energy gamma-rays.

142.527 **Cosmic γ ray line observations.** G. Vedrenne.
Non-solar gamma-rays, (see 012.047), p. 85 - 99 (1980).

The present status of solar and non-solar nuclear gamma-ray line research from several 10's of keV up to several MeV is reviewed. These results demonstrate clearly the existence of solar and cosmic nuclear lines; future programs and especially satellite experiments such as HEAO C and GRO promise to advance our understanding of these phenomena.

142.528 **Cosmic gamma-ray lines: theory.** R. E. Lingenfelter, R. Ramaty.
Non-solar gamma-rays, (see 012.047), p. 103 - 115 (1980).

The authors consider the processes of nuclear excitation, radiative capture, positron annihilation and cyclotron radiation, which may produce gamma-ray line emission from such diverse sources as the interstellar medium, novae, supernovae, pulsars, accreting compact objects, the galactic nucleus and the nuclei of active galaxies. The significance of the relative intensities, widths and frequency shifts of the lines are also discussed. Particular emphasis is placed on understanding those gamma-ray lines that have already been observed from astrophysical sources.

142.529 **Observational properties of gamma-ray bursts.** K. Hurley.
Non-solar gamma-rays, (see 012.047), p. 123 - 139 (1980).

The present status of gamma-ray burst astronomy is reviewed, with emphasis on the progress made recently by dedicated gamma burst experiments. Results on both the statistical properties of gamma-ray bursts (log N-log S and spatial distributions) and on their individual properties (time histories and energy spectra) are discussed.

142.530 **Gamma ray burst observations by the 3 satellite Signe network.** G. Chambon, K. Hurley, M. Niel, G. Vedrenne, V. M. Zenchenko, A. V. Kuznetsov, I. V. Estulin (*Ehstulin*).
Non-solar gamma-rays, (see 012.047), p. 141 - 144 (1980).

The Signe network, consisting of French experiments aboard the Soviet Venus probes Venera 11 and 12, and aboard the Soviet eccentric earth orbit Prognoz 7 satellite, has detected 16 gamma-ray bursts in the period September 1978 - March 1979. The experiments and the Venera and Prognoz missions are described, and time profiles for two November events are presented.

142.531 **The effect of the diffuse gamma radiation on pulsars.** V. Radhakrishnan.
Non-solar gamma-rays, (see 012.047), p. 163 - 174 (1980).

The possible role of the diffuse background gamma radiation in triggering the polar gap discharges in pulsars is examined. Several characteristics of pulsars that were previously not understood appear now to find a rational explanation on the basis of triggering by gamma rays.

142.532 **Gamma rays and the local origin of cosmic rays.** M. Cassé, J. A. Paul.
Non-solar gamma-rays, (see 012.047), p. 187 - 190 (1980).

A supersonic wind must inevitably trigger a shock wave in the surrounding interstellar medium (ISM). The shock wave generated at the boundary between the stellar wind and the

ISM would be able (in certain conditions) to accelerate pre-injected particles. The close proximity of a young (mass-losing) star to the Rho Oph molecular cloud is invoked to explain the enhanced emission of this region inferred from the COS-B satellite observations. The local character of the observed cosmic rays is quite naturally explained in this context by the high concentration of OB stars (potential cosmic-ray sources) in the Gould Belt.

142.533 **Gamma-ray emission under gas accretion onto a neutron star.**
G. S. Bisnovatyj-Kogan, M. Yu. Khlopov, V. M. Chechetkin, R. A. Ehramzhyan.
Astron. Zh., Tom 57, 1242 - 1252 (1980). In Russian.
English translation in Soviet Astron., Vol. 24, No. 6.

Possible mechanisms of γ-ray emission from a neutron star due to the accretion of matter onto its surface are discussed. The suggested mechanism of birth of π^0-mesons in the collisions of nuclei leads to a significant enhancement of the expected γ-ray emission.

142.534 **Gamma ray spectroscopy.** L. Peterson.
X-ray and gamma-ray astronomy in the 1980's, (see 012.054), p. 31 (1979).

Catalogue of cosmic gamma-ray bursts from data of the "Konus" experiment. Part 3. See Abstr. 002.073.

"The Europhysics Study Conference on Gamma-Ray Astronomy after COS-B" Erice, Italy, 17-23 May 1979. See Abstr. 011.036.

The NASA gamma ray observatory (G.R.O.). See Abstr. 032.573.

The India/Soviet Gamma Ray Astronomy Programme. See Abstr. 032.574.

Description of the EXUV, GRO and AXAF satellites. See Abstr. 032.579.

Ruimte-onderzoek en sterrekunde (2). See Abstr. 051.026.

One dimensional hydrodynamics of asteroid-neutron star collisions. See Abstr. 062.122.

Supernovae gamma astronomy? See Abstr. 125.008.

Intensity and spectrum of the continuum gamma ray emission from supernovae. See Abstr. 125.011.

Gamma ray emission from supernova remnants. See Abstr. 125.077.

COS-B observation of high-energy gamma-ray emission from the Orion cloud complex. See Abstr. 131.145.

Low energy γ rays from PSR 1822–09. See Abstr. 141.503.

Hard X-ray and gamma-ray pulsed emission from the Crab pulsar. See Abstr. 141.514.

Detailed characteristics of the high-energy gamma radiation from PSR 0833–45 measured by COS-B. See Abstr. 141.525.

The boundary of a pulsar magnetosphere. Gamma-ray pulsars. See Abstr. 141.534.

Implications of UHE γ-radiation from Cyg X-3. See Abstr. 142.020.

Two X-ray pulsars: 2S 1145–619 and 1E1145.1–6141. See Abstr. 142.037.

2S1417-624: a variable galactic X-ray source near CG312-1. See Abstr. 142.052.

The periodic component of high-energy gamma-radiation of the source Cyg X-3. See Abstr. 142.075.

Cosmic rays in the galactic centre. See Abst.143.001

Galactic X-ray and gamma-ray emission and the nature of the interstellar electron spectrum. See Abstr. 143.041.

Dynamics of the Galaxy. See Abstr. 151.090.

The galactic and extragalactic gamma radiation. See Abstr. 157.017.

Astrophysical gamma-ray production by inverse Compton interactions of relativistic electrons. III Cutoff effect for inverse Compton spectra applied to the case of the hard X-ray and gamma-ray emission of NGC 4151. See Abstr. 158.095.

The Penrose Photoproduction Scenario for NGC 4151; (PCS–SSC). A black hole γ-ray emission mechanism for active galactic nuclei and Seyfert galaxies. See Abstr. 158.105.

Balloon-borne observations of NGC 4151 using the MISO telescope. See Abstr. 158.233.

Erratum

142.901 **Erratum: "High-resolution spectroscopy of two gamma-ray bursts in 1978 November"**
[Astrophys. J., Lett., Vol. 236, L67 - L70 (1980)].
B. J. Teegarden, T. L. Cline.
Astrophys. J., Lett., Vol. 240, L175 (1980). – See Abstr. 27.142.504.

143 Cosmic Radiation

143.001 **Cosmic rays in the galactic centre.**
M. Yoshimori.
Australian J. Phys., Vol. 33, 115 - 120 (1980).

The cosmic ray flux in the galactic centre region is predicted from the observed data for high energy γ rays, γ-ray lines and massive molecular clouds. The predicted cosmic ray fluxes above 1 GeV and below 100 MeV are two and four orders of magnitude respectively larger than the value in the neighbourhood of the solar system. The corresponding energy density of cosmic rays is estimated to be 100 eV cm^{-3}. Such a concentrated stream of cosmic rays could accelerate the dense and massive molecular clouds by transfer of their momentum.

143.002 **Particle trapping and acceleration during the August 1972 event.** X. Moussas.
Sol. Phys., Vol. 67, 163 - 180 (1980).

Several features of the August 1972 events are studied using neutron monitor data together with solar wind streamlines calculated on the basis of an approximate kinematic approach.The relative cosmic ray increase (3–7 UT, 5 August), is attributed to trapping and acceleration of particles between two shock waves. The extremely large solar wind velocities during the main phase of the event are not only due to the large energy of the flare but also to the fact that the ambient solar wind was already almost empty because of the sweeping action of previous shock waves.

143.003 **Large amplitude wave-trains of cosmic-ray intensity.**
H. Mavromichalaki.
Astrophys. Space Sci., Vol. 71, 101 - 110 (1980).

The large amplitude wave-trains of cosmic-ray intensity observed during June, July and August, 1973, were analysed. During these days the phase of the enhanced diurnal anisotropy is shifted to a point earlier than either the corotation direction or the anti-garden-hose direction. For this analysis the author used data from high- and middle-latitude neutron monitors and from the satellites HEOS-2, IMP-7 and IMP-8. The diurnal variation of these days is well understood in terms of a radially outward convective vector and a field-aligned inward diffusive vector yielding a diurnal anisotropy vector along about 1600 h in space.

143.004 **Cosmic ray studies with a gas Cherenkov counter in association with an ionization spectrometer.**
V. K. Balasubrahmanyan, J. F. Ormes, J. F. Arens, F. Siohan, G. B. Yodh, M. Simon, H. Spiegelhauer.
Astrophys. Space Sci., Vol. 71, 135 - 145 (1980).

The results from a ballon-borne gas Cherenkov counter (threshold 16.5 GeV nuc^{-1}) and an ionization spectrometer are presented. The gas Cherenkov counter provides an absolute energy calibration for the response of the calorimeter for $5 \leqslant Z \leqslant 26$ nuclei of cosmic rays. The contribution of scintillation to the gas Cherenkov pulse height has been obtained by independently selecting particles below the gas Cherenkov threshold using the ionization spectrometer. Energy spectra were derived by minimizing the χ^2 between a Monte Carlo simulated data and flight data. Best fit power laws were determinated for C, N, O, Ne, Mg, and Si.

143.005 **The problem of modulation of galactic cosmic rays in the solar system using the results of radioactivity investigation in meteorites.**
A. K. Lavrukhina, G. K. Ustinova.
Yaderna Energ., No. 11, p. 6 - 25 (1980). In Bulgarian.
Abstr. in Phys. Abstr., Vol. 83, Abstr. 72728 (1980).

143.006 **Energy spectra of cosmic-ray nuclei to above 100 GeV per nucleon.**
M. Simon, H. Spiegelhauer, W. K. H. Schmidt, F. Siohan, J. F. Ormes, V. K. Balasubrahmanyan, J. F. Arens.
Astrophys. J., Vol. 239, 712 - 724 (1980).

Energy spectra of cosmic-ray nuclei boron to iron have been measured from 2 GeV per nucleon to beyond 100 GeV per nucleon. The data were obtained using an ionization calorimeter flown on a balloon from Palestine, Texas. The data are consistent with the leaky box model of cosmic-ray propagation.

143.007 **The effect of convection on the propagation of relativistic galactic electrons.** I. Lerche, R. Schlickeiser.
Astrophys. J., Vol. 239, 1089 - 1106 (1980).

The authors present exact analytical solutions describing the steady-state transport, in a direct perpendicular to the galactic plane, of relativistic electrons subject to diffusion, convection, and also radiation losses. The diffusion coefficient is both spatially and energy dependent, as is the radiative loss term. The bulk convection velocity is taken to be constant and directed outward from the galactic plane. The results are derived for arbitrary spatial and energy dependences of the source of relativistic electrons.

143.008 **A cosmic-ray age based on the abundance of ^{10}Be.**
M. E. Wiedenbeck, D. E. Greiner.
Astrophys. J., Lett., Vol. 239, L139 - L142 (1980).

A measurement of the isotopic composition of galactic cosmic-ray beryllium (60–185 MeV amu^{-1}) has been made with high resolution ($\sigma_M < 0.2$ amu). The isotope fractions of ^{7}Be, ^{9}Be, and ^{10}Be obtained are 0.546 ± 0.029, 0.390 ± 0.029, and 0.064 ± 0.015, respectively. The abundance of the radioactive isotope ^{10}Be ($T_{1/2} = 1.6 \times 10^6$ yr) is used to deduce a cosmic-ray confinement time of $8.4\ (+4.0, -2.4) \times 10^6$ yr and an interstellar gas density of 0.33 (+0.13, −0.11) atoms per cm^3 in the confinement volume. These uncertainties are due only to the measurement errors (primarily statistical). Errors in the propagation model parameters, which result in comparable uncertainties, are discussed.

143.009 **The neutron-rich isotopes of cosmic-ray neon and magnesium.**
P. S. Freier, J. S. Young, C. J. Waddington.
Astrophys. J., Lett., Vol. 240, L53 - L58 (1980).

A balloon-borne detector flown during the period of minimum activity in the last solar cycle has been used to observe the isotopes of Ne, Na, Mg and Al in the cosmic radiation, over an energy range ~ 390–530 MeV/n. While the Al and Na nuclei are consistent with essentially pure ^{23}Na and ^{27}Al beams, both Ne and Mg show appreciable abundances of neutron-rich isotopes. The authors find relative abundances corrected to the top of the atmosphere for ^{20}Ne: ^{21}Ne: ^{22}Ne of 57:17:26 and for ^{24}Mg: ^{25}Mg: ^{26}Mg of 59:21:20. The neon abundances are in good agreement with those found from three different satellite experiments. The abundances of magnesium are more controversial, and these measurements add evidence that magnesium is also neutron rich. These abundances, corrected for interstellar propagation, appear significantly different from those of solar-system material, and reflect the special nature of the cosmic-ray sources.

143.010 **Kinetic theory of modulation of galactic cosmic rays by an interplanetary magnetic piston.**
L. I. Dorman, V. Kh. Shogenov.
Geomagn. Aehron., Tom 20, 588 - 594 (1980). In Russian.

143.011 **Rigidity dependence of cosmic ray scintillations in the 50- to 300-GV range.**
L. Bergamasco, A. R. Osborne, S. Alessio, G. Cini.
J. Geophys. Res., Vol. 85, 4288 - 4294 (1980).

The authors present results of the spectral analysis of the fluctuations of cosmic rays recorded by different underground muon detectors (h = 7–76 meters water equivalent). The experimental power spectra, compared with the predictions of diffusion theory in the nonlinear closure approximation, suggest the persistence of solar modulation up to rigidities of ~300 GV and the existence of nonlinear interactions between particles and the interplanetary magnetic field.

143.012 **Energy changes of cosmic rays.**
L. J. Gleeson, G. M. Webb.
Proc. Astron. Soc. Australia, Vol. 3, 233 - 234 (1978).

143.013 **The Forbush decrease of February 1978.**
A. G. Fenton, K. B. Fenton, J. E. Humble.
Proc. Astron. Soc. Australia, Vol. 3, 241 (1978).

143.014 **Sidereal variations in cosmic rays at 365 hg cm^{-2}.**
A. G. Fenton, K. B. Fenton, J. E. Humble.
Proc. Astron. Soc. Australia, Vol. 3, 262 - 263 (1978).

143.015 **Do energetic heavy nuclei penetrate deeply into Earth's atmosphere?**
P. B. Price, F. Askary, G. Tarle.
Proc. Natl. Acad. Sci. USA, Vol. 77, No. 1, p. 44 - 48 (1980). Abstr. in Phys. Abstr., Vol. 83, Abstr. 85826 (1980).

143.016 **Heavy particles in the cosmic radiation.**
P. C. M. Yock.
Phys. Rev. D, Vol. 22, 61 - 64 (1980). – Abstr. in Phys. Abstr., Vol. 83, Abstr. 90300 (1980).

143.017 **Detection of heavy cosmic rays aboard Cosmos-936.**
J. Tripier, M. Debeauvais.
Nucl. Instrum. Methods, Vol. 173, 237 - 240 (1980). – Abstr. in Phys. Abstr., Vol. 83, Abstr. 94663 (1980).

143.018 **Charge and LET distributions of cosmic heavy ions measured on COSMOS 690, 782, and 936.**
D. Hasegan, V. E. Dudkin, A. M. Marenny.
Nucl. Tracks Methods Instrum. Appl., Vol. 4, No. 1, p. 27 - 32 (1980). – Abstr. in Phys. Abstr., Vol. 83, Abstr. 94664 (1980).

143.019 **Antiprotons from the Galaxy?**
Eh. A. Bogomolov, V. A. Romanov.
Priroda, 1980, No. 10, p. 110 - 111. In Russian.

143.020 **The problem of modulation of galactic cosmic rays in the solar system using results of radioactivity investigations of meteorites.**
A. K. Lavrukhina, G. K. Ustinova.
Yader. ehnerg., 1980, No. 11, p. 6 - 25. In Russian. – Abstr. in Ref. zh., 51. Astron., 8.51.333 (1980).

143.021 **Cosmic ray effects due to the general magnetic field of the sun.** Yu. I. Stozhkov.
XIth seminar on cosmophysics, (see 012.035), p. 316 - 330 (1979). In Russian. – Abstr. in Ref. zh., 51. Astron., 8.51.394 (1980).

143.022 **Magnetic cycle and anomalous component of galactic cosmic rays. Taking into account outside modulation.** M. B. Krajnev.
XIth seminar on cosmophysics, (see 012.035), p. 331 - 352 (1979). In Russian. – Abstr. in Ref. zh., 51. Astron., 8.51.395 (1980).

143.023 **Expected fluctuations of the heliosphere and long-term cosmic ray variations.**
L. I. Dorman, V. S. Ptuskin.
XIth seminar on cosmophysics, (see 012.035), p. 367 - 378 (1979). In Russian. – Abstr. in Ref. zh., 51. Astron., 8.51.396 (1980).

143.024 **Peculiarities of zonal modulation of cosmic rays.**
A. N. Charakhch'yan, T. N. Charakhch'yan.
XIth seminar on cosmophysics, (see 012.035), p. 379 - 394 (1979). In Russian. – Abstr. in Ref. zh., 51. Astron., 8.51.397 (1980).

143.025 **Galactic cosmic ray currents in high-velocity recurrent fluxes of the solar wind and long-term changes of cosmic ray anisotropy.**
I. S. Samsonov, V. G. Grigor'ev, Z. N. Samsonova, N. P. Chirkov.
Kosm. luchi sverkhvysok. ehnerg. Yakutsk, 1979, p. 85 - 96. In Russian. – Abstr. in Ref. zh., 51. Astron., 8.51.402 (1980).

143.026 **The source distribution of cosmic-ray electrons.**
E. Massaro, B. Sacco, G. Manzo.
Astron. Astrophys., Vol. 90, 140 - 145 (1980).

A diffusion-loss transfer equation for the cosmic-ray electron component is solved for gaussian and exponential distributions of sources within the Galaxy. Spectral shapes and electron densities at various distances from the Galactic centre are given.

143.027 **Remarkable anisotropic cosmic-ray intensity oscillation.** S. P. Duggal, M. A. Pomerantz, C. H. Tsao.
Geophys. Res. Lett., Vol. 7, 613 - 615 (1980).

A large cosmic ray intensity wave in which a single oscillation of the galactic flux with a peak-to-peak amplitude of about 4% and a period of approximately 27 hours, occured on October 27-28, 1977. This unusual intensity profile was not associated with a classical Forbush decrease. Harmonic analysis in Universal Time of data from a global distribution of stations revealed that in each case the diurnal vector on October 27 differs significantly from those on other days before and after the event. The mechanism which produced this unusual oscillation is not yet understood.

143.028 **Distribution of cosmic-ray electrons in the Galaxy.**
M. R. Issa, P. A. Riley, A. W. Strong, A. W. Wolfendale.
Nature, Vol. 287, 810 - 812 (1980).

Although it seems that cosmic-ray electrons in the range 100–300 MeV are mostly generated in sources within the Galaxy their distribution is not known. The authors examine here the distribution of the electron flux from studies of the γ-ray flux produced by electrons interacting in the interstellar medium by way of the dominant bremsstrahlung mechanism.

143.029 **Cosmic-ray confinement in the Galaxy.**
C. J. Cesarsky.
Annu. Rev. Astron. Astrophys., Vol. 18, (see 003.004), 289 - 319 (1980).

Contents: Physical mechanisms governing cosmic-ray confinement and escape, interpretation of the observations of cosmic rays at the earth, cosmic rays elsewhere: synchrotron and gamma-ray diffuse galactic emission, very high energy cosmic rays.

143.030 **A nonrandom component in cosmic rays of energy $\geqslant 10^{14}$ eV.** C. L. Bhat, M. L. Sapru, C. L. Kaul.
Nature, Vol. 288, 146 - 149 (1980).

The authors have been studying (1979) the distribution of the arrival times of atmospheric Cerenkov light pulses, initiated by cosmic rays of energy $\geqslant 10^{14}$ eV. They have now detected a nonrandom component for time separations

<40 s, and here they discuss the possibility that it has a point-source origin.

143.031 **On proton energy variations in their motion towards the sun.**
E. V. Gorchakov, T. I. Morozova, G. A. Timofeev.
Geomagn. Aehron., Tom 20, 785 - 789 (1980). In Russian.

143.032 **Cosmophysical aspects of cosmic ray investigation. International seminar.**
E. V. Kolomeets (Editor).
Kazakhsk. univ., Alma-Ata. 103 pp. Price 70 Kop. (1980). In Russian. – From Ref. zh., 62. Issled. kosm. prostranstva, 10.62.58 (1980).

143.033 **Isotope anomalies in meteorites and the origin of the galactic cosmic rays.**
J. Audouze, J.-P. Chièze, E. Vangioni-Flam.
Astron. Astrophys., Vol. 91, 49 - 52 (1980).
Some isotopic anomalies have recently been discovered in carbonaceous meteorites while significant progress has been made in the determination of the composition of the galactic cosmic rays. These new data such as the possible relation between the supernova outbursts and the physics of the interstellar medium are briefly reviewed. A scenario trying to locate where the galactic cosmic rays originate and are accelerated is presented.

143.034 **Abundances of cosmic ray nuclei groups with $Z \geq 50$ by studying the fossil tracks in meteoritic olivines.** M. Haiduc, D. Hasegan, D. Lhagvasuren (*Lkhagvasurehn*), A. Marin, O. Otgonsuren (*Otgonsurehn*), P. Pellas, V. P. Perelyghin (*Perelygin*), S. G. Stetsenko.
Rev. Roumaine Phys., Vol. 25, 353 - 365 (1980). – Abstr. in Phys. Abstr., Vol. 83, Abstr. 105139 (1980).

143.035 **Cosmic-ray electrons in the diffusion-convection model of particle propagation.**
V. A. Dogel', V. M. Kovalenko, V. L. Prishchep.
Pis'ma Astron. Zh., Tom 6, 696 - 699 (1980). In Russian. English translation in Soviet Astron. Lett., Vol. 6.
The one-dimensional diffusion-convection equation for the propagation of cosmic rays with synchrotron and adiabatic losses is solved. The asymptotic behaviour of electron spectra at Earth in various energy regions is analysed. From comparison of electron and radioemission spectra with observations it is concluded that convection (galactic wind) does not seem to be dominating in carrying cosmic rays out of the Galaxy and propagation of the particles in the halo is rather of diffusion character.

143.036 **Twenty-two year modulation of cosmic rays associated with polarity reversal of polar magnetic field of the Sun.** I. Morishita.
Proc. Cosmic-Ray Res. Lab. Nagoya Univ., Vol. 23, No. 1, p. 30 - 66 (1980). In Japanese. – From Phys. Abstr., Vol. 84, Abstr. 9337 (1981).

143.037 **The acceleration of interstellar grains and the composition of the cosmic rays.** R. I. Epstein.
Mon. Not. R. Astron. Soc., Vol. 193, 723 - 729 (1980).
It is shown that if strong interstellar shock waves provide the main energy source for the galactic cosmic rays, then these same shock waves efficiently accelerate the nuclei contained in the interstellar grains up to relativistic energies. This may explain the relatively high abundances of the refractory elements in the cosmic rays and possibly the unusual isotropic ratios of the cosmic ray neon.

143.038 **Energy spectra and charge states of low energy cosmic rays in the Skylab experiment.**
N. Durgaprasad, V. S. Venkatavaradan, S. Sarkar, S. Biswas.
Space Research, Vol. XX, (see 012.043), 259 - 262 (1980).
The authors have studied the energy spectra, the charge composition and ionisation states of ions in the energy range 8-100 MeV/amu within the magnetosphere, using solid state dielectric detectors of Lexan Polycarbonate, with a view to obtain clues as to their origin. They compare these spectra with the anomalous component of cosmic rays observed in interplanetary space.

143.039 **Cosmic ray effects in solar system objects.**
N. Bhandari, M. N. Rao.
Space Research, Vol. XX, (see 012.043), 263 - 266 (1980).
Two important effects caused by the nucleonic component of cosmic rays in solar system objects are nuclear reactions and radiation damage. These can be quantitatively studied to determine the intensity, energy spectra and the composition of the cosmic rays. The purpose of the paper is to focus on some of the important results obtained from a study of moon and meteorite samples.

143.040 **Comments on stochastic acceleration of cosmic rays.**
R. Cowsik.
Astrophys. J., Vol. 241, 1195 - 1198 (1980).
The spectrum of particles subjected to second-order Fermi acceleration is studied, and the spectra of both the primary particles such as C and O and their nuclear secondaries such as Li, Be, and B are derived explicitely. The second-order term is seen to be of prime importance in the acceleration process. The secondary-to-primary ratio is shown to increase logarithmically with energy contrary to observation. These considerations are extended to arbitrary stochastic processes, and it is concluded that cosmic rays are not accelerated in the interstellar medium.

143.041 **Galactic X-ray and gamma-ray emission and the nature of the interstellar electron spectrum.**
R. J. Protheroe, A. W. Wolfendale.
Astron. Astrophys., Vol. 92, 175 - 181 (1980).
An analysis is made of all available data, both direct and indirect, on the energy spectrum of cosmic ray electrons. It is shown that the data are consistent with an injection spectrum having constant exponent, $\gamma = 2.1 \pm 0.1$, over a wide range of energy: $10 - 10^9$ MeV. Attention is drawn to the role of a possible deficit of sources in reducing the intensity of local electrons both above 10 GeV and below a few hundred MeV.

143.042 **Acceleration and propagation of cosmic rays.**
C. Fransson, R. I. Epstein.
Astrophys. J., Vol. 242, 411 - 415 (1980).
Two general categories of cosmic ray models are discussed, concomitant acceleration and propagation (CAP) models and sequential acceleration and propagation (SAP) models. For the CAP models it is found that the ratio of the predominantly secondary nuclei to the predominantly primary nuclei varies by less than a factor of 1.5 between 1 and 100 GeV per nucleon. It appears that the evolution of cosmic rays is best described by SAP models.

143.043 **Study of intensity fluctuations in cosmic rays during Forbush-decreases on the basis of the data obtained with the Izmiran scintillation supertelescope.**
L. I. Dorman, I. Ya. Libin, O. V. Gulinsky (*Gulinskij*).
Astrophys. Space Sci., Vol. 73, 337 - 347 (1980).
The one-minute, five-minute, and hourly values of cosmic-ray intensity obtained with a scintillation telescope at Izmiran have been used to carry out the correlational and spectral analysis of large Forbush-decreases of cosmic rays in 1978. The power spectra obtained for the periods of interplanetary disturbances are characterized by individual spectral lines whose amplitude exceeds at least a 95% interval of reliability.

143.044 **A method to study the atmospheric influence on the isotopic composition of primary cosmic rays applied to the elements carbon and oxygen.**
C. Bjarle, N.-Y. Herrström, G. Jönsson.
Phys. Scr., Vol. 22, 551 - 554 (1980).

143.045 **The interactions of energetic particles with the solar wind.** L. A. Fisk.
Solar system plasma physics, Vol. 1, (see 003.010), 177 - 247 (1979).

The author reviews the status of studies of the interaction of energetic particles with the solar wind. He discusses the theory of the solar modulation of galactic cosmic rays, which has come under some criticism in recent years. He reviews theories for the propagation of energetic particles in the solar wind and theories for their acceleration.

143.046 **Effective sizes of the region of cosmic ray modulation.** A. G. Zusmanovich, L. F. Churunova.
Ionos. i soln.-zemn. svyazi. Alma-Ata, 1980, p. 114 - 122. In Russian. – Abstr. in Ref. zh., 51. Astron., 11.51.472 (1980).

143.047 **Longitudinal run of cosmic rays under the radiation belts of the earth.**
Yu. A. Aleksandrov, S. N. Kuznetsov, S. P. Ryumin.
Kosm. Issled., Tom 18, 946 - 947 (1980). In Russian.

143.048 **Cosmic rays.** P. Meyer.
X-ray and gamma-ray astronomy in the 1980's, (see 012.054), p. 26 - 28 (1979).

143.049 **Cosmic ray modulation by a shock wave taking into account adiabatic cooling and transversal diffusion (automodel solution).**
M. B. Bagdasaryan, L. I. Dorman.
Geomagn. Aehron., Tom 20, 990 - 996 (1980). In Russian.

143.050 **On the nature of increase of cosmic ray intensity before the Forbush effect.** N. S. Kaminer.
Geomagn. Aehron., Tom 20, 1097 - 1099 (1980). In Russian.

143.051 **On near-weakly variations of the daily anisotropy of cosmic rays.**
R. M. Arslanova, N. G. Ptitsyna, M. I. Tyasto.
Geomagn. Aehron., Tom 20, 1099 - 1100 (1980). In Russian.

143.052 **Solar polar coronal holes and galactic cosmic ray intensities.** S. P. Agrawal, L. J. Lanzerotti, D. Venkatesan, R. T. Hansen.
J. Geophys. Res., Vol. 85, 6845 - 6852 (1980).

The authors present a comparison between the areas of solar polar coronal holes and north-south (N-S) cosmic ray asymmetries. The observed N-S asymmetries over a 2-year period (1973 - 1974) are found to be related to asymmetries in the sizes of selected solar polar holes; the hemisphere with the smaller hole has a larger cosmic ray flux. An analysis is also made of the N-S asymmetries for intervals when there exists a relatively symmetric coronal region (±40°) about the solar equator; the N-S asymmetry tends to be small for this situation of comparable-sized polar coronal holes. The authors discuss various mechanisms which can contribute to a N-S asymmetry.

143.053 **On the three-dimensional nature of the modulation of galactic cosmic rays.**
D. Venkatesan, S. P. Agrawal, L. J. Lanzerotti.
J. Geophys. Res., Vol. 85, 6893 - 6894 (1980).

Recent results, on two different time scales, of the influence of solar polar coronal holes on the galactic cosmic ray intensity appear on the surface to be inconsistent. The authors show, in the context of present knowledge of the effects of high-speed solar wind streams on energetic particles, that the results are indeed mutually consistent and they support the view that coronal holes are influential in determining the three-dimensional modulation of galactic cosmic rays in the solar system.

Where do cosmic rays come from? See Abstr. 011.025.

Cosmophysical aspects of cosmic ray investigations. International seminar. See Abstr. 012.048.

On high-energy astrophysics. See Abstr. 013.017.

Ion states of low energy cosmic rays: the Indian experiment on the first Space Shuttle-Spacelab mission. See Abstr. 032.554.

Cosmic ray experiment (E 6). See Abstr. 032.560.

Cosmic ray experiment (E 7). See Abstr. 032.561.

A new upper limit on optical bursts from primordial black hole explosions. See Abstr. 066.175.

Solar polar field reversals and secular variation of cosmic ray intensity. See Abstr. 072.028.

Using the method of the heliolatitude-longitude index of solar activity for an analysis of the 27-day cosmic ray variations. See Abstr. 072.058.

Periodic structure of solar activity and cosmic ray intensity at the 1964 - 1965 minimum. See Abstr. 072.068.

Display of solar activity on the branch of growth of the 21st cycle in cosmic rays from data of the artificial earth satellite Meteor. See Abstr. 072.069.

Two classes of fast solar wind streams: their origin and influence on the galactic cosmic ray intensity. See Abstr. 074.068.

The importance of energetic particle precipitation on the chemical composition of the middle atmosphere. See Abstr. 082.057.

Background interplanetary magnetic fields in the earth's orbit and anisotropy of galactic cosmic rays. See Abstr. 106.020.

Energetic particles in space. See Abstr. 106.053.

Variations of interplanetary parameters and cosmic-ray intensities. See Abstr. 106.063.

The propagation of cosmic-rays in the interplanetary region (the theory). See Abstr. 106.075.

Gamma rays and the local origin of cosmic rays. See Abstr. 142.532.

Dynamics of the Galaxy. See Abstr. 151.090.

Gamma radiation, cosmic rays, and galactic structure. See Abstr. 157.003.

Galactic gamma rays and the origin of cosmic rays. See Abstr. 157.015.

Formation of radio halos in clusters of galaxies from cosmic-ray protons. See Abstr. 160.025.

Stellar Systems, Galaxy, Extragalactic Objects, Cosmology

151 Stellar Systems (Kinematics, Dynamics, Evolution)

151.001 **A correlation between the lengths of bars and the sizes of bulges.** E. Athanassoula, L. Martinet.
Astron. Astrophys., Vol. 87, L10 - L11 (1980).

A correlation between the length of the bar and the size of the bulge component of barred galaxies is found. This is in agreement with numerical simulation results. The significance and consequences of this correlation are discussed.

151.002 **Nonlinear effects near the particle resonance.** J. Palouš.
Astron. Astrophys., Vol. 87, 361 - 364 (1980).

The nonlinear behaviour of stellar orbits near the particle resonance is described. For the rotation speed of the spiral structure the author uses the value of 20 km s^{-1} kpc^{-1} which was derived elsewhere (Palouš et al., 1977). This new value draws the attention to the orbits approaching the corotation from the distant peripheries of the stellar system. These orbits exhibit near the particle resonance damped a "leap-frogging motion". The maximum of the response density is near the points $L1$ and $L2$ and it is slightly shifted in the direction opposite to that of galactic rotation.

151.003 **Stability of spherical stellar systems for aspherical modes.** D. Gillon.
C. R. Acad. Sci. Paris, Tome 290, Sér. B, 545 - 547 (1980). In French.

The author considers spherically symmetric stellar systems with distribution function F depending only on the energy ϵ. Using the second variation of energy and the corresponding eigenvalue equation, he shows that the condition $dF/d\epsilon < 0$ is a sufficient condition of stability for aspherical modes.

151.004 **Tunnel effect on density waves of galaxies at corotation circle and the switch character of the "waser" mechanism.** J. Xu.
Sci. Sinica, Vol. 23, 992 - 1005 (1980).

A nonlinear complex eigenvalue problem posed in galactic dynamics is studied. The uniformly valid asymptotic solutions and the quantum condition are obtained. In terms of these solutions the author has analyzed the wave propagation near the corotation circle. The tunnel effect on density waves through the potential barrier at corotation circle and the "switch character" of a certain type possessed by the "waser" mechanism are explained.

151.005 **Contributions to the theory of spiral structure. I. Energy and lifetime of density waves and the classification of spiral galaxies.**
J. V. Feitzinger, T. Schmidt-Kaler.
Astron. Astrophys., Vol. 88, 41 - 51 (1980).

In order to understand better the finite lifetime of spiral density waves, the transport of energy by these waves and the interpretation of the Hubble galaxy classification system, the authors investigate a sample of 25 well-known galaxies with known rotation curve and luminosity classification. They calculate the lifetime and energy density of the spiral wave for all these galaxies considering the group velocity and the energy dissipation process. The authors use axisymmetric models of disk galaxies and a gas dynamical approach. They make the hypothesis $Q=1=$ const over the whole disk, the pattern velocity Ω_p is chosen so that corotation R_{co} is at the end of the observed spiral structure.

151.006 **The formation of super-rings.** G. Tenorio-Tagle.
Astron. Astrophys., Vol. 88, 61 - 65 (1980).

The author has calculated the collision of a small neutral cloud (surface density $\sim 10^{19}$ cm^{-2}) with a constant density galactic disk. Through the collision, a large amount of energy is deposited in a small volume of the galaxy, resulting in a supersonic expansion of very hot ($10^6 - 10^7$ K) gas into the galaxy and out of the galactic disk. The expansion generates a large cavity (a super-ring) with physical characteristics (diameter, velocity of expansion, etc.) in agreement with the observations, and a large volume of hot low-density gas with properties similar to those of the observed coronal gas.

151.007 **The dynamics of the spiral galaxy M81. I. Axisymmetric models and the stellar density wave.**
H. C. D. Visser.
Astron. Astrophys., Vol. 88, 149 - 158 (1980).

In this paper the author starts the analysis with the discussion of axisymmetric models, constructed on the basis of the rotation curve; the effects of a density wave can then be considered as perturbations of this axisymmetric state. He discusses the requirements that the theoretical spiral pattern be consistent with the observed spiral pattern and he treats the amplitude of the spiral potential perturbation as computed on the basis of surface photometry. He summarizes the conclusions that can be drawn from this paper.

151.008 **The dynamics of the spiral galaxy M81. II. Gas dynamics and neutral-hydrogen observations.**
H. C. D. Visser.
Astron. Astrophys., Vol. 88, 159 - 174 (1980).

The author describes the construction of the gas-flow models and relates the galactic shock to observations (dust lanes and H II regions). Detailed comparisons of the theoretical gas flow with observed hydrogen density and velocity fields at three angular resolutions (25″, 50″, and 2′) are made. The effect of the density wave on the rotation curve is discussed. He draws the more general conclusions with respect to Papers I and II and enumerates the more specific results of this paper.

151.009 **Bar-driven spiral structure.** E. Athanassoula.
Astron. Astrophys., Vol. 88, 184 - 193 (1980).

In the present paper the author studies the gaseous and stellar spirals driven by a growing, symmetric and directly rotating bar in a disk with no inner Lindblad resonance. The self-potential of the spirals is taken into account and that of the stars is found to be particularly important in determining the form of the arms. The response was never found to be bar-like between corotation and outer Lindblad resonance. He discusses the form of the spirals as a function of the disk and bar parameters.

151.010 **Can elliptical galaxies be equilibrium systems?** R. Caimmi.
Astrophys. Space Sci., Vol. 71, 75 - 85 (1980).

The author finds that equilibrium models such as Emden-

Chandrasekhar polytropes and Roche polytropes with $n = 0$ can account for the main part of observations relative to the ratio of maximum rotational velocity to central velocity dispersion in elliptical systems. More complex models involving, for example, massive halos could lead to a more complete agreement. Models that are a good fit to the observed data are characterized by an inner component (where most of the mass is concentrated) and a low-density outer component. A comparison is performed between some theoretical density distributions and the density distribution observed by Young et al. (1978) in NGC 4473. Alternative models, such as triaxial oblate non-equilibrium configurations with coaxial shells, involve a number of problems which are briefly discussed. The author concludes that spheroidal oblate models describing elliptical galaxies cannot be ruled out.

151.011 **Numerical simulations of collisions in Keplerian systems.** K. A. Hämeen-Anttila, J. Lukkari.
Astrophys. Space Sci., Vol. 71, 475 - 497 (1980).

Computer simulations which were carried out for Keplerian collisional systems of 250 frictionless particles with a ratio of particle radius to mean semi-major axis of 0.001, confirm the theoretically predicted evolution very well until the thickness of the system is a few times the particle radius and the mass-point approximation becomes invalidated. Before this happens, the collisional contraction of denser regions can be observed. The local dispersions of the perihelia and ascending nodes diminish if the local mean orbit is not too close to a circle with zero inclination. When the mass-point approximation ceases to be valid, the system begins to expand, but with parameter values of our standard system this process is much slower than the simultaneously observed evolution toward grazing collisions which do not affect the orbital elements. Therefore, such systems are not dispersed into the space.

151.012 **On the stability of an inhomogeneous stellar disk.**
A. G. Morozov.
Astron. Zh., Tom 57, 681 - 686 (1980). In Russian.
English translation in Soviet Astron., Vol. 24, No. 4.

The dispersion equation describing the dynamics of non-axisymmetric disturbances in the plane of a differentially rotating stellar disk with inhomogeneous density and velocity dispersion of stars has been obtained. The laws of dispersion of non-axisymmetric Jeans and gradient disturbances are investigated and the boundaries of stability of these disturbances are revealed.

151.013 **On the excitation of warps in galaxy disks.**
G. Bertin, J. W.-K. Mark.
Astron Astrophys., Vol. 88, 289 - 297 (1980).

Recent observations have revived new interest in the kinematical and dynamical interpretation of large scale warps in galaxy disks. Here the authors outline a simple excitation mechanism for warping structures which is based on the transfer of angular momentum between the flat and the spheroidal components of a disk galaxy. A moderate spheroidal (bulge-halo) component is sufficient to provide the additional amplification needed by Hunter and Toomre in order to interpret the warps in the Milky Way as due to the passage of the Large Magellanic Cloud. In addition, the authors suggest a new mechanism (similar to that successfully used for the spiral structure problem) which may give self-excited and self-sustained warping modes; these modes may be helpful in understanding the observations in galaxies without close companions. They argue that warps may provide important clues as to the size and structure of galaxy halos.

151.014 **The gravothermal catastrophe of stellar systems.**
S. Inagaki.
Publ. Astron. Soc. Japan, Vol. 32, 213 - 227 (1980).

The mechanism of the gravothermal catastrophe of stellar systems is examined by using the Fokker–Planck equation in energy space. The gravothermal catastrophe is induced by the flow of stars in energy space into the low-energy region. It occurs in such a situation that the distribution function in μ-space decreases when the number of stars in the low-energy region increases by the above-mentioned flow. This decrease of the distribution function in μ-space is due to the increase of the phase-space volume, which is eventually ascribed to the deepening of the gravitational potential. Thus the presence of self-gravity is the prime cause of the onset of the gravothermal catastrophe.

151.015 **Qualitative properties of motion of stars on the periphery of a galaxy.**
Kh. B. Ibragimova, M. Kh. Khasanova.
Pis'ma Astron. Zh., Tom 6, 405 - 407 (1980). In Russian.
English translation in Soviet Astron. Lett., Vol. 6.

Differential-geometrical properties of star motion in the gravitational field of a uniformly rotating galaxy are discussed.

151.016 **Viscous effects in the gas flow in barred spirals.**
G. D. van Albada, W. W. Roberts, Jr.
Bull. American Astron. Soc., Vol. 12, 444 (1980). – Abstract.

151.017 **Hydrodynamic simulation of galactic disks.**
P. R. Woodward.
Bull. American Astron. Soc., Vol. 12, 444 (1980). – Abstract.

151.018 **Galaxy warps and the quest for properties of the unseen halo – dynamical considerations.**
G. Bertin, J. W.-K. Mark.
Bull. American Astron. Soc., Vol. 12, 444 (1980). – Abstract.

151.019 **Galaxy warps and the quest for properties of the unseen halo – preliminary results for three galaxies.**
G. Lake, J. W.-K. Mark.
Bull. American Astron. Soc., Vol. 12, 444 - 445 (1980).
Abstract.

151.020 **Density scaling of the specific angular momentum (J/M) in the universe.**
L. Carrasco, M. Roth, A. Serrano.
Bull. American Astron. Soc., Vol. 12, 445 (1980). – Abstract.

151.021 **The formation of massive galactic halos.**
C. P. Pryor, M. Lecar.
Bull. American Astron. Soc., Vol. 12, 470 - 471 (1980).
Abstract.

151.022 **Translational rotation.** T. W. Noonan.
Bull. American Astron. Soc., Vol. 12, 524 (1980).
Abstract.

151.023 **A dispersion relation for open spiral galaxies.**
G. Contopoulos.
J. Astrophys. Astron., Vol. 1, 79 - 95 (1980).

The Lin-Shu dispersion relation is applicable in the (asymptotic) case of tight spirals (large wave number). The author reconsiders the various steps leading to the Lin-Shu dispersion relation in higher approximation, under the assumption that the wave number is not large and derives a new dispersion relation. This is valid for open spiral waves and bars. He proves that this dispersion relation is the appropriate limit of the nonlinear self-consistency condition in the case where the linear theory is applicable.

151.024 **Star clusters containing massive, central black holes. III. Evolution calculations.**
A. B. Marchant, S. L. Shapiro.
Astrophys. J., Vol. 239, 685 - 704 (1980).

The authors present detailed, two-dimensional simulations of star cluster evolution. A Monte Carlo method which they have presented previously for steady-state problems is adapted to simulate the development with time of isolated star clusters. They treat clusters which evolve on relaxation time scales with and without central black holes. They first apply the method to follow the development and core collapse of an initial Plummer-model cluster without a central black hole. They present three calculations of cluster reexpansion, each beginning with the insertion of a black hole at the center of a highly collapsed cluster core. Each case is characterized by a different value of initial black hole mass or black hole accretion efficiency for the consumption of debris from disrupted stars.

151.025 **Potential in the central bar structure.**
V. Szebehely, J. Lundberg, W. J. McGahee.
Astrophys. J., Vol. 239, 880 - 881 (1980).

The figure-eight orbits obtained by Miller and Smith inside the central bar structure of galaxies are used to establish possible potential functions which result in such orbits. It is shown that r^{-6} type potentials are special cases of distance and angle-dependent potential functions.

151.026 **Galactic spiral shocks: vertical structure, thermal phase effects, and self-gravity.** A. D. Tubbs.
Astrophys. J., Vol. 239, 882 - 892 (1980).

By use of two-dimensional, time-dependent, hydrodynamical calculations the author has determined the steady-state vertical gas structure of galactic shocks. A variety of energy densities of gas, magnetic field, and relativistic cosmic rays was assumed. The midplane shock structure is remarkably similar to that found in one-dimensional calculations, with no significant decrease in shock strength and no sudden vertical motion of the postshock gas. The shocks are straight and nearly perpendicular to the galactic plane for almost any realistic galactic parameters. Two-phase calculations, utilizing a time-dependent model of the heating and cooling of the interstellar medium, show no evidence for so-called accretion fronts. The clouds or high-density regions which form in these calculations are shown to form only in the midplane of the galaxy. The author has also briefly considered the self-gravity of the gas and shows it to be important for average midplane gas densities of over 0.5 particles cm^{-3}. The relevance of these findings to the observations is discussed.

151.027 **A comment on the formation of spiral galaxies.**
M. De Robertis, R. N. Henriksen.
J. R. Astron. Soc. Canada, Vol. 74, 189 - 202 (1980).

The authors develop the hypothesis that disc systems are secondary structures, resulting from the accretion of a metal-enriched intracluster (inter-galactic) medium by an elliptical (spheroidal) system. The original elliptical will have by now become the spiral bulge/halo component. This hypothesis is able to account for the metal-rich, high-specific-angular-momentum material currently found within the disc components, with reasonable parameter values. It is further suggested that the intracluster gas with near-solar metal abundances, that is now observed in rich clusters, originated in the primary systems during the epoch of galaxy formation. Conditions required for the subsequent "successful" accretion of this material are given.

151.028 **On approximate analytical solutions of the Poisson equation for inner regions of galaxy disks with finite thickness.** J. Xu.
Kexue Tongbao, Vol. 25, 748 - 754 (1980).

The thickness parameter ϵ_0 and another small parameter ϵ_* are adopted for solving the Poisson equation of galaxy disks of finite thickness. The meaning of the parameter ϵ_0 is the maximum thickness of disk, and the physical significance of the parameter ϵ_* is the rate of variation of the wavelength of the density wave along the radius or the characteristic value of the wavelength. The author obtains by an iterative method the approximate solutions of the zero order, first order and second order for the parameters ϵ_0 and ϵ_* which are applicable to rather loosely wound spiral cases as well as to tightly wound spiral cases.

151.029 **Formation and rotation of disc galaxies with haloes.**
S. M. Fall, G. Efstathiou.
Mon. Not. R. Astron. Soc., Vol. 193, 189 - 206 (1980).

The authors consider a picture in which disc galaxies formed from collapsing gas in extended haloes of dark material. They have developed some models that determine the structure of the haloes in terms of the observable properties of the discs. A comparison of the models with a sample of 25 galaxies is made.

151.030 **The mass–angular momentum density relation for spiral galaxies.**
G. Vettolani, B. Marano, G. Zamorani, R. Bergamini.
Mon. Not. R. Astron. Soc., Vol. 193, 269 - 276 (1980).

A relation between optical luminosity and angular momentum density for spiral galaxies is found. It is interpreted as a relationship between mass and angular momentum density and is compared with theoretical predictions from galaxy formation models. Finally, the authors show that no evidence can be found that shows that the sequence of morphological types (at least from Sbc to later types) is a sequence of angular momentum density.

151.031 **The equilibrium of a galactic bar.**
P. O. Vandervoort.
Astrophys. J., Vol. 240, 478 - 487 (1980).

It is shown that, for values of the polytropic index not less than 0.5, every uniformly rotating, gaseous polytrope has an exact stellar-dynamical counterpart. Like gaseous polytropes, uniformly rotating polytropic stellar systems have axisymmetric configurations of equilibrium and, provided that the polytropic index is less than 0.808, nonaxisymmetric configurations of equilibrium as well.

151.032 **Transient annular structures in barred galaxies.**
J. L. Sérsic, J. H. Calderón.
Bull. Astron. Inst. Czechoslovakia, Vol. 31, 253 - 256 (1980).

The existence of ring-like structures in barred galaxies has been proposed by the authors. They develop a dynamical model which requires a variable-mass field in order to give an interpretation of these features. The geometrical properties of the ring – such as the inclination with respect to the bar – shed light on the sense of the mass variation. All observed cases suggest that there is accretion.

151.033 **On the rotation of star clusters and nebulae as determined from radial velocities.** T. W. Noonan.
Astrophys. J., Vol. 240, 803 (1980).

The fictitious apparent rotation due to a body's tangential velocity is unimportant for Galactic nebulae, but it may be important for globular clusters and the high-velocity clouds.

151.034 **On the rotation of star clusters.** T. W. Noonan.
Astron. J., Vol. 85, 1274 - 1275 (1980).

A cluster of stars with a translational motion in the plane of the sky will appear to rotate with an angular velocity equal to the cluster's overall proper motion. The effect is evaluated for Cudworth's method of proper-motion dispersions.

151.035 **Computer simulations of close encounters between single stars and hard binaries.**
J. G. Hills, L. W. Fullerton.
Astron. J., Vol. 85, 1281 - 1291 (1980).

The authors have simulated 5175 close encounters between hard binaries and single stars having masses M_3 ranging from 0.01 to 100 times the mass M_1 of the binary components. The authors use the results of the computer simulations to compute improved cross sections governing the rate at which binaries feed kinetic energy into their parent star clusters. They compare these to earlier work.

151.036 **Population synthesis of giant ellipticals. A composite approach.** J. M. Alvarez-Falcon.
Astron. Astrophys., Vol. 89, 291 - 295 (1980).

The population in the nucleus of a mean elliptical galaxy is synthesized by a composite approach. Evolutionary tracks parametrize the population and a static technique based on linear minimization is applied to narrow-band spectrophotometry in order to find the best values of the parameters. Some deficiencies in the theory of post-main-sequence evolution are avoided. A small "plateau" in the initial luminosity function extending over $M_v \sim +3$ is found and explained. The model gives a value of 0.9 for the slope of the initial mass function, if a power law parametrization is assumed. The turnoff group is well determined at G0–5 V. Some difficulties with the giant branch are discussed, and large values for the corresponding lifetimes set are proposed. Luminosity evolution and cosmological implications are also discussed.

151.037 **Galaxy models with live halos.** J. A. Sellwood.
Astron. Astrophys., Vol. 89, 296 - 307 (1980).

Computer models of galaxies are described, in which the disc stars and the halo stars are both treated fully self consistently. These were used to test the validity of the "rigid halo" approximation usually employed when studying the instabilities of a disc of stars. The models show that very little interaction between the populations occurs while the disc remains nearly axisymmetric, but a strong bar is able to transfer angular momentum from the disc to the halo. The global bar stability of the disc is not greatly affected by the interaction between the two stellar populations, and it is found that a large fraction of the total galactic mass is still required in the spheroidal component if bar formation is to be prevented.

151.038 **Dispersion einer Gruppe von Sternen im Gravitationsfeld einer Spiralgalaxie.**
B. Fuchs, K. O. Thielheim.
Mitt. Astron. Ges., Nr. 48, (see 012.015), p. 161 - 167 (1980).

151.039 **Die stellare Antwort auf eine anwachsende balkenförmige Störung in einer Scheibengalaxie.**
K. O. Thielheim, H. Wolff.
Mitt. Astron. Ges., Nr. 48, (see 012.015), p. 168 - 172 (1980).

151.040 **Why aren't all galaxies barred?** J. Sellwood.
Messenger, No. 21, p. 27 - 29 (1980).

151.041 **A model of the formation of spherical galaxies.**
W. K. Brown.
Astrophys. Space Sci., Vol. 72, 15 - 31 (1980).

Galactic mass distributions produced by the fragmentation model (in addition to those published previously) are examined. This is done by allowing the cubical fragment to assume any orientation in space. All distributions examined closely resemble observed galactic luminosity profiles, and, among the variety produced, was found the special $\log I$ versus $r^{1/4}$ behavior of de Vaucouleurs' empirical law as well as the exponential behavior of spirals. Because of the apparent success of the model, the initial conditions and analytical methods are re-examined in detail.

151.042 **Protogalactic explosions and intracluster chemical enrichment.** A. Di Fazio, F. Vagnetti,
J. R. Wilson.
Astrophys. Space Sci., Vol. 72, 223 - 231 (1980).

Using a hydrodynamic model of protogalactic evolution, the authors explain the amount of iron in rich clusters, deduced from X-ray observations. The calculations show that a strong shock wave originates in the first violent collapse phase and leaves the protogalaxy, carrying out a substantial fraction of its mass, with roughly solar metallicity and high temperature (10^7 - 10^8 K). The authors also show that hot material ejected from proto-cD galaxies can probably explain the observed X-ray emission in clusters.

151.043 **On the dynamics of locally equilibrium stellar systems.** V. G. Gurzadyan, A. G. Kechek.
Fiz. inst. AN SSSR. Prepr., 1980, No. 38, 13 pp. In Russian.
Abstr. in Ref. zh., 51. Astron., 9.51.664 (1980).

151.044 **Evolution of homogeneous ellipsoids and the structure of rich clusters of galaxies.** A. A. Klypin.
Astron. Zh., Tom 57, 913 - 925 (1980). In Russian. – English translation in Soviet Astron., Vol. 24, No. 5.

The evolution of systems consisting of approximately $\sim 10^4$ gravitating point masses is numerically investigated. The initial configurations of all systems are close to homogeneous oblate ellipsoids. In the course of violent relaxation the particles are divided into some energetic groups, and universal density and velocity dispersion profiles are established. The final configurations are oblate in spite of absence of rotation.

151.045 **Binuclear model of an SB galaxy.**
E. M. Nezhinskij.
Astron. Zh., Tom 57, 926 - 935 (1980). In Russian. – English translation in Soviet Astron., Vol. 24, No. 5.

To explain the cause of the origin of the observed peculiarities of SB galaxies a self-consistent stable model of an SB galaxy is constructed. The evolutionary scheme of origin of this model is described. The model consists of an extensive spherical corona, of two point nuclei and of a cloud of particles of infinitely small mass, the latter being bar-shaped and situated in the neighbourhood of the nuclei.

151.046 **Decay instability of a spiral wave in a stellar disk.**
V. G. Lapin, M. A. Raevskij.
Astron. Zh., Tom 57, 991 - 996 (1980). In Russian. – English translation in Soviet Astron., Vol. 24, No. 5.

Nonlinear interaction of density spiral waves in a plane galaxy is considered. A thin stellar disk is taken as a galaxy model. The possibility of resonance interaction of wave triplets is analysed. An expression for the increment of the nonlinear instability of a moderate-amplitude wave has been obtained. The timescale of the development of this instability for the spiral structure of our Galaxy in the solar vicinity is several periods of revolution of the sun around the galactic center.

151.047 **On the dynamical evolution of clusters of non-point gravitating bodies.** V. A. Churkin.
Pis'ma Astron. Zh., Tom 6, 648 - 650 (1980). In Russian.
English translation in Soviet Astron. Lett., Vol. 6.

An applicability criterion for the material-points model while considering the approaches of non-point bodies is given. It is shown that the dynamical evolution of the cluster may be considered in the frame of the point-mass model if the number of bodies in the cluster is large enough. For some clusters of galaxies the time scale for galaxies to prove their non-point nature is comparable with the ages of clusters.

151.048 **Vår åldrande galax.** G. Lyngå.
Astron. Tidsskr., Årg. 13, 119 - 126 (1980).

151.049 **A numerical model for a triaxial stellar system in dynamical equilibrium. II. Some dynamical features of the model.** D. Merritt.
Astrophys. J., Suppl. Ser., Vol. 43, 435 - 455 (1980).

The first and second velocity moments of the stellar distribution function were calculated numerically for the trixial model of paper I, (see abstract 26.151.009) in two extreme cases: first, in such a way that the mean motions everywhere canceled, and second, such that they took on their maximum permitted value. The ratio of kinetic energy in mean motions to potential energy is about 3% in the maximal case. The pressure is roughly isotropic in the center but becomes increasingly anisotropic in the outer regions, in such a way that the greatest dispersion at any point tends to lie along a radial axis. The radial dispersion peaks near the center, falling off at both large and small radii. Projected rotation and velocity dispersion curves are presented for various orientations.

151.050 **On the interpretation of colors of faint galaxies.** G. Bruzual A., R. G. Kron.
Astrophys. J., Vol. 241, 25 - 40 (1980).

The authors present new calculations for evolving light in galaxies which allow the color distribution expected for faint field galaxies to be computed. The authors normalize the expected counts to data in catalogs of bright galaxies, and find that an excellent fit to Kron's faint photometry can be achieved with a Friedmann model and no other special assumptions.

151.051 **On the interpretation of galaxy counts.** B. M. Tinsley.
Astrophys. J., Vol. 241, 41 - 53 (1980).

New models are presented for the interpretation of recent counts of galaxies to 24th magnitude, and predictions are shown to 28th magnitude for future comparison with data from the Space Telescope. These results supersede earlier, more schematic models by the author. Tyson and Jarvis found in their counts a "local" density enhancement at 17th magnitude, on comparison with the earlier models; the excess is no longer significant when a more realistic mixture of galaxy colors is used. Bruzual and Kron's conclusion that Kron's counts show evidence for evolution at faint magnitudes is confirmed, and it is predicted that some 23d magnitude galaxies have redshifts greater than unity.

151.052 **Application of the virial theorem to clusters of galaxies with unseen mass.** H. Smith, Jr.
Astrophys. J., Vol. 241, 63 - 66 (1980).

Application of the virial theorem to N-body systems with various distributions of unseen mass and a reasonable set of initial conditions gives mass values that are overestimates, both for cylindrical samples and for the entire system. The overestimation in the latter case is as much as a factor 3.5-4 and is implicit in earlier work by Limber.

151.053 **Properties of solutions to integral master equations for closed and open stellar systems.**
H. E. Kandrup.
Astrophys. J., Vol. 241, 334 - 342 (1980).

Integral master equations for the evolution of one-particle probability densities have recently been applied to various problems in stellar dynamics. The purpose of this paper is to discuss those mathematical properties of these equations that are relevant to their applications in stellar dynamics.

151.054 **On normal modes of gas sheets and discs.** L. O'C. Drury.
Mon. Not. R. Astron. Soc., Vol. 193, 337 - 343 (1980).

A method is described for calculating the reflection and transmission coefficients characterizing normal modes of the Goldreich–Lynden-Bell gas sheet. Two families of gas discs without self-gravity for which the normal modes can be found analytically are given and used to illustrate the validity of the sheet approximation.

151.055 **Dynamical theory of collisionless relaxation.** G. Severne, M. Luwel.
Astrophys. Space Sci., Vol. 72, 293 - 313 (1980).

In his theory of violent relaxation, Lynden-Bell gave a rigorous derivation of the equilibrium distribution, but only a qualitative discussion of the manner in which equilibrium is attained. The authors present a fully explicit dynamical theory of collisionless relaxation towards Lynden-Bell equilibrium.

151.056 **Orbits near a 2/3 resonance.** P. Michaelidis.
Astron. Astrophys., Vol. 91, 165 - 174 (1980).

The author studies orbits in a galactic-type potential $V = 1/2\,(Ax^2 + By^2) - \epsilon xy^2$ in the near resonance case $\sqrt{A}/\sqrt{B} \simeq 2/3$. He finds periodic orbits, empirically and by means of the "third integral". Two types of characteristics are found. The regular characteristics which are generated directly or through intermediate characteristics from the periodic orbits of the unperturbed problem and irregular ones which are independent of the first. The author finds orbits which intersect the x-axis up to 14 times perpendicularly. The theoretical periodic orbits were found by using the Lie transforms or by a computer program that gives the "third integral" up to 14th degree. Finally a comparison between the experimental and theoretical shapes of the characteristics is made.

151.057 **Relativistic star clusters with high central redshift.** M. C. Durgapal, P. S. Rawat, R. Banerji.
Pramāṇa, Vol. 15, No. 1, p. 53 - 63 (1980). – Abstr. in Phys. Abstr., Vol. 83, Abstr. 105293 (1980).

151.058 **Theoretical considerations on the dynamics of normal galactic nuclei.** R. H. Sanders.
Highlights of Astronomy, Vol. 5, (see 012.023), 197 - 204 (1980).

The author discusses the origin of non-circular gas motions observed in the nuclei of normal spiral galaxies and the possibility that recurring violent activity in normal nuclei excites such motion.

151.059 **Tri-axial dynamics in the core of normal galaxies.** M. Schwarzschild.
Highlights of Astronomy, Vol. 5, (see 012.023), 205 - 208 (1980).

151.060 **Potential-density relations for flat galaxies.** C. Hunter.
Bull. American Astron. Soc., Vol. 12, 738 (1980). Abstract.

151.061 **N-body simulations of spiral wave amplification.** T. A. Zang.
Bull. American Astron. Soc., Vol. 12, 738 (1980). – Abstract.

151.062 **Evolution of spiral structure in a model galaxy.** J. C. Haass.
Bull. American Astron. Soc., Vol. 12, 739 (1980). – Abstract.

151.063 **Stability of stellar orbits in star clusters and the zero velocity surfaces.** D. W. Keenan.
Bull. American Astron. Soc., Vol. 12, 739 (1980). – Abstract.

151.064 **Collisions of galaxies with massive halos.** R. H. Miller, B. F. Smith.
Bull. American Astron. Soc., Vol. 12, 739 (1980). – Abstract.

151.065 **Stability of Schwarzschild's triaxial model.** B. F. Smith, R. H. Miller.
Bull. American Astron. Soc., Vol. 12, 739 (1980). – Abstract.

151.066 **Slowly rotating elliptical galaxies.** B. F. Smith, R. H. Miller.
Bull. American Astron. Soc., Vol. 12, 749 (1980). – Abstract.

151.067 **Warping of galaxies.** K. A. Papp, K. A. Innanen.
Bull. American Astron. Soc., Vol. 12, 749 - 750 (1980). – Abstract.

151.068 **A driving mechanism for galactic spirals.** K. A. Innanen, K. A. Papp.
Bull. American Astron. Soc., Vol. 12, 750 (1980). – Abstract.

151.069 **An accretion theory of spiral structure.** J. Jaaniste.
Spiral structure of the Galaxy (see 012.034), Abastumansk. Astrofiz. Obs. Byull., No. 52, p. 51 - 54 (1980). In Russian.

A possible mechanism of generation of spiral structure – the accretion of intergalactic gas clouds from a thin layer – is studied. It is shown that such infall can generate a spiral structure and explain the correlation between the rotation curve and the morphological type of the galaxy.

151.070 **A generalized model of the large-scale gravitational field of galaxies.** S. A. Kutuzov, L. P. Osipkov.
Spiral structure of the Galaxy (see 012.034), Abastumansk. Astrofiz. Obs. Byull., No. 52, p. 93 - 108 (1980). In Russian.

151.071 **Hydrodynamical theory of three-dimensional density waves for spiral structure of galaxies (II) – global mode solution and effect of thickness.** J. Xu.
Sci. Sinica, Vol. 23, 1545 - 1558 (1980).

The paper presents the solution of the equation governing density wave propagation in a galaxy disk with finite thickness, its global uniformly valid asymptotic solutions and the dispersion relation. By means of these solutions the influences of disk's thickness, the spiral arm's inclination and other physical factors at the corotation circle of the spiral galaxy are investigated. Results show that with other conditions kept invariant, the thicker the disk, the lower becomes the growth rate of the mode solution. From this it follows that in a thick lenticular or an elliptical galaxy, no spiral structure exists.

151.072 **Bar instability and rotation curves.** J. A. Sellwood.
ESO Sci. Prepr. No. 107, 49 pp. (1980). – Submitted to Astron. Astrophys.

151.073 **Invariant surfaces and orbital behaviour in dynamical systems with 3 degrees of freedom.**
L. Martinet, P. Magnenat.
ESO Sci. Prepr. No. 120, 27 pp. (1980). – Submitted to Astron. Astrophys.

151.074 **Dynamics with variable masses.** J. M. Whittaker.
Proc. R. Soc. London, Ser. A, Vol. 372, 485 - 487 (1980). – Abstr. in Phys. Abstr., Vol. 84, Abstr. 55 (1981).

151.075 **Bisymmetric open-spiral configuration of magnetic fields in disk galaxies in differential rotation.**
T. Sawa, M. Fujimoto.
Publ. Astron. Soc. Japan, Vol. 32, 551 - 566 (1980).

Local induction analyses are made of magnetic fields in a galactic gaseous disk of finite thickness which is surrounded with a huge halo of low-density gas and magnetic fields. If magnetic fields in the disk are transported randomly by interstellar turbulence and diffuse outward across the surface of the disk, and if magnetic fields in the halo can diffuse into the disk simultaneously, the field lines in the disk are in open-spiral configuration without being twisted by differential rotation. The steady-state configuration of this kind is considered as achieved by the circulation of magnetic fields through the halo and disk.

151.076 **Spiral condensation of gas in disk galaxies by bisymmetric twisted magnetic fields: two-dimensional case.** M. Fujimoto, M. Tosa.
Publ. Astron. Soc. Japan, Vol. 32, 567 - 580 (1980).

On the basis of the recent discovery of a bisymmetric twisted magnetic field in the spiral galaxies M51, M81, M33, and possibly in the Galaxy, the authors examine the hydromagnetic motion of the conducting gas in the disk by using a linear approximation. Two spiral condensations are found. The non-gravitational spiral condensation of the gas would contribute to exciting and sustaining the spiral density waves.

151.077 **Recent developments in the mathematical investigation of the initial value problem of stellar dynamics and plasma physics.** J. Batt.
Ann. Nucl. Energy, Vol. 7, 213 - 217 (1980). – Abstr. in Phys. Abstr., Vol. 84, Abstr. 6932 (1981).

151.078 **The morphology of galaxies in dense clusters – nature or nurture?** R. Farouki.
News Lett. Astron. Soc. N. Y., Vol. 1, No. 7, p. 23 - 24 (1980). Abstract.

151.079 **Triaxiality in elliptical galaxies.**
L. Benacchio, G. Galletta.
Mon. Not. R. Astron. Soc., Vol. 193, 885 - 894 (1980).

The existence of a triaxial shape for elliptical galaxies has been considered in recent years to explain the kinematical and geometrical findings. A simple geometrical model of elliptical galaxies having shells with different axial ratios c/a, b/a has been produced to interpret three fundamental key-features of elliptical galaxies: (1) the distribution of the maximum flattening observed; (2) the percentage of ellipticals showing twisting; and (3) the correlation between maximum twisting and maximum flattening. The model has been compared with observational data for 348 elliptical systems. The authors find that a triaxial ellipsoid with coaxial shells having axial ratios c/a and b/a mutually dependent in a linear way can satisfy the observations.

151.080 **Computer simulations of environmental influences on galaxy evolution in dense clusters. I. Ram-pressure stripping.** R. Farouki, S. L. Shapiro.
Astrophys. J., Vol. 241, 928 - 945 (1980).

This paper describes N-body computer simulations of ram-pressure stripping of flat galaxies that the authors have performed to address the S0 problem. Their galaxy models consist of 1000 softened point stars and 100 diffuse gas clouds rotating as an equilibrium disk in the spherical potential of a stabilizing "dark" halo. The validity and behaviour of these models are investigated in depth to reliably interpret their "environmental impact" results. It is found that the ejection of a substantial gas component induces appreciable disruption and thickening in the outer galaxy disk. However, the tightly bound nuclear regions suffer little effect. The simulations indicate that ram-pressure stripping alone is not a viable mechanism for transforming spiral galaxies into S0 types.

151.081 ***N*-body simulations of slow tidal encounters and the formation of galactic halos.**
A. Dekel, M. Lecar, J. Shaham.
Astrophys. J., Vol. 241, 946 - 964 (1980).

Slow hyperbolic encounters, between a 250-body spherical system (a "halo") and a few-body perturber of comparable total mass, are studied by means of N-body simulations. The tidal effect on a system which initially has a decreasing $M(R)/R$ profile ($\rho \propto R^{-3}$) is to produce, or extend, an inner region of constant M/R, to encompass 50 - 70% of the bound mass. The rates of mass loss and energy exchange are investigated.

151.082 **Stability theory of the orbit-averaged Boltzmann equation.** J. R. Ipser, H. E. Kandrup.
Astrophys. J., Vol. 241, 1141 - 1147 (1980).

The question is raised whether the "thermal runaway" apparent in recent calculations of the evolution of stellar systems can be explained in terms of the occurrence of linear instabilities associated with the basic evolution equation. It is shown here that the answer is yes for spherical stellar systems obeying the orbit-averaged Boltzmann equation. The linear stability theory of this equation is developed for its probably unique equilibrium solutions, the isothermal configurations confined by spherical boxes. It is shown that these equilibria become unstable to small spherical perturbations at the critical turning point where the Boltzmann entropy ceases to be a local maximum.

151.083 **Orbits in weak and strong bars.**
G. Contopoulos, T. Papayannopoulos.
Astron. Astrophys., Vol. 92, 33 - 46 (1980).

The authors study plane orbits in simple bar models embedded in an axisymmetric background when the bar density is about 1% (weak), 10% (intermediate) or 100% (strong bar) of the axisymmetric density. Most orbits follow the stable periodic orbits. The basic families of periodic orbits are described.

151.084 **A fluid dynamical flow model for the central peak in the rotation curve of disk galaxies.**
T. Bhattacharyya, B. Basu.
Astrophys. Space Sci., Vol. 73, 395 - 410 (1980).

Solving hydrodynamical equations of motion, a flow model has been derived which imitates very closely the actually observed linear rotational velocity, followed by the falling branch of the rotation curve to minimum. The theoretiical flow model has been compared with observed results for nine galaxies. The agreement obtained is extremely encouraging. The distance of the primary peak of the rotation curve from the galactic centre has been shown to be correlated with the angular velocity in the linear part of the rotation curve. It is concluded that the distance of the primary peak from the centre also speaks of the potential capability of the nucleus of the galaxy for repeating explosions.

151.085 **Does each spiral galaxy house a slowly growing elliptical one?** K. O. Thielheim.
Astrophys. Space Sci., Vol. 73, 499 - 502 (1980).

N-body simulations performed by the author suggest a mechanism for the generation of spiral waves in galaxies in which a mutual quasi-ellipsoidal rotating equilibrium configuration increasing slowly by accretion from the surrounding disk influences the density distribution of stars in the disk such as to give rise to a trailing spiral density wave. Interaction of the spiral wave with the viscous interstellar gas and mutual gravitation between the stars in the disk are believed to influence the form of the spiral. Nevertheless the basic assumption of conventional density wave theory according to which the mutual interaction of stars in the disk is essential for the formation of spirals may not be true.

151.086 **Dynamical theory of binaries in clusters.**
D. C. Heggie.
Globular clusters, (see 003.008), p. 281 - 299 (1980).

The author summarizes some important facts about the effect of encounters between binaries and single stars. Then he uses this information to study the implications for the dynamics of star clusters and the binaries within them.

151.087 **Globular clusters as survivors.**
S. M. Fall.
Globular clusters, (see 003.008), p. 309 - 314 (1980).

151.088 **Late core collapse in star clusters and the gravothermal instability.** H. Cohn.
Astrophys. J., Vol. 242, 765 - 771 (1980).

Numerical Fokker-Planck computations of core collapse in a one-component star cluster are presented. The evolution of the cluster has been followed to the point where the central density has increased by a factor of 10^{20}. During the late stages of the core collapse, nonisothermal self-similar structure develops in the region which lies between the rapidly shrinking isothermal core and the halo. In this region, the radial profiles of the stellar density, the gravitational potential, and the velocity dispersion are characterized by power laws. The results provide strong new evidence for the identification of the late phase of core collapse with the gravothermal instability of Lynden-Bell and Wood.

151.089 **Tidal effects on the mass profile of galactic haloes.**
A. Dekel, M. Lecar, J. Shaham.
Nature, Vol. 286, 135 - 136 (1980).

The authors suggest tidal interactions between galactic haloes, or possibly between their smaller building blocks, while in the hierarchical gravitational clustering process. In this process typical tidal encounters are slow, such that relative orbital velocities of the interacting systems are comparable to the internal velocities of the stars in each system. The two systems are slightly unbound.

151.090 **Dynamics of the Galaxy.** S. A. Stephens.
Non-solar gamma-rays, (see 012.047), p. 191 - 202 (1980).

A large part of the observed gamma rays seems to be resulting from the interaction of cosmic rays with interstellar gas. In this paper the author restricts himself to the theoretical results concerning the structure and dynamics of the Galaxy, relevant to Gamma Ray Astronomy. The first part deals with the spiral pattern of the Galaxy and the implications of density wave theory. The second one concerns with the dynamics of the gas-field system with specific reference to cosmic rays.

151.091 **The evolutionary tendency of binary and multiple galaxies.** J.-x. Rong.
Acta Astron. Sinica, Vol. 21, 361 - 367 (1980). In Chinese.

The following three problems are discussed: (1) The distribution deviation of the numbers of binary and multiple galaxies from the distribution in the ideal statistical equilibrium. (2) The ratio of the numbers of quadrilateral multiple galaxies to all multiple galaxies. (3) The total energy of multiple galaxies. The results of the above three problems are different from Ambartsumian's.

151.092 **The effects of environment on the evolution of elliptical and disk galaxies.** K. M. Strom, S. E. Strom.
Photometry, kinematics and dynamics of galaxies, (see 012.051), p. 37 - 52 (1979).

A variety of observations have raised the possibility that differences in the distribution of Hubble-types and other fundamental structural features, may reflect not only relatively recent environmental interactions but also the conditions which applied during the epoch of galaxy formation. The authors review recent observational work which delineates the observed differences among galaxies located in a variety of environmental settings.

151.093 **Dynamical models of elliptical galaxies.**
G. Monnet, F. Simien.
Photometry, kinematics and dynamics of galaxies, (see 012.051), p. 191 - 194 (1979).

151.094 **Kinematics of spiral and irregular galaxies.**
P. Pişmiş.
Photometry, kinematics and dynamics of galaxies, (see 012.051), p. 235 - 247 (1979).

151.095 **Galaxy structure in terms of distinct components in the mass distribution.** J. Kormendy.
Photometry, kinematics and dynamics of galaxies, (see 012.051), p. 341 - 355 (1979).

Galaxies are considered to be composed of a small number of building blocks, the distinct components in the mass distribution. The aim is to identify these components, and to explore galaxy evolution by studying their dynamics, their origin, and especially their mutual interactions.

151.096 **The stellar dynamics of elliptical systems.**
J. Binney.
Photometry, kinematics and dynamics of galaxies, (see 012.051), p. 357 - 364 (1979).

151.097 **Formation of an elliptical galaxy.**
R. H. Miller, B. F. Smith.
Photometry, kinematics and dynamics of galaxies, (see 012.051), p. 365 - 368 (1979).

151.098 **Elliptical galaxies: oblate or prolate?**
D. O. Richstone.
Photometry, kinematics and dynamics of galaxies, (see 012.051), p. 375 - 378 (1979).

151.099 **Triaxial models of galaxies.**
P. O. Vandervoort.
Photometry, kinematics and dynamics of galaxies, (see 012.051), p. 379 - 380 (1979).

A new family of triaxial stellar systems is the subject of this report. The key to the construction of these new models is provided by the remark that every uniformly rotating gaseous polytrope of polytropic index greater than or equal to 0.5 has an exact stellar-dynamical counterpart.

151.100 **A dynamical test to determine the intrinsic shape of galaxies.** G. Lake.
Photometry, kinematics and dynamics of galaxies, (see 012.051), p. 381 - 386 (1979).

151.101 **Galaxy collisions and mergers.**
S. D. M. White.
Photometry, kinematics and dynamics of galaxies, (see 012.051), p. 389 - 391 (1979).

151.102 **Recent developments in the global density wave theory of disk galaxies.** J. W-K. Mark.
Photometry, kinematics and dynamics of galaxies, (see 012.051), p. 393 - 403 (1979).

Recent studies of global spiral modes have provided one resolution of the problem of long term maintenance of large-scale spiral structure. It appears likely that several spiral modes coexist in many disk galaxies. General mode properties, physical mechanisms and simplified formulae are cited for aid in comparison with observations. Illustration by numerical N-body experiments is noted.

151.103 **Dynamical models of axisymmetric galaxies.**
M. Miyamoto, C. Satoh.
Photometry, kinematics and dynamics of galaxies, (see 012.051), p. 405 - 406 (1979).

151.104 **Global instability in polytropic disk galaxies.**
S. Aoki, M. Noguchi, M. Iye.
Photometry, kinematics and dynamics of galaxies, (see 012.051), p. 411 - 413 (1979).

151.105 **Non-linear corrugation waves in disc galaxies.**
A. H. Nelson, T. Matsuda.
Photometry, kinematics and dynamics of galaxies, (see 012.051), p. 415 - 418 (1979).

The authors report the first results of a numerical simulation of non-linear galactic corrugation waves.

151.106 **Stellar dynamics of barred spirals.**
G. Contopoulos.
Photometry, kinematics and dynamics of galaxies, (see 012.051), p. 425 - 434 (1979).

The stellar dynamics of barred galaxies includes three main problems: 1) Find the main forms of the orbits. 2) Find the density response, i.e. the density due to the concentration of the orbits in a given galaxy, and 3) Find possible self-consistent solutions, i.e. models where the response density equals the imposed bar density. The author ignores the effects of gas and/or any departure from bar-symmetry. Furthermore he considers the galaxies as flat, i.e. he studies the effects on a plane perpendicular to the axis of rotation.

151.107 **Models of barred spiral galaxies.**
E. Athanassoula.
Photometry, kinematics and dynamics of galaxies, (see 012.051), p. 441 - 444 (1979).

The author studies the spiral structure driven by a growing, symmetric and directly rotating bar in a disk with no inner Lindblad resonance. The model galaxy is composed of gas and stars and a spiral structure is found in both these components.

151.108 **Correlation of galaxy rotation curves and the morphology of spiral structure.**
J. Kormendy, C. A. Norman.
Photometry, kinematics and dynamics of galaxies, (see 012.051), p. 445 - 447 (1979).

From a survey of all published rotation curves, the authors find that a galaxy shows a global spiral pattern if it has either nearly solid-body rotation throughout the extent of the spiral arms or a density-wave driving mechanism in the form of a bar or a companion.

151.109 **Warps in the gaseous disks of spirals.**
A. D. Tubbs, R. H. Sanders.
Photometry, kinematics and dynamics of galaxies, (see 012.051), p. 449 - 452 (1979).

The authors propose that long lived warps of the gaseous component exist in regions essentially outside of the massive stellar disk, where the gravitational field is becoming highly spherically symmetric. In a nearly spherically symmetric field the rate of differential precession is slow, and warps may persist for many rotation times.

151.110 **On the bar instability of model galaxies.**
J. A. Sellwood.
Photometry, kinematics and dynamics of galaxies, (see 012.051), p. 457 - 459 (1979).

151.111 **Gas dynamics in ordinary and barred spirals.**
W. W. Roberts, Jr.
Photometry, kinematics and dynamics of galaxies, (see 012.051), p. 461 - 473 (1979).

The dynamics of the gas in ordinary and barred spirals are considered from the standpoint of recent gas dynamical studies, both steady-state and time-evolutionary. Such systems are complex; the gaseous component consists of multiple phases; many diverse phenomena on the small scale contribute to the chaotic nature of the turbulent interstellar medium. A degree of order can often be found on the large-scale. This review focuses on dynamical factors which are believed to contribute to the order and on others which tend to break it up.

151.112 **Stochastic star formation and spiral galaxies.**
H. Gerola, P. E. Seiden.
Photometry, kinematics and dynamics of galaxies, (see 012.051), p. 475 - 478 (1979).

151.113 **Spiral waves and the decomposition of luminosity profiles.** H. C. D. Visser.
Photometry, kinematics and dynamics of galaxies, (see 012.051), p. 479 - 482 (1979).

151.114 **Self-gravitating spiral waves in barred spiral galaxies.** J. M. Huntley.
Photometry, kinematics and dynamics of galaxies, (see 012.051), p. 483 - 486 (1979).

151.115 **Gaseous spiral arms in disk galaxies.** S. A. Sørensen.
Photometry, kinematics and dynamics of galaxies, (see 012.051), p. 489 - 492 (1979).

151.116 **Warping of galaxies.** K. A. Papp, K. A. Innanen. J. R. Astron. Soc. Canada, Vol. 74, 362 (1980).
Abstract.

151.117 **A driving mechanism for galactic spirals.** K. A. Innanen, K. A. Papp.
J. R. Astron. Soc. Canada, Vol. 74, 362 - 363 (1980).
Abstract.

151.118 **The generation of spiral waves in galaxies.** K. O. Thielheim.
Mitt. Astron. Ges., Nr. 50, p. 161 - 164 (1980).

151.119 **Numerical models of star clusters with a central black hole. I. Adiabatic models.** P. Young.
Astrophys. J., Vol. 242, 1232 - 1237 (1980).

Numerical models of star clusters containing a massive black hole are computed for the case of a black hole which grows adiabatically in the cluster center. The growth of the hole is assumed to be at a rate longer than the cluster dynamical time scale but shorter than the relaxation time scale. The angular momentum and radial action of each star in the cluster are conserved during the adiabatic variations. This leads to the invariance of the distribution function in (E, J) space which is used to facilitate the numerical calculations.

Lectures on the density wave theory. See Abstr. 003.096.

Automated graphical plots for the study of the gravitational N-body problem. See Abstr. 021.043.

Periodic solution of the generalized Hill problem. See Abstr. 042.045.

Approximations of higher order resonances with an application to Contopoulos' model problem. See Abstr. 042.063.

Higher order fluid equations for multicomponent nonequilibrium stellar (plasma) atmospheres and star clusters. See Abstr. 062.017.

The nonaxisymmetric configurations of uniformly rotating polytropes. See Abstr. 062.067.

The equilibria of rotating stars and stellar systems. See Abstr. 065.019.

Stellar age and mass determinations from kinematic data. See Abstr. 065.073.

The Jeans instability in an expanding medium. See Abstr. 066.171.

Stellar tidal disruption by a massive binary black hole. See Abstr. 066.172.

The first stars. See Abstr. 131.003.

Molecular clouds and star formation. See Abstr. 131.023.

The collision of clouds with a galactic disk. See Abstr. 131.262.

Orbital dynamics of the radio galaxy 3C 129. II. Internal tail structure. See Abstr. 141.074.

The kinematics and dynamics of the galactic globular cluster system. See Abstr. 154.008.

The tidal radius of a globular cluster. See Abstr. 154.013.

Random gravitational encounters and the evolution of spherical systems. VIII. Clusters with an initial distribution of binaries. See Abstr. 154.022.

Globular clusters and galaxy evolution. See Abstr. 154.036.

On the dynamics of globular clusters. See Abstr. 154.039.

On the relation between local kinematics and galactic structure. See Abstr. 155.016.

Can galaxy warps be used to provide constraints on halo properties? See Abstr. 155.017.

High resolution observations of the neutral hydrogen in the galaxy NGC 925. See Abstr. 158.003.

Dynamical models of the gas flow in the barred spiral galaxy NGC 1300. See Abstr. 158.067.

Hat die Scheibenkomponente von galaktischen Massenmodellen ein zentrales Loch? See Abstr. 158.106.

Ring galaxies. See Abstr. 158.109.

The kinematics of the nuclear spiral of the barred galaxy NGC 1512. See Abstr. 158.188.

The dynamical evolution of NGC 5128. See Abstr. 158.207.

The systematics of galactic disks. See Abstr. 158.254.

Kinematics of early type galaxies: methods and results. See Abstr. 158.268.

Rotation and dispersion profiles in elliptical galaxies. See Abstr. 158.269.

Velocity dispersions in the bulges of spiral galaxies. See Abstr. 158.270.

Kinematics of blue compact galaxies. See Abstr. 158.271.

Rotation of the bulge components of disk galaxies. See Abstr. 158.273.

Large scale distribution and kinematics of neutral hydrogen in elliptical galaxies. See Abstr. 158.275.

Rotation of early-type galaxies from H I line observations. See Abstr. 158.276.

Spectroscopic observation of velocity fields in barred spiral galaxies. See Abstr. 158.282.

The interaction between M81, M82, and NGC 3077. See Abstr. 158.284.

H I line widths and H-band magnitudes: an improved version of the Tully-Fisher relation. See Abstr. 158.288.

The measurement of rotation curves and velocity dispersion profiles for elliptical galaxies: results and implications. See Abstr. 158.294.

Spectral analysis of the spiral pattern of M51. See Abstr. 158.295.

Studies of disk galaxies. See Abstr. 158.296.

Ring structures in barred spirals. See Abstr. 158.298.

A dynamical model of NGC 1097. See Abstr. 158.299.

Observations of the kinematics of barred spiral galaxies. I. NGC 1300. See Abstr. 158.322.

The Magellanic Stream and the Galaxy with a massive halo. See Abstr. 159.018.

Does the binding energy of binaries masquerade as missing mass? See Abstr. 160.033.

NGC 5457 (M101), NGC 5194 (M51) and NGC 5055 (M63) as an interacting group of galaxies. See Abstr. 160.059.

NGC 5236 (M83), NGC 5128 and NGC 4945 as an interacting group of galaxies. See Abstr. 160.060.

Galactic cannibalism. IV. The evidence – correlations between dynamical time scales and Bautz-Morgan type. See Abstr. 160.062.

X-ray emission from clusters and the formation of galaxies. See Abstr. 160.083.

Dynamical models and the mass of the Virgo cluster. See Abstr. 160.086.

Errata

151.901 **Erratum: "Ring galaxies – a review"** [Bull Astron. Soc. India, Vol. 7, 32 - 37 (1979)].
T. K. Chatterjee.
Bull. Astron. Soc. India, Vol. 8, 30 (1980).
See Abstr. 26.151.071.

151.902 **Erratum: 'Density wave-star interaction of a differentially rotating spiral system in a scalar field'** [Astrophys. Space Sci., Vol. 54, 467 - 478 (1978)].
E. Evangelidis.
Astrophys. Space Sci., Vol. 72, 255 (1980). – See Abstr. 21.151.090.

152 Stellar Associations

152.001 **On the initial mass function: the mass spectrum of young OB associations.**
M. Claudius, P. J. Grosbøl.
Astron. Astrophys., Vol. 87, 339 - 342 (1980).

uvbyβ data for the young associations Orion OBI, NGC 2264, α Persei, Scorpius, and Centaurus have been used to derive the initial mass function (IMF) for the mass range $2.2m_{\odot}-10m_{\odot}$. This was done by calculating individual stellar masses, combining the position in the HR-diagram and theoretical isochrones. The authors have found that the upper IMF for all the associations could be approximated by a power law with the slope of 1.9.

152.002 **On the sequential formation of subgroups in OB associations.** P. J. Bedijn, G. Tenorio-Tagle.
Astron. Astrophys., Vol. 88, 58 - 60 (1980).

Sequential formation of subgroups in OB associations as proposed by Elmegreen and Lada (1977) is investigated, taking into account the fact that subgroups move towards the parent cloud from which they were formed. A model association is compared with observations.

152.003 **Association membership for the 20 day Cepheid RU Scuti.** D. G. Turner.
Astrophys. J., Vol. 240, 137 - 144 (1980).

New photoelectric *UBV* photometry and MK spectral types are presented for luminous stars in the vicinity of the 19ᵈ7 Cepheid RU Scuti. Two distinct groupings of stars are delineated: a nearby group which is associated in terms of distance, reddening and evolutionary status with the cluster Trumpler 35, and a distant group consisting of stars of larger reddening and younger evolutionary status. Based upon the arguments of apparent agreement in reddening and evolutionary status, RU Scuti is a likely member of the nearer Trumpler 35 association, which may consist solely of coronal members of this cluster and nearby Dolidze 32.

152.004 **The effect of mass loss on the age-determination of young clusters, with an application to the Orion OB-association.** F. B. S. Paerels, H. J. G. L. M. Lamers, C. de Loore.
Astron. Astrophys., Vol. 90, 204 - 206 (1980).

The effect of mass loss on the isochrones and the main-sequence turn-up for young clusters is studied. Assuming reasonable mass loss rates the authors find that the effect on the determination of the nuclear age of young clusters is very small. The results are applied to the Orion OB-association. Mass loss cannot explain the discrepancy of about a factor of two between the nuclear age and the shorter kinematic age of young clusters.

152.005 **Multicolour *UBVRI* photometry of stars in M 17.** R. Chini, H. Elsässer, T. Neckel.
Astron. Astrophys., Vol. 91, 186 - 193 (1980).

Multicolour *UBVRI* photometry of stars towards M 17 demonstrate the existence of a young stellar group with partly heavily obscured O and B stars in this region. The photometric distance of this cluster is found to be 2.2 ± 0.2 kpc. The photometric data can only be interpreted by assuming an abnormal reddening law with $R = 4.2$ inside the dark cloud of M 17. The problem of the energy balance of M 17 is discussed by means of the observed early type stars.

152.006 **Luminous stars beyond the solar circle: investigation of a galactic field at $l = 231°$.**
M. P. FitzGerald, A. F. J. Moffat.
Mon. Not. R. Astron. Soc., Vol. 193, 761 - 774 (1980) = Contrib. Univ. Waterloo Obs., Ontario, Canada, No. 81.

Fifty-one candidate luminous field stars from the SLS catalogue, 30 stars in the young open cluster NGC 2414, 13 in a new mini-cluster Wat 7, and six additional stars, all within ~1° of the H II regions S299/300 at $l \sim 231°.0$, $b \sim 1°.5$, have been observed spectroscopically and photometrically. Nine new members were found in NGC 2414 for a total of 19, and eight members identified in Wat 7. Twenty-nine OB stars appear to lie in an association at a distance of 4.2 ± 0.2 kpc, the same as NGC 2414. The remaining stars appear to lie along the line-of-sight in the Orion arm.

152.007 **A 2.3 GHz radio continuum map of the upper Scorpio region.**
E. E. Baart, G. de Jager, P. I. Mountfort.
Astron. Astrophys., Vol. 92, 156 - 162 (1980).

An area between Dec. −15° and −35° and R. A. 15h43m and 16h40m was observed at a wavelength of 13 cm with a resolution of 20′ and a sensitivity better than 30 mK. A large region of emission (12 × 10°) has been found enclosing most of the stars of the Sco OB2 association. Smaller radio emission features are found to correlate with Hα features excited by early B type stars. The strongest source observed coincides with the H II region Sharpless 9. Emission also occurs near areas of high obscuration which contain early B type stars.

152.008 **A survey of emission-line stars in the Per OB2 dark cloud.** C.-p. Liu, C.-s. Zhang, H. Kimura.
Acta Astron. Sinica, Vol. 21, 354 - 360 (1980). In Chinese.

The results of an objective prism survey for Hα emission-line stars in the 5° × 5° region around NGC 1333 are presented. The authors have identified 25 stars as those having the Hα-line in emission. These are suspected to be mainly T Tauri stars.

152.009 **21-cm observations of the Cep IV star-formation region.** E. J. Grayzeck.
Astron. J., Vol. 85, 1631 - 1637 (1980).

The Cep IV star-formation region has been mapped at 21 cm. The resulting H I distribution indicates that the available gas forms a broken ring coincident with the optical nebulosity. Comparison of this feature with other kinematic data for both stars and gas suggests that sequential star formation is an ongoing process in this region.

152.010 **A detailed investigation of the R association containing the 1ᵈ95 cepheid SU Cassiopeiae.**
D. G. Turner, N. R. Evans.
J. R. Astron. Soc. Canada, Vol. 74, 359 - 360 (1980).
Abstract.

Radio emission from Cyg OB 2 No. 12.
See Abstr. 116.024.

Evidence for shocked interstellar gas toward the Perseus OB2 association. See Abstr. 131.045.

Studies of ultraviolet interstellar extinction with the Sky-Survey telescope of the TD-1 satellite. Results for different OB-associations. See Abstr. 131.110.

Comparison of submillimeter and CO brightness in Orion and Mon R2. See Abstr. 131.187.

Interstellar line spectra of a dense cloud: the VI Cygni association. See Abstr. 131.206.

Giant molecular complexes and OB associations. I. The Rosette molecular complex. See Abstr. 131.251.

Velocity structure in the Canis Major R1 molecular clouds. See Abstr. 131.288.

Two-dimensional radiation-hydrodynamics calculations of the formation of O-B associations in dense molecular clouds. See Abstr. 131.309.

Observations of radio emission in the 18-cm hydroxyl lines in the direction of Herbig-Haro objects and reflection nebulae. See Abstr. 131.321.

Dispersion einer Gruppe von Sternen im Gravitationsfeld einer Spiralgalaxie. See Abstr. 151.038.

The open cluster NGC 3293 and the OB complex in Carina. See Abstr. 153.001.

The young open cluster NGC 3293 and its relation to Car OB1 and the Carina Nebula complex.
See Abstr. 153.008.

The young open cluster Ruprecht 55.
See Abstr. 153.024.

Structure of regions of star formation. II. An observational study of some regions. 1. Spatial extension, mass and age of RSF Ori 1. See Abstr. 153.027.

153 Open Clusters

153.001 **The open cluster NGC 3293 and the OB complex in Carina.** A. Feinstein, H. G. Marraco.
Publ. Astron. Soc. Pacific, Vol. 92, 266 - 274 (1980).

Photoelectric measures in *UBVRI* of 46 stars in and around the open cluster NGC 3293 are presented. A distance modulus of $V_0 - M_v = 12.1 \pm 0.15$ (s.d.), corresponding to 2.6 kpc, was derived. The age of the cluster is about 7×10^6 years.

153.002 **Spectral types in the open cluster NGC 6231.** H. Levato, S. Malaroda.
Publ. Astron. Soc. Pacific, Vol. 92, 323 - 327 (1980).

The authors have determined accurate MK types for 26 of the brightest stars in the field of NGC 6231. They found one emission-line star with a weak shell, one star with C III strong (already known), four stars with double lines, (two of them previously known). Fitting their H-R diagram with the faint portion studied by Garrison and Schild provides an accurate H-R diagram for this young cluster.

153.003 ***uvbyβ*** **photometry of the young southern cluster NGC 3293 and comparison with other young clusters.** R. R. Shobbrook.
Mon. Not. R. Astron. Soc., Vol. 192, 821 - 840 (1980).

Strömgren *uvby* photometry has been obtained for 42 members and β photometry for 37 members of the young southern galactic cluster NGC 3293. The distance obtained is 3.55 kpc. Comparison of the colour/colour and the HR diagrams of NGC 3293 with those of five other young northern and southern clusters reveals large differences between the clusters. The fainter stars in the southern clusters appear to be an average of 0.7 mag brighter than those in the northern clusters, but it is not certain at present how much of this difference is due to possible systematic errors in the β index zero point between the northern and southern hemispheres.

153.004 **Membership in the open cluster NGC 6494. Astrometry with a PDS microdensitometer.**
W. L. Sanders, R. Schröder.
Astron. Astrophys., Vol. 88, 102 - 107 (1980).

An image position finding procedure for PDS microdensitometer measures of astrometric plates is described and applied to the open cluster NGC 6494. Probabilities of membership based on relative proper motions for 304 stars in the cluster field yield a cluster proper motion dispersion (m.e.) of 0."0005 and 169 probable members.

153.005 **Galactic distribution of the oldest open clusters.** S. van den Bergh, R. D. McClure.
Astron. Astrophys., Vol. 88, 360 - 362 (1980).

The oldest known open clusters are found to be strongly concentrated in the outer part of the galactic disk. It is suggested that the relatively high survival rate of clusters in the outer disk might possibly be due to the fact that such objects suffer relatively few disruptive encounters with giant molecular clouds which are mainly located in the inner disk of the Galaxy.

153.006 **The metallicity of M67.** J. P. Norris, R. A. Bell, D. Butler, D. Deming.
Bull. American Astron. Soc., Vol. 12, 458 (1980). – Abstract.

153.007 **The beat Cepheid V367 Scuti and NGC 6649.** S. L. Barrell.
Astrophys. J., Vol. 240, 145 - 148 (1980).

A program of radial velocity measurements of the open cluster NGC 6649 is reported, and it is asserted that the beat Cepheid V367 Sct is indeed a member of the cluster on the basis of these observations and previous evidence. Pulsation and evolution masses are calculated for V367 Sct, and it is shown that the beat Cepheid has a mass typical of a population I Cepheid with no significant difference between pulsation and evolution masses.

153.008 **The young open cluster NGC 3293 and its relation to Car OB1 and the Carina Nebula complex.**
D. G. Turner, G. R. Grieve, W. Herbst, W. E. Harris.
Astron. J., Vol. 85, 1193 - 1206 (1980).

From new photometric and spectroscopic observations, the authors find that stars in NGC 3293 are fairly typical of members of moderately young open clusters. The resulting reddening-free H-R diagram for NGC 3293 exhibits a very tight main-sequence band, and leads the authors to estimate an age of $5(\pm 2) \times 10^6$ yr and a distance of 2.50 ± 0.15 kpc for this cluster. It is clearly physically related to other objects belonging to the Carina Nebula complex.

153.009 ***RGU*** **photographic photometry of the open cluster NGC 2194.** G. del Rio.
Astron. Astrophys., Suppl. Ser., Vol. 42, 189 - 192 (1980).

The galactic cluster NGC 2194 has been studied by means of three-color photometry in the *RGU* system. For this cluster a distance of 2740 pc and a reddening $E(G\text{-}R) = 0.58$ were found. From the existence of a giant branch in both color-magnitude diagrams, an evolutionary effect can be attributed to this cluster.

153.010 **The distribution of main-sequence stars in the open cluster NGC 7789.** U. Hopp.
Astrophys. Space Sci., Vol. 72, 233 - 236 (1980).

By star counts on four plates with different limiting magnitudes, it was found that the relative frequency of main-sequence stars increases from the centre of the cluster to the boundary. It is shown that the stars of NGC 7789 are separated according to their masses.

153.011 **Observations of pre-main-sequence stars in the Pleiades.** J. R. Stauffer.
Astron. J., Vol. 85, 1341 - 1353 (1980).

New photometric and spectroscopic observations of faint stars in the Pleiades cluster are presented. Most of the stars observed were chosen on the basis of their proper motions; many are also flare stars. Results of the *BVRI* photometry show a pre-main-sequence turn-on at about $M_v = +9^{m}.4$ (spectral type dM2 in Joy and Abt's system), from which theoretical pre-main-sequence evolutionary models allow to infer a contraction age for the cluster of $\tau_c = 2.2 \times 10^8$ yr. IDS spectroscopic observations of many of the program stars are used to provide approximate spectral types and to study the emission-line characteristics of these stars. The spectroscopic observations are also used in combination with infrared observations of a few of the program stars (Zappala 1974) to show that circumstellar dust shells are unlikely to be present to a significant degree around these low-mass Pleiades stars.

153.012 **Are binaries concentrated toward the centers of open clusters?** H. A. Abt.
Astrophys. J., Vol. 241, 275 - 278 (1980).

The distances in the plane of the sky from the centers of eight open clusters have been derived for 48 spectroscopic and visual binaries and 113 constant-velocity stars. These show no evidence for a concentration of binaries toward the centers of young clusters (ages $< 10^8$ yr) but 2 σ evidence for a binary concentration toward the centers of old clusters (ages $\geq 10^8$ yr) or 2.4 σ evidence from spectroscopic binaries alone. The detailed radial distributions of binaries in old clusters confirm this concentration, and there is some evidence to show that the binaries near the cluster centers are harder (or of shorter periods).

153.013 **Photoelectric *UBV* and DDO photometry of NGC 5138.** J. J. Clariá.
Astrophys. Space Sci., Vol. 72, 347 - 357 (1980).

Results of *UBV* photoelectric photometry in NGC 5138 are presented for 50 stars brighter than 14.0 mag. In addition, four probable red giants were also observed in the DDO system. Sixteen stars previously considered members by Lindoff (1972), were found not to be physically connected with the cluster. NGC 5138 is located 1.80 kpc from the Sun and the visual interstellar absorption determined from the reddened *B* stars amounts to $A_v = 0.75$ mag. Three of the four red stars observed in the DDO system were found to be cluster members. The mean cyanogen anomaly implies that NGC 5138 is richer in CN than the field K giants in the solar neighbourhood, but poorer than the Hyades giants. The cluster age is estimated to be $\sim 1.5 \times 10^8$ yr.

153.014 **Über den Aufbau der Stern- und Galaxienhaufen.** W. Lohmann.
Astrophys. Space Sci., Vol. 72, 439 - 446 (1980).

In star clusters and in galaxy clusters the exponent n of the generalized law of Schuster statistically decreases with increasing masses M of the clusters or increasing numbers L of their members. In most cases n(gal. cl.) is smaller than n(st. cl.). It is confirmed, that in larger galaxy clusters the brighter members are somewhat more concentrated at the center than the fainter members.

153.015 **The mass function for stars in a cluster: a theoretical derivation.** S. K. Bhattacharjee, I. P. Williams.
Astron. Astrophys., Vol. 91, 85 - 89 (1980).

A theoretical mass function has been derived for stars in a cluster under the assumption that the initial fragmentation of a gas cloud is into similar massed stellar nuclei, the fragmentation being triggered off in a sequence due to the passage of a density wave. The stellar nuclei thus form at different times, the maximum spread in the formation time being the shock crossing time. Subsequently, these nuclei increase in mass by accreting matter from the surrounding space. The final mass of the star is determined by the time for which accretion proceeds. Assuming the rate of fragment formation to be dependent on the geometry of the cloud, the mass distribution for stars in a model cluster has been determined for a number of different cases.

153.016 **Etoiles doubles dans les amas proches.** M. Bischoff.
Bull. Inf. Cent. Données Stellaires, No. 19, p. 11 - 13 (1980).

153.017 **The iron abundance in the Hyades cluster.** D. Branch, D. L. Lambert, J. Tomkin.
Astrophys. J., Lett., Vol. 241, L83 - L87 (1980).

Low-noise Reticon spectra of the Moon and two Hyades G dwarfs are used to derive the metallicity of the Hyades cluster. The relative temperatures of the Sun and the Hyades stars are determined from the wings of their Hα profiles, and iron abundances are based on weak Fe I lines. The spectroscopic result is [Fe/H] = 0.20 ± 0.1, in good agreement with photometric determinations of the Hyades metallicity. The $B - V$ color of the Sun is inferred to be 0.64 ± 0.02.

153.018 **A ghost image of Sirius as a hiding place for a new star cluster.**
G. Auner, J. Dengel, H. Hartl, R. Weinberger.
Publ. Astron. Soc. Pacific, Vol. 92, 422 - 425, (1980).

A hitherto unreported open star cluster of small angular extent (~ 2.5 arc min), located within the "ghost" image of Sirius, was identified on the Palomar Sky Survey print no. 1343. The cluster was also examined on a copy of an ESO (B)-plate and is found to be rich ($n \sim 140$ members), slightly reddened ($A_V \sim 1^m$), and distant ($d \gtrsim 4$ kpc).

153.019 **A *UBV* photometric study of the open cluster NGC 654.** R. C. Stone.
Publ. Astron. Soc. Pacific, Vol. 92, 426 - 431 (1980).

Accurate photographic *UBV* magnitudes and colors have been determined for 83 stars in the region of the young open cluster NGC 654. One-third of these stars lie outside of the cluster nuclear region, indicating the presence of a stellar corona. Over the extended region, the reddening varies from $E(B - V) = 0.77$ to 1.13 magnitudes with several stars having larger reddenings, and a variable-extinction analysis gives an averaged absorption ratio of $R = 2.9 \pm 0.6$ (s.e.). The cluster H-R diagram shows a well-defined main sequence and includes possibly as many as five B-F supergiant members. Moreover, a study of the nuclear and coronal cluster regions was made, and the cluster luminosity function was compared with the Initial Luminosity Function of Sandage (1957).

153.020 **The Puppis open cluster pair Czernik 29 and Haffner 10.** M. P. Fitzgerald, A. F. J. Moffat.
Publ. Astron. Soc. Pacific, Vol. 92, 489 - 492 (1980) = Contrib. Univ. Waterloo Obs. No. 80.

Star counts to $B = 18^m$ show each cluster to have central projected star number density 2 - 3 times that of the surrounding field. With radii ~ 2′ and centers separated by only 3′.2 the cluster overlap in the sky. However, photographic *UBV* photometry to $V = 15.1$ indicates the clusters to have different ages. Cz 29 is located 2.9 ± 0.5 kpc from the sun at $l \sim 231°$, $b \sim +1°$ where previous work has already shown the existence of a weak concentration of stars in the galactic disk. Haf 10 has significantly fainter stars but may be located at nearly the same distance.

153.021 **Five open clusters with suspected Wolf-Rayet type members.** I. Lundström, B. Stenholm.
Rep. Obs. Lund, No. 16, 31 pp. (1980).

Photometric observations of five open clusters with suspected Wolf-Rayet type members are presented. The photometric system used is very similar to that of Smith (1968). Memberships of Wolf-Rayet stars in the observed clusters are discussed. Intrinsic colors and absolute magnitudes are derived for the Wolf-Rayet type members. Transformations between the present photometric system and the UBV system are derived.

153.022 **The Einstein central Hyades survey: a progress report.** R. Stern, J. H. Underwood,
M.-C. Zolcinski, S. Antiochos.
Smithsonian Astrophys. Obs., Spec. Rep. 389, (see 012.038), p. 127 - 131 (1980).

153.023 **The Pleiades.** T. Arny, K. J. Gordon.
Mercury, Vol. 9, 113 - 114, 129 (1980).

153.024 **The young open cluster Ruprecht 55.** R. J. Dodd, L. A. Ellery.
Mon. Not. R. Astron. Soc., Vol. 193, 895 - 900 (1980).

UBV photographic photometry of 303 stars, to a limiting magnitude of $V = 17.2$ is used to investigate the young open cluster Ruprecht 55. A distance of 5 kpc with a foreground colour excess of $E_{B-V} = 0.58$ mag and an age of $< 10^7$ yr is estimated for the cluster. It is suggested that Ruprecht 55 may be part of a more extensive association of OB stars.

153.025 **The blue stragglers in NGC 7789 and the binary post-main-sequence mass-exchange hypothesis.**
B. J. McNamara.
Publ. Astron. Soc. Pacific, Vol. 92, 682 - 687 (1980).

Using a recently completed proper-motion membership study of NGC 7789 and photographic magnitudes and colors, 29 blue stragglers are identified. The data suggest that the blue stragglers have masses appropriate to their observed main-sequence positions. Based on a model suggested by Strom and Strom (1970) of the binary origin of blue stragglers, the predicted number of cluster blue stragglers and the number observed are found to be in good agreement. However, this agreement rests on the untested assumption of the similarity of the Hyades and NGC 7789 binary frequency. Five blue stragglers are also identified which are estimated to have masses above the limiting binary value of $M_{BS} \lesssim 2 M_{TO}$. This results, if substantiated, casts severe doubt on the validity of the binary mass-exchange hypothesis of blue stragglers.

153.026 **UBV photometry of the highly reddened open cluster Berkeley 87.** D. Forbes.
J. R. Astron. Soc. Canada, Vol. 74, 360 - 361 (1980).
Abstract.

153.027 **Structure of regions of star formation. II. An observational study of some regions. 1. Spatial extension, mass and age of RSF Ori 1.** V. S. Shevchenko.
Astron. Zh., Tom 57, 1162 - 1173 (1980). In Russian.
English translation in Soviet Astron., Vol. 24, No. 6.

The H-R diagrams of the open clusters Cr 69, Cr 70, NGC 1976, NGC 2024 and M 78 are compared. Photoelectric photometry and quantitative spectral classification have been carried out for the members of the cluster Cr 69 (λ Ori). All the clusters and O-T-associations in the region represent a single complex RSF Ori 1. The extension of RSF Ori 1 in the south-north direction is 190 pc and in the west-east one, as well as in the line of sight, is 90 pc. The age of RSF Ori 1 as a whole is more than 10^7 years. The mass of stars in RSF Ori 1 is $7 \times 10^3 M_\odot$.

Catalogue of masses and ages of stars in 68 open clusters. See Abstr. 002.042.

Webb Society deep-sky observer's handbook, Vol. 3: Open and globular clusters. See Abstr. 003.061.

A new method to restore the spatial distribution of stars in globular clusters and its application to flare stars in the Pleiades. See Abstr. 031.515.

The evolution of mixed long-lived stars. See Abstr. 065.094.

The sun among the stars. II. Solar color, Hyades metal content, and distance. See Abstr. 080.006.

Recent results of BVRI photometry of late-type dwarf stars in nearby clusters and in the solar neighborhood. See Abstr. 113.008.

Wide- and narrow-band photometry of stars in a field around Collinder 135. See Abstr. 113.026.

Photometry and polarimetry of the extreme blue straggler K1211 in NGC 7789. See Abstr. 113.027.

Calibration of photometric abundance determinations. See Abstr. 113.045.

A note on the existence of W UMa-type systems in the cluster Cr 359. See Abstr. 117.063.

Reply to the note of Ruciński on the existence of W UMa-type systems in the cluster Cr 359. See Abstr. 117.064.

RY Cancri is not a member of Praesepe. See Abstr. 119.001.

Die Hülle der Wolf-Rayet-Komponente des Doppelsterns HD 152270. See Abstr. 120.019.

A cepheid variable in NGC 6067. See Abstr. 122.124.

Flare stars in the Pleiades. See Abstr. 122.164.

The Cepheid luminosity scale. See Abstr. 122.189.

Variability of cool carbon stars situated near or in intermediate-age open clusters. See Abstr. 123.009.

Variable star in the galactic cluster NGC 7654. See Abstr. 123.013.

Photographische Beobachtungen an veränderlichen Sternen des Praesepe-Haufens. See Abstr. 123.015.

A photometric search for white dwarfs in IC 2602. See Abstr. 126.007.

A search for variability in white dwarfs in the region of the Hyades. See Abstr. 126.031.

The interstellar spectrum of the central object of NGC 3603 (HD 97950). See Abstr. 131.009.

Computer simulations of close encounters between single stars and hard binaries. See Abstr. 151.035.

The effect of mass loss on the age-determination of young clusters, with an application to the Orion OB-association. See Abstr. 152.004.

Luminous stars beyond the solar circle: investigation of a galactic field at $l = 231°$. See Abstr. 152.006.

Luminosities and temperatures of the reddest stars in three LMC clusters. See Abstr. 159.006.

154 Globular Clusters

154.001 **Pal 12 – a metal-rich globular cluster in the outer halo.**
J. G. Cohen, J. A. Frogel, S. E. Persson, R. Zinn.
Astrophys. J., Vol. 239, 74 - 77 (1980).

New optical and infrared observations of several stars in the distant globular cluster Pal 12 show that they have CO strengths and heavy element abundances only slightly less than in M71, one of the more metal-rich globular clusters. Pal 12 thus has a metal abundance near the high end of the range over which globular clusters exist and lies in the outer galactic halo. Its red horizontal branch is not anomalous in view of the abundance the authors have found.

154.002 **The dying globular cluster E3.**
S. van den Bergh, S. Demers, W. E. Kunkel.
Astrophys. J., Vol. 239, 112 - 120, plates 2 - 4 (1980).

Photoelectrically calibrated CTIO 4 m plates suggest that E3 is probably a very star-poor globular cluster that is reddened by $E_{B-V} \approx 0.28$. The cluster is located at a distance $R = 8.1$ kpc from which $\varpi = 9.4$ kpc and $Z = -2.6$ kpc. The cluster has $M_V = -4.2$, which makes it one of the faintest known globular clusters. The cluster E3 has a core radius $r_c = 4.4$ pc and a tidal radius $r_t = 26.4$ pc. This indicates that E3 may have been very severely truncated by tidal forces. Loss of stars resulting from tidal effects might be partially responsible for the present low luminosity of E3. The evolved star region of the cluster color-magnitude diagram shows an unusually large scatter. This scatter may be due to the fact that this cluster has preferentially retained binary systems which are more resistant to tidal stripping than are single stars.

154.003 **The composition of the anomalous metal rich globular cluster NGC 288.**
C. A. Pilachowski, C. Sneden.
Bull. American Astron. Soc., Vol. 12, 502 - 503 (1980).
Abstract.

154.004 **Accurate equatorial positions for the centers of six X-ray globular clusters.**
S. J. Shawl, R. E. White.
Astrophys. J., Lett., Vol. 239, L61 - L63, plates L3, L4 (1980).

The central positions of the blue photographic image of six X-ray globular clusters, NGC 1851, 6440, 6441, 6624, 6712, and 7078 (M15), have been determined in the equatorial coordinate system to an accuracy on the order of $\pm 1''$. This two-orders-of-magnitude improvement in cluster central positions is a by-product of a larger program to determine axial ratios and orientations for 120 galactic globular clusters.

154.005 **Color-magnitude photometry to the main sequence for the "anomalous" globular cluster Palomar 12.**
W. E. Harris, R. Canterna.
Astrophys. J., Vol. 239, 815 - 838 (1980) = Contrib. Louisiana State Univ. Obs., No. 156.

New photoelectric and photographic color-magnitude photometry to $V = 22$ is presented for the globular cluster Palomar 12. The C-M diagram morphology most closely resembles 47 Tucanae, and all combined evidence suggests that Pal 12 is a moderately metal-rich cluster in agreement with the recent red-giant abundance studies by Cohen. The authors derive a preliminary age estimate of 11×10^9 years for Pal 12 by fitting their main-sequence data to the Yale isochrones in the $(M_V, B-V)$-plane.

154.006 **Photoelectric photometry of globular clusters in the Andromeda nebula and its companion NGC 147. VI.**
A. S. Sharov, V. M. Lyutyj, V. F. Esipov.
Pis'ma Astron. Zh., Tom 6, 571 - 573 (1980). In Russian.
English translation in Soviet Astron. Lett., Vol. 6.

Results of photoelectric *UBV* observations of 18 globular clusters in M31 and its companion NGC 147 obtained in 1979 are presented.

154.007 **Red stars in Magellanic Cloud globular clusters.**
T. Lloyd Evans.
Mon. Not. R. Astron. Soc., Vol. 193, 87 - 96 (1980).

Direct photographs of six SMC and 36 LMC globular clusters have been taken in the *V* and *I* wavebands. These have revealed 235 red stars, more than half of them close enough to the cluster centre to be considered probable members. Spectroscopic observations by the author and others show that the redder stars are either M stars or carbon stars, which cannot be distinguished by the present observations alone. Only five of the LMC clusters appear to be comparable to galactic globular clusters.

154.008 **The kinematics and dynamics of the galactic globular cluster system.** C. S. Frenk, S. D. M. White.
Mon. Not. R. Astron. Soc., Vol. 193, 295 - 311 (1980).

The authors formulate a general method for testing kinematical models of the globular cluster system in our Galaxy against observational data. Their data are consistent with a systemic rotation velocity for the cluster population independent both of position and of cluster metallicity and taking the value 60±26 km s^{-1} for an assumed galactic rotation velocity of 220 km s^{-1} on the solar circle. The observed velocity dispersion of the cluster system increases significantly with galactocentric distance and is inconsistent at the 95 per cent confidence level with any isothermal or polytropic distribution function. All acceptable models predict a low value for the circular velocity at the solar radius and the authors' most plausible models suggest 200 km $s^{-1} \lesssim v_{\odot} \lesssim 225$ km s^{-1}. The globular cluster data thus demonstrate the existence of a massive halo in the Milky Way.

154.009 **The extended giant branches of intermediate age globular clusters in the Magellanic Clouds.**
J. Mould, M. Aaronson.
Astrophys. J., Vol. 240, 464 - 477 (1980).

Vidicon spectra and infrared *JHK* photometry are presented for stars near the tip of the giant branch in a sample of red globular clusters in the Magellanic Clouds. The coverage extends considerably that of the authors' earlier spectroscopic survey. They again find numerous carbon stars, and some M stars, whose luminosities place them on the upper asymptotic giant branch (i.e., above the luminosity of the helium flash). From the IR photometry, the mean bolometric magnitude of carbon stars in the LMC and SMC clusters is found to be -5.02 ± 0.10 mag and -4.69 ± 0.10 mag, respectively. Effective temperatures based on $J-K$ colors are given for all stars. A simplified theory of asymptotic giant branch evolution is applied to calculate ages for the full sample of clusters studied to date.

154.010 **The bright globular cluster in NGC 5128 and its distance.** G. de Vaucouleurs.
Astrophys. J., Lett., Vol. 240, L93 - L94 (1980).

The bright cluster discovered in NGC 5128 by Graham and Phillips satisfies the general relation between the magnitude of the brightest cluster and of the spheroidal component in galaxies. Its colors agree closely with the observed total colors of NGC 5128, $(B-V)_T = 0.98$, $(U-B)_T = 0.41$; the colors corrected for galactic extinction, $(B-V)_T{}^0 = 0.86$, $(U-B)_T{}^0 = 0.30$, are consistent with spectral type G5.5.

154.011 **Photometric studies of composite stellar systems. IV. Infrared photometry of globular clusters in M31 and a comparison with early-type galaxies.**
J. A. Frogel, S. E. Persson, J. G. Cohen.
Astrophys. J., Vol. 240, 785 - 802 (1980).

The results of an infrared photometric investigation of 40 globular clusters in and around M31 are presented. A comparison of the $(V-K)_0$ colors of the M31 globulars with those of galactic globulars allows an independent derivation of the metallicities of individual M31 globulars. The broad-band infrared data are compared with predictions from integrated light models based on the Ciardullo and Demarque isochrones. The agreement is quite good for models with an initial mass function of slope $\lesssim$ the Salpeter value independent of metallicity, thus ruling out the possibility that a late-type dwarf component is making a significant contribution to the infrared light. CO and H_2O indices measured for eight and seven of the clusters, respectively, give the same result. Early-type galaxies are seen to have much redder broad-band colors and stronger CO and H_2O indices than the most metal-rich M31 or galactic globulars observed. The luminosity functions for the M31 and the galactic globulars are examined with the aid of models to investigate the possibility that metal-enhanced star formation or variations in the initial mass function can be detected in integrated light. Two appendices present new infrared data for a faint dE galaxy in the Virgo cluster, and a recalibration of the integrated light models presented by Aaronson et al.

154.012 **Time-resolved imaging and spectral studies of an X-ray burst from the globular cluster Terzan 2.**
J. E. Grindlay, H. L. Marshall, P. Hertz, A. Soltan, M. C. Weisskopf, R. F. Elsner, P. Ghosh, W. Darbro, P. G. Sutherland.
Astrophys. J., Lett., Vol. 240, L121 - L125, plate L11 (1980).

The first image of an X-ray burst was recorded with the HRI detector at the Einstein Observatory while observing the globular cluster Terzan 2. The burst was coincident with a persistent X-ray source located near the center of the cluster. After a rapid rise to peak luminosity, a double-peaked spectral variation was observed over the next ~ 20 s with anticorrelated changes in the apparent emission region radius and temperature derived from blackbody (and "modified" blackbody) spectral fits. A shell or disk geometry, which undergoes adiabatic expansion and contraction, may be implied for the burst emission region. Alternatively, Comptonization is required. The authors also show that the peak burst luminosity must exceed the Eddington limit.

154.013 **The tidal radius of a globular cluster.**
D. W. Keenan.
Mitt. Astron. Ges., Nr. 48, (see 012.015), p. 175 (1980).

154.014 **Infrared bursts from Liller I/MXB 1730-333.**
K. M. V. Apparao, S. M. Chitre.
Astrophys. Space Sci., Vol. 72, 127 - 132 (1980).

The observation of infrared bursts from the globular cluster Liller I has been reported by Kulkarni et al. (1979) and confirmed by Jones et al. (1980). The infrared bursts which resemble Type I X-ray bursts in their characteristics are plausibly attributed to a cyclotron maser instability operating a few tens of neutron star radii above the poles of a magnetized neutron star in a binary system. It is suggested that similar infrared bursts should in general be observable from Type I X-ray burst sources.

154.015 **The main sequence of the metal-poor globular cluster M30 (NGC 7099).** G. Alcaíno, W. Liller.
Astron. J., Vol. 85, 1330 - 1340 (1980).

The authors present photographic photometry for 673 stars in the metal-poor globular cluster M30 (NGC 7099). The Racine wedge was used with the CTIO 1-m Yale telescope ($\Delta m = 3^m\!.60$), the CTIO 4-m telescope ($\Delta m = 6^m\!.83$), and the ESO 3.6-m telescope ($\Delta m = 4^m\!.12$) to extend the photoelectric limit from $V \cong 16.3$ to $V \cong 20.4$. For the main-sequence turn-off, the authors have determined its position to lie at $V = 18.4 \pm 0.1$ (m. e.) and $B-V = 0.49 \pm 0.03$ (m. e.). From these values, they calculate the intrinsic values. For the cluster as a whole, the authors derive the distance modulus and reddening and deduce the cluster's age.

154.016 **Infrared photometry of the globular cluster associated with NGC 5128.** J. A. Frogel.
Astrophys. J., Lett., Vol. 241, L41 - L42 (1980).

Broad-band infrared photometry of a diffuse object discovered by Graham and Phillips in the vicinity of the peculiar galaxy NGC 5128 is presented. These data support their contention that this object is a moderately metal-rich globular cluster. At 2.2 μm, its luminosity is comparable to, or as much as one magnitude brighter than, the brightest globulars observed in M31; the exact value depends on the distance modulus to NGC 5128.

154.017 **Analysis of low dispersion spectra of globular cluster stars.** R. G. Gratton.
Mon. Not. R. Astron. Soc., Vol. 193, 533 - 535 (1980).

Some low-dispersion spectra of stars at the tip of the giant branch of the globular clusters M71 and M10 are roughly analysed by comparing them with the spectra of some stars of the old galactic cluster NGC 6819. There are indications of a relative sodium deficiency in M71.

154.018 **Gli ammassi globulari nelle Galassie.**
G. Clementini.
Coelum, Anno 50, 177 - 194 (1980).

154.019 **Globular clusters.** W. Iwanowska.
Postępy Astron., Tom 28, 185 - 196 (1980). In Polish.

A review of basic data on globular clusters is given with particular attention to their chemical composition.

154.020 **Red giants in globular clusters and large-scale variations in the Galaxy.** B. E. J. Pagel.
Highlights of Astronomy, Vol. 5, (see 012.030), 817 - 826 (1980).

This review concerns recent work on the determination of overall metallicities [Fe/H] in a number of globular clusters and the systematics of mixing effects displayed (usually) by weak CH and strong CN. Special attention is given to the globular cluster ω Centauri, where both metal abundance variations and mixing effects occur and are closely intertwined. Recent observations carried out at the Anglo-Australian Telescope by E. A. Mallia and D. C. Watts have revealed large variations in the strength of metallic lines across the red giant branch of this cluster.

154.021 **The globular cluster system of the Galaxy. II. The spatial and metallicity distributions, the second parameter phenomenon, and the formation of the cluster system.** R. Zinn.
Astrophys. J., Vol. 241, 602 - 617 (1980).

The metal abundance measurements that were collected for 84 globular clusters in paper I are used to describe the cluster system. The ranking of the clusters by metallicity has been calibrated by a new [Fe/H] scale. The most significant properties of the cluster system are: (1) there is a wide range in metal abundance among the clusters in the zone $9 \lesssim R < 40$ kpc, but no evidence of a gradient with R or with distance from the galactic plane, $|Z|$; (2) among the clusters with $R < 9$ kpc, there is a metal abundance gradient with $|Z|$; and (3) the magnitude of the second parameter effect increases with R, and if age is the second parameter, then over the range $0 < R < 40$ kpc the mean cluster age declines by ~3 Gyr and the scatter in age increases from less than 1 Gyr to ~2 Gyr. To explain these properties, a very crude model of the formation of the Galaxy is proposed.

154.022 **Random gravitational encounters and the evolution of spherical systems. VIII. Clusters with an initial distribution of binaries.** L. Spitzer, Jr., R. D. Mathieu.
Astrophys. J., Vol. 241, 618 - 636 (1980).

The effects of an initial binary population on the evolution of an isolated globular cluster have been explored with the Monte Carlo techniques previously developed. Rate coefficients for reactions between single stars and binaries, analyzed by Heggie and by Hills, have been extended to binary-binary reactions with a theory valid when one binary is much harder (more tightly bound) than the other; each such reaction disrupts the less hard binary. Values assumed for the initial mass fraction in binaries were 20 and 50%; all stars were taken to have the same mass.

154.023 **A color-magnitude diagram for the globular cluster NGC 6535.** M. H. Liller.
Astron. J., Vol. 85, 1480 - 1485 (1980).

Photometry in B and V of 103 stars in the region of the globular cluster NGC 6535 yields a color-magnitude diagram with a moderately steep giant branch, a steeply sloping blue horizontal branch, and an absence of a red horizontal branch. The derived apparent distance modulus $(m-M)_V$ is 15.2 ± 0.3 mag, and the distance from the sun is 6.3 ± 0.9 kpc. The color-magnitude diagram morphology indicates moderately low metallicity. The two RR Lyrae variables in the field have magnitudes inappropriate for cluster membership.

154.024 **Star counts in M15 on *U, B* and *V* plates.**
M. Calvani, L. Nobili, R. Turolla.
Astrophys. Space Sci., Vol. 73, 187 - 192 (1980).

The authors present new counts of stars in M15, using plates in B, V and U. They are able to explore relatively close to the central parts of the cluster (0.1 pc) and they derive the best fitting parameters for the star distribution.

154.025 **Abundances in globular cluster red giants. III. M71, M67, and NGC 2420.** J. G. Cohen.
Astrophys. J., Vol. 241, 981 - 1000 (1980).

High-dispersion spectra of four stars in the metal-rich globular cluster M71, four giants in the old open cluster M67, and two members of the metal-poor old open cluster NGC 2420 are analyzed with the aid of model atmospheres. The derived abundances are [Fe/H] = −1.27, −0.39, and −0.61 dex with regard to the Sun for M71, M67, and NGC 2420, respectively.

154.026 **Ellipticities of globular clusters of the Large Magellanic Cloud.** D. Geisler, P. Hodge.
Astrophys. J., Vol. 242, 66 - 73 (1980).

The LMC old populous clusters are found to be markedly more elliptical than the globular clusters of our Galaxy. The mean ellipticity, averaged for 25 clusters, is $\epsilon = 0.22$. The distribution of true ellipticities, corrected for projection effects, suggests that very few LMC clusters are spherical, most having true ellipticities in the range 0.2 - 0.4.

154.027 **Implications of the new globular-cluster metal-abundance scale for the helium abundance in the galactic halo.** P. Demarque, R. D. McClure.
Astrophys. J., Lett., Vol. 242, L5 - L7 (1980).

Adopting the new abundance scale for globular clusters (Cohen; Pilachowski, Sneden, and Canterna), the authors point out a difficulty in fitting main sequences of 47 Tucanae, M3, M5, and M13. In order to circumvent this problem, a substantial difference in helium abundance must be invoked. They discuss consequences of this helium-abundance variation.

154.028 **Terminology and fundamental data on globular clusters.** B. F. Madore.
Globular clusters, (see 003.008), p. 21 - 63 (1980).

Contents: Introduction. The colour-magnitude diagram. Distances to globular clusters. Variable stars in globular clusters. Metallicity: spectral types and other parameters. Structural properties and dynamics.

154.029 **Stellar evolution and globular clusters.**
V. Castellani.
Globular clusters, (see 003.008), p. 65 - 86 (1980).

This paper provides an overview of stellar evolution theory in the context of globular star clusters.

154.030 **Abundance anomalies in globular clusters.**
R. P. Kraft.
Globular clusters, (see 003.008), p. 87 - 101 (1980).

154.031 **Populations in globular clusters.**
K. C. Freeman.
Globular clusters, (see 003.008), p. 103 - 111 (1980).

This review concentrates on the structural aspects of populations in clusters. Much of this review is about ω Cen and 47 Tuc; they are massive and nearby, so the stellar statistics are good and they have been studied in some detail.

154.032 **The correlation of cyanogen, calcium, and the heavy elements on the giant branch of Omega Centauri.** J. Norris.
Globular clusters, (see 003.008), p. 113 - 123 (1980).

It is commonly accepted that calcium variations are almost certainly primordial: variations in cyanogen, strontium and barium, on the other hand, are generally taken to indicate mixing. This paper describes in preliminary form some efforts to place constraints on the relative importance of these two processes. It appears that variations in the primordial indicators are inextricably connected with those in the mixing indicators.

154.033 **Nucleosynthesis in evolved globular clusters.**
I. Iben, Jr.
Globular clusters, (see 003.008), p. 125 - 142 (1980).

154.034 **Infrared observations of red giants in globular clusters.** S. E. Persson, J. A. Frogel.
Globular clusters, (see 003.008), p. 143 - 157 (1980).

Contents: Introduction. Results from broad-band photometry. Results from narrow-band observations. Infrared observations of giants in ω Cen.

154.035 **Infrared observations of globular clusters in M31 and a comparison with galactic globulars and elliptical galaxies.** J. A. Frogel, S. E. Persson, J. G. Cohen.
Globular clusters, (see 003.008), p. 159 - 173 (1980).

The authors present preliminary results of a study of the infrared photometric properties of the brightest M31 clusters. This work is being carried out with the 5-metre Hale telescope. The authors compare these clusters with galactic globulars (Aaronson et al. 1978) and with early-type galaxies (Frogel et al. 1978) to investigate what new conclusions can be drawn from the infrared data.

154.036 **Globular clusters and galaxy evolution.**
S. van den Bergh.
Globular clusters, (see 003.008), p. 175 - 189 (1980) = Dominion Astrophys. Obs. Contrib. No. 383 = NRC No. 16865.

Contents: Introduction. The early evolution of galaxies. The radial density distribution of clusters. The metallicity of globular cluster families. The nuclei of galaxies.

154.037 **Globular clusters as extragalactic distance indicators.** D. A. Hanes.

Globular clusters, (see 003.008), p. 213 - 247 (1980).

The author's conclusions are: (1) Globular clusters within galaxies seem to be photometrically similar on average, and similar in at least the range of metallicities observed. (2) The total globular cluster population associated with a given galaxy is to first approximation directly proportional to the mass of that galaxy. (3) There is strong evidence that the luminosity functions of globular clusters are closely similar in the galaxies of the Local Group and that these functions are likewise closely similar within the elliptical galaxies of the Virgo cluster. There is now strong circumstantial evidence that the luminosity functions of globular clusters in spiral galaxies and elliptical galaxies are closely similar. (4) The apparent distance modulus of the Virgo cluster of galaxies is 30.7±0.3 mag. The present value of the Hubble constant is found to be 80±11 km/s/Mpc. Globular clusters seem at last to be realizing their full potential as extragalactic distance indicators.

154.038 **Luminosity distributions and density profiles.** I. R. King.
Globular clusters, (see 003.008), p. 249 - 269 (1980).

154.039 **On the dynamics of globular clusters.** J. E. Gunn.
Globular clusters, (see 003.008), p. 271 - 279 (1980).

154.040 **Some thoughts concerning the origin of globular clusters.** J. E. Gunn.
Globular clusters, (see 003.008), p. 301 - 307 (1980).

The author considers some topics relevant to the origin of globular clusters by considering the properties of protoclusters which can be inferred from the present state of clusters and the application of simple dynamical and nucleosynthetic theory.

154.041 **Summary of contemporary research on globular clusters.** K. C. Freeman.
Globular clusters, (see 003.008), p. 361 - 368 (1980).

154.042 **Chemical abundances in the globular clusters M3, M13, and NGC 6752.**
R. A. Bell, R. J. Dickens.
Astrophys. J., Vol. 242, 657 - 672 (1980).

The abundances of iron, carbon, nitrogen, and oxygen have been investigated in red giant stars in the globular clusters M3, M13, and NGC 6752. The results are based on application of spectrum synthesis and theoretical colors to observed spectra, DDO colors, and infrared CO measurements.

154.043 **The main sequence of the globular cluster NGC 288.** G. Alcaíno, W. Liller.
Astron. J., Vol. 85, 1592 - 1603 (1980).

The authors present photographic photometry for 911 stars in the globular cluster NGC 288. They have determined that the position of the main-sequence turn-off lies at $V = 18.8 \pm 0.1$ (m. e.) and $B - V = 0.50 \pm 0.03$ (m. e.). After deriving an interstellar reddening $E(B - V) = 0.02 \pm 0.02$, the authors calculate the following intrinsic turn-off point: $M_v = 4.10$ and $(B - V)_0 = 0.48$. For the cluster as a whole, they derive a distance modulus $(m - M)_v = 14.70 \pm 0.15$. They deduce the cluster's age to be $(14.5 \pm 2.5) \times 10^9$ yr. The horizontal branch is populated only by blue stars, and its red boundary is unusually blue at $(B - V)_0 = 0.06$. There is a striking deficiency of stars over ~0.7 mag of the subgiant branch.

154.044 **A bright ultraviolet object in the center of the X-ray globular cluster M15.** N. M. Spasova.
Astron. Tsirk., No. 1101, p. 4 - 6 (1980). In Russian.

154.045 **Parameters of the orbits of galactic globular clusters.** A. S. Rastorguev, V. G. Surdin.
Astron. Tsirk., No. 1102, p. 3 - 6 (1980). In Russian.

Corrections to the catalog of open clusters (7022). See Abstr. 002.038.

Milestones in globular cluster research: a guide to the literature. See Abstr. 002.064.

Webb Society deep-sky observer's handbook, Vol. 3: Open and globular clusters. See Abstr. 003.061.

A new statistical test with application to globular cluster X-ray source masses. See Abstr. 021.020.

Possibilities of photometric investigation of globular clusters by means of photographic equidensity curves. See Abstr. 031.511.

A new method to restore the spatial distribution of stars in globular clusters and its application to flare stars in the Pleiades. See Abstr. 031.515.

Picture processing of the globular cluster NGC 6712. See Abstr. 031.520.

Interactive approach to integrated photometry of globular clusters. See Abstr. 031.587.

SEC Vidicon photometry of the main sequence of ω Centauri. See Abstr. 034.001.

Current problems on horizontal-branch (HB) stars. IV. The influence of rotation on the expected properties of RR Lyrae pulsators. See Abst. 065.061.

A reinvestigation of the standard model for the dynamics of a massive black hole in a globular cluster. See Abstr. 066.176.

Horizontal branch stars and globular clusters with IUE. See Abstr. 114.033.

Period changes in RR Lyrae stars in the globular cluster NGC 6934. See Abstr. 122.065.

The period of V154 in NGC 5272 (M3). See Abstr. 122.127.

Beobachtungen von V 154 im Kugelhaufen M3. See Abstr. 122.142.

Variable stars in the globular cluster NGC 6284. See Abstr. 122.203.

The Blazhko effect of RR Lyr-type stars in globular clusters. 1. V14 in M5 (NGC 5904). See Abstr. 122.210.

Ninety years of variable stars in globular clusters. See Abstr. 122.213.

The variable stars of the globular cluster NGC 6284. See Abstr. 122.214.

The Blazhko effect of RR Lyrae-type stars in globular clusters. II. V 63 in M5 (NGC 5904). See Abstr. 122.218.

The Blazhko effect of RR Lyrae-type stars in globular clusters. III. Slow irregular amplitude variations of V 79 in M3 (NGC 5272). See Abstr. 122.219.

Mehrfarben-Beobachtungen von V 95 im Kugelhaufen M3. See Abstr. 123.017.

Neue veränderliche Sterne im Kugelhaufen M3. See Abstr. 123.018.

The discovery of an O subdwarf in the globular cluster NGC 6712. See Abstr. 126.019.

Radio observations of globular clusters and galactic bulge X-ray sources. See Abstr. 141.024.

Faint QSO candidates in the field of the globular cluster NGC 6752. See Abstr. 141.171.

Ultraviolet emission from strong X-ray sources as observed with IUE. See Abstr. 142.080.

X-ray burst sources in globular clusters and the galactic bulge. See Abstr. 142.117.

X-rays from globular clusters: steady emission. See Abstr. 142.118.

The gravothermal catastrophe of stellar systems. See Abstr. 151.014.

On the rotation of star clusters and nebulae as determined from radial velocities. See Abstr. 151.033.

On the rotation of star clusters. See Abstr. 151.034.

Dynamical theory of binaries in clusters. See Abstr. 151.086.

Globular clusters as survivors. See Abstr. 151.087.

The overall distribution of mass in our Galaxy. See Abstr. 155.020.

Galaxy-QSO associations: a gravitational lens effect of globular clusters. See Abstr. 158.068.

The first bright globular cluster in NGC 5128. See Abstr. 158.077.

Search for (globular) clusters in M31. I: Candidates in a 70′ square field centered on M31. See Abstr. 158.164.

The dwarf spheroidal galaxies. See Abstr. 158.224.

A classification of star clusters in the Magellanic Clouds. See Abstr. 159.007.

155 Galaxy (Structure, Evolution)

155.001 **O stars and the interstellar medium.** B. T. Lynds.
Astron. J., Vol. 85, 1046 - 1052 (1980).

About 400 O stars covering all galactic longitudes have been classified with respect to the presence or absence of bright nebulosity associated with the stars. Stellar spectrophotometric distances were used to plot the galactic distribution of the O stars and the associated nebulosity. The O stars define relatively thick spiral arms within which thinner arcs of bright nebulosity, indicating regions of high interstellar density, are located. It is found that 57% of the O stars are located inside bright H II regions. The earliest O stars are usually found near the galactic plane and are associated with bright nebulosity. The later-type O stars are often found outside bright nebulae and at larger z distances. The influence of the O stars on low-density regions of the interstellar medium is discussed.

155.002 **The primary source and the fates of galactic positrons.** P. N. Okeke.
Astrophys. Space Sci., Vol. 71, 371 - 375 (1980).

The author calculates the number of positrons emitted per second from the evaporating primordial black hole (PBH) in the various mass ranges. He estimates the number of those positrons from PBH which may likely annihilate within the Galaxy to give rise to the 0.511 MeV line radiation and those which may be degraded or leak out of the Galaxy. The author then shows that the PBH positrons are likely to be not only the primary source of low energy positrons but also high energy positrons which are energetic enough to leak out of the Galaxy.

155.003 **The local density enhancement or local system: a new approach to determining the density function in the galactic plane.** W. Herbst, D. L. Sawyer.
Bull. American Astron. Soc., Vol. 12, 443 (1980). – Abstract.

155.004 **Simple model of the Galaxy.** W. L. H. Shuter.
Bull. American Astron. Soc., Vol. 12, 444 (1980). – Abstract.

155.005 **Surface density, warping and thickness of the galactic H I layer beyond the solar circle.**
A. P. Henderson, P. D. Jackson.
Bull. American Astron. Soc., Vol. 12, 456 - 457 (1980). Abstract.

155.006 **Further high resolution H I observations of the Southern Milky Way and presentation of a complete longitude – velocity map for the galactic equator.**
F. J. Kerr, P. D. Jackson, P. F. Bowers, A. P. Henderson.
Bull. American Astron. Soc., Vol. 12, 457 (1980). – Abstract.

155.007 **The Parkes completely sampled survey of neutral hydrogen in the Southern Milky Way.**
P. D. Jackson, P. F. Bowers, F. J. Kerr.
Bull. American Astron. Soc., Vol. 12, 457 (1980). – Abstract.

155.008 **H I observations towards the Puppis Window of the Galaxy.** J. G. Stacy, P. D. Jackson.
Bull. American Astron. Soc., Vol. 12, 458 (1980). – Abstract.

155.009 **Galactic spiral structure in a cloudy, supernova-dominated interstellar medium.**
F. H. Levinson, W. W. Roberts, Jr.
Bull. American Astron. Soc., Vol. 12, 469 - 470 (1980). Abstract.

155.010 **A dynamical explanation for the tilted disk at the galactic center.** J. W.-K. Mark, L. Blitz, R. P. Sinha.
Bull. American Astron. Soc., Vol. 12, 497 (1980). – Abstract.

155.011 **The distribution of atomic hydrogen in the outer Galaxy.** S. Kulkarni, L. Blitz, C. Heiles.
Bull. American Astron. Soc., Vol. 12, 523 (1980). – Abstract.

155.012 **A deep near-infrared objective-prism survey for carbon stars toward the galactic center and anticenter.** F. J. Fuenmayor.
Bull. American Astron. Soc., Vol. 12, 525 (1980). – Abstract.

155.013 **Zur Erforschung des Milchstraßensystems (II).** H. -E. Fröhlich.
Astron. Schule, 17. Jahrg., 76 - 79 (1980).

155.014 **Molecular clouds and galactic spiral structure.** R. S. Cohen, H. Cong, T. M. Dame, P. Thaddeus.
Astrophys. J., Lett., Vol. 239, L53 - L56, plate L2 (1980).

Two large-scale 2.6 mm CO surveys of the galactic plane, one in the first quadrant (l = 12° to 60°, b = –1° to +1°), the other in the second (l = 105° to 139°, b = –3° to +3°), have provided evidence that, contrary to previous findings, molecular clouds constitute a highly specific tracer of spiral structure. Molecular counterparts of five of the classical 21 cm spiral arms have been identified: the Perseus arm, the local arm (including Lindblad's local expanding ring), the Sagittarius arm, the Scutum arm, and the 4-kpc arm.

155.015 **Brown and black dwarfs: their structure, evolution and contribution to the missing mass.**
D. J. Stevenson.
Proc. Astron. Soc. Australia, Vol. 3, 227 - 229 (1978).

155.016 **On the relation between local kinematics and galactic structure.** P. O. Lindblad.
Mitt. Astron. Ges., Nr. 48, (see 012.015), p. 151 - 159 (1980).

The information about the structure and dynamics of the Galaxy that can be obtained from a study of local kinematics is discussed. Expressions for the velocity field and the galactic rotation constants are given for the cases of an expanding local system and for a linear density wave.

155.017 **Can galaxy warps be used to provide constraints on halo properties?** G. Lake, J. W.-K. Mark.
Nature, Vol. 287, 705 - 706 (1980).

The authors explore the consequence of separating disk and halo mass distributions using a recently developed theory of bending waves. In this theory, the height of bending allows the functional form of the disk density distribution to be determined. With proper normalization this determination allows the halo density to be inferred from the rotation curve. The authors discuss the warps in our Galaxy and in galaxies NGC 2841 and M 33.

155.018 **A new method of determination of the distance to the galactic center.** B. G. Surdin.
Astron. Zh., Tom 57, 959 - 967 (1980). In Russian. – English translation in Soviet Astron., Vol. 24, No. 5.

The average metallicity of globular clusters increases with decreasing galactocentric distance. If the system of globular clusters is axially symmetric and the value of R_0 is correct, then the correlation between metallicity of globular clusters and their galactocentric azimuth must vanish. The author obtains $R_0 = 9.9 \pm 0.3$ kpc for Kukarkin's system of distance moduli of globular clusters and $R_0 = 10.3 \pm 0.6$ kpc for Harris' system.

155.019 **Investigation of the structure of the Orion spiral arm by the statistical modelling method.**
T. S. Basharina, E. D. Pavlovskaya, A. A. Filippova.
Astron. Zh., Tom 57, 975 - 984 (1980). In Russian. – English translation in Soviet Astron., Vol. 24, No. 5.

A method of investigation of the spiral structure based on statistical modelling methods is suggested. This method is used for the study of the Orion spiral arm. The maxima of density and the widths of the Orion arm in the direction of the areas considered for the longitude interval 55° - 187° are defined under the assumption of normal distribution of stars across the arm. The sun is shown to be at the inner edge of the arm.

155.020 **The overall distribution of mass in our Galaxy.**
M. Miyamoto, C. Satoh, M. Ohashi.
Astron. Astrophys., Vol. 90, 215 - 223 (1980).

In order to survey the overall mass distribution of our Galaxy, motions of eighty-one globular clusters and five members of the local group of galaxies are investigated. Examined are two extreme but possible space velocities of the satellites in three gravitational fields of the plausible galactic models A with the total mass $M = 2.7 \times 10^{11} M_\odot$, B with $M = 1.4 \times 10^{11} M_\odot$, and C with $M = 9.4 \times 10^{11} M_\odot$, which differ largely in their mass distribution exterior to the sun. The authors conclude that model A is one of the most plausible descriptions of the overall mass distribution in the Galaxy, while model C with a massive halo seems to be no more than a formal compromise for binding all the galactic satellites.

155.021 **Observations of the motion and distribution of the ionized gas in the central parsec of the Galaxy. II.**
J. H. Lacy, C. H. Townes, T. R. Geballe, D. J. Hollenbach.
Astrophys. J., Vol. 241, 132 - 146 (1980).

Observations of infrared fine-structure line emission from compact clouds of ionized gas within Sgr A West are presented. The authors draw conclusions from the data regarding the distribution of mass, the lifetimes of the clouds, the gas density in the clouds, and the requirements on the spectrum of ionizing radiation. A most probable mass distribution is derived which includes a central pointlike mass of several $\times 10^6 M_\odot$ in addition to several $\times 10^6 M_\odot$ of stars within 1 pc of the center. However, the small number of clouds (14) makes the mass determination quite uncertain.

155.022 **Spiral structure of our Galaxy.** A. Strobel.
Postępy Astron., Tom 28, 197 - 214 (1980). In Polish.

Observational data concerning the existence of the probable spiral structure of our Galaxy are presented.

155.023 **Radio continuum observations of the nucleus of our Galaxy and other normal galaxies.** R. D. Ekers.
Highlights of Astronomy, Vol. 5, (see 012.023), 143 - 148 (1980).

The radio properties of the nuclei of normal galaxies have been reviewed by Ekers (1974, 1978) and of the galactic centre by Oort (1977). In this report the author concentrates on material available since then.

155.024 **Atomic and molecular gas in the inner regions of the Milky Way and other galaxies.** H. S. Liszt.
Highlights of Astronomy, Vol. 5, (see 012.023), 149 - 161 (1980).

The author reviews the rather limited group of observations of carbon monoxide in the inner regions of other galaxies. This is meant to serve as an introduction to the more complex situation seen locally and to provide a gross standard for judging the extent to which our own system may be considered typical. Relatively recent observations of H I and molecular gas in the galactic center region are then dealt with at greater length, in an effort to describe the most general organization of the gaseous material there.

155.025 **Infrared observations of the galactic nucleus.**
J. H. Lacy.
Highlights of Astronomy, Vol. 5, (see 012.023), 163 - 169 (1980).

Infrared observations of the galactic nucleus and conclusions regarding the nature of the objects present there are reviewed. Observations of three sources of infrared radiation are discussed: near-infrared emission from cool stars, mid- and far-infrared emission from dust, and line emission from ionized gas. These observations provide information about the mass distribution, the stellar population, and the origin and ionization of the compact mid-infrared sources. The possibility of the existence of a massive central black hole is discussed.

155.026 **An anisotropic gas jet from the nucleus of the Galaxy.** Y. Fukui.
Highlights of Astronomy, Vol. 5, (see 012.023), 171 (1980).

155.027 **The galactic rotation curve to R = 18 kpc.**
L. Blitz, M. Fich, A. A. Stark.
Interstellar molecules, (see 012.033), p. 213 - 220 (1980).

155.028 **Some problems of the spiral structure of the Galaxy.**
V. I. Voroshilov, N. B. Kalandadze.
Spiral structure of the Galaxy (see 012.034), Abastumansk. Astrofiz. Obs. Byull., No. 52, p. 5 - 14 (1980). In Russian.

The results of investigations of the spiral structure of the Galaxy carried out in Abastumani and Kiev were generalised from the point of their agreement with present-day data. The point of relation of the B3-B5 stars, as well as the dust and gaseous components to the spiral arms was considered. The relation of some members of the gaseous component to the spiral arms, as well as some observational problems of the spiral structure were investigated.

155.029 **Three-dimensional structure of spiral arms.**
L. N. Kolesnik, I. P. Vedenicheva.
Spiral structure of the Galaxy (see 012.034), Abastumansk. Astrofiz. Obs. Byull., No. 52, p. 21 - 22 (1980). In Russian.

The stellar distribution relative to the galactic plane is studied for 3600 O - B - A stars on the basis of photoelectric magnitudes and MK spectral types determined by various authors.

155.030 **The structure of the Milky Way in the area between the local and Sagittarius spiral arms.**
V. I. Kuznetsov, N. G. Guseva, M. D. Metreveli.
Spiral structure of the Galaxy (see 012.034), Abastumansk. Astrofiz. Obs. Byull., No. 52, p. 23 - 28 (1980). In Russian.

The two areas of the Milky Way with their centres coincident with the open stellar clusters NGC 6823 and NGC 6802 are studied. The conclusion is drawn that the area between the spiral arms is characterised by low density for all spectral type stars and by high density for dark matter.

155.031 **Investigation of the galactic structure in the direction $l = 80°$.** I. A. Dubyago.
Spiral structure of the Galaxy (see 012.034), Abastumansk. Astrofiz. Obs. Byull., No. 52, p. 29 - 32 (1980). In Russian.

Interstellar absorption of light and the spatial distribution of stars of early spectral types have been investigated in the areas of 2.8 square degrees in Cygnus on the basis of the Catalogue of UBV-magnitudes of 4126 stars. A density change of dust matter has been calculated in the direction of the z-coordinate.

155.032 **The rotation law and density distribution of neutral hydrogen in the inner regions of the galactic plane.**
I. V. Petrovskaya, G. V. Korzin.

Spiral structure of the Galaxy (see 012.034), Abastumansk. Astrofiz. Obs. Byull., No. 52, p. 33 - 36 (1980). In Russian.

The rotation curve of the neutral hydrogen subsystem in the fourth quarter of the galactic plane is obtained. A map of the neutral hydrogen density distribution is drawn for the galactic plane region 300° ≤ l ≤ 340° with distances from the galactic centre $R<R_0$, where R_0 is the distance of the sun from the galactic centre.

155.033 **Uncertainty of estimating the radial velocity dispersion of stars in the solar neighbourhood.**
K. F. Ogorodnikov, L. P. Osipkov.
Spiral structure of the Galaxy (see 012.034), Abastumansk. Astrofiz. Obs. Byull., No. 52, p. 37 - 42 (1980). In Russian.

It is shown that the presence of few high-velocity stars at the solar neighbourhood causes a significant uncertainty of the radial velocity dispersion σ_ρ^2. Depending on including or neglecting such stars the value of σ_ρ changes twice as much.

155.034 **Kinematical method of galactic mapping.**
G. B. Anisimova, T. G. Malysheva, R. B. Shatsova.
Spiral structure of the Galaxy (see 012.034), Abastumansk. Astrofiz. Obs. Byull., No. 52, p. 43 - 50 (1980). In Russian.

155.035 **"Quasi-" and "semiempirical" lower mass-limits for the three-color-photometrically defined halo-population.** R. P. Fenkart.
Astron. Astrophys., Vol. 91, 352 - 355 (1980).

With the help of three-colour photometrically observed density functions starting at the galactic centre (Becker, 1980) lower mass-limits for the halo population are derived and compared for different models of the halo density distribution assumed to be symmetrical with respect to the galactic equator and to the galactic rotation axis. The derived mass values is low: $4.6\times10^8\ M_\odot$ for model I, $8.1\times10^8\ M_\odot\ (x=3.5)$ and $4.5\times10^8\ M_\odot\ (x=5.5)$ for model II.

155.036 **Influence of perturbed gravitational fields on galactic shock waves.** W. Hu.
Kexue Tongbao, Vol. 25, 900 - 905 (1980).

155.037 **The chemical evolution of the solar neighborhood. I. A bias-free reduction technique and data sample.**
B. A. Twarog.
Astrophys. J., Suppl. Ser., Vol. 44, 1 - 29 (1980).

It is shown that the use of a field star sample chosen on the basis of effective temperature, the most commonly used technique of measuring of the age-metallicity relation, introduces a bias which results in a monotonic increase in the metal abundance of the disk with time. It is concluded that a sample selected and analyzed through the use of *uvby* and Hβ photometry in conjunction with a self-consistent set of theoretical isochrones provides the least biased, most accurate estimate of the age-metallicity relation for the disk. New data for 1007 stars which were chosen to avoid selection effects and observed in the *uvby* and Hβ systems, are presented.

155.038 **The universe at faint magnitudes. I. Models for the Galaxy and the predicted star counts.**
J. N. Bahcall, R. M. Soneira.
Astrophys. J., Suppl. Ser., Vol. 44, 73 - 110 (1980).

A detailed model is constructed for the disk and spheroid components of the Galaxy from which the distribution of visible stars and mass in the Galaxy is calculated. The application of star counts to the determination of galactic structure parameters is demonstrated. The possibility of detecting a halo component with the aid of star counts is also investigated quantitatively.

155.039 **Background starlight at the north and south celestial, ecliptic, and galactic poles.** G. Toller.
News Lett. Astron. Soc. N. Y., Vol. 1, No. 7, p. 9 (1980). Abstract.

155.040 **Structure of the local spiral arm from the 21-cm hydrogen line profiles.** G. Chlewicki.
Acta Astron., Vol. 30, 299 - 314 (1980).

The results of the study of the 21-cm line profiles for the local gaseous arm by means of the model-making method are presented. The use of this method leads to a relatively low value of the pitch angle of the local arm (from 1.0° to 5.0°) with the internal edge of the arm lying at a distance of 300 - 400 pc from the Sun in the direction of the galactic centre. Noncircular motions in the local arm are briefly discussed and the simple model of longitudinal wave-like flow is suggested.

155.041 **An indirect measurement of the galactic center distance.** R. J. Quiroga.
Astron. Astrophys., Vol. 92, 186 - 188 (1980).

The Sun-galactic center distance R_0 is calculated using observational data in neutral hydrogen, OB stars and H II regions with their exciting stars. The method consists in the calibration of the scale of the kinematic distances to hydrogen features by comparisons with the photometric distances of OB stars coexisting with these features. A value $R_0=8.4$ kpc is obtained from comparisons between kinematic distances of H II regions and photometric distances of their exciting stars.

155.042 **Carbon monoxide in the inner Galaxy: the 3 kiloparsec arm and other expanding features.**
T. M. Bania.
Astrophys. J., Vol. 242, 95 - 111 (1980).

A new survey of the latitude distribution of ^{12}CO emission in the inner Galaxy shows that much of the dense molecular gas is found in several large-scale features spanning tens of degrees in l. The most prominent are the rotating nuclear disk, the 3 kpc arm, and the +135 km s^{-1} feature. These objects, which are all present in the H I data as well, together constitute the largest (50 - 180 km s^{-1}) deviations from circular motion observed for molecular gas in the Galaxy. The (l, v) loci of the 3 kpc arm and + 135 km s^{-1} feature can be described by simple kinematic ring structures. Because the features cross $l=0°$ at −53 and + 135 km s^{-1}, respectively, neither symmetric explosions nor on-axis rotating bars can account for the observed kinematics.

155.043 **The chemical evolution of the solar neighborhood. II. The age-metallicity relation and the history of star formation in the galactic disk.** B. A. Twarog.
Astrophys. J., Vol. 242, 242 - 259 (1980).

The age-metallicity relation for the disk in the neighborhood of the Sun is derived from four-color and Hβ photometry of a large sample of southern F dwarfs, analyzed in combination with theoretical isochrones. It is found that the mean metallicity of the disk increased by about a factor of 5 between 12 and 5 billion years ago and has increased only slightly since then: After taking into account stellar evolution and disk scale height corrections, the star formation rate (SFR) for the disk as a function of time is obtained from the number count data. An upper limit for the ratio of the average past SFR to the present SFR of 2.5 is derived. It is concluded that the disk has collapsed by about a factor of 6 perpendicular to the galactic plane during its lifetime, with most of the collapse occurring 13 to 7 billion years ago. It is shown that the derived SFR requires that the disk was subject to the infall of gaseous material from outside the disk at a rate equal to approximately half the SFR. The age-metallicity relation predicted by the theoretical models in a galaxy with the above characteristics is in excellent agreement with the observed age-metallicity relation.

155.044 **The possible nature of the high-velocity OB stars: hot UV-bright stars in the galactic disk.**
L. Carrasco, G. F. Bisiacchi, C. Cruz-Gonzáles, C. Firmani, R. Costero.
Astron. Astrophys., Vol. 92, 253 - 259 (1980).

The authors propose that many of the high-velocity, low surface gravity, early-type stars are old disk population objects in the same evolutionary phase as the hot UV-bright stars in globular clusters. They analyze the kinematics of the sample of OB stars divided into low-velocity, $|V_{rp}|<20$ km s^{-1}, and high-velocity stars, $|V_{rp}|>45$ km s^{-1}. The results for the solar motion and mean peculiar velocity of the groups are the expected ones for the extreme Population I objects in the case of the low-velocity group, and an asymmetric drift of 5.8±2.2 km s^{-1} for the high-velocity stars. This lag behind circular motion is interpreted as the result of a mixture of stellar population in the sample of high-velocity stars.

155.045 **A new look at the North Polar Spur.**
C. Heiles, Y.-H. Chu, R. J. Reynolds, I. Yegingil, T. H. Troland.
Astrophys. J., Vol. 242, 533 - 540 (1980).

The North Polar Spur (NPS) seems to consist of distinct shells. On the outside there is the H I shell (the H I NPS) and interior to this is the radio continuum shell (the RC NPS), which contains ionized gas. The authors attempt to derive densities, temperatures, and magnetic fields in the RC NPS and the H I NPS. They discuss these results in terms of a model that considers the H I shell as the cooled gas that was produced behind the spherical shock, which by now has dissipated.

155.046 **Peculiarities of local spiral arm features of the Galaxy.** M. V. Dolidze.
Pis'ma Astron. Zh., Tom 6, 745 - 749 (1980). In Russian. English translation in Soviet Astron. Lett., Vol. 6.

From data on possible supernova remnants it is found that close to the sun parts of two local details of the Orion spiral arm of the Galaxy are situated in thin layers tilted to the galactic plane at different angles. The cloud-, ring- and shell-like structures are characteristic of the distribution of the objects in these details.

155.047 **Abundances of chemical elements in the spiral branches of the Galaxy.**
N. S. Komarov, A. N. Shcherbak.
Astron. Tsirk., No. 1104, p. 6 - 8 (1980). In Russian.

155.048 **The gas kinematics in the inner Galaxy.**
W. B. Burton, H. S. Liszt.
Photometry, kinematics and dynamics of galaxies, (see 012.051), p. 287 - 295 (1979).

The authors describe some of the more straightforward observational characteristics of the kinematics and distribution of gas lying within 2 kpc of the galactic center. These characteristics can be accounted for by a model involving confinement of the gas to a bar-like structure. Within the tilted bar the gas is smoothly distributed; its motions are along closed elliptical paths and thus involve no net outflow of material. The model accounts for the appearance of an assortment of observational characteristics, including apparently isolated features, without requiring any local kinematic or density perturbations and without requiring ejection of material from the galactic nucleus.

155.049 **The dynamics of the molecular gas in the galactic center.** R. Güsten, D. Downes.
Mitt. Astron. Ges., Nr. 50, p. 76 - 79 (1980).

155.050 **Luminosity function and space density of the nearest stars.** V. A. Zakhozhaj.
Astrometr. Astrofiz., Vyp. (No.) 42, p. 64 - 69 (1980). In Russian.

The nearest stars. See Abstr. 002.022.

The result of a treatment of the tape version of the [Fe/H] catalogue by Morel et al. See Abstr. 002.036.

Galaxies. See Abstr. 003.044.

Early galactic structure. See Abstr. 005.016.

Oort's work reflected in current studies of galactic CO. See Abstr. 013.016.

Synthesis of light metals in the Galaxy. Aluminium abundances in cool halo stars. See Abstr. 061.012.

Relations between nucleosynthesis rates and the metal abundance. See Abstr. 061.013.

The chemical evolution in the Galaxy and isotopic ratios in the solar system. See Abstr. 061.036.

The origin of the elements. See Abstr. 061.045.

The most massive stars in the Galaxy and the LMC: quasi-homogeneous evolution, time-averaged mass loss rates and mass limits. See Abstr. 065.068.

Dispersion inhomogeneities of nearby stars velocities along the direction $l = 330°$, $b = 0°$.
See Abstr. 111.005.

The impact of star parallaxes and very accurate proper motions on galactic structure and dynamics studies.
See Abstr. 111.015.

Recent results of BVRI photometry of late-type dwarf stars in nearby clusters and in the solar neighborhood.
See Abstr. 113.008.

A search for Hα-emission objects in a region in Ara.
See Abstr. 113.012.

***UBV* sequences in three northern Milky Way regions and a comment on the interstellar extinction around $l = 90°$.**
See Abstr. 113.022.

Low-dispersion spectral classification and *UBV* photographic photometry of Hα-emission objects in the Coalsack region. See Abstr. 114.069.

Abundance of chemical elements in the sun's neighbourhood. See Abstr. 114.080.

The chemical compositions of 26 distant late-type supergiants and the metallicity gradient in the galactic disk.
See Abstr. 114.083.

On the space distribution of semi-regular variables.
See Abstr. 122.071.

A different view of variable stars.
See Abstr. 122.072.

Molecule formation in shock waves from the galactic nucleus. See Abstr. 131.026.

Two CO surveys of the first galactic quadrant.
See Abstr. 131.047.

A search for ultracold molecular gas in our Galaxy. See Abstr. 131.067.

Contributions to the theory of spiral structure. IV. The propagation of sound waves in an inhomogeneous interstellar medium. See Abstr. 131.084.

The distribution of the interstellar dust in the galactic plane within 3 kpc. See Abstr. 131.085.

Evidence for the existence of a low-velocity molecular cloud near Sgr A. See Abstr. 131.096.

Dust and gas correlations in the region of the South Celestial Pole. See Abstr. 131.097.

Formaldehyde in the galactic center region: observations. See Abstr. 131.111.

The spatial distribution of the interstellar extinction. See Abstr. 131.112.

Statistical evidence of absorption at high latitudes. See Abstr. 131.127.

HCO^+ emission in the galactic center region. I. Observations. See Abstr. 131.133.

Star formation and ionization in the 3 kiloparsec arm. See Abstr. 131.139.

On the variation of the colour excess in the Carina-Crux-Centaurus-Norma region of the Milky Way. See Abstr. 131.147.

Formaldehyde in L1551, L134 and the galactic centre. See Abstr. 131.178.

Columbia CO survey: molecular clouds and spiral structure. See Abstr. 131.202.

Molecular fan of 360-pc radius in the galactic center region. See Abstr. 131.203.

Distribution of the sources of maser radio emission in the spiral arms of the Galaxy. See Abstr. 131.249.

Intermediate velocity clouds in the region of the south celestial pole. See Abstr. 131.293.

Radio continuum interferometry of dark clouds. I. A search for newly formed H II regions. See Abstr. 132.001.

Radio continuum interferometry of dark clouds. II. A study of the physical properties of local newly formed H II regions. See Abstr. 132.002.

H I shells in the Galaxy. See Abstr. 132.010.

2 μ spectra of sources in the galactic center. See Abstr. 133.006.

The oxygen enrichment of the Galaxy. See Abstr. 135.008.

Cosmic rays in the galactic centre. See Abst. 143.001

Cosmic-ray confinement in the Galaxy. See Abstr. 143.029.

On the excitation of warps in galaxy disks. See Abstr. 151.013.

Decay instability of a spiral wave in a stellar disk. See Abstr. 151.046.

Luminous stars beyond the solar circle: investigation of a galactic field at $l = 231°$. See Abstr. 152.006.

Galactic distribution of the oldest open clusters. See Abstr. 153.005.

The kinematics and dynamics of the galactic globular cluster system. See Abstr. 154.008.

Red giants in globular clusters and large-scale variations in the Galaxy. See Abstr. 154.020.

The globular cluster system of the Galaxy. II. The spatial and metallicity distributions, the second parameter phenomenon, and the formation of the cluster system. See Abstr. 154.021.

Abundances in globular cluster red giants. III. M71, M67, and NGC 2420. See Abstr. 154.025.

2.4 μm emission from the galactic centre region. See Abstr. 156.007.

The nature of the UV radiation background. See Abstr. 157.002.

Hot gas in galactic haloes and winds. See Abstr. 158.147.

The numbers of red supergiants and WR stars in galaxies: an extremely sensitive indicator of chemical composition. See Abstr. 159.013.

Erratum

155.901 **Erratum: "High-resolution far-infrared observations of the galactic center"** [Astrophys. J., Lett., Vol. 205, L69 - L73 (1976)]. P. M. Harvey, M. F. Campbell, W. F. Hoffmann.
Astrophys. J., Lett., Vol. 241, L183, (1980). – See Abstr. 17.155.032.

156 Galaxy (Magnetic Field, Radio and Infrared Radiation)

156.001 **2.2-μm polarization mapping of the galactic center.** Y. Kobayashi, K. Kawara, T. Kozasa, S. Sato, H. Okuda.
Publ. Astron. Soc. Japan, Vol. 32, 291 - 294 (1980).

A polarization map at the *K*-band has been made for the central 7′×7′ region of the galactic center. Polarizations are almost uniform throughout the observed region in their amplitudes as well as in their directions, indicating that they are generated outside the galactic core region.

156.002 **Medium resolution IR-maps of the Cygnus X region.** S. D. Price.
Bull. American Astron. Soc., Vol. 12, 468 (1980). – Abstract.

156.003 **Submillimeter sky survey of the galactic plane.** L. H. Cheung, D. Y. Gezari, M. G. Hauser, T. Kelsall, J. C. Mather, R. F. Silverberg, M. T. Stier.
Bull. American Astron. Soc., Vol. 12, 482 (1980). – Abstract.

156.004 **A recombination line survey of the Milky Way.** F. J. Lockman.
Bull. American Astron. Soc., Vol. 12, 523 (1980). – Abstract.

156.005 **Far-infrared survey of the galactic plane.** T. Nishimura, F. J. Low, R. F. Kurtz.
Astrophys. J., Lett., Vol. 239, L101 - L106 (1980).

The galactic plane has been mapped from $l = 352°$ to 45° in the band 100–300 μm with beam size of 30′. Latitudinal and longitudinal distribution of diffuse far-infrared (FIR) emission are given in addition to identifications of the prominent discrete sources with H II and CO emission. FIR luminosities are calculated for the nucleus and the surrounding disk.

156.006 **A 21 cm radio continuum survey of the galactic plane between $l = 93°$ and $l = 162°$.**
E. Kallas, W. Reich.
Astron. Astrophys., Suppl. Ser., Vol. 42, 227 - 243 (1980).

A 21 cm radio continuum survey of the galactic plane of an area between $93° \leq l \leq 162°$ and $|b| \leq 4°$ has been carried out with the Effelsberg 100 m radio telescope. Contour maps are presented with a sensitivity in temperature of 0.2 K T_B and an angular resolution of 9′ (HPBW). A list of 236 radio sources stronger than 0.3 Jy is given.

156.007 **2.4 μm emission from the galactic centre region.** A. W. Harris, D. Lemke, W. Hofmann.
Astrophys. Space Sci., Vol. 72, 111 - 116 (1980).

It is shown that the 2.6 mm CO emission profile in the regions of two unidentified 2.4 μm features observed near the galactic centre is consistent with an explanation of these features in terms of inhomogeneities in interstellar extinction. From their observations the authors estimate the mass to luminosity ratio of the galactic central bulge to be $M/L_v = 4$.

156.008 **Far-infrared [O III] line emission from the galactic center.** D. M. Watson, J. W. V. Storey, C. H. Townes, E. E. Haller.
Astrophys. J., Lett., Vol. 241, L43 - L46 (1980).

The [O III] 51.8 μm fine-structure line has been detected in Sgr A West. It appears that the emission arises from the same compact clouds within the central parsec of the Galaxy which are observed in [Ne II]. The line intensity is used to derive an effective temperature of 32,000 - 40,000 K for the radiation field that ionizes the clouds. The authors also report an upper limit for the [O III] 88.4 μm fine-structure line in Sgr A West.

156.009 **Numerical models of the galactic dynamo.** M. P. White.
Astron. Nachr., Band 299, 209 - 216 (1978).

Axisymmetric $\alpha\omega$-dynamo models are investigated numerically for the galactic dynamo. Contrasting to the eigenvalue formulation of the problem by Stix (1976), an initial-value formulation is developed in a manner which is a generalization of the approach to the solar dynamo by Jepps (1975). It is found, for Stix's model, that the critical dynamo numbers, P_c, obtained by this approach do not agree with those obtained by Stix. A comparison is made between predictions of the dynamo models and the observed radiation of certain external galaxies which provides insight into the nature of the intergalactic medium.

156.010 **Further observations of radio recombination lines from the direction of the Galactic Centre.**
L. Hart, A. Pedlar.
Mon. Not. R. Astron. Soc., Vol. 193, 781 - 792 (1980).

The hydrogen recombination lines H351α, H252α and H166α have been used to investigate the properties of low-density ionized gas which lies in the direction of the Galactic Centre. A limit of < 5K antenna temperature was found for the H351α line, which implies that the electron density of the region is $\gtrsim 5$ cm^{-3}. H166α and H252α observations show that the low density gas extends over approximately 2°.5. The authors conclude that the ionized gas arises from evolved H II regions which partially overlap in front of the Galactic Centre.

156.011 **Rotation measures and the galactic magnetic field.** M. Simard-Normandin, P. P. Kronberg.
Astrophys. J., Vol. 242, 74 - 94 (1980).

Analysis of a new, large sample of extragalactic source rotation measures (RMs) provides firm evidence for large scale prevailing magnetic fields in the Galaxy. Several regions of strong galactic RM are identified, and their boundaries in (l, b) are well delineated for the first time. Various models of the prevailing magnetic field in the plane of the Galaxy are discussed. The authors conclude that there are no large scale reversals of prevailing magnetic field outside the Sun's radius on our side of the Galaxy. They determine the thickness of the magnetoionic plane of the Galaxy to be ~1.4 kpc. They find no evidence for large scale intergalactic or metagalactic magnetic fields.

156.012 **The H166α radio line from the Galaxy's periphery?** V. I. Ariskin.
Astron. Tsirk., No. 1112, p. 1 - 3 (1980). In Russian.

156.013 **The galactic centre and Aquila regions in the near infrared.** A. W. Harris, W. Hofmann, D. Lemke.
Mitt. Astron. Ges., Nr. 50, p. 75 - 76 (1980).

156.014 **Far-infrared balloon-borne observations.** S. Drapatz, L. Haser, R. Hofmann, H. Rothermel.
Mitt. Astron. Ges., Nr. 50, p. 109 - 115 (1980).

An infrared balloon gondola has been flown successfully (pointing stability achieved: ± 15 arcsec). Photometric observations (λ > 50 μm) of the central region of the Milky Way (Sgr B2) and of another galaxy (M 101) have been carried through with high spatial resolution (beam width 1.4 arcmin obtained by the observation of Saturn).

156.015 **The galactic center: 16–30 micron observations and the 18 micron extinction.**
J. F. McCarthy, W. J. Forrest, D. A. Briotta, Jr., J. R. Houck.
Astrophys. J., Vol. 242, 965 - 975 (1980).

The authors have obtained low-resolution 16–30 μm

spectra of three positions in the Sgr A West region. They have mapped the area in two colors (19 and 28 μm) with a spatial resolution of 30″(1.5 pc at 10 kpc); these results are presented. Upper limits to the 18.71 μm [S III] flux are given and interpreted in terms of the dust extinction and the ionization structure and elemental abundances of the ionized gas. The authors discuss the continuum emission and propose models to explain it. An 18 μm absorption feature is observed and identified as a resonance of silicate minerals present in dust grains. They examine the results of the continuum models, determine the consequences of extending the models to other wavelengths, and discuss the relative strengths of the 9.7 μm and 18 μm silicate features and the shape of the silicate spectrum for $\lambda > 16$ μm.

The hydrostatic equilibrium of interstellar gas and magnetic fields in the 6 kpc region of the Galaxy.
See Abstr. 131.327.

Infrared observations of the galactic nucleus.
See Abstr. 155.025.

Longitudinal distribution of gamma rays and the variation of magnetic field in the Galaxy.
See Abstr. 157.016.

157 Galaxy (UV, X, Gamma Radiation)

157.001 **A galactic component of the diffuse X-ray flux in the range 2–7 keV.**
R. J. Protheroe, A. W. Wolfendale, J. Wdowczyk.
Mon. Not. R. Astron. Soc., Vol. 192, 445 - 454 (1980).

An analysis has been made of the spatial distribution of the 2–7 keV X-ray background as measured by Uhuru and reported by Schwartz. The latitude distribution above about 10° is consistent with a uniform isotropic component comprising the bulk of the radiation plus a galactic part varying from about 3 per cent at $[b] = 20°$ to 1 per cent at $[b] = 90°$. The galactic component is readily interpreted in terms of synchroton and Inverse Compton radiation from cosmic ray electrons in the ISM.

157.002 **The nature of the UV radiation background.**
M. Maucherat-Joubert, J. M. Deharveng, P. Cruvellier.
Astron. Astrophys., Vol. 88, 323 - 328 (1980).

The discrepancy between recent measurements of the far UV diffuse radiation leads to different interpretations as to its dominant contributor, either extragalactic light or galactic plane light scattered off dust grains. This situation has stimulated a review of existing measurements and a more detailed analysis of the UV survey performed by the ELZ photometer on board the D2B satellite. A correlation between far UV brightness and hydrogen column density is found. Two other intensity measurements of the UV diffuse radiation at high galactic latitudes at 2200 and 3100 Å are discussed. The sky background spectrum obtained is strikingly similar to the spectrum of the Coalsack dark nebula. These results suggest a significant contribution of diffuse galactic light in the UV observed background but do not rule out an extragalactic component.

157.003 **Gamma radiation, cosmic rays, and galactic structure.** D. A. Kniffen, C. E. Fichtel.
Bull. American Astron. Soc., Vol. 12, 447 - 448 (1980). Abstract.

157.004 **Observations of 1 - 30 MeV gamma rays from the galactic center.**
E. Zanrosso, J. L. Long, A. D. Zych, R. S. White.
Bull. American Astron. Soc., Vol. 12, 449 (1980). – Abstract.

157.005 **A search for gamma ray spectral features from the galactic plane and Cas-A.**
W. S. Paciesas, B. J. Teegarden, T. L. Cline, R. C. Haymes, W. K. H. Schmidt.
Bull. American Astron. Soc., Vol. 12, 449 (1980). – Abstract.

157.006 **Far-ultraviolet studies. VII. The spectrum and latitude dependence of the local interstellar radiation field.** R. C. Henry, R. C. Anderson, W. G. Fastie.
Astrophys. J., Vol. 239, 859 - 866 (1980).

A direct measurement has been made of the spectrum (1180–1680 Å) and Gould-latitude dependence of the local interstellar radiation field, over about one-third of the sky. The result is corrected to give expected values for the entire sky. The average local 1180–1680 Å energy density is 5.8×10^{-17} ergs cm^{-3} Å^{-1}. The surface brightness falls off toward high latitudes much more steeply than published models predict.

157.007 **Gamma-ray lines and continuum radiation from the galactic center direction.**
M. Leventhal, C. J. MacCallum, A. F. Huters, P. D. Stang.
Astrophys. J., Vol. 240, 338 - 343 (1980).

The authors' balloon-borne germanium γ-ray telescope has been reflown over Alice Springs, Australia. The detection of a sharp 511 keV electron-positron annihilation line from the galactic center (GC) direction has been confirmed at a flux level compatible with a previous result. No evidence for several other candidate lines was found, indicating that they were statistical fluctuations in the previous data. Evidence for a significant change in the GC inverse power-law continuum has been obtained, suggesting that its origin lies in the many unresolved and highly variable pointlike X-ray sources in the field of view.

157.008 **The nature of the spatial structure of diffuse soft X-rays in the direction of the North Polar Spur.**
J. P. Morrison.
Bull. American Astron. Soc., Vol. 12, 748 (1980). – Abstract.

157.009 **Search for gamma-ray emission from the vicinity of the galactic center.**
G. R. Riegler, A. S. Jacobson, J. C. Ling, W. A. Mahoney.
Bull. American Astron. Soc., Vol. 12, 749 (1980). – Abstract.

157.010 **X-ray characteristics of Loop I and the local interstellar medium.**
J. Davelaar, J. A. M. Bleeker, A. J. M. Deerenberg.
Astron. Astrophys., Vol. 92, 231 - 237 (1980).

The authors present some new rocket borne soft X-ray observations of Loop I (North Polar Spur). The observed position of the X-ray ridge of Loop I relative to the radio ridge indicates that one is dealing with a supernova remnant in the adiabatic phase of its evolution with an age of

~75000 yrs which has exploded in a region of anomalously low gas density (~10^{-2} cm^{-3}).

157.011 **50 keV - 3 MeV high resolution observation of the galactic center region.** P. Durouchoux, D. Boclet, R. Rocchia, F. Albernhe, J. F. Leborgne, G. Vedrenne, J. Marques da Costa.
Non-solar gamma-rays, (see 012.047), p. 117 - 121 (1980).

In this paper the authors present their results obtained with a high-resolution gamma-ray spectrometer. The main objective was the observation of the galactic-center region, which has been performed during more than 8 hours.

157.012 **Observations of diffuse galactic gamma rays.** G. A. Simpson.
Non-solar gamma-rays, (see 012.047), p. 147 - 158 (1980).

Observations of 35 MeV to several GeV diffuse gamma radiation from the Galaxy are available from the SAS-2 and COS-B instruments. This paper gives a brief review of the experimental problem, a discussion of the high-latitude (local) galactic component, and a study of the more distant low-latitude emission from the galactic plane. Finally, the emerging observations in other energy ranges are mentioned.

157.013 **Large-scale structure of the Milky Way, as observed by COS-B in high-energy gamma rays.** R. D. Wills.
Non-solar gamma-rays, (see 012.047), p. 159 - 161 (1980).

ESA's satellite COS-B carries a single spark-chamber experiment for gamma-ray astronomy. In the period August 1975 to February 1978 it devoted about 23 months to observations of the region of the sky along the galactic disc. These observations constitute the first complete detailed gamma-ray survey of the Milky Way.

157.014 **Origin of diffuse galactic gamma rays.** S. Hayakawa.
Non-solar gamma-rays, (see 012.047), p. 175 - 186 (1980).

Theoretical interpretations of the diffuse component of galactic gamma rays with energies around 100 MeV are reviewed, on account of that these gamma rays are generated by three mechanisms, the 2γ-decay of neutral pions produced by the collision of the nuclear component of cosmic rays with interstellar matter, the bremsstrahlung of cosmic ray electrons, and the Compton scattering of cosmic-ray electrons by cosmic microwaves and starlight. The contribution of localized sources to the diffuse component is discussed in connection with the star formation in dense molecular clouds.

157.015 **Galactic gamma rays and the origin of cosmic rays.** R. Cowsik.
Non-solar gamma-rays, (see 012.047), p. 205 - 221 (1980).

In this paper the author argues that cosmic rays leaking from gamma-ray sources must contribute significantly to the density of cosmic rays in the Galaxy. Conversely, the sources of cosmic rays, whatever they may be, are also likely to be the gamma ray sources. He discusses several possible models for the sources.

157.016 **Longitudinal distribution of gamma rays and the variation of magnetic field in the Galaxy.** S. A. Stephens.
Non-solar gamma-rays, (see 012.047), p. 223 - 226 (1980).

An analysis has been carried out to understand the longitudinal distribution of gamma rays and radio emission. It is inferred from the analysis that the large scale variation of the density of cosmic rays in the Galaxy is of the form $\rho_{CR}(w) = \rho_{CR}(\theta) \exp(1-w/w_\theta)$ upto w_θ and the density remains nearly constant in the outer parts. It is also found that the variation of magnetic field is exponential in nature with a scale length of about 2 w_θ over the entire extent of the Galaxy. There is no evidence for the proton to electron ratio in cosmic rays to vary over the Galaxy.

157.017 **The galactic and extragalactic gamma radiation.** C. E. Fichtel.
X-ray and gamma-ray astronomy in the 1980's, (see 012.054), p. 28 - 30 (1979).

A "clustering" method applied to the analysis of sky maps in gamma-ray astronomy. See Abstr. 031.592.

The spectrum and latitude dependence of the local interstellar radiation field. See Abstr. 131.041.

Galactic and extragalactic contributions to the far-ultraviolet background. See Abstr. 142.053.

High-energy gamma-ray sources observed by COS-B. See Abstr. 142.523.

158 Single and Multiple Galaxies, Peculiar Objects

158.001 **Photometric properties of bright early-type spiral galaxies. I. The data: multiaperture *UBV* photometry for 251 galaxies.** D. Griersmith.
Astron. J., Vol. 85, 789 - 800 (1980).

The details of a large observing program consisting of photometry of 251 bright, mainly early-type spiral galaxies are described. New multiaperture *UBV* data are presented for these galaxies.

158.002 **Mass-to-light ratios for elliptical galaxies.** P. L. Schechter.
Astron. J., Vol. 85, 801 - 811 (1980).

New velocity dispersion measurements for 32 galaxies, mostly ellipticals, have been obtained using the Mt. Hopkins Photon Counting Reticon. The rms fractional difference for ten galaxies in common with Faber and Jackson is 0.09, with no zero-point difference. The relationship between absolute magnitude M and the log of the velocity dispersion σ is determined with a simultaneous solution for the amplitude of the Virgocentric perturbation to the Hubble flow. The author finds that galaxy luminosity L varies as $\sigma^{5.4 \pm 1.0}$, and that the local amplitude of the Virgocentric flow is 190 ± 130 km s^{-1}. Mass-to-light ratios, properly corrected for seeing effects, are calculated for the elliptical galaxies studied by King.

158.003 **High resolution observations of the neutral hydrogen in the galaxy NGC 925.** S. T. Gottesman.
Astron. J., Vol. 85, 824 - 835 (1980) = Contrib. No. 13 Dep. Astron. Univ. Florida, Gainesville, Florida.

The late-type barred spiral NGC 925 has been observed with the line interferometer of the National Radio Astronomy Observatory. An angular resolution of 62″ × 31″ was achieved with a velocity resolution of 24.5 km/s. The neutral hydrogen distribution is found to correlate well with the optical structure. The bright star formation rate is found to be proportional to the H I surface density raised to the 2.35 power. Noncircular motions are observed in the direction of the spiral arms of the galaxy and are compatible with density wave expectations.

158.004 **Spectroscopic survey of southern compact and bright-nucleus galaxies – III.** A. P. Fairall.
Mon. Not. R. Astron. Soc., Vol. 192, 389 - 397 (1980).

80 galaxies, selected from Schmidt telescope sky surveys, have been observed spectroscopically. Most of the galaxies show conventional absorption line spectra, but 15 emission line galaxies have been discovered – these include five 'near Seyferts' (four of them have Fe II emission lines). There is also a case of a possible blueshift of 1500 km s^{-1}.

158.005 **The dynamics of three radio galaxies – NGC 741, 1316 and 7626.** C. R. Jenkins, P. A. G. Scheuer.
Mon. Not. R. Astron. Soc., Vol. 192, 595 - 600 (1980).

The rotation axes of three early-type radio galaxies, radio sources of Fanaroff–Riley class 1, have been determined, using absorption lines so as to obtain information on the dynamics of the stars. In two of the three, the rotation axes are quite different from the minor axes of the images on Palomar Sky Survey prints, thus showing that the assumption of rotation about the minor axis made in early studies is invalid. For all three, the rotation axes bear no obvious relation to the major axes of the radio structures, and in this respect the present work confirms the conclusion of earlier studies. If the radio emission is a consequence of accretion on to a black hole in the galactic nucleus, this finding may indicate that the direction of the angular momentum changes with time, either because of merging with other galaxies, or by accretion of gas from intergalactic space.

158.006 **Radio continuum observations of galaxies at 11 cm.** J. Pfleiderer, C. Durst, K.-H. Gebler.
Mon. Not. R. Astron. Soc., Vol. 192, 635 - 640, Microfiche MN 192/1 (1980).

1616 galaxies, mostly from the Reference Catalogue of Bright Galaxies and the Uppsala General Catalogue of Galaxies, have been observed at 11 cm with the 100-m telescope at Effelsberg and 296 radio sources were detected that are closer than about 2.5 arcmin to the centres of 323 galaxies. The authors give individual detection limits for the 1293 undetected galaxies. 136 additional radio sources found near the galaxies are also listed. The catalogue is compared to the Arecibo 13-cm survey by Dressel & Condon.

158.007 **Further evidence for a large central mass in M 87.** C. R. Jenkins.
Mon. Not. R. Astron. Soc., Vol. 192, 41P - 45 P (1980).

Observations of gas velocities and velocity dispersions, measured from the [O II] λ 3737 blend in the nucleus of M87, are discussed. Long-slit spectroscopy with good spatial resolution shows a steep rise in the velocity dispersion of gas in the nucleus, which, if the material is gravitationally bound, implies a nuclear mass of $\sim 10^{10}\ M_\odot$ and a mass-to-light ratio of ~ 60 (in solar units).

158.008 **Extended radio sources and elliptical galaxies. IV. Structures of 40 resolved sources.**
E. B. Fomalont, J. J. Palimaka, A. H. Bridle.
Astron. J., Vol. 85, 981 - 994 (1980).

Partial-synthesis maps are presented for a sample of extended radio galaxies at 2.7 and 8.1 GHz. These maps were used to determine the overall sizes, orientations, and gross morphologies of the sources during a study of the relative orientations of extended radio sources and their parent elliptical galaxies. The optical identifications of 13 sources have been confirmed by the detection of small-diameter radio components within their parent galaxies.

158.009 **Extended radio sources and elliptical galaxies. V. Optical positions for 40 identified sources.**
J. J. Palimaka, A. H. Bridle, E. B. Fomalont.
Astron. J., Vol. 85, 995 - 1002 (1980).

Optical positions generally accurate to ~0.4 arcsec are given for the centers of 40 radio galaxies. The authors have examined the distribution of angular offsets of the radio cores from the centroids of the extended structures. The distribution confirms the result of paper I (Bridle and Fomalont 1978) that > 90% of the identifications of clearly bifurcated radio structures are within 0.15 (LAS=*largest angular size*) of their radio centroids.

158.010 **The dynamics of some binary galaxies.** J. R. Dickel, H. J. Rood.
Astron. J., Vol. 85, 1003 - 1009 (1980).

Seven binary galaxies have been studied in the 21-cm hydrogen line to determine individual masses of each galaxy and the pair masses. It appears that all the mass in the pairs is within the individual galaxies as measured by their rotational curves.

158.011 **A continuum radio survey of isolated galaxies.** M. T. Adams, E. B. Jensen, J. T. Stocke.
Astron. J., Vol. 85, 1010 - 1026 (1980).

The results of an extensive continuum radio survey of isolated galaxies drawn from the Zwicky catalog by Karachentseva show that both the spirals and the ellipticals in the sample are extremely deficient in radio sources. The results are particularly striking for the early-type systems (E/S0), for which only one galaxy was detected out of the ~120 observed, compared to 10 ± 3 expected detections. A clear correlation is found between frequency of radio emission and local galaxy density. Such a correlation suggests that gas falling into a galaxy triggers the radio emission.

158.012 **Survey of late-type and irregular southern galaxies on plates taken with the UK 1.2-m Schmidt telescope. III.**
H. G. Corwin, Jr., A. de Vaucouleurs, G. de Vaucouleurs.
Astron. J., Vol. 85, 1027 - 1045 (1980).

This third installment of a survey of late-type southern galaxies is based on 129 UK Schmidt film copies or original plates. It lists morphological types, luminosity classes, and inner and outer diameters for 437 galaxies of type Scd ($T = 6$) and later, including some peculiar and irregular systems. Tables give the data for 38 galaxies in the Second Reference Catalogue of Bright Galaxies, and for 399 mainly anonymous objects including many new dwarf systems, some of which have fairly large apparent diameters and are therefore nearby.

158.013 **Radio observations of a complete sample of spiral galaxies at 408 MHz.** I. M. Gioia, L. Gregorini.
Astron. Astrophys., Suppl. Ser., Vol. 41, 329 - 334 (1980).

Radio observations at 408 MHz are presented for 91 spiral galaxies selected from the Uppsala General Catalogue of Galaxies and lying within the declination range $+20° < \delta < +60°$. Optically the sample is complete down to 12th apparent photographic magnitude. The flux density radio limit of the observations is $S \gtrsim 0.07$ Jy. Radiomaps are presented for a number of galaxies.

158.014 **On the dust content of nearby galaxies.**
I. A. M. Issa.
Astron. Nachr., Band 301, 177 - 180 (1980).

The frequency distribution of the sizes of dark cloud complexes in the four nearby galaxies NGC 3031, 5128, 5194 and 5457 is derived and the total amount of dust in these systems is estimated. The frequency distribution of the clouds is nearly the same in all the considered galaxies and may be approximated by $n(R) \sim e^{-kR}$, $k \approx 0.050$ pc^{-1}. The total amount of dust yields to roughly $10^6 M_{\odot}$.

158.015 **Possible variations in the spectrum of the extreme Seyfert galaxy Fairall 9 (=ESO 113-IG45).**
A. P. Fairall.
Mon. Notes Astron. Soc. South. Africa, Vol. 39, 11 - 13 (1980).

Spectroscopic observations made in 1979 December suggest that the strength of the [O III] lines may be nearly 40% greater than in 1977 - 1978. Such variation would indicate that the diameter of the forbidden emission line region, which presumably surrounds the nucleus proper, is less than ~0.5 pc.

158.016 **On the role of interstellar gas in spiral galaxies.**
G.- x. Song.
Acta Astron. Sinica, Vol. 21, 189 - 195 (1980). In Chinese.

The paper discusses the role played by the interstellar gas in spiral galaxies. To represent a disk-shaped galaxy the author adopts a two-gas-disk model (one for the interstellar gas disk and the other for the stellar disk).

158.017 **Analysis of the mass density and oblateness of disc galaxies.** R.- l. Liu.
Acta Astron. Sinica, Vol. 21, 196 - 199 (1980). In Chinese.

Using the data of inclination, radius and mass of disc galaxies, the mass density and oblateness of 93 disc galaxies have been obtained. The results show that the mass density of disc galaxies increases an order of magnitude from the early-type disc galaxies to the later ones, oblateness decreases to minimum at Sd,m, and then increases.

158.018 **Statistics of the structure and rotation of disk galaxies.** F.- x. Hu.
Acta Astron. Sinica, Vol. 21, 200 - 207 (1980). In Chinese.

The data of the optical and radio rotation curves of 57 galaxies are analyzed. The dynamical quantities of structure and rotation of disk galaxies have a systematic variation from early type to later type along the Hubble sequence. The dispersions of the dynamical quantities in each morphological type are related to the masses of galaxies.

158.019 **The complex radio galaxy 4C47.51 (1919+479).**
J. G. Robertson.
Nature, Vol. 286, 579 - 580 (1980).

An observation of the source 1919+479 with the Westerbork Synthesis Radio Telescope at $\lambda = 21$ cm has shown it to have a remarkable morphology, probably that of a very unusual and very large wide angle tail radio galaxy. It is situated in a cluster of galaxies at an estimated redshift of ~0.1. The observed structure of 1919+479 is shown to differ in several respects from previously known cluster sources.

158.020 **The redshift-magnitude relation for bright galaxies at low redshifts.** I. E. Segal.
Mon. Not. R. Astron. Soc., Vol. 192, 755 - 767 (1980).

Recent observations on Shapley - Ames and similar bright galaxies at low redshifts are compared with predictions based on the Hubble law and the redshift-distance square law. The phenomenological indications are unfavourable to the Hubble law, but favourable to the square law.

158.021 **The UV spectrum of the narrow emission line X-ray emitting nucleus of the galaxy NGC 7582.**
J. Clavel, P. Benvenuti, A. Cassatella, A. Heck, M. V. Penston, P. L. Selvelli, F. Beeckmans, F. Macchetto.
Mon. Not. R. Astron. Soc., Vol. 192, 769 - 777 (1980).

UV spectra of the narrow emission line, X-ray emitting nucleus of the galaxy NGC 7582 obtained with the IUE instrument show a steep featureless continuum, obeying a power-law of spectral index $\alpha = 3.4 \pm 0.4$. The total energy distribution is rather well represented by a power-law spectrum of spectral index $\alpha = 2.1$. From the presence of a jump in the continuum near 3600 Å, the authors estimate that hot stars must contribute approximately 30 per cent of the flux at visible wavelengths.

158.022 **Two radio-quiet BL Lac type objects?**
B. K. McIlwrath, D. Stannard.
Mon. Not. R. Astron. Soc., Vol. 192, 79P - 81P (1980).

The optical objects lying close to the positions of the radio sources 1210+121 and 1620+103 may be the first examples of radio-quiet BL Lac type objects.

158.023 **Neutral hydrogen study of 40 Sa spiral galaxies.**
L. Bottinelli, L. Gouguenheim, G. Paturel.
Astron. Astrophys., Vol. 88, 32 - 40 (1980).

Forty Sa spiral galaxies have been investigated in the 21-cm line of neutral hydrogen with the Nançay radiotelescope. Among them, 38 have been detected and 32 for the first time. They appear to form a homogeneous class of galaxies, with a relative H I content smaller than that of later spirals. Their indicative total mass to luminosity ratio is higher than for later type galaxies belonging to the same luminosity range; so this ratio does not exhibit a constant value along the entire type sequence as previously suggested. No correlation is found between the hydrogen mass to luminosity ratio or the mean H I projected density and the luminosity.

158.024 **A UV image of M 31.**
J. M. Deharveng, P. Jakobsen, B. Milliard, M. Laget.
Astron. Astrophys., Vol. 88, 52 - 57 (1980).

A UV(2000 Å) picture of the Andromeda galaxy (M 31) was obtained from a balloon-borne experiment. In addition to the central region of the galaxy, the general spiral structure is seen as a ring which coincides with the distribution of the OB associations and H II regions. The elliptical companion NGC 205 is also barely detected. The fluxes of this galaxy NGC 205 and of NGC 206, a bright association in the spiral structure of M 31, are found compatible with their known stellar content. For the central part of M 31, the flux is in agreement with previous determinations. Detailed analysis confirms the diffuse aspect of this central region and reveals a well peaked luminosity distribution. Possible origins for the central UV radiation are also discussed.

158.025 **On the nature of VV 493=UGC 07910.**
J. W. Sulentic.
Astron. Astrophys., Vol. 88, 94 - 96 (1980).

Spectroscopic observations have been made on the extragalactic object VV 493. It has been suggested that this object is a chain of galaxies in the process of formation by fragmentation. While this interpretation cannot be ruled out, the morphology and spectroscopic observations favor the identification of this system as a single unusually fragmented, spiral galaxy.

158.026 **Neutral hydrogen in the field of the elliptical galaxy NGC 1052.**
L. Bottinelli, L. Gouguenheim.
Astron. Astrophys., Vol. 88, 108 - 112 (1980).

The neutral hydrogen has been mapped in the field of the elliptical galaxy NGC 1052 with the Nançay radiotelescope. An H I emission has been detected at ten positions, including the centres of NGC 1035, 1042, 1047, and 1052.

158.027 **The mass – luminosity relation for active nuclei of galaxies.** Eh. A. Dibaj.
Astron. Zh., Tom 57, 677 - 680 (1980). In Russian.
English translation in Soviet Astron., Vol. 24, No. 4.

The masses of active nuclei of 37 type I Seyfert galaxies and 12 nearby quasars are estimated using the virial theorem approach. The above mentioned masses are compared with bolometric luminosities (optical + infrared + X-ray). The luminosities of active nuclei are close to the corresponding Eddington limit for such masses.

158.028 **An optical analysis of dust complexes in spiral galaxies.** D. M. Elmegreen.
Astrophys. J., Suppl. Ser., Vol. 43, 37 - 56, plates 1 - 4 (1980).

The author presents a method for quantitatively investigating properties of dust regions in external galaxies. The technique involves matching radiative transfer models (with absorption plus scattering) to multicolor photographic and photometric observations. This technique was used to study dust complexes in the late-type spiral galaxies NGC 628 (M74), NGC 5194 (M51), NGC 5457 (M101), and NGC 7793. Most of the features in the prominent dust lanes were found to have internal visual extinctions corresponding to 2 - 3 mag. A different type of feature that was found to be common to all program galaxies is a high latitude (~ 150 - 300 pc) cloud with an internal extinction of some 2 mag. A few moderately dense clouds (~ 150 cm^{-3}) were also observed in some of the galaxies. These features were determined to be in the midplane, and to have masses of ~ $10^6 - 10^7 M$.

158.029 **The nuclear continuum of the Seyfert galaxy NGC 4151.** N. Kaneko.
Publ. Astron. Soc. Japan, Vol. 32, 185 - 195 (1980).

The luminosity profile of NGC 4151 is shown to follow the $r^{1/4}$-law with an effective radius 17″. Starlight from the underlying galaxy is subtracted from published UBV and scanner data. The results show that the nuclear continuum radiation of NGC 4151 has an essentially flat spectrum and changes its intensity from nearly zero ($\ll 10^{-28}$ W m^{-2} Hz^{-1}) to about 3×10^{-28} W m^{-2} Hz^{-1}. This flat nuclear continuum does not explain the infrared emission.

158.030 **Surface photometry of edge-on galaxies. II. NGC 4565.** M. Hamabe, K. Kodaira, S. Okamura, B. Takase.
Publ. Astron. Soc. Japan, Vol. 32, 197 - 212 (1980).

Detailed surface photometry of the edge-on spiral galaxy NGC 4565 was carried out by a computerized digital method. The luminosity distribution was analyzed in the standard manner and then decomposed into sub-components with the help of a spheroidal-shell model and an exponential-disk model. The results of the present analysis in the standard manner are in good agreement with those of previous investigations.

158.031 **Velocity field of the galaxy NGC 7541.**
G. A. Kyazumov.
Pis'ma Astron. Zh., Tom 6, 398 - 401 (1980). In Russian.
English translation in Soviet Astron. Lett., Vol. 6.

Results of spectral investigation of the spiral galaxy NGC 7541 are given. The curve of rotation is obtained and used to construct a three-component model of mass distribution.

158.032 **Rotation and mass of the inner 5 kiloparsecs of the S0 galaxy NGC 3115.**
V. C. Rubin, C. J. Peterson, W. K. Ford, Jr.
Astrophys. J., Vol. 239, 50 - 53, plate 1 (1980).

NGC 3115 is an isolated field galaxy of type S0. It has a small disk embedded in a large flattened halo. Observed velocities rise steeply to 200 km s^{-1} at r = 450 pc and then are flat at V = 267 km s^{-1} from r = 700 to r = 4.7 kpc. Based on a simple spheroid model, the mass is greater than $7 \times 10^{10} M_{\odot}$ interior to r = 5 kpc. By its kinematic properties, NGC 3115 resembles a rapidly rotating high-density disk galaxy rather than a slowly rotating elliptical galaxy. There is no evidence for a variation in $(M/L)_B$ from 1 to 5 kpc.

158.033 **A photometric and kinematic study of the barred spiral galaxy NGC 253. I. Detailed surface photometry.** W. D. Pence.
Astrophys. J., Vol. 239, 54 - 64 (1980).

Detailed isophotometry of NGC 253 in the Sculptor Group in blue light to a limiting surface brightness of 29 mag $arcsec^{-2}$ has been obtained from a combination of photographic and photoelectric photometry. The standard photometric parameters have been derived; in particular, the total magnitude is B_T = 8.05, and the maximum detected dimensions are 60′ × 26′, or 44 × 19 kpc at the assumed distance of 2.5 Mpc. The absolute luminosity of NGC 253, corrected for galactic and internal extinction,is $L_T = 2.65 \times 10^{10} L_{\odot}$. A three component luminosity model of NGC 253 is derived consisting of a $r^{1/4}$ spheroid, an exponential disk, and the spiral arms, which contribute, respectively, 15, 59, and 26% of the total B luminosity.

158.034 **X-ray observations of the Seyfert 1 galaxies AKN 120 and MCG 8-11-11.**
R. F. Mushotzky, F. E. Marshall.
Astrophys. J., Lett., Vol. 239, L5 - L9 (1980).

The authors identify a new X-ray source, H0523–00, with the optically variable Seyfert 1 galaxy AKN 120. The source has a 2–10 keV X-ray flux of $\sim 2 \times 10^{-11}$ ergs cm^{-2} s^{-1} which corresponds to a 2-10 keV X-ray luminosity of $L_x \sim 1 \times 10^{44}$ ergs s^{-1}. X-ray observations over a 1.5 year time span combined with contemporaneous optical photometry (Miller) show a decrease in the optical with no corresponding decrease in the X-ray. In contrast, similar observations of MCG 8-11-11 show a contemporaneous decrease in optical and X-ray fluxes. The authors note that the infrared and X-ray spectral slopes for these two objects are similar, with the optical being steeper by roughly one unit.

158.035 **Massive black hole binaries in active galactic nuclei.**
M. C. Begelman, R. D. Blandford, M. J. Rees.
Nature, Vol. 287, 307 - 309 (1980).

The authors explore the possibility that some active nuclei may contain two massive black holes in orbit about each other. This hypothesis suggests a new interpretation for the observed bending and apparent precession of radio jets emerging from these objects and may indeed be verified through detection of the direct consequences of orbital motion.

158.036 **The galactic distribution of angular momentum density and star formation.**
L. Carrasco, A. Serrano, M. Roth.
Bull. American Astron. Soc., Vol. 12, 445 (1980). – Abstract.

158.037 **A study of star formation and spiral structure in the Sc galaxy M33.** M. Kaufman.
Bull. American Astron. Soc., Vol. 12, 460 (1980). – Abstract.

158.038 **Visual and far red surface photometry of I Zw 1727+50.**
D. Weistrop, H. J. Reitsema, D. B. Shaffer, B. A. Smith.
Bull. American Astron. Soc., Vol. 12, 462 (1980). – Abstract.

158.039 **Detection of X-ray emission from Markarian 180.**
D. J. Hutter, S. L. Mufson.
Bull. American Astron. Soc., Vol. 12, 486 (1980). – Abstract.

158.040 **Models of central sources in active galactic nuclei.**
M. Kafatos.
Bull. American Astron. Soc., Vol. 12, 487 (1980). – Abstract.

158.041 **X-ray variability of active galactic nuclei.**
C. J. Hailey, D. J. Helfand, W. H.-M. Ku.
Bull. American Astron. Soc., Vol. 12, 488 (1980). – Abstract.

158.042 **Detection of a red halo in NGC 3877.**
E. D. Loh, D. T. Wilkinson.
Bull. American Astron. Soc., Vol. 12, 490 (1980). – Abstract.

158.043 **Search for luminous halos of spiral galaxies.**
P. R. Saulson, B. H. Siebers, Jr., E. D. Loh, D. T. Wilkinson.
Bull. American Astron. Soc., Vol. 12, 490 (1980). – Abstract.

158.044 **Surface photometry of NGC 4656/7.**
L. Stayton, R. Angione, F. Talbert.
Bull. American Astron. Soc., Vol. 12, 490 (1980). – Abstract.

158.045 **Dimensionless systematics of inner ring structures in galaxies.** G. de Vaucouleurs, R. Buta.
Bull. American Astron. Soc., Vol. 12, 490 (1980). – Abstract.

158.046 **Correlation of the dark mass in galaxies with Hubble type.** B. M. Tinsley.
Bull. American Astron. Soc., Vol. 12, 491 (1980). – Abstract.

158.047 **Using gas disks to probe the 3-D structure of elliptical galaxies.**
J. E. Tohline, G. Simonson, C. N. Caldwell.
Bull. American Astron. Soc., Vol. 12, 491 (1980). – Abstract.

158.048 **A spectroscopic study of the galaxy IC5063.**
C. N. Caldwell, M. M. Phillips.
Bull. American Astron. Soc., Vol. 12, 491 (1980). – Abstract.

158.049 **A mass model for the Sombrero galaxy.**
T. Y. Steiman-Cameron.
Bull. American Astron. Soc., Vol. 12, 491 - 492 (1980). Abstract.

158.050 **Ultraviolet images of M101 and M33.**
T. P. Stecher, R. C. Bohlin.
Bull. American Astron. Soc., Vol. 12, 492 (1980). – Abstract.

158.051 **Velocity dispersions in M31 and M32.**
B. C. Whitmore.
Bull. American Astron. Soc., Vol. 12, 492 (1980). – Abstract.

158.052 **Intermediate band surface photometry of NGC 3379 and NGC 4406.** R. J. Boyle.
Bull. American Astron. Soc., Vol. 12, 492 (1980). – Abstract.

158.053 **Velocity dispersions and mass to light ratios in cD galaxies.** E. M. Malumuth, R. P. Kirshner.
Bull. American Astron. Soc., Vol. 12, 492 - 493 (1980). Abstract.

158.054 **The dynamical evolution of NGC 5128.**
A. D. Tubbs.
Bull. American Astron. Soc., Vol. 12, 493 (1980). – Abstract.

158.055 **An observed change in the radio absorption spectrum of the BL Lac object AO 0235+164.**
M. M. Davis, A. M. Wolfe.
Bull. American Astron. Soc., Vol. 12, 494 (1980). – Abstract.

158.056 **Observations of the ultraviolet energy distribution of BL Lacertae using the IUE.**
K. R. Hackney, R. L. Hackney, W. M. Kinzel, R. L. Scott, J. T. Pollock, A. J. Pica, R. J. Leacock, A. G. Smith.
Bull. American Astron. Soc., Vol. 12, 494 (1980). – Abstract.

158.057 **Profiles of prominent emission lines in type 1 Seyfert galaxies.**
C.-C. Wu, A. Boggess, T. R. Gull.
Bull. American Astron. Soc., Vol. 12, 495 (1980). – Abstract.

158.058 **The origin of the forbidden line region in Seyfert galaxies.** A. S. Wilson, J. S. Ulvestad.
Bull. American Astron. Soc., Vol. 12, 496 (1980). – Abstract.

158.059 **Radio structures of Seyfert galaxies.**
J. S. Ulvestad, A. S. Wilson, R. A. Sramek.
Bull. American Astron. Soc., Vol. 12, 496 (1980). – Abstract.

158.060 **Spectral characteristics of emission-line galaxies.**
J. M. Shuder.
Bull. American Astron. Soc., Vol. 12, 496 (1980). – Abstract.

158.061 **Evolutionary implications of a rotating Kerr black hole model for active Seyfert galaxies.** D. Leiter.
Bull. American Astron. Soc., Vol. 12, 496 (1980). – Abstract.

158.062 **The bivariate luminosity function of E and S0 galaxies.** L. L. Dressel.
Bull. American Astron. Soc., Vol. 12, 496 - 497 (1980). Abstract.

158.063 **Internal tail structure and orbit dynamics of the radio galaxy 3C129.** G. G. Byrd, M. J. Valtonen.
Bull. American Astron. Soc., Vol. 12, 503 (1980). – Abstract.

158.064 **Detection of ^{12}CO in the dwarf elliptical galaxy NGC 185.** D. W. Johnson, S. T. Gottesman.
Bull. American Astron. Soc., Vol. 12, 503 (1980). – Abstract.

158.065 **Comparing infrared fluxes and emission line widths for galactic nuclei.**
D. Weedman, V. Balzano, F. Feldman, L. Ramsey.
Bull. American Astron. Soc., Vol. 12, 504 - 505 (1980). Abstract.

158.066 **Theory of dwarf galaxies.**
H. Gerola, P. E. Seiden, L. S. Schulman.
Bull. American Astron. Soc., Vol. 12, 528 (1980). – Abstract.

158.067 **Dynamical models of the gas flow in the barred spiral galaxy NGC 1300.** J. M. Huntley.
Bull. American Astron. Soc., Vol. 12, 529 (1980). – Abstract.

158.068 **Galaxy-QSO associations: a gravitational lens effect of globular clusters.**
J. M. Barnothy, M. F. Barnothy.
Bull. American Astron. Soc., Vol. 12, 537 (1980). – Abstract.

158.069 **Neutrino emission from galaxies, and mechanisms for producing radio lobes.**
R. Silberberg, M. M. Shapiro.
Bull. American Astron. Soc., Vol. 12, 537 - 538 (1980). Abstract.

158.070 **HEAO-1 X-ray observations of time variability in active galactic nuclei.**
A. Tennant, R. Mushotzky, E. Boldt.
Bull. American Astron. Soc., Vol. 12, 544 (1980). – Abstract.

158.071 **Toward a physical understanding of the Hubble sequence.** S. E. Strom.
Bull. American Astron. Soc., Vol. 12, 566 (1980). – Abstract of a paper from the 155th meeting of the American Astronomical Society. – See Abstr. 27.010.002.

158.072 **Photometry of a complete sample of faint galaxies.**
R. G. Kron.
Astrophys. J., Suppl. Ser., Vol. 43, 305 - 325 (1980).

This paper develops techniques for the automatic measurement of the colors and magnitudes of a complete sample of very faint galaxies. The observations consist of photographic surface photometry in two wavebands for ~20000 galaxies in two widely separated, high-latitude fields, each having an area of ~1080 arcmin^2. The photometric zero points are derived directly from faint stars in the same fields measured photoelectrically by other observers. The photometry of the galaxy sample is complete to a limit about 10 times fainter than has been systematically investigated before. The sample is characterized by a population of galaxies which becomes bluer with increasing faintness and which increases steeply in number with increasing faintness in the blue waveband. The interpretation of these results will be attempted in a later paper.

158.073 **Double galaxies as indicators of large scale structure.**
W. G. Tifft.
Astrophys. J., Vol. 239, 445 - 462 (1980).

Northern-hemisphere double galaxies are shown to contain large-scale clumping in the form of clouds associated into four or five major large-scale filamentary structures. The filaments are on the order of 10 Mpc thick and 100 Mpc long. Position angles of the doubles are aligned regionally, and the alignment varies in a regular manner within the large-scale structure. Clusters within the structures appear to have elongations which fit the orientation pattern defined by doubles.

158.074 **Three new cases of galaxies with large discrepant redshifts.** H. Arp.
Astrophys. J., Vol. 239, 469 - 474, plates 6 - 8 (1980).

Three new cases where smaller galaxies appear to be connected by luminous filaments to larger galaxies are presented.

158.075 **Large- and small-scale structure in the continuum energy distributions of quasi-stellar objects and Seyfert 1 galaxies.** S. A. Grandi, M. M. Phillips.
Astrophys. J., Vol. 239, 475 - 482 (1980).

Moderate-resolution spectrophotometric observations of more than 40 Seyfert 1 galaxies and QSOs have been analyzed to study large- and small-scale structure in the continuum energy distributions of such objects.

158.076 **Southern galaxies. VIII. Surface photometry of the Sd spiral NGC 7793.**
G. de Vaucouleurs, E. Davoust.
Astrophys. J., Vol. 239, 783 - 802, plates 10, 11 (1980).

The authors report on detailed surface photometry of NGC 7793, the prototype of morphological type Sd and the faintest of the five major members of the Sculptor group which is at an estimated distance of 2.5 Mpc. This study will also contribute to two other programs: (1) monographs of the members of the Sculptor group which are important to extent the extragalactic distance scale beyond the Local Group in the southern hemisphere and (2) detailed analyses of the velocity fields in hydrogen-rich galaxies by Fabry-Perot interferometry.

158.077 **The first bright globular cluster in NGC 5128.**
J. A. Graham, M. M. Phillips.
Astrophys. J., Lett., Vol. 239, L97 - L99, plate L5 (1980).

A 17th magnitude, slightly diffuse object has been found in the halo of the radio galaxy NGC 5128. Radial velocity and spectrophotometric measurements suggest that the object is a bright globular cluster, the first to be associated with this unusual galaxy.

158.078 **The optical spectra of narrow-line X-ray galaxies.**
J. M. Shuder.
Astrophys. J., Vol. 240, 32 - 40 (1980) = Lick Obs. Bull., No. 865.

Spectrophotometric results of narrow-emission-line galaxies known to be X-ray sources are discussed. In many respects, the spectra of these galaxies are similar to Seyfert type 2 and other narrow-emission-line galaxies. However, evidence for a high level of ionization is indicated in two objects (NGC 2110 and NGC 2992) by the presence of [Fe X] λ6374, and in four of the five survey objects a broad Hα component has been found. The FWZI of Hα in these objects ranges from 2300 to 5700 km s^{-1}, contrasted with the forbidden lines, which have a FWZI of 1000 to 1500 km s^{-1}. This finding is consistent with Seyfert type 1 and broad-line radio galaxies, which show an association between a broad-line component (at least at Hα) and the emission of X-rays.

158.079 **Galaxies with the spectra of giant H II regions.**
H. B. French.
Astrophys. J., Vol. 240, 41 - 59 (1980) = Lick Obs. Bull., No. 863.

Line fluxes in the region 3700–7100 Å are presented for 14 galaxies with strong, sharp, H II region-like emission lines. Ten of these galaxies are low-luminosity objects ($M > -17$); the others have $M \lesssim -20$. Ratios of the line fluxes are used to derive electron temperatures and densities, and the abundances of helium, oxygen, nitrogen, neon, and sulfur relative to hydrogen. The nature of these galaxies is uncertain. All of the available data are consistent with their being young objects, at least in the sense of having only recently formed most of their stars.

158.080 **Detection of the CO $J = 2 \rightarrow 1$ line in M82 and IC 342.** G. R. Knapp, T. G. Phillips, P. J. Huggins, R. B. Leighton, P. G. Wannier.
Astrophys. J., Vol. 240, 60 - 64 (1980).

The authors have searched for the CO(2 → 1) line at 230 GHz in four galaxies and have detected it in two, IC 342 and M82. Comparison with data for the CO(1 → 0) line in these galaxies shows that the molecular gas in IC 342 appears to be optically thick; that in M82, however, is of low optical depth. A comparison of the gas and stellar content of the central region of M82 suggests that much of the current star formation is taking place as high-mass stars.

158.081 **The spectrum of the central luminosity spike in M87.** A. Dressler.
Astrophys. J., Lett., Vol. 240, L11 - L16 (1980).

A spectrum of the central luminosity spike in M87 has been obtained in 0''75 seeing through a 1" × 1" aperture using the du Pont Reticon Spectrograph at Las Campanas Observatory. A comparison of this spectrum with one taken at $r = 13''$ indicates that less than 20% of the visible and red light comes from a nonthermal component. A Fourier transform analysis of Mg b and Na D in the nuclear spectra yields a velocity dispersion $\Delta V = 350 \pm 32$ at an average distance $r \lesssim 0''35$. These measurements are inconsistent with the 5×10^9 black hole model of Young et al. and Sargent et al.

158.082 **Spectra of some double nucleus Markarian galaxies.** A. R. Petrosyan, K. A. Saakyan, Eh. E. Khachikyan.
Pis'ma Astron. Zh., Tom 6, 552 - 553 (1980). In Russian. English translation in Soviet Astron. Lett., Vol. 6.

From spectra obtained with the 6-m telescope radial velocities of six Markarian galaxies are determined for the first time. For two of them, Mark 608 and 708, the apparent double nucleus structure turned out to be caused by a projection of a star on the galaxy. For Mark 104, 786, 799 and 1027 in addition to radial velocities of both nuclei the equivalent widths of the Hα line as well as the intensity ratios of [N II] and [S II] lines to Hα are determined.

158.083 **Preliminary spectrophotometric investigation of the nucleus of the galaxy NGC 5929.**
V. K. Golev, I. M. Yankulova, G. T. Petrov.
Pis'ma Astron. Zh., Tom 6, 554 - 558 (1980). In Russian. English translation in Soviet Astron. Lett., Vol. 6.

Components of emission lines are found in the spectrum of the galaxy NGC 5929 nucleus from 6-m and 1.25-m telescopes observations. Assuming the temperature of the emitting gas $T_e = 10^4$ K the characteristic parameters are estimated. The relative abundances of some ions are determined. The Ly_c radiation of young hot stars appears to be the source of ionization of the gas in the galaxy nucleus.

158.084 **On the asymmetry of permitted line profiles in the spectra of Seyfert galaxies.** S. N. Fabrika.
Pis'ma Astron. Zh., Tom 6, 559 - 563 (1980). In Russian. English translation in Soviet Astron. Lett., Vol. 6.

The asymmetry of permitted line profiles in the spectra of Seyfert galaxies is supposed to be due to self-absorption in the envelope. In this case a direct estimate of the filling factor is possible. For the galaxies NGC 4151 and NGC 3516 the filling factor is estimated. A simple method of determination of the gas velocity field in the nucleus is suggested.

158.085 **The distribution of faint galaxies in a field of 15 square degrees near the South Galactic Pole.**
H. T. MacGillivray, R. J. Dodd.
Mon. Not. R. Astron. Soc., Vol. 193, 1 - 6 (1980).

Measurements are made to examine the distribution of galaxies down to $B \sim 22.0$ in a field of 15 deg^2 near the South Galactic Pole. The mean multiplicity of galaxies is 3.6, in good agreement with analyses of the Lick and Jagellonian galaxy counts, indicating no strong evolutionary effect in the clustering of galaxies at least out to redshifts $z^* \sim 0.4$–0.7.

158.086 **Lopsided galaxies.**
J. E. Baldwin, D. Lynden-Bell, R. Sancisi.
Mon. Not. R. Astron. Soc., Vol. 193, 313 - 319 (1980).

Attention is drawn to the large-scale asymmetries in a number of nearby galaxies. A mechanism is proposed which is partially successful in explaining the long persistence of these asymmetries.

158.087 **Iron line emission and variability in the X-ray spectrum of the Seyfert galaxy NGC 5548.**
M. J. C. Hayes, J. L. Culhane, R. J. Blissett, P. Barr, S. J. Bell Burnell.
Mon. Not. R. Astron. Soc., Vol. 193, 15P - 20P (1980).

The MSSL (*Mullard Space Science Laboratory*) spectrometer on the Ariel V satellite has detected a variable iron emission feature in the X-ray spectrum of NGC 5548. Comparison with other observations indicates that both the slope and intensity of the continuum spectrum are variable. It is suggested that the iron emission is due to fluorescent excitation of the gas in the broad emission-line region surrounding the X-ray emitting nucleus of the galaxy.

158.088 **La relation diamètre-luminosité pour les galaxies elliptiques et lenticulaires.**
J. H. Bigay, G. Paturel.
Astron. Astrophys., Suppl. Ser., Vol. 42, 69 - 79 (1980).

From photoelectric observations and isophotometry of a Schmidt plate, total B magnitudes and isophotal diameters are obtained for 61 E and S0 galaxies of the Coma cluster central region. These data are used to find the exponent q in the diameter-luminosity relation. The q value is not significantly different from the divergent value $q = 2$ at the 0.01 probability level.

158.089 **The radio continuum emission from spiral galaxies in double systems.** E. Hummel.
Astron. Astrophys., Vol. 89, L1 - L2 (1980).

The radio continuum properties of interacting and isolated spiral galaxies are compared. It is found that the disk emissivity is similar but that the central sources in interacting galaxies are on average a factor 2 to 3 stronger than those in isolated spirals.

158.090 **The warped Sb-galaxy NGC 4565.**
W. K. Huchtmeier, J. H. Seiradakis, G. A. Tammann.
Astron. Astrophys., Vol. 89, 95 - 99 (1980).

The distribution of the neutral hydrogen (H I) of the nearly edge-on-Sb-galaxy NGC 4565 has been observed with the 100 m radiotelescope. The H I extends to about twice the optical dimension being asymmetric like the optical galaxy. In the outer part a considerable warp is observed reaching about 18 kpc at a distance of ~100 kpc from the centre. A flat rotation curve is compatible with the observations leading to a total mass of $\sim 10^{12} M_\odot$, the relative hydrogen content being 4%. Thus the Sb-galaxy NGC 4565 is among the largest (extent) and most massive of the relatively near spiral galaxies.

158.091 **High-redshift objects near the companion galaxies to NGC 2859.** H. Arp.
Astrophys. J., Vol. 240, 415 - 420, plate 4 (1980).

As a continuation of a systematic search around companions to bright galaxies, the environs of NGC 2859 have been studied. There are four smaller galaxies that are within 30′ of NGC 2859. Each of these four companions has a quasar or compact object within about 1′. The redshifts of the four compact objects are $z = 2.25$, 1.46, 0.23, and 0.027. The probability of any one of the three bona fide quasars being an accidental projection of a background object is $p \leq 0.01$. The probability that all four higher-redshift objects are accidental projections is around $p \leq 10^{-8}$.

158.092 **Radio structures of Seyfert galaxies. I.**
A. S. Wilson, A. G. Willis.
Astrophys. J., Vol. 240, 429 - 441 (1980).

Radio maps of 10 Seyfert and Seyfert-like galaxies have been made with the Very Large Array at 4.885 GHz. This paper presents the first results. The methods of observation and reduction are explained and the maps on a source by source basis are described. A discussion of the source sizes and morphologies, magnetic fields and cosmic ray densities, the origin of the radio emission, and possible implications for models of superluminally expanding radio sources is presented.

158.093 ***IUE* spectra of the jet and the nucleus of M87.** G. C. Perola, M. Tarenghi.
Astrophys. J., Vol. 240, 447 - 454 (1980).

IUE spectra from 1300 to 3000 Å of the jet and the nucleus of M87 are presented. The flux detected with the 10″ × 20″ aperture positioned on the jet is consistent with the extrapolation of the optical power law spectrum ($\lambda^{-0.3}$) of the bright complex of knots A + J + B, plus a contribution of diffuse light from M87.

158.094 **On the abundance of carbon monoxide in galaxies: a comparison of spiral and Magellanic irregular galaxies.** B. G. Elmegreen, D. M. Elmegreen, M. Morris.
Astrophys. J., Vol. 240, 455 - 463 (1980).

CO emission has been sought in 52 directions toward six irregular, Magellanic-type galaxies, with null results at a typical detection limit of 0.04 K for a line width of 10 km s^{-1}. The positions searched correspond to those of bright OB star clusters, H II regions, H I peaks, and dark clouds. In two cases, one-third of the optical galaxy was covered by the observations. CO has been detected at 2.6 mm toward the nucleus of the low-mass SA(s)dm galaxy NGC 7793.

158.095 **Astrophysical gamma-ray production by inverse Compton interactions of relativistic electrons. III. Cutoff effect for inverse Compton spectra applied to the case of the hard X-ray and gamma-ray emission of NGC 4151.**
R. Schlickeiser.
Astrophys. J., Vol. 240, 636 - 641 (1980).

The recently reported spectral features of NGC 4151 in the hard X-ray and γ-ray energy range are analyzed. A two-component emission model is suggested in which the ultraviolet and soft X-ray photons ($E < 20$ keV) are scattered by relativistic electrons via the inverse Compton process into the hard X-ray and γ-ray regime. It is shown that the rather flat, hard X-ray spectrum can be explained if a low-energy cutoff in the relativistic electron distribution is postulated. The concept of a low-energy cutoff in the electron distribution is not inconsistent with the observed overall continuum emission of NGC 4151.

158.096 **Neutral hydrogen in isolated galaxies: first results for five early-type systems.**
M. P. Haynes, R. Giovanelli.
Astrophys. J., Lett., Vol. 240, L87 - L91 (1980).

The neutral hydrogen properties of the first five early-type galaxies observed in the authors' survey of isolated galaxies are presented. The galaxies are found in regions of very low density and hence are "isolated" at least with respect to the majority of galaxies. Among this sample, early-type systems are extremely rare. The five isolated early-type galaxies presented here contain substantial masses of neutral hydrogen.

158.097 **Photographic *UBV* surface photometry and gross structure of the halo of M 82.**
W. Bronkalla, P. Notni, H. Tiersch.
Astron. Nachr., Band 301, 217 - 232, with a correction p. 335 (1980).

The authors present a surface photometry of M 82 of medium spatial resolution. The results are discussed in terms of the distribution of intensity, colour, reddening, free parameter Q and colour excess E_{B-V}. The population index Q of the inner halo is bluer than that of the disc everywhere. Two sources must be responsible for the illumination of the halo; one of them is responsible for the light from the great disturbances near the minor axis. Both may be located near the center but must be partly shadowed. A major contribution of the disc to the illumination is less probable because the population index is different in most places of the halo from that of the disc. The inclination of the disc is estimated to lie between 80° and 90°.

158.098 **Optical line studies of the nuclei of NGC 4945 and 5128.** J. B. Whiteoak, F. F. Gardner.
Proc. Astron. Soc. Australia, Vol. 3, 319 - 321 (1979).

158.099 **The spectrum and polarization of the nucleus of NGC 4151.** G. D. Schmidt, J. S. Miller.
Astrophys. J., Vol. 240, 759 - 767 (1980) = Lick Obs. Bull., No. 862.

Simultaneous spectropolarimetric and spectrophotometric observations of the nucleus of NGC 4151 are presented. In addition to the strong features characteristic of a Seyfert type 1 galaxy, the spectrum records a host of weak emission features, including broad Fe II and many narrow lines of both Fe II and [Fe II]. The polarization data disclose that the narrow Balmer line cores and forbidden lines are polarized alike, but that the broad-line emission is unpolarized. The results therefore confirm the distinction often made in these objects between a high-density, broad-line emission region and the lower-density material responsible for the permitted-line cores and forbidden lines. The continuum polarization varies smoothly across the spectrum, but the position angle is constant. The wavelength dependence is consistent with the presence of a uniformly polarized synchrotron component. The observed NGC 4151 continuum can be reproduced by a combination of this synchrotron component and a galactic stellar population in proportions ~1:1 at visual wavelengths.

158.100 **Iron-peak abundances in the nuclear regions of M31, M81, and M94.** C. Pritchet, B. Campbell.
Astrophys. J., Vol. 240, 768 - 778 (1980).

The authors have obtained high signal-to-noise ratio near-infrared spectrophotometry for the nuclear regions of the galaxies M31 (NGC 224), M81 (NGC 3031), and M94 (NGC 4736), and for a number of standard stars possessing high-quality curve-of-growth analyses. Blends of weak lines (due primarily to iron-peak elements) have been identified in G, K, and M stars in the spectral region near λ7400; these blends are found to retain their visibility in the velocity-broadened spectra of galaxies. An index which measures the integrated strengths of these blends has been calibrated against effective temperature and metal abundance in the standard stars. The results indicate near-solar abundances of iron-peak elements for super-metal-rich stars. The authors have employed theoretical isochrones and luminosity functions to predict the abundance sensitivity of near-infrared line-blends in the integrated spectra of galaxies. A comparison of predicted blend strength with the observations yields solar abundances of iron-peak elements for the nuclear regions of the galaxies M31, M81, and M94. This is in contrast to several other abundance determinations (based primarily on damped lines of Mg and Na), which have inferred moderate overabundances of heavy elements in the nuclei of M31 and M81.

158.101 **Infrared photometry of the semistellar nucleus of M31.**
S. E. Persson, J. G. Cohen, K. Sellgren, J. Mould, J. A. Frogel.
Astrophys. J., Vol. 240, 779 - 784 (1980).

New broad-band infrared *JHK* data and narrow-band CO and H_2O indices for the semistellar nucleus of M31 are presented. The data were obtained specifically to test a prediction of a recent synthesis model by Faber and French in which the ratio of dwarf-to-giant light increases strongly in going from the bulge to the nucleus of M31. The new infrared data do not support such a model. Some alternative explanations for the behavior of the various indices are given.

158.102 **Neutral-hydrogen observations of smooth-arm spiral galaxies.** M. S. Wilkerson.

Astrophys. J., Lett., Vol. 240, L115 - L119, plates L9 - L10 (1980).

A class of galaxies has been identified whose members show spiral arms but not the clumpy regions that are indicative of recent star formation in those arms. Neutral-hydrogen observations have been made of five of these smooth-arm spirals in three galaxy clusters, using the 305 m Arecibo telescope. Four nonsmooth spirals in two of the same clusters were also observed. Three of the four nonsmooth galaxies were detected and found to be deficient in neutral hydrogen with respect to "field" galaxies. One of the smooth-arm spirals was definitely detected, and a second one was detected at a low confidence level. Upper limits substantially below the neutral-hydrogen levels for actively star-forming spirals were established for the other three smooth spirals. These observations argue that smooth-arm spirals are gas deficient even when compared to their clumpy spiral neighbours in rich clusters.

158.103 **Photometric properties of bright early-type spiral galaxies. II.Fully corrected colors and standard magnitudes.** D. Griersmith.
Astron. J., Vol. 85, 1135 - 1154 (1980).

Multiaperture *UBV* data have been compiled for a large homogeneous sample of early-type spiral galaxies. Corrections have been applied to the data for aperture effect, *K*-dimming, Galactic reddening, internal absorption, and color-absolute magnitude effect. The correction for aperture effect is used to estimate surface color gradients in early-type spirals. Fully corrected colors and magnitudes are presented for 269 early-type spirals. This sample includes 207, or 84%, of the Shapley-Ames galaxies with revised Hubble types in the range $T=-1$ to 2.

158.104 **The velocity field of the Seyfert galaxy NGC 7469.** B. A. M. Westin.
Astron. Astrophys., Vol. 89, L11 - L12 (1980).

Spectrographic observations of the type 1 Seyfert galaxy NGC 7469 have been made. Measurements have shown that the forbidden lines are blue-shifted, with respect to the permitted lines, thus confirming the results obtained by Barbieri et al. (1977). The tilt of the Hα line differs from that of the [N II] line, λ6584. Those two lines also extend relatively far out from the nuclear region. With one exception, the red shift of these lines increases with distance from the nucleus.

158.105 **The Penrose Photoproduction Scenario for NGC 4151; (PCS–SSC). A black hole γ-ray emission mechanism for active galactic nuclei and Seyfert galaxies.**
D. Leiter.
Astron. Astrophys., Vol. 89, 370 - 376 (1980).

On the basis of general arguments, it has been suggested (Bignami et al., 1979) that a steepening of the spectrum between X-ray and γ-ray energies may be a general γ-ray characteristic of Seyfert galaxies, if the diffuse γ-ray spectrum is considered to be a superposition of unresolved contributions from one or more classes of extragalactic objects. The author shows that the above suggestion can be given a consistent theoretical interpretation in the context of the Penrose Photoproduction Scenario (PCS–SSC). Specifically in the case of NGC 4151, the dominant process is shown to be Penrose Compton Scattering PCS in the ergosphere of an $M \geqslant 10^8 M_{\odot}$ Kerr black hole, assumed in its nucleus.

158.106 **Hat die Scheibenkomponente von galaktischen Massenmodellen ein zentrales Loch?**
J. Kreitschmann, K. Rohlfs.
Mitt. Astron. Ges., Nr. 48, (see 012.015), p. 173 - 175 (1980).

158.107 **Observations of radio galaxies.** R. A. E. Fosbury.
Messenger, No. 21, p. 11 - 14 (1980).

158.108 **Smaller galaxies.** H. Arp.
Messenger, No. 21, p. 25 - 27 (1980).

158.109 **Ring galaxies.** M. Dennefeld, J. Materne.
Messenger, No. 21, p. 29 - 32 (1980).

158.110 **New high resolution radio observations of NGC 4258. II.NGC 4258 as a spiral galaxy.**
G. D. van Albada.
Astron. Astrophys., Vol. 90, 123 - 133 (1980).

It is demonstrated that the H I in the large, nearby spiral NGC 4258 behaves very much as expected for a normal SAB galaxy, notwithstanding the presence of a pair of anomalous arms indicating some kind of violent activity. The large angular dimensions of NGC 4258 and the good signal-to-noise ratio of the observations allow a detailed study of the properties of the underlying, unperturbed part of the galaxy. Strong evidence for the presence of a bar is found, as well as indications of a high pattern speed, putting corotation well inside the optically observable disk.

158.111 **Criterion $M_n - M_g$.** S. G. Iskudaryan.
Dokl. AN ArmSSR, Vol. 70, 50 - 55 (1980). In Russian. – Abstr. in Ref. zh., 51. Astron., 9.51.684 (1980).

158.112 **On the kinematics of high-velocity gas in the envelopes of the nuclei of type 1 Seyfert galaxies and quasars.** S. N. Fabrika.
MGU, Moskva, 1980. 7 pp. In Russian. – Abstr. in Ref. zh., 51. Astron., 9.51.734 (1980).

158.113 **Ion abundance and chemical composition in the nuclei of type 2 Seyfert galaxies and radio galaxies with narrow lines.** G. T. Petrov.
Dokl. AN ArmSSR, Vol. 70, 46 - 49 (1980). In Russian. Abstr. in Ref. zh., 51. Astron., 9.51.739 (1980).

158.114 **Photometry and structure of lenticular galaxies. II. NGC 4111 and NGC 4762.** V. Tsikoudi.
Astrophys. J., Suppl. Ser., Vol. 43, 365 - 377, plates 5 - 8 (1980).

Surface photometry of the edge-on lenticular galaxies NGC 4111 and NGC 4762 and a study of their structure are presented. The brightness distributions, when decomposed into the contributing luminosity components, reveal: (*a*) a spheroidal luminosity component due to the nuclear bulge, fitted best by the $r^{1/4}$ luminosity law and contributing about 43% of the light in the galaxy; (*b*) a lens, which is quite prominent and substantial in both objects and exhibits a Gaussian distribution; (*c*) an exponential luminosity component in the outer regions, due to both the disk and the envelope along the galactic plane, but due only to the envelope in all other directions; the disk is thin and Gaussian in directions perpendicular to the equatorial plane; (*d*) additional, weaker light contributions along the galactic plane, which possibly represent "rings" and/or "relic" spiral arms.

158.115 **Colors and magnitudes predicted for high redshift galaxies.**
G. D. Coleman, C.-C. Wu, D. W. Weedman.
Astrophys. J., Suppl. Ser., Vol. 43, 393 - 416 (1980).

Ultraviolet observations of nearby galaxies with the *ANS* are used to derive ultraviolet spectra for different galaxy types. These spectra are used with existing visible spectrophotometry to calculate *K*-corrections, and to predict colors and magnitudes for various galaxy types as a function of redshift, to $z = 2$. No evolutionary effects are considered. It appears that the first-ranked cluster galaxies on blue emulsions should be spirals for $z \gtrsim 0.5$.

158.116 **The size distribution of dark clouds as a new method to determine the distances of galaxies and the amount of dust.** I. A. Issa.

J. Astron. Soc. Egypt. Vol. 1, 53 - 66 (1979).

The areas of many dark clouds in the galaxies NGC 3031, NGC 5128, NGC 5194 and NGC 5457 were measured. The absorptions of the clouds were estimated. The amount of dust in these galaxies was found to be of the order of 10^6 $M_\odot$. A new method was tested concerning the determination of the distances of galaxies from the size distribution function of the apparent radii of the dark clouds.

158.117 **Evolution of stellar composition and some parameters of galaxies and their nuclei.**
A. V. Tutukov, E. Krügel.
Astron. Zh., Tom 57, 942 - 952 (1980). In Russian. – English translation in Soviet Astron., Vol. 24, No. 5.

The stellar composition of galaxies and their nuclei was studied on the basis of stellar evolution theory. Analytical approximations for some important observable parameters of galaxies were found. The influence of mass loss and accretion on the galaxies evolution was explored. The importance of the lower mass limit of forming stars for the rate of galaxy evolution is pointed out.

158.118 **Photometric properties of bright early-type spiral galaxies. III. The color-absolute magnitude relation and the analysis of fully corrected colors.** D. Griersmith.
Astron. J., Vol. 85, 1295 - 1311 (1980).

New results concerning (a) the color-absolute magnitude (CM) relation for early-type spiral galaxies and (b) colors corrected for the CM effect (i. e., fully corrected colors) are presented. The results are based upon analysis of the corrected *UBV* data given in Paper II (Griersmith 1980).

158.119 **101 spiral galaxies: neutral hydrogen distribution.**
N. Krumm, E. E. Salpeter.
Astron. J., Vol. 85, 1312 - 1324 (1980).

This is a first report on a survey of the H I distributions in a large sample of spiral galaxies, using the 21-cm spectral-line system at Arecibo Observatory. One hundred and thirty galaxies have been observed, and 101 have been detected and roughly mapped in H I. The data have been used to determine systemic velocity, H I content, gravitational mass, and an approximate measure of the diameter of the H I distribution.

158.120 **The location of star-forming regions in barred Magellanic-type galaxies.**
D. M. Elmegreen, B. G. Elmegreen.
Astron. J., Vol. 85, 1325 - 1327 (1980).

Approximately half of all barred Magellanic irregular and spiral galaxies larger than 4′ have their largest H II regions near the ends of their bars. Magellanic irregular galaxies also appear to have solid-body rotation throughout most of their barred regions. Since enhanced star formation and solid-body rotation are associated with the bars in massive barred spiral galaxies, the authors suggest that the gas dynamics is similar in all barred galaxies regardless of mass. Star formation in 30 Doradus may have been triggered by the large-scale compression of gas moving around the bar in the Large Magellanic CLoud.

158.121 **Accurate optical positions for Markarian objects 701 - 797.**
C. B. Foltz, B. M. Peterson, T. A. Boroson.
Astron. J., Vol. 85, 1328 - 1329 (1980).

Optical positions of Markarian objects 701 - 797 are reported with an accuracy of a few arcseconds.

158.122 **Brightening of a hot spot in NGC 2903.**
P. Laques, J.-L. Nieto, J.-L. Vidal, A. Augé, R. Despiau.
Nature, Vol. 288, 145 - 146 (1980).

The nuclear region of the Sc galaxy NGC 2903 exhibits a very chaotic structure with eight knots. During high resolution–broad band and narrow band–photometric studies of the central region of NGC 2903, the authors have observed and report here that one of the knots appears 1.3 mag brighter on plates taken in February 1980 than on a plate taken in January 1978, which agrees with other observations taken in 1966–68.

158.123 **The continuous spectrum of Markarian 421 during periods of X-ray satellite observations.**
S. L. Mufson, W. Z. Wisniewski, K. Wood, D. P. McNutt, D. J. Yentis, J. F. Meekins, E. T. Byram, T. A. Chubb, H. Friedman.
Astrophys. J., Vol. 241, 74 - 80 (1980) = Goethe Link Publ., No. 202.

New *UBVRI* photometry of Mrk 421 obtained during periods of X-ray satellite observations are presented. An X-ray light curve for 1977 November from the *HEAO* A-1 experiment is also given. The decomposition of the *UBVR* fluxes into a compact nonthermal component and an extended galactic component shows that there are coordinated variations in the optical nonthermal and X-ray emission. The authors' data are consistent with the hypothesis that the mini-BL Lac object is emitting by the synchrontron–self-Compton process. The host galaxy of this composite source has properties like those of a giant elliptical.

158.124 **The stellar content of dwarf spheroidal galaxies.**
A. W. Hirshfeld.
Astrophys. J., Vol. 241, 111 - 124 (1980).

The stellar content of dwarf spheroidal galaxies has been investigated with an emphasis on determining the nature of the anomalous Cepheid variables. Models of extremely metal-poor horizontal branch stars in the mass range 0.70–1.60 $M_\odot$ have been constructed. It is found that the models with masses of 1.30–1.60 $M_\odot$ spend considerable time within the instability strip and successfully predict the observational characteristics of the anomalous Cepheids. The results support binary mass transfer as the origin of the anomalous Cepheids. Age is held to be the second parameter affecting the horizontal branch distribution in the color-magnitude diagrams of dwarf spheroidal galaxies. An initial helium abundance $Y = 0.24 \pm 0.02$ and a metal abundance $0.2 \times 10^{-4} \lesssim Z \lesssim 3 \times 10^{-4}$ is indicated for the stars in these systems.

158.125 **The recent evolutionary history of the galaxies NGC 6822 and IC 1613.** P. W. Hodge.
Astrophys. J., Vol. 241, 125 - 131 (1980).

The pattern of star formation during the recent past in the galaxies NGC 6822 and IC 1613 is derived from color-magnitude diagrams of their stellar associations and from the brightest stars of star clusters. The pattern moves about irregularly in the galaxies, and there is evidence for sporadic bursts of enhanced star formation. Other features of the stellar associations, including luminosity functions, are described as are the implied formation rates for star clusters, OB associations, and star-formation bursts.

158.126 **X-ray spectral constraints on the broad-line cloud geometry of NGC 4151.** S. S. Holt, R. F. Mushotzky, R. H. Becker, E. A. Boldt, P. J. Serlemitsos, A. E. Szymkowiak, N. E. White.
Astrophys. J., Lett., Vol. 241, L13 - L17 (1980).

X-ray spectral data from NGC 4151 taken with the Einstein Solid-State Spectrometer (SSS) and the HEAO 1 A-2 experiment cannot be simply reconciled with absorption from a uniform column of cold gas. The SSS data can, however, be explained in terms of a clumped absorber with approximately 10% uncovered fraction and factor-of-two overabundances in $Z \geq 14$ elements relative to solar oxygen. The authors suggest that the lack of significant X-ray absorption observed from much higher luminosity Seyferts and quasars is a natural consequence of the picture for NGC 4151.

158.127 **The variability of 3C 390.3.** P. Barr, G. Pollard, P. W. Sanford, J. C. Ives, M. Ward, R. G. Hine, M. S. Longair, M. V. Penston, A. Boksenberg, C. Lloyd.
Mon. Not. R. Astron. Soc., Vol. 193, 549 - 562 (1980).

Continuum observations at X-ray, optical and radio frequencies of 3C 390.3 over a period of 10 yr, and also optical and X-ray spectra, are presented. The authors compare the structure of the light curves, and use variations in the optical line-strengths and profiles to examine the physical conditions and geometry of the nuclear regions. Estimates of the extinction of the narrow-line region are made. The ratios of Lα/Hα and Hα/Hβ are not consistent with simple reddening, suggesting that self-absorption processes are important. Variations of the [Fe X] emission line imply that it originates in regions of relatively high density for both collisional or photoionization models.

158.128 **New optical and radio observations of the X-ray galaxies NGC 7582 and NGC 2992.**
M. Ward, M. V. Penston, J. C. Blades, A. J. Turtle.
Mon. Not. R. Astron. Soc., Vol. 193, 563 - 582 (1980).

The authors present radio maps and spectrophotometric observations of two X-ray emitting emission-line galaxies, NGC 7582 and 2992, which suggest that their nuclear emission has Seyfert-like characteristics. The authors contrast these with the properties of the companion galaxies NGC 7552 and 2993 which, although exhibiting emission-line spectra, are thought to be dominated by thermal processes. Estimates of the extinctions appropriate to the emission-line and continuum-emitting regions are made. The total energy distributions of NGC 7582 and 2992 are very similar and may be explained by simple Synchrotron Self-Compton theory.

158.129 **Gas structure in the nucleus of the Seyfert galaxy NGC 1275 revealed by the study of emission lines variability.** I. I. Pronik.
Izv. Krymskoj Astrofiz. Obs., Tom 61, 131 - 144 (1980). In Russian.

An analysis of relative intensities of emission lines in the spectrum of the Seyfert galaxy NGC 1275 nucleus has been carried out. The relative intensities of forbidden lines showed two kinds of time variations with scales of 6.5 and 1.0 years. These variations are probably due to the changes of gas electron temperatures under the influences of variable UV emission and some unknown source connected with radio bursts in the nucleus.

158.130 **Multicolour photometry of the Markarian galaxy 40.** L. P. Metik.
Izv. Krymskoj Astrofiz. Obs., Tom 61, 145 - 149 (1980). In Russian.

Multicolour photometry of the Markarian galaxy 40 has been carried out. This is a compact, symmetric and spheroidal galaxy with a diffuse band containing two condensations. Stellar magnitudes of the central body of the galaxy and its brighter condensation have been obtained. Colour characteristics of the central part of the galaxy are rather similar to those of the central parts of Seyfert galaxies and to synchrotron emission, while the halo of the galaxy and the brighter condensation are alike to that of the central parts of normal galaxies.

158.131 **Bursts of star formation in the central region of the hot-spot nucleus galaxy NGC 4314.**
K.-i. Wakamatsu, M. T. Nishida.
Publ. Astron. Soc. Japan, Vol. 32, 389 - 404 (1980).

NGC 4314 (SBa pec) is one of the galaxies of the most peculiar appearance owing to the presence of tightly wound inner spiral arms (called nuclear hot spots) located at the center of the bar. Spectrograms (59 Å mm^{-1} or 39 Å mm^{-1}) have been obtained, and the dynamical mass and the absolute Hα luminosity of the nuclear region are evaluated. The authors find that (1) the hot spots are a transient structure, (2) they are the result of bursts of star formation recently occurred in the central region of this galaxy, and (3) their star formation has been triggered by galactic shock waves due to the presence of inner stellar arms. From these facts, they conclude that gases were recently supplied into the nuclear region.

158.132 **Number-magnitude count for galaxies at the South Galactic Pole.** H. T. MacGillivray, R. J. Dodd.
Astrophys. Space Sci., Vol. 72, 315 - 318 (1980).

COSMOS measures on a deep UK Schmidt Telescope Plate have been used to obtain the number-magnitude count for galaxies in a field of 14.6 square degrees near the South Galactic Pole. The results are in excellent agreement with data for the North Galactic Pole for galaxies fainter than $B = 18.0$, indicating no large-scale differences between north and south. A deficiency in numbers is observed for galaxies with $B \sim 16.0$. This is comparable to the deficiency at $B \sim 17.5$ for counts at the North Galactic Pole and supports the suggested asymmetry of the bright galaxy distribution between north and south galactic poles.

158.133 **On the M/L ratios in elliptical galaxies.** R. Michard.
Astron. Astrophys., Vol. 91, 122 - 128 (1980).

The available measurements of velocity dispersion in elliptical galaxies are discussed in connection with photometric data, i.e. absolute magnitudes $M_T(B)$ and effective radii r_e, with the purpose to check whether or not there exist correlations between the M/L ratios and luminosities or colors. The derivation of representative M/L ratios for galaxies obeying the $r^{1/4}$ law is rediscussed and a correction to the classical formula suggested. The known correlations $L \propto \sigma^4$ and $L \propto r_e^{\alpha}$ (with $\alpha < 2$ according to all recent studies) imply an increase of M/L with L in elliptical galaxies. A significant correlation is found between the representative M/L ratios and $U - V$ colors.

158.134 **Search for H_2O maser emission in nearby galaxies.** W. K. Huchtmeier, O.-G. Richter, A. Witzel, I. Pauliny-Toth.
Astron. Astrophys., Vol. 91, 259 - 260 (1980).

The authors report on the search for H_2O maser emission in a complete sample of H II regions in the galaxy M33 and on the search in a number of Local Group galaxies of different morphological type.

158.135 **The structure of galactic nuclei: recent observations.** S. M. Faber.
Highlights of Astronomy, Vol. 5, (see 012.023), 135 - 142 (1980).

The author finds the structure of the inner region of M31 appealing as a model for other galaxies. In M31, there exists a diffuse, quasi-isothermal bulge. Imbedded in this bulge, there is a second, denser, quasi-isothermal core which is usually called the semi-stellar nucleus. If M31 is a valid model, the marginally resolved core radii in ellipticals represent diffuse bulge components. Semi-stellar nuclei within the bulges might then be the source of excess central light. If this picture holds, mass-to-light ratio estimates of elliptical bulges would remain substantially correct, but estimates of the innermost mass density and velocity dispersion could be seriously in error.

158.136 **Infrared emission from spiral galaxy nuclei.** C. G. Wynn-Williams.
Highlights of Astronomy, Vol. 5, (see 012.023), 173 - 175 (1980).

158.137 **Radio continuum emission from the nuclei of normal galaxies.** J. M. van der Hulst.
Highlights of Astronomy, Vol. 5, (see 012.023), 177 - 184 (1980).

The author describes recent results from the Westerbork Synthesis Radio Telescope. He discusses the radio morphology, the luminosities, and the spectra of nuclear sources of galaxies. He comments upon the possible implications for the physical processes in the nuclei that are responsible for the radio emission.

158.138 **Activity in the nuclei of normal galaxies.** T. M. Heckman.
Highlights of Astronomy, Vol. 5, (see 012.023), 185 - 190 (1980).

The principal motivation was to define the nature of a "typical" galactic nucleus and thereby determine whether such nuclei differ from active nuclei in a qualitative or only quantitative sense. Such a study would yield clues as to the origin and duration of nuclear activity and the effects such activity might have on the structure and evolution of the surrounding galaxy.

158.139 **Kinematics of nuclei of Sc galaxies.** V. C. Rubin.
Highlights of Astronomy, Vol. 5, (see 012.023), 191 (1980).

158.140 **Optical spectra of nuclei of early type galaxies.** G. F. O. Schnur, W. A. Sherwood.
Highlights of Astronomy, Vol. 5, (see 012.023), 193 (1980). Abstract.

158.141 **The stellar content of the nuclei of spiral galaxies.** D. Crampton.
Highlights of Astronomy, Vol. 5, (see 012.023), 195 (1980). Abstract.

158.142 **Star formation in the nuclei of normal galaxies.** P. Biermann.
Highlights of Astronomy, Vol. 5, (see 012.023), 209 - 222 (1980).

The author defines all those galaxies as normal whose nuclear regions are not known to be dominated by a compact source. He includes galaxies like M81 and M82 which have weak compact nuclei. This review is organized in three parts: (1) tracers of star formation, (2) star formation in the nuclear regions of observed galaxies, and (3) the theoretical attempts to interpret the observational material.

158.143 **Centers of star formation in the nuclei of galaxies.** V. I. Pronik.
Highlights of Astronomy, Vol. 5, (see 012.023), 223 - 225 (1980).

158.144 **Ultraviolet observations of normal galaxies.** F. Bertola.
Highlights of Astronomy, Vol. 5, (see 012.024), 311 - 315 (1980).

The UV spectrum of galaxies provides us with the unique chance to detect those hot stellar components, which normally do not appear in the visible spectrum. Such components have been found in spectra of early-type galaxies. All these spectra exhibit an UV excess shortward of 3000 A with respect to the energy distribution of a K-type star. UV excess is present in spectra of late-type galaxies. The emission lines in elliptical galaxies are found much stronger than one would expect if the physical conditions are similar to those of H II regions. Direct images of galaxies allow to trace the spiral structure defined by hot stars and to study the UV luminosity profile in the nuclear region.

158.145 **Observations of Seyfert galaxies with the International Ultraviolet Explorer.** M.-H. Ulrich.
Highlights of Astronomy, Vol. 5, (see 012.024), 317 - 323 (1980).

The observations with IUE of the $L\alpha/H\beta$ ratio and the non-detection of the Fe II resonance lines, together with recent theoretical calculations of the hydrogen line ratios and of the Fe II line intensities, all converge to indicate that collisional processes rather than photoionization play the major role in the excitation of the lines in the broad-line region and that the particle density in this region is of the order of 3×10^{10} cm^{-3}.

158.146 **Extragalactic astronomy and IUE.** M. S. Longair.
Highlights of Astronomy, Vol. 5, (see 012.024), 325 - 329 (1980).

The author deals first with normal galaxies and then the emission and absorption line spectra and continuum emission of active galaxies and quasars.

158.147 **Hot gas in galactic haloes and winds.** L. L. Cowie.
Highlights of Astronomy, Vol. 5, (see 012.025), 411 - 417 (1980).

158.148 **X-ray properties of galactic nuclei.** D. W. Weedman.
Highlights of Astronomy, Vol. 5, (see 012.028), 623 - 630 (1980).

158.149 **Optical spectrum variability of Seyfert galaxy nuclei.** A. G. de Bruyn.
Highlights of Astronomy, Vol. 5, (see 012.028), 631 - 639 (1980).

158.150 **Recent X-ray observations of Seyferts and emission-line galaxies.** R. E. Griffiths.
Highlights of Astronomy, Vol. 5, (see 012.028), 641 - 651 (1980).

Timing observations have shown that the X-ray emission from Seyfert galaxies arises in their nuclei. This is confirmed by the high resolution images from the Einstein Observatory. Einstein Observatory data on several unusual emission-line galaxies have indicated X-ray luminosities $L_X > 10^{40}$ ergs s^{-1}. In the case of the peculiar galaxy M 82, the emission is observed to come mainly from many discrete sources or a diffuse source extending into the region of the filaments. For the other emission-line galaxies, it is not yet clear whether the X-ray source is principally nuclear or arises from multiple components.

158.151 **X-ray variability in active galaxy nuclei and quasars in less than one day.** M. Elvis, E. Feigelson, R. E. Griffiths, J. P. Henry, H. Tananbaum.
Highlights of Astronomy, Vol. 5, (see 012.028), 653 - 656 (1980).

158.152 **The active region in galactic nuclei: an outline.** F. Pacini.
Highlights of Astronomy, Vol. 5, (see 012.028), 663 - 666 (1980).

158.153 **X-ray observations of M87's halo.** D. Fabricant, M. Lecar, P. Gorenstein.
Highlights of Astronomy, Vol. 5, (see 012.028), 689 - 693 (1980).

The authors describe the soft X-ray image of M87 obtained with the Einstein Observatory. These data provide further strong evidence for the existence of a massive halo of dark matter surrounding M87 and allow a much more precise determination of its mass.

158.154 **Observations of three radio galaxies with the Einstein X-ray Observatory.** E. D. Feigelson, E. J. Schreier.
Highlights of Astronomy, Vol. 5, (see 012.028), 695 - 697 (1980).

The authors present early results from the Einstein X-ray Observatory on three radio galaxies: Centaurus A, NGC 315 = DW0055+30, and Cygnus A = 3C405.

158.155 **Active galaxies.** Eh. E. Khachikyan.
Zemlya Vselennaya, 1980, No. 5, p. 31 - 35. In Russian.

158.156 **Rotation and mass of the spiral galaxies NGC 7339 and NGC 7537.** G. A. Kyazumov.
Pis'ma Astron. Zh., Tom 6, 687 - 690 (1980). In Russian. English translation in Soviet Astron. Lett., Vol. 6.

Results are presented of spectral observations of two galaxies which are members of pairs. The mass distribution and mass-to-luminosity ratio are found in the frame of the model of a flat disc galaxy.

158.157 **Physical conditions in the nucleus of the Markarian galaxy 534.**
I. M. Yankulova, V. K. Golev, G. T. Petrov.
Pis'ma Astron. Zh., Tom 6, 691 - 695 (1980). In Russian. English translation in Soviet Astron. Lett., Vol. 6.

A spectrophotometric investigation of the nucleus of Mark 534 (NGC 7679) is carried out. The electron density, the electron temperature of the emitting gas, the flux, as well as the luminosity are estimated. The relative abundances of ions are determined.

158.158 **Optical and theoretical studies of giant clouds in spiral galaxies.** B. G. Elmegreen, D. M. Elmegreen.
Interstellar molecules, (see 012.033), p. 191 - 196 (1980).

An optical study of four spiral galaxies, combined with radiative transfer models for transmitted and scattered light, has led to a determination of the opacities and masses of numerous dark patches and dust lanes that outline spiral structure. The observed compression factors for the spiral-like dust lanes are in accord with the authors' expectations from the theory of gas flow in spiral density waves. Several low density (10^2 cm^{-3}) clouds containing 10^6 to 10^7 $M_\odot$ were also studied. The authors discuss these results in terms of recent theoretical models of cloud and star formation in spiral galaxies.

158.159 **Molecule formation in the Seyfert galaxy NGC 1068.** W. J. Carlson, C. B. Foltz.
Interstellar molecules, (see 012.033), p. 471 - 472 (1980).

The authors have constructed models of relatively high-density radiation-bounded filaments near the nucleus of a Seyfert 2 galaxy. The amount of molecular hydrogen predicted by the models for reasonable values of the physical parameters is consistent with observations of the infrared continuum and the quadrupole rotation-vibration lines of H_2 in NGC 1068.

158.160 **Further observations of the H_2O emission from NGC 4945.** J. R. D. Lépine, P. Marques dos Santos.
Interstellar molecules, (see 012.033), p. 599 - 601 (1980).

The authors report the results of new observations of the H_2O maser emission from the galaxy NGC 4945 made in 1979 July. The main purpose of the new observations was to obtain a more accurate position for the source and to look for variability.

158.161 **A large 10-μm source near the center of M51 (NGC 5194).** C. M. Telesco, P. D. Owensby.
Bull. American Astron. Soc., Vol. 12, 750 (1980). – Abstract.

158.162 **Spiral structure and superassociations of young stars in the peculiar pair NGC 4038/39.**
O. A. Dobrodij, I. I. Pronik.
Spiral structure of the Galaxy (see 012.034), Abastumansk. Astrofiz. Obs. Byull., No. 52, p. 15 - 20 (1980). In Russian.

The results of multicolour photometry in the spectral range 3600 - 7400 Å of the pair of galaxies NGC 4038/39 combined with data published by other authors are discussed. It is concluded that not only the morphology of these galaxies is peculiar, but the characteristics of their superassociations and the physical conditions of interstellar matter are peculiar too. It is supposed that there are gas streams along the spiral branches of NGC 4038 and 4039.

158.163 **Centres of star formation in galaxies having no spiral structure.** L. P. Metik, I. I. Pronik.
Spiral structure of the Galaxy (see 012.034), Abastumansk. Astrofiz. Obs. Byull., No. 52, p. 59 - 64 (1980). In Russian.

The photometric structures of the Markarian galaxies NN 34, 42, 69, 205, 279, 290, 298 and the Seyfert galaxy NGC 1275 have been investigated using narrow-band color filters with λ_{eff} 3600, 3730, 4400, 4680, 5090, 5280, 6090, 6600 and 7400 Å. The possibility that the condensations contain the centres of star formation is discussed.

158.164 **Search for (globular) clusters in M31. I: Candidates in a 70′ square field centered on M31.**
P. Battistini, F. Bònoli, A. Braccesi, F. Fusi Pecci, M. L. Malagnini, B. Marano.
Astron. Astrophys., Suppl. Ser., Vol. 42, 357 - 374 (1980).

B and *V* plates for 19 fields covering a 4° × 4° area centered on M31 have been obtained with the Loiano 152 cm F/8 Ritchey-Chrétien telescope (17″/mm scale). The search has yielded three classes of candidates with different degrees of confidence. A cross check against available lists of H II regions, planetary nebulae, miscellaneous objects has been made to limit contamination of the sample. The two best classes obtained include 240 objects. The third class lists 48 objects, mostly faint, whose degree of confidence as globular cluster candidates is lower. The limiting magnitude down to which this survey is estimated to be complete is $V \simeq 18$.

158.165 **X-ray measurements of the mass of M87.**
D. Fabricant, M. Lecar, P. Gorenstein.
Astrophys. J., Vol. 241, 552 - 560 (1980).

The authors have mapped the 0.7–3.0 keV surface brightness of M87 to a radial distance exceeding 50′ where it falls 2.5 orders of magnitude below its peak value. The temperature profile of the hot (~2.5 keV) gas responsible for the X-ray emission has been determined from the X-ray spectral data. The gas temperature is approximately constant between radii of 6′ and 20′, but decreases at radii less than 6′. Beyond 20′, a negative temperature gradient is excluded. Because this gas responds to the gravitational potential of M87, the X-ray observations may be used to measure the radial mass distribution in M87. The well-supported hypothesis of hydrostatic equilibrium relates this mass distribution directly to the density and temperature profiles of the gas. As suggested by previous authors, they find that M87 possesses a dark halo. The authors estimate that the mass of this halo lies between $1.7 \times 10^{13} M_\odot$ and $4.0 \times 10^{13} M_\odot$ within a radius of 50′ or 230 kpc.

158.166 **Photoelectric color measurements of outer rings in galaxies.** J. S. Gallagher, A. Wirth.
Astrophys. J., Vol. 241, 567 - 572 (1980).

Beam-switched photoelectric measurements of *UBV* colors have been obtained for faint outer rings in four representative galaxies. The data show that rings exist over a range of colors and that the most common rings, which are found in RSB0/a galaxies like NGC 2859, consist largely of older stars. The surface brightness increase within such rings is therefore most plausibly attributed to an increase in the projected stellar density associated with the nonaxisymmetric structure of barred and oval galaxies. The outer ring in the oval galaxy NGC 4736 has a relatively blue $B - V$ color, and this provides an example of an outer ring in which young stars are important. A brief discussion of possible formation mechanisms for outer rings is presented.

158.167 **H II regions in NGC 628. III. Hα luminosities and the luminosity function.**

R. C. Kennicutt, P. W. Hodge.
Astrophys. J., Vol. 241, 573 - 586 (1980).

A complete photometric mapping of NGC 628 in the light of Hα is used to measure the luminosities of 593 H II regions. Isophotometry of a few individual objects is used to derive the basic properties of the H II region population. Both the luminosity and the diameter functions are well represented by steep power laws. The H II region luminosity function is largely independent of radius in NGC 628 and is the same in or out of large clumps of H II regions. There exists a marked difference, however, between the populations in or out of spiral arms. The H II region function is used to derive the approximate form of the OB association mass function. Comparison of the total flux of the resolved H II regions with the integrated equivalent width of Hα in NGC 628 indicates that most, if not all, of the Hα emission in this galaxy arises from discrete H II regions.

158.168 **Studies of luminous stars in nearby galaxies. VI. The brightest supergiants and the distance to M33.**
R. M. Humphreys.
Astrophys. J., Vol. 241, 587 - 597, plates 5, 6 (1980).

Eleven normal early type supergiants and 11 M type supergiants have been spectroscopically confirmed in M33. The visual extinction for the supergiants, determined from spectroscopic and color data and from the neutral hydrogen column density, is about 0.5 mag greater ($\bar{A}_v$ =0.8 mag) than the foreground reddening ($\bar{A}_v$ =0.3 mag) from field stars. Internal reddening is a serious problem is external galaxies, and corrections must be applied for it when using stars as distance indicators. The luminosity calibration, $M_v \approx -8$ mag, observed for the brightest M supergiants in other Local Group galaxies is adopted to derive a new and more reliable true distance modulus of $(m-M)_0$=23.9±0.2 mag [$(m-M)_v$= 24.7 mag] for M33. The luminosities, color-magnitude diagrams, and evolution of the brightest stars are also discussed and compared with the most luminous galactic and LMC supergiants.

158.169 **Studies of luminous stars in nearby galaxies. VII. The brightest blue stars in the spiral galaxies M101 and NGC 2403.** R. M. Humphreys.
Astrophys. J., Vol. 241, 598 - 601, plates 7, 8 (1980).

The first spectra of individual stars outside our own Local Group of galaxies have been obtained for suspected supergiants in the fields of M101 and NGC 2403. Four stars in NGC 2403 and three in M101 are confirmed to be members. The visually brightest stars in both galaxies are A-type supergiants, and their spectral characteristics are consistent with the luminosities derived from their membership in M101 and NGC 2403. The visual luminosities of the A-type supergiants are −9.4 mag for the one in NGC 2403 and −10.3 mag and −10.1 mag for the two in M101. These two supergiants in M101 are the visually brightest normal stars yet known in any galaxy.

158.170 **Why extended radio doubles are found in elliptical galaxies.** L. S. Sparke, F. H. Shu.
Astrophys. J., Lett., Vol. 241, L65 - L68 (1980).

Elliptical galaxies rotate very slowly, and this rotation affects the way in which gas lost from stars finds its way into the nucleus. The authors suggest that the dynamical nature of this flow is the reason why extended double radio sources occur in elliptical galaxies, rather than in spirals. They present a specific computed example to illustrate the principle, and discuss some observational consequences.

158.171 **Extended 20 micron emission from the center of NGC 1068.** C. M. Telesco, E. E. Becklin, C. G. Wynn-Williams.
Astrophys. J., Lett., Vol. 241, L69 - L72 (1980).

The authors report multiaperture photometry of the Seyfert galaxy NGC 1068 which demonstrates that significant 20 μm emission originates at positions located more than 3″, or 260 pc, from the nucleus. These observations strongly support arguments that most of the infrared flux is thermal emission from dust. It is argued that the dust giving rise to this extended emission cannot be heated solely by a compact nuclear object. The authors speculate that there is a powerful energy-generation mechanism, possibly an enormous burst of star formation, operating on a scale much larger than that identified with the visible nucleus.

158.172 **H I observations and star formation in the blue compact galaxy IZw 18.**
J. Lequeux, F. Viallefond.
Astron. Astrophys., Vol. 91, 269 - 275 (1980).

21-cm line observations of IZw 18 with the Westerbork Synthesis Radiotelescope have resolved it into a complex structure. A possible model consists in about 6 hydrogen clouds with hydrogen masses between 3 and 30×10^6 $M_\odot$. Some H I components and the H II regions have a particularly high mean gas density. The authors suggest that IZw 18 is a galaxy presently in the process of formation by merging of primordial clouds which have been previously supported against gravitational collapse by quiet formation of stars of intermediate masses. The present burst of star formation is probably not older than 2×10^7 years.

158.173 **Radial velocities of some interacting galaxies.**
V. L. Afanasiev (*Afanas'ev*), I. D. Karachentsev, V. P. Arkhipova, V. A. Dostal (*Dostal'*), V. G. Metlov.
Astron. Astrophys., Vol. 91, 302 - 304 (1980).

Radial velocities of 41 interacting systems of galaxies have been observed at the 6-m telescope. For 23 systems masses and *M/L* ratios are determined.

158.174 **Bisymmetric open-spiral configuration of magnetic fields in the galaxies M 51 and M 81.**
Y. Sofue, T. Takano, M. Fujimoto.
Astron. Astrophys., Vol. 91, 335 - 340 (1980).

Positional variations are studied of the Faraday rotation measure (*RM*) of linearly polarized radio waves from the spiral galaxies M 51 and M 81. A double periodicity is found in the *RM* distribution round the center of the galaxy, which indicates the presence of a bisymmetric and open-spiral magnetic field in the galactic plane. The field configuration suggests a primordial origin of the galactic magnetic fields. The field strength is estimated as 5×10^{-6} G in the disk of M 51, and 3×10^{-6} G in M 81.

158.175 **Late-type galaxies with extended envelopes of neutral hydrogen.**
W. K. Huchtmeier, J. H. Seiradakis, J. Materne.
Astron. Astrophys., Vol. 91, 341 - 351 (1980).

Extended H I-envelopes are found in three hydrogen-rich galaxies or systems of galaxies of late type, these are NGC 3109 and the pairs NGC 4490/4485 and NGC 4618/4625. All three galaxies are of luminosity type III or fainter and of morphological type Scd and later. They are comparable in size to Magellanic type galaxies. All three galaxies show indications of deformations like warps in their H I-distribution. Virial calculations of the NGC 4631 group of galaxies reveal a significant difference between the virial mass and the sum of the luminous masses of the twelve brightest member galaxies.

158.176 **Radio survey of Markarian galaxies at 6 and 11 cm.**
G. Kojoian, H. M. Tovmassian (*G. M. Tovmasyan*), D. F. Dickinson, A. St. Clair Dinger.
Astron J., Vol. 85, 1462 - 1467 (1980) = Contrib. No. 2, Casey Obs.

One hundred and fifty-one objects from Markarian's lists 6 and 7 were observed at 6 cm with a 3σ detection limit of about 30 mJy. Eight Markarian objects were detected, six of which were also detected at 11 cm. Forty-five others were negative at this wavelength. Two of the detections, numbers 533 and 668, are Seyfert galaxies. Additionally, UB1 was detected at 6 cm and NGC 7715 and III Zw 2 were detected at 11 cm.

158.177 **On the metal abundance range in the Draco dwarf galaxy.** R. Zinn.
Astron. J., Vol. 85, 1468 - 1479 (1980).

This paper examines the evidence for and against a range in metal abundance in the Draco dwarf spheroidal galaxy. It is shown that (1) Zinn's scanner observations of 17 red giants in Draco provide strong evidence in favor of a significant dispersion in metallicity; (2) the scatter in $B - V$ color of the subgiant branch stars (i. e., $20.0 \leqslant V \leqslant 20.5$) in Stetson's color-magnitude diagram is consistent with the dispersion in metallicity that is inferred from Zinn's observations, but the possibility that the errors in the measurements alone are responsible for the scatter cannot be ruled out; and (3) the scatter in $B - V$ of a brighter sample of red giants (i. e., $19.2 \leqslant V < 20.0$) is also consistent with Zinn's observations, and in this case the scatter is clearly larger than that expected from the observational errors. It is concluded that there is no evidence against there being a dispersion in metallicity in Draco and that there is some substantial evidence in favor of this hypothesis.

158.178 **1308+32.**
IAU Circ., No. 3500 (1980).

158.179 **Reuzenschillen rond elliptische sterrenstelsels.**
G. W. E. Beekman.
Zenit, 7e Jaarg., 502 - 503 (1980).

158.180 **Mass-luminosity ratio in NGC 5236.** E. L. Agüero.
Astrophys. Space Sci., Vol. 73, 193 - 197 (1980).

The mass-to-light ratio in NGC 5236 was studied in terms of four spheroids. The resulting values are quite small and increasing from the centre outwards, suggesting that Population I stars are rather uniformly distributed over its body with a certain concentration towards the central region.

158.181 **Observations of the kinematics of the excited gas in the late-type spiral galaxy NGC 4945.**
C. J. Peterson.
Publ. Astron. Soc. Pacific, Vol. 92, 397 - 408 (1980).

From a spectroscopic investigation of the late-type galaxy NGC 4945 (de Vaucouleurs classification SB(s)cd:), the author derives the following conclusions: (1) Over the major part of the optical image of the galaxy, the mean rotation curve is linear, with the center of rotation displaced 45″= 1.1 kpc northeast of the apparent optical center of the galaxy. (2) On the southwest side of the galaxy, velocity data in positions offset from the major axis show strong departures from purely circular motions. This is most easily interpreted in terms of an infall of gas into the galaxy. (3) With regard to the interpretation (de Vaucouleurs 1964) of NGC 4945 as an asymmetrical barred spiral seen nearly edge-on, there is no evidence for the outward streaming of gas along the bar which has been reported in other galaxies of similar morphological class. If a bar does exist in the galaxy, it can be only a minor perturbation in the mass distribution. Radial velocities and photometry of 15 background galaxies in the vicinity of NGC 4945 are reported in an appendix.

158.182 **The miniature spiral NGC 3928 = Markarian 190.**
S. van den Bergh.
Publ. Astron. Soc. Pacific, Vol. 92, 409 - 410 (1980).

A 3.6-m plate taken in excellent seeing shows that NGC 3928 is not an E0 or S0 galaxy but a miniature spiral galaxy.

158.183 **The ionised gas in NGC 5128: evidence for a shock-heated component.** M. M. Phillips.
Anglo-Australian Obs. Prepr. No. 139 (1980).

New spectrophotometric and kinematic data for the ionised gas in the main body of the giant radio galaxy NGC 5128 are presented. An analysis of the emission-line spectra shows that two basic types of gas are present: 1) discrete and diffuse low-ionisation H II regions which are photoionised by the radiation of normal O and B stars, and 2) a turbulent, diffuse component which emits a peculiar spectrum more typical of shock-ionised nebulae. Circumstantial evidence suggests that the peculiar gas covers an even greater area of the elliptical component and may be connected with the extended X-ray emission that has recently been identified. Alternatively, the two components of ionised gas may have both naturally resulted from the infall of a gas-rich spiral galaxy.

158.184 **Radiostrahlung ausgedehnter Galaxien – ein Entfaltungsproblem.** J. Pfleiderer, A. von Kap-herr.
Anz. Math.-Naturwiss. Kl., Österreichische Akad. Wiss., Jahrg. 78, Nr. 10, p. 225 - 229 (1978) = Mitt. Sternw. Innsbruck Nr. 48.

158.185 **A spectroscopic survey of emission-line objects in two fields.**
D. Kunth, W. L. W. Sargent, C. Kowal.
ESO Sci. Prepr. No. 99, 18 pp. (1980). – Submitted to Astron. Astrophys., Suppl. Ser.

158.186 **The peculiar Seyfert galaxy ESO 012-G21.**
R. M. West, P. Grosbøl, C. Sterken.
ESO Sci. Prepr. No. 101, 23 pp. (1980). – Submitted to Astron. Astrophys.

158.187 **High-resolution spectrophotometry of the "low-excitation" X-ray galaxies NGC 1672 and NGC 6221.** M. P. Véron, P. Véron, E. J. Zuiderwijk.
ESO Sci. Prepr. No. 109, 10 pp. (1980). – Submitted to Astron. Astrophys.

158.188 **The kinematics of the nuclear spiral of the barred galaxy NGC 1512.** P. O. Lindblad, S. Jörsäter.
ESO Sci. Prepr. No. 112, 18 pp. (1980). – Submitted to Astron. Astrophys.

158.189 **Spectroscopic measures of galaxies, their companions, and peculiar galaxies in the southern hemisphere.** H. Arp.
ESO Sci. Prepr. No. 113, 44 pp. (1980). – Submitted to Astrophys. J., Suppl. Ser.

158.190 **Characteristics of companion galaxies.** H. Arp.
ESO Sci. Prepr. No. 114, 58 pp. (1980). – Submitted to Astrophys. J.

158.191 **The dynamics of the S0 galaxy IC 5063.**
I. J. Danziger, W. M. Goss, K. J. Wellington.
ESO Sci. Prepr. No. 117, 24 pp. (1980). – Submitted to Mon. Not. R. Astron. Soc.

158.192 **Far UV study on the non-thermal activity in the narrow line galaxies NGC 4507 and NGC 5506.**
J. Bergeron, T. Maccacaro, C. Perola.
ESO Sci. Prepr. No. 119, 25 pp. (1980). – Submitted to Astron. Astrophys.

158.193 **MK 1055.**
IAU Circ., No. 3557 (1980).

158.194 **Evolution effects in the distribution of distant galaxies?** G. Dautcourt, N. Richter.

Astron. Nachr., Band 299, 171 - 176 (1978).

The amplitudes of the two-point correlation function for galaxies are compared for galaxy catalogues extending to different depths. Strong evolution effects in the pattern of galaxy distribution seem to be present suggesting a secular increase of the clustering amplitude. Other explanations would involve large errors in the identification of faint objects as galaxies or a considerable contamination of the counts by very young highly redshifted galaxies as recently proposed by Tinsley.

158.195 **Galaxy distribution in a 36 □° field near the north galactic pole.**
G. Dautcourt, K. Kempe, L. Richter, N. Richter.
Astron. Nachr., Band 299, 177 - 192 (1978).

More than 26000 galaxies on four blue plates (covering the whole region) and nearly 18000 galaxies on two red plates (covering half the region) have been counted in $5' \times 5'$ cells. The discussion includes a first determination of the two-point correlation function and a dispersion curve analysis. The results are not very different from those obtained for the Jagiellonian field. It is suggested that counting errors may influence the conclusions drawn for the space distribution of galaxies and its possible secular change.

158.196 **The dynamics of the giant dumbell galaxy IC2082.** D. Carter, G. Efstathiou, R. S. Ellis, I. Inglis, J. Godwin.
Anglo-Australian Obs., Prepr. No. 144, 11 pp. (1980). Submitted to Mon. Not. R. Astron. Soc.

158.197 **The dwarf blue compact galaxies.** J. Audouze, M. Dennefeld, D. Kunth.
Messenger, No. 22, p. 1 - 4 (1980).

158.198 **The density of the broad-line emission region in Seyfert 1 galaxies.** M. P. Véron, P. Véron.
Messenger, No. 22, p. 13 - 14 (1980).

158.199 **Optical and ultraviolet spectroscopy of the nuclei of Seyfert galaxies.** H. Schleicher, H. W. Yorke.
Messenger, No. 22, p. 14 - 16 (1980).

158.200 **The dynamical importance of magnetic fields in beam models of radio galaxies.**
G. V. Bicknell, R. N. Henriksen.
Astrophys. Lett., Vol. 21, 29 - 34, with a correction p. 81 (1980).

Attention is drawn to the dynamical importance of magnetic fields in beam models of the jets emerging from active galaxies. It is shown that lateral expansion of a jet can generate an azimuthal field which is capable of partially collimating the plasma flow. Long wavelength variations in the NGC 315 beam are interpreted as oscillations induced by the competing magnetic stress and thermal pressure.

158.201 **The black hole in the nucleus of galaxies.** G.-z. Xie, J.-f. Lu.
Wuli, Vol. 9, 159 - 162 (1980). In Chinese. – Abstr. in Phys. Abstr., Vol. 84, Abstr. 9674 (1981).

158.202 **VLBI observations of galactic nuclei.** D. Jones. News Lett. Astron. Soc. N. Y., Vol. 1, No. 7, p. 15 - 16 (1980). – Abstract.

158.203 **Radio observations of three supernova remnants in M33.** W. M. Goss, R. D. Ekers, I. J. Danziger, F. P. Israel.
Mon. Not. R. Astron. Soc., Vol. 193, 901 - 909 (1980).

Discrete sources in the southern spiral arm of M33 have been observed with the Westerbork Synthesis Radio Telescope at 49 and 21 cm and the Very Large Array at 6 cm. The three optically confirmed supernova remnants were found to have non-thermal spectral indices. If one assumes that these M33 SNR are similar to galactic SNR, the surface-brightness–diameter relationship proposed by Caswell & Lerche indicates a distance of 860±200 kpc for M33.

158.204 **Spectroscopic observations of three elliptical galaxies.** G. Efstathiou, R. S. Ellis, D. Carter.
Mon. Not. R. Astron. Soc., Vol. 193, 931 - 946 (1980).

The authors report on spectroscopic observations of three elliptical galaxies, NGC 4472, NGC 5813 and IC 4296. Rotation curves, velocity dispersion profiles and line-strength profiles have been obtained. NGC 4472 and 5813 are found to rotate slowly. IC 4296 rotates quite rapidly, consistent with isotropic oblate models. Spectra obtained along the minor axis of each galaxy show little evidence for minor axis rotation, except possibly for NGC 5813. Mass to light ratios have been determined for each galaxy and also for 10 galaxies in Schechter & Gunn's sample. The authors find $(M/L)_B \sim 7$ in the inner regions ($r \approx 5$ kpc) of elliptical galaxies (assuming $H_0 = 50$ km s^{-1} Mpc^{-1}).

158.205 **The dynamics of the broad-line-emitting regions of active galactic nuclei and quasars. I. Broad-line profiles.** E. Capriotti, C. Foltz, P. Byard.
Astrophys. J., Vol. 241, 903 - 909 (1980).

The line profiles resulting from various kinematical and dynamical models for the broad-line-emitting gas in Seyfert 1 galaxies and quasars are calculated. It is shown that profiles resulting from spherical systems in radial motion, for several different cases of acceleration and optical thickness to the ionizing continuum, are logarithmic. Models with rotation and/or expansion confined to a disk seem to be excluded on the basis ofobserved profile shapes. A reasonable ballistic model predicts line profiles given by the first exponential integral.

158.206 **Dust-to-gas ratio in the outer regions of spiral galaxies.** M. Jura.
Astrophys. J., Vol. 241, 965 - 968 (1980).

Some spiral galaxies have extended clouds of neutral hydrogen around them. If there is dust associated with this gas, the dust should appear as a faint extended reflection nebula illuminated by the light from the entire galaxy. While there is not enough information currently available with which to come to any definite conclusions, it seems that observations of the surface brightness of the outer regions of M83 and M101 could be worthwhile, since these two galaxies are bright and have extended surrounding gas clouds. Such studies will be of interest with regard to (1) the question of whether there are faint extended halos of stars around galaxies; (2) the extent of the metallicity gradients in galaxies; and (3) the value of the dust-to-gas ratio in quasar absorption-line systems.

158.207 **The dynamical evolution of NGC 5128.** A. D. Tubbs.
Astrophys. J., Vol. 241, 969 - 980 (1980).

It is shown that the giant radio galaxy NGC 5128 (Centaurus A) may be the partially relaxed product of the collision of a nearly prolate elliptical galaxy with a small gas-rich galaxy or intergalactic cloud. A numerical simulation of the postcollision disk of gas and dust surrounding NGC 5128 demonstrates that its observed distortion results from an interaction with the prolate gravitational field. The gravitational accelerations and potential for an arbitrary (centrally condensed) prolate mass distribution are given.

158.208 **Detection of the 3.3 micron feature in the Seyfert galaxy NGC 4151.** R. M. Cutri, R. J. Rudy.
Astrophys. J., Lett., Vol. 241, L141 - L144 (1980).

The emission feature at 3.3 μm has been detected in the near-infrared spectrum of the Seyfert galaxy NGC 4151. Although not yet identified, this feature is seen in many galactic thermal sources, and its presence suggests that at least a large fraction of the infrared emission in NGC 4151 is due to thermal radiation by dust.

158.209 **Rotation and mass of NGC 2976.**
N. Carozzi-Meyssonnier.
Astron. Astrophys., Vol. 92, 189 - 195 (1980). In French.

NGC 2976 is a peculiar edge-on galaxy: blue and red photographs do not show any arm but H II regions scattered in the disc. Spectra taken with nebular spectrographs give a linear rotation curve, slightly steeper for the NW region than for the SE one, leading to a very small mass of $2 \times 10^9 M_\odot$. This mass and the low luminosity ($4.2 \times 10^8 L_\odot$) classify this galaxy among the latest types of spiral.

158.210 **Kompakte Galaxien – Erkenntnisse und Probleme.**
N. Richter.
Sterne, 56. Band, 341 - 357 (1980).

158.211 **Variability in emission lines in Seyferts. Observational data.** I. I. Pronik.
Variability in stars and galaxies, (see 012.044), p. C.1.1 - 1.10 (1980).

158.212 **Line variations in Seyfert galaxies.**
S. Collin-Souffrin.
Variability in stars and galaxies, (see 012.044), p. C.1.11 - 1.24 (1980).

The author briefly discusses the observations. Some theoretical aspects linked with this problem are mentioned.

158.213 **Short term variability (and a characteristic timescale?) in the X-ray emission of active galaxies.**
K. A. Pounds.
Variability in stars and galaxies, (see 012.044), p. C.5.1 - 5.17 (1980).

Observations by the Ariel V Sky Survey Instrument extending over a 5 year period have been used to investigate the variability of X-ray emission from 31 active galaxies. These results, together with recent studies with the Einstein observatory suggest ~1 day may be a characteristic timescale in the X-ray emission from a wide range of active galaxies. This possibility is briefly discussed in relation to the massive black hole model widely favoured as the power source in galactic nuclei.

158.214 **BL Lac objects.** M. H. Ulrich.
Variability in stars and galaxies, (see 012.044), p. C.6.1 - 6.5 (1980).

The intensity, the spectrum, and the polarization of the flux emitted by BL Lac objects are variable. This paper gives a review of the main properties of the variability, with emphasis on the most recent results obtained in the optical range.

158.215 **Lyman alpha fluxes of Seyfert galaxies and low-redshift quasars.**
C.-C. Wu, A. Boggess, T. R. Gull.
Astrophys. J., Vol. 242, 14 - 17 (1980).

The authors present new Lα flux measures for 15 objects, together with the published data of four others. The available data indicate that, for most objects, the intrinsic spectrum emerging from the emitting regions has Lα/Hβ < 20 and Hα/H$\beta \gtrsim 2.8$. These ratios are further modified by external dust, both within the parent galaxy and in our own Galaxy.

158.216 **Velocity fields in late-type galaxies from Hα Fabry-Perot interferometry. II. Kinematics and dynamics of the Sd spiral NGC 7793.**
E. Davoust, G. de Vaucouleurs.
Astrophys. J., Vol. 242, 30 - 52, plates 1 - 4 (1980).

The velocity field in the SA(s)d spiral NGC 7793 is derived from 3822 velocities measured on 21 Hα interferograms. A catalog of 132 H II regions is presented. Isovelocity maps at 12″ and 24″ resolution are presented. Analysis of the velocity field gives an inclination $i = 53°$ and a position angle of the line of nodes $\theta_0 = 108°$. The center of rotation coincides with the optical nucleus and the systemic velocity is $V_s = 221 \pm 1$ km s^{-1}. The maximum rotation velocity is $V_M = 95$ km s^{-1} at $R_M = 4'1 = 3.7$ kpc. The masses estimated by fitting the observed rotation curve to various conventional mass distribution models – neglecting velocity dispersion – are in the range $8 < M_T < 16 \times 10^9$ solar units, of which 2 - 4% is in a small spheroidal component. The velocity dispersion at the center is $\sigma_r = 40$ km s^{-1}. The mass-to-blue luminosity ratio is 2.74, with 6.59 for the bulge component and 2.68 for the disk component. There is no clear indication of significant noncircular motions except perhaps in the northeast outer regions of the disk.

158.217 **Velocity dispersions in M31 and M32.**
B. C. Whitmore.
Astrophys. J., Vol. 242, 53 - 62 (1980).

Velocity dispersion have been obtained in the central regions of M31 and M32. The author finds the velocity dispersion in the nucleus of M31 is $\sigma_{nuc} = 181 \pm 12$ km s^{-1}, and in the bulge $\sigma_{bulge} = 151 \pm 16$ km s^{-1}. He estimates the ratio of the intrinsic dispersions, $\sigma_{bulge}/\sigma_{nuc} = 0.83 \pm 0.12$. This value is inconsistent with the model of Ruiz. The author finds the velocity dispersion in the nucleus of M32 is $\sigma_{nuc} = 75 \pm 7$ km s^{-1}. This value is consistent with the $L \approx \sigma^4$ relationship found in elliptical galaxies, even though M32 appears to be a tidal remnant of a once larger system.

158.218 **CCD camera observations of the Local Group galaxy LGS-3.** R. Schild.
Astrophys. J., Vol. 242, 63 - 65, plates 6 - 7 (1980).

From CCD observations of LGS-3 with an R filter the author finds that the R magnitude of the brightest resolved stars is $R = 18.3$ mag, and the total integrated R magnitude is 14.2 mag. With allowances for interstellar absorption and the assumption of an old stellar population he determines an absolute visual magnitude of $M_V = -7.1$ mag and numerous other physical parameters. If the galaxy is at the distance of M33 as suggested previously, it must contain a young stellar population.

158.219 **Radio observations at 408 MHz of E and S0 nearby galaxies.** L. Feretti, G. Giovannini.
Astron. Astrophys., Vol. 92, 296 - 301 (1980).

The authors present radio observations of a complete sample of E and S0 galaxies selected from the Humason et al. (1956) catalogue with an absolute photographic magnitude $M_{pg} \lesssim -20$, in the declination range $18° \lesssim \delta \lesssim 60°$, plus a sample of E and S0 galaxies belonging to the Virgo cluster. They present and briefly discuss the distribution of the linear size of detected sources, and estimate the radio luminosity function in the range of low radio powers.

158.220 **NGC 4507: a weak Seyfert 1 and X-ray galaxy.**
P. Véron, M. P. Véron, E. J. Zuiderwijk.
ESO Sci. Prepr. No. 127, 9 pp. (1980). – Submitted to Astron. Astrophys.

158.221 **OJ 287: polarization and photometric behaviour during 1971–76.** V. A. Hagen-Thorn.
Astrophys. Space Sci., Vol. 73, 263 - 277, 279 - 294 (1980). In English and Russian.

The results are given of polarimetric and photometric observations of BL Lacertae-type object OJ 287 for 1972–76. The variations of time-scales from several years to several hours are noted. The variability is caused by the flaring up and fading of separate sources (hot spots) of polarized (synchrotron) radiation. The existence of a preferable direction of polarization ($\theta_0 = 80°$) is an indication of a stable magnetic field. It may be used as an argument in favour of the single-body hypothesis of Lacertids.

158.222 **Differential *UBV* photometry in the central regions of spiral galaxies.** H. A. Dottori, C. M. Bevilaqua. Astrophys. Space Sci., Vol. 73, 327 - 331 (1980).

The differential color indices in the central bulk of spiral 'SA, SAB and SB galaxies were analyzed. A comparison of the data derived for the three families does not show any systematical difference among them, within the most probable error of the sample.

158.223 **New radial velocities for 301 pairs of galaxies.** I. D. Karachentsev. Astrophys. J., Suppl. Ser., Vol. 44, 137 - 149 (1980).

A list of radial velocities is presented for 600 galaxies from the "Catalogue of Isolated Pairs of Galaxies." Spectral observations were made using the Soviet 6 meter telescope.

158.224 **The dwarf spheroidal galaxies.** R. Zinn. Globular clusters, (see 003.008), p. 191 - 212 (1980).

The author presents the major properties of the dwarf spheroidal galaxies that have been discussed in the recent literature and emphasizes the differences and similarities between the dwarf spheroidal galaxies and globular clusters.

158.225 **The velocity field of bright nearby galaxies. III. The distribution in space of galaxies within 80 megaparsecs: the north galactic density anomaly.** A. Yahil, A. Sandage, G. A. Tammann. Astrophys. J., Vol. 242, 448 - 468, plates 9 - 11 (1980).

The authors inquire into the three-dimensional distribution of the nearby galaxies in the Revised Shapley-Ames Catalog, as derived from their positions and radial velocities. The aim is to map the density contrasts in various directions as a prelude to the calculation of the expected velocity perturbations caused by the north galactic density anomaly.

158.226 **X-ray observations of Seyfert galaxies with the Einstein Observatory.** G. A. Kriss, C. R. Canizares, G. R. Ricker. Astrophys. J., Vol. 242, 492 - 501 (1980).

The authors have observed 37 Seyfert galaxies with the Imaging Proportional Counter on the Einstein Observatory. They have detected X-ray emission from 20 Seyfert 1 galaxies, three of which have been previously detected, and from four Seyfert 2 galaxies.

158.227 **NGC 7385: generation of an H II region by thermal instability associated with the creation of a radio source.** P. E. Hardee, J. A. Eilek, F. N. Owen. Astrophys. J., Vol. 242, 502 - 510 (1980).

The authors have mapped the radio source NGC 7385 with the VLA at high resolution. The radio map suggests ejection of material from the galactic nucleus in directions parallel and antiparallel to the motion of the galaxy through the intergalactic medium. An optical knot detected by Simkin and Ekers is associated with the leading radio component. The observations place the optical knot which is probably an H II region toward one edge of the radio lobe. The authors show that, for this particular galaxy, it is quite possible that an H II region may be produced in the interstellar medium (ISM) by interaction between the ISM and a jet ejected from the galactic nucleus, and thermal instabilities in the ISM. The authors also investigate the conditions for production of H II regions by thermal instability from a jet containing mixed relativistic and thermal material.

158.228 **On the origin of the radio, optical emission-line, and X-ray structure of M87.** D. S. De Young, J. J. Condon, H. Butcher. Astrophys. J., Vol. 242, 511 - 516 (1980).

The X-ray emission from M87 is quite symmetric about the center of the galaxy, but both the optical emission-line data and the radio emission show a marked north-south asymmetry. Neither transonic head-tail radio source dynamics nor inhomogeneous accretion or outflow can account for these observations. The authors find that they can be reconciled by a model which couples radiative accretion with very slow subsonic motion of M87 to the north.

158.229 **Theory of dwarf galaxies.** H. Gerola, P. E. Seiden, L. S. Schulman. Astrophys. J., Vol. 242, 517 - 527 (1980).

A striking aspect of dwarf galaxies is that they are characterized by very diverse behavior, ranging from unremarkable low surface brightness objects to the high surface brightness blue compact objects. The dispersion in their properties is much greater than for normal spirals. The stochastic self-propagating star formation model predicts that small systems can easily experience a fluctuation large enough to completely halt propagation of star formation.

158.230 **Properties of spurs in spiral galaxies.** D. M. Elmegreen. Astrophys. J., Vol. 242, 528 - 532, plates 12 - 13 (1980).

Spiral arm spurs were analyzed on *B* and *I* photographs of NGC 628, NGC 2403, NGC 3031, NGC 4321, NGC 5194, NGC 5236, and NGC 5457. The inclination-corrected pitch angles, lengths, and widths of the spurs, and separations between adjacent parallel spurs, were measured. Time scales for significant shearing of the spurs due to differential rotation and to spiral density waves were determined from published rotation curves. Spurs appear to be long-lived features with pitch angles of about 60° and widths of some 560 pc.

158.231 **Medium-resolution spectra of M82 and NGC 1068 from 16 to 30 microns.** J. R. Houck, W. J. Forrest, J. F. McCarthy. Astrophys. J., Lett., Vol. 242, L65 - L68 (1980).

Spectra of M82 are presented over the range of 16–40 μm. The [S III] 18.7 μm emission line is clearly observed at a flux level of $(4.3 \pm 0.6) \times 10^{-18}$ W cm^{-2} in a 30″ diameter beam. The abundance of S III in M82 is approximately half the abundance of sulfur in the solar neighborhood. The continuum shows an absorption dip at ~19 μm. A two-layer model fit implies a 19 μm optical depth $1.2 \lesssim \tau_{18.7} \lesssim 2.4$. A low-resolution spectrum of NGC 1068 is presented. The continuum spectrum appears quite smooth, and no emission lines are observed.

158.232 **Astrophysical statistics of compact galaxies near the galactic north pole. Part II. Blue galaxies in the Tautenburg catalogue of compact galaxies.** N. Richter. Astron. Nachr., Band 301, 301 - 304 (1980).

From the Tautenburg catalogue of 745 compact galaxies in 4 fields round M3 blue objects are selected by help of the color index $U - B \leqslant -0.40$ mag and by spectroscopic inspection on objective prism plates. There are found 45 objects as blue or suspected to have emission.

158.233 **Balloon-borne observations of NGC 4151 using the MISO telescope.** F. Perotti, A. Della-Ventura, G. Sechi, G. Villa, G. Di Cocco, R. E. Baker, R. C. Butler, A. J. Dean, S. J. Martin, D. Ramsden. Non-solar gamma-rays, (see 012.047), p. 67 - 70 (1980).

The Seyfert Galaxy NGC 4151 was observed from balloon altitudes using the MISO low energy gamma-ray telescope on 23rd May 1977. The spectrum was measured over the range 0.04 to 20 MeV and a change in the spectral index was observed at about 2 MeV. The data at low energies is in good agreement with other hard X-ray data but at higher energies the data could be consistent with an exponential spectrum having an e-folding energy of 2.1 MeV.

158.234 **Photometry of the galaxy NGC 1023 on composite photographs referred to brightness distribution in the image of surrounding stars.** V. G. Khristich.
Vestn. LGU, 1980, No. 7, p. 109 - 116. In Russian. – Abstr. in Ref. zh., 51. Astron., 11.51.720 (1980).

158.235 **Shape parameters of the photometric standard galaxy NGC 3379.** P. W. Hodge.
Astron. J., Vol. 85, 1582 - 1586 (1980).

Photographic surface isophotometry of NGC 3379 gives an ellipticity curve from $r = 17$ to 184 arcsec, with a mean value of $\epsilon = 0.15 \pm 0.04$. The orientation angle derived in these studies varies from 76° for $r < 60$ to 93° for $r > 60$. These data are compared with the values found by de Vaucouleurs and Capaccioli (1979) for this standard galaxy, as well as with data from several other sources. The comparison shows that these parameters are still quite uncertain.

158.236 **The structure of the Sculptor system.**
S. Demers, W. E. Kunkel, A. Krautter.
Astron. J., Vol. 85, 1587 - 1591 (1980).

The structure of Sculptor is determined from counts of over 22 000 stars on prints made from a CTIO 4-m plate. The ellipticities of the isophotes of constant density vary from 0.0 at the center to ~0.3 for $15' < r < 25'$. The lack of a measured value of the foreground stellar density prevents from obtaining a definitive tidal radius, but a lower limit of $r_t = 75'$ is indicated. Sculptor is much bigger than previously believed.

158.237 **Multicolour photometry of bright patches in the galaxy NGC 4303.** N. B. Grigor'eva.
Izv. Krymskoj Astrofiz. Obs., Tom 62, 39 - 43 (1980). In Russian.

To detect regions of H II complexes in the central part of NGC 4303 (M 61) a photometric study of its plates has been carried out. 40 bright patches and among them 29 complexes of H II regions have been detected.

158.238 **Primeval galaxies: a new look in red light.**
D. C. Koo, R. G. Kron.
Publ. Astron. Soc. Pacific, Vol. 92, 537 - 545 (1980).

The authors describe a new search in the far red for high-redshift objects. The technique employs prime-focus slitless spectroscopy with CCD and photographic detectors; the authors look for evidence for either emission lines or continuum breaks in the first-order spectra. So far no convincing candidate has been found, although the sensitivity is good enough to reveal Lyman α at $z \sim 5$ if the line were as bright as in typical QSOs observed at $z \sim 2.5$. The upper limits they derive are useful for constraining models of galaxies in an early luminous phase.

158.239 **Spectra of additional predicted and suspected Seyfert galaxies.** D. E. Osterbrock, R. Stoughton.
Publ. Astron. Soc. Pacific, Vol. 92, 548 - 549 (1980). Abstract.

158.240 **Circular and noncircular motions in the SAB(s)bc galaxy NGC 1566.** C. J. Peterson.
Publ. Astron. Soc. Pacific, Vol. 92, 549 (1980).

158.241 **Spectra of some Markarian galaxies with multicomponent structure.** A. R. Petrosyan.
Astron. Tsirk., No. 1096, p. 1 - 3 (1980). In Russian.

158.242 **Average characteristics of type 1 and 2 Seyfert galaxies.** Eh. A. Dibaj, Z. I. Tsvetanov.
Astron. Tsirk., No. 1102, p. 1 - 3 (1980). In Russian.

158.243 **The first half century of galaxy photometry: methods and results.** G. de Vaucouleurs.
Photometry, kinematics and dynamics of galaxies, (see 012.051), p. 1 - 12 (1979).

158.244 **NGC 3379 as a luminosity distribution standard.**
M. Capaccioli, G. de Vaucouleurs.
Photometry, kinematics and dynamics of galaxies, (see 012.051), p. 13 - 14 (1979). – Abstract.

158.245 **Luminosities, diameters and colors of early-type galaxies.** R. Michard.
Photometry, kinematics and dynamics of galaxies, (see 012.051), p. 15 (1979). – Abstract.

158.246 **CM effect in early type galaxies.**
N. Visvanathan.
Photometry, kinematics and dynamics of galaxies, (see 012.051), p. 17 - 21 (1979).

158.247 **Picture processing analysis of NGC 5128.**
R. J. Dufour.
Photometry, kinematics and dynamics of galaxies, (see 012.051), p. 23 - 26 (1979).

The spatial luminosity and color structure of the peculiar elliptical radio galaxy NGC 5128 (Centaurus A) was studied using photoelectrically calibrated photometry of six UBV plates. Color-coded maps of U–B, B–V, and Q = (U–B) – 0.72(B–V) over a 15.5 x 15.5 arcmin region centered on the nucleus of NGC 5128 were constructed and analyzed with picture processing systems. The results are summarized.

158.248 **Photometry of the outer envelope of NGC 5128.**
R. D. Cannon.
Photometry, kinematics and dynamics of galaxies, (see 012.051), p. 27 - 29 (1979).

158.249 **UBV surface photometry of N6181 calibrated using standard aperture-magnitude relationships.**
G. F. Benedict, T. C. Talley.
Photometry, kinematics and dynamics of galaxies, (see 012.051), p. 31 - 35 (1979).

158.250 **Contamination gradients in elliptical galaxies.**
B. F. Madore.
Photometry, kinematics and dynamics of galaxies, (see 012.051), p. 53 - 55 (1979).

158.251 **Carbon monoxide observations of early type galaxies.** D. W. Johnson, S. T. Gottesman.
Photometry, kinematics and dynamics of galaxies, (see 012.051), p. 57 - 61 (1979).

The authors have reported observations of the ^{12}CO molecule in early type galaxies. Their negative results eliminate the possibility of large molecular clouds but still permit molecular material to be present if spread over sufficient velocities. The authors report a marginal detection of CO emission from the nucleus of NGC 185, a dwarf elliptical companion of M31.

158.252 **CCD photometry of field ellipticals.**
R. W. Leach.
Photometry, kinematics and dynamics of galaxies, (see 012.051), p. 75 - 79 (1979).

The author presents data on standard stars and elliptical galaxies taken with the Smithsonian Astrophysical Observatory's CCD camera. The sample consists of a magnitude-limited list of nearby elliptical galaxies both to reproduce established photometric data and to study them in their own right.

158.253 **The existence of thick disks in spirals and ellipticals.**
D. Burstein.

Photometry, kinematics and dynamics of galaxies, (see 012.051), p. 81 - 84 (1979).

158.254 **The systematics of galactic disks.**
K. C. Freeman.
Photometry, kinematics and dynamics of galaxies, (see 012.051), p. 85 - 91 (1979).

158.255 **The Palomar-Westerbork survey of bright northern-hemisphere galaxies.**
P. C. van der Kruit, L. Searle.
Photometry, kinematics and dynamics of galaxies, (see 012.051), p. 93 - 95 (1979).

The project aims at three-colour optical surface photometry to faint levels on deep Palomar-Schmidt plates combined with mapping of the H I distribution and velocity field with the Westerbork Synthesis Radio Telescope for a sample of about 20 spiral galaxies in the northern hemisphere. The authors give a short summary of the presently reduced optical surface photometry.

158.256 **Multicolor surface photometry of M83.**
R. J. Talbot, Jr., E. B. Jensen.
Photometry, kinematics and dynamics of galaxies, (see 012.051), p. 97 - 99 (1979).

158.257 **Electronographic photometry of spirals.**
P. J. Grosbøl.
Photometry, kinematics and dynamics of galaxies, (see 012.051), p. 101 - 103 (1979).

The aim of the paper is to discuss the common features of the three late-type spiral galaxies NGC 157, 628 and 4535 which can be used in the comparison with theoretical models.

158.258 **Surface photometry of edge-on galaxies.**
M. Hamabe, K. Kodaira, S. Okamura, B. Takase.
Photometry, kinematics and dynamics of galaxies, (see 012.051), p. 109 - 111 (1979).

158.259 **A detailed photometric study of the edge-on spiral NGC 4565.** E. B. Jensen, T. X. Thuan.
Photometry, kinematics and dynamics of galaxies, (see 012.051), p. 113 - 117 (1979).

158.260 **A measurement of the color gradient in the halo surrounding NGC 4565.**
D. J. Hegyi, G. L. Gerber.
Photometry, kinematics and dynamics of galaxies, (see 012.051), p. 119 - 124 (1979).

158.261 **Integrated magnitudes and colors of galaxies.**
H. G. Corwin, Jr.,
Photometry, kinematics and dynamics of galaxies, (see 012.051), p. 125 - 133 (1979).

158.262 **Ten colour photometry of galaxies.**
J.-E. Solheim.
Photometry, kinematics and dynamics of galaxies, (see 012.051), p. 139 - 141 (1979).

158.263 **Preliminary photometry of M 104.**
M. S. Burkhead.
Photometry, kinematics and dynamics of galaxies, (see 012.051), p. 143 - 145 (1979).

158.264 **Infrared photometry applied to some traditional astronomical problems.** M. Aaronson.
Photometry, kinematics and dynamics of galaxies, (see 012.051), p. 147 - 150 (1979).

158.265 **Low surface brightness spiral galaxies.**
W. Romanishin, S. E. Strom.
Photometry, kinematics and dynamics of galaxies, (see 012.051), p. 151 - 154 (1979).

158.266 **The optical properties of H I-rich early-type galaxies.** T. G. Hawarden, A. J. Longmore, W. M. Goss, U. Mebold, S. B. Tritton.
Photometry, kinematics and dynamics of galaxies, (see 012.051), p. 155 - 159 (1979).

Most early-type spirals and lenticulars which contain very large amounts of neutral hydrogen seem to have unusually faint discs. It is suggested that in those systems with revised type earlier than Sb and with $M_H/L_B \gtrsim 0.5\ M_\odot/L_\odot$ the conversion of gas into disc stars has largely failed to occur or has been suppressed.

158.267 **Quantitative analysis of M87 and its jet.**
J.-L. Nieto, G. de Vaucouleurs.
Photometry, kinematics and dynamics of galaxies, (see 012.051), p. 161 - 163 (1979).

158.268 **Kinematics of early type galaxies: methods and results.** M. Capaccioli.
Photometry, kinematics and dynamics of galaxies, (see 012.051), p. 165 - 176 (1979).

The purpose of this paper is to review the recent work on kinematics of elliptical galaxies made with optical telescopes. S0 galaxies have also been included because they present the same observational problems as ellipticals.

158.269 **Rotation and dispersion profiles in elliptical galaxies.** J. Fried, G. Illingworth.
Photometry, kinematics and dynamics of galaxies, (see 012.051), p. 177 - 178 (1979).

158.270 **Velocity dispersions in the bulges of spiral galaxies.**
B. C. Whitmore.
Photometry, kinematics and dynamics of galaxies, (see 012.051), p. 179 - 182 (1979).

158.271 **Kinematics of blue compact galaxies.**
R. W. O'Connell.
Photometry, kinematics and dynamics of galaxies, (see 012.051), p. 183 - 185 (1979).

158.272 **Isophote twists in elliptical galaxies – photometry and spectroscopy of NGC 596.**
T. B. Williams.
Photometry, kinematics and dynamics of galaxies, (see 012.051), p. 187 - 190 (1979).

158.273 **Rotation of the bulge components of disk galaxies.**
J. Kormendy, G. Illingworth.
Photometry, kinematics and dynamics of galaxies, (see 012.051), p. 195 - 196 (1979).

158.274 **Kinematics of the nuclei of the barred galaxies NGC 1512 and 1365.** S. Jörsäter.
Photometry, kinematics and dynamics of galaxies, (see 012.051), p. 197 - 200 (1979).

158.275 **Large scale distribution and kinematics of neutral hydrogen in elliptical galaxies.** L. Gouguenheim.
Photometry, kinematics and dynamics of galaxies, (see 012.051), p. 201 - 207 (1979).

The 21-cm line of neutral hydrogen has been detected so far in 11 elliptical galaxies. These observations are described and theoretical interpretations are given.

158.276 **Rotation of early-type galaxies from H I line observations.** N. Krumm, E. E. Salpeter.
Photometry, kinematics and dynamics of galaxies, (see 012.051), p. 209 - 212 (1979).

158.277 **H I syntheses of early-type galaxies.**
G. S. Shostak, H. van Woerden, U. J. Schwarz.
Photometry, kinematics and dynamics of galaxies, (see 012.051), p. 213 - 217 (1979).

158.278 **Preliminary results of an H I survey of southern low surface brightness galaxies.**
A. J. Longmore, T. G. Hawarden, B. L. Webster.
Photometry, kinematics and dynamics of galaxies, (see 012.051), p. 223 - 226 (1979).

Preliminary results of an H I survey of a new list of southern low surface brightness galaxies are given. Some problems of converting optical and H I parameters to a standard form are discussed. Extension of the Tully-Fisher relation between H I line width and absolute magnitude to galaxies of $M_B < -13$ is supported by this sample, for which all optical parameters have been derived independently of previous catalogues.

158.279 **An H I survey of blue compact galaxies.**
D. Gordon, S. T. Gottesman.
Photometry, kinematics and dynamics of galaxies, (see 012.051), p. 227 - 229 (1979).

158.280 **Velocity field of the SAB spiral NGC 253 from Hα interferometry.** W. D. Pence.
Photometry, kinematics and dynamics of galaxies, (see 012.051), p. 253 (1979). – Abstract.

158.281 **Photometry and kinematics of NGC 7793 from Hα Fabry-Perot interferometry.**
E. Davoust, G. de Vaucouleurs.
Photometry, kinematics and dynamics of galaxies, (see 012.051), p. 255 - 258 (1979).

158.282 **Spectroscopic observation of velocity fields in barred spiral galaxies.** C. J. Peterson.
Photometry, kinematics and dynamics of galaxies, (see 012.051), p. 259 - 262 (1979).

158.283 **The structure and nuclear activity of Seyfert galaxies.** S. M. Simkin, H. J. Su, M. P. Schwartz.
Photometry, kinematics and dynamics of galaxies, (see 012.051), p. 263 - 266 (1979).

158.284 **The interaction between M81, M82, and NGC 3077.**
D. J. Killian, S. T. Gottesman.
Photometry, kinematics and dynamics of galaxies, (see 012.051), p. 267 - 270 (1979).

158.285 **Asymmetries in the central part of M33.**
J. Boulesteix, J. Colin, E. Athanassoula, G. Monnet.
Photometry, kinematics and dynamics of galaxies, (see 012.051), p. 271 - 274 (1979).

158.286 **The z-dispersion velocity of the H I in the spiral galaxy NGC 3938.** P. C. van der Kruit, G. S. Shostak, T. S. van Albada.
Photometry, kinematics and dynamics of galaxies, (see 012.051), p. 277 - 281 (1979).

158.287 **Extended neutral hydrogen in some spiral and I0 galaxies.** L. Bottinelli.
Photometry, kinematics and dynamics of galaxies, (see 012.051), p. 283 - 285 (1979).

158.288 **H I line widths and H-band magnitudes: an improved version of the Tully-Fisher relation.**
J. R. Mould.
Photometry, kinematics and dynamics of galaxies, (see 012.051), p. 297 - 300 (1979).

The relation between the luminosity and rotational velocity of spiral galaxies is discussed. For application to the distance scale the advantages of measuring infrared luminosities are stressed.

158.289 **High resolution observations of the neutral hydrogen in the galaxy NGC 925.**
S. T. Gottesman.
Photometry, kinematics and dynamics of galaxies, (see 012.051), p. 301 - 305 (1979).

158.290 **Asymmetric rotation of the barred spiral galaxy NGC 1313.** M. Marcelin.
Photometry, kinematics and dynamics of galaxies, (see 012.051), p. 307 - 309 (1979).

For its type (SBd), this galaxy has a rapidly rotating bar (50 km s^{-1} kpc^{-1}) which, furthermore, is very eccentric (1.5 kpc). This could explain its curious morphology; the southern satellite regions may have been formed by separation of the tip of the main arm as is observed in the northeast.

158.291 **The nearby blue compact galaxy VII Zw 403.**
R. B. Tully, A. M. Boesgaard, W. V. Schempp.
Photometry, kinematics and dynamics of galaxies, (see 012.051), p. 325 - 328 (1979).

VII Zw 403 is a dwarf irregular galaxy which is vigorously forming stars at the present epoch. This report provides a summary of photographic, spectroscopic, infrared photometric and 21-cm line observations of this interesting object.

158.292 **Computer enhanced photography of M 51.**
M. S. Burkhead, W. Matuska.
Photometry, kinematics and dynamics of galaxies, (see 012.051), p. 333 - 336 (1979).

158.293 **VLA observations of the 6 cm radio continuum emission of galactic nuclei.** J. M. van der Hulst.
Photometry, kinematics and dynamics of galaxies, (see 012.051), p. 337 - 340 (1979).

The author briefly describes results from a 6 cm VLA survey of 81 nuclei in galaxies of various morphological type. He discusses the radio morphology and the radio luminosities.

158.294 **The measurement of rotation curves and velocity dispersion profiles for elliptical galaxies: results and implications.** R. L. Davies.
Photometry, kinematics and dynamics of galaxies, (see 012.051), p. 369 - 373 (1979).

158.295 **Spectral analysis of the spiral pattern of M51.**
M. Iye, M. Hamabe, M. Watanabe, S. Okamura.
Photometry, kinematics and dynamics of galaxies, (see 012.051), p. 407 - 410 (1979).

On the assumption that the observed spiral pattern of a galaxy consists of an infinite number of logarithmic spiral components, a quantitative analysis of the surface brightness distribution of M51 to decompose the pattern into respective spiral components has been carried out.

158.296 **Studies of disk galaxies.** A. Bosma.
Photometry, kinematics and dynamics of galaxies, (see 012.051), p. 419 - 420 (1979).

158.297 **The peculiar spiral galaxy NGC 4258.**
G. D. van Albada.
Photometry, kinematics and dynamics of galaxies, (see 012.051), p. 421 - 423 (1979).

158.298 **Ring structures in barred spirals.**
J. L. Sérsic, J. H. Calderón.
Photometry, kinematics and dynamics of galaxies, (see 012.051), p. 439 - 440 (1979).

158.299 **A dynamical model of NGC 1097.**
W. V. Schempp, R. D. Wolstencroft.
Photometry, kinematics and dynamics of galaxies, (see 012.051), p. 453 - 456 (1979).

The authors derive a dynamical model for the barred spiral NGC 1097 and compare it with observations of the velocity field. Their method is to construct a rotation curve, to deduce from it an axisymmetric mass distribution and then to introduce a non-axisymmetric component. The parameters defining the non-axisymmetric mass are varied until the best agreement with observation is produced.

158.300 **Evidence for diffusion of relativistic electrons in the interstellar medium of the elliptical galaxy NGC 3862.** J. P. Vallée.
J. R. Astron. Soc. Canada, Vol. 74, 357 (1980). – Abstract.

158.301 **The surface-brightness profiles of six supergiant galaxies.** D. English, G. A. Welch.
J. R. Astron. Soc. Canada, Vol. 74, 361 (1980). – Abstract.

158.302 **Ultraviolet photometry of spiral and irregular galaxies with the Orbiting Astronomical Observatory.**
G. A. Welch.
J. R. Astron. Soc. Canada, Vol. 74, 361 (1980). – Abstract.

158.303 **Non-circular motions in the spiral galaxy NGC 157.**
G. A. Kyazumov.
Astron. Tsirk., No. 1108, p. 1 - 2 (1980). In Russian.

158.304 **Spectrophotometry of spiral galaxies.**
M. A. Smirnov, G. A. Kyazumov.
Astron. Tsirk., No. 1108, p. 2 - 3 (1980). In Russian.

158.305 **Comparative characteristics of Seyfert galaxies of types 1 and 2.**
Eh. A. Dibaj, Z. I. Tsvetanov.
Astron. Zh., Tom 57, 1143 - 1152 (1980). In Russian.
English translation in Soviet Astron., Vol. 24, No. 6.

The bolometric luminosities (IR+optical+X-ray) are calculated for 35 Sy 1 and 15 Sy 2 galaxies. The parameters of the gas in the nuclei (volume, mass, kinetic energy and typical size) are estimated in broad-line (H II) and narrow-line (O III) regions. The average values for these regions are given. The established data are compared with theoretical models.

158.306 **Isolated galaxies: analysis of the criterion.**
V. E. Karachentseva.
Astron. Zh., Tom 57, 1153 - 1161 (1980). In Russian.
English translation in Soviet Astron., Vol. 24, No. 6.

Using the results of simulation of the apparent distribution of galaxies an analysis of the criterion accepted in compiling the catalogue of isolated galaxies is carried out. It is shown that the criterion applied selected from the Zwicky catalogue a homogeneous sample of galaxies isotropically distributed in the local supercluster. The criterion is not selective with respect to the absolute characteristics of galaxies and their belonging to systems of different multiplicity. At the same time, it is shown that the catalogue contains members of systems together with single isolated galaxies.

158.307 **Flugzeugbeobachtungen von Galaxien im fernen IR-Bereich.**
D. Rouan, F. Viallefond, S. Drapatz, P. Lena, J. L. Puget.
Mitt. Astron. Ges., Nr. 50, p. 13 - 17 (1980).

158.308 **Magnetfelder in M31.** R. Beck.
Mitt. Astron. Ges., Nr. 50, p. 18 - 20 (1980).

158.309 **Active galaxies.** K. A. Pounds.
X-ray and gamma-ray astronomy in the 1980's, (see 012.054), p. 31 - 32 (1979).

158.310 **On gas sweeping from central regions of galaxies with active nuclei.** S. A. Silich, P. I. Fomin.
Inst. teor. fiz. AN USSR. Prepr., 1980, No. 27, 18 pp. In Russian. – Abstr. in Ref. zh., 51. Astron., 12.51.639 (1980).

158.311 **X-ray emission from active galaxies.**
K. A. Pounds.
X-ray astronomy, (see 012.055), p. 273 - 290 (1980).

Contents: Seyfert galaxies. High excitation, narrow emission line galaxies (HEXELG's). X-radiation from QSO's. Radio galaxies, etc.

158.312 **Einstein observations of active galaxies.**
H. Tananbaum.
X-ray astronomy, (see 012.055), p. 291 - 310 (1980).

The author presents a detailed discussion of the Einstein observations of Centaurus A (NGC5128). He then briefly describes the Einstein imaging results on Seyfert galaxies. He concludes with a discussion of his observations of time variations in Seyfert galaxies and quasars.

158.313 **X-ray spectra of active galactic nuclei.**
S. S. Holt.
X-ray astronomy, (see 012.055), p. 327 - 337 (1980).

Contents: Active galactic nuclei. Seyfert I spectra. BL Lac spectra. Quasar spectra. The diffuse background.

158.314 **X-ray emission from galactic nuclei.**
M. J. Rees.
X-ray astronomy, (see 012.055), p. 339 - 354 (1980).

158.315 **Gas close to the radiative continuum source(s) in active nuclei.** J. Bergeron.
X-ray astronomy, (see 012.055), p. 355 - 362 (1980).

158.316 **Central regions of Sérsic-Pastoriza galaxies: a photographic study.** T. P. Prabhu.
J. Astrophys. Astron., Vol. 1, 129 - 154 (1980).

A classification scheme is proposed for the central regions of Sérsic-Pastoriza galaxies based on high resolution photographs of 50 objects in integrated light (4000 Å - 8700 Å). Structures of two different linear scales are recognized: (1) nucleus ($\lesssim$ 1 kpc) and (2) perinuclear formation (~ 1.5 kpc). Equal intensity contours and luminosity profiles are presented for the central regions of 27 galaxies. A comparison of their axial ratios with those of the parent galaxies indicates that the perinuclear formations are prolate or barlike. The dependence of the peak surface brightness of the central formation on the size of the bar is investigated as also the dependence of the central surface brightness of the bar on the size of the bar.

158.317 **NGC 4650 A: a nearly edge-on ring galaxy?**
S. Laustsen, R. M. West.
J. Astrophys. Astron., Vol. 1, 177 - 187 (1980).

The peculiar galaxy NGC 4650 A has been studied by means of direct and spectral observations with the ESO 3.6-m telescope. It is interpreted as a prolate, elliptical galaxy surrounded by a warped ring of H II regions, dust and stars. The distance is 47 Mpc (H_0 = 55 km s^{-1} Mpc^{-1}). The ring is seen nearly edge-on and it rotates. It has a diameter of about 21 kpc and is bluer than the elliptical galaxy for which the (M/L_V) ratio is ~ 12 in solar units. The observed configuration may be the result of interaction with the nearby galaxy, NGC 4650.

158.318 **On the stellar content and structure of the spiral galaxy M33.** R. M. Humphreys, A. Sandage.
Astrophys. J., Suppl. Ser., Vol. 44, 319 - 381, plates 3 - 12 (1980).

The authors discuss the search for the brightest red and blue stars in M33 and the identification of the associations;

the photometry and the photographic catalogs; the identification of the brightest stars and their absolute magnitude calibration; the spiral pattern from the associations, H II regions, and dust; the ages of the associations and the blue to red supergiant ratio; and the nature of the arm system and the evolutionary fate of M33 as an Sc galaxy.

158.319 **Galaxies and their nuclei.** M. J. Rees.
The state of the Universe, (see 003.013), p. 16 - 39 (1980).

The author attempts to outline what galaxies are, alludes to some physical processes that might determine their size and shape and control their evolution, and place them in their cosmological context. He briefly mentions how galaxies might have formed, and how the study of galaxies may aid in answering such basic cosmological questions as whether the Universe is destined to expand for ever or tò recollapse.

158.320 **Strong radio sources in bright spiral galaxies.** J. J. Condon.
Astrophys. J., Vol. 242, 894 - 902 (1980).

High-resolution maps of all spirals with $R > 50$ from the Arecibo survey were made with the NRAO Very Large Array (VLA) at 4885 MHz. Several other spiral or irregular galaxies known to contain compact radio sources were observed as well. The VLA observations and reductions are described. The radio maps of several sources of individual interest are presented. The author discusses the group properties of strong radio sources in a statistically complete magnitude-limited sample of spiral galaxies, the relative contributions of synchrotron radiation and free-free emission, and the energy sources responsible for these radio sources.

158.321 **Are "anemic" spirals deficient in neutral hydrogen?** G. D. Bothun, W. T. Sullivan III.
Astrophys. J., Vol. 242, 903 - 912 (1980).

The anemic class of galaxies would become an especially useful designation if it could be quantiatively demonstrated that these galaxies are gas-poor relative to normal spirals and thus represent a class of galaxies in which star formation is not vigorously proceeding. In this paper the authors study this question by comparing the H I and optical properties of anemics and normal spirals. In addition, they attempt to establish whether or not the RDDO system distinguishes between gas-rich and gas-poor galaxies better than the Hubble-Sandage system.

158.322 **Observations of the kinematics of barred spiral galaxies. I. NGC 1300.**
C. J. Peterson, J. M. Huntley.
Astrophys. J., Vol. 242, 913 - 930 (1980).

The observed emission-line velocity field of the SBb galaxy NGC 1300 shows strong departures from the pattern expected for purely circular motions. A comparison of the observed data with the kinematics predicted by the Huntley barred galaxy model is given. The deviations from circular motion are due to the response of the gaseous disk to the rotation of the stellar bar. The mean stellar velocities have been measured along the bar. The stellar rotation curve shows a turnover to constant velocity at a radius of about 13″. The model implies a dynamical mass of $M \leqslant 1.5 \times 10^{11} M_\odot$; the mass-luminosity ratio is therefore $\langle M/L \rangle \leqslant 7$ interior to a radius of 11.6 kpc. The systemic velocity of the galaxy determined in this study is V_H = 1566±2 (s.d.m.) km s^{-1}.

158.323 **Neutral hydrogen in elliptical galaxies: a bimodal distribution.** R. H. Sanders.
Astrophys. J., Vol. 242, 931 - 937 (1980).

A sample of 46 elliptical galaxies with published neutral hydrogen upper limits or detections is analyzed by a maximum likelihood technique. The sample is most likely drawn from a parent distribution which is bimodal in the ratio of hydrogen mass to blue luminosity ($M_{\rm H\,I}/L_B$). The character of this bimodal distribution is such that about 70% of all ellipticals are undetectable at the sensitivities of present surveys ($M_{\rm H\,I}/L_B < 0.003$) and the remainder contain neutral hydrogen at the level of $M_{\rm H\,I}/L_B \approx 0.03$. The significance of this bimodal distribution is high even if galaxies in the core of the Virgo cluster are excluded from the sample.

158.324 **A new relation for estimating the intrinsic luminosities of spiral galaxies.**
V. C. Rubin, D. Burstein, N. Thonnard.
Astrophys. J., Lett., Vol. 242, L149 - L152 (1980).

For 21 Sc galaxies which span a large range in luminosity, the slope of the conventional Tully-Fisher relationship between blue absolute magnitude and maximum rotation velocity is shown to be $M_B \propto (13 \pm 1.5) \log V_{\rm max}$. A family of correlations between the parameters of the rotation curves and intrinsic luminosities of Sc's exists. The ratio of the radial distance to the isophotal radius is an equally useful luminosity discriminant. The authors expect that this family of relations, once calibrated, will extend the usefulness of the original Tully-Fisher relation.

158.325 **The 21 centimeter line width as an extragalactic distance indicator.** L. Bottinelli, L. Gouguenheim, G. Paturel, G. de Vaucouleurs.
Astrophys. J., Lett., Vol. 242, L153 - L156 (1980).

The Tully-Fisher relation between 21 cm line width and absolute magnitude of spiral galaxies has been the subject of much interest and controversy. The authors report here on the main results of a new statistical analysis of a larger sample of well-documented spiral galaxies, leading to a tighter correlation and still more precise distance scale.

A revised optical catalog of quasi-stellar objects
See Abstr. 002.013.

THE ESO/Uppsala survey of the ESO (B) atlas of the southern sky – VIII. See Abstr. 002.057.

An update of the status of the Revised 3C Catalog of Radio Sources; 22 new galaxy redshifts.
See Abstr. 002.066.

Galaxies. See Abstr. 003.044.

Ten years of discovery with Oort's synthesis radio telescope. See Abstr. 013.015.

Two dimensional Fourier analysis of galactic images.
See Abstr. 031.522.

On the shape of the innermost isophotes of galaxies.
See Abstr. 031.589.

An analytical approximation to the problem of spatial deprojection of rotation curves of elliptical galaxies.
See Abstr. 031.590.

Analysis of ellipticity and twisting of the isophotes of some bright galaxies in Virgo using the INMP interactive numerical mapping package. See Abstr. 031.591.

The measurement of faint galaxy magnitudes.
See Abstr. 031.611.

A new technique for eliminating errors in computerised surface photometry. See Abstr. 031.612.

Velocity fields in late-type galaxies by Hα Fabry-Perot interferometry. See Abstr. 031.613.

Color photography of galaxies. See Abstr. 031.614.

Velocity fields in late-type galaxies from Hα Fabry-Perot interferometry. I. Instrumentation and data reduction. See Abstr. 034.047.

Galaxy electrography at McDonald Observatory. See Abstr. 034.070.

Astronomical consequences of the neutrino rest mass. III. The nonlinear stage of evolution of perturbations and the hidden mass. See Abstr. 061.002.

On the supersonic dynamics of magnetized jets of thermal gas in radio galaxies. See Abstr. 062.091.

Monte Carlo simulation of relativistic Comptonization. See Abstr. 063.037.

Giant flares on supermassive accretion disks: polarization properties and energetics. See Abstr. 064.063.

Inhomogeneous spherical accretion onto massive black holes as a model for active galactic nuclei. See Abstr. 066.170.

Carbon stars in the Fornax dwarf spheroidal galaxy. See Abstr. 114.063.

Stellar atmospheres and chemical compositions of galaxies. See Abstr. 114.091.

Spectrophotometry of supernova remnants in M31. See Abstr. 125.017.

Optical and radio studies of supernova remnants in the Local Group galaxy M33. See Abstr. 125.038.

Radial distribution of supernovae in elliptical galaxies. See Abstr. 125.083.

Star formation and activity in the nuclei of barred galaxies. See Abstr. 131.020.

Empirical information on star formation in galaxies. See Abstr. 131.022.

Star formation in spiral galaxies. See Abstr. 131.247.

Collapse of ionized gas in galactic nuclei. See Abstr. 131.314.

The largest H II regions in M101. See Abstr. 132.011.

On the composition of H II regions in southern galaxies – II. NGC 6822 and 1313. See Abstr. 132.015.

Radio observations of H II regions in external galaxies. III. Thermal emission, H II regions and star formation in 14 late-type galaxies. See Abstr. 132.025.

Spectrophotometry and abundances in H II regions in external galaxies. See Abstr. 132.042.

Neutral hydrogen in the radiogalaxy Centaurus A. See Abstr. 132.043.

Identification of a nebulous blue object near Maffei 1. See Abstr. 134.040.

Spectroscopic observations of galactic nebulae and galaxies with the Imaging Photon Counting System (IPCS). See Abstr. 135.024.

1610 – 771: a QSO with a steep optical spectrum. See Abstr. 141.006.

***JHK* observations of two $z = 3$ QSOs.** See Abstr. 141.007.

Multifrequency observations of extended radio galaxies III: 3C 465. See Abstr. 141.019.

The Broad-Line Region in active nuclei and quasars: correlations with luminosity and radio emission. See Abstr. 141.023.

VLA observations of the M87 jet at 6 and 2 centimeters. See Abstr. 141.029.

The polarization spectra of radio outbursts in extragalactic variable sources. See Abstr. 141.032.

Long term optical behavior of 114 extragalactic sources. See Abstr. 141.036.

Jet formation in asymmetric radio galaxies. See Abstr. 141.037.

High-resolution radio observations at 6 and 20 cm of the jet in NGC 6251. See Abstr. 141.038.

Discovery of low-redshift X-ray selected quasars: new clues to the QSO phenomenon. See Abstr. 141.056.

High-resolution observations of the neutral hydrogen absorption and radio continuum emission of the radio source 3C 178. See Abstr. 141.058.

Radio variability in the nuclei of double radio galaxies and quasars. See Abstr. 141.069.

A study of the 5′ halo of 3C84. See Abstr. 141.071.

Orbital dynamics of the radio galaxy 3C 129. II. Internal tail structure. See Abstr. 141.074.

Physical processes in the envelopes of quasars and nuclei of Seyfert galaxies. See Abstr. 141.094.

Optical and infrared polarization of active extragalactic objects. See Abstr. 141.099.

On collimation of relativistic jets from quasars. See Abstr. 141.105.

Observations of M87 at 15.4 GHz with the 5-km telescope. See Abstr. 141.107.

Statistical analysis of the optical variability of QSOs. See Abstr. 141.114.

Optical observations of radio jets. See Abstr. 141.118.

X-ray studies of quasars and active galaxies with the Einstein Observatory. See Abstr. 141.119.

Structure of the compact nuclear radio source in M82. See Abstr. 141.130.

B2 1141+37: a giant radio galaxy with remarkable radio and optical properties. See Abstr. 141.132.

Radio galaxies and quasars.
See Abstr. 141.139.

Optical continuum and emission-line luminosity of active galactic nuclei and quasars. See Abstr. 141.147.

Collimation of the radio jets in 3C 31.
See Abstr. 141.150.

Variability in QSOs and active galactic nuclei.
See Abstr. 141.151.

Variabilité du rayonnement continu dans les domaines optique et ultraviolet. See Abstr. 141.152.

Optical polarimetry of quasi-stellar and BL Lac objects. See Abstr. 141.165.

Comparisons of the orientations of double-lobed radio sources and their associated elliptical galaxies.
See Abstr. 141.169.

Comments on the relativistic ejection of QSOs from galaxies. See Abstr. 141.170.

On the nature of the faint ($B \simeq 20$) ultraviolet excess objects and the problem of the X-ray background.
See Abstr. 142.063.

X-ray observations of six BL Lacertae fields.
See Abstr. 142.115.

Discovery of soft X-ray emission from a BL Lacertae object. See Abstr. 142.121.

Study of high luminosity X-ray sources in external galaxies (M 31). See Abstr. 142.133.

A correlation between the lengths of bars and the sizes of bulges. See Abstr. 151.001.

Contributions to the theory of spiral structure. I. Energy and lifetime of density waves and the classification of spiral galaxies. See Abstr. 151.005.

The dynamics of the spiral galaxy M81. I. Axisymmetric models and the stellar density wave.
See Abstr. 151.007.

The dynamics of the spiral galaxy M81. II. Gas dynamics and neutral-hydrogen observations.
See Abstr. 151.008.

Bar-driven spiral structure.
See Abstr. 151.009.

Can elliptical galaxies be equilibrium systems?
See Abstr. 151.010.

On the excitation of warps in galaxy disks.
See Abstr. 151.013.

Viscous effects in the gas flow in barred spirals.
See Abstr. 151.016.

Hydrodynamic simulation of galactic disks.
See Abstr. 151.017.

Galaxy warps and the quest for properties of the unseen halo – dynamical considerations.
See Abstr. 151.018.

Galaxy warps and the quest for properties of the unseen halo – preliminary results for three galaxies.
See Abstr. 151.019.

A dispersion relation for open spiral galaxies.
See Abstr. 151.023.

Potential in the central bar structure.
See Abstr. 151.025.

Galactic spiral shocks: vertical structure, thermal phase effects, and self-gravity. See Abstr. 151.026.

Formation and rotation of disc galaxies with haloes.
See Abstr. 151.029.

The mass–angular momentum density relation for spiral galaxies. See Abstr. 151.030.

Transient annular structures in barred galaxies.
See Abstr. 151.032.

Why aren't all galaxies barred?
See Abstr. 151.040.

Binuclear model of an SB galaxy.
See Abstr. 151.045.

On the interpretation of colors of faint galaxies.
See Abstr. 151.050.

On the interpretation of galaxy counts.
See Abstr. 151.051.

Theoretical considerations on the dynamics of normal galactic nuclei. See Abstr. 151.058.

Tri-axial dynamics in the core of normal galaxies.
See Abstr. 151.059.

Slowly rotating elliptical galaxies.
See Abstr. 151.066.

Warping of galaxies. See Abstr. 151.067.

A driving mechanism for galactic spirals.
See Abstr. 151.068.

Bisymmetric open-spiral configuration of magnetic fields in disk galaxies in differential rotation.
See Abstr. 151.075.

Spiral condensation of gas in disk galaxies by bisymmetric twisted magnetic fields: two-dimensional case.
See Abstr. 151.076.

Triaxiality in elliptical galaxies.
See Abstr. 151.079.

***N*-body simulations of slow tidal encounters and the formation of galactic halos.** See Abstr. 151.081.

A fluid dynamical flow model for the central peak in the rotation curve of disk galaxies.
See Abstr. 151.084.

The effects of environment on the evolution of elliptical and disk galaxies. See Abstr. 151.092.

Dynamical models of elliptical galaxies.
See Abstr. 151.093.

Kinematics of spiral and irregular galaxies.
See Abstr. 151.094.

Galaxy structure in terms of distinct components in the mass distribution. See Abstr. 151.095.

The stellar dynamics of elliptical systems.
See Abstr. 151.096.

Formation of an elliptical galaxy.
See Abstr. 151.097.

Elliptical galaxies: oblate or prolate?
See Abstr. 151.098.

Triaxial models of galaxies.
See Abstr. 151.099.

A dynamical test to determine the intrinsic shape of galaxies. See Abstr. 151.100.

Gas dynamics in ordinary and barred spirals.
See Abstr. 151.111.

Stochastic star formation and spiral galaxies.
See Abstr. 151.112.

Spiral waves and the decomposition of luminosity profiles. See Abstr. 151.113.

Self-gravitating spiral waves in barred spiral galaxies.
See Abstr. 151.114.

Photoelectric photometry of globular clusters in the Andromeda nebula and its companion NGC 147. VI.
See Abstr. 154.006.

The bright globular cluster in NGC 5128 and its distance. See Abstr. 154.010.

Photometric studies of composite stellar systems. IV. Infrared photometry of globular clusters in M31 and a comparison with early-type galaxies. See Abstr. 154.011.

Infrared observations of globular clusters in M31 and a comparison with galactic globulars and elliptical galaxies. See Abstr. 154.035.

Globular clusters and galaxy evolution.
See Abstr. 154.036.

Can galaxy warps be used to provide constraints on halo properties? See Abstr. 155.017.

Radio continuum observations of the nucleus of our Galaxy and other normal galaxies. See Abstr. 155.023.

Atomic and molecular gas in the inner regions of the Milky Way and other galaxies. See Abstr. 155.024.

Far-infrared balloon-borne observations.
See Abstr. 156.014.

Galaxien vom Magellanschen Typ.
See Abstr. 159.004.

On the distribution of radio emission in the X-ray cluster of galaxies Abell 401. See Abstr. 160.001.

The distribution of spiral galaxies in the direction of the Coma/A1367 supercluster. See Abstr. 160.012.

New redshifts in the Virgo cluster.
See Abstr. 160.035.

The Westerbork survey of rich clusters of galaxies.
See Abstr. 160.039.

X-ray imaging studies of NGC 1275 and the core of the Perseus cluster. See Abstr. 160.043.

The luminosity function of Coma cluster galaxies.
See Abstr. 160.065.

Very faint blue objects in the Virgo cluster region.
See Abstr. 160.068.

Photographic surface photometry of the brightest galaxies in poor clusters. See Abstr. 160.072.

CCD camera photometry of cD galaxies in poor clusters. See Abstr. 160.073.

Neutral hydrogen observations of spiral galaxies in clusters. See Abstr. 160.074.

High resolution X-ray spectroscopy of the gas surrounding M87 and NGC1275: emission line detection and evidence for radiatively regulated accretion.
See Abstr. 161.005.

Stars, galaxies, cosmos: the past decade, the next decade. See Abstr. 162.002.

Background light from galaxies as a cosmological probe. See Abstr. 162.098.

Errata

158.901 **Erratum: "The dwarf spheroidal galaxy in Draco. I. New *BV* photometry"** [Astron. J., Vol. 84, 1149 - 1166 (1979)]. P. B. Stetson.
Astron. J., Vol. 85, 1134 (1980). – See Abstr. 26.158.049.

158.902 **Erratum: "Surface photometry of elliptical galaxies"** [Astrophys. J., Vol. 222, 1 - 13 (1978)].
I. R. King.
Astrophys. J., Vol. 241, 474 (1980). – See Abstr. 21.158.119.

158.903 **Erratum: "Possible multiple imaging by spherical galaxies"** [Astrophys. J., Lett., Vol. 238, L67 - L70 (1980)]. C. C. Dyer, R. C. Roeder.
Astrophys. J., Lett., Vol. 242, L53 (1980). – See Abstr. 27.158.325.

159 Magellanic Clouds

159.001 **Calcium abundances of F supergiants in the Magellanic Clouds.** H. A. Smith.
Astron. J., Vol. 85, 848 - 852, with a correction p. 1425 (1980).

Low-dispersion image tube spectra have been obtained for six supergiant stars in the Large Magellanic Cloud and for four supergiants in the Small Magellanic Cloud. These stars are of spectral type F, though differences from galactic MK standards make precise classification difficult. The observed calcium K-line and Balmer-line strengths have been compared with simple synthetic spectra based on the model atmospheres of Kurucz. For the LMC stars, [Ca/H] = –0.2 ± 0.1; for the SMC stars, [Ca/H] = –0.6 ± 0.1.

159.002 **The giant and supergiant shells of the Magellanic Clouds.** J. Meaburn.
Mon. Not. R. Astron. Soc., Vol. 192, 365 - 375 (1980).

A process of high contrast printing and unsharp masking has been applied to two long exposure Hα + [N II] plates of the LMC and one of the SMC. Thirty-two new giant (20–260 pc diameter) interstellar shells have been discovered in the LMC. Also two new supergiant (≃ 1000 pc diameter) shells have been certainly revealed and three tentatively identified in the LMC. In all, nine supergiant shells in the LMC and one in the SMC could exist. The formation and statistics of all these phenomena are discussed and their relationship to the young spiral arms of the LMC considered.

159.003 **IUE observations of Large Magellanic Cloud members.** K. Nandy, D. H. Morgan.
Mon. Not. R. Astron. Soc., Vol. 192, 905 - 916 (1980).

Spectra and flux distributions in the wavelength range 1150 - 3100 Å are presented for six supergiants ranging from O to late B in the Large Magellanic Cloud.

159.004 **Galaxien vom Magellanschen Typ.**
J. V. Feitzinger.
Space Sci. Rev., Vol. 27, 35 - 105 (1980).

In the first part of this paper the morphological structure of Magellanic type galaxies (Irr I) is investigated. The galaxies of Magellanic type present a basic pattern consisting of a disk, a bar, stellar arms, rudimentary or well developed, spiral filaments and condensations in the disk. With the help of this pattern a well-defined classification scheme is set up. The subgroup of Irr II-systems consists of normal galaxies which are more or less tidally disturbed. Bursts of star formation have a great influence on structure and colour of irregular galaxies. 580 galaxies of Magellanic type (out of a sample of 3187 galaxies) were classified. In the second part of the paper the kinematics and dynamics of the Large Magellanic Cloud as the nearest and best-known example of a galaxy of Magellanic type is investigated.

159.005 **Highly luminous stars in the Magellanic Clouds.**
W. Buscombe, M. A. Bosko.
Bull. American Astron. Soc., Vol. 12, 521 (1980). – Abstract.

159.006 **Luminosities and temperatures of the reddest stars in three LMC clusters.**
J. A. Frogel, S. E. Persson, J. G. Cohen.
Astrophys. J., Vol. 239, 495 - 501 (1980).

Infrared observations in the 1.2–2.2 μm region are presented for 12 of the reddest stars in the Large Magellanic Cloud (LMC) clusters NGC 1783, 1846, and 1978. Bolometric magnitudes and temperatures are derived from the infrared data.

159.007 **A classification of star clusters in the Magellanic Clouds.**
L. Searle, A. Wilkinson, W. G. Bagnuolo.
Astrophys. J., Vol. 239, 803 - 814 (1980).

Four-color photometry of the integrated light from 61 star clusters in the Magellanic Clouds, together with a rediscussion of Danziger's multivariate spectrophotometric data for some of these same clusters, have made possible for the first time an intelligible classification scheme for their integrated spectra. It is found that the integrated spectra of populous star clusters in the Magellanic Clouds may be arranged in a one-dimensional sequence in such a way, that on traversing the sequence, all spectral features behave regularly – not necessarily monotonically, but as smoothly as can be expected from the magnitude of observational error.

159.008 **Photoelectric photometry of stars in the Small Magellanic Cloud.** A. Ardeberg.
Astron. Astrophys., Suppl. Ser., Vol. 42, 1 - 7 (1980).

Results from photoelectric photometry on the *UBV* system are reported for 105 stars found to be members of the Small Magellanic Cloud. Variability in the *V* magnitude is suspected for 20% of the stars, randomly distributed in magnitude.

159.009 **The dynamics of the giant filamentary shell, N51D, in the LMC.** J. Meaburn, D. L. Terrett.
Astron. Astrophys., Vol. 89, 126 - 131 (1980).

The profiles of the [O II] emission lines have been observed at many positions over the giant filamentary shell N51D. Evidence is presented which suggests that this nebula could either be an expanding spherical shell driven by energetic stellar winds or more likely a supernova remnant approaching fossilisation.

159.010 **Die Eigenbewegung der Magellanschen Wolken: Vorhersagen für Astrometriesatelliten.**
J. V. Feitzinger.
Mitt. Astron. Ges., Nr. 48, (see 012.015), p. 148 (1980).

159.011 **IUE and La Silla observations of mass loss in the Magellanic Clouds.** F. Macchetto.
Messenger, No. 21, p. 4 - 6 (1980).

159.012 **Highly ionized species in the spectra of Small Magellanic Cloud stars.** L. Prévot, C. Laurent, J. Paul, A. Vidal-Madjar, J. Audouze, R. Ferlet, J. Lequeux, M. Maucherat-Joubert, M.-L. Prévot-Burnichon, B. Rocca-Volmerange.
Astron. Astrophys., Vol. 90, L13 - L16 (1980).

The authors present preliminary results of high-resolution IUE observations of the star SK 159 in the Small Magellanic Cloud (SMC), and of low-resolution observations of this star and of 6 other SMC stars. A study of C IV and Si IV circumstellar absorption suggests a stellar mass loss rate smaller than in the Galaxy and the Large Magellanic Cloud. Interstellar C IV and Si IV lines formed in the SMC are weak or absent in the spectrum of SK 159. The authors thus do not confirm the existence of a hot halo in the SMC proposed by de Boer and Savage (1980).

159.013 **The numbers of red supergiants and WR stars in galaxies: an extremely sensitive indicator of chemical composition.** A. Maeder, J. Lequeux, M. Azzopardi.
Astron. Astrophys., Vol. 90, L17 - L20 (1980).

It is shown that the ratio N_R/N_{WR} of the numbers of red supergiants to the number of WR stars varies very strongly with galactocentric distance, by a factor of about 90 between zones located at 7-9 kpc and 11-13 kpc from the galactic center. However, the ratio of the sum of the numbers of red supergiants and WR stars to that of blue supergiants, $(N_R+N_{WR})/N_B$, remains almost constant with galactocentric distance. A relation, also satisfied by the LMC and SMC, is obtained between the N_R/N_{WR} ratio and the metal content Z.

159.014 **Ultraviolet studies of the Magellanic Clouds. II. Internal extinction, formation of massive stars, comparison with other galaxies.** E. Vangioni-Flam, J. Lequeux, M. Maucherat-Joubert, B. Rocca-Volmerange.
Astron. Astrophys., Vol. 90, 73 - 82 (1980).

The absolute integrated UV fluxes of the Magellanic Clouds measured by the D 2 B-Aura satellite (Maucherat-Joubert et al., 1980, Paper I) are corrected from interstellar extinction. For the purpose of this correction, the authors give a detailed discussion of the internal extinction in these galaxies.

159.015 **Five-colour photometry of blue stars in the Magellanic-Cloud region.** W. Wamsteker.
ESO Sci. Prepr. No. 108, 26 pp. (1980). – Submitted to Astron. Astrophys., Suppl. Ser.

159.016 **UV and optical observations of X-ray sources in the Magellanic Clouds.** M. Tarenghi, E. G. Tanzi, A. Treves, W. M. Glencross, I. Howarth, G. Hammerschlag-Hensberge, E. P. J. Van den Heuvel, H. J. G. L. M. Lamers, M. Burger, P. A. Whitelock.
ESO Sci. Prepr. No. 116, 19 pp. (1980). – Submitted to Astron. Astrophys., Suppl. Ser.

159.017 **The dynamics of giant filamentary shells in the Large Magellanic Cloud – III N59A (DEM 241).**
J. Meaburn, D. L. Terrett, J. C. Blades.
Anglo-Australian Obs., Prepr. No. 143, 31 pp. (1980).
Submitted to Mon. Not. R. Astron. Soc.

159.018 **The Magellanic Stream and the Galaxy with a massive halo.** T. Murai, M. Fujimoto.
Publ. Astron. Soc. Japan, Vol. 32, 581 - 603 (1980).

A number of series of orbits are obtained for the Large and Small Magellanic Clouds (LMC and SMC) revolving around a model Galaxy. The authors can reproduce the main characteristics of the Magellanic Stream, if the following dynamically permissible parameters are assumed: (1) the orbital plane of the LMC is perpendicular to the galactic plane, (2) the perigalactic distance of the LMC is 50 kpc, (3) the sense of the revolution around the Galaxy is counterclockwise as seen from the present position of the sun, and (4) the SMC approached the LMC as close as 3 kpc from its center about 200 Myr ago. The distance of the model stream is in the range of 30 to 60 kpc measured from the sun. The total hydrogen mass of the Stream is, therefore, $10^8\ M_\odot$ or slightly less.

159.019 **Ultraviolet studies of the Magellanic Clouds – I. Interstellar lines in the spectra of FD 70 and SK-71-45.** P. M. Gondhalekar, A. J. Willis, D. H. Morgan, K. Nandy.
Mon. Not. R. Astron. Soc., Vol. 193, 875 - 883 (1980).

High dispersion ultraviolet spectra of the two LMC members, HD 38282 (FD 70) and HD 269676 (SK-71-45), have been obtained with IUE. Velocity profiles have been presented of the numerous interstellar lines observed in the spectra of these stars and associated with the absorption in both the Galaxy and the LMC. The extended blue asymmetric absorption observed in the LMC interstellar Ca II *k* line absorption is also observed in both ultraviolet interstellar line profiles in FD 70 and SK-71-45. This result favours an origin for these features associated with gas in a halo around the LMC.

159.020 **On the inclination of the Large Magellanic Cloud.** G. de Vaucouleurs.
Publ. Astron. Soc. Pacific, Vol. 92, 576 - 578 (1980).

The inclination of the Large Magellanic Cloud recently derived from residuals of the cepheid period-luminosity relation is shown to be in good agreement with the previous determinations by this and other methods. The weighted mean value of four different methods, $i = 27\overset{\circ}{.}2 \pm 2\overset{\circ}{.}4$, confirms the previously adopted value. The near side of the Cloud is the east side and there is no evidence for a large-scale gradient of galactic extinction across the central 10° field.

159.021 **On the distance modulus of the Large Magellanic Cloud.** G. de Vaucouleurs.
Publ. Astron. Soc. Pacific, Vol. 92, 579 - 586 (1980).

The corrected modulus $\langle\mu_0\rangle = 18.31 \pm 0.15$ (external m.e.) previously derived from five independent primary distance indicators is confirmed by nine different secondary and tertiary indicators which give $\langle\mu_0\rangle = 18.20 \pm 0.10$ (internal mean error) on the same system of zero points and extinction corrections. Several recent, independent determinations from M supergiants, cepheids, and RR Lyrae variables are in poor agreement. The total galactic and internal extinction in the LMC field derived by six independent methods of $E(B-V) = 0.13 \pm 0.012$ and $A_V = 0.41 \pm 0.04$ (m.e.) of which $\sim 0\overset{m}{.}3$ is galactic and $\sim 0\overset{m}{.}1$ is internal. An apparent correlation between extinction corrections and absolute magnitudes of M supergiants in the LMC, recently presented by Humphreys, is discussed. The low extinction value derived from cepheids by Martin, White, and Feast is discussed. The zero point of the cepheid P-L-C relation is discussed.

159.022 **A note on interstellar absorption in the Magellanic Clouds.** J. B. Hutchings.
Publ. Astron. Soc. Pacific, Vol. 92, 592 - 595 (1980).

Optical spectroscopy is presented of diffuse interstellar absorption in spectra of reddened stars in the Magellanic Clouds. The λ4430 absorption is definitely present and λλ5780, 6284 probably present in some spectra, but the correlation with E_{B-V} is poor. The strongest λ4430 absorption is seen in the star S-69-108 whose UV extinction is most like the galactic. No correlation of interstellar absorption with positions in the Clouds is yet evident.

159.023 **Kinematics and dynamics of the Large Magellanic Cloud.** J. V. Feitzinger.
Photometry, kinematics and dynamics of galaxies, (see 012.051), p. 435 - 438 (1979).

159.024 **Carbon and late M-type stars in the Magellanic Clouds.**
V. M. Blanco, M. F. McCarthy, B. M. Blanco.
Astrophys. J., Vol. 242, 938 - 964 (1980).

Identification charts, coordinates, and *R* and *I* photometric measurements are presented for 320 carbon stars and 107 giant M stars later than type M5 found in five sample (0.12 square degrees) areas in the two Magellanic Clouds. The carbon stars are found to be far more abundant relative to the M giants in the SMC than in the LMC, and they show a single-mode luminosity distribution with a mean *I* magnitude of –4.6. The differences in mean *I* magnitudes between the clouds suggest a distance modulus difference of 0.51±0.03.

The most massive stars in the Galaxy and the LMC: quasi-homogeneous evolution, time-averaged mass loss rates and mass limits. See Abstr. 065.068.

Circumstellar absorption and intrinsic colours of massive stars. See Abstr. 112.019.

Spectral classification of carbon stars in Magellanic Cloud clusters. See Abstr. 114.047.

Spectroscopy of the Small Magellanic Cloud emission line star Hen S 18.
See Abstr. 114.112.

A large Magellanic Cloud member intermediate between Of and WN7. See Abstr. 114.124.

Discovery of the first SC star in the Magellanic Clouds. See Abstr. 114.133.

HV 11417: a peculiar M supergiant in the Small Magellanic Cloud. See Abstr. 114.134.

A period–luminosity relation for supergiant red variables in the Large Magellanic Cloud. See Abstr. 115.006.

New photoelectric observations of the Wolf-Rayet star HD5980 in the Small Magellanic Cloud.
See Abstr. 119.035.

IUE and ground based observations of the LMC star S Doradus. See Abstr. 122.024.

Metal abundances of Magellanic Cloud variable stars.
See Abstr. 122.044.

Spectra of red supergiant variables in the SMC.
See Abstr. 122.075.

Multicolor photoelectric photometry of Magellanic Cloud Cepheids. III: BVI observations of ten LMC Cepheids.
See Abstr. 122.144.

Ultra-short period Cepheids in the LMC.
See Abstr. 122.188.

X-ray line emission from LMC supernova remnants.
See Abstr. 125.020.

A new oxygen-rich supernova remnant in the Large Magellanic Cloud. See Abstr. 125.069.

Observations of supernova remnants in the Large Magellanic Cloud with the Einstein Observatory.
See Abstr. 125.087.

The galactic foreground reddening in the direction of the Magellanic Clouds. See Abstr. 131.062.

The galactic foreground reddening in the direction of the Magellanic Clouds. See Abstr. 131.317.

The motions in N51D, a bubble-like nebula in the Large Magellanic Cloud. See Abstr. 132.008.

CO (J = 2 – 1) observations of the Carina nebula and G333.6–0.2 and a search for CO in LMC and SMC.
See Abstr. 134.031.

Ring nebulae associated with Wolf-Rayet stars in the Large Magellanic Cloud. See Abstr. 135.041.

Observations of outbursts from the recurrent X-ray transient A0538 – 66 and LMC X-4. See Abstr. 142.054.

Optical candidate for LMC X-1.
See Abstr. 142.101.

The 1979 March 5 gamma ray transient reviewed: its source location in N49 within the LMC and its characteristics as evidence for a vibrating neutron star.
See Abstr. 142.504.

On the origin of the March 5, 1979 gamma ray transient: a vibrating neutron star in the Large Magellanic Cloud. See Abstr. 142.507.

Red stars in Magellanic Cloud globular clusters.
See Abstr. 154.007.

The extended giant branches of intermediate age globular clusters in the Magellanic Clouds.
See Abstr. 154.009.

Ellipticities of globular clusters of the Large Magellanic Cloud. See Abstr. 154.026.

160 Groups of Galaxies, Clusters of Galaxies, Superclusters

160.001 **On the distribution of radio emission in the X-ray cluster of galaxies Abell 401.**
J. O. Burns, M. P. Ulmer.
Astron. J., Vol. 85, 773 - 779 (1980).

The authors report 4885-MHz VLA observations of four galaxies in the rich cluster Abell 401. They produced ~ 2-arcsec-resolution maps of the fields near the sources 4C13.17A, 4C13.17B, and 14W12, as well as the cD in A401. 4C13.17A (total flux ≅ 120 mJy) has an ~ 20-arcsec structure aligned along the minor axis of its associated optical galaxy. A portion of this radio structure is very highly polarized–48±4%. The apparent location of high polarization (near the nucleus of the radio galaxy) is probably due to projection effects.

160.002 **Scale covariant gravitation: virial masses of groups of galaxies.** F. R. Klinkhamer.
Astron. Astrophys., Vol. 87, 354 - 356 (1980).

Scale-covariant theory of gravitation introduces an extra term in the virial theorem. Mass-to-light ratios of groups of galaxies are reduced to typically 20–70 solar units. An interpretation is given of the observed correlations of mass-to-light ratios with velocity dispersions and virial radii.

160.003 **A radio survey of clusters of galaxies. III. 6.2 cm observations, radio spectra and optical identifications of sources in 29 Abell clusters.**
H. Andernach, H. Waldthausen, R. Wielebinski.
Astron. Astrophys., Suppl. Ser., Vol. 41, 339 - 394 (1980).

The authors mapped twenty-nine Abell clusters of galaxies at 4 850 MHz (λ 6.2 cm) using the 100-m radio telescope. The radio maps were superimposed onto the optical fields of the Palomar Observatory Sky Survey red (E) prints. This revealed a large fraction of empty fields or optical objects significantly separated from the radio peak. Radio spectra of most of the detected sources were obtained. The authors find a trend for cluster source spectra to become steeper with decreasing distance from the cluster centre.

160.004 **The two-point covariance function of galaxy clustering for extreme values of its argument.**
G. Dautcourt.
Astron. Nachr., Band 301, 155 - 156 (1980).

The two-point covariance function of galaxy clustering is discussed for small and large values of its argument.

160.005 **Investigation of the galaxy distribution in a field at the boundary between Virgo and Serpens Caput.**
F. W. Baier.
Astron. Nachr., Band 301, 165 - 175 (1980).

The number density distributions for further 13 clusters of galaxies are derived by counting galaxies on the red Palomar Sky Survey prints. For these clusters the radial number density distributions and the radial cumulative galaxy distribution are derived.

160.006 **The most compact nest of galaxies VV 644.**
B. A. Vorontsov-Vel'yaminov, V. A. Dostal', V. G. Metlov.
Pis'ma Astron. Zh., Tom 6, 394 - 397 (1980). In Russian. English translation in Soviet Astron. Lett., Vol. 6.

Photographic and spectral data for the close blue group of interacting galaxies VV 644 are given. The conclusion is put forward that VV 644 is a nest of galaxies consisting of 6 compact components.

160.007 **A distance scale from the infrared magnitude/H I velocity-width relation. III. The expansion rate outside the Local Supercluster.** M. Aaronson, J. Mould, J. Huchra, W. T. Sullivan III, R. A. Schommer, G. D. Bothun.
Astrophys. J., Vol. 239, 12 - 37 (1980).

Infrared magnitudes and 21 cm H I velocity widths are presented for galaxies in the Pegasus I cluster, the Cancer cluster, cluster Zwicky 1400.4+0949 (Z74-23), and the Perseus supercluster. The data are used to determine redshift-independent distances from which values of the Hubble ratio can be derived.

160.008 **The range of $V-R$ colors for a cluster of E and S0 galaxies as a function of redshift.**
K. DeGioia-Eastwood, G. L. Grasdalen.
Astrophys. J., Lett., Vol. 239, L1 - L3 (1980).

The expected $(V-R)$ color distribution for a centrally condensed, relaxed cluster of E and S0 galaxies has been calculated as a function of redshift. Because of the differences in the ultraviolet spectra of E and S0 galaxies, which are correlated with absolute magnitude, the spread in $(V-R)$ colors for such a cluster becomes increasingly wide for increasing redshift. This effect becomes pronounced for redshifts of 0.4 and beyond. Thus evidence for the color evolution of cluster galaxies will be seen as an additional broadening of the color distribution.

160.009 **The foreground of the Virgo cluster.**
C. Ftaclas, M. Fanelli, M. Struble, M. Zuber.
Bull. American Astron. Soc., Vol. 12, 471 (1980). – Abstract.

160.010 **Cosmic rays in clusters of galaxies and the formation of extended radio halos.** B. Dennison.
Bull. American Astron. Soc., Vol. 12, 471 (1980). – Abstract.

160.011 **Observations of clusters of galaxies at meter wavelengths.**
H. V. Cane, W. C. Erickson, R. J. Hanisch, P. J. Turner.
Bull. American Astron. Soc., Vol. 12, 471 (1980). – Abstract.

160.012 **The distribution of spiral galaxies in the direction of the Coma/A1367 supercluster.**
B. A. Williams, F. J. Kerr.
Bull. American Astron. Soc., Vol. 12, 472 (1980). – Abstract.

160.013 **Further searches for velocity-inclination correlations in clusters of galaxies.**
M. F. Struble, C. Ftaclas, M. Zuber, M. Fanelli.
Bull. American Astron. Soc., Vol. 12, 472 (1980). – Abstract.

160.014 **On the magnetic field in clusters of galaxies as inferred from Faraday rotation observations.**
J. M. Lawler, B. K. Dennison.
Bull. American Astron. Soc., Vol. 12, 472 (1980). – Abstract.

160.015 **HEAO-1 observations of the Perseus cluster above 15 keV.** F. A. Primini, E. Basinska, A. M. Levine, W. H. G. Lewin, R. Rothschild, J. L. Matteson, L. E. Peterson.
Bull. American Astron. Soc., Vol. 12, 472 (1980). – Abstract.

160.016 **A search for Z=9 neutral hydrogen emission from primordial protoclusters of galaxies.**
J. T. Bonnell, P. D. Jackson, I. F. Mirabel.
Bull. American Astron. Soc., Vol. 12, 472 (1980). – Abstract.

160.017 **Dynamics of the supercluster 1451 +22: evidence for an open universe.**
R. J. Harms, H. C. Ford, F. Bartko, R. Ciardullo, E. Eason.
Bull. American Astron. Soc., Vol. 12, 472 - 473 (1980). Abstract.

160.018 **The HEAO A-2 survey of Abell clusters and the X-ray luminosity function.**
J. D. McKee, R. F. Mushotzky, E. A. Boldt, S. S. Holt, F. E. Marshall, S. H. Pravdo, P. J. Serlemitsos.
Bull. American Astron. Soc., Vol. 12, 486 (1980). – Abstract.

160.019 **X-ray properties of poor clusters with dominant galaxies.** G. A. Kriss, C. R. Canizares.
Bull. American Astron. Soc., Vol. 12, 487 (1980). – Abstract.

160.020 **Discovery of a bright X-ray cluster of galaxies at low galactic latitude.**
M. D. Johnston, R. E. Doxsey, F. E. Marshall, D. A. Schwartz.
Bull. American Astron. Soc., Vol. 12, 487 (1980). – Abstract.

160.021 **X-ray properties of clusters of galaxies.**
W. H.-M. Ku, F. Abramopoulos, K. S. Long.
Bull. American Astron. Soc., Vol. 12, 487 - 488 (1980). Abstract.

160.022 **Observations of distant clusters with a CCD spectrometer/imager.** M. Bautz, S. Meyer.
Bull. American Astron. Soc., Vol. 12, 488 (1980). – Abstract.

160.023 **Diffuse radio emission in the Coma cluster and Abell 1367 – observations at 430 and 1400 MHz.**
R. J. Hanisch, G. D. Holman.
Bull. American Astron. Soc., Vol. 12, 493 - 494 (1980). Abstract.

160.024 **The heating of gas in clusters of galaxies by relativistic electrons: collective effects.** J. S. Scott, G. D. Holman, J. A. Ionson, K. Papadopoulos.
Astrophys. J., Vol. 239, 769 - 773 (1980).

The authors show that the rate at which gas is heated in X-ray clusters of galaxies by streaming relativistic electrons can be much greater than the Coulomb heating rate because of the stimulated growth of a high level of electrostatic turbulence and its subsequent collapse to shorter wavelengths. This enhanced heating (and, hence, energy loss) rate allows the X-ray emitting gas to be heated by those particles which are observable through their synchrotron emission at low radio frequencies and yields a radio source size consistent with the observed radio halo sizes in the Coma cluster. The heating of gas in clusters of galaxies by relativistic electrons will significantly affect the cluster gas dynamics.

160.025 **Formation of radio halos in clusters of galaxies from cosmic-ray protons.** B. Dennison.
Astrophys. J., Lett., Vol. 239, L93 - L96 (1980).

Relativistic protons produced in radio galaxies in a cluster can diffuse over distances ~1 Mpc in the intracluster medium before suffering inelastic collisions with thermal protons. This results in a population of relativistic secondary electrons and positrons formed in situ on this scale. It is shown that these secondary particles can be entirely responsible for the observed radio halo in Coma, if it is assumed that the ratio of the energy-dependent production rate of primary protons to primary electrons is the same as that apparent from cosmic rays in our Galaxy, and that the intracluster magnetic field is in approximate equipartition (~2 microgauss) with relativistic particles.

160.026 **Radio emission of near clusters of galaxies at 102.5 MHz.**
A. G. Gubanov, R. D. Dagkesamanskij, V. A. Rudenko.
Pis'ma Astron. Zh., Tom 6, 548 - 551 (1980). In Russian. English translation in Soviet Astron. Lett., Vol. 6.

Observations of near Abell clusters of galaxies have been made at 102.5 MHz. The cluster radio luminosity increases on the average with the cluster richness.

160.027 **Broadband 21-cm H I emission from Stephan's Quintet.** W. T. Sullivan III.
Astron. Astrophys., Vol. 89, L3 - L5 (1980).

The H I profile at the position of NGC 7318a/b in Stephan's Quintet reveals the presence of broadband emission at a level of ~8 mJy over the entire range of redshifts from 5600 to 6800 km s^{-1}. It is argued that this emission gives further evidence for a high degree of interaction among NGC 7318a, 7318b, and 7319. The galaxies appear to be enveloped in a huge cloud of H I.

160.028 **A Westerbork survey of clusters of galaxies. XIII. Deep 610 MHz source counts from the Cancer cluster field.** E. A. Valentijn.
Astron. Astrophys., Vol. 89, 234 - 238 (1980).

A catalogue of 220 radio sources detected at 610 MHz in the Cancer cluster field and complete down to a map flux density of 2.5 mJy is presented. The catalogue is used to derive a deep source count down to 2.5 mJy. It is concluded that the 610 MHz dN/dS distribution fits very well with a single power law with a slope of −1.66 in the flux density range 2.5–158 mJy.

160.029 **Kelvin-Helmholtz instability in clusters of galaxies.** M. Livio, O. Regev, G. Shaviv.
Astrophys. J., Lett., Vol. 240, L83 - L86 (1980).

The motion of galaxies through the hot intracluster gas is examined. It is found that a Kelvin-Helmholtz instability develops at the interface between the moving galaxy and the gas. Its effect is gas stripping from the moving galaxy: the rate of stripping is estimated. The role of viscosity in this process is discussed.

160.030 **Isothermal clusters of galaxies.** B. M. Lewis.
Proc. Astron. Soc. Australia, Vol. 3, 286 - 287 (1978).

160.031 **The group of galaxies NGC 2805–2814–2820-Markarian 108.** A. Bosma, C. Casini, J. Heidmann, J. M. van der Hulst, H. van Woerden.
Astron. Astrophys., Vol. 89, 345 - 352 (1980).

Radio-continuum, H I-line and optical observations of the group of galaxies NGC 2805–2814–2820-Markarian 108 are presented. The authors discuss the properties of the galaxies and of the group and present their conclusions.

160.032 **3-D nearest neighbours test for Abell clusters of galaxies.** M. Kalinkov, V. N. Dermendjiev, I. F. Kuneva.
C. R. Acad. Bulgare Sci., Vol. 33, No. 1, p. 7 - 10 (1980). Abstr. in Phys. Abstr., Vol. 83, Abstr. 90545 (1980).

160.033 **Does the binding energy of binaries masquerade as missing mass?** P. S. Wesson.
Astron. Astrophys., Vol. 90, 1 - 7 (1980).

A fresh examination is given of the idea that substructure (especially binary subclustering) in clusters of galaxies contributes significantly to the binding energy, and that substructured clusters obey the virial theorem without the need to appeal to missing or hidden mass to balance the virial equation. In particular, the criticism of the binary model of Wesson and Lermann by Ozernoy and Reinhardt is replied to.

160.034 **Radio and X-ray observations of Abell 754.**
D. E. Harris, C. H. Costain, R. G. Strom, F. J. Pineda, J. P. Delvaille, H. W. Schnopper.
Astron. Astrophys., Vol. 90, 283 - 289 (1980).

The cluster of galaxies, Abell 754, has been observed at 610 MHz, and at 1420 MHz, and in the interval 2–10 keV with the rotating modulation collimators on SAS-3. A complete source list for the field is given together with radio contour diagrams of extended sources associated with cluster galaxies. No X-rays were detected.

160.035 **New redshifts in the Virgo cluster.**
J. W. Sulentic.
Astrophys. J., Vol. 241, 67 - 73 (1980).

Spectra have been obtained for 23 galaxies near the center of the Virgo cluster. Almost all of these galaxies are fainter than m_{pg} = 15.0. Nine of the galaxies have redshifts $V_0 \gg 2500$ km s^{-1} and, consequently, are background objects. A significant grouping has become evident at about + 7500 km s^{-1}, but the region between the Virgo cluster and this grouping (~80 Mpc) appears to contain very few galaxies. With 182 galaxies within 6° of the cluster center now having measured redshifts, a brief reanalysis of the mean redshift of the sample as a function of morphology, radius, and apparent magnitude is presented.

160.036 **Stephan's Quintet: H I distribution at high redshifts.**
S. D. Peterson, G. S. Shostak.
Astrophys. J., Lett., Vol. 241, L1 - L5, plate L1 (1980).

Highly sensitive 21 cm observations between 5450 and 6950 km s^{-1} have measured H I emission from the Stephan's Quintet region on a uniformly spaced grid. A composite spectrum presenting a total profile of the region reveals weak emission at 5700 km s^{-1} and features centered at 6000 and 6600 km s^{-1}. All three neutral hydrogen features appear to be spatially extended, with dimensions ~ 100 kpc and masses ~ $10^{10} M_\odot$ (assuming D_{HI} = 88 Mpc). The H I emission regions are offset from each other and the optical galaxies in the quintet.

160.037 **Structure of superclusters and supercluster formation.** J. Einasto, M. Jôeveer, E. Saar.
Mon. Not. R. Astron. Soc., Vol. 193, 353 - 375 (1980).

The study of the spatial distribution of galaxies and clusters of galaxies in the southern galactic hemisphere is continued using Zwicky clusters as principal tracers of the large-scale structure. The whole picture resembles cells. All clusters of galaxies and most galaxies are located in superclusters. Neighbouring superclusters are in contact and contain common elements. The space inside the cell is void of clusters and almost void of galaxies. Arguments have been given suggesting that superclusters formed prior to the formation of galaxies or simultaneously with them.

160.038 **Clusters of galaxies.** N. A. Bahcall.
Highlights of Astronomy, Vol. 5, (see 012.028), 699 - 714 (1980).

This paper summarizes the present understanding of the observed cluster sequence, its dynamical evolution, correlations of optical and X-ray properties, structural cluster parameters, and the global cluster luminosity functions in the optical and X-ray bands.

160.039 **The Westerbork survey of rich clusters of galaxies.**
E. A. Valentijn.
Highlights of Astronomy, Vol. 5, (see 012.028), 715 - 721 (1980).

160.040 **Clustering of galaxies about extragalactic radio sources – implications for observations with the Einstein X-ray Observatory.** M. S. Longair, M. Seldner.
Highlights of Astronomy, Vol. 5, (see 012.028), 723 - 726 (1980).

160.041 **Review of recent observations of cluster X-ray sources using the Einstein Observatory.**
S. S. Murray.
Highlights of Astronomy, Vol. 5, (see 012.028), 727 - 734 (1980).

160.042 **X-ray spectra of clusters of galaxies.**
R. F. Mushotzky, B. W. Smith.
Highlights of Astronomy, Vol. 5, (see 012.028), 735 - 740 (1980).

160.043 **X-ray imaging studies of NGC 1275 and the core of the Perseus cluster.**
J. Grindlay, G. Branduardi, A. Fabian.
Highlights of Astronomy, Vol. 5, (see 012.028), 741 - 745 (1980).

The Einstein X-ray Observatory has been used to study the X-ray emission from the center of the Perseus cluster, including the active galaxy NGC 1275. The central 40′ X 40′ region of the Perseus cluster around NGC 1275 displays an interesting temperature and surface brightness distribution. Simple hydrostatic isothermal sphere models do not well describe the cluster emission. The surface brightness of the high resolution image of NGC 1275 can be fit with a constant-pressure but centrally-cooling gas which suggests a radiative cooling accretion flow onto NGC 1275.

160.044 **X-ray studies of clusters of galaxies with the Einstein Observatory.**
D. J. Helfand, W. H.-M. Ku, F. Abramopoulos.
Highlights of Astronomy, Vol. 5, (see 012.028), 747 - 751 (1980).

In this report on cluster observations with the Einstein Observatory the authors present three comments concerning the following areas: (1) The correlation of X-ray brightness with cluster morphology, (2) the detection of a spiral-rich cluster at z ~ 0.4, and (3) detailed observations of Coma, the nearest rich cluster.

160.045 **X-ray observations of clusters of galaxies.**
F. Abramopoulos.
News Lett. Astron. Soc. N. Y., Vol. 1, No. 7, p. 19 (1980). Abstract.

160.046 **Westerbork observations of B2 radio sources in Abell clusters of galaxies.**
D. E. Harris, C. Lari, J. P. Vallée, A. S. Wilson.
Astron. Astrophys., Suppl. Ser., Vol. 42, 319 - 329 (1980).

Short observations at 1 415 MHz have been made of 120 Abell clusters. This sample contains all apparent coincidences between sources from the Bologna (B2) catalogue with $S(408) > 0.2$ Jy, $24° < \delta < 42°$, and Abell clusters of distance class $\leqslant 5$. The authors present the results for those clusters (58) containing B2 sources which are probably associated with the cluster (i.e. identified with a galaxy of optical magnitude corresponding to cluster membership). Positions, flux densities, and optical identifications are listed for these clusters in a table, and contour diagrams for some of the extended sources are also included.

160.047 **The photometric properties of brightest cluster galaxies. I. Absolute magnitudes in 116 nearby Abell clusters.** J. G. Hoessel, J. E. Gunn, T. X. Thuan.
Astrophys. J., Vol. 241, 486 - 492 (1980).

Two-color aperture photometry of the brightest galaxies in a complete sample of nearby Abell clusters is presented. The results are used to anchor the bright end of the Hubble diagram; essentially the entire formal error for this method is then due to the sample of distant clusters used. New determinations of the systematic trend of galaxy absolute magnitude with the cluster properties of richness and Bautz-Morgan type are derived. When these new results are combined with the Gunn and Oke data on high-redshift clusters, a formal value of $q_0 = -0.55 \pm 0.45$ is found.

160.048 **The photometric properties of brightest cluster galaxies. II. SIT and CCD surface photometry.**
J. G. Hoessel.

Astrophys. J., Vol. 241, 493 - 506, plate 4 (1980).

Surface photometry of the first-ranked galaxy in 108 Abell clusters, obtained with SIT vidicon and CCD detectors, is presented. Galaxy structure, as parametrized by simple Hubble law models is found to correlate with galaxy absolute magnitude and cluster structure. All these structure data support and are interpreted on the dynamical friction evolution model. Average magnitude and structure evolution rates are derived from the data. When this result is combined with the expected rate of evolution of the stellar population in the galaxies, a corrected value of q_0 near the formal value is derived. The correlation of absolute magnitudes and galaxy structure may be used to eliminate the magnitude dependence on cluster richness and Bautz-Morgan type and remove the effect of bias introduced into cluster samples by selection procedures.

160.049 **On the luminosity-segregation in rich clusters of galaxies. Application to Coma.**
H. V. Capelato, D. Gerbal, G. Mathez, A. Mazure, E. Salvador-Sole, H. Sol.
Astrophys. J., Vol. 241, 521 - 527 (1980).

Various methods for analysis of luminosity segregation in rich clusters of galaxies are given. Application to the central parts of the Coma cluster reveals a definite luminosity segregation. It appears effective under a critical magnitude (about $m_c \sim 16$). A dynamical dimension of the cluster of about 1.7 Mpc is also derived.

160.050 **Groups of galaxies with large crossing times.**
F. R. Klinkhamer.
Astron. Astrophys., Vol. 91, 365 - 368 (1980).

Virial masses of groups of galaxies from the sample of Rood and Dickle (1978) drop at least a factor 20 if the crossing times are larger than 1.0×10^{10} yr. These groups appear to have virial masses equal to those expected if the motions of the member galaxies are caused by the Hubble expansion only. One criterion for a group of galaxies to be bound thus appears to be a crossing-time less than 1.0×10^{10} yr, i. e. half the Hubble time.

160.051 **Lokale groep blijft groeien.**
Zenit, 7e Jaarg., 169 (1980).

160.052 **Redshifts of southern clusters of galaxies.**
R. M. West, S. Frandsen.
ESO Sci. Prepr. No. 110, 22 pp. (1980). – Submitted to Astron. Astrophys., Suppl. Ser.

160.053 **Observations of galaxies in the southern cluster CA 0340-538.**
G. Chincarini, M. Tarenghi, C. Bettis.
ESO Sci. Prepr. No. 123, 20 pp. (1980). – Submitted to Astron. Astrophys.

160.054 **The mass of Coma Cluster.** B. M. Lewis.
New Zealand J. Sci., Vol. 22, 371 - 372 (1979) = Carter Obs. Repr. Ser. 2, No. 12.

The central velocity dispersion of Coma Cluster is shown to be larger than would be expected for a King-type model. The causes are examined, and the mass estimate obtained from the King model is refined.

160.055 **The X-ray luminosity function of galaxy clusters.**
K.-H. Schmidt.
Astron. Nachr., Band 299, 193 - 195 (1978).

From published data on X-ray sources identified with clusters of galaxies the X-ray luminosity function of the clusters was derived to $\varphi_x \sim L_x^{-0.8}$ in the range $10^{43} \leq L_x \leq (2\text{–}3) \times 10^{45}$ ergs/s. For $L_x > 3 \times 10^{45}$ ergs/s the function decreases abruptly.

160.056 **Investigation of nine clusters of galaxies.**
F. W. Baier, W. Mai.
Astron. Nachr., Band 299, 197 - 207 (1978).

The number-density distributions of further 9 clusters of galaxies were derived by counting galaxies on the red Palomar Sky Survey prints. For all the clusters the radial number density distribution and the radial cumulative galaxy distribution were calculated.

160.057 **On the observability of young X-ray clusters of galaxies.** S. Ikeuchi, Y. Hirayama.
Prog. Theor. Phys., Vol. 64, 81 - 93 (1980). – Abstr. in Phys. Abstr., Vol. 84, Abstr. 4987 (1981).

160.058 **Counts of ultraviolet-bright galaxies and their distributions in clusters of galaxies.** B. Takase.
Publ. Astron. Soc. Japan, Vol. 32, 605 - 612 (1980).

Some 1,100 ultraviolet-bright galaxies were detected by means of the *UGR* three-image method. The surveyed area covers about 650 square degrees and the limiting magnitude is estimated to be 17–18 mag. The ratio *f* of the number of ultraviolet-bright galaxies to that of all galaxies is 0.25 on average. This *f* value is smaller in clusters than in the field, and also it decreases towards the denser central part of the cluster.

160.059 **NGC 5457 (M101), NGC 5194 (M51) and NGC 5055 (M63) as an interacting group of galaxies.**
K. A. Papp, K. A. Innanen.
Bull. American Astron. Soc., Vol. 12, 739 (1980). – Abstract.

160.060 **NGC 5236 (M83), NGC 5128 and NGC 4945 as an interacting group of galaxies.**
K. A. Innanen, K. A. Papp.
Bull. American Astron. Soc., Vol. 12, 739 (1980). – Abstract.

160.061 **The X-ray structure of two rich galaxy clusters and its implications for the detectability of microwave diminutions.** S. D. M. White, J. Silk.
Astrophys. J., Vol. 241, 864 - 874 (1980).

From their study of two rich clusters the authors draw the following general conclusions: 1) Even rich centrally condensed clusters with smooth X-ray emission may not always be in dynamical equilibrium. 2) In at least one cluster the core radius of the distribution of gravitating matter appears to be much larger than that inferred from galaxy counts. 3) In general there is no physical justification for a polytropic equation of state. 4) Evidence for a microwave decrement due to the Syunyaev-Zel'dovich effect of 10^{-3} K in A576 should be viewed with suspicion.

160.062 **Galactic cannibalism. IV. The evidence – correlations between dynamical time scales and Bautz-Morgan type.** T. A. McGlynn, J. P. Ostriker.
Astrophys. J., Vol. 241, 915 - 924 (1980).

If the luminosity of supergiant cD galaxies in particular, and the Bautz-Morgan sequence of galaxy types in general, is produced by dynamical evolutionary processes, then one expects to find a correlation between dynamical times and ΔM_{12}, the magnitude difference between first and second brightest cluster members. Analysis of the data from 15 Abell clusters gives moderate support for the hypothesis.

160.063 **On the morphology of the magnetic field in galaxy clusters.** W. Jaffe.
Astrophys. J., Vol. 241, 925 - 927 (1980).

Faraday rotation measurements of radio sources in rich clusters indicate that the general cluster magnetic field is probably highly tangled. Using dimensional analysis the author investigates the role of the turbulent wakes behind cluster galaxies in creating this field. He concludes that the wakes are sufficient to explain both the strength and the morphology of the observed field.

160.064 **On the derivation of higher order correlation functions.** S. A. Bonometto, N. A. Sharp.
Astron. Astrophys., Vol. 92, 222 - 224 (1980).

The authors investigate the method suggested by Bonometto and Lucchin (1980) of estimating the four-point function, by using the Zwicky catalogue of galaxies. Although noisy, the results seem reliable, and clearly encourage the application of this method to the type of survey for which it was designed.

160.065 **The luminosity function of Coma cluster galaxies.** L. A. Thompson, S. A. Gregory.
Astrophys. J., Vol. 242, 1 - 7 (1980).

Separate differential luminosity functions are presented for elliptical, S0, and spiral + irregular galaxies in the Coma cluster. The data span a range of more than five magnitudes. The authors find evidence that the ellipticals taken separately do not fit the simple Schechter function over the entire magnitude interval. Ellipticals are dominant among the brightest galaxies, yet S0's quickly overcome the deficit in numbers at intermediate brightnesses. A comparison of the Coma spiral + irregular galaxies with a local sample indicates a drastic deficiency of faint spirals in Coma. In a related discussion the authors question the reliability of using M^* as a cosmological distance indicator.

160.066 **Galaxy content of the Hydra I and Fornax clusters of galaxies.** A. Wirth, J. S. Gallagher.
Astrophys. J., Vol. 242, 469 - 485 (1980).

A detailed study of morphological types for galaxies in the centrally concentrated Hydra I and irregular Fornax clusters of galaxies provides the basis for a determination of the fractional population of spirals and ellipticals within each cluster. The spiral content of Hydra I is similar to that derived from color studies of distant concentrated clusters. The less dense Fornax cluster has both a larger spiral content and more normal spirals. As both clusters are nearby, an examination of the low surface-brightness and dwarf members also proved possible.

160.067 **X-ray emission – an ageing effect of galaxy clusters.** K.-H. Schmidt.
Astron. Nachr., Band 301, 297 - 299 (1980) = Mitt. Univ. Sternw. Jena, Nr. 145.

The observed X-ray luminosity and temperature of clusters of galaxies are correlated with the dynamical evolution modulus, W, defined by von Hoerner (1976). Obviously X-ray emission is an ageing effect of galaxy clusters.

160.068 **Very faint blue objects in the Virgo cluster region.** F. Börngen.
Astron. Nachr., Band 301, 305 - 310 (1980).

Finding charts, coordinates, $U-B$ color estimates and approximate B magnitudes are given for 95 very faint blue objects located near the centre of the Virgo cluster of galaxies.

160.069 **Diffuse radio emission in the Coma cluster and Abell 1367: observations at 430 and 1400 MHz.**
R. J. Hanisch.
Astron. J., Vol. 85, 1565 - 1576 (1980).

Two rich clusters of galaxies, Abell 1656 (the Coma cluster) and Abell 1367, have been mapped at both 430 and 1400 MHz. The diffuse radio sources in these clusters have been isolated by correcting the observed maps for point sources measured in interferometer surveys at Cambridge and Westerbork. The author has investigated the implications of the results on the various theories concerning particle streaming and the heating of the intracluster gas by relativistic electrons.

160.070 **The infrared color-magnitude relations for the Virgo and Coma clusters: new evidence for intermediate age effects?** M. Aaronson.
Publ. Astron. Soc. Pacific, Vol. 92, 547 (1980). – Abstract.

160.071 **Some parameters of the Coma cluster of galaxies (A1656).** A. A. Klypin.
Astron. Tsirk., No. 1105, p. 3 - 5 (1980). In Russian.

160.072 **Photographic surface photometry of the brightest galaxies in poor clusters.**
T. X. Thuan, W. Romanishin.
Photometry, kinematics and dynamics of galaxies, (see 012.051), p. 63 - 67 (1979).

160.073 **CCD camera photometry of cD galaxies in poor clusters.** R. E. Schild, T. C. Weekes.
Photometry, kinematics and dynamics of galaxies, (see 012.051), p. 69 - 74 (1979).

The authors have determined the cutoff radii for five cD galaxies in poor groups from CCD camera images. They find that the cutoff radii, a useful description of the galaxy sizes, are around 75 kpc. These are approximately 1/10 the value derived for the cD galaxies in rich clusters.

160.074 **Neutral hydrogen observations of spiral galaxies in clusters.** W. T. Sullivan, III, R. A. Schommer, G. D. Bothun.
Photometry, kinematics and dynamics of galaxies, (see 012.051), p. 231 - 234 (1979).

The authors' motivation is to compare the H I gas content and related properties of spirals in different types of clusters with galaxies outside clusters.

160.075 **Clusters of galaxies.** R. Giacconi.
X-ray and gamma-ray astronomy in the 1980's, (see 012.054), p. 33 - 35 (1979).

160.076 **3-d nearest neighbours test for Abell clusters of galaxies.**
M. Kalinkov, V. N. Dermendjiev, I. F. Kuneva.
Dokl. Bolg. AN, Vol. 33, No. 1, p. 7 - 10 (1980). – Abstr. in Ref. zh., 51. Astron., 12.51.634 (1980).

160.077 **The structure and evolution of X-ray clusters of galaxies.** C. Jones.
X-ray astronomy, (see 012.055), p. 153 - 170 (1980).

The Einstein observations of nearby X-ray clusters have shown a surprising complexity and variety in their surface brightness distribution. Much of this variety can be ordered by associating the different types of cluster morphology with various stages in their dynamic evolution.

160.078 **The X-ray spectra of clusters of galaxies.** R. Mushotzky.
X-ray astronomy, (see 012.055), p. 171 - 179 (1980).

Spectral observations of clusters of galaxies have demonstrated that the X-ray emission is due to a hot evolved intracluster gas with ~ 10% of the virial mass of the cluster. In low luminosity clusters there is an indication that there is thermal structure that may be related to the X-ray spatial structure. In the Perseus cluster there is evidence for cooling at the core and therefore for evolution of the cluster gas.

160.079 **Einstein observations of the Virgo cluster.** W. Forman.
X-ray astronomy, (see 012.055), p. 181 - 195 (1980).

The author discusses first the X-ray emission from M87 and then proceeds to discuss the individual galaxies of the Virgo cluster which he observes and their interaction with the diffuse cluster gas.

160.080 **Cluster parameters.** G. Chincarini.
X-ray astronomy, (see 012.055), p. 197 - 216 (1980).
The author discusses the cluster environment and the parameters which can be derived using the isothermal, or King, profile. He gives structural parameters derived from some of the published data. The appendix deals with the richness parameter.

160.081 **Models of X-ray emission from clusters of galaxies.** A. Cavaliere.
X-ray astronomy, (see 012.055), p. 217 - 237 (1980).

160.082 **X-ray structure of the Coma cluster from optical data.** J. Binney.
X-ray astronomy, (see 012.055), p. 239 - 243 (1980).

160.083 **X-ray emission from clusters and the formation of galaxies.** J. Binney.
X-ray astronomy, (see 012.055), p. 245 - 251 (1980).

160.084 **The HEAO A-2 survey of Abell clusters and the X-ray luminosity function.**
J. D. McKee, R. F. Mushotzky, E. A. Boldt, S. S. Holt, F. E. Marshall, S. H. Pravdo, P. J. Serlemitsos.
Astrophys. J., Vol. 242, 843 - 856 (1980).
The authors have examined the HEAO A-2 all-sky data base for 2–10 keV X-ray emission from the 225 Abell clusters of galaxies listed in Abell's (1958) catalog. They report on twelve new identifications of clusters with X-ray sources as well as on the correlation of the cluster X-ray luminosities with Bautz-Morgan type and richness, and present a new X-ray luminosity function for Abell clusters.

160.085 **The revised HEAO A-2 X-ray luminosity function for Abell clusters.**
P. Hintzen, J. S. Scott, J. D. McKee.
Astrophys. J., Vol. 242, 857 - 860, plates 19 - 20 (1980).
Spectroscopic redshifts are determined for Abell clusters in the HEAO A-2 X-ray sample. These data are used to correct the cluster X-ray luminosity function of McKee et at.

160.086 **Dynamical models and the mass of the Virgo cluster.** G. L. Hoffman, D. W. Olson, E. E. Salpeter.
Astrophys. J., Vol. 242, 861 - 878 (1980).
Dynamical model calculations for a supercluster of galaxies are carried out, assuming spherical symmetry and ignoring two-body collisions. Observational data are analyzed for the dispersion of systemic galaxy velocities in and near the Virgo cluster. This velocity dispersion decreases with increasing angular distance θ from the Virgo center. Allowing the possibility that the mass-to-light ratio may vary with θ, the observational data are fitted against numerical models with Ω (the present mass density at spatial infinity divided by the closure density) ranging from 0.03 to 0.7. The derived gravitational mass of the Virgo I cluster is almost independent of Ω and is well determined.

X-ray astronomical spectroscopy.
See Abstr. 031.567.

Astronomical consequences of the neutrino rest mass. III. The nonlinear stage of evolution of perturbations and the hidden mass. See Abstr. 061.002.

Flow past a massive object and the gravitational drag. See Abstr. 062.030.

Microwave background radiation in the directions to clusters of galaxies. See Abstr. 066.016.

Angular distribution of the microwave background and its intensity in the directions of clusters of galaxies. See Abstr. 066.052.

A radio continuum survey at 1.4 GHz of the galaxies in the Virgo region. See Abstr. 141.013.

Analysis of quasars found in the CTIO Curtis Schmidt survey in the −40° zone. See Abstr. 141.054.

Distorted radio sources in Abell 2255: evidence of intergalactic gas 2.5 to 5 megaparsecs from the cluster center. See Abstr. 141.057.

3C 206: a resolved quasar in a cluster of galaxies. See Abstr. 141.062.

Quasars, isotropy of H_0 and the Local Supercluster of galaxies. See Abstr. 141.088.

Ca II absorption lines in the spectrum of the quasar PKS 2020-370 due to galactic material in the group Klemola 31. See Abstr. 141.140.

Evidence for the location of quasars in superclusters. See Abstr. 141.158.

Ca II absorption lines in the spectrum of the quasar PKS 2020–370 due to galactic material in the group Klemola 31. See Abstr. 141.182.

Protogalactic explosions and intracluster chemical enrichment. See Abstr. 151.042.

Evolution of homogeneous ellipsoids and the structure of rich clusters of galaxies. See Abstr. 151.044.

On the dynamical evolution of clusters of non-point gravitating bodies. See Abstr. 151.047.

Application of the virial theorem to clusters of galaxies with unseen mass. See Abstr. 151.052.

The morphology of galaxies in dense clusters – nature or nurture? See Abstr. 151.078.

Computer simulations of environmental influences on galaxy evolution in dense clusters. I. Ram-pressure stripping. See Abstr. 151.080.

Über den Aufbau der Stern- und Galaxienhaufen.
See Abstr. 153.014.

Globular clusters as extragalactic distance indicators. See Abstr. 154.037.

Double galaxies as indicators of large scale structure.
See Abstr. 158.073.

Three new cases of galaxies with large discrepant redshifts. See Abstr. 158.074.

Neutral-hydrogen observations of smooth-arm spiral galaxies. See Abstr. 158.102.

Colors and magnitudes predicted for high redshift galaxies. See Abstr. 158.115.

Late-type galaxies with extended envelopes of neutral hydrogen. See Abstr. 158.175.

Evolution effects in the distribution of distant galaxies? See Abstr. 158.194.

CCD camera observations of the Local Group galaxy LGS-3. See Abstr. 158.218.

Hot gas in clusters of galaxies. See Abstr. 161.004.

High resolution X-ray spectroscopy of the gas surrounding M87 and NGC 1275: emission line detection and evidence for radiatively regulated accretion. See Abstr. 161.005.

On the existence of (nonluminous) intergalactic H I clouds in the Virgo cluster. See Abstr. 161.006.

Hubble constant in the local region. See Abstr. 162.026.

The problem of the Einasto and Ostriker team hypothesis on a model of galaxies. See Abstr. 162.062.

Revisions in the extragalactic distance scale. See Abstr. 162.085.

161 Intergalactic Matter

161.001 **Absorption effects due to intergalactic long whiskers of pyrolytic graphite and the cosmic microwave background.** N. C. Rana.
Astrophys. Space Sci., Vol. 71, 123 - 133 (1980).

Earlier (Rana, 1979, 1980) it was shown that an intergalactic medium containing natural graphite whiskers could not adequately thermalize the ambient radiations to generate 3 K microwave background. Now the author has carried out a similar investigation with whiskers of pyrolytic graphite. Provided the abundance is about 10^{-34} g cm^{-3}, the model is capable of thermalizing the background. Some of the observational consequences have been studied with reference to extragalactic astronomy and quasistellar objects. No conflicting evidence has been found so far.

161.002 **The effects of X-ray absorption on the spectra of distant objects.** P. R. Shapiro, J. N. Bahcall.
Astrophys. J., Vol. 241, 1 - 24 (1980).

The authors have calculated in detail the X-ray absorption spectrum above 0.1 keV that would be introduced into the continuous X-ray spectrum of a quasar by an intervening uniform, hot ($T \gtrsim 10^6$ K), intergalactic gas with a small admixture of atoms of C, N, O, Ne, Mg, Si, S and Fe. The results indicate that soft X-ray absorption can be appreciable for all quasar X-ray sources for a significant range of IGM temperatures, densities, heavy-element abundances, and observed photon energies. A brief comparison with the preliminary results of the Einstein Observatory quasar sample is made. A table of minimum detectable densities and abundances for each $T \gtrsim 10^6$ K and for certain T's for each ion is calculated for a representative broad-band soft X-ray measurement (between 0.1 and 6 keV) of a quasar at $z = 3$. The authors also consider the possibility that gas "clumped" on noncosmological scales produces observable absorption lines and/or edges in quasar X-ray spectra.

161.003 **The intergalactic medium.** G. B. Field.
Highlights of Astronomy, Vol. 5, (see 012.025), 375 - 386 (1980).

Observations of Lyman-α absorption show that any gas present in the space between clusters of galaxies must be hot, and X-ray observations indicate that there may be present gas whose temperature at low redshifts is 4×10^8 K and whose density is about half the critical value for cosmology. The interpretation of these observations is reviewed, and difficulties with proposed heating mechanisms are emphasized.

161.004 **Hot gas in clusters of galaxies.** J. L. Culhane.
Highlights of Astronomy, Vol. 5, (see 012.025), 387 - 396 (1980).

The author discusses the evidence for the presence of hot plasma in clusters of galaxies and describes its properties. He examines the available models for the origin and heating of the gas and discusses the evidence for the presence of a gas of lower temperature associated with certain galaxies. He discusses briefly the interactions between the hot plasma with the cluster galaxies.

161.005 **High resolution X-ray spectroscopy of the gas surrounding M87 and NGC 1275: emission line detection and evidence for radiatively regulated accretion.**
C. R. Canizares, C. Berg, G. Clark, J. G. Jernigan, G. Kriss, T. H. Markert, M. Schattenburg, P. F. Winkler.
Highlights of Astronomy, Vol. 5, (see 012.028), 657 - 662 (1980).

The authors have detected several X-ray emission lines from the vicinity of M87 in the Virgo Cluster and NGC 1275 in Perseus. The lines are indicative of material which is cooler than the bulk of the intracluster gas. This material is most likely accreting onto the central galaxy with the accretion rate controlled by the rate of radiative cooling.

161.006 **On the existence of (nonluminous) intergalactic H I clouds in the Virgo cluster.**
C. Wetherill, W. T. Sullivan III, T. Heckman.
Publ. Astron. Soc. Pacific, Vol. 92, 551 (1980). – Abstract.

161.007 **Intergalactic neutral hydrogen: tidal debris and discrete clouds.** M. P. Haynes.
Photometry, kinematics and dynamics of galaxies, (see 012.051), p. 219 - 222 (1979).

161.008 **Distribution of the interstellar scattering medium in the Galaxy.** V. I. Altunin.
Astron. Zh., Tom 57, 1174 - 1186 (1980). In Russian.
English translation in Soviet Astron., Vol. 24, No. 6.

It is shown that the scattering medium discovered through the scintillations of the radio emission of galactic and extragalactic sources is distributed along the radius of the Galaxy inhomogeneously. The space distribution, the magnitude of electron density fluctuations, a large dispersion of the velocities of scattering knots suggest that a strongly turbulent medium is concentrated in H II regions and supernova remnants.

Ultraviolet spectroscopy of interstellar and intergalactic matter. See Abstr. 032.571.

Flow past a massive object and the gravitational drag. See Abstr. 062.030.

The quasar 3C351: VLA maps and a deep search for optical emission in the outer lobes. See Abstr. 141.011.

Observations of quasars with the International Ultraviolet Explorer satellite. See Abstr. 141.055.

Distorted radio sources in Abell 2255: evidence of intergalactic gas 2.5 to 5 megaparsecs from the cluster center. See Abstr. 141.057.

QSO evolution and the intergalactic medium. See Abstr. 141.086.

Turbulence-related morphology in extragalactic radio sources. See Abstr. 141.104.

Quasar Lα absorbers: are precise conclusions possible? See Abstr. 141.146.

Galactic and extragalactic contributions to the far-ultraviolet background. See Abstr. 142.053.

Kα-lines in the X-ray background spectrum and interstellar gas in galaxies. See Abstr. 142.084.

Protogalactic explosions and intracluster chemical enrichment. See Abstr. 151.042.

The dynamical evolution of NGC 5128. See Abstr. 158.207.

The heating of gas in clusters of galaxies by relativistic electrons: collective effects. See Abstr. 160.024.

Broadband 21-cm H I emission from Stephan's Quintet. See Abstr. 160.027.

Kelvin-Helmholtz instability in clusters of galaxies. See Abstr. 160.029.

On the morphology of the magnetic field in galaxy clusters. See Abstr. 160. 063.

X-ray emission from clusters and the formation of galaxies. See Abstr. 160.083.

162 Universe (Structure, Evolution)

162.001 **Two-dimensional simulation of the gravitational system dynamics and formation of the large-scale structure of the Universe.** A. G. Doroshkevich, E. V. Kotok, I. D. Novikov, A. N. Polyudov, S. F. Shandarin, Yu. S. Sigov.
Mon. Not. R. Astron. Soc., Vol. 192, 321 - 337 (1980).

The results of a numerical experiment are given that describe the non-linear stages of the development of perturbations in gravitating matter density in the expanding Universe. This process simulates the formation of the large-scale structure of the Universe from an initially almost homogeneous medium. In the one- and two-dimensional cases of this numerical experiment the evolution of the system from 4096 point masses that interact gravitationally only was studied with periodic boundary conditions (simulation of the infinite space). The initial conditions were chosen that resulted from the theory of the evolution of small perturbations in the expanding Universe. The results of numerical experiments are systematically compared with the approximate analytic theory.

162.002 **Stars, galaxies, cosmos: the past decade, the next decade.** V. C. Rubin.
Science, Vol. 209, 63 - 71 (1980).

162.003 **Cosmology with a material vacuum.** S. V. M. Clube.
Irish Astron. J., Vol. 14, 51 - 54 (1979).

162.004 **Cosmological fluctuations produced near a singularity.** Ya. B. Zeldovich (*Zel'dovich*).
Mon. Not. R. Astron. Soc., Vol 192, 663 - 667 (1980).

The perturbations of a uniform Friedmannian universe, leading to galaxy formation, are explained by the strings formed during the symmetry loss of vacuum of a complex Higgs field with mass characteristic of grand unification. Difficulties are pointed out inherent to phase transition, particle decay and black hole evaporation as sources of growing perturbations.

162.005 **Fluctuations in the cosmic background radiation produced by evolving hierarchical cosmologies.** C. J. Hogan.
Mon. Not. R. Astron. Soc., Vol. 192, 891 - 903 (1980).

Thorough investigation of anisotropies in the cosmic background radiation over a wide range of angles and wavelengths might reveal a rich structure which contains many clues about the course of cosmic evolution during epochs which have so far been inaccessible to direct observation, $20 \lesssim z \lesssim 1000$. Such complex fluctuations are a general feature of cosmological models where matter is hot and inhomogeneous at high z, but the magnitude of the effect depends on the amount of radiative activity and on the amount of intervening scattering. A positive measurement of fluctuations increasing toward small θ, or varying with λ, would be unambiguous evidence for pregalactic activity.

162.006 **Non-standard cosmologies.** J. V. Narlikar, A. K. Kembhavi.
Fundam. Cosmic Phys., Vol. 6, 1 - 186 (1980).

Contents: The general theory of relativity and the standard cosmology. Newtonian cosmology. The steady state theory. The Brans-Dicke theory. Dirac cosmology. The Hoyle-Narlikar cosmologies. Matter-antimatter cosmologies. An assorted collection of non-standard cosmologies. The observational tests.

162.007 **The growth of density perturbations in Friedmann model universes with a decoupled radiation field.** A. G. Emslie.
Astrophys. Space Sci., Vol. 71, 363 - 370 (1980).

The author examines the linear growth of density perturbations in homogeneous isotropic (Friedmann) model universes, including the effect of a decoupled radiation pressure field in the modelling. Amplification factors for density perturbations in all models are derived numerically, and it is shown that the effect of radiation pressure is to decelerate the growth of such condensations, thus requiring larger inhomogeneities to be produced at radiation decoupling in order to produce protogalaxies.

162.008 **Evolution of plane-symmetric self-similar space-times containing perfect fluid.** I. S. Shikin.
Gen. Relativ. Gravitation, Vol. 11, 433 - 451 (1979).

162.009 **On Wheeler's "rule of unanimity" in quantum cosmology.** J. Demaret.
Gen. Relativ. Gravitation, Vol. 11, 453 - 463 (1979).

Counterexamples are given to Wheeler's "rule of unanimity", which implies that every quantum solution for the problem of a closed universe within the framework of Einstein's field equations leads to a singularity.

162.010 **Application of the propagator method to pair production in the Robertson-Walker metric.** M. B. Mensky, O. Yu. Karmanov.
Gen. Relativ. Gravitation, Vol. 12, 267 - 277 (1980).

162.011 **Entropy perturbations in low-density universe models.** K. Tomita.
Publ. Astron. Soc. Japan, Vol. 32, 179 - 184 (1980).

In low-density universe models the gravitational growth of density perturbations are slower than that in the Einstein–de Sitter critical model, and larger amplitudes are required at the decoupling time, in order that protoclusters may form to the present epoch. In the paper it is shown that, if the thermal model without secondary heating is assumed, the required amplitudes of entropy perturbations exceed the observational limit derived from the high degree of isotropy of the cosmic background radiation, as well as those of adiabatic perturbations.

162.012 **Astronomical consequences of the neutrino rest mass. I. The universe.** Ya. B. Zel'dovich, R. A. Syunyaev.
Pis'ma Astron. Zh., Tom 6, 451 - 456 (1980). In Russian.
English translation in Soviet Astron. Lett., Vol. 6.

The measurements of the tritium decay electrons spectrum (Lyubimov et al., 1980) give evidence for the existence of the electronic neutrino rest energy $m_\nu c^2 \approx 30$ eV. In this case the relic neutrinos determine the average matter density of the universe, and our universe turns out to be closed. If further reactor and accelerator experiments lead to the discovery of a rest mass of muonic and tau neutrinos of the same order of magnitude, the age of Friedmann universe would become less than 10^{10} years. This contradicts the age of the oldest stars of the Galaxy. Introducing the cosmological constant one can eliminate this contradiction.

162.013 **Astronomical consequences of the neutrino rest mass. II. Spectrum of density perturbations and small-scale fluctuations of the microwave background.** A. G. Doroshkevich, Ya. B. Zel'dovich, R. A. Syunyaev, M. Yu. Khlopov.
Pis'ma Astron. Zh., Tom 6, 457 - 464 (1980). In Russian.
English translation in Soviet Astron. Lett., Vol. 6.

The neutrino rest mass determines in the universe the

scale of the order of the distance between clusters of galaxies. The amplitude of the neutrino density perturbations of smaller scales decreases as a high degree of the perturbation wavelength. The evolution of adiabatic and isothermal density perturbations is considered.

162.014 **Tenacious myths about cosmological perturbations larger than the horizon size.**
W. H. Press, E. T. Vishniac.
Astrophys. J., Vol. 239, 1 - 11 (1980).

The authors review the linear perturbation theory of the Einstein–de Sitter (k = 0, Friedmann) big-bang cosmology in synchronous gauge, taking particular care to distinguish physical perturbations, which are locally measurable, from pure-gauge perturbations, which correspond to an unperturbed spacetime written in gauge-perturbed coordinates. Some new results are obtained about the growth of physical perturbations at early times, while they are still outside their horizon; and some commonly accepted rules for estimating the growth and decay of perturbations are shown to be false.

162.015 **Uniformity of the radial distribution of quasars in the chronometric cosmology and the X-ray background.** I. E. Segal, J. Loncaric, W. Segal.
Astrophys. J., Vol. 239, 38 - 41 (1980).

The claim by Green and Schmidt for cosmology independence in their determination of strong evolution in the radial distribution of quasars requires revision.

162.016 **Isotropy of the Universe and equation of state of vacuum.** R. Farbri.
Nuovo Cimento B, Ser. 11, Vol. 58B, 125 - 136 (1980).
Abstr. in Phys. Abstr., Vol. 83, Abstr. 73083 (1980).

162.017 **A proposed optical test of preferred frame cosmologies.** M. P. Haugan, M. O. Scully, K. Just.
Phys. Lett. A, Vol. 77A, 88 - 90 (1980). – Abstr. in Phys. Abstr., Vol. 83, Abstr. 77531 (1980).

162.018 **Spontaneous symmetry breaking and the expansion rate of the early universe.**
E. W. Kolb, S. Wolfram.
Astrophys. J., Vol. 239, 428 - 432 (1980).

Gauge theories for weak interactions which employ the Higgs mechanism for spontaneous symmetry breakdown imply that there should exist a large vacuum energy associated with the Higgs scalar field condensate. A cosmological term in Einstein's field equations can be arranged to remove the unobserved gravitational effect of this vacuum energy in the present universe. However, in the early universe, the spontaneously broken symmetry should have been restored, leaving the cosmological term uncanceled. In this paper the authors investigate the conditions necessary for the uncanceled cosmological term to be dynamically important in the early universe.

162.019 **Primordial nucleosynthesis and Dirac's Large Numbers Hypothesis.** V. Canuto, S.-H. Hsieh.
Astrophys. J., Lett., Vol. 239, L91 (1980).

It is shown that a recent analysis concerning the amount of primordial helium produced within the scale covariant cosmology is based on two invalid extrapolations.

162.020 **The Hubble diagram of QSOs by statistical method.**
S. Cao, X. Xiao, Y. Liu, F. Cheng, L. Yang.
Kexue Tongbao, Vol. 25, 743 - 747 (1980).

The authors construct a Hubble diagram by means of the redshift and apparent magnitude data of 1022 QSOs. They conclude that the redshifts are of cosmological nature.

162.021 **Chaotic cosmologies and the topology of the universe.** J. R. Gott III.
Mon. Not. R. Astron. Soc., Vol. 193, 153 - 169 (1980).

Topologically multiply connected universes with finite volume naturally produce the special initial conditions required by the Rees chaotic cosmology model. One can set lower limits on the proper radius of the fundamental volume in such a model by our failure to find multiple images of galaxy clusters in the Shane-Wirtanen sample, i.e. $R_H >$ 400 $(H_0/50 \text{ km s}^{-1} \text{ Mpc}^{-1})^{-1}$Mpc. For $\Omega_0 \neq 1$ models topological constraints dictate even larger values of R_H. Because of these observational and topological constraints, multiply connected universes with chaotic initial conditions are not capable of thermalizing the cosmic black-body radiation with normal thermalization processes.

162.022 **Entropy perturbations and cosmogonic processes in the hot universe.** A. D. Chernin, A. S. Zentsova.
Astron. Astrophys., Vol. 89, 1 - 5 (1980).

The authors discuss the evolution of primordial structure in the early universe and concentrate on the specific properties of entropy perturbations in the hot cosmic medium. This study is stimulated by Parijskij's recent report on observational constraints for possible temperature fluctuations in the microwave background radiation. The authors explore the implications of these data and show that these data together with gasdynamic considerations lead one to favour entropy perturbations as preferred elements of the primordial cosmic structure.

162.023 **Cosmological gravitational waves: their origin and consequences.** B. J. Carr.
Astron. Astrophys., Vol. 89, 6 - 21 (1980).

The author considers the origin and consequences of a cosmological background of gravitational radiation. Such a background might be primordial, in the sense that it goes back to the beginning of the Universe or was produced by quantum processes at the Planck time. Alternatively, it might have been generated at some finite time in the past by, for example, the formation of a population of black holes. Cosmological gravitational waves can be described by their density (Ω_g) and a characteristic period (P_0), in which case these parameters uniquely determine their cosmological effects and observational consequences. Primordial waves would have had an important effect on the dynamics of the early Universe even if Ω_g is small and they may have rendered it chaotic. The author discusses various scenarios for the generation of a gravitational wave background and finds that, in principle, Ω_g could be as large as 10^{-2} over periods in the range 10^{-3} s to 10^5 s.

162.024 **The effect of variability on the V/V_{MAX}- test.**
J. N. Bahcall.
Astrophys. J., Vol. 240, 377 - 383 (1980).

Knowledge of the distribution of quasars in space has been obtained primarily from applications of the V/V_{MAX}-test to complete flux-limited samples of sources. The intrinsic luminosities of many quasars are variable. Other objects, such as X-ray or γ-ray sources, for which one might wish to use the V/V_{MAX}-test are also variable. However, previous formulations of the V/V_{MAX}-test have not included the effects of variations. The purpose of the paper is to describe and estimate the effects of source variability (or random photometric errors) on the V/V_{MAX}-test. The approximate sizes of the effects considered are determined by solving analytically some specific examples, as well as by exploiting simple lemmas.

162.025 **Time dependence of the cosmological redshift in Friedmann universes.** R. Rüdiger.
Astrophys. J., Vol. 240, 384 - 386 (1980).

The author generalizes a formula of Ebert and Trümper which gives the time dependence of the cosmological redshift for certain special Friedmann universes to all Friedmann universes.

162.026 **Hubble constant in the local region.**
N. Visvanathan.
Proc. Astron. Soc. Australia, Vol. 3, 309 - 311 (1979).

162.027 **Some perfect fluid cosmological models of plane symmetry with incident magnetic field.**
S. R. Roy, O. P. Tiwari.
Indian J. Pure Appl. Math., Vol. 11, 609 - 617 (1980).
Abstr. in Phys. Abstr., Vol. 83, Abstr. 82556 (1980).

162.028 **Space-time metrical fluctuations induced by cosmic turbulence.** G. Rosen.
Nuovo Cimento B, Ser. 11, Vol. 57B, 125 - 130 (1980).
Abstr. in Phys. Abstr., Vol. 83, Abstr. 82557 (1980).

162.029 **Baryon asymmetry of the Universe versus left-right symmetry.**
V. A. Kuzmin (*Kuz'min*), M. E. Shaposhnikov.
Phys. Lett. B, Vol. 92B, 115 - 118 (1980). – Abstr. in Phys. Abstr., Vol. 83, Abstr. 82559 (1980).

162.030 **Are the protons to stay with us for ever?**
D. Falik.
Phys. Lett. B, Vol. 93B, 74 (1980). – Abstr. in Phys. Abstr., Vol. 83, Abstr. 82560 (1980).

162.031 **Local inhomogeneities in a Robertson-Walker background. I. General framework.** K. Lake.
Astrophys. J., Vol. 240, 744 - 750 (1980).
A complete generalization of the "Swiss cheese" type of locally inhomogeneous cosmologies is given. Neither the explicit form of the spherically symmetric interior metric nor the spatial curvature and equation of state of the Robertson-Walker background is restricted a priori. The history, mass, and mass growth rate of any timelike inhomogeneity is developed in terms of a single function characteristic of the inhomogeneity. Recent results which have generalized the standard "Swiss cheese" case to a Vaidya interior metric follow immediately from the framework given in the paper.

162.032 **Cosmological solution with matter in a new theory of gravitation.**
G. Kunstatter, J. W. Moffat, P. Savaria.
Canadian J. Phys., Vol. 58, 729 - 736 (1980). – Abstr. in Phys. Abstr., Vol. 83, Abstr. 86021 (1980).

162.033 **Some recent tests of the chronometric cosmology.**
I. E. Segal.
Proc. Natl. Acad. Sci. USA, Vol. 77, 10 - 13 (1980). – Abstr. in Phys. Abstr., Vol. 83, Abstr. 86022 (1980).

162.034 **Neutrinos in cosmological models.** J. R. Ray.
Prog. Theor. Phys., Vol. 63, 1213 - 1216 (1980).
Abstr. in Phys. Abstr., Vol. 83, Abstr. 86023 (1980).

162.035 **Nonhomogeneous distributions in the Brans-Dicke cosmology.** A. B. Batista.
Phys. Rev. D, Vol. 21, 2119 - 2121 (1980). – Abstr. in Phys. Abstr., Vol. 83, Abstr. 86024 (1980).

162.036 **Quantum effects in the early universe. III. Dissipation of anisotropy by scalar particle production.**
J. B. Hartle, B. L. Hu.
Phys. Rev. D, Vol. 21, 2756 - 2769 (1980). – Abstr. in Phys. Abstr., Vol. 83, Abstr. 86025 (1980).

162.037 **Classical predictive electrodynamics in a conformally flat universe. The case of the Einstein-de Sitter universe.** V. Iranzo, R. Lapiedra.
Phys. Rev. D, Vol. 21, 3299 - 3304 (1980). – Abstr. in Phys. Abstr., Vol. 83, Abstr. 86175 (1980).

162.038 **Cosmological model with gravitational, electromagnetic, and scalar waves.** Ch. Charach, S. Malin.
Phys. Rev. D, Vol. 21, 3284 -3294 (1980). – Abstr. in Phys. Abstr., Vol. 83, Abstr. 86188 (1980).

162.039 **Cosmological amount of baryons.**
T. Yanagida, M. Yoshimura.
Nucl. Phys. B, Vol. B168, 534 - 548 (1980). – Abstr. in Phys. Abstr., Vol. 83, Abstr. 86640 (1980).

162.040 **Are grand unified theories compatible with standard cosmology?** M. B. Einhorn, D. L. Stein, D. Toussaint.
Phys. Rev. D, Vol. 21, 3295 - 3298 (1980). – Abstr. in Phys. Abstr., Vol. 83, Abstr. 86647 (1980).

162.041 **Cosmological f-g fields relevant to quark confinement.** M. Gurses.
J. Phys. A, Vol. 13, L223 - L225 (1980). – Abstr. in Phys. Abstr., Vol. 83, Abstr. 86659 (1980).

162.042 **Particle creation and vacuum polarization in an isotropic universe.** A. A. Grib, S. G. Mamayev (*Mamaev*), V. M. Mostepanenko.
J. Phys. A, Vol. 13, 2057 - 2065 (1980). – Abstr. in Phys. Abstr., Vol. 83, Abstr. 90582 (1980).

162.043 **Massive particle production in anisotropic space-times.** N. D. Birrell, P. C. W. Davies.
J. Phys. A, Vol. 13, 2109 - 2120 (1980). –Abstr. in Phys. Abstr., Vol. 83, Abstr. 90583 (1980).

162.044 **Symmetries and metrics of homogeneous cosmologies.** M. A. Melvin.
J. Math. Phys., Vol. 21, 1938 - 1951 (1980). – Abstr. in Phys. Abstr., Vol. 83, Abstr. 90584 (1980).

162.045 **Spatially homogeneous neutrino cosmologies.**
T. R. Michalik, M. A. Melvin.
J. Math. Phys., Vol. 21, 1952 - 1964 (1980). – Abstr. in Phys. Abstr., Vol. 83, Abstr. 90585 (1980).

162.046 **Cosmic matter-antimatter asymmetry and gravitational force.** J. P. Hsu.
Nuovo Cimento, Lett., Ser.2, Vol. 28, 128 - 132 (1980).
Abstr. in Phys. Abstr., Vol. 83, Abstr. 90586 (1980).

162.047 **On a possibility to close the Universe.**
T. Grabinska, M. Zabierowski.
Nuovo Cimento, Lett., Ser. 2, Vol. 28, 139 - 140 (1980).
Abstr. in Phys. Abstr., Vol. 83, Abstr. 90587 (1980).

162.048 **Time variation of atomic masses and expansion of the Universe.** T. L. Chow.
Nuovo Cimento, Lett., Ser. 2, Vol. 28, 158 - 160 (1980).
Abstr. in Phys. Abstr., Vol. 83, Abstr. 90588 (1980).

162.049 **Cosmological baryon production in a 'superconducting' early Universe.**
R. N. Mohapatra, G. Senjanovic.
Phys. Rev. D, Vol. 21, 3470 - 3473 (1980). – Abstr. in Phys. Abstr., Vol. 83, Abstr. 90592 (1980).

162.050 **Vacuum stress-energy tensor and particle creation in isotropic cosmological models.**
A. A. Grib, S. G. Mamayev (*Mamaev*), V. M. Mostepanenko.
Fortschr. Phys., Vol. 28, 173 - 199 (1980). In German.
Abstr. in Phys. Abstr., Vol. 83, Abstr. 94901 (1980).

162.051 **Polarization effects in cosmological models with anisotropic curvature.** R. Fabbri, R. A. Breuer.

Nuovo Cimento B, Ser. 11, Vol. 58B, 113 - 122 (1980). Abstr. in Phys. Abstr., Vol. 83, Abstr. 94903 (1980).

162.052 **Qualitative analysis of homogeneous universes.** M. Novello, R. A. Araujo.
Phys. Rev. D, Vol. 22, 260 - 266 (1980). – Abstr. in Phys. Abstr., Vol. 83, Abstr. 94904 (1980).

162.053 **Kinetic modelling of gravitating systems in the expanding universe.**
A. G. Doroshkevich, Eh. V. Kotok, I. D. Novikov, A. N. Polyudov, Yu. S. Sigov, S. F. Shandarin.
Chisl. modelir. kollektivn. protsessov v plazme. Moskva, 1980, p. 224 - 256. In Russian. – Abstr. in Ref. zh., 51. Astron. 8.51.760 (1980).

162.054 **Models of the universe compatible with present observations.** R. Bouigue.
Astrophys. Space Sci., Vol. 72, 87 - 96 (1980).

The author investigates the various models of the universe which fit the results obtained, or suggested, by recent observations and which allow a Planck distribution for the residual cosmic radiation.

162.055 **Tetrad vector fields and the formation of protogalaxies.** F. I. Mikhail, M. I. Wanas.
J. Astron. Soc. Egypt, Vol. 1, 1 - 6 (1979).

The possibility of forming spherical condensation in a world model is being studied. The world model used is a tetrad analogue of Hoyle-Narlikar's model (1963). Condensations formed can be considered as protogalaxies in the model.

162.056 **The gravitational fragmentation of primordial gas clouds.** J. E. Tohline.
Astrophys. J., Vol. 239, 417 - 427 (1980).

Analytic stability analyses performed by other investigators on collapsing, pressure-free spheres seem to indicate that gas clouds are susceptible to gravitational fragmentation on a free-fall time scale. In this paper, time scale arguments are used to show that clouds collapsing from configurations near the Jeans limit are not susceptible to gravitational fragmentation in the manner described by these earlier analyses; the self-gravity of perturbations alone cannot drive the fragmentation process to completion in a single free-fall time. Serious objections are raised to models of "opacity limited star formation" insofar as they attempt to predict a lower mass limit to stars which form in a dynamically collapsing gas cloud. In light of these objections to earlier models, a reevaluation of the fragmentation process in primordial gas clouds is made.

162.057 **Origin of the structure of the Universe.**
G. S. Bisnovatyi-Kogan (*Bisnovatyj-Kogan*), V. N. Lukash, I. D. Novikov.
Variability in stars and galaxies, (see 012.044), p. G.1.1 - 1.20 (1980).

The origin of structure in the homogeneous isotropic Universe from quantum fluctuations of the matter density is discussed. A theory of gravitational instability of the hot Universe is analyzed when collisionless neutrinos with the nonzero rest mass are taken into account. The problem of relic radiation anisotropy is considered.

162.058 **Cosmology with non-zero neutrino rest mass.**
G. S. Bisnovatyj-Kogan, I. D. Novikov.
Astron. Zh., Tom 57, 899 - 902 (1980). In Russian. – English translation in Soviet Astron., Vol. 24, No. 5.

If the rest mass of the neutrino would be equal to $m_{\nu 0} \approx 30$ eV, then the average density of the universe would be determined by the neutrino and would be close to the critical one. The evolution of perturbations is considered in the neutrino component and in the matter. Development of the perturbations leads to the formation of galactic clusters and a neutrino halo. The characteristic visible and hidden mass of galactic clusters are naturally explained in this scheme.

162.059 **A chaotic universe, Friedmannian in the mean. I.**
L. S. Marochnik.
Astron. Zh., Tom 57, 903 - 912 (1980). In Russian. – English translation in Soviet Astron., Vol. 24, No. 5.

A chaotic universe, homogeneous and isotropic at a scale $L \gg \bar{L}$ ($\bar{L}$ being the scale of averaging) is considered. From Einstein's equations the equations for correlation functions describing a statistically completely chaotic model with fluctuations of arbitary amplitude are obtained.

162.060 **Microwave background radiation as a probe of the contemporary structure and history of the Universe.**
R. A. Sunyaev (*Syunyaev*), Ya. B. Zel'dovich.
Annu. Rev. Astron. Astrophys., Vol. 18, (see 003.004), 537 - 560 (1980).

162.061 **Neutrinos and the age of the Universe.**
E. M. D. Symbalisty, J. Yang, D. N. Schramm.
Nature, Vol. 288, 143 - 145 (1980).

The age of the Universe should be calculable by independent methods with similar results. Previous calculations using nucleochronometers, globular clusters and dynamical measurements coupled with Friedmann models and nucleosynthesis constraints have given different values of the age. The authors report here a consistent age whose implications for the constituent mass density are very interesting and are affected by the existence of a third neutrino flavour, τ, and by allowing the possibility that neutrinos may have a non-zero rest mass.

162.062 **The problem of the Einasto and Ostriker team hypothesis on a model of galaxies.**
M. Zabierowski.
Mem. Soc. Astron. Italiana, Vol. 51, 233 - 246 (1980).

The author presents some arguments in favour of the hypothesis that the giant galaxies may be embedded into massive invisible coronae and that a closed world model cannot be ruled out.

162.063 **The exact dynamical evolution of spherical inhomogeneous perturbations to a Friedmann background of dust.** R. N. Henriksen, M. De Robertis.
Astrophys. J., Vol. 241, 54 - 62 (1980).

The authors give an exact, simple, and self-contained method for dealing with the dynamical evolution of spherically symmetric inhomogeneities in a Friedmann background. The authors apply these techniques to the development of an initially acoustic perturbation. The behavior is shown to be homologous on all mass scales. The authors find that the formation of the core of the perturbation is followed by the growth of mass shells proceeding sequentially outward behind a rarefaction front. They demonstrate also how limits may be set on the formation redshift of a galaxy from evidence concerning its maximum linear extent.

162.064 **The application of dimensional analysis to cosmology** (or, how to make cosmology simple by using dimensional conspiracy). P. S. Wesson.
Space Sci. Rev., Vol. 27, 109 - 153 (1980).

Applied to Einstein's general relativity, the Conspiracy Hypothesis (CH) discussed here yields a simple cosmological model consisting of static clusters of galaxies with inverse-square density profiles embedded in an expanding, homogeneous background. This model agrees well with the observed Universe insofar as the latter can be described by general relativity. The CH can also be applied to other theories of gravity, especially those in which the gravitational parameter G is variable, and can also in itself be taken as a basis for gravitational theory.

162.065 **Was ist ein gekrümmter Raum?** V. Kasten.
Sterne Weltraum, Jahrg. 19, 368 - 373 (1980).

162.066 **Cosmological spaces and closed conformal infinitesimal transformations.** P. Pigeaud, M. Sakoto.
C. R. Acad. Sci. Paris, Tome 291, Sér. B, 181 - 184 (1980). In French.

The class of cosmologically defined fluid universes which admit closed time-like conformal infinitesimal transformation is determined.

162.067 **The Hagedorn temperature and Dirac's large numbers hypothesis.**
T. Grabinska, M. Zabierowski.
Acta Phys. Polonica B, Vol. B11, 471 - 474 (1980). – Abstr. in Phys. Abstr., Vol. 83, Abstr. 105410 (1980).

162.068 **Cosmological implications of the primeval nucleosynthesis.** M. Ostrowski.
Postępy Astron., Tom 28, 215 - 225 (1980). In Polish.

A review is given of numerous cosmological implications and constraints derived from fitting of observed and theoretical light elements abundances.

162.069 **High energy astrophysics and cosmology.** L. Woltjer.
Highlights of Astronomy, Vol. 5, (see 012.028), 753 - 761 (1980).

A brief review is given of cosmological tests based on quasars, of source evolution and of the X-ray background. High energy astrophysics impacts on cosmology in several ways: quasars or radio galaxies may be used for classical cosmological tests; counts of these objects at different redshifts give information on evolution in the universe; the background at X- and γ-ray wavelengths contains information on faint sources, hot gas and energetic particles in the universe.

162.070 **New determination of the Hubble constant.** V. G. Surdin.
Zemlya Vselennaya, 1980, No. 5, p. 14. In Russian.

162.071 **A "Copernican" evaluation of the Hubble constant.** D. J. Westpfahl.
Bull. American Astron. Soc., Vol. 12, 747 (1980). – Abstract.

162.072 **Gravitational radiation dominated cosmologies.** R. L. Zimmerman, R. W. Hellings.
Astrophys. J., Vol. 241, 475 - 485 (1980).

Robertson-Walker cosmologies with matter, radiation, and nonzero cosmological constant are considered to see how much high-frequency gravitational radiation may be present at the current epoch without violating observations. The models are represented as points in a three-dimensional parameter space (q, σ_r, σ_m) and classified according to the scheme of Stabell and Refsdal. Evolutionary limits due to a maximum redshift requirement, minimum age requirement, and magnitude-redshift relation are then used to rule out most M_2 models and to restrict singular models to those satisfying acceleration parameter, matter density, and radiation density limits of $-4.4<q_0<5.6$, $\sigma_{m0}<4.7$, $\sigma_{r0}<3.4$. These limits are then compared with direct limits from various experimental searches for a cosmic gravitational radiation background, indicating that several experiments are very close to a significant sensitivity.

162.073 **Dynamics of the universe and spontaneous symmetry breaking.** D. Kazanas.
Astrophys. J., Lett., Vol. 241, L59 - L63 (1980).

It is shown that the presence of a phase transition early in the history of the universe, associated with spontaneous symmetry breaking (believed to take place at very high temperatures at which the various fundamental interactions unify), significantly modifies its dynamics and evolution. This is due to the energy "pumping" during the phase transition from the vacuum to the substance, rather than the gravitating effects of the vacuum. The expansion law of the universe then differs substantially from the $R \propto t^{1/2}$ relation considered so far for the very early time expansion. In particular it is shown that under certain conditions this expansion law is exponential. It is further argued that under reasonable assumptions for the mass of the associated Higgs boson this expansion stage could last long enough to potentially account for the observed isotropy of the universe.

162.074 **The unitary Bogoliubov transformation for a quantized scalar field in a Friedman space-time.**
S. A. Bonometto, M. D. Pollock.
Gen. Relativ. Gravitation, Vol. 12, 511 - 520 (1980).

162.075 **The Robertson-Walker metrics expressible in static form.** P. S. Florides.
Gen. Relativ. Gravitation, Vol. 12, 563 - 574 (1980).

It is shown that there are six, and only six, Robertson-Walker metrics which can be expressed in static form. They are precisely those Robertson-Walker metrics whose spacetime curvature is constant. The coordinate transformations which transform these metrics into their static form are also given.

162.076 **Global rotation.** K. Rosquist.
Gen. Relativ. Gravitation, Vol. 12, 649 - 664 (1980)

Global rotation in cosmological models is defined on an observational basis. A theorem is proved saying that, for rigid motion, the global rotation is equal to the ordinary local vorticity. The global rotation is calculated in the space-time homogeneous class III models, with Gödel's model as a special case. It is shown that, with the exception of Gödel's model, the rotation in these models becomes infinite for finite affine parameter values. The physical interpretation of the infinite rotation is discussed, and a comparison with the behavior of the area distance at conjugate points is given.

162.077 **Open debate.** M. G. Edmunds.
Nature, Vol. 288, 431 - 432 (1980).

162.078 **An exact Bianchi-type II cosmological model with matter and an electromagnetic field.**
D. Lorenz.
Phys. Lett. A, Vol. 79A, 19 - 20 (1980). – Abstr. in Phys. Abstr., Vol. 83, Abstr. 105560 (1980).

162.079 **Variation of parameters in cosmology.** R. T. Jantzen.
Ann. Physics, Vol. 127, 302 - 309 (1980). – Abstr. in Phys. Abstr., Vol. 83, Abstr. 109568 (1980).

162.080 **An elliptic space temporal closed Universe.** L. M. Georgiev.
Bulgarian J. Phys., Vol. 7, 233 - 238 (1980). – Abstr. in Phys. Abstr., Vol. 83, Abstr. 109570 (1980).

162.081 **Baryon number generation in the early Universe.** E. W. Kolb, S. Wolfram.
Nucl. Phys. B, Vol. B172, 224 - 284 (1980). – Abstr. in Phys. Abstr., Vol. 83, Abstr. 109571 (1980).

162.082 **Dipole and quadrupole anisotropies in homogeneous cosmological models.** R. Fabbri.
Phys. Lett., A, Vol. 79A, 21 - 22 (1980). – Abstr. in Phys. Abstr., Vol. 83, Abstr. 109572 (1980).

162.083 **Decay of fluctuations in the early universe and high energy particle interaction.** H. Sato.

Prog. Theor. Phys., Vol. 63, 1971 - 1983 (1980). – Abstr. in Phys. Abstr., Vol. 83, Abstr. 109573 (1980).

162.084 **An observational test of cosmological particle production theories.** T. Dannehold.
Phys. Lett. B, Vol. 94B, 450 - 452 (1980). – Abstr. in Phys. Abstr., Vol. 83, Abstr. 109574 (1980).

162.085 **Revisions in the extragalactic distance scale.** D. A. Hanes.
Anglo-Australian Obs., Prepr. No. 140 (1980).

The conventional chain of reasoning in the determination of the extragalactic distance scale (Tammann, Sandage & Yahil 1979) is carefully reviewed. Various inconsistencies are identified. The presence of obscuration internal to galaxies is recognized as a potential source of error in the brightest star correlation. Strictly independent tests suggest that the first steps outside the Local Group have been overestimated by ~0.7 mag in distance modulus. The Virgo cluster true distance modulus is 30.8 ± 0.2 mag, and the Hubble constant at Virgo is 76 ± 12 km/sec/Mpc. The far-field value of the Hubble constant may be ~100 km/sec/Mpc.

162.086 **The cosmic asymmetry between matter and antimatter.** F. Wilczek.
Sci. American, Vol. 243, No. 6, p. 60 - 68 (1980).

It seems the universe today is almost entirely the former. Evidence from both cosmology and particle physics (the study of the universe on the largest scale and the smallest) now suggests an explanation.

162.087 **Problemi di cosmologia.** R. Gallino, A. Masani.
Rend. Semin. Fac. Sci. Univ. Cagliari, Vol. 48, 87 - 142 (1978)= Pubbl. Stn. Astron. Int. Latitudine, Carloforte-Cagliari, Nuova Ser., N. 65.

162.088 **Phase transitions in cosmological models: temperature vs. curvature.**
H. Fleming, V. L. R. da Silveira.
Nuovo Cimento B, Ser. 11, Vol. 58B, 208 - 214 (1980). Abstr. in Phys. Abstr., Vol. 84, Abstr. 5008 (1981).

162.089 **Entropy generation in the early Universe by dissipative processes near the Higgs phase transition.**
L. Z. Fang.
Phys. Lett. B, Vol. 95B, 154 - 156 (1980). – Abstr. in Phys. Abstr., Vol. 84, Abstr. 5009 (1981).

162.090 **Cosmogenesis and the origin of the fundamental length scale.** R. Brout, F. Englert, J.-M. Frere, E. Gunzig, P. Nardone, C. Truffin.
Nucl. Phys. B, Vol. B170, 228 - 264 (1980). – Abstr. in Phys. Abstr., Vol. 84, Abstr. 9748 (1981).

162.091 **Thermalization of baryon asymmetry.** S. B. Treiman, F. Wilczek.
Phys. Lett. B, Vol. 95B, 222 - 226 (1980). – Abstr. in Phys. Abstr., Vol. 84, Abstr. 9749 (1981).

162.092 **Canonical quantization of diagonal Bianchi V models filled with a perfect fluid.** J. Demaret.
Phys. Lett. B, Vol. 95B, 413 - 418 (1980). – Abstr. in Phys. Abstr., Vol. 84, Abstr. 9750 (1981).

162.093 **Space-time singularities and microwave background radiation.** P. S. Joshi.
Pramāṇa, Vol. 15, 225 - 230 (1980). – Abstr. in Phys. Abstr., Vol. 84, Abstr. 9751 (1981).

162.094 **On the effect of a variable vacuum energy density upon the spectrum of the cosmological microwave background radiation.** M. D. Pollock.
Mon. Not. R. Astron. Soc., Vol. 193, 825 - 831 (1980).

The cosmological term Λ admitted by the Einstein equations represents the energy density of the vacuum. Kirzhnits & Linde have pointed out that if the gauge theories with broken symmetry are applicable to the Universe, then this cosmological term is not a constant. Rather, it must vary with temperature. The vacuum assumes a dynamical role, and interacts with the observable matter, via the gravitational field. One consequence of this interaction is that the spectrum of equilibrium radiation which is decoupled from other matter cannot maintain an initially Planckian shape as the Universe evolves. The magnitude of this effect is determined by the relative size of the cosmological term and how it depends upon temperature. The authors find that it is possible to reproduce the distortion to the cosmological microwave background radiation observed by Woody & Richards, if it is assumed that variation of the cosmological term is accompanied by non-conservation of photon number. A time-dependence of photon number is equivalent to a time-dependence of the gravitational constant *G*, whose value at the epoch of helium production would be approximately twice its value today.

162.095 **Possible determination of $\bar{q}_0$ using lunar occultations and laser ranging observations.**
V. Canuto, S. - H. Hsieh, J. R. Owen.
Astrophys. J., Vol. 241, 886 - 888 (1980).

Using recent determinations of the atomic and tidal lunar acceleration, the authors propose a model to evaluate the deceleration parameter $\bar{q}_0$. Their conclusion is that the universe is open for values of H_0 close to 50 km s^{-1} Mpc^{-1}.

162.096 **A deep search for ghost images in the universe.** F. Biraud, S. Mavrides.
Astron. Astrophys., Vol. 92, 128 - 131 (1980).

In a closed cosmological model, the antipole of the observer can be detected if it is located within his horizon: a source lying near the pole can be observed from two diametrically opposite directions. Using the Nançay radiotelescope, the authors looked for a correlation between drifts carried out on opposite regions of the sky: 43 pairs of opposite fields, each one of 1°20′, have been observed. They found no correlation. This negative result is discussed as well as the resulting limits to the cosmological parameters.

162.097 **Galaxy formation, clustering and the initial cosmological fluctuations.** M. J. Rees.
Variability in stars and galaxies, (see 012.044), p. G.2.1 - 2.12 (1980).

The general question the author addresses is: how and when did the contents of the Universe condense from gaseous form into the aggregates of stars – the galaxies – which are the most prominent large scale entities in the sky? He discusses three aspects of the problem. (1) The unseen mass (or 'missing light'). (2) Gravitational clustering. (3) The nature of the initial irregularities.

162.098 **Background light from galaxies as a cosmological probe.** R. Stabell, P. S. Wesson.
Astrophys. J., Vol. 242, 443 - 447 (1980).

Galaxies emit radiation that forms a background of intergalactic light. The authors have calculated the expected energy density of the extragalactic background light for (1) a range of galaxy formation epochs; (2) a range of luminosity-evolution parameters for galaxies (taking into account differing types of evolution for ellipticals and spirals); (3) a range of Robertson-Walker universe models (with and without the cosmological constant).

162.099 **Neutrinos and the universe.** V. Pohánka.
Kozmos, Vol. 11, 141 - 142 (1980). In Slovak.

162.100 **Neutrinos and the properties of the universe.** V. Vanýsek.
Vesmír, Vol. 59, 325 - 326 (1980). In Czech.

162.101 **A new determination of the age of the universe.** Z. Mikulášek.
Říše hvězd, Vol. 61, 246 - 247 (1980). In Czech.

162.102 **The wave equation in a curved space-time: on the consistency of the perturbation expansion about Minkowski metric.** R. W. John.
Astron. Nachr., Band 301, 277 - 283 (1980).

162.103 **Peculiar velocities of clusters of galaxies and average matter density in the universe.**
Ya. B. Zel'dovich, R. A. Syunyaev.
Pis'ma Astron. Zh., Tom 6, 737 - 741 (1980). In Russian.
English translation in Soviet Astron. Lett., Vol. 6.

Formulae describing the velocity perturbation evolution during the expansion of the universe are found in the framework of the linear theory of matter density perturbation growth. The type of evolution strongly depends on the average matter density of the universe.

162.104 **Do neutrino rest masses affect cosmological helium production?**
S. L. Shapiro, S. A. Teukolsky, I. Wasserman.
Phys. Rev. Lett., Vol. 45, 669 - 672 (1980).

It is shown that the possibility of nonzero neutrino rest masses $m_\nu \approx 10$ eV/c^2 does not alter previous predictions of the standard big-bang model for the primordial ^{4}He abundance. Although both left- and right-handed neutrinos could then be present, reactions mediated by the standard Weinberg-Salam-Glashow model extended to include Dirac neutrinos with nonzero rest mass cannot maintain ν_R in thermal equilibrium.

162.105 **Galactic neutrinos and UV astronomy.** A. De Rújula, S. L. Glashow.
Phys. Rev. Lett., Vol. 45, 942 - 944 (1980).

Slowly moving massive neutrinos may be responsible for the invisible mass in galactic halos and the missing mass of the universe. Massive neutrinos are expected to decay into lighter neutrinos and UV photons, with lifetimes long on the Hubble scale. The possible detection of these neutrino-decay photons is discussed.

162.106 **Massive neutrinos and the large-scale structure of the Universe.**
J. R. Bond, G. Efstathiou, J. Silk.
Phys. Rev. Lett., Vol. 45, 1980 - 1984 (1980).

162.107 **Hierarchy of cosmological baryon generation.** J. N. Fry, K. A. Olive, M. S. Turner.
Phys. Rev. Lett., Vol. 45, 2074 - 2077 (1980).

162.108 **Cellular structure of the universe and modern theory of formation of galaxies.**
A. G. Doroshkevich, Ya. B. Zel'dovich, S. F. Shandarin.
Inst. prikl. mat. AN SSSR. Prepr., 1980, No. 67, 29 pp. In Russian. – Abstr. in Ref. zh., 51. Astron., 11.51.770 (1980).

162.109 **Sound wave creation in an isotropic universe.** V. N. Lukash.
Inst. kosm. issled. AN SSSR. Prepr., 1980, No. 559, 24 pp. In Russian. – Abstr. in Ref. zh., 51. Astron., 11.51.779 (1980).

162.110 **On the influence of quantized fields on space-time metrics in cosmology.** V. M. Mostepanenko.
Inst. teor. fiz. AN USSR. Prepr., 1980, No. 24R, 16 pp. In Russian. – Abstr. in Ref. zh., 51. Astron., 11.51.780 (1980).

162.111 **Hubble diagram at relativistic velocities: implications for world geometry.** B. M. P. Trivedi.
Publ. Astron. Soc. Pacific, Vol. 92, 551 (1980). – Abstract.

162.112 **Local inhomogeneities in a Robertson-Walker background.** K. Lake.
J. R. Astron. Soc. Canada, Vol. 74, 357 (1980). – Abstract.

162.113 **A chaotic universe, Friedmannian on the average. II.** L. S. Marochnik.
Astron. Zh., Tom 57, 1129 - 1142 (1980). In Russian.
English translation in Soviet Astron., Vol. 24, No. 6.

The cosmological solutions are found for the equations for correlators describing a statistically chaotic universe, Friedmannian on the average in which δ-correlated fluctuations with amplitudes $h \ll 1$ are excited. The contribution of quantum fluctuations and of short-wave parts of the spectrum of classical fluctuations to the expansion law is considered. The restrictions are obtained for the degree of chaos (the spectrum characteristics) compatible with the observed helium abundance.

162.114 **On quantum gravitational effects in an anisotropic universe.** V. A. Bejlin, G. M. Vereshkov, Yu. S. Grishkan, N. M. Ivanov, V. A. Nesterenko, A. N. Poltavtsev.
Zh. ehksperim. i teor. fiz., Vol. 78, 2081 - 2098 (1980). In Russian. – Abstr. in Ref. zh., 51. Astron., 12.51.680 (1980).

162.115 **Quantum evaporation of black holes and baryon asymmetry of the universe.** A. D. Dolgov.
Zh. ehksperim. i teor. fiz., Vol. 79, 337 - 349 (1980). In Russian. – Abstr. in Ref. zh., 51. Astron., 12.51.883 (1980).

162.116 **Some consequences of the theory of interacting fields in cosmology and astrophysics.**
V. G. Krechet, V. M. Nikolaenko, G. N. Shikin.
Gorenie i vzryv v kosmose i na Zemle. Moskva, 1980, p. 3 - 19. In Russian. – Abstr. in Ref. zh., 51. Astron., 12.51.690 (1980).

162.117 **The universe as a self-organizing cybernetic system.** E. A. Sedov.
Zh. Vses. khim. o-va im. D. I. Mendeleeva, Vol. 25, 440 - 443 (1980). In Russian. – Abstr. in Ref. zh., 51. Astron., 12.51.691 (1980).

162.118 **The origin of the Universe.** D. W. Sciama.
The state of the Universe, (see 003.013). p. 3 - 15 (1980).

The author presents the observational and theoretical evidence in favour of the Big Bang origin of the Universe, and the question whether an actually singular origin of the Universe can be avoided.

162.119 **Exact spatially homogeneous cosmologies.** C. B. Collins, E. N. Glass, D. A. Wilkinson.
Gen. Relativ. Gravitation, Vol. 12, 805 - 823 (1980).

The authors consider perfect fluid spatially homogeneous cosmological models. Starting with a new exact solution of Bianchi type VIII, they study generalizations which lead to new classes of exact solutions. These new solutions are discussed and classified in several ways. In the original type VIII solution, the ratio of matter shear to expansion is constant, and the authors present a theorem which delimits those space-times for which this condition holds.

162.120 **Poisson's equation in de Sitter space-time.** E. Pessa.
Gen. Relativ. Gravitation, Vol. 12, 913 - 916 (1980).

The form of "gravitation" law in "projective relativity" is examined, based on a suitable generalization of Poisson's equation in de Sitter space-time; it is found that, in the interior case, a small difference with the customary Newtonian law arises. This difference, of a repulsive character, can be very important in cosmological problems.

162.121 **Machian effects in nonasymptotically flat space-times.** S. M. Lewis.
Gen. Relativ. Gravitation, Vol. 12, 917 - 924 (1980).

162.122 **Nonhomogeneity of the early universe and the origin of primordial black holes.**
N. A. Zabotin, L. S. Marochnik, P. D. Nasel'skij.
Inst. kosm. issled. AN SSSR. Prepr., 1980, No. 564, 37 pp. In Russian. – Abstr. in Ref. zh., 51. Astron., 1.51.746 (1981).

162.123 **Nonlinear influence of small-scale fluctuations on the instability of long-wavelength perturbations in the expanding universe.**
L. S. Marochnik, P. D. Nasel'skij.
Inst. kosm. issled. AN SSSR. Prepr., 1980, No. 565, 22 pp. In Russian. – Abstr. in Ref. zh., 51. Astron., 1.51.748 (1981).

162.124 **Sound wave origin and galaxy formation in the expanding universe.** V. N. Lukash.
Inst. kosm. issled. AN SSSR. Prepr., 1980, No. 300, 15 pp. In Russian. – Abstr. in Ref. zh., 51. Astron., 1.51.749 (1981).

162.125 **Cosmological models of the universe with reversal of time arrow.** A. D. Sakharov.
Zh. ehksperim. i teor. fiz., Vol. 79, 689 - 693 (1980). In Russian. – Abstr. in Ref. zh., 51. Astron., 1.51.754 (1981).

162.126 **Local inhomogeneities in a Robertson-Walker background. II. Flux conditions at boundary surfaces.**
K. Lake.
Astrophys. J., Vol. 242, 1238 - 1242 (1980).
Energy flux conditions imposed on spherical boundary surfaces are examined. The zero flux restriction, which is the hallmark of the standard "Swiss cheese" type construction, is relaxed. The author discusses a class of locally inhomogeneous exact solutions to the Einstein equations which admit an effectively Newtonian accretion mode.

Bulletin GRG, No. 40. List of publications.
See Abstr. 002.014.

Spacetime, geometry, cosmology.
See Abstr. 003.028.

Other worlds: space, superspace and the quantum universe. See Abstr. 003.034.

Stardoom. See Abstr. 003.035.

A theory of cosmology. See Abstr. 003.046.

Relativistic cosmology. See Abstr. 003.055.

Steady-state cosmology re-visited.
See Abstr. 003.059.

Physics and the physical universe.
See Abstr. 003.075.

Theoretical cosmology. See Abstr. 003.093.

Foundations of cosmology.
See Abstr. 003.106.

Gravity, particles and astrophysics.
See Abstr. 003.111.

The paradox of the dark night sky.
See Abstr. 004.021.

The elementary universe. See Abstr. 011.006.

Mathematical theories and philosophical insights in cosmology. See Abstr. 015.004.

Die philosophische Relevanz der Kosmologie.
See Abstr. 015.005.

Cosmology confronts particle physics.
See Abstr. 022.027.

On relativistic kinetic theory of transport processes, in particular of neutrino systems. See Abstr. 022.029.

Cosmological limits on photon splitting.
See Abstr. 022.033.

Vacuum instability, cosmology and constraints on particle masses in the Weinberg-Salam model.
See Abstr. 022.046.

Vacuum structure in gauge theories. The problem of strong CP violation and cosmology. See Abstr. 022.057.

Conformal-symmetry breaking and cosmological particle creation in $\lambda\phi^4$ theory. See Abstr. 022.061.

Possible production of 'collapsed' hadronic matter in very-high-energy nucleon-nucleon collisions.
See Abstr. 022.150.

Helium synthesis, neutrino flavors, and cosmological implications. See Abstr. 061.005.

On the propagation of electromagnetic waves in a cosmological neutrino sea. See Abstr. 061.011.

On the neutrino number and isotropy of the Universe in grand unified theories. See Abstr. 061.014.

L'evoluzione chimica nell'universo.
See Abstr. 061.024.

Oszillierende Neutrinos und die Materiedichte des Weltalls. See Abstr. 061.026.

Heavy neutrinos with nonzero rest mass and astrophysical consequences of their discovery.
See Abstr. 061.041.

Cosmological and astrophysical implications of heavy Majorana particles. See Abstr. 061.042.

Have massive cosmological neutrinos already been detected? See Abstr. 061.043.

The origin of the elements. See Abstr. 061.045.

Some effects of magnetic fields in spatially homogeneous universes with conductivity. See Abstr. 062.050.

Observational tests of the cosmic turbulence theory.
See Abstr. 066.002.

Shear hell holes and anisotropic universes.
See Abstr. 066.007.

High-energy gravity and the very early universe.
See Abstr. 066.009.

Integrability conditions for a gravitational theory with nonmetric connection. See Abstr. 066.011.

Radiation and the structure of space-time. See Abstr. 066.012.

The role of general relativity in astronomy: retrospect and prospect. See Abstr. 066.021.

Symmetric vectors and algebraic classification. See Abstr. 066.022.

Search for the intermediate-scale anisotropy of the cosmological background radiation. See Abstr. 066.045.

On caustics and singularities in general relativity. See Abstr. 066.056.

Where has the fifth dimension gone? See Abstr. 066.059.

On the problem of the singularities in the general cosmological solution of the Einstein equations. See Abstr. 066.068.

Hamiltonian formalism for perfect fluids in general relativity. See Abstr. 066.070.

Electromagnetic fields in space-times with local rotational symmetry. See Abstr. 066.073.

Cosmic microwave background spectrum and G-varying cosmology. See Abstr. 066.080.

Gauge theories, time-dependence of the gravitational constant and antigravity in the early Universe. See Abstr. 066.091.

Conformal covariance of general relativity. See Abstr. 066.092.

Geometric quantization and gravitational collapse. See Abstr. 066.095.

A spherical cavity in an Einstein universe. See Abstr. 066.105.

Quantum many-particle systems in curved spacetime. See Abstr. 066.107.

The material vacuum. See Abstr. 066.111.

Primordial black holes and the deuterium abundance. See Abstr. 066. 112.

The role of general relativity in astronomy: retrospect and prospect. See Abstr. 066.122.

Bimetric general relativity and cosmology. See Abstr. 066.124.

Supergravity with and without superspace. See Abstr. 066.139.

Gravitational bounce. See Abstr. 066.145.

Two-soliton waves in anisotropic cosmology. See Abstr. 066.148.

The gravitational constant at time zero. See Abstr. 066.155.

Conformal invariance, microscopic physics, and the nature of gravitation. See Abstr. 066.158.

Scalar field generalizations of electromagnetic Bianchi models of types II, VIII and IX. See Abstr. 066.162.

On the determination of the degree of cosmological Compton distortions and the temperature of the cosmic black-body radiation. See Abstr. 066.168.

Interacting gravitational shocks in vacuum plane-symmetric cosmologies. See Abstr. 066.169.

On the description of a black hole in the expanding universe. See Abstr. 066.189.

Algebraic isometric embeddings of charged spherically symmetric space-times. See Abstr. 066.195.

Motion of primordial black holes in the early universe and their likely distribution today. See Abstr. 066.199.

Newtonian analogs of Szekeres' space-times. See Abstr. 066.200.

Molecular clouds and star formation. See Abstr. 131.023.

On the density of star formation in the universe. See Abstr. 131.294.

An analysis of the cosmological evolution of radio sources. I. Spectral-index dependent counts of sources and spectral index distributions at 1400 MHz. See Abstr. 141.012.

Anomalous redshifts of quasi-stellar objects. See Abstr. 141.072.

QSO evolution and the intergalactic medium. See Abstr.141.086.

Quasars, isotropy of H_0 and the Local Supercluster of galaxies. See Abstr. 141.088.

Is C IV λ 1549 a standard cosmic candle? See Abstr. 141.110.

Models of the cosmological evolution of extragalactic radio sources – I. The 408-MHz source count. See Abstr. 141.145.

Quasar Lα absorbers: are precise conclusions possible? See Abstr. 141.146.

A new investigation of the redshift–angular-diameter relation for quasars. See Abstr. 141.155.

Spatial distribution of quasars and cosmology. See Abstr. 141.176.

Galactic and extragalactic contributions to the far-ultraviolet background. See Abstr. 142.053.

On observations of the cosmic radiation background. See Abstr. 142.109.

X-ray astronomy and cosmology. See Abstr. 142.130.

The X-ray background as a probe of the matter distribution on very large scales. See Abstr. 142.137.

The diffuse gamma-ray background in the Hoyle-Narlikar cosmology. See Abstr. 142.517.

Vår åldrande galax. See Abstr. 151.048.

On the interpretation of colors of faint galaxies. See Abstr. 151.050.

On the interpretation of galaxy counts. See Abstr. 151.051.

Globular clusters as survivors. See Abstr. 151.087.

Globular clusters as extragalactic distance indicators. See Abstr. 154.037.

The universe at faint magnitudes. I. Models for the Galaxy and the predicted star counts. See Abstr. 155.038.

The redshift-magnitude relation for bright galaxies at low redshifts. See Abstr. 158.020.

Colors and magnitudes predicted for high redshift galaxies. See Abstr. 158.115.

Primeval galaxies: a new look in red light. See Abstr. 158.238.

Galaxies and their nuclei. See Abstr. 158.319.

A distance scale from the infrared magnitude/H I velocity-width relation. III. The expansion rate outside the Local Supercluster. See Abstr. 160.007.

A search for Z=9 neutral hydrogen emission from primordial protoclusters of galaxies. See Abstr. 160.016.

Dynamics of the supercluster 1451 +22: evidence for an open universe. See Abstr. 160.017.

The photometric properties of brightest cluster galaxies. I. Absolute magnitudes in 116 nearby Abell clusters. See Abstr. 160.047.

The photometric properties of brightest cluster galaxies. II. SIT and CCD surface photometry. See Abstr. 160.048.

The mass of Coma Cluster. See Abstr. 160.054.

The luminosity function of Coma cluster galaxies. See Abstr. 160.065.

Dynamical models and the mass of the Virgo cluster. See Abstr. 160.086.

The effects of X-ray absorption on the spectra of distant objects. See Abstr. 161.002.

Author Index

The authors are listed in alphabetical order

according to the initial letter following the first names.

Subject Index

Starting with Volume 18 of Astronomy and Astrophysics Abstracts, some alterations concerning formation, arrangement, and versatility of the key words have been made. In order to provide an adequate description of a paper, specific key words are used as frequently as possible. References to a whole subject category are suppressed now. The user, therefore, has to refer to the contents at the beginning of each volume.

Whenever possible, the key words are formed in such a way that there are two different supplementary terms, e.g. the pair

interstellar matter
  molecules.

An effort is made to choose preferably terms which can be inverted in order to increase the usefulness of this index. In the example given there are the two entries

interstellar matter
  molecules

and

molecules
  interstellar matter.

Exceptions to the rule of inversion of terms are given in all cases where the second key word is either a very specific one (e.g. Urca processes) or a general one (e.g. history). The use of substantives is preferred. In order to obtain the possibility to extend a one-term key word in a two-term one, combinations as

Mars
  atmosphere

or

sun
  active regions

are changed into

Mars atmosphere and solar active regions,

respectively. The number of cross references indicating such slightly different entries is reduced drastically. In previous volumes combinations like

close binaries and binaries
  close binaries,

peculiar A stars and A stars
  peculiar,

groups of galaxies and galaxies
  groups

have been used. Now only the specific key words

close binaries
peculiar A stars
groups of galaxies

have to be considered as a substitute.

The user is requested to look for more specialized entries,as further references to this topic might exist elsewhere in the index under another current astronomical term.

ASTRONOMY AND ASTROPHYSICS ABSTRACTS

A Publication of the Astronomisches Rechen-Institut Heidelberg

Member of the Abstracting Board
of the International Council of Scientific Unions

Editors: S. Böhme, W. Fricke, I. Heinrich, W. Hofmann,
D. Krahn, D. Rosa, L. D. Schmadel, G. Zech

Volume 1	Literature 1969, Part 1	X + 435 pp. (1969)
Volume 2	Literature 1969, Part 2	X + 516 pp. (1970)
Volume 3	Literature 1970, Part 1	X + 490 pp. (1970)
Volume 4	Literature 1970, Part 2	X + 562 pp. (1971)
Volume 5	Literature 1971, Part 1	X + 505 pp. (1971)
Volume 6	Literature 1971, Part 2	X + 560 pp. (1972)
Volume 7	Literature 1972, Part 1	X + 526 pp. (1972)
Volume 8	Literature 1972, Part 2	X + 594 pp. (1973)
Volume 9	Literature 1973, Part 1	X + 610 pp. (1973)
Volume 10	Literature 1973, Part 2	X + 661 pp. (1974)
Volume 11	Literature 1974, Part 1	X + 579 pp. (1974)
Volume 12	Literature 1974, Part 2	X + 699 pp. (1975)
Volume 13	Literature 1975, Part 1	X + 632 pp. (1975)
Volume 14	Literature 1975, Part 2	X + 747 pp. (1976)
Volume 15/16	Author and Subject Indexes to Volumes 1–10	
	Literature 1969–1973	VII + 655 pp. (1976)
Volume 17	Literature 1976, Part 1	XII + 645 pp. (1976)
Volume 18	Literature 1976, Part 2	X + 859 pp. (1977)
Volume 19	Literature 1977, Part 1	X + 732 pp. (1977)
Volume 20	Literature 1977, Part 2	X + 786 pp. (1978)
Volume 21	Literature 1978, Part 1	X + 834 pp. (1978)
Volume 22	Literature 1978, Part 2	X + 849 pp. (1979)
Volume 23/24	Author and Subject Indexes to Volumes 11–14 and 17–22	
	Literature 1974–1978	VIII +1127 pp. (1979)
Volume 25	Literature 1979, Part 1	X + 872 pp. (1979)
Volume 26	Literature 1979, Part 2	X + 794 pp. (1980)
Volume 27	Literature 1980, Part 1	X + 939 pp. (1980)
Volume 28	Literature 1980, Part 2,	X + 841 pp. (1981)

Published for Astronomisches Rechen-Institut by
Springer-Verlag Berlin Heidelberg New York

The manufacturer's authorised representative in the EU is Springer Nature Customer Service Centre GmbH, Europaplatz 3, 69115 Heidelberg, Germany. If you have any concerns regarding our products, please contact ProductSafety@springernature.com

Printed and bound by CPI Group (UK) Ltd, Croydon, CR0 4YY
15/07/2026
02167621-0011